A SYSTEMATIC CATALOGUE OF THE DIASPIDIDAE (ARMOURED SCALE INSECTS) OF THE WORLD, SUBFAMILIES ASPIDIOTINAE, COMSTOCKIELLINAE AND ODONASPIDINAE

A SYSTEMATIC CATALOGUE OF THE DIASPIDIDAE (ARMOURED SCALE INSECTS) OF THE WORLD, SUBFAMILIES ASPIDIOTINAE, COMSTOCKIELLINAE AND ODONASPIDINAE

Yair Ben-Dov and **Victoria German**

*Department of Entomology, Agricultural Research Organization,
The Volcani Center, Bet Dagan, 50250 ISRAEL*

in collaboration with
Douglass R. Miller
*Systematic Entomology Laboratory, Agricultural Research Service,
U.S. Department of Agriculture, Beltsville, Maryland 20705 U.S.A.*
and
Gary. A. P. Gibson
*Biological Resources Program, Eastern Cereal and Oilseed Research Center,
Agriculture and Agri-Food Canada, Ottawa, Ontario, CANADA*

Intercept Ltd
Andover

British Library CIP data available

ISBN 1 898298 93 9

Published by Intercept Limited, P.O. Box 716, Andover, Hants SP10 1YG UK
Email: intercept@andover.co.uk
Website: http://www.intercept.co.uk

Cover photograph: Scale covers of female, male and nymphs of
Aonidiella aurantii (Maskell) (Hemiptera: Coccoidea: Diaspididae: Aspidiotinae)

Printed in Great Britain

CONTENTS

DEDICATION

This volume is dedicated to Victor Antoine Signoret (1816-1889), Maria E. Fernald (1839-1919) and Nikolay Sergeyevitch Borchsenius (1906-1965), whose early catalogues on scale insects (Signoret, 1869, 1877; Fernald, 1903b; Borchsenius, 1966) were significant milestones along the road to the publication of this catalogue.

ABSTRACT

This catalogue on armoured scale insects, subfamilies Aspidiotinae, Comstockiellinae and Odonaspidinae of the family Diaspididae (Hemiptera: Coccoidea) includes data on 864 species that are placed among 118 genera. The family of the armoured scale insects includes some of the most destructive pests of agricultural crops such as citrus, coconut, date palm, manihot, deciduous fruit trees, forest trees and ornamentals. This book is a synthesis and catalogue of all of the information published on these genera and species worldwide up to December 2002, and gives information on their correct scientific name, taxonomy, common names, synonyms, host plants, distribution, natural enemies, biology, economic importance, and published references. The following nomenclatural changes are introduced for the first time in this catalogue: *Sadaotakagia* Ben-Dov **replacement name** for *Takagia* Tang, 1984, homonym of *Takagia* Matsumura, 1951; *Targionidea* MacGillivray, 1921, **n. syn.** of *Aonidia* Targioni Tozzetti, 1868; *Aspidiotus tangfangtehi* Ben-Dov **replacement name** for the homonym *Aspidiotus theae* Tang, 1977; *Furcaspis biformis taquarae* Fonseca, 1969, raised to species level under *Furcaspis taquarae* Fonseca; *Hemiberlesia lataniae laciniata* Gómez-Menor Ortega, 1965, raised to species level under *Hemiberlesia laciniata* Gómez-Menor Ortega; *Rhizaspidiotus artemisiae adiscus* Gómez-Menor Ortega, 1968, raised to species level under *Rhizaspidiotus adiscus* Gómez-Menor Ortega; *Aspidiotus sinensis* (Ferris, 1952) **n. comb.**, transferred from *Temnaspidiotus; Aspidiotus taraxacus* (Tang, 1984) **n. comb.**, transferred from *Temnaspidiotus; Diaspidiotus centrafricanus* (Balachowsky & Ferrero, 1967) **n. comb.** transferred from *Quadraspidiotus; Diaspidiotus cotoneastri* (Takagi, 1975) **n. comb.**, transferred from *Quadraspidiotus; Diaspidiotus fabernii* (Houser, 1918), **n. comb.**, transferred from *Quadraspidiotus; Diaspidiotus inusitatus* (Munting), **n. comb.**, transferred from *Quadraspidiotus; Diaspidiotus socialis* (Hoke), **n. comb.**, transferred from *Quadraspidiotus; Diaspidiotus taxodii* (Ferris), **n. comb.**, *Diaspidiotus tillandsiae* (Takagi & Tippins, 1972) **n. comb.**, transferred from *Quadraspidiotus; Dynaspidiotus apacheca* (Ferris, 1941) **n. comb.**, transferred from *Nuculaspis; Dynaspidiotus piceae* (Tang, Hao, Shi & Tang, 1991) **n. comb.**, transferred from *Tsugaspidiotus; Octaspidiotus yunnanensis* (Tang & Chu), **n. comb.**, transferred from *Metaspidiotus*.

Key words

Hemiptera Coccoidea Diaspididae Aspidiotinae Comstockiellinae Odonaspidinae armoured scale insects   world catalogue   systematics   distribution   host plants natural enemies economic importance

PREFACE

In 1995, Yair Ben-Dov (Bet Dagan, Israel) and Dug Miller (Beltsville, Maryland, USA) initiated a project to catalogue the scale insects (Hemiptera: Coccoidea) of the world. The project was designed around three components: 1) development of a database for each of the 21 scale-insect families; 2) making this information available on a queriable web site on the Internet; and 3) publishing hard-copy catalogues for each of the families.

The foundation of this initiative was the development of a computerized database by Yair Ben-Dov, for the soft scales (Coccidae), and the mealybugs (Pseudococcidae), which subsequently were the bases for the Coccidae Catalogue (Ben-Dov. 1993) and the Pseudococcidae Catalogue (Ben-Dov. 1994). The inception of this project was motivated by the above-mentioned data, along with the bibliographic information on the Coccoidea that had been published to cover the scale-insect from 1758 to 1985 (Morrison & Renk, 1957: Morrison & Morrison, 1965: Russell et al., 1974: Kosztarab & Kosztarab, 1988) and a bibliographic database that is kept up-to-date by the Systematic Entomology Laboratory, USDA and the Virginia Polytechnic Institute and State University in Blacksburg, Virginia.

This project was realized through the financial support from two grants from United States-Israel Binational Agricultural Research and Development Fund (BARD) (Grant Numbers IS-2423-94 and IS-2605-95) which empowered us to carry out this important project.

A database system was chosen that was developed by Gary Gibson, Biological Resources Program, Eastern Cereal and Oilseed Research Centre, Agriculture and Agri-Food Canada, Ottawa, Ontario, Canada and later enhanced by Gary Gibson and Jennifer Read. The database program called **BASIS** is an elegant, user-friendly system that has a series of tools that greatly facilitate the data-entry process. At its inception, the primary purpose of the database system was to produce hard-copy catalogues and it is this output that is the source of this publication. Later, it was slightly modified so that it could be used as an interface with the Internet. From 1995 to 2000 the web site named **ScaleNet** - placed at the following URL http://www.sel.barc.usda.gov.il/scalenet/scalenet.htm - through a contract with Richard Carson & Associates, Advanced Research Division. Reston, Virginia, USA. A team of Jane Lemmer, Robert Todd, and Arek Grantham developed a series of programs to interface with data from the BASIS database. The ScaleNet site allows users to obtain information through a series of personalized requests.

This volume is the second in a series of publications that will result from the above-discussed project. The first publication was the Eriococcidae Catalogue (Miller & Gimpel, 2000),

ACKNOWLEDGMENTS

Many of our colleagues kindly and willingly responded to our requests and queries by providing information and data on material in their collections, sending photocopies of publications and translating various texts. We thank all of them from the bottom of heart. Our apology is extended to any colleague, who we might have overlooked, and is not thanked in the following list:

Songbi, Chen, Academy of Agricultural Sciences, Beijing, CHINA
Lucia E. Claps, Instituto Miguel Lillo, S.M. de Tucumán, ARGENTINA
Vivian Crawford (Mr.), The Institute of Jamaica, Kingston, JAMAICA
Evelyna Danzig, Zoological Institute, St. Petersburg, RUSSIA
John Donaldson, Queensland Dept of Agriculture, Inddoroopilly, AUSTRALIA
Imre Foldi, Muséum National d'Histoire Naturelle, Paris, FRANCE
José Carlos Franco, Instituto Superior de Agronomia, Lisbon, PORTUGAL
Saul Frommer, University of California, Riverside, California, USA
Carl-Axel Gertsson, Lund, SWEDEN
Ray J. Gill, California Department of Agriculture, Sacramento, California, USA
Maren E. Gimpel, Sys. Ent. Lab., USDA, Beltsville, Maryland, USA
Roberto Gonzalez, University of Chile, Santiago, CHILE
Penny J. Gullan, University of California, Davis, California, USA
Emilio Guerrieri, Istituto per la Protezione delle Piante, Portici, ITALY
Avas Hamon, Florida Department of Agriculture, Gainesville, Florida, USA
Rosa Henderson, Landcare Research, Auckland, NEW ZEALAND
Maurice Jansen, Plantenziektenkundige Dienst, Wageningen, NETHERLANDS
Panayotis Katsoyannos, Benaki Phytopathological Institute, Athens, GREECE
Michael Kosztarab, VPI & State University, Blacksburg, Virginia. USA
Jan Koteja, Cracow Agricultural University, Cracow, POLAND
Paris Lambdin, The University of Tennessee, Knoxville, Tennessee, USA
Ireno L. Lit, Jr. Museum of Natural History, Los Baños, Laguna, PHILIPPINES
Ligun, Lu, (Ph.D. student at ARO), Wuhan, Hubei Province, CHINA
The late Salvatore Marrota, Universita Della Basilicata, Potenza, ITALY
Jon Martin, The Natural History Museum, London, UK
Ian Millar, National Collection of Insects, Pretoria, SOUTH AFRICA
John Noyes, The Natural History Museum, London, UK
Giuseppina Pellizzarri, Universita di Padova, Padova, ITALY
Gerhard Prinsloo, National Collection of Insects, Pretoria, SOUTH AFRICA
Salvatore Ragusa Di Chiara, Universita di Palermo, Palermo, ITALY
Kondo Takumasa, (Demian), Auburn University, Auburn, Alabama, USA
Fang-the, Tang, Shanxi Agricultural University, Taigu, Shanxi, CHINA
The late Charles Chia-chu Tao, Agricultural Research Institute, TAIWAN
Elizabeth O. Wangari, UNESCO - World Heritage Centre, Paris, FRANCE
Gillian M. Watson, The Natural History Museum, London, UK
Vera Regina Dos Santos Wolff, FEPAGRO, Porto Alegre, RS, BRAZIL

Sanan, Wu, Beijing Forestry University, Beijing, CHINA
Meron Zalucki, University of Queensland, St. Lucia, Queensland, AUSTRALIA

We would like to give special thanks to Karen Veilleux, Virginia Polytechnic Institute and State University, Blacksburg, Virginia, USA. Karen is responsible for the Reference table of BASIS and has meticulously entered or validated the input of more than 17,400 bibliographic entries, and all the references cited in this catalogue.

We are also very grateful to Jennifer Read, Biological Resources Program, Eastern Cereal and Oilseed Research Centre, Agriculture and Agri-Food Canada, Ottawa, Ontario, Canada for her unending care, patience, and persistence to solve computer problems that we encountered with BASIS and the catalogue preparation.

Sadao Takagi, Systematic Entomology, Faculty of Agriculture, Hokkaido University, Sapporo, Japan, with his authoritative knowledge of the Diaspididae, was available for discussions and comments on taxonomic issues of the group, provided photocopies of papers in Japanese with a translation into English, and finally reviewed the manuscript of this catalogue. Thank you!!!

We very much appreciate the kind help of Douglas J. Williams, The Natural History Museum, London, UK, who reviewed the manuscript, and willingly shared with us his vast experience and knowledge of scale insects.

We also would like to give thanks to Chris Hodgson (National Museum of Wales, Cardiff, Wales, UK), who consented to review the manuscript and provided comments and criticisms that improved its final presentation.

Special thanks are due to Danièle Matile-Ferrero, Laboratoire d'Entomologie, Muséum national d'Histoire naturelle, Paris, France. She was extremely helpful in providing the correct comprehension of French texts, in providing reprints of difficult-to-obtain papers and, in addition, she agreed to correct the spelling of paper titles in French, that are cited in the References chapter. We are responsible for any misspellings that might still be found in the References.

The authorship of this publication is Ben-Dov and German. Dug Miller and Gary Gibson are not to be considered as authors since they did not contribute to the actual data entry or information synthesis. However, their contributions were very important and therefore are acknowledged in a special category as collaborators.

Last but not least, Yair Ben-Dov gives thanks to his wife Yehudith Ben-Dov, for her patience, support and understanding.

INTRODUCTION

Suprageneric Classification within the family

The description of the genus *Diaspis* Costa O.G., 1828 - the type genus of the family Diaspididae - and of *Aspidiotus* Bouché, 1833, retrospectively should be considered as the first attempts to dismember the Linnaean genus *Coccus* into more natural subdivisions among scale insects. However, the beginning in the process of establishing the armoured scale insects as a separate group among scale insects formally is to be credited to Targioni Tozzetti (1868) and Signoret (1870).

From that time to the present, several systems were proposed for the classification and subdivisions within the family, and these are summarized in Tables 1,2 and 3.

Targioni Tozzetti (1868)	Signoret (1870)	Atkinson (1886)	Maskell (1895a)	Leonardi (1897b)	Fernald (1903b)
Diaspites	Diaspides	Diaspina Aspidiotaria Leucaspiaria	Diaspidinae	Diaspiti Aspidioti Leucaspides Diaspides Mytilaspides Parlatoriae	Diaspinae

Table 1. Classification of suprageneric subdivisions within the family Diaspididae from 1868 to 1903.

MacGillivray (1921)	Ferris (1942)	Balachowsky (1954e)	Brown & McKenzie (1962)	Borchsenius (1966)
Diaspidinae	Diaspididae	Diaspidinae	Diaspididae	Diaspididae
Aspidiotini	Aspidiotini	Aspidiotini	Aspidiotini	Aspidiotinae
Diaspidini Lepidosaphidini Fioriniini	Diaspidini	Diaspidini Diaspidina Lepidosaphedina	Diaspidini	Diaspidinae Antakaspidini Lepidosaphidini Ancepaspidini Chionaspidini Fioriniini Diaspidini
Parlatoriini Leucaspidini		Parlatoriini Leucaspidina Parlatorina	Parlatoriini	Leucaspidinae
		Odonaspidini Rugaspidiotina Odonaspidina	Odonaspidini	Odonaspidinae
	Odonaspidini			

Table 2. Classification of suprageneric subdivisions within the family Diaspididae from 1948 to 1966.

Takagi (1969a)	Chou (1982)	Tang (1984)	Danzig (1993)	Takagi (2002)
Diaspididae Aspidiotini Parlatorini Leucaspidini Odonaspidini Diaspidini Lepidosaphedini Rugaspidiotini	Diaspididae Aspidiotinae Parlatoriinae Leucaspinae Odonaspinae Diaspinae Fioriniinae Lepidosaphinae Chionaspinae	Diaspididae Aspidiotinae Parlatoriinae Odonaspidinae Diaspidinae Lepidosaphedinae	Diaspididae Aspidiotinae Parlatoriini Leucaspidini Odonaspidini Aspidiotini Diaspidinae Lepidosaphini Diaspidini	Diaspididae Aspidiotinae Parlatoriini Aspidiotini Odonaspidini Smilacicolini Leucaspidini Thysanaspidini Diaspidinae Lepidosaphidini Diaspidini Ulucoccinae

Table 3. Classification of suprageneric subdivisions within the family Diaspididae from 1969 to 2002.

The data presented in Tables 1, 2 and 3 clearly demonstrate that the following are far from approaching general agreement: the suprageneric classification within the family, the hierarchic level of subdivisions (subfamilial or tribal), the limits between suprageneric units and the taxa contained within a subdivision. We hope that sounder and more thorough decisions will be taken in the future, based on more diversified studies, such as on first-instar morphology, male morphology, symbionts, and molecular-phylogenetic studies.

Until such studies are undertaken, we divide the family in this catalogue into six subfamilies, namely Aspidiotinae, Leucaspidinae, Odonaspidinae, Diaspidinae, Comstockiellinae and Ulococcinae.

Diaspidid Statistics

The descriptive activity, from 1758 to 2002, in the Aspidiotinae, Comstockiellinae and Odonaspidinae subfamilies of the Diaspididae, is presented in Fig. 1, Fig. 2 and Table 4.

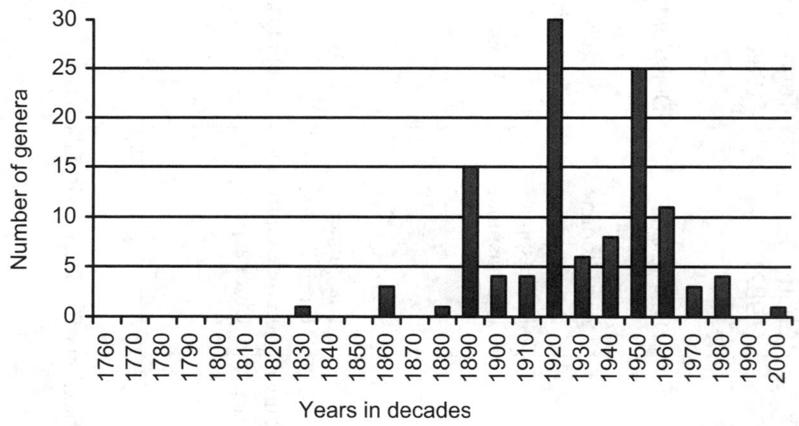

Fig. 1. Number of genera in the Aspidiotinae, Comstockiellinae and Odonaspidinae described between 1758 to 2002.

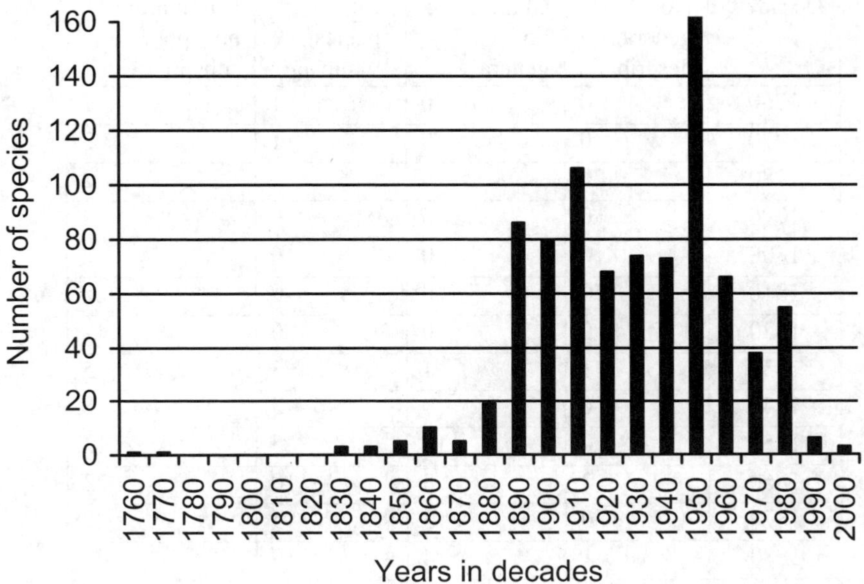

Fig. 2.   Number of species in the Aspidiotinae, Comstockiellinae and Odonaspidinae described between 1758 to 2002.

During this period, a total of 175 generic names (including synonyms, homonyms, replacement names and emendation of names) were described or proposed. Through further studies and revisions, the number of accepted valid genera was reduced to 118, constituting 67% of the proposed and/or described names.

At the species level, 1240 names of species and subspecies (including synonyms, homonyms, replacement names and emendation of names) were introduced. Only 69% of the 1240 names are currently accepted valid species names, numbering 864.

Decade	No. genera described	Cumulative no. genera	No. species. & subspp.	Cumulative no. spp. & subspp.
1760	0	0	1	1
1770	0	0	1	2
1780	0	0	0	2
1790	0	0	0	2
1800	0	0	0	2
1810	0	0	0	2
1820	0	0	0	2
1830	1	1	3	5
1840	0	0	3	8
1850	0	0	5	13
1860	2	3	10	23
1870	0	3	5	28
1880	1	4	19	47
1890	16	20	86	133
1900	4	24	80	213
1910	4	28	106	319
1920	30	58	68	387
1930	6	64	74	461
1940	9	73	73	534
1950	25	98	162	696
1960	11	109	66	762
1970	3	112	38	800
1980	5	117	55	855
1990	0	117	6	861
2000	1	118	3	864
TOTAL	**118**	**118**	**864**	**864**

Table 4. Number of species and genera in the Aspidiotinae, Comstockiellinae and Odonaspidinae, described in each decade between 1758 to 2002.

From the above data, it is clear that the establishment of genera started to increase from 1880 through 1910, mainly by authors such as Signoret, Maskell, Cockerell and Green. The conspicuous peak in the 1920's is attributable to MacGillivray (1921) who introduced 51 generic names, in the subfamilies of this catalogue; 30 of them are currently accepted as valid. During the period 1930-1970 the studies of Ferris, Mamet, Balachowsky, Borchsenius and D.J. Williams) 53 new genera were described.

The apparent decrease in descriptive activity since 1990 (see Table 4) is misleading, because 57 species and 26 genera were described in all subfamilies of Diaspididae during this period.

Family	Number species & subspp. described	Number Genera described	Median number species per genus	Genera with 10 or more species	Monotypic genera
Diaspididae (3 subfamilies)	864	118	2	23 (20%)	47 (40%)
Coccidae	1098	152	2	24 (16%)	62 (41%)
Eriococcidae	551	67	2	5 (7%)	32 (48%)
Pseudococcidae	1923	268	2	32 (12%)	126 (47%)
Ortheziidae	115	10	4	3 (30%)	3 (30%)

Table 5. Several statistical parameters for 5 families of scale insects, Coccoidea. Data taken from Ben-Dov et al. (2000), Miller & Gimpel (2000) and this catalogue.

When comparing statistics of three subfamilies of this catalogue with four other families of scale insects (Table 5), it appears that the percentage of monotypic genera is similar to that in the Coccidae, but remarkably lower than in the Eriococcidae and Pseudococcidae.

CATALOGUE FORMAT

Cutoff date

The cutoff date of literature survey for this catalogue was December 2002. However, a few publications that were issued in 2003 were also included.

Languages

We believe that we managed to obtain translations of pertinent portions from publications in languages other than English. The junior author, Viktoria German is fluent in Russian. For other languages - Chinese, French, German, Japanese, Portuguese, and Spanish - we have been assisted by our colleagues, who are acknowledged in the Acknowledgements section.

Transferred taxa

The species that were included in any of the three subfamilies of this catalogue at one time, but have been transferred to another family, are listed at the end of the Catalogue of Genera and Species.

Citation codes

In this catalogue, there are four sections, namely Host Plants, Distribution, Keys and Citations, in which citation codes are used. These codes are unique combinations of letters and digits that identify a specific publication. In its simplest form, it is composed of the first six letters of the last name of the author (fewer if the author's last name contains less than six letters) and the four digits of the year of publication. For example, the reference code for Miller 1993 is Miller1993. No hyphens, diacritic marks, or spaces between multiple words names of authors are included. For example, the publication by De Lotto 1965, is coded DeLott1965. If the author published more than one paper in a year, ascending letters in lower case are placed after the date of the second, third, etc. publications. The first publication of an author in a certain year has no letter. For example, if Hall published two papers in 1925, they are coded Hall1925 and Hall1925a. For a multi-authored publication, the code is composed of up to six letters of the first author's last name, and the first two letters of the last name of the next two authors. For example, the publication by Berry, J.A., Morales, C.F., Hill, M.G., Lofroth, B.J. & Allan, D.J. 1989, would be coded BerryMoHi1989. If these authors published two papers in 1989, the second paper has a citation code BerryMoHi1989a. If there are two different authors with the same last name who published in the same year, the same process listed above is used but it is necessary to also place their first initials after the date in upper case. For example, in 1977 D.J. Williams, J.R. Williams, and M.L. Williams each published at least one paper; the codes are Willia1977DJ. Willia1977JR, and Willia1977ML. Occasionally, the process of truncating author's names to form a code produces identical codes. For example, the papers by Kosztarab & Kosztarab (1988) and Kosztarab & Kozar (1988) end up with the same codes (KosztaKo1988). This problem is solved by adding the first initials of one of the authors in capital letters after the date, e.g., KosztaKo1988F and KosztaKo1988MP. Finally, papers by societies are given special codes with all capital letters and are placed as the first reference for a letter in the references cited section. If there is still uncertainty about a particular citation code, it is easy to find the appropriate reference since the citation code is given at the end of each reference in the "References" section.

SPECIES CATALOGUE

Species Header. Each species entry is headed by the currently accepted valid name, with the author's name in parentheses if that species was first described in a genus other than the currently accepted generic placement. The next section does not have a heading, and it provides the nomenclatural history of the species. It lists all available and unavailable names (according to the rules of the international Code of Zoological Nomenclature) that have been used in the literature for the species, including junior synonyms, changes of combination, misidentifications, nomina nuda, misspellings, and others. Records are listed chronologically, with the originally cited genus-species combination given along with the author, year (letter), and page number in the publication. The following information is given for each cited-name record that has type data: type locality and host: type category (e.g., holotype, syntypes, lectotype): method of type designation (e.g., original or subsequent) and, if by subsequent designation, the name of the author of the subsequent designation and the reference, type sex/stage (e.g., female, nymph), and museum where the type is deposited. The following information may be given for any cited-name record, the author of synonymy or discoverer of taxonomic or nomenclatural "errors" such as nomina nuda, homonyms, and misidentifications; and notes relating to the specific record. A complete list of references is given in the "Citations" section.

Common Names. This section gives all of the common names, which have been proposed or used in publications, of the species, together with at least one of the sources of the common name as a citation code. Common Names, in alphabetical order, are included in all languages that use Latin letters, as well as trans-latinized common names in Russian.

Systematics. This section is present for some of the species. It provides brief notes on the taxonomy and nomenclature of the taxon.

Scale Cover. This section includes information on the external appearance of the scale cover, as described in the original description and, if available, as described by later authors, together with the reference to this information. The descriptions taken from publications in English are cited with only minor modifications of the text. Descriptions from publications in other languages were translated by our colleagues who are acknowledged in the Acknowledgements Section.

Host Plants are listed alphabetically by plant family and within the family alphabetically by genus, species, and subspecies. A question mark preceding any host record indicates that it is a questionable record. Each host record has at least one literature reference in the form of a citation code as a validation source. We have not attempted to update or verify host names but have used the names given in the literature. For host family names we have followed the system of Willis (1988). We also found valuable sources of information in several internet sites, such as:

Germplasm Resources Information Network, Agricultural Research Service, U.S. Department of Agriculture (http://www.ars-grin.gov/npgs/tax/taxgenfrom.html): Natural Resources Conservation Service, U.S. Department of Agriculture (http://plants.usda.gov/plants/fr_enter.cgi?earl=fr_qurymenu), Missouri Botanical Garden (http://mobot.inobot.org/pick/search/pick.htinl): Kew Gardens, England (http://www.rbgkew.org.uk/web.dbs/webdbsintro.html).

Natural Enemies. This section provides names of natural enemies (arthropods and fungi) that were mentioned in the scale-insect literature, followed by Citations Codes. We are not experts in these areas, but we hope that the list is as up-to-date as possible for the diversity of natural enemies, their currently accepted names, and synonyms.

Distribution. Records are given alphabetically by biogeographic realm (Afrotropical, Antarctic, Australasian, Nearctic. Neotropical, Oriental, and Palaearctic) and within each realm alphabetically by country and subpolitical unit (state or province). A question mark preceding any country or subpolitical unit indicates that the record is doubtful. Each distribution record is validated by at least one citation code between square brackets. Country names listed in this catalogue are as current as possible. The link between 'old' and current country names was ascertained in the 1958 Oxford Atlas (Lewis & Campbell, 1958). Our information sources have been the Hammond Atlas of the World (1995) and several web sites on the Internet, such as: Public National Imagery and Mapping Agency, Department of Defense http://164.214.2.59/gns/html/index/htinl; Getty Information Institute. Thesaurus of Geographic Names (http://www.ahip.getty.edu/tgn_browser/).

Biology section includes brief information on life history as reported in the literature, with appropriate references.

Economic Importance provides a summary of the damage inflicted by the species as a pest to agricultural crops. We have not included in this section information on chemical or biological control methods. However, we have surveyed numerous papers that deal with chemical and biological control, and these were listed in the Citations section. These will be easily traced by the appropriate key words.

General. In this section we provide a list, arranged chronologically, of references that present taxonomic descriptions of the species. For some species, the description and illustration may be poor and almost inadequate for proper identification of the species, but it is the only information available. This section may include any comments that did not fit in other categories of the catalogue,

The **Keys** section gives references to all published keys that include the species in question. Also given is the species name used in the key if different from the currently accepted valid name, the life stage keyed, and the geographical scope of the key.

Citations consist of an alphabetical listing of citation codes of all publications that we have surveyed that cite the species name or its various forms, followed by several 'keyword' descriptors that summarize the general contents of each paper, and the relevant page numbers within the publication for the species. The keywords are: behaviour, biological control, catalogue, chemical control, chemistry, description, distribution, ecology, economic importance, host, illustration, life history, pheromone, physiology, structure and taxonomy.

GENUS CATALOGUE

Genus catalogues are similar in structure and format to species catalogues, except that the former consists only of the following Sections: Header, Synonymy, Systematics, General, Keys and Citations. The **Synonymy** section provides information on the type species of the currently accepted valid genus and its synonyms. The **Systematics** section gives the main distinguishing characters of a genus. In some genera, with unique nomenclatural history, this is discussed. The **Keys** section lists all the publications that include keys to the species of the genus, and/or for the genus within the family.

REFERENCES

The **References** section provides complete bibliographic information for all citations in the main body of the catalogue. Journal titles are spelled out in full when the correct spelling of all components is known. In a few cases, abbreviations are given because the exact spelling of the journal title has not been ascertained. Also, it should be noted that the references listed for an author are not necessarily in order. For example, *I. Foldi* published five papers in 1990 and they were labelled, in the References database of **ScaleNet**, Foldi 1990 to Foldi 1990d. However, only three of these contain information on the Diaspididae. For the sake of consistency throughout the Coccoidea cataloguing system, the same paper codes were retained here, even though only three publications by Foldi, namely Foldi 1990, Foldi 1990c and Foldi 1990d, are listed in the References section of this Diaspididae volume.

INDICES

The catalogue concludes with an Index to Genera and an Index to Species that will help the user to trace the status and catalogue of any of the names used in the Aspidiotinae, Comstockiellinae and Odonaspidinae. Each of the Indices is arranged in two columns. The left column of each Index provides the name, followed by the name of an author, and the status of that name, i.e. accepted valid name, junior synonym, homonym, misspelling, replacement name (that is justified or unjustified), nomen nudum that is placed, nomen nudum that is unplaced, change of combination and change of rank. The right column provides the valid name for the taxon in the left column. For more information on a certain name, the user may check the list of synonyms or the valid name for details. Since the catalogue is arranged in alphabetical order, it will be easy to locate the valid name in the text of the catalogue. Due to technical reasons, it was pertinent to use, in part of indices entries, the following **abbreviation** in the status description of some taxa:

acc. - accepted	**agree.** - agreement
& - and	**cha.** - change
comb. - combination	**epit.** - epithet
gend. - gender	**jun.** - junior
misspel. - misspelling	**nom.** - nomen
nud. - nudum	**replace.** - replacement
spe. - species	**syn.** - synonym
val. -　valid	

CATALOGUE OF GENERA AND SPECIES

Subfamily Aspidiotinae

Abgrallaspis Balachowsky

Abgrallaspis Balachowsky, 1948b: 306. Type species: *Aspidiotus cyanophylli* Signoret, by original designation.
Adgrallaspis; Chou, 1985: 283. Misspelling of genus name.
Abrallaspis; Dziedzicka, 1989: 96. Misspelling of genus name.
Agbrallaspis; Dutta & Singh, 1990: 1. Misspelling of genus name.
SYSTEMATICS: The genus is closely related to *Hemiberlesia* Cockerell, from which it is distinguished (Balachowsky, 1956) mainly by the size and position of the anus, which in *Abgrallaspis* lies at a distance from the apex more than twice the diameter, whereas in *Hemiberlesia* the space is less than the anus diameter (Balachowsky, 1948b; Komosinska, 1969). Takagi (1969a, 1974, and in personal communication, July 20 2002, to Yair Ben-Dov) holds the view that *Hemiberlesia*, as understood by Balachowsky and others, and *Abgrallaspis* are close, and do not seem to be clearly distinguishable from each other. The type species of *Hemiberlesia* is characterized by the enormously large anal opening, but between the type species of the two genera, there are various species that connect these genera concerning the anal opening as well as the pygidial margin and the dorsal ducts (Takagi, 1969a).
GENERAL: Definition and characters by Balachowsky (1948b, 1956), Gómez-Menor Guerrero (1962), Davidson (1964), Almeida (1969), Komosinska (1969), Williams & Watson (1988) and by Yaşar (1995).
KEYS: Colon-Ferrer & Medina-Gaud 1998: 28-32 (female) [Genera of Puerto Rico]; Gill 1997: 24-26 (female) [Genera of California]; Gill 1997: 32 (female) [Species of California]; Kosztarab 1996: 406-407, 410-411 (female) [Northeastern North America]; Blay Goicoechea 1993: 474 (female) [Spain]; Wolff & Corseuil 1993: 29 (female) [Brazil, Rio Grande do Sul]; Zahradník 1990b: 74 (female) [Czech Republic]; Williams & Watson 1988: 20 (female) [Tropical South Pacific]; Chou 1985: 283 (female) [Genera of China]; Chou 1985: 306 (female) [Species of China]; Komosinska 1969: 50 (female) [*Abgrallaspis* group]; Danzig 1964: 646 (female) [Europe]; Davidson 1964: 639-640 (female) [species North America]; Gómez-Menor Guerrero 1962: 157 (female) [Canary Islands]; 1959: 65-67 (female); Zahradník 1959a: 548 (female) [Czech Republic]; Zahradník 1959: 66 (female) [Czech Republic]; Balachowsky 1958b: 230 (female) [*Aspidiotina* of Africa]; Ezzat 1958: 237-239 (female) [Egypt]; Balachowsky 1951: 600-601 (female) [Mediterranean].
CITATIONS: Almeid1969 [taxonomy, description: 153]; Balach1948b [taxonomy, description: 306-307]; Balach1950b [taxonomy: 535,545]; Balach1951 [taxonomy: 600-601]; Balach1953k [taxonomy: 113]; Balach1956 [taxonomy, description: 14-

15]; Balach1958b [taxonomy: 158,230]; BenDov1990h [taxonomy: 82]; BlayGo1993 [taxonomy, description: 496]; Borchs1966 [catalogue: 313]; Chou1985 [taxonomy, description: 283,305-306]; ColonFMe1998 [taxonomy, description: 32]; Danzig1964 [taxonomy: 652]; Davids1964 [taxonomy, description: 639]; DuttaSi1990 [taxonomy: 1]; Dziedz1989 [taxonomy: 96]; Ezzat1958 [taxonomy: 239]; Ghauri1962 [taxonomy, description: 211]; Gill1997 [taxonomy: 32]; GomezM1962 [taxonomy, description: 162]; Hadzib1983 [taxonomy: 232]; Kawai1980 [taxonomy: 216]; Komosi1969 [taxonomy, description: 78-80]; Koszta1996 [taxonomy, description: 406,410-411]; Lindin1957 [taxonomy: 543]; McKenz1937a [taxonomy: 176]; McKenz1938 [taxonomy: 2,5]; McKenz1950 [taxonomy: 99]; MorrisMo1966 [taxonomy, catalogue: 1]; Schmut1959 [taxonomy, description: 66-67]; Takagi1969a [taxonomy: 76-77]; Takagi1974 [taxonomy: 24]; Tao1999 [taxonomy: 67]; WilliaWa1988 [taxonomy, description: 22]; WolffCo1993 [taxonomy: 29]; Yasar1995a [taxonomy, description: 36].

Abgrallaspis colorata (Cockerell)
Aspidiotus uvae coloratus Cockerell, 1893ff: 571. Type data: USA: New Mexico, Las Cruces, altitude 3800 feet, on *Chilopsis* sp. Syntypes, female. Type depository: Washington: United States National Entomological Collection, U.S. National Museum of Natural History, District of Columbia, USA.
Aspidiotus coloratus; Cockerell, 1897a: 14. Change of status.
Aspidiotus (Diaspidiotus) coloratus; Cockerell, 1897i: 20. Change of combination.
Aspidiotus (Evaspidiotus) coloratus; Leonardi, 1898a: 76. Change of combination.
Diaspidiotus coloratus; Cockerell, 1905b: 202. Change of combination.
Hemiberlesia colorata; Ferris, 1938a: 234. Change of combination.
Aspidiotus (Hemiberlesia) coloratus; Merrill, 1953: 17. Change of combination.
Abgrallaspis colorata; Balachowsky, 1953k: 113. Change of combination.
COMMON NAME: colorata scale [Dekle1965c].
SCALE COVER: Female scale whitish, flat, circular, exuviae central; that of the male slightly elongate oval, with the exuvia subcentral (Ferris, 1938a).
HOST PLANTS: **Acanthaceae**: *Chilopsis* [Cocker1893ff], *Chilopsis linearis* [Ferris1938a, Komosi1969]. **Aceraceae**: *Acer* [Dekle1965c]. **Bignoniaceae**: *Bignonia capreolata* [Merril1953, Dekle1965c]. **Fagaceae**: *Quercus obtusiloba* [Ferris1938a, Komosi1969].
DISTRIBUTION: **Nearctic**: United States of America (Florida [Ferris1938a, Merril1953, Dekle1965c], New Mexico [Cocker1893ff], North Carolina [Ferris1938a], South Carolina [Ferris1938a], Texas [Ferris1938a, McDani1969]). **Neotropical**: Guatemala [Ferris1938a].
BIOLOGY: Occurring on underside of leaves; causes formation of a slight pit, especially in case of male (Ferris, 1938a).
GENERAL: Description and illustration of adult female by Ferris (1938a) and by Komosinska (1969).
KEYS: Komosinska 1969: 76-78 (female) [World]; McDaniel 1969: 101 (female) [U.S.A.: Texas]; Davidson 1964: 639-640 (female) [North America]; Ferris 1942: 34 (female) [North America]; Cockerell 1905b: 202 (female) [Colorado]; Newell 1899: 4-5 (female) [North America].

CITATIONS: Balach1953k [taxonomy: 113]; Borchs1966 [catalogue: 313-314]; Cocker1893ff [taxonomy, description, host, distribution: 571-572]; Cocker1896b [distribution: 334]; Cocker1897a [taxonomy, host, distribution: 14]; Cocker1897i [taxonomy, description, host, distribution: 20]; Cocker1905b [taxonomy: 202]; Comsto1883 [taxonomy, description, host, distribution: 80-81]; Davids1964 [taxonomy: 639]; Dekle1965c [taxonomy, description, host, distribution: 67]; Fernal1903b [catalogue: 254]; Ferris1938a [taxonomy, description, illustration, host, distribution: 234]; Ferris1941e [taxonomy: 42]; Ferris1942 [taxonomy: 34]; Inserr1966a [host, distribution, life history: 1-26]; Komosi1969 [taxonomy, description, illustration, host, distribution: 54-56]; Leonar1897 [taxonomy: 285]; Leonar1898a [taxonomy: 76]; Leonar1898c [taxonomy, description, illustration, host, distribution: 65-67]; Lindin1909c [taxonomy, host, distribution: 448]; Lindin1910c [taxonomy: 438]; MacGil1921 [taxonomy, description, host, distribution: 397]; Maskel1896 [taxonomy: 14]; McDani1969 [taxonomy, illustration, host, distribution: 101-102]; McKenz1944 [taxonomy: 54]; Merril1953 [taxonomy, description, host, distribution: 17]; Nakaha1982 [host, distribution: 1]; Newell1899 [taxonomy, description, host, distribution: 4,11-12].

Abgrallaspis corporifusca Chiesa Molinari
Abgrallaspis corporifuscus Chiesa Molinari, 1963: 160. Type data: ARGENTINA: San Juan, on *Laurus nobilis*. Holotype female. Type depository: Manfredi: Instituto Nacional de Tecnología Agropecuaria, Córdoba, Argentina.
SCALE COVER: Female scale, 1.5-2.0 mm long, 0.75-0.9 mm wide; oval; white; larval exuviae centrally; male scale similar, slightly oval (Chiesa Molinari, 1963).
HOST PLANTS: **Lauraceae**: *Laurus nobilis* [Chiesa1963, ClapsWoGo2001], *Phoebe porphyria* [ClapsWoGo2001]. **Sapindaceae**: *Cupania vernalis* [ClapsWoGo2001]. **Solanaceae**: *Solanum* [ClapsWoGo2001]. **Ulmaceae**: *Celtis tala* [Chiesa1963, ClapsWoGo2001].
DISTRIBUTION: **Neotropical**: Argentina (San Juan [Chiesa1963, ClapsWoGo2001], Tucuman [ClapsWoGo2001]).
GENERAL: Description and illustration of adult female by Chiesa Molinari (1963).
CITATIONS: Chiesa1963 [taxonomy, description, illustration, host, distribution: 159-162]; ClapsWoGo2001 [host, distribution: 240].

Abgrallaspis cyanophylli (Signoret)
Aspidiotus cyanophylli Signoret, 1869: 850. Nomen nudum.
Aspidiotus cyanophylli Signoret, 1869a: 119. Type data: FRANCE: Paris, Luxembourg Garden, on *Cyanophyllum magnificum*, a plant introduced from Venezuela; collected Me. Riviere, Chief Gardener, Luxembourg Garden. Syntypes, female. Type depository: Vienna: Naturhistorisches Museum Wien, Austria.
Aspidiotus cyanophyli; Comstock, 1883: 57. Misspelling of species name.
Aspidiotus (*Aspidiotus*) *cyanophylli*; Cockerell, 1897i: 29. Change of combination.
Aspidiotus (*Evaspidiotus*) *cyanophylli*; Leonardi, 1898c: 53. Change of combination.
Furcaspis cyanophylli; MacGillivray, 1921: 407. Change of combination.
Furcaspis (*Aspidiotus*) *cyanophylli*; Borchsenius, 1935a: 30. Change of combination.

Hemiberlesia cyanophylli; Ferris, 1938a: 237. Change of combination.
Abgrallaspis cyanophylli; Balachowsky, 1948b: 308. Change of combination.
Diaspidiotus cyanophylli; Bodenheimer, 1952: 337. Change of combination.
Aspidiotus (*Hemiberlesia*) *cyanophylli*; Merrill, 1953: 18. Change of combination.
Abgrallaspis cynophylli; Gerson & Zor, 1973: 514. Misspelling of species name.
Aspidiotus cyanophlli; Chou, 1985: 306. Misspelling of species name.
COMMON NAMES: Cyanophyllum scale [McKenz1956, Borchs1966].
SYSTEMATICS: This species was reported to have uniparental as well as biparental populations (Gerson & Zor, 1973).
SCALE COVER: Scale of the female flat, elongate oval, whitish or grey, exuviae central; that of the male similar in form, the exuvia near one end (Ferris, 1938a). Colour photograph by Gill (1997) and by Wong *et al.* (1999).
HOST PLANTS: **Acanthaceae**: *Brochosiphon acerifolium* [GomezM1962]. **Agavaceae**: *Cordyline* [Leonar1920], *Cordyline indivisa* [GomezM1962], *Cordyline terminalis* [WilliaWa1988], *Yucca* [BesheaTiHo1973]. **Aizoaceae**: *Mesembryanthemum* [Green1931a]. **Amaranthaceae**: *Alternanthera* [Almeid1973b]. **Anacardiaceae**: *Mangifera* [McKenz1956], *Mangifera indica* [Takaha1933, Takagi1969a, WolffCo1993a], *Schinus molle* [DeLott1967a]. **Annonaceae**: *Annona* [MerrilCh1923, Merril1953, McKenz1956, GomezM1962], *Annona muricata* [Leonar1920, GomezM1962, WilliaWa1988], *Annona reticulata* [Mamet1943a, Mamet1949, Borchs1966], *Annona squamosa* [Cohic1958, WilliaWa1988]. **Apocynaceae**: *Carissa grandiflora* [Merril1953], *Nerium* [Lepage1938], *Plumeria acutifolia* [Mamet1943a, Mamet1949, Borchs1966, WilliaWa1988], *Thevetia nereifolia* [Merril1953]. **Aquifoliaceae**: *Ilex latifolia* [BesheaTiHo1973]. **Araceae**: *Anthurium* [McKenz1956, GomezM1962], *Spathiphyllum* [Merril1953]. **Araliaceae**: *Aralia* [Leonar1920, GomezM1962]. **Asclepiadaceae**: *Stephanotis floribunda* [Merril1953]. **Barringtoniaceae**: *Barringtonia* [WilliaWa1988]. **Bombacaceae**: *Ceiba pentandra* [WilliaWa1988]. **Bromeliaceae**: *Billbergia* [Merril1953], *Billbergia thyrsoidea* [MerrilCh1923], *Nidularium* [Merril1953]. **Buxaceae**: *Buxus* [Lepage1938]. **Cactaceae** [Bodenh1952], *Cactus* [Green1931a], *Cereus* [Balach1932d, McKenz1956, GomezM1962], *Cereus giganteus* [Komosi1969], *Echinocactus* [GomezM1962], *Mammillaria* [GomezM1962], *Mammillaria seitziana* [Balach1932d], *Opuntia* [Balach1932d, GomezM1962], *Opuntia cereiformis* [Mamet1954, Borchs1966], *Opuntia rafinesquei* [Mamet1954, Borchs1966]. **Caprinaceae**: *Ostrya virginiana* [BesheaTiHo1973]. **Chenopodiaceae**: *Chenopodium album* [Mamet1943a, Mamet1949, Borchs1966]. **Combretaceae**: *Combretum* [Merril1953], *Terminalia arjuna* [Merril1953]. **Compositae**: *Berlandiera* [Merril1953]. **Convolvulaceae**: *Ipomoea* [Wilson1917, Leonar1920, MerrilCh1923, GomezM1962]. **Cornaceae**: *Cornus* [BesheaTiHo1973]. **Cycadaceae**: *Cycas* [Green1896e, Green1937, Lepage1938], *Cycas revoluta* [Wilson1917, Mamet1943a, Mamet1949, Merril1953, GomezM1962, Borchs1966]. **Dioscoreaceae**: *Dioscorea alata* [WilliaWa1988]. **Ehretiaceae**: *Ehretia laevis* [Merril1953]. **Ericaceae**: *Leucothoe* [BesheaTiHo1973], *Rhododendron* [McKenz1956], *Vaccinium* [Merril1953]. **Euphorbiaceae**: *Acalypha* [GomezM1962], *Acalypha hispida* [WilliaWa1988], *Euphorbia bajeri* [Balach1932e, GomezM1962], *Euphorbia splendens*

[Merril1953], *Hevea brasiliensis* [Green1911], *Jatropha curcas* [WilliaWa1988], *Manihot esculenta* [WilliaWa1988], *Manihot glazioui* [Balach1956, EtiennMa1993], *Phyllanthus* [Hall1922]. **Fagaceae**: *Quercus* [BesheaTiHo1973]. **Flacourtiaceae**: *Aberia caffra* [Hall1923], *Aphloia theaeformis* [Matile1978]. **Guttiferae**: *Gardenia* [Takaha1934, Takagi1969a], *Haronga madagascariensis* [Mamet1954, Borchs1966]. **Hernandiaceae**: *Hernandia bivalvis* [MerrilCh1923, Merril1953]. **Labiatae**: *Coleus* [WilliaWa1988]. **Lauraceae**: *Cinnamomum zeylanicum* [Leonar1920, WilliaWa1988], *Laurus* [Wilson1917, MerrilCh1923, Lepage1938], *Machilus kusanoi* [Takagi1969a, Tang1984], *Persea* [BesheaTiHo1973], *Persea americana* [Mamet1943a, Mamet1949, Dekle1965c, WilliaWa1988], *Persea gratissima* [DeLott1967a], *Tamala pubescens* [Merril1953]. **Leguminosae**: *Acacia rubra* [Almeid1973], *Bauhinia* [WilliaWa1988], *Cassia spectabilis* [MatileNo1984], *Ceratonia siliqua* [Hall1922], *Inga* [Ferris1942], *Wisteria* [Zimmer1948]. **Liliaceae**: *Anthericum* [Merril1953, McKenz1956], *Liriope muscari* [BesheaTiHo1973], *Mondo* [McKenz1956]. **Magnoliaceae**: *Magnolia* [Ferris1938a, McKenz1956, GomezM1962]. **Malvaceae**: *Hibiscus syriacus* [WilliaWa1988]. **Marantaceae**: *Calathea vandenheckei* [Merril1953]. **Melastomaceae**: *Cyanophyllum* [Wilson1917, MerrilCh1923], *Cyanophyllum magnificum* [Signor1869b, Green1896e, Cocker1899n, Leonar1920, DelageMaBa1972], *Miconia magnifica* [MerrilCh1923, Merril1953, McKenz1956]. **Meliaceae**: *Cedrela toona* [WilliaWa1988], *Swietenia macrophylla* [WilliaWa1988]. **Moraceae**: *Artocarpus altilis* [Beards1966, WilliaWa1988], *Ficus* [Green1896e, Kuwana1933, Lepage1938, GomezM1962, WilliaWa1988], *Ficus indica* [Wilson1917], *Ficus inga* [McKenz1956], *Ficus laurifolia* [Wilson1917], *Ficus sycomorus* [Hall1923]. **Musaceae**: *Musa* [Green1915c, GomezM1962, WilliaBu1987, WilliaWa1988], *Musa sapientum* [Cohic1958]. **Myrsinaceae**: *Maesa* [Takaha1932a, Takaha1933, Takagi1969a]. **Myrtaceae**: *Eucalyptus* [MerrilCh1923, McKenz1956, GomezM1962], *Eugenia* [Ferris1942, McKenz1956, WilliaWa1988], *Eugenia jambos* [Zimmer1948, DeLott1967a], *Psidium guajava* [Houser1918, Hall1922, Cohic1958, WilliaWa1988]. **Nyctaginaceae**: *Bougainvillea* [MerrilCh1923]. **Oleaceae**: *Jasminum* [Hall1923], *Ligustrum* [MerrilCh1923, Merril1953]. **Orchidaceae** [Balach1932d, Balach1937c]. **Palmae** [Green1896e, Green1937, BesheaTiHo1973], *Chrysalidocarpus* [Mamet1954, Borchs1966], *Chrysalidocarpus lutescens* [McKenz1956], *Cocos* [Green1916, Balach1932d, Lepage1938, McKenz1956, GomezM1962], *Cocos nucifera* [Leonar1920, Takaha1932a, Takaha1933, Cohic1958, Takagi1969a, WilliaWa1988], *Hyphaene crinita* [Merril1953], *Kentia* [GomezM1962], *Phoenix* [GomezM1962], *Pritchardia filifera* [Wilson1917, MerrilCh1923], *Sabal* [BesheaTiHo1973]. **Pandanaceae**: *Pandanus* [McKenz1956, Cohic1958, Dekle1965c]. **Piperaceae**: *Piper methysticum* [WilliaWa1988]. **Pittosporaceae**: *Pittosporum* [Ferris1938a, McKenz1956, GomezM1962]. **Polygonaceae**: *Coccoloba uvifera* [Takagi1969a, Tang1984, WilliaWa1988]. **Polypodiaceae**: *Polypodium* [Leonar1920]. **Proteaceae**: *Macadamia tetraphylla* [WilliaWa1988]. **Rosaceae**: *Eriobotrya japonica* [Mamet1959a, Borchs1966, WilliaWa1988], *Rosa* [Houser1918]. **Rubiaceae**: *Cinchona* [Green1896e, Wilson1917, MerrilCh1923, Green1937, Lepage1938], *Coffea* [Ferris1938a, McKenz1956, GomezM1962], *Coffea arabica* [Cohic1958], *Guettarda speciosa*

[WilliaWa1988], *Morinda citrifolia* [Merril1953], *Psychotria serpens* [Takaha1934, Takagi1969a], *Serissa* [Merril1953]. **Scrophulariaceae**: *Russelia* [Ferris1942, McKenz1956]. **Smilacaceae**: *Smilax* [McKenz1956]. **Solanaceae**: *Capsicum ovatum* [WilliaWa1988], *Solanum* [Ferris1942, McKenz1956]. **Sterculiaceae**: *Brachychiton acerifolium* [Leonar1920], *Theobroma cacao* [Green1904a, Lepage1938, WilliaWa1988]. **Strelitziaceae**: *Strelitzia reginae* [McKenz1956]. **Theaceae**: *Camellia japonica* [Takagi1969a], *Camellia sinensis* [Green1896e, Takagi1969a, WilliaWa1988], *Camellia thea* [Lepage1938], *Thea* [Ferris1938a, GomezM1962], *Thea japonica* [Takaha1933], *Thea sinensis* [Takaha1929, Takaha1932a, Borchs1934, Borchs1936, Merril1953]. **Ulmaceae**: *Celtis sinensis* [Takaha1932a, Takagi1969a], *Celtis sinensis* [Takaha1933]. **Verbenaceae**: *Clerodendrum* [WilliaWa1988]. **Viscaceae**: *Phoradendron* [Lepage1938]. **Zamiaceae**: *Zamia* [Merril1953]. **Zingiberaceae**: *Elettaria cardamomum* [WilliaWa1988]. **Zygophyllaceae**: *Guaiacum officinale* [Leonar1908a, Leonar1920, GomezM1962], *Porliera angustifolia* [Ferris1938a, McKenz1956].

NATURAL ENEMIES: COLEOPTERA **Coccinellidae**: *Chilocorus nigritus* (F.) [PonsonCo2000], *Telsimia nitida gemmosa* Chazeau [Chazea1984]. HYMENOPTERA **Aphelinidae**: *Aphytis chrysomphali* (Mercet) [Zimmer1948], *Ap. costalimai* (Gomes) [RosenDe1979], *Ap. maculicornis* (Masi) [AbdRab2001a], *Ap. margaretae* DeBach & Rosen [RosenDe1979], *Aspidiotiphagus citrinus* (Craw) [Zimmer1948], *Encarsia citrina* (Craw) [AbdRab2001a], *Encarsia lounsburyi* (Berlese & Paoli) [AbdRab2001a]. **Signiphoridae**: *Signiphora aspidioti* Ashmead [Zimmer1948], *Signiphora fax* Girault [Woolle1990], *Signiphora flavella* [Woolle1990], *Signiphora prepauca* Girault [Woolle1990].

DISTRIBUTION: **Afrotropical**: Angola [Almeid1973b]; Cameroon [Vayssi1913, MatileNo1984]; Comoros Islands [Matile1978]; Kenya [DeLott1967a]; Madagascar [Mamet1950, Mamet1954, Mamet1959a, Borchs1966]; Mauritius [Mamet1943a, Mamet1949, Borchs1966]; Mozambique [Almeid1973]; Réunion [Mamet1957]; Saint Helena [Matile1976]; Senegal [EtiennMa1993]; Seychelles [Mamet1943a, Borchs1966]; Uganda [Newste1910c, Newste1917b]; Zanzibar [Green1916]. **Australasian**: Bonin Islands (= Ogasawara-Gunto) [Kawai1987]; Cook Islands [WilliaWa1988]; Fiji [Green1915c, WilliaWa1988]; French Polynesia (Society Islands [WilliaWa1988], Tahiti [WilliaWa1988]); Hawaiian Islands (Hawaii [Zimmer1948]); Kiribati [WilliaWa1988]; Marshall Islands [Beards1966]; New Caledonia [Cohic1958]; Papua New Guinea [WilliaWa1988]; Tonga [WilliaWa1988]; Tuvalu [WilliaWa1988]; Vanuatu [WilliaBu1987, WilliaWa1988]; Western Samoa [Laing1927, WilliaWa1988]. **Nearctic**: United States of America (California [McKenz1956], Florida [Wilson1917, MerrilCh1923, Ferris1938a, Merril1953, Dekle1965c], Georgia [BesheaTiHo1973], Mississippi [Herric1911], Missouri [Hollin1923], Texas [Ferris1938a]). **Neotropical**: Argentina (Cordoba [Lizery1936]); Brazil [WolffCo1993a] (Bahia [Lepage1938], Distrito federal (Brasilia) [Lepage1938], Rio de Janeiro [Lepage1938, RosenDe1979], São Paulo [Hempel1900a, Lepage1938]); Colombia [Kondo2001]; Cuba [Houser1918]; Guadeloupe [Balach1957c]; Martinique [Balach1957c]; Panama [Ferris1942]; Peru [VasqueDeCo2002]; Puerto Rico & Vieques Island (Puerto Rico [ColonFMe1998]); U.S. Virgin Islands [Nakaha1983]. **Oriental**: India [Green1919c]; Indonesia (Java

[Green1904a], Sumatra [Green1937]); Philippines [Velasq1971] (Luzon [Velasq1971]); Sri Lanka [Green1896e, Green1911, Ramakr1921a, Green1937]; Taiwan [Takaha1929, Takaha1932a, Takagi1969a, WongChCh1999]. **Palaearctic**: Canary Islands [GomezM1962, MatileOr2001]; China (People's Republic) (Henan (Honan) [Shen1993]); Czech Republic [Zahrad1953, Zahrad1977, Zahrad1990b]; Egypt [Hall1922, Hall1923, Ezzat1958]; France [Balach1932d, Balach1932e, Balach1937c, Ferris1938a]; Georgia (Adzhar ASSR [Borchs1934, Borchs1936]); Germany (United) [Lindin1909b]; Greece [Korone1934]; Ireland [Green1934d]; Israel [GersonZo1973]; Italy [Leonar1908a, Leonar1920, LongoMaPe1995]; Japan [Kuwana1907, Kuwana1917a, Kuwana1933, Green1937, Kawai1980]; Poland [Komosi1969, Dziedz1989]; Turkey [Bodenh1949, Bodenh1952].

ECONOMIC IMPORTANCE: The cyanophyllum scale is widely distributed in tropical and subtropical regions. It is highly polyphagous, causing damage to various ornamentals (Davidson & Miller, 1990). Economic importance on avocado trees in Israel discussed by Gerson & Zor (1973). Lever (1945) recorded this species doing damage to *Psidium guajava* in Fiji. Greve & Ismay (1983) and Williams & Watson (1988) reported damage to tea in Western Highlands Province (W.H.P.), Papua New Guinea.

GENERAL: Description and illustration of adult female by Kuwana (1933), Ferris (1938a), Balachowsky (1948b, 1956), Zimmerman (1948), McKenzie (1956), Komosinska (1969), Takagi (1969a), Velasquez (1971), Chou (1985, 1986), Dziedzicka (1989), Zahradník (1990b), Danzig (1993), Yaşar (1995a), Kosztarab (1996), Gill (1997) and by Colon-Ferrer & Medina-Gaud (1998).

KEYS: Gill 1997: 32 (female) [Species of California]; Kosztarab 1996: 410-411 (female) [Northeastern North America]; Danzig 1993: 169 (female) [Europe]; Chou 1985: 306 (female) [Species of China]; Gerson & Zor 1973: 516 (female) [Israel]; Velasquez 1971: 110 (female) [Philippines]; Komosinska 1969: 76-78 (female) [World]; McDaniel 1969: 101 (female) [U.S.A.: Texas]; Beardsley 1966: 520 (female) [Federated States of Micronesia]; Davidson 1964: 639-640 (female) [North America]; Ezzat 1958: 240 (female) [Egypt]; Balachowsky 1956: 16 (female) [Africa]; McKenzie 1956: 26 (female) [U.S.A.: California]; Balachowsky 1948b: 308 (female) [Mediterranean]; Lupo 1948: 139 (female) [Italy]; Zimmerman 1948: 358 (female) [Hawaii]; Ferris 1942: 34 (female) [North America]; Kuwana 1933: 3 (female) [Japan]; Kuwana 1933b: 49 (female) [Japan]; Fullaway 1932: 97, 107 (female) [Hawaii]; Britton 1923: 371 (female) [U.S.A.: Connecticut]; Hollinger 1923: 7-8 (female) [U.S.A.: Missouri]; Leonardi 1920: 29-30 (female) [Italy]; Lawson 1917: 217 (female) [U.S.A.: Kansas]; Dietz & Morrison 1916a: 289-290 (female) [U.S.A.: Indiana]; Newstead 1901b: 82 (female) [England]; Green 1896e: 40 (female) [Sri Lanka]; Comstock 1883: 55-57 (female) [North America].

CITATIONS: AbdRab2001a [host, distribution, biological control: 175,176]; AbouEl2001 [host, distribution, biological control: 185-195]; Almeid1973 [host, distribution: 3]; Almeid1973b [host, distribution: 7]; Archan1937 [taxonomy, description, illustration, host, distribution: 99,105]; Balach1932e [host, distribution: 236]; Balach1937c [host, distribution: 2]; Balach1948b [taxonomy, description, illustration, host, distribution: 322-325]; Balach1953k [taxonomy: 113]; Balach1956 [taxonomy, description, illustration, host, distribution: 16-18]; Balach1957c [host, distribution: 199]; Beards1966 [host, distribution: 520]; BeardsDaHo1976 [economic importance: 103]; BesheaTiHo1973 [host, distribution: 4]; BiezanSe1940 [host, distribution: 67-68]; Bodenh1949 [taxonomy, description, illustration, host, distribution: 50-52]; Bodenh1952 [host, distribution, structure: 337]; Bondar1930 [host, distribution, economic importance: 343-348]; Bondar1930a [host, distribution, economic importance: 17-19]; Borchs1934 [host, distribution: 28]; Borchs1935a [taxonomy: 30]; Borchs1936 [host, distribution: 131]; Borchs1937 [taxonomy, description, illustration, host, distribution: 125-126]; Borchs1937a [taxonomy, host, distribution: 46,47]; Borchs1939 [taxonomy: 9, 22]; Borchs1950b [taxonomy, description, illustration, host, distribution: 219,224]; Borchs1966 [catalogue: 314-315]; Brick1912 [host, distribution: 1-22]; Britto1923 [taxonomy, description, host, distribution: 371,373]; BurgerUl1990 [economic importance: 313-327]; CanaleVa1999 [biological control]; Castel1951a [biological control: 95-98]; Chazea1984 [host, distribution, biological control: 9-10]; Chiesa1948 [host, distribution, economic importance]; Chou1985 [taxonomy, description, host, distribution: 306-307]; Chou1986 [taxonomy, illustration: 687]; ChuaWo1990 [host, distribution, economic importance: 543-552]; ClapsWoGo2001a [taxonomy, host, distribution: 10]; Cocker1896b [distribution: 334]; Cocker1897i [taxonomy, description, host, distribution: 20,27,29]; Cocker1897q [taxonomy: 703]; Cocker1899d [host, distribution: 168]; Cocker1899n [host, distribution: 21-22]; Cohic1958 [host, distribution: 12]; ColonFMe1998 [taxonomy, description, illustration, host, distribution: 32-34]; Comsto1883 [taxonomy, description, illustration, host, distribution: 59-61]; Costan1938 [host, distribution: 25-44]; Danzig1964 [taxonomy, host, distribution: 652]; Danzig1972 [taxonomy, host, distribution, economic importance: 206]; Danzig1993 [taxonomy, description, illustration, host, distribution, economic importance: 170-171]; Danzig1995 [taxonomy, life history, structure: 19-24]; DanzigPe1998 [catalogue: 271]; Davids1964 [taxonomy: 639]; DavidsMi1990 [host, distribution, economic importance: 603-632]; DeBachRo1976a [host, distribution, biological control: 541-545]; DEDAC1923 [host, distribution]; Dekle1965c [taxonomy, description, host, distribution: 68]; Dekle1976 [taxonomy, description, host, distribution, economic importance: 22]; DeLott1967a [host, distribution: 114]; DeSant1979 [biological control]; DicksoFl1955 [host, distribution: 614-615]; DietzMo1916a [taxonomy, description, illustration, host, distribution: 294-295]; Dozier1926 [host, distribution, biological control: 267-277]; Dzhash1971 [host, distribution: 325-326]; Dziedz1989 [taxonomy, description, illustration, host, distribution: 95-96]; Ehrhor1913 [host, distribution: 101]; EtiennMa1993 [host, distribution: 257]; Ezzat1958 [distribution: 240]; EzzatNa1987 [distribution: 86]; FDACSB1983 [host, distribution: 6-8]; FDACSB1988 [host, distribution: 5-6]; Fernal1903b [catalogue: 255]; Ferris1938a [taxonomy, description, illustration, host, distribution: 237]; Ferris1941e [taxonomy: 42]; Ferris1942 [host, distribution: 445:10;446:33]; Foldi2001 [distribution: 303-308]; FonsecAu1932a [host, distribution: 202-214]; Fullaw1932 [taxonomy: 97,107]; Gaprin1975 [host, distribution, economic importance, biological control: 29-33]; Garcia1930 [host, distribution, biological control]; Gavalo1936 [host, distribution: 80]; Gentry1965 [host, distribution, economic importance]; Gerson1990 [taxonomy: 130,132]; GersonZo1973 [taxonomy, life history, host, distribution, economic importance: 513-533]; Ghauri1962 [taxonomy, description, host, distribution: 98,211]; Gill1997 [host, distribution, taxonomy, description, illustration, economic importance: 33,35]; GomezM1962 [taxonomy, description, host, distribution: 163-165]; Gowdey1913 [host, distribution: 247-249]; Gowdey1917 [host, distribution: 189]; GrandpCh1899 [taxonomy, description, host, distribution: 24-25]; Green1896e [taxonomy, description, illustration, host, distribution: 51-52]; Green1904a [host, distribution: 208]; Green1907 [host, distribution: 203]; Green1911 [host, distribution: 28]; Green1915c [host, distribution: 44]; Green1916 [host, distribution: 376]; Green1919c [host, distribution: 439]; Green1931a [host, distribution: 105]; Green1934d [distribution: 114]; Green1937 [host, distribution: 329]; GreenMa1907 [taxonomy, distribution: 343]; Hadzib1983 [taxonomy, description, illustration, host, distribution, life history, biological control, economic importance: 232-234]; Halber1996 [host, distribution: 4-10]; Hall1922 [taxonomy, description, host, distribution: 24-25]; Hall1923 [host, distribution: 42]; Hargre1937 [host, distribution, economic importance: 505-520]; HeBaLu1998 [life history, ecology: 1-3]; HeCuBa1991 [life history: 58-60]; Hempel1900a [taxonomy, description, host, distribution: 498]; Herric1911 [taxonomy, description, illustration, host, distribution: 9,15,48]; Hinckl1963 [host, distribution, biological control]; Hollin1923 [taxonomy, description, host, distribution: 10-11]; Houser1918 [host, distribution: 165]; Hunt1939 [host, distribution: 548-566]; Kawai1980 [taxonomy, description, host, distribution: 220]; Kawai1987 [host, distribution: 78]; Kiritc1932a [taxonomy: 245]; Komosi1961 [taxonomy, description, illustration, host, distribution: 224-226]; Komosi1969 [taxonomy, description, illustration, host, distribution: 56-58]; Kondo2001 [taxonomy, host, distribution: 43]; Korone1934 [taxonomy, description, illustration, host, distribution: 2-3]; Koszta1996 [taxonomy, description, illustration, host, distribution: 412-413]; Kuwana1907 [host, distribution: 195]; Kuwana1917a [taxonomy, distribution: 174]; Kuwana1933 [taxonomy, description, illustration, host, distribution: 3,7-8]; Laing1927 [host, distribution: 39]; Lawson1917 [taxonomy, description, illustration, host, distribution: 222-223]; Leonar1897 [taxonomy: 285]; Leonar1898a [taxonomy: 75]; Leonar1898c [taxonomy, description, illustration, host, distribution: 53-55]; Leonar1908a [taxonomy, host, distribution: 188]; Leonar1920 [taxonomy, description, illustration, host, distribution: 44-47]; Lepage1938 [catalogue: 393-394]; Lepesm1947 [taxonomy, description, host, distribution, life history: 208-210]; LiLi1990 [host, biological control: 68-70]; Lindin1909b [host, distribution: 149]; Lindin1909c [taxonomy, host, distribution: 449]; Lindin1910b [host, distribution: 34]; Lindin1912b [taxonomy, description, host, distribution: 119,240]; Lindin1935 [taxonomy: 128,147]; Lindin1957 [taxonomy: 543]; Lizery1936 [host, distribution: 113]; LongoMaPe1995 [distribution: 125]; Lupo1948 [taxonomy, description, illustration, host, distribution: 139,169-173]; MacGil1921 [taxonomy, description, host, distribution: 397,407]; Maleno1916a [taxonomy, description, illustration, host, distribution: 326]; Mamet1943a [catalogue: 160]; Mamet1949 [catalogue: 59,60]; Mamet1950 [host, distribution: 22]; Mamet1954 [host, distribution: 17]; Mamet1957 [distribution: 369]; Mamet1959a [host, distribution: 387]; Mansfi1920 [host, distribution: 145-155]; Maskew1916 [host, distribution: 308-309]; Matile1976 [host, distribution: 310-311]; Matile1978 [host, distribution: 60]; MatileNo1984 [host, distribution: 64]; MatileOr2001 [host, distribution: 189]; McDani1969 [taxonomy, illustration, host, distribution: 103]; McKenz1943 [taxonomy: 153]; McKenz1946 [taxonomy: 30]; McKenz1956 [taxonomy, description, illustration, host, distribution: 65-68]; Mead1983 [host, distribution: 1-5]; Merril1953 [taxonomy, description, host, distribution: 17-20]; MerrilCh1923 [taxonomy, description, host, distribution, economic importance: 199]; MetcalMe1993 [economic importance, host, distribution, control]; Miller1983 [host, distribution: 4-6]; MillerDa1990 [host, distribution, economic importance: 300]; Moreir1921b [host, distribution: 182]; MorrisMo1966 [taxonomy: 1]; MoutiaMa1947 [distribution]; Muraka1970 [host, distribution: 69]; NagarkSa1990 [host, distribution, economic importance, biological control: 553-562]; Nakaha1982 [host, distribution: 1]; Nakaha1983 [host, distribution: 8]; Newste1901b [taxonomy, description, illustration, host, distribution: 82,124-126]; Newste1910c [host, distribution: 198-199]; Newste1914 [host, distribution: 307]; Newste1917b [host, distribution: 131]; Peleka1962 [host, distribution: 62]; PonsonCo2000 [life history, biological control: 295-310]; PriesnHo1940 [biological control: 58-70]; Ramakr1919a [taxonomy, description, host, distribution: 18]; Ramakr1921a [host, distribution: 356]; Ramakr1930 [taxonomy, host, distribution: 22-23]; RosenDe1979 [host, distribution, biological control: 244-247,545-548]; Saakya1954 [host, distribution, economic importance]; Sander1904a [taxonomy, description, illustration, host, distribution: 55,59]; Schmut1957a [host, distribution: 134-136]; Schmut1959 [taxonomy, description, host, distribution: 59]; Schmut1969 [host, distribution: 116]; SchmutKlLu1957 [host, distribution, economic importance: 477]; SchmutKlLu1959 [taxonomy: 374]; Shen1993 [host, distribution: 57]; Shiao1979 [taxonomy, host, distribution, life history, ecology: 267-276]; Signor1869 [taxonomy: 850]; Signor1869b [taxonomy, description, illustration, host, distribution: 119]; Silva1944 [host, distribution: 8-14]; Soares1942 [host, distribution, economic importance: 54-56]; Soares1945 [host, distribution, economic importance: 49-81]; Strong1922 [host, distribution: 775-780]; Sugimo1994 [host, distribution: 115-121]; SugimoKaTa1996 [host, distribution: 99-101]; Swirsk1976b [host, distribution, economic importance: 555-559]; SwirskWyIz2002 [taxonomy, host, distribution, life history, economic importance, biological control: 101-102]; Takagi1969a [taxonomy, description, illustration, host, distribution: 78-79,101]; Takaha1929 [host, distribution: 79]; Takaha1932a [host, distribution: 103]; Takaha1933 [host, distribution: 27]; Tang1984 [taxonomy, description, illustration, host, distribution: 52-53]; Tao1999 [taxonomy, host, distribution: 67]; VasqueDeCo2002 [host, distribution: 331]; Vayssi1913 [host, distribution: 430]; Velasq1971 [taxonomy, description, illustration, host, distribution: 114-117]; WilliaBu1987 [host, distribution: 94]; WilliaGr1990 [host, distribution, economic importance, biological control: 563-578]; WilliaWa1988 [taxonomy, illustration, host, distribution, economic importance: 8,22-24]; Wilson1917 [taxonomy, description, host, distribution: 20-21]; WolffCo1993a [host, distribution: 153]; WongChCh1999 [taxonomy, description, host, distribution: 18,57]; Woolle1990 [biological control: 167-176]; Wysoki1997 [host, distribution, economic importance: 805-811]; Yasar1995a [taxonomy, description, illustration, host, distribution: 37-39]; Zahrad1953 [taxonomy, description, illustration, host, distribution: 131,134-136]; Zahrad1977 [taxonomy, distribution: 119]; Zahrad1990b [taxonomy, description, illustration, host, distribution: 76-78]; Zimmer1948 [taxonomy, description, illustration, host, distribution, biological control: 358,360].

Abgrallaspis flabellata (Ferris)
Hemiberlesia flabellata Ferris, 1938a: 239. Type data: MEXICO: near Acapulco, Hacienda La Providencia, on *Pinus* sp. Holotype female. Type depository: Davis: The Bohart Museum of Entomology, University of California, California, USA.
Abgrallaspis flabellata; Davidson, 1964: 639. Change of combination.
SCALE COVER: Scale of female circular or oval, rather thick and but slightly convex, white; exuviae central; that of male elongate oval, white; exuvia central (Ferris, 1938a).
HOST PLANTS: **Pinaceae**: *Pinus* [Ferris1938a].
DISTRIBUTION: **Nearctic**: Mexico [Ferris1938a].
BIOLOGY: Occurring on the outside of the needles beneath the basal sheath (Ferris, 1938a).
GENERAL: Description and illustration of adult female by Ferris (1938a) and by Komosinska (1969).
KEYS: Komosinska 1969: 76-78 (female) [World]; Davidson 1964: 639-640 (female) [North America]; Balachowsky 1956: 105-108 (female) [Africa]; Balachowsky 1953k: 114-115 (female) [World]; Ferris 1942: 35 (female) [North America].
CITATIONS: Balach1953k [taxonomy: 115]; Balach1956 [taxonomy: 107]; Borchs1966 [catalogue: 315]; Davids1964 [taxonomy: 639]; Ferris1938a [taxonomy, description, illustration, host, distribution: 239]; Ferris1942 [taxonomy: 446:35]; Komosi1969 [taxonomy, description, illustration, host, distribution: 62-63].

Abgrallaspis flavida De Lotto
Abgrallaspis flavida De Lotto, 1957: 225. Type data: KENYA: Nairobi, on small branches and leaves of *Elaeodendron stuhlmanni*. Holotype female. Type depository: London: The Natural History Museum, England, UK.
SCALE COVER: Scale of female circular, slightly convex; exuviae central; colour evenly yellow; diameter up to 1.4 mm. Scale of male larger, elongate, with exuvia subcentral; length up to 1.7 mm (De Lotto, 1957).
HOST PLANTS: **Celastraceae**: *Elaeodendron* [DeLott1957], *Elaeodendron stuhlmanni* [DeLott1957]. **Flacourtiaceae**: *Aberia caffra* [DeLott1957].
DISTRIBUTION: **Afrotropical**: Kenya [DeLott1957].
GENERAL: Description and illustration of adult female by De Lotto (1957) and by Komosinska (1969).
KEYS: Komosinska 1969: 76-78 (female) [World].
CITATIONS: Borchs1966 [catalogue: 315]; DeLott1957 [taxonomy, description, illustration, host, distribution: 225-226]; Komosi1969 [taxonomy, description, illustration, host, distribution: 63-64].

Abgrallaspis fraxini (McKenzie)

Hemiberlesia fraxini McKenzie, 1944: 53. Type data: U.S.A.: California, Riverside County, property at Ondio (near La Quinta), on *Fraxinus velutina*. Holotype female. Type depository: Davis: The Bohart Museum of Entomology, University of California, California, USA.

Abgrallaspis fraxini; Balachowsky, 1953k: 113. Change of combination.

Aspidiotus fraxini; Lindinger, 1957: 543. Change of combination.

Aspidiotus fraxinorum Lindinger, 1957: 543. Unjustified emendation; discovered by Borchsenius, 1966: 315.

COMMON NAMES: Ash scale [McKenz1956].

SCALE COVER: Female circular, about 1.5 mm diameter, 0.8 mm wide (McKenzie, 1944).

HOPST PLANTS: Oleaceae: *Fraxinus velutina* [McKenz1944, McKenz1956].

DISTRIBUTION: Nearctic: Mexico [Nakaha1982]; United States of America (Arizona [Nakaha1982], California [McKenz1944, McKenz1956]).

ECONOMIC IMPORTANCE: Recorded from leaves of ash, *Fraxinus velutina*, Arizona, where it caused noticeable injury by removing chlorophyll from leaf tissue (McKenzie, 1944).

GENERAL: Description and illustration of adult female by McKenzie (1944, 1956), Komosinska (1969) and by Gill (1997).

KEYS: Gill 1997: 32 (female) [Species of California]; Komosinska 1969: 76-78 (female) [World]; Davidson 1964: 639-640 (female) [North America]; McKenzie 1956: 26 (female) [U.S.A.: California].

CITATIONS: Balach1953k [taxonomy: 113]; Borchs1966 [catalogue: 315]; Davids1964 [taxonomy: 639]; Gill1997 [host, distribution, taxonomy, description, illustration, economic importance: 34,37]; Komosi1969 [taxonomy, description, illustration, host, distribution: 64-66]; Lindin1957 [taxonomy: 543]; McKenz1944 [taxonomy, description, illustration, host, distribution: 53-54,58]; McKenz1956 [taxonomy, description, illustration, host, distribution: 69-71]; MillerDa1990 [economic importance: 300]; Nakaha1982 [host, distribution: 2]; SchmutKlLu1957 [host, distribution, economic importance: 476].

Abgrallaspis furcillae (Brain)

Aspidiotus furcillae Brain, 1918: 119. Type data: SOUTH AFRICA: Transvaal, Pretoria, on *Acacia horrida*; collected by C.P. Lounsbury, September 20, 1914. Lectotype female, by subsequent designation Munting, 1970a: 38. Type depository: Pretoria: South African National Collection of Insects, South Africa; type no. 209/1.

Octaspidiotus furcillae; MacGillivray, 1921: 396. Change of combination.

Abgrallaspis furcillae; Balachowsky, 1956: 16. Change of combination.

SCALE COVER: Female scale small, about 1mm diameter, circular, roundly arched, sordid white in fresh material, but when older appearing dark brown to blackish brown with rich red brown exuviae. True colour of scale rarely seen, as nearly all specimens partly or wholly covered with outer layers of bark of host plant. Exuviae central or nearly so (Brain, 1918).

HOST PLANTS: Leguminosae: *Acacia horrida* [Brain1918], *Berlinia grobiflora* [Hall1929, Balach1958b].

DISTRIBUTION: **Afrotropical**: South Africa [Brain1918]; Zimbabwe [Hall1929, Balach1958b].

GENERAL: Description and illustration of adult female by Brain (1918) and by Balachowsky (1958b).

KEYS: Balachowsky 1956: 16 (female) [Africa]; Brain 1918: 118 (female) [South Africa].

CITATIONS: Balach1948b [taxonomy: 272]; Balach1956 [taxonomy, description, illustration, host, distribution: 16-18]; Balach1958b [taxonomy, description, illustration, host, distribution: 223,225]; Borchs1966 [catalogue: 315]; Brain1918 [taxonomy, description, illustration, host, distribution: 119-120]; Ferris1941e [taxonomy: 43]; Hall1929 [taxonomy, description, host, distribution: 347]; MacGil1921 [taxonomy, description, host, distribution: 396]; Muntin1970a [taxonomy: 38].

Abgrallaspis ithacae (Ferris)

Aspidiotus abietis Comstock, 1883: 57. Type data: USA: New York, Ithaca, on lower surface of leaves of hemlock, *Abies canadensis*. Syntypes, female. Type depository: Washington: United States National Entomological Collection, U.S. National Museum of Natural History, District of Columbia, USA. Homonym of *Aspidiotus abietis* Schrank, 1776.

Aspidaspis ithacae Ferris, 1938a: 185. Replacement name for *Aspidiotus abietis* Comstock, 1883.

Abgrallaspis ithacae; Davidson, 1964: 638. Change of combination.

Gonaspidiotus ithacae; Borchsenius, 1966: 301. Change of combination.

Abgrallaspis ithicae; Beardsley et al., 1976: 105. Misspelling of species name.

Abgrallaspis ithaeae; Nakahara, 1982: 2. Misspelling of species name.

COMMON NAME: hemlock scale [Comsto1883, Stimme2000].

SCALE COVER: Female scale resembles that of *Aspidiotus pini* except that usually more nearly circular; colour dark grey, often approaching black, with margin lighter, and sometimes with a bluish, brownish, or purple tinge; 1.3 - 2 mm long, width about 9/10 of length (Comstock, 1883). Scale of female circular, flat, blackish centrally and paler about margins; exuviae central; that of male elongate oval, blackish; exuvia near one end (Ferris, 1938a).

HOST PLANTS: **Pinaceae**: *Abies canadensis* [Comsto1883, Ferris1938a], *Abies grandis* [Ferris1942], *Picea pungens* [Ferris1942], *Pseudotsuga taxifolia* [Ferris1942], *Tsuga canadensis* [BesheaTiHo1973, StoetzDa1974], *Tsuga heterophylla* [Ferris1942].

DISTRIBUTION: **Nearctic**: United States of America (Connecticut [Koszta1996], Georgia [BesheaTiHo1973], Idaho [Ferris1942], Indiana [Koszta1996], Maryland [StoetzDa1974a, Koszta1996], Montana [Ferris1942], New York [Ferris1938a, Koszta1996], Ohio [Koszta1996], Oregon [Ferris1942], Pennsylvania [Koszta1996, Stimme2000], Tennessee [BesheaTiHo1973], Virginia [Koszta1996], Washington [Ferris1942], West Virginia [Koszta1996]).

BIOLOGY: Occurring on the needles (Ferris, 1938a).

ECONOMIC IMPORTANCE: Regarded a pest of hemlock (Miller & Davidson, 1990).

GENERAL: Description and illustration of adult female by Ferris (1938a) and by Kosztarab (1996). Description and illustration of female, male nymphs, male pupa and prepupa by Stoetzel & Davidson (1974a).
KEYS: Kosztarab 1996: 410-411 (female) [Northeastern North America]; Davidson 1964: 639-640 (female) [North America]; Ferris 1942: 30 (female) [North America].
CITATIONS: Balach1948b [taxonomy: 339]; BeardsDaHo1976 [economic importance: 105]; BesheaTiHo1973 [host, distribution: 4]; Borchs1966 [catalogue: 301]; Comsto1883 [taxonomy, description, illustration, host, distribution: 57-58]; Davids1964 [taxonomy: 639]; Ferris1938a [taxonomy, description, illustration, host, distribution: 185,251]; Ferris1942 [host, distribution: 9]; Ferris1942 [taxonomy: 30]; Koszta1996 [taxonomy. description, illustration, host, distribution: 410-411,415-416]; Lindin1935 [taxonomy: 128]; Marlat1908c [taxonomy: 14]; MillerDa1990 [host, distribution, economic importance: 300]; Nakaha1982 [taxonomy, host, distribution: 2]; Reh1903 [taxonomy, description, host, distribution: 465]; Schran1776 [taxonomy, description, host, distribution: 48]; Signor1882b [taxonomy, description, host, distribution: clxxxiv]; Stimme2000 [taxonomy, host, distribution, life history, economic importance, control: 15-17]; StoetzDa1974 [taxonomy, life history: 138-140]; StoetzDa1974a [taxonomy, description, illustration, host, distribution, life history: 484-485].

Abgrallaspis latastei (Cockerell)
Aspidiotus latastei Cockerell, 1894n: 35. Type data: CHILE: Banos de Cauquenos, on under and upper side of pale green smooth leaves, about 60 mm long. Syntypes, female. Type depository: Washington: United States National Entomological Collection, U.S. National Museum of Natural History, District of Columbia, USA.
Aspidiotus (*Evaspidiotus*) *latastei*; Leonardi, 1898c: 55. Change of combination.
Aspidiella latastei; MacGillivray, 1921: 405. Change of combination.
Hemiberlesia latastei; Lizer y Trelles, 1942: 79. Change of combination.
Abgrallaspis latastei; Komosinska, 1969: 67. Change of combination.
SCALE COVER: Female scale circular or ovate, diameter 1.5 mm; sometimes elongate ovate; dusky white and semi-transparent; pellicles yellow; secretionary covering thin, transparent. Male scale elongate, oval, dusky white, semi-transparent with yellow pellicle.; length 1 mm; breadth 1-1.5 mm (Hall, 1923).
HOST PLANTS: **Aextoxicaceae**: *Aextoxicon punctatum* [ClapsWoGo2001]. **Celastraceae**: *Euonymus japonica* [ClapsWoGo2001]. **Ephedraceae**: *Ephedra andina* [ClapsWoGo2001]. **Lauraceae**: *Laurus* [ClapsWoGo2001], *Persea americana* [ClapsWoGo2001]. **Moraceae**: *Ficus benjamini* [ClapsWoGo2001]. **Myrtaceae**: *Callistemon pinnifolius* [ClapsWoGo2001]. **Oleaceae**: *Olea europaea* [ClapsWoGo2001]. **Proteaceae**: *Lomatia hirsuta* [ClapsWoGo2001]. **Rhamnaceae**: *Colletia cruciata* [ClapsWoGo2001]. **Rosaceae**: *Prunus domestica* [ClapsWoGo2001]. **Winteraceae**: *Drymis winterii* [ClapsWoGo2001].
DISTRIBUTION: **Neotropical**: Argentina (Buenos Aires [ClapsWoGo2001]); Chile [Cocker1894n, GonzalCh1968, ClapsWoGo2001].
GENERAL: Description of adult female by Cockerell (1894n), Hall (1923) and by Komosinska (1969).
KEYS: Komosinska 1969: 76-78 (female) [World].

CITATIONS: Borchs1966 [catalogue: 308]; Chiesa1938 [host, distribution, taxonomy, description, control: 5-13]; Chiesa1938a [host, distribution: 1-21]; Chiesa1948 [host, distribution, economic importance]; Chiesa1948a [host, distribution, economic importance]; ClapsWoGo2001 [host, distribution: 245]; Cocker1894n [taxonomy, description, host, distribution: 35-36]; Cocker1896b [distribution: 334]; Cocker1897i [taxonomy, description, host, distribution: 24]; Fernal1903b [catalogue: 267]; Ferris1941e [taxonomy: 45]; GonzalCh1968 [distribution: 110]; Komosi1969 [taxonomy, description, illustration, host, distribution: 67-69]; Leonar1898a [taxonomy: 75]; Leonar1898c [taxonomy, description, illustration, host, distribution: 55-56]; Lindin1935 [taxonomy: 129]; Lindin1957 [taxonomy: 545]; Lizery1942 [host, distribution: 79]; MacGil1921 [taxonomy, description, host, distribution: 405]; Morris1923 [taxonomy, host, distribution: 125-126]; Willia1985a [taxonomy: 235].

Abgrallaspis liriodendri Miller & Howard
Abgrallaspis liriodendri Miller & Howard, 1981: 164. Type data: U.S.A.: Louisiana, West of Bogalusa, on *Liriodendron tulipifera*. Holotype female. Type depository: Washington: United States National Entomological Collection, U.S. National Museum of Natural History, District of Columbia, USA.
SCALE COVER: Female scale about 1 mm in diameter, circular or subcircular, thin, flat, yellow; exuviae central; female occurring in a pocket on underside of leaf and forming a gall-like deformity on upper surface (Miller & Howard, 1981).
HOST PLANTS: **Magnoliaceae**: *Liriodendron tulipifera* [MillerHo1981].
DISTRIBUTION: **Nearctic**: United States of America (Louisiana [MillerHo1981]).
GENERAL: Description and illustration of adult female by Miller & Howard (1981).
CITATIONS: DuttaSi1990 [taxonomy: 1]; Larew1990 [ecology, life history, structure: 293-300]; MillerHo1981 [taxonomy, description, illustration, host, distribution: 164-166].

Abgrallaspis mammillaris (Lindinger)
Aspidiotus mammillaris Lindinger, 1910b: 37. Type data: ETHIOPIA: near Harrar, on *Aloe eru*; collected 27.04.1909. Syntypes, female. Type depository: Hamburg, Zoologisches Institut und Zoologishces Museum, Universität von Hamburg, Germany.
Hemiberlesia mammillaris; MacGillivray, 1921: 437. Change of combination.
Gonaspidiotus mammillaris; Borchsenius, 1966: 302. Change of combination.
Abgrallaspis mammillaris; Komosinska, 1969: 69. Change of combination.
SCALE COVER: Female scale white grey, circular, 1-2 mm in diameter; exuviae brownish, central (Lindinger, 1910b). Scale of female grey, circular, somewhat convex; exuviae brown placed centrally (Balachowsky, 1956; Komosinska, 1969).
HOST PLANTS: **Liliaceae**: *Aloe eru* [Balach1956], *Aloe percrassa* [Komosi1969].
DISTRIBUTION: **Afrotropical**: Eritrea [DeLottNa1955]; Ethiopia [Balach1956].
GENERAL: Description and illustration of adult female by Lindinger (1910b), Balachowsky (1956) and by Komosinska (1969).

KEYS: Komosinska 1969: 76-78 (female) [World]; Balachowsky 1956: 107 (female) [Africa].
CITATIONS: Balach1956 [taxonomy, description, illustration, host, distribution: 112-113]; Borchs1966 [catalogue: 302]; DeLottNa1955 [host, distribution: 53-60]; Ferris1941e [taxonomy: 45]; Komosi1969 [taxonomy, description, illustration, host, distribution: 69-70]; Lindin1910b [taxonomy, description, host, distribution: 37]; MacGil1921 [taxonomy, description, host, distribution: 437]; Sassce1912 [taxonomy, host, distribution: 93]; WeidneWa1968 [taxonomy: 173].

Abgrallaspis mendax (McKenzie)

Hemiberlesia mendax McKenzie, 1943: 152. Type data: GUATEMALA: on orchid. Holotype female. Type depository: Davis: The Bohart Museum of Entomology, University of California, California, USA.
Abgrallaspis mendax; Davidson, 1964: 639. Change of combination.
SCALE COVER: Scale of female averages 1 mm in diameter, reddish-brown, with black subcentral exuvium. Scale of male unknown (McKenzie, 1943).
HOST PLANTS: **Orchidaceae** [McKenz1943], *Orchis* [Komosi1969].
DISTRIBUTION: **Neotropical**: Guatemala [McKenz1943, Komosi1969].
BIOLOGY: Occurring on leaves (McKenzie, 1943).
GENERAL: Description and illustration of adult female by McKenzie (1943) and by Komosinska (1969).
KEYS: Komosinska 1969: 76-78 (female) [World]; Davidson 1964: 639-640 (female) [North America].
CITATIONS: Borchs1966 [catalogue: 316]; Davids1964 [taxonomy: 639]; Komosi1969 [taxonomy, description, illustration, host, distribution: 70-71]; McKenz1943 [taxonomy, description, illustration, host, distribution: 152-153,160].

Abgrallaspis mitchelli (Marlatt)

Aspidiotus (*Hemiberlesia*) *mitchelli* Marlatt, 1908c: 22. Type data: SOUTH AFRICA: Cape Province, Mitchell's Pass, on undetermined host plant; collected by C.P. Lounsbury, 29 January 1897. Syntypes, female. Type depository: Washington: U.S. National Entomological Collection, U.S. National Museum of Natural History, District of Columbia, USA; type no. 7695.
Aspidiotus (*Hemiberlesia*) *mitchelli*; Sanders, 1909a: 53. Change of combination.
Hemiberlesia mitchelli; MacGillivray, 1921: 435. Change of combination.
Abgrallaspis mitchelli; Borchsenius, 1966: 316. Change of combination.
SCALE COVER: Scale of female, 1.5 mm long; subcircular to broad oval, strongly convex, and of the general *Camelliae* type; secretionary matter rather dense; colour dull yellowish, due chiefly to the extraneous matter taken up from the surface; exuviae yellowish brown, near the anterior end; ventral scale a distinct white flocculent patch, thinnest at center (Marlatt, 1908c).
HOST PLANTS: **Aizoaceae**: *Mesembryanthemum edule* [Balach1956]. **Celastraceae**: *Cassine maritimum* [Muntin1965b].
DISTRIBUTION: **Afrotropical**: South Africa [Balach1956, Muntin1965b].
GENERAL: Description and illustration of adult female by Marlatt (1908c), Brain (1918) and by Balachowsky (1958b).

KEYS: Balachowsky 1956: 105-108 (female) [Africa].
CITATIONS: Balach1956 [taxonomy, host, distribution: 107,114]; Balach1958b [taxonomy, description, illustration, host, distribution: 226,229]; Borchs1966 [catalogue: 316]; Brain1918 [taxonomy, description, host, distribution: 129-130]; Ferris1941e [taxonomy: 46]; Laing1929a [taxonomy, description, illustration, host, distribution: 490]; MacGil1921 [taxonomy, description, host, distribution: 435]; Mamet1941 [taxonomy: 28]; Marlat1908c [taxonomy, description, illustration, host, distribution: 22-23]; Muntin1965b [host, distribution: 193]; Sander1909a [taxonomy, host, distribution: 53].

Abgrallaspis narainus Dutta & Singh

Abgrallaspis narainus Dutta & Singh, 1990: 1. Type data: INDIA: Uttar Pradesh, District Firozabad, Shikohabad, from *Musa paradisiaca* and *Carica papaya*; collected 2.ix.1986. Holotype female. Type depository: London: The Natural History Museum, England, UK.
Agbrallaspis narainus; Dutta & Singh, 1990: 1. Misspelling of genus name.
SYSTEMATICS: Dutta & Singh (1990) stated that holotype was selected but did not indicate its host plant.
SCALE COVER: Scale of female roughly circular, 1.4 mm long, 1.2 mm wide, flat, pinkish or light brown with first and second larval subcentral exuviae (Dutta & Singh, 1990).
HOST PLANTS: **Caricaceae**: *Carica papaya* [DuttaSi1990]. **Musaceae**: *Musa paradisiaca* [DuttaSi1990].
DISTRIBUTION: **Oriental**: India (Uttar Pradesh [DuttaSi1990]).
GENERAL: Description and illustration of adult female by Dutta & Singh (1990).
CITATIONS: DuttaSi1990 [taxonomy, description, illustration, host, distribution: 1-6].

Abgrallaspis oxycoccus (Woglum)

Aspidiotus oxycoccus Woglum, 1906: 73. Type data: U.S.A.: New Jersey, locality not indicated, on cranberry *Vaccinium oxycoccus*; collected by J.B. Smith, 1891. Syntypes, female. Type depository: Washington: United States National Entomological Collection, U.S. National Museum of Natural History, District of Columbia, USA.
Aspidiella oxycoccus; MacGillivray, 1921: 404. Change of combination.
Aspidiotus oxycocci; Lindinger, 1932f: 196. Emendation that is unjustified.
Aspidaspis oxycoccus; Ferris, 1938a: 186. Change of combination.
Abgrallaspis oxycoccus; Davidson, 1964: 638. Change of combination.
Gonaspidiotus oxycoccus; Borchsenius, 1966: 302. Change of combination.
SCALE COVER: Scale of female almost flat, and very variable in shape; usually circular, or nearly so, but may be elongate with sides parallel; scale on upper surface of leaves black, those on under surface dirty grey to dark brown, usually a lighter colour; exuviae central to sub-central (Woglum, 1906). Scale of female flat, circular, grey; exuviae subcentral; that of male elongate oval, grey; exuvia toward one end (Ferris, 1938a).
HOST PLANTS: **Ericaceae**: *Vaccinium oxycoccus* [Ferris1938a].

NATURAL ENEMIES; HYMENOPTERA **Aphelinidae**: *Aphytis diaspidis* (Howard) [Gordh1979].
DISTRIBUTION: **Nearctic**: United States of America (Massachusetts [Koszta1996], New Jersey [Sander1906, Woglum1906, Ferris1938a, Koszta1996], Oregon [Ferris1938a], Washington [Ferris1938a]).
BIOLOGY: Occurring on leaves and fruit (Ferris, 1938a).
GENERAL: Description and illustration of adult female by Ferris (1938a) and by Kosztarab (1996).
KEYS: Kosztarab 1996: 410-411 (female) [Northeastern North America]; Davidson 1964: 639-640 (female) [North America]; Ferris 1942: 29 (female) [North America].
CITATIONS: Borchs1966 [catalogue: 302]; Davids1964 [taxonomy: 639]; Ferris1938a [taxonomy, description, illustration, host, distribution: 186]; Ferris1941e [taxonomy: 46]; Ferris1942 [taxonomy: 446:29]; Gordh1979 [biological control: 894]; Koszta1996 [taxonomy, description, illustration, host, distribution: 410-411,417-418]; Lindin1932f [taxonomy: 196]; MacGil1921 [taxonomy, description, host, distribution: 404]; Nakaha1982 [host, distribution: 2]; Sander1906 [taxonomy, host, distribution: 14]; Woglum1906 [taxonomy, description, illustration, host, distribution: 73-74].

Abgrallaspis paucitatis (McKenzie)

Aonidiella paucitatis McKenzie, 1942b: 143. Type data: PANAMA: Chiriqui Province, Armuelles on *Carludovica palmata*. Holotype female. Type depository: Davis: Bohart Museum of Entomology, University of California, California, USA.
Hemiberlesia paucitatis; McKenzie, 1946: 29. Change of combination.
Abgrallaspis paucitatis; Borchsenius, 1966: 316. Change of combination.
SCALE COVER: Only slide-mounted females available for description by McKenzie (1942b).
HOST PLANTS: **Cyclanthaceae**: *Carludovia palmata* [McKenz1942b].
DISTRIBUTION: **Neotropical**: Panama [McKenz1942b].
BIOLOGY: Known only from leaves (McKenzie, 1942b).
GENERAL: Description and illustration of adult female by McKenzie (1942b).
KEYS: McKenzie 1942b: 144-145 (female) [World].
CITATIONS: Borchs1966 [catalogue: 316]; McKenz1942b [taxonomy, description, illustration, host, distribution: 143-147]; McKenz1946 [taxonomy: 29-30].

Abgrallaspis perseae Davidson

Abgrallaspis perseus Davidson, 1964: 641. Type data: U.S.A.: Texas, Brownsville, intercepted on avocado from Mexico. Holotype. Type depository: Washington: United States National Entomological Collection, U.S. National Museum of Natural History, District of Columbia, USA.
SCALE COVER: Female circular, light brown, exuviae subcentral (Davidson, 1964).
HOST PLANTS: **Lauraceae**: *Persea americana* [Davids1964].
DISTRIBUTION: **Nearctic**: United States of America (Texas [Davids1964]).
GENERAL: Description and illustration of adult female by Davidson (1964) and by Komosinska (1969).

KEYS: Komosinska 1969: 76-78 (female) [World]; Davidson 1964: 639-640 (female) [North America]
CITATIONS: Davids1964 [taxonomy, description, illustration, host, distribution: 641-642]; Komosi1969 [taxonomy, description, illustration, host, distribution: 72-73].

Abgrallaspis ruebsaameni (Cockerell)
Cryptophyllaspis rübsaameni Cockerell, 1902a: 26. Type data: PAPUA NEW GUINEA: Bismarck Archipelago, on leaves of *Codiaeum* sp.; received from Mr. E.H. Rubsaamen. Syntypes, female. Type depository: Washington: United States National Entomological Collection, U.S. National Museum of Natural History, District of Columbia, USA.
Cryptophyllaspis ruebsaameni; Borchsenius, 1966: 273. Justified emendation.
Abgrallaspis ruebsaameni; Williams & Watson, 1988: 17. Change of combination.
SYSTEMATICS: Williams & Watson (1988) indicated that a few original specimens were available for study, but were in poor condition. The species seems to be a member of *Abgrallaspis* or a genus close to it (Williams & Watson, 1988).
SCALE COVER: Cockerell (1902a) noted that this species was found in small galls, cylindrical, about 2 mm long; thickly clustered on leaves of *Codiaeum*.
HOST PLANTS: **Euphorbiaceae**: *Codiaeum* [WilliaWa1988].
BIOLOGY: Causing small sub cylindrical galls (Cockerell, 1902a).
CITATIONS: Borchs1966 [catalogue: 273]; Cocker1902a [taxonomy, description, host, distribution: 26]; Cocker1902c [taxonomy, description, host, distribution: 75]; Fernal1903b [catalogue: 282]; Larew1990 [ecology, life history, structure: 293-300]; MacGil1921 [taxonomy, description, host, distribution: 428]; Rubsaa1907 [taxonomy, description, host, distribution: 9]; Willia1985a [taxonomy: 238]; WilliaWa1988 [taxonomy, host, distribution: 17].

Acanthaspidiotus Borchsenius & Williams

Acanthaspidiotus Borchsenius & Williams, 1963: 381. Type species: *Aspidiotus (Evaspidiotus) pustulans* Green, by monotypy and original designation.
SYSTEMATICS: Genus close to *Aspidiotus* Bouché and *Metaspidiotus* Takagi (1957) but differs from both in possessing slender ducts, large spine-like marginal setae and poorly developed second and third lobes (Borchsenius & Williams, 1963). In possessing a small anal opening situated towards the apex of the pygidium; the genus *Acanthaspidiotus* resembles *Monaonidiella* MacGillivray (Borchsenius & Williams, 1963).
GENERAL: Definition and characters by Borchsenius & Williams (1963).
CITATIONS: Borchs1966 [catalogue: 276]; BorchsWi1963 [taxonomy, description: 381]; Kawai1980 [taxonomy: 229]; MorrisMo1966 [taxonomy, catalogue: 1].

Acanthaspidiotus pustulans (Green)

Aspidiotus (*Evaspidiotus*) *pustulans* Green, 1905: 31. Type data: INDONESIA: Java, on *Erythrina lithosperma*. Syntypes, female. Type depository: London: The Natural History Museum, England, UK.

Aspidiotus pustulans; MacGillivray, 1921: 398. Change of combination.

Acanthaspidiotus pustulans; Borchsenius & Williams, 1963: 381. Change of combination.

SCALE COVER: Female scale irregularly circular, moderately convex, brownish-fulvous. Pellicles concolorous, inconspicuous. Surface dull and roughened. Diameter, 1 to 1.50 mm. Male scale not observed (Green, 1905).

HOST PLANTS: **Leguminosae**: *Erythrina lithosperma* [Green1905, Sander1906].

DISTRIBUTION: **Oriental**: Indonesia (Java [Green1905, Sander1906]).

GENERAL: Description and illustration of adult female by Borchsenius & Williams (1963).

CITATIONS: Borchs1966 [catalogue: 276]; BorchsWi1963 [taxonomy, description, illustration: 380-381]; Ferris1941e [taxonomy: 47]; Green1905 [taxonomy, description, illustration, host, distribution: 31]; MacGil1921 [taxonomy, description, host, distribution: 398]; Sander1906 [taxonomy, host, distribution: 14].

Achorophora Brimblecombe

Achorophora Brimblecombe, 1957: 273. Type species: *Achorophora obliqua* Brimblecombe, by original designation.

SYSTEMATICS: Brimblecombe (1957) stated that *Achorophora* resembles several genera such as *Pseudaonidia* because female has a constricted thorax, a chitinized body, and perispiracular pores near anterior spiracles, but differs from all other genera in *Pseudaonidia* group in having oblique duct orifices.

GENERAL: Definition and characters by Brimblecombe (1957).

CITATIONS: Borchs1966 [catalogue: 240]; Brimbl1957 [taxonomy, description: 273-275]; MorrisMo1966 [taxonomy, catalogue: 2].

Achorophora divergens Brimblecombe

Achorophora divergens Brimblecombe, 1957: 275. Type data: AUSTRALIA: Queensland, Marmor, on *Casuarina glauca*; collected October 1955. Holotype female. Type depository: Brisbane: Queensland Museum, Queensland, Australia; type no. T5634.

SCALE COVER: Female scale oval, length 1.8 mm, width 1.3 mm; sometimes partly curved around branchlet; dark reddish brown or brownish black; exuviae dark orange, placed near anterior end of scale (Brimblecombe, 1957).

HOST PLANTS: **Casuarinaceae**: *Casuarina glauca* [Brimbl1957].

DISTRIBUTION: **Australasian**: Australia (Queensland [Brimbl1957]).

GENERAL: Description and illustration of adult female by Brimblecombe (1957).

CITATIONS: Borchs1966 [catalogue: 240]; Brimbl1957 [taxonomy, description, illustration, host, distribution: 275-277].

Achorophora obliqua Brimblecombe
Achorophora obliqua Brimblecombe, 1957: 274. Type data: AUSTRALIA: Queensland, Drillham, on *Casuarina luehmannii*; collected by J. Mann, April 1953. Holotype female. Type depository: Brisbane: Queensland Museum, Queensland, Australia; type no. T5630.
SCALE COVER: Insects on branches, single or several adjacent to each other, with anterior end close to or partly under whorl of leaflets; scale oval with exuviae dark orange (Brimblecombe, 1957).
HOST PLANTS: **Casuarinaceae**: *Casuarina luehmannii* [Brimbl1957].
DISTRIBUTION: **Australasian**: Australia (Queensland [Brimbl1957]).
GENERAL: Description and illustration of adult female by Brimblecombe (1957).
CITATIONS: Borchs1966 [catalogue: 240]; Brimbl1957 [taxonomy, description, illustration, host, distribution: 274-275].

Acontonidia **Brimblecombe**

Acontonidia Brimblecombe, 1957: 285. Type species: *Acontonidia triangularis* Brimblecombe, by monotypy and original designation.
SYSTEMATICS: Genus *Acontonidia* Brimblecombe (1957) has some resemblance to *Aspidonymus* Brimblecombe (1957), differing in shape of lobes, plates, paraphyses and of abdominal margin (Brimblecombe, 1957).
GENERAL: Definition and characters by Brimblecombe (1957).
CITATIONS: Borchs1966 [catalogue: 240]; Brimbl1957 [taxonomy, description: 285-287]; MorrisMo1966 [taxonomy, catalogue: 2].

Acontonidia triangularis Brimblecombe
Acontonidia triangularis Brimblecombe, 1957: 286. Type data: AUSTRALIA: Queensland, Beenleigh, on *Dissilaria baloghioides*; collected May 1956. Holotype female. Type depository: Brisbane: Queensland Museum, Queensland, Australia; type no. T5654.
SCALE COVER: Insects single and sparse, embedded under cork tissue on twigs (Brimblecombe, 1957).
HOST PLANTS: **Euphorbiaceae**: *Dissilaria baloghioides* [Brimbl1957].
DISTRIBUTION: Australasian: Australia (Queensland [Brimbl1957]).
GENERAL: Description and illustration of adult female by Brimblecombe (1957).
CITATIONS: Borchs1966 [catalogue: 240]; Brimbl1957 [taxonomy, description, illustration, host, distribution: 286-287].

Acutaspis **Ferris**

Acutaspis Ferris, 1941d: 328. Type species: *Aspidiotus perseae* Comstock, by original designation.
Acutaspi; Balachowsky, 1959: 356. Misspelling of genus name.
SYSTEMATICS: Genus closely related to *Melanaspis* Cockerell, from which it differs in having a narrow pygidium, and large anal opening, being twice or more length of median lobes (Ferris, 1941d; McKenzie, 1937).

GENERAL: Definition and characters by Ferris (1941d), McKenzie (1947) and by Balachowsky (1951).

KEYS: Colon-Ferrer & Medina-Gaud 1998: 28-32 (female) [Genera of Puerto Rico]; Gill 1997: 24-26 (female) [Genera of California]; Wolff & Corseuil 1993: 29 (female) [Brazil, Rio Grande do Sul]; Tereznikova 1986: 83 (female) [Ukraine]; McDaniel 1968: 202 (female) [species U.S.A.: Texas]; Danzig 1964: 646 (female) [Europe]; McKenzie 1956: 22 (female) [U.S.A.: California]; Borchsenius 1950b: 167 (female) [USSR]; Ferris 1942: 446:27 (female) [North America]; Ferris 1942: 446:28 (female) [species North America].

CITATIONS: Balach1951 [taxonomy, description: 593-594]; Balach1958b [taxonomy: 164]; Balach1959 [taxonomy: 356,360]; Borchs1950b [taxonomy, description: 167,221-222]; Borchs1966 [catalogue: 354]; ColonFMe1998 [taxonomy, description: 34]; Danzig1964 [taxonomy: 651]; Danzig1993 [taxonomy, description: 240]; DanzigPe1998 [catalogue: 174]; Ferris1941d [taxonomy, description: 328]; Ferris1942 [taxonomy: 446:27]; Koszta1996 [taxonomy, description: 419]; Kozar1990f [distribution: 143]; McKenz1947 [taxonomy, description: 32-33]; McKenz1950 [taxonomy: 99]; McKenz1956 [taxonomy: 22]; MorrisMo1966 [taxonomy, catalogue: 3]; Schmut1959 [taxonomy, description: 354]; WolffCo1993 [taxonomy: 29].

Acutaspis acuta (Mamet)

Melanaspis acuta Mamet, 1951: 248. Type data: MADAGASCAR: Foulpointe, on upper surface of leaves of *Anacardium occidentale*. Holotype. Type depository: Paris: Muséum national d'Histoire naturelle, France.

Acutaspis acuta; Borchsenius, 1966: 354. Change of combination.

SCALE COVER: Female scale dark brown, almost black; surface roughened by concentric laminations, obscured by brownish to greyish, thick, shiny secretion; brittle, subcircular and produced to one side, flat; exuviae jet black, to one side, sometimes obscured with white to reddish-brown secretion; ventral scale brown to dark-brown, thick; attached to host plant. Male scale not observed (Mamet, 1951).

DISTRIBUTION: Afrotropical: Madagascar [Mamet1951, Borchs1966].

GENERAL: Description and illustration of adult female by Mamet (1951).

CITATIONS: Borchs1966 [catalogue: 354]; Lindin1957 [taxonomy: 550]; Mamet1951 [taxonomy, description, illustration, host, distribution: 229,248].

Acutaspis agavis (Townsend & Cockerell)

Aspidiotus agavis Townsend & Cockerell, 1898: 178. Type data: MEXICO: Toluca, on leaves of *Agave* sp. Syntypes, female. Type depository: Washington: United States National Entomological Collection, U.S. National Museum of Natural History, District of Columbia, USA.

Chrysomphalus agavis; Cockerell, 1899a: 396. Change of combination.

Acutaspis agavis; Ferris, 1941d: 329. Change of combination.

Melanaspis agaves Lindinger, 1943a: 147. Unjustified emendation.

Melanaspis agaves; Lindinger, 1943a: 147. Change of combination.

Melanaspis agaves Lindinger, 1943a: 147. Unjustified emendation.

Acutaspis agavis; Borchsenius, 1966: 354. Revived combination.

SCALE COVER: Scale of female quite convex, circular with one side somewhat produced; exuviae black beneath a film of wax and surrounded by a band that is yellowish brown; margins of scale whitish. Scale of male oval, yellowish brown (Ferris, 1941d).

HOST PLANTS: **Agavaceae**: *Agave* [Ferris1941d], *Agave lecheguilla* [Ferris1941d, McDani1968].

NATURAL ENEMIES: HYMENOPTERA **Aphelinidae**: *Aphytis diaspidis* (Howard) [RosenDe1979], *Aphytis mytilaspidis* (Le Baron) [MyartsRu2000].

DISTRIBUTION: **Nearctic**: Mexico [Cocker1899n, MyartsRu2000] (Colima [Ferris1941d], Mexico State [Ferris1941d], Puebla [Ferris1941d]); United States of America (Arizona [Nakaha1982], Florida [Wilson1917], Texas [Ferris1941d, McDani1968]). **Neotropical**: Costa Rica [Nakaha1982]; Trinidad and Tobago (Trinidad [Nakaha1982]); Venezuela [Nakaha1982].

BIOLOGY: Occurring exposed on leaves (Ferris, 1941d).

GENERAL: Description and illustration of adult female by Ferris (1941d).

KEYS: McDaniel 1968: 202 (female) [U.S.A.: Texas]; Ferris 1942: 29 (female) [North America].

CITATIONS: Borchs1966 [catalogue: 354]; Castel1951a [biological control: 95-98]; Cocker1899a [taxonomy: 396]; Cocker1899n [host, distribution: 26]; Cocker1905 [taxonomy, host, distribution: 46]; DeSant1979 [biological control]; Fernal1903b [catalogue: 285]; Ferris1937c [taxonomy, illustration: 51,79]; Ferris1941d [taxonomy, description, illustration, host, distribution: 329]; Ferris1941e [taxonomy: 40]; Ferris1942 [taxonomy: 29]; Gordh1979 [biological control: 894]; Leonar1900 [taxonomy, host, distribution: 342]; Lindin1943a [taxonomy: 147]; Lindin1957 [taxonomy: 544]; MacGil1921 [taxonomy, description, host, distribution: 421]; McDani1968 [taxonomy, illustration, host, distribution: 202-204]; McKenz1939 [taxonomy: 53]; McKenz1947 [taxonomy: 34]; MyartsRu2000 [distribution, biological control: 7-33]; Nakaha1982 [host, distribution: 3]; RosenDe1979 [host, distribution, biological control: 405-409]; TownseCo1898 [taxonomy, description, host, distribution: 178-179]; Willia1985a [taxonomy: 231]; Wilson1917 [taxonomy, description, host, distribution: 15].

Acutaspis albopicta (Cockerell)

Aspidiotus (*Chrysomphalus*) *albopictus* Cockerell, 1898j: 433. Type data: MEXICO: Cuernavaca, on leaves of orange; collected December 8, 1897. Syntypes, female. Type depository: Washington: United States National Entomological Collection, U.S. National Museum of Natural History, District of Columbia, USA.

Aspidiotus albopictus leonis Townsend & Cockerell, 1898: 179. Type data: MEXICO: Nuevo Leon, Linares, on orange leaves. Syntypes, female. Type depository: Washington, D.C.: U.S. National Entomological Collection, U.S. National Museum of Natural History, USA. Synonymy by Ferris, 1941e: 45.

Aspidiotus koebelei Townsend & Cockerell, 1898: 179. Type data: MEXICO: Oaxaca State, Oaxaca, on leaves of orange. Syntypes, female. Type depository: Washington, D.C.: U.S. National Entomological Collection, U.S. National Museum of Natural History, District of Columbia, USA. Synonymy by Ferris, 1941e: 44.

Chrysomphalus albopictus; Cockerell, 1899a: 396. Change of combination.

Chrysomphalus albopictus leonis; Cockerell, 1899a: 396. Change of combination.
Chrysomphalus koebelei; Cockerell, 1899a: 396. Change of combination.
Aspidiotus albopictus; Cockerell, 1905: 46. Change of combination.
Aspidiotus leonis; McKenzie, 1939: 53. Change of combination and rank.
Acutaspis albopicta; Ferris, 1941d: 329. Change of combination requiring emendation of species name for agreement in gender.

COMMON NAME: albopicta scale [McKenz1956].

SCALE COVER: Scale of female of type common to genus, circular or slightly oval, dark brown; that of male similar (Ferris, 1941d).

HOST PLANTS: **Apocynaceae**: *Tabernaemontana* [McKenz1956]. **Araceae**: *Aglaonema* [Gill1997], *Philodendron* [McKenz1956, Gill1997]. **Bromeliaceae**: *Tillandsia* [Gill1997]. **Ebenaceae**: *Brayodendron texanum* [McDani1968]. **Lauraceae**: *Persea americana* [McKenz1956]. **Leguminosae**: *Inga* [Ferris1941d, McKenz1956]. **Menispermaceae**: *Hyperbaena denticulata* [Ferris1941d, McKenz1956]. **Musaceae**: *Musa paradisiaca sapientum* [McKenz1956]. **Oleaceae**: *Ligustrum vulgare* [McDani1968]. **Palmae**: *Cocos nucifera* [Ferris1941d, McKenz1956]. **Rubiaceae**: *Gardenia jasminoides* [McKenz1956]. **Rutaceae**: *Citrus* [Ferris1941d, McKenz1956]. **Tiliaceae**: *Jacquinia* [Ferris1941d, McKenz1956].

NATURAL ENEMIES: HYMENOPTERA **Aphelinidae**: *Aphytis acutaspidis* Rosen & DeBach [RosenDe1979]

DISTRIBUTION: **Nearctic**: Mexico [Cocker1899n] (Colima [Ferris1941d], Guerrero [Ferris1941d], Morelos [Ferris1941d], Nuevo Leon [TownseCo1898, Ferris1941d], Oaxaca [TownseCo1898, Ferris1941d], Sinola [Ferris1941d]); United States of America (California [McKenz1956], Texas [McDani1968]). **Neotropical**: Brazil (Rio de Janeiro [RosenDe1979]); Costa Rica [Nakaha1982]; Ecuador [YustCe1956]; Guatemala [Nakaha1982]; Honduras [Nakaha1982]; Panama [Ferris1941d]; Peru [VasqueDeCo2002].

BIOLOGY: Occurring on leaves (Ferris, 1941d).

GENERAL: Description and illustration of adult female by Ferris (1941d), McKenzie (1956) and by Gill (1997).

KEYS: McKenzie 1956: 23 (female) [U.S.A.: California]; Ferris 1942: 29 (female) [North America]; Cockerell 1905: 45-46 (female) [Mexico].

CITATIONS: Borchs1966 [catalogue: 354]; Cocker1898j [taxonomy, description, host, distribution: 433-434]; Cocker1899a [taxonomy: 396]; Cocker1899d [host, distribution: 170]; Cocker1899n [host, distribution: 26]; Cocker1905 [taxonomy: 46]; CoronaRuMo1997 [host, distribution: 38-41]; DicksoFl1955 [host, distribution: 614-615]; Fernal1903b [catalogue: 285-286,291]; Ferris1941d [taxonomy, description, illustration, host, distribution: 330]; Ferris1941e [taxonomy: 40,44,45]; Ferris1942 [taxonomy: 29]; Gill1997 [host, distribution, taxonomy, illustration: 40,41]; Leonar1900 [taxonomy, host, distribution: 342]; Lindin1957 [taxonomy: 544]; Lindin1957 [taxonomy: 545]; MacGil1921 [taxonomy, description, host, distribution: 415,418]; McDani1968 [taxonomy, illustration, host, distribution: 204-205]; McKenz1939 [taxonomy: 53-54]; McKenz1956 [taxonomy, description, illustration, host, distribution: 36-37]; Nakaha1982 [host, distribution: 3]; RosenDe1979 [host, distribution, biological control: 248-249]; TownseCo1898 [taxonomy, description, host, distribution: 179]; VasqueDeCo2002 [host, distribution: 331]; Willia1985a [taxonomy: 232,235]; YustCe1956 [host, distribution: 425-442].

Acutaspis aliena (Newstead)

Aspidiotus alienus Newstead, 1901a: 81. Type data: UNITED KINGDOM: England, London, on *Cattleya skinneri*. Syntypes, female. Type depository: London: The Natural History Museum, England, UK.

Chrysomphalus alienus; Fernald, 1903b: 286. Change of combination.

Pseudischnaspis alienus; Houser, 1918: 169. Change of combination.

Insaspidiotus alienus; McKenzie, 1939: 53. Change of combination.

Melanaspis aliena; Ferris, 1941d: 348. Change of combination requiring emendation of species name for agreement in gender.

Acutaspis aliena; Deitz & Davidson, 1986: 11. Change of combination.

COMMON NAME: Alien Scale [MerrilCh1923].

SCALE COVER: Scale of female circular, rather flat, brown; exuviae central, with a quite distinct, black ventral scale. Scale of male somewhat elongate, brown (Ferris, 1941d).

HOST PLANTS: **Agavaceae**: *Yucca gloriosa* [Houser1918, MerrilCh1923]. **Burseraceae**: *Bursera* [Ferris1941d]. **Guttiferae**: *Mammea* [MerrilCh1923]. **Leguminosae**: *Cassia obtusifolia* [Houser1918, MerrilCh1923]. **Orchidaceae** [Ferris1941d], *Cattleya skinneri* [Newste1901a, MerrilCh1923, Ferris1941d]. **Polygonaceae**: *Muehlenbeckia platyclada* [Houser1918, MerrilCh1923]. **Proteaceae**: *Grevillea robusta* [Houser1918]. **Rosaceae**: *Rosa* [Houser1918]. **Salicaceae**: *Salix babylonica* [Houser1918, MerrilCh1923]. **Solanaceae**: *Datura arborea* [MerrilCh1923]. **Verbenaceae**: *Clerodendron* [Houser1918, MerrilCh1923].

DISTRIBUTION: **Australasian**: Hawaiian Islands (Hawaii [Nakaha1982]). **Nearctic**: Mexico (Oaxaca [Ferris1941d], Veracruz [Ferris1941d]); United States of America (Florida [MerrilCh1923, Ferris1941d, Dekle1965c]). **Neotropical**: Brazil [Nakaha1982]; Cuba [Ferris1941d]; Ecuador [Nakaha1982]; Guatemala [Ferris1941d]; Jamaica [Nakaha1982]; Puerto Rico & Vieques Island [Nakaha1982]. **Palaearctic**: United Kingdom (England [Newste1901a, Ferris1941d]).

GENERAL: Description and illustration of the female by Newstead (1901a), Ferris (1941d) and by Balachowsky (1951).

KEYS: Danzig 1993: 238 (female) [world]; Balachowsky 1951: 579 (female) [Mediterranean]; Ferris 1943: 64 (female) [North America]; Ferris 1942: 36 (female) [North America].

CITATIONS: Balach1951 [taxonomy, description, illustration, host, distribution: 586-589]; Borchs1966 [catalogue: 346]; ClapsWoGo2001a [taxonomy, host, distribution: 10]; Danzig1993 [taxonomy: 238]; DanzigPe1998 [catalogue: 174-175]; DeitzDa1986 [taxonomy: 11,76]; Dekle1965c [taxonomy, description, host, distribution: 86]; Dekle1976 [taxonomy, description, host, distribution, economic importance: 107]; Fernal1903b [catalogue: 286]; Ferris1941d [taxonomy, description, illustration, host, distribution: 348]; Ferris1941e [taxonomy: 40]; Ferris1942 [taxonomy: 446:36]; Ferris1943 [taxonomy: 64]; Houser1918 [host, distribution: 169]; Lindin1935 [taxonomy: 149]; Lindin1957 [taxonomy: 550];

MacGil1921 [taxonomy, description, host, distribution: 418]; McKenz1939 [taxonomy: 53]; McKenz1947 [taxonomy: 33]; MerrilCh1923 [taxonomy, description, host, distribution, economic importance: 250-251]; MillerDaSt1984 [taxonomy: 95]; Nakaha1982 [host, distribution: 53]; Newste1901a [taxonomy, description, illustration, host, distribution: 81].

Acutaspis arbelaezi Balachowsky

Acutaspis arbelaezi Balachowsky, 1959: 358. Type data: COLOMBIA: Valle, Cauca Valley, 15 km north east of Cali, on *Pithecellobium dulce*. Syntypes, female. Type depository: Paris: Muséum national d'Histoire naturelle, France.
SCALE COVER: Female subcircular, very flat, colour brown-black; larval exuviae subcentral, black; diameter 2-2.2 mm. Male scale unknown (Balachowsky, 1959).
HOST PLANTS: **Leguminosae**: *Pithecellobium dulce* [Balach1959].
DISTRIBUTION: **Neotropical**: Colombia [Balach1959].
GENERAL: Description and illustration of adult female by Balachowsky (1959).
CITATIONS: Balach1959 [taxonomy, description, illustration, host, distribution: 358-359]; Borchs1966 [catalogue: 354].

Acutaspis decorosa Ferris

Acutaspis decorosa Ferris, 1941d: 331. Type data: MEXICO: off the west coast of Mexico, Maria Madre Island, on *Tillandsia fasciculata*. Holotype female. Type depository: Davis: The Bohart Museum of Entomology, University of California, California, USA.
SCALE COVER: Scale of female more or less buried under epidermis of leaf, of a pale brown colour. Scale of male not recognized (Ferris, 1941d).
HOST PLANTS: **Bromeliaceae**: *Tillandsia fasciculata* [Ferris1941d].
DISTRIBUTION: **Nearctic**: Mexico [Ferris1941d] (Colima [Ferris1941d]).
BIOLOGY: Occurring on leaves (Ferris, 1941d).
GENERAL: Description and illustration of the female by Ferris (1941d).
KEYS: Ferris 1942: 28 (female) [North America].
CITATIONS: Borchs1966 [catalogue: 354]; Ferris1942 [taxonomy: 28].

Acutaspis erythraspidis (Newstead)

Aspidiotus (*Chrysomphalus*) *erythraspidis* Newstead, 1917: 372. Type data: GUYANA: Turkeyn, on *Erythraspis [=Erythrina] glauca*. Syntypes, female. Type depository: London: The Natural History Museum, England, UK.
Chrysomphalus erythraspidis; MacGillivray, 1921: 420. Change of combination.
Pseudischnaspis erythraspidis; Lindinger, 1937: 194. Change of combination.
Aspidiotus erythraspidis; McKenzie, 1939: 54. Change of combination.
Acutaspis erythraspidis; Borchsenius, 1966: 354. Change of combination.
SCALE COVER: Female scale more or less circular or irregularly ovate, moderately convex, surface more or less roughened by fibers of plant upon which fixed; pale brownish buff or greyish buff, sometimes with slightly darker lines of growth; underside blackish; margin similar in colour to exterior; exuviae central, subcentral or submarginal, black; secretionary covering greyish, but usually absent from larval exuviae; maximum diameter 1.7-2.1 mm. Male scale dark brown or

brownish black, margin paler; larval exuviae black, with a narrow, but sharply defined, concentric ring of white secretion (Newstead, 1917).
HOST PLANTS: Leguminosae: *Erythrina glauca* [Newste1917].
DISTRIBUTION: **Neotropical**: Guyana [Newste1917].
GENERAL: Description and illustration of female by Newstead (1917). This species described from material off *Erythraspis glauca*, but there is no plant genus *Erythraspis*, Newstead must have mistaken the plant for *Erythrina glauca* (Leguminosae), often used as a shade tree in cacao plantations.
CITATIONS: Borchs1966 [catalogue: 354]; Ferris1941e [taxonomy: 43]; Lindin1937 [taxonomy: 194]; MacGil1921 [taxonomy, description, host, distribution: 354]; McKenz1939 [taxonomy: 54]; Newste1917 [taxonomy, description, illustration, host, distribution: 372-373].

Acutaspis litorana Lepage
Acutaspis litorana Lepage, 1942: 180. Type data: BRAZIL: São Paolo State, Alcatrazes Island, on undetermined plant. Syntypes, female. Type depository: Curitiba: Departamento de Zoologia, Setor de Ciencias Biologicas, Universidade Federal do Parana, Brazil.
Pseudischnaspis litorana; Lindinger, 1957: 544. Change of combination.
Acutaspis litorana; Borchsenius, 1966: 355. Revived combination.
SCALE COVER: Female circular, diameter 2 mm; "marron escura" [=dark reddish brown]; exuviae subcentral. Male scale short, elongate, exuviae terminal (Lepage, 1942).
DISTRIBUTION: **Neotropical**: Brazil (São Paulo [Lepage1942, ClapsWoGo2001]).
GENERAL: Description and illustration of adult female by Lepage (1942).
CITATIONS: Borchs1966 [catalogue: 355]; Claps1993 [taxonomy: 9]; ClapsWoGo2001 [host, distribution: 240]; Lepage1942 [taxonomy, description, illustration, host, distribution: 179-181]; Lindin1957 [taxonomy: 544].

Acutaspis morrisonorum Kosztarab
Acutaspis morrisonorum Kosztarab, 1963: 18. Type data: U.S.A.: Ohio, Wooster, Secrest Arboretum, on *Tsuga caroliniana*. Holotype female. Type depository: Washington: United States National Entomological Collection, U.S. National Museum of Natural History, District of Columbia, USA.
COMMON NAME: round conifer scale [Dekle1965c, Koszta1996].
SCALE COVER: Scale of adult female slightly oval, flat, yellowish-brown, lighter on margin, with exuviae central. First nymphal exuviae with white margin. Length 1.5 mm, width 1.125 mm (Kosztarab, 1963).
HOST PLANTS: **Cupressaceae**: *Chamaecuparis thyoides* [Dekle1965c], *Cupressus doclouxiana* [Koszta1963], *Juniperus* [Koszta1963, Dekle1965c], *Juniperus bermudiana* [Koszta1963], *Juniperus virginiana* [Koszta1963, TippinBe1970, BesheaTiHo1973], *Juniperus virginiana glauca* [Koszta1963], *Thuja* [Dekle1965c]. **Pinaceae**: *Abies concolor* [Koszta1963], *Picea pungens moerheim* [Koszta1963], *Pinus* [Koszta1963], *Pinus elliottii* [Dekle1965c], *Pinus taeda* [Koszta1963], *Tsuga* [Koszta1963], *Tsuga canadensis* [Koszta1963,

BesheaTiHo1973], *Tsuga caroliniana* [Koszta1963]. **Taxaceae**: *Torreya taxifolia* [Koszta1963, Dekle1965c].

DISTRIBUTION: **Nearctic**: Canada (Quebec [Koszta1963, Koszta1996]); United States of America (Alabama [Koszta1963, USDAAP1978], Arkansas [Koszta1963], Florida [Koszta1963, Dekle1965c], Georgia [TippinBe1970, BesheaTiHo1973], Louisiana [Koszta1963, Koszta1996], Massachusetts [Koszta1963, Koszta1996], Michigan [Koszta1996], North Carolina [Nakaha1982], Ohio [Koszta1996], Pennsylvania [Koszta1963, Koszta1996], Tennessee [Nakaha1982], Virginia [Koszta1963, Koszta1996]). **Neotropical**: Bermuda [Koszta1996]; Puerto Rico & Vieques Island (Puerto Rico [ColonFMe1998]).

GENERAL: Description and illustration of adult female by Kosztarab (1963, 1993, 1996) and by Colon-Ferrer & Medina-Gaud (1998).

CITATIONS: BesheaTiHo1973 [host, distribution: 4]; Borchs1966 [catalogue: 355]; ColonFMe1998 [taxonomy, host, distribution: 3]; ColonFMe1998 [taxonomy, description, illustration, host, distribution: 34]; Dekle1965c [taxonomy, description, host, distribution: 15]; Dekle1976 [taxonomy, description, host, distribution, economic importance: 25]; FDACSB1987 [host, distribution: 4-7]; Koszta1963 [taxonomy, description, illustration, host, distribution: 18-20]; Koszta1996 [taxonomy, description, illustration, host, distribution: 420-421]; Nakaha1982 [host, distribution: 4]; TippinBe1970 [host, distribution: 7]; USDAAP1978 [host, distribution: 1-4,6].

Acutaspis oliveirae (Lepage & Giannotti)

Melanaspis oliveirae Lepage & Giannotti, 1942: 446. Type data: BRAZIL: Joaquim Tavora, on *Licania rigida*. Syntypes, female. Type depositories: São Paulo: Instituto Biologico de São Paulo, Brazil, and Curitiba: Departamento de Zoologia, Setor de Ciencias Biologicas, Universidade Federal do Parana, Brazil.

Acutaspis oliveirae; Borchsenius, 1966: 355. Change of combination.

Acutaspis oliveirai; Claps, 1993: 6. Misspelling of species name.

SCALE COVER: Female scale grey, circular; exuviae black, placed centrally. Male scale not observed (Lepage & Giannotti, 1942).

HOST PLANTS: **Chrysobalanaceae**: *Licania rigida* [LepageGi1942, ClapsWoGo2001]. **Oleaceae**: *Olea* [ClapsWoGo2001].

DISTRIBUTION: **Neotropical**: Brazil [LepageGi1942] Cear [ClapsWoGo2001], São Paulo [ClapsWoGo2001].

GENERAL: Description and illustration of adult female by Lepage & Giannotti (1942).

CITATIONS: Borchs1966 [catalogue: 355]; Claps1993 [taxonomy: 6,9]; ClapsWoGo2001 [host, distribution: 240]; LepageGi1942 [taxonomy, description, illustration, host, distribution: 3-4]; McKenz1947 [taxonomy: 32-33].

Acutaspis paulista (Hempel)

Aspidiotus (*Chrysomphalus*) *paulistus* Hempel, 1900a: 504. Type data: BRAZIL: São Paulo, Ypiranga, on leaves of *Laurus* sp. Syntypes, female. Type depository: São Paulo: Museu de Zoologia, Universidade de São Paulo, Brazil; type no. 203.

Chrysomphalus paulistus; Fernald, 1903b: 292. Change of combination.

Pseudischnaspis paulista; Lindinger, 1937: 194. Change of combination requiring emendation of species name for agreement in gender.

Melanaspis paulistus; McKenzie, 1939: 54. Change of combination.

Melanaspis palustris; Trjapitzin, 1989: 312. Misspelling of species name.

Acutaspis paulista; Claps, 2000: 91. Change of combination.

SCALE COVER: Scale of female subcircular, flattish, opaque, dull brown, above blackish beneath; exuviae blackish usually disposed eccentrically; diameter 2 to 2.5 mm (Green, 1930b).

HOST PLANTS: **Anacardiaceae**: *Mangifera* [ClapsWoGo2001], *Mangifera indica* [Lepage1938, WolffCo1993a], *Spondias* [ClapsWoGo2001]. **Annonaceae**: *Annona* [ClapsWoGo2001], *Annona muricata* [ClapsWoGo2001]. **Apocynaceae**: *Aspidosperma quebracho* [Claps2000], *Aspidosperma quebracho-blanco* [ClapsWoGo2001], *Nerium* [ClapsWoGo2001]. **Aquifoliaceae**: *Ilex* [ClapsWoGo2001]. **Araliaceae**: *Hedera* [Claps2000, ClapsWoGo2001]. **Begoniaceae**: *Begonia* [ClapsWoGo2001]. **Buxaceae**: *Buxus sempervirens* [Claps2000]. **Celastraceae**: *Euonymus* [Claps2000], *Maytenus* [Claps2000], *Maytenus chilensis* [Claps2000], *Maytenus viscifolia* [Claps2000, ClapsWoGo2001]. **Chrysobalanaceae**: *Moquilea* [ClapsWoGo2001], *Moquilea tomentosa* [Lepage1938]. **Lauraceae**: *Laurus* [Hempel1900a, Lepage1938, Claps2000, ClapsWoGo2001], *Laurus nobilis* [Lizery1916c, Claps2000]. **Loranthaceae**: *Phoradendron* [ClapsWoGo2001]. **Moraceae**: *Ficus* [Lepage1938, ClapsWoGo2001]. **Musaceae**: *Musa* [ClapsWoGo2001]. **Myrsinaceae**: *Myrsine* [ClapsWoGo2001]. **Myrtaceae**: *Eucalyptus microrys* [Claps2000], *Psidium* [Lepage1938, ClapsWoGo2001]. **Oleaceae**: *Ligustrum* [Claps2000, ClapsWoGo2001], *Olea* [ClapsWoGo2001], *Olea europaea* [Lizery1916c, Claps2000]. **Pittosporaceae**: *Pittosporum* [Claps2000]. **Rosaceae**: *Rosa* [ClapsWoGo2001]. **Rutaceae**: *Citrus* [Claps2000, ClapsWoGo2001]. **Salicaceae**: *Populus* [Claps2000]. **Simaroubaceae**: *Castela* [Claps2000]. **Solanaceae**: *Brunfelsia australis* [Claps2000, ClapsWoGo2001]. **Theaceae**: *Camellia* [ClapsWoGo2001], *Camellia japonica* [Lepage1938].

NATURAL ENEMIES: HYMENOPTERA **Aphelinidae**: *Aphytis costalimai* [Fidalg1983], *Prospaltella ectophaga* [Fidalg1983]. **Encyrtidae**: *Zaomma lambinus* (Walker) [Trjapi1989]. **Signiphoridae**: *Signiphora fax* Girault [Woolle1990].

DISTRIBUTION: **Neotropical**: Argentina [Lizery1916c] (Buenos Aires [ClapsWoGo2001], Catamarca [ClapsWoGo2001], La Rioja [ClapsWoGo2001], Tucuman [ClapsWoGo2001]); Brazil [WolffCo1993a] (Distrito federal (Brasilia) [Lepage1938, ClapsWoGo2001], Minas Gerais [Lepage1938, ClapsWoGo2001], Rio Grande do Sul [Lepage1938, BertelBa1966, ClapsWoGo2001], Rio de Janeiro [Lepage1938, ClapsWoGo2001], São Paulo [Hempel1900a, Lepage1938, ClapsWoGo2001]).

ECONOMIC IMPORTANCE: This polyphagous species (see Host Plant), is considered a pest of olive in South America (Chiesa Molinari, 1948; Schmutterer et al., 1957).

GENERAL: Description and illustration of adult female by Green (1930b) and by Claps (2000).

CITATIONS: Argyri1990 [host, distribution, economic importance: 579-583]; Bertel1956 [host, distribution, description, life history, biological control]; BertelBa1966 [host, distribution: 17-46]; BiezanFr1939 [host, distribution: 1-18]; BiezanSe1940 [host, distribution: 67-68]; Borchs1966 [catalogue: 351]; Bustsh1958 [taxonomy, description, host, distribution: 219,234]; Chiesa1938a [taxonomy, description, illustration, host, distribution, life history: 1-21]; Chiesa1948 [host, distribution, economic importance]; Chiesa1948a [host, distribution, economic importance]; Claps1993 [taxonomy: 5]; Claps2000 [taxonomy, description, illustration, host, distribution: 91-92,94]; ClapsWoGo2001 [host, distribution: 247]; CostaL1949 [host, distribution, biological control: 65-87]; DeSant1941a [host, distribution, biological control: 21-24]; Fernal1903b [catalogue: 292]; Ferris1941e [taxonomy: 46]; Fidalg1983 [host, distribution, biological control: 119-125]; Fonsec1963 [host, distribution: 32-35]; Fonsec1964 [host, distribution: 515]; Green1930b [taxonomy, description, illustration, host, distribution: 217-219]; Haywar1939 [host, distribution, control: 1]; Haywar1944 [host, distribution: 1-32]; Hempel1900a [taxonomy, description, illustration, host, distribution: 504-505]; Hempel1901a [taxonomy, description, host, distribution: 107]; Lepage1938 [catalogue: 399-400]; Lindin1937 [taxonomy: 194]; Lizery1916c [host, distribution: 432-433]; Lizery1942 [host, distribution: 80]; MacGil1921 [taxonomy, description, host, distribution: 416,420]; McKenz1939 [taxonomy: 54]; MillerDa1990 [host, distribution, economic importance: 303]; Monte1943 [host, distribution: 134]; Newste1920 [taxonomy: 197]; SchmutKlLu1957 [host, distribution, economic importance: 492]; Trjapi1989 [biological control: 312]; Vernal1957 [taxonomy: 20]; WolffCo1993a [host, distribution: 153]; Woolle1990 [biological control: 167-176].

Acutaspis perseae (Comstock)

Aspidiotus perseae Comstock, 1881a: 305. Type data: USA: Florida, Cedar Keys, on leaves of red bay, *Persea carolinensis*. Syntypes, female. Type depository: Washington: United States National Entomological Collection, U.S. National Museum of Natural History, District of Columbia, USA.

Chrysomphalus perseae; Leonardi, 1897: 286. Change of combination.

Aspidiotus (*Chrysomphalus*) *perseae*; Cockerell, 1897i: 22. Change of combination.

Pseudischnaspis perseae; Lindinger, 1911: 355. Change of combination.

Acutaspis perseae; Ferris, 1941d: 332. Change of combination.

Chrysomphalus (*Acutaspis*) *perseae*; Merrill, 1953: 38. Change of combination.

COMMON NAME: red bay scale [Comsto1881a, MerrilCh1923, Merril1953].

SCALE COVER: Scale of female flat, rather thin, circular; exuviae central, colour dark reddish brown; that of male not recognized (Ferris, 1941d).

HOST PLANTS: **Aquifoliaceae**: *Ilex* [MerrilCh1923]. **Araceae**: *Anthurium* [MerrilCh1923, Ferris1941d], *Anthurium harrisii* [Morgan1889a]. **Cactaceae**: *Harrisia* [MerrilCh1923]. **Caprifoliaceae**: *Viburnum* [MerrilCh1923, Ferris1941d]. **Celastraceae**: *Euonymus* [Lepage1938, Ferris1941d]. **Ericaceae**: *Ampelothamnus phillyreifolius* [Merril1953], *Lyonia ferruginea* [Merril1953], *Lyonia lucida* [TippinBe1970], *Xolisma* [Ferris1941d], *Xolisma fruticosa* [MerrilCh1923]. **Lauraceae**: *Persea* [Ferris1941d], *Persea borbonia* [BesheaTiHo1973], *Persea carolinensis* [Comsto1881a, Wilson1917, McDani1968], *Persea gratissima*

[Lepage1938, Ferris1941d]. **Loranthaceae**: *Phoradendron flavescens* [Ferris1941d]. **Magnoliaceae**: *Magnolia* [Lepage1938, Ferris1941d, McDani1968], *Magnolia grandiflora* [Wilson1917, Dekle1965c]. **Oleaceae**: *Olea* [Ferris1941d], *Osmanthus* [Merril1953]. **Orchidaceae**: *Bletia* [Ferris1941d], *Laelia* [Ferris1941d]. **Palmae**: *Cocos nucifera* [Ferris1941d], *Sabal* [Ferris1941d], *Serenoa repens* [BesheaTiHo1973]. **Pinaceae**: *Pinus* [Ferris1941d, Balach1951]. **Theaceae**: *Gordonia* [MerrilCh1923]. **Zamiaceae**: *Zamia* [Ferris1941d], *Zamia floridana* [Dekle1965c], *Zamia pumila* [MerrilCh1923].

DISTRIBUTION: **Nearctic**: Mexico [Cocker1899n] (Oaxaca [Ferris1941d], Sinola [Ferris1941d], Veracruz [Ferris1941d]); United States of America (Alabama [Nakaha1982], Florida [Wilson1917, MerrilCh1923, Dekle1965c, Ferris1941d, Merril1953], Georgia [TippinBe1970, BesheaTiHo1973], Louisiana [Nakaha1982], Mississippi [Herric1911, Ferris1941d], Missouri [Hollin1923], Oklahoma [Nakaha1982], South Carolina [Nakaha1982], Tennessee [Nakaha1982], Texas [McDani1968]). **Neotropical**: Brazil (Rio Grande do Sul [Lepage1938, Ferris1941d], Rio de Janeiro [Lepage1938, Ferris1941d]); Cuba [Nakaha1982]; Trinidad and Tobago (Trinidad [Nakaha1982]); Venezuela [Ferris1941d] [Nakaha1982]. **Palaearctic**: Ukraine (Kiev Oblast [Danzig1993]); United Kingdom (England [Morgan1889a]).

BIOLOGY: Occurring characteristically on leaves (Ferris, 1941d).

ECONOMIC IMPORTANCE: Considered a pest of red bay, *Persea barbonia* (Miller & Davidson, 1990), and in Bermuda of *Juniperus bermudiana* (Schmutterer et al., 1957). Of no economic importance in Florida (Dekle, 1976).

GENERAL: Description and illustration of adult female by Ferris (1941d) and by Tereznikova (1986).

KEYS: McDaniel 1968: 202 (female) [U.S.A.: Texas]; Ferris 1942: 28 (female) [North America]; Hollinger 1923: 29 (female) [U.S.A.: Missouri]; Cockerell 1905: 45-46 (female) [Mexico]; Newstead 1901b: 82 (female) [England]; Comstock 1883: 55-57 (female) [North America].

CITATIONS: Balach1951 [taxonomy, description, illustration, host, distribution: 594-597]; BeardsDaHo1976 [economic importance: 106]; BesheaTiHo1973 [host, distribution: 4]; Borchs1937 [taxonomy, description, illustration, host, distribution: 120]; Borchs1939 [taxonomy, description, host, distribution: 11,44]; Borchs1950b [taxonomy, description, illustration, host, distribution: 222]; Borchs1966 [catalogue: 355]; ClapsWoGo2001a [taxonomy, host, distribution: 10]; Cocker1896b [distribution: 334]; Cocker1896f [taxonomy, description, host, distribution: 33-34]; Cocker1897i [taxonomy, description, host, distribution: 22]; Cocker1897k [taxonomy, description, host, distribution: 90]; Cocker1899n [host, distribution: 25]; Cocker1905 [taxonomy: 46]; Comsto1881a [taxonomy, description, illustration, host, distribution: 305-306]; Comsto1883 [taxonomy, host, distribution: 65]; Danzig1964 [taxonomy, host, distribution: 651-652]; Danzig1993 [taxonomy, description, illustration, host, distribution: 239-240]; DanzigPe1998 [catalogue: 175]; Dekle1965c [taxonomy, description, host, distribution: 16]; Dekle1976 [taxonomy, description, host, distribution, economic importance: 26]; Fernal1903b [catalogue: 292]; Ferris1941d [taxonomy, description, illustration, host, distribution: 332]; Ferris1941e [taxonomy: 47]; Ferris1942 [taxonomy: 446:28]; Foldi2001 [distribution: 303-308]; Green1930b [taxonomy: 219]; Herric1911 [taxonomy, description, illustration, host, distribution: 11,33-34,68]; Hollin1923 [taxonomy, description, host, distribution: 31]; Leonar1897 [taxonomy: 286]; Leonar1899 [taxonomy: 199,200,223]; Lepage1938 [catalogue: 400]; Lindin1909c [taxonomy, host, distribution: 449]; Lindin1911 [taxonomy: 355]; Lindin1921 [host, distribution: 428]; Lindin1935 [taxonomy: 149]; Lindin1936c [host, distribution: 155]; Lindin1957 [taxonomy: 544]; LongoMaPe1995 [distribution: 125]; MacGil1921 [taxonomy, description, host, distribution: 420]; Maranh1946 [taxonomy: 164-179]; McDani1968 [taxonomy, illustration, host, distribution: 204]; McKenz1939 [taxonomy: 54]; Merril1953 [taxonomy, description, host, distribution: 38-40]; MerrilCh1923 [taxonomy, description, host, distribution, economic importance: 225-226]; MillerDa1990 [host, distribution, economic importance: 300]; Morgan1889a [taxonomy, host, distribution: 350]; Nakaha1982 [host, distribution: 4]; Newste1901b [taxonomy, description, illustration, host, distribution: 82,112-114]; Newste1920 [taxonomy, description, host, distribution: 197]; Schmut1959 [taxonomy, description, host, distribution: 45]; SchmutKlLu1957 [host, distribution, economic importance: 492]; Terezn1986 [taxonomy, description, illustration, host, distribution: 83-84]; TippinBe1970 [host, distribution: 8]; Wilson1917 [host, distribution: 54-55].

Acutaspis ramirezi Balachowsky

Acutaspis ramirezi Balachowsky, 1959: 356. Type data: COLOMBIA: Sabana de Bogotá, near Zipaquira, 2600 m altitude, on *Acacia melanoxylon*. Holotype female. Type depository: Paris: Muséum national d'Histoire naturelle, France.
SCALE COVER: Female scale subcircular, flat, larval exuviae central, colour brown black; secretion of adult brown; diameter 2.2 mm. Male scale unknown (Balachowsky, 1959).
HOST PLANTS: **Leguminosae**: *Acacia melanoxylon* [Balach1959].
DISTRIBUTION: **Neotropical**: Colombia [Balach1959, Kondo2001].
GENERAL: Description and illustration of adult female by Balachowsky (1959).
CITATIONS: Balach1959 [taxonomy, description, illustration, host, distribution: 356-357]; Borchs1966 [catalogue: 355]; Kondo2001 [taxonomy, host, distribution: 43].

Acutaspis reniformis (Cockerell)

Aspidiotus reniformis Cockerell, 1897u: 265. Type data: MEXICO: Michoacan, Tehuantepec City, on undetermined plant, on undersides of entire, lanceolate leaves ,about 60 mm long; collected by Townsend. Syntypes, female. Type depository: Washington: United States National Entomological Collection, U.S. National Museum of Natural History, District of Columbia, USA; type no. 7196.
Aspidiotus (*Chrysomphalus*) *reniformis*; Cockerell, 1897i: 24. Change of combination.
Chrysomphalus reniformis; Cockerell, 1899a: 396. Change of combination.
Acutaspis reniformis; Ferris, 1941d: 333. Change of combination.
SCALE COVER: Female scale circular, diameter 2 mm, flat, pale reddish-brown; exuviae concolour or slightly darker, both skins very distinctly visible, large, laterad

of middle. First skin when rubbed shining coppery (Cockerell, 1897u). Scale of female circular, flat, pale reddish brown or straw colour; that of male slightly elongate, similar in colour to that of female (Ferris, 1941d).

DISTRIBUTION: **Nearctic**: Mexico [Cocker1899n] (Michoacan [Cocker1897u]).

BIOLOGY: Occurring on leaves (Ferris, 1941d).

GENERAL: Description and illustration of adult female by Ferris (1941d).

KEYS: Ferris 1942: 29 (female) [North America]; Cockerell 1905: 45-46 (female) [Mexico].

CITATIONS: Borchs1966 [catalogue: 355]; ClapsWoGo2001a [taxonomy, host, distribution: 10-11]; Cocker1897i [taxonomy, description, host, distribution: 24-25]; Cocker1897u [taxonomy, description, host, distribution: 265-266]; Cocker1899a [taxonomy: 396]; Cocker1899n [host, distribution: 25]; Cocker1905 [taxonomy: 46]; Fernal1903b [catalogue: 293]; Ferris1941d [taxonomy, description, illustration, host, distribution: 333]; Ferris1941e [taxonomy: 47]; Ferris1942 [taxonomy: 29]; Ferris1942 [taxonomy: 29]; Leonar1900 [taxonomy, host, distribution: 343]; Lindin1957 [taxonomy: 544]; MacGil1921 [taxonomy, description, host, distribution: 417]; McKenz1939 [taxonomy: 54].

Acutaspis scutiformis (Cockerell)

Aspidiotus scutiformis Cockerell, 1893o: 48. Type data: MEXICO: on *Persea americana*. Syntypes, female. Type depository: Washington: United States National Entomological Collection, U.S. National Museum of Natural History, District of Columbia, USA.

Aspidiotus (*Chrysomphalus*) *scutiformis*; Cockerell, 1897i: 25. Change of combination.

Chrysomphalus scutiformis; Berlese & Leonardi, 1898a: 116. Change of combination.

Acutaspis scutiformis; Ferris, 1941d: 207. Change of combination.

Insaspidiotus scutiformis; Costa Lima, 1942: 289. Change of combination.

Acutaspis scutiformis; Borchsenius, 1966: 355. Revived combination.

SCALE COVER: Scale of female very large, reaching a diameter of nearly 3 mm, circular, flat, thick, dark brown; exuviae yellowish. Scale of male not recognized (Ferris, 1941d).

HOST PLANTS: **Chrysobalanaceae**: *Moquilea tomentosa* [Lepage1938]. **Lauraceae**: *Laurus* [Lepage1938, Ferris1941d], *Persea* [Ferris1941d], *Persea americana* [Cocker1893o], *Persea gratissima* [Lepage1938]. **Leguminosae**: *Pithecellobium flexicaule* [McDani1968]. **Moraceae**: *Artocarpus integrifolia* [Lepage1938], *Ficus* [Lepage1938], *Ficus benjamina* [WolffCo1993]. **Rutaceae**: *Citrus* [Lizery1936, Lepage1938], *Citrus limon* [WolffCo1993]. **Zygophyllaceae**: *Porlieria angustifolia* [McDani1968].

NATURAL ENEMIES: HYMENOPTERA **Aphelinidae**: *Aphytis hispanicus* (Mercet) [RosenDe1979]. **Signiphoridae**: *Signiphora prepauca* Girault [Woolle1990].

DISTRIBUTION: **Nearctic**: Mexico [Cocker1893o] (Guerrero [Ferris1941d], Nuevo Leon [Ferris1941d], Tamaulipas [Ferris1941d], Veracruz [Ferris1941d]); United States of America (Texas [McDani1968]). **Neotropical**: Argentina

(Corrientes [Lizery1936], Entre Rios [Lizery1936]); Brazil [RosenDe1979] (Minas Gerais [Hempel1900a, Lepage1938], Parana [Lepage1938], Rio Grande do Sul [WolffCo1993], Rio de Janeiro [Lepage1938], São Paulo [Lepage1938]); Colombia [Nakaha1982]; Guatemala [Nakaha1982].

BIOLOGY: Occurring on leaves (Ferris, 1941d).

ECONOMIC IMPORTANCE: Reported as a pest of citrus in South America (Bondar, 1914; Chiesa Molinari, 1948, 1948a; Ebeling, 1959) and of banana in Central America (Chua & Wood, 1990).

GENERAL: Description and illustration of adult female by Cockerell (1893o) and by Ferris (1941d).

KEYS: McDaniel 1968: 202 (female) [U.S.A.: Texas]; Ferris 1942: 29 (female) [North America]; Cockerell 1905: 45-46 (female) [Mexico].

CITATIONS: BerlesLe1898a [taxonomy: 116]; BiezanFr1939 [host, distribution: 1-18]; Bondar1914 [host, distribution, economic importance: 1064-1106]; Bondar1915 [host, distribution, economic importance: 44-47]; Borchs1966 [catalogue: 355-356]; Chiesa1938a [host, distribution: 1-21]; Chiesa1948 [host, distribution, economic importance]; Chiesa1948a [host, distribution, economic importance]; ChuaWo1990 [host, distribution, economic importance: 543-552]; ClapsWoGo2001a [taxonomy, host, distribution: 11]; Cocker1893o [taxonomy, description, host, distribution: 48-49]; Cocker1896b [distribution: 334]; Cocker1897i [taxonomy, description, host, distribution: 25]; Cocker1899n [host, distribution: 26]; Cocker1900k [taxonomy: 350]; Cocker1905 [taxonomy: 46]; CoronaRuMo1997 [host, distribution: 38-41]; CostaL1942 [taxonomy, description, host, distribution: 289]; Ebelin1949 [host, distribution, life history, control: 268,279]; Fernal1903b [catalogue: 293]; Ferris1941d [taxonomy, description, illustration, host, distribution: 334]; Ferris1941e [taxonomy: 48]; Ferris1942 [taxonomy: 446:29]; Figuer1952 [host, distribution: 208]; FonsecAu1932a [host, distribution: 202-214]; Haywar1939 [host, distribution, control: 1]; Hempel1900a [taxonomy, description, host, distribution: 503-504]; Kondo2001 [taxonomy, host, distribution: 43]; Leonar1899 [taxonomy: 199-200,222]; Lepage1938 [catalogue: 401]; Lindin1910c [taxonomy: 440]; Lindin1935 [taxonomy: 149]; Lindin1957 [taxonomy: 544]; Lizery1936 [host, distribution: 113]; MacGil1921 [taxonomy, description, host, distribution: 418]; McDani1968 [taxonomy, illustration, host, distribution: 207-208]; McKenz1939 [taxonomy: 54]; MillerDa1990 [host, distribution, economic importance: 300]; Nakaha1982 [host, distribution: 4]; Pace1939 [host, distribution: 664-665]; RosenDe1979 [host, distribution, biological control: 390-394]; Sassce1918 [host, distribution: 125-129]; Sassce1923 [host, distribution: 152-158]; SchmutKlLu1957 [host, distribution, economic importance: 493]; Sefer1961 [host, distribution: 23]; Strong1922 [host, distribution: 775-780]; Vernal1957 [taxonomy, description, host, distribution: 28-30]; WolffCo1993 [taxonomy, description, illustration, host, distribution: 29-31]; Woolle1990 [biological control: 167-176].

Acutaspis subnigra McKenzie
Acutaspis subnigra McKenzie, 1947: 33. Type data: PERU: Tulumayo-Tingo Maria, at the Delicias Plantacion, on *Persea americana*. Holotype female. Type depository: Davis: Bohart Museum of Entomology, University of California, California, USA.
Pseudischnaspis subnigra; Lindinger, 1957: 544. Change of combination.
Acutaspis subnigra; Borchsenius, 1966: 356. Revived combination.
SCALE COVER: Scale of female practically circular, averaging 2mm in diameter, chocolate brown (from whence name was derived); exuvium subcentral; scale of male unknown (McKenzie, 1947).
HOST PLANTS: Lauraceae: *Persea americana* [McKenz1947].
DISTRIBUTION: **Neotropical**: Peru [McKenz1947].
BIOLOGY: Occurring on the leaves of avocado (McKenzie, 1947).
GENERAL: Description and illustration of adult female by McKenzie (1947).
CITATIONS: Borchs1966 [catalogue: 356]; Lindin1957 [taxonomy: 544]; McKenz1947 [taxonomy, description, illustration, host, distribution: 33-35].

Acutaspis tingi McKenzie
Acutaspis tingi McKenzie, 1947: 34. Type data: MEXICO: on *Cocos nucifera*. Holotype female. Type depository: Davis: The Bohart Museum of Entomology, University of California, California, USA.
Pseudischnaspis tingi; Lindinger, 1957: 544. Change of combination.
Acutaspis tingi; Borchsenius, 1966: 356. Revived combination.
SCALE COVER: Only slide-mounted specimens available for description (McKenzie, 1947).
HOST PLANTS: **Moraceae**: *Ficus reclinata* [Balach1959]. **Palmae**: *Cocos nucifera* [McKenz1947].
DISTRIBUTION: Nearctic: Mexico [McKenz1947].
GENERAL: Description and illustration of adult female by McKenzie (1947).
CITATIONS: Balach1959 [host, distribution: 360]; Borchs1966 [catalogue: 356]; Lindin1957 [taxonomy: 544]; McKenz1947 [taxonomy, description, illustration, host, distribution: 34-36].

Acutaspis umbonifera (Newstead)
Aspidiotus (*Chrysomphalus*) *umboniferus* Newstead, 1920: 196. Type data: GUYANA: Ayaria Creek, Fssequibo, on *Lecythis* sp. Syntypes, female. Type depository: London: The Natural History Museum, England, UK.
Chrysomphalus umboniferus; MacGillivray, 1921: 417. Change of combination.
Acutaspis umbonifera; Ferris, 1941d: 335. Change of combination requiring emendation of species name for agreement in gender.
Pseudischnaspis umbonifera; Lindinger, 1957: 544. Change of combination.
Acutaspis umbonifera; Borchsenius, 1966: 356. Revived combination.
SCALE COVER: Scale of female flat, circular, thin, of a reddish brown colour; that of male similar in colour and texture, but slightly elongate (Ferris, 1941d).
HOST PLANTS: **Araceae**: *Anthurium* [Ferris1941d]. **Cactaceae**: *Pereskia* [Ferris1941d]. **Heliconiaceae**: *Heliconia* [Ferris1941d]. **Lecythidaceae**: *Lecythis* [Newste1920, Ferris1941d]. **Palmae**: *Attalea* [Ferris1941d].

NATURAL ENEMIES: HYMENOPTERA **Aphelinidae**: *Aphytis chrysomphali* (Mercet) [Gordh1979].
DISTRIBUTION: **Nearctic**: United States of America (New York [Ferris1941d]). **Neotropical**: Colombia [Balach1959, Kondo2001]; Guyana [Newste1920, Ferris1941d]; Panama [Ferris1941d]; Paraguay [Ferris1941d].
BIOLOGY: Occurring exposed upon the leaves of the host (Ferris, 1941d).
GENERAL: Description and illustration of adult female by Newstead (1920) and by Ferris (1941d).
KEYS: Ferris 1942: 28 (female) [North America].
CITATIONS: Balach1959 [host, distribution: 360]; Borchs1966 [catalogue: 356]; ClapsWoGo2001a [taxonomy, host, distribution: 11]; Ferris1941d [taxonomy, description, illustration, host, distribution: 335]; Ferris1941e [taxonomy: 49]; Ferris1942 [taxonomy: 446:28]; Figuer1952 [host, distribution: 209]; Gordh1979 [biological control: 893]; Kondo2001 [taxonomy, host, distribution: 43]; Lindin1937 [taxonomy: 194]; Lindin1957 [taxonomy: 544]; MacGil1921 [taxonomy, description, host, distribution: 417]; McKenz1939 [taxonomy: 55]; Newste1920 [taxonomy, description, illustration, host, distribution: 196-197]; Swezey1945 [taxonomy, host, distribution: 388].

Africonidia McKenzie

Africonidia McKenzie, 1947b: 110. Type species: *Africonidia halli* McKenzie (= *Gymnaspis africana* Newstead), by monotypy and original designation.
Hallaspidiotus Mamet, 1951: 217. Type species: *Gymnaspis africana* Newstead. Synonymy by Balachowsky, 1954c: 77.
SYSTEMATICS: *Africonidia* was established by McKenzie (1947b) for the species *Africonidia halli*. However, Balachowsky (1954c; 1958b) showed that *A. halli* was a junior synonym of *Gymnaspis africana* Newstead, 1913, and transferred the latter to *Africonidia*. Consequently, the genus *Hallaspidiotus* Mamet, 1951 (type-species: *Gymnaspis africana* Newstead), became a junior objective synonym of *Africonidia*. Borchsenius (1966) did not accept the validity of *Africonidia* and resurrected the genus *Varicaspis* MacGillivray, 1921 (type-species: *Aspidiotus fiorineides* Newstead, 1920). Borchsenius' interpretation will doubtlessly be acceptable as soon as it can be shown that *A. fiorineides* - which at present is known only from inadequate type material (Balachowsky, 1958b) and a poor description - is in fact congeneric with *Africonidia africana*. Until these two genera have been revised, *Varicaspis* is restricted to its type species, while five species are placed in *Africonidia*, namely *africana, carreti, macdanieli, mkuzensis, subsimplex*.
GENERAL: Definition and characters by McKenzie (1947), Mamet (1951) (as *Hallaspidiotus*) and by Balachowsky (1954c; 1958b).
KEYS: Ben-Dov 1974c: 22 (female) [World]; Beardsley 1966: 502-504 (female) [Federated States of Micronesia]; Balachowsky 1958b: 150, 228 (female) [*Aspidiotina* of Africa].
CITATIONS: Balach1954c [taxonomy, description: 77-80]; Balach1956 [taxonomy: 24]; Balach1958b [taxonomy, description: 149-150]; Beards1966 [taxonomy: 505]; BenDov1974c [taxonomy: 19,22]; Borchs1966 [taxonomy,

catalogue: 316]; Mamet1951 [taxonomy, description: 217-218]; Mamet1959a [taxonomy: 386]; McKenz1947b [taxonomy, description: 110-111]; MorrisMo1966 [taxonomy, catalogue: 4].

Africonidia africana (Newstead)
Gymnaspis africana Newstead, 1913: 78. Type data: UGANDA: Tero Forest, on undetermined plant. Syntypes, female. Type depository: London: The Natural History Museum, England, UK.
Cryptaspidiotus africanus; Lindinger, 1913: 73. Change of combination.
Neosignoretia africana; MacGillivray, 1921: 425. Change of combination.
Africonidia halli McKenzie, 1947b: 111. Type data: SOUTH AFRICA: Durban, on *Trichilia* sp. Holotype. Type depository: Davis: The Bohart Museum of Entomology, University of California, California, USA. Synonymy by Borchsenius, 1966: 316.
Hallaspidiotus africana; Mamet, 1951: 218. Change of combination.
Africonidia africana; Balachowsky, 1954c: 78. Change of combination.
SCALE COVER: Female scale nude, except for a small central area covered with larval exuviae; ventral vellum very thin, circular in outline, about one-fourth diameter of scale and occupying a central position; in form highly convex and attenuated posteriorly; length 0.9-1 mm (Newstead, 1913). Scale covering of female, oval, measuring approximately 1.1 mm long, 0.8 mm wide, deep red with blackish exuvium subcentral (McKenzie, 1947b).
HOST PLANTS: **Annonaceae**: *Annona muricata* [Almeid1973b]. **Euphorbiaceae**: *Aleurites moluccana* [DeLott1967a], *Manihot* [Lindin1913, Borchs1966]. **Leguminosae**: *Inga* [MatileNo1984]. **Meliaceae**: *Trichilia* [Lindin1913, McKenz1947b, Borchs1966]. **Monimiaceae**: *Tambourissa* [Matile1978].
DISTRIBUTION: **Afrotropical**: Angola [Almeid1973b]; Cameroon [MatileNo1984]; Comoros Islands [Matile1978]; Kenya [DeLott1967a]; Madagascar [Mamet1950, Mamet1951, Mamet1959a, Borchs1966]; South Africa [McKenz1947b, Borchs1966]; Tanzania [Balach1958b]; Uganda [Newste1913].
BIOLOGY: Occurring on leaves (McKenzie, 1947b).
GENERAL: Description and illustration of adult female by Newstead (1913), McKenzie (1947b) (as *A. halli*), Mamet (1951) and by Balachowsky (1958b).
KEYS: Ben-Dov 1974c: 22 (female) [World]; Balachowsky 1958b: 150 (female) [Africa].
CITATIONS: Almeid1973b [host, distribution: 8]; Balach1954c [taxonomy: 78]; Balach1958b [taxonomy, description, illustration, host, distribution: 150-152]; BenDov1974c [taxonomy: 19,22]; Borchs1966 [taxonomy, catalogue: 316]; DeLott1967a [host, distribution: 112]; Hall1946a [taxonomy: 520,536,548]; Lindin1913 [taxonomy: 73]; Lindin1957 [taxonomy: 544]; MacGil1921 [taxonomy: 255,425]; Mamet1950 [host, distribution: 22]; Mamet1951 [taxonomy, description, distribution: 216-217]; Mamet1959a [host, distribution : 386]; Matile1978 [host, distribution: 67]; MatileNo1984 [host, distribution: 67]; McKenz1947b [taxonomy, description, illustration, host, distribution: 111-114]; Newste1913 [taxonomy, description, illustration, host, distribution: 78-79]; Sassce1915 [taxonomy, host, distribution: 35].

Africonidia carreti Balachowsky

Africonidia carreti Balachowsky, 1954c: 78. Type data: CAMEROON: near falls of river Lobe, 10 km south of Kribi, on branches of undetermined bush. Holotype female. Type depository: Paris: Muséum national d'Histoire naturelle, France.

Varicaspis carreti; Borchsenius, 1966: 316. Change of combination.

SCALE COVER: Female scale circular or subcircular, convex; colour snow-white; larval exuviae placed centrally or subcentrally; rounded, colour pale grey (Balachowsky, 1954c).

HOST PLANTS: Leguminosae: *Erythrina caffra* [Balach1958b].

DISTRIBUTION: **Afrotropical**: Cameroon [Balach1954c]; Tanzania [Balach1958b].

GENERAL: Description and illustration of adult female by Balachowsky (1954c; 1958b).

KEYS: Ben-Dov 1974c: 22 (female) [World]; Balachowsky 1958b: 150 (female) [Africa].

CITATIONS: Balach1954c [taxonomy, description, illustration, host, distribution: 78-80]; Balach1958b [taxonomy, description, illustration, host, distribution: 152-154]; BenDov1974c [taxonomy: 19,22]; Borchs1966 [taxonomy, catalogue: 316].

Africonidia macdanieli Beardsley

Africonidia macdanieli Beardsley, 1966: 505. Type data: FEDERATED STATES OF MICRONESIA: Palau Is., Ulebsehel (Auluptagel), on unknown vine. Holotype female. Type depository: Honolulu: Bernice P. Bishop Museum, Department of Entomology Collection, Hawaii, USA.

SCALE COVER: Female scale circular, light brown; second exuvium central (Beardsley, 1966).

DISTRIBUTION: Australasian: Federated States of Micronesia [Beards1966].

GENERAL: Description and illustration of adult female by Beardsley (1966).

KEYS: Ben-Dov 1974c: 22 (female) [World].

CITATIONS: Beards1966 [taxonomy, description, illustration, host, distribution: 505-506]; BenDov1974c [taxonomy: 19,22].

Africonidia mkuzensis Ben-Dov

Africonidia mkuzensis Ben-Dov, 1974c: 19. Type data: SOUTH AFRICA: Natal, Lower Mkuze, on *Euclea crispa*. Holotype female. Type depository: Pretoria: South African National Collection of Insects, South Africa.

SCALE COVER: Scale of female (secreted part) greyish white; circular; 0.8-1 mm in diameter; flat; larval exuviae central. Scale of male similar in colour to that of female, oval, 0.8 - 1 mm long, 0.5-0.6 mm wide; larval exuviae placed about middle of longitudinal axis of scale (Ben-Dov, 1974c).

HOST PLANTS: Ebenaceae: *Euclea crispa* [BenDov1974c].

DISTRIBUTION: Afrotropical: South Africa [BenDov1974c].

GENERAL: Description and illustration of adult female by Ben-Dov (1974c).

KEYS: Ben-Dov 1974c: 22 (female) [World].

CITATIONS: BenDov1974c [taxonomy, description, illustration, host, distribution: 19-20,22].

Africonidia subsimplex (Hall)

Aonidia subsimplex Hall, 1931: 285. Type data: ZIMBABWE: Umtali and Mtoroshanga Pass, Umvukwes, on *Acacia karroo* and *Acacia* sp. Syntypes, female. Type depository: London: The Natural History Museum, England, UK.
Africonidia subsimplex; Balachowsky, 1954c: 78. Change of combination.
Varicaspis subsimplex; Borchsenius, 1966: 317. Change of combination.
SCALE COVER: Scale of adult female circular in outline, capsular; consisting of two plates composed of dorsal and ventral portions of nymphal exuviae. These plates more or less flat, highly chitinised, and overlap sandwiched adult female. Male scale smaller than that of adult female; narrowly oval. Exuviae black or very dark, with posterior portion pale brown; secretionary appendix dirty white (Hall, 1931).
HOST PLANTS: **Leguminosae**: *Acacia* [Hall1931, DeLott1967a], *Acacia karroo*.
DISTRIBUTION: **Afrotropical**: Kenya [DeLott1967a]; South Africa [Hall1931, Balach1958b, BenDov1974c]; Zimbabwe [Hall1931].
GENERAL: Description and illustration of adult female by Hall (1931) and by Balachowsky (1958b).
KEYS: Ben-Dov 1974c: 22 (female) [World]; Balachowsky 1958b: 150 (female) [Africa].
CITATIONS: Balach1954c [taxonomy: 78]; Balach1958b [taxonomy, description, illustration, host, distribution: 154-156]; BenDov1974c [taxonomy, host, distribution: 19,22]; Borchs1966 [taxonomy, catalogue: 317]; DeLott1967a [host, distribution: 112]; Hall1931 [taxonomy, description, illustration, host, distribution: 285-286]; Lindin1943b [taxonomy: 218].

Anaspidiotus Borchsenius & Williams

Anaspidiotus Borchsenius & Williams, 1963: 381. Type species: *Aspidiotus* (*Hemiberlesia*) *immaculatus* Green, by monotypy and original designation.
SYSTEMATICS: Genus differs from *Aspidiotus* Bouché in possessing large dorsal ducts with swollen inner ends, and from *Hemiberlesia* Cockerell in position of anal opening and absence of paraphyses (Borchsenius & Williams, 1963).
GENERAL: Definition and characters by Borchsenius & Williams (1963).
CITATIONS: BorchsWi1963 [taxonomy, description: 381]; MorrisMo1966 [taxonomy, catalogue: 9].

Anaspidiotus immaculatus (Green)

Aspidiotus (*Hemiberlesia*) *immaculatus* Green, 1904: 65. Type data: AUSTRALIA: Victoria, Shepperton, on stems of *Styphelia virgata*. Holotype female. Type depository: London: The Natural History Museum, England, UK.
Aspidiotus immaculatus; Sanders, 1906: 13. Change of combination.
Neosignoretia immaculata; MacGillivray, 1921: 424. Change of combination.
Anaspidiotus immaculatus; Borchsenius & Williams, 1963: 381. Change of combination.
SCALE COVER: Female scale snowy-white; exuviae completely concealed - both above and below - by white secretionary covering, but indicated by presence of a

raised disc above first larval skin; form strongly convex; apex tilted over towards anterior extremity; diameter 1.25-1.5 mm (Green, 1904).
HOST PLANTS: **Epacridaceae**: *Styphelia virgata* [Green1904, Sander1906, Frogga1914].
DISTRIBUTION: **Australasian**: Australia [Sander1906] (Victoria [Green1904, Frogga1914]).
GENERAL: Description and illustration of adult female by Green (1904) and by Borchsenius & Williams (1963).
CITATIONS: Borchs1966 [catalogue: 275]; BorchsWil1963 [taxonomy, description, illustration: 381-382]; Ferris1941e [taxonomy: 44]; Frogga1914 [taxonomy, description, host, distribution: 315]; Frogga1915 [taxonomy, description, host, distribution: 19]; Green1904 [taxonomy, description, illustration, distribution: 65-68]; MacGil1921 [taxonomy, description, host, distribution: 424]; Sander1906 [taxonomy, host, distribution: 13].

Anastomoderma Beardsley

Anastomoderma Beardsley, 1966: 507. Type species: *Anastomoderma palauensis* Beardsley, by original designation.
SYSTEMATICS: *Anastomoderma* is allied to *Aspidiotus* Bouché, but prosoma of latter membranous at maturity. Sclerotized prosoma, and distinctive reticulated venter distinguish female *Anastomoderma* distinguish from other known genera in the Aspidiotinae (Beardsley, 1966).
GENERAL: Definition and characters by Beardsley (1966).
KEYS: Beardsley 1966: 502-504 (female) [Federated States of Micronesia].
CITATIONS: Beards1966 [taxonomy, description: 507].

Anastomoderma palauensis Beardsley
Anastomoderma palauensis Beardsley, 1966: 507. Type data: FEDERATED STATES OF MICRONESIA: Palau Is., Koror, on *Premna integrifolia*. Holotype female. Type depository: Honolulu: Bernice P. Bishop Museum, Department of Entomology Collection, Hawaii, USA.
SCALE COVER: Female scale brownish, roughly circular, living in shallow pits in gall-like callous growth on twigs of hosts (Beardsley, 1966).
HOST PLANTS: Verbenaceae: *Premna integrifolia* [Beards1966].
DISTRIBUTION: **Australasian**: Federated States of Micronesia [Beards1966].
GENERAL: Description and illustration of adult female by Beardsley (1966).
CITATIONS: Beards1966 [taxonomy, description, illustration, host, distribution: 507-508]; Larew1990 [ecology, life history, structure: 293-300].

Annulaspis Ferris

Annulaspis Ferris, 1938a: 154. Type species: *Annulaspis polygona* Ferris, by original designation.
SYSTEMATICS: Ferris (1938a) assigned *Annulaspis* to the Odonaspidini, and his interpretation was accepted by McKenzie (1963) and Borchsenius (1966). Ben-Dov (1988b) concluded that the genus did not belong to the Odonaspidinae, but did not

indicate a subfamily placement. Superficially the species placed in *Annulaspis* by Ferris (1938a) and by McKenzie (1963) resemble other Odonaspidine species in the absence of lobes and plates on the pygidial margin, but differ remarkably in the absence of crenulae on abdominal sternites. Moreover, it is evident that *A. polygona* and *A. singularis*, are not congeneric. More work is required to place them in appropriate higher-level categories. Until this challenge has been met, *Annulaspis* is temporarily placed in the Aspidiotinae.

GENERAL: Description and definition by Ferris (1938a, 1938b).

CITATIONS: Balach1949 [taxonomy: 109]; Balach1953g [taxonomy: 728]; BenDov1988b [taxonomy: 6]; Borchs1966 [catalogue: 227]; Ferris1938a [taxonomy, description: 154]; Ferris1938b [taxonomy, description: 71,73]; Ferris1942 [taxonomy: 64]; McKenz1963 [taxonomy: 29]; MorrisMo1966 [taxonomy, catalogue: 10].

Annulaspis polygona Ferris

Annulaspis polygona Ferris, 1938a: 155. Type data: U.S.A.: Texas, Chisos Mountains, on undetermined, small, perennial grass. Syntypes, female. Type depository: Davis: Bohart Museum of Entomology, University of California, USA. *Dycryptaspis polygona*; Lindinger, 1957: 544. Change of combination.

SYSTEMATICS: *Annulaspis polygona* the type species of *Annulaspis* superficially resemble species of Odonaspidinae, in the absence of lobes and plates on the pygidial margin, but remarkably differs in the absence of crenulae on abdominal sternites. It is evident that *A. polygona* and *A. singularis*, are not congeneric. More work is required to place them in appropriate higher-level categories. Until this challenge will be met, both species are retained in *Annulaspis*.

SCALE COVER: Scale of female white, elongate, slender; exuviae apical, a strong ventral scale being formed. Male scale similar in form, colour and texture (Ferris, 1938a).

HOST PLANTS: Gramineae [Ferris1938a].

DISTRIBUTION: Nearctic: United States of America (Texas [Ferris1938a]).

BIOLOGY: Occurring on stems, protected by bases of leaves (Ferris, 1938a).

GENERAL: Description and illustration of adult female by Ferris (1938a).

CITATIONS: BenDov1988b [taxonomy: 6]; Borchs1966 [catalogue: 227]; Ferris1938a [taxonomy, description, illustration, host, distribution: 155]; Ferris1938b [taxonomy: 71,73]; Lindin1957 [taxonomy: 544].

Annulaspis singularis McKenzie

Annulaspis singularis McKenzie, 1963: 29. Type data: MEXICO: Baja California, 100 miles north of La Paz, Loreto, on undetermined species of Chenopodiaceae. Holotype female. Type depository: Davis: The Bohart Museum of Entomology, University of California, California, USA.

COMMON NAME: Unique scale [McKenz1963].

SYSTEMATICS: Based on original description of *A. polygona* and *A. singularis*, it is evident that they are not congeneric. Until the appropriate higher-level placement of this genus will be revised, *A. singularis* is temporarily placed in the *Annulaspis*.

SCALE COVER: Scale of female and male quite similar in size and colour, and described as resembling "little brown seeds" (McKenzie, 1963).
HOST PLANTS: **Chenopodiaceae** [McKenz1963].
DISTRIBUTION: **Nearctic**: Mexico (Baja California [McKenz1963]).
BIOLOGY: In bark cracks (McKenzie, 1963).
GENERAL: Description and illustration of adult female by McKenzie (1963).
CITATIONS: Borchs1966 [catalogue: 227]; McKenz1963 [taxonomy, description, illustration, host, distribution: 29-31].

Aonidia Targioni Tozzetti

Aonidia Targioni Tozzetti, 1868: 735. Type species: *Aonidia purpurea* Targioni Tozzetti (= *Aspidiotus lauri* Bouché), by monotypy.
Targionidea MacGillivray, 1921: 393. Type species: *Aspidiotus campylanthi* Lindinger, by original designation. New Synonym.
Cupressaspis Borchsenius, 1962b: 866. Type species: *Cupressaspis isfarensis* Borchsenius, by original designation. Synonymy by Danzig, 1993: 214.
SYSTEMATICS: Genus *Aonidia* includes 31 pupillarial species. Related to *Cryptaspidiotus* Lindinger, differing from latter in that the diameter of anal opening equals width of median lobes (whereas about 1/5 of width on *Cryptaspidiotus*).
GENERAL: Definition and characters by Froggatt (1914), Kuwana (1933), Ferris (1938a), Balachowsky (1951, 1958b), Lupo (1957), Borchsenius (1962b), Bazarov & Shmelev (1971), Danzig (1993) and by Yaşar (1995).
KEYS: Gill 1997: 24-26 (female) [Genera of California]; Danzig 1993: 215 (female) [species Europe]; Zahradník 1990b: 74 (female) [Czech Republic]; Tereznikova 1986: 83 (female) [Ukraine]; Chou 1985: 324 (female) [Genera of China]; Bazarov & Shmelev 1971: 186 (female) [Central Asia]; Munting 1965b: 189 (female) [South Africa]; Danzig 1964: 646 (female) [Europe]; Borchsenius 1962b: 869 (female) [species Palaearctic region]; Zahradník 1959a: 546 (female) [Czech Republic]; Balachowsky 1958b: 232 (female) [*Aspidiotina* of Africa]; Ezzat 1958: 237-239 (female) [Egypt]; McKenzie 1956: 22 (female) [U.S.A.: California]; Balachowsky 1951: 603 (female) [Mediterranean]; Borchsenius 1950b: 168 (female) [USSR]; Gómez-Menor Ortega 1946: 59-61 (female) [Spain]; Ruiz Castro 1944: 57 (female) [Spain]; Ferris 1942: 25 (female) [North America]; Ferris 1942: 29 (female) [species North America]; Archangelskaya 1937: 94 (female) [Middle Asia]; Borchsenius 1937: 100 (female) [USSR]; Borchsenius 1937a: 32-33 (female) [Palaearctic Region]; Kuwana 1933a: 43-45 (female) [Japan]; Archangelskaya 1929: 189 (female) [Palaearctic Region]; Leonardi 1920: 26 (female) [Italy]; Lawson 1917: 206 (female) [U.S.A.: Kansas]; Hempel 1900a: 496-497 (female) [Brazil]; Green 1896e: 37 (female) [Sri Lanka].
CITATIONS: Archan1929 [taxonomy: 189]; Archan1937 [taxonomy, description: 94,112]; Ashmea1891 [taxonomy: 102]; Balach1948b [taxonomy: 269]; Balach1951 [taxonomy, description: 603]; Balach1958b [taxonomy, description: 232-233]; BazaroSh1971 [taxonomy, description: 223]; BerlesLe1898a [taxonomy, description: 10-11]; BlayGo1993 [taxonomy, description: 387,393]; Bodenh1949 [taxonomy, description: 26,38-39]; Bodenh1952 [taxonomy: 329]; Borchs1937

[taxonomy, description: 100]; Borchs1937a [taxonomy, description: 33,68]; Borchs1950b [taxonomy, description: 168,234]; Borchs1962b [taxonomy, description: 866-868,871]; Borchs1966 [catalogue: 252,360,361]; Brain1918 [taxonomy: 116]; Brain1919 [taxonomy: 214]; Chou1985 [taxonomy, description: 325]; Cocker1896b [taxonomy: 338]; Cocker1897i [taxonomy: 31]; Cocker1899a [taxonomy: 396]; Comsto1883 [taxonomy: 128]; Danzig1964 [taxonomy: 654]; Danzig1993 [taxonomy, description: 214]; DanzigPe1998 [catalogue: 179]; Ezzat1958 [taxonomy: 237]; Fernal1903b [catalogue: 301]; Ferris1921b [taxonomy: 94]; Ferris1937c [taxonomy: 50,52]; Ferris1938a [taxonomy, description: 175]; Ferris1942 [taxonomy: 446:25]; Frogga1914 [taxonomy, description: 599]; Frogga1915 [taxonomy, description: 25]; Gill1997 [taxonomy: 44]; GomezM1937 [taxonomy, description: 99-100]; GomezM1946 [taxonomy: 60]; Green1896e [taxonomy, description: 37,68]; Hadzib1983 [taxonomy: 243]; Hempel1900a [taxonomy: 496]; HowellTi1990 [taxonomy: 57]; Kozar1990f [distribution: 143]; Kuwana1933 [taxonomy, description: 39-40,43]; Lawson1917 [taxonomy, description: 208]; Leonar1897 [taxonomy: 284-286]; Leonar1897b [taxonomy: 109,111]; Leonar1899 [taxonomy: 204]; Leonar1900 [taxonomy, description: 320-323]; Leonar1903 [taxonomy: 3-4]; Leonar1920 [taxonomy, description: 26,85-86]; Lindin1906 [taxonomy: 4,15,16]; Lindin1908b [taxonomy: 98]; Lindin1924 [taxonomy: 173]; Lindin1937 [taxonomy: 197]; Lindin1943b [taxonomy: 206]; Lupo1957 [taxonomy, description: 77-78]; MacGil1921 [taxonomy, description: 393,395,449,462-463]; Maskel1887a [taxonomy, description: 40]; Maskel1895b [taxonomy, description: 42-43]; McKenz1956 [taxonomy, description: 22]; Miller1990 [taxonomy: 169-178]; MorrisMo1966 [taxonomy, catalogue: 13,52,193]; Muntin1965b [taxonomy: 189]; Muntin1969 [taxonomy: 119]; Ramakr1930 [taxonomy: 13]; RuizCa1944 [taxonomy: 57]; Schmut1959 [taxonomy, description: 125]; Signor1869 [taxonomy: 860]; Signor1869b [taxonomy, description: 99]; Signor1870 [taxonomy: 102-103]; Tao1999 [taxonomy: 72]; Targio1868 [taxonomy, description: 735]; Targio1888 [taxonomy, description: 419,422]; Willia1969a [taxonomy: 319]; Yasar1995a [taxonomy, description: 40].

Aonidia atlantica Ferris

Aonidia atlantica Ferris, 1942: 424. Type data: U.S.A.: Florida, Tampa, on *Juniperus* sp. Holotype female. Type depository: Davis: The Bohart Museum of Entomology, University of California, California, USA.
Cryptaspidiotus atlanticius; Lindinger, 1957: 544. Change of combination.
Cupressaspis atlantica; Borchsenius, 1966: 360. Change of combination.
Aonidia atlantica; Danzig, 1993: 214. Revived combination.
COMMON NAME: juniper needle scale [Dekle1965c].
SCALE COVER: Female scale elongate oval, composed of enlarged second exuvia, covered with a thin film of wax; yellowish brown in colour. First exuvia placed toward anterior end of scale. Scale of male not recognized (Ferris, 1942).
HOST PLANTS: **Cupressaceae**: *Chamaecyparis thyoides* [BesheaTiHo1973], *Juniperus* [Ferris1941d, Merril1953, Dekle1965c], *J. virginiana* [BesheaTiHo1973].
DISTRIBUTION: **Nearctic**: United States of America (Alabama [BesheaTiHo1973], Florida [Ferris1942, Dekle1965c], Georgia [BesheaTiHo1973]).

BIOLOGY: Occurring on the inner surface of the needles (Ferris, 1942).
GENERAL: Description and illustration of adult female by Ferris (1942).
KEYS: Ferris 1942: 29 (female) [North America].
CITATIONS: BesheaTiHo1973 [host, distribution: 9]; Borchs1966 [catalogue: 360]; Dekle1965c [taxonomy, description, host, distribution: 18]; Dekle1976 [taxonomy, description, host, distribution, economic importance: 29]; Ferris1942 [taxonomy, description, illustration, host, distribution: 29,424]; Lindin1957 [taxonomy: 544]; Merril1953 [taxonomy, description, host, distribution: 12-13]; Nakaha1982 [host, distribution: 6].

Aonidia badia Brain
Aonidia badia Brain, 1919: 217. Type data: SOUTH AFRICA: Zeerust, on *Rhus* sp.; collected May 1915. Lectotype female, by subsequent designation Munting, 1970a: 36. Type depository: Pretoria: South African National Collection of Insects, South Africa; type no. 291/1.
Targionia badia; Lindinger, 1957: 544. Change of combination.
Aonidia badia; Borchsenius, 1966: 362. Revived combination.
SCALE COVER: Scale of adult female almost circular, flat, about 1.2 mm in diameter, consisting of thickened second stage *plus* a very thin, transparent layer of secretion, only noticeable where it projects at margins (Brain, 1919).
HOST PLANTS: **Anacardiaceae**: *Rhus* [Brain1919, Muntin1965a].
DISTRIBUTION: **Afrotropical**: South Africa [Brain1919, Muntin1965a].
GENERAL: Description and illustration of adult female by Brain (1919) and by Balachowsky (1958b).
KEYS: Munting 1965b: 189 (female) [South Africa].
CITATIONS: Balach1958b [taxonomy, description, illustration, host, distribution: 234-235]; Borchs1966 [catalogue: 362]; Brain1919 [taxonomy, description, illustration, host, distribution: 217-218]; Lindin1957 [taxonomy: 544]; Muntin1965a [taxonomy, description, host, distribution: 181,183,189]; Muntin1970a [taxonomy: 36-37].

Aonidia banksiae Fuller
Aonidia banksiae Fuller, 1897b: 1346. Type data: AUSTRALIA: Western Australia, on *Banksia attenuata, B. menziesii, B. prionotes* and *B. ilicifolia*. Syntypes, female.
Aonidia banksiae; Fuller, 1897c: 11. Notes: Described again as n. sp.
Aonidia banksiae; Fuller, 1899: 473. Notes: Described again as n. sp.
SCALE COVER: Female scale grey; appearing orange-red owing to colour of large second exuvia; circular, very convex; diameter 1/15 of an inch (Fuller, 1897, 1897c).
HOST PLANTS: **Proteaceae**: *Banksia attenuata* [Fuller1897b, Frogga1914], *Banksia ilicifolia* [Fuller1897b, Frogga1914], *Banksia menziesii* [Fuller1897b, Frogga1914], *Banksia prionotes* [Fuller1897b, Frogga1914].
DISTRIBUTION: **Australasian**: Australia (Western Australia [Fuller1897b, Frogga1914]).
GENERAL: Description of adult female by Fuller (1897, 1897c).
CITATIONS: Borchs1966 [catalogue: 362]; Cocker1899a [taxonomy: 396]; Fernal1903b [catalogue: 302]; Frogga1914 [taxonomy, description, host,

distribution: 599]; Frogga1915 [taxonomy, description, host, distribution: 25]; Fuller1897b [taxonomy, description, host, distribution: 1346]; Fuller1897c [taxonomy, description, host, distribution: 11]; Fuller1899 [taxonomy, description, host, distribution: 473]; Lindin1911 [taxonomy: 173]; MacGil1921 [taxonomy, description, host, distribution: 462].

Aonidia biafrae Lindinger

Aonidia biafrae Lindinger, 1909e: 40. Type data: CAMEROON: Bipinde, Urwaldgebiet, on *Crudia zenkeri* and *Schotia humboldtioides*. Syntypes, female. Type depository: Hamburg: Zoologisches Institut und Zoologishces Museum, Universität von Hamburg, Germany.
Greeniella biafrae; MacGillivray, 1921: 459. Change of combination.
Aonidia biafrae; Borchsenius, 1966: 362. Revived combination.
SCALE COVER: Pupillarial species; scale thin; colour dark yellow-brown (Lindinger, 1909e).
HOST PLANTS: **Leguminosae**: *Crudia zenkeri* [Lindin1909e, Balach1958b], *Cynometra* [Balach1958b], *Schotia humboldtioides* [Lindin1909e, Balach1958b].
DISTRIBUTION: **Afrotropical**: Cameroon [Lindin1909e, Balach1958b].
GENERAL: Description and illustration of adult female by Lindinger (1909e).
CITATIONS: Balach1958b [host, distribution: 240]; Borchs1966 [catalogue: 362]; Lindin1909e [taxonomy, description, illustration, host, distribution: 40-42]; MacGil1921 [taxonomy, description, host, distribution: 362]; Sassce1911 [taxonomy: 70]; Vayssi1913 [host, distribution: 431]; WeidneWa1968 [taxonomy: 171].

Aonidia bullata Green

Aonidia bullata Green, 1896e: 72. Type data: SRI LANKA: Punduloya, on an undetermined tree. Holotype female. Type depository: London: The Natural History Museum, England, UK.
SCALE COVER: Female and male scale skillfully illustrated and described in great detail by Green (1896e).
HOST PLANTS: **Anacardiaceae**: *Nothopegia colebrookiana* [Green1900a, Ramakr1921a].
DISTRIBUTION: **Oriental**: Sri Lanka [Green1896e, Green1900a, Ramakr1921a].
GENERAL: Description and illustration of adult female by Green (1896e).
KEYS: Green 1896e: 68 (female) [Sri Lanka].
CITATIONS: DEDAC1923 [host, distribution]; Green1896e [taxonomy, description, illustration, host, distribution: 72-73]; Green1900a [taxonomy, description, host, distribution: 73]; Green1922 [taxonomy: 460]; Johnst1915 [host, distribution, biological control: 1-33]; Ramakr1921a [host, distribution: 358].

Aonidia campylanthi (Lindinger)

Targionia ? campylanthi Lindinger, 1911a: 25. Type data: CANARY ISLANDS: Tenerife, between Santa Cruz and San Andres, on *Campylanthus salsoloides*. Syntypes, female. Type depository: Hamburg: Zoologisches Institut und Zoologishces Museum, Universität von Hamburg, Germany.

Aspidiotus campylanthi; MacGillivray, 1921: 393. Change of combination.
Targionidea campylanthi; MacGillivray, 1921: 449. Change of combination.
Aonidia campylanthi; Balachowsky, 1946: 211. Change of combination.
Targionidea campylanthi; Borchsenius, 1966: 252. Revived combination.
Aonidia campylanthi; Danzig & Pellizzari, 1998: 179. Revived combination.
SCALE COVER: Material available for original description very poor condition. Female scale whitish. Second instar not available (Lindinger, 1911a).
HOST PLANTS: **Scrophulariaceae**: *Campylanthus salsoides* [Lindin1911a, Balach1946].
DISTRIBUTION: **Palaearctic**: Canary Islands [Balach1946, MatileOr2001].
GENERAL: Description and illustration of adult female by Lindinger (1911a).
CITATIONS: Balach1946 [host, distribution: 211]; Borchs1966 [catalogue: 252]; DanzigPe1998 [catalogue: 179-180]; Ferris1937c [taxonomy, illustration: 52,100]; Ferris1943a [taxonomy: 85]; Lindin1911 [taxonomy, description, illustration, host, distribution: 25-26]; Lindin1912b [taxonomy, description, host, distribution: 93]; MacGil1921 [taxonomy, description, host, distribution: 393,449]; MatileOr2001 [host, distribution: 190]; Sassce1912 [taxonomy, host, distribution: 94]; WeidneWa1968 [taxonomy: 179].

Aonidia chaetachmeae Brain
Aonidia chaetachmeae Brain, 1919: 215. Type data: SOUTH AFRICA: Natal, Durban, on "umkavoti", *Chaetachme [=Chaetacme] aristata*; collected by C. Fuller, October 12, 1914. Syntypes, female. Type depository: Pretoria: South African National Collection of Insects, South Africa.
Aonidia chaetachmes Lindinger, 1932f: 196. Unjustified emendation.
SCALE COVER: Scale of female small, about 1-1.3 mm long, pyriform, with irregularly crenulate edges, dull pitch black in colour, with a scanty white layer of waxy secretion, most noticeable around margins. Larval exuviae central, raised, black, with dark brown margins (Brain 1919).
HOST PLANTS: **Ulmaceae**: *Chaetacme aristata* [Brain1919, Balach1958b, Muntin1965a].
DISTRIBUTION: **Afrotropical**: South Africa [Brain1919, Balach1958b, Muntin1965a].
GENERAL: Description and illustration of adult female by Brain (1919) and by Munting (1965a).
KEYS: Munting 1965b: 189 (female) [South Africa].
CITATIONS: Balach1958b [host, distribution: 240]; Borchs1966 [catalogue: 362]; Brain1919 [taxonomy, description, illustration, host, distribution: 215]; Lindin1932f [taxonomy: 196]; Muntin1965a [taxonomy, description, illustration, host, distribution: 182-183,189].

Aonidia crenulata Green
Aonidia ebeni Leonardi, 1899: 205. Nomen nudum.
Aonidia crenulata Green, 1900a: 74. Type data: SRI LANKA: Peradeniya, near Kandy, Royal Botanic Gardens, on *Memecylon umbellatum*. Syntypes, female. Type depository: London: The Natural History Museum, England, UK.

Aonidia ebeni Leonardi, 1900: 329. Type data: SRI LANKA: on *Diospyros* sp. Syntypes, female. Type depository: Portici: Dipartimento de Entomologia e Zoologia Agraria di Portici, Università di Napoli Federico II, Italy. Synonymy by Sanders, 1906: 16. Notes: Incorrect citation of "Green" as author.
Gymnaspis ebeni; Lindinger, 1909: 149. Change of combination.
Gymnaspis crenulata; MacGillivray, 1921: 257. Change of combination.
Aonidia crenulata; Borchsenius, 1966: 362. Revived combination.
SCALE COVER: Female scale circular, diameter 1 mm; moderately convex; dull pale reddish or yellowish-brown, with darker zones caused by transmitted colour of underlying pellicle. Male scale similar in size and colour to that of female, but flatter and slightly oval (Green, 1900a).
HOST PLANTS: **Dipterocarpaceae**: *Vatica lanceifolia* [Green1919c, Ramakr1921a]. **Ebenaceae**: *Diospyros* [Leonar1900]. **Memecylaceae**: *Memecylon* [Green1905a], *Memecylon umbellatum* [Green1900a, Green1905a, Green1922, Green1937].
NATURAL ENEMIES: FUNGI **Ascomycotina**: *Nectria diploa* [EvansPr1990].
DISTRIBUTION: **Oriental**: India [Ramakr1921a, Green1937] (Assam [Green1919c]); Sri Lanka [Green1900a, Leonar1900, Green1905a, Green1937].
GENERAL: Description and illustration of adult female by Green (1900a).
CITATIONS: Borchs1966 [catalogue: 362]; Cocker1899a [taxonomy: 396]; Cocker1922a [taxonomy: 149]; EvansPr1990 [biological control: 3-17]; Fernal1903b [catalogue: 302-303]; Ferris1941e [taxonomy: 43]; Green1900a [taxonomy, description, illustration, host, distribution: 74]; Green1905a [taxonomy, host, distribution: 348]; Green1919c [host, distribution: 441]; Green1922 [host, distribution: 463]; Green1937 [host, distribution: 336]; Johnst1915 [host, distribution, biological control: 1-33]; Leonar1899 [taxonomy: 205]; Leonar1900 [taxonomy, description, illustration, host, distribution: 329-330]; Leonar1903a [taxonomy: 6]; Lindin1909 [taxonomy: 149]; Lindin1911 [taxonomy: 12]; MacGil1921 [taxonomy, description, host, distribution: 257]; Petch1921a [biological control: 89-167]; Ramakr1921a [host, distribution : 359]; Sander1906 [taxonomy, host, distribution: 16].

Aonidia echinata Green
Aonidia echinata Green, 1905a: 347. Type data: SRI LANKA: Anaradhapura, on *Hemicyclia sepiaria*. Syntypes, female. Type depository: London: The Natural History Museum, England, UK.
SCALE COVER: Female scale dull reddish-brown (yellowish when immature), roughened, with innumerable slender curved spines that are firmly attached to nymphal pellicle and persist after treatment with caustic potash; circular; strongly convex; diameter 0.35 mm (Green, 1905a).
HOST PLANTS: **Euphorbiaceae**: *Hemicyclia sepiaria* [Green1905a, Sander1906, Ramakr1921a, Green1922, Green1937].
DISTRIBUTION: **Oriental**: Sri Lanka [Green1905a, Sander1906, Ramakr1921a, Green1922, Green1937].
GENERAL: Description and illustration of adult female by Green (1905a).

CITATIONS: Borchs1966 [catalogue: 363]; DEDAC1923 [host, distribution]; Green1905a [taxonomy, description, illustration, host, distribution: 347]; Green1922 [host, distribution: 463]; Green1937 [host, distribution: 337]; MacGil1921 [taxonomy, description, host, distribution: 463]; Ramakr1921a [host, distribution: 359]; Sander1906 [taxonomy, host, distribution: 16].

Aonidia elaeagna Maskell

Aonidia elaeagnus Maskell, 1897a: 241. Type data: JAPAN: on *Elaeagnus macrophylla*. Syntypes, female and first instar. Type depository: Auckland: New Zealand Arthropod Collection, Landcare Research, New Zealand.
Aonidia eleagnus; Leonardi, 1900: 241. Incorrect synonymy. Notes: See Systematics below.
Aonidia eleagnus; Leonardi, 1900: 327. Misspelling of species name.
SYSTEMATICS: Leonardi (1900: 327) regarded *Aonidia elaeagnus* Maskell, 1897a, a synonym of *Aonidia lauri* Bouché, 1833, whereas Borchsenius (1966) retained *A. elaeagna* as an unrecognizable, but valid species.
SCALE COVER: Female scale very small, reddish-brown, very slightly convex; exuviae yellow (Maskell, 1897a).
HOST PLANTS: **Elaeagnaceae**: *Elaeagnus macrophylla* [Kuwana1933].
DISTRIBUTION: **Palaearctic**: Japan [Kuwana1917a, Kuwana1933].
GENERAL: Description and illustration of adult female by Maskell (1897a) and by Kuwana (1933).
KEYS: Kuwana 1933b: 50 (female) [Japan].CITATIONS: Borchs1966 [catalogue: 363]; Cocker1899a [taxonomy: 396]; DanzigPe1998 [catalogue: 180]; DeitzTo1980 [taxonomy: 36]; Fernal1903b [catalogue: 302]; Kuwana1917a [taxonomy, distribution: 176]; Kuwana1933 [taxonomy, description, illustration, host, distribution: 40]; Maskel1897a [taxonomy, description, host, distribution: 241]; Maskel1898 [taxonomy, description, host, distribution: 227-228].

Aonidia formosana Takahashi

Aonidia tentaculata formosana Takahashi, 1935: 35. Type data: TAIWAN: Habon, near Musha, on *Cinnamomum* sp. Syntypes, female. Type depository: Taichung: Entomology Collection, Taiwan Agricultural Research Institute, Wu-feng, Taichung, Taiwan.
Aonidia formosana; Borchsenius, 1966: 363. Change of status.
SCALE COVER: Scale of adult female yellowish brown, shining, with a large longitudinal black patch at median part; flattened, smooth (Takahashi, 1935).
HOST PLANTS: **Lauraceae**: *Cinnamomum* [Takaha1935, Takagi1970].
DISTRIBUTION: **Oriental**: Taiwan [Takaha1935, Takagi1970].
GENERAL: Description and illustration of adult female by Takahashi (1935).
CITATIONS: Borchs1966 [catalogue: 363]; Chou1985 [taxonomy: 325]; Chou1985 [taxonomy, description, host, distribution]; Takagi1970 [taxonomy, host, distribution: 131]; Takaha1935 [taxonomy, description, illustration, host, distribution: 35-36]; Tao1999 [taxonomy, host, distribution: 72].

Aonidia ilicitana Gómez-Menor Ortega
Aonidia ilicitana Gómez-Menor Ortega, 1968: 542. Type data: SPAIN: Elche, Pantano de Vinalapo, Alicante and Valencia, Benejama, on *Pinus*. Syntypes, female. Type depository: Madrid: Museo Nacional de Ciencias Naturales, Spain.
SCALE COVER: Pupillarial species. Female scale 0.8-0.9 mm long, 0.85-0.95 mm wide; elliptical; ventral scale white, attached to host plant. Male scale white; 1.1-1.2 mm long, 0.6 mm wide; ventral scale white (Gómez-Menor Ortega, 1986).
HOST PLANTS: **Pinaceae**: *Pinus* [GomezM1968].
DISTRIBUTION: **Palaearctic**: Spain [GomezM1968].
GENERAL: Description and illustration of adult female by Gómez-Menor Ortega (1968).
CITATIONS: DanzigPe1998 [catalogue: 180]; GomezM1968 [taxonomy, description, illustration, host, distribution: 542-546]; Martin1983 [taxonomy, host, distribution: 60].

Aonidia isfarensis (Borchsenius)
Cryptaspidiotus mediterraneus; Archangelskaya, 1937: 112. Misidentification; discovered by Borchsenius, 1966: 360.
Cryptaspidiotus mediterraneus; Borchsenius, 1950b: 234. Misidentification; discovered by Borchsenius, 1966: 360.
Cupressaspis isfarensis Borchsenius, 1962b: 870. Type data: TADZHIKISTAN: Turkestan mountain ridge, south of Voruh, on *Juniperus* sp. Syntypes, female. Type depository: St. Petersburg: Zoological Museum, Academy of Science, Russia.
Aonidia isfarensis; Danzig, 1993: 215. Change of combination.
SCALE COVER: Female scale almost circular; exuviae not discernible from top, covered with white wax secretion; diameter 0.8-0.9 mm (Borchsenius, 1962b).
HOST PLANTS: **Cupressaceae**: *Juniperus* [Borchs1962b], *Juniperus seravschanica* [BazaroSh1971].
DISTRIBUTION: **Palaearctic**: Tajikistan (=Tadzhikistan) [Borchs1962b].
GENERAL: Description and illustration of adult female by Borchsenius (1962b), Bazarov & Shmelev (1971) and by Danzig (1993).
KEYS: Danzig 1993: 215 (female) [Europe]; Borchsenius 1962b: 869 (female) [Palaearctic region].
CITATIONS: Archan1937 [taxonomy: 112]; BazaroSh1971 [taxonomy, description, illustration, host, distribution: 223-226]; Borchs1950b [taxonomy: 234]; Borchs1962b [taxonomy, description, illustration, host, distribution: 868-870]; Borchs1966 [catalogue: 360]; Danzig1993 [taxonomy, description, illustration, host, distribution, economic importance: 215-217]; Sassce1911 [taxonomy: 71].

Aonidia laticornis Balachowsky
Aonidia laticornis Balachowsky, 1949b: 114. Type data: MOROCCO: on *Cupressus sempervirens*. Syntypes, female. Type depository: Paris: Muséum national d'Histoire naturelle, France.
SCALE COVER: Female scale subcircular, flat, white; larval exuviae bright yellow, subcentral; 0.8-1.2 mm; pupillarial species in which adult female enclosed in nymphal exuviae (Balachowsky, 1951).

HOST PLANTS: **Cupressaceae**: *Cupressus sempervirens* [Balach1949b, Balach1951].
DISTRIBUTION: **Palaearctic**: Morocco [Balach1949b, Balach1951].
GENERAL: Description and illustration of adult female by Balachowsky (1949, 1951).
KEYS: Balachowsky 1951: 604 (female) [Mediterranean].
CITATIONS: Balach1949b [taxonomy, description, illustration, host, distribution: 113-115]; Balach1951 [taxonomy, description, illustration, host, distribution: 612-615]; Borchs1962b [taxonomy, host, distribution: 868]; Borchs1966 [catalogue: 363]; DanzigPe1998 [catalogue: 180].

Aonidia lauri (Bouché)
Aspidiotus lauri Bouché, 1833: 52. Type data: GERMANY: Berlin, in greenhouse, on laurel [=*Laurus nobilis*]. Syntypes, both sexes. Notes: Type material lost (Sachtleben, 1944).
Chermes lauri; Boisduval, 1867: 340. Change of combination.
Diaspis aonidum Targioni-Tozzetti, 1867: 78. Synonymy by Borchsenius, 1966: 363.
Aonidia purpurea Targioni Tozzetti, 1868: 735. Unjustified replacement name for *Aspidiotus lauri* Bouché.
Aonidia lauri; Signoret, 1870: 327. Change of combination.
COMMON NAMES: cochenille du laurier [SchmutKlLu1957]; laurel scale [McKenz1956]; Lorbeerschildlaus [SchmutKlLu1957].
SCALE COVER: Scale of female somewhat oval, rather convex, brown, first exuvia near one end; that of male similar in colour, oval; exuvia at one end; pupillarial species (Ferris, 1938a).
HOST PLANTS: **Lauraceae**: *Apollonias canariensis* [Bodenh1949, BlayGo1993], *Laurus* [Bodenh1928, Bodenh1937, Balach1951, Bodenh1952], *Laurus canariensis* [Bodenh1949, BlayGo1993], *Laurus nobilis* [Bouche1833, Bouche1834, Balach1927, Balach1931a, Balach1932d, BachmaGe1950, Bodenh1952, Bachma1953], *Laurus nobilis* [SengonUyKa1998, UygunSeEr1998, ErlerTu2001], *Ocotea foetens* [Balach1951].
NATURAL ENEMIES: ACARI **Hemisarcoptidae**: *Hemisarcoptes malus* (Shimer) [GersonOcHo1990]. FUNGI **Ascomycotina**: *Myriangium duriaei* [EvansPr1990], *Nectria flammea* [EvansPr1990]. HYMENOPTERA **Aphelinidae**: *Aphytis aonidiae* (Mercet) [RosenDe1979], *Aphytis chilensis* Howard [RosenDe1979], *Aphytis mytilaspidis* (Le Baron) [RosenDe1979], *Hispaniella lauri* Mercet [Viggia1990a], *Pteroptrix lauri* (Mercet) [SengonUyKa1998, ErlerTu2001].
DISTRIBUTION: **Nearctic**: United States of America (California [McKenz1956]). **Neotropical**: Brazil [Hempel1900a]. **Palaearctic**: Algeria [Balach1927, Balach1932d]; Canary Islands [MatileOr2001]; Corsica [Balach1931a, Balach1932d]; Czech Republic [Zahrad1977, Zahrad1990b]; Egypt [Ezzat1958]; France [Signor1870, Balach1932d, Balach1937c]; Georgia [Hadzib1983]; Germany (United) [Bouche1833]; Greece [Bodenh1928, Korone1934, ArgyriStMo1976, RosenDe1979]; Israel [Borchs1966]; Italy [Leonar1920, LongoMaPe1995]; Morocco [Balach1932d]; Poland [Komosi1968, Dziedz1989]; Portugal

[Seabra1942]; Spain [GomezM1954, GomezM1958c, RosenDe1979, Martin1983, BlayGo1993]; Switzerland [BachmaGe1950]; Tunisia [Balach1932d]; Turkey [Bodenh1949, Bodenh1952, SengonUyKa1998, UygunSeEr1998]; United Kingdom (England [Malump1997]); Yugoslavia [Bachma1953].

BIOLOGY: Occurring on twigs or leaves.

ECONOMIC IMPORTANCE: Widely distributed in Mediterranean basin, almost exclusively on *Laurus nobilis* and other species of *Laurus*. Dense populations may develop on laurel, causing leaf drop and twig die-back (Balachowsky, 1951).

GENERAL: Description and illustration of adult female by Ferris (1938a), Balachowsky (1951), McKenzie (1956), Dziedzicka (1989), Tereznikova (1986), Zahradník (1990b), Danzig (1993) and by Yaşar (1995a). *Aspidiotus lauri* was also described as n.sp. by Bouché (1834) page 16.

KEYS: Danzig 1993: 215 (female) [Europe]; Ezzat 1958: 240 (female) [Egypt]; McKenzie 1956: 23 (female) [U.S.A.: California]; Balachowsky 1951: 604 (female) [Mediterranean]; Ferris 1942: 29 (female) [North America].

CITATIONS: Apstei1915 [taxonomy: 119]; ArgyriStMo1976 [host, distribution, biological control: 24]; Bachma1953 [host, distribution: 182]; BachmaGe1950 [host, distribution: 119]; Balach1927 [host, distribution: 178]; Balach1931a [host, distribution: 98]; Balach1932d [taxonomy, host, distribution: XII, XLVIII]; Balach1937c [host, distribution: 3]; Balach1951 [taxonomy, description, illustration, host, distribution: 604-607]; BaldanGaVi1999 [biological control: 209-215]; BenDov1990c [taxonomy: 116]; BeneGaLa2000 [chemistry, chemical control: 43-61]; Blanch1840 [taxonomy: 214]; BlayGo1993 [taxonomy, description, illustration, host, distribution: 388-392]; Bodenh1928 [host, distribution: 191]; Bodenh1935 [host, distribution: 247]; Bodenh1937 [host, distribution: 217]; Bodenh1949 [taxonomy, description, illustration, host, distribution: 84-86]; Bodenh1952 [taxonomy, host, distribution: 351]; Boisdu1867 [taxonomy, description, host, distribution: 340]; Borchs1937 [taxonomy, description, illustration, host, distribution: 137]; Borchs1937a [taxonomy, description, host, distribution: 69-70]; Borchs1950b [taxonomy, description, host, distribution: 235]; Borchs1966 [catalogue: 363]; Bouche1833 [taxonomy, description, host, distribution: 52]; Bouche1834 [taxonomy: 16]; Brown1963 [taxonomy, structure: 360-406]; BurgerUl1990 [economic importance: 313-327]; Burmei1835 [taxonomy: 68]; Comsto1883 [taxonomy, description, host, distribution: 129]; Costan1938 [host, distribution: 25-44]; Danzig1964 [taxonomy, host, distribution: 654]; Danzig1972 [taxonomy, host, distribution, economic importance: 206]; Danzig1993 [taxonomy, description, illustration, host, distribution: 219-220]; DanzigPe1998 [catalogue: 180]; DeBach1964d [biological control: 5-18]; Dingle1924 [taxonomy: 371]; Dziedz1989 [taxonomy, description, illustration, host, distribution: 97-98]; ErlerTu2001 [host, distribution, biological control: 299-305]; EvansPr1990 [biological control: 3-17]; Ezzat1958 [distribution: 240]; EzzatNa1987 [distribution: 86]; Fernal1903b [catalogue: 302]; Ferris1937c [taxonomy, illustration: 50,59]; Ferris1938a [taxonomy, description, illustration, host, distribution: 176]; Ferris1941e [taxonomy: 45]; Ferris1942 [taxonomy: 29]; Foldi2001 [distribution: 303-308]; Foldi2002 [host, distribution: 246]; Garcia1930 [host, distribution, biological control]; GersonOcHo1990 [biological control: 77-97]; Gill1997 [host, distribution, taxonomy, illustration: 43,44]; GomezM1937 [taxonomy, description, illustration, host, distribution: 100-104]; GomezM1954 [host, distribution: 120]; GomezM1958a [host, distribution: 7]; GomezM1958c [host, distribution: 406]; Green1937 [host, distribution: 338]; Hadzib1983 [taxonomy, host, distribution, economic importance: 244]; Hempel1900a [taxonomy, description, host, distribution: 508]; Iperti1961 [economic importance: 14-30]; Komosi1968 [host, distribution: 205-208]; Korone1934 [taxonomy, description, illustration, host, distribution: 26-27]; Koteja1990b [life history, structure, anatomy: 233-242]; KozarHi1996 [host, distribution: 91-96]; LandiDe1994 [host, distribution, life history: 33-45]; Leonar1897 [taxonomy: 286]; Leonar1900 [taxonomy, description, illustration, host, distribution: 327-329]; Leonar1903a [taxonomy: 4]; Leonar1920 [taxonomy, description, illustration, host, distribution: 86-89]; Lindin1909a [taxonomy: 324]; Lindin1912b [taxonomy, description, host, distribution: 70,197,255]; Lindin1935 [taxonomy: 127]; LongoMaPe1995 [distribution: 125]; Lupo1957 [taxonomy, description, illustration, host, distribution: 78-84]; MacGil1921 [taxonomy, description, host, distribution: 464]; Malump1997 [host, distribution: 195-198]; Martin1983 [taxonomy, host, distribution: 61]; MatileOr2001 [host, distribution: 189]; McKenz1956 [taxonomy, description, illustration, host, distribution: 36-38]; Melis1949 [host, distribution: 17-25]; MillerDa1990 [host, distribution, economic importance: 300]; Nakaha1982 [host, distribution: 6]; Nur1990a [taxonomy, structure, chromosomes: 184-185]; Peleka1962 [host, distribution: 62]; Porcel1995 [structure: 25-45]; Priore1964 [host, distribution: 131-178]; Priore1965 [host, distribution: 101-145]; Pulsel1927 [biological control: 300-327]; RosenDe1979 [host, distribution, biological control: 349-354,464-473,]; Ruhl1913 [host, distribution: 79-80]; Ryan1946 [host, distribution: 124-125]; Saakya1954 [host, distribution, economic importance]; Schmut1957a [host, distribution: 137]; Schmut1959 [taxonomy: 125]; SchmutKlLu1957 [host, distribution, economic importance: 493]; Seabra1942 [distribution: 2]; SengonUyKa1998 [host, distribution, biological control: 128-131]; Signor1869 [taxonomy: 860,868]; Signor1870 [taxonomy, description, illustration, host, distribution: 103-105]; Targio1867 [taxonomy: 78]; Targio1868 [taxonomy: 735]; Terezn1986 [taxonomy, description, illustration, host, distribution: 84-85]; UygunSeEr1998 [host, distribution: 183-191]; Viggia1990a [biological control: 128]; Yasar1995a [taxonomy, description, illustration, host, distribution: 40-42]; Zahrad1977 [taxonomy, distribution: 119]; Zahrad1990b [taxonomy, description, illustration, host, distribution: 78-80].

Aonidia longa Lindinger

Aonidia longa Lindinger, 1911: 172. Type data: NEW CALEDONIA: Mont Humboldt, on *Podocarpus gnidioides*; collected 16.xi.1902. Syntypes, female. Type depository: Hamburg: Zoologisches Institut und Zoologishces Museum, Universität von Hamburg, Germany; type no. MP156.

Greeniella longa; MacGillivray, 1921: 462. Change of combination.

Aonidia longa; Borchsenius, 1966: 363. Revived combination.

SYSTEMATICS: Williams & Watson (1988) examined specimens (deposited in Zoologisches Institut und Zoologisches Museum, Universität Hamburg) of this interesting pupillarial species, and indicated that it does not belong to *Aonidia* as presently understood, but the specimens available were not adequate for illustration and further description.

SCALE COVER: Not available for description (Lindinger, 1911).

HOST PLANTS: **Podocarpaceae**: *Podocarpus gnidioides* [Lindin1911, Laing1933].

DISTRIBUTION: **Australasian**: New Caledonia [Lindin1911, Laing1933].

GENERAL: Description and illustration of adult female by Lindinger (1911).

CITATIONS: Borchs1966 [catalogue: 363]; Laing1933 [host, distribution: 676]; Lindin1911 [taxonomy, description, illustration, host, distribution: 172-174]; MacGil1921 [taxonomy, description, host, distribution: 462]; Sassce1912 [taxonomy, host, distribution: 92]; WeidneWa1968 [taxonomy: 171]; WilliaWa1988 [taxonomy: 17].

Aonidia loranthi Green

Aonidia loranthi Green, 1896e: 74. Type data: SRI LANKA: on *Loranthus* sp. Holotype female. Type depository: London: The Natural History Museum, UK.

SCALE COVER: Female scale skillfully illustrated and described in great detail by Green (1896e).

HOST PLANTS: **Loranthaceae**: *Loranthus* [Green1896e, Leonar1900, Ramakr1921a, Green1937].

DISTRIBUTION: **Oriental**: Sri Lanka [Green1896e, Leonar1900, Green1937].

GENERAL: Description and illustration of adult female by Green (1896e)

KEYS: Green 1896e: 68 (female) [Sri Lanka].

CITATIONS: Borchs1966 [catalogue: 363]; Cocker1899a [taxonomy: 396]; Fernal1903b [catalogue: 363]; Green1896e [taxonomy, description, illustration, host, distribution: 68,74-75]; Green1937 [host, distribution: 336]; Larew1990 [ecology, life history, structure: 293-300]; Leonar1900 [taxonomy, description, illustration, host, distribution: 335-336]; Leonar1903a [taxonomy: 6]; Lindin1943b [taxonomy: 264]; MacGil1921 [taxonomy, description, host, distribution: 464]; Ramakr1921a [host, distribution: 358].

Aonidia marginalis Brain

Aonidia marginalis Brain, 1919: 216. Type data: SOUTH AFRICA: Transvaal, Zeerust, on stems of *Rhus* sp.; collected by A. Kelly, May 1915. Lectotype female, by subsequent designation Munting, 1970a: 39. Type depository: Pretoria: South African National Collection of Insects, South Africa; type no. 292/1.

SCALE COVER: Scale of adult female about 1.5 mm long, composed of large second exuviae, with a thin layer of buff-coloured secretion. Exuviae often showing through, black (Brain 1919).

HOST PLANTS: **Anacardiaceae**: *Rhus* [Brain1919, Balach1958b, Muntin1965a].

DISTRIBUTION: **Afrotropical**: South Africa [Brain1919, Balach1958b, Muntin1965a].

GENERAL: Description and illustration of adult female by Brain (1919) and by Munting (1965a).

KEYS: Munting 1965b: 189 (female) [South Africa].

CITATIONS: Balach1958b [host, distribution: 240]; Borchs1966 [catalogue: 363-364]; Brain1919 [taxonomy, description, illustration, host, distribution: 216-217]; Muntin1965a [taxonomy, description, illustration, host, distribution: 184-185,189]; Muntin1970a [taxonomy: 39].

Aonidia maroccana Balachowsky

Aonidia maroccana Balachowsky, 1949b: 115. Type data: MOROCCO: Moyen Atlas, 12 km northwest of Ksiba, altitude 1500 m, on *Laurus nobilis*. Syntypes, female. Type depository: Paris: Muséum national d'Histoire naturelle, France.
Cryptaspidiotus maroccanus; Lindinger, 1957: 545. Change of combination.
Aonidia maroccana; Borchsenius, 1966: 364. Revived combination.
SCALE COVER: Pupillarial species; female scale subcircular, light brown, sometimes covered with white, fine secretion; diameter 0.6-0.8 mm. Male scale of similar structure, more oval, elongated, 0.5 mm (Balachowsky, 1949b, 1951).
HOST PLANTS: **Lauraceae**: *Laurus nobilis* [Balach1949b, Balach1951].
DISTRIBUTION: **Palaearctic**: Morocco [Balach1949b, Balach1951].
GENERAL: Description and illustration of adult female by Balachowsky (1949b, 1951).
KEYS: Balachowsky 1951: 604 (female) [Mediterranean].
CITATIONS: Balach1949b [taxonomy, description, illustration, host, distribution: 115-116]; Balach1951 [taxonomy, description, illustration, host, distribution: 610-612]; Borchs1966 [catalogue: 364]; DanzigPe1998 [catalogue: 180]; Lindin1957 [taxonomy: 545].

Aonidia mediterranea (Lindinger)

Cryptaspidiotus mediterraneus Lindinger, 1910c: 437. Type data: ALGERIA: on *Juniperus phoenicea*. Holotype female. Type depository: Hamburg: Zoologisches Institut und Zoologishces Museum, Universität von Hamburg, Germany.
Aonidia mediterranea; Ferris, 1942: 424. Change of combination.
Cryptaspidiotus juniperi Borchsenius, 1949b: 352. Type data: ARMENIA: Megri, on *Juniperus* sp. Lectotype female, by subsequent designation Danzig, 1993: 219. Type depository: St. Petersburg: (= Leningrad) Zoological Museum, Academy of Science, Russia. Synonymy by Danzig, 1993: 215.
Cupressaspis juniperi; Borchsenius, 1962b: 869. Change of combination.
Cupressaspis mediterranea; Borchsenius, 1962b: 869. Change of combination.
Aonidia mediterranea; Danzig, 1993: 215. Revived combination.
SCALE COVER: Female scale small, 0.9 mm; subcircular, flat; larval exuviae central or subcentral, faint yellow; adult secretion white (Balachowsky, 1951).
HOST PLANTS: **Cupressaceae**: *Callitris articulata* [Balach1951], *Ca. quadrivalvis* [Lindin1910c, Rungs1935], *Cupressus sempervirens* [Balach1951], *Juniperus* [Borchs1949b], *Ju. oxycedrus* [GomezM1960O], *Ju. phoenicea* [GomezM1960O, Zahrad1972], *Thuja orientalis* [Balach1951, Zahrad1972].
DISTRIBUTION: **Palaearctic**: Algeria [Lindin1910c]; Armenia [Borchs1949b]; France [Foldi2002]; Greece [Korone1934]; Morocco [Rungs1935]; Sicily [NucifoWa2001]; Spain [GomezM1960O, Martin1983, BlayGo1993]; Turkey [ErlerKoTu1996]; Uzbekistan [Balach1951].

GENERAL: Description and illustration of adult female by Lindinger (1910c), Borchsenius (1949b), Balachowsky (1951) and by Danzig (1993).
KEYS: Danzig 1993: 215 (female) [Europe]; Borchsenius 1962b: 869, 871 (female) [Palaearctic region]; Borchsenius 1962b: 869 (female) [Palaearctic region]; Balachowsky 1951: 604 (female) [Mediterranean].
CITATIONS: Archan1937 [taxonomy, description, host, distribution: 112]; Balach1951 [taxonomy, description, illustration, host, distribution: 607-609]; BlayGo1993 [taxonomy, description, illustration, host, distribution: 394-397]; Borchs1949b [taxonomy, description, illustration, host, distribution: 352]; Borchs1949d [taxonomy, description, host, distribution: 251]; Borchs1950b [taxonomy, description, host, distribution: 234]; Borchs1962b [taxonomy: 869,871]; Borchs1966 [catalogue: 360]; Danzig1993 [taxonomy, description, illustration, host, distribution: 215-219]; ErlerKoTu1996 [host, distribution: 53-59]; Ferris1937c [taxonomy: 51,68,106]; Ferris1941e [taxonomy: 45]; Ferris1942 [taxonomy: 424]; Foldi2002 [host, distribution: 246]; GomezM1960O [taxonomy, description, illustration, host, distribution: 174-177]; Korone1934 [taxonomy, description, illustration, host, distribution: 25-26]; Lindin1910c [taxonomy, description, host, distribution: 437]; Lindin1912b [taxonomy, description, host, distribution: 89,189-190]; Lindin1935 [taxonomy: 132]; MacGil1921 [taxonomy, description, host, distribution: 426]; Martin1983 [taxonomy, host, distribution: 63]; NucifoWa2001 [host, distribution: 207-209]; Rungs1935 [host, distribution: 270]; TerGri1962 [taxonomy, description, host, distribution: 151-152]; WeidneWa1968 [taxonomy: 174]; Zahrad1972 [host, distribution: 440].

Aonidia mesembryanthemae Brain
Aonidia mesembryanthemae Brain, 1919: 216. Type data: SOUTH AFRICA: Natal, Durban, on fleshy leaves of *Mesembryanthemum edule*; collected 20.iv.1915. Lectotype female, by subsequent designation Munting, 1970a: 39. Type depository: Pretoria: South African National Collection of Insects, South Africa; type no. 290/1.
Aonidia mesembryanthemi Lindinger, 1932f: 196. Unjustified emendation.
SCALE COVER: Female dorsal scale about 1.5 mm long and 1 mm broad, white, thin, translucent, very delicate with buff-coloured larval exuviae central. Male scale about 1 mm long, slightly more elongate and more convex than female scale, white to buff in colour, with yellowish larval exuviae (Brain, 1919).
HOST PLANTS: **Aizoaceae**: *Mesembryanthemum edule* [Brain1919, Balach1958b].
DISTRIBUTION: **Afrotropical**: South Africa [Brain1919, Balach1958b].
BIOLOGY: Developing scale induces formation of pits on fleshy leaves of *Mesembryanthemum edule* (Brain, 1919).
GENERAL: Description and illustration of adult female by Brain (1919) and by Balachowsky (1958b).
KEYS: Munting 1965b: 189 (female) [South Africa].
CITATIONS: Balach1958b [taxonomy, description, illustration, host, distribution: 234-237]; Borchs1966 [catalogue: 364]; Brain1919 [taxonomy, description, illustration, host, distribution: 216]; Lindin1932f [taxonomy: 196]; Muntin1965b [taxonomy: 189]; Muntin1970a [taxonomy: 39].

Aonidia mimusopis Green

Aonidia mimusopis Green, 1922a: 1009. Type data: SRI LANKA: Paradeniya, on the smaller branches of *Diospyros thwaitesii*. Syntypes, female. Type depository: London: The Natural History Museum, England, UK.
SCALE COVER: Female scale small, inconspicuous; pale greyish, centre darker; broadly ovate, flat. Exuviae covered by secretionary covering (Green, 1922a).
HOST PLANTS: **Ebenaceae**: *Diospyros thwaitesii* [Green1922a]. **Sapotaceae**: *Mimusops hexandra* [Green1937].
DISTRIBUTION: **Oriental**: Sri Lanka [Green1922a, Green1937].
GENERAL: Description and illustration of adult female by Green (1922a).
CITATIONS: Borchs1966 [catalogue: 364]; Green1922a [taxonomy, description, illustration, host, distribution: 1009-1010]; Green1937 [host, distribution: 338].

Aonidia nullispina Munting

Aonidia nullispina Munting, 1969: 119. Type data: SOUTH AFRICA: Cape Province, Tosca, on *Olea* sp.; collected 8.iii.1967. Holotype female. Type depository: Pretoria: South African National Collection of Insects, South Africa; type no. 2713/14.
SCALE COVER: Scale of adult female oval, brown, with a pale white secreted covering, about 1 mm in length. Male scale oval, cream coloured, with a yellow subcentral exuviae; about 1 mm long (Munting, 1969).
HOST PLANTS: **Capparidaceae**: *Boscia albitrunca* [Muntin1969], *Boscia foetida* [Muntin1969]. **Oleaceae**: *Olea* [Muntin1969].
DISTRIBUTION: **Afrotropical**: Namibia (Southwest Africa) [Muntin1969]; South Africa [Muntin1969].
GENERAL: Description and illustration of adult female by Munting (1969).
CITATIONS: Muntin1969 [taxonomy, description, illustration, host, distribution: 119-120,145].

Aonidia obscura Green

Aonidia obscura Green, 1896e: 76. Type data: SRI LANKA: Punduloya, on *Loranthus* sp. Syntypes, female. Type depository: London: The Natural History Museum, England, UK.
SCALE COVER: Female scale skillfully illustrated and described in great detail by Green (1896e).
HOST PLANTS: **Loranthaceae**: *Loranthus* [Green1896e, Ramakr1921a].
DISTRIBUTION: **Oriental**: Sri Lanka [Green1896e, Ramakr1921a, Green1937].
GENERAL: Description and illustration of adult female by Green (1896e) and by Leonardi (1900).
KEYS: Ferris 1943: 64 (female) [North America]; Ferris 1942: 36 (female) [North America]; Green 1896e: 76 (female) [Sri Lanka].
CITATIONS: BeardsDaHo1976 [economic importance: 105]; Borchs1966 [catalogue: 364]; Cocker1899a [taxonomy: 396]; Fernal1903b [catalogue: 303]; Ferris1942 [taxonomy: 446:36]; Green1896e [taxonomy, description, illustration, host, distribution: 76]; Green1937 [host, distribution: 336]; Leonar1900 [taxonomy, description, illustration, host, distribution: 336-337]; Leonar1903a [taxonomy: 6];

MacGil1921 [taxonomy, description, host, distribution: 464]; Ramakr1921a [host, distribution: 358].

Aonidia obtusa Green & Laing

Aonidia obtusa Green & Laing, 1921: 126. Type data: SEYCHELLES: on *Verschaffeltia splendida*. Syntypes, female. Type depository: London: The Natural History Museum, England, UK.

SCALE COVER: Female scale consisting almost entirely of enlarged nymphal exuvia; transversely oval, flat, or very slightly convex; a narrow marginal area ornamented with sutures running irregularly inwards and intertwining; colour from pale to dark brown, often thinly coated with white powdery secretion over a wide marginal area, leaving only centre bare (Green & Laing, 1921).

HOST PLANTS: **Palmae**: *Verschaffeltia splendida* [GreenLa1921, Mamet1943a].

DISTRIBUTION: **Afrotropical**: Seychelles [GreenLa1921, Mamet1943a, Balach1958b, Borchs1966].

GENERAL: Description and illustration of adult female by Green & Laing (1921).

CITATIONS: Balach1958b [host, distribution: 240]; Borchs1966 [catalogue: 364]; GreenLa1921 [taxonomy, description, illustration, host, distribution: 126-127]; Mamet1943a [catalogue: 156].

Aonidia oleae Leonardi

Aonidia oleae Leonardi, 1913c: 66. Type data: ERITREA: on olive [=*Olea europaea*]. Holotype female. Type depository: Portici: Dipartimento de Entomologia e Zoologia Agraria di Portici, Università di Napoli Federico II, Italy.

Aonidiae oleae; Borchsenius, 1966: 409. Misspelling of genus name.

SCALE COVER: Female and male scale illustrated by Leonardi (1913c).

HOST PLANTS: **Oleaceae**: *Olea europaea* [Leonar1913c].

DISTRIBUTION: **Afrotropical**: Eritrea [Leonar1913c].

GENERAL: Description and illustration of adult female by Leonardi (1913c).

CITATIONS: Balach1958b [taxonomy: 249]; Leonar1913c [taxonomy, description, illustration, host, distribution: 66-68]; Lindin1928 [taxonomy: 106]; Lindin1931a [taxonomy: 89]; Lindin1943b [taxonomy: 221].

Aonidia operta De Lotto

Aonidia operta De Lotto, 1957: 225. Type data: KENYA: Nairobi, on branches of *Ficus hochstetteri*. Holotype female. Type depository: London: The Natural History Museum, England, UK.

SCALE COVER: Female scale represented by exuviae only, moderately convex; black; diameter up to 2.5 mm.; covered by a thin film of whitish wax. Scale of male elongate with subapical exuvia; dirty white; length up to 1.4 mm (De Lotto, 1957).

HOST PLANTS: **Moraceae**: *Ficus hochstetteri* [DeLott1957].

DISTRIBUTION: **Afrotropical**: Kenya [DeLott1957].

GENERAL: Description and illustration of adult female by De Lotto (1957).

CITATIONS: Borchs1966 [catalogue: 364]; DeLott1957 [taxonomy, description, illustration, host, distribution: 225-227].

Aonidia paradoxa (Lindinger)
Aonidia ? paradoxa Lindinger, 1911: 173. Type data: AUSTRALIA: South Australia, Mount Lyndhurst, on *Casuarina glauca*; collected x.1899. Syntypes, female. Type depository: Hamburg: Zoologisches Institut und Zoologishces Museum, Universität von Hamburg, Germany; type no. MP158.
Greeniella paradoxa; MacGillivray, 1921: 460. Change of combination.
Aonidia paradoxa; Borchsenius, 1966: 364. Revived combination.
SCALE COVER: Female scale white, slightly elongate (Lindinger, 1911).
HOST PLANTS: **Casuarinaceae**: *Casuarina glauca* [Lindin1911].
DISTRIBUTION: **Australasian**: Australia (South Australia) [Lindin1911].
GENERAL: Description and illustration of adult female by Lindinger (1911).
CITATIONS: Borchs1966 [catalogue: 364]; Lindin1911 [taxonomy, description, illustration, host, distribution: 173]; MacGil1921 [taxonomy, description, host, distribution: 460]; Sassce1912 [taxonomy, host, distribution: 92]; WeidneWa1968 [taxonomy: 171].

Aonidia perplexa Green
Aonidia perplexa Green, 1900a: 252. Type data: SRI LANKA: Paradeniya, Botanic Gardens, on under surface of leaves of *Mesua ferrea*. Syntypes, female. Type depository: London: The Natural History Museum, England, UK.
Fiorinia ? perplexa; Lindinger, 1943b: 220. Change of combination.
Aonidia perplexa; Borchsenius, 1966: 364. Revived combination.
SCALE COVER: Female scale oval, flat; secreted area greenish-grey, semi-transparent, completely covering pellicles except over a circular spot in centre of first, where surface of pellicles is exposed; scale 1.75 mm long; 1.25 mm wide. Male scale greyish, semi-transparent. Pellicle very pale yellow: a circular spot in centre exposed. Length 1.50 mm. Breadth 1 mm (Green, 1900a).
HOST PLANTS: **Guttiferae**: *Mesua ferrea* [Green1900a, Sander1909a, Ramakr1921a, Green1922, Green1937].
DISTRIBUTION: **Oriental**: Sri Lanka [Green1900a, Sander1909a, Ramakr1921a, Green1922, Green1937].
GENERAL: Description and illustration of adult female by Green (1900a).
CITATIONS: Borchs1966 [catalogue: 364]; DEDAC1923 [host, distribution]; Green1900a [taxonomy, description, illustration, host, distribution: 252]; Green1922 [host, distribution: 463]; Green1937 [host, distribution: 337]; Lindin1909b [taxonomy: 109]; Lindin1943b [taxonomy: 220]; MacGil1921 [taxonomy, description, host, distribution: 462]; Ramakr1921a [host, distribution: 358]; Ruther1915a [taxonomy, description, host, distribution: 108].

Aonidia planchonioides Green
Aonidia planchonioides Leonardi, 1899: 205. Nomen nudum.
Aonidia planchonioides Green, 1900a: 252. Type data: SRI LANKA: Paradeniya, Botanic Gardens, on leaves of *Ficus* sp. Syntypes, female. Type depository: London: The Natural History Museum, England, UK.
Aonidia planchonioides; Leonardi, 1900: 333. Notes: Correct citation of "Green" as author.

Aonidia planchonoides; Ramakrishna, 1921a: 358. Misspelling of species name.
SCALE COVER: Female scale oval: very pale yellow, transparent, revealing adult insect enclosed within large second pellicle (Green, 1901a).
HOST PLANTS: **Moraceae**: *Ficus* [Green1901a, Green1922, Green1937].
DISTRIBUTION: **Oriental**: Sri Lanka [Green1901a, Leonar1900, Ramakr1921a, Green1922, Green1937].
GENERAL: Description and illustration of adult female by Green (1901a).
CITATIONS: Borchs1966 [catalogue: 364]; Cocker1899a [taxonomy: 396]; DEDAC1923 [host, distribution]; Fernal1903b [catalogue: 303]; Green1901a [taxonomy, description, illustration, host, distribution: 252-253]; Green1905a [taxonomy: 348]; Green1937 [host, distribution: 337]; Leonar1899 [taxonomy: 205]; Leonar1900 [taxonomy, description, illustration, host, distribution: 333-334]; Leonar1903a [taxonomy: 6]; MacGil1921 [taxonomy, description, host, distribution: 462]; Ramakr1921a [host, distribution: 358].

Aonidia pusilla Green
Aonidia pusilla Green, 1905a: 347. Type data: SRI LANKA: Northern Province, Elephant Pass, on *Carissa spinarum*. Syntypes, female. Type depository: London: The Natural History Museum, England, UK.
SCALE COVER: Female scale oval, yellow; obscured - in very fresh examples - by a thin covering of whitish secretion which, in older examples, persists only as a marginal fringe, leaving yellow nymphal pellicle exposed. Total length 0.50 mm. Male scale oval; somewhat larger, but much less convex: pellicle pale-yellow, occupying anterior two-thirds of scale: secretionary area whitish, translucent. Length 0.65 mm (Green, 1905a).
HOST PLANTS: **Apocynaceae**: *Carissa spinarum* [Green1905a, Green1922].
DISTRIBUTION: **Oriental**: Sri Lanka [Green1905a, Ramakr1921a, Green1922, Green1937].
GENERAL: Description and illustration of adult female by Green (1905a).
CITATIONS: Borchs1966 [catalogue: 364]; Green1905a [taxonomy, description, illustration, host, distribution: 347-348]; Green1922 [host, distribution: 463]; Green1937 [host, distribution: 337]; MacGil1921 [taxonomy, description, host, distribution: 463]; Ramakr1921a [host, distribution: 359]; Sander1906 [taxonomy, host, distribution: 16].

Aonidia rageaui nomen nudum
Aonidia rageaui Cohic, 1958: 12. Nomen nudum.
Aonidia rageaui Borchsenius, 1966: 376. Nomen nudum.
Aonidia rageaui Williams & Watson, 1988: 17. Nomen nudum.

Aonidia rarasana Takahashi
Aonidia rarasana Takahashi, 1934: 31. Type data: TAIWAN: Rarasan, near Urai, on undetermined host. Syntypes, female. Type depository: Taichung: Entomology Collection, Taiwan Agricultural Research Institute, Wu-feng, Taichung, Taiwan.
SCALE COVER: Takahashi (1934) did not describe scale cover.
DISTRIBUTION: **Oriental**: Taiwan [Takaha1934, Takagi1970].

GENERAL: Description and illustration of adult female by Takahashi (1934).
CITATIONS: Takagi1970 [taxonomy, host, distribution: 131]; Takaha1934 [taxonomy, description, illustration, host, distribution: 31-32].

Aonidia relicta Balachowsky

Aonidia relicta Balachowsky, 1958b: 236. Type data: ERITREA: on *Juniperus procera*. Holotype female. Type depository: Portici: Dipartimento de Entomologia e Zoologia Agraria di Portici, Università di Napoli Federico II, Italy.
SCALE COVER: Female scale oval, small, moderately convex; larval exuviae yellow, slightly covered with white secretion; 0.8 mm long. Male scale unknown (Balachowsky, 1958b).
HOST PLANTS: **Cupressaceae**: *Juniperus procera* [Balach1958b].
DISTRIBUTION: **Afrotropical**: Eritrea [Balach1958b].
GENERAL: Description and illustration of adult female by Balachowsky (1958b).
CITATIONS: Balach1958b [taxonomy, description, illustration, host, distribution: 236,238-239]; Borchs1962b [taxonomy: 868]; Borchs1966 [catalogue: 364].

Aonidia rhusae Brain

Aonidia simplex; Brain, 1919: 214. Misidentification; discovered by Munting, 1965a: 185.
Aonidia rhusae Brain, 1919: 215. Type data: SOUTH AFRICA: Cape Town, on *Rhus* sp.; collected by C. Fuller 1898. Syntypes, female. Type depository: Pretoria: South African National Collection of Insects, South Africa.
Cryptaspidiotus rhois Lindinger, 1931a: 43. Unjustified emendation; discovered by Borchsenius, 1966: 364.
Cryptaspidiotus rhois; Lindinger, 1931a: 43. Change of combination.
Aonidia rhusae; Balachowsky, 1958b: 238. Revived combination.
SCALE COVER: Female scale ca. 1.5 mm long, brown, or blackish brown, owing to black second stage skin showing through brownish dorsal scale. Dorsal scale large and robust, brown, probably from admixture of epidermal tissues. Exuviae central, dark brown. Male scale circular to elongate oval, somewhat conical, smooth, parchment-like, buff, or brownish, with dark brown central exuviae (Brain, 1919).
HOST PLANTS: **Anacardiaceae**: *Rhus* [Brain1919, Balach1958b, Muntin1965a], *Rhus longispina* [Muntin1965a]. **Boraginaceae**: *Ehretia hottentottica* [Brain1919], *Ehretia rigida* [Muntin1965a]. **Oleaceae**: *Olea africana* [Muntin1969].
NATURAL ENEMIES: HYMENOPTERA **Encyrtidae**: *Metaphycus* sp. [Prinsl1983].
DISTRIBUTION: **Afrotropical**: South Africa [Brain1919, Balach1958b, Muntin1969].
GENERAL: Description and illustration of adult female by Brain (1919) and by Balachowsky (1958b).
KEYS: Munting 1965b: 189 (female) [South Africa].
CITATIONS: Balach1958b [taxonomy, description, illustration, host, distribution: 238,241]; Borchs1966 [catalogue: 364-365]; Brain1919 [taxonomy, description, illustration, host, distribution: 215-216]; Lindin1931a [taxonomy: 43,90]; Muntin1965a [taxonomy, host, distribution: 185,187,189]; Muntin1969 [host,

distribution: 120]; Muntin1970a [taxonomy: 39-40]; Prinsl1983 [distribution, biological control: 26].

Aonidia sclerosa Munting

Aonidia sclerosa Munting, 1969: 120. Type data: NAMIBIA: on escarpment of Weissrand, 7 miles from Gibeon to Witbooisvlei, on *Zygophyllum suffruticosum*; collected 28.viii.1967. Holotype female. Type depository: Pretoria: South African National Collection of Insects, South Africa; type no. 3194/14.

SCALE COVER: Female scale oval or slightly pyriform, covered with whitish secretion, 1.3 mm in length; when secretory matter rubbed off, golden brown second-instar exuviae exposed. Male scale oval, whitish, about 1.3 mm in length (Munting, 1969).

HOST PLANTS: **Zygophyllaceae**: *Zygophyllum suffruticosum* [Muntin1969].

DISTRIBUTION: **Afrotropical**: Namibia (Southwest Africa) [Muntin1969].

GENERAL: Description and illustration of adult female by Munting (1969).

CITATIONS: Muntin1969 [taxonomy, description, illustration, host, distribution: 120-121].

Aonidia shastae (Coleman)

Aspidiotus coniferarum shastae Coleman, 1903: 67. Type data: USA: California, Shasta County, Clear Creek, near Shasta P.O., on *Cupressus macnabiana*; collected August 29, 1901. Syntypes, female. Type depository: Davis: The Bohart Museum of Entomology, University of California, California, USA.

Aonidia juniperi Marlatt, 1908c: 24. Type data: USA: Utah, Logan, on fruit of *Juniperus* sp.; collected by E.G. Titus. Syntypes, female. Type depository: Washington: United States National Entomological Collection, U.S. National Museum of Natural History, District of Columbia, USA; type no. 14123. Synonymy by Ferris, 1943a: 86.

Aspidiotus shastae; Ferris, 1920b: 54. Change of status.

Gonaspidiotus shastae; MacGillivray, 1921: 432. Change of combination.

Targionia juniperi; MacGillivray, 1921: 448. Change of combination.

Aonidia shastae; Ferris, 1938a: 177. Change of combination.

Aspidiotus juniperi; Ferris, 1943a: 86. Change of combination.

Cupressaspis shastae; Borchsenius, 1966: 361. Change of combination.

Aonidia shastae; Danzig, 1993: 214. Revived combination.

COMMON NAMES: red-wood scale [Borchs1966]; redwood scale [McKenz1956].

SCALE COVER: Outer female scale thin, transparent, brownish-white, about 1 mm in diameter; minute yellowish exuvia at apex; beneath shell a thick, opaque, reddish-brown skin present, enclosing insect; very thin white ventral scale present; outer scale very similar to little drops of exuded gum of host tree, *Cupressus macnabiana* (Coleman, 1903). Female scale elongate oval, composed of second exuvia which in undisturbed specimens, covered with wax of a grey colour; in rubbed specimens, exuvia shows as yellow or red. Scale of male oval, exuvia nearly central (Ferris, 1938a).

HOST PLANTS: **Cupressaceae**: *Juniperus* [Marlat1908c, Ferris1938a, McKenz1956], *Ju. virginiana* [Ferris1938a], *Libocedrus decurrens* [Ferris1938a,

McKenz1956]. **Ericaceae**: *Macnabia* [McKenz1956]. **Pinaceae**: *Cupressus goveniana* [Ferris1938a, McKenz1956], *Cu. macnabiana* [Colema1903, McDani1968]. **Taxodiaceae**: *Sequoia sempervirens* [Ferris1920b, McKenz1956].
DISTRIBUTION: **Nearctic**: U.S.A. (Arizona [Nakaha1982], California [Colema1903, McKenz1956], Kansas [Ferris1938a], Missouri [Nakaha1982], New Jersey [Nakaha1982], Oklahoma [Nakaha1982], Texas [McDani1968], Utah [Marlat1908c, Ferris1938a]).
BIOLOGY: Occurring on needles.
GENERAL: Description and illustration of adult female by Marlatt (1908c) (as *Aonidia juniperi*), Ferris (1920b, 1938a) and by McKenzie (1956).
KEYS: McKenzie 1956: 23 (female) [U.S.A.: California]; Ferris 1942: 29 (female) [North America].
CITATIONS: Borchs1966 [catalogue: 361]; Colema1903 [taxonomy, description, host, distribution: 67]; Ferris1920b [taxonomy, description, illustration, host, distribution: 54-55]; Ferris1921b [taxonomy: 94]; Ferris1938a [taxonomy, description, illustration, host, distribution: 177]; Ferris1941e [taxonomy: 48]; Ferris1942 [taxonomy: 446:29]; Ferris1943a [taxonomy: 86]; FurnisCa1977 [host, distribution: 111]; Koehle1964 [host, distribution, control]; Lawson1917 [taxonomy, description, illustration, host, distribution: 208-209]; MacGil1921 [taxonomy, description, host, distribution: 431-432,448]; Marlat1908c [taxonomy, description, illustration, host, distribution: 24]; McDani1968 [taxonomy, illustration, host, distribution: 207,209]; McKenz1956 [taxonomy, description, illustration, host, distribution: 23,38-39]; MillerDa1990 [economic importance: 301]; Nakaha1982 [host, distribution: 6]; Nur1990a [taxonomy, structure, chromosomes: 184-185]; RahmanAn1941 [taxonomy, description: 825]; Sander1906 [taxonomy, host, distribution: 13]; Sander1909a [taxonomy, host, distribution: 56]; SchmutKlLu1957 [host, distribution, economic importance: 493].

Aonidia simplex Leonardi

Aonidia simplex Leonardi, 1914: 209. Type data: SOUTH AFRICA: Pretoria, on an undetermined plant. Syntypes, female. Type depository: Portici: Dipartimento de Entomologia e Zoologia Agraria di Portici, Università di Napoli Federico II, Italy.
SCALE COVER: Female scale of irregular shape, slightly wider than long; not highly convex; exuviae not central; dorsal secreted part of scale dirty grey; ventral vellum fairly robust, white (Leonardi, 1914).
DISTRIBUTION: **Afrotropical**: South Africa [Leonar1914, Balach1958b].
GENERAL: Description and illustration of adult female by Leonardi (1914).
CITATIONS: Balach1958b [host, distribution: 240]; Borchs1966 [catalogue: 365]; Brain1919 [taxonomy: 214]; Leonar1914 [taxonomy, description, illustration, host, distribution: 209-210]; Muntin1965a [taxonomy: 185].

Aonidia spatulata Green

Aonidia spatulata Green, 1900a: 73. Type data: SRI LANKA: Pundaluoya, on upper surface of leaves of *Psychotria thwaitesii*. Syntypes, female. Type depository: London: The Natural History Museum, England, UK.

SCALE COVER: Female scale hemispherical, consisting chiefly of enlarged second pellicle, which completely encloses adult insect; diameter 1.75 mm. Male scale circular, flattish, pellicle central; colour reddish-yellow, pellicle pale clear yellow; diameter equal to that of female, 0.75 mm (Green, 1900a).
HOST PLANTS: **Rubiaceae**: *Psychotria* [Ramakr1921a, Green1922], *Psychotria thwaitesii* [Green1900a, Green1937].
DISTRIBUTION: **Oriental**: Sri Lanka [Green1900a, Green1922, Green1937].
GENERAL: Description and illustration of adult female by Green (1900a).
CITATIONS: Borchs1966 [catalogue: 365]; DEDAC1923 [host, distribution]; Fernal1903b [catalogue: 303]; Green1900a [taxonomy, description, illustration, host, distribution: 73-74]; Green1905a [taxonomy: 348]; Green1922 [host, distribution: 463]; Green1937 [host, distribution: 337]; MacGil1921 [taxonomy, description, host, distribution: 463]; Ramakr1921a [host, distribution: 358].

Aonidia truncata Green & Laing

Aonidia truncata Green & Laing, 1923: 128. Type data: AUSTRALIA: Queensland, Magnetic Island, on leaves of an unknown plant. Syntypes, female. Type depository: London: The Natural History Museum, England, UK.
Greeniella truncata; Brimblecombe, 1958: 83. Change of combination.
Aonidia truncata; Borchsenius, 1966: 365. Revived combination.
SCALE COVER: Female scale circular or subcircular, highly convex; shining black. Ventral scale thin, adhering to the leaf. Diameter averaging 0.5 mm (Green & Laing, 1923). Insects in sparse on leaves; female scale circular, 0.6 mm in diameter; second exuviae black, covering adult body; first exuviae an orange colour with a grey margin; pale secreted filaments arising from first exuviae (Brimblecombe, 1958).
DISTRIBUTION: **Australasian**: Australia (Queensland [GreenLa1923, Brimbl1958]).
GENERAL: Description and illustration of adult female by Green & Laing (1923) and by Brimblecombe (1958).
CITATIONS: Borchs1966 [catalogue: 365]; Brimbl1958 [taxonomy, description, illustration, host, distribution: 83-85]; GreenLa1923 [taxonomy, description, illustration, host, distribution: 129-130].

Aonidia visci Hall

Aonidia visci Hall, 1931: 286. Type data: ZIMBABWE: Bulawayo Park, on *Viscum* sp. Syntypes, female. Type depository: London: The Natural History Museum, UK.
SCALE COVER: Adult female entirely enclosed within nymphal exuviae; nymphal exuviae pyriform, relatively flat, dark reddish brown in colour and coated with a thin semi-transparent dirty-white film, which extends somewhat beyond margin; length, 1.25 mm ; breadth, 1 mm. Male scale relatively large, parallel-sided and rounded at either extremity. Length, 1.4 mm; breadth, 0.6 mm (Hall, 1931).
HOST PLANTS: **Viscaceae**: *Viscum* [Hall1931, Balach1958b].
DISTRIBUTION: **Afrotropical**: Zimbabwe [Hall1931, Balach1958b].
GENERAL: Description and illustration of adult female by Hall (1931) and by Balachowsky (1958b).

CITATIONS: Balach1958b [taxonomy, description, illustration, host, distribution: 235,238,240]; Borchs1966 [catalogue: 365]; Ferris1941e [taxonomy: 41]; Hall1931 [taxonomy, description, illustration, host, distribution: 286-288]; Lindin1943b [taxonomy: 218].

Aonidia yomae Munting
Aonidia yomae Munting, 1965a: 187. Type data: SOUTH AFRICA: Cape Province. Ysterfontein, on *Mesembryanthemum* sp.; collected 21.v.1962. Holotype female. Type depository: Pretoria: South African National Collection of Insects, South Africa; type no. 487/1.
SCALE COVER: Scale of adult female oval, covered in a cream-coloured secretion, occurring in cracks in bark, about 1.5 mm long; male scale oval, felty-white, about 1 mm in length, with exuvium at one end (Munting, 1965a).
HOST PLANTS: **Aizoaceae**: *Mesembryanthemum* [Muntin1965a].
DISTRIBUTION: **Afrotropical**: South Africa [Muntin1965a].
GENERAL: Description and illustration of adult female by Munting (1965a).
KEYS: Munting 1965b: 189 (female) [South Africa].
CITATIONS: Muntin1965a [taxonomy, description, illustration, host, distribution: 186-187,189].

Aonidia zizyphi Rahman & Ansari
Aonidia zizyphi Rahman & Ansari, 1941: 824. Type data: INDIA: Punjab, Bannu (Nartahal Ram) on *Ziziphus jujube*. Holotype female. Type depository: Faisalabad: Entomological Laboratory, University of Agriculture, Pakistan.
SCALE COVER: Female scale 1.1-1.2 mm long, 0.9-1.1 mm broad; whitish grey; almost circular, convex dorsally; exuviae yellow to dark orange, slightly towards the margin (Rahman & Ansari, 1941)
HOST PLANTS: **Rhamnaceae**: *Ziziphus jujube* [RahmanAn1941].
DISTRIBUTION: **Oriental**: India (Punjab [RahmanAn1941]).
GENERAL: Description and illustration of adult female by Rahman & Ansari (1941).
CITATIONS: Borchs1966 [catalogue: 365]; RahmanAn1941 [taxonomy, description, illustration, host, distribution: 824-825].

Aonidiella Berlese & Leonardi
Aonidiella Berlese & Leonardi, 1896: 77. Type species: *Aspidiotus aurantii* Maskell, by monotypy.
Chrysomphalus (*Aonidiella*); Cockerell, 1897i: 9. Change of status.
Aonidiella; Fernald, 1903b: 285. Revived rank.
Heteraspis Leonardi, 1914: 197. Type species: *Aspidiotus replicatus* Lindinger, by monotypy and original designation. Synonymy by Ferris, 1937c: 55. Homonym of *Heteraspis* Chevrolat, 1836, in Coleoptera.
Abnidiella; Chou, 1985: 283. Misspelling of genus name.
SYSTEMATICS: *Aonidiella* Berlese & Leonardi has been synonymized with *Chrysomphalus* by early authors. *Aonidiella* differs from *Chrysomphalus* in having

cephalothorax sclerotized and reniform, prepygidial lobes curving posteriorly to pygidium, and paraphyses shorter than adjacent lobes (Takagi, 1969a; Williams & Watson, 1988).

GENERAL: Definition and characters by McKenzie (1938, 1939, 1956), Borchsenius (1950b), Lupo (1954a), Balachowsky (1948b, 1956), Gómez-Menor Guerrero (1962), Takagi (1969a), Velasquez (1971), Williams & Watson (1988) and by Yaşar (1995).

KEYS: Colon-Ferrer & Medina-Gaud 1998: 28-32 (female) [genera of Puerto Rico]; Gill 1997: 24-26 (female) [Genera of California]; Gill 1997: 44 (female) [Species of California]; Blay Goicoechea 1993: 473 (female) [Spain]; Danzig 1993: 159-160 (female) [species Europe]; Wolff & Corseuil 1993: 29 (female) [Brazil, Rio Grande do Sul]; Williams & Watson 1988: 19 (female) [Tropical South Pacific]; Tereznikova 1986: 83 (female) [Ukraine]; Chou 1985: 283 (female) [Genera of China]; Chou 1985: 292 (female) [Species of China]; Paik 1978: 291 (female) [species South Korea]; Velasquez 1971: 118 (female) [Philippines]; McDaniel 1968: 215 (female) [species U.S.A.: Texas]; Beardsley 1966: 502-504 (female) [Federated States of Micronesia]; Beardsley 1966: 509 (female) [species Federated States of Micronesia]; Danzig 1964: 645 (female) [Europe]; Gómez-Menor Guerrero 1962: 157 (female) [Canary Islands]; Zahradník 1959a: 548 (female) [Czech Republic]; Balachowsky 1958b: 226 (female) [*Aspidiotina* of Africa]; Ezzat 1958: 237-239 (female) [Egypt]; Gómez-Menor Ortega 1956: 7-8 (female) [Spain]; McKenzie 1956: 22 (female) [U.S.A.: California]; Bodenheimer 1952: 330 (female) [Turkey]; Balachowsky 1951: 600 (female) [Mediterranean]; Borchsenius 1950b: 167 (female) [USSR]; Zimmerman 1948: 351 (female) [Hawaii]; Gómez-Menor Ortega 1946: 59-61 (female) [Spain]; McKenzie 1946: 29 (female) [World]; Ruiz Castro 1944: 57 (female) [Spain]; Ferris 1942: 26 (female) [North America]; Ferris 1942: 29 (female) [North America]; McKenzie 1938: 17-18 (female) [species World]; Archangelskaya 1937: 94 (female) [Middle Asia]; Borchsenius 1937: 99 (female) [USSR]; Borchsenius 1937a: 32-33 (female) [Palaearctic Region]; Gómez-Menor Ortega 1937: 43 (female) [Spain]; Leonardi 1920: 26 (female) [Italy]; Leonardi 1920: 75 (female) [Species of Italy].

CITATIONS: Archan1937 [taxonomy: 94,95]; Balach1948b [taxonomy, description: 359-360]; Balach1951 [taxonomy: 600]; Balach1956 [taxonomy, description: 22-25]; Balach1958b [taxonomy: 149,226]; Beards1966 [taxonomy: 507]; BenDov1990h [taxonomy: 81]; Berles1895c [taxonomy: 206]; BerlesLe1896 [taxonomy, description: 77]; BlayGo1993 [taxonomy, description: 517]; Bodenh1949 [taxonomy, description: 37]; Bodenh1952 [taxonomy: 330]; Borchs1937 [taxonomy, description: 99,121]; Borchs1937a [taxonomy, description: 33,62]; Borchs1949d [taxonomy, description: 194,232]; Borchs1950b [taxonomy, description: 220]; Borchs1966 [catalogue: 291]; Brimbl1962a [taxonomy: 403-404]; Bustsh1958 [taxonomy, description: 239]; Chou1985 [taxonomy, description: 283,291-292]; Cocker1897i [taxonomy, description: 9,13]; Cocker1899a [taxonomy: 396]; ColonFMe1998 [taxonomy, description: 35]; Danzig1964 [taxonomy: 651]; Danzig1993 [taxonomy, description: 159]; DanzigPe1998 [catalogue: 181]; Ezzat1958 [taxonomy: 238]; Ferris1937c [taxonomy: 50,51,53,55,60]; Ferris1938a [taxonomy, description: 178]; Ferris1942 [taxonomy: 446:26]; Ghauri1962 [taxonomy: 210]; Gill1997 [taxonomy: 44]; GomezM1937 [taxonomy, description: 43,84]; GomezM1946 [taxonomy: 60]; GomezM1956 [taxonomy, description: 41-42]; GomezM1962 [taxonomy, description: 186]; Hadzib1983 [taxonomy: 226]; Kawai1980 [taxonomy: 211]; Kozar1990f [distribution: 143]; Leonar1897 [taxonomy: 284-286]; Leonar1897a [taxonomy: 375]; Leonar1897b [taxonomy: 109,111]; Leonar1899 [taxonomy: 174]; Leonar1914 [taxonomy, description: 197]; Leonar1920 [taxonomy, description: 26,74-75]; Lepage1938 [taxonomy: 391]; Lepesm1947 [taxonomy, description: 203-]; Lindin1949 [taxonomy: 211]; Lupo1954a [taxonomy, description: 40-41]; MacGil1921 [taxonomy, description: 392,442]; Mamet1949 [taxonomy: 52-53]; McKenz1937 [taxonomy: 324]; McKenz1937a [taxonomy, description: 176-177]; McKenz1938 [taxonomy, description: 2-3]; McKenz1939 [taxonomy, description: 64-65]; McKenz1939 [taxonomy: 53]; McKenz1942b [taxonomy: 141-142]; McKenz1944 [taxonomy: 55]; McKenz1946 [taxonomy: 29]; McKenz1947b [taxonomy: 111-112]; McKenz1950 [taxonomy: 99]; McKenz1956 [taxonomy: 22]; Miller1990 [taxonomy: 169-178]; MorrisMo1922 [taxonomy, description: 96]; MorrisMo1966 [taxonomy, catalogue: 8,92]; Nel1933 [taxonomy: 417-419]; RuizCa1944 [taxonomy: 57]; Takagi1969a [taxonomy, description: 81-82]; Tao1999 [taxonomy: 71]; ThiemGe1934a [taxonomy, description: 232]; Velasq1971 [taxonomy, description: 117-118]; WilliaWa1988 [taxonomy, description: 35]; WolffCo1993 [taxonomy: 29]; Yasar1995a [taxonomy, description: 42-43]; Zimmer1948 [taxonomy: 361].

Aonidiella abietina Hall & Williams

Aonidiella abietina Hall & Williams, 1962: 35. Type data: PAKISTAN: Murree, on the needles of *Abies pindrow*; collected 1.XI.1961. Holotype female. Type depository: London: The Natural History Museum, England, UK.

SCALE COVER: Female circular, translucent, body of underlying female showing through; exuviae subcentral, pale brown or reddish brown. Diameter about 1.5 mm. Male scale not seen (Hall & Williams, 1962).

HOST PLANTS: **Pinaceae**: *Abies pindrow* [HallWi1962].

DISTRIBUTION: **Oriental**: Pakistan [HallWi1962].

GENERAL: Description and illustration of adult female by Hall & Williams (1962).

CITATIONS: Borchs1966 [catalogue: 292]; DanzigPe1998 [catalogue: 181]; HallWi1962 [taxonomy, description, illustration, host, distribution: 35-36].

Aonidiella araucariae Costa Lima

Aonidiella araucariae Costa Lima, 1951: 174. Type data: BRAZIL: Santa Catarina State, Canoinhas, Tries Barras, Parque Florestal do Instituto Nacional do Pinho, on *Araucaria angustifolia brasiliensis*. Holotype female. Type depository: Rio de Janeiro: Colecao Entomologica, Escola de Postgraduacao em Parasitologia Animal, Universidade Federal Rural de Rio de Janeiro, Brazil; type no. 10.250.

SCALE COVER: Female scale circular, flat, bright ash; slightly transparent; exuviae central, yellow; diameter 1.25 mm. Male scale (colour and structure) similar to female; exuviae eccentric; 1 mm long, 0.75 mm wide (Costa Lima, 1951).

HOST PLANTS: **Araucariaceae**: *Araucaria angustifolia brasiliensis* [Newste1920], *Araucaria angustifolia* [ClapsWoGo2001].
DISTRIBUTION: **Neotropical**: Brazil [CostaL1951] (Rio Grande do Sul [ClapsWoGo2001], Santa Catarina [ClapsWoGo2001], São Paulo [ClapsWoGo2001]).
GENERAL: Description and illustration of adult female by Costa Lima (1951).
CITATIONS: Borchs1966 [catalogue: 292]; ClapsWoGo2001 [host, distribution: 240]; CostaL1951 [taxonomy, description, illustration, host, distribution: 174-175].

Aonidiella atlantorum Matile-Ferrero & Balachowsky

Aonidiella atlantorum Matile-Ferrero & Balachowsky, 1972: 110. Type data: CANARY ISLANDS: Tenerife, Puertito de Guimar, on *Euphorbia canariensis*. Holotype female. Type depository: Paris: Muséum national d'Histoire naturelle, France.
SCALE COVER: Female scale circular, flat; larval exuviae central or subcentral; colour bright brown, semi-transparent, with exuviae darker; the whole scale covered with a white layer (Matile-Ferrero & Balachowsky, 1972).
HOST PLANTS: **Euphorbiaceae**: *Euphorbia canariensis* [MatileBa1972].
DISTRIBUTION: **Palaearctic**: Canary Islands [MatileBa1972, MatileOr2001].
GENERAL: Description and illustration of adult female by Matile-Ferrero & Balachowsky (1972).
CITATIONS: DanzigPe1998 [catalogue: 181]; MatileBa1972 [taxonomy, description, illustration, host, distribution: 110-112]; MatileOr2001 [host, distribution: 189].

Aonidiella aurantii (Maskell)

Aspidiotus aurantii Maskell, 1879: 199. Type data: NEW ZEALAND: Governors Bay, on oranges and lemons imported to New Zealand from Sydney. Syntypes, female. Type depository: Auckland: New Zealand Arthropod Collection, Landcare Research, New Zealand.
Aonidia gennadii Targioni Tozzetti, 1881: 151. Type data: ITALY: on undetermined host plant. Syntypes, female. Synonymy by Leonardi, 1920: 75. Notes: Type material probably lost; G. Pellizzari-Scaltriti, 1998, information to Yair Ben-Dov.
Aspidiotus coccineus Gennadius, 1881: 189. Type data: GREECE: Chio Island, on *Citrus* sp., *Euonymus japonicus*, *Ficus elastica*, *Pistacia lentiscus* and *Vitis* sp. Syntypes, female. Synonymy by Lindinger, 1914: 157. Notes: Depository of type material unknown.
Aspidiotus citri Comstock, 1881: 8. Type data: U.S.A.: California, southern California, on *Citrus* sp. Syntypes, female. Type depository: Washington: United States National Entomological Collection, U.S. National Museum of Natural History, District of Columbia, USA. Synonymy by Cockerell, 1896b: 334.
Aonidia aonidum Targioni Tozzetti, 1884: 386. Unjustified replacement name for *Aonidiella aurantii* (Maskell).
Aonidiella aurantii; Berlese, 1895a: 125. Change of combination.
Aspidiotus (Aonidiella) aurantii; Cockerell, 1897i: 29. Change of combination.

Chrysomphalus (*Aonidiella*) *aurantii*; Cockerell, 1899a: 396. Change of combination.

Aspidiotus (*Chrysomphalus*) *aurantii*; Kuwana, 1902: 70. Change of combination.

Chrysomphalus citri; Lindinger, 1935: 132. Change of combination.

Aonidiella coccineus; McKenzie, 1939: 54. Change of combination.

Aonidiella gennadi; McKenzie, 1939: 54. Change of combination.

Chrysomphalus coccineus; Lindinger, 1949: 211. Change of combination.

COMMON NAMES: California red scale [Merril1953, McKenz1956, Borchs1966]; escama roja de California [CoronaRuMo1997]; escama roja de los citricos [Gonzal1989]; krasnaya pomeranzevaya shitovka [Borchs1936]; pijo rojo de California [Lloren1990]; pinta-vermelha [CarvalAg1997]; pou rouge de Californie [Smirno1951a]; red-scale of California [Comsto1881a].

SYSTEMATICS: For a detailed discussion on the differences between *Aonidiella aurantii* and *Aonidiella citrina* refer to Nel (1933); McKenzie, (1937); Compere (1961), and to Remarks under *Aonidiella citrina*.

SCALE COVER: Scale of female circular, flat, exuviae central; thin and pale, permitting red-brown colour of heavily sclerotized adult female to show through. Scale of male elongate oval, colour paler than female, exuvia slightly towards one end (Ferris, 1938a). Colour photograph by Gonzalez (1989), Katsoyannos (1996), Carvalho & Aguiar (1997), Gill (1997) and by Wong *et al.* (1999).

HOST PLANTS: **Acanthaceae**: *Libonia* [Brain1919], *Maranta* [Brain1919]. **Aceraceae**: *Acer* [Brain1919]. **Agavaceae**: *Agave* [Green1896, RahmanAn1941, Balach1948b, McKenz1956], *Agave americana* [Green1896e, RahmanAn1941, Green1937], *Agave variegata* [RahmanAn1941], *Cordyline* [Brain1919], *Dracaena* [Brain1919]. **Amaryllidaceae**: *Clivia* [McKenz1946a], *Doryanthes* [Brain1919]. **Anacardiaceae**: *Mangifera indica* [Hall1923, Takaha1933, RahmanAn1941, WolffCo1993a, KinjoNaHi1996], *Pistacia lentiscus* [Bodenh1924, Bodenh1928]. **Apocynaceae**: *Alstonia scholaris* [RahmanAn1941], *Nerium odorum* [RahmanAn1941], *Nerium oleander* [RahmanAn1941], *Nerium oleander* [UygunSeEr1998], *Rhyncospermum* [Wilson1917]. **Araliaceae**: *Aralia* [Wilson1917, Brain1919], *Hedera helix* [Hall1923], *Heptapleurum octophyllum* [Takaha1929]. **Araucariaceae**: *Araucaria* [Brain1919]. **Aucubaceae**: *Aucuba* [Brain1919, Balach1948b]. **Balsaminaceae**: *Impatiens* [Brain1919]. **Berberidaceae**: *Berberis* [Brain1919], *Mahonia* [McKenz1946a]. **Bignoniaceae**: *Bignonia* [Brain1919], *Jacaranda* [Brain1919], *Tecoma* [RahmanAn1941]. **Bombacaceae**: *Bombax malabaricum* [RahmanAn1941]. **Boraginaceae**: *Benthamia* [Brain1919]. **Buxaceae**: *Buxus* [McKenz1946a]. **Cactaceae**: *Cactus* [Laing1929]. **Cannaceae**: *Canna indica* [RahmanAn1941]. **Caprifoliaceae**: *Viburnum* [Brain1919, McKenz1946a]. **Caricaceae**: *Carica papaya* [Takaha1935]. **Casuarinaceae**: *Casuarina* [Brain1919], *Casuarina equisetifolia* [Mamet1943a, Mamet1949, Borchs1966]. **Celastraceae**: *Euonymus* [McKenz1946a, Bodenh1952, McKenz1956, McDani1968, UygunSeEr1998], *Euonymus japonica* [Bodenh1928, RahmanAn1941]. **Chenopodiaceae**: *Chenopodium ambrosiodes* [McKenz1946a]. **Compositae**: *Ambrosia artemisiifolia* [McDani1968], *Calendula officinalis* [Hall1923], *Gynura* [Brain1919], *Leptilon canadensis* [McDani1968], *Zinnia* [Brain1919, McKenz1946a]. **Cornaceae**: *Cornus* [Brain1919]. **Crassulaceae**:

Sedum [McKenz1946a]. **Cruciferae**: *Brassica nigra* [McKenz1946a]. **Cucurbitaceae**: *Cucurbita* [McKenz1946a], *Cucurbita pepo* [WilliaWa1988]. **Cupressaceae**: *Cupressus* [Brain1919], *Thuja* [Brain1919]. **Cycadaceae**: *Cycas* [Green1904a, MerrilCh1923, Balach1948b, McKenz1956, Cohic1958]. **Cyperaceae**: *Cyperus* [Brain1919]. **Ebenaceae**: *Diospyros kaki* [Leonar1909, Hall1923], *Diospyros montana* [RahmanAn1941]. **Ehretiaceae**: *Cordia myxa* [RahmanAn1941]. **Elaeagnaceae**: *Elaeagnus* [Brain1919]. **Ericaceae**: *Arbutus unedo* [McKenz1946a]. **Erythropalaceae**: *Mackaya* [Brain1919]. **Euphorbiaceae**: *Aleurites* [Lepage1938], *Bridelia ovata* [Takaha1934], *Gelonium aequoreum* [Takaha1933], *Phyllanthus* [Hall1922], *Poinsettia* [Brain1919, RahmanAn1941], *Ricinus communis* [Hall1922, Hall1923, Bodenh1924, Lepage1938, WilliaWa1988]. **Fagaceae**: *Ilex aquifolium* [McKenz1946a], *Ilex opaca* [McDani1968], *Quercus rubra* [Balach1956]. **Flacourtiaceae**: *Scolopia oldhami* [Takaha1932a, Takaha1933]. **Goodeniaceae**: *Scaevola frutescens* [Takaha1932a, Takaha1933]. **Gramineae**: *Bambusa* [McKenz1946a], *Bambusa vulgaris* [WilliaWa1988], *Cydonia vulgaris* [Green1923b]. **Greyaceae**: *Greyia* [Brain1919]. **Hamamelidaceae**: *Trichocladus* [Brain1919]. **Hydrangeaceae**: *Forsythia* [Brain1919], *Hydrangea* [Brain1919]. **Iridaceae** [Mamet1954, Borchs1966]. **Juglandaceae**: *Juglans* [McKenz1946a], *Juglans regia* [McKenz1946a]. **Labiatae**: *Salvia* [Brain1919]. **Lauraceae**: *Camphora officinalis* [Green1904a], *Cinnamomum* [McKenz1946a], *Laurus nobilis* [Hall1923, MerrilCh1923, Bodenh1924, Almeid1973b], *Persea americana* [McKenz1946a, GersonZo1973], *Persea gratissima* [DeLott1967a], *Umbellularia californica* [McKenz1946a]. **Leguminosae**: *Acacia* [Balach1948b, McKenz1956, RosenDe1979, UygunSeEr1998], *Ac. farnesiana* [Bodenh1924], *Ac. longifolia* [Bodenh1924], *Bauhinia* [Brain1919, McDani1968], *Bauhinia alba* [RahmanAn1941], *Bauhinia racemosa* [RahmanAn1941], *Cassia* [RahmanAn1941], *Cassia fistula* [RahmanAn1941], *Ceratonia siliqua* [Hall1922, Bodenh1924, McKenz1946a, InserrCa1987], *Erythrina* [Brain1919, Hall1923], *Genista* [McKenz1946a], *Gleditsia* [Brain1919], *Grevillea* [Green1900a, Brain1919], *Grevillea robusta* [RahmanAn1941], *Kennedia* [Brain1919, MerrilCh1923], *Pongamia glabra* [RahmanAn1941], *Robinia* [Brain1919], *Sesbania cannabina* [McDani1968], *Sophora* [Brain1919]. **Liliaceae**: *Aloe vera* [RahmanAn1941], *Asparagus* [Takaha1933], *Asparagus officinalis* [Hall1923], *Asparagus plumosus* [McDani1968], *Convallaria* [Brain1919], *Yucca* [Brain1919, DeLott1967a]. **Lythraceae**: *Lagerstroemia* [Brain1919], *Lagerstroemia indica* [RahmanAn1941]. **Magnoliaceae**: *Illicium randaiense* [Takaha1935], *Liriodendron* [Brain1919], *Liriodendron lucidum* [Merril1953], *Michelia champaca* [Green1900a]. **Malvaceae**: *Abutilon* [Brain1919], *Hibiscus* [RahmanAn1941], *Hibiscus mutabilis* [McDani1968]. **Meliaceae**: *Cedrela* [Brain1919], *Cedrela odorata* [Wilson1917], *Melia* [Brain1919], *Melia azedarach* [McDani1968], *Trichilia* [Brain1919]. **Menispermaceae**: *Cocculus* [Merril1953, Dekle1965c]. **Moraceae**: *Artocarpus* [MerrilCh1923, McKenz1956, Beards1966], *Artocarpus altilis* [WilliaWa1988], *Broussonetia papyrifera* [RahmanAn1941, WilliaWa1988], *Ficus* [Lepage1938, RahmanAn1941, Balach1948b, Dekle1965c], *Ficus benghalensis* [RahmanAn1941], *Ficus carica* [Hall1923, Almeid1971], *Ficus elastica* [Wilson1917,

RahmanAn1941], *Ficus infectoria* [Hall1923], *Ficus religiosa* [RahmanAn1941], *Ficus roxburghii* [RahmanAn1941], *Morus* [Hall1922, Ferris1953], *Morus alba* [Lepage1938, RahmanAn1941, Balach1948b]. **Musaceae**: *Musa sapientum* [Bodenh1924, RahmanAn1941, Cohic1958]. **Myrsinaceae**: *Ardisia* [Takaha1929], *Maesa tenera* [Takagi1969a]. **Myrtaceae**: *Callistemon* [Brain1919, McKenz1946a], *Eucalyptus* [Wilson1917, Brain1919, RahmanAn1941], *Eugenia* [Brain1919], *Eugenia jambolana* [RahmanAn1941], *Myrtus* [McKenz1946a, Bodenh1952], *Psidium guajava* [RahmanAn1941, Matile1984c], *Tristania* [Brain1919]. **Nyctaginaceae**: *Bougainvillea* [Brain1919]. **Oleaceae**: *Jasminum* [Hall1923, McKenz1946a], *Jasminum humile* [McDani1968], *Jasminum pubescens* [RahmanAn1941], *Jasminum sambac* [RahmanAn1941], *Ligustrum* [McKenz1946a, McKenz1956, RosenDe1979], *Ligustrum japonicus* [McDani1968], *Ligustrum lucidum* [Merril1953, Dekle1965c, McDani1968], *Olea* [Hall1922, Bodenh1952], *Olea europaea* [McKenz1946a], *Osmanthus* [Brain1919]. **Orchidaceae**: *Cypripedium* [Laing1929]. **Palmae**: *Areca alicaceae* [Wilson1917], *Areca triandra* [Wilson1917], *Chamaedorea* [Wilson1917], *Cocos* [McKenz1956], *Cocos nucifera* [WilliaBu1987, WilliaWa1988], *Cocos plumosa* [Wilson1917], *Dictyosperma album* [Wilson1917], *Dictyosperma rubrum* [Wilson1917], *Dypsis madagascariensis* [Wilson1917], *Hyophorbe* [Wilson1917], *Kentia fostercana* [Wilson1917], *Latania* [Wilson1917], *Phoenix* [McKenz1946a], *Phoenix canariensis* [Wilson1917], *Phoenix dactylifera* [Hall1922], *Roscheria melanochaetes* [Wilson1917], *Seaforthia* [Wilson1917], *Washingtonia robusta* [Wilson1917]. **Pandanaceae**: *Pandanus* [Wilson1917, Brain1919], *Pandanus tectorius* [Takaha1935]. **Philadelphaceae**: *Deutzia* [Brain1919]. **Phytolaccaceae**: *Phytolacca dioica* [Brain1919]. **Pinaceae**: *Pinus* [Hall1923], *Pinus thunbergii* [Takaha1932a, Takaha1933, Takagi1969a]. **Pittosporaceae**: *Pittosporum* [McKenz1946a]. **Plumbaginaceae**: *Statice* [Brain1919]. **Podocarpaceae**: *Podocarpus* [MerrilCh1923], *Podocarpus chinensis* [Kuwana1902], *Podocarpus macrophyllus* [Takagi1969a], *Podocarpus neriifolius* [Balach1930a, Balach1932d, Borchs1934, Borchs1936]. **Proteaceae**: *Hakea* [Brain1919]. **Rhamnaceae**: *Rhamnus* [Brain1919], *Ziziphus* [RahmanAn1941]. **Rosaceae**: *Cotoneaster* [McKenz1946a], *Co. horizontalis* [McKenz1946a], *Co. microphylla* [McKenz1946a], *Heteromeles* [RosenDe1979], *Lauro-Cerasus* [McKenz1946a], *Lauro-Cerasus officinalis* [McKenz1946a], *Prunus domestica* [Hall1922, Bodenh1926], *Pr. tomentosa* [McKenz1946a], *Pyrus* [DeLott1967a], *Pyrus communis* [Hall1923, Almeid1971], *Pyrus cydonia* [Hall1922], *Pyrus malus* [Hall1922], *Rosa* [Bodenh1937, Lepage1938, RahmanAn1941, Mamet1943a, McKenz1946a, Mamet1949, McKenz1956, Borchs1966], *Rosa indica* [WilliaWa1988], *Spiraea* [Brain1919]. **Rubiaceae**: *Bouvardia* [Brain1919], *Coprosma* [Brain1919]. **Ruscaceae**: *Ruscus hypoglossum* [Balach1930a], *Ruscus hypophyllus* [Hall1923]. **Rutaceae**: *Aegle marmelos* [RahmanAn1941], *Choisya* [Brain1919, McKenz1946a], *Citrus* [Mamet1943a, Mamet1949, Mamet1954, Mamet1954a, Borchs1966, UygunSeEr1998, Ferris1921a, Takaha1929, Bodenh1930, Lepage1938, Takaha1939b, Takaha1942d, McKenz1946a, Bodenh1952, MerrilCh1923, SengonUyKa1998], *Ci. aurantium* [Bodenh1924, Bodenh1926, Bodenh1928, Borchs1934, Takagi1975, WilliaWa1988], *Ci.*

decumana [Green1896e], *Ci. grandis* [McKenz1946a, NakaoTaTa1977, WilliaWa1988], *Ci. limon* [Green1929, McKenz1946a, Almeid1973b, Martin1983, WilliaWa1988], *Ci. maxima* [WilliaWa1988], *Ci. nobilis unshiu* [Borchs1934], *Ci. nobilis* [Borchs1934], *Ci. paradisi* [WilliaWa1988], *Ci. pomela* [Green1896], *Ci. reticulata* [WilliaWa1988], *Ci. sinensis* [McKenz1946a, Takagi1969a], *Murraya exotica* [RahmanAn1941]. **Salicaceae**: *Salix* [Bodenh1924, Bodenh1937], *Salix babylonica* [Hall1923], *Salix discolor* [McKenz1946a]. **Sapotaceae**: *Chrysophyllum* [Brain1919]. **Scrophulariaceae**: *Penstemon* [Brain1919], *Veronica* [Brain1919]. **Solanaceae**: *Capsicum frutescens* [WilliaWa1988], *Cestrum* [Brain1919], *Solanum* [Hall1923]. **Sterculiaceae**: *Dombeya* [Brain1919], *Dombeya acutangula* [RahmanAn1941], *Sterculia* [Brain1919]. **Strelitziaceae**: *Strelitzia* [Brain1919], *Strelitzia reginae* [McKenz1946a]. **Taxaceae**: *Taxus baccata* [Borchs1936]. **Taxodiaceae**: *Cryptomeria* [Brain1919], *Taxodium* [Brain1919]. **Theaceae**: *Camellia* [McKenz1946a], *Thea* [Balach1956]. **Verbenaceae**: *Clerodendrum* [McKenz1946a], *Duranta* [Brain1919]. **Vitaceae**: *Ampelopsis* [Brain1919], *Vitis* [RahmanAn1941, McKenz1946a, Bodenh1952], *Vitis vinifera* [Hall1923, RahmanAn1941].

NATURAL ENEMIES: ACARI **Camerobiidae**: *Neophyllobius ambulans* [Meyer1962]. **Hemisarcoptidae**: *Hemisarcoptes coccophagus* [Meyer1962], *Hemisarcoptes cooremani* Thomas [HouckOc1996, LuckJiHo1999], *Hemisarcoptes malus* (Shimer) [GersonOcHo1990, HouckOc1996]. **Phytoseiidae**: *Euseius delhiensis* (Naryanan & Kaur) [ErlerTu2001], *Euseius finlandicus* (Quedemans) [ErlerTu2001]. COLEOPTERA **Coccinellidae**: *Chilocorus bipustulatus* L. [NadelBi1964, BenDovRo1969, RosenDe1978], *Ch. bisugus infernalis* Mulsant [RosenDe1978], *Ch. bivulnerus* Mulsant [Comper1961], *Ch. circumdatus* Gyllenhal [RosenDe1978, SmithSmLi1999], *Ch. discoideus* Crotch [RosenDe1978], *Ch. distigma* Klug [RosenDe1978], *Ch. hauseri* Weise [RosenDe1978], *Ch. kuwanae* Silvestri [RosenDe1978], *Ch. nigritus* (Fabricius) [RosenDe1978, Samway1989], *Ch. rubidus tristis* Fald. [RosenDe1978], *Ch. stigma* (Say) [RosenDe1978], *Ch. wahlbergi* Mulsant [RosenDe1978], *Coccidophilus citricola* Brethes [Elguet1932, Flande1936a, Clause1952, RosenDe1978], *Cryptolaemus montrouzieri* Mulsant [SmithSmLi1999], *Exochomus flavipes* Thunberg [RosenDe1978], *Lindorus lophanthae* (Blaisdell) [Comper1961, RosenDe1978], *Lotis neglecta* Mulsant [RosenDe1978], *Lotis niggerima* Casey [RosenDe1978], *Microweisia coccidivora* (Ashmead) [RosenDe1978], *Nephus includens* Kirsch [ErlerTu2001], *Orcus australasiae* Boisduval [RosenDe1978], *Orcus chalybeus* (Boisduval) [Comper1961, RosenDe1978], *Paropsis caseyi* [Fursch1985], *Pentilia nigella* Weise [RosenDe1978], *Pharoscymnus* [RosenDe1978], *Pharoscymnus exiguus* Weise [RosenDe1978], *Pharoscymnus horni* (Weise) [Flande1934a, RosenDe1978], *Rhyzobius satellus* Blacb. [RosenDe1978], *Rhyzobius ventralis* [Pope1981], *Scymnus* [RosenDe1978, ErlerTu2001], *Scymnus marginicollis* Mann. [RosenDe1978], *Scymnus notescens* Blackb. [RosenDe1978], *Spiliconis picticornis* Banks [RosenDe1978], *Telsimia* [RosenDe1978], *Telsimia emarginata* Chapin [Flande1934a, Comper1961, RosenDe1978]. **Nitidulidae**: *Cybocephalus binotatus* Grouvelle [BlumbeSw1974], *Cy. fodori-minor* Enrody-Younga [ErlerTu2001], *Cy. micans* Reitter [Blumbe1971, Blumbe1973a], *Cy. nigriceps nigriceps* (Sahlberg)

[Blumbe1971, Blumbe1973a, BlumbeSw1974]. DIPTERA **Cecidomyiidae**: *Cecidomyia coccidarum* (Cockerell) [Harris1990], *Lestodiplosis aonidiellae* Harris [Harris1990, Harris1997]. FUNGI **Ascomycotina**: *Myriangium duriaei* [EvansPr1990], *Nectria aurantiicola* [EvansPr1990], *Nectria flammea* [EvansPr1990]. **Deuteromycotina**: *Fusarium* [EvansPr1990]. HETEROPTERA **Isometopidae**: *Isomethopus mirificus* Mulsant & Rey [ErlerTu2001]. HYMENOPTERA **Aphelinidae**: *Apytis africanus* Quednau [RosenDe1978, RosenDe1979], *Ap. aonidiae* (Mercet) [RosenDe1979], *Ap. capillatus* (Howard) [RosenDe1979], *Ap. chilensis* Howard [RosenDe1979], *Ap. chrysomphali* (Mercet) [BartleFi1950, Flande1953a, RosenDe1978, RosenDe1979, ErlerTu2001], *Ap. coheni* DeBach [RosenDe1979], *Ap. comperei* DeBach & Rosen [RosenDe1979, MyartsRu2000], *Ap. desantisi* DeBach & Rosen [RosenDe1979], *Ap. diaspidis* (Howard) [RosenDe1979], *Ap. fisheri* DeBach [RosenDe1979], *Ap. hispanicus* (Marcet) [Gordh1979], *Ap. holoxanthus* DeBach [RosenDe1978], *Ap. immaculatus* Compere [Gordh1979], *Ap. lingnanensis* Compere [DeBachLaWh1962, RosenDe1978, RosenDe1979, MyartsRu2000], *Ap. maculicornis* (Masi) [ErlerTu2001], *Ap. mandalayensis* Rosen & DeBach [RosenDe1979], *Ap. melinus* DeBach [DeBachLaWh1962, RosenDe1978, RosenDe1979, HareMo2000], *Ap. mytilaspidis* (Le Baron) [RosenDe1979], *Ap. paramaculicornis* DeBach & Rosen [RosenDe1979], *Ap. philippiensis* DeBach & Rosen [RosenDe1979], *Ap. pinnaspidis* Rosen & DeBach [RosenDe1979, MyartsRu2000], *Ap. proclia* (Walker) [RosenDe1979, SengonUyKa1998], *Ap. yasumatsui* Azim [RosenDe1979], *Aspidiotiphagus citrinus* (Craw) [RosenDe1978, Blumbe1997], *Aspidiotiphagus lounsburyi* (Berlese & Paoli) [RosenDe1978], *Azotus* sp. [SengonUyKa1998], *Coccophagoides moeris* (Walker) [ErlerTu2001], *Encarsia citrina* (Craw) [MyartsRu2000, AbdRab2001a], *Encarsia lounsburyi* (Berlese & Paoli) [MyartsRu2000, AbdRab2001a], *Encarsia perniciosi* (Tower) [Rosen1987, StouthLu1991, MyartsRu2000], *Marietta carnesi* (Howard) [RosenDe1978], *Physcus debachi* Compere [RosenDe1978], *Prospaltella aurantii* (Howard) [Gordh1979, Blumbe1997], *Prospaltella perniciosi* Tower [RosenDe1978], *Pteroptrix chinensis* (Howard) [ComperSm1927, FlandeGrFi1958, RosenDe1978, Blumbe1997], *Pteroptrix wanhsiensis* (Compere) [RosenDe1978]. **Encyrtidae**:*Adelencyrtus sarawaki* Trjapitzin & Myartseva [TrjapiMy2001]; *Aphycus immaculatus* Howard [Gordh1979], *Comperiella* [ErlerTu2001], *Comperiella bifasciata* Howard [ComperSm1927, Coy1938, RosenDe1978, BlumbeLu1990, Blumbe1997, SengonUyKa1998], *Comperiella unifasciata* Ishii [Trjapi1989], *Habrolepis rouxi* Compere [Comper1936a, Comper1961, Flande1962, RosenDe1978, BlumbeDe1979, Blumbe1997], *Pseudhomalopoda prima* Girault [Gordh1979]. **Signiphoridae**: *Signiphora fax* Girault [Woolle1990], *Signiphora flavella* [Woolle1990], *Signiphora flavopalliata* Ashmead [Gordh1979, Woolle1990], *Signiphora merceti* [Woolle1990], *Signiphora occidentalis* Howard [Gordh1979], *Signiphora prepauca* Girault [Woolle1990]. NEUROPTERA **Chrysopidae**: *Chrysopa californica* [Comper1961], *Chrysoperla plorabunda* (Fitch) [Drea1990]. **Coniopterygidae**: *Conwentzia psociformis* (Curtis) [Drea1990]. THYSANOPTERA **Phlaeothripidae**: *Aleurodothrips fasciapennis* (Franklin) [PalmerMo1990].

DISTRIBUTION: **Afrotropical**: Angola [Almeid1973b]; Guinea [Balach1956]; Kenya [Balach1956, DeLott1967a]; Madagascar [Mamet1954, Borchs1966]; Mauritius [GrandpCh1899, Mamet1949, Borchs1966]; Mozambique [Almeid1971]; Rodrigues Island [Mamet1954a, Borchs1966]; Réunion [Mamet1957]; Saint Helena [Matile1976]; South Africa [BrainKe1917, Newste1917b, Brain1919, McKenz1937, RosenDe1978, Bedfor1989a]; Sudan [Balach1956]; Tanzania [Newste1911a]; Uganda [Balach1956]; Zaire [Ghesqu1932]; Zanzibar [Balach1956]; Zimbabwe [Newste1917b, Hall1928, Balach1956]. **Australasian**: Australia [Mamet1943a] (New South Wales [Laing1929, RosenDe1978], Queensland [Brimbl1962a], Victoria [Laing1929]); Bonin Islands (= Ogasawara-Gunto) [Kuwana1909a, Mamet1943a]; Cook Islands [WilliaWa1988]; Federated States of Micronesia (Caroline Islands [Takaha1942d]); Fiji [WilliaWa1988]; Marshall Islands [Beards1966]; New Caledonia [Cohic1958]; New Zealand [Spille1952]; Niue [WilliaWa1988]; Palau [Takaha1939b]; Papua New Guinea [WilliaWa1988]; Solomon Islands [WilliaWa1988]; Tonga [WilliaWa1988]; Vanuatu [WilliaBu1987, WilliaWa1988] [WilliaBu1987, WilliaWa1988]; Western Samoa [WilliaWa1988]. **Nearctic**: Mexico [MyartsRu2000] (Tamaulipas [LunaSaMa1995]); United States of America (California [Comsto1881a, McKenz1956, RosenDe1978, RosenDe1979], Florida [Wilson1917, MerrilCh1923, Merril1953, Dekle1965c], Mississippi [Herric1911], Texas [McKenz1937, McDani1968, RosenDe1978]). **Neotropical**: Argentina [McKenz1937, Mamet1943a, ClapsTe2001]; Brazil [WolffCo1993a] (Alagoas [WolffCo1993], Cear [WolffCo1993], Distrito federal (Brasilia) [Lepage1938, WolffCo1993], Maranhao [WolffCo1993], Paraiba [WolffCo1993], Parana [WolffCo1993], Pernambuco [WolffCo1993], Rio Grande do Norte [WolffCo1993], Rio Grande do Sul [Lepage1938, WolffCo1993], Rio de Janeiro [Lepage1938, WolffCo1993], Santa Catarina [WolffCo1993], São Paulo [WolffCo1993]); Chile [McKenz1937, GonzalCh1968, Gonzal1989]; Puerto Rico & Vieques Island (Puerto Rico [Martor1976, ColonFMe1998]); Uruguay [AsplanGa2001]. **Oriental**: Burma (= Myanmar) [RosenDe1979]; China (People's Republic) (Yunnan [Ferris1953]); Hong Kong [RosenDe1979]; India [Green1900a, McKenz1937, RosenDe1979] (Bihar [Ali1968], Punjab [RahmanAn1941], West Bengal [Nath1972]); Indonesia (Java [Green1904a]); Mongolia [DanzigKo1990]; Nepal [Takagi1975]; Pakistan [Janjua1959]; Philippines [VelasqRi1969] (Luzon [Velasq1971], Palawan [Velasq1971]); Ryukyu Islands (= Nansei Shoto) [KinjoNaHi1996]; Sri Lanka [Green1896, Green1896e, Ramakr1921a]; Taiwan [Ferris1921a, Takaha1929, Takaha1932a, Takaha1934, Takagi1969a, WongChCh1999]; Thailand [Takaha1942b, NakaoTaTa1977]. **Palaearctic**: Afghanistan [Siddiq1966]; Canary Islands [PerezGCa1987, MatileOr2001]; China (People's Republic) [McKenz1937] [Tang1984] (Henan (Honan) [Shen1993]); Crete [Ayouta1940]; Cyprus [Balach1932d, RosenDe1978]; Egypt [Hall1922, Balach1932d, Ezzat1958]; France [Balach1930a, Balach1932d]; Georgia (Abkhaz ASSR [Borchs1934, Borchs1936], Adzhar ASSR [Borchs1934], Georgia [Borchs1934]); Greece [Bodenh1928, Korone1934, RosenDe1978]; Iran [Bodenh1944b, Kaussa1955]; Israel [Bodenh1924, Bodenh1930, Balach1932d, Bodenh1937, GersonZo1973, RosenDe1979]; Italy [Leonar1909, Leonar1920, Lupo1936, Viggia1970a, LongoMaPe1995]; Japan [Kuwana1902, Sakai1939,

Kawai1980]; Lebanon [Bodenh1926]; Madeira Islands [Green1923b, CarvalAg1997]; Morocco [Balach1932d, RosenDe1978]; Saudi Arabia [Beccar1971, Matile1984c]; Sicily [RosenDe1978, InserrCa1987]; Spain [GomezM1937, Martin1983, BlayGo1993]; Syria [Balach1932d]; Tunisia [Balach1932d]; Turkey [Bodenh1949, Bodenh1952, SengonUyKa1998, UygunSeEr1998, ErlerTu2001].

BIOLOGY: California red scale (CRS) is biparental and ovoviviparous infesting all above-ground parts of host plants (Ferris, 1938a; Bartlett, 1978). Crawlers responded to various solutions behind artificial membranes by secreting wax threads that form scale cover. This response was used to evaluate gustatory stimuli responsible for crawler settling (Walker & Bendar, 1986). Sex pheromone of CRS has been identified and synthesized (Roelofs et al., 1978). Millar & Hare (1993) isolated and identified a kairomone from scale cover, which functions as an oviposition stimulant.

ECONOMIC IMPORTANCE: Most important pest of citrus pest in most citrus-growing areas of world (Quayle, 1911a, 1938a; Bodenheimer, 1951; Ebeling, 1959; Rosen & DeBach, 1978). Besides extensive research on chemical control (see Citations) there have been extensive studies on biological control of this pest (Compere, 1961; Rosen & DeBach, 1978).

GENERAL: Description and illustration of adult female by Leonardi (1910), Kuwana (1933), Ferris (1938a), McKenzie (1938, 1956), Balachowsky (1948b, 1956), Takagi (1969a), Velasquez (1971), Chou (1985, 1986), Tereznikova (1986), Williams & Watson (1988), Yaşar (1995a), Gill (1997) and by Colon-Ferrer & Medina-Gaud (1998).

KEYS: Claps & Teran 2001: 392 (female) [South Africa]; Gill 1997: 44 (female) [Species of California]; Danzig 1993: 159 (female) [Europe]; Williams & Watson 1988: 35 (female) [Tropical South Pacific]; Tereznikova 1986: 85 (female) [Ukraine]; Chou 1985: 292 (female) [Species of China]; Kawai 1980: 211 (female) [Japan]; Gerson & Zor 1973: 516 (female) [Israel]; Velasquez 1971: 118 (female) [Philippines]; McDaniel 1968: 210 (female) [U.S.A.: Texas]; Beardsley 1966: 509 (female) [Federated States of Micronesia]; Ezzat 1958: 240 (female) [Egypt]; Balachowsky 1956: 25 (female) [Africa]; McKenzie 1956: 24 (female) [U.S.A.: California]; Lupo 1954a: 41 (female) [Italy]; McKenzie 1953: 36 (female) [World]; Balachowsky 1948b: 362 (female) [Mediterranean]; McKenzie 1946: 33 (female) [World]; Ferris 1942: 29 (female) [North America]; McKenzie 1942b: 144-145 (female) [World]; McKenzie 1938: 17-18 (female) [World]; McKenzie 1937a: 178 (female) [World]; McKenzie 1937: 330 (female) [World]; Lupo 1936: 261 (female) [World]; Kuwana 1933: 26 (female) [Japan]; Kuwana 1933b: 49 (female) [Japan]; Fullaway 1932: 95-97, 107 (female) [Hawaii]; Britton 1923: 376 (female) [U.S.A.: Connecticut]; Leonardi 1920: 75 (female) [Italy]; Brain 1919: 198 (female) [South Africa]; Lawson 1917: 210 (female) [U.S.A.: Kansas]; Robinson 1917: 24 (female) [Philippines]; Dietz & Morrison 1916a: 289-290, 307 (female) [U.S.A.: Indiana]; Newstead 1901b: 81 (female) [England]; Green 1896e: 40 (female) [Sri Lanka]; Comstock 1883: 55-57 (female) [North America].

CITATIONS: Abbas1992 [host, distribution, life history: 477-485]; AbdelFElDa1978a [host, distribution, life history, economic importance: 74-78];

AbdElKDaKo1988 [chemical control, biological control: 270-275]; Abdelr1973 [economic importance, chemical control, biological control: 119-133]; Abdelr1974 [life history, ecology, biological control: 203-212]; Abdelr1974a [life history, ecology, biological control: 213-220]; Abdelr1974b [life history, ecology, biological control: 231-247]; AbdRab2001a [host, distribution, biological control: 174-176]; AblesRi1981 [biological control, economic importance: 273]; AbouEl2001 [host, distribution, biological control: 185-195]; Aldric1996 [life history, physiology, chemistry: 201-204]; Alexan1980 [host, distribution, life history, economic importance: 555-560]; Alexan1983 [host, distribution, life history: 831-838]; AlexanMi1982 [host, distribution, life history: 639-644]; Alfier1929 [host, distribution: 7-9]; Ali1968 [host, distribution: 132]; Almeid1971 [host, distribution: 6,8]; Almeid1973b [host, distribution: 8]; Amin1981 [host, distribution, life history: 1-11]; AndersAdCh1980 [chemistry, pheromone: 2229-2236]; Andrie1932 [host, distribution, economic importance, control]; Anneck1963 [host, distribution, life history, economic importance: 195-225]; AnneckIn1971 [host, distribution, biological control: 14]; Archan1937 [taxonomy, description, illustration, host, distribution: 95-96]; ArgovZcRo1995 [life history, biological control: 315-320]; Argyri1969 [biological control: 817-822]; Argyri1970 [host, distribution, biological control: 57-65]; Argyri1974 [host, distribution, economic importance, biological control: 89-94]; Argyri1979a [host, distribution, economic importance, biological control: 517-520]; Argyri1986 [host, distribution, biological control: 545-548]; Argyri1990 [host, distribution, economic importance: 579-583]; ArgyriMo1981 [host, distribution, economic importance: 623-627]; ArgyriStMo1976 [host, distribution, biological control: 24-25]; AsplanGa1998 [host, distribution, life history, biological control: 637-646]; AsplanGa2001 [host, distribution, life history, ecology: 54-67]; AsquitCrHo1980 [control, chemistry, economic importance]; Atkins1977 [host, distribution, life history, ecology: 65-87]; Atkins1983 [host, distribution, life history, biological control: 239-258]; Atkins1983a [host, distribution, life history, biological control: 417-426]; AungLeJe2001 [host, distribution, chemical control: 93-100]; AvidovBaGe1970 [host, distribution, life history, biological control: 191-207]; AvidovRoGe1963 [biological control, economic importance, chemical control, host, distribution: 205-212]; Ayouta1940 [host, distribution: 2-4]; AytasYuMa2001 [host, distribution, life history, biological control: 653-664]; Azeved1923A [host, distribution: 86-90]; Baker1976 [biological control, life history: 1-25]; Balach1930a [host, distribution: 179]; Balach1932d [taxonomy, host, distribution: XII, XLVIII]; Balach1948b [taxonomy, description, illustration, host, distribution: 366-370]; Balach1950e [chemical control, biological control: 220-223]; Balach1956 [taxonomy, description, illustration, host, distribution: 29-32]; Ballou1912 [host, distribution, economic importance, control]; Ballou1912a [host, distribution: 412-425]; Ballou1913 [host, distribution: 61-65]; Banks1990 [physiology, chemistry: 267-274]; Barany1953 [life history, chemistry, physiology, structure: 202-209]; Barber1893 [host, distribution: 50-51]; Barnes1930 [biological control: 319-329]; Bartle1953a [chemical control, biological control: 565-569]; Bartle1963 [chemical control, biological control: 694-698]; Bartle1964 [chemical control, biological control: 489-511]; BartleFi1950 [biological control: 802-806]; BarZakPeHe1989 [life history, physiology: 1228-1231]; BasuNaCh1969 [economic importance, host, distribution: 169-178]; BatraSaSo1989 [host, distribution, economic importance: 161-162]; BazaroSh1970 [host, distribution, taxonomy: 109-111]; Beards1966 [host, distribution: 509]; BeardsDaHo1976 [economic importance: 103]; BeardsGo1975 [economic importance: 49]; Beatti1984 [host, distribution, ecology, biological control: 25-29]; BeattiGe1983 [host, distribution: 220-226]; BeattiHu1988 [host, distribution, life history, biological control: 386]; Beccar1971 [host, distribution: 193]; Bedfor1968a [biological control, economic importance: 1-15]; Bedfor1968b [host, distribution, chemical control, biological control: 1-14]; Bedfor1969 [chemical control, biological control: 3-10]; Bedfor1973 [host, distribution, chemical control, biological control: 5-11]; Bedfor1981 [host, distribution, chemical control, biological control: 1-6]; Bedfor1989a [economic importance, life history, host, distribution, biological control, chemical control: 1-4]; Bedfor1990 [host, distribution, economic importance, biological control: 507-513]; Bedfor1998 [host, distribution, chemical control, biological control, economic importance, life history: 132-144]; BedforBo1981 [host, distribution, chemical control, biological control: 1-15]; BedforCi1994 [host, distribution, biological control: 143-179]; BedforDe1984 [host, distribution, chemical control, biological control: 1-7]; BedforDo1977 [host, distribution, chemical control, biological control: 1-10]; BedforGe1978 [host, distribution, economic importance, life history, chemical control, biological control: 109-242]; BedforGr1981 [host, distribution, chemical control, biological control: 616-620]; BedforVaDe1998 [host, distribution, economic importance, chemical control, biological control]; BedforVeDe1985 [host, distribution, chemical control, biological control: 1-15]; BedforVeKo1992 [host, distribution, chemical control, biological control: 1-112]; Bellow2001 [biological control: 199-205]; Benass1957 [biological control: 283-291]; Benass1959b [host, distribution, biological control, chemical control: 867-872]; Benass1961b [host, distribution, ecology: 1-157]; Benass1965a [host, distribution, chemical control, biological control, economic importance, life history: 112-125]; Benass1969 [biological control: 793-799]; BenassAlLi1984 [host, distribution: 325-327]; BenassBi1967 [host, distribution: 247-256]; BenassBi1974 [distribution, biological control: 39-50]; BenassEu1966 [host, distribution, economic importance, biological control: 19-26]; BenassEu1967 [biological control: 449-459]; BenassEu1968 [host, distribution, biological control: 1-60]; BenassEu1970 [biological control: 95-100]; BenassEu1970a [host, distribution, biological control: 357-372]; BenassViRo1979 [biological control: 281-287]; BenDavGr2000 [chemical control: 423-425]; BenDov1990a [taxonomy: 89]; BenDovRo1969 [biological control, life history, economic importance: 1057-1060]; BenfatCa1996 [host, distribution, chemical control: 38,73-76]; BenfatCa2002 [host, distribution, life history: 73-75]; BenfatCoTu1998 [biological control, chemical control: 217-222]; BennetRoCo1976 [biological control, economic importance: 359-395]; BerlesLe1899 [taxonomy, description, host, distribution: 265-273]; BiezanFr1939 [host, distribution: 1-18]; BiezanSe1940 [host, distribution: 67-68]; BilogOMo2000 [host, distribution, biological control: 137-147]; BlayGo1993 [taxonomy, description, illustration, host, distribution: 518-522]; BlissBrWa1931 [host, distribution, life history, ecology: 1222-1229]; Blumbe1971 [economic importance, biological control: 434-440]; Blumbe1973a [host, distribution, biological control: 125-131]; Blumbe1976 [biological control: 131-139]; Blumbe1990 [structure, biological control: 221-228]; Blumbe1997 [biological control, ecology: 225-236]; BlumbeDe1979 [life history, biological control: 299-306]; BlumbeIsGo1994 [chemical control, biological control: 434-440]; BlumbeIsGo1994 [chemical control, biological control: 434-440]; BlumbeLu1990 [life history, biological control: 591-597]; BlumbeSw1974 [host, distribution, biological control: 437-443]; BlumbeSw1974a [biological control: 3-11]; BlumbeSw1982a [biological control: 67-76]; BoboyeCa1975 [chemical control: 473-476]; Bodenh1924 [taxonomy, description, host, distribution: 34-36]; Bodenh1926 [host, distribution: 42]; Bodenh1928 [host, distribution: 191]; Bodenh1930 [host, distribution: 4]; Bodenh1934 [life history, taxonomy, ecology, host: 139-148]; Bodenh1935 [host, distribution: 247]; Bodenh1936 [taxonomy: 4]; Bodenh1937 [host, distribution: 217]; Bodenh1938 [taxonomy, host, distribution, life history, biological control, economic importance: 6-7]; Bodenh1944b [host, distribution: 93]; Bodenh1949 [taxonomy, description, illustration, host, distribution: 72-75]; Bodenh1951 [structure: 133]; Bodenh1951a [taxonomy, host, distribution, life history, economic importance, chemical control, biological control: 202-274]; Bodenh1952 [host, distribution: 345]; Bodkin1925 [host, distribution, chemical control: 143-149]; BogranHeCi2002 [biological control: 653-668]; Bondar1914 [host, distribution, economic importance: 1064-1106]; Bondar1915 [host, distribution, economic importance: 44-47]; Borchs1934 [host, distribution, economic importance: 30-31]; Borchs1937 [taxonomy, description, illustration, host, distribution: 121]; Borchs1937a [taxonomy, description, illustration, host, distribution: 63-65]; Borchs1939 [taxonomy, description, host, distribution: 8,18]; Borchs1949d [taxonomy, description, host, distribution: 234]; Borchs1950b [taxonomy, description, illustration, host, distribution: 219,221]; Borchs1966 [catalogue: 292-294]; Borer2002 [life history, ecology, biological control: 957-965]; BorianNi1995 [chemical control: 43]; BourijBo1982 [life history, ecology: 303-315]; Boyce1928 [chemical control, physiology: 715-720]; Boyce1948 [host, distribution, economic importance, control]; Boyce1950 [host, distribution, economic importance: 741-766]; BoyceKaPe1939 [chemical control: 432-450]; BoyeroAnMo2000 [control: 673-688]; Brain1919 [taxonomy, description, illustration, host, distribution: 199-200]; BrainKe1917 [distribution: 184]; Brewer1971 [host, distribution, life history, biological control: 53-63]; Brick1912 [host, distribution: 1-22]; Brimbl1962 [host, distribution, economic importance: 219]; Brimbl1962a [taxonomy, description, illustration, host, distribution: 404-408]; Britto1923 [taxonomy, description, host, distribution: 376-377]; Brock1925 [chemical control: 349,366]; BrownPo1990 [biological control: 527-533]; BruwerDe1988 [host, distribution, biological control: 12-16]; BurgerUl1990 [economic importance: 313-327]; Burke1930 [host, distribution, biological control: 783-785]; Bustsh1958 [taxonomy, description, host, distribution: 220,245]; Buxton1920 [host, distribution: 287-303]; CABI1968 [host, distribution: 1-2]; CABI1996 [host, distribution: 1-5]; Calkin1983 [distribution, economic importance: 321-359]; Caltag1985 [taxonomy, biological control: 189-200]; Caltag1999 [ecology, life history: 217]; CameroGaSo1969 [host, distribution, life history, biological control: 694-696]; Campbe1972 [host, distribution, biological control:

43]; Campbe1975 [host, distribution, chemical control, biological control: 161-164]; Campbe1976MM [host, distribution, biological control: 659-668]; Cangar1960 [host, distribution, chemical control: 69-77]; Carman1948 [chemical control: 1]; Carman1953 [chemical control: 307-308]; Carman1956 [chemical control: 103-111]; Carman1981 [host, distribution: 7,9,11]; CarmanElEw1954 [host, distribution, chemical control: 1-11]; CarmanEw1950 [chemical control: 15A-16A]; CarmanEw1950a [chemical control]; CarmanEwJe1951 [host, distribution, chemical control: 1-16]; CarmanEwJe1956 [chemical control]; CarmanEwJe1957 [chemical control]; CarmanEwJe1958 [chemical control]; CarmanEwJe1959 [chemical control]; CarmanEwJe1960 [chemical control]; CarmanEwJe1961 [chemical control]; CarmanEwJe1962 [chemical control]; CarmanLi1956 [chemical control: 534-539]; Carmin1936 [biological control: 173-175]; CarminSc1934 [host, distribution, economic importance, control: 242-274]; Carnes1907 [taxonomy, host, distribution: 214]; Carrol1979 [host, distribution, life history, ecology: 1-145]; CarrolLu1984 [host, distribution, life history, ecology, biological control: 847-853]; CarrolLu1984a [host, distribution, life history, ecology, biological control: 179-183]; CarvalAg1997 [life history, description, economic importance, biological control, host, distribution: 249-257]; CasasNiSw2000 [biological control: 185-193]; Castel1951a [biological control: 95-98]; CasuOnGe1999 [host, distribution, pheromone, control: 69-72]; Catlin1971 [host, distribution, biological control: 393-411]; Catlin1971a [host, distribution, biological control: 5-9]; Cendan1937 [host, biological control: 337-339]; Charle1998 [distribution, economic importance, biological control: 51N,]; CharmoGe1921 [host, distribution: 188]; Chen1936 [taxonomy: 211]; Chiesa1948 [host, distribution, economic importance]; Chiesa1948a [host, distribution, economic importance]; Chorle1939 [host, distribution]; Chou1947a [chemical control: 34]; Chou1985 [taxonomy, description, host, distribution: 292-295]; Chou1986 [taxonomy, illustration: 446]; Chou1986 [taxonomy, illustration: 676]; ChuaWo1990 [host, distribution, economic importance: 543-552]; ClapsTe2001 [taxonomy, description, illustration, host, distribution, economic importance: 392-394]; ClapsWoGo2001a [taxonomy, host, distribution: 11]; ClarkFr1932 [economic importance, life history, host, distribution, chemical control, biological control: 1-35]; Clause1956 [host, distribution, economic importance, biological control]; Clause1958 [economic importance, biological control: 291-310]; Clause1958a [host, distribution, biological control: 443-447]; CliftBe1993 [life history, chemical control: 470-472]; Cocker1893cc [host, distribution: 101]; Cocker1896b [taxonomy, distribution: 334]; Cocker1897i [taxonomy, description, host, distribution: 13,29]; Cocker1899a [taxonomy: 396]; Cocker1900k [taxonomy: 350]; Cohen1969 [biological control, economic importance: 769-772]; Cohen1975 [host, distribution, economic importance, biological control: 38-41]; CohenPoEl1994 [chemical control, biological control: 183-190]; Cohic1958 [host, distribution: 12]; Collar1918 [host, distribution: 154-162]; Collie1995 [life history, biological control, host, distribution: 206-214]; CollieMuNi1994 [life history, biological control, host, distribution: 299-306]; Collin1950 [host, distribution, economic importance: 158-160]; CollinLaBo1994 [host, distribution, physiology, chemical control: 325-326]; ColonFMe1998 [taxonomy, description, illustration, host, distribution: 35-36]; Comper1926 [host, distribution, biological control: 33-50]; Comper1936 [host, distribution, biological control]; Comper1936a [biological control, host, distribution: 493-496]; Comper1953 [biological control: 35-46]; Comper1955 [biological control: 271-319]; Comper1961 [taxonomy, host, distribution, biological control, economic importance: 173-278]; Comper1961a [biological control: 17-71]; Comper1969 [biological control: 755-764]; Comper1969a [host, distribution, economic importance, biological control: 5-10]; ComperAn1961 [host, distribution, biological control: 17]; ComperFlSm1941 [biological control: 291,301]; ComperSm1927 [biological control: 63-73]; Comsto1881 [taxonomy, description, host, distribution: 8-9]; Comsto1881a [taxonomy, description, illustration, host, distribution: 293-295]; Comsto1883 [taxonomy, host, distribution: 59]; Cook1909 [taxonomy, description, host, distribution, chemical control: 15]; Coquil1888 [chemical control: 123-133]; Coquil1890a [chemical control: 9-17]; Coquil1891a [chemical control, taxonomy: 19-36]; CoronaRuMo1997 [host, distribution: 38-41]; CostaL1942 [taxonomy: 284]; CostaL1949 [host, distribution, biological control: 65-87]; CostaLRa1922 [host, distribution: 1101]; Costan1956a [host, distribution, economic importance: 74-79]; CostilOsBa1970 [host, distribution, life history, ecology: 57-65]; Cottie1956 [host, distribution, economic importance, chemical control, biological control: 209-215]; Coy1938 [biological control: 445-446]; Craw1906 [host, distribution: 139-158]; Creigh1942 [host, distribution, control: 219-233]; Cressm1941 [chemical control: 859]; Cressm1943 [host, distribution, chemical control: 17-26]; Cressm1943a [chemical control: 413-419]; Cressm1954 [chemical control: 100-102]; Cressm1955 [chemical control: 216-217]; Cressm1956 [chemical control: 406,408]; Cressm1957 [chemical control: 593-595]; Cressm1958 [chemical control: 911-912]; CressmBr1943 [chemical control: 439-441]; CressmBr1944 [chemical control: 809-813]; CressmBr1953 [chemical control: 907]; CressmBrMu1954 [chemical control: 379,392,395]; CressmBrMu1957 [chemical control: 1-7]; CressmMuBr1949 [chemical control: 332,350,351]; CressmMuBr1950 [chemical control: 610-614]; CressmMuBr1953 [chemical control: 1070-1074]; CressmMuBr1954 [chemical control: 424-440,441,444]; Crouze1971 [biological control: 200]; Crouze1973 [host, distribution, biological control: 15-39]; CrouzeBiZa1973 [host, distribution, biological control: 251-318]; DahmsSm1994 [host, distribution, biological control: 245-255]; DaneelMeJa1994 [host, distribution: 72-74]; Danzig1972 [taxonomy, host, distribution, economic importance: 206-207]; Danzig1993 [taxonomy, description, illustration, host, distribution, economic importance: 161-162]; DanzigKo1990 [host, distribution: 45]; DanzigPe1998 [catalogue: 181-182]; DarvasFaKo1985 [chemical control: 347-350]; DarvasVa1990 [chemical control: 393-408]; DavidsDiFl1991 [chemical control: 1-47]; DavidsMi1990 [host, distribution, economic importance: 603-632]; DaviesMc1977 [physiology, chemical control, biological control: 323-328]; Dean1955 [biological control: 444-447]; DeBach1948 [host, distribution, biological control: 985]; DeBach1951a [biological control: 443-447]; DeBach1958 [host, distribution, biological control, ecology: 187-194]; DeBach1958a [biological control: 759-768]; DeBach1958b [host, distribution, ecology: 187-194]; DeBach1960 [host, distribution, biological control: 701-705]; DeBach1962a [distribution, economic importance, biological control: 17-25]; DeBach1964 [biological control]; DeBach1964b [biological control: 673-713]; DeBach1964c [host, distribution, biological control, ecology, life history, economic importance: 221-224]; DeBach1964d [biological control: 5-18]; DeBach1965 [biological control, ecology: 848-863]; DeBach1966 [host, distribution, biological control, ecology, life history : 183-212]; DeBach1969 [biological control: 801-815]; DeBach1969a [biological control: 11-28]; DeBach1971 [biological control: 293-307]; DeBach1972 [biological control: 211-233]; DeBach1974 [biological control]; DeBachAr1967 [host, distribution, biological control: 325-342]; DeBachBa1951 [chemical control, biological control: 372-383]; DeBachBa1964 [biological control: 402-426]; DeBachDiFl1949 [biological control: 546-547]; DeBachDiFl1950 [biological control: 783-802]; DeBachDiFl1951 [biological control: 347-348]; DeBachFiLa1955 [biological control, life history: 742-753]; DeBachFlDi1949 [biological control: 12,14]; DeBachHeRo1978 [host, distribution, life history, ecology, biological control: 1-35]; DeBachHu1971 [biological control: 113-140]; DeBachHuMa1976 [biological control: 255-285]; DeBachLa1959 [biological control: 301-304]; DeBachLaJe1959 [biological control: 12,15]; DeBachLaWh1955 [biological control, economic importance, host, distribution: 254,271-275]; DeBachLaWh1962 [biological control, economic importance: 2-3]; DeBachRo1976a [host, distribution, biological control: 541-545]; DeBachRo1991 [biological control]; DeBachRoKe1971 [biological control: 165-194]; DeBachSi1960 [biological control: 153-160]; DeBachSu1963 [biological control: 105-166]; DeBachWh1960 [life history, biological control: 4-58]; DEDAC1923 [host, distribution]; DeitzTo1980 [taxonomy: 33]; Dekle1965c [taxonomy, description, host, distribution: 19]; Dekle1976 [taxonomy, description, host, distribution, economic importance: 30]; DeLott1967a [host, distribution: 112]; Delucc1965 [host, distribution, economic importance, life history, biological control: 739-788]; Delucc1969 [biological control: 871-874]; DeluccRoSc1976 [taxonomy, biological control: 81-91]; deOngKnCh1927 [host, distribution, chemical control: 351-384]; DeSant1940 [biological control: 29-44]; DeSant1941a [host, distribution, biological control: 21-24]; DeSant1979 [biological control]; DeStef1910 [host, distribution: 189-196]; Dickso1941 [chemical control, physiology, structure: 515-522]; Dickso1951 [structure, life history: 596-602]; DicksoLi1947 [host, distribution, economic importance: 6-7,34]; DicksoLi1947a [host, distribution, economic importance, chemical control: 524,542-544]; Dietri1981 [biological control, economic importance, host, distribution: 151-160]; DietzMo1916a [taxonomy, description, illustration, host, distribution: 289,307-308]; Dikov1965 [host, distribution, control: 8-9]; Dingle1924 [taxonomy, description, host, distribution, life history: 368]; Dohani1937 [host, distribution, biological control: 243-247]; Doutt1959 [biological control: 161-182]; Dowson1935 [host, distribution: 225]; Dozier1933 [host, distribution, biological control: 85-100]; Drea1990 [biological control: 51-59]; DreistClFl1994 [taxonomy, life history, economic importance, control]; Dunkel1999 [chemistry, life history: 251-276]; duToit1996 [host, distribution, biological control, chemical control: 526-529]; Dutta1990 [host, distribution: 152-163]; DziedzKa1990 [host, distribution: 39-43]; Early1984 [host, distribution, biological control: 271-308]; Ebelin1931 [chemical control: 669-672]; Ebelin1932 [chemical control: 1007-1012]; Ebelin1933 [host, distribution, life history, ecology: 851-854]; Ebelin1934 [biological control: 362-363]; Ebelin1936 [chemical control, life history: 95-25]; Ebelin1940 [chemical control: 92-102]; Ebelin1945 [chemical control: 556-563]; Ebelin1947 [chemical control: 619-632]; Ebelin1949 [host, distribution, life history, control]; EbelinPe1953 [host, distribution, economic importance: 1-35]; Edward1936 [host, distribution, economic importance: 335-337]; Efimof1937 [host, distribution]; EHG1897 [host, distribution: 67-85]; Ehrhor1926c [host, distribution: 20]; EhrhorFuSw1913 [distribution: 295-300]; Elguet1932 [biological control: 85]; ElirazRo1978 [biological control: 96-101]; ElmerBr1982 [host, distribution, life history, economic importance: 699-700]; ElmerBrEw1980 [host, distribution, economic importance: 20-21]; ElMinsElHa1974 [taxonomy, description, illustration, host, distribution: 223-232]; ErichsSaHa1991 [biological control: 493-498]; ErlerTu2001 [host, distribution, biological control: 299-305]; Ervin1982 [host, distribution, economic importance: 1-174]; ErvinGaBa1986 [host, distribution, economic importance: 71-77]; Esaki1940a [host, distribution: 274-280]; Essig1949 [host, distribution: 673-677]; EssigCh1909 [host, distribution, taxonomy: 1-14]; Euvert1967 [biological control: 59-100]; Evans1942 [host, distribution, taxonomy, chemical control: 156-159]; Evans1943 [host, distribution, taxonomy, chemical control]; EvansPr1990 [biological control: 3-17]; Ewart1969 [chemical control: 879-880]; EwartCaJe1954 [chemical control: 1-11]; EwingYaBa2002 [host, distribution, ecology, biological control: 35-50]; Ezzat1958 [distribution: 240]; EzzatNa1987 [distribution: 86]; EzzatRa1966 [control: 1-46]; FargerMo1974 [life history, chemistry, biological control: 26-28]; Fawcet1948 [biological control: 627-664]; Fernal1903b [catalogue: 287-288]; FernanWa1997 [biological control: 137-144]; FernanWa1999 [biological control, life history: 416-425]; Ferris1921a [host, distribution: 219]; Ferris1937c [taxonomy: 50,60]; Ferris1938a [taxonomy, description, illustration, host, distribution: 179]; Ferris1941e [taxonomy: 41-43]; Ferris1942 [taxonomy: 446:29]; Ferris1953 [host, distribution: 65]; FinneyFi1964 [biological control: 328-355]; Fisher1964 [biological control: 305-325]; FisherDe1976 [biological control: 43-50]; Flande1934a [biological control: 723-724]; Flande1936a [biological control: 1023-1024]; Flande1940b [biological control: 245-253]; Flande1942c [biological control, life history: 834-835]; Flande1943a [biological control: 233-235]; Flande1943b [biological control: 78]; Flande1944a [biological control: 365-371]; Flande1945a [biological control: 711-712]; Flande1945b [life history, biological control: 122-141]; Flande1947 [host, biological control: 746-747]; Flande1949 [biological control: 160-162]; Flande1949a [biological control: 257-274]; Flande1950a [biological control: 719-720]; Flande1951b [biological control: 93-98]; Flande1953 [biological control: 10-28]; Flande1953a [host, distribution, biological control, economic importance: 266-269]; Flande1954 [biological control: 343-352]; Flande1958a [biological control: 579-584]; Flande1959b [biological control: 125-142]; Flande1962 [biological control: 1133-1147]; Flande1966 [biological control: 79-82]; Flande1969 [biological control: 29-33]; Flande1971 [biological control, life history: 857-872]; FlandeGrFi1958 [biological control, economic importance: 65-91]; Flesch1960 [biological control: 183-208]; Fletch1951 [host, distribution, chemical control: 1-24]; FlintVa1981 [biological control: 1]; Foldi1982 [structure, taxonomy: 317-330];

Foldi1990 [structure: 43-54]; Foldi1990d [life history, structure: 257-265]; Foldi2001 [distribution: 303-308]; FoldiPe1978 [structure: 321-337]; FonsecAu1932a [host, distribution: 202-214]; ForsteLu1996 [host, distribution, biological control: 504-507]; ForsteLuGr1995 [host, distribution, life history, biological control]; FoxWil1939 [host, distribution, economic importance: 2296]; Frogga1914 [taxonomy, description, host, distribution: 132-133]; Frogga1915 [taxonomy, description, host, distribution: 9]; Fryer1936 [host, distribution, economic importance]; Fullaw1932 [taxonomy: 97,107]; Fuller1897c [host, distribution: 3]; Fuller1907 [taxonomy, description, host, distribution, economic importance, control: 1031-1055]; FultonBu1943 [chemical control: 628-629]; FurnesBuGe1983 [host, distribution, chemical control, biological control: 199-212]; Fursch1985 [biological control: 223-231]; Gaprin1954 [biological control: 587-597]; Garcia1930 [host, distribution, biological control]; Garcia1931a [host, distribution, biological control: 659-669]; Garcia1931a [host, distribution, biological control]; GarciaRo1995 [host, distribution, life history, chemical control: 118-125]; GardneErMo1983 [economic importance, life history, biological control: 601-604]; Gavalo1931 [host, distribution: 8]; Gavalo1936 [host, distribution: 81-82]; Gennad1881 [taxonomy, description, host, distribution: 189-192]; Georga1963a [economic importance, control: 5,7]; Georga1964a [chemical control: 3-15]; Georga1967 [chemical control: 3,5]; Georga1984 [economic importance, chemical control: 13-18]; Georga1984a [chemical control: 9-12]; GeorgaBuHo1972 [chemical control: 16-22]; GeorghMe1983 [life history, chemical control: 1-46]; GeorghTa1976 [chemical control: 759]; Gerson1967c [host, distribution, biological control: 632-638]; GersonIz1997 [biological control: 33-42]; GersonOcHo1990 [biological control: 77-97]; GersonOcHo1990 [biological control: 77-97]; GersonZo1973 [taxonomy, life history, host, distribution, economic importance: 513-533]; Ghabbo1995 [taxonomy: 379-387]; Ghauri1962 [taxonomy, structure: 211]; Ghesqu1927 [host, distribution: 310-316]; Ghesqu1932 [host, distribution: 59]; Giesel1990 [life history, chemistry: 221-224]; Giesel1990a [chemistry: 225-232]; GieselHeAn1980 [life history, chemistry, pheromone: 179-182]; GieselRi1990 [life history, ecology, chemistry: 349-352]; Gill1997 [host, distribution, taxonomy, description, illustration, economic importance: 7,44-47]; Glover1935 [host, distribution, chemical control, biological control: 151-153]; GomezC1950 [host, distribution, biological control : 1-18]; GomezM1937 [taxonomy, description, illustration, host, distribution: 84-88]; GomezM1956 [taxonomy, description, illustration, host, distribution, biological control: 42-49]; GomezM1957 [host, distribution: 48]; GomezM1958a [host, distribution: 8]; Gonzal1969 [biological control: 839-848]; Gonzal1989 [taxonomy, description, host, distribution, economic importance: 93]; GonzalCh1968 [distribution: 110]; GonzalRo1967 [biological control, distribution: 138]; Gordh1977 [host, distribution, biological control: 125-148]; Gordh1979 [biological control: 893-896,900,907,911,]; Gordh1994 [biological control: 188-205]; Gowdey1921 [taxonomy, host, distribution: 31]; GraebnMoBa1984 [host, distribution, biological control, ecology, life: 27-33]; Grafto1994 [chemical control: 7-9]; GraftoMoOC2000 [host, distribution, control]; GraftoOu1993 [chemical control: 21-29]; GraftoOuSa1998 [host, distribution, economic importance, chemical control, life history, physiology: 812-819];

GraftoOuSt2001 [distribution, chemical control: 20-25]; GraftoRe1995 [host, distribution, life history, biological control, chemical control: 1717-1725]; GraftoStOu1996 [host, distribution, biological control, chemical control: 553-555]; GraftoVe1995 [host, distribution, life history, biological control, chemical control: 495-504]; GrahamSt1996 [host, distribution, chemical control: 640-651]; GrandpCh1899 [taxonomy, description, host, distribution: 22-23]; GrayKi1929 [chemical control: 878-892]; GreanyViLe1984 [biological control: 690-696]; Greath1971 [host, distribution, biological control]; Greath1976 [biological control, economic importance]; Greath1986 [biological control: 289-318]; Greath1989 [biological control: 28-37]; Greath1990 [life history, ecology: 305-308]; Green1896 [host, distribution: 4]; Green1896e [taxonomy, description, illustration, host, distribution: 40,58-59]; Green1900a [host, distribution: 71]; Green1900c [host, distribution: 2]; Green1904a [host, distribution: 208]; Green1907 [host, distribution: 203]; Green1923b [host, distribution: 89]; Green1929 [host, distribution: 377]; Green1937 [host, distribution: 331]; Greig1944 [host, distribution, control]; GressiFl1949 [biological control: 150]; GriffiHoLu1985 [host, distribution, biological control: 1-6]; GroutDuHo1989 [host, distribution, life history, ecology, biological control: 793-798]; GroutRi1989 [host, distribution, life history, ecology, biological control: 277-283]; GroutRi1989a [life history, biological control, chemistry: 11-13]; GroutRi1991 [host, distribution, chemical control: 1802-1805]; GroutRi1991a [host, distribution, life history, biological control: 20-27]; GroutRi1992 [host, distribution, chemical control: 1-7]; GroutRiSt1992 [chemical control: 34-36]; GumusUy1992 [host, distribution, life history: 209-216]; Haas1934 [chemistry, chemical control, physiology: 477-492]; HabibSaAm1971 [host, distribution, life history: 318-330]; HabibSaAm1972 [host, distribution, life history: 324-338]; HabibSaAm1972a [host, distribution, life history: 378-385]; Hadzib1983 [host, distribution: 229]; HakkonPi1984 [biological control: 1109-1121]; Hall1922 [taxonomy, description, host, distribution: 32-33]; Hall1923 [host, distribution: 46]; Hall1924a [host, distribution, economic importance: 4-5]; Hall1928 [host, distribution: 276]; Hall1969 [economic importance: 823-826]; HallFo1933 [host, distribution, economic importance: 1-55]; HammadKaRa1981 [host, distribution, economic importance: 252-268]; HanksDe1998 [life history, ecology: 239-262]; HardieMi1999 [chemistry, life history]; Hardis1941 [control: 283,308-309]; HardmaCr1941 [chemical control, physiology: 187]; Hare1996 [host, chemistry, life history, biological control: 263-269]; HareLu1991 [life history, biological control, host: 1576-1585]; HareMiLu1993 [biological control, chemistry, ecology: 92-94]; HareMo1997 [host, distribution, biological control: 207-214]; HareMo2000 [life history, ecology, biological control, chemistry: 509-519]; HareMoNg1997 [life history, biological control: 73-81]; HareYuLu1990 [host, life history: 1451-1460]; HarpazRo1971 [biological control: 458-468]; Harris1990 [biological control: 61-66]; Harris1997 [host, distribution, biological control: 65]; Hart1990 [life history, ecology, distribution, economic importance: 353-356]; Hassan1999 [chemical control: 23-29]; HassanSu1997 [chemical control: 415-418]; HasselWa1984 [biological control, ecology: 89-114]; Hattin1996 [chemical control, biological control: 14-17]; HattinSa1990 [biological control: 385-390]; HattinSa1991 [host, distribution, biological control: 169-174]; HattinSa1991a [host, distribution, life history, biological control: 143-148]; HattinTa1995 [host, distribution, biological control, chemical control: 489-493]; HattinTa1996 [host, distribution, biological control, chemical control: 523-525]; HattinTa1996a [host, distribution, biological control, chemical control: 560-563]; HavronKeRo1991 [host, distribution, biological control: 229-235]; HavronRo1992 [chemical control, biological control: 984-987]; HavronRo1994 [biological control: 209-220]; HavronRoRu1995 [chemical control, biological control: 309-313]; Hawkin1994 [biological control: 3]; Hayat1989 [host, distribution, biological control: 1-3]; Haywar1939 [host, distribution, control: 1]; Hecht1936 [host, distribution, life history, biological control: 299-236]; HeCuBa1991 [life history: 58-60]; HefetzKrPe1988 [host, distribution, life history, biological control: 1121-1127]; HeimpeRoKa1997 [life history, biological control: 305-315]; HelmyHaHa1992 [chemical control: 763-771]; HelmyHiHa1997 [chemical control: 601-609]; Hempel1904 [host, distribution: 322]; Hempel1937 [taxonomy: 27]; HenrikCaAn1982 [chemistry, pheromone: 27-60]; HeratySc1998 [biological control: 1227-1244]; Herric1911 [taxonomy, description, illustration, host, distribution: 11,30-31,48]; Herric1925 [host, distribution, description, life history, economic importance]; Hewitt1943 [host, distribution: 266-274]; Hill1975 [host, distribution, economic importance, control]; Hill1989a [host, distribution, economic importance, biological control: 177-182]; HiroseNaTa1990 [biological control: 171-183]; Hodgso1969c [economic importance, life history, biological control, chemical control: 18-19]; HoffmaKe1985 [life history: 19-20]; HoleSa1997 [chemical control: 12-13]; HoleSa1998 [host, distribution, life history: 199-200]; HoleSa1999 [host, distribution, life history: 93-102]; HondaLu1995 [life history, biological control: 441-450]; Honiba1975 [biological control: 217-220]; HonibaGiRa1979 [biological control, economic importance: 17-18]; Horn1988 [host, distribution, chemical control]; Horton1918 [host, distribution, life history, economic importance: 1-74]; Hosny1968 [host, distribution, ecology: 179-182]; HosnyAmEl1972 [host, distribution, economic importance: 286-296]; Houck1999 [life history, biological control: 97-118]; HouckJeMo1989 [chemical control: 287-292]; HouckOc1996 [life history, ecology, biological control: 667-682]; Howard1894 [taxonomy, biological control: 228]; Howard1895e [biological control: 1-44]; Howard1908 [host, distribution: 265-277]; Howard1911 [host, distribution, biological control: 276-277]; Hoy1990 [biological control: 441-451]; HoyHe1985 [biological control]; Hsu1935 [host, distribution: 578-590]; Huffak1990 [biological control: 205-220]; HuffakGu1990a [biological control, economic importance: 473-492]; HuffakMeDe1971 [biological control: 16-67]; HuffakSiLa1976 [biological control: 41-78]; HuffakSm1980 [chemical control, biological control, chemistry, host, distribution: 1-24]; HuffakSt1971 [biological control: 333-350]; IFAC1962d [host, distribution: 140]; Imms1931 [biological control: 98-102]; Inserr1966 [host, distribution, biological control: 176-186]; Inserr1966a [host, distribution, life history, economic importance: 1-26]; Inserr1968 [host, distribution, biological control: 45-77]; Inserr1969 [host, distribution, life history, economic importance, control: 1-26]; Inserr1970a [host, distribution, biological control: 39-46]; InserrCa1987 [host, distribution: 93]; IntegrPeMa1991 [taxonomy, life history, biological control: 1]; Ishaay1971 [chemistry, physiology: 935-943]; IshaayBlYa1989 [chemical control: 1003-1008]; IshaayMeBl1992 [chemical control, biological control: 67-71]; IshaaySw1970 [host, chemistry, physiology: 37-42]; IshaaySw1970a [chemistry, physiology: 1599-1605]; IshaaySw1990 [physiology, chemistry: 353-356]; IshaaySwNe1980 [chemistry, physiology: 212]; Ishii1923 [host, distribution, biological control: 69]; Ishii1926 [biological control: 31-36]; Ishii1928 [host, distribution, biological control: 79]; Ishii1932a [host, distribution, biological control: 161]; JamesStOM1997 [host, distribution, biological control: 257-259]; Janjua1959 [distribution: 231-264]; Jenkin1945 [host, distribution, life history, taxonomy, chemical control, biological control: 10-16]; Jeppso1969 [economic importance, chemical control, physiology: 917-921]; JiIzGa1996 [life history, biological control: 503-509]; Jimene1969 [biological control: 781-784]; JiYa1990 [biological control: 134-136]; Johnst1915 [host, distribution, biological control: 1-33]; Jones1936 [host, distribution, life history, chemical control, biological control: 11-52]; Jourdh1979 [biological control: 75-79]; KansuUy1979 [host, distribution, biological control: 565-567]; Kapur1942 [biological control: 49-66]; Karaca1998 [host, distribution, biological control, life history: 101-108]; KaracaSeUy1987 [life history, ecology: 129-138]; KaracaUy1990 [host, distribution, biological control: 97-108]; KaracaUy1992 [host, distribution, life history: 9-19]; KaracaUyEl2002 [host, distribution, life history, biological control: 313-317]; Katsoy1993 [host, distribution, biological control: 177-198]; Katsoy1996 [life history, economic importance, host, distribution, chemical control, biological control: 15-17,61-63]; KattarHeOd1999 [biological control: 640-644]; Kaussa1955 [host, distribution: 15]; Kawai1980 [taxonomy, description, host, distribution: 213]; KeetchWh1972 [biological control, chemical control: 253-263]; KehatBlGr1968 [economic importance, chemical control: 28-29]; KfirRo1981 [biological control: 141-150]; KiddJe1996 [biological control: 293]; KingLe1984 [biological control: 1]; KingMo1984 [biological control: 206-222]; KinjoNaHi1996 [host, distribution: 125-127]; Kiritc1932a [taxonomy: 250]; KiriukTa1947 [host, distribution: 1-4]; Klein1935 [host, distribution, life history, ecology: 71-73,115-116]; Klein1940 [host, distribution, life history, economic importance, control]; Knight1932 [life history: 1]; Koebel1890 [host, distribution, biological control: 1-32]; Koebel1893 [host, distribution, biological control: 1-39]; Koehle1964 [host, distribution, control]; KollerCoCo1997 [host, distribution, biological control: 122]; Komosi1964 [host, distribution, taxonomy, description, illustration: 210-212]; KonarGh1994 [host, distribution, ecology: 9-14]; Korone1934 [taxonomy, description, illustration, host, distribution: 22-23]; Koszta1990 [structure, biological control: 307-311]; Kotins1909 [host, distribution: 97]; Kozar1990a [life history, economic importance: 341-347]; KozarKi1979 [host, distribution: 246-250]; Krambi1977 [host, distribution, biological control: 351-353]; KrishnRa1995 [chemical control, biological control: 71-79]; KrishnRa1998 [chemical control, biological control: 83-88]; KrishnRa1999 [biological control: 86-87]; Kuwana1902 [host, distribution: 70]; Kuwana1907 [host, distribution: 196]; Kuwana1909a [host, distribution: 160]; Kuwana1917a [taxonomy, distribution: 175]; Kuwana1927 [host, distribution: 72]; Kuwana1933 [taxonomy, description, illustration, host, distribution: 26,29-30]; Laing1929 [host, distribution: 25]; Lawson1917 [taxonomy, description, host, distribution: 210]; LawsonHa1984 [ecology, life history: 451]; Lenter1994 [biological control, life history: 13-39]; Leonar1897 [taxonomy: 286];

Leonar1899 [taxonomy, description, host, distribution: 175,176,184]; Leonar1907a [taxonomy, description, host, distribution: 122]; Leonar1910 [taxonomy, description, illustration, host, distribution: 5-10]; Leonar1920 [taxonomy, description, illustration, host, distribution: 75-81]; Lepage1938 [catalogue: 392]; Lepesm1947 [taxonomy, description, host, distribution, life history: 204-207]; Lever1969 [host, distribution]; LevitiCo1998 [chemical control, physiology, chemistry: 115-121]; LiLi1990 [host, biological control: 68-70]; Lindgr1938 [chemical control: 213-225]; Lindgr1941 [chemical control: 499-511]; LindgrBo1944 [chemical control: 123-124]; LindgrDi1942 [chemical control: 827-829]; LindgrDi1945 [chemical control, physiology: 296-299]; LindgrGe1947 [chemical control: 680-682]; LindgrLaDi1945 [chemical control: 567-572]; LindgrLaDo1944 [chemical control: 6,7,30]; LindgrLaDo1944a [chemical control: 180-181]; LindgrSi1944 [chemical control, physiology: 303-315]; Lindin1909c [taxonomy, host, distribution: 449]; Lindin1910b [taxonomy, description, host, distribution: 40]; Lindin1912b [taxonomy, description, host, distribution: 49,108,153]; Lindin1914 [taxonomy: 157]; Lindin1935 [taxonomy: 132]; Lindin1943a [taxonomy: 145]; Lindin1943b [taxonomy: 218]; Lindin1949 [taxonomy: 211]; Liotta1970 [host, distribution, economic importance, biological control: 35-36]; LiottaMi1974 [chemical control, host, distribution: 381-385]; LizzioSiLo1998 [host, distribution, life history, ecology: 165-183]; Lloren1990 [taxonomy, illustration, life history, host, distribution, biological control, life history: 41-49]; Lobdel1937 [taxonomy: 78]; LongoMaPe1995 [distribution: 125]; LongoMaRu1995a [host, distribution: 126-129]; LongoMaRu1996 [host, distribution, economic importance: 126-129]; Lorbee1971 [biological control, host, distribution: 199-201]; Lounsb1898 [host, distribution: 35-38]; Lounsb1906 [host, distribution: 80-91]; Lounsb1914 [host, distribution: 1]; Lounsb1916 [host, distribution: 83-103]; Lounsb1921 [host, distribution: 35-38]; Lounsb1922 [host, distribution: 205-210]; Lu1989 [host, distribution, life history, biological control: 207-217]; Lu1989a [host, distribution, life history, ecology, biological control: 218-223]; Luck1986 [biological control: 69-84]; Luck1986a [host, distribution, life history, biological control: 355-363]; LuckAlBa1980 [host, distribution, biological control, ecology, economic importance: 365-396]; LuckFoMo1996 [host, distribution, ecology, economic importance: 499-503]; LuckJiHo1999 [host, distribution, economic importance, life history, biological control: 173-183]; LuckPo1985 [life history, ecology, biological control: 904-913]; LuckPoKf1982 [life history, biological control: 397-408]; LuckUy1986 [life history, chemistry, biological control: 129-136]; LuckVaGa1977 [chemical control, economic importance, host, distribution: 606-611]; LunaSaMa1995 [host, distribution: 265]; Lupo1936 [taxonomy, description, illustration, host, distribution: 249-255]; Lupo1954a [taxonomy, description, illustration, host, distribution: 41-49]; MacGil1921 [taxonomy, description, host, distribution: 443]; Mackie1936 [host, distribution: 455]; MagaguSa2000 [biological control, chemical control, host, distribution, economic importance: 47-56]; MahmouHaHe2001 [chemistry, structure: 441]; Maksim1978a [life history: 46]; MalipaDuSm2000 [host, distribution, biological control]; Mallam1954 [distribution: 24-60]; MaltbyJiDe1968 [biological control: 1086-1088]; Mamet1943a [catalogue: 156]; Mamet1949 [catalogue: 53]; Mamet1954 [host, distribution: 15]; Mamet1954a [host, distribution: 265]; Mamet1957 [distribution: 369]; Marco1959 [host, distribution, biological control: 25-30]; Marlat1903 [host, distribution, economic importance, control: 24]; Malat1915 [host, distribution]; Martin1983 [taxonomy, host, distribution: 60]; Martor1976 [host, distribution: 63,77,176,224]; Maskel1879 [taxonomy, description, illustration, host, distribution: 199]; Maskel1884 [taxonomy, description, host, distribution: 120-121]; Maskel1887a [taxonomy, description, host, distribution: 42]; Maskel1892 [taxonomy, host, distribution: 12-13]; Maskel1893b [taxonomy, description, host, distribution: 206]; Maskel1895b [taxonomy, description, host, distribution: 40-41]; Matile1976 [host, distribution: 311]; Matile1984c [host, distribution: 220]; MatileOr2001 [host, distribution: 189]; Maxwel1903 [taxonomy, description, host, distribution: 37]; MayneGh1934 [host, distribution: 3-38]; MazzeoLoBe2002 [host, distribution, economic importance, biological control: 485-488]; McBethAl1941 [chemical control: 310-311]; McClur1990b [taxonomy, host, distribution, ecology: 285-288]; McClur1990c [ecology, host: 289-291]; McClur1990e [life history, ecology: 309-314]; McClur1990f [life history: 315-318]; McClur1990g [taxonomy, host, distribution, ecology: 319-330]; McClur1990h [life history, ecology: 331-337]; McDani1968 [taxonomy, illustration, host, distribution: 210-211]; McKenz1935 [host, distribution, life history, control: 1-48]; McKenz1937 [taxonomy, description, illustration, host, distribution: 324-327,331]; McKenz1938 [taxonomy, description, illustration, host, distribution: 6-7,23]; McKenz1939 [taxonomy: 53-54]; McKenz1942b [taxonomy: 145]; McKenz1946 [taxonomy: 33]; McKenz1946a [taxonomy, host, distribution: 95-99]; McKenz1953 [taxonomy: 36]; McKenz1956 [taxonomy, description, illustration, host, distribution: 38-40]; McLare1971 [life history, ecology: 189-204]; McLareBu1973 [host, distribution, biological control: 111-117]; Mead1983 [host, distribution: 1-5]; Mead1987 [host, distribution: 2-4]; Melis1949 [host, distribution: 17-25]; MendelBlIs1994 [chemical control, biological control: 199-209]; MendelPoRo1990 [biological control: 289-290]; Merkel1938 [host, distribution: 88-99]; Merril1953 [taxonomy, host, distribution: 13]; MerrilCh1923 [taxonomy, description, host, distribution, economic importance: 221]; MessenBiVa1976 [biological control: 543-565]; MessenVa1971 [biological control: 68-92]; MessenWiWh1976 [biological control: 209]; Metcal1982 [chemical control: 217-277]; MetcalCaMu1949 [chemical control: 5,12]; MetcalMe1993 [economic importance, host, distribution, control]; Meyer1962 [biological control: 411-417]; MichaePr1960 [chemical control: 389-391]; MillarHa1993 [life history, chemistry, biological control: 1721-1736]; MillerDa1990 [host, distribution, economic importance: 300]; Miyosh1926 [host, distribution: 303-326]; MoffitBa1990 [ecology, economic importance: 357-362]; MohammGhTa2001 [host, distribution, life history, biological control: 413-418]; Monast1958 [economic importance, control: 131-165]; Monte1930 [host, distribution: 3-36]; Moore1933 [chemical control: 1140-1161]; Moore1934 [chemical control: 1042-1055]; Moore1936 [chemical control: 65-78]; Moreno1972 [structure, anatomy, chemistry, pheromone: 1283-1286]; Moreno1983 [pheromone, control: 77-79]; MorenoFa1975 [structure, anatomy, pheromone: 425-428]; MorenoFaSh1973 [life history, pheromone, control: 1333]; MorenoFaSh1976 [chemical control, life history: 292-

297]; MorenoKe1983 [economic importance, life history, chemistry, biological control: 687-689]; MorenoKe1985 [host, distribution, life history, ecology: 1-9]; MorenoLu1992 [host, distribution, biological control, economic importance: 1112-1119]; MorenoRiCa1972 [life history, chemistry, pheromone: 698-701]; MorganHa1997 [biological control: 679-684]; MorganHa1998 [life history, ecology, biological control: 463-479]; MorganHa1998a [host, distribution, biological control, life history: 235-245]; Morley1909 [host, distribution, biological control: 254-257]; MorrisKi1977 [host, distribution, biological control: 183-217]; MorrisKi1977 [host, biological control: 183-217]; MorrisMo1922 [taxonomy: 96]; MorseArMo1985 [control: 8-10]; MoutiaMa1947 [distribution]; Moznet1922 [host, distribution]; Munger1948 [life history, physiology: 422-423]; MungerCr1948 [life history, physiology: 424-427]; Muraka1970 [host, distribution, life history: 69-70]; Murdoc1990 [biological control: 1-24]; MurdocBrNi1996 [biological control: 807-826]; MurdocChCh1985 [biological control: 344]; MurdocLuSw1995 [host, distribution, life history, ecology, biological control: 206-217]; MurdocLuWa1989 [host, distribution, life history, biological control, ecology: 1707-1714]; MurdocNiLu1992 [host, distribution, life history, biological control, ecology: 533-541]; MurdocSwLu1996 [host, distribution, life history, biological control, ecology: 424-444]; MyartsRu2000 [distribution, biological control: 7-33]; NadelBi1964 [biological control: 195-206]; Nafus1996 [host, distribution: 1]; NagarkSa1990 [host, distribution, economic importance, biological control: 553-542]; Nakaha1982 [host, distribution: 6]; NakaoTaTa1977 [host, distribution: 65]; NandakReSh1988 [chemical control: 275-277]; Nath1972 [host, distribution: 1]; Nel1933 [taxonomy, life history, anatomy, host, distribution: 416-466]; NelDeVa1979 [chemical control, physiology: 275-281]; Newste1901b [taxonomy, description, host, distribution: 81,88-91]; Newste1911a [host, distribution: 168]; Newste1917b [host, distribution: 130-131]; NourElRi1970 [host, distribution: 123-127]; Noyes1990a [biological control: 155]; NRC1969 [taxonomy, economic importance, ecology, biological control, chemical control]; NSWDAE1963 [host, distribution, taxonomy, economic importance]; Nur1990a [taxonomy, structure, chromosomes: 182,186]; OfekHuYz1997 [host, distribution, chemical control, biological control: 212-218]; OhgushNiTa1967 [host, distribution, life history: 118-121]; OmerCoJoWh1946 [biological control: 154]; OmerCoWh1950 [host, distribution, biological control: 3,12]; OmerCoWhGl1946 [biological control: 5-8]; Onder1982 [host, distribution, life history, economic importance: 1-171]; Orphan1983 [life history, host, distribution, biological control: 203-212]; Orphan1984 [host, distribution, life history, ecology: 195-209]; Orphan1984a [biological control, host, distribution: 275-281]; Orphan1996 [host, distribution, biological control: 53-57]; OrtuAc1999 [life history, host, distribution: 127-131]; Pace1939 [host, distribution: 664-665]; Painte1951 [economic importance, control]; Paoli1927a [host, distribution: 382-387]; Paolo1920 [biological control: 31-38]; PappasCa1952 [chemical control]; Peleg1982 [chemical control: 27-32]; Peleg1983 [chemical control: 367-372]; Peleg1988 [chemical control: 88-92]; PelegBa1995 [economic importance, host, distribution: 261-264]; PelegGo1981 [chemical control: 124-126]; Peleka1962 [host, distribution: 61]; Peleka1974 [host, distribution, biological control: 14-20]; Penzig1887 [host, distribution: 3]; PeralL1968 [host, biological control: 22-29];

Perez1972 [host, distribution, life history, biological control: 227-245]; PerezGCa1987 [host, distribution: 128]; Perkin1982 [economic importance, chemical control, biological control: 5]; PessonFo1978 [structure: 389-399]; Petch1921 [biological control: 18-40]; Petch1921a [biological control: 89-167]; PicartMa2000 [host, distribution: 16]; Pickel1928 [taxonomy: 55]; PietriBiCo1969 [chemical control: 909-915]; PietriBiCo1969 [chemical control: 909-915]; Pope1981 [biological control: 19-31]; Popova1962 [host, distribution, biological control: 147-175]; Porcel1995 [structure: 25-45]; PowellHo1979 [host, distribution]; Price1984 [biological control: 19]; PriesnHo1940 [biological control: 58-70]; Prinsl1983 [distribution, biological control: 26]; Prinsl1984 [taxonomy, biological control: 25-35]; Priore1964 [host, distribution: 131-178]; Priore1965 [host, distribution: 101-145]; PruthiBa1960 [host, distribution, economic importance,: 1-113]; PruthiMa1945 [host, distribution, life history, control: 1-42]; PuttarCh1953a [biological control: 87-95]; Quayle1910 [biological control: 398-401]; Quayle1911 [life history: 301-306]; Quayle1911a [host, distribution, description, economic importance, life history, biological control: 99-150]; Quayle1911d [host, distribution, description, economic importance, life history, biological control: 443-512]; Quayle1911e [biological control: 510-515]; Quayle1916 [life history, ecology: 486-492]; Quayle1916a [chemical control, physiology: 333-334,358]; Quayle1920 [chemical control: 188,189,193]; Quayle1922 [chemical control, host, distribution: 400-405]; Quayle1927 [chemical control, physiology: 667-673]; Quayle1932 [life history, chemical control: 1-87]; Quayle1938 [chemical control, physiology: 183-210]; Quayle1938a [host, distribution, life history, economic importance, chemical control, biological control]; Quayle1942 [chemical control, physiology: 813-816]; QuayleEb1934 [chemical control: 1-22]; QuayleRo1934 [chemical control: 1083-1095]; Quedna1964 [biological control, life history: 335-340]; Quedna1964a [host, distribution, biological control: 521-530]; Quedna1964b [biological control: 86-116]; Quedna1965 [life history, biological control: 43-56]; QuednaAn1963 [economic importance, biological control: 11-18]; Quilis1935 [life history, ecology: 621-633]; QuisumKy1989 [life history, chemistry: 149-171]; RabbDeKe1984 [ecology, biological control: 697]; RacitiSa2001 [host, distribution, biological control: 39-40]; RacitiTuMa1996 [host, distribution, biological control: 652-658]; RagusaRu1989 [host, distribution: 71-74]; RahmanAn1941 [taxonomy, description, host, distribution: 817-818]; Ramakr1919 [host, distribution, economic importance: 623]; Ramakr1919a [taxonomy, description, host, distribution: 20]; Ramakr1921a [host, distribution: 356]; Ramakr1930 [taxonomy, description, host, distribution: 25]; Ramakr1938a [host, distribution: 341-351]; RangelGo1945 [distribution, description: 1-44]; Rao1969 [biological control: 785-792]; RaoCh1950 [taxonomy, host, distribution: 14,27]; RauchLh1964 [chemical control, host, distribution, economic importance: 61-73]; ReeveMu1985 [biological control: 797-816]; RiceMo1969 [life history, chemistry: 558-560]; RiceMo1969a [life history, host, chemistry, pheromone: 958]; RiceMo1970 [life history: 91-96]; Richar1960AM [host, distribution: 693-698]; RidgwaKiCa1977 [host, distribution, biological control: 379-416]; Riehl1969 [chemical control: 897-907]; Riehl1990 [chemical control: 365-392]; RiehlBrMc1980 [chemical control, biological control, economic importance: 319-363]; RiehlCa1953 [chemical control: 1007-1013]; RiehlGaLa1964 [chemical control: 522-525]; RiehlLa1950 [chemical control: 29-43]; RiehlLa1952 [chemical control: 25-36]; RiehlLaRo1958 [chemical control: 193-195]; RiehlLaRo1959 [chemical control: 857-860]; RiehlLaRo1965 [chemical control: 907-909]; Rivnay1945 [biological control: 117-122]; Rivnay1968 [host, distribution, economic importance, biological control: 1-156]; RobbCoBe2001 [host, distribution, taxonomy, control]; Robert1971 [chemistry, physiology: 449-456]; Robert1973 [chemistry, physiology: 313-323]; RobertSl1974 [chemistry, physiology: 163-175]; Robins1917 [taxonomy, description, host, distribution: 24-25]; RodrigGa1990 [host, distribution, life history: 25-35]; RodrigGa1992 [host, distribution, life history: 31-44]; RodrigGa1994 [host, distribution: 151-164]; RodrigTrGa1996 [host, distribution, biological control: 77-94]; RoelofGiCa1977 [chemistry: 698-699]; RoelofGiCa1978 [chemistry, life history: 211-224]; RoelofGiMo1982 [chemistry, pheromone: 348]; RongGr1998 [biological control: 43-63]; Rose1990a [biological control, host: 263-287]; Rose1990b [biological control, economic importance: 433-440]; Rose1990c [distribution, economic importance: 535-541]; Rose1990d [host, biological control, economic importance: 357-365]; RoseDe1990a [biological control: 461-472]; Rosen1965 [host, distribution, biological control: 388-396]; Rosen1966 [taxonomy, biological control: 43-79]; Rosen1967b [host, distribution, biological control: 1422-1427]; Rosen1969 [biological control: 45-53]; Rosen1987 [taxonomy, biological control: 191]; Rosen1990 [biological control: 413-415]; Rosen1990a [biological control: 497-498]; Rosen1993 [biological control: 411-416]; RosenDe1973 [biological control, taxonomy: 215-222]; RosenDe1978 [economic importance, biological control, distribution, life history: 79-91]; RosenDe1979 [host, distribution, biological control: 262-268,349-354,]; RosenhHe1994 [life history, biological control: 41-78]; RosenhRo1991 [life history, biological control: 873-893]; RosenhRo1992 [life history, biological control: 263-272]; RossleRo1990 [biological control: 519-526]; RSEA1915 [host, distribution, description, life history, economic importance, control: 1]; Ruhl1913 [host, distribution: 79-80]; Rungs1970 [host, distribution, economic importance: 91-94]; Russo1959 [economic importance, chemical control: 11-12]; Ruther1948 [economic importance, biological control: 55,68]; Sailer1983 [distribution, economic importance: 15-38]; Sakai1939 [taxonomy, description, illustration, host, distribution : 45-62]; SalamaAmHa1972 [host, distribution, life history, ecology: 395-405]; SalamaSa1984 [host, distribution, life history, chemistry: 393-398]; SalasGo2001 [life history, control: 27-30]; SalemEl1979 [host, distribution, life history, pheromone: 129-138]; SalemSa1992 [host, distribution, life history: 209-216]; Samway1981 [host, distribution, life history, biological control: 663-670]; Samway1981a [biological control: 1]; Samway1985 [host, distribution, life history, biological control: 379-393]; Samway1986 [host, distribution, life history, biological control, chemical control: 671-683]; Samway1986a [host, distribution, life history, biological control, chemical control: 265-274]; Samway1989 [biological control: 345-351]; SamwayGrPr1998 [host, distribution, economic importance, biological control: 234-242]; SamwayMa1983 [biological control: 4-6]; SamwayNePr1982 [biological control: 155-157]; SamwayOsHa1999 [biological control: 795-812]; Sander1904a [taxonomy, description, illustration, host, distribution: 70-71]; Sankar1984 [host, distribution, biological control]; SchildSc1928 [biological control]; SchlinDo1964 [taxonomy, biological control: 247-280]; Schmut1969 [taxonomy, description, host, distribution, life history, economic importance: 108-109]; Schmut2001 [host, distribution: 339-345]; SchmutKlLu1957 [host, distribution, economic importance, distribution: 482-483]; SchmutKlLu1959 [taxonomy: 374]; SchoonGi1982 [chemical control, biological control: 261-273]; Searle1964 [host, distribution: 1-18]; SearleAnWi1963 [economic importance, chemical control, biological control: 3-9]; SekeroUyKa1989 [host, distribution, life history, ecology: 147-152]; SelhimBr1977 [host, distribution, biological control: 475-478]; SenalKaUn2002 [biological control: 427-433]; SengonUyKa1998 [host, distribution, biological control: 128-131]; Sharon1980 [host, distribution, life history, biological control: 1-59]; ShawMoFa1971 [life history, control: 1305-1306]; ShawMoHe1971 [life history, control: 67-69]; ShawSuMo1973 [life history, chemical control, pheromone: 1062-1063]; Shen1993 [host, distribution: 57]; ShiLi1991 [host, distribution: 165]; SibbetVaFe2000 [host, distribution, control]; Siddiq1966 [host, distribution, economic importance: 4-5]; Siddiq1981 [economic importance, host, distribution: 172-180]; Silves1926a [host, distribution, control: 97-101]; Silves1929 [host, distribution: 897-904]; Simant1962 [chemical control: 99-103]; Simmon1944 [host, distribution, biological control, ecology: 38-39]; Simmon1969a [biological control: 765-767]; SimmonGr1977 [host, distribution, ecology: 109-124]; Singh1964 [host, distribution, economic importance: 213]; SiscarLoLi1999 [host, distribution, economic importance, life history, biological control: 41-52]; Smetni1991 [chemistry: 92-129]; Smirno1950a [biological control: 190-194]; Smirno1951a [economic importance, host, distribution, chemical control: 47-55]; Smirno1952 [host, distribution, biological control: 63-69]; Smirno1957a [host, distribution, economic importance, control: 47-55]; Smit1964 [host, distribution, economic importance, biological control, chemical control]; Smith1926 [biological control: 294-302]; Smith1941 [biological control: 76-77]; Smith1948 [biological control: 597]; Smith1948a [economic importance: 813]; Smith1957 [host, distribution, life history, biological control: 219-230]; SmithBeBr1997 [host, distribution, description, life history, biological control, chemical control]; SmithCo1931a [biological control: 328]; SmithEsFa1933 [economic importance: 1]; SmithFl1948 [biological control: 17-20]; SmithFl1950 [biological control: 362,376,378]; SmithFrPa1996 [biological control]; SmithMa1986 [life history, biological control, host, distribution: 452-434]; SmithSmLi1999 [biological control, chemical control: 995-1000]; SokoloKl1941 [biological control: 40-41]; SokoloKl1942 [biological control: 187-198]; SokoloKl1943 [biological control: 6-7]; Spille1952 [life history, ecology, host, distribution: 483-487]; SpolleHo1993 [chemical control, biological control: 16-19]; SpolleHo1993a [chemical control, biological control: 195-204]; SpolleHo1993b [chemical control, biological control: 87-94]; SpolleRoHo1994 [biological control: 189-208]; Sproul1981 [economic importance, host, chemical control, biological control: 50]; StarleRi1977 [host, distribution, biological control: 431-448]; Starne1897 [taxonomy: 23]; Statha2001b [life history, biological control: 113-116]; Stehr1974 [biological control: 124-136]; SteinbPoRo1994 [life history, biological control, ecology: 79-91]; Steine1987 [host, distribution, description, economic importance, control: 1-7]; SternAdBe1976 [biological control: 593]; Sternl1973a [host, distribution, biological control: 513-519]; Steyn1951 [life history, ecology: 165-170]; Steyn1954 [life history, biological control, ecology: 252-264]; Steyn1954a [host, distribution, life history, biological control, ecology, economic importance: 1-96]; Steyn1955 [life history, biological control, ecology, host, distribution: 93-105]; Steyn1958 [host, distribution, biological control: 589-594]; Steyn1959 [host, distribution, biological control: 899-902]; Stofbe1937a [host, distribution, life history, ecology: 1-24]; StouthLu1991 [biological control: 150-157]; StreibFrKa1994 [chemical control: 23-30]; Sumaro1967 [biological control: 179-185]; Sweetm1958 [biological control, economic importance: 449-459]; Swirsk1976b [host, distribution, economic importance: 555-559]; Swirsk1989 [biological control: 11-44]; SwirskAr1971 [host, distribution: 12-15]; SwirskWyIz2002 [taxonomy, host, distribution, life history, economic importance, biological control: 105-110]; Takagi1969a [taxonomy, description, illustration, host, distribution: 82,102]; Takagi1975 [taxonomy, host, distribution: 11]; TakagiRo1981 [host, distribution, biological control: 314-321]; Takaha1929 [host, distribution: 80-81]; Takaha1932a [host, distribution: 104]; Takaha1933 [host, distribution: 25-34]; Takaha1934 [host, distribution: 34]; Takaha1935 [host, distribution: 3-4]; Takaha1936c [host, distribution: 118]; Takaha1939b [host, distribution: 270]; Takaha1942b [host, distribution: 47]; Takaha1942d [host, distribution: 358]; Takaha1953a [taxonomy, host, distribution: 10-13]; Talhou1950 [host, distribution, economic importance: 133-141]; Talhou1969 [host, distribution, economic importance: 107-109]; Talhou1975 [economic importance: 25]; Tamaki1997 [structure, chemistry]; Tanaka1966 [biological control: 1-42]; Tang1984 [taxonomy, description, host, distribution: 37]; Tao1999 [taxonomy, host, distribution: 71]; Targio1881 [taxonomy, description, illustration, host, distribution: 151,153]; Targio1884 [taxonomy: 386-388]; Targio1885 [taxonomy, description, host, distribution: 108-109]; Tashir1966 [host, biological control: 604-608]; TashirBe1968 [life history: 1009-1014]; TashirBeMo1969 [life history, taxonomy: 279-280]; TashirCh1967 [life history, pheromone: 1166-1170]; TashirChMo1969 [life history, pheromone: 935-940]; TashirMo1968 [life history: 1014-1020]; TawfikGh2001 [host, distribution, life history, biological control: 267-272]; Terezn1986 [taxonomy, description, illustration, host, distribution: 85-87]; Thorpe1930WH [taxonomy: 177]; TianCh1991 [biological control: 64-66]; Timber1916 [host, distribution, biological control]; Torres1922 [host, distribution: 72]; Toumey1895 [host, distribution, economic importance, taxonomy, description, life history, chemical control: 48-56]; TranfaVi1987a [economic importance, control: 215-221]; TremblRo1975 [pheromone: 195-200]; Trjapi1989 [biological control: 296]; TrjapiMy2001[biological control: 163-165]; TronchRoGa1992 [host, distribution, biological control: 11-30]; Trujil1942 [host, distribution, economic importance]; Tryon1889 [taxonomy: 129]; Tschor1939 [host, distribution: 90]; TsolakSi1995 [chemical control: 223-229]; Tsukam1983 [chemical control, life history,: 71-98]; TumminAmCo2001 [host, distribution, chemical control, biological control: 31-36]; TumminCoSa1996 [life history, ecology, host, distribution, biological control: 493-503]; Tuncyu1970 [host, distribution, economic importance: 30-52]; Tuncyu1970a [host, distribution, economic importance: 67-80]; Tuncyu1976 [host, distribution, biological control: 32-45]; Ullyet1946 [host, distribution, biological control: 333-337]; UygunEl1998 [host, distribution, biological control,

life history: 153-162]; UygunEl1998 [host, life history, biological control: 153-162];
UygunKaSe1992 [host, distribution, economic importance: 171-182];
UygunKaSe1995 [life history, economic importance, host, distribution, biological
control: 239-246]; UygunKaUl1996 [host, distribution, biological control: 171-183];
UygunSeEr1998 [host, distribution: 183-191]; Vacant1985a [host, distribution, life
history, biological control: 749-758]; Valent1963 [biological control: 6-13];
Valent1967 [biological control: 1100]; VandenTe1964 [ecology, biological control:
459-488]; VanDij1998 [host, distribution, biological control, economic importance:
23-28]; VehrsGr1994 [host, distribution, chemical control: 1046-1057]; Velasq1971
[taxonomy, description, illustration, host, distribution: 120-123]; VelasqRi1969
[host, distribution: 195-208]; Viggia1970a [host, distribution, economic importance:
51]; Viggia1984 [biological control: 257-276]; Viggia1989 [economic importance,
host, distribution: 63-64]; Vinson1977 [host, distribution, biological control,
chemistry: 237-279]; Vinson1990 [biological control, chemistry: 453-459];
WadhiBa1964 [host, distribution: 227-260]; WaldeLuYu1989 [host, distribution, life
history, biological control, ecology: 1700-1706]; WalkerAi1996 [host, distribution,
biological control, chemical control: 175-180]; WalkerAiOC1990 [host, distribution,
chemical control: 189-196]; WalkerBe1986 [behaviour, life history, physiology:
549-553]; WalkerZaAr1999 [biological control, economic importance: 47-58];
WalkerZaAr1999a [economic importance, biological control: 906-914];
WarePa1996 [host, distribution, biological control: 511-514]; WartheRuMo1970
[pheromone, life history: 2207-2209]; WashinWa1990 [structure, histology: 939-
948]; Watson1918 [host, distribution]; Watson1926 [host, distribution];
WatsonBe1937 [host, distribution, control]; WatsonDuLi2000 [life history,
biological control: 351-357]; Wentze1970 [host, biological control, life history: 195-
198]; Wester1918 [host, distribution, economic importance: 5-57]; Whitna1959
[host, distribution, economic importance: 5-9]; Willar1972 [host, distribution, life
history: 37-47]; Willar1972a [host, distribution, life history: 49-65]; Willar1973
[host, distribution, life history: 217-229]; Willar1973a [host, distribution, life
history: 567-573]; Willar1974 [host, distribution, life history, ecology: 531-548];
Willar1976 [host, distribution, life history, ecology: 7-11]; WilliaBu1987 [host,
distribution: 94]; WilliaWa1988 [taxonomy, description, illustration, host,
distribution, economic importance: 8,36-37]; Wilson1917 [taxonomy, description,
host, distribution: 12]; WilsonGo1962 [host, distribution, economic importance,
control: 41-61]; Woglum1923 [chemical control, economic importance: 1-59];
Woglum1925 [chemical control, physiology: 593-597]; Woglum1925b [chemical
control, physiology: 178]; Woglum1926 [chemical control: 723-733]; Woglum1942
[chemical control: 1-3]; WoglumLa1926 [chemical control, physiology: 396-397];
WolffCo1993 [taxonomy, description, illustration, host, distribution: 32-33];
WolffCo1993a [host, distribution: 153]; WongChCh1999 [taxonomy, description,
host, distribution: 18-19,58]; Wood1962 [distribution, biological control: 8-11];
Woodhi1928 [chemical control: 561]; WoodwaEvEa1970 [distribution];
Woolle1990 [biological control: 167-176]; Wysoki1997 [host, distribution,
economic importance: 805-811]; YamamoOg1989 [chemistry: 123-148]; Yan1981
[life history]; YanIs1986 [life history, ecology: 971-975]; YanIs1986 [life history:
971-975]; YaromBlIs1988 [chemical control: 1581-1585]; Yasar1995a [taxonomy,

description, illustration, host, distribution: 44-46];Yasnos1987 [economic importance: 229-234]; YerushCo2002 [chemistry, chemical control: 133-141]; YuLu1988 [life history, host, taxonomy, biological control: 154-161]; YuLuMu1990 [host, distribution, life history, biological control, ecology: 469-480]; Yust1941 [taxonomy: 785]; Yust1943 [life history: 868-872]; YustBuHo1942 [chemical control: 521-524]; YustBuNe1942 [chemical control, physiology: 816-820]; YustFuNe1951 [chemical control, physiology: 833-838]; YustHo1942 [chemical control: 821-824]; YustNeBu1942 [chemical control: 339-342]; YustNeBu1943 [chemical control: 744-749]; YustNeBu1943a [chemical control: 872-874]; YustSh1952 [chemical control, physiology: 220-228]; Zahrad1959b [host, distribution: 60]; ZchoriFaZe1995 [life history, structure: 173-178]; ZchoriRoRo1994 [life history, structure, biological control: 169-172]; ZhangGu1994 [host, distribution, biological control: 103-105]; ZhangGuZh1992 [life history, biological control: 1,45]; Zhao1990 [host, distribution, life history, biological control: 1-245].

Aonidiella bruni Balachowsky

Aonidiella bruni Balachowsky, 1952a: 101. Type data: GUINEA: on slopes of Gangan Ridge, north of Kindia, at altitude of 1000 meters, on shoots of *Vitis* sp.; collected 13.xii.1951. Syntypes, female. Type depository: Paris: Muséum national d'Histoire naturelle, France.

SCALE COVER: Female scale circular or subcircular, very flat, colour brown; larval exuviae central, covered with slight white, powdery matter in fresh specimens; 1.8-2.2 mm. Male scale white, exuviae central, greenish; form variable, but generally oval; 1.1-1.2 mm (Balachowsky, 1952a, 1956)

HOST PLANTS: **Vitaceae**: *Vitis* [Balach1952a].

DISTRIBUTION: **Afrotropical**: Guinea [Balach1952a, Balach1956].

GENERAL: Description and illustration of adult female by Balachowsky (1952a, 1956).

KEYS: Balachowsky 1956: 26 (female) [Africa].

CITATIONS: Balach1952a [taxonomy, description, illustration, host, distribution: 101-104]; Balach1956 [taxonomy, description, illustration, host, distribution: 31-32,34]; Borchs1966 [catalogue: 295].

Aonidiella citrina (Coquillett)

Aspidiotus citrinus Coquillett, 1891a: 29. Type data: U.S.A.: California, San Gabriel Valley, on *Citrus*. Syntypes, female. Type depository: Washington: United States National Entomological Collection, U.S. National Museum of Natural History, District of Columbia, USA.

Aspidiotus citrinus; Craw, 1891: 10. Notes: The species was listed as "*Aspidiotus citrinus* Coquillett".

Aspidiotus aurantii citrinus; Howard, 1894: 228. Change of status.

Aspidiotus citrinus; Cockerell, 1896b: 334. Incorrect synonymy. Notes: Incorrect synonymy with *Aonidiella aurantii* (Maskell).

Aspidiotus (*Aonidiella*) *aurantii citrinus*; Cockerell, 1897i: 29. Change of combination and rank.

Aonidiella aurantii citrina; Leonardi, 1899: 175. Change of combination requiring emendation of species name for agreement in gender.
Chrysomphalus aurantii citrinus; Fernald, 1903b: 288. Change of combination.
Chrysomphalus citrinus; Lindinger, 1914: 118. Change of status.
Aonidiella aurantii; Kiritchenko, 1929: 173. Misidentification; discovered by Borchsenius, 1966: 295.
Aonidiella citrina; Nel, 1933: 417. Change of combination.
Aonidiella citrina; Danzig, 1993: 160. Notes: Incorrect citation of "Craw" as author.
COMMON NAMES: escama amarilla [CoronaRuMo1997]; jeltaya pomeranzevaya shitovka [Borchs1936]; yellow scale [MerrilCh1923, McKenz1956, Borchs1966].
SYSTEMATICS: The California red scale, *Aonidiella aurantii* (Maskell, 1879) was first introduced into USA, California in 1879 from Australia. The yellow scale, *Aonidiella citrina* (Coquillett, 1891) was present in USA, California at the San Gabriel district as early as 1872. During the early 1880's, field entomologists, taxonomists and biological control experts in California observed evident differences in injuriousness, physiology and ecology of above two species, but failed to find reliable morphological distinguishing features. Morphological basis for separating these species was first provided by McKenzie (1937). The authorship of *Aonidiella citrina* generally credited to Coquillett (1891) (Fernald, 1903; McKenzie, 1937). However, Danzig (1993: 160) suggested that this name was first published by Craw (1891). Here, we follow interpretation of Nel (1933: 418) who stated that "the first description of *Aonidiella citrina* (Coq.) was made by Coquillett in 1891, who was conducting some spray experiments on scale insects at the San Gabriel Valley, California. He [Coquillett] referred to it as follows: "The orange tree experimented upon was infested with the yellow scale (*Aspidiotus citrinus*)". Coquillett (1891) followed binominal format, presented a brief description, locality and host plant, therefore making *Aspidiotus citrinus* Coquillett a valid name. Moreover, Craw (1891, 1891a) clearly credited this species to Coquillett. For a more detailed discussion refer to Nel (1933) and Compere (1961).
SCALE COVER: Scale of female circular, flat, exuviae subcentral; scale thin and pale, permitting yellow colour of the sclerotized female to show through. Scale of male elongate oval, colour similar to female, exuvia near one end (Ferris, 1938a). Colour photograph by Gill (1997).
HOST PLANTS: **Agavaceae**: *Agava* [Hadzib1983]. **Anacardiaceae**: *Mangifera indica* [KondoKa1995a]. **Apocynaceae**: *Carissa grandiflora* [Merril1953, Dekle1965c]. **Aquifoliaceae**: *Ilex aquifolium* [McKenz1956], *Ilex colchica* [Hadzib1983]. **Araliaceae**: *Aralia* [McKenz1956], *Hedera helix* [McKenz1956]. **Aucubaceae**: *Aucuba* [MerrilCh1923]. **Berberidaceae**: *Mahonia aquifolium* [McKenz1956]. **Buxaceae**: *Buxus colchica* [Hadzib1983], *Buxus sempervirens* [McKenz1956]. **Caprifoliaceae**: *Viburnum* [McKenz1956]. **Celastraceae**: *Euonymus* [MerrilCh1923, Merril1953, McKenz1956, Dekle1965c]. **Elaeagnaceae**: *Elaeagnus* [McKenz1956]. **Ericaceae**: *Arbutus unedo* [McKenz1956]. **Lauraceae**: *Laurus cerasus* [McKenz1956], *Persea americana* [McKenz1956]. **Liliaceae**: *Aspidistra* [McKenz1956]. **Loranthaceae**: *Helicia formosana* [Takagi1969a]. **Moraceae**: *Ficus elastica* [McKenz1956, TakahaTa1956], *Ficus indica* [MerrilCh1923], *Ficus retusa* [KawaiMaUm1971]. **Myrsinaceae**: *Ardisia sieboldii*

[Takagi1969a], *Maesa tenera* [Takagi1969a]. **Myrtaceae**: *Eugenia jambolana* [RahmanAn1941], *Leptospermum* [McKenz1956], *Myrtus* [McKenz1956], *Psidium guajava* [DanzigKo1990]. **Nyctaginaceae**: *Ceodes umbellifera* [KawaiMaUm1971]. **Oleaceae**: *Jasminum* [McKenz1956], *Ligustrum* [McKenz1956], *Ligustrum lucidum* [Merril1953, Dekle1965c], *Ligustrum medium* [KawaiMaUm1971], *Olea europaea* [McKenz1956], *Olea fragrans* [Borchs1936], *Osmanthus* [McKenz1956]. **Palmae**: *Livistona subglobosa* [Takagi1969a], *Phoenix dactylifera* [McDani1968]. **Rosaceae**: *Photinia arbutifolia* [McKenz1956], *Prunus ilicifolia* [McKenz1956], *Prunus laurocerasus* [Borchs1936, McKenz1956], *Prunus lyoni* [McKenz1956], *Rosa* [McKenz1956]. **Rutaceae**: *Choisya ternata* [McKenz1956], *Citrus* [Kuwana1933, Borchs1936, RahmanAn1941, Merril1953, KawaiMaUm1971, Danzig1972c, BesheaTiHo1973], *Citrus* [UygunSeEr1998], *Citrus limon* [Hadzib1983], *Citrus paradisi* [Hadzib1983], *Citrus sinensis* [McKenz1956, Takagi1969a, RosenDe1979, Hadzib1983], *Citrus unshiu* [TakahaTa1956, Hadzib1983], *Poncirus trifoliata* [Hadzib1983]. **Smilacaceae**: *Smilax* [Takagi1969a]. **Theaceae**: *Camellia* [McKenz1956], *Camellia japonica* [Hadzib1983], *Cleyera japonica morii* [Takagi1969a], *Schima mertensiana* [KawaiMaUm1971]. **Thymelaeaceae**: *Daphne* [MerrilCh1923].

NATURAL ENEMIES: ACARI **Cheyletidae**: *Hemicheyletia bakeri* (Ehara) [GersonOcHo1990]. **Hemisarcoptidae**: *Hemisarcoptes dzhashii* Dzhibladze [GersonOcHo1990]. COLEOPTERA **Coccinellidae**: *Coccidophilus citricola* [AguileMeVa1984]. HYMENOPTERA **Aphelinidae**: *Aphytis aonidiae* (Mercet) [RosenDe1979, MyartsRu2000], *Ap. chrysomphali* (Mercet) [RosenDe1978, RosenDe1979], *Ap. citrinus* Compere [RosenDe1978], *Ap. coheni* DeBach [RosenDe1979], *Ap. holoxanthus* DeBach [AnneckIn1971], *Ap. lingnanensis* Compere [RosenDe1979], *Ap. mazalae* DeBach & Rosen [RosenDe1979], *Aphytis melinus* DeBach [RosenDe1979], *Ap. mytilaspidis* (Le Baron) [Hadzib1983], *Aspidiotiphagus citrinus* (Craw) [RosenDe1978, Hadzib1983], *Prospaltella aurantii* (Howard) [RosenDe1978, Gordh1979]. **Encyrtidae**: *Comperiella bifasciata* Howard [Coy1938, Flande1953a, RosenDe1978, Rosen1987], *Comperiella unifasciata* Ishii [Trjapi1989]. **Signiphoridae**: *Signiphora flavopalliata* Ashmead [Gordh1979, Woolle1990], *Signiphora occidentalis* Howard [Gordh1979]. NEUROPTERA **Coniopterygidae**: *Coniopteryx* sp. [Drea1990], *Conwentzia barretti* (Banks) [Drea1990], *Conwentzia nigrans* Carpenter [Drea1990].

DISTRIBUTION: **Afrotropical**: Benin [CABI1997a]; Cameroon [CABI1997a]; Congo [CABI1997a]; Côte d'Ivoire (=Ivory Coast) [CABI1997a]; Ethiopia [CABI1997a]; Gabon [CABI1997a]; Guinea [CABI1997a]; Madagascar [CABI1997a]; Mali [CABI1997a]; Mauritius [CABI1997a]; Niger [CABI1997a]; Saint Helena [Matile1976, CABI1997a]; Senegal [CABI1997a]; Tanzania [CABI1997a]; Yemen (North Yemen [CABI1997a], South Yemen [CABI1997a], South Yemen [Nakaha1982]); Zimbabwe [Balach1956]. **Australasian**: Australia (New South Wales [CABI1997a], South Australia [CABI1997a], Victoria [CABI1997a], Western Australia [CABI1997a]); Bonin Islands (= Ogasawara-Gunto) [KawaiMaUm1971, Kawai1987]; Fiji [Nakaha1982]; Papua New Guinea [Nakaha1982]; Western Samoa [Nakaha1982]. **Nearctic**: Mexico [CABI1997a, MyartsRu2000]; United States of America (California [McKenz1956, Takagi1969a,

RosenDe1978, RosenDe1979], Florida [Wilson1917, MerrilCh1923, Merril1953, Dekle1965c, BesheaTiHo1973, CABI1997a], Texas [McKenz1937, McDani1968]). **Neotropical**: Argentina [CABI1997a]; Chile [GonzalCh1968]; Trinidad and Tobago [CABI1997a]. **Oriental**: Bangladesh [CABI1997a, Nakaha1982]; China (People's Republic) (Fujian (Fukien) [CABI1997a], Guangdong (Kwangtung) [CABI1997a], Guangxi (Kwangsi) [CABI1997a], Hubei (Hupei) [CABI1997a], Hunan (Hunan) [CABI1997a], Jiangsu (Kiangsu) [CABI1997a], Jiangxi (Kiangsi) [CABI1997a], Sichuan (Szechwan) [CABI1997a], Yunnan [CABI1997a]); Hong Kong [CABI1997a]; India [McKenz1937] (Andhra Pradesh [CABI1997a], Delhi [CABI1997a], Karnataka [CABI1997a], Maharashtra [CABI1997a], Punjab [RahmanAn1941], Uttar Pradesh [CABI1997a]); Indonesia [Nakaha1982]; Malaysia (Sabah [CABI1997a], Sarawak [CABI1997a]); Mongolia [DanzigKo1990]; Nepal [Takagi1975]; Pakistan [RosenDe1979, Nakaha1982, CABI1997a]; Philippines [CABI1997a] (Luzon [Velasq1971]); Taiwan [TakahaTa1956, Takagi1969a]; Thailand [CABI1997a]. **Palaearctic**: Afghanistan [Siddiq1966, Danzig1972c, CABI1997a]; Armenia [Nakaha1982]; Azerbaijan [Hadzib1983, CABI1997a] [Nakaha1982]; China (People's Republic) (Hebei (Hopei) [CABI1997a], Henan (Honan) [Shen1993], Qinghai (Chinghai) [CABI1997a], Xizang (Tibet) [CABI1997a]); France [GermaiBe2002]; Georgia (Abkhaz ASSR [Borchs1936], Adzhar ASSR [Borchs1936], Georgia [Borchs1936, CABI1997a]); Iran [CABI1997a]; Italy [CABI1997a]; Japan [Kuwana1917a, Sakai1939, Kawai1980] (Honshu [TakahaTa1956, CABI1997a], Kyushu [TakahaTa1956], Shikoku [TakahaTa1956]); Libya [CABI1997a]; Saudi Arabia [CABI1997a]; South Korea [CABI1997a]; Turkey [Tuncyu1970a, CABI1997a, UygunSeEr1998].

BIOLOGY: Biparental and ovoviviparous, infesting chiefly leaves and fruit, rarely or not at all on bark (Ferris, 1938a). Female sex pheromone identified (Gieselmann et al., 1979).

ECONOMIC IMPORTANCE: Polyphagous, and a serious pest of citrus in certain citrus-growing areas of world, such as California, Texas, Florida, China, Japan, India, Iran, Australia (Ebeling, 1959; Rosen & DeBach, 1978). Recent distribution map (CABI, 1997a) indicates that *A. citrina* is more widely distributed in other territories of Africa, South America, Mediterranean Basin, and Oriental region.

GENERAL: Description and illustration of adult female by McKenzie (1938, 1956), Ferris (1938a), Balachowsky (1956), Takagi (1969a), Velasquez (1971), Chou (1985, 1986), Tereznikova (1986), Danzig (1993), Yaşar (1995a) and by Gill (1997).

KEYS: Gill 1997: 44 (female) [Species of California]; Danzig 1993: 159 (female) [Europe]; Tereznikova 1986: 85 (female) [Ukraine]; Chou 1985: 292 (female) [Species of China]; Kawai 1980: 211 (female) [Japan]; Velasquez 1971: 118 (female) [Philippines]; McDaniel 1968: 210 (female) [U.S.A.: Texas]; Balachowsky 1956: 25 (female) [Africa]; McKenzie 1956: 24 (female) [U.S.A.: California]; McKenzie 1953: 36 (female) [World]; McKenzie 1946: 33 (female) [World]; Ferris 1942: 29 (female) [North America]; McKenzie 1942b: 144-145 (female) [World]; McKenzie 1938: 17-18 (female) [World]; McKenzie 1937a: 178 (female) [World]; McKenzie 1937: 330 (female) [World]; Lupo 1936: 261 (female) [World]; Kuwana 1933b: 49 (female) [Japan].

CITATIONS: AguileMeVa1984 [life history, biological control: 47-54]; AlamSa1965 [host, distribution, life history: 161-171]; Aldric1996 [life history, physiology, chemistry: 201-204]; AndersHe1979 [chemistry, life history: 773-779]; AsquitCrHo1980 [control, chemistry, economic importance]; Balach1956 [taxonomy, description, illustration, host, distribution: 29,34]; Bartle1959 [biological control: 1-2]; BeardsDaHo1976 [economic importance: 106]; BeardsGo1975 [economic importance: 49]; BeattiGe1983 [host, distribution: 220-226]; BenDov1990a [taxonomy: 89]; BesheaTiHo1973 [host, distribution: 4]; Blumbe1990 [structure, biological control: 221-228]; Borchs1935a [taxonomy: 36]; Borchs1936 [host, distribution, life history, economic importance, biological control: 135-137]; Borchs1937 [taxonomy, description, illustration, host, distribution: 121]; Borchs1937a [taxonomy: 63,66]; Borchs1939 [taxonomy, description, host, distribution: 8,20]; Borchs1949d [taxonomy, description, host, distribution: 234]; Borchs1950b [taxonomy, description, illustration, host, distribution: 212,221]; Borchs1966 [catalogue: 295]; Boyce1948 [host, distribution, economic importance, control]; Boyce1950 [host, distribution, economic importance: 741-766]; Brooks1969 [chemical control: 923-932]; BrooksBu1966 [host, distribution, chemical control: 185-188]; Browni1994a [biological control: 27-49]; BurgerUl1990 [economic importance: 313-327]; Bustsh1958 [taxonomy, description, host, distribution: 220,245]; CABI1975 [host, distribution: 1-3]; CABI1997a [host, distribution: 1-3]; Calkin1983 [distribution, economic importance: 321-359]; Caltag1985 [taxonomy, biological control: 189-200]; Carman1953 [chemical control: 307-308]; CarmanElEw1954 [host, distribution, chemical control : 1-11]; CarmanEw1950 [chemical control: 15A-16A]; CarmanEwJe1951 [host, distribution, chemical control: 1-16]; CarmanEwJe1956 [chemical control]; CarmanEwJe1957 [chemical control]; CarmanEwJe1958 [chemical control]; CarmanEwJe1959 [chemical control]; CarmanEwJe1960 [chemical control]; CarmanEwJe1961 [chemical control]; CarmanEwJe1962 [chemical control]; CarmanLi1956 [chemical control: 534-539]; Carnes1907 [taxonomy, host, distribution: 216]; Chou1985 [taxonomy, description, host, distribution: 295-296]; Chou1986 [structure, anatomy: 441,443]; Chou1986 [taxonomy, illustration: 677]; ClapsWoGo2001a [taxonomy, host, distribution: 11]; Clause1956 [host, distribution, economic importance, biological control]; Clause1958 [economic importance, biological control: 291-310]; Clause1958a [host, distribution, biological control: 443-447]; Cocker1896b [taxonomy, distribution: 334]; Cocker1897i [taxonomy, description, host, distribution: 29]; Comper1961 [taxonomy, host, distribution, biological control, economic importance: 173-278]; Comper1969 [biological control: 755-764]; Cook1909 [taxonomy, description, host, distribution, chemical control: 19]; Coquil1891a [taxonomy, description, host, distribution, chemical control: 19-26]; CoronaRuMo1997 [host, distribution: 38-41]; Coy1938 [biological control: 445-446]; Craw1891 [taxonomy, description, illustration, host, distribution, life history: 10]; Danzig1964 [taxonomy, host, distribution: 651]; Danzig1972 [taxonomy, host, distribution, economic importance: 207]; Danzig1972c [host, distribution: 583]; Danzig1993 [taxonomy, description, illustration, host, distribution, economic importance: 160-161]; DanzigKo1990 [host, distribution: 45]; DanzigPe1998 [catalogue: 182]; DavidsMi1990 [host, distribution, economic importance: 603-632]; Dean1955 [biological control: 444-447]; DeBach1958 [host, distribution, biological control, ecology: 187-194]; DeBach1958a [biological control: 759-768]; DeBach1958b [host, distribution: 187-194]; DeBach1964 [biological control]; DeBach1966 [host, distribution, biological control, ecology, life history: 183-212]; DeBach1969 [biological control: 801-815]; DeBach1972 [biological control: 211-233]; DeBach1974 [biological control]; DeBachBa1951 [chemical control, biological control: 372-383]; DeBachDiFl1951 [biological control: 347-348]; DeBachHeRo1978 [host, distribution, life history, ecology, biological control: 1-35]; DeBachRo1991 [biological control]; Dekle1965c [taxonomy, description, host, distribution: 20]; Dekle1976 [taxonomy, description, host, distribution, economic importance: 31]; DeSant1979 [biological control]; Drea1990 [biological control: 51-59]; Dunkel1999 [chemistry, life history]; Ebelin1949 [host, distribution, life history, control]; Ehler1996 [host, distribution, biological control: 337-342]; ElirazRo1978 [biological control: 96-101]; ElmerEwCa1951 [economic importance, biological control, host, distribution: 593-597]; Essig1949 [host, distribution: 673-677]; Ewart1969 [chemical control: 879-880]; EwartCa1951 [host, distribution, chemical control: 1-10]; EwartCaJe1952 [host, distribution, chemical control: 1-6]; EwartCaJe1953 [host, distribution, chemical control: 1-7]; FargerMo1974 [life history, chemistry, biological control: 26-28]; FDACSB1983 [host, distribution: 6-8]; Fernal1903b [catalogue: 288]; Ferris1938b [taxonomy, description, illustration, host, distribution: 179]; Ferris1941e [taxonomy: 42]; Ferris1942 [taxonomy: 29]; FisherDe1976 [biological control: 43-50]; Flande1934 [host, distribution: 145-150]; Flande1944a [biological control: 365-371]; Flande1945a [host, distribution, biological control: 711-712]; Flande1948a [host, distribution, economic importance, biological control: 56,76-77]; Flande1953 [biological control: 10-28]; Flande1953a [host, distribution, biological control, economic importance: 266-269]; Flande1958a [biological control: 579-584]; Flande1966 [biological control: 79-82]; Flande1971 [biological control, life history: 857-872]; Flesch1960 [biological control: 183-208]; FlintVa1981 [biological control: 1]; Gaprin1954 [biological control: 587-597]; Gaprin1956 [host, distribution: 103-137]; GermaiBe2002 [taxonomy, host, distribution: 49-51]; GersonOcHo1990 [biological control: 77-97]; Ghabbo1995 [taxonomy: 379-387]; Giesel1990 [life history, chemistry: 221-224]; Giesel1990a [chemistry: 225-232]; GieselMoFa1979 [chemistry, pheromone: 27-33]; GieselRi1990 [life history, ecology, chemistry: 349-352]; GieselRiJo1979 [chemistry, pheromone: 891-900]; Gill1997 [host, distribution, taxonomy, description, illustration, economic importance: 47-51]; Gonzal1969 [biological control: 839-847]; GonzalCh1968 [host, distribution: 110]; Gordh1979 [biological control: 893-895,900,907,911,]; Grafto1994 [chemical control: 7-9]; GraftoMiOC2000 [life history, chemistry, ecology, control, pheromone: 75-88]; GraftoMoOC2000 [host, distribution, control]; GraftoOu1993 [chemical control: 21-29]; GraftoOuSt2001 [distribution, chemical control: 20-25]; GraftoVe1995 [host, distribution, life history, biological control, chemical control: 495-504]; GriffiTh1947 [chemical control: 386-388]; GriffiTh1949 [life history: 101-109]; GriffiTh1949a [taxonomy, description: 1-13]; GriffiTh1957 [host, distribution: 5-30]; Haas1934 [chemistry, chemical control, physiology: 477-492]; Hadzib1983 [taxonomy, description, illustration, host, distribution, life history, biological control, economic importance: 226-228]; Hamble1947 [host, distribution: 949-956]; HardieMi1999 [chemistry, life history]; Hewitt1943 [host, distribution: 266-274]; Hinckl1963 [host, distribution, biological control]; Howard1894 [taxonomy, biological control: 228]; Howard1895e [biological control: 1-44]; HoyHe1985 [biological control]; HoyHe1985 [biological control]; HuffakMeDe1971 [biological control: 16-67]; JiYa1990 [biological control: 134-136]; KattarHeOd1999 [biological control: 640-644]; Kawai1980 [taxonomy, description, host, distribution: 211]; Kawai1987 [host, distribution: 78]; KawaiMaUm1971 [host, distribution: 18-19]; Kiritc1929 [taxonomy: 173]; KiriukTa1947 [host, distribution: 1-4]; Koehle1964 [host, distribution, control]; KondoKa1995a [host, distribution: 97-98]; Kuwana1909 [host, distribution: 154]; Kuwana1917a [taxonomy, distribution: 175]; Kuwana1933 [taxonomy, description, illustration, host, distribution: 30-31]; Laport1948 [host, distribution, biological control: 35-37]; LawsonHa1984 [ecology: 451]; Leonar1899 [taxonomy: 175,176,187]; Lindin1912b [taxonomy: 357-358]; Lindin1914 [taxonomy: 118]; Lindin1943b [taxonomy: 145]; LongoMaPa2002 [host, distribution, life history, economic importance: 508-509]; LongoMaRu1994a [host, distribution, economic importance: 19-25]; LongoMaRu1995a [host, distribution: 126-129]; Lorbee1971 [biological control, host, distribution: 199-201]; Lupo1936 [taxonomy, description, illustration, host, distribution: 255-257]; MacGil1921 [taxonomy, description, host, distribution: 443]; Maranh1946 [taxonomy: 164-179]; Maskel1895b [taxonomy, description, host, distribution: 41]; Matile1976 [host, distribution: 311]; McDani1968 [taxonomy, illustration, host, distribution: 212]; McKenz1937 [taxonomy, description, illustration, host, distribution: 324-327,331]; McKenz1937a [taxonomy: 178]; McKenz1938 [taxonomy, description, illustration, host, distribution: 7-8,24]; McKenz1939 [taxonomy: 54]; McKenz1942b [taxonomy: 145]; McKenz1946 [taxonomy: 33]; McKenz1953 [taxonomy: 36]; McKenz1956 [taxonomy, description, illustration, host, distribution: 41-42]; McLare1971 [life history, ecology: 189-204]; Merril1953 [taxonomy, description, host, distribution: 13-14]; MerrilCh1923 [taxonomy, description, host, distribution, economic importance: 221-223]; MessenBiVa1976 [biological control: 453-565]; MessenVa1971 [biological control: 68-92]; MessenWiWh1976 [biological control: 209]; MetcalMe1993 [economic importance, host, distribution, control]; MillerDa1990 [host, distribution, economic importance: 300]; Moreno1972 [structure, anatomy, chemistry, pheromone: 1283-1286]; Moreno1983 [pheromone, control: 77-79]; MorenoCaFa1974 [life history: 15-20]; MorenoCaRi1972 [life history, chemistry, pheromone: 443-446]; MorenoFa1975 [structure, anatomy, pheromone: 425-428]; MorenoRiCa1972 [life history, chemistry, pheromone: 698-701]; Mori1981 [chemistry, pheromone: 41-53]; MoriKu1982 [chemistry, pheromone: 521-525]; Muma1948 [biological control: 193-194]; Muma1971 [host, distribution, biological control: 139-150]; MumaSeDe1961 [biological control, host, distribution: 1-39]; Muraka1970 [host, distribution: 71]; MyartsRu2000 [distribution, biological control: 7-33]; Nakaha1982 [host, distribution: 7]; Nel1933 [taxonomy, life history, anatomy, host, distribution: 416-466]; NSWDAE1963 [host, distribution, taxonomy, economic importance]; Onder1982 [host, distribution, life history, economic importance: 1-171]; Perkin1982 [economic importance, chemical control, biological control: 5]; Pratt1958 [taxonomy, illustration, distribution]; Quayle1911e [biological control: 510-515]; Quayle1932 [life history, chemical control: 1-87]; Quayle1938a [host, distribution, life history, economic importance, chemical control, biological control]; Quedna1964b [biological control: 86-116]; RahmanAn1941 [taxonomy, description, host, distribution: 818-819]; RiehlBrMc1980 [chemical control, biological control, economic importance: 319-363]; RoelofGiMo1982 [chemistry: 348]; Rose1990 [biological control: 229]; Rose1990a [biological control, host: 263-287]; Rose1990c [distribution, economic importance: 535-542]; Rosen1987 [taxonomy, biological control: 191]; RosenDe1973 [biological control, taxonomy: 215-222]; RosenDe1978 [economic importance, biological control, life history, distribution: 91-93]; RosenDe1979 [host, distribution, biological control: 476-483,533-542,]; Sakai1939 [taxonomy, description, illustration, host, distribution: 45-62]; SchmutKlLu1957 [host, distribution, economic importance: 482]; SchmutKlLu1959 [taxonomy: 374]; SelhimBr1977 [host, distribution, biological control: 475-478]; Shen1993 [host, distribution: 57]; Siddiq1966 [host, distribution, economic importance: 4-5]; Siddiq1981 [economic importance, host, distribution: 172-180]; Simant1960a [host, distribution, life history: 49-57]; Simant1962a [biological control: 105-112]; Simant1969 [biological control, economic importance: 889-896]; Simant1969a [life history, economic importance, life history: 889-896]; Simant1976 [host, distribution, chemical control: 135-164]; Smetni1991 [chemistry: 92-129]; Smith1941 [biological control: 76-77]; Smith1948 [biological control: 597]; SmithBeBr1997 [host, distribution, description, life history, biological control, chemical control]; SmithCo1931a [biological control: 328]; Sweetm1958 [biological control, economic importance: 449-458]; Takagi1969a [taxonomy, description, illustration, host, distribution: 82-83,100]; Takagi1975 [taxonomy, host, distribution: 11]; TakagiRo1981 [host, distribution, biological control: 314-321]; Takaha1953a [taxonomy, host, distribution: 10-13]; TakahaTa1956 [host, distribution: 16]; Tao1999 [taxonomy, host, distribution: 71]; Terezn1986 [taxonomy, description, illustration, host, distribution: 87-89]; Trembl1999 [economic importance: 19-28]; TremblRo1975 [pheromone: 195-200]; Trjapi1989 [biological control: 296]; Tuncyu1970 [host, distribution, economic importance: 30-52]; Tuncyu1970a [host, distribution, economic importance: 67-80]; Tuncyu1976 [host, distribution, biological control: 32-45]; UygunKaUl1996 [host, distribution, biological control: 171-183]; UygunSeEr1998 [host, distribution: 183-191]; Velasq1971 [taxonomy, description, illustration, host, distribution: 124-127]; Wilson1917 [host, distribution: 64]; Woglum1923 [chemical control, economic importance: 1-59]; Woglum1925a [chemical control, physiology: 2]; WoglumLaLa1947 [biological control, chemical control: 818-820]; WoodwaEvEa1970 [distribution]; Woolle1990 [biological control: 167-176]; YamamoOg1989 [chemistry: 123-148]; Yasar1995a [taxonomy, description, illustration, host, distribution: 47-49]; Yasnos1987 [economic importance: 229-234]; Yasnos1994 [host, distribution, biological control: 317-333]; Yust1941 [structure: 785]; YustNeBu1942a [chemical control: 825-826].

Aonidiella comperei McKenzie

Aonidiella comperei McKenzie, 1937: 327. Type data: INDIA: Bombay, Calaba, on *Citrus* sp. Holotype female. Type depository: Davis: The Bohart Museum of Entomology, University of California, California, USA.
Chrysomphalus comperei; Lindinger, 1957: 545. Change of combination.
Aonidiella comperei; Borchsenius, 1966: 295. Revived combination.
COMMON NAMES: Compere scale [VelasqRi1969]; escama falsa amarilla [CoronaRuMo1997]; false yellow scale [McKenz1937, Velasq1971].
SCALE COVER: Scale of female smooth, circular, flat, yellow, hard, and brittle, 1.5 to 1.75 mm in diameter. Scale of male not identified (McKenzie, 1937).
HOST PLANTS: **Annonaceae**: *Annona muricata* [Balach1958b, Takagi1962c, Takagi1970]. **Compositae**: *Pluchea odorata* [WilliaWa1988]. **Cucurbitaceae**: *Cucurbita maxima* [Velasq1971]. **Ebenaceae**: *Diospyros* [WilliaWa1988]. **Euphorbiaceae**: *Annesijoa* [WilliaWa1988]. **Moraceae**: *Ficus* [WilliaWa1988]. **Musaceae**: *Musa* [Takagi1970, WilliaWa1988]. **Palmae**: *Cocos nucifera* [McKenz1946, Takagi1970]. **Rubiaceae**: *Morinda citrifolia* [WilliaWa1988]. **Rutaceae**: *Citrus* [McKenz1937, Takagi1962c, Takagi1970], *Citrus aurantifolia* [McKenz1946], *Citrus grandis* [McKenz1946]. **Vitaceae**: *Vitis* [McKenz1946].
DISTRIBUTION: **Australasian**: Federated States of Micronesia (Caroline Islands [Beards1966], Truk Islands [Beards1966], Yap [Beards1966]); Kiribati [WilliaWa1988]; Marshall Islands [Beards1966]; Palau [Beards1966]; Papua New Guinea [WilliaWa1988]. **Neotropical**: Dominica [McKenz1946]; Guadeloupe [Balach1957c]; Guatemala [McKenz1946]; Haiti [McKenz1946]; Martinique [Balach1957c]; Puerto Rico & Vieques Island (Puerto Rico [Takagi1962c, Takagi1970, Martor1976]); U.S. Virgin Islands [Nakaha1983]. **Oriental**: India [McKenz1937, Takagi1970]; Mongolia [DanzigKo1990]; Philippines [VelasqRi1969] (Cebu [Velasq1971]); Taiwan [Takagi1970]; Thailand [McKenz1946]. **Palaearctic**: China (People's Republic) [Takagi1962c]; Japan [Takagi1970].
BIOLOGY: Occurring on leaves, twigs, and larger branches (McKenzie, 1937).
ECONOMIC IMPORTANCE: A polyphagous species, recorded from South Pacific, Far East and Central America (see Distribution and Host Plants) but not considered a pest.
GENERAL: Description and illustration of adult female by McKenzie (1937), Balachowsky (1958b), Velasquez (1971), Chou (1985, 1986), Williams & Watson (1988) and by Colon-Ferrer & Medina-Gaud (1998).
KEYS: Williams & Watson 1988: 35 (female) [Tropical South Pacific]; Velasquez 1971: 118 (female) [Philippines]; Beardsley 1966: 509 (female) [Federated States of Micronesia]; McKenzie 1946: 33 (female) [World]; McKenzie 1942b: 144-145 (female) [World]; McKenzie 1938: 17-18 (female) [World]; McKenzie 1937a: 178 (female) [World]; McKenzie 1937: 330 (female) [World].
CITATIONS: Balach1957c [host, distribution: 199]; Balach1958b [taxonomy, description, illustration, host, distribution: 223-224,227]; Beards1966 [host, distribution: 509]; Borchs1966 [catalogue: 295-296]; Chou1986 [taxonomy, illustration: 681]; ClapsWoGo2001a [taxonomy, host, distribution: 12]; ColonFMe1998 [taxonomy, description, illustration, host, distribution: 37];

CoronaRuMo1997 [host, distribution: 38-41]; DanzigKo1990 [host, distribution: 45]; Ebelin1949 [host, distribution, life history, control]; Hunt1939 [host, distribution: 548-566]; Lindin1957 [taxonomy: 545]; Martor1976 [host, distribution: 14]; McKenz1937 [taxonomy, description, illustration, host, distribution: 327-330]; McKenz1937a [taxonomy: 178]; McKenz1938 [taxonomy, description, illustration, host, distribution: 8-9,17,25]; McKenz1942b [taxonomy: 144]; McKenz1946 [taxonomy, host, distribution: 31-33]; MillerDa1990 [host, distribution, economic importance: 300]; Nakaha1983 [host, distribution: 8]; PruthiMa1945 [host, distribution, life history, control: 1-42]; Sander1909a [taxonomy, host, distribution: 52]; SchmutKlLu1957 [host, distribution, economic importance: 484]; Takagi1962c [host, distribution: 52]; Takagi1970 [taxonomy, host, distribution: 131]; Tao1999 [taxonomy, host, distribution: 71]; Velasq1971 [taxonomy, description, illustration, host, distribution: 118-119]; VelasqRi1969 [host, distribution: 195-208]; WilliaWa1988 [taxonomy, description, illustration, host, distribution: 37-40].

Aonidiella crenata Balachowsky
Aonidiella crenata Balachowsky, 1953b: 77. Type data: CONGO: on *Eriocoelum microspermum*. Syntypes, female. Type depository: Paris: Muséum national d'Histoire naturelle, France.
SCALE COVER: Female scale circular, colour uniform red; glassy aspect formed by shiny, central exuviae; marginal crenulations visible beneath scale. Ventral scale flat, white, completely sealing scale below; 1.8-2.1 mm. Male scale unknown (Balachowsky, 1953b, 1956).
HOST PLANTS: **Euphorbiaceae**: *Cleistanthus polystahyus* [Almeid1973b]. **Sapindaceae**: *Eriocoelum microspermum* [Balach1953b].
DISTRIBUTION: **Afrotropical**: Angola [Almeid1973b]; Zimbabwe [Balach1953b].
GENERAL: Description and illustration of adult female by Balachowsky (1953b, 1956).
KEYS: Balachowsky 1956: 25 (female) [Africa].
CITATIONS: Almeid1973b [host, distribution: 8]; Balach1953b [taxonomy, description, illustration, host, distribution: 77-80]; Balach1956 [taxonomy, description, illustration, host, distribution: 33-36]; Borchs1966 [catalogue: 296].

Aonidiella ensifera McKenzie
Aonidiella ensifera McKenzie, 1942b: 142. Type data: CHILE: Valparaiso, on *Hedera helix*; collected May 14, 1935. Holotype female. Type depository: Davis: The Bohart Museum of Entomology, University of California, California, USA.
SCALE COVER: Scale of female thin; pale reddish-brown colour; area of second exuvia slightly paler than remainder; no male scale found (McKenzie, 1942b).
HOST PLANTS: **Araliaceae**: *Hedera helix* [McKenz1942b, ClapsWoGo2001].
NATURAL ENEMIES: HYMENOPTERA **Signiphoridae**: *Signiphora flavella* [Woolle1990].
DISTRIBUTION: **Neotropical**: Chile [GonzalCh1968, ClapsWoGo2001] (Valparaiso [McKenz1942b]).
BIOLOGY: Occurring on leaves (McKenzie, 1942b).

GENERAL: Description and illustration of adult female by McKenzie (1942b).

KEYS: McKenzie 1946: 33 (female) [World]; McKenzie 1942d: 144-145 (female) [World].

CITATIONS: Borchs1966 [catalogue: 296]; ClapsWoGo2001 [host, distribution: 240]; GonzalCh1968 [distribution: 110]; McKenz1942b [taxonomy, description, illustration, host, distribution: 142-143,147]; McKenz1946 [taxonomy: 33]; Woolle1990 [biological control: 167-176].

Aonidiella eremocitri McKenzie

Aonidiella eremocitri McKenzie, 1937a: 177. Type data: AUSTRALIA: Queensland, Marmor, on *Eremocitrus glauca*. Holotype female. Type depository: Davis: Bohart Museum of Entomology, University of California, California, USA.
Chrysomphalus eremocitri; Lindinger, 1943b: 207. Change of combination.
Aonidiella eremocitri; Borchsenius, 1966: 296. Revived combination.
SCALE COVER: Scale of female smooth, circular, flat, yellow, hard, and brittle, 1.5 to 1.75 mm in diameter. Male scale not identified (McKenzie, 1937a).
HOST PLANTS: **Anacardiaceae**: *Campnosperma brevipetiolata* [Beards1966]. **Barringtoniaceae**: *Barringtonia* [WilliaWa1988]. **Celastraceae**: *Celastrus bilocularis* [Brimbl1962a]. **Euphorbiaceae**: *Glochidion* [Beards1966]. **Meliaceae**: *Owenia venosa* [Brimbl1962a]. **Orchidaceae**: *Coelogyne·asperata* [McKenz1946]. **Palmae**: *Cocos nucifera* [McKenz1946, Beards1966, WilliaBu1987, WilliaWa1988]. **Rutaceae**: *Citrus* [WilliaWa1988], *Eremocitrus glauca* [McKenz1937a].
NATURAL ENEMIES: HYMENOPTERA **Encyrtidae**: *Comperiella bifasciata* Howard [Flande1944a].
DISTRIBUTION: **Australasian**: Australia (Queensland [McKenz1937a, Brimbl1962a]); Fiji [WilliaWa1988]; Palau [Beards1966]; Papua New Guinea [WilliaWa1988]; Solomon Islands [WilliaWa1988]; Vanuatu [WilliaBu1987, WilliaWa1988] [WilliaBu1987, WilliaWa1988]. **Oriental**: Thailand [McKenz1946].
BIOLOGY: Occurring on leaves, twigs and larger branches (McKenzie, 1937a).
ECONOMIC IMPORTANCE: Considered a pest of oil palm and coconut (Mariau, 1998).
GENERAL: Description and illustration of adult female by McKenzie (1937a, 1938) and by Williams & Watson (1988).
KEYS: Williams & Watson 1988: 35 (female) [Tropical South Pacific]; Beardsley 1966: 509 (female) [Federated States of Micronesia]; McKenzie 1946: 33 (female) [World]; McKenzie 1942b: 144-145 (female) [World]; McKenzie 1938: 17-18 (female) [World]; McKenzie 1937a: 178 (female) [World].
CITATIONS: Balach1958b [taxonomy: 224]; Beards1966 [host, distribution: 510]; Borchs1966 [catalogue: 296]; Brimbl1962a [taxonomy, description, illustration, host, distribution: 408-409]; Flande1944a [biological control: 365-371]; Lindin1943b [taxonomy: 207]; Lindin1957 [taxonomy: 545]; Mariau1998 [host, distribution, economic importance: 269-277]; McKenz1937a [taxonomy, description, illustration, host, distribution: 177-179]; McKenz1938 [taxonomy, description, illustration, host, distribution: 9,26]; McKenz1942b [taxonomy: 144]; McKenz1946 [taxonomy, host, distribution: 32-33]; WilliaBu1987 [host,

distribution: 94]; WilliaWa1988 [taxonomy, description, illustration, host, distribution: 39-40].

Aonidiella eugeniae (Hempel)

Aspidiotus eugeniae Hempel, 1937: 25. Type data: BRAZIL: São Paolo, Santos, on *Eugenia* sp. Holotype female. Type depository: São Paulo: Instituto Biologico de São Paulo, Brazil; type no. 718.

Aonidiella eugeniae; McKenzie, 1938: 9. Change of combination.

Chrysomphalus eugeniae; Lindinger, 1957: 545. Change of combination.

SCALE COVER: Female scale slender, flat, approximately circular, about 2 mm in diameter (Hempel, 1937).

HOST PLANTS: **Melastomataceae**: *Miconia* [ClapsWoGo2001]. **Myrtaceae**: *Eugenia* [Hempel1937, Lepage1938, ClapsWoGo2001], *Eugenia cauliflora* [McKenz1946], *Eugenia dombeyi* [ClapsWoGo2001], *Myrica jaboticaba* [ClapsWoGo2001], *Paivaea langsdorffii* [ClapsWoGo2001].

DISTRIBUTION: **Neotropical**: Brazil (São Paulo [Hempel1937, Lepage1938, McKenz1946, ClapsWoGo2001]).

GENERAL: Description and illustration of adult female by Hempel (1937) and by McKenzie (1938).

KEYS: McKenzie 1946: 33 (female) [World]; McKenzie 1942b: 144-145 (female) [World]; McKenzie 1938: 17-18 (female) [World].

CITATIONS: Borchs1966 [catalogue: 296]; Claps1993 [taxonomy: 6]; ClapsWoGo2001 [host, distribution: 241]; Ferris1941e [taxonomy: 43]; Hempel1937 [taxonomy, description, illustration, host, distribution: 25-27]; Lepage1938 [catalogue: 394]; Lindin1943b [taxonomy: 207]; Lindin1957 [taxonomy: 545]; McKenz1938 [taxonomy, description, illustration, host, distribution: 9,27]; McKenz1942b [taxonomy: 145]; McKenz1944 [taxonomy, description, illustration: 55,59]; McKenz1946 [taxonomy, host, distribution: 32-33]; Rungs1941 [taxonomy: 29].

Aonidiella godfreyi Takahashi

Aonidiella godfreyi Takahashi, 1942b: 48. Type data: THAILAND: Chiengmai, on an undetermined tree. Syntypes, female. Type depository: Taichung: Entomology Collection, Taiwan Agricultural Research Institute, Wu-feng, Taichung, Taiwan.

SCALE COVER: Scale circular, slightly convex, pale yellowish brown, a little greyish, 1.5 mm in diameter (Takahashi, 1942b).

DISTRIBUTION: **Oriental**: Thailand [Takaha1942b].

GENERAL: Description and illustration of adult female by Takahashi (1942b).

CITATIONS: Borchs1966 [catalogue: 296]; Takaha1942b [taxonomy, description, illustration, host, distribution: 48-49].

Aonidiella gracilis (Balachowsky)

Aspidiotus gracilis Balachowsky, 1929: 141. Type data: ZAIRE: Sankuru, on leaves of cacao; collected by Ghesqière, January 1925. Syntypes, female. Type depository: Paris: Muséum national d'Histoire naturelle, France.

Aspidiotus gracillimus Lindinger, 1937: 180. Unjustified replacement name for *Aspidiotus gracilis* Balachowsky.
Aonidiella gracilis; McKenzie, 1938: 9. Change of combination.
Chrysomphalus gracillimus; Lindinger, 1943b: 218. Change of combination.
Aonidiella gracillima; Schmutterer et al., 1957: 484. Change of combination.
Aonidiella gracilis; Borchsenius, 1966: 296. Revived combination.
SCALE COVER: Female scale, circular, small 1.4-1.8 mm; sometimes posterior part projecting giving a pyriform shape; moderately convex, larval exuviae central; scale brown. Male scale same colour, oval, 0.8-1 mm (Balachowsky, 1956).
HOST PLANTS: **Palmae**: *Elaeis guineensis* [Balach1956]. **Sterculiaceae**: *Theobroma cacao* [Balach1929, Liegeo1944].
DISTRIBUTION: **Afrotropical**: Zaire [Balach1929, Ghesqu1932, MayneGh1934, Liegeo1944]; Zimbabwe [Balach1956].
ECONOMIC IMPORTANCE: Recorded as cocoa pest in Congo (Liegeois, 1944).
GENERAL: Description and illustration of adult female by McKenzie (1938) and by Balachowsky (1956).
KEYS: Balachowsky 1956: 26 (female) [Africa]; McKenzie 1946: 33 (female) [World]; McKenzie 1942b: 144-145 (female) [World]; McKenzie 1938: 17-18 (female) [World].
CITATIONS: Balach1929 [taxonomy, description, illustration, host, distribution: 141-145]; Balach1956 [taxonomy, description, illustration, host, distribution: 35-38]; Borchs1966 [catalogue: 296]; Ferris1941e [taxonomy: 44]; Ghesqu1932 [host, distribution: 59]; Liegeo1944 [host, distribution, economic importance: 164-165]; Lindin1937 [taxonomy: 180]; Lindin1943b [taxonomy: 218]; MayneGh1934 [host, distribution: 3-38]; McKenz1938 [taxonomy, description, illustration, host, distribution: 9-10,28]; McKenz1942b [taxonomy: 145]; McKenz1946 [taxonomy: 33]; MillerDa1990 [host, distribution, economic importance: 300]; SchmutKlLu1957 [host, distribution, economic importance: 484].

Aonidiella inornata McKenzie

Chrysomphalus aurantii; Robinson, 1917: 24. Misidentification; discovered by McKenzie, 1938: 10.
Aonidiella inornata McKenzie, 1938: 10. Type data: PHILIPPINES: Luzon, Los Banos, on *Astronia* sp.; collected April 1915, by C.F. Baker. Holotype female. Type depository: Davis: The Bohart Museum of Entomology, University of California, California, USA.
Aonidiella inormata; Borchsenius, 1966: 296. Misspelling of species name.
Aonidiella inormata; Chou, 1986: 682. Misspelling of species name.
COMMON NAME: papaya red scale [LeeWe1977].
SCALE COVER: Female scale yellow-brown, translucent, with body outline showing through; "waxy margin of scale tending to be abruptly descending at the margin of the body of the insect" [sic]. Male scale yellow-brown, elongate (McKenzie, 1938).
HOST PLANTS: **Agavaceae**: *Cordyline terminalis* [McKenz1946, Zimmer1948]. **Anacardiaceae**: *Campnosperma brevipetiolata* [Takaha1941b, McKenz1946], *Mangifera indica* [McKenz1946]. **Apocynaceae**: *Allemanda* [WilliaWa1988],

Nerium oleander [McKenz1938], *Ochrosia* [Beards1966], *Plumeria acuminata* [Takaha1939b, Beards1966], *Plumeria rubra* [WilliaWa1988]. **Barringtoniaceae**: *Barringtonia* [WilliaWa1988]. **Bischofiaceae**: *Bischofia javanica* [WilliaWa1988]. **Caricaceae**: *Carica papaya* [Takaha1939c, LeeWe1977]. **Casuarinaceae**: *Casuarina* [Beards1966]. **Cycadaceae**: *Cycas* [Takagi1969a]. **Euphorbiaceae**: *Annesijoa* [WilliaWa1988], *Euphorbia* [WilliaWa1988]. **Hippocrateaceae**: *Salacea* [WilliaWa1988]. **Leguminosae**: *Cassia* [WilliaWa1988]. **Moraceae**: *Artocarpus alticis* [Beards1966]. **Musaceae**: *Musa* [WilliaBu1987, WilliaWa1988]. **Myrtaceae**: *Melaleuca* [WilliaWa1988]. **Oleaceae**: *Jasminum sambac* [McKenz1946, Zimmer1948, Beards1966]. **Palmae** [McKenz1938, Takagi1969a], *Areca catechu* [McKenz1946], *Cocos nucifera* [Beards1966, Takagi1969a], *Nipa fruticans* [Beards1966]. **Pandanaceae**: *Pandanus odoratissimus* [WilliaWa1988]. **Piperaceae**: *Piper* [Zimmer1948], *Piper aduncum* [WilliaWa1988], *Piper betle* [McKenz1938, McKenz1946], *Piper methysticum* [WilliaBu1987, WilliaWa1988]. **Polygonaceae**: *Polygonum* [WilliaWa1988]. **Potaliaceae**: *Fagraea cambageana* [Brimbl1962a]. **Rhizophoraceae**: *Rhizophora mucronata* [Beards1966]. **Rubiaceae**: *Hedyotis ocutangulus* [McKenz1938], *Platanocephalus morindaefolius* [WilliaWa1988]. **Rutaceae**: *Astronia* [McKenz1938, Takagi1969a], *Citrus* [McKenz1938, Beards1966], *Citrus paradisi* [McKenz1946], *Citrus reticulata* [McKenz1946]. **Vitaceae**: *Vitis vinifera* [WilliaWa1988]. **Zingiberaceae**: *Elettaria cardamomum* [WilliaWa1988].

DISTRIBUTION: **Afrotropical**: Guinea [McKenz1946]. **Australasian**: Australia (Queensland [Brimbl1962a]); Federated States of Micronesia (Caroline Islands [Takaha1939b, Takaha1941b, Beards1966], Kosare (=Kusiae) [Beards1966], Ponape Island [Beards1966], Truk Islands [Beards1966], Yap [Beards1966]); Fiji [WilliaWa1988]; Hawaiian Islands (Hawaii [McKenz1938, McKenz1946, Zimmer1948, Takagi1969a]); Kiribati [WilliaWa1988]; Marshall Islands [Beards1966]; Palau [Takaha1939b, Beards1966]; Papua New Guinea [WilliaWa1988]; Vanuatu [WilliaBu1987, WilliaWa1988] [WilliaBu1987, WilliaWa1988]; Wake Island [Beards1966]; Western Samoa [WilliaWa1988]. **Nearctic**: United States of America (Texas [Nakaha1982]). **Neotropical**: Dominican Republic [Nakaha1982]; Ecuador [Nakaha1982]; Puerto Rico & Vieques Island (Puerto Rico [ColonFMe1998]). **Oriental**: Hong Kong [McKenz1938, Takagi1969a]; India (Uttar Pradesh [GuptaSi1988]); Indonesia (Irian Jaya [WilliaWa1988]); Philippines [McKenz1946] (Luzon [McKenz1938, Takagi1969a]); Taiwan [Takaha1939c, Takagi1969a, LeeWe1977]; Thailand [Nakaha1982]. **Palaearctic**: China (People's Republic) [Tang1984] (Henan (Honan) [Shen1993]); Japan [Sakai1939, Kawai1980].

ECONOMIC IMPORTANCE: Pest of papaya in Taiwan (Lee & Wen, 1977) and mango in the Philippines and Puerto Rico (Chua & Wood, 1990).

GENERAL: Description and illustration of adult female by McKenzie (1938), Zimmerman (1948), Takagi (1969a), Velasquez (1971), Tang (1984), Chou (1985, 1986), Williams & Watson (1988) and by Colon-Ferrer & Medina-Gaud (1998).

KEYS: Williams & Watson 1988: 35 (female) [Tropical South Pacific]; Chou 1985: 292 (female) [Species of China]; Kawai 1980: 211 (female) [Japan]; Velasquez 1971: 118 (female) [Philippines]; McDaniel 1968: 210 (female) [U.S.A.: Texas];

Beardsley 1966: 509 (female) [Federated States of Micronesia]; McKenzie 1946: 33 (female) [World]; McKenzie 1942b: 144-145 (female) [World]; McKenzie 1938: 17-18 (female) [World].

CITATIONS: Beards1966 [host, distribution: 510-511]; Borchs1966 [taxonomy, catalogue: 296]; Brimbl1962a [taxonomy, description, illustration, host, distribution: 410]; Chou1985 [taxonomy, description, host, distribution: 298-299]; Chou1986 [taxonomy, illustration: 682]; ChuaWo1990 [host, distribution, economic importance: 543-552]; ColonFMe1998 [taxonomy, description, illustration, host, distribution: 37-38]; DanzigPe1998 [catalogue: 182]; Esaki1940a [host, distribution: 274-280]; GuptaSi1988 [host, distribution, control: 357-361]; Hinckl1963 [host, distribution, biological control]; Hunt1939 [host, distribution: 548-566]; Kawai1980 [taxonomy, description, host, distribution: 211-212]; LeeWe1977 [economic importance, life history, host, distribution, chemical control: 196-201]; Lindin1943b [taxonomy: 207]; McClur1990e [taxonomy, host, distribution, ecology: 309-314]; McDani1968 [taxonomy, illustration, host, distribution: 212-215]; McKenz1938 [taxonomy, description, illustration, host, distribution: 10-11,29-30]; McKenz1942b [taxonomy: 145]; McKenz1946 [taxonomy, host, distribution: 32-33]; Nafus1996 [host, distribution: 1]; Nakaha1982 [host, distribution: 7]; Robins1917 [taxonomy, host, distribution: 24-25]; Sakai1939 [taxonomy, description, illustration, host, distribution: 45-62]; Schmut2001 [host, distribution: 339-345]; Shen1993 [host, distribution: 58]; Takagi1969a [taxonomy, description, illustration, host, distribution: 83-84,103]; Takaha1939b [taxonomy, host, distribution: 270]; Takaha1939c [taxonomy, host, distribution: 88]; Takaha1941b [host, distribution: 220]; Tang1984 [taxonomy, description, illustration, host, distribution: 39-40]; Tao1999 [taxonomy, host, distribution: 71]; Velasq1971 [taxonomy, description, illustration, host, distribution: 128-130]; VelasqRi1969 [host, distribution: 195-208]; WilliaBu1987 [host, distribution: 94]; WilliaWa1988 [taxonomy, description, illustration, host, distribution: 41-42]; Zimmer1948 [taxonomy, description, illustration, host, distribution: 361,364,366].

Aonidiella lauretorum (Lindinger)

Aspidiotus lauretorum Lindinger, 1911a: 15. Type data: CANARY ISLANDS: Gran Canaria, on *Dracaena draco*; Tenerife, on *Gymnosporia cassinoides, Ilex canariensis, I. platyphylla, Oreodaphne foetens, Picconia excelsa, Smilax canariensis, Hedera helix canariensis, Appolonias canariensis*. Syntypes, female. Type depository: Hamburg: Zoologisches Institut und Zoologisches Museum, Universität von Hamburg, Germany.

Aonidiella lauretorum; McKenzie, 1938: 11. Change of combination.

Aonidiella mimeuri Rungs, 1941: 27. Type data: PORTUGAL: Cintra, botanic garden, on *Dicksonia squarrosa*; collected by M.J.M. Mimeur, 19.ix.1935. Syntypes, female. Type depository: Paris: Muséum national d'Histoire naturelle, France; type no. 1416. Synonymy by Borchsenius, 1966: 296.

SCALE COVER: Female scale white or bright yellow-brown, transparent; exuviae more or less central; 2.5-3 mm long, 2-2.5 wide. Male scale white, elongate, 1-1.2 mm long 0.7-0.8 mm wide; exuviae eccentric (Lindinger, 1911a). Female scale circular or subcircular, very flat, without any projections; colour red-yellow or yellow, transparent; larval exuviae central or subcentral of similar colour; 1.8-2.1 mm (Balachowsky, 1948b).

HOST PLANTS: **Aquifoliaceae**: *Ilex canariensis* [Lindin1911a, GomezM1962]. **Araliaceae**: *Hedera canariense* [Balach1938a], *Hedera helix canariensis* [GomezM1962], *Hedera helix* [Balach1948b], *Hedera helix canariensis* [Lindin1911a]. **Celastraceae**: *Gymnosporia* [Balach1948b], *Gymnosporia cassinoides* [Lindin1911a]. **Cyatheaceae**: *Dicksonia squarrosa* [Rungs1941, Balach1948b]. **Fagaceae**: *Ilex platyphylla* [Lindin1911a, GomezM1962]. **Gentianaceae**: *Wildpretina viscosa* [Balach1948b]. **Lauraceae**: *Apollonias* [Balach1948b], *Apollonias canariensis* [Lindin1911a, GomezM1962], *Laurus canariensis* [Lindin1911a, GomezM1962, GomezM1967O], *Ocotea foetens* [Balach1946, GomezM1962, GomezM1967O], *Oreodaphne foetens* [Lindin1911a, Balach1938a], *Persea indica* [Balach1948b]. **Liliaceae**: *Dracaena* [Lindin1912b], *Dracaena draco* [Balach1948b]. **Myrsinaceae**: *Heberdenia excelsa* [Lindin1911a, GomezM1962]. **Oleaceae**: *Picconia excelsa* [Lindin1911a, GomezM1962]. **Ruscaceae**: *Semele androgyna* [Balach1948b]. **Smilacaceae**: *Smilax* [Balach1948b], *Smilax canariensis* [Lindin1911a]. **Theaceae**: *Visnea mocanera* [Lindin1911a, Balach1948b, GomezM1962].

DISTRIBUTION: **Palaearctic**: Canary Islands [Lindin1911a, Balach1946, Balach1948b, GomezM1962, GomezM1967O, MatileOr2001]; Madeira Islands [Balach1938a]; Portugal [Rungs1941].

GENERAL: Description of adult female by Lindinger (1911a), McKenzie (1938), Balachowsky (1948b) and by Gómez-Menor Guerrero (1962).

KEYS: Gómez-Menor Guerrero 1962: 187 (female) [Canary Islands]; Balachowsky 1948b: 361 (female) [Mediterranean]; McKenzie 1946: 33 (female) [World]; McKenzie 1942b: 144-145 (female) [World]; McKenzie 1938: 17-18 (female) [World].

CITATIONS: Balach1938a [host, distribution: 151-152]; Balach1946 [host, distribution: 212]; Balach1948b [taxonomy, description, illustration, host, distribution: 373-376]; Borchs1966 [catalogue: 296-297]; Bustsh1958 [taxonomy, description, host, distribution: 220,242]; DanzigPe1998 [catalogue: 182]; Ferris1941e [taxonomy: 45]; GomezM1962 [taxonomy, description, host, distribution: 196-198]; GomezM1967O [host, distribution: 132]; Lindin1911a [taxonomy, description, illustration, host, distribution: 15-17]; Lindin1912b [taxonomy, description, host, distribution: 69-70,135,174,175,]; MacGil1921 [taxonomy, description, host, distribution: 402]; MatileOr2001 [host, distribution: 189]; McKenz1938 [taxonomy, description, illustration, host, distribution: 11-12,31]; McKenz1942b [taxonomy: 145]; McKenz1946 [taxonomy: 31,33]; Rungs1941 [taxonomy, description, illustration, host, distribution: 27-30]; Sassce1912 [taxonomy, host, distribution: 93]; WeidneWa1968 [taxonomy: 172].

Aonidiella longicorna McKenzie

Aonidiella longicorna McKenzie, 1939: 65. Type data: ETHIOPIA: Nefasit, on undetermined, ornamental plant. Holotype female. Type depository: Davis: The Bohart Museum of Entomology, University of California, California, USA.
Aspidiotus longicornis; Lindinger, 1943b: 207. Change of combination.

Aonidiella longicorna; Borchsenius, 1966: 297. Revived combination.
SCALE COVER: Scale of female grey-brown, translucent, with body outline showing through; waxy margin tending to be considerably extended beyond body of insect; scale of male grey, elongate-oval, with exuvia at one end (McKenzie, 1939).
DISTRIBUTION: Afrotropical: Ethiopia [McKenz1939].
BIOLOGY: Occurring on the leaves (McKenzie, 1939).
GENERAL: Description and illustration of adult female by McKenzie (1939) and by Balachowsky (1956).
KEYS: Balachowsky 1956: 25 (female) [Africa]; McKenzie 1946: 33 (female) [World]; McKenzie 1942b: 144-145 (female) [World].
CITATIONS: Balach1956 [taxonomy, description, illustration, host, distribution: 37-40]; Borchs1966 [catalogue: 297]; Lindin1943b [taxonomy: 207]; McKenz1939 [taxonomy, description, illustration, host, distribution: 65,76]; McKenz1942b [taxonomy: 145]; McKenz1946 [taxonomy: 33].

Aonidiella marginipora McKenzie
Aonidiella marginipora McKenzie, 1946: 30. Type data: UGANDA: Kampala, on *Hydnocarpus* sp. Holotype female. Type depository: Davis: The Bohart Museum of Entomology, University of California, California, USA.
SCALE COVER: Scale of female approximately 1.2 mm in diameter, circular, pale brown, thin, with a series of puffy wax layers forming a somewhat convex mound. Scale of male elongate, light yellowish (McKenzie, 1946).
HOST PLANTS: Flacourtiaceae: *Hydnocarpus* [McKenz1946].
DISTRIBUTION: Afrotropical: Sierra Leone [Balach1956]; Uganda [McKenz1946].
BIOLOGY: Occurring on the leaves (McKenzie, 1946).
GENERAL: Description and illustration of adult female by McKenzie (1946) and by Balachowsky (1956).
KEYS: Balachowsky 1956: 26 (female) [Africa]; McKenzie 1946: 33 (female) [World].
CITATIONS: Balach1956 [taxonomy, description, illustration, host, distribution: 39-42]; Borchs1966 [catalogue: 297]; McKenz1946 [taxonomy, description, illustration, host, distribution: 30,33,35].

Aonidiella messengeri McKenzie
Aonidiella messengeri McKenzie, 1953: 35. Type data: RYUKYU: Miyako Islands, on *Calophyllum inophyllum*; collected September 5 1951, by A.P. Messenger. Holotype female. Type depository: Davis: The Bohart Museum of Entomology, University of California, California, USA.
SCALE COVER: Female scale about 1.5 mm in diameter, circular, thin and translucent with reddish body of insect showing through; scale develops a wide, greyish-white marginal zone much exceeding body of insect. Male scale elongate, and with body colour similar to female (McKenzie, 1953).
HOST PLANTS: Cycadaceae: *Cycas revoluta* [Takaha1955f]. **Daphniphyllaceae**: *Daphniphyllum glaucescens* [TakahaTa1956, Takagi1970]. **Guttiferae**: *Calophyllum inophyllum* [Takagi1970]. **Moraceae**: *Ficus retusa* [Takaha1955f].

Myrsinaceae: *Bladhia siebaldi* [Takagi1970]. **Palmae**: *Phoenix roebelenii* [TakahaTa1956, Takagi1970].
DISTRIBUTION: **Oriental**: Ryukyu Islands (= Nansei Shoto) [McKenz1953, TakahaTa1956, Takagi1970]; Taiwan [McKenz1953, TakahaTa1956, Takagi1970]. **Palaearctic**: Japan [Takaha1955f, Takagi1970, Kawai1980] (Honshu [TakahaTa1956], Shikoku [TakahaTa1956]).
GENERAL: Description and illustration of adult female by McKenzie (1953).
KEYS: Kawai 1980: 211 (female) [Japan]; McKenzie 1953: 36 (female) [World].
CITATIONS: DanzigPe1998 [catalogue: 183]; Kawai1980 [taxonomy, description, host, distribution: 213]; McKenz1953 [taxonomy, description, illustration, host, distribution: 35-38]; Muraka1970 [host, distribution: 71]; Takagi1970 [taxonomy, host, distribution: 131]; TakahaTa1956 [taxonomy, host, distribution: 15-16]; Tao1999 [taxonomy, host, distribution: 71].

Aonidiella misrae Laing
Aonidiella misrae Laing, 1929a: 491. Type data: INDIA: Pusa, on leaves of *Tamarindus indica*. Syntypes, female. Type depository: London: The Natural History Museum, England, UK.
Melanaspis misrai Lindinger, 1943b: 147. Unjustified emendation; discovered by Borchsenius, 1966: 297.
SCALE COVER: Female scale low convex, subcircular, pale greyish with a few narrow, brown concentric rings, greyish brown, or brownish fuscous, with a slight mealy deposit; exuviae whitish grey, subcentral; diameter 1.4 mm (Laing, 1929a).
HOST PLANTS: **Leguminosae**: *Tamarindus indica* [Laing1929a].
DISTRIBUTION: **Oriental**: India [Laing1929a].
GENERAL: Description and illustration of adult female by Laing (1929a).
CITATIONS: Borchs1966 [catalogue: 297]; Laing1929a [taxonomy, description, illustration, host, distribution: 491-492]; Lindin1943b [taxonomy: 147].

Aonidiella orientalis (Newstead)
Aspidiotus orientalis Newstead, 1894c: 26. Type data: INDIA: Andhra Pradesh, Seven Pagodas, on *Panicum* sp. Syntypes. Type depository: London: The Natural History Museum, England, UK.
Aspidiotus osbeckiae Green, 1896: 4. Type data: SRI LANKA: Punduloya, on *Osbeckia* sp. Holotype female. Type depository: London: The Natural History Museum, England, UK. Synonymy by Green, 1922: 459.
Aspidiotus (*Diaspidiotus*) *osbeckiae*; Cockerell, 1897i: 28. Change of combination.
Aspidiotus (*Evaspidiotus*) *orientalis*; Leonardi, 1898a: 76. Change of combination.
Aspidiotus (*Evaspidiotus*) *osbechiae*; Leonardi, 1898a: 76. Change of combination.
Aspidiotus (*Evaspidiotus*) *osbechiae*; Leonardi, 1898a: 76. Misspelling of species name. Notes: Misspelling of "*osbechiae*" for "*osbeckiae*".
Aspidiotus (*Chrysomphalus*) *pedronis* Green, 1905a: 341. Type data: SRI LANKA: Pedrotalagalla, altitude of about 8000 feet, on an undetermined tree. Syntypes, female. Type depository: London: The Natural History Museum, England, UK. Synonymy by Ferris, 1941e: 46.

Aspidiotus (*Aonidiella*) *taprobanus* Green, 1905a: 344. Type data: SRI LANKA: Paradeniya, on *Phyllanthus myrtifolius*. Syntypes, female. Type depository: London: The Natural History Museum, England, UK. Synonymy by McKenzie, 1939: 65.

Chrysomphalus pedronis; Sanders, 1906: 15. Change of combination.

Chrysomphalus taprobanus; Sanders, 1906: 15. Change of combination.

Aspidiotus (*Aonidiella*) *cocotiphagus* Marlatt, 1908c: 14. Type data: CUBA: Santiago de las Vegas, on *Cocos nucifera*; collected by W.T. Horne, August 1914. Holotype female. Type depository: Washington: United States National Entomological Collection, U.S. National Museum of Natural History, District of Columbia, USA; type no. 14136. Synonymy by Lindinger, 1908: 241.

Aspidiotus cocotiphagus; Lindinger, 1908e: 241. Change of combination.

Chrysomphalus orientalis; Lindinger, 1913: 73. Change of combination.

Chrysomphalus pedroniformis Cockerell & Robinson, 1915: 107. Type data: PHILIPPINES: Luzon, Bataan on *Eriodendron anfractuosum*; Laguna, Los Banos, on *Vitis vinifera*. Syntypes, female. Type depository: Washington: United States National Entomological Collection, U.S. National Museum of Natural History, District of Columbia, USA. Synonymy by Borchsenius, 1966: 297.

Furcaspis cocotiphaga; MacGillivray, 1921: 408. Change of combination.

Furcaspis orientalis; MacGillivray, 1921: 408. Change of combination.

Aonidiella taprobana; MacGillivray, 1921: 444. Change of combination requiring emendation of species name for agreement in gender.

Aspidiotus orientalis cocotiphagus; Merrill & Chaffin, 1923: 204. Change of combination and rank.

Aspidiotus (*Furcaspis*) *orientalis*; Glover, 1933: 1. Change of combination.

Aonidiella orientalis; McKenzie, 1938: 12. Change of combination.

Aonidiella cocotiphagus; Ferris, 1938a: 190. Change of combination.

Aonidiella pedroniformis; McKenzie, 1939: 54. Change of combination.

Aonidiella pedronis; McKenzie, 1939: 54. Change of combination.

COMMON NAMES: escama amarilla oriental [CoronaRuMo1997]; Oriental Scale [Merril1953, Dekle1965c].

SCALE COVER: As on coconut, scale of female quite thick, flat, circular, ranging in colour from almost white to a very light brown; exuviae central, second exuvia dark brown. Scale of male slightly elongate oval; yellowish exuvia near one end (Ferris, 1938a).

HOST PLANTS: **Acanthaceae**: *Adhatoda vasica* [RahmanAn1941], *Barleria cristata* [RahmanAn1941]. **Aceraceae**: *Acer oblongum* [RahmanAn1941], *Acer pictum* [RahmanAn1941]. **Amaryllidaceae**: *Agave* [RahmanAn1941], *Agave americana* [RahmanAn1941], *Agave sisalana* [Balach1948b, Beards1966, DeLott1967a, Matile1978], *Agave variegata* [RahmanAn1941]. **Anacardiaceae**: *Mangifera indica* [RahmanAn1941, McKenz1946, KinjoNaHi1996, OfekHuYz1997], *Spondias cytherea* [McKenz1946]. **Annonaceae**: *Annona* [Houser1918, Merril1953], *Annona glabra* [MerrilCh1923], *Annona squamosa* [McKenz1946], *Polyalthia korihthi* [Balach1948b], *Polyathia* [Ramakr1919a], *Rollinia emarginata* [Merril1953]. **Apocynaceae**: *Alstonia* [RahmanAn1941], *Alstonia scholaris* [RahmanAn1941], *Carissa* [Ramakr1919a, Ramakr1921a, Merril1953], *Carissa carandas* [MerrilCh1923, RahmanAn1941], *Nerium*

[RahmanAn1941], *Nerium oleander* [RahmanAn1941, Bodenh1944a, Beards1966], *Plumeria* [MerrilCh1923, Beards1966], *Tabernaemontana* [MerrilCh1923], *Tabernaemontana coronaria* [RahmanAn1941]. **Araliaceae**: *Hedera* [Dekle1965c]. **Aristolochiaceae**: *Aristolochia* [RahmanAn1941]. **Asclepiadaceae**: *Calotropis* [Ramakr1919a, Ramakr1921a], *Calotropis procera* [RahmanAn1941, Balach1948b]. **Averrhoaceae**: *Averrhoa carambola* [RahmanAn1941]. **Bignoniaceae**: *Bignonia radicans* [RahmanAn1941], *Bignonia vinusta* [RahmanAn1941], *Kigelia pinnata* [RahmanAn1941], *Oroxylum indicum* [RahmanAn1941], *Tecoma australis* [RahmanAn1941], *Tecoma stans* [RahmanAn1941], *Tecoma undulata* [RahmanAn1941]. **Bischofiaceae**: *Bischofia javanica* [RahmanAn1941]. **Bombacaceae**: *Adansonia* [Merril1953], *Bombax malabaricum* [RahmanAn1941], *Eriodendron anfractuosum* [CockerRo1915]. **Burseraceae**: *Boswellia serrata* [RahmanAn1941], *Bursera serrata* [RahmanAn1941]. **Buxaceae**: *Buxus sempervirens* [RahmanAn1941]. **Cactaceae**: *Cactus* [RahmanAn1941], *Opuntia* [RahmanAn1941]. **Cannaceae**: *Canna indica* [RahmanAn1941]. **Capparidaceae**: *Crataeva religiosa* [RahmanAn1941]. **Caprifoliaceae**: *Lonicera chinensis* [RahmanAn1941]. **Caricaceae**: *Carica papaya* [Balach1948b, Brimbl1962a, WilliaWa1988], *Papaya carica* [Green1914c]. **Celastraceae**: *Catha edulis* [MerrilCh1923], *Celastrus paniculata* [RahmanAn1941]. **Chenopodiaceae**: *Bassia latifolia* [RahmanAn1941]. **Combretaceae**: *Quisqualis indica* [RahmanAn1941], *Terminalia arjuna* [RahmanAn1941], *Terminalia belerica* [RahmanAn1941]. **Convolvulaceae**: *Calonyction roxburghii* [RahmanAn1941], *Ipomoea* [RahmanAn1941], *Porana paniculata* [RahmanAn1941]. **Cunoniaceae**: *Weinmannia* [Lepage1938]. **Cycadaceae**: *Cycas* [MerrilCh1923, McKenz1937], *Cycas revoluta* [Green1900c, Houser1918]. **Ebenaceae**: *Diospyros* [Green1937, Balach1948b], *Diospyros embryopteris* [RahmanAn1941], *Diospyros montana* [RahmanAn1941]. **Ehretiaceae**: *Cordia* [RahmanAn1941], *Cordia myxa* [RahmanAn1941, Bodenh1944a], *Cordia obliqua* [RahmanAn1941], *Cordia rothii* [RahmanAn1941], *Ehretia serrata* [RahmanAn1941]. **Elaeagnaceae**: *Elaeagnus pungens* [Merril1953]. **Euphorbiaceae**: *Acalypha* [Matile1984c], *Croton tiglium* [RahmanAn1941], *Mallotus philippinensis* [RahmanAn1941], *Phyllanthus myrtifolius* [Green1905a, Sander1906, Ramakr1921a, Green1922, Green1937], *Poinsettia* [RahmanAn1941], *Putranjiva roxburghii* [RahmanAn1941], *Ricinus* [Green1937, Balach1948b, Beards1966], *Ricinus communis* [RahmanAn1941, Beards1966], *Sapium sebiferum* [RahmanAn1941]. **Flindersiaceae**: *Chloroxylon swietenia* [Ramakr1919a]. **Gramineae**: *Panicum* [Mamet1951, Borchs1966], *Psidium guajava* [RahmanAn1941, Matile1984c]. **Heliconiaceae**: *Heliconia* [Velasq1971]. **Leguminosae**: *Acacia cyanophila* [Matile1984c], *Albizia* [RahmanAn1941], *Albizia lebbek* [RahmanAn1941], *Atylosia* [Ramakr1921a], *Atylosia candollii* [Green1900a, Ramakr1919a], *Bauhinia* [Green1937, Balach1948b], *Ba. alba* [RahmanAn1941], *Ba. purpurea* [RahmanAn1941], *Ba. racemosa* [RahmanAn1941], *Ba. vahlii* [RahmanAn1941], *Ba. variegata* [RahmanAn1941], *Butea frondosa* [RahmanAn1941], *Caesalpinia bonducella* [RahmanAn1941], *Cassia* [Ramakr1919a, Ramakr1921a, Green1937, Balach1948b], *Cassia auriculata* [RahmanAn1941], *Cassia fistula* [RahmanAn1941], *Ceratonia siliqua*

[RahmanAn1941], *Dalbergia* [Green1908a, Green1937, Balach1948b], *Dalbergia lanceolaria* [RahmanAn1941], *Dalbergia sissoo* [RahmanAn1941, Bodenh1943, Bodenh1944a, McKenz1946], *Erythrina crista galli* [RahmanAn1941], *Inga dulcis* [RahmanAn1941], *Poinciana regia* [Matile1984c], *Pongamia glabra* [RahmanAn1941], *Saraca indica* [RahmanAn1941], *Tamarindus* [Green1908a, Green1937, Balach1948b], *Tamarindus indica* [RahmanAn1941], *Tephrosia* [Ramakr1919a]. **Liliaceae**: *Aloe vera* [RahmanAn1941], *Asparagus* [RahmanAn1941], *Asparagus sprengeri* [MerrilCh1923, Merril1953]. **Lythraceae**: *Lagerstroemia indica* [RahmanAn1941], *Lawsonia inermis* [Matile1984c]. **Magnoliaceae**: *Magnolia grandiflora* [RahmanAn1941]. **Malpighiaceae**: *Hiptage madablota* [RahmanAn1941]. **Malvaceae**: *Hibiscus* [MerrilCh1923]. **Melastomataceae**: *Osbeckia* [Green1896, Green1896e, Ramakr1921a, Green1937, Balach1948b], *Wrightia coccinea* [RahmanAn1941]. **Meliaceae**: *Azedarach indica* [Schmut1998], *Cedrela toona* [RahmanAn1941], *Melia* [Green1937, Balach1948b], *Melia azadirachta* [RahmanAn1941], *Melia composita* [RahmanAn1941], *Melia indica* [RahmanAn1941], *Melia volkensii* [Schmut1990a], *Swietenia mahagoni* [RahmanAn1941]. **Menispermaceae**: *Cocculus laurifolius* [RahmanAn1941]. **Moraceae**: *Broussonetia papyrifera* [RahmanAn1941], *Ficus* [MerrilCh1923, Green1937, RahmanAn1941, Bodenh1944b, McKenz1946, Balach1948b, Merril1953, Danzig1972c], *Fi. benghalensis* [RahmanAn1941], *Fi. carica* [RahmanAn1941], *Fi. elastica* [RahmanAn1941], *Fi. glomerata* [RahmanAn1941], *Fi. infectoria* [RahmanAn1941], *Fi. nitida* [Matile1984c], *Fi. orbicularis* [Green1916e, Balach1948b], *Fi. palmata* [RahmanAn1941], *Fi. religiosa* [RahmanAn1941, Matile1984c], *Fi. retusa* [RahmanAn1941], *Fi. roxburghii* [RahmanAn1941], *Fi. salicifolia* [Balach1948b], *Maclura aurantiaca* [RahmanAn1941], *Morus* [RahmanAn1941], *Morus alba* [RahmanAn1941], *Morus laevigata* [RahmanAn1941]. **Moringaceae**: *Moringa pterygosperma* [RahmanAn1941]. **Musaceae**: *Musa* [Green1937, Balach1948b], *Musa sapientum* [RahmanAn1941]. **Myrtaceae** [Lepage1938], *Callistemon rigidus* [RahmanAn1941], *Eucalyptus* [RahmanAn1941, Matile1984c], *Eugenia* [Green1937, Balach1948b], *Eugenia jambolana* [RahmanAn1941], *Myrrhinium rubriflorum* [Lepage1938], *Myrtus communis* [RahmanAn1941]. **Naucleaceae**: *Stephegyne parviflora* [RahmanAn1941]. **Nyctaginaceae**: *Bougainvillea* [RahmanAn1941], *Mirabilis jalapa* [RahmanAn1941], *Nyctaginia* [RahmanAn1941]. **Ochnaceae**: *Ochna squarrosa* [RahmanAn1941]. **Oleaceae**: *Jasminum* [RahmanAn1941], *Olea europaea* [Bodenh1943, Danzig1972c, Matile1984c]. **Orchidaceae** [Lepage1938]. **Palmae** [Houser1918], *Cocos* [Green1937, Balach1948b], *Cocos nucifera* [Marlat1908c, Houser1918, McKenz1946, Dekle1965c, BesheaTiHo1973], *Inodes neglecta* [MerrilCh1923], *Loroma amethystiora* [MerrilCh1923, Merril1953], *Phoenix* [Houser1918], *Phoenix* [RahmanAn1941], *Phoenix dactylifera* [McKenz1946], *Roystonea regia* [McKenz1946]. **Pistaciaceae**: *Pistacia integerrima* [RahmanAn1941]. **Podocarpaceae**: *Podocarpus lamberti* [Lepage1938], *Podocarpus neriifolius* [Balach1948b]. **Polygonaceae**: *Antigonon leptopus* [RahmanAn1941]. **Proteaceae**: *Grevillea robusta* [RahmanAn1941]. **Punicaceae**: *Punica granatum* [RahmanAn1941]. **Ranunculaceae**: *Clematis paniculatus* [RahmanAn1941].

Rhamnaceae: *Rhamnus persicus* [RahmanAn1941], *Ziziphus* [Green1937, RahmanAn1941, Balach1948b], *Zi.jujuba* [RahmanAn1941], *Zi. oenoplia* [RahmanAn1941]. **Rhizophoraceae**: *Bruguiera sexangulata* [Beards1966], *Rhizophora mucronata* [Beards1966]. **Rosaceae**: *Eriobotrya japonica* [RahmanAn1941], *Prunus armeniaca* [Matile1984c], *Pyrus sinensis* [RahmanAn1941], *Rosa* [Green1937, RahmanAn1941, Balach1948b, Matile1984c]. **Rutaceae**: *Aegle* [Green1937], *Aegle marmelos* [RahmanAn1941], *Casimiroa* [MerrilCh1923, Merril1953], *Citrus* [RahmanAn1941, Bodenh1944a, Bodenh1944b, Balach1948b, Matile1976], *Ci. aurantium* [Bodenh1943], *Ci. limon* [DeLott1967a], *Ci. trifoliata* [Houser1918, MerrilCh1923], *Feronia elephantum* [RahmanAn1941], *Limonia* [Green1937], *Murraya exotica* [MerrilCh1923, RahmanAn1941]. **Salicaceae**: *Populus alba* [RahmanAn1941], *Po. euphratica* [Bodenh1943], *Salix tetrasperma* [RahmanAn1941]. **Sambucaceae**: *Sambucus javanica* [Velasq1971]. **Santalaceae**: *Santalum album* [Balach1948b]. **Sapindaceae**: *Dodonaea viscosa* [RahmanAn1941, Matile1984c], *Litchi chinensis* [McKenz1946, Balach1948b], *Nephelium litchi* [RahmanAn1941], *Sapindus detergens* [RahmanAn1941]. **Sapotaceae**: *Mimusops elengi* [RahmanAn1941], *Mi. kauki* [RahmanAn1941]. **Simaroubaceae**: *Ailanthus aladulosa* [RahmanAn1941]. **Solanaceae**: *Solanum arundo* [Balach1948b], *So. melongena* [Beards1966]. **Sterculiaceae**: *Pterospermum acerifolium* [RahmanAn1941], *Sterculia* [RahmanAn1941], *St. alata* [RahmanAn1941]. **Theaceae**: *Camellia* [Green1937, Balach1948b], *Thea* [Green1937, Balach1948b]. **Thunbergiaceae**: *Thunbergia grandiflora* [RahmanAn1941]. **Tiliaceae**: *Grewia asiatica* [RahmanAn1941]. **Ulmaceae**: *Celtis* [RahmanAn1941], *Celtis australis* [RahmanAn1941], *Ulmus* [RahmanAn1941], *Ulmus integrifolia* [RahmanAn1941]. **Verbenaceae**: *Callicarpa macrophyla* [RahmanAn1941], *Citharexylum subserratum* [RahmanAn1941], *Clerodendrum phlomoides* [RahmanAn1941], *Duranta ellisi* [RahmanAn1941], *Du. plumieri* [Matile1984c], *Gmelina arborea* [RahmanAn1941], *Nyctanthes arbor-tristis* [RahmanAn1941], *Vitex negundo* [RahmanAn1941]. **Vitaceae**: *Vitis vinifera* [CockerRo1915, RahmanAn1941]. **Zamiaceae**: *Zamia* [Merril1953]. **Zingiberaceae**: *Alpinia nutans* [RahmanAn1941].

NATURAL ENEMIES: ACARI **Hemisarcoptidae**: *Hemisarcoptes coccophagus* Meyer [OfekHuYz1997]. COLEOPTERA **Coccinellidae**: *Chilocorus baileyi* (Blackburn) [ElderBe1998], *Ch. bipustulatus* Linnaeus [OfekHuYz1997, SwirskWyIz2002], *Ch. circumdatus* Gyllenhal [ElderBe1998], *Stethorus gilvifrons* Mulsant [OfekHuYz1997]. HYMENOPTERA **Aphelinidae**: *Ablerus guadrii* Agarwal [OfekHuYz1997], *Ap. aonidiae* (Mercet) [RosenDe1979], *Ap. lingnanensis* Compere [OfekHuYz1997], *Ap. melinus* DeBach [OfekHuYz1997, ElderSmBe1998], *Ap. mytilaspidis* (Le Baron) [RosenDe1979], *Ap. peculiaris* (Girault) [RosenDe1979], *Ap. riyadhi* DeBach [DeBach1979], *Encarsia citrina* Craw [ElderSmBe1998], *Marietta javensis* Howard [OfekHuYz1997]. **Encyrtidae**: *Comperiella bifasciata* Howard [SwirskWyIz2002], *Co. lemniscata* Compere & Annecke [ElderGuSm1997, ElderSmBe1998], *Co. unifasciata* Ishii [Trjapi1989], *Habrolepis aspidiotii* Compere & Annecke [OfekHuYz1997, MohammGhTa2001, SwirskWyIz2002], *Ha. obscura* Compere & Annecke [Prinsl1983]. NEUROPTERA **Chrysopidae**: *Chrysoperla carnea* Stephens [OfekHuYz1997].

DISTRIBUTION: **Afrotropical**: Comoros Islands [Balach1956, Matile1978]; Kenya [Newste1917b, DeLott1967a, Schmut1998]; Madagascar [Balach1956, Borchs1966]; Saint Helena [Matile1976]; Somalia [Maleno1916a, Balach1956, Borchs1966, DeLott1967a]; South Africa [Balach1956, Borchs1966]; Tanzania [DeLott1967a]. **Australasian**: Australia (Northern Territory [Green1914c, Green1916e], Queensland [Brimbl1962a, Borchs1966]); Federated States of Micronesia (Yap [Beards1966]); Papua New Guinea [WilliaWa1988]. **Nearctic**: United States of America (Florida [MerrilCh1923, Merril1953, Dekle1965c, Borchs1966, BesheaTiHo1973]). **Neotropical**: Bahamas [McKenz1946]; Brazil [Borchs1966] (Rio Grande do Sul [Lepage1938], Rio de Janeiro [Lepage1938], Santa Catarina [Lepage1938]); Colombia [Kondo2001]; Cuba [Marlat1908c, Houser1918, McKenz1946, Borchs1966]; Jamaica [Borchs1966]; Panama [McKenz1946, Borchs1966]; Puerto Rico & Vieques Island (Puerto Rico [McKenz1946, Borchs1966, Martor1976, ColonFMe1998]); U.S. Virgin Islands [Nakaha1983].

Oriental: China (People's Republic) (Guangdong (Kwangtung) [Borchs1966]); India [Green1908a, Ramakr1919a, Ramakr1921a, Green1937, McKenz1937, McKenz1946] (Andhra Pradesh [Mamet1951, Borchs1966], Bihar [Ali1968], Karnataka [UsmanPu1955], Punjab [RahmanAn1941]); Philippines [McKenz1946, Borchs1966, VelasqRi1969] (Luzon [Velasq1971]); Ryukyu Islands (= Nansei Shoto) [KinjoNaHi1996]; Sri Lanka [Green1896, Green1896e, Green1900a, Green1905a, Ramakr1921a, Borchs1966]. **Palaearctic**: China (People's Republic) [McKenz1946]; Iran [Lindin1909b, Bodenh1944a, Bodenh1944b, Borchs1966, RosenDe1979]; Iraq [Bodenh1943, Borchs1966]; Israel [BenDovVe1983, BenDov1985]; Saudi Arabia [Beccar1971, Matile1984c].

BIOLOGY: Usually on foliage (Ferris, 1938a).

ECONOMIC IMPORTANCE: Highly polyphagous (see Host Plants) and widely distributed in tropics and subtropics (CABI, 1978b, and see Distribution). Recorded as a pest of several agricultural crops: Citrus (Rose, 1990c); Tea (Nagarkatti & Sankaran, 1990); date palm, India (Glover, 1933; Rajagopal & Krishnamoorty, 1996); date palms, Saudi Arabia (Hammad et al., 1981; Moussa, 1986); palms and ornamentals, Florida (Dekle, 1976); papaya, Queensland, Australia (Elder et al., 1998); mango, Israel (Ben-Dov & Wysoki, 1990; Swirski et al., 2002).

GENERAL: Description and illustration of adult female by McKenzie (1938), Ferris (1938a), Balachowsky (1948b, 1956), Velasquez (1971), Tang (1984), Chou (1985, 1986), Williams & Watson (1988), Colon-Ferrer & Medina-Gaud (1998).

KEYS: Danzig 1993: 160 (female) [Europe]; Williams & Watson 1988: 35 (female) [Tropical South Pacific]; Chou 1985: 292 (female) [Species of China]; Velasquez 1971: 118 (female) [Philippines]; Beardsley 1966: 509 (female) [Federated States of Micronesia]; Balachowsky 1956: 26 (female) [Africa]; Balachowsky 1948b: 360 (female) [Mediterranean]; McKenzie 1946: 33 (female) [World]; Ferris 1942: 29 (female) [North America]; McKenzie 1942b: 144-145 (female) [World]; McKenzie 1938: 17-18 (female) [World]; McKenzie 1937a: 178 (female) [World]; McKenzie 1937: 330 (female) [World]; Robinson 1917: 24 (female) [Philippine Islands]; Green 1896e: 40 (female) [Sri Lanka].

-CITATIONS: AbouEl2001 [host, distribution, biological control: 185-195]; Ali1968 [host, distribution: 132-133]; BadawiAl1990 [host, distribution, life history: 81-89]; BaghelDu2001 [taxonomy, host, distribution: 76-78]; Balach1948b [taxonomy, description, illustration, host, distribution: 362-365]; Balach1956 [taxonomy, description, illustration, host, distribution: 41-42]; Beards1966 [host, distribution: 511]; BeardsDaHo1976 [economic importance: 106]; Beccar1971 [host, distribution: 193]; BenDov1985 [host, distribution: 186-187]; BenDovVe1983 [host, distribution: 124-125]; BenDovWy1990 [economic importance, host, distribution: 1242]; BesheaTiHo1973 [host, distribution: 4]; BindraVa1972 [host, distribution, economic importance: 14-24]; Bodenh1943 [taxonomy, description, illustration, host, distribution: 3-4]; Bodenh1944a [host, distribution: 81]; Bodenh1944b [host, distribution: 94]; Borchs1966 [catalogue: 297-298]; Brimbl1962a [taxonomy, description, illustration, host, distribution: 410]; CABI1978b [taxonomy, distribution, host: 1-2]; Chou1985 [taxonomy, description, host, distribution: 296-298]; Chou1986 [taxonomy, illustration: 678]; ChuaWo1990 [host, distribution, economic importance: 453-552]; ClapsWoGo2001a [taxonomy, host, distribution: 12]; Cocker1896b [distribution: 334]; Cocker1897i [taxonomy, description, host, distribution: 28]; CockerRo1915 [taxonomy, description, host, distribution: 107]; ColonFMe1998 [taxonomy, description, illustration, host, distribution: 38-39]; Comper1961a [biological control: 17-71]; ComperAn1961 [host, distribution, biological control: 17]; CoronaRuMo1997 [host, distribution: 38-41]; DahmsSm1994 [host, distribution, biological control: 245-255]; DaneelMeJa1994 [host, distribution: 72-74]; Danzig1972 [taxonomy, host, distribution, economic importance: 207]; Danzig1972c [host, distribution: 583]; DanzigPe1998 [taxonomy, description, host, distribution: 183-184]; DavidsMi1990 [host, distribution, economic importance: 603-632]; DeBach1969a [biological control: 11-28]; DeBach1979 [host, distribution, biological control: 131-138]; DEDAC1923 [host, distribution]; Dekle1965c [taxonomy, description, host, distribution: 21]; Dekle1976 [taxonomy, description, host, distribution, economic importance: 32]; DeLott1967a [host, distribution: 112]; DonliBuAm1998 [host, distribution, life history, economic importance: 1-10]; DonliGoAl1998 [host, , life history, ecology: 39-44]; DonliSuAm1998 [host, distribution: 23-31]; DonliWaAm1998 [host, ecology, life history: 55-61]; Dutta1990 [host, distribution: 152-163]; DuttaBa1991 [taxonomy, description, host, distribution, economic importance: 31-35]; Ebelin1949 [host, distribution, life history, control]; ElderBe1998 [host, distribution, biological control: 362-365]; ElderGuSm1997 [host, distribution, biological control: 299-301]; ElderSm1995 [host, distribution, biological control: 253-254]; ElderSmBe1998 [host, distribution, biological control, economic importance: 74-79]; Elwan2000 [host, distribution: 653-664]; Fernal1903b [catalogue: 268]; Ferris1938a [taxonomy, description, illustration, host, distribution: 180]; Ferris1941e [taxonomy: 42,46,48]; Ferris1942 [taxonomy: 29]; Flande1971 [biological control, life history: 857-872]; Ghabbo1995 [taxonomy: 379-387]; Ghauri1962 [taxonomy, description, host, distribution: 82,211]; Glover1933 [economic importance, host, distribution, life history: 1-23]; Green1896 [taxonomy, host, distribution: 4]; Green1896e [taxonomy, description, illustration, host, distribution: 47-48]; Green1900a [taxonomy, host, distribution: 69];

Green1900c [host, distribution: 3]; Green1905a [taxonomy, description, illustration, host, distribution: 341-342,344]; Green1908a [host, distribution: 33]; Green1914c [host, distribution: 232]; Green1916e [host, distribution: 53]; Green1922 [taxonomy, host, distribution: 459,462-463]; Green1937 [host, distribution: 329-332]; Halber2000 [host, distribution: 3-4]; HammadKaRa1981 [host, distribution, economic importance: 252-268]; Hayat1989 [host, distribution, biological control: 1-3]; Houser1918 [host, distribution: 162]; Howard1991 [host, distribution, life history, ecology, control: 217-225]; Kapur1942 [biological control: 49-66]; Kaussa1955 [host, distribution: 15]; KhalafSo1993 [host, life history, biological control: 11-12]; KinjoNaHi1996 [host, distribution: 125-127]; Kondo2001 [taxonomy, host, distribution: 43]; KonstaGu1987 [host, distribution: 161]; Laing1929a [taxonomy, description, host, distribution: 490]; Lale1998 [host, distribution, economic importance, life history: 191-197]; Leonar1898a [taxonomy: 76]; Leonar1898c [taxonomy, description, illustration, host, distribution: 77-80]; Lepage1938 [catalogue: 396]; Lindin1908e [taxonomy: 241]; Lindin1909b [host, distribution: 108]; Lindin1909c [taxonomy, host, distribution: 449]; Lindin1913 [taxonomy, host, distribution: 73]; Lindin1943b [taxonomy: 146]; MacGil1921 [taxonomy, description, host, distribution: 408,419,444]; Maleno1916a [taxonomy, description, illustration, host, distribution: 326-329]; Mamet1951 [distribution: 225]; ManiKr1996 [host, distribution, biological control: 273-274]; Marlat1908c [taxonomy, description, illustration, host, distribution: 14-15]; Martor1976 [host, distribution: 61,108,177,200268]; Matile1976 [host, distribution: 311-312]; Matile1978 [host, distribution: 60]; Matile1984c [host, distribution: 220-221]; McKenz1937 [taxonomy: 327]; McKenz1938 [taxonomy, description, illustration, host, distribution: 12-13,32]; McKenz1939 [taxonomy, host, distribution: 54-55,65-66]; McKenz1942b [taxonomy: 145]; McKenz1946 [taxonomy, host, distribution: 32-33]; Merril1953 [taxonomy, description, host, distribution: 14-15]; MerrilCh1923 [taxonomy, description, host, distribution, economic importance: 204-205]; MillerDa1990 [host, distribution, economic importance: 300]; MohammGhTa2001 [host, distribution, biological control: 413-418]; Montgo1921 [host, distribution: 41-55]; Moussa1986 [host, distribution, biological control, life history, ecology: 227-234]; NagarkSa1990 [host, distribution, economic importance, biological control: 553-542]; Nair1964 [host, distribution, economic importance: 72]; NairMe1963 [host, distribution: 139-147]; Nakaha1982 [host, distribution: 8]; Nakaha1983 [host, distribution: 8]; Newste1894c [taxonomy, description, illustration, host, distribution: 26-27]; Newste1917b [host, distribution: 132]; OfekHuYz1997 [host, distribution, life history, biological control, economic importance: 212-218]; OfekHuYz1997 [host, distribution, chemical control, 101biological control: 212-218]; Ponnam1999 [host, distribution, chemical control: 445-451]; Prinsl1983 [distribution, biological control: 26]; PruthiBa1960 [host, distribution, economic importance: 1-113]; PruthiMa1945 [host, distribution, life history, control: 1-42]; RahmanAn1941 [taxonomy, description, host, distribution: 819-821]; RajagoKr1996 [host, distribution, economic importance, control: 139-146]; Ramakr1919a [taxonomy, description, host, distribution: 17]; Ramakr1921a [host, distribution: 357,359]; Ramakr1930 [taxonomy, host, distribution: 21]; Rao1969 [biological control: 785-792]; RaoCh1950 [taxonomy: 28]; Robins1917 [taxonomy, description, host, distribution: 24]; Rose1990c [distribution, economic importance: 535-542]; RosenDe1979 [host, distribution, biological control: 238-241,464-484]; Ruther1915a [taxonomy, description, host, distribution: 106-107]; Sander1906 [taxonomy, host, distribution: 15]; Sander1909a [taxonomy, host, distribution: 51]; Sankar1984 [host, distribution, biological control]; Sassce1923 [host, distribution: 152-158]; Schmut1969 [host, distribution: 116]; Schmut1990a [economic importance, host, distribution: 392,398]; Schmut1998 [economic importance, host, distribution: 37]; Schmut2001 [host, distribution: 339-345]; SchmutKlLu1957 [host, distribution, economic importance: 484]; Shalab1961 [host, distribution: 211-228]; Singh1964 [host, distribution, economic importance: 213]; SinhaDi1984 [host, distribution, biological control: 7-13]; Soares1942 [host, distribution, economic importance: 54-56]; Soares1945 [host, distribution, economic importance: 49-81]; SwirskWyIz2002 [taxonomy, host, distribution, life history, economic importance, biological control: 110-111]; Tang1984 [taxonomy, description, illustration, host, distribution: 39,41]; Tao1999 [taxonomy, host, distribution: 71]; Trjapi1989 [biological control: 296]; UsmanPu1955 [host, distribution: 48]; Velasq1971 [taxonomy, description, illustration, host, distribution: 119-120]; VelasqRi1969 [host, distribution: 195-208]; WadhiBa1964 [host, distribution, economic importance: 227]; WadhiBa1964 [host, distribution: 227-260]; Wester1918 [host, distribution, economic importance: 5-57]; Wester1920 [host, distribution]; Willia1985a [taxonomy: 237]; WilliaWa1988 [taxonomy, description, illustration, host, distribution: 42-44]; WysokiIsRo1995 [life history, biological control: 267].

Aonidiella pini Young

Aonidiella pini Young *in*: Young & Lu, 1988: 189. Type data: CHINA: Sichuan, Zi Gong, Xinglong Xiang, on *Pinus elliottii* or *Pinus taeda*; host plant of holotype not indicated; collected September 1988 by Dehui Lu. Holotype female. Type depository: Shanghai: Shanghai Institute of Entomology, China.

SCALE COVER: Female scale circular, flat, translucent, pale brownish; exuviae subcentral. Male scale elongate, pale brownish (Young & Lu, 1988).

HOST PLANTS: **Pinaceae**: *Pinus elliottii* [YoungLu1988], *Pinus taeda* [YoungLu1988].

DISTRIBUTION: **Oriental**: China (People's Republic) (Sichuan (Szechwan) [YoungLu1988]).

GENERAL: Description and illustration of adult female by Young & Lu (1988).

CITATIONS: Tao1999 [taxonomy, host, distribution: 71]; YoungLu1988 [taxonomy, description, illustration, host, distribution: 189-192].

Aonidiella pothi Rutherford

Aonidiella pothi Rutherford, 1914: 262. Type data: SRI LANKA: Paradeniya, on *Pothos scandens* and on *Loranthus* sp. Syntypes, female. Type depository: London: The Natural History Museum, England, UK.

Aonidiella pothi; MacGillivray, 1921: 445. Notes: Incorrect citation of "Green" as author.

Aspidiotus (*Aonidiella*) *pothi*; Ferris, 1941e: 47. Change of combination.

Aonidiella pothi; Borchsenius, 1966: 298. Revived combination.

SCALE COVER: Female scale very dark brown and shining; first exuvium with reddish tinge; whole scale slightly elongated, with exuviae towards end; when removed, a faint white ventral scale remains (Rutherford, 1914).
HOST PLANTS: **Araceae**: *Pothos scandens* [Ruther1914, Ramakr1921a, Green1922, Green1937]. **Loranthaceae**: *Loranthus* [Ruther1914, Green1922, Green1937].
DISTRIBUTION: **Oriental**: Sri Lanka [Ruther1914, Ramakr1921a, Green1922, Green1937].
GENERAL: Description of adult female by Rutherford (1914)
CITATIONS: Borchs1966 [catalogue: 298]; DEDAC1923 [host, distribution]; Ferris1941e [taxonomy: 47]; Green1922 [host, distribution: 462]; MacGil1921 [taxonomy, description, host, distribution: 445]; McKenz1938 [taxonomy: 4]; Ramakr1921a [host, distribution: 358]; Ruther1914 [taxonomy, description, host, distribution: 262].

Aonidiella replicata (Lindinger)
Aspidiotus replicatus Lindinger, 1909e: 17. Type data: CAMEROON: Bipinde, Urwaldgebiet, on *Ehretia cymosa*, Anacardiaceae, *Illigera pentaphylla*; Viktoria, on *Mitragyne macrophylla*. Syntypes, female. Type depository: Zoologisches Institut und Zoologishces Museum, Universität von Hamburg, Germany.
Heteraspis replicatus; Leonardi, 1914: 197. Change of combination.
Comstockaspis replicata; MacGillivray, 1921: 439. Change of combination.
Aonidiella replicata; McKenzie, 1938: 13. Change of combination.
SCALE COVER: Scale of female approximately 1.2 mm in diameter; very similar with that of *A. marginipora*; generally circular, pale brown, thin and with a series of puffy wax layers forming a somewhat convex mound. Scale of male elongate, light yellowish (McKenzie, 1946).
HOST PLANTS: **Anacardiaceae** [Lindin1909e, Balach1956]. **Ehretiaceae**: *Ehretia cymosa* [Lindin1909e, Leonar1914, Balach1956, MatileNo1984]. **Euphorbiaceae**: *Manihot glaziovii* [Balach1956]. **Flacourtiaceae**: *Hydnocarpus* [Mamet1951, Borchs1966]. **Hernandiaceae**: *Illigera pentaphylla* [Lindin1909e, Leonar1914, Balach1956]. **Lauraceae**: *Laurus camphora* [Mamet1954, Borchs1966]. **Leguminosae**: *Acrocarpus fraxinifolius* [DeLott1967a]. **Palmae**: *Neodypsis* [Mamet1951, Borchs1966]. **Rubiaceae**: *Mitragyna macrophylla* [Lindin1909e, Leonar1914, Balach1956]. **Ulmaceae**: *Chaetacme aristata* [DeLott1967a]. **Urticaceae**: *Myrianthus arboreus* [DeLott1967a].
DISTRIBUTION: **Afrotropical**: Cameroon [Lindin1909e, Balach1956, MatileNo1984]; Guinea [Balach1956]; Kenya [DeLott1967a]; Madagascar [Mamet1951, Mamet1954, Borchs1966]; South Africa [Leonar1914]; Tanzania [Balach1956]; Uganda [McKenz1946].
BIOLOGY: Occurring on leaves (McKenzie, 1946).
ECONOMIC IMPORTANCE: Recorded from cocoa in Ghana, but of no economic importance (Chua & Wood, 1990).
GENERAL: Description and illustration of adult female by Lindinger (1909e), McKenzie (1946) and by Balachowsky (1956).

KEYS: Balachowsky 1956: 25 (female) [Africa]; McKenzie 1946: 33 (female) [World]; McKenzie 1942b: 144-145 (female) [World]; McKenzie 1938: 17-18 (female) [World].

CITATIONS: Balach1956 [taxonomy, description, illustration, host, distribution: 43-46]; Borchs1966 [catalogue: 298]; ChuaWo1990 [host, distribution, economic importance: 549]; DeLott1967a [host, distribution: 113]; Ferris1937c [taxonomy: 51,53,55]; Ferris1938a [taxonomy: 178]; Ferris1941e [taxonomy: 47]; Leonar1914 [taxonomy, host, distribution: 197]; Lindin1909e [taxonomy, description, illustration, host, distribution: 17-19]; MacGil1921 [taxonomy, description, host, distribution: 439]; Mamet1951 [host, distribution: 225]; Mamet1954 [host, distribution: 15]; Mamet1959a [host, distribution: 386]; MatileNo1984 [host, distribution: 64]; McKenz1938 [taxonomy, description, host, distribution: 13]; McKenz1942b [taxonomy: 145]; McKenz1946 [taxonomy, description, illustration, host, distribution: 30-31,33,36]; Sassce1911 [taxonomy: 70]; Vayssi1913 [host, distribution: 430]; WeidneWa1968 [taxonomy: 173].

Aonidiella rex Balachowsky

Aonidiella rex Balachowsky, 1953b: 78. Nomen nudum.

Aonidiella rex Balachowsky, 1954d: 296. Type data: ZAIRE: Mont Hawa, near d'Aru, on a plant of Euphorbiaceae. Holotype female. Type depository: Tervuren: Musée Royal de l'Afrique Centrale, Section d'Entomologie, Belgium.

SCALE COVER: Female scale circular, flat, formed of three envelopes of white secretion; larval exuviae black, subcircular, central, white; second-instar exuviae bivalvate: dorsal, subcircular, black and ventral, flat, circular, attached to venter of adult female; diameter 1.8-2.1 mm. Male scale oval; with larval exuviae black, circular; nymphal secretion white; 1.2-1.3 mm long (Balachowsky, 1956).

HOST PLANTS: **Euphorbiaceae** [Balach1954d], *Euphorbia candelabrum* [Balach1956]. **Palmae**: *Elaeis guineensis* [Balach1956].

DISTRIBUTION: **Afrotropical**: Zaire [Balach1954d].

GENERAL: Description and illustration of adult female by Balachowsky (1954d, 1956).

KEYS: Balachowsky 1956: 25 (female) [Africa].

CITATIONS: Balach1953b [taxonomy: 78]; Balach1954d [taxonomy, description, illustration, host, distribution: 296-298]; Balach1956 [taxonomy, description, illustration, host, distribution: 45-48]; Borchs1966 [catalogue: 298].

Aonidiella schoutedeni Balachowsky

Aonidiella schoutedeni Balachowsky, 1954d: 298. Type data: ZAIRE: massif forestier central, Eala s/Ruki, on *Macrolobium dewevrei*. Holotype female. Type depository: Tervuren: Musée Royal de l'Afrique Centrale, Section d'Entomologie, Belgium.

SCALE COVER: Female scale subcircular or slightly oval, slightly convex; secretion of adult large, flat; colour red, shiny, translucent; a distinct, circular mark retained on leaves after removal of scale; diameter 1.4-1.6 mm. Many samples were collected but male scale not observed (Balachowsky, 1956).

HOST PLANTS: **Leguminosae**: *Macrolobium dewevrei* [Balach1954d].

DISTRIBUTION: **Afrotropical**: Zaire [Balach1954d].
GENERAL: Description and illustration of adult female by Balachowsky (1954d, 1956).
KEYS: Balachowsky 1956: 26 (female) [Africa].
CITATIONS: Balach1954d [taxonomy, description, illustration, host, distribution: 298-300]; Balach1956 [taxonomy, description, illustration, host, distribution: 47-48]; Borchs1966 [catalogue: 298].

Aonidiella simplex (Grandpré & Charmoy)
Aspidiotus aloes simplex Grandpré & Charmoy, 1899: 20. Type data: MAURITIUS: on *Furcraea gigantea*. Syntypes. Notes: Type-material lost (Mamet, 1941).
Aspidiotus hederae simplex; Fernald, 1903b: 264. Change of status.
Aspidiotus andersoni Laing, 1929a: 489. Type data: KENYA: on unknown host plant; collected by T.J. Anderson. Holotype female. Type depository: London: The Natural History Museum, England, UK. Synonymy by Borchsenius, 1966: 298.
Aonidiella andersoni; McKenzie, 1938: 6. Change of combination.
Hemiberlesia simplex; Mamet, 1941: 25. Change of combination.
Aonidiella simplex; Balachowsky, 1958b: 224. Change of combination.
SYSTEMATICS: Balachowsky (1956: 28) suggested that *Aspidiotus andersoni* Laing, 1929 was identical with *Aonidiella orientalis* (Newstead), differing from the latter only in the absence of perivulvar pores. Later, Balachowsky (1958b: 224) concluded, following a study of type material, that *Aspidiotus andersoni* was a synonym of *Aspidiotus aloes simplex* (Grandpre & Charmoy, 1899: 20). Balachowsky (1958b: 224) re-iterated that *Aonidiella simplex* (Grandpre & Charmoy, 1899) was a viviparous form of *Aonidiella orientalis* (Newstead), but interpreted each of them as a distinct species.
SCALE COVER: Scale of female of *Aspidiotus aloes simplex* convex, white greyish; exuviae central, brown-reddish or yellow (Charmoy, 1899). Scale of female of synonym *Aspidiotus andersoni* subcircular, greyish white; exuviae subcentral, very dark, almost a reddish brown; diameter 1.2 mm (Laing, 1929a).
HOST PLANTS: **Agavaceae**: *Agave amaniensis* [Mamet1949, Borchs1966], *Furcraea gigantea* [GrandpCh1899, Mamet1943a, Mamet1949, Borchs1966]. **Apocynaceae**: *Nerium oleander* [Laing1929, McKenz1938, Balach1956]. **Cleomaceae**: *Cleome viscosa* [Mamet1943a, Mamet1949, Borchs1966]. **Euphorbiaceae**: *Ricinus communis* [Mamet1941, Mamet1943a, Mamet1949, Borchs1966]. **Solanaceae**: *Solanum melongena* [Mamet1941, Mamet1943a, Mamet1949, Borchs1966].
DISTRIBUTION: **Afrotropical**: Kenya [Laing1929, Balach1956]; Mauritius [GrandpCh1899, Mamet1941, Mamet1943a, Mamet1949, Mamet1953a, Borchs1966]; Tanzania [Laing1929, Balach1956].
GENERAL: Description and illustration of adult female by Laing (1929a) (as *A. andersoni*, McKenzie (1938), Balachowsky (1956) (as *A. andersoni)* and by Tang (1984).
KEYS: Balachowsky 1956: 26 (female) [Africa]; McKenzie 1946: 33 (female) [World]; McKenzie 1942b: 145 (female) [World]; McKenzie 1938: 17-18 (female) [World].

CITATIONS: Balach1948b [taxonomy: 360]; Balach1953k [taxonomy: 114]; Balach1956 [taxonomy, description, illustration, host, distribution: 28,41]; Balach1958b [taxonomy, host, distribution: 224]; Borchs1966 [catalogue: 298-299]; Fernal1903b [catalogue: 264]; Ferris1941e [taxonomy: 40,48]; GrandpCh1899 [taxonomy, description, host, distribution: 20]; Green1907 [host, distribution: 203]; Laing1929 [taxonomy, description, illustration, host, distribution: 489-490]; Lindin1957 [taxonomy: 545]; MacGil1921 [taxonomy, description, host, distribution: 400]; Mamet1941 [taxonomy, description, illustration, host, distribution: 25-26]; Mamet1943a [catalogue: 161]; Mamet1949 [catalogue: 61]; Mamet1953a [taxonomy: 152]; McKenz1938 [taxonomy, description, illustration, host, distribution: 6,22]; McKenz1942b [taxonomy: 145]; McKenz1946 [taxonomy: 33]; MoutiaMa1947 [distribution]; Tang1984 [taxonomy, description, illustration, host, distribution: 42-43]; Tao1999 [taxonomy, host, distribution: 71]; UsmanPu1955 [host, distribution: 48].

Aonidiella sotetsu (Takahashi)
Chrysomphalus sotetsu Takahashi, 1933: 57. Type data: TAIWAN: Ishigaki, Miyako, on *Cycas revoluta* and *Ficus retusa*. Syntypes, female. Type depository: Taichung: Entomology Collection, Taiwan Agricultural Research Institute, Wu-feng, Taichung, Taiwan.
Aonidiella sotetsu; McKenzie, 1938: 13. Change of combination.
SCALE COVER: Scale of adult female pale brown, semi-transparent, thin, nearly circular, about 1.3 mm in diameter (Takahashi, 1933).
HOST PLANTS: **Cycadaceae**: *Cycas revoluta* [Takaha1933]. **Moraceae**: *Ficus retusa* [Takaha1933].
DISTRIBUTION: **Oriental**: Taiwan [Takaha1933]; Thailand [Takaha1942b]. **Palaearctic**: China (People's Republic) (Henan (Honan) [Shen1993]); Japan [Sakai1939, Kawai1980].
GENERAL: Description and illustration of adult female by Takahashi (1933), McKenzie (1938) and by Chou (1985, 1986).
KEYS: Kawai 1980: 211 (female) [Japan]; McKenzie 1946: 33 (female) [World]; McKenzie 1942b: 144-145 (female) [World]; McKenzie 1938: 17-18 (female) [World].
CITATIONS: Borchs1966 [catalogue: 299]; Chou1986 [taxonomy, illustration: 679]; DanzigPe1998 [catalogue: 184]; Kawai1980 [taxonomy, description, host, distribution: 211-212]; Lindin1957 [taxonomy: 545]; McKenz1938 [taxonomy, description, illustration, host, distribution: 13-14,33]; McKenz1939 [taxonomy: 55]; McKenz1942b [taxonomy: 145]; McKenz1946 [taxonomy: 33]; Muraka1970 [host, distribution: 71]; Sakai1939 [taxonomy, description, illustration, host, distribution: 45-62]; Shen1993 [host, distribution: 58]; Takaha1933 [taxonomy, description, illustration, host, distribution: 57-58]; Takaha1942b [host, distribution: 47]; Tao1999 [taxonomy, host, distribution: 72]; WuPaZh1998 [host, distribution, life history: 51-57]; YoungLu1988 [taxonomy: 190,191].

Aonidiella taorensis (Lindinger)

Aspidiotus taorensis Lindinger, 1911a: 17. Type data: CANARY ISLANDS: Gran Canaria, Barranco Guinguada, near Las Palmas, on *Euphorbia aphylla* and *E. regisjubae*; Baia del Confital, on *E. aphylla*; Tenerife, Valle de Taoro on *E. regis-juba* and *E. aphylla*. Syntypes, female. Type depository: Hamburg: Zoologisches Institut und Zoologisches Museum, Universität von Hamburg, Germany.

Hemiberlesia taorensis; MacGillivray, 1921: 436. Change of combination.

Aonidiella taorensis; McKenzie, 1938: 14. Change of combination.

SCALE COVER: Female scale circular, white; larval exuviae yellow, central. Male scale unknown.

HOST PLANTS: **Euphorbiaceae**: *Euphorbia aphylla* [Lindin1911a, Balach1948b, GomezM1962, MatileBa1972], *Euphorbia canariensis* [GomezM1962], *Euphorbia regis-jubae* [Lindin1911a, Balach1948b, GomezM1962, MatileBa1972].

DISTRIBUTION: **Palaearctic**: Canary Islands [Lindin1911a, Balach1948b, MatileBa1972, MatileOr2001].

BIOLOGY: This species was found to induce the formation of pit-like pseudo-galls on the leaves of Euphorbiaceae (Balachowsky, 1948b).

GENERAL: Description and illustration of adult female by Lindinger (1911a), Balachowsky (1948b) and by Gómez-Menor Guerrero (1962).

KEYS: Gómez-Menor Guerrero 1962: 187 (female) [Canary Islands]; Balachowsky 1948b: 361 (female) [Mediterranean]; McKenzie 1946: 33 (female) [World]; McKenzie 1942b: 144-145 (female) [World]; McKenzie 1938: 17-18 (female) [World].

CITATIONS: Balach1946 [host, distribution: 211]; Balach1948b [taxonomy, description, illustration, host, distribution: 376-379]; Borchs1966 [catalogue: 299]; DanzigPe1998 [catalogue: 184]; Ferris1941e [taxonomy: 48]; GomezM1962 [taxonomy, description, illustration, host, distribution: 191-196]; Houard1913 [host, distribution: 1]; Larew1990 [ecology, life history, structure: 293-300]; Lindin1911a [taxonomy, description, illustration, host, distribution: 17-18]; Lindin1912b [taxonomy, description, host, distribution: 148-149]; MacGil1921 [taxonomy, description, host, distribution: 436]; MatileBa1972 [host, distribution: 112-113]; MatileOr2001 [host, distribution: 189]; McKenz1938 [taxonomy, description, illustration, host, distribution: 14]; McKenz1942b [taxonomy: 145]; McKenz1946 [taxonomy: 33]; Sassce1912 [taxonomy, host, distribution: 94]; WeidneWa1968 [taxonomy: 173].

Aonidiella taxus Leonardi

Aonidiella aurantii; Berlese & Leonardi, 1895: 23. Misidentification.

Aonidiella taxus Leonardi, 1906: 1. Type data: ITALY: Portici, on *Taxus baccata*. Syntypes, female. Type depository: Portici: Dipartimento de Entomologia e Zoologia Agraria di Portici, Università di Napoli Federico II, Italy.

Chrysomphalus taxus; Sanders, 1909a: 55. Change of combination.

Aonidiella taxa; MacGillivray, 1921: 443. Change of combination requiring emendation of species name for agreement in gender.

Aspidiotus britannicus; Balachowsky, 1928a: 138. Misidentification; discovered by Borchsenius, 1966: 299.

Aspidiotus taxus; Lindinger, 1935: 132. Revived combination.

Aspidiotus taxus; Lindinger, 1935: 132. Change of combination requiring emendation of species name for agreement in gender.

Aonidiella taxus; McKenzie, 1938: 15. Revived combination.

COMMON NAME: Asiatic red scale [McKenz1956].

SCALE COVER: Scale of female flat, circular, thin and translucent, with red-brown body showing through it. Diameter of body of adult female much smaller than diameter of scale, and a conspicuous, pale, marginal zone thus appears. Scale of male pale grey or slightly yellowish; exuvia towards one end (Ferris, 1942).

HOST PLANTS: **Cephalotaxaceae**: *Cephalotaxus drupacea* [TakahaTa1956], *Cephalotaxus fortunei* [Hadzib1983]. **Podocarpaceae**: *Podocarpus* [Takaha1955f, Takagi1970, BesheaTiHo1973], *Podocarpus andina* [McKenz1937, Ferris1942, McKenz1956], *Podocarpus elongata* [Balach1948b], *Podocarpus halli* [Hadzib1983], *Podocarpus macrophyllus* [McKenz1946, McKenz1956, TakahaTa1956, Dekle1965c, Takagi1970, RosenDe1979], *Podocarpus nageia* [Ferris1942, McKenz1956], *Podocarpus neriifolia* [Ferris1942, Balach1948b, McKenz1956], *Podocarpus sinensis* [McKenz1937, Ferris1942, McKenz1956]. **Taxaceae**: *Taxus* [Takagi1970], *Taxus adpressa* [McKenz1946], *Taxus baccata* [Leonar1906, Sander1909a, Leonar1920, Balach1927, McKenz1937, Ferris1942, Balach1948b, Tang1984].

NATURAL ENEMIES: HYMENOPTERA **Aphelinidae**: *Aphytis japonicus* DeBach & Azim [RosenDe1979], *Ap. vandenboschi* DeBach & Rosen [RosenDe1979], *Ap. yasumatsui* Azim [RosenDe1979], *Marietta carnesi* (Howard) [Viggia1990b]. **Encyrtidae**: *Comperiella bifasciata* Howard [Flande1944a].

DISTRIBUTION: **Nearctic**: United States of America (Alabama [Nakaha1982], California [McKenz1946, Ferris1942, McKenz1956], Florida [Dekle1965c], Georgia [BesheaTiHo1973], Louisiana [Ferris1942]). **Neotropical**: Argentina [Ferris1942] (Buenos Aires [McKenz1937]); Brazil [Nakaha1982]. **Oriental**: Philippines [Velasq1971]; Taiwan [Takagi1970]. **Palaearctic**: Algeria [Balach1927]; China (People's Republic) [Tang1984] (Henan (Honan) [Shen1993]); Georgia [Hadzib1983]; Italy [Leonar1906, Leonar1920, McKenz1937, Ferris1942, LongoMaPe1995]; Japan [McKenz1937, Ferris1942, Takaha1955f, RosenDe1979, Kawai1980] (Honshu [TakahaTa1956], Kyushu [TakahaTa1956], Shikoku [TakahaTa1956]); Spain [BlayGo1993].

BIOLOGY: Occurring on leaves (Ferris, 1942).

ECONOMIC IMPORTANCE: Distributed in southern USA, South America, some Mediterranean countries, Japan and China, damaging *Taxus* and *Podocarpus* trees (Schmutterer, 1957; Miller & Davidson, 1990).

GENERAL: Description and illustration of adult female by Ferris (1942), McKenzie (1938, 1956), Velasquez (1971), Tang (1984), Chou (1985, 1986) and by Gill (1997).

KEYS: Gill 1997: 44 (female) [Species of California]; Danzig 1993: 159 (female) [Europe]; Tereznikova 1986: 85 (female) [Ukraine]; Chou 1985: 292 (female) [Species of China]; Kawai 1980: 211 (female) [Japan]; McKenzie 1956: 23 (female) [U.S.A.: California]; Lupo 1954a: 41 (female) [Italy]; McKenzie 1953: 36 (female) [World]; Balachowsky 1948b: 361 (female) [Mediterranean]; McKenzie 1946: 33

(female) [World]; Ferris 1942: 29 (female) [North America]; McKenzie 1942b: 144-145 (female) [World]; McKenzie 1938: 17-18 (female) [World]; McKenzie 1937a: 178 (female) [World]; McKenzie 1937: 330 (female) [World]; Lupo 1936: 261 (female) [World]; Leonardi 1920: 75 (female) [Italy].

CITATIONS: Balach1927 [host, distribution: 178]; Balach1928a [taxonomy, host, distribution: 138,142]; Balach1948b [taxonomy, description, illustration, host, distribution: 370-373]; BeardsDaHo1976 [economic importance: 103]; BesheaTiHo1973 [host, distribution: 4]; BlayGo1993 [taxonomy, description, illustration, host, distribution: 523-526]; Borchs1939 [taxonomy, description, host, distribution: 8,21]; Borchs1950b [taxonomy, description, illustration, host, distribution: 212,221]; Borchs1966 [catalogue: 299]; Bustsh1958 [taxonomy, description, host, distribution: 220,248]; Caltag1985 [taxonomy, biological control: 189-200]; Chou1985 [taxonomy, description, host, distribution: 296]; Chou1986 [taxonomy, illustration: 680]; ClapsWoGo2001a [taxonomy, host, distribution: 12]; ComperFlSm1941 [biological control: 291,301]; CoronaRuMo1997 [host, distribution: 38-41]; Danzig1993 [taxonomy, description, illustration, host, distribution, economic importance: 162-163]; DanzigPe1998 [catalogue: 184]; DavidsMi1990 [host, distribution, economic importance: 603-632]; Dekle1957 [host, distribution: 71-72]; Dekle1965c [taxonomy, description, host, distribution: 22]; Dekle1976 [taxonomy, description, host, distribution, economic importance: 33]; Ebelin1949 [host, distribution, life history, control]; Ferris1941e [taxonomy: 48]; Ferris1942 [taxonomy, description, illustration, host, distribution: 425,446:29]; Flande1944a [biological control: 365-371]; Foldi2001 [distribution: 303-308]; Gill1997 [host, distribution, taxonomy, illustration: 50-51]; Hadzib1983 [taxonomy, description, host, distribution, life history, biological control, economic importance: 227-229]; HoyHe1985 [biological control]; HoyHe1985 [biological control]; Kawai1980 [taxonomy, description, host, distribution: 211]; Leonar1906 [taxonomy, description, illustration, host, distribution: 1-5]; Leonar1907a [taxonomy, description, host, distribution: 131]; Leonar1920 [taxonomy, description, illustration, host, distribution: 81-83]; Lindin1912b [taxonomy: 358]; Lindin1935 [taxonomy: 132]; LongoMaPe1995 [distribution: 125]; Lupo1936 [taxonomy, description, illustration, host, distribution: 257-260]; Lupo1954a [taxonomy, description, illustration, host, distribution: 49-55]; MacGil1921 [taxonomy, description, host, distribution: 443]; McKenz1937 [taxonomy, description, illustration, host, distribution: 327-328,331]; McKenz1938 [taxonomy, description, illustration, host, distribution: 15,34]; McKenz1939 [taxonomy: 55]; McKenz1942b [taxonomy: 145]; McKenz1946 [taxonomy, host, distribution : 32-33]; McKenz1953 [taxonomy: 36]; McKenz1956 [taxonomy, description, illustration, host, distribution: 42-43]; Mead1983 [host, distribution: 1-5]; MillerDa1990 [host, distribution, economic importance: 300]; Muraka1970 [host, distribution: 71]; Nakaha1982 [host, distribution: 8]; QinTaFe1997 [host, distribution, life history: 109-113]; Robiso1990 [structure, anatomy: 213]; RosenDe1979 [host, distribution, biological control: 400-402,558-562,]; Ryan1946 [host, distribution: 124-125]; Sakai1939 [taxonomy, description, illustration, host, distribution: 45-62]; Sander1909a [taxonomy, host, distribution: 55]; SchmutKlLu1957 [host, distribution, economic importance: 484]; Shen1993 [host, distribution: 58];

StoetzDa1974 [taxonomy, life history: 138-140]; Takagi1970 [taxonomy, host, distribution: 131]; Takagi1990b [taxonomy, structure: 7,17]; TakagiRo1981 [host, distribution, biological control: 314-321]; Takaha1955f [host, distribution: 242]; TakahaTa1956 [host, distribution: 15]; Tang1984 [taxonomy, description, illustration, host, distribution: 37-38]; Tao1999 [taxonomy, host, distribution: 72]; Terezn1986 [taxonomy, description, illustration, host, distribution: 88-89]; Uemats1972 [host, distribution, life history, biological control: 187-192]; Uemats1974 [host, distribution, life history, biological control: 177-182]; Uemats1978 [host, distribution, life history, ecology: 25-31]; Uemats1979 [host, distribution, life history, ecology, biological control: 79-86]; Velasq1971 [taxonomy, description, illustration, host, distribution: 127-128]; Viggia1990b [biological control: 180]; Zahrad1990 [host, distribution, description: 642].

Aonidiella tectaria (Lindinger)
Aspidiotus tectarius Lindinger, 1909e: 20. Type data: CAMEROON: Bipinde, Urwaldgebiet, on *Euphorbia* sp. Syntypes, female. Type depository: Zoologisches Institut und Zoologishces Museum, Universität von Hamburg, Germany.
Neosignoretia tectaria; MacGillivray, 1921: 425. Change of combination requiring emendation of species name for agreement in gender.
Aonidiella tectaria; McKenzie, 1938: 16. Change of combination.
SCALE COVER: Lindinger (1909e) did not describe scale cover.
HOST PLANTS: **Euphorbiaceae**: *Euphorbia* [Lindin1909e, Balach1956].
DISTRIBUTION: **Afrotropical**: Cameroon [Lindin1909e, Balach1956].
GENERAL: Description and illustration of adult female by Lindinger (1909e) and by Balachowsky (1956).
KEYS: Balachowsky 1956: 25 (female) [Africa]; McKenzie 1946: 33 (female) [World]; McKenzie 1942b: 145 (female) [World]; McKenzie 1938: 17-18 (female) [World].
CITATIONS: Balach1956 [taxonomy, description, illustration, host, distribution: 25,43,49]; Borchs1966 [catalogue: 299]; Ferris1941e [taxonomy: 48]; Hall1946a [taxonomy: 536]; Lindin1909e [taxonomy, description, illustration, host, distribution: 20]; MacGil1921 [taxonomy, description, host, distribution: 425]; McKenz1938 [taxonomy, description, host, distribution: 16]; McKenz1942b [taxonomy: 145]; McKenz1946 [taxonomy: 33]; Sassce1912 [taxonomy, host, distribution: 94]; Vayssi1913 [host, distribution: 430]; WeidneWa1968 [taxonomy: 173].

Aonidiella tinerfensis (Lindinger)
Aspidiotus tinerfensis Lindinger, 1911a: 18. Type data: CANARY ISLANDS: Tenerife, Valle de Taoro, on *Dracaena draco*. Syntypes, female. Type depository: Hamburg: Zoologisches Institut und Zoologishces Museum, Universität von Hamburg, Germany.
Hemiberlesia tinerfensis; MacGillivray, 1921: 437. Change of combination.
Aonidiella tinerfensis; McKenzie, 1938: 16. Change of combination.
SCALE COVER: Female scale circular or subcircular, slightly convex; scale of fresh specimens entirely covered with white secretion; larval exuviae brown red;

larval exuviae central or subcentral, 2-2.2 mm diameter. Male scale oval, elongate, white; larval exuviae brown-red, central; 1.3-1.5 mm long (Balachowsky (1948b).
HOST PLANTS: **Liliaceae**: *Dracaena* [Lindin1912b, Balach1948b, Fernan1992], *Dracaena draco* [Lindin1911a, Balach1946, GomezM1962, GomezM1967O].
DISTRIBUTION: **Palaearctic**: Canary Islands [Lindin1911a, Balach1948b, GomezM1962, GomezM1967O, MatileOr2001]; Portugal [Fernan1992].
GENERAL: Description and illustration of adult female by Lindinger (1911a), McKenzie (1938), Balachowsky (1948b) and by Gómez-Menor Guerrero (1962).
KEYS: Gómez-Menor Guerrero 1962: 187 (female) [Canary Islands]; Balachowsky 1948b: 361 (female) [Mediterranean]; McKenzie 1946: 33 (female) [World]; McKenzie 1942b: 144-145 (female) [World]; McKenzie 1938: 17-18 (female) [World].
CITATIONS: Balach1946 [host, distribution: 212]; Balach1948b [taxonomy, description, illustration, host, distribution: 379-382]; Borchs1966 [catalogue: 299]; Bustsh1958 [taxonomy, description, host, distribution: 220,242]; DanzigPe1998 [catalogue: 184-185]; Fernan1992 [host, distribution: 60]; Ferris1941e [taxonomy: 49]; Garcia1930 [host, distribution, biological control]; GomezM1962 [taxonomy, description, illustration, host, distribution: 187-191]; GomezM1967O [host, distribution: 132]; Lindin1911a [taxonomy, description, illustration, host, distribution: 18-20]; Lindin1912b [taxonomy, description, host, distribution: 136]; MacGil1921 [taxonomy, description, host, distribution: 437]; MatileOr2001 [host, distribution: 189]; McKenz1938 [taxonomy, description, illustration, host, distribution: 16,35]; McKenz1942b [taxonomy: 145]; McKenz1946 [taxonomy: 33]; Sassce1912 [taxonomy, host, distribution: 94]; Seabra1942 [distribution: 2]; WeidneWa1968 [taxonomy: 173].

Aonidiella tsugae Takagi
Aonidiella tsugae Takagi, 1969a: 84. Type data: TAIWAN: Tung-pu, on the leaf of *Tsuga chinensis* var. *formosana*. Holotype female. Type depository: Sapporo: Entomological Institute, Faculty of Agriculture, Hokkaido University, Japan.
Aomidiella tsugae; Chou, 1985: 401. Misspelling of genus name.
Aonidiella tsagae; Chou, 1985: 401. Misspelling of species name.
SCALE COVER: Female scale dark brown, nipple-like in shape (Takagi, 1969a).
HOST PLANTS: **Pinaceae**: *Tsuga chinensis formosana* [Takagi1969a].
DISTRIBUTION: **Oriental**: Taiwan [Takagi1969a].
GENERAL: Description and illustration of adult female by Takagi (1969a) and by Chou (1985, 1986).
KEYS: Kosztarab 1996: 543 (female) [Northeastern North America].
CITATIONS: BeardsDaHo1976 [economic importance: 106]; Chou1985 [taxonomy, description, host, distribution: 401-402]; Chou1986 [taxonomy, illustration: 683]; Takagi1969a [taxonomy, description, illustration, host, distribution: 84-85]; Tao1999 [taxonomy, host, distribution: 72].

Aspidaspis Ferris

Aspidaspis Ferris, 1938: 45. Nomen nudum.

Aspidaspis Ferris, 1938a: 181. Type species: *Aspidiotus densiflorae* Bremner, by original designation.

SYSTEMATICS: Ferris (1938a) introduced *Aspidaspis* as an attempt to create some satisfactory groupings of the numerous species assigned to *Aspidiotus*. It is very close to *Diaspidiotus* and *Abgrallaspis*, from which it differs in the absence of paraphyses or intersegmental scleroses on the pygidium margin.

GENERAL: Definition and characters by Ferris (1938a) and by Balachowsky (1950b, 1958b).

KEYS: Gill 1997: 24-26 (female) [Genera of California]; Gill 1997: 58-59 (female) [Species of California]; Ezzat & Afifi 1966: 371-372 (female) [Egypt]; Balachowsky 1958b: 228 (female) [*Aspidiotina* of Africa]; Ezzat 1958: 237-239 (female) [Egypt]; McKenzie 1956: 23 (female) [U.S.A.: California]; Balachowsky 1951: 599 (female) [Mediterranean]; Ferris 1942: 28 (female) [North America]; Ferris 1942: 29 (female) [species North America].

CITATIONS: Balach1948b [taxonomy: 307]; Balach1950b [taxonomy, description: 534-536]; Balach1951 [taxonomy: 599]; Balach1958b [taxonomy, description: 156-158,228]; Borchs1966 [catalogue: 302]; ColonFMe1998 [taxonomy, description: 39]; DanzigPe1998 [catalogue: 185]; Ezzat1958 [taxonomy: 238]; Ferris1938 [taxonomy: 45]; Ferris1938a [taxonomy, description: 181]; Ferris1938b [taxonomy, description: 65-66]; Ferris1942 [taxonomy: 446:28]; Gill1997 [taxonomy: 58]; Kozar1990f [distribution: 142]; McKenz1939 [taxonomy: 53]; McKenz1956 [taxonomy: 23]; MorrisMo1966 [taxonomy, catalogue: 16].

Aspidaspis arctostaphyli (Cockerell & Robbins)

Aspidiotus arctostaphyli Cockerell & Robbins, 1909a: 104. Type data: U.S.A.: California, Tehama Co., Red Bluff, on *Arctostaphylos viscida*. Syntypes, female. Type depository: Washington: United States National Entomological Collection, U.S. National Museum of Natural History, District of Columbia, USA.

Aspidiella arctostaphyli; MacGillivray, 1921: 404. Change of combination.

Aspidaspis arctostaphyli; Ferris, 1938a: 182. Change of combination.

Aspidaspis arctoshaphyli; Borchsenius, 1966: 383. Misspelling of species name.

COMMON NAME: Arctostaphylis scale [McKenz1956].

SCALE COVER: Scale of female quite flat, circular, rather brownish, darker centrally; exuviae central or subcentral; scale of male elongate oval, quite dark (Ferris, 1938a). Colour photograph by Gill (1997).

HOST PLANTS: **Ericaceae**: *Arbutus menziesii* [Ferris1938a, McKenz1956], *Arctostaphylos* [Ferris1920b], *Arctostaphylos glauca* [Ferris1938a, McKenz1956], *Arctostaphylos manzanita* [Ferris1938a, McKenz1956], *Arctostaphylos viscida* [CockerRo1909aWW, Ferris1938a, McKenz1956]. **Gramineae**: *Stenotaphrum secundatum* [Martor1976]. **Lauraceae**: *Umbellularia californica* [Ferris1938a].

DISTRIBUTION: **Nearctic**: United States of America (California [Ferris1920b, McKenz1956], Oregon [Ferris1938a]). **Neotropical**: Puerto Rico & Vieques Island (Puerto Rico [Martor1976, ColonFMe1998]).

BIOLOGY: Occurring either on twigs or leaves, apparently more commonly on the latter (Ferris, 1938a).

GENERAL: Description and illustration of adult female by Cockerell & Robbins (1909a), Ferris (1920b, 1938a), McKenzie (1956), Gill (1997) and by Colon-Ferrer & Medina-Gaud (1998).

KEYS: Gill 1997: 58-59 (female) [Species of California]; McKenzie 1956: 24 (female) [U.S.A.: California]; Ferris 1942: 30 (female) [North America].

CITATIONS: Borchs1966 [catalogue: 302-303,383]; CockerRo1909aWW [taxonomy, description, host, distribution: 104-105]; ColonFMe1998 [taxonomy, description, illustration, host, distribution: 40]; Ferris1920b [taxonomy, description, illustration, host, distribution: 48-49]; Ferris1938a [taxonomy, description, illustration, host, distribution: 182]; Ferris1941e [taxonomy : 41]; Ferris1942 [taxonomy, host, distribution: 445:9;446:30]; Gill1997 [host, distribution, taxonomy, description, illustration, economic importance: 59-60]; Lindin1957 [taxonomy: 545]; Martor1976 [host, distribution: 250]; MacGil1921 [taxonomy, description, host, distribution: 404]; McKenz1956 [taxonomy, description, illustration, host, distribution: 43-45]; Nakaha1982 [host, distribution: 11]; Sassce1911 [taxonomy: 69]; SchuhMo1948 [host, distribution, control]; Willia1985a [taxonomy: 232].

Aspidaspis densiflorae (Bremner)

Aspidiotus densiflorae Bremner, 1907: 366. Type data: USA: California, Mendocino County, on *Quercus densiflora*. Syntypes, female. Type depository: Davis: The Bohart Museum of Entomology, University of California, California, USA.

Aspidiella densiflorae; MacGillivray, 1921: 405. Change of combination.

Aspidaspis densiflorae; Ferris, 1938a: 183. Change of combination.

COMMON NAME: tan oak scale [McKenz1956].

SCALE COVER: Female scale snow-white in colour, varying in form from round to sub-oval, according to position on leaf; slightly convex; exuviae situated a little to one side of centre; first-larval skin light-yellow, second nearly white; length 1.5-2 mm. Male scale much smaller than female (1 mm), snow white in colour and oval in form (Bremner, 1907).

HOST PLANTS: **Anacardiaceae**: *Rhus integrifolia* [McKenz1956]. **Berberidaceae**: *Mahonia aquifolium* [McKenz1956]. **Fagaceae**: *Lithocarpus densiflorus* [McKenz1956], *Quercus agrifolia* [Ferris1938a, McKenz1956], *Qu. chrysolepis* [Ferris1920b, McKenz1956], *Qu. densiflora* [Bremne1907], *Qu. tomentella* [Ferris1921, McKenz1956], *Qu. wislizeni* [McKenz1956].

NATURAL ENEMIES: HYMENOPTERA **Aphelinidae**: *Aphytis aonidiae* (Mercet) [RosenDe1979].

DISTRIBUTION: **Nearctic**: Mexico (Baja California [Ferris1921]); United States of America (California [Bremne1907, McKenz1956]).

BIOLOGY: Confined to leaves (Ferris, 1938a).

GENERAL: Description and illustration of adult female by Bremner (1907), Ferris (1920b, 1938a), McKenzie (1956) and by Gill (1997).

KEYS: Gill 1997: 58-59 (female) [Species of California]; McKenzie 1956: 24 (female) [U.S.A.: California]; Ferris 1942: 30 (female) [North America].

CITATIONS: Borchs1966 [catalogue: 303]; Bremne1907 [taxonomy, description, illustration, host, distribution: 366-367]; CSCSH1914 [host, distribution: 189]; Essig1928 [host, distribution: 76-78]; Ferris1920b [taxonomy, description, illustration, host, distribution: 50]; Ferris1921 [host, distribution: 126]; Ferris1938a [taxonomy, description, illustration, host, distribution: 183]; Ferris1938b [illustration: 66]; Ferris1941e [taxonomy: 42]; Ferris1942 [taxonomy: 446:30]; Gill1997 [host, distribution, taxonomy, description, illustration, economic importance: 59,62]; Lindin1932f [taxonomy: 196]; Lobdel1937 [taxonomy: 78]; MacGil1921 [taxonomy, description, host, distribution: 405]; McKenz1956 [taxonomy, description, illustration, host, distribution: 45-46]; Nakaha1982 [host, distribution: 11]; RosenDe1979 [taxonomy, description, biological control: 476-483]; Sander1909a [taxonomy, host, distribution: 52].

Aspidaspis dentiloba Kaussari & Balachowsky

Aspidaspis dentilobus Kaussari & Balachowsky, 1953a: 99. Type data: IRAN: Djahrom-Fars, on *Atraphaxis spinosa*. Syntypes, female. Type depository: Paris: Muséum national d'Histoire naturelle, France.
Aspidaspis dentiloba; Borchsenius, 1966: 303. Emendation of species name for agreement in gender.
SCALE COVER: Female scale irregularly circular; slightly convex; white; exuviae yellow, central; diameter 1.5-1.8 mm (Kaussari & Balachowsky, 1953a).
HOST PLANTS: **Polygonaceae**: *Atraphaxis spinosa* [KaussaBa1953a].
DISTRIBUTION: **Palaearctic**: Iran [KaussaBa1953a].
GENERAL: Description and illustration of adult female by Kaussari & Balachowsky (1953a).
CITATIONS: Borchs1966 [catalogue: 303]; DanzigPe1998 [catalogue: 185]; Ferris1941e [taxonomy: 42]; Kaussa1955 [host, distribution: 16]; KaussaBa1953a [taxonomy, description, illustration, host, distribution: 99-102].

Aspidaspis florenciae (Coleman)

Aspidiotus florenciae Coleman, 1903: 66. Type data: U.S.A.: California, Pine Ridge, on *Pinus ponderosa*; collected from herbarium specimens. Syntypes, female. Type depository: Davis: The Bohart Museum of Entomology, University of California, California, USA.
Aspidaspis florenciae; Ferris, 1938a: 184. Change of combination.
COMMON NAME: florence scale [McKenz1956].
SCALE COVER: Female scale of rectangular shape with rounded corners, nearly semi-cylindrical, about 3 mm long, 1 mm wide; colour light slaty blue, paler at the margin; exuviae bright red, usually situated near one end, but sometimes middle (Coleman, 1903). Scale of female somewhat elongate, colour variable, some specimens being almost white, others with dark center and pale margins, exuviae subcentral. Scale of male not observed (Ferris, 1938a).
HOST PLANTS: **Pinaceae**: *Pinus coulteri* [Ferris1942, McKenz1956], *Pinus ponderosa* [Colema1903, Sander1906, McKenz1956].
DISTRIBUTION: **Nearctic**: United States of America (Arizona [Nakaha1982], California [Colema1903, Sander1906, McKenz1956]).

BIOLOGY: Occurring on needles (Ferris, 1938a).
GENERAL: Description and illustration of adult female by Ferris (1938a) and by McKenzie (1956).
KEYS: Gill 1997: 58-59 (female) [Species of California]; McKenzie 1956: 24 (female) [U.S.A.: California]; Ferris 1942: 30 (female) [North America].
CITATIONS: Borchs1966 [catalogue: 303]; Colema1903 [taxonomy, description, host, distribution: 66]; Ferris1938a [taxonomy, description, illustration, host, distribution: 185]; Ferris1941e [taxonomy: 43]; Ferris1942 [taxonomy: 30]; Gill1997 [host, distribution, taxonomy, description, illustration, economic importance: 63,64]; MacGil1921 [taxonomy, description, host, distribution: 399]; McKenz1956 [taxonomy, description, illustration, host, distribution: 45-46]; Nakaha1982 [host, distribution: 12]; Sander1906 [taxonomy, host, distribution: 13].

Aspidaspis gainesi McDaniel
Aspidaspis gainesi McDaniel, 1968: 215. Type data: U.S.A.: Texas, Cameron Co., southern tip of Padre Island, on *Antirrhinum* sp. Holotype female. Type depository: Washington: United States National Entomological Collection, U.S. National Museum of Natural History, District of Columbia, USA.
SCALE COVER: McDaniel (1968) did not describe scale cover.
HOST PLANTS: **Scrophulariaceae**: *Antirrhinum* [McDani1968].
DISTRIBUTION: **Nearctic**: United States of America (Texas [McDani1968]).
GENERAL: Description and illustration of adult female by McDaniel (1968).
CITATIONS: McDani1968 [taxonomy, description, illustration, host, distribution: 215-216]; Nakaha1982 [host, distribution: 12].

Aspidaspis longiloba (Hall)
Targionia longiloba Hall, 1923: 28. Type data: EGYPT: Upper Egypt, Armant, on *Tamarix* sp. Syntypes, female. Type depository: London: The Natural History Museum, England, UK.
Aspidaspis longiloba; Balachowsky, 1950b: 542. Change of combination.
SCALE COVER: Scale of female small, diameter 0.8 mm, circular, highly convex, dusky white to dark brown in colour. Pellicles brownish, thinly coated with secretionary matter. Ventral scale well developed. Male scale elongate, ovate, only slightly smaller than female scale, white, with yellow pellicle (Hall, 1923).
HOST PLANTS: **Polygonaceae**: *Atraphaxis spinosa* [Balach1958a]. **Tamaricaceae**: *Tamarix* [Hall1923, Balach1950b, Balach1958b, EzzatAf1966].
DISTRIBUTION: **Palaearctic**: Egypt [Hall1923, Balach1950b, Balach1958b, Ezzat1958].
GENERAL: Description and illustration of adult female by Balachowsky (1950b, 1958b).
KEYS: Ezzat 1958: 240 (female) [Egypt]; Balachowsky 1950b: 536 (female) [Mediterranean].
CITATIONS: Balach1950b [taxonomy, description, illustration, host, distribution: 542-544]; Balach1958a [host, distribution: 35]; Balach1958b [taxonomy, description, illustration, host, distribution: 157-158]; Borchs1966 [catalogue: 303]; DanzigPe1998 [catalogue: 185]; Ezzat1958 [distribution: 240]; EzzatAf1966

[taxonomy, description, illustration, host, distribution: 372-374]; EzzatNa1987 [distribution: 86]; Ferris1943a [taxonomy: 86]; Hall1923 [taxonomy, description, illustration, host, distribution: 28-29]; KaussaBa1953a [taxonomy: 102]; Lindin1957 [taxonomy: 545].

Aspidaspis tubulifera (Balachowsky)

Aspidiotus (*Evaspidiotus*) *tubuliferus* Balachowsky, 1933b: 245. Type data: MOROCCO: Moyen Atlas, Midelt, on *Vella glabrescens*; collected by Maire. Syntypes, female. Type depository: Paris: Muséum national d'Histoire naturelle, France.

Aspidiotus tubulifera; Balachowsky, 1950b: 540. Change of combination an emendation of species name for agreement in gender.

SCALE COVER: Female scale circular or slightly subcircular, convex, flat; colour varies from bright grey to rose; in fresh specimens covered with white powdery wax; exuviae brown; secreted part rough, with fine concentric striation; ventral vellum thick, attached to plant; diameter 1.4-1.8 mm. Male scale similar to that of female, but elongate; 1.1 mm long (Balachowsky, 1933b).

HOST PLANTS: **Cruciferae**: *Vella glabrescens* [Balach1933b].

DISTRIBUTION: **Palaearctic**: Morocco [Balach1933b].

GENERAL: Description and illustration of adult female by Balachowsky (1933b, 1950b).

KEYS: Balachowsky 1950b: 536 (female) [Mediterranean].

CITATIONS: Balach1933b [taxonomy, description, illustration, host, distribution: 245-248]; Balach1950b [taxonomy, description, illustration, host, distribution: 540-542]; Borchs1966 [catalogue: 303]; DanzigPe1998 [catalogue: 185]; Ferris1941e [taxonomy: 49]; Rungs1936 [taxonomy: 53,55].

Aspidiella Leonardi

Aspidiotus (*Aspidiella*) Leonardi, 1898a: 50, 60. Type species: *Aspidiotus sacchari* Cockerell, by original designation.

Aspidiella; MacGillivray, 1921: 387. Change of status.

Aspidiclla; Chou, 1985: 254. Misspelling of genus name.

SYSTEMATICS: This genus is closely related to *Aspidiotus* and its relatives, but differs in having only 2 pairs of well-developed lobes, while the third pair represented as short points (Williams & Watson, 1988).

GENERAL: Definition and characters by Ferris (1938a) and by Balachowsky (1958b).

KEYS: Colon-Ferrer & Medina-Gaud 1998: 28-32 (female) [Genera of Puerto Rico]; Williams & Watson 1988: 20 (female) [Tropical South Pacific]; Chou 1985: 254 (female) [China]; Komosinska 1969: 50 (female) [*Abgrallaspis* group]; Beardsley 1966: 502-504 (female) [Federated States of Micronesia]; Balachowsky 1958b: 282 (female) [*Targionina* of Africa]; Ferris 1942: 28 (female) [North America]; Ferris 1942: 30 (female) [species North America]; Leonardi 1920: 26-27 (female) [Italy].

CITATIONS: Balach1958b [taxonomy, description: 282-283]; Beards1966 [taxonomy: 512]; Borchs1966 [catalogue: 243]; Brimbl1958 [taxonomy: 74]; Chou1985 [taxonomy, description: 254]; Cocker1899a [taxonomy: 395]; ColonFMe1998 [taxonomy, description: 41]; DanzigPe1998 [catalogue: 186]; DuttaSi1990 [taxonomy: 1]; Fernal1903b [taxonomy: 251]; Ferris1937c [taxonomy: 50]; Ferris1938a [taxonomy, description: 187]; Ferris1941f [taxonomy: 22]; Ferris1942 [taxonomy: 28]; Kawai1980 [taxonomy: 230]; Leonar1898a [taxonomy, description: 60-62]; MacGil1921 [taxonomy: 387,403]; Mamet1949 [taxonomy: 53]; MorrisMo1966 [taxonomy, catalogue: 16]; Tao1999 [taxonomy: 72]; Willia1969a [taxonomy: 320]; WilliaWa1988 [taxonomy: 44].

Aspidiella agalegae Mamet

Aspidiella agalegae Mamet, 1974: 166. Type data: AGALEGA ISLANDS: South Island, on *Musa sapientum*. Holotype female. Type depository: Paris: Muséum national d'Histoire naturelle, France.
SCALE COVER: Female scale circular, pale brown (Mamet, 1974).
HOST PLANTS: **Musaceae**: *Musa sapientum* [Mamet1974].
DISTRIBUTION: **Afrotropical**: Agalega Islands [Mamet1974].
GENERAL: Description and illustration of adult female by Mamet (1974).
CITATIONS: ChuaWo1990 [host, distribution, economic importance: 548]; Mamet1974 [taxonomy, description, illustration, host, distribution: 166-168].

Aspidiella dentata Borchsenius

Aspidiella dentata Borchsenius, 1958: 166. Type data: CHINA: Kwangtung Province, near Sinchou, on stem of undetermined grass (Gramineae). Syntypes, female. Type depository: St. Petersburg: (= Leningrad) Zoological Museum, Academy of Science, Russia.
Aspidiclla dentata; Chou, 1985: 255. Misspelling of genus name.
SCALE COVER: Female scale oval, 2 mm long, 1.5 mm wide; pale greyish-brown; dorsal and ventral scale thick and hard (Borchsenius, 1958).
HOST PLANTS: **Gramineae** [Borchs1958].
DISTRIBUTION: **Oriental**: China (People's Republic) (Guangdong (Kwangtung) [Borchs1958]).
GENERAL: Description and illustration of adult female by Borchsenius (1958) and by Chou (1985, 1986).
KEYS: Chou 1985: 254 (female) [China].
CITATIONS: Borchs1958 [taxonomy, description, illustration, host, distribution: 166-167]; Borchs1966 [catalogue: 243]; Chou1985 [taxonomy, description, host, distribution: 255-256]; Chou1986 [taxonomy, illustration: 656]; DanzigPe1998 [catalogue: 186]; Tao1999 [taxonomy, host, distribution: 72].

Aspidiella hartii (Cockerell)

Aspidiotus hartii Cockerell, 1895w: 7. Type data: TRINIDAD: in great numbers on tubers of yam. Syntypes, female. Type depository: Washington: United States National Entomological Collection, U.S. National Museum of Natural History, District of Columbia, USA.

Aspidiotus hartii luntii Hart, 1896: 156. Nomen nudum.

Aspidiotus hartii luntii Cockerell, 1896k: v. Type data: TRINIDAD: on "stems of some plant"; collected by Mr. Lunt, August 1895. Syntypes, female. Type depository: Washington: U.S. Entomological Collection, U.S. National Museum of Natural History, District of Columbia, USA. Synonymy by Borchsenius, 1966: 301.

Aspidiotus luntii Cockerell, 1896k: v. Type data: TRINIDAD: on an undetermined plant. Syntypes, female. Type depository: Washington: United States National Entomological Collection, U.S. National Museum of Natural History, District of Columbia, USA. Synonymy by Ferris, 1942: 445-11.

Aspidiotus (*Aspidiotus*) *hartii*; Cockerell, 1897i: 24. Change of combination.

Aspidiotus (*Aspidiella*) *hartii*; Leonardi, 1898a: 67. Change of combination.

Aspidiotus curcumae Kasargode, 1914: 134. Nomen nudum.

Aspidiella hartii; MacGillivray, 1921: 404. Change of combination.

Aspidiotus curcumae Ferris, 1941e: 42. Nomen nudum. Notes: Authorship credited to "Fletcher".

Aspidiotus luntii; Ferris, 1941e: 45. Change of status.

Aspidiotus luntii Ferris, 1941e: 45. Synonymy by Ferris, 1941e: 45.

Gonaspidiotus hartii; Borchsenius, 1966: 300. Change of combination.

Aspidiotus curcumae Borchsenius, 1966: 376. Nomen nudum.

Aspidiella hartii; Williams & Butcher, 1987: 94. Revived combination.

Aspidiotus curcumae Kotikal & Kulkarni, 2000: 52. Nomen nudum. Notes: Authorship credited to "Gr."

COMMON NAMES: ubi scale [VelasqRi1969]; Yam scale [Borchs1966].

SYSTEMATICS: The binomen *Aspidiotus curcumae* is a nomen nudum that is placed as a 'synonym' of *Aspidiella hartii*. The *nomen nudum* was published by Kasargode (1914: 134), Fletcher (1919: 232), Ramakrishna Ayyar (1920a: 326), Ramakrishna Ayyar (1930), Ferris (1941e: 42), Borchsenius (1966: 376) and Kotikal & Kulkarmi (2000: 52).

SCALE COVER: Scale of female brownish with pale center; circular, quite flat; that of male similar in colour; oval (Ferris, 1938a).

HOST PLANTS: **Araceae**: *Colocasia sculenta* [GomezM1941]. **Convolvulaceae**: *Ipomoea batatas* [WilliaWa1988]. **Cyperaceae**: *Cyperus odoratus* [GomezM1941]. **Dioscoreaceae**: *Dioscorea* [Green1915c, Mamet1943a, Mamet1949, Balach1958b, Borchs1966, WilliaWa1988, Beards1966], *Dioscorea alata* [Balach1958b, Cohic1958, WilliaBu1987, WilliaWa1988]. **Gramineae**: *Tripsacum laxum* [GomezM1941]. **Zingiberaceae**: *Curcuma* [Green1919c, Ramakr1921a], *Curcuma longa* [KotikaKu2000], *Zingiber officinale* [WilliaWa1988].

DISTRIBUTION: **Afrotropical**: Côte d'Ivoire (=Ivory Coast) [CouturMaRi1985]; Mauritius [Mamet1943a, Mamet1949, Balach1958b, Borchs1966]; Sierra Leone [Hargre1937]. **Australasian**: Federated States of Micronesia [Beards1966]; Fiji [Green1915c, WilliaWa1988]; New Caledonia [Cohic1958]; Papua New Guinea [WilliaWa1988]; Solomon Islands [WilliaWa1988]; Tonga [WilliaWa1988]; Vanuatu [WilliaBu1987, WilliaWa1988] [WilliaBu1987, WilliaWa1988]. **Neotropical**: Dominican Republic [GomezM1941]; Puerto Rico & Vieques Island (Puerto Rico [Martor1976, ColonFMe1998]); Saint Croix [Beatty1944]; Trinidad and Tobago (Trinidad [Ferris1938a]); U.S. Virgin Islands [Nakaha1983]. **Oriental**:

India [Ramakr1921a] (Maharashtra [Green1919c], Tamil Nadu [Ramakr1921]); Philippines [VelasqRi1969].

BIOLOGY: Occurring on tubers (Ferris, 1938a).

ECONOMIC IMPORTANCE: Tropicopolitan (see CABI, 1966, and Distribution); a pest of yams of the genus *Dioscorea*, both in field and particularly in storage (Schmutterer et al., 1957; Williams & Watson, 1988; Chua & Wood, 1990; Morse et al., 2000).

GENERAL: Description and illustration of adult female by Ferris (1938a), Balachowsky (1958b), Williams & Watson (1988) and by Colon-Ferrer & Medina-Gaud (1998).

KEYS: Williams & Watson 1988: 44 (female) [Tropical South Pacific]; Beardsley 1966: 512 (female) [Federated States of Micronesia]; Balachowsky 1958b: 283 (female) [Africa]; Ferris 1942: 30 (female) [North America].

CITATIONS: AkinloKo1988 [host, distribution, chemical control: 21-23]; Balach1958b [taxonomy, description, illustration, host, distribution: 283-285]; Ballou1915 [host, distribution: 121]; Beards1966 [host, distribution: 512]; Beatty1944 [host, distribution: 114-172]; Borchs1966 [catalogue: 300-301]; BurgerUl1990 [economic importance: 313-327]; CABI1966 [host, distribution: 1-2]; ChuaWo1990 [host, distribution, economic importance: 551]; Cocker1895w [taxonomy, description, host, distribution: 7]; Cocker1895y [host, distribution, economic importance: 85]; Cocker1896b [distribution: 334]; Cocker1896k [taxonomy, description, host, distribution: v]; Cocker1897i [taxonomy, description, host, distribution: 10,24]; Cocker1897l [taxonomy: 151]; Cohic1958 [host, distribution: 12]; ColonFMe1998 [taxonomy, description, illustration, host, distribution: 41-42]; CouturMaRi1985 [host, distribution: 278]; DevasaKoVe1998 [host, distribution: 157-164]; Ehrhor1925g [host, distribution: 101]; Fernal1903b [catalogue: 260]; Ferris1938a [taxonomy, description, illustration, host, distribution: 188]; Ferris1941e [taxonomy: 44,45]; Ferris1942 [taxonomy: 445:11;446:30]; Fletch1919 [host, distribution: 232]; FouaBi1982 [host, distribution, economic importance, life history: 265-273]; Frogga1914 [taxonomy, description, host, distribution: 314]; Frogga1915 [taxonomy, description, host, distribution: 17-18]; GomezM1941 [taxonomy, illustration, host, distribution: 126-128]; Gowdey1921 [taxonomy, description, host, distribution: 29]; Green1915c [host, distribution: 44]; Green1919c [host, distribution: 439]; Hargre1937 [host, distribution, economic importance: 505-520]; Hart1896 [taxonomy: 156]; Hinckl1963 [host, distribution, biological control]; KotikaKu2000 [host, distribution: 51-52]; Leonar1898a [taxonomy, description, illustration, host, distribution: 67-69]; Lindin1907a [taxonomy: 19]; MacGil1921 [taxonomy, description, host, distribution: 404]; Mamet1943a [catalogue: 156]; Mamet1949 [catalogue: 53,54]; Martor1976 [host, distribution: 101]; Maskew1914a [host, distribution: 446-447]; Maxwel1903 [taxonomy, description, host, distribution: 40]; MillerDa1990 [host, distribution, economic importance: 300]; MorseAcMc2000 [economic importance, chemical control]; MoutiaMa1947 [distribution]; Nakaha1983 [host, distribution: 8]; PatilThMo1988 [host, distribution: 8-9]; Ramakr1921 [host, distribution: 38]; Sassce1923 [host, distribution: 152-158]; SchmutKlLu1957 [host, distribution, economic importance: 494]; VelasqRi1969 [host, distribution: 195-208];

Willia1985a [taxonomy: 235]; WilliaBu1987 [host, distribution: 94]; WilliaWa1988 [taxonomy, description, illustration, host, distribution, economic importance: 8,44-47]; Wilson1921 [host, distribution: 20-34].

Aspidiella panici (Rutherford)

Aspidiotus panici Rutherford, 1915: 113. Type data: SRI LANKA: Paradeniya, on *Panicum incinatum*; collected July 1914. Syntypes, female. Type depository: London: The Natural History Museum, England, UK.
Aspidiella panici; Borchsenius, 1966: 243. Change of combination.
SCALE COVER: Female scale pinkish-grey and slightly elongated, narrower at one end than at other; exuviae golden-yellow, situated towards broader end and partly covered by secretion; ventral scale complete (Rutherford, 1915).
HOST PLANTS: Gramineae: *Panicum uncinatum* [Ruther1915, Green1922].
DISTRIBUTION: Oriental: Sri Lanka [Ruther1915, Green1922].
GENERAL: Description of adult female given by Rutherford (1915).
CITATIONS: Borchs1966 [catalogue: 243]; Ferris1941e [taxonomy: 46]; Green1922 [host, distribution: 462]; Ruther1915 [taxonomy, description, host, distribution: 113].

Aspidiella phragmitis (Takahashi)

Aspidiotus phragmitis Takahashi, 1931: 4. Type data: TAIWAN: Suo, Daichikko near Daibu, on *Phragmites* sp. Syntypes, female. Type depository: Taichung: Ent. Collection, Taiwan Agricultural Research Institute, Wu-feng, Taichung, Taiwan.
Aspidiotus miscanthii Kuwana, 1931b: 169. Type data: RYUKU ISLANDS: Amami-Oshima, Nase, on *Miscanthus* sp. Syntypes, female. Type depository: Ibaraki-ken: Insect Taxonomy Laboratory, National Institute of Agricultural Environmental Sciences, Kannon-dai, Yatabe, Tsukuba-shi, (I. Kuwana), Japan. Synonymy by Takahashi, 1933: 54.
Aspidiotus mithcanthii; Kuwana, 1933: 2,14. Misspelling of species name.
Aspidiella phragmitis; Ferris, 1941e: 47. Change of combination.
Chortinaspis phragmitis; Borchsenius, 1966: 280. Change of combination.
Aspidiella phragmitis; Danzig & Pellizzari, 1998: 186. Revived combination.
SCALE COVER: Scale of adult female white, somewhat greyish, not transparent, scarcely convex, about 1.5 mm in diameter. Larval skins yellowish brown, shining, covered with white secretion (Takahashi, 1931).
HOST PLANTS: Gramineae: *Miscanthus* [Kuwana1931b, Kuwana1933, Takagi1958, Takagi1970], *Phragmites* [Takaha1931, Takaha1932a, Takaha1933, Takagi1970], *Thysanolaena maxima* [Takaha1933, Takagi1970].
DISTRIBUTION: Oriental: Ryukyu Islands (= Nansei Shoto) [Kuwana1931b]; Taiwan [Takaha1931, Takaha1932a, Takaha1933, Takagi1970]. **Palaearctic**: Japan [Kuwana1933, Takagi1958, Kawai1980].
GENERAL: Description and illustration of adult female by Takahashi (1931b), Kuwana (1933) and by Takagi (1958).
KEYS: Chou 1985: 279 (female) [Species of China]; Kuwana 1933: 2 (female) [Japan]; Kuwana 1933b: 49 (female) [Japan].

CITATIONS: Borchs1966 [catalogue: 280]; Chou1985 [taxonomy, description, host, distribution: 280-281]; DanzigPe1998 [catalogue: 186]; Ferris1941e [taxonomy: 46-47]; Ferris1955c [taxonomy: 32]; Kawai1980 [taxonomy, description, host, distribution: 230]; Komemo1982 [host, distribution: 26-30]; Kuwana1931b [taxonomy, description, illustration, host, distribution: 169-170]; Kuwana1933 [taxonomy, description, illustration, host, distribution: 14-15]; Lindin1957 [taxonomy: 546]; Muraka1970 [host, distribution: 71]; Takagi1958 [taxonomy, description, illustration, host, distribution: 121-122]; Takagi1970 [taxonomy, host, distribution: 132]; Takaha1931 [taxonomy, description, illustration, host, distribution: 4-5]; Takaha1932a [host, distribution: 104]; Takaha1933 [taxonomy, description, host, distribution: 25-34,54,62]; Tao1999 [taxonomy, host, distribution: 80].

Aspidiella rigida Ferris

Aspidiella rigida Ferris, 1941d: 336. Type data: PANAMA: Chiriqui Province, Boquete, on undetermined tree. Holotype female. Type depository: Davis: The Bohart Museum of Entomology, University of California, California, USA.
Rhizaspidiotus rigidus; Ferris, 1942: 433. Change of combination requiring emendation of species name for agreement in gender.
Rhizaspidiotus rigida; Ferris, 1943a: 243. Change of combination.
Aspidiella rigida; Borchsenius, 1966: 243. Revived combination.
Aspidiella rigina; Borchsenius, 1966: 416. Misspelling of species name.
SCALE COVER: Scale of female quite flat, roughly circular or somewhat elongate, varying with its position, whitish in colour, usually partially covered by or mingled with soft bark tissues of host, exuvia subcentral; with a white and slightly cottony ventral scale. Scale of male tending to be more exposed than that of female, white, elongate, with exuvia at one end (Ferris, 1941d).
DISTRIBUTION: **Neotropical**: Panama [Ferris1941d].
BIOLOGY: Occurring on the tree trunk, buried under bark flakes (Ferris, 1941d).
GENERAL: Description and illustration of adult female by Ferris (1941d).
KEYS: Ferris 1942: 40 (female) [North America].
CITATIONS: Borchs1966 [catalogue: 243-244]; Ferris1941d [taxonomy, description, illustration, host, distribution: 336]; Ferris1942 [taxonomy: 433;446:40]; Ferris1943a [taxonomy: 99].

Aspidiella sacchari (Cockerell)

Aspidiotus sacchari Cockerell, 1893j: 255. Type data: JAMAICA: Kingston, on sugar-cane. Syntypes, female. Type depository: Washington: U.S National Entomological Collection, U.S. National Museum of Natural History, District of Columbia, USA.
Aspidiotus (Aspidiotus) sacchari; Cockerell, 1897i: 25. Change of combination.
Aspidiotus (Aspidiella) sacchari; Leonardi, 1898a: 72. Change of combination.
Aspidiella sacchari; MacGillivray, 1921: 405. Change of combination.
Targionia sacchari; Merrill & Chaffin, 1923: 253-254. Change of combination.
Targionia (Aspidiella) sacchari; Merrill, 1953: 81. Change of combination.
Aspidiella sacchari; Borchsenius, 1966: 244. Revived combination.

COMMON NAMES: sugar cane root scale [SchmutKlLu1957]; sugar-cane scale; sugarcane scale [Dekle1965c].

SCALE COVER: Scale of female pale brown, circular, thin, flat; exuviae central; that of male elongate; exuvia apical (Ferris, 1938a).

HOST PLANTS: **Araceae**: *Alocasia macrorhiza* [WilliaWa1988]. **Gramineae** [GomezM1941, BesheaTiHo1973, WilliaWa1988], *Andropogon citratus* [Martor1976], *Brachiaria mutica* [WilliaWa1988], *Coix lacryma* [Houser1918], *Cymbopogon martini* [Matile1978], *Cynodon dactylon* [Takaha1941b, Mamet1949, Dekle1965c], *Eremochloa ophiuroides* [Dekle1965c, BesheaTiHo1973], *Gynerium sagittatum* [Balach1959a], *Ischaemum* [WilliaWa1988], *Panicum* [Laing1929a], *Pa. molle* [Houser1918], *Paspalum millegrana* [Martor1976], *Pas. vaginatum* [Mamet1941a], *Pennisetum purpureum* [Balach1959a], *Saccharum* [Dekle1965c], *Sa. officinarum* [Cocker1893j, Ferris1955c, WilliaWa1988], *Sporobolus* [Beards1966], *Stenotaphrum* [Mamet1957], *Stenotaphrum dimidiatum* [Mamet1949, Borchs1966], *St. secundatum* [Dekle1965c], *Stenotaphrum subulatum* [Mamet1941a, Borchs1966].

NATURAL ENEMIES: THYSANOPTERA **Phlaeothripidae**: *Podothrips semiflavus* Hood [PalmerMo1990].

DISTRIBUTION: **Afrotropical**: Comoros Islands [Matile1978]; Liberia [Nakaha1982]; Madagascar [Nakaha1982]; Mauritius [Mamet1941a, Mamet1949]; Nigeria [Nakaha1982]; Rodrigues Island [Nakaha1982]; Réunion [Mamet1957]; Sierra Leone [Hargre1937]. **Australasian**: Cook Islands [WilliaWa1988]; Federated States of Micronesia [Takaha1936c, Takaha1941b]; Fiji [Nakaha1982]; Guam [Nakaha1982]; Marshall Islands [Beards1966]; Palau [Beards1966]; Papua New Guinea [WilliaWa1988]; Solomon Islands [WilliaWa1988]; Western Samoa [WilliaWa1988]. **Nearctic**: Mexico [Nakaha1982]; United States of America (Florida [MerrilCh1923, Ferris1938a, Dekle1965c, BesheaTiHo1973], Texas [McDani1968]). **Neotropical**: Bahamas [Nakaha1982]; Colombia [Balach1959a, Kondo2001]; Cuba [Houser1918]; Guadeloupe [Balach1957c]; Guyana [Nakaha1982]; Jamaica [Cocker1893j, Laing1929a, Ferris1938a]; Martinique [Balach1957c]; Panama [Ferris1942, Ferris1955c]; Puerto Rico & Vieques Island (Puerto Rico [VanDin1913, Martor1976]); Saint Croix [Beatty1944]; U.S. Virgin Islands [Nakaha1983]. **Oriental**: China (People's Republic) (Yunnan [Ferris1955c]); Indonesia [Nakaha1982]; Malaysia [Nakaha1982]; Pakistan [Nakaha1982]; Sri Lanka [Ferris1938a, Green1922].

BIOLOGY: Occurring on the stems and beneath the sheathing bases of the leaves (Ferris, 1938a).

ECONOMIC IMPORTANCE: Distributed in many sugarcane-growing areas of world (see Distribution), and considered a pest of sugarcane (Schmutterer et al., 1957; Williams & Watson, 1988; Williams & Greathead, 1990).

GENERAL: Description and illustration of adult female by Laing (1929a), Ferris (1938a), Balachowsky (1958b), Chou (1985, 1986), Williams & Watson (1988) and by Colon-Ferrer & Medina-Gaud (1998).

KEYS: Williams & Watson 1988: 44 (female) [Tropical South Pacific]; Chou 1985: 254 (female) [China]; Beardsley 1966: 512 (female) [Federated States of Micronesia]; Ferris 1942: 30 (female) [North America].

CITATIONS: Balach1957c [host, distribution: 200]; Balach1958b [taxonomy, description, illustration, host, distribution: 284,286-287]; Balach1959a [host, distribution: 362]; Ballou1912 [host, distribution, economic importance, control]; Beards1966 [host, distribution: 512]; BeardsDaHo1976 [economic importance: 106]; Beatty1944 [host, distribution: 114-172]; Bodkin1913 [host, distribution: 29-32]; Borchs1966 [catalogue: 244]; Bourne1921 [host, distribution, economic importance]; Box1953 [host, distribution, biological control: 51]; Chou1985 [taxonomy, description, host, distribution: 254-255]; Chou1986 [taxonomy, illustration: 655]; Cocker1893j [taxonomy, description, host, distribution: 255-256]; Cocker1896b [distribution: 334]; Cocker1897i [taxonomy, description, host, distribution: 25]; Cocker1897l [host, distribution: 150]; ColonFMe1998 [taxonomy, description, illustration, host, distribution: 42-43]; DanzigPe1998 [catalogue: 186]; Dekle1965c [taxonomy, description, host, distribution: 26]; Dekle1976 [taxonomy, description, host, distribution, economic importance: 38]; FDACSB1982 [host, distribution: 5-11]; Fernal1903b [catalogue: 278]; Ferris1937c [taxonomy, illustration: 50,61]; Ferris1938a [taxonomy, description, illustration, host, distribution: 189]; Ferris1941e [taxonomy: 48]; Ferris1942 [taxonomy, host, distribution: 446:9;446:30]; Ferris1943a [taxonomy: 86]; Ferris1955c [taxonomy, host, distribution: 33]; GomezM1941 [host, distribution: 128]; Gowdey1921 [taxonomy, host, distribution: 30]; Green1922 [host, distribution: 463]; Hargre1937 [host, distribution, economic importance: 505-520]; Houser1918 [host, distribution: 167]; Hutson1916 [host, distribution: 426]; Jayant1999 [host, distribution: 76]; Kondo2001 [taxonomy, host, distribution: 43]; Laing1929a [taxonomy, description, illustration, host, distribution: 490-491]; Leonar1897 [taxonomy: 285]; Leonar1898a [taxonomy, description, illustration, host, distribution: 72-73]; Lepesm1947 [host, distribution: 214]; Lindin1909c [taxonomy, host, distribution: 449]; MacGil1921 [taxonomy, description, host, distribution: 405]; Mamet1941a [host, distribution: 40]; Mamet1943a [catalogue: 156]; Mamet1949 [catalogue: 54]; Mamet1957 [host, distribution: 369,375]; Martor1976 [host, distribution: 13,192]; Matile1978 [host, distribution: 61]; Maxwel1903 [taxonomy, host, distribution: 41]; McDani1968 [taxonomy, illustration, host, distribution: 216-217]; Merril1953 [taxonomy, description, illustration, host, distribution: 81]; Merril1953 [taxonomy, description, illustration, host, distribution]; MerrilCh1923 [taxonomy, description, host, distribution, economic importance: 253-254]; MillerDa1990 [host, distribution, economic importance: 300]; Moore1915 [host, distribution: 305-310]; MoutiaMa1947 [distribution]; Nakaha1982 [host, distribution: 12]; Nakaha1983 [host, distribution: 9]; PalmerMo1990 [biological control: 67-76]; RiherdCh1952 [host, distribution, economic importance, control: 1-5]; Sassce1923 [host, distribution: 152-158]; SchmutKlLu1957 [host, distribution, economic importance: 494]; Takaha1931 [taxonomy, host, distribution: 5]; Takaha1936c [host, distribution: 118]; Takaha1939b [taxonomy: 272]; Takaha1940a [taxonomy: 332]; Takaha1941b [host, distribution: 220]; Tao1999 [taxonomy, host, distribution: 72]; VanDin1913 [host, distribution: 251-257]; WilliaGr1990 [host, distribution, economic importance, biological control: 563-578]; WilliaWa1988 [taxonomy, description, illustration, host, distribution: 46-47]; Wilson1921 [host, distribution: 20-34]; WoodruBeSk1998 [distribution].

Aspidiella zingiberi Mamet

Aspidiotus subterraneus Mamet, 1939b: 580. Type data: MAURITIUS: Rose Hill, on *Zingiber officinale*. Holotype. Type depository: Paris: Muséum national d'Histoire naturelle, France. Homonym of *Aspidiotus subterraneus* Lindinger, 1935.

Aspidiella zingiberi Mamet, 1942: 36. Replacement name for *Aspidiotus subterraneus* Mamet, 1939.

Aspidiella subterranea; Takahashi, 1942: 47. Change of combination requiring emendation of species name for agreement in gender.

COMMON NAME: luya scale [VelasqRi1969].

SCALE COVER: Female scale brownish-buff in colour; slightly conical, irregularly circular; diameter 1.9-2.3 mm; exuviae subcentral; larval exuvium yellow and shiny; nymphal exuvium darker, obscured by layer of pale buff secretion; ventral scale represented as a whitish scar on surface of rhizomes of host plant (Mamet, 1939b).

HOST PLANTS: **Zingiberaceae**: *Curcuma longa* [Mamet1949, Borchs1966], *Zingiber officinale* [Mamet1939b, Mamet1943a, Mamet1949, Borchs1966].

NATURAL ENEMIES: HYMENOPTERA **Aphelinidae**: *Aphytis acrenulatus* DeBach & Rosen [RosenDe1979].

DISTRIBUTION: **Afrotropical**: Mauritius [Mamet1939b, Mamet1943a, Mamet1949, Borchs1966]. **Oriental**: Philippines [VelasqRi1969].

GENERAL: Description and illustration of adult female by Mamet (1939).

CITATIONS: Balach1958b [taxonomy, host, distribution: 283,284]; Borchs1966 [catalogue: 244]; DeBachRo1976a [host, distribution, biological control: 541-545]; Ferris1941e [taxonomy: 48]; Mamet1939b [taxonomy, description, illustration, host, distribution: 580]; Mamet1942 [taxonomy: 36]; Mamet1943a [catalogue: 156]; Mamet1949 [catalogue: 54]; MoutiaMa1947 [distribution]; RosenDe1979 [host, distribution, biological control: 419-421]; Takaha1942 [taxonomy, host, distribution: 47]; VelasqRi1969 [host, distribution: 195-208].

Aspidioides **MacGillivray**

Aspidioides MacGillivray, 1921: 387. Type species: *Aspidiotus corokiae* Maskell, by original designation.

Aspidoides; MacGillivray, 1921: 407, 477. Misspelling of genus name.

SYSTEMATICS: Distinctive features of genus are presence of three pairs of lobes, median lobes with well developed basal scleroses, plates fringed, dorsal pygidial ducts small and very slender, ventral surface with a few microducts only, and presence of only anterolateral groups of perivulvar pores. This genus appears to resemble *Monaonidiella* MacGillivray, but differs from it in possessing three pairs of lobes (Borchsenius & Williams, 1963).

GENERAL: Definition and characters by Borchsenius & Williams (1963).

CITATIONS: Borchs1966 [catalogue: 276]; BorchsWi1963 [taxonomy, description: 384]; Ferris1937c [taxonomy: 50]; Ferris1941e [taxonomy: 42]; Kozar1990f [distribution: 142]; Lindin1937 [taxonomy: 179]; MacGil1921 [taxonomy, description: 387,407,477]; MorrisMo1966 [taxonomy, catalogue: 16-17].

Aspidioides corokiae (Maskell)

Aspidiotus corokiae Maskell, 1891: 2. Type data: NEW ZEALAND: South Island, Reefton, on *Corokia cotoneaster*. Syntypes, female. Type depository: Auckland: New Zealand Arthropod Collection, Landcare Research, New Zealand.

Aspidiotus (Selenaspis) corokiae; Leonardi, 1898a: 53. Change of combination.

Aspidiotus (Selenaspis) corokiae; Leonardi, 1898a: 53. Misspelling of genus name.

Selenaspis corockiae; Leonardi, 1898a: 54. Misspelling of genus and species names.

Aspidioides corokiae; MacGillivray, 1921: 406. Change of combination.

Aspidoides corokiae; MacGillivray, 1921: 406. Misspelling of genus name.

SCALE COVER: Female scale circular, rather solid, slightly convex, with exuviae in centre; colour varying from yellow to (less frequently) white; exuviae yellow; diameter about 1/24 inch (Maskell, 1891).

HOST PLANTS: **Cornaceae**: *Corokia* [Maskel1891].

DISTRIBUTION: **Australasian**: New Zealand (South Island [Maskel1891]).

GENERAL: Description and illustration of adult female by Maskell (1891) and by Borchsenius & Williams (1963).

CITATIONS: Borchs1966 [catalogue: 276-277]; BorchsWi1963 [taxonomy, description, illustration: 384-385]; Cocker1896b [distribution: 335]; Cocker1897i [taxonomy, description, host, distribution: 25]; DeitzTo1980 [taxonomy: 35]; Fernal1903b [catalogue: 255]; Ferris1937c [taxonomy: 50]; Leonar1898a [taxonomy, description, illustration, host, distribution: 53-55]; MacGil1921 [taxonomy, description, host, distribution: 387,406]; Maskel1891 [taxonomy, description, illustration, host, distribution: 2].

Aspidiotus Bouché

Aspidiotus Bouché, 1833: 52. Type species: *Aspidiotus nerii* Bouché. Subsequently designated by Leonardi, 1897: 285. Notes: *Aspidiotus* was also described as new by Bouché (1834) page 9.

Aspidiotus (Evaspidiotus) Leonardi, 1898a: 74. Type species: *Aspidiotus hederae* (= *Aspidiotus nerii* Bouché), by original designation. Synonymy Fernald, 1903b: 251.

Brainaspis MacGillivray, 1921: 390. Type species: *Aspidiotus kellyi* Brain, by monotypy and original designation. Synonymy by Lindinger, 1937: 181.

Temnaspidiotus MacGillivray, 1921: 387. Type species: *Aspidiotus excisus* Green, by original designation. Synonymy by Takagi, 1969a: 62.

Aspidicotus; Green, 1930c: 281. Misspelling of genus name.

Genaspidiotus; Bodenheimer, 1949: 33. Misspelling of genus name.

Evaspidiotus; Morrison & Morrison, 1966: 75. Change of status.

SYSTEMATICS: The modern concept of the diagnostic characters of *Aspidiotus* was defined by Ferris (1938a), Balachowsky (1948b, 1956), Ferris (1952a), Takagi (1957, 1969a), Williams & Watson (1988) and by Danzig (1993). Although the genus *Brainaspis* (type species: *Aspidiotus kellyi* Brain) was synonymized with *Aspidiotus* by Lindinger, 1937, Dr. Sadao Takagi (in personal communication, 8 January 2003, to Yair Ben-Dov) suggested that *Aspidiotus kellyi* Brain and *Aspidiotus sinensis* (Ferris) may belong to a separate genus, for which the name *Brainaspis* MacGillivray, 1921, is available.

GENERAL: Definition and characters by Froggatt (1914), Dietz & Morrison (1916a), Brain (1919), Green (1928), Fullaway (1932), Ferris (1938a), Balachowsky (1948b, 1956), Borchsenius (1950b), Ferris (1952a), Gómez-Menor Guerrero (1962), Takagi (1957, 1969a), Bazarov & Shmelev (1971), Velasquez (1971), Danzig (1980b), Williams & Watson (1988), Danzig (1993) and by Yaşar (1995).

KEYS: Colon-Ferrer & Medina-Gaud 1998: 28-32 (female) [Genera of Puerto Rico]; Gill 1997: 24-26 (female) [Genera of California]; Gill 1997: 64 (female) [Species of California]; Kosztarab 1996: 406-407 (female) [Northeastern North America]; Danzig 1993: 140-141 (female) [species Europe]; Wolff & Corseuil 1993: 29 (female) [Brazil, Rio Grande do Sul]; Zahradník 1990b: 74 (female) [Czech Republic]; Williams & Watson 1988: 20 (female) [Tropical South Pacific]; Tereznikova 1986: 83 (female) [Ukraine]; Chou 1985: 260 (female) [Genera of China]; Danzig 1980b: 296 (female) [Far East of USSR]; Paik 1978: 297 (female) [species South Korea]; Bazarov & Shmelev 1971: 186 (female) [Central Asia]; Velasquez 1971: 100 (female) [Philippines]; Beardsley 1966: 502-504 (female) [Federated States of Micronesia]; Ezzat & Afifi 1966: 371-372 (female) [Egypt]; Danzig 1964: 645 (female) [Europe]; Gómez-Menor Guerrero 1962: 157 (female) [Canary Islands]; Zahradník 1959a: 547 (female) [Czech Republic]; Balachowsky 1958b: 228 (female) [*Aspidiotina* of Africa]; Ezzat 1958: 237-239 (female) [Egypt]; Takagi 1957: 31-32 (female) [species Japan]; Gómez-Menor Ortega 1956: 7-8 (female) [Spain]; McKenzie 1956: 23 (female) [U.S.A.: California]; Balachowsky 1951: 599 (female) [Mediterranean]; Borchsenius 1950b: 167 (female) [USSR]; Bodenheimer 1949: 45-46 (female) [Turkey]; Balachowsky 1948b: 274-275 (female) [species Mediterranean]; Zimmerman 1948: 351 (female) [Hawaii]; Gómez-Menor Ortega 1946: 59-61 (female) [Spain]; McKenzie 1946: 11 (female) [Canary Islands]; Ruiz Castro 1944: 56-57 (female) [Spain]; Ferris 1942: 28 (female) [North America]; Ferris 1942: 30 (female) [species North America]; Ferris 1941e: 60-61 (female) [species World]; Archangelskaya 1937: 94 (female) [Middle Asia]; Borchsenius 1937: 99 (female) [USSR]; Borchsenius 1937a: 32-33 (female) [Palaearctic Region]; Fullaway 1932: 97-98 (female) [Hawaii]; Archangelskaya 1929: 89 (female) [Palaearctic Region]; Balachowsky 1928b: 157 (female) [Africa]; Hollinger 1923: 6-7 (female) [U.S.A.: Missouri]; Leonardi 1920: 29-30 (female) [Species of Italy]; Leonardi 1920: 26 (female) [Italy]; Brain 1918: 115 (female) [South Africa]; Lawson 1917: 206 (female) [U.S.A.: Kansas]; Lawson 1917: 217 (female) [species U.S.A.: Kansas]; Robinson 1917: 16-17 (female) [Philippines]; Cockerell 1905b: 201 (female) [U.S.A.: Colorado]; Hempel 1904a: 496-497 (female) [Brazil]; Green 1896e: 37 (female) [Sri Lanka]; Comstock 1883: 55-57 (female) [species U.S.A.]; Comstock 1883: 54 (female) [North America].

CITATIONS: Archan1929 [taxonomy: 189]; Archan1937 [taxonomy, description: 94,98]; Ashmea1891 [taxonomy, description: 101]; Atkins1886 [taxonomy, description: 272]; Baeren1849 [taxonomy, description: 165]; Balach1928a [taxonomy: 157]; Balach1928b [taxonomy: 157]; Balach1942 [taxonomy: 47]; Balach1948b [taxonomy, description: 273-274]; Balach1950b [taxonomy, description: 545]; Balach1951 [taxonomy, description: 599]; Balach1956 [taxonomy, description: 49-51,132]; Balach1958b [taxonomy, description: 228]; BazaroSh1971 [taxonomy, description: 186-187]; Beards1966 [taxonomy: 513];

Berles1896 [taxonomy, description: 78]; Berles1896b [taxonomy: 207]; BerlesLe1896 [taxonomy, description: 345-352]; BerlesLe1898a [taxonomy: 131]; Blanch1840 [taxonomy: 214]; BlayGo1993 [taxonomy, description: 427-428,467]; Bodenh1924 [taxonomy: 21]; Bodenh1927c [taxonomy: 25-44]; Bodenh1949 [taxonomy, description: 32-33,45-46]; Bodenh1952 [taxonomy, description: 330]; Borchs1937 [taxonomy, description: 32,33]; Borchs1937a [taxonomy, description: 99]; Borchs1938 [taxonomy, description: 138]; Borchs1949d [taxonomy, description: 194,235]; Borchs1950b [taxonomy, description: 167,214]; Borchs1966 [catalogue: 258,269,300]; Bouche1833 [taxonomy, description: 52]; Bouche1834 [taxonomy, description: 9]; Bouche1844 [taxonomy, description: 294]; Brain1918 [taxonomy, description: 115,117]; Brain1919 [taxonomy, description: 214]; Brimbl1968 [taxonomy: 39-42]; Britto1923 [taxonomy, description: 360,371]; Burmei1835 [taxonomy: 66]; Bustsh1958 [taxonomy, description: 234]; Chou1985 [taxonomy, description: 260-263]; Chou1985 [taxonomy, description: 273]; Cocker1896b [taxonomy: 333]; Cocker1897i [taxonomy, description: 3-31]; Cocker1897l [taxonomy: 150]; Cocker1899a [taxonomy: 395]; Cocker1905b [taxonomy: 200,201]; ColonFMe1998 [taxonomy, description: 43]; Comsto1881a [taxonomy, description: 292]; Comsto1883 [taxonomy, description: 54,55]; Danzig1964 [taxonomy: 650]; Danzig1980b [taxonomy, description: 336]; Danzig1993 [taxonomy, description: 140]; DanzigPe1998 [catalogue: 187,238,362]; DietzMo1916a [taxonomy, description: 263,288]; Dougla1886 [taxonomy: 245]; Ezzat1958 [taxonomy: 238]; Fernal1903b [taxonomy: 251]; Ferris1921b [taxonomy: 94]; Ferris1937c [taxonomy: 50-53,62,106]; Ferris1938 [taxonomy: 43,56,65,67]; Ferris1938a [taxonomy, description: 190]; Ferris1938b [taxonomy, description: 65,67]; Ferris1941e [taxonomy, description: 33,37]; Ferris1942 [taxonomy, description: 446]; Ferris1943 [taxonomy, description: 82]; Ferris1952a [taxonomy, description: 8-9]; FrankKr1900 [taxonomy, description: 40]; Frogga1914 [taxonomy, description: 131]; Frogga1915 [taxonomy, description: 7]; Fullaw1932 [taxonomy, description: 98,106-107]; Fuller1897c [taxonomy: 3]; Ghauri1962 [taxonomy, description: 210]; Gill1997 [taxonomy: 64]; Goethe1884 [taxonomy: 113]; GomezM1937 [taxonomy, description: 44-46]; GomezM1946 [taxonomy: 60]; GomezM1956 [taxonomy, description: 8]; GomezM1959 [taxonomy, description: 157-158]; GomezM1962 [taxonomy, description: 158]; Gowdey1921 [taxonomy, description: 28]; GrandpCh1899 [taxonomy, description: 1]; Green1896e [taxonomy, description: 39]; Green1927 [taxonomy: 3]; Green1928 [taxonomy, description: 8]; Green1928c [taxonomy: 374-376]; Green1928c [taxonomy: 376]; Green1930c [taxonomy: 281]; Hadzib1983 [taxonomy: 217-218]; Hempel1900a [taxonomy, description: 496]; Hempel1904 [taxonomy, description: 319]; Hempel1920 [taxonomy: 140]; Hollin1923 [taxonomy, description: 7,67,68]; Jorgen1934 [taxonomy, description: 278]; Kawai1980 [taxonomy, description: 226]; Kozar1990f [distribution: 143]; Kuwana1933 [taxonomy, description: 2]; Lawson1917 [taxonomy, description: 206,216-217]; Leonar1897 [taxonomy: 283-286]; Leonar1897a [taxonomy: 375]; Leonar1897b [taxonomy: 102-134]; Leonar1898a [taxonomy: 48-78]; Leonar1903a [taxonomy: 3]; Leonar1920 [taxonomy, description: 26,28-29]; Lepage1938 [catalogue: 393]; Lepesm1947 [taxonomy, description: 185]; Lidget1898a [taxonomy: 13]; Lindin1908b [taxonomy: 98]; Lindin1913 [taxonomy: 64]; Lindin1924 [taxonomy: 171]; Lindin1932f [taxonomy: 182]; Lindin1934e [taxonomy: 160]; Lindin1937 [taxonomy: 181,197]; Lindin1957 [taxonomy: 543]; Low1882c [taxonomy: 521]; Lupo1948 [taxonomy, description: 137]; MacGil1921 [taxonomy, description: 387,390,396,403,431]; Mamet1949 [taxonomy: 54-55]; Maskel1879 [taxonomy, description: 192,197]; Maskel1887a [taxonomy: 39,40]; Maskel1889 [taxonomy: 102]; Maskel1891a [taxonomy: 59]; Maskel1898 [taxonomy: 221]; McKenz1938 [taxonomy: 2,5]; McKenz1956 [taxonomy, description: 23]; Miller1990 [taxonomy: 169-178]; Morgan1888a [taxonomy: 47]; MorrisMo1966 [taxonomy, catalogue: 17,65,75,85,194]; Muntin1971a [taxonomy: 305]; Myers1925 [taxonomy, description: 166]; Nel1933 [taxonomy: 417-419]; Newell1899 [taxonomy, description: 1]; Newste1901b [taxonomy, description: 78,80]; Ramakr1930 [taxonomy: 12]; Robins1917 [taxonomy, description: 17,29]; RuizCa1944 [taxonomy: 56-57]; Sachtl1944 [taxonomy: 69]; Sander1904 [taxonomy: 30,55]; Savesc1982 [taxonomy, description: 302]; Schmut1959 [taxonomy, description: 48]; Signor1869b [taxonomy, description: 98,113]; Silves1902 [taxonomy, description: 98]; Takagi1957 [taxonomy, description: 31,38]; Takagi1969a [taxonomy, description: 62-64]; Tao1999 [taxonomy: 73,119]; Targio1881 [taxonomy: 149]; Targio1888 [taxonomy: 419,420]; ThiemGe1934 [taxonomy, description: 529]; ThiemGe1934a [taxonomy, description: 130]; Velasq1971 [taxonomy, description: 99-100]; Westwo1840 [taxonomy, description: 118]; Willia1969a [taxonomy: 320]; WilliaWa1988 [taxonomy, description: 49]; Wolff1911 [taxonomy: 80]; WolffCo1993 [taxonomy: 29]; Yasar1995a [taxonomy, description: 49-50]; Zimmer1948 [taxonomy: 352].

Aspidiotus abietoides nomen nudum
Aspidiotus abietoides Cockerell, 1894: 32. Nomen nudum.
Aspidiotus abietoides Ferris, 1941e: 40. Nomen nudum.
Aspidiotus abietoides Borchsenius, 1966: 376. Nomen nudum.

Aspidiotus aldabricus nomen nudum
Aspidiotus aldabricus Dupont, 1917: 1. Nomen nudum.
Aspidiotus aldabricus Ferris, 1941e: 40. Nomen nudum.
Aspidiotus aldabricus Borchsenius, 1966: 376. Nomen nudum.

Aspidiotus anningensis Tang & Chu
Aspidiotus cryptomeriae; Ferris, 1953: 65. Misidentification; discovered by Tang & Chu, 1983: 302.
Aspidiotus anningensis Tang & Chu, 1983: 302. Type data: CHINA: Yunnan, Kunming, Anning, on *Keteleeria evelyniana*. Holotype female. Type depository: Shanxi: Entomological Institute, Shanxi Agricultural University, Taigu, China.
SYSTEMATICS: Tang & Chu (1983: 302, 305) regarded record of *Aspidiotus cyptomeriae* Kuwana, by Ferris (1953: 65) (off needles of *Keteleeria evelyniana* in Yunnan, China) as a misidentification of *Aspidiotus anningensis* Tang & Chu, 1983.
SCALE COVER: Female scale circular, about 2 mm in diameter, thin, yellow-white in colour; exuviae pale yellow (Tang & Chu, 1983).

HOST PLANTS: **Pinaceae**: *Keteleeria evelyniana* [TangCh1983].
DISTRIBUTION: **Palaearctic**: China (People's Republic) [TangCh1983, Tang1984]; Japan [Kawai1980].
GENERAL: Description and illustration of adult female by Tang & Chu (1983) and by Tang (1984).
CITATIONS: DanzigPe1998 [catalogue: 187]; Ferris1953 [host, distribution, taxonomy: 65]; Kawai1980 [taxonomy, description, host, distribution: 227-228]; Tang1984 [taxonomy, description, illustration, host, distribution: 18-19]; TangCh1983 [taxonomy, description, illustration, host, distribution: 302-303]; Tao1999 [taxonomy: 73].

Aspidiotus artus Munting

Aspidiotus artus Munting, 1971a: 305. Type data: SOUTH AFRICA: Cape Province, Witzenberg, on *Protea laurifolia*; collected 1.ix.1965, by D.J. Rust. Holotype female. Type depository: Pretoria: South African National Collection of Insects, South Africa; type no. 1927/3.
SCALE COVER: Scale of adult female subcircular, white with golden yellow subcentral exuviae, about 3 mm in diameter. Male scale similar but about 1.3 mm in diameter (Munting, 1971a).
HOST PLANTS: **Proteaceae**: *Protea barbigera* [Muntin1971a], *Protea laurifolia* [Muntin1971a].
DISTRIBUTION: **Afrotropical**: South Africa [Muntin1971a].
GENERAL: Description and illustration of adult female by Munting (1971a).
CITATIONS: Muntin1971a [taxonomy, description, illustration, host, distribution: 305-307].

Aspidiotus atomarius (Hall)

Aonidiella atomaria Hall, 1946: 55. Type data: ZAIRE: Dilolo, on undersurface of leaves of unidentified plant. Syntypes, female. Type depository: London: The Natural History Museum, England, UK.
Aonidiella atomariae; Borchsenius, 1966: 292. Misspelling of species name.
Aspidiotus atomarius; Munting, 1971a: 307. Change of combination requiring emendation of species name for agreement in gender.
SCALE COVER: Scale of adult female circular, diameter 1.50-1.75 mm; low convex, white but, although thin, sufficiently opaque to mask outline of body; thinner at the extreme margin and semi translucent; exuviae central, very dark brown in fully developed scales but paler in early stages; ventral scale very thin and transparent, remaining adherent to host plant. Male scale smaller than that of female, oval and opaque white with pale yellow exuvia (Hall, 1946).
HOST PLANTS: **Theaceae**: *Thea* [Muntin1971a].
NATURAL ENEMIES: HYMENOPTERA **Encyrtidae**: *Habrolepis* sp. [Prinsl1983].
DISTRIBUTION: **Afrotropical**: Kenya [Muntin1971a]; Zaire [Hall1946, Balach1956].
GENERAL: Description and illustration of adult female by Hall (1946), Balachowsky (1956) and by Munting (1971a).

KEYS: Balachowsky 1956: 26 (female) [Africa].
CITATIONS: Balach1956 [taxonomy, description, illustration, host, distribution: 27-30]; Borchs1966 [catalogue: 292]; Hall1946 [taxonomy, description, illustration, host, distribution: 55-56]; Muntin1971a [taxonomy, description, illustration, host, distribution: 307-309]; NagarkSa1990 [host, distribution, economic importance, biological control: 553-542]; Prinsl1983 [distribution, biological control: 26]; Sudoi1995 [host, distribution, chemical control: 119-123].

Aspidiotus atripileus Munting

Aspidiotus atripileus Munting, 1971a: 309. Type data: SOUTH AFRICA: Cape Province, near Kuruman, on *Euclea crispa*; collected 10.iii.1967. Holotype female. Type depository: Pretoria: South African National Collection of Insects, South Africa; type no. 2721/5.
SCALE COVER: Female scale subcircular, about 2 mm in diameter, with dark brown almost black exuviae surrounded by white secreted matter. Male scale similar but about 1 mm in diameter (Munting, 1971a).
HOST PLANTS: **Ebenaceae**: *Euclea crispa* [Muntin1971a].
DISTRIBUTION: **Afrotropical**: South Africa [Muntin1971a].
GENERAL: Description and illustration of adult female by Munting (1971a).
CITATIONS: Muntin1971a [host, distribution, taxonomy, illustration, description: 309-311].

Aspidiotus beilschmiediae Takagi

Aspidiotus beilschmiediae Takagi, 1969a: 67. Type data: TAIWAN: Fen-chi-hu, on *Beilschmiedia erythrophloia*. Holotype female. Type depository: Sapporo: Entomological Institute, Faculty of Agriculture, Hokkaido University, Japan.
Temnaspidiotus beilschmiediae; Chou, 1985: 399. Change of combination.
SCALE COVER: Takagi (1969a) did not describe scale cover.
HOST PLANTS: **Lauraceae**: *Beilschmiedia erythrophloia* [Takagi1969a].
DISTRIBUTION: **Oriental**: Taiwan [Takagi1969a]. **Palaearctic**: Japan [Tachik1973, Kawai1980].
GENERAL: Description and illustration of adult female by Takagi (1969a) and by Chou (1985, 1986).
CITATIONS: Chou1985 [taxonomy, description, host, distribution: 399-400]; Chou1986[taxonomy, illustration: 665]; Kawai1980 [taxonomy, description, host, distribution: 228]; Tachik1973 [host, distribution, biological control: 137]; Takagi1969a [taxonomy, description, illustration, host, distribution: 66-67,99]; Tao1999 [taxonomy, host, distribution: 119].

Aspidiotus bicarinatus Walker
See *Aspidiotus bicarinatus* in the section "Taxa transferred from family" at end of catalogue.

Aspidiotus brachystegiae Hall
Aspidiotus brachystegiae Hall, 1928: 271. Type data: ZIMBABWE: Mazoe, on smaller branches of *Brachystegia flagristipulata*. Syntypes, female. Type depository: London: The Natural History Museum, England, UK.
SCALE COVER: Scale of female circular and semi-transparent; colour dull brown, paler at margin, with a reddish brown tinge in some examples; diameter 1.0-1.75 mm. Male scale pale brown in colour and of usual form; exuviae yellow, with conspicuous broad median longitudinal reddish brown stripe (Hall, 1928).
HOST PLANTS: **Leguminosae**: *Brachystegia flagristipulata* [Hall1928].
DISTRIBUTION: **Afrotropical**: Zimbabwe [Hall1928, Balach1956].
GENERAL: Description and illustration of adult female by Hall (1928) and by Balachowsky (1956).
KEYS: Balachowsky 1956: 51 (female) [Africa].
CITATIONS: Balach1932f [taxonomy, description, host, distribution: 231]; Balach1956 [taxonomy, description, illustration, host, distribution: 53-56]; Borchs1966 [catalogue: 260]; Ferris1941e [taxonomy: 41]; Hall1928 [taxonomy, description, illustration, host, distribution: 272-273].

Aspidiotus capensis Newstead
Aspidiotus (*Evaspidiotus*) *fimbriatus capensis* Newstead, 1917: 373. Type data: SOUTH AFRICA: Port Elizabeth, on cycads; collected by de Charmoy, 1914. Syntypes, female. Type depository: London: The Natural History Museum, England, UK.
Aspidiotus fimbriatus capensis; Brain, 1918: 121. Change of combination.
Aspidiotus capensis; Balachowsky, 1955: 391. Change of status.
SCALE COVER: Scale of female circular, 2.5 - 3 mm in diameter, moderately convex, pure opaque white. Exuviae central and normally covered with a thin layer of secretion, through which appear slightly brownish. Male scale similar in shape, but small, semi-transparent, with yellow exuviae (Brain, 1918).
HOST PLANTS: **Cycadaceae** [Newste1917, Balach1956], *Cycas* [RosenDe1979], *Encephalartos* [Brain1918, Balach1956].
NATURAL ENEMIES: HYMENOPTERA **Aphelinidae**: *Aphytis taylori* Quednau [RosenDe1979].   **Encyrtidae**: *Habrolepis algoensis* Annecke & Mynhardt [Prinsl1983].
DISTRIBUTION: **Afrotropical**: South Africa [BrainKe1917, Newste1917, Brain1918, Balach1956, RosenDe1979].
GENERAL: Description and illustration of adult female by Balachowsky (1956).
KEYS: Balachowsky 1956: 52 (female) [Africa]; Brain 1918: 117 (female) [South Africa].
CITATIONS: Balach1932f [taxonomy, host, distribution: 230]; Balach1955 [taxonomy: 391]; Balach1956 [taxonomy, description, illustration, host, distribution: 55-58]; Borchs1966 [catalogue: 260]; Brain1918 [taxonomy, description, illustration, host, distribution: 121]; BrainKe1917 [distribution: 183]; MacGil1921 [taxonomy, description, host, distribution: 402]; Muntin1971a [taxonomy: 307]; Newste1917 [taxonomy, description, illustration, host, distribution: 373]; Prinsl1983 [distribution, biological control: 26]; Quedna1964b [biological control: 86-116];

RosenDe1979 [host, distribution, biological control: 494-496]; Sander1906 [taxonomy, host, distribution: 13].

Aspidiotus capsulatus nomen nudum

Aspidiotus capsulatus Green, 1905a: 343. Nomen nudum.
Aspidiotus capsulatus Ferris, 1941e: 41. Nomen nudum.
Aspidiotus capsulatus Borchsenius, 1966: 376. Nomen nudum.

Aspidiotus cerasi Fitch

Aspidiotus cerasi Fitch, 1857a: 368. Type data: U.S.A.: New York State, on bark of choke cherry; No. 15. Syntypes, female. Type depositories: Washington: United States National Entomological Collection, U.S. National Museum of Natural History, District of Columbia, USA, and Albany: New York State Museum Insect Collection, New York, USA.
Aspidiotus cerasi; Fernald, 1903b: 217. Incorrect synonymy. Notes: With *Chionaspis furfura* (Fitch); see Borchsenius, 1966: 368.
SYSTEMATICS: Fernald (1903b: 217) incorrectly synonymized *Aspidiotus cerasi* Fitch with *Chionaspis furfura* (Fitch), whereas Borchsenius (1966: 368) regarded the former a valid species.
SCALE COVER: Fitch (1857a) described this species as:" In winter, on the bark of the choke cherry, little roundish white wax-like blisters, scarcely perceptible to the naked eye, containing beneath them in an open cavity a cluster of minute dull red or resin-like eggs."
HOST PLANTS: **Rosaceae**: *Cerasus* [Fitch1857a].
DISTRIBUTION: **Nearctic**: United States of America (New York [Fitch1857a].
CITATIONS: Borchs1966 [taxonomy: 368]; Fernal1903b [taxonomy: 217]; Ferris1941e [taxonomy: 41]; Fitch1857a [taxonomy: 368]; McCabeJo1980 [taxonomy: 7]; Signor1870 [taxonomy: 107].

Aspidiotus chamaeropsis Signoret

Aspidiotus chamaeropsis Signoret, 1869: 848. Nomen nudum.
Aspidiotus chamaeropsis Signoret, 1869b: 118. Type data: FRANCE: apparently Paris, on *Chamaerops australis*; collected by Mr. Boisduval. Syntypes, female. Type depository: Vienna: Naturhistorisches Museum Wien, Austria.
Aspidiotus (*Aspidiotus*) *chamaeropsis*; Cockerell, 1897i: 29. Change of combination.
Aspidiotus chamaeropis Lindinger, 1907a: 20. Unjustified emendation.
SCALE COVER: Female scale elongate, transparent; exuviae bright yellow, placed near margin (Signoret, 1869b).
HOST PLANTS: **Palmae**: *Chamaerops australis* [Signor1869b].
DISTRIBUTION: **Palaearctic**: France [Signor1869b].
GENERAL: Description and illustration of adult female by Signoret (1869b).
CITATIONS: Borchs1966 [catalogue: 368]; Cocker1896b [distribution: 334]; Cocker1897i [taxonomy, description, host, distribution: 29]; Comsto1883 [taxonomy: 74-75]; Fernal1903b [catalogue: 254]; Ferris1941e [taxonomy: 42]; Lepesm1947 [host, distribution: 195]; Lindin1907a [taxonomy: 20]; Lindin1935

[taxonomy: 368]; Lindin1957 [taxonomy: 543]; SchmutKlLu1959 [taxonomy: 374]; Signor1869 [taxonomy: 848]; Signor1869b [taxonomy, description, illustration, host, distribution: 118]; Targio1892 [taxonomy: 81].

Aspidiotus chinensis Kuwana & Muramatsu

Aspidiotus chinensis Kuwana & Muramatsu, 1931: 335. Type data: CHINA: Shanghai, intercepted at Japan, Nagasaki and Yokohama, on *Cymbidium faberi* [= Kyuka-ran]. Syntypes, female. Type depository: Ibaraki-ken: Insect Taxonomy Laboratory, National Institute of Agricultural Environmental Sciences, Kannon-dai, Yatabe, Tsukuba-shi, (I. Kuwana), Japan.

SCALE COVER: Female scale nearly oblong, flat, more or less delicate in texture, greyish dark in colour; exuviae nearly central, pale yellow; length 2 mm, width 1 mm (Kuwana & Muramatsu, 1931).

HOST PLANTS: **Orchidaceae**: *Cymbidium faberi* [KuwanaMu1931].

DISTRIBUTION: **Oriental**: China (People's Republic) (Shanghai [KuwanaMu1931]).

GENERAL: Description and illustration of adult female by Kuwana & Muramatsu (1931).

CITATIONS: Borchs1966 [catalogue: 260]; Chou1985 [taxonomy, description, host, distribution: 398]; DanzigPe1998 [catalogue: 187]; Ferris1941e [taxonomy: 42]; KuwanaMu1931 [taxonomy, description, illustration, host, distribution: 335-339]; Lindin1936b [taxonomy: 286]; Takagi1957 [taxonomy: 35]; TangQiWa1990 [host, distribution, life history, chemical control: 187-194]; Tao1999 [taxonomy, description, host, distribution: 73].

Aspidiotus circularis Fitch

Aspidiotus circularis Fitch, 1857b: 426. Type data: USA: New York, gardens of the city of Albany, on currant [=*Ribes*] stalks. Syntypes, female. Type depository: New York: American Museum of Natural History, Department of Entomology Collection, New York, USA.

SCALE COVER: Fitch (1857b) described this species as follows: "On the bark of currant stalks in gardens of the city of Albany early in the spring, I have observed a minute circular flat scale only 0.03 inch diameter, similar to a species named *Aspidiotus nerii* but differently coloured, being of the same blackish brown hue with the surrounding bark and having in the centre a smooth round wart-like elevation of a pale yellow colour".

HOST PLANTS: **Saxifragaceae**: *Ribes* [Fitch1857b].

DISTRIBUTION: **Nearctic**: United States of America (New York [Fitch1857b]).

CITATIONS: Borchs1966 [taxonomy: 368]; Fernal1903b [taxonomy: 252]; Ferris1941e [taxonomy: 41]; Fitch1857b [taxonomy, description, host, distribution: 426]; Marlat1900a [taxonomy: 590]; Marlat1908b [taxonomy: 309]; Sander1910 [taxonomy: 61]; Signor1877 [taxonomy: 601].

Aspidiotus cochereaui Matile-Ferrero & Balachowsky
Aspidiotus cochereaui Matile-Ferrero & Balachowsky, 1973: 241. Type data: NEW
CALEDONIA: Mont Koghi, on undetermined plant; collected 2.XI.1966. Holotype
female. Type depository: Paris: Muséum national d'Histoire naturelle, France.
SCALE COVER: Female scale circular, slightly convex; colour dark brown; scale
partly or completely covered with white powdery secretion; diameter 1.8-2 mm
(Matile-Ferrero & Balachowsky, 1973).
HOST PLANTS: **Epacridaceae**: *Dracophyllum* [WilliaWa1988], *Dracophyllum
ramosum* [WilliaWa1988]. **Fagaceae**: *Nothofagus codonandra* [WilliaWa1988].
DISTRIBUTION: **Australasian**: New Caledonia [MatileBa1973, WilliaWa1988].
GENERAL: Description and illustration of adult female by Matile-Ferrero &
Balachowsky (1973) and by Williams & Watson (1988).
KEYS: Williams & Watson 1988: 49 (female) [Tropical South Pacific].
CITATIONS: MatileBa1973 [taxonomy, description, illustration, host, distribution:
241-243]; WilliaWa1988 [taxonomy, description, illustration, host, distribution:
48,51-53].

Aspidiotus combreti Hall
Aspidiotus combreti Hall, 1928: 273. Type data: ZIMBABWE: Mazoe, on
Combretum apiculatum and *Combretum* sp. Syntypes, female. Type depository:
London: The Natural History Museum, England, UK.
SCALE COVER: Female scale highly convex, usually circular, but crowded
examples oval in outline; exuviae more or less central; diameter 1.25-1.75 mm.
Male scale not seen (Hall, 1928).
HOST PLANTS: **Combretaceae**: *Combretum* [Hall1928, Balach1956],
Combretum apiculatum [Hall1928, Balach1956].
DISTRIBUTION: **Afrotropical**: Mauritania [BalachMa1970]; Zimbabwe
[Hall1928, Balach1956].
GENERAL: Description and illustration of adult female by Hall (1928) and by
Balachowsky (1956).
KEYS: Balachowsky 1956: 52 (female) [Africa].
CITATIONS: Balach1932f [taxonomy, host, distribution: 230-231]; Balach1955
[taxonomy, distribution: 391]; Balach1956 [taxonomy, description, illustration, host,
distribution: 57-60]; BalachMa1970 [host, distribution: 1080]; Borchs1966
[catalogue: 260]; Ferris1941e [taxonomy: 42]; Hall1928 [taxonomy, description,
illustration, host, distribution: 273-274].

Aspidiotus commelinae nomen nudum
Aspidiotus commelinae Lindinger, 1957: 545. Nomen nudum. Notes: This name was
listed by Lindinger (1957: 545) as follows: "*A. cammelinae* Seabra (1920) = *A.
obliquus*". No publication by Seabra (1920) was found (Y. Ben-Dov, December,
2002).
Aspidiotus commelinae Borchsenius, 1966: 368. Nomen nudum.

Aspidiotus comorensis Mamet

Aspidiotus comorensis Mamet, 1960: 159. Type data: COMOROS ISLANDS: "Coeur de Boeuf", on undetermined plant. Holotype female. Type depository: Paris: Muséum national d'Histoire naturelle, France.
SCALE COVER: Female scale straw-coloured, low convex, circular; exuviae subcentral. Male scale not observed (Mamet, 1960).
HOST PLANTS: **Annonaceae**: *Annona reticulata* [Matile1978].
DISTRIBUTION: **Afrotropical**: Comoros Islands [Mamet1960, Matile1978].
GENERAL: Description and illustration of adult female by Mamet (1960) and by Matile-Ferrero (1978).
CITATIONS: Borchs1966 [catalogue: 260]; Mamet1960 [taxonomy, description, illustration, host, distribution: 159-160]; Matile1978 [taxonomy, description, illustration, host, distribution: 61-62].

Aspidiotus comperei Marlatt

Aspidiotus comperei Marlatt, 1908c: 12. Type data: AUSTRALIA: West Australia, Raventhorpe, on *Hakea* sp.; collected by George Compere, 1902. Holotype female. Type depository: Washington: U.S. National Entomological Collection, U.S. National Museum of Natural History, District of Columbia, USA; type no. 14129.
Aspidoides comperei; MacGillivray, 1921: 406. Change of combination.
Aspidiotus comperei; Borchsenius, 1966: 268. Revived combination.
SCALE COVER: Female scale 1 mm in diameter, strongly convex, circular; exuviae covered; covering secretion easily rubbed off, exposing lemon-yellow to brown exuviae; secretion covering larval exuviae sometimes more dense and adherent, thus sometimes giving scale an annulated appearance due to second exuvia showing through section surrounding larval skin; secretionary supplement normally white, sometimes slightly yellowed; ventral scale inconsiderable. Male scale oval, about half size of female (Marlatt, 1908c).
HOST PLANTS: **Proteaceae**: *Hakea* [Marlat1908c, Frogga1914].
DISTRIBUTION: **Australasian**: Australia (Western Australia [Frogga1914]).
GENERAL: Description and illustration of adult female by Marlatt (1908c).
CITATIONS: Borchs1966 [catalogue: 268-269]; Ferris1941e [taxonomy: 42]; Frogga1914 [taxonomy, description, host, distribution: 136]; Frogga1915 [taxonomy, description, host, distribution: 13]; MacGil1921 [taxonomy, description, host, distribution: 406]; Marlat1908c [taxonomy, description, illustration, host, distribution: 12-13].

Aspidiotus congolensis Balachowsky

Aspidiotus destructor congolensis Balachowsky, 1956: 64. Type data: ZAIRE: Bambesa, on *Musa* sp. Holotype female. Type depository: Tervuren: Musée Royal de l'Afrique Centrale, Section d'Entomologie, Belgium.
Temnaspidiotus congolensis; Borchsenius, 1966: 270. Change of combination and rank.
Aspidiotus congolensis; Williams & Watson, 1988: 49. Revived combination.
SYSTEMATICS: Balachowsky (1956) described *Aspidiotus destructor congolensis* noting that it differed from typical African form in constantly possessing a

continuous row of short microducts along submargin from abdominal segment III to head.

SCALE COVER: General appearance identical to that of typical African form (Balachowsky, 1956).

HOST PLANTS: **Musaceae**: *Musa* [Balach1956].

DISTRIBUTION: **Afrotropical**: Zaire [Balach1956].

GENERAL: Description and illustration of adult female by Balachowsky (1956).

KEYS: Balachowsky 1956: 51 (female) [Africa].

CITATIONS: Balach1956 [taxonomy, description, illustration, host, distribution: 63-64]; Borchs1966 [catalogue: 270].

Aspidiotus corticalis nomen nudum

Aspidiotus corticalis Cockerell, 1894: 376. Nomen nudum.

Aspidiotus corticalis Ferris, 1941e: 42. Nomen nudum.

Aspidiotus corticalis Borchsenius, 1966: 376. Nomen nudum.

Aspidiotus corticalis Gordh, 1979: 94. Nomen nudum.

Aspidiotus coryphae Cockerell & Robinson

Aspidiotus coryphae Cockerell & Robinson, 1915a: 425. Type data: PHILIPPINES: Los Banos, on *Corypha elata*. Syntypes, female. Type depository: Washington: United States National Entomological Collection, U.S. National Museum of Natural History, District of Columbia, USA.

COMMON NAME: corypha scale [Velasq1971].

SCALE COVER: Female scale subcircular, flat, thin, dull white or creamish and about 2 mm diameter; exuviae oval, subcentral, darker than rest of scale. Male unknown (Velasquez, 1971).

HOST PLANTS: **Cycadaceae**: *Cycas revoluta* [Velasq1971]. **Palmae**: *Coccus nucifera* [Velasq1971], *Corypha elata* [CockerRo1915a, Muntin1971a], *Roystonea elata* [Velasq1971].

DISTRIBUTION: **Oriental**: Philippines [CockerRo1915a, VelasqRi1969, Velasq1971] (Luzon [Muntin1971a, Velasq1971]).

GENERAL: Description and illustration of adult female by Cockerell & Robinson (1915a), Munting (1971a) and by Velasquez (1971). Description of adult female by Ferris (1941e).

KEYS: Velasquez 1971: 100 (female) [Philippines]; Robinson 1917: 29 (female) [Philippines].

CITATIONS: Borchs1966 [catalogue: 260]; CockerRo1915a [taxonomy, description, illustration: 425]; Ferris1941e [taxonomy, description, host, distribution: 42,51]; Lepesm1947 [host, distribution: 195]; MacGil1921 [taxonomy, description, host, distribution: 401]; Muntin1971a [taxonomy, description, illustration, host, distribution: 315-316]; Robins1917 [taxonomy, description, host, distribution: 31]; Sugimo1994 [host, distribution: 115-121]; Velasq1971 [taxonomy, description, illustration, host, distribution: 103-104]; VelasqRi1969 [host, distribution: 195-208].

Aspidiotus crenulatus (Pampaloni)

Diaspis crenulata Pampaloni, 1902: 130. Type data: ITALY: Sicily, Syracuse Province, half an hour walk to northeast of Melilli, along Fiuminello torrent, in a form of 'brown coal' (lignite), from the Miocene age. Syntypes, female. Notes: Type of this fossil not traced by Jan Koteja (Koteja & Ben-Dov, 2003).

Aspidiotus proteus; Koteja, 2000c: 208. Misspelling of species name. Note: *Aspidiotus proteus* Koteja is *lapsus calami* for *Aspidiotus crenulatus* Pampaloni.

Aspidiotus crenulatus; Koteja & Ben-Dov, 2003: 165. Change of combination.

DISTRIBUTION: **Palaearctic**: Sicily [Pampal1902, Pampal1903].

GENERAL: Fossil scale insect found in 'brown coal' (lignite) from the Miocene period (Pampaloni, 1902).

CITATIONS: Koteja2000c [taxonomy, distribution: 208]; KotejaBe2003 [taxonomy: 165-166]; Pampal1902 [taxonomy, illustration, distribution: 130]; Pampal1903 [taxonomy: 253].

Aspidiotus cryptomeriae Kuwana

Aspidiotus cryptomeriae Kuwana, 1902: 69. Type data: JAPAN: Honshu, Gifu-Ken, on *Cryptomeria japonica*. Syntypes, female. Type depository: Ibaraki-ken: Insect Taxonomy Laboratory, National Institute of Agricultural Environmental Sciences, Kannon-dai, Yatabe, Tsukuba-shi, (I. Kuwana), Japan.

COMMON NAME: Cryptomeria Scale [Koszta1996].

SCALE COVER: Female scale elliptical, flatly convex; length 1.1-2.0 mm, width 1 mm; usual colour greyish sub-transparent; exuviae usually a little to one side of center; straw colour; first skin usually showing segmentation distinctly, length 0.4 mm; second skin covered with secretion, length about 0.65 mm; ventral scale a mere film applied to bark of plant. Scale of male elongated, with larval skin nearly central; colour greyish, same as female in texture; larval skin straw colour. Length about 1.0 mm , width 0.6 mm (Kuwana, 1902).

HOST PLANTS: **Cephalotaxaceae**: *Cephalotaxus* [Koszta1996], *Cephalotaxus harringtonia* [Takagi1969a]. **Cupressaceae**: *Chamaecyparis* [Takagi1957, Danzig1978, Danzig1980b, Koszta1996], *Chamaecyparis obtusa* [Kuwana1933, Takagi1969a], *Chamaecyparis obtusa breviramea* [Kuwana1933], *Chamaecyparis pisifera* [Kuwana1933, Takagi1969a], *Juniperus* [Danzig1980b], *Juniperus chinensis* [Takagi1969a, Tang1984]. **Pinaceae**: *Abies* [Takagi1957, Danzig1978, Koszta1996], *Abies firma* [Kuwana1933, TakahaTa1956, Takagi1969a, Kawai1977, Tang1984], *Abies halophylla* [Danzig1980b], *Abies nephrolepis* [Danzig1980b], *Abies sachalinensis* [Takagi1969a, Danzig1980b], *Cedrus* [Koszta1996], *Keteleeria* [Danzig1978, Danzig1980b], *Keteleeria evelyniana* [Takagi1969a, Tang1984], *Picea* [Danzig1978], *Picea ajanensis* [Danzig1980b], *Picea excelsa* [Takagi1969a], *Picea glehnii* [Danzig1980b], *Pinus* [Takagi1957], *Pinus koraiensis* [Danzig1978, Danzig1980b], *Pinus thunbergii* [Kuwana1933], *Pseudotsuga* [Koszta1996], *Pseudotsuga japonica* [Takagi1969a], *Tsuga canadensis* [StoetzDa1974a], *Tsuga sieboldii* [Takagi1969a]. **Taxaceae**: *Taxus* [Koszta1996], *Taxus cuspidata* [Takagi1957, Takagi1969a, Danzig1978, Danzig1980b, Tang1984], *Torreya* [Danzig1978, Danzig1980b], *Torreya nucifera* [Takagi1957, Takagi1969a, Kawai1977]. **Taxodiaceae**: *Cryptomeria* [Takagi1957, Danzig1978, Danzig1980b],

Cryptomeria japonica [Kuwana1902, Kuwana1933, TakahaTa1956, Takagi1969a, Tang1984].
NATURAL ENEMIES: COLEOPTERA **Coccinellidae**: *Chilocorus kuwanae* [Muraka1970], *Pseudoscymnus hareja* [Muraka1970]. HYMENOPTERA **Aphelinidae**: *Aspidiotiphagus citrinus* (Craw) [Muraka1970]. **Encyrtidae**: *Comperiella bifasciata* Howard [Trjapi1989], *Comperiella indica* Ramakrishna Ayyar [Trjapi1989].
DISTRIBUTION: **Nearctic**: United States of America (Connecticut [Nakaha1982], Delaware [Nakaha1982], Indiana [Nakaha1982], Maryland [StoetzDa1974a], New York [Nakaha1982], Pennsylvania [Stimme1986]). **Oriental**: China (People's Republic) (Yunnan [Takagi1969a]); Taiwan [Takagi1969a]. **Palaearctic**: China (People's Republic) [Tang1984] (Shandong (Shantung) [Danzig1980b]); Japan [Kuwana1917a, Takagi1969a] (Hokkaido [TakahaTa1956, Takagi1957], Honshu [Kuwana1902, TakahaTa1956, Takagi1957], Kyushu [TakahaTa1956], Shikoku [TakahaTa1956]); Russia (Khabarovsk Kray [Danzig1980b, Danzig1988], Primor'ye Kray [Danzig1980b, Danzig1988], Sakhalin Oblast [TakahaTa1956, Danzig1980b, Danzig1988]); South Korea [TakahaTa1956, Takagi1969a].
BIOLOGY: Infests leaves.
ECONOMIC IMPORTANCE: A pest of conifers in Maryland and Pennsylvania, USA (Stimmel, 1986; Davidson & Raupp, 1999; Gardosik, 2001), Japan (Kawai, 1977) and China (Schmutterer et al., 1957).
GENERAL: Description and illustration of adult female by Kuwana (1902), Kuwana (1933), Ferris (1953), Takagi (1957, 1969a), Danzig (1980b), Tang (1984) and by Kosztarab (1996). Description and illustration of female and male nymphs, and male pupa and prepupa by Stoetzel & Davidson (1974a). This species exhibits a remarkable host-induced variation in Japan in following parameters: variation in Esterase isozymes (Miyanoshita *et al.*, 1991); adult male and female morphology (Miyanoshita *et al.*, 1993; Miyanoshita & Tatsuki, 1995); and role of sex pheromone (Miyanoshita & Tatsuki, 2001). Morphological differences and host preference of two forms of *Aspidiotus cryptomeriae* Kuwana associated with *Cryptomeria japonica* D. Don and *Torreya nucifera* Sieb. et Zucc. examined in Kanto district, Japan. Adult males and females showed statistically significant differences in some characters. However, impossible to divide specimens perfectly by using any one character. Linear discriminant analysis using combinations of several characteristics (>99%) discriminated them. Two sympatric populations of adult females collected at same stand correctly classified by linear discriminant functions. Test of forced host preference to five conifers indicated that each population monophagous. Studies by Miyanoshita et al. suggest strongly that these two forms are distinct host races. If same holds true in other parts of Japan, these races might be considered to have differentiated into distinct species (Miyanoshita et al., 1993).
KEYS: Kosztarab 1996: 427 (female) [Northeastern North America]; Danzig 1993: 40-41 (female) [Europe]; Chou 1985: 262-263 (female) [Species of China]; Takagi 1957: 32 (female) [Japan]; Kuwana 1933: 3 (female) [Japan]; Kuwana 1933b: 49 (female) [Japan].
CITATIONS: Borchs1966 [catalogue: 260]; Brick1912 [host, distribution: 1-22]; Brown1960a [taxonomy, structure, chromosomes: 135-160]; Bustsh1958

[taxonomy: 219,239]; Chou1985 [taxonomy, description, host, distribution: 267-268]; Danzig1977b [taxonomy: 57]; Danzig1978 [host, distribution: 18]; Danzig1980b [taxonomy, description, illustration, host, distribution: 336-338]; Danzig1988 [taxonomy, host, distribution: 725]; DanzigPe1998 [catalogue: 187-188]; DavidsRa1999 [economic importance, chemical control, biological control: 1]; DonMiTa1995 [host, distribution, taxonomy: 159-162]; Fernal1903b [catalogue: 255]; Ferris1941e [taxonomy: 42]; Ferris1953 [taxonomy, description, illustration, host, distribution: 65]; Gardos2001 [host, distribution, economic importance, control, taxonomy: 23-25]; HolmesDa1984 [biological control: 65-70]; JohnsoLy1991 [host, distribution, economic importance: 5]; Kawai1977 [host, distribution, economic importance: 157]; Koszta1996 [taxonomy, description, illustration, host, distribution, biological control, economic importance: 427-429]; Kuwana1902 [taxonomy, description, illustration, host, distribution: 69]; Kuwana1907 [host, distribution: 196]; Kuwana1917a [taxonomy, distribution: 174]; Kuwana1933 [taxonomy, description, illustration, host, distribution: 4-5]; Lindin1909c [taxonomy, host, distribution: 449]; MacGil1921 [taxonomy, description, host, distribution: 400]; MillerDa1990 [host, distribution, economic importance: 300]; MiyanoKaFu1993 [taxonomy, description, illustration, host, distribution: 71-80]; MiyanoTa1995 [taxonomy, description, host, distribution: 159-162]; MiyanoTa2001 [taxonomy, chemistry, life history, host, distribution: 199-201]; MiyanoTaKu1991 [taxonomy, chemistry, host, distribution: 317-321]; Muraka1970 [host, distribution, biological control: 71-72]; Nakaha1982 [host, distribution: 13]; SchmutKlLu1957 [host, distribution, economic importance: 475]; Stimme1986 [host, distribution, description, life history, economic importance, control: 21-22]; StoetzDa1974 [taxonomy, life history: 138-140]; StoetzDa1974a [taxonomy, description, illustration, host, distribution, life history: 491-494]; Takagi1957 [taxonomy, description, illustration, host, distribution: 32-33]; Takagi1969a [taxonomy, description, illustration, host, distribution: 64-65,99]; Takagi1990 [taxonomy, structure: 100]; TakahaTa1956 [taxonomy, host, distribution: 14]; Tang1984 [taxonomy, description, illustration, host, distribution: 16-17]; Tao1999 [taxonomy, host, distribution: 73]; Trjapi1989 [biological control: 296,297]; Zahrad1990 [host, distribution, description: 641].

Aspidiotus cryptus Signoret
See *Aspidiotus cryptus* in the section "Taxa transferred from family" at end of catalogue.

Aspidiotus cymbidii Bouché
Aspidiotus cymbidii Bouché, 1844: 296. Type data: GERMANY: Berlin, in greenhouse, on *Cymbidium chinense* introduced from China. Syntypes, both sexes. Notes: Type material lost (Sachtleben, 1944).
Diaspis cymbidii; Signoret, 1869d: 436. Change of combination.
Aulacaspis cymbidii; Cockerell, 1893k: 548. Change of combination.
Aspidiotus cymbidii; Borchsenius, 1966: 369. Revived combination.
SCALE COVER: Oval, flat, with eccentric "absatzen"; brown at top (Bouché, 1844).

HOST PLANTS: **Orchidaceae**: *Cymbidium* [Kuwana1927], *Cymbidium chinense* [Bouche1844].
DISTRIBUTION: **Palaearctic**: China (People's Republic) [Kuwana1927]; Germany (United) [Bouche1844].
GENERAL: Bouché (1844) noted that this species similar to *Aspidiotus nerii* Bouché, but scale different.
CITATIONS: Borchs1966 [catalogue: 369]; Bouche1844 [taxonomy, description, host, distribution: 296]; Cocker1893k [taxonomy, host, distribution: 548]; Cocker1899n [taxonomy: 30]; Comsto1883 [taxonomy, description, host, distribution: 369]; Fernal1903b [catalogue: 324]; Ferris1941e [taxonomy: 42]; Kuwana1927 [host, distribution: 71]; Lindin1934e [taxonomy, description, host, distribution: 160]; Sachtl1944 [taxonomy, description, host, distribution: 71]; Signor1869 [taxonomy: 851]; Signor1869d [taxonomy, description, host, distribution: 436-437]; Tao1999 [taxonomy, host, distribution: 101].

Aspidiotus dallonii Balachowsky
Aspidiotus (*Evaspidiotus*) *dallonii* Balachowsky, 1932f: 228. Type data: CHAD: Massif du Tibesti, oasis de Goumeur, altitude 1000-1200m., on *Ficus salicifolia* var. *teloucat*. Syntypes, female. Type depository: Paris: Muséum national d'Histoire naturelle, France.
Octaspidiotus dallonii; Balachowsky, 1948b: 271. Change of combination.
Aspidiotus dallonii; Borchsenius, 1966: 260. Revived combination.
SCALE COVER: Female scale circular, white; exuviae central, yellow, sometimes covered with white secretion of adult; slightly convex; without ventral vellum; margin of scale thin; diameter 1.8-2 mm. Male scale subcircular, slightly oval, white; exuviae yellow, subcentral; posterior margin thin and slightly cottony; 1.2-1.3 mm long (Balachowsky, 1948b).
HOST PLANTS: **Apocynaceae**: *Arduina edulis* [Balach1956]. **Leguminosae**: *Bauhinia* [Balach1956], *Tamarindus indicus* [Balach1956]. **Moraceae**: *Ficus salicifolia* var. *teloukat* [Balach1932f, Balach1948b], *Ficus thonningii* [Balach1956]. **Ochnaceae**: *Lophira alata* [Balach1956, Balach1958a].
DISTRIBUTION: **Afrotropical**: Central African Republic [Balach1958a]; Chad [Balach1932f, Balach1948b]; Eritrea [Balach1956]; Guinea [Balach1956, Balach1958a]; Sudan [Balach1956]; Zaire [Balach1956].
GENERAL: Description and illustration of adult female by Balachowsky (1932f, 1948b, 1956).
KEYS: Balachowsky 1956: 52 (female) [Africa].
CITATIONS: Balach1932f [taxonomy, description, illustration, host, distribution: 228-231]; Balach1948b [taxonomy, description, illustration, host, distribution: 271-273]; Balach1954f [host, distribution: 99]; Balach1955 [taxonomy, host, distribution: 391]; Balach1956 [taxonomy, description, illustration, host, distribution: 59-61]; Balach1958 [host, distribution: 23]; Balach1958a [host, distribution: 34]; Borchs1966 [catalogue: 260-261]; Ferris1941e [taxonomy: 42]; Lindin1957 [taxonomy: 550].

Aspidiotus destructor Signoret

Aspidiotus destructor Signoret, 1869: 851. Nomen nudum.

Aspidiotus destructor Signoret, 1869a: 120. Type data: RÉUNION ISLAND: on coconut palm and other palms; collected by Dr. Vinson. Syntypes, female. Type depository: Vienna: Naturhistorisches Museum Wien, Austria.

Aspidiotus transparens Green, 1890: 20. Type data: SRI LANKA: on tea plant, [*Thea* sp.]. Lectotype, by subsequent designation Williams & Watson, 1988: 53. Type depository: London: The Natural History Museum, England, UK. Synonymy by Green, 1910a: 201.

Aspidiotus cocotis Newstead, 1893d: 186. Type data: GUYANA: Demerara, Botanic Garden, on *Cocos nucifera*. Lectotype female, by subsequent designation Williams & Watson, 1988: 53. Type depository: London: The Natural History Museum, England, UK. Synonymy by Leonardi, 1898c: 62.

Aspidiotus fallax Cockerell, 1893j: 252. Type data: ANTIGUA: on mango and *Terminalia catappa*. Syntypes, female. Type depository: Washington: United States National Entomological Collection, U.S. National Museum of Natural History, District of Columbia, USA. Synonymy by Ferris, 1941e: 43.

Aspidiotus vastatrix Leroy, 1896: xx. Type data: Unknown. Syntypes, female and first instar. Synonymy by Ferris, 1941e: 49. Notes: The publication by Leroy, as cited by Ferris (1941e: 45), in which this name was listed, could not be traced (Yair Ben-Dov, December 2002).

Aspidiotus destructor fallax; Cockerell, 1896b: 334. Change of status.

Aspidiotus lataniae; Green, 1896e: 40. Misidentification; discovered by Borchsenius, 1966: 270.

Aspidiotus (Aspidiotus) transparens; Cockerell, 1897i: 28. Change of combination.

Aspidiotus (Aspidiotus) destructor; Cockerell, 1897i: 29. Change of combination.

Aspidiotus (Evaspidiotus) destructor; Leonardi, 1898a: 76. Change of combination.

Aspidiotus (Evaspidiotus) lataniae; Leonardi, 1898a: 76. Misidentification; discovered by Borchsenius, 1966: 270.

Aspidiotus transparens simillimus Cockerell, 1898m: 27. Type data: AUSTRALIA: New South Wales, Sydney, on a palm; collected by Mr. Alex Craw, December 1897. Syntypes, female. Type depository: Washington: United States National Entomological Collection, U.S. National Museum of Natural History, District of Columbia, USA. Synonymy by Cockerell & Robinson, 1915: 106.

Aspidiotus simillimus translucens Cockerell *in* Fernald, 1903b: 278. Replacement name for *Aspidiotus (Evaspidiotus) transparens* Cockerell, 1899q.

Aspidiotus translucens; Cockerell & Robinson, 1915: 106. Change of status.

Aspidiotus destructor transparens; Green, 1915c: 44. Change of status.

Aspidiotus translucens; Ferris, 1941e: 49. Change of combination.

Aspidiotus destructor; Williams & Watson, 1988: 53. Revived combination.

Aspidiotus coccotis; Danzig, 1993: 141. Misspelling of species name.

Aspidiotus transparens simillimus; Danzig, 1993: 145. Incorrect synonymy. Notes: Incorrect synonymy with *Aspidiotus nerii*.

Aspidiotus vastatrex; Tao, 1999: 73. Misspelling of species name.

COMMON NAMES: coconut scale [MoutiaMa1946, Merril1953, McKenz1956, RosenDe1978]; razrushaushaya shitovka [Borchs1936].

SYSTEMATICS: Ferris (1938a, 1941e) and Balachowsky (1956) discussed variation in relative size of median and second lobes on pygidium of adult female. Balachowsky (1957c: 199) named two forms in populations of *Aspidiotus destructor*, namely, *A. destructor* forma *africane* (with second lobes more developed than median, occurring on coconut) and *A. destructor* forma *americane* (in which second lobes less developed than median, occurring on avocado). Williams & Watson (1988) elucidated this variation in four illustrations from different regions and host plants, but all were named *Aspidiotus destructor*. Takagi (1969a) noted that *Aspidiotus watanabei* is close to *A. destructor*. Williams & Watson (1988) stated that "... the range of variation of *A. destructor* from the South Pacific area encompasses that of *A. watanabei*...", but did not synonymize latter. Danzig (1993) listed *A. watanabei* as a synonym of *A. destructor*. Reyne (1947, 1948) described subspecies *Aspidiotus destructor rigidus* from Indonesia, which was distinguished from typical *destructor* mainly by biological features, as well as difference in scale structure.

SCALE COVER: Female scale somewhat straw-coloured, flat circular very thin and delicate, exuviae central and quite pale, that of male slightly elongate, similar to that of female in colour and texture (Ferris, 1938a). See colour photograph in Wong *et al.* (1999).

HOST PLANTS: **Anacardiaceae**: *Anacardium occidentale* [Lepage1938, Balach1956, Beards1966], *Mangifera indica* [Cocker1893j, Takaha1933, Lepage1938, RahmanAn1941, McKenz1956, Takagi1969a, Velasq1971, Matile1984c]. **Annonaceae**: *Annona* [Green1904a, Takaha1932a, Takaha1933, Merril1953], *Annona cherimolia* [Balach1956, Matile1984c], *Annona muricata* [GomezM1941, Almeid1973b], *Annona reticulata* [Mamet1956, Borchs1966, Velasq1971, WilliaWa1988], *Annona senegalensis* [Balach1956], *Annona squamosa* [Lepage1938, WilliaWa1988], *Anomianthus heterocarpus* [Green1904a]. **Apocynaceae**: *Allemanda hendersoni* [WilliaWa1988], *Carissa* [Ramakr1921a], *Plumeria acuminata* [Takaha1939b], *Plumeria acutifolia* [Mamet1943a, Mamet1949, Borchs1966, WilliaWa1988], *Plumeria rubra* [WilliaWa1988], *Trachelospermum asiaticum* [TakahaTa1956]. **Aquifoliaceae**: *Ilex colchicum* [Hadzib1983]. **Araceae**: *Alocasia* [Green1900a], *Colocasia esculenta* [WilliaWa1988], *Xanthosoma sagittifolium* [WilliaWa1988]. **Arecaceae** [WilliaWa1988]. **Asclepiadaceae**: *Calotropis* [Ramakr1921a]. **Avicenniaceae**: *Avicennia* [Balach1956]. **Bixaceae**: *Bixa orellana* [Green1904a]. **Bombacaceae**: *Ceiba pentandra* [Beards1966, WilliaWa1988]. **Cannaceae**: *Canna indica* [Almeid1973b, WilliaWa1988]. **Caprifoliaceae**: *Lonicera japonica* [Takaha1934]. **Caricaceae**: *Carica papaya* [Takaha1929, Takaha1941b, DeLott1967a, Brimbl1968, Almeid1973b, WilliaWa1988]. **Celastraceae**: *Euonymus radicans* [Lindin1911]. **Chrysobalanaceae**: *Chrysobalanus* [Leonar1914]. **Clusiaceae**: *Calophyllum inophyllum* [Newste1906a, Merril1953, WilliaWa1988]. **Combretaceae**: *Combretum erythrophyllum* [Merril1953], *Terminalia* [Lepage1938, Takaha1942d, MatileNo1984], *Terminalia catappa* [Cocker1893j, MerrilCh1923, GomezM1941, Almeid1973b]. **Cruciferae**: *Brassica asiatica* [WilliaWa1988], *Brassica chinensis* [WilliaWa1988], *Brassica napus* [WilliaWa1988], *Brassica oleracea* [WilliaWa1988], *Raphanus sativus*

[WilliaWa1988]. **Cucurbitaceae** [WilliaWa1988], *Cucumis sativus* [WilliaWa1988]. **Cycadaceae**: *Cycas* [Green1930b, WilliaWa1988], *Cycas revoluta* [Lindin1911]. **Dilleniaceae**: *Dillenia biflora* [WilliaWa1988]. **Dioscoreaceae**: *Dioscorea nummularia* [WilliaWa1988]. **Dipterocarpaceae**: *Dipterocarpus* [Takaha1942b]. **Euphorbiaceae**: *Aleurites moluccana* [WilliaWa1988], *Aleurites triloba* [WilliaWa1988], *Breynia disticha* [WilliaBu1987, WilliaWa1988], *Euphorbia* [WilliaWa1988], *Euphorbia pulcherrima* [Mamet1949, Beards1966, Borchs1966, WilliaWa1988], *Hevea ˙ brasiliensis* [Green1904a, Newste1917b, WilliaWa1988], *Macaranga seemannii* [WilliaWa1988], *Manihot* [Green1900a], *Manihot glazioui* [Lindin1910b], *Poinsettia* [Mamet1943a, Borchs1966], *Sapium sebiferum* [Takaha1934]. **Flacourtiaceae**: *Scolopia oldhami* [Takaha1935]. **Geraniaceae**: *Pelargonium* [Mamet1943a, Mamet1949, Borchs1966]. **Gnetaceae**: *Gnetum pirifolium* [Lindin1911]. **Gramineae**: *Saccharum officinarum* [McKenz1956, WilliaWa1988]. **Heliconiaceae**: *Heliconia bahai* [WilliaWa1988]. **Hernandiaceae**: *Hernandia* [Beards1966]. **Lardizabalaceae**: *Stauntonia obovatitolia* [Takaha1934]. **Lauraceae**: *Cinnamomum camphora* [Takaha1932a, Takaha1933], *Cinnamomum zeylanicum* [WilliaWa1988], *Laurus nobilis* [Borchs1934, Borchs1936], *Litsea vitiensis* [WilliaWa1988], *Persea americana* [Takaha1941b, Mamet1943a, Mamet1949, Borchs1966, WilliaWa1988], *Persea gratissima* [Brain1918, Lepage1938, GomezM1941]. **Lecythidaceae**: *Barringtonia* [Beards1966, WilliaWa1988], *Barringtonia asiatica* [WilliaWa1988]. **Leguminosae**: *Albizia lebbek* [WilliaWa1988], *Bauhinia* [Mamet1943a, Mamet1949, Borchs1966], *Cassia* [Ramakr1921a, WilliaWa1988], *Cassia nodosa* [WilliaWa1988], *Cassia occidentalis* [WilliaWa1988], *Cassia tora* [WilliaWa1988], *Crotalaria mucronata* [WilliaWa1988], *Crotalaria saltiana* [WilliaWa1988], *Dalbergia* [Ramakr1921a], *Dalbergia championi* [Green1900a], *Inocarpus fagifer* [WilliaWa1988], *Vigna unguiculata* [WilliaWa1988]. **Liliaceae**: *Asparagus sprengeri* [Merril1953], *Ophiopogon japonicus* [Hadzib1983]. **Loranthaceae**: *Loranthus* [Balach1956]. **Lythraceae**: *Lagerstroemia indica* [GomezM1941, Mamet1943a, Mamet1949, Borchs1966]. **Magnoliaceae**: *Michelia alba* [Takaha1932a, Takaha1933], *Michelia champaca* [Velasq1971]. **Malvaceae**: *Hibiscus* [WilliaWa1988]. **Meliaceae**: *Swietenia mahagoni* [GomezM1941]. **Moraceae**: *Artocarpus altilis* [Beards1966, WilliaBu1987, WilliaWa1988], *Artocarpus communis* [Beards1966], *Artocarpus incisa* [WilliaWa1988], *Ficus* [Takaha1929, Balach1956, MatileNo1984, WilliaWa1988], *Ficus foveolata* [Takaha1935], *Ficus retusa* [Takaha1932a, Takaha1933]. **Musaceae**: *Musa* [Lepage1938, Takaha1939b, Almeid1971, Matile1984c, WilliaWa1988], *Musa acuminata* [Brimbl1968], *Musa paradisiaca sapientum* [McKenz1956], *Musa paradisiaca* [GomezM1941, WilliaWa1988], *Musa sapientum* [GomezM1941, Mamet1943a, Mamet1949, Borchs1966, Velasq1971]. **Myrsinaceae**: *Maesa* [Takaha1935], *Maesa indica* [Green1900a]. **Myrtaceae**: *Caryophyllus* [Leonar1914, Balach1956], *Decaspermum fruticosum* [Takaha1934], *Eucalyptus deglupta* [WilliaBu1987, WilliaWa1988], *Eugenia* [Merril1953, WilliaWa1988], *Eugenia eucalyptoides* [Merril1953], *Eugenia jambolana* [Merril1953], *Eugenia malaccensis* [WilliaWa1988], *Psidium* [McKenz1956, WilliaWa1988], *Psidium guajava* [GomezM1941, RahmanAn1941, Mamet1943a, Beards1966, Borchs1966,

Velasq1971, WilliaWa1988]. **Ochnaceae**: *Lophira* [Balach1956]. **Oleaceae**: *Jasminum* [RahmanAn1941, WilliaWa1988], *Jasminum officinale* [WilliaWa1988], *Jasminum sambac* [WilliaWa1988], *Ligustrum japonicum* [Takaha1932a, Takaha1933], *Osmanthus asiaticus* [Takagi1969a]. **Orchidaceae** [WilliaWa1988]. **Oxalidaceae**: *Averrhoa carambola* [WilliaWa1988]. **Palmae** [Cocker1898m, Takagi1969a], *Areca catechu* [Takaha1939b, Mamet1943a, Mamet1949, Beards1966, Borchs1966], *Chrysalidocarpus lutescens* [Takaha1935, Mamet1943a, Mamet1949, Borchs1966, Velasq1971], *Cocos* [Lepage1938], *Cocos nucifera* [Signor1869b, Mamet1949, Mamet1959a, Castel1963, Dekle1965c, Borchs1966, WilliaBu1987, WilliaWa1988], *Dictyosperma alba* [Mamet1949, Borchs1966], *Elaeis guineensis* [Castel1963, Almeid1973b, WilliaWa1988], *Lithocarpus* [Takaha1942b], *Neodypsis* [Mamet1951, Borchs1966], *Nipa fruticans* [Beards1966], *Phoenix* [WilliaWa1988], *Phoenix canariensis* [Beards1966], *Raphia ruffia* [Mamet1949, Borchs1966], *Trachycarpus excelsus* [Takaha1929]. **Pandanaceae**: *Pandanus* [McKenz1956, Dekle1965c], *Pandanus boninensis* [Takaha1929], *Pandanus odoratissimus* [Green1916e], *Pandanus tectorius* [Takaha1933, Beards1966], *Pandanus utilis* [GomezM1941, Mamet1943a, Mamet1949, Borchs1966]. **Passifloraceae**: *Passiflora quadrangularis* [WilliaWa1988]. **Pedaliaceae**: *Uncaria gambir* [Green1904a]. **Piperaceae**: *Piper* [Green1937, WilliaBu1987, WilliaWa1988], *Piper macgillivrayi* [WilliaWa1988], *Piper methysticum* [WilliaBu1987, WilliaWa1988], *Piper puberulum* [WilliaWa1988], *Piper subpeltatum* [Newste1911a]. **Proteaceae**: *Grevillea robusta* [MerrilCh1923]. **Rhamnaceae**: *Ziziphus jujuba* [GomezM1941]. **Rhizophoraceae**: *Rhizophora* [Balach1956]. **Rosaceae**: *Prunus persica* [Mamet1949, Borchs1966]. **Rubiaceae**: *Catesbaea parviflora* [Merril1953], *Platanocephalus morindaefolius* [WilliaWa1988], *Psychotria* [Green1900a], *Psychotria elliptica* [Takaha1932a, Takaha1933]. **Rutaceae**: *Citrus* [Takaha1932a, Matile1984c], *Citrus grandis* [WilliaWa1988], *Citrus maxima* [WilliaWa1988], *Citrus sinensis* [Takagi1969a]. **Solanaceae** [WilliaWa1988], *Capsicum* [WilliaWa1988], *Capsicum annuum* [WilliaWa1988], *Capsicum frutescens* [Velasq1971, WilliaWa1988], *Capsicum minimum* [WilliaWa1988], *Lycopersicon esculentum* [WilliaWa1988], *Physalis lanceolata* [WilliaWa1988], *Physalis peruviana* [WilliaWa1988], *Solanum melongena* [WilliaWa1988]. **Sonneratiaceae**: *Sonneratia caseolaris* [Beards1966]. **Sterculiaceae**: *Theobroma cacao* [Green1904a, Lepage1938, WilliaWa1988]. **Strelitziaceae**: *Ravenala madagascariensis* [Mamet1943a, Mamet1949, Borchs1966], *Strelitzia reginae* [Brimbl1968]. **Theaceae**: *Camellia japonica* [TakahaTa1956, Hadzib1983], *Camellia sasanqua* [Hadzib1983], *Eurya* [Takaha1935], *Eurya japonica* [Hadzib1983], *Thea* [Borchs1934], *Thea sinensis* [Takaha1933, Borchs1936, TakahaTa1956, Takagi1957, Hadzib1983]. **Thymelaeaceae**: *Daphne odora* [TakahaTa1956]. **Ulmaceae**: *Celtis occidentalis* [MerrilCh1923, Lepage1938], *Chaetacme aristata* [Brain1918]. **Urticaceae**: *Lapotrea photiniphylla* [WilliaWa1988]. **Verbenaceae**: *Lantana camara* [WilliaWa1988]. **Viscaceae**: *Viscum taenioides* [Lindin1910b]. **Vitaceae**: *Vitis* [Green1904a, Takaha1933, Takaha1934, Takaha1935], *Vitis vinifera* [Mamet1943a, Mamet1949, Borchs1966, Matile1984c]. **Zingiberaceae**: *Alpinia nutans* [WilliaWa1988], *Zingiber officinale* [Mamet1949, Borchs1966].

NATURAL ENEMIES: ACARI **Hemisarcoptidae**: *Hemisarcoptes malus* (Shimer) [GersonOcHo1990].   COLEOPTERA **Coccinellidae**: *Azya trinitatis* Marshall [RosenDe1978], *Chilocorus cacti* (L.) [RosenDe1978], *Chilocorus nigritus* Mulsant [RosenDe1978, Kinawy1991], *Chilocorus politus* Mulsant [Beards1970, RosenDe1978], *Chilocorus wahlbergi* [Almeid1973b], *Chnoodes nr. cinctipennis* Gorham [RosenDe1978], *Cleothera* [RosenDe1978], *Cryptognatha nodiceps* Marshall [Castel1959, Beards1970, RosenDe1978, Fernan1987a], *Cryptognatha simillima* Sicard FUNGI **Ascomycotina**: *Myriangium duriaei* [EvansPr1990], *Nectria diploa* [EvansPr1990], *Nectria flammea* [EvansPr1990], *Torrubiella luteorostrata* [EvansPr1990], *Torrubiella tenuis* [EvansPr1990]. **Deuteromycotina**: *Fusarium* [EvansPr1990], *Verticillium lecanii* [EvansPr1990]. HYMENOPTERA **Aphelinidae**: *Aphytis chrysomphali* (Mercet) [Beards1970, RosenDe1978, RosenDe1979], *Ap. equatorialis* Rosen & DeBach [RosenDe1979], *Ap. margaretae* DeBach & Rosen [RosenDe1979, MyartsRu2000], *Ap. melinus* DeBach [RosenDe1979], *Ap. proclia* (Walker) [Gordh1979], *Aspidiotiphagus citrinus* (Craw) [RosenDe1978], *Encarsia citrina* [LiLoCa1997], *Pteroptrix parvipennis* (Gahan) [RosenDe1978].   **Encyrtidae**: *Apterencyrtus microphagus* (Mayr) [RosenDe1978], *Comperiella* [Ramakr1930], *Comperiella bifasciata* Howard [Flande1944a], *Comperiella unifasciata* Ishii [RosenDe1978], *Spaniopterus crucifer* Gahan [RosenDe1978, Noyes1990a], *Thomsonisca amathus* (Walker) [Noyes1990a].   **Signiphoridae**: *Signiphora borinquensis* Quezada [Gordh1979], *Signiphora fax* [RosenDe1978], *Delphastus* [RosenDe1978], *Exoplectra dubia* Crotch [RosenDe1978], *Hyperaspis* [RosenDe1978], *Lindorus lophanthae* (Blaisdell) [Cocher1965, Beards1970], *Pentilia castanea* Mulsant [RosenDe1978], *Pentilia insidiosa* Mulsant [RosenDe1978], *Prodilis* [RosenDe1978], *Pseudoscymnus anomalus* Chapin [Beards1970], *Rhizobius pulchellus* Montrouzier [Cocher1969], *Scymnus* [RosenDe1978], *Scymnus (Nephus) aeneipennis* Sicard [RosenDe1978], *Telsimia nitida* Chapin [Beards1970, RosenDe1978]. Girault [Woolle1990].   NEUROPTERA **Chrysopidae**: *Chrysopa* [Drea1990]. THYSANOPTERA **Phlaeothripidae**: *Aleurodothrips fasciapennis* (Franklin) [Beards1970, RosenDe1978, PalmerMo1990].

DISTRIBUTION: **Afrotropical**: Angola [Balach1956, Almeid1973b]; Benin [Leonar1914]; Cameroon [Balach1956, MatileNo1984]; Cape Verde Islands [Fernan1972]; Côte d'Ivoire (=Ivory Coast) [RosenDe1979]; Eritrea [Lindin1910b, Balach1956]; Ethiopia [Balach1956]; Ghana [Newste1917b]; Guinea-Bissau [Fernan1987a]; Kenya [Newste1911a, Newste1917b, DeLott1967a]; Madagascar [Mamet1951, Mamet1954, Borchs1966]; Mauritius [Mamet1943a, Mamet1949, Borchs1966]; Mozambique [Almeid1971]; Niger [Vayssi1913]; Nigeria [AisagbNwAg1985]; Réunion [Mamet1943a, Mamet1954a, Mamet1957, Borchs1966]; São Tomé and Príncipe [Castel1963]; Senegal [Leonar1914]; Sierra Leone [Hargre1927]; Somalia [Balach1956]; South Africa [BrainKe1917, Newste1917b]; Tanzania [Newste1911a]; Togo [Newste1906a, Newste1908b]; Uganda [Newste1914, Gowdey1917, Newste1917b, Green1937]; Zaire [Balach1956]; Zanzibar [Green1916, Newste1917b, Mamet1956, Borchs1966]; Zimbabwe [Balach1956]. **Australasian**: American Samoa [WilliaWa1988]; Australia [Frogga1914] (New South Wales [Cocker1898m], Northern Territory

[Green1914c, Green1916e], Queensland [Brimbl1968]); Bonin Islands (= Ogasawara-Gunto) [Kawai1987]; Federated States of Micronesia [Takaha1936c, Takaha1942d, RosenDe1978] (Caroline Islands [Takaha1939b, Takaha1941b], Ponape Island [Beards1966], Truk Islands [Beards1966], Yap [Beards1966]); Fiji [Green1937, RosenDe1978, WilliaWa1988]; French Polynesia [DoaneHa1909, Ferris1935, WilliaWa1988] (Tahiti [RosenDe1979]); Hawaiian Islands (Hawaii [RosenDe1979]); New Caledonia [Cohic1958]; Palau [Takaha1939b, Beards1966]; Papua New Guinea [RosenDe1978, WilliaWa1988]; Solomon Islands [WilliaWa1988]; Vanuatu [WilliaBu1987, WilliaWa1988] [WilliaBu1987, WilliaWa1988]; Western Samoa [Laing1927]. **Nearctic**: Mexico [MyartsRu2000] (Baja California Sur [RosenDe1979]); United States of America (California [McKenz1956], Florida [MerrilCh1923, Merril1953, Dekle1965c, RosenDe1978], Georgia [Nakaha1982]). **Neotropical**: Antigua and Barbados (Antigua [Cocker1893j]); Brazil [RosenDe1979, WolffCo1993a] (Bahia [Lepage1938], Distrito federal (Brasilia) [Lepage1938], Rio de Janeiro [Hempel1904, Lepage1938], São Paulo [Green1930b, Lepage1938], Sergipe [Lepage1938]); Colombia [Balach1959a, Kondo2001]; Costa Rica [RosenDe1979]; Dominican Republic [Russo1929, GomezM1941, RosenDe1978, RosenDe1979]; Ecuador [YustCe1956]; Galapagos Islands [PeckHeLa1998]; Guadeloupe [Balach1957c]; Guyana [Newste1893d, Newste1914]; Martinique [Balach1957c]; Panama [RosenDe1979]; Peru [Wolcot1958]; Puerto Rico & Vieques Island [RosenDe1978, RosenDe1979] (Puerto Rico [Martor1976, ColonFMe1998]); Saint Croix [Beatty1944]; Trinidad and Tobago (Trinidad [RosenDe1979]); U.S. Virgin Islands [Nakaha1983]. **Oriental**: British Indian Ocean Territory (Chagas) [Mamet1943a]; India [Green1908a, Ramakr1919a, Ramakr1921a] (Andhra Pradesh [MaheswPu1999], Bihar [Ali1968], Karnataka [UsmanPu1955], Punjab [Ansari1942]); Indonesia [RosenDe1978] (Irian Jaya [WilliaWa1988], Java [Green1904a]); Kampuchea (Cambodia) [Takaha1942b]; Mongolia [DanzigKo1990]; Pakistan [RosenDe1979]; Philippines [Cocker1905f, VelasqRi1969]; Ryukyu Islands (= Nansei Shoto) [KinjoNaHi1996]; Sri Lanka [Green1900a, Ramakr1921a, Green1922]; Taiwan [Takaha1929, Takaha1932a, TakahaTa1956, Takagi1969a, WongChCh1999]; Thailand [Takaha1942b]. **Palaearctic**: Azerbaijan (Azerbaijan [Borchs1936]); China (People's Republic) [Kuwana1927] (Henan (Honan) [Shen1993]); Egypt [Hall1922, Ezzat1958]; Georgia [Dzhash1988] (Abkhaz ASSR [Borchs1934, Hadzib1983], Adzhar ASSR [Borchs1934, Borchs1936, Hadzib1983]); Iran [Kaussa1955]; Japan [Lindin1911, Kuwana1933, Kawai1980] (Honshu [TakahaTa1956, Takagi1957], Kyushu [TakahaTa1956, Takagi1957], Shikoku [TakahaTa1956, Takagi1957]); Russia (Caucasus [Borchs1936]); Saudi Arabia [Matile1984c].

BIOLOGY: Usually on underside of leaves, but in heavy infestation also on upper side (Taylor, 1935; Ferris, 1938a).

ECONOMIC IMPORTANCE: A polyphagous insect, mainly distributed in tropical countries (CABI, 1966a); mainly a pest of coconut and banana (Taylor, 1935; Ebeling, 1959; Rosen & De Bach, 1978; Williams & Watson, 1988).

GENERAL: Description and illustration of adult female by Brain (1918), Kuwana (1933), Ferris (1938a, 1941e), Balachowsky (1948b, 1956), McKenzie (1956),

Takagi (1969a), Beardsley (1970), Williams & Watson (1988), Danzig (1993), Gill (1997) and by Colon-Ferrer & Medina-Gaud (1998).
KEYS: Gill 1997: 64 (female) [Species of California]; Danzig 1993: 140-141 (female) [Europe]; Williams & Watson 1988: 49 (female) [Tropical South Pacific]; Chou 1985: 262-263 (female) [Species of China]; Beardsley 1970: 508 (female) [Hawaii]; Beardsley 1966: 513 (female) [Federated States of Micronesia]; Ezzat 1958: 240 (female) [Egypt]; De Lotto 1957: 228 (female) [Africa]; Takagi 1957: 32 (female) [Japan]; Balachowsky 1956: 51 (female) [Africa]; McKenzie 1956: 24 (female) [U.S.A.: California]; Balachowsky 1948b: 275 (female) [Mediterranean]; Ferris 1942: 30 (female) [North America]; Ferris 1941e: 61 (female) [World]; Kuwana 1933: 3 (female) [Japan]; Kuwana 1933b: 49 (female) [Japan]; Brain 1918: 118 (female) [South Africa]; Robinson 1917: 29 (female) [Philippines].
CITATIONS: AisagbNwAg1985 [host, distribution, economic importance, life history, ecology: 24-32]; Alenca2000 [host, distribution: 1-12]; Ali1968 [host, distribution: 133]; Almeid1971 [host, distribution: 8]; Almeid1973b [host, distribution: 8]; Alvara1939 [host, distribution: 3]; Ansari1942 [host, distribution, economic importance: 233]; Apstei1915 [taxonomy: 119]; Archan1929 [taxonomy: 190]; Azeved1929 [host, distribution: 113-115]; Azeved1929a [host, distribution: 126-128]; Balach1948b [taxonomy, description, illustration, host, distribution: 275-280]; Balach1956 [taxonomy, description, illustration, host, distribution: 61-64]; Balach1957c [host, distribution: 199]; Balach1959a [host, distribution: 362]; Ballou1915 [host, distribution: 121]; Ballou1922a [host, distribution: 239]; Banks1906b [host, distribution, taxonomy: 211-228]; Banks1990 [chemistry: 272]; Beards1966 [host, distribution: 513]; Beards1970 [taxonomy, host, distribution, life history, biological control, economic importance: 505-508]; BeardsDaHo1976 [economic importance: 103]; BeardsGo1975 [economic importance: 49]; Beatty1944 [host, distribution: 114-172]; Beccar1971 [host, distribution: 193]; Benass1961b [host, distribution, ecology: 1-157]; BennetRoCo1976 [biological control, economic importance: 359-395]; BilogOMo2000 [host, distribution, biological control: 137-147]; Borchs1934 [host, distribution: 27]; Borchs1936 [host, distribution, life history, biological control, economic importance: 128-130]; Borchs1937 [taxonomy, description, illustration, host, distribution: 124]; Borchs1949d [taxonomy, description, illustration, host, distribution: 236]; Borchs1950b [taxonomy, description, illustration, host, distribution: 215,219]; Borchs1966 [catalogue: 270-271]; Bordag1914 [distribution]; Bourne1923 [host, distribution, economic importance: 1]; Box1953 [host, distribution, biological control: 51]; Brain1918 [taxonomy, description, illustration, host, distribution: 120]; BrainKe1917 [distribution: 183,184]; Brick1912 [host, distribution: 1-22]; Brimbl1968 [taxonomy, host, distribution: 43]; BurgerUl1990 [economic importance: 313-327]; Butani1974 [host, distribution, biological control: 689-691]; CABI1966a [host, distribution: 1-2]; CarpenEl1978 [host, distribution, economic importance, taxonomy, life history, control: 1-42]; Castel1951a [biological control: 95-98]; Castel1956a [host, distribution, economic importance, control: 225-238]; Castel1959 [host, distribution, biological control: 235-239]; Castel1963 [taxonomy, description, illustration, host, distribution: 131-139]; ChatteBo1934 [biological control: 1-10]; Chazea1981 [host, distribution, economic importance, biological

control: 11-22]; Chou1938 [host, distribution, taxonomy, description: 240-249]; Chou1947a [chemical control: 34]; Chou1985 [taxonomy, description, host, distribution: 269-272]; Chou1986 [taxonomy, illustration: 439,440,444,445]; ChuaWo1990 [host, distribution, economic importance: 543-552]; ClapsWoGo2001a [taxonomy, host, distribution: 27]; Clause1940 [biological control]; Clause1956 [host, distribution, economic importance, biological control]; Clause1958 [economic importance, biological control: 291-310]; Cocher1965 [host, distribution, biological control: 318-321]; Cocher1965a [host, distribution, economic importance, biological control: 507-512]; Cocher1969 [host, distribution, biological control: 57-100]; Cocher1972 [host, distribution, economic importance, biological control: 89-104]; Cock1985a [biological control: 3]; Cocker1893j [taxonomy, description, host, distribution: 255]; Cocker1896b [taxonomy, distribution: 334]; Cocker1897i [taxonomy, description, host, distribution: 9,21,28,29]; Cocker1897q [taxonomy, description, host, distribution: 703]; Cocker1898m [taxonomy, description, host, distribution: 27]; Cocker1899a [taxonomy: 395]; Cocker1899q [host, distribution: 93]; Cocker1905f [taxonomy, host, distribution: 133]; CockerRo1915 [taxonomy, description, host, distribution: 106]; Cohic1958 [host, distribution: 13]; ColonFMe1998 [taxonomy, description, illustration, host, distribution: 44-45]; Comsto1883 [taxonomy: 75-76]; Corbet1932 [host, distribution]; CoronaRuMo1997 [host, distribution: 38-41]; Danzig1972 [taxonomy, host, distribution, economic importance: 207]; Danzig1993 [taxonomy, description, illustration, host, distribution: 141-143]; DanzigKo1990 [host, distribution: 46]; DanzigPe1998 [catalogue: 188]; DavidsMi1990 [host, distribution, economic importance: 603-632]; Davis1972 [economic importance, biological control: 187-190]; DeBach1964 [biological control]; DeBach1964b [biological control: 673-713]; DeBach1969a [biological control: 11-28]; DeBach1971 [biological control: 303]; DeBach1974 [biological control]; DeBachRo1976a [host, distribution, biological control: 541-545]; DeBachRo1991 [biological control]; DEDAC1923 [host, distribution]; Dekle1965c [taxonomy, description, host, distribution: 27]; Dekle1976 [taxonomy, description, host, distribution, economic importance: 39]; DeLott1957 [taxonomy: 228]; DeLott1967a [host, distribution: 113]; DeSant1940 [biological control: 29-44]; DeSant1979 [biological control]; Devasa1992 [host, distribution, life history: 153]; DevasaKo1993 [host, distribution: 101-102]; DevasaKoVe1998 [host, distribution: 157-164]; Dhilee1996 [host, distribution, biological control: 64-74]; DicksoFl1955 [host, distribution: 614-615]; Dinthe1958 [host, distribution, economic importance: 421]; Dixon1997 [biological control: 205]; DoaneHa1909 [host, distribution: 297]; Dohani1937 [biological control, host, distribution: 243-247]; Dozier1926 [host, distribution, biological control: 267-277]; Dozier1933 [host, distribution, biological control: 85-100]; Dozier1937 [host, distribution, biological control: 121-135]; Drea1990 [biological control: 51-59]; Dupont1931 [host, distribution: 1-18]; Dzhash1971 [host, distribution: 325-326]; Dzhash1988 [economic importance, life history, host, distribution: 134-143]; Dzhash1988 [host, distribution, life history, economic importance: 134-143]; Ebelin1949 [host, distribution, life history, control]; Esaki1940a [host, distribution: 274-280]; EvansPr1990 [biological control: 3-17]; Ezzat1958 [distribution: 240]; EzzatNa1987 [distribution: 88]; FDACSB1982 [host, distribution: 5-11]; FDACSB1987 [host, distribution: 4-7]; FergusF11991 [host, distribution, biological control: 260-261]; Fernal1903b [catalogue: 257,278]; Fernan1972 [taxonomy, description, host, distribution: 11-12]; Fernan1987a [taxonomy, host, distribution: 32]; Ferris1935 [host, distribution: 131]; Ferris1938a [taxonomy, description, illustration, host, distribution: 191]; Ferris1941e [taxonomy, description, illustration, host, distribution: 42,43,48-53,61]; Ferris1942 [taxonomy: 446:30]; Ferris1946 [taxonomy: 44]; Ferris1953 [taxonomy: 65]; Figuer1952 [host, distribution: 208]; Fisher1964 [biological control: 305-325]; Flande1936b [biological control: 251-255]; Flande1937 [biological control: 401-422]; Flande1944a [biological control: 365-371]; Flande1966 [biological control: 79-82]; Flande1969 [biological control: 29-33]; Flande1971 [biological control, life history: 857-872]; Foldi1990 [structure: 43-54]; Fonsec1963 [host, distribution: 32-35]; Fonsec1964 [host, distribution: 515]; Frogga1914 [taxonomy, description, host, distribution: 311,318]; Frogga1915 [taxonomy, description, host, distribution: 14]; Gahan1925 [host, distribution, biological control: 1-23]; Gahan1927a [host, distribution, biological control: 149-153]; Gaprin1954 [biological control: 587-597]; Gaprin1956 [host, distribution: 103-137]; Gaprin1975 [host, distribution, economic importance, biological control: 29-33]; Garcia1930 [host, distribution, biological control]; Gavalo1936 [host, distribution: 79-80]; Gentry1965 [host, distribution, economic importance: 3]; GersonOcHo1990 [biological control: 77-97]; Gill1997 [host, distribution, taxonomy, description, illustration, economic importance: 64,66]; GiraldRu1962 [life history, behaviour, host, economic importance: 121-144]; Goberd1962 [host, distribution, economic importance: 49-70]; GomezM1941 [host, distribution: 128]; Goot1928 [host, distribution: 1]; Gordh1979 [biological control: 896,900,910]; Gordon1978 [biological control: 205-218]; Gordon1980 [biological control: 149]; Gowdey1913 [host, distribution: 247-249]; Gowdey1917 [host, distribution: 189]; Greath1971 [host, distribution, biological control]; Greath1973 [biological control: 29-33]; Greath1986 [biological control: 289-318]; Greath1989 [biological control: 28-37]; GreathGr1992 [host, distribution, biological control: 61-68]; Green1890 [taxonomy, description, host, distribution: 20]; Green1900a [taxonomy, description, illustration, host, distribution: 69-70]; Green1904a [host, distribution: 208]; Green1907 [host, distribution: 203]; Green1908a [host, distribution: 33]; Green1910a [taxonomy: 201-202]; Green1914c [host, distribution: 232]; Green1915c [host, distribution: 44]; Green1915e [host, distribution: 608-636]; Green1916 [host, distribution: 376]; Green1922 [host, distribution: 462]; Green1930b [host, distribution: 214]; Green1937 [host, distribution: 329]; GreenMa1907 [taxonomy, distribution: 343-344]; Gressi1958 [host, distribution, economic importance: 395-398]; GuptaSi1988 [host, distribution: 357-361]; Hadzib1983 [taxonomy, description, illustration, host, distribution, life history, economic importance: 219-221]; Hagen1974 [biological control: 25-44]; HagenBoMc1976 [biological control: 93]; HakkonPi1984 [biological control: 1109-1121]; Halber1996a [host, distribution: 4-8]; Halber2000 [host, distribution: 4-8]; Hall1922 [taxonomy, description, host, distribution: 26]; HandaDa1999 [host, distribution, chemical control: 112-114]; HanksDe1998 [life history, ecology: 239-262]; Hargre1927 [host, distribution, economic importance: 113-128]; Hargre1937 [host, distribution, economic importance: 505-520]; Hargre1948 [host, distribution];

Harris1937a [host, distribution: 88-94]; Hempel1904 [host, distribution: 319]; Hill1975 [host, distribution, economic importance, control]; Hinckl1963 [host, distribution, biological control]; Houck1999 [life history, biological control: 97-118]; Houser1918 [host, distribution: 165]; Howard1991 [host, distribution, life history, ecology, control: 217-225]; HuffakCa1986 [biological control, economic importance: 95-107]; HuffakMeDe1971 [biological control: 16-67]; HuffakSiLa1976 [biological control: 93]; HuffakSt1971 [biological control: 333-350]; Hunt1939 [host, distribution: 548-566]; Hutson1933 [host, distribution, economic importance, taxonomy, life history: 254-256]; Ishii1932a [host, distribution, biological control: 161]; JalaluMo1989 [chemical control: 203-206]; JalaluMo1989a [biological control, chemical control: 199-202]; JalaluMoSu1992 [host, distribution, life history: 5-7]; JalaluThMo1991 [host, distribution: 17]; JalaluSaMa2001 [host, distribution: 347-348]; Jannon1940 [host, distribution: 241-253]; JannonCaBi1962 [life history, behaviour, host, economic importance, taxonomy, distribution: 5-126]; JiYa1990 [biological control: 134-136]; Jourdh1979 [biological control: 75-79]; Kalsho1981 [taxonomy, host, distribution, economic importance: 166-170]; Kamath1979 [host, distribution, economic importance, biological control: 55-72]; Kathir1993 [host, distribution, economic importance, chemical control: 1026]; Kaussa1955 [host, distribution: 15]; Kawai1980 [taxonomy, description, host, distribution: 228]; Kawai1987 [host, distribution: 78]; Kinawy1991 [host, distribution, biological control: 387-389]; KinjoNaHi1996 [host, distribution: 125-127]; Kobakh1965 [biological control: 323-330]; Kondo2001 [taxonomy, host, distribution: 43]; KondoKa1995 [host, distribution: 57-58]; KondoKa1995a [host, distribution: 97-98]; KoyaDeSe1996 [host, distribution: 129-136]; Kuwana1917a [taxonomy, distribution: 174]; Kuwana1927 [host, distribution: 71]; Kuwana1933 [taxonomy, description, illustration, host, distribution: 9-10,18-19]; Laing1927 [host, distribution: 40]; Leefma1929 [host, distribution: 1]; Leonar1897 [taxonomy: 285]; Leonar1898a [taxonomy: 76]; Leonar1898c [taxonomy, description, illustration, host, distribution: 62-64]; Leonar1914 [taxonomy, host, distribution: 196,197]; Leonar1920 [taxonomy, description, illustration, host, distribution: 29,40-41]; Lepage1938 [catalogue: 394,397]; Lepesm1947 [taxonomy, description, host, distribution, life history, biological control: 189-193]; Lever1969 [host, distribution]; LiLoCa1997 [host, distribution, biological control: 225-226]; Lindin1908b [taxonomy: 106]; Lindin1909b [host, distribution: 150]; Lindin1909c [taxonomy, host, distribution: 449]; Lindin1910b [taxonomy, description, illustration, host, distribution: 34-35,38]; Lindin1911 [host, distribution: 88]; Lindin1924 [taxonomy: 174]; Lindin1932g [taxonomy: 223]; Lindin1957 [taxonomy: 545,546]; LongoMaPe1995 [distribution: 125]; MacGil1921 [taxonomy, description, host, distribution: 396,401]; MaChZh1995 [host, distribution: 117-119]; MaheswPu1999 [host, distribution, economic importance: 17]; MahmooMo1986 [host, distribution, economic importance: 1-11]; Maleno1916a [taxonomy, description, illustration, host, distribution: 321-326]; Mallam1954 [distribution: 24-60]; Mamet1943a [catalogue: 157]; Mamet1949 [catalogue: 55]; Mamet1951 [host, distribution: 226]; Mamet1954 [host, distribution: 15]; Mamet1954a [distribution: 264]; Mamet1956 [host, distribution: 136]; Mamet1957 [distribution: 369]; Mamet1959a [host, distribution: 386];

Mansfi1920 [host, distribution: 145-155]; Mariau1998 [host, distribution, economic importance: 269-277]; MariauJu1977 [life history, economic importance, chemical control, biological control: 217-224]; Martor1976 [host, distribution: 13,21,101,165,177,200]; Maskel1892 [host, distribution: 12]; Matile1984c [host, distribution: 221]; MatileNo1984 [host, distribution: 64]; MayneGh1934 [host, distribution: 3-38]; McClur1990g [taxonomy, host, distribution, ecology: 319-330]; McKenz1956 [taxonomy, description, illustration, host, distribution: 47-48]; Merril1953 [taxonomy, description, host, distribution: 19-20]; MerrilCh1923 [taxonomy, description, host, distribution, economic importance: 200]; MillerDa1990 [host, distribution, economic importance: 300]; MoutiaMa1946 [host, distribution, economic importance, life history, biological control: 457-462]; MoutiaMa1947 [distribution]; MunozG1937 [host, distribution: 3-9]; Muntin1971a [taxonomy: 313]; Muraka1970 [host, distribution: 72]; MyartsRu2000 [distribution, biological control: 7-33]; Nafus1996 [host, distribution: 1]; NagarkSa1990 [host, distribution, economic importance, biological control: 553-542]; Nair1964 [host, distribution, economic importance: 72]; NairMe1963 [host, distribution: 139-147]; Nakaha1982 [host, destructor: 13]; Nakaha1983 [host, distribution: 9]; Newell1923 [host, distribution: 263-266]; Newste1893d [taxonomy, description, host, distribution: 186]; Newste1906a [host, distribution: 73]; Newste1908b [host, distribution: 33-34]; Newste1910a [taxonomy: 68]; Newste1911a [host, distribution: 167-168]; Newste1914 [host, distribution: 307]; Newste1917b [host, distribution: 131]; NotzP1974 [host, distribution, biological control: 127-143]; Noyes1990a [biological control: 155,156]; ObraRe2000 [life history, biological control: 137-147]; OilPaCo1981 [host, distribution, taxonomy, description, life history, economic importance, control: 168-228]; Ordish1967 [economic importance, biological control]; PalmerMo1990 [biological control: 67-76]; PeckHeLa1998 [host, distribution: 219-237]; PruthiBa1960 [host, distribution, economic importance,: 1-113]; PruthiMa1945 [host, distribution, life history, control: 1-42]; Quelch1890 [host, distribution: 369-370]; RahmanAn1941 [taxonomy, description, host, distribution: 824]; Ramakr1919 [host, distribution, economic importance: 623]; Ramakr1919a [taxonomy, description, host, distribution: 16-17]; Ramakr1921a [host, distribution: 356,359]; Ramakr1930 [taxonomy, host, distribution, biological control: 24]; Ramakr1938a [host, distribution: 341-351]; Ramos1946 [host, distribution: 1]; RaoGhSa1971 [host, distribution, biological control]; Reboul1976 [host, distribution, economic importance]; Robert1958 [host, distribution, economic importance: 411-415]; Robins1917 [taxonomy, description, host, distribution: 29,31-32]; Rose1990c [distribution, economic importance: 535-542]; Rosen1973 [biological control: 47-54]; Rosen1990 [biological control: 413-415]; Rosen1990a [biological control: 499]; RosenDe1978 [economic importance, biological control, host, distribution, life history, distribution: 93-98]; RosenDe1979 [host, distribution, biological control: 533-539,545-548,]; Rubtso1952a [biological control: 96-106]; Russo1929 [host, distribution, economic importance: 2-3]; Sadaka1993 [host, distribution, biological control: 12-13]; Schmut1964 [host, distribution, economic importance: 102-106]; Schmut2001 [host, distribution: 339-345]; SchmutKlLu1957 [host, distribution, economic importance: 475]; Schrea1970 [host, distribution, control]; Schrei1989 [biological control: 57-69]; Sefer1961 [host, distribution: 23];

SelvakKaDe1996 [host, distribution, life history, biological control: 79-83]; ShahJhPa1988 [chemical control: 19-22]; Shen1993 [host, distribution: 60]; Shirom1969 [host, distribution: 283]; Signor1869 [taxonomy: 851]; Signor1869b [taxonomy, description, illustration, host, distribution: 120-121]; Silva1944 [host, distribution: 8-14]; Simmon1958b [host, distribution, biological control: 475-478]; Simmon1959a [host, distribution, biological control: 1099-1102]; Simmon1960 [host, distribution, biological control, economic importance: 223-237]; SinhaDi1984 [host, distribution, biological control: 7-13]; Smirno1952 [host, distribution, biological control: 63-69]; SugimoKaTa1996 [host, distribution: 99-101]; SureshMo1995 [host, distribution: 429-430]; SureshMo1995 [host, distribution: 429-430]; Sweetm1958 [biological control, economic importance: 449-458]; SzentI1963 [host, distribution: 67-71]; TabibuGa1973 [host, distribution, life history: 409-426]; Takagi1957 [taxonomy, host, distribution: 32-34,40]; Takagi1969a [taxonomy, description, illustration, host, distribution: 65-66,69,99]; TakagiRo1981 [host, distribution, biological control: 314-321]; Takaha1929 [host, distribution: 79]; Takaha1932a [host, distribution: 103-105]; Takaha1933 [host, distribution: 25-34]; Takaha1934 [taxonomy, host, distribution: 33-37]; Takaha1935 [host, distribution: 3,4]; Takaha1936c [host, distribution: 109,118]; Takaha1939b [host, distribution: 269]; Takaha1941b [host, distribution: 219]; Takaha1942b [host, distribution: 47]; Takaha1942d [host, distribution: 357]; Takaha1953a [taxonomy, host, distribution: 10-13]; TakahaTa1956 [host, distribution: 13-14]; Tao1999 [taxonomy, host, distribution: 73,120]; Taylor1935 [host, distribution, economic importance, biological control: 1-102]; Thomps1958 [host, distribution, biological control, economic importance: 479-482]; TianCh1991 [biological control: 64-66]; Tourne1970 [host, distribution, biological control: 97-107]; Trjapi1989 [biological control: 296]; Tuncyu1970 [host, distribution, economic importance: 30-52]; UsmanPu1955 [host, distribution: 48]; Valles1965 [biological control: 259-279]; Vayssi1913 [host, distribution: 430]; Vayssi1932a [economic importance, biological control: 629-648]; Velasq1971 [taxonomy, description, illustration, host, distribution: 107-109]; VelasqRi1969 [host, destructor: 195-208]; VeseyF1953 [host, distribution, biological control: 405-413]; WadhiBa1964 [host, distribution, economic importance: 227]; WadhiBa1964 [host, distribution: 227-260]; Wester1918 [host, distribution, economic importance: 5-57]; Wester1920 [host, distribution]; Whitne1933 [host, distribution: 59-70]; Willia1985a [taxonomy: 234,238]; WilliaBu1987 [host, distribution: 94]; WilliaGr1990 [host, distribution, economic importance, biological control: 563-578]; WilliaWa1988 [taxonomy, description, illustration, host, distribution, economic importance: 8,50-56]; Wilson1921 [host, distribution: 20-34]; Wolcot1958 [host, distribution, biological control, economic importance: 511-513]; WolffCo1993a [host, distribution: 153]; WongChCh1999 [taxonomy, description, host, distribution: 19,58-59]; WoodruBeSk1998 [distribution]; Woolle1990 [biological control: 167-176]; YustCe1956 [host, distribution: 425-442]; Zagain1956 [distribution: 85-90]; ZhouZoPe1993 [host, distribution, life history: 18-20].

Aspidiotus elaeidis Marchal

Aspidiotus elaeidis Marchal, 1909c: 69. Type data: BENIN [=DAHOMEY]: Porto-Novo, on leaves of *Elaeis guineensis*. Syntypes, female. Type depository: Paris: Muséum national d'Histoire naturelle, France.

Aspidiotus oppugnatus Silvestri, 1915: 258. Type data: ERITREA: Nefasit, on *Olea chrysophilla*. Syntypes, female. Type depository: Portici: Dipartimento de Entomologia e Zoologia Agraria di Portici, Università di Napoli Federico II, Italy. Synonymy by Borchsenius, 1966: 261.

Aspidiotus transparens Brain, 1918: 120. Misidentification. Discovered by Borchsenius, 1966: 261.

SCALE COVER: Female scale large, diameter 2.8-3.1 mm, circular, moderately convex, white in typical form, may be brown in several populations; exuviae central, pale yellow. Male scale white, oval, 1.2-1.3 mm long (Balachowsky, 1956).

HOST PLANTS: **Apocynaceae**: *Arduina edulis* [Balach1956], *Funtumia africana* [Balach1956]. **Chrysobalanaceae**: *Chrysobalanus* [Balach1956]. **Ebenaceae**: *Diospyros batocana* [Balach1956, Almeid1973b], *Diospyros mespiliformis* [Almeid1973b]. **Euphorbiaceae**: *Bridelia* [Balach1956], *Manihot glaziovii* [Balach1956]. **Leguminosae**: *Carissa edulis* [Almeid1973b]. **Meliaceae**: *Trichilia emetica* [Almeid1971], *Xylocarpus obovatus* [Balach1956]. **Ochnaceae**: *Lophira alata* [Balach1956]. **Oleaceae**: *Olea chrysophylla* [Silves1915, Balach1956]. **Palmae**: *Cocos nucifera* [Balach1956], *Elaeis guineensis* [Marcha1909c, Sander1909a, Leonar1914, Balach1956]. **Rosaceae**: *Prunus domestica* [Almeid1971]. **Viscaceae**: *Viscum taenioides* [Balach1956].

NATURAL ENEMIES: HYMENOPTERA **Aphelinidae**: *Aphytis erythraeus* (Silvestri) [Balach1956, RosenDe1979]. **Encyrtidae**: *Habrolepis oppugnati* Silvestri [Balach1956].

DISTRIBUTION: **Afrotropical**: Angola [Almeid1973b]; Benin [Marcha1909c, Leonar1914]; Cameroon [Balach1956]; Côte d'Ivoire (=Ivory Coast) [Balach1956]; Eritrea [Silves1915, Balach1956, RosenDe1979]; Guinea [Balach1956]; Mozambique [Almeid1971]; Niger [Vayssi1913]; Senegal [Balach1956]; Sierra Leone [Balach1956]; Somalia [Balach1956]; Tanzania [Balach1956]; Zaire [Balach1956].

ECONOMIC IMPORTANCE: Recorded as pest of oil palm, *Elaeis guineensis* in Congo (Chua & Wood, 1990) and Sierra Leone (Hargreaves, 1927, 1937; Lepesme, 1947).

GENERAL: Description and illustration of adult female by Marchal (1909c), Silvestri (1915) and by Balachowsky (1956).

KEYS: Balachowsky 1956: 52 (female) [Africa].

CITATIONS: Almeid1971 [host, distribution: 8]; Almeid1973b [host, distribution: 8-9]; Balach1932f [taxonomy, host, distribution: 231]; Balach1956 [taxonomy, description, illustration, host, distribution, biological control: 64-66]; Borchs1966 [catalogue: 261]; Brain1918 [taxonomy, description, host, distribution: 120]; ChuaWo1990 [host, distribution, economic importance: 543-552]; DanzigPe1998 [catalogue: 188-189]; Ferris1941e [taxonomy: 43]; Hargre1927 [host, distribution, economic importance: 113-128]; Hargre1937 [host, distribution, economic importance: 505-520]; Leonar1914 [taxonomy, host, distribution: 196]; Lepesm1947 [taxonomy, description, host, distribution, life history, biological control: 193-194]; Lindin1914 [taxonomy: 244]; Lindin1928 [taxonomy: 106]; MacGil1921 [taxonomy, description, host, distribution: 397]; Marcha1909c [taxonomy, description, host, distribution: 69]; Muntin1971a [taxonomy, host, distribution, description, illustration: 311-313]; Prinsl1983 [distribution, biological control: 26]; RosenDe1979 [host, distribution, biological control: 674-678]; Sander1909a [taxonomy, host, distribution: 52]; Silves1915 [taxonomy, description, illustration, host, distribution: 258-260]; Vayssi1913 [host, distribution: 430]; Viggia1978 [taxonomy: 351].

Aspidiotus excisus Green

Aspidiotus excisus Green, 1896e: 53. Type data: SRI LANKA: Punduloya, on *Cyanotis pilosa*. Lectotype female, by subsequent designation Williams & Watson, 1988: 56. Type depository: London: The Natural History Museum, England, UK.

Aspidiotus (*Aspidiotus*) *excisus*; Cockerell, 1897i: 27. Change of combination.

Aspidiotus (*Evaspidiotus*) *excisus*; Leonardi, 1898c: 46. Change of combination.

Temnaspidiotus excisus; MacGillivray, 1921: 403. Change of combination.

Aspidiotus excisus; Williams & Watson, 1988: 56. Revived combination.

COMMON NAME: cyanotis scale [VelasqRi1969].

SCALE COVER: Female scale convex, of irregular outline, margin often lobed, thin, and semi-transparent, whitish or very pale ochreous. Exuviae yellow, approximately central. Male scale smaller and more oblong (Ferris, 1941e). Colour photograph in Wong *et al.* (1999).

HOST PLANTS: **Anacardiaceae**: *Rhus semialata* [Takaha1932a, Takaha1933, Takagi1969a]. **Araceae**: *Aglaonema commutatum* [Velasq1971]. **Asclepiadaceae**: *Hoya carnosa* [Takaha1932a, Takaha1933, Takagi1969a]. **Boraginaceae**: *Tournefortia* [Beards1966, Takagi1969a, Tang1984], *Tournefortia augentea* [Takaha1942d]. **Caprifoliaceae**: *Viburnum* [Takaha1932a, Takaha1933, Takagi1969a]. **Caricaceae**: *Carica papaya* [Takagi1969a, WilliaWa1988]. **Commelinaceae**: *Cyanotis* [Green1937], *Cyanotis pilosa* [Green1896, Green1896e, Ramakr1921a, Takagi1969a, Tang1984]. **Compositae**: *Elephantopus mollis* [Takaha1929, Takagi1969a, Tang1984]. **Convolvulaceae**: *Ipomoea* [Green1900a, Green1937, Takagi1969a, Tang1984]. **Ericaceae**: *Rhododendron* [Takaha1935, Takagi1969a]. **Euphorbiaceae**: *Euphorbia* [WilliaWa1988], *Glochidion hongkongense* [Takagi1969a]. **Malvaceae**: *Thespesia* [Beards1966, Takagi1969a], *Thespesia populnea* [Beards1966], *Urena lobata* [Takaha1932a, Takaha1933, Takagi1969a]. **Musaceae**: *Musa* [Takagi1969a]. **Orchidaceae** [Velasq1971]. **Palmae**: *Cocos nucifera* [Takaha1941b, Takagi1969a]. **Piperaceae**: *Piper* [Green1937, Takagi1969a, Tang1984]. **Rutaceae**: *Citrus* [Takagi1969a, WilliaWa1988], *Citrus aurantifolia* [WilliaWa1988]. **Verbenaceae**: *Clerodendron inerme* [Takaha1929, Takaha1936b, Green1937, Ferris1941e], *Clerodendrum neriifolium* [Takagi1969a].

DISTRIBUTION: **Australasian**: Federated States of Micronesia (Caroline Islands [Takaha1941b, Beards1966]); Fiji [Green1937]; Palau [Takaha1942d, Beards1966]; Papua New Guinea [WilliaWa1988]. **Nearctic**: Mexico [Nakaha1982]; United States of America (Florida [Dekle1976]). **Neotropical**: Antigua and Barbados

(Antigua [Nakaha1982]); Colombia [Kondo2001]; Costa Rica [Nakaha1982]; Dominican Republic [Nakaha1982]; Ecuador [Nakaha1982]; El Salvador [Nakaha1982]; Grenada [Nakaha1982]; Guatemala [Nakaha1982]; Guyana [Nakaha1982]; Honduras [Nakaha1982]; Jamaica [Nakaha1982]; Martinique [Nakaha1982]; Panama [Nakaha1982]; Puerto Rico & Vieques Island (Puerto Rico [ColonFMe1998]); Saint Croix [Nakaha1982]; Suriname [Nakaha1982]; Trinidad and Tobago (Trinidad [Nakaha1982]); U.S. Virgin Islands [Nakaha1983]; Venezuela [Nakaha1982]. **Oriental**: India [Nakaha1982]; Indonesia [Nakaha1982]; Mongolia [DanzigKo1990]; Pakistan [Nakaha1982]; Philippines (Luzon [VelasqRi1969, Velasq1971], Samar [Velasq1971]); Singapore [Nakaha1982]; Sri Lanka [Green1896e, Green1900a, Ramakr1921a, Green1937, Takagi1969a]; Taiwan [Takaha1929, Takaha1932a, Takaha1936b, Green1937, Takagi1969a, WongChCh1999]; Thailand [Takaha1942b, Takagi1969a]. **Palaearctic**: China (People's Republic) [Tang1984]; Japan [Kawai1980].

ECONOMIC IMPORTANCE: Distribution (see Distribution) disjunctive from North and South America, Far East, Asia, and Pacific Islands. Considered a pest of ornamentals (Dekle, 1976; Davidson & Miller, 1990).

GENERAL: Description and illustration of adult female by Ferris (1941e), Velasquez (1971), Tang (1984), Chou (1985, 1986), Williams & Watson (1988) and by Colon-Ferrer & Medina-Gaud (1998).

KEYS: Chou 1985: 274 (female) [Species of China]; Velasquez 1971: 100 (female) [Philippines]; Beardsley 1966: 513 (female) [Federated States of Micronesia]; Ferris 1946: 43 (female) [World]; Ferris 1941e: 61 (female) [World]; Green 1896e: 40 (female) [Sri Lanka].

CITATIONS: Beards1966 [host, distribution: 514-515]; BeardsDaHo1976 [economic importance: 103]; Borchs1966 [catalogue: 271-272]; Chou1985 [taxonomy, description, host, distribution: 274-275]; Chou1986 [taxonomy, illustration: 664]; ChuaWo1990 [host, distribution, economic importance: 548]; Cocker1897i [taxonomy, description, host, distribution: 27]; Cocker1899a [taxonomy: 395]; ColonFMe1998 [taxonomy, description, illustration, host, distribution: 45-46]; CoronaRuMo1997 [host, distribution: 38-41]; DanzigKo1990 [host, distribution: 46]; DanzigPe1998 [catalogue: 362]; DavidsMi1990 [host, distribution, economic importance: 603-632]; Dekle1966 [taxonomy, host, distribution, biological control: 1-2]; Dekle1976 [taxonomy, description, host, distribution, economic importance: 40]; Fernal1903b [catalogue: 258]; Ferris1937c [taxonomy: 52]; Ferris1941e [taxonomy, description, illustration, host, distribution: 43,53-54,63]; Ferris1946 [taxonomy: 43]; Flande1971 [biological control, life history: 857-872]; FoxWil1939 [host, distribution, economic importance: 2296]; Green1896e [taxonomy, description, illustration, host, distribution: 53]; Green1900a [host, distribution: 71]; Green1937 [host, distribution: 330]; Kawai1980 [taxonomy, description, host, distribution: 229]; Kondo2001 [taxonomy, host, distribution: 43]; Leonar1898c [taxonomy, description, illustration, host, distribution: 46-48]; MacGil1921 [taxonomy, description, host, distribution: 403]; MillerDa1990 [host, distribution, economic importance: 300]; Nakaha1982 [host, distribution: 13]; Nakaha1983 [host, distribution: 9]; Ramakr1921a [host, distribution: 357]; Sugimo1994 [host, distribution: 115-121]; Takagi1969a [taxonomy, description, illustration, host, distribution: 70-72,100]; Takaha1929 [host, distribution: 79]; Takaha1932a [host, distribution: 104-105]; Takaha1933 [host, distribution: 25-34]; Takaha1935 [host, distribution: 4]; Takaha1936b [taxonomy, host, distribution: 426]; Takaha1941b [host, distribution: 220]; Takaha1942b [host, distribution: 47]; Takaha1942d [host, distribution: 358]; Tang1984 [taxonomy, description, illustration, host, distribution: 20-21]; Tao1999 [taxonomy, host, distribution: 119]; Velasq1971 [taxonomy, description, illustration, host, distribution: 100-103]; VelasqRi1969 [host, distribution: 195-208]; WilliaWa1988 [taxonomy, description, illustration, host, distribution: 58-60]; WongChCh1999 [taxonomy, description, host, distribution: 35,79]; WoodruBeSk1998 [distribution].

Aspidiotus flavus nomen nudum
Aspidiotus flavus Ferris, 1941e: 43. Nomen nudum. Notes: Ferris (1941e: 43) credited this binomen to Green (1899), but no *Aspidiotus flavus* was listed by Green (1889).
Aspidiotus flavus Borchsenius, 1966: 369. Nomen nudum.

Aspidiotus fularum Balachowsky
Aspidiotus fularum Balachowsky, 1956: 68. Type data: GUINEA: Dalaba (Fouta-Djalon), altitude 1300 meters, on undetermined plant. Holotype female. Type depository: Tervuren: Musée Royal de l'Afrique Centrale, Section d'Entomologie, Belgium.
SCALE COVER: Female scale subcircular, flat, bright brown, translucent at margin; exuviae central, very transparent; diameter 1.8-2 mm. Male scale dark brown, oval, 1 mm long (Balachowsky, 1956).
HOST PLANTS: **Ebenaceae**: *Diospyros mespiliformis* [Balach1956]. **Palmae**: *Cocos* [Balach1956].
DISTRIBUTION: **Afrotropical**: Guinea [Balach1956]; Sierra Leone [Balach1956]; Tanzania [Balach1956].
GENERAL: Description and illustration of adult female by Balachowsky (1956).
KEYS: De Lotto 1957: 228 (female) [Africa]; Balachowsky 1956: 52 (female) [Africa].
CITATIONS: Balach1956 [taxonomy, description, illustration, host, distribution: 67-70]; Borchs1966 [catalogue: 261]; DeLott1957 [taxonomy: 228].

Aspidiotus furcraeicola Lindinger
Aspidiotus furcraeicola Lindinger, 1910b: 36. Type data: TANZANIA [=Deutsch-Ostafrica]: Tanga, on *Furcraea gigantea*. Syntypes, female. Type depository: Hamburg: Zoologisches Institut und Zoologishces Museum, Universität von Hamburg, Germany.
Spinaspidiotus furcraeicolus; MacGillivray, 1921: 429. Change of combination requiring emendation of species name for agreement in gender.
Aspidiotus furcraeicola; Ferris, 1941e: 43. Revived combination.
SCALE COVER: Female scale circular, 1.5-2 mm in diameter, brown. Male scale elongate, 1.5 mm long, 1 mm wide, white, exuviae brown yellow, situated towards cephalic end (Lindinger, 1910b).

HOST PLANTS: **Agavaceae**: *Furcraea gigantea* [Lindin1910b].
DISTRIBUTION: **Afrotropical**: Tanzania [Lindin1910b].
GENERAL: Description and illustration of adult female by Lindinger (1910b).
CITATIONS: Borchs1966 [catalogue: 269]; Ferris1941e [taxonomy: 43]; Lindin1910b [taxonomy, description, illustration, host, distribution: 36]; MacGil1921 [taxonomy, description, host, distribution: 429]; Sassce1912 [taxonomy, host, distribution: 93].

Aspidiotus gossypii Fitch
See *Aspidiotus* gossypii in the section "Taxa transferred from family" at end of catalogue.

Aspidiotus gymnosporiae Lindinger
Aspidiotus gymnosporiae Lindinger, 1911a: 13. Type data: CANARY ISLANDS: Tenerife, Puerto de la Cruz, Botanical Garden, on *Gymnosporia cassinoides*; Palma: Barranco del Rio, on *Gymnosporia cassinoides*. Syntypes, female. Type depository: Hamburg: Zoologisches Institut und Zoologishces Museum, Universität von Hamburg, Germany.
SCALE COVER: Female scale white, with yellow, central exuviae; flat, circular, 2 mm in diameter. Male scale elongate, 1.3 mm long, 1 mm wide, exuviae situated slightly towards cephalic end (Lindinger, 1911a).
HOST PLANTS: **Celastraceae**: *Gymnosporia cassinoides* [Lindin1911a].
DISTRIBUTION: **Palaearctic**: Canary Islands [Lindin1911a, MatileOr2001].
GENERAL: Description and illustration of adult female by Lindinger (1911a).
CITATIONS: Borchs1966 [catalogue: 269]; DanzigPe1998 [catalogue: 189]; Ferris1941e [taxonomy: 44]; Lindin1911a [taxonomy, description, illustration, host, distribution: 13-14]; Lindin1912b [taxonomy, description, host, distribution: 174]; MacGil1921 [taxonomy, description, host, distribution: 403]; MatileOr2001 [host, distribution: 189]; Sassce1912 [taxonomy, host, distribution: 93]; WeidneWa1968 [taxonomy: 172].

Aspidiotus hedericola Leonardi
Aspidiotus hedericola Leonardi, 1918: 188. Nomen nudum. Notes: Leonardi (1918: 188) credited authorship to "Lindinger".
Aspidiotus hedericola Leonardi, 1920: 36. Type data: ITALY: Liguria, Bordighera, on *Hedera*, and YUGOSLAVIA: Dalmatia, Ragusa, on *Hedera*. Syntypes, female. Type depository: Portici: Dipartimento de Entomologia e Zoologia Agraria di Portici, Università di Napoli Federico II, Italy. Notes: Leonardi (1920: 36) incorrectly credited authorship to "Lindinger".
Aspidiotus hedericola; Koroneos, 1934: 7. Notes: Incorrect citation of "Lindinger" as author.
Aspidiotus hedericola; Balachowsky, 1948b: 45. Notes: Incorrect citation of "Lindinger" as author. The citation (p.45) "O. Jaap Cocc. Saml., no. 209, 1912" does not refer to a publication but to specimens in Jaap collection.
Aspidiotus hedericola; Bodenheimer, 1949: 55. Notes: Incorrect citation of "Lindinger" as author.

SCALE COVER: Female scale flat, white, circular, exuvia central, and yellowish. Scale of male slightly elongate, exuvium toward one end (Ferris, 1946).
HOST PLANTS: **Araliaceae**: *Hedera* [Leonar1918], *Hedera helix* [Leonar1920, Ferris1941e, Lupo1948, Balach1948b, Bodenh1949, Bachma1953, Bodenh1952].
NATURAL ENEMIES: HYMENOPTERA **Aphelinidae**: *Aphytis chilensis* Howard [RosenDe1979], *Aphytis mytilaspidis* (Le Baron) [RosenDe1979].
DISTRIBUTION: **Palaearctic**: Greece [Korone1934, ArgyriStMo1976]; Italy [Leonar1918, Leonar1920, LongoMaPe1995]; Sicily [LongoMaPe1995]; Spain [RosenDe1979]; Turkey [Bodenh1952]; Yugoslavia [Balach1948b, Bachma1953].
BIOLOGY: Occurring on leaves (Ferris, 1946).
GENERAL: Description and illustration of adult female by Ferris (1946), Balachowsky (1948b) and by Yaşar (1995a).
KEYS: Danzig 1993: 141 (female) [Europe]; Balachowsky 1948b: 275 (female) [Mediterranean]; Lupo 1948: 138 (female) [Italy]; Ferris 1946: 43 (female) [World]; Leonardi 1920: 29-30 (female) [Italy].
CITATIONS: ArgyriStMo1976 [host, distribution, biological control: 25]; Bachma1953 [host, distribution: 177]; Balach1948b [taxonomy, description, illustration, host, distribution: 285-288]; Bodenh1949 [taxonomy, description, illustration, host, distribution: 55]; Bodenh1952 [host, distribution, structure: 338]; Borchs1966 [catalogue: 261]; DanzigPe1998 [catalogue: 189]; DeBach1964d [biological control: 5-18]; Ferris1941e [taxonomy, host, distribution: 44,56]; Ferris1946 [taxonomy, description, illustration, host, distribution: 42-43,49]; Korone1934 [taxonomy, description, illustration, host, distribution: 7]; Leonar1918 [host, distribution: 188]; Leonar1920 [taxonomy, description, illustration, host, distribution: 36-37]; Lepesm1947 [host, distribution: 189]; LongoMaPe1995 [distribution: 125]; Lupo1948 [taxonomy, description, illustration, host, distribution: 150-154]; Lupo1953 [taxonomy: 39]; Mamet1954 [taxonomy: 52]; RosenDe1979 [host, distribution, biological control: 349-354,464-473]; Ulgent1996 [host, distribution: 541-548]; Viggia1987 [host, distribution, biological control: 121-123]; WeidneWa1968 [taxonomy: 172]; Yasar1995a [taxonomy, description, illustration, host, distribution: 51-52].

Aspidiotus hordeolum Walker
See *Aspidiotus* **hordeolum in the section "Taxa transferred from family" at end of catalogue.**

Aspidiotus hoyae Takagi
Aspidiotus hoyae Takagi, 1969a: 70. Type data: TAIWAN: Southeastern Tai-pei Hsien, on *Hoya carnosa*. Holotype female. Type depository: Sapporo: Entomological Institute, Faculty of Agriculture, Hokkaido University, Japan.
Temnaspidiotus hoyae; Chou, 1985: 400. Change of combination.
SCALE COVER: Takagi (1969a) did not describe scale cover.
HOST PLANTS: **Asclepiadaceae**: *Hoya carnosa* [Takagi1969a].
DISTRIBUTION: **Oriental**: Taiwan [Takagi1969a].
GENERAL: Description and illustration of adult female by Takagi (1969a) and by Chou (1985, 1986).

CITATIONS: Chou1985 [taxonomy, description, host, distribution: 400]; Chou1986 [taxonomy, illustration: 665]; Takagi1969a [taxonomy, description, illustration, host, distribution: 70-71,100]; Tao1999 [taxonomy, host, distribution: 119-120].

Aspidiotus hybridum nomen nudum
Aspidiotus hybridum Jarvis, 1911: 72. Nomen nudum.
Aspidiotus hybridum Ferris, 1941e: 44. Nomen nudum.
Aspidiotus hybridum Borchsenius, 1966: 376. Nomen nudum.

Aspidiotus japonicus (Takagi)
Temnaspidiotus japonicus Takagi, 1957: 38. Type data: JAPAN: Honsyu, Sizuoka-ken, Amagi-san, on *Camellia* sp. Syntypes, female. Type depository: Sapporo: Entomological Institute, Faculty of Agriculture, Hokkaido University, Japan.
Aspidiotus japonica; Danzig, 1993: 141. Incrrect emendation of species name.
SCALE COVER: Scale of female irregularly circular, flat or slightly convex, pale brown; in male slightly elongate (Takagi, 1957).
HOST PLANTS: **Theaceae**: *Camellia* [Takagi1957].
DISTRIBUTION: **Palaearctic**: China (People's Republic) (Henan (Honan) [Wu1999b]); Japan [Kawai1980] (Honshu [Takagi1957]).
BIOLOGY: On underside of leaves.
GENERAL: Description and illustration of adult female by Takagi (1957).
KEYS: Danzig 1993: 140-141 (female) [Europe].
CITATIONS: Borchs1966 [catalogue: 272]; DanzigPe1998 [catalogue: 362]; Kawai1980 [taxonomy, description, host, distribution: 229]; Muraka1970 [host, distribution: 78]; Takagi1957 [taxonomy, description, illustration, host, distribution: 38-40]; Wu1999b [host, distribution: 234].

Aspidiotus juglandis Colvée
Aspidiotus juglandis Colvée, 1881d: clxv. Type data: SPAIN: Catalonia, near Tarragone, on "noyer" [=*Juglans* sp.]. Syntypes, female. Notes: Depository of type material unknown.
Aspidiotus (*Diaspidiotus*) *juglandis*; Cockerell, 1897i: 18. Change of combination.
Aspidiotus iuglandis; Leonardi, 1898c: 40. Misspelling of species name.
Aspidiella juglandis; MacGillivray, 1921: 405. Change of combination.
Aspidiotus juglandis; Lindinger, 1935: 128. Revived combination.
SYSTEMATICS: Leonardi (1898c: 39) regarded *Aspidiotus tiliae* Bouché, 1851, a synonym of *Aspidiotus juglans-regiae* Comstock, whereas Borchsenius (1966) retained it as a valid species.
SCALE COVER: Female scale small, reddish; exuviae central (Colvée, 1881d).
HOST PLANTS: **Juglandaceae**: *Juglans regia* [Colvee1881d, BlayGo1993]. **Salicaceae**: *Populus tremula* [Martin1983].
DISTRIBUTION: **Palaearctic**: Spain [Colvee1881d, GomezM1937, Martin1983, BlayGo1993].
GENERAL: Description of adult female by Colvée (1881d).

CITATIONS: BlayGo1993 [taxonomy, description, illustration, host, distribution: 429-431]; Borchs1966 [catalogue: 369]; Cocker1896b [distribution: 333,334]; Cocker1897i [taxonomy, description, host, distribution: 18]; Colvee1881d [taxonomy, description, host, distribution: clxv-clxvi]; Colvee1882 [taxonomy, description, host, distribution: 5-7]; DanzigPe1998 [catalogue: 189]; Fernal1903b [catalogue: 265]; Ferris1941e [taxonomy: 44]; GomezM1937 [taxonomy, description, illustration, host, distribution: 57-59]; Lindin1912b [taxonomy, description, host, distribution: 187]; Lindin1935 [taxonomy: 128]; Lindin1957 [taxonomy: 545]; MacGil1921 [taxonomy, description, host, distribution: 405]; Martin1983 [taxonomy, host, distribution: 61].

Aspidiotus kellyi Brain
Aspidiotus kellyi Brain, 1918: 122. Type data: SOUTH AFRICA: Transavaal, Pretoria, on *Andropogon amplectens*; collected by A. Kelly, 13.x.1913. Lectotype female, by subsequent designation Munting, 1970a: 39. Type depository: Pretoria: South African National Collection of Insects, South Africa; type no. 188/1.
Brainaspis kellyi; MacGillivray, 1921: 427. Change of combination.
Temnaspidiotus kellyi; Balachowsky, 1956: 132. Change of combination.
Temnaspidiotus kelleyi; Balachowsky, 1956: 132. Misspelling of species name.
Aspidiotus kellyi; Williams & Watson, 1988: 49. Revived combination.
SYSTEMATICS: *Aspidiotus kellyi* Brain is type species of the genus *Brainaspis*, synonymized with *Aspidiotus* by Lindinger, 1937. Dr. Sadao Takagi (in personal communication to Yair Ben-Dov, 8 January 2003) suggested that *Aspidiotus kellyi* Brain and *Aspidiotus sinensis* (Ferris) may belong to a separate genus.
SCALE COVER: Female scale about 2 mm in diameter, circular or slightly elongate, flat to slightly convex, rather robust, faintly buff or brownish in colour, with almost central exuviae, which are covered but, in rubbed specimens appear metallic yellow to bronze in colour. Seen from below, second exuviae yellow. Male scale flat, about 1 mm long, somewhat elongate, often with ends slightly pointed, dull light brown in colour with paler margins. Exuviae covered yellowish (Brain, 1918).
HOST PLANTS: **Gramineae**: *Andropogon amplectens* [Brain1918, Balach1956], *Saccharum officinalis* [PruthiRa1942, Box1953].
DISTRIBUTION: **Afrotropical**: South Africa [Brain1918, Balach1956]. **Oriental**: India [PruthiRa1942].
ECONOMIC IMPORTANCE: Pruthi & Rao (1942) recorded this species from sugarcane in India.
GENERAL: Description and illustration of adult female by Brain (1918) and by Balachowsky (1956).
KEYS: Brain 1918: 118 (female) [South Africa].
CITATIONS: AgarwaSi1964 [host, distribution, economic importance: 149]; Balach1956 [taxonomy, description, illustration, host, distribution: 134-135]; Borchs1966 [catalogue: 272]; Brain1918 [taxonomy, description, illustration, host, distribution: 122-123]; Ferris1937c [taxonomy: 50]; Ferris1938b [illustration: 67]; Ferris1941e [taxonomy: 44]; MacGil1921 [taxonomy, description, host, distribution: 427]; Muntin1970a [taxonomy: 39]; PruthiRa1942 [host, distribution: 87-88];

WilliaGr1990 [host, distribution, economic importance, biological control: 563-578].

Aspidiotus kennedyae (Boisduval)

Chermes kennedyae Boisduval, 1867: 326. Type data: FRANCE: on "glycine" [=*Kennedia*], imported from Nouvelle-Hollande [=AUSTRALIA]. Syntypes, female. Notes: Type material probably lost; Daniele Matile-Ferrero, 1998, personal communication to Yair Ben-Dov.

Aspidiotus kennedyae; Signoret, 1869b: 124. Change of combination.

Aspidiotus (*Aspidiotus*) *kennedyae*; Cockerell, 1897i: 29. Change of combination.

Aspidiotus kennedyae; Borchsenius, 1966: 369. Revived combination.

SCALE COVER: Boisduval (1867) described this species as follows: "This insect is a pest of certain climbing ornamentals from New Holland [=New York] currently placed under the name *Kennedya* [=*Kennedia*]. It resembles very much that of oleander, but it is slightly orange brown. It could be simply a variety of this. It infests, like that insect of the date palm, on the lower surface of leaves, and later both surfaces. This is the reason that certain gardeners abandoned *Kennedia* culture because it is very susceptible to plant sucking insects. It would be necessary, according to the example of Bouché, to start to study especially on the biology of scale insects and study the male, in order to establish that this [*Chermes kennedyae*] constitutes a distinct species."

HOST PLANTS: **Leguminosae**: *Kennedy* [Boisdu1867, Frogga1914].

DISTRIBUTION: **Australasian**: Australia [Frogga1914].

CITATIONS: Boisdu1867 [taxonomy, description, host, distribution: 326-327]; Borchs1966 [catalogue: 369]; Cocker1896b [distribution: 334]; Cocker1897i [taxonomy, description, host, distribution: 29]; Comsto1883 [taxonomy: 78]; Fernal1903b [catalogue: 266]; Ferris1941e [taxonomy: 44]; Frogga1914 [taxonomy, host, distribution: 315]; Frogga1915 [taxonomy, host, distribution: 19]; Signor1869 [taxonomy: 858]; Signor1869b [taxonomy, host, distribution: 124].

Aspidiotus lectularis nomen nudum

Aspidiotus lectularis French, 1907: 184. Nomen nudum.

Aspidiotus lectularius Sanders, 1909a: 53. Nomen nudum.

Aspidiotus lectularis Ferris, 1941e: 45. Nomen nudum.

Aspidiotus lectularis Borchsenius, 1966: 376. Nomen nudum.

Aspidiotus ligusticus Leonardi

Aspidiotus ligusticus Leonardi, 1918: 189. Type data: ITALY: Ventimiglia, on vine. Syntypes, female. Type depository: Portici: Dipartimento de Entomologia e Zoologia Agraria di Portici, Università di Napoli Federico II, Italy.

SCALE COVER: Female scale circular or almost circular; diameter about 1 mm; slightly convex, formed of whitish fine texture; covered by epidermis of bark of host plant; exuviae light yellow, subcentral (Leonardi, 1920).

HOST PLANTS: **Apocynaceae**: *Nerium oleander* [Goux1945]. **Chenopodiaceae**: *Atriplex* [Lupo1948]. **Leguminosae**: *Mimosa* [Lupo1948]. **Vitaceae**: *Vitis vinifera* [Leonar1918, Leonar1920, Ferris1941e].

DISTRIBUTION: **Afrotropical**: Eritrea [Lupo1948]. **Palaearctic**: France [Goux1945]; Italy [Leonar1918, Leonar1920, Lupo1948]; Sicily [Costan1938].
GENERAL: Description and illustration of adult female by Leonardi (1918, 1920).
KEYS: Lupo 1948: 138 [Italy]; Leonardi 1920: 29 [Italy].
CITATIONS: Borchs1966 [catalogue: 261]; Costan1938 [host, distribution: 25-44]; DanzigPe1998 [catalogue: 189]; Ferris1941e [taxonomy: 45]; Goux1945 [taxonomy, host, distribution: 36]; KozarWa1985 [taxonomy: 82]; Leonar1918 [taxonomy, description, host, distribution: 189-192]; Leonar1920 [taxonomy, description, illustration, host, distribution: 29,41-43]; Lindin1957 [taxonomy: 546]; Lupo1948 [taxonomy, description, illustration, host, distribution: 138,145-150].

Aspidiotus luzulae Dufour
See *Aspidiotus* luzulae in the section "Taxa transferred from family" at end of catalogue.

Aspidiotus macfarlanei Williams & Watson
Aspidiotus macfarlanei Williams & Watson, 1988: 58. Type data: SOLOMON ISLANDS: Guadalcanal, Kukum, on *Cocos nucifera*; collected 1.ii.1956. Holotype female. Type depository: London: The Natural History Museum, England, UK.
SCALE COVER: Female scale circular, fawn, with slightly yellow subcentral exuviae. Male scale oval with subcentral exuviae, same colour as female scale (Williams & Watson, 1988).
HOST PLANTS: **Caricaceae**: *Carica papaya* [WilliaWa1988]. **Palmae**: *Cocos nucifera* [WilliaWa1988].
DISTRIBUTION: **Australasian**: Solomon Islands [WilliaWa1988].
GENERAL: Description and illustration of adult female by Williams & Watson (1988).
KEYS: Williams & Watson 1988: 51 (female) [Tropical South Pacific].
CITATIONS: WilliaWa1988 [taxonomy, description, illustration, host, distribution: 58-60].

Aspidiotus maddisoni Williams & Watson
Aspidiotus maddisoni Williams & Watson, 1988: 60. Type data: WESTERN SAMOA: Upolu, Utumapu, on *Asplenium nidus*; collected 7.i.1977. Holotype female. Type depository: London: The Natural History Museum, England, UK.
SCALE COVER: Female scale dirty white with brown exuviae. Male scale not seen (Williams & Watson, 1988).
HOST PLANTS: **Aspleniaceae**: *Asplenium nidus* [WilliaWa1988].
DISTRIBUTION: **Australasian**: Western Samoa [WilliaWa1988].
GENERAL: Description and illustration of adult female by Williams & Watson (1988).
KEYS: Williams & Watson 1988: 51 (female) [Tropical South Pacific].
CITATIONS: WilliaWa1988 [taxonomy, description, illustration, host, distribution: 60-62].

Aspidiotus madecassus Mamet

Aspidiotus madecassus Mamet, 1954: 50. Type data: MADAGASCAR: Périnet, on "Vahim davenona", and Tsinjoarivo, on undetermined plant. Syntypes. Type depository: Paris: Muséum national d'Histoire naturelle, France.

SCALE COVER: Female scale flat, pale-straw coloured, transparent, circular; adult female at centre of scale; exuviae more opaque, yellowish, subcentral. Male scale similar to that of female, but a little more elongate (Mamet, 1954).

DISTRIBUTION: **Afrotropical**: Madagascar [Mamet1954, Borchs1966].

BIOLOGY: On under surface of leaves (Mamet, 1954).

GENERAL: Description and illustration of adult female by Mamet (1954).

CITATIONS: Borchs1966 [catalogue: 261]; Mamet1954 [taxonomy, description, illustration, host, distribution: 15,50].

Aspidiotus maderensis Lindinger

Aspidiotus maderensis Lindinger, 1912b: 189. Type data: MADEIRA ISLANDS: on *Juniperus cedrus*. Holotype female. Type depository: Hamburg: Zoologisches Institut und Zoologishces Museum, Universität von Hamburg, Germany.

SCALE COVER: Female scale more or less circular, diameter 1.5 mm, flat to slightly convex; yellow-white and yellow-brown in center (Lindinger, 1912b).

HOST PLANTS: **Cupressaceae**: *Juniperus cedrus* [Lindin1912b, Green1923b].

DISTRIBUTION: **Palaearctic**: Madeira Islands [Lindin1912b].

GENERAL: Description of adult female by Lindinger (1912b).

CITATIONS: Borchs1966 [catalogue: 269]; DanzigPe1998 [catalogue: 189-190]; Ferris1941e [taxonomy: 45]; Green1923b [host, distribution: 89]; Lindin1912b [taxonomy, description, host, distribution: 189]; WeidneWa1968 [taxonomy: 173].

Aspidiotus marisci Tippins & Beshear

Aspidiotus marisci Tippins & Beshear, 1971: 85. Type data: U.S.A.: Georgia, Camden County, Cumberland Island, on *Mariscus jamaicensis*. Holotype female. Type depository: Washington: United States National Entomological Collection, U.S. National Museum of Natural History, District of Columbia, USA.

SCALE COVER: Female scale circular, ca. 2 mm in diameter, flat, papery, brownish, with exuviae central. Male scale similar but elongate oval (Tippins & Beshear, 1971).

HOST PLANTS: **Cyperaceae**: *Mariscus jamaicensis* [TippinBe1971, BesheaTiHo1973].

DISTRIBUTION: **Nearctic**: United States of America (Alabama [Nakaha1982], Florida [BesheaTiHo1973], Georgia [TippinBe1971, BesheaTiHo1973]).

GENERAL: Description and illustration of adult female by Tippins & Beshear (1971).

CITATIONS: BesheaTiHo1973 [host, distribution: 5]; Dekle1976 [taxonomy, description, host, distribution, economic importance: 41]; Nakaha1982 [host, distribution: 14]; TippinBe1971 [taxonomy, description, illustration, host, distribution: 85-86].

Aspidiotus mespili nomen nudum
Aspidiotus mespili Del Guercio, 1894: 152. Nomen nudum.
Aspidiotus mespili Lindinger, 1936: 153. Nomen nudum.
Aspidiotus mespili Ferris, 1941e: 45. Nomen nudum.
Aspidiotus mespili Borchsenius, 1966: 376. Nomen nudum.

Aspidiotus minutus Cockerell
Aspidiotus minutus Cockerell, 1892b: 333. Type data: JAMAICA: Montego Bay, on coconut palm; collected by Dr. Sinclair. Syntypes, female and first instar. Type depository: Washington: United States National Entomological Collection, U.S. National Museum of Natural History, District of Columbia, USA.
SYSTEMATICS: Cockerell (1892b: 33) briefly described this species, as follows: "*Aspidiotus minutus* Ckll. MS. On coconut palm, near Montego Bay. Collected by Dr. Sinclair. Not yet studied sufficiently; seems to be mature. Occurs with *A. rapax* var., but the young of that species, when of the size of *minutus*, are black with slight, pale rim". Ferris (1941e: 45) regarded this species a *nomen nudum*, whereas Borchsenius (1966) listed it among the *incertae sedis* species. We interpret original description, although very meager, as providing enough characters to validate this species.
HOST PLANTS: **Palmae**: *Cocos nucifera* [Cocker1892b].
DISTRIBUTION: **Neotropical**: Jamaica.
CITATIONS: Borchs1966 [catalogue: 369]; Cocker1892b [taxonomy, description, host, distribution: 333]; Ferris1941e [taxonomy: 45].

Aspidiotus moreirai Hempel
Aspidiotus moreirai Hempel, 1904: 320. Type data: BRAZIL: Rio de Janeiro, Itatiya, on leaves of *Drimys* sp. Syntypes, female. Type depository: Curitiba: Departamento de Zoologia, Setor de Ciencias Biologicas, Universidade Federal do Parana, Brazil; type no. 81-106.
SCALE COVER: Female scale more or less circular, flat, colour white, or sometimes transparent; 2 mm in diameter; exuviae light yellow, placed more or less centrally (Hempel, 1904).
HOST PLANTS: **Winteraceae**: *Drimys winterii* [Hempel1904, Lepage1938, ClapsWoGo2001], *Drimys* [Sander1906].
DISTRIBUTION: **Neotropical**: Brazil (Rio de Janeiro [Hempel1904, Sander1906, Lepage1938, ClapsWoGo2001]).
GENERAL: Description of adult female by Hempel (1904).
CITATIONS: Borchs1966 [catalogue: 269]; Claps1993 [taxonomy: 9]; ClapsWoGo2001 [host, distribution: 241]; Ferris1941e [taxonomy: 46]; Hempel1904 [taxonomy, description, host, distribution: 320-321]; Lepage1938 [catalogue: 395-396]; Lindin1957 [taxonomy: 546]; MacGil1921 [taxonomy, description, host, distribution: 400]; Sander1906 [taxonomy, host, distribution: 14].

Aspidiotus msolonus Hall

Aspidiotus msolonus Hall, 1929: 348. Type data: ZIMBABWE: Mazoe, on *Pseudolachnostylis maprounaefolia*; Lomagundi, Sipolilo, on *Pseudolachnostylis* sp. Syntypes, female. Type depository: London: The Natural History Museum, England, UK.

SCALE COVER: Female scale pure white, more or less circular, diameter 1.25-1.75 mm; highly convex; exuviae shiny brown green, pale at margin; nymphal exuviae brown, also paler at margin; exuviae usually a little to one side; colour of exuviae masked by a semi-opaque thin white secretionary film; that part covering larval exuviae occasionally wanting, having presumably been knocked off; secretionary film thinner in vicinity of margin of nymphal exuviae, since pale brown belt round margin often seen through film as a pale brown ring; ventral scale extremely thin and delicate, remaining attached to host plant. Male scale of normal form, white with brown exuviae (Hall, 1929).

HOST PLANTS: **Euphorbiaceae**: *Pseudolachnostylis* [Hall1929], *Pseudolachnostylis maprounaefolia* [Balach1956].

DISTRIBUTION: **Afrotropical**: Zimbabwe [Hall1929, Balach1956].

GENERAL: Description and illustration of adult female by Hall (1929) and by Balachowsky (1956).

KEYS: Balachowsky 1956: 51 (female) [Africa].

CITATIONS: Balach1955 [taxonomy, host, distribution: 391]; Balach1956 [taxonomy, description, illustration, host, distribution: 70-72]; Borchs1966 [catalogue: 261]; Ferris1941e [taxonomy: 46]; Hall1929 [taxonomy, description, illustration, host, distribution: 348-350].

Aspidiotus musae Williams & Watson

Aspidiotus musae Williams & Watson, 1988: 62. Type data: PAPUA NEW GUINEA: Morobe P., Lasagna Is., on *Musa* sp.; collected 18.ix.1979. Holotype female. Type depository: London: The Natural History Museum, England, UK.

SCALE COVER: Williams & Watson (1988) did not describe scale cover.

HOST PLANTS: **Musaceae**: *Musa* [WilliaWa1988].

DISTRIBUTION: **Australasian**: Papua New Guinea [WilliaWa1988].

GENERAL: Description and illustration of adult female by Williams & Watson (1988).

KEYS: Williams & Watson 1988: 51 (female) [Tropical South Pacific].

CITATIONS: SugimoKaTa1996 [host, distribution: 99-101]; WilliaWa1988 [taxonomy, description, illustration, host, distribution: 62-64].

Aspidiotus myoporii Lidgett

Aspidiotus myoporii Lidgett, 1898a: 13. Type data: AUSTRALIA: Victoria, Myrniong, on *Myoporum deserti*; collected by H. Lidgett. Syntypes, both sexes.
Aspidiotus yoporii; Lidgett, 1898a: 14. Misspelling of species name.
Chrysomphalus yoporii; Leonardi, 1900: 343. Change of combination.
Chrysomphalus yoporii; Leonardi, 1900: 343. Misspelling of species name.
Aspidiotus myoporii; Borchsenius, 1966: 369. Revived combination.

SYSTEMATICS: On page 14 of original description, species name misspelled *Aspidiotusm yoporii*.

SCALE COVER: Female scale circular, diameter about 1/12 inch; slightly convex, dark brown, of a lighter tinge towards edge; in some cases, colour almost black, and scale forms quite conspicuous objects on green leaves of the host plant; exuviae yellowish, forming a slight depression. Male scale somewhat elongated, light brown in colour; exuviae at one end; average length 1/16 inch (Lidgett, 1898a).

HOST PLANTS: **Myoporaceae**: *Myoporum deserti* [Lidget1898a, Frogga1914].

DISTRIBUTION: **Australasian**: Australia (Victoria [Lidget1898a, Frogga1914]).

GENERAL: Description of adult female by Lidgett (1898a).

CITATIONS: Borchs1966 [catalogue: 369]; Fernal1903b [catalogue: 267]; Ferris1941e [taxonomy: 46]; Frogga1914 [taxonomy, description, host, distribution: 315-316]; Frogga1915 [taxonomy, description, host, distribution: 20]; Leonar1900 [taxonomy, host, distribution: 343]; Lidget1898 [taxonomy, description, host, distribution: 13-14]; Lindin1907a [taxonomy: 20].

Aspidiotus myrthi Bouché

Aspidiotus myrthi Bouché, 1851: 112. Type data: GERMANY: Berlin, on *Myrthus communis* [=*Myrtus communis*]. Syntypes, female. Notes: Type material lost (Sachtleben, 1944).

Chionaspis? myrthi; Signoret, 1869d: 445. Change of combination.

Mytilaspis myrthi; Cockerell, 1901c: 93. Change of combination.

Lepidosaphes myrthi; Fernald, 1903b: 311. Change of combination.

Aspidiotus myrti; Lindinger, 1935: 128. Misspelling of species name.

Aspidiotus myrthii; Borchsenius, 1966: 369. Revived combination.

SYSTEMATICS: Borchsenius (1966) listed this species among the species *incertae sedis*.

SCALE COVER: Female scale elongate, brown (Bouché, 1851).

HOST PLANTS: **Myrtaceae**: *Myrtus communis* [Bouche1851].

DISTRIBUTION: **Palaearctic**: Germany (United) [Bouche1851].

GENERAL: Description of adult female by Bouché (1851).

CITATIONS: Borchs1966 [catalogue: 369]; Bouche1851 [taxonomy, description, host, distribution: 112]; Cocker1901c [taxonomy: 93]; Fernal1903b [catalogue: 220]; Ferris1937a [taxonomy: 5]; Ferris1938a [taxonomy: 229]; Ferris1941e [taxonomy: 46]; Lindin1912b [taxonomy: 372]; Lindin1928 [taxonomy: 106]; Lindin1934e [taxonomy, description, host, distribution: 166]; Lindin1935 [taxonomy: 128]; Lindin1957 [taxonomy: 545]; Signor1869d [taxonomy, description, host, distribution: 445].

Aspidiotus mytiliformis nomen nudum

Aspidiotus mytiliformis Amerling, 1858a: 103. Nomen nudum.

Aspidiotus mytiliformis Borchsenius, 1966: 376. Nomen nudum.

Aspidiotus nerii Bouché

Diaspis obliquum Costa, 1829: 2. Type data: ITALY:. Syntypes, female. Synonymy by Ben-Dov & Marotta, 2001a: 191. Notes: Ben-Dov & Marotta (2001a, 2001b) proved that this species was a *nomen oblitum,* and placed it as a synonym of *Aspidiotus nerii* Bouché, 1833. Type material lost (Giuseppina Pellizzari, personal communication to Yair Ben-Dov, 1999).

Aspidiotus nerii Bouché, 1833: 52. Type data: GERMANY: Berlin, in greenhouse, on *Nerium, Arbutus, Magnolia* and *Acacia.* Syntypes, both sexes. Type depository: Eberswalde: Institut fur Pflanzenschutzforschung, Germany.

Aspidiotus genistae Westwood, 1840: 118. Nomen nudum; discovered by Fernald, 1903b: 261.

Diaspis bouchei Targioni Tozzetti, 1867: 13. Nomen nudum; discovered by Fernald, 1903b: 261.

Chermes aloes Boisduval, 1867: 327. Type data: FRANCE: apparently Paris, in greenhouse, on *Aloe umbellata.* Syntypes, female. Synonymy by Fernald, 1903b: 261. Notes: Type material probably lost; Daniele Matile-Ferrero, 1998, personal communication to Yair Ben-Dov.

Chermes ericae Boisduval, 1867: 330. Type data: FRANCE: Vincennes and Montreuil, on *Erica.* Syntypes, female. Synonymy by Fernald, 1903b: 261. Notes: Type material probably lost; Daniele Matile-Ferrero, 1998; personal communication to Yair Ben-Dov.

Chermes cycadicola Boisduval, 1867: 345. Type data: FRANCE: apparently Paris, on *Cycas revoluta.* Syntypes, female. Synonymy by Fernald, 1903b: 261. Notes: Type material probably lost; Daniele Matile-Ferrero, 1998; personal communication to Yair Ben-Dov.

Chermes nerii; Boisduval, 1868: 281. Change of combination.

Aspidiotus affinis Targioni Tozzetti, 1868: 736. Type data: ITALY: on leaves of *Rusci aculeati [=Ruscus aculeatus].* Syntypes, female. Synonymy by Fernald, 1903b: 261. Notes: Type material probably lost; Giuseppina Pellizzari, 1990, personal communication to Yair Ben-Dov.

Aspidiotus bouchei Targioni Tozzetti, 1868: 736. Unjustified replacement name for *Aspidiotus nerii* Bouché, 1833; discovered by Fernald, 1903b: 261.

Aspidiotus caldesii Targioni Tozzetti, 1868: 736. Type data: ITALY: on leaves of *Daphne collinae.* Syntypes, female. Synonymy by Borchsenius, 1966: 262. Notes: Type material probably lost; Giuseppina Pellizzari, 1990, personal communication to Yair Ben-Dov.

Aspidiotus denticulatus Targioni Tozzetti, 1868: 736. Type data: ITALY: on leaves of *Rubia peregrinae.* Syntypes, female. Synonymy by Fernald, 1903b: 261. Notes: Type material probably lost; Giuseppina Pellizzari, 1990, personal communication to Yair Ben-Dov.

Aspidiotus villosus Targioni Tozzetti, 1868: 736. Type data: ITALY: on leaves of *Olea europaea.* Syntypes, female. Synonymy by Fernald, 1903b: 261. Notes: Type material probably lost; Giuseppina Pellizzari, 1990, personal communication to Yair Ben-Dov.

Aspidiotus aloes; Signoret, 1869: 843. Change of combination.

Aspidiotus budleiae Signoret, 1869: 845. Nomen nudum.

Aspidiotus limonii Signoret, 1869: 860. Nomen nudum.

Aspidiotus vrieseiae Signoret, 1869: 876. Nomen nudum.

Aspidiotus budleiae Signoret, 1869b: 115. Type data: FRANCE: Paris, Luxembourg Garden, on *Buddleia salicina*. Holotype female. Type depository: Vienna: Naturhistorisches Museum Wien, Austria. Synonymy by Fernald, 1903b: 261.

Aspidiotus ceratoniae Signoret, 1869b: 118. Type data: FRANCE: Nice, on carob [=*Ceratonia siliqua*]. Syntypes, female. Type depository: Vienna: Naturhistorisches Museum Wien, Austria. Synonymy by Fernald, 1903b: 261.

Aspidiotus cycadicola; Signoret, 1869b: 119. Change of combination.

Aspidiotus ericae; Signoret, 1869b: 121. Change of combination.

Aspidiotus gnidii Signoret, 1869b: 122. Type data: FRANCE: Le Midi [=South-East France], on *Daphne gnidium*. Syntypes, female. Type depository: Vienna: Naturhistorisches Museum Wien, Austria. Synonymy by Fernald, 1903b: 261.

Aspidiotus ilicis Signoret, 1869b: 123. Type data: FRANCE: Le Midi [=South-East France], on *Quercus ilicis*. Syntypes, female. Type depository: Vienna: Naturhistorisches Museum Wien, Austria. Synonymy by Fernald, 1903b: 261.

Aspidiotus limonii Signoret, 1869b: 125. Type data: FRANCE: Provence, on "citron" [=*Citrus limon*]. Syntypes, both sexes. Type depository: Vienna: Naturhistorisches Museum Wien, Austria. Synonymy by Fernald, 1903b: 261.

Aspidiotus myricinae Signoret, 1869b: 125. Type data: FRANCE: Paris, Luxembourg Garden, in greenhouse, on *Myricinia retusa [= probably Myristica retusa]*. Syntypes, female. Type depository: Vienna: Naturhistorisches Museum Wien, Austria. Synonymy by Fernald, 1903b: 125.

Aspidiotus capparis Signoret, 1869b: 129. Type data: FRANCE: probably Dijon, on caprier [=*Capparis* sp.]. Syntypes, female. Type depository: Vienna: Naturhistorisches Museum Wien, Austria. Synonymy by Signoret, 1877: 663. Notes: Authorship incorrectly credited to "Vallot, 1829".

Aspidiotus ulicis Signoret, 1869b: 132. Type data: FRANCE: Le Midi [=South East France], on "genet epineux" [=*Ulex* sp.]. Syntypes, female and first instar. Type depository: Vienna: Naturhistorisches Museum Wien, Austria. Synonymy by Signoret, 1877: 675.

Aspidiotus vriesciae Signoret, 1869b: 134. Type data: FRANCE: apparently Paris, in greenhouse, on *Vriescia [= Vriesea] splendens* imported from Cayenne. Syntypes, female. Type depository: Vienna: Naturhistorisches Museum Wien, Austria. Synonymy by Fernald, 1903b: 262.

Aspidiotus genistae Signoret, 1869c: 122. Type data: FRANCE: Cannes, on "genets" [=*Genista* sp.]. Syntypes, female. Type depository: Vienna: Naturhistorisches Museum Wien, Austria. Synonymy by Fernald, 1903b: 261. Notes: Species incorrectly credited to Westwood.

Aspidiotus hederae Signoret, 1869c: 122. Unavailable name.

Aspidiotus palmarum; Signoret, 1869c: 131. Misidentification; discovered by Danzig, 1993: 145.

Aspidiotus epidendri Signoret, 1869c: 145. Type data: FRANCE: Paris, Luxembourg garden, on *Epidendrum* sp. Syntypes, both sexes. Type depository: Vienna: Naturhistorisches Museum Wien, Austria. Synonymy by Borchsenius, 1966: 263. Notes: Authorship incorrectly credited to Bouché.

Aspidiotus lentisci Signoret, 1877: 601. Type data: FRANCE: Le Midi [=South-East France], on "lentisque" [=*Pistacia lentiscus*]; also from ALGERIA, on same host; collected by Bigot. Syntypes, female. Type depository: Vienna: Naturhistorisches Museum Wien, Austria. Synonymy by Fernald, 1903b: 262.

Aspidiotus ? osmanthi Signoret, 1877: 621. Type data: FRANCE: Locality not indicated, on *Olea fragrans* [=*Osmanthus fragrans*]. Syntypes, female. Type depository: Vienna: Naturhistorisches Museum Wien, Austria. Synonymy by Borchsenius, 1966: 264. Notes: Authorship incorrectly credited to "Vallot, 1829".

Aspidiotus atherospermae Maskell, 1879: 198. Type data: NEW ZEALAND: on *Atherosperma novae-zealandiae*. Syntypes, female. Type depository: Auckland: New Zealand Arthropod Collection, Landcare Research, New Zealand. Synonymy by Borchsenius, 1966: 264.

Aspidiotus budlaei; Maskell, 1879: 198. Misspelling of species name. Notes: Misspelling of *Aspidiotus budleiae* Signoret, 1869b.

Aspidiotus dysoxyli Maskell, 1879: 198. Type data: NEW ZEALAND: on *Dysoxylum* sp. Lectotype female, by subsequent designation Henderson, 2001a: 89. Type depository: Auckland: New Zealand Arthropod Collection, Landcare Research, New Zealand. Synonymy by Henderson, 2001a: 89.

Aspidiotus oleae Colvée, 1880: 39. Type data: SPAIN: on olive. Syntypes, female. Synonymy by Fernald, 1903b: 261. Notes: Depository of type material unknown. Original description by Colvée (1880) included material of both *Diaspis oleae* (=*Parlatoria oleae* (Colvée)) and *Aspidiotus oleae* (=*Aspidiotus nerii* (Bouché)).

Aspidiotus corynocarpi Colvée, 1881: 39. Type data: SPAIN: on *Corynocarpus* sp. Syntypes, female. Synonymy by Borchsenius, 1966: 264. Notes: Depository unknown.

Aspidiotus oleastri Colvée, 1882: 12. Type data: SPAIN: Cambrils, on olive, *Olea europaea*. Syntypes, female. Synonymy by Borchsenius, 1966: 264. Notes: Depository of type material unknown.

Aspidiotus offinis; Comstock, 1883: 72. Misspelling of species name. Notes: Misspelling of *Aspidiotus affinis* Targioni Tozzetti, 1868

Aspidiotus hederae; Comstock, 1883: 77. Misidentification; discovered by Ben-Dov & Matile-Ferrero, 1999: 5.

Aspidiotus myrsinae; Comstock, 1883: 79. Misspelling of species name. Notes: Misspelling of *Aspidiotus myrcinae* Signoret.

Aspidiotus sophorae Maskell, 1884: 121. Type data: NEW ZEALAND: on *Sophora tetraptera*. Syntypes, female and first instar. Type depository: Auckland: New Zealand Arthropod Collection, Landcare Research, New Zealand. Synonymy by Ferris, 1941e: 48.

Aspidiotus carpodeti Maskell, 1885a: 21. Type data: NEW ZEALAND: on *Carpodetus serratus* and *Vitex littoralis*. Syntypes, female and first instar. Type depository: Auckland: New Zealand Arthropod Collection, Landcare Research, New Zealand. Synonymy by Borchsenius, 1966: 264.

Aspidiotus budlaeiae; Maskell, 1887a: 40. Misspelling of species name. Notes: Misspelling of *Aspidiotus budleiae* Signoret, 1869b.

Aspidiotus epidendri; Cockerell, 1896b: 334. Notes: Incorrect citation of "Bouché" as author.

Aspidiotus myrsinae; Cockerell, 1896b: 334. Misspelling of species name. Notes: Misspelling of *Aspidiotus myrcinae* Signoret.

Aspidiotus nerii limonii; Cockerell, 1896b: 334. Change of combination and rank.

Aspidiotus (*Aspidiotus*) *affinis*; Cockerell, 1897i: 18. Change of combination.

Aspidiotus (*Aspidiotus*) *caldesii*; Cockerell, 1897i: 18. Change of combination.

Aspidiotus (*Aspidiotus*) *ceratoniae*; Cockerell, 1897i: 18. Change of combination.

Aspidiotus (*Aspidiotus*) *denticulatus*; Cockerell, 1897i: 18. Change of combination.

Aspidiotus (*Aspidiotus*) *ericae*; Cockerell, 1897i: 18. Change of combination.

Aspidiotus (*Aspidiotus*) *genistae*; Cockerell, 1897i: 18. Change of combination.

Aspidiotus (*Aspidiotus*) *gnidii*; Cockerell, 1897i: 18. Change of combination.

Aspidiotus (*Aspidiotus*) *hederae*; Cockerell, 1897i: 18. Misidentification; discovered by Ben-Dov & Matile-Ferrero, 1999: 5.

Aspidiotus (*Aspidiotus*) *ilicis*; Cockerell, 1897i: 18. Change of combination.

Aspidiotus (*Aspidiotus*) *lentisci*; Cockerell, 1897i: 18. Change of combination.

Aspidiotus (*Diaspidiotus*) *villosus*; Cockerell, 1897i: 19. Change of combination.

Aspidiotus (*Aspidiotus*) *carpodeti*; Cockerell, 1897i: 25. Change of combination.

Aspidiotus (*Aspidiotus*) *aloes*; Cockerell, 1897i: 29. Change of combination.

Aspidiotus (*Aspidiotus*) *buddleiae*; Cockerell, 1897i: 29. Change of combination.

Aspidiotus (*Aspidiotus*) *cycadicola*; Cockerell, 1897i: 29. Change of combination.

Aspidiotus (*Aspidiotus*) *epidendri*; Cockerell, 1897i: 29. Change of combination.

Aspidiotus (*Aspidiotus*) *myrsinae*; Cockerell, 1897i: 30. Misspelling of species name. Notes: Misspelling of *Aspidiotus myrcinae* Signoret.

Aspidiotus (*Aspidiotus*) *nerii*; Cockerell, 1897i: 30. Change of combination.

Aspidiotus (*Aspidiotus*) *nerii limonii*; Cockerell, 1897i: 30. Change of combination.

Aspidiotus osmanthi; Cockerell, 1897i: 30. Notes: Incorrect citation of "Vallot" as author.

Aspidiotus (*Aspidiotus*) *vriesciae*; Cockerell, 1897i: 30. Change of combination.

Aspidiotus (*Evaspidiotus*) *hederae*; Leonardi, 1898a: 76. Change of combination.

Aspidiotus (*Evaspidiotus*) *hederae*; Leonardi, 1898c: 71. Misidentification; discovered by Morrison & Morrison, 1966: 17. Notes: Cited as *Aspidiotus hederae* (Vallot, 1829).

Aspidiotus (*Evaspidiotus*)) *nerii*; Leonardi, 1898c: 71. Incorrect synonymy. Notes: Incorrect synonymy of *Aspidiotus nerii* Bouché, 1833 with *Aspidiotus hederae* (Vallot, 1829); see Morrison & Morrison, 1966: 17.

Aspidiotus hederae nerii; Hunter, 1899: 11. Change of status.

Aspidiotus hederae carpodeti; Cockerell & Parrott, 1899: 276. Change of combination and rank.

Aspidiotus capparis; Cockerell, 1899a: 395. Notes: Incorrect citation of "Vallot" as author.

Aspidiotus corinocarpi; Cockerell, 1899a: 395. Misspelling of species name. Notes: Misspelling of "Aspidiotus corynocarpi".

Aspidiotus epidendri; Cockerell, 1899a: 395. Notes: Incorrect citation of "Bouché" as author.

Aspidiotus hederae; Cockerell, 1899a: 395. Misidentification; discovered by Ben-Dov & Matile-Ferrero, 1999: 5.

Aspidiotus myrsinae; Cockerell, 1899a: 395. Misspelling of species name. Notes: Misspelling of "Aspidiotus myrciniae" Signoret.

Aspidiotus nerii; Cockerell, 1899a: 395. Incorrect synonymy. Notes: Incorrect synonymy of *Aspidiotus nerii* Bouché, 1833 with *Aspidiotus hederae* (Vallot, 1829); see Ben-Dov & Matile-Ferrero, 1999: 5.

Aspidiotus vagabundus Cockerell, 1899n: 20. Type data: MEXICO: Mexico City, on bark of ash. Syntypes, female. Type depository: Washington: United States National Entomological Collection, U.S. National Museum of Natural History, District of Columbia, USA. Synonymy by Ferris, 1941e: 49.

Aspidiotus hederae limonii; Cockerell, 1900: 350. Change of combination.

Aspidiotus epidendri; Newstead, 1901: 120. Notes: Incorrect citation of "Bouché" as author.

Aspidiotus hederae; Newstead, 1901b: 120. Misidentification; discovered by Ben-Dov & Matile-Ferrero, 1999: 5.

Aspidiotus epidendri; Fernald, 1903b: 261. Notes: Incorrect citation of "Bouché" as author.

Aspidiotus osmanthi; Fernald, 1903b: 268. Notes: Incorrect citation of "Vallot" as author.

Aspidiotus nerii; Fernald, 1903b: 269. Incorrect synonymy. Notes: Incorrect synonymy of *Aspidiotus nerii* Bouché, 1833, with *Aspidiotus hederae* (Vallot, 1829); see Ben-Dov & Matile-Ferrero, 1999: 5.

Aspidiotus simillimus; Fernald, 1903b: 278. Change of status.

Aspidiotus transparens rectangulatus Lindinger, 1913: 97. Type data: KENYA: Mombassa, on *Sansevieria* sp. Syntypes, female. Type depository: Hamburg: Zoologisches Institut und Zoologishces Museum, Universität von Hamburg, Germany. Synonymy by Ferris, 1941e: 47.

Aspidiotus confusus Froggatt, 1914: 136. Type data: AUSTRALIA: New South Wales, at Narara, on *Eucalyptus* sp. Syntypes, female. Type depository: Canberra: Australian National Insect Collection, CSIRO Entomology, Australia. Synonymy by Borchsenius, 1966: 265.

Aspidiotus transvaalensis Leonardi, 1914: 198. Type data: SOUTH AFRICA: Pretoria, on *Nerium oleander*. Syntypes, female. Type depository: Portici: Dipartimento de Entomologia e Zoologia Agraria di Portici, Università di Napoli Federico II, Italy. Synonymy by Ferris, 1941e: 49.

Aspidiotus confusus; Froggatt, 1915: 13. Notes: Described again as "n. sp.".

Aspidiotus tasmaniae Green, 1915d: 50. Type data: AUSTRALIA: Tasmania, Launceston, on *Ribes* sp. and *Ampelopsis* sp.; Victoria, on *Eucalyptus* sp., *Acacia* sp., and *Cytisus* sp. Syntypes, female. Type depository: London: The Natural History Museum, England, UK. Synonymy by Ferris, 1941e: 48.

Aspidiotus hederae; Dietz & Morrison, 1916a: 296. Misidentification; discovered by Ben-Dov & Matile-Ferrero, 1999: 5.

Aspidiotus epidendri; Leonardi, 1920: 31. Notes: Incorrect citation of "Bouché" as author.

Aspidiotus guidii; Leonardi, 1920: 31. Misspelling of species name. Notes: Misspelling of *Aspidiotus gnidii* Signoret, 1869b.

Aspidiotus hederae; Leonardi, 1920: 31. Misidentification. Notes: Cited as *Aspidiotus hederae* (Vallot, 1829).

Aspidiotus myrsinae; Leonardi, 1920: 31. Misspelling of species name. Notes: Misspelling of *Aspidiotus myrcinae* Signoret.

Aspidiotus viresciae; Leonardi, 1920: 31. Misspelling of species name. Notes: Misspelling of *Aspidiotus vriesciae* Signoret, 1869b.

Octaspidiotus atherospermae; MacGillivray, 1921: 395. Change of combination.

Aspidiotus hederae; MacGillivray, 1921: 400. Misidentification; discovered by Ben-Dov & Matile-Ferrero, 1999: 5. Notes: Cited as *Aspidiotus hederae* (Vallot, 1829).

Aspidiotus hederae urenae Hall, 1923: 19. Type data: EGYPT: Giza, garden of the Horticultural Section of the Ministry of Agriculture, on *Urena lobata*. Syntypes, female. Type depository: London: The Natural History Museum, England, UK. Synonymy by Ferris, 1941e: 49.

Aspidiotus hederae unipectinata Carimini, 1930: 121. Type data: ITALY: Livorno, in several gardens of Castglioncello, on *Acacia dealbata*. Syntypes, female. Synonymy by Ferris, 1941e: 49. Notes: Depository of type series unknown.

Aspidiotus nederae; Archangelskaya, 1930: 85. Misspelling of species name.

Aspidiotus (*Dynaspidiotus*) *hederae*; Thiem & Gerneck, 1934: 131. Change of combination.

Chermes hederae; Ferris, 1937c: 62. Change of combination.

Aspidiotus hederae; Ferris, 1938a: 192. Misidentification; discovered by Ben-Dov & Matile-Ferrero, 1999: 5.

Aspidiotus hederae; Ferris, 1938a: 192. Incorrect synonymy. Notes: Incorrect synonymy of *Aspidiotus nerii* Bouché, 1833, with *Aspidiotus hederae* (Vallot, 1829); see Ben-Dov & Matile-Ferrero, 1999: 5.

Aspidiotus hederale; Tscorbadjiew, 1939: 89. Misspelling of species name.

Chermes genistae; Ferris, 1941e: 43. Change of combination.

Aspidiotus osmanthi; Ferris, 1941e: 46. Notes: Incorrect citation of "Vallot" as author.

Chermes osmanthi; Ferris, 1941e: 46. Change of combination.

Aspidiotus rectangulatus; Ferris, 1941e: 47. Change of combination and rank.

Aspidiotus unipectinatus; Ferris, 1941e: 49. Change of combination and rank.

Aspidiotus urenae; Ferris, 1941e: 49. Change of combination and rank.

Aspidiotus fonsecai Giannotti, 1942: 214. Type data: BRAZIL: São Paulo, on undetermined host; collected by J.P. da Fonseca, December 1930. Holotype female. Type depositories: São Paulo: Instituto Biologico de São Paulo, Brazil, and Curitiba: Departamento de Zoologia, Setor de Ciencias Biologicas, Universidade Federal do Parana, Brazil. Synonymy by Munting, 1971a: 313.

Octaspidiotus anthospermae; Balachowsky, 1948b: 272. Misspelling of species name.

Octaspidiotus athospermae; Balachowsky, 1948b: 272. Change of combination.

Aspidiotus hederae hederae; Schmutterer, 1952: 566. Change of combination.

Aspidiotus hederae unisexualis Schmutterer, 1952: 566. Type data: GERMANY: in greenhouse, on *Phoenix canariensis*, *Chamaerops humilis*, *Asparagus sprengeri*, *Aloe* sp. and *Nerium oleander*. Syntypes, female. Type depository: Wetlenberg: The Schmutterer Collection, Germany. Synonymy by Borchsenius, 1966: 265.

Aspidiotus hederae; Balachowsky, 1956: 70. Misidentification; discovered by Morrison & Morrison, 1966: 17. Notes: Cited as *Aspidiotus hederae* (Vallot, 1829).

Aspidiotus heredae; Balachowsky, 1956: 70. Misspelling of species name.

Aspidiotus nerii; Morrison & Morrison, 1966: 17. Revived status.

Aspidiotus nerii; Borchsenius, 1966: 261. Revived status.

Aspidiotus nerii; Gerson & Zor, 1973: 516. Notes: Author incorrectly cited as "Vallot, 1829".

Aspidiotus paranerii Gerson *in* Gerson & Hazan, 1979: 281. Type data: ISRAEL: Rehovot, laboratory culture on potato [*Solanum tuberosum*] tubers; laboratory culture initiated from population collected on *Pittosporum undulatum* in Rehovot. Holotype female. Type depository: Bet Dagan: Department of Entomology, The Volcani Center, Israel. Synonymy by Danzig, 1993: 145.

Aspidiotus atherospirmae; Chou, 1985: 264. Misspelling of species name.

Aspidiotus nereii; Foldi, 1990c: 199. Misspelling of species name.

Aspidiotus anthospermae; Tao, 1999: 73. Misspelling of species name.

Aspidiotus atherospinmae; Tao, 1999: 73. Misspelling of species name.

Aspidiotus budlaei; Tao, 1999: 73. Misspelling of species name.

Aspidiotus budlei; Tao, 1999: 73. Misspelling of species name.

Aspidiotus transvalensis; Tao, 1999: 73. Misspelling of species name.

Aspidiotus vagobundus; Tao, 1999: 73. Misspelling of species name.

COMMON NAMES: escama blanca de la hiedra [Gonzal1989]; escama hiedra [CoronaRuMo1997]; ivy scale [MerrilCh1923, McKenz1956, Dekle1965c]; Oleander scale [Merril1953, Borchs1966]; pinta-branca [CarvalAg1997]; piojo blanco [Lloren1990]; plushevaya shitovka [Borchs1936].

SYSTEMATICS: *Aspidiotus nerii* Bouché, 1833 is senior synonym of oleander scale. Taxonomic interpretation of this species has gradually evolved through studies of Signoret (1869b), Comstock (1883), Newstead (1901) and Dietz & Morrison (1916). Currently, its taxonomy is widely recognized and established in accordance with taxonomic facies elucidated by Ferris (1938), Balachowsky (1956) and Borchsenius (1966); see detailed discussion in Ben-Dov & Matile-Ferrero (1999). *Diaspis obliquum* was described by O.G. Costa from Italy, in his *Fauna del Regno di Napoli, famiglia de' coccinigliferi, o de'gallinsetti*. For more than a century, publication date of Costa's book accepted as 1835 and therefore synonymy of *Diaspis obliquum* with *Aspidiotus nerii* was nomenclaturally acceptable. In 1983, The ICZN ruled (ICZN, 1983) date of publication of Costa's book was 1829. Consequently, *Diaspis obliquum* Costa, 1829, predated *Aspidiotus nerii* Bouché, 1833. Ben-Dov & Marotta (2001), in order to stabilize senior synonym status of *Aspidiotus nerii* Bouché, 1833, regarded *Diaspis obliquum* Costa, 1829 as a *nomen oblitum* and placed it as a junior synonym of *Aspidiotus nerii*. Signoret (1869b, p. 122, and Plate IV, Figs. c,e,f) redescribed a species which he named *Aspidiotus hederae* and incorrectly credited the authorship to Vallot. Signoret illustrated (Plate IV, Fig. c) a tubular duct which is more than 5 times as long as wide, and therefore *Aspidiotus hederae* Signoret, 1869b, is clearly different from *Aspidiotus nerii* Bouché, see discussion in Ben-Dov & Matile-Ferrero (1999). Synonym *Aspidiotus epidendri* first described by Signoret (1869c: 121) who erroneously credited it to Bouché (1844: 293). However, Bouché did not describe this species and therefore, the author is Signoret. Signoret's error was perpetuated by Newstead (1901: 120), Fernald (1903b: 261) and Leonardi (1920: 31). Borchsenius (1966: 261) regarded *Aspidiotus ligusticus* Leonardi, 1918, as a distinct species. *Aspidiotus urenae* Hall, 1923 was regarded as a valid species, but Borchsenius (1966: 265) placed it as a synonym of *Aspidiotus nerii*. Danzig (1993: 143) regarded *Aspidiotus urenae* a synonym of *Aspidiotus nerii*. *Aspidiotus nerii* was reported to have uniparental as well as biparental populations (Brown, 1965).

SCALE COVER: Female scale white or pale grey, circular, flat, exuviae subcentral. Male scale similar in colour, slightly oval, exuvia subcentral (Ferris, 1938a). Colour photograph of scale cover and general appearance see Gonzalez (1986, 1989), Carvalho & Aguiar (1997) and Gill (1997).

HOST PLANTS: **Actinidiaceae**: *Actinidia chinensis* [Gonzal1986, Gonzal1989a, GonzalCu1994]. **Agavaceae**: *Agave* [McKenz1956, Hadzib1983, Martin1983], *Agave americana* [Balach1927, Balach1932d, GomezM1962, Martin1983], *Agave palmeri* [MerrilCh1923], *Agave sisal* [Bodenh1924], *Dracaena* [Brain1918], *Furcraea gigantea* [Balach1932d, GomezM1962], *Sansevieria* [Lindin1913, Ferris1941e], *Yucca* [Brain1918, Borchs1934, Borchs1936, Bodenh1952, McDani1968], *Yucca gloriosa* [Borchs1934, Martin1983]. **Aizoaceae**: *Mesembryanthemum acinaciforme* [Balach1932d], *Mesembryanthemum edule* [Balach1932d]. **Anacardiaceae**: *Pistacia mutica* [Bodenh1924], *Pistacia lentiscus* [Signor1877, Balach1932d, Balach1935b, Martin1983], *Pleiogynium cerasiferum* [Brimbl1968], *Rhodosphaera rhodanthema* [Brimbl1968], *Schinus* [Hall1923], *Schinus terebinthifolius* [Balach1932d]. **Annonaceae**: *Annona squamosa* [DeLott1967a]. **Apocynaceae**: *Acokanthera schimperi* [DeLott1967a], *Carissa ovata* [Brimbl1968], *Nerium* [Bodenh1937, Bachma1953, Balach1956], *Nerium oleander* [UygunSeEr1998, ErlerTu2001], *Nerium oleandrum* [Leonar1914, Bodenh1926, Balach1932d, Borchs1934, Bodenh1944b, Bodenh1952, GomezM1962, Muntin1969], *Oleander* [Bodenh1928], *Plumeria alba* [Cohic1958], *Plumeria rubra* [Brimbl1968], *Vinca* [DietzMo1916, Bodenh1952], *Vinca major* [Bodenh1952, Merril1953, GomezM1962, Martin1983], *Vinca rosea* [Martin1983]. **Aquifoliaceae**: *Ilex aquifolium* [McKenz1956]. **Araceae**: *Acorus gramineus* [Martin1983]. **Araliaceae**: *Aralia* [Brain1918], *Aralia sieboldi* [Martin1983], *Hedera* [Signor1869b, Borchs1936, Martin1983], *Hedera helix* [DietzMo1916, Balach1932d, Bodenh1952, McKenz1956, Hadzib1983, Martin1983, UygunSeEr1998, Foldi2000], *Hedera pastuchovii* [Hadzib1983], *Meryta* [Green1929]. **Aristolochiaceae**: *Aristolochia* [Hall1923, Balach1932d], *Aristolochia baetica* [GomezM1946, Martin1983], *Aristolochia elegans* [BesheaTiHo1973]. **Asclepiadaceae**: *Stapelia variegata* [Martin1983]. **Aspleniaceae**: *Asplenium lucidum* [TachikVa1969], *Asplenium nidus* [WilliaWa1988]. **Aucubaceae**: *Aucuba* [Brain1918, Green1928, Merril1953, McKenz1956], *Aucuba japonica* [Bodenh1952, Bachma1953, BesheaTiHo1973, RosenDe1979, Martin1983]. **Betulaceae**: *Alnus rhombifolia* [Ferris1920b]. **Bignoniaceae**: *Bignonia* [Balach1932d], *Tecoma* [Balach1932d]. **Bromeliaceae**: *Ananas comosus* [Cohic1958], *Billbergia rutas* [Hadzib1983], *Vriesea splendens* [Signor1869b, MerrilCh1923]. **Buddlejaceae**: *Buddleja* [McKenz1956], *Buddleja madagascariensis* [Balach1932d], *Buddleja salicina* [Signor1869b]. **Buxaceae**:

Buxus [Signor1869b], *Buxus sempervirens* [Balach1932d]. **Cactaceae**: *Phyllocactus* [GomezM1962], *Rhipsalis* [GomezM1962]. **Cannaceae**: *Canna* [Bodenh1924]. **Capparidaceae**: *Capparis lucida* [Brimbl1968], *Capparis nobilis* [Brimbl1968], *Capparis spinosa* [Bodenh1928, GomezM1948, Martin1983]. **Caprifoliaceae**: *Lonicera* [GomezM1946, Bodenh1952, Martin1983], *Lonicera caprifolium* [Lepage1938], *Lonicera implexa* [Balach1932d], *Lonicera japonica* [Brimbl1968], *Lonicera meisneri* [Martin1983], *Lonicera pentademia* [Martin1983], *Viburnum tinus* [Martin1983]. **Caryophyllaceae**: *Dianthus* [Balach1932d], *Dianthus cariofilus* [Martin1983], *Petrocoptis lagascae* [Martin1983]. **Casuarinaceae**: *Casuarina glauca* [Brimbl1968]. **Celastraceae**: *Celastrus* [Brimbl1968], *Euonymus japonica* [Bachma1953], *Euonymus japonicum* [Balach1927, Balach1932d], *Gymnosporia baetica* [Martin1983]. **Cistaceae**: *Cistus* [Martin1983], *Cistus heterophyllus* [Balach1932d], *Cistus salviaefolius* [Balach1927, Balach1932d], *Halimium lasianthum* [Martin1983], *Helianthemum* [Martin1983]. **Cneoraceae**: *Cneorum* [Martin1983]. **Compositae**: *Carduus* [Balach1932d], *Chrysanthemum coronarium* [Hall1923], *Inula viscosa* [Balach1932d, Balach1933e, Foldi2000], *Senecio* [Martin1983]. **Convolvulaceae**: *Calistegia sepium* [Martin1983], *Ipomoea mexicana* [Balach1932d]. **Corynocarpaceae**: *Corynocarpus laevigatus* [Martin1983]. **Crassulaceae**: *Sedum* [Martin1983]. **Cruciferae**: *Iberis pruitis* [Martin1983], *Iberis sempervirens* [Martin1983]. **Cupressaceae**: *Callitris* [Brain1918], *Callitris glauca* [Brimbl1968], *Cupressus sempervirens* [Brain1918], *Thuja* [Brain1918]. **Cycadaceae**: *Cycas* [Merril1953, McKenz1956, Hadzib1983], *Cycas media* [Brimbl1968], *Cycas revoluta* [Boisdu1867, Signor1869b, DietzMo1916, Balach1932d, Borchs1934, Dekle1965c, Martin1983]. **Cyperaceae**: *Cyperus* [Bodenh1924, Borchs1934], *Cyperus alternifolius* [Balach1932d, Martin1983]. **Dioscoreaceae**: *Dioscorea* [Brimbl1968], *Tamus communis* [Balach1932d]. **Ebenaceae**: *Diospyros ferrea* [Zimmer1948], *Diospyros kaki* [Borchs1934, Brimbl1968, WilliaWa1988, TomkinWiTh2000], *Diospyros lotus* [Borchs1934], *Royena lucida* [Martin1983]. **Elaeagnaceae**: *Elaeagnus* [Brain1918, Borchs1934, BachmaGe1950], *Elaeagnus ombellata* [Balach1932d], *Elaeagnus pungens* [Merril1953], *Elaeagnus reflexa* [Balach1932d]. **Ephedraceae**: *Ephedra* [Bodenh1949, Bodenh1952]. **Ericaceae** [BesheaTiHo1973], *Arbutus* [Brain1918], *Arbutus menziesii* [Ferris1920b], *Arbutus unedo* [Balach1932d, Balach1933e, Martin1983, Foldi2000], *Arctostaphylos* [Ferris1920b, McKenz1956], *Erica* [Boisdu1867, MerrilCh1923], *Erica arborea* [Bodenh1949, Bodenh1952], *Erica verticiallata* [UygunSeEr1998], *Rhododendron ponticum* [Hadzib1983], *Vaccinium* [BesheaTiHo1973]. **Escalloniaceae**: *Carpodetus serratus* [Maskel1885a, MerrilCh1923]. **Euphorbiaceae**: *Aleurites fordii* [Hadzib1983], *Aleurites moluccana* [DeLott1967a], *Baloghia lucida* [Brimbl1968], *Cluytia pulchella* [GomezM1946, Martin1983], *Codiaeum variegatum* [Brimbl1968], *Colmeiroa buxifolia* [Martin1983], *Croton tiglium* [Brimbl1968], *Euphorbia* [Green1896e, Green1923b, Balach1932d, GomezM1954, GomezM1956b], *Eu. aphylla* [Martin1983], *Eu. characias* [Balach1933e, Foldi2000], *Eu. pithyusa* [Balach1930, Foldi2000], *Eu. terracina* [Balach1932d], *Eu. wulpheni* [GomezM1946, Martin1983], *Jatropha multifida* [Wilson1917, MerrilCh1923, Merril1953], *Mallotus claoxyloides* [Brimbl1968], *Poinsettia* [Hall1923], *Ricinus communis*

[Bodenh1924, Balach1927, Balach1932d]. **Fagaceae**: *Quercus* [Borchs1934, McKenz1956], *Quercus coccifera* [Balach1932d], *Quercus ilex* [Signor1869b, MerrilCh1923, Balach1932d], *Quercus suber* [Balach1927, Balach1932d]. **Flacourtiaceae**: *Casearia multinervosa* [Brimbl1968]. **Flindersiaceae**: *Flindersia australis* [Brimbl1968], *Flindersia bennettiana* [Brimbl1968], *Flindersia xanthoxyla* [Brimbl1968]. **Garryaceae**: *Garrya elliptica* [TachikVa1969], *Garrya macrophyla* [Balach1932d]. **Grossulariaceae**: *Ribes* [Green1915d]. **Guttiferae**: *Hypericum* [McKenz1956], *Hypericum ritchteri* [Martin1983], *Mammea africana* [MerrilCh1923], *Ochrocarpos africanus* [MerrilCh1923, Merril1953]. **Hippocastanaceae**: *Aesculus hippocastani* [Bachma1953, Zahrad1972]. **Icacinaceae**: *Pennantia cunninghamii* [Brimbl1968]. **Iridaceae**: *Gladiolus* [Brimbl1968], *Iris germanica* [Balach1932d, Martin1983], *Rochea* [Martin1983]. **Juglandaceae**: *Carya* [Borchs1934]. **Labiatae**: *Rosmarinus officinalis* [Balach1932d, Martin1983], *Salvia* [Hall1922, Bodenh1949], *Sideritis subatlantica* [Balach1932d], *Stachys circinnata* [Newste1897a, Balach1927, Balach1932d], *Teucrium polium* [GomezM1958c], *Teucrium scorodonia* [Martin1983]. **Lauraceae**: *Cinnamomum camphorae* [Brimbl1968], *Endiandra pubens* [Brimbl1968], *Laurus* [McKenz1956], *Laurus nobilis* [Balach1932d, GomezM1962, Martin1983], *Persea* [Wilson1917], *Persea americana* [McKenz1956, Brimbl1968, GersonZo1973], *Umbellularia californica* [Ferris1920b]. **Leguminosae**: *Acacia* [Green1915d, MerrilCh1923, Bodenh1937, BachmaGe1950, McKenz1956, Martin1983, SengonUyKa1998], *Ac. aulacocarpa* [Brimbl1968], *Ac. cunninghamii* [Brimbl1968], *Ac. cyanophylla* [Balach1927, Balach1932d, Matile1984c, UygunSeEr1998, KaracaSeCo2001], *Ac. dealbata* [Carimi1930, Balach1932d, Borchs1934], *Ac. decurrens* [Hall1922], *Ac. dodonefolia* [Martin1983], *Ac. floribunda* [Balach1932d, Martin1983], *Ac. harpophylla* [Brimbl1968], *Ac. longifolia* [Bodenh1924, Bodenh1949, Bodenh1952], *Ac. melanoxylon* [Borchs1934], *Ac. mollissima* [DeLott1967a], *Ac. myrtifolia* [Laing1929], *Ac. nematophila* [Martin1983], *Ac. provissima* [Borchs1934], *Ac. xilocarpa* [Martin1983], *Adenocarpus foliosus gomerae* [GomezM1962], *Albizia neumaniana* [Martin1983], *Amorpha nana* [Martin1983], *Anthyllis cytisoides* [Martin1983], *Bauhinia* [Brain1918], *Caesalpinia bonduc* [DeLott1967a], *Cajanus cajan* [DeLott1967a], *Cajanus indica* [Hadzib1983], *Calycotome spinosa* [Newste1897a, Balach1927, Balach1932d, Bodenh1952], *Calycotome villosa* [Bodenh1926, Martin1983], *Cassia didymobotrya* [DeLott1967a], *Centrosema brasilianum* [Martin1983], *Ceratonia siliqua* [Signor1869b, MerrilCh1923, Bodenh1926, Balach1932d, Bodenh1952, Bachma1953, McKenz1956, Martin1983], *Cerceris* [MerrilCh1923], *Cercis siliquastrum* [Balach1932d, GomezM1946, ErlerTu2001], *Coronilla glauca* [GomezM1946, Martin1983], *Coronilla juncea* [Martin1983], *Cytisus* [Green1915d, Balach1932d], *Cytisus prolifer palmensis* [GomezM1962], *Dalbergia* [Green1896], *Genista* [MerrilCh1923, Balach1931a, Balach1932d], *Genista linifolia* [Balach1932d], *Gleditsia* [Borchs1934, Bodenh1952], *Mimosa* [Martin1983], *Platylobium* [Frogga1914], *Robinia pseudacacia* [Newste1897a, Balach1932d, Borchs1934, Martin1983], *Sarothamnus scoparius* [Green1923b], *Sophora tetrapora* [Maskel1884], *Spartium junceum* [Newste1897a, Balach1927, Balach1932d], *Tipuana speciosa* [Balach1932d], *Ulex* [Signor1869b, GomezM1946,

Martin1983], *Ulex boeticus* [Martin1983], *Ulex europaeus* [Martin1983], *Ulex spectabilis* [Balach1932d]. **Liliaceae**: *Aloe umbellata* [Signor1869b, MerrilCh1923], *Asparagus* [Green1930b, Balach1935b, Bodenh1952, Savesc1982, Hadzib1983, Martin1983], *Asparagus albus* [Balach1932d], *Asparagus aphyllus* [Balach1932d], *Asparagus horridus* [Martin1983], *Asparagus plumosus* [Balach1932d, Bodenh1944b, Martin1983], *Asparagus racemosus* [Brimbl1968], *Asparagus sprengieri* [Green1923b, Balach1932d, Merril1953, McDani1968], *Asphodelus* [Balach1932d], *Aspidistra* [Laing1929], *Aspidistra elatior* [Martin1983], *Chlorophytum sapense* [Brimbl1968], *Haworthia retusa* [Martin1983], *Haworthia tesselata* [GomezM1946, Martin1983]. **Loganiaceae**: *Gelsemium semperviren* [GomezM1962]. **Loranthaceae**: *Amyema congener* [Brimbl1968], *Amyema ferruginiflora* [Brimbl1968], *Amyema pendula* [Brimbl1968], *Loranthus* [Green1896], *Phrygilanthus bidwillii* [Brimbl1968]. **Lythraceae**: *Bergenia crassifolia* [McKenz1956]. **Magnoliaceae**: *Magnolia* [Hall1922], *Magnolia fuscata* [GomezM1962], *Magnolia grandiflora* [Balach1932d, Martin1983]. **Malpighiaceae**: *Stigmatophyllon ciliatum* [Zimmer1948]. **Malvaceae**: *Hoheria* [TachikVa1969], *Malope malachoides stipulacea* [Balach1932d], *Urena lobata* [Hall1923]. **Meliaceae**: *Cedrela toona* [Brimbl1968], *Cedrella fissilis* [Lepage1938], *Dysoxylum* [Comsto1916a, Brimbl1968, Hender2001a], *Melia* [Brain1918, MerrilCh1923], *Melia azedarach* [Hall1922, Bodenh1924, Hall1928, Balach1932d, Martin1983, UygunSeEr1998]. **Monimiaceae**: *Hedicarya arborea* [TachikVa1969], *Wilkiea huegeliana* [Brimbl1968]. **Moraceae**: *Ficus carica* [Balach1932d], *Morus* [Hall1922], *Morus alba* [Balach1932d, Bodenh1952], *Morus nigra* [Martin1983]. **Musaceae**: *Musa* [GomezM1962], *Musa sapientum* [Cohic1958]. **Myoporaceae**: *Eremophila mitchelli* [Frogga1914, Brimbl1968], *Myoporum* [Balach1932d], *Myoporum acuminatum* [Muntin1969], *Myoporum deserti* [Brimbl1968], *Myoporum pictum* [Martin1983]. **Myrsinaceae**: *Aegiceras corniculatum* [Brimbl1968], *Myrsine africana* [Martin1983], *Myrsine retusa* [Signor1869b, MerrilCh1923]. **Myrtaceae**: *Eucalyptus* [Frogga1914, Green1915d], *Eugenia smithii* [Frogga1914], *Melaleuca viridiflora* [Brimbl1968], *Myrtus communis* [Martin1983], *Psidium guajava* [Brimbl1968], *Tristania conferta* [Brimbl1968], *Tristania suaveolens* [Brimbl1968]. **Nandinaceae**: *Nandina domestica* [Brimbl1968]. **Oleaceae**: *Fraxinus* [Balach1932d, Borchs1934, Borchs1936], *Fraxinus berlandieri* [Balach1932d], *Fraxinus oxyphylla* [Balach1932d], *Jasminum* [Hall1922, Bodenh1952], *Jasminum officinalis* [Martin1983], *Jasminum primulinum* [Merril1953], *Ligustrum* [McKenz1956], *Ligustrum caricana* [GomezM1946, Martin1983], *Ligustrum coriaceum* [Martin1983], *Ligustrum sinensis* [Balach1932d], *Olea* [Bodenh1937], *Olea europaea* [Targio1868, Bodenh1927b, Lepage1938, Bodenh1949, Bodenh1952, McKenz1956, Almeid1973b, Martin1983], *Olea fragrans* [Wilson1917], *Osmanthus* [Brain1918], *Osmanthus aquifolium* [Balach1932d], *Phyllirea media* [Newste1897a, Balach1927, Balach1932d], *Picconia excelsa* [GomezM1962], *Syringa* [Borchs1934, Bodenh1952], *Syringa vulgaris* [Borchs1934, Martin1983]. **Orchidaceae**: *Cymbidium* [McKenz1956]. **Paeoniaceae**: *Paeonia officinalis* [GomezM1946, Martin1983]. **Palmae** [Green1896, DietzMo1916, Green1929, BesheaTiHo1973], *Areca* [Wilson1917], *Chamaerops excelsa* [Balach1932d],

Chamaerops humilis [Newste1897a, Balach1927, Balach1932d, Martin1983], *Cocos nucifera* [Cohic1958, WilliaWa1988], *Cocos plumosa* [Wilson1917], *Cocos romanzoffiana* [Balach1932d], *Howeia belmoreana* [StoetzDa1974a], *Howeia forsteriana* [Balach1932d], *Kentia balmoreana* [Martin1983], *Livistona sinensis* [Balach1932d], *Phoenix* [McKenz1956], *Phoenix canariensis* [Wilson1917, Bodenh1924, Balach1927, Bodenh1928, Balach1932d, Balach1937c, Bodenh1952], *Phoenix dactylifera* [Bodenh1924, Martin1983], *Phoenix reclinata* [Balach1927, Balach1932d], *Pritchardia filifera* [Balach1932d], *Rhopalostylis baueri* [PicartMa2000], *Trachycarpus excelsa* [Borchs1934, GomezM1962]. **Pandanaceae**: *Pandanus* [DietzMo1916, Cohic1958]. **Passifloraceae**: *Passiflora* [Balach1932d], *Passiflora coerulea* [Balach1932d], *Tacsonia manicata* [Green1923b]. **Phytolaccaceae**: *Phytolacca dioica* [Balach1932d]. **Pinaceae**: *Picea* [Martin1983], *Pinus halepensis* [Balach1932d]. **Pittosporaceae**: *Pittosporum* [Fernan1992], *Pittosporum tobira* [Martin1983], *Pittosporum undulatum* [GersonHa1979, Martin1983]. **Platanaceae**: *Platanus* [Martin1983], *Platanus orientalis* [Balach1932d, Borchs1934]. **Podocarpaceae**: *Podocarpus elatus* [Brimbl1968]. **Polygonaceae**: *Coccoloba* [Wilson1917], *Muehlenbeckia platyclados* [Martin1983], *Owenia venosa* [Brimbl1968]. **Proteaceae**: *Grevillea* [GomezM1946, Martin1983], *Grevillea preissei* [Balach1932d], *Grevillea robusta* [Newste1914, Balach1932d, Brimbl1968], *Hakea* [Brain1918], *Macadamia integrifolia* [Brimbl1968], *Persoonia cornifolia* [Brimbl1968], *Stenocarpus sinuatus* [Brimbl1968]. **Punicaceae**: *Punica granatum* [Hall1923, Balach1932d]. **Ranunculaceae**: *Clematis cirhosa* [Balach1932d], *Clematis flammula* [Newste1897a, Balach1927, Balach1932d]. **Rhamnaceae**: *Ceanothus* [Ferris1920b, McKenz1956], *Colletia spinosa* [Balach1932d], *Rhamnus alaternus* [Balach1932d], *Rhamnus californica* [McKenz1956], *Ziziphus lotus* [Martin1983]. **Rosaceae**: *Crataegus azarolus* [Newste1897a, Balach1927, Balach1932d], *Malus sylvestris* [Brimbl1968], *Prunus avium* [Brimbl1968], *Prunus japonica* [Balach1932d], *Prunus pissardi* [Brain1918], *Pyrophorum pictum* [Martin1983], *Pyrus communis* [Brimbl1968], *Pyrus cydonia* [Balach1932d], *Rosa* [Lepage1938], *Rubus cuesus* [UygunSeEr1998], *Spiraea* [Brain1918]. **Rubiaceae**: *Canthium coprosmoides* [Brimbl1968], *Coprosma* [Brain1918], *Galium* [Balach1935b, Martin1983], *Gardenia* [Lepage1938], *Putoria calabrica* [Bachma1953], *Randia fitzalani* [Brimbl1968], *Rubia peregrina* [Targio1868, MerrilCh1923]. **Ruscaceae**: *Ruscus* [Bodenh1952, Martin1983], *Ruscus aculeatus* [Targio1868, MerrilCh1923, Bodenh1926, Martin1983]. **Rutaceae**: *Acronychia baueri* [Brimbl1968], *Citrus* [Howard1908, Balach1932d, McKenz1956, WilliaWa1988, CarvalAg1997], *Citrus* [MerrilCh1923], *Citrus aurantium* [Balach1932d, Bodenh1926], *Citrus limon* [Signor1869b, Balach1932d, Bodenh1926, Brimbl1968, Hadzib1983], *Citrus sinensis* [Brimbl1968], *Eriostemon queenslandicus* [Brimbl1968], *Evodia microcarpa* [Brimbl1968], *Geijera salicifolia* [Brimbl1968], *Microcitrus australasica* [Brimbl1968], *Platydesma* [Zimmer1948], *Ruta angustifolia* [Balach1932d, Balach1933e, Foldi2000], *Ruta graveolens* [Green1923b, Balach1927, Balach1932d, Martin1983], *Xanthoxylum americanum* [Balach1932d]. **Salicaceae**: *Populus alba* [Balach1932d], *Populus nigra* [Balach1932d]. **Sambucaceae**: *Sambucus nigra* [ErlerTu2001]. **Santalaceae**: *Osyris alba* [Newste1897a, Balach1927, Balach1932d, Bodenh1952], *Santalum haleakalae*

[Zimmer1948], *Santalum lanceolatum* [Brimbl1968]. **Sapindaceae**: *Alectryon coriaceus* [Brimbl1968], *Atalaya hemiglauca* [Brimbl1968], *Diploglottis australe* [Brimbl1968], *Dodonaea triquerta* [Brimbl1968], *Dodonaea viscosa* [GomezM1946, Martin1983, Matile1984c], *Guioa semiglauca* [Brimbl1968], *Harpullia pendula* [Brimbl1968], *Heterodendrum oleaefolium* [Frogga1914, Laing1929]. **Sapotaceae**: *Argania sideroxylon* [Balach1932d], *Bumelia tenax* [Martin1983]. **Saxifragaceae**: *Saxifraga* [Hadzib1983]. **Scrophulariaceae**: *Antirrhinum* [Bodenh1924], *Antirrhinum majus* [Newste1897a, Balach1927, Balach1932d, Martin1983], *Digitalis obscura* [Martin1983], *Paulownia* [Balach1932d], *Penstemon* [Brain1918], *Silvia* [Bodenh1952], *Veronica* [Brain1918]. **Smilacaceae**: *Smilax* [McDani1968], *Smilax aspera* [Newste1897a, Balach1927, Balach1932d], *Smilax australe* [Brimbl1968]. **Solanaceae**: *Datura alba* [Hall1922], *Datura arborea* [Martin1983], *Nicotiana glauca* [Bodenh1924, Martin1983], *Solanum coagulans* [Bodenh1924], *Solanum jasminoides* [Balach1932d], *Solanum lycopersicum* [Hall1923], *Solanum macrocarpum* [DeLott1967a], *Solanum sodomaeum* [Balach1932d, Martin1983], *Solanum tuberosum* [GersonHa1979], *Withania frutescens* [Martin1983]. **Siphonodontaceae:** *Siphonodon australe* [Brimbl1968]. **Strelitziaceae**: *Strelitzia augusta* [Balach1932d], *Strelitzia ovata* [Martin1983], *Strelitzia reginae* [Balach1932d, Martin1983]. **Taxaceae**: *Taxus* [McKenz1956], *Taxus iberica* [Martin1983], *Taxus polium* [Martin1983]. **Taxodiaceae**: *Sequoia sempervirens* [Ferris1920b]. **Theaceae**: *Camellia* [McKenz1956], *Camellia japonica* [MerrilCh1923], *Thea* [Green1896]. **Thymelaeaceae**: *Daphne* [GomezM1954, Martin1983], *Daphne collina* [Targio1868], *Daphne gnidium* [Signor1869b, MerrilCh1923, Balach1932d, Balach1933e, Martin1983, Foldi2000], *Thymelaea hirsuta* [Bodenh1924, Balach1927, Balach1932d, Martin1983], *Thymelaea lythroides* [Balach1932d]. **Tmesipteridaceae**: *Tmesipteris tannensis* [WilliaWa1988]. **Ulmaceae**: *Celtis australis* [Balach1932d], *Ulmus* [Balach1932d]. **Umbelliferae**: *Foeniculum vulgaris* [Martin1983], *Washingtonia* [Bodenh1924, Bodenh1944b], *Washingtonia sonorae* [Martin1983]. **Verbenaceae**: *Lantana camara* [Brimbl1968], *Vitex littoralis* [Maskel1885a, MerrilCh1923]. **Vitaceae**: *Ampelopsis* [Green1915d], *Cayratia acris* [Brimbl1968], *Vitis* [Green1923b, McKenz1956], *Vitis vinifera* [Leonar1918, Leonar1920, Balach1932d, Bodenh1924, Ferris1941e, Bodenh1949, Bodenh1952, Cohic1958]. **Zamiaceae**: *Bowenia serrulata* [Brimbl1968], *Macrozamia* [Frogga1914], *Macrozamia moorei* [Brimbl1968], *Zamia floridana* [Dekle1965c]. **NATURAL ENEMIES**: ACARI **Hemisarcoptidae**: *Hemisarcoptes malus* (Shimer) [GersonOcHo1990]. COLEOPTERA **Coccinellidae**: *Chilocorus bipustulatus* L. [Balach1948b, Inserr1970a, Zahrad1972, KaracaSeCo2001], *Chilocorus circumdatus* Gyllenhal [Housto1991], *Chilocorus undecimpunctata* L. [Housto1991], *Coccidophilus citricola* [AguileMeVa1984], *Exochomus quadrimaculatus* [Balach1948b], *Exochomus quadripustulatus* L. [Inserr1970a, ErlerTu2001], *Lindorus lophantae* (Blaisdell) [Smirno1950a], *Nephus caneparii* Fursch & Uygun [ErlerTu2001], *Pharoscymnus setulosus* Mulsant [Balach1948b], *Rhyzobius lophanthae* (Blaisdell) [ErlerTu2001, KaracaSeCo2001]. **Nitidulidae**: *Cybocephalus rufifrons* Reitter [Balach1948b]. FUNGI **Ascomycotina**: *Nectria flammea* [EvansPr1990]. **Deuteromycotina**: *Verticillium lecanii* [EvansPr1990].

HYMENOPTERA **Aphelinidae**: *Aphytis aonidiae* (Mercet) [RosenDe1979], *Ap. capillatus* (Howard) [RosenDe1979], *Ap. chilensis* Howard [RosenDe1979, Hadzib1983, MyartsRu2000], *Ap. chrysomphali* (Mercet) [RosenDe1979], *Ap. coheni* DeBach [RosenDe1979], *Ap. diaspidis* Howard [Balach1948b, RosenDe1979], *Ap. fisheri* DeBach [RosenDe1979], *Ap. fuscipennis* Howard [Balach1948b], *Ap. hispanicus* (Mercet) [RosenDe1979], *Ap. holoxanthus* DeBach [AnneckIn1971, RosenDe1979], *Ap. lingnanensis* Compere [Inserr1970a, RosenDe1979], *Ap. longiclavae* Mercet [Balach1948b, Zahrad1972], *Ap. maculicornis* (Masi) [Balach1948b, RosenDe1979, MyartsRu2000], *Ap. melinus* DeBach [Inserr1970a, RosenDe1979, SengonUyKa1998, KaracaSeCo2001], *Ap. mytilaspidis* (Le Baron) [RosenDe1979], *Ap. notialis* De Santis [RosenDe1979], *Ap. proclia* (Walker) [RosenDe1979], *Aspidiotiphagus citrinus* Craw [Balach1948b, Zahrad1972, Liotta1974a, Hadzib1983], *Aspidiotiphagus latipennis* Compere [AnneckIn1971], *Azotus capensis* Howard [AnneckIn1970], *Coccophagus scutellaris* Dalman [Balach1948b], *Encarsia citrina* (Craw) [SengonUyKa1998, AbdRab2001a], *Encarsia lounsburyi* (Berlese & Paoli) [AbdRab2001a], *Prospaltella aurantii* (Howard) [Gordh1979]. **Encyrtidae**: *Adelencyrtus inglisiae* Compere & Annecke [Prinsl1983], *Aphycus punctipes* Dalman [Balach1948b], *Euaphycus flavus* [Zahrad1972], *Euaphycus hederaceus* (Westwood) [Balach1948b, Zahrad1972], *Habrolepis rouxi* Compere [BlumbeDe1979], *Metaphycus eurhinus* Annecke & Mynhardt [Prinsl1983], *Metaphycus nigripectus* Annecke & Mynhardt [Prinsl1983], *Zelaphycus aspidioti* (Tachikawa & Valentine) [TachikVa1969]. **Signiphoridae**: *Chartocerus mexicanus* (Ashmead) [Gordh1979], *Signiphora aspidioti* Ashmead [Woolle1990], *Signiphora merceti* Malenotti [Woolle1990], *Signiphora prepauca* Girault [Woolle1990], *Signiphora thoeauini* Girault [Gordh1979]. NEUROPTERA **Chrysopidae**: *Chrysoperla carnea* (Stephens) [Drea1990]. THYSANOPTERA **Phlaeothripidae**: *Aleurodothrips fasciapennis* (Franklin) [Beshea1975, PalmerMo1990], *Karnyothrips flavipes* (Jones) [PalmerMo1990].

DISTRIBUTION: **Afrotropical**: Angola [Almeid1973b]; Eritrea [DeLottNa1955, DeLott1967a]; Kenya [Lindin1913, Ferris1941e, DeLott1967a]; Malawi [Newste1914]; Namibia (Southwest Africa) [Muntin1969]; South Africa [Leonar1914, BrainKe1917, Brain1918, Balach1956]; Tanzania [Newste1911a, Balach1956]; Uganda [Newste1911, Balach1956]; Zaire [Balach1956]; Zimbabwe [Hall1928, Balach1956]. **Australasian**: Australia [Frogga1914] (Queensland [Brimbl1968], Tasmania [Green1915d], Victoria [Green1915d, Laing1929]); Hawaiian Islands (Hawaii [Zimmer1948]); Lord Howe Island [WilliaWa1988]; New Caledonia [Laing1933, WilliaWa1988]; New Zealand [Maskel1884, Maskel1885a, Green1929]; Norfolk Island [WilliaWa1988]. **Nearctic**: Mexico [RosenDe1979, MyartsRu2000]; United States of America (Alabama [BesheaTiHo1973], California [Comsto1881a, McKenz1956], Florida [Wilson1917, MerrilCh1923, Merril1953, Dekle1965c], Georgia [TippinBe1970, BesheaTiHo1973], Indiana [DietzMo1916], Kansas [Hunter1899], Maryland [StoetzDa1974a], Mississippi [Herric1911], Missouri [Hollin1923], Texas [McDani1968]). **Neotropical**: Argentina [ClapsTe2001] (Buenos Aires [RosenDe1979]); Brazil (Minas Gerais [Lepage1938, WolffCo1993], Rio Grande do Sul [Lepage1938, WolffCo1993], Rio de Janeiro

[WolffCo1993], São Paulo [Green1930b, Gianno1942, Muntin1971a, WolffCo1993, ClapsWoGo2001]); Chile [GonzalCh1968, RosenDe1979, Gonzal1986, Gonzal1989, Gonzal1989a]; Colombia [Balach1959a, Kondo2001]; Cuba [Houser1918]; Puerto Rico & Vieques Island (Puerto Rico [Martor1976, ColonFMe1998]). **Oriental**: Mongolia [DanzigKo1990]; Sri Lanka [Green1896]. **Palaearctic**: Algeria [Signor1877, Newste1897a, Balach1932d]; Canary Islands [GomezM1962, GomezM1967O, MatileOr2001]; China (People's Republic) (Henan (Honan) [Shen1993]); Corsica [Balach1931a, Balach1932d]; Crete [Ayouta1940]; Czech Republic [Zahrad1977, Zahrad1990b]; Egypt [Hall1922, Hall1923, Ezzat1958]; France [Boisdu1867, Signor1869b, Balach1930, Balach1932d, Balach1933e, Foldi2000]; Georgia (Abkhaz ASSR [Borchs1934, Borchs1936, Hadzib1983], Adzhar ASSR [Borchs1934, Borchs1936, Hadzib1983], Georgia [Borchs1936, Hadzib1983]); Germany (United) [Bouche1844, Lindin1909b]; Greece [Bodenh1928, Korone1934, ArgyriStMo1976, RosenDe1979]; Iran [Bodenh1944b, Kaussa1955]; Israel [Bodenh1924, Bodenh1927b, Bodenh1937, GersonZo1973, GersonHa1979, RosenDe1979]; Italy [Targio1868, Leonar1920, Carimi1930, Ferris1941e, Viggia1970a]; Japan [Kuwana1917a, Kuwana1933, Kawai1980]; Lebanon [Bodenh1926]; Madeira Islands [Green1923b, Balach1938a, CarvalAg1997]; Malta [Borg1919]; Morocco [Vayssi1920, Balach1927, Balach1932d, Rungs1970]; Poland [Dziedz1989]; Portugal [Seabra1930, Seabra1941, Fernan1992]; Romania [Savesc1982]; Russia (Caucasus [Borchs1934]); Saudi Arabia [Beccar1971, Matile1984c]; Spain [Balach1935b, GomezM1946, GomezM1956b, GomezM1958c, Martin1983, BlayGo1993]; Syria [Hariri1971]; Turkey [Bodenh1949, Bodenh1952, UygunSeEr1998, ErlerTu2001, KaracaSeCo2001]; United Kingdom (Channel Islands [Green1925b], England [Newste1901b, Green1916, Green1928]); Yugoslavia [Bachma1953, RosenDe1979].

BIOLOGY: On bark or leaves (Ferris, 1938a). Comstock (1883: 63) reared this species in USA, from specimens taken on lemons imported from Mediterranean. Sex pheromone emitted by the female, *Aspidiotus nerii* (Homoptera, Diaspididae), isolated and characterized as (1R, 2S)-cis-2isopropenyl-1-(4'-methyl-4'-penten-1'-yl) cyclobutaneethanol acetate by using advanced MS and NMR spectroscopic methods, as well as a variety of microderivatization sequences. Structure confirmed by stereo- and enetioselective synthesis of four possible stereoisomers. Absolute configuration determined by comparison of activity of cis (1S,2R) and (1R,2S) enantiomers with that exhibited by natural material in greenhouse bioassays and field tests. The structure of this sesquitrpenoid pheromone is new in the Coccoidea and in the pheromone field in general.

ECONOMIC IMPORTANCE: Among most highly polyphagous scale insects (see Host Plants) and widely distributed in almost all zoogeographical regions (CABI, 1970, and see Distribution). Unisexual and bisexual populations reported (e.g. Schmutterer, 1952; DeBach & Fisher (1956). A pest of citrus in several Mediterranean countries (Bodenheimer, 1951); a serious pest of olive in Mediterranean basin (Argyriou, 1990); damaging kiwifruit in Chile (Gonzalez, 1989a); pest of avocado in Israel (Gerson & Zor (1973); and troublesome pest of many ornamental plants (e.g. Gill, 1997).

GENERAL: DeBach & Fisher (1956) studied unisexual and bisexual populations, which had both been identified as *Aspidiotus hederae* and concluded that these "races" were biologically distinct. Description and illustration of adult female by Green (1915d) [as *A. tasmaniae*], Leonardi (1918), Dietz & Morrison (1916) [as *A. hederae* (Vallot)], Kuwana (1933), Ferris (1938a, 1941e) [as *A. hederae* (Vallot)], Balachowsky (1948b, 1956) [as *A. hederae* (Vallot)], Zimmerman (1948), McKenzie (1956), Gómez-Menor Guerrero (1962), Bazarov & Shmelev (1971), Chou (1985, 1986), Tereznikova (1986), Williams & Watson (1988), Dziedzicka (1989), Danzig (1993), Yaşar (1995a), Kosztarab (1996), Gill (1997) and by Colon-Ferrer & Medina-Gaud (1998). Description and illustration of female nymph, male nymph, male pupa and prepupa by Stoetzel & Davidson (1974a).

KEYS: Claps & Teran 2001: 392 (female) [South America]; Gill 1997: 64 (female) [Species of California]; Kosztarab 1996: 427 (female) [Northeastern North America]; Danzig 1993: 140-141 (female) [Europe]; Williams & Watson 1988: 51 (female) [Tropical South Pacific]; Chou 1985: 262-263 (female) [Species of China]; Gerson & Zor 1973: 516 (female) [Israel]; Beardsley 1970: 508 (female) [Hawaii]; Ezzat 1958: 241 (female) [Egypt]; Balachowsky 1956: 51 (female) [Africa]; McKenzie 1956: 24 (female) [U.S.A.: California]; Mamet 1954: 52 (female) [Madagascar]; Balachowsky 1948b: 275 (female) [Mediterranean]; Lupo 1948: 138 (female) [Italy]; Zimmerman 1948: 355 (female) [Hawaii]; Ferris 1946: 43 (female) [World]; Ferris 1942: 30 (female) [North America]; Ferris 1941e: 61 (female) [World]; Archangelskaya 1937: 99 (female) [Middle Asia]; Borchsenius 1935: 127-128 (female) [Former USSR]; Kuwana 1933: 3 (female) [Japan]; Kuwana 1933b: 49 (female) [Japan]; Fullaway 1932: 95-97, 107 (female) [Hawaii]; Archangelskaya 1929: 190 (female) [Palaearctic Region]; Britton 1923: 371 (female) [U.S.A.: Connecticut]; Hollinger 1923: 7-8 (female) [U.S.A.: Missouri]; Leonardi 1920: 29-30 (female) [Italy]; Brain 1918: 117 (female) [South Africa]; Lawson 1917: 217 (female) [U.S.A.: Kansas]; Dietz & Morrison 1916a: 289-290 (female) [U.S.A.: Indiana]; Cockerell 1905: 45-46 (female) [Mexico]; Cockerell 1905b: 201 (female) [U.S.A.: Colorado]; Comstock 1883: 55-57 (female) [North America].

CITATIONS: AbdElKDaKo1988 [chemical control, biological control: 270-275]; AbdElKKo1988a [life history, host: 55-60]; AbdelR1995 [life history, economic importance: 1553-1564]; AbdRab2001a [host, distribution, biological control: 175,176]; AbouEl2001 [host, distribution, biological control: 185-195]; AguileDiGr1981 [host, distribution, life history, economic importance: 175-178]; AguileMeVa1984 [life history, biological control: 47-54]; Alam1957 [biological control: 421-466]; Alexan1990 [host, distribution, life history, ecology, biological control: 339-342]; AlexanBe1979 [host, distribution, economic importance, life history: 49-56]; AlexanBe1981 [host, distribution, biological control: 13-25]; AlexanBe1982 [host, distribution, life history, ecology: 843-850]; AlexanNe1979 [host, distribution, economic importance, biological control,: 171-184]; AlexanNe1980 [host, distribution, life history, biological control: 61-71]; Alfier1929 [host, distribution: 7-9]; Almeid1973b [host, distribution: 9]; AnneckIn1970 [host, distribution, biological control: 240]; AnneckIn1971 [host, distribution, biological control: 28,29]; Archan1929 [taxonomy: 190]; Archan1930 [host, distribution: 85]; Archan1937 [taxonomy, description, illustration, host, distribution: 106-107];

Argyri1969 [biological control: 817-822]; Argyri1970 [host, distribution, biological control: 57-65]; Argyri1976 [host, distribution, life history, biological control: 209-218]; Argyri1979a [host, distribution, economic importance, biological control: 517-520]; Argyri1990 [host, distribution, economic importance, biological control: 579-583]; Argyri1990 [host, distribution, economic importance: 579-583]; ArgyriKo1980 [host, distribution, life history, economic importance, biological control: 633-638]; ArgyriKo1981 [host, distribution, life history, economic importance, chemical control: 65-72]; ArgyriMo1981 [host, distribution, economic importance: 623-627]; ArgyriStMo1976 [host, distribution, biological control: 25]; Autran1907 [catalogue: 165]; Ayouta1940 [host, distribution: 2-4]; Bachma1953 [host, distribution: 177]; BaetaN1947 [host, distribution: 128]; Baker1976 [biological control, life history: 1-25]; Balach1927 [host, distribution: 176-177]; Balach1928d [biological control: 284,289,293,295]; Balach1930 [host, distribution: 312]; Balach1931a [host, distribution: 97]; Balach1932d [taxonomy, host, distribution: I-III; XLV]; Balach1933e [host, distribution: 3]; Balach1935b [host, distribution: 256]; Balach1937c [host, distribution: 2]; Balach1938a [host, distribution: 148]; Balach1948b [taxonomy, description, illustration, host, distribution, biological control: 272,280-285]; Balach1951 [taxonomy: 697-698]; Balach1956 [taxonomy, description, illustration, host, distribution: 69-70]; Balach1959a [host, distribution: 362]; BaldanGaVi1999 [biological control: 209-215]; Banti1893 [taxonomy, description, host, distribution: 12]; Barber1893 [host, distribution: 50-51]; BartleDe1952 [biological control: 16-17]; Bartra1977 [host, distribution, life history: 43-48]; BazaroSh1970 [host, distribution, taxonomy: 109-111]; BazaroSh1971 [taxonomy, description, illustration, host, distribution: 187-188]; BeardsDaHo1976 [economic importance: 105]; BeardsGo1975 [economic importance: 49]; Beccar1971 [host, distribution: 193]; Bedfor1978 [host, distribution, economic importance: 129-133]; Beffa1937 [host, distribution, economic importance: 63-70]; BeglyaSm1977 [host, distribution, biological control : 283-328]; Benass1958d [biological control: 179-181]; Benass1965a [host, distribution, chemical control, biological control, economic importance, life history: 112-125]; BenDov1990a [taxonomy: 85]; BenDov1990c [taxonomy: 111]; BenDovMa1999 [taxonomy: 5-8]; BenDovMa2001Sa [taxonomy: 191-192]; BenDovMa2001Sb [taxonomy: 426-427]; BentleBa1931 [host, distribution, economic importance, control: 1-61]; Berles1895c [taxonomy, description, illustration, host, distribution: 229-232]; Berles1911 [biological control: 436-461]; BerlesLe1898 [taxonomy, description, host, distribution: 97]; BerlesLe1899 [taxonomy, description, illustration, host, distribution: 262-264]; BerlinSePo1999 [host, distribution, life history: 1113-1119]; Berro1927 [host, distribution: 55-59]; Berry1991 [taxonomy: 323-341]; BerryMoHi1989 [host, distribution, economic importance: 182-186]; Beshea1975 [host, distribution, biological control: 223-224]; BesheaTiHo1973 [host, distribution: 5]; BiezanFr1939 [host, distribution: 1-18]; BiezanSe1940 [host, distribution: 67-68]; BindraVa1972 [host, distribution, economic importance: 14-24]; Blanch1922 [host, distribution: 387-398]; BlankGiDo1997 [host, distribution, economic importance, life history: 293-297]; BlankGiDo1999 [host, distribution, economic importance, life history: 1-12]; BlankOl1990a [chemical control: 243-246]; BlayGo1993 [taxonomy, description, illustration, host, distribution: 432-444]; BlumbeDe1979 [life history, biological control: 299-306]; BlumbeSw1974 [biological control: 437-443]; BlumbeSw1974a [biological control: 3-11]; Bodenh1924 [taxonomy, description, illustration, host, distribution: 25-30]; Bodenh1926 [host, distribution: 42]; Bodenh1927b [host, distribution: 79]; Bodenh1928 [host, distribution: 191]; Bodenh1935 [host, distribution: 246]; Bodenh1937 [host, distribution: 216]; Bodenh1944b [host, distribution: 93]; Bodenh1949 [taxonomy, description, illustration, host, distribution: 52-55]; Bodenh1951 [structure: 133]; Bodenh1951a [taxonomy, host, distribution, economic importance: 199-201]; Bodenh1952 [host, distribution: 337-338]; Bohm1976 [host, distribution: 41-42]; Boisdu1867 [taxonomy, description, illustration, host, distribution: 330,339,345]; Boisdu1868 [taxonomy, description, host, distribution: 281]; Borchs1934 [host, distribution: 27]; Borchs1935a [taxonomy, description, host, distribution: 3,27]; Borchs1936 [host, distribution: 130]; Borchs1937 [taxonomy, description, illustration, host, distribution: 124-125]; Borchs1949d [taxonomy, description, host, distribution: 236]; Borchs1950b [taxonomy, description, illustration, host, distribution: 214,219]; Borchs1965 [taxonomy: 212]; Borchs1966 [catalogue: 261-266,269]; Borg1919 [taxonomy, description, host, distribution: 24-27]; Borg1922 [host, distribution]; BorianNi1995 [chemical control: 43]; Bouche1833 [taxonomy, description, host, distribution: 52]; Bouche1834 [taxonomy, description, host, distribution: 12-14]; Bouche1844 [taxonomy, description, host, distribution: 293]; BouhelDeDe1932 [host, distribution, control: 1-60]; Boyce1948 [host, distribution, economic importance, control]; BoyerDu1999 [chemistry, pheromone: 29-33]; BoyerDu1999a [chemistry, pheromone: 1201-1211]; Brain1918 [taxonomy, description, illustration, host, distribution: 118-119]; BrainKe1917 [distribution: 183]; BrandtBo1948 [taxonomy: 3]; Brick1912 [host, distribution: 1-22]; Brimbl1962 [host, distribution, economic importance: 220]; Brimbl1968 [taxonomy, description, illustration, host, distribution: 43-47]; Britto1923 [taxonomy, description, host, distribution: 371,373-374]; Buchne1953 [taxonomy, structure: 217-218]; BurgerUl1990 [economic importance: 313-327]; Burmei1835 [taxonomy, description, host, distribution: 67]; Bustsh1958 [taxonomy, description, host, distribution: 219,234]; Buxton1920 [host, distribution: 287-303]; CABI1970 [host, distribution: 1-2]; Caltag1985 [taxonomy, biological control: 189-200]; CanaleVa1999 [biological control]; Carimi1930 [taxonomy, description, illustration, host, distribution: 121-123]; Carnes1907 [host, distribution: 206]; CarvalAg1997 [life history, description, economic importance, biological control, host, distribution: 258-261]; Castel1951a [biological control: 95-98]; Chambe1925EL [host, distribution, control: 149-165]; Charle1998 [distribution, economic importance, biological control: 51N]; CharleHiAl1995 [host, distribution, life history, biological control: 319-324]; CheahIr1997 [host, distribution]; Chiesa1938a [host, distribution: 1-21]; Chiesa1948 [host, distribution, economic importance]; Chiesa1948a [host, distribution, economic importance]; Chou1947a [chemical control: 33]; Chou1985 [taxonomy, description, host, distribution: 263-267]; Chou1986 [taxonomy, illustration: 659]; ChuaWo1990 [host, distribution, economic importance: 543-552]; Cirio1979 [host, distribution, biological control : 297-303]; CividaGu1996 [biological control, life history: 257-266]; Claps1993 [taxonomy: 6,9]; ClapsTe2001 [taxonomy, description, illustration, host, distribution, economic importance: 392,394]; ClapsWoGo2001 [host, distribution: 241]; ClapsWoGo2001a [taxonomy, host, distribution: 12-13]; Claus1864 [life history, structure: 42-54]; Cocker1893j [taxonomy, host, distribution: 255]; Cocker1893k [taxonomy, description, host, distribution: 548]; Cocker1896b [taxonomy, distribution: 333-335]; Cocker1897i [taxonomy, description, host, distribution: 18,19,25,29,30]; Cocker1898m [taxonomy, description, host, distribution: 27]; Cocker1899a [taxonomy: 395]; Cocker1899n [taxonomy, description, illustration, host, distribution: 20-21]; Cocker1900k [taxonomy: 350]; Cocker1905 [taxonomy: 46]; Cocker1905b [taxonomy: 201]; CockerPa1899 [taxonomy, host, distribution: 276]; Cohen1969 [biological control, economic importance: 769-772]; ColonFMe1998 [taxonomy, description, illustration, host, distribution: 46-47]; Colvee1880 [taxonomy: 39]; Colvee1881 [taxonomy, description, host, distribution: 16,39]; Colvee1882 [taxonomy, description, host, distribution: 12-14]; Compto1924 [chemical control: 222-225]; Comsto1881a [taxonomy, description, illustration, host, distribution: 301-303]; Comsto1883 [taxonomy, host, distribution: 63,72-79,83,89]; Comsto1916a [taxonomy, description, host, distribution: 537]; CoronaRuMo1997 [host, distribution: 38-41]; CostaL1949 [host, distribution, biological control: 65-87]; Costan1938 [host, distribution: 25-44]; Costan1956a [host, distribution, economic importance: 74-79]; Cotte1912 [host, distribution: 81]; Coutin1997 [host, distribution: 1-4]; Crouze1971 [biological control: 200]; Crouze1973 [host, distribution, biological control: 15-39]; CuiHa1987 [host, distribution, life history: 332-335]; Curtis1843d [taxonomy, host, distribution: 588]; DahmsSm1994 [host, distribution, biological control: 245-255]; Danzig1964 [taxonomy, host, distribution: 651]; Danzig1972 [taxonomy, host, distribution, economic importance: 208]; Danzig1977 [taxonomy: 101]; Danzig1993 [taxonomy, description, illustration, host, distribution, life history, economic importance: 144-147]; Danzig1995 [taxonomy, life history, structure: 19-24]; DanzigKo1990 [host, distribution: 46]; DanzigPe1998 [catalogue: 190-193]; DarvasFaKo1985 [chemical control: 347-350]; DarvasVi1983 [chemical control: 455-463]; DavidsDiFl1991 [chemical control: 1-47]; DavidsMi1990 [host, distribution, economic importance: 603-632]; DeBach1958 [host, distribution, biological control, ecology: 187-194]; DeBach1958b [host, distribution, ecology: 187-194]; DeBach1964 [biological control]; DeBach1964d [biological control: 5-18]; DeBach1969 [biological control: 801-815]; DeBach1971 [biological control: 293-307]; DeBach1974 [biological control]; DeBachAr1967 [host, distribution, biological control: 325-342]; DeBachFi1956 [taxonomy, life history: 235-239]; DeBachRo1976a [host, distribution, biological control: 541-545]; DeBachRo1991 [biological control]; DeBachWh1960 [life history, biological control: 4-58]; DEDAC1923 [host, distribution]; DeitzTo1980 [taxonomy: 33,34,36,42]; Dekle1965c [taxonomy, description, host, distribution: 28]; Dekle1976 [taxonomy, description, host, distribution, economic importance: 42]; DelGue1906 [host, distribution : 257-263]; DeLott1967a [host, distribution: 113]; DeLottNa1955 [host, distribution: 53-60]; DeSant1941a [host, distribution, biological control: 21-24]; DeSant1979 [biological control]; DeStef1910 [host, distribution: 189-196]; DietzMo1916 [taxonomy, description, illustration, host, distribution: 296-297]; DietzMo1916a [taxonomy, description, illustration, host, distribution: 296-297];

Doane1931 [host, distribution, control]; Dougla1912 [taxonomy: 203]; Dowson1935 [host, distribution: 225]; Drea1990 [biological control: 51-59]; DreistClFl1994 [taxonomy, life history, economic importance, control]; Dunkel1999 [chemistry, life history: 251-276]; DymockHo1996 [host, distribution: 249-257]; Dziedz1989 [taxonomy, description, illustration, host, distribution: 98-99]; Ebelin1949 [host, distribution, life history, control]; Ebelin1975 [host, distribution, economic importance]; EbelinPe1953 [host, distribution, economic importance: 1-35]; EHG1897 [host, distribution: 67-85]; Ehrhor1913 [host, distribution: 101]; EinhorGuDu1998 [chemistry, life history, physiology, pheromone: 9867-9871]; ElirazRo1978 [biological control: 96-101]; ElMinsElHa1974 [taxonomy, description, illustration, host, distribution: 223-232]; ErichsSaHa1991 [biological control: 493-498]; ErlerTu2001 [host, distribution, biological control: 299-305]; Evans1943 [host, distribution, taxonomy, chemical control]; EvansPr1990 [biological control: 3-17]; Ezzat1958 [distribution: 241]; EzzatAf1966 [taxonomy, description, illustration, host, distribution: 374-377]; EzzatNa1987 [distribution: 86]; Fawcet1948 [biological control: 627]; Fawcet1948 [biological control: 627-664]; Felt1901 [taxonomy, description, host, distribution, biological control, chemical control: 333-334]; Fernal1903b [catalogue: 253,254,258,260-268,]; Ferrie1927 [biological control: 55-67]; Ferris1920b [host, distribution: 51]; Ferris1937c [taxonomy, illustration: 50,62]; Ferris1938a [taxonomy, description, illustration, host, distribution: 192]; Ferris1941e [taxonomy, description, illustration, host, distribution: 40-49,54-57]; Ferris1942 [taxonomy, host, distribution: 445:5;446:30]; Ferris1946 [taxonomy: 43]; Figuer1952 [host, distribution: 208]; FinneyFi1964 [biological control: 328-355]; FisherDe1976 [biological control: 43-50]; Fjeldd1996 [host, distribution: 4-24]; Flande1971 [biological control, life history: 857-872]; Flesch1960 [biological control: 183-208]; Foldi1990 [structure: 43-54]; Foldi2000 [host, distribution: 82]; Foldi2001 [distribution: 303-308]; Foldi2002 [host, distribution: 246]; FoldiSo1989 [host, distribution: 411]; FoxWil1939 [host, distribution, economic importance: 2296]; Frogga1914 [taxonomy, description, host, distribution: 133,136,314,315]; Frogga1915 [taxonomy, description, host, distribution: 10,13,18,22]; Fullaw1932 [taxonomy: 97,107]; Fuller1897c [host, distribution: 3]; Fuller1897c [host, distribution: 4]; Fuller1907 [taxonomy, description, host, distribution, economic importance, control: 1031-1055]; Gabrit1923 [life history, structure, physiology: 295-332]; Gahan1925 [host, distribution, biological control: 1-23]; Gaprin1954 [biological control: 587-597]; Gaprin1956 [host, distribution: 103-137]; Garcia1916 [host: 776-788]; Garcia1922 [host, distribution, biological control: 196-200]; Garcia1930 [host, distribution, biological control]; Gavalo1931 [host, distribution: 7]; Gavalo1936 [host, distribution: 78-79]; Gentry1965 [host, distribution, economic importance]; Gerson1990 [taxonomy, structure, life history: 130-132]; GersonHa1979 [taxonomy, description, host, distribution: 281-283]; GersonOcHo1990 [biological control: 77-97]; GersonZo1973 [taxonomy, life history, host, distribution, economic importance: 513-533]; Ghauri1962 [taxonomy, description, host, distribution: 65,211]; Gianno1942 [taxonomy, description, illustration, host, distribution: 214-216]; GibsonRo1922 [host, distribution: 1]; Gill1997 [host, distribution, taxonomy, description, illustration, economic importance: 65,67,68]; GomesCRe1947 [taxonomy, description, host, distribution: 66]; Gomez1936 [host, distribution: 42-43]; GomezC1950 [host, distribution, biological control, economic importance: 1-18]; GomezM1937 [taxonomy, description, illustration, host, distribution: 47-53]; GomezM1946 [host, distribution: 61]; GomezM1948 [host, distribution: 74]; GomezM1954 [host, distribution: 119]; GomezM1956 [taxonomy, description, illustration, host, distribution, biological control: 8-12]; GomezM1956b [host, distribution: 482]; GomezM1957 [host, distribution: 39-40]; GomezM1958c [host, distribution: 406]; GomezM1960O [host, distribution: 158]; GomezM1962 [taxonomy, description, illustration, host, distribution: 158-162]; GomezM1965 [host, distribution: 88]; GomezM1967O [host, distribution: 131]; GomezM1968 [host, distribution: 541]; Gonzal1986 [host, distribution, economic importance: 23]; Gonzal1989 [taxonomy, description, host, distribution, economic importance: 94]; Gonzal1989a [life history, economic importance, chemical control, host, distribution: 35-43]; GonzalCh1968 [distribution: 110]; GonzalCu1994 [host, distribution, life history, economic importance, chemical control: 5-20]; GonzalRo1967 [biological control, distribution: 138]; Gordh1979 [biological control: 893-896,900,907,911,]; Gowdey1921 [taxonomy, description, host, distribution: 29]; Greath1976 [biological control, economic importance]; GreaveToWi1992 [host, distribution, chemical control: 79-83]; Green1896 [host, distribution: 4]; Green1915d [taxonomy, description, illustration, host, distribution: 50]; Green1916 [taxonomy, host, distribution: 29]; Green1923b [host, distribution: 89]; Green1925b [host, distribution: 518]; Green1928 [taxonomy, description, host, distribution: 8]; Green1929 [host, distribution: 377]; Green1930b [host, distribution: 214]; GreenMa1907 [taxonomy, distribution: 344]; Hadzib1957a [host, distribution: 102]; Hadzib1983 [taxonomy, description, host, distribution, life history, biological control, economic importance: 218-219]; Hall1922 [taxonomy, description, host, distribution: 26-27]; Hall1923 [taxonomy, description, illustration, host, distribution: 19,43]; Hall1924a [host, distribution, economic importance: 11]; Hall1928 [host, distribution: 274]; HallFo1933 [host, distribution, economic importance: 1-55]; HardieMi1999 [chemistry, life history]; HareMo1997 [host, distribution, biological control: 207-214]; Hariri1971 [distribution: 48]; HaseyOlVa1999 [host, distribution, control]; HattinSa1992 [life history, biological control: 327-334]; HattinSa1993 [life history, biological control: 13-20]; Haywar1939 [host, distribution, control: 1]; Haywar1944 [host, distribution: 1-32]; HeimpeRo1995 [host, life history, biological control: 153-167]; Hellen1921 [host, distribution: 120-128]; Hender2001a [taxonomy, host, distribution: 89-90]; Herric1911 [taxonomy, description, illustration, host, distribution: 10,16-17,50]; Herric1925 [host, distribution, description, life history, economic importance]; Hill1989a [host, distribution, economic importance, biological control: 177-182]; Hofer1903 [taxonomy, host, distribution: 479]; Hollin1923 [taxonomy, description, host, distribution: 7,12,68]; Holzap1932 [host, distribution: 325]; HondaLu1995 [life history, biological control: 441-450]; Hosny1939 [taxonomy, description, host, distribution: 14-15]; HouckOc1996 [life history, ecology, biological control: 667-682]; Houser1918 [host, distribution: 166-167]; Housto1991 [host, distribution, biological control: 341-342]; Howard1898b [biological control: 133-139]; Howard1907 [host, distribution, biological control: 69-88]; Howard1908 [host, distribution: 265-277]; HowellTi1990 [taxonomy, structure, description, illustration: 30]; HoyHe1985 [biological control]; HoyHe1985 [biological control]; Hsu1935 [host, distribution: 578-590]; Huffak1990 [biological control: 205-220]; Hunter1899 [taxonomy, description, host, distribution: 11-12]; Inserr1966 [biological control: 176-186]; Inserr1970a [host, distribution, biological control: 39-45]; Iperti1961 [economic importance: 14-30]; IzraylGe1993 [host, distribution, biological control: 861-875]; IzraylGe1993a [host, distribution, biological control: 877-888]; IzraylGe1995 [host, distribution, biological control: 439-446]; IzraylGe1995b [host, distribution, biological control: 235-240]; IzraylGeHa1996 [life history, ecology, biological control: 390-395]; IzraylHaGe1995 [life history, ecology, biological control: 138-145]; JamiesDoCa2002 [host, distribution: 354-360]; Jancke1955 [taxonomy: 304]; Jannon1940 [host, distribution: 241-253]; JaszaiDa1983 [chemical control: 198-202]; JiYa1990 [biological control: 134-136]; John1930 [host, distribution: 3-7]; Jorgen1934 [taxonomy, host, distribution: 279]; Jourdh1979 [biological control: 75-79]; KaracaSeCo1998 [host, distribution, life history, biological control: 23]; KaracaSeCo2001 [host, distribution, economic importance, life history, biological control: 407-412]; KaracaUy1993 [host, distribution, life history, ecology: 217-224]; Katsoy1992 [host, distribution, economic importance, biological control]; Kaussa1955 [host, distribution: 15]; Kawai1980 [taxonomy, description, host, distribution: 227]; King1899d [taxonomy: 255]; King1899h [host, distribution: 350]; KingLe1984 [biological control: 1]; KingMo1984 [biological control: 206-222]; Kiritc1932a [taxonomy: 245]; Koebel1893 [host, distribution, biological control: 1-39]; Koehle1964 [host, distribution, control]; Komosi1964 [host, distribution, taxonomy, description, illustration: 212-214]; Kondo2001 [taxonomy, host, distribution: 43]; KondoKa1995a [host, distribution: 97-98]; Konsta1976 [host, distribution, economic importance: 49-50]; Korone1934 [taxonomy, description, illustration, host, distribution: 3-6]; Koszta1996 [taxonomy, description, illustration, host, distribution, life history: 429-431]; Koteja1990b [life history, structure, anatomy: 233-242]; Koteja1990c [life history: 243-254]; Kozar1990a [life history, economic importance: 341-347]; KozarHi1996 [host, distribution: 91-96]; KozarKi1979 [host, distribution: 246-250]; Krassi1893 [life history: 69-76]; Ksiazk1980 [structure, anatomy: 149-157]; Ksiazk1984 [structure, anatomy: 1-6]; Kuwana1917a [taxonomy, distribution: 174]; Kuwana1933 [taxonomy, description, illustration, host, distribution: 10-11]; Lagows1995 [biological control: 5-10]; LagowsFi1995 [host, distribution: 375-378]; Laing1929 [taxonomy, host, distribution: 23]; Laing1933 [host, distribution: 676]; Laport1948 [host, distribution, biological control: 35-37]; Larew1990 [ecology, life history, structure: 293-300]; Lawson1917 [taxonomy, description, illustration, host, distribution: 217,223-224]; LenterDe1981 [life history, biological control: 504-532]; Leonar1897 [taxonomy: 285]; Leonar1898c [taxonomy, description, illustration: 71-77]; Leonar1914 [taxonomy, description, illustration, host, distribution: 198-199]; Leonar1920 [taxonomy, description, illustration, host, distribution: 31-35]; Lepage1938 [catalogue: 395]; Lepesm1947 [taxonomy, description, host, distribution, life history, biological control: 185-189]; Lidget1902 [host, distribution: 43-45]; Lindbl1938 [host, distribution: 1]; Lindin1909a [taxonomy: 324]; Lindin1909b [host, distribution: 150-151]; Lindin1909c [taxonomy, host, distribution: 449]; Lindin1912b [taxonomy, description, host, distribution: 50,55,58,63,68,]; Lindin1913b [taxonomy: 97]; Lindin1924 [taxonomy: 174]; Lindin1931 [taxonomy: 114]; Lindin1935 [taxonomy: 128]; Lindin1939 [host, distribution: 37]; Lindin1957 [taxonomy: 545,546]; Lintne1895 [host, distribution: 263-305]; Liotta1970 [host, distribution, economic importance: 33]; Liotta1974 [host, distribution, biological control: 175-186]; Liotta1974a [biological control, chemical control: 187-194]; Liotta1974b [biological control: 83-88]; Liotta1980 [host, distribution, life history, biological control: 695-699]; LiottaBiLo1985 [host, distribution, life history, ecology: 591-598]; LiottaBuMo1985 [host, distribution, economic importance, biological control: 833-840]; LiottaMa1974 [chemical control: 207-213]; Lloren1990 [taxonomy, illustration, life history, host, distribution, biological control, life history: 51-60]; LoBl1989 [host, distribution: 1-4]; LoganTh2002 [biological control, life history, ecology: 361-367]; LongoMaPe1995 [distribution: 125]; LongoMaRu1995a [host, distribution: 126-129]; LongoMaRu1996 [host, distribution, economic importance: 126-129]; Lounsb1898 [host, distribution: 35-58]; Lounsb1906 [host, distribution: 80-91]; Lounsb1914 [host, distribution: 1]; Lozzia1985 [host, distribution: 122-124]; LuckUy1986 [life history, chemistry, biological control: 129-136]; Lugger1900 [host, distribution: 208-245]; Lupo1948 [taxonomy, description, illustration, host, distribution: 139-145]; Lyne1921 [distribution: 146-148]; MacGil1921 [taxonomy, description: 395,398,400,402]; MalleaMaGa1974 [host, distribution, life history: 145-147]; Mamet1954 [taxonomy: 52]; Mark1877 [structure, anatomy,: 31-81]; Marlat1903 [host, distribution, economic importance, control: 25]; Martel1913 [chemical control: 1-28]; Martin1958 [host, distribution, economic importance: 120-123]; Martin1983 [taxonomy, host, distribution: 61]; Martor1976 [host, distribution: 63]; Maskel1879 [taxonomy, description, host, distribution: 197-199]; Maskel1882 [taxonomy, description, host, distribution: 217]; Maskel1884 [taxonomy, description, host, distribution: 121]; Maskel1885a [taxonomy, description, illustration, host, distribution: 21]; Maskel1887 [taxonomy, description, host, distribution: 43]; Maskel1887a [taxonomy: 40,44]; Matile1984c [host, distribution: 221]; MatileOr2001 [host, distribution: 189]; Matta1979 [host, distribution, biological control: 231-242]; MayneGh1934 [host, distribution: 3-38]; McClur1990a [taxonomy, host, distribution, ecology: 165-168]; McClur1990c [taxonomy, host, distribution, ecology: 289-291]; McClur1990d [host: 301-303]; McClur1990f [taxonomy, host, distribution, ecology: 315-318]; McClur1990g [taxonomy, host, distribution, ecology: 319-330]; McDani1968 [taxonomy, illustration, host, distribution: 218-219]; McKenz1935 [host, distribution, life history, control: 1-48]; McKenz1956 [taxonomy, description, illustration, host, distribution: 24,47-49]; Melis1930 [distribution: 81]; Melis1949 [host, distribution: 17-25]; Merkel1938 [host, distribution: 88-99]; Merril1953 [taxonomy, description, host, distribution: 20-21]; MerrilCh1923 [taxonomy, description, host, distribution, economic importance: 201-202]; MessenVa1971 [biological control: 68-92]; MessenWiWh1976 [biological control: 209]; MetcalMe1993 [economic importance, host, distribution, control]; Metsch1866 [life history, structure: 389-500]; Michel1990 [host, distribution, economic importance: 38-40]; MillerDa1990 [host, distribution, economic importance: 300]; MillerDa1998 [taxonomy: 197];

MonastZa1960 [host, distribution: 169-236]; Monte1930 [host, distribution: 3-36]; Morgan1888b [taxonomy, description: 118-120]; Morgan1889a [taxonomy: 350]; MorrisKi1977 [host, biological control: 183-217]; MorrisMo1966 [taxonomy: 17]; MouradMeFa2001 [host, distribution: 571-580]; Muntin1969 [host, distribution: 121]; Muntin1971a [taxonomy, host, distribution: 313]; MyartsRu2000 [distribution, biological control: 7-33]; Myers1925 [taxonomy, description: 164-167]; NagarkSa1990 [host, distribution, economic importance, biological control: 553-542]; Nakaha1982 [host, distribution: 14]; Nel1933 [taxonomy: 418]; NeuensMiAl1977 [host, distribution, life history, ecology: 418-427]; Newste1897a [host, distribution: 93-94]; Newste1901b [taxonomy, description, illustration, host, distribution: 120-124]; Newste1911 [distribution: 85]; Newste1911a [host, distribution: 168]; Nonell1932 [host, distribution: 183-190]; Nonell1935 [host, distribution: 281-287]; NourElRi1970 [host, distribution: 123-127]; NSWDAE1963 [host, distribution, taxonomy, economic importance]; Nur1990b [taxonomy, life history: 191-197]; OrdoghTa1983 [chemical control: 417-419]; Paglia1929 [host, distribution, economic importance: 274-307]; Paglia1929 [host, distribution: 274]; Paik1972 [host, distribution: 1-4]; PalmerMo1990 [biological control: 67-76]; Paoli1915 [host, distribution: 259]; Peleka1962 [host, distribution: 62]; Peleka1974 [host, distribution, biological control: 14-20]; Penzig1887 [host, distribution: 3]; PeralL1968 [biological control: 22-29]; PeralL1968 [host, biological control: 22-29]; PetschBoGu2000 [chemistry: 1691-1695]; PetschPaBo1999 [chemistry: 3299-3309]; PicartMa2000 [host, distribution: 44-46]; PietriBiCo1969 [chemical control: 909-915]; PietriBiCo1969 [chemical control: 909-915]; Pizzam1977 [host, distribution, life history, ecology, biological control: 1-96]; Porcel1995 [structure: 25-45]; Poutie1928 [biological control: 267-270]; PriesnHo1940 [biological control: 58-70]; Prinsl1983 [distribution, biological control: 26]; Priore1964 [host, distribution: 131-178]; Priore1965 [host, distribution: 101-145]; Proven2002 [taxonomy: 512]; Pulsel1927 [biological control: 300-327]; Quayle1911d [host, distribution, description, economic importance, life history, biological control: 443-512]; Quedna1964b [biological control: 86-116]; Quilis1935 [life history, ecology: 621-633]; QuirogArAr1991 [host, distribution: 469-472]; RagusaRu1989 [host, distribution: 71-74]; Reh1903 [taxonomy, description, host, distribution: 466]; RobbCoBe2001 [host, distribution, taxonomy, control]; Rose1990b [biological control: 437]; Rose1990c [distribution, economic importance: 535-542]; Rose1990d [host, biological control, economic importance: 357-365]; RosenDe1979 [host, distribution, biological control: 262-268,349-354,]; RosenhRo1991 [life history, biological control: 873-893]; RosenhRo1992 [life history, biological control: 263-272]; RothscEuRe1973 [chemistry, biological control: 89-90]; RSEA1915 [host, distribution, description, life history, economic importance, control: 1]; Rubtso1949 [biological control: 74-75]; Rubtso1952a [biological control: 96-106]; Ruhl1913 [host, distribution: 79-80]; Rungs1970 [host, distribution, economic importance: 92]; Russo1959 [economic importance, chemical control: 8-9]; Ruther1915a [taxonomy, description, host, distribution: 111]; Saakya1954 [host, distribution, economic importance]; SalamaHa1974a [host, distribution, life history, biological control: 138-139]; Salaza1989 [host, distribution]; SalazaSo1990 [host, distribution, life history, biological control: 135-137]; SamwayMa1983 [biological control: 4-6];

Sander1904a [taxonomy, description, host, distribution: 55,62]; Sassce1915 [taxonomy, host, distribution: 33]; Savesc1982 [taxonomy, description, host, distribution, biological control: 303-304]; ScheurRu1974 [chemical control: 218-222]; SchlinDo1964 [taxonomy, biological control: 247-280]; Schmid1883 [life history, anatomy, structure: 169-200]; Schmut1952 [taxonomy, description, illustration, host, distribution: 567]; Schmut1957a [host, distribution: 134,135]; Schmut1957b [taxonomy: 148]; Schmut1959 [taxonomy, description, host, distribution: 49]; SchmutKlLu1957 [host, distribution, economic importance: 476]; Schrad1929a [taxonomy, life history: 232-236]; SchuhMo1948 [host, distribution, control]; Scott1984a [host, distribution: 11-31]; Seabra1930 [taxonomy, host, distribution: 134-135]; Seabra1930a [host, distribution: 143-148]; Seabra1941 [distribution: 8]; SengonUyKa1998 [host, distribution, biological control: 128-131]; Shalab1961 [host, distribution: 23]; Shen1993 [host, distribution: 58]; ShiLi1991 [host, distribution: 165]; Shoema1980 [host, distribution, biological control, chemical control: 26-49]; ShoemaHuKe1978 [biological control: 16-17]; SibbetVaFe2000 [host, distribution, control]; Siddiq1981 [economic importance, host, distribution: 172-180]; Signor1869 [taxonomy: 842-846,851-862,872,]; Signor1869b [taxonomy, description, illustration, host, distribution: 114-129,132-134]; Signor1877 [taxonomy: 601,621-622,663]; Silves1902 [taxonomy, description, host, distribution: 98]; Silves1921 [host, distribution, economic importance: 1-11]; Smirno1950a [biological control: 190-194]; Smirno1952 [host, distribution, biological control: 63-69]; SmithFrPa1996 [biological control]; SmithSmSm1998 [biological control, chemical control: 136-139]; SouissPa1999 [host, distribution, life history: 87-92]; Staffo1915 [taxonomy, structure: 72]; Statha2000 [life history, biological control: 203-211]; Statha2000a [life history, biological control: 439-451]; Statha2001b [life history, biological control: 113-116]; StathaEl2001 [host, distribution, biological control: 125-133]; StavraArYa1979 [host, distribution, biological control: 574-577]; Steine1987 [host, distribution, description, economic importance, control: 1-6]; Steini1938 [host, distribution, life history, economic importance: 160-163]; Steinw1948 [host, distribution, taxonomy, chemical control: 105-111]; StevenMcBl1997 [chemical control: 288-292]; StevenRe1999 [host, distribution, economic importance, biological control, chemical control: 345-354]; StevenVaGo1997 [host, distribution: 773-777]; StoetzDa1974 [taxonomy, life history: 138-140]; StoetzDa1974a [taxonomy, description, illustration, host, distribution, life history: 494-495]; Swirsk1976b [host, distribution, economic importance: 555-559]; Swirsk1989 [biological control: 11-44]; SwirskAr1971 [host, distribution: 12-15]; SwirskWyIz2002 [taxonomy, host, distribution, life history, economic importance, biological control: 102-103]; SzklarBi1995 [anatomy, structure: 23-29]; Szulcz1926 [host, distribution: 137-143]; TachikVa1969 [host, distribution, biological control: 535-540]; Talhou1950 [host, distribution, economic importance: 133-141]; Talhou1969 [host, distribution, life history, economic importance: 106-107]; Talhou1975 [economic importance: 25]; Tao1999 [taxonomy, host, distribution: 73-74]; Targio1867 [taxonomy: 13]; Targio1868 [taxonomy: 736]; Targio1881 [taxonomy: 150]; Targio1884 [taxonomy: 389-390]; Terezn1986 [taxonomy, description, illustration, host, distribution: 89-90]; ThiemGe1934a [taxonomy, description, host, distribution: 130-158,208-238];

Timber1924 [host, distribution, biological control]; Tomkin1992 [chemical control: 151-155]; TomkinAlTh1997 [host, distribution, life history, ecology: 791-795]; TomkinGrWi1992 [chemical control: 146-150]; TomkinGrWi1996 [chemical control: 12-16]; TomkinThWi1992 [host, distribution, life history: 58-63]; TomkinWiTh2000 [host, distribution, economic importance: 211-215]; TranfaVi1987a [economic importance: 215-221]; Trimbl1929 [host, distribution, economic importance, description, control: 1-21]; Trujil1942 [host, distribution, economic importance]; Tschor1939 [host, distribution: 89]; Tullgr1906 [taxonomy: 83]; Tuncyu1970 [host, distribution, economic importance: 30-52]; Tuncyu1976 [host, distribution, biological control: 32-45]; UygunEl1998 [host, life history, biological control: 153-162]; UygunKaUl1996 [host, distribution, biological control: 171-183]; UygunSeEr1998 [host, distribution: 183-191]; Vacant1985a [host, distribution, life history, biological control: 749-758]; VacantGe1987 [host, distribution, life history, biological control: 385-401]; Valent1963 [biological control: 6-13]; Valent1967 [biological control: 1100]; VandenTe1964 [ecology, biological control: 459-488]; Vayssi1913 [host, distribution: 430]; Vayssi1920 [host, distribution: 257]; Viggia1970a [host, distribution, economic importance: 50]; Viggia1978 [host, distribution, biological control: 30-38]; Viggia1984 [biological control: 257-276]; Viggia1987 [host, distribution, biological control: 121-123]; WeidneWa1968 [taxonomy: 173]; Wester1920 [host, distribution]; Westwo1840 [taxonomy, description, host: 118]; WhitinHoCo1998 [host, distribution, control, economic importance: 211-215]; WilliaWa1988 [taxonomy, description, illustration, host, distribution: 64-66]; Wilson1917 [host, distribution: 43-44]; WilsonGo1962 [host, distribution, economic importance, control: 41-61]; Wolff1911 [taxonomy, description, host, distribution: 75]; WolffCo1993 [taxonomy, description, illustration, host, distribution: 33-36]; WoodruBeSk1998 [distribution]; WoodwaEvEa1970 [distribution]; Woolle1990 [biological control: 167-176]; Wysoki1977 [host, distribution: 187-188]; Yasar1995a [taxonomy, description, illustration, host, distribution: 53-57]; Yasnos1987 [economic importance: 228-234]; Yasnos1994 [host, distribution, biological control: 317-333]; Zagain1956 [distribution: 85-90]; Zahrad1959b [host, distribution: 60]; Zahrad1972 [host, distribution, biological control: 432]; Zahrad1977 [taxonomy, distribution: 119]; Zahrad1990 [host, distribution, description: 640-641]; Zahrad1990a [host, distribution, description: 649]; Zahrad1990b [taxonomy, description, illustration, host, distribution: 80-83]; Zimmer1948 [taxonomy, description, illustration, host, distribution: 355-356].

Aspidiotus niger Signoret

Aspidiotus niger Signoret, 1869: 862. Nomen nudum.

Aspidiotus niger Signoret, 1869c: 130. Type data: FRANCE: Paris, on banks of the Seine, on "saule" [=*Salix*]. Syntypes, female. Type depository: Vienna: Naturhistorisches Museum Wien, Austria.

Aspidiotus (*Diaspidiotus*) *niger*; Cockerell, 1897i: 19. Change of combination.

Aspidiotus niger; Borchsenius, 1966: 369. Revived combination.

SYSTEMATICS: Ferris (1941e: 46) regarded this species as "indeterminate, but not *Aspidiotus*". Borchsenius (1966) listed it among *species incertae sedis*, whereas

Leonardi (1900: 305), regarded *Aspidiotus niger* Signoret, 1869c, a synonym of *Diaspidiotus distinctus* Leonardi, 1900. Original description of *Aspidiotus niger* Signoret definitely validates species, and here it is retained in *Aspidiotus*.

SCALE COVER: Female scale circular, brownish, with a yellow central point; male scale elongate (Signoret, 1869c: 130).

HOST PLANTS: Salicaceae: *Salix* [Signor1869b].

DISTRIBUTION: Palaearctic: France [Signor1869b].

KEYS: Balachowsky 1958b: 258 (female) [Africa].

CITATIONS: Borchs1966 [catalogue: 369-370]; Cocker1896b [distribution: 333]; Cocker1897i [taxonomy, description, host, distribution: 19]; Comsto1883 [taxonomy, host, distribution: 79-80]; Ferris1941e [taxonomy: 46]; Ferris1943a [taxonomy: 86]; Signor1869 [taxonomy: 862]; Signor1869b [taxonomy, description, illustration, host, distribution: 130].

Aspidiotus ophiopogonus Kuwana

Aspidiotus ophiopogonus Kuwana in: Kuwana & Muramatsu, 1932a: 96. Type data: JAPAN: Oita, on *Ophiopogon japonica*. Syntypes, female. Type depository: Ibaraki-ken: Insect Taxonomy Laboratory, National Institute of Agricultural Environmental Sciences, Kannon-dai, Yatabe, Tsukuba-shi, (I. Kuwana), Japan.

Aspidiotus pavlovskii Borchsenius, 1955b: 247. Type data: NORTH KOREA: South Hamgen, at Sea of Japan coast, near Tezo, between Hamhin and Pukchen, on *Festuca* sp. Syntypes, female. Type depository: St. Petersburg: (= Leningrad) Zoological Museum, Academy of Science, Russia. Synonymy by Borchsenius, 1966: 266.

SCALE COVER: Female scale subcircular, greyish dark brown, exuviae nearly central, yellow (Kuwana & Muramatsu, 1932a). Female scale, of junior synonym *Aspidiotus pavlovskii* Borchsenius, 1955, elliptical, 1.6 mm long, 0.9 mm wide; flat; light brown; exuviae brown and shiny, placed centrally (Borchsenius, 1955b).

HOST PLANTS: Cyperaceae: *Carex* [Takagi1957, Takagi1958]. **Gramineae**: *Festuca* [Borchs1955b, Takagi1957, Takagi1958]. **Liliaceae**: *Ophiopogon* [Takagi1958], *Ophiopogon japonicus* [KuwanaMu1932a, Takagi1957]. **Polypodiaceae** [Takagi1958].

DISTRIBUTION: Palaearctic: Japan [KuwanaMu1932a, Takagi1958, Kawai1980] (Hokkaido [Takagi1957], Honshu [Takagi1957]); North Korea [Borchs1955b, Takagi1957].

BIOLOGY: A grass-infesting species, found on leaves.

GENERAL: Description and illustration of adult female by Kuwana & Muramatsu (1932a), Borchsenius (1955b) and by Takagi (1957).

KEYS: Danzig 1993: 140-141 (female) [Europe]; Takagi 1957: 31 (female) [Japan].

CITATIONS: Borchs1955b [taxonomy, description, illustration, host, distribution: 247-249]; Borchs1966 [catalogue: 266]; DanzigPe1998 [catalogue: 193]; Ferris1941e [taxonomy: 46]; Kawai1980 [taxonomy, description, host, distribution: 227]; KuwanaMu1932a [taxonomy, description, illustration, host, distribution: 95-100]; Muraka1970 [host, distribution: 72]; Takagi1957 [taxonomy, description, illustration, host, distribution: 34-35]; Takagi1958 [taxonomy, host, distribution: 122].

Aspidiotus pacificus Williams & Watson
Aspidiotus pacificus Williams & Watson, 1988: 66. Type data: SOLOMON ISLANDS: Guadalcanal, Lungga, on *Cocos nucifera*; collected 18.VII.1956. Holotype female. Type depository: London: The Natural History Museum, UK.
SCALE COVER: Appearance of scale cover unknown (Williams & Watson, 1988).
HOST PLANTS: **Palmae**: *Cocos nucifera* [WilliaWa1988].
DISTRIBUTION: **Australasian**: American Samoa [WilliaWa1988]; Solomon Islands [WilliaWa1988].
GENERAL: Description and illustration of adult female by Williams & Watson (1988).
KEYS: Williams & Watson 1988: 49, 51 (female) [Tropical South Pacific].
CITATIONS: WilliaWa1988 [taxonomy, description, illustration, host, distribution: 66-69].

Aspidiotus palmarum Bouché
Aspidiotus palmarum Bouché, 1834: 17. Type data: GERMANY: Berlin, in greenhouse of Botanical Garden, on tropical palms. Syntypes, both sexes. Notes: Type material lost (Sachtleben, 1944).
Chermes palmarum Boisduval, 1868: 281. Change of combination.
Aspidiotus (Aspidiotus) palmarum; Cockerell, 1897i: 30. Change of combination.
Aspidiotus palmarum; Fernald, 1903b: 260. Incorrect synonymy. Notes: Incorrect synonymy with *Aspidiotus nerii* Bouché; see Borchsenius, 1966: 370.
Aspidiotus palmarum; Leonardi, 1920: 31. Incorrect synonymy. Notes: Incorrect synonymy with *Aspidiotus nerii* Bouché; see Borchsenius, 1966: 370.
Diaspis palmarum; Lindinger, 1934e: 159. Change of combination.
Aspidiotus palmarum; Borchsenius, 1966: 370. Revived combination.
SYSTEMATICS: *Aspidiotus palmarum* Bouché, 1834 was synonymized with *Aspidiotus nerii* Bouché by Fernald (1903b) and by Lindinger (1957), but the synonymy rejected by Borchsenius (1966).
SCALE COVER: Female scale circular, about 1/2 line; flat, white. Male scale elongate, length 1/2 line; white (Bouché, 1834).
HOST PLANTS: **Palmae** [Bouche1834], *Chamaerops* [Signor1869b].
DISTRIBUTION: **Palaearctic**: France [Signor1869b]; Germany (United) [Bouche1834].
CITATIONS: Blanch1840 [taxonomy, description, host, distribution: 215]; Borchs1966 [catalogue: 370]; Bouche1834 [taxonomy, description, illustration, host, distribution: 17-18]; Burmei1835 [taxonomy: 69]; Cocker1893j [taxonomy: 255]; Cocker1896b [distribution: 334]; Cocker1897i [taxonomy, description, host, distribution: 30]; Comsto1883 [taxonomy, description, host, distribution: 81]; Fernal1903b [taxonomy: 260]; Ferris1941e [taxonomy: 46,54]; Lindin1957 [taxonomy: 546]; Signor1869 [taxonomy: 863]; Signor1869b [taxonomy, description, illustration, host, distribution: 131-132]; Targio1892 [taxonomy: 81].

Aspidiotus pandani (Boisduval)

Chermes pandani Boisduval, 1868a: 301. Type data: FRANCE: Paris, in greenhouse, on *Pandanus utilis*; collected by Mr. Burel. Syntypes, female. Notes: Type material probably lost; Daniele Matile-Ferrero (1999) personal communication to Yair Ben-Dov.

Aspidiotus pandani Signoret, 1869: 863. Nomen nudum.

Aspidiotus pandani; Signoret, 1869b: 131. Change of combination.

Aspidiotus pandani; Cockerell, 1869b: 334. Notes: Incorrect citation of "Signoret" as author.

SYSTEMATICS: Signoret (1869: 863) referred to this species as "Signoret nov. spec.", but later (Signoret, 1869b: 131) correctly credited it to Boisduval (1868).

SCALE COVER: Female scale resembling a "chapeau chinois"; brown, circular - oval; sharply terminating in a rust-colour prominence on top (Boisduval, 1868a).

HOST PLANTS: **Pandanaceae**: *Pandanus utilis* [Boisdu1868a, Signor1869b].

DISTRIBUTION: **Palaearctic**: France [Boisdu1868a, Signor1869b].

CITATIONS: Boisdu1868a [taxonomy, description, host, distribution: 301]; Borchs1966 [catalogue: 370]; Cocker1896b [taxonomy, distribution: 334]; Cocker1897i [taxonomy, description, host, distribution: 30]; Fernal1903b [catalogue: 270]; Ferris1941e [taxonomy: 46]; Leonar1897 [taxonomy: 285]; Signor1869 [taxonomy: 863]; Signor1869b [taxonomy, description, illustration, host, distribution: 131].

Aspidiotus paolii Balachowsky

Aspidiotus paolii Balachowsky, 1956: 72. Type data: ERITREA: Asmara, on *Dodonaea viscosa*. Holotype female. Type depository: Tervuren: Musée Royal de l'Afrique Centrale, Section d'Entomologie, Belgium.

SCALE COVER: Female scale circular, convex; exuviae red orange, central, covered with white secretion; diameter 1.5-1.6 mm. Male scale oval, of similar colour (Balachowsky, 1956).

HOST PLANTS: **Sapindaceae**: *Dodonaea viscosa* [Balach1956].

DISTRIBUTION: **Afrotropical**: Eritrea [Balach1956].

GENERAL: Description and illustration of adult female by Balachowsky (1956).

KEYS: Balachowsky 1956: 51 (female) [Africa].

CITATIONS: Balach1956 [taxonomy, description, illustration, host, distribution: 72-74]; Borchs1966 [catalogue: 267].

Aspidiotus philippinensis Velasquez

Aspidiotus philippinensis Velasquez, 1971: 104. Type data: PHILIPPINES: Luzon, Laguna, Luisiana, on *Pandanus tectorius*; collected by M.P. Mariano. Holotype. Type depository: Los Banos: Entomological Museum, Museum of Natural History, University of the Philippines at Los Banos, College, Laguna, Luzon, Philippines.

COMMON NAMES: Pandan scale [Velasq1971].

SYSTEMATICS: Lit & Rimando (1993) discussed the status and depository of the type series.

SCALE COVER: Female scale circular, slightly convex, moderately thin, soft, brown; 0.5-2 mm in diameter; exuviae circular, lighter than rest of scale. Male unknown (Velasquez, 1971).

HOST PLANTS: **Pandanaceae**: *Pandanus tectorius* [Velasq1971].

DISTRIBUTION: **Oriental**: Philippines (Luzon [Velasq1971, LitRi1993]).

GENERAL: Description and illustration of adult female by Velasquez (1971).

CITATIONS: LitRi1993 [taxonomy: 163-167]; Velasq1971 [taxonomy, description, illustration, host, distribution: 104-107].

Aspidiotus phormii Signoret

Aspidiotus phormii Signoret, 1869: 865. Nomen nudum. Notes: Citation of "Bremi" as author.

Aspidiotus phormii Signoret, 1869b: 130. Type data: FRANCE: Midi, locality not indicated, on leaves of *Phormium tenax*; collected by Mayr. Syntypes, female. Notes: Incorrect citation of "de Breme" as author.

Aspidiotus phormii; Signoret, 1877: 671. Notes: Incorrect citation of "Brem." as author.

Aspidiotus phormii; Comstock, 1883: 80. Notes: Incorrect citation of "Breme." as author.

Aspidiotus phormii; Cockerell, 1896b: 334. Notes: Incorrect citation of "de Breme" as author.

Aspidiotus (*Aspidiotus*) *phormii*; Cockerell, 1897i: 30. Change of combination. Notes: Incorrect citation of "de Breme" as author.

Aspidiotus phormii; Fernald, 1903b: 276. Notes: Incorrect citation of " "Breme," Sign." as author.

Aspidiotus phormii; Ferris, 1941e: 47. Notes: Incorrect citation of "de Breme" as author.

Aspidiotus phormii; Ferris, 1941e: 47. Notes: Incorrect citation of "de Breme" as author.

Aspidiotus phormii; Borchsenius, 1966: 370. Notes: Correct citation of "Signoret" as author.

SCALE COVER: Female scale white, circular, exuviae central; male scale elongate (Signoret, 1869b).

HOST PLANTS: **Liliaceae**: *Phormium tenax* [Signor1869b].

DISTRIBUTION: **Palaearctic**: France [Signor1869b].

GENERAL: History of nomenclature *Aspidiotus phormii* was as follows: 1. Bremy (1847) presented a general discussion on scale insects in Switzerland, using only generic names, but no species names. 2. Signoret (1868: 520) listed above publication as by Bremi Wolff (1847), Ueber Schildlause, fait l'enumeration des espèces qui se trouve en Suisse. (Verhandlungen der Schweizer naturforch, Geselschaft, Schafhausen, 1847: 41-45). 3. Signoret (1869: 865) listed *phormii* Bremi - *Aspidiotus* ... Suisse. Signoret (1869b: 130) described *Aspidiotus phormii* (incorrectly credited to de Breme), from *Phormium tenax*, collected by Mayr in France. This is the original description of *Aspidiotus phormii* Signoret. 4. Signoret (1877: 671) listed *phormii* Brem. - *Aspidiotus*. 5. Comstock (1883: 80) discussed *Aspidiotus phormii* Brem. 6. Cockerell (1896b: 334) listed *Aspidiotus phormii* de

Breme. 7. Fernald (1903b: 276) listed *Aspidiotus phormii* "Breme", Signoret. 8. Ferris (1941e) listed *Aspidiotus phormii* de Breme. 9. Borchsenius (1966: 370) catalogued *Aspidiotus phormii* and correctly credited authorship to Signoret.

CITATIONS: Borchs1966 [catalogue: 370]; Cocker1895w [taxonomy: 8]; Cocker1896b [taxonomy: 334]; Cocker1897i [taxonomy, description: 30]; Comsto1883 [taxonomy, host, distribution: 80]; Fernal1903b [catalogue: 276]; Ferris1941e [taxonomy: 47]; Lindin1912b [taxonomy: 359]; MacGil1921 [taxonomy: 387]; Signor1869 [taxonomy: 865]; Signor1869b [taxonomy, description, host, distribution: 130]; Signor1877 [taxonomy: 671].

Aspidiotus populi Baerensprung

Aspidiotus populi Baerensprung, 1849: 167. Type data: GERMANY: Berlin, on "Pappeln und Linden". Syntypes, both sexes. Notes: Depository unknown

SCALE COVER: Female scale circular or slightly elongate, white, with a central or subcentral 'navel'. Male scale white, parallel-sided (Baerensprung, 1849).

HOST PLANTS: **Salicaceae**: *Populus* [Baeren1849]. **Tiliaceae**: *Tilia* [Baeren1849].

DISTRIBUTION: **Palaearctic**: Germany [Baeren1849].

GENERAL: Description of adult female and male by Baerensprung (1849).

CITATIONS: Baeren1849 [taxonomy, description, host, distribution: 167]; Borchs1966 [catalogue: 370]; Comsto1883 [taxonomy: 109-110]; Fernal1903b [catalogue: 370]; Ferris1941e [taxonomy: 47]; Kaussa1955 [host, distribution: 16]; Signor1869 [taxonomy: 446].

Aspidiotus pothos Takagi

Aspidiotus pothos Takagi, 1969a: 69. Type data: TAIWAN: Southeastern Tai-pei Hsien, on *Pothos seemannii*. Holotype female. Type depository: Sapporo: Entomological Institute, Faculty of Agriculture, Hokkaido University, Japan.

Aspidiotus potos; Danzig, 1993: 141, 143. Misspelling of species name.

Aspidiotus potos; Danzig & Pellizzari, 1998: 193. Misspelling of species name.

SCALE COVER: Takagi (1969) did not describe scale cover.

HOST PLANTS: **Araceae**: *Pothos seemannii* [Takagi1969a]. **Liliaceae**: *Ophiopogon japonicus* [Danzig1993]. **Theaceae**: *Thea* [Danzig1993].

DISTRIBUTION: **Oriental**: Taiwan [Takagi1969a]. **Palaearctic**: Georgia (Adzhar ASSR [Danzig1993]); Russia [Danzig1993].

GENERAL: Description and illustration of adult female by Takagi (1969a), Chou (1985, 1986), and by Danzig (1993).

KEYS: Danzig 1993: 140-141 (female) [Europe].

CITATIONS: Chou1985 [taxonomy, description, host, distribution: 398-399]; Chou1986 [taxonomy, illustration: 661]; Danzig1993 [taxonomy, description, illustration, host, distribution: 143-144]; DanzigPe1998 [catalogue: 193]; Takagi1969a [taxonomy, description, illustration, host, distribution: 68-70,100]; Tao1999 [taxonomy, host, distribution: 120].

Aspidiotus putearius Green

Aspidiotus putearius Green, 1896e: 54. Type data: SRI LANKA: Punduloya, on *Strobilanthes viscosus*. Syntypes, female. Type depository: London: The Natural History Museum, England, UK.

Hemiberlesia putearia; Leonardi, 1897b: 131. Change of combination requiring emendation of species name for agreement in gender.

Aspidiotus (*Aspidiotus*) *putearius*; Cockerell, 1897i: 28. Change of combination.

Cryptophyllaspis putearia; MacGillivray, 1921: 427. Change of combination requiring emendation of species name for agreement in gender.

Aspidiotus (*Hemiberlesia*) *putearius*; Green, 1937: 333. Change of combination.

Aspidiotus putearius; Borchsenius, 1966: 267. Revived combination.

SCALE COVER: Female scale round, flat or slightly concave, forming operculum to pit-like depression in which insect rests; colour very pale brownish ochreous, semi-opaque; minutely rugose, with concentric lines of growth; exuviae central, pale yellow; second exuvia slightly concave, first slightly convex; ventral scale obsolete, a mere powdery film lining cavity below insect; diameter 1.5 mm. Male scale broadly oval; similar in texture to that of female; median area convex where it covers insect; size 1 x 1.2mm (Green, 1896e). Colour illustration of scale cover by Green (1896e).

HOST PLANTS: **Acanthaceae**: *Strobilanthes* [Ramakr1921a], *Strobilanthes viscosus* [Green1896e, Green1937].

DISTRIBUTION: **Oriental**: Sri Lanka [Green1896e, Ramakr1921a, Green1937].

BIOLOGY: Forming small pits on undersides of leaves (Ferris, 1941e).

GENERAL: Description and illustration of adult female by Green (1896e) and by Ferris (1941e).

KEYS: Ferris 1946: 43 (female) [World]; Ferris 1941e: 61 (female) [World]; Green 1896e: 40 (female) [Sri Lanka].

CITATIONS: Borchs1966 [catalogue: 267]; Cocker1897i [taxonomy, description, host, distribution: 28]; Cocker1899a [taxonomy: 395]; DEDAC1923 [host, distribution]; Fernal1903b [catalogue: 276]; Ferris1941e [taxonomy, description, illustration, host, distribution: 47,58,66]; Ferris1946 [taxonomy: 43]; Green1896e [taxonomy, description, illustration, host, distribution: 54-55]; Green1937 [host, distribution: 333]; Larew1990 [ecology, life history, structure: 283-300]; Leonar1897b [taxonomy, description, illustration, host, distribution: 119,131-132]; MacGil1921 [taxonomy, description, host, distribution: 427]; Ramakr1921a [host, distribution: 357].

Aspidiotus queenslandicus Brimblecombe

Aspidiotus queenslandicus Brimblecombe, 1959: 121. Type data: AUSTRALIA: Queensland, Beenleigh, on *Acronychia laevis*; collected 29.v.1956. Holotype female. Type depository: Brisbane: Queensland Museum, Queensland, Australia; type no. T5692.

SCALE COVER: Female scale circular, 2.5 mm diameter, dark fawn in colour. Pellicles orange coloured (Brimblecombe, 1959).

HOST PLANTS: **Rutaceae**: *Acronychia laevis* [Brimbl1959, Muntin1971a].

DISTRIBUTION: **Australasian**: Australia (Queensland [Brimbl1959, Muntin1971a]).

GENERAL: Description and illustration of adult female by Brimblecombe (1959) and by Munting (1971a).

CITATIONS: Borchs1966 [catalogue: 267]; Brimbl1959 [taxonomy, description, illustration, host, distribution: 121-123]; Brimbl1968 [taxonomy, illustration, host, distribution: 47]; Muntin1971a [taxonomy, description, illustration, host, distribution: 313-314].

Aspidiotus remaudierei Balachowsky

Aspidiotus remaudierei Balachowsky, 1956: 74. Type data: NIGER: Oasis de Myrriah, 25 km from Zinder, on *Psidium guajava*. Holotype female. Type depository: Tervuren: Musée Royal de l'Afrique Centrale, Section d'Entomologie, Belgium.

Aspidiotus semaudierei; Borchsenius, 1966: 417. Misspelling of species name.

SCALE COVER: Female scale white, circular, slightly convex; exuviae brown, central; diameter 2.2-2.3 mm. Male scale white, oval, 1.5 mm long (Balachowsky, 1956).

HOST PLANTS: **Myrtaceae**: *Psidium guajava* [Balach1956]. **Palmae**: *Elaeis guineensis* [Balach1956].

DISTRIBUTION: **Afrotropical**: Niger [Balach1956]; Zaire [Balach1956].

GENERAL: Description and illustration of adult female by Balachowsky (1956).

KEYS: Balachowsky 1956: 52 (female) [Africa].

CITATIONS: Balach1956 [taxonomy, description, illustration, host, distribution: 74-76]; Borchs1966 [catalogue: 267,417]; Muntin1971a [taxonomy: 309].

Aspidiotus rhododendri nomen nudum

Aspidiotus rhododendri Wailes, 1858: 85. Nomen nudum.

Aspidiotus rhododendri Signoret, 1870: 109. Nomen nudum. Notes: Signoret (1870: 109) credited this Nomen Nudum to G. Wailes (1859).

Aspidiotus rhododendri Fernald, 1903b: 324. Nomen nudum.

Aspidiotus rhododendri Ferris, 1941e: 47. Nomen nudum.

Aspidiotus rhododendri Borchsenius, 1966: 376. Nomen nudum.

Aspidiotus rigidus Reyne

Aspidiotus destructor rigidus Reyne, 1947: 294. Type data: INDONESIA: Sangi Island, half way to Sulawesi [=Celebes], and PHILIPPINES: Mindanao, on coconut palm [=*Cocos nucifera*]. Syntypes, female. Type depository: Amsterdam: Institut voor Taxonomische Zoologie, The Netherlands.

Aspidiotus rigidus; Borchsenius, 1966: 267. Change of status.

SYSTEMATICS: Species first described as a subspecies of *Aspidiotus destructor* by Reyne (1947, 1948), who distinguished it from typical *destructor* mainly by biological features, as well as difference in scale structure.

SCALE COVER: Female scale 1.8-2.1 mm in diameter, height 0.1-0.2 mm, more or less smoke-coloured, fibrous, sometimes with an irregular outline, caused by one or more incisions; exuviae of first- and second-larval stage 0.45 x 030 and 0.7 x 0.6

mm respectively (Reyne, 1948). Reyne (1948) also noted that, in typical *Aspidiotus destructor,* scale is generally more transparent, less fibrous, and sometimes provided with a faint radial striation.

HOST PLANTS: **Palmae**: *Cocos nucifera* [Reyne1947, Reyne1948].

NATURAL ENEMIES: COLEOPTERA **Coccinellidae**: *Chilocorus nigritus* F. [Reyne1948], *Nephus luteus* Sicard [Reyne1948], *Telsimia nitida* Chap. [Reyne1948].   HYMENOPTERA **Encyrtidae**: *Comperiella unifasciata* Ishii [Reyne1947].

DISTRIBUTION: **Oriental**: Indonesia (Sulawesi (Celebes) [Reyne1947]).

ECONOMIC IMPORTANCE: A serious pest of coconut palm in Sangi island, near Sulawesi, Indonesia (Reyne, 1947, 1948).

GENERAL: Description and illustration of adult female by Reyne (1947, 1948).

CITATIONS: Borchs1966 [catalogue: 267]; Kalsho1981 [taxonomy, host, distribution, economic importance: 166-170]; MillerDa1990 [host, distribution, economic importance: 300]; Reyne1947 [taxonomy, description, host, distribution, economic importance, biological control: 294-302]; Reyne1948 [taxonomy, description, illustration, host, distribution, life history, economic importance, biological control: 83-123]; SchmutKlLu1957 [host, distribution, economic importance: 475]; WilliaWa1988 [taxonomy: 56].

Aspidiotus riverae Cockerell

Aspidiotus riverae Cockerell, 1905d: 161. Type data: CHILE: Province of Arauco, on stems of *Chusquea* sp. Syntypes, female. Type depository: Washington, D.C.: U.S. Entomological Collection, U.S. National Museum of Natural History, USA.

Comstockiella riverae; MacGillivray, 1921: 423. Change of combination.

Dycryptaspis (?) riverai; Lindinger, 1937: 184. Change of combination.

Aspidiotus riverai Lindinger, 1957: 546. Unjustified emendation.

Aspidiotus riverae; Borchsenius, 1966: 370. Revived combination.

SCALE COVER: Female scale 3 mm long, oval, moderately convex, greyish brown, with large uncovered ochreous exuviae near one end (Cockerell, 1905d).

HOST PLANTS: **Gramineae**: *Chusquea* [Cocker1905d, ClapsWoGo2001].

DISTRIBUTION: **Neotropical**: Chile [Cocker1905d, Sander1906, GonzalCh1968, ClapsWoGo2001].

GENERAL: Description of adult female by Cockerell (1905d).

CITATIONS: Borchs1966 [catalogue: 370]; ClapsWoGo2001 [host, distribution: 241]; Cocker1905d [taxonomy, description, host, distribution: 161]; Ferris1941e [taxonomy: 47]; GonzalCh1968 [distribution: 110]; Lindin1907a [taxonomy: 19]; Lindin1937 [taxonomy: 180]; Lindin1957 [taxonomy: 546]; MacGil1921 [taxonomy, description, host, distribution: 423].

Aspidiotus robiniae nomen nudum

Aspidiotus robiniae Targioni Tozzetti, 1879a: 26. Nomen nudum.

Aspidiotus robiniae Lindinger, 1949: 210. Nomen nudum.

Aspidiotus robiniae Borchsenius, 1966: 376. Nomen nudum.

Aspidiotus ruandensis Balachowsky

Aspidiotus ruandensis Balachowsky, 1955: 391. Type data: RWANDA: Nyanza territory, Gitarama, 1800 meters altitude, on *Euphorbia tommingi*. Holotype female. Type depository: Paris: Muséum national d'Histoire naturelle, France.

SCALE COVER: Female scale circular, slightly convex; colour dark brown; exuviae dark, situated centrally; secretion of second stage brown; diameter 2-2.2 mm. Male scale unknown (Balachowsky, 1956).

HOST PLANTS: **Euphorbiaceae**: *Euphorbia tommingi* [Balach1955]. **Leguminosae**: *Cassia* [MatileNo1984]. **Myrtaceae**: *Melaleuca leucodendron* [Balach1955, MatileNo1984].

DISTRIBUTION: **Afrotropical**: Cameroon [Balach1956, MatileNo1984]; Guinea [Balach1956]; Rwanda [Balach1955].

GENERAL: Description and illustration of adult female by Balachowsky (1955, 1956).

KEYS: Balachowsky 1956: 52 (female) [Africa].

CITATIONS: Balach1955 [taxonomy, description, illustration, host, distribution: 391-393]; Balach1956 [taxonomy, description, illustration, host, distribution: 76-78]; Borchs1966 [catalogue: 267]; MatileNo1984 [host, distribution: 64]; Muntin1971a [taxonomy: 313].

Aspidiotus saliceti Bouché

Aspidiotus saliceti Bouché, 1851: 111. Type data: GERMANY: Berlin, on twigs of *Salix holosericea*. Syntypes, female. Notes: Type material lost (Sachtleben, 1944).

Mytilaspis saliceti; Targioni-Tozzetti, 1868: 737. Change of combination.

Lepidosaphes saliceti; Lindinger, 1934: 63. Change of combination.

Aspidiotus saliceti; Borchsenius, 1966: 370. Revived combination.

SYSTEMATICS: Borchsenius (1966) listed species as *incertae sedis* species.

SCALE COVER: Female scale "schinkenmuschelformig", light brown with darker basis (Bouché, 1851).

HOST PLANTS: **Salicaceae**: *Salix holosericea* [Bouche1851].

DISTRIBUTION: **Palaearctic**: Germany (United) [Bouche1851].

GENERAL: Description of male and female by Bouché (1851).

CITATIONS: Borchs1966 [catalogue: 370]; Bouche1851 [taxonomy, description, host, distribution: 111]; Comsto1883 [taxonomy, description, host, distribution: 126]; Fernal1903b [catalogue: 224]; Ferris1941e [taxonomy: 48]; Lindin1934 [taxonomy: 164]; Lindin1935 [taxonomy: 139]; Targio1868 [taxonomy: 737].

Aspidiotus selangorensis Hall & Williams

Aspidiotus selangorensis Hall & Williams, 1962: 35. Type data: MALAYSIA: Malaya, Kuala Lumpur, on *Adiantum fergusoni*; collected 1. VI. 1926. Holotype female. Type depository: London: The Natural History Museum, England, UK.

SCALE COVER: Characteristics of scale were unknown to Hall & Williams (1962).

HOST PLANTS: **Adiantaceae**: *Adiantum fergusoni* [HallWi1962].

DISTRIBUTION: **Australasian**: Hawaiian Islands (Hawaii [Beards1965]). **Oriental**: Malaysia (Malaya [HallWi1962]).

GENERAL: Description and illustration of adult female by Hall & Williams (1962).
KEYS: Beardsley 1970: 508 (female) [Hawaii].
CITATIONS: Beards1965 [host, distribution: 12]; Borchs1966 [catalogue: 267]; HallWi1962 [taxonomy, description, illustration, host, distribution: 35,37-38].

Aspidiotus serratus Froggatt

Aspidiotus serrata Froggatt, 1914: 318. Type data: AUSTRALIA: New South Wales, Darling River, growing at Pera Bore, on *Acacia cambagei*. Syntypes, female. Type depository: Brisbane: Queensland Museum, Queensland, Australia.
Aspidiotus serratus; Ferris, 1941e: 48. Emendation of species name for agreement in gender.
SCALE COVER: Female scale almost circular, very convex; diameter about 1/35 inch; outer surface greyish brown, inner surface white; exuviae light yellow; centre of exuviae sometimes slightly depressed at apex (Froggatt, 1914).
HOST PLANTS: **Leguminosae**: *Acacia cambagei* [Frogga1914].
DISTRIBUTION: **Australasian**: Australia (New South Wales [Frogga1914]).
GENERAL: Description of adult female by Froggatt (1914).
CITATIONS: Borchs1966 [catalogue: 370]; Ferris1941e [taxonomy: 48]; Sassce1915 [taxonomy, host, distribution: 34].

Aspidiotus simmondsi nomen nudum

Aspidiotus simmondsi Green & Laing in Simmonds, 1925: 1. Nomen nudum.
Aspidiotus simmondsi Ferris, 1941e: 48. Nomen nudum.
Aspidiotus simmondsi Lepesme, 1947: 195. Nomen nudum.
Aspidiotus simmondsi Borchsenius, 1966: 376. Nomen nudum.

Aspidiotus simulans De Lotto

Aspidiotus simulans De Lotto, 1957: 228. Type data: KENYA: Nairobi, on leaves of *Ficus vallis-choudae*. Holotype female. Type depository: London: The Natural History Museum, England, UK.
SCALE COVER: Scale of female extremely thin and transparent; dirty white or pale brown in colour; low convex; exuviae golden yellow, central; diameter up to 2 mm. Scale of male elongate, white, up to 1.2 mm in length (De Lotto, 1957).
HOST PLANTS: **Apocynaceae**: *Acokanthera longiflora* [DeLott1957], *Acokanthera schimperi* [DeLott1957], *Nerium oleander* [DeLott1957]. **Bignoniaceae**: *Markhamia platycalyx* [DeLott1957]. **Canellaceae**: *Warburgia stuhlmanni* [DeLott1957]. **Celastraceae**: *Elaeodendron* [DeLott1957, BrownDe1959], *Gymnosporia* [BrownDe1959]. **Ehretiaceae**: *Ehretia silvatica* [DeLott1957]. **Flacourtiaceae**: *Aberia caffra* [DeLott1957]. **Leguminosae**: *Ceratonia siliqua* [DeLott1957]. **Moraceae**: *Ficus vallis-choudae* [DeLott1957]. **Musaceae**: *Musa paradisiaca* [DeLott1957]. **Rubiaceae**: *Rytigynia schumannii* [DeLott1957].
DISTRIBUTION: **Afrotropical**: Kenya [DeLott1957, BrownDe1959].
GENERAL: Description and illustration of adult female by De Lotto (1957).
KEYS: De Lotto 1957: 228 (female) [Africa].

CITATIONS: Borchs1966 [catalogue: 267]; BrownDe1959 [taxonomy, host, distribution, structure, chromosome: 369-379]; DeLott1957 [taxonomy, description, illustration, host, distribution: 227-228]; Nur1990a [taxonomy, structure, chromosomes: 186]; Trembl1990a [anatomy, structure: 275-283].

Aspidiotus sinensis (Ferris), new combination

Temnaspidiotus sinensis Ferris, 1952a: 9. Type data: CHINA: Yunnan Province, near Kunming, at An-lin-wen-chian, on a small, undetermined grass; collected by G.F. Ferris, April 30, 1949. Syntypes, female. Type depository: Davis: The Bohart Museum of Entomology, University of California, California, USA.
SYSTEMATICS: *Temnaspidiotus sinensis* Ferris, 1952 is transferred here to *Aspidiotus*. Dr. Sadao Takagi (in personal communication, 8 January 2003, to Yair Ben-Dov) suggested that *Aspidiotus kellyi* Brain and *Aspidiotus sinensis* (Ferris) may belong to a separate genus, for which the name *Brainaspis* MacGillivray, 1921, is available.
SCALE COVER: Scale of the female white, oval, the exuvia entirely or nearly covered by secretion. Scale of the male similar to that of the female but smaller (Ferris, 1952a).
HOST PLANTS: **Gramineae** [Ferris1952a].
DISTRIBUTION: **Oriental**: China (People's Republic) (Yunnan [Ferris1952a]).
BIOLOGY: Occurring on the leaves and stems (Ferris, 1952a).
GENERAL: Description and illustration of adult female by Ferris (1952a) and by Chou (1985, 1986).
KEYS: Chou 1985: 274 (female) [Species of China].
CITATIONS: Borchs1966 [catalogue: 272]; Chou1985 [taxonomy, description, host, distribution: 275]; Chou1986 [taxonomy, illustration: 666]; DanzigPe1998 [catalogue: 362]; Ferris1952a [taxonomy, description, illustration, host, distribution: 9,15]; Tao1999 [taxonomy, host, distribution: 120].

Aspidiotus spurcatus Signoret

Aspidiotus spurcatus Signoret, 1869b: 138. Type data: FRANCE: locality not indicated, probably Paris, on *Populus*. Syntypes, both sexes. Type depository: Vienna: Naturhistorisches Museum Wien, Austria.
Aspidiotus (Diaspidiotus) spurcatus; Cockerell, 1897i: 19. Change of combination.
Aspidiotus spurcatus; Leonardi, 1898c: 38. Incorrect synonymy. Notes: See Systematics below.
Aspidiotus spurcatus; Fernald, 1903b: 269. Incorrect synonymy. Notes: See Systematics below.
Aspidiotus spurcatus; Borchsenius, 1966: 371. Revived combination.
SYSTEMATICS: *Aspidiotus spurcatus* Signoret, 1869c was synonymized with *Quadraspidiotus ostreaeformis* (Curtis), by Leonardi (1898c) and by Fernald (1903b), but Borchsenius (1966) regarded the former as a valid species.
SCALE COVER: Female scale circular, brown and light yellow in centre. Male scale elongate, brown-red (Signoret, 1869c).
HOST PLANTS: **Salicaceae**: *Populus* [Signor1869b].
DISTRIBUTION: **Palaearctic**: France [Signor1869b].

GENERAL: Description and illustration of adult female and male by Signoret (1869c).
CITATIONS: Borchs1950b [taxonomy, description, illustration, host, distribution: 225,227,231]; Borchs1966 [catalogue: 371]; Chumak1961 [host, distribution, biological control: 313-338]; Cocker1896b [distribution: 333]; Cocker1897i [taxonomy, description, host, distribution: 19]; Comsto1883 [taxonomy, description, host, distribution: 82]; Fernal1903b [taxonomy: 269]; Ferris1941e [taxonomy: 48]; Leonar1897 [taxonomy: 285]; Signor1869c [taxonomy, description, illustration, host, distribution: 138]; Smetni1991 [chemistry: 92-129].

Aspidiotus suvaensis nomen nudum
Aspidiotus suvaensis Williams & Watson, 1988: 17. Nomen nudum. Notes: Recorded as "*Aspidiotus suvaensis* Green & Laing".

Aspidiotus symplocos Ramakrishna Ayyar
Aspidiotus calophylli symplocos Ramakrishna Ayyar, 1924: 340. Nomen nudum.
Aspidiotus calophylli symplocos Ramakrishna Ayyar, 1926: 456. Nomen nudum.
Aspidiotus calophylli symplocos Ramakrishna Ayyar, 1930: 27. Type data: INDIA: Coonnor, altitude 5500 feet, on *Symplocos*. Syntypes, female. Notes: Depository of type material unknown.
Aspidiotus symplocos; Ferris, 1941e: 48. Change of status.
Aspidiotus symlpocos; Borchsenius, 1966: 371. Misspelling of species name.
SCALE COVER: Scale white, though covered with earthly deposits (Ramakrishna Ayyar, 1930).
HOST PLANTS: Symplocaceae: *Symplocos* [Ramakr1930].
DISTRIBUTION: Oriental: India [Ramakr1930].
GENERAL: Described for first time as *Aspidiotus calophylli* var. *symplocos* by Ramakrishna Ayyar (1930: 27), who referred to an earlier description ("B.J. xxviii, p. 1008" 1922). It is very likely that the latter is a reference to page 1008 in Green (1922a), in which *Aspidiotus calophylli* Green was described, but not the subspecies *symplocos* .
CITATIONS: Borchs1966 [catalogue: 371]; Ferris1941e [taxonomy: 48]; Ramakr1924 [taxonomy: 340]; Ramakr1926 [taxonomy: 456]; Ramakr1930 [taxonomy, description, host, distribution: 27].

Aspidiotus tafiranus Lindinger
Aspidiotus tafiranus Lindinger, 1912b: 229. Type data: CANARY ISLANDS: Gran Canaria, on *Olea europaea*. Syntypes, female. Type depository: Zoologisches Institut und Zoologishces Museum, Universität von Hamburg, Germany.
Aspidioides tafiranus; MacGillivray, 1921: 406. Change of combination.
Aspidiotus tafiranus; Borchsenius, 1966: 269. Revived combination.
SCALE COVER: Female scale white or grey white, circular or slightly elongate, convex, diameter 2 mm; exuviae brown yellow, subcentral (Lindinger, 1912b).
HOST PLANTS: Oleaceae: *Olea europaea* [Lindin1912b].
DISTRIBUTION: Palaearctic: Canary Islands [Lindin1912b].
GENERAL: Description of adult female by Lindinger (1912b).

CITATIONS: Borchs1966 [catalogue: 269]; DanzigPe1998 [catalogue: 194]; Ferris1941e [taxonomy: 48]; Lindin1912b [taxonomy, description, host, distribution: 229]; MacGil1921 [taxonomy, description, host, distribution: 406]; Sassce1915 [taxonomy, host, distribution: 34]; WeidneWa1968 [taxonomy: 173].

Aspidiotus tangfangtehi Ben-Dov, replacement name

Aspidiotus theae Tang, 1977: 236. Type data: CHINA: Guangxi Province, Guizhou, on bark and leaves of tea plant. Syntypes, female. Type depository: Shanxi: Entomological Institute, Shanxi Agricultural University, Taigu, Shanxi, China. Junior homonym of *Aspidiotus theae* Green, 1890.

SCALE COVER: Female scale, circular, about 2.5 mm in diameter; thin; semitransparent, light yellow; exuviae subcentral. Male scale oval (Tang, 1977).

HOST PLANTS: **Theaceae**: *Thea sinensis* [Tang1977].

DISTRIBUTION: **Oriental**: China (People's Republic) (Guangxi (Kwangsi) [Tang1977]).

GENERAL: Description and illustration of adult female by Tang (1977).

CITATIONS: Chou1985 [taxonomy, description, host, distribution: 399]; Tang1977 [taxonomy, description, illustration, host, distribution: 236-237].

Aspidiotus taraxacus (Tang), new combination

Temnaspidiotus taraxacus Tang, 1984: 22. Type data: CHINA: Hainan Islands, on *Taraxacum koksagyz*. Holotype female. Type depository: Shanxi: Entomological Institute, Shanxi Agricultural University, Taigu, Shanxi, China.

SCALE COVER: Tang (1984) did not describe scale cover.

HOST PLANTS: **Compositae**: *Taraxacum koksagyz* [Tang1984].

DISTRIBUTION: **Oriental**: China (People's Republic) (Hainan [Tang1984]).

GENERAL: Description and illustration of adult female by Tang (1984).

CITATIONS: Tang1984 [taxonomy, description, illustration, host, distribution: 22,24]; Tao1999 [taxonomy, host, distribution: 121].

Aspidiotus targionii Del Guercio

Aspidiotus targionii Del Guercio, 1894: 148. Type data: ITALY: Sicily, Messina Province, on *Mespilus germanica*; collected 1892. Syntypes, female. Notes: Depository of type material unknown.

SYSTEMATICS: Del Guercio (1894) gave a detailed description with partial illustration of adult female. Borchsenius (1966) placed this species among *incertae sedis* taxa.

SCALE COVER: Female scale oval, white; exuviae yellow, oval, situated between center and margin (Del Guercio, 1894).

HOST PLANTS: **Rosaceae**: *Mespilus germanica* [DelGue1894].

DISTRIBUTION: **Palaearctic**: Sicily [DelGue1894].

GENERAL: Description and illustration of adult female by Del Guercio (1894).

CITATIONS: Borchs1966 [catalogue: 371]; DelGue1894 [taxonomy, description, illustration, host, distribution: 148-158]; Fernal1903b [catalogue: 320]; Ferris1941e [taxonomy: 48]; McKenz1945 [taxonomy: 54].

Aspidiotus taverdeti Balachowsky

Aspidiotus taverdeti Balachowsky, 1956: 78. Type data: CAMEROON: between Tibati and N'Gaoundére, 100 km south of the latter, on *Syzygium guineense*. Holotype female. Type depository: Tervuren: Musée Royal de l'Afrique Centrale, Section d'Entomologie, Belgium.

SCALE COVER: Female scale circular, slightly convex; exuviae brown, central or subcentral; secreted part of scale white, slightly pink; diameter 2.2-2.3 mm. Male scale oval, 1.6 mm long (Balachowsky, 1956).

HOST PLANTS: **Myrtaceae**: *Syzygium guineense* [Balach1956].

DISTRIBUTION: **Afrotropical**: Cameroon [Balach1956].

GENERAL: Description and illustration of adult female by Balachowsky (1956).

KEYS: Balachowsky 1956: 52 (female) [Africa].

CITATIONS: Balach1956 [taxonomy, description, illustration, host, distribution: 78-81]; Borchs1966 [catalogue: 268]; Muntin1971a [taxonomy: 309].

Aspidiotus tiliae Bouché

Aspidiotus tiliae Bouché, 1851: 111. Type data: GERMANY: Berlin, on twigs of *Tilia* sp. and *Aleus* sp. Syntypes, both sexes. Notes: Type material lost (Sachtleben, 1944).

SYSTEMATICS: Leonardi (1898c: 38) and Fernald (1903b: 268) regarded *Aspidiotus tiliae* Bouché, 1851, a synonym of *Quadraspidiotus ostreaeformis* Curtis, whereas Borchsenius (1966) listed it among *incertae sedis* species.

SCALE COVER: Female scale elongate, narrow at base, white yellow (Bouché, 1851).

HOST PLANTS: *Aleus* [Bouche1851]. **Tiliaceae**: *Tilia* [Bouche1851].

GENERAL: Description of adult female and male by Bouché (1851).

CITATIONS: Borchs1966 [catalogue: 371]; Bouche1851 [taxonomy, description, host, distribution: 111]; Comsto1881a [taxonomy, description, host, distribution: 83]; Comsto1883 [taxonomy, description, host, distribution: 83]; Comsto1916a [taxonomy, description, host, distribution: 544]; Fernal1903b [taxonomy: 268]; Ferris1941e [taxonomy: 49]; Lindin1932f [taxonomy: 200]; Signor1869c [taxonomy, description, host, distribution: 137].

Aspidiotus tridentifer Ferris

Aspidiotus tridentifer Ferris, 1941d: 337. Type data: MEXICO: State of Oaxaca, at Chivela, on *Smilax* sp. Holotype female. Type depository: Davis: The Bohart Museum of Entomology, University of California, California, USA.

SCALE COVER: Scale of adult female quite convex, roughly circular with posterior end somewhat produced; texture thick and surface rather rough; exuviae displaced somewhat toward one side, brown except for a whitish area over exuvia. Scale of male oval, white; exuvia close to one end (Ferris, 1941d).

HOST PLANTS: **Smilacaceae**: *Smilax* [Ferris1941d].

DISTRIBUTION: **Nearctic**: Mexico (Oaxaca [Ferris1941d]).

BIOLOGY: Occurring on the stems of the host (Ferris, 1941d).

GENERAL: Good description and illustration of adult female by Ferris (1941d, 1941e).

KEYS: Ferris 1946: 43 (female) [World]; Ferris 1941e: 60 (female) [World].
CITATIONS: Borchs1966 [catalogue: 268]; Ferris1941d [taxonomy, description, illustration, host, distribution: 337]; Ferris1941e [taxonomy, description, illustration, host, distribution: 49,60,69]; Ferris1942 [taxonomy: 446:30]; Ferris1946 [taxonomy: 43].

Aspidiotus tripinnatus nomen nudum

Aspidiotus tripinnatus Ramakrishna Ayyar, 1924: 340. Nomen nudum.
Aspidiotus tripinnatus Ramakrishna Ayyar, 1930: 27. Nomen nudum.
Aspidiotus tripinnatus McKenzie, 1939: 55. Nomen nudum. Notes: McKenzie (1939) credited this name to Green.
Aspidiotus tripinnatus Ferris, 1941e: 49. Nomen nudum.
Aspidiotus tripinnatus Borchsenius, 1966: 377. Nomen nudum.

Aspidiotus undulatus Lindinger

Aspidiotus undulatus Lindinger, 1909e: 20. Type data: CAMEROON: Bipinde, Urwaldgebiet, on *Acioa pallescens* and on *Strychnos cinnabarina*. Syntypes, female. Type depository: Hamburg: Zoologisches Institut und Zoologishces Museum, Universität von Hamburg, Germany.
Gonaspidiotus undulatus; MacGillivray, 1921: 432. Change of combination.
Gonaspidiotus ungulatus; Borchsenius, 1966: 313. Misspelling of species name.
Aspidiotus undulatus; Borchsenius, 1966: 269. Revived combination.
SCALE COVER: Female scale similar to that of *Aspidiotus maendrius* Lindinger, 1909e, but smaller (Lindinger, 1909e).
HOST PLANTS: **Chrysobalanaceae**: *Acioa pallescens* [Lindin1909e].
Strychnaceae: *Strychnos cinnabarina* [Lindin1909e].
DISTRIBUTION: **Afrotropical**: Cameroon [Lindin1909e].
GENERAL: Description and illustration of adult female by Lindinger (1909e).
CITATIONS: Borchs1966 [catalogue: 269]; Ferris1941e [taxonomy: 49]; Lindin1909e [taxonomy, description, illustration, host, distribution: 20-22]; MacGil1921 [taxonomy, description, host, distribution: 432]; McKenz1938 [taxonomy: 5]; Sassce1911 [taxonomy: 70]; Vayssi1913 [host, distribution: 431]; WeidneWa1968 [taxonomy: 174].

Aspidiotus varians Lindinger

Aspidiotus varians Lindinger, 1910b: 39. Type data: MADAGASCAR: on *Cocos nucifera*. Syntypes. Type depository: Hamburg: Zoologisches Institut und Zoologishces Museum, Universität von Hamburg, Germany.
Aspidiotus varianus; Borchsenius, 1966: 423. Misspelling of species name.
SCALE COVER: Female scale circular, up to 2 mm in diameter, brown grey; exuviae yellow, central (Lindinger, 1910b).
HOST PLANTS: **Palmae**: *Cocos nucifera* [Lindin1910b, Mamet1943a].
DISTRIBUTION: **Afrotropical**: Madagascar [Lindin1910b, Mamet1943a, Borchs1966]; Tanzania [Lindin1910b, Borchs1966].
GENERAL: Description and illustration of adult female by Lindinger (1910b).

CITATIONS: Borchs1966 [catalogue: 269]; Ferris1941e [taxonomy: 49]; Lepesm1947 [host, distribution: 195]; Lindin1910b [taxonomy, description, illustration, host, distribution: 39]; MacGil1921 [taxonomy, description, host, distribution: 397]; Mamet1943a [catalogue: 157]; Sassce1912 [taxonomy, host, distribution: 94]; WeidneWa1968 [taxonomy: 174].

Aspidiotus vernoniae Hall

Aspidiotus vernoniae Hall, 1929: 350. Type data: ZIMBABWE: Embeza, on branches of *Vernonia podocoma*. Syntypes, female. Type depository: London: The Natural History Museum, England, UK.

SCALE COVER: Female scale circular, low convex; pale brown. Secretionary covering semi-transparent white and thick. Exuviae pale brown; larval exuvia with a median longitudinal carina; nymphal exuvia not apparent owing to secretionary covering and uniform colouration of exuvia and secretionary area. Ventral scale thin and poorly developed, remaining adherent to host plant. Diameter 2 mm. Male scale of normal shape, pale brown, covered with a somewhat opaque white secretionary film, which masks pale brown colour of exuvia and secretionary area (Hall, 1929).

HOST PLANTS: **Compositae**: *Vernonia lasiopus* [DeLott1967a], *Ve. podocoma* [Hall1929, Balach1956]. **Euphorbiaceae**: *Cluytia lanceolata* [Balach1956].

DISTRIBUTION: **Afrotropical**: Guinea [Balach1956]; Kenya [Balach1956, DeLott1967a]; Zimbabwe [Hall1929, Balach1956].

GENERAL: Description and illustration of adult female by Hall (1929) and by Balachowsky **(1956)**.

KEYS: Balachowsky 1956: 51 (female) [Africa].

CITATIONS: Balach1955 [taxonomy, host, distribution: 391]; Balach1956 [taxonomy, description, illustration, host, distribution: 80-83]; Borchs1966 [catalogue: 268]; DeLott1967a [host, distribution: 113]; Ferris1941e [taxonomy: 49]; Hall1929 [taxonomy, description, illustration, host, distribution: 350-351].

Aspidiotus watanabei Takagi

Aspidiotus watanabei Takagi, 1969a: 67. Type data: TAIWAN: Fen-chi-hu, on *Viburnum arboricolum*. Holotype female. Type depository: Sapporo: Entomological Institute, Faculty of Agriculture, Hokkaido University, Japan.

Temnaspidiotus watanabei; Chou, 1985: 400. Change of combination.

SYSTEMATICS: Takagi (1969a) noted *Aspidiotus watanabei* as close to *A. destructor*. Williams & Watson (1988) stated "... range of variation of *A. destructor* from the South Pacific area encompasses that of *A. watanabei*...", but did not synonymize latter. Danzig (1993) listed *A. watanabei* as a synonym of *A. destructor*.

SCALE COVER: Takagi (1969a) did not describe scale cover.

HOST PLANTS: **Caprifoliaceae**: *Viburnum arboricolum* [Takagi1969a].

DISTRIBUTION: **Oriental**: Taiwan [Takagi1969a].

GENERAL: Description and illustration of adult female by Takagi (1969a) and by Chou (1985, 1986).

CITATIONS: Chou1985 [taxonomy, description, host, distribution: 400-401]; Chou1986 [taxonomy, illustration: 667]; Takagi1969a [taxonomy, description, illustration, host, distribution: 67-69,99].

Aspidiotus zizyphi Hall
Aspidiotus combreti zizyphi Hall, 1929: 346. Type data: ZIMBABWE: Mazoe, on smaller branches of *Ziziphus jujuba*; Umvukwes, on smaller branches of *Acacia* sp. Syntypes, female. Type depository: London: The Natural History Museum, UK.
Aspidiotus combreti ziziphi; Balachowsky, 1956: 60. Misspelling of species name.
Aspidiotus zizyphi; Borchsenius, 1966: 268. Change of status.
SCALE COVER: Female scale not as white as *Aspidiotus combreti* Hall, 1928; white or dull brown owing to the incorporation of foreign matter (Hall, 1929).
HOST PLANTS: **Leguminosae**: *Acacia* [Hall1929, Balach1956]. **Rhamnaceae**: *Ziziphus jujuba* [Hall1929, Balach1956].
DISTRIBUTION: **Afrotropical**: Zimbabwe [Hall1929, Balach1956].
GENERAL: Description and illustration of adult female by Hall (1929).
CITATIONS: Balach1932f [taxonomy, host, distribution: 230-231]; Balach1956 [taxonomy, description, host, distribution: 60]; Borchs1966 [catalogue: 268]; Hall1929 [taxonomy, description, illustration, host, distribution: 346-347].

Aspidonymus Brimblecombe
Aspidonymus Brimblecombe, 1957: 283. Type species: *Aspidonymus woodwardi* Brimblecombe, by monotypy and original designation.
SYSTEMATICS: The genus *Aspidonymus* resembles *Pseudaonidia* in having a constricted thorax and similarly-shaped pygidial lobes. It differs in having more slender lobes and plates, a smaller and more slender body, more slender ducts and pygidial chitinization much finer (Brimblecombe, 1957).
GENERAL: Definition and characters by Brimblecombe (1957).
CITATIONS: Borchs1966 [catalogue: 240]; Brimbl1957 [taxonomy, description: 283-285]; MorrisMo1966 [taxonomy, taxonomy: 18].

Aspidonymus woodwardi Brimblecombe
Aspidonymus woodwardi Brimblecombe, 1957: 284. Type data: AUSTRALIA: Queensland, Beenleigh, on *Dissilaria baloghioides*; collected May 1956. Holotype female. Type depository: Brisbane: Queensland Museum, Queensland, Australia; type no. T5650.
SCALE COVER: Insects sparse on undersurface of leaves; scale circular, 1.5 mm diameter, pale to light brown; exuviae light yellow to orange (Brimblecombe, 1957).
HOST PLANTS: **Euphorbiaceae**: *Dissilaria baloghioides* [Brimbl1957].
DISTRIBUTION: **Australasian**: Australia (Queensland [Brimbl1957]).
GENERAL: Description and illustration of adult female by Brimblecombe (1957).
CITATIONS: Borchs1966 [catalogue: 240]; Brimbl1957 [taxonomy, description, illustration, host, distribution: 284-285].

Avidovaspis Gerson & Davidson
Avidovaspis Gerson & Davidson, 1974: 159. Type species: *Avidovaspis phoenicis* Gerson & Davidson, by monotypy and original designation.
SYSTEMATICS: *Avidovaspis* is close to *Greenoidea*, *Melanaspis*, *Crenulaspidiotus* and *Pseudomelanaspis,* but differs in having a series of

intermittent marginal ducts and plates on segment V, and unique median paraphyses (Gerson & Davidson, 1974).
GENERAL: Definition and characters by Gerson & Davidson (1974).
CITATIONS: DanzigPe1998 [catalogue: 202]; GersonDa1974 [taxonomy, description: 159]; KosztaBeKo1986 [taxonomy, catalogue: 3].

Avidovaspis phoenicis Gerson & Davidson
Avidovaspis phoenicis Gerson & Davidson, 1974: 159. Type data: EGYPT: Sinai Peninsula, Wadi Feiran, on pinnae of date palm; collected July 17, 1968. Holotype female. Type depository: Bet Dagan: Department of Entomology, The Volcani Center, Israel.
SCALE COVER: Female scale circular, 0.85-0.95 mm in diameter, convex; subcentral exuviae shiny black, surrounded by greyish fluffy film. Male scale elongate about 0.9 mm long, 0.75 mm wide, exuviae subterminal, dark-brown, rest of shield brownish (Gerson & Davidson, 1974).
HOST PLANTS: **Palmae**: *Phoenix dactylifera* [GersonDa1974].
DISTRIBUTION: **Palaearctic**: Egypt [GersonDa1974].
BIOLOGY: Male and female scales occur in large numbers on both sides of date palm pinnae, where they settle along veins (Gerson & Davidson, 1974).
GENERAL: Description and illustration of adult female by Gerson & Davidson (1974).
CITATIONS: DanzigPe1998 [catalogue: 202]; GersonDa1974 [taxonomy, description, illustration, host, distribution: 159-162].

Chentraspis Leonardi

Chentraspis Leonardi, 1897: 284. Type species: *Aspidiotus unilobis* Maskell. Subsequently designated by Fernald, 1903b: 251.
Aspidiotus (*Chentraspis*); Cockerell, 1899a: 395. Change of status.
Neglectaspis Lindinger, 1937: 190. Type species: *Aspidiotus unilobis* Maskell, by monotypy and original designation. Synonymy by Ferris, 1937c: 51.
SYSTEMATICS: *Chentraspis* is related to *Aspidiotus* in absence of pygidial paraphyses, but differs in having median lobes fused into one lobe. Morrison & Morrison (1922: 93) erroneously alleged that only *Aspidiotus unilobis* Maskell was included in *Chentraspis* at the time it was established. Actually *Aspidiotus extensus* Maskell was also assigned there. Lindinger (1937: 181) designated *A. extensus* as type species of *Chentraspis*, and ignored the prior designation by Leonardi (1897). Ferris (1937e: 528), Morrison & Morrison (1966: 34) and Borchsenius (1966: 303) rejected Lindinger's 1937 type selection, and maintained it as *Aspidiotus unilobis* Maskell.
GENERAL: Description and definition by Leonardi (1897b), Morrison & Morrison (1922), Ferris (1937c) and by Brimblecombe (1955).
CITATIONS: BerlesLe1898a [taxonomy: 131]; Borchs1966 [catalogue: 303]; Brimbl1955 [taxonomy: 39-42]; Cocker1897i [taxonomy: 31]; Cocker1899a [taxonomy: 395]; Ferris1937c [taxonomy: 51,53,55]; Ferris1937e [taxonomy: 528]; Ferris1938 [taxonomy: 46]; Ferris1941f [taxonomy, description: 22-23];

Leonar1897 [taxonomy, description: 284-286]; Leonar1897b [taxonomy: 109-111]; Lindin1937 [taxonomy: 181,190]; Lindin1943b [taxonomy: 217]; MacGil1921 [taxonomy, description: 391,434]; MorrisMo1922 [taxonomy, description: 93]; MorrisMo1966 [taxonomy, catalogue: 34,128-129].

Chentraspis uniloba (Maskell)
Aspidiotus unilobis Maskell, 1895b: 40. Type data: AUSTRALIA: New South Wales, Mount Victoria, on *Acacia* sp. Syntypes, female. Type depositories: Auckland: New Zealand Arthropod Collection, Landcare Research, New Zealand, and Washington: United States National Entomological Collection, U.S. National Museum of Natural History, District of Columbia, USA.
Chentraspis uniloba; Leonardi, 1897: 286. Change of combination requiring emendation of species name for agreement in gender.
Aspidiotus (*Chentraspis*) *unilobis*; Cockerell, 1897i: 27. Change of combination.
Neglectaspis unilobis; Lindinger, 1937: 190. Change of combination.
Chentraspis uniloba; Brimblecombe, 1955: 39. Revived combination.
SCALE COVER: Female scale whitish, generally covered by so much dense black fungus that it seems black, and is very difficult to distinguish; form circular, slightly convex; exuviae orange, central, forming a minute boss; diameter about 1/20 inch. Male scale white, elongated, not carinated; length about 1/25 inch (Maskell, 1895b).
HOST PLANTS: **Leguminosae**: *Acacia* [Maskel1895b, Frogga1914]. **Myrtaceae**: *Callistemon salignus* [Brimbl1955], *Callistemon viminalis* [Brimbl1955], *Melaleuca leucadendra* [Green1916e, Brimbl1955].
DISTRIBUTION: **Australasian**: Australia (New South Wales [Maskel1895b, Frogga1914], Northern Territory [Green1916e, Brimbl1955]).
GENERAL: Description and illustration of adult female Maskell (1895b), Morrison & Morrison (1922) and by Brimblecombe (1955).
CITATIONS: Borchs1966 [catalogue: 303]; Brimbl1955 [taxonomy, description, illustration, host, distribution: 39-42]; Cocker1896b [distribution: 335]; Cocker1897i [taxonomy, description, host, distribution: 27]; DeitzTo1980 [taxonomy: 44]; Fernal1903b [catalogue: 280]; Ferris1937c [taxonomy: 51]; Ferris1938 [taxonomy: 43]; Ferris1941e [taxonomy: 49]; Ferris1941f [illustration: 22]; Frogga1914 [taxonomy, description, host, distribution: 319]; Frogga1915 [taxonomy, description, host, distribution: 23]; Green1916e [host, distribution: 53]; Leonar1897 [taxonomy: 286]; Leonar1897b [taxonomy, description, illustration, host, distribution: 111-112]; Lindin1937 [taxonomy: 190]; Lobdel1937 [taxonomy: 78]; MacGil1921 [taxonomy, description, host, distribution: 434]; Maskel1895b [taxonomy, description, illustration, host, distribution: 40]; MorrisMo1922 [taxonomy, description, illustration, host, distribution: 93-96].

Chinaspis Gómez-Menor Ortega

Chinaspis Gómez-Menor Ortega, 1954: 122. Type species: *Chinaspis vellae* Gómez-Menor, by monotypy and original designation.
SYSTEMATICS: This genus is very close to *Lindingaspis* from which it differs in absence of a thoracic tubercle (Gómez-Menor Ortega, 1954).

GENERAL: Description and definition by Gómez-Menor Ortega (1954).
KEYS: Blay Goicoechea 1993: 474 (female) [Spain].
CITATIONS: BlayGo1993 [taxonomy, description: 626]; Borchs1966 [taxonomy, catalogue: 353]; DanzigPe1998 [catalogue: 205]; GomezM1954 [taxonomy, description: 122-125]; MorrisMo1966 [taxonomy, catalogue: 35].

Chinaspis vellae Gómez-Menor Ortega

Chinaspis vellae Gómez-Menor Ortega, 1954: 123. Type data: SPAIN: Madrid province, Aranjuez, on *Vella pseudocytisus*. Lectotype female, by subsequent designation Blay Goicoechea, 1993: 627. Type depository: Madrid: Museo Nacional de Ciencias Naturales, Spain.
Aspidiotus vellae Gómez-Menor Ortega, 1956b: 482. Synonymy by Borchsenius, 1966: 353. Notes: Type data exactly the same as that of *Chinaspis vellae* Gómez-Menor Ortega, 1954.
SYSTEMATICS: Collection data of type series of *Chinaspis vellae* Gómez-Menor Ortega, 1954 and of *Aspidiotus vellae* Gómez-Menor Ortega, 1956b identical.
SCALE COVER: Female scale elliptical, slightly longer than wide, colour creamy white, sometimes slightly pink; convex; 0.8-0.9 mm; exuviae subcentral, white; ventral vellum fine, white (Gómez-Menor Ortega, 1954).
HOST PLANTS: **Cruciferae**: *Vella pseudocytisus* [GomezM1954, Martin1983, BlayGo1993].
DISTRIBUTION: **Palaearctic**: Spain [GomezM1954, GomezM1956b, Martin1983, **BlayGo1993**].
GENERAL: Description and illustration of adult female by Gómez-Menor Ortega (1954).
CITATIONS: BlayGo1993 [taxonomy, description, illustration, host, distribution: 627-631]; Borchs1966 [catalogue: 353-354]; DanzigPe1998 [catalogue: 205]; GomezM1954 [taxonomy, description, illustration, host, distribution: 123-125]; GomezM1956b [taxonomy, description, host, distribution: 482-484]; GomezM1958a [host, distribution: 7]; Martin1983 [taxonomy, host, distribution: 62].

Chortinaspis Ferris

Chortinaspis Ferris, 1938a: 194. Type species: *Aspidiotus chortinus* Ferris, by original designation.
SYSTEMATICS: Genus differs from other genera in Aspidiotina, Aspidiotinae in having long and pointed lateral pygidial plates (Balachowsky, 1958b).
GENERAL: Definition and characters by Ferris (1938a), Borchsenius (1950b), Balachowsky (1948b, 1958b) and by Danzig (1993).
KEYS: Gill 1997: 24-26 (female) [Genera of California]; Danzig 1993: 176 (female) [species Europe]; Tereznikova 1986: 83 (female) [Ukraine]; Chou 1985: 260 (female) [Genera of China]; Chou 1985: 279 (female) [Species of China]; Kosztarab & Kozár 1978: 144-147 (female) [Hungary]; Danzig 1964: 645 (female) [Europe]; Balachowsky 1958b: 230 (female) [*Aspidiotina* of Africa]; McKenzie 1956: 23 (female) [U.S.A.: California]; Balachowsky 1951: 599 (female)

[Mediterranean]; Borchsenius 1950b: 167 (female) [USSR]; Ferris 1942: 28 (female) [North America]; Ferris 1942: 30 (female) [species North America].
CITATIONS: Balach1948b [taxonomy, description: 382-383]; Balach1951 [taxonomy: 599]; Balach1958b [taxonomy, description: 160]; Borchs1949d [taxonomy, description: 194,238]; Borchs1950b [taxonomy, description: 216]; Borchs1966 [catalogue: 279]; Chou1985 [taxonomy, description: 279]; Danzig1964 [taxonomy: 651]; Danzig1993 [taxonomy, description: 175-176]; DanzigPe1998 [catalogue: 212-213]; Ferris1938 [taxonomy: 46]; Ferris1938a [taxonomy, description: 194]; Ferris1938b [taxonomy, description: 65,68]; Ferris1942 [taxonomy: 446:28]; Ferris1946 [taxonomy, description: 37-38]; Gill1997 [taxonomy: 95]; KosztaKo1978 [taxonomy, description: 154]; McKenz1956 [taxonomy: 23]; MorrisMo1966 [taxonomy, catalogue: 35]; Tao1999 [taxonomy: 80].

Chortinaspis biloba (Maskell)
Aspidiotus bilobis Maskell, 1898: 225. Type data: HONG KONG: on grass; collected by Koebele, Num. 1518. Syntypes, female and first instar. Type depository: Auckland: New Zealand Arthropod Collection, Landcare Research, New Zealand.
Hemiberlesia bilobis; Leonardi, 1900: 338. Change of combination.
Chortinaspis bilobis; Ferris, 1946: 38. Change of combination.
Chortinaspis biloba; Borchsenius, 1966: 279. Emendation of species name for agreement in gender.
SCALE COVER: Colour very variable, apparently changing with age. Scale of some specimens above soil surface white, in others almost straw coloured; those beneath soil surface and subjected to staining action of soil and perhaps to fungi black. Scale of the female quite thick, with a thick ventral portion; quite convex, irregularly circular or slightly oval. Scale of male slightly elongate, white or straw coloured (Ferris, 1946).
HOST PLANTS: **Gramineae** [Ferris1946].
DISTRIBUTION: **Oriental**: Hong Kong [Ferris1955c, Takagi1970]; Taiwan [Ferris1955c, Takagi1970]. **Palaearctic**: China (People's Republic) [Takagi1970].
BIOLOGY: At base of leaves, on rootstock and even on roots (Ferris, 1946).
GENERAL: Description and illustration of adult female by Ferris (1946) and by Chou (1985, 1986).
KEYS: Chou 1985: 279 (female) [Species of China]; Ferris 1946: 41 (female) [World].
CITATIONS: Borchs1966 [catalogue: 279]; Chou1985 [taxonomy, description, host, distribution: 279-280]; Chou1986 [taxonomy, illustration: 669]; Cocker1899a [taxonomy: 395]; DeitzTo1980 [taxonomy: 33]; Fernal1903b [catalogue: 253]; Ferris1936 [taxonomy, description, illustration, host, distribution: 9-10]; Ferris1941e [taxonomy: 41]; Ferris1946 [taxonomy, description, illustration, host, distribution: 38-39,45]; Ferris1955c [taxonomy, host, distribution: 32]; Leonar1900 [taxonomy, host, distribution: 338]; Maskel1898 [taxonomy, description, host, distribution: 225]; Takagi1958 [taxonomy: 122]; Takagi1970 [taxonomy, host, distribution: 133]; Tao1999 [taxonomy, host, distribution: 80].

Chortinaspis chortina (Ferris)

Aspidiotus chortinus Ferris, 1921: 123. Type data: MEXICO: Baja California, San Jose del Cabo, on *Chaetochloa caudata*. Holotype female. Type depository: Davis: The Bohart Museum of Entomology, University of California, California, USA.

Epidiaspis chortina; Lindinger, 1932f: 189. Change of combination requiring emendation of species name for agreement in gender.

Chortinaspis chortina; Ferris, 1938a: 195. Change of combination.

Morganella chortinus; Lindinger, 1957: 545. Change of combination.

Chortinaspis chortina; Borchsenius, 1966: 279. Revived combination.

SCALE COVER: Scale of female elongate oval, 1.5-2 mm in diameter; tapering posteriorly, brown, thick and heavy, with a thick ventral scale. Male scale similar in colour but small and slender, with apical exuvia (Ferris, 1921, 1938a).

HOST PLANTS: **Gramineae**: *Chaetochloa caudata* [Ferris1921].

DISTRIBUTION: **Nearctic**: Mexico (Baja California [Ferris1921]).

BIOLOGY: On stems (Ferris, 1938a).

GENERAL: Description and illustration of adult female by Ferris (1921, 1938a).

KEYS: Ferris 1946: 41 (female) [World]; Ferris 1942: 31 (female) [North America].

CITATIONS: Borchs1966 [catalogue: 279]; Ferris1921 [taxonomy, description, illustration, host, distribution: 123-124]; Ferris1938a [taxonomy, description, illustration, host, distribution: 195]; Ferris1938b [taxonomy: 65,68]; Ferris1942 [taxonomy: 446:31]; Ferris1946 [taxonomy, host, distribution: 39,41]; Laing1929a [taxonomy: 487]; Lindin1932f [taxonomy: 189]; Lindin1957 [taxonomy: 545].

Chortinaspis consolidata Ferris

Chortinaspis consolidata Ferris, 1941d: 338. Type data: U.S.A.: California, San Bernandino County, near Old Woman Springs, on a grass *Hilaria rigida*. Holotype female. Type depository: Davis: The Bohart Museum of Entomology, University of California, California, USA.

COMMON NAME: grass scale [McKenz1956].

SCALE COVER: Scale of female white, somewhat convex, oval, with the exuviae toward one end, ventral scale thin. Scale of male white, elongate, with apical exuvia (Ferris, 1941d).

HOST PLANTS: **Gramineae**: *Hilaria* [Ferris1946, McKenz1956], *Hilaria rigida* [Ferris1941d, McKenz1956].

DISTRIBUTION: **Nearctic**: United States of America (Arizona [Ferris1946], California [Ferris1941d, McKenz1956]).

BIOLOGY: Occurring on the stems, concealed beneath the coating of tomentum (Ferris, 1941d).

GENERAL: Description and illustration of adult female by Ferris (1941d), McKenzie (1956) and by Gill (1997).

KEYS: McKenzie 1956: 24 (female) [U.S.A.: California]; Ferris 1946: 41 (female) [World]; Ferris 1942: 30 (female) [North America].

CITATIONS: Borchs1966 [catalogue: 279]; Ferris1941d [taxonomy, description, illustration, host, distribution: 338]; Ferris1942 [taxonomy: 446:30]; Ferris1946 [taxonomy, host, distribution: 39]; Gill1997 [host, distribution, taxonomy, description, illustration, economic importance: 94-95]; McKenz1956 [taxonomy, description, illustration, host, distribution: 50-51]; Nakaha1982 [distribution: 21].

Chortinaspis cottami McDaniel

Chortinaspis cottami McDaniel, 1968: 221. Type data: U.S.A.: Texas, San Patricio Co., on *Stenotaphrum secundatum*. Holotype female. Type depository: Washington: U.S. National Museum of Natural History, District of Columbia, USA.
SCALE COVER: McDaniel (1968) did not describe scale cover.
HOST PLANTS: **Gramineae**: *Stenotaphrum secundatum* [McDani1968].
DISTRIBUTION: **Nearctic**: United States of America (Texas [McDani1968]).
GENERAL: Description and illustration of adult female by McDaniel (1968).
KEYS: McDaniel 1968: 221 (female) [U.S.A.: Texas].
CITATIONS: McDani1968 [taxonomy, description, illustration, host, distribution: 221-223]; Nakaha1982 [host, distribution: 21].

Chortinaspis decorata Ferris

Chortinaspis decorata Ferris, 1952a: 8. Type data: CHINA: Yunnan Province, near Kunming, at An-lin-wen-chian, on undetermined small, perennial grass; collected by G.F. Ferris, April, 29, 1949. Syntypes, female. Type depository: Davis: The Bohart Museum of Entomology, University of California, California, USA.
SCALE COVER: Scale of female black, almost circular, quite convex, rather rough, marked with concentric lines; exuvia to one side; a distinct ventral scale is formed. Scale of male not recognized (Ferris, 1952a).
HOST PLANTS: **Gramineae** [Ferris1952a].
DISTRIBUTION: **Oriental**: China (People's Republic) (Yunnan [Ferris1952a]).
BIOLOGY: Among bases of roots, mostly below level of ground (Ferris, 1952a).
GENERAL: Description and illustration of adult female by Ferris (1952a) and by Chou (1985, 1986).
KEYS: Chou 1985: 279 (female) [Species of China].
CITATIONS: Borchs1966 [catalogue: 279]; Chou1985 [taxonomy, description, host, distribution: 281]; Chou1986 [taxonomy, illustration: 670]; DanzigPe1998 [catalogue: 212]; Ferris1952a [taxonomy, description, illustration, host, distribution: 8,14]; Ferris1955c [taxonomy, host, distribution: 32-33]; Tao1999 [taxonomy, host, distribution: 80].

Chortinaspis divaricata Ferris

Chortinaspis divaricata Ferris, 1946: 39. Type data: U.S.A.: Florida, Lake Gem, on *Eleusine indica*; collected by H.W. Fogg, August, 31, 1923. Syntypes, female. Type depository: Davis: The Bohart Museum of Entomology, University of California, California, USA.
COMMON NAME: wiregrass scale [Dekle1965c].
SCALE COVER: Ferris (1946) indicated that only slide-mounted specimens were available for description.

HOST PLANTS: **Gramineae**: *Aristida* [TippinBe1972, BesheaTiHo1973], *Eleusine indica* [Ferris1946, Merril1953, Dekle1965c].
DISTRIBUTION: **Nearctic**: United States of America (Florida [Ferris1946, Merril1953, TippinBe1972, BesheaTiHo1973], Georgia [BesheaTiHo1973]).
GENERAL: Description and illustration of adult female by Ferris (1946).
KEYS: Ferris 1946: 41 (female) [World].
CITATIONS: BesheaTiHo1973 [host, distribution: 5]; Borchs1966 [catalogue: 279]; Dekle1965c [taxonomy, description, host, distribution: 40]; Dekle1976 [taxonomy, description, host, distribution, economic importance: 57]; Ferris1946 [taxonomy, description, illustration, host, distribution: 39,46]; Merril1953 [taxonomy, description, host, distribution: 34]; Nakaha1982 [host, distribution: 21]; TippinBe1972 [host, distribution: 287].

Chortinaspis fissurella Hall & Williams

Chortinaspis fissurella Hall & Williams, 1962: 38. Type data: PAKISTAN: Murree, on *Imperata cylindrica*. Holotype female. Type depository: London: The Natural History Museum, England, UK.
SCALE COVER: Female scale subcircular, moderately convex, dark brown; exuviae of similar colour, set marginally; ventral scale thin but well developed; diameter of scale about 1.0 mm. Male scale not observed (Hall & Williams, 1962).
HOST PLANTS: **Gramineae**: *Imperata cylindrica* [HallWi1962].
DISTRIBUTION: **Oriental**: Pakistan [HallWi1962].
GENERAL: Description and illustration of adult female by Hall & Williams (1962).
CITATIONS: Borchs1966 [catalogue: 279]; DanzigPe1998 [catalogue: 213]; HallWi1962 [taxonomy, description, illustration, host, distribution: 38-39].

Chortinaspis frankliniana Ferris

Aspidiotus graminellus; Ferris, 1919a: 65. Misidentification; discovered by Ferris, 1938a: 196.
Chortinaspis frankliniana Ferris, 1938a: 196. Type data: U.S.A.: Texas, near El Paso, on Mt. Franklin, on *Hilaria cenchroides*. Holotype female. Type depository: Davis: Bohart Museum of Entomology, University of California, California, USA.
SCALE COVER: Scale of female white, elongate oval, rather thin; exuvia apical. Scale of male elongate, slightly darker than that of female; exuvia apical (Ferris, 1938a).
HOST PLANTS: **Gramineae**: *Hilaria cenchroides* [Ferris1938a, McDani1968].
DISTRIBUTION: **Nearctic**: United States of America (New Mexico [Ferris1938a], Texas [Ferris1938a, McDani1968]).
BIOLOGY: On tomentose stems of host and almost concealed by silvery hairs (Ferris, 1938a).
GENERAL: Description and illustration of adult female by Ferris (1938a).
KEYS: McDaniel 1968: 221 (female) [U.S.A.: Texas]; Ferris 1946: 41 (female) [World]; Ferris 1942: 31 (female) [North America].
CITATIONS: Borchs1966 [catalogue: 279]; Ferris1919a [taxonomy, description, illustration, host, distribution: 65-66]; Ferris1938a [taxonomy, description,

illustration, host, distribution: 196]; Ferris1941e [taxonomy: 44]; Ferris1942 [taxonomy: 446:31]; Ferris1946 [taxonomy, host, distribution: 39-40]; McDani1968 [taxonomy, illustration, host, distribution: 223-224]; Nakaha1982 [host, distribution: 21].

Chortinaspis graminella (Cockerell)

Aspidiotus graminellus Cockerell, 1901d: 333. Type data: U.S.A.: New Mexico, Las Vegas, on leaves of grass. Syntypes, female. Type depository: Washington: U.S. National Museum of Natural History, District of Columbia, USA.

Targionia graminella; Fernald, 1903b: 297. Change of combination requiring emendation of species name for agreement in gender.

Gonaspidiotus graminellus; MacGillivray, 1921: 432. Change of combination.

Chortinaspis graminella; Ferris, 1938a: 197. Change of combination.

SCALE COVER: Scale of female elongate oval, white, quite thin; exuvia apical; male scale white, elongate; distinct purple blotching present associated with this species on *Bouteloua* in Colorado, USA (Ferris, 1938a).

HOST PLANTS: **Gramineae**: *Bouteloua* [Ferris1919a, Ferris1946], *Bouteloua bromoides* [Ferris1946], *Eragrostis* [TippinBe1978], *Monanthochloe* [Ferris1946, McDani1968].

DISTRIBUTION: **Nearctic**: United States of America (Arizona [Ferris1946], Colorado [Ferris1919a, Ferris1946], Florida [Dekle1976], Georgia [Nakaha1982], New Mexico [Cocker1901d, Ferris1946], Texas [Ferris1946, McDani1968]).

BIOLOGY: Occurring on the leaves (Ferris, 1938a).

GENERAL: Description and illustration of adult female by Cockerell (1901d) and by Ferris (1919a, 1938a).

KEYS: McDaniel 1968: 221 (female) [U.S.A.: Texas]; Ferris 1946: 41 (female) [World]; Ferris 1942: 31 (female) [North America]; Cockerell 1905b: 201 (female) [U.S.A.: Colorado].

CITATIONS: Borchs1966 [catalogue: 279-280]; Cocker1901d [taxonomy, description, host, distribution: 333]; Cocker1905b [taxonomy: 201]; Dekle1976 [taxonomy, description, host, distribution, economic importance: 58]; Fernal1903b [catalogue: 297]; Ferris1919a [taxonomy, host, distribution: 68]; Ferris1921b [taxonomy: 94]; Ferris1938a [taxonomy, description, illustration, host, distribution: 197]; Ferris1941e [taxonomy: 44]; Ferris1942 [taxonomy: 446:30]; Ferris1943a [taxonomy: 86]; Ferris1946 [taxonomy, host, distribution: 40]; MacGil1921 [taxonomy, description, host, distribution: 432]; McDani1968 [taxonomy, illustration, host, distribution: 223-224]; Nakaha1982 [host, distribution: 22]; TippinBe1978 [host, distribution: 14]; Willia1985a [taxonomy: 234].

Chortinaspis inyangae Hall

Chortinaspis inyangae Hall, 1941: 222. Type data: ZIMBABWE: Inyanga, Pungwe Falls, on undetermined grass. Syntypes, female. Type depository: London: The Natural History Museum, England, UK.

SCALE COVER: Female scale dark grey, almost black, coated with thin semitransparent greyish secretionary film which shows as narrow pale area marginally and, in some cases, gives a somewhat frosted appearance to dorsal scale;

diameter 1.25-1.50 mm. Male scale resembling that of female in colour and texture but more elongate oval, with exuvia at one end (Hall, 1941).
HOST PLANTS: **Gramineae** [Balach1958b].
DISTRIBUTION: **Afrotropical**: Zimbabwe [Hall1941, Balach1958b].
GENERAL: Description and illustration of adult female by Hall (1941) and by Balachowsky (1958b).
KEYS: Ferris 1946: 41 (female) [World].
CITATIONS: Balach1958b [taxonomy, description, illustration, host, distribution: 159-160]; Borchs1966 [catalogue: 280]; Ferris1946 [taxonomy, host, distribution: 40]; Hall1941 [taxonomy, description, illustration, host, distribution: 222-223].

Chortinaspis iridis Balachowsky

Chortinaspis iridis Balachowsky, 1941: 9. Type data: SYRIA: on *Iris* sp. and ISRAEL: on *Iris* sp. and *Iris nazarina*. Syntypes, female. Type depository: Paris: Muséum national d'Histoire naturelle, France.
Aspidiotus iridis; Lindinger, 1943b: 217. Change of combination.
Chortinaspis iridis; Borchsenius, 1966: 280. Revived combination.
SCALE COVER: Female scale circular, grey; exuviae central, brown; ventral scale not distinct; diameter 1.6 mm. Male scale of similar structure, grey, elongate, 1.3 mm (Balachowsky, 1941, 1948b).
HOST PLANTS: **Iridaceae**: *Iris* [Balach1941, Ferris1946], *Iris nazarina* [Balach1941, Ferris1946].
DISTRIBUTION: **Palaearctic**: Israel [Balach1941, Balach1948b]; Syria [Balach1941, Ferris1946].
BIOLOGY: On subterranean, herbaceous parts but not on rhizomes (Balachowsky, 1941).
GENERAL: Description and illustration of adult female by Balachowsky (1941, 1948b).
KEYS: Danzig 1993: 176 (female) [Europe]; Balachowsky 1948b: 383 (female) [Mediterranean]; Ferris 1946: 41 (female) [World].
CITATIONS: Balach1941 [taxonomy, description, illustration, host, distribution: 9-11]; Balach1948b [taxonomy, description, illustration, host, distribution: 386-388]; Borchs1966 [catalogue: 280]; DanzigPe1998 [catalogue: 213]; Ferris1946 [taxonomy, host, distribution: 40]; Lindin1943b [taxonomy: 217-218].

Chortinaspis salavatiani Balachowsky & Kaussari

Chortinaspis salavatiani Balachowsky & Kaussari, 1951: 2. Type data: IRAN: on *Pennisetum* sp. Syntypes, female. Type depository: Paris: Muséum national d'Histoire naturelle, France.
SCALE COVER: Female scale circular, 1.8-2,1 mm; slightly convex; white, matt; with concentric zones; exuviae central, brown; generally covered with white secretion (Balachowsky & Kaussari, 1951).
HOST PLANTS: **Gramineae**: *Pennisetum* [BalachKa1951].
DISTRIBUTION: **Palaearctic**: Iran [BalachKa1951, Kaussa1955].
GENERAL: Description and illustration of adult female by Balachowsky & Kaussari (1951).

KEYS: Danzig 1993: 176 (female) [Europe].
CITATIONS: BalachKa1951 [taxonomy, description, illustration, host, distribution: 2-3,9]; Borchs1966 [catalogue: 280]; DanzigPe1998 [catalogue: 213]; Kaussa1955 [host, distribution: 15].

Chortinaspis senapirensis Ben-Dov

Chortinaspis senapirensis Ben-Dov, 1976: 205. Type data: EGYPT: Sinai Peninsula, Senapir Island, on *Panicum turgidum*; collected by Y. Ben-Dov, 31.V.1968. Holotype female. Type depository: Bet Dagan: Department of Entomology, The Volcani Center, Israel.
SCALE COVER: Female scale circular, 1.0-1.3 mm in diameter; larval exuviae brownish-yellow placed subcentrally; excreted portion of scale reddish-brown, turning brighter along the margin. Male scale oval in outline (1.2 X 2.0 mm); resembling in colours to female scale (Ben-Dov, 1976).
HOST PLANTS: **Gramineae**: *Panicum turgidum* [BenDov1976].
DISTRIBUTION: **Palaearctic**: Egypt [BenDov1976].
GENERAL: Description and illustration of adult female by Ben-Dov (1976).
KEYS: Danzig 1993: 176 (female) [Europe].
CITATIONS: BenDov1976 [taxonomy, description, illustration, host, distribution: 205-207]; DanzigPe1998 [catalogue: 213].

Chortinaspis subchortina (Laing)

Aspidiotus subchortinus Laing, 1929a: 486. Type data: JAMAICA: Hope Laboratory, on water-grass [*Panicum*] sp. Holotype female. Type depository: London: The Natural History Museum, England, UK.
Chortinaspis subchortina; Ferris, 1941d: 339. Change of combination requiring emendation of species name for agreement in gender.
HOST PLANTS: Female scale pale brown, area over first exuvia white, moderately convex, oval, exuvia to one side; ventral scale thin. Scale of male somewhat elongate, with exuvia at one end; colour as in female (Ferris, 1941d).
HOST PLANTS: **Gramineae**: *Panicum* [Laing1929a, Ferris1941d, McDani1968].
DISTRIBUTION: **Australasian**: Hawaiian Islands (Hawaii [Nakaha1982]).
Nearctic: Mexico (Morelos [Ferris1942], Veracruz [Ferris1942]); United States of America (Arizona [Nakaha1982], Florida [Dekle1976], Mississippi [Nakaha1982], Texas [McDani1968]). **Neotropical**: Bahamas [Nakaha1982]; Colombia [Nakaha1982]; Jamaica [Laing1929a, Ferris1946, Nakaha1982]; Panama [Ferris1941d] [Nakaha1982]; Peru [Nakaha1982].
BIOLOGY: Occurring in the material at hand on the smaller stems of the host (Ferris, 1941d).
GENERAL: Description and illustration of adult female by Laing (1929a) and by Ferris (1941d).
KEYS: McDaniel 1968: 221 (female) [U.S.A.: Texas]; Ferris 1946: 41 (female) [World]; Ferris 1942: 30 (female) [North America].
CITATIONS: Borchs1966 [catalogue: 280]; Dekle1976 [taxonomy, description, host, distribution, economic importance: 59]; Ferris1941d [taxonomy, description, illustration, host, distribution: 339]; Ferris1941e [taxonomy: 48]; Ferris1942

[taxonomy, host, distribution: 446:11;446:30]; Ferris1946 [taxonomy, host, distribution: 40]; Laing1929a [taxonomy, description, illustration, host, distribution: 486-487]; McDani1968 [taxonomy, illustration, host, distribution: 223-226]; Nakaha1982 [host, distribution: 22]; RiherdCh1952 [host, distribution, economic importance, control: 1-5].

Chortinaspis subterranea (Lindinger)

Epidiaspis subterranea Lindinger, 1912b: 174. Type data: FRANCE: Montpellier, on grass. Syntypes, female. Type depository: Hamburg: Zoologisches Institut und Zoologishces Museum, Universität von Hamburg, Germany.

Aspidiotus (*Hemiberlesia*) *provincialis* Vayssière, 1914a: 207. Type data: FRANCE: Bouches-du-Rhone, Carry-le-Rouet, on a common grass, probably *Psamma arenaria*. Syntypes, female. Type depository: Paris: Muséum national d'Histoire naturelle, France. Synonymy by Ferris, 1943a: 38.

Hemiberlesia provincialis; Sasscer, 1915: 35. Change of combination.

Hemiberlesia subterranea; Leonardi, 1918: 192. Change of combination.

Aspidiotus subterraneus; Lindinger, 1935: 134. Change of combination requiring emendation of species name for agreement in gender.

Chortinaspis subterraneus; Balachowsky, 1941: 11. Change of combination.

Separaspis subterraneus; Lupo, 1954: 2. Change of combination.

Chortinaspis subterranea; Borchsenius, 1966: 280. Revived combination.

SCALE COVER: Female scale convex, circular or slightly elongate, 1.5 mm in diameter, black brown with darker center; exuvia subcentral, sometimes central; brown yellow; ventral scale grey or yellow brown (Lindinger, 1912b).

HOST PLANTS: **Gramineae** [Balach1932e], *Agropyrum* [Balach1932d], *Agropyrum intermedium* [Leonar1918, Leonar1920, Ferris1946, Balach1948b, Lupo1954], *Agropyrum repens* [Leonar1918, Leonar1920, Borchs1936, Ferris1946, Balach1948b, Bachma1953, Lupo1954], *Ammophila arenaria* [Vayssi1914a, Balach1948b, Lupo1954, Foldi2000], *Festuca* [Ferris1946]. **Zingiberaceae**: *Curcuma longa* [Takaha1942b].

DISTRIBUTION: **Oriental**: Thailand [Takaha1942b]. **Palaearctic**: France [Vayssi1914a, Balach1932d, Balach1932e, Lupo1954, Foldi2000]; Georgia (Georgia [Borchs1936, Ferris1946]); Italy [Leonar1918, Leonar1920, Ferris1946, LongoMaPe1995]; Ukraine [Ferris1946] (Krym (= Crimea) Oblast [Ferris1946]); Yugoslavia [Ferris1946, Bachma1953, Lupo1954].

GENERAL: Description and illustration of adult female by Ferris (1946), Balachowsky (1948b), Lupo (1954), Tereznikova (1986) and by Danzig (1993).

KEYS: Danzig 1993: 176 (female) [Europe]; Kosztarab & Kozár 1978: 155 (female) [Hungary]; Balachowsky 1948b: 383 (female) [Mediterranean]; Ferris 1946: 41 (female) [World]; Leonardi 1920: 90 (female) [Italy].

CITATIONS: Bachma1953 [host, distribution: 177]; Balach1932d [taxonomy, host, distribution: XLVII]; Balach1932e [taxonomy, host, distribution: 236]; Balach1941 [taxonomy: 11]; Balach1948b [taxonomy, description, illustration, host, distribution: 383-386]; Borchs1935a [taxonomy, description, host, distribution: 35]; Borchs1936 [host, distribution: 135]; Borchs1937 [taxonomy, description, illustration, host, distribution: 123]; Borchs1937a [taxonomy, description, host, distribution: 61,62];

Borchs1939 [taxonomy, description, host, distribution: 9,25]; Borchs1950b [taxonomy, description, host, distribution: 216-217]; Borchs1966 [catalogue: 280]; Danzig1964 [taxonomy, host, distribution: 651]; Danzig1972 [taxonomy, host, distribution, economic importance: 208]; Danzig1993 [taxonomy, description, illustration, host, distribution: 176-177]; DanzigPe1998 [catalogue: 213-214]; Ferris1941e [taxonomy: 47]; Ferris1946 [taxonomy, description, illustration, host, distribution: 40-41,47]; Foldi1990 [structure: 43-54]; Foldi2000 [host, distribution: 84]; KosztaKo1978 [taxonomy, description, host, distribution: 154]; Kozar1990a [life history, economic importance: 341-347]; Leonar1918 [taxonomy, host, distribution: 192]; Leonar1920 [taxonomy, description, illustration, host, distribution: 94-95]; Lindin1912b [taxonomy, description, host, distribution: 173-174]; Lindin1935 [taxonomy: 134]; LongoMaPe1995 [distribution: 126]; Lupo1954 [taxonomy, description, illustration, host, distribution: 2-7]; MacGil1921 [taxonomy, description, host, distribution: 436]; MillerDa1990 [host, distribution, economic importance: 301]; Sassce1915 [taxonomy, host, distribution: 35]; Takaha1942b [host, distribution: 47]; Terezn1986 [taxonomy, description, illustration, host, distribution: 92-93]; Vayssi1914a [taxonomy, description, host, distribution: 207-208]; Vayssi1915 [taxonomy, description, host, distribution: 298]; WeidneWa1968 [taxonomy: 175].

Chrysomphalus Ashmead

Chrysomphalus Ashmead, 1880: 267. Type species: *Chrysomphalus ficus* Ashmead, by monotypy.
Chrysamphalus; Chou, 1985: 283. Misspelling of genus name.
Chrisomphalus; Yasnosh, 1995: 248. Misspelling of genus name.
SYSTEMATICS: *Chrysomphalus* Ashmead closely related to *Aonidiella*, but differs from latter, in the cephalothorax never being reniform, and prepygidial lobes not curving posteriorly to pygidium; furthermore, paraphyses in *Chrysomphalus* as long as or longer than lobes, but much shorter in *Aonidiella* (Takagi, 1969a; Williams & Watson, 1988).
GENERAL: Definition and characters by Dietz & Morrison (1916a), Robinson (1917), Fullaway (1932), Kuwana (1933), Ferris (1938a), Lupo (1953), Balachowsky (1948b, 1956), Borchsenius (1950b), Gómez-Menor Guerrero (1962), Almeida (1969), Takagi (1969a), Velasquez (1971), Bazarov & Shmelev (1971), Williams & Watson (1988), Danzig (1993), Yaşar (1995) and by Kosztarab (1996).
KEYS: Colon-Ferrer & Medina-Gaud 1998: 28-32 (female) [Genera of Puerto Rico]; Gill 1997: 24-26 (female) [Genera of California]; Gill 1997: 96 (female) [Species of California]; Kosztarab 1996: 406-407 (female) [Northeastern North America]; Blay Goicoechea 1993: 473 (female) [Spain]; Danzig 1993: 164 (female) [species Europe]; Wolff & Corseuil 1993: 29 (female) [Brazil, Rio Grande do Sul]; Zahradník 1990b: 74 (female) [Czech Republic]; Williams & Watson 1988: 20 (female) [Tropical South Pacific]; Tereznikova 1986: 83 (female) [Ukraine]; Chou 1985: 283 (female) [Genera of China]; Chou 1985: 284 (female) [Species of China]; Paik 1978: 311 (female) [species South Korea]; Bazarov & Shmelev 1971: 186 (female) [Central Asia]; Velasquez 1971: 131 (female) [Philippines]; Beardsley

1966: 502-504 (female) [Federated States of Micronesia]; Danzig 1964: 645 (female) [Europe]; Gómez-Menor Guerrero 1962: 157 (female) [Canary Islands]; Zahradník 1959a: 548 (female) [Czech Republic]; Balachowsky 1958b: 230 (female) [*Aspidiotina* of Africa]; Ezzat 1958: 237-239 (female) [Egypt]; Gómez-Menor Ortega 1956: 7-8 (female) [Spain]; McKenzie 1956: 22-23 (female) [U.S.A.: California]; Balachowsky 1951: 600 (female) [Mediterranean]; Borchsenius 1950b: 167 (female) [USSR]; Zimmerman 1948: 351 (female) [Hawaii]; Gómez-Menor Ortega 1946: 59-61 (female) [Spain]; Ruiz Castro 1944: 57 (female) [Spain]; Ferris 1942: 27 (female) [North America]; Ferris 1942: 31 (female) [species North America]; Archangelskaya 1937: 94 (female) [Middle Asia]; Borchsenius 1937: 99 (female) [USSR]; Borchsenius 1937a: 32-33 (female) [Palaearctic Region]; Kuwana 1933a: 43-45 (female) [Japan]; Fullaway 1932: 97-98 (female) [Hawaii]; Archangelskaya 1929: 189 (female) [Palaearctic Region]; Balachowsky 1928b: 157 (female) [Africa]; Britton 1923: 360 (female) [U.S.A.: Connecticut]; Hollinger 1923: 6-7 (female) [U.S.A.: Missouri]; Leonardi 1920: 26 (female) [Italy]; Leonardi 1920: 65 (female) [Species of Italy]; Brain 1919: 198 (female) [World]; Lawson 1917: 206 (female) [U.S.A.: Kansas]; Lawson 1917: 210 (female) [species U.S.A.: Kansas]; Robinson 1917: 16-17 (female) [Philippines]; Robinson 1917: 24 (female) [species Philippines]; Dietz & Morrison 1916a: 263 (female) [U.S.A.: Indiana]; Lindinger 1913: 64 (female) [Africa].

CITATIONS: Almeid1969 [taxonomy: 154-156]; Archan1929 [taxonomy: 189]; Archan1937 [taxonomy, description: 94]; Ashmea1880 [taxonomy, description: 267]; Balach1928b [taxonomy: 157]; Balach1948b [taxonomy, description: 345-346]; Balach1951 [taxonomy: 590,600]; Balach1956 [taxonomy, description: 82-85]; Balach1958b [taxonomy: 230]; BazaroSh1971 [taxonomy, description: 193]; Beards1966 [host, distribution: 515]; BenDov1990h [taxonomy: 81]; BerlesLe1896 [taxonomy, description: 347]; BerlesLe1898a [taxonomy: 131]; BlayGo1993 [taxonomy, description: 502]; Bodenh1924 [taxonomy: 21]; Bodenh1949 [taxonomy, description: 26,37]; Bodenh1952 [taxonomy: 329]; Borchs1937 [taxonomy, description: 94]; Borchs1937a [taxonomy, description: 32,48]; Borchs1949d [taxonomy, description: 194,231]; Borchs1950b [taxonomy, description: 167,217]; Borchs1966 [taxonomy, catalogue: 283-284]; Brain1918 [taxonomy: 116]; Brain1919 [taxonomy: 166,198]; Brimbl1962a [taxonomy: 411-412]; Britto1923 [taxonomy, description: 360,376]; Bustsh1958 [taxonomy: 229]; Chou1947 [taxonomy, description: 9-24]; Chou1985 [taxonomy, description: 283-284]; Cocker1897i [taxonomy: 9,12,31]; Cocker1899a [taxonomy: 396]; Cocker1899n [taxonomy: 25]; Cocker1905b [taxonomy: 200]; ColonFMe1998 [taxonomy, description: 48]; Danzig1964 [taxonomy: 651]; Danzig1993 [taxonomy, description: 163-164]; DanzigPe1998 [catalogue: 214]; DietzMo1916a [taxonomy, description: 263,306]; Ezzat1958 [taxonomy: 239]; Fernal1903b [catalogue: 285]; Ferris1937c [taxonomy: 50,53,54,64]; Ferris1937d [taxonomy: 105]; Ferris1938a [taxonomy, description: 198]; Ferris1942 [taxonomy: 446:27]; Fullaw1932 [taxonomy, description: 97,107]; Ghauri1962 [taxonomy: 210]; Gill1997 [taxonomy: 95]; GomezM1937 [taxonomy, description: 43,88-89]; GomezM1946 [taxonomy: 60]; GomezM1956 [taxonomy, description: 31-32]; GomezM1962 [taxonomy, description: 179]; Gowdey1921 [taxonomy: 31]; Hadzib1983 [taxonomy: 223-224]; Hempel1920 [taxonomy: 140]; Hollin1923 [taxonomy: 7,67,68]; HosnyEz1957 [taxonomy: 332]; Kawai1980 [taxonomy: 209]; Koszta1996 [taxonomy, description: 477]; Kuwana1933 [taxonomy, description: 26,43]; Lawson1917 [taxonomy, description: 206,209]; Lawson1917 [taxonomy, description: 209]; Leonar1897 [taxonomy: 284]; Leonar1897a [taxonomy: 375]; Leonar1897b [taxonomy: 111]; Leonar1899 [taxonomy: 198]; Leonar1920 [taxonomy, description: 26,64-65]; Lepage1938 [taxonomy: 398]; Lepesm1947 [taxonomy, description: 196]; Lindin1908b [taxonomy: 98]; Lindin1910a [taxonomy: 440]; Lindin1910b [taxonomy: 39]; Lindin1911 [taxonomy: 355]; Lindin1924 [taxonomy: 171]; Lindin1937 [taxonomy: 182]; Lupo1953 [taxonomy, description: 23]; MacGil1921 [taxonomy: 388,414-421]; Mamet1949 [taxonomy: 55-56]; McKenz1939 [taxonomy, description: 51-53]; McKenz1943 [taxonomy: 148]; McKenz1956 [taxonomy: 23]; MorrisMo1966 [taxonomy, catalogue: 37]; Nel1933 [taxonomy: 417-419]; Robins1917 [taxonomy, description: 16,23]; RuizCa1944 [taxonomy: 57]; Savesc1982 [taxonomy, description: 304]; Schmut1959 [taxonomy, description: 47,53]; Takagi1969a [taxonomy, description: 85-86]; Tao1999 [taxonomy: 80]; ThiemGe1934a [taxonomy: 232]; Velasq1971 [taxonomy, description: 131]; WilliaWa1988 [taxonomy, description: 90,93]; WolffCo1993 [taxonomy: 29]; Yasar1995a [taxonomy, description: 57-58]; Yasnos1995 [taxonomy: 248]; Zimmer1948 [taxonomy: 351,368].

Chrysomphalus aberrans Mamet

Chrysomphalus aberrans Mamet, 1951: 245. Type data: MADAGASCAR: North Tamatave, Tampolo, on lower surface of leaves of tea plant. Holotype. Type depository: Paris: Muséum national d'Histoire naturelle, France.
SCALE COVER: Female scale very small, circular, more or less conical, dark reddish-brown to rust brown in colour, sometimes with a clearer marginal zone. Male scale not observed (Mamet, 1951).
HOST PLANTS: Theaceae: *Thea* [Mamet1951, Borchs1966].
DISTRIBUTION: Afrotropical: Madagascar [Mamet1951].
GENERAL: Description and illustration of adult female by Mamet (1951).
CITATIONS: Borchs1966 [catalogue: 284]; Mamet1951 [taxonomy, description, illustration, host, distribution: 226,245-246].

Chrysomphalus ansei (Green)

Aspidiotus (*Chrysomphalus*) *ansei* Green, 1916f: 193. Type data: SEYCHELLES: Anse aux Pins, crowded on fronds of *Cocos nucifera*. Syntypes, female. Type depository: London: The Natural History Museum, England, UK.
Chrysomphalus ansei; MacGillivray, 1921: 415. Change of combination.
Chrysomphalus anseyi; Lepesme, 1947: 203. Misspelling of species name.
SCALE COVER: Female scale irregularly circular or broadly ovate, diameter averaging 1.45 mm; flattish or moderately convex; very pale brownish ochreous, semitransparent; exuviae darker, central. Male scale smaller and more distinctly ovate; exuviae nearer one extremity; length 1 mm (Green, 1916f). Scale of female thin and transparent, yellow brown, with exuviae darker brown; that of male similar in colour, somewhat elongate, exuvia toward one end (McKenzie, 1939).

HOST PLANTS: **Lauraceae**: *Litsea glutinosa* [Mamet1943a, Borchs1966].
Palmae: *Cocos nucifera* [Green1916f, McKenz1939, Mamet1943a, Borchs1966].
DISTRIBUTION: **Afrotropical**: Seychelles [Green1916f, McKenz1939, Mamet1943a, Borchs1966].
GENERAL: Description and illustration of adult female by Green (1916f) and by McKenzie (1939).
KEYS: McKenzie 1943: 150 (female) [World]; McKenzie 1939: 64 (female) [World].
CITATIONS: Borchs1966 [catalogue: 284]; Dupont1931 [host, distribution: 1-18]; Ferris1941e [taxonomy: 40]; Green1916f [taxonomy, description, host, distribution: 193]; Lepesm1947 [host, distribution, taxonomy: 203]; Lindin1943b [taxonomy: 207]; MacGil1921 [taxonomy, description, host, distribution: 415]; Mamet1943a [catalogue: 157]; McKenz1939 [taxonomy, description, illustration, host, distribution: 53,56-57,68]; McKenz1943 [taxonomy: 150]; MillerDa1990 [host, distribution, economic importance: 301]; SchmutKlLu1957 [distribution, economic importance: 478]; VeseyF1953 [host, distribution, biological control: 405-413].

Chrysomphalus aonidum (Linnaeus)

Coccus aonidum Linnaeus, 1758: 455. Type data: ASIA: on perennial fruit trees and on *Camellia*. Syntypes. Type depository: London: The Linnean Society of London, England.

Chrysomphalus ficus Ashmead, 1880: 267. Type data: U.S.A.: Florida, Orange Co., Orlando, on *Ficus nitida*. Syntypes, female. Type depository: Washington: United States National Entomological Collection, U.S. National Museum of Natural History, District of Columbia, USA. Synonymy by Cockerell, 1899n: 25.

Aspidiotus ficus; Comstock, 1881a: 296. Notes: Incorrect citation of "Riley MSS" as author.

Aspidiotus ficus; Comstock, 1883: 61. Notes: Incorrect citation of Riley as author.

Aspidiotus ficus; Morgan, 1889a: 350. Notes: Incorrect citation of "Riley" as author.

Aspidiotus (*Chrysomphalus*) *ficus*; Berlese, 1895a: 83. Change of combination.

Aspidiotus (*Chrysomphalus*) *ficus*; Grandpre & Charmoy, 1899: 25. Notes: Incorrect citation of "Riley" as author.

Chrysomphalus aonidum; Cockerell, 1899n: 25. Change of combination.

Aspidiotus (*Chrysomphalus*) *aonidum*; Hempel, 1900a: 502. Change of combination.

Aspidiotus aonidum; Cockerell, 1905: 46. Change of combination.

Chrysomphalus adonidum; Ferris & Kelly, 1923: 318. Misspelling of species name.

Chrysomphalus aonidum; McKenzie, 1939: 53. Revived combination.

COMMON NAMES: circular black scale [Brimbl1962]; citrus black scale [Bodenh1951a]; cochonilha-purpura [CarvalAg1997]; Egyptian black scale [Bodenh1951a]; escama roja de Florida [CoronaRuMo1997]; Florida red scale [Merril1953, McKenz1956, Dekle1965c, Koszta1996]; la cochenille de l'oranger [PicartMa2000]; pou de Floride [SchmutKlLu1957].

SCALE COVER: Scale of female flat, circular, somewhat variable in colour but tending to be quite dark; centrally placed exuviae somewhat paler than other parts; scale of male somewhat elongate oval, exuvia near one end (Ferris, 1938a). Colour photograph by Carvalho & Aguiar (1997), Gill (1997) and by Wong *et al.* (1999).

HOST PLANTS: **Agavaceae**: *Cordyline* [McKenz1956], *Furcraea gigantea* [Mamet1943a, Mamet1949, Borchs1966], *Sansevieria* [McKenz1956, Cohic1958, WilliaWa1988]. **Anacardiaceae**: *Anacardium occidentale* [DeLott1967a], *Mangifera indica* [Takaha1933, Lepage1938, Mamet1943a, Mamet1949, McKenz1956, Borchs1966, Ali1967a], *Protorhus thouarsii* [Mamet1954, Borchs1966], *Sclerocarya caffra* [Mamet1959a], *Spondias lutea* [Morgan1889a, Lepage1938]. **Annonaceae**: *Annona* [Lepage1938, Balach1948b], *Annona muricata* [Mamet1943a, Mamet1949, Borchs1966, Cohic1958, WilliaWa1988], *Annona reticulata* [Mamet1956, Borchs1966, Cohic1958, WilliaWa1988], *Annona squamosa* [Hall1922, Cohic1958, WilliaWa1988], *Artabotrys odoratissima* [Leonar1920, Takaha1929]. **Apocynaceae**: *Acokanthera* [Hall1923], *Carissa bispinosa* [MerrilCh1923, Balach1932d], *Carissa carandas* [Takaha1929], *Carissa edulis* [Hall1923], *Nerium* [Lepage1938], *Nerium oleander* [Balach1927, Balach1932d, McKenz1956, GonzalCh1968, WilliaWa1988], *Ochrosia elliptica* [Merril1953], *Plumeria* [WilliaWa1988], *Plumeria acutifolia* [Mamet1943a, Mamet1949, Borchs1966], *Plumeria rubra* [WilliaWa1988], *Rhynchospermum* [MerrilCh1923], *Tabernaemontana* [MerrilCh1923], *Thevetia* [Takaha1929], *Trachelospermum* [Merril1953], *Vinca major* [Merril1953]. **Aquifoliaceae**: *Ilex* [Green1937, Lepage1938, Dekle1965c], *Ilex cornuta* [McKenz1956], *Ilex latifolia* [Kuwana1902]. **Araceae**: *Anthurium* [Zimmer1948], *Pothos aureus* [Merril1953, McKenz1956]. **Araliaceae**: *Aralia* [MerrilCh1923, Balach1948b], *Aralia papyrifera* [Balach1932d], *Hedera* [Lepage1938], *Hedera helix* [McKenz1956, GomezM1962, BesheaTiHo1973], *Schefflera* [Merril1953, BesheaTiHo1973], *Trevesia palmata* [Merril1953]. **Araucariaceae**: *Agathis lanceolata* [WilliaWa1988], *Agathis moorei* [WilliaWa1988], *Araucaria bidwelli* [MerrilCh1923]. **Arecaceae**: *Latania commersonii* [Cohic1958, WilliaWa1988]. **Begoniaceae**: *Begonia* [Hall1923, Lepage1938], *Begonia magnifica* [Balach1948b]. **Berberidaceae**: *Mahonia japonica* [Takaha1929]. **Bischofiaceae**: *Bischoffia* [Merril1953]. **Brexiaceae**: *Brexia* [Balach1932d]. **Buxaceae**: *Buxus japonica* [Takaha1929]. **Cactaceae**: *Opuntia* [Balach1932d]. **Calycanthaceae**: *Calycanthus* [Merril1953], *Chimonanthus praecox* [Merril1953]. **Caprifoliaceae**: *Viburnum* [Merril1953, McKenz1956]. **Caryophyllaceae**: *Dianthus caryophyllus* [Hall1923]. **Celastraceae**: *Elaeodendron* [Mamet1949, Borchs1966], *Euonymus* [McKenz1956, McDani1968], *Euonymus japonicum* [MerrilCh1923, Balach1927, Balach1932d, Kuwana1933]. **Combretaceae**: *Terminalia* [Almeid1971]. **Compositae**: *Calendula officinalis* [Hall1923], *Calostemma* [Hall1923], *Gerbera* [WilliaWa1988], *Gerbera jamesoni* [Cohic1958, WilliaWa1988]. **Crassulaceae**: *Cotyledon orbiculata* [Mamet1954, Borchs1966]. **Cycadaceae**: *Cycas* [MerrilCh1923, Lepage1938, Mamet1959a, Matile1978, WilliaWa1988], *Cycas circinalis* [GomezM1941], *Cycas revoluta* [Houser1918, Ferris1921a, Takaha1929, Mamet1943a, Mamet1949, Borchs1966], *Cycas thouarsi* [Mamet1954, Borchs1966]. **Cyclanthaceae**: *Carludovica palmata* [GomezM1941]. **Ebenaceae**: *Diospyros kaki* [Hall1922]. **Elaeagnaceae**: *Elaeagnus* [MerrilCh1923]. **Epacridaceae**: *Leucopogon* [Cohic1958, WilliaWa1988]. **Ericaceae**: *Rhododendron* [Ramakr1921a, Ali1967a], *Rhododendron arboreum* [Green1896e, Green1900a, Ramakr1919a, Green1937, Lepage1938]. **Euphorbiaceae**: *Aleurites moluccana* [GomezM1941, Zimmer1948, Cohic1958,

WilliaWa1988], *Croton* [Green1904a], *Euphorbia* [WilliaWa1988], *Phyllanthus* [Hall1922], *Poinsettia* [Hall1923], *Ricinus communis* [Balach1927, Balach1932d, WilliaWa1988]. **Fagaceae**: *Nothofagus aequilateralis* [WilliaWa1988], *Quercus robur* [Hall1923]. **Flacourtiaceae**: *Hydnocarpus wightiana* [Mamet1943a, Mamet1949, Borchs1966]. **Gramineae**: *Bambusa* [Green1937]. **Guttiferae**: *Calophyllum coloba* [Houser1918], *Calophyllum inophyllum* [Takaha1929, WilliaWa1988], *Garcinia* [Ali1967a], *Garcinia spicata* [Takaha1929, Takaha1955f], *Gardenia* [McKenz1956, Cohic1958, WilliaWa1988], *Mammea americana* [Houser1918]. **Heliconiaceae**: *Heliconia* [WilliaWa1988]. **Illiciaceae**: *Illicium floridanum* [Merril1953]. **Iridaceae**: *Belamcanda* [Merril1953], *Gladiolus* [Cohic1958, WilliaWa1988], *Gladiolus illyricus* [Mamet1943a, Mamet1949, Borchs1966], *Moraea iridioides* [Merril1953]. **Lauraceae**: *Beilschmiedia elliptica* [RosenDe1979], *Cinnamomum camphora* [Takaha1929], *Laurus* [Lepage1938], *Laurus camphora* [Mamet1954, Borchs1966], *Laurus nobilis* [Balach1932d, Balach1948b, Cohic1958, WilliaWa1988], *Machilus thunbergii* [Kuwana1902, Takaha1929], *Persea americana* [McKenz1956, GersonZo1973], *Persea gratissima* [Hall1922, Balach1932d]. **Lecythidaceae**: *Barringtonia asiatica* [Cohic1958, WilliaWa1988], *Barringtonia asiatica* [Cohic1958, WilliaWa1988]. **Leguminosae**: *Acacia arabica* [Hall1923], *Acacia decurrens* [Hall1922], *Acacia longifolia* [Bodenh1924], *Acacia melanoxylon* [RosenDe1979], *Bauhinia* [Hall1922], *Bauhinia purpurea* [McDani1968], *Bauhinia speciosa* [GomezM1941], *Bauhinia variegata* [Cohic1958, WilliaWa1988], *Cassia auriculata* [RahmanAn1941], *Cassia occidentalis* [RahmanAn1941], *Castanospermum australis* [Frogga1914], *Ceratonia siliqua* [Hall1922], *Erythrina crista-galli* [McKenz1956], *Robinia* [McKenz1956], *Ruppelia grata* [Leonar1920], *Tamarindus indica* [GomezM1941, Cohic1958, WilliaWa1988]. **Liliaceae**: *Aloe* [Hall1922], *Asparagus plumosus* [Kuwana1902], *Aspidistra* [McKenz1956, Cohic1958, WilliaWa1988], *Aspidistra lurida* [Kuwana1902, MerrilCh1923], *Danielia?* [Mamet1959a], *Dracaena draco* [Balach1948b], *Dracaena marginata* [PicartMa2000], *Mondo* [McKenz1956]. **Loganiaceae**: *Fagraea* [WilliaWa1988]. **Lythraceae**: *Lawsonia alba* [Hall1922]. **Magnoliaceae**: *Magnolia grandiflora* [Houser1918, Hall1923]. **Malvaceae**: *Gossypium* [Hall1922], *Hibiscus mutabilis* [GomezM1941], *Hibiscus rosa* [Hall1923], *Hibiscus rosa sinensis* [GomezM1941]. **Meliaceae**: *Azadirachta indica* [Schmut1998], *Cedrela odorata* [MerrilCh1923]. **Moraceae**: *Artocarpus altilis* [WilliaWa1988], *Artocarpus communis* [Mamet1943a, Mamet1949, Borchs1966, Ali1967a], *Ficus* [Hall1922, MerrilCh1923, Balach1927, Lepage1938, Cohic1958, Ali1967a, WilliaWa1988], *Ficus carica* [Hall1923, GomezM1962], *Ficus doescherii* [McKenz1956], *Ficus elastica* [Hall1923, Balach1932d, Kuwana1933, McKenz1956], *Ficus foveolata* [Kuwana1933], *Ficus macrophylla* [Balach1948b], *Ficus morica* [FerrisKe1923], *Ficus nitida* [Ashmea1880, Balach1932d, GomezM1941, Balach1948b], *Ficus pumila* [WilliaWa1988], *Ficus religiosa* [Houser1918, Hall1923, GomezM1941], *Ficus retusa* [Takaha1929, Kuwana1933, HabibEzAt1960], *Ficus sycomorus* [Hall1922], *Morus* [Hall1923]. **Musaceae**: *Musa* [Lepage1938, Balach1948b, WilliaWa1988], *Musa cavendishi* [Balach1932d], *Musa paradisiaca* [GomezM1941], *Musa sapientum* [Bodenh1924, Takaha1929, Mamet1949, Cohic1958, Borchs1966, WilliaWa1988]. **Myristicaceae**: *Myristica* *heterophylla* [Takaha1929]. **Myrtaceae**: *Decaspermum fruticosum* [Takaha1934], *Eucalyptus* [MerrilCh1923, Balach1948b], *Eucalyptus globolus* [Bodenh1924], *Eugenia* [Ramakr1919a, Ramakr1921a, MerrilCh1923, Lepage1938, Ali1967a, WilliaWa1988], *Eugenia cumini* [WilliaWa1988], *Eugenia jambolana* [RahmanAn1941, Zimmer1948], *Eugenia jambusa* [Hall1922], *Eugenia vaccinifolia* [Mamet1954, Borchs1966], *Feijoa* [Merril1953], *Melaleuca* [WilliaWa1988], *Melaleuca leucadendron* [WilliaWa1988], *Melaleuca leucadendron viridiflora* [Cohic1958], *Melaleuca quinquenervia* [WilliaWa1988], *Myrtus* [Merril1953], *Myrtus communis* [Hall1922, Martin1983], *Myrtus hilli* [Frogga1914], *Psidium* [Lepage1938], *Psidium guajava* [DoaneHa1909, Balach1932d]. **Naucleaceae**: *Adina cordifolia* [Hall1922]. **Nyctaginaceae**: *Bougainvillea* [WilliaWa1988]. **Oleaceae**: *Jasminum* [Hall1922, Merril1953, McKenz1956, WilliaWa1988], *Jasminum humila* [MerrilCh1923], *Jasminum sambac* [Cohic1958, WilliaWa1988], *Ligustrum* [MerrilCh1923], *Ligustrum japonicum* [Kuwana1902, Hall1923, Takaha1929], *Olea* [Hall1922], *Olea europaea* [Bodenh1924, Takaha1932a, Lepage1938], *Osmanthus* [Merril1953], *Osmanthus fragrans* [MerrilCh1923, Takaha1929, Kuwana1933, McKenz1956]. **Onagraceae**: *Ludwigia octovalvis* [WilliaWa1988]. **Orchidaceae**: *Coelogyne cristata* [Morgan1889a], *Cypripedium* [McKenz1956], *Dendrobium* [Zimmer1948], *Odontoglossum* [McKenz1956], *Oncidium* [GomezM1941], *Vanilla* [WilliaWa1988]. **Palmae** [Green1930b, Mamet1959a, Almeid1971], *Areca* [McKenz1956], *Areca catechu* [Green1908a, Takaha1929, Ali1967a], *Chamaeodorea* [MerrilCh1923], *Cocos* [Lepage1938, Balach1948b], *Cocos nucifera* [Houser1918, Takaha1929, Mamet1949, Mamet1954, McKenz1956, Takagi1962b, Ali1967a, WilliaWa1988], *Dictyosperma album* [Cocker1899n, Lepage1938], *Dypsis madagascariensis* [MerrilCh1923], *Howeia selloviana* [Balach1932d], *Kentia* [Balach1948b], *Kentia fosteri* [McKenz1956], *Latania* [Hall1922, Balach1932d, Zimmer1948], *Martinezia caryotaefolia* [Houser1918], *Oreodoxa carribea* [GomezM1941], *Oreodoxa regia* [Mamet1959a], *Phoenicophorium sechellarum* [MerrilCh1923], *Phoenix* [Green1908a, Ramakr1921a, Balach1948b, Ali1967a], *Phoenix canariensis* [Balach1927, Balach1932d], *Phoenix dactylifera* [Hall1922, GomezM1941, McDani1968], *Phoenix humilis loureiri* [McKenz1956], *Pritchardia* [Hall1923], *Roscheria melanochaetes* [MerrilCh1923], *Washingtonia* [Bodenh1924]. **Pandanaceae**: *Pandanus* [MerrilCh1923, Laing1933, Mamet1957, Cohic1958, McDani1968, WilliaWa1988], *Pandanus odoratissimus* [Takaha1929, WilliaWa1988], *Pandanus utilis* [Mamet1943a, Mamet1949, Borchs1966], *Pandanus vetichi* [McKenz1956]. **Pinaceae**: *Pinus* [Takaha1929], *Pinus caribaea* [WilliaWa1988]. **Podocarpaceae**: *Podocarpus* [Takagi1969a], *Podocarpus macrophylla* [Takaha1929]. **Punicaceae**: *Punica granatum* [Hall1922]. **Rhamnaceae**: *Rhamnus alaternus* [Balach1932d]. **Rhizophoraceae**: *Kandelia rheedii* [Takaha1932a, Takaha1933, Takaha1936d]. **Rosaceae**: *Cerasus* [Zimmer1948], *Eriobotrya japonica* [Hall1922, Balach1932d, Cohic1958, WilliaWa1988], *Photinia serrata* [Balach1927, Balach1932d], *Photinia serrulata* [Merril1953], *Prunus domestica* [Hall1922], *Prunus laurocerasus* [Balach1932d], *Prunus persica* [Hall1923], *Pyrus malus* [Hall1922], *Rosa* [Takaha1929, Lepage1938, Mamet1943a, Mamet1949, Cohic1958, Borchs1966, WilliaWa1988].

Rubiaceae: *Morinda citrifolia* [WilliaWa1988]. **Rutaceae**: *Atlantia buxifolia* [Frogga1914], *Citrus* [Houser1918, Takaha1929, Green1930b, Kuwana1933, Lepage1938, RahmanAn1941, Mamet1949, Mamet1951], *Citrus* [MerrilCh1923, Merril1953], *Citrus acida* [Green1914c], *Citrus aurantifolia* [WilliaWa1988], *Citrus aurantium* [Bodenh1924, Balach1928b, McKenz1939, GomezM1962, WilliaWa1988], *Citrus aurantium bigaradia* [Matile1978], *Citrus bigaradia* [Balach1928b], *Citrus decumana* [Houser1918], *Citrus grandis* [WilliaWa1988], *Citrus japonica* [Balach1928b], *Citrus limon* [Bodenh1924, Balach1928b, WilliaWa1988], *Citrus myrthifolia* [Balach1928b], *Citrus nobilis* [Balach1928b], *Citrus paradisi* [Houser1918, Beards1966, WilliaWa1988], *Citrus reticulata* [WilliaWa1988], *Citrus sinensis* [McKenz1956, McDani1968], *Citrus trifoliata* [Houser1918], *Citrus triptera* [Balach1928b], *Evodia roxburghiana* [Takaha1929], *Feronia elephantum* [RahmanAn1941], *Glycosmis pentaphylla* [Merril1953], *Xanthoxylum americanum* [Balach1932d]. **Salicaceae**: *Salix babylonica* [Hall1923]. **Santalaceae**: *Santalum austrocaledonicum* [Cohic1958, WilliaWa1988]. **Sapindaceae**: *Blighia sapida* [GomezM1941], *Dodonaea viscosa* [Cohic1958, WilliaWa1988]. **Sapotaceae**: *Palaquium formosanum* [Takaha1929]. **Strelitziaceae**: *Ravenala madagascariensis* [Mamet1959a], *Strelitzia augusta* [Balach1927, Balach1932d], *Strelitzia reginae* [Zimmer1948]. **Theaceae**: *Camellia japonica* [MerrilCh1923, McKenz1956], *Cleyera ochnacea* [Takaha1929], *Thea* [Mamet1951, Borchs1966], *Thea chinensis* [Takaha1929], *Thea japonica* [Takaha1929]. **Verbenaceae**: *Duranta plumieri* [Hall1923, Mamet1943a, Mamet1949, Borchs1966], *Premna* [WilliaWa1988]. **Vitaceae**: *Vitis* [Balach1948b], *Vitis vinifera* [Hall1922, GomezM1941, Almeid1971]. **Zamiaceae**: *Zamia* [MerrilCh1923, Green1914d, Mamet1943a, Merril1953, Borchs1966], *Zamia pumila* [Houser1918].

NATURAL ENEMIES: ACARI **Eupalopsellidae**: *Saniosulus nudus* Summers [GersonOcHo1990]. **Hemisarcoptidae**: *Hemisarcoptes coccophagus* Meyer [GersonOcHo1990], *Hemisarcoptes malus* (Shimer) [GersonOcHo1990]. **Phytoseiidae**: *Amblyseius gossipi* El Badry [YousefEl1982]. COLEOPTERA **Coccinellidae**: *Chilocorus bipustulatus* [Balach1928b], *Chilocorus circumdatus* Sch. [DasBoGo1988], *Chilocorus distigma* Klug [RosenDe1978], *Curinus coeruleus* (Mulsant) [Zimmer1948], *Lindorus lophantae* (Blaisdell) [Smirno1950a], *Sukunahikana popei* Vazirani [Vazira1982], *Telsimia nitida gemmosa* Chazeau [Chazea1984]. **Nitidulidae**: *Cybocephalus binotatus* Grouvelle [BlumbeSw1974], *Cybocephalus micans* Reitter [BlumbeSw1974], *Cybocephalus nigriceps nigriceps* (Sahlberg) [BlumbeSw1974]. **Tenebrionidae**: *Epitragus tomentosus* Leconte [Drea1990]. FUNGI **Ascomycotina**: *Nectria aurantiicola* [EvansPr1990], *Nectria diploa* [EvansPr1990], *Podonectria coccicola* [EvansPr1990]. **Deuteromycotina**: *Hirsutella* [EvansPr1990]. **Mastigomycotina**: *Myiophagus* [EvansPr1990]. HYMENOPTERA **Aphelinidae**: *Ablerus perspeciosus* Girault [Gordh1979], *Aphytis africanus* Quednau [RosenDe1979], *Ap. australiensis* DeBach & Rosen [RosenDe1979], *Ap. chrysomphali* Mercet [Balach1948b], *Ap. chrysomphali* (Mercet) [Zimmer1948, RosenDe1979, MyartsRu2000], *Ap. columbi* (Girault) [RosenDe1979], *Ap. comperei* DeBach & Rosen [RosenDe1979], *Ap. costalimai* (Gomes) [RosenDe1979], *Ap. diaspidis* (Howard) [RosenDe1979, MyartsRu2000],

Ap. holoxanthus DeBach [QuezadCoDi1972, RosenDe1978, RosenDe1979, SteinbPoRo1987, Koszta1996], *Ap. lingnanensis* Compere [RosenDe1979], *Ap. mazalae* DeBach & Rosen [RosenDe1979], *Ap. merceti* Compere [RosenDe1979], *Ap. philippinensis* DeBach & Rosen [RosenDe1979], *Ap. proclia* (Walker) [MyartsRu2000], *Aspidiotiphagus citrinus* (Craw) [Balach1948b, Zimmer1948], *Aspidiotiphagus latipennis* Compere [AnneckIn1971], *Aspidiotiphagus lounsburyi* Berlese & Paoli [Balach1948b], *Azotus separaspidis* Annecke & Insley [AnneckIn1970], *Encarsia aurantii* (Howard) [MyartsRu2000], *Encarsia lounsburyi* (Berlese & Paoli) [AbdRab2001a], *Marietta marchali* Mercet [AnneckIn1971], *Prospaltella elongata* [FlandeGrDe1950], *Prospaltella elongata* Dozier [RosenDe1978], *Pteroptrix chinensis* (Howard) [ComperSm1927], *Pteroptrix smithi* (Compere) [RosenDe1978, SteinbPoRo1987]. **Encyrtidae**: *Adelencyrtus ficusae* Risbec [AnneckIn1971], *Comperiella bifasciata* Howard [ComperSm1927], *Habrolepis aspidioti* Compere & Annecke [Prinsl1983], *Habrolepis fanari* Delucchi & Traboulsi [DeluccTr1965], *Habrolepis pascuorum* Mercet [Trjapi1989], *Habrolepis rouxi* Compere [AnneckIn1971], *Pseudhomalopoda prima* Girault [Muma1959, RosenDe1978], *Zaomma lambinus* (walker) [Trjapi1989]. **Signiphoridae**: *Chartocerus niger* (Ashmead) [Gordh1979], *Signiphora aleyrodis* Ashmead [Gordh1979], *Signiphora fax* Girault [Woolle1990], *Signiphora merceti* [Woolle1990], *Signiphora prepauca* Girault [Woolle1990]. NEUROPTERA **Chrysopidae**: *Ceraeochrysa cubana* Hagen) [Drea1990], *Ceraeochrysa sanchezi* (Navas) [Drea1990], *Ceraeochrysa valida* (Banks) [Drea1990], *Chrysopa* [Drea1990], *Chrysoperla carnea* (Stephens) [Drea1990], *Chrysoperla plorabunda* (Fitch) [Drea1990], *Chrysoperla rufilabris* (Burmeister) [Drea1990], *Nodita pavida* (Hagen) [Drea1990]. **Coniopterygidae**: *Semidalis vicina* (Hagen) [Drea1990]. THYSANOPTERA **Phlaeothripidae**: *Aleurodothrips fasciapennis* (Franklin) [Beshea1975, PalmerMo1990, WatsonDuLi2000a], *Haplothrips cahirensis* (Trybom) [PalmerMo1990].

DISTRIBUTION: **Afrotropical**: Comoros Islands [Matile1978]; Guinea [Leonar1914, Balach1956]; Kenya [DeLott1967a, Schmut1998]; Madagascar [Mamet1951, Mamet1954, Mamet1959a, Borchs1966]; Mauritius [GrandpCh1899, Mamet1943a, Mamet1949]; Mozambique [Almeid1971]; Rodrigues Island [Mamet1954a, Borchs1966]; Réunion [Mamet1957]; Seychelles [Green1907, Green1914d, Mamet1943a]; South Africa [BrainKe1917, Brain1919, Balach1956, RosenDe1979, Bedfor1989]; Tanzania [Balach1956, Mamet1956, Borchs1966]; Uganda [DeLott1967a]; Zaire [Balach1956]; Zanzibar [Green1916]; Zimbabwe [Hall1928, Balach1956]. **Australasian**: Australia (Northern Territory [Green1914c], Queensland [Frogga1914, Brimbl1962a]); Bonin Islands (= Ogasawara-Gunto) [Kuwana1909a]; Federated States of Micronesia (Caroline Islands [Takaha1941b], Ponape Island [Beards1966]); Fiji [WilliaWa1988]; French Polynesia (Tahiti [DoaneHa1909]); Hawaiian Islands (Hawaii [Zimmer1948]); Kiribati [WilliaWa1988]; New Caledonia [Cohic1958, WilliaWa1988]; Papua New Guinea [WilliaWa1988]; Tuvalu [WilliaWa1988]; Western Samoa [Green1923, Laing1927]. **Nearctic**: Mexico [Cocker1899n, MyartsRu2000] (Baja California [Ferris1921, FerrisKe1923], Tamaulipas [LunaSaMa1995]); United States of America (Alabama [BesheaTiHo1973], California [McKenz1956, Dekle1965c], Florida [Ashmea1880,

Comsto1881a, MerrilCh1923, McKenz1939, Merril1953, BesheaTiHo1973], Georgia [BesheaTiHo1973], Mississippi [Herric1911], Missouri [Hollin1923], Texas [McDani1968]). **Neotropical**: Brazil [Green1937, Lepage1938, RosenDe1979] (Amazonas [WolffCo1993], Bahia [WolffCo1993], Maranhao [WolffCo1993], Mato Groso [WolffCo1993], Minas Gerais [WolffCo1993], Paraiba [WolffCo1993], Parana [WolffCo1993], Par [WolffCo1993], Pernambuco [WolffCo1993], Rio Grande do Norte [WolffCo1993], Rio Grande do Sul [WolffCo1993], Rio de Janeiro [WolffCo1993], Santa Catarina [WolffCo1993, HickelDu1995], São Paulo [Hempel1900a, Green1930b]); Chile [GonzalCh1968]; Colombia [Balach1959a, Kondo2001]; Cuba [Houser1918]; Dominican Republic [GomezM1941, RosenDe1979]; El Salvador [QuezadCoDi1972]; Panama [Cocker1899n]; Peru [Beingo1969d]; Puerto Rico & Vieques Island (Puerto Rico [Martor1976, ColonFMe1998]); U.S. Virgin Islands [Nakaha1983]. **Oriental**: British Indian Ocean Territory (Chagas) [Mamet1943a]; India [Green1908a, Ramakr1921a] (Bihar [Ali1967a], Karnataka [UsmanPu1955], West Bengal [Nath1972]); Indonesia (Java [Green1904a]); Mongolia [DanzigKo1990]; Philippines [VelasqRi1969, RosenDe1979] (Luzon [Velasq1971]); Sri Lanka [Green1896e, Green1900a]; Taiwan [Ferris1921a, Takaha1929, Takaha1932a, Takaha1934, Takagi1969a, RosenDe1979, WongChCh1999]. **Palaearctic**: Algeria [Balach1928b, Balach1932d]; Belgium [Balach1948b]; Canary Islands [Balach1948b, PerezGCa1987, MatileOr2001]; China (People's Republic) (Henan (Honan) [Shen1993]); Denmark [Balach1948b]; Egypt [Hall1922, Hall1923, Ezzat1958]; Germany (United) [Lindin1909b]; Greece [Korone1934, ArgyriStMo1976]; Israel [Bodenh1924, Bodenh1930, Bodenh1937, GersonZo1973]; Italy [Leonar1920, LongoMaPe1995]; Japan [Kuwana1902, Kuwana1917a, Kuwana1933, Takaha1936d, Takaha1955f, Kawai1980] (Kyushu [Takagi1962b]); Lebanon [Bodenh1926]; Madeira Islands [Balach1938a, CarvalAg1997]; Morocco [Balach1932d]; Poland [Dziedz1989]; Saudi Arabia [Beccar1971, Matile1984c]; Spain [Martin1983]; Turkey [Bodenh1949, Bodenh1952]; United Kingdom [Green1931a] (England [Morgan1889a]); Yugoslavia [Bachma1953].

BIOLOGY: A biparental species that infests leaves and fruits.

ECONOMIC IMPORTANCE: Widely distributed in many tropical and subtropical regions in North and South America, Africa, the Mediterranean Basin, the Far East, Pacific Islands and Australia (CABI, 1988). Has been recorded as a serious pest of citrus in Florida, Texas, Brazil, Mexico, Lebanon, Egypt and Israel (Quayle, 1938; Bodenheimer, 1951; Ebeling, 1959). In recent years also damaging bananas in Central America and coconut palm in the Philippines (Rosen & DeBach, 1978).

GENERAL: Description and illustration of adult female by Kuwana (1933), McKenzie (1939, 1956), Balachowsky (1948b, 1956), Zimmerman (1948), Habib et al. (1960), Velasquez (1971), Chou (1985, 1986), Williams & Watson (1988), Dziedzicka (1989), Yaşar (1995a), Kosztarab (1996), Gill (1997) and by Colon-Ferrer & Medina-Gaud (1998).

KEYS: Claps & Teran 2001: 392 (female) [South Africa]; Gill 1997: 96 (female) [Species of California]; Kosztarab 1996: 476 (female) [Northeastern North America]; Danzig 1993: 164 (female) [Europe]; Williams & Watson 1988: 93

(female) [Tropical South Pacific]; Tereznikova 1986: 92-93 (female) [Ukraine]; Chou 1985: 284 (female) [Species of China]; Kawai 1980: 209-210 (female) [Japan]; Gerson & Zor 1973: 516 (female) [Israel]; Velasquez 1971: 131 (female) [Philippines]; McDaniel 1968: 227 (female) [U.S.A.: Texas]; Beardsley 1966: 515 (female) [Federated States of Micronesia]; Gómez-Menor Guerrero 1962: 180 (female) [Canary Islands]; Ezzat 1958: 241 (female) [Egypt]; Balachowsky 1956: 86 (female) [Africa]; McKenzie 1956: 24 (female) [U.S.A.: California]; Lupo 1953: 24 (female) [Italy]; Balachowsky 1948b: 346 (female) [Mediterranean]; Zimmerman 1948: 368-369 (female) [Hawaii]; McKenzie 1943: 150 (female) [World]; Ferris 1942: 31 (female) [North America]; McKenzie 1939: 63 (female) [World]; Kuwana 1933: 26 (female) [Japan]; Kuwana 1933b: 49 (female) [Japan]; Fullaway 1932: 95-97, 107 (female) [Hawaii]; Balachowsky 1928b: 157 (female) [North Africa]; Britton 1923: 376 (female) [U.S.A.: Connecticut]; Hollinger 1923: 29 (female) [U.S.A.: Missouri]; Leonardi 1920: 65 (female) [Italy]; Brain 1919: 198 (female) [South Africa]; Lawson 1917: 210 (female) [U.S.A.: Kansas]; Robinson 1917: 24 (female) [Philippines]; Dietz & Morrison 1916a: 307 (female) [U.S.A.: Indiana]; Cockerell 1905: 45-46 (female) [Mexico]; Cockerell 1905b: 201 (female) [U.S.A.: Colorado]; Green 1896e: 39 (female) [Sri Lanka]; Comstock 1883: 55-57 (female) [North America].

CITATIONS: Abbas1992 [host, distribution, life history: 477-485]; AbdElKDaKo1988 [chemical control, biological control: 270-275]; AbdRab2001a [host, distribution, biological control: 174,176]; AbouEl2001 [host, distribution, biological control: 185-195]; Aguila1980 [host, distribution, biological control: 83-91]; AguilaSaNu1980 [host, distribution, biological control: 97-100]; Alfier1929 [host, distribution: 7-9]; Ali1967a [host, distribution: 40]; Almeid1971 [host, distribution: 9]; Alvara1939 [host, distribution: 3]; AminRiSa2001 [host, distribution, biological control, economic importance: 441]; Andrie1932 [host, distribution, economic importance, control]; Anneck1963 [host, distribution, life history, economic importance: 195-225]; Anneck1969 [economic importance, host, distribution, biological control, chemical control: 849-854]; AnneckIn1970 [host, distribution, biological control: 241-242]; AnneckIn1971 [host, distribution, biological control: 2,14,29,35]; Argyri1970 [host, distribution, biological control: 57-65]; Argyri1979a [host, distribution, economic importance, biological control: 517-520]; Argyri1990 [host, distribution, economic importance: 579-583]; ArgyriMo1981 [host, distribution, economic importance: 623-627]; ArgyriStMo1976 [host, distribution, biological control: 25]; Ashmea1880 [taxonomy, description, host, distribution: 267]; Avidov1960 [host, distribution, life history: 17-31]; Avidov1970 [host, distribution, life history, biological control]; AvidovGa1960 [host, distribution, economic importance, life history: 5-16]; Azeved1925 [host, distribution: 85-86]; Azeved1929 [host, distribution: 113-115]; Azeved1929a [host, distribution: 126-128]; Bachma1953 [host, distribution: 178]; Baker1976 [biological control, life history: 1-25]; Balach1927 [host, distribution: 178]; Balach1928b [taxonomy, description, host, distribution, biological control, economic importance, chemical control, life history: 156-180]; Balach1928d [biological control: 296]; Balach1932b [ecology: 517-522]; Balach1932d [taxonomy, host, distribution: XI]; Balach1938a [host, distribution, economic importance: 151]; Balach1948b [taxonomy, description, illustration, host, distribution, biological control: 346-351]; Balach1956 [taxonomy, description, illustration, host, distribution: 86-88]; Balach1957c [host, distribution: 200]; Ballou1912 [host, distribution, economic importance, control]; Banks1990 [chemistry: 272]; Barrit1929 [host, distribution, economic importance, chemical control: 44]; BartleVa1964 [biological control: 283-304]; BasuNaCh1969 [economic importance, host, distribution: 169-178]; Beards1966 [host, distribution: 516]; BeardsDaHo1976 [economic importance: 105]; BeardsGo1975 [economic importance: 49]; BeattiGe1983 [host, distribution: 220-226]; Beccar1971 [host, distribution: 194]; Bedfor1968b [host, distribution, chemical control, biological control: 1-14]; Bedfor1969 [chemical control, biological control: 3-10]; Bedfor1973 [host, distribution, chemical control, biological control: 4-11]; Bedfor1989 [economic importance, host, distribution, biological control, chemical control: 1-16]; BedforCi1994 [host, distribution, biological control: 143-179]; BedforGr1981 [host, distribution, chemical control, biological control: 616-620]; BedforVaDe1998 [host, distribution, economic importance, chemical control, biological control]; Beingo1967 [host, distribution, biological control: 67-81]; Beingo1969d [biological control: 827-838]; Benass1959b [host, distribution, biological control, chemical control: 867-872]; Benass1965a [host, distribution, chemical control, biological control, economic importance, life history: 112-125]; BenassViRo1979 [biological control: 281-287]; BenDov1990e [host, distribution: 656]; BennetRoCo1976 [biological control, economic importance: 359-395]; BentleBa1931 [host, distribution, economic importance, control: 1-61]; Berger1921 [host, distribution, biological control: 141-154]; Berger1921a [host, distribution, biological control: 333-334]; Berger1932 [host, distribution, biological control: 131-136]; Berles1895a [taxonomy, description, illustration, structure: 83,112,195]; Bertel1956 [host, distribution, description, life history, biological control]; BertelBa1966 [host, distribution: 17-46]; Beshea1975 [host, distribution, biological control: 223-224]; BesheaTiHo1973 [host, distribution: 5]; BiezanFr1939 [host, distribution: 1-18]; BiezanSe1940 [host, distribution: 67-68]; BilogOMo2000 [host, distribution, biological control: 137-147]; Blanch1922 [host, distribution: 387-398]; BlumbeSw1974 [biological control: 437-443]; BlumbeSw1974a [biological control: 3-11]; BlumbeSwWy1995 [biological control: 33-44]; Bodenh1924 [taxonomy, description, host, distribution: 36-37]; Bodenh1926 [host, distribution: 42]; Bodenh1930 [host, distribution: 4]; Bodenh1935 [host, distribution: 246]; Bodenh1937 [host, distribution: 217]; Bodenh1938 [taxonomy, host, distribution, life history, biological control, economic importance: 7-8]; Bodenh1949 [taxonomy, description, illustration, host, distribution: 74-77]; Bodenh1951a [taxonomy, host, distribution, life history, economic importance, chemical control, biological control: 274-301]; Bodenh1952 [taxonomy, host, distribution, structure: 345-346]; Bodkin1925 [host, distribution, chemical control: 143-149]; Bonafo1979 [structure: 505-512]; Bonafo1981 [life history, ecology: 157-170]; Bondar1914 [host, distribution, economic importance: 1064-1106]; Borchs1937a [taxonomy, description, host, distribution: 49,51]; Borchs1939 [taxonomy, description, host, distribution: 8,17]; Borchs1949d [taxonomy, description, host, distribution: 232]; Borchs1950b [taxonomy, description, illustration, host, distribution: 217,219];

Borchs1966 [catalogue: 283-286]; Bourne1923 [host, distribution, economic importance: 1]; Boyce1948 [host, distribution, economic importance, control]; Boyce1950 [host, distribution, economic importance: 741-766]; Brain1919 [taxonomy, description, illustration, host, distribution: 198,200-201]; BrainKe1917 [distribution: 184]; Brick1912 [host, distribution: 1-22]; Brimbl1962 [host, distribution, economic importance: 220]; Brimbl1962a [taxonomy, description, illustration, host, distribution: 414-416]; Britto1923 [taxonomy, description, host, distribution: 376-377]; Broodr1964 [host, distribution, economic importance, biological control: 7-13]; Brooks1977 [host, distribution, life history, economic importance, biological control]; BrooksTh1963 [chemical control: 279-284]; Browni1994a [biological control: 27-49]; Bruner1930 [host, distribution, biological control: 11-18]; BrusseBh1970 [host, distribution, life history, control: 64-76]; Bryan1915 [host, distribution]; BurditSi1963 [chemical control: 223-227]; BurgerUl1990 [economic importance: 313-327]; Bustsh1958 [taxonomy, description, host, distribution: 229]; Buxton1920 [host, distribution: 287-303]; CABI1988 [host, distribution, taxonomy: 1-3]; CaiLuWe2001 [chemical control: 28-32]; Caltag1985 [taxonomy, biological control: 189-200]; Carnes1907 [taxonomy, host, distribution: 213]; CarvalAg1997 [life history, description, economic importance, biological control, host, distribution: 262-264]; Castel1951a [biological control: 95-98]; CatchiWh1924 [host, distribution, life history, ecology: 604-606]; CaveMa1994 [biological control: 3-8]; CaveMa1994 [host, distribution, biological control: 3-8]; Chambe1925EL [host, distribution, control: 149-165]; CharmoGe1921 [host, distribution: 188]; Chazea1984 [host, distribution, biological control: 9-10]; Chen1936 [taxonomy: 222]; Chiesa1938a [host, distribution: 1-21]; Chiesa1948 [host, distribution, economic importance]; Chou1947 [taxonomy, description, illustration, host, distribution: 17-24]; Chou1947a [chemical control: 33]; Chou1985 [taxonomy, description, host, distribution: 284-287]; Chou1986 [taxonomy, illustration: 672]; ChuaWo1990 [host, distribution, economic importance: 543-552]; Cillie1971 [host, distribution, life history, biological control: 269-284]; Cillie1978b [host, distribution, biological control, life history, economic importance, chemical control: 119-122]; Cillie1998a [host, distribution, life history, economic importance, biological control, chemical control: 145-149]; ClancySeMu1963 [biological control, economic importance, host, distribution: 603-605]; ClapsTe2001 [taxonomy, description, illustration, host, distribution, economic importance: 392,395-396]; ClapsWoGo2001a [taxonomy, host, distribution: 14-15]; Clause1940 [biological control]; Cocker1892a [host, distribution: 54]; Cocker1892b [host, distribution: 333]; Cocker1893cc [host, distribution: 101]; Cocker1896b [distribution: 334]; Cocker1897i [taxonomy, description, host, distribution: 23-24]; Cocker1899j [taxonomy, description, host, distribution: 273]; Cocker1899n [host, distribution: 25]; Cocker1900k [taxonomy: 350]; Cocker1905 [taxonomy: 46]; Cocker1905b [taxonomy: 201]; Cohen1969 [biological control, economic importance: 769-772]; Cohen1975 [host, distribution, economic importance, biological control: 38-41]; CohenPoEl1987 [life history, physiology, chemical control: 303-307]; CohenPoEl1988 [chemical control, biological control: 91-95]; CohenPoEl1994 [chemical control, biological control: 183-190]; Cohic1958 [host, distribution: 13]; ColonFMe1998 [taxonomy, description, illustration, host, distribution: 49-50]; Comper1926 [host, distribution, biological control: 33-50]; Comper1953 [biological control: 35-46]; Comper1961a [biological control: 17-71]; Comper1969 [biological control: 755-764]; ComperAn1961 [host, distribution, biological control: 17]; ComperSm1927 [host, distribution: 63-73]; Compto1924 [chemical control: 222-225]; Comsto1881 [taxonomy: 9]; Comsto1881a [taxonomy, description, illustration, host, distribution, life history: 296-300]; Comsto1883 [taxonomy: 55,61]; CoronaRuMo1997 [host, distribution: 38-41]; CostaL1949 [host, distribution, biological control: 65-87]; CostaLRa1922 [host, distribution: 1101]; Craw1891 [taxonomy, description, illustration, host, distribution, economic importance: 10-12]; Craw1896 [taxonomy, description, host, distribution: 34]; Craw1906 [host, distribution: 139-158]; Creigh1942 [host, distribution, control: 219-233]; Cressm1943 [chemical control: 17-26]; Crouze1971 [biological control: 200]; Crouze1973 [host, distribution, biological control: 15-39]; DahmsSm1994 [host, distribution, biological control: 245-255]; DaneelMeJa1994 [host, distribution: 72-74]; Danzig1972 [taxonomy, host, distribution, economic importance: 209]; Danzig1993 [taxonomy, description, illustration, host, distribution, economic importance: 164-166]; DanzigKo1990 [host, distribution: 46]; DanzigPe1998 [catalogue: 214-215]; DarlinJo1984 [host, distribution, biological control: 555]; Das1988 [life history, biological control: 44-45]; DasBoGo1988 [biological control, economic importance: 15-18]; DavidsMi1990 [host, distribution, economic importance: 603-632]; Dean1955 [biological control: 444-447]; Dean1982 [host, distribution, life history, biological control, economic importance: 147-149]; DeBach1960 [host, distribution, biological control: 701-705]; DeBach1964 [biological control]; DeBach1964b [biological control: 674-713]; DeBach1972 [biological control: 211-233]; DeBach1974 [biological control]; DeBachBa1951 [chemical control, biological control: 372-383]; DeBachRo1976a [host, distribution, biological control: 541-545]; DeBachRo1991 [biological control]; DeBachRoKe1971 [biological control: 165-194]; DebnatHa1991 [host, distribution: 39-40]; DEDAC1923 [host, distribution]; Dekle1954 [host, distribution, economic importance: 226-228]; Dekle1965c [taxonomy, description, host, distribution: 42]; Dekle1976 [taxonomy, description, host, distribution, economic importance: 60]; DeLott1967a [host, distribution: 113-114]; Delucc1969 [biological control: 871-874]; DeluccRoSc1976 [taxonomy, biological control: 81-91]; DeluccTr1965 [host, distribution, biological control: 495-500]; DeSant1979 [biological control]; Dhilee1991 [host, distribution: 94-99]; Dhilee1996 [host, distribution, biological control: 64-74]; DicksoFl1955 [host, distribution: 614-615]; DietzMo1916a [taxonomy, description, illustration, host, distribution: 312-313]; Dinthe1958 [host, distribution, economic importance: 421]; Doane1931 [host, distribution, control]; Dougla1912 [taxonomy: 213]; DouttAnTr1976 [biological control, life history: 143]; Dowson1935 [host, distribution: 225]; Dozier1933 [host, distribution, biological control: 85-100]; Dozier1937 [host, distribution, biological control: 121-135]; Drea1990 [biological control: 41-49]; Dupont1931 [host, distribution: 1-18]; Dziedz1989 [taxonomy, description, illustration, host, distribution: 101-102]; Ebelin1936 [chemical control: 95]; Ebelin1949 [host, distribution, life history, control]; Edward1936 [host, distribution, economic importance: 335-337]; Efimof1937 [host, distribution]; Ehler1996 [host, distribution, biological control: 337-342]; Ehrhor1913 [host, distribution: 101]; EhrhorFuSw1913 [distribution: 295-300]; ElMinsElHa1974 [taxonomy, description, illustration, host, distribution: 223-232]; EnglisTu1940 [host, distribution, economic importance, life history, control: 3-18]; EvansPr1990 [biological control: 3-17]; Ezzat1958 [distribution: 241]; EzzatNa1987 [distribution: 87]; Fawcet1948 [biological control: 627-664]; Fernal1903b [catalogue: 286-287]; Ferris1921 [host, distribution: 129]; Ferris1921a [host, distribution: 220]; Ferris1937c [taxonomy, illustration: 50,64]; Ferris1938a [taxonomy, description, illustration, host, distribution: 201]; Ferris1941e [taxonomy: 40,43]; Ferris1942 [taxonomy: 446:31]; FerrisKe1923 [host, distribution: 318]; Figuer1952 [host, distribution: 209]; Filho1931 [host, distribution, control: 606]; Fisher1950 [biological control: 305-309]; FisherDe1976 [biological control: 43-50]; FisherGr1950 [chemical control: 712-718]; FisherThGr1949 [biological control: 1-11]; Fjeldd1996 [host, distribution: 4-24]; Flande1959b [biological control: 125-142]; Flande1969 [biological control: 29-33]; Flande1971 [biological control, life history: 857-872]; FlandeGrDe1950 [biological control: 254-255]; Foldi1990 [structure: 43-54]; Foldi1990c [structure, anatomy: 199-204]; Foldi2001 [distribution: 303-308]; Fonsec1963 [host, distribution: 32-35]; FonsecAu1932a [host, distribution: 202-214]; FoxWil1939 [host, distribution, economic importance: 2296]; Frogga1914 [taxonomy, description, host, distribution: 313]; Frogga1915 [taxonomy, description, host, distribution: 16-17]; Fullaw1932 [taxonomy: 97,107]; Fuller1907 [taxonomy, description, host, distribution, economic importance, control: 1031-1055]; GanLiZh1993 [host, distribution, life history, ecology, biological control: 347-348]; Garcia1930 [host, distribution, biological control]; Garcia1931a [host, distribution, biological control: 659-666]; Garcia1931a [host, distribution, biological control]; GeorghMe1983 [life history, chemical control: 1-46]; GeorghTa1976 [chemical control: 759]; GerlinBa1971 [host, distribution, life history, ecology, biological control: 37-44]; Gerson1967c [host, distribution, biological control: 632-638]; GersonOcHo1990 [biological control: 77-97]; GersonZo1973 [taxonomy, life history, host, distribution, economic importance: 513-533]; Ghauri1962 [taxonomy, structure: 93,210]; GibsonRo1922 [host, distribution: 1]; Gill1997 [host, distribution, taxonomy, description, illustration, economic importance: 96,98]; GillClDu1999 [host, distribution, control]; GirolaAr1925 [host, distribution]; GomesC1940 [host, distribution, chemical control, biological control: 329-354]; GomezM1937 [taxonomy, description, illustration, host, distribution: 97-98]; GomezM1941 [host, distribution: 130]; GomezM1956 [taxonomy, description, illustration, host, distribution, biological control: 32,39-41]; GomezM1958a [host, distribution: 7]; GomezM1962 [taxonomy, description, host, distribution: 184-186]; GonzalCh1968 [host, distribution: 111]; GonzalHeSi1991 [host, distribution, biological control: 433-441]; Gordh1979 [biological control: 894,896,899,900,910,]; Gough1914 [chemical control: 17-19]; Gowdey1921 [taxonomy, description, host, distribution: 31]; GrandpCh1899 [taxonomy, description, host, distribution: 25-26]; GravenFo1979 [host, distribution, life history: 109-113]; Greath1971 [host, distribution, biological control]; Green1896e [taxonomy, description, illustration, host, distribution: 39,43-44]; Green1900a [taxonomy, host, distribution: 69]; Green1904a [host, distribution: 208]; Green1907 [host, distribution: 202];

Green1908a [host, distribution: 33]; Green1914c [taxonomy: 232]; Green1914d [host, distribution: 47]; Green1915e [host, distribution: 608-636]; Green1916 [host, distribution: 376]; Green1930b [host, distribution: 214]; Green1931a [host, distribution: 105]; Green1937 [host, distribution: 331]; Griffi1951 [economic importance, chemical control: 464-468]; GriffiFi1949 [chemical control: 829-833]; GriffiSt1947 [chemical control: 2-8]; GriffiTh1949a [taxonomy, description, host, distribution: 8]; GriffiTh1957 [host, distribution: 5-30]; HabibAt1960 [host, distribution, life history, ecology: 353-365]; HabibElRa1973 [host, distribution, life history: 133-141]; HabibEzAt1960 [taxonomy, description, illustration, host, distribution: 329-336]; HabibSaAm1971 [host, distribution, life history: 318-330]; HafezTaRa1970 [host, distribution, life history, ecology: 3-16]; HakkonPi1984 [biological control: 1109-1121]; Halber1995 [host, distribution: 6-9]; Hall1922 [taxonomy, description, host, distribution: 30-32]; importance: 823-826]; HallFo1933 [host, distribution, economic importance: 1-55]; Hamble1947 [host, distribution: 949-956]; Hamlen1974 [host, distribution: 6-8]; HanksDe1998 [life history, ecology: 239-262]; HavronRo1992 [chemical control, biological control: 984-987]; HavronRo1994 [biological control: 209-220]; HavronRoPr1991 [host, distribution, biological control: 221-228]; Haywar1939 [host, distribution, control: 1]; Haywar1944 [host, distribution: 1-32]; Hecht1936 [host, distribution, life history, biological control: 299-326]; Hempel1900a [taxonomy, description, host, distribution: 502-503]; Hempel1904 [taxonomy, host, distribution: 321-322]; Hempel1922 [host, distribution: 133-144]; Herric1911 [taxonomy, description, illustration, host, distribution: 11,29-30,65]; Herric1925 [host, distribution, description, life history, economic importance]; HickelDu1995 [host, distribution: 665-668]; Hill1975 [host, distribution, economic importance, control]; HindiAmRo1964 [chemical control: 27-43]; Hollin1923 [taxonomy, description, host, distribution: 29-30]; HorticInNo1923 [host, distribution: 236-239]; Houser1918 [host, distribution: 168]; Howard1895e [biological control: 1-44]; Howard1908 [host, distribution: 265-277]; Howard1991 [host, distribution, life history, ecology, control: 217-225]; HoyHe1985 [biological control]; Hsu1935 [host, distribution: 578-590]; Huffak1990 [biological control: 205-220]; HuffakSt1971 [biological control: 333-350]; Hughes1957 [taxonomy, structure: 709-718]; HunterWo2001 [life history, biological control: 251-290]; IpertiBr1969 [host, biological control: 149-157]; IpertiLa1968 [host, biological control: 543-552]; Ishaay1971 [chemistry, physiology: 935-943]; IshaaySw1970 [host, chemistry, physiology: 37-42]; IshaaySw1970a [chemistry, physiology: 1599-1605]; IshaaySw1990 [physiology, chemistry: 353-356]; IshaaySwNe1980 [chemistry, physiology: 212]; Ishii1928 [host, distribution, biological control: 79]; Ishii1932a [host, distribution, biological control: 161]; Jackso1913 [taxonomy: 1-48]; Jimene1969 [biological control: 781-784]; JiYa1990 [biological control: 134-136]; Johnst1915 [host, distribution, biological control: 1-33]; Jourdh1979 [biological control: 75-79]; Kawai1980 [taxonomy, description, host, distribution: 210]; KfirRo1981 [biological control: 141-150]; KhalifHa1958 [host, distribution, life history, ecology: 449-459]; Kiritc1932a [taxonomy: 251]; Klein1940 [host, distribution, life history, economic importance, control: 1-25]; Koebel1893 [host, distribution, biological control: 1-39]; KoebelMa1897 [host, distribution, biological control: 65-67]; Komosi1964 [host, distribution, taxonomy, description, illustration: 216-218]; Kondo2001 [taxonomy, host, distribution: 43]; KondoKa1995a [host, distribution: 97-98]; Korone1934 [taxonomy, description, illustration, host, distribution: 21-22]; Koszta1990 [structure, biological control: 307-311]; Koszta1996 [taxonomy, description, illustration, host, distribution, life history, biological control, economic importance: 477-478]; Koteja1990b [life history, structure, anatomy: 233-242]; Kotins1909 [host, distribution: 97]; Kotter1977 [host, distribution: 1807-1815]; KozarKi1979 [host, distribution: 246-250]; Kuwana1902 [host, distribution: 71]; Kuwana1907 [host, distribution: 196]; Kuwana1909a [host, distribution: 160]; Kuwana1917a [taxonomy, distribution: 175]; Kuwana1927 [host, distribution: 72]; Kuwana1933 [taxonomy, description, illustration, host, distribution: 27-28]; Labano1999 [host, distribution, life history, chemical control: 14-15]; Laing1933 [host, distribution: 676]; Laport1948 [host, distribution, biological control: 35-37]; Laport1949 [host, distribution, life history, biological control: 150-158]; Larter1937 [host, distribution: 65-72]; Lawson1917 [taxonomy, description, illustration, host, distribution: 214-216]; Leonar1897 [taxonomy: 286]; Leonar1899 [taxonomy, description, host, distribution: 199,211]; Leonar1907 [taxonomy, description, host, distribution: 119]; Leonar1914 [host, distribution: 205]; Leonar1920 [taxonomy, description, illustration, host, distribution: 65-69]; Lepage1938 [catalogue: 398-399]; LepageFi1947 [taxonomy, description, host, distribution: 39]; Lepesm1947 [taxonomy, description, host, distribution, life history: 196-198]; Lever1969 [host, distribution]; LiLi1990 [host, biological control: 68-70]; Lindin1909b [host, distribution: 222]; Lindin1909c [taxonomy, host, distribution: 449]; Lindin1910b [host, distribution: 40]; Lindin1912b [taxonomy, description, host, distribution: 58,72,154,175,358,]; Lindin1924 [taxonomy: 175]; Lindin1935 [taxonomy: 132]; Linnae1758 [taxonomy, description, host, distribution: 455]; Lobdel1937 [taxonomy: 78,79]; LongoMaPe1995 [distribution: 126]; Lounsb1898 [host, distribution: 35-58]; Lounsb1914 [host, distribution: 1]; Lounsb1916 [host, distribution: 83-103]; Lu1989a [host, distribution, life history, ecology, biological control: 218-223]; Lugger1900 [host, distribution: 208-245]; LunaSaMa1995 [host, distribution: 265]; Lupo1953 [taxonomy, description, illustration, host, distribution: 24-31]; MacGil1921 [taxonomy, description, host, distribution: 415-416]; MalipaDuSm2000 [host, distribution, biological control]; MaltbyJiDe1968 [biological control: 1086-1088]; Mamet1943a [catalogue: 158]; Mamet1949 [catalogue: 56,57]; Mamet1951 [host, distribution: 226]; Mamet1954 [host, distribution: 16]; Mamet1954a [host, distribution: 265]; Mamet1956 [host, distribution: 138]; Mamet1957 [host, distribution: 369,375]; Mamet1959a [host, distribution: 386,387]; Mansfi1920 [host, distribution: 145-155]; Mariau1998 [host, distribution, economic importance: 269-277]; Marlat1903 [host, distribution, economic importance, control: 23]; Marlat1915 [host, distribution]; Martin1983 [taxonomy, host, distribution: 62]; Martor1976 [host, distribution: 1-303]; Maskel1895b [host, distribution: 39]; Maskew1914 [host, distribution: 193-194]; Maskew1914a [host, distribution: 446-447]; Maskew1916 [host, distribution: 308-309]; Mathis1941 [host, distribution, life history, biological control: 1-5]; Mathis1947 [host, distribution, life history, biological control: 13-35]; Matile1978 [host, distribution: 63]; Matile1984c [host, distribution: 221]; MatileOr2001 [host, distribution: 189]; Matsud1927 [life history, structure: 391-417]; Matsud1929 [host, distribution, life history, structure: 1-79]; Matsud1935 [host, distribution, life history: 29-36]; McClur1990c [ecology, host: 289-291]; McClur1990e [life history, ecology: 309-314]; McClur1990g [taxonomy, host, distribution, ecology: 319-330]; McCoy1985 [biological control: 481-499]; McDani1968 [taxonomy, illustration, host, distribution: 230-231]; McKenz1939 [taxonomy, description, illustration, host, distribution: 53,59-60,63,71]; McKenz1943 [taxonomy: 150]; McKenz1956 [taxonomy, description, illustration, host, distribution: 24,54-57]; Merkel1938 [host, distribution: 88-99]; Merril1953 [taxonomy, description, host, distribution: 34-35]; MerrilCh1923 [taxonomy, description, host, distribution, economic importance: 219-221]; MetcalMe1993 [economic importance, host, distribution, control]; Miller1937RL [economic importance, control: 100-106]; MillerDa1990 [host, distribution, economic importance: 301]; Miyosh1926 [host, distribution: 303-326]; MonrreCoRu1997 [host, distribution, economic importance, life history, biological control: 43-48]; Monte1930 [host, distribution: 3-36]; Morgan1889a [taxonomy, host, distribution: 350-351]; Morley1909 [host, distribution, biological control: 254-257]; MoutiaMa1947 [distribution]; Muma1948 [biological control: 193-194]; Muma1955 [biological control, economic importance: 432-438]; Muma1959 [host, distribution, life history, biological control: 577-586]; Muma1969 [biological control: 863-870]; Muma1971 [host, distribution, biological control: 139-150]; MumaCl1961 [host, distribution, life history, biological control: 29-32]; MumaSeDe1961 [biological control, host, distribution: 1-39]; MunozG1937 [host, distribution: 3-9]; Muraka1970 [host, distribution: 73]; MurakaAbCo1984 [host, distribution, biological control: 237-244]; MyartsRu2000 [distribution, biological control: 7-33]; Myers1927LE [taxonomy, description: 341-346]; NagarkSa1990 [host, distribution, economic importance, biological control: 553-542]; Nair1964 [host, distribution, economic importance: 72]; Nakaha1982 [host, distribution: 22]; Nakaha1983 [host, distribution: 10]; Nath1972 [host, distribution: 1-2]; Nel1933 [taxonomy: 418]; Newste1901b [taxonomy, description, host, distribution: 82,104]; Newste1917b [host, distribution: 131]; NotzP1974 [host, distribution, biological control: 127-143]; NourElRi1970 [host, distribution: 123-127]; Noyes1990a [biological control: 155]; NRC1969 [taxonomy, economic importance, ecology, biological control, chemical control]; Nur1990a [taxonomy, structure, chromosomes: 181,183,185]; Osburn1939 [chemical control: 688-690]; OsburnMa1944 [chemical control: 516-519]; OsburnMa1946 [life history, ecology, host, distribution: 571-574]; OsburnMa1948 [chemical control: 454-456]; OsburnSp1938 [chemical control: 731-732]; Otero1935 [host, distribution: 1-26]; OteroCaMo1996 [host, distribution, biological control: 530-535]; Pace1939 [host, distribution: 664-665]; Paik1972 [host, distribution: 1-4]; Painte1951 [economic importance, control]; PalmerMo1990 [biological control: 67-76]; Peleg1982 [chemical control: 27-32]; Peleg1983 [chemical control: 367-372]; PelegGo1981 [chemical control: 124-126]; PelegNa1966 [life history, chemistry, biological control: 97-98]; Pelliz1987 [host, distribution: 121]; Penzig1887 [host, distribution: 3]; PeralL1968 [biological control: 22-29]; PeralL1968 [host, biological control: 22-29]; PerezGCa1987 [host, distribution: 128]; Petch1921a [biological control: 89-167]; PicartMa2000 [host, distribution: 44-46]; Ponnam1999 [host, distribution, chemical control: 445-451]; Porath1969 [economic importance, biological control: 41-43]; Pratt1956 [life history, ecology, economic importance: 87-93]; Pratt1958 [taxonomy, illustration, distribution]; Priesn1931 [host, distribution, life history, economic importance, control: 1-19]; PriesnHo1940 [biological control: 58-70]; Prinsl1983 [biological control: 26]; Prinsl1984 [taxonomy, biological control: 38-43]; PrunaAl1973 [host, distribution: 1-12]; PruthiMa1945 [host, distribution, life history, control: 1-42]; Pulsel1927 [biological control: 300-327]; Quayle1938a [host, distribution, life history, economic importance, chemical control, biological control]; Quedna1964b [biological control: 86-16]; QuezadCoDi1972 [host, distribution, biological control, economic importance: 12-14]; RahmanAn1941 [taxonomy, description, host, distribution: 816,821-822]; Ramakr1919 [host, distribution, economic importance: 623]; Ramakr1919a [taxonomy: 21]; Ramakr1921a [host, distribution: 356]; Ramakr1930 [taxonomy: 25]; Ramakr1938a [host, distribution: 341-351]; RangelGo1945 [distribution, description: 1-44]; Rao1969 [biological control: 785-792]; RaoGhSa1971 [host, distribution, biological control]; Rehman1996 [biological control, chemical control: 1-96]; RehmanBrNi1999 [biological control, chemical control: 252-257]; RehmanBrNi2000 [host, distribution, chemical control, biological control: 87-93]; RehmanBrSa1999a [biological control, chemical control: 28-33]; ReiderKo1994 [host, distribution: 423-427]; RiehlBrMc1980 [chemical control, biological control, economic importance: 319-363]; Rivnay1968 [host, distribution, economic importance, biological control: 1-156]; Robert1958 [host, distribution, economic importance: 411-415]; Robins1917 [taxonomy, description, host, distribution: 25]; Ronna1928 [host, distribution: 1]; Rose1990c [distribution, economic importance: 535-542]; Rose1990d [host, biological control, economic importance: 357-365]; Rose1992 [biological control]; Rosen1965 [host, distribution, biological control: 388-396]; Rosen1966 [taxonomy, biological control: 43-79]; Rosen1967b [host, distribution, biological control: 1422-1427]; Rosen1969 [biological control: 45-53]; Rosen1973 [biological control: 47-54]; Rosen1979 [host, distribution, biological control: 289-292]; Rosen1987 [taxonomy, biological control: 191]; Rosen1990 [biological control: 413-415]; Rosen1990a [biological control: 502]; Rosen1993 [biological control: 411-416]; RosenDe1973 [biological control: 215-222]; RosenDe1978 [economic importance, biological control, life history, host, distribution: 98-100,102-104]; RosenDe1979 [host, distribution, biological control: 244-247,272-274,]; RosenhHe1994 [life history, biological control: 41-78]; RossleRo1990 [biological control: 519-526]; Russo1929 [distribution, economic importance: 2-3]; Ruther1915a [taxonomy, description, host, distribution: 107]; Sailer1973 [biological control: 15-27]; Sailer1976 [biological control: 195-209]; Sailer1981 [biological control, economic importance: 419-432]; Sailer1983 [distribution, economic importance : 15-38]; SalamaSa1984 [host, chemistry, ecology: 393-398]; Sander1904a [taxonomy, description, illustration, host, distribution: 70]; SassceWe1923 [chemical control: 84-87]; Schmut1959 [taxonomy, description, host, distribution: 54,57]; Schmut1969 [taxonomy, description, host, distribution, life history, economic importance: 109-110]; Schmut1990a [host, distribution: 393]; Schmut1998 [economic importance, host, distribution: 37]; Schmut2001 [host, distribution: 339-345]; SchmutKlLu1957 [host, distribution, economic importance, distribution: 480-481]; SchuhMo1948 [host, distribution, control]; SchweiGr1936 [host, distribution, economic importance, life history, control: 677-714]; Searle1964 [host, distribution: 1-18]; Sefer1961 [host, distribution: 23]; SelhimBr1977 [host, distribution, biological control: 475-478]; SelhimMuSi1969 [biological control: 954-955]; ShafikHu1938 [chemical control: 357-395]; Shalab1961 [host, distribution: 211-228]; Sharon1980 [host, distribution, life history, biological control: 1-59]; Shen1993 [host, distribution: 58]; ShiLi1991 [host, distribution: 165]; Shirak1912 [host, distribution, economic importance]; Signor1869 [taxonomy: 843]; Signor1870 [taxonomy, description, host, distribution: 109]; Silves1929 [host, distribution: 897-904]; Simant1960a [host, distribution, life history: 49-57]; Simant1962a [economic importance: 105-112]; Simant1962b [economic importance: 1182]; Simant1969 [biological control, economic importance: 889-896]; Simant1969a [life history, economic importance, life history: 889-896]; Simant1976 [host, distribution, chemical control: 135-164]; Simmon1969a [biological control: 765-767]; SimmonBe1976 [biological control: 460]; Smirno1950a [biological control: 190-194]; Smirno1952 [host, distribution, biological control: 63-69]; Smit1964 [host, distribution, economic importance, biological control, chemical control]; Smith1948a [economic importance: 813]; Smith1978a [host, distribution, biological control: 373-377]; SmithBeBr1997 [host, distribution, description, life history, biological control, chemical control]; SpenceOs1938 [host, distribution: 728-730]; Staffo1915 [taxonomy, structure: 69-70]; Starne1897 [taxonomy: 22]; StathaElKo2002 [biological control: 105-109]; SteinbPoRo1987 [host, distribution, biological control: 299-310]; SteinbPoRo1987a [host, distribution, biological control: 199-204]; SteinbPoRo1994 [life history, biological control, ecology: 79-91]; Steine1987 [host, distribution, description, economic importance, control: 1-6]; Strong1922 [host, distribution: 775-780]; Sulliv1930 [host, distribution: 51-59]; SurisC1993 [host, distribution, description: 121-127]; Sweetm1958 [biological control, economic importance: 449-458]; Swirsk1976b [host, distribution, economic importance: 555-559]; Swirsk1989 [biological control: 11-44]; SwirskWyIz2002 [taxonomy, host, distribution, life history, economic importance, biological control: 111-113]; Takagi1962b [taxonomy, host, distribution: 52]; Takagi1969a [taxonomy, description, host, distribution: 86-87]; TakagiRo1981 [host, distribution, biological control: 314-321]; Takaha1929 [host, distribution: 81]; Takaha1932a [host, distribution: 104]; Takaha1934 [host, distribution: 34]; Takaha1936d [taxonomy, description, host, distribution: 7-8]; Takaha1941b [host, distribution: 220]; Takaha1953a [taxonomy, host, distribution: 10-13]; Takaha1955f [host, distribution: 242]; Talhou1950 [host, distribution, economic importance: 133-141]; Talhou1975 [economic importance: 25]; Tang1984 [taxonomy, description, host, distribution: 35]; Tao1999 [taxonomy, host, distribution: 80]; Targio1884 [taxonomy: 388-389]; Terezn1986 [taxonomy, description, illustration, host, distribution: 96-97]; Thomps1935 [chemical control: 1-38]; Thomps1938 [chemical control: 109-114]; Thomps1938a [host, distribution, control: 51-59]; Thomps1939 [chemical control: 782-789]; Thomps1939a [economic importance, chemical control, host, distribution: 104-111]; Thomps1942a [host, distribution, economic importance, control: 51-59]; Thomps1950 [chemical control: 155]; ThompsGr1949 [taxonomy, description, host, distribution, life history, economic importance, chemical control, biological control:

1-40]; TianCh1991 [biological control: 64-66]; Toledo1940 [host, distribution, life history: 559-578]; Tower1911 [host, distribution, control: 1]; Trabut1911 [host, distribution: 61]; Trimbl1929 [host, distribution, economic importance, description, control: 1-21]; Trjapi1989 [biological control: 296,312]; Trujil1942 [host, distribution, economic importance]; Tryon1889 [taxonomy, description, host, distribution: 131]; Tschor1939 [host, distribution: 90]; Tuncyu1970 [host, distribution, economic importance: 30-52]; UsmanPu1955 [host, distribution: 48]; VandenTe1964 [ecology, biological control: 459-488]; Vayssi1913 [host, distribution: 431]; Vayssi1930 [host, distribution]; Vazira1982 [host, distribution, biological control: 29-32]; Velasq1971 [taxonomy, description, illustration, host, distribution: 131-133]; VelasqRi1969 [host, distribution: 195-208]; VeseyF1940 [host, distribution, economic importance, life history: 253-285]; VeseyF1941 [host, distribution, biological control: 161]; VeseyF1953 [host, distribution, biological control: 405-413]; Viggia1984 [biological control: 257-276]; Viggia1990a [biological control: 131]; WadhiBa1964 [host, distribution: 227-260]; WatanaTaCo2000 [host, distribution, life history, biological control: 49-64]; Watson1918 [host, distribution]; Watson1926 [host, distribution]; WatsonBe1937 [host, distribution, control]; WatsonDuLi2000 [host, distribution, life history, biological control: 45-61]; WatsonDuLi2000a [host, distribution, life history, biological control: 31-37]; Webber1897 [chemical control, biological control: 53-58]; WeigelBr1924 [chemical control: 386-389]; Wester1918 [host, distribution, economic importance: 5-57]; Whitco1974 [biological control: 150-169]; Willia1970DJ [host, distribution: 5]; WilliaWa1988 [taxonomy, description, illustration, host, distribution, economic importance: 9,91,93-94]; Wilson1917 [taxonomy, description, host, distribution: 24-26]; Wilson1921 [host, distribution: 20-34]; WilsonGo1962 [host, distribution, economic importance, control: 41-61]; Wolfen1955 [host, distribution, economic importance, chemical control: 27-34]; WolffCo1993 [taxonomy, description, illustration, host, distribution: 37-39]; WongChCh1999 [taxonomy, description, host, distribution: 21,61]; WoodruBeSk1998 [distribution]; WoodwaEvEa1970 [distribution]; Woodwo1909 [host, distribution: 359-360]; Woolle1990 [biological control: 167-176]; Wysoki1997 [host, distribution, economic importance: 805-811]; Yasar1995a [taxonomy, description, illustration, host, distribution: 59-61]; Yinon1969 [life history, biological control: 139-146]; Yother1917 [host, distribution, life history, ecology: 30-40]; Yother1922 [chemical control: 63-67]; YotherMi1933 [chemical control: 38-52]; YousefEl1982 [life history, biological control: 113-117]; Yu1986 [biological control: 143-144]; Zhang2000 [host, distribution, chemical control, economic importance: 6,23]; Zimmer1948 [taxonomy, description, illustration, host, distribution, biological control: 369,372].

Chrysomphalus bifasciculatus Ferris

Chrysomphalus ficus; Ferris, 1937c: 64. Misidentification.

Chrysomphalus bifasciculatus Ferris, 1938a: 199. Type data: U.S.A.: California, Pasadena, Huntington Gardens, on *Aspidistra* sp. Holotype female. Type depository: Davis: Bohart Museum of Entomology, University of California, California, USA.

Chrysomphalus bifascicularis; Lindinger, 1943b: 218. Misspelling of species name.

Chrysomphalus bifasciatus; Murakami, 1970: 72. Misspelling of species name.
Chrysophalus bifasciatus; Wu, 1999b: 234. Misspelling of genus and species names.
COMMON NAME: bifasciculate scale [McKenz1956].
SCALE COVER: Scale of female, circular, quite flat, of a dark chocolate brown colour, with region of subcentrally placed exuviae slightly paler; that of male similar in colour, oval; exuvia towards one end (Ferris, 1938a). Colour photograph by Gill (1997).
HOST PLANTS: **Anacardiaceae**: *Rhus* [McKenz1944, McKenz1956]. **Apocynaceae**: *Nerium oleander* [McKenz1944, McKenz1956]. **Aquifoliaceae**: *Ilex* [Kawai1977], *Ilex aquifolium* [McKenz1944, McKenz1956]. **Araliaceae**: *Aralia* [McKenz1944, McKenz1956], *Fatsia japonica* [TakahaTa1956], *Gilibertia trifida* [TakahaTa1956], *Hedera* [Koszta1996], *Hedera helix* [McKenz1944, McKenz1956], *Hedera rhombea* [TakahaTa1956]. **Aucubaceae**: *Aucuba japonica* [McKenz1944, McKenz1956, TakahaTa1956]. **Buxaceae**: *Buxus microphylla suffruticosa* [TakahaTa1956]. **Caprifoliaceae**: *Viburnum* [Kawai1977], *Viburnum odoratissimum* [TakahaTa1956]. **Celastraceae**: *Catha edulis* [McKenz1944, McKenz1956], *Euonymus* [Borchs1950b, McKenz1956, Kawai1977], *Euonymus japonicus* [Ferris1938a, McKenz1939, McKenz1944, TakahaTa1956, Takagi1969a, RosenDe1979]. **Cycadaceae**: *Cycas* [Takaha1940, Takagi1969a], *Cycas revoluta* [McKenz1944, McKenz1956, TakahaTa1956]. **Cyperaceae**: *Cyperus alternifolius* [McKenz1944, McKenz1956]. **Elaeagnaceae**: *Elaeagnus* [McKenz1944, McKenz1956]. **Fagaceae**: *Castanopsis cuspidata* [Kawai1977], *Lithocarpus edulis* [RosenDe1979], *Quercus glauca* [TakahaTa1956], *Quercus phillyraeoides* [TakahaTa1956]. **Illiciaceae**: *Illicium religiosum* [RosenDe1979]. **Iridaceae**: *Iris japonica* [TakahaTa1956]. **Lauraceae**: *Cinnamomum camphora* [McKenz1944, McKenz1956], *Laurus nobilis* [McKenz1944, McKenz1956, TakahaTa1956]. **Liliaceae**: *Aspidistra* [Ferris1938a, McKenz1939, McKenz1944, McKenz1956, McDani1968, Takagi1969a, BesheaTiHo1973], *Aspidistra elatior* [TakahaTa1956, RosenDe1979], *Dianella ensifolia* [Takagi1969a], *Libertia* [McKenz1944, McKenz1956], *Mondo* [McKenz1956], *Ophiopogon* [McKenz1944]. **Moraceae**: *Ficus* [McKenz1944, McKenz1956]. **Oleaceae**: *Ligustrum* [Kawai1977], *Ligustrum japonicum* [McKenz1944, McKenz1956, TakahaTa1956], *Olea* [McKenz1944, McKenz1956], *Osmanthus* [Kawai1977], *Osmanthus fortunei* [TakahaTa1956]. **Palmae**: *Phoenix canariensis* [McKenz1944, McKenz1956]. **Pandanaceae**: *Pandanus* [Takaha1940, Takagi1969a], *Pandanus odoratissimus* [Takagi1969a]. **Pittosporaceae**: *Pittosporum* [McKenz1944, McKenz1956]. **Proteaceae**: *Hakea saligna* [McKenz1944, McKenz1956]. **Rosaceae**: *Lauro-Cerasus officinalis* [McKenz1944], *Prunus ilicifolia* [McKenz1944, McKenz1956], *Prunus laurocerasus* [McKenz1956], *Raphiolepis* [McKenz1956]. **Rutaceae**: *Citrus* [McKenz1944, McKenz1956]. **Strelitziaceae**: *Strelitzia reginae* [McKenz1944, McKenz1956]. **Theaceae**: *Camellia* [Koszta1996], *Camellia japonica* [McKenz1944, McKenz1956, TakahaTa1956]. **Vitaceae**: *Ampelopsis tricuspidata* [McKenz1944, McKenz1956].
NATURAL ENEMIES: COLEOPTERA **Coccinellidae**: *Chilocorus kuwanae* Silvestri [Kato1968]. HYMENOPTERA **Aphelinidae**: *Aphytis japonicus* DeBach & Azim [RosenDe1979], *Aphytis vandenboschi* DeBach & Rosen [RosenDe1979],

Aphytis yasumatsui Azim [RosenDe1979], *Marietta carnesi* [Uemats1972, Uemats1974, Uemats1976]. **Encyrtidae**: *Comperiella bifasciata* Howard [Trjapi1989].

DISTRIBUTION: **Australasian**: Hawaiian Islands (Hawaii [Ferris1938a, McKenz1939, Zimmer1948, Takagi1969a]). **Nearctic**: Mexico [Nakaha1982]; United States of America (Alabama [BesheaTiHo1973], California [Ferris1938a, McKenz1939, McKenz1956, Takagi1969a], Delaware [Nakaha1982], Georgia [Nakaha1982], Louisiana [Nakaha1982], Maryland [Nakaha1982], New Jersey [Nakaha1982], North Carolina [Nakaha1982], Oklahoma [Nakaha1982], South Carolina [Nakaha1982], Texas [McDani1968], Virginia [Nakaha1982]). **Oriental**: Mongolia [DanzigKo1990]; Taiwan [Takaha1940, TakahaTa1956, Takagi1969a, Nakaha1982]. **Palaearctic**: China (People's Republic) [Ferris1938a, Takagi1969a] (Henan (Honan) [Shen1993, Wu1999b]); Japan [Ferris1938a, Takagi1969a, Kawai1977, RosenDe1979, Kawai1980] (Honshu [TakahaTa1956], Kyushu [TakahaTa1956], Shikoku [TakahaTa1956]); South Korea [Nakaha1982]; Ukraine (Odessa Oblast [Borchs1950b]).

BIOLOGY: On leaves (Ferris, 1938a).

ECONOMIC IMPORTANCE: McKenzie (1956) reported this species as damaging *Euonymus* and *Hedera* in USA, California.

GENERAL: Description and illustration of adult female by Ferris (1938a), McKenzie (1939, 1956), Takagi (1969a), Chou (1985, 1986), Tereznikova (1986), Danzig (1993), Kosztarab (1996) and by Gill (1997).

KEYS: Gill 1997: 96 (female) [Species of California]; Kosztarab 1996: 476 (female) [Northeastern North America]; Danzig 1993: 164 (female) [Europe]; Tereznikova 1986: 92-93 (female) [Ukraine]; Chou 1985: 284 (female) [Species of China]; Kawai 1980: 209-210 (female) [Japan]; McDaniel 1968: 227 (female) [U.S.A.: Texas]; Balachowsky 1956: 88 (female) [Africa]; McKenzie 1956: 24 (female) [U.S.A.: California]; Zimmerman 1948: 368-369 (female) [Hawaii]; McKenzie 1943: 150 (female) [World]; Ferris 1942: 31 (female) [North America]; McKenzie 1939: 63 (female) [World].

CITATIONS: Azim1961 [host, distribution, biological control: 97-109]; Balach1948b [taxonomy: 350]; Balach1956 [taxonomy: 88]; BeardsDaHo1976 [economic importance: 103]; BesheaTiHo1973 [host, distribution: 5]; Borchs1950b [taxonomy, description, host, distribution: 218]; Borchs1966 [catalogue: 286-287]; Caltag1985 [taxonomy, biological control: 189-200]; Chou1985 [taxonomy, description, host, distribution: 284,287-288]; Chou1986 [taxonomy, illustration: 675]; ComperFlSm1941 [biological control: 291,301]; Danzig1964 [taxonomy, host, distribution: 651]; Danzig1993 [taxonomy, description, illustration, host, distribution, economic importance: 166-167]; DanzigKo1990 [host, distribution: 46]; DanzigPe1998 [catalogue: 215]; DavidsMi1990 [host, distribution, economic importance: 603-632]; Ferris1937c [taxonomy: 64]; Ferris1938a [taxonomy, description, illustration, host, distribution: 199]; Ferris1942 [taxonomy: 446:31]; Flande1969 [biological control: 29-33]; Flande1971 [biological control, life history: 857-872]; Gill1997 [host, distribution, taxonomy, description, illustration, economic importance: 96-97,99]; Hewitt1943 [host, distribution: 266-274]; HoyHe1985 [biological control]; Kato1968 [host, distribution, life history, biological control: 29-

38]; Kawai1977 [host, distribution, economic importance: 158,160-162]; Kawai1980 [taxonomy, description, host, distribution: 210]; Koszta1996 [taxonomy, description, illustration, host, distribution, biological control, life history, economic importance: 479-480]; LiLi1990 [host, biological control: 68-70]; Lindin1943b [taxonomy: 218]; McDani1968 [taxonomy, illustration, host, distribution: 227-228]; McKenz1939 [taxonomy, description, illustration, host, distribution: 53,57,69]; McKenz1943 [taxonomy: 150]; McKenz1944 [taxonomy, host, distribution : 56]; McKenz1956 [taxonomy, description, illustration, host, distribution: 24,51-53]; MillerDa1990 [host, distribution, economic importance: 301]; Muraka1970 [host, distribution, life history: 72-73]; Nakaha1982 [host, distribution: 22]; RosenDe1979 [host, distribution, biological control: 400-402,558-561,]; SchmutKlLu1957 [host, distribution, economic importance: 478]; Shen1993 [host, distribution: 58]; ShiLi1991 [host, distribution: 165]; Takagi1969a [taxonomy, description, host, distribution: 87]; TakagiRo1981 [host, distribution, biological control: 314-321]; Takaha1940 [taxonomy, host, distribution: 28]; Takaha1953a [taxonomy, host, distribution: 12]; TakahaTa1956 [host, distribution: 16]; Tanaka1966 [biological control: 1-42]; Tao1999 [taxonomy, host, distribution: 80]; Terezn1986 [taxonomy, description, illustration, host, distribution: 93-94]; Trjapi1989 [biological control: 296]; Uemats1972 [biological control: 187-192]; Uemats1974 [host, distribution, life history, biological control: 177-182]; Uemats1976 [host, distribution, life history, biological control: 115-119]; Uemats1978a [host, distribution, life history, biological control: 135-140]; Wu1999b [taxonomy, host, distribution: 234]; Zimmer1948 [taxonomy, description, illustration, host, distribution: 369-370].

Chrysomphalus dictyospermi (Morgan)

Aspidiotus dictyospermi Morgan, 1889a: 352. Type data: GUYANA: Demerara, on *Dictyospermum album*; collected by McIntire. Syntypes, female. Type depository: London: The Natural History Museum, England, UK.

Aspidiotus dictyospermi arecae Newstead, 1893d: 185. Type data: GUYANA: Demerara, Botanic Garden, on leaves of *Areca triandra*. Syntypes, female. Type depository: London: The Natural History Museum, England, UK. Synonymy by McKenzie, 1939: 287.

Aspidiotus mangiferae Cockerell, 1893j: 255. Type data: JAMAICA: on mango [=*Mangifera indica*]. Syntypes, female. Type depository: Washington: United States National Entomological Collection, U.S. National Museum of Natural History, District of Columbia, USA. Synonymy by McKenzie, 1939: 57.

Aspidiotus arecae; Cockerell, 1894: 129. Change of status.

Aspidiotus dictyospermi jamaicensis Cockerell, 1894g: 129. Type data: JAMAICA: King's House, on *Cycas* sp. Syntypes, female. Type depository: Washington: United States National Entomological Collection, U.S. National Museum of Natural History, District of Columbia, USA. Synonymy by Borchsenius, 1966: 287.

Chrysomphalus dictyospermi; Maskell, 1895b: 44. Change of combination.

Chrysomphalus minor Berlese in Berlese & Leonardi, 1896: 346. Type data: ITALY: Florence, Botanical Garden, on *Pandanus graminifoliae*. Syntypes, female and first instar. Type depository: Portici: Dipartimento de Entomologia e Zoologia Agraria di Portici, Università di Napoli Federico II, Italy. Synonymy by Borchsenius, 1966: 288.

Aspidiotus minor; Cockerell, 1896b: 334. Change of combination.

Aspidiotus (*Chrysomphalus*) *dictyospermi*; Cockerell, 1897i: 23. Change of combination.

Aspidiotus (*Chrysomphalus*) *dictyospermi arecae*; Cockerell, 1897i: 23. Change of combination.

Aspidiotus (*Chrysomphalus*) *dictyospermi jamaicensis*; Cockerell, 1897i: 23. Change of combination.

Aspidiotus (*Chrysomphalus*) *mangiferae*; Cockerell, 1897i: 24. Change of combination.

Aspidiotus (*Chrysomphalus*) *minor*; Cockerell, 1897i: 30. Change of combination.

Chrysomphalus mangiferae; Leonardi, 1899: 199. Change of combination.

Chrysomphalus dictyospermi arecae; Leonardi, 1899: 218. Change of combination.

Chrysomphalus dictyospermi mangiferae; Cockerell, 1899a: 396. Change of combination and rank.

Chrysomphalus dictyospermi minor; Marchal, 1904: 246. Change of combination and rank.

Chrysomphalus dictyospermi; Lindinger, 1912: 92. Revived combination.

Aspidiotus agrumincola De Gregorio, 1915: 125, 164. Type data: ITALY: Sicily, Palermo, on *Citrus*. Syntypes, female. Synonymy by Ferris, 1941e: 40. Notes: Type material very probably lost; personal communication from S. Ragusa di Chiara (Palermo, Sicily) to Yair Ben-Dov, September 2000.

Chrysomphalus dictyospermi agrumicola De Gregorio, 1915: 161. Type data: ITALY: Sicily, on *Citrus*. Syntypes, female and first instar. Synonymy by McKenzie, 1939: 55. Notes: Depository of type material unknown.

Chrysomphalus arecae; Malenotti, 1916c: 114. Change of combination.

Chrysomphalus jamaicensis; Malenotti, 1917: 114. Change of combination and rank.

Chrysomphalus castigatus Mamet, 1936: 94. Type data: MAURITIUS: Rose Hill and Ebéne, on leaves of *Furcraea gigantea*, in association with *Chrysomphalus ficus* Maskell, *Pinnaspis minor* Maskell and *Lecanium mauritiense* sp. n. Syntypes. Type depository: Paris: Muséum national d'Histoire naturelle, France. Synonymy by McKenzie, 1939: 58.

Chrysomphalus dictyospermi; Ferris, 1938a: 200. Revived combination.

Aspidiotus agrumincola; Ferris, 1941e: 40. Notes: Incorrect citation of "Gregorio" as author.

Aspidiotus jamaicensis; Ferris, 1941e: 44. Change of combination and rank.

Chrysomphalus dictyospermatis Lindinger, 1949: 211. Unjustified emendation; discovered by Borchsenius, 1966: 288.

Chrysomphalus dictiospermi; Quezada et al., 1972: 14. Misspelling of species name.

Chrysomphalus jamaucebsis; Chou, 1985: 288. Misspelling of species name.

Chrisomphalus dictyospermi; Yasnosh, 1995: 248. Misspelling of genus name.

COMMON NAMES: Bianca-Rossa [Maleno1918]; Dictyospermum scale [Merril1953, McKenz1956, Dekle1965c, RosenDe1978]; escama anaranjada [Gonzal1989]; escama dictiosperma [CoronaRuMo1997]; korichnevaya shitovka [Borchs1936]; pinta-amarela [CarvalAg1997]; piojo rojo [Lloren1990]; pou rouge des orangers [SchmutKlLu1957]; Spanish red scale [Katsoy1996].

SCALE COVER: Female scale greyish-white; exuviae central, depressed, elongate oval; about 1.2 mm longest diameter; centre of larval skin dark orange, whilst exuviae light yellow (Morgan, 1899a). Scale of female rather thin, circular, flat, light, brown or yellowish, exuviae central; that of male elongate oval, similar in colour, exuvia toward one end (Ferris, 1938a). Colour photograph of scale cover and general appearance by Gonzalez (1989), Katsoyannos (1996), Gill (1997), Carvalho & Aguiar (1997) and by Wong *et al.* (1999).

HOST PLANTS: **Acanthaceae**: *Sciaphyllum brownii* [Martin1983]. **Aceraceae**: *Acer palmatum* [McKenz1956]. **Agavaceae**: *Agave* [MerrilCh1923, Matile1978], *Dracaena* [Balach1956], *Dracaena draco* [GomezM1962], *Furcraea gigantea* [Mamet1936, Mamet1943a, Mamet1949, Borchs1966], *Pincenectitia tuberculata* [Hall1926a], *Yucca* [Balach1932d], *Yucca elephantipes* [Balach1932d]. **Amaryllidaceae**: *Doryanthes palmeri* [Merril1953]. **Anacardiaceae**: *Anacardium occidentale* [WilliaWa1988], *Lithraea arhoeirihna* [Balach1932d], *Mangifera indica* [Green1916, Takaha1933, Lepage1938, Takaha1939b, Takagi1970, Almeid1973b, WilliaWa1988, KinjoNaHi1996], *Rhus succudenae* [Borchs1936], *Schinus terebinthifolius* [Balach1932d], *Spondias dulcis* [WilliaWa1988], *Spondias mangifera* [Lepage1938]. **Annonaceae**: *Annona* [Lepage1938], *Annona cherimola* [McKenz1956], *Annona muricata* [Balach1956]. **Apocynaceae**: *Allamanda* [MerrilCh1923], *Carissa* [MerrilCh1923], *Nerium oleander* [Balach1927, Balach1932d, McKenz1956, Martin1983], *Plumeria* [WilliaWa1988], *Trachelospermum* [Merril1953], *Vinca major* [Merril1953]. **Araceae**: *Anthurium* [MerrilCh1923, Lepage1938, Zimmer1948], *Epipremnum pinnatum* [WilliaWa1988], *Monstera deliciosa* [Balach1932d], *Xanthosoma* [WilliaWa1988]. **Araliaceae**: *Aralia sieboldi* [Martin1983], *Hedera chrysocarpa* [Martin1983], *Hedera helix* [Cocker1900h, Balach1932d, Green1937, Bodenh1952, McKenz1956, Hadzib1983, Martin1983], *Meryta macrophylla* [WilliaWa1988], *Schefflera* [WilliaWa1988]. **Araucariaceae**: *Araucaria braziliana* [Martin1983]. **Arecaceae** [WilliaWa1988], *Elaeis guineensis* [Almeid1973b, WilliaWa1988], *Howeia forsteriana* [WilliaWa1988], *Roystonea regia* [WilliaWa1988], *Veitchia joannis* [WilliaWa1988]. **Berberidaceae**: *Berberis aquifolium* [Martin1983]. **Bromeliaceae**: *Cryptanthus* [Merril1953]. **Buxaceae**: *Buxus* [Leonar1920], *Buxus balearica* [Balach1927, Balach1932d, GomezM1946, Martin1983], *Buxus sempervirens* [Green1923b, Balach1932d, Borchs1936, Hadzib1983]. **Cactaceae**: *Cactus* [McKenz1956], *Opuntia* [Ramakr1919a, Ramakr1921a], *Opuntia cochinellifera* [Green1905a]. **Capparidaceae**: *Capparis spinosa* [GomezM1946, Martin1983]. **Caprifoliaceae**: *Lonicera caprifolium* [Martin1983], *Lonicera implexa* [Balach1932d]. **Caricaceae**: *Carica papaya* [WilliaWa1988]. **Casuarinaceae**: *Casuarina* [WilliaWa1988], *Casuarina stricta* [Merril1953]. **Celastraceae**: *Elaeodendron* [Brimbl1962a], *Euonymus* [McKenz1956], *Euonymus japonicum* [Balach1932d, GomezM1948], *Euonymus japonicus* [Bodenh1952, Takagi1958,

Hadzib1983, Martin1983]. **Clusiaceae**: *Calophyllum* [Ramakr1919a, Ramakr1921a, Green1937], *Calophyllum inophyllum* [WilliaWa1988]. **Combretaceae**: *Terminalia catappa* [WilliaWa1988]. **Compositae**: *Bahia fastigata* [MerrilCh1923]. **Cornaceae**: *Cornus mas* [Hadzib1983]. **Cupressaceae**: *Cupressus macrocarpa* [Brain1919], *Thuja occidentalis* [Almeid1973b]. **Cycadaceae**: *Cycas* [Cocker1894g, Ramakr1921a, MerrilCh1923, Lepage1938, WilliaWa1988], *Cycas circinalis* [GomezM1941, Martin1983], *Cycas revoluta* [Takaha1929, Balach1932d, Almeid1973b, Takagi1970, Martin1983, Dziedz1989]. **Cyperaceae**: *Cyperus alternifolius* [Balach1932d]. **Daphniphyllaceae**: *Daphniphyllum humile* [Merril1953]. **Ebenaceae**: *Diospyros* [Green1904a], *Diospyros lotus* [Hadzib1983]. **Ehretiaceae**: *Patagonula americana* [Martin1983]. **Elaeagnaceae**: *Elaeagnus* [MerrilCh1923, Merril1953, McKenz1956, Dekle1965c, RosenDe1979], *Elaeagnus reflexa* [Balach1932d]. **Ericaceae**: *Arbutus unedo* [Balach1932d, Martin1983], *Enkianthus* [Takagi1958]. **Euphorbiaceae**: *Aleurites* [Borchs1936], *Euphorbia* [Merril1953], *Euphorbia regie-jubae* [MatileBa1972], *Macaranga* [WilliaWa1988], *Manihot esculenta* [WilliaWa1988], *Manihot palmata* [Martin1983]. **Fagaceae**: *Quercus lusitanica* [Balach1932d], *Quercus mirbeckii* [Balach1932d], *Quercus suber* [Balach1932d]. **Gramineae**: *Bambusa vulgaris* [Dziedz1989]. **Guttiferae**: *Mammea* [MerrilCh1923], *Mammea africana* [Merril1953], *Rheedia aristata* [Wilson1917, MerrilCh1923]. **Heliconiaceae**: *Heliconia rex* [McKenz1956]. **Iridaceae**: *Iris germanica* [Balach1932d]. **Lauraceae**: *Benzoin* [Merril1953], *Cinnamomum* [Ferris1938a], *Cinnamomum camphora* [Green1923b, Borchs1936, Hadzib1983, Martin1983], *Cinnamomum zeylanicum* [Houser1918, Laing1932], *Endiandra palmerstoni* [Brimbl1962a], *Laurus maderensis* [Martin1983], *Laurus nobilis* [Balach1932d, Borchs1936, McKenz1956, Almeid1973b, BenDov1980, Martin1983], *Ocotea foetens* [GomezM1962], *Persea americana* [Newste1914, Houser1918, McKenz1956, Dekle1965c, WilliaWa1988], *Persea gratissima* [Martin1983]. **Lecythidaceae**: *Barringtonia* [WilliaWa1988], *Barringtonia racemosa* [WilliaWa1988]. **Leguminosae**: *Acacia* [MerrilCh1923, Balach1932d], *Acacia cyanophylla* [Balach1932d], *Albizia* [MerrilCh1923], *Baikiaea minor* [Ghesqu1932], *Caesalpinia sappan* [Merril1953], *Cassia spectabilis* [MatileNo1984], *Ceratonia siliqua* [Balach1932d, Martin1983], *Cercis siliquastrum* [Balach1932d, Martin1983], *Crotalaria spectabilis* [Merril1953], *Erythrina indica* [Maskel1898, Kuwana1927, Lepage1938], *Sarothamnus scoparius* [Green1923b], *Sophora japonica* [Balach1932d], *Sophora secundiflora* [Balach1932d, Martin1983], *Spartium junceum* [Martin1983]. **Liliaceae**: *Aloe ciliaris* [Brimbl1962a], *Aloe purpurascens* [GomezM1946, Martin1983], *Aloe zeyberi* [Lepage1938], *Anthericum* [McKenz1956], *Asparagus plumosus* [WilliaWa1988], *Asparagus sprengeri* [Green1923b], *Aspidistra* [McKenz1956], *Dianella intermedia* [WilliaWa1988], *Gasteria* [Martin1983], *Ophiopogon* [McKenz1956, GomezM1962], *Ophiopogon japonicus* [Balach1932d]. **Lomariopsidaceae**: *Arthrobotrya odoratisima* [GomezM1962]. **Lythraceae**: *Lagerstroemia flos-reginae* [WilliaWa1988], *Lagerstroemia indica* [GomezM1941]. **Magnoliaceae**: *Magnolia* [Bodenh1952, GomezM1962]. **Malpighiaceae**: *Malpighia glabra* [Wilson1917]. **Malvaceae**: *Hibiscus syriacus* [Hadzib1983]. **Marantaceae**: *Calathea* [WilliaWa1988]. **Meliaceae**: *Swietenia macrophylla* [WilliaWa1988].

Menispermaceae: *Cocculus* [Merril1953]. **Moraceae**: *Artocarpus altilis* [WilliaWa1988], *Artocarpus heterophyllus* [WilliaWa1988], *Artocarpus incisa* [WilliaWa1988], *Artocarpus integra* [Takaha1939b], *Ficus* [Ferris1938a, McKenz1956, GomezM1962, Savesc1982, WilliaWa1988], *Ficus* [MerrilCh1923, Merril1953, Dekle1965c], *Ficus altissima* [Wilson1917], *Ficus benjamina* [Martin1983], *Ficus carica* [Bodenh1952, GomezM1954], *Ficus corynocali* [Martin1983], *Ficus eburnia* [Wilson1917], *Ficus elastica* [Wilson1917, Balach1927, Balach1932d, Bodenh1952, McKenz1956, TakahaTa1956, Martin1983], *Ficus indica* [Martin1983], *Ficus macrophylla* [Balach1927, Balach1932d, Martin1983], *Ficus nitida* [Balach1927, Balach1932d, Martin1983], *Ficus populnea* [Martin1983], *Ficus pumila* [MatileNo1984], *Ficus tinctoria* [WilliaWa1988]. **Musaceae**: *Musa* [WilliaWa1988], *Musa cavendishi* [Green1923b, Balach1932d], *Musa sapientum* [WilliaWa1988]. **Myoporaceae**: *Myoporum* [Balach1932d], *Myoporum desertum* [Martin1983], *Myoporum loetum* [Martin1983], *Myoporum pictum* [GomezM1948, Martin1983]. **Myristicaceae**: *Myristica* [WilliaWa1988], *Myristica fragrans* [Green1904a]. **Myrtaceae**: *Agonis flexuosa* [Brimbl1962a], *Eucalyptus* [MerrilCh1923, Balach1932d], *Eucalyptus cinerifolia* [Martin1983], *Eucalyptus corynocali* [Martin1983], *Eucalyptus gunni* [Martin1983], *Eugenia* [MerrilCh1923], *Eugenia jambo* [Balach1932d, DeLott1967a], *Eugenia malaccensis* [WilliaWa1988], *Eugenia micheli* [Martin1983], *Eugenia vaccinifolia* [Mamet1954, Borchs1966], *Feijoa selloviana* [Balach1932d], *Feijoa sellowiana* [Merril1953], *Myrtus* [Merril1953, McKenz1956, GomezM1962], *Psidium guajava* [Hall1925, Balach1932d, Lepage1938, WilliaWa1988], *Spermolepis gummifera* [Cohic1958, WilliaWa1988], *Syzygium* [Merril1953]. **Ochnaceae**: *Schuurmansia henningsii* [WilliaWa1988]. **Oleaceae**: *Jasminum* [McKenz1956], *Jasminum* [Merril1953], *Jasminum officinalis* [Wilson1917, Martin1983], *Ligustrum* [Leonar1920, MerrilCh1923, RosenDe1979], *Ligustrum coriaceum* [Martin1983], *Ligustrum kellerianus* [Martin1983], *Ligustrum sinensis* [Balach1932d], *Olea* [Bodenh1952], *Olea europaea* [Balach1932d, Martin1983], *Olea fragrans* [Borchs1934, Borchs1936]. **Orchidaceae**: *Bachia mayor* [MerrilCh1923], *Coclogyne cristata* [Cocker1922a], *Cymbidium* [Green1931a], *Cypripedium* [Lepage1938], *Dendrobium* [Ramakr1919a, Ramakr1921a, Lepage1938], *Epidendrum tampens* [McKenz1956], *Odontoglossum* [McKenz1956], *Vanilla* [WilliaWa1988], *Vanilla fragrans* [WilliaWa1988]. **Paeoniaceae**: *Paeonia* [GomezM1948, Martin1983]. **Palmae**: *Areca* [WilliaWa1988], *Areca catechu* [WilliaWa1988], *Areca sajuda* [GomezM1962], *Areca triandra* [Newste1893d, Lepage1938], *Chamaerops humilis* [Balach1932d, Martin1983], *Cocos* [Lepage1938, McKenz1956], *Cocos nucifera* [Mamet1943a, Mamet1954, Borchs1966, Almeid1973b, WilliaWa1988], *Cocos romanzoffiana* [Balach1932d], *Dictyosperma* [McKenz1939, McKenz1956, McDani1968], *Dictyosperma album* [Cocker1899n, Morgan1889a, Lepage1938, Matile1978], *Drymophloeus robustus* [Lepage1938], *Howeia selloviana* [Balach1932d], *Hyphaene* [Mamet1959a, Borchs1966], *Hyphaene thebaica* [DeLott1967a], *Kentia* [Ferris1938a, Balach1937c, McKenz1956], *Kentia belmoreana* [Wilson1917, Martin1983], *Latania* [Hempel1900a, Martin1983], *Latania borbonica* [Martin1983], *Phoenix* [Green1923b, Balach1937c, Balach1938a], *Phoenix* *canariensis* [Balach1927, Balach1932d, Bachma1953, McKenz1956, Martin1983], *Phoenix dactylifera* [GomezM1948, Martin1983], *Phoenix reclinata* [Balach1932d], *Sabal andasoni* [Balach1932d], *Sabal blackburniana* [Wilson1917], *Sabal havanensis* [Wilson1917], *Verschaffeltia splendida* [Zimmer1948]. **Pandanaceae**: *Pandanus* [Wilson1917, MerrilCh1923, McKenz1956], *Pandanus graminifolia* [BerlesLe1896]. **Passifloraceae**: *Passiflora coerulea* [Balach1932d]. **Phytolaccaceae**: *Phytollaca americana* [Hadzib1983]. **Pinaceae**: *Pinus caribaea* [WilliaWa1988], *Pinus thunbergii* [Takaha1932a, Takaha1933, Takagi1970]. **Pistaciaceae**: *Pistacia lentiscus* [Balach1927, Balach1932d, Balach1933e, Foldi2000], *Pistacia mutica* [Bodenh1944b]. **Pittosporaceae**: *Pittosporum tobira* [Balach1932d]. **Platanaceae**: *Platanus orientalis* [Balach1932d]. **Polygonaceae**: *Muehlenbeckia platyclados* [Martin1983]. **Proteaceae**: *Banksia* [Balach1932d], *Macadamia tetraphylla* [WilliaWa1988]. **Ranunculaceae**: *Caltha edulis* [Martin1983]. **Rosaceae**: *Cerasus lauro-cerassus* [Martin1983], *Cotoneaster* [Balach1932d], *Crataegus azarolus* [GomezM1946, Martin1983], *Eriobotrya japonica* [Balach1932d], *Fragaria vesca* [GomezM1946, Martin1983], *Laurocerasus officinalis* [Hadzib1983], *Malus pumila* [PerezGCa1987], *Malus sylvestris* [WilliaWa1988], *Photinia* [Martin1983], *Photinia serrulata* [Merril1953], *Prunus laurocerasus* [Balach1932d, Borchs1934, Borchs1936, Martin1983], *Prunus spinosa* [Martin1983], *Pyrus communis* [Brimbl1962a], *Pyrus cydonia* [Balach1932d], *Rosa* [Houser1918, Green1923b, Lepage1938, McKenz1956], *Rosa centifolia* [Martin1983], *Rosa indica* [WilliaWa1988]. **Rubiaceae**: *Randia fitzalani* [Brimbl1962a]. **Ruscaceae**: *Ruscus* [Merril1953], *Ruscus aculeatus* [Balach1932d], *Ruscus hypoglossum* [Balach1932d]. **Rutaceae**: *Citrus* [Cocker1900h, Balach1932d, Lepage1938, Bodenh1952, Merril1953, Almeid1973b, Takagi1970, Martin1983], *Ci. aurantifolia* [WilliaWa1988], *Ci. aurantium* [Bodenh1928, Hall1929, Martin1983], *Ci. deliciosa* [DeLott1967a], *Ci. limetta* [Martin1983], *Ci. limon* [Hadzib1983, Martin1983], *Ci. maxima* [WilliaWa1988], *Ci. nobilis unshiu* [Borchs1934], *Ci. paradisi* [Hadzib1983], *Ci. sinensis* [Hadzib1983, WilliaWa1988], *Ci. unschiu* [Hadzib1983], *Ruta halepensis* [Hall1926a]. **Salicaceae**: *Populus alba* [Balach1932d], *Populus nigra* [Balach1932d], *Salix* [McKenz1956]. **Sapotaceae**: *Mimusops elenga* [Ramakr1919a, Ramakr1921a], *Palaquium* [Green1904a]. **Solanaceae**: *Solanum melongena* [WilliaWa1988]. **Sterculiaceae**: *Brachychiton* [Balach1932d], *Brachychiton populneus* [Martin1983], *Sterculia diversifolia* [Martin1983]. **Strelitziaceae**: *Strelitzia* [Bodenh1952], *Strelitzia augusta* [Balach1927, Balach1932d], *Strelitzia ericolai* [Martin1983], *Strelitzia reginae* [Balach1932d, McKenz1956]. **Taxaceae**: *Taxus* [McKenz1956], *Taxus baccata* [Martin1983]. **Theaceae**: *Camellia* [MerrilCh1923], *Camellia japonica* [Wilson1917, McKenz1956, Martin1983], *Camellia sinensis* [WilliaWa1988], *Camellia thea* [Lepage1938], *Cleyera japonica* [BesheaTiHo1973], *Thea sinensis* [Hadzib1983]. **Ulmaceae**: *Chaetacme aristata* [Brain1919]. **Vitaceae**: *Vitis* [Bodenh1952, Merril1953], *Vitis vinifera* [Martin1983]. **Zamiaceae**: *Zamia* [MerrilCh1923]. **Zingiberaceae**: *Alpinia nutans* [WilliaWa1988], *Phaeomeria speciosa* [WilliaWa1988], *Zingiber* [WilliaWa1988]. **NATURAL ENEMIES**: ACARI **Hemisarcoptidae**: *Hemisarcoptes dzhashii* Dzhibladze [GersonOcHo1990], *Hemisarcoptes malus* (Shimer) [Coorem1951].

COLEOPTERA **Coccinellidae**: *Chilocorus bipustulatus* L. [Balach1948b, Inserr1970a], *Chilocorus renipustulatus* [Hadzib1983], *Exochomus quadripustulatus* L. [Balach1948b, Inserr1970a, Hadzib1983], *Lindorus lophanthae* (Blaisdell) [Smirno1950a, RosenDe1978], *Orcus chalybeus* (Boisduval) [RosenDe1978], *Rhizobius ventralis* (Erich.) [RosenDe1978]. **Nitidulidae**: *Cybocephalus flaviceps* Reitter [Balach1948b]. FUNGI **Ascomycotina**: *Nectria diploa* [EvansPr1990]. **Deuteromycotina**: *Fusarium* sp. [EvansPr1990], *Verticillium* sp. [EvansPr1990]. HYMENOPTERA **Aphelinidae**: *Aphytis africanus* Quednau & Annecke [AnneckIn1971], *Ap. anomalus* Compere [RosenDe1979], *Ap. boselli* Malenotti [Balach1948b, GomezM1946], *Ap. chilensis* Howard [RosenDe1979], *Ap. chrysomphali* (Mercet) [Balach1948b, RosenDe1979, Hadzib1983], *Ap. coheni* DeBach [Gordh1979], *Ap. diaspidis* (Howard) [RosenDe1979], *Ap. hispanica* (Mercet) [RosenDe1979], *Ap. holoxanthus* DeBach [AnneckIn1971], *Ap. lingnanensis* Compere [Inserr1970a, RosenDe1978, RosenDe1979], *Ap. longiclava* Mercet [Balach1948b], *Ap. maculicornis* Masi [Balach1948b, GomezM1946], *Ap. melinus* DeBach [Inserr1970a, RosenDe1978, RosenDe1979], *Ap. mytilaspidis* (Le Baron) [RosenDe1979], *Ap. proclia* (Walker) [QuezadCoDi1972, RosenDe1979, MyartsRu2000], *Aspidiotiphagus citrinus* (Craw) [Balach1948b, Gordh1979, Hadzib1983], *Aspidiotiphagus lounsburyi* (Berlese & Paoli) [Balach1948b, AnneckIn1971, RosenDe1978], *Encarsia citrina* (Craw) [AbdRab2001a], *Prospaltella aurantii* (Howard) [Gordh1979, Hadzib1983], *Prospaltella fasciata* Malenotti [GomezM1946, Balach1948b]. **Encyrtidae**: *Comperiella bifasciata* Howard [Flande1944a], *Comperiella lemniscata* Compere & Annecke [Garonn1991, GaronnVi1993], *Habrolepis pascuorum* Mercet [Antong1937, Balach1948b]. **Signiphoridae**: *Signiphora aleyrodis* Ashmead [Gordh1979], *Signiphora flavella* Girault [Woolle1990], *Signiphora flavopalliata* Ashmead [Gordh1979], *Signiphora merceti* Malenotti [Maleno1916d, Gordh1979, Woolle1990], *Signiphora prepauca* Girault [Woolle1990].

DISTRIBUTION: **Afrotropical**: Angola [Almeid1969, Almeid1973b]; Cape Verde Islands [SchmutPiKl1978, Fernan1999]; Comoros Islands [Matile1978]; Eritrea [DeLott1967a]; Guinea [Balach1956]; Kenya [DeLott1967a]; Madagascar [Mamet1943a, Mamet1954, Mamet1959a, Borchs1966]; Mauritius [Mamet1936, Mamet1949, Borchs1966]; Mozambique [Almeid1971]; Nigeria [Newste1914, Balach1956]; Réunion [Mamet1957]; Seychelles [Mamet1943a, Borchs1966]; South Africa [BrainKe1917, Newste1917b, Brain1919]; Tanzania [Balach1956]; Uganda [Gowdey1917, Newste1917, Balach1956]; Zaire [Laing1932, Ghesqu1932, Balach1956]; Zanzibar [Green1916]; Zimbabwe [Hall1929, Balach1956]. **Australasian**: Australia (Queensland [Brimbl1962a]); Bonin Islands (= Ogasawara-Gunto) [Kawai1987]; Cook Islands [WilliaWa1988]; Federated States of Micronesia (Truk Islands [Beards1966]); Fiji [RosenDe1979, WilliaWa1988]; French Polynesia (Society Islands [WilliaWa1988], Tahiti [WilliaWa1988]); Kiribati [WilliaWa1988]; Marshall Islands [Beards1966]; New Caledonia [Cohic1958]; Niue [WilliaWa1988]; Palau [Takaha1939b, Beards1966]; Papua New Guinea [WilliaWa1988]; Solomon Islands [WilliaWa1988]; Tonga [WilliaWa1988]; Tuvalu [WilliaWa1988]; Western Samoa [WilliaWa1988]. **Nearctic**: Mexico [Cocker1899n, MyartsRu2000] (Veracruz [RosenDe1979]); United States of America (California [McKenz1956,

RosenDe1978], Colorado [Cocker1922a], Florida [Wilson1917, MerrilCh1923, Merril1953, Dekle1965c], Georgia [BesheaTiHo1973], Louisiana [Ferris1938a], Mississippi [Herric1911, Ferris1938a], Missouri [Hollin1923], Texas [McDani1968]). **Neotropical**: Argentina [RosenDe1979, ClapsTe2001]; Brazil [RosenDe1979, WolffCo1993a] (Bahia [WolffCo1993], Distrito federal (Brasilia) [Lepage1938], Par [WolffCo1993], Rio Grande do Sul [WolffCo1993], Rio de Janeiro [Lepage1938], São Paulo [Hempel1900a, Lepage1938]); Chile [GonzalCh1968, Gonzal1989]; Colombia [Kondo2001]; Cuba [Houser1918]; Dominican Republic [GomezM1941]; El Salvador [QuezadCoDi1972]; Guyana [Morgan1889a, Newste1893d, McKenz1939]; Jamaica [Cocker1894g]; Panama [Cocker1899n]; Peru [Beingo1969d]; Puerto Rico & Vieques Island (Puerto Rico [Martor1976, ColonFMe1998]); U.S. Virgin Islands [Nakaha1983]. **Oriental**: India (Karnataka [UsmanPu1955]); Indonesia (Irian Jaya [WilliaWa1988], Java [Green1904a]); Mongolia [DanzigKo1990]; Philippines [VelasqRi1969] (Luzon [Velasq1971]); Ryukyu Islands (= Nansei Shoto) [KinjoNaHi1996]; Sri Lanka [Green1900a, Green1905a, Ramakr1921a]; Taiwan [Takaha1929, Takaha1932a, Takagi1970, WongChCh1999]; Thailand [Takaha1942b]. **Palaearctic**: Algeria [Balach1927, Balach1932d]; Armenia [Borchs1936]; Azerbaijan (Azerbaijan [Borchs1936]); Azores [Fernan1981]; Canary Islands [GomezM1962, GomezM1967O, MatileBa1972, PerezGCa1987, MatileOr2001]; China (People's Republic) [Maskel1898, Kuwana1927] (Henan (Honan) [Shen1993]); Corsica [Balach1931a, Balach1932d]; Crete [Ayouta1940]; Czech Republic [Zahrad1977]; Egypt [Hall1925, Ezzat1958]; France [Cocker1900h, Balach1932d, Balach1933e, Foldi2000]; Georgia (Abkhaz ASSR [Borchs1934, Borchs1936], Adzhar ASSR [Borchs1934], Georgia [Borchs1936, RosenDe1979, Hadzib1983]); Greece [Bodenh1928, ArgyriStMo1976, RosenDe1978]; Iran [Bodenh1944b, Kaussa1955]; Ireland [Green1934d]; Israel [BenDov1980]; Italy [BerlesLe1896, Leonar1920, Viggia1970a, RosenDe1978, RosenDe1979, LongoMaPe1995]; Japan [Kawai1980] (Honshu [TakahaTa1956], Kyushu [Takagi1958], Shikoku [TakahaTa1956]); Libya [Green1937, Martin1954]; Madeira Islands [Green1923b, Balach1938a, CarvalAg1997]; Malta [Borg1919]; Morocco [Vayssi1920, Balach1932d, Rungs1970, RosenDe1979]; Poland [Dziedz1989]; Portugal [Seabra1941]; Romania [Savesc1982]; Russia (Caucasus [BazaroSh1971], Kabardino-Balkarian AR [Borchs1936], Krasnodar Kray [Hadzib1983]); Spain [Balach1935b, GomezM1954, RosenDe1979, Martin1983, BlayGo1993]; Tunisia [Balach1932d]; Turkey [Bodenh1949, Bodenh1952, Tuncyu1970a, RosenDe1979, SengonUyKa1998, UygunSeEr1998]; United Kingdom (England [Green1931a, Ferris1938a]); Yugoslavia [Bachma1953, RosenDe1979].

BIOLOGY: Reported to have uniparental as well as biparental populations in USA (Brown, 1965).

ECONOMIC IMPORTANCE: Widely distributed in subtropical regions of the world (CABI, 1951). Very polyphagous (see Host Plants), and recorded as a most serious pest of citrus in Western Mediterranean Basin, Greece and Iran. A minor pest in Mexico, South America and Republic of Georgia (Rosen & DeBach, 1978).

GENERAL: Description and illustration of adult female by Leonardi (1910), Kuwana (1933), McKenzie (1939, 1956), Balachowsky (1948b, 1956), Zimmerman (1948), Gómez-Menor Guerrero (1962), Bazarov & Shmelev (1971), Velasquez (1971), Chou (1985, 1986), Tereznikova (1986), Dziedzicka (1989), Zahradník (1990b), Danzig (1993), Yaşar (1995a), Kosztarab (1996), Gill (1997) and by Colon-Ferrer & Medina-Gaud (1998).

KEYS: Claps & Teran 2001: 392 (female) [South Africa]; Gill 1997: 96 (female) [Species of California]; Kosztarab 1996: 476 (female) [Northeastern North America]; Danzig 1993: 164 (female) [Europe]; Williams & Watson 1988: 93 (female) [Tropical South Pacific]; Tereznikova 1986: 92-93 (female) [Ukraine]; Chou 1985: 284 (female) [Species of China]; Kawai 1980: 209-210 (female) [Japan]; Velasquez 1971: 131 (female) [Philippines]; McDaniel 1968: 227 (female) [U.S.A.: Texas]; Beardsley 1966: 515 (female) [Federated States of Micronesia]; Gómez-Menor Guerrero 1962: 180 (female) [Canary Islands]; Ezzat 1958: 241 (female) [Egypt]; Balachowsky 1956: 86 (female) [Africa]; McKenzie 1956: 24 (female) [U.S.A.: California]; Lupo 1953: 24 (female) [Italy]; Balachowsky 1948b: 346 (female) [Mediterranean]; Zimmerman 1948: 368-369 (female) [Hawaii]; McKenzie 1943: 150 (female) [World]; Ferris 1942: 31 (female) [North America]; McKenzie 1939: 63 (female) [World]; Kuwana 1933: 26 (female) [Japan]; Kuwana 1933b: 49 (female) [Japan]; Fullaway 1932: 95-97, 107 (female) [Hawaii]; Balachowsky 1928b: 157 (female) [North Africa]; Britton 1923: 376 (female) [U.S.A.: Connecticut]; Hollinger 1923: 29 (female) [U.S.A.: Missouri]; Leonardi 1920: 65 (female) [Italy]; Brain 1919: 198 (female) [South Africa]; Lawson 1917: 210 (female) [U.S.A.: Kansas]; Dietz & Morrison 1916a: 307 (female) [U.S.A.: Indiana]; Cockerell 1905: 45-46 (female) [Mexico]; Cockerell 1905b: 201 (female) [U.S.A.: Colorado]; Newstead 1901b: 82 (female) [England].

CITATIONS: AbdRab2001a [host, distribution, biological control: 176]; Alfier1929 [host, distribution: 7-9]; Almeid1969 [taxonomy, description, host, distribution, biological control: 156-157]; Almeid1971 [host, distribution: 9]; Almeid1973b [host, distribution: 9]; Andrie1932 [host, distribution, economic importance, control]; AnneckIn1971 [host, distribution, biological control: 28]; Antong1937 [biological control: 44-46]; Archan1937 [taxonomy, description, illustration, host, distribution: 94-95]; Argyri1969 [biological control: 817-822]; Argyri1970 [host, distribution, biological control: 57-65]; Argyri1974 [host, distribution, economic importance, biological control: 89-94]; Argyri1977a [host, distribution, biological control: 630-634]; Argyri1979a [host, distribution, economic importance, biological control: 517-520]; Argyri1986 [host, distribution, biological control: 545-548]; Argyri1990 [host, distribution, economic importance: 579-583]; ArgyriMo1981 [host, distribution, economic importance: 623-627]; ArgyriStMo1976 [host, distribution, biological control: 25-26]; Ayouta1940 [host, distribution: 2-4]; Azeved1925 [host, distribution: 85-86]; Azeved1929a [host, distribution: 126-128]; Bachma1953 [host, distribution: 177]; Balach1927 [host, distribution: 178]; Balach1928b [taxonomy: 157]; Balach1928d [biological control: 292,293,296]; Balach1931a [host, distribution: 97]; Balach1932b [ecology: 517-522]; Balach1932d [taxonomy, host, distribution: IX-X, XLVII]; Balach1933e [host, distribution: 3]; Balach1935b [host, distribution: 260]; Balach1937c [host, distribution: 2]; Balach1938a [host, distribution: 148-149]; Balach1948b [taxonomy, description, illustration, host, distribution, biological control: 351-355]; Balach1956 [taxonomy, description, illustration, host, distribution: 85-86]; BazaroSh1971 [taxonomy, description, illustration, host, distribution: 193-195]; Beards1966 [host, distribution: 516]; Beards1966 [host, distribution: 524]; BeardsDaHo1976 [economic importance: 103]; BeardsGo1975 [economic importance: 49]; Beffa1937 [host, distribution, economic importance: 63-70]; Beingo1969d [biological control: 827-838]; Bellio1932 [life history, ecology, physiology: 1-22]; Benass1965a [host, distribution, chemical control, biological control, economic importance, life history: 112-165]; Benass1977 [host, distribution, life history: 1-20]; BenassBi1967 [host, distribution: 247-256]; BenassBi1974 [host, distribution, life history, economic importance: 247-256]; BenassEu1967a [host, distribution, life history, ecology: 95-111]; BenassEu1970a [host, distribution, biological control: 357-372]; BenassSo1964 [host, distribution, life history, ecology: 193-222]; BenassViRo1979 [biological control: 281-287]; BenDov1980 [taxonomy, host, distribution: 264-265]; BennetRoCo1976 [biological control, economic importance: 359-395]; BerlesLe1896 [taxonomy, description, illustration, host, distribution: 346-347]; BerlesPa1916 [host, distribution, biological control: 305-307]; Berro1927 [host, distribution: 55-59]; Bertel1956 [host, distribution, description, life history, biological control]; BertelBa1966 [host, distribution: 17-46]; BesheaTiHo1973 [host, distribution: 5]; BiezanFr1939 [host, distribution: 1-18]; BiezanSe1940 [host, distribution: 67-68]; BindraVa1972 [host, distribution, economic importance: 14-24]; BlayGo1993 [taxonomy, description, illustration, host, distribution: 503-511]; Bodenh1928 [host, distribution: 191]; Bodenh1944b [host, distribution: 94]; Bodenh1949 [taxonomy, description, illustration, host, distribution: 77-78]; Bodenh1952 [host, distribution, structure: 346]; Borchs1934 [taxonomy, host, distribution, economic importance, life history: 29-30]; Borchs1935a [taxonomy, description, host, distribution: 30]; Borchs1936 [host, distribution, life history, economic importance, biological control: 133-135]; Borchs1937 [taxonomy, description, illustration, host, distribution: 119-120]; Borchs1939 [taxonomy: 16]; Borchs1949d [taxonomy, description, host, distribution: 232]; Borchs1950b [taxonomy, description, illustration, host, distribution: 218-219]; Borchs1966 [catalogue: 287-289]; Borg1919 [taxonomy, description, host, distribution: 27-28]; Borg1922 [host, distribution]; BorianNi1995 [chemical control: 43]; Bouhel1935 [host, distribution, biological control: 17-20]; BouhelDeDe1932 [host, distribution, control: 1-60]; Boyce1948 [host, distribution, economic importance, control]; Brain1919 [taxonomy, description, illustration, host, distribution: 203-204]; BrainKe1917 [distribution: 184]; BrandtBo1948 [taxonomy: 3]; Brick1912 [host, distribution: 1-22]; Brimbl1962 [host, distribution, economic importance: 220]; Brimbl1962a [taxonomy, description, illustration, host, distribution: 412-414]; Britto1923 [taxonomy, description, host, distribution: 377]; Bunzli1935 [host, distribution]; BurgerUl1990 [economic importance: 313-327]; Bustsh1958 [taxonomy, description, host, distribution: 220,231]; CABI1951 [host, distribution: 1-2]; Cabido1949 [host, distribution, description, life history, taxonomy, ecology: 374-465]; CarvalAg1997 [life history, description, economic importance, biological control, host, distribution: 265-273]; Castel1951a [biological control: 95-98];

CatchiWh1924 [host, distribution, life history, ecology: 604-606]; Chambe1927 [taxonomy, description, illustration, host, distribution: 484-491]; Chen1936 [taxonomy, description, host, distribution: 211,222]; Chiesa1938a [taxonomy, description, illustration, host, distribution, life history: 1-21]; Chiesa1948 [host, distribution, economic importance]; Chiesa1948a [host, distribution, economic importance]; Chou1947 [taxonomy, description, illustration, host, distribution: 13-16]; Chou1947a [chemical control: 30-32]; Chou1985 [taxonomy, description, host, distribution: 284,288-291]; Chou1986 [taxonomy, illustration: 673]; ChuaWo1990 [host, distribution, economic importance: 543-552]; ClapsTe2001 [taxonomy, description, illustration, host, distribution, economic importance: 392,396-397]; ClapsWoGo2001a [taxonomy, host, distribution: 15]; Cocker1893j [taxonomy, description, host, distribution: 255]; Cocker1894g [taxonomy, description, host, distribution: 128-129]; Cocker1896b [taxonomy, distribution: 334]; Cocker1897i [taxonomy, description, host, distribution: 23,24,30]; Cocker1899a [taxonomy: 396]; Cocker1899n [host, distribution: 25]; Cocker1900h [taxonomy, host, distribution: 157]; Cocker1905 [taxonomy: 46]; Cocker1905b [taxonomy: 201]; Cocker1922a [host, distribution: 149]; Cohic1958 [host, distribution: 13]; ColonFMe1998 [taxonomy, description, illustration, host, distribution: 50-51]; Comper1961 [biological control: 210,236]; Coorem1951 [host, distribution, biological control: 30-34]; CoronaRuMo1997 [host, distribution: 38-41]; CostaL1949 [host, distribution, biological control: 65-87]; Costan1938 [host, distribution: 25-44]; Costan1956a [host, distribution, economic importance: 74-79]; Creigh1942 [host, distribution, control: 219-233]; Cressm1933 [host, distribution, control, taxonomy: 696-706]; Crouze1971 [biological control: 200]; Crouze1973 [host, distribution, biological control: 15-39]; Danzig1964 [taxonomy, host, distribution: 651]; Danzig1972 [taxonomy, host, distribution, economic importance: 209]; Danzig1993 [taxonomy, description, illustration, host, distribution, economic importance: 167-168]; Danzig1993 [taxonomy, description, illustration, host, distribution, economic importance: 167-168]; DanzigKo1990 [host, distribution: 47]; DanzigPe1998 [catalogue: 215-217]; DavidsMi1990 [host, distribution, economic importance: 603-632]; DeBach1962b [biological control: 29]; DeBach1964d [biological control: 5-18]; DeBach1969 [biological control: 801-815]; DeBach1972 [biological control: 211-233]; DeBach1974 [biological control]; DeBachAr1967 [host, distribution, biological control: 325-342]; DeBachRo1991 [biological control]; DEDAC1923 [host, distribution]; DeGreg1915 [taxonomy, description, host, distribution, biological control: 125-190]; Dekle1954 [host, distribution, economic importance: 226-228]; Dekle1965c [taxonomy, description, host, distribution: 41]; Dekle1976 [taxonomy, description, host, distribution, economic importance: 61]; DelGueMa1915 [host, distribution, life history, control: 1-127]; DeLott1967a [host, distribution: 113]; Delucc1969 [biological control: 871-874]; DeSant1935 [host, distribution, biological control: 262-271]; DeSant1941a [host, distribution, biological control: 21-24]; DeSant1979 [biological control]; DicksoFl1955 [host, distribution: 614-615]; DietzMo1916a [taxonomy, description, illustration, host, distribution: 307,310-312]; Dinthe1958 [host, distribution, economic importance: 421]; Dougla1912 [taxonomy: 219]; Dupont1931 [host, distribution: 1-18]; Dziedz1989 [taxonomy, description, illustration, host, distribution: 101];

DziedzKa1990 [host, distribution: 39-43]; Ebelin1949 [host, distribution, life history, control]; Ebelin1959 [host, distribution, economic importance]; EbelinPe1953 [host, distribution, economic importance: 1-35]; Eblako1938 [biological control: 75-77]; Efimof1937 [host, distribution]; Esaki1940a [host, distribution: 274-280]; Essig1916a [host, distribution: 192-197]; EvansPr1990 [biological control: 3-17]; Ezzat1958 [distribution: 241]; EzzatNa1987 [distribution: 87]; FAO1976 [host, distribution: 6-8]; Fawcet1948 [biological control: 627-664]; FDACSB1982 [host, distribution: 5-11]; FDACSB1988 [host, distribution: 5-6]; Fernal1903b [catalogue: 289-290]; Fernan1981 [host, distribution: 48]; Fernan1999 [host, distribution: 86]; Ferris1938a [taxonomy, description, illustration, host, distribution: 200]; Ferris1941d [taxonomy: 330]; Ferris1941e [taxonomy: 40-42,44-45]; Ferris1942 [taxonomy: 446:31]; Fjeldd1996 [host, distribution: 4-24]; Flande1944a [biological control: 365-371]; Flande1971 [biological control, life history: 857-872]; Fleury1934 [host, distribution: 483-500]; Foldi1990 [structure: 43-54]; Foldi2000 [host, distribution: 84]; Foldi2001 [distribution: 303-308]; Fonsec1963 [host, distribution: 32-35]; Fonsec1964 [host, distribution: 515]; FonsecAu1932a [host, distribution: 202-214]; FontenRoSu1987 [host, distribution, life history, ecology: 1-28]; Fullaw1932 [taxonomy: 97,107]; Fursch1987 [host, distribution, biological control: 387-394]; Gaprin1954 [biological control: 587-597]; Gaprin1956 [host, distribution: 103-137]; Garcia1916 [host, biological control: 776-788]; Garcia1921 [host, distribution, biological control]; Garcia1930 [host, distribution, biological control]; Garcia1949 [host, distribution, taxonomy, life history, ecology: 374-465]; Garonn1991 [host, distribution, life history, biological control: 363-366]; GaronnVi1989 [host, distribution, life history, biological control: 523-527]; GaronnVi1993 [life history, biological control: 117-124]; Gavalo1931 [host, distribution: 8]; Gavalo1936 [host, distribution: 80-81]; GerharLi1952 [host, distribution, life history, control: 874-877]; Gerson1990 [taxonomy: 130]; GersonOcHo1990 [biological control: 77-97]; Ghesqu1932 [host, distribution: 59]; Gill1997 [host, distribution, taxonomy, description, illustration, economic importance: 97,100]; GirolaAr1925 [host, distribution]; GomesCRe1947 [taxonomy, description, host, distribution: 66]; GomezC1943 [taxonomy, description, host, distribution: 308]; GomezC1950 [host, distribution, economic importance, biological control: 1-18]; GomezC1954a [host, distribution, biological control, economic importance: 19-35]; GomezM1937 [taxonomy, description, illustration, host, distribution: 89-96]; GomezM1941 [host, distribution: 131]; GomezM1946 [host, distribution: 62]; GomezM1948 [host, distribution: 74]; GomezM1954 [host, distribution: 121]; GomezM1956 [taxonomy, description, illustration, host, distribution, biological control: 32-39]; GomezM1962 [taxonomy, description, illustration, host, distribution: 180-184]; GomezM1965 [host, distribution: 88]; GomezM1967O [host, distribution: 131]; GomezM1968 [host, distribution: 542]; Gonzal1969 [biological control: 839-848]; Gonzal1989 [taxonomy, description, host, distribution, economic importance: 94]; GonzalCh1968 [distribution: 110]; Gordh1979 [biological control: 893-896,900,907,911]; Gowdey1917 [host, distribution: 189]; Gowdey1921 [taxonomy, description, host, distribution: 32]; Greath1971 [host, distribution, biological control]; Greath1973 [biological control: 29-33]; Greath1976 [biological control, economic importance]; Green1900a [taxonomy, description, illustration, host, distribution: 68-69]; Green1900c [taxonomy, host, distribution: 2-3]; Green1904a [host, distribution: 208]; Green1905a [taxonomy, host, distribution: 345]; Green1916 [host, distribution: 376]; Green1922 [host, distribution: 462]; Green1923b [host, distribution: 89,97]; Green1931a [host, distribution: 105]; Green1934d [distribution: 114]; Green1937 [host, distribution: 332]; GreenMa1907 [taxonomy, distribution: 344]; GriffiTh1957 [host, distribution: 5-30]; Hadzib1983 [taxonomy, description, illustration, host, distribution, life history, biological control, economic importance: 223-226]; HakkonPi1984 [biological control: 1109-1121]; Halber1996 [host, distribution: 4-10]; Hall1923 [taxonomy: 54]; Hall1925 [taxonomy, host, distribution: 14]; Hall1926a [host, distribution: 34]; Hall1929 [host, distribution: 356]; HallFo1933 [host, distribution, economic importance: 1-55]; Haywar1939 [host, distribution, control: 1]; Haywar1944 [host, distribution: 1-32]; Hempel1900a [taxonomy, description, host, distribution: 505-506]; Herric1911 [taxonomy, description, illustration, host, distribution: 11,31-32,66]; Hinckl1963 [host, distribution, biological control]; Hodek1973 [biological control]; Hodgki1904 [host, distribution, description, life history, control: 97-106]; Hollin1923 [taxonomy, description, host, distribution: 30]; HorticInNo1923 [host, distribution: 236-239]; Houser1918 [host, distribution: 168-169]; Hsu1935 [host, distribution: 578-590]; HuffakGu1990a [biological control, economic importance: 473-492]; Hunt1939 [host, distribution: 548-566]; Hussai1974 [host, distribution, economic importance]; IFAC1962d [host, distribution: 140]; Inserr1966 [biological control: 176-186]; Inserr1970a [host, distribution, biological control: 39-45]; Iperti1961 [economic importance: 14-30]; Jannon1940 [host, distribution: 241-253]; Jarray1970 [host, distribution, economic importance: 86]; JiYa1990 [biological control: 134-136]; John1930 [host, distribution: 3-7]; Jourdh1979 [biological control: 75-79]; Katsoy1996 [life history, economic importance, host, distribution, chemical control, biological control: 15-18,75-78]; Kaussa1955 [host, distribution: 15]; Kawai1980 [taxonomy, description, host, distribution: 210-211]; Kawai1987 [host, distribution: 78]; KinjoNaHi1996 [host, distribution: 125-127]; Kiritc1932a [taxonomy: 250-251]; KiriukTa1947 [host, distribution: 1-4]; Kobakh1965 [biological control: 323-330]; Komosi1964 [host, distribution, taxonomy, description, illustration: 214-216]; Kondo2001 [taxonomy, host, distribution: 43]; KondoKa1995 [host, distribution: 57-58]; KondoKa1995a [host, distribution: 97-98]; Korone1934 [taxonomy, description, illustration, host, distribution: 19-21]; Koszta1996 [taxonomy, description, illustration, host, distribution, life history, biological control, economic importance: 479-482]; Kozar1990a [life history, economic importance: 341-347]; KozarKi1979 [host, distribution: 246-250]; Kuwana1927 [host, distribution: 71-72]; Kuwana1933 [taxonomy, description, illustration, host, distribution: 31-32]; Laing1932 [host, distribution: 67]; Laport1948 [host, distribution, biological control: 35-37]; Lawson1917 [taxonomy, description, illustration, host, distribution: 210-214]; Leonar1897 [taxonomy: 286]; Leonar1899 [taxonomy, description, host, distribution: 199,218]; Leonar1910 [taxonomy, description, illustration, host, distribution: 11-13]; Leonar1920 [taxonomy, description, illustration, host, distribution: 69-72]; Lepage1938 [catalogue: 399]; Lepesm1947 [taxonomy, description, host, distribution, life history: 199-201]; Lepine1928 [host, distribution, economic importance: 313-321]; Lindin1909b [host, distribution: 108,222]; Lindin1909c [taxonomy, host, distribution: 449]; Lindin1910b [host, distribution: 40]; Lindin1912b [taxonomy, description, host, distribution: 49,57,58,65,69,71,]; Lindin1924 [taxonomy: 175]; Lindin1935 [taxonomy: 132]; Lindin1949 [taxonomy: 211]; Liotta1970 [host, distribution, economic importance: 33]; Lloren1990 [taxonomy, illustration, life history, host, distribution, biological control, life history: 61-68]; LongoMaRu1995a [host, distribution: 126-129]; LongoMaRu1996 [host, distribution, economic importance: 126-129]; Lounsb1914 [host, distribution: 1]; Lounsb1916 [host, distribution: 83-103]; Lounsb1922 [host, distribution: 205-210]; Lupo1953 [taxonomy, description, illustration, host, distribution: 31-38]; MacGil1921 [taxonomy, description, host, distribution: 415,416]; Mackie1931 [host, distribution, economic importance: 419-441]; Maleno1916c [taxonomy, host, distribution: 109-123]; Maleno1916d [host, distribution, biological control: 181-182]; Maleno1917 [biological control: 195-196]; Maleno1918 [biological control: 17-53]; Maleno1927 [taxonomy: 53]; MaltbyJiDe1968 [biological control: 1086-1088]; Mamet1936 [taxonomy, description, illustration, host, distribution: 96]; Mamet1942 [taxonomy: 35]; Mamet1943a [catalogue: 157]; Mamet1949 [catalogue: 56]; Mamet1954 [host, distribution: 16]; Mamet1957 [distribution: 369]; Mamet1959a [host, distribution: 386]; Mansfi1920 [host, distribution: 145-155]; Marcha1904 [taxonomy, life history, host, distribution: 246-249]; Marlat1921a [host, distribution: 1-36]; Martel1913 [chemical control: 1-28]; Martin1954 [host, distribution, economic importance: 113-116]; Martin1983 [taxonomy, host, distribution: 62-63]; Martor1976 [host, distribution: 91,166,216,224,252]; Maskel1895b [taxonomy, description, host, distribution: 44]; Maskel1898 [taxonomy, description, host, distribution: 224]; Matile1978 [host, distribution: 64]; MatileBa1972 [host, distribution: 113]; MatileNo1984 [host, distribution: 65]; MatileOr2001 [host, distribution: 189]; MayneGh1934 [host, distribution: 3-38]; McDani1968 [taxonomy, illustration, host, distribution: 227-229]; McKenz1935 [host, distribution, life history, control: 1-48]; McKenz1939 [taxonomy, description, illustration, host, distribution: 53,57-58,70]; McKenz1943 [taxonomy: 150]; McKenz1956 [taxonomy, description, illustration, host, distribution: 24,52-55]; Mead1983 [host, distribution: 1-5]; Mead1987 [host, distribution: 2-4]; Melis1949 [host, distribution: 17-25]; Merkel1938 [host, distribution: 88-99]; Merril1953 [taxonomy, description, host, distribution: 36-37]; MerrilCh1923 [taxonomy, description, host, distribution, economic importance: 223-224]; MetcalMe1993 [economic importance, host, distribution, control]; Miller1983 [host, distribution: 4-6]; MillerDa1990 [host, distribution, economic importance: 301]; Monast1958 [economic importance, control: 131-165]; MonastZa1960 [host, distribution: 169-236]; Morgan1889a [taxonomy, description, illustration, host, distribution: 350,352-353]; Moulto1931 [host, distribution: 745]; MouradMeFa2001 [host, distribution: 571-580]; MoutiaMa1947 [distribution]; Moznet1920 [host, distribution, life history: 5-11]; Moznet1922a [host, distribution, life history, chemical control: 1-31]; Muraka1970 [host, distribution: 73]; MyartsRu2000 [distribution, biological control: 7-33]; NagarkSa1990 [host, distribution, economic importance, biological control: 553-542]; Nakaha1982 [host, distribution: 23]; Nakaha1983 [host, distribution: 10]; Newste1893d [taxonomy, description, host, distribution: 185-186]; Newste1901b [taxonomy, description, illustration, host, distribution: 82,107-110]; Newste1917 [taxonomy, host, distribution: 371-372]; Newste1917b [host, distribution: 131]; Nonell1932 [host, distribution: 183-190]; Nonell1935 [host, distribution: 281-287]; NRC1969 [taxonomy, economic importance, ecology, biological control, chemical control]; Nur1990b [taxonomy, life history: 192]; Paglia1929 [host, distribution, economic importance: 274-307]; Paglia1929 [host, distribution: 274]; Painte1951 [economic importance, control]; Paoli1922 [biological control, host, distribution: 407-416]; Paoli1927a [host, distribution: 382-387]; Peleka1962 [host, distribution: 62]; Peleka1974 [host, distribution, biological control: 14-20]; Pelot1950 [host, distribution: 17]; PerezGCa1987 [host, distribution: 128]; PietriBiCo1969 [chemical control: 909-915]; Pope1981 [biological control: 19-31]; Poutie1922 [biological control: 3-28]; Poutie1924 [host, distribution, biological control: 490-496]; Poutie1928 [biological control: 267-270]; Pratt1958 [taxonomy, illustration, distribution]; Priore1964 [host, distribution: 131-178]; Priore1965 [host, distribution: 101-145]; PruthiMa1945 [host, distribution, life history, control: 1-42]; Quayle1938 [host, distribution, life history, economic importance, control]; Quedna1964b [biological control: 86-116]; QuezadCoDi1972 [host, distribution, biological control, economic importance: 14]; Ramakr1919a [taxonomy, description, host, distribution: 19-20]; Ramakr1921a [host, distribution: 356]; Ramakr1930 [taxonomy, host, distribution: 24]; RangelGo1945 [distribution, description: 1-44]; Reboul1976 [host, distribution, economic importance]; Rose1990c [distribution, economic importance: 535-542]; RosenDe1978 [economic importance, biological control, life history, host, distribution: 104-106]; RosenDe1979 [host, distribution, biological control: 275-276,349-354,]; RSEA1915 [host, distribution, description, life history, economic importance, control: 1]; Rubtso1952a [biological control: 96-106]; Rungs1950 [host, distribution, biological control: 9-11]; Rungs1970 [host, distribution, economic importance: 91-94]; Russo1951 [distribution, biological control: 86-97]; Russo1959 [economic importance, chemical control: 4-7]; RzaevaYa1985 [biological control: 55-58]; Saakya1954 [host, distribution, economic importance]; Salama1970 [host, distribution, life history: 427-430]; Sander1904a [taxonomy, description, illustration, host, distribution: 70,71]; SassceWe1923 [chemical control: 84-87]; Savast1930 [host, distribution, taxonomy, life history: 1-77]; Savesc1982 [taxonomy, description, host, distribution, biological control: 304-305]; SchildSc1928 [biological control]; Schmut1957a [host, distribution: 136]; Schmut1957b [taxonomy: 148]; Schmut1959 [taxonomy, description, host, distribution: 54]; Schmut1969 [host, distribution: 116]; Schmut1990a [host, distribution: 393]; Schmut2001 [host, distribution: 339-345]; SchmutKlLu1957 [host, distribution, economic importance, distribution: 478-480]; SchmutKlLu1959 [taxonomy: 374]; SchmutPiKl1978 [host, distribution, economic importance: 330]; Seabra1930a [host, distribution: 143-148]; Seabra1941 [distribution: 8]; SengonUyKa1998 [host, distribution, biological control: 128-131]; Shen1993 [host, distribution: 58]; ShiLi1991 [host, distribution: 165]; Sigwal1971 [host, distribution: 5-15]; Silves1921 [host, distribution, economic importance: 1-11]; Silves1926a [host, distribution, control: 97-101]; Silves1929 [host, distribution: 897-904]; Simant1960a [host, distribution, life history]; Simant1962a [biological control: 105-112]; Singh1964 [host, distribution, economic importance: 213];

Smirno1950a [biological control: 190-194]; Smirno1951 [host, distribution, description, life history, biological control, chemical control: 3-7]; Smirno1951a [host, distribution, economic importance, life history, biological control, chemical control: 1-29]; Smirno1952 [host, distribution, biological control: 63-69]; Smith1948 [biological control: 597]; Smith1948a [economic importance: 813]; SmithEsFa1933 [economic importance: 1]; Soares1942 [host, distribution, economic importance: 54-56]; Soares1945 [host, distribution, economic importance: 49-81]; SoriaEsVi1996 [host, distribution: 241-249]; Steine1987 [host, distribution, description, economic importance, control: 1-7]; Steinw1948 [host, distribution, taxonomy, chemical control: 105-111]; Swirsk1989 [biological control: 11-44]; Takagi1958 [taxonomy, host, distribution: 123]; Takagi1970 [taxonomy, host, distribution: 133-134]; Takaha1929 [host, distribution: 82]; Takaha1932a [host, distribution: 104]; Takaha1939b [host, distribution: 270]; Takaha1942b [host, distribution: 49]; Tao1999 [taxonomy, host, distribution: 80]; Terezn1986 [taxonomy, description, illustration, host, distribution: 93,95-96]; Trabut1911 [taxonomy: 62]; Trimbl1929 [host, distribution, economic importance, description, control: 1-21]; Trjapi1989 [biological control: 293,296]; Tuncyu1970 [host, distribution, economic importance: 30-52]; Tuncyu1970a [host, distribution, economic importance: 67-80]; Tuncyu1976 [host, distribution, biological control: 32-45]; Tzalev1964b [host, distribution: 15-20]; UsmanPu1955 [host, distribution: 48]; UygunKaUl1996 [host, distribution, biological control: 171-183]; UygunSeEr1998 [host, distribution: 183-191]; Vayssi1913 [host, distribution: 431]; Vayssi1920 [host, distribution: 258]; Vayssi1932a [economic importance, biological control: 629-648]; Velasq1971 [taxonomy, description, illustration, host, distribution: 134-136]; VernalGaRo1970 [host, distribution: 41-43]; VeseyF1940 [host, distribution, life history, ecology, economic importance, control: 253-285]; VeseyF1953 [host, distribution, biological control: 405-413]; Viggia1970a [host, distribution, economic importance: 50-51]; Viggia1974 [host, distribution, biological control : 117-120]; Viggia1987 [host, distribution, biological control: 121-123]; ViggiaIa1973 [biological control, life history, distribution, host: 104-116]; Wang1936 [host, distribution: 382-389]; Watson1918 [host, distribution]; Watson1926 [host, distribution]; WatsonBe1937 [host, distribution, control]; Wester1920 [host, distribution]; Wille1941 [host, distribution: 3-26]; WilliaWa1988 [taxonomy, description, illustration, host, distribution: 92,94,97]; Wilson1917 [taxonomy, description, host, distribution: 21-22]; WilsonGo1962 [host, distribution, economic importance, control: 41-61]; Wolfen1951 [host, distribution, economic importance, chemical control: 54-58]; Wolfen1955 [host, distribution, economic importance, chemical control: 27-34]; WolfeToSt1934 [host, distribution, economic importance: 1]; WolffCo1993 [taxonomy, description, illustration, host, distribution: 40-42]; WolffCo1993a [host, distribution: 153]; WongChCh1999 [taxonomy, description, host, distribution: 21-22,61]; Woolle1990 [biological control: 167-176]; Yasar1995a [taxonomy, description, illustration, host, distribution: 61-64]; Yasnos1987 [economic importance: 229-234]; Yasnos1994 [host, distribution, biological control: 317-333]; Yasnos1995 [taxonomy, economic importance, host, distribution: 248]; Zagain1956 [distribution: 85-90]; Zahrad1957 [host, distribution: 48]; Zahrad1959b [host, distribution: 60]; Zahrad1977 [taxonomy, distribution:

120]; Zahrad1990 [host, distribution, description: 642]; Zahrad1990b [taxonomy, description, illustration, host, distribution: 86-88]; Zimmer1948 [taxonomy, description, illustration, host, distribution: 369,371].

Chrysomphalus diversicolor (Green)

Aspidiotus (*Chrysomphalus*) *pinnulifera diversicolor* Green, 1923b: 96. Type data: MADEIRA: on *Citrus* sp., *Musa* sp., *Buxus* sp., *Psidium* sp., *Phoenix* sp., *Asparagus* sp., and on *Dracaena* sp. Syntypes. Type depository: London: The Natural History Museum, England, UK.

Chrysomphalus diversicolor; McKenzie, 1939: 59. Change of status.

SCALE COVER: Scale of female differing from that of typical *pinnulifera* in diverse colouring of secretionary appendix, which varies from blackish or purplish, through various shades of brown, to almost pure white; exuviae always reddish (Green, 1923b). Scale of female extremely variable in colour, ranging from blackish or purplish through various shades of brown to almost grey; central exuviae in all specimens reddish (McKenzie, 1939).

HOST PLANTS: **Apocynaceae**: *Mascarenhasia arborescens* [Mamet1951, Borchs1966]. **Araliaceae**: *Hedera helix* [McKenz1939, GomezM1967O]. **Araucariaceae**: *Araucaria* [Mamet1951, Borchs1966]. **Bignoniaceae**: *Tecoma radicans* [Mamet1950, Borchs1966]. **Buxaceae**: *Buxus* [Green1923b, McKenz1939, Borchs1966], *Buxus sempervirens* [Green1923b]. **Cycadaceae**: *Cycas revoluta* [Green1923b, Mamet1949, Borchs1966]. **Euphorbiaceae**: *Bridelia talasmeana* [Mamet1954, Borchs1966]. **Lauraceae**: *Laurus camphora* [Mamet1951, Borchs1966], *Persea gratissima* [Mamet1950, Borchs1966, GomezM1967O]. **Leguminosae**: *Bauhinia* [Mamet1949, Borchs1966]. **Liliaceae**: *Asparagus* [Green1923b, McKenz1939, Borchs1966], *Dracaena* [Green1923b, McKenz1939]. **Moraceae**: *Ficus* [Mamet1950, Borchs1966]. **Musaceae**: *Musa* [Green1923b, McKenz1939, Borchs1966], *Musa cavendishi* [Green1923b]. **Myrtaceae**: *Callistemon rigidum* [Mamet1950, Borchs1966], *Eugenia* [Mamet1950, Borchs1966], *Eugenia condensata* [Mamet1954, Borchs1966], *Feijoa* [Mamet1951, Borchs1966], *Psidium* [Green1923b, McKenz1939, Borchs1966], *Psidium guajava* [Mamet1950, Borchs1966], *Psidium littorale* [Green1923b]. **Oleaceae**: *Fraxinus berlandieri* [Mamet1950, Borchs1966]. **Orchidaceae**: *Cymbidium* [McKenz1939], *Dendrobium dalhousianum* [Mamet1949, Borchs1966], *Dendrobium nobile* [Mamet1949, Borchs1966], *Vanda suavis* [Mamet1949, Borchs1966], *Vanilla* [Mamet1950, Borchs1966]. **Palmae** [Hall1928], *Neodypsis baroni* [Mamet1954, Borchs1966], *Phoenix* [Green1923b, Borchs1966], *Phoenix canariensis* [McKenz1939], *Phoenix dactylifera* [Green1923b, McKenz1939]. **Pandanaceae**: *Pandanus* [Mamet1951, Borchs1966]. **Proteaceae**: *Grevillea* [Mamet1954, Borchs1966]. **Rubiaceae**: *Cephalanthus spathelliferus* [Mamet1951, Borchs1966], *Cinchona ledgeriana* [Mamet1951, Borchs1966], *Cinchona succirubra* [Mamet1951, Borchs1966]. **Rutaceae**: *Citrus* [Green1923b, McKenz1939, Mamet1951, Borchs1966], *Citrus aurantium* [Hall1928]. **Salicaceae**: *Populus amiana* [Mamet1954, Borchs1966]. **Theaceae**: *Camellia* [Mamet1949, Borchs1966]. **Urticaceae**: *Pilea urticifolia* [Mamet1949, Borchs1966].

DISTRIBUTION: **Afrotropical**: Madagascar [Mamet1950, Mamet1951, Mamet1954, Borchs1966]; Mauritius [Mamet1943a, Mamet1949, Borchs1966]; South Africa [Green1923, McKenz1939, Borchs1966]. **Oriental**: Sri Lanka [McKenz1939]. **Palaearctic**: Canary Islands [GomezM1967O]; Madeira Islands [Green1923, Green1923b, McKenz1939, Borchs1966]; Portugal [Seabra1942].
BIOLOGY: Occurring on the leaves (McKenzie, 1939).
GENERAL: Description and illustration of adult female by Green (1923b). Description of adult female by McKenzie (1939).
KEYS: McKenzie 1939: 63, 64; 1943: 150; (female) [World].
CITATIONS: Balach1948b [taxonomy: 355]; Borchs1966 [catalogue: 289]; Chambe1927 [taxonomy: 289]; Ferris1941e [taxonomy: 43]; GomezM1967O [host, distribution: 131]; Green1923b [taxonomy, description, illustration, host, distribution: 89,96,97]; Hall1928 [host, distribution: 276]; Mamet1943a [catalogue: 158]; Mamet1949 [catalogue: 56]; Mamet1950 [host, distribution : 22]; Mamet1951 [host, distribution: 226]; Mamet1954 [host, distribution: 16]; McKenz1939 [taxonomy, description, host, distribution: 54,59,64]; McKenz1943 [taxonomy: 150]; Seabra1942 [distribution: 2].

Chrysomphalus fodiens (Maskell)
Aspidiotus fodiens Maskell, 1892: 10. Type data: AUSTRALIA: on *Acacia* sp.; collected by Mr. French. Syntypes, female. Type depository: Auckland: New Zealand Arthropod Collection, Landcare Research, New Zealand.
Aspidiotus (Chrysomphalus) fodiens; Cockerell, 1897i: 26. Change of combination.
Chrysomphalus fodiens; Leonardi, 1899: 198. Change of combination.
Chrysomphalus fodiens albus Laing, 1929: 26. Type data: AUSTRALIA: Northern Territory, Darwin; collected by G.F. Hill. Syntypes, female. Type depository: London: The Natural History Museum, England, UK. Synonymy by Borchsenius, 1966: 290.
Chrysomphalus albus McKenzie, 1939: 53. Synonymy by Borchsenius, 1966: 290.
SCALE COVER: Scale of female yellowish brown, with a darker, reddish brown band surrounding the pale; centrally placed exuviae; scale of male similar in colour to that of female, oval; exuvia near one end (McKenzie, 1939).
HOST PLANTS: **Acanthaceae**: *Strobilanthes viscosus* [Green1911]. **Leguminosae**: *Acacia* [Maskel1892, Frogga1914, McKenz1939], *Pithecelobium moniliferum* [McKenz1939]. **Myrtaceae**: *Melaleuca leucadendron* [Green1916e].
DISTRIBUTION: **Australasian**: Australia [Maskel1892, McKenz1939] (Northern Territory [Green1911, Green1916e, Laing1929], Victoria [Frogga1914]).
ECONOMIC IMPORTANCE: Chua & Wood (1990) listed this species as a pest of banana in northern Australia.
GENERAL: Description and illustration of adult female by McKenzie (1939).
KEYS: McKenzie 1939: 64 (female) [World].
CITATIONS: Borchs1966 [catalogue: 290]; ChuaWo1990 [host, distribution, economic importance: 548]; Cocker1896b [distribution: 335]; Cocker1897i [taxonomy, description, host, distribution: 26]; DeitzTo1980 [taxonomy: 37]; Fernal1903b [catalogue: 259]; Ferris1941e [taxonomy: 43]; Frogga1914 [taxonomy, description, host, distribution: 312-313]; Frogga1915 [taxonomy, description, host,

distribution: 16]; Green1914c [taxonomy, description, illustration, host, distribution: 231]; Green1916e [host, distribution: 53]; Laing1929 [taxonomy, host, distribution: 26]; Larew1990 [ecology, life history, structure: 293-300]; Leonar1899 [taxonomy, description, host, distribution: 198-200]; MacGil1921 [taxonomy, description, host, distribution: 417]; Maskel1892 [taxonomy, description, illustration, host, distribution: 10]; McKenz1939 [taxonomy, description, illustration, host, distribution: 53,54,60-61,72].

Chrysomphalus greeni Leonardi

Chrysomphalus greeni Leonardi, 1914: 206. Type data: GUINEA: Conakry, on undetermined plant. Syntypes, female. Type depository: Portici: Dipartimento de Entomologia e Zoologia Agraria di Portici, Università di Napoli Federico II, Italy.
Melanaspis greeni; Lindinger, 1943a: 147. Change of combination.
Chrysomphalus greeni; Borchsenius, 1966: 290. Revived combination.
SCALE COVER: Female scale circular, slightly convex, robust; exuviae central; colour black, excluding a marginal area which is less compact and chestnut-coloured; dorsal part of scale covered by epidermis of host plant; 1.1 mm in diameter (Leonardi, 1914).
DISTRIBUTION: **Afrotropical**: Guinea [Leonar1914].
GENERAL: Description and illustration of adult female by Leonardi (1914).
CITATIONS: Borchs1966 [catalogue: 290]; Leonar1914 [taxonomy, description, illustration, host, distribution: 206-207]; Lindin1943a [taxonomy: 147]; McKenz1939 [taxonomy: 54].

Chrysomphalus minutus Kotinsky

Chrysomphalus minutus Kotinsky, 1908: 170. Type data: SINGAPORE: on undetermined plant. Holotype female. Type depository: Davis: The Bohart Museum of Entomology, University of California, California, USA; type no. 853.
Aonidiella minuta; MacGillivray, 1921: 444. Change of combination requiring emendation of species name for agreement in gender.
Aspidiotus pigmaeus Lindinger, 1937: 180. Unjustified replacement name for *Chrysomphalus minutus* Kotinsky.
SCALE COVER: Scale of adult female greenish yellow, subcircular, convex, diameter 0.65 mm; exuviae sub-central, 1st orange, 2nd dark brown within white circle; ventral scale complete except central opening, tough (Kotinsky, 1908).
DISTRIBUTION: **Oriental**: Singapore [Kotins1908].
GENERAL: Description and illustration of adult female by Kotinsky (1908).
CITATIONS: Borchs1966 [catalogue: 371-372]; Ferris1941e [taxonomy: 45]; Kotins1908 [taxonomy, description, illustration, host, distribution: 170-171]; Lindin1937 [taxonomy: 180]; MacGil1921 [taxonomy, description, host, distribution: 444]; McKenz1938 [taxonomy: 4]; McKenz1939 [taxonomy: 54]; Sander1909a [taxonomy, host, distribution: 55].

Chrysomphalus mume Tang

Chrysomphalus mume Tang, 1984: 35. Type data: CHINA: Yunnan Province, Gejiu, on *Prunus mume*. Holotype female. Type depository: Shanxi: Entomological Institute, Shanxi Agricultural University, Taigu, Shanxi, China.
SCALE COVER: Scales of adult females yellowish turning white, 1.9 mm in diameter (Tang, 1984).
HOST PLANTS: **Rosaceae**: *Prunus mume* [Tang1984].
DISTRIBUTION: **Oriental**: China (People's Republic) (Yunnan [Tang1984]).
GENERAL: Description and illustration of adult female by Tang (1984).
CITATIONS: DanzigPe1998 [catalogue: 217]; Tang1984 [taxonomy, description, illustration, host, distribution: 35-36]; Tao1999 [taxonomy, host, distribution: 81].

Chrysomphalus nulliporus McKenzie

Chrysomphalus nulliporus McKenzie, 1939: 76. Type data: PHILIPPINES: on an orchid *Dendrobium lyonis*. Holotype female. Type depository: Davis: The Bohart Museum of Entomology, University of California, California, USA.
SCALE COVER: McKenzie (1939) did not describe scale cover.
HOST PLANTS: **Orchidaceae**: *Dendrobium lyonis* [McKenz1939].
DISTRIBUTION: **Oriental**: Philippines [McKenz1939, Velasq1971].
GENERAL: Description and illustration of adult female by McKenzie (1939) and by Velasquez (1971).
KEYS: Velasquez 1971: 131 (female) [Philippines]; McKenzie 1943: 150 (female) [World].
CITATIONS: Borchs1966 [catalogue: 290]; Mamet1951 [taxonomy: 246]; McKenz1939 [taxonomy, description, illustration, host, distribution: 76-77]; McKenz1943 [taxonomy: 150]; Velasq1971 [taxonomy, description, illustration, host, distribution: 136-137].

Chrysomphalus pallens nomen nudum

Chrysomphalus pallens McKenzie, 1939: 54. Nomen nudum. Notes: Name credited to Green.
Chrysomphalus pallens Borchsenius, 1966: 377. Nomen nudum.

Chrysomphalus pinnulifer (Maskell)

Diaspis pinnulifera Maskell, 1891: 4. Type data: FIJI: on undetermined plant. Syntypes. Type depositories: Auckland: New Zealand Arthropod Collection, Landcare Research, New Zealand, and Washington: U.S. National Entomological Collection, U.S. National Museum of Natural History, District of Columbia, USA.
Chrysomphalus dictyospermi pinnulifera; Cockerell, 1900h: 157. Change of combination and rank.
Diaspis pinnulifera; Leonardi, 1920: 69. Incorrect synonymy. Notes: Incorrect synonymy with *Chrysomphalus dictyospermi* (Moragan).
Chrysomphalus pinnuliferus; MacGillivray, 1921: 416. Change of combination requiring emendation of species name for agreement in gender.
Aspidiotus (*Chrysomphalus*) *pinnulifera*; Green, 1923b: 96. Change of combination.
Chrysomphalus pinnulifer; Chamberlin, 1927: 486. Change of combination.

Chrysomphalus pinnulifer diversicolor; Chamberlin, 1927: 486. Change of combination.

Chrysomphalus pinnulifera; Balachowsky, 1928c: 276. Change of combination.

Aspidiotus pinnuliferus; Ferris, 1941e: 47. Change of combination.

Chrysomphalus diversicolor; Balachowsky, 1948b: 355. Incorrect synonymy. Notes: Incorrect synonymy with *Chrysomphalus pinnulifer*; see Borchsenius, 1966: 289.

Chrysomphalus pinnulifer diversicolor; Balachowsky, 1948b: 355. Incorrect synonymy. Notes: Incorrect synonymy with *Chrysomphalus pinnulifer*; see Borchsenius, 1966: 289.

Chrysomphalus pinnulifera; Gómez-Menor Ortega, 1957: 46. Misspelling of species name.

COMMON NAME: pinta-amarela [CarvalAg1997].

SCALE COVER: Female scale circular or subcircular, diameter about 1/16 inch; reddish-brown, flat; exuviae subcentral, yellow. Male scale elongated, white, distinctly carinated, length about 1/20 inch (Maskell, 1891). Scale of female reddish-brown, with margin somewhat lighter; rather thin, circular, exuviae central; scale of male similar in colour and texture to that of female, oval, exuvia near one end (McKenzie, 1939). Colour photograph of scale cover and general appearance see Carvalho & Aguiar (1997).

HOST PLANTS: **Agavaceae**: *Agave sisalana* [DeLott1967a], *Dracaena* [Balach1948b], *Phormium tenax* [Matile1976]. **Anacardiaceae**: *Mangifera indica* [Balach1948b, DeLott1967a]. **Annonaceae**: *Annona* [Balach1948b]. **Apocynaceae**: *Nerium oleander* [Mamet1954, Borchs1966], *Plumeria acutifolia* [Mamet1954, Borchs1966]. **Araceae**: *Monstera deliciosa* [Mamet1954, Borchs1966]. **Araliaceae**: *Hedera* [Green1923b], *Hedera helix* [Lepage1938, Balach1948b]. **Buxaceae**: *Buxus* [Balach1948b], *Buxus sempervirens* [GomezM1957, Martin1983]. **Caprifoliaceae**: *Viburnum* [Bodenh1952]. **Celastraceae**: *Euonymus* [Martin1983], *Euonymus japonicus* [Balach1948b]. **Combretaceae** [Lepage1938]. **Cycadaceae**: *Cycas* [Mamet1959a, Borchs1966]. **Ericaceae**: *Arbutus unedo* [Balach1932d]. **Euphorbiaceae**: *Aleurites mollucana* [DeLott1967a]. **Flacourtiaceae**: *Aphloia theaeformis* [Mamet1959a, Borchs1966], *Aphloia theaeformis* [Mamet1954, Borchs1966]. **Iridaceae**: *Gladiolus* [DeLott1967a]. **Lauraceae** [Takaha1942b], *Laurus camphora* [Mamet1954, Borchs1966], *Laurus nobilis* [Lizery1916a, Balach1948b, GomezM1957, Martin1983], *Persea gratissima* [DeLott1967a]. **Leguminosae**: *Acrocarpus fraxinifolius* [DeLott1967a], *Bauhinia* [Mamet1950, Borchs1966], *Ceratonia siliqua* [Balach1948b], *Ceratonia siliqua* [DeLott1967a]. **Moraceae**: *Artocarpus* [Mamet1954, Borchs1966], *Ficus carica* [Bodenh1952]. **Musaceae**: *Musa* [Almeid1971]. **Myrtaceae**: *Callistemon lanceolatus* [DeLott1967a], *Eucalyptus* [Mamet1959a, Borchs1966], *Eugenia* [Mamet1950, Mamet1959a, Borchs1966], *Eugenia caryophylata* [Mamet1943a, Borchs1966], *Eugenia jambolona* [Mamet1954, Borchs1966], *Eugenia jambos* [DeLott1967a], *Myrtus communis* [Mamet1954, Borchs1966], *Psidium guajava* [Balach1948b, Mamet1950, Borchs1966]. **Oleaceae**: *Fraxinus berlandieri* [Mamet1950, Borchs1966], *Jasminum* [Mamet1943a, Balach1948b, Borchs1966], *Ligustrum* [Matile1976]. **Orchidaceae**: *Angraecum* [Mamet1950, Borchs1966], *Cymbidium* [Balach1948b], *Vanilla* [Mamet1950, Borchs1966]. **Palmae**: *Cocos nucifera*

[Mamet1943a, Borchs1966], *Kentia* [Hall1928, Balach1956], *Neodyspis baroni* [Mamet1954, Borchs1966], *Seaforthia* [Hall1928, Balach1956]. **Pandanaceae**: *Pandanus* [Mamet1943a, Balach1948b, Borchs1950b, Borchs1966]. **Pittosporaceae**: *Pittosporum tobira* [Balach1948b]. **Proteaceae**: *Grevillea robusta* [DeLott1967a]. **Rosaceae**: *Crataegus mexicana* [Mamet1954, Borchs1966], *Prunus capuli* [Mamet1954, Borchs1966]. **Rubiaceae**: *Cinchona ledgeriana* [Mamet1951, Borchs1966], *Thunbergia* [Mamet1943a, Borchs1966], *Thunbergia grandiflora* [Mamet1950, Borchs1966]. **Ruscaceae**: *Ruscus hypoglossum* [Balach1932d]. **Rutaceae** [Balach1928c], *Citrus* [Balach1932d, Mamet1950, Mamet1959a, Borchs1966, WilliaWa1988, CarvalAg1997], *Citrus aurantium* [DeLott1967a, Almeid1973b], *Citrus limon* [Lizery1916a], *Citrus mitis* [Mamet1943a, Borchs1966]. **Salicaceae**: *Salix babylonica* [Mamet1954, Borchs1966]. **Strelitziaceae**: *Strelitzia* [Balach1948b], *Strelitzia augusta* [Balach1938a], *Strelitzia regine* [Balach1938a]. **Taxaceae**: *Taxus* [Bodenh1952]. **Theaceae**: *Camellia* [Mamet1959a, Borchs1966], *Thea sinensis* [DeLott1967a]. **Verbenaceae**: *Duranta plumieri* [Mamet1950, Borchs1966]. **Vitaceae**: *Vitis vinifera* [Lizery1916a].

NATURAL ENEMIES: HYMENOPTERA **Aphelinidae**: *Aspidiotiphagus lounsburyi* Berlese & Paoli [Balach1948b].

DISTRIBUTION: **Afrotropical**: Angola [Almeid1973b]; Kenya [Balach1956, DeLott1967a]; Madagascar [Mamet1950, Mamet1951, Mamet1954, Mamet1959a, Borchs1966]; Mozambique [Almeid1971]; Réunion [Mamet1957]; Saint Helena [Matile1976]; Seychelles [Mamet1943a, Borchs1966]; South Africa [Balach1956]; Zimbabwe [Hall1928, Balach1956]. **Australasian**: Papua New Guinea [WilliaWa1988]. **Neotropical**: Argentina (Corrientes [Lizery1916a]); Brazil (São Paulo [Lepage1938]). **Oriental**: Thailand [Takaha1942b]. **Palaearctic**: Algeria [Balach1928c, Balach1932d]; Canary Islands [MatileOr2001]; Madeira Islands [Green1923b, McKenz1939, CarvalAg1997]; Sicily [DeStef1910, Martel1913]; Spain [GomezM1957, GomezM1958c, BlayGo1993]; Turkey [Bodenh1949, Bodenh1952].

BIOLOGY: On leaves (McKenzie, 1939).

ECONOMIC IMPORTANCE: Recorded as tea pest in Kenya and India (Nagarkatti & Sankaran, 1990). Recently considered a potential international invader of citrus-growing regions (Rose, 1990c).

GENERAL: Description and illustration of adult female by McKenzie (1939), Balachowsky (1956), Williams & Watson (1988) and by Yaşar (1995a).

KEYS: Danzig 1993: 164 (female) [Europe]; Williams & Watson 1988: 93 (female) [Tropical South Pacific]; Balachowsky 1956: 86 (female) [Africa]; Balachowsky 1948b: 346 (female) [Mediterranean]; McKenzie 1943: 150 (female) [World]; McKenzie 1939: 63 (female) [World]; Balachowsky 1928b: 157 (female) [North Africa].

CITATIONS: Almeid1971 [host, distribution: 9]; Almeid1973b [host, distribution: 9]; Andrie1932 [host, distribution, economic importance, control]; Balach1928b [taxonomy: 157]; Balach1928c [host, distribution: 279]; Balach1928d [biological control: 296]; Balach1932d [taxonomy, host, distribution: X-XI]; Balach1938a [host, distribution: 149-150]; Balach1948b [taxonomy, description, illustration, host, distribution, biological control: 355-358]; Balach1956 [taxonomy, description, illustration, host, distribution: 88-90]; Bedfor1978 [host, distribution, economic importance: 129-133]; BlayGo1993 [taxonomy, description, illustration, host, distribution: 512-516]; Bodenh1949 [taxonomy, description, illustration, host, distribution: 78-79]; Bodenh1952 [taxonomy, host, distribution, structure: 346]; Borchs1950b [taxonomy, description, illustration, host, distribution: 212,218-219]; Borchs1966 [catalogue: 290]; CarvalAg1997 [life history, description, economic importance, biological control, host, distribution: 274-275]; Castel1951a [biological control: 95-98]; Chambe1927 [taxonomy: 486]; Chorle1939 [host, distribution]; ClapsWoGo2001a [taxonomy, host, distribution: 16]; Cocker1900h [taxonomy, host, distribution: 157]; CoronaRuMo1997 [host, distribution: 38-41]; DanzigPe1998 [catalogue: 217]; DeitzTo1980 [taxonomy: 41]; DeLott1967a [host, distribution: 114]; DeStef1910 [host, distribution: 189-196]; Fernal1903b [catalogue: 290]; Ferris1941e [taxonomy: 47]; GomezM1957 [host, distribution: 46]; GomezM1958c [host, distribution: 406]; Green1923b [host, distribution: 89,96]; Hall1928 [host, distribution: 276]; HallFo1933 [host, distribution, economic importance: 1-55]; Leonar1907a [taxonomy: 119]; Lepage1938 [catalogue: 400]; Lepesm1947 [taxonomy, description, host, distribution, life history: 201-202]; Lizery1916a [host, distribution: 177]; MacGil1921 [taxonomy, description, host, distribution: 416]; Maleno1916c [taxonomy: 109]; Mamet1943a [catalogue: 158]; Mamet1950 [host, distribution: 22]; Mamet1951 [host, distribution: 226]; Mamet1954 [host, distribution: 16]; Mamet1957 [host, distribution: 370,375]; Mamet1959a [host, distribution: 387]; Martel1913 [host, distribution, economic importance, chemical control: 1-28]; Martin1983 [taxonomy, host, distribution: 63]; Maskel1891 [taxonomy, description, illustration, host, distribution: 4]; Maskel1893b [taxonomy, host, distribution: 208]; Maskel1895b [taxonomy, description, host, distribution: 44]; Matile1976 [host, distribution: 312]; MatileOr2001 [host, distribution: 189]; McKenz1939 [taxonomy, description, illustration, host, distribution: 54,61-62,73]; McKenz1943 [taxonomy: 150]; MillerDa1990 [host, distribution, economic importance: 301]; Mottar1915 [host, distribution, life history: 29-31]; NagarkSa1990 [host, distribution, economic importance, biological control: 553-542]; Neves1935 [host, distribution: 59-61]; Nur1990b [taxonomy, life history: 196]; Rose1990c [host, economic importance: 539]; SchmutKlLu1957 [host, distribution, economic importance: 480]; Stepha1909 [taxonomy, description, host, distribution: 189]; Sudoi1995 [host, distribution, chemical control: 119-123]; TakagiRo1981 [host, distribution, biological control: 314-321]; Takaha1942b [taxonomy, host, distribution: 49]; WilliaWa1988 [taxonomy, description, illustration, host, distribution: 95-97]; WilsonGo1962 [host, distribution, economic importance, control: 41-61]; Yasar1995a [taxonomy, description, illustration, host, distribution: 64-66].

Chrysomphalus propsimus Banks

Chrysomphalus propsimus Banks, 1906: 230. Type data: PHILIPPINES: Manila, San Miguel de Mayumo, Province of Bulaean, on leaves of *Cocos nucifera*. Holotype female. Type depository: Manila: Entomological Collection, Bureau of Science, Philippines; type no. 10164.

Chrysomphalus calami Malenotti, 1916: 317. Type data: MALAYSIA: on *Calamus spectabilis*; collected by Prof. Odorado Beccari, 1898. Syntypes, female. Synonymy by Lindinger, 1943b: 218. Notes: Depository of type material unknown.

Chrysomphalus proposimus; MacGillivray, 1921: 415. Misspelling of species name.

SCALE COVER: Scale of female dark chocolate colour, flat and circular, exuviae subcentral; male scale similar in colour, oval, exuvia towards one end (McKenzie, 1939).

HOST PLANTS: **Palmae**: *Calamus spectabilis* [Maleno1916, McKenz1939, Zimmer1948], *Cocos nucifera* [Banks1906, Sander1909a, McKenz1939, WilliaWa1988], *Corypha elata* [McKenz1939, Zimmer1948]. **Pandanaceae**: *Pandanus* [Zimmer1948], *Pandanus odoratissimus* [WilliaWa1988].

DISTRIBUTION: **Australasian**: Hawaiian Islands (Hawaii [Zimmer1948]); Kiribati [WilliaWa1988]; Tuvalu [WilliaWa1988]. **Oriental**: Indonesia (Sumatra [Maleno1916, McKenz1939]); Malaysia [Maleno1916]; Philippines [McKenz1939] (Luzon [Banks1906]).

BIOLOGY: On leaves (McKenzie, 1939).

GENERAL: Description and illustration of adult female by Banks (1906), Malenotti (1916), McKenzie (1939) and by Williams & Watson (1988).

KEYS: Williams & Watson 1988: 93 (female) [Tropical South Pacific]; Balachowsky 1956: 88 (female) [Africa]; Zimmerman 1948: 368-369 (female) [Hawaii]; McKenzie 1943: 150 (female) [World]; McKenzie 1939: 64 (female) [World].

CITATIONS: Balach1948b [taxonomy: 350]; Balach1956 [taxonomy: 88]; Banks1906 [taxonomy, description, illustration, host, distribution: 230-231]; Borchs1966 [catalogue: 290]; Lepesm1947 [taxonomy, description, host, distribution, life history: 202-203]; Lindin1943b [taxonomy: 218]; LitRi1993 [distribution: 163-167]; Maleno1916 [taxonomy, description, illustration, host, distribution: 317-319]; McKenz1939 [taxonomy, description, illustration, host, distribution: 53,54,62-63,74]; McKenz1943 [taxonomy: 150]; Wester1918 [host, distribution, economic importance: 5-57]; WilliaWa1988 [taxonomy, description, illustration, host, distribution: 96,98]; Zimmer1948 [taxonomy, description, host, distribution: 373].

Chrysomphalus rubribullatus (Froggatt)

Aspidiotus perniciosus eucalypti Fuller, 1897b: 1344. Type data: AUSTRALIA: Western Australia, on Tasmanian gum, *Eucalyptus globulus*. Syntypes, female. Type depository: Canberra: Australian National Insect Collection, CSIRO Entomology, Australia. Homonym of *Aspidiotus eucalypti* Maskell, 1889. Notes: Secondary homonym of *Aspidiotus eucalypti* Maskell, 1889, through subsequent elevation of *Aspidiotus perniciosus eucalypti* Froggatt, 1914 to specific rank by Ferris (1941e: 45).

Aspidiotus (Diaspidiotus) perniciosus? eucalypti; Cockerell, 1899a: 396. Change of combination.

Aspidiotus (Aspidiella) rubribullata Froggatt, 1914: 317. Type data: AUSTRALIA: West Australia, near Perth, on *Eucalyptus* sp., and New South Wales, Trangie, on *Eucalyptus* sp. Syntypes, female. Type depository: Canberra: Australian National

Insect Collection, CSIRO Entomology, Australia. Junior synonym and Replacement Name. Notes: *Aspidiotus perniciosus eucalypti* Fuller, 1897b became a secondary homonym of *Aspidiotus eucalypti* Maskell, 1889, through subsequent elevation of this subspecies to specific rank by Ferris (1941e: 45). In retrospect, the description of *Aspidiotus (Aspidiella) rubribullata* Froggatt, 1914, which is identical with *Aspidiotus perniciosus eucalypti* Fuller, also provided the Replacement Name.

Aspidiotus (Aspidiella) rubribullata; Froggatt, 1915: 22. Notes: Described again as n. sp.

Aspidiotus (Aonidiella) miniatae Green, 1916e: 53. Type data: AUSTRALIA: Northern Australia, on twigs of *Eucalyptus miniata*. Syntypes, female. Type depository: London: The Natural History Museum, England, UK. Synonymy by Ferris, 1941e: 45.

Neosignoretia miniatae; MacGillivray, 1921: 424. Change of combination.

Aonidiella eucalypti Lindinger, 1932c: 204. Synonymy by Borchsenius, 1966: 291.

Aspidiotus eucalypti; Ferris, 1941e: 43. Change of combination. Homonym of *Aspidiotus eucalypti* Maskell, 1887.

Aspidiotus miniatae; Ferris, 1941e: 43. Change of combination.

Aspidiotus (Aonidiella) rubribullatus; Ferris, 1941e: 48. Change of combination requiring emendation of species name for agreement in gender.

Quadraspidiotus rubribullatus; Ferris, 1941e: 48. Change of combination.

Chrysomphalus miniatae; Lindinger, 1943: 291. Change of combination.

Chrysomphalus rubribullatus; Brimblecombe, 1958: 69. Change of combination.

SCALE COVER: Female scale circular, diameter up to 1/20 inch; convex; chocolate brown; exuviae large, forming a regular boss in centre, deep orange red, sometimes clouded with white secretion (Froggatt, 1914). Illustration of scale cover by Froggatt (1914). Insects numerous on small branches; scale convex and mostly circular; about 1.25 mm diameter; dark brown in colour with a fawn margin; first and second exuviae central, with a pale grey flaky covering weathering to expose the dark orange colour of the exuviae (Brimblecombe, 1958).

HOST PLANTS: **Myrtaceae**: *Eucalyptus* [Frogga1914, Brimbl1958], *Eu. camaldulensis* [Brimbl1962a], *Eu. crebra* [Brimbl1962a], *Eu. globulus* [Fuller1897b], *Eu. macrocarpa* [Brimbl1958], *Eu. melanophloia* [Brimbl1962a], *Eu. miniata* [Green1916e, Brimbl1958], *Eu. transcontinentalis* [Brimbl1958].

DISTRIBUTION: **Australasian**: Australia (New South Wales [Frogga1914], Northern Territory [Green1916e], Queensland [Brimbl1962a], Victoria [Brimbl1958], Western Australia [Fuller1897b, Brimbl1958]).

BIOLOGY: Insects numerous on small branches (Brimblecombe, 1958).

GENERAL: Description and illustration of adult female by Froggatt (1914) and by Brimblecombe (1958).

CITATIONS: Borchs1966 [catalogue: 291]; Brimbl1958 [taxonomy, description, illustration, host, distribution: 69-73]; Brimbl1962a [taxonomy, description, illustration, host, distribution: 416-418]; Cocker1899a [taxonomy: 396]; Ferris1941e [taxonomy: 43,45,48]; Frogga1914 [taxonomy, description, host, distribution: 317-318]; Frogga1915 [taxonomy, description, host, distribution: 22]; Fuller1897b [taxonomy, description, host, distribution: 1344]; Fuller1897c [taxonomy, description, host, distribution: 4]; Fuller1899 [taxonomy, description, host, distribution: 465]; Green1916e [taxonomy, description, illustration, host, distribution: 53]; Lindin1932c [taxonomy: 204]; Lindin1943b [taxonomy: 207]; MacGil1921 [taxonomy: 424,425]; McKenz1938 [taxonomy: 4]; Sassce1915 [taxonomy, host, distribution: 34].

Chrysomphalus silvestrii Chou
Chrysomphalus silvestrii Chou, 1946: 3. Type data: CHINA: Yunnan, on undetermined Leguminosae. Syntypes, female. Notes: Depository of type material unknown.
Chrysomphalus silvesteii; Chou, 1985: 291. Misspelling of species name.
SCALE COVER: Chou (1946, 1947) illustrated female and male scales.
HOST PLANTS: **Leguminosae** [Chou1946].
DISTRIBUTION: **Oriental**: China (People's Republic) (Yunnan [Chou1946, Chou1947, Borchs1966]).
GENERAL: Description and illustration of adult female by Chou (1946, 1947, 1985, 1986).
KEYS: Chou 1985: 284 (female) [Species of China].
CITATIONS: Borchs1966 [catalogue: 291]; Chou1946 [taxonomy, description, illustration, host, distribution: 3-8]; Chou1947 [taxonomy, description, illustration, host, distribution: 21-34]; Chou1985 [taxonomy, description, host, distribution: 284,291]; Chou1986 [taxonomy, illustration: 674]; DanzigPe1998 [catalogue: 217]; Tao1999 [taxonomy, host, distribution: 81].

Chrysomphalus trifasciculatus Brimblecombe
Chrysomphalus trifasciculatus Brimblecombe, 1959: 123. Type data: AUSTRALIA: Queensland, South Queensland, on *Eucalyptus* sp.; collected March 1945. Holotype female. Type depository: Brisbane: Queensland Museum, Queensland, Australia; type no. T5696.
SCALE COVER: Female scale circular, 1.5-2.0 mm diameter, light fawn, exuviae orange coloured (Brimblecombe, 1959).
HOST PLANTS: **Myrtaceae**: *Eucalyptus* [Brimbl1959, Brimbl1962a].
DISTRIBUTION: **Australasian**: Australia (New South Wales [Brimbl1962a], Queensland [Brimbl1959, Brimbl1962a]).
GENERAL: Description and illustration of adult female by Brimblecombe (1959).
CITATIONS: Borchs1966 [catalogue: 291]; Brimbl1959 [taxonomy, description, illustration, host, distribution: 123-125]; Brimbl1962a [taxonomy, description, illustration, host, distribution: 416,418].

Chrysomphalus variabilis McKenzie
Chrysomphalus variabilis McKenzie, 1943: 148. Type data: AUSTRALIA: West Australia, Norseman, on Quandang or "wild peach of Australia" *Santalum acuminatum*. Holotype female. Type depository: Davis: The Bohart Museum of Entomology, University of California, California, USA.
SCALE COVER: Female scale ca. 1.25 mm diameter, yellow-brown, with dark brown band surrounding the pale, yellowish-brown central exuvium; male scale similar in colour to female, oval, exuviae near one end (McKenzie, 1943).

HOST PLANTS: **Santalaceae**: *Santalum acuminatum* [McKenz1943].
DISTRIBUTION: **Australasian**: Australia (Western Australia [McKenz1943]).
BIOLOGY: On leaves (McKenzie, 1943).
GENERAL: Description and illustration of adult female by McKenzie (1943).
CITATIONS: Borchs1966 [catalogue: 291]; McKenz1943 [host, distribution, taxonomy, description, illustration: 148-149].

Clavaspidiotus Takagi & Kawai

Clavaspidiotus Takagi & Kawai, 1966: 115. Type species: *Clavaspidiotus abietis* Takagi & Kawai, by original designation.
SYSTEMATICS: This genus comes close to *Quadraspidiotus* MacGillivray and *Clavaspis* MacGillivray, but differs from *Quadraspidiotus* in presence of a fourth lobe, presence of bifid marginal spines between third and fourth lobe, and extreme development of paraphyses 2a. It differs from *Clavaspis* in less acute pygidium, presence of lateral lobes (L2, L3, L4) and bifid marginal spines between third and fourth lobes (Takagi & Kawai, 1966).
GENERAL: Definition and characters by Takagi & Kawai (1966).
CITATIONS: DanzigPe1998 [catalogue: 217]; Kawai1980 [taxonomy: 213]; Takagi1974 [taxonomy, description: 10-12]; TakagiKa1966 [taxonomy, description: 115].

Clavaspidiotus abietis Takagi & Kawai

Clavaspidiotus abietis Takagi & Kawai, 1966: 115. Type data: JAPAN: Tokyo and Tusima, Okayama-ken, on *Abies firma*. Syntypes, female. Type depository: Sapporo: Ent. Institute, Faculty of Agriculture, Hokkaido University, Japan.
SCALE COVER: Scale cover not described by Takagi & Kawai (1966).
HOST PLANTS: **Pinaceae**: *Abies firma* [TakagiKa1966, Kawai1977].
DISTRIBUTION: **Palaearctic**: Japan [TakagiKa1966, Kawai1977, Kawai1980].
GENERAL: Description and illustration of adult female by Takagi & Kawai (1966).
KEYS: Kawai 1980: 214 (female) [Japan].
CITATIONS: DanzigPe1998 [catalogue: 218]; Kawai1977 [host, distribution, economic importance: 157]; Kawai1980 [taxonomy, description, host, distribution: 214]; Muraka1970 [host, distribution: 73]; Takagi1974 [taxonomy, description, illustration, host, distribution: 15-17]; TakagiKa1966 [taxonomy, description, illustration, host, distribution: 115,117].

Clavaspidiotus apicalis Takagi

Clavaspidiotus apicalis Takagi, 1974: 17. Type data: INDONESIA: Java, on shaddock tree, intercepted at U.S.A. Holotype female. Type depository: Washington: United States National Entomological Collection, U.S. National Museum of Natural History, District of Columbia, USA.
SYSTEMATICS: Type series includes specimens intercepted at ports of entry in the U.S.A. The holotype is from Java but paratypes are from the Philippines, Singapore, Penang and Egypt (Takagi, 1974).

SCALE COVER: Scale cover not described by Takagi (1974).
HOST PLANTS: **Flacourtiaceae**: *Pangium* [Takagi1974]. **Lauraceae**: *Cylicodaphne sebifera* [Takagi1974]. **Rutaceae**: *Citrus* [Takagi1974], *Citrus grandis* [Takagi1974], *Citrus limon* [Takagi1974], *Citrus paradisi* [Takagi1974].
DISTRIBUTION: **Oriental**: Indonesia (Java [Takagi1974]); Philippines [Takagi1974]; Singapore [Takagi1974]. **Palaearctic**: Egypt [Takagi1974].
GENERAL: Description and illustration of adult female by Takagi (1974).
CITATIONS: DanzigPe1998 [catalogue: 218]; Takagi1974 [taxonomy, description, illustration, host, distribution: 17-20].

Clavaspidiotus tayabanus (Cockerell)
Aspidiotus tayabanus Cockerell, 1905f: 133. Type data: PHILIPPINES: Lucban, Tayabas, on cultivated plant called "rosal" or "campopot". Syntypes, female. Type depository: Washington: United States National Entomological Collection, U.S. National Museum of Natural History, District of Columbia, USA.
Aonidiella tayabana; MacGillivray, 1921: 445. Change of combination requiring emendation of species name for agreement in gender.
Chrysomphalus tayabanus; Takahashi, 1933: 56. Change of combination.
Clavaspis tayabanus; McKenzie, 1939: 55. Change of combination.
Aspidiotus tabayanus; Takagi, 1970: 134. Misspelling of species name.
Clavaspidiotus tayabanus; Takagi, 1974: 15. Change of combination.
COMMON NAME: rosal scale [VelasqRi1969].
SYSTEMATICS: Takagi & Kawai (1966) suggested that this species possibly belongs to *Clavaspidiotus* Takagi & Kawai.
SCALE COVER: Female scale crowded on bark, not distinctly separable, flat, dark ferruginous; exuviae marked by a distinct dot and ring in grey or yellowish-white but, on rubbing, the second skin appears, bright orange-ferruginous or orange-chestnut; a thin whitish ventral scale (Cockerell, 1905f).
HOST PLANTS: **Guttiferae**: *Gardenia* [Takagi1970]. **Oleaceae**: *Jasminum* [Takagi1970]. **Rosaceae**: *Pyracantha koidzumii* [Takaha1933, Takagi1970].
DISTRIBUTION: **Oriental**: Philippines [Cocker1905f, Sander1906, Takagi1970]; Taiwan [Takaha1933, Takagi1970]. **Palaearctic**: Japan [Kawai1980].
GENERAL: Description of adult female by Cockerell (1905f).
KEYS: Kawai 1980: 214 (female) [Japan]; Robinson 1917: 29 (female) [Philippines].
CITATIONS: Borchs1966 [catalogue: 319]; Chou1985 [taxonomy, distribution: 310]; Cocker1905f [taxonomy, description, host, distribution: 133-134]; DanzigPe1998 [catalogue: 218]; Ferris1941e [taxonomy: 48]; Kawai1980 [taxonomy, description, host, distribution: 214]; MacGil1921 [taxonomy, description, host, distribution: 445]; McKenz1939 [taxonomy: 55]; Robins1917 [taxonomy, description, host, distribution: 29,32]; Sander1906 [taxonomy, host, distribution: 15]; Takagi1970 [taxonomy, host, distribution: 134]; Takagi1974 [taxonomy, description, illustration, host, distribution: 15-17]; Takaha1933 [taxonomy, illustration, host, distribution: 32,56-57]; Tao1999 [taxonomy, host, distribution: 81]; VelasqRi1969 [host, distribution: 195-208].

Clavaspis MacGillivray

Clavaspis MacGillivray, 1921: 391. Type species: *Aspidiotus subsimilis v. anonae* Houser, by original designation.

SYSTEMATICS: *Clavaspis* is close to *Diaspidiotus* and *Diclavaspis,* differing from *Diclavaspis* in possessing 3 pairs of lobes, and from *Diaspidiotus* in presence of clavate paraphyses on pygidium (Balachowsky, 1956; Munting, 1969).

GENERAL: Definition and characters by Ferris (1938a), Lepage (1940), Balachowsky (1956), Williams & Watson (1988) and by Kosztarab (1996).

KEYS: Colon-Ferrer & Medina-Gaud 1998: 28-32 (female) [Genera of Puerto Rico]; Gill 1997: 24-26 (female) [Genera of California]; Gill 1997: 101 (female) [Species of California]; Williams & Watson 1988: 19 (female) [Tropical South Pacific]; Chou1985 1985: 283 (female) [Genera of China]; Balachowsky 1958b: 226-231 (female) [*Aspidiotina* of Africa]; McKenzie 1956: 22-23 (female) [U.S.A.: California]; Ferris 1942: 446:26 (female) [North America]; Ferris 1942: 31-32 (female) [species North America].

CITATIONS: Balach1950b [taxonomy: 490]; Balach1956 [taxonomy, description: 90-91]; Balach1958b [taxonomy: 230]; Borchs1966 [catalogue: 317]; Brimbl1955 [taxonomy: 42]; Chou1985 [taxonomy, description: 283,309-310]; ColonFMe1998 [taxonomy, description: 51]; DanzigPe1998 [catalogue: 218]; Ferris1921b [taxonomy: 94]; Ferris1937c [taxonomy: 50]; Ferris1938 [taxonomy: 46]; Ferris1938a [taxonomy, description, illustration, host, distribution: 202]; Ferris1942 [taxonomy: 446:26]; Gill1997 [taxonomy: 101]; Koszta1996 [taxonomy, description: 482]; Lepage1940 [taxonomy, description: 73]; Lindin1937 [taxonomy: 182]; MacGil1921 [taxonomy, description: 391,441]; Mamet1949 [taxonomy: 57]; McKenz1939 [taxonomy: 53]; McKenz1956 [taxonomy: 22]; MorrisMo1966 [taxonomy, catalogue: 39]; Muntin1969 [taxonomy: 125]; Tao1999 [taxonomy: 81]; WilliaWa1988 [taxonomy, description: 98].

Clavaspis barbigera Ferris

Clavaspis barbigera Ferris, 1954: 43. Type data: U.S.A.: Florida, at Miami, on mastic [=*Bursera simaruba?*]; collected by O.D. Link, February 25, 1953. Syntypes, female. Type depository: Davis: The Bohart Museum of Entomology, University of California, California, USA.

COMMON NAME: barbigera scale [Dekle1965c].

SCALE COVER: Only slide-mounted specimens available to Ferris (1954).

HOST PLANTS: **Burseraceae**: *Bursera simaruba* [Balach1932d]. **Sapotaceae**: *Sideroxylon* [Dekle1965c].

DISTRIBUTION: **Nearctic**: United States of America (Florida [Balach1932d, Dekle1965c]). **Neotropical**: Trinidad and Tobago (Trinidad [Nakaha1982]).

GENERAL: Description and illustration of adult female by Ferris (1954).

CITATIONS: Borchs1966 [catalogue: 317]; Dekle1965c [taxonomy, description, host, distribution: 44]; Dekle1976 [taxonomy, description, host, distribution: 63]; Ferris1954 [taxonomy, description, illustration, host, distribution: 43]; Nakaha1982 [host, distribution: 23].

Clavaspis coursetiae (Marlatt)

Aspidiotus (*Diaspidiotus*) *coursetiae* Marlatt, 1908c: 20. Type data: MEXICO: Hormosillo, on *Coursetia glandulosa*; collected by Albert Koebele, 23 & 24 April 1897. Syntypes, female. Type depository: Washington: United States National Entomological Collection, U.S. National Museum of Natural History.

Aspidiotus coursetiae; Sanders, 1909a: 52. Change of combination.

Diaspidiotus coursetiae; MacGillivray, 1921: 413. Change of combination.

Clavaspis coursetiae; Ferris, 1938a: 203. Change of combination.

Aspidiotus (*Clavaspis*) *coursetiae*; Merrill, 1953: 17. Change of combination.

Clavaspis coursetiae; Borchsenius, 1966: 317. Revived combination.

COMMON NAME: pigeon plum scale [Dekle1965c].

SCALE COVER: Female scale nearly circular, 1.5 mm in diameter; of medium density, depressed; colour greyish, more or less soiled by adhering extraneous matter; exuviae sublateral, covered. Male scale similar, of normal shape (Marlatt, 1908c). Scales grey or whitish, that of female circular, flat, exuvia subcentral; that of male elongate oval, exuvia near one end (Ferris, 1938a).

HOST PLANTS: **Burseraceae**: *Bursera simarouba* [Dekle1965c]. **Ericaceae**: *Arctostaphylos* [Ferris1942]. **Euphorbiaceae**: *Ricinella vaseyi* [Ferris1938a]. **Leguminosae**: *Acacia* [Ferris1938a], *Acacia constricta* [Ferris1938a, McDani1968], *Cercidium floridanum* [Ferris1938a], *Coursetia glandulosa* [Marlat1908c, Sander1909a, MerrilCh1923, Ferris1938a], *Pithecellobium brevifolium* [Ferris1938a]. **Malpighiaceae**: *Byrsonima crassifolia* [Ferris1942]. **Polygonaceae**: *Coccoloba diversifolia* [Dekle1965c], *Coccoloba laurifolia* [MerrilCh1923, Ferris1938a]. **Rhamnaceae**: *Karwinskia humboltiana* [Ferris1938a]. **Rutaceae**: *Amyris parvifolia* [McDani1968], *Ptelea tomentosa* [Ferris1938a], *Xanthoxylum pterota* [Ferris1938a]. **Scrophulariaceae**: *Leucophyllum texanum* [Ferris1938a]. **Solanaceae**: *Lycium fremontii* [Ferris1938a]. **Ulmaceae**: *Celtis pallida* [Ferris1938a]. **Zygophyllaceae**: *Porlieria angustifolia* [Ferris1938a].

DISTRIBUTION: **Nearctic**: Mexico (Sonora [Ferris1938a]); U.S.A. (Arizona [Ferris1938a], Florida [MerrilCh1923, Dekle1965c], Texas [Ferris1938a). **Neotropical**: Mexico (Chihuahua [Ferris1938a]); Panama [Ferris1942].

BIOLOGY: Occurring on bark (Ferris, 1938a).

GENERAL: Description and illustration of adult female by Marlatt (1908c) and by Ferris (1938a).

KEYS: McDaniel 1968: 231-232 (female) [U.S.A.: Texas]; Ferris 1942: 32 (female) [North America].

CITATIONS: Borchs1966 [catalogue: 317]; Dekle1965c [taxonomy, description, host, distribution: 45]; Dekle1976 [taxonomy, description, host, distribution: 64]; Ferris1921b [taxonomy: 94]; Ferris1938a [taxonomy, description, illustration, host, distribution: 203]; Ferris1941e [taxonomy: 42]; Ferris1942 [taxonomy, host, distribution: 445:10; 446:31]; MacGil1921 [taxonomy, description, host, distribution: 413]; Marlat1908c [taxonomy, description, illustration, host, distribution: 20]; McDani1968 [taxonomy, illustration, host, distribution: 232-233]; Merril1953 [taxonomy, description, host, distribution: 17-18]; MerrilCh1923 [taxonomy, description, host, distribution: 198]; Nakaha1982 [host, distribution: 23-24]; Sander1909a [taxonomy, host, distribution: 52].

Clavaspis covilleae (Ferris)

Aspidiotus covilleae Ferris, 1919a: 64. Type data: U.S.A.: Arizona, east of Phoenix, Mormon Flat, on *Covillea glutinosa*. Syntypes, female. Type depository: Davis: The Bohart Museum of Entomology, University of California, California, USA.
Ferrisaspis covilleae; MacGillivray, 1921: 388. Change of combination.
Clavaspis covilleae; Ferris, 1938a: 204. Change of combination.
COMMON NAMES: Covillea scale [McKenz1956].
SCALE COVER: Scale of female flat, circular, whitish, exuviae subcentral; that of male elongate oval with exuvia near one end (Ferris, 1938a).
HOST PLANTS: **Anacardiaceae**: *Rhus trilobata* [McKenz1956]. **Capparidaceae**: *Forchammeria* [Ferris1942, McKenz1956]. **Celastraceae**: *Euonymus* [Ferris1938a, McKenz1956, McDani1968]. **Compositae**: *Pluchea sericea* [McKenz1956]. **Leguminosae**: *Acacia paucispina* [Ferris1942, McKenz1956]. **Oleaceae**: *Olea europaea* [Ferris1938a, McKenz1956]. **Rosaceae**: *Prunus amygdalus* [McKenz1956], *Pyrus communis* [McKenz1956]. **Solanaceae**: *Lycium fremontii* [Ferris1942]. **Zygophyllaceae**: *Covillea glutinosa* [Ferris1919a, McKenz1956].
DISTRIBUTION: **Nearctic**: Mexico (Sonora [Ferris1942]); United States of America (Arizona [Ferris1919a, Ferris1938a], California [McKenz1956], New Mexico [Ferris1942], Texas [McDani1968]).
BIOLOGY: Type series taken from beneath loose bark on exposed roots of *Covillea glutinosa*. Ferris (1938a) also recorded it from twigs of olive.
GENERAL: Description and illustration of adult female by Ferris (1919a, 1938a), McKenzie (1956) and by Gill (1997).
KEYS: Gill 1997: 101 (female) [Species of California]; McDaniel 1968: 232 (female) [U.S.A.: Texas]; McKenzie 1956: 24 (female) [U.S.A.: California]; Ferris 1942: 32 (female) [North America].
CITATIONS: Borchs1966 [catalogue: 317-318]; Ferris1919a [taxonomy, description, illustration, host, distribution: 64-65]; Ferris1921b [taxonomy: 94]; Ferris1937c [taxonomy, illustration: 54,72]; Ferris1938a [taxonomy, description, illustration, host, distribution: 204]; Ferris1942 [taxonomy, host, distribution: 445:10; 446:31]; Gill1997 [host, distribution, taxonomy, description, illustration, economic importance: 101,102]; MacGil1921 [taxonomy, description, host, distribution: 422-423]; McDani1968 [taxonomy, illustration, host, distribution: 232-234]; McKenz1939 [taxonomy: 54]; McKenz1956 [taxonomy, description, illustration, host, distribution: 54,57]; Nakaha1982 [host, distribution: 24].

Clavaspis crypta Howell & Tippins

Clavaspis crypta Howell & Tippins, 1975: 338. Type data: U.S.A.: Georgia, Clinch County, near Cargo, on *Carya pecan*. Holotype female. Type depository: Washington: United States National Entomological Collection, U.S. National Museum of Natural History, District of Columbia, USA.
SCALE COVER: Female scale circular, ca. 2 mm in diameter, but may appear elongate because of irregularities in bark; light grey; exuviae subcentral, gold-coloured; scale covered with bark flakes and nearly same colour as bark; gold exuviae often the only indication of its presence. Scale of male not seen (Howell & Tippins, 1975).

HOST PLANTS: **Juglandaceae**: *Carya pecan* [HowellTi1975].
DISTRIBUTION: **Nearctic**: United States of America (Georgia [HowellTi1975]).
BIOLOGY: Occurring on bark (Howell & Tippins, 1975).
GENERAL: Description and illustration of adult female by Howell & Tippins (1975).
CITATIONS: HowellTi1975 [taxonomy, description, illustration, host, distribution: 338-340]; Nakaha1982 [host, distribution: 24].

Clavaspis dentata Ferris

Clavaspis dentata Ferris, 1942: 427. Type data: PANAMA: Chiriqui Province, between Dolega and David, on an undetermined tree. Holotype female. Type depository: Davis: The Bohart Museum of Entomology, University of California, California, USA.
SCALE COVER: Scale of female roughly circular, brown, quite hard and brittle. Scale of male not recognized (Ferris, 1942).
DISTRIBUTION: **Neotropical**: Panama [Ferris1942].
BIOLOGY: Occurring on trunk, concealed beneath bark flakes (Ferris, 1942).
GENERAL: Description and illustration of adult female by Ferris (1942).
KEYS: Ferris 1942: 31 (female) [North America].
CITATIONS: Borchs1966 [catalogue: 318]; Ferris1942 [taxonomy, description, illustration, host, distribution: 427; 446:31]; Lindin1957 [taxonomy: 547].

Clavaspis disclusa Ferris

Clavaspis disclusa Ferris, 1938a: 205. Type data: U.S.A.: California, Ventura County, Camarillo, on "Persian walnut" [=*Juglans* sp.]. Holotype female. Type depository: Davis: The Bohart Museum of Entomology, University of California, California, USA.
Clavaspis diclusa; Nakahara, 1982: 24. Misspelling of species name.
COMMON NAME: decluse scale [McKenz1956].
SCALE COVER: Scale of female dark grey, flat, circular; exuviae central almost entirely concealed; scale of male not seen (Ferris, 1938a).
HOST PLANTS: **Cornaceae**: *Cornus* [McKenz1956]. **Fagaceae**: *Quercus* [McKenz1956]. **Juglandaceae**: *Carya illinoensis* [McKenz1956], *Juglans* [Ferris1938a], *Juglans californica* [McKenz1956], *Juglans regia* [McKenz1956]. **Oleaceae**: *Fraxinus* [McKenz1956]. **Rosaceae**: *Prunus persica* [McKenz1956].
DISTRIBUTION: **Nearctic**: U.S.A. (California [Ferris1938a, McKenz1956]).
BIOLOGY: Occurring on the bark (Ferris, 1938a).
GENERAL: Description and illustration of adult female by Ferris (1938a), McKenzie (1956) and by Gill (1997).
KEYS: Gill 1997: 101 (female) [Species of California]; McKenzie 1956: 24 (female) [U.S.A.: California]; Ferris 1942: 31 (female) [North America].
CITATIONS: Borchs1966 [catalogue: 318]; Ferris1938a [taxonomy, description, illustration, host, distribution: 205]; Ferris1942 [taxonomy: 446:31]; Gill1997 [host, distribution, taxonomy, description, illustration, economic importance: 101,103,106]; Lindin1957 [taxonomy: 547]; McKenz1956 [taxonomy, description, illustration, host, distribution: 56-57]; Nakaha1982 [host, distribution: 24].

Clavaspis herculeana (Cockerell & Hadden)

Aspidiotus herculeanus Cockerell & Hadden, *in*: Doane & Hadden, 1909: 298. Type data: FRENCH POLYNESIA: Society Islands, on bark of undetermined plant. Syntypes, female. Type depository: Davis: The Bohart Museum of Entomology, University of California, California, USA. Notes: The credit of authorship to Cockerell & Hadden is clearly indicated by Doane & Hadden (1909: 296).

Aspidiotus subsimilis anonae Houser, 1918: 163. Type data: CUBA: Havana, University of Havana Botanic Garden, on 11 host plants; host of holotype not indicated. Holotype female. Type depository: Washington: United States National Entomological Collection, U.S. National Museum of Natural History, District of Columbia, USA. Synonymy by Ferris, 1920a: 64.

Clavaspis anonae MacGillivray, 1921: 441. Change of combination and rank.

Clavaspis herculeana; MacGillivray, 1921: 441. Change of combination requiring emendation of species name for agreement in gender.

Aspidiotus anonae; Ferris, 1921b: 94. Change of combination.

Aspidiotus symbioticus Hempel, 1932: 334. Type data: BRAZIL: São Paulo State, S. Roque, Cascata, Itatiba and Morungaba, on "pereiras cultivadas" [=cultivated pear trees] and *Annona*. Syntypes, female. Type depository: São Paulo: Instituto Biologico de São Paulo, Brazil. Synonymy by Lepage, 1940: 74.

Chrysomphalus alluaudi Mamet, 1936: 93. Type data: MAURITIUS: on *Rosa* sp.; Rose Hill on *Spondias dulcis*. Syntypes. Type depository: Paris: Muséum national d'Histoire naturelle, France. Synonymy by Mamet, 1942: 35.

Clavaspis alluaudi; McKenzie, 1939: 53. Change of combination.

Aspidiotus herculeanus; Ferris, 1941e: 44. Notes: Incorrect citation of "Hadden" as author.

Clavaspis herculeana; Ferris, 1942: 31. Notes: Incorrect citation of "Hadden" as author.

Clavaspis symbioticus; Vernalha, 1953: 169. Change of combination.

Aspidiotus (Clavaspis) herculeanus; Merrill, 1953: 21. Change of combination.

Clavaspis herculeana; Borchsenius, 1966: 318. Revived combination. Notes: Incorrect citation of "Doane & Hadden" as authors.

COMMON NAMES: cassia bark scale [VelasqRi1969]; clavate scale [Brimbl1962]; herculeana scale [Dekle1965c].

SCALE COVER: Ferris (1938a) reported that scales of type-material were so concealed beneath and so mingled with bark scales and epidermis of host that it was impossible to tell much about them; scale seems to be grey or white and circular.

HOST PLANTS: **Anacardiaceae**: *Mangifera* [Balach1956], *Mangifera cambodiana* [Houser1918, MerrilCh1923, Merril1953], *Mangifera indica* [WilliaWa1988], *Metopium toxiferum* [MerrilCh1923], *Spondias cytherea* [Mamet1943a, Mamet1949, Borchs1966], *Spondias dulcis* [Mamet1936, Borchs1966], *Spondias purpurea* [Houser1918, MerrilCh1923]. **Annonaceae**: *Annona* [Houser1918, Hempel1932], *Annona muricata* [Dekle1965c], *Annona squamosa* [Dekle1965c]. **Avicenniaceae**: *Avicennia nitida* [BesheaTiHo1973]. **Bignoniaceae**: *Tabebuia heterophylla* [Martor1976]. **Bombacaceae**: *Eriodendron* [Balach1956], *Eriodendron anfractuosum* [Mamet1951, Borchs1966]. **Caricaceae**: *Carica papaya* [Brimbl1955]. **Cochlospermaceae**: *Cochlospermum vitifolium*

[Martor1976]; **Euphorbiaceae**: *Aleurites* [Almeid1973], *Aleurites fordii* [Balach1956]. **Lauraceae**: *Cinnamomum* [Balach1956], *Cinnamomum zeylanicum* [Houser1918]. **Leguminosae**: *Acacia farnesiana* [McDani1968], *Acacia flexicaulis* [McDani1968], *Cassia fistula* [Merril1953], *Delonix regia* [MerrilCh1923, Dekle1965c], *Erythrina* [Brimbl1955], *Erythrina caffra* [DeLott1967a], *Erythrina indica* [WilliaWa1988], *Lonchocarpus* [Balach1956], *Lonchocarpus latifolius* [Houser1918], *Mimosa pudica* [WilliaWa1988], *Pithecellobium saman* [WilliaWa1988]. **Magnoliaceae**: *Magnolia grandiflora* [Houser1918]. **Malvaceae**: *Gossypium* [WilliaWa1988]. **Meliaceae**: *Cedrela toona* [Brimbl1955]. **Moraceae**: *Ficus* [BesheaTiHo1973], *Ficus capensis* [Merril1953], *Ficus carica* [Brimbl1955], *Ficus religiosa* [Merril1953], *Ficus roxburgii* [Merril1953], *Maclura tinctoria* [Houser1918]. **Myristicaceae**: *Myristica hypargyraea* [WilliaWa1988]. **Myrtaceae**: *Eugenia* [MerrilCh1923]. **Naucleaceae**: *Cephalanthus occidentalis* [BesheaTiHo1973]. **Rhizophoraceae**: *Rhizophora* [Balach1956]. **Rosaceae**: *Eriobotrya japonica* [Houser1918, Brimbl1955], *Pyrus* [Mamet1943a, Mamet1949, Borchs1966], *Rosa* [Houser1918, Mamet1936, Mamet1949, Borchs1966]. **Rutaceae**: *Citrus* [Ferris1955b]. **Salicaceae**: *Populus* [Almeid1973]. **Styracaceae**: *Halesia* [BesheaTiHo1973].

DISTRIBUTION: **Afrotropical**: Guinea [Balach1956]; Kenya [DeLott1967a]; Madagascar [Mamet1951, Mamet1954, Borchs1966]; Malawi [Nakaha1982]; Mauritius [Mamet1936, Mamet1943a, Mamet1949, Borchs1966]; Mozambique [Almeid1973]; Zimbabwe [Nakaha1982]. **Australasian**: Australia (Queensland [Brimbl1955]); Cook Islands [WilliaWa1988]; Fiji [WilliaWa1988]; French Polynesia (Society Islands [DoaneHa1909], Tahiti [Balach1956]); Hawaiian Islands (Hawaii [Nakaha1982]); New Caledonia [Nakaha1982]; Tonga [WilliaWa1988]; Western Samoa [WilliaWa1988]. **Nearctic**: Mexico (Guerrero [Ferris1955b], San Luis Potosi [Ferris1955b], Veracruz [Ferris1955b]); United States of America (Florida [MerrilCh1923, Merril1953, Balach1956, Dekle1965c, BesheaTiHo1973], Texas [McDani1968]). **Neotropical**: Brazil [Lepage1940] (São Paulo [Hempel1932, Lepage1938]); Colombia [Kondo2001]; Cuba [Houser1918]; Panama [Ferris1942]; Puerto Rico & Vieques Island (Puerto Rico [Martor1976, ColonFMe1998]); U.S. Virgin Islands [Nakaha1983]. **Oriental**: Philippines [VelasqRi1969]; Singapore [Nakaha1982].

BIOLOGY: Scales concealed beneath and mingled with bark scales and epidermis of host (Ferris, 1938a). Associated with fungus *Septobasidium saccardinum* (Lepage, 1940; Ferris, 1955b).

ECONOMIC IMPORTANCE: Brimblecombe (1955) considered infestation on fig and Papaya in Australia, Queensland, as evidence of pest potential.

GENERAL: Description and illustration of adult female by Ferris (1938a), Lepage (1940), Balachowsky (1956), Williams & Watson (1988) and by Colon-Ferrer & Medina-Gaud (1998).

KEYS: McDaniel 1968: 230-231 (female) [U.S.A.: Texas]; Balachowsky 1956: 91 (female) [Africa]; Ferris 1942: 31 (female) [North America].

CITATIONS: Almeid1973 [host, distribution: 3-4]; Balach1956 [taxonomy, description, illustration, host, distribution: 94-96]; BeardsDaHo1976 [economic importance: 105]; BesheaTiHo1973 [host, distribution: 5]; Borchs1966 [catalogue:

318]; Brimbl1955 [taxonomy, description, illustration, host, distribution: 42-45]; Brimbl1962 [host, distribution: 220]; ClapsWoGo2001a [taxonomy, host, distribution: 14]; ColonFMe1998 [taxonomy, description, illustration, host, distribution: 51-52]; Dekle1965c [taxonomy, description, host, distribution: 46]; Dekle1976 [taxonomy, description, host, distribution, economic importance: 65]; DeLott1967a [host, distribution: 114]; DoaneHa1909 [taxonomy, description, host, distribution: 298-299]; Ferris1920a [taxonomy: 64]; Ferris1921b [taxonomy: 94]; Ferris1937c [taxonomy, illustration: 50,65]; Ferris1938a [taxonomy, description, illustration, host, distribution: 206]; Ferris1941e [taxonomy: 40,44,48]; Ferris1942 [taxonomy, host, distribution: 445:10; 446:31]; Ferris1955b [host, distribution, life history: 25]; Hall1943 [taxonomy: 2]; Hempel1932 [taxonomy, description, host, distribution: 334-335]; Hinckl1963 [host, distribution, biological control]; Houser1918 [taxonomy, description, illustration, host, distribution: 163-165]; Kondo2001 [taxonomy, host, distribution: 43]; Lepage1938 [catalogue: 397]; Lepage1940 [taxonomy, description, illustration, host, distribution: 74-76]; Lindin1957 [taxonomy: 547]; Lizery1942c [taxonomy, description, host, distribution: 235-236]; MacGil1921 [taxonomy, description, host, distribution: 441]; Mamet1936 [taxonomy, description, illustration, host, distribution: 93]; Mamet1942 [taxonomy: 35]; Mamet1943a [catalogue: 158]; Mamet1949 [catalogue: 57,58]; Mamet1951 [host, distribution: 227]; Mamet1954 [host, distribution: 17]; Martor1976 [host, distribution: 74,248,253]; McDani1968 [taxonomy, illustration, host, distribution: 234-236]; McKenz1939 [taxonomy: 53]; Mead1983 [host, distribution: 1-5]; Merril1953 [taxonomy, description, host, distribution: 21-23]; MerrilCh1923 [taxonomy, description, host, distribution, economic importance: 210]; MillerDa1990 [host, distribution, economic importance: 301]; Montgo1921 [host, distribution: 41-55]; MoutiaMa1947 [distribution]; Nakaha1982 [host, distribution: 24-25]; Nakaha1983 [host, distribution: 10]; Newell1923 [host, distribution: 263-266]; Sander1909a [taxonomy, host, distribution: 52]; Sassce1923 [host, distribution: 152-158]; VelasqRi1969 [host, distribution: 195-208]; Vernal1953 [taxonomy, host, distribution: 169]; WilliaWa1988 [taxonomy, description, illustration, host, distribution: 98-100].

Clavaspis kalaharica Munting

Clavaspis kalaharica Munting, 1969: 121. Type data: SOUTH AFRICA: Kalahari Gemsbok National Park, Union Ends, on *Acacia giraffae*; collected 22.viii.1967. Holotype female. Type depository: Pretoria: South African National Collection of Insects, South Africa; type no. 3323/1.
SCALE COVER: Scale structure not examined by Munting (1969).
HOST PLANTS: **Leguminosae**: *Acacia giraffae* [Muntin1969].
DISTRIBUTION: **Afrotropical**: South Africa [Muntin1969].
GENERAL: Description and illustration of adult female by Munting (1969).
CITATIONS: Muntin1969 [taxonomy, description, illustration, host, distribution: 121-122,147].

Clavaspis mori (Herrick)

Aspidiotus mori Herrick, 1910: 22. Type data: U.S.A.: Texas, Banks of Brazos River, six miles from the Agricultural and Mechanical College, College Station, on underside of branches of native red mulberry, *Morus rubra*. Syntypes, female. Type depository: Washington: United States National Entomological Collection, U.S. National Museum of Natural History, District of Columbia, USA.

Diaspidiotus mori; MacGillivray, 1921: 413. Change of combination.

Clavaspis mori; Ferris, 1938a: 207. Change of combination.

SCALE COVER: Scale of female circular or subcircular, flat; male scale similar to that of the female but smaller and more elongated (Ferris, 1938a).

HOST PLANTS: **Moraceae**: *Morus rubra* [Herric1910, McDani1968].

DISTRIBUTION: **Nearctic**: United States of America (Texas [Herric1910, Herric1911, McDani1968]).

BIOLOGY: Occurring on branches (Ferris, 1938a).

GENERAL: Description and illustration of adult female by Ferris (1938a).

KEYS: McDaniel 1968: 230-231 (female) [U.S.A.: Texas]; Ferris 1942: 31 (female) [North America].

CITATIONS: Borchs1966 [catalogue: 318]; Ferris1938a [taxonomy, description, illustration, host, distribution: 207]; Ferris1941e [taxonomy: 46]; Ferris1942 [taxonomy: 446:31]; Herric1910 [taxonomy, description, illustration, host, distribution: 22]; Herric1911 [taxonomy, description, illustration, host, distribution: 10,19-20,53]; Lindin1957 [taxonomy: 547]; MacGil1921 [taxonomy, description, host, distribution: 413]; McDani1968 [taxonomy, illustration, host, distribution: 236-237]; Nakaha1982 [host, distribution: 25]; Sassce1911 [taxonomy: 69].

Clavaspis pedilanthi (Ferris)

Aspidiotus pedilanthi Ferris, 1921: 127. Type data: MEXICO: Baja California, near Pescadero, on *Pedilanthus macrocarpa* ("candelilla"). Holotype female. Type depository: Davis: The Bohart Museum of Entomology, University of California, California, USA.

Clavaspis pedilanthi; Ferris, 1938a: 208. Change of combination.

SCALE COVER: Scale of female white; circular, about 1.5 mm in diameter; flat, grey; exuviae subcentral. Male scale slightly elongate with exuvia near middle (Ferris, 1921, 1938a).

HOST PLANTS: **Compositae**: *Franseria* [Ferris1921]. **Euphorbiaceae**: *Pedilanthus macrocarpa* [Ferris1921, McDani1968]. **Malvaceae**: *Horsfordia* [Ferris1921]. **Salicaceae**: *Populus* [Ferris1921]. **Solanaceae**: *Lycium* [Ferris1938a].

DISTRIBUTION: **Nearctic**: Mexico (Baja California [Ferris1921]); United States of America (Texas [McDani1968]).

BIOLOGY: Type specimens taken off stem base and roots, although material collected from Texas was collected on stems (Ferris, 1938a).

GENERAL: Description and illustration of adult female by Ferris (1921, 1938a).

KEYS: Ferris 1942: 32 (female) [North America].

CITATIONS: Borchs1966 [catalogue: 318]; Ferris1921 [taxonomy, description, illustration, host, distribution: 127-128]; Ferris1938a [taxonomy, description, illustration, host, distribution: 208]; Ferris1941e [taxonomy: 46]; Ferris1942 [taxonomy: 446:32]; Lindin1957 [taxonomy: 547]; McDani1968 [taxonomy, illustration, host, distribution: 236,238]; Nakaha1982 [host, distribution: 25].

Clavaspis perplexa Munting

Clavaspis perplexa Munting, 1971: 120. Type data: SOUTH AFRICA: Transvaal, Chuniespoort, on *Euphorbia* sp.; collected by H.K. Munro, 8.xii.1965. Holotype female. Type depository: Pretoria: South African National Collection of Insects, South Africa; type no. 2022/12.
SCALE COVER: Female scale circular, 1.3 mm in diameter, greyish-brown in colour. Male scale oval, 0.8 mm long, grey with a whitish apex (Munting, 1971).
HOST PLANTS: **Euphorbiaceae**: *Euphorbia* [Muntin1971].
DISTRIBUTION: **Afrotropical**: South Africa [Muntin1971].
GENERAL: Description and illustration of adult female by Munting (1971).
CITATIONS: Muntin1971 [taxonomy, description, illustration, host, distribution: 120-121].

Clavaspis pituranthi Williams

Clavaspis pituranthi Williams, 1962a: 128. Type data: SOUTH AFRICA: Twee Rivieren (Kalahari Gemsbok National Park), on *Pituranthos aphyllus*. Holotype female. Type depository: London: The Natural History Museum, England, UK.
SCALE COVER: Scale of adult female slightly convex, subcircular, of a pale straw colour; exuviae of immature stages, subcentral pale brown. Diameter about 1.2 mm. Male scale same shape and colour as female; exuviae pale brown, diameter about 0.75 mm (Williams, 1962a).
HOST PLANTS: **Umbelliferae**: *Pituranthos aphyllus* [Willia1962a].
NATURAL ENEMIES: HYMENOPTERA **Aphelinidae**: *Azotus capensis* Howard [AnneckIn1970]. **Encyrtidae**: *Adelencyrtus inglisiae* Compere & Annecke [AnneckIn1971], *Comperiella ponticula* Prinsloo & Annecke [PrinslAn1976], *Comperiella ponticula* Prinsloo & Annecke [Prinsl1983], *Metaphycus ustulatus* Annecke & Mynhardt [Prinsl1983].
DISTRIBUTION: **Afrotropical**: South Africa [Willia1962a].
BIOLOGY: Occurring on stems (Williams, 1962a).
GENERAL: Description and illustration of adult female by Williams (1962a).
CITATIONS: AnneckIn1970 [host, distribution, biological control: 240]; AnneckIn1971 [host, distribution, biological control: 2]; Borchs1966 [catalogue: 318]; Prinsl1983 [distribution, biological control: 26]; PrinslAn1976 [host, distribution, biological control: 186-188]; Willia1962a [taxonomy, description, illustration, host, distribution: 128-130].

Clavaspis quadriloba Brimblecombe

Clavaspis quadriloba Brimblecombe, 1959: 125. Type data: AUSTRALIA: Queensland, Yarraman, on *Owenia venosa*; collected May 1947. Holotype female. Type depository: Brisbane: Queensland Museum, Queensland, Australia; type no. T5700.
SCALE COVER: Insects sparse on twigs of host (Brimblecombe, 1959). Details of scale were not available to Brimblecombe (1959).

HOST PLANTS: Meliaceae: *Owenia venosa* [Brimbl1959].
DISTRIBUTION: Australasian: Australia (Queensland [Brimbl1959]).
GENERAL: Description and illustration of adult female by Brimblecombe (1959).
CITATIONS: Borchs1966 [catalogue: 318-319]; Brimbl1959 [taxonomy, description, illustration, host, distribution: 125-127].

Clavaspis subcuticularis (Green)

Aspidiotus (*Aonidiella*) *subcuticularis* Green, 1916e: 54. Type data: AUSTRALIA: Northern Australia, on *Ficus orbicularis*. Syntypes, female. Type depository: London: The Natural History Museum, England, UK.
Aonidiella subcuticularis; MacGillivray, 1921: 444. Change of combination.
Clavaspis subcuticularis; Brimblecombe, 1958: 63. Change of combination.
SCALE COVER: Female scale buried under cuticle of leaf; reddish-brown larval exuviae partially exposed; rest of scale closely adherent to and difficult to separate from superimposed cuticle. Presence indicated by very inconspicuous blister-like swellings on surface of leaf. These vesicles with a diameter of 2.0-2.5 mm. Nymphal exuviae with a denser median area (Green, 1916e). Female scale single and sparse, embedded within leaf tissue with only apex of dark brown exuviae exposed; scale dense and brittle (Brimblecombe, 1958).
HOST PLANTS: Moraceae: *Ficus orbicularis* [Green1916e, Brimbl1958].
DISTRIBUTION: Australasian: Australia (Northern Territory [Green1916e].
GENERAL: Description and illustration of adult female by Green (1916e) and by Brimblecombe (1958).
CITATIONS: Borchs1966 [catalogue: 319]; Brimbl1958 [taxonomy, description, illustration, host, distribution: 63-65]; Ferris1941e [taxonomy: 48]; Green1916e [taxonomy, description, illustration, host, distribution: 54-55]; MacGil1921 [taxonomy, description, host, distribution: 444]; McKenz1938 [taxonomy: 4].

Clavaspis subfervens (Green)

Aspidiotus (*Targionia*) *subfervens* Green, 1904: 66. Type data: AUSTRALIA: Victoria, on *Acacia* sp. Holotype female. Type depository: London: The Natural History Museum, England, UK.
Aspidiotus (*Tagionia*) *subfervens*; Green, 1904: 66. Misspelling of genus name.
Aspidiotus subfervens; Sanders, 1906: 14. Change of combination.
Monaonidiella subfervens; MacGillivray, 1921: 446. Change of combination.
Targionia subfervens; Laing, 1929: 28. Change of combination.
Clavaspis subfervens; Brimblecombe, 1958: 61. Change of combination.
SCALE COVER: Female scale circular, diameter 1.25 mm; strongly convex; dull blackish brown, thickly dusted with greyish scurfy secretion; exuviae brown, obscured by grey secretion; below deep chocolate-brown; exuviae fiery red. Male scale not observed (Green, 1904). Insects scattered on twigs; female scale blackish brown; convex, 1.25 mm diameter; exuviae reddish brown (Brimblecombe, 1958).
HOST PLANTS: Leguminosae: *Acacia* [Green1904, Frogga1914], *Ac. harpophylla* [Brimbl1958], *Ac. melanoxylon* [Laing1929, Brimbl1958], *Ac. pycnantha* [Laing1929]. **Myrtaceae:** *Eucalyptus* [Laing1929]. **Rhamnaceae:** *Pomaderris* [Sander1906, Frogga1914].

DISTRIBUTION: **Australasian**: Australia [Sander1906] (Queensland [Brimbl1958], Victoria [Green1904, Frogga1914, Laing1929]).
GENERAL: Description and illustration of adult female by Green (1904), Laing (1929) and by Brimblecombe (1958).
CITATIONS: Borchs1966 [catalogue: 319]; Brimbl1958 [taxonomy, description, illustration, host, distribution: 61-63]; Ferris1941e [taxonomy: 48]; Ferris1943a [taxonomy: 86]; Frogga1914 [taxonomy, description, host, distribution: 318]; Frogga1915 [taxonomy, description, host, distribution: 22]; Green1904 [taxonomy, description, illustration, host, distribution: 66-67]; Laing1929 [host, distribution: 28-29]; MacGil1921 [taxonomy, description, host, distribution: 446]; Sander1906 [taxonomy, host, distribution: 14].

Clavaspis subsimilis (Cockerell)
Aspidiotus subsimilis Cockerell, 1899d: 168. Type data: MEXICO: Cuautla, on a leafless tree; Hermosillo, on *Caesalpinia palmeri*. Syntypes, female. Type depository: Washington: United States National Entomological Collection, U.S. National Museum of Natural History, District of Columbia, USA.
Aspidiotus (*Diaspidiotus*) *subsimilis*; Cockerell, 1899a: 396. Change of combination.
Aonidiella subsimilis; Leonardi, 1900: 342. Change of combination.
Hendaspidiotus subsimilis; MacGillivray, 1921: 440. Change of combination.
Clavaspis subsimilis; Ferris, 1938a: 209. Change of combination.
SCALE COVER: Cockerell (1899d) described scale: "Female scale circular, about 1.5 mm in diameter; flat, thin; pale grey to whitish, or tinged with brown; exuviae covered, inconspicuous, marked by a whitish boss; this scale is very like that of *Aspidiotus perniciosus*, but there is no distinct dot and ring. Male scale oval, slightly stained with blackish; exuviae yellowish".
HOST PLANTS: **Burseraceae**: *Bursera microphylla* [RosenDe1979]. **Ebenaceae**: *Diospyros texana* [McDani1968]. **Leguminosae**: *Acacia constricta* [McDani1968], *Caesalpinia palmeri* (?) [Cocker1899d, Leonar1900, McDani1968]. **Rutaceae**: *Ptelea tomentosa* [McDani1968].
NATURAL ENEMIES: HYMENOPTERA **Aphelinidae**: *Aphytis melanostictus* Compere [RosenDe1979, MyartsRu2000]. **Signiphoridae**: *Signiphora pulchra* Girault [Woolle1990].
DISTRIBUTION: **Nearctic**: Mexico [Cocker1899n, MyartsRu2000] (Baja California Sur [RosenDe1979], Morelos [Cocker1899d, Ferris1938a]); United States of America (Texas [McDani1968]). **Neotropical**: Guatemala [Nakaha1982].
BIOLOGY: Occurring on bark (Ferris, 1938a).
GENERAL: Description and illustration of adult female by Ferris (1938a).
KEYS: McDaniel 1968: 232 (female) [U.S.A.: Texas]; Ferris 1942: 32 (female) [North America]; Newell 1899: 4-5 (female) [North America].
CITATIONS: Borchs1966 [catalogue: 319]; Chiesa1948 [host, distribution, economic importance]; ClapsWoGo2001a [taxonomy, host, distribution: 14]; Cocker1899a [taxonomy: 396]; Cocker1899d [taxonomy, description, host, distribution: 168]; Cocker1899n [host, distribution, illustration: 21]; Fernal1903b [catalogue: 279]; Ferris1938a [taxonomy, description, illustration, host, distribution:

209]; Ferris1941e [taxonomy: 48]; Ferris1942 [taxonomy: 446:32]; Leonar1900 [taxonomy, host, distribution: 342]; Lindin1957 [taxonomy: 547]; MacGil1921 [taxonomy, description, host, distribution: 440]; McDani1968 [taxonomy, illustration, host, distribution: 237-241]; MyartsRu2000 [distribution, biological control: 7-33]; Nakaha1982 [host, distribution: 25]; Newell1899 [taxonomy, description, host, distribution: 14]; RosenDe1979 [host, distribution, biological control: 291-294]; Sassce1923 [host, distribution: 125-129]; Willia1985a [taxonomy: 239]; Woolle1990 [biological control: 167-176].

Clavaspis texana Ferris

Clavaspis texana Ferris, 1938a: 210. Type data: U.S.A.: Texas, Chisos Mountain, on *Diospyros (or Brayodendron) texanum*. Holotype female. Type depository: Davis: The Bohart Museum of Entomology, University of California, California, USA.
SCALE COVER: Scales white, that of female irregularly circular, quite flat; exuviae subcentral, that of male elongate, exuviae central (Ferris, 1938a).
HOST PLANTS: **Ebenaceae**: *Diospyros texanum* [Ferris1938a, McDani1968]. **Rhamnaceae**: *Ziziphus lycioides* [Ferris1938a].
DISTRIBUTION: **Nearctic**: Mexico (San Luis Potosi [Ferris1955b]); United States of America (Texas [Ferris1938a, McDani1968]).
BIOLOGY: Occurring on bark, usually covered with a film of epidermis of host (Ferris, 1938a). Associated with fungus *Septobasidium* sp. (Ferris, 1955b).
GENERAL: Description and illustration of adult female by Ferris (1938a).
KEYS: McDaniel 1968: 232 (female) [U.S.A.: Texas]; Ferris 1942: 32 (female) [North America].
CITATIONS: Borchs1966 [catalogue: 319]; Ferris1938a [taxonomy, description, illustration, host, distribution: 210]; Ferris1942 [taxonomy: 446:32]; Ferris1955b [host, distribution, life history: 25]; Lindin1957 [taxonomy: 547]; McDani1968 [taxonomy, illustration, host, distribution: 240,242]; McKenz1939 [taxonomy: 43]; Nakaha1982 [host, distribution: 25].

Clavaspis ulmi (Johnson)

Aspidiotus ulmi Johnson, 1896: 152. Type data: U.S.A.: Illinois, Champaign, University campus, on *Ulmus americana*. Syntypes, female. Type depository: Washington: United States National Entomological Collection, U.S. National Museum of Natural History, District of Columbia, USA.
Aspidiotus (Hemiberlesia) ulmi; Cockerell, 1899a: 396. Change of combination.
Aonidiella ulmi; Leonardi, 1900: 342. Change of combination.
Hendaspidiotus ulmi; MacGillivray, 1921: 440. Change of combination.
Clavaspis ulmi; Ferris, 1938a: 211. Change of combination.
COMMON NAME: elm Aspidiotus [Johnso1896].
SCALE COVER: Scale of female white or grey, circular, quite convex; with exuviae submarginal; male scale not available (Ferris, 1938a). Colour photograph by Gill (1997).
HOST PLANTS: **Aceraceae**: *Acer* [Ferris1938a, BesheaTiHo1973]. **Bignoniaceae**: *Catalpa* [Ferris1938a]. **Celastraceae**: *Euonymus* [Ferris1938a]. **Cornaceae**: *Cornus* [Ferris1938a]. **Cycadaceae**: *Cycas revoluta* [Kuwana1902].

Hippocastanaceae: *Aesculus* [Ferris1938a]. **Juglandaceae**: *Juglans* [Ferris1938a]. **Leguminosae**: *Robinia* [Ferris1938a], *Robinia pseudacacia* [BesheaTiHo1973]. **Tiliaceae**: *Tilia* [Koszta1996]. **Ulmaceae**: *Celtis* [Ferris1938a], *Ulmus* [Koszta1996], *Ulmus americana* [Johnso1896, Leonar1900, McDani1968], *Ulmus fulva* [Ferris1938a, McDani1968].

NATURAL ENEMIES: ACARI **Oribatulidae**: *Zygoribatula pyrostigmata* [Koszta1996]. HYMENOPTERA **Signiphoridae**: *Signiphora pulchra* Girault [Woolle1990].

DISTRIBUTION: **Nearctic**: Canada (Ontario [FletchGi1908]); United States of America (California [Nakaha1982], Colorado [Nakaha1982], Connecticut [Ferris1938a], Georgia [BesheaTiHo1973], Illinois [Johnso1896, Leonar1900], Indiana [Ferris1938a], Kansas [Hunter1899, Ferris1938a], Louisiana [Nakaha1982], Maryland [Nakaha1982], Mississippi [Nakaha1982], Missouri [Hollin1923, Ferris1938a], New Jersey [Nakaha1982], New York [Ferris1938a], Ohio [Nakaha1982], South Carolina [Nakaha1982], Texas [Ferris1938a, McDani1968], Virginia [Nakaha1982], West Virginia [Nakaha1982], Wisconsin [Nakaha1982]). **Palaearctic**: Japan [Kuwana1902, Kuwana1917a].

BIOLOGY: Occurring ordinarily on rough bark (Ferris, 1938a).

ECONOMIC IMPORTANCE: Recorded from many states in USA (see Distribution), but not reported to cause damage (Gill, 1997).

GENERAL: Description and illustration of adult female by Ferris (1938a), Kosztarab (1996) and by Gill (1997).

KEYS: Gill 1997: 101 (female) [Species of California]; McDaniel 1968: 232 (female) [U.S.A.: Texas]; Ferris 1942: 32 (female) [North America]; Kuwana 1933b: 49 (female) [Japan]; Britton 1923: 371 (female) [U.S.A.: Connecticut]; Hollinger 1923: 7-8 (female) [U.S.A.: Missouri]; Lawson 1917: 217 (female) [U.S.A.: Kansas]; Dietz & Morrison 1916a: 289-290 (female) [U.S.A.: Indiana]; Newell 1899: 25 (female) [North America].

CITATIONS: BeardsDaHo1976 [economic importance: 103]; BesheaTiHo1973 [host, distribution: 5]; Borchs1966 [catalogue: 319]; Britto1923 [taxonomy, description, host, distribution: 371,375]; Cocker1899a [taxonomy: 396]; DanzigPe1998 [catalogue: 218-219]; DietzMo1916a [taxonomy, description, host, distribution: 293-294]; Fernal1903b [catalogue: 280]; Ferris1937c [taxonomy, illustration: 51,78]; Ferris1938a [taxonomy, description, illustration, host, distribution: 211]; Ferris1941e [taxonomy: 49]; Ferris1942 [taxonomy: 446:32]; FletchGi1908 [host, distribution: 113-133]; Gill1997 [host, distribution, taxonomy, description, illustration, economic importance: 104,106]; Harned1928 [host, distribution: 23-24]; Hollin1923 [taxonomy, description, host, distribution: 17]; Hunter1899 [taxonomy, host, distribution: 6]; Johnso1896 [taxonomy, description, host, distribution: 152]; Kaston1938 [host, distribution: 235-242]; Koszta1996 [taxonomy, description, illustration, host, distribution, life history, biological control: 482-484]; Kuwana1902 [host, distribution: 68]; Kuwana1907 [host, distribution: 195]; Kuwana1917a [taxonomy, distribution: 175]; Lawson1917 [taxonomy, description, illustration, host, distribution: 239-241]; Leonar1900 [taxonomy, host, distribution: 342]; Lindin1957 [taxonomy: 547]; MacGil1921 [taxonomy, description, host, distribution: 440]; McDani1968 [taxonomy, illustration, host, distribution: 241-242]; MillerDa1990 [host, distribution, economic importance: 301]; Nakaha1982 [host, distribution: 25-26]; Newell1899 [taxonomy, description, host, distribution: 25,28-29]; Sander1904a [taxonomy, description, illustration, host, distribution: 56,67]; Takagi1990b [taxonomy, structure: 11]; Woolle1990 [biological control: 167-176].

Crassaspidiotus Takagi

Crassaspidiotus Takagi, 1969a: 89. Type species: *Crassaspidiotus takahashii* Takagi, by original designation.
Crassiaspidiotus; Tao, 1999: 82. Misspelling of genus name.
SYSTEMATICS: *Crassaspidiotus* Takagi is characterized by derm being sclerotized throughout, all pygidial lobes about same shape and size, submarginal dorsal macroducts numerous and disposed in three rows on each side of pygidium, and anal opening situated at center of pygidium. May be close to *Metaspidiotus*, from which it differs in having marginal setae of pygidium all normal in shape, not thickened and lanceolate (Takagi, 1969a).
GENERAL: Definition and characters by Takagi (1969a).
CITATIONS: Takagi1969a [taxonomy, description: 89]; Tao1999 [taxonomy: 82].

Crassaspidiotus takahashii Takagi
Crassaspidiotus takahashii Takagi, 1969a: 89. Type data: TAIWAN: Tung-pu, on leaves of *Tsuga chinensis* var. *formosana*. Holotype female. Type depository: Sapporo: Entomological Institute, Faculty of Agriculture, Hokkaido University, Japan.
Dynaspidiotus takahashii; Chou, 1985: 401. Change of combination.
Crassaspidiotus takahashii; Tao, 1999: 82. Revived combination.
SCALE COVER: Female scale elliptical, moderately convex dorsally, thin, and pale brownish in colour (Takagi, 1969a).
HOST PLANTS: **Pinaceae**: *Tsuga chinensis formosana* [Takagi1969a].
DISTRIBUTION: **Oriental**: Taiwan [Takagi1969a].
GENERAL: Description and illustration of adult female by Takagi (1969a) and by Chou (1985, 1986).
CITATIONS: Chou1985 [taxonomy, description, host, distribution: 401]; Chou1986 [taxonomy, illustration: 671]; Takagi1969a [taxonomy, description, illustration, host, distribution: 89-91]; Tao1999 [taxonomy, host, distribution: 82].

Crenulaspidiotus MacGillivray

Crenulaspidiotus MacGillivray, 1921: 389. Type species: *Chrysomphalus (Melanaspis) portoricensis* Lindinger.
Grenulaspidiotus; Kozár, 1990f: 143. Misspelling of genus name.
SYSTEMATICS: The genus *Crenulaspidiotus* is unique among Aspidiotine genera in having dorsum of pygidium divided into several sclerotized areas. It resembles *Melanaspis* from which it differs in having long pygidial paraphyses (Miller & Davidson, 1981).
GENERAL: Definition and characters by Miller & Davidson (1981).

KEYS: Colon-Ferrer & Medina-Gaud 1998: 28-32 (female) [Genera of Puerto Rico].
CITATIONS: Balach1958 [taxonomy: 191]; BenDov1990h [taxonomy: 82]; Borchs1966 [catalogue: 359]; ColonFMe1998 [taxonomy, description: 52-53]; Ferris1937c [taxonomy: 51]; Ferris1941d [taxonomy: 347]; Kozar1990f [taxonomy, distribution: 143]; Lindin1937 [taxonomy: 182]; MacGil1921 [taxonomy, description: 389,392,426-427]; MillerDa1981 [taxonomy, description: 534-555]; MorrisMo1966 [taxonomy, catalogue: 47].

Crenulaspidiotus anticheir Miller & Davidson
Crenulaspidiotus anticheir Miller & Davidson, 1981: 556. Type data: JAMAICA: on *Coccoloba uvifera*; collected March 28, 1920. Holotype female. Type depository: Washington: United States National Entomological Collection, U.S. National Museum of Natural History, District of Columbia, USA.
SCALE COVER: Scale structure not available to Miller & Davidson (1981).
HOST PLANTS: **Polygonaceae**: *Coccoloba barbadensis* [MillerDa1981], *Co. diversiflora* [MillerDa1981], *Co. diversifolia* [MillerDa1981], *Co. jamaicensis* [MillerDa1981], *Co. krugii* [MillerDa1981], *Co. longiflora* [MillerDa1981], *Co. longifolia* [MillerDa1981], *Co. longipes* [MillerDa1981], *Co. plumieri* [MillerDa1981], *Co. uvifera* [MillerDa1981].
DISTRIBUTION: **Neotropical**: Costa Rica [MillerDa1981]; Honduras [MillerDa1981]; Jamaica [MillerDa1981]; U.S. Virgin Islands [MillerDa1981].
GENERAL: Description and illustration of adult female and nymphs by Miller & Davidson (1981).
KEYS: Miller & Davidson 1981: 536, 538-539 (female) [World].
CITATIONS: MillerDa1981 [taxonomy, description, illustration, host, distribution: 539,550,555-559].

Crenulaspidiotus cyrtus Miller & Davidson
Crenulaspidiotus cyrtus Miller & Davidson, 1981: 559. Type data: TAIWAN: on *Coccoloba* sp. Holotype female. Type depository: Washington, D.C.: U.S.s National Entomological Collection, U.S. National Museum of Natural History, USA.
SCALE COVER: Scale structure not available to Miller & Davidson (1981).
HOST PLANTS: **Polygonaceae**: *Coccoloba* [MillerDa1981].
DISTRIBUTION: **Oriental**: Taiwan [MillerDa1981].
GENERAL: Description and illustration of adult female by Miller & Davidson (1981).
KEYS: Miller & Davidson 1981: 537 (female) [World].
CITATIONS: MillerDa1981 [taxonomy, description, illustration, host, distribution: 540,559-560].

Crenulaspidiotus dicentron Miller & Davidson
Crenulaspidiotus dicentron Miller & Davidson, 1981: 560. Type data: PUERTO RICO: on *Coccoloba pirifolia*; collected May 5, 1940. Holotype female. Type depository: Washington: United States National Entomological Collection, U.S. National Museum of Natural History, District of Columbia, USA.

SCALE COVER: Female dorsal scale dark, convex, circular (Miller & Davidson, 1981).

HOST PLANTS: **Polygonaceae**: *Coccoloba diversifolia* [MillerDa1981], *Co. pyrifolia* [MillerDa1981], *Co. sintenisii* [MillerDa1981], *Co. swartzii urbaniana* [MillerDa1981].

DISTRIBUTION: **Neotropical**: Puerto Rico & Vieques Island (Puerto Rico [MillerDa1981]).

BIOLOGY: Found predominantly on upper surface of leaves, and often under outer epidermis of leaf tissue (Miller & Davidson, 1981).

GENERAL: Description and illustration of adult female and nymphs by Miller & Davidson (1981).

KEYS: Miller & Davidson 1981: 537-538 (female) [World].

CITATIONS: MillerDa1981 [taxonomy, description, illustration, host, distribution: 541,551-552,560-563].

Crenulaspidiotus greeneri Miller & Davidson

Crenulaspidiotus greeneri Miller & Davidson, 1981: 563. Type data: ARGENTINA: Concordia; host plant unknown; from H.L. Parker, October 31, 1940. Holotype female. Type depository: WashingtonD.C.: U.S. National Entomological Collection, U.S. National Museum of Natural History, USA.

SCALE COVER: Scale structure not available to Miller & Davidson (1981).

DISTRIBUTION: **Neotropical**: Argentina [MillerDa1981].

GENERAL: Description and illustration of adult female by Miller & Davidson (1981).

KEYS: Miller & Davidson 1981: 537 (female) [World].

CITATIONS: MillerDa1981 [taxonomy, description, illustration, host, distribution: 542,563-564].

Crenulaspidiotus lahillei (Lizer y Trelles)

Chrysomphalus obscurus lahillei Lizer y Trelles, 1917b: 242. Type data: ARGENTINA: Buenos Aires, on *Vitis vinifera, Juglans regia* and on *Populus* sp. Syntypes, female. Type depository: Buenos Aires: Museo Argentino de Ciencias Naturales, Division Entomologia, Argentina.

Chrysomphalus lahillei; McKenzie, 1939: 54. Change of status.

Greenoidea lahillei; Lizer y Trelles, 1939: 202. Change of combination.

Melanaspis lahillei; Lindinger, 1943a: 147. Change of combination.

Crenulaspidiotus lahillei; Borchsenius, 1966: 359. Change of combination.

SYSTEMATICS: Borchsenius (1966) included this species in *Crenulaspidiotus*. Miller & Davidson (1981) stressed that it was clearly not a species of *Crenulaspidiotus*, but were uncertain to which genus it should be assigned.

SCALE COVER: Lizer y Trelles (1917b) did not describe scale cover.

HOST PLANTS: **Juglandaceae**: *Juglans regia* [Lizery1917b, ClapsWoGo2001]. **Lauraceae**: *Ocotea acutifolia* [ClapsWoGo2001]. **Leguminosae**: *Vachellia astringens* [ClapsWoGo2001]. **Salicaceae**: *Populus* [Lizery1917b, ClapsWoGo2001]. **Vitaceae**: *Vitis vinifera* [Lizery1917b, ClapsWoGo2001].

DISTRIBUTION: **Neotropical**: Argentina (Buenos Aires [Lizery1917b, ClapsWoGo2001], Entre Rios [ClapsWoGo2001]).
GENERAL: Description and illustration of adult female by Lizer y Trelles (1917b), Davidson (1970) and by Miller & Davidson (1981).
CITATIONS: Blanch1940 [host, distribution, biological control: 106-128]; Borchs1966 [catalogue: 359]; Chiesa1948 [host, distribution, economic importance]; ClapsWoGo2001 [host, distribution: 241-242]; DeSant1979 [biological control]; Lindin1943a [taxonomy: 147]; Lizery1917b [taxonomy, description, illustration, host, distribution: 242-244]; Lizery1939 [taxonomy: 202]; McKenz1939 [taxonomy: 54].

Crenulaspidiotus maurellae (Laing)
Aonidiella maurellae Laing, 1929a: 492. Type data: COLOMBIA: Tucurinca, on bark of unspecified plant. Lectotype female, by subsequent designation Miller & Davidson, 1981: 566. Type depository: London: The Natural History Museum, England, UK.
Melanaspis maurellae; Lindinger, 1943a: 147. Change of combination.
Aonidiella maurellae; Borchsenius, 1966: 359. Incorrect synonymy. Notes: See Systematics below.
Crenulaspidiotus maurellae; Miller & Davidson, 1981: 564. Change of combination.
SYSTEMATICS: Borchsenius (1966: 359) regarded *Aonidiella maurellae* Laing, 1929 as synonym of *Crenulaspidiotus portoricensis* (Lindinger), but Miller & Davidson (1981: 564) regarded the former a distinct species.
SCALE COVER: Female subcircular, highly convex, concentrically ridged, dull black, obscured marginally by flecks of greyish-white deposit from bark of plant; ventral scale entire, strong, inner side with broad reddish-brown border and whitish centre; exuviae paler than scale itself, subcentral; diameter 1.4 mm (Laing, 1929a).
HOST PLANTS: **Polygonaceae**: *Coccoloba acuminata* [MillerDa1981], *Co. barbadensis* [MillerDa1981], *Co. caracasana* [MillerDa1981], *Co. densifrons* [MillerDa1981], *Co. humboldtii* [MillerDa1981], *Co. lehmannii* [MillerDa1981], *Co. marginata* [MillerDa1981], *Co. padiformis* [MillerDa1981], *Co. paraguariensis* [MillerDa1981], *Co. ramosissima* [MillerDa1981], *Co. uvifera* [MillerDa1981], *Co. venosa* [MillerDa1981].
DISTRIBUTION: **Nearctic**: Mexico [MillerDa1981] (Michoacan [MillerDa1981], San Luis Potosi [MillerDa1981]). **Neotropical**: Argentina [MillerDa1981]; Colombia [Laing1929a, MillerDa1981]; El Salvador [MillerDa1981]; Guatemala [MillerDa1981]; Guyana [MillerDa1981]; Honduras [MillerDa1981]; Mexico (Chiapas [MillerDa1981]); Nicaragua [MillerDa1981]; Panama [MillerDa1981]; Panama Canal Zone [MillerDa1981]; Paraguay [MillerDa1981]; Venezuela [MillerDa1981].
GENERAL: Description and illustration of adult female by Laing (1929a) and by Miller & Davidson (1981).
KEYS: Miller & Davidson 1981: 537-538 (female) [World].
CITATIONS: Laing1929a [taxonomy, description, illustration, host, distribution: 492-493]; Lindin1957 [taxonomy: 545]; MillerDa1981 [taxonomy, description, illustration, host, distribution: 543,550,552,564-567].

Crenulaspidiotus mini Davidson

Crenulaspidiotus mini Davidson, 1970a: 500. Type data: ARIZONA: on *Prosopis* sp.; collected July 15, 1969. Holotype female. Type depository: Washington: United States National Entomological Collection, U.S. National Museum of Natural History, District of Columbia, USA.

SCALE COVER: Illustration of scale cover by Davidson (1970a). Female scale circular to slightly elongate, 0.7-1 mm in diameter; black; highly convex; tilted up at one side at maturity, often surrounded by bark flakes; exuvium subcentral; ventral scale present. Male scale elongate oval, 0.4-0.6 mm wide, 0.8-1.2 mm long; flattened, greyish; exuvium at one end and ventral scale present (Davidson, 1970a)

HOST PLANTS: **Leguminosae**: *Prosopis* [Davids1970a, MillerDa1981].

DISTRIBUTION: **Nearctic**: United States of America (Arizona [Davids1970a, MillerDa1981], California [MillerDa1981]).

GENERAL: Description and illustration of adult female by Davidson (1970a) and by Miller & Davidson (1981).

KEYS: Miller & Davidson 1981: 537-539 (female) [World].

CITATIONS: Davids1970a [taxonomy, description, illustration, host, distribution: 500-503]; MillerDa1981 [taxonomy, description, illustration, host, distribution: 544,550,552,567-569].

Crenulaspidiotus monocentron Miller & Davidson

Crenulaspidiotus monocentron Miller & Davidson, 1981: 570. Type data: JAMAICA: Mt. Diabolo, on *Coccoloba longifolia*; collected May 25-27, 1904. Holotype female. Type depository: Washington: U.S. Entomological Collection, U.S. National Museum of Natural History, District of Columbia, USA.

SCALE COVER: Scale structure not described by Miller & Davidson (1981).

HOST PLANTS: **Polygonaceae**: *Coccoloba* [MillerDa1981], *Co. longifolia* [MillerDa1981], *Co. venosa* [MillerDa1981].

DISTRIBUTION: **Neotropical**: Jamaica [MillerDa1981].

GENERAL: Description and illustration of adult female by Miller & Davidson (1981).

KEYS: Miller & Davidson 1981: 536, 538 (female) [World].

CITATIONS: MillerDa1981 [taxonomy, description, illustration, host, distribution: 545,550,552,570-572].

Crenulaspidiotus portoricensis (Lindinger)

Chrysomphalus (*Melanaspis*) *portoricensis* Lindinger, 1910c: 441. Type data: PUERTO RICO: Cayey, near La Cruz, on *Coccoloba excoriata*. Lectotype female, by subsequent designation Miller & Davidson, 1981: 575. Type depository: Hamburg: Zoologisches Institut und Zoologishces Museum, Universität von Hamburg, Germany.

Aspidiotus portoricensis; MacGillivray, 1921: 389. Change of combination.

Crenulaspidiotus portoricensis; MacGillivray, 1921: 427. Change of combination.

Melanaspis portoricensis; Lindinger, 1931a: 27. Change of combination.

Chrysomphalus portoricensis; Ferris, 1937c: 51. Change of combination.

Crenulaspidiotus portoricensis; Borchsenius, 1966: 359. Revived combination.

SCALE COVER: Female scale on twigs of host; black, circular, high convex, with exuviae subcentral. Scale of male oval, with exuvia at one end (Ferris, 1941d).

HOST PLANTS: **Polygonaceae**: *Coccoloba buchii* [MillerDa1981], *Co. costata* [MillerDa1981], *Co. diversifolia* [MillerDa1981], *Co. excoriata* [MillerDa1981], *Co. krugii* [MillerDa1981], *Co. microstachya* [MillerDa1981], *Co. obtusifolia* [MillerDa1981], *Co. pubescens* [MillerDa1981], *Co. pyrifolia* [MillerDa1981], *Co. tenuifolia* [MillerDa1981], *Co. uvifera* [MillerDa1981], *Co. venosa* [MillerDa1981].

DISTRIBUTION: **Nearctic**: Mexico (Michoacan [Ferris1941d]). **Neotropical**: Colombia [Ferris1941d]; Haiti [MillerDa1981]; Puerto Rico & Vieques Island (Puerto Rico [Martor1976, MillerDa1981]); U.S. Virgin Islands [MillerDa1981].

GENERAL: Description and illustration of adult female by Ferris (1941d), Miller & Davidson (1981) and by Colon-Ferrer & Medina-Gaud (1998).

KEYS: Miller & Davidson 1981: 537-539 (female) [World]; Ferris 1943: 64 (female) [North America]; Ferris 1942: 36 (female) [North America].

CITATIONS: Borchs1966 [catalogue: 359]; ColonFMe1998 [taxonomy, description, illustration, host, distribution: 53]; Ferris1937c [taxonomy: 51]; Ferris1941d [taxonomy, description, illustration, host, distribution: 364]; Ferris1942 [taxonomy: 446:36]; Ferris1943 [taxonomy: 64]; Lindin1910c [taxonomy, description, illustration, host, distribution: 441]; Lindin1931a [taxonomy: 27]; MacGil1921 [taxonomy, description, host, distribution: 389,426-427]; Martor1976 [host, distribution: 71-74; McKenz1939 [taxonomy: 54]; MillerDa1981 [taxonomy, description, illustration, host, distribution: 546,551,553,572-576]; Sassce1911 [taxonomy: 70]; WeidneWa1968 [taxonomy: 177].

Crenulaspidiotus russellae Miller & Davidson

Crenulaspidiotus russellae Miller & Davidson, 1981: 576. Type data: HAITI: Vicinity of Ennery, on *Coccoloba incrassata*; collected January 14, 1926. Holotype female. Type depository: Washington: United States National Entomological Collection, U.S. National Museum of Natural History, District of Columbia, USA.

SCALE COVER: Scale structure not described by Miller & Davidson (1981).

HOST PLANTS: **Polygonaceae**: *Co. diversifolia* [MillerDa1981], *Co. fuertesii* [MillerDa1981], *Co. incrassata* [MillerDa1981].

DISTRIBUTION: **Neotropical**: Dominican Republic [MillerDa1981]; Haiti [MillerDa1981].

GENERAL: Description and illustration of adult female by Miller & Davidson (1981).

KEYS: Miller & Davidson 1981: 537-538 (female) [World].

CITATIONS: MillerDa1981 [taxonomy, description, illustration, host, distribution: 547,551,553,576-579].

Crenulaspidiotus sinuatus (Ferris)

Melanaspis sinuata Ferris, 1941d: 365. Type data: PANAMA: Chiriqui Province, near David, Pedregal, on an undetermined tree. Lectotype female, by subsequent designation Miller & Davidson, 1981: 580. Type depository: Davis: The Bohart Museum of Entomology, University of California, California, USA.

Crenulaspidiotus sinuata; Borchsenius, 1966: 579. Change of combination.

SCALE COVER: Scale of female of type common to the genus, that of male not recognized (Ferris, 1941d).

DISTRIBUTION: **Neotropical**: Panama [Ferris1941d, MillerDa1981].

BIOLOGY: Concealed beneath bark flakes and actually within soft bark of host (Ferris, 1941d).

GENERAL: Description and illustration of adult female by Ferris (1941d) and Miller & Davidson (1981).

KEYS: Miller & Davidson 1981: 537-538 (female) [World]; Ferris 1943: 64 (female) [North America]; Ferris 1942: 36 (female) [North America].

CITATIONS: Borchs1966 [catalogue: 359]; Ferris1941d [taxonomy, description, illustration, host, distribution: 365]; Ferris1942 [taxonomy: 446:36]; Ferris1943 [taxonomy: 64]; MillerDa1981 [taxonomy, description, illustration, host, distribution: 548,551,553,579-581].

Crenulaspidiotus truncus Miller & Davidson

Crenulaspidiotus truncus Miller & Davidson, 1981: 581. Type data: CUBA: Oriente, Saetia, on *Coccoloba uvifera*. Holotype female. Type depository: Washington: United States National Entomological Collection, U.S. National Museum of Natural History, District of Columbia, USA.

SCALE COVER: Scale structure not described by Miller & Davidson (1981).

HOST PLANTS: **Polygonaceae**: *Coccoloba* [MillerDa1981], *Co. buchii* [MillerDa1981], *Co. costata* [MillerDa1981], *Co. diversifolia* [MillerDa1981], *Co. leoganensis* [MillerDa1981], *Co. nodosa* [MillerDa1981], *Co. reflexiflora* [MillerDa1981], *Co. retusa* [MillerDa1981], *Co. saxicola* [MillerDa1981], *Co. uvifera* [MillerDa1981], *Co. wrightii* [MillerDa1981].

DISTRIBUTION: **Neotropical**: Cuba [MillerDa1981]; Dominican Republic [MillerDa1981].

GENERAL: Description and illustration of adult female and nymphs by Miller & Davidson (1981).

KEYS: Miller & Davidson 1981: 536 (female) [World].

CITATIONS: MillerDa1981 [taxonomy, description, illustration, host, distribution: 549,551,553,581-584].

Cryptaspidiotus Lindinger

Cryptaspidiotus Lindinger, 1910: 156. Type species: *Chrysomphalus barbusano* Lindinger, by monotypy.

SYSTEMATICS: The genus *Cryptaspidiotus* includes three pupillarial species. It differs from *Aonidia* Targioni Tozzetti in circular shape of females, pygidium less projecting less and the first lobe wider than projecting (Balachowsky, 1951).

GENERAL: Definition and characters by Borchsenius (1950b), Balachowsky (1951) and by Gómez-Menor Guerrero (1962).

KEYS: Gómez-Menor Guerrero 1962: 157 (female) [Canary Islands]; Balachowsky 1958b: 232 (female) [*Aspidiotina* of Africa]; Balachowsky 1951: 603 (female) [Mediterranean]; Borchsenius 1950b: 168 (female) [USSR]; Archangelskaya 1937: 94 (female) [Middle Asia].

CITATIONS: Archan1937 [taxonomy, description: 94,111]; Balach1948b [taxonomy: 269]; Balach1951 [taxonomy, description: 624-625]; Balach1958b [taxonomy: 232]; Borchs1949d [taxonomy, description: 195,250]; Borchs1950b [taxonomy, description: 234]; Borchs1966 [catalogue: 361]; Brain1919 [taxonomy: 197]; DanzigPe1998 [catalogue: 221]; Ferris1937c [taxonomy: 51]; Ferris1937d [taxonomy: 106]; GomezM1960O [taxonomy, description: 173-174]; GomezM1962 [taxonomy, description: 198]; Hall1946a [taxonomy: 520]; HowellTi1990 [taxonomy: 57]; Lindin1910 [taxonomy, description: 156]; Lindin1910b [taxonomy: 41]; Lindin1937 [taxonomy: 179,182]; MacGil1921 [taxonomy, description: 389,426]; McKenz1939 [taxonomy: 53]; Miller1990 [taxonomy: 169-178]; MorrisMo1966 [taxonomy, catalogue: 48].

Cryptaspidiotus aonidioides Lindinger
Cryptaspidiotus aonidioides Lindinger, 1911a: 21. Type data: CANARY ISLANDS: Tenerife, between Içod and Garachico, on *Laurus canariensis*; Palma, Barranco del Rio, on *Apollonias canariensis*. Syntypes, female. Type depository: Zoologisches Institut und Zoologishces Museum, Universität von Hamburg, Germany.
Hemiberlesia aonidioides; MacGillivray, 1921: 438. Change of combination.
Cryptaspidiotus aonidioides; Borchsenius, 1966: 361. Revived combination.
SCALE COVER: Female scale small, "kapselartig" [=cap-like], 1 mm long, 0.5-07 mm wide. Male scale elliptical, 0.75 mm long, 0.4 mm wide; brown; exuviae towards cephalic end (Lindinger, 1911a).
HOST PLANTS: **Lauraceae**: *Apollonias canariensis* [Lindin1911a, Lindin1912b, Balach1946], *Laurus canariensis* [Lindin1911a, GomezM1962, GomezM1967O], *Ocotea foetens* [Balach1951].
DISTRIBUTION: **Palaearctic**: Canary Islands [Lindin1911a, Balach1946, GomezM1962, GomezM1967O, MatileOr2001]; Madeira Islands [Balach1951].
GENERAL: Description and illustration of adult female by Lindinger (1911a), Balachowsky (1951) and by Gómez-Menor Guerrero (1962).
KEYS: Gómez-Menor Guerrero 1962: 198 (female) [Canary Islands]; Balachowsky 1951: 625 (female) [Mediterranean].
CITATIONS: Balach1946 [host, distribution: 211]; Balach1951 [taxonomy, description, illustration, host, distribution: 628-630]; Borchs1966 [catalogue: 361]; DanzigPe1998 [catalogue: 221]; GomezM1962 [taxonomy, description, illustration, host, distribution: 199-201]; GomezM1967O [host, distribution: 132]; Lindin1911a [taxonomy, description, illustration, host, distribution: 21-23]; Lindin1912b [taxonomy, description, host, distribution: 70-71,198]; MacGil1921 [taxonomy, description, host, distribution: 438]; MatileOr2001 [host, distribution: 189]; Sassce1912 [taxonomy, host, distribution: 92]; WeidneWa1968 [taxonomy: 174].

Cryptaspidiotus austroafricanus Lindinger

Cryptaspidiotus austro-africanus Brick, 1910: 499. Nomen nudum.
Cryptaspidiotus austro-africanus Lindinger, 1910b: 41. Type data: SOUTH AFRICA: Natal, Mariann Hill, on *Euphorbia* sp.; collected 26.8.1909. Syntypes, female. Type depository: Hamburg: Zoologisches Institut und Zoologishces Museum, Universität von Hamburg, Germany; type no. MP112.
Greeniella austro-africana; MacGillivray, 1921: 461. Change of combination.
Cryptaspidiotus austroafricanus; Borchsenius, 1966: 361. Revived combination and justified emendation to species name.
SCALE COVER: Female scale elongate or circular, white-grey; exuviae dark yellow; 1 mm long, 0.8-1 mm wide (Lindinger, 1910b). Female scale flat, circular, 1 mm and whitish grey or yellowish; exuviae central, dark yellow (Brain, 1919).
HOST PLANTS: **Euphorbiaceae**: *Euphorbia* [Brain1919].
DISTRIBUTION: **Afrotropical**: South Africa [Brain1919].
GENERAL: Description and illustration of adult female by Lindinger (1910b) and by Brain (1919).
CITATIONS: Balach1951 [taxonomy: 625]; Borchs1966 [catalogue: 361]; Brain1919 [taxonomy, description, illustration, host, distribution: 197-198]; Brick1910 [taxonomy: 499]; Lindin1910b [taxonomy, description, illustration, host, distribution: 41]; MacGil1921 [taxonomy, description, host, distribution: 461]; Sassce1912 [taxonomy, host, distribution: 92]; WeidneWa1968 [taxonomy: 174].

Cryptaspidiotus barbusano (Lindinger)

Chrysomphalus barbusano Lindinger, 1908b: 101. Type data: CANARY ISLANDS: Tenerife, Fagana, Cumbre, altitude 900 meters, on *Apollonias canariensis* [=*Phoebe barbusano*]. Syntypes, female. Type depository: Hamburg: Zoologisches Institut und Zoologishces Museum, Universität von Hamburg, Germany.
Cryptaspidiotus barbusano; Lindinger, 1910: 156. Change of combination.
Targionia barbusano; MacGillivray, 1921: 448. Change of combination.
Cryptaspidiotus barbusano; Balachowsky, 1951: 626. Revived combination.
SCALE COVER: Female scale circular, with only one exuvia; female enclosed within exuvia of second instar (Lindinger, 1908b).
HOST PLANTS: **Lauraceae**: *Apollonias canariensis* [Lindin1908b, Balach1946, GomezM1962], *Laurus canariensis* [GomezM1962], *Phoebe barbusano* [Lindin1909b, Sander1909a, Balach1951].
DISTRIBUTION: **Palaearctic**: Canary Islands [Lindin1909b, Sander1909a, Balach1946, GomezM1962, MatileOr2001].
GENERAL: Description and illustration of adult female by Lindinger (1908b; 1911a), Balachowsky (1951) and by Gómez-Menor Guerrero (1962).
KEYS: Gómez-Menor Guerrero 1962: 198 (female) [Canary Islands]; Balachowsky 1951: 625 (female) [Mediterranean].
CITATIONS: Balach1946 [host, distribution: 211]; Balach1951 [taxonomy, description, illustration, host, distribution: 626-628]; Borchs1966 [catalogue: 361]; DanzigPe1998 [catalogue: 222]; Ferris1937d [taxonomy: 106]; Ferris1943a [taxonomy: 85]; GomezM1962 [taxonomy, description, illustration, host, distribution: 202-204]; Lindin1908b [taxonomy, description, host, distribution: 101-

102]; Lindin1909b [taxonomy, description, illustration, host, distribution: 105-107]; Lindin1910 [taxonomy, host, distribution: 156,192]; Lindin1911a [taxonomy: 23]; Lindin1912b [taxonomy, description, host, distribution: 69]; MacGil1921 [taxonomy, description, host, distribution: 448]; MatileOr2001 [host, distribution: 189]; McKenz1939 [taxonomy: 53]; Sander1909a [taxonomy, host, distribution: 55]; WeidneWa1968 [taxonomy: 174].

Cryptophyllaspis Cockerell

Aspidiotus (*Cryptophyllaspis*) Cockerell, 1897i: 14. Type species: *Aspidiotus occultus* Green, by monotypy and original designation.
Cryptophyllaspis; Fernald, 1903b: 281. Change of status.
SYSTEMATICS: Three species are here included in this genus, which is close to *Aspidiotus*. The type species is a gall-forming form, and Ferris (1938) doubted whether this feature by itself was a good character to distinguish this genus.
GENERAL: Definition and characters by Cockerell (1897i, 1899a) and by MacGillivray (1921).
CITATIONS: Borchs1966 [catalogue: 272]; Cocker1897i [taxonomy, description: 14,31]; Cocker1899a [taxonomy: 396]; DanzigPe1998 [catalogue: 224]; Fernal1903b [catalogue: 281]; Ferris1937c [taxonomy: 51]; Ferris1938 [taxonomy: 43]; Leonar1897a [taxonomy: 375]; Lindin1937 [taxonomy: 183]; MacGil1921 [taxonomy, description: 390,426]; MorrisMo1966 [taxonomy, catalogue: 51].

Cryptophyllaspis bornmuelleri (Lindinger)
Cryptophyllaspis bornmülleri Rubsaamen, 1902: 62. Nomen nudum; discovered by Lindinger, 1911a: 9.
Aspidiotus bornmülleri Lindinger, 1911a: 9. Type data: CANARY ISLANDS: Tenerife, Barrano de San Andre, on *Globularia salicina*; collected 30.v.1901. Syntypes, female. Type depository: Hamburg: Zoologisches Institut und Zoologishces Museum, Universität von Hamburg, Germany; type no. MP226.
Cryptophyllaspis bornmuelleri; Borchsenius, 1966: 272. Justified emendation.
SCALE COVER: Female scale white, flat, elongate; exuviae subcentral. Male scale linear, white, 1.1 mm long, 0.3 mm wide (Lindinger, 1911a).
HOST PLANTS: **Globulariaceae**: *Globularia salicina* [Rubsaa1902, Sander1906].
DISTRIBUTION: **Palaearctic**: Canary Islands [Sander1906, MatileOr2001]; Madeira Islands [Rubsaa1902, Sander1906].
BIOLOGY: Causing leaf galls on *Globularia salicina* (Lindinger, 1911a; Rubsaamen, 1902).
GENERAL: Description and illustration of adult female by Lindinger (1911a).
CITATIONS: Borchs1966 [catalogue, taxonomy: 272]; DanzigPe1998 [catalogue: 224]; Larew1990 [ecology, life history, structure: 293-300]; Lindin1911a [taxonomy, description, illustration, host, distribution: 9]; Lindin1912b [taxonomy, description, host, distribution: 163-164]; MacGil1921 [taxonomy, description, host, distribution: 428]; MatileOr2001 [host, distribution: 189]; Rubsaa1902 [taxonomy, description, host, distribution: 62]; Sander1906 [taxonomy, host, distribution: 15]; WeidneWa1968 [taxonomy: 171].

Cryptophyllaspis elongata (Green)

Aspidiotus (*Cryptophyllaspis*) *occultus elongatus* Green, 1905a: 345. Type data: SRI LANKA: on *Grewia* sp. Syntypes, female. Type depository: London: The Natural History Museum, England, UK.

Cryptophyllaspis occultus elongatus; Sanders, 1906: 15. Change of combination.

Cryptophyllaspis elongata; MacGillivray, 1921: 428. Change of status.

Cryptophyllaspis elongata; MacGillivray, 1921: 428. Change of combination requiring emendation of species name for agreement in gender.

Aspidiotus elongatus; Ferris, 1941e: 43. Change of combination requiring emendation of species name for agreement in gender.

Cryptophyllaspis elongata; Borchsenius, 1966: 273. Revived combination.

SCALE COVER: Female scale consisting of a delicate film lining cavity of gall; exuviae forming an operculum at its base. Male scale not observed, but probably occupying shallow depressions on the surface of leaf as in the type (Green, 1905a).

HOST PLANTS: **Tiliaceae**: *Grewia* [Green1905a, Sander1906, Green1922, Green1937, Muntin1971a].

DISTRIBUTION: **Oriental**: Sri Lanka [Green1905a, Sander1906, Ramakr1921a, Green1922, Muntin1971a].

BIOLOGY: Forming pit galls on under surface of leaves; opening of the galls being closed by scale (Ferris, 1941e).

GENERAL: Illustration of the female pygidium given by Munting (1971a).

CITATIONS: Borchs1966 [catalogue: 273]; Ferris1941e [taxonomy, description, host, distribution: 43,53]; Green1905a [taxonomy, description, illustration, host, distribution: 345]; Green1922 [host, distribution: 462]; Green1937 [host, distribution: 333]; Larew1990 [ecology, life history, structure: 293-300]; MacGil1921 [taxonomy, description, host, distribution: 428]; Muntin1971a [taxonomy, illustration, host, distribution: 315,317]; Ramakr1921a [host, distribution: 357]; Sander1906 [taxonomy, host, distribution: 15].

Cryptophyllaspis occulta (Green)

Aspidiotus occultus Green, 1896: 4. Type data: SRI LANKA: Punduloya, on *Grewia orientalis*. Holotype female. Type depository: London: The Natural History Museum, England, UK.

Hemiberlesia occulta; Leonardi, 1897b: 129. Change of combination requiring emendation of species name for agreement in gender.

Aspidiotus (*Cryptophyllaspis*) *occultus*; Cockerell, 1897i: 28. Change of combination.

Cryptophyllaspis occultus; Cockerell, 1899a: 396. Change of combination.

Aspidiotus (*Cryptophyllaspis*) *occultus*; Ramakrishna Ayyar, 1921a: 357. Change of combination.

Cryptophyllaspis occulta; Borchsenius, 1966: 273. Revived combination.

SCALE COVER: Female scale represented externally by an ochreous scale closing aperture of the gall; a whitish film lining gall cavity may be taken to represent ventral scale; exuviae do not appear on surface, but lie within gall; second exuvia is very large, sometimes as large as, or even larger, than adult insect, in this respect approaching the genus *Aonidia*; the aperture of gall, opening on under surface of leaf, is surrounded by a prominent irregularly lobed rim; diameter of gall from 0.5-0.75 mm. Male scale occupying shallow depressions on under surface of leaf; oval, flat; transparent, very pale reddish yellow; exuviae bright yellow; size about 1 mm long, 0.75 mm wide (Green, 1896e). Colour illustration of gall by Green (1896e).

HOST PLANTS: **Tiliaceae**: *Grewia orientalis* [Green1896, Green1896e, Ramakr1921a, Green1937].

DISTRIBUTION: **Oriental**: Sri Lanka [Green1896, Green1896e, Ramakr1921a, Green1937].

BIOLOGY: Very small species, forming minute rounded galls on upper surface of leaves of *Grewia orientalis*; although galls appear on upper surface, insects in reality placed on under surface; at first forming depressions as in *Aspidiotus putearius*, but in *Cryptophyllaspis occulta* the process is carried further by the formation of a cavity, which almost completely encloses insect; occasionally two individuals occupy a single cavity (Green, 1896e).

GENERAL: Description and illustration of adult female by Ferris (1941e).

KEYS: Ferris 1946: 43 (female) [World]; Ferris 1941e: 61 (female) [World]; Green 1896e: 40 (female) [Sri Lanka].

CITATIONS: Borchs1966 [catalogue: 273]; Cocker1896b [distribution: 334]; Cocker1897i [taxonomy, description, host, distribution: 28]; Cocker1899a [taxonomy: 396]; Fernal1903b [catalogue: 281-282]; Ferris1937c [taxonomy: 51]; Ferris1938 [taxonomy: 43,47]; Ferris1941e [taxonomy, description, illustration, host, distribution: 46,57,65]; Ferris1946 [taxonomy: 43]; Green1896 [taxonomy, host, distribution: 4]; Green1896e [taxonomy, description, illustration, host, distribution: 56-57]; Green1937 [host, distribution: 333]; Larew1990 [ecology, life history, structure: 293-300]; Leonar1897b [taxonomy, description, host, distribution: 129-130]; MacGil1921 [taxonomy, description, host, distribution: 427]; Ramakr1921a [host, distribution: 357].

Cryptoselenaspidus Lindinger

Cryptoselenaspidus Lindinger, 1910: 259. Type species: *Cryptoselenaspidus serra* Lindinger, by original designation.

SYSTEMATICS: Available information on this genus is confined to original description. Ferris (1937c: 101) and Borchsenius (1966: 372) placed this genus among unplaced genera of the Aspidiotinae. The genus was not mentioned by Mamet (1958) among genera of the *Selenaspidus* complex, nor by Balachowsky (1958b) in the monograph of aspidiotine genera from Central Africa.

CITATIONS: Borchs1966 [catalogue: 372]; Ferris1937c [taxonomy: 101]; Lindin1910 [taxonomy, description: 259]; Lindin1937 [taxonomy: 183]; MorrisMo1966 [taxonomy, description: 51].

Cryptoselenaspidus serra Lindinger

Cryptoselenaspidus serra Lindinger, 1910: 259. Type data: CAMEROON: locality and host plant not indicated. Syntypes, female. Type depository: Hamburg: Zoologisches Institut und Zoologishces Museum, Universität von Hamburg, Germany.

SYSTEMATICS: Ferris (1937c: 101) and Borchsenius (1966: 372) listed this species among unplaced species of the Aspidiotinae. Mamet (1958) did not mention it among species of the *Selenaspidus* complex, nor did Balachowsky (1958b) in the monograph of aspidiotine species from Central Africa.

SCALE COVER: Lindinger (1910) noted that the adult female remains enclosed within the second stage, and that the second stage shows similarity to the adult female of *Selenaspidus silvaticus* Lindinger.

DISTRIBUTION: **Afrotropical**: Cameroon [Lindin1910].

CITATIONS: Borchs1966 [catalogue: 372]; Ferris1937c [taxonomy: 101]; Lindin1910 [taxonomy, description, illustration, host, distribution: 259]; Lindin1937 [taxonomy: 183]; MorrisMo1966 [taxonomy: 51]; WeidneWa1968 [taxonomy: 175].

Diaonidia Takahashi

Diaonidia Takahashi, 1956b: 25. Type species: *Aonidia yabunikkei* Kuwana, by monotypy and original designation.

SYSTEMATICS: Genus *Diaonidia* includes two pupillarial species. It is very close to *Aonidia*, but differs from latter in having well-developed lobes and plates (Takagi, 1969a).

GENERAL: Definition and characters by Takahashi (1956b) and by Takagi (1969a).

KEYS: Chou 1985: 324 (female) [Genera of China].

CITATIONS: Borchs1966 [catalogue: 360]; Chou1985 [taxonomy, description: 324]; DanzigPe1998 [catalogue: 225]; HowellTi1990 [taxonomy: 57]; Kawai1980 [taxonomy: 226]; MorrisMo1966 [taxonomy, catalogue: 56]; Takagi1969a [taxonomy, description: 95-96]; Takaha1956b [taxonomy, description: 25]; Tao1999 [taxonomy, host, distribution: 82].

Diaonidia cinnamomi (Takahashi)

Gymnaspis cinnamomi Takahashi, 1936: 82. Type data: TAIWAN: Chushinron near Rokki, Takao Prefecture, on *Cinnamomum* sp. Syntypes, female. Type depository: Taichung: Entomology Collection, Taiwan Agricultural Research Institute, Wu-feng, Taichung, Taiwan.
Diaonidia cinnamomi; Borchsenius, 1966: 360. Change of combination.

SCALE COVER: Scale of adult female reddish brown, shining, with no secretion evident; flattened, nearly circular (Takahashi, 1936).

HOST PLANTS: **Lauraceae**: *Cinnamomum* [Takaha1936, Takagi1969a], *Machilus kusanoi* [Takagi1969a].

DISTRIBUTION: **Oriental**: Taiwan [Takaha1936, Takagi1969a].

GENERAL: Description and illustration of adult female by Takahashi (1936), Takagi (1969a) and by Chou (1985, 1986).

CITATIONS: Borchs1966 [catalogue: 360]; Chou1985 [taxonomy, description, host, distribution: 324-325]; Chou1986 [taxonomy, illustration: 698]; Takagi1969a [taxonomy, description, illustration, host, distribution: 96-98,104]; Takaha1936 [taxonomy, description, illustration, host, distribution: 82-83]; Tao1999 [taxonomy, host, distribution: 82].

Diaonidia yabunikkei (Kuwana)

Aonidia yabunikkei Kuwana, 1933: 41. Type data: JAPAN: Kyushu, near Moji, Mutsure-jima, on the leaf of *Cinnamomum pedunculatum*. Syntypes, female. Type depository: Ibaraki-ken: Insect Taxonomy Laboratory, National Institute of Agricultural Environmental Sciences, Kannon-dai, Yatabe, Tsukuba-shi, (I. Kuwana), Japan.

Gymnaspis yabunikkei; Takahashi, 1936: 83. Change of combination.

Diaonidia yabunikkei; Takahashi, 1956: 26. Change of combination.

SCALE COVER: Scale of female circular or subcircular; reddish brown. First exuvia exposed, black, slightly convex, subcentral; second exuvia completely enclosing adult insect, more or less convex. Diameter about 0.7 mm. Scale of male similar in size and form to that of female, but smaller (Kuwana, 1933).

HOST PLANTS: **Lauraceae**: *Cinnamomum* [Takaha1955f], *Cinnamomum pedunculatum* [Kuwana1933, Takaha1956b], *Machilus* [Takaha1955f].

DISTRIBUTION: **Palaearctic**: Japan [Kuwana1933, Takaha1955f, Takaha1956b, Kawai1980] (Kyushu [Kuwana1933]).

GENERAL: Description and illustration of adult female by Kuwana (1933).

KEYS: Kuwana 1933: 40 (female) [Japan]; Kuwana 1933b: 50 (female) [Japan].

CITATIONS: Borchs1966 [catalogue: 360]; DanzigPe1998 [catalogue: 226]; Kawai1980 [taxonomy, description, host, distribution: 226]; Kuwana1933 [taxonomy, description, illustration, host, distribution: 40,41-42]; Lindin1943b [taxonomy: 206]; Lindin1957 [taxonomy: 548]; Muraka1970 [host, distribution: 73]; Takaha1936 [taxonomy: 83]; Takaha1955f [host, distribution: 242]; Takaha1956b [taxonomy, description, illustration, host, distribution: 26].

Diaphoraspis Brimblecombe

Diaphoraspis Brimblecombe, 1957: 277. Type species: *Diaphoraspis orbata* Brimblecombe, by original designation.

SYSTEMATICS: Genus *Diaphoraspis* resembles *Achorophora* in having a constricted thorax, possessing perispiracular disc pores, and sutures on dorsal pygidial chitinization. It differs in shape and arrangement of lobes, and in having duct orifices not oblique (Brimblecombe, 1957).

GENERAL: Definition and characters by Brimblecombe (1957).

CITATIONS: Borchs1966 [catalogue: 239]; Brimbl1957 [taxonomy, description: 277-279]; MorrisMo1966 [taxonomy, catalogue: 56].

Diaphoraspis compacta Brimblecombe

Diaphoraspis compacta Brimblecombe, 1957: 281. Type data: AUSTRALIA: Queensland, Forest Hill, on *Casuarina cunninghamiana*; collected January 1953. Holotype female. Type depository: Brisbane: Queensland Museum, Queensland, Australia; type no. T-5646.

SCALE COVER: Scale broadly oval to subcircular, 1.5 mm long, fawn to grey; exuviae towards anterior end, orange to dark yellow (Brimblecombe, 1957).

HOST PLANTS: **Casuarinaceae**: *Casuarina cunninghamiana* [Brimbl1957].

DISTRIBUTION: **Australasian**: Australia (Queensland [Brimbl1957]).

BIOLOGY: Insects scattered on twigs or in axils of leaflet whorls (Brimblecombe, 1957).

GENERAL: Description and illustration of adult female by Brimblecombe (1957).

CITATIONS: Borchs1966 [catalogue: 239]; Brimbl1957 [taxonomy, description, illustration, host, distribution: 281-283].

Diaphoraspis incisa Brimblecombe

Diaphoraspis incisa Brimblecombe, 1957: 279. Type data: AUSTRALIA: Queensland, Moggill, on *Casuarina cunninghamiana*; collected July 1953. Holotype female. Type depository: Brisbane: Queensland Museum, Queensland, Australia; type no. T-5642.

SCALE COVER: Scale subcircular, 1.5 mm diameter, convex, hard, black with a brown loose suffusion; exuviae central, reddish black (Brimblecombe, 1957).

HOST PLANTS: **Casuarinaceae**: *Casuarina cunninghamiana* [Brimbl1957].

DISTRIBUTION: **Australasian**: Australia (Queensland [Brimbl1957]).

BIOLOGY: Insects single on twigs mostly in axils of branchlets or leaflet whorls (Brimblecombe, 1957).

GENERAL: Description and illustration of adult female by Brimblecombe (1957).

CITATIONS: Borchs1966 [catalogue: 239]; Brimbl1957 [taxonomy, description, illustration, host, distribution: 279-281].

Diaphoraspis orbata Brimblecombe

Diaphoraspis orbata Brimblecombe, 1957: 278. Type data: AUSTRALIA: Queensland, Moggill, on *Casuarina cunninghamiana*; collected July 1953. Holotype female. Type depository: Brisbane: Queensland Museum, Queensland, Australia; type no. T-5638.

SCALE COVER: Scale subcircular, 1.2 mm diameter, or slightly oval; dark to blackish brown, exuviae brown (Brimblecombe, 1957).

HOST PLANTS: **Casuarinaceae**: *Casuarina cunninghamiana* [Brimbl1957], *Casuarina luehmannii* [Brimbl1957].

DISTRIBUTION: **Australasian**: Australia (Queensland [Brimbl1957]).

BIOLOGY: Insects single on twigs mostly in axils of branchlets (Brimblecombe, 1957).

GENERAL: Description and illustration of adult female by Brimblecombe (1957).

CITATIONS: Borchs1966 [catalogue: 239-240]; Brimbl1957 [taxonomy, description, illustration, host, distribution: 278-279].

Diaspidiotus Berlese

Aspidiotus (*Diaspidiotus*) Berlese *in*: Berlese & Leonardi, 1896: 350. Type species: *Aspidiotus* (*Diaspidiotus*) *patavinus* Berlese, *in*: Berlese & Leonardi, 1896 (= *Aspidiotus pyri* Lichtenstein), by monotypy.

Aspidiotus (*Diaspidiotus*); Leonardi, 1898a: 50. Notes: Incorrect citation of "Berlese & Leonardi" as authors.

Aspidiotus (*Diaspidiotus*); Cockerell, 1899a: 395.

Aspidiotus (*Diaspidiotus*); Fernald, 1903b: 251. Notes: Incorrect citation of "Leonardi" as author.

Affirmaspis MacGillivray, 1921: 393. Type species: *Aspidiotus socotranus* Lindinger, by original designation. Synonymy by Morrison & Morrison, 1966: 4.

Chemnaspidiotus MacGillivray, 1921: 391. Type species: *Cryptophyllaspis liquidambaris* Kotinsky, by monotypy and original designation. Synonymy by Ferris, 1938a: 223.

Comstockaspis MacGillivray, 1921: 391. Type species: *Aspidiotus perniciosus* Comstock, by original designation. Synonymy by Ferris, 1937a: 53.

Diaspidiotus; MacGillivray, 1921: 388. Notes: Incorrect citation of "Leonardi" as author.

Diaspidiotus; MacGillivray, 1921: 388. Change of status.

Ferrisaspis MacGillivray, 1921: 388. Type species: *Aspidiotus covilleae* Ferris, by original designation. Synonymy by Ferris, 1937c: 54.

Forbesaspis MacGillivray, 1921: 388. Type species: *Aspidiotus forbesi* Johnson, by monotypy and original designation. Synonymy by Ferris, 1938a: 255.

Hendaspidiotus MacGillivray, 1921: 391. Type species: *Aspidiotus ulmi* Johnson, by original designation. Synonymy by Ferris, 1937a: 55.

Quadraspidiotus MacGillivray, 1921: 388. Type species: *Aspidiotus ostreaeformis* Curtis, by original designation. Synonymy by Danzig, 1993: 177.

Euraspidiotus Thiem & Gerneck, 1934: 231. Type species: *Aspidiotus ostreaeformis* Curtis, by original designation. Synonymy by Ferris, 1937a: 54.

Paraspidiotus Thiem & Gerneck, 1934: 231. Type species: *Aspidiotus viticola* Leonardi, by original designation. Synonymy by Ferris, 1941e: 41.

Diaspidiotus; Lindinger, 1937: 183. Notes: Incorrect citation of "Leonardi" as author.

Diaspidiotus; Ferris, 1937c: 69. Notes: Incorrect citation of "Berlese & Leonardi" as authors.

Diaspidiotus; Ferris, 1938a: 214. Notes: Incorrect citation of "Berlese & Leonardi" as authors.

Archaspis Bodenheimer, 1943: 25. Type species: *Archaspis ephedrae* Bodenheimer (= *Quadraspidiotus cecconii* (Leonardi)), by monotypy. Synonymy by Ben-Dov, 1980: 264.

Diaspidiotus; Lupo, 1948: 194. Notes: Incorrect citation of "Leonardi" as author.

Dinaspidiotus; Bodenheimer, 1949: 67. Misspelling of genus name.

Diaspidiotus; Balachowsky, 1950b: 488. Notes: Incorrect citation of "Leonardi" as author.

Diaspidiotus; Borchsenius, 1950b: 224. Notes: Incorrect citation of "Berlese & Leonardi" as authors.

Diaspidiotus; Schmutterer, 1959: 100. Notes: Incorrect citation of "Leonardi" as author.

Diaspidiotus; De Lotto, 1963: 144.

Diaspidiotus; Borchsenius, 1966: 320. Notes: Incorrect citation of "Cockerell" as author.

Diaspidiotus; Morrison & Morrison, 1966: 58.

Qiadraspidiotus; Borchsenius, 1966: 403. Misspelling of genus name.

Comstockiaspis; Chou, 1985: 310. Misspelling of genus name.

Diaspidiotus; Blay Goicoechea, 1993: 593. Notes: Incorrect citation of "Cockerell" as author.

Diaspidiotus; Yaşar, 1995: 66. Notes: Incorrect citation of "Berlese & Leonardi" as authors.

SYSTEMATICS: De Lotto (1963) first pointed out that the authorship of *Diaspidiotus* Berlese, 1896 has been incorrectly credited to Leonardi, or to Berlese & Leonardi. The issue was further discussed by Morrison & Morrison (1966) and by Danzig (1993). In this catalogue we accept De Lotto's (1963) interpretation as follows: *Diaspidiotus* Berlese, in: Berlese & Leonardi (1896: 350), type species: *Aspidiotus* (*Diaspidiotus*) *patavinus* Berlese, in: Berlese & Leonardi (1896: 350). *Quadraspidiotus* MacGillivray was accepted until the early 1990's a distinct genus, to which Borchsenius (1966) assigned about 50 species. In this catalogue we follow the view of Danzig (1993) and of Dr. Sadao Takagi (personal communication, 17 October 2002, to Yair Ben-Dov) that there is no good basis to distinguish *Quadraspidiotus* from *Diaspidiotus*.

GENERAL: Definition and characters by Ferris (1938a), Balachowsky (1950b), Borchsenius (1950b), Zahradník (1952, 1972), Gómez-Menor Guerrero (1962), Bazarov & Shmelev (1971), Stoetzel & Davidson (1974a), Kosztarab & Kozár (1978), Danzig (1980b, 1993), Yaşar (1995) and by Kosztarab (1996).

KEYS: Colon-Ferrer & Medina-Gaud 1998: 28-32 (female) [Genera of Puerto Rico]; Gill 1997: 24-26 (female) [Genera, California]; Gill 1997: 113, 243 (female) [Species, California]; Kosztarab 1996: 406-407 (female) [Northeastern North America]; Blay Goicoechea 1993: 475 (female) [Spain]; Blay Goicoechea 1993: 475, 528-529 (female) [Spain]; Danzig 1993: 179-182 (female) [species Europe]; Wolff & Corseuil 1993: 29 (female) [Brazil]; Tereznikova 1986: 83 (female) [Ukraine]; Chou 1985: 283 (female) [Genera, China]; Chou 1985: 311 (female) [Species, China]; Danzig 1980b: 296, 340 (female) [species Far East USSR]; Kosztarab & Kozár 1978: 144-147 (female) [Hungary]; Kosztarab & Kozár 1978: 144-147 (female) [Hungary]; Bazarov & Shmelev 1971: 186 (female) [Central Asia]; McDaniel 1970: 429 (female) [species U.S.A.: Texas]; Komosinska 1969: 50 (female) [*Abgrallaspis* group]; McDaniel 1969: 89-91 (female) [species U.S.A.: Texas]; Ezzat & Afifi 1966: 371-372 (female) [Egypt]; Danzig 1964: 646 (female) [Europe]; Gómez-Menor Guerrero 1962: 157 (female) [Canary Islands]; Zahradník 1959a: 549 (female) [Czech Republic]; Zahradník 1959a: 549 (female) [Czech Republic]; Balachowsky 1958b: 230 (female) [*Aspidiotina* of Africa]; Ezzat 1958: 237-239 (female) [Egypt]; Gómez-Menor Ortega 1956: 7-8 (female) [Spain]; Gómez-Menor Ortega 1956: 7-8 (female) [Spain]; McKenzie 1956: 22 (female) [U.S.A.: California]; McKenzie 1956: 22 (female) [U.S.A.: California]; Balachowsky 1951: 601 (female) [Mediterranean]; Balachowsky 1951: 601 (female) [Mediterranean]; Borchsenius 1950b: 168 (female) [USSR]; Gómez-Menor Ortega 1946: 59-61 (female) [Spain]; Gómez-Menor Ortega 1946: 59-61 (female) [Spain]; Ruiz Castro 1944: 56-57 (female) [Spain]; Ferris 1942: 26 (female) [North America]; Ferris 1942: 32-33; 39-40 (female) [species North America]; Ferris 1942: 26 (female) [North America]; Borchsenius 1937a: 32-33 (female) [Palaearctic Region]; Borchsenius 1937a: 32-33 (female) [Palaearctic Region]; Brain 1918: 117

(female) [South Africa]; Newell 1899: 3 (female) [North America].
CITATIONS: Balach1948b [taxonomy: 259]; Balach1950b [taxonomy, description: 397-398,480-490,535]; Balach1951 [taxonomy: 601]; Balach1958b [taxonomy, description: 158,210,230]; BazaroSh1971 [taxonomy, description: 197-198]; BenDov1980 [taxonomy: 264]; BenDov1981c [taxonomy: 144]; BerlesLe1896 [taxonomy, description: 350]; BlayGo1993 [taxonomy, description: 527,593]; Bodenh1943 [taxonomy, description: 25]; Bodenh1949 [taxonomy, description: 26,32-34,67]; Bodenh1952 [taxonomy, description: 330,340-341]; Borchs1937 [taxonomy, description: 32-33,42,53,55]; Borchs1937a [taxonomy, description: 32,40]; Borchs1949d [taxonomy, description: 195,240]; Borchs1950b [taxonomy, description: 168,224]; Borchs1966 [catalogue: 273,312,320,327,368]; Brain1918 [taxonomy: 117]; Brimbl1962a [taxonomy: 418]; Brimbl1968 [taxonomy: 47-48]; Bustsh1958 [taxonomy, description: 248]; Chou1985 [taxonomy, description: 283,310-311]; Cocker1897i [taxonomy, description: 9,11]; Cocker1899a [taxonomy: 396]; Cocker1905b [taxonomy: 201]; ColonFMe1998 [taxonomy, description: 79]; Danzig1964 [taxonomy: 652-653]; Danzig1972 [taxonomy: 209]; Danzig1980b [taxonomy, description: 333-334,339-340]; Danzig1993 [taxonomy, description: 177-179]; DanzigPe1998 [catalogue: 226]; DeLott1963 [taxonomy: 144-145]; Ezzat1958 [taxonomy: 239]; Fernal1903b [taxonomy, catalogue: 251]; Ferris1921b [taxonomy: 94]; Ferris1937a [taxonomy: 50-56,66,95]; Ferris1937c [taxonomy: 50-55,69]; Ferris1938a [taxonomy, description: 214,255]; Ferris1942 [taxonomy: 446:26]; Ferris1943a [taxonomy, description: 84,95]; Ghauri1962 [taxonomy, description: 210]; Ghauri1962 [taxonomy, description: 211]; Gill1997 [taxonomy: 113,243]; GomezM1937 [taxonomy, description: 43,80]; GomezM1946 [taxonomy: 60,66]; GomezM1956 [taxonomy, description: 7,13,21]; GomezM1959 [taxonomy, description: 151-152]; GomezM1962 [taxonomy, description: 165-166,173]; Hadzib1983 [taxonomy: 234]; Kawai1980 [taxonomy: 215-216]; Koszta1996 [taxonomy, description: 484,573-575]; KosztaKo1978 [taxonomy, description, host, distribution: 154-155,166-167]; Kozar1990f [distribution: 142,144]; Leonar1897a [taxonomy: 375]; Leonar1898a [taxonomy: 55-56]; Lindin1937 [taxonomy: 178,181-183,185,195]; Lupo1948 [taxonomy: 174,194]; Lupo1954a [taxonomy, description: 55-56]; MacGil1921 [taxonomy, description: 388,391,422-423,]; Mamet1954b [taxonomy: 193]; McKenz1939 [taxonomy: 53]; McKenz1943a [taxonomy: 96]; McKenz1947b [taxonomy: 111-112]; McKenz1951 [taxonomy: 80]; McKenz1956 [taxonomy: 22]; McKenz1963 [taxonomy: 31-32]; MorrisMo1966 [taxonomy, catalogue: 4,14-15,34,41,58-59,]; Muntin1969 [taxonomy: 139]; Muntin1971 [taxonomy: 119-143]; Newell1899 [taxonomy, description: 3]; RuizCa1944 [taxonomy: 55-57]; Sander1904a [taxonomy, description, illustration, host, distribution: 55,56]; Savesc1982 [host, distribution: 305]; Schmut1959 [taxonomy: 48,76,100]; Silves1902 [taxonomy: 99]; StoetzDa1974a [taxonomy, description: 495]; Takagi1975 [taxonomy: 11-13]; TakagiKa1966 [taxonomy: 116]; TakagiTi1972 [taxonomy: 182]; Tao1999 [taxonomy: 82-83,114]; ThiemGe1934a [taxonomy, description: 231]; WolffCo1993 [taxonomy: 29]; Yasar1995a [taxonomy, description: 66]; Zahrad1952 [taxonomy, description: 98,113,142-143]; Zahrad1972 [taxonomy, description: 439].

Diaspidiotus acutus (Borchsenius)

Quadraspidiotus acutus Borchsenius, 1964: 164. Type data: NORTH KOREA: Choanche province, vicinity of Sarivon, on *Castanea* sp. Syntypes, female. Type depository: St. Petersburg: Zoological Museum, Academy of Science, Russia.
Diaspidiotus acutus; Danzig & Pellizzari, 1998: 227. Change of combination.
SCALE COVER: Female scale circular, diameter about 2 mm; moderately convex; secreted part grey, but lighter at margin; exuviae light brown, placed centrally. Male scale similar to that of female, about 1.5 mm long (Borchsenius, 1964).
HOST PLANTS: **Fagaceae**: *Castanea* [Borchs1964].
DISTRIBUTION: **Palaearctic**: North Korea [Borchs1965].
GENERAL: Description and illustration of adult female by Borchsenius (1964).
CITATIONS: Borchs1964 [taxonomy, description, illustration, host, distribution: 164,166,168]; Borchs1966 [catalogue: 328]; DanzigPe1998 [catalogue: 226-227]; ShiLi1991 [host, distribution: 166].

Diaspidiotus aesculi (Johnson)

Aspidiotus aesculi Johnson, 1896: 152. Type data: U.S.A.: California, Santa Clara County, on buckeye, *Aesculus californica*. Syntypes, female. Type depository: Washington: United States National Entomological Collection, U.S. National Museum of Natural History, District of Columbia, USA.
Aspidiotus (*Diaspidiotus*) *aesculi*; Cockerell, 1897i: 20. Change of combination.
Chrysomphalus aesculi; Leonardi, 1900: 342. Change of combination.
Diaspidiotus aesculi; MacGillivray, 1921: 414. Change of combination.
COMMON NAMES: buckeye Aspidiotus [Johnso1896]; Buckeye scale [McKenz1956].
SCALE COVER: Scale of female grey, circular, flat; exuviae subcentral, that of male elongate, grey; exuvia near one end (Ferris, 1938a). Colour photograph by Gill (1997).
HOST PLANTS: **Aceraceae**: *Acer negundo* [Ferris1920b, McKenz1956]. **Betulaceae**: *Alnus* [Ferris1938a, McKenz1956]. **Ericaceae**: *Arbutus menziesii* [Ferris1938a, McKenz1956]. **Euphorbiaceae**: *Adelia neomexicana* [Ferris1938a, McKenz1956]. **Fagaceae**: *Quercus* [McKenz1956]. **Hippocastanaceae**: *Aesculus californica* [Johnso1896, Leonar1900, Fernal1903b, McKenz1956]. **Juglandaceae**: *Juglans californica* [McKenz1956], *Juglans regiae* [McKenz1956]. **Oleaceae**: *Fraxinus* [McKenz1956], *Syringa* [McKenz1956]. **Salicaceae**: *Populus* [Ferris1920b, McKenz1956], *Populus fremontii* [Ferris1938a, McKenz1956], *Populus trichocarpa* [Ferris1938a, McKenz1956], *Salix argentea* [McKenz1956]. **Sambucaceae**: *Sambucus glauca* [McKenz1956].
NATURAL ENEMIES: HYMENOPTERA **Aphelinidae**: *Prospaltella fasciaventris* Girault [Gordh1979].
DISTRIBUTION: **Nearctic**: United States of America (Arizona [Ferris1938a], California [Johnso1896, Fernal1903b, Ferris1920b, Ferris1938a, McKenz1956], Colorado [Nakaha1982], Idaho [Nakaha1982], Montana [Nakaha1982], Nevada [Ferris1938a], New Mexico [Nakaha1982], North Dakota [Nakaha1982], Oregon [Nakaha1982], Texas [Nakaha1982], Utah [Ferris1920b, Ferris1938a], Washington [Nakaha1982]).

BIOLOGY: Occurring on bark. Causing pitting in bark of host (Ferris, 1938a).
ECONOMIC IMPORTANCE: Reported to cause pitting on twigs of host plants, but not of economic importance (Ferris, 1938a; Gill, 1997).
GENERAL: Description and illustration of adult female by Ferris (1938a), McKenzie (1956), Kosztarab (1996) and by Gill (1997).
KEYS: Gill 1997: 113 (female) [Species of California]; Kosztarab 1996: 484-485 (female) [Northeastern North America]; McKenzie 1956: 25 (female) [U.S.A.: California]; Ferris 1942: 32 (female) [North America]; Newell 1899: 4-5 (female) [North America].
CITATIONS: Borchs1966 [catalogue: 320-321]; Cocker1896b [distribution: 343]; Cocker1897i [taxonomy, description, host, distribution: 20]; Cocker1899a [taxonomy: 396]; Fernal1903b [catalogue: 251-252]; Ferris1920b [taxonomy, description, illustration, host, distribution: 48]; Ferris1938a [taxonomy, description, illustration, host, distribution: 215]; Ferris1941e [taxonomy: 40]; Ferris1942 [taxonomy: 445:10; 446:32]; FletchGi1908 [host, distribution: 113-133]; Gordh1979 [biological control: 908]; Harned1928 [host, distribution: 23-24]; IPMW1987 [economic importance, biological control]; Johnso1896 [taxonomy, description, host, distribution: 152,386]; Koszta1996 [taxonomy, description, illustration, host, distribution, life history: 485-487]; Leonar1900 [taxonomy, host, distribution: 342]; MacGil1921 [taxonomy, description, host, distribution: 414]; McKenz1939 [taxonomy: 53]; McKenz1956 [taxonomy, description, illustration, host, distribution: 25,58-61]; Nakaha1982 [host, distribution: 27]; Newell1899 [taxonomy, description, host, distribution: 13]; Zahrad1990a [host, distribution, description: 649].

Diaspidiotus africanus (Marlatt)

Aspidiotus (*Diaspidiotus*) *africanus* Marlatt, 1908c: 15. Type data: SOUTH AFRICA: Orange Free State, Bloemfontein, on *Gleditsia triacanthos*, *Schinus molle*, fig, almond and quince; collected by W.J. Palmer, December 1907. Syntypes, female. Type depository: Washington: United States National Entomological Collection, U.S. National Museum of Natural History, District of Columbia, USA.
Aspidiotus africanus; Sanders, 1909a: 51. Change of combination.
Aspidiotus pectinatus Lindinger, 1909e: 43. Type data: SOUTH AFRICA: Cape Province, on pear. Syntypes, female. Type depository: Hamburg: Zoologisches Institut und Zoologishces Museum, Universität von Hamburg, Germany. Synonymy by Munting, 1971: 120.
Aspidiotus (*Diaspidiotus*) *pectinatus*; Brain, 1918: 126. Change of combination.
Furcaspis pectinata; MacGillivray, 1921: 408. Change of combination requiring emendation of species name for agreement in gender.
Hendaspidiotus africanus; MacGillivray, 1921: 439. Change of combination.
Aspidiotus mashonae Hall, 1929: 347. Type data: ZIMBABWE: Mazoe, on leaves of *Combretum* sp. Syntypes, female. Type depository: London: The Natural History Museum, England, UK. Synonymy by Munting, 1971: 120.
Aspidiotus (*Diaspidiotus*) *mazoeensis* Hall, 1929: 351. Type data: ZIMBABWE: Mazoe, on branches of unknown plant. Syntypes, female. Type depository: London: The Natural History Museum, England, UK. Synonymy by Munting, 1971: 120.

Diclavaspis mashonae; Balachowsky, 1956: 102. Change of combination.

Clavaspis africanus; Balachowsky, 1956: 92. Change of combination.

Clavaspis mazoeensis; Balachowsky, 1956: 96. Change of combination.

Clavaspis pectinatus; Balachowsky, 1956: 98. Change of combination.

Clavaspis betrokensis Mamet, 1959a: 468. Type data: MADAGASCAR: Betroka, on *Sclerocarya caffra*. Holotype female. Type depository: Paris: Muséum national d'Histoire naturelle, France. Synonymy by Munting, 1971: 120.

Diaspidiotus africanus; Borchsenius, 1966: 321. Change of combination.

Diaspidiotus pectinatus; Borchsenius, 1966: 325. Change of combination.

COMMON NAME: grey scale [Muntin1971].

SCALE COVER: Female scale dull opaque brown, circular, 1.5 mm in diameter; moderately convex, with fairly prominent central nipple, and slightly annulated as in *perniciosus*; when rubbed, resinous exuviae appear, and when old and massed the ashen appearance of *perniciosus* is seen (Marlatt, 1908c). Female scale circular to broadly oval, moderately convex; exuviae more or less central and brown in colour; secretionary area a dirty white or dull brown; ventral scale, extremely thin, remaining attached to host plant as a thin white film (Hall, 1929).

HOST PLANTS: **Aceraceae**: *Acer* [Brain1918]. **Anacardiaceae**: *Rhus* [Brain1918], *Schinus molle* [Balach1956], *Sclerocarya caffra* [Mamet1959a, Borchs1966]. **Apocynaceae**: *Nerium oleander* [Muntin1971]. **Berberidaceae**: *Berberis* [Brain1918]. **Boraginaceae**: *Ehretia rigida* [Muntin1971]. **Combretaceae**: *Combretum* [Hall1929, Muntin1971], *Combretum erythrophyllum* [Muntin1971]. **Lauraceae**: *Laurus* [Muntin1971]. **Leguminosae**: *Acacia* [Balach1956, Muntin1971], *Acacia giraffae* [Muntin1971], *Acacia horrida* [Brain1918], *Acacia karroo* [Muntin1971], *Acacia nigrescens* [Muntin1971], *Ceratonia* [Brain1918], *Gleditsia* [Brain1918], *Gleditsia triacanthos* [Muntin1971], *Psoralea* [Muntin1971], *Robinia pseudoacacia* [Brain1918, Balach1956]. **Moraceae**: *Ficus* [Muntin1971], *Ficus carica* [Balach1956]. **Oleaceae**: *Olea europaea* [Muntin1965b, Muntin1971]. **Rosaceae**: *Cotoneaster* [Brain1918], *Crataegus* [Brain1918], *Cydonia* [Muntin1971], *Prunus communis* [Muntin1971], *Prunus domesticus* [Muntin1971], *Prunus persica* [Muntin1971], *Pyrus communis* [Muntin1971], *Pyrus malus* [Muntin1971]. **Tiliaceae**: *Grewia occidentalis* [Muntin1971]. **Verbenaceae**: *Clerodendron glabrum* [Muntin1971].

NATURAL ENEMIES: HYMENOPTERA **Aphelinidae**: *Azotus capensis* Howard [AnneckIn1970]. **Encyrtidae**: *Adelencyrtus* sp. [Prinsl1983], *Metaphycus* sp. [Prinsl1983].

DISTRIBUTION: **Afrotropical**: Madagascar [Mamet1959a, Borchs1966, Muntin1971]; South Africa [Marlat1908c, Leonar1914, Brain1918, Balach1956, Muntin1965b, Muntin1971]; Zimbabwe [Hall1929, Balach1956, Muntin1971].

ECONOMIC IMPORTANCE: A pest of deciduous fruit trees in the Western Cape and Langkloof Valley in Cape Province of South Africa (Munting, 1971).

GENERAL: Marlatt (1908c) studied intraspecific variation of some taxonomic characters in adult female. Munting (1971) extensively studied intraspecific variation of taxonomic characters in adult female, and showed that several were host-induced variations. Description and illustration of adult female by Marlatt (1908c), Brain (1918), Hall (1929), Balachowsky (1956) and by Munting (1971).

KEYS: Balachowsky 1956: 92, 100 (female) [Africa]; Brain 1918: 124 (female) [South Africa].

CITATIONS: AnneckIn1970 [host, distribution, biological control: 240]; Balach1956 [taxonomy, description, illustration, host, distribution: 92-99,102-104]; Borchs1966 [catalogue: 315-317,321,324,325]; Brain1918 [taxonomy, description, illustration, host, distribution: 125-127]; Ferris1941e [taxonomy: 40,45-46]; Hall1929 [taxonomy, description, illustration, host, distribution : 347-348,351-352]; Kozar1990c [life history, economic importance, host, distribution: 593-602]; Leonar1914 [taxonomy, host, distribution: 196-197]; Lindin1909e [taxonomy, description, illustration, host, distribution: 43]; Lounsb1914 [host, distribution: 1]; Lounsb1916 [host, distribution: 83-103]; MacGil1921 [taxonomy, description, host, distribution: 408,439]; Mamet1959a [taxonomy, description, illustration, host, distribution: 386,468]; Marlat1908c [taxonomy, description, illustration, host, distribution: 15-20]; MillerDa1990 [host, distribution, economic importance: 301]; Muntin1965b [host, distribution: 190]; Muntin1971 [taxonomy, description, illustration, host, distribution: 120-134]; Prinsl1983 [distribution, biological control: 26]; Sander1909a [taxonomy, host, distribution: 51]; Sassce1911 [taxonomy: 70]; SchmutKlLu1957 [host, distribution, economic importance: 491]; WeidneWa1968 [taxonomy: 173].

Diaspidiotus alni (Marchal)

Targionia alni Marchal, 1909: 872. Type data: FRANCE: Var, Agay, on *Alnus glutinosa*. Syntypes, female. Type depository: Paris: Muséum national d'Histoire naturelle, France.

Aspidiotus (*Hemiberlesiella*) *alni*; Thiem & Gerneck, 1934a: 132. Change of combination.

Aspidiotus alni; Lindinger, 1935: 128. Change of combination.

Diaspidiotus alni; Borchsenius, 1950b: 229. Change of combination.

Hemiberlesiella alni; Lindinger, 1957: 549. Change of combination.

Diaspidiotus alni; Danzig, 1993: 212. Revived combination.

SCALE COVER: Female scale circular, 1.5 mm, relatively flat; covered by epidermis of host plant; slender, grey; larval exuviae dark yellow (Marchal, 1909).

HOST PLANTS: **Betulaceae**: *Alnus* [Balach1950b], *Alnus glutinosa* [Marcha1909, Balach1932d, Kaweck1935, Kozar1999a].   **Carpinaceae**: *Carpinus betulus* [Zahrad1972]. **Fagaceae**: *Fagus sylvatica* [Zahrad1972], *Quercus* [Zahrad1972], *Quercus robur* [PellizCa1991a]. **Salicaceae**: *Populus* [Balach1950b].

NATURAL ENEMIES: HYMENOPTERA **Aphelinidae**: *Pteroptrix dimidiata* Westwood [Zahrad1972].

DISTRIBUTION: **Palaearctic**: France [Marcha1909, Sander1909a, Balach1932d]; Germany (United) [Balach1950b]; Hungary [Kozar1999a]; Italy [PellizCa1991a, LongoMaPe1995]; Poland [Kaweck1935]; Ukraine [Balach1950b].

GENERAL: Description and illustration of adult female by Balachowsky (1950b) and by Tereznikova (1986).

KEYS: Danzig 1993: 179-182 (female) [Europe]; Tereznikova 1986: 97-98 (female) [Ukraine]; Kosztarab & Kozár 1978: 155-156 (female) [Hungary]; Balachowsky 1950b: 491 (female) [Mediterranean].

CITATIONS: Balach1932d [taxonomy, host, distribution: XLVIII]; Balach1950b [taxonomy, description, illustration, host, distribution: 529-531]; Borchs1939 [taxonomy, description, host, distribution: 10,34]; Borchs1950b [taxonomy, description, illustration, host, distribution: 229,231]; Borchs1966 [catalogue: 321]; Bustsh1958 [taxonomy, description, host, distribution: 220,250]; Danzig1964 [taxonomy, host, distribution: 653]; Danzig1993 [taxonomy, description, illustration, host, distribution: 212-214]; DanzigPe1998 [catalogue: 227]; Ferris1941e [taxonomy: 40]; Ferris1943a [taxonomy: 85]; Foldi2001 [distribution: 303-308]; Kaweck1935 [host, distribution: 76]; KosztaKo1978 [taxonomy, description, host, distribution: 155-156]; Kozar1999a [host, distribution: 141]; Lagows2002a [host, distribution: 184]; Lindin1912b [taxonomy, description, host, distribution: 63]; Lindin1935 [taxonomy: 128]; Lindin1957 [taxonomy: 549]; LongoMaPe1995 [distribution: 126]; MacGil1921 [taxonomy, description, host, distribution: 448]; Marcha1909 [taxonomy, description, host, distribution: 872]; PellizCa1991a [taxonomy, host, distribution: 199]; Sander1909a [taxonomy, host, distribution: 55]; Schmut1959 [taxonomy, description, host, distribution: 100,103]; Terezn1986 [taxonomy, description, illustration, host, distribution: 98-99]; ThiemGe1934a [taxonomy, description, host, distribution: 130-158,208-238]; Zahrad1972 [host, distribution, biological control: 439].

Diaspidiotus anatolicus (Bodenheimer)

Forbesaspis anatolica Bodenheimer, 1949: 68. Type data: TURKEY: Nigde, on *Amygdalus communis*; collected 12.vii.1939. Syntypes, female. Type depository: Paris: Muséum national d'Histoire naturelle, France. Notes: Bodenheimer (1952: 341) again referred to this species as n.sp.

Quadraspidiotus anatolica; Borchsenius, 1966: 328. Change of combination.

Diaspidiotus anatolica; Danzig & Pellizzari, 1998: 227. Change of combination.

SCALE COVER: Scale of female: almost circular. Exuviae subcentral. Exact position of first exuvia not recognisable, as damaged. Colour dark grey, first exuvia yellow-brown, second exuvia orange. Diameter 1 mm (Bodenheimer, 1952).

HOST PLANTS: **Rosaceae**: *Amygdalus communis* [Bodenh1949, Bodenh1952].

DISTRIBUTION: **Palaearctic**: Turkey [Bodenh1949, Bodenh1952].

GENERAL: Description and illustration of adult female by Bodenheimer (1952) and by Yaşar (1995a).

CITATIONS: Bodenh1949 [taxonomy, description, illustration, host, distribution: 68-70]; Bodenh1952 [taxonomy, description, illustration, host, distribution: 341-342]; Borchs1966 [catalogue: 328]; DanzigPe1998 [catalogue: 227]; Yasar1995a [taxonomy, description, illustration, host, distribution: 106-108].

Diaspidiotus ancylus (Putnam)

Diaspis ancylus Putnam, 1878: 323. Type data: U.S.A.: Iowa, Davenport, on *Acer dasycarpum*. Syntypes, female. Type depository: Washington: United States National Entomological Collection, U.S. National Museum of Natural History, District of Columbia, USA.

Diaspis aneglus; Putnam, 1878: 321. Misspelling of species name.

Aspidiotus convexus Comstock, 1881a: 295. Type data: U.S.A.: California, locality not indicated, on bark of trunk and limbs of native willows. Syntypes, both sexes. Type depository: Washington: U.S. Entomological Collection, U.S. National Museum of Natural History, District of Columbia, USA. Synonymy by Marlatt, 1899: 208.

Aspidiotus howardi Cockerell, 1895b: 16. Type data: U.S.A.: Colorado, Canon City, on plum (*Prunus* sp.). Syntypes, female. Type depository: Washington: United States National Entomological Collection, U.S. National Museum of Natural History, District of Columbia, USA. Synonymy by Stannard, 1965: 575.

Aspidiotus comstocki Johnson, 1896: 151. Type data: U.S.A.: Illinois, Mt. Carmel, on *Acer saccharinum*. Syntypes, female. Type depository: Washington: United States National Entomological Collection, U.S. National Museum of Natural History, District of Columbia, USA. Synonymy by Stannard, 1965: 575.

Aspidiotus ancylus; Cockerell, 1896b: 334. Change of combination.

Aspidiotus comstocki; Cockerell, 1896b: 334. Change of combination.

Aspidiotus convexus; Cockerell, 1896b: 334. Incorrect synonymy. Notes: Incorrect synonymy with *Aspidiotus rapax* Comstock; see Borchsenius, 1966: 322.

Aspidiotus townsendi Cockerell, 1896h: 20. Type data: MEXICO: Piedras Negras, on leaves of undetermined plant; collected by Townsend. Syntypes, female. Type depository: Washington: United States National Entomological Collection, U.S. National Museum of Natural History, District of Columbia, USA. Synonymy by Ferris, 1938: 190.

Aspidiotus (*Diaspidiotus*) *ancylus*; Cockerell, 1897i: 20. Change of combination.

Aspidiotus (*Diaspidiotus*) *comstocki*; Cockerell, 1897i: 20. Change of combination.

Aspidiotus (*Hemiberlesia*) *convexus*; Cockerell, 1897i: 20. Change of combination.

Aspidiotus (*Diaspidiotus*) *howardi*; Cockerell, 1897i: 21. Change of combination.

Aspidiotus (*Diaspidiotus*) *townsendi*; Cockerell, 1897i: 22. Change of combination.

Aspidiotus circularis; Cockerell, 1897i: 4. Misidentification; discovered by Borchsenius, 1966: 322.

Aspidiotus ancylus serratus Newell & Cockerell in Osborn, 1898: 229. Type data: U.S.A.: Iowa, on native willows. Syntypes, female. Type depository: Washington: United States National Entomological Collection, U.S. National Museum of Natural History, District of Columbia, USA. Synonymy by Ferris, 1941e: 48.

Aspidiotus (*Aspidiella*) *townsendi*; Leonardi, 1898a: 70. Change of combination.

Aspidiotus (*Evaspidiotus*) *howardi*; Leonardi, 1898a: 76. Change of combination.

Aspidiotus (*Evaspidiotus*) *convexus*; Leonardi, 1898c: 50. Change of combination.

Aspidiotus aesculi solus Hunter, 1899: 12. Type data: USA: Kansas, Lawrence, on *Juglans nigra*. Syntypes, female. Type depositories: Manhattan: Kansas State University Collections, Kansas, USA, and Washington: United States National Entomological Collection, U.S. National Museum of Natural History, District of Columbia, USA. Synonymy by Ferris, 1938a: 190.

Aspidiotus ancylus latilobis Newell, 1899: 9. Type data: U.S.A.: Iowa, Ames, on "Mountain ash" [=*Pyrus americana*]; collected 2 May, 1899. Syntypes, female. Type depository: Washington D.C.: U.S. Entomological Collection, U.S. National Museum of Natural History, USA. Synonymy by Ferris, 1938a: 190.

Aspidiotus (*Diaspidiotus*) *ancylus ornatus*; Leonardi, 1900: 339. Change of combination and rank.

Aspidiotus (*Aspidiella*) *comstocki*; Leonardi, 1900: 339. Change of combination.

Aspidiotus (*Diaspidiotus*) *ohioensis* York, 1905: 325. Type data: USA: Ohio, Ohio State University Campus, on *Aesculus glabra*. Syntypes, female. Type depository: Washington: United States National Entomological Collection, U.S. National Museum of Natural History, District of Columbia, USA. Synonymy by Borchsenius, 1966: 322.

Diaspidiotus ancylus; Cockerell, 1905b: 202. Change of combination.

Diaspidiotus howardi; Cockerell, 1905b: 202. Change of combination.

Diaspidiotus townsendi; Cockerell, 1905b: 202. Change of combination.

Aspidiotus ohioensis; Sanders, 1906: 14. Change of combination.

Aspidiotus pseudospinosus Woglum, 1906: 75. Type data: USA: Florida, on Saw Palmetto; collected by W.H. Fields, 1882. Syntypes, female. Type depository: Washington: United States National Entomological Collection, U.S. National Museum of Natural History, District of Columbia, USA. Synonymy by Ferris, 1938a: 190.

Aspidiotus (*Hemiberlesia*) *epigaeae* Marlatt, 1908c: 21. Type data: USA: Virginia, Chain Bridge, near Washington, D.C., on *Epigaea repens*; collected by J.G. Sanders, 6 May 1906. Syntypes, female. Type depository: Washington: United States National Entomological Collection, U.S. National Museum of Natural History, District of Columbia, USA; type no. 14135. Synonymy by Ferris, 1938a: 190.

Aspidiella pseudospinosa; MacGillivray, 1921: 404. Change of combination requiring emendation of species name for agreement in gender.

Aspidiella comstocki; MacGillivray, 1921: 405. Change of combination.

Quadraspidiotus townsendi; MacGillivray, 1921: 409. Change of combination.

Quadraspidiotus epigaeae; MacGillivray, 1921: 410. Change of combination.

Diaspidiotus ancylus latilobis; MacGillivray, 1921: 411. Change of combination.

Diaspidiotus ancylus serratus; MacGillivray, 1921: 411. Change of combination.

Diaspidiotus ohioensis; MacGillivray, 1921: 412. Change of combination.

Diaspidiotus solus; MacGillivray, 1921: 414. Change of combination and rank.

Aspidiotus toxycrataegi Hollinger, 1923: 15. Type data: U.S.A.: Missouri, Lawrence County, under exfoliating bark of hawthorn and osage orange or bois d'arc. Syntypes, female. Type depository: Washington: United States National Entomological Collection, U.S. National Museum of Natural History, District of Columbia, USA. Synonymy by Ferris, 1941e: 49.

Aspidiotus toxycrataegi; Hollinger, 1923: 8, 15. Misspelling of species name.

Aspidiotus solus; Lindinger, 1937: 180. Change of combination.

Aspidiotus serratus; Ferris, 1938: 190. Change of combination and rank.

Aspidiotus epigeae; Ferris, 1938a: 190. Misspelling of species name.

Aspidiotus epigeae; Ferris, 1938a: 190. Change of combination.

Aspidiotus latilobis; Ferris, 1938a: 190. Change of combination and rank.

Hemiberlesia comstocki; Ferris, 1938a: 235. Change of combination.

Hemiberlesia howardi; Ferris, 1938a: 240. Change of combination.

Aspidiotus oxycrataegi; Ferris, 1942: 4. Justified emendation.

Aspidiotus (*Hemiberlesia*) *howardi*; Merrill, 1953: 23. Change of combination.

Abgrallaspis comstocki; Balachowsky, 1953k: 113. Change of combination.
Abgrallaspis howardi; Balachowsky, 1953k: 113. Change of combination.
Diaspidiotus ancylus; McKenzie, 1956: 25. Revived combination.
Gonaspidiotus comstocki; Borchsenius, 1966: 300. Change of combination.
COMMON NAMES: falsa escama de San Jose [Gonzal1989]; Howard scale [McKenz1956]; maple bark louse [Putnam1878]; maple leaf Aspidiotus [Johnso1896]; Putnam scale [Comsto1883, MerrilCh1923, Merril1953, McKenz1956, Dekle1965c, Stimme1976]; putnam's scale [Comsto1883].
SYSTEMATICS: Putnam (1878 p. 321) printed the species name *Diaspis aneglus*, which is very likely a misspelling of *Diaspis ancylus* that was printed on page 323 (Putnam, 1878). Stannard (1965) synonymized *Aspidiotus comstocki* Johnson and *Aspidiotus howardi* Cockerell with *Diaspidiotus ancylus* (Putnam), while Kosztarab (1996) noted that more studies are required to confirm the synonymy, and regarded each of these as distinct species.
SCALE COVER: Female scale of variable form, depending upon it's position in relation to leaf veins; pale grey or white, exuviae submarginal; that of male grey, elongate, exuvia near one end (Ferris, 1938a). Colour photograph by Gonzalez (1989).
HOST PLANTS: **Aceraceae**: *Acer* [Ferris1938a, BesheaTiHo1973], *Acer dasycarpum* [Ferris1938a], *Acer saccharum* [Johnso1896, McKenz1956]. **Agavaceae**: *Yucca aloifolia* [Dekle1965c]. **Annonaceae**: *Asimina* [BesheaTiHo1973]. **Betulaceae**: *Betula* [McKenz1956], *Betula nigra* [BesheaTiHo1973]. **Bignoniaceae**: *Catalpa* [Ferris1938a, McKenz1956, McDani1969]. **Buxaceae**: *Pachysandra terminalis* [StoetzDa1974a]. **Caprifoliaceae**: *Viburnum* [McKenz1956]. **Compositae**: *Eupatorium* [Dekle1965c]. **Elaeagnaceae**: *Elaeagnus pungens* [TippinBe1970, BesheaTiHo1973]. **Ericaceae**: *Andromeda* [Wilson1917, MerrilCh1923], *Epigaea repens* [Marlat1908c, Ferris1938a, McKenz1956], *Kalmia latifolia* [BesheaTiHo1973], *Vaccinium* [Dekle1965c], *Vaccinium corymbosum* [PolavaDaMi2000]. **Fagaceae**: *Fagus* [Ferris1938a, McKenz1956], *Fagus grandifolia* [BesheaTiHo1973], *Ilex* [Ferris1938a], *Ilex cornuta* [McKenz1956], *Ilex opaca* [McKenz1956, StoetzDa1974a], *Quercus* [Ferris1938a, Dekle1965c], *Quercus alba* [BesheaTiHo1973], *Quercus phellos* [BesheaTiHo1973], *Quercus virginiana* [BesheaTiHo1973], *Quercus wislizeni* [McKenz1956], *Quercus wrightii* [Cocker1896m]. **Gramineae**: *Avena japoneza* [Lepage1938]. **Grossulariaceae**: *Ribes* [McKenz1956]. **Hippocastanaceae**: *Aesculus glabra* [York1905, Sander1906]. **Hydrangeaceae**: *Hydrangea quercifolia* [McKenz1956, BesheaTiHo1973]. **Juglandaceae**: *Carya* [Ferris1938a], *Carya illinoensis* [McKenz1956, Brimbl1962, Dekle1965c, Brimbl1968, BesheaTiHo1973], *Carya pecan* [McKenz1956], *Juglans nigra* [Hunter1899, Fernal1903b, McKenz1956, BesheaTiHo1973], *Juglans regiae* [McKenz1956]. **Leguminosae**: *Acacia* [Comsto1881a], *Robinia* [Ferris1938a, McDani1969], *Robinia hispida* [McKenz1956], *Robinia pseudacacia* [Muntin1971]. **Liliaceae**: *Aloe* [Dekle1965c]. **Magnoliaceae**: *Liriodendron tulipifera* [McKenz1956], *Magnolia* [Ferris1938a, McKenz1956], *Magnolia grandiflora* [BesheaTiHo1973]. **Marantaceae**: *Maranta* [McKenz1956]. **Moraceae**: *Maclura aurantiaca* [Ferris1942], *Maclura pomifera*

[Ferris1938a, McKenz1956]. **Naucleaceae**: *Cephalanthus occidentalis* [Ferris1938a, McKenz1956, McDani1969]. **Oleaceae**: *Fraxinus* [Ferris1938a, McKenz1956], *Olea europaea* [McKenz1956]. **Pinaceae**: *Tsuga canadensis* [BesheaTiHo1973]. **Rosaceae**: *Armeniaca vulgaris* [Ferris1938a], *Crataegus* [Ferris1938a, McKenz1956, McDani1969], *Cydonia* [McKenz1956], *Cydonia oblonga* [Ferris1938a], *Malus* [Ferris1938a, McKenz1956], *Malus pumila* [McKenz1956], *Prunus* [Cocker1895b, Ferris1938a, McKenz1956], *Prunus amygdalus* [McKenz1956], *Prunus armeniaca* [McKenz1956], *Prunus persica* [Ferris1938a, McKenz1956], *Pyracantha* [McKenz1956], *Pyrus americana* [Newell1899, Ferris1938a], *Pyrus communis* [Ferris1938a, McKenz1956], *Pyrus malus* [Ferris1938a], *Sorbus americana* [McKenz1956]. **Salicaceae**: *Populus candicans* [McKenz1956], *Salix* [Ferris1938a, McKenz1956]. **Sapotaceae**: *Bumelia* [Davids1964, McDani1969], *Bumelia lanugenosa* [Ferris1938a]. **Smilacaceae**: *Smilax* [BesheaTiHo1973]. **Styracaceae**: *Halesia* [BesheaTiHo1973]. **Tiliaceae**: *Tilia* [Ferris1938a, McKenz1956]. **Ulmaceae**: *Celtis* [Ferris1938a, McDani1969], *Celtis occidentalis* [McKenz1956], *Ulmus* [McKenz1956].

NATURAL ENEMIES: FUNGI **Ascomycotina**: *Myriangium duriaei* [EvansPr1990]. HYMENOPTERA **Aphelinidae**: *Aspidiotiphagus citrinus* (Craw) [Gordh1979], *Coccophagus varicornis* Howard [Felt1901], *Physcus varicornis* (Howard) [Gordh1979], *Prospaltella aurantii* (Howard) [Gordh1979].

DISTRIBUTION: **Afrotropical**: South Africa [Muntin1971]. **Australasian**: Australia (Queensland [Brimbl1962, Brimbl1968]). **Nearctic**: Mexico [Cocker1896h, Cocker1896m, Ferris1938a]; United States of America (Alabama [BesheaTiHo1973], Arizona [Cocker1899n, Ferris1938a], California [Comsto1881a, McKenz1956], Colorado [Cocker1895b, Ferris1938a], Connecticut [Ferris1938a], Delaware [Nakaha1982], District of Columbia [Comsto1881a, Ferris1938a], Florida [Woglum1906, Wilson1917, MerrilCh1923, Merril1953, Dekle1965c, BesheaTiHo1973], Georgia [TippinBe1970, BesheaTiHo1973], Illinois [Johnso1896], Indiana [Ferris1938a], Iowa [Comsto1881a, Newell1899], Kansas [Hunter1899, Fernal1903b], Kentucky [Koszta1996], Louisiana [Ferris1938a], Maryland [StoetzDa1974a], Mississippi [Herric1911], Missouri [Hollin1923, Ferris1938a], Montana [Nakaha1982], New Jersey [PolavaDaMi2000], New York [Comsto1881a, Ferris1938a], North Carolina [Nakaha1982], Ohio [York1905, Sander1906, Ferris1938a], Oklahoma [Davids1964], Pennsylvania [Stimme1976, Koszta1996], Rhode Island [Nakaha1982], South Carolina [Nakaha1982], Tennessee [Ferris1938a], Texas [Herric1911, Ferris1938a, McDani1969], Utah [Ferris1938a], Virginia [Marlat1908c, BesheaTiHo1973], West Virginia [Koszta1996]). **Neotropical**: Brazil (São Paulo [Lepage1938]); Chile [GonzalCh1968, Gonzal1989]. **Palaearctic**: Japan [Kuwana1917a]; Portugal [Seabra1941]; Spain [RuizCa1944, GomezM1954, Martin1983].

BIOLOGY: Occurring on bark, leaves or fruit. Occurring characteristically on leaves although on plums occasionally occurring on fruit (Ferris, 1938a).

ECONOMIC IMPORTANCE: A pest of elm, walnut, cranberry, blueberry and ornamentals in several states in USA (McKenzie, 1947; English & Decker, 1954; Gill, 1997).

GENERAL: Description and illustration of adult female by Ferris (1938a), McKenzie (1956), Munting (1971)(based on material from South Africa), Gill (1997) and by Polavarapu *et al.* (2000). Description and illustration of female and male nymphs and male pupa and prepupa given by Stoetzel & Davidson (1974a).

KEYS: Gill 1997: 32, 113 (female) [Species of California]; Kosztarab 1996: 410-411, 484-485 (female) [Northeastern North America]; McDaniel 1969: 90-91, 101 (female) [U.S.A.: Texas]; Davidson 1964: 639-640 (female) [North America]; McKenzie 1956: 25 (female) [U.S.A.: California]; McKenzie 1956: 25 (female) [U.S.A.: California]; Ferris 1942: 33-35 (female) [North America]; Britton 1923: 371 (female) [U.S.A.: Connecticut]; Hollinger 1923: 7-8 (female) [U.S.A.: Missouri]; Lawson 1917: 217 (female) [U.S.A.: Kansas]; Dietz & Morrison 1916a: 289-290 (female) [U.S.A.: Indiana]; Cockerell 1905: 45-46 (female) [Mexico]; Cockerell 1905b: 202 (female) [U.S.A.: Colorado]; Newell 1899: 4-5 (female) [North America]; Comstock 1883: 55-57 (female) [North America].

CITATIONS: AranciSaCh1990 [host, distribution, life history: 106-108]; Balach1953k [taxonomy: 113]; BeardsDaHo1976 [economic importance: 106]; Beckwi1945 [economic importance, control: 43-54]; BerlesLe1898a [taxonomy: 107]; BesheaTiHo1973 [host, distribution: 4,6]; Borchs1937a [taxonomy, description, host, distribution: 41-42]; Borchs1966 [catalogue: 300-301,321-322]; Boynto1901 [taxonomy: 347,351]; Brick1912 [host, distribution: 1-22]; Brimbl1962 [host, distribution, economic importance: 221]; Brimbl1968 [taxonomy, illustration, host, distribution: 48-49]; Britto1923 [taxonomy, description, host, distribution: 371-373]; Britto1923b [host, distribution]; ClapsWoGo2001a [taxonomy, host, distribution: 16]; Cocker1894l [taxonomy, description, host, distribution: 191]; Cocker1895b [taxonomy, description, host, distribution: 16-18]; Cocker1896b [taxonomy, distribution: 334]; Cocker1896f [taxonomy, description, host, distribution: 32]; Cocker1896h [taxonomy, description, host, distribution: 20]; Cocker1896m [host, distribution: 226]; Cocker1897i [taxonomy, description, host, distribution: 5,8,20-22,25]; Cocker1899a [taxonomy: 396]; Cocker1899n [host, distribution: 21]; Cocker1905 [taxonomy: 46]; Cocker1905b [taxonomy: 202]; Comsto1881a [taxonomy, description, illustration, host, distribution: 292-295]; Comsto1883 [host, distribution: 56-59]; Danzig1972 [taxonomy, host, distribution, economic importance: 209-210]; Danzig1995 [taxonomy, life history, structure: 19-24]; Davids1964 [taxonomy, description, illustration, host, distribution: 640-641]; Dekle1965c [taxonomy, description, host, distribution: 48,70]; Dekle1976 [taxonomy, description, host, distribution, economic importance: 21-24,67]; DeLott1963 [taxonomy: 144-145]; DeSant1940 [biological control: 29-44]; DietzMo1916a [taxonomy, description, illustration, host, distribution: 298-299,303-305]; Dougla1912 [taxonomy: 193]; Drake1935 [host, distribution, control: 83-91]; EnglisDe1954 [host, distribution, chemical control: 624-627]; EvansPr1990 [biological control: 3-17]; Fawcet1948 [biological control: 627]; Fawcet1948 [biological control: 627-664]; FDACSB1983 [host, distribution: 6-8]; Felt1901 [taxonomy, description, illustration, host, distribution, life history, biological control: 326-328]; Felt1924 [host, distribution, economic importance, life history, control]; Fernal1903b [catalogue: 252-254,264,279]; Ferris1937c [taxonomy, illustration: 51,69]; Ferris1938a [taxonomy, description, illustration, host, distribution: 190,216,235,240]; Ferris1941e [taxonomy: 42-49]; Ferris1942 [taxonomy, host, distribution: 445:4; 446:33;]; Fitch1857 [taxonomy, description, host, distribution: 426]; FrankKr1900 [taxonomy: 72]; Garcia1931a [host, distribution, biological control: 659-669]; Gauthi1993 [host, distribution, life history, chemical control: 4-5]; Gill1997 [host, distribution, taxonomy, illustration: 34,38,114-115,118]; GilletLi1918 [host, distribution: 3]; GilletLi1923 [host, distribution: 12-18]; GomezM1954 [taxonomy, description, illustration, host, distribution: 132-134]; Gonzal1989 [taxonomy, description, host, distribution, economic importance: 94-95]; GonzalCh1968 [distribution: 110]; Gordh1979 [biological control: 900,907]; Hall1969 [economic importance: 823-826]; Hamilt1936 [chemistry, chemical control, physiology: 150-160]; Harned1928 [host, distribution: 23-24]; Herric1911 [taxonomy, description, illustration, host, distribution: 9,14-15,47]; Herric1925 [host, distribution, description, life history, economic importance]; Hollin1923 [taxonomy, description, host, distribution: 9-10,15-16]; Houser1908 [host, distribution: 173-181]; Howard1895e [biological control: 1-44]; HowardAs1895 [biological control: 633]; Hunter1899 [taxonomy, description, host, distribution: 12-14]; IPMW1987 [economic importance, biological control]; Johnso1896 [taxonomy, description, host, distribution: 151-152]; Johnso1898 [description, life history: 82-83]; Jorgen1934 [taxonomy: 279]; Kaston1938 [host, distribution: 235-242]; King1903b [taxonomy, host, distribution: 195]; KonstaGu1987 [host, distribution: 161]; Koszta1996 [taxonomy, description, illustration, host, distribution: 411-419,487-489]; Kozar1990c [life history, economic importance, host, distribution: 593-602]; Kuwana1917a [taxonomy, distribution: 174]; Larew1990 [ecology, life history, structure: 293-300]; Lawson1917 [taxonomy, description, illustration, host, distribution: 217,233-236]; Leonar1897 [taxonomy: 286]; Leonar1898a [taxonomy, description, illustration, host, distribution: 56-58,69-71,76]; Leonar1898c [taxonomy, description, illustration, host, distribution: 50-51]; Leonar1900 [taxonomy, host, distribution: 339]; Lepage1938 [catalogue: 393]; Lindin1909c [taxonomy, host, distribution: 448,449]; Lindin1909e [taxonomy: 44]; Lindin1936b [taxonomy: 286]; Lindin1937 [taxonomy: 180]; Lindin1943b [taxonomy: 207]; Lindin1957 [taxonomy: 546]; Lobdel1937 [taxonomy: 78]; Lord1922 [host, distribution: 1]; Lugger1900 [host, distribution: 208-245]; MacGil1921 [taxonomy, description, host, distribution: 300,404,409-414]; Mackie1936 [host, distribution: 455]; Marlat1899b [taxonomy: 208,211]; Marlat1902a [biological control: 155-174]; Marlat1908b [taxonomy, description, host, distribution: 309]; Marlat1908c [taxonomy, description, illustration, host, distribution: 21-22]; Martin1983 [taxonomy, host, distribution: 63]; Marucc1966 [host, distribution, control: 99-235]; McClur1990a [taxonomy, host, distribution, ecology: 165-168]; McDani1969 [taxonomy, illustration, host, distribution: 91-93,102-107]; McKenz1947 [taxonomy, economic importance: 31]; McKenz1956 [taxonomy, description, illustration, host, distribution: 69-70]; Mead1983 [host, distribution: 1-5]; Mead1987 [host, distribution: 2-4]; Meerwa1900 [taxonomy: 3]; Merkel1938 [host, distribution: 88-99]; Merril1953 [taxonomy, description, host, distribution: 16]; MerrilCh1923 [taxonomy, description, host, distribution, economic importance: 197-198,208]; MichelOr1958 [host, distribution, taxonomy, economic importance, biological control, chemical control: 46-57];

MilholMe1984 [host, distribution: 1-33]; MillerDa1990 [host, distribution, economic importance: 300-301]; Muntin1971 [taxonomy, description, illustration, host, distribution: 134-135]; Nakaha1982 [host, distribution: 1,2]; Nakaha1982 [host, distribution: 27-28]; Neiswa1966 [host, distribution, taxonomy, economic importance: 1-54]; Newell1899 [taxonomy, description, illustration, host, distribution: 7-11,13,22,25-26]; NewellRo1908 [host, distribution: 150-155]; Osborn1898 [taxonomy, description, host, distribution: 229]; Pitt1945 [host, distribution, economic importance, control: 3-42]; PolavaDaMi2000 [taxonomy, description, illustration, host, distribution, life history: 549-560]; PrittsHa1992 [host, distribution, control: 3]; Putnam1878 [taxonomy, description, host, distribution, biological control: 317-324]; Putnam1880 [description, life history: 346]; Reh1900a [host, distribution, taxonomy: 271]; RuizCa1944 [taxonomy, distribution: 65]; Salaza1989 [host, distribution]; SalazaSo1990 [host, distribution, life history, biological control: 135-137]; Sander1904a [taxonomy, description, illustration, host, distribution: 57-58]; Sander1906 [taxonomy, host, distribution: 12,14]; Sander1909a [taxonomy, host, distribution: 52]; SchmutKlLu1957 [host, distribution, economic importance: 476-477,491]; SchuhMo1948 [host, distribution, control]; Seabra1941 [distribution: 8]; Staffo1915 [taxonomy, structure: 70]; Stanna1965 [taxonomy, description, illustration, host, distribution: 574-576]; Stimme1976 [host, distribution, description, life history, economic importance, control: 19-20]; StoetzDa1974 [taxonomy, life history: 138-140]; StoetzDa1974a [taxonomy, description, illustration, host, distribution, life history: 486-490]; Sulliv1930 [host, distribution: 51-59]; Takaha1952a [taxonomy: 15]; TippinBe1970 [host, distribution: 7]; Wilson1917 [taxonomy, description, host, distribution: 6,52]; Woglum1906 [taxonomy, description, host, distribution: 75]; York1905 [taxonomy, description, illustration, host, distribution: 325-326]; Zahrad1990a [host, distribution, description: 649].

Diaspidiotus armenicus (Borchsenius)

Aspidiotus armenicus Borchsenius, 1935: 132. Type data: ARMENIA: Erivanskii region, Erevan, on *Populus* sp. Lectotype female, by subsequent designation Danzig, 1993: 198. Type depository: St. Petersburg: (= Leningrad) Zoological Museum, Academy of Science, Russia.
Diaspidiotus armenicus; Borchsenius, 1949d: 240. Change of combination.
Quadraspidiotus armeniacus Borchsenius, 1949d: 240. Unjustified emendation.
Quadraspidiotus armeniacus Balachowsky, 1950b: 472. Unjustified emendation.
Quadraspidiotus armenicus; Balachowsky, 1950b: 472. Change of combination.
Diaspidiotus armenicus; Danzig & Pellizzari, 1998: 228. Revived combination.
SCALE COVER: Scale of adult female circular, slightly convex, yellowish-white in colour. Exuviae situated in the central part of scale, orange in colour. Diameter of scale 2.0-2.2 mm. Males unknown (Borchsenius, 1935).
HOST PLANTS: **Salicaceae**: *Populus* [Borchs1935, Borchs1936, Balach1950b], *Salix* [Balach1950b].
DISTRIBUTION: **Palaearctic**: Armenia [Borchs1935, Borchs1936, Balach1950b]; Iran [Balach1950b, Kaussa1955]; Turkey [Yasar1995a].

GENERAL: Description and illustration of adult female by Borchsenius (1935), Balachowsky (1950b), Danzig (1993) and by Yaşar (1995a).
KEYS: Danzig 1993: 179-182 (female) [Europe]; Balachowsky 1950b: 402 (female) [Mediterranean]; Borchsenius 1935: 127-128 (female) [Former USSR].
CITATIONS: Balach1950b [taxonomy, description, illustration, host, distribution: 472-475]; Borchs1935 [taxonomy, description, illustration, host, distribution: 132]; Borchs1936 [host, distribution: 131]; Borchs1937 [taxonomy, description, illustration, host, distribution: 128-129]; Borchs1939 [taxonomy, description, host, distribution: 10,39]; Borchs1949d [taxonomy, description, host, distribution: 240]; Borchs1950b [taxonomy, description, illustration, host, distribution: 225,227,230]; Borchs1966 [catalogue: 328]; Danzig1993 [taxonomy, description, illustration, host, distribution: 198-199]; DanzigPe1998 [catalogue: 228]; Ferris1941e [taxonomy: 41]; Kaussa1955 [host, distribution: 16]; MillerDa1990 [host, distribution, economic importance: 305]; TerGri1962 [taxonomy, host, distribution: 148]; Yasar1995a [taxonomy, description, illustration, host, distribution: 108-110].

Diaspidiotus arroyoi (Balachowsky)

Quadraspidiotus arroyoi Balachowsky, 1968: 75. Type data: CANARY ISLANDS: Tenerife, Las Canadas, 2100 meters altitude, near Pico de Teide, on *Spartocytisus nubigenus*. Holotype female. Type depository: Paris: Muséum national d'Histoire naturelle, France.
Diaspidiotus arroyoi; Danzig & Pellizzari, 1998: 228. Change of combination.
SCALE COVER: Female scale circular, highly convex; larval exuviae subcentral, red, slightly covered by wax secretion of adult; adult secretion white; 1-1.5 to 2 mm. Male scale unknown (Balachowsky, 1968).
HOST PLANTS: Leguminosae: *Spartocytisus nubigenus* [Balach1968].
DISTRIBUTION: Palaearctic: Canary Islands [Balach1968, MatileOr2001].
GENERAL: Description and illustration of adult female by Balachowsky (1968).
CITATIONS: Balach1968 [taxonomy, description, illustration, host, distribution: 75-79]; DanzigPe1998 [catalogue: 228]; MatileOr2001 [host, distribution: 190]; Muntin1969 [taxonomy: 139].

Diaspidiotus baiati (Kaussari)

Quadraspidiotus baiati Kaussari, 1958: 232. Type data: IRAN: near road to Shiraz-Kazeroun, altitude 2000 meters, on twigs of *Daphne* sp. Syntypes, female. Type depository: Paris: Muséum national d'Histoire naturelle, France.
Diaspidiotus baiati; Danzig & Pellizzari, 1998: 228. Change of combination.
SCALE COVER: Female scale circular, slightly convex; grey yellowish; thinner at margin; exuviae central or subcentral, brown, and covered with white, wax secretion; diameter 2.5-3 mm (Kaussari, 1958).
HOST PLANTS: Thymelaeaceae: *Daphne* [Kaussa1958].
DISTRIBUTION: Palaearctic: Iran [Kaussa1958].
GENERAL: Description and illustration of adult female by Kaussari (1958).
CITATIONS: Borchs1966 [catalogue: 328]; DanzigPe1998 [catalogue: 228]; Kaussa1958 [taxonomy, description, illustration, host, distribution: 232-234].

Diaspidiotus bavaricus (Lindinger)

Aspidiotus (*Diaspidiotus*) *bavaricus* Lindinger, 1912: 31. Type data: GERMANY: near Harburg, on *Calluna vulgaris* and *Erica tetralix*; Bayern, Hessen-Nassau, Steiermark and Norwegen on *Calluna vulgaris*. Syntypes, female. Type depository: Hamburg: Zoologisches Institut und Zoologishces Museum, Universität von Hamburg, Germany; type no. 116.

Aspidiotus bavaricus; Sasscer, 1912: 92. Change of combination.

Quadraspidiotus bavaricus; MacGillivray, 1921: 410. Change of combination.

Aspidiotus (*Euraspidiotus*) *bavaricus*; Thiem & Gerneck, 1934: 131. Change of combination.

Diaspidiotus bavaricus; Borchsenius, 1950b: 228. Change of combination.

SCALE COVER: Female scale small, convex, about 2 mm in diameter; black greyish; exuviae light brown, central or subcentral (Lindinger, 1912).

HOST PLANTS: **Ericaceae**: *Arbutus unedo* [Balach1950b], *Calluna vulgaris* [Lindin1912, Green1928, Balach1937c, Zahrad1952, Danzig1962, Pelliz1987, Foldi2000], *Erica* [Martin1983], *Erica ciliaris* [GomezM1957], *Er. cinerea* [Green1925b, Green1928, Balach1937c, Balach1950b], *Er. tetralix* [Lindin1912], *Er. umbellata* [GomezM1957, Martin1983], *Vaccinium baccata* [Balach1950b].

NATURAL ENEMIES: HYMENOPTERA **Encyrtidae**: *Epitetracnemus zetterstedtii* (Westwood) [Trjapi1989], *Metaphycus nadius* (Walker) [GuerriNo2000], *Trichomasthus bavarici* Hoffer [Trjapi1989].

DISTRIBUTION: **Palaearctic**: Belgium [Balach1950b]; Czech Republic [Zahrad1952, Zahrad1977]; Finland [Vikber1991]; France [Balach1937c, Balach1950b, Foldi2000]; Germany (United) [Lindin1912]; Italy [Balach1950b, Pelliz1987, LongoMaPe1995]; Netherlands [Balach1950b]; Poland [Kaweck1935, Kaweck1948, LagowsKo1996, Koteja2000a]; Russia (St. Petersberg (= Leningrad) Oblast [Danzig1962]); Spain [Martin1983, BlayGo1993]; Sweden [Gertss2001]; Switzerland [Balach1950b]; United Kingdom [Green1928] (Channel Islands [Green1925b]).

GENERAL: Description and illustration of adult female by Balachowsky (1950b), Zahradník (1952), Tereznikova (1986) and by Danzig (1993).

KEYS: Danzig 1993: 179-182 (female) [Europe]; Tereznikova 1986: 97-98 (female) [Ukraine]; Kosztarab & Kozár 1978: 155 (female) [Hungary]; Zahradník 1952: 143 (female) [Czech Republic]; Balachowsky 1950b: 493 (female) [Mediterranean].

CITATIONS: Alam1957 [biological control: 421-466]; Balach1932g [taxonomy, description, host, distribution: 103]; Balach1937c [host, distribution: 2]; Balach1950b [taxonomy, description, illustration, host, distribution: 500-503]; BlayGo1993 [taxonomy, description, illustration, host, distribution: 588-592]; Boraty1953 [taxonomy, description, illustration, host, distribution: 461-463]; Borchs1935a [taxonomy, description, host, distribution: 27]; Borchs1937a [taxonomy, host, distribution: 43-44]; Borchs1950b [taxonomy, description, illustration, host, distribution: 228,231]; Borchs1966 [catalogue: 328-329]; Danzig1962 [host, distribution: 22]; Danzig1964 [taxonomy, host, distribution: 653]; Danzig1993 [taxonomy, description, illustration, host, distribution: 201-202]; DanzigPe1998 [catalogue: 228-229]; Ferris1941e [taxonomy: 41]; Foldi2000 [host, distribution: 84]; Foldi2001 [distribution: 303-308]; Gertss2001 [distribution: 123-

130]; Ghauri1962 [taxonomy, description, host, distribution: 114,210]; GomezM1957 [taxonomy, description, illustration, host, distribution: 40-45]; GomezM1958a [host, distribution: 7]; Goux1951 [taxonomy, host, distribution: 14]; Green1925b [host, distribution: 518]; Green1928 [host, distribution: 8]; GuerriNo2000 [host, distribution, biological control: 157-159]; Jaap1914 [host, distribution: 135-142]; Kaweck1935 [host, distribution: 76]; Kaweck1948 [host, distribution: 5-6]; Killin1936 [host, distribution: 119]; KosztaKo1978 [taxonomy, description, host, distribution: 155]; Koteja2000a [distribution: 172]; LagowsKo1996 [host, distribution: 32]; Lindin1912 [taxonomy, description, host, distribution: 31]; Lindin1912b [taxonomy, description, host, distribution: 89]; Lindin1932g [taxonomy, host, distribution: 219]; Lindin1935 [taxonomy: 128]; Lindin1957 [taxonomy: 545-546]; LongoMaPe1995 [distribution: 126]; MacGil1921 [taxonomy, description, host, distribution: 410]; Martin1983 [taxonomy, host, distribution: 66]; Pelliz1987 [host, distribution: 122]; PolavaDaMi2000 [taxonomy: 558]; Reyne1957 [taxonomy, host, distribution: 25-26,33]; Sassce1912 [taxonomy, host, distribution: 92]; Schmut1959 [taxonomy, description, host, distribution: 45]; Terezn1986 [taxonomy, description, illustration, host, distribution: 98,100]; Theron1958 [taxonomy: 17,41,53]; ThiemGe1934 [taxonomy, description, host, distribution, life history: 537]; ThiemGe1934a [taxonomy, description, host, distribution: 130-158,208-238]; Trjapi1989 [biological control: 187,292]; Vikber1991 [host, distribution: 4]; WeidneWa1968 [taxonomy: 171]; Zahrad1951 [taxonomy: 198]; Zahrad1952 [taxonomy, description, illustration, host, distribution: 143-147]; Zahrad1977 [taxonomy, distribution: 120].

Diaspidiotus botanicus (Gómez-Menor Ortega)

Aspidiotus botanicus Gómez-Menor Ortega, 1927a: 295. Type data: SPAIN: Madrid, Botanic Garden, on *Buxus sempervirens*. Lectotype female, by subsequent designation Blay Goicoechea, 1993: 605. Type depository: Madrid: Museo Nacional de Ciencias Naturales, Spain.

Diaspidiotus botanicus; Balachowsky, 1950b: 506. Change of combination.

SCALE COVER: Female scale more or less circular, diameter 1.3-1.5 mm; convex; exuviae central, yellow gold. Male scale elliptical, slightly convex; anterior part wider than posterior; exuviae yellow, subcentral, rest of scale grey, white at margin; ventral scale white (Gómez-Menor Ortega, 1927a).

HOST PLANTS: **Buxaceae**: *Buxus sempervirens* [GomezM1927a, Balach1935b, GomezM1946, Martin1983, BlayGo1993].

DISTRIBUTION: **Palaearctic**: Spain [GomezM1927a, Balach1935b, GomezM1946, Martin1983, BlayGo1993].

GENERAL: Description and illustration of adult female by Gómez-Menor Ortega (1927a), (Balachowsky (1950b) and by Blay Goicoechea (1993).

KEYS: Balachowsky 1950b: 493 (female) [Mediterranean].

CITATIONS: Balach1935b [host, distribution: 257]; Balach1950b [taxonomy, description, illustration, host, distribution: 506-509]; BlayGo1993 [taxonomy, description, illustration, host, distribution: 605-609]; Borchs1966 [catalogue: 322]; DanzigPe1998 [catalogue: 229]; Ferris1941e [taxonomy: 41]; GomezM1927a [taxonomy, description, host, distribution: 295-298]; GomezM1937 [taxonomy, description, illustration, host, distribution: 67-70]; GomezM1946 [host, distribution: 61]; GomezM1958a [host, distribution: 7]; Lindin1957 [taxonomy: 545]; Martin1983 [taxonomy, host, distribution: 63].

Diaspidiotus braunschvigi (Rungs)

Aspidiotus braunschvigi Rungs, 1937: 329. Type data: MOROCCO: East Morocco, Oudja, on leaves of *Pistacia atlantica*; collected by Coinder, 18 August, 1936. Syntypes, female. Type depository: Paris: Muséum national d'Histoire naturelle, France.
Aspidaspis braunschvigi; Ferris, 1942: 426. Change of combination.
Quadraspidiotus braunschvigi; Balachowsky, 1950b: 421. Change of combination.
Aspidiotus braunschwigi; Lindinger, 1957: 545. Misspelling of species name.
Diaspidiotus braunschvigi; Danzig & Pellizzari, 1998: 229. Change of combination.
COMMON NAME: braunschvigi scale [McKenz1956].
SCALE COVER: Scale of female flat, more or less circular, translucent, frequently appearing like oiled paper; dirty grey or grey yellow. Exuviae red or brown, frequently eccentric. Scale of male smaller, more elongate, clearer, sometimes whitish (Rungs, 1937).
HOST PLANTS: **Moraceae**: *Ficus* [Ferris1942], *Ficus carica* [McKenz1956]. **Pistaciaceae**: *Pistacia atlantica* [Rungs1937, Balach1950b, McKenz1956].
DISTRIBUTION: **Nearctic**: United States of America (California [Ferris1942, McKenz1956]). **Palaearctic**: Morocco [Rungs1937, Ferris1942, Balach1950b].
ECONOMIC IMPORTANCE: Apparently imported from the Mediterranean onto D'Orrigo Ranch in San Jose, Santa Clara County, California, USA in 1939 where it was collected on common fig. Probably no longer present in California (Gill, 1997).
GENERAL: Description and illustration of adult female by Ferris (1942), Balachowsky (1950b), McKenzie (1956) and by Gill (1997).
KEYS: Gill 1997: 58-59 (female) [Species of California]; McKenzie 1956: 24 (female) [U.S.A.: California]; Balachowsky 1950b: 400 (female) [Mediterranean]; Ferris 1942: 30 (female) [North America].
CITATIONS: Balach1950b [taxonomy, description, illustration, host, distribution: 421-424]; Borchs1966 [catalogue: 329]; DanzigPe1998 [catalogue: 229]; Ferris1941e [taxonomy: 41]; Ferris1942 [taxonomy, description, illustration, host, distribution: 426;446:30]; Gill1997 [taxonomy, description, illustration, host, distribution: 59,61]; Lindin1957 [taxonomy: 545]; McKenz1956 [taxonomy, description, illustration, host, distribution: 24,44-45]; Nakaha1982 [host, distribution: 77]; Rungs1937 [taxonomy, description, illustration, host, distribution: 329-332].

Diaspidiotus bumeliae Ferris

Diaspidiotus bumeliae Ferris, 1938a: 217. Type data: U.S.A.: Texas, Corsicana, on *Bumelia lanuginosa*. Holotype female. Type depository: Davis: The Bohart Museum of Entomology, University of California, California, USA.
SCALE COVER: Female scale dark grey or brown, circular, highly convex, exuviae toward one side giving scale a tipped-over appearance; Male scale not identified (Ferris, 1938a).

HOST PLANTS: **Moraceae**: *Maclura aurantiaca* [Ferris1938a], *Maclura pomifera* [McDani1969]. **Rosaceae**: *Crataegus* [Ferris1938a, McDani1969]. **Sapotaceae**: *Bumelia lanuginosa* [Ferris1938a, McDani1969].

DISTRIBUTION: **Nearctic**: United States of America (Kansas [Nakaha1982], Oklahoma [Nakaha1982], Texas [Ferris1938a, McDani1969]).

BIOLOGY: Occurring on twigs (Ferris, 1938a).

GENERAL: Description and illustration of adult female by Ferris (1938a).

KEYS: McDaniel 1969: 90-91 (female) [U.S.A.: Texas]; Ferris 1942: 33 (female) [North America].

CITATIONS: Borchs1966 [catalogue: 322]; Ferris1938a [taxonomy, description, illustration, host, distribution: 217]; Ferris1942 [taxonomy: 446:33]; McDani1969 [taxonomy, illustration, host, distribution: 93-94]; Nakaha1982 [host, distribution: 28].

Diaspidiotus buxii nomen nudum

Diaspidiotus buxii Hadzibejli, 1957: 102. Nomen nudum.

Diaspidiotus buxii Borchsenius, 1966: 377. Nomen nudum.

Diaspidiotus caryae Kosztarab

Diaspidiotus caryae Kosztarab, 1963: 25. Type data: U.S.A.: Ohio, on barks of twigs of "hickory" *Carya* sp. Syntypes, female. Type depository: Washington: United States National Entomological Collection, U.S. National Museum of Natural History, District of Columbia, USA.

COMMON NAME: Hicory Scale [Koszta1996].

SCALE COVER: Scale of female circular, flat, grey, darker in center; about 2.0 mm in diameter; exuviae subcentral (Kosztarab, 1963).

HOST PLANTS: **Juglandaceae**: *Carya* [Koszta1963].

DISTRIBUTION: **Nearctic**: United States of America (Georgia [Nakaha1982], Ohio [Koszta1963]).

GENERAL: Description and illustration of adult female by Kosztarab (1963, 1996).

KEYS: Kosztarab 1996: 484-485 (female) [Northeastern North America].

CITATIONS: Borchs1966 [catalogue: 322-323]; Koszta1963 [taxonomy, description, illustration, host, distribution: 27-29]; Koszta1996 [taxonomy, description, illustration, host, distribution, life history: 489-490]; Nakaha1982 [host, distribution: 28].

Diaspidiotus caucasicus (Borchsenius)

Aspidiotus caucasicus Borchsenius, 1935: 130. Type data: GEORGIA: Eastern Georgia, Rodnikovskaya station (Azovo-Chernomorskii kray), on *Populus* sp. Lectotype female, by subsequent designation Danzig, 1993: 202. Type depository: St. Petersburg: (= Leningrad) Zoological Museum, Academy of Science, Russia.

Diaspidiotus caucasicus; Borchsenius, 1949b: 245. Change of combination.

Aspidiotus caucasicus; Lindinger, 1957: 545. Revived combination.

Aspidiotus caucasius; Lindinger, 1957: 545. Misspelling of species name.

Diaspidiotus caucasicus; Lindinger, 1957: 548. Incorrect synonymy. Notes: Incorrect synonymy with *Diaspidiotus lenticularis*

Diaspidiotus caucasicus; Borchsenius, 1966: 323. Revived combination.
SCALE COVER: Female scale circular, diameter 1.5 mm; slightly convex. Exuviae eccentric, situated close to anterior margin. First exuviae brown, second brownish; scale in central part over exuviae greyish-green, other part white. Male scale of form common to this genus, greyish-green , length about 0.9 mm (Borchsenius, 1935).
HOST PLANTS: Salicaceae: *Populus* [Borchs1935, Borchs1936, Balach1950b].
DISTRIBUTION: Palaearctic: Georgia (Georgia [Borchs1935, Borchs1936, Balach1950b]); Russia (Caucasus [Borchs1936]); Turkey [Yasar1995a].
ECONOMIC IMPORTANCE: Schmutterer et al., (1957) noted that this species occasionally is a minor pest of *Populus*.
GENERAL: Description and illustration of adult female by Borchsenius (1935), Balachowsky (1950b), Tereznikova (1986), Danzig (1993) and by Yaşar (1995a).
KEYS: Danzig 1993: 179-182 (female) [Europe]; Tereznikova 1986: 97-98 (female) [Ukraine]; Balachowsky 1950b: 494 (female) [Mediterranean]; Borchsenius 1935: 127-128 (female) [Former USSR].
CITATIONS: Balach1950b [taxonomy, description, illustration, host, distribution: 509-512]; Borchs1935 [taxonomy, description, illustration, host, distribution: 128,130-131]; Borchs1936 [host, distribution: 131]; Borchs1937 [taxonomy, description, illustration, host, distribution: 132-133]; Borchs1939 [taxonomy, description, host, distribution: 133]; Borchs1949d [taxonomy, description, host, distribution: 245]; Borchs1950b [taxonomy, description, illustration, host, distribution: 227-228]; Borchs1966 [catalogue: 323]; Danzig1964 [host, distribution: 653]; Danzig1993 [taxonomy, description, illustration, host, distribution, economic importance: 202-203]; DanzigPe1998 [catalogue: 229]; Ferris1941e [taxonomy: 41]; KaussaBa1953 [taxonomy: 26]; Lindin1957 [taxonomy: 545]; MillerDa1990 [host, distribution, economic importance: 301]; RzaevaYa1985 [biological control: 55-58]; SchmutKlLu1957 [host, distribution, economic importance: 491]; Terezn1986 [taxonomy, description, illustration, host, distribution: 99-101]; Yasar1995a [taxonomy, description, illustration, host, distribution: 67-69].

Diaspidiotus cecconii (Leonardi)
Hemiberlesia cecconii Leonardi, 1908a: 188. Type data: ITALY: Sardinia, Aggius, on *Osyris alba*. Syntypes, female. Type depository: Portici: Dipartimento de Entomologia e Zoologia Agraria di Portici, Università di Napoli Federico II, Italy.
Aspidiotus trabuti; Sanders, 1909: 53. Change of combination.
Aspidiotus cecconii; Sanders, 1909a: 51. Change of combination.
Aspidiotus (*Hemiberlesia*) *trabuti* Marchal, 1909b: 59. Type data: ALGERIA: Oran, on *Ephedra altissima*. Syntypes, female. Type depository: Paris: Muséum national d'Histoire naturelle, France. Synonymy by Borchsenius, 1966: 329.
Hemiberlesia trabuti; Malenotti, 1916: 312. Change of combination.
Hemiberlesia cecconii; Leonardi, 1920: 97. Revived combination.
Neosignoretia cecconii; MacGillivray, 1921: 425. Change of combination.
Aspidiotus herzlianus Bodenheimer, 1924: 30. Type data: ISRAEL: Jordan Valley, Ghor, on *Asparagus aphyllus*; collected by F.S. Bodenheimer. Syntypes, female. Type depository: Bet Dagan: Department of Entomology, The Volcani Center, Israel. Synonymy by Lindinger, 1957: 545.

Hemiberlesea trabuti; Gómez-Menor Ortega, 1937: 114. Misspelling of genus name.
Hemiberlesia jourdani Rungs, 1939: 231. Type data: MOROCCO: Ljoukak (Grand-Atlas), altitude 1200 meters, on *Asparagus stipularis*; collected M.L. Jourdan, September 1937. Syntypes, female. Type depository: Paris: Muséum national d'Histoire naturelle, France. Synonymy by Balachowsky, 1950b: 449.

Archaspis ephedrae Bodenheimer, 1943: 26. Type data: IRAQ: Shuatra, on *Ephedra alte*, 11.X.1942. Lectotype first-instar nymph, by subsequent designation Ben-Dov, 1980: 264. Type depository: Bet Dagan: Department of Entomology, The Volcani Center, Israel. Synonymy by Ben-Dov, 1980: 264.

Diaspidiotus herzlianus; Bodenheimer, 1943: 4. Change of combination.

Quadraspidiotus jourdani; Rungs, 1948: 110. Change of combination.

Quadraspidiotus cecconii; Balachowsky, 1950b: 449. Change of combination.

Quadraspidiotus ceconi; Gómez-Menor Ortega, 1965: 90. Misspelling of species name.

Diaspidiotus cecconii; Danzig & Pellizzari, 1998: 229. Change of combination.

SCALE COVER: Female scale circular, diameter 1.5 mm; highly convex; robust; exuviae small, subcentral, yellow; secreted part brown; ventral vellum white, remains on host plant after removal of insect (Leonardi, 1908a).

HOST PLANTS: **Crassulaceae**: *Sedum* [GomezM1948, GomezM1958c], *Sedum album* [Balach1932d], *Sedum altissimum* [Martin1983]. **Ephedraceae**: *Ephedra* [Bodenh1937, Bodenh1943, Bachma1953], *Ephedra alte* [BenDov1980], *Ephedra altissima* [Marcha1909b, Sander1909a, Balach1927, Balach1932d], *Ephedra campylopoda* [Hall1927], *Ephedra fragilis* [Rungs1948], *Ephedra nabrodensis* [Leonar1918, Leonar1920]. **Leguminosae**: *Retama sphaerocarpa* [Martin1983], *Ulex* [GomezM1937, GomezM1946, Martin1983]. **Liliaceae**: *Asparagus* [Bodenh1937], *Asparagus aphyllus* [Bodenh1924, Balach1950b], *Asparagus horridus* [Balach1932d, Martin1983], *Asparagus stipularis* [Rungs1939, Rungs1948]. **Pistaciaceae**: *Pistacia lentiscus* [InserrCa1987]. **Santalaceae**: *Osyris alba* [Leonar1908a, Leonar1920, Balach1932e, Bachma1953]. **Umbelliferae**: *Pituranthos tortuosus* [Balach1950b].

NATURAL ENEMIES: HYMENOPTERA **Encyrtidae**: *Metaphycus ibericus* (Mercet) [GuerriNo2000].

DISTRIBUTION: **Palaearctic**: Algeria [Marcha1909b, Sander1909a, Balach1927, Balach1932d]; Egypt [Hall1926a, Balach1950b]; France [Balach1932d, Balach1932e]; Greece [Korone1934]; Iran [Kaussa1955]; Israel [Bodenh1924, Bodenh1937]; Italy [Leonar1918, Balach1950b, LongoMaPe1995]; Morocco [Rungs1935, Rungs1939, Rungs1948]; Portugal [Seabra1942]; Sardinia [Leonar1908a, Leonar1920]; Sicily [Leonar1920, InserrCa1987]; Spain [GomezM1937, GomezM1946, GomezM1948, Martin1983, BlayGo1993]; Yugoslavia [Bachma1953].

GENERAL: Description and illustration of adult female by Leonardi (1908a), Hall (1926a), Rungs (1939) and by Balachowsky (1950b).

KEYS: Lupo 1953a: 75-76 (female) [Italy]; Balachowsky 1950b: 404 (female) [Mediterranean]; Balachowsky 1928a: 132 (female) [North Africa]; Leonardi 1920: 90 (female) [Italy].

CITATIONS: Bachma1953 [host, distribution: 178]; Balach1927 [host, distribution: 177]; Balach1928a [taxonomy: 132]; Balach1932d [taxonomy, host, distribution: VI; XLVII]; Balach1950b [taxonomy, description, illustration, host, distribution: 449-453]; BenDov1980 [taxonomy: 264]; BlayGo1993 [taxonomy, description, illustration, host, distribution: 561-566]; Bodenh1924 [taxonomy, description, illustration, host, distribution: 30-32]; Bodenh1935 [host, distribution: 246]; Bodenh1935c [taxonomy, distribution: 1155]; Bodenh1937 [host, distribution: 217]; Bodenh1943 [taxonomy, description, illustration, host, distribution: 4,26]; Bodenh1944 [taxonomy: 5]; Borchs1966 [catalogue: 329,368]; DanzigPe1998 [catalogue: 229-230]; Ferris1941e [taxonomy: 41,44,49]; Foldi2001 [distribution: 303-308]; GomezM1937 [taxonomy, description, illustration, host, distribution: 114-116]; GomezM1946 [host, distribution: 66]; GomezM1948 [host, distribution: 74]; GomezM1957 [host, distribution: 48]; GomezM1958a [host, distribution: 7]; GomezM1958c [host, distribution: 406]; GomezM1960O [host, distribution, illustration: 168-169]; GomezM1965 [taxonomy, host, distribution: 90]; GomezM1968 [host, distribution: 541]; GuerriNo2000 [host, distribution, biological control: 162]; Hall1926a [taxonomy, description, illustration, host, distribution: 21-22]; Hall1927 [host, distribution: 107-108]; Hall1927b [taxonomy, description, host, distribution: 144-145]; InserrCa1987 [host, distribution: 94]; Kaussa1955 [host, distribution: 16]; Korone1934 [taxonomy, description, illustration, host, distribution: 15-17]; Leonar1908a [taxonomy, description, illustration, host, distribution: 188-199]; Leonar1918 [host, distribution: 192]; Leonar1920 [taxonomy, description, illustration, host, distribution: 90,97-100]; Lindin1912b [taxonomy, description, host, distribution: 139,237-238]; Lindin1957 [taxonomy: 545]; LongoMaPe1995 [distribution: 128]; Lupo1953a [taxonomy, description, illustration, host, distribution: 76,86-91]; MacGil1921 [taxonomy, description, host, distribution: 425,437]; Maleno1916 [taxonomy, description, illustration, host, distribution: 312-313]; Marcha1909b [taxonomy, description, host, distribution: 59]; Martin1983 [taxonomy, host, distribution: 66]; Rungs1935 [host, distribution: 270-271]; Rungs1939 [taxonomy, description, illustration, host, distribution: 231-233]; Rungs1948 [host, distribution: 110]; Sander1909a [taxonomy, host, distribution: 51,53]; Seabra1942 [distribution: 2].

Diaspidiotus centrafricanus (Balachowsky & Ferrero), new combination
Quadraspidiotus centrafricanus Balachowsky & Ferrero, 1967c: 218. Type data: CENTRAL AFRICAN REPUBLIC: Savane de Bebe, near Maboke, on Gramineae; collected 29 December 1965. Holotype female. Type depository: Paris: Muséum national d'Histoire naturelle, France; type no. 3081.
SCALE COVER: Illustration of scale cover by Balachowsky & Matile-Ferrero (1967c). Female scale subcircular, small size, 1.2-1.3 mm in diameter; colour grey; exuviae central or slightly eccentric, brown acajou in colour. Male scale unknown (Balachowsky & Matile-Ferrero, 1967c).
HOST PLANTS: **Gramineae** [BalachFe1967c].
DISTRIBUTION: **Afrotropical**: Central African Republic [BalachFe1967c].
GENERAL: Description and illustration of adult female by Balachowsky & Ferrero (1967c).

CITATIONS: BalachFe1967c [taxonomy, description, illustration, host, distribution: 218-220]; Muntin1969 [taxonomy: 139].

Diaspidiotus coniferarum (Cockerell)

Aspidiotus (*Diaspidiotus*) *coniferarum* Cockerell, 1898e: 201. Type data: U.S.A.: New Mexico, Organ Mts., on a small pine tree, (*Pinus ponderosa* var. *scopulorum*). Syntypes, female. Type depository: Washington: United States National Entomological Collection, U.S. National Museum of Natural History, District of Columbia, USA.

Aspidiotus conirerarum; Newell, 1899: 21. Misspelling of genus name.

Hemiberlesia coniferarum; Leonardi, 1900: 338. Change of combination.

Diaspidiotus coniferarum; Cockerell, 1905b: 202. Change of combination.

Comstockaspis coniferarum; MacGillivray, 1921: 439. Change of combination.

Diaspidiotus conferarum; McKenzie, 1956: 62. Misspelling of species name.

Diaspidiotus coniferarum; Borchsenius, 1966: 323. Revived combination.

COMMON NAME: conifer scale [McKenz1956, Dekle1965c].

SCALE COVER: Female scale whitish or grey, usually nearly bark colour, subcircular, flat; exuviae subcentral. Male scale not observed (Ferris, 1938a). Colour photograph by Gill (1997).

HOST PLANTS: **Cupressaceae**: *Cupressus* [Ferris1938a], *Cu. guadelupensis* [Ferris1920b], *Cu. macrocarpa* [McKenz1956], *Cu. sargenti* [Ferris1938a, McKenz1956], *Juniperus* [Ferris1938a, Koszta1996], *Ju. monosperma* [Ferris1938a, McKenz1956, McDani1969], *Ju. occidentalis* [McKenz1956], *Ju. pachyphloea* [Ferris1938a, McKenz1956, McDani1969], *Ju. virginiana* [Ferris1938a, McKenz1956, Dekle1965c, TippinBe1970, BesheaTiHo1973], *Libocedrus* [Koszta1996], *Libocedrus decurrens* [Ferris1920b, McKenz1956]. **Pinaceae**: *Abies* [Koszta1996], *Pinus* [Ferris1920b, Koszta1996], *Pi. ponderosa scopulorum* [Cocker1898e, Ferris1938a, McKenz1956], *Tsuga canadensis* [BesheaTiHo1973].

DISTRIBUTION: **Nearctic**: Mexico (Baja California [Ferris1938a]); United States of America (Arizona [Nakaha1982], California [Ferris1920b, McKenz1956], Colorado [Nakaha1982], Florida [Dekle1965c], Georgia [TippinBe1970, BesheaTiHo1973], Indiana [Nakaha1982], Kansas [Nakaha1982], Louisiana [Nakaha1982], Mississippi [Ferris1938a], New Mexico [Cocker1898e, Leonar1900, Ferris1920b], Oregon [Nakaha1982], South Carolina [Nakaha1982], Texas [Ferris1938a, McDani1969], Utah [Ferris1938a], Virginia [Nakaha1982]).

BIOLOGY: Occurring on bark of smaller limbs, never on foliage (Ferris, 1938a).

GENERAL: Description and illustration of adult female by Ferris (1920b, 1938a), McKenzie (1956) and by Gill (1997).

KEYS: Gill 1997: 113 (female) [Species of California]; Kosztarab 1996: 484-485 (female) [Northeastern North America]; McDaniel 1969: 90-91 (female) [U.S.A.: Texas]; McKenzie 1956: 25 (female) [U.S.A.: California]; Ferris 1942: 33 (female) [North America]; Cockerell 1905b: 202 (female) [U.S.A.: Colorado]; Newell 1899: 4-5 (female) [North America].

CITATIONS: BesheaTiHo1973 [host, distribution: 6]; Borchs1966 [catalogue: 323]; Cocker1898e [taxonomy, description, host, distribution: 201]; Cocker1899a [taxonomy: 396]; Cocker1905b [taxonomy: 202]; CockerPa1899 [taxonomy, description, host, distribution: 284]; Dekle1965c [taxonomy, description, host, distribution: 49]; Dekle1976 [taxonomy, description, host, distribution, economic importance: 68]; FDACSB1982 [host, distribution: 5-11]; Fernal1903b [catalogue: 255]; Ferris1920b [taxonomy, description, illustration, host, distribution: 49-50]; Ferris1938a [taxonomy, description, illustration, host, distribution: 218]; Ferris1941e [taxonomy: 42]; Ferris1942 [taxonomy: 446:33]; Gill1997 [host, distribution, taxonomy, description, illustration, economic importance: 115,119]; Koszta1996 [taxonomy, description, illustration, host, distribution, life history: 490-491]; Leonar1900 [taxonomy, host, distribution: 338]; MacGil1921 [taxonomy, description, host, distribution: 439]; McDani1969 [taxonomy, illustration, host, distribution: 93,95]; McKenz1956 [taxonomy, description, illustration, host, distribution: 60,62]; Nakaha1982 [host, distribution: 28]; Newell1899 [taxonomy, description, host, distribution: 21]; Takagi1958 [taxonomy: 124]; TippinBe1970 [host, distribution: 8]; Willia1985a [taxonomy: 233]; Zahrad1990 [host, distribution, description: 643].

Diaspidiotus convexus Goux

Diaspidiotus convexus Goux, 1951: 10. Type data: FRANCE: Bouches-du-Rhône, near Aix-en-Provence, on *Lithospermum fruticosum*. Holotype female. Type depository: Paris: Muséum national d'Histoire naturelle, France.
SCALE COVER: Female scale oval, 1.2 mm long, 0.8 mm wide; dark grey-yellow; convex, area containing exuviae very prominent; exuviae subcentral (Goux, 1951).
HOST PLANTS: **Boraginaceae**: *Lithospermum fruticosum* [Goux1951].
DISTRIBUTION: **Palaearctic**: France [Goux1951].
GENERAL: Description and illustration of adult female by Goux (1951).
CITATIONS: Borchs1966 [catalogue: 323]; DanzigPe1998 [catalogue: 230]; Foldi2001 [distribution: 303-308]; Goux1951 [taxonomy, description, illustration, host, distribution: 10-13].

Diaspidiotus cotoneastri (Takagi), new combination

Quadraspidiotus cotoneastri Takagi, 1975: 11. Type data: NEPAL: Tukucha, altitude 2600 meters, on the branches of *Cotoneaster* sp. Holotype female. Type depository: Sapporo: Entomological Institute, Faculty of Agriculture, Hokkaido University, Japan.
SCALE COVER: Takagi (1975) did not describe scale cover.
HOST PLANTS: **Rosaceae**: *Cotoneaster* [Takagi1975].
DISTRIBUTION: **Oriental**: Nepal [Takagi1975].
GENERAL: Description and illustration of adult female by Takagi (1975).
CITATIONS: Takagi1975 [taxonomy, description, illustration, host, distribution: 11-13].

Diaspidiotus crescentiae Ferris

Diaspidiotus crescentiae Ferris, 1938a: 219. Type data: MEXICO: near Mazatlan, Hacienda de Barron, on *Crescentia alata* ("tecomate"); collected Ferris, 1925. Holotype female. Type depository: Davis: The Bohart Museum of Entomology, University of California, California, USA.

SCALE COVER: Scale of female circular, rather highly convex, white, composed of brittle and rather crystalline wax; exuviae central, ventral scale strongly developed. Scale of male elongate, white, exuvia at one end (Ferris, 1938a).

HOST PLANTS: **Agavaceae**: *Alibertia* [Ferris1942]. **Bignoniaceae**: *Crescentia alata* [Ferris1938a].

DISTRIBUTION: **Nearctic**: Mexico [Ferris1938a]. **Neotropical**: Panama [Ferris1942].

BIOLOGY: Occurring on bark (Ferris, 1938a).

GENERAL: Description and illustration of adult female by Ferris (1938a).

KEYS: Ferris 1942: 33 (female) [North America].

CITATIONS: Borchs1966 [catalogue: 323]; Ferris1938a [taxonomy, description, illustration, host, distribution: 219]; Ferris1942 [host, distribution: 445:10; 446:33].

Diaspidiotus cryptoxanthus (Cockerell)

Aspidiotus (*Diaspidiotus*) *cryptoxanthus* Cockerell, 1900f: 71. Type data: JAPAN: on *Quercus glandulifera*. Syntypes, female. Type depository: Washington, D.C.: U. S. Entomological Collection, U.S. National Museum of Natural History, USA.

Aspidiotus cryptoxanthus; Fernald, 1903b: 255. Change of combination.

Aspidiotus crytoxanthus; Kuwana, 1917a: 176. Misspelling of species name.

Quadraspidiotus cryptoxanthus; MacGillivray, 1921: 411. Change of combination.

Diaspidiotus cryptoxanthus; Danzig & Pellizzari, 1998: 230. Change of combination.

SCALE COVER: Female scale on bark of twigs, almost invisible; colour similar to bark; about 1.2 mm diameter; circular to suboval; often massed; slightly convex; exuviae orange-red, very conspicuous when exposed by rubbing; young scales with a dot and ring; scales removed from bark leave a whitish patch (Cockerell, 1900f).

HOST PLANTS: **Fagaceae**: *Castanea crenata* [Takagi1958], *Quercus* [Kawai1977], *Qu. glandulifera* [Cocker1900f], *Qu. serrata* [Kuwana1933, Takagi1958, Takagi1974].

DISTRIBUTION: **Palaearctic**: Japan [Cocker1900f, Kuwana1917a, Kawai1980].

GENERAL: Description and illustration of adult female by Kuwana (1933) and by Takagi (1958, 1974).

KEYS: Chou 1985: 311 (female) [Species of China]; Kuwana 1933: 3 (female) [Japan]; Kuwana 1933b: 49 (female) [Japan].

CITATIONS: Borchs1966 [catalogue: 329]; Chou1985 [taxonomy, description, host, distribution: 314-315]; Cocker1900f [taxonomy, description, illustration, host, distribution: 71]; DanzigPe1998 [catalogue: 230]; Fernal1903b [catalogue: 255]; Ferris1941e [taxonomy: 42]; HorticInNo1923 [host, distribution: 236-239]; HorticInNo1923a [host, distribution: 237-239]; Kawai1977 [host, distribution, economic importance: 158]; Kawai1980 [taxonomy, description, host, distribution: 219]; Kuwana1917a [taxonomy, distribution: 174]; Kuwana1933 [taxonomy, description, illustration, host, distribution: 6-7]; MacGil1921 [taxonomy, description, host, distribution: 411]; Muraka1970 [host, distribution: 77]; ShiLi1991 [host, distribution: 166]; Takagi1958 [taxonomy, description, illustration, host, distribution: 126-127]; Takagi1974 [taxonomy, description, illustration, host, distribution: 4-9]; Tao1999 [taxonomy, host, distribution: 114].

Diaspidiotus cryptus (Ferris)

Quadraspidiotus cryptus Ferris, 1953: 67. Type data: CHINA: Yunnan Province, in compound of the American Consulate at Kunming, on a species of *Juniperus*; collected by G.F. Ferris, April 23, 1949. Syntypes, female. Type depository: Davis: The Bohart Museum of Entomology, University of California, California, USA.

Diaspidiotus cryptus; Danzig & Pellizzari, 1998: 230. Change of combination.

SCALE COVER: Scale of female quite flat, roughly circular and white or grey. Scale of male slightly elongate and darker than female (Ferris, 1953).

HOST PLANTS: **Cupressaceae**: *Juniperus* [Ferris1953].

DISTRIBUTION: **Oriental**: China (People's Republic) (Yunnan [Ferris1953]).

BIOLOGY: Occurring mostly on the inner face of the more or less closely oppressed needles of the host, especially near the tips of the twigs (Ferris, 1953).

GENERAL: Description and illustration of adult female by Ferris (1953) and by Chou (1985, 1986).

KEYS: Chou 1985: 311 (female) [Species of China].

CITATIONS: Borchs1966 [catalogue: 329]; Chou1985 [taxonomy, description, host, distribution: 315-316]; Chou1986 [taxonomy, illustration: 692]; DanzigPe1998 [catalogue: 230]; Ferris1953 [taxonomy, description, illustration, host, distribution: 67]; Tao1999 [taxonomy, host, distribution: 114].

Diaspidiotus crystallinus Ferris

Diaspidiotus crystallinus Ferris, 1938a: 220. Type data: U.S.A.: Texas, Woodville, on "ironwood" [probably either *Olneya tesota* or *Acacia flexicaulis*, both of which trees are commonly known as ironwood]. Holotype female. Type depository: Davis: The Bohart Museum of Entomology, University of California, California, USA.

SCALE COVER: Scales white, composed of thick and crystalline-appearing wax, that of female elongate oval, with exuviae near one end and with a quite thick ventral scale; that of male similar in form and with exuvia apical (Ferris, 1938a).

HOST PLANTS: **Leguminosae**: *Acacia flexicaulis* [Ferris1938a], *Olneya tesota* [Ferris1938a].

DISTRIBUTION: **Nearctic**: United States of America (Texas [Ferris1938a, McDani1969]).

BIOLOGY: Occurring in cracks in bark (Ferris, 1938a).

GENERAL: Description and illustration of adult female by Ferris (1938a).

KEYS: McDaniel 1969: 90-91 (female) [U.S.A.: Texas]; Ferris 1942: 33 (female) [North America].

CITATIONS: Borchs1966 [catalogue: 323]; Ferris1938a [taxonomy, description, illustration, host, distribution: 220]; Ferris1942 [taxonomy: 446:33]; McDani1969 [taxonomy, illustration, host, distribution: 94-96]; Nakaha1982 [host, distribution: 28].

Diaspidiotus danzigae Kuznetsov

Diaspidiotus danzigae Kuznetsov, 1976: 77. Type data: UKRAINE: Crimea, Ai-Petri plateau, altitude 1400 meters, on *Juniperus depressa*. Holotype female. Type depository: St. Petersburg: (= Leningrad) Zoological Museum, Academy of Science, Russia.

SCALE COVER: Female scale circular, 1.6 mm in diameter; convex; yellow brown, white in center of scale. Male scale similar in shape and colour to that of female (Kuznetsov, 1976).
HOST PLANTS: **Cupressaceae**: *Juniperus depressa* [Kuznet1976].
DISTRIBUTION: **Palaearctic**: Russia (Karachay-Cherkessia AR [Danzig1985]); Ukraine (Krym (= Crimea) Oblast [Kuznet1976]).
GENERAL: Description and illustration of adult female by Kuznetsov (1976) and by Danzig (1993).
KEYS: Danzig 1993: 179-182 (female) [Europe].
CITATIONS: Danzig1985 [distribution: 112]; Danzig1993 [taxonomy, description, illustration, host, distribution: 195-196]; DanzigPe1998 [catalogue: 230]; Kuznet1976 [taxonomy, description, illustration, host, distribution: 77-79].

Diaspidiotus distinctus (Leonardi)
Targionia distincta Leonardi, 1900: 305. Type data: ITALY: on *Quercus robur*. Syntypes, female. Type depository: Portici: Dipartimento de Entomologia e Zoologia Agraria di Portici, Università di Napoli Federico II, Italy.
Diaspidiotus distinctus; Balachowsky, 1948: 14. Change of combination requiring emendation of species name for agreement in gender.
Chorizaspidiotus distinctus; Lupo, 1954: 29. Change of combination.
Diaspidiotus slovenicus Bachmann, 1956: 99. Type data: YUGOSLAVIA: Opatje village, on *Quercus pubescens*. Holotype female. Synonymy by Danzig, 1993: 211. Notes: Depository of type material unknown.
Diaspidiotus slavenicus; Borchsenius, 1966: 326. Misspelling of species name.
Diaspidiotus alni; Tereznikova, 1986: 211. Misidentification; discovered by Danzig, 1993: 211.
Diaspidiotus distinctus; Gómez-Menor Ortega, 1960: 160. Revived combination.
SCALE COVER: Female scale circular, convex; about 1 mm in diameter; white-grey or yellow-brown; underneath flakes of bark; exuviae subcentral; dark red; ventral scale delicate, white, attached to bark (Leonardi, 1920).
HOST PLANTS: **Compositae**: *Matricaria officinalis* [GomezM1960O, Martin1983, BlayGo1993]. **Corylaceae**: *Corylus avellana* [Zahrad1972]. **Fagaceae**: *Quercus* [PellizCa1991a], *Qu. cerris* [Zahrad1972], *Qu. coccifera* [Bodenh1924], *Qu. lusitanica* [Bodenh1924, Balach1932d, Balach1950b], *Qu. pubescens* [BachmaGe1950, Bachma1956, Zahrad1972], *Qu. ruber* [Leonar1900, Leonar1920, Zahrad1972]. **Leguminosae**: *Gonocytisus* [UygunSeEr1998].
DISTRIBUTION: **Palaearctic**: Algeria [Balach1932d]; Czech Republic [Zahrad1977]; France [Balach1950b]; Israel [Bodenh1924]; Italy [Leonar1900, Leonar1920, Balach1950b, Pelliz1987, PellizCa1991a]; Morocco [Balach1950b]; Spain [Balach1950b, GomezM1960O, BlayGo1993]; Switzerland [BachmaGe1950]; Turkey [UygunSeEr1998]; Yugoslavia [Bachma1956].
BIOLOGY: Recorded from roots of *Matricaria officinalis* in Spain (Gómez-Menor Ortega, 1960).
GENERAL: Description and illustration of adult female by Balachowsky (1948, 1950b), Bachmann (1956) and Danzig (1993).

KEYS: Danzig 1993: 179-182 (female) [Europe]; Leonardi 1920: 104-105 (female) [Italy].

CITATIONS: Bachma1956 [taxonomy, description, illustration, host, distribution: 99-102]; BachmaGe1950 [host, distribution: 119]; Balach1932d [taxonomy, host, distribution: XIII-XIV]; Balach1948 [taxonomy, description, illustration, host, distribution: 14-18]; Balach1950b [taxonomy, description, illustration, host, distribution: 523-525]; BlayGo1993 [taxonomy, description, illustration, host, distribution: 610-613]; Bodenh1924 [taxonomy, description, host, distribution: 38]; Bodenh1935 [host, distribution: 247]; Borchs1966 [catalogue: 323,326]; Danzig1993 [taxonomy, description, illustration, host, distribution: 211-212]; DanzigPe1998 [catalogue: 230-231]; Fernal1903b [catalogue: 298]; Ferris1941e [taxonomy: 43]; Ferris1943a [taxonomy: 85]; Foldi2001 [distribution: 303-308]; GomezM1960O [taxonomy, description, host, distribution: 160-162]; Leonar1900 [taxonomy, description, illustration, host, distribution: 305-307]; Leonar1920 [taxonomy, description, illustration, host, distribution: 105,114-115]; Lindin1912b [taxonomy, description, host, distribution: 278,302]; Lindin1935 [taxonomy: 146]; Lindin1957 [taxonomy: 548]; LongoMaPe1995 [distribution: 126]; Lupo1954 [taxonomy, description, illustration, host, distribution: 29-33]; Martin1983 [taxonomy, host, distribution: 64]; Pelliz1987 [host, distribution: 121-122]; PellizCa1991a [taxonomy, host, distribution: 199-200]; UygunSeEr1998 [host, distribution: 183-191]; Zahrad1962 [taxonomy, description, host, distribution: 92-93]; Zahrad1972 [host, distribution: 439-440]; Zahrad1977 [taxonomy, distribution: 120]; Zahrad1990a [host, distribution, description: 649].

Diaspidiotus ehrhorni (Coleman)

Aspidiotus (*Diaspidiotus*) *ehrhorni* Coleman, 1903: 68. Type data: USA: California, Siskiyou County, near Sisson Mt. Shasta, concealed among and underneath the lichens on the bark of *Abies concolor*; collected by E.M. Ehrhorn, September 4, 1901. Syntypes, female. Type depository: Davis: The Bohart Museum of Entomology, University of California, California, USA.

Aspidiotus ehrhorni; Sanders, 1906: 13. Change of combination.

Diaspidiotus ehrhorni; MacGillivray, 1921: 413. Change of combination.

COMMON NAME: Ehrhorn scale [McKenz1956].

SCALE COVER: Female scale nearly circular, about 2 mm in diameter; very slightly convex; dark grey, with light yellow exuviae at apex; covered with minute granules and resembling the lichens under which it is found; beneath this outer scale is a dark reddish-brown skin enclosing the insect; ventral scale very thin, transparent and white (Coleman, 1903). Scale of the female grey, flat, circular, exuviae subcentral; that of the male somewhat elongate oval, the exuvia near one end (Ferris, 1938a). Colour photograph by Gill (1997).

HOST PLANTS: **Cupressaceae**: *Libocedrus decurrens* [Sander1906, Ferris1938a, McKenz1956]. **Pinaceae**: *Abies concolor* [Colema1903, Sander1906, McKenz1956], *Pinus cembroides parryana* [McKenz1956], *Pinus sabiniana* [McKenz1956], *Pseudotsuga taxifolia* [Ferris1920b, Ferris1938a, McKenz1956].

DISTRIBUTION: **Nearctic**: United States of America (California [Colema1903, Sander1906, McKenz1956], Colorado [Nakaha1982], Washington [Nakaha1982]).

BIOLOGY: Occurring on the bark, frequently concealed beneath lichens (Ferris, 1938a).

GENERAL: Description and illustration of adult female by Ferris (1920b, 1938a), McKenzie (1956) and by Gill (1997).

KEYS: Gill 1997: 113 (female) [Species of California]; McKenzie 1956: 25 (female) [U.S.A.: California]; Ferris 1942: 32 (female) [North America].

CITATIONS: Borchs1966 [catalogue: 323]; Colema1903 [taxonomy, description, host, distribution: 68]; Ferris1920b [taxonomy, description, illustration, host, distribution: 51]; Ferris1938a [taxonomy, description, illustration, host, distribution: 221]; Ferris1941e [taxonomy: 43]; Ferris1942 [catalogue: 446:32]; FurnisCa1977 [host, distribution: 111]; Gill1997 [host, distribution, taxonomy, description, illustration, economic importance: 115,120]; MacGil1921 [taxonomy, description, host, distribution: 413]; McKenz1956 [taxonomy, description, illustration, host, distribution: 60,62]; Nakaha1982 [host, distribution: 29]; Sander1906 [taxonomy, host, distribution: 13].

Diaspidiotus elaeagni (Borchsenius)

Aspidiotus elaeagni Borchsenius, 1939: 35. Type data: KYRGYZSTAN: 18 km from Oktyabr' station, near Kugartka river, on *Elaeagnus* sp. Lectotype female, by subsequent designation Danzig, 1993: 202. Type depository: St. Petersburg: (= Leningrad) Zoological Museum, Academy of Science, Russia.

Aspidiotus eleagni; Ferris, 1941e: 43. Misspelling of species name.

Diaspidiotus elaeagni; Borchsenius, 1950b: 230. Change of combination.

COMMON NAME: kruglaya djigdovaya shitovka [BazaroSh1971].

SCALE COVER: Female scale very flat; circular, diameter 1.8-2 mm; black; exuviae red, central or subcentral, always covered with white wax secretion; surface of scale with small, punctiform areas, that do not exist in other palearctic species. Male scale oval; structure similar to that of female, 1.2 mm (Balachowsky, 1950b).

HOST PLANTS: **Elaeagnaceae**: *Elaeagnus* [Borchs1939, Balach1950b], *Elaeagnus angustifolia* [BazaroSh1971], *Hippophae rhamnoides* [BazaroSh1971].

DISTRIBUTION: **Palaearctic**: Armenia [BazaroSh1971]; China (People's Republic) [Tang1984]; Georgia (Georgia [Hadzib1983, BazaroSh1971]); Kazakhstan [BazaroSh1971]; Kyrgyzstan (= Kirgizia) [Borchs1939, Balach1950b]; Tajikistan (=Tadzhikistan) [Balach1950b]; Turkmenistan [Balach1950b].

ECONOMIC IMPORTANCE: Schmutterer et al. (1957) reported that this species damaged *Elaeagnus* sp. in Central Asia (Turkmenistan, Tadzhikistan, Uzbekistan).

GENERAL: Description and illustration of adult female by Balachowsky (1950b), Bazarov & Shmelev (1971), Tang (1984) and by Danzig (1993).

KEYS: Danzig 1993: 179-182 (female) [Europe]; Bazarov & Shmelev 1971: 198 (female) [Central Asia]; Balachowsky 1950b: 494 (female) [Mediterranean]; Kuwana 1933: 40 (female) [Japan].

CITATIONS: Balach1950b [taxonomy, description, illustration, host, distribution: 520-523]; BazaroSh1971 [taxonomy, description, illustration, host, distribution, life history: 198-200]; Borchs1939 [taxonomy, description, illustration, host, distribution: 10,35]; Borchs1950b [taxonomy, description, illustration, host, distribution: 230-231]; Borchs1966 [catalogue: 323-324]; BurgerU11990 [economic importance: 321]; Danzig1972 [taxonomy, host, distribution, economic importance: 210]; Danzig1993 [taxonomy, description, illustration, host, distribution: 203-205]; DanzigPe1998 [catalogue: 231]; Ferris1941e [taxonomy: 43]; Hadzib1983 [taxonomy, host, distribution, life history, biological control, economic importance: 234-235]; KaussaBa1953 [taxonomy: 26]; MillerDa1990 [host, distribution, economic importance: 301]; SchmutKlLu1957 [host, distribution, economic importance: 491]; Tang1984 [taxonomy, description, illustration, host, distribution: 68-69]; Tao1999 [taxonomy: 83].

Diaspidiotus eurotiae (Bazarov)

Rhizaspidiotus eurotiae Bazarov, 1967: 57. Type data: KYRGYZSTAN: Alaiskii ridge, near Akbasaga, on *Eurotia ceratoides*. Syntypes, female. Type depository: St. Petersburg: (= Leningrad) Zoological Museum, Academy of Science, Russia.
Diaspidiotus eurotiae; Bazarov & Shmelev, 1971: 207. Change of combination.
Diaspidiotus halophilus Danzig, 1983: 521. Type data: KAZAKHSTAN: Central Kazakhstan, Dzhezkazgan region, Koksengir, 40 km south Jan-Ark, on *Halocnemum strobilaceum*. Holotype female. Type depository: St. Petersburg: (= Leningrad) Zoological Museum, Academy of Science, Russia. Synonymy by Danzig, 1993: 208.
COMMON NAME: tereskenovaya shitovka [BazaroSh1971].
SCALE COVER: Female scale elliptical, broad, almost circular, 1.2-1.4 mm long, 0.9-1.1 mm wide; slightly convex; white grey; exuviae placed centrally (Bazarov & Shmelev, 1971).
HOST PLANTS: **Chenopodiaceae**: *Atriplex cana* [Danzig1983], *Atriplex verrucifera* [Danzig1983], *Camphorosma monspeliacum* [Danzig1983], *Eurotia ceratoides* [BazaroSh1971], *Halocnemum strobilaceum* [Danzig1983], *Suaeda physophora* [Danzig1983].
DISTRIBUTION: **Palaearctic**: Kazakhstan (Dzhezkazgan Oblast [Danzig1983]); Kyrgyzstan (= Kirgizia) [Borchs1939, BazaroSh1971].
GENERAL: Description and illustration of adult female by Bazarov & Shmelev (1971), Danzig (1983, 1993).
KEYS: Danzig 1993: 179-182 (female) [Europe]; Bazarov & Shmelev 1971: 208 (female) [Central Asia].
CITATIONS: BazaroSh1971 [taxonomy, description, illustration, host, distribution: 207-208]; Danzig1983 [taxonomy, description, illustration, host, distribution: 521-522]; Danzig1993 [taxonomy, description, illustration, host, distribution: 208-209]; DanzigPe1998 [catalogue: 231].

Diaspidiotus fabernii (Houser), new combination

Aspidiotus fabernii Houser, 1918: 165. Type data: CUBA: Havana, Jardin Botanica del Instituto Segunda Ensenada de la Habana, on *Faberia* sp. Holotype female. Type depository: Washington: United States National Entomological Collection, U.S. National Museum of Natural History, District of Columbia, USA.
Hemiberlesia fabernii; MacGillivray, 1921: 438. Change of combination.
Aspidiotus fabernae Lindinger, 1957: 545. Unjustified emendation.

Aspidiotus fabernii; Borchsenius, 1966: 305. Incorrect synonymy. Notes: Incorrect synonymy with *Hemiberlesia diffinis* Newstead; see Miller & Davidson, 1998: 197.
Quadraspidiotus fabernii; Miller & Davidson, 1998: 197. Change of combination.
SYSTEMATICS: Borchsenius (1966: 305) listed this species as a synonym of *Hemiberlesia diffinis* (Newstead) but Miller & Davidson (1998) did not accept it.
SCALE COVER: Female scale usually circular, 1.25 mm in diameter, though sometimes slightly elongate; strongly convex; exuviae central, covered, but granular covering easily rubbed off, leaving exposed the yellow or orange exuviae; ventral scale white, conspicuous (Houser 1918).
HOST PLANTS: **Compositae**: *Faberia* [Houser1918].
DISTRIBUTION: **Neotropical**: Cuba [Houser1918].
BIOLOGY: Scales piled one upon another so that bark appears to bear scattered nodules (Houser, 1918).
GENERAL: Description and illustration of adult female by Houser (1918).
CITATIONS: Ferris1941e [taxonomy: 43]; Houser1918 [taxonomy, description, host, distribution: 165-166]; Lindin1957 [taxonomy: 545]; MacGil1921 [taxonomy, description, host, distribution: 438]; MillerDa1998 [taxonomy: 197]; Newell1923 [host, distribution: 263-266].

Diaspidiotus farahbakhchi Kaussari
Diaspidiotus farahbakhchi Kaussari, 1955a: 235. Type data: IRAN: province of Mazaderan, near Ramsar, on indigenous *Quercus* of *robur* type. Syntypes, female. Type depository: Paris: Muséum national d'Histoire naturelle, France.
Diaspidiotus farahbakhehi; Danzig & Pellizzari, 1998: 231. Misspelling of species name.
SCALE COVER: Female scale subcircular, slightly convex; exuviae central or subcentral, colour brown reddish; secreted part, dark grey, matt; 1.8-2.1 mm (Kaussari, 1955a).
HOST PLANTS: **Fagaceae**: *Quercus* [Kaussa1955a, Kaussa1957, Danzig1993].
DISTRIBUTION: **Palaearctic**: Armenia [Danzig1993]; Iran [Kaussa1955a, Kaussa1957].
GENERAL: Description and illustration of adult female by Kaussari (1955a) and by Danzig (1993).
KEYS: Danzig 1993: 179-182 (female) [Europe].
CITATIONS: Borchs1966 [catalogue: 324]; Danzig1993 [taxonomy, description, illustration, host, distribution: 190]; DanzigPe1998 [catalogue: 231]; Kaussa1955a [taxonomy, description, illustration, host, distribution: 235-237]; Kaussa1957 [host, distribution: 1].

Diaspidiotus forbesi (Johnson)
Aspidiotus forbesi Johnson, 1896: 151. Type data: U.S.A.: Illinois, locality not indicated, on cherry, currant, apple, peach, pear and honey locust. Syntypes, female. Type depository: Washington: United States National Entomological Collection, U.S. National Museum of Natural History, District of Columbia, USA.
Aspidiotus (*Diaspidiotus*) *forbesi*; Cockerell, 1897i: 21. Change of combination.
Aspidiotus (*Aspidiella*) *forbesi*; Leonardi, 1898a: 63. Change of combination.

Aspidiotus (*Diaspidiotus*) *fernaldi hesperius* Cockerell, 1902: 450. Type data: U.S.A.: Arizona, Prescott, on bark of an undetermined bush; collected May 1902. Syntypes, female. Type depository: Washington: United States National Entomological Collection, U.S. National Museum of Natural History, District of Columbia, USA. Synonymy by Borchsenius, 1966: 330.

Aspidiotus fernaldi hesperius; Fernald, 1903b: 259. Change of combination.

Diaspidiotus forbesi; Cockerell, 1905b: 202. Change of combination.

Forebesaspis forbesi; MacGillivray, 1921: 388. Change of combination.

Forbesaspis (*Aspidiotus*) *forbesi*; Borchsenius, 1935a: 31. Change of combination.

Aspidiotus hesperius; Ferris, 1938a: 190. Change of combination and rank.

Quadraspidiotus forbesi; Ferris, 1938a: 255. Change of combination.

Aspidiotus (*Quadraspidiotus*) *forbesi*; Merrill, 1953: 20. Change of combination.

Diaspidiotus fobesi; Borchsenius, 1966: 330. Misspelling of species name.

Diaspidiotus forbesi; Borchsenius, 1966: 330. Revived combination.

COMMON NAMES: cherry Aspidiotus [Johnso1896]; cherry scale [Brain1918, Hollin1923, MerrilCh1923]; Forbes scale [McKenz1956, Dekle1965c].

SCALE COVER: Scale of female grey, elongate oval, high convex; exuviae toward one end; scale of male elongate, exuvia toward one end (Ferris, 1938a).

HOST PLANTS: **Aceraceae**: *Acer pseudoplatanus* [Wilson1917, MerrilCh1923]. **Aquifoliaceae**: *Ilex* [BesheaTiHo1973], *Ilex aquifolium* [McKenz1956], *Ilex opaca* [BesheaTiHo1973]. **Betulaceae**: *Betula nigra* [BesheaTiHo1973]. **Caprifoliaceae**: *Viburnum* [BesheaTiHo1973]. **Cornaceae**: *Cornus* [Dekle1965c, BesheaTiHo1973]. **Ericaceae**: *Oxydendrum arboreum* [BesheaTiHo1973]. **Juglandaceae**: *Carya ilinoensis* [McKenz1956], *Hicoria pecan* [Ferris1938a]. **Leguminosae**: *Gleditsia triacanthos* [BesheaTiHo1973]. **Nyssaceae**: *Nyssa sylvatica* [BesheaTiHo1973]. **Oleaceae**: *Chionanthus virginicus* [BesheaTiHo1973], *Jasminum* [Martor1976], *Ligustrum* [Ferris1938a, McKenz1956]. **Rhamnaceae**: *Rhamnus* [Ferris1938a, McKenz1956]. **Rosaceae** [BesheaTiHo1973], *Crataegus* [Wilson1917, MerrilCh1923, McDani1970], *Cydonia* [Ferris1938a], *Cydonia oblonga* [McKenz1956], *Malus pumila* [McKenz1956], *Prunus* [Ferris1938a, McKenz1956, Balach1958b, Dekle1965c, BesheaTiHo1973], *Prunus cerasus* [Ferris1938a, McKenz1956], *Prunus domestica* [Brain1918, Ferris1938a], *Prunus persica* [Ferris1938a], *Pyrus* [Dekle1965c], *Pyrus americana* [RosenDe1979], *Pyrus malus* [Ferris1938a]. **Ulmaceae**: *Celtis* [Ferris1938a, McKenz1956].

NATURAL ENEMIES: ACARI: **Acaridae**: *Thyreophagus entomophagus* [Koszta1996]. **Cymbaeremaeidae**: *Scapheremaeus marginalis* [Koszta1996]. **Hemisarcoptidae**: *Hemisarcoptes malus* (Shimer) [SummerHa1951, GersonOcHo1990]. **Oribatulidae**: *Zygoribatula pyrostigmata* [Koszta1996]. HYMENOPTERA **Aphelinidae**: *Ablerus clisiocampae* (Ashmead) [Gordh1979], *Aphytis diaspidis* (Howard) [Koszta1996], *Aphytis proclia* (Walker) [RosenDe1979], *Marietta pulchella* (Howard) [Gordh1979], *Prospaltella aurantii* (Howard) [Felt1901, Gordh1979], *Prospaltella fasciaventris* Girault [Gordh1979], *Prospaltella forbesi* Dozier [Gordh1979], *Prospaltella murtfeldtiae* (Howard) [Felt1901, Gordh1979]. **Encyrtidae**: *Arrhenophagus chionaspidis* Aurivillius [Gordh1979]. **Signiphoridae**: *Signiphora pulchra* Girault [Woolle1990].

DISTRIBUTION: **Afrotropical**: South Africa [BrainKe1917, Brain1918, Balach1958b]. **Nearctic**: Canada [Nakaha1982]; Mexico [Ferris1938a]; United States of America (Arizona [Cocker1902k, Ferris1938a], California [McKenz1956], Florida [Wilson1917, MerrilCh1923, Merril1953, Dekle1965c], Georgia [TippinBe1970, BesheaTiHo1973], Illinois [Johnso1896], Kansas [Hunter1899], Mississippi [Herric1911], Missouri [Hollin1923], Ohio [RosenDe1979], Texas [Herric1911]). **Neotropical**: Puerto Rico & Vieques Island (Puerto Rico [Martor1976, ColonFMe1998]).

BIOLOGY: Occurring on the bark (Ferris, 1938a).

ECONOMIC IMPORTANCE: Pest of forest and fruit trees in North America (Davidson & Miller, 1990).

GENERAL: Description and illustration of adult female by Brain (1918), Ferris (1938a), McKenzie (1956), Balachowsky (1958b), Kosztarab (1996), Gill (1997) and by Colon-Ferrer & Medina-Gaud (1998).

KEYS: Gill 1997: 243 (female) [Species of California]; Kosztarab 1996: 575 (female) [Northeastern North America]; McKenzie 1956: 26 (female) [U.S.A.: California]; Ferris 1942: 39 (female) [North America]; Britton 1923: 371 (female) [U.S.A.: Connecticut]; Hollinger 1923: 7-8 (female) [U.S.A.: Missouri]; Brain 1918: 124 (female) [South Africa]; Lawson 1917: 217 (female) [U.S.A.: Kansas]; Dietz & Morrison 1916a: 289-290 (female) [U.S.A.: Indiana]; Cockerell 1905b: 202 (female) [U.S.A.: Colorado]; Newell 1899: 4-5 (female) [North America].

CITATIONS: Balach1958b [taxonomy, description, illustration, host, distribution: 210-212]; BeardsDaHo1976 [economic importance: 105]; BeardsGo1975 [economic importance: 49]; BesheaTiHo1973 [host, distribution: 7-8]; Borchs1935a [taxonomy: 31]; Borchs1937a [taxonomy, description, illustration, host, distribution: 53-54]; Borchs1966 [catalogue: 329-330]; Boynto1901 [taxonomy, description, illustration, host, distribution: 347]; Brain1918 [taxonomy, description, illustration, host, distribution: 124]; BrainKe1917 [distribution: 183]; Brick1912 [host, distribution: 1-22]; Britto1923 [taxonomy, description, host, distribution: 371,373]; Britto1923b [host, distribution]; Chandl1950 [host, distribution, chemical control, economic importance: 398]; Cocker1896b [distribution: 334]; Cocker1897i [taxonomy, description, host, distribution: 5,21]; Cocker1902k [taxonomy, description, host, distribution: 450]; Cocker1905b [taxonomy: 202]; ColonFMe1998 [taxonomy, description, illustration, host, distribution: 79-80]; DavidsMi1990 [host, distribution, economic importance: 603-632]; Dekle1965c [taxonomy, description, host, distribution: 124]; Dekle1976 [taxonomy, description, host, distribution, economic importance : 143]; DietzMo1916a [taxonomy, description, illustration, host, distribution: 302-303]; Drake1935 [host, distribution, control: 83-91]; Felt1901 [taxonomy, description, illustration, host, distribution, biological control, life history: 330-331]; Fernal1903b [catalogue: 259-260]; Ferris1937c [taxonomy, illustration: 51,73]; Ferris1938a [taxonomy, description, illustration, host, distribution: 190,256]; Ferris1941e [taxonomy: 43-44]; Ferris1942 [taxonomy: 446:39]; Garcia1931a [host, distribution, biological control: 659-669]; GersonOcHo1990 [biological control: 77-97]; Gill1997 [host, distribution, taxonomy, description, illustration, economic importance: 243,247]; Gordh1979 [biological control: 897,899,907,908,929]; Harned1928 [host, distribution: 23-24];

Headle1929 [host, distribution: 125]; Herric1911 [taxonomy, description, illustration, host, distribution: 9,15-16,49]; Herric1925 [host, distribution, description, life history, economic importance]; Hollin1923 [taxonomy, description, host, distribution: 11-12]; Hunter1899 [taxonomy, description, host, distribution: 3-4]; Jarvis1908TD [taxonomy: 59]; JiYa1990 [biological control: 134-136]; Johnso1896 [taxonomy, description, host, distribution: 151]; Johnso1898 [taxonomy, host, distribution, life history: 82-83]; Koszta1996 [taxonomy, description, illustration, host, distribution, life history, biological control, economic importance: 575-577]; Kozar1990c [life history, economic importance, host, distribution: 593-602]; Lawson1917 [taxonomy, description, illustration, host, distribution: 226-228]; Leonar1898 [taxonomy, description, illustration, host, distribution: 63-64]; Lyne1921 [distribution: 146-148]; MacGil1921 [taxonomy, description, host, distribution: 388,422]; Mackie1934 [host, distribution: 396]; McDani1970 [taxonomy, illustration, host, distribution : 429-431]; Martor1976 [host, distribution: 151]; McKenz1956 [taxonomy, description, illustration, host, distribution: 79-81]; Merril1953 [taxonomy, description, host, distribution: 20-21]; MerrilCh1923 [taxonomy, description, host, distribution, economic importance: 201]; MetcalMe1993 [economic importance, host, distribution, control]; MillerDa1990 [host, distribution, economic importance: 305]; Nakaha1982 [host, distribution: 77]; Newell1899 [taxonomy, description, host, distribution: 14-16]; NewellRo1908 [host, distribution: 150-155]; RosenDe1979 [host, distribution, biological control: 377-383]; Sander1904a [taxonomy, description, illustration, host, distribution: 57,60]; SchmutKlLu1957 [host, distribution, economic importance: 484]; Staffo1915 [taxonomy, structure: 71]; Sulliv1930 [host, distribution: 51-59]; SummerHa1951 [host, distribution, biological control: 818]; Takagi1958 [taxonomy: 127]; TippinBe1970 [host, distribution: 11]; Wilson1917 [taxonomy, description, host, distribution: 14-15]; Woolle1990 [biological control: 167-176].

Diaspidiotus gigas (Thiem & Gerneck)

Aspidiotus populi Glaser, 1877: 49. Type data: GERMANY: Rhein region, vicinity of Worms, on "Schwartzpappel" [=*Populus*]. Syntypes, female. Synonymy by Lindinger, 1936: 152. Homonym of *Aspidiotus populi* Baerensprung, 1849. Notes: Depository of type material unknown.

Aspidiotus (*Euraspidiotus*) *gigas* Thiem & Gerneck, 1934a: 131. Type data: GERMANY: host plant not indicated. Syntypes, female and first instar. Notes: Type depository unknown. Junior synonym and Replacement Name.

Aspidiotus multiglandulatus Borchsenius, 1935: 46. Type data: RUSSIA: Dagestan, Levashinskii region, Hadjal-Mekhi, on *Populus* sp. Lectotype female, by subsequent designation Danzig, 1993: 193. Type depository: St. Petersburg: (= Leningrad) Zoological Museum, Academy of Science, Russia. Synonymy by Ferris, 1941e: 46.

Aspidiotus gigas; Ferris, 1941e: 43. Change of combination.

Quadraspidiotus gigas; Balachowsky, 1948: 18. Change of combination.

Diaspidiotus gigas; Borchsenius, 1949d: 243. Change of combination.

Quadraspidiotus gigans; Bachmann, 1956: 102. Misspelling of species name.

Diaspidiotus gigas; Danzig, 1980: 340. Revived combination.

Aspidiotus multigrandulatus; Chou, 1985: 311. Misspelling of species name.

COMMON NAMES: topolevaya shitovka [BazaroSh1971]; Willow Scale [Koszta1996].

SCALE COVER: Female scale circular or subcircular, 2.4-2.8 mm; very flat; colour dark brown; mingles with bark; exuviae central, red orange. Male scale of similar structure, oval; exuviae subcentral, dark brown; 1.8-2 mm (Balachowsky, 1950b).

HOST PLANTS: **Betulaceae**: *Alnus* [Zahrad1972]. **Salicaceae**: *Populus* [Borchs1935, Borchs1936, Balach1950b, Danzig1962, BazaroSh1971, Danzig1980b], *Populus alba* [Balach1950b, Zahrad1952, Zahrad1972], *Populus berolinensis* [Zahrad1972], *Populus canadensis* [Zahrad1972], *Populus canescens* [Zahrad1972], *Populus deltoides* [Zahrad1972], *Populus euramericana* [Zahrad1972], *Populus nigra* [Zahrad1972], *Populus piramidalis* [Balach1950b, Bachma1953, Zahrad1972], *Populus tremula* [Zahrad1952, GomezM1960O, Zahrad1972], *Populus trichocarpae* [Zahrad1972], *Populus trimola* [Bachma1953, Martin1983], *Salix* [Balach1950b, Bachma1953, BazaroSh1971, Zahrad1972, Danzig1978a, Danzig1980b], *Salix acutifolia* [Balach1950b, Zahrad1972], *Salix alba* [Zahrad1972], *Salix aurita* [Zahrad1972], *Salix caprea* [Zahrad1972], *Salix daphnoides* [Balach1950b, Zahrad1972], *Salix fragilis* [Balach1950b, Zahrad1972], *Salix pedicellata* [Balach1950b], *Salix viminalis?* [BachmaGe1950]. **Tiliaceae**: *Tilia parvifolia* [Balach1950b, Zahrad1972].

NATURAL ENEMIES: ACARI **Hemisarcoptidae**: *Hemisarcoptes budensis* Fain & Ripka [FainRi1998]. COLEOPTERA **Coccinellidae**: *Chilocorus bipustulatus* L. [Zahrad1972], *Chilocorus kuwanae* Silvestri [MaLiLi1997, MaLiLi1997a], *Coccinella bipunctata* L. [Zahrad1972]. HETEROPTERA **Anthocoridae**: *Ectemnus nigriceps* [Zahrad1972], *Temnostethus longirostris* Horvath [Zahrad1972]. **Microphysidae**: *Loricula elengatula* Baerensprung [Zahrad1972], *Loricula pselaphiformis* Curt. [Zahrad1972]. HYMENOPTERA **Aphelinidae**: *Aphytis diaspidioti* Tshumakova [RosenDe1979], *Aphytis mytiliaspidis* Le Baron [Bachma1956], *Aphytis quadraspidioti* Li, C.D. [Li1996], *Archenomus maritimus* Nikolskaya [Zahrad1972], *Aspidiotiphagus citrinus* Crawford [Bachma1956, Zahrad1972], *Azotus* sp. [Bachma1956], *Azotus matritensis* Mercet [Zahrad1972], *Encarsia gigas* [LiLoCa1997], *Prospaltella perniciosi* Tower [Bachma1956], *Pteroptrix dimidiata* Westwood [Zahrad1972]. **Encyrtidae**: *Comperiella bifasciata* Howard [Trjapi1989], *Metaphycus duplus* (Tshumakova) [Trjapi1989], *Metaphycus nadius* (Walker) [GuerriNo2000], *Trichomasthus dissimilis* (Tshumakova) [Trjapi1989].

DISTRIBUTION: **Nearctic**: Canada (New Brunswick [Nakaha1982], Nova Scotia [Nakaha1982], Ontario [Nakaha1982]); United States of America (Idaho [Nakaha1982], Montana [Nakaha1982], New Mexico [Nakaha1982], Ohio [Nakaha1982], Oregon [Nakaha1982], Pennsylvania [Nakaha1982], Rhode Island [Nakaha1982], Utah [Nakaha1982], Washington [Nakaha1982], Wisconsin [Nakaha1982], Wyoming [Nakaha1982]). **Palaearctic**: Algeria [Balach1950b]; Azerbaijan (Azerbaijan [Borchs1935]); Bulgaria [Balach1950b]; China (People's Republic) [Tang1984] (Henan (Honan) [Shen1993]); Czech Republic [Zahrad1952, Zahrad1977]; France [Balach1950b]; Georgia (Georgia [Borchs1935, Borchs1936]); Germany (United) [Balach1950b]; Hungary [Nakaha1982]; Italy [Balach1950b];

Kazakhstan (Alma Ata Oblast [BazaroSh1971]); Netherlands [Reyne1957, Jansen2001]; Poland [Komosi1968a]; Romania [Savesc1982]; Russia (Caucasus [BazaroSh1971], Dagestan AR [Danzig1993], Irkutsk Oblast [Danzig1980b], Novosibirsk Oblast [Danzig1980b], Primor'ye Kray [BazaroSh1971], St. Petersberg (= Leningrad) Oblast [Danzig1962, Danzig1962b], Yakutia-Sakha (= Yakut) AR [Danzig1978a, Danzig1980b]); Slovak Republic [Zahrad1952]; Spain [GomezM1960O, Martin1983, BlayGo1993]; Switzerland [BachmaGe1950]; Turkey [Balach1950b]; Ukraine [BazaroSh1971]; Yugoslavia [Bachma1953].

ECONOMIC IMPORTANCE: Schmutterer et al. (1957) considered it a pest of *Populus* spp. in several countries in Europe and Central Asia.

GENERAL: Description and illustration of adult female by Balachowsky (1948, 1950b), Zahradník (1952, 1972), Bazarov & Shmelev (1971), Tang (1984), Tereznikova (1986), Danzig (1980b, 1993), Yaşar (1995a) and by Kosztarab (1996).

KEYS: Kosztarab 1996: 575 (female) [Northeastern North America]; Danzig 1993: 179-182 (female) [Europe]; Tereznikova 1986: 97-98 (female) [Ukraine]; Chou 1985: 311 (female) [Species of China]; Danzig 1980b: 340 (female) [Far East of USSR]; Kosztarab & Kozár 1978: 172 (female) [Hungary]; Bazarov & Shmelev 1971: 211 (female) [Central Asia]; Reyne 1957: 33 (female) [Netherlands]; Zahradník 1952: 114-115 (female) [Czech Republic]; Balachowsky 1950b: 404 (female) [Mediterranean]; Borchsenius 1935: 127-128 (female) [Former USSR]; Danzig 1935: 127 (female) [Former USSR].

CITATIONS: Bachma1953 [host, distribution: 178]; Bachma1956 [biological control: 102]; BachmaGe1950 [host, distribution: 117]; Balach1948 [taxonomy, description, illustration, host, distribution: 18-23]; Balach1950b [taxonomy, description, illustration, host, distribution: 465-469]; BazaroSh1971 [taxonomy, description, illustration, host, distribution: 213-215]; BeardsDaHo1976 [economic importance: 106]; BlayGo1993 [taxonomy, description, illustration, host, distribution: 567-571]; Borchs1935 [taxonomy, description, illustration, host, distribution: 132-133]; Borchs1936 [host, distribution: 131]; Borchs1937 [taxonomy, description, illustration, host, distribution: 130-131]; Borchs1939 [taxonomy, description, host, distribution: 9,29]; Borchs1949d [taxonomy, description, host, distribution: 243]; Borchs1950b [taxonomy, description, illustration, host, distribution: 226,231]; Borchs1966 [catalogue: 330]; Bustsh1958 [taxonomy, description, host, distribution: 220,262]; ChiMiQu1997 [host, distribution, chemical control, biological control: 10-14]; ChiZhHu1997 [host, distribution, life history, biological control: 15-21]; Chou1985 [taxonomy, description, host, distribution: 311-312]; Chumak1957 [host, distribution, biological control: 533-547]; Chumak1961 [host, distribution, biological control: 313-338]; Danzig1962 [host, distribution: 22]; Danzig1962b [taxonomy, distribution: 25]; Danzig1964 [taxonomy, host, distribution: 653]; Danzig1977b [taxonomy: 57]; Danzig1978a [host, distribution: 78]; Danzig1980b [taxonomy, description, illustration, host, distribution: 340-341]; Danzig1993 [taxonomy, description, illustration, host, distribution: 193-194]; DanzigKo1991 [distribution: 1-15]; DanzigPe1998 [catalogue: 231-232]; DavidsMi1990 [host, distribution, economic importance: 603-632]; FainRi1998 [host, distribution, biological control: 33-39]; Ferris1941e [taxonomy: 43,46]; Foldi2001 [distribution: 303-308]; FoldiDe1998 [taxonomy, host, distribution: 201]; FreyFr1995 [taxonomy, chemistry: 777-780]; FreyFr1995a [chemistry, structure: 100]; Glaser1877 [taxonomy, description, host, distribution: 48-49]; GomezM1960O [taxonomy, description, host, distribution: 166-168]; GuerriNo2000 [host, distribution, biological control: 157-159]; Huba1960 [taxonomy: 39-50]; HuDaHu1982 [host, distribution, economic importance, chemical control, biological control: 160-169]; Jansen2001 [host, distribution: 197-206]; Komosi1968a [host, distribution, taxonomy, description, illustration: 51-54]; Komosi1974a [host, distribution, life history, ecology: 1-84]; Komosi1986 [host, distribution: 3-12]; Komosi1986a [host, distribution: 13-20]; Koszta1990 [structure, biological control: 307-311]; Koszta1996 [taxonomy, description, illustration, host, distribution, life history, biological control, economic importance: 577-578]; KosztaKo1978 [taxonomy, description, host, distribution: 172]; Koteja1990c [life history: 243-254]; Kozar1995b [taxonomy: 51]; KozarHiMa1996 [taxonomy, description: 433-437]; Lagows1998a [host, distribution: 63-71]; Lellak1963 [taxonomy, description, host, distribution, life history: 611-648]; Lellak1966 [taxonomy, host, distribution: 297]; Li1996 [host, distribution, biological control: 98-101]; LiLoCa1997 [host, distribution, biological control: 225-226]; Lindin1957 [taxonomy: 546]; LiuLiYa1997 [host, distribution, economic importance, life history, ecology: 5-9]; LongoMaPe1995 [distribution: 128]; MaLiLi1997 [host, distribution, life history, biological control: 64-67]; MaLiLi1997a [life history, biological control: 59-61]; Martin1983 [taxonomy, host, distribution: 66]; MillerDa1990 [host, distribution, economic importance: 305]; Nakaha1982 [host, distribution: 77]; Reyne1957 [taxonomy, host, distribution: 26,33]; RosenDe1979 [host, distribution, biological control: 731]; RzaevaYa1985 [biological control: 55-58]; Savesc1982 [taxonomy, description, host, distribution, life history, biological control: 323-325]; Schmut1959 [taxonomy, description, host, distribution: 77,97]; SchmutKlLu1957 [p. 484]; Shen1993 [host, distribution: 60]; Smetni1991 [chemistry: 92-129]; Tang1984 [taxonomy, description, illustration, host, distribution: 62,64]; Tao1999 [taxonomy, host, distribution: 114-115]; Terezn1986 [taxonomy, description, illustration, host, distribution: 101-103]; ThiemGe1934a [taxonomy, description, host, distribution: 130-158,208-238]; Trjapi1989 [biological control: 188,244,295]; Xie1998 [taxonomy, description, illustration, host, distribution: 103-107]; XieXuLi1995 [host, distribution, life history, ecology: 114-118]; Yasar1995a [taxonomy, description, illustration, host, distribution: 110-112]; Yasnos1994 [host, distribution, biological control: 317-333]; Zahrad1952 [taxonomy, description, illustration, host, distribution: 136-138]; Zahrad1972 [taxonomy, description, illustration, host, distribution, biological control: 434-435]; Zahrad1977 [taxonomy, distribution: 121]; Zahrad1990a [host, distribution, description: 650-651].

Diaspidiotus hunteri (Newell)

Aspidiotus hunteri Newell, 1899: 10. Type data: U.S.A.: Iowa, Alton, on currant [*Ribes* sp.]; imported from Texas, in 1897. Syntypes, female. Type depository: Washington: United States National Entomological Collection, U.S. National Museum of Natural History, District of Columbia, USA.
Aspidiotus (*Diaspidiotus*) *hunteri*; Leonardi, 1900: 340. Change of combination.

Diaspidiotus hunteri; MacGillivray, 1921: 412. Change of combination.
SCALE COVER: Female scale round, 1.3-1.4 mm in diameter; light grey in colour; portion immediately surrounding exuviae dark; exuviae sublateral in position, dark orange in colour (Newell, 1899).
HOST PLANTS: **Fagaceae**: *Quercus laurifolia* [TippinBe1970]. **Grossulariaceae**: *Grossularia leptantha* [Ferris1938a].
DISTRIBUTION: **Nearctic**: United States of America (Georgia [TippinBe1970], New Mexico [Ferris1938a], Ohio [Sander1904, Ferris1938a], Texas [Ferris1938a]).
GENERAL: Description and illustration of adult female by Newell (1899). According to Nakahara (1982: 29) and in correspondence to Yair Ben-Dov (25 February, 2003) *Diaspidiotus hunteri* (Newell) was interpreted by Ferris (1938a: 222) as a misidentification of *Diaspidiotus piceus* (Sanders).
KEYS: Ferris 1942: 33 (female) [North America]; Newell 1899: 4-5 (female) [North America].
CITATIONS: Borchs1966 [catalogue: 324]; Fernal1903b [catalogue: 265]; Ferris1941e [taxonomy: 44,46]; Ferris1942 [taxonomy: 446:33]; Leonar1900 [taxonomy, host, distribution: 340]; Lindin1957 [taxonomy: 545]; MacGil1921 [taxonomy, description, host, distribution: 412,413]; Newell1899 [taxonomy, description, illustration, host, distribution: 10]; TippinBe1970 [host, distribution: 9].

Diaspidiotus hydrangeae Takagi
Diaspidiotus hydrangeae Takagi, 1974: 20. Type data: JAPAN: Honsyu, Yamanasi-ken, Masutomi, on *Hydrangea paniculata*. Holotype female. Type depository: Sapporo: Ent. Institute, Faculty of Agriculture, Hokkaido University, Japan.
SCALE COVER: Takagi (1974) did not describe scale cover.
HOST PLANTS: **Hydrangeaceae**: *Hydrangea paniculata* [Takagi1974].
DISTRIBUTION: **Palaearctic**: Japan [Kawai1980] (Honshu [Takagi1974]).
GENERAL: Description and illustration of adult female by Takagi (1974).
CITATIONS: DanzigPe1998 [catalogue: 232]; Kawai1980 [taxonomy, description, host, distribution: 222]; Takagi1974 [taxonomy, description, illustration, host, distribution: 20-22,31].

Diaspidiotus inusitatus (Munting), new combination
Quadraspidiotus inusitatus Munting, 1969: 138. Type data: NAMIBIA: Witbooisvlei, on *Boscia foetida*; collected by J. Munting, 24.viii.1967. Holotype female. Type depository: Pretoria: South African National Collection of Insects, South Africa; type no. 2907/1.
SCALE COVER: Scale cover not observed by Munting (1969).
HOST PLANTS: **Capparidaceae**: *Boscia foetida* [Muntin1969].
DISTRIBUTION: **Afrotropical**: Namibia (Southwest Africa) [Muntin1969].
GENERAL: Description and illustration of adult female by Munting (1969).
CITATIONS: Muntin1969 [taxonomy, description, illustration, host, distribution: 138-139,161].

Diaspidiotus iranicus Kaussari & Balachowsky
Diaspidiotus iranicus Kaussari & Balachowsky, 1953: 24. Type data: IRAN: Kerman province, Chahdad, on *Tamarix* sp. Syntypes, female. Type depository: Paris: Muséum national d'Histoire naturelle, France.
SCALE COVER: Female scale subcircular, 1.7-2 mm; flat or slightly convex; colour dark brown, uniform, central area sometimes darker than margin; exuviae central or subcentral, brown, covered by secreted material of adult (Kaussari & Balachowsky, 1953).
HOST PLANTS: **Tamaricaceae**: *Tamarix* [KaussaBa1953].
DISTRIBUTION: **Palaearctic**: Iran [KaussaBa1953, Kaussa1955].
GENERAL: Description and illustration of adult female by Kaussari & Balachowsky (1953).
CITATIONS: Borchs1966 [catalogue: 324]; DanzigPe1998 [catalogue: 232]; Kaussa1955 [host, distribution: 16]; KaussaBa1953 [taxonomy, description, illustration, host, distribution: 24-26].

Diaspidiotus jaapi (Leonardi)
Targionia iaapi; Leonardi, 1918: 193. Nomen nudum. Listed as *Targionia iaapi* (Lindinger).
Targionia jaapi Leonardi, 1920: 115. Type data: ITALY: Liguria, Sestri-Lavante, on *Genista pilosa*. Syntypes, female. Type depository: Portici: Dipartimento de Entomologia e Zoologia Agraria di Portici, Università di Napoli Federico II, Italy. Notes: Leonardi incorrectly credited the authorship to Lindinger.
Aspidiotus jaapi; Lindinger, 1936: 152. Notes: Incorrect citation of "Lindinger" as author.
Targionia jaapi; Gómez-Menor Ortega, 1937: 122. Notes: Incorrect citation of "Lindinger" as author.
Quadraspidiotus jaapi; Ferris, 1943a: 86. Change of combination.
Gonaspidiotus jaapi; Lupo, 1954: 19. Change of combination.
Quadraspidiotus jaapi; Martin Mateo, 1983: 66. Notes: Incorrect citation of "Lindinger y Leonardi" as authors.
Diaspidiotus jaapi; Danzig & Pellizzari, 1998: 232. Change of combination.
SYSTEMATICS: Leonardi (1920: 115) incorrectly credited the authorship to Lindinger, as follows "*Aspidiotus jaapi* Linding., in Jaap, Cocciden-Sammlung, No. 173". This error was also published by Lindinger (1936: 152). Lindinger did not describe a species named *Aspidiotus jaapi* and Leonardi (1920) was the first to publish and describe it.
SCALE COVER: Female scale almost circular, diameter 1 mm; margin irregular; slightly convex; exuviae yellow orange, central or subcentral; secreted part of scale thin to robust, yellow; ventral vellum robust, white or white grey, remains attached to host plant (Leonardi, 1920). Scale of female circular, slightly convex, brown (Ferris, 1943a).
HOST PLANTS: **Chenopodiaceae**: *Salicornia* [Martin1983], *Salsola longifolia* [Martin1983]. **Compositae**: *Santolina chamacyparisus* [GomezM1946, Lupo1954, Martin1983]. **Leguminosae**: *Cytisus laburnum* [GomezM1937, Lupo1954], *Genista cinerea* [Balach1932d], *Genista pilosa* [Leonar1918, Leonar1920, Balach1930a,

Ferris1943a, Lupo1954], *Genista scorpionis* [Martin1983], *Retama sphaerocarpa* [GomezM1937, Lupo1954, Martin1983], *Ulex* [Lupo1954, GomezM1954, GomezM1956b, Martin1983], *Ulex boeticus* [GomezM1957, BlayGo1993].
Oleaceae: *Olea europaea*? [GomezM1946, Martin1983].
NATURAL ENEMIES: HYMENOPTERA **Aphelinidae**: *Physcus testaceus* Masi [GomezM1937].
DISTRIBUTION: **Palaearctic**: France [Balach1930a, Balach1932d, Ferris1943a]; Italy [Leonar1920, Ferris1943a, LongoMaPe1995]; Spain [GomezM1946, GomezM1956b, GomezM1957, Martin1983, BlayGo1993].
BIOLOGY: Occurring on the bark (Ferris, 1943a).
GENERAL: Description and illustration of adult female by Leonardi (1920), Ferris (1943a), Balachowsky (1950b) and by Lupo (1954).
KEYS: Blay Goicoechea 1993: 528-529 (female) [Spain]; Lupo 1954: 8-9 (female) [Italy]; Balachowsky 1950b: 401 (female) [Mediterranean]; Leonardi 1920: 104-105 (female) [Italy].
CITATIONS: Balach1930a [host, distribution: 179]; Balach1932d [taxonomy, host, distribution: XLIX]; Balach1950b [taxonomy, description, illustration, host, distribution: 443-446]; Balach1958b [taxonomy: 212]; BlayGo1993 [taxonomy, description, illustration, host, distribution: 554-558]; Borchs1966 [catalogue: 330]; Danzig1972 [taxonomy, host, distribution, economic importance: 210]; DanzigPe1998 [catalogue: 232]; Ferris1941e [taxonomy: 44]; Ferris1943a [taxonomy, description, illustration, host, distribution: 86,96,107]; Foldi2001 [distribution: 303-308]; Garcia1930 [host, distribution, biological control]; GomezM1937 [taxonomy, description, illustration, host, distribution: 122-124]; GomezM1946 [host, distribution: 73]; GomezM1954 [host, distribution: 122]; GomezM1956b [host, distribution: 482]; GomezM1957 [host, distribution: 48]; GomezM1958a [host, distribution: 8]; GomezM1958c [host, distribution: 406]; GomezM1960O [host, distribution: 162]; GomezM1965 [host, distribution: 90]; Leonar1918 [host, distribution: 193]; Leonar1920 [taxonomy, description, illustration, host, distribution: 105,115-116]; Lindin1936 [taxonomy: 152]; LongoMaPe1995 [distribution: 128]; Lupo1954 [taxonomy, description, illustration, host, distribution: 9,19-23]; Martin1983 [taxonomy, host, distribution: 66]; Rungs1933 [taxonomy: 116]; WeidneWa1968 [taxonomy: 172].

Diaspidiotus juglansregiae (Comstock)
Aspidiotus juglans-regiae Comstock, 1881a: 300. Type data: U.S.A.: California, Los Angeles, on bark of the larger limbs of English walnut, *Juglans regia*. Syntypes, female. Type depository: Washington: United States National Entomological Collection, U.S. National Museum of Natural History, District of Columbia, USA.
Aspidiotus juglans-regiae pruni Cockerell, 1894g: 131. Type data: U.S.A.: New Mexico, Las Cruses, on twigs of plum. Syntypes, female. Type depository: Washington: United States National Entomological Collection, U.S. National Museum of Natural History, District of Columbia, USA. Synonymy by Ferris, 1941e: 47.

Aspidiotus juglans-regiae albus Cockerell, 1894g: 132. Type data: U.S.A.: New Mexico, Mesila, on bark of pear trees. Syntypes, female. Type depository: Washington: United States National Entomological Collection, U.S. National Museum of Natural History, District of Columbia, USA. Synonymy by Ferris, 1938a: 190.

Aspidiotus iuglans-regiae; Leonardi, 1897: 285. Misspelling of species name.

Aspidiotus (*Diaspidiotus*) *juglans-regiae*; Cockerell, 1897i: 21. Change of combination.

Aspidiotus (*Diaspidiotus*) *juglans-regiae pruni*; Cockerell, 1897i: 21. Change of combination.

Aspidiotus iuglans-regiae; Leonardi, 1898c: 40. Misspelling of species name.

Aspidiotus (*Evaspidiotus*) *juglans-regiae*; Leonardi, 1898c: 40. Change of combination.

Aspidiotus iuglans-regiae pruni; Leonardi, 1898c: 42. Misspelling of species name.

Aspidiotus (*Evaspidiotus*) *juglans-regiae pruni*; Leonardi, 1898c: 42. Change of combination.

Aspidiotus (*Evaspidiotus*) *iuglans-regiae albus*; Leonardi, 1898c: 43. Change of combination.

Aspidiotus iuglans-regiae albus; Leonardi, 1898c: 43. Misspelling of species name.

Aspidiotus (*Evaspidiotus*) *juglans-regiae albus*; Leonardi, 1898c: 43. Change of combination.

Aspidiotus fernaldi Cockerell, 1898q: 323. Type data: U.S.A.: Massachusetts, Boston, Charlesbank Park, on *Gleditsia triacanthos*. Syntypes, female. Type depository: Washington: United States National Entomological Collection, U.S. National Museum of Natural History, District of Columbia, USA. Synonymy by Borchsenius, 1966: 331.

Aspidiotus fernaldi cockerelli Parrott, 1899: 10. Type data: U.S.A.: Kansas, Manhattan, on rough bark on trunk of maples; collected September 18, 1898. Syntypes, female. Type depository: Washington: United States National Entomological Collection, U.S. National Museum of Natural History, District of Columbia, USA. Synonymy by Borchsenius, 1966: 331.

Aspidiotus fernaldi albiventer Hunter, 1899: 6. Type data: USA: Kansas, Lawrence, on trunk of maple. Syntypes, female. Type depositories: Manhattan: Kansas State University Collections, Kansas, USA, and Washington: United States National Entomological Collection, U.S. National Museum of Natural History, District of Columbia, USA. Synonymy by Ferris, 1938a: 190.

Aspidiotus (*Diaspidiotus*) *glanduliferus* Cockerell, 1902l: 287. Type data: USA: Ohio, on *Pinus sylvestris*. Syntypes, female. Type depository: Washington: United States National Entomological Collection, U.S. National Museum of Natural History, District of Columbia, USA. Synonymy by Borchsenius, 1966: 331.

Aspidiotus glanduliferus; Fernald, 1903b: 260. Change of combination.

Diaspidiotus juglans-regiae albus; Cockerell, 1905b: 202. Change of combination.

Diaspidiotus juglans-regiae pruni; Cockerell, 1905b: 202. Change of combination.

Aspidiotus glandulifer Lindinger, 1907a: 20. Unjustified emendation.

Aspidiotus juglandis-regiae Lindinger, 1907a: 22. Unjustified emendation.

Aspidiotus (*Diaspidiotus*) *juglans-regiae albus*; Cockerell, 1907i: 21. Change of combination.

Furcaspis juglans-regiae; MacGillivray, 1921: 409. Change of combination.

Furcaspis juglans-regiae pruni; MacGillivray, 1921: 409. Change of combination.

Quadraspidiotus fernaldi; MacGillivray, 1921: 411. Change of combination.

Quadraspidiotus glanduliferus; MacGillivray, 1921: 411. Change of combination.

Aspidiotus albiventer; Ferris, 1938a: 190. Change of combination.

Aspidiotus albus; Ferris, 1938a: 190. Change of status.

Aspidiotus cockerelli Ferris, 1938a: 190. Synonymy by Borchsenius, 1966: 331.

Quadraspidiotus juglans-regiae; Ferris, 1938a: 257. Change of combination.

Aspidiotus pruni; Ferris, 1941e: 47. Change of status.

Aspidiotus (*Quadraspidiotus*) *juglans-regiae*; Merrill, 1953: 23. Change of combination.

Diaspidiotus juglans-regiae; Bustshik, 1958: 219. Change of combination.

Diaspidiotus juglansregiae; Borchsenius, 1966: 331. Justified emendation.

COMMON NAMES: English walnut scale [Comsto1881a, MerrilCh1923, Merril1953, Dekle1965c]; walnut scale [McKenz1956].

SCALE COVER: Female scale circular, 3 mm in diameter; flat, exuviae laterad of center; pale greyish brown; exuviae covered with secretion; position of first skin is indicated by pink or reddish brown prominence; ventral scale a mere film which adheres to bark (Comstock, 1881a). Colour photograph by Gill (1997).

HOST PLANTS: **Aceraceae**: *Acer* [Parrot1899], *Acer barbatum* [Ferris1938a], *Acer negundo* [McDani1970], *Acer saccharum* [MerrilCh1923]. **Altingiaceae**: *Liquidambar styraciflua?* [BesheaTiHo1973]. **Aquifoliaceae**: *Ilex* [Dekle1965c], *Ilex crenata* [BesheaTiHo1973], *Ilex glabra* [BesheaTiHo1973]. **Avicenniaceae**: *Avicennia nitida* [BesheaTiHo1973]. **Betulaceae**: *Betula* [McKenz1956], *Betula nigra* [Ferris1938a]. **Caprifoliaceae**: *Viburnum tinus* [Merril1953]. **Cornaceae**: *Cornus* [McKenz1956, McDani1970, BesheaTiHo1973]. **Euphorbiaceae**: *Adelia neomexicana* [Ferris1938a, McKenz1956]. **Hippocastanaceae**: *Aesculus* [BesheaTiHo1973]. **Juglandaceae**: *Juglans* [Ferris1938a, McKenz1956], *Juglans regia* [Comsto1881a, McKenz1956, McDani1970], *Juglans ruprestis* [McDani1970]. **Leguminosae**: *Cercis canadensis* [BesheaTiHo1973], *Gleditsia* [McDani1970], *Gleditsia triacanthos* [Cocker1898q, Ferris1938a], *Robinia pseudacacia* [BesheaTiHo1973]. **Magnoliaceae**: *Liriodendron tulipifera* [BesheaTiHo1973], *Magnolia virginiana* [BesheaTiHo1973]. **Oleaceae**: *Chionanthus virginicus* [BesheaTiHo1973], *Fraxinus* [Ferris1938a, McKenz1956]. **Pinaceae**: *Pinus* [Ferris1938a, McKenz1956, BesheaTiHo1973], *Pinus sylvestris* [Cocker1902l, Ferris1938a]. **Rosaceae**: *Photinia arbutifolia* [McKenz1956], *Photinia serrulata* [Merril1953], *Prunus* [Dekle1965c], *Prunus americana* [Dekle1965c], *Prunus cerasus* [Comsto1883], *Prunus persica* [Ferris1938a, McKenz1956, Dekle1965c], *Prunus serotina* [BesheaTiHo1973], *Pyracantha crenulata* [Merril1953], *Pyrus malus* [Comsto1883], *Rosa* [Ferris1938a, McKenz1956]. **Rutaceae**: *Zanthoxylum* [Ferris1938a, McKenz1956]. **Salicaceae**: *Salix* [Ferris1938a, McKenz1956, McDani1970]. **Tiliaceae**: *Tilia* [BesheaTiHo1973]. **Ulmaceae**: *Ulmus* [Ferris1938a, McKenz1956], *Ulmus crassifolia* [McDani1970]. **Vitaceae**: *Vitis* [Ferris1938a, McKenz1956].

NATURAL ENEMIES: ACARI **Acaridae**: *Thyreophagus entomophagus* [Koszta1996]. **Hemisarcoptidae**: *Hemisarcoptes* [GersonOcHo1990]. COLEOPTERA **Coccinellidae**: *Adalia bipunctata* [Koszta1996], *Chilocorus bivulnerus* [Koszta1996], *Chilocorus stigma* [Koszta1996]. HYMENOPTERA **Aphelinidae**: *Aphytis aonidiae* (Mercet) [RosenDe1979], *Ap. diaspidis* (Howard) [RosenDe1979], *Ap. maculicornis* (Masi) [RosenDe1979], *Ap. melanostictus* Compere [RosenDe1979], *Ap. melinus* DeBach [RosenDe1979], *Ap. mytilaspidis* (Le Baron) [RosenDe1979], *Ap. proclia* (Walker) [Gordh1979], *Marietta pulchella* (Howard) [Gordh1979], *Physcus varicornis* (Howard) [Gordh1979], *Prospaltella aurantii* (Howard) [Gordh1979], *Prospaltella fasciaventris* Girault [Gordh1979]. **Encyrtidae**: *Apterencyrtus microphagus* (Mayr) [Gordh1979, Koszta1996], *Coccidencyrtus ensifer* (Howard) [Gordh1979, Koszta1996], *Zaomma lambinus* (Walker) [Trjapi1989]. **Signiphoridae**: *Signiphora occidentalis* Howard [Gordh1979].

DISTRIBUTION: **Nearctic**: Canada [Nakaha1982]; Mexico [Nakaha1982]; United States of America (Alabama [BesheaTiHo1973], Arizona [Cocker1902k], California [Comsto1881a, McKenz1956, RosenDe1979], District of Columbia [Comsto1883], Florida [Wilson1917, Merril1953, Dekle1965c], Georgia [BesheaTiHo1973], Kansas [Parrot1899, Hunter1899], Massachusetts [Cocker1898q, Ferris1938a], Mississippi [Herric1911], Missouri [Hollin1923], New York [Comsto1883], Ohio [Cocker1902l, Ferris1938a], Texas [Herric1911, Ferris1938a, McDani1970], Virginia [BesheaTiHo1973]).

BIOLOGY: Occurring on bark (Ferris, 1938a).

ECONOMIC IMPORTANCE: A pest of walnut, *Juglans* sp. in North America (Ebeling, 1959; Gill, 1997; Davidson & Miller, 1990).

GENERAL: Description and illustration of adult female by Ferris (1938a), McKenzie (1956), Kosztarab (1996) and by Gill (1997).

KEYS: Gill 1997: 243 (female) [Species of California]; Kosztarab 1996: 575 (female) [Northeastern North America]; McDaniel 1970: 429 (female) [U.S.A.: Texas]; McKenzie 1956: 26 (female) [U.S.A.: California]; Ferris 1942: 40 (female) [North America]; Britton 1923: 371 (female) [U.S.A.: Connecticut]; Hollinger 1923: 7-8 (female) [U.S.A.: Missouri]; Lawson 1917: 217 (female) [U.S.A.: Kansas]; Dietz & Morrison 1916a: 289-290 (female) [U.S.A.: Indiana]; Cockerell 1905b: 202 (female) [U.S.A.: Colorado]; Newell 1899: 4-5 (female) [North America]; Comstock 1883: 55-57 (female) [North America].

CITATIONS: Balach1950b [taxonomy: 408]; Barnes1930 [biological control: 319-329]; BeardsDaHo1976 [economic importance: 103]; BeardsGo1975 [economic importance: 49]; BentleCoHa2000 [host, distribution, control]; BesheaTiHo1973 [host, distribution: 8]; Borchs1966 [catalogue: 331-332]; BoyceKaPe1939 [chemical control: 432-450]; Britto1923 [taxonomy, description, host, distribution: 371,374]; Bustsh1958 [taxonomy, description, host, distribution: 219,262]; Carnes1907 [taxonomy, host, distribution: 208]; ChuaWo1990 [host, distribution, economic importance: 550]; Cocker1894g [taxonomy, description, host, distribution: 131-132,286,354]; Cocker1896b [distribution: 334]; Cocker1897i [taxonomy, description, host, distribution: 5,8,16,17,21]; Cocker1898d [taxonomy: 65]; Cocker1898q [taxonomy, description, host, distribution: 323]; Cocker1899a [taxonomy: 396]; Cocker1902k [taxonomy, description, host, distribution: 450]; Cocker1902l [taxonomy, description, host, distribution: 287-288]; Cocker1905b [taxonomy: 202]; Comper1961a [biological control: 17-71]; Comsto1881a [taxonomy, description, illustration, host, distribution: 300-301]; Comsto1883 [host, distribution: 61-62]; DavidsMi1990 [host, distribution, economic importance: 603-632]; Dekle1965c [taxonomy, description, host, distribution: 125]; Dekle1976 [taxonomy, description, host, distribution, economic importance: 144]; DeSant1940 [biological control: 29-44]; DietzMo1916a [taxonomy, description, illustration, host, distribution: 301-302]; Dougla1912 [taxonomy: 204]; Drake1935 [host, distribution, control: 83-91]; Ebelin1949 [host, distribution, life history, control]; Ebelin1959 [host, distribution, economic importance: 343]; Felt1924 [host, distribution, economic importance, life history, control]; Fernal1903b [catalogue: 258,259,260,265]; Ferris1921b [taxonomy: 94]; Ferris1938a [taxonomy, description, illustration, host, distribution: 257]; Ferris1941e [taxonomy: 43,44]; Ferris1942 [taxonomy: 446:40]; Garcia1930 [host, distribution, biological control]; Garcia1931a [host, distribution, biological control: 659-666]; GersonOcHo1990 [biological control: 77-97]; Gill1997 [host, distribution, taxonomy, description, illustration, economic importance: 244-245,248]; Gordh1979 [biological control: 895-897,907,908,911,]; GordonPo1988 [host, distribution, life history, biological control: 1181-1185]; Hamilt1936 [chemistry, chemical control, physiology: 150-160]; Herric1911 [taxonomy, description, illustration, host, distribution: 10,17-18,51]; Hollin1923 [taxonomy, description, host, distribution: 12-13]; Howard1895e [biological control: 1-44]; Hunter1899 [taxonomy, description, illustration, host, distribution: 6-8]; IPMW1987 [economic importance, biological control]; Jarvis1908TD [taxonomy: 61]; KnowltSm1936 [host, distribution: 263-267]; Koehle1964 [host, distribution, control]; Koszta1996 [taxonomy, description, illustration, host, distribution, life history, biological control, economic importance: 578-580]; Kozar1990c [life history, economic importance, host, distribution: 593-602]; Lawson1917 [taxonomy, description, illustration, host, distribution: 217-221]; Leonar1897 [taxonomy: 285]; Leonar1898a [taxonomy: 75]; Leonar1898c [taxonomy, description, illustration, host, distribution: 40-44]; Lindin1907a [taxonomy: 22]; Lindin1957 [taxonomy: 545]; MacGil1921 [taxonomy, description, host, distribution: 409-411]; Maranh1946 [taxonomy: 164-179]; McKenz1956 [taxonomy, description, illustration, host, distribution: 26,81-82]; Merril1953 [taxonomy, description, host, distribution: 23-24]; MerrilCh1923 [taxonomy, description, host, distribution, economic importance: 202-203]; MichelOr1958 [host, distribution, taxonomy, biological control, chemical control: 46-57]; MillerDa1990 [host, distribution, economic importance: 305]; Morgan1894 [host, distribution: 982-1006]; Nakaha1982 [host, distribution: 78]; Newell1899 [taxonomy, description, host, distribution: 5,18-21]; NewellRo1908 [host, distribution: 150-155]; Parrot1899 [taxonomy, description, illustration, host, distribution: 10-11]; RosenDe1979 [host, distribution, biological control: 291-294,383-387,]; Sander1904a [taxonomy, description, illustration, host, distribution: 57,61,62]; SchmutKlLu1957 [host, distribution, economic importance: 484]; SchuhMo1948 [host, distribution, control]; SeveriSe1909 [taxonomy: 298]; Starne1897 [taxonomy: 23]; StoetzDa1974 [taxonomy, life history: 138-140]; Trjapi1989 [biological

control: 312]; Wilson1917 [taxonomy, description, host, distribution: 22-23];
Zahrad1990a [host, distribution, description: 651].

Diaspidiotus kafkai (Cockerell)

Aspidiotus juglans-regiae kafkae Cockerell, 1898d: 65. Type data: AUSTRIA:
Vienna, on bark of *Fraxinus excelsior*; collected by Karl L. Kafka. Syntypes,
female. Type depository: Washington: United States National Entomological
Collection, U.S. National Museum of Natural History, District of Columbia, USA.
Aspidiotus (*Diaspidiotus*) *juglans-regiae kafkae*; Cockerell, 1899a: 396. Change of
combination.
Aspidiotus (*Evaspidiotus*) *iuglans-regiae kafkae*; Leonardi, 1900: 341. Change of
combination.
Aspidiotus (*Evaspidiotus*) *iuglans-regiae kafkae*; Leonardi, 1900: 341. Misspelling
of species name.
Aspidiotus (*Evaspidiotus*) *juglansregiae kafkae*; Leonardi, 1900: 341. Justified
emendation.
Aspidiotus (*Evaspidiotus*) *juglansregiae kafkae*; Leonardi, 1900: 341. Change of
combination.
Aspidiotus juglans-regiae kafkai; Lindinger, 1907a: 19. Emendation that is justified.
Notes: The species was named after Karl L. Kafka.
Aspidiotus juglans-regiae kafkae; Lindinger, 1912b: 359. Incorrect synonymy.
Notes: Incorrect synonymy with *Quadraspidiotus pyri*; see Borchsenius, 1966: 332.
Aspidiotus kafkae; Ferris, 1941e: 44. Change of status.
Quadraspidiotus kafkae; Borchsenius, 1966: 332. Change of combination.
Quadraspidiotus kafkae; Borchsenius, 1966: 332. Change of combination.
Diaspidiotus kafkae; Danzig & Pellizzari, 1998: 232. Change of combination.
SCALE COVER: Female scale circular, 2.5 mm diameter, flat, dark grey or
blackish, same colour as bark; exuviae subcentral, covered, orange-reddish. Male
scale small and elongate, about twice as long as broad (Cockerell, 1898d).
HOST PLANTS: **Oleaceae**: *Fraxinus excelsior* [Cocker1898d].
DISTRIBUTION: **Palaearctic**: Austria [Cocker1898d].
CITATIONS: Borchs1966 [catalogue: 332]; Cocker1898d [taxonomy, description,
host, distribution: 65-66]; Cocker1899a [taxonomy: 396]; DanzigPe1998 [catalogue:
232-233]; Fernal1903b [catalogue: 266]; Ferris1941e [taxonomy: 44]; Leonar1900
[taxonomy, host, distribution: 341]; Lindin1907a [taxonomy: 19]; Lindin1912b
[taxonomy: 359].

Diaspidiotus kaussari Balachowsky

Diaspidiotus kaussari Balachowsky, 1950b: 494. Type data: IRAN: Abe-Ali, near
Teheran, on branches of *Salix* sp. Syntypes, female. Type depository: Paris:
Muséum national d'Histoire naturelle, France.
SCALE COVER: Female scale subcircular, very flat; diameter 2 mm; exuviae light
yellow, central; secreted part white, grey in center. Male scale of similar structure,
oval, grey, exuviae darker, 1.6 mm (Balachowsky, 1950b).
HOST PLANTS: **Salicaceae**: *Salix* [Balach1950b].
DISTRIBUTION: **Palaearctic**: Iran [Balach1950b, Kaussa1955].

GENERAL: Description and illustration of adult female by Balachowsky (1950b).
KEYS: Balachowsky 1950b: 492 (female) [Mediterranean].
CITATIONS: Balach1950b [taxonomy, description, illustration, host, distribution: 494-497]; Borchs1966 [catalogue: 324]; DanzigPe1998 [catalogue: 233]; Kaussa1955 [host, distribution: 16].

Diaspidiotus kuwanai Takahashi

Diaspidiotus kuwanai Takahashi, 1952a: 14. Type data: JAPAN: Tokyo, Hikawa near Tokyo, Hakone, on *Quercus glandulifera*. Syntypes, female. Type depository: Sapporo: Entomological Institute, Faculty of Agriculture, Hokkaido University, Japan.
Hemiberlesia kuwanai; Borchsenius, 1966: 306. Change of combination.
Diaspidiotus kuwanai; Danzig & Pellizzari, 1998: 233. Revived combination.
SCALE COVER: Female scale subcircular or oval, 1-1.2 mm long; convex; greyish pale brown; exuviae pale brown, at margin of scale (Takahashi, 1952a).
HOST PLANTS: **Fagaceae**: *Castanea crenata* [Takagi1974], *Quercus* [Kawai1977], *Quercus glandulifera* [Takaha1952a], *Quercus serrata* [Takagi1974].
DISTRIBUTION: **Palaearctic**: Japan [Takaha1952a, Takagi1974, Kawai1977, Kawai1980].
GENERAL: Description and illustration of adult female by Takahashi (1952a) and by Takagi (1974).
CITATIONS: Borchs1966 [catalogue: 306]; DanzigPe1998 [catalogue: 233]; Kawai1980 [taxonomy, description, host, distribution: 219]; Muraka1970 [host, distribution: 74]; Takagi1974 [taxonomy, description, illustration, host, distribution: 4-9]; Takaha1952a [taxonomy, description, illustration, host, distribution: 14-15].

Diaspidiotus labiatarum (Marchal)

Aspidiotus (*Evaspidiotus*) *labiatarum* Marchal, 1909: 872. Type data: FRANCE: Corsica, Cort, on *Stachys glutinosa* and *Teucrium capitatum*. Syntypes, female. Type depository: Paris: Muséum national d'Histoire naturelle, France.
Aspidiotus privignus Lindinger, 1909b: 151. Type data: ITALY: Eremo di S. Zeno, Baldi (M. Baldo), altitude 800-1000 meters, on *Hypericum coris*; GREECE: Attika, Mount Pentelikon, on *Thymelaea tartonraira*. Syntypes, female. Type depository: Hamburg: Zoologisches Institut und Zoologishces Museum, Universität von Hamburg, Germany. Synonymy by Ferris, 1941e: 47.
Aspidiotus labiatarum; Lindinger, 1912b: 315. Change of combination.
Quadraspidiotus labiatarum; Balachowsky, 1932e: 234. Change of combination.
Aspidiotus thymi; Šulc, 1934: 19. Misspelling of species name. Notes: Typing error or misspelling of *Aspidiotus thymi* rather than *Aspidiotus baudysi* Šulc.
Aspidiotus baudysi Šulc, 1934: 9. Type data: CZECH REPUBLIC: near Brno, on *Thymus serphyllum*. Syntypes, female. Type depository: Brno: K. Šulc Collection, Moravian Museum, Czech Republic. Synonymy by Lindinger, 1936: 152.
Aspidiotus (*Euraspidiotus*) *labiatarum*; Thiem & Gerneck, 1934a: 131. Change of combination.
Diaspidiotus labiatarum; Rungs, 1948: 114. Change of combination.

Rhizaspidiotus labiatarum; Lupo, 1948: 199. Change of combination.

Aspidiotus baudsysi; Blay Goicoechea, 1993: 572. Misspelling of species name. Notes: Misspelling of *Aspidiotus baudysi* Šulc, 1934.

Diaspidiotus labiatarum; Danzig & Pellizzari, 1998: 233. Revived combination.

SCALE COVER: Female scale elongated, 1.2-1.5 mm long, 0.8-1.2 mm wide; white; ventral scale thick, remains attached to host plant (Marchal, 1909).

HOST PLANTS: **Anacardiaceae**: *Rhus pentaphylla* [Balach1950b]. **Caryophyllaceae**: *Spergula* [Balach1950b], *Spergula arvensis* [Foldi2000], *Spergula fimbriata* [Balach1950b], *Spergula fimbriata var. condensata* [Rungs1948]. **Chenopodiaceae**: *Salicornia fructicosa* [Balach1937]. **Cistaceae**: *Fumana spachii* [Balach1950b], *Fumana thymifolia* [Balach1950b]. **Crassulaceae**: *Sedum* [Balach1950b, Bachma1953]. **Euphorbiaceae**: *Euphorbia characias* [Balach1950b]. **Globulariaceae**: *Globularia cordifolia* [Leonar1920, Balach1950b]. **Guttiferae**: *Hypericum coris* [Lindin1909b, Balach1950b]. **Labiatae**: *Prasium maius* [Balach1950b, Bachma1953], *Salvia* [Balach1950b, Bachma1953], *Stachys glutinosa* [Marcha1909, Sander1909a, Leonar1920, Balach1929a, Balach1932d], *Teucrium* [Bachma1953, GomezM1957, Martin1983], *Teucrium capitatum* [Marcha1909, Sander1909a, Leonar1920, Balach1929a], *Teucrium montanum* [Zahrad1952], *Thymus mastichinus* [GomezM1960O, Martin1983], *Thymus serphyllum* [Balach1931a, Balach1933a, Sulc1934, Zahrad1952]. **Leguminosae**: *Cytisus tridentatus var. riphaeus* [Rungs1948], *Cytisus tridentaus* [Balach1950b], *Genista hispanica* [Balach1932e, Balach1950b]. **Pistaciaceae**: *Pistacia lentiscus* [Foldi2000]. **Thymelaeaceae**: *Thymelaea* [Bodenh1937], *Thymelaea hirsuta* [Lindin1909b, Korone1934, Bodenh1935].

NATURAL ENEMIES: HYMENOPTERA **Encyrtidae**: *Metaphycus hubai* Hoffer [Trjapi1989, GuerriNo2000].

DISTRIBUTION: **Palaearctic**: Algeria [Balach1929a]; Corsica [Sander1909a, Balach1931a, Balach1932d]; Czech Republic [Sulc1934, Zahrad1952, Zahrad1977]; France [Balach1932e, Balach1933a, Foldi2000]; Greece [Lindin1909b, Korone1934]; Italy [Lindin1909b, Leonar1920, Balach1950b, LongoMaPe1995]; Morocco [Rungs1948]; Slovak Republic [Zahrad1952]; Spain [Balach1935b, GomezM1937, Martin1983, BlayGo1993]; Switzerland [Balach1950b]; Yugoslavia [Balach1950b, Bachma1953].

GENERAL: Description and illustration of adult female by Balachowsky (1950b), Zahradník (1952) and by Danzig (1993).

KEYS: Danzig 1993: 179-182 (female) [Europe]; Kosztarab & Kozár 1978: 171 (female) [Hungary]; Zahradník 1952: 114-115 (female) [Czech Republic]; Balachowsky 1950b: 403 (female) [Mediterranean]; Leonardi 1920: 29-30 (female) [Italy].

CITATIONS: Bachma1953 [host, distribution: 178]; Balach1929a [host, distribution: 317]; Balach1931a [host, distribution: 97]; Balach1932d [taxonomy, host, distribution: XLV-XLVI]; Balach1932e [host, distribution: 234,236]; Balach1933a [taxonomy, illustration, host, distribution: 37]; Balach1935b [host, distribution: 257]; Balach1937 [host, distribution: 339]; Balach1950b [taxonomy, description, illustration, host, distribution: 478-482]; BlayGo1993 [taxonomy, description, illustration, host, distribution: 572-576]; Bodenh1935 [host, distribution: 246]; Bodenh1935c [taxonomy, distribution: 1156]; Bodenh1937 [host, distribution: 216]; Borchs1966 [catalogue: 332]; Danzig1993 [taxonomy, description, illustration, host, distribution: 199-200]; DanzigPe1998 [catalogue: 233]; Ferris1941e [taxonomy: 41,45-46]; Foldi2000 [host, distribution: 84]; Foldi2001 [distribution: 303-308]; Foldi2002 [host, distribution: 246]; GomezM1937 [taxonomy, description, illustration, host, distribution: 78-79]; GomezM1957 [taxonomy, host, distribution: 46]; GomezM1958a [host, distribution: 8]; GomezM1960O [host, distribution: 162]; GuerriNo2000 [host, distribution, biological control: 159-161]; Korone1934 [taxonomy, description, illustration, host, distribution: 10-11,27]; KosztaKo1978 [taxonomy, description, host, distribution: 171]; Leonar1920 [taxonomy, description, illustration, host, distribution: 59-61]; Lindin1909b [taxonomy, description, illustration, host, distribution: 151-152,220]; Lindin1912b [taxonomy, description, host, distribution: 164,183,315]; Lindin1935 [taxonomy: 128]; Lindin1936 [taxonomy: 152]; LongoMaPe1995 [distribution: 128]; Lupo1948 [taxonomy, description, illustration, host, distribution: 199-203]; MacGil1921 [taxonomy, description, host, distribution: 398,399]; Marcha1909 [taxonomy, description, illustration, host, distribution: 872]; Martin1983 [taxonomy, host, distribution: 66-67]; Rungs1948 [host, distribution: 114]; Sander1909a [taxonomy, host, distribution: 52,53]; Schmut1959 [taxonomy, description, host, distribution: 77,95]; Sulc1934 [taxonomy, description, illustration, host, distribution: 9-17,20-21]; ThiemGe1934a [taxonomy, description, host, distribution: 130-158,208-238]; Trjapi1989 [biological control: 244]; WeidneWa1968 [taxonomy: 173]; Zahrad1952 [taxonomy, description, illustration, host, distribution: 139-142]; Zahrad1959b [host, distribution: 60]; Zahrad1977 [taxonomy, distribution: 121].

Diaspidiotus laperrinei (Balachowsky)

Aspidiotus (*Hemiberlesia*) *laperrinei* Balachowsky, 1929a: 314. Type data: ALGERIA: Hoggar, Mountain Amezzereei, 2500 meters altitude), on *Olea laperrinei*. Holotype female. Type depository: Paris: Muséum national d'Histoire naturelle, France. Notes: See discussion on spelling of species epithet in Systematics below.

Aspidiotus lapperinei; Lindinger, 1932f: 200. Misspelling of species name.

Parlatoreopsis lapperinei; Lindinger, 1932f: 200. Change of combination. Notes: Lindinger (1932f: 200) misspelled species name as *Aspidiotus lapperinei*.

Aspidaspis laperrinei; Balachowsky, 1950b: 536. Change of combination.

Quadraspidiotus laperrinei; Balachowsky, 1958b: 212. Change of combination.

Diaspidiotus laperrinei; Danzig & Pellizzari, 1998: 233. Change of combination.

SYSTEMATICS: Balachowsky (1929a) in heading (p. 314) misspelled the species epithet as *Aspidiotus (Hemiberlesia) lapperrinei*, while in caption to Fig. 25 (p. 315) it was spelled *laperrinei*. On page 317, Balachowsky (1929a) clearly indicated that the species was named after général Laperrine. Therefore, the species epithet *laperrinei* is here accepted.

SCALE COVER: Female scale circular or subcircular, small, diameter 0.7-0.8 mm; slightly convex; irregular along margin; all scale covered with white waxy secretion; exuviae yellow gold; ventral vellum poorly developed (Balachowsky, 1929a).

HOST PLANTS: **Apocynaceae**: *Nerium oleander* [Balach1950b]. **Myrtaceae**: *Myrtus nivellii* [Balach1950b, Balach1958a]. **Oleaceae**: *Olea laperrinei* [Balach1929a, Balach1932d].

DISTRIBUTION: **Palaearctic**: Algeria [Balach1929a, Balach1932d, Balach1934d, Kaussa1955].

GENERAL: Description and illustration of adult female by Balachowsky (1929a, 1950b, 1958b).

KEYS: Balachowsky 1950b: 536 (female) [Mediterranean].

CITATIONS: Balach1929a [taxonomy, description, illustration, host, distribution: 314-316]; Balach1932d [taxonomy, host, distribution: VIII]; Balach1934d [host, distribution: 146]; Balach1950b [taxonomy, description, illustration, host, distribution: 536-539]; Balach1954f [host, distribution: 99]; Balach1958 [host, distribution: 23]; Balach1958a [host, distribution: 36]; Balach1958b [taxonomy, description, illustration, host, distribution: 212-214]; Borchs1966 [catalogue: 332]; DanzigPe1998 [catalogue: 233-234]; Ferris1941e [taxonomy: 45]; Kaussa1955 [host, distribution: 16]; Lindin1932f [taxonomy: 200]; Lindin1957 [taxonomy: 545]; Quilis1935 [life history, ecology: 621-633].

Diaspidiotus laurinus (Lindinger)

Targionia laurina Lindinger, 1912b: 198. Type data: MADEIRA: on *Oreodaphne (=Ocotea) foetens*. Syntypes, female. Type depository: Zoologisches Institut und Zoologishces Museum, Universität von Hamburg, Germany.

Gonaspidiotus laurinus; MacGillivray, 1921: 432. Change of combination requiring emendation of species name for agreement in gender.

Cryptaspidiotus laurinus; Lindinger, 1935: 146. Change of combination.

Quadraspidiotus laurinus; Ferris, 1943a: 86. Change of combination.

Diaspidiotus laurinus; Danzig & Pellizzari, 1998: 234. Change of combination.

SCALE COVER: Female scale dark brown, elongated, 1-1.5 mm long, 0.5-1 mm wide; exuviae brown, subcentral, broadly elliptical to circular (Lindinger, 1912b). Scale of female quite convex, dark brown or reddish, circular but produced at one side into a sort of spout. Scale of male very small, oval, pale (Ferris, 1943a).

HOST PLANTS: **Lauraceae**: *Laurus canariensis* [Green1923b, Ferris1943a], *Ocotea foetens* [Lindin1912b, Green1923b, Ferris1943a], *Oreodaphne foetens* [Balach1938a]. **Leguminosae**: *Spartocytisus nubigenus* [GomezM1962]. **Zygophyllaceae**: *Zygophyllum fontanesi* [GomezM1962].

DISTRIBUTION: **Palaearctic**: Canary Islands [GomezM1962, MatileOr2001]; Madeira Islands [Lindin1912b, Balach1938a, Ferris1943a].

BIOLOGY: Infesting leaves and bark (Ferris, 1943a).

GENERAL: Description and illustration of adult female by Ferris (1943a) and by Balachowsky (1950b).

KEYS: Balachowsky 1950b: 399 (female) [Mediterranean].

CITATIONS: Balach1938a [host, distribution: 152]; Balach1950b [taxonomy, description, illustration, host, distribution: 437-440]; Borchs1966 [catalogue: 332]; DanzigPe1998 [catalogue: 234]; Ferris1943a [taxonomy, description, illustration, host, distribution: 86,97-98,109]; GomezM1962 [taxonomy, description, host, distribution: 173-175]; Green1923b [host, distribution: 89]; Lindin1912b [taxonomy, description, host, distribution: 198,227]; Lindin1935 [taxonomy: 146]; MacGil1921 [taxonomy, description, host, distribution: 432]; MatileOr2001 [host, distribution: 190]; Sassce1915 [taxonomy, host, distribution: 35]; WeidneWa1968 [taxonomy: 179].

Diaspidiotus leguminosum (Archangelskaya)

Aspidiotus leguminosum Archangelskaya, 1937: 109. Type data: UZBEKISTAN: Fergana, on *Robinia pseudacacia*. Lectotype female, by subsequent designation Danzig, 1993: 205. Type depository: St. Petersburg: (= Leningrad) Zoological Museum, Academy of Science, Russia.

Diaspidiotus leguminosum; Borchsenius, 1950b: 229. Change of combination.

COMMON NAME: bobovaya shitovka [BazaroSh1971].

SCALE COVER: Female scale more or less circular, 1.5 mm in diameter; slightly convex, but not flat; colour white or yellowish; exuviae dark yellow or yellow brown, central (Archangelskaya, 1937).

HOST PLANTS: **Leguminosae**: *Caragana* [Balach1950b, BazaroSh1971], *Halimodendron* [Balach1950b], *Halimodendron argenteum* [Balach1950b], *Halimodendron halodendron* [BazaroSh1971], *Robinia pseudacacia* [Archan1937, Balach1950b, BazaroSh1971].

DISTRIBUTION: **Palaearctic**: Kazakhstan (Karaganda Oblast [BazaroSh1971]); Tajikistan (=Tadzhikistan) [BazaroSh1971]; Turkmenistan [Balach1950b, BazaroSh1971]; Uzbekistan (Fergana Oblast [Archan1937], Samarkand Oblast [Balach1950b]).

ECONOMIC IMPORTANCE: Schmutterer et al. (1957) reported that it occasionally damage *Caragana*, *Halimodendron* and *Robinia* in Tajikistan, Turkmenistan and Uzbekistan.

GENERAL: Description and illustration of adult female by Balachowsky (1950b), Bazarov & Shmelev (1971) and by Danzig (1993).

KEYS: Danzig 1993: 179-182 (female) [Europe]; Bazarov & Shmelev 1971: 198 (female) [Central Asia]; Balachowsky 1950b: 494 (female) [Mediterranean].

CITATIONS: Archan1937 [taxonomy, description, host, distribution: 99,109-110]; Balach1950b [taxonomy, description, illustration, host, distribution: 531-534]; BazaroSh1971 [taxonomy, description, illustration, host, distribution, life history: 201-203]; Borchs1950b [taxonomy, description, illustration, host, distribution: 229,231]; Borchs1966 [catalogue: 324]; Danzig1993 [taxonomy, description, illustration, host, distribution, life history: 205-206]; DanzigPe1998 [catalogue: 234]; Ferris1941e [taxonomy: 45]; KaussaBa1953 [taxonomy: 26]; MillerDa1990 [host, distribution, economic importance: 301]; SchmutKlLu1957 [host, distribution, economic importance: 491].

Diaspidiotus lenticularis (Lindinger)

Aspidiotus lenticularis Lindinger, 1912b: 149. Type data: CROATIA: Arbe, on *Euphorbia wulfeni*. Syntypes, female. Type depository: Hamburg: Zoologisches Institut und Zoologishces Museum, Universität von Hamburg, Germany.

Targionidea lenticularis; MacGillivray, 1921: 449. Change of combination.

Aspidiotus lenticularis marocanus Green, 1928c: 374. Type data: MOROCCO: on *Olea europaea*; collected by A. Balachowsky. Syntypes, female and first instar. Type depositories: London: The Natural History Museum, England, UK, and Paris: Muséum national d'Histoire naturelle, France. Synonymy by Danzig, 1993: 200.

Aspidiotus ostreiformis anactenus Koroneos, 1934: 13. Notes: Koroneos (1934: 13) credited this *nomen nudum* to Malenotti.

Diaspidiotus anactenus Ferris, 1941e: 40. Nomen nudum.

Aspidiotus marocanus; Ferris, 1941e: 45. Change of status.

Quadraspidiotus lenticularis; Ferris, 1943a: 96. Change of combination.

Diaspidiotus lenticularis; Borchsenius, 1950b: 228. Change of combination.

Aspidiotus ostreiformis anactenus Borchsenius, 1966: 332. Nomen nudum.

Quadraspidiotus marocanus; Borchsenius, 1966: 334. Change of combination.

Qiadraspidiotus lenticularis; Borchsenius, 1966: 403. Misspelling of genus name.

Aspidiotus ostreiformis anactenus Danzig, 1993: 200. Nomen nudum.

Aspidiotus ostreaeformis anactenus Danzig & Pellizzari, 1998: 234. Nomen nudum.

Diaspidiotus lenticularis; Danzig & Pellizzari, 1998: 234-235. Revived combination.

SYSTEMATICS: Danzig (1993: 200) synonymised *Aspidiotus lenticularis marocanus* with *Quadraspidiotus lenticularis* while Borchsenius (1966: 332) regarded *Aspidiotus lenticularis marocanus* as a valid species: *Quadraspidiotus marocanus*.

SCALE COVER: Scale of female quite flat, circular, grey, or slightly brown; exuviae subcentral. Scale of male oval, pale, exuvia subcentral (Ferris, 1943a).

HOST PLANTS: **Betulaceae**: *Betula alba* [Balach1950b, Zahrad1972]. **Euphorbiaceae**: *Euphorbia wulfeni* [Lindin1912b, Ferris1943a, Bachma1953]. **Fagaceae**: *Quercus* [BachmaGe1950, Balach1950b, Bachma1953, Zahrad1972], *Quercus petraea* [Zahrad1972], *Quercus ruber* [Balach1950b, Zahrad1972], *Quercus suber* [GomezM1959, GomezM1960O, Martin1983]. **Leguminosae**: *Ceratonia siliqua* [InserrCa1987]. **Moraceae**: *Ficus carica* [Balach1928a, Balach1932d, Ferris1943a], *Ficus silvatica* [Zahrad1972]. **Oleaceae**: *Fraxinus excelsior* [BachmaGe1950, Zahrad1972], *Olea* [Balach1950b], *Olea europaea* [Balach1928a, Green1928c, Balach1932d, Ferris1943a, GomezM1960O]. **Pinaceae**: *Pinus* [Martin1983]. **Pistaciaceae**: *Pistacia lentiscus* [Leonar1918, Leonar1920, Ferris1943a, Balach1950b, Bachma1953], *Pistacia vera* [InserrCa1987]. **Rhamnaceae**: *Rhamnus alaternus* [Korone1934, Ferris1943a]. **Rosaceae**: *Prunus* [Balach1950b], *Prunus domestica* [BachmaGe1950, BrookeHu1968], *Pyrus communis* [BrookeHu1968]. **Salicaceae**: *Populus alba* [Zahrad1972], *Populus tremula* [Leonar1918, Leonar1920, Ferris1943a, Balach1950b, Zahrad1972].

DISTRIBUTION: **Australasian**: Australia (South Australia [BrookeHu1968]). **Palaearctic**: Canary Islands [Ferris1943a]; Croatia [Lindin1912b, Ferris1943a, Bachma1953]; France [Balach1932d, Foldi2000]; Greece [Korone1934, Ferris1943a]; Italy [Leonar1918, Leonar1920, Balach1950b, LongoMaPe1995]; Morocco [Green1928c, Balach1928a, Balach1932d, Ferris1943a]; Sicily [InserrCa1987]; Spain [GomezM1959, GomezM1960O, Martin1983, BlayGo1993]; Sweden [Gertss2001]; Switzerland [Ferris1943a, BachmaGe1950]; Turkey [Yasar1995a]; Ukraine (Krym (= Crimea) Oblast [Ferris1943a]).

BIOLOGY: Occurring on twigs of host plant (Ferris, 1943a).

GENERAL: Description and illustration of adult female by Ferris (1943a), Balachowsky (1950b), Brookes & Hudson (1968), Tereznikova (1986), Danzig (1993) and by Yaşar (1995a). Description of adult female by Gómez-Menor Ortega (1959). Description and illustration of second instar by Brookes & Hudson (1968).

KEYS: Blay Goicoechea 1993: 528-529 (female) [Spain]; Danzig 1993: 179-182 (female) [Europe]; Tereznikova 1986: 97-98 (female) [Ukraine]; Brookes & Hudson 1968: 91 (larva) [Australia]; Brookes & Hudson 1968: 91 (female) [Australia]; Balachowsky 1950b: 401 (female) [Mediterranean]; Lupo 1948: 175 (female) [Italy]; Leonardi 1920: 29-30 (female) [Italy].

CITATIONS: Argyri1990 [host, distribution, economic importance: 579-583]; Bachma1953 [host, distribution: 178]; BachmaGe1950 [host, distribution: 118]; Balach1928a [host, distribution: 138]; Balach1930a [host, distribution: 178]; Balach1932b [ecology: 517-522]; Balach1932d [taxonomy, host, distribution, economic importance: V; XLVI]; Balach1950b [taxonomy, description, illustration, host, distribution: 433-437]; BlayGo1993 [taxonomy, description, illustration, host, distribution: 545-548]; Borchs1937 [taxonomy, description, illustration, host, distribution: 133]; Borchs1950b [taxonomy, description, illustration, host, distribution: 228,231]; Borchs1966 [catalogue: 332-334]; BrookeHu1968 [taxonomy, description, illustration, host, distribution: 92-93]; Danzig1964 [taxonomy, host, distribution: 652]; Danzig1972 [taxonomy, host, distribution, economic importance: 210]; Danzig1993 [taxonomy, description, illustration, host, distribution: 200-201]; DanzigPe1998 [catalogue: 234-235]; Ferris1941e [taxonomy: 40,45]; Ferris1943a [taxonomy, description, illustration, host, distribution: 96-97,108]; Foldi2000 [host, distribution : 84]; Foldi2001 [distribution: 303-308]; FoldiDe1998 [taxonomy, host, distribution: 201-202]; Gertss2001 [distribution: 123-130]; GomezM1958a [host, distribution: 7]; GomezM1959 [taxonomy, description, host, distribution: 155-157]; GomezM1960O [taxonomy, description, illustration, host, distribution: 163-166]; GomezM1968 [host, distribution: 541]; Green1928c [taxonomy, description, illustration, host, distribution: 374-376]; InserrCa1987 [host, distribution: 94]; Kiritc1932a [taxonomy: 246]; Korone1934 [taxonomy, description, illustration, host, distribution: 13]; KosztaKo1978 [taxonomy, description, host, distribution: 171]; Koteja1990b [life history, structure, anatomy: 233-242]; Leonar1918 [host, distribution: 189]; Leonar1920 [taxonomy, description, illustration, host, distribution: 61-62]; Lindin1912b [taxonomy, description, illustration, host, distribution: 173]; Lindin1935 [taxonomy: 129]; Lindin1957 [taxonomy: 546]; LongoMaPe1995 [distribution: 128]; Lupo1948 [taxonomy, description, illustration, host, distribution: 189-193]; MacGil1921 [taxonomy, description, host, distribution: 449]; Martin1983 [taxonomy, host, distribution: 67]; MathysGuSt1965 [taxonomy: 65-67]; MillerDa1990 [host, distribution, economic importance: 305]; Sassce1915 [taxonomy, host, distribution: 34]; Schmut1959 [taxonomy, description, host, distribution: 76,77,92]; Takagi1956b [taxonomy: 89]; Terezn1986 [taxonomy, description, illustration, host, distribution: 103-104]; WeidneWa1968 [taxonomy: 173]; Yasar1995a [taxonomy, description, illustration, host, distribution: 112-114]; Zahrad1972 [taxonomy, host, distribution: 435]; Zahrad1990 [host, distribution, description: 651].

Diaspidiotus lepineyi (Balachowsky)
Quadraspidiotus lepineyi Balachowsky, 1950b: 415. Type data: MOROCCO: Rabat, on branches of *Olea europaea*. Syntypes, female. Type depository: Paris: Muséum national d'Histoire naturelle, France.
Diaspidiotus lepineyi; Danzig & Pellizzari, 1998: 235. Change of combination.
SCALE COVER: Female scale circular, light grey, exuviae brown, central; diameter 2.2 mm. Male scale unknown (Balachowsky, 1950b).
HOST PLANTS: **Oleaceae**: *Olea europaea* [Balach1950b].
DISTRIBUTION: **Palaearctic**: France [Foldi2001]; Morocco [Balach1950b].
GENERAL: Description and illustration of adult female by Balachowsky (1950b).
KEYS: Balachowsky 1950b: 400 (female) [Mediterranean].
CITATIONS: Balach1950b [taxonomy, description, illustration, host, distribution: 415-417]; Borchs1966 [catalogue: 333]; DanzigPe1998 [catalogue: 235]; Foldi2001 [distribution: 303-308].

Diaspidiotus liaoningensis (Tang)
Quadraspidiotus liaoningensis Tang, 1984: 65. Type data: CHINA: Liaoning Province, Gaixian County, on *Populus* sp. Holotype female. Type depository: Shanxi: Entomological Institute, Shanxi Agricultural University, Taigu, Shanxi, China.
Diaspidiotus liaoningensis; Danzig & Pellizzari, 1998: 235. Change of combination.
Diaspidiotus liaoningense; Tao, 1999: 115. Misspelling of species name.
SCALE COVER: Female scale circular; greyish black in colour (Tang, 1984).
HOST PLANTS: **Salicaceae**: *Populus* [Tang1984].
DISTRIBUTION: **Palaearctic**: China (People's Republic) (Liaoning [Tang1984]).
GENERAL: Description and illustration of adult female by Tang (1984).
CITATIONS: DanzigPe1998 [catalogue: 235]; Tang1984 [taxonomy, description, illustration, host, distribution: 65-67]; Tao1999 [taxonomy, host, distribution: 115]; Xie1998 [taxonomy, description, illustration, host, distribution: 113].

Diaspidiotus liquidambaris (Kotinsky)
Cryptophyllaspis liquidambaris Kotinsky, 1903: 149. Type data: U.S.A.: Georgia, Atlanta; Washington, D.C., on *Liquidambar styraciflua*. Syntypes, female. Type depository: Davis: The Bohart Museum of Entomology, University of California, California, USA.
Chemnaspidiotus liquidambaris; MacGillivray, 1921: 439. Change of combination.
Diaspidiotus liquidambaris; Ferris, 1938a: 223. Change of combination.
Cryptophyllaspis (*Diaspidiotus*) *liquidambaris*; Merrill, 1953: 41. Change of combination.
COMMON NAME: sweet gum scale [MerrilCh1923, McKenz1956, Dekle1965c].
SCALE COVER: Female scale flat, white; exuviae central. Male scale exposed upon leaves, white, elongate oval; exuvia toward one end (Ferris, 1938a). Colour photograph by Gill (1997).
HOST PLANTS: **Aceraceae**: *Acer* [McKenz1956], *Acer rubrum* [McKenz1956].
Altingiaceae: *Liquidambar styraciflua* [Kotins1903, MerrilCh1923, Merril1953, McKenz1956, Dekle1965c, McDani1969, TippinBe1970].

DISTRIBUTION: **Nearctic**: United States of America (Alabama [BesheaTiHo1973], California [McKenz1956], Connecticut [Nakaha1982], Delaware [Nakaha1982], District of Columbia [Kotins1903, MerrilCh1923, Merril1953], Florida [MerrilCh1923, Merril1953], Georgia [Kotins1903, MerrilCh1923, Merril1953, TippinBe1970, BesheaTiHo1973], Illinois [Nakaha1982], Indiana [Nakaha1982], Louisiana [Merril1953], Maryland [StoetzDa1974a], Mississippi [Herric1911, MerrilCh1923, Merril1953, BesheaTiHo1973], Missouri [Nakaha1982], New Jersey [Nakaha1982], New York [Nakaha1982], North Carolina [Nakaha1982], Ohio [MerrilCh1923, Merril1953], Oklahoma [Nakaha1982], Pennsylvania [Nakaha1982], South Carolina [Nakaha1982], Tennessee [Nakaha1982], Texas [Merril1953, McDani1969], Virginia [BesheaTiHo1973]).

BIOLOGY: Causing formation of pit galls on leaves; each gall being occupied by a single female the scale of which fills opening of gall (Ferris, 1938a).

ECONOMIC IMPORTANCE: Considered a pest of *Liquidambar* in Florida (Dekle, 1976), while a minor pest in California (Gill, 1997).

GENERAL: Description and illustration of adult female by Ferris (1938a), McKenzie (1956), Kosztarab (1996) and by Gill (1997). Description and illustration of female and male nymphs and male pupa and prepupa by Stoetzel & Davidson (1974a).

KEYS: Gill 1997: 113 (female) [Species of California]; Kosztarab 1996: 485 (female) [Northeastern North America]; McDaniel 1969: 90-91 (female) [U.S.A.: Texas]; McKenzie 1956: 25 (female) [U.S.A.: California]; Ferris 1942: 33 (female) [North America].

CITATIONS: BeardsDaHo1976 [economic importance: 106]; BesheaTiHo1973 [host, distribution: 6]; Borchs1966 [catalogue: 312]; Dekle1965c [taxonomy, description, host, distribution: 50]; Dekle1976 [taxonomy, description, host, distribution, economic importance: 69]; Fernal1903b [catalogue: 281]; Ferris1937c [taxonomy, illustration: 50,63]; Ferris1938a [taxonomy, description, illustration, host, distribution: 223]; Ferris1942 [taxonomy: 446:33]; Gill1997 [host, distribution, taxonomy, description, illustration, economic importance: 116,121]; Herric1911 [taxonomy, description, illustration, host, distribution: 12,35-36]; Koszta1990 [structure, biological control: 307-311]; Koszta1996 [taxonomy, description, illustration, host, distribution, life history, economic importance: 492-493]; Kotins1903 [taxonomy, description, host, distribution: 149-150]; Larew1990 [ecology, life history, structure: 293-300]; Lobdel1937 [taxonomy: 78]; MacGil1921 [taxonomy, description, host, distribution: 439]; McDani1969 [taxonomy, illustration, host, distribution: 96-98]; McKenz1943 [taxonomy, life history: 153]; McKenz1956 [taxonomy, description, illustration, host, distribution: 62-64]; Merril1953 [taxonomy, description, host, distribution: 41-42]; MerrilCh1923 [taxonomy, description, host, distribution, economic importance: 227-228]; MillerDa1990 [host, distribution, economic importance: 301]; Nakaha1982 [host, distribution: 29]; Neiswa1966 [host, distribution, taxonomy, economic importance: 1-54]; StoetzDa1974 [taxonomy, life history: 138-140]; StoetzDa1974a [taxonomy, description, illustration, host, distribution, life history: 495-498]; TippinBe1970 [host, distribution: 9].

Diaspidiotus macroporanus (Takagi)

Quadraspidiotus macroporanus Takagi, 1956b: 86. Type data: JAPAN: Sapporo, Hokkaido, on *Cercidiphyllum japonicum*. Holotype female. Type depository: Sapporo: Ent. Institute, Faculty of Agriculture, Hokkaido University, Japan.

Comstockaspis macroporana; Hirayama & Nogami, 1975: 2. Change of combination.

Diaspidiotus macroporanus; Danzig & Pellizzari, 1998: 325. Change of combination.

SCALE COVER: Scale of female flat or slightly convex; dull brown; that of male not identified (Takagi, 1956b).

HOST PLANTS: **Cercidiphyllaceae**: *Cercidiphyllum japonicum* [Takagi1956b]. **Tiliaceae**: *Tilia japonica* [Takagi1956b].

DISTRIBUTION: **Palaearctic**: Japan [Kawai1980] (Hokkaido [Takagi1956b]).

BIOLOGY: Type-material collected on branches of host plant (Takagi, 1956b).

GENERAL: Description and illustration of adult female by Takagi (1956b).

KEYS: Kawai 1980: 215 (female) [Japan].

CITATIONS: Borchs1966 [catalogue: 333]; DanzigPe1998 [catalogue: 235]; HirayaNo1975 [host, distribution, economic importance, biological control: 2-6]; Kawai1980 [taxonomy, description, host, distribution: 215]; Lindin1957 [taxonomy: 551]; Muraka1970 [host, distribution: 77]; Takagi1956b [taxonomy, description, illustration, host, distribution: 86-88]; Takagi1961 [taxonomy, description, host, distribution: 42].

Diaspidiotus mairei (Balachowsky)

Hemiberlesia mairei Balachowsky, 1928a: 128. Type data: MOROCCO: South Taroudant, Adar-Quaman (Anti-Atlas), on *Laburnum platycarpum*; collected 1922. Holotype female. Type depository: Paris: Muséum national d'Histoire naturelle, France.

Targionia regnieri Rungs, 1933: 114. Type data: MOROCCO: Mamora Forest, on *Ulex spectabilis*; collected by Ch. Rungs, 28.iii.1933. Holotype female. Type depository: Paris: Muséum national d'Histoire naturelle, France. Synonymy by Borchsenius, 1966: 333.

Hemiberlesea mairei; Balachowsky, 1935b: 257. Misspelling of genus name.

Aspidiotus mairei; Lindinger, 1936: 157. Change of combination.

Hemiberlesea mairei; Gómez-Menor Ortega, 1937: 118. Misspelling of genus name.

Hemiberlesia regnieri; Ferris, 1941e: 47. Change of combination.

Quadraspidiotus mairei; Balachowsky, 1950b: 440. Change of combination.

Diaspidiotus mairei; Danzig & Pellizzari, 1998: 235. Change of combination.

SCALE COVER: Female scale circular, diameter 1.8-2 mm; highly convex; colour grey; exuviae central or subcentral brown, covered with white secretion; ventral vellum well developed (Balachowsky, 1928a). Male scale flat, oval; white, with dark yellow exuviae, 1.2-1.4 mm (Balachowsky, 1950b).

HOST PLANTS: **Compositae**: *Adenocarpus bacquei* [Rungs1935]. **Crassulaceae**: *Sedum album* [Balach1932d]. **Leguminosae**: *Calycotome intermedia* [Rungs1948], *Cytisus* [Balach1932d], *Cytisus fontanesii* [Rungs1935], *Cytisus grandiflorus* [Rungs1935], *Cytisus triflorus* [Rungs1935], *Genista* [Balach1932d, GomezM1937,

Martin1983], *Genista ferox microphylla* [Balach1928a, Balach1932d], *Genista ferox* [Rungs1948], *Genista tricuspidata* [Rungs1948], *Laburnum anagyroides* [GomezM1937, GomezM1946, Martin1983], *Laburnum platycarpum* [Balach1928a, Balach1932d], *Retama* [Balach1950b], *Retama monosperma* [Rungs1948], *Retama sphaerocarpa* [Balach1935b, GomezM1937, Martin1983], *Ulex spectabilis* [Rungs1933, Rungs1935]. **Liliaceae**: *Asparagus stipularis* [Rungs1948]. **Oleaceae**: *Olea europaea* [Rungs1935]. **Sapotaceae**: *Argania spinosa* [Balach1932d].

DISTRIBUTION: **Palaearctic**: Morocco [Balach1928a, Balach1932d, Rungs1933, Rungs1948]; Spain [Balach1935b, GomezM1937, GomezM1946, Martin1983].

GENERAL: Description and illustration of adult female by Balachowsky (1928a, 1950b).

KEYS: Blay Goicoechea 1993: 528-529 (female) [Spain]; Balachowsky 1950b: 399 (female) [Mediterranean]; Balachowsky 1928a: 132 (female) [North Africa].

CITATIONS: Balach1928a [taxonomy, description, illustration, host, distribution: 128-129]; Balach1929a [host, distribution: 316]; Balach1932 [taxonomy, host, distribution: VI-VII]; Balach1935b [host, distribution: 257]; Balach1950b [taxonomy, description, illustration, host, distribution: 440-443]; BlayGo1993 [taxonomy, description, illustration, host, distribution: 549-553]; Borchs1966 [catalogue: 333]; DanzigPe1998 [catalogue: 235]; Ferris1941e [taxonomy: 45,47]; Ferris1943a [taxonomy: 86]; GomezM1937 [taxonomy, description, illustration, host, distribution: 118-121]; GomezM1946 [host, distribution: 62]; GomezM1958a [host, distribution: 8]; Lindin1936 [taxonomy: 157]; Martin1983 [taxonomy, host, distribution: 67]; Rungs1933 [taxonomy, description, illustration, host, distribution: 114-117]; Rungs1935 [host, distribution: 272-274]; Rungs1936 [taxonomy: 53]; Rungs1948 [host, distribution: 110].

Diaspidiotus makii (Kuwana)

Aspidiotus makii Kuwana, 1932: 51. Type data: JAPAN: on *Pinus luchuensis*. Syntypes, female. Type depository: Ibaraki-ken: Insect Taxonomy Laboratory, National Institute of Agricultural Environmental Sciences, Kannon-dai, Yatabe, Tsukuba-shi, (I. Kuwana), Japan.

Diaspidiotus makii; Borchsenius, 1966: 324. Change of combination.

SCALE COVER: Scale of female elongate, oval, convex; white or dirty white; exuviae subcentral, orange yellow; usually covered with a white waxy substance; ventral scale very thin; when removed, insect leaves a white patch on host; length, about 1.5 mm, width, about 1 mm (Kuwana, 1932).

HOST PLANTS: **Pinaceae**: *Abies* [Takagi1974], *Pinus* [Takagi1958, Takagi1974], *Pinus luchuensis* [Kuwana1932, Kuwana1933]. **Podocarpaceae**: *Podocarpus* [Takagi1974].

DISTRIBUTION: **Oriental**: Taiwan [Takagi1974]. **Palaearctic**: Japan [Kuwana1932, Kuwana1933, Takagi1958, Takagi1974, Kawai1980]; South Korea [Takagi1974].

GENERAL: Description and illustration of adult female by Kuwana (1932, 1933) and by Takagi (1958, 1974).

KEYS: Kuwana 1933: 3 (female) [Japan]; Kuwana 1933b: 49 (female) [Japan].

CITATIONS: Borchs1966 [catalogue: 324]; DanzigPe1998 [catalogue: 235-236]; Ferris1941e [taxonomy: 45]; Kawai1980 [taxonomy, description, host, distribution: 218]; Kuwana1932 [taxonomy, description, illustration, host, distribution: 51-52]; Kuwana1933 [taxonomy, description, illustration, host, distribution: 13-14]; Muraka1970 [host, distribution: 74]; Takagi1958 [taxonomy, description, illustration, host, distribution: 123-124]; Takagi1974 [taxonomy, description, illustration, host, distribution: 9-10,29].

Diaspidiotus malenconi (Rungs)

Aspidiotus malenconi Rungs, 1936: 50. Type data: MORROCO: Oued Ziz valley, Amzoudj, on *Antirrhinum ramosissimum*; collected by G. Malencon, 30.i.1934. Syntypes, female. Type depository: Paris: Muséum national d'Histoire naturelle, France.

Quadraspidiotus malenconi; Balachowsky, 1950b: 459. Change of combination.

Diaspidiotus malenconi; Danzig & Pellizzari, 1998: 236. Change of combination.

SCALE COVER: Female scale circular, convex; colour varies from grey-blue to grey-rose; exuviae central, brown, covered with white secretion; 1.2-1.8 mm. Male scale light white, elongate, 1-1.2 X 0.7-0.9 mm (Rungs, 1936).

HOST PLANTS: **Scrophulariaceae**: *Antirrhinum ramosissimum* [Rungs1936].

DISTRIBUTION: **Palaearctic**: Morocco [Rungs1936].

GENERAL: Description and illustration of adult female by Rungs (1936) and by Balachowsky (1950b).

KEYS: Balachowsky 1950b: 403 (female) [Mediterranean].

CITATIONS: Balach1950b [taxonomy, description, illustration, host, distribution: 459-462]; Balach1958a [host, distribution: 36-37]; Borchs1966 [catalogue: 333]; DanzigPe1998 [catalogue: 236]; Ferris1941e [taxonomy: 45]; Rungs1936 [taxonomy, description, illustration, host, distribution: 50].

Diaspidiotus maleti (Vayssière)

Aspidiotus (Aonidiella) maleti Vayssière, 1920: 257. Type data: MOROCCO: between Meknes and Fez, on leaves and twigs of olive; collected April 1919. Syntypes, female. Type depository: Paris: Muséum national d'Histoire naturelle, France.

Aonidiella maleti; Balachowsky, 1932d: xii. Change of combination.

Aspidiotus maleti; Lindinger, 1932f: 199. Change of combination.

Quadraspidiotus maleti; Balachowsky, 1950b: 462. Change of combination.

Diaspidiotus maleti; Danzig & Pellizzari, 1998: 236. Change of combination.

SCALE COVER: Female scale 1.5-2 mm in diameter; convex; exuviae central or subcentral, colour yellow reddish; ventral vellum almost completely closed (Vayssière, 1920).

HOST PLANTS: **Oleaceae**: *Olea europaea* [Vayssi1920, Balach1927].

DISTRIBUTION: **Palaearctic**: Algeria [Balach1927, Balach1932d]; Morocco [Vayssi1920].

ECONOMIC IMPORTANCE: Ddamages olive in Morocco and Algeria (Schmutterer et al., 1957).

GENERAL: Description and illustration of adult female by Balachowsky (1950b).

KEYS: Balachowsky 1950b: 404 (female) [Mediterranean].
CITATIONS: Argyri1990 [host, distribution, economic importance: 579-583]; Balach1927 [host, distribution: 178]; Balach1932d [taxonomy, host, distribution: XII]; Balach1950b [taxonomy, description, illustration, host, distribution: 462-465]; Benass1967 [host, distribution, life history, biological control: 1133-1139]; Borchs1966 [catalogue: 333]; BouhelDeDe1932 [host, distribution, control: 1-60]; DanzigPe1998 [catalogue: 236]; Ferris1941e [taxonomy: 45]; Lindin1932f [taxonomy: 199]; McKenz1938 [taxonomy: 3]; MillerDa1990 [host, distribution, economic importance: 305]; SchmutKlLu1957 [host, distribution, economic importance: 484]; Vayssi1920 [taxonomy, description, host, distribution: 257-258].

Diaspidiotus marani (Zahradník)

Quadraspidiotus schneideri Bachmann, 1952: 144. Nomen nudum; discovered by Danzig, 1993: 186.

Quadraspidiotus marani Zahradník, 1952a: 449. Type data: CZECH REPUBLIC: Prague, on *Pyrus communis* ssp. *sativa*; collected 2.iii.1951. Syntypes, female. Type depository: Prague: National Museum (Natural History), Department of Entomology, Czech Republic.

Quadraspidiotus schneideri Bachmann, 1952a: 357. Type data: SWITZERLAND: on *Malus pumila, Pyrus communis, Prunus domestica* and *Prunus spinosa*. Syntypes, female. Synonymy by Danzig, 1993: 186. Notes: Type material presumably kept by Bachmann.

Aspidiotus marani; Lindinger, 1957: 546. Change of combination.

Aspidiotus schneideri; Lindinger, 1957: 546. Change of combination.

Aspidiotus morani; Borchsenius, 1966: 333. Misspelling of species name.

Diaspidiotus marani; Hadzibejli, 1983: 241. Change of combination.

Diaspidiotus marani; Danzig & Pellizzari, 1998: 236. Revived combination.

SYSTEMATICS: Borchsenius (1966: 333) placed *Quadraspidiotus marani olivicola* Pegazzano, 1955, as a synonym of *Q. marani*, while Danzig (1993: 187) synonymised *Quadraspidiotus marani olivicola* with *Diaspidiotus perieri* Goux, 1949.

SCALE COVER: Colour photograph of scale cover by Porcelli & Pizza (1993).

HOST PLANTS: **Oleaceae**: *Fraxinus excelsior* [Zahrad1952a], *Olea europaea* [PorcelPi1993]. **Rosaceae**: *Crataegus* [Zahrad1952a], *Prunus domestica* [Zahrad1952a], *Pyrus* [Zahrad1952], *Pyrus communis sativa* [Zahrad1952a].

NATURAL ENEMIES: COLEOPTERA **Coccinellidae**: *Chilocorus bipustulatus* (L.) [Zahrad1972, PorcelPi1993]. **Nitidulidae**: *Cybocephalus politus* (Germ.) [Zahrad1972, PorcelPi1993]. HYMENOPTERA **Aphelinidae**: *Aphytis mytilaspidis* (Le Baron) [Bachma1953a, PorcelPi1993], *Aphytis proclia* Walker [Zahrad1972, PorcelPi1993], *Archenomus bicolor* Howard [Zahrad1972, PorcelPi1993]. **Encyrtidae**: *Cheiloneurus microphagus* Mayr [Zahrad1972], *Metaphycus notatus* Hoffer [Zahrad1972, PorcelPi1993], *Zaomma lambinus* (Walker) [Trjapi1989].

DISTRIBUTION: **Palaearctic**: Czech Republic [Zahrad1952, Zahrad1977]; Georgia [Hadzib1983]; Italy [PorcelPi1993, LongoMaPe1995]; Netherlands [Jansen2001]; Romania [Savesc1982]; Slovak Republic [Zahrad1952]; Turkey [Yasar1995a].

BIOLOGY: Develops one annual generation on olive in Italy. A biparental species. Adult females occur from December to April-May, and oviposit in may-June (Porcelli & Pizza, 1993).
ECONOMIC IMPORTANCE: Pest of deciduous fruit trees in Europe (Bachmann, 1953a; Schmutterer et al., 1957; Kozár, 1990c; Porcelli & Pizza, 1993).
GENERAL: Description and illustration of adult female by Zahradník (1952), Tereznikova (1986), Danzig (1993), Porcelli & Pizza (1993) and by Yaşar (1995a).
KEYS: Danzig 1993: 179-182 (female) [Europe]; Tereznikova 1986: 97-98 (female); Kosztarab & Kozár 1978: 170 (female) [Hungary]; Zahradník 1952: 114-115 (female) [Czech Republic].
CITATIONS: Bachma1952a [taxonomy, description, host, distribution: 357]; Bachma1953 [taxonomy: 181-182]; Bachma1953a [taxonomy, description, illustration, host, distribution, life history, chemical control, biological control: 357-404]; Bachma1955 [taxonomy: 122-123]; Bohm1954 [host, distribution: 55-57]; Borchs1966 [catalogue: 333-334]; Danzig1964 [taxonomy, host, distribution: 652]; Danzig1972 [taxonomy, host, distribution, economic importance: 210]; Danzig1993 [taxonomy, description, illustration, host, distribution: 186-187]; DanzigPe1998 [catalogue: 236]; Duskov1952 [taxonomy, description, host, distribution: 452-455]; Duskov1953a [host, distribution, taxonomy, life history: 229-250]; Foldi2001 [distribution: 303-308]; FreyFr1995 [taxonomy, chemistry: 777-780]; FreyFr1995a [structure, chemistry: 100]; Hadzib1983 [taxonomy, host, distribution, life history, biological control, economic importance: 241]; Huba1960 [taxonomy: 39-50]; HuffakDo1965 [biological control: 61-63]; Jansen2001 [host, distribution: 197-206]; Komosi1974a [host, distribution, life history, ecology: 1-84]; KosztaKo1978 [taxonomy, description, host, distribution: 170]; Kozar1990 [life history, economic importance: 335-340]; Kozar1990c [life history, economic importance, host, distribution: 593-602]; Kozar1995b [taxonomy: 51]; KozarGuBa1994 [host, distribution: 151-161]; KozarHiMa1996 [taxonomy, description: 433-437]; Krzysz1957 [taxonomy, description, host, distribution, life history: 233]; Lagows1998a [host, distribution: 63-71]; Lellak1965 [taxonomy, description, host, distribution, life history: 202-209]; Lindin1957 [taxonomy: 546]; LongoMaPe1995 [distribution: 128]; ManiHiSc1993 [life history, economic importance, host, distribution: 299-302]; MillerDa1990 [host, distribution, economic importance: 305]; MuelleEi1954 [taxonomy, description: 151-153]; Pegazz1955 [taxonomy, description, host, distribution: 311-314]; PorcelPi1993 [taxonomy, description, illustration, host, distribution, life history, economic importance, biological control: 13-16]; Savesc1982 [taxonomy, description, host, distribution, life history, biological control: 321-323]; Schern1954 [taxonomy: 225]; Schmut1957b [taxonomy: 149]; Schmut1959 [taxonomy, description, host, distribution: 76,89]; SchmutKlLu1957 [host, distribution, economic importance: 484]; Terezn1986 [taxonomy, description, illustration, host, distribution: 104-105]; Trjapi1989 [biological control: 312]; TzalevVa1965 [host, distribution: 22-25]; Yasar1995a [taxonomy, description, illustration, host, distribution: 114-116]; Zahrad1952 [taxonomy, description, illustration, host, distribution: 121-124]; Zahrad1952a [taxonomy, description, illustration, host, distribution: 449-451]; Zahrad1955 [taxonomy: 88]; Zahrad1955a [taxonomy: 125]; Zahrad1972 [taxonomy, host,

distribution, biological control: 435]; Zahrad1977 [taxonomy, distribution: 121]; Zahrad1990a [host, distribution, description: 651].

Diaspidiotus mccombi McKenzie

Diaspidiotus mccombi McKenzie, 1963: 32. Type data: U.S.A.: Maryland, Prince Georges County, University of Maryland Campus, College Park, on *Pinus mugo mughus*. Holotype female. Type depository: Davis: The Bohart Museum of Entomology, University of California, California, USA.

COMMON NAMES: McComb pine scale [McKenz1963].

SCALE COVER: Female scale blackish; with somewhat lightened, central exuviae, appearing rather elongate, probably due to colour of needle upon which insect feeds. Male scale lighter, elongate oval, exuvia central (McKenzie, 1963).

HOST PLANTS: **Pinaceae**: *Pinus* [McKenz1963, BesheaTiHo1973, Koszta1996], *Pinus clausa* [BesheaTiHo1973], *Pinus glabra* [BesheaTiHo1973], *Pinus mugo mughus* [McKenz1963, StoetzDa1974a], *Pinus nigra* [McKenz1963], *Pinus resinosa* [McKenz1963], *Pinus virginiana* [McKenz1963].

DISTRIBUTION: **Nearctic**: United States of America (Alabama [USDAAP1978], District of Columbia [Nakaha1982], Florida [BesheaTiHo1973], Georgia [BesheaTiHo1973], Louisiana [Nakaha1982], Maryland [McKenz1963, StoetzDa1974a], Mississippi [Nakaha1982], North Carolina [Nakaha1982], Pennsylvania [Nakaha1982], South Carolina [McKenz1963], Virginia [Nakaha1982]).

BIOLOGY: Occurring on needles, usually hidden at base within needle sheath (McKenzie, 1963).

GENERAL: Description and illustration of adult female by McKenzie (1963) and by Kosztarab (1996). Description and illustration of female and male nymphs and male pupa and prepupa by Stoetzel & Davidson (1974a).

KEYS: Kosztarab 1996: 484-485 (adult female) [Northeastern North America].

CITATIONS: BesheaTiHo1973 [host, distribution: 6]; Borchs1966 [catalogue: 324]; Dekle1976 [taxonomy, description, host, distribution, economic importance: 70]; Koszta1996 [taxonomy, description, illustration, host, distribution, life history, biological control: 492-495]; Mead1987 [host, distribution: 2-4]; Nakaha1982 [host, distribution: 29]; StoetzDa1974 [taxonomy, life history: 138-140]; StoetzDa1974a [taxonomy, description, illustration, host, distribution, life history: 498-500]; USDAAP1978 [host, distribution: 1-4,6].

Diaspidiotus naracola Takagi

Diaspidiotus naracola Takagi, 1956b: 83. Type data: JAPAN: Honsyu, Toyama-ken, Kamidaki, on *Quercus serrata*. Holotype female. Type depository: Sapporo: Entomological Institute, Faculty of Agriculture, Hokkaido University, Japan.

SCALE COVER: Scale of the type common to genus, being grey (Takagi, 1956b).

HOST PLANTS: **Caprifoliaceae**: *Viburnum wrightii* [Takagi1956b]. **Fagaceae**: *Castanea crenata* [Takagi1974], *Quercus serrata* [Takagi1956b, Takagi1974].

DISTRIBUTION: **Palaearctic**: Japan [Takagi1974] (Hokkaido [Takagi1962b], Honshu [Takagi1956b, Takagi1962b]).

BIOLOGY: Types were collected on bark of host plant (Takagi, 1956b).

GENERAL: Description and illustration of adult female by Takagi (1956b, 1974).
CITATIONS: Borchs1966 [catalogue: 324]; DanzigPe1998 [catalogue: 236]; Kawai1980 [taxonomy, description, host, distribution: 219]; Muraka1970 [host, distribution: 74]; Takagi1956b [taxonomy, description, illustration, host, distribution: 83-84]; Takagi1962b [host, distribution: 52]; Takagi1974 [taxonomy, description, illustration, host, distribution: 4-9].

Diaspidiotus nitrariae (Marchal)

Aspidiotus (*Hemiberlesia*) *nitrariae* Marchal, 1911a: 150. Type data: TUNISIA: Medenin, on leaves of *Nitraria* sp. Syntypes, female. Type depository: Paris: Muséum national d'Histoire naturelle, France.
Aspidiotus nitrariae; Lindinger, 1912b: 226. Change of combination.
Hemiberlesia nitrariae; Sasscer, 1915: 35. Change of combination.
Quadraspidiotus nitrariae; Balachowsky, 1950b: 456. Change of combination.
Diaspidiotus nitrariae; Danzig & Pellizzari, 1998: 236. Change of combination.
SCALE COVER: Female scale circular, 0.5-0.6 mm, white, small and transparent; glassy; exuviae central, flat; colour yellow; depressed into tissue of lower surface of leaves, to form a pseudo-gall (Marchal, 1911a; Balachowsky, 1950b).
HOST PLANTS: **Zygophyllaceae**: *Nitraria* [Marcha1911a, Balach1930c, Balach1932d, Bodenh1937], *Nitraria tridentata* [Bodenh1927b, Balach1950b].
DISTRIBUTION: **Palaearctic**: Algeria [Balach1930c, Balach1932d]; Israel [Bodenh1927b, Bodenh1937, Balach1950b]; Tunisia [Marcha1911a].
BIOLOGY: Insects depressed into tissue of lower surface of leaves, forming a pseudo-gall (Marchal, 1911a; Balachowsky, 1950b).
GENERAL: Description and illustration of adult female by Marchal (1911a) and by Balachowsky (1950b).
KEYS: Balachowsky 1950b: 402 (female) [Mediterranean].
CITATIONS: Balach1930c [host, distribution: 119]; Balach1932d [taxonomy, host, distribution: VIII]; Balach1950b [taxonomy, description, illustration, host, distribution: 456-459]; Bodenh1927b [host, distribution: 79-80]; Bodenh1935 [host, distribution: 246]; Bodenh1937 [host, distribution: 217]; Borchs1966 [catalogue: 334]; DanzigPe1998 [catalogue: 236-237]; Ferris1941e [taxonomy: 46]; Houard1913 [host, distribution: 1]; Larew1990 [ecology, life history, structure: 293-300]; Lindin1912b [taxonomy, description, host, distribution: 226]; MacGil1921 [taxonomy, description, host, distribution: 437]; Marcha1911a [taxonomy, description, host, distribution: 150]; Sassce1915 [taxonomy, host, distribution: 35].

Diaspidiotus osborni (Newell & Cockerell)

Aspidiotus osborni Newell & Cockerell, in: Osborn, 1898: 229. Type data: U.S.A: Iowa, Ames, on white oak. Syntypes, female. Type depository: Washington: United States National Entomological Collection, U.S. National Museum of Natural History, District of Columbia, USA.
Diaspis snowii Hunter, 1899: 14. Type data: U.S.A.: Kansas, Douglass County, on *Salix nigra*. Holotype female. Type depository: Washington: United States National Entomological Collection, U.S. National Museum of Natural History, District of Columbia, USA. Synonymy by Borchsenius, 1966: 325.

Aspidiotus (*Diaspidiotus*) *osborni*; Cockerell, 1899a: 369. Change of combination.

Aspidiotus (*Diaspidiotus*) *osborni*; Leonardi, 1900: 340. Change of combination.

Aspidiotus yulupae Bremner, 1907: 367. Type data: USA: California, Sonoma County, Yulupa Valley, on *Quercus lobata*. Syntypes, female. Type depository: Davis: The Bohart Museum of Entomology, University of California, California, USA. Synonymy by Ferris, 1920b: 51.

Neosignoretia yulupae; MacGillivray, 1921: 424. Change of combination.

Diaspidiotus osborni; MacGillivray, 1921: 413. Change of combination.

COMMON NAME: Osborn scale [McKenz1956, Dekle1965c].

SCALE COVER: Female scale small, oval, 1-1.25 mm in length, 0.5-0.75 mm in breadth; irregularly margined; dark, spotted minutely; of general scurfy appearance; exuviae dark brown, submarginal, small; ventral vellum a mere white film (Osborn, 1898). Female scale grey or whitish, usually nearly colour of bark; circular, flat; that of male grey, elongate oval, exuvia near one end (Ferris, 1938a).

HOST PLANTS: **Aceraceae**: *Acer* [MerrilCh1923]. **Altingiaceae**: *Liquidambar styraciflua* [BesheaTiHo1973]. **Betulaceae**: *Betula* [Koszta1996]. **Carpinaceae**: *Carpinus* [Koszta1996], *Ostrya virginica* [Newell1899, MerrilCh1923]. **Cornaceae**: *Cornus nutalli* [McKenz1956]. **Ebenaceae**: *Diospyros* [McKenz1956], *Diospyros virginiana* [Ferris1938a]. **Fagaceae**: *Castanea* [BesheaTiHo1973, Koszta1996], *Fagus* [Koszta1996], *Fagus sylvatica* [McKenz1956], *Quercus* [Ferris1938a, McKenz1956, BesheaTiHo1973, PellizCa1991a], *Quercus agrifolia* [Ferris1920b, McKenz1956], *Quercus alba* [Hunter1899, Newell1899, MerrilCh1923, Ferris1938a, McKenz1956, StoetzDa1974a], *Quercus brandegei* [Ferris1921, McKenz1956], *Quercus falcata* [BesheaTiHo1973], *Quercus laevis* [BesheaTiHo1973], *Quercus laurifolia* [BesheaTiHo1973], *Quercus lobata* [Bremne1907, Sander1909a, McKenz1956], *Quercus michauxii* [BesheaTiHo1973], *Quercus nigra* [McKenz1956, TippinBe1970, BesheaTiHo1973], *Quercus robur* [Muntin1971], *Quercus undulata* [Ferris1938a, McKenz1956, McDani1969], *Quercus virginiana* [Ferris1938a, McDani1969, BesheaTiHo1973]. **Juglandaceae**: *Carya illinoensis* [Ferris1938a, McKenz1956, Dekle1965c, BesheaTiHo1973], *Juglans* [Ferris1938a, McKenz1956]. **Leguminosae**: *Robinia* [Koszta1996]. **Moraceae**: *Morus* [McKenz1956]. **Oleaceae**: *Fraxinus* [Koszta1996]. **Rosaceae**: *Crataegus* [McKenz1956], *Prunus* [Ferris1938a], *Prunus caroliniana* [BesheaTiHo1973], *Prunus pissardi* [McKenz1956]. **Tiliaceae**: *Tilia* [Koszta1996], *Tilia americana* [StoetzDa1974a]. **Vitaceae**: *Vitis* [McKenz1956, Dekle1965c], *Vitis vinifera* [Ferris1938a].

NATURAL ENEMIES: HYMENOPTERA **Aphelinidae**: *Ablerus clisiocampae* (Ashmead) [Koszta1996].

DISTRIBUTION: **Afrotropical**: South Africa [Muntin1971]. **Nearctic**: Canada (Ontario [FletchGi1908]); Mexico (Baja California [Ferris1921]); United States of America (Alabama [Nakaha1982], California [Bremne1907, Sander1909a, McKenz1956], Connecticut [Ferris1938a], Delaware [Nakaha1982], District of Columbia [Nakaha1982], Florida [MerrilCh1923, Ferris1938a, Merril1953, Dekle1965c], Georgia [Ferris1938a, TippinBe1970, BesheaTiHo1973], Illinois [Nakaha1982], Indiana [Nakaha1982], Iowa [Leonar1900, Ferris1938a], Kansas [Hunter1899, Ferris1938a], Kentucky [Nakaha1982], Louisiana [Ferris1938a],

Maryland [StoetzDa1974a], Massachusetts [Nakaha1982], Michigan [Nakaha1982], Mississippi [Ferris1938a, BesheaTiHo1973], Missouri [Hollin1923, Ferris1938a], New Hampshire [Nakaha1982], New Jersey [Nakaha1982], New Mexico [Nakaha1982], New York [Nakaha1982], North Carolina [Nakaha1982], Ohio [Ferris1938a], Oklahoma [Nakaha1982], Pennsylvania [Nakaha1982], South Carolina [Nakaha1982], Tennessee [Nakaha1982], Texas [Ferris1938a, McDani1969], Vermont [Nakaha1982], Virginia [Nakaha1982], West Virginia [Nakaha1982], Wisconsin [Nakaha1982]). **Palaearctic**: Italy [PellizCa1991a, LongoMaPe1995]; Turkey [Yasar1995a].

BIOLOGY: Occurring on bark of host (Ferris, 1938a).

GENERAL: Description and illustration of adult female by Ferris (1920b, 1938a), McKenzie (1956), Munting (1971) (based on material from South Africa), Yaşar (1995, 1995a), Kosztarab (1996) and by Gill (1997). Description and illustration of female and male nymphs and male pupa and prepupa by Stoetzel & Davidson (1974a).

KEYS: Gill 1997: 113 (female) [Species of California]; Kosztarab 1996: 484-485 (female) [Northeastern North America]; McDaniel 1969: 90-91 (female) [U.S.A.: Texas]; McKenzie 1956: 25 (female) [U.S.A.: California]; Ferris 1942: 33 (female) [North America]; Britton 1923: 371 (female) [U.S.A.: Connecticut]; Hollinger 1923: 7-8 (female) [U.S.A.: Missouri]; Lawson 1917: 217 (female) [U.S.A.: Kansas]; Newell 1899: 4-5 (female) [North America].

CITATIONS: Amos1933a [taxonomy, description, host, distribution: 210]; BeardsDaHo1976 [economic importance: 106]; BesheaTiHo1973 [host, distribution: 6]; Borchs1966 [catalogue: 325]; Bremne1907 [taxonomy, description, illustration, host, distribution: 367-368]; Britto1923 [taxonomy, description, host, distribution: 371,374]; Cocker1899a [taxonomy: 396]; ColesTa1976 [host, distribution, life history: 481-488]; Couch1931 [host, distribution, life history: 383-437]; DanzigPe1998 [catalogue: 237]; Dekle1965c [taxonomy, description, host, distribution: 51]; Dekle1976 [taxonomy, description, host, distribution: 71]; FDACSB1982 [host, distribution: 5-11]; Fernal1903b [catalogue: 268]; Ferris1920b [taxonomy, description, illustration, host, distribution: 51-52]; Ferris1921 [host, distribution: 126]; Ferris1938a [taxonomy, description, illustration, host, distribution: 224]; Ferris1941e [taxonomy: 46,49]; Ferris1942 [taxonomy: 445:5; 446:33]; Gill1997 [host, distribution, taxonomy, description, illustration, economic importance: 116,122,124]; Hollin1923 [taxonomy, description, host, distribution: 13-14]; Hunter1899 [taxonomy, description, illustration, host, distribution: 5-6,14-15]; Koszta1996 [taxonomy, description, illustration, host, distribution, life history, biological control: 495-497]; Lawson1917 [taxonomy, description, illustration, host, distribution: 232-233]; Leonar1900 [taxonomy, host, distribution: 340]; Lindin1957 [taxonomy: 546]; LongoMaPe1995 [distribution: 126]; MacGil1921 [taxonomy, description, host, distribution: 413,424]; McDani1969 [taxonomy, illustration, host, distribution: 98-99]; McKenz1956 [taxonomy, description, illustration, host, distribution: 63-64]; Merril1953 [taxonomy, description, host, distribution: 25]; MerrilCh1923 [taxonomy, description, host, distribution, economic importance: 205-206]; MillerDa1990 [host, distribution, economic importance: 301]; Muntin1971 [taxonomy, description, illustration, host, distribution: 134-137];

Nakaha1982 [host, distribution: 29]; Newell1899 [taxonomy, description, illustration, host, distribution: 5-7]; NewellCo1898 [taxonomy, description, host, distribution: 229]; Osborn1898 [host, distribution: 224]; PellizCa1991a [taxonomy, host, distribution: 199]; Sander1909a [taxonomy, host, distribution: 53]; StoetzDa1974 [taxonomy, life history: 138-140]; StoetzDa1974a [taxonomy, description, illustration, host, distribution, life history: 500-501]; Takagi1956 [taxonomy: 84]; TippinBe1970 [host, distribution: 9]; Yasar1995 [taxonomy, description, illustration, host, distribution: 50-51]; Yasar1995a [taxonomy, description, illustration, host, distribution: 69-71].

Diaspidiotus ostreaeformis (Curtis)
Aspidiotus ostreaeformis Curtis, 1843c: 805. Type data: ENGLAND: locality not indicated, on bark of pear tree. Syntypes, both sexes. Type depository: Abbotsford: Department of Entomology, Museum of Victoria, Victoria, Australia.
Aspidiotus betulae Baerensprung, 1849: 167. Type data: GERMANY: on *Betula* sp. Syntypes, female. Synonymy by Borchsenius, 1966: 334. Notes: Depository of type material unknown.
Mytilococcus ellipticus Amerling, 1858: 103. Nomen nudum; discovered by Borchsenius, 1966: 378.
Diaspis ostraeformis; Signoret, 1869: 854, 862, 863. Misspelling of species name.
Aspidiotus hyppocastani Signoret, 1869: 857. Nomen nudum.
Aspidiotus oxyacanthae Signoret, 1869: 863. Nomen nudum.
Aspidiotus hippocastani Signoret, 1869b: 136. Type data: FRANCE: apparently Paris, on bark of "marronier d'Indie" [=*Aesculus hippocastanum*]. Syntypes, female. Type depository: Vienna: Naturhistorisches Museum Wien, Austria. Synonymy by Leonardi, 1898c: 38.
Aspidiotus oxyacanthae Signoret, 1869b: 137. Type data: FRANCE: apparently Paris, on "aubepine" [=*Crataegus oxyacanthae*]. Syntypes, both sexes. Type depository: Vienna: Naturhistorisches Museum Wien, Austria. Synonymy by Leonardi, 1898c: 38.
Aspidiotus tiliae Signoret, 1869b: 137. Type data: FRANCE: on "tilleul" [=*Tilia*]. Syntypes, female. Type depository: Vienna: Naturhistorisches Museum Wien, Austria. Synonymy by Leonardi, 1898c: 38. Homonym of *Aspidiotus tiliae* Bouché, 1851.
Diaspis ostreaeformis; Signoret, 1869d: 439. Change of combination.
Diaspis ostraeformis; Lichtenstein, 1881: li. Change of combination.
Diaspis ostraeformis; Lichtenstein, 1881: li. Misspelling of species name.
Diaspis ostreaeformis; Goethe, 1884: 114. Change of combination.
Aspidiotus (*Diaspidiotus*) *betulae*; Cockerell, 1897i: 18. Change of combination.
Aspidiotus (*Diaspidiotus*) *hippocastani*; Cockerell, 1897i: 18. Change of combination.
Aspidiotus (*Diaspidiotus*) *ostreaeformis*; Cockerell, 1897i: 19. Change of combination.
Aspidiotus (*Diaspidiotus*) *oxyacanthae*; Cockerell, 1897i: 19. Change of combination.
Aspidiotus (*Evaspidiotus*) *betulae*; Leonardi, 1898c: 38. Change of combination.

Aspidiotus ostreaeformis oblongus Goethe, 1899: 16. Type data: GERMANY: Geisenheim am Rhein, on plums. Syntypes, both sexes. Synonymy by Ferris, 1941e: 46. Notes: Depository of type material unknown.

Aspidiotus ostreaeformis magnus Goethe, 1899: 17. Type data: GERMANY: Geisenheim am Rhein, on oaks. Syntypes, both sexes. Synonymy by Ferris, 1941e: 45. Notes: Depository of type material unknown.

Aspidiotus scutiformis; Goethe, 1899: 18. Misidentification; discovered by Borchsenius, 1966: 335.

Aspidiotus ortraeformis; Leonardi, 1909: 124. Misspelling of species name.

Aspidiotus ostreiformis; Lindinger, 1909c: 449. Misspelling of species name.

Aspidiotus ostreiformis; Lindinger, 1912b: 49. Misspelling of species name.

Quadraspidiotus ostreaeformis; MacGillivray, 1921: 410. Change of combination.

Aspidiotus ostraeiformis; Bodenheimer, 1924: 33. Misspelling of species name.

Aspidiotus ostreiformis; Koroneos, 1934: 11. Misspelling of species name.

Aspidiotus (Evaspidiotus) ostreaeformis; Thiem & Gerneck, 1934: 532. Change of combination.

Aspidiotus alma-atensis Borchsenius, 1935: 128. Type data: KAZAKHSTAN: Alma-Ata, on *Malus* sp. and *Crataegus* sp. Syntypes, female. Type depository: St. Petersburg: (= Leningrad) Zoological Museum, Academy of Science, Russia. Synonymy by Danzig, 1993: 182.

Aspidiotus ostreiformis; Kawecki, 1935: 76. Misspelling of species name.

Aspidiotus (Quadraspidiotus) ostreaeformis; Borchsenius, 1935a: 6. Change of combination.

Aspidiotus magnus; Ferris, 1941e: 45. Change of combination and rank.

Aspidiotus oblongus; Ferris, 1941e: 46. Change of combination and rank.

Quadraspidiotus ostrasformis; Bodenheimer, 1949: 57. Misspelling of species name.

Diaspidiotus ostreaeformis; Borchsenius, 1949d: 224. Change of combination.

Diaspidiotus alma-atensis; Borchsenius, 1950b: 228. Change of combination.

Quadraspidiotus alma-atensis; Balachowsky, 1950b: 469. Change of combination.

Quadraspidiotus ostreaefornis; Bodenheimer, 1952: 338. Misspelling of species name.

Quadraspidiotus ostraeformis; Gómez-Menor Ortega, 1956: 21. Misspelling of species name.

Aspidiotus ostreiformis; Gómez-Menor Ortega, 1957: 46. Misspelling of species name.

Quadraspidiotus ostreiformis; Gómez-Menor Ortega, 1957: 46. Misspelling of species name.

Aspidiotus ostreiformis; Lindinger, 1957: 545. Misspelling of species name.

Aspidiotus ostreiformis; Lindinger, 1957: 548. Misspelling of species name.

Quadraspidiotus williamsi Takagi, 1958: 127. Type data: JAPAN: Sapporo, Hokkaido, on a species of Rosaceae. Holotype female. Type depository: Sapporo: Entomological Institute, Faculty of Agriculture, Hokkaido University, Japan. Synonymy by Danzig, 1993: 182.

Quadraspidiotus almaatensis; Borchsenius, 1966: 328. Justified emendation.

Diaspidiotus ostraeformis; Hadzibejli, 1983: 238. Misspelling of species name.

Aspidiotus ostreaeformis obiongus; Chou, 1985: 312. Misspelling of species name.
Quadraspidiotus ostraeformis; Chou, 1985: 312. Misspelling of species name.
Aspidiotus hippocastini; Blay Goicoechea, 1993: 577. Misspelling of species name.
Quadraspidiotus ostraeformis; Aleksidze, 1995: 187. Misspelling of species name.
Quadraspidiotus ostraeformis; Frey & Frey, 1995a: 100. Misspelling of species name.

Diaspidiotus ostreaeformis; Danzig & Pellizzari, 1998: 237-238. Change of combination.

Aspidiotus hippocaastani; Tao, 1999: 115. Misspelling of species name.

COMMON NAMES: almaatinskaya shitovka [BazaroSh1971]; Austernformige schildlaus [SchmutKlLu1957]; European fruit scale [Kozar1990c]; oyster-shell scale [Green1928]; pear-tree oyster scale [Curtis1843c]; ustrizevidnaya ili lojnokaliforniiskya shitovka [Borchs1936]; zitronenfarbene Austernschildlaus [SchmutKlLu1957].

SYSTEMATICS: Ferris (1941e: 48) suggested that record of *Aspidiotus scutiformis* Cockerell by Goethe (1899), was a misidentification of *Aspidiotus juglansregiae*.

SCALE COVER: Female scale grey, circular, moderately convex, exuviae subcentral; scale of male elongate, exuvia near one end (Ferris, 1938a). Female scale circular, slightly convex; greyish-brown in colour, dark in central part and turning light towards margin; margin sometimes with a white border; exuviae eccentric; first exuvia light-brown in colour, second brown; diameter of scale about 1.5 mm. Male scale oval, greyish-green in colour, almost white on margins. Length about 0.6-0.8 mm (Borchsenius, 1935). Female scale circular, somewhat convex, dark grey; in male smaller and slightly elongate (Takagi, 1958).

HOST PLANTS: **Aceraceae**: *Acer* [Balach1950b], *Acer campestre* [Zahrad1972], *Acer platanoides* [Zahrad1972]. **Betulaceae**: *Alnus* [Balach1950b], *Alnus glutinosa* [Zahrad1972], *Alnus incana* [Balach1950b, Zahrad1972], *Betula* [Green1928, Balach1950b, Zahrad1952, Danzig1978], *Be. alba* [Balach1950b], *Be. dahurica* [Danzig1980b], *Be. manshurica* [Danzig1980b], *Be. pendula* [Zahrad1972], *Be. platyphylla* [Danzig1978a], *Be. pubescens* [Zahrad1972], *Be. tauschii* [Danzig1978, Danzig1980b]. **Bignoniaceae**: *Catalpa bignonioides* [Zahrad1972]. **Caprinaceae**: *Ostrya* [Balach1950b, Zahrad1972]. **Carpinaceae**: *Carpinus* [Balach1950b], *Carpinus betulus* [Bachma1953, Zahrad1972]. **Corylaceae**: *Corylus* [Balach1950b], *Cor. avellana* [Zahrad1972, RosenDe1979]. **Ericaceae**: *Calluna vulgaris* [Foldi2000], *Ledum macrophyllum* [Danzig1978, Danzig1980b], *Le. palustre* [Takagi1958], *Vaccinium uligimosum* [Danzig1978, Danzig1980b]. **Fagaceae**: *Fagus* [Balach1950b], *Fagus sylvatica* [Zahrad1972], *Quercus* [Balach1950b, Zahrad1972], *Qu. mongolica* [Danzig1978, Danzig1980b], *Qu. pubescens* [Balach1932d], *Qu. robur* [Zahrad1952, Zahrad1972]. **Fraxinaceae**: *Fraxinus* [Balach1950b], *Fr. excelsior* [Balach1935b, Zahrad1972, Martin1983]. **Grossulariaceae**: *Ribes aureum* [Zahrad1952], *Ri. rubrum* [Balach1950b]. **Hippocastanaceae**: *Aesculus* [Green1928, Balach1950b], *Ae. hippocastanum* [Signor1869b, Bachma1953, Zahrad1972], *Ae. pavia* [Zahrad1972]. **Juglandaceae**: *Juglans* [Bodenh1937], *Juglans regia* [Bodenh1924, Zahrad1972]. **Leguminosae**: *Caragana arborescens* [RosenDe1979], *Cytisus* [Balach1950b], *Gleditsia* [Balach1950b]. **Moraceae**: *Ficus* [Balach1950b, Bodenh1952], *Ficus carica*

[Balach1932d]. **Myricaceae**: *Myrica tomentosa* [Danzig1978, Danzig1980b]. **Oleaceae**: *Olea* [Balach1950b], *Olea europaea* [Leonar1909], *Syringa amurensis* [Danzig1978, Danzig1980b], *Syringa vulgaris* [Bachma1953, Takagi1958]. **Palmae**: *Phoenix* [Balach1950b]. **Pinaceae**: *Abies* [Borchs1938, Balach1950b], *Abies alba* [Kaweck1935]. **Platanaceae**: *Platanus* [Balach1932d, Zahrad1972], *Platanus occidentalis* [GomezM1957, Martin1983], *Platanus orientalis* [Balach1931a, Balach1932d]. **Rhamnaceae**: *Rhamnus* [Balach1950b, Danzig1980b]. **Rosaceae** [Borchs1938, Takagi1958], *Amygdalus bucharica* [BazaroSh1971], *Amygdalus communis* [Martin1983], *Crataegus* [Borchs1935, Balach1950b], *Cr. chlorosarca* [Danzig1978, Danzig1980b], *Cr. oxyacanthae* [Signor1869b], *Cydonia* [Balach1950b], *Malus* [Borchs1935, Balach1950, Zahrad1952], *Malus manshurica* [Danzig1978, Danzig1980b], *Prunus* [Borchs1934, Borchs1936, Balach1950b, Zahrad1952], *Pr. cerasus* [Zahrad1952], *Pr. domestica* [Green1928, Ferris1938a, Bachma1953], *Pr. spinosa* [Zahrad1952], *Pr. ussuriensis* [Danzig1978, Danzig1980b], *Pyrus* [Curtis1843c, Bodenh1937], *Py. communis* [Balach1932d], *Py..communis sativa* [Zahrad1952], *Py. malus* [Signor1869d, Bodenh1924, Green1928, Balach1950b], *Sorbaria sorbifolia* [Danzig1980b], *Sorbus* [Balach1950b], *Sorbus terminalis* [Zahrad1972], *Spiraea salicifolia* [Danzig1978, Danzig1980b]. **Salicaceae**: *Populus* [Green1928, Borchs1934, Balach1950b, Bodenh1952, Zahrad1972], *Populus alba* [Zahrad1972], *Populus canescens* [Zahrad1972], *Populus nigra* [Zahrad1972], *Populus pyramidalis* [Borchs1934, Zahrad1972], *Salix* [Bodenh1944a, Bodenh1952, Danzig1978a], *Salix alba* [Balach1928a, Balach1932d], *Salix viminalis* [Zahrad1972]. **Tiliaceae**: *Tilia* [Balach1950b, Zahrad1952], *Tilia cordata* [Zahrad1972], *Tilia platyphyllos* [Zahrad1972]. **Ulmaceae**: *Ulmus* [Balach1950b, Zahrad1972], *Ulmus propinqua* [Danzig1980b].

NATURAL ENEMIES: FUNGI : *Fusarium larvarum* [HornokKo1984]. **Ascomycotina**: *Nectria aurantiicola* [EvansPr1990]. HYMENOPTERA **Aphelinidae**: *Aphytis aonidiae* (Mercet) [RosenDe1979], *Ap. bovelli* Malenotti [Zahrad1972], *Ap. mytilaspidis* Le Baron [Zahrad1972], *Ap. testaceus* Tshumakova [RosenDe1979], *Archenomus longiclava* (Girault) [Viggia1990a], *Azotus marchalli* Howard [Balach1950b, Zahrad1972], *Az. matritensis* Mercet [Zahrad1972], *Az. pinifoliae* Mercet [Zahrad1972], *Coccophagoides similis* (Masi) [Zahrad1972, Gordh1979], *Prospaltella aurantii* Howard [Zahrad1972], *Pteroptrix dimidiata* Westwood [Novits1961], *Pt. dimidiata* Westwood [Balach1950b, Zahrad1972], *Pteroptrix maritimus* Nikolskaya [Zahrad1972]. **Encyrtidae**: *Anabrolepis zetterstedtii* (Westwood) [Gordh1979], *Anagyrus schmuttereri* Ferriere [Zahrad1972], *An. schonherri* Westwood [Zahrad1972], *Chiloneurinus microphagus* Mayr [Zahrad1972], *Epitetracnemus zetterstedtii* (Westwood) [Trjapi1989], *Habrolepis pascuorum* Mercet [Trjapi1989], *Zaomma lambinus* (Walker) [Trjapi1989]. **Signiphoridae**: *Thysanus ater* Haliday [Balach1950b, Woolle1990]. THYSANOPTERA **Phlaeothripidae**: *Haplothrips subtilissimus* (Haliday) [PalmerMo1990].

DISTRIBUTION: **Australasian**: Australia (South Australia [BrookeHu1968], Tasmania [BrookeHu1968], Victoria [BrookeHu1968]); New Zealand [Nakaha1982]. **Nearctic**: Canada [Nakaha1982]; United States of America

(Colorado [Nakaha1982], Connecticut [Nakaha1982], Idaho [Nakaha1982], Iowa [Nakaha1982], Kansas [Nakaha1982], Maine [Nakaha1982], Massachusetts [Nakaha1982], Michigan [Nakaha1982], Montana [Ferris1938a], New Hampshire [Nakaha1982], New Mexico [Nakaha1982], New York [Nakaha1982], Ohio [Nakaha1982], Oregon [Nakaha1982], Pennsylvania [Nakaha1982], Rhode Island [Nakaha1982], South Dakota [Nakaha1982], Utah [Ferris1938a], Washington [Nakaha1982], Wisconsin [Nakaha1982], Wyoming [Nakaha1982]). **Neotropical**: Argentina (Mendoza [Lizery1936]). **Palaearctic**: Algeria [Balach1928a, Balach1932d]; Azerbaijan (Azerbaijan [Borchs1936]); China (People's Republic) [Danzig1980b, Tang1984]; Corsica [Balach1931a, Balach1932d]; Czech Republic [Zahrad1952, Zahrad1977]; France [Signor1869b, Balach1932d, Balach1933a, Balach1933e, Foldi2000]; Georgia (Abkhaz ASSR [Borchs1934, Borchs1936], Adzhar ASSR [Borchs1934, Borchs1936], Georgia [Borchs1936, Hadzib1983]); Germany (United) [Lindin1909b, Lindin1909c, Balach1950b]; Greece [Korone1934]; Hungary [Kozar1999a]; Iran [Bodenh1944a, Kaussa1955]; Iraq [Dowson1935]; Ireland [Green1934d, Balach1950b]; Israel [Bodenh1924, Bodenh1937]; Italy [Leonar1909, Leonar1920, LongoMaPe1995]; Japan [Kawai1980] (Hokkaido [Takagi1958]); Kazakhstan (Alma Ata Oblast [Borchs1935, BazaroSh1971]); Kyrgyzstan (= Kirgizia) [Balach1950b]; Morocco [Balach1932d]; Netherlands [Reyne1957, Jansen2001]; North Korea [Danzig1980b]; Poland [Kaweck1935, Koteja2000a]; Portugal [Seabra1941]; Romania [Savesc1982]; Russia (Caucasus [Borchs1936], Karachay-Cherkessia AR [Danzig1985], Primor'ye Kray [Danzig1980b], Sakhalin Oblast [Danzig1980b], St. Petersberg (= Leningrad) Oblast [Danzig1962b], Yakutia-Sakha (= Yakut) AR [Danzig1978a]); Slovak Republic [Zahrad1952]; Spain [Balach1935b, GomezM1937, BlayGo1993]; Sweden [Balach1950b, Gertss2001]; Switzerland [Balach1950b]; Tajikistan (=Tadzhikistan) [BazaroSh1971]; Turkey [Bodenh1949, Bodenh1952]; United Kingdom [Green1928] (England [Curtis1843c]); Uzbekistan (Samarkand Oblast [BazaroSh1971]); Yugoslavia [Bachma1953].

BIOLOGY: Occurring on bark (Ferris, 1938a).

ECONOMIC IMPORTANCE: Widely distributed in Palearctic region, and has been introduced to other regions of the world (see Distribution). Pest of deciduous fruit trees, mainly Rosaceae (Balachowsky, 1950b; Schmutterer et al., 1957; Argyriou, 1990; Kozár, 1990c).

GENERAL: Description and illustration of adult female by Ferris (1938a), Balachowsky (1948a, 1950b), Zahradník (1952), Takagi (1958), Brookes & Hudson (1968), Bazarov & Shmelev (1971), Tang (1984), Chou (1985, 1986), Tereznikova (1986), Danzig (1980b, 1993), Yaşar (1995a) and by Kosztarab (1996). Description and illustration of second instar-nymph by Brookes & Hudson (1968).

KEYS: Kosztarab 1996: 575 (female) [Northeastern North America]; Danzig 1993: 179-182 (female) [Europe]; Tereznikova 1986: 97-98 (female) [Ukraine]; Chou 1985: 311 (female) [Species of China]; Danzig 1980b: 340 (female) [Far East of USSR]; Kosztarab & Kozár1978: 173-174 (female) [Hungary]; Bazarov & Shmelev 1971: 211 (female) [Central Asia]; Brookes & Hudson 1968: 91 (larva) [Australia]; Brookes & Hudson 1968: 91 (female) [Australia]; Reyne 1957: 33 (female) [Netherlands]; Zahradník 1952: 114-115 (female) [Czech Republic]; Balachowsky

1950b: 404-405 (female) [Mediterranean]; Balachowsky 1948a: 90 (female) [World]; Lupo 1948: 174 (female) [Italy]; Ferris 1942: 40 (female) [North America]; Borchsenius 1938: 142 (female) [Far East of USSR]; Borchsenius 1935: 127-128 (female) [Former USSR]; Britton 1923: 371 (female) [U.S.A.: Connecticut]; Leonardi 1920: 29-30 (female) [Italy]; Lawson 1917: 217 (female) [U.S.A.: Kansas]; Newstead 1901b: 81 (female) [England]; Newell 1899: 4-5 (female) [North America].

CITATIONS: Aleksi1995 [host, distribution, economic importance, biological control: 187-190]; AngeriLo1985 [host, distribution, chemical control: 31-35]; Archan1937 [taxonomy, description, illustration, host, distribution: 99,101]; Argyri1990 [host, distribution, economic importance: 579-583]; Bachma1953 [host, distribution: 178]; Baeren1849 [taxonomy, description, host, distribution: 165-167]; Balach1928a [host, distribution: 138]; Balach1931a [host, distribution: 97]; Balach1932b [ecology: 517-522]; Balach1932d [taxonomy, host, distribution: IV-V; XLVI]; Balach1933a [host, distribution: 38]; Balach1933e [host, distribution: 3]; Balach1935b [host, distribution: 256]; Balach1937c [host, distribution: 2]; Balach1948a [taxonomy, description, illustration, host, distribution: 91-93]; Balach1950b [taxonomy, description, illustration, host, distribution, biological control, economic importance: 405-409,469-472]; Banks1990 [physiology, chemistry: 267-274]; BazaroSh1971 [taxonomy, description, illustration, host, distribution: 211-217]; BeardsDaHo1976 [economic importance: 105]; BeardsGo1975 [economic importance: 49]; BlayGo1993 [taxonomy, description, illustration, host, distribution: 577-582]; Bodenh1924 [taxonomy, host, distribution: 33]; Bodenh1935 [host, distribution: 246]; Bodenh1935c [taxonomy, distribution: 1156]; Bodenh1937 [host, distribution: 217]; Bodenh1944b [host, distribution: 93]; Bodenh1949 [taxonomy, description, illustration, host, distribution: 57-60]; Bodenh1952 [host, distribution, structure: 338]; Bonnem1936 [taxonomy, description: 230-243]; Boraty1953 [taxonomy, description, host, distribution: 463-465]; Borchs1934 [host, distribution: 28]; Borchs1935 [taxonomy, description, illustration, host, distribution: 128-129]; Borchs1936 [host, distribution: 132]; Borchs1937 [taxonomy, description, illustration, host, distribution: 129,130]; Borchs1937a [taxonomy, description, illustration, host, distribution: 43-44]; Borchs1938 [host, distribution: 143]; Borchs1939 [taxonomy: 10,31]; Borchs1939a [taxonomy, distribution: 43]; Borchs1949d [taxonomy, description, host, distribution: 244]; Borchs1950b [taxonomy, description, illustration, host, distribution: 226-228,231]; Borchs1966 [catalogue: 328,334-335,341]; BrandtBo1948 [taxonomy: 3]; Britto1923 [taxonomy, description, host, distribution: 371,374-375]; BrookeHu1968 [taxonomy, description, illustration, host, distribution: 93-95]; BrookeHu1969 [host, distribution: 228-233]; BurgerUl1990 [economic importance: 313-327]; Bustsh1958 [taxonomy, description, host, distribution: 220,255]; Buxton1920 [host, distribution: 287-303]; Calkin1983 [distribution, economic importance: 321-359]; Charle1998 [distribution, economic importance, biological control: 52N]; Chou1986 [taxonomy, illustration: 691]; Chumak1957 [host, distribution, biological control: 533-547]; Chumak1961 [host, distribution, biological control: 313-338]; ClapsWoGo2001a [taxonomy, host, distribution: 26]; Cocker1895b [taxonomy: 16]; Cocker1896b [distribution: 333];

Cocker1897i [taxonomy, description, host, distribution: 15,18,19]; Collin1950 [host, distribution, economic importance: 158-160]; Comsto1883 [taxonomy: 73,77-78,80]; Curtis1843c [taxonomy, description, illustration, host, distribution: 805]; Danzig1959 [taxonomy, host, distribution: 450,459]; Danzig1962b [taxonomy, distribution: 27]; Danzig1964 [taxonomy, host, distribution: 653]; Danzig1972 [taxonomy, host, distribution, economic importance: 210]; Danzig1977b [taxonomy: 57]; Danzig1978 [host, distribution: 19-20]; Danzig1978a [host, distribution: 78]; Danzig1980b [taxonomy, description, illustration, host, distribution: 341-342]; Danzig1985 [distribution: 112]; Danzig1988 [taxonomy, host, distribution: 725]; Danzig1993 [taxonomy, description, illustration, host, distribution, economic importance: 182-184]; DanzigKo1991 [distribution: 1-15]; DanzigPe1998 [catalogue: 237-238]; DavidsMi1990 [host, distribution, economic importance: 603-632]; Dougla1887 [taxonomy: 239]; Dowson1935 [host, distribution: 225]; DumasVa1950 [chemical control: 235-245]; Duskov1953a [host, distribution, taxonomy, life history: 229-250]; Evans1942 [host, distribution, taxonomy, chemical control: 156-159]; Evans1943 [host, distribution, taxonomy, chemical control]; EvansPr1990 [biological control: 3-17]; Felt1901 [taxonomy: 347,352]; Fernal1903b [catalogue: 268-270]; Ferris1937c [taxonomy, illustration: 50-52,95]; Ferris1938a [taxonomy, description, illustration, host, distribution: 258]; Ferris1941e [taxonomy: 40,44-46]; Ferris1942 [taxonomy: 446:40]; Foldi1990c [structure, anatomy: 199-204]; Foldi2000 [host, distribution: 84]; Foldi2001 [distribution: 303-308]; FoldiDe1998 [taxonomy, host, distribution: 202]; FrankKr1898 [taxonomy: 397]; FrankKr1900 [taxonomy: 41]; FreyFr1995 [taxonomy, chemistry: 777-780]; FreyFr1995a [taxonomy, structure, chemistry: 100]; Garcia1930 [host, distribution, biological control]; Gavalo1931 [host, distribution: 7]; Gavalo1932b [host, distribution: 1]; Gavalo1936 [host, distribution: 75-76]; Gertss2001 [distribution: 123-130]; Ghauri1962 [taxonomy, description, host, distribution: 109,211]; Goethe1884 [taxonomy, description, host, distribution: 114]; Goethe1899 [taxonomy, description, host, distribution: 16-19]; GolanLaJa2001 [taxonomy, host, distribution: 229-249]; GomezM1937 [taxonomy, description, illustration, host, distribution: 59-64]; GomezM1956 [taxonomy, description, illustration, host, distribution, biological control: 21-24]; GomezM1957 [taxonomy, host, distribution: 46]; GomezM1958c [host, distribution: 406]; Gordh1979 [biological control: 895,901,944]; Goux1949 [taxonomy: 32]; Green1916 [host, distribution: 29]; Green1928 [taxonomy, description, host, distribution: 8]; Green1934d [distribution: 114]; Griswo1926 [biological control: 331-334]; Hadzib1983 [taxonomy, host, distribution, life history, biological control, economic importance: 238-240]; Heriot1934 [structure, life history: 602-612]; Herric1925 [host, distribution, description, life history, economic importance]; Hill1989a [host, distribution, economic importance, biological control: 177-182]; HippeScMa1995 [host, distribution, economic importance, chemical control, biological control: 4,84-85]; HornokKo1984 [host, distribution, biological control: 9-11]; Horvat1897 [taxonomy: 95]; Huba1960 [taxonomy: 39-50]; HunterWo2001 [life history, biological control: 251-290]; Jaap1914 [host, distribution: 135-142]; Jansen2001 [host, distribution: 197-206]; Jarvis1908TD [taxonomy: 57]; Jorgen1934 [taxonomy, distribution: 279]; Jura1959 [taxonomy, description, structure: 17-34]; Kaussa1955 [host, distribution: 16]; Kawai1980 [taxonomy, description, host, distribution: 221]; Kaweck1935 [taxonomy, host, distribution: 76]; Kiritc1932a [taxonomy: 246]; Komosi1974a [host, distribution, life history, ecology: 1-84]; Komosi1986 [host, distribution: 3-12]; Komosi1986a [host, distribution: 13-20]; Komosi1987 [host, distribution: 95-103]; Komosi1987a [host, distribution: 105-116]; Konsta1976 [host, distribution, economic importance: 49-50]; Koreck1974 [structure, anatomy: 85-93]; Korone1934 [taxonomy, description, illustration, host, distribution: 11-12]; Koszta1996 [taxonomy, description, illustration, host, distribution, life history, biological control, economic importance: 581-582]; KosztaKo1978 [taxonomy, description, host, distribution: 173-174]; Koteja1990b [life history, structure, anatomy: 233-242]; Koteja1990b [anatomy, life history: 233-242]; Koteja1990c [life history: 243-254]; Koteja2000a [distribution: 172]; Kozar1990 [life history, economic importance: 335-340]; Kozar1990c [host, distribution, life history, economic importance: 593-602]; Kozar1995b [taxonomy: 51]; Kozar1999a [host, distribution: 141]; KozarGuBa1994 [host, distribution: 151-161]; KozarHiMa1996 [taxonomy, description: 433-437]; KozarzKo1972 [host, distribution: 42-46]; Krzysz1957 [taxonomy: 223]; Lagows1998a [host, distribution: 63-71]; LagowsGo1998 [host, distribution, economic importance: 21-23]; LagowsKo1996 [host, distribution: 32, 35]; Lawson1917 [taxonomy, description, illustration, host, distribution: 217,224-225]; Leonar1898a [taxonomy: 75]; Leonar1898c [taxonomy, description, illustration, host, distribution: 38-40]; Leonar1909 [host, distribution: 124]; Leonar1920 [taxonomy, description, illustration, host, distribution: 30,50-54]; Lichte1881 [taxonomy, description: li-lii]; Lindin1907 [taxonomy: 6]; Lindin1909b [host, distribution: 151]; Lindin1909c [host, distribution: 449]; Lindin1909e [taxonomy: 44]; Lindin1912b [taxonomy, description, host, distribution: 48,49,57,83,151,163,]; Lindin1935 [taxonomy: 129]; Lindin1957 [taxonomy: 545,546,548]; Lizery1936 [host, distribution: 113]; LongoMaPe1995 [distribution: 128]; Lupo1948 [taxonomy, description, illustration, host, distribution: 174-180]; MacGil1921 [taxonomy, description, host, distribution: 410]; ManiHiSc1993 [life history, economic importance, host, distribution: 299-302]; Marlat1899d [taxonomy, description, host, distribution: 76-82]; Martin1983 [taxonomy, host, distribution: 67]; Masi1934 [host, distribution, biological control: 97-102]; May1899a [taxonomy: 151]; McClur1990d [taxonomy, host, distribution, ecology: 301-303]; McClur1990e [taxonomy, host, distribution, ecology: 309-314]; McLare1989 [host, distribution, chemical control: 221-227]; McLare1989a [host, distribution, life history, ecology: 215-219]; McLareFr1992 [life history, ecology: 21]; Meerwa1900 [taxonomy: 3-15]; Merkel1938 [host, distribution: 88-99]; MillerDa1990 [host, distribution, economic importance: 305]; Morgan1888b [taxonomy, description: 45,118-120,350]; Morgan1889a [taxonomy, host, distribution: 350]; Morgan1890 [taxonomy: 42-44]; Morgan1967 [host, distribution, life history, economic importance: 650-659]; MuelleEi1954 [taxonomy, description: 151-153]; Muraka1970 [host, distribution: 78]; Nakaha1982 [host, distribution: 78]; Newell1899 [taxonomy, description, host, distribution: 17-18]; Newste1901b [taxonomy, description, illustration, host, distribution: 81,97,99-104]; Noel1894 [taxonomy, description, host, distribution, life history, chemical control: 67-72]; Novits1961 [biological control: 193-194]; PalmerMo1990 [biological control: 67-

76]; Penman1984 [host, distribution, biological control: 33-50]; Pesson1950 [structure, taxonomy, life history: 566-570]; Pettit1900 [host, distribution, taxonomy, description: 1]; PolavaDaMi2000 [taxonomy: 558]; Reh1900a [taxonomy: 259]; Reh1900b [taxonomy: 497]; Reh1903 [taxonomy: 468]; Reh1933 [life history, physiology: 109-112]; Reyne1949 [taxonomy, host, distribution: 32]; Reyne1957 [taxonomy, host, distribution: 26,33]; Richar1960AM [host, distribution: 693-698]; RosenDe1979 [host, distribution, biological control: 411-414,476-483]; RSEA1915 [host, distribution, description, life history, economic importance, control: 1]; Ruhl1913 [host, distribution: 79-80]; RzaevaYa1985 [biological control: 55-58]; Sander1904a [taxonomy, description, illustration, host, distribution: 57,64]; Savesc1953 [host, distribution, taxonomy, description, economic importance, control: 3-46]; Savesc1982 [taxonomy, description, host, distribution, life history, biological control: 306-308]; Schmut1959 [taxonomy, description, host, distribution: 77]; SchmutKlLu1957 [host, distribution, economic importance: 484,485]; SchuhMo1948 [host, distribution, control]; Seabra1941 [distribution: 8]; SeveriSe1909 [taxonomy: 298]; ShiLi1991 [host, distribution: 166]; Signor1869 [taxonomy: 844,854,857,862,863]; Signor1869b [taxonomy, description, host, distribution: 136-137]; Signor1869c [taxonomy, description, host, distribution: 115]; Signor1869d [taxonomy, description, illustration, host, distribution: 439-441]; Signor1877 [taxonomy, description: 603]; Smetni1991 [chemistry: 92-129]; SzklarBi1995 [anatomy, structure: 23-29]; Szulcz1926 [host, distribution: 137-143]; Szulcz1949 [distribution: 219-224]; Takagi1958 [taxonomy, description, illustration, host, distribution: 127-129]; Tang1984 [taxonomy, description, illustration, host, distribution: 65-66]; Tao1999 [taxonomy, host, distribution: 115]; Targio1868 [taxonomy: 44]; Terezn1986 [taxonomy, description, illustration, host, distribution: 104-106]; TerGri1956 [taxonomy, description, host, distribution, life history: 55]; TerGri1962 [taxonomy, description, host, distribution: 149-150]; ThiemGe1934 [taxonomy, description, host, distribution, life history: 532]; ThiemGe1934a [taxonomy, description, host, distribution: 130-158,208-238]; TranfaVi1987a [economic importance: 215-221]; Trembl1990a [anatomy, structure: 275-283]; Trjapi1989 [biological control: 292,293,312]; Tschor1939 [host, distribution: 90]; Valent1963 [biological control: 6-13]; Valent1967 [biological control: 1100]; Viggia1985 [host, distribution, biological control: 3-9]; Viggia1987 [host, distribution, biological control: 121-123]; Viggia1990a [biological control: 124]; Weglar1962 [structure, anatomy: 267-294]; Weglar1962a [structure, anatomy: 41-68]; Weglar1966 [structure, anatomy: 59-98]; Weglar1968 [structure, anatomy: 63-82]; WoodwaEvEa1970 [distribution]; Woolle1990 [biological control: 167-176]; Xie1998 [taxonomy, description, host, distribution: 113]; Yasar1995a [taxonomy, description, illustration, host, distribution: 116-118]; Yasnos1994 [host, distribution, biological control: 317-333]; Zahrad1952 [taxonomy, description, illustration, host, distribution: 115-119]; Zahrad1959b [host, distribution: 60]; Zahrad1972 [taxonomy, host, distribution, biological control: 435-436]; Zahrad1977 [taxonomy, distribution: 121]; Zahrad1990a [host, distribution, description, taxonomy: 651-652].

Diaspidiotus paraphyses (Takagi)

Quadraspidiotus paraphyses Takagi, 1956b: 88. Type data: JAPAN: Honsyu, Toyama-ken, Toyama, on *Castanopsis cuspidata*. Holotype female. Type depository: Sapporo: Entomological Institute, Faculty of Agriculture, Hokkaido University, Japan.

Diaspidiotus paraphyses; Danzig & Pellizzari, 1998: 238. Change of combination.

SCALE COVER: Female scale of the type common to genus, being dark brown; that of male not identified (Takagi, 1956b).

HOST PLANTS: **Fagaceae**: *Castanopsis cuspidata* [Takagi1956b].

DISTRIBUTION: **Palaearctic**: Japan [Kawai1980] (Honshu [Takagi1956b]).

BIOLOGY: The types were collected on branches of the host plant (Takagi, 1956b).

GENERAL: Description and illustration of adult female by Takagi (1956b).

KEYS: Kawai 1980: 215 (female) [Japan].

CITATIONS: Borchs1966 [catalogue: 335]; DanzigPe1998 [catalogue: 238]; Kawai1980 [taxonomy, description, host, distribution: 215]; Lindin1957 [taxonomy: 551]; Muraka1970 [host, distribution: 77]; Takagi1956b [taxonomy, description, illustration, host, distribution: 88-89].

Diaspidiotus perieri (Goux)

Quadraspidiotus perieri Goux, 1949: 27. Type data: FRANCE: Marseille, pinede du Lycée Perier, on leaves of *Olea europaea*; collected by L. Goux, 1939. Holotype female. Type depository: Paris: Muséum national d'Histoire naturelle, France.

Quadraspidiotus marani olivicola Pegazzano, 1955: 314. Type data: ITALY: Florence, on leaves of *Olea europaea*. Syntypes, female. Type depository: Florence: Istituto Sperimentale per la Zoologia, Italy. Synonymy by Danzig & Pellizzari, 1998: 238.

Diaspidiotus perieri; Danzig & Pellizzari, 1998: 238. Change of combination.

SYSTEMATICS: Borchsenius (1966: 333) placed *Quadraspidiotus marani olivicola* Pegazzano, 1955, as a synonym of *Q. marani*, while Danzig (1993: 187) synonymised *Quadraspidiotus marani olivicola* with *Diaspidiotus perieri* Goux, 1949.

SCALE COVER: Female scale circular, about 2 mm in diameter; flat; placed beneath epidermis of the leaves; colour bright, yellow, similar to that of *Aspidiotus nerii*; exuviae central or subcentral. Male scale elongated, about 1 mm long; colour bright; exuviae subcentral (Goux, 1949).

HOST PLANTS: **Oleaceae**: *Olea europaea* [Goux1949, Pegazz1955].

DISTRIBUTION: **Palaearctic**: France [Goux1949]; Italy [Pegazz1955].

BIOLOGY: Develops mainly on underside of olive leaves (Pegazzano, 1955).

GENERAL: Description and illustration of adult female by Goux (1949), Pegazzano (1955) and by Danzig (1993).

KEYS: Danzig 1993: 179-182 (female) [Europe].

CITATIONS: Borchs1966 [catalogue: 336]; Danzig1993 [taxonomy, description, illustration, host, distribution: 187-188]; DanzigPe1998 [catalogue: 238]; Foldi2001 [distribution: 303-308]; Goux1949 [taxonomy, description, illustration, host, distribution: 27-32]; Pegazz1955 [taxonomy, description, illustration, host, distribution: 311-324].

Diaspidiotus perniciabilus Wang & Zhang

Diaspidiotus perniciabilus Wang & Zhang, 1994a: 326. Type data: CHINA: Anhui Province, host plant not indicated. Holotype female. Type depository: Beijing: Institute of Entomology, Academy of Sciences, China.

Diaspidiotus pemiciabilus; Tang & Zhang, 1994a: 327. Misspelling of species name.

SCALE COVER: Only slide-mounted females available for original description (Wang & Zhang, 1994a).

DISTRIBUTION: **Palaearctic**: China (People's Republic) (Anhui (Anhwei) [WangZh1994a]).

GENERAL: Description and illustration of adult female by Wang & Zhang (1994a).

CITATIONS: Tao1999 [taxonomy, host, distribution: 83]; WangZh1994a [taxonomy, description, illustration, host, distribution: 326-328].

Diaspidiotus perniciosus (Comstock)

Aspidiotus perniciosus Comstock, 1881a: 304. Type data: U.S.A.: California, Santa Clara County, on apple, pear, plum, and other trees. Syntypes, female. Type depository: Washington: United States National Entomological Collection, U.S. National Museum of Natural History, District of Columbia, USA.

Aonidia fusca Maskell, 1895b: 43. Type data: AUSTRALIA: New South Wales, Bulga, on peach, *Persica vulgaris*, sent by Mr. French. Syntypes, female and first instar. Type depository: Auckland: New Zealand Arthropod Collection, Landcare Research, New Zealand. Synonymy by Maskell, 1896: 14.

Aspidiotus albopunctatus Cockerell, 1896h: 20. Type data: JAPAN: on twigs of orange seedlings; collected Craw. Syntypes, female. Type depository: Washington: United States National Entomological Collection, U.S. National Museum of Natural History, District of Columbia, USA. Synonymy by Danzig, 1993: 191.

Aonidiella fusca; Leonardi, 1897: 286. Change of combination.

Aonidiella perniciosa; Leonardi, 1897: 286. Change of combination requiring emendation of species name for agreement in gender.

Aspidiotus (*Diaspidiotus*) *andromelas* Cockerell, 1897i: 20. Type data: JAPAN: on "Phaetenia glauca". Syntypes, female. Type depository: Washington: United States National Entomological Collection, U.S. National Museum of Natural History, District of Columbia, USA. Synonymy by Danzig, 1993: 191.

Aspidiotus (*Diaspidiotus*) *perniciosus albopunctatus*; Cockerell, 1897i: 20. Change of combination and rank.

Aspidiotus (*Diaspidiotus*) *perniciosus*; Cockerell, 1897i: 30. Change of combination.

Aonidiella perniciosa; Berlese & Leonardi, 1899: 255. Change of combination.

Diaspidiotus perniciosus andromelas; Cockerell, 1899a: 376. Change of combination and rank.

Diaspidiotus perniciosus; Cockerell, 1899a: 396. Change of combination.

Aonidiella andromelas; Leonardi, 1900: 341. Change of combination.

Aspidiotus perniciosus albopunctatus; Fernald, 1903b: 275. Change of combination.

Aspidiotus perniciosus andromelas; Fernald, 1903b: 276. Change of combination and rank.

Aspidiotus (*Diaspidiotus*) *perniciosus*; Brain, 1918: 125. Change of combination.

Comstockaspis perniciosa; MacGillivray, 1921: 438. Change of combination.

Aspidiotus (*Hemiberlesiana*) *perniciosus*; Thiem & Gerneck, 1934a: 132. Change of combination.

Aspidiotus (*Comstockaspis*) *perniciosus*; Borchsenius, 1935a: 33. Change of combination.

Quadraspidiotus perniciosus; Ferris, 1938a: 337. Change of combination.

Aspidiotus fuscus; Ferris, 1941e: 43. Change of combination requiring emendation of species name for agreement in gender.

Aspidiotus (*Quadraspidiotus*) *perniciosus*; Merrill, 1953: 25. Change of combination.

Hemiberlesiana perniciosa; Lindinger, 1957: 549. Change of combination.

Quadraspidiotus perniciosus; Borchsenius, 1966: 337. Revived combination.

Aspidiotus albopunctatus; Borchsenius, 1966: 368. Revived combination.

Aspidiotus andromelas; Borchsenius, 1966: 368. Revived combination.

Aspidiotus perniciasus; Chou, 1985: 310. Misspelling of species name.

Diaspidiotus perniciosus; Danzig & Pellizzari, 1998: 238. Change of combination.

COMMON NAMES: California scale [Borchs1936]; Chinese scale [Xie1998]; cocciniglia San Iose [Pegazz1948]; escama de San Jose [Gonzal1989]; kaliforniiskaya ili vrednaya shitovka [Borchs1936]; pernicious scale [Comsto1881a]; Pou de St-Jose [Balach1926]; round pear scale [Xie1998]; San José scale [McKenz1956, Dekle1965c, Borchs1966, RosenDe1978].

SYSTEMATICS: Danzig (1993: 191) synonymised *Aspidiotus albopunctatus* Cockerell and *Aspidiotus andromelas* Cockerell with *Q. perniciosus* while Borchsenius (1966:368) regarded these as *species incertae sedis*. Cockerell (1897i) noted a great similarity of *A. andromelas* to *Quadraspidiotus perniciosus*.

SCALE COVER: Scale of female grey, circular; slightly convex; exuviae subcentral; scale of male grey, slightly elongate; exuvia toward one end (Ferris, 1938a). Colour photograph by Gonzalez (1986, 1989), Gill (1997) and by Wong *et al.* (1999).

HOST PLANTS: **Aceraceae**: *Acer* [MerrilCh1923], *Ac. campestre* [Zahrad1972], *Ac. japonica* [Zahrad1972], *Ac. platonoides* [Zahrad1972], *Ac. saccharinum* [Zahrad1972]. **Actinidiaceae**: *Actinidia chinensis* [Gonzal1986, Gonzal1989a]. **Apocynaceae**: *Nerium* [McKenz1956]. **Araliaceae**: *Hedera* [McKenz1956]. **Betulaceae**: *Alnus* [Zahrad1972], *Alnus nepalennis* [RahmanAn1941], *Alnus nitida* [RahmanAn1941], *Betula alba* [Balach1950b], *Be. pendula* [Zahrad1972], *Be. pubescens* [Zahrad1972]. **Bignoniaceae**: *Catalpa bignoniodes* [Zahrad1972]. **Buxaceae**: *Buxus* [McKenz1956]. **Cannabidaceae**: *Cannabis sativa* [RahmanAn1941]. **Cannaceae**: *Canna indica* [RahmanAn1941]. **Caprifoliaceae**: *Symphoricarpos* [Merril1953], *Symphoricarpos racemosus* [Balach1950b], *Viburnum lantana* [Balach1950b]. **Carpinaceae**: *Carpinus betulus* [Zahrad1972]. **Cercidiphyllaceae**: *Cercidiphyllum japonicum* [MerrilCh1923]. **Cornaceae**: *Cornus alba sibirica* [Balach1950b], *Cornus baileyi* [Balach1950b], *Cornus sanguinea* [Balach1950b]. **Corylaceae**: *Corylus avenallana* [GomezM1946, Balach1950b, Zahrad1972], *Corylus tubulosa* [Zahrad1972]. **Ebenaceae**: *Diospyros kaki* [GomezM1946, Balach1950b]. **Elaeagnaceae**: *Elaeagnus* [MerrilCh1923]. **Ericaceae**: *Azalea* [McKenz1956]. **Euphorbiaceae**: *Aleurites fordii* [Balach1950b].

Fagaceae: *Castanea americana* [Zahrad1972], *Castanea sativa* [RahmanAn1941, Balach1950b, Zahrad1972], *Fagus sylvatica purpurea* [Balach1950b], *Fagus sylvatica* [Zahrad1972], *Quercus* [Muntin1971], *Quercus dilatata* [RahmanAn1941]. **Grossulariaceae**: *Ribes* [Muntin1971], *Ribes aureum* [Balach1950b], *Ribes nigrum* [Balach1950b], *Ribes oxycantha* [Balach1950b], *Ribes rubrum* [Balach1950b]. **Guttiferae**: *Hypericum* [MerrilCh1923, Merril1953]. **Hippocastanaceae**: *Aesculus hippocastanum* [Zahrad1972]. **Juglandaceae**: *Juglans* [McKenz1956], *Juglans regia* [RahmanAn1941, Balach1950b, Zahrad1972], *Juglans sieboldiana* [Balach1950b, Zahrad1972]. **Lardizabalaceae**: *Akebia* [Balach1950b], *Akebia quinata* [MerrilCh1923, Merril1953]. **Lauraceae**: *Persea* [McKenz1956]. **Leguminosae**: *Acacia* [MerrilCh1923, McKenz1956], *Gleditsia triacanthos* [Muntin1971], *Robinia pseudacacia* [Zahrad1972]. **Liliaceae**: *Aloe* [McKenz1956]. **Malvaceae**: *Hibiscus* [MerrilCh1923]. **Moraceae**: *Maclura aurantiaca* [Balach1950b], *Toxylon pomiferum* [Balach1950b]. **Myrtaceae**: *Eucalyptus* [Zahrad1972]. **Nyssaceae**: *Nyssa* [MerrilCh1923]. **Oleaceae**: *Fraxinus* [Muntin1971], *Fraxinus excelsior* [Balach1950b, Zahrad1972], *Ligustrum vulgare* [Balach1950b], *Olea* [McKenz1956], *Olea europaea* [TippinBe1970, BesheaTiHo1973], *Syringa amurensis* [Borchs1938], *Syringa persica* [Balach1950b], *Syringa vulgaris* [Balach1950b]. **Pinaceae**: *Pinus* [McKenz1956]. **Platanaceae**: *Platanus* [Zahrad1972]. **Ranunculaceae**: *Sieboldia* [MerrilCh1923]. **Rhamnaceae**: *Ceanothus* [MerrilCh1923]. **Rosaceae**: *Amelanchier canadensis* [Balach1950b], *Cerasus vulgaris* [RahmanAn1941], *Chaenomelis* [BesheaTiHo1973], *Chaenomelis japonica* [Muntin1971], *Chaenomelis lagenaria moorloosei* [Muntin1971], *Cotoneaster* [Balach1950b, McKenz1956], *Cotoneaster microphyla* [Muntin1971], *Cotoneaster vulgaris* [Balach1950b], *Crataegus* [MerrilCh1923, RahmanAn1941, McKenz1956], *Cr. coccinea* [Balach1950b], *Cr. cordata* [Balach1950b], *Cr. crus-galli* [Balach1950b], *Cr. monogyna* [Kozar1999a], *Cr. oxyacantha* [Balach1950b], *Cydonia japonica* [Balach1950b, McDani1970], *Cydonia oblonga* [Muntin1971], *Cydonia vulgaris* [RahmanAn1941, GomezM1946, Balach1950b], *Eriobotrya japonica* [UygunSeEr1998], *Malus communis* [GomezM1946, Martin1983, BlayGo1993, SengonUyKa1998, UygunSeEr1998], *Persica vulgaris* [Maskel1895b], *Photinia* [Merril1953, McDani1970], *Photinia glauca* [Cocker1897i, Leonar1900], *Prunus* [Cocker1897i, Lepage1938, RahmanAn1941, Dekle1965c, BesheaTiHo1973], *Pr. amygdalus* [RahmanAn1941, GomezM1946, Balach1950b], *Pr. armeniaca* [RahmanAn1941, GomezM1946, Balach1950b], *Salix* [Borchs1936, RahmanAn1941, Balach1950b, BesheaTiHo1973], *Pr. avium* [Balach1950b], *Pr. carolina* [Wilson1917], *Pr. cerasifera astropurpurea* [Balach1950b], *Pr. cerasus* [Muntin1971], *Pr. divaricata* [RahmanAn1941], *Pr. domesticus* [Lepage1938, GomezM1946, Balach1950b, Muntin1971, BlayGo1993], *Pr. germanica* [Lepage1938], *Pr. hortulana* [Balach1950b], *Pr. japonica* [Balach1950b], *Pr. laurocerasus* [Balach1950b], *Pr. mahaleb* [Kozar1999a], *Pr. maritima* [Balach1950b], *Pr. persica* [RahmanAn1941, GomezM1946, Balach1950b, Muntin1971, ErlerTu2001], *Pr. pumila* [Balach1950b], *Pr. serotina* [Balach1950b, BesheaTiHo1973], *Pr. spinosa* [Kozar1999a], *Pr. triflora* [Balach1950b], *Pr. virginiana* [Balach1950b], *Pyracantha* [McKenz1956, Dekle1965c], *Pyracantha rogersii* [Muntin1971], *Pyrus*

[RahmanAn1941, Dekle1965c], *Pyrus baccata* [Balach1950b], *Pyrus communis* [Lepage1938, Iren1970, Muntin1971, Almeid1973b, ErlerTu2001], *Pyrus malus* [Lepage1938, RahmanAn1941, Iren1970, Muntin1971, Almeid1973b], *Pyrus simonii* [TakahaTa1956], *Rosa* [Borchs1936, RahmanAn1941, GomezM1946, Dekle1965c, Muntin1971], *Rosa carolina* [Balach1950b], *Rosa lucida* [Balach1950b], *Rosa rugosa* [Balach1950b], *Rosa viriginiana* [Balach1950b], *Sorbus* [Balach1950b], *Sorbus americana* [Balach1950b, Zahrad1972], *Sorbus aria* [Zahrad1972], *Sorbus aucuparia* [Balach1950b, Zahrad1972], *Sorbus domestica* [Zahrad1972], *Sorbus melanocarpa* [Balach1950b, Zahrad1972]. **Rutaceae**: *Citrus* [Cocker1896h, Cocker1897i, McKenz1956], *Citrus trifoliata* [Wilson1917, MerrilCh1923, Balach1950b], *Citrus unshiu* [TakahaTa1956], *Poncirus trifoliata* [Borchs1936, GomezM1946, Merril1953], *Ptelea triflora* [Balach1950b], *Pyrus sinensis* [Balach1950b]. **Salicaceae**: *Populus* [Muntin1971], *Po. deltoides* [Balach1950b, Zahrad1972], *Po. italica* [Balach1950b], *Po. nigra* [Balach1950b, Zahrad1972], *Po. nigra italica* [Zahrad1972], *Po. trimula* [Zahrad1972], *Salix acutifolia* [Zahrad1972], *Sa. alba* [Zahrad1972], *Sa. babylonica* [Balach1950b, Zahrad1972], *Sa. bumilis* [Zahrad1972], *Sa. caprea* [Zahrad1972], *Sa. discolor* [BesheaTiHo1973], *Sa. elaeagnus* [Zahrad1972], *Sa. humilis* [Balach1950b], *Sa. incana* [Balach1950b], *Sa. lucida* [Balach1950b, Zahrad1972], *Sa. pentendra* [Balach1950b, Zahrad1972], *Sa. repens* [Zahrad1972], *Sa. vitellina* [Balach1950b]. **Sambucaceae**: *Sambucus* [Balach1950b]. **Tamaricaceae**: *Tamarix* [McKenz1956]. **Theaceae**: *Camellia* [Muntin1971]. **Tiliaceae**: *Tilia* [Balach1950b], *Tilia americana* [Zahrad1972], *Tilia platyphyllos* [Zahrad1972]. **Ulmaceae**: *Ulmus* [RahmanAn1941, McKenz1956], *Ulmus americana* [Balach1950b, Zahrad1972], *Ulmus campestris* [Balach1950b, Zahrad1972], *Ulmus procera* [Zahrad1972]. **Vitaceae**: *Vitis* [McKenz1956], *Vitis vinifera* [Balach1950b].

NATURAL ENEMIES: ACARI **Hemisarcoptidae**: *Hemisarcoptes malus* (Shimer) [Andre1942, Zahrad1972, GersonOcHo1990, Koszta1996]. COLEOPTERA **Alleculidae**: *Omophlus proteus* Kirsch [Drea1990]. **Cantharidae**: *Cantharis rustica* Fallen [Drea1990]. **Coccinellidae**: *Chilocorus bijugus* Mulsant [RawatThPa1988], *Ch. bipustulatus* L. [Balach1950b, Zahrad1972, Koszta1996, ErlerTu2001], *Ch. bivulnerus* [Marlat1902a], *Ch. circumdata* Schon. [RosenDe1978], *Ch. kuwanae* Silvestri [RosenDe1978], *Ch. quadripustulatus* L. [Zahrad1972], *Ch. renipustulatus* Scriba [Balach1950b, Zahrad1972], *Ch. renipustulatus inornatus* Weise [RosenDe1978], *Ch. similis* (Rossi) [Nakaya1912, RosenDe1978], *Ch. stigma* Say [RosenDe1978], *Coccinella bipunctata* L. [Zahrad1972]. *Exochomus quadrimaculatus* L. [Zahrad1972, ErlerTu2001], *Lindorus lophanthae* (Blaisdell) [Smirno1950a, RosenDe1978], *Pharoscymnus flexibilis* Mulsant [RawatThPa1988], *Rhyzobius lindi* Blackburn [RosenDe1978], *Rhyzobius ventralis* [Pope1981], *Scymnus marginicollis* Mann. [RosenDe1978]. **Nitidulidae**: *Cybocephalus* sp. [RosenDe1978], *Cybocephalus fodori* [Koszta1996], *Cybocephalus fodori-minor* Enrody-Younga [ErlerTu2001], *Cybocephalus politus* Germ. [Balach1950b, Zahrad1972]. **Tenebrionidae**: *Epitragus similis* Steinheil [Drea1990]. FUNGI **Ascomycotina**: *Myriangium duriaei* [EvansPr1990], *Nectria aurantiicola* [EvansPr1990], *Nectria coccophila* [Muraka1970], *Nectria diploa* [EvansPr1990], *Nectria flammea* [EvansPr1990], *Podonectria coccicola* [EvansPr1990].

Deuteromycotina: *Fusarium* [EvansPr1990], *Fusarium aspidioti* [Muraka1970], *Fusarium larvarum* [HornokKo1984], *Peziotrichum saccardinum* [EvansPr1990]. HETEROPTERA **Anthocoridae**: *Temnostethus dacicus* Puton [ErlerTu2001], *Temnostethus longirostis* (Horvath) [ErlerTu2001]. **Miridae**: *Deraeocoris ruber* L. [Zahrad1972]. HYMENOPTERA **Aphelinidae**: *Ablerus clisiocampae* (Ashmead) [Balach1950b, Gordh1979], *Anaphes gracilis* Howard [Balach1950b], *Aphytis aonidiae* (Mercet) [GomezM1946, GulmahDe1978a, RosenDe1979], *Ap. diaspidis* (Howard) [Flande1960, RosenDe1979], *Ap. fuscipennis* Howard [Balach1950b], *Ap. hispanicus* (Mercet) [SengonUyKa1998, MyartsRu2000], *Ap. melinus* DeBach [RosenDe1979], *Ap. mytilaspidis* (Le Baron) [Balach1950b, Flande1960, Zahrad1972, RosenDe1979], *Ap. paramaculicornis* DeBach & Rosen [RosenDe1979], *Ap. proclia* Walker [Zahrad1972, RosenDe1978], *Ap. vandenboschi* DeBach & Rosen [RosenDe1979], *Archenomus bicolor* Howard [Zahrad1972], *Aspidiotiphagus citrinus* Howard [Balach1950b, Zahrad1972], *Azotus americanus* Dozier [Gordh1979], *Azotus elegantulus* Silvestri [AnneckIn1970], *Azotus marchali* Howard [Balach1950b, Zahrad1972], *Azotus separaspidis* Annecke & Insley [AnneckIn1970], *Coccophagoides kuwanae* (Silvestri) [Flande1960, Gordh1979], *Coccophagoides similis* (Masi) [Gordh1979], *Coccophagus immaculatus* Howard [Gordh1979], *Encarsia perniciosi* (Tower) [Rosen1987, StouthLu1991, PaloukNa1995, Koszta1996, MyartsRu2000], *Hispaniella lauri* Mercet [Zahrad1972], *Marietta carnesi* (Howard) [Gordh1979], *Marietta mexicana* (Howard) [MyartsRu2000], *Marietta pulchella* (Howard) [Gordh1979], *Physcus varicornis* (Howard) [Gordh1979], *Prospaltella aurantii* (Howard) [Balach1950b, Gordh1979], *Prospaltella berlesei* (Howard) [Benass1954], *Prospaltella diaspidicola* Silvestri [Gordh1979], *Prospaltella fasciaventris* Girault [Gordh1979], *Prospaltella perniciosi* Tower [Tower1913, Flande1960, BenassBi1960, Zahrad1972, RosenDe1978], *Pteroptrix dimidiata* Westwood [Balach1950b, Novits1961, Zahrad1972]. **Encyrtidae**: *Acerophagus citrinus* (Howard) [Gordh1979], *Adelencyrtus inglisiae* Compere & Annecke [AnneckIn1971], *Anabrolepis zetterstedtii* (Westwood) [Gordh1979], *Apterencyrtus microphagus* (Mayr) [Gordh1979], *Arrhenophagus chionaspidis* Aurivillius [Gordh1979], *Coccidencyrtus ensifer* (Howard) [Gordh1979], *Coccidencyrtus steinbergi* Tshumakova & Trjapitzin [Trjapi1989], *Epitetracnemus zetterstedtii* (Westwood) [Trjapi1989], *Euussuria shutovae* Trjapitzin [RosenDe1978], *Habrolepis aspidioti* Compere & Annecke [RosenDe1978], *Habrolepis obscura* Compere & Annecke [AnneckIn1971], *Metaphycus nadius* (Walker) [GuerriNo2000], *Thomsonisca pallipes* (Tshumakova) [Trjapi1989], *Zaomma lambinus* (Walker) [Trjapi1989]. **Mymaridae**: *Polynema fulmeki* Soyka [Zahrad1972]. **Signiphoridae**: *Chartocerus pulcher* (Girault) [Gordh1979], *Signiphora flavella* [Woolle1990], *Signiphora pulchra* Girault [Woolle1990], *Signiphora townsendi* Ashmead [Gordh1979], *Thysanus ater* Haliday [Zahrad1972, Woolle1990]. NEUROPTERA **Chrysopidae**: *Chrysopa* sp. [Drea1990]. THYSANOPTERA **Phlaeothripidae**: *Leptothrips mali* Fitch [PalmerMo1990].
DISTRIBUTION: **Afrotropical**: Angola [Almeid1973b]; South Africa [BrainKe1917, Brain1918, Muntin1971, CABI1986]; Zaire [Ghesqu1932, CABI1986]; Zimbabwe [Muntin1971, CABI1986]. **Australasian**: Australia (New

South Wales [CABI1986], Queensland [Brimbl1962a, CABI1986], South Australia [CABI1986], Tasmania [Borchs1966, BrookeHu1968, CABI1986], Victoria [CABI1986], Western Australia [CABI1986]); Hawaiian Islands (Hawaii [Borchs1966]); New Zealand [Borchs1966, EmmsMc1984]. **Nearctic**: Canada [Borchs1966] (British Columbia [CABI1986], Nova Scotia [CABI1986], Ontario [CABI1986], Quebec [CABI1986]); Mexico [Borchs1966] (District federal [CABI1986]); United States of America (Alabama [CABI1986], Alaska [CABI1986], Aleutian Islands [CABI1986], Arizona [CABI1986], Arkansas [CABI1986], California [CABI1986], Colorado [CABI1986], Connecticut [CABI1986], Delaware [CABI1986], District of Columbia [CABI1986], Florida [Wilson1917, MerrilCh1923, Merril1953, Dekle1965c, CABI1986], Georgia [TippinBe1970, BesheaTiHo1973, CABI1986], Idaho [CABI1986], Illinois [CABI1986], Indiana [CABI1986], Iowa [CABI1986], Kansas [Hunter1899, CABI1986], Kentucky [CABI1986], Louisiana [CABI1986], Maryland [CABI1986], Massachusetts [CABI1986], Michigan [CABI1986], Minnesota [CABI1986], Mississippi [Herric1911, CABI1986], Missouri [Hollin1923, CABI1986], Montana [CABI1986], Nebraska [CABI1986], Nevada [CABI1986], New Hampshire [CABI1986], New Jersey [CABI1986], New Mexico [CABI1986], New York [CABI1986], North Carolina [CABI1986], Ohio [CABI1986], Oklahoma [CABI1986], Oregon [CABI1986], Pennsylvania [CABI1986], Rhode Island [CABI1986], South Carolina [CABI1986], Tennessee [CABI1986], Texas [Herric1911, McDani1970, CABI1986], Utah [CABI1986], Vermont [CABI1986], Virginia [CABI1986], Washington [CABI1986], West Virginia [CABI1986], Wisconsin [CABI1986]). **Neotropical**: Argentina [Borchs1966, CABI1986] (Buenos Aires [CABI1986]); Bolivia [CABI1986]; Brazil [Borchs1966] (Minas Gerais [Lepage1938, CABI1986], Parana [Lepage1938, CABI1986], Rio Grande do Sul [Lepage1938, BertelBa1966, CABI1986], Rio de Janeiro [CABI1986], Santa Catarina [Lepage1938, CABI1986], São Paulo [Lepage1938, CABI1986]); Chile [Borchs1966, GonzalCh1968, Gonzal1981, Gonzal1989, Gonzal1989a]; Cuba [CABI1986]; Peru [CABI1986]; Uruguay [CABI1986]; Venezuela [CABI1986]. **Oriental**: China (People's Republic) (Guangdong (Kwangtung) [CABI1986], Hubei (Hupei) [CABI1986], Jiangsu (Kiangsu) [CABI1986], Jiangxi (Kiangsi) [CABI1986], Sichuan (Szechwan) [CABI1986], Zhejiang (Chekiang) [CABI1986]); India [Borchs1966] (Andhra Pradesh [CABI1986], Assam [CABI1986], Delhi [CABI1986], Himachal Pradesh [CABI1986], Jammu & Kashmir [CABI1986], Karnataka [UsmanPu1955, CABI1986], Maharashtra [CABI1986], Orissa [CABI1986], Punjab [RahmanAn1941, CABI1986], Tamil Nadu [CABI1986], Uttar Pradesh [CABI1986], West Bengal [CABI1986]); Mongolia [DanzigKo1990]; Nepal [CABI1986]; Pakistan [RosenDe1978, RosenDe1979, CABI1986]; Taiwan [WongChCh1999]. **Palaearctic**: Afghanistan [Siddiq1966, CABI1986]; Algeria [Borchs1966, CABI1986]; Armenia [Babaia1987]; Austria [Borchs1966]; Azerbaijan (Azerbaijan [Borchs1966]); Bulgaria [Borchs1966]; Canary Islands [GomezM1967O, CABI1986, PerezGCa1987, MatileOr2001]; China (People's Republic) [Borchs1966] (Anhui (Anhwei) [CABI1986], Hebei (Hopei) [CABI1986], Heilongliang (HeilungKiang) [CABI1986], Henan (Honan) [CABI1986], Henan (Honan) [Shen1993], Jilin (Kirin) [CABI1986], Liaoning [CABI1986], Nei

Monggol (Inner Mongolia) [CABI1986], Shandong (Shantung) [CABI1986], Shanxi (Shensi) [CABI1986], Xingiang Uygur (Sinkiang) [CABI1986]); Czech Republic [Zahrad1952, Borchs1966, Zahrad1977]; France [Borchs1966]; Georgia [Borchs1966] (Abkhaz ASSR [Borchs1936], Adzhar ASSR [Borchs1936], Georgia [Borchs1936]); Germany (United) [Borchs1966]; Greece [CABI1986, PaloukNa1995]; Hungary [Borchs1966, Kozar1999a]; Iran [CABI1986]; Iraq [Borchs1966, CABI1986]; Italy [Borchs1966, LongoMaPe1995]; Japan [Cocker1896h, Cocker1897i, Sasaki1901, Kuwana1902, Kuwana1917a, Kuwana1933, Borchs1966, Kawai1980] (Hokkaido [TakahaTa1956], Honshu [TakahaTa1956, CABI1986], Kyushu [TakahaTa1956, CABI1986], Shikoku [TakahaTa1956, CABI1986]); Kazakhstan [CABI1986]; Madeira Islands [CABI1986]; Moldavia [Borchs1966]; North Korea [Borchs1966]; Poland [Borchs1966]; Portugal [Seabra1941, Borchs1966]; Romania [Borchs1966, Savesc1982]; Russia (Amur Oblast [CABI1986], Dagestan AR [Borchs1966], Kabardino-Balkarian AR [CABI1986], Khabarovsk Kray [Borchs1966, CABI1986], Krasnodar Kray [CABI1986], Kurile Islands [CABI1986], North Ossetia AR [CABI1986], Primor'ye Kray [Borchs1966, CABI1986], Sakhalin Oblast [Borchs1966], Stavrapol Oblast [CABI1986]); South Korea [Borchs1966, CABI1986]; Spain [Balach1935b, GomezM1937, Borchs1966, Martin1983, BlayGo1993]; Sweden [Gertss2001]; Switzerland [Borchs1966]; Tajikistan (=Tadzhikistan) [Borchs1966, BazaroSh1971]; Turkey [Iren1970, CABI1986, UygunSeEr1998, ErlerTu2001]; Turkmenistan [Borchs1966]; Ukraine [Borchs1966] (Krym (= Crimea) Oblast [Borchs1966]); United Kingdom (England [Borchs1966]); Uzbekistan (Fergana Oblast [CABI1986]); Yugoslavia [Borchs1966].

BIOLOGY: Generally on bark, although it may occur on fruit of apples (Ferris, 1938a). Symbiosis of this species with the fungi *Septobasidium burtii* and *S. mariani* on apple and pear trees in Turkey studied by Iren (1970). Female sex pheromone identified (Gieselmann et al., 1979).

ECONOMIC IMPORTANCE: The San Jose scale is a major pest of deciduous fruit trees in many regions of the world. Origin believed to be in north of the Oriental region, north China, Far East Russia. First recorded in USA, California about 1870, from where it was originally described in 1881. It spread rapidly throughout main fruit-growing regions of the United States, from west to east. It has since spread to Canada, Central and South America, Europe (from Spain to Caucasus), Japan, India, South Africa, Zimbabwe and Australia (Ebeling, 1959; Rosen & DeBach, 1978; Kozár, 1990). A pest of kiwifruit in Chile (Gonzalez, 1989a). The plant growth regulator heteroauxin was most effective in controlling *Quadraspidiotus perniciosus*, (Ivanova & Pavlyuchuk, 1988).

GENERAL: Description and illustration of adult female by Brain (1918), Ferris (1938a), Zahradník (1952), Balachowsky (1950b, 1958b), McKenzie (1956), Brookes & Hudson (1968), Bazarov & Shmelev (1971), Chou (1985, 1986), Tereznikova (1986), Danzig (1993), Yaşar (1995a), Kosztarab (1996) and by Gill (1997). Description and illustration of second instar by Brookes & Hudson (1968).

KEYS: Gill 1997: 243 (female) [Species of California]; Kosztarab 1996: 575 (female) [Northeastern North America]; Blay Goicoechea 1993: 528-529 (female) [Spain]; Danzig 1993: 179-182 (female) [Europe]; Tereznikova 1986: 97-98 (female) [Ukraine]; Chou 1985: 311 (female) [Species of China]; Danzig 1980b: 340 (female) [Far East of USSR]; Kawai 1980: 215 (female) [Japan]; Kosztarab & Kozár 1978: 167-169 (female) [Hungary]; Bazarov & Shmelev 1971: 211 (female) [Central Asia]; McDaniel 1970: 429 (female) [U.S.A.: Texas]; Brookes & Hudson 1968: 91 (larva) [Australia]; Brookes & Hudson 1968: 91 (female) [Australia]; McKenzie 1956: 26 (female) [U.S.A.: California]; Zahradník 1952: 114-115 (female) [Czech Republic]; Balachowsky 1950b: 401 (female) [Mediterranean]; Ferris 1942: 40 (female) [North America]; Borchsenius 1938: 143 (female) [Far East of USSR]; Borchsenius 1935: 128 (female) [Former USSR]; Kuwana 1933: 3 (female) [Japan]; Kuwana 1933b: 49 (female) [Japan]; Britton 1923: 371 (female) [U.S.A.: Connecticut]; Hollinger 1923: 7-8 (female) [U.S.A.: Missouri]; Brain 1918: 124 (female) [South Africa]; Lawson 1917: 217 (female) [U.S.A.: Kansas]; Dietz & Morrison 1916a: 289-290 (female) [U.S.A.: Indiana]; Cockerell 1905b: 202 (female) [U.S.A.: Colorado]; Newell 1899: 4-5 (female) [North America]; Comstock 1883: 55-57 (female) [North America].

CITATIONS: Abbott1926 [chemical control: 858-860]; AbbottCuMo1926 [chemical control: 1-26]; AbdElKDaKo1988 [chemical control, biological control]; AblesRi1981 [biological control, economic importance: 273]; Ackerm1923 [economic importance, host, distribution, chemical control: 1-18]; AhmadGh1966 [life history, biological control: 101-106]; Alden1925 [chemical control: 253-257]; AlderdSpWe1984 [chemistry, pheromone: 1643-1646]; Aldric1899 [taxonomy, description, illustration, host, distribution: 4]; Aldric1996 [life history, physiology, chemistry: 201-204]; Aleksi1995 [host, distribution, economic importance, biological control: 187-190]; Almeid1973b [host, distribution: 11]; AlvareVa1998 [biological control: 130-136]; Alwood1896 [host, distribution, economic importance: 33-44]; AndersChGi1979 [chemistry, life history: 919-927]; AndersGiCh1981 [chemistry, pheromone: 695-706]; Andre1942 [biological control: 173-180]; Angeri1990 [life history, chemical control: 409-411]; AngeriGaLo1986 [host, distribution, economic importance: 493-497]; AngeriLo1985 [host, distribution, chemical control: 31-35]; AngeriLo1986 [host, distribution, life history, biological control: 767-774]; AnneckIn1970 [host, distribution, biological control: 241-242,246]; AnneckIn1971 [host, distribution, biological control: 2]; AnneckIn1971 [host, distribution, biological control: 13,14]; Anthon1960 [host, distribution, economic importance, chemical control: 1085-1087]; Antong1954 [chemical control: 1-23]; AntropFaPa1990 [host, distribution, chemical control: 65-67]; Apstei1915 [taxonomy: 119]; Archan1937 [taxonomy, description, illustration, host, distribution: 99,107-108]; Argyri1981 [host, distribution, biological control: 125-129]; ArmourScan1991 [host, distribution, economic importance: 31-33]; AsquitCrHo1980 [control, ecology, economic importance: 249-317]; Babaia1987 [host, distribution, economic importance: 136]; BabenkYa1997 [chemical control: 64-68]; Bachma1974 [economic importance, chemical control: 755-760]; BadeneZaBe2002 [chemistry, ecology, life history: 545-549]; BadeneZaBe2002a [host, distribution, chemical control,: 291-296]; Baerg1921 [chemical control: 177]; BaggioGeMa1951 [host, distribution, economic importance: 931-937]; Baicu1982 [chemical control: 1]; BaioccGa1991 [economic importance, biological control, chemical control: 104-106]; Balach1926 [taxonomy, host, distribution, economic importance: 5-6];   Balach1932j [economic importance: 171-176]; Balach1935b [host, distribution, economic importance: 258]; Balach1950b [taxonomy, description, illustration, host, distribution, biological control, economic importance: 424-433]; Balach1958b [taxonomy, description, illustration, host, distribution: 214-216]; Banks1990 [physiology, chemistry: 267-274]; Bardia1967 [host, taxonomy, life history, control: 3-23]; Baroff1993 [life history, biological control: 371-378]; Baroff1997 [life history, biological control: 323-333]; BassinPeTi1975 [biological control, host, distribution: 21-23,26-42]; BazaroSh1971 [taxonomy, description, illustration, host, distribution, life history: 220-223]; BeardsDaHo1976 [economic importance: 106]; BeardsGo1975 [economic importance: 49]; Beauch1964 [host, distribution, chemical control: 45-49]; BeglyaSm1977 [host, distribution, biological control: 283-328]; Benass1954 [biological control: 528-530]; Benass1957 [biological control, chemical control: 867-872]; Benass1958b [life history, ecology, biological control: 93-108]; Benass1958d [biological control: 179-181]; Benass1959b [host, distribution, biological control, chemical control: 867-872]; Benass1961a [host, distribution, life history, biological control: 1-165]; Benass1961b [host, distribution, ecology: 1-157]; Benass1969 [biological control: 33-39]; Benass1969a [biological control: 33-39]; Benass1970 [biological control: 45-50]; Benass1971 [biological control, economic importance: 9-23]; BenassBi1960 [biological control, economic importance: 165-191]; BenassBiMi1960a [host, distribution, biological control: 29-36]; BenassBiMi1962 [biological control: 59-68]; BenassBiMi1964 [host, distribution, biological control: 271-280]; BenassBiMi1964a [host, distribution, biological control: 457-472]; BenassBiMi1964b [host, distribution, chemical control, biological control: 27-37]; BenassBiMi1967 [biological control: 1-5]; BenassBiMi1967a [biological control: 241-255]; BenassBiMi1971 [host, distribution, biological control: 411-420]; BenassMaNe1968 [host, distribution, biological control: 4-28]; BenassMaNe1968 [biological control, host, distribution: 1-28]; BenDov1990a [taxonomy: 85]; BenDov1990e [host, distribution: 656]; BenDov2001 [distribution, economic importance: 141-142]; Bentle1913 [host, distribution, economic importance, life history, chemical control, biological control: 39-59]; BentleCoHa2000 [host, distribution, control]; BentleHeDu2001 [host, distribution, control: 12-19]; BentleRiBr2000 [host, biological control, chemical control]; BentleRiDa2000 [host, distribution, control]; Beran1962 [chemical control: 135-136]; Beran1963 [host, distribution, chemical control, biological control: 95-96]; Berger1921 [host, distribution, biological control: 141-154]; Berger1921a [host, distribution, biological control: 333-334]; Berger1942a [host, distribution, biological control: 26-29]; BerlesLe1899 [taxonomy, description, illustration, host, distribution, economic importance: 253,255-260]; BerlesLe1899 [taxonomy: 250]; Bertel1956 [host, distribution, description, life history, biological control]; BertelBa1966 [host, distribution: 17-46]; BesheaTiHo1973 [host, distribution: 8]; BichinYa1985 [taxonomy, life history, host, distribution, economic importance, control: 1-96]; BichnaVa1985 [host, distribution, chemical control, chemistry: 31]; BiezanFr1939 [host, distribution: 1-18]; BiezanSe1940 [host, distribution: 67-68]; Biliot1977 [host, distribution, biological control: 341-347]; BiliotBeBi1960 [host, distribution, biological control: 707-711]; Blahut1990 [biological control: 1-98]; BloescSt1992 [host, distribution, life history, pheromone: 305-310]; Bohm1951 [host, distribution, life history, economic importance: 66-76]; Bohm1955 [host, distribution, economic importance, chemical control, biological control: 245-267]; Bohm1963 [host, distribution, biological control: 95]; Bohm1964 [host, distribution, biological control: 147-148]; Bohm1965 [distribution, biological control : 104]; Bohm1976 [host, distribution: 41-42]; Bonfan1999 [host, distribution, economic importance, control: 24-26]; Bonnem1936 [taxonomy, description: 230-243]; Borchs1935 [taxonomy, description, illustration: 127-128]; Borchs1935a [taxonomy, description, host, distribution: 5,9,33]; Borchs1936 [host, distribution: 132]; Borchs1937 [taxonomy, description, illustration, host, distribution: 134]; Borchs1937a [taxonomy, description, illustration, host, distribution: 55-56]; Borchs1938 [host, distribution: 143]; Borchs1939 [taxonomy, description, host, distribution: 9,27]; Borchs1949d [taxonomy, description, host, distribution: 249]; Borchs1950b [taxonomy, description, illustration, host, distribution: 227,229,231]; Borchs1960b [taxonomy, host, distribution: 217,218]; Borchs1966 [catalogue: 336-339,368]; BorgesViSa1986 [distribution]; BorianNi1995 [chemical control: 43]; BoveyGe1946 [host, distribution, control: 201-211]; Bower1987 [host, distribution, economic importance, chemical control: 55-58]; Bower1989 [host, distribution, life history: 239-245]; BowerPeDo1993 [economic importance, chemical control: 57-62]; BoyceKaPe1939 [chemical control: 432-450]; Boynto1901 [taxonomy: 347,349]; Brain1918 [taxonomy, description, illustration, host, distribution: 124-126]; BrainKe1917 [distribution: 183]; BrandtBo1948 [taxonomy: 14-18]; Brick1912 [host, distribution: 1-22]; Brimbl1962 [host, distribution, economic importance: 222-223]; Brimbl1962a [taxonomy, description, illustration, host, distribution: 418-422]; Britto1901 [host, distribution, taxonomy, control: 3-14]; Britto1901a [host, distribution, chemical control: 1-12]; Britto1909 [host, distribution, economic importance, chemical control: 1-4]; Britto1919 [host, distribution: 255-261]; Britto1920a [host, distribution: 119-125]; Britto1920b [host, distribution: 140-143]; Britto1923 [taxonomy, description, host, distribution: 371,375]; Britto1923a [chemical control: 329-331]; Britto1924 [host, distribution: 387-404]; Britto1924a [host, distribution]; Britto1924b [host, distribution: 156-170]; Britto1926 [host, distribution: 156]; Britto1928 [host, distribution: 198]; Britto1929 [host, distribution: 669]; Britto1930 [host, distribution: 490]; Britto1931 [host, distribution: 461]; Britto1932 [host, distribution: 499]; Britto1937 [host, distribution: 293]; Britto1938 [host, distribution: 137]; BrittoWa1903 [host, distribution, chemical control: 1-26]; BrittoWa1904 [host, distribution, chemical control: 1-32]; BrittoZa1927a [host, distribution: 1]; BrittoZa1929a [host, distribution: 689]; BrittoZa1930 [host, distribution: 502]; BrockFl1919 [chemical control: 1-4]; BrookeHu1968 [taxonomy, description, illustration, host, distribution: 97-99]; BrookeHu1969 [host, distribution: 228-233]; BrownPo1990 [biological control: 527-533]; Brunet1943 [chemical control: 221-233]; Buckle1987 [life history, ecology: 53-85]; BuhrooChMa2000 [life history, biological control: 117-123]; BurgerUl1990 [economic importance: 313-327]; Burke1930 [host, distribution, biological control: 783-785]; Bustsh1958 [taxonomy, description, host, distribution: 220,250]; CABI1986 [host, distribution: 1-3]; Caesar1914 [host, distribution, taxonomy: 1-30]; CaprilVaPi2000 [host, distribution, control];

CarbonBr1975 [host, distribution, economic importance, life history: 1-11]; Carnes1907 [taxonomy, host, distribution: 209]; Castel1951a [biological control: 95-98]; CevikOkSo1996 [chemical control: 28-29,61-62]; Chambe1926 [host, distribution: 55]; ChandeKa1994 [chemical control: 222-223]; ChangDaYu1989 [host, distribution, life history, ecology: 29-30]; ChapmaAvPe1944 [chemical control: 305-307]; Charle1998 [distribution, economic importance, biological control: 51N]; CharleAlWe1998 [host, distribution, biological control: 93-98]; CharleHiAl1995 [host, distribution, biological control: 319-324]; CharleWa1981 [chemical control, host, distribution: 252-253]; CheahIr1997 [host, distribution]; CherkaDu2000 [chemical control: 26]; Chiesa1939 [taxonomy, description, host, distribution, control: 27-44]; Chiesa1948 [host, distribution, economic importance]; Chou1985 [taxonomy, description, host, distribution: 310,316-318]; Chou1986 [taxonomy, illustration: 690]; Chu1992 [host, life history, control: 361-369]; ChuaWo1990 [host, distribution, economic importance: 550,551]; Chumak1957 [host, distribution, biological control: 533-547]; Chumak1961 [host, distribution, biological control: 313-338]; Chumak1964 [host, distribution, biological control: 535-552]; Chumak1969 [life history, ecology, biological control: 247-254]; ChumakGo1963 [biological control: 178-181]; ChumakMu1975 [biological control: 174-184]; ClapsWoGo2001a [taxonomy: 26]; Clarke1906 [host, distribution, economic importance, control: 1-8]; Clause1956 [host, distribution, economic importance, biological control]; Cocker1895b [taxonomy: 16]; Cocker1896b [distribution: 333-334]; Cocker1896h [taxonomy, description, host, distribution: 20-21]; Cocker1897i [taxonomy, description, host, distribution: 3-17,20,30]; Cocker1898bb [host, distribution: 95]; Cocker1899a [taxonomy: 396]; Cocker1905b [taxonomy: 202]; Collin1950 [host, distribution, economic importance: 158-160]; CollyeVa1975 [host, distribution, chemical control, biological control: 101-134]; Comper1961a [biological control: 17-71]; Comsto1881a [taxonomy, description, illustration, host, distribution: 304-305]; Comsto1883 [taxonomy, host, distribution: 56,65]; Conway1951 [host, distribution: 159-164]; Cooley1898b [host, distribution: 61-63]; CoronaRuMo1997 [host, distribution: 38-41]; CostaL1922a [host, distribution: 37-45]; CostaL1942 [taxonomy: 279]; CostaL1949 [host, distribution, biological control: 65-87]; Cottie1956 [host, distribution, economic importance, chemical control, biological control: 209-215]; Cox1942 [chemical control, biological control: 698-701]; Craved2000 [host, distribution, control: 177-197]; CravedMo1993 [control, pheromone: 170-173]; CravedMo1995 [control, pheromone: 115-121]; Craw1896 [taxonomy: 33]; Creigh1942 [host, distribution, control: 219-233]; Cressm1943 [chemical control: 17-26]; CressmDa1936 [chemical control: 865-878]; Croft1982 [host, distribution, economic importance, chemical control, biological control: 465-498]; CroftBo1983 [chemical control: 219-270]; CroftHu1983 [host, distribution, economic importance: 19-42]; DahmsSm1994 [host, distribution, biological control: 245-255]; Danzig1964 [taxonomy, host, distribution: 653]; Danzig1972 [taxonomy, host, distribution, economic importance: 210-212]; Danzig1977b [taxonomy: 57]; Danzig1978 [host, distribution: 20]; Danzig1980b [taxonomy, description, illustration, host, distribution: 342-343]; Danzig1988 [taxonomy, host, distribution: 725]; Danzig1993 [taxonomy, description, illustration, host, distribution, life history, economic importance: 191-

193]; DanzigKo1990 [host, distribution: 47]; DanzigKo1991 [distribution: 1-15]; DanzigPe1998 [catalogue: 238-239]; DarvasFaKo1985 [chemical control: 347-350]; DarvasSz1987 [host, distribution, chemical control: 343-351]; DarvasVa1990 [chemical control: 393-408]; DavidsDiFl1991 [chemical control: 1-47]; DavidsMi1990 [host, distribution, economic importance: 603-632]; DavidsRa1999 [economic importance, control: 1]; Davis1924 [economic importance, life history, ecology: 192-195]; Davis1924a [chemical control: 285-289]; Davis1929 [host, distribution: 2299]; Davis1935 [host, distribution: 198]; Davis1936 [host, distribution: 257]; Davis1937 [host, distribution: 230]; DeBach1960 [life history, biological control: 4-58]; DeBach1964 [biological control]; DeBach1964b [biological control: 673-713]; DeBach1971 [biological control: 302]; DeBach1974 [biological control]; DeBachRo1976a [host, distribution, biological control: 541-545]; DeBachRo1991 [biological control]; Decaux1899 [host, distribution, chemical control, biological control: 158-184]; DeitzTo1980 [taxonomy: 37]; Dekle1965c [taxonomy, description, host, distribution: 126]; Dekle1976 [taxonomy, description, host, distribution, economic importance: 145]; Delins1989 [life history, ecology: 102-105]; DeluccRoSc1976 [taxonomy, biological control: 81-91]; DengNa2001 [host, distribution, economic importance: 37-40]; DeSant1940 [biological control: 29-44]; DeSant1941a [host, distribution, biological control: 21-24]; DeSant1979 [biological control]; DietzMo1916a [taxonomy, description, illustration, host, distribution: 290-291]; DinabaBh1981 [host, distribution, life history: 63-67]; Dissel1954 [life history, structure: 109-155]; Doane1931 [host, distribution, control]; DowninLo1977 [life history, control: 1249-1252]; Dozier1933 [host, distribution, biological control: 85-100]; Drake1935 [host, distribution, control: 83-91]; Drea1990 [biological control: 41-49]; DreistClFl1994 [taxonomy, life history, economic importance, control]; DrgonHuBe1962 [chemical control: 221-243]; DumasVa1950 [chemical control: 235-245]; Dunkel1999 [chemistry, life history: 251-276]; Durr1951 [chemical control: 200-201]; Duskov1953a [host, distribution, taxonomy, life history: 229-250]; Duzgun1969 [host, distribution, economic importance]; Early1984 [host, distribution, biological control: 271-308]; Edland1990 [economic importance: 19-22]; Edmund1918 [chemical control: 1-16]; Ehrhor1913 [host, distribution: 101]; Ehrhor1925 [host, distribution: 18-20]; ElkinsVaVa2000 [host, distribution, control]; Emblet1902 [biological control: 219-229]; EmmsMc1984 [host, distribution: 81-82]; EMPPO1955 [economic importance: 1-19]; EMPPO1962 [host, distribution, economic importance, biological control: 7-27]; EMPPO1965 [economic importance: 1-8]; ErlerTu2001 [host, distribution, biological control: 299-305]; Evans1942 [host, distribution, taxonomy, chemical control: 156-159]; Evans1943 [host, distribution, taxonomy, chemical control]; Evans1975 [taxonomy: xxvii]; EvansPr1990 [biological control: 3-17]; FariFa1935 [chemical control: 339]; Fawcet1948 [biological control: 627]; Fawcet1948 [biological control: 627-664]; FDACSB1988 [host, distribution: 5-6]; Felt1901 [taxonomy, description, illustration, host, distribution, life history, biological control, chemical control: 304-316]; Felt1924 [host, distribution, economic importance, life history, control]; Felt1929 [biological control: 43-48]; Fenton1939 [host, distribution: 71-78]; FergusFl1991 [host, distribution, biological control: 260-261]; Fernal1903b [catalogue: 271-276]; Ferrar1976 [host, life history, control, ecology: 391-400]; Ferrie1927 [biological control: 55-67]; Ferris1920b [taxonomy: 52]; Ferris1937c [taxonomy, illustration: 51,66]; Ferris1938a [taxonomy, description, illustration, host, distribution: 259]; Ferris1941e [taxonomy: 40,43,46]; Ferris1942 [taxonomy: 446:40]; FinneyFi1964 [biological control: 328-355]; Fisher1950 [biological control: 305-309]; Flande1944b [economic importance, biological control: 105]; Flande1945c [biological control: 168-169]; Flande1947 [host, biological control: 746-747]; Flande1950a [biological control: 719-720]; Flande1960 [economic importance, distribution, biological control: 757-758]; Flande1971 [biological control, life history: 857-872]; Flesch1960 [biological control: 183-208]; Flint1920 [chemical control: 339-343]; Flint1923 [chemical control: 209-215]; FlintVa1981 [biological control: 1]; Foldi2001 [distribution: 303-308]; FoldiDe1998 [taxonomy, host, distribution: 202]; Folkin1978 [biological control: 301]; Folkin1980 [host, distribution, life history, economic importance, control: 24-31]; Forbes1898 [host, distribution, taxonomy, life history: 1-25]; Forbes1915 [host, distribution, chemical control: 545-561]; Forbes1915a [host, distribution, economic importance, chemical control: 63-79]; FotidaRa1936 [host, distribution, life history, biological control: 692-693]; FrankKr1898 [taxonomy: 393]; FrankKr1900 [taxonomy: 58]; Franz1958 [host, distribution, economic importance, biological control: 461-465]; Freita1962 [host, distribution: 21-32]; Freita1966 [life history, host, distribution: 289-235]; Freita1975 [host, distribution, life history, biological control: 235-285]; FreyFr1995 [taxonomy, chemistry, host, distribution: 777-780]; FreyFr1995a [structure, chemistry: 100]; Frogga1914 [taxonomy, description, host, distribution: 316-317]; Frogga1915 [taxonomy, description, host, distribution: 20-21]; Fuller1897c [host, distribution: 4]; FurnisCa1977 [host, distribution: 111]; Gahan1925 [host, distribution, biological control: 1-23]; Gambar1947 [host, distribution, life history: 45-58]; Gambar1955 [host, distribution, life history: 255-272]; Gambar1963a [host, distribution, biological control: 403-406]; Gambar1965 [host, distribution, biological control: 373-376]; Gamzae2002 [biological control: 27-28]; Garcia1930 [host, distribution, biological control]; Garcia1931a [host, distribution, biological control: 659-666]; Gatina1973 [host, distribution, economic importance, control: 121-192]; Gavalo1936 [host, distribution, life history, economic importance: 76-78]; Geier1950 [life history, control: 329-336]; Gendri1999 [biological control: 1-6]; GentilSu1958 [host, distribution, life history, ecology: 269-285]; GeorghTa1976 [chemical control: 759]; Gerasi1938 [host, distribution, life history, ecology: 47-65]; GersonOcHo1990 [biological control: 77-97]; Gertss2001 [distribution: 123-130]; Ghesqu1932 [host, distribution: 58]; Giesel1990 [life history, chemistry: 221-224]; Giesel1990a [chemistry: 225-232]; GieselRi1990 [life history, ecology, chemistry: 349-352]; GieselRiJo1979 [life history, chemistry, pheromone: 891-900]; Gill1990 [economic importance, control: 331]; Gill1997 [host, distribution, taxonomy, description, illustration, economic importance: 245-246,249]; GilletLi1918 [host, distribution: 3-52]; GilletLi1922 [host, distribution: 1]; GilletLi1923 [host, distribution: 12-18]; Giraul1911a [host, distribution, biological control: 175-189]; Glenn1915 [host, distribution, life history, control: 87-106]; Glenn1931 [life history, ecology: 167-180]; Glick1922 [host, distribution, economic importance: 55-77]; GomesCRe1947 [taxonomy, host, distribution: 229]; GomezM1937 [taxonomy, description, illustration, host, distribution: 64-67]; GomezM1946 [taxonomy, description, illustration, host, distribution: 66-73]; GomezM1956 [taxonomy, description, illustration, host, distribution, biological control: 21,24-31]; GomezM1960O [host, distribution: 162]; GomezM1967O [host, distribution: 131]; Gonzal1956 [life history, chemical control:9-13]; Gonzal1973 [economic importance, control: 56-63]; Gonzal1975 [economic importance, control: 135-145]; Gonzal1981 [taxonomy, description, illustration, host, distribution, life history, economic importance, biological control, chemical control: 1-64]; Gonzal1986 [host, distribution, economic importance: 18,23]; Gonzal1989 [taxonomy, description, host, distribution, life history, economic importance: 100-106]; Gonzal1989a [life history, economic importance, chemical control, host, distribution: 35-43]; GonzalCh1968 [distribution: 110]; GonzalCu1994 [host, distribution, economic importance: 9]; GonzalRo1967 [biological control, distribution: 138]; Gordh1977 [host, distribution, biological control: 125-148]; Gordh1979 [biological control: 894-895,897,899-901,]; Goryun1964 [biological control: 40-55]; Greath1971 [host, distribution, biological control]; Greath1976 [biological control, economic importance]; Greath1989 [biological control: 28-37]; Green1896a [taxonomy: 84]; Griffi1919 [chemical control: 14-18]; Gruys1979 [host, distribution, biological control, chemical control : 107-112]; GuerriNo2000 [host, distribution, biological control: 157-159]; GulmahDe1978 [host, distribution, life history, ecology, biological control: 205-238]; GulmahDe1978a [host, distribution, life history, ecology, biological control: 239-256]; Habibi1980 [host, distribution, chemical control: 127-134]; Hadzib1957a [distribution: 102]; HakkonPi1984 [biological control: 1109-1121]; Halber2000 [host, distribution: 3-4]; Hamilt1936 [host, distribution: 150-160]; HanksDe1998 [life history, ecology: 239-262]; Hansen2001 [control: 186-188]; HardieMi1999 [chemistry]; HasemaSu1923 [chemical control: 1-4]; Hayat1989 [host, distribution, biological control: 1-3]; Haywar1939 [host, distribution, control: 1]; Haywar1944 [host, distribution: 1-32]; Headle1927 [host, distribution: 112-113]; HeimpeRo1998 [biological control: 160-168]; HeimpeRoMa1996 [life history, biological control: 2410-2420]; Hempel1922 [host, distribution: 133-144]; HenrikCaAn1982 [chemistry, pheromone: 27-60]; Herber1919a [host, distribution: 333-337]; Herric1911 [taxonomy, description, illustration, host, distribution: 10,20-21,54]; Herric1920 [host, distribution, economic importance]; Herric1925 [host, distribution, description, life history]; Hill1975 [host, distribution, economic importance, control]; Hill1989a [host, distribution, economic importance, biological control: 177-182]; Hillma1895 [host, distribution, life history, control: 1-9]; HippeMa1995 [host, distribution, life history, biological control: 184]; HixPlDe1999 [chemical control: 106-108]; HodgkiPa1914 [biological control: 227-228]; Hollin1923 [taxonomy, description, host, distribution: 14]; Holman2002 [biological control: 41-54]; HornokKo1984 [biological control, host, distribution: 9-11]; Houck1999 [life history, biological control: 97-118]; Hough1932 [chemical control: 613-62]; Hough1933 [chemical control: 470-473]; Houser1908 [host, distribution: 173-181]; Houser1920 [chemical control: 49-51]; Houser1926 [chemical control: 94-95]; Houska1926 [chemical control: 94-95]; Howard1895e [biological control: 1-44]; Howard1897b [host, distribution: 81-82]; Howard1898a [host, distribution, economic importance: 1-31]; HowardMa1899 [host, distribution: 36-39]; HowellGe1984 [host, distribution, chemical control: 534-536]; HoytCa1971 [economic importance, biological control, host, distribution: 395-421]; HoytLeBr1983 [host, distribution, economic importance, life history, control: 93-151]; HoytWeRi1983 [host, distribution, life history, control: 371-375]; Hsu1935 [host, distribution: 578-590]; Huba1957 [host, distribution, biological control: 306-353]; Huba1958 [biological control: 395-403]; Huba1960 [taxonomy: 39-50]; Huba1962 [host, distribution, economic importance: 415-422]; Huffak1990 [biological control: 205-220]; HuffakDo1965 [biological control: 61-63]; HuffakGu1990a [biological control, economic importance: 473-492]; HuffakMeDe1971 [biological control: 16-67]; HuffakSt1971 [biological control: 333-350]; Hunger1944 [host, distribution, economic importance: 52-55]; Hunter1899 [taxonomy, description, host, distribution: 10-11]; Hurt1933 [chemical control: 1-16]; IngramNi1991 [host, distribution, chemical control: 53-58]; InnerhReWe1985 [economic importance, chemical control: 261-268]; Iperti1961 [economic importance: 14-30]; IpertiBr1969 [host, biological control: 149-157]; IPMW1987 [economic importance, biological control: 1-96]; Iren1970 [life history, ecology, host, distribution: 1-21]; Isayev1975 [economic importance: 181-187]; Isely1924 [chemical control: 1-4]; IshcheBiKo1988 [chemistry, pheromone: 99-101]; IvanovPa1988 [biological control, chemical control: 28-29]; JahnPo1999 [biological control, life history, ecology, host, distribution: 201-202]; JalaliSi1995 [chemical control, biological control: 617-620]; Jancke1955 [taxonomy: 307]; Janjua1959 [distribution: 231-264]; Jansen2001 [host, distribution: 197-206]; Jarvis1907 [chemical control: 5-12]; Jarvis1908CD [chemical control: 169-197]; Jarvis1908TD [taxonomy: 52]; Jarvis1927 [economic importance, life history, host, distribution, chemical control, biological control: 513-517]; JenserSh1972 [life history, chemical control: 119-124]; JerminBr1994 [chemical control: 5-6,115-119]; JiYa1990 [biological control: 134-136]; Johnso1898 [description, life history: 82-83]; Johnso1899 [chemistry, life history: 87-88]; Johnso1900 [biological control: 73-75]; Jorgen1934 [taxonomy: 279]; JorgenRiHo1981 [host, distribution, life history, control: 149-159]; Joseph1984 [economic importance: 1-32]; Kagy1936 [chemical control: 393]; Kapur1956 [biological control: 257-274]; KatsoyAr1985 [host, distribution, life history, biological control: 3-11]; KattarHeOd1999 [biological control: 640-644]; Kawai1980 [taxonomy, description, host, distribution: 216]; Kaweck1950 [host, distribution, economic importance: 1-54]; KazandTzSt1995 [life history: 247-252]; KhajurSh1997 [host, distribution, chemical control: 488-489]; Kim1983 [economic importance: 261-319]; Kiritc1932a [taxonomy: 246-247]; Klett1963 [life history, economic importance, biological control: 86-91]; KnightAl1974 [life history, physiology: 73-86]; KnightChUn2001 [ecology, life history, economic importance: 413-428]; Knowlt1932 [host, distribution, economic importance: 79-83]; KnowltSm1936 [host, distribution, economic importance: 263-267]; Kobakh1965 [biological control: 323-330]; KocourBeSt1999 [host, distribution, life history, chemistry, control: 407-412]; Koehle1964 [host, distribution, control]; Konsta1976 [host, distribution, economic importance: 49-50]; Konsta1988 [distribution, life history, economic importance, control: 50-51]; KonstaGu1987 [host, distribution: 162]; KonstaKuDe1985 [chemical control, physiology, chemistry: 29-30]; KonstaMaKo1982 [host, distribution, economic importance, chemical control, biological control: 24-25]; KonstaMo1982 [economic importance, chemical control: 63]; Korcha1987 [host, distribution, economic importance, life history: 58-59]; Koszta1987 [economic importance: 219-220]; Koszta1990 [economic importance: 307-311]; Koszta1996 [taxonomy, description, illustration, host, distribution, life history, biological control, economic importance: 583-584]; KosztaKo1978 [taxonomy, description, host, distribution: 167-169]; Kotins1909 [host, distribution: 97]; Kozar1990 [life history, economic importance: 335-340]; Kozar1990a [life history, economic importance: 341-347]; Kozar1990c [life history, economic importance, host, distribution: 593-602]; Kozar1995b [taxonomy: 51]; Kozar1999 [life history, chemistry, control: 125]; Kozar1999a [host, distribution: 141]; KozarHiMa1996 [taxonomy, description: 433-437]; KozarHiMa1997 [taxonomy: 321-327]; KozarKo1981a [host, distribution: 1981a]; KozarKo1996 [life history, biological control, ecology: 499-506]; KozarzKo1972 [host, distribution: 42-46]; Krause1950 [host, distribution, taxonomy, economic importance: 1-36]; KrnjajInPe1993 [life history, chemistry: 63-71]; Kruger1899 [host, distribution, economic importance: 3]; Krzysz1957 [taxonomy: 223]; Kuhmin1970 [host, distribution, life history, economic importance, biological control: 45-96]; Kuwana1901a [taxonomy: 13]; Kuwana1902 [host, distribution: 67-68]; Kuwana1904 [taxonomy, description, host, distribution: 1-33]; Kuwana1907 [host, distribution: 195]; Kuwana1917a [taxonomy, distribution: 175]; Kuwana1927 [host, distribution: 72]; Kuwana1933 [taxonomy, description, illustration, host, distribution: 15-16]; Kypari1987 [biological control, chemistry: 7-12]; Kypari1987a [life history, ecology, chemistry: 75-80]; Kypari1990 [chemical control, life history, chemistry: 5-9]; Lagows1995a [host, distribution, economic importance: 2,8]; LagowsGo1998 [host, distribution, economic importance: 21-23]; LancheDe1957 [host, distribution, chemical control: 14-15]; LangfoCo1939 [host, distribution: 3]; Laport1947 [host, distribution, biological control: 209-212]; Lawson1917 [taxonomy, description, illustration, host, distribution: 229-230]; Leonar1897 [taxonomy: 286]; Leonar1899 [taxonomy, description, host, distribution: 175-176,189]; Leonar1900 [taxonomy, host, distribution: 341]; Lepage1938 [catalogue: 396]; LiangAiZh1997 [host, distribution, life history, control: 152-153]; LiangAiZh1999 [host, distribution, life history, biological control: 29-30]; LiLoCa1997 [host, distribution, biological control: 225-226]; Lindin1909e [taxonomy: 44]; Lindin1957 [taxonomy: 546,549]; Lindne1954 [taxonomy, structure: 456-458]; Lintne1895 [taxonomy, description, host, distribution: 263-305]; Lintne1896 [host, distribution: 54-61]; Lobdel1937 [taxonomy: 78]; Lochhe1900 [host, distribution, taxonomy, control: 1-48]; LombarWe1986 [chemistry, pheromone: 5555-5558]; LongoMaPe1995 [distribution: 128]; LongoRuSi1989 [host, distribution, life history, ecology, biological control: 3-11]; Lord1922 [host, distribution: 1]; LoucifBo1977 [host, distribution, life history, ecology: 253-261]; Lounsb1914 [host, distribution: 1]; Lounsb1916 [host, distribution: 83-103]; Lounsb1921 [host, distribution: 35-38]; Lourei1992 [host, distribution, economic importance, control: 5-24]; LuckVaGa1977 [economic importance, chemical control, host, distribution: 606-611]; Ludick1950 [host, distribution, chemical control: 17-32]; Lugger1900 [host, distribution: 208-245]; Lupo1954a [taxonomy, description, illustration, host, distribution: 56-63]; LurieFaKl1998 [life history, control: 110-114]; Lyne1921 [distribution: 146-148]; MaLi2001 [biological control: 269-270]; MacGil1921 [taxonomy, description, host, distribution: 438]; Mackie1933 [host, distribution: 457]; MadsenMo1970 [host, distribution, economic importance: 295]; Mague1982 [host, distribution, life history, ecology: 975]; MagueRe1983 [host, distribution, life history, ecology: 692-696]; MagueRe1983a [host, distribution, life history: 717-722]; Maksim1973 [taxonomy, structure, life history: 81-84]; Maksim1973a [life history, ecology: 78-80]; Maksim1974 [life history, ecology: 81-82]; Maksim1976 [life history: 76-78]; Maksim1978 [life history, physiology: 58-61]; Maksim1991 [life history, ecology, chemistry, control: 42]; Maksim1991a [life history, ecology, chemistry, control: 24]; Maleno1946 [life history, economic importance: 3-8]; ManiHiSc1993 [life history, economic importance, host, distribution: 299-302]; ManiScBa1995 [host, distribution, life history, chemical control, biological control: 264-267]; Maranh1946 [taxonomy: 164-179]; Marco1959 [host, distribution, biological control: 25-30]; Marlat1902 [taxonomy, biological control, economic importance: 65-78]; Marlat1902a [host, distribution, biological control, economic importance: 155-174]; Marlat1906 [taxonomy, description, life history, economic importance, biological control, chemical control: 1-89]; Marr1949 [chemical control: 18-25]; Marsha1952 [host, distribution, economic importance: 25-31]; Marsha1953 [host, distribution, economic importance: 7-11]; Marsha1958 [host, distribution, economic importance, chemical control: 249-254]; Martin1983 [taxonomy, host, distribution: 67]; MaShWa1985 [host, distribution, economic importance, life history, ecology, chemical control, biological control: 9-10]; Maskel1895b [taxonomy, description, illustration, host, distribution: 43-44]; Maskel1896 [taxonomy, host, distribution: 14-16]; Maskel1896b [taxonomy, description, host, distribution: 386]; Maskew1914 [host, distribution: 193-194]; MasoodBhKo1989 [chemical control, biological control: 50-52]; MasoodBhSo1995 [chemical control, biological control: 154-156]; MasoodBhSo1996 [host, distribution, biological control: 77-78]; MasoodSoBh1996 [host, biological control: 119-121]; MasoodTrBh1989 [host, distribution, biological control: 39-43]; MasoodTrBh1989a [host, distribution, biological control: 71-73]; Mathys1953 [life history, ecology, economic importance: 981-984]; Mathys1959 [host, distribution, chemical control, biological control: 53-56]; Mathys1966 [host, distribution, economic importance, biological control: 53-64]; MathysGu1961 [biological control, economic importance: 53-56]; MathysGu1962 [host, distribution, biological control: 59-61]; MathysGu1963 [biological control: 96-101]; MathysGu1965 [biological control, host, distribution: 193-220]; MathysGu1967 [biological control, distribution: 223-224]; MathysGu1967a [biological control, host, distribution: 212-222]; MathysGuSt1965 [taxonomy: 65-67]; MatileOr2001 [host, distribution: 190]; Matvie1983 [chemical control, host, distribution: 25]; May1899a [taxonomy, illustration: 152]; Mazzon2001 [life history, ecology: 201-206]; MazzonCr2002 [host, distribution, taxonomy, life history, ecology: 373-383]; MazzonCr2002 [life history, distribution: 201-205]; McClaiRoSt1990 [host, distribution, ecology: 916-925]; McClaiRoSt1990a [host, distribution, life history, biological control: 1396-1402]; McClaiRoWo1990 [host, distribution, life history, ecology, biological control: 926-931]; McClur1990c [ecology, host,: 289-291]; McClur1990d [host:

301-303]; McClur1990e [life history, ecology: 309-314]; McClur1990f [life history: 315-318]; McClur1990h [life history, ecology: 331-337]; McDani1970 [taxonomy, illustration, host, distribution: 433-435]; McKenz1938 [taxonomy: 4]; McKenz1956 [taxonomy, description, illustration, host, distribution: 26,81-83]; Meerwa1900 [taxonomy, description: 1-15]; Meland1914 [chemical control, physiology: 167-173]; Meland1915 [chemical control: 475-481]; Meland1915a [host, distribution, taxonomy, economic importance, control]; Meland1922 [chemical control: 21-22]; Meland1922a [chemical control: 24-26]; Meland1923 [chemical control, physiology: 1-52]; Meland1926 [chemical control, physiology: 7,40]; Melis1943 [taxonomy, description, illustration, host, distribution, life history, biological control, chemical control: 1-170]; Melis1949 [host, distribution: 17-25]; Melis1951 [host, distribution, life history, ecology: 1-91]; Merkel1938 [host, distribution: 88-99]; Merril1953 [taxonomy, description, host, distribution: 25-27]; MerrilCh1923 [taxonomy, description, host, distribution, economic importance: 206-207]; MessenBiVa1976 [biological control: 543-565]; MessenWiWh1976 [biological control: 209]; Metcal1982 [chemical control: 217-277]; MetcalMe1993 [economic importance, host, distribution, control]; MeyerRa1973 [chemical control: 1354]; Milair1958 [host, distribution: 15-18]; Milair1969 [host, distribution, biological control: 77-85]; MilairAu1979 [host, distribution, biological control: 321-355]; Miller1999 [chemical control: 14]; MillerDa1990 [host, distribution, economic importance: 305]; Miyosh1926 [host, distribution: 303-326]; MoiseeIsVe1989 [chemistry, pheromone: 366-368]; Monte1930 [host, distribution: 3-36]; MordkoCh1993 [chemical control: 30]; Moreir1899 [host, distribution, control: 89-90]; Moreir1929 [economic importance: 85-89]; Morgan1967 [host, distribution, life history, economic importance: 650-659]; MorganAn1968 [host, distribution, economic importance: 499-503]; MorganAn1969 [host, distribution, life history, ecology, economic importance: 983-989]; MorganAr1971 [host, distribution, life history, ecology, economic importance: 1-12]; MuelleEi1954 [taxonomy, description: 151-153]; Muntin1971 [taxonomy, host, distribution: 141-142]; Muraka1970 [host, distribution, life history, biological control: 77-78]; MyartsRu2000 [distribution, biological control: 7-33]; NAC2000 [host, distribution, economic importance, control: 1-31]; NagaraHu1968 [biological control: 249-256]; Nakaha1982 [host, distribution: 78-79]; Nakaya1912 [host, distribution, biological control: 932-936]; NavrozZaPa1999 [host, distribution, economic importance, life history, biological control: 695-698]; Neiswa1966 [host, distribution, taxonomy, economic importance: 1-54]; Nekras1938 [chemical control: 481]; Nepveu1943 [life history, economic importance: 461-462]; NepveuVa1946 [host, distribution, life history, ecology: 415-418]; NepveuVa1948 [host, distribution, economic importance: 1-41]; Neuffe1962 [biological control, host, distribution: 97-101]; Neuffe1964 [biological control: 1-11]; Neuffe1964a [host, distribution, biological control: 131-136]; Neuffe1966 [biological control: 383-393]; Neuffe1967 [biological control: 235-239]; Neuffe1968 [host, distribution, biological control: 97-101]; Neuffe1969 [biological control: 49-55]; Neuffe1969a [life history, ecology: 63-68]; Neuffe1990 [host, distribution, economic importance, biological control: 89-96]; NewcomYo1929 [chemical control, host, distribution, life history: 821-822]; NewcomYo1931 [chemical control, host, distribution: 1-12]; Newell1899 [taxonomy, description, host, distribution: 16-17]; NewellRo1908 [host, distribution: 150-155]; Nicola1979 [host, distribution, life history, chemical control: 143-148]; NormanAl1982 [chemistry: 33-40]; Novits1961 [biological control: 193-194]; NRC1969 [taxonomy, economic importance, ecology, biological control, chemical control]; NSWDAE1963 [host, distribution, taxonomy, economic importance]; Nur1990a [taxonomy, structure, chromosomes: 182,183]; ObreteSt1992 [chemical control: 118-123]; OdinokKuZa1989 [chemistry, pheromone: 364-366]; OdyMaHu1969 [structure: 38-45]; OkaneCo1930 [chemical control: 1-15]; OneilHa1952 [life history, control: 170-178]; Orlins1989a [biological control: 54]; Paik1958 [host, distribution, economic importance: 31]; Painte1951 [economic importance, control]; PalmerMo1990 [biological control: 67-76]; Palouk1987 [chemical control: 179-183]; PaloukNa1995 [chemical control, economic importance, host, distribution, biological control: 285-286]; PasquaCi2000 [chemical control: 93-95]; Pegazz1948 [life history, host, distribution: 177-188]; Penman1984 [host, distribution, biological control: 33-50]; PerezGCa1987 [host, distribution: 129]; PernicSaJo1971 [host, distribution, economic importance: 27]; Petch1921a [biological control: 89-167]; Pfeiff1985 [host, distribution, life history, chemical control: 351-353]; Philip1965 [host, distribution, economic importance: 218-219]; PickelBeRi2000 [host, distribution, control]; PickelOlZa2000 [host, distribution, control]; PlessDeSa1995 [host, distribution, chemical control: 94-97]; PokrovUnDe1960 [chemical control: 27]; PollinBa1998 [host, distribution, economic importance, life history, chemical control, biological control: 79-80]; Pope1981 [biological control: 19-31]; Popova1962 [host, distribution, biological control: 147-175]; Popova1964 [biological control: 61-64]; Popova1971 [biological control: 42-43]; Popova1979 [biological control: 538-547]; Poutie1948 [host, distribution, economic importance, control: 23-26]; Powell1986 [biological control, host, distribution: 319-340]; PreeMeMa1985 [host, distribution, chemical control: 19-23]; Prinsl1983 [distribution, biological control: 27]; Prints1965 [host, control: 49-52]; Priore1964 [host, distribution: 131-178]; Priore1964a [host, distribution, chemical control: 179-204]; Priore1965 [host, distribution: 101-145]; PruthiBa1960 [host, distribution, economic importance: 1-113]; PruthiMa1945 [host, distribution, life history, control: 1-42]; PruthiRa1951 [host, distribution, life history, economic importance, chemical control, biological control : 1-48]; Pulsel1927 [biological control: 300-327]; Quaint1909 [chemical control: 1-33]; Quaint1915 [chemical control: 1-27]; QuaintSc1912 [chemical control: 1-48]; Quayle1911e [biological control: 510-515]; Quayle1938 [chemical control, physiology: 183-210]; Quedna1964b [biological control: 86-116]; RabbDeKe1984 [ecology, chemical control: 697]; RafalsNi1988 [host, distribution, chemical control: 40-42]; RahmanAn1941 [host, distribution, host, distribution: 816,822-823]; RahmanKa1940 [life history: 272]; RaoCh1950 [taxonomy: 8,27]; RaoGhSa1971 [host, distribution, biological control]; Rao1943VPa [host, distribution, economic importance]; RawatPa1992 [host, distribution, biological control: 7-10]; RawatSa1993 [biological control: 109-110]; RawatThPa1988 [host, distribution, economic importance, biological control: 1250-1251]; Reh1900 [taxonomy, description, host, distribution: 237-257]; Reh1900a [host, distribution, taxonomy: 259-271]; Reh1900c [taxonomy, structure: 502-504]; Reh1926 [host, distribution:

308-328]; RehmanGhKa1961 [biological control: 165-177]; ReissiWeOn1985 [host, distribution, chemical control: 238-248]; Ribaga1902 [distribution, biological control: 299]; Rice1974 [life history, chemistry, pheromone: 561-562]; RiceAtDe1997 [life history, chemistry, pheromone, control: 151-161]; RiceDiJo1974 [host, distribution, chemical control: 3]; RiceFlJo1982 [life history, ecology: 13-14]; RiceHo1980 [life history, chemistry, pheromone: 190-194]; RiceJo1977 [host, distribution, life history: 1403-1404]; RiceJo1982 [host, distribution, life history, biological control: 876-880]; Richar1960AM [host, distribution: 693-698]; RLER1935 [host, distribution: 60-62]; RobbCoBe2001 [host, distribution, taxonomy, control]; RockMc1990 [host, distribution, life history: 615-621]; RodrigToPo2001 [host, distribution, life history: 195-199]; Rolfs1897 [host, distribution, biological control: 519-536]; Ronna1928 [host, distribution: 1]; Rose1990a [biological control, host: 263-287]; Rose1990b [biological control: 438]; Rose1990d [host, biological control, economic importance: 357-365]; Rosen1986 [taxonomy, biological control: 121-129]; Rosen1987 [taxonomy, biological control: 191]; Rosen1990 [biological control: 413-415]; Rosen1990a [biological control: 502-503]; RosenDe1973 [biological control, taxonomy: 215-222]; RosenDe1978 [economic importance, biological control, life history, host, distribution: 123-128]; RosenDe1979 [host, distribution, biological control: 387-390,400-402,]; RosenhHe1994 [life history, biological control: 41-78]; Rota1953 [host, distribution, taxonomy: 77-84]; Rota1955 [structure, taxonomy: 231-252]; Rubtso1947a [distribution, biological control: 61-83]; Rubtso1952a [biological control: 96-106]; Russo1951 [distribution, biological control: 86-97]; Russo1958 [distribution, biological control: 141-147]; Russo1987 [host, distribution, life history: 203-208]; RzaevaYa1985 [biological control: 55-58]; SahaiJo1965 [host, distribution, economic importance, life history, biological control, ecology: 37-43]; Sailer1983 [distribution, economic importance: 15-38]; Salaza1989 [host, distribution]; SalazaSo1990 [host, distribution, life history, biological control: 135-137]; Sampai1898 [host, distribution: 73-80]; Sander1904a [taxonomy, description, illustration, host, distribution: 57,65]; Sander1908 [economic importance, chemical control: 1-12]; Sasaki1901 [taxonomy, description, illustration, host, distribution: 165-172]; Savesc1953 [host, distribution, taxonomy, description, economic importance, control: 3-46]; Savesc1982 [taxonomy, description, host, distribution, life history, economic importance, biological control: 308-321]; SchaubBlHi1999 [host, distribution, life history: 253-257]; SchaubMaBl1995 [host, distribution, pheromone: 631-636]; Schaus1998 [biological control, life history: 53-56]; SchildSc1928 [biological control]; Schmut1957b [taxonomy: 148]; Schmut1959 [taxonomy, description, host, distribution: 77,80]; SchmutKlLu1957 [host, distribution, economic importance, life history, biological control, chemical control: 485-490]; Schuhm1954 [host, distribution, chemical control: 284-303]; Schuhm1954 [chemical control: 284-303]; SchuhMo1948 [host, distribution, control]; Seabra1941 [distribution: 8]; Seghat1980 [host, distribution, life history: 149-154]; SengonUyKa1998 [host, distribution, biological control: 128-131]; SharmaRaPa1990 [biological control: 11-14]; ShawBrWa2000 [host, distribution, chemical control: 13-17]; Shen1993 [host, distribution: 60]; ShenefBe1955 [host, distribution]; ShetaJe1970 [life history, control: 76-81]; Shidra1990 [economic importance, chemical control: 403-408]; ShiLi1991 [host, distribution: 166]; ShiLiLi1997 [host, distribution, life history, ecology: 161-167]; Siddiq1966 [host, distribution, economic importance: 4-5]; Siddiq1981 [economic importance, host, distribution: 172-180]; SiegleBa1924 [biological control, distribution: 497-499]; SimmonFrSi1976 [biological control: 17]; Singh1963 [taxonomy, description, economic importance, host, distribution, chemical control, biological control: 6-8]; Singh1964 [host, distribution, economic importance: 213]; SkibaPa1989 [economic importance, control: 48-51]; Skorki1938 [life history, economic importance: 147-186]; Smetni1991 [chemistry: 92-129]; SmetniKo1983 [chemistry, biological control: 882]; SmetniKoKo1982 [chemistry, life history, pheromone: 39]; SmetniKoMa1987 [chemistry, life history, chemical control, pheromone: 209-212]; SmetniMaKo1983 [chemistry, physiology, life history, pheromone: 38]; Smirno1950a [biological control: 190-194]; Smit1964 [host, distribution, economic importance, biological control, chemical control]; Smith1898 [host, distribution: 32-39]; SmithEsFa1933 [economic importance: 1]; Staffo1915 [taxonomy, structure: 72]; Stanev1963 [host, distribution, economic importance, biological control, chemical control: 5-27]; StanevDe1989 [host, distribution, life history, ecology: 85-91]; Starne1897 [taxonomy, description, illustration, host, distribution: 7]; StoetzDa1974 [taxonomy, life history: 138-140]; StoliaOkCh1977 [biological control: 54-55]; StouthLu1991 [biological control: 150-157]; StreibFrKa1994 [chemical control: 23-30]; SudhaRRa1960 [biological control: 120-123]; Sulliv1930 [host, distribution: 51-59]; Swider1980 [structure, anatomy: 331-339]; SwinglSn1931 [chemical control, host, distribution, economic importance: 1-48]; SymonsCoBa1911 [chemical control: 221-234]; SymonsWe1907 [chemical control: 139-152]; Tadic1961 [host, distribution, biological control: 130-132]; Tadic1967 [host, distribution, biological control: 147-154]; Takagi1956b [taxonomy: 87]; TakagiRo1981 [host, distribution, biological control: 314-321]; Takaha1953a [taxonomy, host, distribution: 10-13]; TakahaTa1956 [host, distribution: 15]; TanquaHa1920 [chemical control: 3-7]; Tao1999 [taxonomy, host, distribution: 115]; Terezn1986 [taxonomy, description, illustration, host, distribution: 104-109]; Thakur1999 [host, distribution: 866-872]; ThakurPaRa1993 [biological control: 99-101]; ThakurRaPa1988 [chemical control, biological control: 72-73]; ThakurRaPa1989 [host, distribution, biological control: 143-146]; ThiemGe1934 [taxonomy: 541]; ThiemGe1934a [taxonomy, description, host, distribution: 130-158,208-238]; Thorpe1930WH [taxonomy: 177]; Timber1924 [host, distribution, biological control]; Timlin1964a [host, distribution: 531-535]; TippinBe1970 [host, distribution: 11]; Torres1922 [host, distribution: 72]; Toumey1895 [host, distribution, economic importance, taxonomy, description, life history, chemical control: 32-47]; TorresRoAv2001 [chemical control: 207-212]; Tower1913 [biological control: 125-126]; Tower1914 [life history, biological control: 422-432]; TranfaVi1987a [economic importance: 215-221]; Treher1914 [host, distribution: 67-71]; Treher1916a [host, distribution, economic importance: 66]; Trimbl1929 [host, distribution, economic importance, description, control: 1-21]; Trjapi1989 [biological control: 290-297,312]; TrouveVe1942 [host, distribution, economic importance: 1-12]; Trujil1942 [host, distribution, economic importance]; Tuncyu1970 [host, distribution, biological control: 30-52]; Tuncyu1976 [host, distribution, biological control: 32-45]; Ugolin1979 [host, distribution, biological control: 597-605]; Unterb1964 [chemical control: 26-27]; Unterb1964a [chemical control: 34-41]; UsmanPu1955 [host, distribution: 48]; UygunSeEr1998 [host, distribution: 183-191]; Vacant1985 [host, distribution, biological control: 741-747]; Vacant1985a [host, distribution, biological control: 749-758]; Vacant1985a [host, distribution, life history, biological control: 749-758]; Valent1963 [biological control: 6-13]; Valent1967 [biological control: 1100]; VanDer1962 [host, distribution: 328-331]; vanEmd1990 [biological control: 63-80]; Vargas1987 [host, distribution, biological control, economic importance, taxonomy, description: 44-48]; Vasseu1949 [host, distribution, economic importance, chemical control: 47-51]; VasseuBe1953 [distribution, biological control: 283-290]; VasseuBi1949 [chemical control: 280-282]; VasseuBi1953 [host, distribution, chemical control: 76-83]; VasseuBi1953a [economic importance, chemical control: 45-58]; VasseuSc1957 [host, distribution, life history, ecology: 5-66]; VasseuScBi1952 [biological control, chemical control, host, distribution: 339-350]; VasseuScBi1957 [host, distribution, chemical control: 101-110]; Vayssi1932a [economic importance, biological control: 629-648]; VernalSiGa1985 [host, distribution, economic importance: 193-206]; Viel1946 [chemical control: 401-413]; Viggia1984 [biological control: 257-276]; Viggia1987 [host, distribution, biological control: 121-123]; Viggia1990a [biological control: 127-128]; Vogel1960 [taxonomy: 265-267]; Wahl1933 [host, distribution, economic importance: 693-698]; Waldne1988 [chemical control: 186-188]; Waterw1981 [host, distribution, biological control, economic importance: 269-296]; Watson1918 [host, distribution]; Watson1926 [host, distribution]; WatsonBe1937 [host, distribution, control]; Wearin1976 [host, distribution, life history, economic importance: 1-2]; Webste1899 [life history, chemistry: 4]; Webste1900 [host, distribution: 55-60]; Wilson1917 [host, distribution: 57]; Wilson1921 [host, distribution: 20-34]; WongChCh1999 [taxonomy, description, host, distribution: 35,79]; WoodwaEvEa1970 [distribution]; Woolle1990 [biological control: 167-176]; Worthl1932 [chemical control: 22-26]; Xie1998 [taxonomy, description, illustration, host, distribution: 107-111]; YamamoOg1989 [chemistry: 123-148]; Yanovs2001 [host, distribution, life history, behaviour, biological control: 13-14]; YanovsLa2000 [chemical control: 25]; Yasar1995a [taxonomy, description, illustration, host, distribution: 119-121]; Yasnos1994 [host, distribution, biological control: 317-333]; Yother1940 [chemical control, physiology: 890-892]; Zahrad1952 [taxonomy, description, illustration, host, distribution: 114,128-133]; Zahrad1959b [host, distribution: 60]; Zahrad1972 [taxonomy, description, illustration, host, distribution, biological control: 436-438]; Zahrad1977 [taxonomy, distribution: 121]; Zahrad1990 [host, distribution, description: 643]; Zahrad1990a [host, distribution, description: 652]; ZalomVaBe2001 [host, distribution, control]; Zhang1983 [life history, distribution, biological control: 86-88].

Diaspidiotus piceus (Sanders)

Aspidiotus piceus Sanders, 1904: 96. Type data: U.S.A.: Ohio, Lake County, Painesville, on *Liriodendron tulipifera*. Syntypes, female. Type depository: Washington: United States National Entomological Collection, U.S. National Museum of Natural History, District of Columbia, USA.

Diaspidiotus piceus; MacGillivray, 1921: 413. Change of combination.

SCALE COVER: Female scale, diameter 1.8-2 mm, flat, subelliptical to oval, with subcentral exuviae; black shading to dark toward margin, having appearance of pitch covered with dust; raised, shiny black, deciduous first exuvia surrounded by an indistinct ring-like depression; when rubbed second orange exuvia appears (Sanders, 1904).

HOST PLANTS: **Lauraceae**: *Sassafras* [Nakaha1982]. **Magnoliaceae**: *Liriodendron tulipifera* [Sander1904]. **Oleaceae**: *Fraxinus* [Nakaha1982].

DISTRIBUTION: **Nearctic**: United States of America (Missouri [Nakaha1982], Ohio [Nakaha1982], Tennessee [Nakaha1982]).

GENERAL: Description and illustration of adult female by Sanders (1904).

CITATIONS: Borchs1966 [catalogue: 324]; Nakaha1982 [taxonomy, host, distribution: 29-30]; Sander1904 [taxonomy, description, illustration, host, distribution: 96-98]; Sander1904a [taxonomy, description, illustration, host, distribution,: 56,66].

Diaspidiotus prunorum (Laing)

Aspidiotus prunorum Laing, 1931: 99. Type data: PAKISTAN: Quetta, on bark of damson, almond and cherry. Holotype female. Type depository: London: The Natural History Museum, England, UK.

Aspidiotus (Targionidea) prunorum; Borchsenius, 1935: 16. Change of combination.

Targionidea prunorum; Borchsenius, 1937a: 68. Change of combination.

Diaspidiotus prunorum; Borchsenius, 1949d: 247. Change of combination.

COMMON NAME: turanskaya shitovka [Borchs1936].

SCALE COVER: Female scale greyish-white; subcircular, 0.8 mm in diameter; semi-opaque; exuviae subcentral, orange-yellow; second exuvia wholly and the first sometimes partially, obscured with a whitish-grey secretion (Laing, 1931).

HOST PLANTS: **Rosaceae**: *Amygdalus* [Balach1950b], *Cerasus* [Danzig1972c], *Malus* [Borchs1936], *Prunus armeniaca* [Borchs1936], *Pr. divaricatus* [Balach1950b], *Pr. domestica* [Borchs1936], *Pr.persica* [Borchs1936], *Pyrus* [Danzig1972c].

NATURAL ENEMIES: HYMENOPTERA **Encyrtidae**: *Anthemus aspidioti* Nikolskaya [Sharip1980, Trjapi1989], *Epitetracnemus zetterstedtii* (Westwood) [Trjapi1989], *Habrolepis tergrigorianae* Trjapitzin [Trjapi1989], *Zaomma lambinus* (Walker) [Trjapi1989].

DISTRIBUTION: **Oriental**: Pakistan [Laing1931, Balach1950b]. **Palaearctic**: Afghanistan [Danzig1972c]; Armenia [Borchs1936, Babaia1987]; Georgia (Georgia [Borchs1936, Hadzib1983]); Iran [Balach1950b, Kaussa1955, DastghBeBa1988]; Kazakhstan [BazaroSh1971]; Turkey [Yasar1995a]; Turkmenistan [Balach1950b].

ECONOMIC IMPORTANCE: Noted as a minor pest of plum and almond in Central Asia (Kozár, 1990c).

GENERAL: Description and illustration of adult female by Laing (1931), Balachowsky (1950b), Bazarov & Shmelev (1971), Danzig (1993) and by Yaşar (1995a).

KEYS: Danzig 1993: 179-182 (female) [Europe]; Bazarov & Shmelev 1971: 198 (female) [Central Asia]; Balachowsky 1950b: 492 (female) [Mediterranean];

Archangelskaya 1937: 100 (female) [Central Asia]; Borchsenius 1935: 127-128 (female) [Former USSR].

CITATIONS: Archan1937 [taxonomy, description, illustration, host, distribution: 100,110-111]; Babaia1987 [host, distribution, economic importance: 136]; Balach1950b [taxonomy, description, illustration, host, distribution: 515-518]; BazaroSh1971 [taxonomy, description, illustration, host, distribution, life history: 203-205]; Borchs1935 [taxonomy: 128]; Borchs1935a [taxonomy: 16,37]; Borchs1936 [host, distribution: 132-133]; Borchs1937 [taxonomy, description, illustration, host, distribution: 135]; Borchs1937a [taxonomy, description, illustration, host, distribution: 68-69]; Borchs1939 [taxonomy, description, host, distribution: 10,38]; Borchs1939a [taxonomy, distribution: 43]; Borchs1949d [taxonomy, host, distribution: 247]; Borchs1949d [taxonomy, description, host, distribution: 247]; Borchs1950b [taxonomy, description, illustration, host, distribution: 230-231]; Borchs1966 [catalogue: 325-326]; Bustsh1958 [taxonomy, description, host, distribution: 220,265]; Danzig1972 [taxonomy, host, distribution, economic importance: 212]; Danzig1972c [host, distribution: 583]; Danzig1993 [taxonomy, description, illustration, host, distribution, economic importance: 208-211]; DanzigPe1998 [catalogue: 239]; DastghBeBa1988 [host, distribution, economic importance: 31-32]; Ferris1941e [taxonomy: 47]; Gavalo1936 [host, distribution: 78]; Hadzib1983 [taxonomy, host, distribution, life history, biological control, economic importance: 236-238]; Janjua1959 [distribution: 231-264]; Kaussa1955 [host, distribution: 16]; KaussaBa1953 [taxonomy: 26]; Laing1931 [taxonomy, description, illustration, host, distribution: 99-100]; MillerDa1990 [host, distribution, economic importance: 301]; Myarts1984 [host, distribution, biological control: 23-30]; RzaevaYa1985 [biological control: 55-58]; SchmutKlLu1957 [host, distribution, economic importance: 491]; Sharip1980 [host, distribution, biological control: 381-384]; TerGri1962 [taxonomy, description, host, distribution: 150]; Trjapi1989 [biological control: 292,312,378]; Yasar1995a [taxonomy, description, illustration, host, distribution:71-73]; Yasnos1994 [host, distribution, biological control: 317-333].

Diaspidiotus pseudocamelliae (Green)

Aspidiotus pseudocamelliae Ramakrishna Ayyar, 1919a: 20. Nomen nudum.

Aspidiotus pseudocamelliae Green, 1919c: 438. Type data: INDIA: Tamil Nadu, Bellary District, Ittige, on *Capparis stylosa*. Syntypes, female. Type depository: London: The Natural History Museum, England, UK.

Aspidiotus (*Hemiberlesia*) *pseudocamelliae*; Green, 1919c: 438. Change of combination.

Morganella pseudocamelliae; Lindinger, 1957: 546. Change of combination.

Aspidiotus pseudocamelliae; Borchsenius, 1966: 326. Notes: Incorrect citation of "Green" as author.

Diaspidiotus pseudocamelliae; Borchsenius, 1966: 326. Change of combination.

SCALE COVER: Female scale ochreous (on twigs or upper surface of foliage), whitish (on undersurface of foliage): exuviae darker ochreous, occupying greater part of area of scale; form irregularly circular, slightly convex above; diameter 0.75 to 1 mm. Male scale slightly paler in colour: ovate; length 0.75 mm (Green, 1919c).

HOST PLANTS: **Capparidaceae**: *Capparis stylosa* [Green1919c, Ramakr1921a].
DISTRIBUTION: **Oriental**: India [Ramakr1921a] (Tamil Nadu [Green1919c]).
GENERAL: Description and illustration of adult female by Green (1919c).
CITATIONS: Borchs1966 [catalogue: 326]; Ferris1941e [taxonomy: 47]; Green1919c [taxonomy, description, illustration, host, distribution: 438-439]; Lindin1957 [taxonomy: 546]; Ramakr1919a [taxonomy: 20]; Ramakr1921a [host, distribution: 356]; Ramakr1930 [taxonomy, host, distribution: 24].

Diaspidiotus pyri (Lichtenstein)

Aspidiotus pyri Lichtenstein, 1881: lii. Type data: FRANCE: Hérault, Montpellier, la Lironde, on pear [=*Pyrus*]. Syntypes, female. Type depository: Paris: Muséum national d'Histoire naturelle, France.

Aspidiotus (*Diaspidiotus*) *patavinus* Berlese in: Berlese & Leonardi, 1896: 350. Type data: ITALY: Patavi, on trunk of *Prunus cerasus*. Syntypes, female. Type depository: Portici: Dipartimento de Entomologia e Zoologia Agraria di Portici, Università di Napoli Federico II, Italy. Synonymy by Borchsenius, 1966: 339.

Aspidiotus pyri; Cockerell, 1896b: 333. Incorrect synonymy. Notes: incorrect synonymy with *Quadraspidiotus ostreaeformis*.

Aspidiotus patavinus; Cockerell, 1896b: 334. Change of combination.

Aspidiotus (*Evaspidiotus*) *patavinus*; Leonardi, 1898c: 48. Change of combination.

Aspidiotus patavinus; Fernald, 1903b: 270. Change of combination.

Aspidiotus piri; Lindinger, 1909e: 44. Change of combination.

Aspidiotus piri; Lindinger, 1909e: 44. Misspelling of species name.

Aspidiotus piri; Lindinger, 1912b: 158. Misspelling of species name.

Furcaspis patavina; MacGillivray, 1921: 408. Change of combination requiring emendation of species name for agreement in gender.

Aspidiotus aharonii Bodenheimer, 1924: 23. Type data: ISRAEL: mountains near Benjamina [=Binyamina], on branches of *Ceratonia siliqua*. Lectotype female, by subsequent designation Ben-Dov, 2001c: 161. Type depository: London: The Natural History Museum, England, UK. Synonymy by Ben-Dov, 2001c: 161.

Aspidiotus ostreaeformis aegyptiacus Hall, 1925: 12. Type data: EGYPT: Kafr Sheikh (Lower Egypt), on *Ficus carica*. Syntypes, female. Type depository: London: The Natural History Museum, UK. Synonymy by Danzig, 1993: 184.

Aspidiotus sinaiticus Bodenheimer, 1929b: 106. Type data: EGYPT: Sinai, Wadi Ledscha, on *Olea europaea*. Syntypes, female. Type depository: Bet Dagan: Department of Entomology, The Volcani Center, Israel. Synonymy by Borchsenius, 1966: 339.

Aspidiotus piri; Kiritchenko, 1932a: 247. Misspelling of species name.

Aspidiotus (*Euraspidiotus*) *piri*; Thiem & Gerneck, 1934: 131. Misspelling of species name.

Aspidiotus (*Euraspidiotus*) *piri*; Thiem & Gerneck, 1934: 131. Change of combination.

Furcaspis piri; Borchsenius, 1934: 28. Misspelling of species name.

Aspidiotus piri; Borchsenius, 1935: 127. Misspelling of species name.

Aspidiotus piri; Kawecki, 1935: 77. Misspelling of species name.

Aspidiotus (*Furcaspis*) *piri*; Borchsenius, 1935a: 7. Change of combination.

Aspidiotus (*Furcaspis*) *piri*; Borchsenius, 1935a: 7. Misspelling of species name.

Aspidiotus piri; Borchsenius, 1936: 131. Misspelling of species name.

Aspidiotus ostreiformis; Gómez-Menor Ortega, 1937: 59. Misidentification; discovered by Gómez-Menor Ortega, 1960: 166.

Quadraspidiotus aegyptiacus; Ferris, 1941e: 40. Change of combination and rank.

Quadraspidiotus pyri; Lupo, 1948: 175, 185. Change of combination.

Aspidiotus pattarinus; Bodenheimer, 1949: 61. Misspelling of species name.

Diaspidiotus spurcatus; Borchsenius, 1949d: 242. Misidentification; discovered by Danzig, 1993: 184.

Aspidiotus spurcatus; Borchsenius, 1950b: 225. Change of combination.

Quadraspidiotus piri; Zahradník, 1952: 119. Misspelling of species name.

Quadraspidiotus piri; Zahradník, 1952: 119. Change of combination.

Quadraspidiotus pelagijae Bachmann, 1953: 178. Type data: YUGOSLAVIA: Bar, on *Ficus carica* and *Platanus orientalis*. Syntypes, female. Synonymy by Danzig, 1993: 184. Notes: Depository of type material unknown.

Quadraspidiotus piri; Bachmann, 1953: 178. Misspelling of species name.

Quadraspidiotus piri; Duskova, 1953: 229. Misspelling of species name.

Quadraspidiotus piri; Pegazzano, 1955: 322. Misspelling of species name.

Aspidiotus piri; Lindinger, 1957: 546. Misspelling of species name.

Quadraspidiotus piri; Zahradník, 1972: 438. Misspelling of species name.

Diaspidiotus pyri; Hadzibejli, 1983: 240. Change of combination.

Diaspidiotus pyri; Danzig & Pellizzari, 1998: 239. Change of combination.

COMMON NAMES: Apfelsinenfarbene Austernschildlaus [SchmutKlLu1957]; Europaische Pseudo-San-Jose-Schildlaus [SchmutKlLu1957]; jeltaya grushevaya shitovka [Borchs1936]; Pirus-Austrenschildlaus [SchmutKlLu1957].

SYSTEMATICS: Lindinger (1936: 152) regarded *Aspidiotus populi* Glaser (1877) as a synonym of *Aspidiotus gigas* Thiem (1934). Borchsenius (1966: 370) regarded *A. populi* Glaser as *species incertae sedis*. Reported to have uniparental as well as biparental populations (Schmutterer, 1959).

SCALE COVER: Female scale circular, moderately large convex and dark greyish brown; exuviae deep orange, the second exuvia being usually masked by secretionary matter and of a similar colour to rest of scale; on removing scale and sublying female, the ventral scale may be observed as a circular powdery white film on host plant; diameter 1.5-2 mm (Hall, 1925).

HOST PLANTS: **Betulaceae**: *Betula* [Balach1950b, Zahrad1972]. **Carpinaceae**: *Carpinus betulus* [Zahrad1972]. **Hippocastanaceae**: *Aesculus hippocastanum* [Zahrad1972]. **Leguminosae**: *Ceratonia* [Bodenh1937], *Ceratonia siliqua* [Bodenh1924, BenDov2001c]. **Moraceae**: *Ficus carica* [Hall1925, Bachma1953]. **Oleaceae**: *Fraxinus* [Balach1950b], *Fraxinus angustifolia* [Rungs1948, Zahrad1972], *Fraxinus excelsior* [Kaweck1935, Zahrad1972], *Fraxinus ornus* [Zahrad1972], *Olea europaea* [Balach1950b]. **Platanaceae**: *Platanus* [Balach1932d, Zahrad1972], *Platanus orientalis* [Balach1950b, Bachma1953, Zahrad1972]. **Rosaceae**: *Crataegus* [Bodenh1949, Bodenh1952], *Malus pumila* [Bachma1953], *Prunus* [Borchs1934, Balach1950b], *Prunus cerasus* [BerlesLe1896, Leonar1920], *Pyrus* [Lichte1881], *Pyrus communis* [Leonar1909, Zahrad1952, Bachma1953, Zahrad1972], *Sorbus aucuparia* [Zahrad1972]. **Salicaceae**: *Salix* [Zahrad1972].

NATURAL ENEMIES: COLEOPTERA **Coccinellidae**: *Chilocorus bipustulatus* L. [Zahrad1972], *Ch. renipustulatus* Scriba [Zahrad1972]. HYMENOPTERA **Aphelinidae**: *Aphytis moldavicus* Yasnosh [RosenDe1979], *Aphytis mytilaspidis* (Le Baron) [Balach1950b, Bachma1953a, Zahrad1972], *Aspidiotiphagus citrinus* Crawford [Bachma1953a, Zahrad1972], *Azotus marchali* Howard [Balach1950b, Zahrad1972], *Pteroptrix dimidiata* Westwood [Bachma1953a, Zahrad1972]. **Encyrtidae**: *Epitetracnemus zetterstedtii* (Westwood) [Trjapi1989], *Habrolepis dalmani* Westwood [Zahrad1972], *Zaomma lambinus* (Walker) [Trjapi1989].

DISTRIBUTION: **Palaearctic**: Algeria [Balach1928a, Balach1932d]; Armenia [Borchs1936]; Azerbaijan (Azerbaijan [Borchs1936]); Canary Islands [GomezM1967O, MatileOr2001]; Czech Republic [Zahrad1952, Zahrad1977]; Egypt [Hall1925, Ezzat1958]; France [Lichte1881]; Georgia [Hadzib1983]; Hungary [Balach1950b]; Israel [Bodenh1924, Bodenh1937, BenDov2001c]; Italy [BerlesLe1896, Leonar1909, Leonar1920, Balach1950b, LongoMaPe1995]; Morocco [Rungs1948]; Netherlands [Reyne1957, Jansen2001]; Poland [Kaweck1935]; Russia (Caucasus [Borchs1934, Borchs1936], Dagestan AR [Borchs1936]); Spain [Balach1950b, GomezM1960O, BlayGo1993]; Switzerland [Balach1950b]; Turkey [Bodenh1949, Bodenh1952]; Ukraine (Krym (= Crimea) Oblast [Borchs1936]); Yugoslavia [Bachma1953].

BIOLOGY: Found in thick incrustations on branches of *Ceratonia siliqua* together with populations of *Hemiberlesia lataniae* and *Lepidosaphes pinniformis* (Bodenheimer, 1924). Reported to have uniparental as well as biparental populations (Schmutterer, 1959).

ECONOMIC IMPORTANCE: A pest of deciduous fruit trees, mainly pears and plums (Balachowsky, 1950b; Schmutterer et al., 1957), as well as to forest trees (Zahradník, 1990a).

GENERAL: Good description and illustration of adult female by Bodenheimer (1924), Hall (1925), Balachowsky (1948a, 1950b), Zahradník (1952), Bachmann (1953), Brookes & Hudson (1968), Tereznikova (1986), Danzig (1993) and by Yaşar (1995a). Description of second instar-nymph by Brookes & Hudson (1968).

KEYS: Blay Goicoechea 1993: 528-529 (female) [Spain]; Danzig 1993: 179-182 (female) [Europe]; Tereznikova 1986: 97-98 (female) [Ukraine]; Kosztarab & Kozár 1978: 170 (female) [Hungary]; Brookes & Hudson 1968: 91 (larva) [Australia]; Brookes & Hudson 1968: 91 (female) [Australia]; Ezzat 1958: 242 (female) [Egypt]; Reyne 1957: 33 (female) [Netherlands]; Zahradník 1952: 114-115 (female) [Czech Republic]; Balachowsky 1950b: 400 (female) [Mediterranean]; Borchsenius 1950b: 225 (female) [Palaearctic Region]; Balachowsky 1948a: 90 (female) [World]; Lupo 1948: 175 (female) [Italy]; Borchsenius 1935: 127-128 (female) [Former USSR]; Leonardi 1920: 29-30 (female) [Italy].

CITATIONS: Aleksi1995 [host, distribution, economic importance, biological control: 187-190]; Bachma1953 [taxonomy, description, illustration, host, distribution, life history: 178-182]; Bachma1953a [taxonomy, description, illustration, host, distribution, life history, ecology, economic importance, chemical control, biological control: 357-404]; Bachma1955 [taxonomy: 122]; Balach1928a [host, distribution: 138]; Balach1932d [taxonomy, host, distribution: V]; Balach1948a [taxonomy, description, illustration, host, distribution: 89,93-96];

Balach1950b [taxonomy, description, illustration, host, distribution, biological control: 409-415]; BeardsGo1975 [economic importance: 49]; BenDov2001c [taxonomy, host, distribution: 161]; BerlesLe1896 [taxonomy, description, illustration, host, distribution: 350-352]; Berro1927 [host, distribution: 55-59]; BlayGo1993 [taxonomy, description, illustration, host, distribution: 530-534]; Bodenh1924 [taxonomy, description, illustration, host, distribution: 23-24]; Bodenh1929b [taxonomy, description, illustration, host, distribution: 106-107]; Bodenh1935 [host, distribution: 246]; Bodenh1935c [taxonomy, distribution: 1156]; Bodenh1937 [host, distribution: 216]; Bodenh1949 [taxonomy, description, illustration, host, distribution: 57,60-62]; Bodenh1952 [host, distribution: 338]; Borchs1934 [host, distribution: 28]; Borchs1935 [taxonomy: 127]; Borchs1935a [taxonomy: 7,13,29]; Borchs1936 [host, distribution: 131]; Borchs1937 [taxonomy, description, illustration, host, distribution: 131]; Borchs1937a [taxonomy, description, illustration, host, distribution: 46-47]; Borchs1939 [taxonomy, description, host, distribution: 9,28]; Borchs1939a [taxonomy, distribution: 43]; Borchs1949d [taxonomy, description, host, distribution: 242]; Borchs1950b [taxonomy, description, host, distribution: 225]; Borchs1966 [catalogue: 328,336,339-340]; BrandtBo1948 [taxonomy: 3]; BrookeHu1968 [taxonomy, description, illustration, host, distribution: 98-100]; Bustsh1958 [taxonomy, description, host, distribution: 221,258]; Cocker1896b [taxonomy, distribution: 334]; Cocker1897i [taxonomy, description, host, distribution: 19]; Danzig1964 [taxonomy, host, distribution: 652]; Danzig1972 [taxonomy, host, distribution, economic importance: 212]; Danzig1993 [taxonomy, description, illustration, host, distribution, economic importance: 184-186]; DanzigKo1991 [distribution: 1-15]; DanzigPe1998 [catalogue: 187,239-240]; DeLott1963 [taxonomy: 144-145]; DumasVa1950 [chemical control: 235-245]; Duskov1952 [taxonomy, description, host, distribution: 452-455]; Duskov1953 [taxonomy, description, illustration, host, distribution, life history: 8-36]; Duskov1953a [host, distribution, taxonomy, life history: 229-250]; Ezzat1958 [distribution: 242]; EzzatAf1966 [taxonomy, description, illustration, host, distribution: 377-379]; EzzatNa1987 [distribution: 88]; Fernal1903b [catalogue: 270,276]; Ferris1941e [taxonomy: 40,46-48]; Foldi2001 [distribution: 303-308]; FoldiDe1998 [taxonomy, host, distribution: 202]; FreyFr1995 [taxonomy, chemistry, host, distribution: 777-780]; FreyFr1995a [chemistry, structure: 100]; Gavalo1936 [host, distribution: 76]; Gerson1990 [taxonomy: 130]; GomezM1960O [taxonomy, host, distribution: 166]; GomezM1967O [host, distribution: 131]; Goux1949 [taxonomy: 31]; Hadzib1983 [taxonomy, host, distribution, life history, biological control, economic importance: 240-241]; Hall1925 [taxonomy, description, host, distribution: 12]; HippeScMa1995 [host, distribution, economic importance, chemical control, biological control: 4,84-85]; Huba1960 [taxonomy: 39-50]; Jansen2001 [host, distribution: 197-206]; Kaussa1955 [host, distribution: 15]; Kaweck1935 [host, distribution: 77]; Kiritc1932a [taxonomy: 247]; Kislyi1965 [host, distribution, taxonomy: 31-32]; Konsta1976 [host, distribution, economic importance: 49-50]; KosztaKo1978 [taxonomy, description, host, distribution: 170]; Kozar1995b [taxonomy: 51]; KozarGuBa1994 [host, distribution: 161-151]; KozarHiMa1996 [taxonomy, description: 433-437]; Krzysz1957 [taxonomy, description, host, distribution: 223];

LagowsGo1998 [host, distribution, economic importance: 21-23]; Leonar1897 [taxonomy: 285]; Leonar1898a [taxonomy: 75]; Leonar1898c [taxonomy, description, illustration, host, distribution: 48-49]; Leonar1909 [host, distribution: 102]; Leonar1920 [taxonomy, description, illustration, host, distribution: 47-48,57-58]; Lichte1881 [taxonomy, description, host, distribution: lii]; Lindin1909e [taxonomy: 44]; Lindin1912b [taxonomy, description, host, distribution: 158,214,260]; Lindin1932 [taxonomy: 204]; Lindin1935 [taxonomy: 129]; Lindin1935 [taxonomy: 129]; Lindin1936 [taxonomy: 152]; Lindin1957 [taxonomy: 546]; LongoMaPe1995 [distribution: 128]; Lupo1948 [taxonomy, description, illustration, host, distribution: 185-189]; MacGil1921 [taxonomy, description, host, distribution: 408]; ManiHiSc1993 [life history, economic importance, host, distribution: 299-302]; Marlat1900a [taxonomy: 593]; MatileOr2001 [host, distribution: 190]; MillerDa1990 [host, distribution, economic importance: 305]; MuelleEi1954 [taxonomy, description: 151-153]; Pegazz1955 [taxonomy: 322]; Reh1900a [taxonomy: 271]; Reh1900b [taxonomy: 497]; Reh1903 [taxonomy: 468]; Reyne1957 [taxonomy, host, distribution: 26,33]; RosenDe1979 [host, distribution, biological control: 475-476]; Rungs1948 [host, distribution: 110]; RzaevaYa1985 [biological control: 55-58]; Schern1954 [taxonomy: 225]; Schmut1957b [taxonomy: 148-149]; Schmut1959 [taxonomy, description, host, distribution: 76,85,89]; SchmutKlLu1957 [host, distribution, economic importance: 490]; Szelen1936 [host, distribution, life history: 159-170]; Szulcz1926 [host, distribution: 137-143]; Szulcz1931 [host, distribution: 124-135]; Szulcz1949 [distribution: 219-224]; Terezn1986 [taxonomy, description, illustration, host, distribution: 109-110]; TerGri1956 [taxonomy, description, host, distribution: 54-55]; TerGri1962 [taxonomy, description, host, distribution: 149]; ThiemGe1934a [taxonomy, description, host, distribution: 130-158,208-238]; Trembl1990a [anatomy, structure: 275-283]; Trjapi1989 [biological control: 292,312]; Yasar1995a [taxonomy, description, illustration, host, distribution: 122-124]; Yasnos1994 [host, distribution, biological control: 317-333]; Zahrad1952 [taxonomy, description, illustration, host, distribution: 114,119-124]; Zahrad1955a [taxonomy: 125]; Zahrad1959b [host, distribution: 60]; Zahrad1972 [taxonomy, description, illustration, host, distribution, biological control: 436,438]; Zahrad1977 [taxonomy, distribution: 121]; Zahrad1990a [host, distribution, description, economic importance: 652-653].

Diaspidiotus roseni Danzig

Diaspidiotus roseni Danzig, 2000: 287. Type data: ISRAEL: 45 km South of Neve Zohar, on *Nitraria retusa*; collected by D. Gerling, V. Kravchenko and E. Sugonyaev, 25.V.1992. Holotype female. Type depository: St. Petersburg: (= Leningrad) Zoological Museum, Academy of Science, Russia.

SCALE COVER: Female scale white, opaque; exuviae yellow, central. Male scale similar to scale of female, but elongate (Danzig, 2000).

HOST PLANTS: **Zygophyllaceae**: *Nitraria retusa* [Danzig2000].

DISTRIBUTION: **Palaearctic**: Israel [Danzig2000].

BIOLOGY: Females form pit galls on leaves. As typical of gall-forming diaspidids (Beardsley, 1984), insects incompletely covered by plant tissue; top of gall is covered by the scale. Males form no galls, occur not only on leaves but also on

twigs. Ovoviviparous. At end of May, young females, females with crawlers under scale and second-instar larvae were observed. Numerous crawlers were on leaves and twigs. Scales of males were empty (Danzig, 2000).
GENERAL: Description and illustration of the adult female by Danzig (2000).
CITATIONS: Danzig2000 [taxonomy, description, illustration, host, distribution: 287-289].

Diaspidiotus salicis (Lupo)

Hemiberlesia salicis Lupo, 1953a: 91. Type data: ITALY: Salerno, Battipaglia, on *Salix* sp. Syntypes, female. Type depository: Portici: Dipartimento de Entomologia e Zoologia Agraria di Portici, Università di Napoli Federico II, Italy.
Quadraspidiotus salicis; Borchsenius, 1966: 340. Change of combination.
Diaspidiotus salicis; Danzig & Pellizzari, 1998: 240. Change of combination.
SCALE COVER: Female scale almost circular, about 2.1 mm in diameter; convex; colour brown-hazelnut to grey; exuviae orange; ventral vellum white grey, robust; remains attached to host plant. Male scale not observed (Lupo, 1953a).
HOST PLANTS: **Salicaceae**: *Salix* [Lupo1953a, Zahrad1972].
DISTRIBUTION: **Palaearctic**: Italy [Lupo1953a, Zahrad1972, LongoMaPe1995].
GENERAL: Description and illustration of adult female by Lupo (1953a).
KEYS: Lupo 1953a: 75-76 (female) [Italy].
CITATIONS: Borchs1966 [catalogue: 340]; DanzigPe1998 [catalogue: 240]; Dimitr1935 [host, distribution: 220]; LongoMaPe1995 [distribution: 128]; Lupo1953a [taxonomy, description, illustration, host, distribution: 76,91-96]; Zahrad1959b [host, distribution: 60]; Zahrad1972 [host, distribution: 432].

Diaspidiotus slavonicus (Green)

Aspidiotus transcaspiensis; Archangelskaya, 1923: 262. Misidentification; discovered by Danzig, 1993: 197.
Targionia slavonica Green, 1934a: 95. Type data: UZBEKISTAN: Samarkand, on *Populus* sp.; collected by A. Kiritchenko. Syntypes, female. Type depository: London: The Natural History Museum, England, UK.
Aspidiotus slavonica; Archangelskaya, 1937: 99. Change of combination.
Quadraspidiotus populi Bodenheimer, 1943: 3. Type data: IRAQ: Baghdad, on leaves of *Populus euphratica*. Syntypes, female. Type depository: Bet Dagan: Department of Entomology, The Volcani Center, Israel. Synonymy by Borchsenius, 1966: 340.
Quadraspidiotus slavonicus; Ferris, 1943a: 98. Change of combination requiring emendation of species name for agreement in gender.
Diaspidiotus slavonicus; Borchsenius, 1949d: 241. Change of combination.
Diaspidiotus slavonicus; Danzig & Pellizzari, 1998: 240. Revived combination.
COMMON NAME: topolevaya vipuklaya shitovka [BazaroSh1971].
SCALE COVER: Female scale greyish or white, opaque; circular or broadly ovate, diameter 1.7-2 mm; moderately convex; exuviae eccentric, often situated close to anterior margin, bright orange-red; concealed in fresh examples but exposed or older specimens (Green, 1934a). Female scale flat, circular, white, with subcentral exuviae yellow. Male scale as large as female, elongate white (Ferris, 1943a).

HOST PLANTS: **Salicaceae**: *Populus* [Green1934a, Ferris1943a], *Po. bolleana* [BazaroSh1971], *Po. diversifolia* [BazaroSh1971], *Po. euphratica* [Bodenh1943, Bodenh1944a, Balach1950b], *Po. pyramidalis* [BazaroSh1971], *Po. tadschikistanica* [BazaroSh1971], *Salix* [Bodenh1944b, Balach1950b], *Salix australior* [BazaroSh1971].

NATURAL ENEMIES: HYMENOPTERA **Aphelinidae**: *Aphytis neuter* Yasnosh & Myartseva [YasnosMy1971a, RosenDe1979]. **Encyrtidae**: *Anthemus aspidioti* Nikolskaya [Trjapi1989]. **Signiphoridae**: *Thysanus ater* Haliday [Woolle1990].

DISTRIBUTION: **Palaearctic**: China (People's Republic) [Tang1984] (Henan (Honan) [Shen1993]); Iran [Bodenh1944a, Kaussa1955]; Iraq [Bodenh1943, Bodenh1944a]; Kazakhstan [Balach1950b]; Kyrgyzstan [Balach1950b]; Tajikistan (=Tadzhikistan) [Balach1950b, RosenDe1979]; Turkmenistan (Ashkhabad Oblast [Ferris1943a]); Uzbekistan (Samarkand Oblast [Green1934a, Ferris1943a]).

ECONOMIC IMPORTANCE: Pest of *Populus* spp. in Central Asia (Archangelskaya, 1937; Balachowsky, 1950b; Schmutterer et al., 1957).

GENERAL: Description and illustration of adult female by Green (1934a), Bodenheimer (1943), Ferris (1943a), Balachowsky (1950b), Bazarov & Shmelev (1971), Tang (1984) and by Danzig (1993).

KEYS: Danzig 1993: 179-182 (female) [Europe]; Bazarov & Shmelev 1971: 211 (female) [Central Asia]; Balachowsky 1950b: 402 (female) [Mediterranean]; Archangelskaya 1937: 99 (female) [Middle Asia].

CITATIONS: Archan1937 [taxonomy, description, host, distribution: 99,103]; Balach1950b [taxonomy, description, illustration, host, distribution: 475-478]; BazaroSh1971 [taxonomy, description, illustration, host, distribution, life history: 217-220]; Bodenh1943 [taxonomy, description, illustration, host, distribution: 3]; Bodenh1944a [host, distribution: 81]; Bodenh1944b [host, distribution: 93]; Borchs1937 [taxonomy, description, illustration, host, distribution: 133]; Borchs1949d [taxonomy, description, host, distribution: 241]; Borchs1950b [taxonomy, description, illustration, host, distribution: 225,231]; Borchs1966 [catalogue: 340]; Bustsh1958 [taxonomy, description, host, distribution: 221,255]; Danzig1964 [taxonomy, host, distribution: 652]; Danzig1993 [taxonomy, description, illustration, host, distribution: 197-199]; DanzigPe1998 [catalogue: 240]; Ferris1941e [taxonomy: 48]; Ferris1943a [taxonomy, description, illustration, host, distribution: 86,98-99,110]; Green1934a [taxonomy, description, illustration, host, distribution: 95-96]; Lindin1957 [taxonomy: 546]; MillerDa1990 [host, distribution, economic importance: 305]; Myarts1972 [host, distribution: 53-60]; Myarts1984 [host, distribution, biological control: 23-30]; RosenDe1979 [host, distribution, biological control: 490-492]; SchmutKlLu1957 [host, distribution, economic importance: 491]; Sharip1980 [host, distribution, biological control: 381-384]; Shen1993 [host, distribution: 60]; Tang1984 [taxonomy, description, illustration, host, distribution: 60-61]; Tao1999 [taxonomy, host, distribution: 115]; TerGri1962 [taxonomy, description, host, distribution: 148-149]; Trjapi1989 [biological control: 378]; Woolle1990 [biological control: 167-176]; Xie1998 [taxonomy, description, illustration, host, distribution: 111-112]; Yasnos1994 [host, distribution, biological control: 317-333]; YasnosMy1971a [host, distribution, biological control: 35-41].

Diaspidiotus socialis (Hoke), new combination
Aspidiotus socialis Hoke, 1927: 349. Type data: U.S.A.: Mississippi, Aberdeen, on *Quercus* sp. Holotype female. Type depository: Mississippi State (Starkeville): Mississippi Entomological Museum, Mississippi State University, USA..
Quadraspidiotus socialis; Ferris, 1938a: 260. Change of combination.
SCALE COVER: Female scale yellowish brown; moderately convex; circular, 1 mm diameter; exuviae subcentral, covered by secretion. Male scale not observed (Hoke, 1927).
HOST PLANTS: **Fagaceae**: *Quercus* [Hoke1927, Ferris1938a].
DISTRIBUTION: **Nearctic**: United States of America (Georgia [Nakaha1982], Mississippi [Hoke1927, Ferris1938a], Texas [Ferris1938a, McDani1970]).
BIOLOGY: Occurring on bark of host (Ferris, 1938a).
GENERAL: Description and illustration of adult female by Ferris (1938a).
KEYS: McDaniel 1970: 429 (female) [U.S.A.: Texas]; Ferris 1942: 40 (female) [North America].
CITATIONS: BesheaTi1977 [taxonomy: 181]; Borchs1966 [catalogue: 340]; Ferris1938a [taxonomy, description, illustration, host, distribution: 260]; Ferris1941e [taxonomy: 41,48]; Ferris1942 [taxonomy: 446:40]; Hoke1927 [taxonomy, description, illustration, host, distribution: 349-350,356-357]; McDani1970 [taxonomy, illustration, host, distribution: 434-437]; Nakaha1982 [host, distribution: 79]; Takagi1958 [taxonomy: 127].

Diaspidiotus sphaerocarpae (Balachowsky)
Hemiberlesea spaerocarpae Balachowsky, 1934e: 160. Type data: MOROCCO: Moyen-Atlas, Itzer, on *Retama sphaerocarpa*. Syntypes, female. Type depository: Paris: Muséum national d'Histoire naturelle, France.
Hemiberlesea spaerocarpae; Balachowsky, 1934e: 160. Misspelling of genus and species names.
Hemiberlesea sphaerocarpae; Rungs, 1935: 271. Justified emendation.
Hemiberlesea sphaerocarpae; Rungs, 1935: 271. Misspelling of genus name.
Quadraspidiotus sphaerocarpae; Rungs, 1948: 111. Change of combination.
Diaspidiotus sphaerocarpae; Danzig & Pellizzari, 1998: 240. Change of combination.
SCALE COVER: Female scale circular, diameter 1.6-1.8 mm; convex, matt; white or smoky grey; exuviae central, brown chestnut, covered with white secretion. Male scale similar to that of female, oval 1.2 mm long (Balachowsky, 1950b).
HOST PLANTS: **Leguminosae**: *Adenocarpus anagyrifolius loeiocarpus* [Rungs1948], *Retama spharocarpa* [Balach1934e, Rungs1935].
DISTRIBUTION: **Palaearctic**: Morocco [Balach1935b, Rungs1935, Rungs1948]; Sicily [LongoMaPe1995].
GENERAL: Description and illustration of adult female by Balachowsky (1950b).
KEYS: Balachowsky 1950b: 403 (female) [Mediterranean].
CITATIONS: Balach1934e [taxonomy, description, illustration, host, distribution: 160-163]; Balach1950b [taxonomy, description, illustration, host, distribution: 453-456]; Borchs1966 [catalogue: 340]; DanzigPe1998 [catalogue: 240]; LongoMaPe1995 [distribution: 127]; Rungs1935 [host, distribution: 271];

Rungs1936 [taxonomy: 55]; Rungs1948 [host, distribution: 111].

Diaspidiotus spiraspinae Takagi
Diaspidiotus spiraspinae Takagi, 1956b: 85. Type data: JAPAN: Honsyu, Toyama-ken, Toyama, on *Ilex crenata*. Holotype female. Type depository: Sapporo: Entomological Institute, Faculty of Agriculture, Hokkaido University, Japan.
SCALE COVER: Of type common to genus, being grey; tends to be covered by epidermis of host when on *Viburnum* (Takagi, 1969b).
HOST PLANTS: **Caprifoliaceae**: *Viburnum* [Kawai1977], *Viburnum wrightii* [Takagi1956b]. **Cornaceae**: *Cornus* [Kawai1977]. **Fagaceae**: *Ilex* [Kawai1977], *Ilex crenata* [Takagi1956b]. **Oleaceae**: *Ligustrum* [Kawai1977].
DISTRIBUTION: **Palaearctic**: Japan [Kawai1980] (Honshu [Takagi1956b]).
BIOLOGY: Occurs on bark of host (Takagi, 1956b).
GENERAL: Description and illustration of adult female by Takagi (1956b).
CITATIONS: Borchs1966 [catalogue: 326]; DanzigPe1998 [catalogue: 241]; Kawai1977 [host, distribution, economic importance: 160,162]; Kawai1980 [taxonomy, description, host, distribution: 221-222]; Muraka1970 [host, distribution: 74]; Takagi1956b [taxonomy, description, illustration, host, distribution: 85-86].

Diaspidiotus sulci (Balachowsky)
Quadraspidiotus sulci Balachowsky, 1950b: 446. Type data: CZECH REPUBLIC: Moravia, Mohelno, on *Genista pilosa*. Syntypes, female. Type depository: Paris: Muséum national d'Histoire naturelle, France.
Gonaspidiotus sulci; Lupo, 1954: 14. Change of combination.
Diaspidiotus sulci; Danzig & Pellizzari, 1998: 241. Change of combination.
SCALE COVER: Female scale circular, 1.6-1.8 mm diameter; convex; dark almost black; surface not even, matt; exuviae central, dark brown red. Male scale narrowly oval, dark grey; surface not even, matt; 1.4 mm long (Balachowsky, 1950b).
HOST PLANTS: **Leguminosae**: *Genista pilosa* [Balach1950b, Lupo1954], *Genista sagittalis* [Lupo1954].
DISTRIBUTION: **Palaearctic**: Czech Republic [Zahrad1977]; Italy [Lupo1954].
GENERAL: Description and illustration of adult female by Balachowsky (1950b), Zahradník (1952), Lupo (1954) and by Danzig (1993).
KEYS: Danzig 1993: 179-182 (female) [Europe]; Kosztarab & Kozár 1978: 169-170 (female) [Hungary]; Lupo 1954: 8-9 (female) [Italy]; Zahradník 1952: 114-115 (female) [Czech Republic]; Balachowsky 1950b: 401 (female) [Mediterranean].
CITATIONS: Balach1950b [taxonomy, description, illustration, host, distribution: 401,446-449]; Balach1958b [taxonomy: 212]; Borchs1966 [catalogue: 340]; Danzig1993 [taxonomy, description, illustration, host, distribution: 194-195]; DanzigPe1998 [catalogue: 241]; KosztaKo1978 [taxonomy, description, host, distribution: 169-170]; Lindin1957 [taxonomy: 551]; LongoMaPe1995 [distribution: 128]; Lupo1954 [taxonomy, description, illustration, host, distribution: 14-19]; Schmut1959 [taxonomy, description, host, distribution: 77,83]; Zahrad1952 [taxonomy, description, illustration, host, distribution: 133-137]; Zahrad1977 [taxonomy, distribution: 121].

Diaspidiotus taxodii (Ferris), new combination
Quadraspidiotus taxodii Ferris, 1938a: 261. Type data: U.S.A.: Texas, six miles east of Wharton, on *Taxodium distichum*; collected 1921. Holotype female. Type depository: Davis: The Bohart Museum of Entomology, University of California, California, USA.
COMMON NAME: bald cypress scale [Dekle1965c].
SCALE COVER: Female scale circular, flat, white or grey (Ferris, 1938a).
HOST PLANTS: **Taxodiaceae**: *Taxodium ascendens* [TippinBe1970, BesheaTiHo1973], *Ta. distichum* [Ferris1938a, Dekle1965c, McDani1970].
DISTRIBUTION: **Nearctic**: United States of America (District of Columbia [Nakaha1982], Florida [Dekle1965c, BesheaTiHo1973], Georgia [TippinBe1970, BesheaTiHo1973], Louisiana [Ferris1938a], Maryland [Nakaha1982], Pennsylvania [Nakaha1982], Texas [Ferris1938a, McDani1970]).
GENERAL: Description and illustration of adult female by Ferris (1938a) and by Kosztarab (1996).
KEYS: Kosztarab 1996: 575 (female) [Northeastern North America]; McDaniel 1970: 429 (female) [U.S.A.: Texas]; Ferris 1942: 39 (female) [North America].
CITATIONS: BesheaTiHo1973 [host, distribution: 8]; Borchs1966 [catalogue: 340]; Dekle1965c [taxonomy, description, host, distribution: 127]; Dekle1976 [taxonomy, description, host, distribution, economic importance: 146]; Ferris1938a [taxonomy, description, illustration, host, distribution: 261]; Ferris1942 [taxonomy: 446:39]; Koszta1996 [taxonomy, description, illustration, host, distribution, life history: 585-586]; McDani1970 [taxonomy, host, distribution: 437]; Mead1983 [host, distribution: 1-5]; Nakaha1982 [host, distribution: 79]; Nakaha1982 [host, distribution: 79]; TippinBe1970 [host, distribution: 11].

Diaspidiotus ternstroemiae (Ferris)
Quadraspidiotus ternstroemiae Ferris, 1952a: 9. Type data: CHINA: Yunnan Province, near Kunming, at Si-shan, on *Ternstroemia* sp.; collected by G.F. Ferris, May, 8, 1949. Syntypes, female. Type depository: Davis: The Bohart Museum of Entomology, University of California, California, USA.
Diaspidiotus ternstroemiae; Danzig & Pellizzari, 1998: 241. Change of combination.
Diaspidiotus ternstromeiae; Danzig & Pellizzari, 1998: 241. Misspelling of species name.
SCALE COVER: Female scale circular, quite high convex; exuvia apparently toward one side; colour usually concealed by covering bark and debris but apparently grey. Scale of male not recognized (Ferris, 1952a).
HOST PLANTS: **Theaceae**: *Ternstroemia* [Ferris1952a].
DISTRIBUTION: **Oriental**: China (People's Republic) (Yunnan [Ferris1952a]).
BIOLOGY: Occurring on branches almost completely concealed beneath bark flakes and lichens (Ferris, 1952a).
GENERAL: Description and illustration of adult female by Ferris (1952a) and by Chou (1985, 1986).
KEYS: Chou 1985: 311 (female) [Species of China].

CITATIONS: Borchs1966 [catalogue: 340]; Chou1985 [taxonomy, description, host, distribution: 315]; Chou1986 [taxonomy, illustration: 693]; DanzigPe1998 [catalogue: 241]; Ferris1952a [taxonomy, description, illustration, host, distribution: 9,16]; KaussaBa1953 [taxonomy: 26]; Tao1999 [taxonomy, host, distribution: 115].

Diaspidiotus thymbrae (Koroneos)

Aspidiotus thymbrae Koroneos, 1934: 17. Type data: GREECE: Volos - Ghorista, Melissiatica, Latomion, Ano Volos, Khalkis, Colline Karababa, on *Satureja thymbra* and *Thymbra capitala*. Syntypes, female. Notes: Depository of type material unknown (P. Katsoyannos ,Athens, Greece, in message to Yair Ben-Dov, 2000).
Rhizaspidiotus thymbrae; Goux, 1941: 39. Change of combination.
Quadraspidiotus thymbrae; Balachowsky, 1950b: 482. Change of combination.
Gonaspidiotus thymbrae; Lupo, 1954: 23. Change of combination.
Diaspidiotus thymbrae; Danzig & Pellizzari, 1998: 241. Change of combination.
SCALE COVER: Female scale matt, circular or subcircular, 1.8-2.2 mm; convex; exuviae central, brown chestnut; secreted part of scale grey with dark brown strings; whole scale generally covered with white secretion; ventral vellum white, adhered to plant. Male scale narrowly oval; exuviae white, 1.2 mm (Balachowsky, 1950b).
HOST PLANTS: **Labiatae**: *Satureja thymbra* [Korone1934, Lupo1954], *Thymus capitata* [Korone1934, Lupo1954], *Thymus vulgaris* [Balach1950b, Lupo1954].
DISTRIBUTION: **Palaearctic**: France [Lupo1954]; Greece [Korone1934]; Italy [Lupo1954, LongoMaPe1995].
GENERAL: Description and illustration of adult female by Balachowsky (1950b) and by Lupo (1954).
KEYS: Lupo 1954: 8-9 (female) [Italy]; Balachowsky 1950b: 403 (female) [Mediterranean].
CITATIONS: Balach1935 [taxonomy: 236]; Balach1950b [taxonomy, description, illustration, host, distribution: 482-485]; Borchs1966 [catalogue: 340-341]; DanzigPe1998 [catalogue: 241]; Ferris1941e [taxonomy: 49]; Ferris1943a [taxonomy: 99]; Foldi2001 [distribution: 303-308]; Goux1941a [taxonomy: 39]; Korone1934 [taxonomy, description, illustration, host, distribution: 17-18]; Lindin1936 [taxonomy: 153]; LongoMaPe1995 [distribution: 128]; Lupo1954 [taxonomy, description, illustration, host, distribution: 23-28]; Muntin1969 [taxonomy: 138].

Diaspidiotus thymicola (Balachowsky)

Aspidiotus (*Evaspidiotus*) *thymicola* Balachowsky, 1935: 233. Type data: SPAIN: Andalusia, Sierra Nevada, near Trevelez village, altitude 1900m, on *Thymus mastichina*. Syntypes, female. Type depository: Paris: Muséum national d'Histoire naturelle, France.
Quadraspidiotus thymicola; Balachowsky, 1950b: 485. Change of combination.
Diaspidiotus thymicola; Danzig & Pellizzari, 1998: 241. Change of combination.

SCALE COVER: Female scale circular or subcircular, more or less irregular, diameter 1.6-1.8 mm; generally beneath bark fibers; moderately convex, elevated anteriorly; exuviae central or subcentral; colour brown, transparent, wholly covered with white-grey waxy secretion (Balachowsky, 1935).

HOST PLANTS: **Compositae**: *Artemisia herba-alba* [Martin1983]. **Labiatae**: *Corydothymus capitatus* [GomezM1957, Martin1983], *Satureja montana* [GomezM1954, Martin1983], *Teucrium polium* [Martin1983], *Thymus* [Balach1935b, GomezM1954, GomezM1956b, Martin1983], *Thymus mastichina* [Balach1935, Balach1935b, Martin1983], *Thymus vulgaris* [Martin1983].

DISTRIBUTION: **Palaearctic**: Spain [Balach1935, Balach1935b, GomezM1937, Martin1983, BlayGo1993].

GENERAL: Description and illustration of adult female by Balachowsky (1935, 1950b).

KEYS: Balachowsky 1950b: 402 (female) [Mediterranean].

CITATIONS: Balach1935 [taxonomy, description, illustration, host, distribution: 233-236]; Balach1935b [host, distribution: 257]; Balach1950b [taxonomy, description, illustration, host, distribution: 485-488]; BlayGo1993 [taxonomy, description, illustration, host, distribution: 583-587]; Borchs1966 [catalogue: 341]; DanzigPe1998 [catalogue: 241]; Ferris1941e [taxonomy: 49]; GomezM1937 [taxonomy, description, illustration, host, distribution: 72-75]; GomezM1954 [host, distribution: 120]; GomezM1956b [host, distribution: 482]; GomezM1957 [host, distribution: 40]; GomezM1958a [host, distribution: 8]; GomezM1958c [host, distribution: 406]; GomezM1960O [host, distribution: 163]; GomezM1965 [host, distribution: 90]; GomezM1968 [host, distribution: 541]; Lindin1936 [taxonomy: 285]; Lindin1937 [taxonomy: 180]; Martin1983 [taxonomy, host, distribution: 68].

Diaspidiotus tillandsiae (Takagi & Tippins), new combination
Quadraspidiotus tillandsiae Takagi & Tippins, 1972: 182. Type data: U.S.A.: Georgia, on *Tillandsia usneoides*; collected April 5 1969. Holotype female. Type depository: Washington: U.S. National Entomological Collection, U.S. National Museum of Natural History, District of Columbia, USA; type no. HHT5069.

SCALE COVER: Takagi & Tippins (1972) did not describe scale cover.

HOST PLANTS: **Bromeliaceae**: *Tillandsia usneoides* [TakagiTi1972, BesheaTiHo1973].

DISTRIBUTION: **Nearctic**: U.S.A. (Georgia [TakagiTi1972, BesheaTiHo1973]).

GENERAL: Description and illustration of adult female by Takagi & Tippins (1972).

CITATIONS: BesheaTiHo1973 [host, distribution: 8]; TakagiTi1972 [taxonomy, description, illustration, host, distribution: 182-184].

Diaspidiotus transcaspiensis (Marlatt)
Aspidiotus (Diaspidiotus) transcaspiensis Marlatt, 1908c: 21. Type data: TURKMENISTAN: C. Ahnger, Ashkhabad, on bark of *Populus*; collected November 1898. Syntypes, female. Type depository: Washington: U.S. National Museum of Natural History, District of Columbia, USA; type no. 8216.
Aspidiotus transcaspiensis; Sanders, 1909a: 53. Change of combination.

Hendaspidiotus transcaspiensis; MacGillivray, 1921: 440. Change of combination.
Diaspidiotus trancaspiensis; Borchsenius, 1950b: 229. Misspelling of species name.
Diaspidiotus transcaspiensis; Borchsenius, 1950b: 229. Change of combination.
SCALE COVER: Female scale subcircular, 1.5 mm in diameter; yellowish or buff; exuviae orange, exposed when rubbed or old. Male scale of normal oval shape, similar in colour to female (Marlatt, 1908c).
HOST PLANTS: **Elaeagnaceae**: *Elaeagnus angustifolia* [Balach1950b], *Hippophae rhamnoides* [Balach1950b]. **Leguminosae**: *Halimodendron arsenticum* [Balach1950b]. **Salicaceae**: *Populus* [Marlat1908c, Sander1909a, Archan1930, BazaroSh1971], *Populus alba* [Archan1930], *Salix* [Archan1930, Balach1950b].
NATURAL ENEMIES: HYMENOPTERA **Encyrtidae**: *Anthemus aspidioti* Nikolskaya [Trjapi1989].
DISTRIBUTION: **Palaearctic**: Iran [Kaussa1955]; Kazakhstan [Balach1950b]; Tajikistan [Balach1950b, BazaroSh1971]; Turkmenistan (Ashkhabad Oblast [Marlat1908c, Archan1930, BazaroSh1971]); Uzbekistan [Balach1950b].
GENERAL: Description and illustration of adult female by Marlatt (1908c), Balachowsky (1950b), Bazarov & Shmelev (1971) and by Danzig (1993).
KEYS: Danzig 1993: 179-182 (female) [Europe]; Bazarov & Shmelev 1971: 198 (female) [Central Asia]; Balachowsky 1950b: 492 (female) [Mediterranean]; Archangelskaya 1937: 99 (female) [Central Asia].
CITATIONS: Archan1930 [host, distribution: 84-85]; Archan1937 [taxonomy, description, host, distribution: 99,102]; Balach1950b [taxonomy, description, illustration, host, distribution: 512-515]; BazaroSh1971 [taxonomy, description, illustration, host, distribution: 205-207]; Borchs1937 [taxonomy, description, illustration, host, distribution: 134]; Borchs1939 [taxonomy, description, host, distribution: 10,36]; Borchs1950b [taxonomy, description, illustration, host, distribution: 229,231]; Borchs1966 [catalogue: 326]; Bustsh1960 [taxonomy: 181]; Danzig1993 [taxonomy, description, illustration, host, distribution: 206-207]; DanzigPe1998 [catalogue: 242]; Ferris1941e [taxonomy: 49]; Kaussa1955 [host, distribution: 16]; KaussaBa1953 [taxonomy: 26]; MacGil1921 [taxonomy, description, host, distribution: 440]; Marlat1908c [taxonomy, description, illustration, host, distribution: 21]; MillerDa1990 [host, distribution, economic importance: 301]; Sander1909a [taxonomy, host, distribution: 53]; Trjapi1989 [biological control: 378].

Diaspidiotus turanicus (Borchsenius)
Aspidiotus turanicus Borchsenius, 1935: 131. Type data: ARMENIA: Erevanskii region, on *Salix* sp. Lectotype female, by subsequent designation Danzig, 1993: 206. Type depository: St. Petersburg: Zoological Museum, Academy of Science, Russia.
Diaspidiotus turanicus; Borchsenius, 1949d: 246. Change of combination.
Diaspidiotus truanicus; Tao, 1999: 83. Misspelling of species name.
SCALE COVER: Scale of female circular, slightly convex, brownish-grey in central part of scale to white on margins; exuviae situated in central part of scale; first exuviae reddish-brown, second brown; diameter of scale 2.5 mm. Male unknown. (Borchsenius, 1935).

HOST PLANTS: **Salicaceae**: *Salix* [Borchs1935, Borchs1936, Balach1950b, BazaroSh1971, Danzig1972c].

DISTRIBUTION: **Palaearctic**: Afghanistan [Danzig1972c]; Armenia [Borchs1935, Borchs1936, Balach1950b, BazaroSh1971]; China (People's Republic) [Tang1984]; Iran [Kaussa1955]; Tajikistan (=Tadzhikistan) [BazaroSh1971].

GENERAL: Description and illustration of adult female by Borchsenius (1935), Balachowsky (1950b), Bazarov & Shmelev (1971), Tang (1984) and by Danzig (1993).

KEYS: Danzig 1993: 179-182 (female) [Europe]; Bazarov & Shmelev 1971: 198 (female) [Central Asia]; Balachowsky 1950b: 491 (female) [Mediterranean]; Borchsenius 1935: 127-128 (female) [Former USSR].

CITATIONS: Balach1950b [taxonomy, description, illustration, host, distribution : 518-520]; BazaroSh1971 [taxonomy, description, illustration, host, distribution: 208-210]; Borchs1935 [taxonomy, description, illustration, host, distribution: 128,131-132]; Borchs1936 [host, distribution: 131]; Borchs1937 [taxonomy, description, illustration, host, distribution: 131-132]; Borchs1939 [taxonomy, description, host, distribution: 10,37]; Borchs1949d [taxonomy, description, host, distribution: 246]; Borchs1950b [taxonomy, description, illustration, host, distribution: 227-228,231]; Borchs1966 [catalogue: 326]; Bustsh1958 [taxonomy, description, host, distribution: 221,258]; Danzig1972c [host, distribution: 583]; Danzig1993 [taxonomy, description, illustration, host, distribution: 207-208]; DanzigPe1998 [catalogue: 242]; Ferris1941e [taxonomy: 49]; KaussaBa1953 [taxonomy: 26]; MillerDa1990 [host, distribution, economic importance: 301]; Tang1984 [taxonomy, description, illustration, host, distribution: 70-71]; Tao1999 [taxonomy, host, distribution: 83]; TerGri1962 [taxonomy, description, host, distribution: 180].

Diaspidiotus uvae (Comstock)

Aspidiotus uvae Comstock, 1881a: 309. Type data: USA: Indiana, Vevay, on grapevines; collected by Charles G. Boerner. Syntypes, female. Type depository: Washington: United States National Entomological Collection, U.S. National Museum of Natural History, District of Columbia, USA.

Aspidiotus (*Diaspidiotus*) *uvae*; Cockerell, 1897i: 22. Change of combination.

Aspidiotus uvae var. Hunter, 1899: 4. Nomen nudum; discovered by Lindinger, 1937: 180.

Diaspidiotus uvae; Silvestri, 1902: 100. Change of combination.

Diaspidiotus uviae; MacGillivray, 1921: 412. Misspelling of species name.

Aspidiotus uvaspis Lindinger, 1937: 180. Type data: U.S.A.: Kansas, Lawrence, on *Carya alba*. Syntypes, female. Synonymy by Ferris, 1941e: 49. Notes: Lindinger's (1937) action validated the status of *Aspidiotus uvae* var., as described by Hunter (1899: 4).

COMMON NAME: grape scale [Comsto1881a, Merril1953, McKenz1956, Dekle1965c, Koszta1996].

SYSTEMATICS: Ferris (1941e: 49) explained in details the reasons for the synonymy of *Aspidiotus uvaspis* Lindinger, 1937 with *Diaspidiotus uvae*.

SCALE COVER: Female scale flat; nearly circular, diameter 1.6 mm; exuviae covered and more or less upon one side; light yellowish brown, being a little lighter than dry bark of vine; part of scale covering exuviae white, latter are bright yellow; ventral scale thin, white, contains ventral half of molted skins, and adheres to bark (Comstock, 1881a). Scale of female circular, flat, exuviae central or subcentral, colour of scale white but this usually changed into yellowish by a film of epidermis of host; scale of male perhaps slightly darker than that of female, elongate oval, exuvia near one end (Ferris, 1938a).

HOST PLANTS: **Agavaceae**: *Yucca* [McDani1969]. **Betulaceae**: *Betula* [McKenz1956], *Betula nigra* [Ferris1938a]. **Juglandaceae**: *Carya* [McKenz1956], *Carya* [Ferris1938a, Balach1950b], *Carya alba* [Ferris1942]. **Moraceae**: *Maclura pomifera* [McKenz1956]. **Platanaceae**: *Platanus* [Ferris1938a, Balach1950b, McKenz1956, McDani1969]. **Rosaceae**: *Crataegus* [Ferris1938a, Balach1950b, McKenz1956, McDani1969], *Eriobotrya japonica* [Balach1950b, Martin1983], *Pyrus vitis* [BesheaTiHo1973]. **Vitaceae**: *Vitis* [Comsto1881a, Lepage1938, Balach1950b, McKenz1956, Dekle1965c, McDani1969, TippinBe1970], *Vitis vinifera* [Ferris1938a, GomezM1962, Martin1983].

NATURAL ENEMIES: DIPTERA **Cecidomyiidae**: *Dentifibula viburni* (Felt) [Harris1990]. HYMENOPTERA **Aphelinidae**: *Ablerus americanus* Girault [Gordh1979], *Aphytis proclia* (Walker) [Gordh1979], *Azotus marchali* Howard [Gordh1979], *Prospaltella murtfeldtiae* (Howard) [Howard1894c, Gordh1979]. **Signiphoridae**: *Signiphora pulchra* Girault [Woolle1990].

DISTRIBUTION: **Nearctic**: United States of America (Alabama [Nakaha1982], Arkansas [Nakaha1982], California [McKenz1956], Connecticut [Nakaha1982], Delaware [Nakaha1982], District of Columbia [Nakaha1982], Florida [Wilson1917, MerrilCh1923, Ferris1938a, Merril1953, Dekle1965c], Georgia [TippinBe1970, BesheaTiHo1973], Illinois [Nakaha1982], Indiana [Comsto1881a], Kansas [Hunter1899, Ferris1938a, Ferris1942], Kentucky [Nakaha1982], Maryland [Nakaha1982], Mississippi [Herric1911, Ferris1938a], Missouri [Hollin1923, Ferris1938a], New Jersey [Nakaha1982], New York [Nakaha1982], North Carolina [Nakaha1982], Ohio [Ferris1938a], Pennsylvania [Nakaha1982], Tennessee [Nakaha1982], Texas [Herric1911, Ferris1938a, McDani1969], Virginia [Nakaha1982], West Virginia [Nakaha1982]). **Neotropical**: Brazil (Rio de Janeiro [Lepage1938], São Paulo [KimouiCo1987]). **Palaearctic**: Canary Islands [GomezM1962, MatileOr2001]; Italy [Ferris1938a]; Spain [GomezM1937, RuizCa1944, GomezM1956, Martin1983, BlayGo1993].

BIOLOGY: Occur on bark (Ferris, 1938a).

ECONOMIC IMPORTANCE: A pest of grapes, *Vitis* spp. in south central USA (Whitehead, 1963; Miller & Davidson, 1990; Johnson et al., 1999). Not considered a pest in Florida (Dekle, 1976) or in California (Gill, 1997).

GENERAL: Description and illustration of adult female by Comstock (1881a), Ferris (1938a), Balachowsky (1950b), McKenzie (1956), Gómez-Menor Guerrero (1962), Kosztarab (1996) and by Gill (1997).

KEYS: Gill 1997: 113 (female) [Species of California]; Kosztarab 1996: 484-485 (female) [Northeastern North America]; McDaniel 1969: 90-91 (female) [U.S.A.: Texas]; Gómez-Menor Guerrero 1962: 166 (female) [Canary Islands]; McKenzie

1956: 25 (female) [U.S.A.: California]; Balachowsky 1950b: 492 (female) [Mediterranean]; Ferris 1942: 33 (female) [North America]; Britton 1923: 371 (female) [U.S.A.: Connecticut]; Hollinger 1923: 7-8 (female) [U.S.A.: Missouri]; Lawson 1917: 217 (female) [U.S.A.: Kansas]; Dietz & Morrison 1916a: 289-290 (female) [U.S.A.: Indiana]; Newell 1899: 4-5 (female) [North America]; Comstock 1883: 55-57 (female) [North America].

CITATIONS: Balach1950b [taxonomy, description, illustration, host, distribution: 497-500]; Barnes1930 [biological control: 319-326]; BeardsDaHo1976 [economic importance: 105]; BesheaTiHo1973 [host, distribution: 6]; BlayGo1993 [taxonomy, description, illustration, host, distribution: 595-599]; Borchs1966 [catalogue: 326-327]; Britto1923 [taxonomy, description, host, distribution: 371,375-376]; ChuaWo1990 [host, distribution, economic importance: 551]; ClapsWoGo2001a [taxonomy, host, distribution: 16]; Cocker1896b [distribution: 334]; Cocker1897i [taxonomy, description, host, distribution: 22]; Comsto1881a [taxonomy, description, illustration, host, distribution: 309-310]; Comsto1883 [taxonomy, host, distribution: 71]; CostaLRa1922 [host, distribution: 1101]; DanzigPe1998 [catalogue: 242]; Dekle1965c [taxonomy, description, host, distribution: 52]; Dekle1976 [taxonomy, description, host, distribution, economic importance: 72]; DeLott1963 [taxonomy: 144-145]; DeSant1940 [biological control: 29-44]; DeSant1979 [biological control]; DietzMo1916a [taxonomy, description, illustration, host, distribution: 290,305-306]; Dougla1912 [taxonomy: 213]; Essig1920 [taxonomy, description, host, distribution: 37]; Felt1924 [biological control: 458-459]; Felt1925 [biological control: 3]; Fernal1903b [catalogue: 280]; Ferris1938a [taxonomy, description, illustration, host, distribution: 225]; Ferris1941e [taxonomy: 49]; Ferris1942 [taxonomy: 446:33]; Garcia1922 [host, distribution, biological control: 196-200]; Garcia1930 [host, distribution, biological control]; Gill1997 [host, distribution, taxonomy, description, illustration, economic importance: 123-124]; Giraul1911a [host, distribution, biological control: 175-189]; Giraul1916 [host, distribution, biological control: 42-44]; GomezM1937 [taxonomy, description, illustration, host, distribution: 77-78]; GomezM1956 [taxonomy, description, illustration, host, distribution, biological control: 17-21]; GomezM1960O [taxonomy, description, illustration, host, distribution: 158-160]; GomezM1962 [taxonomy, description, illustration, host, distribution: 170-173]; Gordh1979 [biological control: 896,899,900,908]; Gowdey1921 [taxonomy, description, host, distribution: 30]; Harris1990 [biological control: 61-66]; Herric1911 [taxonomy, description, illustration, host, distribution: 10,21-22,55]; Herric1925 [host, distribution, description, life history, economic importance]; Hollin1923 [taxonomy, description, host, distribution: 8,17-18,68]; Howard1894c [host, distribution, biological control: 5-7]; Howard1895e [biological control: 1-44]; Hunter1899 [taxonomy, host, distribution: 4]; JohnsoLeWh1999 [host, distribution, economic importance, life history: 161-170]; KimouiCo1987 [host, distribution, life history, economic importance: 104-106]; KonstaGu1987 [host, distribution: 161]; Koszta1996 [taxonomy, description, illustration, host, distribution, life history, biological control, economic importance: 497-499]; Lawson1917 [taxonomy, description, illustration, host, distribution: 236-237]; Leonar1897 [taxonomy: 286]; Leonar1898a [taxonomy, description, illustration, host, distribution: 58-60];

Lepage1938 [catalogue: 397]; Lindin1937 [taxonomy: 180]; MacGil1921 [taxonomy, description, host, distribution: 412]; Martin1983 [taxonomy, host, distribution: 64]; MatileOr2001 [host, distribution: 189]; McClur1990c [taxonomy, host, distribution, ecology: 289-291]; McDani1969 [taxonomy, illustration, host, distribution: 98-101]; McKenz1956 [taxonomy, description, illustration, host, distribution: 25,63-66]; Merril1953 [taxonomy, description, host, distribution: 28-29]; MerrilCh1923 [taxonomy, description, host, distribution, economic importance: 211]; MillerDa1990 [host, distribution, economic importance: 302]; Nakaha1982 [host, distribution: 30]; Newell1899 [taxonomy, description, host, distribution: 12-13]; ReisMe1984 [economic importance, host, distribution: 68-72]; Ruhl1913 [host, distribution: 79-80]; RuizCa1944 [taxonomy, description, illustration, host, distribution, chemical control: 55-74]; Sander1904a [taxonomy, description, illustration, host, distribution: 56,68]; SchmutKlLu1957 [host, distribution, economic importance: 491]; Silves1902 [taxonomy, description, host, distribution: 100]; Targio1884 [taxonomy: 389-390]; TippinBe1970 [host, distribution: 9]; Whiteh1963 [host, distribution, economic importance, control, life history: 1-98]; Wilson1917 [taxonomy, description, host, distribution: 27-28]; Woolle1990 [biological control: 167-176]; Zimmer1912 [taxonomy, description, host, distribution, biological control, chemical control: 117].

Diaspidiotus viticola (Leonardi)

Aspidiotus viticola Leonardi, 1913b: 61. Type data: ITALY: Napoli, Portici, S. Giorgio a Cremano, on trunk of grapevine [*Vitis vinifera*]. Syntypes, female. Type depository: Portici: Dipartimento de Entomologia e Zoologia Agraria di Portici, Università di Napoli Federico II, Italy.
Paraspidiotus viticola; Thiem & Gerneck, 1934: 231. Change of combination.
Diaspidiotus viticola; Castro, 1944: 63. Change of combination.
SCALE COVER: Female scale oval, 1 mm long, 0.7 mm wide; exuviae central; colour yellowish with a rusty tint (Leonardi, 1913b).
HOST PLANTS: **Vitaceae**: *Vitis vinifera* [Leonar1913b, Leonar1920, GomezM1941, Lupo1948, Balach1950b].
NATURAL ENEMIES: HYMENOPTERA **Aphelinidae**: *Azotus matriensis* Mercet [Viggia1990b], *Coccophagoides moeris* (Walker) [Gordh1979, Viggia1990a].
DISTRIBUTION: **Neotropical**: Dominican Republic [GomezM1941]. **Palaearctic**: Italy [Leonar1913b, Leonar1920, Lupo1948, Balach1950b, LongoMaPe1995].
GENERAL: Description and illustration of adult female by Leonardi (1913b) and by Balachowsky (1950b).
KEYS: Balachowsky 1950b: 493 (female) [Mediterranean]; Leonardi 1920: 29-30 (female) [Italy].
CITATIONS: Balach1950b [taxonomy, description, illustration, host, distribution: 503-506]; Borchs1966 [catalogue: 327]; DanzigPe1998 [catalogue: 242-243]; Ferris1937c [taxonomy: 51]; Ferris1941e [taxonomy: 49]; GomezM1941 [host, distribution: 128]; Gordh1979 [biological control: 901]; Leonar1913b [taxonomy, description, illustration, host, distribution: 61-63]; Leonar1920 [taxonomy, description, illustration, host, distribution: 62-64]; Lindin1928 [taxonomy: 106]; Lindin1957 [taxonomy: 550]; LongoMaPe1995 [distribution: 126]; Lupo1948

[taxonomy, description, illustration, host, distribution: 194-198]; Priore1964 [host, distribution: 131-178]; Priore1965 [host, distribution: 101-145]; RuizCa1944 [taxonomy: 66]; ThiemGe1934a [taxonomy, description, host, distribution: 231]; TranfaVi1987a [economic importance: 215-221]; Viggia1990a [biological control: 124]; Viggia1990b [biological control: 178].

Diaspidiotus wuenni (Lindinger)

Aspidiotus alni; Lindinger, 1911: 245. Misidentification. Discovered by Lindinger, 1923: 152.

Aspidiotus wünni Lindinger, 1923: 147. Type data: AUSTRIA: Vienna, on *Quercus cerris*; collected by Lindinger, 18.9.1909. Syntypes, female. Type depository: Hamburg: Zoologisches Institut und Zoologishces Museum, Universität von Hamburg, Germany.

Diaspidiotus wünni; Balachowsky, 1950b: 526. Change of combination.

Diaspidiotus wuenni; Borchsenius, 1966: 327. Justified emendation.

SCALE COVER: Female scale circular, 1 mm in diameter; convex; grey; exuviae brown red, subcentral (Lindinger, 1923; Balachowsky, 1950b).

HOST PLANTS: **Betulaceae**: *Alnus* [Balach1950b], *Alnus glutinosa* [Balach1950b]. **Fagaceae**: *Castanea sativa* [BachmaGe1950], *Quercus* [Balach1950b, Zahrad1952, Bachma1953], *Qu. cerris* [Balach1950b, Zahrad1972], *Qu. petraea* [Zahrad1972], *Qu. robur* [Zahrad1952, Zahrad1972].

NATURAL ENEMIES: HYMENOPTERA **Aphelinidae**: *Aspidiotiphagus citrinus* Crawford [Zahrad1972].

DISTRIBUTION: **Palaearctic**: Austria [Balach1950b]; Czech Republic [Zahrad1952, Zahrad1977]; Slovak Republic [Zahrad1952]; Switzerland [BachmaGe1950, Balach1950b]; Turkey [Yasar1995a]; Yugoslavia [Bachma1953].

GENERAL: Description and illustration of adult female by Balachowsky (1950b), Zahradník (1952), Tereznikova (1986), Danzig (1993) and by Yaşar (1995a).

KEYS: Danzig 1993: 179-182 (female) [Europe]; Tereznikova 1986: 97-98 (female) [Ukraine]; Kosztarab & Kozár 1978: 156 (female) [Hungary]; Zahradník 1952: 143 (female) [Czech Republic]; Balachowsky 1950b: 492 (female) [Mediterranean].

CITATIONS: Bachma1953 [host, distribution: 182]; Bachma1956 [taxonomy: 101]; Bachma1956 [taxonomy: 101]; BachmaGe1950 [host, distribution: 118]; Balach1950b [taxonomy, description, illustration, host, distribution: 526-528]; Borchs1966 [catalogue: 327]; Danzig1993 [taxonomy, description, illustration, host, distribution: 212-213]; DanzigPe1998 [catalogue: 243]; Ferris1941e [taxonomy: 49]; Ferris1943a [taxonomy: 86]; KosztaKo1978 [taxonomy, description, host, distribution: 156]; Lellak1959 [taxonomy, description, host, distribution: 338]; Lindin1911 [taxonomy, description, host, distribution: 245]; Lindin1912b [taxonomy, description, host, distribution: 63-64,278]; Lindin1923 [taxonomy, description, illustration, host, distribution: 147,152]; Lindin1931a [taxonomy: 90]; Lindin1932a [taxonomy: 79]; Lindin1935 [taxonomy: 128]; Lindin1949 [taxonomy: 211]; MacGil1921 [taxonomy, description, host, distribution: 440]; Schmut1959 [taxonomy, description, host, distribution: 100,105]; Takagi1956b [taxonomy: 84]; Terezn1986 [taxonomy, description, illustration, host, distribution: 110-111]; WeidneWa1968 [taxonomy: 174]; Yasar1995a [taxonomy, description, illustration, host, distribution: 73-75]; Zahrad1952 [taxonomy, description, illustration, host, distribution: 143,147-152]; Zahrad1972 [host, distribution, biological control: 440]; Zahrad1977 [taxonomy, distribution: 120]; Zahrad1990a [host, distribution, description: 649-650].

Diaspidiotus xinjiangensis Tang

Diaspidiotus xinjiangensis Tang, 1984: 70. Type data: CHINA: Xinjiang Province, Korla County, Luntai County, on *Populus* sp. Syntypes, female. Type depository: Shanxi: Ent. Institute, Shanxi Agricultural University, Taigu, Shanxi, China.
Diaspidiotus xingjiangensis; Tao, 1999: 83. Misspelling of species name.
SCALE COVER: Female scale circular, greyish white, 1.5 mm in diameter; that of male ovoid, about 1 mm long, similar in colour and texture to female's (Tang, 1984).
HOST PLANTS: **Salicaceae**: *Populus* [Tang1984].
DISTRIBUTION: **Palaearctic**: China (People's Republic) (Xingiang Uygur (Sinkiang) [Tang1984]).
GENERAL: Description and illustration of adult female by Tang (1984).
CITATIONS: DanzigPe1998 [catalogue: 243]; Tang1984 [taxonomy, description, illustration, host, distribution: 70,72].

Diaspidiotus yomae Munting

Diaspidiotus yomae Munting, 1971: 137. Type data: SOUTH AFRICA: Transvaal, Christiana, on *Diospyros* sp.; collected 11.iii.1967. Holotype female. Type depository: Pretoria: South African National Collection of Insects, South Africa; type no. 2831/1.
SCALE COVER: Female scale small, about 0.75 mm in diameter; hidden in cracks in bark of host and varying in shape accordingly (Munting, 1971).
HOST PLANTS: **Ebenaceae**: *Diospyros* [Muntin1971].
DISTRIBUTION: **Afrotropical**: South Africa [Muntin1971].
GENERAL: Description and illustration of adult female by Munting (1971).
CITATIONS: Muntin1971 [taxonomy, description, illustration, host, distribution: 137-139].

Diaspidiotus zonatus (Frauenfeld)

Aspidiotus zonatus Frauenfeld, 1868: 888. Type data: AUSTRIA: Vienna, Botanic Garden, on underside of leaves of *Quercus montana*. Syntypes, female. Notes: Depository of type material unknown.
Aspidiotus quercus Signoret, 1869: 868. Nomen nudum.
Aspidiotus quercus Signoret, 1869b: 132. Type data: FRANCE: locality not indicated, on trunk of *Quercus* sp. Syntypes, both sexes. Type depository: Vienna: Naturhistorisches Museum Wien, Austria. Synonymy by Signoret, 1877: 673.
Aspidiotus (*Diaspidiotus*) *zonatus*; Cockerell, 1897i: 19. Change of combination.
Aspidiotus (*Aspidiella*) *zonatus*; Leonardi, 1898a: 64. Change of combination.
Furcaspis zonata; MacGillivray, 1921: 409. Change of combination requiring emendation of species name for agreement in gender.
Aspidiotus (*Euraspidiotus*) *zonatus*; Thiem & Gerneck, 1934a: 131. Change of combination.

Aspidiotus (*Furcaspis*) *zonatus*; Borchsenius, 1935a: 8. Change of combination.
Quadraspidiotus zonata; Bodenheimer, 1943: 3. Change of combination.
Diaspidiotus zonatus; Borchsenius, 1949d: 244. Change of combination.
Diaspidiotus hungaricus Kosztarab, 1956: 435. Type data: HUNGARY: Budapest, on *Quercus petraea*. Holotype female. Type depository: Budapest: Hungarian Natural History Museum, Zoological Department, Hungary. Synonymy by Danzig, 1993: 188.

COMMON NAME: dubovaya shitovka [Borchs1936].

SCALE COVER: Illustration of female and male scale by Zahradník (1952). Female scale circular, 2-2.5 mm diameter; little convex; grey; exuviae oval, orange colour, placed centrally or eccentrally. Male scale narrow, elongated, 1-1.5 mm long; circular anteriorly and posteriorly; grey to grey black; exuviae orange, placed on anterior part; ventral scale poorly developed; white (Zahradník, 1952).

HOST PLANTS: **Betulaceae**: *Betula* [Balach1950b, Zahrad1972]. **Ericaceae**: *Vaccinium myrtillus* [Balach1950b, Zahrad1952]. **Fagaceae**: *Fagus* [Zahrad1952], *Fagus orientalis* [Bodenh1949, Bodenh1952], *Fagus sylvatica* [Balach1931a, Balach1932d, Borchs1934, Zahrad1972], *Quercus* [Signor1869b, Newste1893f, Green1928, Bodenh1949, Martin1983, PellizCa1991a, Borchs1936, BachmaGe1950], *Qu. cerris* [GomezM1959, Zahrad1972, Martin1983, Pelliz1987], *Qu. coccifera* [Bodenh1924, Bodenh1926], *Qu. ilex* [BlayGo1993], *Qu. lusitanica* [Bodenh1924, GomezM1959, Zahrad1972, Martin1983], *Qu. montana* [Frauen1868, GomezM1959, Zahrad1972, Martin1983], *Qu. nigra* [GomezM1959, Zahrad1972, Martin1983], *Qu. palustris* [GomezM1959, Zahrad1972, Martin1983], *Qu. pedunculata* [Leonar1920, Borchs1934, GomezM1957, GomezM1959, Martin1983], *Qu. persica* [Bodenh1943, Balach1950b], *Qu. petraea* [Zahrad1952, Zahrad1972, Koszta1956], *Qu. pubescens* [Leonar1920, BachmaGe1950, Bachma1953, GomezM1959, Zahrad1972, Martin1983, Pelliz1987], *Qu. robur* [Balach1937c, BachmaGe1950, Zahrad1952, GomezM1959, Zahrad1972, RosenDe1979, Martin1983], *Qu. sessiliflora* [GomezM1959, Martin1983], *Qu. toza* [BlayGo1993]. **Juglandaceae**: *Juglans regia* [Balach1950b, Zahrad1952, Zahrad1972]. **Loranthaceae**: *Loranthus europeus* [Bodenh1943, Balach1950b]. **Moraceae**: *Ficus* [Bodenh1937], *Ficus carica* [Balach1950b]. **Rhamnaceae**: *Ziziphus* [Bodenh1937], *Ziziphus spina-christi* [Bodenh1924]. **Rosaceae**: *Sorbus* [Balach1950b], *Sorbus aucuparia* [Zahrad1972]. **Salicaceae**: *Salix* [Bodenh1952].

NATURAL ENEMIES: COLEOPTERA **Anthribidae**: *Anthribus nebulosus* (Forster) [Drea1990]. HYMENOPTERA **Aphelinidae**: *Aphytis bovelli* Malenotti [Zahrad1972], *Ap. mytilaspidis* (Le Baron) [Zahrad1972, RosenDe1979], *Ap. proclia* (Walker) [RosenDe1979], *Azotus atomon* Walker [Podsia1995], *Azotus pinifoliae* Mercet [Zahrad1972], *Coccophagoides similis* Masi [Zahrad1972], *Pteroptrix dimidiata* Westwood [Zahrad1972, Podsia1995]. **Encyrtidae**: *Metaphycus nadius* (Walker) [GuerriNo2000], *Zaomma lambinus* (Walker) [Trjapi1989].

DISTRIBUTION: **Palaearctic**: Austria [Frauen1868]; Corsica [Balach1932d]; Czech Republic [Zahrad1977]; France [Balach1937c]; Georgia [Hadzib1983] (Abkhaz ASSR [Borchs1936]); Germany (United) [Lindin1909b, Balach1950b]; Greece [Bodenh1928, Korone1934]; Hungary [Balach1950b, Koszta1956]; Iran

[Kaussa1955]; Iraq [Bodenh1943]; Israel [Bodenh1924, Bodenh1937]; Italy [Leonar1920, LongoMaPe1995]; Lebanon [Bodenh1926]; Netherlands [Jansen2001]; Poland [Kaweck1935]; Portugal [Morgan1889a]; Russia (Caucasus [Borchs1936]); Sicily [Pelliz1987]; Slovak Republic [Zahrad1952]; Spain [GomezM1954, GomezM1959, Martin1983]; Sweden [Gertss2001]; Switzerland [BachmaGe1950, Balach1950b]; Turkey [Bodenh1952, Yasar1995a]; United Kingdom [Green1928] (England [Morgan1889a, RosenDe1979]); Yugoslavia [Bachma1953].

BIOLOGY: Both biparental and univoltine populations were observed in Poland (Podsiadlo, 1995).

GENERAL: Description and illustration of adult female by Balachowsky (1950b), Zahradník (1952), Kosztarab (1956), Gómez-Menor Ortega (1959), Danzig (1993) and by Yaşar (1995a).

KEYS: Blay Goicoechea 1993: 528-529 (female) [Spain]; Danzig 1993: 179-182 (female) [Europe]; Tereznikova 1986: 97-98 (female) [Ukraine]; Kosztarab & Kozár 1978: 155, 170-171 (female) [Hungary]; Zahradník 1952: 114-115 (female) [Czech Republic]; Balachowsky 1950b: 400 (female) [Mediterranean]; Lupo 1948: 175 (female) [Italy]; Borchsenius 1935: 127-128 (female) [Former USSR]; Leonardi 1920: 29-30 (female) [Italy].

CITATIONS: Alam1957 [biological control: 421-466]; Bachma1953 [host, distribution: 178]; BachmaGe1950 [host, distribution: 118]; Balach1931a [host, distribution: 97]; Balach1932d [taxonomy, host, distribution, economic importance: XLVI]; Balach1937c [host, distribution: 2]; Balach1950b [taxonomy, description, illustration, host, distribution: 418-421]; BlayGo1993 [taxonomy, description, illustration, host, distribution: 535-539]; Bodenh1924 [taxonomy, description, host, distribution: 34]; Bodenh1926 [host, distribution: 42]; Bodenh1928 [host, distribution: 191]; Bodenh1935 [host, distribution: 246]; Bodenh1937 [host, distribution: 217]; Bodenh1943 [taxonomy, host, distribution: 3]; Bodenh1949 [taxonomy, description, host, distribution: 57,62-65]; Bodenh1952 [host, distribution, structure: 339]; Boraty1953 [taxonomy, description, illustration, host, distribution: 464-466]; Borchs1934 [host, distribution: 28]; Borchs1935 [taxonomy: 127]; Borchs1936 [host, distribution: 131]; Borchs1937 [taxonomy, description, illustration, host, distribution: 129-130]; Borchs1937a [taxonomy, description, host, distribution: 45-46]; Borchs1939 [taxonomy, description, host, distribution : 10,30]; Borchs1949d [taxonomy, description, host, distribution: 244]; Borchs1950b [taxonomy, description, illustration, host, distribution: 226,231]; Borchs1966 [catalogue: 324,341]; Bustsh1958 [taxonomy, description, host, distribution: 220,253]; Chumak1961 [host, distribution, biological control: 313-338]; Cocker1896b [distribution: 333]; Cocker1897i [taxonomy, description, host, distribution: 19]; Colvee1882 [taxonomy, description, host, distribution: 14-15]; Comsto1883 [taxonomy, description, host, distribution: 81,85]; Danzig1964 [taxonomy, host, distribution: 652-653]; Danzig1993 [taxonomy, description, illustration, host, distribution: 188-190]; DanzigPe1998 [catalogue: 243]; Dougla1886c [taxonomy: 150]; Drea1990 [biological control: 41-49]; Dunkel1999 [chemistry, life history: 251-276]; Fernal1903b [catalogue: 281]; Ferris1941e [taxonomy: 47,49]; Foldi2001 [distribution: 303-308]; Frauen1868 [taxonomy, description, illustration, host, distribution: 888-889]; FreyFr1995 [taxonomy, chemistry, host, distribution: 777-780]; FreyFr1995a [chemistry, structure: 100]; Gavalo1931 [host, distribution: 8]; Gertss2001 [distribution: 123-130]; Ghauri1962 [taxonomy, description, host, distribution: 103,211]; GolanLaJa2001 [taxonomy, host, distribution: 229-249]; GomezM1937 [taxonomy, description, illustration, host, distribution: 70-71]; GomezM1954 [host, distribution: 120]; GomezM1956b [host, distribution: 482]; GomezM1957 [host, distribution: 40]; GomezM1959 [taxonomy, description, illustration, host, distribution: 152-155]; GomezM1960O [host, distribution: 162]; Green1895 [taxonomy: 230]; Green1928 [taxonomy, description, host, distribution: 8]; GuerriNo2000 [host, distribution, biological control: 157-159]; Hadzib1983 [taxonomy, host, distribution, life history, biological control, economic importance: 241-243]; HardieMi1999 [chemistry, life history]; Jaap1914 [host, distribution: 135-142]; Jansen2001 [host, distribution: 197-206]; Kaussa1955 [host, distribution: 16]; Kaweck1935 [host, distribution: 77]; Killin1932 [host, distribution]; Komosi1974a [host, distribution, life history, ecology: 1-84]; Korone1934 [taxonomy, description, illustration, host, distribution: 12-13]; Koszta1956 [taxonomy, description, illustration, host, distribution: 435-438]; KosztaKo1978 [taxonomy, description, host, distribution: 155,170-171]; Kozar1995b [taxonomy: 51]; KozarHiMa1996 [taxonomy, description: 433-437]; Lagows1998a [host, distribution: 63-71]; Leonar1897 [taxonomy: 285]; Leonar1898a [taxonomy, description, illustration, host, distribution: 64-67]; Leonar1920 [taxonomy, description, illustration, host, distribution: 54-57]; Lindin1907 [taxonomy: 6]; Lindin1909b [host, distribution: 220]; Lindin1912b [taxonomy, description, host, distribution: 278-279]; Lindin1957 [taxonomy: 548]; LongoMaPe1995 [distribution: 128]; Lupo1948 [taxonomy, description, illustration, host, distribution: 175,181-184]; MacGil1921 [taxonomy, description, host, distribution: 409]; Martin1983 [taxonomy, host, distribution: 68]; MillerDa1990 [host, distribution, economic importance: 305]; Morgan1888 [taxonomy, description: 205-208]; Morgan1888b [taxonomy, description: 118-120,351]; Morgan1889a [host, distribution: 351]; Newste1893f [taxonomy, description, illustration, host, distribution: 279-281]; Newste1901b [taxonomy, description, host, distribution: 81,94-98]; Nonell1935 [host, distribution: 281-287]; Pelliz1987 [host, distribution: 123]; PellizCa1991a [taxonomy, host, distribution: 200]; Podsia1995 [life history, host, distribution, biological control: 237-238]; Podsia1997 [taxonomy, description: 193-200]; Podsia1998 [taxonomy: 32]; Podsia2000 [taxonomy, description, illustration, host: 397-404]; Podsia2001 [taxonomy, description: 139-140]; Podsia2002 [taxonomy, structure: 159-164]; PolavaDaMi2000 [taxonomy: 558]; RosenDe1979 [host, distribution, biological control : 377-383,474-483]; Schmut1959 [taxonomy, description, host, distribution: 76,90]; Signor1869 [taxonomy: 868]; Signor1869b [taxonomy, description, host, distribution: 132,135]; Signor1877 [taxonomy: 673,676]; Szulcz1926 [host, distribution: 137-143]; Szulcz1931 [host, distribution: 124-135]; Szulcz1949 [distribution: 219-224]; Terezn1986 [taxonomy, description, illustration, host, distribution: 110-113]; TerGri1962 [taxonomy, description, host, distribution: 149]; ThiemGe1934a [taxonomy, description, host, distribution: 130-158,208-238]; Trjapi1989 [biological control: 312]; Yasar1995a [taxonomy, description, illustration, host, distribution:

124-126]; Zahrad1952 [taxonomy, description, illustration, host, distribution: 124-128]; Zahrad1959b [host, distribution: 60]; Zahrad1972 [taxonomy, description, host, distribution, biological control: 439]; Zahrad1977 [taxonomy, distribution: 121]; Zahrad1990a [host, distribution, description: 653].

Diaspidopus Brimblecombe

Diaspidopus Brimblecombe, 1959: 129. Type species: *Diaspidopus distinctus* Brimblecombe, by monotypy and original designation.

SYSTEMATICS: This genus resembles *Pseudotargionia* and related genera, in possessing a deep thoracic constriction and heavy sclerotization, but differs in presence of unique large and long ducts or megaducts (Brimblecombe, 1959).

GENERAL: Definition and characters by Brimblecombe (1959).

CITATIONS: Borchs1966 [catalogue: 258]; Brimbl1959 [taxonomy, description: 129].

Diaspidopus distinctus Brimblecombe

Diaspidopus distinctus Brimblecombe, 1959: 129. Type data: AUSTRALIA: Queensland, Glenmorgan, on *Homoranthus virgatus*; collected July, 1957. Holotype female. Type depository: Brisbane: Queensland Museum, Queensland, Australia; type no. T5705.

SCALE COVER: Scale of adult female circular, white, embedded in mealy filaments, up to 0.8 mm diameter, surmounted by orange coloured pellicles, loose meal up to 2.0 mm diameter (Brimblecombe, 1959).

HOST PLANTS: **Myrtaceae**: *Homoranthus virgatus* [Brimbl1959].

DISTRIBUTION: **Australasian**: Australia (Queensland [Brimbl1959]).

BIOLOGY: Clustered in axils of small leaves amongst loose cottony filaments (Brimblecombe, 1959).

GENERAL: Description and illustration of adult female by Brimblecombe (1959).

CITATIONS: Borchs1966 [catalogue: 258]; Brimbl1959 [taxonomy, description, illustration, host, distribution: 129-131].

Diastolaspis Brimblecombe

Diastolaspis Brimblecombe, 1959: 132. Type species: *Diastolaspis novata* Brimblecombe, by monotypy and original designation.

SYSTEMATICS: The genus *Diastolaspis* is close to *Pseudotargionia* but differs in having long ducts, a toothed pygidial margin, absence of perispiracular disc pores associated with anterior spiracles, contiguous median lobes and four longitudinal dense bands of ducts on venter of pygidium (Brimblecombe, 1959).

GENERAL: Definition and characters by Brimblecombe (1959).

CITATIONS: Borchs1966 [catalogue: 239]; Brimbl1959 [taxonomy, description: 132]; MorrisMo1966 [taxonomy, catalogue: 60].

Diastolaspis novata Brimblecombe

Diastolaspis novata Brimblecombe, 1959: 132. Type data: AUSTRALIA: Victoria, Lake Hattah, on *Eucalyptus* sp.; collected 1916. Holotype female. Type depository: Brisbane: Queensland Museum, Queensland, Australia.

SCALE COVER: Scale of adult female circular, diameter 1.5 mm; dark fawn in colour with a light fawn margin; first pellicle central, dark olive green; second pellicle covered with a pale suffusion (Brimblecombe, 1959).

HOST PLANTS: **Myrtaceae**: *Eucalyptus* [Brimbl1959].

DISTRIBUTION: **Australasian**: Australia (Victoria [Brimbl1959]).

GENERAL: Description and illustration of adult female by Brimblecombe (1959).

CITATIONS: Borchs1966 [catalogue: 239]; Brimbl1959 [taxonomy, description, illustration, host, distribution: 132-134].

Dichosoma **Brimblecombe**

Dichosoma Brimblecombe, 1957: 271. Type species: *Dichosoma convexa* Brimblecombe, by monotypy and original designation.

SYSTEMATICS: The genus *Dichosoma* resembles *Neomorgania* and *Mimeraspis* in having median lobes united. It differs from both genera in absence of perispiracular disc pores, and in having long spines and a great number of prominent plates on pygidium (Brimblecombe, 1957).

GENERAL: Definition and characters by Brimblecombe (1957).

CITATIONS: Borchs1966 [catalogue: 239]; Brimbl1957 [taxonomy, description: 271-273]; MorrisMo1966 [taxonomy, catalogue: 61].

Dichosoma convexa Brimblecombe

Dichosoma convexa Brimblecombe, 1957: 271. Type data: AUSTRALIA: Queensland, Mt. Isa, on *Eucalyptus microtheca*; collected May 1946. Holotype female. Type depository: Brisbane: Queensland Museum, Queensland, Australia; type no. T5626.

SCALE COVER: Insects numerous on leaves; scale convex, circular, 0.5 mm diameter; fawn in colour and centrally surrounded by dark greenish black exuviae (Brimblecombe, 1957).

HOST PLANTS: **Myrtaceae**: *Eucalyptus microtheca* [Brimbl1957].

DISTRIBUTION: **Australasian**: Australia (Queensland [Brimbl1957]).

GENERAL: Description and illustration of adult female by Brimblecombe (1957).

CITATIONS: Borchs1966 [catalogue: 239]; Brimbl1957 [taxonomy, description, illustration, host, distribution: 271-273].

Diclavaspis **Balachowsky**

Diclavaspis Balachowsky, 1956: 100. Type species: *Aspidiotus ehretiae* Brain, by original designation.

SYSTEMATICS: Balachowsky (1956) separated *Diclavaspis* from *Clavaspis* by the possession of 2 or 3 pygidial lobes, whereas the species of *Clavaspis* have only one pair of lobes. In addition, *Diclavaspis* differs from *Diaspidiotus, Abgrallaspis* and *Aspidaspis* in presence of clavate paraphyses on segment VIII.

GENERAL: Definition and characters by Balachowsky (1956).
KEYS: Balachowsky 1958b: 230 (female) [Aspidiotina of Africa].
CITATIONS: Balach1956 [taxonomy, description: 100]; Balach1958b [taxonomy: 230]; Borchs1966 [catalogue: 273]; MorrisMo1966 [taxonomy, catalogue: 61]; Muntin1969 [taxonomy: 125].

Diclavaspis ehretiae (Brain)

Aspidiotus (*Diaspidiotus*) *ehretiae* Brain, 1918: 127. Type data: SOUTH AFRICA: Cape Province, Cookhouse, on *Ehretia hottentotica*; collected by A. Kelly, 13.iii.1915. Lectotype female, by subsequent designation Munting, 1970a: 37. Type depository: Pretoria: South African National Collection of Insects, South Africa; type no. 189/1.
Affirmaspis ehretiae; MacGillivray, 1921: 449. Change of combination.
Diclavaspis ehretiae; Balachowsky, 1956: 100. Change of combination.
SCALE COVER: Female scale 2 mm diameter, circular; greyish buff to brownish grey, often obscured by fragments of bark; margins are depressed and central portion raised, almost conical; highest portion occupied by covered exuviae; portion of scale covering second exuviae smoother than remainder of secreted scale and pale brown exuviae slightly visible; small greyish or slate-coloured area in centre, with a central white dot surrounded by a distinct shining ring of opaque white; ring particularly prominent on young and male scales; ventral scale white, remaining attached to host plant. Male scale about 1 mm long, and 0.6 mm broad, of similar colour to that of female scale (Brain, 1918).
HOST PLANTS: **Ehretiaceae**: *Ehretia hottentottica* [Brain1918, Balach1956]. **Sapindaceae**: *Allophyllus natalensis* [Almeid1971]. **Solanaceae**: *Solanum campylacanthum* [Balach1956].
NATURAL ENEMIES: HYMENOPTERA **Aphelinidae**: *Archenomus aethiopicus* Annecke [Anneck1963], *Azotus capensis* Howard [AnneckIn1970]. **Encyrtidae**: *Habrolepis* sp. [Prinsl1983].
DISTRIBUTION: **Afrotropical**: Kenya [Balach1956]; Mozambique [Almeid1971]; South Africa [Brain1918, Balach1956, Muntin1970a].
GENERAL: Description and illustration of adult female by Brain (1918) and by Balachowsky (1956).
KEYS: Balachowsky 1956: 100 (female) [Africa]; Brain 1918: 124 (female) [South Africa].
CITATIONS: Almeid1971 [host, distribution: 10]; Anneck1963 [host, distribution, biological control: 346-348]; AnneckIn1970 [host, distribution, biological control: 240]; Balach1956 [taxonomy, description, illustration, host, distribution: 100-102]; Borchs1966 [catalogue: 273]; Brain1918 [taxonomy, description, illustration, host, distribution: 127-128]; Ferris1941e [taxonomy: 43]; MacGil1921 [taxonomy, description, host, distribution: 449]; Mamet1959a [host, distribution: 387]; Muntin1970a [taxonomy: 37]; Prinsl1983 [distribution, biological control: 26].

Diclavaspis socotrana (Lindinger)

Aspidiotus socotranus Lindinger, 1913b: 96. Type data: YEMEN: Sokotra Island, taken from herbarium specimens of *Dracaena cinnabari*; collected 1880. Syntypes, female. Type depository: Hamburg: Zoologisches Institut und Zoologishces Museum, Universität von Hamburg, Germany.

Affirmaspis socotrana; MacGillivray, 1921: 449. Change of combination requiring emendation of species name for agreement in gender.

Targionia soccotrana; Lindinger, 1937: 197. Change of combination.

Targionia soccotrana; Lindinger, 1937: 197. Misspelling of species name.

Diclavaspis socotranus; Balachowsky, 1956: 100. Change of combination.

SCALE COVER: Female scale more or less circular; thin; exuviae red-yellow, central (Lindinger, 1913b).

HOST PLANTS: **Agavaceae**: *Dracaena cinnabari* [Lindin1913b].

DISTRIBUTION: **Afrotropical**: Yemen [Lindin1913b].

GENERAL: Description and illustration of adult female by Lindinger (1913b).

CITATIONS: Balach1956 [taxonomy: 100]; Borchs1966 [catalogue: 273]; Ferris1937c [taxonomy: 50]; Ferris1941e [taxonomy: 48]; Ferris1943a [taxonomy: 86]; Lindin1913b [taxonomy, description, illustration, host, distribution: 96-97]; Lindin1937 [taxonomy: 197]; MacGil1921 [taxonomy, description, host, distribution: 449]; WeidneWa1968 [taxonomy: 173].

Diclavaspis xerophila Munting

Diclavaspis xerophila Munting, 1969: 124. Type data: SOUTH AFRICA: Kalahari Gemsbok National Park, Gemsbokplein, on *Rhigozum trichotomum*; collected 21.viii.1967. Holotype female. Type depository: Pretoria: South African National Collection of Insects, South Africa; type no. 3173/6.

SCALE COVER: Scales of males and females hidden under bark of host plant and not discernible, except as small bumps on bark (Munting, 1969).

HOST PLANTS: **Bignoniaceae**: *Rhigozum trichotomum* [Muntin1969].

DISTRIBUTION: **Afrotropical**: South Africa [Muntin1969].

GENERAL: Description and illustration of adult female by Munting (1969).

CITATIONS: Muntin1969 [taxonomy, description, illustration, host, distribution: 124-125,150].

Duplaspidiotus MacGillivray

Duplaspidiotus MacGillivray, 1921: 394. Type species: *Pseudaonidia clavigera* Cockerell, by original designation.

Lattaspidiotus MacGillivray, 1921: 394. Type species: *Aspidiotus* (*Diaspidiotus*) *tesseratus* de Charmoy, by original designation. Synonymy by Ferris, 1937a: 55.

SYSTEMATICS: *Duplaspidiotus* differs from *Pseudaonidia* in presence of well-developed marginal paraphyses on segments VI, VII, VIII of pygidium (Balachowsky, 1958b).

GENERAL: Definition and characters by Ferris (1938a) and by Balachowsky (1958b).

KEYS: Colon-Ferrer & Medina-Gaud 1998: 28-32 (female) [Genera of Puerto Rico]; Williams & Watson 1988: 19 (female) [Tropical South Pacific]; Beardsley 1966: 502-504 (female) [Federated States of Micronesia]; Balachowsky 1958b: 256 (female) [*Pseudaonidina* of Africa]; Zimmerman 1948: 351 (female) [Hawaii]; Ferris 1942: 25 (female) [North America]; Ferris 1942: 33 (female) [species North America].

CITATIONS: Balach1951 [taxonomy: 680]; Balach1953i [taxonomy: 1512]; Balach1958b [taxonomy, description: 257]; Beards1966 [taxonomy: 517]; Borchs1966 [catalogue: 233]; ColonFMe1998 [taxonomy, description: 54]; DanzigPe1998 [catalogue: 250]; Ferris1937c [taxonomy: 51,54-55,70,80]; Ferris1938a [taxonomy, description: 226]; Kawai1980 [taxonomy: 205]; Lindin1937 [taxonomy: 184,187]; MacGil1921 [taxonomy: 394,452-453]; Mamet1949 [taxonomy: 58]; MorrisMo1966 [taxonomy, catalogue: 64]; Tao1999 [taxonomy: 84]; Young1986 [taxonomy: 199]; Zimmer1948 [taxonomy, description: 351-352].

Duplaspidiotus aldabracus (Green & Laing)
Pseudaonidia aldabraca Green & Laing, 1921: 125. Type data: SEYCHELLES: Aldabra Island, on bark of "Bois d'Amande" [=*Pemphis acidula*]. Syntypes, female. Type depository: London: The Natural History Museum, England, UK.
Duplaspidiotus aldabraca; Borchsenius, 1966: 233. Change of combination.
SCALE COVER: Female scale circular, brownish, partly overlaid with greyish-white secretion; diameter about 2 mm; exuviae subcentral (Green & Laing, 1921).
HOST PLANTS: **Lythraceae**: *Pemphis acidula* [GreenLa1921].
DISTRIBUTION: **Afrotropical**: Seychelles (Aldabra Island [GreenLa1921, Mamet1943a, Borchs1966]).
GENERAL: Description and illustration of adult female by Green & Laing (1921).
CITATIONS: Borchs1966 [catalogue: 233]; GreenLa1921 [taxonomy, description, illustration, host, distribution: 125-126]; Mamet1943a [catalogue: 159].

Duplaspidiotus angularis Ferris
Duplaspidiotus angularis Ferris, 1954: 43. Type data: JAMAICA: on "sapodilla", presumably *Achras zapota*. Syntypes, female. Type depository: Davis: The Bohart Museum of Entomology, University of California, California, USA.
SCALE COVER: Only slide-mounted specimens available for original description (Ferris, 1954).
HOST PLANTS: **Sapotaceae**: *Achras zapota* [Ferris1954].
DISTRIBUTION: **Neotropical**: Jamaica [Ferris1954].
GENERAL: Description and illustration of adult female by Ferris (1954).
CITATIONS: Borchs1966 [catalogue: 233]; Ferris1954 [taxonomy, description, illustration, host, distribution: 43-44].

Duplaspidiotus carptellus Brimblecombe
Duplaspidiotus carptellus Brimblecombe, 1956: 115. Type data: AUSTRALIA: Victoria, Mallee, on *Acacia* sp.; collected by J.E. Dixon, 1918. Holotype female. Type depository: Brisbane: Queensland Museum, Queensland, Australia.

SCALE COVER: Insects sparse on leaves; circular, outer part white, gradually darkening to fawn towards centre; second exuviae black but with a whitish suffusion; first exuviae pale yellow-brown (Brimblecombe, 1956).
HOST PLANTS: **Leguminosae**: *Acacia* [Brimbl1956].
DISTRIBUTION: **Australasian**: Australia (Victoria [Brimbl1956]).
GENERAL: Description and illustration of adult female by Brimblecombe (1956).
CITATIONS: Borchs1966 [catalogue: 233]; Brimbl1956 [taxonomy, description, illustration, host, distribution: 115-116].

Duplaspidiotus claviger (Cockerell)
Pseudaonidia clavigera Cockerell, 1901l: 226. Type data: SOUTH AFRICA: Natal, Durban, on twigs of *Camellia* sp. Holotype. Type depository: Washington: United States National Entomological Collection, U.S. National Museum of Natural History, District of Columbia, USA.
Pseudaonidia iota Green & Laing, 1921: 127. Type data: SEYCHELLES: on *Eugenia caryophyllata*. Syntypes, female. Type depository: London: The Natural History Museum, England, UK. Synonymy by Balachowsky, 1958b: 258.
Pseudaonidea clavigera; MacGillivray, 1921: 394. Misspelling of genus name.
Duplaspidiotus claviger; MacGillivray, 1921: 453. Change of combination requiring emendation of specific epithet for agreement in gender.
Pseudaonidia clavagera; Tao, 1999: 84. Misspelling of species name.
COMMON NAMES: Camellia mining scale [Dekle1965c]; duplascale [Brimbl1962].
SCALE COVER: Female scale 2.5 mm in diameter; moderately convex; blackish; entirely covered by epidermis of twig, except small shining, sublateral orange-ferruginous exuviae (Cockerell, 1901l). Female scale broad oval to circular, about 2.5 mm in longest diameter, completely covered by outer layers of host plant stem, but with brownish or resinous coloured exuviae faintly exposed; observed from below when scale is raised flatly convex, and brownish (Brain, 1919).
HOST PLANTS: **Caprifoliaceae**: *Viburnum* [Takagi1963a, Dekle1965c]. **Euphorbiaceae**: *Acalypha* [Green1937]. **Malvaceae**: *Hibiscus* [Zimmer1948]. **Moraceae**: *Ficus* [WilliaWa1988]. **Myrtaceae**: *Eugenia caryophyllata* [GreenLa1921, Mamet1943a, Balach1958b, Takagi1963a, Borchs1966]. **Oleaceae**: *Ligustrum lucidum* [Dekle1965c]. **Proteaceae**: *Macadamia ternifolia* [Zimmer1948], *Macadamia tetraphylla* [WilliaWa1988]. **Punicaceae**: *Punica granatum* [Tang1984]. **Rubiaceae**: *Gardenia* [Green1937, Beards1966, WilliaWa1988], *Gardenia jasminoides* [Beards1966]. **Rutaceae**: *Citrus* [Zimmer1948], *Ci. limon* [WilliaWa1988], *Ci. sinensis* [WilliaWa1988]. **Santalaceae**: *Santalum freycinetianum* [Zimmer1948]. **Sapindaceae**: *Dodonaea viscosa* [WilliaWa1988]. **Theaceae**: *Camellia* [Marlat1908, Brain1919, Balach1958b, BesheaTiHo1973], *Thea* [Takagi1963a], *Th. chinensis* [Takagi1963a].
DISTRIBUTION: **Afrotropical**: Seychelles [GreenLa1921, Mamet1943a, Balach1958b, Takagi1963a]; South Africa [Marlat1908, Brain1919, Green1937, Balach1958b, Takagi1963a]. **Australasian**: Cook Islands [WilliaWa1988]; Fiji [WilliaWa1988]; Guam [Beards1966]; Hawaiian Islands (Hawaii [Zimmer1948]); Niue [WilliaWa1988]; Papua New Guinea [WilliaWa1988]; Western Samoa

[WilliaWa1988]. **Nearctic**: United States of America (Florida [Dekle1965c, BesheaTiHo1973]). **Oriental**: Sri Lanka [Green1937]. **Palaearctic**: China (People's Republic) [Tang1984]; Japan [Kawai1980] (Kyushu [Takagi1963a]).

ECONOMIC IMPORTANCE: Reported to damage *Camellia* spp. in Florida (Dekle & Kuitert, 1975; Dekle, 1976), and considered a pest of tea plants in India (Nagarkatti & Sankaran, 1990)

GENERAL: Description and illustration of adult female by Green & Laing (1921), Zimmerman (1948), Balachowsky (1958b) and by Tang (1984).

KEYS: Beardsley 1966: 517 (female) [Federated States of Micronesia]; Balachowsky 1958b: 258 (female) [Africa]; Zimmerman 1948: 352 (female) [Hawaii]; Fullaway 1932: 96, 109 (female) [Hawaii]; Marlatt 1908: 135 (female) [World].

CITATIONS: Balach1958b [taxonomy, description, illustration, host, distribution: 258-260]; Beards1966 [host, distribution: 517]; BeardsDaHo1976 [economic importance: 103]; BesheaTiHo1973 [host, distribution: 7]; Borchs1966 [catalogue: 233]; Brain1919 [taxonomy, description, illustration, host, distribution: 207]; Brimbl1962 [host, distribution: 221]; Cocker19011 [taxonomy, description, illustration, host, distribution: 226]; DanzigPe1998 [catalogue: 250-251]; DavidsMi1990 [host, distribution, economic importance: 603-632]; Dekle1962 [host, distribution: 1]; Dekle1965c [taxonomy, description, host, distribution: 116]; Dekle1976 [host, distribution, economic importance: 78]; DekleKu1975 [taxonomy, description, host, distribution, life history, control: 1-4]; FDACSB1982 [host, distribution: 5-11]; Fernal1903b [catalogue: 283]; Ferris1937c [taxonomy, illustration: 51,70]; Ferris1941e [taxonomy: 42]; Fullaw1932 [taxonomy: 96,109]; GreenLa1921 [taxonomy, description, illustration, host, distribution: 125]; GreenLa1923 [taxonomy: 127]; Hunt1939 [host, distribution: 548-566]; Kawai1980 [taxonomy, description, host, distribution: 205]; MacGil1921 [taxonomy, description, host, distribution: 453]; Mamet1936 [taxonomy: 92]; Mamet1943a [catalogue: 159]; Marlat1908 [taxonomy, host, distribution: 135,139]; Maskew1914a [host, distribution: 446-447]; Mead1985 [host, distribution: 1-3]; MillerDa1990 [host, distribution, economic importance: 302]; Muraka1970 [taxonomy, host, distribution: 74]; NagarkSa1990 [host, distribution, economic importance, biological control: 553-542]; Nakaha1982 [host, distribution: 33]; Sassce1923 [host, distribution: 152-158]; Strong1922 [host, distribution: 775-780]; Takagi1963a [taxonomy, host, distribution: 123]; Tang1984 [taxonomy, description, illustration, host, distribution: 12-13]; Tao1999 [taxonomy, host, distribution: 84]; Wester1920 [host, distribution]; WilliaWa1988 [taxonomy, illustration, host, distribution: 107-108]; Young1986 [taxonomy: 199]; Zimmer1948 [taxonomy, description, illustration, host, distribution: 352-353].

Duplaspidiotus fossor (Newstead)

Aspidiotus (*Pseudaonidia*) *fossor* Newstead, 1914: 308. Type data: BARBADOS: Queen's Park, on grape vine. Syntypes, female. Type depository: London: The Natural History Museum, England, UK.

Lattaspidiotus fossor; MacGillivray, 1921: 457. Change of combination.

Duplaspidiotus fossor; Ferris, 1941d: 340. Change of combination.

Aspidiotus fossor; Ferris, 1941e: 43. Change of combination.
Duplaspidiotus fossor; Borchsenius, 1966: 234. Revived combination.
SCALE COVER: Female scale subcircular, diameter 2.5 mm; highly convex; black or piceous; very thick and strong, but invariably covered with a superficial layer of bark, so highly protected and inconspicuous; exuviae (not buried beneath bark) placed centrally, colour red, or dull orange red; second exuviae dull castaneous; under surface black, smooth, but with a thin pearly grey secretion; ventral exuvia thick, greyish, with a well-defined black margin and a subcentral circular white patch, with an underlying blackish secretion showing through in fine dark and somewhat concentric lines towards margin (Newstead, 1914).
HOST PLANTS: **Fagaceae**: *Quercus* [Ferris1941d]. **Vitaceae**: *Vitis* [Lepage1938, Ferris1941d], *Vitis vinifera* [Newste1914].
DISTRIBUTION: **Nearctic**: Mexico (Michoacan [Ferris1941d]). **Neotropical**: Barbados [Newste1914, Ferris1941d]; Brazil (Bahia [Lepage1938, Ferris1941d]); Guyana [Newste1917b]; Puerto Rico & Vieques Island (Puerto Rico [ColonFMe1998]). **Oriental**: Sri Lanka [Ramakr1921a].
GENERAL: Description and illustration of adult female by Newstead (1914), Ferris (1941d) and by Colon-Ferrer & Medina-Gaud (1998).
KEYS: Ferris 1942: 33 (female) [North America].
CITATIONS: Borchs1966 [catalogue: 234]; ClapsWoGo2001a [taxonomy, host, distribution: 17]; ColonFMe1998 [taxonomy, description, illustration, host, distribution: 54-55]; Ferris1941d [taxonomy, description, illustration, host, distribution: 340]; Ferris1941e [taxonomy: 43]; Ferris1942 [taxonomy: 446:33]; FoldiSo1989 [host, distribution: 411]; Lepage1938 [catalogue: 417-418]; MacGil1921 [taxonomy, description, host, distribution: 457]; Newste1914 [taxonomy, description, illustration, host, distribution: 308-309]; Newste1917b [host, distribution: 131]; Ramakr1921a [host, distribution: 359]; Young1986 [taxonomy: 199].

Duplaspidiotus incisus Ferris

Duplaspidiotus incisus Ferris, 1941d: 341. Type data: PANAMA: Chiriqui Province, altitude of about 7000 feet on the Volcan de Chiriqui, on undetermined shrub of the family Ericaceae. Holotype female. Type depository: Davis: The Bohart Museum of Entomology, University of California, California, USA.
SCALE COVER: Scale of female, when free, circular, highly convex; exuviae subcentral, reddish brown; scale dark brown or black. Scale of male not recognized (Ferris, 1941d).
HOST PLANTS: **Ericaceae** [Ferris1941d]. **Leguminosae**: *Inga* [Ferris1941d].
DISTRIBUTION: **Neotropical**: Panama [Ferris1941d].
BIOLOGY: Occurring on the bark, with a strong tendency to concealment under bark flakes and in cracks (Ferris, 1941d).
GENERAL: Description and illustration of adult female by Ferris (1941d).
KEYS: Ferris 1942: 33 (female) [North America].
CITATIONS: Borchs1966 [catalogue: 234]; Ferris1941d [taxonomy, description, illustration, host, distribution: 341]; Ferris1942 [taxonomy: 446:33].

Duplaspidiotus koehleri Lizer y Trelles
Duplaspidiotus köhleri Lizer y Trelles, 1954: 193. Type data: ARGENTINA: Province of Mendoza, in neighborhood of Cacheuta, on bark of *Condalia* probably *microphylla*. Holotype female. Type depository: Buenos Aires: Museo Argentino de Ciencias Naturales, Division Entomologia, Argentina.
Duplaspidiotus koehleri; Borchsenius, 1966: 234. Justified emendation.
SCALE COVER: Female scale oval, about 1.5 mm long; thick, bivalvate; upper and lower scales of same consistency and size; dirty white; exuviae subcentral, covered with white secretion; exuviae golden colour; sometimes outer margin of scales slightly separate. Male scale about 1.5 mm , elongate, sides subparallel, same colour as those of female; exuviae at one extremity, also covered with white secretion (Lizer y Trelles, 1954).
HOST PLANTS: **Rhamnaceae**: *Condalia microphylla* [Lizery1954, ClapsWoGo2001].
DISTRIBUTION: **Neotropical**: Argentina (Mendoza [Lizery1954, ClapsWoGo2001], Rio Negro [ClapsWoGo2001]).
GENERAL: Description and illustration of adult female by Lizer y Trelles (1954).
CITATIONS: Borchs1966 [catalogue: 234]; ClapsWoGo2001 [host, distribution: 243]; Lizery1954 [taxonomy, description, illustration, host, distribution: 192-193].

Duplaspidiotus laciniae (Brain)
Pseudaonidia laciniae Brain, 1919: 207. Type data: SOUTH AFRICA: Pietermaritzburg, on stems of *Acacia melanoxylon*; collected by A. Kelly, 13.vii.1915. Lectotype female, by subsequent designation Munting, 1970a: 39. Type depository: Pretoria: South African National Collection of Insects, South Africa; type no. 219a/1.
Lattaspidiotus laciniae; MacGillivray, 1921: 457. Change of combination.
Duplaspidiotus laciniae; Balachowsky, 1958b: 258. Change of combination.
SCALE COVER: Female scale circular, about 1.5 mm diameter, hemispherical, completely covered by bark of host, through which black exuviae show faintly (Brain, 1919).
HOST PLANTS: **Leguminosae**: *Acacia melanoxylon* [Brain1919, Balach1958b], *Acacia xanthophloea* [Almeid1973]. **Rosaceae**: *Prunus* [Almeid1973].
DISTRIBUTION: **Afrotropical**: Mozambique [Almeid1973]; South Africa [Brain1919].
GENERAL: Description and illustration of adult female by Brain (1919) and by Balachowsky (1958b).
KEYS: Balachowsky 1958b: 258 (female) [Africa]; Brain 1919: 206 (female) [South Africa].
CITATIONS: Almeid1973 [host, distribution: 4]; Balach1958b [taxonomy, description, illustration, host, distribution: 258-262]; Borchs1966 [catalogue: 234]; Brain1919 [taxonomy, description, illustration, host, distribution: 207-208]; MacGil1921 [taxonomy, description, host, distribution: 457]; Mamet1936 [taxonomy: 93]; Muntin1970a [taxonomy: 39].

Duplaspidiotus lacinioides (Mamet)
Pseudaonidia lacinioides Mamet, 1936: 92. Type data: MAURITIUS: Rose Hill, on twigs of *Tamarindus indicus*. Holotype. Type depository: Paris: Muséum national d'Histoire naturelle, France.
Duplaspidiotus lacinioides; Mamet, 1949: 58. Change of combination.
SCALE COVER: Female scale circular to subcircular, 1.0-1.5 mm diameter; hemispherical to conical; black; covered by bark of host, through which black exuviae visible; when seen from below scale capsular with its inner walls shiny black; exuviae subcircular; first exuviae black and sometimes surrounded by a ring of dirty white secretion; second exuviae black obscured by black tissue; ventral scale white and film-like. Male scale unknown (Mamet, 1936).
HOST PLANTS: **Casuarinaceae**: *Casuarina equisetifolia* [Mamet1943a, Borchs1966]. **Leguminosae**: *Haematoxylon campechianum* [Mamet1943a, Mamet1949, Borchs1966], *Tamarindus indicus* [Mamet1936, Mamet1949, Borchs1966]. **Rosaceae**: *Crataegus oxyacantha* [Mamet1943a, Mamet1949].
DISTRIBUTION: **Afrotropical**: Mauritius [Mamet1936, Mamet1943a].
GENERAL: Description and illustration of adult female by Mamet (1936).
CITATIONS: Borchs1966 [catalogue: 234]; Mamet1936 [taxonomy, description, illustration, host, distribution: 92]; Mamet1943a [catalogue: 159]; Mamet1949 [taxonomy, host, distribution: 58].

Duplaspidiotus magnus Brimblecombe
Duplaspidiotus magnus Brimblecombe, 1957: 287. Type data: AUSTRALIA: Queensland, Tugun, on *Melichrus urceolatus*; collected January 1950. Holotype female. Type depository: Brisbane: Queensland Museum, Queensland, Australia; type no. T5658.
SCALE COVER: Insects mostly single near ground level on trunk of host; almost embedded in cork tissue; scale circular to slightly oval, 2-3 mm diameter; dark brown to brownish grey; first exuviae light brown (Brimblecombe, 1957).
HOST PLANTS: **Epacridaceae**: *Melichrus urceolatus* [Brimbl1957].
DISTRIBUTION: **Australasian**: Australia (Queensland [Brimbl1957]).
GENERAL: Description and illustration of adult female by Brimblecombe (1957).
CITATIONS: Borchs1966 [catalogue: 234]; Brimbl1957 [taxonomy, description, illustration, host, distribution: 287-289].

Duplaspidiotus merrilli (Cockerell)
Targionia merrilli Cockerell, 1920: 385. Type data: PHILIPPINES: Manila, on leaves of *Rhizophora mucronata*; collected by E.D. Merrill, September 1918. Holotype female. Type depository: Washington: U.S. National Entomological Collection, U.S. National Museum of Natural History, District of Columbia, USA.
Pseudischnaspis merrilli; Lindinger, 1937: 194. Change of combination.
Duplaspidiotus merrilli; Borchsenius, 1966: 234. Change of combination.
COMMON NAME: mangrove scale [VelasqRi1969].
SCALE COVER: Female scale circular, 3-3.5 mm diameter, flattened, somewhat convex; pale grey; first exuvia near margin, appearing as a small black nipple-like prominence (Cockerell, 1920).

HOST PLANTS: **Rhizophoraceae**: *Rhizophora mucronata* [Cocker1920].
DISTRIBUTION: **Oriental**: Philippines [Cocker1920, VelasqRi1969].
GENERAL: Description and illustration by Cockerell (1920).
CITATIONS: Borchs1966 [catalogue: 234]; Cocker1920 [taxonomy, description, illustration, host, distribution: 385-386]; Ferris1943a [taxonomy: 86]; Lindin1937 [taxonomy: 194]; VelasqRi1969 [host, distribution: 195-209].

Duplaspidiotus niger (Brain)

Pseudaonidia nigra Brain, 1919: 211. Type data: SOUTH AFRICA: Natal, Durban, on leaves of undetermined plant; collected by C.P. van der Merwe, 1.viii.1916. Lectotype female, by subsequent designation Munting, 1970a: 39. Type depository: Pretoria: South African National Collection of Insects, South Africa; type no. 257/1.
Duplaspidiotus niger MacGillivray, 1921: 453. Change of combination requiring emendation of specific epithet for agreement in gender.
SCALE COVER: Whole female scale beneath epidermis of leaf and appears as a black blister (Brain, 1919). Female scale circular or subcircular, diameter 2.5 mm; slightly convex; brown black; exuviae central; scale frequently covered by bark of host plant. Male scale unknown (Balachowsky, 1958b).
HOST PLANTS: **Sterculiaceae**: *Cola natalensis* [Muntin1965b].
DISTRIBUTION: **Afrotropical**: South Africa [Brain1919, Balach1958b, Muntin1965b].
GENERAL: Description and illustration of adult female by Balachowsky (1958b).
KEYS: Brain 1919: 206 (female) [South Africa].
CITATIONS: Balach1958b [taxonomy, description, illustration, host, distribution: 262-264]; Borchs1966 [catalogue: 234]; Brain1919 [taxonomy, description, illustration, host, distribution: 211-212]; MacGil1921 [taxonomy, description, host, distribution: 453]; Muntin1965b [host, distribution: 191]; Muntin1970a [taxonomy: 39].

Duplaspidiotus pavettae (Balachowsky)

Pseudaonidia pavettae Balachowsky, 1953i: 1517. Type data: GUINEA: Bena Plateau, 1000 meters altitude, on *Pavetta corymbosa*. Syntypes, female. Type depository: Paris: Muséum national d'Histoire naturelle, France.
Duplaspidiotus pavettae; Balachowsky, 1958b: 258. Change of combination.
SCALE COVER: Illustration of female and male scale cover by Balachowsky (1953i). Female scale subcircular or oval, 2-2.2 mm; light brown; very flat; exuviae central or slightly eccentric, yellow. Male scale oval, narrow; same colour and structure as that of female; brighter at margin; exuviae placed anteriorly (Balachowsky, 1953i).
HOST PLANTS: **Rubiaceae**: *Pavetta corymbosa* [Balach1953i, Balach1958b].
NATURAL ENEMIES: HYMENOPTERA **Encyrtidae**: *Habrolepis guineensis* Ferriere [AnneckIn1971].
DISTRIBUTION: **Afrotropical**: Guinea [Balach1953i, Balach1958b].
GENERAL: Description and illustration of adult female by Balachowsky (1953i, 1958b).
KEYS: Balachowsky 1958b: 258 (female) [Africa].

CITATIONS: AnneckIn1971 [host, distribution, biological control: 13]; Balach1953i [taxonomy, description, illustration, host, distribution: 1517-1522]; Balach1958b [taxonomy, description, illustration, host, distribution: 264-266]; Borchs1966 [catalogue: 234]; Prinsl1983 [distribution, biological control: 26].

Duplaspidiotus quadriclavatus (Green)

Aspidiotus (*Chrysomphalus*) *quadriclavatus* Green, 1905a: 343. Type data: SRI LANKA: Peradeniya, on *Murraya exotica*. Syntypes, female. Type depository: London: The Natural History Museum, England, UK.
Chrysomphalus quadriclavatus; Sanders, 1906: 15. Change of combination.
Pseudaonidia quadriclavatus; McKenzie, 1939: 54. Change of combination.
Duplaspidiotus quadriclavatus; Borchsenius, 1966: 234. Change of combination.
SCALE COVER: Female scale flat, subcircular, very dark chocolate-brown. Diameter 3 mm. Male scale similar in colour and texture to that of female, but smaller and oblong. Length 2 mm. Breadth about 1 mm (Green, 1905a).
HOST PLANTS: **Rutaceae**: *Murraya exotica* [Green1905a, Sander1906, Ramakr1921a, Green1922, Green1937].
DISTRIBUTION: **Oriental**: Sri Lanka [Green1905a, Sander1906, Ramakr1921a, Green1922, Green1937].
GENERAL: Description and illustration of adult female by Green (1905a).
CITATIONS: Borchs1966 [catalogue: 234]; Cocker1920 [taxonomy: 386]; DEDAC1923 [host, distribution]; Ferris1941e [taxonomy: 47]; Green1905a [taxonomy, description, illustration, host, distribution: 343-344]; Green1922 [host, distribution: 463]; Green1937 [host, distribution: 332]; MacGil1921 [taxonomy, description, host, distribution: 414]; McKenz1939 [taxonomy: 54]; Ramakr1921a [host, distribution: 357]; Sander1906 [taxonomy: 15].

Duplaspidiotus spinosus Brimblecombe

Duplaspidiotus spinosus Brimblecombe, 1956: 113. Type data: AUSTRALIA: Queensland, Chincilla, on *Geijera parviflora*; collected by J. Mann, April 1953. Holotype female. Type depository: Brisbane: Queensland Museum, Queensland, Australia; type no. T5524.
SCALE COVER: Female scale 1.5 mm diameter, mostly greyish white, slightly darkening near centre; second exuviae black with a thin pale greyish suffusion; first exuviae yellowish brown (Brimblecombe, 1956).
HOST PLANTS: **Rutaceae**: *Geijera parviflora* [Brimbl1956].
DISTRIBUTION: **Australasian**: Australia (Queensland [Brimbl1956]).
BIOLOGY: Insects single and sparse on twigs (Brimblecombe, 1956).
GENERAL: Description and illustration of adult female by Brimblecombe (1956).
CITATIONS: Borchs1966 [catalogue: 234]; Brimbl1956 [taxonomy, description, illustration, host, distribution: 113-115].

Duplaspidiotus tesseratus (Grandpré & Charmoy)

Aspidiotus (*Diaspidiotus*) *tesseratus* Grandpré & Charmoy, 1899: 23. Type data: MAURITIUS: on grapevine. Syntypes, female. Notes: Type-material lost (Mamet, 1941).

Aspidiotus tesseratus; Cockerell, 1899n: 24. Change of combination.

Pseudaonidia tesserata; Fernald, 1903b: 284. Notes: Incorrect citation of "de Charm." as author.

Pseudaonidia tesserata; Fernald, 1903b: 284. Change of combination requiring emendation of species name for agreement in gender.

Pseudaonidia oreodoxae Rutherford, 1914: 260. Type data: SRI LANKA: Paradeniya, on stem of Cabbage Palm, *Oreodoxa oleracea*, Royal Palm, *Acalypha* sp. and on *Broussonetia papyrifera*. Syntypes, female. Type depository: London: The Natural History Museum, England, UK. Synonymy by Borchsenius, 1966: 235.

Pseudaonidia tesserata; Brain, 1919: 206. Notes: Incorrect citation of "d'Emmerez" as author.

Lattaspidiotus oreodoxae; MacGillivray, 1921: 458. Change of combination.

Lattaspidiotus tesserata; MacGillivray, 1921: 458. Change of combination.

Pseudaonidia subtesserata Green & Laing, 1923: 127. Type data: JAMAICA: Hill Crest, on gungo (Congo) peas. Syntypes, female. Type depository: London: The Natural History Museum, England, UK. Synonymy by Borchsenius, 1966: 235.

Duplaspidiotus tesseratus; Ferris, 1938a: 227. Change of combination.

Duplaspidiotus subtesseratus; Ferris, 1942: 445. Change of combination requiring emendation of species name for agreement in gender.

Pseudaonidia (*Duplaspidiotus*) *tesserata*; Merrill, 1953: 75. Change of combination.

Duplaspidiotus tesseratus; Balachowsky, 1958b: 266. Revived combination.

SCALE COVER: Female scale circular; slightly convex anteriorly, reddish; rest of scale flat, uniformly brown reddish; ventral scale white, remains attached to the host plant twigs (Grandpré & Charmoy, 1899).

HOST PLANTS: **Apocynaceae**: *Allamanda* [Mamet1941, Mamet1949, Borchs1966]. **Dipterocarpaceae**: *Monotes glaber* [Hall1929a, Green1937, Balach1958b]. **Euphorbiaceae**: *Acalypha* [Ruther1914, DeLott1967a], *Acalypha grandis* [Mamet1943a, Mamet1949, Borchs1966]. **Leguminosae**: *Cassia fistula* [Newste1914], *Inga laurina* [Ferris1938a], *Leucaena glauca* [Houser1918, MerrilCh1923, Green1937, Mamet1943a, Mamet1949, Borchs1966]. **Malvaceae**: *Hibiscus* [Zimmer1948, Dekle1965c], *Malvaviscus* [Cocker1899n, Ferris1938a]. **Meliaceae**: *Swietenia mahagoni* [GomezM1941]. **Moraceae**: *Broussonetia* [Green1937], *Broussonetia papyrifera* [Ruther1914], *Ficus* [Ferris1938a]. **Palmae**: *Oreodoxa* [Ramakr1921a, Green1922], *Oreodoxa oleracea* [Ruther1914], *Oreodoxa regia* [Green1937, Balach1958b]. **Passifloraceae**: *Passiflora* [Balach1958b]. **Rosaceae**: *Prunus* [MerrilCh1923], *Rosa* [Ferris1938a]. **Theaceae**: *Camellia* [Dekle1965c]. **Vitaceae**: *Vitis* [GrandpCh1899, Green1937, Lepage1938, Mamet1941, Borchs1966], *Vitis vinifera* [GrandpCh1899, Mamet1943a, Mamet1949, Borchs1966].

DISTRIBUTION: **Afrotropical**: Kenya [DeLott1967a]; Mauritius [GrandpCh1899, Mamet1941, Mamet1943a, Mamet1949, Mamet1953a, Borchs1966]; South Africa [Brain1919, Green1937]; Zimbabwe [Hall1929a, Balach1958b]. **Australasian**: Guam [Beards1966]; Hawaiian Islands (Hawaii [Zimmer1948]). **Nearctic**: Mexico [Ferris1938a] (Veracruz [Cocker1899n, Ferris1938a]); United States of America (Florida [MerrilCh1923, Merril1953, Dekle1965c]). **Neotropical**: Antigua and Barbados (Antigua [Green1907]); Barbados [Newste1914]; Brazil (Rio Grande do

Sul [BertelBa1966], Rio de Janeiro [Lepage1938]); Colombia [Kondo2001]; Cuba [Houser1918]; Dominican Republic [GomezM1941]; Jamaica [Newste1917b, GreenLa1923, Ferris1942]; Puerto Rico & Vieques Island (Puerto Rico [Ferris1938a, Martor1976]); U.S. Virgin Islands [Nakaha1983]. **Oriental**: Indonesia (Java [Ferris1938a]); Sri Lanka [Ruther1914, Ramakr1921a, Green1922].

BIOLOGY: Occurring on bark (Ferris, 1938a).

ECONOMIC IMPORTANCE: Noted as a pest of grapevine in Central and South America Schmutterer et al. (1957).

GENERAL: Description and illustration of adult female by Green & Laing (1923), Ferris (1938a), Zimmerman (1948), Balachowsky (1958b) and by Colon-Ferrer & Medina-Gaud (1998).

KEYS: Beardsley 1966: 517 (female) [Federated States of Micronesia]; Balachowsky 1958b: 258 (female) [Africa]; Zimmerman 1948: 352 (female) [Hawaii]; Ferris 1942: 33 (female) [North America]; Brain 1919: 206 (female) [South Africa]; Marlatt 1908: 135 (female) [World].

CITATIONS: Balach1957c [host, distribution: 200]; Balach1958b [taxonomy, description, illustration, host, distribution: 266-268]; Balach1959a [host, distribution: 362]; Beards1966 [host, distribution: 517]; BeardsDaHo1976 [economic importance: 106]; BertelBa1966 [host, distribution: 17-46]; BiezanSe1940 [host, distribution: 67-68]; Borchs1966 [catalogue: 234-235]; Brain1919 [taxonomy, description, illustration, host, distribution: 206-207]; BurgerUl1990 [economic importance: 313-327]; ClapsWoGo2001a [taxonomy, host, distribution: 17]; Cocker1899n [taxonomy, description, illustration, host, distribution: 24]; Cocker1899r [taxonomy: 900]; ColonFMe1998 [taxonomy, description, illustration, host, distribution: 55-56]; CostaL1942 [taxonomy, host, distribution: 277]; Dekle1965c [taxonomy, description, host, distribution: 60]; Dekle1976 [taxonomy, description, host, distribution, economic importance: 79]; DeLott1967a [host, distribution: 114]; Fernal1903b [catalogue: 284]; Ferris1937c [taxonomy, illustration: 51,80]; Ferris1938a [taxonomy, description, illustration, host, distribution: 227]; Ferris1941e [taxonomy: 49]; Ferris1942 [taxonomy: 446:33]; FoldiSo1989 [host, distribution: 411]; GomezM1941 [host, distribution: 139]; Gowdey1921 [host, distribution: 30]; GrandpCh1899 [taxonomy, description, illustration, host, distribution: 23]; Green1907 [host, distribution: 203]; Green1922 [host, distribution: 462]; Green1937 [host, distribution: 334]; GreenLa1921 [taxonomy: 126]; GreenLa1923 [taxonomy, description, illustration, host, distribution: 127]; Hall1929a [taxonomy, host, distribution: 359]; Houser1918 [host, distribution: 167]; Hunt1939 [host, distribution: 548-566]; Kondo2001 [taxonomy, host, distribution: 44]; Lepage1938 [catalogue: 418]; Lepesm1947 [host, distribution: 196]; MacGil1921 [taxonomy, description, host, distribution: 457-458]; Mamet1941 [taxonomy, description, illustration, host, distribution: 28]; Mamet1943a [catalogue: 159]; Mamet1949 [catalogue: 58-59]; Mamet1953a [taxonomy: 152]; Marlat1908 [taxonomy, host, distribution: 139]; Martor1976 [host, distribution: 14,224]; Merril1953 [taxonomy, description, host, distribution: 75]; MerrilCh1923 [taxonomy, description, host, distribution, economic importance: 249-250]; MillerDa1990 [host, distribution, economic importance: 302]; Montgo1921 [host, distribution: 41-55]; Nakaha1982 [host, distribution: 34];

Nakaha1983 [host, distribution: 11]; Newell1923 [host, distribution: 263-266]; Newste1914 [host, distribution: 309]; Newste1917b [host, distribution: 132]; Ramakr1921a [host, distribution: 359]; RaoCh1950 [taxonomy: 19,27]; Ruther1914 [taxonomy, description, host, distribution: 260-261]; Sassce1923 [host, distribution: 152-158]; SchmutKlLu1957 [host, distribution, economic importance: 494]; Singh1964 [host, distribution, economic importance: 213]; Zimmer1948 [taxonomy, description, illustration, host, distribution: 352-354].

Duplaspidiotus tripartitus (Mamet)
Pseudaonidia tripartita Mamet, 1936: 90. Type data: MAURITIUS: Rose Hill, on *Camellia* twigs. Holotype. Type depository: Paris: Muséum national d'Histoire naturelle, France.
Duplaspidiotus tripartitus; Mamet, 1949: 59. Change of combination requiring emendation of species name for agreement in gender.
SCALE COVER: Female scale oval to circular, 2-2.5 mm in diameter; dark brown; slightly convex; covered by the outer layers of host bark; exuviae brownish; first exuviae darker; ventral vellum represented as a whitish scar on the surface of the bark (Mamet, 1936).
HOST PLANTS: **Theaceae**: *Camellia* [Mamet1936, Mamet1943a, Mamet1949, Borchs1966].
DISTRIBUTION: **Afrotropical**: Mauritius [Mamet1936, Mamet1943a, Mamet1949, Borchs1966].
GENERAL: Description and illustration of adult female by Mamet (1936).
CITATIONS: Borchs1966 [catalogue: 235]; Mamet1936 [taxonomy, description, illustration, host, distribution: 90]; Mamet1943a [catalogue: 159]; Mamet1949 [catalogue: 59].

Duplaspidiotus xishuangensis Young
Duplaspidiotus xishuangensis Young, 1986: 199. Type data: CHINA: Yunnan Province, Xishuangbanna, on branches of undetermined tree; collected April 1957. Syntypes, female. Type depository: Beijing: Institute of Entomology, Academy of Sciences, China.
Duplaspidiotus xishuaensis; Tao, 1999: 84. Misspelling of species name.
SCALE COVER: Young (1986) did not describe scale cover.
DISTRIBUTION: **Oriental**: China (People's Republic) (Yunnan [Young1986]).
GENERAL: Description and illustration of adult female by Young (1986).
CITATIONS: Tao1999 [taxonomy, host, distribution: 84]; Young1986 [taxonomy, description, illustration, host, distribution: 199-200,207].

Dynaspidiotus Thiem & Gerneck

Dynaspidiotus Thiem & Gerneck, 1934a: 231. Type species: *Aspidiotus britannicus* Newstead, by original designation.
Nuculaspis Ferris, 1938a: 250. Type species: *Aspidiotus californicus* Coleman, by original designation. Synonymy by Danzig, 1993: 149.

Ephedraspis Borchsenius, 1949c: 738. Type species: *Aspidiotus ephedrarum* Lindinger, by monotypy and original designation. Synonymy by Danzig, 1993: 149.

Dinaspidiotus; Gómez-Menor Ortega, 1957: 45. Misspelling of genus name.

Tsugaspidiotus Takahashi & Takagi, 1957: 52. Type species: *Aspidiotus* (*Diaspidiotus*) *tsugae* Marlatt, by original designation. Synonymy by Danzig, 1993: 149.

SYSTEMATICS: Danzig (1993) regarded *Tsugaspidiotus* Takahashi & Takagi (1957) as a subjective synonym of *Dynaspidiotus* Thiem & Gerneck. Takahashi & Takagi (1957) indicated the affinity of both genera, but distinguished *Tsugaspidiotus* from *Dynaspidiotus* in the former having large anus, no plates outside third lobe, and in presence of distinct plates.

GENERAL: Definition and characters by Ferris (1938a), Borchsenius (1950b), Zahradník (1952), Balachowsky (1948b, 1958b), Takahashi & Takagi (1957), Bazarov and Shmelev (1971), Kosztarab & Kozár (1978), Danzig (1993) and by Yaşar (1995).

KEYS: Gill 1997: 24-26 (female) [Genera of California]; Kosztarab 1996: 406-407 (female) [Northeastern North America]; Danzig 1993: 150 (female) [species Europe]; Zahradník 1990b: 74 (female) [Czech Republic]; Tereznikova 1986: 83 (female) [Ukraine]; Tereznikova 1986: 83 (female) [Ukraine]; Danzig 1980b: 296 (female) [Far East of USSR]; Kosztarab & Kozár 1978: 144-147 (female) [Hungary]; Paik 1978: 392 (female) [species South Korea]; Bazarov & Shmelev 1971: 186 (female) [Central Asia]; Komosinska 1969: 50 (female) [*Abgrallaspis* group]; Danzig 1964: 645 (female) [Europe]; Danzig 1964: 645 (female) [Europe]; Zahradník 1959a: 548 (female) [Czech Republic]; Balachowsky 1958b: 228 (female) [*Aspidiotina* of Africa]; Ezzat 1958: 237-239 (female) [Egypt]; Takahashi & Takagi 1957: 102 (female) [species Japan]; McKenzie 1956: 23 (female) [U.S.A.: California]; Balachowsky 1951: 599 (female) [Mediterranean]; Borchsenius 1950b: 167 (female) [USSR]; Borchsenius 1950b: 167 (female) [USSR]; Ferris 1942: 28 (female) [North America]; Ferris 1942: 33 (female) [species North America]; Ferris 1942: 27 (female) [North America]; Ferris 1942: 39 (female) [species North America].

CITATIONS: Balach1948b [taxonomy, description: 325-326]; Balach1950b [taxonomy: 545]; Balach1951 [taxonomy: 599]; Balach1956 [taxonomy: 15]; Balach1958b [taxonomy, description: 162]; BazaroSh1971 [taxonomy, description: 188-189]; BlayGo1993 [taxonomy, description: 445,450,456]; Borchs1949c [taxonomy, description: 738]; Borchs1949d [taxonomy, description: 236]; Borchs1950b [taxonomy, description: 167,215,216,218,220]; Borchs1966 [catalogue: 273,274,281,316]; Bustsh1958 [taxonomy: 223]; Chou1985 [taxonomy, description: 281-282]; Danzig1964 [taxonomy: 651]; DanzigPe1998 [catalogue : 251,364]; Ezzat1958 [taxonomy: 238]; Ferris1937c [taxonomy: 51,53,54,71]; Ferris1938a [taxonomy, description: 228,250]; Ferris1938b [taxonomy, description: 65,70,75]; Ferris1942 [taxonomy: 446:27]; Gill1997 [taxonomy: 135]; Gill1997 [taxonomy: 204]; GomezM1957 [taxonomy: 45]; Hadzib1983 [taxonomy: 222]; Kaussa1954 [taxonomy: 80]; Kawai1980 [taxonomy: 223]; Koszta1996 [taxonomy, description: 541]; KosztaKo1978 [taxonomy, description, host, distribution: 156-157]; Kozar1990f [distribution: 143]; Lupo1948 [taxonomy, description: 203];

McKenz1956 [taxonomy: 23]; MorrisMo1966 [taxonomy, catalogue: 65,67,136,199]; Schmut1959 [taxonomy, description: 48,59]; TakahaTa1957 [taxonomy, description: 102]; TangHaSh1991 [taxonomy: 459-460]; Tao1999 [taxonomy: 85]; ThiemGe1934a [taxonomy, description: 231]; Yasar1995a [taxonomy, description: 75]; Zahrad1952 [taxonomy, description: 152].

Dynaspidiotus abieticola (Koroneos)

Aspidiotus abieticola Koroneos, 1934: 9. Type data: GREECE: Ano Lekhonia and Pelion, on *Abies cephalonica*. Syntypes, female. Notes: Depository of type material unknown (S. Katsoyannos, Athens, Greece, in letter to Yair Ben-Dov, 2000).
Dynaspidiotus abieticola; Balachowsky, 1948b: 339. Change of combination.
Nuculaspis abieticola; Borchsenius, 1966: 274. Change of combination.
Dynaspidiotus abieticola; Danzig & Pellizzari, 1998: 252. Revived combination.
SCALE COVER: Female scale subrectangular, margins parallel, contracted laterally, truncated at two apices, 2-2.2 mm; sometimes subcircular; light brown; exuviae central, yellow gold. Male scale unknown (Balachowsky, 1948b).
HOST PLANTS: **Pinaceae**: *Abies cephalonica* [Korone1934, Kaussa1954, Zahrad1972], *Cedrus libanotica libani* [Balach1954b, Kaussa1954].
DISTRIBUTION: **Palaearctic**: Greece [Korone1934, Zahrad1972]; Iran [Kaussa1954]; Lebanon [Balach1954b]; Turkey [Ulgent1996].
GENERAL: Description and illustration of adult female by Balachowsky (1948b) and by Yaşar (1995a).
KEYS: Balachowsky 1948b: 326 (female) [Mediterranean].
CITATIONS: Balach1948b [taxonomy, description, illustration, host, distribution: 339-342]; Balach1954b [host, distribution: 110]; Bodenh1949 [taxonomy: 47]; Bodenh1952 [taxonomy: 334]; Borchs1966 [catalogue: 274]; DanzigKo1991 [distribution: 1-15]; DanzigPe1998 [catalogue: 252]; Ferris1941e [taxonomy: 40]; Kaussa1954 [host, distribution: 80]; Korone1934 [taxonomy, description, illustration, host, distribution: 9]; Smetni1991 [chemistry: 92-129]; Ulgent1996 [host, distribution: 541-548]; Yasar1995a [taxonomy, description, illustration, host, distribution: 97-99]; Zahrad1972 [host, distribution: 433].

Dynaspidiotus abietis (Schrank)

Coccus abietis Schrank, 1776: 48. Type data: AUSTRIA: on *Pinus*. Syntypes, female. Notes: Depository of type material unknown.
Coccus arborum Schrank, 1781: 295. Unjustified replacement name for *Coccus abietis* Schrank, 1776; discovered by Danzig, 1993: 152.
Coccus pineti Schrank, 1801: 146. Unjustified replacement name for *Coccus abietis* Schrank, 1776; discovered by Danzig, 1993: 153.
Coccus flavus Hartig, 1839: 642. [Publication not seen]. Synonymy by Leonardi, 1908a: 187. Notes: Depository of type material unknown.
Aspidiotus flavus; Signoret, 1870: 108. Change of combination.
Aspidiotus (*Aspidiotus*) *abietis*; Cockerell, 1897i: 18. Change of combination.
Aspidiotus (*Evaspidiotus*) *abietis*; Leonardi, 1898c: 67. Change of combination.
Aspidiotus abietis; Koroneos, 1934: 9. Notes: Incorrect citation of "(Schr.) Loew" as author.

Aspidiotus (*Dynaspidiotus*) *abietis*; Thiem & Gerneck, 1934a: 131. Change of combination.

Nuculaspis abietis; Lupo, 1948: 204. Change of combination.

Dynaspidiotus abietis; Balachowsky, 1948b: 336. Change of combination.

Dynaspidiotus abietis; Danzig, 1993: 152. Revived combination.

COMMON NAMES: hemlock scale [Wilson1917, MerrilCh1923]; pihtovaya shitovka [Borchs1936].

SCALE COVER: Female scale suboval, 1.8-2.1 mm; convex; contracted laterally, and truncated at apices; grey, lighter at margins; exuviae central, more or less light brown, sometimes yellow gold. Male scale same structure as that of female, but more elongate and paler; 1.4-1.6 mm (Balachowsky, 1948b).

HOST PLANTS: **Aceraceae**: *Acer* [Bodenh1949], *Acer rubrum* [Leonar1898c]. **Cupressaceae**: *Juniperus* [Abai1995], *Ju. communis* [Balach1933e], *Ju. communis nana* [Balach1948b]. **Pinaceae**: *Abies* [Borchs1934, Bodenh1949, Kaussa1954], *Abies alba* [Zahrad1952], *Ab. bornmulleriana* [Bodenh1952], *Ab. cephalonica* [Korone1934], *Ab. nordmanianna* [Balach1948b], *Ab. pectinata* [Balach1932d], *Ab. sylvestris* [Leonar1898c], *Picea* [Borchs1936, Kaussa1954], *Pic. abies* [Reyne1957], *Pic. excelsa* [Balach1937c, Zahrad1952, Danzig1959], *Pic. omorica* [Bachma1953], *Pic. pungens* [Zahrad1952], *Pinus* [Schran1776, Schran1801, Bodenh1949, Kaussa1954], *Pin. brutia* [Bodenh1952], *Pin. halepensis* [Balach1928a], *Pin. laricio* [Bachma1953], *Pin. mitis* [Targio1884], *Pin. montana* [Balach1948b], *Pin. mugo* [Abai1995], *Pin. nigra* [Bodenh1952, Abai1995], *Pin. rigida* [Targio1884], *Pin. silvestris* [Newste1894, Leonar1908a, Leonar1920, Balach1932d, Balach1935b, Zahrad1952, Martin1983, Abai1995], *Pin. silvestris hammata* [Zahrad1952], *Pseudotsuga taxifolia* [Abai1995]. **Rosaceae**: *Pyrus sylvestris* [Leonar1898c].

NATURAL ENEMIES: HYMENOPTERA **Aphelinidae**: *Aphytis mytilaspidis* Le Baron [Schmut1951], *Prospaltella aspidioticola* Mercet [GomezM1946, Balach1948b], *Prospaltella aurantii* Howard [Schmut1951].

DISTRIBUTION: **Nearctic**: United States of America (Florida [Wilson1917, MerrilCh1923], Georgia [Fernal1903b], Maine [Fernal1903b], Massachusetts [Fernal1903b], Mississippi [Herric1911], New Jersey [Fernal1903b], New York [Fernal1903b]). **Palaearctic**: Algeria [Balach1928a]; Austria [Schran1776, Schran1801]; Corsica [Balach1933e]; Czech Republic [Newste1894, Zahrad1952, Zahrad1977]; France [Balach1932d, Balach1937c]; Georgia (Abkhaz ASSR [Borchs1934, Borchs1936]); Germany (United) [Lindin1909b]; Greece [Korone1934]; Iran [Kaussa1954, Abai1995]; Italy [Leonar1908a, Leonar1920, LongoMaPe1995]; Netherlands [Reyne1957]; Poland [LagowsKo1996, Koteja2000a]; Russia (Karachay-Cherkessia AR [Danzig1985], St. Petersberg (= Leningrad) Oblast [Danzig1959]); Slovak Republic [Zahrad1952]; Spain [Balach1935b, GomezM1937, Martin1983, BlayGo1993]; Sweden [Gertss2000, Gertss2001]; Turkey [Bodenh1949, Bodenh1952]; Yugoslavia [Bachma1953].

ECONOMIC IMPORTANCE: Caused injury to conifers in Central Europe (Schmutterer et al., 1957; Zahradník, 1990).

GENERAL: Description and illustration of adult female by Balachowsky (1937a, 1948b), Zahradník (1952), Danzig (1993) and by Yaşar (1995a).

KEYS: Danzig 1993: 150 (female) [Europe]; Kosztarab & Kozár 1978: 157 (female) [Hungary]; Reyne 1957: 31 (female) [Netherlands]; Balachowsky 1948b: 327 (female) [Mediterranean]; Britton 1923: 371 (female) [U.S.A.: Connecticut]; Leonardi 1920: 29-30 (female) [Italy]; Comstock 1883: 55-57 (female) [North America].

CITATIONS: Abai1995 [host, distribution: 26,108-109]; Bachma1953 [host, distribution: 177]; Balach1928a [host, distribution: 138]; Balach1932d [taxonomy, host, distribution: XLVI]; Balach1934a [host, distribution: 71]; Balach1935b [host, distribution: 257]; Balach1937c [host, distribution: 2]; Balach1948b [taxonomy, description, illustration, host, distribution, biological control: 336-339]; Barbey1925 [host, distribution, taxonomy: 3]; BlayGo1993 [taxonomy, description, illustration, host, distribution: 457-461]; Bodenh1949 [taxonomy, description, illustration, host, distribution: 46-48]; Bodenh1952 [taxonomy, host, distribution, structure: 333-336]; Borchs1934 [host, distribution: 28]; Borchs1935 [taxonomy, description, host, distribution: 23]; Borchs1936 [host, distribution: 131]; Borchs1937 [taxonomy, description, illustration, host, distribution: 126-127]; Borchs1937a [taxonomy: 35,37]; Borchs1939 [taxonomy, description, host, distribution: 9,23]; Borchs1950b [taxonomy, description, host, distribution: 215-216]; Borchs1966 [catalogue: 274-275]; Britto1923 [taxonomy, description, host, distribution: 371,372]; Britto1924 [host, distribution: 387-404]; Britto1938 [host, distribution: 137]; Brown1916 [host, distribution, economic importance: 414-422]; Chumak1961 [host, distribution, biological control: 313-338]; Cocker1894l [taxonomy, host, distribution: 190]; Cocker1896b [distribution: 333]; Cocker1897i [taxonomy, description, host, distribution: 10,18]; Comsto1883 [taxonomy, description, illustration, host, distribution: 57-58]; Danzig1959 [host, distribution: 450]; Danzig1964 [taxonomy, host, distribution: 651]; Danzig1985 [distribution: 112]; Danzig1993 [taxonomy, description, illustration, host, distribution, economic importance: 152-154]; DanzigPe1998 [catalogue: 252]; Felt1924 [host, distribution, economic importance, life history, control]; Fernal1903b [catalogue: 251]; Ferris1941e [taxonomy: 40,46]; Fjeldd1996 [host, distribution: 4-24]; Foldi2001 [distribution: 303-308]; Garcia1930 [host, distribution, biological control]; Garcia1931a [host, distribution, biological control: 659-669]; Gavalo1931 [host, distribution: 7]; Gertss2000 [host, distribution: 152]; Gertss2001 [distribution: 123-130]; Goidan1956 [ecology, structure: 207-224]; Goidan1958 [structure: 387]; GolanLa1998 [host, distribution: 4-6]; GolanLaJa2001 [taxonomy, host, distribution: 229-249]; GomezM1937 [taxonomy, description, illustration, host, distribution: 55-57]; GomezM1946 [host, distribution: 61]; GomezM1958a [host, distribution: 7]; Green1930 [taxonomy, description, distribution: 16]; Hartig1839 [taxonomy, description, host, distribution: 187]; Herric1911 [taxonomy, description, illustration, host, distribution: 9,13,46]; Jaap1914 [host, distribution: 135-142]; John1930 [host, distribution: 3-7]; Kaussa1954 [host, distribution: 80]; Kaweck1935 [host, distribution: 75]; Komosi1974a [host, distribution, life history, ecology: 1-84]; Korone1934 [taxonomy, description, illustration, host, distribution: 9-10]; KosztaKo1978 [taxonomy, description, host, distribution: 157]; Koteja2000a [distribution: 172]; LagowsKo1996 [host, distribution: 32]; Leonar1897 [taxonomy: 285]; Leonar1898c [taxonomy, description, illustration, host, distribution: 67-69]; Leonar1908a [taxonomy, host, distribution: 187]; Leonar1920 [taxonomy, description, illustration, host, distribution: 48-50]; Lindin1907 [taxonomy: 6]; Lindin1909b [host, distribution: 149]; Lindin1912b [taxonomy, description, host, distribution: 48,250,256,328]; Lindin1957 [taxonomy: 550]; Lobdel1937 [taxonomy: 78]; LongoMaPe1995 [distribution: 128]; Low1882 [taxonomy, description, host, distribution: 270-273]; Lupo1948 [taxonomy, description, illustration, host, distribution: 204-208]; MacGil1921 [taxonomy, description, host, distribution: 399]; Martin1983 [taxonomy, host, distribution: 65]; MerrilCh1923 [taxonomy, description, host, distribution: 197]; MillerDa1990 [host, distribution, economic importance: 304]; Nakaha1982 [host, distribution: 60]; Newste1894 [taxonomy, description, illustration, host, distribution: 179-180]; Reyne1957 [taxonomy, host, distribution: 25,31]; Schmut1951 [host, distribution, life history, biological control: 124]; Schmut1959 [taxonomy, description, host, distribution: 60]; SchmutKlLu1957 [host, distribution, economic importance: 478]; Schran1781 [taxonomy, description, host, distribution: 295]; Schran1801 [taxonomy, description, life history, host, distribution: 146]; Signor1870 [taxonomy: 108]; Szulcz1926 [host, distribution: 137-143]; Szulcz1931 [host, distribution: 124-135]; Szulcz1949 [distribution: 219-224]; Targio1884 [taxonomy, host, distribution: 384]; Terezn1986 [taxonomy, description, illustration, host, distribution: 117-119]; ThiemGe1934 [taxonomy, description, host, distribution: 537]; ThiemGe1934a [taxonomy, description, host, distribution: 130-158,208-238]; Ulgent1996 [host, distribution: 541-548]; Wilson1917 [host, distribution: 31-32]; UlgentTo2000 [distribution, biological control: 106-110]; Wolff1911 [taxonomy, description, host, distribution: 80]; Yasar1995a [taxonomy, description, illustration, host, distribution: 99-101]; Yasnos1994 [host, distribution, biological control: 317-333]; Zahrad1952 [taxonomy, description, illustration, host, distribution: 152-157]; Zahrad1959b [host, distribution: 60]; Zahrad1977 [taxonomy, distribution: 120]; Zahrad1990 [host, distribution, description: 641].

Dynaspidiotus amygdalicola (Borchsenius)
Diaspidiotus amygdalicola Borchsenius, 1952: 261. Type data: IRAN: Horosh-Abad, Kuh-i-Sefid, on *Amygdalus* sp. Syntypes, female. Type depository: St. Petersburg: (= Leningrad) Zoological Museum, Academy of Science, Russia.
Dynaspidiotus (*Diaspidiotus*) *amygdalicola*; Kaussari, 1954: 80. Change of combination.
Dynaspidiotus amygdalicola; Borchsenius, 1966: 281. Change of combination.
Abgrallaspis amygdalicola; Komosinska, 1969: 53. Change of combination.
Dynaspidiotus amygdalicola; Danzig & Pellizzari, 1999: 252. Revived combination.
SCALE COVER: Female scale almost circular, up to 2 mm in diameter; flat, brownish-grey or light brown, with brighter marginal part; larval exuviae central, first brown, second yellow (Borchsenius, 1952; Komosinska, 1969).
HOST PLANTS: **Rosaceae**: *Amygdalus* [Borchs1952, Komosi1969].
DISTRIBUTION: **Palaearctic**: Iran [Borchs1952, Komosi1969].
GENERAL: Description and illustration of adult female by Komosinska (1969).
KEYS: Komosinska 1969: 76-78 (female) [World].

CITATIONS: Borchs1952 [taxonomy, description, illustration, host, distribution: 261-262]; Borchs1966 [catalogue: 281]; DanzigPe1998 [catalogue: 252]; Kaussa1954 [host, distribution: 80]; Komosi1969 [taxonomy, description, illustration, host, distribution: 53-54]; Lobdel1937 [taxonomy: 78,79]; Ramakr1919a [taxonomy, host, distribution: 21-22].

Dynaspidiotus andersoni Balachowsky

Dynaspidiotus andersoni Balachowsky, 1958b: 162. Type data: SOUTH AFRICA: North Pretoria, on *Crassula* sp. Holotype female. Type depository: London: The Natural History Museum, England, UK.

SCALE COVER: Female scale circular, 2-2.2 mm in diameter; flat; exuviae red, central (Balachowsky, 1958b).

HOST PLANTS: **Crassulaceae**: *Crassula* [Balach1958b].

NATURAL ENEMIES: HYMENOPTERA **Encyrtidae**: *Metaphycus nigripectus* Annecke & Mynhardt [Prinsl1983].

DISTRIBUTION: **Afrotropical**: South Africa [Balach1958b].

GENERAL: Description and illustration of adult female by Balachowsky (1958b).

CITATIONS: Balach1958b [taxonomy, description, illustration, host, distribution: 161-163]; Borchs1966 [catalogue: 281]; Ferris1941e [taxonomy: 40]; Prinsl1983 [distribution, biological control: 26].

Dynaspidiotus apacheca (Ferris), new combination

Nuculaspis apacheca Ferris, 1941d: 376. Type data: U.S.A.: Arizona, Chiricahua Mountains, Cave Creek, near South Fork Public Camp, on *Pinus apacheca*. Holotype female. Type depository: Davis: The Bohart Museum of Entomology, University of California, California, USA.

SCALE COVER: Scale of female slightly elongate, flat and thin; white, exuviae more or less central; scale of male similar to that of female in form and colour (Ferris, 1941d).

HOST PLANTS: **Pinaceae**: *Pinus apacheca* [Ferris1941d].

DISTRIBUTION: **Nearctic**: Mexico [Nakaha1982]; United States of America (Arizona [Ferris1941d]). **Neotropical**: Guatemala [Nakaha1982].

BIOLOGY: Occurring on inner side of needles (Ferris, 1941d).

GENERAL: Description and illustration of adult female by Ferris (1941d).

KEYS: Ferris 1942: 39 (female) [North America].

CITATIONS: Borchs1966 [catalogue: 275]; Ferris1941d [taxonomy, description, illustration, host, distribution: 376]; Ferris1942 [taxonomy: 446:39]; Lindin1957 [taxonomy: 550]; Nakaha1982 [host, distribution: 61].

Dynaspidiotus atlanticus (Balachowsky)

Hemiberlesia atlantica Balachowsky, 1928a: 125. Type data: MOROCCO: Guelber-Rahal (Moyen Atlas), altitude 1600-2000 meters, on *Buxus balearica*. Holotype female. Type depository: Paris: Muséum national d'Histoire naturelle, France.

Aspidiotus atlanticus; Lindinger, 1936: 157. Change of combination requiring emendation of species name for agreement in gender.

Dynaspidiotus atlanticus; Balachowsky, 1948b: 333. Change of combination.

SCALE COVER: Female scale circular, diameter 2-2.3 mm; flattened laterally, elevated in middle; exuviae central, brown; secretion of adult, white, shiny; transparent; ventral vellum absent. Male unknown (Balachowsky, 1928a).

HOST PLANTS: **Araliaceae**: *Hedera helix* [Balach1948b, Kaussa1954]. **Buxaceae**: *Buxus* [Kaussa1954], *Buxus balearica* [Balach1928a, Balach1932d, Rungs1935, Rungs1948], *Buxus sempervirens* [Rungs1948]. **Salicaceae**: *Populus* [Kaussa1954], *Populus hickeliana* [Rungs1948].

DISTRIBUTION: **Palaearctic**: Iran [Kaussa1954]; Morocco [Balach1928a, Balach1932d, Rungs1935, Rungs1948]; Turkey [Yasar1995a].

GENERAL: Description and illustration of adult female by Balachowsky (1928a) and by Yaşar (1995a).

KEYS: Balachowsky 1948b: 328 (female) [Mediterranean]; Balachowsky 1928a: 132 (female) [North Africa].

CITATIONS: Balach1928a [taxonomy, description, illustration, host, distribution: 125-127]; Balach1929a [host, distribution: 317]; Balach1932d [taxonomy, host, distribution: VIII]; Balach1948b [taxonomy, description, illustration, host, distribution: 333-335]; Borchs1966 [catalogue: 281]; DanzigPe1998 [catalogue: 252]; Ferris1941e [taxonomy: 41]; Kaussa1954 [host, distribution: 80]; Lindin1936 [taxonomy: 157]; Rungs1935 [host, distribution: 271-272]; Rungs1948 [host, distribution: 111]; Yasar1995 [taxonomy, description, illustration, host, distribution: 51-53]; Yasar1995a [taxonomy, description, illustration, host, distribution: 76-78].

Dynaspidiotus britannicus (Newstead)

Aspidiotus hederae; Newstead, 1896: 279. Misidentification; discovered by Newstead, 1898: 94.

Aspidiotus britannicus Newstead, 1898: 93. Type data: ENGLAND: Teddington near London, on holly, *Ilex aquifolium*. Syntypes, female. Type depositories: London: The Natural History Museum, England, UK, and Hamburg: Zoologisches Institut und Zoologishces Museum, Universität von Hamburg, Germany.

Aspidiotus (*Evaspidiotus*) *britannicus*; Leonardi, 1900: 340. Change of combination.

Aspidiotus latastei; Hall, 1923: 20. Misidentification; discovered by Danzig, 1993: 150.

Aspidiotus (*Dynaspidiotus*) *britannicus*; Thiem & Gerneck, 1934a: 156. Change of combination.

Dynaspidiotus britannicus; Ferris, 1938a: 229. Change of combination.

Dinaspidiotus britanicus; Gómez-Menor Ortega, 1957: 45. Misspelling of genus and species names.

COMMON NAME: holly scale [Borchs1966, McKenz1956].

SCALE COVER: Female scale circular or almost so, diameter 0.75-2 mm; moderately convex; colour dusky ochreous, with a broad smoky-brown central zone; exuviae central, or a little to one side, those of larva dark yellow or dull orange; secretionary covering very thin; second secretionary covering smoky-brown (Newstead, 1898). Scale of female grey or brown, circular, flat, exuviae subcentral; that of male grey, oval, exuvia toward one end (Ferris, 1938a). Colour photograph by Gill (1997).

HOST PLANTS: **Anacardiaceae**: *Pistacia* [Bodenh1949]. **Apocynaceae**: *Vinca* [Bodenh1949]. **Aquifoliaceae**: *Ilex* [Green1928, Ferris1938a, Bodenh1949, Kaussa1954], *Ilex aquifolium* [Newste1898, Leonar1900, Balach1932d, Borchs1934, McKenz1956], *Ilex colchica* [Hadzib1983, Danzig1993]. **Araliaceae**: *Hedera colchica* [Danzig1993], *Hedera helix* [Leonar1918, Leonar1920, Borchs1934, Borchs1936, Ferris1938a, Bodenh1949, Bodenh1952, Kaussa1954], *Hedera helix* [BachmaGe1950]. **Berberidaceae**: *Berberis* [Ferris1938a, McKenz1956]. **Buxaceae**: *Buxus* [Green1928, Green1930, Ferris1938a, Bodenh1949, Kaussa1954], *Buxus balearica* [Balach1932d, Martin1983], *Buxus colchica* [Hadzib1983, Danzig1993], *Buxus rotundifolia* [BachmaGe1950], *Buxus sempervirens* [Balach1932d, Borchs1934, Borchs1936, Balach1937c, McKenz1956, Martin1983]. **Caprifoliaceae**: *Viburnum* [Leonar1918, Leonar1920, Ferris1938a, Bodenh1949, McKenz1956], *Viburnum tinus* [Rungs1948, Danzig1993]. **Elaeagnaceae**: *Elaeagnus* [Bodenh1949]. **Ericaceae**: *Arbutus* [Bodenh1949]. **Fagaceae**: *Quercus* [Borchs1934]. **Lauraceae**: *Cinnamomum camphorae* [Hadzib1983, Danzig1993], *Laurus* [Bodenh1937, Ferris1938a, Bodenh1949, Bodenh1952], *Laurus nobilis* [Bodenh1924, Green1925, Green1928, Borchs1934, Bodenh1952, Kaussa1954, McKenz1956, Hadzib1983]. **Leguminosae**: *Ceratonia siliqua* [Bodenh1926, Bodenh1949, Bodenh1952, InserrCa1987], *Sophora* [Balach1932d, Bodenh1949]. **Myrtaceae**: *Myrtus communis* [Bodenh1949, Bodenh1952, GomezM1957, Martin1983]. **Oleaceae**: *Jasminum* [Bodenh1949], *Ligustrum* [Borchs1934, Bodenh1949], *Olea* [Bodenh1924, Bodenh1937], *Olea europaea* [Bodenh1924, Bodenh1924a, Bodenh1926, Bodenh1949, Bodenh1952], *Osmanthus* [Bodenh1949]. **Palmae**: *Livistona* [Bodenh1949], *Phoenix* [Borchs1934, Bodenh1949]. **Pinaceae**: *Pinus* [Balach1932d, Bodenh1949, Zahrad1972], *Pinus brutia* [Zahrad1972, Martin1983], *Pinus halepensis* [Balach1932d, Danzig1993]. **Pistaciaceae**: *Pistacia lentiscus* [Rungs1948, Bodenh1952, InserrCa1987]. **Rhamnaceae**: *Rhamnus* [Ferris1938a, Bodenh1949, McKenz1956], *Rhamnus alaternus* [Leonar1918, Leonar1920], *Ziziphus* [Bodenh1949], *Ziziphus spina-christi* [Hall1923]. **Rosaceae**: *Lauro-Cerasus officinalis* [Hadzib1983, Danzig1993], *Prunus* [Bodenh1949], *Prunus laurocerasus* [Borchs1934, Borchs1936], *Pyrus malus* [Bodenh1949]. **Ruscaceae**: *Ruscus* [Ferris1938a, Lepage1938, McKenz1956], *Ruscus aculeatus* [Balach1948b, Kaussa1954], *Ruscus hypoglossum* [Britto1923, Balach1932d], *Ruscus pontica* [Hadzib1983, Danzig1993], *Ruscus racemosus* [BattagVi1987]. **Rutaceae**: *Citrus* [Bodenh1949], *Citrus nobilis unshui* [Borchs1934]. **Taxaceae**: *Taxus baccata* [Balach1928a, Balach1932d, Zahrad1972, Danzig1993]. **Thymelaeaceae**: *Daphne* [Bodenh1952], *Daphne gnidium* [GomezM1957, Martin1983].

NATURAL ENEMIES: HYMENOPTERA **Aphelinidae**: *Aphytis aonidiae* (Mercet) [RosenDe1979], *Aphytis libanicus* Traboulsi [RosenDe1979, BattagVi1987], *Aphytis mytilaspidis* (LeBaron) [Gordh1979], *Aphytis yasumatsui* Azim [RosenDe1979], *Aspidiotiphagus citrinus* (Craw) [Gordh1979], *Encarsia aspidioticola* (Mercet) [BattagVi1987], *Encarsia citrina* (Craw) [BattagVi1987]. **Encyrtidae**: *Coccidencyrtus dynaspidiotus* Battaglia [Battag1989], *Coccidencyrtus steinbergii* Chumakova & Trjapitzin [BattagVi1987], *Eusemion cornigerum* (Walker) [Trjapi1989].

DISTRIBUTION: **Nearctic**: Canada (British Columbia [Ferris1938a]); United States of America (California [McKenz1956], Connecticut [Ferris1938a], Illinois [Nakaha1982], Indiana [Nakaha1982], Massachusetts [Britto1923, Ferris1938a], Michigan [Nakaha1982], Oregon [Ferris1938a], Pennsylvania [Nakaha1982], Washington [Nakaha1982]). **Neotropical**: Brazil (Rio Grande do Sul [Lepage1938]). **Palaearctic**: Algeria [Balach1928a, Balach1932d]; Corsica [Balach1932d]; Cyprus [Balach1948b]; Czech Republic [Zahrad1977, Zahrad1990b]; Egypt [Ezzat1958]; France [Balach1932d, Balach1937c]; Georgia (Abkhaz ASSR [Borchs1934, Borchs1936, Hadzib1983], Adzhar ASSR [Borchs1934, Borchs1936, Hadzib1983]); Greece [Korone1934, ArgyriStMo1976]; Iran [Kaussa1954]; Israel [Bodenh1924a, Bodenh1937, Balach1948b, RosenDe1979]; Italy [Leonar1918, Leonar1920, LongoMaPe1995]; Latvia [Danzig1993]; Lebanon [Bodenh1926]; Madeira Islands [Balach1938a]; Morocco [Balach1932d, Rungs1948]; Portugal [Seabra1942, Fernan1992]; Russia (Caucasus [Borchs1934]); Sicily [Leonar1918, InserrCa1987]; Spain [Martin1983, BlayGo1993]; Switzerland [BachmaGe1950]; Turkey [Bodenh1949, Bodenh1952]; United Kingdom (England [Newste1898, Green1925, Green1928, Green1930]).
BIOLOGY: Occurring on leaves (Ferris, 1938a).
ECONOMIC IMPORTANCE: Recorded from several countries in the Palearctic and Nearctic regions (see Distribution). Considered a minor pest of olive in Mediterranean countries (Argyriou, 1990), and of palms and ornamentals in various countries (Zahradník, 1990; Gill, 1997).
GENERAL: Description and illustration of adult female by Ferris (1938a), Balachowsky (1948b), McKenzie (1956), Zahradník (1990b), Danzig (1993), Yaşar (1995a) and by Gill (1997).
KEYS: Danzig 1993: 150 (female) [Europe]; Ezzat 1958: 241 (female) [Egypt]; Lupo 1953: 39 (female) [Italy]; Balachowsky 1948b: 327 (female) [Mediterranean]; Lupo 1948: 138 (female) [Italy]; Ferris 1942: 33 (female) [North America]; Britton 1923: 371 (female) [U.S.A.: Connecticut]; Leonardi 1920: 29-30 (female) [Italy]; Dietz & Morrison 1916a: 289-290 (female) [U.S.A.: Indiana]; Lindinger 1907: 5 (female) [Germany]; Newstead 1901a: 82 (female) [England].
CITATIONS: Apstei1915 [taxonomy: 119]; Argyri1990 [host, distribution, economic importance: 579-583]; ArgyriStMo1976 [host, distribution, biological control: 26]; BachmaGe1950 [host, distribution: 117]; Balach1928a [host, distribution: 138]; Balach1932d [taxonomy, host, distribution, economic importance: V; XLV]; Balach1937c [host, distribution: 2]; Balach1938a [host, distribution: 148]; Balach1948b [taxonomy, description, illustration, host, distribution, economic importance: 328-331]; Balach1951 [taxonomy: 697]; BaldanGaVi1999 [biological control: 209-215]; Battag1989 [host, distribution, biological control: 149-165]; BattagVi1985 [host, distribution, biological control: 337-340]; BattagVi1987 [host, distribution, biological control: 139-142]; BeardsDaHo1976 [economic importance: 105]; BlayGo1993 [taxonomy, description, illustration, host, distribution: 451-455]; Bodenh1924 [taxonomy, description, host, distribution: 24-25]; Bodenh1926 [host, distribution: 42]; Bodenh1935 [host, distribution: 246]; Bodenh1937 [host, distribution: 216]; Bodenh1949 [taxonomy, description, illustration, host, distribution: 48-50];

Bodenh1952 [taxonomy, host, distribution: 337]; Borchs1934 [host, distribution: 27-28]; Borchs1935a [taxonomy, description, host, distribution: 24]; Borchs1936 [host, distribution: 130]; Borchs1937 [taxonomy, description, illustration, host, distribution: 127]; Borchs1937a [taxonomy, description, host, distribution: 35,37]; Borchs1939a [taxonomy, distribution: 43]; Borchs1950b [taxonomy, description, illustration, host, distribution: 219-220]; Borchs1966 [catalogue: 281-282]; Britto1923 [taxonomy, description, host, distribution: 371,372]; Bustsh1958 [taxonomy, description, illustration, host, distribution: 219,227]; ClapsWoGo2001a [taxonomy, host, distribution: 17]; Cocker1899a [taxonomy: 395]; Danzig1964 [taxonomy, host, distribution: 651]; Danzig1972 [taxonomy, host, distribution, economic importance: 212-213]; Danzig1993 [taxonomy, description, illustration, host, distribution, economic importance: 150-151]; DanzigPe1998 [catalogue: 253]; DavidsMi1990 [host, distribution, economic importance: 603-632]; DeBach1964d [biological control: 5-18]; DelBen1984 [host, distribution, economic importance, biological control: 323-336]; DietzMo1916a [taxonomy, description, illustration, host, distribution: 289,295-296]; Dingle1924 [taxonomy, description, host, distribution, life history: 174]; Essig1913c [host, distribution: 597]; Ezzat1958 [distribution: 241]; EzzatNa1987 [distribution: 87]; Fernal1903b [catalogue: 253]; Fernan1992 [host, distribution: 60]; Ferris1937c [taxonomy, illustration: 50,51,71]; Ferris1938a [taxonomy, description, illustration, host, distribution: 229]; Ferris1941e [taxonomy: 41]; Ferris1942 [taxonomy: 446:33]; Fjeldd1996 [host, distribution: 4-24]; Fleury1934a [distribution: 278-289]; Fleury1935 [distribution: 504]; Foldi2001 [distribution: 303-308]; Garcia1930 [host, distribution, biological control]; Gavalo1931 [host, distribution: 7]; Gentry1965 [host, distribution, economic importance]; Gill1997 [host, distribution, taxonomy, description, illustration, economic importance: 135-136]; GomezM1937 [taxonomy, description, illustration, host, distribution: 46,53-55]; GomezM1957 [host, distribution: 45]; GomezM1958c [host, distribution: 406]; Gordh1979 [biological control: 895]; Green1925 [host, distribution: 44]; Green1928 [host, distribution: 8]; Green1930 [host, distribution: 16]; Hadzib1983 [taxonomy, description, host, distribution, life history: 222-223]; Hall1923 [taxonomy, description, host, distribution: 18-19]; Hall1926a [taxonomy, host, distribution: 31]; Harris1916 [host, distribution: 172-174]; Houard1913 [host, distribution: 1]; Hulsen1928 [host, distribution, chemical control: 285-315]; Kaussa1954 [host, distribution: 80]; Korone1934 [taxonomy, description, illustration, host, distribution: 7-8]; Larew1990 [ecology, life history, structure: 293-300]; Leonar1900 [taxonomy, host, distribution: 340]; Leonar1918 [host, distribution: 188-189]; Leonar1920 [taxonomy, description, illustration, host, distribution: 37-40]; Lepage1938 [catalogue: 393]; Lindin1907 [taxonomy: 5]; Lindin1909a [taxonomy, description, illustration, host, distribution, life history: 324-328]; Lindin1912b [taxonomy, description, host, distribution: 87,99,176,185,196]; Lindin1924 [taxonomy: 174]; Lindin1935 [taxonomy: 128,147]; LongoMaPe1995 [distribution: 126]; Lupo1948 [taxonomy, description, illustration, host, distribution: 155-159]; Lupo1953 [taxonomy: 39]; MacGil1921 [taxonomy, description, host, distribution: 402]; Mackie1933 [host, distribution: 457]; Martin1983 [taxonomy, host, distribution: 64]; McKenz1956 [taxonomy, description, illustration, host, distribution: 65-66]; MillerDa1990 [host, distribution, economic importance: 302];

Nakaha1982 [host, distribution: 34]; Newste1896 [taxonomy: 279]; Newste1898 [taxonomy, description, illustration, host, distribution: 93-94]; Newste1901a [taxonomy, description, illustration, host, distribution: 82,117-119]; RoafMo1935 [taxonomy, description, host, distribution, life history, chemical control: 1042]; RoafMo1935 [taxonomy, description, life history: 1041-1043]; RosenDe1979 [host, distribution, biological control: 476-485,558-561]; Ruhl1913 [host, distribution: 79-80]; Rungs1948 [host, distribution: 111]; Schmut1959 [taxonomy, description, illustration, host, distribution: 60,63]; SchmutKlLu1957 [host, distribution, economic importance: 478]; SchmutKlLu1959 [taxonomy: 374]; Seabra1942 [distribution: 2]; SmithEsFa1933 [economic importance: 1]; SoriaMoVi2000 [host, distribution: 335-348]; Staffo1915 [taxonomy, structure: 71]; Szulcz1926 [host, distribution: 137-143]; TakagiRo1981 [host, distribution, biological control: 314-321]; Tao1999 [taxonomy, host, distribution: 85]; Terezn1986 [taxonomy, description, illustration, host, distribution: 113]; ThiemGe1934a [taxonomy, description, host, distribution: 130-158,208-238]; TranfaVi1987a [economic importance: 215-221]; Trjapi1989 [biological control: 301]; Viggia1987 [host, distribution, biological control: 121-123]; WeidneWa1968 [taxonomy: 171]; Yasar1995a [taxonomy, description, illustration, host, distribution: 78-80]; Yasnos1994 [host, distribution, biological control: 317-333]; Zagain1956 [distribution: 85-90]; Zahrad1959b [host, distribution: 60]; Zahrad1972 [host, distribution: 432]; Zahrad1977 [taxonomy, distribution: 120]; Zahrad1990 [host, distribution, description: 642]; Zahrad1990b [taxonomy, description, illustration, host, distribution: 95-96].

Dynaspidiotus californicus (Coleman)

Aspidiotus ? pini Comstock, 1881a: 306. Type data: U.S.A.: New York, Ithaca, on leaves of *Pinus rigida* and Georgia, Macon, on *Pinus mitis*. Syntypes, female. Type depository: Washington: United States National Entomological Collection, U.S. National Museum of Natural History, District of Columbia, USA. Homonym of *Aspidiotus pini* Bouché, 1851.

Aspidiotus abietis; Cockerell, 1894l: 190. Misidentification; discovered by Borchsenius, 1966: 275.

Aspidiotus abietis pini; Cockerell, 1896b: 333. Change of status.

Aspidiotus californicus Coleman, 1903: 64. Type data: U.S.A.: California, at nine localities, on *Pinus sabiniana, P. ponderosa, P. lambertiana* and *P. attenuata*. Syntypes, female. Type depository: Davis: The Bohart Museum of Entomology, University of California, California, USA. Junior synonym and Replacement Name.

Nuculaspis californica; Ferris, 1938a: 251. Change of combination requiring emendation of species name for agreement in gender.

Aspidiotus (Nuculaspis) californicus; Merrill, 1953: 16. Change of combination.

Dynaspidiotus californicus; Danzig, 1993: 149. Change of combination.

COMMON NAME: black pine leaf scale [Merril1953, McKenz1956, Dekle1965c].

SYSTEMATICS: *Aspidiotus ? pini* Comstock, 1881a: 306, (described from *Pinus* in USA, New York) is a homonym of *Aspidiotus pini* Bouché, 1851. No Replacement Name has been formally proposed for the former. However, *Aspidiotus californicus* Coleman, 1903 (described from *Pinus* spp. in USA,

California) was found by Ferris (1938a: 190) to be a synonym of *Aspidiotus ? pini* Comstock, 1881a, becoming therefore a Replacement Name.

SCALE COVER: Female scale oval, 2 mm long, 1 mm wide; conical; blackish with pale edges; exuviae central, reddish-brown; size and form variable (Coleman, 1903). Female scale black, with one margin set against edge of needle and therefore scale is elongate; exuviae central; male scale darker, elongate oval, exuvia central (Ferris, 1938a). Colour photograph by Gill (1997).

HOST PLANTS: **Pinaceae**: *Pinus* [Ferris1920b, Dekle1965c, BesheaTiHo1973], *Pin. attenuata* [Colema1903, Sander1906], *Pin. cembroides* [McKenz1956, McDani1970], *Pin. lambertiana* [Colema1903, Sander1906, McKenz1956], *Pin. mitis* [Comsto1881a, McKenz1956], *Pin. murrayana* [Ferris1938a, McKenz1956], *Pin. occidentalis* [GomezM1941], *Pin. palustris* [MerrilCh1923], *Pin. ponderosa* [Colema1903, Sander1906, McKenz1956, McDani1970], *Pin. radiata* [Ferris1938a, McKenz1956], *Pin. rigida* [Comsto1881a, McKenz1956], *Pin. sabiniana* [Colema1903, Sander1906, McKenz1956, McDani1970], *Pseudotsuga taxifolia* [Ferris1920b, MerrilCh1923, Ferris1938a, McKenz1956, McDani1970].

NATURAL ENEMIES: HYMENOPTERA **Aphelinidae**: *Aspidiotiphagus citrinus* (Craw) [Gordh1979], *Physcus howardi* Compere [Gordh1979], *Physcus varicornis* (Howard) [Gordh1979]. NEROPTERA **Raphidiidae**: *Raphidia* [Drea1990].

DISTRIBUTION: **Nearctic**: Canada [Nakaha1982]; Mexico [Ferris1938a]; U.S.A. (California [Sander1906, McKenz1956], Florida [MerrilCh1923, Dekle1965c], Georgia [Comsto1881a, BesheaTiHo1973], New York [Comsto1881a], Texas [Ferris1938a, McDani1970]). **Neotropical**: Dominican Republic [GomezM1941].

BIOLOGY: Occurring on the needles (Ferris, 1938a).

ECONOMIC IMPORTANCE: Considered a serious pest of pines in the West Coast of USA (Gill, 1997).

GENERAL: Description and illustration of adult female by Ferris (1920b; 1938a), McKenzie (1956), Kosztarab (1996) and by Gill (1997).

KEYS: Kosztarab 1996: 543 (female) [Northeastern North America]; McKenzie 1956: 26 (female) [U.S.A.: California]; Ferris 1942: 39 (female) [North America]; Britton 1923: 371 (female) [U.S.A.: Connecticut]; Comstock 1883: 55-57 (female) [North America].

CITATIONS: Alstad1998 [host, distribution, life history, ecology: 3-21]; AlstadEd1983 [host, distribution, life history, ecology: 93-95]; AlstadEd1983a [host, distribution, life history, ecology: 413-426]; AlstadEd1987 [host, distribution, life history, ecology: 652-654]; AlstadEd1989 [host, distribution, life history, ecology: 253-263]; AlstadEdJo1980 [host, distribution, life history, ecology: 665-667]; Amos1933 [taxonomy: 208]; Baker1972 [host, distribution, economic importance, control: 1-642]; BeardsDaHo1976 [economic importance: 103]; BesheaTiHo1973 [host, distribution: 7]; Borchs1966 [catalogue: 275]; Britto1923 [taxonomy, description, host, distribution: 371,372]; BrownEa1967 [host, distribution, economic importance, control: 1-72]; Cocker1894l [taxonomy, description, host, distribution: 190-191]; Cocker1896b [taxonomy, distribution: 333]; Colema1903 [taxonomy, description, host, distribution: 64]; Comper1928 [biological control: 209-230]; Comsto1881a [taxonomy, description, illustration, host, distribution: 306-307]; Comsto1883 [taxonomy: 67]; Dekle1965c [taxonomy, description, host, distribution:

98]; Dekle1976 [taxonomy, description, host, distribution, economic importance: 119]; DeSant1940 [biological control: 29-44]; Drea1990 [biological control: 51-59]; Edmund1973 [host, distribution, life history, ecology, biological control: 765-777]; EdmundAl1958 [host, distribution, life history, ecology: 391-392]; EdmundAl1981 [host, distribution, life history, ecology]; EdmundAl1984 [host, distribution, life history, ecology: 267-268]; EdmundAl1985 [host, distribution, life history, chemical control, ecology: 403-405]; Essig1928 [host, distribution: 76-78]; Ferris1920b [taxonomy, description, illustration, host, distribution: 52-54]; Ferris1938a [taxonomy, description, illustration, host, distribution: 251]; Ferris1938b [taxonomy: 65,70]; Ferris1941e [taxonomy: 41,47]; Ferris1942 [taxonomy: 446:39]; FurnisCa1977 [host, distribution: 111]; Gill1997 [host, distribution, taxonomy, description, illustration, economic importance: 204-205,207]; GomezM1941 [host, distribution: 128]; Gordh1979 [biological control: 907]; HagenVaDa1971 [host, distribution, biological control: 253-293]; Herric1911 [taxonomy, description, illustration, host, distribution: 13]; Howard1895e [biological control: 1-44]; JohnsoYoAl1997 [host, distribution, life history, chemistry, ecology: 1794-1804]; Keen1952 [host, distribution, economic importance: 1-280]; Koszta1996 [taxonomy, description, illustration, host, distribution, life history, biological control, economic importance: 543-544]; Kuwana1932 [taxonomy: 54]; Leonar1897 [taxonomy: 285]; Lindin1957 [taxonomy: 550]; Lobdel1937 [taxonomy: 78]; MacGil1921 [taxonomy, description, host, distribution: 396]; McClur1990a [host, distribution, ecology: 165-168]; McClur1990g [taxonomy, host, distribution, ecology: 319-330]; McDani1970 [taxonomy, illustration, host, distribution: 425-427]; McKenz1939 [taxonomy: 53]; McKenz1956 [taxonomy, description, illustration, host, distribution: 77-80]; Merril1953 [taxonomy, description, host, distribution: 16-17]; MerrilCh1923 [taxonomy, description, host, distribution, economic importance: 207-208]; Miller1983 [host, distribution: 4-6]; MillerDa1990 [host, distribution, economic importance: 304]; Nakaha1982 [host, distribution: 61]; Nur1990a [taxonomy, structure, chromosomes: 185]; Sander1906 [taxonomy, host, distribution: 13]; SchmutKlLu1957 [host, distribution, economic importance: 494]; StrublJo1964 [host, distribution, economic importance: 1-6]; Wilson1917 [taxonomy, description, host, distribution: 31-32]; YoungJoAl1993 [host, distribution, life history, ecology: 497]; Zahrad1990 [host, distribution, description: 641-642].

Dynaspidiotus degeneratus (Leonardi)
Chrysomphalus degeneratus Leonardi in: Berlese & Leonardi, 1896: 345. Type data: ITALY: Portici, on leaves of *Pandanus graminifolia*. Syntypes, female and first instar. Type depository: Portici: Dipartimento de Entomologia e Zoologia Agraria di Portici, Università di Napoli Federico II, Italy.
Aspidiotus degeneratus; Cockerell, 1896b: 334. Change of combination.
Aspidiotus (Chrysomphalus) degeneratus; Cockerell, 1897i: 29. Change of combination.
Aspidiotus degeneratus; Lindinger, 1912b: 358. Incorrect synonymy. Notes: Incorrect synonymy with *Chrysomphalus dictyospermi*; see Borchsenius, 1966: 282.
Aspidiotus degeneratus; Koroneos, 1934: 10. Notes: Incorrect citation of "Fernald" as author.

Hemiberlesia degenerata; McKenzie, 1939: 54. Change of combination requiring emendation of species name for agreement in gender.

Abgrallaspis degeneratus; Balachowsky, 1948b: 317. Change of combination.

Diaspidiotus degeneratus; Borchsenius, 1950b: 225. Change of combination.

Dynaspidiotus degeneratus; Borchsenius, 1966: 282. Change of combination.

Abgrallaspis degenesatus; Tang, 1984: 54. Misspelling of species name.

Abgrallaspis degoneratus; Chou, 1985: 308. Misspelling of species name.

Dynaspidiotus degeneratus; Danzig, 1993: 151. Revived combination.

COMMON NAME: degenerate scale [McKenz1956].

SCALE COVER: Female scale very light brown, becoming white in region of subcentrally placed exuviae; circular, very slightly convex. Male scale similar to female but slightly ovoid (Ferris, 1941d). Colour photograph by Gill (1997).

HOST PLANTS: **Aquifoliaceae**: *Ilex* [Kawai1977], *Ilex colchica* [Danzig1993], *Ilex cornuta* [McKenz1956], *Ilex integra* [Kuwana1933, Ferris1941d]. **Araliaceae**: *Aralia* [McKenz1956]. **Oleaceae**: *Osmanthus* [Kawai1977], *Osmanthus fragrans* [Kuwana1933, Ferris1941d, McKenz1956]. **Rutaceae**: *Citrus limon* [McKenz1956]. **Theaceae**: *Camellia* [Ferris1941d, McKenz1956, Kawai1977], *Camellia japonica* [BerlesLe1896, Korone1934, McKenz1956, TakahaTa1956, Hadzib1983], *Camellia sasanqa* [Hadzib1983], *Eurya japonica* [Kuwana1933, Ferris1941d, McKenz1956, Kawai1977, Danzig1993], *Eurya ochnacea* [Ferris1941d, McKenz1956], *Thea japonica* [Kuwana1933, Ferris1941d], *Thea sinensis* [Hadzib1983].

DISTRIBUTION: **Nearctic**: United States of America (California [Ferris1941d, McKenz1956], Montana [Nakaha1982], Oregon [Nakaha1982]). **Palaearctic**: China (People's Republic) [Nakaha1982]; Georgia [Hadzib1983, Danzig1993]; Greece [Korone1934]; Italy [BerlesLe1896, Leonar1920, LongoMaPe1995]; Japan [Kuwana1917a, Kuwana1933, Kawai1977, Kawai1980] (Honshu [TakahaTa1956], Shikoku [TakahaTa1956]); North Korea [Danzig1993]; Portugal [Nakaha1982].

BIOLOGY: Occurring on leaves (Ferris, 1941d).

ECONOMIC IMPORTANCE: McKenzie (1956) listed it as common nursery pest in California, USA. Gill (1997) reported it to be rare, not found in nurseries and suggested that it was not established in California, after its first introduction in 1930's - 1940's.

GENERAL: Description and illustration of adult female by Kuwana (1933), Ferris (1941d), Balachowsky (1948b), McKenzie (1956), Komosinska (1969), Chou (1985, 1986), Danzig (1993) and by Gill (1997).

KEYS: Gill 1997: 32 (female) [Species of California]; Danzig 1993: 150 (female) [Europe]; Chou 1985: 306 (female) [Species of China]; Komosinska 1969: 76-78 (female) [World]; Davidson 1964: 639-640 (female) [North America]; McKenzie 1956: 26 (female) [U.S.A.: California]; Balachowsky 1948b: 308 (female) [Mediterranean]; Ferris 1942: 34 (female) [North America]; Kuwana 1933: 3 (female) [Japan]; Kuwana 1933b: 49 (female) [Japan]; Leonardi 1920: 65 (female) [Italy].

CITATIONS: Balach1948b [taxonomy, description, illustration, host, distribution: 317-320]; Balach1951 [taxonomy: 697]; Balach1953k [taxonomy: 113]; BerlesLe1896 [taxonomy, description, illustration, host, distribution: 345]; Borchs1950b [taxonomy, description, host, distribution: 225]; Borchs1966 [catalogue: 282]; Chou1985 [taxonomy, description, host, distribution: 308]; Chou1986 [taxonomy, illustration: 688]; Cocker1896b [taxonomy, distribution: 334]; Cocker1897i [taxonomy, description, host, distribution: 29]; Danzig1972 [taxonomy, host, distribution, economic importance: 213]; Danzig1993 [taxonomy, description, illustration, host, distribution: 151-152]; DanzigPe1998 [catalogue: 253]; Davids1964 [taxonomy: 639]; Fernal1903b [catalogue: 257]; Ferris1941d [taxonomy, description, illustration, host, distribution: 342]; Ferris1941e [taxonomy: 42]; Ferris1942 [taxonomy: 446:34]; Fleury1934a [distribution: 278-289]; Gill1997 [taxonomy, description, illustration, host, distribution, economic importance: 33-34,36]; Hadzib1983 [taxonomy, description, host, distribution: 223]; Hewitt1943 [host, distribution: 266-274]; Kawai1977 [host, distribution, economic importance: 160-161]; Kawai1980 [taxonomy, description, host, distribution: 220]; Komosi1969 [taxonomy, description, illustration, host, distribution: 58-60]; Korone1934 [taxonomy, description, illustration, host, distribution: 10]; Kuwana1917a [taxonomy, distribution: 174]; Kuwana1933 [taxonomy, description, illustration, host, distribution: 8-9]; Leonar1897 [taxonomy: 286]; Leonar1899 [taxonomy, description, host, distribution: 199-200,216]; Leonar1920 [taxonomy, description, illustration, host, distribution: 73-74]; Lindin1912b [taxonomy: 358]; LongoMaPe1995 [distribution: 125]; Lupo1953 [taxonomy, description, illustration, host, distribution: 38-43]; MacGil1921 [taxonomy, description, host, distribution: 419]; Mackie1933 [host, distribution: 457]; Maleno1916c [taxonomy: 119]; McKenz1939 [taxonomy: 54]; McKenz1956 [taxonomy, description, illustration, host, distribution: 67-69]; MillerDa1990 [host, distribution, economic importance: 301]; Muraka1970 [host, distribution: 69]; Nakaha1982 [host, distribution: 1]; SchmutKlLu1957 [host, distribution, economic importance: 478]; SchmutKlLu1959 [taxonomy: 374]; SchuhMo1948 [host, distribution, control]; TakahaTa1956 [host, distribution: 14-15]; Tang1984 [taxonomy, description, illustration, host, distribution: 54-55]; Tao1999 [taxonomy, host, distribution: 68].

Dynaspidiotus ephedrarum (Lindinger)

Aspidiotus ephedrarum Lindinger, 1912b: 139. Type data: SARDINIA and SPAIN: on *Ephedra nebrodensis* and *Ephedra scoparia*. Syntypes, female. Type depository: Hamburg, Zoologisches Institut und Zoologishces Museum, Universität von Hamburg, Germany.

Hemiberlesia ephedrarum; Paoli, 1915: 265. Change of combination.

Spinaspidiotus ephedrarum; MacGillivray, 1921: 429. Change of combination.

Hemiberlesea ephedrarum; Gómez-Menor Ortega, 1937: 116. Misspelling of genus name.

Hemiberlesea ephedrarum; Rungs, 1942: 107. Misspelling of genus name.

Diaspidiotus ephedrarum; Bodenheimer, 1943: 4. Change of combination.

Quadraspidiotus ephedrarum; Rungs, 1948: 111. Change of combination.

Abgrallaspis ephedrarum; Balachowsky, 1948b: 309. Change of combination.

Ephedraspis ephedrarum; Borchsenius, 1949d: 238. Change of combination.

Hemiberlesia ephedracum; Gómez-Menor Ortega, 1958a: 7. Misspelling of species name.

Dynaspidiotus ephedrarum; Danzig, 1993: 149. Change of combination.

SCALE COVER: Female scale white or grey white, 1.5 mm in diameter; exuviae subcentral, large, light brown (Lindinger, 1912b). Female scale circular, 2-2.2 mm diameter; convex, robust; white; exuviae, subcentral, dark brown; ventral scale thick. Male scale white, oval, 1.4-1.5 mm (Balachowsky, 1948b).

HOST PLANTS: **Ephedraceae**: *Ephedra* [Balach1930c, Balach1932d, Rungs1935, Rungs1942, Bodenh1943, BazaroSh1971], *Ephedra alata* [Balach1948b], *Ephedra altissima* [Rungs1935], *Ephedra campylopoda* [Korone1934], *Ephedra nebrodensis* [Lindin1912b, Leonar1918, Leonar1920, Balach1930c, Balach1932d, GomezM1937, Martin1983], *Ephedra procera* [Archan1930], *Ephedra rolandii* [Balach1948b, Balach1956], *Ephedra scoparia* [Lindin1912b, GomezM1937, Martin1983], *Ephedra tilhoana* [Balach1956]. **Liliaceae**: *Asparagus* [Balach1956, BazaroSh1971], *Asparagus stipularis* [Rungs1935].

NATURAL ENEMIES: HYMENOPTERA **Aphelinidae**: *Coccobius sybariticus* Pedata [Pedata1999].

DISTRIBUTION: **Afrotropical**: Central African Republic [Balach1956]; Mauritania [Balach1956, BalachMa1970]. **Palaearctic**: Algeria [Balach1930c]; Armenia [BazaroSh1971]; Greece [Korone1934]; Iran [Kaussa1955]; Iraq [Bodenh1943, Balach1948b]; Italy [Leonar1918, Leonar1920, LongoMaPe1995]; Morocco [Balach1932d, Rungs1935, Rungs1948]; Russia (Dagestan AR [Danzig1993]); Sardinia [Maleno1916]; Spain [GomezM1937, GomezM1946, Martin1983, BlayGo1993]; Tajikistan (=Tadzhikistan) [BazaroSh1971]; Turkmenistan [Archan1930, BazaroSh1971]; Uzbekistan [BazaroSh1971]; Western Sahara [Rungs1942].

GENERAL: Description and illustration of adult female by Malenotti (1916), Balachowsky (1948b, 1956), Bazarov & Shmelev (1971), Tang (1984) and by Danzig (1993).

KEYS: Danzig 1993: 150 (female) [Europe]; Bazarov & Shmelev 1971: 189 (female) [Central Asia]; Balachowsky 1956: 16 (female) [Africa]; Lupo 1953a: 75-76 (female) [Italy]; Balachowsky 1948b: 307 (female) [Mediterranean]; Leonardi 1920: 90 (female) [Italy].

CITATIONS: Archan1930 [host, distribution: 85]; Balach1930c [host, distribution: 119]; Balach1932d [taxonomy, host, distribution: VII]; Balach1948b [taxonomy, description, illustration, host, distribution, biological control: 309-311]; Balach1956 [taxonomy, description, illustration, host, distribution: 18-19]; Balach1958a [host, distribution: 34]; BalachMa1970 [host, distribution: 1080]; BazaroSh1971 [taxonomy, description, illustration, host, distribution: 189-191]; BlayGo1993 [taxonomy, description, illustration, host, distribution: 446-449]; Bodenh1943 [host, distribution: 4]; Borchs1937 [taxonomy, description, illustration, host, distribution: 123-124]; Borchs1939 [taxonomy, description, host, distribution: 9,26]; Borchs1949c [taxonomy: 738]; Borchs1949d [taxonomy, description, host, distribution: 238]; Borchs1950b [taxonomy, description, host, distribution: 216]; Borchs1966 [catalogue: 273-274]; Bustsh1958 [taxonomy, description, illustration, host, distribution: 219]; Bustsh1960 [taxonomy, description, host, distribution: 180]; Danzig1993 [taxonomy, description, illustration, host, distribution: 156]; DanzigPe1998 [catalogue: 253-254]; EzzatNa1987 [distribution: 87]; Ferris1941e [taxonomy: 43]; GomezM1937 [taxonomy, description, host, distribution: 116-117];

GomezM1946 [taxonomy, description, illustration, host, distribution: 62-66]; GomezM1958a [host, distribution: 7]; GomezM1962 [taxonomy, description, illustration, host, distribution: 62-66]; Kaussa1955 [host, distribution: 15]; Korone1934 [taxonomy, description, illustration, host, distribution: 14]; Leonar1918 [host, distribution: 192]; Leonar1920 [taxonomy, description, illustration, host, distribution: 96-97]; Lindin1912b [taxonomy: 139]; Lindin1957 [taxonomy: 548]; LongoMaPe1995 [distribution: 126]; Lupo1953a [taxonomy, description, illustration, host, distribution: 81-86]; MacGil1921 [taxonomy, description, host, distribution: 429]; Maleno1916 [taxonomy, description, illustration, host, distribution: 309-311]; Martin1983 [taxonomy, host, distribution: 64]; MillerDa1990 [host, distribution, economic importance: 302]; Paoli1915 [taxonomy, description, host, distribution: 273]; Pedata1999 [host, distribution, life history, biological control: 45-51]; Rungs1935 [host, distribution: 273]; Rungs1942 [taxonomy, host, distribution: 107]; Rungs1948 [host, distribution: 111]; Sassce1915 [taxonomy, host, distribution: 34]; Tang1984 [taxonomy, description, illustration, host, distribution: 33-34]; TerGri1962 [taxonomy, host, distribution: 148]; WeidneWa1968 [host, distribution: 172].

Dynaspidiotus ericarum (Goux)
Aspidiotus ericarum Goux, 1937c: 345. Type data: FRANCE: Var, Tamaris, on *Erica arborea*. Holotype female. Type depository: Paris: Muséum national d'Histoire naturelle, France.
Dynaspidiotus ericarum; Balachowsky, 1948b: 332. Change of combination.
SCALE COVER: Female scale elongate, 1.5-1.6 mm long, 1 mm wide; colour grey-blue; exuviae yellow gold, slightly eccentric; ventral vellum poorly developed (Goux, 1937c).
HOST PLANTS: **Ericaceae**: *Arbutus unedo* [Balach1948b], *Erica arborea* [Goux1937c, Balach1948b, Kaussa1954], *Erica multiflora* [Goux1937c, Balach1948b, Kaussa1954]. **Leguminosae**: *Ceratonia siliqua* [Balach1948b]. **Myrtaceae**: *Myrtus communis* [Balach1948b, Kaussa1954]. **Oleaceae**: *Olea europaea* [Balach1948b]. **Pistaciaceae**: *Pistacia lentiscus* [Balach1948b, Kaussa1954]. **Santalaceae**: *Osyris alba* [Foldi2000].
DISTRIBUTION: **Palaearctic**: France [Goux1937c, Foldi2000]; Iran [Kaussa1954]; Turkey [Balach1948b].
GENERAL: Description and illustration of adult female by Goux (1937c).
KEYS: Balachowsky 1948b: 327 (female) [Mediterranean].
CITATIONS: Balach1948b [taxonomy, description, host, distribution: 322-323]; Balach1951 [taxonomy: 697]; Borchs1966 [catalogue: 282-283]; Bustsh1958 [taxonomy, description, illustration, host, distribution: 220,227]; DanzigPe1998 [catalogue: 254]; Ferris1941e [taxonomy: 43]; Foldi2000 [host, distribution: 84]; Foldi2001 [distribution: 303-308]; Goux1937c [taxonomy, description, illustration, host, distribution: 345-349]; Kaussa1954 [host, distribution: 80]; Lindin1943b [taxonomy: 207]; Lindin1957 [taxonomy: 545].

Dynaspidiotus greeni Balachowsky

Dynaspidiotus greeni Balachowsky, 1951: 695. Type data: CYPRUS: on *Asparagus* sp. Syntypes, female. Type depository: Paris: Muséum national d'Histoire naturelle, France.

SCALE COVER: Female scale circular or subcircular, 1.8-2 mm; black, matt; exuviae brown red, central. Male scale unknown (Balachowsky, 1951).

HOST PLANTS: **Liliaceae**: *Asparagus* [Balach1951].

DISTRIBUTION: **Palaearctic**: Cyprus [Balach1951].

GENERAL: Description and illustration of adult female by Balachowsky (1951).

CITATIONS: Balach1951 [taxonomy, description, illustration, host, distribution: 695-697]; Borchs1966 [catalogue: 283]; DanzigPe1998 [catalogue: 254]; Kaussa1954 [host, distribution: 80].

Dynaspidiotus medicus Kaussari

Dynaspidiotus medicus Kaussari, 1956: 102. Type data: IRAN: Shiraz Province, Dunheh, on *Haplophyllum* sp. Holotype female. Type depository: Paris: Muséum national d'Histoire naturelle, France.

SCALE COVER: Female scale circular, 2.2 mm in diameter; convex; exuviae yellow-gold, central; adult secretion light brown or slightly pink, with concentric areas, darker and variously coloured (Kaussari, 1956).

HOST PLANTS: **Cruciferae**: *Sameraria* [Kaussa1956]. **Rutaceae**: *Haplophyllum* [Kaussa1956].

DISTRIBUTION: **Palaearctic**: Iran [Kaussa1956].

GENERAL: Description and illustration of adult female by Kaussari (1956).

CITATIONS: Balach1958b [taxonomy: 163]; Borchs1966 [catalogue: 283]; DanzigPe1998 [catalogue: 254]; Kaussa1956 [taxonomy, description, illustration, host, distribution: 102-103].

Dynaspidiotus meyeri (Marlatt)

Aspidiotus meyeri Marlatt, 1908c: 13. Type data: CHINA: Wu Tai Shan, near Peking, on *Abies* sp.; collected by F.N. Meyer, April 21 1908. Holotype female. Type depository: Washington: U.S. National Entomological Collection, U.S. National Museum of Natural History, District of Columbia, USA; type no. 14137.

Aspidiella meyeri; MacGillivray, 1921: 404. Change of combination.

Dynaspidiotus meyeri; Borchsenius, 1966: 283. Change of combination.

SCALE COVER: Scale of female elongate oval, 1.5-2 mm long; very convex; exuviae subcentral; larval secretion whitish except as obscured by extraneous matter; main or subsequent secretion dull resinous brown; larval secretion and exuvia easily lost, exposing large, bright orange coloured second exuvia; scale adhering very loosely to leaf and easily dislodged, leaving insect exposed; inner surface of scale covered with snow-white secretion which entirely conceals exuviae. Male scale not certainly identified, but apparently similar to female, except in size and number of exuviae (Marlatt, 1908c).

HOST PLANTS: **Pinaceae**: *Abies* [Marlat1908c, Sander1909a, Kuwana1927].

DISTRIBUTION: **Palaearctic**: China (People's Republic) [Sander1909a, Kuwana1927] (Beijing (Peking) [Marlat1908c]).

GENERAL: Description and illustration of adult female by Marlatt (1908c).
CITATIONS: Borchs1966 [catalogue: 283]; Chou1985 [taxonomy, description, host, distribution: 282]; DanzigPe1998 [catalogue: 254]; Ferris1941e [taxonomy: 45]; Kuwana1927 [host, distribution: 71-72]; Kuwana1932 [taxonomy: 53]; MacGil1921 [taxonomy, description, host, distribution: 404]; Marlat1908c [taxonomy, description, illustration, host, distribution: 13-14]; Sander1909a [taxonomy, host, distribution: 53]; Tao1999 [taxonomy, host, distribution: 85].

Dynaspidiotus piceae (Tang, Hao, Shi & Tang), new combination
Tsugaspidiotus piceae Tang, Hao, Shi & Tang, 1991: 460. Type data: CHINA: Shanxi, Shui-men, on *Picea asparata*; collected by Guan-lu Shi, July 25, 1985. Holotype female. Type depository: Shanxi: Entomological Institute, Shanxi Agricultural University, Taigu, Shanxi, China.
SCALE COVER: Tang et al. (1991) did not describe the scale cover.
HOST PLANTS: **Pinaceae**: *Picea asparata* [TangHaSh1991].
DISTRIBUTION: **Palaearctic**: China (People's Republic) (Shanxi (Shensi) [TangHaSh1991]).
GENERAL: Description and illustration of adult female by Tang et al. (1991).
CITATIONS: TangHaSh1991 [taxonomy, description, illustration, host, distribution: 469-461,463-464].

Dynaspidiotus pseudomeyeri (Kuwana)
Aspidiotus pseudomeyeri Kuwana, 1932: 52. Type data: JAPAN: on *Juniperus chinensis*. Syntypes, female. Type depository: Ibaraki-ken: Insect Taxonomy Laboratory, National Institute of Agricultural Environmental Sciences, Kannon-dai, Yatabe, Tsukuba-shi, (I. Kuwana), Japan.
Tsugaspidiotus pseudomeyeri; Takahashi & Takagi, 1957: 104. Change of combination.
Dynaspidiotus pseudomeyeri; Danzig, 1993: 149. Change of combination.
SCALE COVER: Female scale elongate oval or irregularly circular, convex; dull greenish yellow; older specimens dirty pale yellow with posterior margin brown (Kuwana, 1932).
HOST PLANTS: **Cupressaceae**: *Chamaecyparis* [Kawai1977], *Ch. pisifera* [TakahaTa1957], *Juniperus chinensis* [Kuwana1932, Kuwana1933, TakahaTa1957].
DISTRIBUTION: **Nearctic**: U.S.A. (New York [Nakaha1982], Pennsylvania [Nakaha1982]). **Palaearctic**: Japan [Kuwana1932, TakahaTa1957, Kawai1980] (Hokkaido [TakahaTa1957]).
GENERAL: Description and illustration of adult female by Kuwana (1932, 1933), Takahashi & Takagi (1957) and by Kosztarab (1996).
KEYS: Kosztarab 1996: 543 (female) [Northeastern North America]; Takahashi & Takagi 1957: 102 (female) [Japan]; Kuwana 1933: 3 (female) [Japan]; Kuwana 1933b: 49 (female) [Japan].
CITATIONS: Borchs1966 [catalogue: 316]; Brown1960a [taxonomy, structure: 135-160]; DanzigPe1998 [catalogue: 364]; Ferris1941e [taxonomy: 47]; Kawai1977 [host, distribution, economic importance: 157]; Kawai1980 [taxonomy, description, host, distribution: 223]; Koszta1996 [taxonomy, description, illustration, host,

distribution, life history: 545-546]; Kuwana1932 [taxonomy, description, illustration, host, distribution: 52-54]; Kuwana1933 [taxonomy, description, illustration, host, distribution: 17-18]; Muraka1970 [host, distribution: 78]; Nakaha1982 [host, distribution: 61]; Stimme2002 [host, distribution, life history, economic importance, control: 27-29]; TakahaTa1957 [taxonomy, description, illustration, host, distribution: 104-105]; TangHaSh1991 [taxonomy: 468-469].

Dynaspidiotus regius (Brain)

Aspidiotus regius Brain, 1918: 121. Type data: SOUTH AFRICA: Cape Province, King Williamstown, on *Aloe* sp.; collected by A. Kelly, March 1913. Lectotype female, by subsequent designation Munting, 1971: 139. Type depository: Washington: United States National Entomological Collection, U.S. National Museum of Natural History, District of Columbia, USA; type no. 200/1.

Dynaspidiotus regius; Balachowsky, 1958b: 162. Change of combination.

SCALE COVER: Female scale small, elongate, about 1.4 mm long and 0.7 mm broad, usually widest about middle and narrowed to each end; buff in colour, with brownish exuviae showing through thin layer of secretion; scale somewhat translucent (Brain, 1918).

HOST PLANTS: **Liliaceae**: *Aloe* [Brain1918, Muntin1971].

DISTRIBUTION: **Afrotropical**: South Africa [Brain1918, Muntin1971].

GENERAL: Description and illustration of adult female by Brain (1918) and by Munting (1971).

KEYS: Brain 1918: 117 (female) [South Africa].

CITATIONS: Balach1958b [taxonomy: 162]; Borchs1966 [catalogue: 283]; Brain1918 [taxonomy, description, illustration, host, distribution: 121-122]; Ferris1941e [taxonomy: 47]; MacGil1921 [taxonomy, description, host, distribution: 403]; Muntin1971 [taxonomy, description, illustration, host, distribution: 139-141].

Dynaspidiotus regnieri (Balachowsky)

Hemiberlesia regnieri Balachowsky, 1928a: 123. Type data: MOROCCO: Cedraies d'Azrou (Moyen Atlas), on *Cedrus libanotica* v. *atlantica*; collected 19 April, 1926. Holotype female. Type depository: Paris: Muséum national d'Histoire naturelle, France.

Hemiberlesia regneri; Balachowsky, 1932d: VI. Misspelling of species name.

Dynaspidiotus regneri; Balachowsky, 1948b: 342. Change of combination.

Dynaspidiotus regneri; Balachowsky, 1948b: 342. Misspelling of species name.

Dynaspidiotus regneri; Kaussari, 1954: 80. Misspelling of species name.

Dynaspidiotus regneri; Balachowsky, 1954b: 109. Misspelling of species name.

Nuculaspis regnieri; Borchsenius, 1966: 275. Change of combination.

Dynaspidiotus regnieri; Danzig & Pellizzari, 1998: 254. Revived combination.

SCALE COVER: Illustration of scale cover by Balachowsky (1928a). Female scale oval, elongate, 2.5-2.6 mm long; thin and truncated at apices; highly convex, retracted laterally; exuviae central, yellow gold; secretion of adult white; ventral vellum almost absent. Male scale unknown (Balachowsky, 1928a).

HOST PLANTS: **Pinaceae**: *Cedrus libanotica atlantica* [Balach1928a, Balach1932d, Balach1954b, BlayGo1993].

NATURAL ENEMIES: COLEOPTERA **Coccinellidae**: *Chilocorus bipustulatus* [DelEstSoVi1994].

DISTRIBUTION: **Palaearctic**: Algeria [Balach1954b]; Morocco [Balach1928a, Balach1932d, Balach1954b]; Spain [BlayGo1993].

ECONOMIC IMPORTANCE: Reported as a serious pest of *Cedrus* in Spain (Del Estal *et al.*, 1994).

GENERAL: Description and illustration of adult female by Balachowsky (1928a, 1948b).

KEYS: Balachowsky 1948b: 327 (female) [Mediterranean]; Balachowsky 1928a: 132 (female) [North Africa].

CITATIONS: Balach1928a [taxonomy, description, illustration, host, distribution: 123-125]; Balach1932d [taxonomy, host, distribution, economic importance: VI]; Balach1948b [taxonomy, description, illustration, host, distribution: 342-345]; Balach1954b [host, distribution: 109]; BlayGo1993 [taxonomy, description, illustration, host, distribution: 462-466]; Borchs1966 [catalogue: 275]; DanzigPe1998 [catalogue: 254]; DelEstSoVi1994 [host, distribution, life history, economic importance, biological control: 477-486]; Ferris1941e [taxonomy: 47]; Kaussa1954 [host, distribution: 80].

Dynaspidiotus rhodesiensis (Hall)

Aspidiotus (*Hemiberlesea*) *rhodesiensis* Hall, 1928: 274. Type data: ZIMBABWE: Umtali, Mazoe and Banket, on smaller branches of *Brachystegia flagristipulata*, *Brachystegia* sp., *Berlinia globiflora* and *Parinarium mobola*. Syntypes, female. Type depository: London: The Natural History Museum, England, UK.

Hemiberlesea rhodesiensis; Hall, 1929: 353. Change of combination.

Abgrallaspis rhodesiensis; Balachowsky, 1956: 16. Change of combination.

Ephedraspis rhodesiensis; Borchsenius, 1966: 274. Change of combination.

Dynaspidiotus rhodesiensis; Danzig, 1993: 149. Change of combination.

SCALE COVER: Female scale highly convex and usually circular, except where scales are crowded; exuviae central; diameter 1-1.25 mm. Male scale of usual form; exuviae yellow, with a dark brown medio-dorsal stripe (Hall, 1928).

HOST PLANTS: **Leguminosae**: *Berlinia globiflora* [Hall1928, Balach1956], *Brachystegia* [Hall1928, Balach1956], *Brachystegia fragristipulata* [Hall1928, Balach1956]. **Rosaceae**: *Parinarium mobola* [Hall1928, Balach1956].

DISTRIBUTION: **Afrotropical**: Zimbabwe [Hall1928, Balach1956].

GENERAL: Description and illustration of adult female by Hall (1928) and by Balachowsky (1956).

KEYS: Balachowsky 1956: 16 (female) [Africa].

CITATIONS: Balach1956 [taxonomy, description, illustration, host, distribution: 18-21]; Borchs1966 [catalogue: 274]; Ferris1941e [taxonomy: 47]; Hall1928 [taxonomy, description, illustration, host, distribution: 274-275]; Hall1929 [taxonomy: 352-353].

Dynaspidiotus riograndensis Wolff & Claps

Dynaspidiotus riograndensis Wolff & Claps, 1999: 22. Type data: BRAZIL: Rio Grande do Sul, São Pedro da Serra, on *Araucaria angustifolia*. Holotype female. Type depository: Porto Alegre: Colecao de Entomologia, do Museu de Ciencia e Tecnologia da Pontificia Universidade Catolica do Rio Grande do Sul, Brazil.

SCALE COVER: Male scale elongate, semitransparent, with exuviae eccentric; 0.96 mm long, 0.50 mm wide. Female scale oval; highly convex; with concentric line of growth throughout scale; white, with exuviae in apex, yellow; 2.16 mm long; 2.07 mm wide (Wolff & Claps, 1999). Colour photograph of female and male scale cover by Wolff & Claps (1999).

HOST PLANTS: **Araucariaceae**: *Araucaria angustifolia* [WolffCl1999].

DISTRIBUTION: **Neotropical**: Brazil (Rio Grande do Sul [WolffCl1999, ClapsWoGo2001]).

GENERAL: Description and illustration of adult female by Wolff & Claps (1999).

CITATIONS: ClapsWoGo2001 [host, distribution: 243-244]; WolffCl1999 [taxonomy, description, illustration, host, distribution: 22-25].

Dynaspidiotus sanctadelaidae Lepage

Dynaspidiotus sanctadelaidae Lepage, 1942: 181. Type data: BRAZIL: Chacara Santa Adelaide, do Ginasio das Conegas de Sto. Agostinho, in Campos do Jordao, on *Araucaria brasiliensis*. Syntypes, female. Type depositories: São Paulo: Instituto Biologico de São Paulo, Brazil, and Curitiba: Departamento de Zoologia, Setor de Ciencias Biologicas, Universidade Federal do Parana, Brazil.

Aonidiella sanctadelaidae; Costa Lima, 1951: 173. Change of combination.

Dynaspidiotus sanctadelaidae; Borchsenius, 1966: 283. Revived combination.

SCALE COVER: Female scale white, circular, 1.8 mm diameter; exuviae yellow subcentral. Male scale white, elongate, exuviae terminal (Lepage, 1942).

HOST PLANTS: **Araucariaceae**: *Araucaria angustifolia* [ClapsWoGo2001], *Araucaria brasiliensis* [Lepage1942].

DISTRIBUTION: **Neotropical**: Brazil (São Paulo [Lepage1942, ClapsWoGo2001]).

GENERAL: Description and illustration of adult female by Lepage (1942).

CITATIONS: Borchs1966 [catalogue: 283]; Claps1993 [taxonomy: 6,9]; ClapsWoGo2001 [host, distribution: 244]; CostaL1951 [taxonomy: 173]; Lepage1942 [taxonomy, description, illustration, host, distribution: 181-183].

Dynaspidiotus spartii Kaussari

Dynaspidiotus spartii Kaussari, 1954: 82. Type data: IRAN: Kerman-Shah, on *Spartium junceum*; collected December 1950, by M. Sarkissian. Syntypes, female. Type depository: Paris: Muséum national d'Histoire naturelle, France.

SCALE COVER: Female scale circular; exuviae central, light brown; adult secretion grey brown (Kaussari, 1954).

HOST PLANTS: **Leguminosae**: *Spartium* [Kaussa1957], *Spartium junceum* [Kaussa1954].

DISTRIBUTION: **Palaearctic**: Iran [Kaussa1954, Kaussa1957].

GENERAL: Description and illustration of adult female by Kaussari (1954).

CITATIONS: Borchs1966 [catalogue: 283]; DanzigPe1998 [catalogue: 255]; Kaussa1954 [taxonomy, description, illustration, host, distribution: 82-83]; Kaussa1957 [host, distribution: 1].

Dynaspidiotus tener (Bazarov & Shmelev)

Ephedraspis tenera Bazarov & Shmelev, 1967: 60. Type data: TAJIKISTAN: Western Pamir, Pyandja Valley, Pastruh Village, 20 km east of Vancha, on *Ephedra* sp. Syntypes, female. Type depository: St. Petersburg: (= Leningrad) Zoological Museum, Academy of Science, Russia.

Dynaspidiotus tener; Danzig, 1993: 156. Change of combination requiring emendation of species name for agreement in gender.

COMMON NAME: pamirskaya seraya efedrovaya shitovka [BazaroSh1971].

SCALE COVER: Female scale broadly elliptical, 1.8 mm long, 1.4 mm wide; convex; exuviae central or slightly towards anterior area; first exuvia orange, second bright grey; secreted part dark grey slightly brownish (Bazarov & Shmelev, 1971).

HOST PLANTS: **Ephedraceae**: *Ephedra* [BazaroSh1971, Danzig1993].

DISTRIBUTION: **Palaearctic**: Tajikistan (=Tadzhikistan) [BazaroSh1971993].

GENERAL: Description and illustration of adult female by Bazarov & Shmelev (1971) and by Danzig (1993).

KEYS: Danzig 1993: 150 (female) [Europe]; Bazarov & Shmelev 1971: 189 (female) [Central Asia].

CITATIONS: BazaroSh1971 [taxonomy, description, illustration, host, distribution: 191-193]; Danzig1993 [taxonomy, description, illustration, host, distribution: 156-159]; DanzigPe1998 [catalogue: 255].

Dynaspidiotus tsugae (Marlatt)

Aspidiotus (*Diaspidiotus*) *tsugae* Marlatt, 1911: 385. Type data: JAPAN: on *Tsuga* sp. Holotype female. Type depository: Washington: United States National Entomological Collection, U.S. National Museum of Natural History, District of Columbia, USA; type no. 14.185.

Aspidiotus tsugae; Sasscer, 1912: 94. Change of combination.

Furcaspis tsugae; MacGillivray, 1921: 407. Change of combination.

Tsugaspidiotus tsugae; Takahashi & Takagi, 1957: 103. Change of combination.

Nuculaspis tsugae; McClure, 1990a: 165. Change of combination.

Dynaspidiotus tsugae; Danzig, 1993: 149. Change of combination.

COMMON NAME: Shortneedle Evergreen Scale [Koszta1996].

SCALE COVER: Female scale circular, 1-1.3 mm in diameter; strongly convex; dark brown; pointed or nippled at center; central area covered by secretion; light resinous yellow when rubbed. Male scale oval, smaller than female; secretion covering center or nipple, somewhat ashen, forming a light central spot (Marlatt, 1911).

HOST PLANTS: **Pinaceae**: *Abies* [Danzig1978], *Abies firma* [Kuwana1933, TakahaTa1957], *Abies halophyla* [Danzig1993], *Picea* [Danzig1978], *Picea ajanensis* [Danzig1993], *Picea excelsa* [TakahaTa1957], *Picea glehni* [Danzig1993], *Pinus koraiensis* [Danzig1978, Danzig1993], *Tsuga* [Marlat1911, Danzig1978, Danzig1993], *Tsuga sieboldii* [Kuwana1933, TakahaTa1957].

NATURAL ENEMIES: HYMENOPTERA **Encyrtidae**: *Aspidiotiphagus citrinus* [Koszta1996].

DISTRIBUTION: **Nearctic**: United States of America (Connecticut [McClurFe1977, Nakaha1982], Maryland [Koszta1996], New Jersey [Nakaha1982, Koszta1996], Rhode Island [Nakaha1982, Koszta1996]). **Palaearctic**: Japan [Marlat1911, Kuwana1917a, TakahaTa1957, Danzig1978, Nakaha1982] (Hokkaido [TakahaTa1957]); Russia (Primor'ye Kray [Danzig1978, Danzig1988], Sakhalin Oblast [Danzig1978, Danzig1988]); South Korea [Nakaha1982].

ECONOMIC IMPORTANCE: A serious pest of Eastern hemlock, *Tsuga canadensis* in Northeastern states of USA (McClure & Fergione, 1977; McClure, 1982; Kosztarab, 1996).

GENERAL: Description and illustration of adult female by Marlatt (1911), Kuwana (1933), Takahashi & Takagi (1957), Danzig (1993) and by Kosztarab (1996).

KEYS: Kosztarab 1996: 543 (female) [Northeastern North America]; Danzig 1993: 150 (female) [Europe]; Takahashi & Takagi 1957: 102 (female) [Japan]; Kuwana 1933: 3 (female) [Japan]; Kuwana 1933b: 49 (female) [Japan].

CITATIONS: Borchs1966 [catalogue: 316]; Britto1924 [host, distribution: 387-404]; Britto1938 [host, distribution: 137]; Danzig1977b [taxonomy: 57]; Danzig1978 [host, distribution: 21]; Danzig1988 [taxonomy, host, distribution: 725]; Danzig1993 [taxonomy, description, illustration, host, distribution: 154-156]; DanzigPe1998 [catalogue: 255,364]; DennoMcOt1995 [host, distribution, life history, ecology: 297-331]; Ferris1941e [taxonomy: 49]; Kawai1980 [taxonomy, description, host, distribution: 223]; Koszta1996 [taxonomy, description, illustration, host, distribution, biology, biological control, economic importance: 545,547-548]; Kuwana1917a [taxonomy, distribution: 175]; Kuwana1933 [taxonomy, description, illustration, host, distribution: 19-20]; MacGil1921 [taxonomy, description, host, distribution: 407]; Marlat1911 [taxonomy, description, illustration, host, distribution: 385-387]; McClur1978 [host, distribution, life history, ecology, biological control: 863-870]; McClur1980a [host, distribution, life history, ecology: 1391-1401]; McClur1981 [host, distribution, life history, ecology: 47-54]; McClur1982 [host, distribution, life history, ecology, economic importance: 150-157]; McClur1983b [host, distribution, life history, ecology: 1811-1815]; McClur1985 [host, distribution, life history, ecology: 413-415]; McClur1986a [host, distribution, life history, ecology: 141-142]; McClur1988 [host, distribution, life history, ecology, biological control: 46-65]; McClur1989 [host, distribution, life history, ecology: 291-302]; McClur1990a [taxonomy, host, distribution, ecology: 165-168]; McClur1990b [ecology, host, distribution: 285-288]; McClur1990d [taxonomy, host, distribution, ecology: 301-303]; McClur1990e [life history, ecology: 309-314]; McClur1990g [ecology, host: 313-330]; McClur1990h [life history, ecology: 331-337]; McClur1991 [host, distribution, ecology: 256-270]; McClurFe1977 [host, distribution, life history, economic importance: 807-811]; McClurHa1984 [host, life history, chemistry: 185-193]; MillerDa1990 [host, distribution, economic importance: 304]; Muraka1970 [host, distribution: 78]; Nakaha1982 [host, distribution: 61]; Sassce1912 [taxonomy, host, distribution: 94]; Takaha1955e [taxonomy: 77]; TakahaTa1957 [taxonomy, description, illustration, host, distribution: 103-104]; Zahrad1990 [host, distribution, description: 642].

Dynaspidiotus umtalii (Hall)

Aspidiotus (*Hemiberlesea*) *rhodesiensis umtalii* Hall, 1929: 352. Type data: ZIMBABWE: Umtali, on branches of *Brachystegia* sp. Syntypes, female. Type depository: London: The Natural History Museum, England, UK.

Aspidiotus umtalii; Ferris, 1941e: 49. Change of combination and rank.

Abgrallaspis rhodesiensis umtalii; Balachowsky, 1956: 16. Change of combination and rank.

Ephedraspis umtalii; Borchsenius, 1966: 274. Change of combination and rank.

Dynaspidiotus umtalii; Danzig, 1993: 149. Change of combination.

SCALE COVER: Hall (1929) first introduced this as a variety of *Dynaspidiotus rhodesiensis*, distinguishing the former only on differences in microscopic characters.

HOST PLANTS: **Leguminosae**: *Brachystegia* [Hall1929, Balach1956].

DISTRIBUTION: **Afrotropical**: Zimbabwe [Hall1929, Balach1956].

GENERAL: Description and illustration of adult female by Hall (1929) and by Balachowsky (1956).

KEYS: Balachowsky 1956: 16 (female) [Africa].

CITATIONS: Balach1956 [taxonomy, description, illustration, host, distribution: 20-21]; Borchs1966 [catalogue: 274]; Ferris1941e [taxonomy: 49]; Hall1929 [taxonomy, description, illustration, host, distribution: 352-353].

Entaspidiotus MacGillivray

Entaspidiotus MacGillivray, 1921: 394. Type species: *Selenaspidus magnus* Lindinger, by original designation.

SYSTEMATICS: The genus *Entaspidiotus* differs from other genera of the *Selenaspidus* complex, in the variously shaped third lobe, which is never spur-like. Also related to *Duplaspidiotus* and *Pseudaonidia* but differs in having a constriction between thoracic segments (Mamet, 1958a).

GENERAL: Definition and characters by Mamet (1958a).

KEYS: Ben-Dov 1974c: 28 (female) [World]; Mamet 1958a: 362 (female) [*Selenaspidus* complex]; Mamet 1958a: 414 (female) [world].

CITATIONS: BenDov1974c [taxonomy: 28]; Borchs1966 [catalogue: 252]; Ferris1937c [taxonomy: 51]; Lindin1937 [taxonomy: 184]; MacGil1921 [taxonomy, description: 394,454-456]; Mamet1958a [taxonomy, description: 362,413-414]; MorrisMo1966 [taxonomy, catalogue: 66]; Muntin1969 [taxonomy: 128].

Entaspidiotus giliomeei Williams

Entaspidiotus giliomeei Williams, 1963a: 44. Type data: SOUTH AFRICA: Stellenbosch, on fleshy leaves of *Cheiridopsis* spp.; collected 19.VII.1961. Holotype female. Type depository: London: The Natural History Museum, England, UK.

SCALE COVER: Female scale circular, white and slightly convex; exuviae subcentral, shiny reddish-brown but usually covered with a layer of white secretion. Diameter about 1.5 mm. Male scale elongate, white and uncarinated; exuviae terminal, reddish-brown. Length about 1.2 mm (Williams, 1963).

HOST PLANTS: **Aizoaceae**: *Cheiridopsis* [Willia1963a], *Cheiridopsis crassa* [Willia1963a].
DISTRIBUTION: **Afrotropical**: South Africa [Willia1963a].
GENERAL: Description and illustration of adult female by Williams (1963a).
KEYS: Ben-Dov 1974c: 28 (female) [World]; Williams 1963a: 46 (female) [World].
CITATIONS: BenDov1974c [taxonomy: 28]; Borchs1966 [catalogue: 252]; Willia1963a [taxonomy, description, illustration, host, distribution: 44-45].

Entaspidiotus lounsburyi (Marlatt)

Pseudaonidia (*Selenaspidus*) *lounsburyi* Marlatt, 1908: 136,139. Type data: SOUTH AFRICA: Cape Town, on *Mesembryanthemum edule*; collected by C.P. Lounsbury, June 29, 1897. Holotype female. Type depository: Washington: United States National Entomological Collection, U.S. National Museum of Natural History, District of Columbia, USA; type no. 7693.
Pseudaonidia lounsburyi; Sanders, 1909a: 54. Change of combination.
Aspidiotus (*Selenaspidus*) *lounsburyi*; Brain, 1918: 131. Change of combination.
Aspidiotus (*Selenaspidus*) *lounsburyi var.* Brain, 1918: 131. Type data: SOUTH AFRICA: Orange Free State, Bloemfontein, on *Mesembryanthemum edule*; collected June 3, 1915, by J.C. Faure. Syntypes, female. Type depository: Pretoria: South African National Collection of Insects, South Africa; type no. B221. Synonymy by Borchsenius, 1966: 252.
Entaspidiotus lounsburyi; MacGillivray, 1921: 455. Change of combination.
Entaspidiotus lounsburyi; Mamet, 1958a: 414. Revived combination.
Selenaspidus (*Aspidiotus*) *lounsburgi*; Borchsenius, 1966: 404. Misspelling of species name.
SCALE COVER: Female scale subcircular, 2-2.5 mm in diameter; flat; yellowish white but dense and opaque; exuviae resinous to brown, covered with a very slight secretion; secreted part of scale usually three times diameter of second exuvia; ventral scale thin but distinct, adheres to the leaf. Male scale similar, oval, 1.5 mm long; exuvia brown, near anterior end (Marlatt, 1908). Female scale circular to subcircular, yellowish white to greyish white, flat; exuviae central, pale brown to dark buff, obscured by layer of secretion; diameter about 2.5 mm. Scale of male same colour, but smaller, oval; exuviae towards one end, brownish (Mamet, 1958a). Colour photograph of scale cover by Russo et al. (1999).
HOST PLANTS: **Aizoaceae**: *Carpobrotus* [Muntin1969], *Conophytum bilobum* [Mamet1958a, Willia1963a, Borchs1966], *Mesembryanthemum* [Mamet1958a, Borchs1966], *Mesembryanthemum edule* [Marlat1908, Sander1909a, Brain1918, Mamet1958a, Willia1963a, Borchs1966].
NATURAL ENEMIES: HYMENOPTERA **Encyrtidae**: *Metaphycus aspidiotinorum* Compere [AnneckIn1971].
DISTRIBUTION: **Afrotropical**: Namibia (Southwest Africa) [Muntin1969]; South Africa [Marlat1908, Sander1909a, Brain1918, Mamet1958a, Willia1963a, Borchs1966]. **Palaearctic**: Italy [RussoMaSu1999].
GENERAL: Description and illustration of adult female by Brain (1918) and by Mamet (1958a).

KEYS: Ben-Dov 1974c: 28 (female) [World]; Williams 1963a: 46 (female) [World]; Brain 1918: 130 (female) [South Africa]; Marlatt 1908: 134 (female) [World].
CITATIONS: AnneckIn1971 [host, distribution, biological control: 17]; BenDov1974c [taxonomy: 28]; Borchs1966 [catalogue: 252,404]; Brain1918 [taxonomy, description, illustration, host, distribution: 131]; Comper1940a [host, distribution, biological control: 7-33]; Ferris1941e [taxonomy: 45]; MacGil1921 [taxonomy, description, host, distribution: 455]; Mamet1958a [taxonomy, description, illustration, host, distribution: 414-415]; Marlat1908 [taxonomy, description, illustration, host, distribution: 134-140]; Muntin1969 [host, distribution: 128]; Prinsl1983 [distribution, biological control: 26]; RussoMaSu1999 [taxonomy, description, illustration, host, distribution: 83-87]; Sander1909a [taxonomy, host, distribution: 54]; Willia1963a [taxonomy, host, distribution: 44].

Entaspidiotus magnus (Lindinger)
Selenaspidus magnus Lindinger, 1909d: 9. Type data: ETHIOPIA: Abyssinia, Harrar, on *Euphorbia* sp. Holotype. Type depository: Hamburg: Zoologisches Institut und Zoologishces Museum, Universität von Hamburg, Germany.
Pseudaonidia magna; Sanders, 1909a: 54. Change of combination requiring emendation of species name for agreement in gender.
Entaspidiotus magnus; MacGillivray, 1921: 454. Change of combination.
SCALE COVER: Female scale subcircular, greyish white to pale buff in colour, flat to slightly convex; exuviae subcentral, brownish, obscured by secretion; diameter about 3 mm. Scale of male elongate, greyish white in colour; exuviae towards one end, brownish; length about 1 mm (Mamet, 1958a).
HOST PLANTS: **Euphorbiaceae**: *Euphorbia* [Lindin1909d, Sander1909a, Mamet1958a, Borchs1966], *Euphorbia abyssinica* [Mamet1958a, Willia1963a].
DISTRIBUTION: **Afrotropical**: Eritrea [Mamet1958a, Borchs1966]; Ethiopia [Lindin1909d, Sander1909a, Mamet1958a, Willia1963a, Borchs1966].
GENERAL: Description and illustration of adult female by Mamet (1958a).
KEYS: Ben-Dov 1974c: 28 (female) [World]; Williams 1963a: 46 (female) [World]; Lindinger 1909d: 4-5 (female) [Genus *Selenaspidus*].
CITATIONS: BenDov1974c [taxonomy: 28]; Borchs1966 [catalogue: 252-253]; Ferris1937c [taxonomy: 51]; Lindin1909c [taxonomy, distribution: 451]; Lindin1909d [taxonomy, description, illustration, host, distribution: 9]; Lindin1910b [host, distribution: 42]; MacGil1921 [taxonomy: 454]; Mamet1958a [taxonomy, description, illustration, host, distribution: 416-418]; Sander1909a [taxonomy: 54]; WeidneWa1968 [taxonomy: 178]; Willia1963a [taxonomy, host, distribution: 44].

Entaspidiotus namaquensis Ben-Dov
Entaspidiotus namaquensis Ben-Dov, 1974c: 28. Type data: SOUTH AFRICA: Cape Province, Namaqualand, Annisfontein, on *Jacobsenia* sp. Holotype. Type depository: Pretoria: South African National Collection of Insects, South Africa.
SCALE COVER: Scale of female and male not observed (Ben-Dov, 1974c).
HOST PLANTS: **Aizoaceae**: *Jacobsenia* [BenDov1974c].
DISTRIBUTION: **Afrotropical**: South Africa [BenDov1974c].

GENERAL: Description and illustration of adult female by Ben-Dov (1974c).
KEYS: Ben-Dov 1974c: 28 (female) [World].
CITATIONS: BenDov1974c [taxonomy, description, illustration, host, distribution: 26-28].

Eremiaspis Balachowsky

Eremiaspis Balachowsky, 1951: 671. Type species: *Hemiberlesia balachowskyi* Rungs, by monotypy and original designation.
SYSTEMATICS: The genus *Eremiaspis* differs from *Rhizaspidiotus* MacGillivray and from *Targionia* Signoret by the presence of dorsal, large tubular ducts with an oval opening. The large anal opening, and poorly developed plates separates this genus from *Aspidiella* Leonardi (Balachowsky, 1951, 1958b).
GENERAL: Definition and characters by Balachowsky (1951, 1958b).
KEYS: Balachowsky 1958b: 281 (female) [*Targionina* of Africa]; Balachowsky 1951: 631 (female) [Mediterranean].
CITATIONS: Balach1951 [taxonomy, description: 671]; Balach1958b [taxonomy, description: 286]; BenDov1974b [taxonomy: 315]; Borchs1966 [catalogue: 248]; DanzigPe1998 [catalogue: 257]; MorrisMo1966 [taxonomy, catalogue: 67].

Eremiaspis balachowskyi (Rungs)
Hemiberlesea balachowskyi Rungs, 1936: 53. Type data: MOROCCO: Oued Draa valley, on *Farsetia hamiltonii*; collected C. Rungs, 22.ii.1934. Syntypes, female. Type depository: Paris: Muséum national d'Histoire naturelle, France.
Aspidiotus balachowskyi; Lindinger, 1943b: 207. Change of combination.
Eremiaspis balachowskyi; Balachowsky, 1951: 671. Change of combination.
Rhizaspidiotus balachowskyi; Bustshik, 1958: 221. Change of combination.
Eremiaspis balachowskyi; Borchsenius, 1966: 248. Revived combination.
SCALE COVER: Female scale of small size, circular or slightly eccentric, 1.3-1.7 mm in diameter; highly convex; grey-rose, frequently covered by white secretion; exuviae pointed, brown-red; nymphal secretion brown-rose; adult secretion bright grey, more or less rose, sometimes white; ventral scale white, thin, adhering to plant (Rungs, 1936).
HOST PLANTS: **Cruciferae**: *Farsetia hamiltonii* [Rungs1936, Balach1958b].
DISTRIBUTION: **Palaearctic**: Morocco [Rungs1936, Balach1951, Balach1958b].
GENERAL: Description and illustration of adult female by Rungs (1936) and by Balachowsky (1951, 1958b).
CITATIONS: Balach1951 [taxonomy, description, illustration, host, distribution: 672-674]; Balach1958a [host, distribution: 39]; Balach1958b [taxonomy, description, illustration, host, distribution: 286-288]; Borchs1966 [catalogue: 248]; Bustsh1958 [taxonomy, description, host, distribution: 219,221]; DanzigPe1998 [catalogue: 257]; Lindin1943b [taxonomy: 207]; Rungs1936 [taxonomy, description, illustration, host, distribution: 53-56].

Eremiaspis graminis Ben-Dov

Eremiaspis graminis Ben-Dov, 1974b: 315. Type data: NAMIBIA: 30 km north of Noordoewer, on *Aristida* cf. *brevifolia*. Holotype. Type depository: Pretoria: South African National Collection of Insects, South Africa.

SCALE COVER: Scale of female white, rounded, 1.0-1.5 mm in diameter, flat; with a well-developed ventral velum, white; nymphal exuviae brown, placed centrally (Ben-Dov, 1974b).

HOST PLANTS: **Gramineae**: *Aristida* cf. *brevifolia* [BenDov1974b].

DISTRIBUTION: **Afrotropical**: Namibia (Southwest Africa) [BenDov1974b].

GENERAL: Description and illustration of adult female by Ben-Dov (1974b).

CITATIONS: BenDov1974b [taxonomy, description, illustration, host, distribution: 315-317].

Eulaingia Brimblecombe

Eulaingia Brimblecombe, 1958: 80. Type species: *Pseudaonidia stenophyllae* Laing, by monotypy and original designation.

SYSTEMATICS: Genus belongs to the *Pseudaonidia - Duplaspidiotus* series of Aspidiotine genera, because of constriction between prothorax and mesothorax. It shares with *Duplaspidiotus* MacGillivray presence of clavate paraphyses arising from base of median lobes, but differs in possessing only two pairs of lobes, as compared to three pairs in *Duplaspidiotus*. Other distinguishing features of *Eulaingia* are the poorly developed plates and presence of dorsal reticulation in centre of pygidium (Borchsenius & Williams, 1963).

GENERAL: Definition and characters by Brimblecombe (1958) and by Borchsenius & Williams (1963).

CITATIONS: Borchs1966 [catalogue: 237]; BorchsWi1963 [taxonomy, description: 384]; Brimbl1958 [taxonomy, description: 80-82]; MorrisMo1966 [taxonomy, catalogue: 72].

Eulaingia stenophyllae (Laing)

Pseudaonidia stenophyllae Laing, 1929: 27. Type data: AUSTRALIA: Victoria, Murray River, near Hattah, on *Acacia stenophylla*; collected by J.E. Dixon. Syntypes, female. Type depository: London: The Natural History Museum, England, UK.

Eulaingia stenophyllae; Brimblecombe, 1958: 81. Change of combination.

SCALE COVER: Female scale not observed by Laing (1929). Brimblecombe (1958) described scale: "Insects single and sparse on twigs; female scale, circular, 1.2 mm in diameter and light brown in colour; first exuviae darker".

HOST PLANTS: **Leguminosae**: *Acacia harpophylla* [Brimbl1958], *Acacia melanoxylon* [Brimbl1958], *Acacia stenophyllae* [Laing1929, Brimbl1958].

DISTRIBUTION: **Australasian**: Australia (New South Wales [Brimbl1958], Victoria [Laing1929, Brimbl1958]).

GENERAL: Description and illustration of adult female by Laing (1929) and by Brimblecombe (1958).

CITATIONS: Borchs1966 [catalogue: 237]; BorchsWi1963 [taxonomy, description, illustration: 384,386]; Brimbl1958 [taxonomy, description, illustration, host, distribution: 80-82]; Laing1929 [taxonomy, description, illustration, host, distribution: 27-28].

Felixiella Almeida, D.M.

Felixiella Almeida, D.M., 1973: 5. Type species: *Felixiella infulenensis* Almeida, by monotypy and original designation.
SYSTEMATICS: Type species of this genus is a pupillarial scale insect. Genus assigned to the Aonidina of Aspidiotinae, and affinity was indicated to *Doriopus* Brimblecombe and *Hybridaspis* Green.
GENERAL: Definition and characters by Almeida (1973).
CITATIONS: Almeid1973 [taxonomy, description: 5]; KosztaBeKo1986 [taxonomy, catalogue: 6].

Felixiella infulenensis Almeida, D.M.

Felixiella infulenensis Almeida, D.M., 1973: 5. Type data: MOZAMBIQUE: Infulene, on *Androstachys johnsonii*. Holotype female. Type depository: Lisbon: Coleccoes do Centro de Zoologia do Instituto de Investigacao Cientifica Tropical, Portugal.
SCALE COVER: Female scale oval, 1-1.5 mm; exuvia of second instar yellow; slightly thick; transparent; exuvia of first instar small, light yellow, transparent, eccentric; covered by white secretion. Male scale white, similar to that of female (Almeida, 1973).
HOST PLANTS: **Euphorbiaceae**: *Androstachys johnsonii* [Almeid1973].
DISTRIBUTION: **Afrotropical**: Mozambique [Almeid1973].
GENERAL: Description and illustration of adult female by Almeida (1973).
CITATIONS: Almeid1973 [taxonomy, description, illustration, host, distribution: 5-7].

Furcaspis Lindinger

Furcaspis Lindinger, 1908b: 99. Type species: *Aspidiotus biformis* Cockerell. Subsequently designated by Sanders, 1909a: 54.
Neofurcaspis Green, 1926: 61. Type species: *Neofurcaspis andamanensis* Green, by monotypy and original designation. Synonymy by Beardsley, 1966: 517.
SYSTEMATICS: The genus *Furcaspis,* together with several other genera belongs in sub-tribe Furcaspidina (sensu Balachowsky, 1958b), which differ from other aspidiotine genera in having multisetose antennal tubercle, perispiracular pores near anterior spiracles, and glandular tubercles on venter of abdomen and thorax. Green (1926) distinguished *Neofurcaspis* from *Furcaspis* solely on basis of presence of a double row of perivulvar pores in the type species of the former. Lindinger (1937) considered *Neofurcaspis* as a synonym of *Furcaspis,* and Beardsley (1966) concluded that there was no basis for maintaining *Neofurcaspis* as distinct from *Furcaspis.*

GENERAL: Definition and characters by Lindinger (1908b), Brain (1918), Green (1926), Ferris (1938a) and by Beardsley (1966).
KEYS: Colon-Ferrer & Medina-Gaud 1998: 28-32 (female) [Genera of Puerto Rico]; Williams & Watson 1988: 19 (female) [Tropical South Pacific]; Beardsley 1966: 502-504 (female) [Federated States of Micronesia]; Zimmerman 1948: 351 (female) [Hawaii]; Ferris 1942: 27 (female) [North America]; Ferris 1942: 33 (female) [species North America]; Borchsenius 1937a: 32-33 (female) [Palaearctic Region].
CITATIONS: Balach1958b [taxonomy: 249-250]; Beards1966 [taxonomy: 517-518]; Borchs1937a [taxonomy, description: 32,44]; Borchs1966 [catalogue: 240,242]; Brain1918 [taxonomy, description: 137]; ColonFMe1998 [taxonomy, description: 56]; Ferris1921b [taxonomy: 94]; Ferris1937c [taxonomy: 51]; Ferris1938 [taxonomy: 43]; Ferris1938a [taxonomy, description: 230]; Ferris1938b [taxonomy: 75]; Ferris1942 [taxonomy: 27]; Green1926 [taxonomy, description: 61]; Kozar1990f [distribution: 143]; Lindin1908b [taxonomy, description: 98-99]; Lindin1932f [taxonomy: 193-194]; Lindin1937 [taxonomy: 190]; Lindin1943b [taxonomy: 220]; MacGil1921 [taxonomy, description: 388,406-407]; McKenz1939 [taxonomy: 53]; MorrisMo1966 [taxonomy, catalogue: 82,130]; Sander1909a [taxonomy: 54]; Takaha1939b [taxonomy: 271]; Willia1969a [taxonomy: 326]; WilliaWa1988 [taxonomy: 124]; Zimmer1948 [taxonomy: 366].

Furcaspis andamanensis (Green)
Neofurcaspis andamanensis Green, 1926: 62. Type data: INDIA: Andaman Islands, Port Blair, Ross Island, on *Cocos nucifera*. Syntypes, female. Type depository: London: The Natural History Museum, England, UK.
Furcaspis andamanensis; Lindinger, 1943b: 220. Change of combination.
SCALE COVER: Female scale subcircular to suboval, maximum diameter 1.5 mm; reddish brown, margin paler; exuviae impinging upon margin, outlined with brighter red; that of larva with a small whitish central boss; ventral scale stout; margin of dorsal scale expanded and flattened; with sinuous raised lines on undersurface. Male scale of similar form, but smaller and of a brighter red colour; length diameter 1.5 mm (Green, 1926).
HOST PLANTS: **Palmae**: *Cocos nucifera* [Green1926].
DISTRIBUTION: **Oriental**: Andaman Islands [Green1926].
GENERAL: Description and illustration of adult female by Green (1926).
CITATIONS: Beards1966 [taxonomy: 517-518]; Borchs1966 [catalogue: 243]; Ferris1937c [taxonomy: 51]; Ferris1938 [taxonomy: 43,48]; Green1926 [taxonomy, description, illustration, host, distribution: 62-63]; Lepesm1947 [host, distribution: 214]; Lindin1943b [taxonomy: 220].

Furcaspis biformis (Cockerell)
Aspidiotus biformis Cockerell, 1893j: 255. Nomen nudum.
Aspidiotus biformis Cockerell, 1893k: 548. Type data: JAMAICA: Kingston, on Orchid, and TRINIDAD: Royal Botanic Gardens, on *Oncidium sprucei*. Syntypes, female and first instar. Type depository: Washington: U.S. National Entomological Collection, U.S. National Museum of Natural History, District of Columbia, USA.

Aspidiotus (*Chrysomphalus*) *biformis cattleyae* Cockerell, 1893k: 548. Type data: JAMAICA: Hope Gardens, on *Cattleya bowringiana*. Syntypes, female and first instar. Type depository: Washington: United States National Entomological Collection, U.S. National Museum of Natural History, District of Columbia, USA. Synonymy by Borchsenius, 1966: 241.

Aspidiotus biformis cattleyae; Cockerell, 1896b: 334. Change of combination.

Aspidiotus (*Chrysomphalus*) *biformis*; Cockerell, 1897i: 23. Change of combination.

Aspidiotus (*Chrysomphalus*) *biformis cattleyae*; Cockerell, 1897i: 23. Change of combination.

Aspidiotus (*Evaspidiotus*) *biformis*; Leonardi, 1898a: 76. Change of combination.

Chrysomphalus biformis; Cockerell, 1899n: 26. Change of combination.

Chrysomphalus biformis cattleyae; Fernald, 1903b: 289. Change of combination.

Furcaspis biformis; Lindinger, 1908b: 99. Change of combination.

Aspidiotus bipromineus Kuwana in Kuwana & Muramatsu, 1931a: 653. Type data: INDONESIA: Java, on orchid. Syntypes, female. Type depository: Ibaraki-ken: 5Insect Taxonomy Laboratory, National Institute of Agricultural Environmental Sciences, Kannon-dai, Yatabe, Tsukuba-shi, (I. Kuwana), Japan. Synonymy by Ferris, 1941e: 41.

Aspidiotus cattleyae; Ferris, 1941e: 41. Change of combination and rank.

Targionia biformis; Lindinger, 1943b: 152. Change of combination.

Furcaspis biformis; Zimmerman, 1948: 366. Revived combination.

COMMON NAME: orchid scale [Dekle1965c].

SCALE COVER: Female scale broadly oval or circular, about 2 mm across; convex; brown-black; exuviae nipple-like dark red-brown, placed away from center; surface of scale strongly granulose. Male scale much smaller, elongate with parallel sides; exuviae placed at one end (Cockerell, 1893k). Scale of female dark, red-brown; circular, moderately convex; exuviae central; that of male similar in colour, quite elongate and slender with exuvia close to one end (Ferris, 1938a).

HOST PLANTS: **Bromeliaceae**: *Aechmea magdalenae* [Balach1959a]. **Orchidaceae** [KuwanaMu1931a, WilliaWa1988], *Cattleya* [Zimmer1948], *Cattleya bowringiana* [Cocker1893k], *Cattleya percivaliana* [Ferris1938a], *Miltonia regnelli* [WilliaWa1988], *Oncidium* [WilliaWa1988], *Vanda* [Zimmer1948].

NATURAL ENEMIES: THYSANOPTERA **Phlaeothripidae**: *Aleurodothrips fascipennis* (Franklin) [Beshea1975, PalmerMo1990].

DISTRIBUTION: **Australasian**: Fiji [WilliaWa1988]; French Polynesia (Society Islands [WilliaWa1988], Tahiti [WilliaWa1988]); Hawaiian Islands (Hawaii [Zimmer1948]). **Nearctic**: United States of America (Colorado [Ferris1938a], District of Columbia [Nakaha1982], Florida [Dekle1965c], New Jersey [Nakaha1982], Washington [Nakaha1982]). **Neotropical**: Brazil [Lepage1940]; Colombia [Balach1959a, Kondo2001]; Guyana [Newste1914]; Jamaica [Ferris1938a]; Puerto Rico & Vieques Island (Puerto Rico [Martor1976, ColonFMe1998]); Trinidad and Tobago (Trinidad [Ferris1938a]); U.S. Virgin Islands [Nakaha1983]. **Oriental**: Indonesia (Java [KuwanaMu1931a]).

BIOLOGY: Confined to the plant family Orchidaceae.

ECONOMIC IMPORTANCE: Occasionally troublesome in greenhouses (Schmutterer et al., 1957). Zimmermann (1948) suggested that nicotine spray might be effective in controlling this scale.

GENERAL: Description and illustration of adult female by Kuwana & Muramatsu (1931a) (as *Aspidiotus bipromineus*), Ferris (1938a), Lepage (1940), Zimmerman (1948), Williams & Watson (1988) and by Colon-Ferrer & Medina-Gaud (1998).

KEYS: Ferris 1942: 33 (female) [North America].

CITATIONS: Balach1959a [host, distribution: 362]; BeardsDaHo1976 [economic importance: 106]; Borchs1966 [catalogue: 241]; ClapsWoGo2001a [taxonomy, host, distribution: 18]; Cocker1893j [taxonomy: 255]; Cocker1893k [taxonomy, description, host, distribution: 548]; Cocker1894g [taxonomy, description, host, distribution: 131]; Cocker1896b [distribution: 334]; Cocker1896k [host, distribution: iii-v]; Cocker1897i [taxonomy, description, host, distribution: 23]; Cocker1897l [taxonomy: 151]; Cocker1899n [host, distribution: 26]; Cocker1919 [taxonomy: 116]; Cocker1920a [host, distribution: 241]; ColonFMe1998 [taxonomy, description, illustration, host, distribution: 57]; Dekle1965c [taxonomy, description, host, distribution: 64]; Dekle1976 [taxonomy, description, host, distribution, economic importance: 84]; DeSant1979 [biological control]; Ehrhor1913 [host, distribution: 101]; Fernal1903b [catalogue: 288-289]; Ferris1921b [taxonomy: 94]; Ferris1937c [taxonomy, illustration: 51,74]; Ferris1938a [taxonomy, description, illustration, host, distribution: 231]; Ferris1941e [taxonomy: 41]; Ferris1942 [taxonomy: 446:33]; Ferris1943a [taxonomy: 85]; Gowdey1921 [host, distribution: 32]; Kondo2001 [taxonomy, host, distribution: 44]; KuwanaMu1931a [taxonomy, description, illustration, host, distribution: 653-654,658]; Leonar1897 [taxonomy: 285]; Leonar1898a [taxonomy: 76]; Leonar1898c [taxonomy, description, illustration, host, distribution: 60-62]; Lepage1940 [taxonomy, description, illustration, host, distribution: 76-78]; LepageFi1947 [taxonomy, description, host, distribution: 39]; Lindin1908b [taxonomy: 99]; Lindin1943b [taxonomy: 220]; MacGil1921 [taxonomy, description, host, distribution: 406-407]; Martor1976 [host, distribution: 30,91,184,193,234,265]; Maxwel1903 [taxonomy, description, host, distribution: 38]; McKenz1939 [taxonomy: 53]; Merril1953 [taxonomy, description, host, distribution: 48-49]; MillerDa1990 [host, distribution, economic importance: 302]; Nakaha1982 [host, distribution: 38]; Nakaha1983 [host, distribution: 11]; Newste1914 [host, distribution: 307]; Nur1990a [taxonomy, structure, chromosomes: 184-185]; PalmerMo1990 [biological control: 67-76]; SassceWe1924 [chemical control: 214-222]; SchmutKlLu1957 [host, distribution, economic importance: 495]; Sulliv1930 [host, distribution: 51-59]; WilliaWa1988 [taxonomy, description, illustration, host, distribution: 124-126]; WoodruBeSk1998 [distribution]; Zimmer1948 [taxonomy, description, illustration, host, distribution: 365-368].

Furcaspis bromeliae Hempel

Furcaspis bromeliae Hempel, 1932: 335. Type data: BRAZIL: Santa Catharina State, Laguna, on a plant of Bromeliaceae. Syntypes, female. Type depository: São Paulo: Instituto Biologico de São Paulo, Brazil.

SCALE COVER: Female scale subcircular or slightly oval, 2-2.5 mm in diameter; convex; exuviae near margin; dark; exuviae very dark; ventral scale thin, white (Hempel, 1932).

HOST PLANTS: **Bromeliaceae** [Hempel1932], *Aechma* [ClapsWoGo2001], *Aechma nudicaulis* [ClapsWoGo2001], *Bromelia* [ClapsWoGo2001].

DISTRIBUTION: **Neotropical**: Brazil (Rio de Janeiro [ClapsWoGo2001], Santa Catarina [ClapsWoGo2001], São Paulo [Hempel1932, ClapsWoGo2001]).

GENERAL: Description of adult female by Hempel (1932).

CITATIONS: Borchs1966 [catalogue: 241]; ClapsWoGo2001 [host, distribution: 244]; Hempel1932 [taxonomy, description, host, distribution: 335-337].

Furcaspis charmoyi Brain

Aspidiotus cladii; Grandpré & Charmoy, 1899: 22. Misidentification; discovered by Brain, 1918: 138.

Furcaspis charmoyi Brain, 1918: 138. Type data: MAURITIUS: on palms. Syntypes, female. Type depository: Pretoria: South African National Collection of Insects, South Africa.

Spinaspidiotus charmoyi; MacGillivray, 1921: 430. Change of combination.

Aspidiotus cladii; Mamet, 1946: 241. Misidentification; discovered by Mamet, 1946: 241.

Furcaspis charmoyi; Borchsenius, 1966: 241. Revived combination.

SYSTEMATICS: First recorded from Mauritius by Grandpré & Charmoy (1899) who misidentified it as *Aspidiotus cladii* Maskell. Brain (1918) noted this misidentification, and described *Furcaspis charmoyi*, based on original material of Grandpré & Charmoy (1899) from Mauritius.

SCALE COVER: Female scale yellow-brown to dark brown, often tinged with a greenish colour (Brain, 1918). Female scale rounded to oval, fairly convex, dark reddish-brown. Exuviae near anterior end, rounded. Male scale similar to that of female but smaller (Mamet, 1946).

HOST PLANTS: **Palmae** [GrandpCh1899, Muntin1965], *Dictyosperma alba* [Mamet1943a, Mamet1946, Mamet1949, Borchs1966].

DISTRIBUTION: **Afrotropical**: Mauritius [GrandpCh1899, Mamet1943a, Mamet1946, Mamet1949, Borchs1966]; South Africa [Muntin1965].

GENERAL: Description and illustration of adult female by Mamet (1946) and by Munting (1965).

CITATIONS: Balach1958b [taxonomy: 219,249]; Borchs1966 [catalogue: 241]; Brain1918 [taxonomy: 138]; Cocker1899r [taxonomy: 900]; GrandpCh1899 [taxonomy, description, illustration, host, distribution: 22]; Lepesm1947 [host, distribution: 196]; MacGil1921 [taxonomy, description, host, distribution: 430]; Mamet1943a [catalogue: 160]; Mamet1946 [taxonomy, description, illustration, host, distribution: 241-243]; Mamet1949 [catalogue: 65]; McKenz1939 [taxonomy: 54]; MoutiaMa1947 [distribution]; Muntin1965 [taxonomy, description, illustration, host, distribution: 232-234].

Furcaspis haematochroa Cockerell
Furcaspis haematochroa Cockerell, 1919: 116. Type data: PHILIPPINES: Batabatan Island, Antique Province, Panay, on coconut palm [=*Cocos nucifera*]; collected R.C. McGregor. Syntypes, female. Type depository: Washington: United States National Entomological Collection, U.S. National Museum of Natural History, District of Columbia, USA.
Spinaspidiotus haematochroa; MacGillivray, 1921: 430. Change of combination.
Furcaspis hematochroa; Lepesme, 1947: 214. Misspelling of species name.
Furcaspis haematochroa; Borchsenius, 1966: 241. Revived combination.
SCALE COVER: Female scale deep rich red, suggesting a drop of blood; circular, 2.5 mm in diameter; slightly convex; large circular exuviae to one side, often reaching or slightly overlapping margin; first exuvia nipple-like, prominent, Male scale suboval (Cockerell, 1919).
HOST PLANTS: **Palmae**: *Cocos nucifera* [Cocker1919].
DISTRIBUTION: **Oriental**: Philippines (Panay [Cocker1919]).
GENERAL: Description and illustration of adult female by Cockerell (1919).
CITATIONS: Borchs1966 [catalogue: 241]; Cocker1919 [taxonomy, description, illustration, host, distribution: 116]; Ferris1921b [taxonomy: 94]; Lepesm1947 [taxonomy, host, distribution: 214]; MacGil1921 [taxonomy, description, host, distribution: 430]; Willia1985a [taxonomy: 235].

Furcaspis oceanica Lindinger
Furcaspis oceanica Lindinger, 1909b: 149. Type data: MARSHALL ISLANDS: on *Cocos nucifera*. Syntypes, female. Type depository: Hamburg: Zoologisches Institut und Zoologishces Museum, Universität von Hamburg, Germany.
Spinaspidiotus oceanicus; MacGillivray, 1921: 430. Change of combination requiring emendation of species name for agreement in gender.
Chrysomphalus saipanensis Sasaki, I., 1935: 866. Type data: MICRONESIA: Truk Island, intercepted at Japan, Port Kobe, on a palm. Syntypes, female. Synonymy by Borchsenius, 1966: 241. Notes: Depository of type material unknown.
Furcaspis saipanensis; McKenzie, 1939: 54. Change of combination.
Furcaspis oceanica; Borchsenius, 1966: 241. Revived combination.
COMMON NAMES: oceanic scale [VelasqRi1969]; red coconut scale [LaliMu1996].
SCALE COVER: Lindinger (1909b) noted that female scale is similar to that of *Furcaspis biformis* (Cockerell).
HOST PLANTS: **Palmae** [Sasaki1935], *Bentinckiopsis ponapensis* [Takaha1939b, Takaha1942d], *Cocos nucifera* [Lindin1909b, Takaha1939b, Takaha1941b, Beards1966], *Exorrhiza* [Beards1966], *Exorrhiza ponapensis* [Beards1966], *Nipa fruticans* [Takaha1939b, Beards1966], *Ponapea ledermanniana* [Beards1966]. **Pandanaceae**: *Pandanus* [Takaha1942d, Beards1966].
NATURAL ENEMIES: HYMENOPTERA **Aphelinidae**: *Ablerus palauensis* Doutt [AAEE1925].   **Encyrtidae**: *Adelencyrtus oceanicus* (Doutt) [Doutt1950, MarutaMu1989, LaliMu1996].
DISTRIBUTION: **Australasian**: Federated States of Micronesia (Caroline Islands [Takaha1939b, Takaha1941b, Takaha1942d, Beards1966], Kosare (=Kusiae)

[Beards1966], Truk Islands [Sasaki1935, Beards1966]); Guam [MarutaMu1990]; Marshall Islands [Lindin1909b, Takaha1939b, Beards1966, Nafus1996]; Palau [Takaha1939b, Takaha1942d]. **Oriental**: Philippines [VelasqRi1969].
ECONOMIC IMPORTANCE: Pest of palms in several Pacific islands (Schmutterer et al., 1957; Sweetman, 1958; Lali & Muniappan, 1996).
GENERAL: Description and illustration of adult female by Lindinger (1909b), Sasaki (1935) and by Takahashi (1939b).
CITATIONS: Borchs1966 [catalogue: 241]; Doutt1950 [host, distribution, biological control: 501-507]; Esaki1940a [host, distribution: 274-280]; Ferris1921b [taxonomy: 94]; Ferris1941e [taxonomy: 46]; Gardne1958 [host, distribution, economic importance, biological control: 465-471]; Gressi1958 [host, distribution, economic importance: 395-398]; LaliMu1996 [host, distribution, biological control, economic importance: 15-20]; Lepesm1947 [host, distribution: 196]; Lindin1909b [taxonomy, description, illustration, host, distribution: 149]; MacGil1921 [taxonomy, description, host, distribution: 430]; MarutaMu1989 [host, distribution, biological control: 61-66]; MarutaMu1990 [host, distribution, biological control: 1]; McKenz1939 [taxonomy: 54]; MillerDa1990 [host, distribution, economic importance: 302]; MuniapMa1989 [host, distribution, economic importance, biological control: 17-18]; Nafus1996 [host, distribution: 1]; Noyes1990a [biological control: 153]; Piment1963 [biological control: 786-787]; RaoGhSa1971 [host, distribution, biological control]; Sander1909a [taxonomy, host, distribution: 54]; Sasaki1935 [taxonomy, description, illustration, host, distribution: 866]; SchmutKlLu1957 [host, distribution, economic importance: 495]; Sweetm1958 [biological control, economic importance: 449-458]; Takaha1936c [host, distribution: 109]; Takaha1939b [taxonomy, description, host, distribution: 270-271]; Takaha1941b [host, distribution: 220]; Takaha1942d [host, distribution: 358]; VelasqRi1969 [host, distribution: 195-208]; WeidneWa1968 [taxonomy: 176]; WilliaWa1988 [taxonomy: 124].

Furcaspis plana Hempel
Furcaspis plana Hempel, 1937: 29. Type data: BRAZIL: São Paolo, Santos, on leaves of "grumixama" [=*Eugenia brasiliensis*. Holotype female. Type depositories: São Paulo: Museu de Zoologia, Universidade de São Paulo, Brazil, São Paulo: Instituto Biologico de São Paulo, Brazil, and Curitiba: Departamento de Zoologia, Setor de Ciencias Biologicas, Universidade Federal do Parana, Brazil.
SCALE COVER: Female scale irregularly shaped, asymmetric, approximately circular, flat; orange; with a small central boss, of dark colour; most specimens 3 mm in diameter (Hempel, 1937).
HOST PLANTS: **Magnoliaceae**: *Magnolia*. **Myrtaceae**: *Eugenia* [ClapsWoGo2001], *Eugenia brasiliensis* [Hempel1937], *Eugenia dombeyi* [ClapsWoGo2001], *Syzygium jambos* [ClapsWoGo2001].
DISTRIBUTION: **Neotropical**: Brazil (São Paulo [ClapsWoGo2001]).
GENERAL: Description and illustration of adult female by Hempel (1937).
CITATIONS: Borchs1966 [catalogue: 241]; Claps1993 [taxonomy: 3,6,9]; ClapsWoGo2001 [host, distribution: 244]; Hempel1937 [taxonomy, description, illustration, host, distribution: 29].

Furcaspis rufa Lindinger

Furcaspis rufa Lindinger, 1913b: 97. Type data: RÉUNION: St. Denis, on *Erythroxylon* [=*Erythroxylum*] sp. Syntypes, female. Type depository: Hamburg: Zoologisches Institut und Zoologishces Museum, Universität von Hamburg, Germany.

Tollaspidiotus rufus; MacGillivray, 1921: 426. Change of combination requiring emendation of species name for agreement in gender.

Furcaspis rufa; Borchsenius, 1966: 242. Revived combination.

SCALE COVER: Scale of young female broadly elliptical, parallel-sided, later circular, 1.5-2 mm in diameter; red brown, thick robust; exuviae lighter in colour, eccentric; ventral scale white. Male scale narrow, linear, 1.3-1.5 mm long, 0.5 mm wide; red brown, with yellow exuviae at head apex (Lindinger, 1913b).

HOST PLANTS: Erythroxylaceae: *Erythroxylum* [Lindin1913b, Mamet1943a, Borchs1966].

DISTRIBUTION: Afrotropical: Réunion   [Lindin1913b,   Mamet1943a, Mamet1954a, Mamet1957, Borchs1966].

GENERAL: Description and illustration of adult female by Lindinger (1913b).

CITATIONS: Balach1958b [taxonomy, host, distribution: 249]; Borchs1966 [catalogue: 242]; Hempel1937 [taxonomy: 31]; Lindin1913 [taxonomy, description, host, distribution: 39]; Lindin1913b [taxonomy, description, host, distribution: 97]; Lindin1931a [taxonomy: 89]; MacGil1921 [taxonomy, description, host, distribution: 426]; Mamet1943a [catalogue: 160]; Mamet1954a [distribution: 264]; Mamet1957 [distribution: 369]; Newste1917 [taxonomy, host, distribution: 375]; WeidneWa1968 [taxonomy: 176].

Furcaspis taquarae Fonseca, new status

Furcaspis biformis taquarae Fonseca, 1969: 35. Type data: BRAZIL: São Paolo, Cantareira, on *Merostachys* sp. Holotype female. Type depositories: São Paulo: Museu de Zoologia, Universidade de São Paulo, Brazil, and São Paulo: Instituto Biologico de São Paulo, Brazil.

SYSTEMATICS: Fonseca (1969) separated *Furcaspis biformis taquarae* from *Furcaspis biformis* (Cockerell) by a pronounced convexity of scale cover, distinct sclerotized zones between pygidial lobes, slightly elongated, resembling form of a paraphyse; by absence of perivulvar pores; and spiracle apodemes shorter and forming a narrower semicircular region than in *F. biformis*. Presence or absence of perivulvar pores has been often applied as a reliable taxonomic feature in the Diaspididae, therefore, this sub-species here raised to species level.

SCALE COVER: Female scale convex, more convex than in *Furcaspis biformis* (Cockerell) (Fonseca, 1969).

HOST PLANTS: Gramineae: *Merostachys* [Fonsec1969, ClapsWoGo2001].

DISTRIBUTION: Neotropical: Brazil   (São   Paulo   [Fonsec1969, ClapsWoGo2001]).

GENERAL: Description and illustration of adult female by Fonseca (1969).

CITATIONS: Claps1993 [taxonomy: 3,6]; ClapsWoGo2001 [host, distribution: 244]; Fonsec1969 [taxonomy, description, illustration, host, distribution: 34-35].

Genistaspis Bodenheimer

Genistaspis Bodenheimer, 1949: 38. Type species: *Genistaspis zelihae* Bodenheimer, by monotypy and original designation. Notes: Bodenheimer (1952:348) again referred to this genus and its type species as n. gen. and n. sp., respectively.

SYSTEMATICS: Genus close to *Targionia*, from which it differs in presence of plates and in bilobed second lobes (Bodenheimer, 1949, 1952).

GENERAL: Description and definition by Bodenheimer (1949, 1951, 1952) and by Yaşar (1995).

CITATIONS: Bodenh1949 [taxonomy, description: 26,38]; Bodenh1951 [taxonomy, description: 329]; Bodenh1952 [taxonomy, description: 348-349]; Borchs1966 [catalogue: 367]; DanzigPe1998 [catalogue: 266]; MorrisMo1966 [taxonomy, catalogue: 84]; Yasar1995a [taxonomy, description: 80].

Genistaspis zelihae Bodenheimer

Genistaspis zelihae Bodenheimer, 1949: 83. Type data: TURKEY: Ankara, on *Genista jouberti*; collected 28.v.1941. Syntypes, female. Type depository: Bet Dagan: Department of Entomology, The Volcani Center, Israel. Notes: Bodenheimer (1952: 350) again referred to this species as n. sp.

SCALE COVER: Female scale short oval, 1.0 to 1.5 mm long, 0.5 to 1.0 mm broad; often with irregular margin and usually narrowing towards one or both ends; highly convex; exuviae distinctly eccentric; ground colour grey, covered by brown secretion as are also orange coloured exuviae (Bodenheimer, 1952)

HOST PLANTS: **Leguminosae**: *Genista jouberti* [Bodenh1949]; *Genista jouberti inops* [Bodenh1952].

DISTRIBUTION: **Palaearctic**: Turkey [Bodenh1949, Bodenh1952].

GENERAL: Description and illustration of adult female by Bodenheimer (1949, 1952) and by Yaşar (1995a).

CITATIONS: Bodenh1949 [taxonomy, description, illustration, host, distribution: 83-84]; Bodenh1952 [taxonomy, description, illustration, host, distribution: 349-350]; Borchs1966 [catalogue: 367]; DanzigPe1998 [catalogue: 266]; Yasar1995a [taxonomy, description, illustration, host, distribution: 81-82].

Gomphaspidiotus Borchsenius & Williams

Gomphaspidiotus Borchsenius & Williams, 1963: 384. Type species: *Aspidiotus cuculus* Green, by monotypy and original designation.

SYSTEMATICS: *Gomphaspidiotus* Borchsenius & Williams (1963) is characterized in having a constriction between prothorax and mesothorax, which place it in the *Pseudaonidia* group. In possessing only a single pair of lobes *Gomphaspidiotus* resembles *Neomorgania* MacGillivray, *Diastolaspis* Brimblecombe and *Dichosoma* Brimblecombe (Borchsenius & Williams, 1963).

GENERAL: Definition and characters by Borchsenius & Williams (1963).

CITATIONS: Borchs1966 [catalogue: 240]; BorchsWi1963 [taxonomy, description: 384,389]; MorrisMo1966 [taxonomy, catalogue: 85].

Gomphaspidiotus cuculus (Green)

Aspidiotus cuculus Green, 1905a: 341. Type data: SRI LANKA: on an undetermined plant. Syntypes, female. Type depository: London: The Natural History Museum, England, UK.

Chorizaspidiotus cuculus; MacGillivray, 1921: 433. Change of combination.

Gomphaspidiotus cuculus; Borchsenius & Williams, 1963: 387. Change of combination.

SCALE COVER: Female scale very irregular in form, due to living in cavity; colour dull brown, usually comprising portions of exuviae and derm of former occupant of the gall, *Amorphococcus mesuae* (Green, 1905a).

HOST PLANTS: **Guttiferae**: *Mesua* [Ramakr1921a], *Mesua ferrea* [Green1922, Green1937].

DISTRIBUTION: **Oriental**: Sri Lanka [Green1905a, Green1922, Green1937].

BIOLOGY: Inhabits galls of *Amorphococcus mesuae* Green, 1902 (Asterolecaniidae) on *Mesua* sp. (Green, 1905a).

GENERAL: Description and illustration of adult female by Green (1905a) and by Borchsenius & Williams (1963).

CITATIONS: Borchs1966 [catalogue: 240]; BorchsWi1963 [taxonomy, description, illustration: 384,387]; Ferris1941e [taxonomy: 42]; Green1905a [taxonomy, description, illustration, host, distribution: 341]; Green1922 [host, distribution: 462]; Green1937 [host, distribution: 333]; MacGil1921 [taxonomy, description, host, distribution: 433]; Ramakr1921a [host, distribution: 357]; Ruther1915a [taxonomy, description, host, distribution: 105]; Sander1906 [taxonomy, host, distribution: 13].

Gonaspidiotus MacGillivray

Gonaspidiotus MacGillivray, 1921: 390. Type species: *Aspidiotus minimus* Leonardi, by original designation.

SYSTEMATICS: Ferris (1937d) and Balachowsky (1950b) accepted *Gonaspidiotus* as distinct genus. *Gonaspidiotus* differs from *Aspidaspis* Ferris in possessing two pairs of pygidial lobes. It also differs from *Aspidiotus, Abgrallaspis* and *Dynaspidiotus* in total absence of plates on segment VI of pygidium (Balachowsky, 1950b).

GENERAL: Definition and characters by MacGillivray (1921), Ferris (1937d), Balachowsky (1950b) and by Yaşar (1995).

KEYS: Balachowsky 1951: 599 (female) [Palaearctic Region].

CITATIONS: Balach1950b [taxonomy, description: 545]; Balach1951 [taxonomy: 599]; BerlesLe1896 [taxonomy, description: 349]; DanzigPe1998 [catalogue: 268]; Ferris1937c [taxonomy: 51]; Ferris1937d [taxonomy, description: 106]; Leonar1897 [taxonomy: 284-286]; Lepesm1947 [taxonomy: 211]; Lindin1937 [taxonomy: 180]; Lupo1954 [taxonomy, description: 7-8]; MacGil1921 [taxonomy, description: 390]; TakahaTa1957 [taxonomy: 102]; Yasar1995a [taxonomy, description: 83].

Gonaspidiotus kaussarii (Balachowsky)

Abgrallaspis kaussarii Balachowsky, 1959b: 211. Type data: IRAN: south of Teheran, 35 km west of Shakyund, on undetermined plant; coll. J.J. Dyca. Holotype female. Type depository: Paris: Muséum national d'Histoire naturelle, France.

Gonaspidiotus kaussarii; Borchsenius, 1966: 302. Change of combination.

SCALE COVER: Female scale circular, small, diameter 1.8 mm; conical; exuviae central, very rounded in form; colour bright brown, with a concentric area, darker (Balachowsky, 1959b).

DISTRIBUTION: **Palaearctic**: Iran [Balach1959b].

GENERAL: Description and illustration of adult female by Balachowsky (1959b).

KEYS: Komosinska 1969: 76-78 (female) [World].

CITATIONS: Balach1959b [taxonomy, description, illustration, host, distribution: 211-213]; Borchs1966 [catalogue: 302]; DanzigPe1998 [catalogue: 268].

Gonaspidiotus minimus (Leonardi)

Aspidites minimus Leonardi in: Berlese & Leonardi, 1896: 350. Type data: ITALY: Portici, on leaves of *Quercus ilicis*. Syntypes, female. Type depository: Portici: Dipartimento de Entomologia e Zoologia Agraria di Portici, Università di Napoli Federico II, Italy.

Aspidiotus minimus; Cockerell, 1896b: 334. Change of combination.

Hemiberlesia minima; Leonardi, 1897b: 126. Change of combination requiring emendation of species name for agreement in gender.

Aspidiotus (*Aspidiotus*) *minimus*; Cockerell, 1897i: 18. Change of combination.

Gonaspidiotus minimus; MacGillivray, 1921: 431. Change of combination.

Aspidiotus tyrrhenus Lindinger, 1928: 100. Unjustified replacement name; discovered by Danzig & Pellizzari, 1998: 268.

Aspidiotus (*Aonidiella*) *occidentalis* Balachowsky, 1932: 18. Type data: ALGERIA: Haut-Atlas, Immouzer (altitude 1750 meters), on *Chamaerops humilis*; collected by P. Vayssière, 30.ix.1930. Syntypes, female. Type depository: Paris: Muséum national d'Histoire naturelle, France. Synonymy by Borchsenius, 1966: 302.

Aonidiella occidentalis; Balachowsky, 1932: 66. Change of combination.

Hemiberlesea minima; Balachowsky, 1935b: 257. Misspelling of genus name.

Hemiberlesea minima; Gómez-Menor Ortega, 1937: 117. Misspelling of genus name.

Dinaspidiotus minimus; Bodenheimer, 1949: 67. Misspelling of genus name.

Dinaspidiotus minimus; Bodenheimer, 1949: 67. Change of combination.

Gonaspidiotus minimus; Borchsenius, 1966: 302. Revived combination.

SCALE COVER: Female scale suboval, 0.75 mm long, 0.55 mm wide; convex; exuviae yellow, eccentric; nymphal exuviae comparatively large; scale texture delicate, colour red; scale of certain specimens covered with white secretion. Male scale elongate, almost reniform; exuviae eccentric; colour similar to female, but slightly paler; 0.9-1 mm long (Leonardi, 1920). Female scale small, circular, abruptly convex at center, flat along margin; colour red; exuviae central; scale of certain specimens covered with white secretion. Male scale oval, narrower in front, white, exuviae eccentric, 0.9-1.0 mm long (Balachowsky, 1950b).

HOST PLANTS: **Fagaceae**: *Quercus* [Bodenh1937, Bodenh1949,], *Qu. coccifera* [Balach1928a, Balach1935b, Bachma1953, Lupo1954, Zahrad1972], *Qu. ilex* [Leonar1920, Balach1928a, Balach1932d, GomezM1937, Lupo1954, Zahrad1972, Martin1983], *Qu. ilex fragilis* [Martin1983], *Qu. ilicis* [BerlesLe1896], *Qu. incana* [Balach1928a, Balach1932d], *Qu. ithaburensis* [Bodenh1935], *Qu. suber* [Balach1931a, Balach1932d, Zahrad1972, Martin1983]. **Palmae**: *Chamaerops humilis* [Balach1932, Balach1932d].

NATURAL ENEMIES: HYMENOPTERA **Aphelinidae**: *Pteroptrix dimidiata* Westwood [Novits1961].

DISTRIBUTION: **Palaearctic**: Algeria [Balach1928a, Balach1932, Balach1932d]; Corsica [Leonar1920, Balach1931a, Balach1932d]; France [Balach1930a, Balach1932d]; Greece [Korone1934, Lupo1954]; Israel [Bodenh1935; BytinsSt1967]; Italy [BerlesLe1896, Leonar1920, Lupo1954, LongoMaPe1995]; Morocco [Balach1932d]; Spain [Balach1935b, GomezM1937, GomezM1959, Martin1983, BlayGo1993]; Turkey [Bodenh1949, Bodenh1952]; Yugoslavia [Bachma1953, Lupo1954].

GENERAL: Description and illustration of adult female by Leonardi (1920), Balachowsky (1950b), Lupo (1954), Gómez-Menor Ortega (1959) and by Yaşar (1995a).

KEYS: Lupo 1954: 8-9 (female) [Italy]; Balachowsky 1928a: 132 (female) [North Africa]; Leonardi 1920: 90 (female) [Italy].

CITATIONS: Bachma1953 [host, distribution: 182]; Balach1928a [host, distribution: 138]; Balach1930a [host, distribution: 178-179]; Balach1931a [host, distribution: 97]; Balach1932 [taxonomy, description, illustration, host, distribution: 18-20]; Balach1932d [taxonomy, host, distribution, economic importance: 66,VI,XII,XLVII]; Balach1935b [host, distribution: 257]; Balach1950b [taxonomy, description, illustration, host, distribution: 546-549]; BerlesLe1896 [taxonomy, description, illustration, host, distribution: 350]; BlayGo1993 [taxonomy, description, illustration, host, distribution: 468-472]; Bodenh1935 [host, distribution: 246]; Bodenh1937 [host, distribution: 217]; Bodenh1949 [taxonomy, description, illustration, host, distribution: 67-68]; Bodenh1952 [taxonomy, description, host, distribution: 339-340]; Borchs1966 [catalogue: 302]; BytinsSt1967: 125]; Cocker1896b [taxonomy, distribution: 334]; Cocker1897i [taxonomy, description, host, distribution: 12,18]; DanzigPe1998 [catalogue: 268]; Fernal1903b [catalogue: 267]; Ferris1937c [taxonomy: 51]; Ferris1937d [taxonomy, illustration: 106,122]; Ferris1941e [taxonomy: 45-46,49]; Foldi2001 [distribution: 303-308]; Garcia1930 [host, distribution, biological control]; GomezM1937 [taxonomy, description, host, distribution: 117-118]; GomezM1958a [host, distribution: 8]; GomezM1959 [taxonomy, description, illustration, host, distribution: 158-162]; Korone1934 [taxonomy, description, illustration, host, distribution: 19]; Leonar1897 [taxonomy: 286]; Leonar1897b [taxonomy, description, illustration, host, distribution: 119,126-127]; Leonar1920 [taxonomy, description, illustration, host, distribution: 90,100-101]; Lepesm1947 [taxonomy, description, host, distribution, life history: 211-213]; Lindin1912b [taxonomy, description, host, distribution: 279]; Lindin1928 [taxonomy: 100,106]; Lindin1957 [taxonomy: 546]; LongoMaPe1995 [distribution: 127]; Lupo1954 [taxonomy, description, illustration, host, distribution: 9-13]; MacGil1921 [taxonomy, description, host, distribution: 431]; Martin1983 [taxonomy, host, distribution: 64]; McKenz1938 [taxonomy: 4]; Novits1961 [biological control: 193-194]; Yasar1995a [taxonomy, description, illustration, host, distribution: 84-86]; Zahrad1972 [host, distribution: 440]; Zahrad1990a [host, distribution, description: 649].

Gonaspidiotus seurati (Marchal)

Aspidiotus (*Hemiberlesia*) *seurati* Marchal, 1911: 71. Type data: ALGERIA: South Algeria, Garadaia, on *Antirrhinum ramosissimum*; collected by M. Seurat. Syntypes, female. Type depository: Paris: Muséum national d'Histoire naturelle, France.
Comstockaspis seurati; MacGillivray, 1921: 438. Change of combination.
Hemiberlesia seurati; Balachowsky, 1928a: 132. Change of combination.
Abgrallaspis seurati; Balachowsky, 1948b: 312. Change of combination.
Gonaspidiotus seurati; Borchsenius, 1966: 302. Change of combination.
Abgrallaspis seurati; Komosinska, 1969: 73. Revived combination.
Gonaspidiotus seurati; Danzig & Pellizzari, 1998: 268. Revived combination.
SCALE COVER: Female scale highly convex; circular or oval, 1.5-2 mm in diameter; light grey, central area yellow red; transparent (Marchal, 1911). Female scale circular, conical; exuviae brown, central; colour grey-white; covered with white secretion. Male scale oval, narrower anteriorly; colour similar to female; 1.4 mm long (Balachowsky, 1948b).
HOST PLANTS: **Boraginaceae**: *Trichodesma africana* [Balach1956]. **Cruciferae**: *Antirrhinum ramosissimum* [Marcha1911], *Moricandia arvensis* [Balach1948b, Balach1956], *Zilla* [Rungs1937], *Zilla macroptera* [Balach1927, Balach1932d], *Zilla spinosa* [Balach1956, Matile1988]. **Resedaceae**: *Randonia africana* [Rungs1937].
DISTRIBUTION: **Palaearctic**: Algeria [Marcha1911, Balach1956]; Egypt [Balach1927, Ezzat1958]; Morocco [Rungs1937]; Saudi Arabia [Matile1988]; Turkey [Yasar1995a].
GENERAL: Description and illustration of adult female by Balachowsky (1948b, 1956), Komosinska (1969) and by Yaşar (1995a).
KEYS: Komosinska 1969: 76-78 (female) [World]; Ezzat 1958: 240 (female) [Egypt]; Balachowsky 1956: 16 (female) [Africa]; Balachowsky 1948b: 307 (female) [Mediterranean]; Balachowsky 1928a: 132 (female) [North Africa].
CITATIONS: Balach1927 [host, distribution: 177]; Balach1929a [host, distribution: 301-302]; Balach1932d [taxonomy, host, distribution: V-VI]; Balach1934d [host, distribution: 146-147]; Balach1948b [taxonomy, description, illustration, host, distribution: 312-314]; Balach1956 [taxonomy, description, illustration, host, distribution: 20-23]; Balach1958a [host, distribution: 34-35]; Balach1958b [taxonomy: 163]; Borchs1966 [catalogue: 302]; DanzigPe1998 [catalogue: 268-269]; Ezzat1958 [distribution: 240]; EzzatNa1987 [distribution: 87]; Ferris1941e [taxonomy: 48]; Hall1923 [taxonomy: 53]; Komosi1969 [taxonomy, description, illustration, host, distribution: 73-74]; Lindin1912b [taxonomy, description, host, distribution: 68,345]; MacGil1921 [taxonomy, description, host, distribution: 438]; Marcha1911 [taxonomy, description, host, distribution: 71]; Matile1988 [host, distribution: 25]; Rungs1937 [host, distribution: 332]; Sassce1911

[taxonomy: 70]; Yasar1995a [taxonomy, description, illustration, host, distribution: 86-87] .

Greeniella Cockerell

Greeniella Cockerell, 1897q: 703. Type species: *Aonidia corniger* Green, by original designation.
Greenidiella Lindinger, 1943a: 148. Unjustified emendation.
Decoraspis Ferris, 1955c: 31. Unjustified replacement name for *Greeniella* MacGillivray, 1921.
SYSTEMATICS: The generic name *Greeniella* was used in scale insect nomenclature for two zoologically different entities. *Greeniella* Cockerell, 1897q (type species: *Aonidia cornigera* Green) is a valid genus in Diaspididae. *Greeniella* MacGillivray, 1921 (type species: *Monophlebus stebbingii* Stebbing) is a junior homonym of *Greeniella* Cockerell, and a junior synonym of *Drosicha* Walker in Margarodidae. The genus *Greeniella* Cockerell includes 13 pupillarial species. *Greeniella* is related to *Aonidia*.
KEYS: Chou 1985: 324 (female) [Genera of China].
CITATIONS: Balach1948b [taxonomy: 269]; Borchs1966 [catalogue: 365]; Brimbl1958 [taxonomy: 82]; Chou1985 [taxonomy, description: 325-326]; Cocker1897q [taxonomy: 703]; Cocker1899a [taxonomy: 396]; DanzigPe1998 [catalogue: 270]; Fernal1903b [catalogue: 303]; Ferris1937c [taxonomy: 51,54,75]; Ferris1937e [taxonomy: 529]; Ferris1955c [taxonomy: 31-32]; HowellTi1990 [taxonomy: 57]; Kozar1990f [distribution: 142]; Leonar1899 [taxonomy: 206]; Leonar1900 [taxonomy, description: 317]; Lindin1908b [taxonomy: 98]; Lindin1937 [taxonomy: 186]; Lindin1943a [taxonomy: 148]; MacGil1921 [taxonomy, description: 395,459-462]; MorrisMo1966 [taxonomy, catalogue: 55,87]; Tao1999 [taxonomy: 90].

Greeniella capitata Brimblecombe

Greeniella capitata Brimblecombe, 1959: 137. Type data: AUSTRALIA: Queensland, Brisbane River, on *Hemicyclia australiasica*; collected before 1900. Holotype female. Type depository: Brisbane: Queensland Museum, Queensland, Australia; type no. T5715.
SCALE COVER: Second pellicle circular, 0.75 mm diameter, completely covering the adult body, convex, black with a slight greyish suffusion and a narrow fawn coloured margin; first pellicle central and black (Brimblecombe, 1959).
HOST PLANTS: **Euphorbiaceae**: *Hemicyclia australiasica* [Brimbl1959].
DISTRIBUTION: **Australasian**: Australia (Queensland [Brimbl1959]).
GENERAL: Description and illustration of adult female by Brimblecombe (1959).
CITATIONS: Borchs1966 [catalogue: 365]; Brimbl1959 [taxonomy, description, illustration, host, distribution: 137-140].

Greeniella columnifera (Green)

Aonidia (*Greeniella*) *columnifera* Green, 1922a: 1008. Type data: SRI LANKA: Hakgala, on the under surface of leaves of *Turpinia pomifera*. Syntypes, female. Type depository: London: The Natural History Museum, England, UK.

Greeniella columnifera; Green, 1937: 336. Change of combination.

Aonidia columnifera; Lindinger, 1943b: 218. Change of combination.

Cryptaspidiotus columnifer; Lindinger, 1957: 544. Change of combination requiring emendation of species name for agreement in gender.

Greeniella columnifera; Borchsenius, 1966: 365. Revived combination.

SCALE COVER: Illustration of scale cover by Green (1922a). Female scale flattish, broadly oblate; dull castaneous; exuvia blackish, central area occupied by a cylindrical column of dense, white wax which is often missing in old examples. Male scale very thin and flat; greyish; exuvia as on female (Green, 1922a).

HOST PLANTS: **Staphyleaceae**: *Turpinia pomifera* [Green1922a, Green1937].

DISTRIBUTION: **Oriental**: Sri Lanka [Green1922a, Green1937].

GENERAL: Description and illustration of adult female by Green (1922a).

CITATIONS: Borchs1966 [catalogue: 365]; Green1922a [taxonomy, description, illustration, host, distribution: 1008-1009]; Green1937 [host, distribution: 336]; Lindin1943b [taxonomy: 218]; Lindin1957 [taxonomy: 544].

Greeniella cornigera (Green)

Aonidia corniger Green, 1896: 5. Type data: SRI LANKA: Punduloya, on *Psychotria* sp. and *Litsea* sp. Holotype female. Type depository: London: The Natural History Museum, England, UK.

Greeniella cornigera; Cockerell, 1899: 396. Change of combination requiring emendation of species name for agreement in gender.

Decoraspis cornigera; Ferris, 1955c: 32. Change of combination.

Greeniella cornigera; Borchsenius, 1966: 365. Revived combination.

SCALE COVER: Female scale semicircular, 1.25-1.75 in diameter; flattish or slightly convex; light reddish-brown, minutely mottled with paler specks; first exuvia approximately central; either exposed or bearing horn-shaped processes of young scale; latter is the normal condition, but exuvia being slightly prominent and processes very brittle; these appendages frequently rubbed away; exuvia itself divided up into three (a median and two lateral) series of distinct plates; second exuvia very large and broad, anterior margin straight, posterior extremity pointed, its dorsal surface concealed by a horny secretion extending slightly beyond margin; scale closed beneath by ventral part of second exuvia; if ventral scale be carefully removed, adult female lies within hollow of second exuvia; size of second exuvia about 1 by 1.25 mm. Male scale oblong, 1 by 0.75 mm; similar in appearance to female, but smaller and darker; single exuvia placed transversely across scale, near middle, and usually bearing larval horn-shaped processes; reddish brown with broad groove below for reception of pupa (Green, 1896e). Colour illustration of female and male scale cover by Green (1896e).

HOST PLANTS: **Euphorbiaceae**: *Ostodes zeylanica* [Green1937]. **Lauraceae**: *Litsea* [Green1896], *Litsea zeylanica* [Green1896e, Green1937]. **Rubiaceae**: *Psychotria* [Green1896], *Psy. thwaitesii* [Green1896e, Leonar1900, Green1937].

DISTRIBUTION: **Oriental**: Sri Lanka [Green1896, Green1896e, Leonar1900, Ramakr1921a, Green1937].
GENERAL: Description and illustration of adult female and adult male by Green (1896e).
KEYS: Green 1896e: 68 (female) [Sri Lanka].
CITATIONS: Borchs1965 [taxonomy: 210]; Borchs1966 [catalogue: 365]; Cocker1899a [taxonomy: 396]; DEDAC1923 [host, distribution]; Fernal1903b [catalogue: 304]; Ferris1937c [taxonomy, illustration: 51,75]; Ferris1955c [taxonomy: 32]; Green1896 [taxonomy, description, host, distribution: 5]; Green1896e [taxonomy, description, illustration, host, distribution: 69-71]; Green1922 [taxonomy: 460]; Green1937 [host, distribution: 336]; Hayat1989 [host, distribution, biological control: 1-3]; Howard1907 [host, distribution, biological control: 69-88]; HowardAs1895 [biological control: 633]; Leonar1897 [taxonomy: 286]; Leonar1899 [taxonomy: 205,206]; Leonar1900 [taxonomy, description, illustration, host, distribution: 317-320]; Leonar1903a [taxonomy: 6]; MacGil1921 [taxonomy, description, host, distribution: 459]; Ramakr1921a [host, distribution: 358].

Greeniella dentata (Lindinger)
Aonidia dentata Lindinger, 1911: 12. Type data: INDIA: Kamelekum Hill, on *Walsura piscidia*, 11.1881. Holotype. Type depository: Hamburg: Zoologisches Institut und Zoologishces Museum, Universität von Hamburg, Germany; type no. MP139.
Greeniella dentata; MacGillivray, 1921: 460. Change of combination.
SCALE COVER: Lindinger (1911) did not describe scale cover.
HOST PLANTS: **Meliaceae**: *Walsura piscidia* [Green1919c, Ramakr1921a].
DISTRIBUTION: **Oriental**: India [Green1919c, Ramakr1921a].
GENERAL: Description and illustration of adult female by Lindinger (1911).
CITATIONS: Borchs1966 [catalogue: 365-366]; Green1919c [host, distribution: 441]; Lindin1911 [taxonomy, description, illustration, host, distribution: 12]; MacGil1921 [taxonomy, description, host, distribution: 366]; Ramakr1921a [host, distribution: 359]; Sassce1911 [taxonomy: 70]; WeidneWa1968 [taxonomy: 171].

Greeniella ferreae (Rutherford)
Aonidia ferreae Rutherford, 1914: 265. Type data: SRI LANKA: Paradeniya, on *Mesua ferrea*. Syntypes, female. Type depository: London: The Natural History Museum, England, UK.
Greeniella ferreae; MacGillivray, 1914: 366. Change of combination.
SCALE COVER: Female scale very black, shining; first exuvia usually situated just within margin, sometimes more centrally, black with apex yellowish; it is raised in centre and striated; rest of scale consists of second exuvia; venter of scale white; when scale removed a white wax deposit left on twig (Rutherford, 1914).
HOST PLANTS: **Guttiferae**: *Mesua ferrea* [Ruther1914, Ramakr1921a, Green1922, Green1937].
DISTRIBUTION: **Oriental**: Sri Lanka [Ruther1914, Green1922, Green1937].
GENERAL: Description of adult female by Rutherford (1914).

CITATIONS: Borchs1966 [catalogue: 366]; DEDAC1923 [host, distribution]; Green1922 [host, distribution: 463]; Green1937 [host, distribution: 338]; MacGil1921 [taxonomy, description, host, distribution: 461]; Ramakr1921a [host, distribution: 359]; Ruther1914 [taxonomy, description, host, distribution: 265].

Greeniella fimbriata (Ferris)

Decoraspis fimbriata Ferris, 1955c: 32. Type data: CHINA: Kwangtung Province, Yeung Kong, on undetermined tree. Holotype female. Type depository: Davis: The Bohart Museum of Entomology, University of California, California, USA.
Greeniella fimbriata; Borchsenius, 1966: 366. Change of combination.
SCALE COVER: Female scale oval, black, overlain by a film of white wax. Male scale round and as large as female, very light yellow; exuvia black (Ferris, 1955c).
DISTRIBUTION: **Oriental**: China (People's Republic) (Guangdong (Kwangtung) [Ferris1955c]).
GENERAL: Description and illustration of adult female by Ferris (1955c) and by Chou (1985, 1986).
CITATIONS: Borchs1966 [catalogue: 366]; Chou1985 [taxonomy, description, host, distribution: 326]; Chou1986 [taxonomy, illustration: 699]; DanzigPe1998 [catalogue: 270]; Ferris1955c [taxonomy, description, illustration, host, distribution: 32]; LongoMaPe1995 [distribution: 127]; Tao1999 [taxonomy, host, distribution: 90].

Greeniella javanensis (Green)

Aonidia javanensis Green, 1905: 31. Type data: INDONESIA: Java, on undersurface of leaves of *Myristica fragrans*. Syntypes, female. Type depository: London: The Natural History Museum, England, UK.
Greeniella javanensis; Robinson, 1918: 147. Change of combination.
SCALE COVER: Female scale subcircular, posterior extremity slightly pointed; occupied almost completely by large second exuvia with a very narrow secretionary border; dull reddish-brown; first exuvia outlined with fulvous; diameter, about 1 mm. Male scale larger, paler and flatter; rather broader than long; brownish-ochreous; diameter about 1 mm (Green, 1905).
HOST PLANTS: **Myristicaceae**: *Myristica fragrans* [Green1905, Sander1906]. **Myrtaceae**: *Eugenia* [Robins1918].
DISTRIBUTION: **Oriental**: Indonesia (Java [Green1905, Sander1906]); Philippines (Luzon [Robins1918]).
GENERAL: Description and illustration of adult female by Green (1905).
CITATIONS: Borchs1966 [catalogue: 366]; Cohic1958 [taxonomy: 12]; Green1905 [taxonomy, description, illustration, host, distribution: 31-32]; Laing1933 [taxonomy: 678]; MacGil1921 [taxonomy, description, host, distribution: 465]; Robins1918 [taxonomy, description, illustration, host, distribution: 147]; Sander1906 [taxonomy, host, distribution: 16].

Greeniella lahoarei (Takahashi)

Aonidia (*Greeniella*) *lahoarei* Takahashi, 1931a: 219. Type data: TAIWAN: Sozan, Suisha, on *Eugenia* sp. Syntypes, female. Type depository: Taichung: Entomology Collection, Taiwan Agricultural Research Institute, Wu-feng, Taichung, Taiwan
Greeniella lahoarei; Borchsenius, 1966: 366. Change of combination.
SCALE COVER: Female scale flattened, somewhat convex on middle area of dorsum, mostly shining reddish black, yellowish brown on marginal area except on posterior projecting portion, very narrowly dusky on margin; about 0.6 mm (Takahashi, 1931a).
HOST PLANTS: **Myrtaceae**: *Eugenia* [Takaha1931a, Takaha1933, Takagi1970].
DISTRIBUTION: **Oriental**: Taiwan [Takaha1931a, Takaha1933, Takagi1970].
GENERAL: Description and illustration of adult female by Takahashi (1931a).
CITATIONS: Borchs1966 [catalogue: 366]; Chou1985 [taxonomy, description, host, distribution: 405]; Takagi1970 [taxonomy, host, distribution: 131]; Takaha1931 [taxonomy, description, illustration, host, distribution: 219-220]; Takaha1933 [host, distribution: 28,62]; Tao1999 [taxonomy, host, distribution: 90].

Greeniella mesuae (Green)

Aonidia messuae Leonardi, 1899: 205. Nomen nudum.
Aonidia messuae Leonardi, 1900: 331. Nomen nudum.
Aonidia mesuae Green, 1900a: 74. Type data: SRI LANKA: Peradeniya, Royal Botanic Gardens, on *Mesua ferrea*. Syntypes, female. Type depository: London: The Natural History Museum, England, UK.
Aonidia messuae; Fernald, 1903b: 303. Notes: Incorrect citation of "Green, Leon." as author.
Aonidia messuae; Fernald, 1903b: 303. Misspelling of species name.
Greeniella messuae; MacGillivray, 1921: 460. Change of combination and misspelling of species name.
Greeniella messuae; MacGillivray, 1921: 460. Notes: Incorrect citation of "Leonardi" as author and misspelling of species name.
SCALE COVER: Female scale circular, 1.25 mm in diameter; reddish brown with a paler zone marking position of second exuvia; secretionary area covering and extending beyond second exuvia; first exuvia exposed, blackish, strongly convex, subcentral; second exuvia completely enclosing adult insect, with a median convex area corresponding with position of larval exuviae; extremity of second exuvia with six distinct chitinous lobes followed by about five irregular marginal prominences, and from 3 to 5 stout spiniform squames in each interlobular space; diameter of second exuvia 1 mm. Male scale similar in size and form to female, but without pale scale (Green, 1900a).
HOST PLANTS: **Guttiferae**: *Mesua* [Ramakr1921a], *Mesua ferrea* [Green1900a, Leonar1900, Green1922, Green1937].
DISTRIBUTION: **Oriental**: Sri Lanka [Green1900a, Ramakr1921a, Green1937].
GENERAL: Description and illustration of adult female by Green (1900a).
CITATIONS: Borchs1966 [catalogue: 366]; Cocker1899a [taxonomy: 396]; DEDAC1923 [host, distribution]; Fernal1903b [catalogue: 303]; Green1900a [taxonomy, description, illustration, host, distribution: 74-75]; Green1905a

[taxonomy: 348]; Green1922 [host, distribution: 463]; Green1937 [host, distribution: 337]; Leonar1899 [taxonomy: 205]; Leonar1900 [taxonomy, description, host, distribution: 331-333]; Leonar1903a [taxonomy: 6]; MacGil1921 [taxonomy, description, host, distribution: 459-460]; Ramakr1921a [host, distribution: 358].

Greeniella ornata Brimblecombe

Greeniella ornata Brimblecombe, 1959: 134. Type data: AUSTRALIA: Queensland, Nambour, on *Callistemon salignus*; collected July, 1956. Holotype female. Type depository: Brisbane: Queensland Museum, Queensland, Australia.
SCALE COVER: Scale of adult female circular, 0.9 mm diameter, convex, black with margin dark fawn; second pellicle dark brown to black, completely covering adult female; first pellicle black, central, with a series of pale lateral filaments, and a pair dorsally (Brimblecombe, 1959).
HOST PLANTS: **Myrtaceae**: *Callistemon salignus* [Brimbl1959].
DISTRIBUTION: **Australasian**: Australia (Queensland [Brimbl1959]).
GENERAL: Description and illustration of adult female by Brimblecombe (1959).
CITATIONS: Borchs1966 [catalogue: 366]; Brimbl1959 [taxonomy, description, illustration, host, distribution: 134-136].

Greeniella ramosa Williams

Greeniella ramosa Williams, 1960d: 153. Type data: MALAYSIA: Malaya, Bukit Mertajam, on *Myristica fragrans*; collected 18.VI.1927. Holotype female. Type depository: London: The Natural History Museum, England, UK.
SCALE COVER: Scale composed of exuviae of second stage female enclosing adult female, ovoid but with anterior end more or less straight and with a small triangular-shaped lobe projecting from posterior edge; becoming hard and slightly convex; pale orange; exuviae of first stage dark brown, situated toward anterior end. Male scale not seen (Williams, 1960d).
HOST PLANTS: **Myristicaceae**: *Myristica fragrans* [Willia1960d].
DISTRIBUTION: **Oriental**: Malaysia (Malaya [Willia1960d]).
BIOLOGY: Occurring on the under surfaces of the leaves and with the anterior ends touching the midribs (Williams, 1960d).
GENERAL: Description and illustration of adult female by Williams (1960d).
CITATIONS: Borchs1966 [catalogue: 366]; Willia1960d [taxonomy, description, illustration, host, distribution: 153-154].

Greeniella tentaculata (Green)

Aonidia tentaculata Ramakrishna Ayyar, 1919a: 22. Nomen nudum.
Aonidia tentaculata Green, 1919c: 440. Type data: INDIA: Kerala, Travancore, Quilon, on *Vateria indica*. Syntypes, female. Type depository: London: The Natural History Museum, England, UK.
Greeniella tentaculata; Borchsenius, 1966: 366. Change of combination.
SCALE COVER: Female scale flattish, dull castaneous, consisting of large, naked nymphal exuviae upon which is superimposed smaller larval exuviae of a darker shade of brown (Green, 1919c).
HOST PLANTS: **Dipterocarpaceae**: *Vateria indica* [Green1919c, Ramakr1921a].

DISTRIBUTION: **Oriental**: India [Ramakr1921a] (Kerala [Green1919c]).
BIOLOGY: Green (1919c) noted "Associated with *Websteriella vaieriae*" in original description.
GENERAL: Description and illustration of adult female by Green (1919c).
CITATIONS: Borchs1966 [catalogue: 366]; Green1919c [taxonomy, description, illustration, host, distribution: 440-441]; Ramakr1919a [taxonomy, host, distribution: 22]; Ramakr1921a [host, distribution: 359]; Ramakr1930 [taxonomy, host, distribution: 28].

Greeniella viridis (Lindinger)

Aonidia viridis Lindinger, 1911: 86. Type data: INDIA: Travancore, on *Aglaia minutiflora*, 29.iii.1895. Syntypes, female. Type depository: Zoologisches Institut und Zoologishces Museum, Universität von Hamburg, Germany; type no. mp141.
Greeniella viridis; MacGillivray, 1921: 462. Change of combination.
SCALE COVER: Lindinger (1911) did not describe the scale cover.
HOST PLANTS: **Meliaceae**: *Aglaia* [Ramakr1919a, Ramakr1921a], *Aglaia minutiflora* [Green1919c].
DISTRIBUTION: **Oriental**: India [Ramakr1921a] (Tamil Nadu [Green1919c]).
GENERAL: Description and illustration of adult female by Lindinger (1911).
CITATIONS: Borchs1966 [catalogue: 366]; Cohic1958 [taxonomy: 12]; Green1919c [host, distribution: 441]; Laing1933 [taxonomy: 678]; Lindin1911 [taxonomy, description, illustration, host, distribution: 86]; MacGil1921 [taxonomy, description, host, distribution: 462]; Ramakr1919a [taxonomy, host, distribution: 22]; Ramakr1921a [host, distribution: 359]; Sassce1912 [taxonomy, host, distribution: 92]; WeidneWa1968 [taxonomy: 171].

Greenoidea MacGillivray

Greenoidea MacGillivray, 1921: 392. Type species: *Aspidiotus (Targionia) phyllanthi* Green, by original designation.
Greenoidea; Lindinger, 1937: 182. Incorrect synonymy; discovered by Gerson & Davidson, 1974: 157. Notes: Incorrect synonymy with *Crenulaspidiotus*.
SYSTEMATICS: *Greenoidea* MacGillivray differs from *Pseudomelanaspis* Borchsenius in having pygidial lobes with parallel axes; from *Crenulaspidiotus* MacGillivray in lacking small transverse furrow above second pygidial lobes; and from *Melanaspis* Cockerell in possessing a reduced number of lobe-associated pygidial paraphyses (Gerson & Davidson, 1974).
GENERAL: Definition and characters by Gerson & Davidson (1974).
CITATIONS: Balach1958 [taxonomy: 191]; Ferris1937c [taxonomy: 51]; GersonDa1974 [taxonomy, description: 156-157].

Greenoidea phyllanthi (Green)

Aspidiotus (Targionia) phyllanthi Green, 1905a: 344. Type data: SRI LANKA: Peradeniya, on *Phyllanthus myrtifolius*; collected by E.E. Green, February 1900. Lectotype female, by subsequent designation Davidson, 1975: 82. Type depository: London: The Natural History Museum, England, UK.

Targionia phyllanthi; Sanders, 1906: 16. Change of combination.
Greenoidea phyllanthi; MacGillivray, 1921: 446. Change of combination.
Aspidiotus phyllanthi; Ferris, 1937c: 51. Change of combination.
Melanaspis phyllanthi; Ferris, 1941d: 347. Change of combination.
Crenulaspidiotus phyllanthi; Borchsenius, 1966: 359. Change of combination.
Greenoidea phyllanthi; Gerson & Davidson, 1974: 157. Revived combination.
SYSTEMATICS: Davidson (1975: 82) clarified that the Lectotype designation by Gerson & Davidson (1974: 159) was invalid, and designated the correct Lectotype.
SCALE COVER: Female scale dull black, with a raised whitish disc on larval exuvia: moderately convex: more or less concealed beneath corky outer bark. Diameter 1 to 1.25 mm. Male scale greyish, (a whitish bloom overlying blackish secretionary area). Exuvia very dark shining brown, with a raised whitish circle in centre. Length 1 mm (Green, 1905a).
HOST PLANTS: **Euphorbiaceae**: *Phyllanthus myrtifolius* [Green1905a, Green1922, GersonDa1974]. **Phyllanthaceae**: *Phyllanthus* [Ramakr1921a].
DISTRIBUTION: **Oriental**: Sri Lanka [Green1905a, Green1922, Green1937].
GENERAL: Description and illustration of adult female by Green (1905a) and by Gerson & Davidson (1974).
CITATIONS: Borchs1966 [catalogue: 359]; DEDAC1923 [host, distribution]; Ferris1937c [taxonomy, illustration: 51]; Ferris1941d [taxonomy: 347]; Ferris1941e [taxonomy: 47]; Ferris1943a [taxonomy: 86]; GersonDa1974 [taxonomy, description, illustration, host, distribution: 157-159]; Green1905a [taxonomy, description, illustration, host, distribution: 344-345]; Green1922 [host, distribution: 462]; Green1937 [host, distribution: 332]; MacGil1921 [taxonomy, description, host, distribution: 446]; Ramakr1921a [host, distribution: 357]; Sander1906 [taxonomy, host, distribution: 16].

Helaspis McKenzie

Helaspis McKenzie, 1963: 34. Type species: *Helaspis mexicana* McKenzie, by monotypy and original designation.
SYSTEMATICS: *Helaspis* resembles *Aspidiotus* Bouché more than any other aspidiotine genera. Short length of dorsal pygidial macroducts resembles certain *Aspidiotus* species, but the 'peg-like' plates, with little or no fimbriation, and bilobed-shape of third pygidial lobes, show wide divergence from *Aspidiotus* (McKenzie, 1963).
GENERAL: Definition and characters by McKenzie (1963).
CITATIONS: Borchs1966 [catalogue: 269]; Kozar1990f [distribution: 142]; McKenz1963 [taxonomy, description: 34]; MorrisMo1966 [taxonomy, catalogue: 90].

Helaspis mexicana McKenzie

Helaspis mexicana McKenzie, 1963: 34. Type data: MEXICO: State of Sinaloa, at Mazatlan, on *Roupala* sp. Holotype female. Type depository: Davis: The Bohart Museum of Entomology, University of California, California, USA.
COMMON NAME: Mexican helaspis scale [McKenz1963].

SCALE COVER: Female scale about 1 mm long, 0.85 mm wide; no further characters given in original description (McKenzie, 1963).

HOST PLANTS: **Proteaceae**: *Roupala* [McKenz1963].

DISTRIBUTION: **Nearctic**: Mexico (Sinola [McKenz1963]).

BIOLOGY: This scale was observed mostly on the fruits of its host (McKenzie, 1963).

GENERAL: Description and illustration of adult female by McKenzie (1963).

CITATIONS: Borchs1966 [catalogue: 269]; McKenz1963 [taxonomy, description, illustration, host, distribution: 34-36].

Hemiberlesia Cockerell

Aspidites Berlese & Leonardi, 1896: 349. Type species: *Aspidiotus rapax* Comstock, 1881, by original designation. Homonym of *Aspidites* in Reptilia (1877), Mollusca (1895); discovered by Cockerell in Leonardi, 1897a: 375.

Hemiberlesia Cockerell, 1897i: 12. Replacement name for *Aspidites* Berlese & Leonardi, 1896.

Aspidiotus (*Hemiberlesia*); Cockerell, 1899a: 395. Change of status.

Hemiberlesea Lindinger, 1908b: 96. Unjustified replacement name for *Hemiberlesia* Cockerell.

Hemiberlesea; Brain, 1918: 117. Misspelling of genus name.

Marlattaspis MacGillivray, 1921: 387. Type species: *Aspidiotus implicatus* Maskell, by monotypy and original designation. Synonymy by Ferris, 1937: 82.

Hemiberlesiana Thiem & Gerneck, 1934: 232. Type species: *Aspidiotus camelliae* Signoret, by original designation. Synonymy by Ferris, 1937c: 54.

Hemiberlesea; Gómez-Menor Ortega, 1937: 109. Misspelling of genus name.

Borchseniaspis Zahradník, 1959: 67. Type species: *Aspidiotus palmae* Morgan & Cockerell, 1893, by monotypy and original designation. Synonymy by Williams & Watson, 1988: 134. Notes: The type species is actually *Aspidiotus palmae* Cockerell, 1893.

SYSTEMATICS: Closely related to *Abgrallaspis* Balachowsky, from which it is distinguished mainly by large size of anus (about as wide as a median lobe), and anus separated from base of median lobes at a distance equal to its diameter. In *Abgrallaspis* latter space longer than anus diameter (Balachowsky, 1948b; Komosinska, 1969). Takagi (1969a, 1974, and in personal communication, July 20 2002, to Yair Ben-Dov) believes that *Hemiberlesia*, as understood by Balachowsky and others, and *Abgrallaspis* are close, and not clearly distinguishable from each other. Type species of *Hemiberlesia* characterized by enormously large anal opening, while the type species of *Abgrallaspis* has a small-sized anus. But a gradation in size of anus and in size and number of dorsal ducts is found among species of these two genera (Takagi, 1969a).

GENERAL: Definition and characters by Ferris (1938a), Borchsenius (1950b), Lupo (1953a), Balachowsky (1948b, 1956), Gómez-Menor Guerrero (1962), Takagi (1969a), Velasquez (1971), Bazarov & Shmelev (1971), Stoetzel & Davidson (1974a), Williams & Watson (1988), Danzig (1993), Yaşar (1995) and by Kosztarab (1996).

KEYS: Colon-Ferrer & Medina-Gaud 1998: 28-32 (female) [Genera of Puerto Rico]; Gill 1997: 24-26 (female) [Genera of California]; Gill 1997: 155 (female) [Species of California]; Kosztarab 1996: 406-407 (female) [Northeastern North America]; Blay Goicoechea 1993: 473 (female) [Spain]; Danzig 1993: 169 (female) [species Europe]; Wolff & Corseuil 1993: 29 (female) [Brazil, Rio Grande do Sul]; Zahradník 1990b: 74 (female) [Czech Republic]; Williams & Watson 1988: 20 (female) [Tropical South Pacific]; Tereznikova 1986: 83 (female) [Ukraine]; Chou 1985: 283 (female) [Genera of China]; Chou 1985: 300 (female) [Species of China]; Paik 1978: 324 (female) [species South Korea]; Bazarov & Shmelev 1971: 186 (female) [Central Asia]; Velasquez 1971: 110 (female) [Philippines]; Komosinska 1969: 50 (female) [*Abgrallaspis* group]; Beardsley 1966: 502-504 (female) [Federated States of Micronesia]; Danzig 1964: 646 (female) [Europe]; Gómez-Menor Guerrero 1962: 157 (female) [Canary Islands]; Zahradník 1959: 65-67 (female); Zahradník 1959a: 548 (female) [Czech Republic]; Zahradník 1959: 66 (female) [Czech Republic]; Balachowsky 1958b: 230 (female) [*Aspidiotina* of Africa]; Ezzat 1958: 237-239 (female) [Egypt]; Gómez-Menor Ortega 1956: 7-8 (female) [Spain]; McKenzie 1956: 22 (female) [U.S.A.: California]; Balachowsky 1951: 600 (female) [Mediterranean]; Borchsenius 1950b: 167 (female) [USSR]; Zimmerman 1948: 351 (female) [Hawaii]; Gómez-Menor Ortega 1946: 59-61 (female) [Spain]; McKenzie 1946: 29 (female) [World]; Ruiz Castro 1944: 57 (female) [Spain]; Ferris 1942: 34-35 (female) [species North America]; Ferris 1942: 26 (female) [North America]; Borchsenius 1937: 99 (female) [USSR]; Borchsenius 1937a: 32-33 (female) [Palaearctic Region]; Archangelskaya 1929: 189 (female) [Palaearctic Region]; Leonardi 1920: 27 (female) [Italy]; Leonardi 1920: 90 (female) [Species of Italy]; Brain 1918: 117 (female) [South Africa]; Cockerell 1905b: 201 (female) [U.S.A.: Colorado]; Newell 1899: 3 (female) [North America].

CITATIONS: Archan1929 [taxonomy: 89]; Balach1928a [taxonomy: 132]; Balach1948b [taxonomy, description: 297-298]; Balach1950b [taxonomy: 490]; Balach1951 [taxonomy: 600]; Balach1953k [taxonomy, description: 113-114]; Balach1956 [taxonomy, description: 104-105]; Balach1958b [taxonomy: 230]; BazaroSh1971 [taxonomy, description: 195-196]; Beards1966 [taxonomy: 520]; BerlesLe1896 [taxonomy, description: 349-350]; BerlesLe1898a [taxonomy: 131]; BlayGo1993 [taxonomy, description: 476]; Borchs1937 [taxonomy: 99]; Borchs1937a [taxonomy, description: 33,60]; Borchs1949d [taxonomy, description: 194,239]; Borchs1950b [taxonomy, description: 223]; Borchs1966 [catalogue: 304,313]; Brain1918 [taxonomy: 117,130]; Brimbl1968 [taxonomy: 50]; Chou1985 [taxonomy, description: 283,299-300]; Chou1985 [taxonomy, description: 299-300]; Cocker1897i [taxonomy: 9,12,31]; Cocker1899a [taxonomy: 396]; Cocker1905b [taxonomy: 201]; ColonFMe1998 [taxonomy, description: 58]; Danzig1964 [taxonomy: 652]; Danzig1993 [taxonomy, description: 168-169]; DanzigPe1998 [catalogue: 271]; DuttaSi1990 [taxonomy: 1]; Ezzat1958 [taxonomy: 239]; Ferris1937c [taxonomy: 51]; Ferris1938a [taxonomy, description: 232]; Ferris1942 [taxonomy: 446:26]; Ferris1943a [taxonomy: 95]; Gill1997 [taxonomy: 155]; GomezM1937 [taxonomy, description: 109-110]; GomezM1946 [taxonomy: 60]; GomezM1956 [taxonomy, description: 50]; GomezM1962 [taxonomy, description: 175]; Hadzib1983 [taxonomy: 230]; Kawai1980 [taxonomy: 216]; Koszta1996 [taxonomy, description: 509]; Kozar1990f [distribution: 143]; Leonar1897 [taxonomy: 284]; Leonar1897a [taxonomy: 375]; Leonar1897b [taxonomy: 109,117-119]; Leonar1920 [taxonomy, description: 27,89-90]; Lepage1938 [taxonomy: 407]; Lepesm1947 [taxonomy, description: 207]; Lindin1908b [taxonomy: 96]; Lindin1937 [taxonomy: 186]; Lindin1949 [taxonomy: 210]; Lupo1953a [taxonomy, description: 74-76]; MacGil1921 [taxonomy, description: 387,406]; Mamet1949 [taxonomy: 59]; McKenz1939 [taxonomy: 54]; McKenz1943 [taxonomy: 152]; McKenz1944 [taxonomy: 53]; McKenz1947 [taxonomy: 31]; McKenz1951 [taxonomy: 80]; McKenz1956 [taxonomy: 22]; MillerDa1998 [taxonomy: 193]; MorrisMo1966 [taxonomy, catalogue: 17-18,24,91,115-116]; Muntin1971 [taxonomy: 119]; Newell1899 [taxonomy, description: 3]; RuizCa1944 [taxonomy: 57]; Sander1904a [taxonomy: 55-56]; Schmut1959 [taxonomy, description: 48,65]; Silves1902 [taxonomy: 12]; StoetzDa1974a [taxonomy, description: 501]; Takagi1969a [taxonomy, description: 76-77]; Tao1999 [taxonomy: 90]; ThiemGe1934a [taxonomy: 132,230,232]; Velasq1971 [taxonomy, description: 109-110]; WilliaWa1988 [taxonomy, description: 130]; WolffCo1993 [taxonomy: 29]; Xie1998 [taxonomy, description: 140]; Xie1998 [taxonomy, description: 140]; Yasar1995a [taxonomy, description: 88]; Zahrad1959 [taxonomy, description : 65-67]; Zimmer1948 [taxonomy: 358].

Hemiberlesia camarana (Seabra)
Selenaspidus camarae Seabra, 1921: 99. Nomen nudum.
Aspidiotus camaranus Seabra, 1922: 7. Type data: SÃO TOMÉ: on coffee. Syntypes, female. Notes: Depository of type material unknown.
Aspidiotus camaranus; Seabra, 1925: 31. Notes: Described again as "n. sp.".
Selenaspidus camaranus; Seabra, 1925: 33. Change of combination.
Hemiberlesia camarana; Borchsenius, 1966: 304. Change of combination requiring emendation of species name for agreement in gender.
SCALE COVER: Illustration of scale cover by Seabra (1922, 1925). Female scale circular, colour dark; exuviae brown; resembles scale cover of *Aspidiotus palmae*. Male scale unknown (Seabra, 1922, 1925).
HOST PLANTS: **Musaceae**: *Musa* [Lepage1938]. **Myrtaceae**: *Eucalyptus tereticonis* [Lepage1938]. **Vitaceae**: *Vitis* [Lepage1938].
DISTRIBUTION: **Afrotropical**: São Tomé and Príncipe (São Tomé) [Seabra1925]). **Neotropical**: Brazil (Parana [Lepage1938], Rio Grande do Sul [Lepage1938], Rio de Janeiro [Lepage1938]).
GENERAL: Description and illustration of adult female by Seabra (1922, 1925).
CITATIONS: Borchs1966 [catalogue: 304]; ClapsWoGo2001a [taxonomy, host, distribution: 18]; Ferris1941e [taxonomy: 41]; Lepage1938 [catalogue: 393]; Seabra1922 [taxonomy, description, illustration, host, distribution : 7]; Seabra1925 [taxonomy, description, illustration, host, distribution: 31-32].

Hemiberlesia candidula (Cockerell)

Aspidiotus (*Hemiberlesia*) *candidulus* Cockerell, 1900d: 130. Type data: U.S.A.: Arizona, Tucson, behind University, on leaves and twigs of *Prosopis velutina*. Syntypes, female. Type depository: Washington: U.S. National Entomological Collection, U.S. National Museum of Natural History, District of Columbia, USA.

Aspidiotus candidulus; Fernald, 1903b: 254. Change of combination.

Diaspidiotus candidula; MacGillivray, 1921: 412. Change of combination requiring emendation of species name for agreement in gender.

Hemiberlesia candidula; Ferris, 1938a: 233. Change of combination.

Morganella candidula; Lindinger, 1957: 545. Change of combination.

Hemiberlesia candidula; Borchsenius, 1966: 304. Revived combination.

SCALE COVER: Scale of female pale, flat, circular; exuviae subcentral; that of male white, oval, exuvia near one end (Ferris, 1938a).

HOST PLANTS: **Agavaceae**: *Yucca* [Ferris1921, Ferris1938a]. **Leguminosae**: *Prosopis velutina* [Cocker1900d, Ferris1919a, Ferris1938a].

DISTRIBUTION: **Nearctic**: Mexico (Baja California [Ferris1921, Ferris1938a]); United States of America (Arizona [Cocker1900d, Ferris1919a, Ferris1938a], California [Nakaha1982]).

BIOLOGY: Occurring on underside of leaves (Ferris, 1938a).

GENERAL: Description and illustration of adult female by Cockerell (1900d), Ferris (1919a, 1938a) and by Gill (1997).

KEYS: Gill 1997: 155 (female) [Species of California]; Balachowsky 1956: 105-108 (female) [Africa]; Balachowsky 1953k: 114-115 (female) [World]; Ferris 1942: 34 (female) [North America].

CITATIONS: Balach1948b [taxonomy: 298]; Balach1953k [taxonomy: 114]; Balach1956 [taxonomy: 106]; Borchs1966 [catalogue: 304]; Cocker1900d [taxonomy, description, host, distribution: 130]; Fernal1903b [catalogue: 254]; Ferris1919a [taxonomy, description, illustration, host, distribution: 63-64]; Ferris1921 [host, distribution: 123]; Ferris1938a [taxonomy, description, illustration, host, distribution: 233]; Ferris1941e [taxonomy: 41]; Ferris1942 [taxonomy: 446:34]; Gill1997 [host, distribution, taxonomy, description, illustration, economic importance: 155-156,158]; Lindin1957 [taxonomy: 545]; MacGil1921 [taxonomy, description, host, distribution: 412]; Nakaha1982 [host, distribution: 41]; Willia1985a [taxonomy: 233].

Hemiberlesia caricis (Gómez-Menor Ortega)

Quadraspidiotus caricis Gómez-Menor Ortega, 1954: 125. Type data: SPAIN: Madrid, Botanical Garden, on *Carex elegantisima*; collected 6.v.1925. Lectotype female, by subsequent designation Blay Goicoechea, 1993: 497. Type depository: Madrid: Museo Nacional de Ciencias Naturales, Spain.

Abgrallaspis caricis; Borchsenius, 1966: 313. Change of combination. Notes: Type data the same as in Gómez-Menor Ortega (1954).

Hemiberlesia caricis; Danzig & Pellizzari, 1998: 271. Change of combination.

SYSTEMATICS: *Quadraspidiotus caricis* was again described as n. sp. by Gómez-Menor Ortega, 1956b: 484. The type data in both descriptions is identical.

SCALE COVER: Female scale broadly elliptical, 2 mm long, 0.9 mm wide; white; exuviae yellow, lateral, slightly anterior; without a ventral vellum (Gómez-Menor Ortega, 1954).

HOST PLANTS: **Cyperaceae**: *Carex elegantissima* [GomezM1954, GomezM1956b, Martin1983, BlayGo1993].

DISTRIBUTION: **Palaearctic**: Spain [GomezM1954, GomezM1956b, Martin1983, BlayGo1993].

GENERAL: Description and illustration of adult female by Gómez-Menor Ortega (1954).

CITATIONS: BlayGo1993 [taxonomy, description, illustration, host, distribution: 497-501]; Borchs1966 [catalogue: 313]; DanzigPe1998 [catalogue: 271]; DuttaSi1990 [taxonomy: 1]; GomezM1954 [taxonomy, description, host, distribution: 125-128]; GomezM1956b [taxonomy, description, host, distribution: 484-486]; GomezM1958a [host, distribution: 7]; Leonar1897 [taxonomy: 286]; Martin1983 [taxonomy, host, distribution: 60].

Hemiberlesia chipponsanensis (Takahashi)

Aspidiotus chipponsanensis Takahashi, 1935: 33. Type data: TAIWAN: Taito Province, Chipponsan, on *Rhododendron* sp. Syntypes, female. Type depository: Taichung: Entomology Collection, Taiwan Agricultural Research Institute, Wu-feng, Taichung, Taiwan.

Hemiberlesia chipponsanensis; Borchsenius, 1966: 304. Change of combination.

Hemiberlesia chippomsanensis; Chou, 1985: 402. Misspelling of species name.

SCALE COVER: Female scale whitish, thick, convex dorsally, about 1.6 mm in diameter, usually beneath epidermis of host. Larval skins pale yellowish (Takahashi, 1935).

HOST PLANTS: **Ericaceae**: *Rhododendron* [Takaha1935, Takagi1970].

DISTRIBUTION: **Oriental**: Taiwan [Takaha1935, Takagi1970].

GENERAL: Description and illustration of adult female by Takahashi (1935).

CITATIONS: Borchs1966 [catalogue: 304]; Chou1985 [taxonomy, description, host, distribution: 402]; Ferris1941e [taxonomy: 42]; FoxWil1939 [host, distribution, economic importance: 2296]; Takagi1970 [taxonomy, host, distribution: 131]; Takaha1935 [taxonomy, description, illustration, host, distribution: 4,33-35]; Tao1999 [taxonomy, host, distribution: 90-91].

Hemiberlesia cupressi (Cockerell)

Aspidiotus cupressi Cockerell, 1899d: 168. Type data: MEXICO: Toluca, on twigs of *Cupressus* sp. Syntypes, female. Type depository: Washington: United States National Entomological Collection, U.S. National Museum of Natural History, District of Columbia, USA.

Aspidiotus (*Hemiberlesia*) *cupressi*; Cockerell, 1899a: 396. Change of combination.

Hemiberlesia cupressi; Leonardi, 1900: 338. Change of combination.

Hendaspidiotus cupressi; MacGillivray, 1921: 440. Change of combination.

Hemiberlesia cupressi; Borchsenius, 1966: 304. Revived combination.

SCALE COVER: Female scale about 1 mm in diameter; convex; white; exuviae subcentral to lateral, covered by white film; film often rubbed off, leaving exuviae

exposed, shining yellow, or coppery yellow; young scales round and very white (Cockerell, 1899d). Female scale flat, circular, white, exuviae central; that of male similar to female; elongate, oval, exuvia somewhat toward one end (Ferris, 1938a).
HOST PLANTS: **Cupressaceae**: *Cupressus* [Cocker1899d, Cocker1899n, Leonar1900, Ferris1938a].
DISTRIBUTION: **Nearctic**: Mexico [Cocker1899d, Cocker1899n, Ferris1938a].
BIOLOGY: Occurring on bark and cones of host (Ferris, 1938a).
GENERAL: Description and illustration of adult female by Cockerell (1899d) and by Ferris (1938a).
KEYS: Balachowsky 1956: 105-108 (female) [Africa]; Balachowsky 1953k: 115 (female) [World]; Ferris 1942: 35 (female) [North America]; Newell 1899: 25 (female) [North America].
CITATIONS: Balach1948b [taxonomy: 298]; Balach1953k [taxonomy: 115]; Balach1956 [taxonomy: 107]; Borchs1966 [catalogue: 304-305]; Cocker1899a [taxonomy: 396]; Cocker1899d [taxonomy, description, host, distribution: 168-169]; Cocker1899n [host, distribution: 23]; Fernal1903b [catalogue: 255]; Ferris1938a [taxonomy, description, illustration, host, distribution: 236]; Ferris1942 [taxonomy: 446:35]; Leonar1 900 [taxonomy, host, distribution: 338]; MacGil1921 [taxonomy, description, host, distribution: 440]; Newell1899 [taxonomy, description, host, distribution: 30-31]; Nur1990b [taxonomy, life history: 196]; Willia1985a [taxonomy: 234]; Zahrad1990 [host, distribution, description: 643].

Hemiberlesia diffinis (Newstead)
Aspidiotus affinis Newstead, 1893d: 186. Type data: GUYANA: Demerara, Botanic Garden, on unspecified host plant. Lectotype female and first instar, by subsequent designation Miller & Davidson, 1998: 197. Type depository: London: The Natural History Museum, England, UK. Homonym of *Aspidiotus affinis* Targioni Tozzetti.
Aspidiotus diffinis Newstead, 1893f: 281. Replacement name for *Aspidiotus affinis* Newstead, 1893d.
Hemiberlesia diffinis; Leonardi, 1897b: 133. Change of combination.
Aspidiotus (*Diaspidiotus*) *diffinis*; Cockerell, 1897i: 23. Change of combination.
Aspidiotus jatrophae Townsend & Cockerell, 1898: 178. Type data: MEXICO: Frontera, on *Jatropha*. Lectotype female and first instar, by subsequent designation Miller & Davidson, 1898: 197. Type depository: Washington: United States National Entomological Collection, U.S. National Museum of Natural History, District of Columbia, USA. Synonymy by Marlatt, 1900: 425.
Aspidiotus jatrophae parrotti Newell, 1899: 23. Type data: MEXICO: Frontera, on "Berenjeno chiquito" [= small eggplant]. Lectotype female and first instar, by subsequent designation Miller & Davidson, 1998: 197. Type depository: Washington: United States National Entomological Collection, U.S. National Museum of Natural History, District of Columbia, USA. Described: nymphal stages. Synonymy by Ferris, 1938a: 190.
Aspidiotus (*Diaspidiotus*) *jatrophae*; Cockerell, 1899a: 396. Change of combination.
Hemiberlesia iatrophae; Leonardi, 1900: 339. Misspelling of species name.
Aspidiotus diffinis parrotti; Fernald, 1903b: 258. Change of combination and rank.

Aspidiotus camelliae; Newstead, 1917: 371. Misidentification; discovered by Lindinger, 1957: 545.

Hemiberlesia diffinis parrotti; MacGillivray, 1921: 438. Change of combination.

Aspidiotus parrotti; Ferris, 1938a: 190. Change of combination and rank.

Aspidiotus guianensis Lindinger, 1957: 545. Type data: GUYANA [=BRITISH GUIANA]: Turkeyn, on *Erythraspis* [=*Erythrina*] *glauca*; collected by G.E. Bodkin, 17.ix.1915. Syntypes, female. Type depository: London: The Natural History Museum, England, UK. Synonymy by Ben-Dov & Williams, 2003: 166.

Aspidiotus iatrophae parrotii; Lindinger, 1957: 545. Change of combination.

Aspidiotus iatrophae parrotti; Lindinger, 1957: 545. Misspelling of species name.

Hemiberlesia guianensis; Borchsenius, 1966: 305. Change of combination.

Abgrallaspis diffinis; Komosinska, 1969: 60. Change of combination.

Hemiberlesia diffinis; Miller & Davidson, 1998: 197. Revived combination.

SYSTEMATICS: Newstead (1917) described and illustrated some specimens under the name *Aspidiotus camelliae* Signoret from British Guiana (Guyana). Lindinger (1957) referred to Newstead's record as "*A. [Aspidiotus] guianensis* nom. nov.". This reference to Newstead's illustration and description validates Lindinger's name. Ben-Dov & Williams (2003) studied types of *Aspidiotus guianensis* Lindinger, 1957, and concluded that it was a synonym of *Aspidiotus diffinis* Newstead, 1893, currently known as *Hemiberlesia diffinis* (Newstead).

SCALE COVER: Scale of female grey, somewhat elongate, high convex; exuviae close to anterior end; scale of male similar in form and colour (Ferris, 1938a).

HOST PLANTS: **Agavaceae**: *Dracaena* [MillerDa1998]. **Anacardiaceae**: *Spondias* [MillerDa1998]. **Annonaceae**: *Annona* [MillerDa1998]. **Apocynaceae**: *Plumeria* [MillerDa1998]. **Araceae**: *Philodendron* [MillerDa1998]. **Balsamaceae**: *Balsa* [Balach1959a]. **Burseraceae**: *Bursera* [MillerDa1998]. **Compositae**: *Faberia* [Ferris1938a]. **Cornaceae**: *Cornus* [Ferris1938a]. **Ebenaceae**: *Diospyros kaki* [TippinBe1970]. **Euphorbiaceae**: *Hevea* [MillerDa1998], *Jatropha* [Leonar1900, Ferris1938a, MillerDa1998], *Manihot* [MillerDa1998]. **Guttiferae**: *Mammea* [MillerDa1998]. **Juglandaceae**: *Carya illinoensis* [BesheaTiHo1973]. **Lauraceae**: *Persea* [Dekle1965c, MillerDa1998]. **Lecythidaceae**: *Couroupita* [MillerDa1998]. **Leguminosae**: *Cassia* [MillerDa1998], *Drepanocarpus* [MillerDa1998], *Erythrina* [MillerDa1998], *Erythrina glauca* [Newste1917, BenDovWi2003], *Mimosa* [Dekle1965c]. **Magnoliaceae**: *Liriodendron tulipifera* [Ferris1938a, StoetzDa1974a], *Magnolia* [Ferris1938a]. **Malvaceae**: *Hibiscus* [MillerDa1998]. **Meliaceae**: *Cedrela fissilis* [Lepage1938], *Melia azedarach* [Lepage1938]. **Myrtaceae**: *Psidium* [Ferris1938a, MillerDa1998], *Psidium guajava* [Ferris1921]. **Orchidaceae**: *Oncidium* [MillerDa1998]. **Palmae**: *Cocos* [MillerDa1998]. **Punicaceae**: *Punica* [MillerDa1998]. **Rosaceae**: *Prunus* [MillerDa1998]. **Sterculiaceae**: *Theobroma* [MillerDa1998]. **Tiliaceae**: *Tilia americana* [Ferris1938a]. **Ulmaceae**: *Celtis* [Ferris1938a, McDani1969], *Ulmus* [Ferris1938a, McDani1969, Dekle1965c]. **Zygophyllaceae**: *Porlieria angustifolia* [McDani1969].

DISTRIBUTION: **Nearctic**: Canada [Ferris1938a]; Mexico [Leonar1900, MillerDa1998] (Baja California [Ferris1921, Ferris1938a]); United States of America (District of Columbia [Ferris1938a], Florida [Dekle1965c], Georgia [TippinBe1970, BesheaTiHo1973], Louisiana [Ferris1938a], Maryland

[StoetzDa1974a], Mississippi [Ferris1938a], New Jersey [Ferris1938a], North Carolina [Ferris1938a], Texas [Ferris1938a, McDani1969]). **Neotropical**: Brazil [MillerDa1998] (Rio Grande do Sul [Lepage1938]); Colombia [Balach1959a, Kondo2001]; Costa Rica [MillerDa1998]; Cuba [Ferris1938a]; Dominica [MillerDa1998]; Ecuador [MillerDa1998]; El Salvador [MillerDa1998]; Guatemala [MillerDa1998]; Guyana [Newste1893d, Newste1917, Ferris1938a, MillerDa1998]; Jamaica [MillerDa1998]; Mexico (Tabasco [Ferris1938a]); Nicaragua [MillerDa1998]; Panama [MillerDa1998]; Peru [MillerDa1998].

BIOLOGY: Occurring on bark (Ferris, 1938a).

ECONOMIC IMPORTANCE: Considered to be of no economic importance in Florida, USA (Dekle, 1976).

GENERAL: Description and illustration of adult female (as *Aspidiotus camelliae* Sign., according to Lindinger, 1957) by Newstead (1917). *Aspidiotus guianensis* described from material off *Erythraspis glauca*. There is no plant genus *Erythraspis* and it is believed that Newstead mistook the host for *Erythrina glauca* (Leguminosae), often used as a shade tree in cacao plantations. Description and illustration of adult female by Ferris (1921, 1938a), Komosinska (1969), Kosztarab (1996) and by Miller & Davidson (1998). Description and illustration of female, male nymphs, male pupa and prepupa by Stoetzel & Davidson (1974a).

KEYS: Kosztarab 1996: 509 (female) [Northeastern North America]; Komosinska 1969: 76-78 (female) [World]; McDaniel 1969: 107 (female) [U.S.A.: Texas]; Balachowsky 1956: 105-108 (female) [Africa]; Balachowsky 1953k: 114-115 (female) [World]; Ferris 1942: 35 (female) [North America]; Lawson 1917: 217 (female) [U.S.A.: Kansas]; Cockerell 1905: 45-46 (female) [Mexico]; Newell 1899: 4-5 (female) [North America].

CITATIONS: Balach1948b [taxonomy: 298]; Balach1953k [taxonomy: 115]; Balach1956 [taxonomy: 107]; BeardsDaHo1976 [economic importance: 103]; BenDovWi2003 [taxonomy, host, distribution: 166-167]; BesheaTiHo1973 [host, distribution: 6]; Borchs1966 [catalogue: 305]; ClapsWoGo2001a [taxonomy, host, distribution: 18-19]; Cocker1894g [taxonomy, description, host, distribution: 130-131]; Cocker1896b [distribution: 334]; Cocker1897i [taxonomy, description, host, distribution: 23]; Cocker1899a [taxonomy: 396]; Cocker1899n [taxonomy, host, distribution: 21]; Cocker1905 [taxonomy: 46]; Dekle1965c [taxonomy, description, host, distribution: 69]; Dekle1976 [taxonomy, description, host, distribution, economic importance: 91]; Fernal1903b [catalogue: 257]; Ferris1921 [taxonomy, description, illustration, host, distribution: 125-126]; Ferris1938a [taxonomy, description, illustration, host, distribution: 238]; Ferris1941e [taxonomy: 40,42,44,46]; Ferris1942 [taxonomy: 446:35]; FletchGi1908 [host, distribution: 113-133]; Houser1918 [taxonomy, description, illustration, host, distribution: 165-166]; Komosi1969 [taxonomy, description, illustration, host, distribution: 60-62]; Kondo2001 [taxonomy, host, distribution: 44]; Koszta1996 [taxonomy, description, illustration, host, distribution, life history: 509-511]; Lawson1917 [taxonomy, description, illustration, host, distribution: 221-222]; Leonar1897b [taxonomy, description, illustration, host, distribution: 119,132-134]; Leonar1900 [taxonomy, host, distribution: 339]; Lepage1938 [catalogue: 394]; Lindin1928 [taxonomy: 106]; Lindin1932c [taxonomy: 204]; Lindin1957 [taxonomy: 545]; Lizery1942a [taxonomy, description, host, distribution: 321-322]; Lizery1943a [host, distribution: 319-335]; MacGil1921 [taxonomy, description, host, distribution: 437-438]; Marlat1900 [taxonomy: 425]; McDani1969 [taxonomy, illustration, host, distribution: 107-109]; MillerDa1990 [host, distribution, economic importance: 302]; MillerDa1998 [taxonomy, description, illustration, host, distribution: 193,197-200]; Nakaha1982 [host, distribution: 41]; Newell1899 [taxonomy, description, illustration, host, distribution: 23]; Newste1893d [taxonomy, description, host, distribution: 186-187]; Newste1893f [taxonomy: 281]; Newste1917 [taxonomy, description, illustration, host, distribution: 371]; StoetzDa1974 [taxonomy, life history: 138-140]; StoetzDa1974a [taxonomy, description, illustration, host, distribution, life history: 501-505]; TippinBe1970 [host, distribution: 9]; TownseCo1898 [taxonomy, description, host, distribution: 178]; Willia1985a [taxonomy: 235].

Hemiberlesia elegans (Lindinger)

Aspidiotus elegans Lindinger, 1913: 69. Type data: TANZANIA: Muansa, near Victoria Lake, on *Trichilia* sp. Syntypes, female. Type depository: Hamburg: Zoologisches Institut und Zoologishces Museum, Universität von Hamburg, Germany.
Hemiberlesia elegans; MacGillivray, 1921: 435. Change of combination.
SCALE COVER: Female scale slightly elongated, slightly broadly oval, 1 mm long, 0.8 mm wide; circular in young individuals; flat; grey brown with darker rim; exuviae yellow, subcentral to eccentric (Lindinger, 1913).
HOST PLANTS: **Meliaceae**: *Trichilia* [Lindin1913].
DISTRIBUTION: **Afrotropical**: Tanzania [Lindin1913].
GENERAL: Description and illustration of adult female by Lindinger (1913).
CITATIONS: Borchs1966 [catalogue: 305]; Ferris1941e [taxonomy: 43]; Lindin1913 [taxonomy, description, illustration, host, distribution: 69-70]; MacGil1921 [taxonomy, description, host, distribution: 435]; WeidneWa1968 [taxonomy: 172].

Hemiberlesia gliwicensis (Komosinska)

Abgrallaspis gliwicensis Komosinska, 1965a: 1. Type data: POLAND: Gliwicne, in greenhouse, on *Billbergia nutans*, March 1959. Holotype female. Type depository: Warsaw: Museum of the Institute of Zoology, Polish Academy of Sciences, Poland.
Abrallaspis gliwicensis; Dziedzicka, 1989: 96. Misspelling of genus name.
Hemiberlesia gliwicensis; Danzig & Pellizzari, 1998: 272. Change of combination.
SCALE COVER: Scale of female elongate 2-3 mm long, 1.1-2 mm wide, white, with first larval exuviae yellow, transparent; second exuvia quite black, thick; exuviae central, sometimes close to side of scale; male scale 1.1-1.5 mm long, distinctly elongate, white, exuviae black (Komosinska, 1965a).
HOST PLANTS: **Bromeliaceae**: *Billbergia nutans* [Komosi1965a, Dziedz1989, Danzig1993].
DISTRIBUTION: **Palaearctic**: Poland [Komosi1965a, Dziedz1989].
BIOLOGY: Recorded so far only from Bromeliaceae in greenhouses in Poland (Komosinska, 1965a; Dziedzicka, 1989).

GENERAL: Description and illustration of adult female by Komosinska (1965a, 1969) and by Dziedzicka (1989).
KEYS: Danzig 1993: 169 (female) [Europe]; Komosinska 1969: 76-78 (female) [World].
CITATIONS: Danzig1993 [taxonomy, host, distribution: 171]; DanzigPe1998 [catalogue: 272]; Dziedz1989 [taxonomy, description, illustration, host, distribution: 96-97]; Komosi1965a [taxonomy, description, illustration, host, distribution: 1-6]; Komosi1969 [taxonomy, description, illustration, host, distribution: 66-67].

Hemiberlesia ignobilis Ferris

Hemiberlesia ignobilis Ferris, 1941d: 343. Type data: PANAMA: Chiriqui Province, Armuelles, on *Ficus* sp. Holotype female. Type depository: Davis: The Bohart Museum of Entomology, University of California, California, USA.
SCALE COVER: Scale of female circular, when free, moderately convex; white; with exuviae central. Scale of male not recognized (Ferris, 1941d).
HOST PLANTS: **Moraceae**: *Ficus* [Ferris1941d].
DISTRIBUTION: **Neotropical**: Panama [Ferris1941d].
BIOLOGY: Occurring on bark, beneath bark flakes and in cracks (Ferris, 1941d).
GENERAL: Description and illustration of adult female by Ferris (1941d).
KEYS: Balachowsky 1956: 105-108 (female) [Africa]; Balachowsky 1953k: 114-115 (female) [World]; Ferris 1942: 35 (female) [North America].
CITATIONS: Balach1948b [taxonomy: 298]; Balach1953k [taxonomy: 115]; Balach1956 [taxonomy: 107]; Borchs1966 [catalogue: 305]; Ferris1941d [taxonomy, description, illustration, host, distribution: 343]; Ferris1942 [taxonomy: 446:35].

Hemiberlesia insularis (Balachowsky)

Chrysomphalus insularis Balachowsky, 1937a: 110. Type data: MADEIRA: near Ribeiro frio, 1000 meters altitude, on *Isoplexis (=Digitalis) sceptrum*. Syntypes, female. Type depository: Paris: Muséum national d'Histoire naturelle, France.
Hemiberlesia insularis; McKenzie, 1939: 54. Change of combination.
Abgrallaspis insularis; Balachowsky, 1948b: 314. Change of combination.
Hemiberlesia insularis; Danzig & Pellizzari, 1998: 272. Revived combination.
SCALE COVER: Female scale large, circular, convex, partly covered by epidermis of host plant; colour dark brown; larval exuviae eccentric, brown; ventral scale not distinct; 2.3-2.5 mm; male scale of similar structure, oval, 1.4-1.8 mm long (Balachowsky, 1948b).
HOST PLANTS: **Lauraceae** [Balach1937a, Balach1948b]. **Scrophulariaceae**: *Digitalis sceptrum* [Balach1937a, Balach1938a, Balach1948b].
DISTRIBUTION: **Palaearctic**: Madeira Islands [Balach1937a, Balach1938a, Balach1948b].
GENERAL: Description and illustration of adult female by Balachowsky (1937a, 1948b).
KEYS: Komosinska 1969: 76-78 (female) [World]; Balachowsky 1948b: 308 (female) [Mediterranean].

CITATIONS: Balach1937a [taxonomy, description, illustration, host, distribution: 110-112]; Balach1938a [host, distribution: 150-151]; Balach1948b [taxonomy, description, illustration, host, distribution: 314-316]; Borchs1966 [catalogue: 315]; DanzigPe1998 [catalogue: 272]; McKenz1939 [taxonomy: 54].

Hemiberlesia laciniata Gómez-Menor Ortega, new status

Hemiberlesia lataniae laciniata Gómez-Menor Ortega, 1965: 88. Type data: SPAIN: Alicante, on *Atriplex halimus*. Syntypes, female. Type depository: Madrid: Museo Nacional de Ciencias Naturales, Spain.

SYSTEMATICS: Gómez-Menor Ortega (1968) distinguished *Hemiberlesia lataniae laciniata* from *Hemiberlesia lataniae* in possessing 7-8 plates, as compared to 6 in *H. lataniae* and in having margin anterior to plates bearing 4 distinct elongations placed at equal spaces. Until types of *H. lataniae laciniata* will be studied, this sub-species is here raised to species level.

SCALE COVER: Gómez-Menor Ortega (1965) did not describe the scale cover.

HOST PLANTS: **Chenopodiaceae**: *Atriplex halimus* [GomezM1965].

DISTRIBUTION: **Palaearctic**: Spain [GomezM1965].

GENERAL: Description and illustration of adult female by Gómez-Menor Ortega (1965).

CITATIONS: GomezM1965 [taxonomy, description, illustration, host, distribution: 88-89].

Hemiberlesia lataniae (Signoret)

Aspidiotus lataniae Signoret, 1869: 860. Nomen nudum.

Aspidiotus lataniae Signoret, 1869a: 124. Type data: FRANCE: probably Paris, on leaves of *Latania*. Syntypes, female. Type depository: Vienna: Naturhistorisches Museum Wien, Austria.

Aspidiotus cydoniae Comstock, 1881a: 295. Type data: USA: Florida, on quince. Syntypes, female. Type depository: Washington: United States National Entomological Collection, U.S. National Museum of Natural History, District of Columbia, USA. Synonymy by Green, 1900a: 71.

Aspidiotus punicae Cockerell, 1893j: 255. Type data: JAMAICA: Kingston, on pomegranate, and DOMINICA: on coconut. Syntypes, female. Type depository: Washington: U.S. National Entomological Collection, U.S. National Museum of Natural History, District of Columbia, USA. Synonymy by Ferris, 1938a: 190.

Aspidiotus diffinis lateralis Cockerell, 1894g: 130. Type data: JAMAICA: Kingston, Parade Garden, Kingston, on stems of *Jasminum pubescens*. Syntypes, female. Type depository: Washington, D.C.: U.S. National Entomological Collection, U.S. National Museum of Natural History, USA. Synonymy by Ferris, 1941e: 45.

Aspidiotus cydoniae tecta Maskell, 1897a: 240. Type data: SANDWICH ISLANDS [=HAWAII]: on "Ohia". Syntypes, female. Type depositories: Auckland: New Zealand Arthropod Collection, Landcare Research, New Zealand, and Washington: United States National Entomological Collection, U.S. National Museum of Natural History, District of Columbia, USA. Synonymy by Borchsenius, 1966: 306. Notes: Described again as *Aspidiotus cydoniae*, Comstock, var. nov., by Maskell, 1898: 224.

Aspidiotus implicatus Maskell, 1897a: 241. Type data: TAIWAN: on *Campanula* sp. Syntypes, female. Type depository: Auckland: New Zealand Arthropod Collection, Landcare Research, New Zealand. Synonymy by Lindinger, 1957: 545. Notes: Described again as *Aspidiotus implicatus* sp. nov., Maskell, 1898: 226.

Aspidiotus (*Hemiberlesia*) *cydoniae*; Cockerell, 1897i: 21. Change of combination.

Aspidiotus (*Hemiberlesia*) *crawii* Cockerell, 1897i: 23. Type data: MEXICO: on twigs of grapevine; collected by Alex Craw. Syntypes, female. Type depository: Washington: U.S. National Entomological Collection, U.S. National Museum of Natural History, District of Columbia, USA. Synonymy by Ferris, 1941e: 42.

Aspidiotus (*Diaspidiotus*) *punicae*; Cockerell, 1897i: 24. Change of combination.

Aspidiotus (*Diaspidiotus*) *greenii* Cockerell, 1897i: 27. Type data: MEXICO: Mazatlan, on coconut palm; collected by Alex Craw. Syntypes, female. Type depository: Washington: United States National Entomological Collection, U.S. National Museum of Natural History, District of Columbia, USA. Synonymy by Cockerell, 1898n: 184.

Aspidiotus (*Aspidiotus*) *lataniae*; Cockerell, 1897i: 29. Change of combination.

Aspidiotus (*Evaspidiotus*) *lataniae*; Leonardi, 1898a: 76. Change of combination.

Aspidiotus (*Evaspidiotus*) *punicae*; Leonardi, 1898a: 77. Change of combination.

Aspidiotus (*Evaspidiotus*) *cydoniae*; Leonardi, 1898c: 44. Change of combination.

Aspidiotus greeni; Cockerell, 1898n: 184. Change of combination.

Aspidiotus crawii; Newell, 1899: 25. Change of combination.

Aspidiotus lateralis; Marlatt, 1899: 658. Change of status.

Aspidiotus (*Hemiberlesia*) *cydoniae tectus*; Cockerell, 1899a: 396. Change of combination.

Aspidiotus (*Hemiberlesia*) *greenii*; Cockerell, 1899a: 396. Change of combination.

Hemiberlesia crawii; Cockerell, 1899a: 396. Change of combination.

Aspidiotus (*Evaspidiotus*) *crawii*; Leonardi, 1900: 340. Change of combination.

Aspidiotus (*Evaspidiotus*) *cydoniae tectus*; Leonardi, 1900: 340. Change of combination requiring emendation of species name for agreement in gender.

Aspidiotus (*Evaspidiotus*) *greenii*; Leonardi, 1900: 340. Change of combination.

Aspidiotus (*Evaspidiotus*) *implicatus*; Leonardi, 1900: 341. Change of combination.

Aspidiotus (*Hemiberlesia*) *lataniae*; Hempel, 1900a: 502. Change of combination.

Aspidiotus cydoniae crawii; Fernald, 1903b: 256. Change of combination and rank.

Aspidiotus cydoniae punicae; Fernald, 1903b: 256. Change of combination and rank.

Hemiberlesia lataniae; Cockerell, 1905b: 202. Change of combination.

Aspidiotus cydoniae crawi; Lindinger, 1907a: 19. Change of combination and rank.

Aspidiotus cydoniae greeni; Cockerell & Robinson, 1915: 427. Change of combination and rank.

Aspidiella punicae; MacGillivray, 1921: 405. Change of combination and rank.

Marlattaspis implicata; MacGillivray, 1921: 406. Change of combination requiring emendation of species name for agreement in gender.

Diaspidiotus lataniae; MacGillivray, 1921: 412. Change of combination.

Aspidiotus sydonia; Seabra, 1921: 98. Misspelling of species name. Notes: Misspelling of *Aspidiotus cydoniae* Comstock.

Aspidiotus aspleniae Sasaki, I., 1935: 864. Type data: PHILIPPINES: Manila, intercepted at Japan, Port Kobe, on *Asplenium nidus*. Syntypes, female. Synonymy by Ferris, 1941e: 41. Notes: Depository of type material unknown.
Diaspidiotus (Aspidiotus) lataniae; Borchsenius, 1935a: 25. Change of combination.
Aspidiotus tectus; Ferris, 1941e: 48. Change of combination and rank.
Hemiberlesia implicatus; Balachowsky, 1953k: 114. Change of combination.
Aspidiella punica; Borchsenius, 1966: 307. Misspelling of species name.
Hemiberlesia lataniae; Borchsenius, 1966: 306. Revived combination.
Aspidiotus askleniae; Borchsenius, 1966: 307. Misspelling of species name.
Aspidiotus implocatus; Chou, 1985: 303. Misspelling of species name.
Hemiberlesia latanieae; Chou, 1985: 303. Misspelling of species name.
COMMON NAMES: escama del latano [Gonzal1989]; Latania scale [MerrilCh1923, McKenz1956, Borchs1966].
SYSTEMATICS: *Aspidiotus implicatus* Maskell, 1897 is a junior synonym of *H. lataniae*. However, Chou (1985: 303) listed the former as a synonym, whereas on page 305 it was regarded as a valid species. This species was reported to have uniparental as well as biparental populations (Brown, 1965).
SCALE COVER: Female scale little elongated; yellow; transparent in center and white around exuviae; exuviae large, oval elongate. Male scale not found (Signoret, 1869b). Female scale quite convex, with exuviae displaced toward one side, scale thus having a slightly tilted appearance, colour ranging from white to grey; exuviae described by some authors as yellowish but quite dark in all the specimens examined. Male scale unknown (Ferris, 1938a). Colour photograph by Gonzalez (1986), Gill (1997) and by Wong *et al.* (1999).
HOST PLANTS: **Palmae**: *Livistona australis* [Brimbl1968]. **Acanthaceae**: *Eranthemum pulchellum* [Brimbl1968]. **Actinidiaceae**: *Actinidia chinensis* [Gonzal1986, Gonzal1989a], *Saurauia tristyla oldhamii* [Takagi1969a]. **Agavaceae**: *Agave cantala* [Velasq1971], *Cohnia floribunda* [Mamet1943a, Borchs1966], *Cordyline* [WilliaWa1988], *Cordyline fruticosa* [Velasq1971], *Cordyline neo-caledonica* [WilliaWa1988], *Dracaena draco* [GomezM1962], *Dracaena reflexa* [Mamet1943a, Borchs1966], *Yucca* [McKenz1956, BesheaTiHo1973], *Yucca aloifolia* [Cohic1958, WilliaWa1988], *Yucca gloriosa* [Mamet1943a, Borchs1966, Martin1983]. **Amaranthaceae**: *Amaranthus viridis* [Brimbl1968], *Gomphrena globosa* [WilliaWa1988]. **Amaryllidaceae**: *Calostemma* [Hall1923]. **Anacardiaceae**: *Mangifera indica* [Ferris1921, Mamet1959a, Borchs1966, Takagi1969a, Almeid1971, WolffCo1993a, KinjoNaHi1996], *Pleiogynium cerasiferum* [Brimbl1968], *Rhus tsaratanana* [Mamet1951, Borchs1966], *Schinus molle* [Hall1923, DeLott1967a], *Sclerocarya caffra* [Almeid1971], *Spondias cytherea* [Mamet1943a, Borchs1966], *Spondias mombium* [Almeid1973b]. **Annonaceae**: *Annona reticulata* [Beards1966], *Annona squamosa* [Hall1922]. **Apocynaceae**: *Carissa* [Ramakr1921a], *Carissa carandas* [Ramakr1919a], *Carissa edulis* [Hall1923, DeLott1967a], *Nerium oleander* [Mamet1943a, Borchs1966, Brimbl1968, BesheaTiHo1973, WilliaWa1988], *Plumeria* [Hall1928, MerrilCh1923, Beards1966], *Plumeria acutifolia* [Mamet1943a, Borchs1966], *Plumeria rubra* [WilliaWa1988], *Tabernaemontana* [MerrilCh1923], *Thevetia nerifolia* [Hall1928], *Thevetia peruviana* [Almeid1971]. **Aquifoliaceae**: *Ilex*

[BesheaTiHo1973, StoetzDa1974a], *Ilex cassine* [MerrilCh1923, BesheaTiHo1973], *Ilex glabra* [BesheaTiHo1973]. **Araceae**: *Alocasia porteri* [Velasq1971], *Cyrtosperma chamissonis* [WilliaWa1988], *Monstera deliciosa* [Mamet1943a, Borchs1966], *Philodendron* [McKenz1956], *Synogonium podophyllum* [ImenesBeWo1999]. **Araliaceae**: *Brassaia actinophylla* [Brimbl1968], *Fatsia* [McKenz1956], *Hedera* [BesheaTiHo1973], *Hedera helix* [Mamet1954, McKenz1956, Borchs1966], *Schefflera* [WilliaWa1988], *Schefflera actinophylla* [WilliaWa1988]. **Arecaceae**: *Areca* [Borchs1934], *Areca lutescens* [Leonar1920, Lepage1938], *Balaca* [WilliaWa1988]. **Aspleniaceae**: *Asplenium nidus* [Sasaki1935]. **Averrhoaceae**: *Averrhoa carambola* [Mamet1943a, Borchs1966, Velasq1971, WilliaWa1988]. **Begoniaceae**: *Begonia* [MerrilCh1923, WilliaWa1988]. **Berberidaceae**: *Berberis vulgaris* [DeLott1967a]. **Bignoniaceae**: *Bignonia* [Hall1923], *Tecoma radicans* [Mamet1950, Mamet1951, Borchs1966]. **Bombacaceae**: *Ceiba pentandra* [WilliaWa1988], *Eriodendron* [Balach1932d], *Myristica hypargyraea* [WilliaWa1988]. **Boraginaceae**: *Cordia holstii* [DeLott1967a]. **Buxaceae**: *Buxus japonica* [Kuwana1927]. **Cactaceae** [Lepage1938], *Cactus* [Green1896e], *Opuntia ficus-indica* [Balach1932d], *Opuntia tomentosa* [Balach1932d]. **Campanulaceae**: *Campanula* [Leonar1900, Takaha1929, Kuwana1927]. **Cannaceae**: *Canna* [Mamet1943a, Borchs1966]. **Capparidaceae**: *Forchhammeria watsoni* [Ferris1921]. **Caricaceae**: *Carica papaya* [Brimbl1968]. **Casuarinaceae**: *Casuarina* [Takaha1941b, Mamet1959a, Dekle1965c, Borchs1966, Almeid1973b, WilliaWa1988], *Casuarina equisetifolia* [Mamet1943a, Borchs1966]. **Celastraceae**: *Denhamia pittosporoides* [Brimbl1968], *Euonymus* [MerrilCh1923, McKenz1956]. **Chenopodiaceae**: *Atriplex halimus* [Martin1983], *Beta maritima* [Balach1932d], *Chenopodium album* [Hall1923]. **Chrysobalanaceae**: *Parinari laurinum* [WilliaWa1988]. **Cistaceae**: *Cistus heterophyllus* [Balach1932d], *Cistus salviaefolius* [Balach1932d]. **Combretaceae**: *Terminalia* [Almeid1971]. **Compositae** [BesheaTiHo1973], *Artemisia* [Balach1938a], *Artemisia argentea* [Balach1938a], *Baccharis* [McDani1969, BesheaTiHo1973], *Baccharis emoryi* [McDani1969], *Baccharis neglecta* [McDani1969], *Bellis* [Mamet1943a, Borchs1966], *Borrichia frutescens* [BesheaTiHo1973], *Cassinia quinquefaria* [Brimbl1968], *Chrysanthemum* [CockerRo1915a, WilliaWa1988], *Chrysanthemum coronarium* [Hall1923], *Chrysanthemum indicum* [Brimbl1968], *Chrysanthemum segetum* [Balach1932d], *Elephantopus scaber* [WilliaWa1988], *Eupatorium* [BesheaTiHo1973], *Eupatorium seartinum* [McDani1969], *Eupatorium triplinerve* [Mamet1943a, Borchs1966], *Fitchia* [WilliaWa1988], *Gerbera jamesoni* [Brimbl1968], *Inula viscosa* [Balach1932d], *Phagnalon saxatile* [Balach1938a], *Scalesia* [Leonar1920, MerrilCh1923, Lepage1938], *Scalesia hopkinsi* [Kuwana1902], *Tithonia diversifolia* [Mamet1943a, Borchs1966]. **Crassulaceae**: *Bryophyllum pinnatum* [Mamet1943a, Borchs1966]. **Cruciferae**: *Lobularia maritima* [Brimbl1968]. **Cupressaceae**: *Cupressus arizonica* [Brimbl1968], *Cupressus lusitanicus* [Hall1928], *Juniperus* [Takagi1969a], *Thuja* [GomezM1962], *Thuja orientalis* [Brimbl1968]. **Cycadaceae**: *Cycas* [Green1896e, Cohic1958, Takagi1969a, WilliaWa1988], *Cycas revoluta* [Takaha1929, Takagi1969a, Velasq1971, Almeid1973b]. **Cyperaceae**: *Cyperus radiatus* [Velasq1971]. **Cyrillaceae**: *Cliftonia monophylla* [BesheaTiHo1973]. **Dioscoreaceae**: *Dioscorea*

[WilliaWa1988]. **Ebenaceae**: *Diospyros discolor* [Takagi1969a], *Diospyros kaki* [Hall1923, TomkinWiTh2000], *Royena pallens* [Hall1928]. **Ehretiaceae**: *Cordia* [WilliaWa1988], *Cordia interrupta* [Mamet1943a, Borchs1966], *Cordia myxa* [Hall1923, Mamet1943a, Borchs1966], *Ehretia membranifolia* [Brimbl1968]. **Elaeagnaceae**: *Elaeagnus* [MerrilCh1923]. **Elaeocarpaceae**: *Elaeocarpus tonganus* [WilliaWa1988]. **Ephedraceae**: *Ephedra* [Brimbl1968]. **Ericaceae** [BesheaTiHo1973], *Erica* [Mamet1959a, Borchs1966], *Leucothoe* [BesheaTiHo1973], *Rhododendron* [Takagi1969a]. **Euphorbiaceae**: *Aleurites fordii* [MerrilCh1923, Balach1956, WilliaWa1988], *Aleurites moluccana* [DeLott1967a, WilliaWa1988], *Clutia mollis* [DeLott1967a], *Gelonium procerum* [DeLott1967a], *Manihot esculenta* [WilliaWa1988], *Petalostigma pubescens* [Brimbl1968], *Phyllanthus* [Hall1922], *Sapium sebiferum* [BesheaTiHo1973]. **Fagaceae**: *Quercus* [Zahrad1972]. **Flacourtiaceae**: *Aberia caffra* [Borchs1934, DeLott1967a], *Hydnocarpus wightiana* [Mamet1943a, Borchs1966]. **Ginkgoaceae**: *Ginkgo biloba* [Mamet1943a, Borchs1966]. **Goodeniaceae**: *Scaevola frutescens* [Mamet1943a, Borchs1966]. **Gramineae**: *Chloris* [WilliaWa1988], *Saccharum* [Hall1922], *Thysanolaena agrostis* [Mamet1943a, Borchs1966]. **Guttiferae**: *Garcinia spicata* [Takagi1969a], *Hypericum* [BesheaTiHo1973], *Hypericum calycinum* [Brimbl1968], *Mammea* [MerrilCh1923]. **Heliconiaceae**: *Heliconia* [WilliaWa1988], *Heliconia platystachys* [Velasq1971]. **Hydrophyllaceae**: *Wigandia caracasana* [GomezM1962]. **Iridaceae**: *Gladiolus* [McKenz1956, Brimbl1968, WilliaWa1988, Lit1997b]. **Juglandaceae**: *Carya illinoiensis* [Brimbl1968], *Juglans regia* [Brimbl1968]. **Lauraceae**: *Cinnamomum camphorae* [Mamet1943a, Borchs1966, Brimbl1968], *Endiandra palmerstoni* [Brimbl1968], *Litsea akoensis* [Takagi1969a], *Litsea glutinosa* [Mamet1943a, Borchs1966], *Persea* [McKenz1956], *Persea americana* [Mamet1943a, Borchs1966, Brimbl1968, McDani1969, GersonZo1973, WilliaWa1988], *Persea gratissima* [Hall1923, DeLott1967a], *Ravensara* [Mamet1950, Borchs1966]. **Lecythidaceae**: *Barringtonia* [Takaha1941b, WilliaWa1988], *Barringtonia asiatica* [Beards1966, WilliaWa1988]. **Leguminosae**: *Abrus precatorus* [Beards1966], *Acacia* [Hall1928, McKenz1956, MerrilCh1923, Almeid1973b, WilliaWa1988], *Ac. abyssinica* [DeLott1967a], *Ac. arabica* [Hall1922], *Ac. cunninghamii* [Brimbl1968], *Ac. decurrens* [Hall1922, Hall1928, Brimbl1968], *Ac. eburnea* [Balach1932d], *Ac. farnesiana* [Bodenh1924, Zimmer1948], *Ac. harpophylla* [Brimbl1968], *Ac. molissima* [DeLott1967a], *Ac. saligna* [Pelliz1987], *Albizia falcataria* [WilliaWa1988], *Albizia lebbek* [Hall1923, Mamet1943a, Mamet1951, Borchs1966], *Andira* [MerrilCh1923], *Anthyllis citisoides* [Martin1983], *Bauhinia* [Newste1917b, Balach1956, Almeid1973b, BesheaTiHo1973], *Caesalpinia sepiaria* [Hall1922], *Cajanus cajan* [Mamet1943a, Borchs1966, DeLott1967a, Brimbl1968], *Canavalia* [WilliaWa1988], *Canavalia ensiformis* [Almeid1973b], *Cassia* [WilliaWa1988], *Cassia artemesioides* [Brimbl1968], *Cassia didymobotrya* [DeLott1967a], *Castanospermum australe* [Brimbl1968], *Ceratonia siliqua* [Bodenh1926, InserrCa1987], *Cercis siliquastrum* [Bodenh1924], *Coronilla pentaphylla* [Balach1932d], *Dalbergia* [Green1896e], *Dalbergia championii* [Green1896e, Leonar1920], *Dalbergia sissoo* [Hall1922], *Desmodium frutescens* [Mamet1943a, Borchs1966], *Dichrostachys cinerea* [Almeid1973b], *Dolichos lablab* [Hall1922], *Erythrina caffra* [DeLott1967a],

Erythrina tomentosa [Hall1928], *Indigofera arrecta* [DeLott1967a], *Indigofera hirsuta* [Beards1966], *Jacksonia scoparia* [Brimbl1968], *Leucaena glauca* [Beards1966, WilliaWa1988], *Mimosa pudica* [WilliaWa1988], *Mundulea* [Mamet1959a, Borchs1966], *Parkinsonia* [Hall1928, Balach1932d], *Parkinsonia aculeata* [Mamet1951, Borchs1966, Brimbl1968], *Pithecellobium dulce* [WilliaWa1988], *Pithecellobium saman* [WilliaWa1988], *Poinciana regia* [Ramakr1919a], *Prosopis insularum* [WilliaWa1988], *Robinia* [Zahrad1972, Martin1983], *Robinia pseudacacia* [Green1923b, Hall1923, Bodenh1924, Balach1932d], *Samanea saman* [Takaha1942b], *Spartocytisus nubigens* [GomezM1962]. **Liliaceae**: *Asparagus* [Mamet1954, Borchs1966], *Asparagus officinalis* [Hall1923], *Asparagus plumosa* [Velasq1971], *Asparagus racemosus* [Brimbl1968], *Taetsia neocaledonica* [Cohic1958]. **Loranthaceae**: *Amyema pendula* [Brimbl1968], *Loranthus* [Green1896e, Leonar1920], *Phrygilanthus bidwillii* [Brimbl1968]. **Lythraceae**: *Lagerstroemia* [MerrilCh1923], *Lagerstroemia indica* [MerrilCh1923, Brimbl1968], *Lawsonia* [MerrilCh1923], *Pemphis acidula* [WilliaWa1988]. **Malvaceae**: *Abutilon* [WilliaWa1988], *Abutilon graveolens* [WilliaWa1988], *Astrochlaena malvacea* [Hall1928], *Gossypium* [WilliaWa1988], *Hibiscus* [Mamet1956, Borchs1966], *Hibiscus mutabilis* [McDani1969], *Hibiscus rosa-sinensis* [Takaha1929, Brimbl1968, Takagi1969a], *Urena lobata* [Mamet1959a, Borchs1966]. **Melastomataceae**: *Melastoma* [MerrilCh1923], *Melastoma candidum* [Takagi1969a]. **Meliaceae**: *Azedarach indica* [Schmut1998], *Cedrela toona australis* [Brimbl1968], *Melia azedarach* [MerrilCh1923, Bodenh1926, Balach1932d, Mamet1943a, Borchs1966, McDani1969, WilliaWa1988], *Owenia venosa* [Brimbl1968]. **Moraceae**: *Artocarpus communis* [Mamet1943a, Borchs1966], *Art. heterophyllus* [WilliaWa1988], *Art. integrifolius* [Mamet1943a, Borchs1966], *Ficus benghalensis* [Borchs1966], *Fi. carica* [Green1896e, Bodenh1926, Bodenh1928, Balach1932d, Mamet1943a, GomezM1962, Borchs1966, Martin1983], *Fi. edulis* [Beards1966], *Fi. ferruginea* [Bodenh1924], *Fi. glandulifera* [WilliaWa1988], *Fi. indica* [Green1907], *Fi. repens* [Mamet1943a, Borchs1966], *Morus* [Hall1922, Bodenh1937, GomezM1962, McDani1969], *Morus alba* [Ferris1921a, Bodenh1924, Takaha1929, Takagi1969a], *Morus nigra* [Martin1983], *Morus rubra* [McDani1969]. **Moringaceae**: *Moringa oleifera* [MerrilCh1923]. **Musaceae**: *Musa* [Green1923b, WilliaWa1988], *Musa acuminata* [Brimbl1968], *Musa banksii* [Brimbl1968], *Musa cavendishi* [Balach1932d], *Musa paradisiaca* [WilliaWa1988], *Musa sapientum* [Hall1923, Mamet1943a, Borchs1966, WilliaWa1988]. **Myoporaceae**: *Myoporum* [Balach1932d]. **Myrsinaceae**: *Ardisia sieboldii* [Takagi1969a]. **Myrtaceae**: *Austromyrtus dulcis* [Brimbl1968], *Eucalyptus crebra* [Brimbl1968], *Eucalyptus grandis* [WilliaWa1988], *Eucalyptus pilularis* [Brimbl1968], *Eugenia malaccensis* [WilliaWa1988], *Leptospermum scoparium* [Brimbl1968], *Melaleuca nesophila* [Wilson1917, MerrilCh1923], *Metrosideros* [Zimmer1948], *Myrtus communis* [Martin1983], *Psidium* [Lepage1938], *Psidium cattleianum* [Cohic1958, WilliaWa1988], *Psidium guajava* [GrandpCh1899, Green1923b, Mamet1943a, Borchs1966, Takagi1969a, WilliaWa1988], *Psidium pomiferum* [Mamet1951, Borchs1966], *Rhodomyrtus psidioides* [Brimbl1968], *Tristania suaeveolens* [Brimbl1968]. **Oleaceae**: *Jasminum humile* [Hall1922], *Jasminum lineare*

[Brimbl1968], *Jasminum pubescens* [Cocker1894g], *Ligustrum* [MerrilCh1923], *Olea* [Bodenh1937], *Olea europaea* [Bodenh1924, McKenz1956, Brimbl1968, Almeid1973b, Matile1984c], *Osmanthus* [BesheaTiHo1973]. **Onagraceae**: *Gaura lindheimeria* [Mamet1954, Borchs1966]. **Orchidaceae**: *Cymbidium sinense* [Takaha1929, Takagi1969a], *Epidendrum* [MerrilCh1923]. **Palmae** [Green1896e, Green1937], *Archontophoenix cunninghamiana* [Brimbl1968], *Chrysalidocarpus lutescens* [Mamet1943a, Borchs1966], *Cocos bonneti* [Wilson1917], *Cocos nucifera* [Takaha1929, Lepage1938, Takaha1942d, Mamet1943a, Borchs1966, Almeid1971, Takagi1969a, WilliaWa1988], *Dictyosperma alba* [Mamet1943a, Borchs1966], *Elaeis* [Balach1956], *Elaeis guineensis* [Muntin1969, Almeid1971, Almeid1973b], *Howeia belmoreana* [Hunter1899], *Howeia selloviana* [Balach1932d], *Hydriastele wendlandiana* [Brimbl1968], *Hyophorbe verschaffeltii* [Beards1966], *Kentia* [McKenz1956], *Kentia selloviana* [Balach1927], *Latania* [Signor1869b, Green1896e, Leonar1920, Hall1923, McKenz1956], *Latania commersonii* [Takagi1969a], *Latania verschaffeltii* [Mamet1943a, Borchs1966], *Livistona sinensis* [Balach1927, Balach1932d], *Livistona subglobosa* [TakahaTa1956], *Phoenix* [Green1908a, Borchs1934, Borchs1936, TakahaTa1956, WilliaWa1988], *Phoenix dactylifera* [Hall1923, Martin1983], *Pritchardia martii* [Zimmer1948], *Ptychosperma vitiensis* [Laing1927], *Veitchia joannis* [WilliaWa1988], *Veitchia merrilli* [Velasq1971]. **Pandanaceae**: *Freycinetia gladiolus* [Zimmer1948], *Pandanus* [Mamet1959a, Borchs1966, WilliaWa1988], *Pandanus veitchi* [Leonar1920]. **Passifloraceae**: *Passiflora edulis* [WilliaWa1988]. **Philydraceae**: *Helmholtzia glaberrima* [Brimbl1968]. **Pinaceae**: *Cedrus* [McKenz1956]. **Piperaceae**: *Piper methysticum* [WilliaBu1987]. **Platanaceae**: *Platanus* [Hall1923, Bodenh1937, Zahrad1972], *Platanus orientalis* [Bodenh1924, Balach1932d]. **Plumbaginaceae**: *Statice gummifer* [Martin1983]. **Polygonaceae**: *Antigonon octopus* [Newste1911], *Coccoloba* [BesheaTiHo1973], *Muehlenbeckia platyclada* [Mamet1943a, Zimmer1948, Borchs1966]. **Proteaceae**: *Grevillea* [McKenz1956], *Grevillea robusta* [Mamet1943a, Borchs1966, Brimbl1968], *Macadamia integrifolia* [Balach1956, Brimbl1968], *Macadamia tetraphylla* [Brimbl1968], *Persoonia cornifolia* [Brimbl1968]. **Pteridaceae**: *Pteris* [Almeid1971]. **Punicaceae**: *Punica granatum* [Mamet1943a, Borchs1966]. **Rhamnaceae**: *Condalia obovata* [McDani1969], *Karwinskia humboldtiana* [Ferris1921], *Ziziphus* [Hall1922]. **Rhizophoraceae**: *Rhizophora* [Beards1966], *Rhizophora mucronata* [WilliaWa1988]. **Rosaceae**: *Amygdalus communis* [Martin1983], *Crataegus oxyacantha* [Mamet1943a, Borchs1966], *Cydonia oblonga* [Comsto1881a, Green1896e, Brimbl1968], *Eriobotrya* [Green1923b, Dekle1965c], *Eriobotrya japonica* [Bodenh1924, Mamet1943a, Borchs1966, BesheaTiHo1973, Martin1983], *Malus communis* [Almeid1973b], *Malus sylvestris* [Brimbl1968, WilliaWa1988], *Prunus* [BesheaTiHo1973], *Pr. avium* [Brimbl1968], *Pr. domestica* [Hall1922, Martin1983], *Pr. persica* [Green1908a, WilliaWa1988], *Pr. spinosa* [Martin1983], *Pyracantha* [BesheaTiHo1973], *Pyrus* [Green1923b, Lepage1938], *Pyrus communis* [Hall1923, Brimbl1968, Almeid1971, Martin1983], *Pyrus cydonia* [Hall1922], *Pyrus malus* [Matile1984c], *Pyrus mamorensis* [Balach1932d], *Rosa* [Hall1928, McKenz1956, Dekle1965c, Almeid1971, Almeid1973b], *Rosa indica* [WilliaWa1988], *Rubus* [McKenz1956]. **Rubiaceae**: *Canthium* [Almeid1973b],

Coffea canephora [WilliaWa1988], *Fuchsia* [McKenz1956], *Morinda* [Beards1966], *Morinda citrifolia* [Beards1966, WilliaWa1988], *Randia fitzalani* [Brimbl1968], *Saprosma ceylanica* [Leonar1920], *Timonius* [WilliaWa1988]. **Rutaceae**: *Calodendron capense* [DeLott1967a], *Citrus* [Green1915c, Takaha1939b, Green1937, Lepage1938, Takagi1969a], *Ci. aurantifolia* [WilliaBu1987, WilliaWa1988], *Ci. decumana* [Green1896e], *Ci. limon* [Brimbl1968], *Ci. maxima* [WilliaWa1988], *Ci. paradisi* [WilliaWa1988], *Ci. reticulata* [WilliaWa1988], *Ci. sinensis* [Laing1927, Beards1966, Brimbl1968, WilliaWa1988], *Coleonema* [Brimbl1968], *Eremocitrus glauca* [Brimbl1968]. **Salicaceae**: *Populus* [Hall1923], *Populus alba* [Balach1932d, Zahrad1972], *Populus nigra* [Balach1932d], *Salix* [McKenz1956, McDani1969], *Salix babylonica* [Hall1923, Brimbl1968], *Salix caprea* [Brimbl1968], *Salix nigra* [McDani1969]. **Sambucaceae**: *Sambucus javanica* [Velasq1971]. **Sapindaceae**: *Dodonaea viscosa* [WilliaWa1988], *Euphoria longana* [Mamet1943a, Borchs1966]. **Sapotaceae** [WilliaWa1988], *Achras sapota* [CockerRo1915a], *Chrysophyllum cainito* [WilliaWa1988], *Manilkara zapota* [Velasq1971], *Mimusops bojeri* [Mamet1943a, Borchs1966]. **Saxifragaceae**: *Saxifraga* [Mamet1959a, Borchs1966]. **Scrophulariaceae**: *Leucophyllum texana* [McDani1969], *Veronica* [Green1925b]. **Solanaceae**: *Capsicum frutescens* [WilliaWa1988], *Datura arborea* [Martin1983], *Solanum* [WilliaWa1988], *Solanum campylacanthum* [DeLott1967a], *Solanum coagulans* [Bodenh1924], *Solanum nigrum* [McDani1969], *Solanum sodomaeum* [Brimbl1968]. **Sterculiaceae**: *Assonia* [MerrilCh1923], *Brachychiton* [Bodenh1924], *Sterculia* [Balach1932d]. **Strelitziaceae**: *Ravenala madagascariensis* [Takagi1969a], *Strelitzia* [Bodenh1952], *Strelitzia augusta* [Balach1932d], *Strelitzia reginae* [Balach1932d]. **Theaceae**: *Camellia thea* [Lepage1938], *Thea* [Green1896e, Leonar1920], *Thea sinensis* [DeLott1967a]. **Thymelaeaceae**: *Daphne gnidium* [Pelliz1987], *Thymelaea hirsuta* [Bodenh1924, Bodenh1924a], *Wilkstroemia indica* [Brimbl1968]. **Ulmaceae**: *Chaetachme aristata* [DeLott1967a], *Trema guineensis* [DeLott1967a], *Ulmus caprinifolia* [Zahrad1972]. **Umbelliferae**: *Foeniculum vulgare* [Green1923b, Martin1983]. **Verbenaceae**: *Avicennia nitida* [WilliaWa1988], *Lantana* [Lepage1938], *Lantana camara* [Brimbl1968], *Verbena hybrida* [Brimbl1968]. **Viscaceae**: *Phoradendron flavescens* [BesheaTiHo1973]. **Vitaceae**: *Vitis* [Green1907, Lepage1938, Mamet1943a, Borchs1966, BesheaTiHo1973], *Vitis vinifera* [Green1923b, Takaha1932a, Takaha1933, Balach1932d, Mamet1943a, Borchs1966, Almeid1973b, Takagi1969a,]. **Winteraceae**: *Zygogynum* [Cohic1958, WilliaWa1988]. **Zingiberaceae**: *Amomum sceptrum* [Balach1956].
NATURAL ENEMIES: ACARI **Cheyletidae**: *Cheletomimus berlesei* (Oudemans) [GersonOcHo1990]. **Hemisarcoptidae**: *Hemisarcoptes coccophagus* Meyer [GersonOcHo1990, HillAlHe1993, CharleHiA11995], *Hemisarcoptes malus* (Shimer) [GersonOcHo1990]. COLEOPTERA **Coccinellidae**: *Chilocorus stigma* (Say) [ChuaWo1990], *Coccidophilus citricola* [AguileMeVa1984], *Cycloneda rubripennis* (Casey) [ChuaWo1990], *Lindorus lophanthae* (Blaisdell) [ChuaWo1990], *Scymnus fagus* (Broun) [CharleHiA11995]. **Nitidulidae**: *Cybocephalus micans* Reitter [Blumbe1973]. DIPTERA **Cecidomyiidae**: *Dentifibula obtusilobae* Felt [Harris1990]. FUNGI **Ascomycotina**: *Nectria flammea*

[EvansPr1990]. HYMENOPTERA **Aphelinidae**: *Aphytis chilensis* Howard [RosenDe1979], *Ap. chrysomphali* (Mercet) [RosenDe1979], *Ap. coheni* DeBach [RosenDe1979], *Ap. confusus* DeBach & Rosen [RosenDe1979], *Ap. diaspidis* (Howard) [Zimmer1948, RosenDe1979], *Ap. lingnanensis* Compere [RosenDe1979], *Ap. maculicornis* (Masi) [RosenDe1979], *Ap. melinus* DeBach [RosenDe1979], *Ap. mytilaspidis* (Le Baron) [RosenDe1979], *Ap. paramaculicornis* DeBach & Rosen [RosenDe1979], *Ap. perplexus* Rosen & DeBach [RosenDe1979], *Ap. proclia* (Walker) [RosenDe1979], *Ap. vandenboschi* DeBach & Rosen [RosenDe1979], *Encarsia citrina* (Craw) [CharleHiAl1995], *Encarsia lounsburyi* (Berlese & Paoli) [AbdRab2001a], *Prospaltella bicolor* Timberlake [Zimmer1948]. **Encyrtidae**: *Habrolepis aspidioti* Compere & Annecke [AbdRab2001a], *Plagiomerus diaspidis* Crawford [Trjapi1989]. **Signiphoridae**: *Signiphora aspidioti* Ashmead [Zimmer1948, Woolle1990], *Signiphora fax* Girault [Woolle1990], *Signiphora flavella* Girault [Gordh1979], *Signiphora flavopalliata* Ashmead [Gordh1979], *Signiphora louisianae* (Dozier) [Gordh1979], *Signiphora merceti* Malenotti [CharleHiAl1995], *Signiphora prepauca* Girault [Woolle1990], *Signiphora townsendi* Ashmead [Gordh1979]. NEUROPTERA **Chrysopidae**: *Chrysopa californica* Coquillett [ChuaWo1990]. THYSANOPTERA **Phlaeothripidae**: *Watsoniella flavipes* (Jones) [ChuaWo1990].

DISTRIBUTION: **Afrotropical**: Agalega Islands [Mamet1943a, Borchs1966]; Angola [Muntin1969, Fernan1989]; Cameroon [Balach1956, MatileNo1984]; Cape Verde Islands [Balach1956, Fernan1972, SchmutPiKl1978]; Comoros Islands [Matile1978]; Ghana [Newste1917b, Balach1956]; Guinea [Balach1956]; Kenya [Newste1917b, DeLott1967a, Schmut1998]; Madagascar [Mamet1950, Mamet1951, Mamet1954, Mamet1959a, Borchs1966]; Mauritius [GrandpCh1899, Mamet1943a, Mamet1949, Borchs1966]; Mozambique [Almeid1971]; Réunion [Mamet1957]; Saint Helena [Matile1976]; São Tomé and Príncipe [Seabra1921]; Senegal [EtiennMa1993]; Seychelles [Green1907, Mamet1943a, Borchs1966]; South Africa [BrainKe1917, Balach1956]; Sudan [Balach1956]; Tanzania [Balach1956]; Uganda [Newste1910a, Newste1910c, Newste1911, Gowdey1917, Newste1917b, Balach1956]; Zaire [Balach1956]; Zanzibar [Green1916, Mamet1956, Balach1956, Borchs1966]; Zimbabwe [Hall1928, Balach1956]. **Australasian**: Australia (Queensland [Brimbl1968]); Bonin Islands (= Ogasawara-Gunto) [Kuwana1909a, Beards1966, Kawai1987]; Cook Islands [WilliaWa1988]; Federated States of Micronesia [Takaha1942d] (Caroline Islands [Takaha1941b], Kosare (=Kusiae) [Beards1966], Ponape Island [Beards1966], Truk Islands [Takaha1939b, Beards1966], Yap [Beards1966]); Fiji [Green1915c, WilliaWa1988]; French Polynesia (Society Islands [DoaneHa1909, Ferris1935, WilliaWa1988]); Hawaiian Islands (Hawaii [Maskel1897a, Zimmer1948]); Kiribati [WilliaWa1988]; Marshall Islands [Beards1966]; New Caledonia [Cohic1958, WilliaWa1988]; New Zealand [CharleHiAl1995]; Niue [WilliaWa1988]; Palau [Beards1966]; Papua New Guinea [WilliaWa1988]; Samoa [DoaneFe1916]; Solomon Islands [WilliaWa1988]; Tokelau [WilliaWa1988]; Tonga [WilliaWa1988]; Tuvalu [WilliaWa1988]; Vanuatu [WilliaBu1987, WilliaWa1988] [WilliaBu1987, WilliaWa1988]; Volcano Islands (= Kazan-Reto) [Beards1966]; Wake Island [Beards1966]; Western Samoa [Laing1927, WilliaWa1988]. **Nearctic**: Mexico [Cocker1899n] (Baja California

[Ferris1921]); United States of America (Alabama [BesheaTiHo1973], California [McKenz1956, RosenDe1979], Florida [Comsto1881a, Wilson1917, MerrilCh1923, Merril1953, Dekle1965c, BesheaTiHo1973], Georgia [TippinBe1970, BesheaTiHo1973], Kansas [Hunter1899], Maryland [StoetzDa1974a], Mississippi [Herric1911], Missouri [Hollin1923], Texas [Herric1911, McDani1969], Virginia [BesheaTiHo1973]). **Neotropical**: Brazil [ImenesBeWo1999, WolffCo1993a] (Minas Gerais [Hempel1900a, Lepage1938], Rio de Janeiro [Lepage1938, RosenDe1979], São Paulo [Lepage1938]); Chile [GonzalCh1968, Gonzal1989, Gonzal1989a]; Colombia [Kondo2001]; Dominican Republic [GomezM1941]; Ecuador [YustCe1956]; Galapagos Islands [Kuwana1902a, PeckHeLa1998]; Guyana [Newste1914]; Jamaica [Newste1917b]; Puerto Rico & Vieques Island (Puerto Rico [Martor1976, ColonFMe1998]); U.S. Virgin Islands [Beatty1944, Nakaha1983]. **Oriental**: India [Green1908a, Ramakr1921a] (Karnataka [UsmanPu1955]); Indonesia (Irian Jaya [WilliaWa1988]); Mongolia [DanzigKo1990]; Philippines (Luzon [CockerRo1915a, Lit1997b]); Ryukyu Islands (= Nansei Shoto) [KinjoNaHi1996]; Sri Lanka [Green1896e]; Taiwan [Leonar1900, Ferris1921a, Takaha1929, Takaha1932a, Takagi1969a, WongChCh1999]; Thailand [Takaha1942b]. **Palaearctic**: Algeria [Balach1927, Balach1932d]; Canary Islands [Balach1946, GomezM1962, GomezM1967O, PerezGCa1987, MatileOr2001]; China (People's Republic) [Kuwana1927, Tang1984]; Crete [RosenDe1979]; Cyprus [RosenDe1979]; Czech Republic [Zahrad1953, Zahrad1977, Zahrad1990b]; Egypt [Hall1922, Bodenh1923, Hall1923, Ezzat1958]; France [Signor1869b]; Georgia (Abkhaz ASSR [Borchs1934, Borchs1936], Adzhar ASSR [Borchs1936]); Greece [Bodenh1928, Korone1934, RosenDe1979]; Israel [Bodenh1924, Bodenh1937, GersonZo1973, RosenDe1979]; Italy [Leonar1920, InserrCa1987, Pelliz1987, LongoMaPe1995]; Japan [Kuwana1917a, Kuwana1933, Kawai1980] (Honshu [Kuwana1902, TakahaTa1956], Shikoku [TakahaTa1956]); Lebanon [Bodenh1926]; Madeira Islands [Green1923b, Balach1938a]; Morocco [Balach1932d]; Poland [Dziedz1989]; Portugal [Seabra1941, Seabra1942, Fernan1992]; Saudi Arabia [Matile1984c]; Sicily [InserrCa1987]; Spain [GomezM1937, GomezM1946, Martin1983, BlayGo1993]; Turkey [Bodenh1949, Bodenh1952]; United Kingdom (Channel Islands [Green1925b]).

BIOLOGY: Occurring on any part of host, but perhaps most commonly on bark (Ferris, 1938a). Reported to have uniparental as well as biparental populations (Brown, 1965).

ECONOMIC IMPORTANCE: Highly polyphagous (see Host Plants) and widely distributed in almost all zoogeographical regions (CABI, 1970, and see Distribution). Pest of several agricultural crops and ornamental plants, such as: Kiwi, in New Zealand (Hill, 1989a; Blank et al. 1992) and Chile (Gonzalez, 1989a); Olive in Mediterranean Basin (Argyriou, 1990); avocado in Israel (Gerson & Zor, 1973) and California (Gill, 1997); mango in Malaysia (Chua & Wood, 1990); tea plants (Nagarkatti & Sankaran, 1990); ornamentals in Florida (Dekle, 1976).

GENERAL: Description and illustration of adult female by Brain (1918), Kuwana (1933), Balachowsky (1948, 1956), Zimmerman (1948), McKenzie (1956), Gómez-Menor Guerrero (1962), Takagi (1969a), Velasquez (1971), Tang (1984), Chou (1985, 1986), Tereznikova (1986), Williams & Watson (1988), Dziedzicka (1989),

Zahradník (1990b), Danzig (1993), Yaşar (1995a), Kosztarab (1996), Gill (1997) and by Colon-Ferrer & Medina-Gaud (1998). Description and illustration of adult female, male nymphs, male pupa and male prepupa by Stoetzel & Davidson (1974a).

KEYS: Gill 1997: 155 (female) [Species of California]; Kosztarab 1996: 509 (female) [Northeastern North America]; Danzig 1993: 169 (female) [Europe]; Zahradník 1990b: 104 (female) [Czech Republic]; Williams & Watson 1988: 132 (female) [Tropical South Pacific]; Tereznikova 1986: 113 (female) [Ukraine]; Chou 1985: 300 (female) [Species of China]; Gerson & Zor 1973: 516 (female) [Israel]; Velasquez 1971: 110 (female) [Philippines]; McDaniel 1969: 107 (female) [U.S.A.: Texas]; Beardsley 1966: 520 (female) [Federated States of Micronesia]; Gómez-Menor Guerrero 1962: 166 (female) [Canary Islands]; Ezzat 1958: 241 (female) [Egypt]; Balachowsky 1956: 106 (female) [Africa]; McKenzie 1956: 25 (female) [U.S.A.: California]; Balachowsky 1953k: 114 (female) [World]; Balachowsky 1948b: 299 (female) [Mediterranean]; Lupo 1948: 139 (female) [Italy]; Zimmerman 1948: 358 (female) [Hawaii]; Ferris 1942: 34 (female) [North America]; Archangelskaya 1937: 99 (female) [Central Asia]; Kuwana 1933: 3 (female) [Japan]; Kuwana 1933b: 49 (female) [Japan]; Fullaway 1932: 95-97, 107 (female) [Hawaii]; Archangelskaya 1929: 190 (female) [Palaearctic Region]; Hollinger 1923: 7-8 (female) [U.S.A.: Missouri]; Leonardi 1920: 29-30 (female) [Italy]; Lawson 1917: 217 (female) [U.S.A.: Kansas]; Robinson 1917: 29 (female) [Philippines]; Dietz & Morrison 1916a: 289-290 (female) [U.S.A.: Indiana]; Cockerell 1905: 45-46 (female) [Mexico]; Cockerell 1905b: 202 (female) [U.S.A.: Colorado]; Newell 1899: 4-5, 25 (female) [North America]; Green 1896e: 40 (female) [Sri Lanka]; Comstock 1883: 55-57 (female) [North America].

CITATIONS: AbdRab2001a [host, distribution, biological control: 176]; AbouEl2001 [host, distribution, biological control: 185-195]; AguileDiGr1981 [host, distribution, life history, economic importance: 175-178]; AguileMeVa1984 [life history, biological control: 47-54]; Almeid1971 [host, distribution: 11-12]; Almeid1973b [host, distribution: 9]; Archan1929 [taxonomy: 90]; Archan1937 [taxonomy, description, illustration, host, distribution: 99-100]; Argyri1990 [host, distribution, economic importance: 579-583]; ArgyriStMo1976 [host, distribution, biological control: 26]; Azeved1925 [host, distribution: 85-86]; Baker1976 [biological control, life history: 1-25]; Balach1927 [host, distribution: 177]; Balach1932b [ecology: 517-522]; Balach1932d [taxonomy, host, distribution: III-IV]; Balach1938a [taxonomy, host, distribution: 146-148]; Balach1946 [host, distribution: 212]; Balach1948b [taxonomy, description, illustration, host, distribution, biological control: 302-306]; Balach1953k [taxonomy: 114]; Balach1956 [taxonomy, description, illustration, host, distribution: 108]; Balach1959a [host, distribution: 362]; Barber1893 [host, distribution: 50-51]; BarJosGe1973 [structure, economic importance: 146-147]; Barnes1930 [biological control: 319-326]; BartleDe1952 [biological control: 16-17]; Beards1966 [host, distribution: 520-521]; BeardsDaHo1976 [economic importance: 105]; BeardsGo1975 [economic importance: 49]; Beatty1944 [host, distribution: 114-172]; BenassAlLi1984 [host, distribution: 325-327]; BerryMoHi1989 [host, distribution, economic importance: 182-186]; BesheaTiHo1973 [host, distribution: 6]; BiezanSe1940 [host, distribution: 67-68]; BlankGiDo1997 [host, distribution, economic importance, life history: 293-297]; BlankGiDo1999 [host, distribution, economic importance, life history: 1-12]; BlankGiMc2000 [host, life history, ecology: 1752-1759]; BlankGiOl1994 [host, distribution, life history: 304-309]; BlankOl1989 [chemical control: 191-194]; BlankOl1990a [chemical control: 243-246]; BlankOlCl1993 [host, distribution, chemical control: 80-85]; BlankOlGi1992 [host, distribution, economic importance: 397-405]; BlankOlGi1993 [economic importance: 139-145]; BlankOlGi1994 [host, distribution, life history, ecology: 28-30]; BlankOlTo1994 [chemical control: 195-202]; BlankOlWa1991 [chemical control: 75-79]; BlayGo1993 [taxonomy, description, illustration, host, distribution: 484-490]; Blumbe1973 [host, distribution, biological control: 125-131]; BoboyeCa1975 [chemical control: 473-476]; Bodenh1924 [taxonomy, description, host, distribution : 32-33]; Bodenh1924a [host, distribution: 122]; Bodenh1926 [host, distribution: 42]; Bodenh1928 [host, distribution: 191]; Bodenh1935 [host, distribution: 246]; Bodenh1937 [host, distribution: 217]; Bodenh1944a [taxonomy: 81]; Bodenh1949 [taxonomy, description, illustration, host, distribution: 65]; Bodenh1952 [taxonomy, description, host, distribution, structure: 339]; Bondar1914 [host, distribution, economic importance: 1064-1106]; Bondar1915 [host, distribution, economic importance: 44-47]; Borchs1934 [host, distribution: 28]; Borchs1935a [taxonomy, description, host, distribution: 25]; Borchs1936 [host, distribution: 133]; Borchs1937 [taxonomy, description, illustration, host, distribution: 128]; Borchs1937a [taxonomy, host, distribution: 41]; Borchs1939 [taxonomy, description, host, distribution: 8,14]; Borchs1949d [taxonomy, description, host, distribution: 240]; Borchs1950b [taxonomy, description, illustration, host, distribution: 218,223]; Borchs1966 [catalogue: 305-308]; Box1953 [host, distribution, biological control: 51]; Brain1918 [taxonomy, description, illustration, host, distribution: 129]; BrainKe1917 [distribution: 183]; Brick1912 [host, distribution: 1-22]; Brimbl1962 [host, distribution, economic importance: 221]; Brimbl1968 [taxonomy, illustration, host, distribution: 50-54]; Britto1923b [host, distribution]; BurgerUl1990 [economic importance: 313-327]; Butani1979 [economic importance: 36-40]; CABI1976 [host, distribution: 1-2]; Caltag1985 [taxonomy, biological control: 189-200]; CanaleVa1999 [biological control]; Charle1998 [distribution, economic importance, biological control: 52N]; CharleHiAl1995 [host, distribution, life history, biological control: 319-324]; CharleHiAl1995a [host, distribution, economic importance, biological control: 297-300]; CheahIr1997 [host, distribution]; Chiesa1948 [host, distribution, economic importance]; Chou1985 [taxonomy, description, host, distribution: 302-305]; Chou1986 [taxonomy, illustration: 685]; ChuaWo1990 [host, distribution, economic importance: 543-552]; ClapsWoGo2001a [taxonomy, host, distribution: 19]; Cocker1893cc [host, distribution: 101]; Cocker1893j [taxonomy, description, host, distribution: 255]; Cocker1894g [taxonomy, description, host, distribution: 129-131]; Cocker1895b [taxonomy: 16]; Cocker1896b [distribution: 334]; Cocker1897i [taxonomy, description, host, distribution: 21-24,27,29]; Cocker1897q [taxonomy: 703]; Cocker1897s [taxonomy, host, distribution: 384]; Cocker1898n [taxonomy, host, distribution: 184-185]; Cocker1899a [taxonomy: 395,396]; Cocker1899n [taxonomy, host, distribution: 22,23]; Cocker1900f [taxonomy, host, distribution: 72]; Cocker1900k [taxonomy: 350]; Cocker1905 [taxonomy: 45-46]; Cocker1905b [taxonomy: 202]; CockerRo1915a [taxonomy, host, distribution: 427]; Cohic1958 [host, distribution: 14]; ColonFMe1998 [taxonomy, description, illustration, host, distribution: 59-60]; Comsto1881a [taxonomy, description, illustration, host, distribution: 295-296]; Comsto1883 [taxonomy: 61,78-79]; CostaL1949 [host, distribution, biological control: 65-87]; Costan1938 [host, distribution: 25-44]; Creigh1942 [host, distribution, control: 219-233]; DahmsSm1994 [host, distribution, biological control: 245-255]; DaneelMeJa1994 [host, distribution: 72-74]; Danzig1964 [taxonomy, host, distribution: 652]; Danzig1972 [taxonomy, host, distribution, economic importance: 213]; Danzig1990 [host, distribution: 47]; Danzig1993 [taxonomy, description, illustration, host, distribution, economic importance: 174-175]; DanzigKo1990 [host, distribution: 47]; DanzigPe1998 [catalogue: 272-273]; DavidsDiFl1991 [chemical control: 1-47]; DavidsMi1990 [host, distribution, economic importance: 603-632]; DeBach1964 [biological control]; DeBach1964d [biological control: 8-15]; DeBach1974 [biological control]; DeBachBa1964 [biological control: 402-426]; DeBachHu1971 [biological control: 113-140]; DeBachRo1991 [biological control]; DeBachWh1960 [life history, biological control: 4-58]; DEDAC1923 [host, distribution]; DeitzTo1980 [taxonomy: 38,43]; Dekle1954 [host, distribution, economic importance: 226-228]; Dekle1965c [taxonomy, description, host, distribution: 71]; Dekle1976 [taxonomy, description, host, distribution, economic importance: 92]; DeLott1967a [host, distribution: 114]; DeSant1940 [biological control: 29-44]; DeSant1979 [biological control]; Dhilee1991 [host, distribution: 94-99]; Dhilee1996 [host, distribution, biological control: 64-74]; DicksoFl1955 [host, distribution: 614-615]; DietzMo1916a [taxonomy, description, illustration, host, distribution: 299-301]; Doane1931 [host, distribution, control]; DoaneFe1916 [host, distribution: 400]; DoaneHa1909 [host, distribution: 297]; Doutt1959 [biological control: 161-182]; Dozier1933 [host, distribution, biological control: 85-100]; DreistClFl1994 [taxonomy, life history, economic importance, control]; Dupont1931 [host, distribution: 1-18]; Dziedz1989 [taxonomy, description, illustration, host, distribution: 106-108]; Ebelin1949 [host, distribution, life history, control]; EbelinPe1953 [host, distribution, life history, control: 1-35]; EHG1897 [host, distribution: 67-85]; EhlerEn1984 [host, distribution, biological control, chemical control: 1-47]; Ehrhor1913 [host, distribution: 101]; ElMinsElHa1971a [host, distribution, life history, ecology: 461-467]; ElMinsElHa1974 [taxonomy, description, illustration, host, distribution: 223-232]; EtiennMa1993 [host, distribution: 258]; EvansPr1990 [biological control: 3-17]; Ezzat1958 [distribution: 241]; EzzatNa1987 [distribution: 87]; FangWuXu2001 [host, distribution: 109]; FDACSB1982 [host, distribution: 5-11]; FDACSB1983 [host, distribution: 6-8]; FergusFl1991 [host, distribution, biological control: 260-261]; Fernal1903b [catalogue: 256,265-267]; Fernan1972 [host, distribution: 12]; Fernan1973a [host, distribution, illustration, host, distribution: 252-255]; Fernan1989 [taxonomy, description, host, distribution: 133-134]; Fernan1992 [host, distribution: 60]; Ferris1921 [host, distribution: 126]; Ferris1921a [host, distribution: 219]; Ferris1935 [host, distribution: 131]; Ferris1936 [taxonomy, description, illustration, host, distribution: 10-12]; Ferris1937c [taxonomy, illustration: 51,82]; Ferris1938a [taxonomy, description, illustration, host, distribution: 241]; Ferris1941e [taxonomy: 42-48]; Ferris1942 [taxonomy: 34]; Figuer1952 [host, distribution: 209]; FinneyFi1964 [biological control: 328-355]; Flande1947 [host, biological control: 746-747]; Flande1949a [biological control: 257-274]; Flande1971 [biological control, life history: 857-872]; Foldi1990 [structure: 43-54]; Foldi2001 [distribution: 303-308]; Foldi2002 [host, distribution: 246]; Fonsec1934a [host, distribution: 263]; FonsecAu1932a [host, distribution: 202-214]; Fullaw1932 [taxonomy: 97,107]; Garcia1930 [host, distribution, biological control]; Gentry1965 [host, distribution, economic importance]; Gerson1967c [host, distribution, biological control: 632-638]; Gerson1990 [taxonomy: 130]; GersonOcHo1990 [biological control: 77-97]; GersonZo1973 [taxonomy, host, distribution, life history, economic importance: 514-533]; Gill1997 [host, distribution, taxonomy, description, illustration, economic importance: 156-157,159]; GomezM1937 [taxonomy, description, illustration, host, distribution: 80-84]; GomezM1941 [host, distribution: 128]; GomezM1946 [host, distribution: 61]; GomezM1956 [taxonomy, description, illustration, host, distribution, biological control: 13-16]; GomezM1957 [host, distribution: 45]; GomezM1962 [taxonomy, description, illustration, host, distribution: 166-169]; GomezM1965 [host, distribution: 90]; GomezM1967O [host, distribution: 131]; GomezM1968 [host, distribution: 541]; Gonzal1986 [host, distribution, economic importance: 23]; Gonzal1989 [taxonomy, description, host, distribution, economic importance: 96]; Gonzal1989a [life history, economic importance, chemical control, host, distribution: 35-43]; GonzalCh1968 [distribution: 110]; GonzalCu1994 [host, distribution, economic importance: 9]; Gordh1979 [biological control: 895,911]; Gowdey1913 [host, distribution: 247-249]; Gowdey1917 [host, distribution: 189]; Green1896e [taxonomy, description, illustration, host, distribution: 49-50,62-63]; Green1899c [taxonomy, description, illustration, host, distribution: 181-183]; Green1900a [taxonomy: 71]; Green1907 [host, distribution: 202]; Green1908a [host, distribution: 33]; Green1915c [host, distribution: 44]; Green1916 [host, distribution: 376]; Green1922 [taxonomy: 460]; Green1923b [host, distribution: 89]; Green1925b [host, distribution: 518]; Green1937 [host, distribution: 330]; GreenMa1907 [taxonomy, distribution: 344]; Haas1934 [chemistry, chemical control, physiology: 477-492]; HagenVaDa1971 [host, distribution, biological control: 253-293]; Hall1922 [taxonomy, description, host, distribution: 25-26]; Hall1923 [host, distribution: 42-43]; Hall1924a [host, distribution, economic importance: 9]; Hall1928 [host, distribution: 274]; Hamlen1974 [host, distribution: 6-8]; Hargre1937 [host, distribution, economic importance: 505-520]; Harris1937 [host, distribution: 483-488]; Harris1990 [biological control: 61-66]; HaseyOlVa1999 [host, distribution, control]; HassanHa1985 [life history, biological control: 63-72]; Hayat1989 [host, distribution, biological control: 1-3]; Hempel1900a [taxonomy, description, host, distribution: 502]; Hempel1904 [host, distribution: 319]; Hempel1920 [taxonomy, description, host, distribution: 140]; Herric1911 [taxonomy, description, illustration, host, distribution: 10,18,52]; Hewitt1943 [host, distribution: 266-274]; Hill1989a [host, distribution, economic importance, biological control: 177-182]; HillAlHe1993 [host, distribution, biological control: 369-376]; Hollin1923 [taxonomy, description, host, distribution: 13]; Houser1918 [taxonomy, description, host, distribution: 166-167]; HoyHe1985 [biological control]; HoyHe1985 [biological control]; Hsu1935 [host, distribution: 578-590]; Hunt1939 [host, distribution: 548-566]; Hunter1899 [taxonomy, host, distribution: 11]; ImenesBeWo1999 [host, distribution: 117-119]; InserrCa1987 [host, distribution: 93]; IzraylGe1993 [host, distribution, biological control: 861-875]; IzraylGe1993a [host, distribution, biological control: 877-888]; IzraylGe1995 [host, distribution, biological control: 439-446]; IzraylGe1995b [host, distribution, biological control: 235-240]; IzraylGeHa1996 [life history, ecology, biological control: 390-395]; IzraylHaGe1995 [life history, ecology, biological control: 138-145]; JiYa1990 [biological control: 134-136]; Kawai1980 [taxonomy, description, host, distribution: 222]; Kawai1987 [host, distribution: 78]; King1900b [taxonomy: 215]; KinjoNaHi1996 [host, distribution: 125-127]; Kiritc1932a [taxonomy: 246]; Koehle1964 [host, distribution, control]; Kondo2001 [taxonomy, host, distribution: 44]; KondoKa1995 [host, distribution: 57-58]; Korone1934 [taxonomy, description, illustration, host, distribution: 1-2]; Koszta1996 [taxonomy, description, illustration, host, distribution, life history, biological control, economic importance: 511-513]; Kozar1990a [life history, economic importance: 341-347]; Kuwana1902 [host, distribution: 68]; Kuwana1902a [host, distribution: 31]; Kuwana1907 [host, distribution: 195]; Kuwana1909a [host, distribution: 160]; Kuwana1917a [taxonomy, distribution: 174]; Kuwana1927 [host, distribution: 71]; Kuwana1933 [taxonomy, description, illustration, host, distribution: 12-13]; Laing1927 [host, distribution: 39,40]; Lawson1917 [taxonomy, description, illustration, host, distribution: 230-232]; Leonar1897 [taxonomy: 285-286]; Leonar1898a [taxonomy, description, illustration, host, distribution: 76-78]; Leonar1898c [taxonomy, description, illustration, host, distribution: 44-46,69-71]; Leonar1900 [taxonomy, host, distribution: 340-341]; Leonar1918 [host, distribution: 189]; Leonar1920 [taxonomy, description, illustration, host, distribution: 40-41]; Lepage1938 [catalogue: 395-397]; Lepesm1947 [taxonomy, description, host, distribution, life history: 207-208]; Lindin1907a [taxonomy: 19]; Lindin1909c [taxonomy, host, distribution: 449]; Lindin1910b [host, distribution: 36-37]; Lindin1911a [taxonomy, description, illustration, host, distribution: 14-15]; Lindin1912b [taxonomy, description, host, distribution: 50,153,240]; Lindin1924 [taxonomy, host, distribution: 174]; Lindin1935 [taxonomy: 129]; Lindin1957 [taxonomy: 545,546,549]; Lit1997b [host, distribution: 92]; Lobdel1937 [taxonomy: 78]; LoBl1989 [host, distribution: 1-4]; LongoMaPe1995 [distribution: 127]; Lupo1948 [taxonomy, description, illustration, host, distribution: 164-169]; MacGil1921 [taxonomy, description, host, distribution: 405,406,412]; Mackie1931 [host, distribution: 419]; Mackie1931 [host, distribution, economic importance: 419-441]; Mackie1935 [host, distribution: 403]; Mallam1954 [distribution: 24-60]; Mamet1943a [catalogue: 160,161]; Mamet1949 [catalogue: 60]; Mamet1950 [host, distribution: 23]; Mamet1951 [host, distribution: 228]; Mamet1954 [host, distribution: 17]; Mamet1956 [host, distribution: 138]; Mamet1957 [host, distribution: 369,375]; Mamet1959a [host, distribution: 387]; Marlat1899b [taxonomy: 211]; Marlat1899e [taxonomy: 658]; Marlat1900 [taxonomy: 427]; Marlat1921a [host, distribution: 1-36]; Martin1983 [taxonomy, host, distribution: 65]; Martor1976 [host, distribution: 73,151,213,249]; Maskel1892 [taxonomy, host, distribution: 13-14]; Maskel1897a [taxonomy, description, host, distribution: 240-241]; Maskel1898 [taxonomy, description, host, distribution: 224-227]; Matile1976 [host, distribution: 312-313]; Matile1978 [host, distribution: 64]; Matile1984c [host, distribution: 221]; MatileBa1972 [host, distribution: 113]; MatileNo1984 [host, distribution: 65]; MatileOr2001 [host, distribution: 190]; Matta1979 [host, distribution, biological control: 231-242]; MayneGh1934 [host, distribution: 3-38]; McDani1969 [taxonomy, illustration, host, distribution: 109-111]; McKenz1935 [host, distribution, life history, control: 1-48]; McKenz1956 [taxonomy, description, illustration, host, distribution: 69-72]; Merkel1938 [host, distribution: 88-99]; Merril1953 [taxonomy, description, host, distribution: 24]; MerrilCh1923 [taxonomy, description, host, distribution, economic importance: 203-204]; MetcalMe1993 [economic importance, host, distribution, control]; MillerDa1990 [host, distribution, economic importance: 302]; Miyosh1926 [host, distribution: 303-326]; Monte1930 [host, distribution: 3-36]; Morale1988 [host, distribution, taxonomy: 77-82]; Moreir1921b [host, distribution: 182]; Moulto1931 [host, distribution: 745]; MouradMeFa2001 [host, distribution: 571-580]; MoutiaMa1947 [distribution]; Muraka1970 [host, distribution: 74]; NagarkSa1990 [host, distribution, economic importance, biological control: 553-542]; Nakaha1982 [host, distribution: 42]; Nakaha1983 [host, distribution: 11]; Newell1899 [taxonomy, description, illustration, host, distribution: 24-27]; Newste1910a [host, distribution: 68]; Newste1910c [host, distribution: 199]; Newste1911 [host, distribution: 86]; Newste1914 [host, distribution: 307]; Newste1917a [taxonomy, host, distribution: 266]; Newste1917b [host, distribution: 131]; NourElRi1970 [host, distribution: 123-127]; Nur1990b [taxonomy, life history: 196]; Otero1935 [host, distribution: 1-26]; Pace1939 [host, distribution: 664-665]; PeckHeLa1998 [host, distribution: 219-237]; Pelliz1987 [host, distribution: 122]; Pelot1950 [host, distribution: 17]; PeralL1968 [biological control: 22-29]; PerezGCa1987 [host, distribution: 128]; PietriBiBa1961 [chemical control: 21-46]; PietriBiCo1969 [chemical control: 909-915]; Ponnam1999 [host, distribution, chemical control: 445-451]; PriesnHo1940 [biological control: 58-70]; Quedna1964b [biological control: 86-116]; Ramakr1919a [taxonomy, description, host, distribution: 17,19]; Ramakr1921a [host, distribution: 355,356]; Ramakr1930 [taxonomy, host, distribution: 22,23]; Ramakr1938a [host, distribution: 341-351]; RaoCh1950 [taxonomy: 12,29]; ReisMe1984 [economic importance, host, distribution: 68-72]; RobbCoBe2001 [host, distribution, taxonomy, control]; Robins1917 [taxonomy, description, host, distribution: 30]; RosenDe1979 [host, distribution, biological control: 249-251,349-354,]; RuizCa1944 [taxonomy, distribution: 64]; Saakya1954 [host, distribution, economic importance]; SalamaHa1974 [host, distribution, life history, economic importance: 200-204]; Sander1904a [taxonomy, description, illustration, host, distribution: 56,59]; Sasaki1935 [taxonomy, description, illustration, host, distribution: 864-865]; Schmut1957b [taxonomy: 148]; Schmut1959 [taxonomy, description, host, distribution: 48,71]; Schmut1969 [host, distribution: 116]; Schmut1990a [host, distribution: 393]; Schmut1998 [economic importance, host, distribution: 37]; Schmut2001 [host, distribution: 339-345]; SchmutKlLu1957 [host, distribution, economic importance: 477]; SchmutPiKl1978 [host, distribution, economic importance: 330]; Seabra1921 [taxonomy, host, distribution: 98];

Seabra1941 [distribution: 8]; Seabra1942 [distribution: 2]; ShiLi1991 [host, distribution: 165]; SibbetVaFe2000 [host, distribution, control]; Signor1869 [taxonomy: 860]; Signor1869b [taxonomy, description, illustration, host, distribution: 124-125]; Silves1929 [host, distribution: 897-904]; Singh1964 [host, distribution, economic importance: 213]; Steine1987 [host, distribution, description, economic importance, control: 1-7]; StevenBlTo1991 [host, distribution, life history, chemical control: 84]; StevenMcBl1997 [chemical control: 288-292]; StevenRe1999 [host, distribution, economic importance, biological control, chemical control: 345-354]; StevenToBl1994 [host, distribution, life history, chemical control, biological control: 135-142]; StevenVaGo1997 [host, distribution: 773-777]; StoetzDa1974 [taxonomy, life history: 138-140]; StoetzDa1974a [taxonomy, description, illustration, host, distribution, life history: 505-507]; Strong1922 [host, distribution: 775-780]; Sugimo1994 [host, distribution: 115-121]; SuWa1988 [host, distribution, economic importance, chemical control: 279-288]; Swirsk1976b [host, distribution, economic importance: 555-559]; SwirskAr1971 [host, distribution: 12-15]; SwirskWyIz2002 [taxonomy, host, distribution, life history, economic importance, biological control: 103-104]; Takagi1969a [taxonomy, description, illustration, host, distribution: 78,101]; Takaha1929 [host, distribution: 80]; Takaha1929a [host, distribution: 431]; Takaha1932a [host, distribution: 105]; Takaha1936c [host, distribution: 118]; Takaha1939b [host, distribution: 269]; Takaha1941b [host, distribution: 219]; Takaha1942b [host, distribution: 47]; Takaha1942d [host, distribution: 358]; Takaha1953a [taxonomy, host, distribution: 10-13]; TakahaTa1956 [host, distribution: 15]; Tang1984 [taxonomy, description, illustration, host, distribution: 47-48]; Tao1999 [taxonomy, host, distribution: 91]; Targio1892 [taxonomy: 81]; TawfikMo2002 [host, distribution, life history, biological control: 267-273]; Terezn1986 [taxonomy, description, illustration, host, distribution: 113-115]; Timber1924 [host, distribution, biological control]; TippinBe1970 [host, distribution: 9]; TomkinWiTh2000 [host, distribution: 211-215]; Trjapi1989 [biological control: 295]; UsmanPu1955 [host, distribution: 48]; Vayssi1913 [host, distribution: 430]; Velasq1971 [taxonomy, description, illustration, host, distribution: 110-113]; VelasqRi1969 [host, distribution: 195-208]; VeseyF1953 [host, distribution, biological control: 405-413]; WadhiBa1964 [host, distribution, economic importance: 227]; WadhiBa1964 [host, distribution: 227-260]; WangSu1989 [host, life history: 151-156]; Wester1918 [host, distribution, economic importance: 5-57]; WhitinHoCo1998 [host, distribution, control, economic importance: 211-215]; Willia1985a [taxonomy: 234]; WilliaBu1987 [host, distribution: 95]; WilliaGr1990 [host, distribution, economic importance, biological control: 563-578]; WilliaWa1988 [taxonomy, description, illustration, host, distribution: 131-134]; Wilson1917 [taxonomy, description, host, distribution: 6,33]; Wilson1921 [host, distribution: 20-34]; WolffCo1993a [host, distribution: 153]; WongChCh1999 [taxonomy, description, host, distribution: 25-26,66]; Woolle1990 [biological control: 167-176]; Wysoki1977 [host, distribution: 187-188]; Wysoki1997 [host, distribution, economic importance: 805-811]; Yasar1995a [taxonomy, description, illustration, host, distribution: 89-90]; YustCe1956 [host, distribution: 425-442]; Zagain1956 [distribution: 85-90]; Zahrad1953 [taxonomy, description, illustration, host, distribution: 130-133]; Zahrad1972 [host, distribution:

432]; Zahrad1977 [taxonomy, distribution: 120]; Zahrad1990 [host, distribution, description: 642-643]; Zahrad1990b [taxonomy, description, illustration, host, distribution: 104-106]; Zimmer1948 [taxonomy, description, illustration, host, distribution, biological control: 361-362].

Hemiberlesia loranthi (Laing)
Aspidiotus loranthi Laing, 1929: 21. Type data: AUSTRALIA: Victoria, Eltham, on *Loranthus pendulus*; collected by J.E. Dixon. Syntypes, female. Type depository: London: The Natural History Museum, England, UK.
Hemiberlesia loranthi; Balachowsky, 1948: 58. Change of combination.
Diaspidiotus loranthi; Brimblecombe, 1958: 59. Change of combination.
Hemiberlesia loranthi; Borchsenius, 1966: 308. Revived combination.
SCALE COVER: Female scale subcircular, highly convex, almost conical, dark blackish grey, practically same colour as bark of host-plant; surface roughened and corrugated; exuviae subcentral bright orange-yellow; a subcircular white patch left on bark when scale removed; diameter 0.75 mm (Laing, 1929).
HOST PLANTS: **Leguminosae**: *Jacksonia scoparia* [Brimbl1968]. **Loranthaceae**: *Amyema pendula* [Brimbl1968], *Dendrophthoe falcata* [Brimbl1968], *Loranthus* [Brimbl1958], *Loranthus pendulus* [Laing1929, Brimbl1958]. **Meliaceae**: *Cedrela toona australis* [Brimbl1968]. **Moraceae**: *Ficus carica* [Brimbl1968].
DISTRIBUTION: **Australasian**: Australia (New South Wales [Brimbl1958], Queensland [Brimbl1958], Victoria [Laing1929, Brimbl1958]).
GENERAL: Description and illustration of adult female by Laing (1929) and by Brimblecombe (1958).
KEYS: Balachowsky 1953k: 115 (female) [World].
CITATIONS: Balach1948b [taxonomy: 298]; Balach1953k [taxonomy: 115]; Balach1956 [taxonomy: 107]; Borchs1966 [catalogue: 308]; Brimbl1958 [taxonomy, description, illustration, host, distribution: 59-61]; Brimbl1962 [host, distribution, economic importance: 221]; Brimbl1968 [taxonomy, illustration, host, distribution: 49]; Ferris1941e [taxonomy: 45]; Laing1929 [taxonomy, description, illustration, host, distribution: 21-22].

Hemiberlesia lottoi Balachowsky
Hemiberlesia lottoi Balachowsky, 1956: 110. Type data: KENYA: Nairobi, on *Ficus mallatocarpa*. Holotype female. Type depository: London: The Natural History Museum, England, UK.
SCALE COVER: Female scale subcircular; moderately convex; light brown; exuviae darker, central or subcentral. Male scale unknown (Balachowsky, 1956).
HOST PLANTS: **Moraceae**: *Ficus mallatocarpa* [Balach1956, CouturMaRi1985]. **Sapotaceae**: *Gambeya taiensis* [CouturMaRi1985].
DISTRIBUTION: **Afrotropical**: Côte d'Ivoire (=Ivory Coast) [CouturMaRi1985]; Kenya [Balach1956]; Madagascar [Mamet1959a, Borchs1966].
GENERAL: Description and illustration of adult female by Balachowsky (1956). This species was named after Giovanni De Lotto. Though Balachowsky (1956) misspelled the last name De Lotto, the original species epithet must be retained.
KEYS: Balachowsky 1956: 105-108 (female) [Africa].

CITATIONS: Balach1956 [taxonomy, description, illustration, host, distribution: 110-111]; Borchs1966 [catalogue: 309]; CouturMaRi1985 [host, distribution: 278]; Mamet1959a [host, distribution: 388].

Hemiberlesia malagassa Mamet

Hemiberlesia malagassa Mamet, 1959a: 470. Type data: MADAGASCAR: Lambomakandro, on *Euphorbia enterophora*. Holotype. Type depository: Paris: Muséum national d'Histoire naturelle, France.
SCALE COVER: Female scale greyish, somewhat conical; circular; exuviae central. Male scale not observed (Mamet, 1959a).
HOST PLANTS: **Euphorbiaceae**: *Euphorbia enterophora* [Mamet1959a].
DISTRIBUTION: **Afrotropical**: Madagascar [Mamet1959a, Borchs1966].
GENERAL: Description and illustration of adult female by Mamet (1959a).
CITATIONS: Borchs1966 [catalogue: 309]; Mamet1959a [taxonomy, description, illustration, host, distribution: 470].

Hemiberlesia manengoubae Balachowsky

Hemiberlesia manengoubae Balachowsky, 1953k: 111. Type data: CAMEROON: Volcanic range of Manengouba, 30 km north of N'Konsamba, near "Lac de l'Enfant", 2200 meters altitude, on undetermined plant. Syntypes, female. Type depository: Paris: Muséum national d'Histoire naturelle, France.
SCALE COVER: Female scale circular, 2-2.2 mm in diameter; convex; light brown; exuviae, darker, blackish brown, central or subcentral. Male scale unknown (Balachowsky, 1953k. 1956)
DISTRIBUTION: **Afrotropical**: Cameroon [Balach1953k].
GENERAL: Description and illustration of adult female by Balachowsky (1953k, 1956).
KEYS: Balachowsky 1956: 107 (female) [Africa]; Balachowsky 1953k: 115 (female) [World].
CITATIONS: Balach1953k [taxonomy, description, illustration, host, distribution: 111-113]; Balach1956 [taxonomy, description, illustration, host, distribution: 112-115]; Borchs1966 [catalogue: 309]; Lindin1957 [taxonomy: 549].

Hemiberlesia massonianae Tang

Hemiberlesia massonianae Tang, 1984: 50. Type data: CHINA: Guandong Province, Zhuhai County, on *Pinus massonianae*. Holotype female. Type depository: Shanxi: Entomological Institute, Shanxi Agricultural University, Taigu, Shanxi, China.
SCALE COVER: Scale of adult female circular, yellowish brown. Male scale elongate, similar in colour and texture to that of female (Tang, 1984).
HOST PLANTS: **Pinaceae**: *Pinus massoniae* [Tang1984].
DISTRIBUTION: **Oriental**: China (People's Republic) (Guangdong (Kwangtung) [Tang1984]).
GENERAL: Description and illustration of adult female by Tang (1984).
CITATIONS: Tang1984 [taxonomy, description, illustration, host, distribution: 50-51]; Tao1999 [taxonomy, host, distribution: 91].

Hemiberlesia momicola (Takagi & Kawai)
Abgrallaspis momicola Takagi & Kawai, 1966: 116. Type data: JAPAN: Tokyo and Amagi-san, Sizuoka-ken, on *Abies firma*. Syntypes, female. Type depository: Sapporo: Entomological Institute, Faculty of Agriculture, Hokkaido University, Japan.
Hemiberlesia momicola; Danzig & Pellizzari, 1998: 273. Change of combination.
SCALE COVER: Takagi & Kawai (1966) did not describe scale cover.
HOST PLANTS: **Pinaceae**: *Abies firma* [TakagiKa1966].
DISTRIBUTION: **Palaearctic**: Japan [TakagiKa1966] (Honshu [Muraka1970]).
GENERAL: Description and illustration of adult female by Takagi & Kawai (1966).
CITATIONS: DanzigPe1998 [catalogue: 273]; DuttaSi1990 [taxonomy: 1]; Kawai1980 [taxonomy, description, host, distribution: 218]; Muraka1970 [host, distribution: 69]; TakagiKa1966 [taxonomy, description, illustration, host, distribution: 116-117].

Hemiberlesia musae Takagi & Yamamoto
Hemiberlesia (*Abgrallaspis*) *musae* Takagi & Yamamoto, 1974: 37. Type data: ECUADOR: intercepted at quarantine at Kobe, Japan, on banana. Holotype female. Type depository: Sapporo: Entomological Institute, Faculty of Agriculture, Hokkaido University, Japan.
SCALE COVER: Takagi & Yamamoto (1974) did not describe the scale cover.
HOST PLANTS: **Musaceae**: *Musa* [TakagiYa1974].
DISTRIBUTION: **Neotropical**: Ecuador [TakagiYa1974]; Puerto Rico & Vieques Island (Puerto Rico [ColonFMe1998]).
ECONOMIC IMPORTANCE: Banana pest in Ecuador (Chua & Wood, 1990).
GENERAL: Description and illustration of adult female by Takagi & Yamamoto (1974) and by Colon-Ferrer & Medina-Gaud (1998).
CITATIONS: ChuaWo1990 [host, distribution, economic importance: 548]; ColonFMe1998 [taxonomy, host, distribution: 3]; ColonFMe1998 [taxonomy, description, illustration, host, distribution: 60-61]; TakagiYa1974 [taxonomy, description, illustration, host, distribution: 37-40].

Hemiberlesia neodiffinis Miller & Davidson
Hemiberlesia neodiffinis Miller & Davidson, 1998: 194. Type data: U.S.A.: Illinois, Simpson, on *Liriodendron tulipifera*; collected by J.E. Appleby, August 7, 1969. Holotype female. Type depository: Washington: U.S. National Entomological Collection, U.S. National Museum of Natural History, District of Columbia, USA.
SYSTEMATICS: Miller & Davidson (1998) discovered that since the turn of the 19th century this species has been misidentified as *Hemiberlesia affinis* (Newstead).
SCALE COVER: Miller & Davidson (1998) did not describe scale cover.
HOST PLANTS: **Anacardiaceae**: *Spondias* [MillerDa1998]. **Apocynaceae**: *Nerium oleander* [MillerDa1998]. **Fagaceae**: *Quercus* [MillerDa1998]. **Juglandaceae**: *Carya illinoensis* [MillerDa1998], *Juglans* [MillerDa1998], *Pterocarya steoptera* [MillerDa1998]. **Lauraceae**: *Cinnamomum camphora* [MillerDa1998], *Persea* [MillerDa1998]. **Loranthaceae**: *Phoradendron*

[MillerDa1998]. **Magnoliaceae**: *Liriodendron* [MillerDa1998], *Li. tulipifera* [MillerDa1998], *Magnolia* [MillerDa1998], *Ma. grandiflora* [MillerDa1998], *Ma. virginiana* [MillerDa1998]. **Moraceae**: *Ficus* [MillerDa1998]. **Oleaceae**: *Fraxinus* [MillerDa1998], *Fraxinus uhdei* [MillerDa1998], *Syringa* [MillerDa1998]. **Salicaceae**: *Salix* [MillerDa1998]. **Tiliaceae**: *Tilia americana* [MillerDa1998]. **Ulmaceae**: *Celtis* [MillerDa1998], *Ulmus* [MillerDa1998], *Ul. americana* [MillerDa1998].
DISTRIBUTION: **Nearctic**: Mexico [MillerDa1998]; U.S.A. (Alabama [MillerDa1998], Arkansas [MillerDa1998], District of Columbia [MillerDa1998], Florida [MillerDa1998], Georgia [MillerDa1998], Illinois [MillerDa1998], Louisiana [MillerDa1998], Maryland [MillerDa1998], Mississippi [MillerDa1998], Missouri [MillerDa1998], New Jersey [MillerDa1998], New York [MillerDa1998], North Carolina [MillerDa1998], Ohio [MillerDa1998], South Carolina [MillerDa1998], Tennessee [MillerDa1998], Texas [MillerDa1998]).
GENERAL: Description and illustration of adult female by Miller & Davidson (1998).
CITATIONS: MillerDa1998 [taxonomy, description, illustration, host, distribution: 193-197].

Hemiberlesia nothofagi Williams

Hemiberlesia nothofagi Williams, 1985e: 251. Type data: ARGENTINA: Bariloche, San Roman, on leaves of *Nothofagus antarctica*. Holotype female. Type depository: London: The Natural History Museum, England, UK.
SCALE COVER: Williams (1985e) did not describe scale cover.
HOST PLANTS: **Fagaceae**: *Nothofagus antarctica* [Willia1985e, ClapsWoGo2001].
DISTRIBUTION: **Neotropical**: Argentina (Rio Negro [Willia1985e, ClapsWoGo2001]); Chile [ClapsWoGo2001].
GENERAL: Description and illustration of adult female by Williams (1985e).
CITATIONS: ClapsWoGo2001 [host, distribution: 245]; Willia1985e [taxonomy, description, illustration, host, distribution: 251-253].

Hemiberlesia ocellata Takagi & Yamamoto

Hemiberlesia (*Abgrallaspis*) *ocellata* Takagi & Yamamoto, 1974: 40. Type data: ECUADOR: intercepted at quarantine at Kobe, Japan, on banana. Holotype female. Type depository: Sapporo: Entomological Institute, Faculty of Agriculture, Hokkaido University, Japan.
SCALE COVER: Takagi & Yamamoto (1974) did not describe the scale cover.
HOST PLANTS: **Musaceae**: *Musa* [TakagiYa1974].
DISTRIBUTION: **Neotropical**: Ecuador [TakagiYa1974]. **Palaearctic**: Japan [TakagiYa1974].
ECONOMIC IMPORTANCE: Chua & Wood (1990) listed this species as a pest of banana in Ecuador.
GENERAL: Description and illustration of adult female by Takagi & Yamamoto (1974).

CITATIONS: ChuaWo1990 [host, distribution, economic importance: 548]; SugimoKaTa1996 [host, distribution: 99-101]; TakagiYa1974 [taxonomy, description, illustration, host, distribution: 40-42].

Hemiberlesia palmae (Cockerell)

Aspidiotus rapax palmae Cockerell, 1892b: 333. Nomen nudum.

Aspidiotus palmae Cockerell, 1893e: 39. Type data: JAMAICA: Kingston, on coconut [=*Cocos nucifera*]. Syntypes, female. Type depository: Washington, D.C.: U.S. Entomological Collection, U.S. National Museum of Natural History, USA.

Aspidiotus palmae Morgan, 1893: 40. Type data: JAMAICA: on palms. Syntypes, female. Synonymy by Fernald, 1903b: 270. Junior synonym and Homonym of *Aspidiotus palmae* Cockerell, 1893e. Notes: Depository of type material unknown.

Aspidiotus palmae; Leonardi, 1897: 285. Notes: Incorrect citation of "Morgan & Cockerell" as authors.

Aspidiotus (*Hemiberlesia*) *palmae*; Cockerell, 1897i: 24. Notes: Incorrect citation of "Morg. & Ckll." as authors.

Aspidiotus (*Hemiberlesia*) *palmae*; Cockerell, 1897i: 24. Change of combination.

Aspidiotus (*Evaspidiotus*) *palmae*; Leonardi, 1898c: 51. Change of combination.

Aspidiotus (*Evaspidiotus*) *palmae*; Leonardi, 1898c: 51. Notes: Incorrect citation of "Morgan & Cockerell" as authors.

Aspidiotus palmae; Fernald, 1903b: 270. Notes: Incorrect citation of "Morgan & Cockerell" as authors.

Aspidiotus palmae; Fernald, 1903b: 270. Notes: Incorrect citation of "Morgan & Cockerell" as authors.

Aspidiotus unguiculatus Leonardi, 1914: 199. Type data: Guinea: Conakry, on undetermined plant. Syntypes, female. Type depository: Portici: Dipartimento de Entomologia e Zoologia Agraria di Portici, Università di Napoli Federico II, Italy. Synonymy by Laing, 1929a: 486.

Aspidiotus palmae; Houser, 1918: 167. Notes: Incorrect citation of "Morgan & Cockerell" as authors.

Furcaspis palmae; MacGillivray, 1921: 407. Change of combination.

Selenaspidus palmae; Seabra, 1921: 98. Notes: Incorrect citation of "Morgan & Cockerell" as author.

Aspidicotus palmae; Green, 1930c: 281. Misspelling of genus name.

Aspidiotus javanensis Kuwana in Kuwana & Muramatsu, 1931a: 654. Type data: INDONESIA: Java, Batavia, on palm. Holotype female. Type depository: Ibaraki-ken: Insect Taxonomy Laboratory, National Institute of Agricultural Environmental Sciences, Kannon-dai, Yatabe, Tsukuba-shi, (I. Kuwana), Japan. Synonymy by Ferris, 1941e: 44.

Hemiberlesia palmae; Ferris, 1938a: 242. Change of combination.

Abgrallaspis palmae; Balachowsky, 1948b: 320. Change of combination.

Abgrallaspis palmae; Balachowsky, 1948b: 320. Notes: Incorrect citation of "Morgan & Cockerell" as authors.

Aspidiotus palmae; Zahradník, 1959: 65. Notes: Incorrect citation of "Morgan & Cockerell" as authors.

Borchseniaspis palmae; Zahradník, 1959: 65. Change of combination.

Abgrallaspis palmae; Almeida, 1969: 153. Notes: Incorrect citation of "Morgan & Cockerell" as authors.

Hemiberlesia palmae; Almeida, 1973b: 9. Notes: Incorrect citation of "Morgan & Cockerell" as authors.

Hemiberlesia palmae; Williams & Watson, 1988: 134. Revived combination.

COMMON NAME: tropical palm scale [McKenz1956, Dekle1965c].

SYSTEMATICS: Authorship of this species was confused, since it was described as a new species by both Cockerell and by Morgan in separate articles in same number of the same journal. Consequently, *H. palmae* has been ascribed variously to Cockerell, to Morgan and to Morgan and Cockerell. Under ICZN rules, it is clear that this species should be ascribed to Cockerell alone (see discussion in Ferris, 1938a).

SCALE COVER: Female scale somewhat oval, quite convex, white; exuviae subcentral and very dark; scale of male not seen and apparently unknown (Ferris, 1938a). Colour photograph by Gill (1997).

HOST PLANTS: **Acanthaceae**: *Graptophyllum pictum* [WilliaWa1988]. **Agavaceae**: *Cordyline* [WilliaWa1988]. **Anacardiaceae**: *Anacardium occidentale* [WilliaWa1988], *Mangifera indica* [Balach1959a, WilliaWa1988], *Mangifera odorata* [Green1930]. **Annonaceae**: *Annona* [Almeid1973b, WilliaWa1988], *Annona muricata* [WilliaWa1988], *Annona reticulata* [Mamet1956, WilliaWa1988], *Annona squamosa* [WilliaWa1988]. **Apocynaceae**: *Plumeria* [WilliaWa1988], *Plumeria rubra* [WilliaWa1988]. **Araceae**: *Acorus gramineus variegatus* [Dekle1965c], *Colocasia* [WilliaWa1988], *Epipremnum pinnatum* [WilliaWa1988], *Xanthosoma sagittifolium* [WilliaWa1988]. **Averrhoaceae**: *Averrhoa carambola* [Velasq1971]. **Bignoniaceae**: *Kigelia pinnata* [WilliaWa1988]. **Bixaceae**: *Bixa orellana* [WilliaWa1988]. **Bombacaceae**: *Ceiba pentandra* [Beards1966, WilliaWa1988]. **Bromeliaceae**: *Aechmea weibachia* [McKenz1956], *Billbergia* [McKenz1956, Dekle1965c], *Bromelia* [McKenz1956], *Karatas* [Green1919c]. **Caprifoliaceae**: *Lonicera* [Green1923b]. **Casuarinaceae**: *Casuarina* [WilliaWa1988]. **Clusiaceae**: *Calophyllum* [WilliaWa1988], *Calophyllum inophyllum* [WilliaWa1988]. **Combretaceae**: *Terminalia catappa* [Beards1966]. **Cycadaceae**: *Cycas* [Takaha1939b, Beards1966, WilliaWa1988], *Cycas circinalis* [WilliaWa1988], *Cycas revoluta* [Balach1948b]. **Dioscoreaceae**: *Dioscorea* [WilliaWa1988], *Dioscorea bulbifera* [WilliaWa1988]. **Ehretiaceae**: *Ehretia littoralis* [DeLott1967a]. **Euphorbiaceae**: *Acalypha* [Balach1956], *Acalypha hispida* [WilliaWa1988], *Aleurites moluccana* [WilliaWa1988], *Euphorbia* [Balach1948b], *Euphorbia heterophylla* [WilliaWa1988], *Hevea brasiliensis* [WilliaWa1988], *Jatropha multifida* [Beards1966], *Manihot* [Balach1956], *Manihot glaziovii* [Balach1948b]. **Flacourtiaceae**: *Aphloia theaeformis* [Matile1978]. **Gramineae**: *Miscanthus* [WilliaWa1988], *Tripsacum laxum* [WilliaWa1988]. **Heliconiaceae**: *Heliconia* [WilliaWa1988], *Heliconia bahai* [WilliaWa1988]. **Lauraceae**: *Cinnamomum camphora* [Velasq1971], *Persea americana* [Takaha1941b, WilliaWa1988], *Persea gratissima* [GonzalCh1968]. **Lecythidaceae**: *Barringtonia* [WilliaWa1988], *Barringtonia asiatica* [WilliaWa1988], *Barringtonia racemosa* [WilliaWa1988], *Bertholletia excelsa* [Takaha1941b]. **Leguminosae**: *Bauhinia* [WilliaWa1988], *Bauhinia monandra*

[WilliaWa1988], *Canavalia* [WilliaWa1988], *Cassia spectabilis* [MatileNo1984], *Gliricidia sepium* [WilliaWa1988], *Intsia bijuga* [Beards1966], *Pithecellobium saman* [WilliaWa1988]. **Lythraceae**: *Lagerstroemia indica* [Velasq1971]. **Magnoliaceae**: *Magnolia grandiflora* [Houser1918]. **Malvaceae**: *Hibiscus* [WilliaWa1988], *Hibiscus tiliaceus* [WilliaWa1988]. **Meliaceae**: *Trichilia* [Balach1956]. **Moraceae**: *Artocarpus altilis* [WilliaWa1988], *Artocarpus communis* [Takaha1939b, Takaha1941b], *Artocarpus heterophyllus* [WilliaWa1988], *Ficus* [Ferris1938a, McKenz1956], *Ficus glandulifera* [WilliaWa1988]. **Musaceae**: *Musa* [Green1915c, Green1937, Ferris1938a, MatileNo1984, WilliaWa1988], *Musa paradisiaca sapientum* [McKenz1956], *Musa sapientum* [GonzalCh1968, WilliaWa1988], *Strelitzia* [Almeid1973b]. **Myrtaceae**: *Decaspermum* [WilliaWa1988], *Eugenia malaccensis* [WilliaWa1988], *Psidium guajava* [WilliaWa1988]. **Ochnaceae**: *Schuurmansia henningsii* [WilliaWa1988]. **Oleaceae**: *Olea europaea* [GonzalCh1968, AguileDiGr1981]. **Orchidaceae**: *Grammatophyllum papuanum* [WilliaWa1988], *Vanilla* [WilliaWa1988], *Vanilla planifolia* [WilliaWa1988]. **Palmae** [KuwanaMu1931a, Almeid1969], *Areca catechu* [Takaha1939b], *Chrysalidocarpus lutescens* [Velasq1971], *Cocos nucifera* [Cocker1892b, Ferris1938a, McKenz1956, WilliaWa1988], *Elaeis guineensis* [Almeid1973b], *Nipa fruticans* [Beards1966], *Phoenix* [Balach1956]. **Pandanaceae**: *Pandanus odoratissimus* [WilliaWa1988]. **Passifloraceae**: *Passiflora edulis* [WilliaWa1988]. **Pinaceae**: *Pinus caribaea* [WilliaWa1988]. **Piperaceae**: *Piper* [WilliaWa1988], *Piper aduncum* [WilliaWa1988]. **Proteaceae**: *Macadamia ternifolia* [Velasq1971], *Macadamia tetraphylla* [WilliaWa1988]. **Rhizophoraceae**: *Bruguiera gymnorhiza* [WilliaWa1988]. **Rosaceae**: *Malus* [WilliaWa1988]. **Rubiaceae**: *Coffea canephora* [WilliaWa1988], *Guettarda speciosa* [WilliaWa1988], *Morinda citrifolia* [WilliaWa1988]. **Rutaceae**: *Citrus* [WilliaWa1988], *Citrus limon* [WilliaWa1988]. **Sapotaceae**: *Chrysophyllum cainito* [WilliaWa1988]. **Solanaceae**: *Solanum grandiflorum* [MatileNo1984]. **Sterculiaceae**: *Theobroma cacao* [WilliaWa1988]. **Strelitziaceae**: *Ravenala madagascariensis* [Velasq1971], *Strelitzia reginae* [Velasq1971]. **Theaceae**: *Camellia sinensis* [WilliaWa1988]. **Tiliaceae**: *Grewia* [DeLott1967a]. **Ulmaceae**: *Ulmus* [Dekle1965c]. **Zingiberaceae**: *Alpinia purpurata* [WilliaWa1988], *Curcuma longa* [WilliaWa1988], *Elettaria cardamomum* [WilliaWa1988]. **Zygophyllaceae**: *Guaiacum sanctum* [Balach1948b].

NATURAL ENEMIES: ACARI **Hemisarcoptidae**: *Hemisarcoptes malus* (Shimer) [GersonOcHo1990]. COLEOPTERA **Coccinellidae**: *Coccidophilus citricola* [AguileMeVa1984]. HYMENOPTERA **Signiphoridae**: *Signiphora flavella* [Woolle1990], *Signiphora prepauca* Girault [Woolle1990].

DISTRIBUTION: **Afrotropical**: Angola [Almeid1969, Almeid1973b]; Cameroon [MatileNo1984]; Comoros Islands [Matile1978]; Ghana [Nakaha1982]; Guinea [Leonar1914, Balach1956]; Kenya [DeLott1967a]; São Tomé and Príncipe (Príncipe [Seabra1921, Castel1963, Fernan1993], São Tomé [Nakaha1982]); Seychelles [Nakaha1982]; South Africa [Nakaha1982]; Tanzania [Mamet1956, Balach1956, Borchs1966]; Togo [Nakaha1982]; Zaire [Balach1956]; Zanzibar [Nakaha1982]. **Australasian**: Australia [Nakaha1982]; Federated States of Micronesia (Caroline Islands [Takaha1939b, Takaha1941b], Kosare (=Kusiae) [Beards1966], Ponape

Island [Beards1966], Yap [Beards1966]); Fiji [Green1915c, WilliaWa1988]; Guam [Nakaha1982]; Kiribati [WilliaWa1988]; Marshall Islands [Takaha1939b]; Palau [Takaha1939b, Takaha1941b, Beards1966]; Papua New Guinea [WilliaWa1988]; Solomon Islands [WilliaWa1988] [Nakaha1982]; Tonga [WilliaWa1988]; Western Samoa [WilliaWa1988]. **Nearctic**: Mexico [Nakaha1982]; United States of America (California [McKenz1956], Florida [Merril1953, Dekle1965c]). **Neotropical**: Antigua and Barbados (Antigua [Nakaha1982]); Argentina [ClapsTe2001]; Bolivia [Nakaha1982]; Brazil [Nakaha1982]; Chile [GonzalCh1968, AguileDiGr1981]; Colombia [Balach1959a, Kondo2001]; Cuba [Houser1918]; Dominica [Nakaha1982]; Ecuador [YustCe1956]; Guyana [Nakaha1982]; Jamaica [Cocker1892b, Cocker1893e, Cocker1894g, Ferris1938a]; Panama [Cocker1899n, Ferris1938a]; Panama Canal Zone [Ferris1938a]; Peru [VasqueDeCo2002]; Puerto Rico & Vieques Island (Puerto Rico [Martor1976, ColonFMe1998]); Saint Kitts & Nevis Islands (Saint Kitts [Nakaha1982]); Suriname [Nakaha1982]; Trinidad and Tobago (Trinidad [Ferris1938a]); U.S. Virgin Islands [Nakaha1983]. **Oriental**: India [Nakaha1982]; Indonesia (Java [KuwanaMu1931a], Sumatra [Green1930c]); Malaysia [Nakaha1982]; Mongolia [DanzigKo1990]; Philippines [VelasqRi1969, Nakaha1982]; Singapore [Nakaha1982]; Sri Lanka [Nakaha1982]; Thailand [Nakaha1982]; Vietnam [Nakaha1982]. **Palaearctic**: China (People's Republic) [Tang1984]; Czech Republic [Zahrad1953, Zahrad1977]; Germany (United) [Lindin1909b]; Madeira Islands [Green1923b, Ferris1938a]; Poland [Dziedz1989]; Portugal [Seabra1941]; United Kingdom (England [Green1920]).

BIOLOGY: Occurring on leaves (Ferris, 1938a). Reported to have uniparental as well as biparental populations (Schmutterer, 1959).

ECONOMIC IMPORTANCE: Pest of crops, such as banana, coconut palm, oil palm, manihot and cocoa, in tropics (Schmutterer et al, 1957; Chua & Wood, 1990). Occasionally damages orchids and palms in greenhouses at cooler regions (Schmutterer et al., 1957).

GENERAL: Description and illustration of adult female by Cockerell (1892b), Kuwana & Muramatsu (1931a) (as *Aspidiotus javanensis*), Ferris (1938a), Balachowsky (1948b, 1956), McKenzie (1956), Velasquez (1971), Bazarov & Shmelev (1971), Tang (1984), Chou (1985, 1986), Williams & Watson (1988), Dziedzicka (1989), Zahradník (1990b), Danzig (1993), Gill (1997) and by Colon-Ferrer & Medina-Gaud (1998).

KEYS: Claps & Teran 2001: 392 (female) [South Africa]; Gill 1997: 155 (female) [Species of California]; Danzig 1993: 169 (female) [Europe]; Williams & Watson 1988: 132 (female) [Tropical South Pacific]; Chou 1985: 306 (female) [Species of China]; Velasquez 1971: 110 (female) [Philippines]; Beardsley 1966: 520 (female) [Federated States of Micronesia]; Balachowsky 1956: 106 (female) [Africa]; McKenzie 1956: 26 (female) [U.S.A.: California]; Balachowsky 1953k: 114 (female) [World]; Ferris 1942: 34 (female) [North America]; Newell 1899: 25 (female) [North America].

CITATIONS: AguileDiGr1981 [host, distribution, life history, economic importance: 175-178]; AguileMeVa1984 [life history, biological control: 47-54]; Almeid1969 [taxonomy, description, host, distribution: 153]; Almeid1973b [host, distribution: 9-10]; Balach1948b [taxonomy, description, illustration, host, distribution: 320-321]; Balach1953k [taxonomy: 114]; Balach1956 [taxonomy, description, illustration, host, distribution: 114-117]; Balach1959a [host, distribution: 362]; BazaroSh1971 [taxonomy, description, illustration, host, distribution: 196-197]; Beards1966 [host, distribution: 521]; Borchs1937 [taxonomy, description, illustration, host, distribution: 128]; Borchs1939 [taxonomy, description, host, distribution: 8,16]; Borchs1950b [taxonomy, description, illustration, host, distribution: 224]; Borchs1966 [catalogue: 313]; Brick1912 [host, distribution: 1-22]; BurgerUl1990 [economic importance: 313-327]; Castel1951a [biological control: 95-98]; Castel1963 [distribution: 140]; Chou1985 [taxonomy, description, host, distribution: 308-309]; Chou1986 [taxonomy, illustration: 689]; ChuaWo1990 [host, distribution, economic importance: 543-552]; ClapsTe2001 [taxonomy, description, illustration, host, distribution, economic importance: 392,394]; ClapsWoGo2001a [taxonomy, host, distribution: 13]; Cocker1892b [taxonomy, host, distribution: 333]; Cocker1892e [economic importance, biological control: 5]; Cocker1893e [taxonomy, description, host, distribution: 39-40]; Cocker1896b [distribution: 334]; Cocker1897i [taxonomy, description, host, distribution: 24]; Cocker1897l [taxonomy: 151]; Cocker1899n [host, distribution: 22]; ColonFMe1998 [taxonomy, description, illustration, host, distribution: 61-62]; Corbet1932 [host, distribution]; Danzig1964 [taxonomy, host, distribution : 652]; Danzig1993 [taxonomy, description, illustration, host, distribution: 171-172]; DanzigKo1990 [host, distribution: 47]; DanzigPe1998 [catalogue: 273-274]; Dekle1965c [taxonomy, description, host, distribution: 72]; Dekle1976 [taxonomy, description, host, distribution, economic importance: 93]; DeLott1967a [host, distribution: 115]; DeSant1979 [biological control]; Dhilee1996 [host, distribution, biological control: 64-74]; DicksoFl1955 [host, distribution: 614-615]; Dziedz1989 [taxonomy, description, illustration, host, distribution: 100]; FDACSB1982 [host, distribution: 5-11]; Fernal1903b [catalogue: 270]; Fernan1993 [host, distribution: 112]; Ferris1938a [taxonomy, description, illustration, host, distribution: 242]; Ferris1941e [taxonomy: 44,46,49]; Ferris1942 [taxonomy: 446:34]; Flande1971 [biological control, life history: 857-872]; Gerson1990 [taxonomy: 130]; GersonOcHo1990 [biological control: 77-97]; Gill1997 [host, distribution, taxonomy, description, illustration, economic importance: 157,160]; GonzalCh1968 [host, distribution: 111]; Gowdey1921 [taxonomy, description, host, distribution: 29]; Green1915c [host, distribution: 44]; Green1920 [taxonomy, host, distribution: 129-130]; Green1923b [host, distribution: 89]; Green1930c [taxonomy, host, distribution: 281]; Green1937 [host, distribution: 330]; Hinckl1963 [host, distribution, biological control]; Houser1918 [host, distribution: 167]; Jancke1955 [taxonomy: 305]; Kawai1980 [taxonomy, description, host, distribution: 220]; Komosi1961 [taxonomy, description, illustration, host, distribution: 221-224]; Kondo2001 [taxonomy, host, distribution: 44]; KondoKa1995 [host, distribution: 57-58]; Kotins1907 [host, distribution: 304-308]; KuwanaMu1931a [taxonomy, description, illustration, host, distribution: 654,658]; Laing1929a [taxonomy: 486]; Leonar1897 [taxonomy: 285]; Leonar1898c [taxonomy, description, illustration, host, distribution: 51-53]; Leonar1914 [taxonomy, description, illustration, host, distribution: 199-201]; Lepesm1947 [taxonomy, description, host, distribution, life history: 211]; Lindin1909b [host, distribution: 151]; Lindin1909c [taxonomy, host, distribution: 449]; Lindin1910a [taxonomy: 440]; Lindin1910b [host, distribution: 37]; Lindin1912b [taxonomy, description, host, distribution: 204-205,240]; Lindin1935 [taxonomy: 129]; MacGil1921 [taxonomy, description, host, distribution: 407]; Mallam1954 [distribution: 24-60]; Mamet1956 [host, distribution: 138]; Martor1976 [host, distribution: 1-303]; Matile1978 [host, distribution: 63]; MatileNo1984 [host, distribution: 64-65]; McKenz1956 [taxonomy, description, illustration, host, distribution: 71-73]; Merril1953 [taxonomy, description, host, distribution: 25]; MillerDa1990 [host, distribution, economic importance: 302]; Montei1973 [host, distribution: 3]; Morgan1893 [taxonomy, description, host, distribution: 40-41]; Nakaha1982 [host, distribution: 42]; Nakaha1983 [host, distribution: 11]; Newell1899 [taxonomy, description, host, distribution: 25,28]; Newell1923 [host, distribution: 263-266]; Nur1990b [taxonomy, life history: 196]; Ponnam1999 [host, distribution, chemical control: 445-451]; Sassce1923 [host, distribution: 152-158]; Schmut1957a [host, distribution: 135,136]; Schmut1959 [taxonomy, description, host, distribution: 45]; Schmut1990a [host, distribution: 393]; Schmut2001 [host, distribution: 339-345]; SchmutKlLu1957 [host, distribution, economic importance: 477]; Seabra1921 [taxonomy, distribution: 98]; Seabra1930a [host, distribution: 143-148]; Seabra1941 [distribution: 8]; ShiLi1991 [host, distribution: 165]; Sugimo1994 [host, distribution: 115-121]; Takaha1939b [host, distribution: 269]; Takaha1941b [host, distribution: 220]; Tang1984 [taxonomy, description, illustration, host, distribution: 58-59]; Tao1999 [taxonomy, host, distribution: 68]; VasqueDeCo2002 [host, distribution: 331]; Vayssi1913 [host, distribution: 430]; Velasq1971 [taxonomy, description, illustration, host, distribution: 113-114]; VelasqRi1969 [host, distribution: 195-208]; WilliaWa1988 [taxonomy, description, illustration, host, distribution: 133-135]; Woolle1990 [biological control: 167-176]; YustCe1956 [host, distribution: 425-442]; Zahrad1953 [taxonomy, description, illustration, host, distribution: 137-139]; Zahrad1959 [taxonomy: 65-67]; Zahrad1959b [host, distribution: 60]; Zahrad1977 [taxonomy, distribution: 119]; Zahrad1990b [taxonomy, description, illustration, host, distribution: 83-85].

Hemiberlesia pictor (Williams)

Abgrallaspis pictor Williams, 1971a: 453. Type data: UNITED KINGDOM: England, Devon, Kingsteignton, on leaves of *Cymbidium* sp. (under glass); collected xii.1969. Holotype female. Type depository: London: The Natural History Museum, England, UK.

Hemiberlesia pictor; Danzig, 1993: 175. Change of combination.

SCALE COVER: Scale of adult female yellow-brown, subcircular to oval; exuviae subcentral, slightly darker. Scale of male not observed (Williams, 1971a).

HOST PLANTS: **Orchidaceae**: *Cymbidium* [Willia1971a], *Cymbidium ensifolium* [Willia1971a].

DISTRIBUTION: **Palaearctic**: United Kingdom (England [Willia1971a]).

GENERAL: Description and illustration of adult female by Williams (1971a).

KEYS: Danzig 1993: 169 (female) [Europe].

CITATIONS: Danzig1993 [taxonomy, host, distribution: 175]; DanzigPe1998 [catalogue: 274]; DuttaSi1990 [taxonomy: 1]; NagarkSa1990 [host, distribution, economic importance, biological control: 553-542]; Willia1971a [taxonomy, description, illustration, host, distribution: 453-455].

Hemiberlesia pitysophila Takagi

Hemiberlesia pitysophila Takagi, 1969a: 79. Type data: TAIWAN: Yang-ming Shan and southeastern Tai-pei Hsien, on the needles of *Pinus* spp. Holotype female. Type depository: Sapporo: Entomological Institute, Faculty of Agriculture, Hokkaido University, Japan.

SCALE COVER: Scale white (Takagi, 1969a). Colour photograph of scale cover by Wong *et al.* (1999).

HOST PLANTS: **Pinaceae**: *Pinus* [Takagi1969a], *Pinus massoniana* [Tang1984, Cadahi1989].

NATURAL ENEMIES: FUNGI: *Cladosporium cladosporioides* [PanChLi1989]. HYMENOPTERA **Aphelinidae**: *Aphytis vandenboschi* [Bao1993], *Coccobius azumai* Tachikawa [Tachik1988, GuMu1990].

DISTRIBUTION: **Oriental**: Hong Kong [Cadahi1989]; Taiwan [Takagi1969a, WongChCh1999]. **Palaearctic**: China (People's Republic) [Tang1984, Cadahi1989]; Japan [Takagi1969a, Kawai1980]; South Korea [Takagi1969a].

ECONOMIC IMPORTANCE: Has been introduced into continental China, near Hong Kong and Macao, from Japan or Taiwan. Recorded as a pest of *Pinus massoniana* in China (Pan et al., 1987; Tong et al., 1988; Cadahia, 1989).

GENERAL: Description and illustration of adult female by Takagi (1969a) and by Chou (1985, 1986).

CITATIONS: Bao1993 [host, distribution, biological control: 1-5]; Cadahi1989 [host, distribution, economic importance, chemical control, biological control: 343-363]; ChenHu1998 [host, distribution, life history, biological control: 136-142]; ChiuLiHu1993 [host, distribution, chemical control: 177-184]; Chou1985 [taxonomy, description, host, distribution: 402-403]; Chou1986 [taxonomy, illustration: 686]; DanzigPe1998 [catalogue: 274]; DingPaTa1992 [host, distribution, biological control: 35-42]; DingTaDu1988 [host, distribution, life history: 88]; DuDiTa1991 [host, distribution, chemistry, life history: 45-50]; GuCh1998 [host, distribution, life history, biological control: 4,156-163]; GuMu1990 [host, distribution, life history, biological control: 31-36]; Kawai1980 [taxonomy, description, host, distribution: 218]; Liang1988 [biological control: 62]; LiangCh1990 [host, distribution, biological control: 1-6]; LianTa1985 [host, distribution, economic importance, chemical control: 29-31]; PanChLi1989 [biological control: 22-23]; PanTaCh1989 [host, distribution, life history: 1-6]; PanTaXi1987 [host, distribution, economic importance, life history, ecology: 177-189]; Tachik1988 [host, distribution, biological control: 69-71]; Takagi1969a [taxonomy, description, illustration, host, distribution: 79-81,102]; Tang1984 [taxonomy, description, host, distribution: 50]; Tao1999 [taxonomy, host, distribution: 91]; TongTaPa1988 [host, distribution, economic importance, life history: 6-11]; WongChCh1999 [taxonomy, description, host, distribution: 26,67]; WuChWe1989 [biological control: 28-30]; WuLi1994 [host, chemical control: 28-30]; Xie1998 [taxonomy, description, illustration, host, distribution: 140-142]; XiePaTa1997 [chemical control: 135-144].

Hemiberlesia popularum (Marlatt)

Aspidiotus (*Hemiberlesia*) *popularum* Marlatt, 1908c: 23. Type data: USA: Mew Mexico, Deming, on cottonwood; collected by T.D.A. Cockerell, 9 February 1897. Syntypes, female. Type depository: Washington: United States National Entomological Collection, U.S. National Museum of Natural History, District of Columbia, USA; type no. 8370.

Aspidiotus popularum; Sanders, 1909a: 53. Change of combination.

Diaspidiotus popularum; MacGillivray, 1921: 414. Change of combination.

Hemiberlesia popularum; Ferris, 1938a: 243. Change of combination.

COMMON NAME: poplar scale [McKenz1956].

SCALE COVER: Female scale broadly oval, length 2 mm, convex; exuviae near anterior end, dark brown; larval exuviae nearly black, normally covered; secretionary matter yellowish white, very dense. Male scale oval shape, length 1-1.5 mm; exuviae showing prominently at one end through scanty secretionary covering (Marlatt, 1908c). Scale of female yellowish white, oval, high convex, dark exuviae submarginal; that of male oval, white, exuviae toward one end (Ferris, 1938a).

HOST PLANTS: **Compositae**: *Baccharis viminea* [McKenz1956], *Pluchea sericea* [McKenz1956]. **Oleaceae**: *Fraxinus* [McKenz1956]. **Salicaceae**: *Populus* [Sander1909a, Ferris1938a, McKenz1956, McDani1969], *Salix* [Ferris1938a, McKenz1956, McDani1969]. **Ulmaceae**: *Ulmus pumila* [McKenz1956]. **Viscaceae**: *Viscum album* [McKenz1956].

DISTRIBUTION: **Nearctic**: U.S.A. (Arizona [Marlat1908c, Ferris1938a], California [McKenz1956], Idaho [Nakaha1982], Montana [Nakaha1982], New Mexico [Marlat1908c, Sander1909a,], Texas [Ferris1938a], Utah [Nakaha1982]).

BIOLOGY: Occurring on the bark (Ferris, 1938a).

GENERAL: Description and illustration of adult female by Marlatt (1908c), Ferris (1938a) and by McKenzie (1956).

KEYS: Gill 1997: 155 (female) [Species of California]; McDaniel 1969: 107 (female) [U.S.A.: Texas]; Balachowsky 1956: 105-108 (female) [Africa]; Balachowsky 1953k: 114-115 (female) [World]; Ferris 1942: 34 (female) [North America].

CITATIONS: Balach1948b [taxonomy: 298]; Balach1953k [taxonomy: 114]; Balach1956 [taxonomy: 106]; Borchs1966 [catalogue: 309]; BurgerU11990 [economic importance: 313-327]; Ferris1938a [taxonomy, description, illustration, host, distribution: 243]; Ferris1941e [taxonomy: 47]; Ferris1942 [taxonomy: 446:34]; Gill1997 [host, distribution, taxonomy, description, illustration, economic importance: 157,161]; MacGil1921 [taxonomy, description, host, distribution: 414]; Marlat1908c [taxonomy, description, illustration, host, distribution: 23-24]; McDani1969 [taxonomy, illustration, host, distribution: 111-112]; McKenz1956 [taxonomy, description, illustration, host, distribution: 25,73-74]; Nakaha1982 [host, distribution: 42-43]; Sander1909a [taxonomy, host, distribution: 53].

Hemiberlesia pseudorapax McKenzie

Hemiberlesia pseudorapax McKenzie, 1951: 81. Type data: U.S.A.: New Mexico, Santa Rita, on *Abies* sp. Holotype female. Type depository: Davis: The Bohart Museum of Entomology, University of California, California, USA.

SCALE COVER: Female scale about 1.5 mm in diameter; grey, with a darker, reddish-brown, subcentral exuviae (sometimes a light grey wax film covers exuvia); scale of male similar in colour to that of female, elongate, exuvia near one end (McKenzie, 1951).
HOST PLANTS: **Pinaceae**: *Abies* [McKenz1951].
DISTRIBUTION: **Nearctic**: United States of America (New Mexico [McKenz1951]).
BIOLOGY: Occurring on bark (McKenzie, 1951).
GENERAL: Description and illustration of adult female by McKenzie (1951).
CITATIONS: Balach1953k [taxonomy: 115]; Balach1956 [taxonomy: 108]; Borchs1966 [catalogue: 309]; McKenz1951 [taxonomy, description, illustration, host, distribution: 81-82]; Nakaha1982 [host, distribution: 43].

Hemiberlesia quercicola Ferris

Hemiberlesia quercicola Ferris, 1941d: 344. Type data: U.S.A.: California, San Diego County, at Banner, on *Quercus engelmanni*. Holotype female. Type depository: Davis: The Bohart Museum of Entomology, University of California, California, USA.
COMMON NAME: irregular oak scale [McKenz1956].
SCALE COVER: Scale of female white or straw coloured; very irregular because of its position, but apparently basically circular; exuviae subcentral, pale yellow and normally covered with secretion. Scale of male not recognized (Ferris, 1941d).
HOST PLANTS: **Fagaceae**: *Quercus* [Ferris1941d, McKenz1956], *Quercus engelmanni* [Ferris1941d, McKenz1956]. **Rosaceae**: *Prunus amygdalus* [Gill1997].
DISTRIBUTION: **Nearctic**: United States of America (California [Ferris1941d, McKenz1956], New Mexico [Ferris1941d]).
BIOLOGY: Occurring buried in cracks of the bark and under bark flakes (Ferris, 1941d).
GENERAL: Description and illustration of adult female by Ferris (1941d), McKenzie (1956) and by Gill (1997).
KEYS: Gill 1997: 32 (female) [Species of California]; Davidson 1964: 639-640 (female) [North America]; McKenzie 1956: 26 (female) [U.S.A.: California]; Ferris 1942: 35 (female) [North America].
CITATIONS: Borchs1966 [catalogue: 309]; Davids1964 [taxonomy: 639]; Ferris1941d [taxonomy, description, illustration, host, distribution: 344]; Ferris1942 [taxonomy: 446:35]; Gill1997 [host, distribution, taxonomy, description, illustration, economic importance: 34,39,40]; McKenz1956 [taxonomy, description, illustration, host, distribution: 73-74]; Nakaha1982 [host, distribution: 3]; Takaha1952a [taxonomy: 15].

Hemiberlesia rapax (Comstock)

Aspidiotus camelliae Signoret, 1869c: 117. Misidentification of *Aspidiotus camelliae* Boisduval, 1867. Notes: see discussion below in Systematics for *Hemiberlesia rapax*.

Aspidiotus rapax Comstock, 1881a: 307. Type data: USA: California and Florida, locality not indicated, on trunk, limbs, leaves and fruit of various trees and shrubs. Syntypes, female. Type depository: Washington: U.S. National Entomological Collection, U.S. National Museum of Natural History, District of Columbia, USA.

Aspidiotus acuminatus Targioni Tozzetti, 1881: 151. Type data: ITALY: on *Robinia* sp. Syntypes, female. Synonymy by Leonardi, 1920: 91. Notes: Type material lost (Giuseppina Pellizzari, personal communication to Yair Ben-Dov).

Aspidiotus evonymi Targioni Tozzetti, 1888: 420. Type data: ITALY: Florence and Rome, on *Euonymus* sp. Syntypes, female. Synonymy by Borchsenius, 1966: 310. Notes: Type material lost (Giuseppina Pellizzari, 1999, personal communication to Yair Ben-Dov).

Aspidiotus flavescens Green, 1890: 21. Type data: SRI LANKA: locality not indicated, on tea plant. Syntypes, female. Type depository: London: The Natural History Museum, England, UK. Synonymy by Borchsenius, 1966: 310.

Diaspis circulata Green, 1896: 4. Type data: SRI LANKA: on young and older twigs of tea plants, on *Chinchoana* sp. and on *Osbeckia* sp. Syntypes, female. Type depository: London: The Natural History Museum, England, UK. Synonymy by Cockerell, 1896b: 339.

Aspidiotus napax; Newstead, 1897a: 94. Misspelling of species name.

Hemiberlesia camelliae; Leonardi, 1897b: 124. Change of combination.

Aspidiotus (*Hemiberlesia*) *rapax*; Cockerell, 1897i: 30. Change of combination.

Aspidiotus (*Hemiberlesia*) *tricolor* Cockerell, 1897u: 266. Type data: MEXICO: Oaxaca, Salina Cruz, host plant not indicated; collected by Townsend, May 29; No. 7193. Syntypes, female. Type depository: Washington: United States National Entomological Collection, U.S. National Museum of Natural History, District of Columbia, USA. Synonymy by Ferris, 1941e: 49.

Aspidiotus tricolor; Townsend, 1897: 185. Change of combination.

Aspidiotus (*Hemiberlesia*) *rapax evonymi*; Cockerell, 1899a: 396. Change of combination and rank.

Aspidiotus euonymi; Cockerell, 1899j: 274. Misspelling of species name.

Aspidiotus lucumae Cockerell, 1899n: 22. Type data: MEXICO: Esperanza, crowded on bark of *Mammea sapota* tree. Syntypes, female. Type depository: Washington: United States National Entomological Collection, U.S. National Museum of Natural History, District of Columbia, USA. Synonymy by Ferris, 1938a: 190.

Hemiberlesia tricolor; Leonardi, 1900: 339. Change of combination.

Aspidiotus (*Hemiberlesia*) *camelliae*; Hempel, 1900a: 501. Change of combination.

Hemiberlesia acuminatus; Cockerell, 1905b: 205. Change of combination.

Hemiberlesia argentina Leonardi, 1911: 237. Type data: ARGENTINA: Cacheuta, on *Ophryosporus andinus*. Syntypes, female. Type depository: Portici: Dipartimento de Entomologia e Zoologia Agraria di Portici, Università di Napoli Federico II, Italy. Synonymy by Borchsenius, 1966: 310.

Aspidiotus argentina; Sasscer, 1912: 92. Change of combination.

Hendaspidiotus tricolor; MacGillivray, 1921: 439. Change of combination.

Hemiberlesiana camelliae; Thiem & Gerneck, 1934a: 232. Change of combination.

Hemiberlesea camelliae; Balachowsky, 1935b: 257. Misspelling of genus name.

Hemiberlesea rapax; Balachowsky, 1935b: 3. Misspelling of genus name.

Hemiberlesea camelliae; Gómez-Menor Ortega, 1937: 111. Misspelling of genus name.

Hemiberlesia rapax; Ferris, 1938a: 244. Change of combination.

Diaspidiotus camelliae; Bodenheimer, 1949: 65. Misidentification.

Hemiberlesia repax; Tippins & Beshear, 1972: 287. Misspelling of species name.

Aspidiotus lacumae; Chou, 1985: 300. Misspelling of species name. Notes: Misspelling of *Aspidiotus lucumae* Cockerell.

Aspidiotus argintina; Chou, 1985: 301. Misspelling of species name.

COMMON NAMES: cochonilha-concha [CarvalAg1997]; escama rapax [Gonzal1989]; escama voraz [CoronaRuMo1997]; greedy scale [Merril1953, McKenz1956, Dekle1965c, Borchs1966]; greedy scale insect [Comsto1881a]; mnogoyadnaya shitovka [Borchs1936]; rapacious scale [Brimbl1961].

SYSTEMATICS: Signoret (1869b: 117) misidentified a record taken from cultivated *Camellia* in greenhouses at Luxembourg Garden and Boulogne forest, Paris, as *Aspidiotus camelliae* Boisduval, 1867. However, *Aspidiotus camelliae* Boisduval, 1867 is a Lepidosaphedine species (Ben-Dov & Matile-Ferrero, unpublished data).

SCALE COVER: Female scale highly convex, exuviae near one side and scale thus having a strongly tipped-over appearance, colour grey; scale of male not known (Ferris, 1938a). Colour photograph by Gonzalez (1989), Gill (1997) and by Carvalho & Aguiar (1997).

HOST PLANTS: **Actinidiaceae**: *Actinidia arisanensis* [Takagi1969a], *Actinidia chinensis* [LoveFe1977, GonzalCu1994]. **Anacardiaceae**: *Rhus laurina* [Ferris1921], *Schinus molle* [DeLott1967a]. **Apocynaceae**: *Nerium oleander* [Brimbl1968], *Plumeria* [Wilson1917, MerrilCh1923, Brimbl1968], *Vinca major* [GomezM1962]. **Araliaceae**: *Hedera helix* [Balach1932d]. **Begoniaceae**: *Begonia* [Leonar1920]. **Bignoniaceae**: *Tecoma* [MerrilCh1923, Balach1932d]. **Cactaceae**: *Opuntia tomentosa* [Balach1932d]. **Caprifoliaceae**: *Lonicera caprifolium* [Martin1983]. **Caricaceae**: *Carica* [Borchs1934]. **Casuarinaceae**: *Casuarina* [Leonar1920]. **Celastraceae**: *Celastrus bilocularis* [Brimbl1968], *Euonymus* [Brain1918, MerrilCh1923, Bodenh1924, Borchs1934, Bodenh1937], *Euonymus grandiflora* [Ferris1953], *Euonymus japonicus* [Wilson1917, Borchs1934, Korone1934, Bodenh1952, Bachma1953, Martin1983], *Gymnosporia europaea* [GomezM1958c]. **Cistaceae**: *Cistus heterophyllus* [Balach1927, Balach1932d], *Cistus monspeliensis* [Balach1932d]. **Compositae**: *Baccharis* [Lepage1938], *Brachymerium* [Martin1983], *Brachymerium* [Martin1983], *Chrysanthemum segetum* [Balach1932d], *Inula viscosa* [Balach1932d, Balach1933e], *Ophryosporus andinus* [Leonar1911, ClapsWoGo2001]. **Corylaceae**: *Corylus* [Lepage1938]. **Crassulaceae**: *Sedum* [MerrilCh1923]. **Cucurbitaceae**: *Sechium edule* [Brimbl1968]. **Ebenaceae**: *Diospyros kaki* [TomkinWiTh2000], *Diospyros pentamera* [Brimbl1968]. **Elaeagnaceae**: *Elaeagnus* [Leonar1920, MerrilCh1923], *Elaeagnus edulis* [Martin1983], *Elaeagnus japonica* [Green1929], *Elaeagnus pungens* [Borchs1934], *Elaeagnus reflexa* [Balach1932d]. **Ericaceae**: *Arbutus peninsularis* [Ferris1921], *Philippia* [Mamet1959a, Borchs1966]. **Euphorbiaceae**: *Euphorbia* [Leonar1920], *Mallotus japonicus* [Borchs1934]. **Fagaceae**: *Castanopsis*

[Merril1953], *Ilex* [Takaha1940, Takagi1969a, Kawai1977], *Quercus nigra* [TippinBe1970, BesheaTiHo1973]. **Gramineae**: *Phyllostachys bambusoides* [TippinBe1972, BesheaTiHo1973]. **Guttiferae**: *Hypericum canariense* [GomezM1962], *Hypericum moserianum* [Brimbl1968], *Hypericum reflexum* [GomezM1962]. **Juglandaceae**: *Carya illinoensis* [Dekle1965c], *Juglans regia* [Lepage1938]. **Labiatae**: *Sideritis leucantha* [Martin1983]. **Lauraceae**: *Cinnamomum camphora* [Brimbl1968], *Laurus nobilis* [Balach1927, Balach1932d, Borchs1934, Balach1935b, Martin1983], *Persea americana* [Brimbl1968], *Umbellularia californica* [Comsto1883]. **Leguminosae** [Mamet1954, Borchs1966], *Acacia* [Green1896e, MerrilCh1923, Green1937, Lepage1938], *Acacia dealbata* [Balach1932d], *Acacia heterophylla* [Mamet1957], *Acacia julibrissin* [Borchs1934], *Acacia juniperina* [Frogga1914], *Acacia koa* [Zimmer1948], *Acacia linifolia* [Brimbl1968], *Acacia longifolia* [Frogga1914], *Acacia melanoxylon* [Brain1918], *Acacia mollisima* [Balach1956], *Bauhinia purpurea* [DeLott1967a], *Ceratonia siliqua* [Borchs1934, InserrCa1987], *Cercis* [MerrilCh1923, Lepage1938], *Cercis siliquastrum* [Kuwana1927, Martin1983], *Colophospermum mopane* [Almeid1973b], *Cytisus prolifer* [GomezM1962], *Dolichos* [Balach1932d], *Genista* [MerrilCh1923], *Indigofera* [Green1907], *Parkinsonia* [Balach1932d], *Pithecellobium dulce* [Ferris1921], *Pterolobium exosum* [DeLott1967a], *Pueraria thunbergiana* [DeLott1967a]. **Liliaceae**: *Asparagus* [Leonar1920], *Smilax* [Hall1929]. **Magnoliaceae**: *Magnolia grandiflora* [Wilson1917, BesheaTiHo1973]. **Malvaceae**: *Hoheria* [Green1929], *Lavatera* [MerrilCh1923]. **Melastomataceae**: *Osbeckia* [Green1896e]. **Meliaceae**: *Cedrela toona australis* [Brimbl1968], *Melia azedarach* [Wilson1917, Balach1932d]. **Moraceae**: *Castilloa elastica* [Green1911], *Ficus* [Lepage1938], *Ficus carica* [Balach1932d, Borchs1934, Brimbl1968], *Ficus macrophylla* [Brimbl1968], *Maclura aurantiacea* [Martin1983], *Morus* [Ferris1953], *Morus alba* [Balach1932d]. **Musaceae**: *Musa acuminata* [Brimbl1968]. **Myoporaceae**: *Myoporum* [Green1923b, MerrilCh1923], *Myoporum loetum* [Martin1983]. **Myrtaceae**: *Austromyrtus bidwillii* [Brimbl1968], *Callistemon* [Leonar1920], *Eucalyptus* [Lepage1938, Martin1983], *Eugenia brachyandra* [Brimbl1968], *Metrosideros* [Zimmer1948], *Myrtus* [Leonar1920, Bodenh1924, Bodenh1937, Balach1927], *Myrtus alba* [Balach1932d], *Myrtus communis parvifolia* [GomezM1958c, Martin1983], *Myrtus communis* [Bachma1953, GomezM1962], *Psidium guajava* [Balach1932d, Brimbl1968, DanzigKo1990], *Psidium pomiferum* [Mamet1954, Borchs1966], *Tristania conferta* [Brimbl1968]. **Nyctaginaceae**: *Bougainvillea glabra* [Balach1932d]. **Ochnaceae**: *Maesia* [Mamet1954, Borchs1966]. **Oleaceae**: *Fraxinus* [Borchs1934], *Fraxinus oxyphylla* [Balach1932d], *Ligustrum* [Martin1983], *Olea europaea* [Comsto1883, Korone1934, Lepage1938], *Osmanthus fragans* [Ferris1953, Tang1984]. **Onagraceae**: *Fuchsia* [MerrilCh1923, Lepage1938]. **Orchidaceae**: *Dendrobium bigibbum* [Brimbl1968]. **Palmae**: *Archontophoenix cunninghamiana* [Brimbl1968]. **Phytolaccaceae**: *Phytolacca dioica* [Balach1932d]. **Platanaceae**: *Platanus orientalis* [Balach1932d]. **Pontederiaceae**: *Michelia* [Ramakr1919a]. **Proteaceae**: *Grevillea* [Ramakr1919a], *Hakea acicularis* [Lindin1909b], *Leucodendrum* [Leonar1920], *Stenocarpus sinuatus* [Brimbl1968]. **Rhamnaceae**: *Ceanothus* [Ferris1920b], *Rhamnus crocea* [Lepage1938], *Ziziphus* [Borchs1934,

Bodenh1937], *Ziziphus spina-christi* [Bodenh1924]. **Rosaceae**: *Cydonia vulgaris* [Lepage1938], *Eriobotrya japonica* [Brimbl1968], *Heteromeles arbutifolia* [Ferris1921], *Malus silvestris* [Brimbl1968], *Prunus domestica* [Lepage1938], *Pyrus* [Lepage1938], *Pyrus communis* [Brimbl1968, Martin1983]. **Rubiaceae**: *Cinchona* [Green1896e, Green1937], *Coprosma* [MerrilCh1923, Lepage1938]. **Rutaceae**: *Citrus* [Leonar1920, Lepage1938, Zimmer1948, CarvalAg1997], *Ruta angustifolia* [Balach1930, Balach1932d, Balach1933e]. **Salicaceae**: *Populus* [Borchs1934], *Populus alba* [Balach1932d], *Populus nigra* [Balach1932d], *Salix babylonica* [Balach1932d, Brimbl1968]. **Sapotaceae**: *Argania* [Borchs1934], *Planchonella australis* [Brimbl1968], *Sideroxylon* [Balach1932d]. **Scrophulariaceae**: *Veronica lindleyana* [Martin1983]. **Solanaceae**: *Solanum macrocarpum* [DeLott1967a]. **Sterculiaceae**: *Brachychiton discolor* [Brimbl1968]. **Strelitziaceae**: *Strelitzia* [MerrilCh1923], *Strelitzia augusta* [Balach1932d]. **Theaceae**: *Camellia* [Signor1869b, MerrilCh1923], *Camellia thea* [Lepage1938], *Thea japonica* [Borchs1934], *Thea sinensis* [Mamet1954a, Borchs1966]. **Verbenaceae**: *Citharexylum* [Balach1927, Balach1932d]. **Vitaceae**: *Cissus* [MerrilCh1923], *Vitis* [Borchs1934, Lepage1938], *Vitis rubra* [Borchs1934], *Vitis vinifera* [Balach1932d, Brimbl1968]. **Winteraceae**: *Drimys pauciflora* [WilliaWa1988].

NATURAL ENEMIES: ACARI **Hemisarcoptidae**: *Hemisarcoptes malus* (Shimer) [GersonOcHo1990]. FUNGI **Ascomycotina**: *Myriangium duriaei* [EvansPr1990]. HYMENOPTERA **Aphelinidae**: *Aphytis chilensis* Howard [RosenDe1979, MyartsRu2000], *Ap. chrysomphali* (Mercet) [Zimmer1948], *Ap. diaspidis* (Howard) [RosenDe1979], *Ap. proclia* (Walker) [Gordh1979], *Aspidiotiphagus citrinus* (Craw) [AnneckIn1971], *Prospaltella bicolor* Timberlake [Zimmer1948], *Pseudopteroptrix imitatrix* Fullaway [Zimmer1948], *Thysanus merceti* Malenotti [Balach1948b]. **Encyrtidae**: *Comperiella bifasciata* Howard [Zimmer1948, Trjapi1989]. **Signiphoridae**: *Signiphora flava* Girault [Gordh1979], *Si. flavopalliata* Ashmead [Woolle1990], *Si. merceti* Malenotti [GomezM1946], *Si. prepauca* Girault [Woolle1990].

DISTRIBUTION: **Afrotropical**: Angola [Almeid1973b]; Kenya [Newste1917b, DeLott1967a]; Madagascar [Mamet1954, Mamet1959a, Borchs1966]; Malawi [Nakaha1982]; Mauritius [Mamet1954a, Borchs1966]; Réunion [Mamet1957]; Seychelles [Green1907]; South Africa [BrainKe1917, Brain1918, Balach1956]; Tanzania [Balach1956]; Zimbabwe [Nakaha1982]. **Australasian**: Australia (New South Wales [Frogga1914], Queensland [Brimbl1968]); Bonin Islands (= Ogasawara-Gunto) [Beards1966, Kawai1987] [Nakaha1982]; French Polynesia (Society Islands [DoaneHa1909]); Hawaiian Islands (Hawaii [Zimmer1948, Nakaha1982]; New Caledonia [WilliaWa1988]; New Zealand [Lindin1909b, LoveFe1977]. **Nearctic**: Mexico [Cocker1899n, MyartsRu2000] (Baja California [Ferris1921], Oaxaca [Cocker1897u, Ferris1942]); U.S.A. (Alabama [Nakaha1982], California [Comsto1881a, McKenz1956, RosenDe1979], District of Columbia [Nakaha1982], Florida [Comsto1881a, Wilson1917, MerrilCh1923, Merril1953, Dekle1965c, BesheaTiHo1973], Georgia [TippinBe1970, BesheaTiHo1973], Idaho [Nakaha1982], Illinois [Nakaha1982], Indiana [Nakaha1982], Louisiana [Nakaha1982], Maryland [Nakaha1982], Mississippi [Herric1911], Missouri [Hollin1923], New Jersey [Nakaha1982], New York [Nakaha1982], Ohio

[Nakaha1982], Oregon [Nakaha1982], Pennsylvania [Nakaha1982], South Carolina [Nakaha1982], Texas [Herric1911, McDani1969], Virginia [Nakaha1982], Washington [Nakaha1982]). **Neotropical**: Argentina [Leonar1911] (Mendoza [ClapsWoGo2001]); Bolivia [Nakaha1982]; Brazil (Minas Gerais [Lepage1938, WolffCo1993], Paraiba [WolffCo1993], Rio Grande do Sul [Lepage1938], Rio de Janeiro [Hempel1900a, Lepage1938], São Paulo [Green1930b, Lepage1938]); Chile [GonzalCh1968, Gonzal1989, Gonzal1989a, GonzalCu1994]; Colombia [Nakaha1982]; Costa Rica [Nakaha1982]; Cuba [Nakaha1982]; Ecuador [Nakaha1982]; Guatemala [Nakaha1982]; Guyana [Newste1917]; Peru [Nakaha1982]; Puerto Rico & Vieques Island (Puerto Rico [Martor1976, Nakaha1982, ColonFMe1998]); Uruguay [Nakaha1982]. **Oriental**: China (People's Republic) (Yunnan [Ferris1953]); India [GreenMa1907]; Mongolia [DanzigKo1990]; Nepal [Takagi1975]; Philippines [VelasqRi1969]; Sri Lanka [Green1896e, GreenMa1907, Green1911, Ramakr1921a] [Nakaha1982]; Taiwan [Takaha1940, Takagi1969a]. **Palaearctic**: Algeria [Newste1897a, Balach1927, Balach1932d]; Armenia [ColonFMe1998]; Azores [ColonFMe1998]; Canary Islands [GomezM1962, GomezM1967O, MatileOr2001]; China (People's Republic) [Kuwana1927, Tang1984]; Czech Republic [Zahrad1953, Zahrad1990b]; Egypt [Ezzat1958]; France [Balach1930, Balach1932d, Balach1933e]; Georgia (Abkhaz ASSR [Borchs1934, Borchs1936], Adzhar ASSR [Borchs1934, Borchs1936], Georgia [Borchs1936, Hadzib1983]); Greece [Korone1934]; Iran [Kaussa1955]; Israel [Nakaha1982]; Italy [Leonar1920, LongoMaPe1995]; Japan [Kuwana1917a, Kuwana1933, Kawai1977] [Nakaha1982]; Madeira Islands [Green1923b, CarvalAg1997]; Malta [Borg1919]; Morocco [Balach1932d]; Poland [Dziedz1989]; Portugal [Seabra1941, Fernan1992]; Sicily [InserrCa1987]; Spain [Balach1935b, GomezM1937, Martin1983, BlayGo1993]; Syria [Nakaha1982]; Tunisia [Balach1932d]; Turkey [Bodenh1949, Bodenh1952]; Yugoslavia [Bachma1953].

BIOLOGY: Occurring either on bark or leaves (Ferris, 1938a).

ECONOMIC IMPORTANCE: Pest of several crops in tropical and subtropical regions, such as kiwifruit (Love & Ferguson, 1977; Gonzalez, 1989a), citrus (Rose, 1990; Carvalho & Aguiar, 1997), olive (Argyriou, 1990), mango (Chua & Wood, 1990) and tea plant (Nagarkatti & Sankaran, 1990).

GENERAL: Description and illustration of adult female by Brain (1918), Kuwana (1933), Ferris (1938a), Balachowsky (1948b), Zimmerman (1948), McKenzie (1956), Gómez-Menor Guerrero (1962), Takagi (1969a), Chou (1985, 1986), Tereznikova (1986), Williams & Watson (1988), Dziedzicka (1989), Zahradník (1990b), Danzig (1993), Yaşar (1995a), Kosztarab (1996), Gill (1997) and by Colon-Ferrer & Medina-Gaud (1998).

KEYS: Gill 1997: 155 (female) [Species of California]; Kosztarab 1996: 509 (female) [Northeastern North America]; Danzig 1993: 169 (female) [Europe]; Zahradník 1990b: 104 (female) [Czech Republic]; Williams & Watson 1988: 130 (female) [Tropical South Pacific]; Tereznikova 1986: 113 (female) [Ukraine]; Chou 1985: 300 (female) [Species of China]; McDaniel 1969: 107 (female) [U.S.A.: Texas]; Beardsley 1966: 520 (female) [Federated States of Micronesia]; Ezzat 1958: 241 (female) [Egypt]; Balachowsky 1956: 106 (female) [Africa]; McKenzie 1956: 26 (female) [U.S.A.: California]; Balachowsky 1953k: 114 (female) [World]; Lupo

1953a: 75-76 (female) [Italy]; Balachowsky 1948b: 299 (female) [Mediterranean]; Zimmerman 1948: 358 (female) [Hawaii]; Ferris 1942: 35 (female) [North America]; Archangelskaya 1937: 98-100 (female) [Central Asia]; Gómez-Menor Ortega 1937: 110 (female) [Spain]; Kuwana 1933: 3 (female) [Japan]; Kuwana 1933b: 48 (female) [Japan]; Fullaway 1932: 95-97, 107 (female) [Hawaii]; Balachowsky 1928a: 132 (female) [North Africa]; Hollinger 1923: 7-8 (female) [U.S.A.: Missouri]; Leonardi 1920: 90 (female) [Italy]; Lawson 1917: 217 (female) [U.S.A.: Kansas]; Robinson 1917: 29 (female) [Philippines]; Dietz & Morrison 1916a: 289-290 (female) [U.S.A.: Indiana]; Cockerell 1905: 45-46 (female) [Mexico]; Cockerell 1905b: 202 (female) [U.S.A.: Colorado]; Newstead 1901b: 81 (female) [England]; Newell 1899: 4-5, 25 (female) [North America]; Green 1896e: 40 (female) [Sri Lanka]; Comstock 1883: 55-57 (female) [North America].

CITATIONS: AbouEl2001 [host, distribution, biological control: 185-195]; Almeid1973b [host, distribution: 10]; AnneckIn1971 [host, distribution, biological control: 29]; Archan1937 [taxonomy, description, illustration, host, distribution: 99,108]; Argyri1990 [host, distribution, economic importance: 579-583]; Azeved1923A [host, distribution: 86-90]; Azeved1929a [host, distribution: 126-128]; Bachma1952 [host, distribution: 177]; BaetaN1947 [host, distribution: 128]; Balach1927 [host, distribution: 177]; Balach1928a [taxonomy: 32]; Balach1928d [biological control: 285,305]; Balach1930 [host, distribution: 312]; Balach1932d [taxonomy, host, distribution, economic importance: VII, XLVII]; Balach1933e [host, distribution: 3]; Balach1935b [host, distribution: 257]; Balach1948b [taxonomy, description, illustration, host, distribution, biological control: 299-302]; Balach1953k [taxonomy: 114]; Balach1956 [taxonomy, description, illustration, host, distribution: 116-118]; Beards1966 [host, distribution: 522]; BeardsDaHo1976 [economic importance: 105]; BeardsGo1975 [economic importance: 49]; BenDov1990e [host, distribution: 656]; BerryMoHi1989 [host, distribution, economic importance: 182-186]; BertelBa1966 [host, distribution: 17-46]; BesheaTiHo1973 [host, distribution: 7]; BlankGiDo1997 [host, distribution, economic importance, life history: 293-297]; BlankGiDo1999 [host, distribution, economic importance, life history: 1-12]; BlankGiKe2000 [host, life history, ecology: 934-942]; BlankGiMc2000 [host, life history, ecology: 1752-1759]; BlankGiOl1994 [host, distribution, life history: 304-309]; BlankGiOl1995 [host, distribution, life history: 1569-1575]; BlankGiOl1995a [host, distribution, life history, biological control: 1634-1640]; BlankGiSt2000 [host, distribution, chemical control, economic importance: 205-210]; BlankGiUp1996 [host, distribution, life history: 239-248]; BlankGiUp1996 [host, distribution, life history, ecology: 239-248]; BlankHoGi1995 [host, distribution, life history, chemical control: 13-23]; BlankLoGi1992 [host, distribution, life history, ecology: 174-179]; BlankOl1989 [chemical control: 191-194]; BlankOl1989a [chemical control: 187-190]; BlankOl1990 [host, distribution, chemical control: 240-242]; BlankOl1990a [chemical control: 243-246]; BlankOlBe1987 [host, distribution, life history, ecology: 127-130]; BlankOlGi1992 [host, distribution, economic importance: 397-405]; BlankOlGi1993 [economic importance: 139-145]; BlankOlLo1990 [host, distribution, life history, ecology: 81-87]; BlankOlLo1993 [host, distribution, chemical control: 71-74]; BlankOlTo1994 [chemical control: 195-202];

BlayGo1993 [taxonomy, description, illustration, host, distribution: 478-483]; Bodenh1924 [taxonomy, description, host, distribution: 25]; Bodenh1935 [host, distribution: 246]; Bodenh1937 [host, distribution: 217]; Bodenh1949 [taxonomy, description, illustration, host, distribution: 65-67]; Bodenh1952 [host, distribution: 339]; Bondar1914 [host, distribution, economic importance: 1064-1106]; Bondar1915 [host, distribution, economic importance: 44-47]; Borang1996 [host, distribution: 474-479]; Borchs1934 [host, distribution: 30]; Borchs1935a [taxonomy, description, host, distribution: 35]; Borchs1937 [taxonomy, description, illustration, host, distribution: 122-123]; Borchs1937a [taxonomy, host, distribution: 61-62]; Borchs1939 [taxonomy, description, host, distribution: 8,13]; Borchs1949d [taxonomy, description, host, distribution: 239]; Borchs1950b [taxonomy, description, illustration, host, distribution: 219,223]; Borchs1966 [catalogue: 309-311,368]; Borg1919 [taxonomy, description, host, distribution: 26-27]; BorgesViSa1986 [distribution]; Boyce1948 [host, distribution, economic importance, control]; Brain1918 [taxonomy, description, illustration, host, distribution: 128-129]; BrainKe1917 [distribution: 183]; BrandtBo1948 [taxonomy: 3]; Brick1912 [host, distribution: 1-22]; Brimbl1962 [host, distribution, economic importance: 221]; Brimbl1968 [taxonomy, illustration, host, distribution: 54-55]; Britto1923b [host, distribution]; BurgerUl1990 [economic importance: 313-327]; Burke1930 [host, distribution, biological control: 783-785]; CABI1987a [host, distribution: 1-3]; Carnes1907 [taxonomy, host, distribution: 210]; CarvalAg1997 [life history, description, economic importance, biological control, host, distribution: 276-279]; Charle1998 [distribution, economic importance, biological control: 51N]; CharleHiAl1995 [host, distribution, life history, biological control: 319-324]; CheahIr1997 [host, distribution]; Chiesa1948 [host, distribution, economic importance]; Chiesa1948a [host, distribution, economic importance]; Chou1985 [taxonomy, description, host, distribution: 300-302]; Chou1986 [taxonomy, illustration: 684]; ClapsWoGo2001 [host, distribution: 253]; ClapsWoGo2001a [taxonomy, host, distribution: 19-20]; Cocker1896b [taxonomy, distribution: 334,339]; Cocker1897i [taxonomy, description, host, distribution: 4,9,25,30]; Cocker1897u [taxonomy, description, host, distribution: 266]; Cocker1899a [taxonomy: 396]; Cocker1899j [taxonomy: 274]; Cocker1899n [taxonomy, description, illustration, host, distribution: 22-23]; Cocker1905 [taxonomy: 46]; Cocker1905b [taxonomy: 202]; ColonFMe1998 [taxonomy, description, illustration, host, distribution: 62-63]; Comsto1881a [taxonomy, description, illustration, host, distribution: 307-308]; Comsto1883 [taxonomy, host, distribution: 56,67,74]; CoronaRuMo1997 [host, distribution: 38-41]; Crouze1971 [biological control: 200]; Crouze1973 [host, distribution, biological control: 15-39]; DahmsSm1994 [host, distribution, biological control: 245-255]; Danzig1964 [taxonomy, host, distribution: 652]; Danzig1972 [taxonomy, host, distribution, economic importance: 213]; Danzig1993 [taxonomy, description, illustration, host, distribution, economic importance: 172-174]; DanzigKo1990 [host, distribution: 48]; DanzigPe1998 [catalogue: 274-275]; DavidsDiFl1991 [chemical control: 1-47]; DavidsMi1990 [host, distribution, economic importance: 603-632]; DeBach1964 [biological control]; DeBach1969 [biological control: 801-815]; DeBach1974 [biological control]; DeBachRo1991 [biological control]; DEDAC1923 [host, distribution];

Dekle1965c [taxonomy, description, host, distribution: 73]; Dekle1976 [taxonomy, description, host, distribution, economic importance: 94]; DeSant1941a [host, distribution, biological control: 21-24]; DeSant1979 [biological control]; DicksoFl1955 [host, distribution: 614-615]; DietzMo1916a [taxonomy, description, illustration, host, distribution: 289,291-293]; DoaneHa1909 [host, distribution: 298]; Dougla1912 [taxonomy: 213]; DreistClFl1994; Dziedz1989 [taxonomy, description, illustration, host, distribution: 106-107]; DziedzKa1990 [host, distribution: 39-43]; Early1984 [host, distribution, biological control: 271-308]; Ebelin1949 [host, distribution, life history, control]; Ebelin1975 [host, distribution, economic importance]; EbelinPe1953 [host, distribution, economic importance: 1-35]; Efimof1937 [host, distribution]; Ehrhor1925a [host, distribution: 20-21]; EhrhorFuSw1913 [distribution: 295-300]; Evans1942 [host, distribution, taxonomy, chemical control: 156-159]; Evans1943 [host, distribution, taxonomy, chemical control]; EvansPr1990 [biological control: 3-17]; Ezzat1958 [distribution: 241]; FDACSB1983 [host, distribution: 6-8]; Fergus1974 [taxonomy, description, host, distribution, economic importance, life history, biological control: 20-26]; FergusFl1991 [host, distribution, biological control: 260-261]; FergusSt1978 [host, distribution, economic importance, chemical control: 135-139]; FergusSt1978a [host, distribution, economic importance, chemical control: 140-146]; FergusWa1979 [economic importance, chemical control, host, distribution: 220-224]; Fernal1903b [catalogue: 258,267,276-280]; Ferris1920b [host, distribution: 54]; Ferris1921 [host, distribution: 126,128]; Ferris1937c [taxonomy, illustration: 50,51,77]; Ferris1938a [taxonomy, description, illustration, host, distribution: 190,244]; Ferris1941e [taxonomy: 40-43,47,49]; Ferris1942 [taxonomy, host, distribution: 445:4-5,10; 446:35]; Ferris1953 [host, distribution: 66]; Fjeldd1996 [host, distribution: 4-24]; Flande1971 [biological control, life history: 857-872]; Foldi2001 [distribution: 303-308]; Foldi2002 [host, distribution: 246]; FrankKr1900 [taxonomy, description, host, distribution: 75]; Frogga1914 [taxonomy, description, host, distribution: 133-134]; Frogga1915 [taxonomy, description, host, distribution: 10]; Fullaw1932 [taxonomy: 97,107]; Fuller1897c [host, distribution: 4]; Fuller1907 [taxonomy, description, host, distribution, economic importance, control: 1031-1055]; Gavalo1931 [host, distribution: 8]; Gavalo1936 [host, distribution: 79]; Gentry1965 [host, distribution, economic importance]; GersonOcHo1990 [biological control: 77-97]; Gill1997 [host, distribution, taxonomy, description, illustration, economic importance: 157,162,164]; GomezM1937 [taxonomy, description, illustration, host, distribution: 110-114]; GomezM1946 [host, distribution: 62]; GomezM1956 [taxonomy, description, illustration, host, distribution, biological control: 50-53]; GomezM1957 [host, distribution: 48]; GomezM1958c [host, distribution: 406]; GomezM1962 [taxonomy, description, illustration, host, distribution: 176-179]; GomezM1965 [host, distribution: 90]; GomezM1967O [host, distribution: 132]; GomezM1968 [host, distribution: 542]; Gonzal1989 [taxonomy, description, host, distribution, economic importance: 96,99]; Gonzal1989a [life history, economic importance, chemical control, host, distribution: 35-43]; GonzalCu1994 [host, distribution, life history, economic importance, chemical control: 5-20]; Gordh1979 [biological control: 893,894,896,907,911]; Gowdey1921 [taxonomy, description, host, distribution: 29]; GreaveDaDo1994 [host, distribution, life history: 7-16]; GreaveToWi1992 [host, distribution, chemical control: 79-83]; Green1890 [taxonomy, description, illustration, host, distribution: 19-22]; Green1896 [taxonomy, description, host, distribution: 4]; Green1896e [taxonomy, description, illustration, host, distribution: 40,60-61]; Green1900c [host, distribution: 2]; Green1907 [host, distribution: 202]; Green1911 [host, distribution: 28]; Green1923b [host, distribution: 89]; Green1929 [host, distribution: 377]; Green1930b [host, distribution: 214]; Green1937 [host, distribution: 333]; GreenMa1907 [taxonomy, distribution: 343]; Hadzib1983 [taxonomy, host, distribution, life history, biological control, economic importance: 230-232]; Hall1929 [host, distribution: 352]; HaseyOlVa1999 [host, distribution, control]; Haywar1939 [host, distribution, control: 1]; Haywar1944 [host, distribution: 1-32]; Hempel1900a [taxonomy, description, host, distribution: 501]; Hempel1904 [taxonomy: 319]; Herric1911 [taxonomy, description, illustration, host, distribution: 12,39,73]; Hewitt1943 [host, distribution: 266-274]; Hill1989a [host, distribution, economic importance, biological control: 177-182]; Hollin1923 [taxonomy, description, host, distribution: 9]; Howard1895e [biological control: 1-44]; InserrCa1987 [host, distribution: 93]; Iperti1961 [economic importance: 14-30]; Ishii1923 [host, distribution, biological control: 69]; JamiesDoCa2002 [host, distribution: 354-360]; JiYa1990 [biological control: 134-136]; Johnst1915 [host, distribution, biological control: 1-33]; Kaussa1955 [host, distribution: 15]; Kawai1977 [host, distribution, economic importance: 160]; Kawai1980 [taxonomy, description, host, distribution: 222-223]; Kawai1987 [host, distribution: 78]; Kiritc1932a [taxonomy: 253]; KnowltSm1936 [host, distribution: 263-267]; Koehle1964 [host, distribution, control]; Komosi1964 [host, distribution, taxonomy, description, illustration: 218-220]; KondoKa1995a [host, distribution: 97-98]; Korone1934 [taxonomy, description, illustration, host, distribution: 14]; Koszta1996 [taxonomy, description, illustration, host, distribution, life history, biological control, economic importance: 513-515]; Kotins1907 [host, distribution: 304-308]; Kotins1909 [host, distribution: 97]; Kuwana1907 [host, distribution: 195]; Kuwana1909a [host, distribution: 160]; Kuwana1917a [taxonomy, distribution: 174-175]; Kuwana1927 [host, distribution: 71]; Kuwana1933 [taxonomy, description, illustration, host, distribution: 4]; Lawson1917 [taxonomy, description, illustration, host, distribution: 238-239]; Leonar1897 [taxonomy: 287]; Leonar1897b [taxonomy, description, illustration, host, distribution: 119,124-126]; Leonar1900 [taxonomy, host, distribution: 339]; Leonar1911 [taxonomy, description, illustration, host, distribution: 277-278]; Leonar1911a [taxonomy, description, illustration, host, distribution: 43-44]; Leonar1920 [taxonomy, description, illustration, host, distribution: 90-93]; Lepage1938 [catalogue: 407]; Lepesm1947 [taxonomy, description, host, distribution, life history: 210-211]; Lindin1909a [taxonomy: 324]; Lindin1909b [host, distribution: 148]; Lindin1909c [taxonomy, host, distribution: 449]; Lindin1910b [host, distribution: 38]; Lindin1912b [taxonomy, description, host, distribution: 70,78,92,198,]; Lindin1924 [taxonomy, host, distribution: 174]; Lindin1935 [taxonomy: 129,147]; Lindin1943b [taxonomy: 148]; Lindin1949 [taxonomy: 210]; Lindin1957 [taxonomy: 546]; Lloren1990 [taxonomy, illustration, life history, host, distribution, biological control, life history: 103-105]; Lobdel1937 [taxonomy: 78]; LoBl1989 [host, distribution: 1-4]; LoganTh2002 [biological control, life history, ecology: 361-367]; LongoMaPe1995 [distribution: 127]; Lounsb1906 [host, distribution: 80-91]; Lounsb1921 [host, distribution: 35-38]; LoveFe1977 [host, distribution, economic importance, chemical control: 95-103]; Lugger1900 [host, distribution: 208-245]; Lupo1953a [taxonomy, description, illustration, host, distribution: 75-81]; MaberHoTo1986 [chemical control: 143-147]; MacGil1921 [taxonomy, description, host, distribution: 435,436,439]; Mamet1954 [host, distribution: 18]; Mamet1954a [taxonomy, description, host, distribution: 263]; Mamet1957 [host, distribution: 369,375]; Mamet1959a [host, distribution: 388]; Marlat1899b [taxonomy: 209,211]; Martin1983 [taxonomy, host, distribution: 65]; Martor1976 [host, distribution: 63]; Maskel1885a [taxonomy, host, distribution: 21]; Maskel1887a [taxonomy, description, host, distribution: 41]; Maskel1891 [taxonomy, description, host, distribution: 3]; Maskel1895b [taxonomy, host, distribution: 39]; MatileNo1984 [host, distribution: 65]; MatileOr2001 [host, distribution: 190]; Matta1979 [host, distribution, biological control: 231-242]; Maxwel1903 [taxonomy, description, host, distribution: 39]; McDani1969 [taxonomy, illustration, host, illustration: 111-113]; McKennRe1999 [chemical control: 365-370]; McKenz1935 [host, distribution, life history, control: 1-48]; McKenz1951 [taxonomy: 82]; McKenz1956 [taxonomy, description, illustration, host, distribution: 73-76]; Merkel1938 [host, distribution: 88-99]; Merril1953 [taxonomy, description, host, distribution: 27]; MerrilCh1923 [taxonomy, description, host, distribution, economic importance: 208-209]; MetcalMe1993 [economic importance, host, distribution, control]; MillerDa1990 [host, distribution, economic importance: 302]; Miyosh1926 [host, distribution: 303-326]; Monte1930 [host, distribution: 3-36]; Morale1988 [taxonomy: 77-82]; Morgan1887 [taxonomy: 68,79-82]; Morgan1888b [taxonomy, description: 118-120]; Morgan1889a [taxonomy: 351]; Morgan1893 [taxonomy: 40-41]; Muraka1970 [host, distribution: 74]; MyartsRu2000 [distribution, biological control: 7-33]; NagarkSa1990 [host, distribution, economic importance, biological control: 543-552]; NagarkSa1990 [host, distribution, economic importance, biological control: 553-542]; Nakaha1982 [host, distribution: 43]; Newell1899 [taxonomy, description, host, distribution: 29-30]; Newste1897a [host, distribution: 94]; Newste1901b [taxonomy, description, illustration, host, distribution: 81,91-94]; Newste1917 [taxonomy, description, illustration, host, distribution: 371]; Newste1917b [host, distribution: 131]; Nur1990b [taxonomy, life history: 196]; Osborn1898 [taxonomy, host, distribution, economic importance, life history: 7-8]; Otero1935 [host, distribution: 1-26]; Pace1939 [host, distribution: 664-665]; Peleka1962 [host, distribution: 62]; Priore1964 [host, distribution: 131-178]; Priore1965 [host, distribution: 101-145]; PruthiMa1945 [host, distribution, life history, control: 1-42]; Quayle1911d [host, distribution, description, economic importance, life history, biological control: 443-512]; Quedna1964b [biological control: 86-116]; RagusaRu1989 [host, distribution: 71-74]; Ramakr1919 [host, distribution: 623]; Ramakr1919a [taxonomy, description, host, distribution: 18-19]; Ramakr1921a [host, distribution: 356]; Ramakr1930 [taxonomy, host, distribution: 23]; RaoCh1950 [taxonomy: 10,27]; Richar1960AM [host, distribution: 693-698]; Robins1917 [taxonomy, description, host, distribution: 32-33]; RongGr1998 [biological control: 43-63]; Rose1990c [distribution, economic importance: 535-542]; RosenDe1979 [host, distribution, biological control: 349-

354,405-409]; Ruhl1913 [host, distribution: 79-80]; Salaza1989 [host, distribution]; SalazaSo1990 [host, distribution, life history, biological control: 135-137]; Sander1904a [taxonomy, description, illustration, host, distribution: 56,67]; Sassce1912 [taxonomy, host, distribution: 92]; Schmut1957a [host, distribution: 137]; Schmut1957b [taxonomy: 148]; Schmut1959 [taxonomy: 45]; SchmutKlLu1957 [host, distribution, economic importance: 477]; SchuhMo1948 [host, distribution, control]; Scott1984a [host, distribution: 11-31]; Seabra1930a [host, distribution: 143-148]; Seabra1941 [distribution: 8]; ShiLi1991 [host, distribution: 165]; SibbetVaFe2000 [host, distribution, control]; Signor1869c [taxonomy, description, illustration, host, distribution: 117]; Silves1902 [taxonomy, description, host, distribution: 122]; Singh1964 [host, distribution, economic importance: 213]; Starne1897 [taxonomy: 24]; Steine1987 [host, distribution, description, economic importance, control: 1-7]; StevenBlTo1991 [host, distribution, life history, chemical control: 84]; StevenMcBl1997 [chemical control: 288-292]; StevenRe1999 [host, distribution, economic importance, biological control, chemical control: 345-354]; StevenToBl1994 [host, distribution, life history, chemical control, biological control: 135-142]; StevenVaGo1997 [host, distribution: 773-777]; Sudoi1995 [host, distribution, chemical control: 119-123]; Sugimo1994 [host, distribution: 115-121]; SwirskWyIz2002 [taxonomy, host, distribution, life history, economic importance, biological control: 104-105]; Takagi1969a [taxonomy, description, illustration, host, distribution: 77-78,101]; Takagi1975 [taxonomy, host, distribution: 13]; Takaha1940 [taxonomy, host, distribution: 27-28]; Takaha1953a [taxonomy, host, distribution: 10-13]; Tang1984 [taxonomy, description, illustration, host, distribution: 45-46]; Tao1999 [taxonomy, host, distribution: 91]; Targio1881 [taxonomy, description, host, distribution: 151]; Targio1888 [taxonomy, description, host, distribution: 420-421]; Terezn1986 [taxonomy, description, illustration, host, distribution: 114-116]; ThiemGe1934a [taxonomy: 132,230,232]; ThomsoToWi1996 [host, distribution, biological control, chemical control: 1-5]; Timber1924 [host, distribution, biological control]; TippinBe1970 [host, distribution: 10]; TippinBe1972 [host, distribution: 287]; Tomkin1992 [host, distribution, chemical control: 151-155]; Tomkin1992a [host, distribution, chemical control: 517-522]; TomkinAlWi1996 [host, distribution, life history, economic importance: 785-789]; TomkinFoTh1994 [host, distribution, chemical control: 337-340]; TomkinGrWi1992 [host, distribution, chemical control: 146-150]; TomkinThWi1992 [host, distribution, life history: 58-63]; TomkinThWi1995 [chemical control, biological control: 139-142]; TomkinWiTh2000 [host, distribution: 211-215]; Townse1897 [host, distribution: 185]; Trimbl1929 [host, distribution, economic importance, description, control: 1-21]; Trjapi1989 [biological control: 296]; Valent1963 [biological control: 6-13]; Valent1967 [biological control: 1100]; VelasqRi1969 [host, distribution: 195-208]; WhitinHoCo1998 [host, distribution, control, economic importance: 211-215]; Willia1985a [taxonomy: 236]; WilliaWa1988 [taxonomy, description, illustration, host, distribution: 135-136]; Wilson1917 [taxonomy, description, host, distribution: 28]; WolffCo1993 [taxonomy, description, illustration, host, distribution: 42-44]; Woolle1990 [biological control: 167-176]; Yasar1995a [taxonomy, description, illustration, host, distribution: 91-93]; Zagain1956 [distribution: 85-90]; Zahrad1953 [taxonomy, description, illustration, host, distribution: 127-130]; Zahrad1959b [host, distribution: 60]; Zahrad1990b [taxonomy, description, illustration, host, distribution: 106-108]; Zimmer1948 [taxonomy, description, illustration, host, distribution, biological control: 361,363].

Hemiberlesia securidacae (Hall)

Aspidiotus (*Hemiberlesea*) *zizyphi securidacae* Hall, 1929: 355. Type data: ZIMBABWE: Sinoia, on small branches of *Securidaca longependuculata*; Mazoe, on small branches of *Cassia tettlens* and *Canthium lanciflorum*. Syntypes, female. Type depository: London: The Natural History Museum, England, UK.
Aspidiotus (*Hemiberlesia*) *securidacae*; Ferris, 1941e: 48. Change of status.
Hemiberlesia ziziphi securidacae; Balachowsky, 1956: 105, 108, 120. Misspelling of species name.
SCALE COVER: Hall (1929) indicated that scale cover of this species resembles that of *Hemiberlesia zizyphi* from *Flacourtia hirtiuscula*, being usually partly hidden by surface tissues of the host plant.
HOST PLANTS: **Leguminosae**: *Cassia tettensis* [Hall1929], *Securidaca longepedunculata* [Hall1929, Balach1956]. **Rubiaceae**: *Canthium lanciflorum* [Hall1929, Balach1956]. **Ulmaceae**: *Chaetacme aristata* [DeLott1967a].
DISTRIBUTION: **Afrotropical**: Kenya [DeLott1967a]; Zimbabwe [Hall1929, Balach1956].
GENERAL: Description and illustration of adult female by Hall (1929) and by Balachowsky (1956).
CITATIONS: Balach1956 [taxonomy, description, illustration, host, distribution: 120,123]; Borchs1966 [catalogue: 311]; DeLott1967a [taxonomy, host, distribution: 115]; Ferris1941e [taxonomy: 48]; Hall1929 [taxonomy, description, illustration, host, distribution: 355-356].

Hemiberlesia silvestrii Gómez-Menor Ortega

Hemiberlesia silvestrii Gómez-Menor Ortega, 1956a: 611. Type data: SPAIN: Malaga, Tolox, on *Staehelina boetica*; collected by G. Gómez-Menor Ortega, viii.1954. Lectotype female, by subsequent designation Blay Goicoechea, 1993: 491. Type depository: Madrid: Museo Nacional de Ciencias Naturales, Spain.
SCALE COVER: Illustration of scale cover by Gómez-Menor Ortega (1956a). Female scale dark brown in colour; highly convex; diameter 1.6 mm; exuviae green, eccentric; rest of scale with concentric striations; ventral vellum robust, white, well developed; insect is hidden in bark cracks (Gómez-Menor Ortega, 1956a).
HOST PLANTS: **Compositae**: *Staehelina boetica* [GomezM1956a].
DISTRIBUTION: **Palaearctic**: Spain [GomezM1956a, BlayGo1993].
GENERAL: Description and illustration of adult female by Gómez-Menor Ortega (1956a) and by Blay Goicoechea (1993).
CITATIONS: BlayGo1993 [taxonomy, description, illustration, host, distribution: 491-495]; Borchs1966 [catalogue: 311]; DanzigPe1998 [catalogue: 275]; GomezM1956a [taxonomy, description, illustration, host, distribution: 611-612]; GomezM1958a [host, distribution: 7].

Hemiberlesia sinensis Ferris

Hemiberlesia sinensis Ferris, 1953: 66. Type data: CHINA: Yunnan Province, near Kunming, at An-lin-wen-chian, on an undetermined shrub of family Apocynaceae; collected by G.F. Ferris, April 29, 1949. Syntypes, female. Type depository: Davis: The Bohart Museum of Entomology, University of California, California, USA.

Hemiberlesia sishanensis; Ferris, 1953: 83. Misspelling of species name.

Abgrallaspis sinensis; Komosinska, 1969: 74. Change of combination.

Hemiberlesia sinensis; Danzig & Pellizzari, 1998: 275. Revived combination.

SCALE COVER: Female scale about 1.0 mm in diameter, basically round but subject to considerable modification according to its position; white or somewhat grey. Scale of male not recognized (Ferris, 1953).

HOST PLANTS: **Apocynaceae** [Ferris1953].

DISTRIBUTION: **Oriental**: China (People's Republic) (Yunnan [Ferris1953]).

BIOLOGY: Occurring on upper side of the leaves (Ferris, 1953).

GENERAL: Description and illustration of adult female by Ferris (1953) and by Komosinska (1969).

KEYS: Chou 1985: 300 (female) [Species of China]; Komosinska 1969: 76-78 (female) [World].

CITATIONS: Borchs1966 [catalogue: 311]; Chou1985 [taxonomy, description, host, distribution: 302]; DanzigPe1998 [catalogue: 275]; Ferris1953 [taxonomy, description, illustration, host, distribution: 66-67,83]; Komosi1969 [taxonomy, description, illustration, host, distribution: 74-76]; Tao1999 [taxonomy, host, distribution: 91].

Hemiberlesia tectonae (Lindinger)

Aspidiotus tectonae Lindinger, 1913: 71. Type data: TANZANIA: Tanga, on *Tectona grandis*. Syntypes, female. Type depository: Hamburg: Zoologisches Institut und Zoologishces Museum, Universität von Hamburg, Germany.

Hemiberlesia tectonae; MacGillivray, 1921: 436. Change of combination.

SCALE COVER: Female scale (observed in preserved material) circular pyriform, 1-1.5 mm long, 1 mm wide; highly convex; brown grey; exuviae yellow subcentral or eccentric (Lindinger, 1913).

HOST PLANTS: **Moraceae**: *Ficus mallatocarpa* [DeLott1967a]. **Rubiaceae**: *Coffea robusta* [DeLott1967a]. **Rutaceae**: *Calodendron capense* [DeLott1967a]. **Verbenaceae**: *Tectona grandis* [Lindin1913, Balach1956].

DISTRIBUTION: **Afrotropical**: Kenya [Balach1956]; Tanzania [Lindin1913, Balach1956].

GENERAL: Description and illustration of adult female by Lindinger (1913) and by Balachowsky (1956).

KEYS: Balachowsky 1956: 107 (female) [Africa].

CITATIONS: Balach1956 [taxonomy, description, illustration, host, distribution: 118,121]; Borchs1966 [catalogue: 311]; DeLott1967a [host, distribution: 115]; Ferris1941e [taxonomy: 48]; Lindin1913 [taxonomy, description, illustration, host, distribution: 71-72]; MacGil1921 [taxonomy, description, host, distribution: 436]; WeidneWa1968 [taxonomy: 173].

Hemiberlesia zizyphi (Hall)

Aspidiotus (*Hemiberlesia*) *zizyphi* Hall, 1929: 353. Type data: ZIMBABWE: Mazoe, on branches of *Ziziphus jujuba*; Sinoia, on branches of *Flacourtia hirtuiscula*. Syntypes, female. Type depository: London:Natural History Museum, England, UK.
Aspidiotus zizyphi; Ferris, 1941e: 49. Change of combination.
Hemiberlesia zizyphi; Balachowsky, 1948b: 298. Change of combination.
Hemiberlesia ziziphi; Balachowsky, 1953k: 115. Misspelling of species name.
Hemiberlesia ziziphi; Balachowsky, 1956: 105, 108, 118. Misspelling of species name.
SCALE COVER: Female scale of irregular shape, usually circular, diameter, 1-1.25 mm; exuviae generally to one side; larval exuviae pale, golden brown round margin; nymphal exuviae dark purplish brown; whole scale covered by thin dirty white secretionary film, which masks colour of exuviae in some specimens but not of nymphal exuviae so that it appears as though there is a whitish boss surrounded by the dark colouration of underlying nymphal exuviae; secretionary area very pale brown or dirty white; ventral scale very thin, attached to host plant. Male scale oval to elongate; exuviae very pale, reddish brown at margin; owing to white secretionary covering, they appear to be surmounted by a white boss; secretionary area dirty white (Hall, 1929).
HOST PLANTS: **Flacourtiaceae**: *Flacourtia hirtiuscula* [Hall1929, Balach1956].
Rhamnaceae: *Ziziphus jujuba* [Hall1929, Balach1956].
DISTRIBUTION: **Afrotropical**: Zimbabwe [Hall1929, Balach1956].
GENERAL: Description and illustration of adult female by Hall (1929) and by Balachowsky (1956).
KEYS: Balachowsky 1953k: 114-115 (female) [World].
CITATIONS: Balach1948b [taxonomy: 298]; Balach1953k [taxonomy: 115]; Balach1956 [taxonomy, description, illustration, host, distribution: 118-120,123]; Borchs1966 [catalogue: 312]; Ferris1941e [taxonomy: 49]; Hall1929 [taxonomy, description, illustration, host, distribution: 353-354].

Hemigymnaspis Lindinger

Melanaspis (*Hemigymnaspis*) Lindinger, 1934c: 45. Type species: *Melanaspis* (*Hemigymnaspis*) *eugeniae* Lindinger, by monotypy and original designation.
Hemigymnaspis; Lindinger, 1943b: 221. Change of status.
SYSTEMATICS: *Hemigymnaspis* Lindinger resembles *Furcaspis* Lindinger, from which it differs in having pygidial plates of various shapes, not furcated; paraspircular areas without pores; third lobe normally longer than wide; and interlobular space between lobes 3 and 4 about equal to width of lobe 3 (Davidson & Miller, 1977).
GENERAL: Definition and characters by Davidson & Miller (1977).
KEYS: Colon-Ferrer & Medina-Gaud 1998: 28-32 (female) [Genera of Puerto Rico]; Davidson & Miller 1977: 501-502 (female) [world].
CITATIONS: ColonFMe1998 [taxonomy, description: 63]; DavidsMi1977 [taxonomy, description: 499-502]; Lindin1934c [taxonomy, description: 45]; Lindin1943b [taxonomy: 221].

Hemigymnaspis brayi Davidson & Miller
Hemigymnaspis brayi Davidson & Miller, 1977: 502. Type data: DOMINICA: on trail to Boire Lake, from Fresh Water Lake, on Araceae, probably *Anthurium*; collected by D.F. Bray, February 22, 1964. Holotype female. Type depository: Washington: United States National Entomological Collection, U.S. National Museum of Natural History, District of Columbia, USA.
SCALE COVER: Female cover elongate oval; slightly convex; black; with beige exuviae terminal. Male cover similar in shape and texture to that of female, but smaller; flat; light brown, with yellow-brown exuviae subterminal (Davidson & Miller, 1977).
HOST PLANTS: Araceae: *Anthurium* [DavidsMi1977].
DISTRIBUTION: **Neotropical**: Dominica [DavidsMi1977].
BIOLOGY: Scales located in bark crevices (Davidson & Miller, 1977).
GENERAL: Description and illustration of adult female, adult male and first instar nymph by Davidson & Miller (1977).
KEYS: Davidson & Miller 1977: 501-502 (female) [world].
CITATIONS: DavidsMi1977 [taxonomy, description, illustration, host, distribution: 501-509].

Hemigymnaspis eugeniae (Lindinger)
Melanaspis (*Hemigymnaspis*) *eugeniae* Lindinger, 1934c: 45. Type data: PUERTO RICO: Mount Cienaga, near Adjuntas, on *Eugenia cordata*. Lectotype female, by subsequent designation Davidson, 1972: 318. Type depository: Hamburg: Zoologisches Institut und Zoologishces Museum, Universität von Hamburg, Germany; type no. 588.
Melanaspis eugeniae; Ferris, 1942: 445. Change of combination.
Hemigymnaspis eugeniae; Lindinger, 1943b: 221. Change of combination.
Melanaspis eugeniae; Borchsenius, 1966: 348. Change of combination.
Hemigymnaspis eugeniae; Davidson & Miller, 1977: 509. Revived combination.
SCALE COVER: Illustration of female and male scale cover by Davidson (1972). Female cover beige, circular to oval; exuviae central to subcentral. Male cover beige, elongate, exuviae black and terminal (Davidson, 1972).
HOST PLANTS: Canellaceae: *Canella winterana* [Davids1972]. **Myrtaceae**: *Eugenia boringuensis* [Davids1972], *Eugenia cordata* [Ferris1942, Davids1972].
DISTRIBUTION: **Neotropical**: Puerto Rico & Vieques Island (Puerto Rico [Ferris1942, Davids1972]).
BIOLOGY: Occurring mainly on upper leaf surface (Davidson, 1972).
GENERAL: Description and illustration of adult female by Davidson (1972) and by Colon-Ferrer & Medina-Gaud (1998).
KEYS: Davidson & Miller 1977: 501-502 (female) [world].
CITATIONS: Borchs1966 [catalogue: 348]; ColonFMe1998 [taxonomy, description, illustration, host, distribution: 63-64]; DavidsMi1977 [taxonomy, description: 509]; Ferris1942 [taxonomy, host, distribution: 445]; Lindin1934c [taxonomy, description, host, distribution: 45-46]; Lindin1943b [taxonomy: 221]; WeidneWa1968 [taxonomy: 177].

Hemigymnaspis jessopae Davidson & Miller
Hemigymnaspis jessopae Davidson & Miller, 1977: 509. Type data: USA: Florida, Indian River County, near Orchid, on *Eugenia simpsonii*; collected by D.R. Miller, R.F. Denno & J.A. Davidson, 8.v.1975. Holotype female. Type depository: Washington: United States National Entomological Collection, U.S. National Museum of Natural History, District of Columbia, USA; type no. 2980.
SCALE COVER: Davidson & Miller (1977) did not describe scale cover.
HOST PLANTS: **Myrtaceae**: *Eugenia simpsonii* [DavidsMi1977].
DISTRIBUTION: **Nearctic**: United States of America (Florida [DavidsMi1977]).
BIOLOGY: Occurs on bark just above and below soil surface (Davidson & Miller, 1977).
GENERAL: Description and illustration of adult female by Davidson & Miller (1977).
KEYS: Davidson & Miller 1977: 501-502 (female) [world].
CITATIONS: DavidsMi1977 [taxonomy, description, illustration, host, distribution: 509-511].

Hemigymnaspis orchidicola Davidson & Miller
Hemigymnaspis orchidicola Davidson & Miller, 1977: 512. Type data: VENEZUELA: on orchid leaf; collected 2.viii.1972, Miami No. 4287. Holotype female. Type depository: Washington: United States National Entomological Collection, U.S. National Museum of Natural History, District of Columbia, USA; type no. 4287.
SCALE COVER: Female cover is elongate oval, flattened, light beige, with the brown exuviae located off center (Davidson & Miller, 1977).
HOST PLANTS: **Orchidaceae** [DavidsMi1977].
DISTRIBUTION: **Neotropical**: Venezuela [DavidsMi1977].
BIOLOGY: Most scales were found on the upper surface of the leaf (Davidson & Miller, 1977).
GENERAL: Description and illustration of adult female by Davidson & Miller (1977).
KEYS: Davidson & Miller 1977: 501-502 (female) [world].
CITATIONS: DavidsMi1977 [taxonomy, description, illustration, host, distribution: 512-514].

Hemigymnaspis pimentae Davidson & Miller
Hemigymnaspis pimentae Davidson & Miller, 1977: 514. Type data: DOMINICAN REPUBLIC: on *Pimenta* leaf; collected 19.xi.1975. Holotype female. Type depository: Washington: United States National Entomological Collection, U.S. National Museum of Natural History, District of Columbia, USA; type no. 21574.
SCALE COVER: Female cover oval, flattened, black; black exuviae located subcentrally (Davidson & Miller, 1977).
HOST PLANTS: **Myrtaceae**: *Pimenta* [DavidsMi1977].
DISTRIBUTION: **Neotropical**: Dominican Republic [DavidsMi1977].
BIOLOGY: Most scales were found on the under surface of the leaf (Davidson & Miller, 1977).

GENERAL: Description and illustration of adult female by Davidson & Miller (1977).
KEYS: Davidson & Miller 1977: 501-502 (female) [world].
CITATIONS: DavidsMi1977 [taxonomy, description, illustration, host, distribution: 514-516].

Hypaspidiotus Takahashi

Hypaspidiotus Takahashi, 1956b: 23. Type species: *Aspidiotus jordani* Kuwana, by monotypy and original designation.
SYSTEMATICS: Takahashi (1956b) diagnosed this genus as having well-developed paraphyses, very peculiar in shape, but did not suggest relationships to other aspidiotine genera.
GENERAL: Definition and characters by Takahashi (1956b).
CITATIONS: Borchs1966 [catalogue: 291]; DanzigPe1998 [catalogue: 276]; Kawai1980 [taxonomy: 225]; MorrisMo1966 [taxonomy, catalogue: 94]; Takagi1958 [taxonomy: 125]; Takaha1956b [taxonomy, description: 23].

Hypaspidiotus jordani (Kuwana)

Aspidiotus jordani Kuwana, 1902: 69. Type data: JAPAN: Angio, Saitama-ken, on *Quercus* sp. Holotype female. Type depository: Probably in Ibaraki-ken: Insect Taxonomy Laboratory, National Institute of Agricultural Environmental Sciences, Kannon-dai, Yatabe, Tsukuba-shi, (I. Kuwana), Japan. Notes: Sadao Takagi, in letter (6 May, 2003) indicated that types were lost in earthquake in 1923.
Furcaspis jordani; MacGillivray, 1921: 407. Change of combination.
Hypaspidiotus jordani; Takahashi, 1956: 23. Change of combination.
Hypaspidiotus jordana; Danzig & Pellizzari, 1998: 276. Misspelling of species name.
SCALE COVER: Female scale circular, 1.5-2.5 mm in diameter; flat; general colour dingy brown, conforming usually to colour of leaves on which insect settled; exuviae central, covered with secretion; first skin pale straw colour, 0.4 mm long; second skin orange-yellow, 0.85 mm long. Male scale circular, flat; same colour as that of female; about 1 mm in diameter (Kuwana, 1902).
HOST PLANTS: **Fagaceae**: *Pasania cuspidata* [Kuwana1933], *Quercus* [Kuwana1902, Kuwana1907], *Shiia sieboldii* [Takaha1956b].
DISTRIBUTION: **Palaearctic**: Japan [Kuwana1902, Kuwana1907, Kuwana1917a, Kuwana1933, Takaha1956b, Kawai1980].
GENERAL: Description and illustration of adult female by Kuwana (1902, 1933) and Takahashi (1956b).
KEYS: Kuwana 1933: 3 (female) [Japan]; Kuwana 1933b: 49 (female) [Japan].
CITATIONS: Borchs1966 [catalogue: 291]; DanzigPe1998 [catalogue: 276]; Fernal1903b [catalogue: 265]; Ferris1921b [taxonomy: 94]; Ferris1941e [taxonomy: 44]; Kawai1980 [taxonomy, description, host, distribution: 225]; Kuwana1902 [taxonomy, description, illustration, host, distribution: 69-70]; Kuwana1907 [host, distribution: 196]; Kuwana1917a [taxonomy, distribution: 174]; Kuwana1933 [taxonomy, description, illustration, host, distribution: 11-12]; MacGil1921

[taxonomy, description, host, distribution: 407]; Muraka1970 [host, distribution: 75]; Takagi1958 [taxonomy: 125]; Takaha1956b [taxonomy, description, illustration, host, distribution: 23-24].

Hypaspidiotus phaneraspis Takagi

Hypaspidiotus phaneraspis Takagi, 1958: 124. Type data: JAPAN: Konja and Nase, Amami-Osima, on *Quercus* sp. Holotype female. Type depository: Sapporo: Entomological Institute, Faculty of Agriculture, Hokkaido University, Japan.
SCALE COVER: Scale of female rounded, thin and flat, pale in colour; male scale smaller (Takagi, 1958).
HOST PLANTS: **Fagaceae**: *Quercus* [Takagi1958].
DISTRIBUTION: **Palaearctic**: Japan [Takagi1958].
GENERAL: Description and illustration of adult female by Takagi (1958).
CITATIONS: Borchs1966 [catalogue: 291]; DanzigPe1998 [catalogue: 276]; Muraka1970 [host, distribution: 75]; Takagi1958 [taxonomy, description, illustration, host, distribution: 124-125].

Icaraspidiotus Takagi

Icaraspidiotus Takagi, 2000: 57. Type species: *Icaraspidiotus chaetopterus* Takagi, by monotypy and original designation.
SYSTEMATICS: The first instar nymph of type species distinguished in having very long setae on thorax and abdomen, that are nearly twice as long as antenna (Takagi, 2000).
GENERAL: Definition and characters by Takagi (2000).
CITATIONS: Takagi2000 [taxonomy, description: 57-59].

Icaraspidiotus chaetopterus Takagi

Icaraspidiotus chaetopterus Takagi, 2000: 57. Type data: PHILIPPINES: Palawan Island, Maasin Forest, Brooke's Point, on *Sideroxylon velutinum*; collected 20 August 1993. Holotype female. Type depository: Los Banos: Entomological Museum, Museum of Natural History, University of the Philippines at Los Banos, College, Laguna, Luzon, Philippines; type no. 93PL-92.
SCALE COVER: First instar nymph is distinguished in having very long setae on thorax and abdomen, that are nearly twice as long as antenna (Takagi, 2000). Female scale scallop-shaped rather than circular, flat, smooth, coriaceous, black, with posterior margin brownish; larval exuviae laid on test margin osculating vein; it seems, however, that first exuvia usually lost at an early stage of test formation. Male scale much smaller, slightly elongate, blackish brown, with posterior margin whitish (Takagi, 2000).
HOST PLANTS: **Sapotaceae**: *Sideroxylon velutinum* [Takagi2000].
DISTRIBUTION: **Oriental**: Philippines (Palawan [Takagi2000]).
BIOLOGY: Tests of both sexes occur on upper surface of leaves along lateral veins (Takagi, 2000).
GENERAL: Description and illustration of adult female and first instar nymph by Takagi (2000).

CITATIONS: Takagi2000 [taxonomy, description, illustration, host, distribution: 57-59,83-85].

Leonardianna MacGillivray

Leonardianna MacGillivray, 1921: 393. Type species: *Aspidiotus pimentae* Newstead, by original designation.

SYSTEMATICS: Balachowsky (1953g) and Borchsenius (1966) placed this genus to Odonaspidinae. Here regarded as a member of Aspidiotinae. Among aspidiotine genera the genus is distinguished by presence of conspicuous paraphyses combined with absence of plates and lobes. *Leonardianna* resembles *Odonaspis* Cockerell in absence of plates and lobes, but differs from latter in having multisetose antennal tubercle and absence of intersegmental crenulae.

GENERAL: Definition and characters by Ferris (1941d).

KEYS: Ferris 1942: 25 (female) [North America]; Ferris 1942: 35 (female) [species North America].

CITATIONS: Balach1953g [taxonomy: 727,731]; Balach1958b [taxonomy: 298]; Borchs1966 [catalogue: 228]; Ferris1937c [taxonomy: 51]; Ferris1938b [taxonomy, description: 65,69]; Ferris1941d [taxonomy, description: 345]; Ferris1942 [taxonomy: 25]; Lindin1937 [taxonomy: 188]; MacGil1921 [taxonomy, description: 393,450]; MorrisMo1966 [taxonomy, catalogue: 107].

Leonardianna pimentae (Newstead)

Aspidiotus pimentae Newstead, 1917: 375. Type data: JAMAICA: on *Pimenta officinalis*; collected by A.H. Ritchie. Syntypes, female. Type depository: London: The Natural History Museum, England, UK.

Leonardianna pimentae; MacGillivray, 1921: 450. Change of combination.

SCALE COVER: Female scale subcircular, greatest diameter 1.5 mm; slightly produced posteriorly and uptilted by thick ventral exuvia, which does not extend to margin behind; dorsal exuvia strongly convex or subconical; larval exuvia subcentral (in young forms it is central), more or less nipple-like, generally nude and bright castaneous or piceous; secretionary portion covered with grey epidermal layer of bark; beneath this exuvia is thick and varies from dark castaneous to dark piceous; ventral exuvia low convex externally, fitting closely into pit of depression; central area with a thin, circular, whitish scale, the diameter of which is approximately equal to width of very dense dark border surrounding it (Newstead, 1917).

HOST PLANTS: **Myrtaceae**: *Pimenta officinalis* [Newste1917, Ferris1941d].

NATURAL ENEMIES: FUNGI **Ascomycotina**: *Myriangium duriaei* [EvansPr1990], *Nectria flammea* [EvansPr1990].

DISTRIBUTION: **Neotropical**: Jamaica [Newste1917, Ferris1941d].

BIOLOGY: Scale resting in a well-defined circular pit or depression in bark of plant (Newstead, 1917).

ECONOMIC IMPORTANCE: Causing pits on bark of host (Newstead, 1917). Apparently caused loss of about 2000 pimento trees in Jamaica (Newstead, 1917).

GENERAL: Description and illustration of adult female by Newstead (1917) and by Ferris (1941d).

KEYS: Ferris 1942: 35 (female) [North America].

CITATIONS: Borchs1966 [catalogue: 228]; EvansPr1990 [biological control: 7-13]; Ferris1937c [taxonomy: 51]; Ferris1938b [illustration: 69]; Ferris1941e [taxonomy: 47]; Ferris1942 [taxonomy: 446:35]; MacGil1921 [taxonomy, description, host, distribution: 450]; MillerDa1990 [host, distribution, economic importance: 302]; Newste1917 [taxonomy, description, illustration, host, distribution: 375-377]; SchmutKlLu1957 [host, distribution, economic importance: 494].

Lindingaspis MacGillivray

Lindingaspis MacGillivray, 1921: 388. Type species: *Melanaspis samoana* Lindinger, by monotypy and original designation.

Lindingraspis; Chou, 1986: 695,696. Misspelling of genus name.

SYSTEMATICS: *Lindingaspis* MacGillivray differs from *Chrysomphalus* in possessing marginal series of paraphyses anterior to position of fourth pygidial lobe, these lacking on *Chrysomphalus*, and in serrate appearance of pygidial margin up to fourth abdominal segment, on *Chrysomphalus* this margin not serrated. It is separable from *Melanaspis* mainly by: possession of two quite different sizes of dorsal pygidial ducts; large-size ducts are lacking in *Melanaspis*. *Lindingaspis* differs from *Marginaspis* Hall in lacking well-developed plated anterior to third lobe, and in lacking conspicuous, sclerotized spurs anterior to these plates (McKenzie, 1950).

GENERAL: Definition and characters by Ferris (1938a), McKenzie (1950), Balachowsky (1951, 1958b), Williams (1963b), Takagi (1969a) and Williams & Watson (1988).

KEYS: Gill 1997: 24-26 (female) [Genera of California]; Blay Goicoechea 1993: 474 (female) [Spain]; Wolff & Corseuil 1993: 29 (female) [Brazil, Rio Grande do Sul]; Williams & Watson 1988: 20 (female) [Tropical South Pacific]; Chou 1985: 321 (female) [Species of China]; Chou 1985: 319 (female) [genera of China]; Balachowsky 1958b: 231 (female) [Africa]; Ezzat 1958: 237-239 (female) [Egypt]; McKenzie 1956: 23 (female) [U.S.A.: California]; Balachowsky 1951: 601 (female) [Mediterranean]; Zimmerman 1948: 351 (female) [Hawaii]; Ferris 1942: 27 (female) [North America]; Ferris 1942: 35 (female) [species North America].

CITATIONS: Balach1951 [taxonomy, description: 589-590]; Balach1953c [taxonomy: 110]; Balach1958b [taxonomy, description: 163-164,231]; BenDov1990h [taxonomy: 82]; BlayGo1993 [taxonomy, description: 621]; Borchs1966 [catalogue: 342]; Brimbl1955 [taxonomy: 45]; Chou1985 [taxonomy, description: 320-321]; Chou1986 [taxonomy: 695,696]; DanzigPe1998 [catalogue: 298]; Ezzat1958 [taxonomy: 239]; Ferris1937c [taxonomy: 51]; Ferris1938 [taxonomy: 46]; Ferris1938a [taxonomy, description: 245]; Ferris1942 [taxonomy: 446:27]; Gill1997 [taxonomy: 189]; GomezM1954 [taxonomy, description: 128-129]; HosnyEz1957 [taxonomy: 332]; Kawai1980 [taxonomy: 207]; Kozar1990f [distribution: 143]; MacGil1921 [taxonomy, description: 388,422]; Mamet1949 [taxonomy: 61]; McKenz1939 [taxonomy: 53]; McKenz1943 [taxonomy: 150]; McKenz1950 [taxonomy: 98]; McKenz1956 [taxonomy: 23]; MorrisMo1966 [taxonomy, catalogue: 110]; Takagi1969a [taxonomy, description: 87-88]; Tao1999 [taxonomy: 97]; Willia1963b [taxonomy, description: 3-4]; WilliaWa1988 [taxonomy, description: 170]; WolffCo1993 [taxonomy: 29]; Zimmer1948 [taxonomy: 368].

Lindingaspis benaensis Balachowsky

Lindingaspis benaensis Balachowsky, 1953c: 110. Type data: GUINEA: centre of massif du Bena, 1100 meters altitude, on *Diospyros mespiliformis*. Syntypes, female. Type depository: Paris: Muséum national d'Histoire naturelle, France.
SCALE COVER: Female scale slightly convex; circular, 1.9-2.2 mm diameter; exuviae of first nymph central, black, covered by white powdery secretion; exuviae of second nymph, brown, embedded in female secretion; ventral vellum absent (Balachowsky, 1958b).
HOST PLANTS: **Ebenaceae**: *Diospyros mespiliformis* [Balach1953c].
DISTRIBUTION: **Afrotropical**: Guinea [Balach1953c].
GENERAL: Description and illustration of adult female by Balachowsky (1953c, 1958b).
KEYS: Williams 1974: 218 (female) [World]; Williams 1963b: 10 (female) [World]; Balachowsky 1958b: 165 (female) [Africa].
CITATIONS: Balach1953c [taxonomy, description, illustration, host, distribution: 110-112]; Balach1958b [taxonomy, description, illustration, host, distribution: 167-168]; Borchs1966 [catalogue: 342]; Willia1963b [taxonomy: 10].

Lindingaspis buxtoni (Laing)

Chrysomphalus buxtoni Laing, 1927: 40. Type data: WESTERN SAMOA: Malololelei, on bark of undetermined shrub. Syntypes, female. Type depository: London: The Natural History Museum, England, UK.
Melanaspis buxtoni; Lindinger, 1932e: 224. Change of combination.
Lindingaspis buxtoni; McKenzie, 1939: 53. Change of combination.
SCALE COVER: Female scale deep brown to black; subcircular to elliptical, size 4.5 mm, by 3 mm in elliptical specimens, 3 mm diameter in subcircular ones; flattish around marginal area and gradually rising to very low nipple-like deep black eccentric larval exuvium; surface somewhat irregular and deposited in concentric layers (Laing, 1927).
HOST PLANTS: **Cyperaceae** [Cohic1958].
DISTRIBUTION: **Australasian**: New Caledonia [Cohic1958]; Samoa [Willia1963b]; Western Samoa [Laing1927, WilliaWa1988].
GENERAL: Description and illustration of adult female by Laing (1927) and by Williams (1963b).
KEYS: Williams & Watson 1988: 172 (female) [Tropical South Pacific]; Williams 1963b: 13 (female) [World].
CITATIONS: Borchs1966 [catalogue: 342]; Cohic1958 [host, distribution: 12]; Laing1927 [taxonomy, description, illustration, host, distribution: 40-41]; Lindin1932e [taxonomy: 224]; McKenz1939 [taxonomy: 53]; Willia1963b

[taxonomy, description, illustration, host, distribution: 4-6]; WilliaWa1988 [taxonomy, illustration, host, distribution: 171-172].

Lindingaspis colae (Laing)

Chrysomphalus rossi colae Laing, 1929a: 496. Type data: SIERRA LEONE: Yandu, crowded on upper surface of leaves of *Cola* sp. Holotype female. Type depository: London: The Natural History Museum, England, UK.
Lindingaspis colae; McKenzie, 1939: 54. Change of combination and rank.
Lindingaspis coleae; Balachowsky, 1953c: 109. Misspelling of species name.
SCALE COVER: Female scale varying from greyish white to greyish brown, subcircular; exuviae subcentral, shining black, second wholly covered with a thin whitish or pale greyish deposit, first with a broad belt of white secretion around periphery; inner surface dull back; ventral scale persisting around margin; diameter up to 2.5 mm. Male scale whitish, or brownish white with a whitish margin; exuvia subcentral, black, with a thin white secretion in places; diameter 1.5 mm (Laing, 1929a).
HOST PLANTS: **Sterculiaceae**: *Cola* [McKenz1950].
DISTRIBUTION: **Afrotropical**: Sierra Leone [Laing1929a, McKenz1950].
BIOLOGY: Occurring on upper surface of leaves (McKenzie, 1950).
GENERAL: Description of adult female by McKenzie (1950) and by Balachowsky (1958b).
KEYS: Williams 1963b: 12 (female) [World]; Balachowsky 1958b: 166 (female) [Africa]; McKenzie 1950: 108 (female) [World].
CITATIONS: Balach1953c [taxonomy: 109]; Balach1958b [taxonomy, description, illustration, host, distribution: 169-170]; Borchs1966 [catalogue: 342]; Hargre1937 [host, distribution, economic importance: 505-520]; Laing1929a [taxonomy, description, illustration, host, distribution: 496-497]; McKenz1939 [taxonomy: 54]; McKenz1950 [taxonomy, description, host, distribution: 100]; Willia1963b [taxonomy: 12].

Lindingaspis crocea De Lotto

Lindingaspis crocea De Lotto, 1957: 228. Type data: KENYA: Nairobi, along the midrib of leaves of *Chaetacme aristata*. Holotype female. Type depository: London: The Natural History Museum, England, UK.
SCALE COVER: Female scale flat, elongate; length 1.8-2.5 mm; colour of secretionary area yellow or orange, margin lighter; exuviae black covered by a filmy layer of whitish wax. Scale of male smaller 1.2 to 1.4 mm long (De Lotto, 1957).
HOST PLANTS: **Ulmaceae**: *Chaetacme aristata* [DeLott1957].
DISTRIBUTION: **Afrotropical**: Kenya [DeLott1957].
GENERAL: Description and illustration of adult female by De Lotto (1957).
KEYS: Williams 1974: 220 (female) [World]; Williams 1963b: 10 (female) [World].
CITATIONS: Borchs1966 [catalogue: 342]; DeLott1957 [taxonomy, description, illustration, host, distribution : 228-230]; Willia1963b [taxonomy: 10].

Lindingaspis equipora Munting
Lindingaspis equipora Munting, 1967a: 261. Type data: SOUTH AFRICA: Cape Province, Worcester, Karroo Botanical gardens, on *Cheiridopsis cuprea*; collected 2.xii.1964. Holotype female. Type depository: Pretoria: South African National Collection of Insects, South Africa; type no. 1721/3.
SCALE COVER: Female scale subcircular, 2 to 2.5 mm in diameter, chocolate-brown in colour; exuviae dark brown, subcentral and sometimes circumscribed by a white ring. Male scale similar in colour to that of female, but oval in shape and about 1.5 mm in length (Munting, 1967a).
HOST PLANTS: **Aizoaceae**: *Cheiridopsis cuprea* [Muntin1967a].
DISTRIBUTION: **Afrotropical**: South Africa [Muntin1967a].
GENERAL: Description and illustration of adult female by Munting (1967a).
KEYS: Williams 1974: 220 (female) [World].
CITATIONS: Muntin1967a [taxonomy, description, illustration, host, distribution: 262-263,266].

Lindingaspis ferrisi McKenzie
Lindingaspis ferrisi McKenzie, 1950: 100. Type data: CHINA: Kwangtung Province, Canton, Lingnan University, on monocotyledonous host in a lath house. Holotype female. Type depository: Davis: The Bohart Museum of Entomology, University of California, California, USA.
Lindingraspis ferrisi; Chou, 1986: 696. Misspelling of genus name.
SCALE COVER: Female scale 2 mm in diameter, chocolate brown; exuviae subcentral, blackish. Male scale elongate, oval, lighter brown than female and with a blackish exuvia situated near one end (McKenzie, 1950).
HOST PLANTS: **Guttiferae**: *Calophyllum inophyllum* [Takagi1969a]. **Rutaceae**: *Citrus* [McKenz1950, Takagi1969a].
DISTRIBUTION: **Oriental**: China (People's Republic) (Guangdong (Kwangtung) [McKenz1950]); India [Takagi1969a]; Pakistan [MahmooMo1986]; Taiwan [Takagi1969a]. **Palaearctic**: China (People's Republic) [Takagi1969a].
BIOLOGY: Occurring on leaves (McKenzie, 1950).
ECONOMIC IMPORTANCE: A pest of tea plants in India (Das & Ganguli, 1961).
GENERAL: Description and illustration of adult female by McKenzie (1950), Takagi (1969a), Tang (1984) and by Chou (1985, 1986).
KEYS: Chou 1985: 321 (female) [Species of China]; Williams 1963b: 12 (female) [World]; McKenzie 1950: 108 (female) [World].
CITATIONS: Borchs1966 [catalogue: 342]; Chou1985 [taxonomy, description, host, distribution: 322-323]; Chou1986 [taxonomy, illustration: 696]; DanzigPe1998 [catalogue: 299]; DasGa1961 [host, distribution, life history, economic importance: 245-256]; MahmooMo1986 [host, distribution, economic importance]; McKenz1950 [taxonomy, description, illustration, host, distribution: 100-101,111]; NagarkSa1990 [host, distribution, economic importance, biological control: 553-542]; Takagi1969a [taxonomy, description, illustration, host, distribution: 88-89,103]; Tang1984 [taxonomy, description, illustration, host, distribution: 44-45]; Tao1999 [taxonomy, host, distribution: 97]; Willia1963b [taxonomy: 12].

Lindingaspis floridana Ferris

Lindingaspis floridana Ferris, 1942: 428. Type data: U.S.A.: Florida, West Palm Beach, on *Mangifera indica*. Holotype female. Type depository: Davis: The Bohart Museum of Entomology, University of California, California, USA.
COMMON NAME: floridana scale [Dekle1965c].
SCALE COVER: Scale of female about 2.5 mm in diameter, circular; flat; pale brown or at times almost white; centrally placed exuviae very dark but covered with a wax film; thin, but very hard and brittle; in one sample a large proportion of scales almost pure white, but this possibly due to a surface film since scale beneath surface brown. Male scale similar in form and texture to that of female but smaller and paler (Ferris, 1942).
HOST PLANTS: **Anacardiaceae**: *Mangifera indica* [Ferris1942, HosnyEz1957, Dekle1965c]. **Moraceae**: *Ficus* [Merril1953]. **Oleaceae**: *Olea* [McKenz1943].
NATURAL ENEMIES: HYMENOPTERA **Aphelinidae**: *Aphytis lingnanensis* Compere [AbdRab2001a].
DISTRIBUTION: **Nearctic**: U.S.A. (Florida [Ferris1942, Dekle1965c]). **Neotropical**: Haiti [Nakaha1982]; Jamaica [Nakaha1982]. **Oriental**: India (Uttar Pradesh [McKenz1943]); Pakistan [Nakaha1982]; Philippines [Nakaha1982]; Thailand [Nakaha1982]. **Palaearctic**: Egypt [HosnyEz1957, Ezzat1958].
BIOLOGY: Occurring on leaves (Ferris, 1942).
GENERAL: Description and illustration of adult female by Ferris (1942) and by McKenzie (1950).
KEYS: Williams 1963b: 13 (female) [World]; Ezzat 1958: 241 (female) [Egypt]; McKenzie 1950: 109 (female) [World]; Ferris 1942: 35 (female) [North America].
CITATIONS: AbdRab2001a [distribution, biological control: 175]; Borchs1966 [catalogue: 342-343]; DanzigPe1998 [catalogue: 299]; Dekle1965c [description, host, distribution: 85]; Dekle1976 [description, host, distribution, economic importance: 105]; Ezzat1958 [distribution: 241]; EzzatNa1987 [distribution: 87]; FDACSB1982 [host, distribution: 5-11]; Ferris1942 [taxonomy, description, illustration, host, distribution: 428,446:35]; Halber2000 [host, distribution: 3-4]; HosnyEz1957 [taxonomy, host, distribution: 332]; KondoKa1995a [host, distribution: 97-98]; Lindin1957 [taxonomy: 550]; McKenz1943 [host, distribution: 152]; McKenz1950 [taxonomy, description, illustration, host, distribution: 101,112]; Merril1953 [taxonomy, description, host, distribution: 61]; Nakaha1982 [host, distribution: 51-52]; Willia1963b [taxonomy: 13].

Lindingaspis fusca McKenzie

Lindingaspis fusca McKenzie, 1943: 151. Type data: INDIA: at Poona, on *Cycas circinalis*. Holotype female. Type depository: Davis: The Bohart Museum of Entomology, University of California, California, USA.
SCALE COVER: Scale of female averaging approximately 2 mm in diameter; reddish-brown (species name derived from colour of scale covering) with a black subcentral exuvia (McKenzie, 1943).
HOST PLANTS: **Capparidaceae**: *Capparis moonii* [Willia1963b]. **Cycadaceae**: *Cycas circinalis* [McKenz1943].
DISTRIBUTION: **Oriental**: India [McKenz1943]; Sri Lanka [Willia1963b].

BIOLOGY: Occurring on leaves of sago palm (McKenzie, 1943).
GENERAL: Description and illustration of adult female by McKenzie (1943, 1950).
KEYS: Williams 1963b: 12 (female) [World]; McKenzie 1950: 108 (female) [World].
CITATIONS: Borchs1966 [catalogue: 343]; Hayat1989 [host, distribution, biological control: 1-3]; Lindin1957 [taxonomy: 550]; McKenz1943 [taxonomy, description, illustration, host, distribution: 151-152,160]; McKenz1950 [taxonomy, description, illustration, host, distribution: 101,113]; Willia1963b [taxonomy, host, distribution: 6].

Lindingaspis greeni (Brain & Kelly)
Aspidiotus rossi; Green, 1896e: 45. Misidentification; discovered by Brain & Kelly, 1917: 184.
Chrysomphalus rossi greeni Brain & Kelly, 1917: 184. Type data: SOUTH AFRICA: East London and Durban, on a native tree. Syntypes, female. Type depository: Pretoria: South African National Collection of Insects, South Africa.
Chrysomphalus greeni; McKenzie, 1939: 54. Change of status.
Lindingaspis ? greeni; McKenzie, 1939: 54. Change of combination.
Lindingaspis greeni; McKenzie, 1950: 101. Change of combination.
SCALE COVER: Brain (1919) cited description of scale cover by Green (1896e). However the latter was based on material from Sri Lanka, whereas *L. greeni* was described from South Africa. Balachowsky (1958b) described scale cover "circular, 2 mm in diameter; slightly convex; brown, with black central exuviae. Male scale elongated, colour similar to that of female".
HOST PLANTS: **Capparidaceae**: *Capparis mooni* [Green1896e, McKenz1950]. **Moraceae**: *Ficus nyanzensis* [Balach1958b]. **Ulmaceae**: *Chaetacme* [Balach1958b], *Chaetacme aristata* [Brain1919].
NATURAL ENEMIES: HYMENOPTERA **Aphelinidae**: *Aphytis merceti* Compere [AnneckIn1971]. **Encyrtidae**: *Habrolepis setigera* Annecke & Mynhardt [AnneckIn1971, Prinsl1983].
DISTRIBUTION: **Afrotropical**: South Africa [Balach1958b, Willia1963b]; Uganda [Balach1958b, Willia1963b]. **Oriental**: Sri Lanka [Green1896e].
GENERAL: Description and illustration of adult female by McKenzie (1950) and by Balachowsky (1958b).
KEYS: Williams 1963b: 13 (female) [World]; Balachowsky 1958b: 166 (female) [Africa]; McKenzie 1950: 109 (female) [World].
CITATIONS: AnneckIn1971 [host, distribution, biological control: 14]; Balach1953c [taxonomy: 109]; Balach1958b [taxonomy, description, illustration, host, distribution: 170-172]; Borchs1966 [catalogue: 343]; Brain1919 [taxonomy, description, illustration, host, distribution: 202-203]; BrainKe1917 [taxonomy, host, distribution: 184]; Green1896e [taxonomy, description, illustration, host, distribution: 40,45-46]; KondoKa1995a [host, distribution: 97-98]; Laing1929a [taxonomy, distribution: 493-494]; MacGil1921 [taxonomy, description, host, distribution: 421]; McKenz1939 [taxonomy: 54]; McKenz1950 [taxonomy, description, illustration, host, distribution: 101-102,114]; Muntin1970a [taxonomy:

38]; Prinsl1983 [distribution, biological control: 27]; Willia1963b [taxonomy, distribution: 12-13].

Lindingaspis kenyae Williams

Lindingaspis kenyae Williams, 1963b: 6. Type data: KENYA: Nairobi, on leaves of *Rangaeris brachyceras*; collected 1961. Holotype female. Type depository: London: The Natural History Museum, England, UK.

SCALE COVER: Scale of adult female purple-brown, about 2.0 mm in diameter; exuviae almost black, sub-central. Male scale more elongate but smaller, light purple-brown (Williams, 1963b).

HOST PLANTS: **Orchidaceae**: *Calanthe volkensii* [Willia1963b], *Rangaeris brachyceras* [Willia1963b].

DISTRIBUTION: **Afrotropical**: Kenya [Willia1963b].

GENERAL: Description and illustration of adult female by Williams (1963b).

KEYS: Williams 1963b: 12 (female) [World].

CITATIONS: Borchs1966 [catalogue: 343]; Willia1963b [taxonomy, description, illustration, host, distribution: 6-8].

Lindingaspis mackenziei Williams

Lindingaspis mackenziei Williams, 1963b: 8. Type data: SRI LANKA: Colombo, on leaves of *Cocos nucifera*. Holotype female. Type depository: London: The Natural History Museum, England, UK.

SCALE COVER: Scale of adult female dark chocolate-brown, with sub-central exuviae even darker; about 2.5 mm in diameter. Male scale similar to that of female but smaller and more elongate (Williams, 1963b).

HOST PLANTS: **Anacardiaceae**: *Nothopegia* [Willia1963b]. **Guttiferae**: *Garcinia spicata* [Willia1963b]. **Palmae**: *Cocos nucifera* [Willia1963b].

DISTRIBUTION: **Oriental**: Sri Lanka [Willia1963b].

GENERAL: Description and illustration of adult female by Williams (1963b).

KEYS: Williams 1963b: 12 (female) [World].

CITATIONS: Borchs1966 [catalogue: 343]; Willia1963b [taxonomy, description, illustration, host, distribution: 8-10,12].

Lindingaspis musae (Laing)

Chrysomphalus rossi musae Laing, 1929a: 495. Type data: TANZANIA: Bukoba, Kamachumu, on banana plant. Holotype female. Type depository: London: The Natural History Museum, England, UK.

Chrysomphalus musae; McKenzie, 1939: 54. Change of status.

Lindingaspis rossi musae; McKenzie, 1943: 151. Change of status.

Lindingaspis musae; Borchsenius, 1966: 343. Change of status.

SCALE COVER: Female scale subcircular, very low convex, grey-brown in colour, finely serrated concentrically; exuviae subcentral, shining black; diameter 2.5 mm (Laing, 1929a).

HOST PLANTS: **Musaceae**: *Musa* [McKenz1950, MatileNo1984]. **Palmae**: *Elaeis guineensis* [Balach1958b]. **Rubiaceae**: *Gardenia* [Balach1958b, MatileNo1984].

DISTRIBUTION: **Afrotropical**: Cameroon [MatileNo1984]; Guinea [Balach1958b]; Sierra Leone [Balach1958b]; Tanzania [McKenz1950]; Zaire [Balach1958b].

ECONOMIC IMPORTANCE: Pest of banana in Tanzania (Chua & Wood, 1990).

GENERAL: Description and illustration of adult female by Laing (1929a) and by Balachowsky (1958b). Description of adult female by McKenzie (1950).

KEYS: Williams 1963b: 12 (female) [World]; Balachowsky 1958b: 166 (female) [Africa]; McKenzie 1950: 108 [World].

CITATIONS: Balach1953c [taxonomy: 109]; Balach1958b [taxonomy, description, illustration, host, distribution: 172-174]; Borchs1966 [catalogue: 343]; ChuaWo1990 [host, distribution, economic importance: 548]; Harris1937 [host, distribution: 483-488]; Laing1929a [taxonomy, description, illustration, host, distribution: 495-496]; MatileNo1984 [host, distribution: 66]; McKenz1939 [taxonomy: 54]; McKenz1943 [taxonomy: 151]; McKenz1950 [taxonomy, description, host, distribution: 102]; Willia1963b [taxonomy: 12].

Lindingaspis neorossi McKenzie

Lindingaspis neorossi McKenzie, 1950: 102. Type data: AUSTRALIA: New South Wales, Sydney, on undetermined plant. Holotype female. Type depository: Davis: The Bohart Museum of Entomology, University of California, California, USA.

SCALE COVER: Female scale about 2 mm in diameter; blackish; exuvia centrally situated, chocolate-brown. Male scale smaller and more elongate, generally more reddish-brown and less black than the female (McKenzie, 1950).

HOST PLANTS: **Xanthorrhoeaceae**: *Xanthorrhoea* [Brimbl1955].

DISTRIBUTION: **Australasian**: Australia (New South Wales [McKenz1950]).

BIOLOGY: Occurring on leaves of host (McKenzie, 1950).

GENERAL: Description and illustration of adult female by McKenzie (1950).

KEYS: Williams 1963b: 13 (female) [World]; McKenzie 1950: 109 (female) [World].

CITATIONS: Borchs1966 [catalogue: 343]; Brimbl1955 [taxonomy, host, distribution: 45]; McKenz1950 [taxonomy, description, illustration, host, distribution: 102-103,115]; Willia1963b [taxonomy: 13].

Lindingaspis opima (Silvestri)

Chrysomphalus opimus Silvestri, 1915: 260. Type data: ERITREA: Nefasit, on upper side of leaves of undetermined plant. Syntypes, female. Type depository: Portici: Dipartimento de Entomologia e Zoologia Agraria di Portici, Università di Napoli Federico II, Italy.

Chrysomphalus rossi ferrandii Malenotti, 1916a: 329. Type data: SOMALIA: Lugh, on leaves of *Garcinia somalensis*; collected November 1913. Syntypes, female. Synonymy by Borchsenius, 1966: 343.

Melanaspis opima; Lindinger, 1928: 106. Change of combination requiring emendation of species name for agreement in gender.

Chrysomphalus ferrandii; McKenzie, 1939: 54. Change of status.

Lindingaspis ferrandii; McKenzie, 1939: 54. Change of combination.

Lindingaspis opimus; McKenzie, 1946: 30. Change of combination.

Lindingaspis optimus; Balachowsky, 1951: 593. Misspelling of species name.
SCALE COVER: Female scale about 1.75 mm in diameter; predominantly orange (older scales appear more light brown than orange), with blackish exuvia usually covered with a light brown film. Male scale smaller and more elongate but similar in colour to female (McKenzie, 1950).
HOST PLANTS: **Anacardiaceae**: *Rhus glaucescens* [DeLott1967a]. **Apocynaceae**: *Acokanthera schimperi* [DeLott1967a]. **Flacourtiaceae**: *Hydnocarpus* [McKenz1950]. **Gramineae**: *Pennisetum* [Balach1958b]. **Guttiferae**: *Garcinia somalensis* [Maleno1916a, McKenz1939]. **Leguminosae**: *Bauhinia* [Almeid1973b]. **Palmae**: *Cocos nucifera* [Almeid1973b], *Elaeis guineensis* [Almeid1973b]. **Podocarpaceae**: *Podocarpus milanjianus* [Almeid1973b].
DISTRIBUTION: **Afrotropical**: Angola [Almeid1973b]; Cameroon [Balach1958b]; Eritrea [Silves1915, McKenz1950]; Kenya [DeLott1967a]; Somalia [Maleno1916a, McKenz1950]; Uganda [Balach1958b].
BIOLOGY: Occurring on leaves of host (McKenzie, 1950).
GENERAL: Description and illustration of adult female by Silvestri (1915), Laing (1929a), McKenzie (1950) and by Balachowsky (1958b).
KEYS: Williams 1963b: 10 (female) [World]; Balachowsky 1958b: 165 (female) [Africa]; McKenzie 1950: 108 (female) [World].
CITATIONS: Almeid1973b [host, distribution: 10]; Balach1951 [taxonomy: 593]; Balach1953c [taxonomy: 109]; Balach1958b [taxonomy, description, illustration, host, distribution: 174-176]; Borchs1966 [catalogue: 343]; Laing1929a [taxonomy, description, illustration, host, distribution: 494]; Lindin1928 [taxonomy: 106]; Maleno1916a [taxonomy, description, host, distribution: 329-330]; McKenz1939 [taxonomy: 54]; McKenz1946 [taxonomy: 30]; McKenz1950 [taxonomy, description, illustration, host, distribution: 103,116]; Silves1915 [taxonomy, description, illustration, host, distribution: 260-262]; Viggia1978 [taxonomy: 351]; Willia1963b [taxonomy: 10].

Lindingaspis penniseti Hall
Lindingaspis penniseti Hall, 1946: 55. Type data: UGANDA: Kampala, on *Pennisetum* sp. Syntypes, female. Type depository: London: The Natural History Museum, England, UK.
SCALE COVER: Female scale large, 3-4 mm in diameter, somewhat subcircular; dull grey-brown with exuvia dark brown to almost black; entire surface coated with a grey secretionary material; undersurface a dark bluish-brown. Male scale smaller and more elongate but similar in colour to female (McKenzie, 1950).
HOST PLANTS: **Gramineae**: *Pennisetum* [Hall1946].
DISTRIBUTION: **Afrotropical**: Uganda [Hall1946].
BIOLOGY: Occurring on stems of host (McKenzie, 1950).
GENERAL: Description and illustration of adult female by Hall (1946) and by McKenzie (1950).
KEYS: Williams 1963b: 12 (female) [World]; Balachowsky 1958b: 166 (female) [Africa]; McKenzie 1950: 108 (female) [World].
CITATIONS: Balach1953c [taxonomy: 110]; Balach1958b [taxonomy, description, illustration, host, distribution: 166,176-178]; Borchs1966 [catalogue: 343]; Hall1946

[taxonomy, description, illustration, host, distribution: 55-57]; Lindin1957 [taxonomy: 550]; McKenz1950 [taxonomy, description, illustration, host, distribution: 103-104,117]; Willia1963b [taxonomy: 12].

Lindingaspis picea (Malenotti)

Chrysomphalus piceus Malenotti, 1916a: 331. Type data: SOMALIA: Aden Caboba, on *Cassine schweinfurthiana*; collected October 1913. Syntypes, female.
Lindingaspis piceus; McKenzie, 1939: 54. Change of combination.
Lindingaspis magnifica McKenzie, 1943: 150. Type data: UGANDA: on tea plant. Holotype female. Type depository: Davis: The Bohart Museum of Entomology, University of California, California, USA. Synonymy by Borchsenius, 1966: 344.
Melanaspis picea; Lindinger, 1943a: 147. Change of combination requiring emendation of species name for agreement in gender.
Lindingaspis piceus; McKenzie, 1950: 104. Revived combination.
SCALE COVER: Female scale small, about 0.9 mm long, 0.65 mm wide; oval. Male scale similar in size to female, but slightly rounded at apex; exuviae black; ventral scale white (Malenotti, 1916a). Female scale circular averaging 1 mm in diameter; black; exuviae placed centrally. Male scale similar in colour to female, exuvia near one end (McKenzie, 1943). Illustration of scale cover by Malenotti (1916a).
HOST PLANTS: **Celastraceae**: *Cassine schweinfurthiana* [Maleno1916a, McKenz1950, Balach1958b, MatileNo1984]. **Flacourtiaceae**: *Hydnocarpus* [Balach1958b]. **Magnoliaceae**: *Liriodendron tulipifera* [Sander1906]. **Rutaceae**: *Citrus paradisi* [McKenz1950]. **Theaceae**: *Thea* [McKenz1943].
DISTRIBUTION: **Afrotropical**: Cameroon [MatileNo1984]; Kenya [McKenz1950, Balach1958b]; Somalia [Maleno1916a, McKenz1950, Balach1958b]; Uganda [McKenz1943, Balach1958b]. **Nearctic**: United States of America (Ohio [Sander1906]).
BIOLOGY: Occurring on leaves (McKenzie, 1943).
ECONOMIC IMPORTANCE: Pest of tea plants in Kenya and Uganda (Le Pelley, 1959).
GENERAL: Description and illustration of adult female by Malenotti (1916a), McKenzie (1943, 1950) and by Balachowsky (1958b).
KEYS: Williams 1963b: 12 (female) [World]; Balachowsky 1958b: 165 (female) [Africa]; McKenzie 1950: 108 (female) [World].
CITATIONS: Balach1951 [taxonomy: 593]; Balach1953c [taxonomy: 109]; Balach1958b [taxonomy, description, illustration, host, distribution: 178-180]; Borchs1966 [catalogue: 343-344]; LePell1959 [host, distribution, economic importance: 33-48]; Lindin1943a [taxonomy: 147]; Lindin1957 [taxonomy: 550]; Maleno1916a [taxonomy, description, illustration, host, distribution: 331-333]; MatileNo1984 [host, distribution: 66]; McKenz1939 [taxonomy: 54]; McKenz1943 [taxonomy, description, illustration, host, distribution: 150-151,159]; McKenz1950 [taxonomy, description, illustration, host, distribution: 104,118]; NagarkSa1990 [host, distribution, economic importance, biological control: 553-542]; Sander1906 [taxonomy, host, distribution: 14]; Willia1963b [taxonomy: 12].

Lindingaspis rossi (Maskell)

Aspidiotus rossi Maskell, 1892: 11. Type data: AUSTRALIA: Melbourne, Sydney and Adelaide, on *Nerium oleander*. Lectotype female, by subsequent designation Deitz & Tocker, 1980: 42. Type depository: Auckland: New Zealand Arthropod Collection, Landcare Research, New Zealand. Notes: Maskell (1892: 11) incorrectly credited the authorship to "Crawford".

Chrysomphalus rossi; Leonardi, 1897: 286. Change of combination.

Aspidiotus (*Chrysomphalus*) *rossi*; Cockerell, 1897i: 27. Change of combination.

Melanaspis rossi; Lindinger, 1910b: 41. Change of combination.

Aspidiotus rosei; Kasargode, 1914: 134. Misspelling of species name.

Aonidiella subrossi Laing, 1929: 25. Type data: AUSTRALIA: New South Wales, on *Acacia rubra*; collected by W.W. Froggatt. Syntypes, female. Type depository: London: Natural History Museum, UK. Synonymy by Williams, 1963b: 10.

Chrysomphalus niger Laing, 1929a: 493. Type data: AUSTRALIA: host plant not indicated. Syntypes, female. Synonymy by Borchsenius, 1966: 344. Notes: Type material not available in BMNH (Jon Martin, The Natural History Museum, January 27, 2003, private communication to Yair Ben-Dov); depository of type material unknown.

Lindingaspis rossi; Ferris, 1938a: 246. Change of combination.

Melanaspis subrossi; Lindinger, 1943a: 147. Change of combination.

Lindingraspis rossi; Chou, 1986: 695. Misspelling of genus name.

COMMON NAMES: Araucaria black scale [VelasqRi1969]; Ross scale [SchmutKlLu1957]; Ross's black scale [DeitzTo1980]; round black scale [SchmutKlLu1957].

SCALE COVER: Scale of female flat, circular, very dark brown or black; exuviae almost central and almost pitchy black; scale of male slightly elongate oval, dark brown; exuvia near one end; with thick ventral scale formed by female and this includes ventral portion of second exuvia (Ferris, 1938a). Illustration of female scale by Froggatt (1914). Colour photograph by Gill (1997) and by Wong *et al.* (1999).

HOST PLANTS: **Anacardiaceae**: *Mangifera indica* [Takaha1929, Brimbl1955, Takagi1970], *Schinus terebinthifolius* [WilliaWa1988]. **Apocynaceae**: *Alstonia constricta* [Brimbl1955], *Carissa* [Ramakr1921a], *Carissa carandus* [Green1919c], *Nerium* [Ferris1938a], *Nerium oleander* [Laing1929, Brimbl1955, McKenz1956]. **Araliaceae**: *Hedera helix* [Brimbl1955, McKenz1956]. **Araucariaceae**: *Araucaria* [McKenz1956, Almeid1971], *Ar. bidwilli* [Ferris1920b, Zimmer1948, Brimbl1955], *Ar. brasiliensis* [GomezM1954, Martin1983, WilliaWa1988], *Ar. cooki* [Mamet1943a, Mamet1949], *Ar. cunninghami* [Mamet1943a, Brimbl1955], *Ar. excelsa* [Brimbl1955], *Ar. hunsteini* [Brimbl1955]. **Avicenniaceae**: *Avicennia officinalis* [Cohic1958]. **Barringtoniaceae**: *Barringtonia* [Ramakr1919a], *Barringtonia acutangula* [Ramakr1919a]. **Capparidaceae**: *Capparis* [Frogga1914, Ramakr1921a, Green1937], *Capparis mitchellii* [Brimbl1955], *Capparis moonii* [Green1896e]. **Celastraceae**: *Celastrus bilocularis* [Brimbl1955], *Euonymus* [Laing1929, Brimbl1955], *Euonymus japonicus* [Green1929]. **Compositae**: *Artemisia* [McKenz1956]. **Crassulaceae**: *Bryophyllum calycinum* [Takaha1929, Takagi1970]. **Cunoniaceae**: *Callicoma serratifolia* [Brimbl1955]. **Cycadaceae**: *Cycas* [Ramakr1919a, Ramakr1921a]. **Ebenaceae**: *Diospyros eriantha*

[Takaha1935, Takagi1970], *Royena glabra* [Balach1958b]. **Elaeocarpaceae**: *Elaeodendron curtipendulum* [WilliaWa1988]. **Epacridaceae**: *Styphelia* [Laing1929]. **Euphorbiaceae**: *Cleistanthus cunninghamii* [Brimbl1955], *Croton acronychioides* [Brimbl1955], *Mallotus philippinensis* [Brimbl1955], *Ricinocarpus* [Ferris1938a, McKenz1956]. **Fagaceae**: *Pasania cuspidata* [Kuwana1909]. **Flacourtiaceae**: *Scolopia oldhamii* [Takagi1970]. **Guttiferae**: *Calophyllum inophyllum* [Takaha1929, Takagi1970], *Garcinia* [Green1937]. **Labiatae**: *Hyssopus* [McKenz1956]. **Leguminosae**: *Acacia aulacocarpa* [Brimbl1955], *Ac.. cunninghamii* [Brimbl1955], *Ac. fimbriata* [Brimbl1955], *Ac. penninervis* [Brimbl1955], *Ac. rubra* [Laing1929, Willia1963b], *Ac. saligna* [Balach1958b], *Hovea longifolia* [Brimbl1955]. **Loranthaceae**: *Loranthus* [Balach1958b], *Loranthus alyxifolius* [Brimbl1955]. **Malvaceae**: *Abutilon* [McKenz1956]. **Meliaceae**: *Owenia venosa* [Brimbl1955]. **Moraceae**: *Ficus* [McKenz1956], *Ficus pumila* [Takaha1932a, Takagi1970]. **Myrtaceae**: *Callistemon salignus* [Brimbl1955], *Callistemon viminalis* [Brimbl1955], *Eucalyptus* [Newste1917b, Laing1929, McKenz1956], *Eu. resinifera* [Brimbl1955], *Eu. rostrata* [Laing1929], *Eu. siderophloia* [Brimbl1955], *Eugenia coolminiana* [Brimbl1955], *Leptospermum citratum* [Brimbl1955], *Melaleuca leucadendra* [Brimbl1955], *Tristania conferta* [Brimbl1955], *Tristania suaveolens* [Brimbl1955]. **Oleaceae**: *Olea apetala* [WilliaWa1988], *Olea europaea* [Brimbl1955, McKenz1956], *Olea verrucosa* [WilliaWa1988]. **Orchidaceae**: *Dendrobium kingianum* [Brimbl1955]. **Palmae** [Hall1929, Brimbl1955], *Cocos nucifera* [Laing1927, Takaha1932a, Takaha1933, Ferris1938a, Takagi1970], *Kentia* [RosenDe1979], *Phoenix* [RosenDe1979], *Phoenix canariensis* [RosenDe1979]. **Pandanaceae**: *Pandanus pedunculatus* [Brimbl1955]. **Pinaceae**: *Pinus* [Hall1929]. **Pittosporaceae**: *Pittosporum undulatum* [Laing1929]. **Podocarpaceae**: *Podocarpus elatus* [Brimbl1955]. **Polygonaceae**: *Coccolobis* [McKenz1956]. **Proteaceae**: *Banksia* [Frogga1914], *Banksia integrifolia* [Brimbl1955], *Banksia serrata* [Frogga1914], *Buckinghamia cleissima* [Brimbl1955], *Grevillea robusta* [Hall1929, Balach1958b], *Macadamia ternifolia* [Brimbl1955], *Stenocarpus sinuatus* [Brimbl1955]. **Rhizophoraceae**: *Kandelia candel* [Takagi1970], *Kandelia rheedi* [Takaha1929]. **Rosaceae**: *Malus sylvestris* [Timlin1964a], *Pyracantha koidzumii* [Takaha1935, Takagi1970]. **Rubiaceae**: *Gardenia* [McKenz1956]. **Rutaceae**: *Citrus* [Green1930b, Lepage1938, McKenz1956], *Eremocitrus glauca* [Brimbl1955], *Phebalium* [RosenDe1979], *Pleiococa wilcoxiana* [Brimbl1955]. **Sapindaceae**: *Alectryon connatus* [Brimbl1955], *Cupaniopsis anacardioides* [Brimbl1955], *Diploglottis australis* [Brimbl1955]. **Scrophulariaceae**: *Veronica* [Laing1929]. **Taxodiaceae**: *Sequoia* [Ferris1938a, McKenz1956]. **Theaceae**: *Camellia* [McKenz1956], *Camellia japonica* [Brimbl1955]. **Vitaceae**: *Vitis* [Takagi1970]. **Xanthorrhoeaceae**: *Lomandra longifolia* [Brimbl1955], *Xanthorrhoea* [Maskel1893b, Brimbl1955, McKenz1956].

NATURAL ENEMIES: HYMENOPTERA **Aphelinidae**: *Aphytis africanus* Quednau [RosenDe1979], *Ap. chrysomphali* (Mercet) [Zimmer1948], *Ap. ignotus* Compere [RosenDe1979], *Ap. merceti* Compere [RosenDe1979], *Aspidiotiphagus citrinus* (Craw) [Zimmer1948]. **Encyrtidae**: *Habrolepis obscura* Compere & Annecke [AnneckIn1971, Prinsl1983].

DISTRIBUTION: **Afrotropical**: Angola [Almeid1969, Almeid1973b]; Cameroon [Vayssi1913]; Guinea [Nakaha1982]; Mauritius [Mamet1943a, Mamet1949, Borchs1966]; Mozambique [Almeid1971]; Réunion [Mamet1957]; Sierra Leone [Hargre1937]; South Africa [BrainKe1917, Newste1917b, Brain1919, Balach1958b, RosenDe1979]; Tanzania [Nakaha1982]; Togo [Nakaha1982]; Zimbabwe [Nakaha1982 .[**Australasian**: Australia [Maskel1891, Leonar1914, Mamet1943a, Brimbl1955, Borchs1966] (New South Wales [Laing1929, Willia1963b], Victoria [Laing1929]); Hawaiian Islands (Hawaii [Nakaha1982]); New Caledonia [Cohic1958, WilliaWa1988] [Nakaha1982]; New Zealand [Green1929, Ferris1938a]; Norfolk Island [WilliaWa1988]; Samoa [Ferris1938a]; Western Samoa [Laing1927]. **Nearctic**: Mexico [Nakaha1982]; United States of America (California [Ferris1920b, McKenz1950, McKenz1956]). **Neotropical**: Argentina [Nakaha1982]; Brazil (São Paulo [Green1930b, Lepage1938]); Chile [Nakaha1982]; Peru [Nakaha1982]. **Oriental**: India (Madhya Pradesh [Green1919c], Tamil Nadu [Green1919c]); Philippines [Ferris1938a, VelasqRi1969]; Sri Lanka [Green1896e, Ramakr1921a]; Taiwan [Takaha1929, Takaha1932a, Takaha1935, Takagi1970, WongChCh1999]. **Palaearctic**: China (People's Republic) [Ferris1938a]; Japan [Kuwana1909, Kuwana1917a, Ferris1938a]; Madeira Islands [Balach1938a]; Portugal [Seabra1942]; Sicily [LongoMaPe1995]; Spain [GomezM1954, Martin1983, BlayGo1993].

BIOLOGY: Occurring on leaves (Ferris, 1938a).

ECONOMIC IMPORTANCE: Published records (see Distribution) of this polyphagous species, show a disjunct distribution in Australia, Africa, Mediterranean Basin, North and South America. Considered a pest of *Citrus* in New Zealand (Charles, 1998), USA, California (Gill, 1997), and southern Africa (Schmutterer et al., 1957), of oil palm in Sierra Leone (Schmutterer et al., 1957) and *Araucaria* in California (Gill, 1997)

GENERAL: Description and illustration of adult female by Froggatt (1914), Laing (1929), Ferris (1938a), Balachowsky (1951, 1958b), Brimblecombe (1955), McKenzie (1950, 1956), Chou (1985, 1986) and by Gill (1997).

KEYS: Williams & Watson 1988: 172 (female) [Tropical South Pacific]; Chou 1985: 321 (female) [Species of China]; Williams 1963b: 13 (female) [World]; Balachowsky 1958b: 166 (female) [Africa]; McKenzie 1956: 26 (female) [U.S.A.: California]; McKenzie 1950: 109 (female) [World]; Ferris 1942: 35 (female) [North America]; Kuwana 1933: 26 (female) [Japan]; Kuwana 1933b: 49 (female) [Japan]; Fullaway 1932: 95-97, 107 (female) [Hawaii]; Brain 1919: 198 (female) [South Africa]; Robinson 1917: 24 (female) [Philippines]; Green 1896e: 40 (female) [Sri Lanka].

CITATIONS: Almeid1969 [taxonomy, description, host, distribution, biological control: 31]; Almeid1971 [host, distribution: 13]; Almeid1973b [host, distribution : 10]; AnneckIn1971 [host, distribution, biological control: 14]; Balach1938a [host, distribution: 151]; Balach1951 [taxonomy, description, illustration, host, distribution: 590-593]; Balach1958b [taxonomy, description, illustration, host, distribution: 180-182]; BenDov1990c [taxonomy: 112]; BlayGo1993 [taxonomy, description, illustration, host, distribution: 621-625]; Borchs1966 [catalogue: 344-345]; Brain1919 [taxonomy, description, illustration, host, distribution: 201-202];

BrainKe1917 [distribution: 184]; Brimbl1955 [taxonomy, description, illustration, host, distribution: 45-51]; BrownEa1967 [host, distribution, economic importance, control: 1-72]; BurgerUl1990 [economic importance: 313-327]; Castel1951a [biological control: 95-98]; Charle1998 [distribution, economic importance, biological control: 51N]; Chou1985 [taxonomy, description, host, distribution: 321-322]; Chou1986 [taxonomy, illustration: 695]; ClapsWoGo2001a [taxonomy, host, distribution: 21]; Cocker1896b [distribution: 335]; Cocker1897i [taxonomy, description, host, distribution: 27]; Cocker1899j [host, distribution: 274]; Cohic1958 [host, distribution: 14]; Collar1918 [host, distribution: 154-162]; Conway1951 [host, distribution: 159-164]; Craw1896 [taxonomy, description, host, distribution: 34]; DanzigPe1998 [catalogue: 299-300]; DEDAC1923 [host, distribution]; DeitzTo1980 [catalogue: 40,42]; DeSant1979 [biological control]; DoaneFe1916 [host, distribution: 401]; Ebelin1949 [host, distribution, life history, control]; Fernal1903b [catalogue: 293]; Ferris1920b [taxonomy, description, illustration, host, distribution: 55]; Ferris1938a [taxonomy, description, illustration, host, distribution: 246]; Ferris1941d [taxonomy: 347]; Ferris1941e [taxonomy: 48]; Ferris1942 [taxonomy: 446:35]; Fletch1951 [host, distribution, chemical control: 1-24]; Foldi2001 [distribution: 303-308]; Frogga1914 [taxonomy, description, host, distribution: 317]; Frogga1915 [taxonomy, description, host, distribution: 21-22]; Fullaw1932 [taxonomy: 95,107]; Fuller1897c [host, distribution: 4]; Fuller1907 [taxonomy, description, host, distribution, economic importance, control: 1031-1057]; FurnisCa1977 [host, distribution: 111]; Gill1997 [host, distribution, taxonomy, description, illustration, economic importance: 189,191,193]; GomezM1954 [taxonomy, description, illustration, host, distribution: 129-132]; GomezM1958a [host, distribution: 8]; Green1896e [taxonomy, description, illustration, host, distribution: 45-46]; Green1919c [host, distribution: 439]; Green1929 [host, distribution: 377]; Green1930b [host, distribution: 214]; Green1937 [host, distribution: 331]; Greig1944 [host, distribution, control]; Hall1929 [host, distribution: 356]; Hargre1937 [host, distribution, economic importance: 505-520]; Kasarg1914 [taxonomy, host, distribution: 134]; KondoKa1995a [host, distribution: 97-98]; Kuwana1909 [host, distribution: 154]; Kuwana1917a [taxonomy, distribution: 176]; Kuwana1927 [host, distribution: 72]; Kuwana1933 [taxonomy, description, illustration, host, distribution: 26,32-33]; Laing1927 [host, distribution: 42]; Laing1929 [taxonomy, description, illustration, host, distribution: 25-27]; Laing1929a [taxonomy, description, host, distribution: 493-494]; Leonar1897 [taxonomy: 286]; Leonar1899 [taxonomy: 199-200,202]; Leonar1914 [host, distribution: 207]; Lepage1938 [catalogue: 400]; Lindin1909c [taxonomy, host, distribution: 449]; Lindin1910b [taxonomy, host, distribution: 41]; Lindin1943a [taxonomy: 147]; Lindin1957 [taxonomy: 546]; Lobdel1937 [taxonomy: 78]; LongoMaPe1995 [distribution: 127]; Lounsb1914 [host, distribution: 1]; Lounsb1916 [host, distribution: 83-103]; Lounsb1922 [host, distribution: 205-210]; MacGil1921 [taxonomy, description, host, distribution: 421]; Mallam1954 [distribution: 24-60]; Mamet1943a [catalogue: 163]; Mamet1949 [catalogue: 61]; Mamet1957 [distribution: 369]; Martin1983 [taxonomy, host, distribution: 65]; Maskel1891 [taxonomy, description, distribution: 3]; Maskel1892 [taxonomy, description, host, distribution: 11-12]; Maskel1893b [taxonomy, host, distribution:

207]; Maskel1898 [taxonomy, host, distribution: 224]; Maskew1916 [host, distribution: 308-309]; McKenz1939 [taxonomy: 54]; McKenz1950 [taxonomy, description, illustration, host, distribution: 104-105,119]; McKenz1956 [taxonomy, description, illustration, host, distribution: 75-77]; MillerDa1990 [host, distribution, economic importance: 303]; Nakaha1982 [host, distribution: 52]; Newste1917b [host, distribution: 132]; Prinsl1983 [distribution, biological control: 27]; PruthiMa1945 [host, distribution, life history, control: 1-42]; Ramakr1919a [taxonomy, description, host, distribution: 18]; Ramakr1921a [host, distribution: 355]; Ramakr1930 [taxonomy, host, distribution: 22]; Robins1917 [taxonomy, description, host, distribution: 25-26]; RosenDe1979 [host, distribution, biological control: 341-344,542-545,]; Ruther1915a [taxonomy, description, host, distribution: 107]; SchmutKlLu1957 [host, distribution, economic importance: 492]; Seabra1942 [distribution: 2]; ShiLi1991 [host, distribution: 165]; SwaileAwSh1976 [host, distribution, life history, ecology, biological control: 257-263]; Takagi1970 [taxonomy, host, distribution: 138]; Takaha1929 [host, distribution: 81-82]; Takaha1932a [host, distribution: 103]; Takaha1933 [host, distribution: 25-34]; Takaha1935 [host, distribution: 3-4]; Tao1999 [taxonomy, host, distribution: 97]; Timber1924 [host, distribution, biological control]; Timlin1964a [host, distribution: 531-535]; Valent1963 [biological control: 6-13]; Valent1967 [biological control: 1100]; Vayssi1913 [host, distribution: 431]; VelasqRi1969 [host, distribution: 195-208]; Whitne1933 [host, distribution: 59-70]; Willia1963b [host, distribution: 10,13]; WilliaWa1988 [taxonomy, illustration, host, distribution: 172-173,176]; WongChCh1999 [taxonomy, description, host, distribution: 28,70]; Zahrad1959b [host, distribution: 60]; Zahrad1990 [host, distribution, description: 643]; Zimmer1948 [taxonomy, description, illustration, host, distribution, biological control: 367-368].

Lindingaspis samoana (Lindinger)
Melanaspis samoana Lindinger, 1911: 177. Type data: WESTERN SAMOA: Savaii, on *Myristica hypargyraea*. Lectotype female, by subsequent designation Williams, 1974: 217. Type depository: Hamburg: Zoologisches Institut und Zoologishces Museum, Universität von Hamburg, Germany.
Chrysomphalus samoana; Sasscer, 1912: 93. Change of combination.
Lindingaspis samoana; MacGillivray, 1921: 422. Change of combination.
Melenaspis samoana; Tang, 1984: 43. Misspelling of genus name.
SCALE COVER: Female scale round, about 2.00 mm in diameter, brown, firm, with dark almost blackish-brown central exuviae (Lindinger, 1911).
HOST PLANTS: **Myristicaceae**: *Myristica hypargyraea* [Laing1927, McKenz1950, Willia1974, WilliaWa1988].
DISTRIBUTION: **Australasian**: Western Samoa [Laing1927, McKenz1950, Willia1974, WilliaWa1988].
GENERAL: Description and illustration of adult female by Lindinger (1911) and by Williams (1974).
KEYS: Williams & Watson 1988: 172 (female) [Tropical South Pacific]; Williams 1974: 218 (female) [World].

CITATIONS: Borchs1966 [catalogue: 345]; Ferris1937c [taxonomy: 51]; Laing1927 [host, distribution: 42]; Leonar1899 [taxonomy: 199-200,210]; Lindin1911 [taxonomy, description, illustration, host, distribution: 177]; MacGil1921 [taxonomy, description, host, distribution: 422]; McKenz1950 [taxonomy, host, distribution: 105]; Sassce1912 [taxonomy, host, distribution: 93]; Tang1984 [taxonomy, description, host, distribution: 43]; WeidneWa1968 [taxonomy: 177]; Willia1974 [taxonomy, description, illustration, host, distribution: 217-219]; WilliaWa1988 [taxonomy, illustration, host, distribution: 174-176].

Lindingaspis setiger (Maskell)

Aspidiotus setiger Maskell, 1897: 298. Type data: JAPAN: Honshu, Yokohama, on *Quercus* sp.; sent by Mr. Koebele. Syntypes. Type depository: Auckland: New Zealand Arthropod Collection, Landcare Research, New Zealand.

Chrysomphalus setiger; Leonardi, 1899: 199. Change of combination.

Chrysomphalus (*Melanaspis*) *setiger*; Cockerell, 1899a: 396. Change of combination.

Aspidiotus (*Chrysomphalus*) *kelloggi* Kuwana, 1902: 71. Type data: JAPAN: Kiushiu, Higuma-yama, Chikujo-gun, on undetermined host. Holotype female. Type depository: Ibaraki-ken: Insect Taxonomy Laboratory, National Institute of Agricultural Environmental Sciences, Kannon-dai, Yatabe, Tsukuba-shi, (I. Kuwana), Japan. Synonymy by Lindinger, 1957: 545.

Chrysomphalus kelloggi; Fernald, 1903b: 291. Change of combination.

Lindingaspis setiger; McKenzie, 1950: 105. Change of combination.

SYSTEMATICS: The species epithet *setiger* is both an adjective and a noun. Here it is treated as a noun in apposition, and retained as *setiger*.

SCALE COVER: Maskell (1897) described cover: "Female scale very dark-brown or intense dull-black; circular; convex; diameter about 1/11 inch; larval exuviae central, very small, forming a minute apical shiny-black boss; texture of scale thick and solid; on turning it over the inside is smooth and black, with reddish edge. Male scale light-brown, subelliptical, flattish; length about 1/25 inch; exuviae yellowish". McKenzie (1950) described the cover: "Female scale circular, diameter about 2 mm, convex thick and solid in texture, exuvia subcentral and shiny black or covered by a greyish secretion". Male scale, according to Kuwana, is "light brown, subelliptical flattish, exuvia yellow, length about 1 mm".

HOST PLANTS: **Anacardiaceae**: *Mangifera indica* [KinjoNaHi1996]. **Caprifoliaceae**: *Viburnum* [Kawai1977]. **Celastraceae**: *Euonymus* [Kawai1977]. **Fagaceae**: *Castanopsis cuspidata* [Kawai1977], *Pasania sieboldii* [McKenz1950], *Pasania sieboldii* [Kuwana1933], *Quercus* [Maskel1897], *Quercus myrsinaefolia* [Kuwana1933, McKenz1950]. **Rosaceae**: *Amelanchier asiatica* [TakahaTa1956]. **Ulmaceae**: *Celtis sinensis japonica* [TakahaTa1956], *Zelkova serrata* [Kawai1977].

DISTRIBUTION: **Oriental**: Ryukyu Islands (= Nansei Shoto) [KinjoNaHi1996]. **Palaearctic**: Japan [Kuwana1917a, Kawai1977, Kawai1980] (Honshu [Maskel1897, TakahaTa1956], Kyushu [Kuwana1902], Shikoku [TakahaTa1956]).

BIOLOGY: Presumably on leaves of host (McKenzie, 1950).

GENERAL: Description and illustration of adult female by Kuwana (1902, 1933) and by McKenzie (1950).

KEYS: Williams 1963b: 12 (female) [World]; McKenzie 1950: 108 (female) [World]; Kuwana 1933: 26 (female) [Japan]; Kuwana 1933b: 49 (female) [Japan].
CITATIONS: Borchs1966 [catalogue: 345]; Cocker1899a [taxonomy: 396]; DanzigPe1998 [catalogue: 300]; DeitzTo1980 [taxonomy: 42]; Fernal1903b [catalogue: 291,294]; Ferris1941e [taxonomy: 44,48]; Kawai1977 [host, distribution, economic importance: 158,160,162]; Kawai1980 [taxonomy, description, host, distribution: 207-208]; KinjoNaHi1996 [host, distribution: 125-127]; Kuwana1902 [taxonomy, description, illustration, host, distribution: 71-72]; Kuwana1907 [host, distribution: 196]; Kuwana1917a [taxonomy, distribution: 176]; Kuwana1933 [taxonomy, description, illustration, host, distribution: 26,33-34]; Lidget1898a [taxonomy: 14]; Lindin1943a [taxonomy: 147]; Lindin1957 [taxonomy: 545]; MacGil1921 [taxonomy, description, host, distribution: 414-415,418]; Maskel1897 [taxonomy, description, illustration, host, distribution: 298]; McKenz1939 [taxonomy: 54-55]; McKenz1950 [taxonomy, description, illustration, host, distribution: 105,120]; Muraka1970 [host, distribution: 75]; Takagi1990b [taxonomy, structure: 17]; TakahaTa1956 [host, distribution: 17]; Willia1963b [taxonomy: 12].

Lindingaspis similis McKenzie

Chrysomphalus rossi; Doane & Ferris, 1916: 401. Misidentification; discovered by McKenzie, 1950: 105.
Lindingaspis similis McKenzie, 1950: 105. Type data: SAMOA: Upolu, 1000 feet altitude, on unidentified plant. Holotype female. Type depository: Davis: The Bohart Museum of Entomology, University of California, California, USA.
SCALE COVER: Female scale diameter about 2.25 mm; chocolate brown, with subcentral, blackish, exuvia. Male scale elongate, oval, lighter brown than female and with blackish exuvia situated near one end (McKenzie, 1950).
HOST PLANTS: **Pandanaceae**: *Pandanus* [McKenz1950].
DISTRIBUTION: **Australasian**: Samoa [McKenz1950].
BIOLOGY: Occurring on leaves of host (McKenzie, 1950).
GENERAL: Description and illustration of adult female by McKenzie (1950).
KEYS: Williams & Watson 1988: 172 (female) [Tropical South Pacific]; Williams 1963b: 12 (female) [World]; McKenzie 1950: 109 (female) [World].
CITATIONS: Borchs1966 [catalogue: 345]; DoaneFe1916 [taxonomy: 401]; Fernal1903b [catalogue: 293]; McKenz1950 [taxonomy, description, illustration, host, distribution: 105-106,121]; Willia1963b [taxonomy: 12].

Lindingaspis tingi McKenzie

Lindingaspis tingi McKenzie, 1950: 106. Type data: PHILIPPINES: Manila, on orchid (probably a species of *Phalaenopsis*) sp. Holotype female. Type depository: Davis: Bohart Museum of Entomology, University of California, California, USA.
Lindingaspis tinge; Borchsenius, 1966: 345. Misspelling of species name.
Acutaspis tingi; Velasquez & Rimando, 1969: 198. Misidentification; discovered by Lit & Rimando, 1993: 164.

SCALE COVER: Scale of female of 2.5 to 2.75 mm in diameter; light chocolate brown with darker subcentral exuvia. Scale of male elongate, oval, similar in colour to female (McKenzie, 1950).

HOST PLANTS: **Orchidaceae**: *Phalaenopsis* [McKenz1950], *Phalaenopsis schilleriana* [McKenz1950].

DISTRIBUTION: **Oriental**: Philippines (Luzon [McKenz1950, Balach1958b]).

BIOLOGY: Occurring on the leaves of the host (McKenzie, 1950).

GENERAL: Description and illustration of adult female by McKenzie (1950) and by Balachowsky (1958b).

KEYS: Williams 1963b: 12 (female) [World]; Balachowsky 1958b: 166 (female) [Africa]; McKenzie 1950: 109 (female) [World].

CITATIONS: Balach1958b [taxonomy, description, illustration, host, distribution: 182-184]; Borchs1966 [catalogue: 345]; LitRi1993 [taxonomy, host, distribution: 163-167]; McKenz1950 [taxonomy, description, illustration, host, distribution: 106-107,122]; VelasqRi1969 [taxonomy, host, distribution: 195-208]; Willia1963b [taxonomy: 12].

Lindingaspis tomarum Balachowsky

Lindingaspis tomarum Balachowsky, 1958b: 184. Type data: GUINEA: Seredou, 1200 meters altitude, on lower surface of leaves of "Quinquina" [=*Chinchona*] sp. Holotype female. Type depository: Tervuren: Musée Royal de l'Afrique Centrale, Section d'Entomologie, Belgium.

SCALE COVER: Female scale circular, 2.5-2.6 mm in diameter; slightly convex; colour dark brown; exuviae black, shiny, placed centrally or subcentrally. Male scale not observed (Balachowsky, 1958b).

HOST PLANTS: **Rubiaceae**: *Chinchona* [Balach1958b], *Gardenia* [MatileNo1984].

DISTRIBUTION: **Afrotropical**: Cameroon [MatileNo1984]; Guinea [Balach1958b].

GENERAL: Description and illustration of adult female by Balachowsky (1958b).

KEYS: Williams 1963b: 13 (female) [World]; Balachowsky 1958b: 166 (female) [Africa].

CITATIONS: Balach1958b [taxonomy, description, illustration, host, distribution: 186-187]; Borchs1966 [catalogue: 345]; MatileNo1984 [host, distribution: 66]; Willia1963b [taxonomy, host, distribution: 13].

Lindingaspis victoriae (Cockerell)

Chrysomphalus rossi victoriae Cockerell, 1899k: 89. Type data: AUSTRALIA: Victoria, Bacchus Marsh, Maddingly Park, on *Eucalyptus globulus*. Syntypes, female. Type depository: Washington: United States National Entomological Collection, U.S. National Museum of Natural History, District of Columbia, USA.

Lindingaspis victoriae; McKenzie, 1939: 55. Change of combination and rank.

Chrysomphalus victoriae; Ferris, 1941e: 49. Change of combination.

Lindingaspis victoriae; McKenzie, 1950: 107. Revived combination.

SCALE COVER: Cockerell (1899k) described scale cover "like *rossi* in size and shape, but rougher and paler (more brown), with the exuviae covered by whitish secretion". McKenzie (1950) supplemented this information "Scale of the female, judging from the single damaged female at hand, with a diameter of approximately 1.75 mm; brownish-black in colour. Male scale not represented in loaned material".

HOST PLANTS: **Leguminosae**: *Acacia penninervis* [Brimbl1955]. **Myrtaceae**: *Eucalyptus* [Brimbl1955], *Eucalyptus globosus* [Cocker1899k, McKenz1950].

DISTRIBUTION: **Australasian**: Australia (Victoria [Cocker1899k, McKenz1950, Brimbl1955]).

BIOLOGY: Apparently occurring on leaves of host (McKenzie, 1950).

GENERAL: Description and illustration of adult female by McKenzie (1950) and by Brimblecombe (1955).

KEYS: Williams 1963b: 13 (female) [World]; McKenzie 1950: 109 (female) [World].

CITATIONS: Borchs1966 [catalogue: 345]; Brimbl1955 [taxonomy, description, illustration, host, distribution: 51-53]; Cocker1899k [taxonomy, description, host, distribution: 89]; Fernal1903b [catalogue: 293]; Ferris1941e [taxonomy: 49]; McKenz1939 [taxonomy: 55]; McKenz1950 [taxonomy, description, illustration, host, distribution: 107,109,123]; Willia1963b [taxonomy: 13]; Willia1985a [taxonomy: 238].

Lindingaspis williamsi Balachowsky

Lindingaspis williamsi Balachowsky, 1958b: 186. Type data: SUDAN: White Nile region, Zerat Cuts, on *Cyperus papyrus*. Holotype female. Type depository: London: The Natural History Museum, England, UK.

SCALE COVER: Balachowsky (1958b) did not describe female scale cover. Male scale unknown (1958b).

HOST PLAQNTS: **Cyperaceae**: *Cyperus papyrus* [Balach1958b].

DISTRIBUTION: **Afrotropical**: Sudan [Balach1958b].

GENERAL: Description and illustration of adult female by Balachowsky (1958b).

KEYS: Williams 1963b: 12 (female) [World]; Balachowsky 1958b: 166 (female) [Africa].

CITATIONS: Balach1958b [taxonomy, description, illustration, host, distribution: 186-188]; Borchs1966 [catalogue: 345]; Willia1963b [taxonomy: 12].

Loranthaspis Cockerell & Bueker

Loranthaspis Cockerell & Bueker, 1930: 4. Type species: *Loranthaspis microconcha* Cockerell & Bueker, by monotypy.

SYSTEMATICS: Cockerell & Bueker (1930) noted affinity of this genus to *Aspidiotus, Targionia* and *Pygidiaspis*. The narrow, almost linear, median lobes regarded as a distinct feature of *Loranthaspis*.

GENERAL: Definition and characters by Cockerell & Bueker (1930).

CITATIONS: Borchs1966 [catalogue: 359]; CockerBu1930 [taxonomy, description: 4]; CockerBu1930 [taxonomy, description: 4]; Ferris1937c [taxonomy: 51]; Ferris1938 [taxonomy: 46]; Ferris1941d [taxonomy: 374]; Lindin1937 [taxonomy: 188]; MorrisMo1966 [taxonomy, catalogue: 112].

Loranthaspis microconcha Cockerell & Bueker

Loranthaspis microconcha Cockerell & Bueker, 1930: 4. Type data: INDONESIA: Java, on *Loranthus pentandrus* and *Canarium* sp. Syntypes, female. Type depository: Washington: United States National Entomological Collection, U.S. National Museum of Natural History, District of Columbia, USA.

SCALE COVER: Female scale circular, 0.6-0.9 mm in diameter; highly convex dorsally and ventrally; dark grey; first and second exuviae central to subcentral; first represented by an oval exuvia (0.25 mm long, 0.24 mm wide), light brown; dorsal side with a circular waxy secretion (0.12 mm across); second exuvia oval (0.46 mm long, 0.39 mm wide) dorsally with a waxy secretion; with well-developed ventral scale, so that the female is completely enclosed, but not in the second exuvia as in *Aonidia* (Cockerell & Bueker, 1930).

HOST PLANTS: Burseraceae: *Canarium* [CockerBu1930]. **Loranthaceae:** *Loranthus pentandrus* [CockerBu1930].

DISTRIBUTION: Oriental: Indonesia (Java [CockerBu1930]).

GENERAL: Description and illustration of adult female by Cockerell & Bueker (1930).

CITATIONS: Borchs1966 [catalogue: 359]; CockerBu1930 [taxonomy, description, illustration, host, distribution: 4-5]; Ferris1937c [taxonomy, illustration: 51,81].

Marginaspis Hall

Marginaspis Hall, 1946: 58. Type species: *Marginaspis thevetiae* Hall, by monotypy and original designation.

SYSTEMATICS: The genus *Marginaspis* is related to *Lindingaspis,* differing from the latter in possessing well-developed plates anterior to the third lobe, and in having conspicuous, sclerotized spurs anterior to these plates (Hall, 1946; McKenzie, 1950).

GENERAL: Definition and characters by Hall (1946), McKenzie (1950) and by Balachowsky (1958b).

KEYS: Balachowsky 1958b: 231 (female) [*Aspidiotina* of Africa].

CITATIONS: Balach1953c [taxonomy: 110]; Balach1958b [taxonomy, description: 188]; Borchs1966 [catalogue: 342]; Hall1946 [taxonomy, description: 58]; McKenz1950 [taxonomy, description: 107]; MorrisMo1966 [taxonomy, catalogue: 115].

Marginaspis affinis (Leonardi)

Chrysomphalus affinis Leonardi, 1914: 204. Type data: GUINEA: Kakoulima, on an undetermined plant, and SOUTH AFRICA: Pretoria, on *Nerium* sp. Syntypes, female. Type depository: Portici: Dipartimento de Entomologia e Zoologia Agraria di Portici, Università di Napoli Federico II, Italy.

Marginaspis affinis; Balachowsky, 1953c: 110. Change of combination.

SCALE COVER: Female scale of irregular shape, slightly convex; exuviae subcentral; secreted cover compact and robust; colour black chestnut, becoming less intense from center to margin; ventral vellum well developed, moderately robust; 2-2.5 mm (Leonardi, 1914).

HOST PLANTS: **Apocynaceae**: *Nerium* [Leonar1914].
DISTRIBUTION: **Afrotropical**: Guinea [Leonar1914]; South Africa [Leonar1914].
GENERAL: Description and illustration of adult female by Leonardi (1914).
CITATIONS: Balach1953c [taxonomy: 110]; Balach1958b [taxonomy: 190]; Borchs1966 [catalogue: 342]; Leonar1914 [taxonomy, description, illustration, host, distribution: 204-205]; McKenz1939 [taxonomy: 53].

Marginaspis thevetiae Hall
Marginaspis thevetiae Hall, 1946: 58. Type data: SIERRA LEONE: Njala, on *Thevetia neriifolia*. Syntypes, female. Type depository: London: The Natural History Museum, England, UK.
SCALE COVER: Female scale hemispherical in outline, 3 mm long, 1.50 to 1.75 mm wide; flat; colour chocolate; exuviae black or very dark brown and placed so close to flattened edge as to appear marginal; ventral scale thin but persistent round margin to a greater or lesser degree. Male scale similar to that of female in shape but very much smaller, with only a very narrow fringe of ventral scale persisting round margin; exuvia black or very dark brown (Hall, 1946).
HOST PLANTS: **Apocynaceae**: *Thevetia neriifolia* [Hall1946, McKenz1950], *Thevetia peruviana* [MatileNo1984]. **Moraceae**: *Ficus thonnigii* [Balach1958b]. **Rhizophoraceae**: *Rhizophora racemosa* [Balach1958b].
DISTRIBUTION: **Afrotropical**: Cameroon [Balach1958b, MatileNo1984]; Guinea [Balach1958b]; Sierra Leone [Hall1946, McKenz1950, Balach1958b].
BIOLOGY: Settled on under-surface of leaf, with flattened portion of margin resting against very narrow thickened rim formed by turning under the extreme margin of leaf (Hall, 1946).
GENERAL: Description and illustration of adult female by Hall (1946), McKenzie (1950) and by Balachowsky (1958b).
CITATIONS: Balach1953c [taxonomy: 342]; Balach1958b [taxonomy, description, illustration, host, distribution: 189-191]; Borchs1966 [catalogue: 342]; Hall1946 [taxonomy, description, illustration, host, distribution: 58-60]; MatileNo1984 [host, distribution: 66]; McKenz1950 [taxonomy, description, illustration, host, distribution: 107,124].

Maskellia Fuller

Maskellia Fuller, 1897: 579. Type species: *Maskellia globosa* Fuller, by monotypy and original designation.
Austromaskellia Lindinger, 1943b: 207. Unjustified replacement name; discovered by Borchsenius (1966).
SYSTEMATICS: Adult female of type species of *Maskellia* Fuller, namely *Maskellia globosa* Fuller, reported to inhabit galls formed on twigs of *Eucalyptus gomphocephala*, while male inhabited galls formed upon leaves. Genus established because of the gall formation habit and gall inhabiting (Fuller, 1897).
GENERAL: Definition and characters by Fuller (1897) and by Froggatt (1914). Fuller (1897c) again referred to this genus as gen. nov.

CITATIONS: Borchs1966 [catalogue: 367]; Cocker1899a [taxonomy: 396]; Fernal1903b [catalogue: 282]; Frogga1914 [taxonomy, description: 989]; Frogga1915 [taxonomy, description: 64]; Fuller1897 [taxonomy, description: 579-580]; Fuller1897c [taxonomy: 5]; HowellTi1990 [taxonomy: 57]; Lindin1908b [taxonomy: 98]; Lindin1943b [taxonomy: 207]; MacGil1921 [taxonomy, description: 392,446]; MorrisMo1966 [taxonomy, catalogue: 116].

Maskellia globosa Fuller

Maskellia globosa Fuller, 1897: 579. Type data: AUSTRALIA: Western Australia, Swan River, in galls formed on *Eucalyptus gomphocephala*. Syntypes, female.

SCALE COVER: Induces formation of galls on twigs of *Eucalyptus gomphocephala* in Western Australia, described by Fuller (1897) as follows: "The female galls are green and red or brown, according to their age, and are formed upon the young twigs, great numbers usually growing side by side, producing an uneven, aborted growth. The apex of each gall is conical, and the orifice small. When formed singly the galls are uneven, globose excrescences, somewhat longer than high. Height, 0.18 inch; width, 0.23 inch. Female chamber-contour pyriform; diameter, 0.08 inch. Adult male, unobserved. Male galls, small, elongated, striated longitudinally, green formed upon leaves; length 0.12 inch. The galls protrude upon the opposite side of the leaf, the orifice being at the base instead at the apex." (Fuller, 1897).

HOST PLANTS: **Myrtaceae**: *Eucalyptus gomphocephala* [Fuller1897].

DISTRIBUTION: **Australasian**: Australia (Western Australia [Fuller1897, Frogga1914]).

BIOLOGY: Forms galls on twigs of *Eucalyptus gomphocephala* in Western Australia (Fuller, 1897).

GENERAL: Description and illustration of adult female and of galls by Fuller (1897). Fuller (1897) noted that he had collected Diaspididae that formed galls on an unnamed plant in New South Wales in 1895, which might be similar to *Maskellia globosa*.

CITATIONS: Borchs1966 [catalogue: 367]; Cocker1899a [taxonomy: 396]; Fernal1903b [catalogue: 282]; Ferris1937c [taxonomy: 101]; Foldi1990 [structure: 43-54]; Frogga1914 [taxonomy, description, host, distribution: 989]; Frogga1915 [taxonomy, description, host, distribution: 64]; Fuller1897 [taxonomy, description, illustration, host, distribution: 579-580]; Fuller1897c [taxonomy, description, host, distribution: 5]; Larew1990 [ecology, life history, structure: 293-300]; MacGil1921 [taxonomy, description, host, distribution: 446].

Megaspidiotus Brimblecombe

Megaspidiotus Brimblecombe, 1954: 155. Type species: *Diaspis fimbriata* Maskell, by monotypy and original designation.

SYSTEMATICS: Type species of genus possesses a constriction between prothorax and mesothorax, thus resembling *Pseudaonidia* group of genera, but differs in having constrictions on all prepygidial segments. The genus clearly belongs to the group of genera allied to *Aspidiotus* Bouché, in having three pairs of

well-developed lobes, structure of plates and absence of reticulated area on pygidium (Borchsenius & Williams, 1963).
GENERAL: Definition and characters by Brimblecombe (1954) and by Borchsenius & Williams (1963).
CITATIONS: Borchs1966 [catalogue: 258]; BorchsWi1963 [taxonomy, description: 389]; Brimbl1954 [taxonomy, description: 155-157]; MorrisMo1966 [taxonomy, catalogue: 118].

Megaspidiotus fimbriatus (Maskell)

Diaspis (?) fimbriata Maskell, 1893b: 208. Type data: AUSTRALIA: New South Wales, Sydney, on *Eugenia smithii*; collected by Mr. Koebele. Lectotype female, by subsequent designation Deitz & Tocker, 1980: 37. Type depository: Canberra: Australian National Insect Collection, CSIRO Entomology, Australia.
Aspidiotus fimbriatus; Cockerell, 1894g: 128. Change of combination requiring emendation of species name for agreement in gender.
Aspidiotus virescens Maskell, 1896b: 384. Type data: AUSTRALIA: New South Wales, probably near Sydney, on *Eugenia smithii*; collected by Froggatt. Lectotype, by subsequent designation Deitz & Tocker, 1980: 44. Type depository: Auckland: New Zealand Arthropod Collection, Landcare Research, New Zealand. Synonymy by Ferris, 1941e: 49.
Aspidiotus (Aspidiotus) fimbriatus; Cockerell, 1897i: 26. Change of combination.
Aspidiotus (Evaspidiotus) fimbriatus; Leonardi, 1898a: 76. Change of combination.
Aspidiotus (Evaspidiotus) virescens; Leonardi, 1898a: 76. Change of combination.
Aspidiotus (Evaspidiotus) virescens; Leonardi, 1898c: 64. Change of combination.
Megaspidiotus fimbriatus; Brimblecombe, 1954: 155. Change of combination.
SCALE COVER: Female scale circular, averaging 1/13 inch in diameter; flat; very thin and papery; whitish, grey or brownish; exuvia subcentral, yellow or greenish. Male scale unknown (Maskell, 1893b).
HOST PLANTS: **Myrtaceae**: *Eugenia smithii* [Maskel1896b, Brimbl1954].
DISTRIBUTION: **Australasian**: Australia (New South Wales [Maskel1896b, Frogga1914, Brimbl1954]).
GENERAL: Description and illustration of adult female by Maskell (1893b), Brimblecombe (1954) and by Borchsenius & Williams (1963).
CITATIONS: Borchs1966 [catalogue: 258]; BorchsWi1963 [taxonomy, description, illustration: 388-389]; Brimbl1954 [taxonomy, description, illustration, host, distribution: 155-157]; Cocker1894g [taxonomy: 128]; Cocker1896b [distribution: 335]; Cocker1897i [taxonomy, description, host, distribution: 26-27]; Cocker1899a [taxonomy: 395]; DeitzTo1980 [taxonomy: 37,44]; Fernal1903b [catalogue: 259,280-281]; Ferris1941e [taxonomy: 43,49]; Frogga1914 [taxonomy, description, host, distribution: 312,319]; Frogga1915 [taxonomy, description, host, distribution: 15-16,24]; Leonar1898a [taxonomy: 76]; Leonar1898a [taxonomy, host, distribution: 76]; Leonar1898c [taxonomy, description, illustration, host, distribution: 58-60,64-65]; Lindin1957 [taxonomy: 551]; MacGil1921 [taxonomy, description, host, distribution: 396]; Maskel1893b [taxonomy, description, illustration, host, distribution: 208-209]; Maskel1896b [taxonomy, description, illustration, host, distribution: 384].

Melanaspis Cockerell

Chrysomphalus (*Melanaspis*) Cockerell, 1897i: 9. Type species: *Aspidiotus obscurus* Comstock, by original designation.
Melanaspis; Lindinger, 1910a: 440. Change of status.
Pelomphala MacGillivray, 1921: 392. Type species: *Aspidiotus lilacinus* Cockerell, by original designation. Synonymy by Lindinger, 1937: 192.
Melanasois; Borchsenius, 1966: 352. Misspelling of genus name.
SYSTEMATICS: *Melanaspis* is related to *Mycetaspis* but differs in absence of broad protuberance on cephalic margin. The lobe-associated pygidial paraphyses in *Melanaspis* are numerous, distinguishing it from *Greenoidea*. In *Melanaspis* the pygidial apex forms an obtuse angle, while in *Acutaspis* the apex is more acute.
GENERAL: Definition and characters by Ferris (1941d), Borchsenius (1950b), Lupo (1954a), Balachowsky (1951, 1958b), Deitz & Davidson (1986) and by Yaşar (1995).
KEYS: Colon-Ferrer & Medina-Gaud 1998: 28-32 (female) [Genera Puerto Rico]; Gill 1997: 24-26 (female) [Genera California]; Gill 1997: 194 (female) [Species California]; Blay Goicoechea 1993: 474 (female) [Spain]; Deitz & Davidson 1986: 12-15 (female) [species North America]; Chou 1985: 319 (female) [Genera China]; McDaniel 1970: 411 (female) [species U.S.A.: Texas]; Beardsley 1966: 502-504 (female) [Federated States of Micronesia]; Balachowsky 1958b: 231 (female) [*Aspidiotina* of Africa]; Ezzat 1958: 237-239 (female) [Egypt]; McKenzie 1956: 23 (female) [U.S.A.: California]; Bodenheimer 1952: 330 (female) [Turkey]; Balachowsky 1951: 601 (female) [Mediterranean]; Borchsenius 1950b: 167 (female) [USSR]; Ferris 1942: 27 (female) [North America]; Ferris 1942: 35-38 (female) [species North America]; Bodenheimer 1924: 21 (female) [Israel]; Brain 1919: 198 (female) [World]; Lindinger 1913: 65 (female) [Africa]; Lindinger 1911: 356 (female) [Germany].
CITATIONS: Balach1951 [taxonomy, description: 578-579]; Balach1958b [taxonomy, description: 191-192]; Balach1959 [taxonomy: 360]; BalachKa1953 [taxonomy: 28]; Beards1966 [taxonomy: 522]; BenDov1990h [taxonomy: 82]; BlayGo1993 [taxonomy, description: 614]; Bodenh1924 [taxonomy: 21]; Bodenh1949 [taxonomy, description: 26,37]; Bodenh1952 [taxonomy: 330]; Borchs1949d [taxonomy, description: 194,234]; Borchs1950b [taxonomy, description: 167,222]; Borchs1952 [taxonomy: 262]; Borchs1966 [catalogue: 345-346,352]; Brain1919 [taxonomy: 198]; Bustsh1958 [taxonomy, description: 223]; Chou1985 [taxonomy, description: 323]; Cocker1897i [taxonomy, description: 9,13,31]; Cocker1899a [taxonomy: 396]; ColonFMe1998 [taxonomy, description: 64-65]; Danzig1993 [description: 237-238]; DanzigPe1998 [catalogue: 303]; DeitzDa1986 [taxonomy, description: 10-11]; Ezzat1958 [taxonomy: 239]; Fernal1903b [catalogue: 285]; Ferris1937c [taxonomy: 51,52,54,55,88]; Ferris1938 [taxonomy: 46]; Ferris1938a [taxonomy: 245]; Ferris1941d [taxonomy, description: 347]; Ferris1942 [taxonomy: 446:27]; Ferris1943 [taxonomy: 58]; Gill1997 [taxonomy: 194]; Kawai1980 [taxonomy: 206]; Koszta1996 [taxonomy, description: 532-533]; Kozar1990f [distribution: 144]; Leonar1897a [taxonomy: 375]; LepageGi1942 [taxonomy, description: 445]; Lindin1910 [taxonomy: 440];

Lindin1911 [taxonomy: 356]; Lindin1913 [taxonomy: 65]; Lindin1924 [taxonomy: 172]; Lindin1934c [taxonomy: 45]; Lindin1937 [taxonomy: 189,192]; Lupo1954a [taxonomy, description: 34-35]; MacGil1921 [taxonomy: 392,441-442]; McKenz1939 [taxonomy: 53]; McKenz1944 [taxonomy: 54]; McKenz1947 [taxonomy: 32]; McKenz1950 [taxonomy: 99]; McKenz1956 [taxonomy: 23]; Miller1990 [taxonomy: 169-178]; MorrisMo1966 [taxonomy, catalogue: 118]; Tao1999 [taxonomy: 98]; Willia1969a [taxonomy: 330]; Yasar1995a [taxonomy, description: 93].

Melanaspis araucariae Lepage

Melanaspis araucariae Lepage, 1942: 182. Type data: BRAZIL: near Sta. Adelaide, altitude 1600 metros, at Capivari, Campos do Jordao, on *Araucaria brasiliensis*. Syntypes, female. Type depositories: São Paulo: Museu de Zoologia, Universidade de São Paulo, Brazil, and Curitiba: Departamento de Zoologia, Setor de Ciencias Biologicas, Universidade Federal do Parana, Brazil.

SCALE COVER: Female scale circular, white, diameter 3 mm; exuviae black, central. Male scale not observed (Lepage, 1942).

HOST PLANTS: **Araucariaceae**: *Araucaria angustifolia* [ClapsWoGo2001], *Araucaria brasiliensis* [Lepage1942].

DISTRIBUTION: **Neotropical**: Brazil (São Paulo [Lepage1942, ClapsWoGo2001]).

GENERAL: Description and illustration of adult female by Lepage (1942).

CITATIONS: Borchs1966 [catalogue: 346]; Claps1993 [taxonomy: 9]; ClapsWoGo2001 [host, distribution: 246]; Lepage1942 [taxonomy, description, illustration, host, distribution: 183-185]; Lindin1957 [taxonomy: 550].

Melanaspis aristotelesi Lepage & Giannotti

Melanaspis aristotelesi Lepage & Giannotti, 1944: 300. Type data: BRAZIL: Alagoas State, Marechal Deodoro City, on *Anacardium occidentalis*. Syntypes, female. Type depositories: São Paulo: Instituto Biologico de São Paulo, Brazil, and Curitiba: Departamento de Zoologia, Setor de Ciencias Biologicas, Universidade Federal do Parana, Brazil.

Melanaspis aristotelis; Lindinger, 1957: 550. Misspelling of species name.

SCALE COVER: Female scale circular, 1.4 mm diameter; colour brown chestnut or ash; convex; exuviae black, central or subcentral; ventral scale white (Lepage & Giannotti, 1944).

HOST PLANTS: **Anacardiaceae**: *Anacardium occidentalis* [LepageGi1944, ClapsWoGo2001].

DISTRIBUTION: **Neotropical**: Brazil (Alagoas [LepageGi1944, ClapsWoGo2001]).

GENERAL: Description and illustration of adult female by Lepage & Giannotti (1944).

CITATIONS: Borchs1966 [catalogue: 346]; Claps1993 [taxonomy: 6,9]; ClapsWoGo2001 [host, distribution: 246]; LepageGi1944 [taxonomy, description, illustration, host, distribution: 300-302]; Lindin1957 [taxonomy: 550].

Melanaspis arnaldoi (Costa Lima)

Aonidiella arnaldoi Azevedo Marques, 1923: 422. Nomen nudum.

Aonidiella arnaldoi Costa Lima, 1924: 139. Type data: BRAZIL: on grapevine. Syntypes, female. Type depository: Rio de Janeiro: Colecao Entomologica, Escola de Postgraduacao em Parasitologia Animal, Universidade Federal Rural de Rio de Janeiro, Brazil.

Aonidiella arnoldi; McKenzie, 1938: 3. Misspelling of species name.

Melanaspis arnaldoi; Costa Lima, 1942: 289. Change of combination.

SCALE COVER: Female scale circular, 1 mm diameter; convex; grey or black; exuvia of first instar shiny black; ventral scale well developed, especially marginally, white (Costa Lima, 1924). Photograph of scale cover by Costa Lima (1924).

HOST PLANTS: **Leguminosae**: *Mimosa scabrella* [ClapsWoGo2001]. **Vitaceae**: *Vitis* [Lepage1938, ClapsWoGo2001], *Vitis vinifera* [CostaL1924].

DISTRIBUTION: **Neotropical**: Brazil [CostaL1924] (Rio de Janeiro [Lepage1938, LepageGi1943]).

GENERAL: Description and illustration of adult female by Costa Lima (1924) and by Lepage & Giannotti (1943).

CITATIONS: Azeved1923LA [taxonomy: 422]; Borchs1966 [catalogue: 346]; ClapsWoGo2001 [host, distribution: 246]; CostaL1924 [taxonomy, description, illustration, host, distribution: 139-140]; CostaL1942 [taxonomy: 289]; Lepage1938 [catalogue: 392]; LepageGi1943 [taxonomy, description, illustration, host, distribution: 340-342]; McKenz1938 [taxonomy: 3].

Melanaspis artemisiae Mamet

Melanaspis artemisiae Mamet, 1959a: 472. Type data: MADAGASCAR: Bevoalavo, South of Menarandrana on *Artemisia* sp. Holotype. Type depository: Paris: Muséum national d'Histoire naturelle, France.

SCALE COVER: Female scale of type common in *Melanaspis* (Mamet, 1959a).

HOST PLANTS: **Compositae**: *Artemisia* [Mamet1959a, Borchs1966].

DISTRIBUTION: **Afrotropical**: Madagascar [Mamet1959a, Borchs1966].

GENERAL: Description and illustration of adult female by Mamet (1959a).

CITATIONS: Borchs1966 [catalogue: 346]; Mamet1959a [taxonomy, description, illustration, host, distribution: 388,472].

Melanaspis arundinariae Deitz & Davidson

Melanaspis arundinariae Deitz & Davidson, 1986: 16. Type data: U.S.A.: South Carolina, Anderson Co., Anderson, on *Arundinaria* sp. Holotype female. Type depository: Washington: United States National Entomological Collection, U.S. National Museum of Natural History, District of Columbia, USA.

SCALE COVER: Deitz & Davidson (1986) did not describe scale cover.

HOST PLANTS: **Gramineae**: *Arundinaria* [DeitzDa1986].

DISTRIBUTION: **Nearctic**: United States of America (South Carolina [DeitzDa1986]).

GENERAL: Description and illustration of adult female by Deitz & Davidson (1986).

CITATIONS: DeitzDa1986 [taxonomy, description, illustration, host, distribution: 16-17].

Melanaspis bolivari Balachowsky
Melanaspis bolivari Balachowsky, 1959: 360. Type data: COLOMBIA: Magdalena, Santa Marta, Quinta de San Pedro Alejandrino, on *Quercus ilex*. Syntypes, female. Type depository: Paris: Muséum national d'Histoire naturelle, France.
SCALE COVER: Female scale circular, 2-2.2 mm diameter; colour black, matt; exuviae black. Male scale suboval; colour similar to female, but lighter (Balachowsky, 1959). Illustration of female scale cover by Balachowsky (1959).
HOST PLANTS: **Fagaceae**: *Quercus ilex* [Balach1959].
DISTRIBUTION: **Neotropical**: Colombia [Balach1959].
GENERAL: Description and illustration of adult female by Balachowsky (1959).
CITATIONS: Balach1959 [taxonomy, description, illustration, host, distribution: 360-361]; Borchs1966 [catalogue: 346].

Melanaspis bondari Lepage & Giannotti
Melanaspis bondari Lepage & Giannotti, 1943: 346. Type data: BRAZIL: Bahia, on an undetermined plant. Syntypes, female. Type depositories: São Paulo: Museu de Zoologia, Universidade de São Paulo, Brazil, São Paulo: Instituto Biologico de São Paulo, Brazil, and Curitiba: Departamento de Zoologia, Setor de Ciencias Biologicas, Universidade Federal do Parana, Brazil.
SCALE COVER: Female scale light grey, about 1.9 mm in diameter; exuviae more or less eccentric, black shiny (Lepage & Giannotti, 1943).
DISTRIBUTION: **Neotropical**: Brazil (Bahia [LepageGi1943, ClapsWoGo2001]).
GENERAL: Description and illustration of adult female by Lepage & Giannotti (1943).
 CITATIONS: Borchs1966 [catalogue: 346-347]; Claps1993 [taxonomy: 3,6,9]; ClapsWoGo2001 [host, distribution: 246]; LepageGi1943 [taxonomy, description, illustration, host, distribution: 345-346].

Melanaspis bromiliae (Leonardi)
Aonidiella bromiliae Leonardi, 1899: 177. Type data: ENGLAND: Chester, on pineapple, apparently imported from the Canary Islands. Syntypes, female. Type depository: Portici: Dipartimento di Entomologia e Zoologia Agraria di Portici, Università di Napoli Federico II, Italy.
Aspidiotus bromiliae; Newstead, 1901b: 81. Change of combination.
Chrysomphalus bromeliae Fernald, 1903b: 289. Unjustified emendation; discovered by Deitz & Davidson, 1986: 18.
Chrysomphalus bromeliae; Fernald, 1903b: 289. Change of combination.
Chrysomphalus bromeliae; Fernald, 1903b: 289. Misspelling of species name.
Pseudischnaspis bromeliae; Lindinger, 1912b: 67. Change of combination.
Pseudischnaspis anassarum Lindinger, 1932: 26. Unjustified replacement name for *Aonidiella bromiliae* Leonardi, 1899.
Aonidiella bromiliae; Lindinger, 1932f: 191. Incorrect synonymy. Notes: See Systematics below; see Deitz & Davidson, 1986: 18.

Chrysomphalus bromeliae; Seabra, 1942: 2. Notes: Incorrect citation of "Newstead" as author.

Chrysomphalus bromeliae; Seabra, 1942: 2. Misspelling of species name.

Chrysomphalus (*Melanaspis*) *bromeliae*; Merrill, 1953: 36. Change of combination.

Chrysomphalus (*Melanaspis*) *bromeliae*; Merrill, 1953: 36. Misspelling of species name.

Melanaspis bromeliae; Dekle, 1965c: 87. Misspelling of species name.

Melanaspis bromeliae; Dekle, 1976: 108. Misspelling of species name.

Melanaspis bromiliae; Deitz & Davidson, 1986: 18. Change of combination.

Melanaspis bromeliae; Gill, 1997: 194. Misspelling of species name.

COMMON NAME: ananas scale [Dekle1965c].

SYSTEMATICS: Borchsenius (1966) synonymized *Melanaspis bromiliae* (Leonardi) with Melanaspis *smilacis* (Comstock), while Deitz & Davidson (1986) regarded these as distinct species.

SCALE COVER: Colour photograph by Gill (1997).

HOST PLANTS: **Bromeliaceae**: *Ananas* [Dekle1965c, DeitzDa1986], *Ananas bracteatus* [DeitzDa1986], *Ananas comosus* [DeitzDa1986], *Bromelia* [Lindin1912b], *Neoglaziovia variegata* [DeitzDa1986]. **Palmae**: *Cocos nucifera* [DeitzDa1986]. **Pandanaceae**: *Pandanus* [DeitzDa1986].

DISTRIBUTION: **Afrotropical**: Cameroon [Nakaha1982]; Côte d'Ivoire (=Ivory Coast) [Nakaha1982]; Guinea [Nakaha1982]; Seychelles [DeitzDa1986]; South Africa [Nakaha1982]; Togo [Nakaha1982]. **Australasian**: Federated States of Micronesia (Ponape Island [Beards1966]); Guam [Beards1966]; Hawaiian Islands (Hawaii [DeitzDa1986]). **Nearctic**: Mexico (Oaxaca [DeitzDa1986], Sonora [DeitzDa1986], Veracruz [DeitzDa1986]); United States of America (District of Columbia [DeitzDa1986], Florida [Merril1953, Dekle1965c, DeitzDa1986]). **Neotropical**: Bahamas [DeitzDa1986]; Bermuda [DeitzDa1986]; Brazil [DeitzDa1986]; Colombia [DeitzDa1986]; Costa Rica [Nakaha1982]; Cuba [DeitzDa1986]; Dominican Republic [DeitzDa1986]; Ecuador [DeitzDa1986]; Guatemala [DeitzDa1986]; Haiti [DeitzDa1986]; Honduras [DeitzDa1986]; Jamaica [DeitzDa1986]; Martinique [Nakaha1982]; Panama Canal Zone [DeitzDa1986]; Puerto Rico & Vieques Island (Puerto Rico [DeitzDa1986]). **Oriental**: India [DeitzDa1986] (Rajasthan); Malaysia (Malaya [DeitzDa1986]); Philippines [DeitzDa1986]; Singapore [DeitzDa1986]. **Palaearctic**: Azores [Lindin1912b]; Canary Islands [DeitzDa1986]; Italy [DeitzDa1986]; Japan [Kawai1980, DeitzDa1986]; Netherlands [DeitzDa1986]; Portugal [Seabra1942, DeitzDa1986].

GENERAL: Description and illustration of adult female by Deitz & Davidson (1986), Gill (1997) and by Colon-Ferrer & Medina-Gaud (1998).

KEYS: Gill 1997: 194 (female) [Species of California]; Kawai 1980: 206 (female) [Japan].

CITATIONS: Beards1966 [host, distribution: 522-523]; Borchs1966 [catalogue: 352]; ColonFMe1998 [taxonomy, description, illustration, host, distribution: 65-67]; DanzigPe1998 [catalogue: 303-304]; DeitzDa1986 [taxonomy, description, illustration, host, distribution: 18-19]; Dekle1965c [taxonomy, description, host, distribution: 87]; Dekle1976 [taxonomy, description, host, distribution, economic importance: 108]; Fernal1903b [catalogue: 289]; Gill1997 [host, distribution, taxonomy, description, illustration, economic importance: 194-196]; Kawai1980 [taxonomy, description, host, distribution: 207]; Leonar1899 [taxonomy, description, illustration, host, distribution: 177]; Lindin1912b [taxonomy, description, host, distribution: 67]; Lindin1932f [taxonomy, host, distribution: 191]; LongoMaPe1995 [distribution: 127]; Mead1983 [host, distribution: 1-5]; Merril1953 [taxonomy, description, host, distribution: 36]; Nakaha1982 [host, distribution: 53-54]; Newste1901b [taxonomy : 81,86]; Seabra1942 [distribution, taxonomy: 2].

Melanaspis calura (Cockerell)

Aspidiotus (*Chrysomphalus*) *calurus* Cockerell, 1898j: 440. Type data: MEXICO: Orizaba, on *Crataegus* sp. Lectotype female, by subsequent designation Deitz & Davidson, 1986: 20. Type depository: Washington: United States National Entomological Collection, U.S. National Museum of Natural History, District of Columbia, USA.

Chrysomphalus calurus; Cockerell, 1899n: 27. Change of combination.

Aonidiella calura; Leonardi, 1900: 341. Change of combination requiring emendation of species name for agreement in gender.

Aspidiotus calurus; Cockerell, 1905: 46. Change of combination.

Chrysomphalus calurnus; MacGillivray, 1921: 441. Misspelling of species name.

Pelomphala calurna; MacGillivray, 1921: 441. Change of combination.

Pelomphala calurna; MacGillivray, 1921: 441. Misspelling of species name.

Melanaspis calurus; McKenzie, 1939: 53. Change of combination.

Melanaspis calura; Ferris, 1941d: 350. Justified emendation.

SCALE COVER: Female scale circular, about 1.5 mm in diameter; slightly convex; covered by epidermis of bark, except shining black exuviae; exuviae exposed and very conspicuous; thick ventral scale present (Cockerell, 1898j).

HOST PLANTS: **Anacardiaceae**: *Spondias* [DeitzDa1986], *Spondias mombin* [DeitzDa1986], *Spondias purpurea* [DeitzDa1986]. **Compositae** [Ferris1941d]. **Guttiferae**: *Mammea americana* [Houser1918, DeitzDa1986]. **Rhizophoraceae**: *Rhizophora* [DeitzDa1986]. **Rosaceae**: *Crataegus* [Ferris1941d, DeitzDa1986].

DISTRIBUTION: **Nearctic**: Mexico [Cocker1899n] (Michoacan [Ferris1941d], Oaxaca [Ferris1941d], Sonora [Cocker1899d], Veracruz [Ferris1941d]). **Neotropical**: Costa Rica [DeitzDa1986]; Cuba [Houser1918, DeitzDa1986]; El Salvador [DeitzDa1986]; Guatemala [DeitzDa1986]; Haiti [DeitzDa1986]; Honduras [DeitzDa1986]. **Oriental**: Philippines [DeitzDa1986].

BIOLOGY: Occurring concealed beneath bark flakes and in cracks (Ferris, 1941d).

GENERAL: Description and illustration of adult female by Ferris (1941d) and by Deitz & Davidson (1986).

KEYS: Deitz & Davidson 1986: 12-15 (female) [North America]; Ferris 1943: 65 (female) [North America]; Ferris 1942: 37 (female) [North America]; Cockerell 1905: 45-46 (female) [Mexico].

CITATIONS: Borchs1966 [catalogue: 347]; Cocker1898i [description, host, distribution: 440]; Cocker1899n [host, distribution: 27]; Cocker1905 [taxonomy: 46]; DeitzDa1986 [taxonomy, description, illustration, host, distribution: 20-21]; Fernal1903b [catalogue: 289]; Ferris1941d [taxonomy, description, illustration, host, distribution: 350]; Ferris1941e [taxonomy: 41]; Ferris1942 [taxonomy: 446:37];

Ferris1943 [taxonomy: 65]; Houser1918 [host, distribution: 168]; Leonar1900 [taxonomy, host, distribution: 341]; MacGil1921 [taxonomy, description, host, distribution: 441]; McKenz1939 [taxonomy: 53]; Willia1985a [taxonomy: 233].

Melanaspis casuarinae Mamet

Melanaspis casuarinae Mamet, 1954: 65. Type data: MADAGASCAR: Mananjary, on *Casuarina equisetifolia*. Holotype. Type depository: Paris: Muséum national d'Histoire naturelle, France.

SCALE COVER: Female scale of type common to genus (Mamet, 1954)

HOST PLANTS: **Casuarinaceae**: *Casuarina equisetifolia* [Mamet1954, Borchs1966].

DISTRIBUTION: **Afrotropical**: Madagascar [Mamet1954, Borchs1966].

BIOLOGY: Insects occur among crevices of bark of host (Mamet, 1954).

GENERAL: Description and illustration of adult female by Mamet (1954).

CITATIONS: Borchs1966 [catalogue: 347]; Mamet1954 [taxonomy, description, illustration, host, distribution: 19,65-67].

Melanaspis chrysobalani (Leonardi)

Aonidiella chrysobalani Leonardi, 1914: 207. Type data: SENEGAL: Dakar, on *Chrysobalanus* sp. Syntypes, female. Type depository: Portici: Dipartimento de Entomologia e Zoologia Agraria di Portici, Università di Napoli Federico II, Italy.
Pseudischnaspis chrysobalani; Lindinger, 1937: 194. Change of combination.
Melanaspis gliricidiae Hall, 1946: 60. Type data: SIERRA LEONE: Njala, on *Gliricidia maculata*. Syntypes, female. Type depository: London: The Natural History Museum, England, UK. Synonymy by Balachowsky, 1958b: 194.
Melanaspis chrysobalani; Hall, 1946: 62. Change of combination.

SCALE COVER: Female scale irregularly oval, convex; exuviae black, placed more or less towards margin; secreted part of scale compact, white grey; ventral vellum well developed, robust, white; 1-2 mm long (Leonardi, 1914).

HOST PLANTS: **Chrysobalanaceae**: *Chrysobalanus* [Leonar1914, Balach1953e, Balach1958b]. **Leguminosae**: *Acacia nilotica tomentosa* [Almeid1973b], *Gliricidia maculata* [Hall1946, Balach1953e, Balach1958b].

DISTRIBUTION: **Afrotropical**: Angola [Almeid1973b]; Guinea [Balach1953e]; Senegal [Leonar1914, Balach1958b]; Sierra Leone [Hall1946].

GENERAL: Description and illustration of adult female by Leonardi (1914), Hall (1946) and by Balachowsky (1958b).

KEYS: Balachowsky 1958b: 194 (female) [Africa]; Hall 1946: 62 (female) [Africa].

CITATIONS: Almeid1973b [host, distribution: 10]; Balach1953e [host, distribution: 146,149]; Balach1958b [taxonomy, description, illustration, host, distribution: 194-196]; Borchs1966 [catalogue: 347]; Hall1946 [taxonomy, description, illustration, host, distribution: 60-62]; Leonar1914 [taxonomy, description, illustration, host, distribution: 207-209]; Lindin1937 [taxonomy: 194]; McKenz1938 [taxonomy: 3].

Melanaspis coccolobae Ferris

Melanaspis latipyga; Ferris, 1941d: 356. Misidentification; discovered by Ferris, 1943: 59.

Melanaspis coccolobae Ferris, 1943: 59. Type data: U.S.A.: Florida, Miami, on *Coccoloba laurifolia*. Lectotype female, by subsequent designation Deitz & Davidson, 1986: 22. Type depository: Davis: The Bohart Museum of Entomology, University of California, California, USA.

Melanaspis latipiga; Borchsenius, 1966: 347. Misspelling of species name.

SCALE COVER: Female scale circular, moderately convex, hard, brittle, black slightly brown; area over first exuvia white. Male scale not identified (Ferris, 1943).

HOST PLANTS: **Moraceae**: *Ficus* [DeitzDa1986]. **Polygonaceae**: *Coccoloba acapulcensis* [DeitzDa1986], *Co. acuminata* [DeitzDa1986], *Co. barbadensis* [DeitzDa1986], *Co. belizensis* [DeitzDa1986], *Co. caracasana* [DeitzDa1986], *Co. cornata* [DeitzDa1986], *Co. cozumelensis* [DeitzDa1986], *Co. cujabensis* [DeitzDa1986], *Co. diversifolia* [DeitzDa1986], *Co. excoriata* [DeitzDa1986], *Co. floridana* [Dekle1965c], *Co. goldmannii* [DeitzDa1986], *Co. hondurensis* [DeitzDa1986], *Co. humboldtii* [DeitzDa1986], *Co. laurifolia* [Ferris1943], *Co. liebmanni X lundellii* [DeitzDa1986], *Co. manzanillensis* [DeitzDa1986], *Co. obtusifolia* [DeitzDa1986], *Co. reflexiflora* [DeitzDa1986], *Co. retusa* [DeitzDa1986], *Co. spicata* [DeitzDa1986], *Co. striata* [DeitzDa1986], *Co. swartzii* [DeitzDa1986], *Co. swartzii urbaniana* [DeitzDa1986], *Co. tenuiflora* [DeitzDa1986], *Co. tuerckheimii* [DeitzDa1986], *Co. uvifera* [DeitzDa1986], *Co. venosa* [DeitzDa1986]. **Rubiaceae**: *Ixora acuminata* [DeitzDa1986].

DISTRIBUTION: **Antarctica**: Subantarctic islands (Marion & Prince Edward Island). **Nearctic**: Mexico (Campeche [DeitzDa1986], Colima [Ferris1943, DeitzDa1986], Guerrero [Ferris1943, DeitzDa1986], Morelos [DeitzDa1986], Nayarit [DeitzDa1986], Oaxaca [Ferris1943, DeitzDa1986], Sinola [DeitzDa1986], Veracruz [DeitzDa1986]); United States of America (Florida [Ferris1943, Merril1953, Dekle1965c, DeitzDa1986]). **Neotropical**: Bahamas [DeitzDa1986]; Barbados [DeitzDa1986]; Belize [DeitzDa1986]; Cuba [DeitzDa1986]; Dominican Republic [DeitzDa1986]; El Salvador [DeitzDa1986]; Grenada [DeitzDa1986] [Nakaha1982]; Guatemala [DeitzDa1986]; Haiti [DeitzDa1986]; Honduras [DeitzDa1986]; Jamaica [DeitzDa1986]; Martinique [DeitzDa1986]; Mexico (Chiapas [DeitzDa1986], Yucatan [DeitzDa1986]); Netherlands Antilles (Bonaire [DeitzDa1986], Curaçao [DeitzDa1986]); Panama [DeitzDa1986]; Panama Canal Zone [DeitzDa1986]; Puerto Rico & Vieques Island (Puerto Rico [DeitzDa1986]); Trinidad and Tobago (Tobago [DeitzDa1986]); U.S. Virgin Islands [Nakaha1983]; Venezuela [DeitzDa1986].

BIOLOGY: Occurring on bark (Ferris, 1943).

GENERAL: Description and illustration of adult female by Ferris (1943), Deitz & Davidson (1986) and by Colon-Ferrer & Medina-Gaud (1998).

KEYS: Deitz & Davidson 1986: 12-15 (female) [North America]; Ferris 1943: 65 (female) [North America].

CITATIONS: Borchs1966 [catalogue: 347]; ColonFMe1998 [taxonomy, description, illustration, host, distribution: 67-68]; DeitzDa1986 [taxonomy, description, illustration, host, distribution: 22-23]; Dekle1965c [taxonomy, description, host, distribution: 88]; Dekle1976 [taxonomy, description, host, distribution, economic importance: 109]; Ferris1941d [taxonomy, description, illustration, host, distribution: 356]; Ferris1943 [taxonomy, description, illustration, host, distribution: 59,67]; McKenz1944 [taxonomy: 55]; Merril1953 [taxonomy, description, host, distribution: 61-62]; Nakaha1982 [host, distribution: 54]; Nakaha1983 [host, distribution: 13].

Melanaspis corticosa (Brain)

Chrysomphalus (*Pseudischnaspis*) *corticosus* Brain, 1919: 204. Type data: SOUTH AFRICA: Cape Province, Rosebank, on 'keurboom', *Virgilia capensis*; collected by C.W. Mally, 21.viii.1914. Lectotype female, by subsequent designation Munting, 1970a: 37. Type depository: Pretoria: South African National Collection of Insects, South Africa; type no. 240/1.

Chrysomphalus corticosus; MacGillivray, 1921: 419. Change of combination.

Melanaspis corticosus; McKenzie, 1939: 54. Change of combination.

Melanaspis corticosa; Borchsenius, 1966: 347. Change of combination requiring emendation of species name for agreement in gender.

COMMON NAMES: South African obscure Scale [Brain1919].

SCALE COVER: Female scale varying greatly on different host-plants; on smooth-barked plants very large and flat, reaching 3.2 mm in diameter, brownish to black with the blackish exuviae covered; as a rule, however, scale almost or entirely covered by outer layers of bark of the host-plant; on *Rhus* this habit is usual, and it has been submitted on many occasions as a browning scale; on *Robinia* the scale is greyish in appearance, like bark, but black exuviae are very conspicuous with a greyish white concentric ring; on wild olive, it forms a thick crust of blackish or greyish black scales, which easily flake off; scale itself, without any admixture of tissues black; exuviae of similar colour; seen from below scale is domed and very glossy; ventral scale is delicate and usually remains on host (Brain, 1919).

HOST PLANTS: **Anacardiaceae**: *Schinus molle* [Balach1953e], *Sclerocarya caffra* [Almeid1971]. **Celastraceae**: *Celastrus* [Balach1953e]. **Ebenaceae**: *Royena pallens* [Hall1928, Balach1958b]. **Leguminosae**: *Erythrina caffra* [Brain1919, Balach1953e, Balach1958b], *Robinia* [Balach1953e], *Virgilia capensis* [Brain1919, Muntin1970a, Balach1958b].

NATURAL ENEMIES: HYMENOPTERA **Aphelinidae**: *Aphytis faurei* Annecke [RosenDe1979], *Aphytis merceti* Compere [RosenDe1979], *Azotus plesius* Annecke & Insley [AnneckIn1970]. **Encyrtidae**: *Adelencyrtus inglisiae* Compere & Annecke [AnneckIn1971], *Habrolepis obscura* Compere & Annecke [AnneckIn1971].

DISTRIBUTION: **Afrotropical**: Guinea [Balach1953e]; Mozambique [Almeid1971]; South Africa [Brain1919, Balach1958b, Muntin1970a, RosenDe1979]; Zimbabwe [Hall1928].

GENERAL: Description and illustration of adult female by Brain (1919) and by Balachowsky (1958b).

KEYS: Balachowsky 1958b: 194 (female) [Africa]; Hall 1946: 62 (female) [Africa]; Brain 1919: 198 (female) [South Africa].

CITATIONS: Almeid1971 [host, distribution: 13]; AnneckIn1970 [host, distribution, biological control: 241-242]; AnneckIn1971 [host, distribution, biological control: 2]; Balach1953e [host, distribution: 146]; Balach1958b [taxonomy, description, illustration, host, distribution: 195-198]; Borchs1966 [catalogue: 347]; Brain1919 [taxonomy, description, illustration, host, distribution: 198, 204-205]; Comper1961a [biological control: 17-71]; ComperAn1961 [host, distribution, biological control: 17]; Ferris1941d [taxonomy: 347]; Hall1928 [host, distribution: 276]; Hall1946 [taxonomy: 62]; MacGil1921 [taxonomy, description, host, distribution: 419]; Mamet1954 [taxonomy: 67]; McKenz1939 [taxonomy: 54]; Muntin1970a [taxonomy: 37]; Prinsl1983 [distribution, biological control: 27]; RosenDe1979 [host, distribution, biological control: 341-346].

Melanaspis cubensis Deitz & Davidson

Melanaspis cubensis Deitz & Davidson, 1986: 24. Type data: CUBA: Pinar del Rio Province, vicinity of Coloma, on *Coccoloba retusa*. Holotype female. Type depository: Washington: United States National Entomological Collection, U.S. National Museum of Natural History, District of Columbia, USA.
SCALE COVER: Deitz & Davidson (1986) did not describe scale cover.
HOST PLANTS: **Polygonaceae**: *Coccoloba retusa* [DeitzDa1986].
DISTRIBUTION: **Neotropical**: Cuba [DeitzDa1986].
GENERAL: Description & illustration of adult female by Deitz & Davidson (1986).
KEYS: Deitz & Davidson 1986: 12-15 (female) [North America].
CITATIONS: DeitzDa1986 [taxonomy, description, illustration, host, distribution: 24-25].

Melanaspis deklei Deitz & Davidson

Melanaspis deklei Deitz & Davidson, 1986: 26. Type data: U.S.A.: Florida, Columbia Co., Santa Fe River, on *Sebastiania* sp. Holotype female. Type depository: Washington: United States National Entomological Collection, U.S. National Museum of Natural History, District of Columbia, USA.
SCALE COVER: Deitz & Davidson (1986) did not describe scale cover.
HOST PLANTS: **Agavaceae**: *Agave* [DeitzDa1986]. **Aquifoliaceae**: *Ilex* [DeitzDa1986]. **Caprifoliaceae**: *Viburnum* [DeitzDa1986], *Viburnum obovatum* [DeitzDa1986]. **Compositae** [DeitzDa1986], *Iva* [DeitzDa1986]. **Euphorbiaceae**: *Sebastiania* [DeitzDa1986]. **Lauraceae**: *Persea americana* [DeitzDa1986]. **Malvaceae**: *Sida rhombifolia* [DeitzDa1986]. **Myricaceae**: *Myrica cerifera* [DeitzDa1986]. **Sambucaceae**: *Sambucus* [DeitzDa1986].
DISTRIBUTION: **Nearctic**: Mexico [DeitzDa1986]; United States of America (Florida [DeitzDa1986], Georgia [DeitzDa1986]).
GENERAL: Description and illustration of adult female by Deitz & Davidson (1986).
KEYS: Deitz & Davidson 1986: 12-15 (female) [North America].
CITATIONS: DeitzDa1986 [taxonomy, description, illustration, host, distribution: 26-27].

Melanaspis delicata Ferris
Melanaspis delicata Ferris, 1941d: 351. Type data: U.S.A.: Texas, at Carrizo Springs, on "mesquite", *Prosopis* sp. Lectotype female, by subsequent designation Deitz & Davidson, 1986: 28. Type depository: Davis: The Bohart Museum of Entomology, University of California, California, USA.
SCALE COVER: Scale of female perhaps slightly elongate. Male scale not recognized (Ferris, 1941d).
HOST PLANTS: **Leguminosae**: *Acacia berlandieri* [DeitzDa1986], *Acacia constricta* [DeitzDa1986], *Pithecellobium flexicaule* [DeitzDa1986], *Prosopis* [Ferris1941d, DeitzDa1986], *Prosopis chinensis* [DeitzDa1986].
DISTRIBUTION: **Nearctic**: United States of America (Texas [Ferris1941d, McDani1970, DeitzDa1986]).
BIOLOGY: Occurring under bark scales or in cracks (Ferris, 1941d).
GENERAL: Description and illustration of adult female given by Ferris (1941d) and by Deitz & Davidson (1986).
KEYS: Deitz & Davidson 1986: 12-15 (female) [North America]; McDaniel 1970: 411-412 (female) [U.S.A.: Texas]; Ferris 1943: 64 (female) [North America]; Ferris 1942: 37 (female) [North America].
CITATIONS: Borchs1966 [catalogue: 347]; DeitzDa1986 [taxonomy, description, illustration, host, distribution: 28-29]; Ferris1941d [taxonomy, description, illustration, host, distribution: 351]; Ferris1942 [taxonomy: 446:37]; Ferris1943 [taxonomy: 64]; McDani1970 [taxonomy, illustration, host, distribution: 412-413]; Nakaha1982 [host, distribution: 54].

Melanaspis deliquescens Ferris
Melanaspis deliquescens Ferris, 1941d: 352. Type data: U.S.A.: Texas, Chisos Mountains, on *Acacia constricta*. Lectotype female, by subsequent designation Deitz & Davidson, 1986: 30. Type depository: Davis: The Bohart Museum of Entomology, University of California, California, USA.
SCALE COVER: Scales of the type common to the genus (Ferris, 1941d).
HOST PLANTS: **Leguminosae**: *Acacia* [McDani1970], *Acacia constricta* [Ferris1941d, McDani1970, DeitzDa1986].
DISTRIBUTION: **Nearctic**: Mexico (Durango [DeitzDa1986]); United States of America (Texas [Ferris1941d, McDani1970, DeitzDa1986]).
BIOLOGY: Occurring concealed beneath bark flakes and in cracks (Ferris, 1941d).
GENERAL: Description and illustration of adult female by Ferris (1941d) and by Deitz & Davidson (1986).
KEYS: Deitz & Davidson 1986: 12-15 (female) [North America]; McDaniel 1970: 411-412 (female) [U.S.A.: Texas]; Ferris 1943: 66 (female) [North America]; Ferris 1942: 38 (female) [North America].
CITATIONS: Borchs1966 [catalogue: 347]; DeitzDa1986 [taxonomy, description, illustration, host, distribution: 30-31]; Ferris1941d [taxonomy, description, illustration, host, distribution: 352]; Ferris1942 [taxonomy: 446:38]; Ferris1943 [taxonomy: 66]; McDani1970 [taxonomy, illustration, host, distribution: 412-414]; Nakaha1982 [host, distribution: 54].

Melanaspis eglandulosa (Lindinger)

Aspidiotus eglandulosus Lindinger in: Brick, 1909: 10. Type data: GUATEMALA: on *Cereus* sp. Lectotype female and first instar, by subsequent designation Deitz & Davidson, 1986: 32. Type depository: Hamburg: Zoologisches Institut und Zoologishces Museum, Universität von Hamburg, Germany.

Melanaspis eglandulosa; Lindinger, 1931a: 44. Change of combination requiring emendation of species name for agreement in gender.

SCALE COVER: Scale cover not described by Lindinger *in* Brick (1909) nor by Deitz & Davidson (1986).

HOST PLANTS: **Cactaceae** [Brick1909, Ferris1942, DeitzDa1986], *Cephalocereus* [DeitzDa1986], *Cereus* [Ferris1942, DeitzDa1986], *Lemaireocereus* [DeitzDa1986], *Pseudopilocereus superfloccosa* [DeitzDa1986]. **Solanaceae**: *Solanum punctulatum* [DeitzDa1986].

DISTRIBUTION: **Neotropical**: Guatemala [Ferris1942]; Panama [Brick1909, Ferris1942].

GENERAL: Description and illustration of adult female by Deitz & Davidson (1986).

KEYS: Deitz & Davidson 1986: 12-15 (female) [North America].

CITATIONS: Borchs1966 [catalogue: 348]; Brick1909 [taxonomy, description, host, distribution: 10]; DeitzDa1986 [taxonomy, description, illustration, host, distribution: 32-33]; Ferris1941e [taxonomy: 43]; Ferris1942 [taxonomy: 3]; Lindin1909c [taxonomy, host, distribution: 449]; Lindin1931a [taxonomy: 348]; Sander1909a [host, distribution: 52]; WeidneWa1968 [taxonomy: 172].

Melanaspis elaeagni McKenzie

Melanaspis elaeagni McKenzie, 1957: 218. Type data: U.S.A.: Texas, Harlingen, three-fourths mile south Dilworth Road, on *Elaeagnus* sp., probably *pungens*. Holotype female. Type depository: Washington: United States National Entomological Collection, U.S. National Museum of Natural History, District of Columbia, USA.

COMMON NAME: Black elaegnus scale [McKenz1957].

SCALE COVER: Female scale circular, 1 mm in diameter; dark chocolate brown or black but overlain with a film of white wax, or by epidermis of plant, thus often appearing mottled with pale grey. Male scale elongate, similar in colour, but smaller than female; exuvia near one end (McKenzie, 1957).

HOST PLANTS: **Elaeagnaceae**: *Elaeagnus* [McKenz1957, McDani1970].

DISTRIBUTION: **Nearctic**: United States of America (Louisiana [DeitzDa1986], Texas [McKenz1957, McDani1970]). **Neotropical**: Costa Rica [DeitzDa1986]; Guatemala [DeitzDa1986]; Honduras [DeitzDa1986]; Mexico (Chiapas [DeitzDa1986]).

BIOLOGY: Occurring on the leaves (McKenzie, 1957).

GENERAL: Description and illustration of adult female by McKenzie (1957) and by Deitz & Davidson (1986).

KEYS: Deitz & Davidson 1986: 12-15 (female) [North America]; McDaniel 1970: 411-412 (female) [U.S.A.: Texas].

CITATIONS: Borchs1966 [catalogue: 348]; DeitzDa1986 [taxonomy, description, illustration, host, distribution: 34-35]; McDani1970 [taxonomy, host, distribution: 415]; McKenz1957 [taxonomy, description, illustration, host, distribution: 218-220]; Nakaha1982 [host, distribution: 54].

Melanaspis enceliae (Ferris)

Chrysomphalus enceliae Ferris, 1921: 129. Type data: MEXICO: Baja California, on *Encelia palmeri*. Lectotype female, by subsequent designation Deitz & Davidson, 1986: 36. Type depository: Davis: The Bohart Museum of Entomology, University of California, California, USA.

Melanaspis enceliae; Lindinger, 1931a: 44. Change of combination.

SYSTEMATICS: Lindinger (1957) suggested that *Melanaspis enceliae* is a synonym of *M. eglandulosa*, while Deitz & Davidson (1986) regarded these as distinct species.

SCALE COVER: Scale of female roughly circular, 2-3 mm in diameter; flat, black, hard and brittle. Male scale not identified (Ferris, 1921).

HOST PLANTS: Compositae: *Encelia palmeri* [Ferris1921, DeitzDa1986].

DISTRIBUTION: Nearctic: Mexico (Baja California [Ferris1921, DeitzDa1986]).

BIOLOGY: Scales buried in soft cortex of host (Ferris, 1941d).

GENERAL: Description and illustration of adult female by Ferris (1921, 1941d) and by Deitz & Davidson (1986).

KEYS: Deitz & Davidson 1986: 12-15 (female) [North America]; Ferris 1943: 64 (female) [North America]; Ferris 1942: 36 (female) [North America].

CITATIONS: Borchs1966 [catalogue: 348]; DeitzDa1986 [taxonomy, description, illustration, host, distribution: 36-37]; Ferris1921 [taxonomy, description, illustration, host, distribution: 129-130]; Ferris1941d [taxonomy, description, illustration, host, distribution: 353]; Ferris1942 [taxonomy: 446:36]; Ferris1943 [taxonomy: 64]; Lindin1931a [taxonomy: 44]; Lindin1957 [taxonomy: 550]; McKenz1939 [taxonomy: 54].

Melanaspis figueiredoi Lepage

Melanaspis figueiredoi Lepage, 1942: 184. Type data: BRAZIL: São Paolo State, Santos, on *Eugenia* sp. Syntypes, female. Type depositories: São Paulo: Museu de Zoologia, Universidade de São Paulo, Brazil, São Paulo: Instituto Biologico de São Paulo, Brazil, and Curitiba: Departamento de Zoologia, Setor de Ciencias Biologicas, Universidade Federal do Parana, Brazil.

SCALE COVER: Female scale 2 mm diameter, circular; colour bright; exuviae black placed centrally or subcentrally; ventral scale thin, attached to host plant (Lepage, 1942).

HOST PLANTS: Myrtaceae: *Eugenia* [Lepage1942, ClapsWoGo2001].

DISTRIBUTION: Neotropical: Brazil (São Paulo [Lepage1942, ClapsWoGo2001]).

GENERAL: Description and illustration of adult female by Lepage (1942).

CITATIONS: Borchs1966 [catalogue: 348]; Claps1993 [taxonomy: 6,9]; ClapsWoGo2001 [host, distribution: 246]; Lepage1942 [taxonomy, description, illustration, host, distribution: 184-186]; Vernal1953 [taxonomy: 179].

Melanaspis frankiniae Ferris

Melanaspis frankiniae Ferris, 1941d: 354. Type data: MEXICO: Baja California, Northern District, Miller's Landing, which is near the northern limit of San Sebastian Vizcaino Bay, on *Frankinia* [=*Frankenia*] *palmeri*. Lectotype female, by subsequent designation Deitz & Davidson, 1986: 38. Type depository: Davis: The Bohart Museum of Entomology, University of California, California, USA.

SCALE COVER: Scale of female slightly elongate, but otherwise of form characteristic of genus; that of male not recognized (Ferris, 1941d).

HOST PLANTS: **Frankeniaceae**: *Frankenia palmeri* [Ferris1941d, DeitzDa1986].

DISTRIBUTION: **Nearctic**: Mexico (Baja California [Ferris1941d, DeitzDa1986]).

BIOLOGY: Occurring on bark of twigs (Ferris, 1941d).

GENERAL: Description and illustration of adult female by Ferris (1941d) and by Deitz & Davidson (1986).

KEYS: Deitz & Davidson 1986: 12-15 (female) [North America]; Ferris 1943: 64 (female) [North America]; Ferris 1942: 37 (female) [North America].

CITATIONS: Borchs1966 [catalogue: 348]; DeitzDa1986 [taxonomy, description, illustration, host, distribution: 38-39]; Ferris1941d [taxonomy, description, illustration, host, distribution: 354]; Ferris1942 [taxonomy: 446:37]; Ferris1943 [taxonomy: 64].

Melanaspis fucata Ferris

Melanaspis fucata Ferris, 1941d: 355. Type data: MEXICO: State of Oaxaca, Ayutla, on *Diospyros ebenaster*. Holotype female. Type depository: Davis: The Bohart Museum of Entomology, University of California, California, USA.

SCALE COVER: Female scale flat, circular with slight prolongation to one side; reddish brown around margins, paler in central portion; outer layer of wax film tends to become detached from inner layers and consequently central region of older scales becomes white; thin, black ventral scale present. Scale of male slightly elongate, similar to female in colour (Ferris, 1941d).

HOST PLANTS: **Ebenaceae**: *Diospyros ebenaster* [Ferris1941d].

DISTRIBUTION: **Nearctic**: Mexico (Oaxaca [Ferris1941d]).

BIOLOGY: Occurring on leaves (Ferris, 1941d).

GENERAL: Description and illustration of adult female by Ferris (1941d).

KEYS: Ferris 1943: 64 (female) [North America]; Ferris 1942: 36 (female) [North America].

CITATIONS: Borchs1966 [catalogue: 348]; Ferris1941d [taxonomy, description, illustration, host, distribution: 355]; Ferris1942 [taxonomy: 446:36]; Ferris1943 [taxonomy: 64]; Lindin1957 [taxonomy: 550].

Melanaspis glomerata (Green)

Aspidiotus (*Targionia*) *glomeratus* Green, 1903a: 93. Type data: INDIA: on *Saccharum officinale*; collected by G. Watt. Syntypes, female. Type depository: London: The Natural History Museum, England, UK.

Targionia glomerata; Fernald, 1903b: 297. Change of combination requiring emendation of species name for agreement in gender.

Aonidiella glomerata; MacGillivray, 1921: 445. Change of combination.

Aspidiotus glomeratus; McKenzie, 1938: 3. Change of combination.
Melanaspis glomerata; Lindinger, 1943a: 147. Change of combination.
Aonidiella gromerata; Borchsenius, 1966: 348. Misspelling of species name.
COMMON NAMES: black araucaria scale [McKenz1956]; black scale [NarayaSuKa1957]; sugarcane scale insect [TewariTr1979].
SCALE COVER: Female scales crowded and adhering together so that it is difficult to isolate a single individual; form irregularly circular, diameter 2.5 mm; slightly convex; smoky-brown or greyish-black. Male scale similar to female, but much smaller and more oval; length 1 mm (Green, 1903a).
HOST PLANTS: **Gramineae**: *Saccharum officinalis* [Green1903a, Ali1962].
NATURAL ENEMIES: COLEOPTERA **Coccinellidae**: *Chilocorus nigritus* [RavindSu1995], *Pharoscymnus horni* [RavindSu1995], *Sticholotis cribellata* Sicard [SunilBaRa2001], *Sticholotis madagassa* Weise [ChandrAv1986]. HYMENOPTERA **Aphelinidae**: *Azotus chionaspidis* Howard [WilliaGr1990], *Azotus delhiensis* Lal [NarayaSuKa1957], *Azotus delthiensis* Lal [WilliaGr1990], *Botryoideclava bharatiya* Subba Rao [EaswarDaGa1996], *Marietta cheriani* (Mani) [NarayaSuKa1957], *Marietta leopardina* Motschulsky [WilliaGr1990]. **Ceraphronidae**: *Ceraphron* sp. [NarayaSuKa1957]. **Encyrtidae**: *Adelencyrtus bifasciata* (Ishii) [WilliaGr1990], *Adelencyrtus femoralis* Compere & Annecke [WilliaGr1990], *Adelencyrtus mayurai* (Subba Rao) [NarayaSuKa1957, WilliaGr1990], *Xanthoencyrtus fullawayi* Timberlake [WilliaGr1990]. **Eulophidae**: *Tetrastichus lecanii* (Girault) [WilliaGr1990]. THYSANOPTERA **Phlaeothripidae**: *Podothrips lucasseni* Kruger [PalmerMo1990].
DISTRIBUTION: **Oriental**: India [Ramakr1921a] (Bihar [Ali1962], Delhi [NarayaSuKa1957], Uttar Pradesh [Gupta1978]).
ECONOMIC IMPORTANCE: Causes serious loss to sugarcane crops in India (Williams & Greathead, 1990).
GENERAL: Description and illustration of adult female by Green (1903a).
CITATIONS: Agarwa1956 [taxonomy, description, host, distribution, economic importance, biological control, chemical control: 24]; Agarwa1969 [chemistry, biological control: 767-776]; AgarwaSh1960 [host, distribution, economic importance, biological control: 197-203]; AgarwaShKa1959 [life history, ecology: 462-463]; AgarwaSi1964 [host, distribution, economic importance: 149]; Ali1962 [host, distribution, economic importance, life history: 74-75]; AnsariPaMu1989 [host, distribution, biological control, economic importance: 21-23]; Borchs1966 [catalogue: 348]; Box1953 [host, distribution, biological control: 51]; BurgerUl1990 [economic importance: 321]; ChandrAv1986 [biological control: 194-196]; DavidEa1986 [host, distribution, biological control: 383-421]; DorgeDaPr1972 [biological control: 138-141]; Dutta1992 [host, distribution, life history, control: 337-342]; DuttaDe1987 [taxonomy, description: 30-35]; DuttaDe1987a [biological control, host, distribution: 54-56]; DuttaDe1990 [host, distribution, control: 161-163]; DuttaDe1994 [biological control: 19-21]; DuttaDe1995 [taxonomy, description, biological control: 39-47]; DuttaDe1995a [biological control: 75-78]; Easwar1986 [host, distribution, economic importance, ecology, biological control: 31-67]; Easwar1986a [host, distribution, economic importance, ecology, biological control: 437-457]; EaswarDaBa1994 [chemical control: 14-17]; EaswarDaGa1996

[host, distribution, life history, biological control: 55-64]; EaswarKu1986 [host, distribution, economic importance, life history, chemical control, biological control: 233-258]; EaswarRa1983 [host, distribution, ecology, biological control: 64-65]; Fernal1903b [catalogue: 297]; Ferris1941e [taxonomy: 44]; Ferris1943a [taxonomy: 86]; Greath1973 [biological control: 29-33]; Greath1989 [biological control: 28-37]; Green1903a [taxonomy, description, illustration, host, distribution: 93]; Gupta1978 [host, distribution, life history, economic importance, control: 67-74]; HanksDe1998 [life history, ecology: 239-262]; JayantDa1986 [structure, life history, biological control: 363-381]; JayantDaGo1994 [life history, structure, biological control: 305-314]; JayantDaGo1996 [host, life history, behaviour, ecology: 7-9]; JayantGo2000 [life history, physiology: 11]; JhansiBa2002 [host, distribution, chemical control: 615-618]; Khanna1957 [host, distribution: 19-22]; Lindin1943a [taxonomy: 147]; MaLi2002 [biological control: 21-26]; MacGil1921 [taxonomy, description, host, distribution : 445]; McClur1990g [taxonomy, host, distribution, ecology: 319-330]; McKenz1938 [taxonomy: 3]; NarayaSuKa1957 [biological control, host, distribution: 144-146]; NorrisKo1980 [biological control: 22-61]; Noyes1990a [biological control: 153]; PalmerMo1990 [biological control: 67-76]; ParsanMaKo1994 [host, distribution, life history, ecology: 15-17]; PrasadVeMu1991 [host, distribution, economic importance, chemical control: 516-520]; PruthiRa1942 [host, distribution: 87-88]; RajuRa1983 [host, distribution, chemical control, biological control: 571-572]; RajuSe1982 [host, distribution, life history: 19-20]; Ramakr1921a [host, distribution: 357]; Rao1957 [biological control: 376-390]; RaoRa1990 [chemical control, biological control: 60]; RaoRa1990a [chemical control: 173-175]; RaoRa1992 [life history, chemical control: 125-128]; RaoRaSa1983 [chemical control: 11-18]; RaoSa1969 [host, distribution, economic importance: 325-342]; RavindSu1994 [host, distribution, economic importance, chemical control: 333-343]; RavindSu1995 [host, distribution, biological control: 117-119]; RavindSu1998 [host, distribution, biological control: 159-161]; ShuklaTr1979 [host, distribution, life history: 535-536]; ShuklaTr1981 [host, distribution, life history: 101-102]; ShuklaTr1983 [host, distribution, life history: 160-162]; SunilBaRa2001 [host, distribution, biological control: 26-31]; SunilPoSi2002 [biological control: 21-26]; TewariTr1979 [economic importance, chemical control, biological control: 19-21]; TripatSh1982 [host, distribution, life history, economic importance: 83-84]; TripatTe1984 [host, distribution, life history, economic importance: 68-78]; UpadhyVa1993 [host, distribution, economic importance: 26-29]; Varun1993 [host, distribution, control, economic importance: 219-220]; WilliaGr1990 [host, distribution, economic importance, biological control: 553-578].

Melanaspis indurata (Ferris)

Chrysomphalus induratus Ferris, 1921: 130. Type data: MEXICO: Baja California, La Laguna, on *Pinus cembroides*; collected by G.F. Ferris, August 1919. Lectotype female, by subsequent designation Deitz & Davidson, 1986: 40. Type depository: Davis: Bohart Museum of Entomology, University of California, California, USA.

Melanaspis indurata; Lindinger, 1931a: 44. Change of combination requiring emendation of species name for agreement in gender.

Chrysomphalus induratus; Borchsenius, 1966: 349-350. Incorrect synonymy. Notes: Incorrect synonymy with *Melanaspis mimosae* (Cockerell); see Deitz & Davidson, 1986: 40.

SYSTEMATICS: Lindinger (1957) and Borchsenius (1966) synonymized *Chrysomphalus induratus* Ferris, 1921 with *Melanaspis mimosae* (Comstock). Deitz & Davidson regarded *Melanaspis indurata* a distinct species.

SCALE COVER: Scale of female circular, about 2 mm in diameter; black, flat, hard and brittle; ventral scale very thin. Male scale not identified (Ferris, 1921).

HOST PLANTS: **Pinaceae**: *Pinus cembroides* [Ferris1921, DeitzDa1986].

DISTRIBUTION: **Nearctic**: Mexico (Baja California [DeitzDa1986]).

GENERAL: Description and illustration of adult female by Ferris (1921) and by Deitz & Davidson (1986).

KEYS: Deitz & Davidson 1986: 12-13 (female) [North America].

CITATIONS: Borchs1966 [taxonomy: 349-350]; DeitzDa1986 [taxonomy, description, illustration, host, distribution: 40-41]; Ferris1921 [taxonomy, description, illustration, host, distribution: 130-132]; Lindin1931a [taxonomy: 44]; Lindin1957 [taxonomy: 547]; McKenz1939 [taxonomy: 54].

Melanaspis inopinata (Leonardi)

Aonidiella inopinata Leonardi, 1913b: 63. Type data: ITALY: Sicily, Siracusa Province, on "Mandorlo" and pear, and Foggia Province, on "Peri selvatici". Syntypes, female. Type depository: Portici: Dipartimento de Entomologia e Zoologia Agraria di Portici, Università di Napoli Federico II, Italy.

Aonidiella robusta Grassi & Berlese in: Berlese, 1915: 409. Type data: ITALY: Rome, on pear. Syntypes, female. Type depository: Portici: Dipartimento de Entomologia e Zoologia Agraria di Portici, Università di Napoli Federico II, Italy. Synonymy by Borchsenius, 1966: 348.

Melanaspis inopinata; Bodenheimer, 1924: 37. Change of combination.

Chrysomphalus robusta; Koroneos, 1934: 23. Change of combination.

Aspidiotus (*Aonidiella*) *inopinata*; McKenzie, 1938: 3. Change of combination.

Melanaspis robustus; McKenzie, 1939: 54. Change of combination requiring emendation of species name for agreement in gender.

Chrysomphalus inopinatus; Ferris, 1941d: 347. Change of combination requiring emendation of species name for agreement in gender.

Pelomphala inopinata; Lupo, 1954a: 35. Change of combination.

Melanaspis inopinata; Borchsenius, 1966: 348. Revived combination.

COMMON NAME: steklovidnaya shitovka [Borchs1936].

SCALE COVER: Female scale circular, maximum diameter 2.1 mm; robust, highly convex; exuviae eccentric but not marginal, black; ventral vellum robust, white, remains attached to host plant (Leonardi, 1913b). Female scale very robust; very convex; circular, diameter 2-2.75 mm; greyish brown in colour; exuviae jet black, eccentric, but not marginal; secretionary covering greyish brown often worn off above exuviae; ventral scale well developed, but greater part remains adherent to plant on lifting back dorsal scale (Hall, 1923).

HOST PLANTS: **Ericaceae**: *Arbutus unedo* [Bodenh1949, Bodenh1952]. **Leguminosae**: *Cercis siliquastrum* [Balach1951], *Sophora japonica* [Balach1951].

Oleaceae: *Fraxinus* [Bodenh1944a, Balach1951, Zahrad1972], *Jasminum fruticans* [Bodenh1949, Bodenh1952]. **Pistaciaceae**: *Pistacia* [Bodenh1937], *Pistacia lentiscus* [Balach1951, InserrCa1987], *Pistacia palestina* [Bodenh1935], *Pistacia terebinthus* [Bodenh1924, Bodenh1926, Bodenh1927b, Balach1951], *Pistacia vera* [Balach1951]. **Rhamnaceae**: *Rhamnus rodopea* [Bodenh1949, Bodenh1952]. **Rosaceae**: *Amygdalus orientalis* [Bodenh1949, Bodenh1952], *Cotoneaster* [DeLillPo1993], *Crataegus* [RosenDe1979], *Crataegus monogyna* [Balach1951], *Malus* [Borchs1936, Balach1951], *Prunus* [Balach1951], *Prunus armeniaca* [Hall1923], *Prunus avium* [Borchs1936], *Pyrus* [Balach1951], *Pyrus communis* [Hall1923, Borchs1936, Bodenh1949, Bodenh1952], *Pyrus eleagnifolia* [Bodenh1949, Bodenh1952], *Pyrus malus* [Hall1923], *Pyrus nigra* [ErlerTu2001], *Pyrus silvatica* [Leonar1913b]. **Salicaceae**: *Populus nigra* [ErlerTu2001]. **Ulmaceae**: *Celtis tournefori* [Balach1951].

NATURAL ENEMIES: ACARI **Ascidae**: *Cheletogenes ornatus* (Canestri & Fanzago) [ErlerTu2001]. **Pyemotidae**: *Pyemotes herfsi* (Oudeman) [DeLillPo1993]. COLEOPTERA **Coccinellidae**: *Chilocorus bipustulatus* (Linnaeus) [ErlerTu2001], *Rhyzobius lophanthae* (Blaisdell) [ErlerTu2001]. **Nitidulidae**: *Cybocephalus fodoriminor* Enrody-Younga [ErlerTu2001]. DIPTERA **Cecidomyiidae**: *Lestodiplosis aonidiellae* Harris [ErlerTu2001]. HYMENOPTERA **Aphelinidae**: *Ablerus celsus* (Walker) [ErlerTu2001], *Aphytis chrysomphali* (Mercet) [RosenDe1979], *Coccobius tescaceus* (Masi) [ErlerTu2001], *Coccophagoides moeris* (Walker) [ErlerTu2001]. THYSANOPTERA **Phlaeothripidae**: *Karnyothrips flavipes* (Jones) [PalmerMo1990].

DISTRIBUTION: **Palaearctic**: Armenia [Borchs1936, Balach1951, Danzig1993]; Egypt [Hall1923, Ezzat1958]; Greece [Korone1934, RosenDe1979]; Iran [MehrneAk2001]; Iraq [Bodenh1944a, Balach1951]; Israel [Bodenh1924, Bodenh1927b, Bodenh1937, Balach1951]; Italy [Balach1951, LongoMaPe1995]; Sicily [Leonar1920, InserrCa1987]; Turkey [Bodenh1949, KaydanUlTo2002].

ECONOMIC IMPORTANCE: Pest of fruit trees, mainly of Rosaceae, in southern Europe, Middle East and Armenia (Balachowsky, 1951; Schmutterer et al., 1957).

GENERAL: Description and illustration of adult female by Leonardi (1913b), Balachowsky (1951) and by Yaşar (1995a).

KEYS: Danzig 1993: 238 (female) [world]; Ezzat 1958: 241 (female) [Egypt]; Balachowsky 1951: 579 (female) [Mediterranean]; Hall 1946: 62 (female) [Africa]; Borchsenius 1937a: 62 (female) [Palaearctic Region]; Leonardi 1920: 75 (female) [Italy].

CITATIONS: Balach1951 [taxonomy, description, illustration, host, distribution: 580-583]; Bodenh1924 [taxonomy, description, host, distribution: 37-38]; Bodenh1926 [host, distribution: 42]; Bodenh1927b [host, distribution: 80]; Bodenh1935 [host, distribution: 247]; Bodenh1937 [host, distribution: 217]; Bodenh1944a [host, distribution: 81]; Bodenh1949 [taxonomy, description, illustration, host, distribution: 79-81]; Bodenh1952 [taxonomy, illustration, host, distribution, structure: 346-348]; Borchs1936 [host, distribution: 137-138]; Borchs1937 [taxonomy, description, illustration, host, distribution: 122]; Borchs1937a [taxonomy, host, distribution: 62-63]; Borchs1939a [taxonomy, distribution: 11,42-43]; Borchs1949d [taxonomy, description, host, distribution:

235]; Borchs1950b [taxonomy, description, illustration, host, distribution: 160,222]; Borchs1966 [catalogue: 348-349]; BurgerUl1990 [economic importance: 313-327]; Bustsh1958 [taxonomy, description, host, distribution: 220,223]; Danzig1972 [taxonomy, host, distribution, economic importance: 217]; Danzig1993 [taxonomy, description, illustration, host, distribution: 238-239]; DanzigPe1998 [catalogue: 304]; DeLillPo1993 [biological control, host, distribution: 117-124]; ErlerTu2001 [host, distribution, biological control: 299-305]; Ezzat1958 [distribution: 241]; EzzatNa1987 [distribution: 87]; Ferris1941d [taxonomy: 347]; Gentry1965 [host, distribution, economic importance]; Hall1923 [taxonomy, description, host, distribution: 18]; Hall1946 [taxonomy: 62]; InserrCa1987 [host, distribution: 95]; KatsoySt1997 [host, distribution, life history: 331-332]; Kaussa1955 [host, distribution: 16]; KaydanUlTo2002 [host, distribution: 253-257]; Kiriuk1947 [host, distribution: 21]; Konsta1976 [host, distribution, economic importance: 49-50]; KonstaGu1987 [host, distribution: 162]; Korone1934 [taxonomy, description, illustration, host, distribution: 23]; Leonar1913b [taxonomy, description, illustration, host, distribution: 83-85]; Leonar1920 [taxonomy, description, illustration, host, distribution: 83-85]; LongoMaPe1995 [distribution: 127]; Lupo1954a [taxonomy, description, illustration, host, distribution: 35-40]; Lupo1957a [taxonomy: 422]; Maleno1927 [taxonomy: 52]; McKenz1938 [taxonomy: 3-4]; McKenz1939 [taxonomy: 54]; MehrneAk2001 [host, distribution: 315-322]; MillerDa1990 [host, distribution, economic importance: 303]; PalmerMo1990 [biological control: 67-76]; Porcel1995 [structure: 25-45]; PriesnHo1940 [biological control: 58-70]; RosenDe1979 [host, distribution, biological control: 593-598]; RSEA1915 [host, distribution, description, life history, economic importance, control: 1]; SchmutKlLu1957 [host, distribution, economic importance: 492]; Ulgent1996 [host, distribution: 451-548]; Yasar1995a [taxonomy, description, illustration, host, distribution: 94-96]; Zahrad1972 [host, distribution: 440].

Melanaspis jaboticabae (Hempel)

Aspidiotus jaboticabae Hempel, 1918: 206. Type data: BRAZIL: State of São Paolo, S. Manuel, on *Eugenia jaboticabae*. Holotype female; type no. 17350.
Targionia jaboticabae; Lepage & Giannotti, 1943: 337. Change of combination.
Melanaspis jaboticabae; Borchsenius, 1966: 349. Change of combination.
SCALE COVER: Female scale circular to sub-circular; hard, opaque and variable, some being more convex then others; colour dark brown to black; with inner surface lined with a white secretion (Hempel, 1918).
HOST PLANTS: **Myrtaceae**: *Eugenia jaboticaba* [Hempel1918, Lepage1938, LepageGi1943, ClapsWoGo2001].
DISTRIBUTION: **Neotropical**: Brazil (São Paulo [Hempel1918, Lepage1938, LepageGi1943, ClapsWoGo2001]).
GENERAL: Description and illustration of adult female by Hempel (1918) and by Lepage & Giannotti (1943).
CITATIONS: Borchs1966 [catalogue: 349]; ClapsWoGo2001 [host, distribution: 252]; Ferris1941e [taxonomy: 44]; Hempel1918 [taxonomy, description, host, distribution: 206-208]; Lepage1938 [catalogue: 395]; LepageGi1943 [taxonomy, description, illustration, host, distribution: 337-339].

Melanaspis jamaicensis Davidson

Melanaspis jamaicensis Davidson, 1970: 33. Type data: U.S.A.: Florida, Miami, intercepted from Jamaica, on bromeliad leaves. Holotype female. Type depository: Washington: United States National Entomological Collection, U.S. National Museum of Natural History, District of Columbia, USA.

SCALE COVER: Female scale circular, approximately 1.5 mm in diameter; moderately convex; greyish in colour; exuviae subcentral, black; exuviae and cover overlain with a brownish grey film of plant tissue. Male scale elongate, similar in colour but smaller than female; exuviae near one end (Davidson, 1970). Illustration of scale cover by Davidson (1970).

HOST PLANTS: **Bromeliaceae**: [Davids1970].

DISTRIBUTION: **Nearctic**: United States of America (Florida [Davids1970]).

GENERAL: Description and illustration of adult female by Davidson (1970) and by Deitz & Davidson (1986).

KEYS: Deitz & Davidson 1986: 12-15 (female) [North America].

CITATIONS: Davids1970 [taxonomy, description, illustration, host, distribution: 33-36]; DeitzDa1986 [taxonomy, description, illustration, host, distribution: 42-43].

Melanaspis latipyga Ferris

Melanaspis latipyga Ferris, 1941d: 356. Type data: PANAMA: Chiriqui Province, altitude about 7000 feet, on the Volcan de Chiriqui, on undetermined woody vine. Lectotype female, by subsequent designation Deitz & Davidson, 1986: 44. Type depository: Davis: The Bohart Museum of Entomology, University of California, California, USA.

SCALE COVER: Female scale of type common to genus; male scale not recognized (Ferris, 1941d).

HOST PLANTS: **Compositae** [Ferris1941d], *Senecio* [Ferris1943]. **Oleaceae**: *Fraxinus* [Ferris1941d, McDani1970]. **Piperaceae**: *Piper* [Ferris1941d]. **Salicaceae**: *Salix* [Ferris1941d, McDani1970]. **Ulmaceae**: *Celtis* [Ferris1941d], *Celtis mississippiensis* [McDani1970].

DISTRIBUTION: **Nearctic**: Mexico (Guerrero [Ferris1941d], Michoacan [Ferris1941d], Oaxaca [Ferris1941d], Veracruz [Ferris1941d]); United States of America (Texas [Ferris1941d, McDani1970]). **Neotropical**: Costa Rica [Nakaha1982]; Panama [Ferris1941d].

BIOLOGY: Associated with fungus, *Septobasidium* (Ferris, 1943).

GENERAL: Description and illustration of adult female by Ferris (1941d) and by Deitz & Davidson (1986).

KEYS: Deitz & Davidson 1986: 12-15 (female) [North America]; McDaniel 1970: 411-412 (female) [U.S.A.: Texas]; Ferris 1943: 65 (female) [North America]; Ferris 1942: 37 (female) [North America].

CITATIONS: Borchs1966 [catalogue: 349]; Ferris1941d [taxonomy, description, illustration, host, distribution: 356]; Ferris1942 [taxonomy: 446:37]; Ferris1943 [taxonomy, host, distribution: 58-59]; Lindin1957 [taxonomy: 550]; McDani1970 [taxonomy, illustration, host, distribution: 415-416]; Nakaha1982 [host, distribution: 55].

Melanaspis leivasi (Costa Lima)

Aonidiella leivas Costa Lima, 1924a: 197. Type data: BRAZIL: Rio Grando do Sul, Pelotas, on *Ficus* sp. Holotype female. Type depository: Rio de Janeiro: Colecao Entomologica, Escola de Postgraduacao em Parasitologia Animal, Universidade Federal Rural de Rio de Janeiro, Brazil.

Aspidiotus (*Aonidiella*) *leivasi*; McKenzie, 1938: 3. Change of combination.

Aspidiotus (*Aonidiella*) *leivasi*; McKenzie, 1938: 3. Justified emendation.

Melanaspis calcarata Ferris, 1941d: 349. Type data: MEXICO: State of Guerrero, near Acapulco, La Providencia, on *Bursera* (= *Elaphrium*) sp. Holotype female. Type depository: Davis: The Bohart Museum of Entomology, University of California, California, USA. Synonymy by Borchsenius, 1966: 349.

Melanaspis leivasi; Costa Lima, 1942: 289. Change of combination.

SCALE COVER: Female scale of type common to genus (Ferris, 1941d).

HOST PLANTS: **Anacardiaceae**: *Anacardium excelsum* [DeitzDa1986]. **Burseraceae**: *Bursera* [Ferris1941d, LepageGi1943, DeitzDa1986, ClapsWoGo2001]. **Moraceae**: *Ficus* [CostaL1924, Lepage1938, Ferris1941d, LepageGi1943, DeitzDa1986, ClapsWoGo2001]. **Vitaceae**: *Vitis* [Lepage1938, LepageGi1943].

DISTRIBUTION: **Nearctic**: Mexico (Guerrero [Ferris1941d, LepageGi1943], Michoacan [Ferris1941d, LepageGi1943]). **Neotropical**: Brazil (Bahia [Lepage1938, LepageGi1943, ClapsWoGo2001], Rio Grande do Sul [CostaL1924a, Lepage1938, LepageGi1943, ClapsWoGo2001], Rio de Janeiro [DeitzDa1986]); Colombia [DeitzDa1986]; Guatemala [DeitzDa1986]; Panama [Ferris1941d, LepageGi1943, DeitzDa1986].

BIOLOGY: Occurring on bark (Ferris, 1941d).

GENERAL: Description and illustration of adult female by Ferris (1941d), Lepage & Giannotti (1943) and by Deitz & Davidson (1986).

KEYS: Deitz & Davidson 1986: 12-15 (female) [North America]; Ferris 1942: 36 (female) [North America].

CITATIONS: BiezanSe1940 [host, distribution: 67-68]; Borchs1966 [catalogue: 349]; ClapsWoGo2001 [host, distribution: 246]; CostaL1924a [taxonomy, description, illustration, host, distribution: 197-199]; CostaL1942 [taxonomy: 289]; DeitzDa1986 [taxonomy, description, illustration, host, distribution: 46-47]; Ferris1941d [taxonomy, description, illustration, host, distribution: 349]; Ferris1942 [taxonomy: 446:36]; Lepage1938 [catalogue: 392]; LepageGi1943 [taxonomy, description, illustration, host, distribution: 341-343]; McKenz1938 [taxonomy: 3].

Melanaspis lilacina (Cockerell)

Aspidiotus (*Chrysomphalus*) *lilacinus* Cockerell, 1898m: 26. Type data: U.S.A.: New Mexico, Dripping Spring, Organ Mountains, on bark of oak, a species of the *Quercus undulata* group; collected April 23, 1898. Lectotype female, by subsequent designation Deitz & Davidson, 1986: 48. Type depository: Washington: United States National Entomological Collection, U.S. National Museum of Natural History, District of Columbia, USA.

Chrysomphalus (*Melanaspis*) *lilacinus*; Cockerell, 1899a: 396. Change of combination.

Chrysomphalus lilacinus; Cockerell, 1899d: 170. Change of combination.
Aonidiella lilacina; Leonardi, 1900: 341. Change of combination.
Pelomphala lilacina; MacGillivray, 1921: 442. Change of combination.
Aspidiotus lilacinus; Ferris, 1937c: 52. Change of combination.
Melanaspis lilacinus; McKenzie, 1939: 54. Change of combination.
Aspidiotus lalicinus; Borchsenius, 1966: 349. Misspelling of species name.
Melanaspis lalicina; Borchsenius, 1966: 349. Misspelling of species name.
Pelomphaga lilacina; Borchsenius, 1966: 349. Misspelling of genus name.
COMMON NAME: dark oak scale [McKenz1956].
SCALE COVER: Female scale about 1 mm in diameter; light grey, similar to colour of bark on which it rests; immature scales show a white dot and ring; old scales when rubbed show jet-black exuviae (Cockerell, 1898m).
HOST PLANTS: **Fagaceae**: *Quercus* [Cocker1898m, DeitzDa1986], *Qu. agrifolia* [McKenz1956], *Qu. engelmanni* [Ferris1941d], *Qu. texana* [McKenz1956, McDani1970], *Qu. undulata* [Cocker1899d, Ferris1941d, McKenz1956].
NATURAL ENEMIES: HYMENOPTERA **Signiphoridae**: *Chartocerus fasciatus* (Girault) [Gordh1979], *Signiphora flavopelliata* Ashmead [Woolle1990].
DISTRIBUTION: **Nearctic**: Mexico (Baja California [DeitzDa1986], Sonora [Ferris1941d]); U.S.A.: (Arizona [DeitzDa1986], California [Ferris1941d, DeitzDa1986], New Mexico [Cocker1898m, McDani1970, DeitzDa1986], Oklahoma [Nakaha1982], Texas [Ferris1941d]). **Neotropical**: Mexico (Chihuahua [Ferris1941d]).
BIOLOGY: Occurring on bark, usually not much concealed (Ferris, 1941d).
ECONOMIC IMPORTANCE: Gill (1997) reported this species as common and at times injurious on oaks in Arizona and New Mexico, while rare and of no economic importance in California.
GENERAL: Description and illustration of adult female by Ferris (1941d), McKenzie (1956), Deitz & Davidson (1986) and by Gill (1997).
KEYS: Gill 1997: 194 (female) [Species of California]; Deitz & Davidson 1986: 12-15 (female) [North America]; McDaniel 1970: 411-412 (female) [U.S.A.: Texas]; Ferris 1943: 65 (female) [North America]; Ferris 1942: 37 (female) [North America]; Cockerell 1905b: 201 (female) [U.S.A.: Colorado].
CITATIONS: Borchs1966 [catalogue: 349]; Cocker1898m [taxonomy, description, host, distribution: 26-27]; Cocker1899a [taxonomy: 396]; Cocker1899d [host, distribution: 170]; Cocker1899n [host, distribution: 27]; Cocker1905b [taxonomy: 201]; DeitzDa1986 [taxonomy, description, illustration, host, distribution: 48-49]; Fernal1903b [catalogue: 291]; Ferris1937c [taxonomy, illustration: 52,88]; Ferris1941d [taxonomy, description, illustration, host, distribution: 357]; Ferris1941e [taxonomy: 45]; Ferris1942 [taxonomy: 446:37]; Ferris1943 [taxonomy: 65]; Gill1997 [host, distribution, description, illustration: 195,197]; Gordh1979 [biological control: 912]; Leonar1900 [taxonomy, host, distribution: 341]; MacGil1921 [taxonomy, description, host, distribution: 442]; McDani1970 [taxonomy, illustration, host, distribution: 415-417]; McKenz1939 [taxonomy: 54]; McKenz1956 [taxonomy, description, illustration, host, distribution: 77-78]; Nakaha1982 [host, distribution: 55]; Willia1985a [taxonomy: 236]; Woolle1990 [biological control: 167-176].

Melanaspis longula Ferris

Melanaspis longula Ferris, 1941d: 358. Type data: PANAMA: Chiriqui Province, at Boquete, on undetermined tree. Lectotype female, by subsequent designation Deitz & Davidson, 1986: 50. Type depository: Davis: The Bohart Museum of Entomology, University of California, California, USA.

SCALE COVER: Female scale of texture and colour common to genus but elongate oval with exuviae toward one end; male scale not recognized (Ferris, 1941d).

HOST PLANTS: **Leguminosae** [Ferris1943].

DISTRIBUTION: **Nearctic**: Mexico (San Luis Potosi [Ferris1943, DeitzDa1986], Veracruz [Ferris1941d]). **Neotropical**: Panama [Ferris1941d].

BIOLOGY: Occurring on bark (Ferris, 1941d). Associated with fungus, *Septobasidium* (Ferris, 1943).

GENERAL: Description and illustration of adult female by Ferris (1941d) and by Deitz & Davidson (1986).

KEYS: Deitz & Davidson 1986: 12-15 (female) [North America]; Ferris 1943: 65 (female) [North America]; Ferris 1942: 37 (female) [North America].

CITATIONS: Borchs1966 [catalogue: 349]; DeitzDa1986 [taxonomy, description, illustration, host, distribution: 50-51]; Ferris1941d [taxonomy, description, illustration, host, distribution: 358]; Ferris1942 [taxonomy: 446:37]; Ferris1943 [host, distribution, life history: 63].

Melanaspis louristanus Balachowsky & Kaussari

Melanaspis louristanus Balachowsky & Kaussari, 1953: 28. Type data: IRAN: Louristan province, on *Quercus* sp. Syntypes, female. Type depository: Paris: Muséum national d'Histoire naturelle, France.

Melanaspis lauristanus; Kaussari, 1955: 16. Misspelling of species name.

Melanaspis louristanus; Borchsenius, 1966: 349. Notes: Order of authors cited incorrectly.

SCALE COVER: Female scale circular or subcircular, 2.2-2.6 mm; black or very dark grey; flat or slightly convex; exuviae central, dark brown; ventral vellum attached to host plant (Balachowsky & Kaussari, 1953).

HOST PLANTS: **Fagaceae**: *Quercus* [BalachKa1953].

DISTRIBUTION: **Palaearctic**: Iran [BalachKa1953].

GENERAL: Description and illustration of adult female by Balachowsky & Kaussari (1953).

KEYS: Danzig 1993: 238 (female) [world].

CITATIONS: BalachKa1953 [taxonomy, description, illustration, host, distribution: 28-31]; Borchs1966 [catalogue: 349]; Danzig1993 [taxonomy: 238]; DanzigPe1998 [catalogue: 304]; Kaussa1955 [host, distribution: 16].

Melanaspis madagascariensis Mamet

Melanaspis madagascariensis Mamet, 1951: 250. Type data: MADAGASCAR: road to Majunga, 530 km, on leaf of "Sohy" *Cephalanthus spathelliferus*. Holotype. Type depository: Paris: Muséum national d'Histoire naturelle, France.

SCALE COVER: Female scale subcircular, fairly flat, pale buff coloured, sometimes covered by debris of plant tissues; exuviae subcentral, yellowish. Scale of male not observed (Mamet, 1954, 1958a).
HOST PLANTS: **Naucleaceae**: *Cephalanthus spathelliferus* [Mamet1951, Borchs1966].
DISTRIBUTION: **Afrotropical**: Angola [Almeid1973b]; Madagascar [Mamet1951, Borchs1966, Almeid1973b].
GENERAL: Description and illustration of adult female by Mamet (1954, 1958a).
CITATIONS: Almeid1973b [host, distribution: 11]; Borchs1966 [catalogue: 349]; Mamet1951 [taxonomy, description, illustration, host, distribution: 229,250-251].

Melanaspis marlatti (Parrott)
Aspidiotus (*Targionia*) *marlatti* Parrott, 1899b: 282. Type data: U.S.A.: Kansas, Manhattan, Blue Mont, on stem base of *Andropogon furcatus* and *A. scoparius*; collected by J.B. Norton. Lectotype female, by subsequent designation Deitz & Davidson, 1986: 52. Type depository: Washington: United States National Entomological Collection, U.S. National Museum of Natural History, District of Columbia, USA.
Targionia marlatti; Fernald, 1903b: 297. Change of combination.
Targaspidiotus marlatti; MacGillivray, 1921: 447. Change of combination.
Melanaspis marlatti; Borchsenius, 1966: 352. Incorrect synonymy. Notes: Incorrect synonymy with *Melanaspis smilacis* (Comstock); see Deitz & Davidson, 1986: 52.
Melanaspis marlatti; Deitz & Davidson, 1986: 52. Change of combination.
SYSTEMATICS: Borchsenius (1966) synonymized *Melanaspis marlatti* (Parrott) with *Melanaspis smilacis* (Comstock), while Deitz & Davidson (1986) regarded them as distinct species.
SCALE COVER: Female scale 2 mm in diameter; flat to highly convex; dark reddish-brown, resembling walnut; on margin to a lighter shade at centre; exuviae lateral, large, black, often covered with brownish secretion; ventral scale thin, light reddish-brown, not easily separated from scale; leaves no mark on host plant when detached (Parrott, 1899b).
HOST PLANTS: **Gramineae**: *Andropogon* [TippinBe1972, BesheaTiHo1973], *Andropogon furcatus* [DeitzDa1986], *Andropogon scoparium* [DeitzDa1986].
DISTRIBUTION: **Australasian**: Bonin Islands (= Ogasawara-Gunto) [Kawai1987]. **Nearctic**: United States of America (Florida [DeitzDa1986], Georgia [TippinBe1972, BesheaTiHo1973, DeitzDa1986], Kansas [DeitzDa1986], South Dakota [DeitzDa1986], Texas [DeitzDa1986]). **Palaearctic**: Japan [Kawai1980].
GENERAL: Description and illustration of adult female by Parrott (1899b) and by Deitz & Davidson (1986).
KEYS: Deitz & Davidson 1986: 12-13 (female) [North America]; Kawai 1980: 206 (female) [Japan]; Lawson 1917: 246 (female) [U.S.A.: Kansas].
CITATIONS: BesheaTiHo1973 [host, distribution: 7]; DeitzDa1986 [taxonomy, description, illustration, host, distribution: 52-53]; Fernal1903b [catalogue: 297]; Ferris1941e [taxonomy: 45]; Kawai1980 [taxonomy, description, host, distribution: 207]; Kawai1987 [host, distribution: 78]; Lawson1917 [taxonomy, description, illustration, host, distribution: 247-248]; Lindin1957 [taxonomy: 546]; MacGil1921

[taxonomy, description, host, distribution: 447]; Nakaha1982 [host, distribution: 55]; Parrot1899b [taxonomy, description, illustration, host, distribution: 282]; TippinBe1972 [host, distribution: 287].

Melanaspis martinsi Lepage

Melanaspis martinsi Lepage, 1942: 186. Type data: BRAZIL: São Paolo State, Campinas, on *Rosa* sp. Syntypes, female. Type depositories: São Paulo: Instituto Biologico de São Paulo, Brazil, and Curitiba: Departamento de Zoologia, Setor de Ciencias Biologicas, Universidade Federal do Parana, Brazil.
SCALE COVER: Female scale circular, 2.5 mm diameter, yellow; exuviae black, central or subcentral; ventral scale thin, attached to host plant (Lepage, 1942).
HOST PLANTS: Rosaceae: *Rosa* [Lepage1942, ClapsWoGo2001].
DISTRIBUTION: Neotropical: Brazil (São Paulo [Lepage1942]).
GENERAL: Description and illustration of adult female by Lepage (1942).
CITATIONS: Borchs1966 [catalogue: 349]; Claps1993 [taxonomy: 6,9]; ClapsWoGo2001 [host, distribution: 247]; Lepage1942 [taxonomy, description, illustration, host, distribution: 185-187]; McKenz1947 [taxonomy: 32].

Melanaspis mimosae (Comstock)

Aspidiotus mimosae Comstock, 1883: 62. Type data: MEXICO: Tampico, on twig of *Mimosa*. Lectotype female, by subsequent designation Deitz & Davidson, 1896: 54. Type depository: Washington: United States National Entomological Collection, U.S. National Museum of Natural History, District of Columbia, USA.
Aonidiella mimosae; Leonardi, 1897: 286. Change of combination.
Aspidiotus (*Chrysomphalus*) *mimosae*; Cockerell, 1897i: 24. Change of combination.
Chrysomphalus mimosae; Cockerell, 1899n: 26. Change of combination.
Pseudischnaspis mimosae; Lindinger, 1937: 194. Change of combination.
Melanaspis mimosae; McKenzie, 1939: 54. Change of combination.
Chrysomphalus (*Melanaspis*) *mimosae*; Merrill, 1953: 37. Change of combination.
Melanaspis mimosae; Borchsenius, 1966: 349. Revived combination.
COMMON NAMES: Mimosa scale [Comsto1883, MerrilCh1923, Dekle1965c].
SCALE COVER: Female scale dark grey, agreeing in colour with bark to which attached; convex, exuviae central; protuberance indicating position of exuviae marked with a white dot and concentric ring (Comstock, 1883).
HOST PLANTS: Anacardiaceae: *Spondias* [Ferris1941d, Dekle1965c], *Spondias purpurea* [MerrilCh1923]. **Fagaceae**: *Quercus* [Ferris1941d], *Quercus brandegeei* [Ferris1921]. **Leguminosae**: *Cercidium floridum* [Ferris1941d], *Mimosa* [Comsto1883], *Vachelia farnesiana* [Ferris1921]. **Pinaceae**: *Pinus cembroides* [Ferris1921].
DISTRIBUTION: Nearctic: Mexico [Cocker1899n] (Tamaulipas [DeitzDa1986]); United States of America (Arizona [DeitzDa1986], California [DeitzDa1986], Florida [MerrilCh1923, Ferris1941d, Merril1953, Dekle1965c]).
BIOLOGY: Occurring under shreds of bark and in cracks (Ferris, 1941d).

GENERAL: Description and illustration of adult female by Comstock (1883), Ferris (1921, 1941d) and by Deitz & Davidson (1986).

KEYS: Deitz & Davidson 1986: 12-15 (female) [North America]; Ferris 1943: 64 (female) [North America]; Ferris 1942: 38 (female) [North America]; Cockerell 1905: 45-46 (female) [Mexico]; Comstock 1883: 55-57 (female) [North America].

CITATIONS: Borchs1966 [catalogue: 349-350]; Cocker1896b [distribution: 334]; Cocker1897i [taxonomy, description, host, distribution: 24]; Cocker1899n [host, distribution: 26]; Cocker1905 [taxonomy: 46]; Comsto1883 [taxonomy, description, illustration, host, distribution: 56,62]; DeitzDa1986 [taxonomy, description, illustration, host, distribution: 54-55]; Dekle1965c [taxonomy, description, host, distribution: 89]; Dekle1976 [taxonomy, description, host, distribution, economic importance: 110]; Fernal1903b [catalogue: 291]; Ferris1941d [taxonomy, description, illustration, host, distribution: 359]; Ferris1941e [taxonomy: 45]; Ferris1942 [taxonomy: 446:38]; Ferris1943 [taxonomy: 64]; Leonar1897 [taxonomy: 286]; Leonar1899 [taxonomy: 175-176,181]; Lindin1937 [taxonomy: 194]; Lindin1957 [taxonomy: 550]; MacGil1921 [taxonomy, description, host, distribution: 442-443]; McKenz1939 [taxonomy: 54]; Merril1953 [taxonomy, description, host, distribution: 37-38]; MerrilCh1923 [taxonomy, description, host, distribution, economic importance : 224-225]; Nakaha1982 [host, distribution: 55].

Melanaspis monotes (Hall)

Aspidiotus (*Aonidiella*) *monotes* Hall, 1929: 357. Type data: ZIMBABWE: Trelawney, on *Monotes glaber*. Syntypes, female. Type depository: London: The Natural History Museum, England, UK.

Aspidiotus monotes; Ferris, 1941e: 46. Change of combination.

Chrysomphalus monotes; Lindinger, 1943b: 207. Change of combination.

Melanaspis monotes; Hall, 1946: 62. Change of combination.

SCALE COVER: Female scale circular; highly convex, almost as high as broad; larval exuviae dark purplish brown with a thin covering film of a white or greyish colour - this is frequently knocked off and sometimes the larval exuviae themselves are missing; nymphal exuviae and secretionary area jet black; beyond the exuviae the surface is beset by minute concentric striations or growth lines (Hall, 1929).

HOST PLANTS: **Dipterocarpaceae**: *Monotes glaber* [Hall1929, Balach1953e, Balach1958b].

DISTRIBUTION: **Afrotropical**: Zimbabwe [Hall1929, Balach1953e, Balach1958b].

GENERAL: Description and illustration of adult female by Hall (1929) and by Balachowsky (1958b).

KEYS: Balachowsky 1958b: 194 (female) [Africa]; Hall 1946: 62 (female) [Africa].

CITATIONS: Balach1953e [host, distribution: 147]; Balach1958b [taxonomy, description, illustration, host, distribution: 194,197-200]; Borchs1966 [catalogue: 350]; Ferris1941e [taxonomy: 46]; Hall1929 [taxonomy, description, illustration, host, distribution: 357-358]; Hall1946 [taxonomy: 62]; Lindin1943b [taxonomy: 207]; McKenz1938 [taxonomy: 4].

Melanaspis nigropunctata (Cockerell)

Aspidiotus nigropunctatus Cockerell, 1896h: 20. Type data: MEXICO: San Luis, on bark of undetermined tree; collected by Townsend, October 12, 1894. Lectotype female, by subsequent designation Deitz & Davidson, 1986: 56. Type depository: Washington: United States National Entomological Collection, U.S. National Museum of Natural History, District of Columbia, USA.

Chrysomphalus nigropunctatus; Leonardi, 1897: 286. Change of combination.

Aspidiotus (*Melanaspis*) *nigropunctatus*; Cockerell, 1897i: 24. Change of combination.

Melanaspis nigropunctata; McKenzie, 1939: 54. Change of combination requiring emendation of species name for agreement in gender.

SCALE COVER: Female scale subcircular to suboval, 3 mm in diameter; slightly convex; dirty grey; exuviae sublateral, pitch black, with a narrow reddish margin; exuviae covered by a film of white secretion (Cockerell, 1896h).

HOST PLANTS: **Compositae**: *Baccharis* [Ferris1941d]. **Oleaceae**: *Fraxinus* [Ferris1941d]. **Rubiaceae**: *Randia* [Ferris1941d].

DISTRIBUTION: **Nearctic**: Mexico [Cocker1899n] (Colima [Ferris1941d], District federal [Ferris1941d], Hidalgo [Ferris1943], Nayarit [Ferris1941d], Puebla [Ferris1941d], San Luis Potosi [Ferris1941d]); United States of America (District of Columbia [Nakaha1982], Texas [McDani1970], Virginia [Nakaha1982]). **Neotropical**: Costa Rica [Nakaha1982]; Guatemala [Nakaha1982]; Panama [Ferris1941d]; Puerto Rico & Vieques Island (Puerto Rico [ColonFMe1998]).

BIOLOGY: Occurring exposed upon bark of host, or concealed beneath bark flakes (Cockerell, 1896h; Ferris, 1941d).

GENERAL: Description and illustration of adult female by Ferris (1941d), Deitz & Davidson (1986), Kosztarab (1996) and by Colon-Ferrer & Medina-Gaud (1998).

KEYS: Kosztarab 1996: 533 (female) [Northeastern North America]; Deitz & Davidson 1986: 12-15 (female) [North America]; McDaniel 1970: 411-412 (female) [U.S.A.: Texas]; Ferris 1943: 64 (female) [North America]; Ferris 1942: 36 (female) [North America]; Cockerell 1905: 45-46 (female) [Mexico].

CITATIONS: BalachKa1953 [taxonomy: 30]; Borchs1966 [catalogue: 350]; Cocker1896b [distribution: 334]; Cocker1896f [taxonomy, description, host, distribution: 31-32]; Cocker1896h [taxonomy, description, host, distribution: 20]; Cocker1897i [taxonomy, description, host, distribution: 13,24]; Cocker1899d [host, distribution: 170]; Cocker1899n [host, distribution: 26-27]; Cocker1905 [taxonomy: 46]; ColonFMe1998 [taxonomy, description, illustration, host, distribution: 68-69]; DeitzDa1986 [taxonomy, description, illustration, host, distribution: 56-57]; Fernal1903b [catalogue: 267]; Ferris1941d [taxonomy, description, illustration, host, distribution: 360]; Ferris1941e [taxonomy: 46]; Ferris1942 [taxonomy: 446:336]; Ferris1943 [host, distribution: 63]; Koszta1996 [taxonomy, description, illustration, host, distribution, life history : 533-535]; Leonar1897 [taxonomy: 286]; Leonar1899 [taxonomy: 199,200,202]; MacGil1921 [taxonomy, description, host, distribution: 421]; McDani1970 [taxonomy, illustration, host, distribution: 418-419]; McKenz1939 [taxonomy: 54]; Nakaha1982 [host, distribution: 55-56]; Willia1985a [taxonomy: 236].

Melanaspis obscura (Comstock)

Aspidiotus obscurus Comstock, 1881a: 303. Type data: USA: Washington, D.C., on the bark of the limbs of willow oak, *Quercus phellos*. Lectotype female, by subsequent designation Deitz & Davidson, 1986: 58. Type depository: Washington: United States National Entomological Collection, U.S. National Museum of Natural History, District of Columbia, USA.

Chrysomphalus obscurus; Leonardi, 1897: 286. Change of combination.

Aspidiotus (*Melanaspis*) *obscurus*; Cockerell, 1897i: 21. Change of combination.

Melanaspis obscura; Lindinger, 1911: 356. Change of combination requiring emendation of species name for agreement in gender.

Chrysomphalus (*Melanaspis*) *obscurus*; Merrill, 1953: 38. Change of combination.

Melanaspis obscura; Borchsenius, 1966: 350. Revived combination.

COMMON NAME: obscure scale [Comsto1881a, MerrilCh1923, Merril1953, McKenz1956, Dekle1965c].

SCALE COVER: Female scale irregular in outline, but nearly circular, 3 mm in diameter; very dark grey; slightly convex; exuviae between center and one side, position indicated by a nipple-like prominence, and marked with a white dot and concentric ring of same colour; ventral scale a delicate film of white secretion, and lower half of exuviae attached to bark (Comstock, 1881a). Colour photograph by Gill (1997).

HOST PLANTS: **Aceraceae**: *Acer* [TippinBe1970]. **Anacardiaceae**: *Spondias purpurea* [MerrilCh1923]. **Caprifoliaceae**: *Viburnum arboratum* [Merril1953], *Viburnum obovatum* [Koszta1996]. **Cornaceae**: *Cornus* [Ferris1941d, McKenz1956, Koszta1996]. **Fagaceae**: *Castanea* [Ferris1941d, McKenz1956, DeitzDa1986, Koszta1996], *Castanea pumila* [DeitzDa1986], *Fagus* [Koszta1996], *Fagus grandifolia* [DeitzDa1986], *Quercus* [McKenz1956, Dekle1965c, BesheaTiHo1973, DeitzDa1986], *Qu. alba* [BesheaTiHo1973, DeitzDa1986], *Qu. bicolor* [DeitzDa1986], *Qu. coccinea* [McKenz1956, DeitzDa1986], *Qu. falcata* [DeitzDa1986], *Qu. laurifolia* [TippinBe1970, BesheaTiHo1973], *Qu. macrocarpa* [DeitzDa1986], *Qu. marilandica* [DeitzDa1986], *Qu. nigra* [DeitzDa1986], *Qu. palustris* [DeitzDa1986, HendriWi1992], *Quercus phellos* [Comsto1881a, McKenz1956, DeitzDa1986], *Quercus prinoides* [DeitzDa1986], *Qu. prinus* [DeitzDa1986], *Qu. robur* [DeitzDa1986], *Qu. rubra* [DeitzDa1986], *Qu. stellata* [BesheaTiHo1973, DeitzDa1986], *Qu. tinctoria* [Hunter1899], *Qu. velutina* [DeitzDa1986], *Qu. virginiana* [McDani1970, BesheaTiHo1973, DeitzDa1986]. **Juglandaceae**: *Carya* [Koszta1996], *Carya illinoensis* [McKenz1956], *Carya laciniosa* [DeitzDa1986], *Carya ovata* [DeitzDa1986], *Hicoria* [Ferris1941d], *Hicoria pecan* [Ferris1941d, McKenz1956], *Juglans* [Ferris1941d, McKenz1956, McDani1970, Koszta1996]. **Leguminosae**: *Prosopis* [Ferris1941d, McKenz1956, McDani1970, Koszta1996]. **Oleaceae**: *Fraxinus americana* [DeitzDa1986]. **Rosaceae**: *Prunus reverchoni* [McKenz1956]. **Sapindaceae**: *Sapindus* [Koszta1996]. **Ulmaceae**: *Planera aquatica* [Koszta1996], *Ulmus* [DeitzDa1986, Koszta1996], *Ulmus procera* [DeitzDa1986]. **Vitaceae**: *Vitis* [McKenz1956, Koszta1996].

NATURAL ENEMIES: ACARI **Hemisarcoptidae**: *Hemisarcoptes malus* (Shimer) [GersonOcHo1990]. FUNGI **Ascomycotina**: *Myriangium duriaei* [EvansPr1990],

Nectria aurantiicola [EvansPr1990], *Nectria flammea* [EvansPr1990]. HYMENOPTERA **Aphelinidae**: *Ablerus clisiocampae* (Ashmead) [Gordh1979], *Coccophagoides fuscipennis* (Girault) [Gordh1979], *Physcus varicornis* (Howard) [Gordh1979], *Prospaltella berlesei* (Howard) [Gordh1979]. **Signiphoridae**: *Thysanus ater* Haliday [Woolle1990].

DISTRIBUTION: **Nearctic**: Canada [Nakaha1982]; United States of America (Alabama [DeitzDa1986, HendriWi1992], Arkansas [DeitzDa1986], California [McKenz1956], Connecticut [DeitzDa1986], Delaware [Nakaha1982], District of Columbia [Comsto1881a, DeitzDa1986], Florida [Wilson1917, MerrilCh1923, Dekle1965c], Georgia [TippinBe1970, BesheaTiHo1973, DeitzDa1986], Illinois [DeitzDa1986], Indiana [DeitzDa1986], Iowa [DeitzDa1986], Kansas [Hunter1899, DeitzDa1986], Kentucky [DeitzDa1986], Louisiana [DeitzDa1986], Maryland [DeitzDa1986], Mississippi [Herric1911, DeitzDa1986], Missouri [Hollin1923, DeitzDa1986], New Jersey [DeitzDa1986], New York [Nakaha1982], North Carolina [DeitzDa1986], Ohio [DeitzDa1986], Oklahoma [DeitzDa1986], Pennsylvania [Stimme1980a, DeitzDa1986], South Carolina [DeitzDa1986], Tennessee [DeitzDa1986], Texas [Ferris1941d, McDani1970, DeitzDa1986], Virginia [BesheaTiHo1973, DeitzDa1986], West Virginia [DeitzDa1986]). **Palaearctic**: Japan [Kawai1977, Kawai1980].

BIOLOGY: Occurring on bark of host, with no special tendency to be hidden in cracks or under bark (Ferris, 1941d). Develops one generation a year in Alabama, USA (Hendricks & Williams (1992), Maryland, USA (Stoetzel & Davidson, 1971), Ohio, USA (Kosztarab, 1963) and Louisiana (Baker, 1933).

ECONOMIC IMPORTANCE: A serious pest of oaks, *Quercus* spp. in eastern states of USA (Gill, 1997; Hendricks & Williams, 1992), of shade trees in Maryland, USA (Stoetzel & Davidson, 1971), and of pecan in southern USA (Osburn & Pierce, 1963).

GENERAL: Description and illustration of adult female by Ferris (1941d), McKenzie (1956), Deitz & Davidson (1986), Kosztarab (1996) and by Gill (1997).

KEYS: Gill 1997: 194 (female) [Species of California]; Kosztarab 1996: 533 (female) [Northeastern North America]; Deitz & Davidson 1986: 12-15 (female) [North America]; Kawai 1980: 206 (female) [Japan]; McDaniel 1970: 411-412 (female) [U.S.A.: Texas]; McKenzie 1956: 26 (female) [U.S.A.: California]; Britton 1923: 376 (female) [U.S.A.: Connecticut]; Hollinger 1923: 29 (female) [U.S.A.: Missouri]; Lawson 1917: 210 (female) [U.S.A.: Kansas]; Dietz & Morrison 1916a: 307 (female) [U.S.A.: Indiana]; Comstock 1883: 55-57 (female) [North America].

CITATIONS: Baker1933 [host, distribution, life history, economic importance, control: 1-19]; Berger1942a [host, distribution, biological control: 26-29]; Berger1942a [host, distribution, life history, biological control: 26-29]; BesheaTiHo1973 [host, distribution: 7]; Borchs1966 [catalogue: 350]; Britto1923 [taxonomy, description, host, distribution: 376-377]; ChuaWo1990 [host, distribution, economic importance: 550]; Cocker1896b [distribution: 334]; Cocker1897i [taxonomy, description, host, distribution: 21]; Comsto1881a [taxonomy, description, illustration, host, distribution: 303-304]; Comsto1883 [taxonomy: 64]; DanzigPe1998 [catalogue: 304-305]; DarlinJo1984 [host, distribution, biological control: 555]; DavidsMi1990 [host, distribution, economic importance: 603-632]; DavidsRa1999 [economic importance, control: 1]; DeitzDa1986 [taxonomy, description, illustration, host, distribution: 58-59]; Dekle1965c [taxonomy, description, host, distribution: 90]; Dekle1976 [taxonomy, description, host, distribution, economic importance: 111]; DeSant1940 [biological control: 29-44]; DietzMo1916a [taxonomy, description, illustration, host, distribution: 308-310]; Ebelin1949 [host, distribution, life history, control]; Ehler1995 [host, distribution, life history, economic importance, biological control: 779-795]; Ehler1996 [host, distribution, biological control: 337-342]; Ehler1997 [host, distribution, biological control, economic importance: 29-32]; EvansPr1990 [biological control: 3-17]; FDACSB1982 [host, distribution: 5-11]; Felt1924 [host, distribution, economic importance, life history, control]; Fernal1903b [catalogue: 291]; Ferris1937c [taxonomy, illustration: 51,83]; Ferris1941d [taxonomy, description, illustration, host, distribution: 361]; Ferris1941e [taxonomy: 46]; Ferris1942 [taxonomy: 446:36]; Ferris1943 [taxonomy: 64]; Fleury1934a [distribution: 278-289]; GersonOcHo1990 [biological control: 77-97]; Gill1997 [host, distribution, taxonomy, description, illustration, economic importance: 195,198,200]; Gordh1979 [biological control: 907,908,912]; HendriWi1992 [host, distribution, life history, economic importance: 452-457]; Herric1911 [taxonomy, description, illustration, host, distribution: 11,32,67]; Hewitt1943 [host, distribution: 266-274]; Hollin1923 [taxonomy, description, host, distribution: 30-31]; Hunter1899 [taxonomy, host, distribution: 7]; Kawai1977 [host, distribution, economic importance: 157]; Kawai1980 [taxonomy, description, host, distribution: 206-207]; Koszta1996 [taxonomy, description, illustration, host, distribution, life history, biological control, economic importance: 535-536]; Lawson1917 [taxonomy, description, illustration, host, distribution: 211-213]; Leonar1897 [taxonomy: 286]; Leonar1899 [taxonomy, description, host, distribution: 199-200,205]; Lindin1911 [taxonomy: 356]; Lindin1957 [taxonomy: 550]; Lizery1917b [taxonomy: 243]; Lobdel1937 [taxonomy: 78]; Lord1922 [host, distribution: 1]; Lyle1947 [host, distribution, control: 1-8]; Mackie1934 [host, distribution: 396]; Mackie1934a [host, distribution: 268-269]; Mackie1936 [host, distribution: 455]; McClur1990a [taxonomy, host, distribution, ecology: 165-168]; McDani1970 [taxonomy, illustration, host, distribution: 418-420]; McKenz1939 [taxonomy: 54]; McKenz1956 [taxonomy, description, host, distribution: 26,77]; Merril1953 [taxonomy, description, host, distribution: 38]; MerrilCh1923 [taxonomy, description, host, distribution, economic importance: 225]; Miller1999 [chemical control: 14]; MillerDa1990 [host, distribution, economic importance: 303]; Nakaha1982 [host, distribution: 56]; OsburnPi1963 [host, distribution, chemical control: 1]; PayneMaKe1979 [host, distribution, economic importance: 1-43]; Pierce1938 [chemical control: 722-724]; PotterJeGo1989 [host, distribution, life history, ecology, biological control: 551-555]; RauppHoSa2001 [host, distribution, chemical control: 203-214]; Robiso1990 [structure, anatomy: 213-214]; Ryan1946 [host, distribution: 124-125]; Sander1904a [taxonomy, description, illustration, host, distribution: 70,72]; SchmutKlLu1957 [host, distribution, economic importance: 492]; Sikes1931 [chemical control: 1-40]; Staffo1915 [taxonomy, structure: 69]; Stimme1980a [host, distribution, description, life history, economic importance, control: 11-12]; StoetzDa1971 [taxonomy, host, distribution, life history: 45-50];

StoetzDa1973 [taxonomy, host, distribution, life history: 308-311]; StoetzDa1974 [taxonomy, life history: 138-140]; TippinBe1970 [host, distribution: 10]; Wilson1917 [taxonomy, description, host, distribution: 43]; Woolle1990 [biological control: 167-176]; Zahrad1990a [host, distribution, description: 653].

Melanaspis odontoglossi (Cockerell)

Aspidiotus biformis odontoglossi Cockerell, 1893k: 548. Type data: JAMAICA: Kingston, on *Odontoglossum grande*; collected Cockerell and Dr. Henderson. Lectotype female, by subsequent designation Deitz & Davidson, 1986: 62. Type depository: Washington: United States National Entomological Collection, U.S. National Museum of Natural History, District of Columbia, USA; type no. 5489.

Aspidiotus (Chrysomphalus) biformis odontoglossi; Cockerell, 1897i: 23. Change of combination.

Aspidiotus (Chrysomphalus) smilacis; Kuwana, 1902a: 32. Misidentification; discovered by Borchsenius, 1966: 350.

Chrysomphalus odontoglossi; Fernald, 1903b: 291. Change of combination and rank.

Aspidiotus (Aonidiella) ritchiei Green & Laing, 1923: 125. Type data: JAMAICA: Kingston, on bark of *Cassia fistula*. Syntypes, female. Type depository: London: Natural History Museum, England, UK. Synonymy by Deitz & Davidson, 1986: 60.

Aspidiotus (Aonidiella) multiclavata Green & Laing, 1923: 126. Type data: JAMAICA: Hill Gardens, on redwood tree, *Brythroxylon areolatum*. Syntypes, female. Type depository: London: The Natural History Museum, England, UK. Synonymy by Deitz & Davidson, 1986: 60.

Melanaspis multiclavata; Lindinger, 1932: 224. Change of combination.

Pseudischnaspis ritchiei; Lindinger, 1937: 194. Change of combination.

Pseudischnaspis ritchiiei; Lindinger, 1937: 194. Misspelling of species name.

Aspidiotus odontoglossi; McKenzie, 1939: 54. Change of combination.

Melanaspis odontoglossi; McKenzie, 1939: 54. Change of combination.

Melanaspis obtusa Ferris, 1941d: 362. Type data: MEXICO: State of Oaxaca, Chivela, on *Byrsonima crassifolia*. Holotype female. Type depository: Davis: The Bohart Museum of Entomology, University of California, California, USA. Synonymy by Borchsenius, 1966: 351.

Aspidiotus multiclavata; Ferris, 1941e: 46. Change of combination.

Melanaspis ritchiei; Ferris, 1942: 445. Change of combination.

Aspidiotus (Aonidiella) multiclavata; Borchsenius, 1966: 352. Incorrect synonymy.

Notes: Incorrect synonymy of *Aspidiotus (Aonidiella) odontoglossi* Green & Laing, 1923 with *Melanaspis smilacis* (Comstock); see Deitz & Davidson, 1986: 60.

HOST PLANTS: **Agavaceae**: *Agave* [DeitzDa1986]. **Apocynaceae**: *Vallesia* [DeitzDa1986]. **Bignoniaceae**: *Jacaranda* [DeitzDa1986]. **Cactaceae**: *Gymnathocereus* [DeitzDa1986]. **Compositae**: *Scalesia gummifera* [Ferris1941d]. **Erythroxylaceae**: *Erythroxylum areolatum* [DeitzDa1986]. **Euphorbiaceae**: *Croton scouleri* [Ferris1941d], *Euphorbia pulcherrima* [DeitzDa1986]. **Gramineae**: *Chusquiea* [Ferris1941d]. **Leguminosae**: *Acacia* [DeitzDa1986], *Albizia lebbek* [DeitzDa1986], *Cassia fistula* [GreenLa1923, Ferris1942, DeitzDa1986], *Enterolobium cyclocarpum* [Ferris1941d], *Samanea saman* [DeitzDa1986].

Malpighiaceae: *Byrsonima crassifolia* [Ferris1941d, DeitzDa1986]. **Melastomataceae**: *Medinilla magnifica* [DeitzDa1986]. **Orchidaceae**: *Brassavola cordata* [DeitzDa1986], *Laeliopsis domingensis* [DeitzDa1986], *Lycaste brevispatha* [DeitzDa1986], *Odontoglossum grande* [Cocker1893k, Ferris1942, DeitzDa1986], *Oncidium* [DeitzDa1986], *Oncidium triquetrum* [DeitzDa1986]. **Proteaceae**: *Grevillea robusta* [DeitzDa1986]. **Rosaceae**: *Cydonia* [DeitzDa1986]. **Rubiaceae**: *Cinchona* [DeitzDa1986]. **Sapotaceae**: *Chrysophyllum albidum* [DeitzDa1986], *Pouteria sapota* [DeitzDa1986]. **Solanaceae**: *Solanum punctulatum* [DeitzDa1986]. **Sterculiaceae**: *Pterospermum acerifolium* [Ferris1941d].
DISTRIBUTION: **Nearctic**: Mexico (Oaxaca [Ferris1941d, DeitzDa1986]); United States of America (Florida [DeitzDa1986], New York [Ferris1941d, DeitzDa1986]). **Neotropical**: Ecuador [DeitzDa1986]; Galapagos Islands [Ferris1941d, PeckHeLa1998]; Honduras [DeitzDa1986]; Jamaica [GreenLa1923, Ferris1942, DeitzDa1986]; Panama [Ferris1941d, DeitzDa1986]; Peru [DeitzDa1986]; Puerto Rico & Vieques Island [DeitzDa1986]; Trinidad and Tobago (Trinidad [DeitzDa1986]).
GENERAL: Description and illustration of adult female by Green & Laing (1923) (as *A. ritchiei* and *A. multiclavata*, Ferris (1941d) (as *Melanaspis obtusa* and by Deitz & Davidson (1986).
KEYS: Deitz & Davidson 1986: 12-15 (female) [North America]; Ferris 1943: 66 (female) [North America]; Ferris 1943: 64 (female) [North America]; Ferris 1942: 38 (female) [North America].
CITATIONS: Borchs1966 [catalogue: 350-352]; Cocker1893k [taxonomy, description, host, distribution: 548]; Cocker1896b [distribution: 334]; Cocker1897i [taxonomy, description, host, distribution: 23]; Cocker1897l [taxonomy: 151]; DeitzDa1986 [taxonomy, description, illustration, host, distribution: 60-62]; Fernal1903b [catalogue: 291]; Ferris1941d [taxonomy, description, illustration, host, distribution: 362]; Ferris1941e [taxonomy: 46,47]; Ferris1942 [taxonomy, description, illustration, host, distribution: 430;445:3;446:37-38]; Ferris1943 [taxonomy: 64]; GreenLa1923 [taxonomy, description, illustration, host, distribution: 125-126]; Kuwana1902a [taxonomy, description, host, distribution: 32]; Lindin1937 [taxonomy: 194]; Lindin1957 [taxonomy: 545-546,550]; McKenz1938 [taxonomy: 4]; McKenz1939 [taxonomy: 54]; Nakaha1982 [host, distribution: 56]; PeckHeLa1998 [host, distribution: 219-237].

Melanaspis pedina Munting
Melanaspis pedina Munting, 1969: 129. Type data: NAMIBIA: Spitzkoppe, on *Boscia foetida*; collected 26.viii.1967. Holotype female. Type depository: Pretoria: South African National Collection of Insects, South Africa; type no. 2862/8.
SCALE COVER: Female scale subcircular, about 1.4 mm in diameter; living in small pit in bark of host plant so that top is more or less even with surface of bark; exuviae black, covered with relatively thick white secretion. Male scale similar in colour when undisturbed but without black exuviae; broadly oval; slightly smaller than female; second instar exuviae completely enclosing adult female even at full maturity (Munting, 1969).

HOST PLANTS: **Capparidaceae**: *Boscia albitrunca* [Muntin1969], *Boscia foetida* [Muntin1969].
NATURAL ENEMIES: HYMENOPTERA **Encyrtidae**: *Habrolepis* [Prinsl1983].
DISTRIBUTION: **Afrotropical**: Namibia (Southwest Africa) [Muntin1969]; South Africa [Muntin1969].
GENERAL: Description and illustration of adult female by Munting (1969).
CITATIONS: Muntin1969 [taxonomy, description, illustration, host, distribution: 129-130,154]; Prinsl1983 [distribution, biological control: 27].

Melanaspis phenax (Cockerell)
Chrysomphalus phenax Cockerell, 1901l: 225. Type data: SOUTH AFRICA: Natal, Verulam, on bark of branches of *Mimosa* sp. Syntypes, female. Type depository: Washington: United States National Entomological Collection, U.S. National Museum of Natural History, District of Columbia, USA.
Chrysomphalus (*Melanaspis*) *phenax*; Brain, 1919: 205. Change of combination.
Aonidiella phenax; MacGillivray, 1921: 442. Change of combination.
Pseudischnaspis phenax Lindinger, 1937: 194. Change of combination.
Melanaspis phenax; Hall, 1946: 62. Change of combination.
Melanaspis phoenax; Balachowsky, 1953e: 146. Misspelling of species name.
SCALE COVER: Female scale about 1.5 mm in greatest diameter; convex, capsular; black, but covered with a secretionary layer of yellowish brown material which shows concentric markings; exuviae nearly always nearer to one side and covered with a whitish or greyish layer; in very young scales a faint concentric ring and dot effect present; ventral scale dense, black (Brain, 1919).
HOST PLANTS: **Leguminosae**: *Acacia giraffae* [Muntin1969], *Acacia horrida* [Brain1919, Balach1953e, Balach1958b], *Acacia karroo* [Hall1928, Balach1953e, Balach1958b], *Mimosa* [Cocker1901l].
NATURAL ENEMIES: HYMENOPTERA **Encyrtidae**: *Habrolepis occidua* Annecke & Mynhardt [AnneckIn1971].
DISTRIBUTION: **Afrotropical**: Guinea [Balach1953e]; South Africa [Cocker1901l, Brain1919, Balach1958b, Muntin1969]; Zimbabwe [Hall1928, Balach1958b].
GENERAL: Description and illustration of adult female by Brain (1919) and by Balachowsky (1958b).
KEYS: Balachowsky 1958b: 194 (female) [Africa]; Hall 1946: 62 (female) [Africa]; Brain 1919: 199 (female) [South Africa].
CITATIONS: AnneckIn1971 [host, distribution, biological control: 14]; Balach1958b [taxonomy, description, illustration, host, distribution: 199-201]; Borchs1966 [catalogue: 351]; Brain1919 [taxonomy, description, illustration, host, distribution: 205]; Cocker1901 [taxonomy, description, host, distribution: 225-226]; Fernal1903b [catalogue: 292]; Ferris1941d [taxonomy: 347]; Hall1928 [host, distribution: 276]; Hall1946 [taxonomy: 62]; Lindin1937 [taxonomy: 194]; Lindin1957 [taxonomy: 550]; MacGil1921 [taxonomy, description, host, distribution: 442]; McKenz1939 [taxonomy: 54]; Muntin1969 [host, distribution: 130]; Prinsl1983 [distribution, biological control: 27].

Melanaspis philippiae Mamet
Melanaspis philippiae Mamet, 1959a: 474. Type data: MADAGASCAR: Ambila Lemaitso, on *Philippia* sp. Holotype. Type depository: Paris: Muséum national d'Histoire naturelle, France.
SCALE COVER: Scale of female of type common to genus (Mamet, 1959a).
HOST PLANTS: **Ericaceae**: *Philippia* [Mamet1959a, Borchs1966].
DISTRIBUTION: **Afrotropical**: Madagascar [Mamet1959a, Borchs1966].
BIOLOGY: Occurring on twigs of host (Mamet, 1959a).
GENERAL: Description and illustration of adult female by Mamet (1959a).
CITATIONS: Borchs1966 [catalogue: 351]; Mamet1959a [taxonomy, description, illustration, host, distribution: 474].

Melanaspis pinicola Deitz & Davidson
Melanaspis pinicola Deitz & Davidson, 1986: 62. Type data: MEXICO: Chiapas, Cirrode Boqueron, on *Pinus teocote*; collected by C.A. Purpus, August 1913. Holotype female. Type depository: Washington: United States National Entomological Collection, U.S. National Museum of Natural History, District of Columbia, USA; type no. 6984.
SCALE COVER: Deitz & Davidson (1986) did not describe scale cover.
HOST PLANTS: **Pinaceae**: *Pinus teocote* [DeitzDa1986]. **Rosaceae**: *Prunus salicifolia* [DeitzDa1986].
DISTRIBUTION: **Neotropical**: Guatemala [DeitzDa1986]; Mexico (Chiapas [DeitzDa1986]).
GENERAL: Description and illustration of adult female by Deitz & Davidson (1986).
KEYS: Deitz & Davidson 1986: 12-15 (female) [North America].
CITATIONS: DeitzDa1986 [taxonomy, description, illustration, host, distribution: 62-64].

Melanaspis ponderosa Ferris
Melanaspis ponderosa Ferris, 1941d: 363. Type data: PANAMA: Chiriqui Province, Armuelles, on *Enallagma cucurbitans*. Lectotype female, by subsequent designation Deitz & Davidson, 1986: 64. Type depository: Davis: The Bohart Museum of Entomology, University of California, California, USA.
SCALE COVER: Female scale of type common to genus (Ferris, 1941d).
HOST PLANTS: **Avicenniaceae**: *Avicennia nitida* [Ferris1941d, DeitzDa1986]. **Bignoniaceae**: *Enallagma cucurbitans* [Ferris1941d, DeitzDa1986]. **Leguminosae**: *Enterlobium cyclocarpum* [DeitzDa1986]. **Polygonaceae**: *Coccoloba* [DeitzDa1986]. **Rhizophoraceae**: *Rhizophora* [DeitzDa1986].
DISTRIBUTION: **Nearctic**: Mexico (District federal [DeitzDa1986]). **Neotropical**: Colombia [DeitzDa1986]; Panama [Ferris1941d, DeitzDa1986].
BIOLOGY: Buried in soft bark, concealed under bark flakes (Ferris, 1941d).
GENERAL: Description and illustration of adult female by Ferris (1941d) and by Deitz & Davidson (1986).
KEYS: Deitz & Davidson 1986: 12-15 (female) [North America]; Ferris 1943: 64 (female) [North America]; Ferris 1942: 36 (female) [North America].

CITATIONS: Borchs1966 [catalogue: 351]; DeitzDa1986 [taxonomy, description, illustration, host, distribution: 64-65]; Ferris1941d [taxonomy, description, illustration, host, distribution: 363]; Ferris1942 [taxonomy: 446:36]; Ferris1943 [taxonomy: 64].

Melanaspis pseudoponderosa Deitz & Davidson

Melanaspis pseudoponderosa Deitz & Davidson, 1986: 66. Type data: U.S.A.: Florida, Key West, on "hog plum" [=*Prunus rivularis*]; collected by Warner & Sanford, June 2, 1919. Holotype female. Type depository: Washington: United States National Entomological Collection, U.S. National Museum of Natural History, District of Columbia, USA.
SCALE COVER: Deitz & Davidson (1986) did not describe scale cover.
HOST PLANTS: **Rosaceae**: *Prunus rivularis* [DeitzDa1986].
DISTRIBUTION: **Nearctic**: United States of America (Florida [DeitzDa1986]).
GENERAL: Description and illustration of adult female by Deitz & Davidson (1986).
KEYS: Deitz & Davidson 1986: 12-15 (female) [North America].
CITATIONS: DeitzDa1986 [taxonomy, description, illustration, host, distribution: 66-67].

Melanaspis reticulata Ferris

Melanaspis reticulata Ferris, 1943: 60. Type data: MEXICO: State of Mexico, "road to Toluca", on *Quercus* sp.; collected by J.N. Couch, January 16, 1940. Lectotype female, by subsequent designation Deitz & Davidson, 1986: 68. Type depository: Davis: The Bohart Museum of Entomology, University of California, California, USA; type no. 11188.
SCALE COVER: Female scale circular, moderately convex; of type common to genus. Scale of male not recognized (Ferris, 1943).
HOST PLANTS: **Fagaceae**: *Quercus* [Ferris1943, DeitzDa1986].
DISTRIBUTION: **Nearctic**: Mexico (Mexico State [Ferris1943, DeitzDa1986]).
BIOLOGY: Occurring on bark; associated with and buried beneath a fungus of genus *Septobasidium* (Ferris, 1943).
GENERAL: Description and illustration of adult female by Ferris (1943) and by Deitz & Davidson (1986).
KEYS: Deitz & Davidson 1986: 12-15 (female) [North America]; Ferris 1943: 65 (female) [North America].
CITATIONS: Borchs1966 [catalogue: 351]; DeitzDa1986 [taxonomy, description, illustration, host, distribution: 68-69]; Ferris1943 [taxonomy, description, host, distribution: 60-61,65].

Melanaspis rhizophorae (Cockerell)

Chrysomphalus rhizophorae Cockerell, 1899d: 169. Type data: MEXICO: Tabasco, El Rio Polo, on leaves of mangrove. Syntypes, female. Type depository: Washington: United States National Entomological Collection, U.S. National Museum of Natural History, District of Columbia, USA.
Pseudischnaspis rhizophorae; Lindinger, 1937: 194. Change of combination.

Mycetaspis rhizophorae; McKenzie, 1939: 54. Change of combination.
Melanaspis rhizophorae; Ferris, 1942: 431. Change of combination.
SCALE COVER: Female scale circular to oval, 1.5 mm in diameter; slightly convex; shining sepia-brown, pale coffee-colour, or purplish brown; exuviae black, covered by a dirty white film, leaving only first skin visible (Cockerell, 1899d).
HOST PLANTS: **Rhizophoraceae**: *Rhizophora mangle* [Ferris1942].
DISTRIBUTION: **Neotropical**: Mexico (Tabasco [Cocker1899d, Ferris1942]).
BIOLOGY: Occurring on leaves (Ferris, 1942).
GENERAL: Description and illustration of adult female by Ferris (1942).
KEYS: Ferris 1943: 64 (female) [North America]; Ferris 1942: 36 (female) [North America].
CITATIONS: Borchs1966 [catalogue: 351]; Cocker1899a [taxonomy: 396]; Cocker1899d [taxonomy, description, host, distribution: 169-170]; Cocker1899n [host, distribution: 25]; Cocker1920 [taxonomy: 386]; Fernal1903b [catalogue: 293]; Ferris1942 [taxonomy, description, illustration, host, distribution: 431; 446:36]; Ferris1943 [taxonomy: 64]; Lindin1937 [taxonomy: 194]; Lindin1957 [taxonomy: 550]; MacGil1921 [taxonomy, description, host, distribution: 418-419]; McKenz1939 [taxonomy: 54]; McKenz1947 [taxonomy: 33]; Willia1985a [taxonomy: 238].

Melanaspis rotunda Ferris
Melanaspis rotunda Ferris, 1943: 62. Type data: MEXICO: Michoacan State, Zitacuaro, on *Eurya theoides*, J.N. Couch, March 7, 1940. Lectotype female, by subsequent designation Deitz & Davidson, 1986: 70. Type depository: Davis: Bohart Museum of Entomology, University of California, California, USA; type no. 11488.
SCALE COVER: Female scale of type common to genus; circular, black, quite highly convex. Scale of male not recognized (Ferris, 1943).
HOST PLANTS: **Theaceae**: *Eurya theoides* [Ferris1943, DeitzDa1986].
DISTRIBUTION: **Nearctic**: Mexico (Michoacan [Ferris1943, DeitzDa1986]).
BIOLOGY: Occurring on bark of twigs; associated with and concealed by fungus of genus *Septobasidium* (Ferris, 1943).
GENERAL: Description and illustration of adult female by Ferris (1943) and by Deitz & Davidson (1986).
KEYS: Deitz & Davidson 1986: 12-15 (female) [North America]; Ferris 1943: 65 (female) [North America].
CITATIONS: Borchs1966 [catalogue: 352]; Ferris1943 [taxonomy, description, illustration, host, distribution: 62-63,72].

Melanaspis saccharicola (Costa Lima)
Aonidiella saccharicola Costa Lima, 1934a: 25. Type data: BRAZIL: Guaratiba (Distrito Federal), on sugarcane; collected by Aristotels Silva. Syntypes, female; type no. 2435. Notes: Costa Lima (1934a) stated that syntypes were deposited in the Oswaldo Cruz Institute. Claps et al. (2001) did not trace type material.
Melanaspis saccharicola; Costa Lima, 1942: 289. Change of combination.
Chrysomphalus saccharicola; Lindinger, 1943b: 207. Change of combination.
Melanaspis saccharicola; Borchsenius, 1966: 352. Revived combination.

SCALE COVER: Female scale circular, 1.5-1.8 mm in diameter; robust, colour grey; exuviae eccentric, black; ventral vellum developed (Costa Lima, 1934a). Photograph of scale cover by Costa Lima (1934a).
HOST PLANTS: **Gramineae**: *Pennisetum purpureum* [Lepage1938, LepageGi1943, ClapsWoGo2001], *Saccharum officinarum* [CostaL1934, Lepage1938, LepageGi1943, ClapsWoGo2001].
DISTRIBUTION: **Neotropical**: Brazil (Distrito federal (Brasilia) [Lepage1938, LepageGi1943], Rio de Janeiro [ClapsWoGo2001]).
GENERAL: Description and illustration of adult female by Costa Lima (1934a) and by Lepage & Giannotti (1943).
CITATIONS: Borchs1966 [catalogue: 352]; Box1953 [host, distribution, biological control: 51]; ClapsWoGo2001 [host, distribution: 247]; CostaL1934a [taxonomy, description, illustration, host, distribution: 25]; CostaL1942 [taxonomy: 289]; Lepage1938 [catalogue: 392]; LepageGi1943 [taxonomy, description, illustration, host, distribution: 344-345]; Lindin1943b [taxonomy: 207]; McKenz1938 [taxonomy: 4].

Melanaspis sacculata Ferris

Melanaspis sacculata Ferris, 1943: 62. Type data: MEXICO: State of Guerrero, Taxco, on undetermined host; collected by J.N. Couch, February 5, 1940. Lectotype female, by subsequent designation Deitz & Davidson, 1986: 72. Type depository: Davis: The Bohart Museum of Entomology, University of California, California, USA; type no. 11325.
SCALE COVER: Female scale of type common to genus, circular, moderately convex, black, hard and brittle (Ferris, 1943).
HOST PLANTS: **Euphorbiaceae**: *Euphorbia schlechtendalii* [Ferris1943, DeitzDa1986].
DISTRIBUTION: **Nearctic**: Mexico (Guerrero [Ferris1943]).
BIOLOGY: Occurring on bark in association with fungus of genus *Septobasidium* under which it is entirely concealed (Ferris, 1943).
GENERAL: Description and illustration of adult female by Ferris (1943) and by Deitz & Davidson (1986).
KEYS: Deitz & Davidson 1986: 12-15 (female) [North America]; Ferris 1943: 65 (female) [North America].
CITATIONS: Borchs1966 [catalogue: 352]; DeitzDa1986 [taxonomy, description, illustration, host, distribution: 72-73]; Ferris1943 [taxonomy, description, illustration, host, distribution: 62,71].

Melanaspis sansevii Mamet

Melanaspis sansevii Mamet, 1959a: 474. Type data: MADAGASCAR: Ankilimivony, on "Sansevy" [=Cruciferae sp.]. Holotype. Type depository: Paris: Muséum national d'Histoire naturelle, France.
SCALE COVER: Female scale of type common to genus (Mamet, 1959a).
HOST PLANTS: **Cruciferae** [Mamet1959a].
DISTRIBUTION: **Afrotropical**: Madagascar [Mamet1959a, Borchs1966].
BIOLOGY: Occurring on twigs of hosts (Mamet, 1959a).

GENERAL: Description and illustration of adult female by Mamet (1959a).
CITATIONS: Borchs1966 [catalogue: 352]; Mamet1959a [taxonomy, description, illustration, host, distribution: 474].

Melanaspis santensis Lepage

Melanaspis santensis Lepage, 1942: 187. Type data: BRAZIL: São Paolo State, Santos, Ilha Porchat, on undetermined plant. Syntypes, female. Type depositories: São Paulo: Museu de Zoologia, Universidade de São Paulo, Brazil, and Curitiba: Departamento de Zoologia, Setor de Ciencias Biologicas, Universidade Federal do Parana, Brazil.
SCALE COVER: Female scale of moderate brown colour; exuviae central or subcentral, black (Lepage, 1942).
DISTRIBUTION: **Neotropical**: Brazil (São Paulo [Lepage1942, ClapsWoGo2001]).
GENERAL: Description and illustration of adult female by Lepage (1942).
CITATIONS: Borchs1966 [catalogue: 352]; Claps1993 [taxonomy: 3,9]; ClapsWoGo2001 [host, distribution: 247]; Lepage1942 [taxonomy, description, illustration, host, distribution: 187-189].

Melanaspis sitreana (Hempel)

Chrysomphalus sitreanus Hempel, 1932: 337. Type data: CHILE: Santiago, Cerro San Cristobal, on *Sitrea caustica*; collected by Walther Horn, Berlin, Germany. Syntypes, female and first instar. Type depositories: São Paulo: Instituto Biologico de São Paulo, Brazil, and Curitiba: Departamento de Zoologia, Setor de Ciencias Biologicas, Universidade Federal do Parana, Brazil.
Melanaspis sitreana; Lepage & Giannotti, 1943: 339. Change of combination requiring emendation of species name for agreement in gender.
Melanasois sitreana; Borchsenius, 1966: 352. Misspelling of genus name.
SCALE COVER: Female scale circular or suboval, 2-2.5 mm in diameter; scale colour grey or black; exuviae eccentric, bright grey. Male scale similar in colour and form to that of female, but smaller (Hempel, 1932).
HOST PLANTS: **Aextoxicaceae**: *Aextoxicon punctatum* [ClapsWoGo2001]. **Anacardiaceae**: *Lithraea caustica* [ClapsWoGo2001]. **Lauraceae**: *Beilschmiedia niersii* [ClapsWoGo2001]. **Sterculiaceae**: *Sitella caustica* [Hempel1932].
DISTRIBUTION: **Neotropical**: Chile [GonzalCh1968] (Santiago [Hempel1932, ClapsWoGo2001], Valparaiso [ClapsWoGo2001]).
GENERAL: Description and illustration of adult female by Lepage & Giannotti (1943).
CITATIONS: Aguile1970 [taxonomy, description, host, distribution, life history: 111-133]; Borchs1966 [catalogue: 352]; Claps1993 [taxonomy: 10]; GonzalCh1968 [distribution: 110]; Hempel1932 [taxonomy, description, host, distribution: 337-338]; LepageGi1943 [taxonomy, description, illustration, host, distribution: 339-340]; McKenz1939 [taxonomy: 55].

Melanaspis smilacis (Comstock)

Aspidiotus smilacis Comstock, 1883: 69. Type data: U.S.A.: Massachusetts, Woods Holl, on *Smilax* sp.; collected by Prof. W. Trelease. Lectotype female, by subsequent designation Deitz & Davidson, 1986: 74. Type depository: Washington: United States National Entomological Collection, U.S. National Museum of Natural History, District of Columbia, USA.

Aonidiella smilacis; Leonardi, 1897: 286. Change of combination.

Aspidiotus (*Chrysomphalus*) *smilacis*; Cockerell, 1897i: 22. Change of combination.

Pseudischnaspis smilacis; Lindinger, 1937: 194. Change of combination.

Melanaspis smilacis; McKenzie, 1939: 55. Change of combination.

Melonaspis smilacis; Borchsenius, 1966: 418. Misspelling of genus name.

COMMON NAMES: Smilax scale [Comsto1883, Dekle1965c].

SCALE COVER: Female scale circular; exuviae central, covered with excretion; colour varies from brown to dark grey, almost black; position of exuviae marked with a white dot and concentric ring of same colour (Comstock, 1883). Scale of character common to genus; in specimens at hand from Florida scales massed on small twigs, being so crowded and piled upon each other that individual scales can scarcely be distinguished; scale very thick and hard; slightly oval; dark brown or black but overlain by a film of white wax or of epidermis of plant so that they appear pale grey; ventral scale of type common to genus present. Male scale similar to female in texture but elongate and with exuvia at one end (Ferris, 1941d, 1942).

HOST PLANTS: **Agavaceae**: *Dasylirion* [Ferris1941d, McDani1970, DeitzDa1986, Koszta1996], *Dasylirion longissimum* [DeitzDa1986, Koszta1996], *Dasylirion wheeleri* [DeitzDa1986, Koszta1996], *Nolina* [Ferris1941d, McDani1970, DeitzDa1986, Koszta1996], *Yucca* [DeitzDa1986, Koszta1996]. **Aquifoliaceae**: *Ilex* [Merril1953]. **Bromeliaceae**: *Ananas comosus* [Mamet1943a, Borchs1966, Takagi1970], *Ananas sativa* [Ferris1941d]. **Caprifoliaceae**: *Viburnum* [Merril1953], *Viburnum obovatum* [Ferris1942]. **Erythroxylaceae**: *Erythroxylum areolatum* [GreenLa1923, Ferris1942]. **Gramineae**: *Andropogon* [Balach1951, Balach1958b, DeitzDa1986, Koszta1996], *An. furcatus* [Parrot1899b], *An. scoparius* [Parrot1899b], *Arundinaria* [BesheaTiHo1973, DeitzDa1986, Koszta1996], *Bambusa* [Takagi1970], *Chrysopogon avenaceus* [Ferris1941d], *Panicum virgatum* [Ferris1941d]. **Guttiferae**: *Mammea americana* [DeitzDa1986, Koszta1996]. **Liliaceae**: *Smilax* [Comsto1883, Merril1953, Balach1958b, Dekle1965c, McDani1970, TippinBe1970, BesheaTiHo1973], *Smilax rotundifolia* [Koszta1996]. **Myricaceae**: *Myrica cerifera* [Dekle1965c]. **Myrtaceae**: *Eugenia buxifolia* [Merril1953].

NATURAL ENEMIES: HYMENOPTERA **Encyrtidae**: *Coccidencyrtus* [Koszta1996], *Signiphora fasciata* [Koszta1996].

DISTRIBUTION: **Afrotropical**: Cameroon [Balach1958b]; Côte d'Ivoire (=Ivory Coast) [Balach1958b]; Guinea [Balach1958b]; Seychelles [Mamet1943a, Borchs1966]. **Nearctic**: Mexico (Baja California [DeitzDa1986], San Luis Potosi [DeitzDa1986]); United States of America (Arizona [DeitzDa1986], District of Columbia [DeitzDa1986], Florida [Ferris1942, Merril1953, Dekle1965c, BesheaTiHo1973], Georgia [TippinBe1970, BesheaTiHo1973, DeitzDa1986], Kansas [Parrot1899b], Louisiana [DeitzDa1986], Maryland [DeitzDa1986],

Massachusetts [Comsto1883], Mississippi [DeitzDa1986], New Mexico [DeitzDa1986], New York [DeitzDa1986], North Carolina [DeitzDa1986], South Carolina [DeitzDa1986], Texas [McDani1970, DeitzDa1986], Virginia [DeitzDa1986]). **Neotropical**: Brazil [Balach1951]; Galapagos Islands [Balach1951]; Jamaica [GreenLa1923, Ferris1942]; Martinique [Balach1958b]; Mexico (Chihuahua [DeitzDa1986]); Panama [Balach1951]. **Oriental**: Mongolia [DanzigKo1990]; Taiwan [Takagi1970]. **Palaearctic**: Azores [Ferris1941d]; France [Balach1951]; Spain [BlayGo1993]; United Kingdom (England [Ferris1941d]).

BIOLOGY: Occurring on stems (Ferris, 1941d).

ECONOMIC IMPORTANCE: Recorded from sugarcane in Brazil, but not considered a pest (Williams & Greathead 1990).

GENERAL: Description and illustration of adult female by Green & Laing (1923), Ferris (1941d, 1942), Balachowsky (1951, 1958b), Chou (1985, 1986), Deitz & Davidson (1986) and by Kosztarab (1996).

KEYS: Kosztarab 1996: 533 (female) [Northeastern North America]; Danzig 1993: 238 (female) [world]; Deitz & Davidson 1986: 12-15 (female) [North America]; Deitz & Davidson 1986: 12-15 (female) [North America]; McDaniel 1970: 411-412 (female) [USA: Texas]; Balachowsky 1958b: 194 (female) [Africa]; Balachowsky 1951: 579 (female) [Mediterranean]; Ferris 1943: 65 (female) [North America]; Ferris 1942: 37, 38 (female) [North America]; Comstock 1833: 56 (female) [USA].

CITATIONS: Balach1951 [taxonomy, description, illustration, host, distribution: 583-586]; Balach1958b [taxonomy, description, illustration, host, distribution: 201-204]; BesheaTiHo1973 [host, distribution: 7]; BlayGo1993 [taxonomy, description, illustration, host, distribution: 615-619]; Borchs1937 [taxonomy, description, illustration, host, distribution: 122]; Borchs1950b [taxonomy, description, illustration, host, distribution: 222-223]; Borchs1966 [catalogue: 352-353]; Chou1985 [distribution: 323]; Chou1985 [taxonomy, description, host, distribution: 403-404]; Chou1986 [taxonomy, illustration: 697]; ClapsWoGo2001a [taxonomy, host, distribution: 21]; Cocker1896b [distribution: 334]; Cocker1897i [taxonomy, description, host, distribution: 22]; Comsto1883 [taxonomy, description, illustration, host, distribution: 56,69-70]; Danzig1972 [taxonomy, host, distribution, economic importance: 217]; Danzig1993 [taxonomy: 238]; DanzigKo1990 [host, distribution: 49]; DeitzDa1986 [taxonomy, description, illustration, host, distribution: 74-75]; Dekle1965c [taxonomy, description, host, distribution: 91]; Dekle1976 [taxonomy, description, host, distribution, economic importance: 112]; Fernal1903b [catalogue: 294]; Ferris1941d [taxonomy, description, illustration, host, distribution: 366]; Ferris1941e [taxonomy: 41,45-48]; Ferris1942 [taxonomy, description, illustration, host, distribution: 429;446:38]; Ferris1943 [taxonomy: 63]; Ferris1943a [taxonomy: 86]; GreenLa1923 [taxonomy: 125-126]; Koszta1996 [taxonomy, description, illustration, host, distribution, life history, economic importance: 537-538]; Leonar1897 [taxonomy: 286]; Lindin1932 [taxonomy, description, illustration, host, distribution: 26-27]; Lindin1937 [taxonomy: 194]; MacGil1921 [taxonomy, description, host, distribution: 444-446]; Mamet1943a [catalogue: 163]; McDani1970 [taxonomy, illustration, host, distribution: 421-422]; McKenz1939 [taxonomy: 53,55]; Merril1953 [taxonomy, description, host, distribution: 62]; Nakaha1982 [host, distribution: 56]; SchmutKlLu1957 [host, distribution, economic importance: 492]; StoetzDa1974 [taxonomy, life history: 138-140]; Takagi1970 [taxonomy, host, distribution: 138]; Tao1999 [taxonomy, host, distribution: 98]; TippinBe1970 [host, distribution: 10]; WilliaGr1990 [host, distribution, economic importance, biological control: 563-576].

Melanaspis squamea Ferris

Melanaspis latipyga; Ferris, 1941d: 356. Misidentification; discovered by Ferris, 1943: 59.

Melanaspis squamea Ferris, 1943: 59. Type data: PANAMA: Chiriqui Province, Boquete, on undetermined tree. Lectotype female, by subsequent designation Deitz & Davidson, 1986: 76. Type depository: Davis: The Bohart Museum of Entomology, University of California, California, USA.

Melanaspis latipiga; Borchsenius, 1966: 353. Misspelling of species name.

SCALE COVER: Scale of female circular; convex; black brittle with a thick ventral scale. Scale of male not recognized (Ferris, 1943).

HOST PLANTS: **Cycadaceae** [DeitzDa1986]. **Cyclanthaceae**: *Carludovica* [DeitzDa1986]. **Lauraceae**: *Persea americana* [DeitzDa1986]. **Sapotaceae**: *Pouteria sapota* [DeitzDa1986]. **Zamiaceae**: *Ceratozamia* [DeitzDa1986], *Dioon spinulosum* [DeitzDa1986].

DISTRIBUTION: **Nearctic**: Mexico (Oaxaca [DeitzDa1986], Veracruz [DeitzDa1986]). **Neotropical**: Guatemala [DeitzDa1986]; Panama [DeitzDa1986] [Ferris1943].

BIOLOGY: Occurring on bark, associated with and concealed under a fungus of genus *Septobasidium* (Ferris, 1943).

GENERAL: Description and illustration of adult female by Ferris (1943) and by Deitz & Davidson (1986).

KEYS: Ferris 1943: 65; Deitz & Davidson 1986: 12-15 (female) [North America].

CITATIONS: Borchs1966 [catalogue: 353]; DeitzDa1986 [taxonomy, description, illustration, host, distribution: 76-77]; Ferris1941e [taxonomy, description, illustration, host, distribution: 356]; Ferris1943 [taxonomy, description, illustration, host, distribution: 59-60,68].

Melanaspis sulcata Ferris

Melanaspis latipyga; Ferris, 1941d: 356. Misidentification; discovered by Ferris, 1943: 59.

Melanaspis sulcata Ferris, 1943: 61. Type data: PANAMA: Chiriqui Province, Armuelles, on undetermined plant, with *Septobasidium*. Lectotype female, by subsequent designation Deitz & Davidson, 1986: 78. Type depository: Davis: The Bohart Museum of Entomology, University of California, California, USA.

SCALE COVER: Ferris (1943) did not describe scale cover.

HOST PLANTS: **Anacardiaceae**: *Spondias* [Ferris1943, DeitzDa1986]. **Annonaceae**: *Annona cherimola* [Ferris1943, DeitzDa1986]. **Apocynaceae**: *Plumeria* [DeitzDa1986]. **Burseraceae**: *Bursera* [DeitzDa1986]. **Compositae** [Ferris1943]. **Ebenaceae**: *Diospyros kaki* [DeitzDa1986]. **Euphorbiaceae**: *Manihot esculenta* [DeitzDa1986]. **Lamiaceae** [DeitzDa1986]. **Rutaceae**: *Citrus* [DeitzDa1986]. **Salicaceae**: *Salix* [DeitzDa1986].

DISTRIBUTION: **Nearctic**: Mexico (Colima [Ferris1943], Guerrero [Ferris1943], Nayarit [Ferris1943], Veracruz [Ferris1943]). **Neotropical**: Cuba [DeitzDa1986]; Jamaica [DeitzDa1986]; Panama [Ferris1943]. **Oriental**: Philippines [DeitzDa1986]. **Palaearctic**: Germany [DeitzDa1986]; Japan [DeitzDa1986].

BIOLOGY: Associated with fungi, *Septobasidium* (Ferris, 1943).

GENERAL: Description and illustration of adult female by Ferris (1941d, 1943) and by Deitz & Davidson (1986).

KEYS: Deitz & Davidson 1986: 12-15 (female) [North America]; Ferris 1943: 65 (female) [North America].

CITATIONS: Borchs1966 [catalogue: 353]; DanzigPe1998 [catalogue: 305]; DeitzDa1986 [taxonomy, description, illustration, host, distribution: 78-79]; Ferris1941d [taxonomy, description, illustration, host, distribution: 356]; Ferris1943 [taxonomy, description, illustration, host, distribution: 61,70].

Melanaspis tenax McKenzie

Melanaspis tenax McKenzie, 1944: 54. Type data: GUATEMALA: on *Odontoglossum rossii*. Holotype female. Type depository: Davis: The Bohart Museum of Entomology, University of California, California, USA.

SCALE COVER: Scale of female circular, averaging 1.25 mm in diameter; chocolate brown; exuvium subcentral. Male scale elongate, similar in colour but smaller than female; exuvia near one end (McKenzie, 1944).

HOST PLANTS: **Araceae**: *Monstera* [DeitzDa1986], *Monstera deliciosa* [DeitzDa1986], *Philodendron* [DeitzDa1986], *Philodendron dubium* [DeitzDa1986], *Philodendron lacerum* [DeitzDa1986], *Philodendron polytomum* [DeitzDa1986], *Philodendron warscewiczii* [DeitzDa1986]. **Arecaceae** [DeitzDa1986]. **Euphorbiaceae**: *Euphorbia pulcherrima* [DeitzDa1986]. **Musaceae**: *Musa* [DeitzDa1986]. **Orchidaceae** [DeitzDa1986], *Brassavola cordata* [DeitzDa1986], *Brassavola nodosa* [DeitzDa1986], *Brassia* [DeitzDa1986], *Cattleya* [DeitzDa1986], *Cattleya dowiana aurea* [DeitzDa1986], *Cattleya trianaei* [DeitzDa1986], *Cattleya velutina* [DeitzDa1986], *Cattleya warscewiczii* [DeitzDa1986], *Dendrobium* [DeitzDa1986], *Epidendrum* [DeitzDa1986], *Epidendrum stamfordianum* [DeitzDa1986], *Miltonia warscewiczii* [DeitzDa1986], *Odontoglossum rossii* [McKenz1944, DeitzDa1986], *Oncidium* [DeitzDa1986], *Oncidium sphacelatum* [DeitzDa1986], *Peristeria elata* [DeitzDa1986]. **Rubiaceae**: *Ixora acuminata* [DeitzDa1986]. **Solanaceae**: *Solanum punctulatum* [DeitzDa1986], *Solanum wendlandii* [DeitzDa1986].

DISTRIBUTION: **Nearctic**: Mexico (San Luis Potosi [DeitzDa1986], Veracruz [DeitzDa1986]); United States of America (California [McKenz1944], Florida [DeitzDa1986]). **Neotropical**: Barbados [DeitzDa1986]; Colombia [DeitzDa1986]; Costa Rica [DeitzDa1986]; Cuba [DeitzDa1986]; Dominican Republic [DeitzDa1986]; Ecuador [DeitzDa1986]; Guatemala [DeitzDa1986]; Honduras [DeitzDa1986]; Jamaica [DeitzDa1986]; Mexico (Chiapas [DeitzDa1986]); Nicaragua [DeitzDa1986]; Panama [DeitzDa1986]; Panama Canal Zone [DeitzDa1986]; Puerto Rico & Vieques Island (Puerto Rico [DeitzDa1986]); Trinidad and Tobago (Trinidad [DeitzDa1986]); Venezuela [DeitzDa1986].

BIOLOGY: Occurring on pseudo bulbs of *Odontoglossum rossii* and presumably on leaves of *Cattleya* sp. (McKenzie, 1943).
GENERAL: Description and illustration of adult female by McKenzie (1944), Deitz & Davidson (1986) and by Colon-Ferrer & Medina-Gaud (1998).
KEYS: Deitz & Davidson 1986: 12-15 (female) [North America].
CITATIONS: Borchs1966 [catalogue: 353]; ColonFMe1998 [taxonomy, description, illustration, host, distribution: 69-70]; DeitzDa1986 [taxonomy, description, illustration, host, distribution: 80-81]; McKenz1944 [taxonomy, description, illustration, host, distribution: 54-55,58].

Melanaspis tenebricosa (Comstock)

Aspidiotus tenebricosus Comstock, 1881a: 308. Type data: U.S.A.: Washington, D.C., on bark of trunk and limbs, of red or swamp maple, *Acer rubrum*. Syntypes, female. Type depository: Washington: United States National Entomological Collection, U.S. National Museum of Natural History, District of Columbia, USA.
Aonidiella tenebricosa; Leonardi, 1897: 286. Change of combination requiring emendation of species name for agreement in gender.
Aspidiotus (Chrysomphalus) tenebricosus; Cockerell, 1897i: 22. Change of combination.
Chrysomphalus tenebricosus; Fernald, 1903b: 294. Change of combination.
Aonidiella tenebricoa; MacGillivray, 1921: 443. Misspelling of species name.
Melanaspis tenebricosa; Lindinger, 1931a: 44. Change of combination.
Chrysomphalus (Melanaspis) tenebricosus; Merrill, 1953: 40. Change of combination.
Melanaspis tenebricosa; Borchsenius, 1966: 353. Revived combination.
COMMON NAMES: gloomy scale [Comsto1881a, MerrilCh1923, Merril1953, Dekle1965c, DeitzDa1986, Koszta1996]; red maple scale [SchmutKlLu1957].
SCALE COVER: Female scale 1.5 mm in diameter; dark grey; protuberance indicating position of exuviae marked with a white dot and concentric ring; in rubbed specimens protuberance is smooth and black, in all cases remainder of surface of scale is rough; very convex; exuviae usually between center and one side; ventral scale well developed, especially at margin, where it is much thickened and dark-coloured; central part white and adheres to bark (Comstock, 1881).
HOST PLANTS: **Aceraceae**: *Acer* [Ferris1941d, Dekle1965c, TippinBe1970, BesheaTiHo1973], *Acer negundo* [McDani1970, BesheaTiHo1973], *Acer rubrum* [Comsto1881a, Wilson1917], *Acer saccharinum* [BesheaTiHo1973, DeitzDa1986], *Negundo* [Ferris1941d]. **Apocynaceae**: *Nerium* [DeitzDa1986], *Thevetia* [DeitzDa1986]. **Bignoniaceae**: *Catalpa* [DeitzDa1986]. **Cornaceae**: *Cornus* [McDani1970, DeitzDa1986], *Cornus florida* [DeitzDa1986]. **Fagaceae**: *Quercus* [Ferris1941d]. **Grossulariaceae**: *Ribes* [DeitzDa1986]. **Juglandaceae**: *Carya illinoensis* [DeitzDa1986], *Hicoria* [Ferris1941d], *Juglans regia* [DeitzDa1986]. **Leguminosae**: *Acacia* [DeitzDa1986], *Gleditsia* [DeitzDa1986], *Pithecellobium flexicaule* [DeitzDa1986]. **Magnoliaceae**: *Liriodendron tulipifera* [DeitzDa1986]. **Moraceae**: *Maclura* [Ferris1941d], *Maclura pomifera* [DeitzDa1986], *Morus* [Ferris1941d, McDani1970, DeitzDa1986], *Morus rubra* [BesheaTiHo1973]. **Nyssaceae**: *Nyssa* [Ferris1941d]. **Oleaceae**: *Fraxinus* [Ferris1941d, McDani1970,

DeitzDa1986], *Fraxinus americana* [DeitzDa1986]. **Platanaceae**: *Platanus occidentalis* [DeitzDa1986]. **Rosaceae**: *Malus pumila* [DeitzDa1986], *Prunus persica* [DeitzDa1986]. **Rubiaceae**: *Gardenia* [DeitzDa1986]. **Salicaceae**: *Populus* [Ferris1941d], *Salix* [DeitzDa1986], *Salix nigra* [McDani1970, DeitzDa1986]. **Sapindaceae**: *Sapindus* [Dekle1965c], *Sapindus drummondii* [DeitzDa1986]. **Ulmaceae**: *Celtis* [Ferris1941d, McDani1970], *Celtis laevigata* [DeitzDa1986], *Celtis mississippiensis* [McDani1970], *Celtis occidentalis* [DeitzDa1986], *Celtis reticulata* [DeitzDa1986], *Ulmus* [DeitzDa1986], *Ulmus americana* [DeitzDa1986]. **Vitaceae**: *Vitis* [Dekle1965c, DeitzDa1986].

NATURAL ENEMIES: HYMENOPTERA **Aphelinidae**: *Aspidiotiphagus citrinus* (Craw) [Gordh1979]. **Signiphoridae**: *Chartocerus nigrellus* (Girault) [Gordh1979].

DISTRIBUTION: **Nearctic**: Mexico [Ferris1941d]; United States of America (Alabama [DeitzDa1986], Arkansas [DeitzDa1986], California [DeitzDa1986], Delaware [DeitzDa1986], District of Columbia [Comsto1881a], Florida [Wilson1917, MerrilCh1923, Merril1953, Dekle1965c, DeitzDa1986], Georgia [TippinBe1970, BesheaTiHo1973, DeitzDa1986], Illinois [DeitzDa1986], Kentucky [DeitzDa1986], Louisiana [DeitzDa1986], Maryland [DeitzDa1986], Mississippi [Herric1911, DeitzDa1986], Missouri [Hollin1923, DeitzDa1986], New Jersey [Nakaha1982], New York [Nakaha1982], North Carolina [DeitzDa1986], Ohio [Ferris1941d], Oklahoma [Nakaha1982], Pennsylvania [DeitzDa1986], South Carolina [Nakaha1982], Tennessee [DeitzDa1986], Texas [Herric1911, Ferris1941d, McDani1970], Virginia [DeitzDa1986], West Virginia [Nakaha1982]). **Neotropical**: Mexico (Tabasco [DeitzDa1986]); Panama [DeitzDa1986].

BIOLOGY: Occurring on bark, commonly exposed (Ferris, 1941d).

ECONOMIC IMPORTANCE: A minor pest in southern states of USA, and in Central America (Schmutterer et al., 1957).

GENERAL: Description and illustration of adult female by Ferris (1941d), Deitz & Davidson (1986), Kosztarab (1996) and by Gill (1997).

KEYS: Gill 1997: 194 (female) [Species of California]; Kosztarab 1996: 533 (female) [Northeastern North America]; Deitz & Davidson 1986: 12-15 (female) [North America]; McDaniel 1970: 411-412 (female) [U.S.A.: Texas]; Ferris 1943: 64 (female) [North America]; Ferris 1942: 37 (female) [North America]; Hollinger 1923: 29 (female) [U.S.A.: Missouri]; Comstock 1883: 55-57 (female) [North America].

CITATIONS: BeardsDaHo1976 [economic importance: 105]; BesheaTiHo1973 [host, distribution: 7]; Borchs1966 [catalogue: 353]; Castel1951a [biological control: 95-98]; Cocker1896b [distribution: 334]; Cocker1897i [taxonomy, description, host, distribution: 22]; Comsto1881a [taxonomy, description, illustration, host, distribution: 308-309]; Comsto1883 [taxonomy, host, distribution: 71]; DavidsMi1990 [host, distribution, economic importance: 603-632]; DavidsRa1999 [economic importance, control: 1]; DeitzDa1986 [taxonomy, description, illustration, host, distribution: 82-84]; Dekle1965c [taxonomy, description, host, distribution: 92]; Dekle1976 [taxonomy, description, host, distribution, economic importance: 113]; Fernal1903b [catalogue: 294]; Ferris1941d [taxonomy, description, illustration, host, distribution: 367]; Ferris1941e [taxonomy: 48]; Ferris1942 [taxonomy: 446:37]; Ferris1943 [taxonomy: 64]; Garcia1930 [host, distribution, biological control]; Gill1997 [host, distribution, taxonomy, illustration: 199-200]; Gordh1979 [biological control: 900,912]; HanksDe1998 [life history, ecology: 239-262]; Herric1911 [taxonomy, description, illustration, host, distribution: 11,34-35,69]; Hollin1923 [taxonomy, description, host, distribution: 31-32]; Koszta1996 [taxonomy, description, illustration, host, distribution, life history, biological control: 537-540]; Leonar1897 [taxonomy: 286]; Leonar1899 [taxonomy: 175-178]; Lindin1931a [taxonomy: 44]; Lobdel1937 [taxonomy: 78]; Lord1922 [host, distribution: 1]; MacGil1921 [taxonomy, description, host, distribution: 443]; McDani1970 [taxonomy, illustration, host, distribution: 423-424]; McKenz1938 [taxonomy: 4]; McKenz1939 [taxonomy: 55]; Merril1953 [taxonomy, description, host, distribution: 40-41]; MerrilCh1923 [taxonomy, description, host, distribution, economic importance: 227]; MillerDa1990 [host, distribution, economic importance: 303]; Nakaha1982 [host, distribution: 57]; SchmutKlLu1957 [host, distribution, economic importance: 492]; StoetzDa1974 [taxonomy, life history: 138-140]; Sulliv1930 [host, distribution: 51-59]; TippinBe1970 [host, distribution: 10]; Wilson1917 [taxonomy, description, host, distribution: 27].

Melanaspis tricuspis Ferris
Melanaspis tricuspis Ferris, 1941d: 368. Type data: MEXICO: State of Colima, Manzanillo, on undetermined woody vine. Lectotype female, by subsequent designation Deitz & Davidson, 1986: 84. Type depository: Davis: The Bohart Museum of Entomology, University of California, California, USA.
SCALE COVER: Scale of female circular; quite high convex; black. Male scale not recognized (Ferris, 1941d).
DISTRIBUTION: **Nearctic**: Mexico (Colima [Ferris1941d]).
BIOLOGY: Occurring on branches. Scales buried in bark (Ferris, 1941d).
GENERAL: Description and illustration of adult female by Ferris (1941d) and by Deitz & Davidson (1986).
KEYS: Deitz & Davidson 1986: 12-15 (female) [North America]; Ferris 1943: 65 (female) [North America]; Ferris 1942: 36 (female) [North America].
CITATIONS: Borchs1966 [catalogue: 353]; DeitzDa1986 [taxonomy, description, illustration, host, distribution: 84-85]; Ferris1941d [taxonomy, description, illustration, host, distribution: 368]; Ferris1942 [taxonomy: 446:36]; Ferris1943 [taxonomy: 65].

Melanaspis vilardeboi Balachowsky
Melanaspis vilardeboi Balachowsky, 1953e: 147. Type data: GUINEA: Fulaya-Kindia, on *Parinarum* [=*Parinari*] *excelsa*. Syntypes, female. Type depository: Paris: Muséum national d'Histoire naturelle, France.
SCALE COVER: Female scale large, circular, 2.8-3.1 mm in diameter; brown black, matt; generally covered by bark of host plant; exuviae black, central or subcentral; ventral vellum white, attached to host plant. Male scale grey; oval; bivalvate; exuvia eccentric; 1.6-1.8 mm (Balachowsky, 1958b).
HOST PLANTS: **Chrysobalanaceae**: *Parinari excelsa* [Balach1953e, Balach1958b].
DISTRIBUTION: **Afrotropical**: Guinea [Balach1953e, Balach1958b].

GENERAL: Description and illustration of adult female by Balachowsky (1953e, 1958b).
KEYS: Balachowsky 1958b: 194 (female) [Africa].
CITATIONS: Balach1953e [taxonomy, description, illustration, host, distribution: 147-149]; Balach1958b [taxonomy, description, illustration, host, distribution: 193,204-205]; Borchs1966 [catalogue: 353].

Melanaspis williamsi De Lotto
Melanaspis williamsi De Lotto, 1957: 230. Type data: KENYA: Nairobi, on branches of *Aberia caffra*. Holotype female. Type depository: London: The Natural History Museum, England, UK.
SCALE COVER: Scale of female conical, very thick; colour uniformly black; diameter up to 0.8 mm. Scale of male not seen (De Lotto, 1957).
HOST PLANTS: **Flacourtiaceae**: *Aberia caffra* [DeLott1957].
DISTRIBUTION: **Afrotropical**: Kenya [DeLott1957].
GENERAL: Description and illustration of adult female by De Lotto (1957).
CITATIONS: Borchs1966 [catalogue: 353]; DeLott1957 [taxonomy, description, illustration, host, distribution: 230-231].

Mesoselenaspidus Fonseca

Mesoselenaspidus Fonseca, 1969: 16. Type species: *Mesoselenaspidus andersoni* Fonseca, by original designation.
SYSTEMATICS: Fonseca (1969) distinguished this genus from other genera in *Selenaspidus* Complex by median lobes being fused.
GENERAL: Definition and characters by Fonseca (1969).
CITATIONS: Fonsec1969 [taxonomy, description: 16-18].

Mesoselenaspidus andersoni Fonseca
Mesoselenaspidus andersoni Fonseca, 1969: 16. Type data: BRAZIL: São Paolo, Juqui, on *Nectandra myriantha*. Holotype female. Type depositories: São Paulo: Museu de Zoologia, Universidade de São Paulo, Brazil, and São Paulo: Instituto Biologico de São Paulo, Brazil.
SCALE COVER: Female scale moderately convex; circular or slightly elongate; light-greyish in colour; exuviae placed laterally. Male scale elongate, 1 mm long, 0.5 mm wide (Fonseca, 1969).
HOST PLANTS: **Lauraceae**: *Nectandra myriantha* [Fonsec1969, ClapsWoGo2001].
DISTRIBUTION: **Neotropical**: Brazil (São Paulo [Fonsec1969, ClapsWoGo2001]).
GENERAL: Description and illustration of adult female by Fonseca (1969).
CITATIONS: Claps1993 [taxonomy: 3]; ClapsWoGo2001 [host, distribution: 247-248]; Fonsec1969 [taxonomy, description, illustration, host, distribution: 16-18].

Mimeraspis Brimblecombe

Mimeraspis Brimblecombe, 1957: 261. Type species: *Mimeraspis cuspiloba* Brimblecombe, by original designation.

SYSTEMATICS: Genus *Mimeraspis* resembles *Neomorgania* MacGillivray and *Pseudotargionia* Cockerell in having a thoracic constriction, prominent median lobes, a small anal opening close to lobes, perispiracular disc pores associated with anterior spiracles and dorsal reticulated chitinization on pygidium. It differs from *Neomorgania* in shape of ducts and their orifices. The contiguous median lobes distinguish *Mimeraspis* from *Pseudotargionia* (Brimblecombe, 1957).

GENERAL: Definition and characters by Brimblecombe (1957).

CITATIONS: Borchs1966 [catalogue: 238]; Brimbl1957 [taxonomy, description: 261-263]; MorrisMo1966 [taxonomy, catalogue: 120].

Mimeraspis cuspiloba Brimblecombe

Mimeraspis cuspiloba Brimblecombe, 1957: 261. Type data: AUSTRALIA: Queensland, Boonah, on *Callistemon viminalis*; collected January 1953. Holotype female. Type depository: Brisbane: Queensland Museum, Queensland, Australia; type no. T5068.

Mimeraspis cuspilobis; Brimblecombe, 1957: 261. Misspelling of species name.

SCALE COVER: Female scale circular, 0.8 mm diameter; dark fawn, margin paler; first exuviae dark orange, surrounded by a greyish band (Brimblecombe, 1957).

HOST PLANTS: **Myrtaceae**: *Callistemon viminalis* [Brimbl1957].

DISTRIBUTION: **Australasian**: Australia (Queensland [Brimbl1957]).

GENERAL: Description and illustration of adult female by Brimblecombe (1957).

CITATIONS: Borchs1966 [catalogue: 238-239]; Brimbl1957 [taxonomy, description, illustration, host, distribution: 261-263].

Mimeraspis rotunda Brimblecombe

Mimeraspis rotunda Brimblecombe, 1957: 263. Type data: AUSTRALIA: South Australia, Mann range, on *Melaleuca dissitiflora*; collected August, 1954. Holotype female. Type depository: Brisbane: Queensland Museum, Queensland, Australia; type no. T5611.

SCALE COVER: Insects thinly scattered on leaves; scale convex, circular, 1.0 mm diameter, light fawn in colour; apical suffusion is often rubbed off, revealing orange brown to dark brown exuviae (Brimblecombe, 1957).

HOST PLANTS: **Myrtaceae**: *Melaleuca dissitiflora* [Brimbl1957].

DISTRIBUTION: **Australasian**: Australia (South Australia [Brimbl1957]).

GENERAL: Description and illustration of adult female by Brimblecombe (1957).

CITATIONS: Borchs1966 [catalogue: 239]; Brimbl1957 [taxonomy, description, illustration, host, distribution: 263-265].

Monaonidiella MacGillivray

Monaonidiella MacGillivray, 1921: 392. Type species: *Aspidiotus ceratus* Maskell, by original designation.

SYSTEMATICS: Lindinger (1937) considered this genus not distinct from *Aspidiotus*, while Ferris (1941f) was uncertain as to its synonymy with *Aspidiella*. Brimblecombe (1958, 1959) accepted it and assigned to it species all having only one pair of median lobes.

GENERAL: Definition and characters by MacGillivray (1921) and by Ferris (1941f).

CITATIONS: Borchs1966 [catalogue: 276]; Brimbl1958 [taxonomy: 74]; Ferris1937c [taxonomy: 51]; Ferris1941f [taxonomy, description: 22,24]; Kozar1990f [distribution: 142]; Lindin1937 [taxonomy: 189]; MacGil1921 [taxonomy, description: 392,445-446]; MorrisMo1966 [taxonomy, catalogue: 123].

Monaonidiella bidens (Green)

Aspidiotus (*Hemiberlesia*) *bidens* Green, 1915d: 50. Type data: AUSTRALIA: Victoria, Lake Albacutya, on *Casuarina* sp. Syntypes, female. Type depository: London: The Natural History Museum, England, UK.

Chorizaspidiotus bidens; MacGillivray, 1921: 434. Change of combination.

Aonidia bidens; Lindinger, 1937: 179. Change of combination.

Monaonidiella bidens; Brimblecombe, 1958: 85. Change of combination.

SCALE COVER: Female scale dirty white; irregularly circular, diameter about 1.65; moderately convex; exuviae yellowish, eccentric (Green, 1915d). Insects scattered on twigs; scale circular, diameter 1.0 mm; white; exuviae dark orange (Brimblecombe, 1958).

HOST PLANTS: **Casuarinaceae**: *Casuarina* [Green1915d, Brimbl1958].

DISTRIBUTION: **Australasian**: Australia (Victoria [Green1915d, Brimbl1958]).

GENERAL: Description and illustration of adult female by Green (1915d) and by Brimblecombe (1958).

CITATIONS: Borchs1966 [catalogue: 276]; Brimbl1958 [taxonomy, description, illustration, host, distribution: 85-87]; Ferris1941e [taxonomy: 41]; Green1915d [taxonomy, description, illustration, host, distribution: 50]; Lindin1937 [taxonomy: 179]; MacGil1921 [taxonomy, description, host, distribution: 434].

Monaonidiella cerata (Maskell)

Aspidiotus ceratus Maskell, 1895b: 39. Type data: AUSTRALIA: South Australia, banks of Murray River, on *Acacia stenophylla*; collected by Mr. French. Syntypes, female and first instar. Type depository: Auckland: New Zealand Arthropod Collection, Landcare Research, New Zealand.

Aonidiella cerata; Leonardi, 1899: 175. Change of combination requiring emendation of species name for agreement in gender.

Monaonidiella cerata; MacGillivray, 1921: 446. Change of combination.

SCALE COVER: Female scale snowy-white; circular, about 1/22 inch diameter; convex; two exuviae central, of a faint yellow tinge, covered by white wax. Male scale white, smaller, and more elongate than female (Maskell, 1895b).

HOST PLANTS: **Leguminosae**: *Acacia stenophylla* [Maskel1895a, Frogga1914].
DISTRIBUTION: **Australasian**: Australia (South Australia [Maskel1895a].
GENERAL: Description and illustration of adult female by Maskell (1895b).
CITATIONS: Borchs1966 [catalogue: 276]; Cocker1896b [distribution: 334]; Cocker1897i [taxonomy, description, host, distribution: 26]; Cocker1897i [taxonomy, description, host, distribution: 26]; DeitzTo1980 [taxonomy: 34]; Fernal1903b [catalogue: 254]; Ferris1937c [taxonomy: 51]; Ferris1941e [taxonomy: 42]; Ferris1941f [taxonomy: 22,24]; Frogga1914 [taxonomy, description, host, distribution: 134]; Frogga1915 [taxonomy, description, host, distribution: 11]; Fuller1897c [host, distribution: 3]; Leonar1899 [taxonomy, description, host, distribution: 175-176,191]; Lindin1937 [taxonomy: 180]; MacGil1921 [taxonomy, description, host, distribution : 446]; Maskel1895b [taxonomy, description, illustration, host, distribution: 39-40]; Maskel1897 [taxonomy, host, distribution: 296]; McKenz1938 [taxonomy: 3].

Monaonidiella nivea (Fuller)

Aspidiotus niveus Fuller, 1897b: 1346. Type data: AUSTRALIA: Western Australia, Swan River, on *Acacia pulchella*. Syntypes, female.
Monaonidiella nivea; MacGillivray, 1921: 446. Change of combination requiring emendation of species name for agreement in gender.
SCALE COVER: Female scale very convex, circular, pure white; exuviae hidden by white secretion, central, and pale yellow colour (Fuller, 1897c).
HOST PLANTS: **Leguminosae**: *Acacia pulchella* [Fuller1897b, Frogga1914].
DISTRIBUTION: **Australasian**: Australia (Western Australia [Fuller1897b, Frogga1914]).
GENERAL: Description of adult female by Fuller (1897b, 1897c).
CITATIONS: Borchs1966 [catalogue: 276]; Cocker1899a [taxonomy: 395]; Fernal1903b [catalogue: 268]; Ferris1941e [taxonomy: 46]; Frogga1914 [taxonomy, host, distribution: 316]; Frogga1915 [taxonomy, description, host, distribution: 20]; Fuller1897b [taxonomy, description, host, distribution: 1346]; Fuller1897c [taxonomy, description, host, distribution: 11]; Lindin1911 [taxonomy, description, host, distribution: 175]; MacGil1921 [taxonomy, description, host, distribution: 446].

Monaonidiella parva Brimblecombe

Monaonidiella parva Brimblecombe, 1959: 140. Type data: AUSTRALIA: Queensland, Beerwah, on *Banksia robur*; collected April 1953. Holotype female. Type depository: Brisbane: Queensland Museum, Queensland, Australia; type no. T5719.
SCALE COVER: Scales white, circular, 1.2 mm in diameter (Brimblecombe, 1959).
HOST PLANTS: **Proteaceae**: *Banksia robur* [Brimbl1959].
DISTRIBUTION: **Australasian**: Australia (Queensland [Brimbl1959]).
GENERAL: Description and illustration of adult female by Brimblecombe (1959).
CITATIONS: Borchs1966 [catalogue: 276]; Brimbl1959 [taxonomy, description, illustration, host, distribution: 140-141].

Morganella Cockerell

Morganella Cockerell, 1897i: 22. Type species: *Aspidiotus maskelli* Cockerell, by monotypy.

Aspidiotus (*Morganella*); Cockerell, 1899a: 395. Change of status.

Morganella; MacGillivray, 1921: 389. Revived rank.

Morganelia; Lepage, 1938: 410. Misspelling of genus name.

SYSTEMATICS: *Morganella* Cockerell is characterized by presence of: only median lobes, well-developed plates anterior to lobes, and interlobular paraphyses. *Sudanaspis* Chou, 1985 (type species: *Aspidiotus vuilleti* Marchal, 1909) is related to *Morganella* having been distinguished from the latter by differences in shape of median lobes and of plates.

GENERAL: Definition and characters by Fullaway (1932), Ferris (1938a), Balachowsky (1948b, 1956) and by Kosztarab (1996).

KEYS: Colon-Ferrer & Medina-Gaud 1998: 28-32 (female) [Genera of Puerto Rico]; Wolff & Corseuil 1993: 29 (female) [Brazil, Rio Grande do Sul]; Williams & Watson 1988: 19 (female) [Tropical South Pacific]; Chou 1985: 260 (female) [Genera of China]; Balachowsky 1958b: 228, 230 (female) [*Aspidiotina* of Africa]; Ezzat 1958: 237-239 (female) [Egypt]; Balachowsky 1951: 598 (female) [Mediterranean]; Balachowsky 1948b: 292-293 (female) [species Mediterranean]; Zimmerman 1948: 351 (female) [Hawaii]; Ferris 1942: 26 (female) [North America]; Ferris 1942: 38 (female) [species North America]; Brain 1918: 117 (female) [South Africa].

CITATIONS: Balach1948b [taxonomy, description: 291-292]; Balach1956 [taxonomy, description: 120-124]; Balach1958b [taxonomy, description: 228,230]; Borchs1966 [catalogue: 277]; Brain1918 [taxonomy, description: 117]; Brimbl1955 [taxonomy: 54]; Chou1985 [taxonomy: 278]; Cocker1897i [taxonomy, description: 22]; Cocker1899a [taxonomy: 395]; ColonFMe1998 [taxonomy, description: 70]; DanzigPe1998 [catalogue: 310]; Ezzat1958 [taxonomy, description: 238]; Fernal1903b [catalogue: 282]; Ferris1937c [taxonomy: 51]; Ferris1938a [taxonomy, description: 247]; Ferris1942 [taxonomy: 446:26]; Fullaw1932 [taxonomy, description: 108]; Kawai1980 [taxonomy: 208]; Koszta1996 [taxonomy, description: 540]; Kozar1990f [distribution: 142]; Lepage1938 [taxonomy: 410]; Lindin1937 [taxonomy: 189]; MacGil1921 [taxonomy, description: 389,426]; Mamet1949 [taxonomy: 62]; MorrisMo1966 [taxonomy, catalogue: 125]; Tao1999 [taxonomy, host, distribution: 98-99]; WilliaWa1988 [taxonomy: 182]; WolffCo1993 [taxonomy: 29]; Zimmer1948 [taxonomy: 358].

Morganella acaciae Munting

Morganella acaciae Munting, 1967a: 259. Type data: SOUTH AFRICA: Transvaal, Pretoria, on *Acacia karroo*. Holotype female. Type depository: Pretoria: South African National Collection of Insects, South Africa; type no. 1775/1.

SCALE COVER: Female scale subcircular, about 1 mm in diameter; whitish with brown subcircular exuviae. Male scale similar to female, but oval and about 1.2 mm in length (Munting, 1967a).

HOST PLANTS: **Leguminosae**: *Acacia* [Muntin1969], *Acacia karroo* [Muntin1967a], *Acacia nigriscens* [Muntin1969].

DISTRIBUTION: **Afrotropical**: Namibia (Southwest Africa) [Muntin1969]; South Africa [Muntin1967a, Muntin1969].

BIOLOGY: Insects scattered singly here and there on host plant and partly hidden under scaly bark (Munting, 1967a).

GENERAL: Description and illustration of adult female by Munting (1967a).

CITATIONS: Muntin1967a [taxonomy, description, illustration, host, distribution: 259-261]; Muntin1969 [host, distribution: 130].

Morganella conspicua (Brain)

Diaspis (*Epidiaspis*) *conspicua* Brain, 1919: 228. Type data: SOUTH AFRICA: Pretoria, on privet; Kroonstad on *Acacia* sp.; Kabah, Uitenhage, on *Gardenia fortunei*. Syntypes, female. Type depository: Pretoria: South African National Collection of Insects, South Africa.

Chorizaspidiotus conspicus; MacGillivray, 1921: 433. Change of combination requiring emendation of species name for agreement in gender.

Aspidiotus (*Morganella*) *conspicua*; Hall, 1929: 356. Change of combination.

Epidiaspis conspicua; Lindinger, 1937: 184. Change of combination.

Morganella conspicua; Balachowsky, 1956: 124. Change of combination.

SCALE COVER: Female scale circular, 2 mm in diameter; convex; dark greenish to almost black, with brown exuviae. Male scale whitish, and comparatively large, with parallel sides; particularly conspicuous because they are attached to stem only by their anterior end and project in all directions from among female scales (Brain, 1919). Female scale subcircular, 2 mm in diameter; convex; colour varies from green to black; exuviae central, brown (Balachowsky, 1956).

HOST PLANTS: **Anacardiaceae**: *Sclerocarya caffra* [Almeid1971]. **Leguminosae**: *Acacia* [Brain1919, Balach1956], *Bauhinia* [Hall1929], *Bauhinia macrantha* [Balach1956]. **Oleaceae**: *Ligustrum* [Balach1956]. **Rosaceae**: *Pyrus communis* [Almeid1973b]. **Rubiaceae**: *Canthium lanciflorum* [Hall1929, Balach1956], *Gardenia fortunei* [Brain1919, Balach1956].

DISTRIBUTION: **Afrotropical**: Angola [Almeid1973b]; Mozambique [Almeid1971]; South Africa [Brain1919, Balach1956]; Zimbabwe [Hall1929].

BIOLOGY: Ben-Dov & Matile-Ferrero (1984) discussed the association of this species with ant, *Melissotarsus beccarii* Emery in Africa.

GENERAL: Description and illustration of adult female by Brain (1919) and by Balachowsky (1956).

KEYS: Balachowsky 1956: 124 (female) [Africa].

CITATIONS: Almeid1971 [host, distribution: 13]; Almeid1973b [host, distribution: 10]; Balach1956 [taxonomy, description, illustration, host, distribution: 124-126]; BenDov1990d [life history, ecology, host, distribution: 339-343]; BenDovMa1984 [life history, economic importance: 378]; Borchs1966 [catalogue: 277]; Brain1919 [taxonomy, description, illustration, host, distribution: 228-229]; Foldi1990 [structure: 43-54]; Hall1929 [taxonomy, host, distribution: 356]; Hall1946 [taxonomy: 517,527,549]; Lindin1937 [taxonomy: 184]; MacGil1921 [taxonomy, description, host, distribution: 433].

Morganella cueroensis (Cockerell)

Aspidiotus cueroensis Cockerell, 1899i: 105. Type data: U.S.A.: Texas, Cuero, on rough bark of trunks of *Celtis* sp. Holotype female. Type depository: Washington: United States National Entomological Collection, U.S. National Museum of Natural History, District of Columbia, USA.

Targionia cueroensis; Leonardi, 1900: 343. Change of combination.

Targionia celtis Herrick, 1910a: 373. Type data: U.S.A.: Texas, College Station, on hackberry, *Celtis occidentalis*. Syntypes, female. Type depository: Washington: United States National Entomological Collection, U.S. National Museum of Natural History, District of Columbia, USA. Synonymy by Sasscer, 1911: 71.

Chorizaspidiotus cueroensis; MacGillivray, 1921: 433. Change of combination.

Epidiaspis cueroensis; Lindinger, 1937: 184. Change of combination.

Morganella cueroensis; Ferris, 1938a: 248. Change of combination.

Targionia (*Morganella*) *cueroensis*; Merrill, 1953: 79. Change of combination.

Morganella cueroensis; Borchsenius, 1966: 277. Revived combination.

COMMON NAME: Cuero scale [Dekle1965c].

SCALE COVER: Cockerell (1899i) described "Female scale circular, slightly over 1 mm in diameter; slightly convex above and beneath, with the margin somewhat elevated, like an oyster; very pale grey or greyish-white; exuviae more or less to one side, covered, inconspicuous, but appearing as a dark spot on inside of scale". Ferris (1938a) described "Female scale grey; somewhat elongate; highly convex; exuviae close to anterior end; a thick ventral scale present. Male scale grey or whitish, elongate oval, exuvia toward one end".

HOST PLANTS: **Aceraceae**: *Acer* [Koszta1996]. **Caricaceae**: *Carica papaya* [Koszta1996]. **Hydrangeaceae**: *Decumaria barbara* [Koszta1996]. **Labiatae**: *Leucas aspera* [SureshMo1995]. **Lauraceae**: *Persea* [Koszta1996]. **Leguminosae**: *Acacia flexicaulis* [Ferris1938a, McDani1970, Koszta1996], *Pithecellobium dulce* [Ferris1942, Koszta1996]. **Magnoliaceae**: *Magnolia* [Ferris1938a, Koszta1996], *Magnolia grandiflora* [Dekle1965c]. **Myricaceae**: *Myrica cerifera* [Dekle1965c, Koszta1996]. **Rutaceae**: *Fagara* [Ferris1938a, McDani1970, Koszta1996]. **Ulmaceae**: *Celtis* [Leonar1900, Ferris1938a, McDani1970, Koszta1996], *Celtis occidentalis* [Herric1910, McDani1970]. **Vitaceae**: *Perthenocissus* [Koszta1996].

DISTRIBUTION: **Nearctic**: Mexico (Sinola [Ferris1942]); United States of America (Florida [Merril1953, Dekle1965c], Georgia [Nakaha1982], Louisiana [Nakaha1982], Maryland [Nakaha1982], Mississippi [Ferris1938a], South Carolina [Nakaha1982], Texas [Herric1910, Herric1911, McDani1970]). **Oriental**: India [SureshMo1995].

BIOLOGY: Occurring on the bark (Ferris, 1938a).

GENERAL: Description and illustration of adult female by Ferris (1938a) and by Kosztarab (1996).

KEYS: Ferris 1942: 38 (female) [North America]; Newell 1899: 4-5 (female) [North America].

CITATIONS: Borchs1966 [catalogue: 277]; Cocker1899i [taxonomy, description, host, distribution: 105]; CockerPa1899 [taxonomy, illustration: 278,284]; Dekle1965c [description, host, distribution: 93]; Dekle1976 [description, host, distribution, economic importance: 114]; FDACSB1983 [host, distribution: 6-8];

Fernal1903b [catalogue: 296]; Ferris1938a [taxonomy, description, illustration, host, distribution: 248]; Ferris1941e [taxonomy: 42]; Ferris1942 [taxonomy: 446:38]; Ferris1943a [taxonomy: 85]; Herric1910a [taxonomy, description, illustration, host, distribution: 373-374]; Herric1911 [taxonomy, description, illustration, host, distribution: 13,43-45,77]; Koszta1996 [taxonomy, description, illustration, host, distribution: 541-542]; Leonar1900 [taxonomy, host, distribution: 343]; Lindin1937 [taxonomy: 184]; Lindin1957 [taxonomy: 550]; MacGil1921 [taxonomy, description, host, distribution: 433]; McDani1970 [taxonomy, illustration, host, distribution: 425-426]; Mead1983 [host, distribution: 1-5]; Merril1953 [taxonomy, description, host, distribution: 80]; Nakaha1982 [host, distribution: 57]; Newell1899 [taxonomy, description, host, distribution: 5,21-22]; Sassce1911 [taxonomy: 71]; SureshMo1995 [host, distribution: 429-430]; SureshMo1996 [host, distribution: 233]; Willia1985a [taxonomy: 234].

Morganella longispina (Morgan)
Aspidiotus longispina Morgan, 1889a: 352. Type data: GUYANA: Demerara, on *Cupania supida*; collected by McIntire. Syntypes, female. Type depository: London: The Natural History Museum, England, UK.
Aspidiotus longispina; Maskell, 1895b: 38. Misidentification; discovered by Cockerell, 1897i: 22.
Hemiberlesia longispina; Leonardi, 1897b: 120. Change of combination.
Aspidiotus (*Morganella*) *maskelli* Cockerell, 1897i: 22. Type data: HAWAII [=SANDWICH ISLANDS]: Kailuna, N. Kona, on Ohia tree; sent by W.S. Wait. Syntypes, female. Type depository: Washington: United States National Entomological Collection, U.S. National Museum of Natural History, District of Columbia, USA. Synonymy by Borchsenius, 1966: 277.
Aspidiotus (*Morganella*) *longispinus*; Cockerell, 1897i: 24. Change of combination requiring emendation of species name for agreement in gender.
Aspidiotus longispina ornata Maskell, 1898: 225. Type data: HAWAII: on various trees, collected by Koebele, and MAURITIUS: on undetermined plant, collected by d'Emmerez de Charmoy. Syntypes, female. Synonymy by Ferris, 1941e: 46. Notes: Depository of type material unknown (Deitz & Tocker, 1980).
Aspidiotus (*Morganella*) *maskelli ornatus*; Cockerell, 1899a: 395. Change of status.
Hemiberlesia longispina ornata; Leonardi, 1900: 339. Change of combination.
Hemiberlesia maskellii; Leonardi, 1900: 339. Change of combination.
Hemiberlesia maskellii; Leonardi, 1900: 339. Misspelling of species name.
Aspidiotus maskelli; Kirkaldy, 1902: 107. Change of combination.
Morganella longispina; Fernald, 1903b: 282. Change of combination.
Morganella maskelli; Fernald, 1903b: 282. Change of combination.
Morganella maschelli; Seabra, 1921: 99. Misspelling of species name. Notes: Misspelling of "Morganella maskelli" Cockerell.
Aspidiotus ornatus; Ferris, 1941e: 46. Change of combination and rank.
Aspidiotus (*Morganella*) *longinus*; Tao, 1999: 99. Misspelling of species name.
Aspidiotus ornataus; Tao, 1999: 99. Misspelling of species name.
COMMON NAMES: champaca scale [VelasqRi1969]; plumose scale [Brimbl1962, Hamon1981].

SCALE COVER: Female scale dark; exuviae central, and a small concentric circle in centre of first exuvia; about 0.6 mm in diameter; convex (Morgan, 1889). Female scale almost black; circular, high convex; exuviae central; a thick ventral scale formed; scale of male of similar colour, elongate; exuvia near one end (Ferris, 1938a). Colour photograph of scale cover by Wong *et al.* (1999).

HOST PLANTS: **Anacardiaceae**: *Mangifera indica* [GrandpCh1899, Mamet1943a, Mamet1949, Borchs1966, WilliaWa1988, KinjoNaHi1996]. **Annonaceae**: *Cananga odoratum* [Matile1978]. **Apocynaceae**: *Nerium oleander* [Brimbl1955]. **Bignoniaceae**: *Tecoma stans* [WilliaWa1988]. **Caricaceae**: *Carica papaya* [Lepage1938, Brimbl1955, Balach1956, Cohic1958, WilliaWa1988]. **Elaeagnaceae**: *Elaeagnus* [Takagi1958]. **Lauraceae**: *Cinnamomum zeylanica* [Ruther1915a]. **Leguminosae**: *Bauhinia* [Green1937, Balach1948b, WilliaWa1988], *Bauhinia variegata* [Cohic1958, WilliaWa1988], *Gleditsia delavayi* [Ferris1953], *Stenolobium stans* [Cohic1958]. **Lythraceae**: *Lagerstroemia* [Brain1918, Balach1956], *Lagerstroemia flos-reginae* [WilliaWa1988]. **Magnoliaceae**: *Michelia champeca* [Brain1918, Balach1956], *Michelia flava* [Hempel1900a, Lepage1938, Balach1948b]. **Malvaceae**: *Hibiscus* [Green1937, Lepage1938], *Hibiscus rosa-sinensis* [GrandpCh1899, Mamet1943a, Mamet1949, Borchs1966, WilliaWa1988], *Hibiscus syruacus* [Ferris1953]. **Meliaceae**: *Cedrela toona* [Brimbl1955]. **Moraceae**: *Artocarpus integrifolia* [Green1905a, Green1922, Green1937, Balach1948b], *Broussonetia papyrifera* [Ruther1915a], *Ficus* [Lepage1938, Cohic1958, WilliaWa1988], *Fi. carica* [Balach1926a, Balach1927a, Balach1932d, Mamet1943a, Mamet1949, TakahaTa1956, Borchs1966, WilliaWa1988], *Fi. macrophylla* [Brimbl1955], *Morus* [Ruther1915a]. **Myrtaceae**: *Eugenia* [WilliaWa1988], *Psidium cattleianum* [WilliaWa1988], *Ps. guajava* [DoaneHa1909, WilliaWa1988]. **Oleaceae**: *Fraxinus* [Green1937], *Fraxinus berlandieri* [Balach1927, Balach1932d], *Jasminum* [WilliaWa1988], *Jasminum sambac* [Cohic1958, WilliaWa1988], *Ligustrum* [Balach1948b], *Ligustrum sinense* [Brimbl1955], *Olea europaea* [Balach1932d]. **Oxalidaceae**: *Averrhoa carambola* [Lepage1938, WilliaWa1988]. **Proteaceae**: *Macadamia ternifolia* [Brimbl1955]. **Rutaceae**: *Citrus* [Balach1926a, Lepage1938, Zimmer1948, WilliaWa1988], *Ci. aurantium* [WilliaWa1988], *Ci. grandis* [Balach1956, WilliaWa1988], *Ci. limon* [WilliaWa1988], *Ci. maxima* [WilliaWa1988], *Ci. paradisi* [WilliaWa1988], *Ci. reticulata* [WilliaWa1988]. **Sapindaceae**: *Alectryon connatus* [Brimbl1955], *Cupania* [Green1937], *Cupania supida* [Morgan1889a]. **Theaceae**: *Camellia* [Balach1948b], *Camellia japonica* [Lepage1938]. **Ulmaceae**: *Celtis* [Ferris1953].

NATURAL ENEMIES: FUNGI: *Fusarium juruanum* [WilliaWa1988]. HYMENOPTERA **Aphelinidae**: *Archenomus perkinsi* (Fullaway) [Zimmer1948], *Prospaltella koebelei* Howard [Zimmer1948], *Pteroptrix perkinsi* Fullaway [SchmutKlLu1957].

DISTRIBUTION: **Afrotropical**: Cameroon [Nakaha1982]; Comoros Islands [Matile1978]; Mauritius [GrandpCh1899, Green1907, Mamet1943a, Mamet1949]; Mozambique [Almeid1971]; São Tomé and Príncipe [Nakaha1982]); South Africa [BrainKe1917, Ferris1938a, Balach1956]. **Australasian**: Australia (Queensland [Brimbl1955]); Cook Islands [WilliaWa1988]; Fiji [WilliaWa1988]; French Polynesia (Society Islands [DoaneHa1909, WilliaWa1988], Tahiti [DoaneHa1909,

Ferris1938a, WilliaWa1988]); Hawaiian Islands (Hawaii [Cocker1897i, Zimmer1948, TakahaTa1956]); New Caledonia [Cohic1958]; Papua New Guinea [WilliaWa1988]; Samoa [Nakaha1982]; Tonga [WilliaWa1988]; Western Samoa [Laing1927, WilliaWa1988]. **Nearctic**: Mexico [Balach1948b]; United States of America (Florida [Nakaha1982]). **Neotropical**: Antigua and Barbados [Nakaha1982]; Bahamas [Nakaha1982]; Bermuda [Nakaha1982]; Brazil [Ferris1938a] (Parana [WolffCo1993], Rio Grande do Sul [WolffCo1993], Rio de Janeiro [Lepage1938], Santa Catarina [WolffCo1993], São Paulo [Lepage1938]); Costa Rica [Nakaha1982]; Dominican Republic [Nakaha1982]; Guatemala [Nakaha1982]; Guyana [Morgan1889a, Newste1920]; Haiti [Nakaha1982]; Jamaica [Nakaha1982]; Puerto Rico & Vieques Island (Puerto Rico [Balach1948b, Martor1976]); Trinidad and Tobago [Nakaha1982]. **Oriental**: China (People's Republic) (Yunnan [Ferris1953]); Hong Kong [Nakaha1982]; India [Ramakr1921a, Ferris1938a]; Indonesia [Nakaha1982]; Philippines [VelasqRi1969]; Ryukyu Islands (= Nansei Shoto) [KinjoNaHi1996]; Sri Lanka [Green1905a, Ruther1915a, Green1922]; Taiwan [WongChCh1999]. **Palaearctic**: Algeria [Balach1926a, Balach1927a, Ferris1938a]; China (People's Republic) [Tang1984]; Egypt [Ezzat1958]; Japan [Takagi1958] (Kyushu [TakahaTa1956], Shikoku [TakahaTa1956]).

BIOLOGY: Occurring on bark (Ferris, 1938a).

ECONOMIC IMPORTANCE: Cohic (1955, 1956) and Reboul (1976) recorded damage to various host plants in New Caledonia and French Polynesia. Das (1965) listed this species as a pest of tea plants in India. Williams & Watson (1988) indicated damage to grapefruit, lemon and fig in Tahiti.

GENERAL: Description and illustration of adult female by Balachowsky (1926a, 1948b, 1956), Zimmerman (1948), Ferris (1938a), Brimblecombe (1955), Tang (1984), Chou (1985, 1986) and by Colon-Ferrer & Medina-Gaud (1998).

KEYS: Ezzat 1958: 241 (female) [Egypt]; Balachowsky 1956: 124 (female) [Africa]; Balachowsky 1948b: 292 (female) [Mediterranean]; Ferris 1942: 38 (female) [North America]; Fullaway 1932: 95-97, 108 (female) [Hawaii].

CITATIONS: Almeid1971 [host, distribution: 13]; Balach1926a [taxonomy, description, illustration, host, distribution: 63,65-69]; Balach1927 [host, distribution: 177]; Balach1927a [host, distribution: 71-72]; Balach1932b [ecology: 517-522]; Balach1932d [taxonomy, host, distribution: IX]; Balach1948b [taxonomy, description, illustration, host, distribution: 293-295]; Balach1956 [taxonomy, description, illustration, host, distribution: 126-127]; BiezanFr1939 [host, distribution: 1-18]; BiezanSe1940 [host, distribution: 67-68]; Borchs1966 [catalogue: 277-278]; Brain1918 [taxonomy, description, illustration, host, distribution: 136-137]; BrainKe1917 [distribution: 184]; Brimbl1955 [taxonomy, description, illustration, host, distribution: 54-56]; Brimbl1962 [host, distribution, economic importance: 222]; BurgerUl1990 [economic importance: 313-327]; Castel1963 [distribution: 140]; Chou1985 [taxonomy, description, host, distribution: 278-279]; Chou1986 [taxonomy, illustration: 668]; ClapsWoGo2001a [taxonomy, host, distribution: 21-22]; Cocker1896b [distribution: 334]; Cocker1897i [taxonomy, description, illustration, host, distribution: 22-23]; Cocker1899a [taxonomy: 395]; Cohic1956 [host, distribution, economic importance]; Cohic1958 [host, distribution:

14]; ColonFMe1998 [taxonomy, description, illustration, host, distribution: 70-71]; CoronaRuMo1997 [host, distribution: 38-41]; DanzigPe1998 [catalogue: 310-311]; DEDAC1923 [host, distribution]; DeitzTo1980 [taxonomy: 40]; Dekle1964 [host, distribution: 1]; DoaneHa1909 [host, distribution: 298]; EHG1897 [host, distribution: 67-85]; Ezzat1958 [distribution: 241]; EzzatNa1987 [distribution: 87]; Fernal1903b [catalogue: 282]; Ferris1937c [taxonomy, illustration: 51,84]; Ferris1938a [taxonomy, description, illustration, host, distribution: 249]; Ferris1941e [taxonomy: 45-46]; Ferris1942 [taxonomy: 446:38]; Ferris1953 [host, distribution: 67]; Flande1971 [biological control, life history: 857-872]; Foldi1990 [structure: 43-54]; Fullaw1932 [taxonomy: 96,108]; Gomez1936 [host, distribution: 42-43]; GrandpCh1899 [taxonomy, description, host, distribution: 24]; Green1905a [taxonomy, description, illustration, host, distribution: 340]; Green1907 [host, distribution: 203]; Green1922 [host, distribution: 462]; Green1937 [host, distribution: 332]; Hall1946a [taxonomy: 527]; Hamon1981 [taxonomy, host, distribution, life history, control: 1-2]; Hamon1983 [host, distribution: 45]; Hempel1900a [taxonomy, description, host, distribution: 498-499]; HorticInNo1923 [host, distribution: 236-239]; Kawai1980 [taxonomy, description, host, distribution: 208-209]; KinjoNaHi1996 [host, distribution: 125-127]; Kirka11902 [taxonomy, host, distribution: 107]; Laing1927 [host, distribution: 40]; Leonar1897b [taxonomy, description, illustration, host, distribution: 120-121]; Leonar1900 [taxonomy, host, distribution: 339]; Lepage1938 [catalogue: 410-411]; MacGil1921 [taxonomy, description, host, distribution: 426]; Mamet1943a [catalogue: 164]; Mamet1949 [catalogue: 62]; Martor1976 [host, distribution: 65]; Maskel1895b [taxonomy, description, host, distribution: 38-39]; Maskel1898 [taxonomy, description, host, distribution: 225-226]; Matile1978 [host, distribution: 65]; MillerDa1990 [host, distribution, economic importance: 303]; Morgan1889a [taxonomy, description, illustration, host, distribution: 352]; MoutiaMa1947 [distribution]; Muraka1970 [host, distribution: 75]; NagarkSa1990 [host, distribution, economic importance, biological control: 553-542]; Nakaha1982 [host, distribution: 57]; Newste1920 [taxonomy, description, illustration, host, distribution: 194-195]; PenaDu1999 [host, distribution, chemical control, economic importance: 210-212]; Ramakr1921a [host, distribution: 357]; RangelGo1945 [distribution, description: 1-44]; Reboul1976 [host, distribution, economic importance]; Reboul1976 [host, distribution, economic importance]; Ruther1915a [taxonomy, description, host, distribution: 109]; Sassce1923 [host, distribution: 152-158]; SchmutKlLu1957 [host, distribution, economic importance: 474]; Seabra1921 [taxonomy, host, distribution: 99]; Silves1929 [host, distribution: 897-904]; Takagi1958 [taxonomy, host, distribution: 126]; TakahaTa1956 [host, distribution: 15]; Tang1984 [taxonomy, description, illustration, host, distribution: 56-57]; Tao1999 [taxonomy, host, distribution: 98-99]; Timber1924 [host, distribution, biological control]; Wester1920 [host, distribution]; Willia1985a [taxonomy: 236]; WilliaWa1988 [taxonomy, description, illustration, host, distribution, economic importance: 9,181-183]; WolffCo1993 [taxonomy, description, illustration, host, distribution: 45-46]; WongChCh1999 [taxonomy, description, host, distribution: 28,70-71]; Zimmer1948 [taxonomy, description, illustration, host, distribution, biological control: 358-359].

Morganella pseudospinigera Balachowsky

Morganella pseudospinigera Balachowsky, 1956: 126. Type data: ZAIRE: Stanleyville, on undetermined plant. Holotype female. Type depository: Tervuren: Musée Royal de l'Afrique Centrale, Section d'Entomologie, Belgium.

SCALE COVER: Female scale circular, 2 mm in diameter; convex; colour brown or black; exuviae central, embedded in secretion of adult female, dark. Male scale unknown (Balachowsky, 1956).

HOST PLANTS: **Anacardiaceae**: *Mangifera indica* [DeJeanMo1991]. **Euphorbiaceae**: *Briedelia* [DelageMaBa1972, DeJeanMo1991]. **Leguminosae**: *Bauhinia* [DelageMaBa1972], *Piliostigma thoningii* [DelageMaBa1972]. **Moraceae**: *Ficus* [DeJeanMo1991]. **Rosaceae**: *Malus communis* [Almeid1973b].

DISTRIBUTION: **Afrotropical**: Angola [Almeid1973b]; Cameroon [DeJeanMo1991]; Côte d'Ivoire (=Ivory Coast) [DelageMaBa1972]; Zaire [Balach1956].

BIOLOGY: Ben-Dov & Matile-Ferrero (1984) discussed the association of this species with ant, *Melissotarsus beccarii* Emery in Africa. DeJean & Mony (1991) and Mony et al. (2002) studied the association of this armoured scale with *Melissotarsus beccarii* and *M. weissi* (Santschi) on mango, *Ficus* and *Briedelia* in Yaounde, Cameroon.

GENERAL: Description and illustration of adult female by Balachowsky (1956).

KEYS: Balachowsky 1956: 124 (female) [Africa].

CITATIONS: Almeid1973b [host, distribution: 10]; Balach1956 [taxonomy, description, illustration, host, distribution: 124-128]; BenDov1990d [life history, ecology, host, distribution: 339-343]; BenDovMa1984 [life history, ecology: 378]; Borchs1966 [catalogue: 278]; DeJeanMo1991 [host, distribution, life history, ecology: 179-187]; DelageMaBa1972 [host, distribution, life history, ecology: 2359-2361]; Foldi1990 [structure: 43-54]; Lepesm1947 [host, distribution: 214]; MonyKeOr2002 [host, distribution, life history, ecology: 645-654].

Morganella spinigera (Lindinger)

Aspidiotus spiniger Lindinger, 1909e: 19. Type data: CAMEROON: Bipinde, Urwaldgebiet, on *Strombosiopsis tetranda*. Syntypes, female. Type depository: Hamburg: Zoologisches Institut und Zoologishces Museum, Universität von Hamburg, Germany.

Chorizaspidiotus spiniger; MacGillivray, 1921: 433. Change of combination.

Epidiaspis spinigera; Lindinger, 1937: 184. Change of combination requiring emendation of species name for agreement in gender.

Morganella spinigera; Balachowsky, 1956: 124. Change of combination.

SCALE COVER: Female scale capsule-like, convex (Lindinger, 1909e).

HOST PLANTS: **Olacaceae**: *Strombosiopsis tetranda* [Lindin1909e].

DISTRIBUTION: **Afrotropical**: Cameroon [Lindin1909e, Vayssi1913].

GENERAL: Description and illustration of adult female by Lindinger (1909e).

KEYS: Balachowsky 1956: 124 (female) [Africa].

CITATIONS: Balach1956 [taxonomy: 124,128]; Borchs1966 [catalogue: 278]; Ferris1941e [taxonomy: 48]; Lindin1909e [taxonomy, description, illustration, host, distribution: 19-20]; Lindin1937 [taxonomy: 184]; MacGil1921 [taxonomy, description, host, distribution: 433]; Sassce1911 [taxonomy: 70]; Vayssi1913 [host, distribution: 430]; WeidneWa1968 [taxonomy: 173].

Murataspis Balachowsky & Richardeau

Murataspis Balachowsky & Richardeau, 1942: 100. Type species: *Hemiberlesia megapora* Balachowsky, by original designation.

SYSTEMATICS: This genus is characterized mainly by: expanded and highly-sclerotized prosoma, quadrate shape pygidial lobes, and reduced plates. May resemble genus *Sishanaspis* Ferris, but the latter belongs to paralatoriine genera of Diaspidinae (Balachowsky, 1956).

GENERAL: Definition and characters by Balachowsky & Richardeau (1942) and by Balachowsky (1949a, 1951, 1956).

KEYS: Balachowsky 1958b: 228 (female) [*Aspidiotina* of Africa]; Balachowsky 1951: 598 (female) [Mediterranean].

CITATIONS: Balach1949a [taxonomy, description: 163-164]; Balach1951 [taxonomy, description: 561-562]; Balach1956 [taxonomy, description: 130]; Balach1958b [taxonomy: 228]; BalachRi1942 [taxonomy, description: 100]; Borchs1966 [catalogue: 278]; DanzigPe1998 [catalogue: 311]; MorrisMo1966 [taxonomy, catalogue: 125].

Murataspis claviformis Balachowsky & Richardeau

Murataspis claviformis Balachowsky & Richardeau, 1942: 102. Type data: ALGERIA: Touggourt Territory, Oasis of Mhraier et El-Arfiane, on *Tamarix gallica*. Syntypes, female. Type depository: Paris: Muséum national d'Histoire naturelle, France.

SCALE COVER: Female scale subcircular, 1.5-1.6 mm in diameter; slightly convex in center; exuviae central, yellow gold; secretion of adult irregular, thin, white matt. Male scale same structure, oval; white; exuviae yellow gold (Balachowsky & Richardeau, 1942; Balachowsky, 1951).

HOST PLANTS: **Tamaricaceae**: *Tamarix gallica* [BalachRi1942, Balach1951].

DISTRIBUTION: **Palaearctic**: Algeria [BalachRi1942, Balach1951].

GENERAL: Description and illustration of adult female by Balachowsky & Richardeau (1942) and by Balachowsky (1951).

KEYS: Balachowsky 1951: 562 (female) [Mediterranean]; Balachowsky 1949a: 165 (female) [World]; Balachowsky & Richardeau 1942: 102 (female) [North Africa].

CITATIONS: Balach1949a [taxonomy: 165]; Balach1951 [taxonomy, description, illustration, host, distribution: 565-567]; Balach1956 [taxonomy: 130]; Balach1958a [host, distribution: 35-36]; BalachRi1942 [taxonomy, description, illustration, host, distribution: 100-103]; Borchs1966 [catalogue: 278]; DanzigPe1998 [catalogue: 311]; Lindin1957 [taxonomy: 550]; Lindin1957 [taxonomy: 550].

Murataspis megapora (Balachowsky)

Hemiberlesia megapora Balachowsky, 1928a: 122. Type data: MOROCCO: Igli (Oued Saoura), Mezzer, Beni-Abbes, on branches of *Tamarix articulata*; collected May 1926. Holotype female. Type depository: Paris: Muséum national d'Histoire naturelle, France.

Parlatoreopsis megapora; Lindinger, 1932f: 202. Change of combination.

Murataspis megaporus; Balachowsky & Richardeau, 1942: 101. Change of combination, and incorrect agreement in gender.

SCALE COVER: Female scale circular or subcircular, diameter 0.75-1.1 mm; slightly convex; frequently embedded in bark; exuviae central, rarely subcentral, yellow gold; secretion of adult white, not shiny; ventral vellum poorly developed (Balachowsky, 1928a). Illustration of scale cover by Balachowsky (1928a).

HOST PLANTS: **Salicaceae**: *Populus* [Balach1951]. **Tamaricaceae**: *Tamarix* [Balach1951], *Ta. aphylla* [Rungs1935], *Ta. articulata* [Balach1928a, Balach1932d, Balach1956], *Ta. speciosa* [Balach1951], *Ta. speciosa massaica* [Balach1949a].

DISTRIBUTION: **Palaearctic**: Morocco [Balach1928a, Balach1932d, Rungs1935, Balach1956].

GENERAL: Description and illustration of adult female by Balachowsky (1928a, 1949a, 1951).

KEYS: Balachowsky 1956: 130 (female) [Africa]; Balachowsky 1951: 562 (female) [Mediterranean]; Balachowsky 1949a: 165 (female) [World]; Balachowsky & Richardeau 1942: 102 (female) [North Africa]; Balachowsky 1928a: 132 (female) [North Africa].

CITATIONS: Balach1928a [taxonomy, description, illustration, host, distribution: 122-123]; Balach1932d [taxonomy, host, distribution: 48; viii]; Balach1949a [taxonomy, description, illustration, host, distribution: 164-165]; Balach1951 [taxonomy, description, illustration, host, distribution: 562-565]; Balach1956 [taxonomy, description, illustration, host, distribution: 130-133]; Balach1958a [host, distribution: 35]; BalachRi1942 [taxonomy: 101-102]; Borchs1966 [catalogue: 278-279]; DanzigPe1998 [catalogue: 311]; Lindin1932f [taxonomy: 202]; Rungs1935 [host, distribution: 271].

Mycetaspis Cockerell

Chrysomphalus (*Mycetaspis*) Cockerell, 1897i: 9. Type species: *Aspidiotus personatus* Comstock, by monotypy and original designation.

Mycetaspis; MacGillivray, 1921: 392. Change of status.

SYSTEMATICS: *Mycetaspis* is related to *Melanaspis* but differs in presence of a broad protuberance on cephalic margin (Ferris, 1941d; Balachowsky, 1958b).

GENERAL: Definition and characters by Ferris (1941d) and by Balachowsky (1951, 1958b).

KEYS: Colon-Ferrer & Medina-Gaud 1998: 28-32 (female) [Genera of Puerto Rico]; Wolff & Corseuil 1993: 29 (female) [Brazil, Rio Grande do Sul]; Zahradník 1990b: 74 (female) [Czech Republic]; Chou 1985: 319 (female) [Genera of China]; Zahradník 1959a: 547 (female) [Czech Republic]; Balachowsky 1958b: 230 (female) [*Aspidiotina* of Africa]; Ezzat 1958: 237-239 (female) [Egypt];

Balachowsky 1951: 601 (female) [Mediterranean]; Ferris 1942: 27 (female) [North America]; Ferris 1942: 38 (female) [species North America].
CITATIONS: Balach1951 [taxonomy, description: 574-575]; Balach1958b [taxonomy, description: 205-206]; Borchs1966 [catalogue: 357]; Chou1985 [taxonomy, description: 319]; Cocker1897i [taxonomy, description: 9,13]; Cocker1899a [taxonomy: 396]; ColonFMe1998 [taxonomy, description: 71-72]; Danzig1993 [taxonomy, description: 241]; DanzigPe1998 [catalogue: 312]; Ezzat1958 [taxonomy: 239]; Ferris1937c [taxonomy: 51,55,85]; Ferris1938 [taxonomy: 46]; Ferris1941d [taxonomy, description: 369]; MacGil1921 [taxonomy, description: 392,442]; McKenz1939 [taxonomy: 53]; MorrisMo1966 [taxonomy, catalogue: 125]; Schmut1959 [taxonomy, description: 48,107]; Tao1999 [taxonomy: 99]; WolffCo1993 [taxonomy: 29].

Mycetaspis apicata (Newstead)
Aspidiotus (*Chrysomphalus*) *apicatus* Newstead, 1920: 195. Type data: GUYANA: Enmore Forest, East Coast, Demerara, on *Avicennia nitida*. Holotype female. Type depository: London: The Natural History Museum, England, UK.
Pelomphala apicata; MacGillivray, 1921: 441. Change of combination requiring emendation of species name for agreement in gender.
Chrysomphalus apicatus; Bodkin, 1922: 59. Change of combination.
Aonidiella apicata; Costa Lima, 1924a: 196. Change of combination.
Aspidiotus apicatus; McKenzie, 1938: 3. Change of combination.
Mycetaspis apicata; Ferris, 1941d: 370. Change of combination.
Melanaspis apicata; Lindinger, 1943a: 147. Change of combination.
Mycetaspis apicata; Borchsenius, 1966: 357. Revived combination.
SCALE COVER: Female scale more or less circular, 1.6-1.9 mm in diameter; convex and very thick; covered with a relatively thick epidermal layer of bark; colour, when denuded, opaque black; larval exuviae nude, shining black, forming a well-defined nipple; second exuviae black; ventral surface shining black; ventral scale rather stout, white or dusky white, with a dark brown or blackish periphery (Newstead, 1920). Scale of female only moderately convex, dark brown or black. Scale of male dark brown, oval (Ferris, 1941d).
HOST PLANTS: **Avicenniaceae**: *Avicennia nitida* [Newste1920, Ferris1941d]. **Euphorbiaceae**: *Croton* [Ferris1941d]. **Guttiferae**: *Clusia* [Lepage1938, Ferris1941d], *Vismia* [Ferris1941d]. **Leguminosae**: *Acacia* [Ferris1941d], *Albizia* [Ferris1941d], *Enterolobium* [Ferris1941d], *Inga* [Ferris1941d]. **Magnoliaceae**: *Magnolia* [Lepage1938, Ferris1941d]. **Rutaceae**: *Cyclocarpus* [Ferris1941d].
DISTRIBUTION: **Nearctic**: Mexico (Colima [Ferris1941d], Guerrero [Ferris1941d], Nayarit [Ferris1941d]); United States of America (Texas). **Neotropical**: Brazil (Rio de Janeiro [Lepage1938, Ferris1941d]); Guyana [Newste1920, Ferris1941d]; Panama [Ferris1941d].
BIOLOGY: Occurring on the bark of the limbs and trunks of the hosts. There is a very strong tendency for the scales to be concealed beneath bark flakes, in cracks or under the epidermis of the bark (Ferris, 1941d).
GENERAL: Description and illustration of adult female by Newstead (1920) and by Ferris (1941d).

KEYS: Ferris 1942: 38 (female) [North America].

CITATIONS: Bodkin1922 [taxonomy, distribution: 59]; Borchs1966 [catalogue: 357]; ClapsWoGo2001a [taxonomy, host, distribution: 22]; CostaL1924a [taxonomy, host, distribution: 196]; Ferris1941d [taxonomy, description, illustration, host, distribution: 370]; Ferris1941e [taxonomy: 41]; Ferris1942 [taxonomy: 446:38]; Lepage1938 [catalogue: 391-392]; Lindin1943a [taxonomy: 147]; Lizery1943a [host, distribution: 319-335]; MacGil1921 [taxonomy, description, host, distribution: 441]; McKenz1938 [taxonomy: 3]; McKenz1939 [taxonomy: 53]; Nakaha1982 [host, distribution: 58]; Newste1920 [taxonomy, description, illustration, host, distribution: 195-196]; WilliaGr1990 [host, distribution, economic importance, biological control: 563-578].

Mycetaspis bezerrai Arruda

Mycetaspis bezerrai Arruda, 1972: 13. Type data: BRAZIL: Pernambuco, on leaves of cashew tree (*Anacardium occidentale*). Holotype female. Type depository: Pernambuco: Coleção de coccideos, Instituto de Pesquisas Agronomicas de Pernambuco, Brazil.

SCALE COVER: Female scale circular, 1.2 mm in diameter; convex, as a high cap; colour black; exuviae central (Arruda, 1972).

HOST PLANTS: **Anacardiaceae**: *Anacardium occidentale* [Arruda1972, ClapsWoGo2001].

DISTRIBUTION: **Neotropical**: Brazil (Pernambuco [Arruda1972, ClapsWoGo2001]).

GENERAL: Description and illustration of adult female by Arruda (1972).

CITATIONS: Arruda1972 [taxonomy, description, illustration, host, distribution: 13-17]; ClapsWoGo2001 [host, distribution: 248].

Mycetaspis brasiliensis Hempel

Mycetaspis brasiliensis Hempel, 1932: 338. Type data: BRAZIL: São Paulo State, Praia Grande, on leaves of a tree of Myrtaceae; collected by Jose Pinto da Fonseca, November 1929. Syntypes, female. Type depositories: São Paulo: Instituto Biologico de São Paulo, Brazil, and Curitiba: Departamento de Zoologia, Setor de Ciencias Biologicas, Universidade Federal do Parana, Brazil.

SCALE COVER: Female scale circular, about 1 mm in diameter; convex; bright grey or black; ventral scale well-developed (Hempel, 1932).

HOST PLANTS: **Myrtaceae** [Hempel1932]. **Theaceae**: *Camellia* [ClapsWoGo2001].

DISTRIBUTION: **Neotropical**: Brazil (Rio de Janeiro [ClapsWoGo2001], São Paulo [Hempel1932, ClapsWoGo2001]).

GENERAL: Description and illustration of adult female by Hempel (1932).

CITATIONS: Borchs1966 [catalogue: 357]; Claps1993 [taxonomy: 6,10]; ClapsWoGo2001 [host, distribution: 248]; Ferris1941d [taxonomy: 369]; Hempel1932 [taxonomy, description, host, distribution: 338-339].

Mycetaspis defectopalus Ferris
Mycetaspis defectopalus Ferris, 1941d: 371. Type data: U.S.A.: Texas, at Brownsville, on *Porlieria angustifolia*. Holotype female. Type depository: Davis: The Bohart Museum of Entomology, University of California, California, USA.
COMMON NAME: defecto scale [Dekle1965c].
SCALE COVER: Scale of female almost hemispherical, dark brown; that of male flat, oval, dark brown (Ferris, 1941d).
HOST PLANTS: **Meliaceae**: *Swietenia mahogani* [Dekle1965c]. **Moraceae**: *Chlorophora tinctoria* [Ferris1941d]. **Rutaceae**: *Fagara fagara* [Ferris1941d]. **Sapotaceae**: *Bumelia tenax* [Dekle1965c], *Sideroxylon* [Dekle1965c]. **Zygophyllaceae**: *Porlieria angustifolia* [Ferris1941d].
DISTRIBUTION: **Nearctic**: Mexico [Nakaha1982]; United States of America (Florida [Merril1953, Dekle1965c], Texas [Ferris1941d]). **Neotropical**: Belize [Nakaha1982]; Ecuador [Nakaha1982]; Nicaragua [Nakaha1982]; Panama [Ferris1941d]; Peru [Nakaha1982].
BIOLOGY: Occurring on twigs and trunk (Ferris, 1941d).
GENERAL: Description and illustration of adult female by Ferris (1941d).
KEYS: Ferris 1942: 38 (female) [North America].
CITATIONS: Borchs1966 [catalogue: 357]; Dekle1965c [taxonomy, description, host, distribution: 94]; Dekle1976 [taxonomy, description, host, distribution, economic importance: 115]; Ferris1941d [taxonomy, description, illustration, host, distribution: 371]; Ferris1942 [taxonomy: 446:38]; Merril1953 [taxonomy, description, host, distribution: 62-63]; Nakaha1982 [host, distribution: 58].

Mycetaspis eneideae Arruda
Mycetaspis eneideae Arruda, 1976: 21. Type data: BRAZIL: Pernambuco, Sirinhaem, Nucleo de Recolonizacao, Cucao, on leaves of mango. Holotype female. Type depository: Pernambuco: Coleção de coccideos, Instituto de Pesquisas Agronomicas de Pernambuco, Brazil; type no. 22.
SCALE COVER: Female scale circular; highly convex, like a cap; black; exuviae central (Arruda, 1976).
HOST PLANTS: **Anacardiaceae**: *Mangifera indica* [Arruda1976, ClapsWoGo2001].
DISTRIBUTION: **Neotropical**: Brazil (Pernambuco [Arruda1976, ClapsWoGo2001]).
GENERAL: Description and illustration of adult female by Arruda (1976).
CITATIONS: Arruda1976 [taxonomy, description, illustration, host, distribution: 21-26]; ClapsWoGo2001 [host, distribution: 248].

Mycetaspis juventinae Lepage & Giannotti
Mycetaspis juventinae Lepage & Giannotti, 1944: 299. Type data: BRAZIL: São Paolo, Capital de S. Paolo, Parque D. Pedro II., on *Erythrina* sp. Syntypes, female. Type depositories: São Paulo: Museu de Zoologia, Universidade de São Paulo, Brazil, São Paulo: Instituto Biologicó de São Paulo, Brazil, and Curitiba: Departamento de Zoologia, Setor de Ciencias Biologicas, Universidade Federal do Parana, Brazil.

SCALE COVER: Female scale more or less circular; colour dark brown chestnut; exuviae black, central or subcentral. Male scale small and elongate (Lepage & Giannotti, 1944).
HOST PLANTS: **Leguminosae**: *Erythrina* [LepageGi1944, ClapsWoGo2001].
DISTRIBUTION: **Neotropical**: Brazil (São Paulo [LepageGi1944, ClapsWoGo2001]).
GENERAL: Description and illustration of adult female by Lepage and Giannotti (1944).
CITATIONS: Borchs1966 [catalogue: 357]; Claps1993 [taxonomy: 6,9]; ClapsWoGo2001 [host, distribution: 248]; LepageGi1944 [taxonomy, description, illustration, host, distribution: 299-300].

Mycetaspis personata (Comstock)

Aspidiotus personatus Comstock, 1883: 66. Type data: CUBA: Havana, on leaves of various trees and shrubs in public gardens. Syntypes, female. Type depository: Washington: United States National Entomological Collection, U.S. National Museum of Natural History, District of Columbia, USA.
Aspidiotus personates; Cockerell, 1891: 5. Misspelling of species name.
Aonidiella personata; Leonardi, 1897: 286. Change of combination requiring emendation of species name for agreement in gender.
Aspidiotus (*Mycetaspis*) *personatus*; Cockerell, 1897i: 24. Change of combination.
Chrysomphalus personatus; Fernald, 1903b: 292. Change of combination.
Melanaspis personata; Lindinger, 1921: 427. Change of combination.
Mycetaspis personata; MacGillivray, 1921: 442. Change of combination.
Pseudaonidia personata; Gómez-Menor Ortega, 1941: 139. Change of combination.
Chrysomphalus ((*Mycetaspis*)) *personatus*; Merrill, 1953: 40. Change of combination.
Mycetaspis personata; Borchsenius, 1966: 357. Revived combination.
COMMON NAMES: masked scale [Comsto1883, Merril1953, Dekle1965c]; pinta-negra [CarvalAg1997].
SCALE COVER: Female scale circular; very convex; exuviae central; grey or black, with exuviae shining black; position of exuviae is usually marked with a white dot and a concentric ring of same colour; ventral scale well-developed (Comstock, 1883). Female scale very black, almost hemispherical or somewhat thimble-shaped; with exuviae central; male scale flat, oval, lighter in colour (Ferris, 1941d). Colour photograph of scale cover and general appearance see Carvalho & Aguiar (1997).
HOST PLANTS: **Anacardiaceae**: *Anacardium* [Ferris1941d], *Mangifera* [Ferris1941d], *Mangifera indica* [Morgan1889a, Dekle1965c, WolffCo1993a]. **Annonaceae**: *Annona* [Balach1951]. **Bignoniaceae**: *Catalpa longisiliqua* [GomezM1941]. **Bromeliaceae**: *Tillandsia* [Ferris1941d]. **Lauraceae**: *Persea* [Balach1951]. **Leguminosae**: *Bauhinia* [Merril1953]. **Malpighiaceae**: *Banisteria laurifolia* [Ferris1941d]. **Monimiaceae**: *Tambourissa* [Matile1978]. **Moraceae**: *Ficus* [Ferris1941d, Merril1953], *Ficus retusa* [Balach1958b]. **Musaceae**: *Musa* [Ferris1941d]. **Myrtaceae**: *Eugenia* [Merril1953], *Feijoa selloviana* [Balach1938a]. **Oleaceae**: *Jasminum* [Ferris1941d]. **Palmae**: *Areca* [Ferris1941d], *Cocos*

[Balach1951], *Cocos nucifera* [Ferris1941d], *Latania* [Balach1951], *Phoenix* [Balach1951], *Sabal* [Ferris1941d]. **Rutaceae**: *Citrus* [CarvalAg1997]. **Sapotaceae**: *Achras sapota* [Dekle1965c]. **Theaceae**: *Camellia japonica* [Balach1938a].

NATURAL ENEMIES: HYMENOPTERA **Aphelinidae**: *Aphytis chrysomphali* (Mercet) [AbdRab2001a], *Aphytis equatorialis* Rosen & DeBach [RosenDe1979]. **Signiphoridae**: *Signiphora fax* Girault [Woolle1990].

DISTRIBUTION: **Afrotropical**: Cameroon [RosenDe1979, MatileNo1984]; Comoros Islands [Matile1978]; Côte d'Ivoire (=Ivory Coast) [Nakaha1982]; Guinea [Nakaha1982]; Senegal [Nakaha1982]; Sierra Leone [Nakaha1982]; Togo [Nakaha1982]. **Australasian**: Hawaiian Islands (Hawaii [Nakaha1982]). **Nearctic**: Mexico [Cocker1899n, Ferris1941d] [Nakaha1982]; United States of America (Florida [Merril1953, Dekle1965c]). **Neotropical**: Argentina [ClapsTe2001]; Brazil [Ferris1941d, WolffCo1993a] (Alagoas [WolffCo1993], Pernambuco [WolffCo1993], Rio de Janeiro [WolffCo1993], Santa Catarina [WolffCo1993], São Paulo [WolffCo1993]); Colombia [Kondo2001]; Cuba [Comsto1883]; Dominican Republic [GomezM1941]; Guyana [Morgan1889a]; Jamaica [Ferris1941d]; Panama [Ferris1941d]; Peru [Nakaha1982]; Puerto Rico & Vieques Island (Puerto Rico [Ferris1941d, Martor1976]); U.S. Virgin Islands [Nakaha1983]; Venezuela [Nakaha1982]. **Oriental**: Hong Kong [Chou1985]; Indonesia [Nakaha1982]; Philippines [Nakaha1982]; Sri Lanka [Nakaha1982]. **Palaearctic**: Belgium [Nakaha1982]; Canary Islands [Nakaha1982]; Czech Republic [Zahrad1977, Zahrad1990b]; Egypt [Ezzat1958, Salama1970a]; France [Nakaha1982]; Germany (United) [Nakaha1982]; Madeira Islands [Balach1938a, CarvalAg1997]; Netherlands [Nakaha1982]; Russia (Moscow Oblast [Danzig1993]); United Kingdom (England [Ferris1941d]).

BIOLOGY: Normally on leaves (Ferris, 1941d).

ECONOMIC IMPORTANCE: Polyphagous species; distributed in tropical and subtropical regions; has been recorded as citrus pest in Madeira (Balachowsky, 1938a; Carvalho & Aguiar, 1997) and mango in Egypt (Salama & Saleh, 1972).

GENERAL: Description and illustration of adult female by Ferris (1941d), Balachowsky (1951, 1958b), Chou (1985, 1986), Zahradník (1990b) and by Colon-Ferrer & Medina-Gaud (1998).

KEYS: Claps & Teran 2001: 392 (female) [South Africa]; Ezzat 1958: 242 (female) [Egypt]; Ferris 1942: 38 (female) [North America]; Cockerell 1905: 45-46 (female) [Mexico]; Comstock 1883: 55-57 (female) [North America].

CITATIONS: AbdRab2001a [host, distribution, biological control: 174]; AbouEl2001 [host, distribution, biological control: 185-195]; Balach1938a [host, distribution, economic importance: 151]; Balach1951 [taxonomy, description, illustration, host, distribution: 575-578]; Balach1958b [taxonomy, description, illustration, host, distribution: 203,206-207]; Barber1893 [host, distribution: 50-51]; BiezanFr1939 [host, distribution: 1-18]; Bondar1914 [host, distribution, economic importance: 1064-1106]; Bondar1915 [host, distribution, economic importance: 44-47]; Borchs1966 [catalogue: 357-358]; BurgerUl1990 [economic importance: 313-327]; CarvalAg1997 [life history, description, economic importance, biological control, host, distribution: 295-297]; Chou1985 [taxonomy, description, host, distribution: 319-320]; Chou1986 [taxonomy, illustration: 694]; ChuaWo1990 [host, distribution, economic importance: 543-552]; ClapsTe2001 [taxonomy, description, illustration, host, distribution: 392,397-398]; ClapsWoGo2001a [taxonomy, host, distribution: 22]; Cocker1892b [host, distribution: 333]; Cocker1893cc [host, distribution: 100-101]; Cocker1896b [distribution: 334]; Cocker1897i [taxonomy, description, host, distribution: 24]; Cocker1899n [host, distribution: 23]; Cocker1905 [taxonomy: 45]; ColonFMe1998 [taxonomy, description, illustration, host, distribution: 72-73]; Comsto1883 [taxonomy, description, illustration, host, distribution: 66-67]; Craw1896 [taxonomy, description, host, distribution: 35]; Danzig1993 [taxonomy, description, illustration, host, distribution: 241-242]; DanzigPe1998 [catalogue: 312]; Dekle1965c [taxonomy, description, host, distribution: 95]; Dekle1976 [taxonomy, description, host, distribution, economic importance: 116]; DeSant1979 [biological control]; ElMinsElHa1974 [taxonomy, description, illustration, host, distribution: 223-232]; ElMinsOs1973 [taxonomy, structure: 169-173]; ElMinsOs1974 [host, distribution, life history, ecology, biological control: 152-170]; Ezzat1958 [distribution: 242]; EzzatNa1987 [distribution: 87]; Fernal1903b [catalogue: 292]; Ferris1937c [taxonomy, illustration: 51,85]; Ferris1941d [taxonomy, description, illustration, host, distribution: 372]; Ferris1941e [taxonomy: 47]; Ferris1942 [taxonomy: 446:38]; Foldi1990 [structure: 43-54]; Fonsec1963 [host, distribution: 32-35]; Fonsec1964 [host, distribution: 515]; FonsecAu1932a [host, distribution: 202-214]; Gentry1965 [host, distribution, economic importance]; Gomez1936 [host, distribution: 42-43]; GomezM1941 [taxonomy, illustration, host, distribution: 139]; Gowdey1921 [taxonomy, description, host, distribution: 32]; Green1897 [taxonomy: 69]; Green1915e [host, distribution: 608-636]; Hall1927d [taxonomy: 274]; Hempel1920 [taxonomy, description, host, distribution: 141]; Koebel1907 [host, distribution, biological control: 159-164]; Kondo2001 [taxonomy, host, distribution: 44]; KondoKa1995 [host, distribution: 57-58]; KondoKa1995a [host, distribution: 97-98]; Koszta1955 [host, distribution: 381]; Leefma1929 [host, distribution: 1]; Leonar1897 [taxonomy: 286]; Leonar1899 [taxonomy: 175-176,193-203]; Lindin1921 [host, distribution: 427]; Lindin1936 [taxonomy: 155]; Lindin1937 [taxonomy: 194]; Lindin1939 [taxonomy: 38]; Lounsb1921 [host, distribution: 35-38]; MacGil1921 [taxonomy, description, host, distribution: 442]; Mallam1954 [distribution: 24-60]; Martor1976 [host, distribution: 1-303]; Matile1978 [host, distribution: 65]; MatileNo1984 [host, distribution: 66]; McInti1889 [taxonomy: 23]; McKenz1938 [taxonomy: 4]; McKenz1939 [taxonomy: 54]; Merril1953 [taxonomy, description, host, distribution: 40]; MillerDa1990 [host, distribution, economic importance: 303]; Morgan1889a [host, distribution: 351]; MouradMeFa2001 [host, distribution: 571-580]; Nakaha1982 [host, distribution: 58]; Nakaha1983 [host, distribution: 13]; Newste1901b [taxonomy, description, illustration, host, distribution: 81-84]; NourElRi1970 [host, distribution: 123-127]; Pace1939 [host, distribution: 664-665]; Ponnam1999 [host, distribution, chemical control: 445-451]; PriesnHo1932 [taxonomy, description, host, distribution: 92]; PriesnHo1940 [biological control: 58-70]; RosenDe1979 [host, distribution, biological control: 564-567]; Salama1970a [host, distribution, life history: 42-46]; Salama1972 [host, distribution, life history: 403-407]; SalamaSa1972 [host, distribution, life history, economic importance: 328-331]; Sassce1918 [host, distribution: 125-129]; Schmut1957a [host, distribution: 136]; Schmut1957b [taxonomy: 148]; Schmut1959 [taxonomy, description, host, distribution: 107]; SchmutKlLu1957 [host, distribution, economic importance: 491]; Tao1999 [taxonomy, host, distribution: 99]; Ward1890 [host, distribution: 306]; WatanaTaCo2000 [host, distribution, life history, biological control: 49-64]; Wester1918 [host, distribution, economic importance: 5-57]; Wester1920 [host, distribution]; Willia1970DJ [taxonomy, host, distribution: 33]; WolffCo1993 [taxonomy, description, illustration, host, distribution: 47-49]; WolffCo1993a [host, distribution: 153]; Woolle1990 [biological control: 167-176]; Zahrad1957 [host, distribution: 49]; Zahrad1977 [taxonomy, distribution: 120]; Zahrad1990b [taxonomy, description, illustration, host, distribution: 113-115].

Mycetaspis sphaerioides (Cockerell)

Aspidiotus sphaerioides Cockerell, 1895w: 7. Type data: U.S.A.: Louisiana, exact locality not indicated, on leaves of New Zealand flax [=*Phormium* sp.]; collected by E.M. Ehrhorn. Syntypes, female. Type depository: Washington: United States National Entomological Collection, U.S. National Museum of Natural History, District of Columbia, USA.
Chrysomphalus sphaerioides; Leonardi, 1897: 286. Change of combination.
Aspidiotus (*Chrysomphalus*) *sphaerioides*; Cockerell, 1897i: 30. Change of combination.
Pseudischnaspis sphaerioides; Lindinger, 1937: 194. Change of combination.
Mycetaspis sphaerioides; McKenzie, 1939: 55. Change of combination.
SCALE COVER: Female scale circular, over 1 mm in diameter; individuals on leaves moderately convex; dark reddish-brown, with part covering exuviae indicated by a pale raised ring; when rubbed, exuviae uncovered and appear shining black; with a ventral vellum that remains attached to host plant (Cockerell, 1895w). In the type lot scales of females almost hemispherical in form, dark brown in colour; male scale flat and paler, oval. In material on *Bursera,* scales seem to be flatter, but they are perhaps not fully mature (Ferris, 1941d).
HOST PLANTS: **Agavaceae**: *Phormium* [Ferris1941d], *Phormium tenax* [Ferris1941d]. **Burseraceae**: *Bursera* [Ferris1941d]. **Myrtaceae**: *Jambosa* [Ferris1941d].
DISTRIBUTION: **Nearctic**: Mexico (Oaxaca [Ferris1941d]); United States of America (Louisiana [Cocker1895w]). **Neotropical**: Guatemala [Nakaha1982]; Panama [Nakaha1982]; Venezuela [Ferris1941d].
BIOLOGY: Specimens on *Phormium* occur on leaves, while those on *Bursera* on bark of the trunk (Ferris, 1941d).
GENERAL: Description and illustration of adult female by Ferris (1941d).
KEYS: Ferris 1942: 38 (female) [North America].
CITATIONS: Borchs1966 [catalogue: 358]; Cocker1895w [taxonomy, description, host, distribution: 7-8]; Cocker1896b [distribution: 334]; Cocker1897i [taxonomy, description, host, distribution: 30]; Fernal1903b [catalogue: 294]; Ferris1941d [taxonomy, description, illustration, host, distribution: 373]; Ferris1941e [taxonomy: 48]; Ferris1942 [taxonomy: 446:38]; Kuwana1902 [taxonomy: 71]; Leonar1897 [taxonomy: 286]; Lindin1909c [taxonomy, host, distribution: 449]; MacGil1921

[taxonomy, description, host, distribution: 420]; McKenz1939 [taxonomy: 55]; Nakaha1982 [host, distribution: 59]; Willia1985a [taxonomy: 239].

Myrtophila Brimblecombe

Myrtophila Brimblecombe, 1957: 265. Type species: *Myrtophila curvata* Brimblecombe, by original designation.

Mytrophila; Borchsenius, 1966: 239. Misspelling of genus name.

SYSTEMATICS: This genus is related to *Pseudotargionia*, *Neomorgania* and *Mimeraspis*. It differs from *Neomorgania* and *Mimeraspis* in: separated median lobes, and possession of plates between them. The large dorsal ducts in median area of pygidium, and short uniform paraphyses distinguish *Myrtophila* from *Pseudotargionia* (Brimblecombe, 1957).

GENERAL: Definition and characters by Brimblecombe (1957).

CITATIONS: Borchs1966 [catalogue: 239]; Brimbl1957 [taxonomy, description: 265-267]; MorrisMo1966 [taxonomy, catalogue: 126].

Myrtophila adnatae Brimblecombe

Myrtophila adnatae Brimblecombe, 1957: 268. Type data: AUSTRALIA: Queensland, Yelarbon, on *Melaleuca adnata*; collected October 1954. Holotype female. Type depository: Brisbane: Queensland Museum, Queensland, Australia; type no. T5619.

Mytrophila adnatae; Borchsenius, 1966: 239. Misspelling of genus name.

SCALE COVER: Female scale circular, 0.8 mm diameter; white to light fawn in colour; exuviae almost brown (Brimblecombe, 1957).

HOST PLANTS: **Myrtaceae**: *Melaleuca adnata* [Brimbl1957].

DISTRIBUTION: **Australasian**: Australia (Queensland [Brimbl1957]).

BIOLOGY: Insects mostly single on basal half of upper surface of small leaves (Brimblecombe, 1957).

GENERAL: Description and illustration of adult female by Brimblecombe (1957).

CITATIONS: Borchs1966 [catalogue: 239]; Brimbl1957 [taxonomy, description, illustration, host, distribution: 268-269].

Myrtophila curvata Brimblecombe

Myrtophila curvata Brimblecombe, 1957: 266. Type data: AUSTRALIA: Queensland, Tugun, on *Leptospermum stellatum*; collected January 1950. Holotype female. Type depository: Brisbane: Queensland Museum, Queensland, Australia; type no. T5615.

Mytrophila curvata; Borchsenius, 1966: 239. Misspelling of genus name.

SCALE COVER: Female scale circular, 0.75 mm diameter; brown to dark brown, margin fawn; first exuviae dark brown (Brimblecombe, 1957).

HOST PLANTS: **Myrtaceae**: *Leptospermum flavescens* [Brimbl1957], *Leptospermum stellatum* [Brimbl1957].

DISTRIBUTION: **Australasian**: Australia (Queensland [Brimbl1957]).

BIOLOGY: Insects single and sparse, mostly in leaf axils (Brimblecombe, 1957).

GENERAL: Description and illustration of adult female by Brimblecombe (1957).

CITATIONS: Borchs1966 [catalogue: 239]; Brimbl1957 [taxonomy, description, illustration, host, distribution: 266-267].

Myrtophila pseudadnatae Brimblecombe
Myrtophila pseudadnatae Brimblecombe, 1959: 142. Type data: AUSTRALIA: Queensland, Mt. Isa, on *Melaleuca bracteata*; collected February 1958. Holotype female. Type depository: Brisbane: Queensland Museum, Queensland, Australia; type no. T5721.
Mytrophila pseudadnatae; Borchsenius, 1966: 239. Misspelling of genus name.
SCALE COVER: Female scale circular, 0.6 mm diameter; convex; dull whitish; exuviae dark (Brimblecombe, 1959).
HOST PLANTS: **Myrtaceae**: *Melaleuca bracteata* [Brimbl1959].
DISTRIBUTION: **Australasian**: Australia (Queensland [Brimbl1959]).
GENERAL: Description and illustration of adult female by Brimblecombe (1959).
CITATIONS: Borchs1966 [catalogue: 239]; Brimbl1959 [taxonomy, description, illustration, host, distribution: 142-143].

Myrtophila suticollis Brimblecombe
Myrtophila suticollis Brimblecombe, 1957: 269. Type data: AUSTRALIA: Queensland, Amberley, on *Melaleuca irbyana*; collected July 1954. Holotype female. Type depository: Brisbane: Queensland Museum, Queensland, Australia; type no. T5623.
Mytrophila suticollis; Borchsenius, 1966: 239. Misspelling of genus name.
SCALE COVER: Female scale circular, 1.2 mm diameter; dark fawn; exuviae paler (Brimblecombe, 1957).
HOST PLANTS: **Myrtaceae**: *Melaleuca irbyana* [Brimbl1957].
DISTRIBUTION: **Australasian**: Australia (Queensland [Brimbl1957]).
BIOLOGY: Insects single and sparse on the minute host leaves (Brimblecombe, 1957).
GENERAL: Description and illustration of adult female by Brimblecombe (1957).
CITATIONS: Borchs1966 [catalogue: 239]; Brimbl1957 [taxonomy, description, illustration, host, distribution: 269-271].

Neoclavaspis Brimblecombe

Neoclavaspis Brimblecombe, 1959: 144. Type species: *Neoclavaspis duplex* Brimblecombe, by monotypy and original designation.
SYSTEMATICS: The species of *Neoclavaspis* resemble some species in *Clavaspis* and in *Diaspidiotus*. The genus differs from *Clavaspis* in not having strongly developed paraphyses in first and second interlobular areas, and from *Diaspidiotus* in absence of well-developed second lobes (Brimblecombe, 1959).
GENERAL: Definition and characters by Brimblecombe (1959).
CITATIONS: Borchs1966 [catalogue: 317]; Brimbl1959 [taxonomy, description: 144]; MorrisMo1966 [taxonomy, catalogue: 130].

Neoclavaspis duplex Brimblecombe

Neoclavaspis duplex Brimblecombe, 1959: 144. Type data: AUSTRALIA: Queensland, Texas, on *Eremophila mitchellii*. Holotype female. Type depository: Brisbane: Queensland Museum, Queensland, Australia; type no. T5725.
SCALE COVER: Female scale circular, 1.6 mm diameter, colour obscured by extraneous matter (Brimblecombe, 1959).
HOST PLANTS: **Myoporaceae**: *Eremophila mitchellii* [Brimbl1959].
DISTRIBUTION: **Australasian**: Australia (Queensland [Brimbl1959]).
GENERAL: Description and illustration of adult female by Brimblecombe (1959).
CITATIONS: Borchs1966 [catalogue: 317]; Brimbl1959 [taxonomy, description, illustration, host, distribution: 144-146].

Neoclavaspis nudata (Brimblecombe)

Morganella nudata Brimblecombe, 1959b: 405. Type data: AUSTRALIA: Queensland, Kenmore, on *Callistemon viminalis*; collected April 1958. Holotype female. Type depository: Brisbane: Queensland Museum, Queensland, Australia; type no. T5787.
Neoclavaspis nudata; Borchsenius, 1966: 317. Change of combination.
SCALE COVER: Insects single and sparse, under corky tissue on twigs (Brimblecombe, 1959b).
HOST PLANTS: **Myrtaceae**: *Callistemon viminalis* [Brimbl1959b].
DISTRIBUTION: **Australasian**: Australia (Queensland [Brimbl1959b]).
GENERAL: Description and illustration of adult female by Brimblecombe (1959b).
CITATIONS: Borchs1966 [catalogue: 317]; Brimbl1959b [taxonomy, description, illustration, host, distribution: 405-407].

Neoleonardia MacGillivray

Neoleonardia MacGillivray, 1921: 392. Type species: *Aspidiotus extensus* Maskell, by monotypy and original designation.
SYSTEMATICS: Lindinger (1937) considered *Neoleonardia* a synonym of *Chentraspis* Leonardi, while Ferris (1938) and Brimblecombe (1953) accepted *Neoleonardia* as valid. All three species currently placed in *Neoleonardia* are characterized by median lobes fully fused into a single lobe.
GENERAL: Definition and characters by MacGillivray (1921).
CITATIONS: Borchs1966 [catalogue: 300]; Brimbl1953 [taxonomy: 161]; Ferris1937c [taxonomy: 51]; Ferris1938 [taxonomy: 43]; Kozar1990f [distribution: 142]; Lindin1937 [taxonomy: 190]; MacGil1921 [taxonomy, description: 392,446]; MorrisMo1966 [taxonomy, catalogue: 131].

Neoleonardia alata (Froggatt)

Aspidiotus alatus Froggatt, 1914: 132. Type data: AUSTRALIA: New South Wales, at Dubbo and Wagga, on *Eucalyptus* sp., Victoria, near Swan Hill, Murray River, on *Eucalyptus rostrata*. Syntypes, female. Type depository: Canberra: Australian National Insect Collection, CSIRO Entomology, Australia.
Aspidiotus alatus; Froggatt, 1915: 8. Notes: Described again as "n. sp.".

Neglectaspis alata; Lindinger, 1937: 190. Change of combination requiring emendation of species name for agreement in gender.
Neoleonardia alata; Ferris, 1938: 43. Change of combination.
SCALE COVER: Female scale hardly circular, about 1/8 inch diameter; rising up into a rounded dome, curving round like a mussel-shell; light chocolate brown, clothed with thin, grey shell, giving it a regular ringed structure; exuviae central, blue-black, surrounded with a regular boss, making it very prominent (Froggatt, 1914). Froggatt (1914) illustrated the female scale.
HOST PLANTS: **Myrtaceae**: *Eucalyptus rostrata* [Frogga1914].
DISTRIBUTION: **Australasian**: Australia (Victoria [Frogga1914]).
GENERAL: Description and illustration of adult female by Froggatt (1914).
CITATIONS: Borchs1966 [catalogue: 300]; Brimbl1953 [taxonomy: 161]; Ferris1938 [taxonomy: 43]; Ferris1941e [taxonomy: 40]; Frogga1914 [taxonomy, description, host, distribution: 132]; Frogga1915 [taxonomy, description, host, distribution: 8-9]; Laing1929 [taxonomy, description, illustration: 24]; Lindin1937 [taxonomy: 190]; Sassce1915 [taxonomy, host, distribution: 33]; Takagi1990b [taxonomy, structure: 19].

Neoleonardia aliformis Brimblecombe

Neoleonardia aliformis Brimblecombe, 1953: 164. Type data: AUSTRALIA: Queensland, Brisbane, on *Eucalyptus acmenioides*; collected by A.R. Brimblecombe, 26.i.1947. Holotype female. Type depository: Brisbane: Queensland Museum, Queensland, Australia; type no. T5280.
SCALE COVER: Exuviae dark, hard and brittle (Brimblecombe, 1953). Photograph of scale cover by Brimblecombe (1953).
HOST PLANTS: **Myrtaceae**: *Eucalyptus acmenioides* [Brimbl1953].
DISTRIBUTION: **Australasian**: Australia (Queensland [Brimbl1953]).
BIOLOGY: Insect and scale occurring on the smooth bark of small branches or under the stringy bark of the larger branches and trunk (Brimblecombe, 1953).
GENERAL: Description and illustration of adult female by Brimblecombe (1953).
CITATIONS: Borchs1966 [catalogue: 300]; Brimbl1953 [taxonomy, description, illustration, host, distribution: 164-166].

Neoleonardia chitinosa Brimblecombe

Neoleonardia chitinosa Brimblecombe, 1953: 161. Type data: AUSTRALIA: Queensland, Inglewood, on *Eucalyptus crebra*; collected by M. Muell, September 1940. Holotype female. Type depository: Brisbane: Queensland Museum, Queensland, Australia; type no. T5278.
SCALE COVER: Female scale averaging 1.5 mm; exuviae black, hard and brittle (Brimblecombe, 1953). Photograph of scale cover by Brimblecombe (1953).
HOST PLANTS: **Myrtaceae**: *Eucalyptus crebra* [Brimbl1953].
DISTRIBUTION: **Australasian**: Australia (Queensland [Brimbl1953]).
BIOLOGY: Insects occurring under epidermis of twigs and branches and giving appearance of small round surface swellings (Brimblecombe, 1953)
GENERAL: Description and illustration of adult female by Brimblecombe (1953).

CITATIONS: Borchs1966 [catalogue: 300]; Brimbl1953 [taxonomy, description, illustration, host, distribution: 161-164].

Neoleonardia delicatula (Laing)

Aspidiotus delicatulus Green, 1918: 146. Nomen nudum.
Aspidiotus delicatulus Laing, 1929: 23. Type data: AUSTRALIA: Northern Territory, Koolpinyah, on *Melaleuca leucodendron*; collected by G.F. Hill. Syntypes, female. Type depository: London: Natural History Museum, UK.
Neglectaspis delicatula; Lindinger, 1937: 190. Change of combination requiring emendation of species name for agreement in gender.
Neoleonardia delicatula; Ferris, 1938: 43. Change of combination.
SCALE COVER: Female scale subcircular, flattish, pale grey, rather thin and papery in texture; exuviae eccentric, dull black or very dark brown; diameter approximately 1 mm (Laing, 1929).
HOST PLANTS: **Myrtaceae**: *Melaleuca leucodendron* [Laing1929].
DISTRIBUTION: **Australasian**: Australia (Northern Territory [Laing1929]).
GENERAL: Description and illustration of adult female by Laing (1929).
CITATIONS: Borchs1966 [catalogue: 300]; Brimbl1953 [taxonomy: 161]; DEDAC1923 [host, distribution]; Ferris1938 [taxonomy: 43]; Ferris1941e [taxonomy: 42]; Green1918 [taxonomy: 146]; Laing1929 [taxonomy, description, illustration, host, distribution: 23-24]; Lindin1937 [taxonomy: 190]; Lindin1957 [taxonomy: 545].

Neoleonardia extensa (Maskell)

Aspidiotus extensus Maskell, 1895b: 41. Type data: AUSTRALIA: New South Wales, Bankstown near Sydney, on *Eucalyptus capitellata*. Syntypes, female and first instar. Type depository: Auckland: New Zealand Arthropod Collection, Landcare Research, New Zealand.
Chentraspis extensa; Leonardi, 1897: 286. Change of combination requiring emendation of species name for agreement in gender.
Aspidiotus (*Chentraspis*) *extensus*; Cockerell, 1897i: 26. Change of combination.
Neoleonardia extensa; MacGillivray, 1921: 446. Change of combination.
SCALE COVER: Female scale dull dirty-yellow or brown; frequently obscured by black fungus; subcircular, diameter 1/12 inch at full development; convex, somewhat conical; larval exuvia black, situated at apex of cone; second exuvia very inconspicuous, and difficult to make out its dimensions; close examination shows that it occupies about half scale; surface of scale finely striated. Male scale subcircular, grey or bluish-grey; rather convex, but less so than female; scale less black than female, and not placed centrally; diameter of scale about 1/20 inch (Maskell, 1895b).
HOST PLANTS: **Myrtaceae**: *Eucalyptus capitellata* [Maskel1895b, Frogga1914].
DISTRIBUTION: **Australasian**: Australia (New South Wales [Frogga1914]).
BIOLOGY: Maskell (1895b) reported that female scales were on the twigs, while male scales on the leaves.
GENERAL: Description and illustration of adult female by Maskell (1895b).

CITATIONS: Borchs1966 [catalogue: 300]; Brimbl1953 [taxonomy: 161]; Cocker1896b [distribution: 335]; Cocker1897i [taxonomy, description, host, distribution: 26]; DeitzTo1980 [taxonomy: 37]; Fernal1903b [catalogue: 258]; Ferris1937c [taxonomy: 51]; Ferris1938 [taxonomy, illustration: 43,49]; Ferris1941e [taxonomy: 43]; Frogga1914 [taxonomy, description, host, distribution: 311-312]; Frogga1915 [taxonomy, description, host, distribution: 15]; Leonar1897 [taxonomy: 286]; Leonar1897b [taxonomy, description, illustration, host, distribution: 113-115]; MacGil1921 [taxonomy, description, host, distribution: 446]; Maskel1895b [taxonomy, description, illustration, host, distribution: 41-42].

Neomorgania MacGillivray

Neomorgania MacGillivray, 1921: 394. Type species: *Aspidiotus junctiloba* Marlatt (= *Neomorgania eucalypti* (Maskell)). Subsequently designated by Ferris, 1937c: 55.

SYSTEMATICS: MacGillivray (1921) assigned three species to this genus, namely *Aspidiotus eucalypti* Maskell, *A. acaciae* Morgan and *A. junctiloba* Marlatt. Brimblecombe (1954) concluded that all three represent one species, *Aspidiotus eucalypti* Maskell. The genus appears to be related to *Neoleonardia*, but on latter the median lobes completely fused, whereas in *Neomorgania* inner margins of median lobes very close to each other, but not conjugated.

GENERAL: Definition and characters by MacGillivray (1921) and by Brimblecombe (1954).

CITATIONS: Balach1948b [taxonomy: 269]; Balach1951 [taxonomy: 677]; Borchs1966 [catalogue: 236]; Ferris1937a [taxonomy: 51]; Lindin1937 [taxonomy: 190]; MacGil1921 [taxonomy, description: 394,458]; Miller1990 [taxonomy: 169-178]; MorrisMo1966 [taxonomy, catalogue: 131].

Neomorgania eucalypti (Maskell)

Aspidiotus eucalypti Maskell, 1889: 102. Type data: AUSTRALIA: South Australia, on *Eucalyptus* sp. Lectotype female and first instar, by subsequent designation Deitz & Tocker, 1980: 36. Type depository: Canberra: Australian National Insect Collection, CSIRO Entomology, Australia.

Aspidiotus acaciae Morgan, 1889a: 353. Type data: AUSTRALIA: Tasmania, on *Acacia pycnantha*. Syntypes, female. Type depository: London: The Natural History Museum, England, UK. Synonymy by Brimblecombe, 1954: 150.

Aspidiotus acaciae propinqua Maskell, 1893b: 205. Type data: AUSTRALIA: New South Wales, Mount Victoria, on *Acacia* sp. Syntypes, female and first instar. Synonymy by Borchsenius, 1966: 236. Notes: Depository of type material unknown (Deitz & Tocker, 1980).

Aspidiotus acaciae propinquus; Cockerell, 1897i: 26. Change of combination requiring emendation of species name for agreement in gender.

Targionia acaciae propinqua; Leonardi, 1900: 191. Change of combination.

Targionia acaciae; Leonardi, 1900: 308. Change of combination.

Targionia eucalypti; Leonardi, 1900: 310. Change of combination.

Pseudaonidia eucalypti; Marlatt, 1908: 138. Change of combination.

Pseudaonidia junctiloba Marlatt, 1908: 138. Type data: AUSTRALIA: Victoria, Shepparton, on *Acacia* sp.; collected by Charles French. Syntypes, female. Type depository: Washington: U.S. National Entomological Collection, U.S. National Museum of Natural History, District of Columbia, USA. Synonymy by Brimblecombe, 1954: 150. Notes: Marlatt (1908) referred to this species as a MS name of Green, but the latter did not describe it. However, the characters and collection data in Marlatt (1908: 135,138) validated this species.

Aspidiotus (Targionia) acaciae; Froggatt, 1914: 131. Change of combination.

Aspidiotus (Targionia) eucalypti; Froggatt, 1914: 312. Change of combination.

Aspidiotus junctilobius Froggatt, 1914: 315. Type data: AUSTRALIA: New South Wales, Whitton (south-west of NSW), on Yarran, *Exocarpus aphylla*. Syntypes, female. Type depository: Canberra: Australian National Insect Collection, CSIRO Entomology, Australia. Synonymy by Brimblecombe, 1954: 150.

Aspidiotus junctilobius; Froggatt, 1915: 19. Notes: Described again as "n. sp.".

Neomorgania acaciae; MacGillivray, 1921: 458. Change of combination.

Neomorgania eucalypti; MacGillivray, 1921: 458. Change of combination.

Neomorgania junctiloba; MacGillivray, 1921: 458. Change of combination.

Aspidiotus propinquus; Ferris, 1943a: 86. Change of status.

Neomorgania iunctiloba Lindinger, 1943b: 223. Unjustified emendation; discovered by Borchsenius, 1966: 236.

SCALE COVER: Female scale circular, diameter averaging 1/12 inch; slightly convex; dirty white; exuviae central, very conspicuous. Male scale narrow, elongated, length about 1/18 inch, semi-cylindrical; white, exuviae yellow; not carinated above (Maskell, 1889).

HOST PLANTS: **Casuarinaceae**: *Casuarina* [Leonar1900, Marlat1908, Frogga1914]. **Leguminosae**: *Acacia* [Maskel1893b, Marlat1908, Sander1909a, Laing1929], *Acacia pycnantha* [Morgan1889a, Leonar1900]. **Myrtaceae**: *Eucalyptus* [Maskel1889, Leonar1900, Marlat1908, Frogga1914]. **Proteaceae**: *Hakea saligna* [Frogga1914]. **Santalaceae**: *Exocarpos aphylla* [Frogga1914].

NATURAL ENEMIES: HYMENOPTERA **Aphelinidae**: *Aphytis capillatus* (Howard) [RosenDe1979].

DISTRIBUTION: **Australasian**: Australia (New South Wales [Maskel1892, Maskel1893b, Leonar1900, Marlat1908, Frogga1914], South Australia [Maskel1889], Tasmania [Morgan1889a], Victoria [Marlat1908, Laing1929]).

GENERAL: Description and illustration of adult female by Maskell (1889), Froggatt (1914) (as *Aspidiotus junctilobius)*, Laing (1929) and by Brimblecombe (1954).

KEYS: Marlatt 1908: 135 (female) [World].

CITATIONS: Borchs1966 [catalogue: 236]; Brimbl1954 [taxonomy, description, illustration, host, distribution: 150-153]; Cocker1896b [distribution: 335]; Cocker1897i [taxonomy, description, host, distribution: 26]; DeitzTo1980 [taxonomy: 36,41]; Fernal1903b [catalogue: 295-297]; Ferris1937c [taxonomy, illustration: 51,86]; Ferris1941e [taxonomy: 40,44,46]; Ferris1943a [taxonomy: 85-86]; Frogga1914 [taxonomy, description, host, distribution: 131-132,312,315]; Frogga1915 [taxonomy, description, host, distribution: 7-8,15,19]; Laing1929 [taxonomy, description, illustration, host, distribution: 29]; Leonar1900 [taxonomy, description, illustration, host, distribution: 308-311]; Lindin1932f [taxonomy: 195]; Lindin1943b [taxonomy: 223]; MacGil1921 [taxonomy, description, host, distribution: 458]; Marlat1908 [taxonomy, host, distribution: 135,138]; Maskel1889 [taxonomy, description, host, distribution: 102]; Maskel1892 [host, distribution: 11]; Maskel1893b [taxonomy, description, illustration, host, distribution: 206-207]; Morgan1889a [taxonomy, description, illustration, host, distribution: 353]; RosenDe1979 [host, distribution, biological control: 262-268]; Sander1909a [taxonomy, host, distribution: 54]; Sassce1915 [taxonomy, host, distribution: 34].

Neoselenaspidus Mamet

Neoselenaspidus Mamet, 1958a: 404. Type species: *Selenaspidus silvaticus* Lindinger, by original designation.

SYSTEMATICS: The genus *Neoselenaspidus* closely allied to *Selenaspidus* and related genera by the spur-shaped third lobe. differs from these genera in absence of a constriction in prosoma. Also resembles *Aspidiotus* and related genera in general shape of body, but differs from them in characteristic shape of third lobe (Mamet, 1958a).

GENERAL: Definition and characters by Mamet (1958a).

KEYS: Mamet 1958a: 362 (female) [*Selenaspidus* complex]; Mamet 1958a: 406 (female) [Species of *Neoselenaspidus*].

CITATIONS: Borchs1966 [catalogue: 257]; Mamet1958a [taxonomy, description: 362,404-406]; MorrisMo1966 [taxonomy, catalogue: 133].

Neoselenaspidus aspidiotiformis (Balachowsky)

Selenaspidus aspidiotiformis Balachowsky, 1954: 62. Type data: GUINEA: Western slope of Mt. Kakoulima, altitude 400 meters, on *Syzygium guineense*. Syntypes, female. Type depository: Paris: Muséum national d'Histoire naturelle, France.
Neoselenaspidus aspidiotiformis; Mamet, 1958a: 406. Change of combination.

SCALE COVER: Female scale subcircular, very flat; larval exuviae central or subcentral and flat; straw-yellowish with exuviae slightly darker; 1.8-2 mm in diameter. Scale of male same colour as female, oval, very flat, length about 1mm unknown (Balachowsky, 1954).

HOST PLANTS: **Myrtaceae**: *Syzygium guineense* [Balach1954, Mamet1958a, Borchs1966].

DISTRIBUTION: **Neotropical**: French Guiana [Balach1954, Mamet1958a].

GENERAL: Description and illustration of adult female by Balachowsky (1954) and by Mamet (1958a).

CITATIONS: Balach1954 [taxonomy, description, illustration, host, distribution: 62-63]; Borchs1966 [catalogue: 257]; Mamet1958a [taxonomy, description, illustration, host, distribution: 406-407].

Neoselenaspidus kenyae Mamet

Neoselenaspidus kenyae Mamet, 1958a: 408. Type data: KENYA: Nairobi, on *Euphorbia candelabrum*. Holotype. Type depository: Paris: Muséum national d'Histoire naturelle, France.

SCALE COVER: Female scale circular, whitish to light brown in colour, flat to low convex; exuviae central, light to very dark brown, obscured by a layer of white secretion; diameter about 2 mm. Scale of male not observed (Mamet, 1958a).

HOST PLANTS: **Euphorbiaceae**: *Euphorbia candelabrum* [Mamet1958a, Borchs1966].

DISTRIBUTION: **Afrotropical**: Eritrea [Mamet1958a, Borchs1966]; Kenya [Mamet1958a, Borchs1966]; South Africa [Mamet1958a, Borchs1966].

GENERAL: Description and illustration of adult female by Mamet (1958a).

CITATIONS: Borchs1966 [catalogue: 257]; Mamet1958a [taxonomy, description, illustration, host, distribution: 408-409]; Nur1990b [taxonomy, life history: 196].

Neoselenaspidus sefanae (Mamet)

Selenaspidus sefanae Mamet, 1954: 85. Type data: MADAGASCAR: Périnet, on "Sefana" [=Anacardiaceae sp.]. Holotype. Type depository: Paris: Muséum national d'Histoire naturelle, France.

Neoselenaspidus sefanae; Mamet, 1958a: 408. Change of combination.

SCALE COVER: Scale of female and male not observed (Mamet, 1954, 1958a).

HOST PLANTS: **Anacardiaceae** [Mamet1959a].

DISTRIBUTION: **Afrotropical**: Madagascar [Mamet1954, Mamet1958a, Borchs1966].

GENERAL: Description and illustration of adult female by Mamet (1954, 1958a).

CITATIONS: Borchs1966 [catalogue: 257]; Mamet1954 [taxonomy, description, illustration, host, distribution: 22,85]; Mamet1958a [taxonomy, description, illustration, host, distribution: 408,410-412].

Neoselenaspidus silvaticus (Lindinger)

Selenaspidus silvaticus Lindinger, 1909d: 10. Type data: CAMEROON: on Anacardiaceae, on *Bandeiraea speciosa*, on *Rinorea exappendiculata*; and TANZANIA: on *Ficus indica*. Syntypes. Type depository: Hamburg: Zoologisches Institut und Zoologishces Museum, Universität von Hamburg, Germany.

Pseudaonidia silvatica; Sanders, 1909a: 54. Change of combination.

Aspidiotus (Selenaspidus) silvaticus; Vayssière, 1913: 431. Change of combination.

Pseudoaonidia silvatica; Leonardi, 1914: 203. Misspelling of genus name.

Entaspidiotus silvaticus; MacGillivray, 1921: 455. Change of combination.

Selenaspidus sylvaticus; Seabra, 1921: 98. Misspelling of species name.

Selenaspidus silvaticus; Bellio, 1939: 236. Change of combination.

Neoselenaspidus silvaticus; Mamet, 1950a: 406. Change of combination.

SCALE COVER: Female scale dark brown, circular to broadly oval, 1-1.6 mm long, 0.9-1.1 mm wide; thin; exuviae yellow, placed centrally (Lindinger, 1909d). Female scale circular to slightly oval; very pale buff to dirty yellowish; length about 2.5 mm. Scale of male smaller than female, flat, brownish in colour; exuviae yellowish to brown (Mamet, 1958a).

HOST PLANTS: **Agavaceae**: *Agave sisalana* [Mamet1958a, Borchs1966], *Dracaena australis* [Brain1918]. **Anacardiaceae** [Lindin1909d, Mamet1958a, Borchs1966], *Mangifera indica* [Almeid1973b]. **Apocynaceae**: *Acokanthera schimperi* [Mamet1958a, Borchs1966]. **Berberidaceae**: *Berberis* [Brain1918].

Celastraceae: *Euonymus* [Brain1918, Mamet1958a, Borchs1966]. **Euphorbiaceae**: *Aleurites moluccana* [Mamet1958a, Borchs1966], *Euphorbia* [Bellio1939, Mamet1958a, Borchs1966], *Euphorbia ingens* [Hall1928, Mamet1958a, Borchs1966]. **Flacourtiaceae**: *Aberia caffra* [Mamet1958a, Borchs1966]. **Lauraceae**: *Laurus camphora* [Mamet1954, Borchs1966]. **Leguminosae**: *Bandeiraea speciosa* [Lindin1909d, Sander1909a, Leonar1914, Mamet1958a, Borchs1966], *Cassia* [MatileNo1984], *Cassia fistula* [Mamet1958a, Borchs1966], *Ceratonia siliqua* [Mamet1958a, Borchs1966]. **Meliaceae**: *Trichilia emetica* [Mamet1958a, Borchs1966]. **Moraceae**: *Ficus indica* [Lindin1909d, Sander1909a, Leonar1914, Mamet1958a, Borchs1966]. **Oleaceae**: *Ligustrum japonicum* [Mamet1958a, Borchs1966], *Olea europaea* [Almeid1973b]. **Palmae** [Hall1928, Mamet1958a, Borchs1966], *Phoenix canariensis* [Mamet1958a, Borchs1966]. **Rhamnaceae**: *Scutia myrtina* [Mamet1958a, Borchs1966]. **Rosaceae**: *Eriobotrya japonica* [Mamet1958a, Borchs1966]. **Rubiaceae**: *Gardenia* [MatileNo1984]. **Rutaceae**: *Citrus aurantium* [Mamet1958a, Borchs1966, Almeid1973b]. **Sapotaceae**: *Chrysophyllum megalies-montana* [Brain1918, Mamet1958a, Borchs1966]. **Sterculiaceae**: *Theobroma cacao* [Laing1932]. **Ulmaceae**: *Chaetacme aristata* [Mamet1958a, Borchs1966]. **Violaceae**: *Rinorea exappendiculata* [Lindin1909d, Sander1909a, Leonar1914, Mamet1958a, Borchs1966]. **Vitaceae**: *Vitis vinifera* [Almeid1973b].

NATURAL ENEMIES: HYMENOPTERA **Encyrtidae**: *Habrolepis* [Prinsl1983].

DISTRIBUTION: **Afrotropical**: Angola [Almeid1973b]; Cameroon [Sander1909a, Leonar1914, Mamet1958a, Borchs1966, MatileNo1984]; Eritrea [DeLottNa1955, Mamet1958a, Borchs1966]; Ethiopia [Bellio1939]; Kenya [Mamet1958a, Borchs1966]; Madagascar [Mamet1954, Borchs1966]; São Tomé and Príncipe [Seabra1921, Seabra1925, Mamet1958a, Borchs1966]); South Africa [Brain1918, Mamet1958a, Borchs1966]; Tanzania [Mamet1958a, Borchs1966]; Uganda [Gowdey1917, Newste1917b, Mamet1958a, Borchs1966]; Zaire [Laing1932, Ghesqu1932, Mamet1958a, Borchs1966]; Zimbabwe [Hall1928, Mamet1958a, Borchs1966].

ECONOMIC IMPORTANCE: Recorded as pest of tea plants in Kenya and Malawi (Le Pelley, 1959), and citrus in Zimbabwe (Ebeling, 1959; Wilson & Goldsmid, 1962).

GENERAL: Description and illustration of adult female by Lindinger (1909d), Brain (1918) and by Mamet (1958a).

KEYS: Brain 1918: 131 (female) [South Africa]; Lindinger 1909d: 4-5 (female) [Genus *Selenaspidus*].

CITATIONS: Almeid1973b [host, distribution: 11]; Bellio1939 [taxonomy, description, illustration, host, distribution: 236-239]; Borchs1966 [catalogue: 257]; Brain1918 [taxonomy, description, illustration, host, distribution: 134]; DeLottNa1955 [host, distribution: 53-60]; Ferris1941e [taxonomy: 48]; Ghesqu1932 [host, distribution: 59]; Gowdey1917 [host, distribution: 189]; Hall1928 [host, distribution: 275]; Laing1932 [host, distribution: 67]; Leonar1914 [taxonomy, host, distribution: 203]; Lindin1909d [taxonomy, description, illustration, host, distribution: 5,10]; Lindin1910b [host, distribution: 42]; MacGil1921 [taxonomy, description, host, distribution: 455]; Mamet1954 [host, distribution: 22];

Mamet1958a [taxonomy, description, illustration, host, distribution: 411-412]; MatileNo1984 [host, distribution: 66]; MayneGh1934 [host, distribution: 3-38]; MillerDa1990 [host, distribution, economic importance: 304]; NagarkSa1990 [host, distribution, economic importance, biological control: 553-542]; Newste1917b [host, distribution: 132]; Nur1990b [taxonomy, life history: 196]; Prinsl1983 [distribution, biological control: 27]; Sander1909a [taxonomy, host, distribution: 54]; SchmutKlLu1957 [host, distribution, economic importance: 494]; Seabra1921 [taxonomy, host, distribution: 98-99]; Seabra1925 [taxonomy, description, illustration, host, distribution: 30]; Vayssi1913 [host, distribution: 431]; WeidneWa1968 [taxonomy: 178]; WilsonGo1962 [host, distribution, economic importance, control: 41-61].

Neoselenaspidus triangularis Munting

Neoselenaspidus triangularis Munting, 1969a: 291. Type data: SOUTH AFRICA: Cape Province, Ceres district, Hottentotskloof, on *Othonna coronopifolia*; collected 1.iii.1966. Holotype female. Type depository: Pretoria: South African National Collection of Insects, South Africa; type no. 2109/2.

SCALE COVER: Scale of adult female dull-white, subcircular, flat, about 1.2 mm in diameter. Male scale not seen (Munting, 1969a).

HOST PLANTS: **Compositae**: *Othonna coronopifolia* [Muntin1969a].

DISTRIBUTION: **Afrotropical**: South Africa [Muntin1969a].

GENERAL: Description and illustration of adult female by Munting (1969a).

CITATIONS: Muntin1969a [taxonomy, description, illustration, host, distribution: 291-292].

Nigridiaspis Ferris

Nigridiaspis Ferris, 1941d: 374. Type species: *Nigridiaspis armigera* Ferris, by monotypy and original designation.

SYSTEMATICS: Ferris (1941d) indicated that the type species suggests resemblance to type species of *Pygidiaspis* MacGillivray and of *Loranthaspis* Cockerell & Bueker, mainly by presence of armour-like sclerotization on dorsum of prepygidial abdominal segments.

GENERAL: Definition and characters by Ferris (1941d).

KEYS: Ferris 1942: 27 (female) [North America]; Ferris 1942: 38 (female) [North America].

CITATIONS: Borchs1966 [catalogue: 359]; Ferris1941d [taxonomy, description: 374]; Ferris1942 [taxonomy: 446:42]; MorrisMo1966 [taxonomy, catalogue: 135].

Nigridiaspis armigera Ferris

Nigridiaspis armigera Ferris, 1941d: 375. Type data: PANAMA: Chiriqui Province, at Armuelles, on *Enterolobium cyclocarpum*. Holotype female. Type depository: Davis: Bohart Museum of Entomology, University of California, California, USA.

SCALE COVER: Female scale black, convex; circular, hard and thick; ventral scale thick and hard especially around margin; exuviae subcentral (Ferris, 1941d)

HOST PLANTS: **Leguminosae**: *Enterolobium cyclocarpum* [Ferris1941d].

DISTRIBUTION: **Neotropical**: Panama [Ferris1941d].
BIOLOGY: Occurring on bark, usually exposed (Ferris, 1941d).
GENERAL: Description and illustration of adult female by Ferris (1941d).
KEYS: Ferris 1942: 38 (female) [North America].
CITATIONS: Borchs1966 [catalogue: 359]; Ferris1941d [taxonomy, description, illustration, host, distribution: 375]; Ferris1942 [taxonomy: 446:38].

Obtusaspis MacGillivray

Obtusaspis MacGillivray, 1921: 393. Type species: *Aspidiotus* (*Odonaspis*) *rhizophilus* Newstead, by monotypy and original designation.
Obtuaspis; Ben-Dov, 1974c: 28. Misspelling of genus name.
SYSTEMATICS: Balachowsky (1958b) considered *Obtusaspis* related to genera of the Furchaspidina, but differed by having unisetose antennal tubercle whereas in the Furchaspidina the tubercle is multisetose.
GENERAL: Definition and characters by Balachowsky (1958b).
KEYS: Balachowsky 1958b: 230 (female) [*Aspidiotina* of Africa].
CITATIONS: Balach1958b [taxonomy, description: 207-208,230]; BenDov1974c [taxonomy: 28]; Borchs1966 [catalogue: 281]; Ferris1937c [taxonomy: 101]; Ferris1938b [taxonomy: 75]; Lindin1937 [taxonomy: 191]; MacGil1921 [taxonomy, description: 393,450]; MorrisMo1966 [taxonomy, catalogue: 137].

Obtusaspis cynodontis (Hall)
Furcaspis cynodontis Hall, 1937: 119. Type data: ZIMBABWE: Mazoe, on *Cynodon dactylon*. Syntypes, female. Type depository: London: The Natural History Museum, England, UK.
Obtusaspis cynodontis; Balachowsky, 1958b: 208. Change of combination.
SCALE COVER: Female scale more or less circular in outline, diameter 1.0-1.25 mm; convex; black; larval exuviae pale straw-coloured and overlaid by a film of white secretionary matter; nymphal exuviae very dark brown and obscured by a black secretionary film; exuviae not quite central; secretionary area black with minute but distinct concentric growth lines; ventral scale entire with a dirty white roughly circular area situated towards one side; remainder of ventral scale black with faint concentric growth lines (Hall, 1937).
HOST PLANTS: **Gramineae**: *Cynodon dactylon* [Hall1937, Balach1958b].
DISTRIBUTION: **Afrotropical**: Zimbabwe [Hall1937, Balach1958b].
GENERAL: Description and illustration of adult female by Hall (1937) and by Balachowsky (1958b).
KEYS: Balachowsky 1958b: 208 (female) [Africa].
CITATIONS: Balach1958b [taxonomy, description, illustration, host, distribution: 208-209]; Borchs1966 [catalogue: 281]; Hall1937 [taxonomy, description, illustration, host, distribution: 119-120]; Lindin1943b [taxonomy: 220].

Obtusaspis rhizophila (Newstead)

Aspidiotus (*Odonaspis*) *rhizophilus* Newstead, 1920: 198. Type data: KENYA: Kabete, on roots of *Chloris incompleta*. Syntypes, female. Type depository: London: The Natural History Museum, England, UK.

Obtusaspis rhizophila; MacGillivray, 1921: 450. Change of combination requiring emendation of species name for agreement in gender.

SCALE COVER: Female scale of irregular form, but old examples slightly narrowed and produced posteriorly; length 1.3-1.5 mm; texture rather rough; dense, hard and capsulate, but two halves slightly separated posteriorly; dull black or brownish black; convex dorsally and flat ventrally; larval exuviae generally towards anterior margin, greyish in colour and sometimes fissured; ventral scale with a greyish patch towards anterior margin (Newstead, 1920).

HOST PLANTS: **Gramineae**: *Chloris* [Balach1958b], *Chloris incompleta* [Newste1920].

DISTRIBUTION: **Afrotropical**: Kenya [Newste1920]; Uganda [Balach1958b].

GENERAL: Description and illustration of adult female by Newstead (1920) and by Balachowsky (1958b).

KEYS: Balachowsky 1958b: 207 (female) [Africa].

CITATIONS: Balach1958b [taxonomy, description, illustration, host, distribution: 208-209]; Borchs1966 [catalogue: 281]; Ferris1941e [taxonomy: 47]; MacGil1921 [taxonomy, description, host, distribution: 450]; Newste1920 [taxonomy, description, illustration, host, distribution: 198-199].

Obtusaspis trilobis Ben-Dov

Obtusaspis trilobis Ben-Dov, 1974c: 28. Type data: SOUTH AFRICA: Cape Province, Port St. John's, on *Digitaria* sp. Holotype. Type depository: Pretoria: South African National Collection of Insects, South Africa.

Obtuaspis trilobis; Ben-Dov, 1974c: 28. Misspelling of genus name.

SCALE COVER: Scale of female oval in outline, 1.0-1.3 mm long, 0.5-0.7 mm wide; convex; with a well-developed ventral scale; both scales dark brown; larval exuviae slightly brighter in colour, placed subcentrally; rounded anteriorly and slightly tapering posteriorly (Ben-Dov, 1974c).

HOST PLANTS: **Gramineae**: *Digitaria* [BenDov1974c].

DISTRIBUTION: **Afrotropical**: South Africa [BenDov1974c].

BIOLOGY: The females infest basal nodes very close to, and sometimes covered by soil (Ben-Dov, 1974c).

GENERAL: Description and illustration of adult female by Ben-Dov (1974c).

CITATIONS: BenDov1974c [taxonomy, description, illustration, host, distribution: 27-28].

Oceanaspidiotus Takagi

Oceanaspidiotus Takagi, 1984: 16. Type species: *Octaspidiotus araucariae* Adachi & Fullaway, by original designation.

SYSTEMATICS: Takagi (1984) and Williams & Watson (1988) noted remarkable morphological variation among the five species currently placed in this genus.

Character variation remarkable in size and shape of lateral lobes, number of lobes, presence or absence of perivulvar pores, and position of the anus.

GENERAL: Definition and characters by Takagi (1984), Williams & Watson (1988) and by Yaşar (1995).

KEYS: Gill 1997: 24-26 (female) [Genera of California]; Williams & Watson 1988: 20 (female) [Tropical South Pacific].

CITATIONS: Gill1997 [taxonomy: 207]; KosztaBeKo1986 [taxonomy, catalogue: 12]; Takagi1984 [taxonomy, description: 16]; WilliaWa1988 [taxonomy, description: 183]; Yasar1995a [taxonomy, description: 101].

Oceanaspidiotus araucariae (Adachi & Fullaway)

Octaspidiotus araucariae Adachi & Fullaway, 1953: 89. Type data: HAWAII: Waipio, Oahu, on *Araucaria excelsa*; collected by J.W. Beardsley, 1.ii.1951. Holotype female and first instar. Type depository: Washington D.C.: U.S. National Entomological Collection, U.S. National Museum of Natural History, USA.
Oceanaspidiotus araucariae; Takagi, 1984: 16. Change of combination.

SCALE COVER: Female scale pale yellow to white; somewhat circular, 1 by 0.8 mm; exuviae slightly concentric (Adachi & Fullaway, 1953). Illustration of female scale cover by Adachi & Fullaway (1953).

HOST PLANTS: **Araucariaceae**: *Araucaria* [WilliaWa1988], *Araucaria columnaris* [Cohic1958], *Araucaria cookii* [WilliaWa1988], *Araucaria excelsa* [AdachiFu1953, Beards1966, Takagi1984], *Araucaria heterophylla* [WilliaWa1988].

DISTRIBUTION: **Australasian**: Federated States of Micronesia (Ponape Island [Beards1966]); Hawaiian Islands (Hawaii [AdachiFu1953, Takagi1984]); New Caledonia [Cohic1958, WilliaWa1988]. **Nearctic**: United States of America (Florida [Hamon1985]). **Neotropical**: Puerto Rico & Vieques Island (Puerto Rico [ColonFMe1998]).

ECONOMIC IMPORTANCE: Serious pest on *Araucaria columnaris* (now *A. cookii*) (Cohic, 1959). On extremities of branches, causing desiccation (Williams & Watson, 1988).

GENERAL: Description and illustration of adult female by Adachi & Fullaway (1953), Takagi (1984), Williams & Watson (1988), Zahradník (1990b) and by Colon-Ferrer & Medina-Gaud (1998).

KEYS: Williams & Watson 1988: 184 (female) [Tropical South Pacific].

CITATIONS: AdachiFu1953 [taxonomy, description, illustration, host, distribution: 89-91]; Beards1966 [host, distribution: 523]; Borchs1966 [catalogue: 272]; BurgerUl1990 [economic importance: 313-327]; Cohic1958 [host, distribution: 15]; ColonFMe1998 [taxonomy, description, illustration, host, distribution: 73-74]; Hamon1985 [host, distribution, economic importance, control: 1-2]; Mead1985 [host, distribution: 1-3]; Takagi1984 [taxonomy, description, illustration, host, distribution: 17-18,43]; WilliaWa1988 [taxonomy, description, illustration, host, distribution, economic importance: 9,184-186]; Zahrad1990b [taxonomy, description, illustration, host, distribution: 115-117].

Oceanaspidiotus caledonicus (Matile-Ferrero & Balachowsky)

Octaspidiotus caledonicus Matile-Ferrero & Balachowsky, 1973: 239. Type data: NEW CALEDONIA: Noumea, on *Erythrina fusca fustigiata*. Holotype female. Type depository: Paris: Muséum national d'Histoire naturelle, France.
Oceanaspidiotus caledonicus; Takagi, 1984: 18. Change of combination.
SCALE COVER: Female scale subcircular, diameter 1.8-2 mm; flat; exuviae central; white or slightly brownish (Matile-Ferrero & Balachowsky, 1973). Scale of female circular, flat, whitish with pale brown subcentral exuviae. Male scale similar to female but more elongate (Williams & Watson, 1988).
HOST PLANTS: **Leguminosae**: *Erythrina fastigata* [WilliaWa1988], *Erythrina fusca fustigata* [MatileBa1973]. **Pittosporaceae**: *Pittosporum* [WilliaWa1988].
DISTRIBUTION: **Australasian**: New Caledonia [MatileBa1973, Takagi1984, WilliaWa1988].
GENERAL: Description and illustration of adult female by Matile-Ferrero & Balachowsky (1973), Takagi (1984) and by Williams & Watson (1988).
KEYS: Williams & Watson 1988: 184 (female) [Tropical South Pacific].
CITATIONS: MatileBa1973 [taxonomy, description, illustration, host, distribution: 239-243]; Takagi1984 [taxonomy, description, illustration, host, distribution: 18,45,67]; WilliaWa1988 [taxonomy, description, illustration, host, distribution: 186-188].

Oceanaspidiotus nendeanus Williams & Watson

Oceanaspidiotus nendeanus Williams & Watson, 1988: 188. Type data: SOLOMON ISLANDS: Santa Cruz Island, on sweet potato vine *Ipomoea batatas*; collected iii, 1983. Holotype female. Type depository: London: The Natural History Museum, England, UK.
SCALE COVER: Female scale subcircular; convex; light to mid-brown, with paler secretion covering yellow-brown subcentral exuviae. Male scale similar to female scale, but smaller and elongate-oval (Williams & Watson, 1988).
HOST PLANTS: **Convolvulaceae**: *Ipomoea batatas* [WilliaWa1988].
DISTRIBUTION: **Australasian**: Solomon Islands [WilliaWa1988].
GENERAL: Description and illustration of adult female by Williams & Watson (1988).
KEYS: Williams & Watson 1988: 188 (female) [Tropical South Pacific].
CITATIONS: WilliaWa1988 [taxonomy, description, illustration, host, distribution: 188-189,192].

Oceanaspidiotus pangoensis (Doane & Ferris)

Aspidiotus pangoensis Doane & Ferris, 1916: 400. Type data: AMERICAN SAMOA: Pango Pango, on coconut husks and on unidentified plant. Syntypes, female. Type depository: Davis: The Bohart Museum of Entomology, University of California, California, USA.
Spinaspidiotus pangoenensis; MacGillivray, 1921: 429. Change of combination.
Spinaspidiotus pangoenensis; MacGillivray, 1921: 429. Misspelling of species name.

Oceanaspidiotus pangoensis; Williams & Watson, 1988: 192. Change of combination.

SCALE COVER: Female scale circular or sub-circular, diameter 2 mm; flat and rather thin and chaffy; brownish grey; exuviae central, yellow. Male scale not identified (Doane & Ferris, 1916). Female scale approximately circular, flat, slightly translucent, light to mid-brown; exuviae central, slightly yellower. Male scale similar to female scale but more elongate (Williams & Watson, 1988).

HOST PLANTS: **Boraginaceae**: *Cordia subcordata* [WilliaWa1988]. **Hernandiaceae**: *Hernandia peltata* [WilliaWa1988]. **Moraceae**: *Broussonetia papyrifera* [WilliaWa1988]. **Palmae**: *Cocos nucifera* [DoaneFe1916, GreenLa1923, Laing1927, WilliaWa1988]. **Rhizophoraceae**: *Bruguiera gymnorhiza* [WilliaWa1988]. **Smilacaceae**: *Smilax* [WilliaWa1988].

DISTRIBUTION: **Australasian**: American Samoa [WilliaWa1988]; Fiji [GreenLa1923, GreenLa1923, WilliaWa1988]; Niue [WilliaWa1988]; Samoa [DoaneFe1916]; Tonga [WilliaWa1988]; Western Samoa [Laing1927].

GENERAL: Description and illustration of adult female by Doane & Ferris (1916) and by Williams & Watson (1988).

KEYS: Williams & Watson 1988: 184 (female) [Tropical South Pacific].

CITATIONS: Borchs1966 [catalogue: 266]; DoaneFe1916 [taxonomy, description, illustration, host, distribution: 400-401]; Ferris1921b [taxonomy: 94]; Ferris1941e [taxonomy: 46]; GreenLa1923 [host, distribution: 125]; Hinckl1963 [host, distribution, biological control]; Laing1927 [host, distribution: 40]; Lepesm1947 [host, distribution: 196]; MacGil1921 [taxonomy, description, host, distribution: 429]; WilliaWa1988 [taxonomy, description, illustration, host, distribution: 190-193].

Oceanaspidiotus spinosus (Comstock)

Aspidiotus spinous Comstock, 1883: 70. Type data: U.S.A.: New York, Ithaca, Cornell University, in greenhouse, on leaves and branches of Camellias. Syntypes, female. Type depository: Washington: United States National Entomological Collection, U.S. National Museum of Natural History, District of Columbia, USA. Notes: Species epithet misspelled as *spinous*.

Aspidiotus spinous; Comstock, 1883: 70. Misspelling of species name.

Aspidiotus spinosus; Cockerell, 1895b: 17. Justified emendation.

Aspidiotus cydoniae; Newstead, 1897b: 74. Misidentification; discovered by Borchsenius, 1966: 268.

Aspidiotus (Aspidiotus) spinosus; Cockerell, 1897i: 30. Change of combination.

Aspidiotus (Evaspidiotus) spinosus; Leonardi, 1898c: 56. Change of combination.

Aspidiotus persearum Cockerell, 1898r: 240. Type data: HAWAII: Honolulu, on *Persea gratissima*. Syntypes, female. Type depository: Washington: United States National Entomological Collection, U.S. National Museum of Natural History, District of Columbia, USA. Synonymy by Ferris, 1938a: 190.

Aspidiotus (Evaspidiotus) persearum; Leonardi, 1900: 341. Change of combination.

Acanthaspidiotus borchsenii Takagi & Kawai, 1966: 116. Type data: JAPAN: Tokyo, on *Platanus orientalis*; Iro-saki, Idu Peninsula, on *Ligustrum obtusifolium*; Hatizyo-sima, Idu Islands, on *Hydrangea macrophylla, Hydrangea involucrata* and *Boehmeria tricuspis*. Syntypes, female. Type depository: Sapporo: Entomological Institute, Faculty of Agriculture, Hokkaido University, Japan. Synonymy by Takagi, 1984: 18.

Oceanaspidiotus spinosus; Takagi, 1984: 18. Change of combination.

Acanthaspidiotus spinosa; Kawai, 1987: 78. Change of combination requiring emendation of species name for agreement in gender.

Acanthaspidiotus borchseniusi; Tao, 1999: 101. Misspelling of species name.

COMMON NAMES: spined scale insect [Comsto1883]; spinose scale [McKenz1956, Dekle1965c, GersonZo1973].

SCALE COVER: Female scale circular, exuviae central and covered with secretion; very light brown or dirty white (Comstock, 1883). Female scale whitish or straw coloured, flat, exuviae central; that of the male described as slightly elongate but not observed in material at hand (Ferris, 1938a).

HOST PLANTS: **Agavaceae**: *Nolina recurvata* [Merril1953], *Yucca gloriosa* [BesheaTiHo1973]. **Anacardiaceae**: *Mangifera indica* [Ferris1955b, Beards1966, Takagi1984]. **Annonaceae**: *Asimina* [TippinBe1970, BesheaTiHo1973]. **Aquifoliaceae**: *Ilex colchica* [Hadzib1983]. **Araliaceae**: *Hedera helix* [Bodenh1924]. **Bromeliaceae**: *Bromelia* [Takagi1984]. **Buxaceae**: *Buxus sempervirens* [Hadzib1983]. **Caprifoliaceae**: *Viburnum* [McKenz1956, Takagi1984], *Viburnum tinus* [Bodenh1949, Bodenh1952]. **Celastraceae**: *Euonymus* [McKenz1956, Takagi1984], *Euonymus japonicus* [Merril1953]. **Cycadaceae**: *Encephalarthos* [Balach1956]. **Euphorbiaceae**: *Euphorbia regis-jubae* [MatileBa1972, Takagi1984]. **Hydrangeaceae**: *Hydrangea involucrata* [TakagiKa1966], *Hydrangea macrophylla* [TakagiKa1966, Takagi1984]. **Labiatae**: *Lycopus* [BesheaTiHo1973]. **Lauraceae**: *Cinnamomum* [Ferris1938a, McKenz1956], *Cinnamomum zeylanicum* [Houser1918], *Laurus* [McKenz1956], *Laurus nobilis* [Hadzib1983], *Litsea laurifolia* [Matile1978], *Persea americana* [McKenz1956, GersonZo1973, WilliaWa1988], *Persea borbonia* [TippinBe1970, BesheaTiHo1973], *Persea gratissima* [Cocker1898r, Leonar1900, Ferris1938a], *Sassafras albidum* [BesheaTiHo1973]. **Leguminosae**: *Bauhinia* [BesheaTiHo1973]. **Liliaceae**: *Asparagus* [Ferris1921, McKenz1956]. **Lythraceae**: *Lawsonia inermis* [Merril1953]. **Magnoliaceae**: *Magnolia* [Ferris1938a, McKenz1956], *Magnolia grandiflora* [Borchs1934, Bodenh1949], *Magnolia virginiana* [BesheaTiHo1973]. **Malvaceae**: *Lagunaria* [McKenz1956]. **Moraceae**: *Ficus* [McKenz1956], *Ficus carica* [Ferris1938a]. **Myrsinaceae**: *Maesa chisia* [Takagi1984], *Maesa macrophylla* [Takagi1984]. **Oleaceae**: *Ligustrum obtusifolium* [TakagiKa1966]. **Palmae** [Muntin1965b, BesheaTiHo1973], *Arenga saccharifera* [MerrilCh1923, Balach1948b], *Livistonia chinensis* [Bodenh1949], *Rhapis* [Merril1953, McKenz1956], *Trachycarpus excelsus* [Kuwana1902]. **Platanaceae**: *Platanus* [Borchs1936], *Platanus orientalis* [Borchs1934, Bodenh1949, TakagiKa1966]. **Podocarpaceae**: *Podocarpus* [BesheaTiHo1973]. **Rosaceae**: *Rosa* [Ferris1938a, McKenz1956], *Rubus* [Ferris1938a, McKenz1956]. **Ruscaceae**: *Ruscus* [BesheaTiHo1973]. **Rutaceae**: *Citrus* [Takagi1984]. **Sapindaceae**: *Nephelium* *longana* [Balach1927, Balach1932d, Bodenh1949]. **Smilacaceae**: *Smilax* [TippinBe1970, BesheaTiHo1973], *Smilax stenopelata* [Takagi1958, Takagi1984]. **Taxaceae**: *Taxus* [McKenz1956]. **Theaceae**: *Camellia* [Comsto1883, MerrilCh1923, McKenz1956, Dekle1965c, Takagi1984], *Camellia japonica* [Hadzib1983], *Camellia sasanqua* [Hadzib1983], *Eurya emarginata* [Takagi1984]. **Urticaceae**: *Boehmeria tricuspis* [TakagiKa1966]. **Vitaceae**: *Vitis* [McKenz1956, TippinBe1970, BesheaTiHo1973], *Vitis vinifera* [Ferris1921, Ferris1938a, WilliaWa1988].

NATURAL ENEMIES: HYMENOPTERA **Signiphoridae**: *Signiphora flava* Girault [Gordh1979], *Signiphora flavella* Girault [Woolle1990].

DISTRIBUTION: **Afrotropical**: Comoros Islands [Matile1978]; Madagascar [Nakaha1982]; Mozambique [Nakaha1982]; South Africa [Muntin1965b]; Tanzania [Balach1956] [Nakaha1982]. **Australasian**: Bonin Islands (= Ogasawara-Gunto) [Kawai1987]; Cook Islands [WilliaWa1988]; Federated States of Micronesia (Yap [Beards1966]); Hawaiian Islands (Hawaii [Cocker1898r, Leonar1900, Zimmer1948]). **Nearctic**: Mexico [Takagi1984] (Baja California [Ferris1938a], Colima [Ferris1938a], Veracruz [Ferris1955b]); United States of America (Alabama [Nakaha1982], California [McKenz1956], District of Columbia [Comsto1883], Florida [MerrilCh1923, Merril1953, Dekle1965c, BesheaTiHo1973, Takagi1984], Georgia [TippinBe1970, BesheaTiHo1973], Louisiana [Takagi1984], Mississippi [Nakaha1982], Texas [Takagi1984]). **Neotropical**: Bahamas [Ferris1938a]; Bermuda [Nakaha1982]; Brazil [Takagi1984]; Colombia [Kondo2001]; Costa Rica [Takagi1984]; Cuba [Houser1918]; Dominican Republic [Nakaha1982]; Peru [Nakaha1982]; Puerto Rico & Vieques Island (Puerto Rico [Takagi1984]); Uruguay [Nakaha1982]. **Oriental**: Nepal [Takagi1984]. **Palaearctic**: Algeria [Balach1932d, Ferris1938a]; Azores [Nakaha1982]; Canary Islands [MatileBa1972, Takagi1984, MatileOr2001]; China (People's Republic) [Nakaha1982]; Egypt [AbouEl2001]; Georgia (Abkhaz ASSR [Borchs1934, Borchs1936], Adzhar ASSR [Borchs1936], Georgia [Borchs1936, Hadzib1983]); Israel [Bodenh1924, GersonZo1973]; Italy [Lozzia1985, LongoMaPe1995]; Japan [Kuwana1917a, TakagiKa1966, Kawai1980, Takagi1984] (Kyushu [Kuwana1902, Takagi1958]); Madeira Islands [Nakaha1982]; Portugal [Seabra1941]; Sicily [Lozzia1985]; Spain [Nakaha1982]; Syria [Nakaha1982]; Turkey [Bodenh1949, Bodenh1952, Takagi1984]; United Kingdom (England [Ferris1938a]).

BIOLOGY: Occurring either on leaves or bark (Ferris, 1938a). Associated with fungus *Septobasidium* (Ferris, 1955b). Reported to have uniparental and biparental populations (Gerson & Zor, 1973).

ECONOMIC IMPORTANCE: A minor pest of avocado trees in Israel (Gerson & Zor (1973), and of tea plants in India (Nagarkatti & Sankaran, 1990).

GENERAL: Description and illustration of adult female by Ferris (1921, 1938a, 1941e), Balachowsky (1948b, 1956), Zimmerman (1948), McKenzie (1956), Takagi & Kawai (1966), Takagi (1984), Chou (1985, 1986), Tereznikova (1986), Williams & Watson (1988), Yaşar (1995a), Gill (1997) and by Colon-Ferrer & Medina-Gaud (1998).

KEYS: Danzig 1993: 40-41 (female) [Europe]; Williams & Watson 1988: 184 (female) [Tropical South Pacific]; Tereznikova 1986: 89 (female) [Ukraine]; Chou

1985: 262-263 (female) [Species of China]; Gerson & Zor 1973: 516 (female) [Israel]; Beardsley 1970: 508 (female) [Hawaii]; McDaniel 1968: 218 (female) [U.S.A.: Texas]; Beardsley 1966: 513 (female) [Federated States of Micronesia]; Balachowsky 1956: 52 (female) [Africa]; McKenzie 1956: 24 (female) [U.S.A.: California]; Lupo 1953: 39 (female) [Italy]; Balachowsky 1948b: 275 (female) [Mediterranean]; Lupo 1948: 139 (female) [Italy]; Zimmerman 1948: 355 (female) [Hawaii]; Ferris 1946: 43 (female) [World]; Ferris 1942: 30 (female) [North America]; Ferris 1941e: 60 (female) [World]; Borchsenius 1938: 142 (female) [Far East of USSR]; Kuwana 1933b: 49 (female) [Japan]; Newstead 1901b: 82 (female) [British Isles]; Comstock 1883: 55-57 (female) [North America].

CITATIONS: AbouEl2001 [host, distribution, biological control: 185-195]; Balach1927 [host, distribution: 177]; Balach1932d [taxonomy, host, distribution, economic importance: IV]; Balach1948b [taxonomy, description, illustration, host, distribution: 288-291]; Balach1956 [taxonomy, description, illustration, host, distribution: 78-79]; Beards1966 [host, distribution: 515]; BesheaTiHo1973 [host, distribution: 5]; Bodenh1924 [taxonomy, description, host, distribution: 33-34]; Bodenh1949 [taxonomy, description, illustration, host, distribution: 55-57]; Bodenh1952 [host, distribution: 338]; Borchs1934 [host, distribution: 27]; Borchs1935a [taxonomy, description, host, distribution: 24]; Borchs1936 [host, distribution: 130]; Borchs1937 [taxonomy, description, illustration, host, distribution: 126]; Borchs1937a [taxonomy, description, host, distribution: 37]; Borchs1938 [host, distribution: 143]; Borchs1950b [taxonomy, description, illustration, host, distribution: 215,219]; Borchs1966 [catalogue: 267-268]; Chou1985 [taxonomy, description, host, distribution: 268-269]; Chou1986 [taxonomy, illustration: 660]; ClapsWoGo2001a [taxonomy, host, distribution: 13]; Cocker1895b [taxonomy: 17]; Cocker1896b [distribution: 334]; Cocker1897i [taxonomy, description, host, distribution: 30]; Cocker1898r [taxonomy, description, host, distribution: 240]; Cocker1899a [taxonomy: 395]; ColonFMe1998 [taxonomy, description, illustration, host, distribution: 47-48]; Comsto1883 [taxonomy, description, illustration, host, distribution: 70-71]; Danzig1964 [taxonomy, host, distribution: 651]; DanzigPe1998 [catalogue: 193]; Dekle1965c [taxonomy, description, host, distribution: 29]; Dekle1976 [taxonomy, description, host, distribution, economic importance: 43]; Ehrhor1913 [host, distribution: 101]; EhrhorFuSw1913 [distribution: 295-300]; FDACSB1983 [host, distribution: 6-8]; Fernal1903b [catalogue: 276,279]; Ferris1921 [taxonomy, description, illustration, host, distribution: 128]; Ferris1938a [taxonomy, description, illustration, host, distribution: 193]; Ferris1941e [taxonomy, description, illustration, host, distribution: 47-48,58-59,67]; Ferris1942 [taxonomy: 446:30]; Ferris1946 [taxonomy: 43]; Ferris1955b [host, distribution, life history: 25]; Gerson1990 [taxonomy: 130]; GersonZo1973 [taxonomy, life history, host, distribution, economic importance: 513-533]; Gill1997 [host, distribution, taxonomy, description, illustration, economic importance: 206-207]; Gordh1979 [biological control: 911]; Hadzib1983 [taxonomy, description, host, distribution: 219]; Hewitt1943 [host, distribution: 266-274]; Houser1918 [host, distribution: 167]; Kawai1980 [taxonomy, description, host, distribution: 229]; Kawai1987 [taxonomy, host, distribution: 78]; Kondo2001 [taxonomy, host, distribution: 45]; Kuwana1902 [host, distribution: 65];

Kuwana1917a [taxonomy, distribution: 175]; Leonar1897 [taxonomy: 285]; Leonar1898 [taxonomy: 75]; Leonar1898c [taxonomy, description, illustration, host, distribution: 56-58]; Leonar1900 [taxonomy, host, distribution: 341]; Lepesm1947 [host, distribution: 194]; Lindin1911 [taxonomy: 247]; Lindin1912b [taxonomy, description, host, distribution: 203,239,335]; Lindin1935 [taxonomy: 129]; Lindin1957 [taxonomy: 546]; LongoMaPe1995 [distribution: 125]; Lozzia1985 [host, distribution: 122-124]; Lupo1948 [taxonomy, description, illustration, host, distribution: 159-164]; Lupo1953 [taxonomy: 39]; MacGil1921 [taxonomy, description, host, distribution: 398]; Matile1978 [host, distribution: 63]; MatileBa1972 [host, distribution: 113]; MatileOr2001 [host, distribution: 189]; McDani1968 [taxonomy, illustration, host, distribution: 218-221]; McKenz1956 [taxonomy, description, illustration, host, distribution: 49-51]; Mead1985 [host, distribution: 1-3]; Merril1953 [taxonomy, description, host, distribution: 28]; MerrilCh1923 [taxonomy, description, host, distribution, economic importance: 209]; MillerDa1990 [host, distribution, economic importance: 304]; Muraka1970 [host, distribution: 69,72]; NagarkSa1990 [host, distribution, economic importance, biological control: 553-542]; Nakaha1982 [host, distribution: 15]; Newste1901b [taxonomy, description, illustration, host, illustration: 82,114-116]; Sassce1923 [host, distribution: 152-158]; Seabra1930a [host, distribution: 143-148]; Seabra1941 [distribution: 8]; SwirskWyIz2002 [taxonomy, host, distribution, life history, economic importance, biological control: 103]; Takagi1958 [taxonomy, host, distribution: 122-123]; Takagi1984 [taxonomy, description, illustration, host, distribution: 18-21,37,47,65]; TakagiKa1966 [taxonomy, description, illustration, host, distribution: 116-118]; Tao1999 [taxonomy, host, distribution: 101]; Terezn1986 [taxonomy, description, illustration, host, distribution: 91-92]; TippinBe1970 [host, distribution: 8]; WilliaWa1988 [taxonomy, description, illustration, host, distribution: 193-195]; Woolle1990 [biological control: 167-176]; Wysoki1997 [host, distribution, economic importance: 905-811]; Yasar1995a [taxonomy, description, illustration, host, distribution: 102-103]; Zimmer1948 [taxonomy, description, illustration, host, distribution: 355-357].

Octaspidiotus MacGillivray

Octaspidiotus MacGillivray, 1921: 387. Type species: *Aspidiotus subrubescens* Maskell, by original designation.

Metaspidiotus Takagi, 1957: 35. Type species: *Aspidiotus stauntoniae* Takahashi, by original designation. Synonymy by Takagi, 1984: 3.

SYSTEMATICS: Takagi (1957, 1969a) established *Metaspidiotus* as separate from *Aspidiotus*, differing from latter in lanceolate marginal setae of pygidium. However, Takagi (1984) found that the presence of lanceolate setae on bases of second and third lobes is a common character to species of *Octaspidiotus* and species of *Metaspidiotus*, and therefore synonymized *Metaspidiotus* with *Octaspidiotus*.

GENERAL: Definition and characters by Balachowsky (1948b), Takagi (1969a, 1984) and by Williams & Watson (1988).

KEYS: Colon-Ferrer & Medina-Gaud 1998: 28-32 (female) [Genera of Puerto Rico]; Zahradník 1990b: 74 (female) [Czech Republic]; Williams & Watson 1988:

20 (female) [Tropical South Pacific]; Chou 1985: 260 (female) [Genera of China]; Chou 1985: 276 (female) [Species of China]; Takagi 1984: 13-14 (female) [species World]; Beardsley 1966: 502-504 (female) [Federated States of Micronesia]; Takagi 1957: 36 (female) [species Japan]; Balachowsky 1951: 598 (female) [Mediterranean].

CITATIONS: Balach1948b [taxonomy, description: 270,272]; Balach1951 [taxonomy: 598]; Beards1966 [taxonomy: 523]; Borchs1966 [catalogue: 272,275]; Chou1985 [taxonomy, description: 275-276]; ColonFMe1998 [taxonomy, description: 73]; DanzigPe1998 [catalogue: 315]; Ferris1937c [taxonomy: 51]; Kawai1980 [taxonomy: 224]; Kozar1990f [distribution: 142]; Lindin1937 [taxonomy: 191]; MacGil1921 [taxonomy, description: 387,395-396]; McKenz1939 [taxonomy: 55]; MorrisMo1966 [taxonomy, catalogue: 120,137]; Takagi1957 [taxonomy, description: 35]; Takagi1958 [taxonomy: 125]; Takagi1969a [taxonomy, description: 91]; Takagi1984 [taxonomy, description: 3-6]; Tao1999 [taxonomy: 101]; WilliaWa1988 [taxonomy, description: 195].

Octaspidiotus australiensis (Kuwana)

Aspidiotus australiensis Kuwana in: Kuwana & Muramatsu, 1931a: 652. Type data: AUSTRALIA: Queensland, Thursday Island, on orchid. Syntypes, female. Type depository: Ibaraki-ken: Insect Taxon. Laboratory, National Institute of Agricultural Environmental Sciences, Kannon-dai, Yatabe, Tsukuba-shi, (I. Kuwana), Japan.

Octaspidiotus australiensis; Borchsenius, 1966: 272. Change of combination.

SCALE COVER: Female scale circular, small; exuviae central; greyish (Kuwana & Muramatsu, 1931a). Scale of female broadly oval to subcircular, flat, light brown with slightly darker central exuviae. Male scale similar to female, but smaller and more elongate (Williams & Watson, 1988).

HOST PLANTS: **Araucariaceae**: *Araucaria hunsteinii* [WilliaWa1988]. **Menispermaceae**: *Cocculus laurifolius* [Takagi1984]. **Moraceae**: *Ficus* [WilliaWa1988]. **Myrsinaceae**: *Maesa chisia* [Takagi1984], *Maesa macrophylla* [Takagi1984]. **Orchidaceae** [KuwanaMu1931a, WilliaWa1988], *Dendrobium sanderae* [Takagi1984], *Dendrobium toftii* [Takagi1984], *Vanda* [Takagi1984]. **Sapotaceae** [WilliaWa1988]. **Symplocaceae**: *Symplocos crataegoides* [Takagi1984]. **Theaceae**: *Eurya* [Takagi1984].

DISTRIBUTION: **Australasian**: Australia, Queensland [Takagi1984] [KuwanaMu1931a]); Hawaiian Islands (Hawaii [Takagi1984]); Papua New Guinea [Takagi1984, WilliaWa1988]; Solomon Islands [Takagi1984]. **Nearctic**: U.S.A.: (California [Takagi1984]). **Oriental**: India (Uttar Pradesh [Takagi1984]); Nepal [Takagi1984]; Philippines [Takagi1984].

GENERAL: Description and illustration of adult female by Kuwana & Muramatsu (1931a), Takagi (1984) and by Williams & Watson (1988).

KEYS: Takagi 1984: 14 (female) [World].

CITATIONS: Borchs1966 [catalogue: 272]; Ferris1941e [taxonomy: 41]; KuwanaMu1931a [taxonomy, description, illustration, host, distribution: 652-653,658]; Takagi1984 [taxonomy, description, illustration, host, distribution: 12-13,35,53,55,65]; Takagi1990c [taxonomy, structure: 59-60]; WilliaWa1988 [taxonomy, description, illustration, host, distribution: 195-197].

Octaspidiotus bituberculatus Tang
Octaspidiotus bituberculatus Tang, 1984: 26. Type data: CHINA: Zhejiang Province, Lishui County, on *Sapium sebiferum*. Holotype female. Type depository: Shanxi: Entomological Institute, Shanxi Agricultural University, Taigu, Shanxi, China.
SCALE COVER: Tang (1984) did not describe scale cover.
HOST PLANTS: **Euphorbiaceae**: *Sapium sebiferum* [Tang1984].
DISTRIBUTION: **Oriental**: China (People's Republic) (Zhejiang (Chekiang) [Tang1984]).
GENERAL: Description and illustration of adult female by Tang (1984).
CITATIONS: Tang1984 [taxonomy, description, illustration, host, distribution: 26,29]; Tao1999 [taxonomy, host, distribution: 101].

Octaspidiotus calophylli (Green)
Aspidiotus calophylli Green, 1922a: 1008. Type data: SRI LANKA: Badulla, Namunakuli Hill, on foliage of *Calophyllum walkeri*. Syntypes, female. Type depository: London: The Natural History Museum, England, UK.
Octaspidiotus calophylli; Takagi, 1984: 7. Change of combination.
SCALE COVER: Female scale pale brown, semi translucent; flat; broadly ovate; exuviae yellowish, sub-central (Green, 1922a).
HOST PLANTS: **Guttiferae**: *Calophyllum walkeri* [Green1922a, Green1937].
Lauraceae: *Neolitsea* [Takagi1984]. **Myrsinaceae**: *Maesa* [Takagi1984].
Symplocaceae: *Symplocos laurina* [Takagi1984].
DISTRIBUTION: **Oriental**: India (Tamil Nadu [Takagi1984]); Sri Lanka [Green1922a].
GENERAL: Description and illustration of adult female by Green (1922a) and by Takagi (1984).
KEYS: Takagi 1984: 14 (female) [World].
CITATIONS: Borchs1966 [catalogue: 268]; Ferris1941e [taxonomy: 41]; Green1922a [taxonomy, description, illustration, host, distribution: 1008]; Green1937 [taxonomy, host, distribution: 330]; Takagi1984 [taxonomy, description, illustration, host, distribution: 7-8,33,48,58].

Octaspidiotus corticoides (Green)
Aspidiotus (*Evaspidiotus*) *subrubescens corticoides* Green, 1905b: 3. Type data: AUSTRALIA: Victoria, Myrniong, on *Eucalyptus globulus*. Syntypes, female. Type depository: London: The Natural History Museum, England, UK.
Aspidiotus subrubescens corticoides; Sanders, 1906: 14. Change of combination.
Octaspidiotus subrubescens corticoides; MacGillivray, 1921: 396. Change of combination.
Aspidiotus (*Evaspidiotus*) *corticoides*; Ferris, 1941e: 42. Change of combination and rank.
Octaspidiotus corticoides; Borchsenius, 1966: 272. Change of combination.
SCALE COVER: Female scale large, diameter, 2.5 mm; dark, chocolate-brown, opaque; exuviae concealed (Green, 1905b).
HOST PLANTS: **Myrtaceae**: *Eucalyptus globulus* [Green1905b, Sander1906].

DISTRIBUTION: **Australasian**: Australia [Sander1906] (Victoria [Green1905b]).
GENERAL: Description and illustration of adult female by Green (1905b).
CITATIONS: Balach1948b [taxonomy, host, distribution: 272]; Borchs1966 [catalogue: 272]; Ferris1941e [taxonomy: 42]; Green1905b [taxonomy, description, illustration, host, distribution: 3-4]; MacGil1921 [taxonomy, description, host, distribution: 396]; Sander1906 [taxonomy, host, distribution: 14].

Octaspidiotus cymbidii Tang

Octaspidiotus cymbidii Tang, 1984: 31. Type data: CHINA: Beijing, on the leaves of *Cymbidium* sp. Holotype female. Type depository: Shanxi: Entomological Institute, Shanxi Agricultural University, Taigu, Shanxi, China.
SCALE COVER: Female scale circular, 2 mm diameter; yellowish; exuviae orange yellow; central in position; male scale ovoid in shape, yellowish (Tang, 1984).
HOST PLANTS: **Orchidaceae**: *Cymbidium* [Tang1984].
DISTRIBUTION: **Palaearctic**: China (People's Republic) (Beijing (Peking) [Tang1984]).
GENERAL: Description and illustration of adult female by Tang (1984).
CITATIONS: DanzigPe1998 [catalogue: 315]; Tang1984 [taxonomy, description, illustration, host, distribution: 31-32].

Octaspidiotus machili (Takahashi)

Aspidiotus machili Takahashi, 1931b: 384. Type data: TAIWAN: Sozan, on *Machilus* sp. Syntypes, female. Type depository: Taichung: Entomology Collection, Taiwan Agricultural Research Institute, Wu-feng, Taichung, Taiwan.
Metaspidiotus machili; Takagi, 1969a: 92. Change of combination.
Octaspidiotus machili; Takagi, 1984: 11. Change of combination.
SCALE COVER: Female scale circular; secretion whitish; larval skins dark greenish brown, occupying most of scale (Takahashi, 1931b).
HOST PLANTS: **Araliaceae**: *Schefflera octophylla* [Takagi1969a, Takagi1984].
Lauraceae: *Machilus* [Takaha1931b, Takaha1932a, Takaha1933, Takagi1969a].
DISTRIBUTION: **Oriental**: Taiwan [Takaha1931b, Takaha1933, Takagi1984].
GENERAL: Description and illustration of adult female by Takahashi (1931b), Takagi (1969a, 1984) and by Chou (1985, 1986).
KEYS: Chou 1985: 276 (female) [Species of China]; Takagi 1984: 14 (female) [World].
CITATIONS: Borchs1966 [catalogue: 261]; Chou1985 [taxonomy, description, host, distribution: 276-277]; Chou1986 [taxonomy, illustration: 667]; Takagi1969a [taxonomy, description, illustration, host, distribution: 92-94]; Takagi1984 [taxonomy, description, illustration, host, distribution: 11,51]; Takaha1931b [taxonomy, description, illustration, host, distribution: 384-385]; Takaha1932a [host, distribution: 103]; Takaha1933 [host, distribution: 25-34,62]; Tao1999 [taxonomy, host, distribution: 101].

Octaspidiotus multipori (Takahashi)
Aspidiotus multipori Takahashi, 1956b: 24. Type data: JAPAN: Moji and Kyushu, Unzen, on *Illicium anisatum*. Syntypes, female. Type depository: Sapporo: Entomological Institute, Faculty of Agriculture, Hokkaido University, Japan.
Metaspidiotus multipori; Takagi, 1957: 37. Change of combination.
Octaspidiotus multipori; Takagi, 1984: 11. Change of combination.
SCALE COVER: Female scale circular, about 2 mm in diameter; thick; slightly convex; dark brown, but yellowish red at centre of exuvia (Takahashi, 1956b).
HOST PLANTS: **Illiciaceae**: *Illicium anisatum* [Takaha1956b], *Illicium religiosum* [Takagi1957, Takagi1962b, Takagi1984]. **Rutaceae**: *Skimmia japonica* [Takagi1984].
DISTRIBUTION: **Palaearctic**: Japan [Takaha1956b, Kawai1980] (Honshu [Takagi1962b, Takagi1984], Kyushu [Takagi1957, Takagi1962b, Takagi1984]).
GENERAL: Description and illustration of adult female by Takahashi (1956b) and by Takagi (1984).
KEYS: Takagi 1984: 14 (female) [World]; Takagi 1957: 36 (female) [Japan].
CITATIONS: Borchs1966 [catalogue: 276]; DanzigPe1998 [catalogue: 316]; Kawai1980 [taxonomy, description, host, distribution: 224]; Muraka1970 [host, distribution: 75]; Takagi1957 [taxonomy, host, distribution: 37-38]; Takagi1962b [host, distribution: 52]; Takagi1984 [taxonomy, description, illustration, host, distribution: 11-12,51,57]; Takaha1956b [taxonomy, description, illustration, host, distribution: 24-25].

Octaspidiotus nothopanacis (Ferris)
Aspidiotus nothopanacis Ferris, 1953: 66. Type data: CHINA: Yunnan Province, near Kunming, at Si-shan, on *Nothopanax delavayi*; collected by G.F. Ferris, May 8, 1953. Syntypes, female. Type depository: Davis: The Bohart Museum of Entomology, University of California, California, USA.
Octaspidiotus nothopanacis; Takagi, 1984: 10. Change of combination.
SCALE COVER: Scale of female circular, diameter about 2 mm; very flat, quite thin; pale brown. Scale of male not recognized (Ferris, 1953).
HOST PLANTS: **Araliaceae**: *Nothopanax delavayi* [Ferris1953, Takagi1984]. **Theaceae**: *Ternstroemia* [Takagi1984].
DISTRIBUTION: **Oriental**: China (People's Republic) (Yunnan [Ferris1953, Takagi1984]).
BIOLOGY: Occurring on underside of leaves (Ferris, 1953).
GENERAL: Description and illustration of adult female by Ferris (1953) and by Takagi (1984).
KEYS: Takagi 1984: 14 (female) [World].
CITATIONS: Borchs1966 [catalogue: 266]; DanzigPe1998 [catalogue: 316]; Ferris1953 [taxonomy, description, illustration, host, distribution: 66]; Takagi1984 [taxonomy, description, illustration, host, distribution: 10-11,41]; Tao1999 [taxonomy, host, distribution: 101].

Octaspidiotus pinicola Tang

Octaspidiotus pinicola Tang, 1984: 26. Type data: CHINA: Guanxi, Bobai County, on *Pinus* sp. Holotype female. Type depository: Shanxi: Entomological Institute, Shanxi Agricultural University, Taigu, Shanxi, China.
Octaspieiotus pinicola; Tang, 1984: 26. Misspelling of genus name.
SCALE COVER: Female scale circular or oval, 2 mm in diameter; thin; light yellow. Male scale elongate, about 1 mm long (Tang, 1984).
HOST PLANTS: **Pinaceae**: *Pinus* [Tang1984].
DISTRIBUTION: **Oriental**: China (People's Republic) (Guangxi (Kwangsi) [Tang1984]).
GENERAL: Description and illustration of adult female by Tang (1984).
CITATIONS: Tang1984 [taxonomy, description, illustration, host, distribution: 26,28]; Tao1999 [taxonomy, host, distribution: 101].

Octaspidiotus rhododendronii Tang

Octaspidiotus rhododendronii Tang, 1984: 26. Type data: CHINA: Yunnan Province, Kunming City, on *Rhododendron* sp. Holotype female. Type depository: Shanxi: Entomological Institute, Shanxi Agricultural University, Taigu, Shanxi, China.
Octaspidiotus rhododendroni; Tao, 1999: 101. Misspelling of species name.
SCALE COVER: Scale of adult female circular, 2 mm in diameter; quite thin; light yellow. Male scale elongate, about 1 mm long (Tang, 1984).
HOST PLANTS: **Ericaceae**: *Rhododendron* [Tang1984].
DISTRIBUTION: **Oriental**: China (People's Republic) (Yunnan [Tang1984]).
GENERAL: Description and illustration of adult female by Tang (1984).
CITATIONS: DanzigPe1998 [catalogue: 316]; Tang1984 [taxonomy, description, illustration, host, distribution: 26,30]; Tao1999 [taxonomy, host, distribution: 101].

Octaspidiotus stauntoniae (Takahashi)

Aspidiotus transparens; Kuwana, 1933: 18. Misidentification; discovered by Borchsenius, 1966: 275.
Aspidiotus stauntoniae Takahashi, 1933: 54. Type data: TAIWAN: Shinten, on *Stauntonia obovatifolia*; collected October 16, 1932. Syntypes, female. Type depository: Taichung: Entomology Collection, Taiwan Agricultural Research Institute, Wu-feng, Taichung, Taiwan.
Metaspidiotus stauntoniae; Takagi, 1957: 36. Change of combination.
Metaspidiotus stountoniae; Kawai, 1977: 161. Misspelling of species name.
Octaspidiotus stauntoniae; Takagi, 1984: 8. Change of combination.
SCALE COVER: Female scale grey, semi-transparent, thin, flat, nearly circular, but irregular at margin; diameter about 1.5 mm (Takahashi, 1933).
HOST PLANTS: **Anacardiaceae**: *Mangifera indica* [Takagi1984]. **Araliaceae**: *Dendropanax trifidus* [Takagi1984], *Fatsia japonica* [TakahaTa1956, Takagi1984], *Hedera rhombea* [Takaha1955f, TakahaTa1956, Takagi1957, Takagi1969a]. **Aucubaceae**: *Aucuba* [Takaha1955f], *Aucuba japonica* [Kawai1977, Takagi1984]. **Elaeagnaceae**: *Elaeagnus pungens* [TakahaTa1956, Takagi1969a, Takagi1984]. **Euphorbiaceae**: *Aleurites cordata* [Takagi1984]. **Flacourtiaceae**: *Scolopia*

oldhamii [Takagi1969a]. **Lardizabalaceae**: *Stauntonia hexaphylla* [Takagi1969a], *Stauntonia obovatifoliola* [Takaha1933]. **Lauraceae**: *Lindera obtusiloba* [Takagi1984], *Litsea cubea* [Takagi1984]. **Malvaceae**: *Hibiscus tiliaceus* [Takagi1969a, Takagi1984]. **Moraceae**: *Ficus* [Takagi1969a], *Ficus foveolata* [Takagi1969a], *Ficus retusa* [Takagi1984]. **Myrsinaceae**: *Maesa* [Takagi1969a]. **Myrtaceae**: *Psidium guajava* [Takagi1984]. **Orchidaceae**: *Arundina bambusifolia* [Takagi1984]. **Rutaceae**: *Citrus depressa* [Takaha1955f, Takagi1969a], *Skimmia japonica* [Takagi1984]. **Sapotaceae**: *Chrysophyllum cainito* [Takagi1984]. **Theaceae**: *Eurya* [Takagi1969a]. **Vitaceae**: *Vitis* [Takagi1969a].

DISTRIBUTION: **Australasian**: Hawaiian Islands (Hawaii [Takagi1984]). **Oriental**: Mongolia [Danzig1990]; Philippines [Takagi1984]; Ryukyu Islands (= Nansei Shoto) [TakahaTa1956, Takagi1969a]; Taiwan [TakahaTa1956, Takagi1969a, Takagi1984]. **Palaearctic**: China (People's Republic) [Takagi1984]; Japan [Takaha1955f, Takagi1957, Kawai1977, Kawai1980, Takagi1984] (Honshu [Takagi1984], Kyushu [Takagi1984], Shikoku [Takaha1956, Takagi1984]).

BIOLOGY: On leaves of host plants.

GENERAL: Description and illustration of adult female by Kuwana (1933) and by Takagi (1957, 1969a, 1984).

KEYS: Chou 1985: 276 (female) [Species of China]; Takagi 1984: 14 (female) [World]; Takagi 1957: 36 (female) [Japan].

CITATIONS: Borchs1966 [catalogue: 275-276]; Chou1985 [taxonomy, description, host, distribution: 276]; DanzigKo1990 [host, distribution: 46]; DanzigPe1998 [catalogue: 316]; Ferris1941e [taxonomy: 48]; Kawai1977 [host, distribution, economic importance: 161]; Kawai1980 [taxonomy, description, host, distribution: 224]; Kuwana1933 [taxonomy, description, illustration, host, distribution: 3,18]; Lindin1943b [taxonomy: 207]; Muraka1970 [host, distribution: 75]; Takagi1957 [taxonomy, description, illustration, host, distribution: 36-37]; Takagi1969a [taxonomy, description, illustration, host, distribution: 91-92,103]; Takagi1984 [taxonomy, description, illustration, host, distribution: 8-9,33,50,60]; Takaha1933 [taxonomy, description, illustration, host, distribution: 54-56]; Takaha1934 [taxonomy, description, host, distribution: 33]; Takaha1955f [host, distribution: 242]; Takaha1956b [taxonomy: 25]; TakahaTa1956 [taxonomy, host, distribution: 14]; Tao1999 [taxonomy, host, distribution: 101].

Octaspidiotus subrubescens (Maskell)

Aspidiotus subrubescens Maskell, 1892: 9. Type data: AUSTRALIA: on *Eucalyptus* sp.; collected by Mr. French. Syntypes, female and first instar. Type depositories: Auckland: New Zealand Arthropod Collection, Landcare Research, New Zealand, London: The Natural History Museum, England, UK, and Washington: United States National Entomological Collection, U.S. National Museum of Natural History, District of Columbia, USA.

Aspidiotus (Aspidiotus) subrubescens; Cockerell, 1897i: 27. Change of combination.

Aspidiotus (Evaspidiotus) subrubescens; Leonardi, 1898a: 77. Change of combination.

Octaspidiotus subrubescens; MacGillivray, 1921: 387. Change of combination.

SCALE COVER: Female scale reddish-brown, sub-circular, flat, and smooth; exuviae in centre, small, forming a small slightly elevated boss, which is rather yellower than rest; diameter variable, from 1/23 inch to 1/8 inch. Male scale white, slightly elongated, not carinated. Length about 1/20 inch (Maskell, 1892). Illustration of female scale by Froggatt (1914).

HOST PLANTS: **Apocynaceae**: *Nerium oleander* [Takagi1984]. **Capparidaceae**: *Capparis mitchelli* [Takagi1984]. **Euphorbiaceae**: *Caelebogyne ilicifolia* [Takagi1984]. **Flindersiaceae**: *Flindersia bennettiana* [Takagi1984]. **Gramineae**: *Tristania conferta* [Frogga1914]. **Loranthaceae**: *Loranthus pendulus* [Laing1929]. **Myrtaceae**: *Eucalyptus* [Maskel1892, Takagi1984], *Eucalyptus globulus* [Takagi1984]. **Pinaceae**: *Pinus caribaea* [Takagi1984]. **Polygonaceae**: *Owenia venosa* [Takagi1984]. **Proteaceae**: *Banksia* [Maskel1893b, Frogga1914]. **Rutaceae**: *Citrus australis* [Takagi1984].

DISTRIBUTION: **Australasian**: Australia [Maskel1892, Frogga1914] (Queensland [Takagi1984], Victoria [Laing1929, Takagi1984]).

GENERAL: Description and illustration of adult female by Froggatt (1914) and by Takagi (1984).

KEYS: Takagi 1984: 13 (female) [World].

CITATIONS: Borchs1966 [catalogue: 272]; Cocker1896b [distribution: 335]; Cocker1897i [taxonomy, description, host, distribution: 27]; DeitzTo1980 [taxonomy: 43]; Fernal1903b [catalogue: 279]; Ferris1937c [taxonomy, illustration: 51,87]; Ferris1941e [taxonomy: 48]; Frogga1914 [taxonomy, description, host, distribution: 318-319]; Frogga1915 [taxonomy, description, host, distribution: 23]; Laing1929 [host, distribution: 25]; Leonar1898a [taxonomy: 77]; Leonar1898c [taxonomy, description, illustration, host, distribution: 84-86]; MacGil1921 [taxonomy, description, host, distribution: 387,395]; Maskel1892 [taxonomy, description, illustration, host, distribution: 9-10]; Maskel1893b [taxonomy, description, host, distribution: 207]; McKenz1939 [taxonomy: 55]; Takagi1984 [taxonomy, description, illustration, host, distribution: 6-7,39,57].

Octaspidiotus tamarindi (Green)

Aspidiotus tamarindus Ramakrishna Ayyar, 1919a: 20. Nomen nudum; discovered by Borchsenius, 1966: 268.

Aspidiotus tamarindi Green, 1919c: 439. Type data: INDIA: Tamil Nadu, Coimbatore, on *Tamarindus* sp. Syntypes, female. Type depository: London: The Natural History Museum, England, UK.

Octaspidiotus tamarindi; Takagi, 1984: 9. Change of combination.

SCALE COVER: Female scale oval or subcircular, diameter 2 mm; flattish; colour stramineous, ochreous, or pale castaneous; darker examples situated on upper surface of leaves. Male scale small, oblong oval, length 0.75 mm; slightly narrower behind; colour paler than female scale, occasionally whitish (Green, 1919c).

HOST PLANTS: **Leguminosae**: *Tamarindus* [Green1919c, Ferris1941e], *Tamarindus indica* [Takagi1984].

NATURAL ENEMIES: HYMENOPTERA **Encyrtidae**: *Comperiella bifasciata* Howard [Flande1944a], *Comperiella indica* Ramakrishna Ayyar [Trjapi1989].

DISTRIBUTION: **Oriental**: India [Ramakr1921a, Ferris1941e] (Tamil Nadu [Green1919c, Takagi1984]).
BIOLOGY: Occurring on both surfaces of leaves (Green, 1919c).
ECONOMIC IMPORTANCE: Pest of leaves of *Tamarindus* (Schmutterer et al., 1957).
GENERAL: Description and illustration of adult female by Green (1919c), Ferris (1941e) and by Takagi (1984).
KEYS: Takagi 1984: 14 (female) [World]; Ferris 1946: 43 (female) [World]; Ferris 1941e: 61 (female) [World].
CITATIONS: Borchs1966 [catalogue: 268]; Dutta1990 [host, distribution: 152-163]; Ferris1941e [taxonomy, description, illustration, host, distribution: 48,59-60,68]; Ferris1946 [taxonomy: 43]; Ferris1953 [taxonomy: 65]; Flande1944a [biological control: 365-371]; Green1919c [taxonomy, description, illustration, host, distribution: 439]; MillerDa1990 [host, distribution, economic importance: 304]; Ramakr1919a [taxonomy: 20]; Ramakr1919b [taxonomy, host, distribution: 97]; Ramakr1921a [host, distribution: 356]; SchmutKlLu1957 [host, distribution, economic importance: 476]; Takagi1984 [taxonomy, description, illustration, host, distribution: 9-10,51]; Trjapi1989 [biological control: 297].

Octaspidiotus tripurensis Takagi
Octaspidiotus tripurensis Takagi, 1984: 9. Type data: INDIA: Tripura, Agartala, on *Thevetia peruviana*. Holotype female. Type depository: Sapporo: Entomological Institute, Faculty of Agriculture, Hokkaido University, Japan.
SCALE COVER: Takagi (1984) did not describe scale cover.
HOST PLANTS: **Apocynaceae**: *Thevetia peruviana* [Takagi1984]. **Moraceae**: *Ficus religiosa* [Takagi1984].
DISTRIBUTION: **Oriental**: India (Tripura [Takagi1984]); Thailand [Takagi1984].
GENERAL: Description and illustration of adult female by Takagi (1984).
KEYS: Takagi 1984: 14 (female) [World].
CITATIONS: Takagi1984 [taxonomy, description, illustration, host, distribution: 9,33,51,63].

Octaspidiotus yunnanensis (Tang & Chu), new combination
Metaspidiotus yunnanensis Tang & Chu, 1983: 304. Type data: CHINA: Yunnan, vicinity of Kunming, on *Keteleeria evelyniana*. Holotype female. Type depository: Shanxi: Entomological Institute, Shanxi Agricultural University, Taigu, Shanxi, China.
SYSTEMATICS: This species is here assigned to *Octaspidiotus,* following the synonymy of *Metaspidiotus* with *Octaspidiotus* by Takagi (1984).
SCALE COVER: Female scale circular or ovate, about 2 mm in diameter; yellow white; very thin and flat; exuviae yellowish in colour (Tang & Chu, 1983).
HOST PLANTS:- **Pinaceae**: *Keteleeria evelyniana* [TangCh1983, Tang1984].
DISTRIBUTION: **Oriental**: China (People's Republic) (Yunnan [TangCh1983, Tang1984]).
GENERAL: Description and illustration of adult female by Tang & Chu (1983) and by Tang (1984).

CITATIONS: DanzigPe1998 [catalogue: 316]; Tang1984 [taxonomy, description, illustration, host, distribution: 25-27]; TangCh1983 [taxonomy, description, illustration, host, distribution: 304-305]; Tao1999 [taxonomy, host, distribution: 101].

Operculaspis Laing

Operculaspis Laing, 1925a: 62. Type species: *Operculaspis crinitus* Laing, by monotypy and original designation.

SYSTEMATICS: Affinities of this genus are obscure. Laing (1925a) assigned it to Aspidiotini as "its affinities are entirely with such genera as *Selenaspidus* and *Pseudaonidia*". Ferris (1937a) interpreted that it belonged to Diaspidinae rather than to Aspidiotinae. Pending future study of the type species, *Operculaspis* here retained in Aspidiotinae, as suggested by Hall (1946a) and Borchsenius (1966).

GENERAL: Definition and characters given by Laing (1925a), Ferris (1937a) and by Hall (1946a).

CITATIONS: Borchs1966 [catalogue: 237]; Ferris1937a [taxonomy, description: 5,19]; Ferris1938b [taxonomy: 75]; Hall1946a [taxonomy, description: 527]; Laing1925a [taxonomy, description: 62]; Lindin1937 [taxonomy: 191]; MorrisMo1966 [taxonomy, catalogue: 138].

Operculaspis crinita Laing

Operculaspis crinitus Laing, 1925a: 63. Type data: TANZANIA: Ngerengere, on a forest tree. Syntypes, female. Type depository: London: The Natural History Museum, England, UK.

Operculaspis crinita; Ferris, 1937a: 3. Emendation of species name for agreement in gender.

SCALE COVER: Female scale subcircular to broadly ovate, greatest diameter about 3 mm; greyish white; moderately convex; very firm, rather brittle and thick; ventral scale present, closely adhering to bark of host plant, papery in texture in young stages but hardening with age, and in mature individuals of about same thickness as dorsal scale (Laing, 1925a).

DISTRIBUTION: **Afrotropical**: Tanzania [Laing1925a].

GENERAL: Description and illustration of adult female by Laing (1925a).

CITATIONS: Borchs1966 [catalogue: 237]; Ferris1937a [taxonomy: 3,19]; Laing1925a [taxonomy, description, illustration, host, distribution: 63-64].

Palinaspis Ferris

Palinaspis Ferris, 1941d: 377. Type species: *Targionia quohogiformis* Merrill, by original designation.

SYSTEMATICS: Ferris (1941d) assigned three species to this genus, but was doubtful whether congeneric. There is a faint relationship to *Morganella*, but latter have intersegmental scleroses between median and second lobes greatly reduced.

GENERAL: Definition and characters by Ferris (1941d).

KEYS: Colon-Ferrer & Medina-Gaud 1998: 28-32 (female) [Genera of Puerto Rico]; Ferris 1942: 26 (female) [North America]; Ferris 1942: 39 (female) [species North America].
CITATIONS: Balach1951 [taxonomy: 571]; Borchs1966 [catalogue: 319]; ColonFMe1998 [taxonomy, description: 74]; Ferris1941d [taxonomy, description: 377]; Ferris1942 [taxonomy: 446:26]; MorrisMo1966 [taxonomy, catalogue: 142].

Palinaspis barbata Ferris
Palinaspis barbata Ferris, 1942: 432. Type data: PANAMA: Chiriqui Province, at Boquete, on undetermined tree. Holotype female. Type depository: Davis: The Bohart Museum of Entomology, University of California, California, USA.
SCALE COVER: Scale of female normally circular with posterior end slightly produced and frequently projecting slightly from beneath covering bark; white, wax of a slightly granular or crystalline texture. Male scale not recognized (Ferris, 1942).
HOST PLANTS: **Leguminosae** [Balach1959].
DISTRIBUTION: **Neotropical**: Colombia [Balach1959]; Panama [Ferris1942].
BIOLOGY: Occurring on bark, usually almost completely concealed beneath bark flakes and in cracks (Ferris, 1942).
GENERAL: Description and illustration of adult female by Ferris (1942).
KEYS: Ferris 1942: 39 (female) [North America].
CITATIONS: Balach1959 [host, distribution: 356]; Borchs1966 [catalogue: 320]; Ferris1942 [taxonomy, description, illustration, host, distribution: 432; 446:39].

Palinaspis elisabethae Balachowsky
Palinaspis elisabethae Balachowsky, 1959: 354. Type data: COLOMBIA: Cordoba, 100 km north of Monteria, 5 km of Tukura, Ile de France Ranch, Alto Sinu, on *Gynerium sagittatum*. Syntypes, female. Type depository: Paris: Muséum national d'Histoire naturelle, France.
SCALE COVER: Female scale circular or subcircular, 2-2.2 mm in diameter; rugose; larval exuvia central or subcentral; brown chestnut, exuvia lighter. Male scale suboval, 1.6 mm long; colour lighter than that of female (Balachowsky, 1959).
HOST PLANTS: **Gramineae**: *Gynerium sagittatum* [Balach1959].
DISTRIBUTION: **Neotropical**: Colombia [Balach1959].
GENERAL: Description and illustration of adult female by Balachowsky (1959).
CITATIONS: Balach1959 [taxonomy, description, illustration, host, distribution: 354-355]; Borchs1966 [catalogue: 320].

Palinaspis lobulata Ferris
Palinaspis lobulata Ferris, 1941d: 378. Type data: PANAMA: Chiriqui Province, altitude about 7000 feet, on Volcan de Chiriqui, on undetermined tree. Holotype female. Type depository: Davis: The Bohart Museum of Entomology, University of California, California, USA.
SCALE COVER: Scale of female roughly circular, flat, whitish, exuvia central. No distinct ventral scale formed. Scale of male not recognized (Ferris, 1941d).
DISTRIBUTION: **Neotropical**: Panama [Ferris1941d].
BIOLOGY: Occurring on bark, concealed under bark scales (Ferris, 1941d).

GENERAL: Description and illustration of adult female by Ferris (1941d).
KEYS: Ferris 1942: 39 (female) [North America].
CITATIONS: Borchs1966 [catalogue: 320]; Ferris1941d [taxonomy, description, illustration, host, distribution: 378]; Ferris1942 [taxonomy: 446:39].

Palinaspis quohogiformis (Merrill)

Targionia quohogiformis Merrill, 1923: 167. Type data: U.S.A.: Florida, Miami, on blue trumpet (*Bignonia speciosa*); collected by Jeff Chaffin, January 19, 1920. Syntypes, female. Type depository: Davis: The Bohart Museum of Entomology, University of California, California, USA; type no. 6296.
Palinaspis quohogiformis; Ferris, 1941d: 379. Change of combination.
COMMON NAME: quohog-shaped scale [Merril1923, Dekle1965c].
SCALE COVER: Female scale oval; 1-1.4 mm long; shape similar to a Round Clam, commonly called Quohog; dorsal (upper) and ventral (lower) scales placed more or less laterally, ventral scale being unusually developed and forming with dorsal a bivalve-like arrangement; brownish, upper edges generally lighter; scale surface covered with fine particles of sand-like material; exuviae subcentral and ringed with a whitish secretion (Merrill, 1923). Female scale circular or oval, quite high convex, exuvia displaced towards one side; brownish; surface sprinkled with minute lumps of amorphous wax-like grains of sand or sugar; ventral scale strongly developed. Scale of male not known (Ferris, 1941d). Illustration of scale cover by Merrill (1923).
HOST PLANTS: **Annonaceae**: *Annona* [Ferris1941d]. **Bignoniaceae**: *Bignonia speciosa* [Merril1923, MerrilCh1923, Ferris1941d]. **Euphorbiaceae**: *Croton* [Ferris1941d]. **Leguminosae**: *Bauhinia* [Ferris1941d, Dekle1965c]. **Moraceae**: *Morus* [Ferris1941d]. **Oleaceae**: *Jasminum sambac* [Dekle1965c]. **Proteaceae**: *Grevillea* [Ferris1941d, Dekle1965c]. **Verbenaceae**: *Petrea volubilis* [Merril1923, MerrilCh1923, Ferris1941d].
DISTRIBUTION: **Nearctic**: United States of America (Florida [MerrilCh1923, Ferris1941d]). **Neotropical**: Cuba [Ferris1941d]; Puerto Rico & Vieques Island (Puerto Rico [ColonFMe1998]); Saint Lucia [Ferris1941d].
BIOLOGY: Occurring in leaf buds or "near the junction of a bud and branch" (Ferris, 1941d).
GENERAL: Description and illustration of adult female by Merrill (1923), Ferris (1941d) and by Colon-Ferrer & Medina-Gaud (1998).
KEYS: Ferris 1942: 39 (female) [North America].
CITATIONS: Borchs1966 [catalogue: 320]; ColonFMe1998 [taxonomy, description, illustration, host, distribution: 75-76]; Dekle1965c [taxonomy, description, host, distribution: 102]; Ferris1941d [taxonomy, description, illustration, host, distribution: 379]; Ferris1942 [taxonomy: 445:11; 446:39]; Ferris1943a [taxonomy: 86]; Merril1923 [taxonomy, description, illustration, host, distribution: 167-168]; Merril1953 [taxonomy, description, host, distribution: 65-66]; MerrilCh1923 [taxonomy, description, host, distribution, economic importance: 252-253].

Palinaspis sordidata Ferris
Palinaspis sordidata Ferris, 1941d: 380. Type data: PANAMA: Chiriqui province, Volcan de Chiriqui, altitude about 7000 feet altitude, on undetermined tree. Holotype female. Type depository: Davis: The Bohart Museum of Entomology, University of California, California, USA.
SCALE COVER: Scale of female very flat and thin, apparently composed in part of soft bark tissues of host with which it is so intermingled as to be detectable only by scraping bark; exuviae roughly central; no distinct ventral scale is formed (Ferris, 1941d).
DISTRIBUTION: **Neotropical**: Panama [Ferris1941d].
BIOLOGY: Occurring on trunk of host (Ferris, 1941d).
GENERAL: Description and illustration of the adult female given by Ferris (1941d).
KEYS: Ferris 1942: 39 (female) [North America].
CITATIONS: Borchs1966 [catalogue: 320]; Ferris1941d [taxonomy, description, illustration, host, distribution: 380]; Ferris1942 [taxonomy: 446:39].

Paranewsteadia MacGillivray

Paranewsteadia MacGillivray, 1921: 391. Type species: *Aspidiotus maculatus* Newstead, by monotypy and original designation.
SYSTEMATICS: Distinct characters of type species, *Aspidiotus maculatus* as described by Newstead (1896a): small rounded median lobes and narrow and long second lobes.
GENERAL: Definition and characters by MacGillivray (1921).
CITATIONS: Borchs1966 [catalogue: 367]; Ferris1937c [taxonomy: 52]; Kozar1990f [distribution: 142]; Lindin1937 [taxonomy: 367]; MacGil1921 [taxonomy, description: 391,432]; MorrisMo1966 [catalogue: 146].

Paranewsteadia maculata (Newstead)
Aspidiotus maculatus Newstead, 1896a: 133. Type data: AFRICA: country and host plant not indicated. Syntypes, female. Notes: Depository of type material unknown; no material available at BMNH (Douglas J. Williams and Jon Martin, personal information to Yair Ben-Dov, December, 2002).
Hemiberlesia maculata; Leonardi, 1897b: 122. Change of combination requiring emendation of species name for agreement in gender.
Paranewsteadia maculata; MacGillivray, 1921: 391. Change of combination.
SCALE COVER: Female scale pure white; rather thick; exuviae black, forming a large, conspicuous, central spot (Newstead, 1896a).
GENERAL: Description and illustration of adult female by Newstead (1896a).
CITATIONS: Borchs1966 [catalogue: 367]; Cocker1897i [taxonomy, description, host, distribution: 28]; Cocker1899a [taxonomy: 395]; Fernal1903b [catalogue: 267]; Ferris1937c [taxonomy: 52]; Ferris1941e [taxonomy: 45]; Leonar1897 [taxonomy: 285]; Leonar1897b [taxonomy, description: 122]; MacGil1921[description, host, distribution: 391,432]; Newste1896a [taxonomy, description, illustration, host, distribution: 133]; Vayssi1913 [host, distribution: 430].

Paraonidia MacGillivray

Paraonidia MacGillivray, 1921: 394. Type species: *Aspidiotus malleolus* Green, by monotypy and original designation.
Paraonidea; MacGillivray, 1921: 454. Misspelling of genus name.
SYSTEMATICS: Ferris (1938) considered this genus possibly valid, but indicated that further study of related genera in the *Pseudaonidia* group would be necessary before reaching definite conclusion. Balachowsky (1948b) placed this genus in Pseudaonidina.
GENERAL: Definition and characters MacGillivray (1921).
CITATIONS: Balach1948b [taxonomy: 269]; Borchs1966 [catalogue: 237]; Ferris1937c [taxonomy: 52]; Ferris1938 [taxonomy: 43]; Lindin1937 [taxonomy: 192]; MacGil1921 [taxonomy, description: 394,454]; McKenz1939 [taxonomy: 54]; MorrisMo1966 [taxonomy, catalogue: 146].

Paraonidia malleola (Green)

Aspidiotus (*Chrysomphalus*) *malleolus* Green, 1905a: 342. Type data: SRI LANKA: on *Mimusops hexandra*. Syntypes, female. Type depository: London: The Natural History Museum, England, UK.
Chrysomphalus malleolus; Sanders, 1906: 15. Change of combination.
Pseudaonidia malleolus; Lindinger, 1909b: 148. Change of combination.
Paraonidea malleola; MacGillivray, 1921: 454. Misspelling of genus name.
Paraonidia malleola; MacGillivray, 1921: 454. Change of combination requiring emendation of species name for agreement in gender.
Aspidiotus malleolus; Ferris, 1937c: 52. Change of combination.
Paraonidia malleola; Borchsenius, 1966: 237. Revived combination.
SCALE COVER: Female scale opaque snowy white; dense, broad and flat; irregularly deltoid. Long diameter 4.0-5.5 mm. Male scale similar but very much smaller. Length 2.25 mm (Green, 1905a).
HOST PLANTS: **Sapotaceae**: *Mimusops hexandra* [Green1905a, Sander1906, Ramakr1921a, Green1922, Green1937].
DISTRIBUTION: **Oriental**: Sri Lanka [Green1905a, Sander1906, Green1922].
GENERAL: Description and illustration of adult female by Green (1905a).
CITATIONS: Borchs1966 [catalogue: 237]; DEDAC1923 [host, distribution]; Ferris1937c [taxonomy: 51]; Ferris1938 [taxonomy: 43,50]; Ferris1941e [taxonomy: 45]; Green1905a [taxonomy, description, illustration, host, distribution: 342]; Green1922 [host, distribution: 462]; Green1937 [host, distribution: 334]; Lindin1909b [taxonomy: 148]; MacGil1921 [taxonomy, description, host, distribution: 454]; McKenz1939 [taxonomy: 54]; Ramakr1921a [host, distribution: 357]; Sander1906 [taxonomy, host, distribution: 15]; Sander1909a [taxonomy: 58].

Paraonidiella MacGillivray

Paraonidiella MacGillivray, 1921: 392. Type species: *Aspidiotus cladii* Maskell, by monotypy and original designation.
SYSTEMATICS: MacGillivray (1921) placed this genus in Aspidiotini. Laing (1929) and Lindinger (1937) considered *Paraonidiella* a synonym of *Furcaspis*.

Ferris (1938) indicated necessity for study of *Pseudaonidia* group of genera in order to verify the status of *Paraonidiella* .

GENERAL: Definition and characters by MacGillivray (1921).

CITATIONS: Balach1958b [taxonomy: 249]; Borchs1966 [catalogue: 242]; Ferris1937c [taxonomy: 52]; Ferris1938 [taxonomy: 43]; Kozar1990f [distribution: 142]; Laing1929 [taxonomy, description: 26-27]; Lindin1937 [taxonomy: 192]; MacGil1921 [taxonomy, description: 392,442]; McKenz1939 [taxonomy: 54]; MorrisMo1966 [taxonomy, catalogue: 146].

Paraonidiella cladii (Maskell)

Aspidiotus cladii Maskell, 1891: 3. Type data: AUSTRALIA: on *Cladium* sp. Syntypes, female. Type depository: Auckland: New Zealand Arthropod Collection, Landcare Research, New Zealand.

Aspidiotus (Chrysomphalus) cladii; Cockerell, 1897i: 26. Change of combination.

Aonidiella cladii; Leonardi, 1899: 176. Change of combination.

Chrysomphalus cladii; Fernald, 1903b: 289. Change of combination.

Furcaspis cladii; Cockerell, 1919: 116. Change of combination.

Paraonidiella cladii; MacGillivray, 1921: 442. Change of combination.

SCALE COVER: Illustration of female and male scale cover by Maskell (1891). Female scale circular, about 1/16 inch in diameter; convex; rich dark-brown, margin orange-red and exuviae dark-yellow; exuviae central. Male scale similar in colour, but narrower and elongated; length about 1/20 inch; not carinated (Maskell, 1891).

HOST PLANTS: **Cyperaceae**: *Cladium* [Maskel1891, Frogga1914], *Gahnia* [Laing1929]. **Myrtaceae**: *Leptospermum* [Frogga1914]. **Xanthorrhoeaceae**: *Xerotes* [Frogga1914].

DISTRIBUTION: **Afrotropical**: Mauritius [Green1907]. **Australasian**: Australia (New South Wales [Frogga1914], Victoria [Laing1929]). **Palaearctic**: Japan [Green1907, Kuwana1917a].

GENERAL: Description and illustration of adult female by Maskell (1891) and by Laing (1929).

CITATIONS: Balach1958b [taxonomy: 249,252]; Borchs1966 [catalogue: 242]; Cocker1896b [distribution: 335]; Cocker1897i [taxonomy, description, host, distribution: 26]; Cocker1919 [taxonomy: 116]; DeitzTo1980 [taxonomy: 34]; Fernal1903b [catalogue: 289]; Ferris1937c [taxonomy: 52]; Ferris1941e [taxonomy: 42]; Frogga1914 [taxonomy, description, host, distribution: 135]; Frogga1915 [taxonomy, description, host, distribution: 12-13]; Fuller1897c [host, distribution: 3]; Green1907 [host, distribution: 203]; Kuwana1917a [taxonomy, distribution: 175]; Laing1929 [taxonomy, description, illustration, host, distribution: 26-27]; Leonar1899 [taxonomy: 176,195]; Lindin1943b [taxonomy: 220]; MacGil1921 [taxonomy, description, host, distribution: 442]; Maskel1891 [taxonomy, description, illustration, host, distribution: 3]; McKenz1939 [taxonomy: 54]; McKenz1939 [taxonomy: 54]; Takaha1939b [taxonomy, host, distribution: 271].

Paraselenaspidus Mamet

Paraselenaspidus Mamet, 1958a: 418. Type species: *Selenaspidus madagascariensis* Mamet, by original designation.

SYSTEMATICS: This genus resembles *Selenaspidus* in the spur-shaped third lobe, but differs in having the constriction between prothorax and mesothorax. In *Selenaspidus,* constriction between mesothorax and metathorax (Mamet, 1958a).

GENERAL: Definition and characters by Mamet (1958a).

KEYS: Mamet 1958a: 362 (female) [*Selenaspidus* complex]; Mamet 1958a: 420 (female) [Species of *Paraselenaspidus*.].

CITATIONS: Borchs1966 [catalogue: 256]; Mamet1958a [taxonomy, description: 362,418-420]; MorrisMo1966 [taxonomy, catalogue: 147].

Paraselenaspidus gracilis (Lindinger)

Selenaspidus gracilis Lindinger, 1909d: 10. Type data: CAMEROON: Bipinde, on *Agelaia* [=*Agelaea*] *fragrans*, and on *Tricalysia* sp. Syntypes. Type depository: Zoologisches Institut und Zoologishces Museum, Universität von Hamburg, Germany.

Pseudaonidia gracilis; Sanders, 1909a: 54. Change of combination.

Aspidiotus (*Selenaspidus*) *gracilis*; Vayssière, 1913: 431. Change of combination.

Entaspidiotus gracilis; MacGillivray, 1921: 455. Change of combination.

Paraselenaspidus gracilis; Mamet, 1958a: 420. Change of combination.

SCALE COVER: Female scale slightly elongated, 1 mm long, 0.8 mm wide; brownish; exuviae central (Lindinger, 1909d).

HOST PLANTS: **Connaraceae**: *Agelaea fragrans* [Lindin1909d, Sander1909a, Vayssi1913, Mamet1958a, Borchs1966]. **Rubiaceae**: *Tricalysia* [Lindin1909d, Sander1909a, Vayssi1913, Mamet1958a, Borchs1966].

DISTRIBUTION: **Afrotropical**: Cameroon [Lindin1909d, Sander1909a, Vayssi1913, Mamet1958a, Borchs1966].

GENERAL: Description and illustration of adult female by Lindinger (1909d).

KEYS: Lindinger 1909: 4-5 (female) [Genus *Selenaspidus*].

CITATIONS: Borchs1966 [catalogue: 256]; Ferris1941e [taxonomy: 44]; Lindin1909d [taxonomy, description, illustration, host, distribution: 10]; MacGil1921 [taxonomy, host, distribution: 455]; Mamet1958a [taxonomy, description, host, distribution: 420]; Sander1909a [taxonomy, host, distribution: 54]; Vayssi1913 [host, distribution: 431]; WeidneWa1968 [taxonomy: 176].

Paraselenaspidus madagascariensis (Mamet)

Selenaspidus madagascariensis Mamet, 1954: 81. Type data: MADAGASCAR: Périnet, on "Varongo mainty". Holotype. Type depository: Paris: Muséum national d'Histoire naturelle, France.

Paraselenaspidus madagascariensis; Mamet, 1958b: 420. Change of combination.

SCALE COVER: Female scale subcircular, fairly flat; pale buff; sometimes covered by debris of plant tissues; exuviae subcentral, yellowish. Male scale not observed (Mamet, 1954, 1958a).

HOST PLANTS: **Buxaceae**: *Buxus sempervirens* [Mamet1958a, Borchs1966]. **Erythroxylaceae**: *Erythroxylum corymbosum* [Matile1978]. **Flacourtiaceae**: *Aberia caffra* [Mamet1958a, Borchs1966], *Dovyalis caffra* [RosenDe1979]. **Guttiferae**: *Haronga madagascariensis* [Mamet1954, Mamet1958a, Borchs1966]. **Monimiaceae**: *Tambourissa* [Mamet1954, Mamet1958a, Borchs1966]. **Moraceae**: *Ficus thonningii* [Mamet1958a, Borchs1966]. **Ochnaceae**: *Lophira alata* [Mamet1958a, Borchs1966]. **Palmae** [Mamet1958a, Borchs1966], *Raphia* [Mamet1959a, Borchs1966]. **Rutaceae**: *Citrus* [Mamet1958a, Borchs1966]. **Sapotaceae**: *Pachystela brevipes* [Mamet1958a, Borchs1966]. **Sterculiaceae**: *Cacao* [Mamet1958a, Borchs1966]. **Theaceae**: *Camellia sinensis* [Mamet1958a, Borchs1966], *Thea* [Mamet1958a, Borchs1966].
NATURAL ENEMIES: HYMENOPTERA **Aphelinidae**: *Aphytis anneckei* DeBach & Rosen [RosenDe1979].
DISTRIBUTION: **Afrotropical**: Cameroon [MatileNo1984]; Comoros Islands [Matile1978]; Kenya [Mamet1958a, Borchs1966, RosenDe1979]; Madagascar [Mamet1954, Mamet1958a, Mamet1959a, Borchs1966]; São Tomé and Príncipe [Mamet1958a, Borchs1966]; Uganda [Mamet1958a, Borchs1966]; Zaire [Mamet1958a]. **Neotropical**: French Guiana [Mamet1958a, Borchs1966].
GENERAL: Description and illustration of adult female by Mamet (1954, 1958a).
CITATIONS: Borchs1966 [catalogue: 256-257]; DeBachRo1976a [host, distribution, biological control: 541-545]; Mamet1954 [taxonomy, description, illustration, host, distribution: 22,81]; Mamet1958a [taxonomy, description, illustration, host, distribution: 420]; Mamet1959a [host, distribution: 389]; Matile1978 [host, distribution: 65]; MatileNo1984 [host, distribution: 66]; RosenDe1979 [host, distribution, biological control: 598-601].

Parrottia MacGillivray

Parrottia MacGillivray, 1921: 394. Type species: *Aspidiotus moorei* Green, by monotypy and original designation.
SYSTEMATICS: MacGillivray (1921) assigned this genus to Aspidiotini. Balachowsky (1948b) placed it in subtribe Pseudaonidina. Ferris (1938) suggested that a further study of *Pseudaonidia* series of genera required to clarify status of *Parrottia*.
GENERAL: Definition and characters by MacGillivray (1921).
CITATIONS: Balach1948b [taxonomy: 269]; Borchs1966 [catalogue: 237]; Ferris1937c [taxonomy: 52]; Ferris1938 [taxonomy: 43,52]; Lindin1937 [taxonomy: 192]; MacGil1921 [taxonomy, description: 394,458]; MorrisMo1966 [taxonomy, catalogue: 149].

Parrottia moorei (Green)

Aspidiotus moorei Green, 1896c: 199. Type data: INDIA: Madras, on bark of *Grislea tomentosa*. Syntypes, female. Type depository: London: The Natural History Museum, England, UK.
Targionia moorei; Leonardi, 1900: 312. Change of combination.
Pseudaonidia moorei; Marlatt, 1908: 139. Change of combination.

Parrottia moorei; MacGillivray, 1921: 458. Change of combination.

SCALE COVER: Female scale very inconspicuous; rugose and coloured like bark; circular, diameter 2.50 mm; flattish above, slightly concave below; very solid and opaque; margin thicker than median area; insect lies in a shallow depression excavated in bark, and coated with whitish or greyish secretion deposited in distinct concentric rings. Male scale not observed (Green, 1896c).

HOST PLANTS: **Lythraceae**: *Grislea tomentosa* [Green1896c, Leonar1900, Marlat1908, Ramakr1919a, Ramakr1921a].

DISTRIBUTION: **Oriental**: India (Tamil Nadu [Green1896c, Leonar1900, Marlat1908, Ramakr1921a]).

GENERAL: Description and illustration of adult female by Green (1896c), Leonardi (1900) and by Marlatt (1908).

KEYS: Marlatt 1908: 135 (female) [World].

CITATIONS: Borchs1966 [catalogue: 237]; Cocker1897i [taxonomy, description, host, distribution: 28]; Cocker1899a [taxonomy: 395]; Cocker1920 [taxonomy: 386]; Fernal1903b [catalogue: 298]; Ferris1937c [taxonomy: 52]; Ferris1938 [taxonomy: 43,52]; Ferris1941e [taxonomy: 46]; Ferris1943a [taxonomy: 86]; Green1896c [taxonomy, description, illustration, host, distribution: 199-200]; Leonar1900 [taxonomy, description, illustration, host, distribution: 312-314]; MacGil1921 [taxonomy, description, host, distribution: 458]; Marlat1908 [taxonomy, description, illustration, host, distribution: 135,139]; Ramakr1919a [taxonomy, host, distribution: 20]; Ramakr1921a [host, distribution: 356].

Phaspis Ben-Dov

Phaspis Ben-Dov, 1974b: 317. Type species: *Phaspis lobulata* Ben-Dov, by monotypy and original designation.

SYSTEMATICS: Genus differs from *Melanaspis* Cockerell in possessing only two pairs of pygidial paraphyses; from *Lindingaspis* MacGillivray in presence of dorsal macroducts of one size; from *Clavaspis* MacGillivray in presence of three pairs of pygidial lobes; from *Diclavaspis* Balachowsky in elongate shape of paraphyses and their absence between median lobes, from *Pseudomelanaspis* Borchsenius in squat median lobes, and from *Greenoidea* MacGillivray in presence of only two pairs of pygidial lobes and their absence in median position (Ben-Dov, 1974b).

GENERAL: Definition and characters by Ben-Dov (1974b).

CITATIONS: BenDov1974b [taxonomy, description: 317]; KosztaBeKo1986 [taxonomy, catalogue: 13].

Phaspis lobulata Ben-Dov

Phaspis lobulata Ben-Dov, 1974b: 318. Type data: SOUTH AFRICA: Transvaal, Phalaborwa, on *Acacia nigrescens*; collected by Y. Ben-Dov. Holotype female (examined). Type depository: Pretoria: National Collection of Insects, South Africa.

SCALE COVER: Scale of female oval in outline, about 1.5 mm long, 1.0 mm wide; black with a bright brown margin; nymphal exuviae central; without a ventral velum, other than a thin layer of white wax attached to host (Ben-Dov, 1974b).

HOST PLANTS: **Leguminosae**: *Acacia nigrescens* [BenDov1974b].

DISTRIBUTION: **Afrotropical**: South Africa [BenDov1974b].
GENERAL: Description and illustration of adult female by Ben-Dov (1974b).
CITATIONS: BenDov1974b [taxonomy, description, illustration, host, distribution: 318-320].

Phaulaspis Leonardi

Phaulaspis Leonardi, 1897: 284. Type species: *Aspidiotus hakeae* Maskell, by monotypy.
Aspidiotus (*Phaulaspis*); Cockerell, 1899a: 395. Change of status.
Cryptoaonidia Leonardi, 1900: 323. Type species: *Aspidiotus hakeae* Maskell, by original designation. Synonymy by Borchsenius, 1966: 366.
Phaulaspis; Borchsenius, 1966: 366. Revived status.
SYSTEMATICS: The type species of *Phaulaspis* is a pupillarial. Morrison & Morrison (1922), Ferris (1937c) and Lindinger (1937) accepted genus as valid. Pygidial margin of adult female of type species without lobes and plates, while pygidium of second instar has lobes.
GENERAL: Definition and characters by MacGillivray (1921) and by Morrison & Morrison (1922).
CITATIONS: Balach1948b [taxonomy: 269]; Borchs1966 [catalogue: 366]; Cocker1897i [taxonomy: 31]; Cocker1899a [taxonomy: 395]; Fernal1903b [catalogue: 251]; Ferris1937c [taxonomy: 55,69]; Ferris1938b [taxonomy: 75]; HowellTi1990 [taxonomy: 57]; Leonar1897 [taxonomy, description: 284-286]; Lindin1932f [taxonomy: 189]; Lindin1937 [taxonomy: 192]; MacGil1921 [taxonomy, description: 395,465]; MorrisMo1922 [taxonomy, description: 89-93]; MorrisMo1966 [taxonomy, catalogue: 152].

Phaulaspis hakeae (Maskell)

Aspidiotus hakeae Maskell, 1896b: 383. Type data: AUSTRALIA: New South Wales, Sydney, on *Hakea* sp.; sent by Mr. Olliff. Syntypes, female and first instar. Type depository: Auckland: New Zealand Arthropod Collection, Landcare Research, New Zealand.
Aspidiotus (*Phaulaspis*) *hakeae*; Cockerell, 1897i: 27. Change of combination.
Aonidia hacheae; Leonardi, 1899: 205. Change of combination, and misspelling of species name.
Aonidia (*Cryptoaonidia*) *hackeae*; Leonardi, 1900: 323. Change of combination, and misspelling of species name
Phaulaspis hakeae; MacGillivray, 1921: 465. Change of combination.
SCALE COVER: Female scale female circular, diameter about 1/45 inch; slightly convex; greyish-white; exuviae dark-orange, central; median portion frequently rubbed off, leaving exuviae exposed, with a ring of secretion. Male scale circular, smaller and whiter than female; diameter about 1/65 inch (Maskell, 1896b).
HOST PLANTS: **Proteaceae**: *Hakea* [Maskel1896b, Leonar1900, Frogga1914].
DISTRIBUTION: **Australasian**: Australia [Leonar1900] (New South Wales [Maskel1896b, Frogga1914]).

GENERAL: Description and illustration of adult female by Maskell (1896b), Leonardi (1900) and by Ferris (1937c).
CITATIONS: Borchs1966 [catalogue: 366-367]; Cocker1897i [taxonomy, description, host, distribution: 27]; DeitzTo1980 [taxonomy: 38]; Fernal1903b [catalogue: 260]; Ferris1937c [taxonomy, illustration: 51,52,89]; Ferris1941e [taxonomy: 44]; Frogga1914 [taxonomy, description, host, distribution: 314]; Frogga1915 [taxonomy, description, host, distribution: 17]; Leonar1897 [taxonomy: 286]; Leonar1900 [taxonomy, description, illustration, host, distribution: 323-326]; Maskel1896b [taxonomy, description, illustration, host, distribution: 383-384].

Pseudaonidia Cockerell

Aspidiotus (*Pseudaonidia*) Cockerell, 1897i: 14. Type species: *Aspidiotus duplex* Cockerell, by original designation.
Pseudaonidia; Fernald, 1903b: 283. Change of status.
Pseudoaonidia; Leonardi, 1914: 201. Misspelling of genus name.
Pseudaonidea; MacGillivray, 1921: 394,453. Misspelling of genus name.
Pseudaonidiella MacGillivray, 1921: 394. Type species: *Pseudaonidia duplex var. paeoniae* Cockerell, by monotypy and original designation. Synonymy by Ferris, 1937a: 55.
Pseudoaonidia; Rungs, 1935: 273. Misspelling of genus name.
Pseudoaunidia; Minamikawa, 1959: 41. Misspelling of genus name.
Pseudoaonidia; Chou, 1985: 320. Misspelling of genus name.
SYSTEMATICS: *Pseudaonidia* differs from *Duplaspidiotus* MacGillivray in absence of marginal paraphyses on pygidium (Balachowsky, 1958b).
GENERAL: Definition and characters by Marlatt (1908), Robinson (1917), Brain (1919), Fullaway (1932), Kuwana (1933), Ferris (1938a), Borchsenius (1950b), Lupo (1953), Balachowsky (1951, 1958b), Almeida (1969), Takagi (1969a), Danzig (1993) and by Kosztarab (1996).
KEYS: Colon-Ferrer & Medina-Gaud 1998: 28-32 (female) [Genera of Puerto Rico]; Kosztarab 1996: 406-407 (female) [Northeastern North America]; Danzig 1993: 136 (female) [species Europe]; Williams & Watson 1988: 19 (female) [Tropical South Pacific]; Chou 1985: 248 (female) [China]; Paik 1978: 380 (female) [species South Korea]; McDaniel 1970: 427 (female) [species U.S.A.: Texas]; Beardsley 1966: 502-504 (female) [Federated States of Micronesia]; Balachowsky 1958b: 256 (female) [*Pseudaonidina* of Africa]; Balachowsky 1958b: 256 (female) [*Pseudaonidina* of Africa]; Ezzat 1958: 237-239 (female) [Egypt]; Balachowsky 1951: 675 (female) [Mediterranean]; Borchsenius 1950b: 167 (female) [USSR]; Ferris 1942: 25 (female) [North America]; Ferris 1942: 39 (female) [species North America]; Borchsenius 1937: 100 (female) [USSR]; Borchsenius 1937a: 32-33 (female) [Palaearctic Region]; Kuwana 1933a: 43-45 (female) [Japan]; Fullaway 1932: 97-98 (female) [Hawaii]; Robinson 1917: 16-17 (female) [Philippines]; Lindinger 1913: 65 (female) [World]; Marlatt 1908: 134-135 (female) [World].
CITATIONS: Almeid1969 [taxonomy, description: 160]; Balach1948b [taxonomy: 266]; Balach1951 [taxonomy, description: 680-681]; Balach1953i [taxonomy, description: 1512]; Balach1958b [taxonomy, description: 268-269]; Beards1966

[taxonomy: 524]; Borchs1937 [taxonomy, description: 32,38]; Borchs1949d [taxonomy: 193,220]; Borchs1950b [taxonomy, description: 212]; Borchs1966 [catalogue: 229]; Brain1918 [taxonomy: 116]; Brain1919 [taxonomy, description: 205-206]; Chou1985 [taxonomy, description: 248,320]; Cocker1897i [taxonomy, description: 14]; Cocker1899a [taxonomy: 396]; CockerRo1915 [taxonomy, description: 108-109]; ColonFMe1998 [taxonomy, description: 76]; Danzig1993 [taxonomy, description: 136]; DanzigPe1998 [catalogue: 338]; Ezzat1958 [taxonomy: 237]; Fernal1903b [catalogue: 283]; Ferris1937c [taxonomy: 51]; Ferris1938a [taxonomy, description: 252]; Ferris1942 [taxonomy: 446:25]; Fullaw1932 [taxonomy, description: 109]; Gowdey1921 [taxonomy: 1]; Hadzib1983 [taxonomy: 215]; Hempel1904 [taxonomy, description: 321]; Kawai1980 [taxonomy: 204]; Koszta1996 [taxonomy, description: 562]; Kuwana1933 [taxonomy, description: 20]; Kuwana1933a [taxonomy: 43-45]; Lepage1938 [taxonomy: 417]; Lindin1908b [taxonomy: 98]; Lindin1913 [taxonomy: 65]; Lindin1932f [taxonomy: 194]; Lindin1937 [taxonomy: 194]; Lupo1953 [taxonomy, description: 16-17]; MacGil1921 [taxonomy, description: 394,453-454]; Mamet1949 [taxonomy: 63]; Marlat1908 [taxonomy, description: 131-141]; Miller1990 [taxonomy: 169-178]; MorrisMo1966 [taxonomy, catalogue: 163]; Robins1917 [taxonomy, description: 17,33]; Rungs1935 [taxonomy: 273]; Schmut1959 [taxonomy: 112]; Takagi1969a [taxonomy, description: 98]; Tao1999 [taxonomy, host, distribution: 111]; Willia1969a [taxonomy: 335]; WilliaWa1988 [taxonomy: 219].

Pseudaonidia baikeae (Newstead)
Aspidiotus (*Pseudaonidia*) *baikeae* Newstead, 1914: 307. Type data: UGANDA: Entebbe, on *Baikea* [=*Baikiaea*] *insignis* and an unknown shrub. Syntypes, female. Type depository: London: The Natural History Museum, England, UK.
Lattaspidiotus baikeae; MacGillivray, 1921: 457. Change of combination.
Pseudaonidia baikeae; Laing, 1929a: 498. Change of combination.
SCALE COVER: Female scale form deltoid (owing to arrest of growth by mid-rib or other prominent veins of leaf), length 2.5-3 mm; pure white or partly yellowish-white; low convex or flat; larval exuvia green, margins yellowish; ventral scale very thin adhering to plant (Newstead, 1914).
HOST PLANTS: **Leguminosae**: *Baikiaea insignis* [Newste1914, Balach1958b], *Copaifera copallifera* [Balach1958b]. **Sterculiaceae**: *Cola* [Laing1929a].
DISTRIBUTION: **Afrotropical**: Sierra Leone [Laing1929a]; Uganda [Newste1914, Gowdey1917, Balach1958b].
GENERAL: Description and illustration of adult female by Newstead (1914), Laing (1929a) and by Balachowsky (1958b).
KEYS: Balachowsky 1958b: 269 (female) [Africa].
CITATIONS: Balach1953i [taxonomy: 1513]; Balach1958b [taxonomy, description, illustration, host, distribution: 269-271]; Borchs1966 [catalogue: 230]; CockerRo1915 [taxonomy: 109]; Ferris1941e [taxonomy: 41]; Gowdey1917 [host, distribution: 189]; Laing1929a [taxonomy, description, illustration, host, distribution: 498-499]; MacGil1921 [taxonomy, description, host, distribution: 457]; Newste1914 [taxonomy, description, illustration, host, distribution: 307-308].

Pseudaonidia casuarinae (Maskell)

Aspidiotus casuarinae Maskell, 1894b: 66. Type data: AUSTRALIA: on *Casuarina equisetifolia*; sent by G.H. Brown from Albury. Lectotype female and first instar, by subsequent designation Deitz & Tocker, 1980: 34. Type depository: Canberra: Australian National Insect Collection, CSIRO Entomology, Australia.

Targionia casuarinae; Leonardi, 1900: 314. Change of combination.

Pseudaonidia casuarinae; Marlatt, 1908: 138. Change of combination.

Aspidiotus (*Targionia*) *casuarinae*; Froggatt, 1914: 134. Change of combination.

Targaspidiotus casuarinae; MacGillivray, 1921: 230. Change of combination.

Pseudaonidia casuarinae; Brimblecombe, 1954: 158. Revived combination.

SCALE COVER: Female scale circular, diameter about 1/23 inch; dark yellowish-brown; convex; exuviae yellow. Male scale elongated, length about 1/23 inch; subcylindrical, slightly convex; not carinated; brown, posterior end whitish (Maskell, 1894b).

HOST PLANTS: **Casuarinaceae**: *Casuarina equisetifolia* [Leonar1900, Marlat1908, Brimbl1954]. **Myrtaceae**: *Melaleuca nodosa* [Frogga1914].

DISTRIBUTION: **Australasian**: Australia (New South Wales [Frogga1914, Brimbl1954]).

GENERAL: Description and illustration of adult female by Newstead (1894b) and by Brimblecombe (1954).

KEYS: Marlatt 1908: 135 (female) [World].

CITATIONS: Borchs1966 [catalogue: 230]; Brimbl1954 [taxonomy, description, illustration, host, distribution: 158-160]; Cocker1896b [distribution: 335]; Cocker1897i [taxonomy, description, host, distribution: 26]; DeitzTo1980 [taxonomy: 34]; Fernal1903b [catalogue: 296]; Ferris1941e [taxonomy: 41]; Ferris1943a [taxonomy: 85]; Frogga1914 [taxonomy, description, host, distribution: 134]; Frogga1915 [taxonomy, description, host, distribution: 11]; Leonar1900 [taxonomy, description, host, distribution: 314-315]; MacGil1921 [taxonomy, description, host, distribution: 447]; Marlat1908 [taxonomy, host, distribution: 135,138-139]; Maskel1894b [taxonomy, description, illustration, host, distribution: 66-67].

Pseudaonidia cingulata (Froggatt)

Aspidiotus cingulatus Froggatt, 1914: 135. Type data: AUSTRALIA: Victoria, at Lake Albacutya, on *Casuarina* sp. Syntypes, female. Type depository: Brisbane: Queensland Museum, Queensland, Australia. Notes: Incorrect citation of "Green" as author.

Pseudaonidia cingulata; Brimblecombe, 1958: 65. Change of combination requiring emendation of species name for agreement in gender.

SCALE COVER: Female scales scattered over branchlets and often broader at base than diameter of branchlet, so almost curve round it; dark biscuit brown, with apex often covered with a rounded cap of white secretion, completely hiding large flattened rounded, brown and yellow exuviae; broadly rounded at base, very conical, with apical portion often curving over; diameter 1/12 inch (Froggatt, 1914). Insects singly on branchlets of host; female scale convex to conical, basally curved round branchlet, 2.5 to 3 mm in length; colour fawn; second exuviae whitish, surmounted by yellow first exuviae (Brimblecombe, 1958). Illustration of female scale by Froggatt (1914).

HOST PLANTS: **Casuarinaceae**: *Casuarina* [Frogga1914], *Casuarina cristata* [Brimbl1958].

DISTRIBUTION: **Australasian**: Australia (New South Wales [Brimbl1958], Queensland [Brimbl1958], Victoria [Frogga1914]).

GENERAL: Description and illustration of adult female given by Froggatt (1914) and by Brimblecombe (1958).

CITATIONS: Borchs1966 [catalogue: 230]; Brimbl1958 [taxonomy, description, illustration, host, distribution: 65-67]; Ferris1941e [taxonomy: 42]; Frogga1914 [taxonomy, description, host, distribution: 135]; Frogga1915 [taxonomy, description, host, distribution: 12]; Frogga1933 [taxonomy: 363].

Pseudaonidia corbetti Hall & Williams

Pseudaonidia corbetti Hall & Williams, 1962: 40. Type data: MALAYSIA: Malaya, Balik Pulau, on *Myristica fragrans*; collected 17.VI.1927. Holotype female. Type depository: London: The Natural History Museum, England, UK.

SCALE COVER: Characters of scale unknown to Hall & Williams (1962).

HOST PLANTS: **Myristicaceae**: *Myristica fragrans* [HallWi1962]. **Myrtaceae**: *Eugenia malaccensis* [HallWi1962].

DISTRIBUTION: **Oriental**: Malaysia (Malaya [HallWi1962]).

GENERAL: Description and illustration of adult female by Hall & Williams (1962).

CITATIONS: Borchs1966 [catalogue: 230]; HallWi1962 [taxonomy, description, illustration, host, distribution: 40-41].

Pseudaonidia dentata Brimblecombe

Pseudaonidia dentata Brimblecombe, 1956: 119. Type data: AUSTRALIA: Queensland, Drillham, on *Acacia harpophylla*; J. Mann, April 1953. Holotype female. Type depository: Brisbane: Queensland Museum, Queensland, Australia; type no. T5527.

SCALE COVER: Female scale circular, general colour brownish grey (Brimblecombe, 1956).

HOST PLANTS: **Leguminosae**: *Acacia harpophylla* [Brimbl1956].

DISTRIBUTION: **Australasian**: Australia (Queensland [Brimbl1956]).

BIOLOGY: Insects on twigs (Brimblecombe, 1956).

GENERAL: Description and illustration of adult female by Brimblecombe (1956).

CITATIONS: Borchs1966 [catalogue: 230]; Brimbl1956 [taxonomy, description, illustration, host, distribution: 119-121].

Pseudaonidia dimidiata Brimblecombe

Pseudaonidia dimidiata Brimblecombe, 1956: 117. Type data: AUSTRALIA: Queensland, Gayndah, on *Acacia harpophylla*; collected August 1951. Holotype female. Type depository: Brisbane: Queensland Museum, Queensland, Australia.

SCALE COVER: Female scale circular, up to 2.5 mm diameter, white; second exuviae dark with grey suffusion; first exuviae olive green (Brimblecombe, 1956).

HOST PLANTS: **Leguminosae**: *Acacia harpophylla* [Brimbl1956].
DISTRIBUTION: **Australasian**: Australia (Queensland [Brimbl1956]).
BIOLOGY: Found single and sparse on leaves (Brimblecombe, 1956).
GENERAL: Description and illustration of adult female by Brimblecombe (1956).
CITATIONS: Borchs1966 [catalogue: 230]; Brimbl1956 [taxonomy, description, illustration, host, distribution: 117-118].

Pseudaonidia dryandrae (Fuller)

Aspidiotus dryandrae Fuller, 1897b: 1344. Type data: AUSTRALIA: Western Australia, Swan River, on *Dryandra florabunda*. Holotype female.
Targaspidiotus dryandrae; MacGillivray, 1921: 447. Change of combination.
Pseudaonidia dryandrae; Brimblecombe, 1958: 67. Change of combination.
SYSTEMATICS: Fuller (1897c) again described this species as n. sp.
SCALE COVER: Female scale sub-circular, diameter 0.12 inch Fuller (1897).
HOST PLANTS: **Proteaceae**: *Dryandra florabunda* [Fuller1897b, Brimbl1958].
DISTRIBUTION: **Australasian**: Australia (Western Australia [Fuller1897b, Frogga1914, Brimbl1958]).
GENERAL: Description and illustration of adult female by Brimblecombe (1958).
CITATIONS: Borchs1966 [catalogue: 230]; Brimbl1958 [taxonomy, description, illustration, host, distribution: 67-69]; Cocker1899a [taxonomy: 395]; Fernal1903b [catalogue: 258]; Ferris1941e [taxonomy: 43]; Ferris1943a [taxonomy: 85]; Frogga1914 [taxonomy, description, host, distribution: 311]; Frogga1915 [taxonomy, description, host, distribution: 15]; Fuller1897b [taxonomy, host, distribution: 1344]; Fuller1897c [taxonomy, description, host, distribution: 3]; Fuller1899 [taxonomy: 465]; Lindin1937 [taxonomy: 197]; MacGil1921 [taxonomy, description, host, distribution: 447].

Pseudaonidia duplex (Cockerell)

Aspidiotus theae Maskell, 1891: 6. Nomen nudum; discovered by Deitz & Tocker, 1980: 43.
Aspidiotus theae Maskell, 1891a: 59. Type data: INDIA: Assam, Kangra Valley, on tea plants. Syntypes, female. Type depository: Auckland: New Zealand Arthropod Collection, Landcare Research, New Zealand. Homonym of *Aspidiotus theae* Green, 1890.
Aspidiotus theae; Cockerell, 1896b: 334. Change of combination.
Aspidiotus duplex Cockerell, 1896h: 20. Type data: JAPAN: Tokyo, host plant not indicated, from Takahashi; in Japanese nursery at San Francisco, California, on *Camellia*; and on orange trees from Japan. Syntypes, female. Type depository: Washington: United States National Entomological Collection, U.S. National Museum of Natural History, District of Columbia, USA. Junior synonym replacing a junior homonym.
Aspidiotus (*Pseudaonidia*) *duplex*; Cockerell, 1897i: 20. Change of combination.
Aspidiotus (*Pseudaonidia*) *theae*; Cockerell, 1897i: 28. Change of combination.
Aspidiotus (*Evaspidiotus*) *duplex*; Leonardi, 1898a: 77. Change of combination.
Aspidiotus (*Evaspidiotus*) *theae*; Leonardi, 1898a: 77. Change of combination.

Aspidiotus theae rhododendri Green, 1900a: 67. Type data: SRI LANKA: Nuwara Eliya, on leaves of *Rhododendron arboreum*. Syntypes, female. Type depository: London: The Natural History Museum, England, UK. Synonymy by Borchsenius, 1966: 231.

Pseudaonidia duplex; Fernald, 1903b: 283. Change of combination.

Pseudaonidia rhododendri; Fernald, 1903b: 283. Change of combination.

Pseudaonidia rhododendri thearum Fernald, 1903b: 283. Replacement name for *Aspidiotus theae* Maskell, 1891a: 59; synonymy by Danzig, 1993: 138.

Pseudaonidia trilobitiformis; Borchsenius, 1935a: 19. Misidentification; discovered by Borchsenius, 1966: 231.

Pseudaonidia duplax; Chou, 1985: 249. Misspelling of species name.

COMMON NAMES: camphor scale [Dekle1965c, Borchs1966]; escama alcanfor [CoronaRuMo1997]; Japanese camphor scale [Borchs1966]; Yaponskaya kamphornaya shitovka [Borchs1966].

SCALE COVER: Scale of female dark brown, circular, high convex, exuviae submarginal; scale of male elongate oval, exuvia at one end (Ferris, 1938a). Colour photograph of scale cover by Wong *et al.* (1999).

HOST PLANTS: **Aceraceae**: *Acer palmatum matsumurae* [TakahaTa1956]. **Anacardiaceae**: *Rhus succedanea* [Kuwana1902, Kuwana1907, Marlat1908, Kuwana1933]. **Aquifoliaceae**: *Ilex* [Kawai1977]. **Betulaceae**: *Alnus hirsuta* [TakahaTa1956]. **Ericaceae**: *Rhododendron* [Marlat1908, Ramakr1921a, Green1922, Kuwana1933, Balach1951], *Rhododendron arboreum* [Green1900a, Green1937]. **Fagaceae**: *Castanea pubinervis* [TakahaTa1956], *Castanopsis cuspidata* [Kawai1977], *Quercus* [Kawai1977], *Quercus phillyraeoides* [TakahaTa1956]. **Illiciaceae**: *Illicium religiosum* [TakahaTa1956]. **Lauraceae**: *Cinnamomum camphor* [Kuwana1933, Takagi1970]. **Magnoliaceae**: *Michelia fuscata* [Takaha1929, Takagi1970]. **Moraceae**: *Ficus carica* [TakahaTa1956], *Ficus kingiana* [Takaha1955f]. **Myricaceae**: *Myrica rubra* [Marlat1908, Kuwana1933, TakahaTa1956]. **Oleaceae**: *Ligustrum* [Kawai1977], *Olea fragrans* [Marlat1908], *Osmanthus* [Kawai1977], *Osmanthus fragrans* [Kuwana1933]. **Rosaceae**: *Photinia glabra* [TakahaTa1956], *Pyracantha angustifolia* [TakahaTa1956]. **Rutaceae**: *Citrus* [Kuwana1927, Green1937], *Citrus junos* [TakahaTa1956], *Citrus natsudaidai* [TakahaTa1956]. **Sapindaceae**: *Nephelium litchi* [Takaha1955f]. **Theaceae**: *Camellia* [Green1937, Balach1951, Dekle1965c, McDani1970], *Camellia japonica* [Marlat1908, Borchs1936, TakahaTa1956], *Camellia sinensis* [Marlat1908, Takagi1970], *Eurya japonica* [TakahaTa1956], *Eurya ochracea* [Kuwana1902, Kuwana1907, Marlat1908, Kuwana1933], *Ternstroemia japonica* [TakahaTa1956], *Thea japonica* [Maskel1891a, Green1900c, Kuwana1933, Balach1951], *Thea sasanqua* [Kuwana1933], *Thea sinensis* [Kuwana1933, TakahaTa1956].

NATURAL ENEMIES: COLEOPTERA **Coccinellidae**: *Chilocorus kuwanae* [Muraka1970], *Ch. similis* Rossi [Nakaya1912]. FUNGI **Ascomycotina**: *Nectria aurantiicola* [EvansPr1990], *Ne.flammea* [EvansPr1990]. HYMENOPTERA **Aphelinidae**: *Aphytis cylindratus* Compere [RosenDe1979], *Ap. longicaudus* Rosen & DeBach [RosenDe1979], *Ap. proclia* (Walker) [Gordh1979], *Aspidiotiphagus citrinus* (Craw) [Gordh1979], *Asp. pseudoaonidiae* [Muraka1970], *Azotus perspeciosus* [Muraka1970], *Prospaltella aurantii* (Howard) [Gordh1979], *Pro.*

fasciata Malenotti [Gordh1979]. **Encyrtidae**: *Anabrolepis lindingaspidis* [Muraka1970], *Comperiella bifasciata* Howard [Flande1944a], *Co. unifasciata* Ishii [Trjapi1989]. **Signiphoridae**: *Signiphora flavopalliata* Ashmead [Gordh1979]. LEPIDOPTERA *Platoeceticus gloveri* [Koszta1996]. THYSANOPTERA **Phlaeothripidae**: *Karnyothrips flavipes* (Jones) [PalmerMo1990].

DISTRIBUTION: **Australasian**: Hawaiian Islands (Hawaii [Marlat1908, Green1937]). **Nearctic**: United States of America (Alabama [Ferris1938a], California [Marlat1908], Florida [Dekle1965c], Georgia [Nakaha1982], Louisiana [Ferris1938a], Mississippi [Ferris1938a], Texas [Ferris1938a, McDani1970], Virginia [Nakaha1982]). **Oriental**: India [Green1900c, Marlat1908, Ramakr1921a, Green1937] (Assam [Maskel1891a]); Indonesia (Java [Balach1951]); Sri Lanka [Green1900a, Green1922, Balach1951]; Taiwan [Takaha1929, TakahaTa1956, Takagi1970, WongChCh1999]. **Palaearctic**: China (People's Republic) [Kuwana1927] (Henan (Honan) [Shen1993, Wu1999b]); Georgia (Abkhaz ASSR [Nakaha1982], Adzhar ASSR [Danzig1993, Borchs1936]); Japan [Marlat1908, Kuwana1917a, Kuwana1933, Green1937, Takaha1955f, Takagi1970, Kawai1977, Kawai1980] (Honshu [TakahaTa1956], Kyushu [Kuwana1902, TakahaTa1956], Shikoku [TakahaTa1956]); South Korea [TakahaTa1956].

ECONOMIC IMPORTANCE: Recorded as a pest of ornamental plants (Schmutterer et al., 1957; Beardsley & Gonzalez, 1975; Dekle, 1976; Kosztarab, 1996) and of citrus (Ebeling, 1959; Talhouk, 1975).

GENERAL: Description and illustration of adult female by Maskell (1891a), Leonardi (1898c), Green (1900a), Kuwana (1933), Borchsenius (1937, 1950b), Ferris (1938a), Balachowsky (1951, 1953i), McDaniel (1978), Chou (1985, 1986), Danzig (1993) and by Kosztarab (1996).

KEYS: Kosztarab 1996: 562 (female) [Northeastern North America]; Kosztarab 1996: 562 (female) [Northeastern North America]; Danzig 1993: 136 (female) [Europe]; Kawai 1980: 204 (female) [Japan]; McDaniel 1970: 427 (female) [U.S.A.: Texas]; Balachowsky 1951: 681 (female) [Mediterranean]; Ferris 1942: 39 (female) [North America]; Kuwana 1933: 20-21 (female) [Japan]; Kuwana 1933b: 49 (female) [Japan]; Marlatt 1908: 134 (female) [World].

CITATIONS: Balach1951 [taxonomy, description, illustration, host, distribution: 681-684]; BeardsDaHo1976 [economic importance: 103]; BeardsGo1975 [economic importance: 49]; BenDov1990c [taxonomy: 114]; BerlesLe1898a [taxonomy: 104]; Borchs1935a [taxonomy: 19]; Borchs1936 [host, distribution: 138]; Borchs1937 [taxonomy, description, illustration, host, distribution: 136]; Borchs1939 [taxonomy: 46]; Borchs1949 [taxonomy: 231]; Borchs1950b [taxonomy, description, illustration, host, distribution: 212-214,219]; Borchs1966 [catalogue: 231]; Boyce1948 [host, distribution, economic importance, control]; CatchiWh1924 [host, distribution, life history, ecology: 604-606]; Cendan1937 [host, distribution]; Chen1936 [taxonomy: 211,224]; ChenWo1936 [host, distribution]; Chou1985 [taxonomy, description, host, distribution: 249-250]; Chou1986 [taxonomy, illustration: 652]; ClapsWoGo2001a [taxonomy, host, distribution: 24]; Cocker1896b [distribution: 333,334]; Cocker1896h [taxonomy, description, host, distribution: 20]; Cocker1897i [taxonomy, description, host, distribution: 20,28]; Comper1926 [host, distribution, biological control: 33-50]; Comper1961a [biological control: 17-71]; ComperAn1961 [host, distribution, biological control: 17]; CoronaRuMo1997 [host, distribution: 38-41]; Craw1906 [host, distribution: 139-158]; CressmBlKe1935 [life history, ecology, host: 267-283]; CressmDa1936 [chemical control: 865-878]; CressmKe1933 [life history: 1177-1179]; Danzig1972 [taxonomy, host, distribution, economic importance: 220]; Danzig1993 [taxonomy, description, illustration, host, distribution: 138-139]; DanzigPe1998 [catalogue: 338]; DavidsMi1990 [host, distribution, economic importance: 603-632]; DEDAC1923 [host, distribution]; DeitzTo1980 [taxonomy: 43]; Dekle1964a [host, distribution: 1-2]; Dekle1965c [taxonomy, description, host, distribution: 117]; Dekle1976 [taxonomy, description, host, distribution, economic importance: 134]; Dozier1923 [host, distribution, taxonomy, life history, natural enemies: 1-15]; Ebelin1949 [host, distribution, life history, control]; EHG1897 [host, distribution: 67-85]; EnglisTu1933 [chemical control: 987-989]; EvansPr1990 [biological control: 3-17]; FDACSB1982 [host, distribution: 5-11]; FDACSB1987 [host, distribution: 4-7]; Fernal1903b [catalogue: 283-284]; Ferris1937c [taxonomy, illustration: 52,90]; Ferris1941e [taxonomy: 43,48]; Ferris1942 [taxonomy: 446:39]; Flande1944a [biological control: 365-371]; Flande1971 [biological control, life history: 957-872]; Fleury1934 [host, distribution: 483-500]; FoxWil1939 [host, distribution, economic importance: 2296]; Gordh1979 [biological control: 896,900,907,908,911]; Green1900a [taxonomy, description, illustration, host, distribution: 67-68]; Green1900c [host, distribution: 3]; Green1922 [host, distribution: 463]; Green1937 [host, distribution: 334]; GreenMa1907 [taxonomy, distribution: 344]; HanksDe1998 [life history, ecology: 239-262]; HorticInNo1923a [host, distribution: 237-239]; Kawai1977 [host, distribution, economic importance: 158,161]; Kawai1980 [taxonomy, description, host, distribution: 205]; Koszta1996 [taxonomy, description, illustration, host, distribution, life history, biological control, economic importance: 563-565]; Kuwana1902 [host, distribution: 66]; Kuwana1907 [host, distribution: 194]; Kuwana1917a [taxonomy, distribution: 175]; Kuwana1927 [host, distribution: 71]; Kuwana1933 [taxonomy, description, illustration, host, distribution: 21-23]; Leonar1897 [taxonomy: 285]; Leonar1898a [taxonomy: 77]; Leonar1898c [taxonomy, description, illustration, host, distribution: 80-82]; Leonar1899 [taxonomy: 173]; Lindin1943a [taxonomy: 151]; Lobdel1937 [taxonomy: 78]; MacGil1921 [taxonomy, description, host, distribution: 454]; MaChZh1995 [host, distribution: 117-119]; Marlat1908 [taxonomy, host, distribution: 134,136-137]; Maskel1891 [taxonomy: 6]; Maskel1891a [taxonomy, description, illustration, host, distribution: 59-60]; Maskel1892 [taxonomy: 11]; Maskel1893b [taxonomy, description, host, distribution: 207]; Maskew1914 [host, distribution: 193-194]; McDani1970 [taxonomy, illustration, host, distribution: 427-428]; MillerDa1990 [host, distribution, economic importance: 304]; Miyosh1926 [host, distribution: 303-326]; Muraka1970 [host, distribution, life history: 76]; NagarkSa1990 [host, distribution, economic importance, biological control: 553-542]; Nakaha1982 [host, distribution: 72-73]; Nakaya1912 [host, distribution, biological control: 932-936]; PalmerMo1990 [biological control: 67-76]; Ramakr1921a [host, distribution: 357]; Rose1990c [distribution, economic importance: 535-542]; RosenDe1979 [host, distribution, biological control: 604-607,681-684]; Sassce1923 [host, distribution: 152-158]; SassceWe1923 [chemical control: 84-87]; SassceWe1924 [chemical control: 214-222]; SchildSc1928 [biological control]; SchmutKlLu1957 [host, distribution, economic importance: 493]; Shen1993 [host, distribution: 59]; ShiLi1991 [host, distribution: 166]; Silves1929 [host, distribution: 897-904]; Takagi1970 [taxonomy, host, distribution: 140]; Takagi2000 [taxonomy: 59]; TakagiRo1981 [host, distribution, biological control: 314-321]; Takaha1929 [host, distribution: 79]; Takaha1953a [taxonomy, host, distribution: 10-13]; Takaha1955f [host, distribution: 242]; TakahaTa1956 [host, distribution: 16]; Tanaka1966 [biological control: 1-42]; Tao1999 [taxonomy, host, distribution: 74,111]; Trjapi1989 [biological control: 296]; Watson1926 [host, distribution]; WatsonBe1937 [host, distribution, control]; Whitne1933 [host, distribution: 59-70]; WongChCh1999 [taxonomy, description, host, distribution: 33,76]; Woodwo1909 [host, distribution: 359-360]; Wu1999b [host, distribution: 234].

Pseudaonidia gigantea Balachowsky

Pseudaonidia gigantea Balachowsky, 1953i: 1514. Type data: GUINEA: Bena Plateau, on *Parinarium* [=*Parinari*] *excelsa*. Syntypes, female. Type depository: Paris: Muséum national d'Histoire naturelle, France.

SCALE COVER: Female scale circular, 3.2-4 mm in diameter; size of scale is exceptionally large for Aspidiotini; slightly convex or flat; grey rusty; concentric rings poorly distinct; exuviae subcentral or eccentric, small, yellow. Male scale unknown (Balachowsky, 1953i). Illustration of scale cover by Balachowsky (1953i).

HOST PLANTS: **Chrysobalanaceae**: *Parinari excelsa* [Balach1953i, Balach1958b].

DISTRIBUTION: **Afrotropical**: Guinea [Balach1953i].

GENERAL: Description and illustration of adult female by Balachowsky (1953i, 1958b).

KEYS: Balachowsky 1958b: 269 (female) [Africa].

CITATIONS: Balach1953i [taxonomy, description, illustration, host, distribution: 1514-1517]; Balach1958b [taxonomy, description, illustration, host, distribution: 270-273]; Borchs1966 [catalogue: 231].

Pseudaonidia greeni Marlatt

Pseudaonidia greeni Marlatt, 1908: 138. Type data: INDONESIA: Java, Buitenzorg, on mangosteen (collected Dr. Treub) and at Boro Boedor on *Mangifera indica* (collected by C.L. Marlatt, December 13, 1901). Holotype female. Type depository: Washington: United States National Entomological Collection, U.S. National Museum of Natural History, District of Columbia, USA; type no. 14095.

Entaspidiotus greeni; MacGillivray, 1921: 455. Change of combination.

Furcaspis greeni; Lindinger, 1943b: 220. Change of combination.

Pseudaonidia greeni; Borchsenius, 1966: 231. Revived combination.

SCALE COVER: Scale of adult female long-oval, 3-3.5 by 2 mm; flat; colour opaque, yellowish brown; ventral scale distinct. Male scale similar in structure to female but much smaller (Marlatt, 1908).

HOST PLANTS: **Anacardiaceae**: *Mangifera indica* [Marlat1908].

DISTRIBUTION: **Oriental**: Indonesia (Java [Marlat1908, Sander1909a]).

GENERAL: Description of adult female by Marlatt (1908).
KEYS: Marlatt 1908: 134 (female) [World].
CITATIONS: Borchs1966 [catalogue: 231]; Lindin1943b [taxonomy: 220]; MacGil1921 [taxonomy, description, host, distribution: 455]; Marlat1908 [taxonomy, description, host, distribution: 140-141]; Sander1909a [taxonomy, host, distribution: 54].

Pseudaonidia irrepta Rutherford

Pseudaonidia irrepta Rutherford, 1914: 261. Type data: SRI LANKA: Paradeniya, on branches of undetermined plant (possibly *Acalypha* sp.). Type depository: Horticultural Crop Research and Development Institute, Paradeniya, Sri Lanka Holotype female.
Duplaspidiotus irreptus; MacGillivray, 1921: 453. Change of combination requiring emendation of species name for agreement in gender.
Aspidiotus irrepta; Green, 1922: 462. Change of combination.
Pseudaonidia irrepta; Borchsenius, 1966: 231. Revived combination.
SCALE COVER: Female scale completely concealed under bark of plant (Rutherford, 1914).
HOST PLANTS: **Euphorbiaceae**: *Acalypha* [Ruther1914, Green1922].
DISTRIBUTION: **Oriental**: Sri Lanka [Ruther1914, Ramakr1921a, Green1922].
CITATIONS: Borchs1966 [catalogue: 231]; Ferris1941e [taxonomy: 44]; Green1922 [taxonomy, host, distribution: 462]; Lindin1932f [taxonomy: 196]; MacGil1921 [taxonomy, description, host, distribution: 452-453]; Ramakr1921a [host, distribution, taxonomy: 359]; Ruther1914 [taxonomy, description, host, distribution: 261-262].

Pseudaonidia lycii Brain

Pseudaonidia lycii Brain, 1919: 210. Type data: SOUTH AFRICA: Cape Province, Uitenhage, on *Lycium afrum*; collected by C.P. Lounsbury, 1.viii.1906. Lectotype female, by subsequent designation Munting, 1970a: 39. Type depository: Pretoria: South African National Collection of Insects, South Africa; type no. 155/1.
SCALE COVER: Scale of adult female about 1.6 mm in diameter, circular or somewhat elongate, moderately convex; sordid buff, but usually obscured by outer layers or bark of host; exuviae central, covered; dull yellow in rubbed specimens; ventral scale very delicate and remains attached to host plant (Brain, 1919).
HOST PLANTS: **Solanaceae**: *Lycium afrum* [Brain1919, Muntin1970a].
DISTRIBUTION: **Afrotropical**: South Africa [Brain1919, Muntin1970a].
GENERAL: Description and illustration of adult female by Brain (1919).
KEYS: Brain 1919: 206 (female) [South Africa].
CITATIONS: Brain1919 [taxonomy, description, illustration, host, distribution: 210-211]; Ferris1937c [taxonomy: 50]; Muntin1970a [taxonomy: 39].

Pseudaonidia manilensis Robinson

Pseudaonidia manilensis Robinson, 1918: 146. Type data: PHILIPPINES: Luzon, Manila, on *Samanea saman*. Holotype female. Type depository: Washington, D.C., U.S. Entomological Collection, U.S. National Museum of Natural History, USA.
SCALE COVER: Female scale circular to subcircular, 1.5-1.75 mm diameter; convex; dark brown; exuviae yellowish brown, lateral to subcentral (Robinson, 1918).
HOST PLANTS: **Anacardiaceae**: *Campnosperma* [Beards1966]. **Euphorbiaceae**: *Glochidion* [Beards1966]. **Leguminosae**: *Samanea saman* [Robins1918].
DISTRIBUTION: **Australasian**: Palau [Beards1966]. **Oriental**: Philippines (Luzon [Robins1918]).
GENERAL: Description and illustration of adult female by Robinson (1918).
CITATIONS: Beards1966 [host, distribution: 524]; Borchs1966 [catalogue: 231]; Robins1918 [taxonomy, description, illustration, host, distribution: 146].

Pseudaonidia marquesi Costa Lima

Pseudaonidia marquesi Costa Lima, 1924: 141. Type data: BRAZIL: Rio de Janeiro, on *Averrhoa carambola*. Syntypes, female. Type depository: Rio de Janeiro: Fundacao Instituto Oswaldo Cruz, Rio de Janeiro, Brazil.
Duplaspidiotus marquesi; Costa Lima, 1942: 278. Change of combination.
Pseudaonidia marquesi; Borchsenius, 1966: 231. Revived combination.
SCALE COVER: Female scale circular or almost circular, 1.5-2 mm in diameter; convex; grey or black; dorsal surface rugose, with concentric rings, robust, fragile; larval exuviae situated at apex, colour yellow, shining; ventral vellum complete, robust, attached to host plant, colour similar to dorsal scale (Costa Lima, 1924).
HOST PLANTS: **Averrhoaceae**: *Averrhoa carambola* [CostaL1924, ClapsWoGo2001]. **Sapotaceae**: *Lucuma caimito* [Lepage1938, ClapsWoGo2001]. **Vitaceae**: *Vitis* [Lepage1938, ClapsWoGo2001].
DISTRIBUTION: **Neotropical**: Brazil (Rio de Janeiro [CostaL1924, Lepage1938, ClapsWoGo2001], Santa Catarina [ClapsWoGo2001]).
GENERAL: Description and illustration of adult female by Costa Lima (1924).
CITATIONS: Borchs1966 [catalogue: 231]; ClapsWoGo2001 [host, distribution: 243]; CostaL1924 [taxonomy, description, illustration, host, distribution: 141]; CostaL1942 [taxonomy: 278]; Lepage1938 [catalogue: 418].

Pseudaonidia obsita Cockerell & Robinson

Pseudaonidia obsita Cockerell & Robinson, 1915: 109. Type data: PHILIPPINES: Los Banos, on underside of leaves of *Ficus caudatifolia*. Type depository: United States National Entomological Collection, U.S. National Museum of Natural History, Washington, D.C., USA. Holotype female.
Furcaspis obsita; Lindinger, 1932c: 204. Change of combination.
Pseudaonidia obsita; Borchsenius, 1966: 232. Revived combination.
COMMON NAME: isis scale [VelasqRi1969].

SCALE COVER: Female scale circular, about 2.5 mm in diameter; slightly convex; appearing brownish-black, but actually light brownish-pink; with a thick covering of fungous growth; exuviae yellowish-fulvous, sublateral; ventral scale thick, white; occasionally scales white (Cockerell & Robinson, 1915).
HOST PLANTS: **Moraceae**: *Ficus caudatifolia* [CockerRo1915, Robins1917].
DISTRIBUTION: **Oriental**: Philippines [CockerRo1915, Robins1917].
GENERAL: Description and illustration of adult female by Cockerell & Robinson (1915).
KEYS: Robinson 1917: 33 (female) [Philippines].
CITATIONS: Borchs1966 [catalogue: 232]; CockerRo1915 [taxonomy, description, illustration, host, distribution: 109-110]; Ferris1941e [taxonomy: 46]; Green1918 [taxonomy: 233]; Lindin1932c [taxonomy: 204]; Robins1917 [taxonomy, description, host, distribution: 33]; VelasqRi1969 [host, distribution: 195-208].

Pseudaonidia paeoniae (Cockerell)

Aspidiotus duplex paeoniae Cockerell, 1899i: 105. Type data: U.S.A.: California, San Francisco, on bark of peony, on *Camellia japonica*, and on *Camellia* sp. Syntypes, female. Type depository: Washington: United States National Entomological Collection, U.S. National Museum of Natural History, District of Columbia, USA.
Aspidiotus (*Evaspidiotus*) *duplex peoniae*; Leonardi, 1900: 340. Change of combination.
Aspidiotus (*Evaspidiotus*) *duplex peoniae*; Leonardi, 1900: 340. Misspelling of species name.
Pseudaonidia paeoniae; Fernald, 1903b: 293. Change of combination and rank.
Aspidiotus paconiae; Kuwana, 1907: 197. Misspelling of species name.
Pseudaonidiella paeoniae; MacGillivray, 1921: 454. Change of combination.
Pseudaonidia theae; Borchsenius, 1937a: 39. Misidentification; discovered by Danzig, 1993: 136.
Aspidiotus paeoniae; Ferris, 1937c: 52. Change of combination.
Pseudaonidia poeoniae; Pegazzano, 1949: 233. Misspelling of species name.
COMMON NAMES: Japanese camellia scale [Dzhash1989]; Peony Scale [Merril1953, Dekle1965c, Borchs1966].
SCALE COVER: Cockerell (1899i) did not describe scale cover. Ferris (1938a) described: "Female scale dark grey or brown; circular, high convex; exuviae subcentral; scale of male elongate, exuvia near one end". Takagi (1974) described: "Female scale is circular, but the expected concentric pattern of growth is rather obscure owing to the agglutination of filaments. The inner surface appears more irregular as to the running direction of filaments (which are secreted for lining), but some filaments may be seen running along the margin. The central area is further lined by fine filaments, which may be produced by microducts occurring submedially and submarginally in the prepygidial region. The male test is smaller and elongate, but, when magnified, much the same as the female test in both the dorsal and ventral views".

HOST PLANTS: **Aquifoliaceae**: *Ilex* [Balach1951, Danzig1993], *Ilex colchica* [Hadzib1983], *Ilex latifolia* [Kuwana1902, Marlat1908]. **Buxaceae**: *Buxus microphylla japonica* [TippinClBa1977]. **Caprifoliaceae**: *Viburnum* [Kawai1977]. **Clethraceae**: *Clethra barbineros* [Kuwana1902]. **Cornaceae**: *Cornus* [Kawai1977]. **Ericaceae**: *Azalea* [Marlat1908, Pegazz1949, BesheaTiHo1973], *Rhododendron* [Marlat1908, Danzig1993], *Rhododendron indicum* [Kuwana1902, Dekle1965c]. **Moraceae**: *Morus bombicis* [Kawai1977]. **Oleaceae**: *Osmanthus* [Kawai1977]. **Paeoniaceae**: *Paeonia* [Ferris1938a, McDani1970]. **Solanaceae**: *Solanum tuberosum* [TippinHoBe1981]. **Theaceae**: *Camellia* [Cocker1899i, Balach1951, Dekle1965c, McDani1970, Kawai1977, TippinHoBe1981], *Camellia japonica* [Cocker1899i, Leonar1900, Marlat1908, Pegazz1949, McDani1970, TippinBe1970, BesheaTiHo1973], *Camellia sasanqua* [Takagi1990, Takagi1990a], *Eurya ochracea* [Kuwana1902], *Thea japonica* [Kuwana1902, Kuwana1933, Ferris1938a, Balach1951], *Thea sinensis* [Kuwana1902, Kuwana1933, Hadzib1983]. **Ulmaceae**: *Celtis sinensis* [Kawai1977].

NATURAL ENEMIES: COLEOPTERA **Coccinellidae**: *Chilocorus kuwanae* [Muraka1970], *Chilocorus similis* Rossi [Nakaya1912]. HYMENOPTERA **Encyrtidae**: *Anabrolepis extranea* [Muraka1970], *Encarsia citrina* (Crawford) [Dzhash1989], *Epitetracnemus zetterstedtii* (Westwood) [Trjapi1989].

DISTRIBUTION: **Nearctic**: U.S.A. (Alabama [BesheaTiHo1973], Arkansas [Nakaha1982], California [Cocker1899i], Delaware [Nakaha1982], District of Columbia [Nakaha1982], Florida [Merril1953, Dekle1965c], Georgia [TippinBe1970, BesheaTiHo1973, TippinClBa1977, TippinHoBe1981], Louisiana [Nakaha1982], Maryland [Nakaha1982], Mississippi [Ferris1938a], Missouri [Nakaha1982], New Jersey [Nakaha1982], New York [Nakaha1982], North Carolina [Nakaha1982], Pennsylvania [Nakaha1982], South Carolina [Nakaha1982], Texas [McDani1970], Virginia [Nakaha1982]). **Oriental**: China (People's Republic) (Yunnan [Ferris1953]). **Palaearctic**: China (People's Republic) (Henan (Honan) [Shen1993]); Georgia (Adzhar ASSR [Hadzib1983, Danzig1993], Georgia [Dzhash1989]); Italy [Pegazz1949, LongoMaPe1995]; Japan [Kuwana1902, Kuwana1917a, Kawai1980, Takagi1990, Takagi1990a].

BIOLOGY: Occurring on bark (Ferris, 1938a).

ECONOMIC IMPORTANCE: Pest of ornamental plants in various regions (English & Turnipseed, 1940; Davidson & Miller, 1990; Kosztarab, 1996).

GENERAL: Description and illustration of adult female by Kuwana (1933), Ferris (1938a), Balachowsky (1951), Tippins *et al.* (1981), Chou (1985, 1986), Danzig (1993) and by Kosztarab (1996). Description and illustration of first instar nymph by Takagi (2000).

KEYS: Kosztarab 1996: 564 (female) [Northeastern North America]; Danzig 1993: 136 (female) [Europe]; Kawai 1980: 204 (female) [Japan]; McDaniel 1970: 427 (female) [U.S.A.: Texas]; Balachowsky 1951: 681 (female) [Mediterranean]; Ferris 1942: 39 (female) [North America]; Kuwana 1933b: 49 (female) [Japan]; Marlatt 1908: 135 (female) [World].

CITATIONS: Balach1951 [taxonomy, description, illustration, host, distribution: 687-689]; BeardsDaHo1976 [economic importance: 106]; BesheaTiHo1973 [host, distribution: 7]; Borchs1937 [taxonomy, description, illustration, host, distribution:

136]; Borchs1950b [taxonomy, description, illustration, host, distribution: 214,219]; Borchs1966 [catalogue: 232]; BurgerUl1990 [economic importance: 313-327]; Chou1985 [taxonomy, description, host, distribution: 251-253]; Chou1986 [taxonomy, illustration: 653]; Cocker1899i [taxonomy, description, host, distribution: 106]; Danzig1972 [taxonomy, host, distribution, economic importance: 220]; Danzig1993 [taxonomy, description, illustration, host, distribution: 136-138]; DanzigPe1998 [catalogue: 338-339]; DavidsMi1990 [host, distribution, economic importance: 603-632]; Dekle1965c [taxonomy, description, host, distribution: 118]; Dekle1976 [taxonomy, description, host, distribution, economic importance: 135]; Dzhash1989 [economic importance, host, distribution, chemical control, biological control: 93-101]; EnglisTu1940 [host, distribution, economic importance, life history, control: 3-18]; Fernal1903b [catalogue: 283]; Ferris1937c [taxonomy, illustration: 52,90]; Ferris1938a [taxonomy, description, illustration, host, distribution: 254]; Ferris1941e [taxonomy: 46]; Ferris1942 [taxonomy: 446:39]; Ferris1953 [host, distribution: 67]; FoxWil1939 [host, distribution, economic importance: 2296]; Hadzib1983 [taxonomy, description, host, distribution, economic importance: 215-216]; HowellTi1990 [taxonomy, structure: 38]; Hunt1939 [host, distribution: 548-566]; Kawai1977 [host, distribution, economic importance: 159,161,162]; Kawai1980 [taxonomy, description, host, distribution: 204-205]; Koszta1996 [taxonomy, description, illustration, host, distribution, life history, biological control, economic importance: 564-566]; Kuwana1902 [host, distribution: 66-67]; Kuwana1907 [taxonomy, host, distribution: 194]; Kuwana1917a [taxonomy, distribution: 175]; Kuwana1933 [taxonomy, description, illustration, host, distribution: 23-24]; Leonar1900 [taxonomy, host, distribution: 340]; LongoMaPe1995 [distribution: 128]; Lupo1948 [taxonomy, description, illustration, host, distribution: 17-23]; Lupo1953 [taxonomy, description, illustration, host, distribution: 17-23]; MacGil1921 [taxonomy, description, host, distribution: 454]; Marlat1908 [taxonomy, host, distribution: 138]; Maskew1914 [host, distribution: 193-194]; McDani1970 [taxonomy, host, distribution: 429]; Melis1949 [host, distribution: 17-25]; Merkel1938 [host, distribution: 88-99]; Merril1953 [taxonomy, description, host, distribution: 74-75]; MillerDa1990 [host, distribution, economic importance: 304]; Muraka1970 [host, distribution: 76]; NagarkSa1990 [host, distribution, economic importance, biological control: 553-542]; Nakaha1982 [host, distribution: 73]; Nakaya1912 [host, distribution, biological control: 932-936]; Pegazz1949 [taxonomy, host, distribution: 233-235]; Pegazz1953 [taxonomy, description, host, distribution, biological control: 281-315]; Sassce1923 [host, distribution: 152-158]; SchmutKlLu1957 [host, distribution, economic importance: 493]; Shen1993 [host, distribution: 59]; StoetzDa1974 [taxonomy, life history: 138-140]; Takagi1990 [taxonomy, structure: 82,100,112]; Takagi1990a [taxonomy, structure: 25,79-80]; Takagi1990b [taxonomy, structure: 12,14]; Takagi2000 [taxonomy, structure: 59,86]; Tao1999 [taxonomy, host, distribution: 111]; TippinBe1970 [host, distribution: 11]; TippinClBa1977 [host, distribution, life history: 68-71]; TippinHoBe1981 [taxonomy, description, illustration, host, distribution: 356-361]; Trjapi1989 [biological control: 292]; Willia1985a [taxonomy: 234].

Pseudaonidia tricuspidata Lindinger

Pseudaonidia tricuspidata Lindinger, 1939: 37. Type data: GERMANY: Hamburg, Botanical Gardens, on *Dillenia indica*. Syntypes, female. Notes: No type material of this species listed by Weidner & Wagner (1968) from Hamburg collection.

SCALE COVER: Female scale thick, highly convex. Male scale circular 3 mm in diameter; convex; ventral scale thick, white (Lindinger, 1939).

HOST PLANTS: **Dilleniaceae**: *Dillenia indica* [Lindin1939].

DISTRIBUTION: **Palaearctic**: Germany (United) [Lindin1939].

GENERAL: Description of adult female by Lindinger (1939) and by Schmutterer (1959).

CITATIONS: Borchs1966 [catalogue: 232]; DanzigPe1998 [catalogue: 339]; Lindin1939 [taxonomy, description, host, distribution: 37-38]; Schmut1959 [taxonomy, description, host, distribution: 113].

Pseudaonidia trilobitiformis (Green)

Aspidiotus trilobitiformis Green, 1896: 4. Type data: SRI LANKA: Pundaluoya, on leaves of unidentified tree. Holotype female. Type depository: London: The Natural History Museum, England, UK.

Aspidiotus (*Pseudaonidia*) *trilobitiformis*; Cockerell, 1897i: 28. Change of combination.

Aspidiotus darutyi Charmoy, 1898: 278. Type data: MAURITIUS: on various plants, especially common on mango. Syntypes, female. Synonymy by Borchsenius, 1966: 232.

Aspidiotus (*Evaspidiotus*) *trilobitiformis*; Leonardi, 1898a: 77. Change of combination.

Pseudaonidia trilobitiformis; Cockerell, 1899a: 396. Change of combination.

Pseudaonidia trilobitiformis darutyi; Fernald, 1903b: 284. Change of status.

Aspidiotus trilobiformis; Kuwana, 1907: 194. Misspelling of species name.

Pseudaonidia darutyi; Marlatt, 1908: 137. Change of combination and rank.

Pseudoaonidia trilobitiformis; Leonardi, 1914: 203. Misspelling of genus name.

Aspidiotus daruyi; Borchsenius, 1966: 232. Misspelling of species name.

COMMON NAME: gingging scale [VelasqRi1969].

SCALE COVER: Green (1896e) described scale cover: "Female scale large, 3-4.5 mm in diameter; almost flat; usually semicircular or deltoid from arrest of growth by prominent veins of the leaf, seldom circular; the insect apparently preferring a position close to the midrib of the leaf; colour pale reddish brown; pellicles yellow, the second usually depressed. Male scale unknown". Brain (1919) described scale cover: "Female scale large (may reach 4 mm in diameter), flat, sometimes circular, but more often with one side flattened against a vein of a leaf; colour is usually brown or reddish brown, but in old exposed specimens it is commonly more or less bleached; exuviae are flat and usually yellowish to brownish". Colour illustration of scale cover by Green (1896e). Colour photograph of scale cover by Wong *et al.* (1999).

HOST PLANTS: **Agavaceae**: *Agave mexicana* [Marlat1908], *Cordyline* [WilliaWa1988], *Cordyline neo-caledonyca* [WilliaWa1988], *Dracaena* [MatileNo1984]. **Anacardiaceae**: *Anacardium* [Green1937, MatileNo1984],

Anacardium occidentale [Leonar1914, Mamet1943a, Mamet1950, Almeid1973b, WilliaBu1987, WilliaWa1988], *Mangifera* [Marlat1908, Green1937], *Mangifera indica* [Charmo1898, GrandpCh1899, Marlat1908, Lepage1938, Mamet1943a, Mamet1951, WolffCo1993a], *Nothopegia colebrookiana* [Green1900a, Marlat1908, Ramakr1919a, Ramakr1930, Green1937], *Schinus molle* [Mamet1957], *Schinus terebinthifolius* [Mamet1957, Cohic1958, WilliaWa1988], *Sclerocarya caffra* [Mamet1950, Mamet1959a, Borchs1966]. **Annonaceae**: *Annona* [Lepage1938, WilliaWa1988], *Annona reticulata* [Mamet1956, Borchs1966], *Annona squamosa* [Cohic1958, WilliaWa1988], *Cananga odoratum* [Matile1978]. **Apocynaceae**: *Acokanthera abessinica* [Marlat1908], *Carissa carandas* [Mamet1943a, Borchs1966], *Carissa madagascariensis* [Mamet1950, Borchs1966], *Carissophyllum* [Leonar1914], *Catharanthus roseus* [Cohic1958, WilliaWa1988], *Cerbera oppositifolia* [WilliaWa1988], *Echites* [Ramakr1930], *Nerium* [Leonar1914], *Nerium indicum* [Almeid1973b], *Nerium oleander* [Marlat1908, Lepage1938, Mamet1943a, Cohic1958, Borchs1966, WilliaWa1988], *Ochrosia oppositifolia* [Cohic1958], *Plumeria acutifolia* [Mamet1943a, Borchs1966, WilliaWa1988], *Plumeria rubra* [WilliaBu1987, WilliaWa1988], *Thevetia* [Balach1958b], *Thevetia peruviana* [Almeid1971], *Trachelospermum foetidum* [Takagi1969a]. **Araceae**: *Monstera deliciosa* [Mamet1943a, Borchs1966], *Philodendron* [Seabra1925], *Pothos aureus* [Cohic1958, WilliaWa1988]. **Bignoniaceae**: *Crescentia cujete* [Cohic1958, WilliaWa1988], *Pyrostegia venusta* [Cohic1958, WilliaWa1988]. **Bromeliaceae**: *Ananas sativa* [Balach1958b]. **Caricaceae**: *Carica papaya* [Cohic1958, WilliaWa1988]. **Combretaceae**: *Terminalia arjuna* [Mamet1943a, Borchs1966], *Terminalia catappa* [Mamet1943a, Borchs1966]. **Corylaceae**: *Corylus* [Lepage1938]. **Ebenaceae**: *Diospyros* [Ramakr1930], *Diospyros eriantha* [Takaha1935, Takagi1969a], *Diospyros kaki* [Lepage1938]. **Ehretiaceae**: *Cordia myxa* [Mamet1943a, Borchs1966]. **Euphorbiaceae** [Mamet1959a, Borchs1966], *Aleurites* [Almeid1971], *Aleurites fordi* [Mamet1954, Borchs1966], *Aleurites moluccana* [Cohic1958, WilliaWa1988], *Aleurites montana* [Cohic1958, WilliaWa1988], *Codiaeum* [Laing1933, Cohic1958, WilliaWa1988], *Gelonium lanceolatum* [Ramakr1930], *Hura crepitans* [Mamet1954, Borchs1966], *Jatropha curcas* [Mamet1943a, Borchs1966]. **Fagaceae**: *Quercus* [Takaha1934, Takagi1969a]. **Flacourtiaceae**: *Flacourtia ramontchi* [Mamet1950, Mamet1959a, Borchs1966], *Hydnocarpus wightiana* [Mamet1943a, Borchs1966], *Scolopia oldhamii* [Takagi1969a]. **Fumariaceae**: *Fumaria* [Balach1958b]. **Guttiferae**: *Calophyllum inophyllum* [Takaha1929, Cohic1958, Takagi1969a, WilliaWa1988]. **Hydrangeaceae**: *Hydrangea* [Takaha1933, Takagi1969a]. **Lauraceae**: *Cinnamomum zeylanicum* [Seabra1925], *Laurus nobilis* [Lepage1938, Cohic1958, RosenDe1979, WilliaWa1988], *Machilus* [Takaha1929, Takaha1933, Takagi1969a], *Persea* [WilliaWa1988], *Persea americana* [Cohic1958, WilliaBu1987, WilliaWa1988], *Persea gratissima* [Lepage1938]. **Lecythidaceae**: *Barringtonia asiatica* [Cohic1958, WilliaWa1988]. **Leguminosae**: *Acacia simplicifolia* [Cohic1958, WilliaWa1988], *Acacia spirorbis* [Cohic1958, WilliaWa1988], *Bauhinia* [Mamet1943a, Mamet1959a, Borchs1966, Matile1978], *Bauhinia variegata* [Cohic1958, WilliaWa1988], *Cassia* [Mamet1959a, Borchs1966], *Cassia siamea* [Almeid1971, Almeid1973b], *Cassia* *spectabilis* [MatileNo1984], *Clitoria terneata* [Cohic1958, WilliaWa1988], *Crotalaria* [Mamet1950, Borchs1966], *Dalbergia* [Ramakr1921a], *Dalbergia championii* [Green1896e, Marlat1908, Ramakr1919a, Green1937], *Mucuna bennettii* [Cohic1958, WilliaWa1988]. **Liliaceae**: *Taetsia neocaledonica* [Cohic1958]. **Magnoliaceae**: *Michelia champaca* [Matile1978]. **Malvaceae**: *Hibiscus* [MatileNo1984]. **Marantaceae**: *Maranta* [WilliaWa1988]. **Meliaceae**: *Xylocarpus obovatus* [Balach1958b]. **Moraceae**: *Artocarpus* [Robins1917, WilliaBu1987, WilliaWa1988], *Art. altilis* [Cohic1958, WilliaWa1988], *Art. communis* [Mamet1943a, Borchs1966], *Art. heterophyllus* [Takagi1969a, WilliaWa1988], *Art. incisa* [WilliaWa1988], *Art. integrifolius* [Takaha1929, Lepage1938, Mamet1943a, Mamet1959a, Cohic1958, Borchs1966], *Cudrania cochinchinensis* [Takagi1969a], *Ficus* [GrandpCh1899, Takaha1929, Lepage1938, Mamet1956, Mamet1959a, Borchs1966, Takagi1969a, WilliaWa1988, Takaha1942b], *Fi. awkeotsang* [Takaha1932a, Takaha1933], *Fi. benghalensis* [Mamet1943a, Borchs1966], *Fi. elastica* [Green1916, Takagi1969a], *Fi. pumila* [Takaha1932a, Takaha1933, Cohic1958, Takagi1969a, WilliaWa1988], *Fi. religiosa* [Mamet1943a, Borchs1966], *Fi. repens* [Mamet1943a, Borchs1966], *Fi. retusa* [Takaha1929, Takagi1969a], *Fi. scandens* [Ruther1915a, Marlat1908], *Fi. swinhoei* [Takagi1969a], *Fi. thonningii* [Balach1958b], *Fi. trichoclada* [Mamet1950, Borchs1966], *Fi. wightiana japonica* [Kuwana1933]. **Myrtaceae** [Marlat1908], *Eugenia* [Mamet1950, Mamet1957, Mamet1959a, Borchs1966], *Eugenia jaboticaba* [Lepage1938], *Myrtus* [Balach1951], *Psidium* [WilliaWa1988], *Psidium cattleianum* [Cohic1958, WilliaWa1988], *Psidium guajava* [Lepage1938, Mamet1943a, Cohic1958, Borchs1966, DanzigKo1990]. **Naucleaceae**: *Cephalanthus* [Mamet1959a, Borchs1966]. **Nyctaginaceae**: *Bougainvillea* [Almeid1973b]. **Oleaceae**: *Jasminum* [Green1937]. **Palmae**: *Cocos nucifera* [Mamet1943a, Borchs1966], *Dictyosperma alba* [Mamet1943a, Borchs1966], *Elaeis guineensis* [Almeid1973b], *Hyphaene thebaica* [Balach1958b]. **Passifloraceae**: *Passiflora* [WilliaWa1988], *Passiflora edulis* [Mamet1943a, Cohic1958, Borchs1966, WilliaWa1988], *Passiflora laurifolia* [Cohic1958, WilliaWa1988], *Passiflora quadrangularis* [Cohic1958, WilliaWa1988]. **Pittosporaceae**: *Pittosporum* [Matile1978]. **Punicaceae**: *Punica granatum* [Mamet1943a, Borchs1966]. **Rhamnaceae**: *Ziziphus* [Mamet1959a, Borchs1966], *Ziziphus spina-christi* [Mamet1950, Borchs1966]. **Rosaceae**: *Eriobotrya japonica* [Cohic1958, WilliaWa1988], *Mespilus germanica* [Almeid1971], *Prunus domestica* [Lepage1938], *Pyrus* [Lepage1938], *Rosa* [Mamet1943a, Borchs1966, Almeid1973b]. **Rubiaceae**: *Coffea* [Mamet1943a, Mamet1950, Borchs1966, Almeid1973b], *Cof. arabica* [Mamet1943a, Borchs1966], *Cof. liberica* [Charmo1898], *Ixora* [Ruther1915a, Ramakr1930], *Ixora coccinia* [Green1905a, Ramakr1919a, Ramakr1930]. **Rutaceae**: *Citrus* [Ferris1921a, Takaha1929, Borchs1934, Borchs1936, Lepage1938, Mamet1959a, Borchs1966], *Ci. aurantium* [Borchs1934, WilliaWa1988], *Ci. aurantium bigaradia* [Matile1978], *Ci. bergamia* [Matile1978], *Ci. decumana* [Green1916], *Ci. grandis* [Marlat1908, WilliaWa1988], *Ci. histrix* [Matile1978], *Ci. limon* [Almeid1971, Almeid1973b, WilliaWa1988], *Ci. maxima* [WilliaBu1987, WilliaWa1988], *Ci. nobilis unchiu* [Borchs1934], *Ci. sinensis* [Takagi1969a], *Murraya exotica* [Charmo1898, Marlat1908, Takaha1929,

Mamet1943a, Borchs1966, Takagi1969a, WilliaWa1988]. **Santalaceae**: *Santalum austro-caledonicum* [Cohic1958, WilliaWa1988]. **Sapindaceae**: *Dodonaea viscosa* [Cohic1958, WilliaWa1988], *Euphoria longana* [Charmo1898, Marlat1908, Mamet1943a, Borchs1966], *Litchi sinensis* [Brain1919, Mamet1943a, Borchs1966]. **Sapotaceae**: *Achras sapota* [Lepage1938], *Mimusops* [Mamet1943a, Borchs1966], *Mimusops elengi* [Ramakr1919a, Ramakr1930]. **Solanaceae**: *Capsicum* [WilliaWa1988], *Capsicum annuum* [WilliaWa1988, Cohic1958], *Capsicum frutescens* [Cohic1958, WilliaWa1988]. **Sterculiaceae**: *Theobroma cacao* [Seabra1925, Lepage1938, Matile1978, WilliaBu1987, WilliaWa1988]. **Theaceae**: *Camellia* [Mamet1943a, Borchs1966], *Camellia japonica* [Takagi1969a], *Eurya japonica* [Takagi1969a], *Thea japonica* [Takaha1929]. **Thymelaeaceae**: *Peddiea africana* [Ruther1915a]. **Tiliaceae**: *Grewia* [DeLott1967a]. **Verbenaceae**: *Premna* [WilliaBu1987, WilliaWa1988], *Tectona grandis* [Mamet1943a, Borchs1966]. **Vitaceae**: *Vitis vinifera* [Mamet1943a, Borchs1966].

NATURAL ENEMIES: FUNGI **Ascomycotina**: *Nectria flammea* [EvansPr1990]. HYMENOPTERA **Aphelinidae**: *Aphytis chrysomphali* (Mercet) [RosenDe1979], *Ap. costalimai* (Gomes) [RosenDe1979], *Ap. cylindratus* Compere [RosenDe1979], *Ap. lingnanensis* Compere [RosenDe1979], *Ap. longicaudus* Rosen & DeBach [RosenDe1979]. **Signiphoridae**: *Signiphora flavella* [Woolle1990].

DISTRIBUTION: **Afrotropical**: Angola [Almeid1969, Almeid1973b]; Benin [Balach1958b]; Cameroon [Balach1958b, MatileNo1984]; Central African Republic [Balach1958b]; Chad [Nakaha1982]; Comoros Islands [Matile1978]; Côte d'Ivoire (=Ivory Coast) [Balach1958b]; Guinea [Balach1958b]; Kenya [DeLott1967a]; Liberia [Marlat1908, Balach1958b]; Madagascar [Mamet1943a, Mamet1950, Mamet1951, Mamet1954, Mamet1959a, Borchs1966]; Malawi [Nakaha1982]; Mauritius [Charmo1898, GrandpCh1899, Mamet1943a, Mamet1949, Borchs1966]; Mozambique [Almeid1971]; Nigeria [Nakaha1982]; Rodrigues Island [Nakaha1982]; Réunion [Mamet1957]; São Tomé and Príncipe (Príncipe) [Seabra1921, Castel1963, Fernan1993], São Tomé and Príncipe [Seabra1921, Seabra1925, Balach1958b]); Senegal [Balach1958b]; Seychelles [Marlat1908, Mamet1943a, Borchs1966]; Sierra Leone [Nakaha1982]; Somalia [Balach1958b]; South Africa [BrainKe1917, Brain1919, Balach1958b]; Tanzania [Newste1911a, Mamet1956, Balach1958b, Borchs1966]; Togo [Nakaha1982]; Uganda [Newste1911, Gowdey1917, Newste1917b]; Zaire [Ghesqu1932, Liegeo1944, Balach1958b]; Zanzibar [Balach1958b]. **Australasian**: Australia [Nakaha1982]; Fiji [Nakaha1982]; New Caledonia [Laing1933, Cohic1958, RosenDe1979]; Vanuatu [WilliaBu1987] [WilliaBu1987]. **Nearctic**: United States of America (Florida [Hamon1980d]). **Neotropical**: Argentina [Nakaha1982]; Bahamas [Nakaha1982]; Bolivia [Nakaha1982]; Brazil [Marlat1908, Lepage1938, RosenDe1979] (Bahia [Hempel1900a], Cear [WolffCo1993], Minas Gerais [WolffCo1993], Paraiba [WolffCo1993], Parana [WolffCo1993], Pernambuco [WolffCo1993], Rio Grande do Norte [WolffCo1993], Rio Grande do Sul [WolffCo1993], Rio de Janeiro [Hempel1900a, Cocker1902k, Hempel1904], São Paulo [WolffCo1993]); Colombia [Kondo2001]; Costa Rica [Nakaha1982]; Ecuador [Nakaha1982]; El Salvador [Nakaha1982]; Guadeloupe [Balach1957c]; Guatemala [Nakaha1982]; Guyana [Nakaha1982]; Martinique [Balach1957c]; Peru [VasqueDeCo2002]; Puerto Rico &

Vieques Island (Puerto Rico [ColonFMe1998]); Suriname [Nakaha1982]; Trinidad and Tobago (Trinidad [RosenDe1979]); U.S. Virgin Islands [Nakaha1983]; Uruguay [Nakaha1982]; Venezuela [Nakaha1982]. **Oriental**: Hong Kong [RosenDe1979]; India [Ramakr1919a, Ramakr1921a, Ramakr1930] (Bihar [Ali1968], Tamil Nadu [Newste1917b]); Indonesia (Java [Marlat1908]); Kampuchea (Cambodia) [Takaha1942b]; Mongolia [DanzigKo1990]; Philippines (Luzon [Robins1917]); Sri Lanka [Green1896, Green1896e, Green1905a, Marlat1908, Ruther1915a, Borchs1966]; Taiwan [Ferris1921a, Takaha1929, Takaha1932a, Takaha1934, Takagi1969a, WongChCh1999]; Thailand [Takaha1942b]. **Palaearctic**: Egypt [Ezzat1958]; Georgia (Abkhaz ASSR [Borchs1934, Borchs1936]); Japan [Kuwana1902, Kuwana1907, Marlat1908, Kuwana1933, Kawai1980].

ECONOMIC IMPORTANCE: Pest of cocoa in Congo (Liegeois, 1944).

GENERAL: Description and illustration of adult female by Marlatt (1908), Brain (1919), Kuwana (1933), Balachowsky (1951, 1958b), Takagi (1969a), Chou (1985, 1986) and by Colon-Ferrer & Medina-Gaud (1998).

KEYS: Kawai 1980: 204 (female) [Japan]; Balachowsky 1958b: 269 (female) [Africa]; Ezzat 1958: 242 (female) [Egypt]; Balachowsky 1951: 681 (female) [Mediterranean]; Kuwana 1933: 20-21 (female) [Japan]; Kuwana 1933b: 49 (female) [Japan]; Brain 1919: 206 (female) [South Africa]; Robinson 1917: 33 (female) [Philippines]; Marlatt 1908: 134 (female) [World]; Green 1896e: 39 (female) [Sri Lanka].

CITATIONS: Ali1968 [host, distribution: 133-134]; Almeid1969 [taxonomy, description, host, distribution, biological control: 160-161]; Almeid1971 [host, distribution: 15]; Almeid1973b [host, distribution: 11]; Azeved1923A [host, distribution: 86-90]; Azeved1929 [host, distribution: 113-115]; Azeved1929a [host, distribution: 126-128]; Balach1951 [taxonomy, description, illustration, host, distribution: 684-686]; Balach1953i [taxonomy, description, illustration, host, distribution: 1512,1520]; Balach1957c [host, distribution: 200]; Balach1958b [taxonomy, description, illustration, host, distribution: 272,274-276]; BenDov1990a [taxonomy: 87]; Bondar1914 [host, distribution, economic importance: 1064-1106]; Bondar1915 [host, distribution, economic importance: 44-47]; Borchs1934 [host, distribution: 31]; Borchs1936 [host, distribution: 138]; Borchs1966 [catalogue: 232-233]; Brain1919 [taxonomy, description, illustration, host, distribution: 209-210]; BrainKe1917 [distribution: 184]; Brick1912 [host, distribution: 1-22]; Castel1963 [distribution: 140]; Charmo1899 [taxonomy: 1,6]; CharmoGe1921 [host, distribution: 188,189]; Chen1936 [host, distribution: 211,224]; ChenWo1936 [host, distribution]; Chou1985 [taxonomy, description, host, distribution: 250-251]; Chou1986 [taxonomy, illustration: 654]; ClapsWoGo2001a [taxonomy, host, distribution: 24-25]; Cocker1896b [distribution: 334]; Cocker1897i [taxonomy, description, host, distribution: 28]; Cocker1899a [taxonomy: 396]; Cocker1899p [taxonomy: 312]; Cocker1902k [host, distribution: 456]; CockerRo1915 [host, distribution: 109]; Cohic1958 [host, distribution : 15]; ColonFMe1998 [taxonomy, description, illustration, host, distribution: 76-77]; CoronaRuMo1997 [host, distribution: 38-41]; CostaL1942 [taxonomy, host, distribution: 277]; CostaLRa1922 [host, distribution: 1101]; DanzigKo1990 [host, distribution: 50]; DeLott1967a [host, distribution: 115]; Dupont1931 [host, distribution: 1-18]; Ebelin1949 [host, distribution, life history, control]; EvansPr1990 [biological control: 7-13]; Ezzat1958 [distribution: 242]; FDACSB1982 [host, distribution: 5-11]; FDACSB1983 [host, distribution: 6-8]; Fernal1903b [catalogue: 284]; Fernan1993 [host, distribution: 112]; Ferris1921a [host, distribution: 220]; Ferris1941e [taxonomy: 42,49]; Flande1971 [biological control, life history: 957-872]; Fonsec1963 [host, distribution: 32-35]; Fonsec1964 [host, distribution: 515]; FonsecAu1932a [host, distribution: 202-214]; Gavalo1936 [host, distribution: 82]; Ghesqu1927 [host, distribution: 310-316]; Ghesqu1932 [host, distribution: 60]; Gowdey1913 [host, distribution: 247-249]; Gowdey1917 [host, distribution: 189]; GrandpCh1899 [taxonomy, description, host, distribution: 26-27]; Green1896 [taxonomy, description, host, distribution: 4]; Green1896e [taxonomy, description, illustration, host, distribution: 41-42]; Green1900a [taxonomy, description, host, distribution: 66-67]; Green1905a [host, distribution: 346]; Green1907 [host, distribution: 202]; Green1916 [host, distribution: 376]; Green1937 [host, distribution: 333]; Hamble1947 [host, distribution: 949-956]; Hamon1980d [host, distribution: 38-39]; Hempel1900a [taxonomy, description, host, distribution: 499-500]; Hempel1904 [host, distribution: 321]; HorticInNo1923 [host, distribution: 236-239]; Hunt1939 [host, distribution: 548-566]; Kawai1980 [taxonomy, description, host, distribution: 205]; Kondo2001 [taxonomy, host, distribution: 45]; KondoKa1995 [host, distribution: 57-58]; KondoKa1995a [host, distribution: 97-98]; Kuwana1902 [host, distribution: 66]; Kuwana1907 [taxonomy, host, distribution: 194]; Kuwana1917a [taxonomy, distribution: 175]; Kuwana1933 [taxonomy, description, illustration, host, distribution: 24-25]; Laing1933 [host, distribution: 676]; Leonar1898a [taxonomy: 77]; Leonar1898c [taxonomy, description, illustration, host, distribution: 82-84]; Leonar1914 [taxonomy, host, distribution: 203]; Lepage1938 [catalogue: 418]; Liegeo1944 [host, distribution, economic importance: 165]; Lindin1909b [host, distribution: 148]; Lindin1910b [host, distribution: 47]; Lindin1913 [host, distribution: 79]; Lindin1932 [taxonomy: 194]; MacGil1921 [taxonomy, description, host, distribution: 453]; Mallam1954 [distribution: 24-60]; Mamet1943a [catalogue: 166]; Mamet1949 [catalogue: 63]; Mamet1950 [host, distribution: 23]; Mamet1951 [host, distribution: 230]; Mamet1954 [host, distribution: 20]; Mamet1956 [host, distribution: 138]; Mamet1957 [host, distribution: 369,377]; Mamet1959a [host, distribution: 389]; Mansfi1920 [host, distribution: 145-155]; Marlat1908 [taxonomy, description, illustration, host, distribution: 134,137,141]; Maskew1916 [host, distribution: 308-309]; Matile1978 [host, distribution: 66-67]; MatileNo1984 [host, distribution: 67]; MayneGh1934 [host, distribution: 3-38]; Miller1983 [host, distribution: 4-6]; MillerDa1990 [host, distribution, economic importance: 304]; Monte1930 [host, distribution: 3-36]; MoutiaMa1947 [distribution]; Muraka1970 [host, distribution: 76]; NagarkSa1990 [host, distribution, economic importance, biological control: 553-542]; Nakaha1982 [host, distribution: 73]; Nakaha1983 [host, distribution: 14]; Newste1911 [host, distribution: 85]; Newste1911a [host, distribution: 168-169]; Newste1917b [host, distribution: 132]; Ramakr1919a [taxonomy, host, distribution: 21-22]; Ramakr1921a [host, distribution: 356]; Ramakr1930 [taxonomy, host, distribution: 26]; RangelGo1945 [distribution, description: 1-44]; Robins1917 [taxonomy, description, host, distribution: 33-34]; RosenDe1979 [host, distribution, biological control: 244-247,533-539,]; Ruther1915a [taxonomy, description, host, distribution: 108]; Sassce1923 [host, distribution: 152-158]; Schmut1969 [taxonomy, description, host, distribution, life history, economic importance: 115]; SchmutKlLu1957 [host, distribution, economic importance: 494]; Seabra1921 [host, distribution: 99]; Seabra1925 [taxonomy, description, illustration, host, distribution: 28-29]; Silva1944 [host, distribution: 8-14]; Strong1922 [host, distribution: 775-780]; Takagi1969a [taxonomy, description, illustration, host, distribution: 98-105]; Takagi2000 [taxonomy: 59]; Takaha1929 [host, distribution: 78-79]; Takaha1932a [host, distribution: 103]; Takaha1933 [host, distribution: 25-34]; Takaha1934 [host, distribution: 36]; Takaha1935 [host, distribution: 3]; Takaha1942b [host, distribution: 50]; Tao1999 [taxonomy, host, distribution: 111]; Torres1922 [host, distribution: 72]; VasqueDeCo2002 [host, distribution: 331]; Vayssi1913 [host, distribution: 431]; VelasqRi1969 [host, distribution: 195-208]; Wester1920 [host, distribution]; WilliaBu1987 [host, distribution: 95]; WilliaWa1988 [taxonomy, host, distribution: 219-220]; WolffCo1993 [taxonomy, description, illustration, host, distribution: 49-51]; WolffCo1993a [host, distribution: 153]; WongChCh1999 [taxonomy, description, host, distribution: 33,76-77]; WoodruBeSk1998 [distribution]; Woolle1990 [biological control: 167-176].

Pseudischnaspis Hempel

Pseudischnaspis Hempel, 1900a: 506. Type species: *Pseudischnaspis linearis* Hempel, by monotypy and original designation.

SYSTEMATICS: *Pseudischnaspis* Cockerell is related to *Melanaspis* Cockerell and to *Acutaspis* Ferris, differing mainly in elongate shape of adult female (Ferris, 1941d).

GENERAL: Definition and characters by Ferris (1941d) and by Miller, Davidson & Stoetzel (1984).

KEYS: Colon-Ferrer & Medina-Gaud 1998: 28-32 (female) [Genera of Puerto Rico]; Miller, Davidson & Stoetzel 1984: 96 (female) [World]; Ferris 1942: 27 (female) [North America]; Ferris 1942: 39 (female) [species North America]; Brain 1919: 198 (female) [South Africa].

CITATIONS: Balach1951 [taxonomy: 594]; Borchs1966 [catalogue: 356]; Cocker1901 [taxonomy: 64]; ColonFMe1998 [taxonomy, description: 77-78]; Fernal1903b [catalogue: 294]; Ferris1937c [taxonomy: 52,55,92]; Ferris1938b [taxonomy: 75]; Ferris1941d [taxonomy, description: 381]; Ferris1942 [taxonomy: 446:27]; Gowdey1921 [taxonomy, description: 32]; Hempel1900a [taxonomy, description: 496,506]; Hempel1901a [taxonomy: 108]; Lindin1911 [taxonomy: 355]; Lindin1924 [taxonomy: 171]; Lindin1937 [taxonomy: 194]; MacGil1921 [taxonomy, description: 388,421-422]; McKenz1939 [taxonomy: 53]; McKenz1950 [taxonomy: 99]; MillerDaSt1984 [taxonomy, description: 94-109]; MorrisMo1966 [taxonomy, catalogue: 164].

Pseudischnaspis acephala Ferris

Pseudischnaspis acephala Ferris, 1941d: 382. Type data: PANAMA: Chiriqui Province, Boquete, on *Cavendishia*. Lectotype female, by subsequent designation Miller, Davidson & Stoetzel, 1984: 97. Type depository: Davis: The Bohart Museum of Entomology, University of California, California, USA.

COMMON NAME: flathead scale [MillerDaSt1984].

SCALE COVER: Female scale black, very elongate and slender; exuviae at anterior extremity, second exuvia broader than scale itself. A very thick ventral scale formed, of same colour and texture as dorsal scale and these two scales so closely fused that adult insect is enclosed as within a flattened tube. Male scale similar to that of female, but short (Ferris, 1941d).

HOST PLANTS: **Amaryllidaceae**: *Narcissus* [MillerDaSt1984]. **Anacardiaceae**: *Anacardium* [MillerDaSt1984], *Mangifera indica* [MillerDaSt1984]. **Ericaceae**: *Cavendishia* [Ferris1941d]. **Lauraceae**: *Persea* [MillerDaSt1984]. **Palmae** [MillerDaSt1984], *Chamaedorea* [MillerDaSt1984], *Cocos* [MillerDaSt1984]. **Rubiaceae**: *Coffea* [MillerDaSt1984]. **Rutaceae**: *Citrus aurantifolia* [MillerDaSt1984], *Citrus medica* [MillerDaSt1984].

DISTRIBUTION: **Nearctic**: Mexico [MillerDaSt1984]. **Neotropical**: Colombia [MillerDaSt1984, Kondo2001]; El Salvador [MillerDaSt1984]; Nicaragua [MillerDaSt1984]; Panama [Ferris1941d, MillerDaSt1984]; Peru [MillerDaSt1984].

BIOLOGY: Occurring on underside of leaves (Ferris, 1941d).

GENERAL: Description and illustration of adult female by Ferris (1941d) and by Miller et al. (1984). Description of first instar nymph by Miller et al. (1984).

KEYS: Miller, Davidson & Stoetzel 1984: 96 (female) [world]; Ferris 1942: 39 (female) [North America].

CITATIONS: Borchs1966 [catalogue: 356]; Ferris1941d [taxonomy, description, illustration, host, distribution: 382]; Ferris1942 [taxonomy: 446:39]; Kondo2001 [taxonomy, host, distribution: 45]; MillerDaSt1984 [taxonomy, description, illustration, host, distribution: 96-99].

Pseudischnaspis bowreyi (Cockerell)

Aspidiotus bowreyi Cockerell, 1893bb: 383. Type data: JAMAICA: Hope, on *Agave rigida*; collected by Bowrey and Cockerell. Lectotype female, by subsequent designation Miller, Davidson & Stoetzel, 1984: 100. Type depository: Washington: United States National Entomological Collection, U.S. National Museum of Natural History, District of Columbia, USA; type no. 7831. Notes: Also described as n. sp. in by Cockerell 1894b.

Aspidiotus (*Chrysomphalus*) *bowreyi*; Cockerell, 1897i: 23. Change of combination.

Aspidiotus (*Chrysomphalus*) *longissimus* Cockerell, 1898j: 439. Type data: MEXICO: Frontera, Tab., on mango; collected June 28, 1897. Lectotype female, by subsequent designation Miller, Davidson & Stoetzel, 1984: 100. Type depository: Washington, D.C.: U.S. Entomological Collection, U.S. National Museum of Natural History, USA; type no. 7973. Synonymy by Miller, Davidson & Stoetzel, 1984: 99.

Chrysomphalus bowreyi; Leonardi, 1899: 220. Change of combination.

Chrysomphalus longissimus; Cockerell, 1899a: 396. Change of combination.

Pseudischnaspis linearis Hempel, 1900a: 506. Type data: BRAZIL: Ypiranga, on leaves of *Myrica* sp.; collected by Hempel, April 28, 1900. Lectotype female, by subsequent designation Miller, Davidson & Stoetzel, 1984: 100. Type depository: Washington: United States National Entomological Collection, U.S. National Museum of Natural History, District of Columbia, USA; type no. 29. Synonymy by Ferris, 1941e: 45.

Pseudischnaspis bowreyi; Cockerell, 1901: 64. Change of combination.

Pseudischnaspis longissima; Cockerell, 1901: 64. Change of combination requiring emendation of species name for agreement in gender.

Aspidiotus longissimus; Cockerell, 1905: 45. Change of combination.

Aspidiotus linearis; Ferris, 1941e: 45. Change of combination.

COMMON NAMES: Bowrey scale [Dekle1965c, MillerDaSt1984].

SCALE COVER: Female scale elongate oblong, 2-3 mm long, 1-1.25 mm in wide; sides approximately parallel; moderately convex; well defined curved ridges extend crosswise over secretionary cover; blackish but covered with a purplish brown or bluish grey powder-like secretion; exuviae marginal or submarginal; brownish black; first exuviae crater-like surrounded by a greyish concentric ring; ventral vellum well developed. Male scale similar to female; about 1 mm in length; exuviae marginal or submarginal, brown or black (Dekle, 1965c). Photograph of scale cover by Dekle (1965c).

HOST PLANTS: **Agavaceae**: *Agave* [Newste1914, MillerDaSt1984], *Agave rigida* [Ferris1941d], *Beaucarnea* [MillerDaSt1984], *Dracaena* [MillerDaSt1984], *Yucca aloifolia* [MillerDaSt1984], *Yucca decipients* [MillerDaSt1984]. **Anacardiaceae**: *Mangifera indica* [MillerDaSt1984], *Spondias* [MillerDaSt1984]. **Annonaceae**: *Annona* [MillerDaSt1984]. **Apocynaceae**: *Nerium* [MillerDaSt1984]. **Araliaceae**: *Hedera helix* [Ferris1941d]. **Bombacaceae**: *Ceiba* [MillerDaSt1984]. **Bromeliaceae**: *Bromelia* [MillerDaSt1984], *Tillandsia* [MillerDaSt1984]. **Cactaceae**: *Hylocerus* [MillerDaSt1984]. **Guttiferae**: *Vismia ferruginea* [Ferris1941d]. **Juglandaceae**: *Carya* [MillerDaSt1984]. **Lauraceae:** *Persea* [MillerDaSt1984]. **Leguminosae**: *Euphorbia* [MillerDaSt1984], *Holocalyx* [MillerDaSt1984], *Poinciana* [MillerDaSt1984]. **Liliaceae**: *Aloe* [MillerDaSt1984]. **Lythraceae**: *Lagerstroemia lanceolata* [MillerDaSt1984]. **Malvaceae**: *Hibiscus* [MillerDaSt1984]. **Moraceae**: *Ficus* [MillerDaSt1984]. **Musaceae**: *Musa* [MillerDaSt1984]. **Myricaceae**: *Myrica* [Hempel1900a]. **Myrtaceae**: *Eucalyptus* [MillerDaSt1984], *Psidium guajava* [MillerDaSt1984]. **Oleaceae**: *Jasminum* [MillerDaSt1984]. **Orchidaceae** [MillerDaSt1984], *Cattleya* [MillerDaSt1984]. **Palmae**: *Phoenix* [MillerDaSt1984]. **Passifloraceae**: *Passiflora* [MillerDaSt1984]. **Polygonaceae**: *Coccoloba uvifera* [MillerDaSt1984]. **Rosaceae**: *Prunus* [MillerDaSt1984], *Pyrus* [MillerDaSt1984], *Rosa* [Dekle1965c, MillerDaSt1984]. **Rutaceae**: *Citrus* [MillerDaSt1984]. **Sterculiaceae**: *Theobroma* [MillerDaSt1984]. **Theaceae**: *Camellia* [Ferris1941d]. **Vitaceae**: *Vitis* [MillerDaSt1984].

DISTRIBUTION: **Nearctic**: Mexico [Cocker1899n, Ferris1941d, MillerDaSt1984] (Guerrero [Ferris1941d]); United States of America (Florida [Dekle1965c, MillerDaSt1984], Missouri [MillerDaSt1984], New York [Ferris1941d]). **Neotropical**: Argentina [Nakaha1982]; Barbados [Newste1914, MillerDaSt1984]; Belize [MillerDaSt1984]; Bermuda [MillerDaSt1984]; Bolivia [Nakaha1982]; Brazil

(Santa Catarina [Ferris1941d], São Paulo [Hempel1900a]); Colombia [MillerDaSt1984, Kondo2001]; Costa Rica [MillerDaSt1984]; Cuba [MillerDaSt1984]; Dominican Republic [Nakaha1982]; Ecuador [MillerDaSt1984]; Guatemala [MillerDaSt1984]; Honduras [MillerDaSt1984]; Jamaica [Ferris1941d, MillerDaSt1984]; Mexico (Tabasco [Ferris1941d]); Nicaragua [MillerDaSt1984]; Panama [MillerDaSt1984]; Peru [MillerDaSt1984]; Puerto Rico & Vieques Island [Ferris1941d, MillerDaSt1984]); Trinidad and Tobago (Trinidad [Ferris1941d, MillerDaSt1984]); U.S. Virgin Islands [Nakaha1983]; Venezuela [MillerDaSt1984].

BIOLOGY: Occurring on the leaves (Ferris, 1941d).

ECONOMIC IMPORTANCE: Occasionally a serious pest on rose in Key West, Florida, USA (Dekle, 1965c).

GENERAL: Description and illustration of adult female by Ferris (1941d) and by Miller, Davidson & Stoetzel (1984). Description and illustration of first-instar nymph and adult male by Miller, Davidson & Stoetzel (1984).

KEYS: Miller, Davidson & Stoetzel 1984: 96 (female) [world]; Ferris 1942: 39 (female) [North America]; Ferris 1942: 39 (female) [North America]; Cockerell 1905: 45-46 (female) [Mexico].

CITATIONS: Borchs1966 [catalogue: 356-357]; Cocker1893bb [taxonomy, description, host, distribution: 383]; Cocker1894b [taxonomy, description, host, distribution: 59-60]; Cocker1896b [distribution: 334]; Cocker1897i [taxonomy, description, host, distribution: 23]; Cocker1898j [taxonomy, description, host, distribution: 439]; Cocker1899a [taxonomy: 396]; Cocker1899n [host, distribution: 26]; Cocker1901 [taxonomy: 64]; Cocker1905 [taxonomy: 45]; ColonFMe1998 [taxonomy, description, illustration, host, distribution: 78-79]; Dekle1965c [taxonomy, description, host, distribution: 121]; Dekle1976 [taxonomy, description, host, distribution, economic importance: 139]; Fernal1903b [catalogue: 295]; Ferris1937c [taxonomy, illustration: 52,92]; Ferris1941d [taxonomy, description, illustration, host, distribution: 383-384]; Ferris1941e [taxonomy: 41,45]; Ferris1942 [taxonomy: 446:39]; Gowdey1921 [taxonomy, description, host, distribution: 32]; Hempel1900a [taxonomy, description, illustration, host, distribution: 506-508]; Hempel1901a [taxonomy: 108]; Houser1918 [taxonomy, description: 171]; Kondo2001 [taxonomy, host, distribution: 45]; Leonar1899 [taxonomy: 220]; Leonar1900 [taxonomy, host, distribution: 342]; Lindin1909c [taxonomy, host, distribution: 449]; MacGil1921 [taxonomy, description, host, distribution: 421-422]; McKenz1939 [taxonomy: 53]; MillerDa1990 [host, distribution, economic importance: 304]; Nakaha1982 [host, distribution: 75]; Nakaha1983 [host, distribution: 15]; Newste1914 [host, distribution: 307]; Wester1920 [host, distribution]; Willia1985a [taxonomy: 233].

Pseudoselenaspidus Fonseca

Pseudoselenaspidus Fonseca, 1962: 26. Type species: *Pseudoselenaspidus inermis* Fonseca, by monotypy and original designation.

SYSTEMATICS: *Pseudoselenaspidus* Fonseca differs from other genera in *Selenaspidus* complex in having third lobes of pygidial margin of regular shape, not spiriform.

GENERAL: Definition and characters by Fonseca (1962).
CITATIONS: Borchs1966 [catalogue: 252]; Fonsec1962 [taxonomy, description: 26]; MorrisMo1966 [taxonomy, catalogue: 168].

Pseudoselenaspidus inermis Fonseca

Pseudoselenaspidus inermis Fonseca, 1962: 26. Type data: BRAZIL: São Paolo, Serra da Cantareira, on indigenous plant. Holotype female. Type depositories: São Paulo: Museu de Zoologia, Universidade de São Paulo, Brazil, and São Paulo: Instituto Biologico de São Paulo, Brazil.
SCALE COVER: Female scale circular, flat, slightly transparent; yellow-straw; exuviae placed centrally, yellow (Fonseca, 1962).
DISTRIBUTION: **Neotropical**: Brazil (São Paulo [Fonsec1962, ClapsWoGo2001]).
GENERAL: Description and illustration of adult female by Fonseca (1962). The illustration of this species, in Fonseca (1962), must be restricted to Fig. 5 - A,B,C,D on page 24, and Fig. 6 - F,G,H on page 25; whereas Fig 6 - J on page 25, represents a species of Diaspidinae.
CITATIONS: Borchs1966 [catalogue: 252]; Claps1993 [taxonomy: 3,6]; ClapsWoGo2001 [host, distribution: 252]; Fonsec1962 [taxonomy, description, illustration, host, distribution: 26-27].

Pseudotargionia Lindinger

Pseudotargionia Lindinger, 1912b: 50. Type species: *Aonidia glandulosa* Newstead, by monotypy.
SYSTEMATICS: Balachowsky (1948b, 1951) placed this genus to sub-tribe Pseudaonidina, and distinguished it from *Neomorgania* MacGillivray by separated median lobes, while in the latter the lobes are fused.
GENERAL: Definition and characters by Balachowsky (1951, 1958b).
KEYS: Ben-Dov 1974b: 324 (female) [Africa]; Munting 1969: 136 (female) [Africa]; Balachowsky 1958b: 256 (female) [*Pseudaonidina* of Africa]; Ezzat 1958: 237-239 (female) [Egypt]; Balachowsky 1951: 675 (female) [Mediterranean]; Bodenheimer 1924: 237 (female) [Egypt].
CITATIONS: Balach1948b [taxonomy: 269]; Balach1951 [taxonomy, description: 676-677]; Balach1953i [taxonomy: 1512]; Balach1958b [taxonomy, description: 276]; BenDov1974b [taxonomy: 320]; Bodenh1924 [taxonomy: 237]; Borchs1966 [catalogue: 237]; DanzigPe1998 [catalogue: 347]; Ezzat1958 [taxonomy: 237]; Ferris1937c [taxonomy: 52]; Lepage1938 [taxonomy: 420]; Lindin1912b [taxonomy: 50]; Lindin1937 [taxonomy: 194]; MacGil1921 [taxonomy, description: 389,426]; MorrisMo1966 [taxonomy, catalogue: 168]; Muntin1969 [taxonomy: 136]; Sassce1915 [taxonomy: 35].

Pseudotargionia anareolae Ben-Dov

Pseudotargionia anareolae Ben-Dov, 1974b: 320. Type data: SOUTH AFRICA: Transvaal, Phalaborwa on *Acacia nigrescens*. Holotype female (examined). Type depository: Pretoria: South African National Collection of Insects, South Africa.

SCALE COVER: Female scale oval, 1.2 mm long, 0.9 mm wide; greyish white; nymphal exuviae brown, placed at anterior apex (Ben-Dov, 1974b).
HOST PLANTS: **Leguminosae**: *Acacia nigrescens* [BenDov1974b].
DISTRIBUTION: **Afrotropical**: South Africa [BenDov1974b].
GENERAL: Description and illustration of adult female by Ben-Dov (1974b).
KEYS: Ben-Dov 1974b: 324 (female) [Africa].
CITATIONS: BenDov1974b [taxonomy, description, illustration, host, distribution: 320-322].

Pseudotargionia asymmetrica Brimblecombe

Pseudotargionia asymmetrica Brimblecombe, 1959: 146. Type data: AUSTRALIA: Queensland, Winton, on *Eucalyptus camaldulensis*; collected April 1954. Holotype female. Type depository: Brisbane: Queensland Museum, Queensland, Australia; type no. T5728.
SCALE COVER: Few male scales on leaf surface; oval; dark fawn; exuvia dark (Brimblecombe, 1959).
HOST PLANTS: **Myrtaceae**: *Eucalyptus camaldulensis* [Brimbl1959].
DISTRIBUTION: **Australasian**: Australia (Queensland [Brimbl1959]).
GENERAL: Description and illustration of adult female by Brimblecombe (1959).
CITATIONS: Borchs1966 [catalogue: 237]; Brimbl1959 [taxonomy, description, illustration, host, distribution: 146-148].

Pseudotargionia comata (Maskell)

Aspidiotus eucalypti comatus Maskell, 1896b: 385. Type data: AUSTRALIA: Victoria, Melbourne, on *Eucalyptus viminalis*. Lectotype female and first instar, by subsequent designation Deitz & Tocker, 1980: 35. Type depository: Canberra: Australian National Insect Collection, CSIRO Entomology, Australia.
Targionia eucalypti comata; Leonardi, 1900: 303. Change of combination requiring emendation of species name for agreement in gender.
Targionia eucalypti comatus; Leonardi, 1900: 311. Change of combination.
Aspidiotus comatus; Ferris, 1941e: 42. Change of combination and rank.
Pseudotargionia comatus; Brimblecombe, 1954: 153. Change of combination.
Pseudotargionia comata; Borchsenius, 1966: 237. Emendation of species name for agreement in gender.
SCALE COVER: Female scale circular, greyish-white, slightly convex. Male scale narrow, subelliptical, white, not carinated (Maskell, 1896b).
HOST PLANTS: **Myrtaceae**: *Eucalyptus viminalis* [Maskel1896b, Leonar1900].
DISTRIBUTION: **Australasian**: Australia (Victoria [Maskel1896b, Leonar1900]).
GENERAL: Description and illustration of adult female by Brimblecombe (1954).
CITATIONS: Borchs1966 [catalogue: 237-238]; Brimbl1954 [taxonomy, description, illustration, host, distribution: 153-154]; Cocker1897i [host, distribution: 26]; Cocker1899a [taxonomy: 395]; DeitzTo1980 [taxonomy: 35]; Fernal1903b [catalogue: 297]; Ferris1941e [taxonomy: 42]; Ferris1943a [taxonomy: 85]; Leonar1900 [taxonomy, description, illustration, host, distribution: 311-312]; Maskel1896b [taxonomy, description, illustration, host, distribution: 385-386].

Pseudotargionia cordata Brimblecombe

Pseudotargionia cordata Brimblecombe, 1956: 107. Type data: AUSTRALIA: Queensland, Carbrook, on *Melaleuca nodosa*. Holotype female. Type depository: Brisbane: Queensland Museum, Queensland, Australia; type no. T5517.
SCALE COVER: Scale and overlying cork tissue thin; under-surface of scale whitish (Brimblecombe, 1956).
HOST PLANTS: **Myrtaceae**: *Melaleuca nodosa* [Brimbl1956].
DISTRIBUTION: **Australasian**: Australia (Queensland [Brimbl1956]).
BIOLOGY: Insects single and sparse; under the cork tissue in surface depressions of twigs (Brimblecombe, 1956).
GENERAL: Description and illustration of adult female by Brimblecombe (1956).
CITATIONS: Borchs1966 [catalogue: 238]; Brimbl1956 [taxonomy, description, illustration, host, distribution: 107-109].

Pseudotargionia crenulata Brimblecombe

Pseudotargionia crenulata Brimblecombe, 1956: 111. Type data: AUSTRALIA: Queensland, Ormiston, on *Melaleuca leucadendra*; collected September 1950. Holotype female. Type depository: Brisbane: Queensland Museum, Queensland, Australia; type no. T5505.
SCALE COVER: Under-surface of scale whitish; exuviae yellow (Brimblecombe, 1956).
HOST PLANTS: **Myrtaceae**: *Melaleuca leucadendra* [Brimbl1956].
DISTRIBUTION: **Australasian**: Australia (Queensland [Brimbl1956]).
BIOLOGY: Insects single and sparse on twigs under thin layer of cork tissue or in bark depressions of twigs (Brimblecombe, 1956).
GENERAL: Description and illustration of adult female by Brimblecombe (1956).
CITATIONS: Borchs1966 [catalogue: 238]; Brimbl1956 [taxonomy, description, illustration, host, distribution: 11-113].

Pseudotargionia damara Munting

Pseudotargionia damara Munting, 1969: 136. Type data: NAMIBIA: Daan Viljoen Game Park, on *Acacia hebeclada*. Holotype. Type depository: Pretoria: South African National Collection of Insects, South Africa.
SCALE COVER: Scale of adult female pale brown; subcircular, about 1.5 mm in diameter. Male scale white elongate, with greenish-yellow apical exuviae, 1-1.2 mm in length (Munting, 1969).
HOST PLANTS: **Leguminosae**: *Acacia hebeclada* [Muntin1969, BenDov1974b].
NATURAL ENEMIES: HYMENOPTERA **Encyrtidae**: *Habrolepis occidua* Annecke & Mynhardt [AnneckIn1971].
DISTRIBUTION: **Afrotropical**: Namibia (Southwest Africa) [Muntin1969, BenDov1974b].
GENERAL: Description and illustration of adult female by Munting (1969).
KEYS: Ben-Dov 1974b: 324 (female) [Africa]; Munting 1969: 136 (female) [Africa].
CITATIONS: AnneckIn1971 [host, distribution, biological control: 14]; BenDov1974b [taxonomy, host, distribution: 322]; Muntin1969 [taxonomy,

description, illustration, host, distribution: 136-137,159]; Prinsl1983 [distribution, biological control: 27].

Pseudotargionia glandulosa (Newstead)

Aonidia glandulosa Newstead, 1911: 103. Type data: EGYPT: Upper Egypt, on "Sunt", *Acacia arabica*. Lectotype fossil (examined), by subsequent designation Ben-Dov, 1974b: 323. Type depository: London: The Natural History Museum, England, UK.

Pseudotargionia glandulosa; Lindinger, 1912b: 50. Change of combination.

Pseudaonidia quadriareolata Malenotti, 1916a: 334. Type data: SOMALIA: Allengo, on *Acacia asak*; collected September 1913. Syntypes, female. Synonymy by Ferris, 1941e: 238.

Pseudaonidia glandulosa; Brain, 1919: 208. Change of combination.

Pseudoaonidia glandulosa occitalis Rungs, 1935: 273. Type data: MOROCCO: Ourika N'Tamsift and Ouled Youb (Oued Draa) on *Acacia raddiana*, Assif el Mal (Grand Atlas 1100 m) on *Acacia gummifera* and Merkala (Hammada du Draa) on *Acacia seyal*. Syntypes, female and first instar. Type depository: Paris: Muséum national d'Histoire naturelle, France. Synonymy by Borchsenius, 1966: 238.

Pseudaonidia glandulosa occitalis; Rungs, 1935: 273. Misspelling of species name. Notes: Misspelling of *occitalis* for *occidentalis*.

Pseudotargionia glandulosa occidentalis; Rungs, 1942: 107. Change of combination.

Pseudotargionia glandulosa occidentalis; Rungs, 1942: 107. Justified emendation.

Pseudotargionia quadriareolata; Balachowsky & Kaussari, 1951: 3. Change of combination.

SCALE COVER: Newstead (1911) described scale cover as: "Female scale circular, diameter 1.25 mm; straw-coloured or ochreous buff with faint patches of dull orange-yellow; highly convex, sometimes obconical; highest portion towards the anterior margin; margins thin and sometimes rounded; larval pellicle yellow, generally completely hidden; second pellicle invariably covered with secretion; ventral surface white, with an external zone of pale yellow white; second pellicle large, bright orange-yellow, nude; ventral scale white, thick at the margin, thin and semi-transparent centrally. Male scale with the cephalic segment strongly defined and distinctly articulated; anterior margin very broadly rounded". Brain (1919) described scale cover as: "Female scale elongate, very occasionally almost circular, about 1.25 to 1.5 mm in diameter, white at first, sometimes faintly buff, with dark brown or resinous exuviae. Male scale similar, but smaller, with pale exuviae".

HOST PLANTS: **Leguminosae**: *Acacia arabica* [Hall1922, Balach1951], *Ac. asak* [Maleno1916a, Balach1951], *Ac. gemmifera* [Rungs1942], *Ac. horrida* [Brain1919], *Ac. raddiana* [Rungs1942, Rungs1948, Balach1951], *Ac. scorpioides* [Balach1951], *Ac. seyal* [Rungs1942], *Ac. tortilis* [Hall1926a, Balach1951]. **Vitaceae**: *Vitis* [Lepage1938], *Vitis vinifera* [Balach1958b].

NATURAL ENEMIES: COLEOPTERA **Coccinellidae**: *Pharoscymnus* [BalachMa1970]. **Nitidulidae**: *Cybocephalus* [BalachMa1970]. HYMENOPTERA **Encyrtidae**: *Habrolepis occidua* Annecke & Mynhardt [AnneckIn1971].

DISTRIBUTION: **Afrotropical**: Mauritania [Balach1958b, BalachMa1970]; Namibia (Southwest Africa) [BenDov1974b]; Senegal [Balach1958b]; Somalia [Maleno1916a]; South Africa [Brain1919, Muntin1969, BenDov1974b]; Sudan [Bodenh1935c]; Zimbabwe [Balach1958b]. **Neotropical**: Brazil (Rio de Janeiro [Lepage1938, Balach1958b]). **Palaearctic**: Egypt [Newste1911, Hall1922, Hall1923, Ezzat1958, BenDov1974b]; Israel [Bodenh1935, Balach1951]; Libya [BenDov1974b]; Morocco [Rungs1948]; Western Sahara [Bodenh1935c, Rungs1942].

GENERAL: Description and illustration of adult female by Newstead (1911), Brain (1919), Balachowsky (1951, 1958b). Data on the intraspecific variation in the number of macroducts on the female's pygidium by Ben-Dov (1974).

KEYS: Ben-Dov 1974b: 324 (female) [Africa]; Munting 1969: 136 (female) [Africa]; Ezzat 1958: 242 (female) [Egypt]; Brain 1919: 206 (female) [South Africa].

CITATIONS: AnneckIn1971 [host, distribution, biological control: 14]; Balach1951 [taxonomy, description, illustration, host, distribution: 677-679]; Balach1958a [host, distribution: 39]; Balach1958b [taxonomy, description, illustration, host, distribution: 276-279]; BalachKa1951 [taxonomy: 3,5]; BalachMa1970 [host, distribution, biological control: 1081-1082]; BenDov1974b [taxonomy, description, host, distribution: 322-324]; Bodenh1924 [taxonomy, description, host, distribution: 62-63]; Bodenh1935 [host, distribution: 249]; Bodenh1935c [host, distribution: 1156]; Brain1919 [taxonomy, description, illustration, host, distribution: 208-209]; ClapsWoGo2001a [taxonomy, host, distribution: 26]; DanzigPe1998 [catalogue: 347]; Ezzat1958 [distribution: 242]; EzzatNa1987 [distribution: 88]; Ferris1937c [taxonomy, illustration: 52,93]; Ferris1941e [taxonomy: 47]; Ferris1943a [taxonomy: 86]; GomesCRe1947 [host, distribution: 228]; Hall1922 [taxonomy, description, host, distribution: 24]; Hall1923 [host, distribution: 42]; Hall1926a [host, distribution : 30]; Hall1927b [taxonomy, description, host, distribution: 169-170]; Lepage1938 [catalogue: 420]; Lindin1912b [taxonomy, description, host, distribution: 50]; Lindin1935 [taxonomy: 145]; MacGil1921 [taxonomy, description, host, distribution: 426]; Maleno1916a [taxonomy, description, illustration, host, distribution: 334-339]; Muntin1969 [host, distribution: 137]; Newste1911 [taxonomy, description, illustration, host, distribution: 103]; PriesnHo1940 [biological control: 58-70]; Prinsl1983 [distribution, biological control: 27]; Rungs1935 [taxonomy, description, host, distribution: 273-275]; Rungs1942 [taxonomy, host, distribution: 107-108]; Rungs1948 [host, distribution: 111-112]; Sander1909a [taxonomy, host, distribution: 56]; Sassce1912 [taxonomy, host, distribution: 91].

Pseudotargionia inconspicua Brimblecombe

Pseudotargionia inconspicua Brimblecombe, 1957: 289. Type data: AUSTRALIA: Queensland, Rocklea, on *Melaleuca leucadendra*; collected H. Tryon, September 1914. Holotype female. Type depository: Brisbane: Queensland Museum, Queensland, Australia; type no. T5662.

SCALE COVER: Scale beneath a thin layer of cork tissue, circular, 1.6 mm diameter; undersurface whitish; exuviae brown (Brimblecombe, 1957).

HOST PLANTS: **Myrtaceae**: *Melaleuca leucadendra* [Brimbl1957].
DISTRIBUTION: **Australasian**: Australia (Queensland [Brimbl1957]).
BIOLOGY: Insects scattered on twigs; scale beneath a thin layer of cork tissue (Brimblecombe, 1957).
GENERAL: Description and illustration of adult female by Brimblecombe (1957).
CITATIONS: Borchs1966 [catalogue: 238]; Brimbl1957 [taxonomy, description, illustration, host, distribution: 289-291].

Pseudotargionia isaensis Brimblecombe
Pseudotargionia isaensis Brimblecombe, 1959: 148. Type data: AUSTRALIA: Queensland, Mt. Isa, on *Amyema sanguinea*. Holotype female. Type depository: Brisbane: Queensland Museum, Queensland, Australia; type no. T5731.
SCALE COVER: Female scale circular, 1.0 mm diameter; light to dark fawn, margin paler; second exuviae covered with light fawn secretion; first exuvia central, dark greenish brown (Brimblecombe, 1959).
HOST PLANTS: **Loranthaceae**: *Amyema sanguinea* [Brimbl1959].
DISTRIBUTION: **Australasian**: Australia (Queensland [Brimbl1959]).
GENERAL: Description and illustration of adult female by Brimblecombe (1959).
CITATIONS: Borchs1966 [catalogue: 238]; Brimbl1959 [taxonomy, description, illustration, host, distribution: 148-150].

Pseudotargionia kalaharica Munting
Pseudotargionia kalaharica Munting, 1969: 137. Type data: SOUTH AFRICA: Cape Province, Kalahari Gemsbok National park, on *Acacia giraffae*. Holotype. Type depository: Pretoria: South African National Collection of Insects, South Africa; type no. 2867/3.
SCALE COVER: Female scale subcircular, white with brown exuviae near margin; about 1.5 mm in diameter. Male scale oval, white, with golden yellow exuviae at one end; about 1 mm long (Munting, 1969).
HOST PLANTS: **Leguminosae**: *Acacia giraffae* [Muntin1969, BenDov1974b].
DISTRIBUTION: **Afrotropical**: Namibia (Southwest Africa) [Muntin1969, BenDov1974b]; South Africa [Muntin1969, BenDov1974b].
GENERAL: Description and illustration of adult female by Munting (1969).
KEYS: Ben-Dov 1974b: 324 (female) [Africa]; Munting 1969: 136 (female) [Africa].
CITATIONS: BenDov1974b [taxonomy, host, distribution: 323]; Muntin1969 [taxonomy, description, illustration, host, distribution: 137-138,160].

Pseudotargionia marginata Brimblecombe
Pseudotargionia marginata Brimblecombe, 1956: 109. Type data: AUSTRALIA: Queensland, Ayr, on *Melaleuca viridifolia*; collected by W.A. Smith, February 1953. Holotype female. Type depository: Brisbane: Queensland Museum, Queensland, Australia; type no. t5509.
SCALE COVER: Insects single and sparse, on twigs beneath cork tissue; undersurface of scale greyish white (Brimblecombe, 1956).
HOST PLANTS: **Myrtaceae**: *Melaleuca viridifolia* [Brimbl1956].

DISTRIBUTION: **Australasian**: Australia (Queensland [Brimbl1956]).
GENERAL: Description and illustration of adult female by Brimblecombe (1956).
CITATIONS: Borchs1966 [catalogue: 238]; Brimbl1956 [taxonomy, description, illustration, host, distribution: 109-111].

Pseudotargionia orientalis Balachowsky & Kaussari

Pseudotargionia orientalis Balachowsky & Kaussari, 1951: 3. Type data: IRAN: Province de Saravan, on *Stocksia brahuica*. Holotype female. Type depository: Paris: Muséum national d'Histoire naturelle, France.
SCALE COVER: Female scale circular or subcircular, 2-2.3 mm; slightly convex; snow-white; exuviae central or subcentral black; generally covered with white secretion; scale covered slightly by bark (Balachowsky & Kaussari, 1951).
HOST PLANTS: **Sapindaceae**: *Stocksia brahuica* [BalachKa1951].
DISTRIBUTION: **Palaearctic**: Iran [BalachKa1951, Kaussa1955].
GENERAL: Description and illustration of adult female by Balachowsky & Kaussari (1951).
CITATIONS: BalachKa1951 [taxonomy, description, illustration, host, distribution: 3-5,11]; Borchs1966 [catalogue: 238]; DanzigPe1998 [catalogue: 347]; Kaussa1955 [host, distribution: 17].

Pseudotargionia subcorticis Ben-Dov

Pseudotargionia subcorticis Ben-Dov, 1972: 309. Type data: SOUTH AFRICA: Transvaal, Mara railway station, on *Combretum imberbe*; collected by Y. Ben-Dov. Holotype female. Type depository: Pretoria: South African National Collection of Insects, South Africa.
SCALE COVER: White female scales almost concealed under thin cork layer of twig, except for yellowish exuviae that can be noticed on superficial observation (Ben-Dov, 1972).
HOST PLANTS: **Combretaceae**: *Combretum imberbe* [BenDov1972].
DISTRIBUTION: **Afrotropical**: South Africa [BenDov1972].
BIOLOGY: Adult females found on twigs only (Ben-Dov, 1972).
GENERAL: Description and illustration of adult female by Ben-Dov (1972).
KEYS: Ben-Dov 1974b: 324 (female) [Africa].
CITATIONS: BenDov1972 [taxonomy, description, illustration, host, distribution: 309-312]; BenDov1974b [taxonomy, host, distribution: 324].

Pygidiaspis MacGillivray

Pygidiaspis MacGillivray, 1921: 392. Type species: *Aspidiotus* (*Targionia*) *cedri* Green, by monotypy and original designation.
SYSTEMATICS: Lindinger (19370) synonymized this genus with *Targionia*. Ferris (1937c) suggested some relation to *Loranthaspis*. The narrow, almost linear, median lobes on *Loranthaspis* distinguish it from *Pygidiaspis*.
GENERAL: Definition and characters by MacGillivray (1921) and by Ferris (1937c).

CITATIONS: Borchs1966 [catalogue: 359]; Ferris1937c [taxonomy: 52]; Ferris1938b [taxonomy: 75]; Ferris1941d [taxonomy: 374]; Lindin1937 [taxonomy: 194]; MacGil1921 [taxonomy, description: 392,447]; MorrisMo1966 [taxonomy, catalogue: 170].

Pygidiaspis cedri (Green)

Aspidiotus (*Targionia*) *cedri* Green, 1915d: 51. Type data: AUSTRALIA: Queensland, on cedar logs [= *Cedrus* sp.]. Syntypes, female. Type depository: London: The Natural History Museum, England, UK.
Pygidiaspis cedri; MacGillivray, 1921: 447. Change of combination.
Targionia cedri; Lindinger, 1937: 197. Change of combination.
Aspidiotus cedri; Ferris, 1937c: 52. Change of combination.
Pygidiaspis cedri; Borchsenius, 1966: 360. Revived combination.
SCALE COVER: Female scale circular, diameter, 1.50 to 1.65 mm; flattish; very dense; very dark blackish brown, inner surface sometimes whitish; exuviae concealed, position of larval exuvia marked by a small raised boss (Green, 1915d).
HOST PLANTS: **Pinaceae**: *Cedrus* [Green1915d].
DISTRIBUTION: **Australasian**: Australia (Queensland [Green1915d]).
GENERAL: Description and illustration of adult female by Green (1915d).
CITATIONS: Borchs1966 [catalogue: 360]; Ferris1937c [taxonomy, illustration: 52,94]; Ferris1941e [taxonomy: 41]; Ferris1943 [taxonomy: 85]; Green1915d [taxonomy, description, illustration, host, distribution: 51]; Lindin1937 [taxonomy: 197]; MacGil1921 [taxonomy, description, host, distribution: 447].

Reclavaspis Komosinska

Reclavaspis Komosinska, 1965: 1. Type species: *Reclavaspis australicus*, by original designation.
SYSTEMATICS: Komosinska (1965) noted that *Reclavaspis* differed from *Diaspidiotus* in having longer and differently-shaped paraphyses on pygidium between segments 8 and 7. *Reclavaspis* also distinguished from *Clavaspis* by reduced size of inner paraphyses between segments 8 and 7.
GENERAL: Definition and characters by Komosinska (1965).
CITATIONS: Komosi1965 [taxonomy, description: 1].

Reclavaspis australicus Komosinska

Reclavaspis australicus Komosinska, 1965: 1. Type data: AUSTRALIA: C. Australia, on *Ficus platypoda*. Syntypes, female. Type depository: London: The Natural History Museum, England, UK.
SCALE COVER: Scale cover not observed by Komosinska (1965).
HOST PLANTS: **Moraceae**: *Ficus platypoda* [Komosi1965].
DISTRIBUTION: **Australasian**: Australia [Komosi1965].
GENERAL: Description and illustration of adult female by Komosinska (1965).
CITATIONS: Komosi1965 [taxonomy, description, illustration, host, distribution: 1-4].

Reclavaspis evexa (Brimblecombe)

Diaspidiotus evexus Brimblecombe, 1959: 127. Type data: AUSTRALIA: Northern Territory, Alice Springs, on *Eremophila sturtii*; collected May 1954. Holotype female. Type depository: Brisbane: Queensland Museum, Queensland, Australia; type no. T5702.

Reclavaspis evexus; Komosinska, 1965: 4. Change of combination.

SCALE COVER: Female scale circular, 0.85 mm diameter; dull white to light fawn; exuviae central, dark orange (Brimblecombe, 1959).

HOST PLANTS: **Myoporaceae**: *Eremophila sturtii* [Brimbl1959].

DISTRIBUTION: **Australasian**: Australia (Northern Territory [Brimbl1959]).

GENERAL: Description and illustration of adult female by Brimblecombe (1959).

CITATIONS: Borchs1966 [catalogue: 324]; Brimbl1959 [taxonomy, description, illustration, host, distribution: 127-129]; Komosi1965 [taxonomy: 4].

Remotaspidiotus MacGillivray

Remotaspidiotus MacGillivray, 1921: 391. Type species: *Aspidiotus* (*Targionia*) *chenopodii* Marlatt, by original designation.

Remataspidiotus; Chou, 1985: 256. Misspelling of genus name.

SYSTEMATICS: *Remotaspidiotus* MacGillivray regarded by Ferris (1937a) a synonym of *Rhizaspidiotus* MacGillivray, but resurrected by Brimblecombe (1958). Borchsenius & Williams (1963) also accepted *Remotaspidiotus* as a valid genus.

GENERAL: Definition and characters by Borchsenius & Williams (1963).

CITATIONS: Borchs1966 [catalogue: 244]; BorchsWi1963 [taxonomy, description: 389,392]; Brimbl1958 [taxonomy, description: 74]; Chou1985 [taxonomy, description: 256]; DanzigPe1998 [catalogue: 348]; Ferris1921b [taxonomy: 94]; Ferris1937a [taxonomy: 33-34,42]; Ferris1937e [taxonomy: 528]; Ferris1943 [taxonomy: 99]; Kozar1990f [distribution: 142]; Lindin1937 [taxonomy: 195]; MacGil1921 [taxonomy, description: 391,434]; MorrisMo1966 [taxonomy, catalogue: 172]; Tao1999 [taxonomy: 115].

Remotaspidiotus albus Brimblecombe

Remotaspidiotus albus Brimblecombe, 1959: 150. Type data: AUSTRALIA: Queensland, Eulo, on *Santalum lanceolatum*; collected October 1954. Holotype female. Type depository: Brisbane: Queensland Museum, Queensland, Australia; type no. T5734.

SCALE COVER: Female scales white, convex, circular, 1.5 mm diameter; second exuvia fawnish-yellow with a white suffusion; first exuvia light olive green (Brimblecombe, 1959).

HOST PLANTS: **Santalaceae**: *Santalum lanceolatum* [Brimbl1959].

DISTRIBUTION: **Australasian**: Australia (Queensland [Brimbl1959]).

GENERAL: Description and illustration of adult female by Brimblecombe (1959).

CITATIONS: Borchs1966 [catalogue: 244]; Brimbl1959 [taxonomy, description, illustration, host, distribution: 150-152].

Remotaspidiotus bossieae (Maskell)

Aspidiotus bossieae Maskell, 1892: 10. Type data: AUSTRALIA: on *Bossiaea procumbens*; sent by Mr. French. Syntypes, female. Type depository: Auckland: New Zealand Arthropod Collection, Landcare Research, New Zealand.

Hemiberlesia bossieae; Leonardi, 1897b: 122. Change of combination.

Aspidiotus bossiaeae Lindinger, 1907: 20. Unjustified emendation.

Aspidiotus bossicae; Kuwana, 1927: 71. Misspelling of species name.

Aspidiella bossieae; Ferris, 1941e: 41. Change of combination.

Aspidiotus bossiae; Ferris, 1941e: 41. Misspelling of species name.

Remotaspidiotus bossieae; Brimblecombe, 1958: 78. Change of combination.

SCALE COVER: Maskell (1897) described scale cover as: "Female scale circular, convex, about 1/18 inch in diameter; colour varying from dirty-white to yellow, and sometimes to dark-brown; texture soft and wooly-looking; exuviae central, very small and inconspicuous, yellow". Brimblecombe (1958) described scale cover as: "Insects in small groups on leaves of host; scale of adult female white, circular, convex, 1.75 mm diameter; exuviae central pale orange with white suffusion".

HOST PLANTS: Leguminosae: *Bossiaea procumbens* [Maskel1892, Brimbl1958].

DISTRIBUTION: Australasian: Australia (Victoria [Maskel1892, Frogga1914]). **Palaearctic**: China (People's Republic) [Kuwana1927].

BIOLOGY: Insects in small groups on leaves of host (Brimblecombe, 1958)

GENERAL: Description and illustration of adult female by Brimblecombe (1958).

CITATIONS: Borchs1966 [catalogue: 244]; Brimbl1958 [taxonomy, description, illustration, host, distribution: 78-80]; Chou1985 [taxonomy, description, host, distribution: 256-257]; Cocker1896b [distribution: 335]; Cocker1897i [taxonomy, description, host, distribution: 26]; DanzigPe1998 [catalogue: 348]; DeitzTo1980 [taxonomy: 33]; Fernal1903b [catalogue: 253]; Ferris1941e [taxonomy: 41]; Frogga1914 [taxonomy, description, host, distribution: 133]; Frogga1915 [taxonomy, description, host, distribution: 9]; Kuwana1927 [host, distribution: 71]; Leonar1897b [taxonomy, description, host, distribution: 122]; Lindin1907a [taxonomy: 20]; Lindin1957 [taxonomy: 545]; Maskel1892 [taxonomy, description, host, distribution: 10-11]; Tao1999 [taxonomy, host, distribution: 115-116].

Remotaspidiotus cassiniae (Brimblecombe)

Rhizaspidiotus cassiniae Brimblecombe, 1956: 121. Type data: AUSTRALIA: Queensland, Inglewood, on *Cassinia laevis*; collected February 1954. Holotype female. Type depository: Brisbane: Queensland Museum, Queensland, Australia; type no. T5513.

Remotaspidiotus cassiniae; Brimblecombe, 1958: 74. Change of combination.

SCALE COVER: Scale dirty, probably due to sooty mould from other insects (Brimblecombe, 1956).

HOST PLANTS: Compositae: *Cassinia laevis* [Brimbl1956].

DISTRIBUTION: Australasian: Australia (Queensland [Brimbl1956]).

GENERAL: Description and illustration of adult female by Brimblecombe (1956).

CITATIONS: Borchs1966 [catalogue: 244-245]; Brimbl1956 [taxonomy, description, illustration, host, distribution: 121-122]; Brimbl1958 [taxonomy, description, illustration, host, distribution: 75-76].

Remotaspidiotus chenopodii (Marlatt)

Aspidiotus (*Targionia*) *chenopodii* Marlatt, 1908c: 24. Type data: AUSTRALIA: New South Wales, Coolabah, on *Chenopodium*; collected by J.G. Smith. Syntypes, female. Type depository: Washington, D.C.: U.S. National Entomological Collection, U.S. National Museum of Natural History, USA; type no. 14143.
Remotaspidiotus chenopodii; MacGillivray, 1921: 434. Change of combination.
Rhizaspidiotus chenopodii; Ferris, 1943a: 99. Change of combination.
Remotaspidiotus chenopodii; Brimblecombe, 1958: 17. Revived combination.
SCALE COVER: Marlatt (1908) described scale cover as: "Female scale subcircular, 1.5-2 mm in diameter, convex; light buff in colour; the secretion covering the larval exuvia whitish; with the loss of the larval exuvia, the light-orange second exuvia appears as a conspicuous spot; ventral scale attached to the bark, white, rather abundant. Male scale elongate, sides nearly parallel, length about 1 mm; same general characters as the female, except that the lower secretion remains attached to the upper, forming a definite flattish sac or cocoon which easily separates from the plant". Brimblecombe (1958) described scale cover as: "Insects scattered or clustered on branches or host; scale of adult female white, subcircular, convex, 1.2 to 1.75 mm diameter; exuviae pale yellow, partially covered with white suffusion".
HOST PLANTS: **Chenopodiaceae**: *Bassia quinquicuspis villosa* [Brimbl1958], *Chenopodium* [Marlat1908c, Frogga1914, Brimbl1958].
DISTRIBUTION: **Australasian**: Australia (New South Wales [Marlat1908c, Frogga1914], Queensland [Brimbl1958]).
BIOLOGY: Insects scattered or clustered on branches of host (Brimblecombe, 1958).
GENERAL: Description and illustration of adult female by Brimblecombe (1958) and by Borchsenius & Williams (1963).
CITATIONS: Borchs1966 [catalogue: 245]; BorchsWil963 [taxonomy, description, illustration: 389-392]; Ferris1921b [taxonomy: 94]; Ferris1937a [taxonomy, illustration: 33,42]; Ferris1941e [taxonomy: 41]; Ferris1943a [taxonomy: 85,99]; Frogga1914 [taxonomy, description, host, distribution: 135]; Frogga1915 [taxonomy, description, host, distribution: 11]; MacGil1921 [taxonomy, description, host, distribution: 434]; Marlat1908c [taxonomy, description, illustration, host, distribution: 24-25]; Sander1909a [taxonomy, host, distribution: 55].

Remotaspidiotus coralinus (Froggatt)

Aspidiotus (*Targionia*) *coralinus* Froggatt, 1914: 136. Type data: AUSTRALIA: New South Wales, Darling River, near Bourke, on *Eremophila sturtii*. Syntypes, female. Type depository: Canberra: Australian National Insect Collection, CSIRO Entomology, Australia. Notes: Brimblecombe (1958) did not locate type material. Peter Gillespie (New South Wales, Agriculture) informed Yair Ben-Dov that syntypes are available in ANIC.
Aspidiotus (*Targionia*) *coralinus*; Froggatt, 1915: 14. Notes: Described again as "n. sp.".

Targionia carolina; Sasscer, 1915: 35. Change of combination and misspelling of species name.

Neglectaspis corallina; Lindinger, 1937: 190. Change of combination and misspelling of species name.

Remotaspidiotus coralinus; Brimblecombe, 1958: 76. Change of combination.

SCALE COVER: Froggatt (1914) illustrated female scale, and described it as: "... Female scale white. conical, circular, not more than 1/40 inch in diameter, with the apex truncate, forming a ring with a depression in the centre above the dull yellow exuviae". Brimblecombe (1958) described scale as: "Insects numerous on leaves and twigs; scale circular, 1.2 mm in diameter, white, convex; exuviae orange yellow".

HOST PLANTS: **Myoporaceae**: *Eremophila sturtii* [Frogga1914, Brimbl1958].

DISTRIBUTION: **Australasian**: Australia (New South Wales [Frogga1914, Brimbl1958], Queensland [Brimbl1958]).

BIOLOGY: On young foliage and branchlets; often clustered together in little patches, causing foliage to become very sticky (Froggatt, 1914).

GENERAL: Description and illustration of adult female by Froggatt (1914) and by Brimblecombe (1958).

CITATIONS: Borchs1966 [catalogue: 245]; Brimbl1958 [taxonomy, description, illustration, host, distribution: 76-78]; Ferris1941e [taxonomy: 42]; Ferris1943a [taxonomy: 85]; Frogga1914 [taxonomy, description, host, distribution: 136]; Frogga1915 [taxonomy, description, host, distribution: 14]; Lindin1937 [taxonomy: 190]; Sassce1915 [taxonomy, host, distribution: 35].

Remotaspidiotus gidgei (Froggatt)

Aspidiotus gidgei Froggatt, 1914: 313. Type data: AUSTRALIA: New South Wales, Darling River, at Pera Bore, on *Acacia cambagei*. Syntypes, female. Type depository: Canberra: Australian National Insect Collection, CSIRO Entomology, Australia.

Aspidiotus gidgei; Froggatt, 1915: 17. Notes: Described again as "n. sp.".

Remotaspidiotus gidgei; Borchsenius, 1966: 245. Change of combination.

SCALE COVER: Female scale almost circular, very convex; diameter 1/35 inch; outer surface greyish brown, but when secretion peeled off almost white; exuviae light yellow, circular, and sometimes slightly depressed in centre (Froggatt, 1914).

HOST PLANTS: **Leguminosae**: *Acacia cambagei* [Frogga1914], *Acacia stenophylla* [Laing1929].

DISTRIBUTION: **Australasian**: Australia (New South Wales [Frogga1914], Victoria [Laing1929]).

GENERAL: Description and illustration of adult female by Froggatt (1914) and by Laing (1929).

CITATIONS: Borchs1966 [catalogue: 245]; Ferris1941e [taxonomy: 43]; Ferris1943a [taxonomy: 99]; Frogga1914 [taxonomy, description, host, distribution: 313]; Frogga1915 [taxonomy, description, host, distribution: 17]; Laing1929 [taxonomy, description, illustration, host, distribution: 21-22]; Sassce1915 [taxonomy, host, distribution: 34].

Remotaspidiotus reconditus Brimblecombe

Remotaspidiotus reconditus Brimblecombe, 1959: 152. Type data: AUSTRALIA: Queensland, Marmor, on *Eremocitrus glauca*; collected October 1955. Holotype female. Type depository: Brisbane: Queensland Museum, Queensland, Australia; type no. T5738.

SCALE COVER: Female scale circular, 1.3 mm diameter; fawn or greyish white; exuviae dark orange (Brimblecombe, 1959).

HOST PLANTS: **Rutaceae**: *Eremocitrus glauca* [Brimbl1959].

DISTRIBUTION: **Australasian**: Australia (Queensland [Brimbl1959]).

GENERAL: Description and illustration of adult female by Brimblecombe (1959).

CITATIONS: Borchs1966 [catalogue: 245]; Brimbl1959 [taxonomy, description, illustration, host, distribution: 152-154].

Remotaspidiotus squamosus Brimblecombe

Remotaspidiotus squamosus Brimblecombe, 1959: 154. Type data: AUSTRALIA: Queensland, Texas, on *Eremophila mitchelli*. Holotype female. Type depository: Brisbane: Queensland Museum, Queensland, Australia; type no. T5742.

SCALE COVER: Female scales whitish, circular, 1.0 mm diameter; exuviae yellow (Brimblecombe, 1959).

HOST PLANTS: **Myoporaceae**: *Eremophila mitchellii* [Brimbl1959].

DISTRIBUTION: **Australasian**: Australia (Queensland [Brimbl1959]).

GENERAL: Description and illustration of adult female by Brimblecombe (1959).

CITATIONS: Borchs1966 [catalogue: 245]; Brimbl1959 [taxonomy, description, illustration, host, distribution: 154-156].

Rhizaspidiotus MacGillivray

Rhizaspidiotus MacGillivray, 1921: 390. Type species: *Aspidiotus* (*Targionia*) *helianthi* Parrott, by monotypy and original designation.

Chorizaspidiotus MacGillivray, 1921: 391. Type species: *Aspidiotus* (*Targionia*) *gutierreziae* Cockerell & Parrott, by original designation. Synonymy by Ferris, 1937: 34.

Thymaspis Šulc, 1934: 2. Type species: *Thymaspis fusca* Šulc, by monotypy and original designation. Synonymy by Gómez-Menor Ortega, 1954: 120.

Hemiberlesiella Thiem & Gerneck, 1934a: 132. Type species: *Aspidiotus canariensis* Lindinger, by original designation. Synonymy by Ferris, 1943a: 99.

Arundaspis Borchsenius, 1949c: 737. Type species: *Arundaspis secreta* Borchsenius, by monotypy & original designation. Synonymy by Danzig, 1993: 227.

SYSTEMATICS: *Rhizaspidiotus* MacGillivray close to *Targionia*, differing in form and distribution of dorsal macroducts on pygidium. In *Rhizaspidiotus,* ducts short and distributed sporadically, not in well-defined furrows as on *Targionia* (Balachowsky, 1951, 1958b; Danzig, 1993).

GENERAL: Definition and characters by Ferris (1938a, 1943a), Lupo (1948, 1954, 1957), Borchsenius (1949c, 1950b), Balachowsky (1951, 1958b), Zahradník (1952), Borchsenius & Williams (1963), Bazarov & Shmelev (1971), Kosztarab & Kozár (1978), Yaşar (1995) and by Kosztarab (1996).

KEYS: Gill 1997: 24-26 (female) [Genera of California]; Danzig 1993: 228 (female) [Palearctic]; Tereznikova 1986: 78 (female) [Ukraine]; Danzig 1980b: 296 (female) [Far East of USSR]; Kosztarab & Kozár 1978: 144-147 (female) [Hungary]; Bazarov & Shmelev 1971: 171 (female) [Central Asia]; Bazarov & Shmelev 1971: 171 (female) [Central Asia]; Ezzat & Afifi 1966: 371-372 (female) [Egypt]; Danzig 1964: 646 (female) [Europe]; Zahradník 1959a: 546 (female) [Czech Republic]; Balachowsky 1958b: 283 (female) [*Targionina* of Africa]; Ezzat 1958: 237-239 (female) [Egypt]; McKenzie 1956: 23 (female) [U.S.A.: California]; Balachowsky 1951: 631 (female) [Mediterranean]; Borchsenius 1950b: 167 (female) [USSR]; Borchsenius 1950b: 168 (female) [USSR]; Ferris 1942: 28 (female) [North America]; Ferris 1942: 40 (female) [species North America].

CITATIONS: Balach1948b [taxonomy: 268]; Balach1951 [taxonomy, description: 650-651]; Balach1953g [taxonomy: 727]; Balach1958b [taxonomy, description: 288-290]; BazaroSh1971 [taxonomy, description: 171-172,179]; BlayGo1993 [taxonomy, description: 414]; Bodenh1949 [taxonomy, description: 34]; Bodenh1952 [taxonomy, description: 342-343]; Borchs1949c [taxonomy, description: 737]; Borchs1949d [taxonomy: 195,247]; Borchs1950b [taxonomy, description: 211-212,232]; Borchs1966 [catalogue: 245,248]; BorchsWi1963 [taxonomy, description: 381-384]; Brimbl1958 [taxonomy: 74]; Bustsh1958 [taxonomy: 221]; Danzig1964 [taxonomy: 653]; Danzig1988 [taxonomy: 724]; Danzig1993 [taxonomy, description: 227-228]; DanzigPe1998 [catalogue: 348-349]; Ezzat1958 [taxonomy: 238]; Ferris1921b [taxonomy: 94]; Ferris1937a [taxonomy: 33-34,42]; Ferris1937e [taxonomy: 528]; Ferris1938a [taxonomy, description: 262]; Ferris1942 [taxonomy: 21]; Ferris1943a [taxonomy, description: 84,99]; Ghauri1962 [taxonomy: 200]; Gill1997 [taxonomy: 251]; GomezM1954 [taxonomy, description: 120]; KaussaBa1953b [taxonomy: 276]; Koszta1996 [taxonomy, description: 587]; KosztaKo1978 [taxonomy, description: 174-175]; Kozar1990f [distribution: 142,143]; Lindin1937 [taxonomy: 195]; Lupo1948 [taxonomy, description: 198-199]; Lupo1954 [taxonomy, description: 28]; Lupo1957 [taxonomy, description : 66-67]; MacGil1921 [taxonomy, description: 390-391,430-431]; McKenz1956 [taxonomy: 23]; Miller1990 [taxonomy: 169-178]; MorrisMo1966 [taxonomy, catalogue: 15,91,173]; Schmut1959 [taxonomy: 121]; Sulc1934 [taxonomy, description: 1-3]; TangHaSh1991 [taxonomy: 458]; Tao1999 [taxonomy: 116]; ThiemGe1934a [taxonomy, description: 132,230,232]; Yasar1995a [taxonomy, description:127]; Zahrad1952 [taxonomy, description: 157].

Rhizaspidiotus adiscus Gómez-Menor Ortega, new status

Rhizaspidiotus artemisiae adiscus Gómez-Menor Ortega, 1968: 546. Type data: SPAIN: Manga del Mar Menor, Murcia, on *Artemisia coerulescens*. Syntypes, female. Type depository: Madrid: Museo Nacional de Ciencias Naturales, Spain.

SYSTEMATICS: Gómez-Menor Ortega (1968) distinguished *Rhizaspidiotus artemisiae adiscus* from *Rhizaspidiotus artemisiae artemisiae* only by absence of perivulvar disc pores. Balachowsky (1951) showed that this character varies in *Rhizaspidiotus artemisiae* and these pores may be absent in individuals of same population. Until types of *Rhizaspidiotus artemisiae adiscus* have been studied, this sub-species is here raised to species level and retained as a distinct species.

SCALE COVER: Scale cover not described by Gómez-Menor Ortega (1968).
HOST PLANTS: **Compositae**: *Artemisia coerulescens* [GomezM1968].
DISTRIBUTION: **Palaearctic**: Spain [GomezM1968].
CITATIONS: GomezM1968 [taxonomy, description, host, distribution: 546].

Rhizaspidiotus albatus Borchsenius

Rhizaspidiotus albatus Borchsenius, 1949b: 351. Type data: TURKMENISTAN: Kopet-dag, urochishe Robertovskoe, Firuzinskoe gorge, on *Ephedra* sp. Syntypes, female. Type depository: St. Petersburg: (= Leningrad) Zoological Museum, Academy of Science, Russia.
Rhizaspidiotus albutus; Danzig, 1972b: 346. Misspelling of species name.
COMMON NAME: bol'shaya ephedrovaya shitovka [BazaroSh1971].
SCALE COVER: Female scale circular or broadly elliptical, diameter 1.5-2 mm; convex; white; exuviae yellow, placed centrally (Borchsenius, 1949b).
HOST PLANTS: **Ephedraceae**: *Ephedra* [Borchs1949b, BazaroSh1971, Danzig1993], *Ephedra przewalskii* [Danzig1972b].
DISTRIBUTION: **Oriental**: Mongolia [Danzig1972b, Danzig1993]. **Palaearctic**: Armenia [BazaroSh1971, Danzig1993]; Kazakhstan (Vostochno Kazakhstan Oblast [BazaroSh1971, Danzig1993]); Tajikistan (=Tadzhikistan) [BazaroSh1971]; Turkmenistan [Borchs1949b, Danzig1993].
GENERAL: Description and illustration of adult female by Borchsenius (1949b) and by Bazarov & Shmelev (1971).
KEYS: Danzig 1993: 228 (female) [Palearctic]; Bazarov & Shmelev 1971: 172 (female) [Central Asia].
CITATIONS: Balach1951 [taxonomy: 654]; BazaroSh1971 [taxonomy, description, illustration, host, distribution: 176-177]; Borchs1949b [taxonomy, description, illustration, host, distribution: 351-352]; Borchs1950b [taxonomy, description, host, distribution: 232-233]; Borchs1966 [catalogue: 246]; Bustsh1960 [taxonomy, host, distribution: 181]; Danzig1972b [host, distribution: 346-347]; Danzig1993 [taxonomy, description, illustration, host, distribution: 228,233]; DanzigPe1998 [catalogue: 349]; TerGri1962 [taxonomy, description, host, distribution: 150].

Rhizaspidiotus amoiensis Tang

Rhizaspidiotus amoiensis Tang, 1984: 14. Type data: CHINA: Fukien Province, Amoy, on *Imperata* sp. Holotype female. Type depository: Shanxi: Entomological Institute, Shanxi Agricultural University, Taigu, Shanxi, China.
SCALE COVER: Adult female scale nearly circular, convex, black brown; exuviae subcentral, ventral scale thick (Tang, 1984).
HOST PLANTS: **Gramineae**: *Imperata* [Tang1984].
DISTRIBUTION: **Oriental**: China (People's Republic) (Fujian (Fukien) [Tang1984]).
GENERAL: Description and illustration of adult female by Tang (1984).
CITATIONS: Tang1984 [taxonomy, description, illustration, host, distribution: 14-15]; Tao1999 [taxonomy, host, distribution: 116].

Rhizaspidiotus balachowskyi Kozár & Matile-Ferrero

Rhizaspidiotus balachowskyi Kozár & Matile-Ferrero, 1983: 392. Type data: HUNGARY: Szarsomlyo, on *Chrysopogon gryllus*. Holotype female. Type depository: Budapest: Hungarian Natural History Museum, Zoological Department, Hungary.

SCALE COVER: Female scale broad-oval, diameter, 1.5-2 mm; strongly convex; black; exuviae subcentral; ventral scale thick, uniting with dorsal scale to form an almost closed capsule. Male test of same shape and colour as female, but smaller (Kozár & Matile-Ferrero, 1983).

HOST PLANTS: **Gramineae**: *Chrysopogon gryllus* [KozarMa1983, Danzig1993].

DISTRIBUTION: **Palaearctic**: Hungary [KozarMa1983, Danzig1993].

BIOLOGY: Occurring on roots (Kozár & Matile-Ferrero, 1983).

GENERAL: Description and illustration of adult female by Kozár & Matile-Ferrero (1983).

KEYS: Danzig 1993: 228 (female) [Palearctic].

CITATIONS: Danzig1993 [taxonomy, host, distribution: 228,234-235]; DanzigPe1998 [catalogue: 349]; KozarMa1983 [taxonomy, description, illustration, host, distribution: 392-395]; Pelliz1994a [host, distribution: 271-274].

Rhizaspidiotus bivalvatus Goux

Targionia festucae Borchsenius, 1937a: 183. Nomen nudum.

Rhizaspidiotus bivalvatus Goux, 1937c: 341. Type data: FRANCE: Rhône, Courzieu, on *Festuca ovina*. Holotype female. Type depository: Paris: Muséum national d'Histoire naturelle, France.

Rhizaspidiotus festucae Kiritchenko, 1940: 117. Nomen nudum

Pseudodiaspis bivalvata; Lindinger, 1943b: 264. Change of combination requiring emendation of species name for agreement in gender.

Rhizaspidiotus bivalvatus; Borchsenius, 1966: 246. Revived combination.

SCALE COVER: Female scale bivalvate, thick, brown; larval exuviae darker placed eccentrically; scale wrinkled, formed of concentric zones; ventral and dorsal valves of similar structure and colour; 1.8-2.2 mm. Male unknown (Goux, 1937c).

HOST PLANTS: **Gramineae**: *Festuca* [Balach1951, Danzig1993], *Festuca arundinacae* [Foldi2000], *Festuca ovina* [Balach1951].

DISTRIBUTION: **Palaearctic**: France [Foldi2000]; Switzerland [KozarHi1996]; Tunisia [Balach1953j]; Ukraine (Krym (=Crimea) Oblast [Balach1951, Danzig1993]).

GENERAL: Description and illustration of adult female by Goux (1937c), Balachowsky (1951), Tereznikova (1986) and by Danzig (1993).

KEYS: Danzig 1993: 234 (female) [Palearctic]; Tereznikova 1986: 78 (female) [Ukraine]; Balachowsky 1951: 652 (female) [Mediterranean].

CITATIONS: Balach1951 [taxonomy, description, illustration, host, distribution: 665-668]; Balach1953j [taxonomy, host, distribution: 230-231]; Borchs1937a [taxonomy: 183]; Borchs1950b [taxonomy, description, host, distribution: 231,233]; Borchs1966 [catalogue: 246]; Danzig1964 [taxonomy, host, distribution: 653]; Danzig1993 [taxonomy, description, illustration, host, distribution: 228,234]; DanzigPe1998 [catalogue: 349]; Ferris1943a [taxonomy: 85,99]; Foldi2000 [host, distribution: 84]; Foldi2001 [distribution: 303-308]; Foldi2002 [host, distribution: 247]; Goux1937c [taxonomy, description, illustration, host, distribution: 341-345]; Goux1941a [taxonomy: 40]; Kiritc1940 [taxonomy: 117]; KozarHi1996 [host, distribution: 91-96]; Lindin1943b [taxonomy: 264]; Terezn1986 [taxonomy, description, illustration, host, distribution: 81].

Rhizaspidiotus canariensis (Lindinger)

Aspidiotus canariensis Lindinger, 1911a: 12. Type data: CANARY ISLANDS: Tenerife, near Santa Cruz; Gran Canaria; Gomera, near San Sebastian, on *Argyranthemum frutescens*. Syntypes, female. Type depository: Zoologisches Institut und Zoologishces Museum, Universität von Hamburg, Germany.

Hemiberlesia canariensis; Lindinger, 1918: 192. Change of combination.

Chorizaspidiotus canariensis; MacGillivray, 1921: 434. Change of combination.

Aspidiotus artemisiae Hall, 1926a: 20. Type data: EGYPT: between the 4th and 5th Towers Suez Road, on *Artemisia monosperma*. Syntypes, female. Type depository: London: Natural History Museum, England, UK. Synonymy by Danzig, 1970: 1021.

Aspidiotus kiritchenkoi Laing, 1929a: 487. Type data: UKRAINE: Odessa, on *Teucrium* sp. Holotype female. Type depository: London: The Natural History Museum, England, UK. Synonymy by Lindinger, 1957: 545.

Aspidiotus kiritshenkoi; Kiritchenko, 1931: 319. Misspelling of species name.

Epidiaspis canariensis; Lindinger, 1932: 202. Change of combination.

Thymaspis fusca Šulc, 1934: 3. Type data: CZECH REPUBLIC: near Brno, on *Thymus serphyllum*. Syntypes, female. Type depository: Brno: K. Šulc Collection, Moravian Museum, Czech Republic. Synonymy by Danzig, 1970: 1021.

Pseudodiaspis canariensis; Lindinger, 1935: 128. Change of combination.

Aspidiotus kiritshenkoi; Borchsenius, 1936: 131. Misspelling of species name.

Rhizaspidiotus artemisiae; Ferris, 1943a: 99. Change of combination.

Rhizaspidiotus canariensis; Ferris, 1943a: 99. Change of combination.

Rhizaspidiotus fusca; Ferris, 1943a: 99. Change of combination.

Thymaspis artemisiae; Bodenheimer, 1949: 70. Change of combination.

Rhizaspidiotus kiritshenkoi; Borchsenius, 1950b: 232. Change of combination.

Rhizaspidiotus pavlovskii Borchsenius, 1955b: 249. Type data: RUSSIA: Primorskii kray, Suchan (Partizansk), host plant not indicated. Lectotype female, by subsequent designation Danzig, 1993: 229. Type depository: St. Petersburg: Zoological Museum, Academy of Science, Russia. Synonymy by Danzig, 1970: 1021.

Hemiberlesiella canariensis; Lindinger, 1957: 549. Change of combination.

Rugaspidiotus artemisiae; Lindinger, 1957: 552. Change of combination.

Rhizaspidiotus canariensis; Danzig, 1993: 228. Revived combination.

COMMON NAMES: polinnaya shitovka [BazaroSh1971]; shitovka Kirichenko [Borchs1936].

SYSTEMATICS: Balachowsky (1951) and Borchsenius (1966) regarded *Rhizaspidiotus canariensis* (Lindinger) and *R. kiritchenkoi* (Laing) as distinct species. Danzig (1993) regarded both as belonging to one species that exhibited great intraspecific variation, but recognized in it five geographic morphs.

SCALE COVER: Lindinger (1911a) described scale cover as: "Female scale circular, up to 2.5 mm in diameter; white grey or brown grey; exuviae yellow, subcentral. Male scale narrow, linear, 1 mm long; white grey; exuviae yellow, towards cephalic end. Hall (1926a) described scale cover as: "Female scale highly convex; approximately circular in outline, diameter 1.25-1.5 mm; exuviae usually eccentric pale green in colour this colour being obscured by a film of white secretionary matter; colour very pale green. Male scale white with exuviae greenish obscured by white secretionary matter.

HOST PLANTS: **Compositae**: *Achillea* [Laing1929a, Bodenh1937, Balach1951, Danzig1970, Danzig1980b], *Ac. fragrantissima* [Hall1926a, Bodenh1935, Bodenh1949], *Ac. gerberi* [Kiritc1931], *Argyranthemum frutescens* [Lindin1911a], *Artemisia* [Bodenh1949, Borchs1955b, GomezM1956b, Danzig1970, Danzig1980b, Martin1983], *Ar. absinthium* [Kiritc1931], *Ar. austriaca* [Kiritc1931], *Ar. campestris* [Balach1930a, Kiritc1931, Balach1932d], *Ar. fragrans* [Bodenh1952], *Ar. glutinosa* [Balach1935b, Martin1983], *Ar. herba-alba* [Rungs1948, Martin1983], *Ar. judaica* [Hall1926a, Bodenh1935], *Ar. monosperma* [Hall1926a], *Ar. rutifolia* [BazaroSh1971], *Ar. valentina* [Martin1983], *Aster* [Borchs1955b, Danzig1970, Danzig1980b], *Aster amellus* [Lagows1990], *Centaurea* [Leonar1918, Leonar1920, Balach1951, Lupo1957], *Chrysanthemum frutescens* [Balach1951, Lupo1957], *Helichrysum* [Balach1951], *Santolina* [Martin1983], *Tanacetum* [Danzig1970, Danzig1980b]. **Convolvulaceae**: *Convolvulus trabutianus* [Rungs1942]. **Crassulaceae**: *Sedum* [Danzig1993]. **Euphorbiaceae**: *Euphorbia* [Danzig1970]. **Labiatae**: *Teucrium* [Laing1929a, Balach1951, Danzig1970], *Te. chamaedrys* [Kiritc1931], *Te. polium* [Kiritc1931], *Thymus communis* [Balach1932d], *Thy. hirtus* [Balach1935b, Martin1983], *Thy. mastichinus* [Martin1983], *Thy. serphyllum* [Sulc1934, Balach1951, Zahrad1952, Pelliz1987]. **Umbelliferae**: *Bupleurum lateriflorum* [Rungs1937, Balach1951, Balach1958b].

NATURAL ENEMIES: HYMENOPTERA **Encyrtidae**: *Paraschedius bicolor* Myartseva [Trjapi1989], *Paraschedius jasnoshae* Myartseva & Trjapitzin [Trjapi1989].

DISTRIBUTION: **Palaearctic**: Canary Islands [Lindin1911a, Leonar1920, Balach1946, Lupo1957, MatileOr2001]; Czech Republic [Sulc1934, Balach1951, Zahrad1952, Zahrad1977]; Egypt [Hall1926a, Bodenh1935c, Ezzat1958]; France [Balach1930a, Balach1932d]; Israel [Bodenh1935, Bodenh1937]; Italy [Leonar1918, Leonar1920, Lupo1957, Pelliz1987]; Kazakhstan (Alma Ata Oblast [BazaroSh1971]); Morocco [Rungs1937, Rungs1948]; Poland [Lagows1990]; Russia (Caucasus [Borchs1936], Karachay-Cherkessia AR [Danzig1985], Primor'ye Kray [Borchs1955b, Danzig1980b, Danzig1988]); Spain [Balach1935b, GomezM1937, GomezM1954, GomezM1956b, BlayGo1993]; Tajikistan (=Tadzhikistan) [BazaroSh1971]; Turkey [Balach1951, Bodenh1952]; Turkmenistan [Bustsh1960]; Ukraine (Krym (= Crimea) Oblast [Laing1929a, Balach1951, Terezn1986], Odessa Oblast [Balach1951, Terezn1986]); Western Sahara [Rungs1942].

GENERAL: Description and illustration of adult female by Lindinger (1911a), Hall (1926a), Laing (1929a), Šulc (1934), Zahradník (1952), Balachowsky (1951, 1958b), Bodenheimer (1952), Borchsenius (1955b), Bazarov & Shmelev (1971), Tereznikova (1986), Danzig (1980b, 1993) and by Yaşar (1995a). Description and illustration of first-instar nymph by Šulc (1934).

KEYS: Danzig 1993: 228 (female) [Palaearctic]; Tereznikova 1986: 78 (female) [Ukraine]; Kosztarab & Kozár 1978: 175 (female) [Hungary]; Bazarov & Shmelev 1971: 172 (female) [Central Asia]; Ezzat 1958: 242 (female) [Egypt]; Lupo 1957: 67 (female) [Italy]; Zahradník 1952: 157 (female) [Czech Republic]; Balachowsky 1951: 652 (female) [Mediterranean]; Leonardi 1920: 90 (female) [Italy].

CITATIONS: Balach1930a [host, distribution: 178]; Balach1932d [taxonomy, host, distribution: XLVI]; Balach1935b [host, distribution: 256-257]; Balach1946 [host, distribution: 212]; Balach1951 [taxonomy, description, illustration, host, distribution: 655-660]; Balach1958b [taxonomy, description, illustration, host, distribution: 290]; BalachMa1970 [host, distribution: 1081]; BazaroSh1971 [taxonomy, description, illustration, host, distribution: 172-176]; BlayGo1993 [taxonomy, description, illustration, host, distribution: 416-420]; Bodenh1929b [taxonomy, host, distribution: 106]; Bodenh1935 [host, distribution: 246]; Bodenh1935c [taxonomy, distribution: 1156]; Bodenh1937 [host, distribution: 216]; Bodenh1949 [taxonomy, description, illustration, host, distribution: 70-71]; Bodenh1952 [taxonomy, description, illustration, host, distribution: 344-345]; Borchs1936 [host, distribution: 131]; Borchs1937 [taxonomy, description, illustration, host, distribution: 132]; Borchs1949d [taxonomy, host, distribution: 248]; Borchs1950b [taxonomy, description, illustration, host, distribution: 231-232]; Borchs1955b [taxonomy, description, illustration, host, distribution: 249-250]; Borchs1966 [catalogue: 246-248]; Bustsh1960 [taxonomy, description, illustration, host, distribution: 181]; Danzig1964 [taxonomy, host, distribution: 653]; Danzig1970 [taxonomy, host, distribution: 1021-1022]; Danzig1977b [taxonomy: 57]; Danzig1980b [taxonomy, description, illustration, host, distribution: 334-335]; Danzig1985 [distribution: 112]; Danzig1988 [taxonomy, host, distribution: 724]; Danzig1993 [taxonomy, description, illustration, host, distribution: 228-232]; DanzigPe1998 [catalogue: 349-350]; Ezzat1958 [distribution: 242]; EzzatAf1966 [taxonomy, description, illustration, host, distribution: 379-381]; EzzatNa1987 [distribution: 88]; Ferris1937c [taxonomy, illustration: 51-52]; Ferris1937d [taxonomy: 132]; Ferris1941e [taxonomy: 41,44]; Ferris1943a [taxonomy: 99]; Foldi2001 [distribution: 303-308]; GomezM1937 [taxonomy, description, illustration, host, distribution: 75-76]; GomezM1954 [host, distribution: 120]; GomezM1956b [host, distribution: 482]; GomezM1957 [host, distribution: 46]; GomezM1958a [host, distribution: 8]; GomezM1958c [host, distribution: 406]; GomezM1965 [host, distribution: 95]; GomezM1968 [host, distribution: 546]; Hall1926a [taxonomy, description, illustration, host, distribution: 20-21]; Hall1927b [taxonomy, description, host, distribution: 142-144]; Hosny1939 [taxonomy, host, distribution: 14]; Kiritc1931 [taxonomy, description, host, distribution: 319-320]; KosztaKo1978 [taxonomy, description, host, distribution: 175]; Lagows1990 [host, distribution: 261-264]; LagowsKo1996 [host, distribution: 32,35]; Laing1929a [taxonomy, description, illustration, host, distribution: 487-489]; Leonar1918 [host, distribution: 192]; Leonar1920 [taxonomy, description, illustration, host, distribution: 102-104]; Lindin1911a [taxonomy, description, illustration, host, distribution: 12-13]; Lindin1912b [taxonomy, description, host, distribution: 103]; Lindin1932 [taxonomy: 202]; Lindin1935 [taxonomy: 128]; Lindin1936 [taxonomy: 152]; Lindin1943b [taxonomy: 264]; Lindin1957 [taxonomy: 545,552]; LongoMaPe1995 [distribution: 128]; Lupo1957 [taxonomy, description, illustration, host, distribution: 72-77]; MacGil1921 [taxonomy, description, host, distribution: 434]; Martin1983 [taxonomy, host, distribution: 68]; MatileOr2001 [host, distribution: 190]; MillerDa1990 [host, distribution, economic importance: 305]; Myarts1982 [host, distribution, biological control: 39-46]; Pelliz1987 [host, distribution: 123]; Rungs1937 [host, distribution: 332-333]; Rungs1942 [host, distribution: 106-107]; Rungs1948 [host, distribution: 114]; Sassce1912 [taxonomy, host, distribution: 92]; Schmut1959 [taxonomy, host, distribution: 122]; Sulc1934 [taxonomy, description, illustration, host, distribution: 3-9,18-19]; TangHaSh1991 [taxonomy: 459]; Tao1999 [taxonomy, host, distribution: 116]; Terezn1986 [taxonomy, description, illustration, host, distribution: 78-79]; WeidneWa1968 [taxonomy: 172]; Yasar1995a [taxonomy, description, illustration, host, distribution: 128-129]; Zahrad1952 [taxonomy, description, illustration, host, distribution: 157-161]; Zahrad1977 [taxonomy, distribution: 121].

Rhizaspidiotus caraganae (Kiritchenko)

Targionia caraganae Kiritchenko, 1940: 115. Type data: UKRAINE: near Odessa, Holodnaya Balka; Popovka; Grosulovo; Zahar'evka on *Caragana frutescens*. Syntypes, female. Type depository: St. Petersburg: (= Leningrad) Zoological Museum, Academy of Science, Russia.

Rhizaspidiotus caraganae; Ferris, 1943a: 99. Change of combination.

COMMON NAME: chiligovaya shitovka [BazaroSh1971].

SCALE COVER: Female scale of medium size; circular elongated; convex; white grey or light brown; scale of fully grown females white; exuviae subcentral; first exuviae grey or brown, smaller than second exuviae; secreted part of scale about twice size of second exuviae (Kiritchenko, 1940).

HOST PLANTS: **Leguminosae**: *Caragana arborescens* [Balach1951 BazaroSh1971, Danzig1993], *Ca. frutescens* [Kiritc1940], *Ca. grandiflora* [Danzig1993], *Ca. zandifora* [BazaroSh1971], *Halimodendron halodendron* [Danzig1993].

DISTRIBUTION: **Palaearctic**: Georgia [Danzig1993] (Georgia [Balach1951, BazaroSh1971]); Kazakhstan (Karaganda Oblast [BazaroSh1971], Karaganda Oblast [Danzig1993], Vostochno Kazakhstan Oblast [Danzig1993]); Ukraine (Odessa Oblast [Kiritc1940, Ferris1943a, BazaroSh1971]).

GENERAL: Description and illustration of adult female by Kiritchenko (1940), Balachowsky (1951), Bazarov & Shmelev (1971), Tereznikova (1986) and by Danzig (1993).

KEYS: Danzig 1993: 228 (female) [Palearctic]; Tereznikova 1986: 78 (female) [Ukraine]; Bazarov & Shmelev 1971: 172 (female) [Central Asia]; Balachowsky 1951: 651 (female) [Mediterranean].

CITATIONS: Balach1951 [taxonomy, description, illustration, host, distribution: 663-665]; BazaroSh1971 [taxonomy, description, illustration, host, distribution: 177-179]; Borchs1949 [taxonomy, description, host, distribution: 248]; Borchs1950b [taxonomy, description, illustration, host, distribution: 231,233]; Borchs1966 [catalogue: 246-247]; Danzig1964 [taxonomy, host, distribution: 653]; Danzig1993 [taxonomy, description, illustration, host, distribution: 228,232-233]; DanzigPe1998 [catalogue: 350]; Ferris1943a [taxonomy, host, distribution: 85,88,99]; Kiritc1940 [taxonomy, description, illustration, host, distribution: 115-117]; Terezn1986 [taxonomy, description, illustration, host, distribution: 79-81].

Rhizaspidiotus dearnessi (Cockerell)

Aspidiotus dearnessi Cockerell, 1898s: 267. Type data: CANADA: Ontario, shore of Lake Huron, on *Arctostaphylos uva-ursi*. Syntypes, female. Type depository: Washington: United States National Entomological Collection, U.S. National Museum of Natural History, District of Columbia, USA.

Aspidiotus (Targionia) gutierreziae Cockerell & Parrott, 1899: 277. Type data: U.S.A.: New Mexico, Mesilla Valley, near the Agricultural College, on stems of *Gutierrezia lucida*. Syntypes, female. Type depository: Washington: United States National Entomological Collection, U.S. National Museum of Natural History, District of Columbia, USA. Synonymy by Ferris, 1943a: 99.

Aspidiotus (Targionia) helianthi Parrott, 1899a: 176. Type data: U.S.A.: Kansas, Wabawnsee County, near Hackberry Glen, on roots of a sunflower, *Helianthus annuus*. Syntypes, female. Type depository: Washington: United States National Entomological Collection, U.S. National Museum of Natural History, District of Columbia, USA. Synonymy by Ferris, 1941e: 44.

Aspidiotus (Targionia) dearnessi; Cockerell, 1899a: 395. Change of combination.

Targionia dearnessi; Leonardi, 1900: 343. Change of combination.

Targionia gutierreziae; Leonardi, 1900: 343. Change of combination.

Targionia helianthi; Fernald, 1903b: 297. Change of combination.

Rhizaspidiotus helianthi; MacGillivray, 1921: 431. Change of combination.

Chorizaspidiotus guterriziae; MacGillivray, 1921: 432. Change of combination.

Chorizaspidiotus guterriziae; MacGillivray, 1921: 432. Misspelling of species name. Notes: Misspelling of *Aspidiotus (Targionia) gutierreziae*.

Remotaspidiotus dearnessi; MacGillivray, 1921: 434. Change of combination.

Pseudodiaspis helianthi; Lindinger, 1937: 194. Change of combination.

Aspidiotus gutierreziae; Ferris, 1937a: 33. Change of combination.

Rhizaspidiotus dearnessi; Ferris, 1938a: 263. Change of combination.

Targionia (Rhizaspidiotus) dearnessi; Merrill, 1953: 80. Change of combination.

Rhizaspidiotus dearnessi; McKenzie, 1956: 26. Revived combination.

COMMON NAME: dearness scale [McKenz1956, Dekle1965c].

SCALE COVER: Cockerell (1898s) described as: "Female scale suboval, 2 mm in diameter; moderately convex; pale grey, concentrically ridged; the orange-yellow, partly exposed exuviae to one side; ventral scale thick, distinct; the scale resemble minute oyster-shells". Ferris (1938a) described as: "Female scale grey or brownish, circular, high convex, exuviae central, a strong ventral scale formed; male scale elongate, brown, exuvia at one end". Colour photograph by Gill (1997).

HOST PLANTS: **Chenopodiaceae**: *Salicornia* [BesheaTiHo1973]. **Compositae** [Ferris1942, BesheaTiHo1973], *Ambrosia* [Ferris1938a, McKenz1956, Dekle1965c, McDani1970], *Aplopappus heterophyllus* [McKenz1956], *Aplopappus laricifolius* [McKenz1956], *Aplopappus linearifolius* [McKenz1956], *Artemisia* [Ferris1938a], *Artemisia filifolia* [Ferris1938a, McKenz1956, McDani1970], *Aster* [Ferris1938a, McKenz1956], *Baccharis pilularis* [McKenz1956], *Chrysoma laricifolia* [Ferris1938a, McDani1970], *Composite* [McKenz1956], *Corethrogyne* [Ferris1920b, McKenz1956], *Eriophyllum confertiflorum* [Ferris1920b], *Franseria dumosa* [Ferris1938a, McKenz1956], *Grindelia cuneifolia* [Ferris1920b, McKenz1956], *Gutierrezia* [Ferris1938a, McDani1970], *Gutierrezia lucida* [Ferris1938a, McKenz1956], *Gymnolomia tenuifolia* [Ferris1938a, McDani1970], *Helianthus* [Ferris1938a, McKenz1956], *Helianthus annuus* [Parrot1899a], *Isocoma wrightii* [Ferris1938a], *Parthenium incanum* [Ferris1938a, McKenz1956, McDani1970], *Solidago* [BesheaTiHo1973], *Viguiera tenuifolia* [McKenz1956]. **Empetraceae**: *Ceratiola ericoides* [BesheaTiHo1973]. **Ericaceae**: *Arctostaphylos uva-ursi* [Cocker1898s, Leonar1900, MerrilCh1923, Ferris1938a, McKenz1956]. **Labiatae** [BesheaTiHo1973]. **Leguminosae**: *Kuhnistera pinnata* [MerrilCh1923], *Petalostemon corymbosus* [Dekle1965c]. **Polygonaceae**: *Eriogonum fasciculatum* [Ferris1938a, McKenz1956], *Thysanella fimbriata* [Merril1953, Dekle1965c]. **Rhamnaceae**: *Ceanothus americanus* [Ferris1938a, McKenz1956, McDani1970]. **Verbenaceae**: *Verbena* [Dekle1965c]. **Zygophyllaceae**: *Larrea* [McKenz1956].

NATURAL ENEMIES: HYMENOPTERA **Encyrtidae**: *Ceraptroceroideus cinctipes* Girault [Gordh1979], *Coccidencyrtus ensifer* (Howard) [Gordh1979].

DISTRIBUTION: **Nearctic**: Canada (Ontario [Cocker1898s, Ferris1938a]); Mexico [Ferris1942]; United States of America (Alabama [Nakaha1982], Arizona [Ferris1938a], California [Ferris1920b, McKenz1956], Colorado [Nakaha1982], Florida [MerrilCh1923, Merril1953, Dekle1965c, BesheaTiHo1973], Georgia [BesheaTiHo1973], Indiana [Nakaha1982], Kansas [Parrot1899a, Ferris1938a], Maryland [Nakaha1982], Massachusetts [Nakaha1982], Missouri [Hollin1923], Nebraska [Nakaha1982], New Jersey [Nakaha1982], New Mexico [CockerPa1899, Ferris1938a], New York [Ferris1938a], North Carolina [Nakaha1982], Oklahoma [Nakaha1982], South Carolina [Nakaha1982], Texas [Ferris1938a, McDani1970], Utah [Ferris1938a], Utah [Nakaha1982], Virginia [Nakaha1982], Wisconsin [Nakaha1982]). **Neotropical**: Cuba [Nakaha1982].

BIOLOGY: Occurring on crowns and stems, at times below ground surface (Ferris, 1938a).

GENERAL: Description and illustration of adult female by Ferris (1920b, 1938a), McKenzie (1956), Kosztarab (1996) and by Gill (1997).

KEYS: McKenzie 1956: 26 (female) [U.S.A.: California]; Ferris 1942: 40 (female) [North America]; Lawson 1917: 246 (female) [U.S.A.: Kansas]; Cockerell 1905b: 201 (female) [U.S.A.: Colorado].

CITATIONS: Balach1951 [taxonomy, description, illustration, host, distribution: 653-654]; BeardsDaHo1976 [economic importance: 103]; BesheaTiHo1973 [host, distribution: 8]; Borchs1966 [catalogue: 247]; Cocker1898s [taxonomy, description, host, distribution: 266-267]; Cocker1899a [taxonomy: 395]; Cocker1905b [taxonomy: 201]; CockerPa1899 [taxonomy, description, host, distribution: 277-

278]; Danzig1993 [taxonomy: 227]; Dekle1965c [taxonomy, description, host, distribution: 130]; Dekle1976 [taxonomy, description, host, distribution, economic importance: 149]; FDACSB1987 [host, distribution: 4-7]; Fernal1903b [catalogue: 296-297]; Ferris1920b [taxonomy, description, illustration, host, distribution: 56]; Ferris1921b [taxonomy: 94]; Ferris1937a [taxonomy, illustration: 33,42]; Ferris1938a [taxonomy, description, illustration, host, distribution: 263]; Ferris1941e [taxonomy: 42,44]; Ferris1942 [taxonomy: 446:40]; Ferris1943a [taxonomy: 85,86,99]; Gill1997 [host, distribution, taxonomy, description, illustration, economic importance: 251-253]; Gordh1979 [biological control: 944]; Hollin1923 [taxonomy, description, host, distribution: 36-37]; Koszta1996 [taxonomy, description, illustration, host, distribution, life history: 589-590]; Lacroi1926 [taxonomy, description, host, distribution, life history, economic control: 250]; Lawson1917 [taxonomy, description, illustration, host, distribution: 246-247]; Leonar1900 [taxonomy, host, distribution: 247,343]; Lindin1937 [taxonomy: 194]; Lobdel1937 [taxonomy: 78]; MacGil1921 [taxonomy, description, host, distribution: 430-434]; McDani1970 [taxonomy, illustration, host, distribution: 437-439]; McKenz1956 [taxonomy, description, illustration, host, distribution: 83-84]; Merril1953 [taxonomy, description, host, distribution: 80-81]; MerrilCh1923 [taxonomy, description, host, distribution, economic importance: 252-253]; MillerDa1990 [host, distribution, economic importance: 305]; Nakaha1982 [host, distribution: 80-81]; Parrot1899 [taxonomy, description, illustration, host, distribution: 176]; TangHaSh1991 [taxonomy: 459].

Rhizaspidiotus donacis (Leonardi)

Targionia donacis Leonardi, 1920: 108. Type data: CROATIA: Dalmazia, Lacroma, on *Arundo donax*. Syntypes, female. Type depository: Portici: Dipartimento de Entomologia e Zoologia Agraria di Portici, Università di Napoli Federico II, Italy. Notes: Incorrect citation of Lindinger as author.

Targionia donacis; Gómez-Menor Ortega, 1937: 127. Notes: Incorrect citation of "Lindinger" as author.

Rhizaspidiotus donacis; Ferris, 1943a: 99. Change of combination.

SCALE COVER: Female scale almost circular, about 2 mm in diameter; with margin wavy in various forms; slightly convex; larval exuviae, yellow, placed at top of scale; nymphal exuviae black; scale colour black grey; surface of scale rugose; scale robust and fragile; ventral vellum complete, robust. Male scale oval elongated, margin parallel-sided; larval exuviae yellow; ventral vellum well developed; 0.9-0.95 mm long (Leonardi, 1920). Illustration of female and male scale cover by Leonardi (1920).

HOST PLANTS: **Gramineae**: *Arundo donax* [Leonar1920, Balach1928c, Balach1932d, Balach1932e, Ferris1943a, Bachma1953, Martin1983], *Phragmites australis* [SengonUyKa1998, UygunSeEr1998].

NATURAL ENEMIES: HYMENOPTERA **Aphelinidae**: *Aphytis acrenulatus* Rosen & DeBach [Garonn1994]. **Eulophidae**: *Euderus* [SengonUyKa1998].

DISTRIBUTION: **Palaearctic**: Algeria [Balach1928c, Balach1932d]; Croatia [Leonar1920]; France [Balach1930a, Balach1932d, Balach1932e]; Italy [LongoMaPe1995]; Spain [GomezM1937]; Turkey [UygunSeEr1998].

BIOLOGY: Beneath leaf sheaths, especially at nodes (Ferris, 1943a).
GENERAL: Description and illustration of adult female by Ferris (1943a) and by Balachowsky (1951).
KEYS: Danzig 1993: 228 (female) [Palearctic]; Lupo 1957: 67 (female) [Italy]; Balachowsky 1951: 651 (female) [Mediterranean]; Leonardi 1920: 104-105 (female) [Italy].
CITATIONS: Bachma1953 [host, distribution: 177]; Balach1928c [host, distribution: 279]; Balach1930a [host, distribution: 179]; Balach1932d [taxonomy, host, distribution, economic importance: XIII, XLIX]; Balach1932e [host, distribution: 237]; Balach1933e [host, distribution: 3]; Balach1935b [host, distribution: 260]; Balach1951 [taxonomy, description, illustration, host, distribution: 660-663]; BlayGo1993 [taxonomy, description, illustration, host, distribution: 421-424]; Borchs1966 [catalogue: 247]; DanzigPe1998 [catalogue: 350-351]; Ferris1943a [taxonomy, description, illustration, host, distribution: 85,100,111]; Foldi2001 [distribution: 303-308]; Garonn1994 [host, distribution, biological control: 53-59]; GomezM1937 [taxonomy, description, illustration, host, distribution: 127-128]; GomezM1958a [host, distribution: 7]; Leonar1920 [taxonomy, description, illustration, host, distribution: 108-111]; LongoMaPe1995 [distribution: 128]; Lupo1957 [taxonomy, description, illustration, host, distribution: 67-72]; Martin1983 [taxonomy, host, distribution: 68]; Rungs1933 [taxonomy: 116]; SengonUyKa1998 [host, distribution, biological control: 128-131]; UygunSeEr1998 [host, distribution: 183-191]; WeidneWa1968 [taxonomy: 179].

Rhizaspidiotus graminis Borchsenius
Rhizaspidiotus graminis Borchsenius, 1955b: 250. Type data: RUSSIA: Primorskii kray, Michailovskii region, on *Arundinella* sp. Syntypes, female. Type depository: St. Petersburg: (= Leningrad) Zoological Museum, Academy of Science, Russia.
SCALE COVER: Female scale circular or subcircular, diameter 1.5-1.7 mm; convex; brown black with thin layer of white secretion; exuviae dark brown, almost black, shiny; ventral scale thick, grey-brown, lighter in centre (Borchsenius, 1955b).
HOST PLANTS: **Gramineae**: *Arundinella* [Borchs1955b, Danzig1980b, Danzig1988, Danzig1993].
DISTRIBUTION: **Palaearctic**: Russia (Primor'ye Kray [Borchs1955b, Danzig1980b, Danzig1988, Danzig1993]).
GENERAL: Description and illustration of adult female by Borchsenius (1955b) and by Danzig (1980b, 1993).
KEYS: Danzig 1993: 228 (female) [Palearctic].
CITATIONS: Borchs1955b [taxonomy, description, illustration, host, distribution: 250-252]; Borchs1966 [catalogue: 247]; Danzig1977b [taxonomy: 57]; Danzig1980b [taxonomy, description, illustration, host, distribution: 335-336]; Danzig1988 [taxonomy, host, distribution: 724]; Danzig1993 [taxonomy, description, illustration, host, distribution: 228,235]; DanzigPe1998 [catalogue: 351].

Rhizaspidiotus marginalis Hall & Williams

Rhizaspidiotus marginalis Hall & Williams, 1962: 40. Type data: MALAYSIA: Kepong, on fruits of *Calamus* sp. Holotype female. Type depository: London: The Natural History Museum, England, UK.

Rhizaspidiotus pavlovskii; Borchsenius, 1966: 248. Misspelling of species name. Notes: *Rhizaspidiotus pavlovskii* in Borchsenius, 1966 (p. 248, line 7) is a misspelling for *R. marginalis*.

SCALE COVER: Scale of adult female rather thick, more or less circular, white and low convex. Exuviae marginal, golden yellow or pale brown, coated with a film of white secretionary matter. Ventral scale well developed, often remaining attached to host plant. Diameter about 1.9 mm. Male scale white and narrowly elongate oval (Hall & Williams, 1962).

HOST PLANTS: **Palmae**: *Calamus* [HallWi1962].

DISTRIBUTION: **Oriental**: Malaysia (Malaya [HallWi1962]).

GENERAL: Description and illustration of adult female by Hall & Williams (1962).

CITATIONS: Borchs1966 [catalogue: 248]; HallWi1962 [taxonomy, description, illustration, host, distribution: 40-41].

Rhizaspidiotus secretus (Borchsenius)

Arundaspis secretus Borchsenius, 1949c: 738. Type data: TAJIKISTAN: South Tajikistan, near Shaartuz, on *Arundo* sp. Syntypes, female. Type depository: St. Petersburg: (= Leningrad) Zoological Museum, Academy of Science, Russia.

Rhizaspidiotus secretus; Balachowsky, 1951: 668. Change of combination.

Arundaspis secreta; Borchsenius & Williams, 1963: 381. Revived combination.

Rhizaspidiotus secretus; Danzig, 1993: 236. Revived combination.

SCALE COVER: Female scale oval or circular, large up to 3.5 mm; white; larval exuviae covered with white secretion (Borchsenius, 1949c).

HOST PLANTS: **Gramineae**: *Arundo* [Borchs1949c, Borchs1950b], *Arundo donax* [BazaroSh1971, Danzig1993], *Phragmites communis* [Danzig1972c].

DISTRIBUTION: **Palaearctic**: Afghanistan [Danzig1972c, Danzig1993]; Tajikistan (=Tadzhikistan) [Borchs1949c, Borchs1950b, BazaroSh1971, Danzig1993].

GENERAL: Description and illustration of adult female by Balachowsky (1951), Borchsenius & Williams (1963), Bazarov & Shmelev (1971) and by Danzig (1993).

KEYS: Balachowsky 1951: 652 (female) [Mediterranean].

CITATIONS: Balach1951 [taxonomy, description, illustration, host, distribution: 668-670]; BazaroSh1971 [taxonomy, description, illustration, host, distribution: 179-181]; Borchs1949c [taxonomy, description, host, distribution: 738]; Borchs1950b [taxonomy, description, host, distribution: 212]; Borchs1966 [catalogue: 248]; BorchsWi1963 [taxonomy, description, illustration: 381,383]; Danzig1972c [host, distribution: 583]; Danzig1993 [taxonomy, description, illustration, host, distribution: 228,236-237]; DanzigPe1998 [catalogue: 351]; Lindin1957 [taxonomy: 545].

Rhizaspidiotus taiyuensis Tang, Hao, Shi & Tang
Rhizaspidiotus taiyuensis Tang, Hao, Shi & Tang, 1991: 458. Type data: CHINA: Shanxi, Taiyue Mountain, on *Artemisia argyii*; collected by Guang-lu Shi, August 19, 1985. Holotype female. Type depository: Shanxi: Entomological Institute, Shanxi Agricultural University, Taigu, Shanxi, China.
SCALE COVER: Female dorsal scale convex; ventral scale present; white; exuviae yellowish, placed centrally. Male scale elongate but similar in texture and colour to female (Tang et al., 1991).
HOST PLANTS: **Compositae**: *Artemisia argyii* [TangHaSh1991].
DISTRIBUTION: **Palaearctic**: China (People's Republic) (Shanxi (Shansi) [TangHaSh1991]).
GENERAL: Description and illustration of adult female by Tang et al. (1991).
CITATIONS: TangHaSh1991 [taxonomy, description, illustration, host, distribution: 458-459,463].

Rungaspis Balachowsky

Rungaspis Balachowsky, 1949c: 74. Type species: *Rungaspis trabuti* Balachowsky, by monotypy and original designation.
Sinaidiaspis Bodenheimer, 1951: 329. Type species: *Diaspis capparidis* Bodenheimer, by monotypy and original designation. Synonymy by Borchsenius, 1966: 280.
Sinaidaspis; Borchsenius, 1966: 280. Misspelling of genus name.
SYSTEMATICS: *Rungaspis* Balachowsky is characterized mainly by presence of median lobes only, and the considerable reduction in size of pygidial plates. Resembles *Palinaspis* Ferris, but differs in absence of pygidial paraphyses on segments 7, 8 (Balachowsky, 1949c, 1951, 1958b).
GENERAL: Definition and characters by Balachowsky (1949c, 1951, 1958b).
KEYS: Ben-Dov 1980: 268 (female) [world]; Balachowsky 1958b: 228 (female) [Palearctic]; Balachowsky 1951: 598 (female) [Mediterranean].
CITATIONS: Balach1949c [taxonomy, description: 74]; Balach1951 [taxonomy, description: 571]; Balach1958b [taxonomy, description: 216]; BenDov1980 [taxonomy, description: 265-268]; Bodenh1951 [taxonomy, description: 280]; Borchs1966 [catalogue: 280]; DanzigPe1998 [catalogue: 351]; MorrisMo1966 [taxonomy, catalogue: 177,184].

Rungaspis arcuata Munting
Rungaspis arcuata Munting, 1967a: 263. Type data: SOUTH AFRICA: Cape Province, Tulbagh district, 10 miles N.E. of Gydo Pass near Prince Alfred Hamlet, on *Elytropappus rhinocerotis*; collected by J. Munting, 1.iii.1966. Holotype female. Type depository: Pretoria: South African National Collection of Insects, South Africa; type no. 2156/9.
SCALE COVER: Female scale subcircular, about 1.5 mm in length; covered with dust and particles of sand; apparently brownish. Male scale whitish, of usual aspidiotine shape and about 1.2 mm in length (Munting, 1967a).
HOST PLANTS: **Compositae**: *Elytropappus rhinocerotis* [Muntin1967a].

DISTRIBUTION: **Afrotropical**: South Africa [Muntin1967a].

GENERAL: Description and illustration of adult female by Munting (1967a).

KEYS: Ben-Dov 1980: 268 (female) [world].

CITATIONS: BenDov1980 [taxonomy: 268]; Muntin1967a [taxonomy, description, illustration, host, distribution: 263-266].

Rungaspis capparidis (Bodenheimer)

Diaspis capparidis Bodenheimer, 1929b: 108. Type data: EGYPT: Sinai Peninsula, Wadi Isle, on *Capparis galeata*. Syntypes, female. Type depository: Bet Dagan: Department of Entomology, The Volcani Center, Israel.

Rungaspis trabuti Balachowsky, 1949c: 74. Type data: MOROCCO: on *Convolvulus trabutianus*. Syntypes, female. Type depository: Paris: Muséum national d'Histoire naturelle, France. Synonymy by Ben-Dov, 1980: 265.

Sinaidiaspis capparidis; Bodenheimer, 1951: 329. Change of combination.

Rungaspis capparidis; Borchsenius, 1966: 280. Change of combination.

SCALE COVER: Female scale circular; light brown in; with eccentric rings (Bodenheimer, 1929). Female scale subcircular, flat, larval exuviae central or subcentral bright yellow; secreted scale white, 1.6 mm (Balachowsky, 1951).

HOST PLANTS: **Asclepiadaceae**: *Calotropis procera* [Balach1958b]. **Capparidaceae**: *Capparis cartilanginea* [BenDov1980], *Capparis galeata* [Bodenh1929b]. **Convolvulaceae**: *Convolvulus trabutianus* [Balach1949c, Balach1951]. **Polygonaceae**: *Calligonum comosum* [Balach1951]. **Zygophyllaceae**: *Zygophyllum dimosum* [BenDov1980].

DISTRIBUTION: **Palaearctic**: Algeria [Balach1949c, Balach1951]; Egypt [Bodenh1929b, BenDov1980]; Iran [Balach1958b]; Israel [BenDov1980]; Morocco [Balach1951]; Sicily [Nucifo1991, LongoMaPe1995].

GENERAL: Description and illustration of adult female by Balachowsky (1949c, 1951, 1958b).

KEYS: Ben-Dov 1980: 268 (female) [world].

CITATIONS: Balach1949c [taxonomy, description, illustration, host, distribution: 74-76]; Balach1951 [taxonomy, description, illustration, host, distribution: 572-574]; Balach1958a [host, distribution: 36]; Balach1958b [taxonomy, description, illustration, host, distribution: 216-218]; BenDov1980 [taxonomy, host, distribution: 265-267]; Bodenh1929b [taxonomy, description, host, distribution: 108]; Bodenh1951 [taxonomy: 329]; Borchs1966 [catalogue: 280,281]; DanzigPe1998 [catalogue: 351-352]; LongoMaPe1995 [distribution: 128]; Nucifo1991 [host, distribution: 533-536].

Rungaspis macrolobis Kaussari

Rungaspis macrolobis Kaussari, 1958: 229. Type data: IRAN: Bandar Abbas, on *Anabasis aphylla*. Lectotype female, by subsequent designation Ben-Dov, 1980: 267. Type depository: Paris: Muséum national d'Histoire naturelle, France.

SCALE COVER: Female scale pale, placed within bark cracks of host plant. Male scale unknown (Kaussari, 1958). Female scale circular or slightly oval, about 1.5 mm long; white; exuviae placed centrally; scales concealed under thin, dry cork layers of twigs. Males present (Ben-Dov, 1980).

HOST PLANTS: **Chenopodiaceae**: *Anabasis aphylla* [BenDov1980], *Arthrocnemum macrostachyum* [BenDov1980], *Seidlitzia* [BenDov1980].
DISTRIBUTION: **Palaearctic**: Iran [BenDov1980]; Israel [BenDov1980].
GENERAL: Description and illustration of adult female by Ben-Dov (1980).
KEYS: Ben-Dov 1980: 268 (female) [world].
CITATIONS: BenDov1980 [taxonomy, description, illustration, host, distribution: 266-268]; Borchs1966 [catalogue: 280]; DanzigPe1998 [catalogue: 352]; Kaussa1958 [taxonomy, description, illustration, host, distribution: 229-230].

Sadaotakagia **Ben-Dov** replacement name

Takagia Tang, 1984: 10. Type species: *Takagia sishanensis* Tang, by monotypy and original designation. Junior homonym of *Takagia* Matsumura, 1951 (*Insecta Matsumrana* 16: 83) in the Cercopoidea.
NAME DERIVATION: The generic name *Sadaotakagia* is derived from the name of Dr. Sadao Takagi, Hokkaido University, Sapporo, Japan.
SYSTEMATICS: Tang (1984) assigned this genus to subtribe Pseudaonidina of Aspidiotinae. It is related to *Dichosoma* Brimblecombe, *Neomorgania* MacGillivray and *Mimeraspis* Brimblecombe, in the contiguous median lobes, but differs in possessing three pairs of pygidial lobes, and presence of spine-like plates anterior to third lobes.
GENERAL: Definition and characters by Tang (1984).
CITATIONS: DanzigPe1998 [catalogue: 358]; KosztaBeKo1986 [taxonomy, catalogue: 16]; Tang1984 [taxonomy, description: 10]; Tao1999 [taxonomy: 119].

Sadaotakagia sishanensis Tang new combination

Takagia sishanensis Tang, 1984: 10. Type data: CHINA: Yunnan Province, near Kunming, Sishan, on an undetermined woody plant. Syntypes, female. Type depository: Shanxi: Entomological Institute, Shanxi Agricultural University, Taigu, Shanxi, China.
Takagia sishanensia; Tang, 1984: 10. Misspelling of species name.
Takagia sishanensia; Tao, 1999: 119. Misspelling of species name.
SCALE COVER: Female scale circular; highly convex; deep brown; exuviae golden yellow (Tang, 1984).
DISTRIBUTION: **Oriental**: China (People's Republic) (Yunnan [Tang1984]).
GENERAL: Description and illustration of adult female by Tang (1984).
CITATIONS: DanzigPe1998 [catalogue: 358]; Tang1984 [taxonomy, description, illustration, host, distribution: 10-11]; Tao1999 [taxonomy, host, distribution: 119].

Saharaspis **Balachowsky**

Saharaspis Balachowsky, 1951: 567. Type species: *Hemiberlesia ceardi* Balachowsky, by monotypy and original designation.
SYSTEMATICS: Balachowsky (1951) indicated an affinity of this genus to *Murataspis* Balachowsky. The dorsal macroducts on *Saharaspis* thin and long, and disposed in longitudinal lines, whereas in *Murataspis* the ducts short and wide (Balachowsky, 1951).

GENERAL: Definition and characters by Balachowsky (1951).
KEYS: Balachowsky 1951: 599 (female) [Mediterranean].
CITATIONS: Balach1951 [taxonomy, description: 567]; Borchs1966 [catalogue: 279]; DanzigPe1998 [catalogue: 352]; Kozar1990f [distribution: 142]; MorrisMo1966 [taxonomy, distribution: 178].

Saharaspis ceardi (Balachowsky)

Hemiberlesia ceardi Balachowsky, 1928a: 129. Type data: MOROCCO: Colomb-Bechar (Sud oranais), on *Ficus carica*. Holotype female. Type depository: Paris: Muséum national d'Histoire naturelle, France.
Aspidiotus ceardi; Lindinger, 1936: 157. Change of combination.
Saharaspis ceardi; Balachowsky, 1951: 568. Change of combination.
SCALE COVER: Female scale subcircular, 1.8-2.2 mm; convex in middle, flat at margin; larval exuviae placed centrally or subcentrally; brown; scale covered with white secreted material (Balachowsky, 1951).
HOST PLANTS: **Leguminosae**: *Ceratonia siliqua* [InserrCa1987]. **Moraceae**: *Ficus carica* [Balach1928a, Balach1932d, Rungs1935, InserrCa1987], *Morus alba* [Balach1951]. **Oleaceae**: *Olea europaea* [Balach1928a, Balach1932d]. **Pistaciaceae**: *Pistacia lentiscus* [InserrCa1987], *Pistacia vera* [InserrCa1987]. **Rhamnaceae**: *Ziziphus lotus* [Balach1932d]. **Vitaceae**: *Vitis vinifera* [Balach1928a, Balach1932d].
DISTRIBUTION: **Palaearctic**: Algeria [Balach1932d]; Morocco [Balach1928a, Balach1932d]; Sicily [InserrCa1987, LongoMaPe1995].
ECONOMIC IMPORTANCE: Minor pest of grapevine and fig in several oases in Sahara (Balachowsky, 1951).
GENERAL: Description and illustration of adult female Balachowsky (1928a, 1951).
KEYS: Balachowsky 1928a: 132 (female) [North Africa].
CITATIONS: Balach1928a [taxonomy, description, illustration, host, distribution: 129-131]; Balach1932b [ecology: 517-522]; Balach1932d [taxonomy, host, distribution: VIII]; Balach1951 [taxonomy, description, illustration, host, distribution: 568-570]; Balach1958a [host, distribution: 36]; Borchs1966 [catalogue: 279]; DanzigPe1998 [catalogue: 352]; Ferris1941e [taxonomy: 279]; InserrCa1987 [host, distribution: 94]; Lindin1936 [taxonomy: 157]; Lindin1957 [taxonomy: 545]; LongoMaPe1995 [distribution: 129]; MillerDa1990 [host, distribution, economic importance: 305]; Rungs1935 [host, distribution: 271]; SchmutKlLu1957 [host, distribution, economic importance: 491].

Sakalavaspis Mamet

Sakalavaspis Mamet, 1954: 74. Type species: *Sakalavaspis perineti* Mamet, by original designation.
SYSTEMATICS: Mamet (1954) assigned this genus to "Diaspididae with 'two-barred' ducts", and indicated that it was "non-pupillarial", but no affinities suggested. Lindinger (1957) transferred both species, described by Mamet in *Sakalavaspis*, to *Aonidia*. However, all species currently placed in *Aonidia* are pupillarial.

GENERAL: Definition and characters by Mamet (1954).
CITATIONS: Borchs1966 [catalogue: 358]; Mamet1954 [taxonomy, description: 74]; MorrisMo1966 [taxonomy, catalogue: 178].

Sakalavaspis bilobis Mamet
Sakalavaspis bilobis Mamet, 1954: 76. Type data: MADAGASCAR: Périnet, on "Tavolo fohiravina". Holotype. Type depository: Paris: Muséum national d'Histoire naturelle, France.
Aonidia mameti Lindinger, 1957: 552. Unjustified replacement name.
SCALE COVER: Female scale rounded, much smaller than that of *Sakalavaspis perineti* Mamet, but similarly tilted to one side; black, obscured by bands of buff secretion; larval exuviae completely obscured by clear buff secretion; nymphal exuviae black, occupying almost whole of discal portion of scale; obscured by buff secretion (Mamet, 1954). Illustration of scale cover by Mamet (1954).
DISTRIBUTION: **Afrotropical**: Madagascar [Mamet1954, Borchs1966].
BIOLOGY: Female scales occur on under surface of leaves (Mamet, 1954).
GENERAL: Description and illustration of adult female by Mamet (1954).
CITATIONS: Borchs1966 [catalogue: 358]; Lindin1957 [taxonomy: 552]; Mamet1954 [taxonomy, description, illustration, host, distribution: 21,76-78].

Sakalavaspis perineti Mamet
Sakalavaspis perineti Mamet, 1954: 74. Type data: MADAGASCAR: Périnet, on "Tavolo malama". Holotype. Type depository: Paris: Muséum national d'Histoire naturelle, France.
Aonidia perineti; Lindinger, 1957: 522. Change of combination.
SCALE COVER: Female scale rounded, tilted to one side; secreted matter black, with a covering of white waxy secretion; larval exuviae black, with a yellowish border; nymphal exuviae occupying nearly whole of discal portion of scale, black; covered with a thin layer of buff secretion, with a yellowish border; scale very brittle; ventral vellum appearing like a flange around the scale (Mamet, 1954). Illustration of scale cover by Mamet (1954).
DISTRIBUTION: **Afrotropical**: Madagascar [Mamet1954, Borchs1966].
BIOLOGY: Female scales occur on upper surface of leaves (Mamet, 1954).
GENERAL: Description and illustration of adult female by Mamet (1954).
CITATIONS: Borchs1966 [catalogue: 358]; Lindin1957 [taxonomy: 552]; Mamet1954 [taxonomy, description, illustration, host, distribution: 21,74-76].

Schizaspis Cockerell & Robinson

Schizaspis Cockerell & Robinson, 1915a: 423. Type species: *Schizaspis lobata* Cockerell & Robinson, by monotypy.
Schizaspidiotus; MacGillivray, 1921: 456. Misspelling of genus name.
SYSTEMATICS: Originally this genus was assigned to Diaspidinae, but from later studies (Ferris, 1937c) it is clear that the genus belongs to Aspidiotinae.
GENERAL: Definition and characters by Cockerell & Robinson (1915a) and by Robinson (1917).

KEYS: Robinson 1917: 16-17 (female) [Philippines].
CITATIONS: Borchs1966 [catalogue: 258]; CockerRo1915a [taxonomy, description: 423]; Ferris1937c [taxonomy: 52,56,96]; Ferris1938b [taxonomy: 75]; Lindin1937 [taxonomy: 195]; MacGil1921 [taxonomy, description: 394,456]; MorrisMo1966 [taxonomy, catalogue: 180]; Robins1917 [taxonomy, description: 16,26].

Schizaspis lobata Cockerell & Robinson

Schizaspis lobata Cockerell & Robinson, 1915a: 423. Type data: PHILIPPINES: Luzon, Los Banos, on *Ficus nota*. Syntypes, female. Type depository: Washington: United States National Entomological Collection, U.S. National Museum of Natural History, District of Columbia, USA.
Schizaspidiotus lobata; MacGillivray, 1921: 456. Misspelling of genus name.
Aspidiotus lobatus; Lindinger, 1943a: 152. Change of combination requiring emendation of species name for agreement in gender.
Schizaspis lobata; Borchsenius, 1966: 258. Revived combination.
COMMON NAME: tibig scale [VelasqRi1969].
SCALE COVER: Female scale about 0.75 mm in diameter, nearly circular; flat; yellowish brown, the surface beaded with little prominences in concentric rows; exuviae large, sublateral or central, dull golden yellow, broad pyriform (Cockerell & Robinson, 1915a).
HOST PLANTS: **Moraceae**: *Ficus nota* [CockerRo1915a].
DISTRIBUTION: **Oriental**: Philippines (Luzon [CockerRo1915a]).
GENERAL: Description and illustration of adult female by Cockerell & Robinson (1915a).
CITATIONS: Borchs1966 [catalogue: 258]; CockerRo1915a [taxonomy, description, illustration, host, distribution: 423-424]; Ferris1937c [taxonomy, illustration: 52,96]; Lindin1943a [taxonomy: 152]; MacGil1921 [taxonomy, description, host, distribution: 456]; Robins1917 [taxonomy, description, host, distribution: 26]; VelasqRi1969 [host, distribution: 195-208]; Willia1985a [taxonomy: 236].

Schizentaspidus Mamet

Schizentaspidus Mamet, 1958a: 421. Type species: *Schizentaspidus loranthi* Mamet, by original designation.
Schizentaspidiotus; Borchsenius, 1966: 253. Misspelling of genus name.
SYSTEMATICS: Genus is closely allied to *Entaspidiotus* in presence of spur-shaped third lobe, and constriction between mesothorax and metathorax. Readily distinguished from *Entaspidiotus* in deep constriction and articulation between metathorax and first abdominal segment (Mamet, 1958a).
GENERAL: Definition and characters by Mamet (1958a) and by Williams & Watson (1988).
KEYS: Williams & Watson 1988: 20 (female) [Tropical South Pacific]; Mamet 1958a: 362 (female) [*Selenaspidus* complex].

CITATIONS: Borchs1966 [catalogue: 253]; Mamet1958a [taxonomy, description: 362,421-422]; MorrisMo1966 [taxonomy, catalogue: 180]; WilliaWa1988 [taxonomy, description: 237-239].

Schizentaspidus loranthi Mamet

Schizentaspidus loranthi Mamet, 1958a: 422. Type data: INDONESIA: Mollucas [=Maluka Province], Amboina, on *Loranthus amboinicum*. Holotype. Type depository: Paris: Muséum national d'Histoire naturelle, France.
Schizentaspidiotus loranthi; Borchsenius, 1966: 253. Misspelling of genus name.
SCALE COVER: Scale cover of female and male not observed (Mamet, 1958a).
HOST PLANTS: **Loranthaceae**: *Loranthus amboinicum* [Mamet1958a].
DISTRIBUTION: **Oriental**: Indonesia [Mamet1958a, Borchs1966].
GENERAL: Description and illustration of adult female by Mamet (1958a).
CITATIONS: Borchs1966 [catalogue: 253]; Mamet1958a [taxonomy, description, illustration, host, distribution: 422-423].

Schizentaspidus silvicola Williams & Watson

Schizentaspidus silvicola Williams & Watson, 1988: 239. Type data: PAPUA NEW GUINEA: Morobe P. coast, Buso, lower garden area, on *Myristica* sp. gall; collected 5.xi.1979. Holotype female. Type depository: London: The Natural History Museum, England, UK.
SCALE COVER: Female scale circular, with central exuviae. Male scale not known (Williams & Watson, 1988).
HOST PLANTS: **Araliaceae**: *Schefflera* [WilliaWa1988]. **Hippocrateaceae**: *Salacea* [WilliaWa1988]. **Myristicaceae**: *Myristica* [WilliaWa1988]. **Pandanaceae**: *Pandanus* [WilliaWa1988]. **Sapotaceae** [WilliaWa1988].
DISTRIBUTION: **Australasian**: Papua New Guinea [WilliaWa1988]; Solomon Islands [WilliaWa1988].
GENERAL: Description and illustration of female by Williams & Watson (1988).
CITATIONS: WilliaWa1988 [taxonomy, description, illustration, host, distribution: 238-239,241].

Selenaspidopsis Nakahara

Selenaspidopsis Nakahara, 1984: 935. Type species: *Selenaspidopsis browni* Nakahara, by monotypy and original designation.
SYSTEMATICS: The genus *Selenaspidopsis* is closely related to *Pseudoselenaspidus* differing only in position of thoracic constriction. On *Selenaspidopsis* it is between meso- and prothorax, whereas on *Pseudoselenaspidus* constriction between meso- and mesothorax (Nakahara, 1984).
GENERAL: Definition and characters by Nakahara (1984).
KEYS: McKenzie 1956: 23 (female) [U.S.A.: California].
CITATIONS: KosztaBeKo1986 [taxonomy, catalogue: 14-15]; Nakaha1984 [taxonomy, description: 935].

Selenaspidopsis browni Nakahara

Selenaspidopsis browni Nakahara, 1984: 936. Type data: MEXICO: Papantla, Veracruz, on *Chamaedorea* sp. Holotype female. Type depository: Washington: United States National Entomological Collection, U.S. National Museum of Natural History, District of Columbia, USA.
SCALE COVER: Scale cover not described by Nakahara (1984).
HOST PLANTS: **Palmae**: *Chamaedorea* [Nakaha1984].
DISTRIBUTION: **Nearctic**: Mexico (Veracruz [Nakaha1984]).
GENERAL: Description and illustration of adult female by Nakahara (1984).
CITATIONS: Nakaha1984 [taxonomy, description, illustration, host, distribution: 936-939].

Selenaspidopsis mexicana Nakahara

Selenaspidopsis mexicana Nakahara, 1984: 939. Type data: MEXICO: Santiago, Tuxtla, Veracruz, on *Chamaedorea* sp. Holotype female. Type depository: Washington: United States National Entomological Collection, U.S. National Museum of Natural History, District of Columbia, USA.
SCALE COVER: Scale cover not described by Nakahara (1984).
HOST PLANTS: **Palmae**: *Chamaedorea* [Nakaha1984].
DISTRIBUTION: **Nearctic**: Mexico (Veracruz [Nakaha1984]).
GENERAL: Description and illustration of adult female by Nakahara (1984).
CITATIONS: Nakaha1984 [taxonomy, description, illustration, host, distribution: 939-941].

Selenaspidus Cockerell

Aspidiotus (*Selenaspidus*) Cockerell, 1897i: 14. Type species: *Aspidiotus articulatus* Morgan, by monotypy and original designation.
Selenaspis; Leonardi, 1897a: 375. Misspelling of genus name.
Selenaspis Leonardi, 1898a: 51. Unjustified emendation.
Selenaspidus; Fernald, 1903b: 284. Change of status.
Selenaspidiotus Thiem & Gerneck, 1934a: 230. Synonymy by Borchsenius, 1966: 253.
Lelenaspidus; Kozár, 1990f: 142. Misspelling of genus name.
SYSTEMATICS: *Selenaspidus* Cockerell differs from other genera in the *Selenaspidus* complex, in having spur-shaped third lobe, prosoma not produced latero-posteriorly in form of well-developed lobes, and prosoma constricted between meso- and metathorax (Mamet, 1958a).
GENERAL: Definition and characters by Robinson (1917), Brain (1918), Ferris (1938a), Balachowsky (1951) and by Mamet (1958a).
KEYS: Colon-Ferrer & Medina-Gaud 1998: 28-32 (female) [Genera of Puerto Rico]; Gill 1997: 24-26 (female) [Genera of California]; Gill 1997: 256 (female) [Species of California]; Wolff & Corseuil 1993: 29 (female) [Brazil, Rio Grande do Sul]; Williams & Watson 1988: 20 (female) [Tropical South Pacific]; Chou 1985: 257 (female) [China]; Mamet 1962: 158 (female) [species Afrotropical]; Mamet 1958a: 362 (female) [*Selenaspidus* complex]; Mamet 1958a: 365-366 (female)

[Species of *Selenaspidus*]; Ferris 1942: 27 (female) [North America]; Ferris 1942: 40 (female) [species North America]; Robinson 1917: 16-17 (female) [Philippines]; Lindinger 1913: 64 (female) [Africa].

CITATIONS: Balach1951 [taxonomy, description: 690-691]; Balach1957 [taxonomy: 57]; BenDov1990h [taxonomy: 32]; Borchs1966 [catalogue: 253]; Brain1918 [taxonomy, description: 130]; Chou1985 [taxonomy, description: 257]; Cocker1897i [taxonomy, description: 14,31]; Cocker1899a [taxonomy: 395]; ColonFMe1998 [taxonomy, description: 80]; DanzigPe1998 [catalogue: 354]; Fernal1903b [catalogue: 284]; Ferris1937c [taxonomy: 52]; Ferris1938a [taxonomy, description: 264]; Ferris1942 [taxonomy: 27]; Gill1997 [taxonomy: 256]; Gowdey1921 [taxonomy: 30]; Hall1946a [taxonomy: 527]; Kozar1990f [taxonomy, distribution: 142]; Leonar1897a [taxonomy: 375]; Leonar1898a [taxonomy: 51]; Lepage1938 [taxonomy: 420]; Lindin1909d [taxonomy, description: 1-2]; Lindin1913 [taxonomy: 64]; Lindin1932f [taxonomy: 194]; Lindin1937 [taxonomy: 195]; MacGil1921 [taxonomy, description: 394,451-452]; Mamet1949 [taxonomy: 64]; Mamet1958a [taxonomy, description: 362-366]; McKenz1956 [taxonomy, description: 23]; Miller1990 [taxonomy: 169-178]; MorrisMo1966 [taxonomy, catalogue: 182]; Robins1917 [taxonomy, description: 17,28]; Schmut1959 [taxonomy, distribution: 45]; Tao1999 [taxonomy: 116]; ThiemGe1934a [taxonomy, description: 230,231]; WilliaWa1988 [taxonomy: 241]; WolffCo1993 [taxonomy: 29].

Selenaspidus albus McKenzie

Selenaspidus albus McKenzie, 1953a: 53. Type data: U.S.A.: California, Riverside County, Norco, on *Euphorbia Caput-Medusae*. Holotype female. Type depository: Washington: United States National Entomological Collection, U.S. National Museum of Natural History, District of Columbia, USA.

COMMON NAME: white euphorbia scale [McKenz1956].

SCALE COVER: Scale of adult female approximately 1.75 mm diameter, circular, white, with a yellowish brown, subcentral exuviae. Male scale not observed (McKenzie, 1953a). Colour photograph by Gill (1997).

HOST PLANTS: **Euphorbiaceae**: *Euphorbia* [Mamet1958a, Borchs1966], *Euphorbia ?atrispina* [Mamet1958a, Borchs1966], *Eu. abyssinica* [Mamet1958a, Borchs1966], *Eu. captiosa* [McKenz1956, Mamet1958a, Borchs1966], *Eu. caput-medusae* [McKenz1953a, McKenz1956, Mamet1958a, Borchs1966], *Eu. fimbriata* [McKenz1956, Mamet1958a, Borchs1966], *Eu. horrida* [Mamet1958a, Borchs1966], *Euphorbia polygona* [McKenz1956, Mamet1958a, Borchs1966], *Eu. pulvinata* [McKenz1956], *Eu. submammilaris* [McKenz1956, Mamet1958a, Borchs1966], *Eu. truncata* [McKenz1956, Mamet1958a, Borchs1966].

DISTRIBUTION: **Afrotropical**: Eritrea [Mamet1958a]; Namibia (Southwest Africa) [Nakaha1982]; South Africa [Mamet1958a, Borchs1966]. **Nearctic**: United States of America (California [McKenz1953a, McKenz1956, Mamet1958a]).

ECONOMIC IMPORTANCE: Gill (1997) recorded heavy populations on Euphorbias in California that may require control.

GENERAL: Description and illustration of adult female by McKenzie (1953a, 1956), Mamet (1958a) and by Gill (1997).

KEYS: Gill 1997: 256 (female) [Species of California]; Mamet 1958a: 365-366 (female) [World]; McKenzie 1956: 26 (female) [U.S.A.: California].
CITATIONS: Borchs1966 [catalogue: 253]; DanzigPe1998 [catalogue: 354]; Gill1997 [host, distribution, taxonomy, description, illustration, economic importance: 256-257,260]; KonstaGu1987 [host, distribution: 161]; LongoMaPe1995 [distribution: 129]; Mamet1958a [taxonomy, description, illustration, host, distribution: 366-367]; McKenz1953a [taxonomy, description, illustration, host, distribution: 53-56]; McKenz1956 [taxonomy, description, illustration, host, distribution: 83-85]; Nakaha1982 [host, distribution: 81]; Schmut1959 [taxonomy, description, host, distribution: 114-115].

Selenaspidus ambanjae Mamet

Selenaspidus ambanjae Mamet, 1951: 253. Type data: MADAGASCAR: Ambanja, on leaf of unknown plant. Holotype. Type depository: Paris: Muséum national d'Histoire naturelle, France.
SCALE COVER: Scale of female flat, whitish-grey, not translucent, circular; exuviae subcentral, light buff. Scale of male not observed (Mamet, 1951).
DISTRIBUTION: **Afrotropical**: Madagascar [Mamet1951, Mamet1958a].
GENERAL: Description and illustration of adult female by Mamet (1951, 1958a).
KEYS: Mamet 1962: 158 (female) [Afrotropical]; Mamet 1958a: 365-366 (female) [World].
CITATIONS: Borchs1966 [catalogue: 253]; Mamet1951 [taxonomy, description, illustration, host, distribution: 253]; Mamet1958a [taxonomy, description, illustration, host, distribution: 368-369]; Mamet1959a [host, distribution: 390]; Mamet1962 [taxonomy, description, host, distribution: 157-158].

Selenaspidus antsingyi Mamet

Selenaspidus antsingyi Mamet, 1950: 37. Type data: MADAGASCAR: Antsingy Nord, 63 km. East of Maintirano, on inner basis of leaves of *Dracaena* sp. Holotype. Type depository: Paris: Muséum national d'Histoire naturelle, France.
SCALE COVER: Female scale, subcircular, flat, white to very pale ochreous, transparent; exuviae subcentral, pale golden yellow, obscured by secretion. Male scale not observed (Mamet, 1950).
HOST PLANTS: **Agavaceae**: *Dracaena* [Mamet1950, Mamet1958a, Borchs1966].
DISTRIBUTION: **Afrotropical**: Madagascar [Mamet1950, Mamet1958a, Borchs1966].
GENERAL: Description and illustration of adult female by Mamet (1950, 1958a).
KEYS: Mamet 1958a: 365-366 (female) [World].
CITATIONS: Borchs1966 [catalogue: 253]; Mamet1950 [taxonomy, description, illustration, host, distribution: 37-38]; Mamet1958a [taxonomy, description, illustration, host, distribution: 368-370].

Selenaspidus articulatus (Morgan)

Aspidiotus articulatus Morgan, 1889a: 352. Type data: GUYANA: Demerara, on *Dictyospermum album*; coll. McIntire. Syntypes, female. Type depository: London: The Natural History Museum, England, UK.

Aspidiotus rufescens Cockerell, 1891: 5. Type data: JAMAICA: Kingston, on Heus and Olive trees. Syntypes, female. Synonymy by Cockerell, 1892i: x. Notes: Type material probably lost; no type material found in BMNH, USNM and UCDC; information from curators to Y. Ben-Dov, December 2002.

Aspidiotus refescens; Cockerell, 1891: 5. Misspelling of species name.

Aspidiotus articulatus rufescens; Cockerell, 1892: 5. Change of status.

Aspidiotus (*Selenaspidus*) *articulatus*; Cockerell, 1897i: 14, 23. Change of combination.

Aspidiotus (*Selenaspis*) *articulatus*; Leonardi, 1898a: 51. Misspelling of genus name.

Aspidiotus (*Selenaspidus*) *articulatus simplex* Grandpré & Charmoy, 1899: 21. Type data: MAURITIUS: on *Jasmin* [=*Jasminum*] sp. and on avocado. Syntypes, female. Synonymy by Mamet, 1958a: 370. Notes: Type-material lost (Mamet, 1941).

Selenaspidus articulatus; Fernald, 1903b: 284. Change of combination.

Selenaspidus articulatus simplex; Fernald, 1903b: 285. Change of combination.

Selenaspidus articulatus simplex; Fernald, 1903b: 285. Notes: Incorrect citation of "de Charm." as author.

Pseudaonidia articulatus; Marlatt, 1908: 132. Change of combination.

Aspidiotus articulatus simplex; Lindinger, 1909d: 3. Change of combination.

Selenaspidus articulatus; Lindinger, 1909d: 5. Revived combination.

Pseudaonidia articulata; Malenotti, 1916a: 339. Change of combination requiring emendation of species name for agreement in gender.

Selenaspidus (*Pseudaonidia*) *articulatus*; Seabra, 1922: 5. Change of combination.

Aspidiotus simplex; Ferris, 1941e: 48. Change of status.

Selenaspidus articulatus; Balachowsky, 1959: 362. Notes: Incorrect citation of "Borgan" as author.

Selenaspidus articulatus; Borchsenius, 1966: 253. Revived combination.

Selenaspidus rufescens; Borchsenius, 1966: 256. Change of combination.

COMMON NAMES: escama "rufous" [CoronaRuMo1997]; rufous scale [MerrilCh1923, Merril1953, McKenz1956, Dekle1965c]; West Indian red scale [SchmutKlLu1957].

SCALE COVER: Female scale greyish-white, depressed, more or less circular; exuviae in centre (Morgan, 1889a). Female scale, flat, circular, white; centrally placed exuviae darker; male scale almost white, oval, exuvia subcentral (Ferris, 1938a). Scale of female, circular, flat, semi-transparent, white to yellowish-brown; exuviae subcentral, yellowish; diameter about 2.3 mm. Scale of male smaller than female, elongate-oval, white to brownish; exuviae towards one end (Mamet, 1958a).

HOST PLANTS: **Agavaceae**: *Cordyline* [Lepage1938], *Cordyline terminalis* [Hempel1900a, Marlat1908, MerrilCh1923, Balach1951]. **Anacardiaceae**: *Anacardium occidentale* [WilliaWa1988], *Mangifera* [Balach1951], *Mangifera indica* [Marlat1908, Mamet1943a, Mamet1958a, Mamet1949, Borchs1966, WilliaWa1988], *Schinus molle* [Mamet1958a, Borchs1966], *Spondias dulcis* [WilliaWa1988], *Spondias purpurea* [MerrilCh1923]. **Annonaceae**: *Annona* [Houser1918], *Annona muricata* [WilliaWa1988], *Annona squamosa* [WilliaWa1988]. **Apocynaceae**: *Carissa edulis* [Brain1918, Mamet1958a, Borchs1966], *Carissa grandiflora* [Brain1918], *Mascarenhasia arborescens*

[Mamet1951, Mamet1958a, Borchs1966], *Nerium oleander* [Marlat1908, Mamet1954, McKenz1956, Mamet1958a, Borchs1966], *Plumeria acutifolia* [Mamet1943a, Mamet1958a, Mamet1949, Borchs1966], *Plumeria rubra* [WilliaWa1988], *Tabernaemontana* [MerrilCh1923]. **Araceae**: *Xanthosoma sagittifolium* [WilliaWa1988]. **Araliaceae**: *Hedera helix* [Mamet1958a, Borchs1966]. **Atherospermataceae**: *Laurelia indica* [MerrilCh1923]. **Averrhoaceae**: *Averrhoa bilimbi* [WilliaWa1988], *Averrhoa carambola* [Mamet1943a, Mamet1949, Mamet1958a, Borchs1966, WilliaWa1988]. **Bignoniaceae**: *Tecoma radicans* [Mamet1950, Mamet1958a, Borchs1966]. **Celastraceae**: *Celastrus laurinus* [Marlat1908]. **Cycadaceae**: *Cycas circinalis* [McKenz1956]. **Cyperaceae**: *Stenophyllus* [Mamet1958a, Borchs1966]. **Euphorbiaceae**: *Acalypha* [Mamet1958a, Borchs1966], *Acalypha tricolor* [WilliaWa1988], *Aleurites moluccana* [GomezM1941], *Codiaeum* [WilliaWa1988], *Codiaeum variegatum* [Dekle1965c, WilliaWa1988], *Croton* [McKenz1956], *Excoecaria* [WilliaWa1988], *Hevea brasiliensis* [WilliaWa1988]. **Flacourtiaceae**: *Aberia caffra* [Mamet1958a, Borchs1966]. **Gramineae**: *Saccharum officinarum* [WilliaWa1988]. **Guttiferae**: *Calophyllum calaba* [Houser1918], *Calophyllum inophyllum* [WilliaWa1988], *Garcinia ovalifolia* [Mamet1958a, Borchs1966], *Mammea* [MerrilCh1923]. **Lauraceae**: *Persea* [McKenz1956], *Persea americana* [Mamet1949, Borchs1966, WilliaWa1988]. **Lecythidaceae**: *Barringtonia asiatica* [WilliaWa1988]. **Leguminosae**: *Bauhinia* [Mamet1950, Mamet1958a, Borchs1966, Almeid1973b], *Bauhinia purpurea* [Mamet1958a, Borchs1966], *Bauhinia variegata* [WilliaWa1988], *Cassia alata* [WilliaWa1988], *Ceratonia siliqua* [Mamet1958a, Borchs1966], *Gliricidia sepium* [WilliaWa1988], *Phaseolus* [Mamet1958a, Borchs1966], *Tamarindus* [Marlat1908, WilliaWa1988], *Tamarindus indica* [Dekle1965c, WilliaWa1988]. **Lythraceae**: *Lagerstroemia indica* [WilliaWa1988]. **Magnoliaceae**: *Magnolia grandiflora* [Houser1918]. **Malpighiaceae**: *Malpighia urens* [GomezM1941]. **Malvaceae**: *Hibiscus* [Mamet1958a, Balach1959a, Borchs1966], *Hibiscus syriacus* [WilliaWa1988], *Thespesia propulnea* [WilliaWa1988]. **Marantaceae**: *Calathea* [WilliaWa1988]. **Meliaceae**: *Swietenia mahagoni* [GomezM1941], *Trichilia emetica* [Almeid1971]. **Monimiaceae**: *Tambourissa* [Matile1978]. **Moraceae**: *Artocarpus* [Marlat1908, Mamet1954, Mamet1958a, Borchs1966], *Ar. heterophyllus* [WilliaWa1988], *Ficus hochstetten* [Mamet1958a, Borchs1966], *Fi. nitida* [Merril1953, Balach1959a], *Fi. verrueocarpa* [Mamet1958a, Borchs1966]. **Musaceae**: *Musa* [WilliaWa1988], *Musa sapientum* [GomezM1941]. **Myrsinaceae**: *Ardisia crenata* [Mamet1943a, Mamet1949, Mamet1958a, Borchs1966]. **Myrtaceae**: *Decaspermum* [WilliaWa1988], *Eugenia* [WilliaWa1988], *Psidium guajava* [Marlat1908]. **Oleaceae**: *Jasminum* [GrandpCh1899, Mamet1943a, Borchs1966], *Olea* [McKenz1956], *Olea europaea* [Marlat1908, Mamet1943a, Mamet1949, Mamet1958a, Borchs1966]. **Orchidaceae**: *Arundina bambusifolia* [WilliaWa1988]. **Palmae** [Houser1918], *Chrysalidocarpus lutescens* [Mamet1949, Mamet1958a, Borchs1966], *Cocos nucifera* [Marlat1908, McKenz1956, Mamet1958a, Borchs1966, WilliaWa1988], *Dictyospermum album* [Morgan1889a, Marlat1908, Lepage1938, McKenz1956], *Elaeis* [Mamet1958a, Borchs1966], *Elaeis guineensis* [Mamet1958a, Borchs1966, Almeid1973b], *Hyphaene thebiaca* [Mamet1958a,

Borchs1966], *Neodypsis* [Mamet1951, Borchs1966], *Neodypsis decaryi* [Mamet1954, Mamet1958a, Borchs1966], *Phoenix dactylifera* [Houser1918, WilliaWa1988]. **Pandanaceae**: *Pandanus* [Marlat1908, MerrilCh1923, Lepage1938, Balach1951]. **Passifloraceae**: *Passiflora edulis* [WilliaWa1988]. **Polygonaceae**: *Muehlenbeckia calophyllum* [Houser1918], *Muehlenbeckia platyctoda* [Houser1918]. **Rosaceae**: *Rosa* [Marlat1908, McKenz1956]. **Rubiaceae**: *Coffea arabica* [Mamet1943a, Lepage1938, Mamet1949, Mamet1958a, Borchs1966, Muntin1969, WilliaWa1988], *Cof. canephora* [WilliaWa1988], *Cof. liberica* [Mamet1958a, Borchs1966], *Cof. macrocarpa* [Mamet1958a, Mamet1949, Borchs1966], *Cof. robusta* [Mamet1958a, Borchs1966], *Gardenia* [Cocker1899n, Marlat1908, MerrilCh1923, Lepage1938, McKenz1956, WilliaWa1988], *Gardenia scadens* [WilliaWa1988], *Ixora coccinea* [Newste1901a, Mamet1958a, Borchs1966, WilliaWa1988]. **Rutaceae**: *Chalcas* [Merril1953], *Citrus aurantifolia* [WilliaWa1988], *Ci. aurantium* [Mamet1958a, Borchs1966, WilliaWa1988], *Ci. decumana* [Mamet1958a, Borchs1966], *Ci. limon* [Houser1918, Almeid1971, WilliaWa1988], *Ci. maxima* [WilliaWa1988], *Ci. paradisi* [Houser1918], *Ci. reticulata* [WilliaWa1988], *Ci. sinensis* [Houser1918, WilliaWa1988], *Fortunella japonica* [WilliaWa1988]. **Sapindaceae**: *Litchi chinensis* [WilliaWa1988], *Nephelium* [WilliaWa1988]. **Sapotaceae**: *Chrysophyllum* [Marlat1908], *Chrysophyllum argyrophyllum* [Mamet1958a, Borchs1966], *Mimusops* [Almeid1971], *Mimusops chapelieri* [Mamet1954, Mamet1958a, Borchs1966]. **Solanaceae**: *Brunfelsia nitida* [MerrilCh1923]. **Theaceae**: *Camellia sinensis* [Mamet1958a, Borchs1966, WilliaWa1988]. **Urticaceae**: *Pilea urticifolia* [Mamet1958a, Mamet1949, Borchs1966]. **Verbenaceae**: *Lantana camara* [WilliaWa1988]. **Vitaceae**: *Vitis vinifera* [Marlat1908, WilliaWa1988].

NATURAL ENEMIES: COLEOPTERA **Coccinellidae**: *Chilocorus cacti* (L.) [QuezadCoDi1972], *Pentilia egena* Mulsant [BenvenGrBo1997], *Scymnus*? [QuezadCoDi1972]. FUNGI : *Aschersonia aleyrodis* Webber [QuezadCoDi1972]. **Ascomycotina**: *Nectria diploa* [EvansPr1990]. **Deuteromycotina**: *Fusarium* [EvansPr1990]. HYMENOPTERA **Aphelinidae**: *Aphytis chrysomphali* Mercet [QuezadCoDi1972], *Ap. diaspidis* (Howard) [RosenDe1979], *Ap. holoxanthus* DeBach [RosenDe1979], *Ap. lingnanensis* Compere [RosenDe1979], *Ap. roseni* DeBach & Gordh [DeBachGo1974, RosenDe1979], *Aspidiotiphagus citrinus agilior* Berlese [AnneckIn1971], *Azotus stylatus* Mercet [AnneckIn1971], *Physcus paolii* Mercet [AnneckIn1971]. **Encyrtidae**: *Adelencyrtus ficusae* Risbec [AnneckIn1971]. **Signiphoridae**: *Signiphora lutea* Rust [Woolle1990].

DISTRIBUTION: **Afrotropical**: Angola [Almeid1973b]; Cameroon [Mamet1958a]; Chad [Mamet1958a]; Comoros Islands [Matile1978]; Côte d'Ivoire (=Ivory Coast) [Mamet1958a]; Eritrea [DeLottNa1955, Mamet1958a]; Ghana [Mamet1958a]; Guinea [Mamet1958a]; Kenya [Mamet1958a]; Madagascar [Mamet1950, Mamet1951, Mamet1954, Mamet1958a, Mamet1959a, Borchs1966]; Mauritius [GrandpCh1899, Mamet1943a, Mamet1949, Mamet1953a, Mamet1958a, Borchs1966]; Mozambique [Mamet1958a, Almeid1971]; Namibia (Southwest Africa) [Muntin1969]; Réunion [Mamet1957, Mamet1958a]; São Tomé and Príncipe (São Tomé) [Seabra1921, Seabra1925, Mamet1958a, Castel1963]); South Africa [Marlat1908, BrainKe1917, Brain1918, Mamet1958a]; Tanzania

[Mamet1958a]; Togo [Mamet1958a]; Uganda [Gowdey1917, Mamet1958a, Borchs1966]; Zaire [Mamet1958a]; Zanzibar [Mamet1958a]; Zimbabwe [Hall1929, Mamet1958a]. **Australasian**: Australia [Mamet1958a]; Fiji [WilliaWa1988]; Solomon Islands [WilliaWa1988]. **Nearctic**: Mexico [Cocker1899n, Marlat1908]; United States of America (Florida [Marlat1908, Morril1911, Wilson1917, MerrilCh1923, Merril1953, Dekle1965c]). **Neotropical**: Barbados [Marlat1908, Mamet1958a]; Brazil [Hempel1900a, Marlat1908]; Colombia [Balach1959a, Kondo2001]; Costa Rica [Mamet1958a]; Cuba [Houser1918]; Ecuador [YustCe1956]; El Salvador [QuezadCoDi1972]; French Guiana [Mamet1958a]; Galapagos Islands [PeckHeLa1998]; Grenada [Mamet1958a]; Guadeloupe [Balach1957c]; Guyana [Mamet1958a]; Haiti [Mamet1958a]; Jamaica [Cocker1891, Cocker1892i, Newste1917b]; Martinique [Balach1957c]; Montserrat [Mamet1958a]; Nicaragua [Marlat1908]; Panama [Cocker1899n, Marlat1908]; Peru [Beingo1969a, Beders1969]; Puerto Rico & Vieques Island (Puerto Rico [Mamet1958a, Martor1976, ColonFMe1998]); Suriname [Mamet1958a]; Trinidad and Tobago (Trinidad [Mamet1958a, RosenDe1979]); U.S. Virgin Islands [Nakaha1983]; Venezuela [Mamet1958a]. **Oriental**: India [Mamet1958a]; Philippines [VelasqRi1969]; Sri Lanka [Mamet1958a]; Taiwan [Mamet1958a]. **Palaearctic**: Japan [Mamet1958a]; United Kingdom (England [Newste1901a, Marlat1908, Mamet1958a]).

ECONOMIC IMPORTANCE: Polyphagous, tropicopolitan species (see Distribution, Host plants and CABI1981b). Mainly a pest of citrus in several regions (Boyce, 1948; Schmutterer et al., 1957; Ebeling, 1959; Beingolea, 1963d; Beardsley & Gonzalez, 1975; Talhouk, 1975; Rose, 1990c). Recorded also a pest of other crops such as coffee, cocoa, avocado, mango, banana and palms (Schmutterer et al., 1957).

GENERAL: Description and illustration of adult female by Marlatt (1908), Brain (1918), Ferris (1938a), Balachowsky (1951, 1959a), McKenzie (1956), Mamet (1958a), Chou (1985, 1986), Gill (1997) and by Colon-Ferrer & Medina-Gaud (1998).

KEYS: Gill 1997: 256 (female) [Species of California]; Mamet 1962: 158 (female) [Afrotropical]; Mamet 1958a: 365-366 (female) [World]; McKenzie 1956: 26 (female) [U.S.A.: California]; Brain 1918: 131 (female) [South Africa]; Lindinger 1909d: 4-5 (female) [Genus *Selenaspidus*]; Marlatt 1908: 134 (female) [World].

CITATIONS: Aguila1980 [host, distribution, biological control: 83-91]; Almeid1971 [host, distribution: 15]; Almeid1973b [host, distribution: 11]; AnneckIn1971 [host, distribution, biological control: 2,29,35]; Balach1951 [taxonomy, description, illustration, host, distribution: 691-694]; Balach1957c [host, distribution: 200]; Balach1959a [host, distribution: 362]; Ballou1912a [host, distribution: 412-425]; Ballou1913 [host, distribution: 61-65]; Ballou1922a [host, distribution: 239]; Barber1893 [host, distribution: 50-51]; Bartra1974 [host, distribution, taxonomy, life history: 60-68]; BeardsDaHo1976 [economic importance: 106]; BeardsGo1975 [economic importance: 49]; Beders1969 [distribution, host, chemical control: 933-940]; Beingo1969a [taxonomy, description, illustration, host, distribution, life history: 119-129]; Beingo1969d [biological control: 827-838]; Beingo1977 [host, distribution, biological control, economic importance: 1-2]; BenDov1990c [taxonomy: 113]; Bennet1974 [biological control: 87-96]; BenvenGrBo1997 [biological control: 105-106]; BergmaStBr1988 [host, distribution: 27-28]; Bondar1914 [host, distribution, economic importance: 1064-1106]; Bondar1915 [host, distribution, economic importance: 44-47]; Borchs1966 [catalogue : 253-254,256]; Bourne1921 [host, distribution, economic importance]; Boyce1948 [host, distribution, economic importance, control]; Brain1918 [taxonomy, description, illustration, host, distribution: 135]; BrainKe1917 [distribution: 184]; Bunzli1935 [host, distribution]; BurgerUl1990 [economic importance: 313-327]; Buxton1920 [host, distribution: 287-303]; CABI1981b [host, distribution: 1-2]; CanaleVa1999 [biological control]; Castel1963 [distribution: 140]; CaveMa1994 [host, distribution, biological control: 3-8]; Chou1985 [taxonomy, description, host, distribution: 257-259]; Chou1986 [taxonomy, illustration: 657]; ChuaWo1990 [host, distribution, economic importance: 548]; ClapsWoGo2001a [taxonomy, host, distribution: 27]; Cocker1891 [taxonomy, description, host, distribution: 5]; Cocker1892b [taxonomy, description, host, distribution: 333]; Cocker1892c [taxonomy, host, distribution, life history, control: 380-382]; Cocker1892i [taxonomy, host, distribution: x]; Cocker1893cc [host, distribution: 100]; Cocker1896b [distribution: 334]; Cocker1897i [taxonomy, description, host, distribution: 14,23]; Cocker1899n [host, distribution: 24]; Cocker1900k [taxonomy: 350]; ColonFMe1998 [taxonomy, description, illustration, host, distribution: 80-82]; CoronaRuMo1997 [host, distribution: 38-41]; CorreaAnLi1997 [host, distribution, economic importance, biological control: 312-313]; Costa1997 [host, distribution, economic importance, chemical control: 294]; Danzig1972 [taxonomy, host, distribution, economic importance: 221]; DanzigPe1998 [catalogue: 355]; DavidsMi1990 [host, distribution, economic importance: 603-632]; DeBachGo1974 [host, distribution, biological control: 259-265]; deFreiMo1998 [life history, ecology: 19-24]; Dekle1965c [taxonomy, description, host, distribution: 131]; Dekle1976 [taxonomy, description, host, distribution, economic importance: 150]; DeLottNa1955 [host, distribution: 53-60]; DeSant1940 [biological control: 29-44]; DeSant1979 [biological control]; DicksoFl1955 [host, distribution: 614-615]; DomeniBaSc1997 [host, distribution, economic importance, life history: 227]; Dumble1954 [host, distribution, economic importance: 1]; DziedzKa1990 [host, distribution: 39-43]; Ebelin1949 [host, distribution, life history, control]; Edward1936 [host, distribution, economic importance: 335-337]; EvansPr1990 [biological control: 3-17]; Fawcet1948 [biological control: 627-664]; Fernal1903b [catalogue: 284, 285]; Ferris1937c [taxonomy, illustration: 51,97]; Ferris1938a [taxonomy, description, illustration, host, distribution: 265]; Ferris1941e [taxonomy: 41]; Ferris1942 [taxonomy: 446:40]; FisherDe1976 [biological control: 43-50]; Fleury1934 [host, distribution: 483-500]; FonsecAu1932a [host, distribution: 202-214]; FontenRoSu1987 [host, distribution, life history, ecology: 1-28]; Garcia1930 [host, distribution, biological control]; GarciaPiGr1997 [host, distribution, chemical control: 160]; Gentry1965 [host, distribution, economic importance]; Gill1997 [host, distribution, taxonomy, description, illustration, economic importance: 258,260]; GomezM1941 [host, distribution: 142]; GonzalHeSi1991 [host, distribution, biological control: 433-441]; Gowdey1917 [host, distribution: 189]; Gowdey1921 [taxonomy, host, distribution:

31]; GrandpCh1899 [taxonomy, description, host, distribution: 21]; GravenFeSa1992 [host, distribution, economic importance, chemical control: 215-222]; GravenLeMo1988 [host, distribution, life history, chemical control, biological control: 209-218]; GravenPaYa1993 [chemical control: 16-25]; GravenYaFe1992 [chemical control: 101-111]; Green1907 [host, distribution: 203]; Hall1929 [host, distribution: 356]; Hall1939 [taxonomy, description, host, distribution: 100]; Hamble1947 [host, distribution: 949-956]; Hargre1927 [host, distribution, economic importance: 113-128]; Hargre1937 [host, distribution, economic importance: 505-520]; Hempel1900a [taxonomy, description, host, distribution: 499]; Herrer1964 [host, distribution, life history: 1-8]; Hinckl1963 [host, distribution, biological control]; Houck1999 [life history, biological control: 97-118]; Houser1918 [host, distribution: 167-168]; Hoz1983 [host, distribution, life history, economic importance: 21-33]; HuffakGu1990a [biological control, economic importance: 473-492]; Hunt1939 [host, distribution: 548-566]; Ibarra1990 [host, distribution: 207-231]; Jeppso1969 [economic importance, chemical control, physiology: 917-921]; Johnst1915 [host, distribution, biological control: 1-33]; Jorgen1934 [taxonomy, host, distribution: 278]; Kerveg1932 [host, distribution, economic importance: 1653]; Kondo2001 [taxonomy, host, distribution: 45]; KondoKa1995 [host, distribution: 57-58]; KonstaGu1987 [host, distribution: 162]; Larter1937 [host, distribution: 65-72]; Leonar1897 [taxonomy: 286]; Leonar1898a [taxonomy, description, illustration, host, distribution: 51-53]; Leonar1914 [taxonomy, host, distribution: 203]; Lepage1938 [catalogue: 420-421]; Lepesm1947 [host, distribution: 214]; Lindin1909c [taxonomy, host, distribution: 451]; Lindin1909d [taxonomy, description, host, distribution: 3,5-7,11]; Lindin1910b [host, distribution: 42]; Lindin1932f [taxonomy: 204]; MacGil1921 [taxonomy, description, host, distribution: 452]; Maleno1916a [taxonomy, description, illustration, host, distribution: 339-340]; Mallam1954 [distribution: 24-60]; Mamet1941 [taxonomy: 24]; Mamet1943a [catalogue: 167]; Mamet1949 [catalogue: 64]; Mamet1950 [host, distribution: 23]; Mamet1951 [host, distribution: 230]; Mamet1953a [taxonomy: 152]; Mamet1954 [host, distribution: 21]; Mamet1957 [distribution: 369]; Mamet1958a [taxonomy, description, illustration, host, distribution: 370-374]; Mamet1959a [host, distribution: 390]; Maranh1946 [taxonomy: 164-179]; Marlat1908 [taxonomy, description, illustration, host, distribution: 132-136]; Martor1976 [host, distribution: 1-255]; Maskew1916 [host, distribution: 308-309]; Matile1978 [host, distribution: 67]; MatileNo1984 [host, distribution: 67]; McKenz1956 [taxonomy, description, illustration, host, distribution: 85-86]; Merril1953 [taxonomy, description, host, distribution: 79-80]; MerrilCh1923 [taxonomy, description, host, distribution, economic importance: 248-249]; MillerDa1990 [host, distribution, economic importance: 305]; Moreir1921a [host, distribution: 124-126]; Moreir1921b [host, distribution: 182]; Moreir1929a [host, distribution, life history, economic importance: 3]; Morgan1889a [taxonomy, description, illustration, host, distribution: 352]; Morril1911 [host, distribution: 277]; MoutiaMa1947 [distribution]; Muntin1969 [host, distribution: 142]; NagarkSa1990 [host, distribution, economic importance, biological control: 553-542]; Nakaha1982 [host, distribution: 82]; Nakaha1983 [host, distribution: 15]; Newell1923 [host, distribution: 263-266]; Newste1896a [host, distribution: 133]; Newste1901a [host, distribution: 81-82,127]; Newste1901b [p. 284]; Newste1917b [host, distribution: 133]; NotzP1974 [host, distribution, biological control: 127-143]; OteroCaMo1996 [host, distribution, biological control: 530-535]; PaivaPiSi1997 [host, distribution, chemical control: 161]; PalaroSiPa1997 [host, distribution, chemical control: 161]; PeckHeLa1998 [host, distribution: 219-237]; PereirSaAs1997 [host, distribution, life history: 223]; PerrusCa1993 [host, distribution, life history, economic importance: 401-404]; PerrusCa1997 [host, distribution, life history: 321-326]; PerrusCaSo1997 [host, distribution, life history, ecology: 241]; Petch1921a [biological control: 89-167]; PintoBuGr1997 [host, distribution, life history, biological control: 239]; Prinsl1983 [distribution, biological control: 27]; PrunaAl1973 [host, distribution: 1-12]; PruthiMa1945 [host, distribution, life history, control: 1-42]; QuezadCoDi1972 [host, distribution, biological control, economic importance: 9-11]; Robins1917 [taxonomy, description, host, distribution: 28-29]; RodrigGoAz1997 [host, distribution, biological control, life history: 132]; Rose1990c [distribution, economic importance: 535-542]; Rosen1987 [taxonomy, biological control: 191]; Rosen1990 [biological control: 413-415]; Rosen1990a [biological control: 503]; Rosen1993 [biological control: 411-416]; RosenDe1979 [host, distribution, biological control: 405-409,533-539,]; SantosGr1995 [host, distribution, chemical control: 411-414]; Sassce1923 [host, distribution: 152-158]; Schmut1969 [taxonomy, description, host, distribution, life history, economic importance: 115-116]; Schmut1990a [host, distribution: 393]; Schmut2001 [host, distribution: 339-345]; SchmutKlLu1957 [host, distribution, economic importance: 494]; Seabra1921 [host, distribution: 98]; Seabra1925 [taxonomy, description, illustration, host, distribution: 29-30]; SenoGa1999 [host, distribution, economic importance, life history, chemical control: 715-720]; SilvaPaPi1997 [host, distribution, chemical control: 161]; Silves1915 [host, distribution: 262]; SoaresMaCa1998 [host, distribution, biological control, life history: 124-129]; Strong1922 [host, distribution: 775-780]; Sudoi1995 [host, distribution, chemical control: 119-123]; Suris1999 [host, distribution, economic importance, life history: 23-24]; SurisC1984 [host, distribution, economic importance, life history: 110]; SurisC1993 [host, distribution, description: 121-127]; SurisCa1987 [host, distribution: 27-31]; SurisHeVa1989 [host, distribution, life history, ecology: 1-8]; SurisVa1988 [host, distribution, life history, ecology: 38-44]; Tao1999 [taxonomy, host, distribution: 116]; Vayssi1913 [host, distribution: 431]; VelasqRi1969 [host, distribution: 195-208]; Viggia1984 [biological control: 257-276]; Watana1997 [host, distribution, biological control: 103]; WatanaTaCo2000 [host, distribution, life history, biological control: 49-64]; WatanaTaNa2000 [host, distribution, life history, ecology: 81-97]; WatanaYo1992 [biological control: 63-65]; WatanaYoSi1994 [host, distribution, biological control: 193-200]; Watson1918 [host, distribution]; WatsonBe1937 [host, distribution, control]; Wester1918 [host, distribution, economic importance: 5-57]; Wester1920 [host, distribution]; Whitne1933 [host, distribution: 59-70]; Wille1941 [host, distribution: 3-26]; Willia1970DJ [taxonomy, host, distribution: 33]; WilliaWa1988 [taxonomy, illustration, host, distribution, economic importance: 11,240-241]; Wilson1917 [host, distribution: 56-57]; Wilson1921 [host, distribution: 20-34]; WilsonGo1962 [host, distribution, economic importance, control: 41-61]; WoodruBeSk1998 [distribution]; Woolle1990 [biological control: 167-176]; XavierDeSc1997 [host, distribution, biological control: 135]; YustCe1956 [host, distribution: 425-442].

Selenaspidus aubrevillei Balachowsky & Ferrero

Selenaspidus aubrevillei Balachowsky & Ferrero, 1965: 711. Type data: CENTRAL AFRICAN REPUBLIC: Saint Felice Reserve, on leaves of *Randia nilotica*. Holotype female. Type depository: Paris: Muséum national d'Histoire naturelle, France.

SCALE COVER: Female scale circular, 2-2.2 mm in diameter; flat; slightly prominent in center; larval exuviae central, brown red; secreted part white. Male scale oval elongate, 1.6-1.7 mm long, 0.8 mm wide; similar in colour and structure to that of female (Balachowsky & Ferrero, 1965).

HOST PLANTS: **Rubiaceae**: *Randia nilotica* [BalachFe1965].

DISTRIBUTION: **Afrotropical**: Central African Republic [BalachFe1965].

GENERAL: Description and illustration of adult female by Balachowsky & Ferrero (1965).

CITATIONS: BalachFe1965 [taxonomy, description, illustration, host, distribution: 711-714].

Selenaspidus celastri (Maskell)

Aspidiotus articulatus celastri Maskell, 1897: 297. Type data: SOUTH AFRICA: Cape Province, Cape of Good Hope, on *Celastrus laurinus*; specimens sent to Maskell by Lounsbury from Cape Town Herbarium, collected 1825. Syntypes, female. Type depository: Auckland: New Zealand Arthropod Collection, Landcare Research, New Zealand.

Selenaspidus articulatus celastri; Fernald, 1903b: 285. Change of combination.

Pseudaonidia articulatus celastri; Marlatt, 1908: 136. Change of combination.

Selenaspidus celastri; Lindinger, 1909d: 8. Change of combination and rank.

Aspidiotus (Selenaspidus) celastri; Brain, 1918: 136. Change of combination.

Stringaspidiotus celastri; MacGillivray, 1921: 451. Change of combination.

Aspidiotus celastri; Ferris, 1941e: 41. Change of status.

Selenaspidus celastri; Borchsenius, 1966: 254. Revived combination.

SCALE COVER: Female scale about 3 mm in diameter, almost circular, ochreous fawn to straw-coloured, often paler towards margin; semi-glassy and smooth; with a dark brown to black blotch in centre of second exuviae. Exuviae central, flat. Male scale smaller, somewhat elongate; pale buff; semi-transparent, thin and delicate, with exuviae slightly darker (Brain, 1918).

HOST PLANTS: **Celastraceae**: *Celastrus laurinus* [Maskel1897, Marlat1908, Brain1918, Mamet1958a, Borchs1966]. **Oleaceae**: *Ligustrum* [Mamet1958a].

NATURAL ENEMIES: HYMENOPTERA **Encyrtidae**: *Habrolepis aspidioti* Compere & Annecke [Prinsl1983, Trjapi1989].

DISTRIBUTION: **Afrotropical**: South Africa [Maskel1897, Marlat1908, Brain1918, Mamet1958a, Borchs1966].

GENERAL: Description and illustration of adult female by Brain (1918) and by Mamet (1958a).

KEYS: Mamet 1962: 158 (female) [Afrotropical]; Mamet 1958a: 365-366 (female) [World]; Brain 1918: 131 (female) [South Africa]; Lindinger 1909d: 4-5 (female) [Genus *Selenaspidus*]; Marlatt 1908: 134 (female) [World].

CITATIONS: Borchs1966 [catalogue: 254]; Brain1918 [taxonomy, description, illustration, host, distribution: 136]; DeitzTo1980 [taxonomy: 34]; Fernal1903b [catalogue: 285]; Ferris1941e [taxonomy: 41]; Lindin1909d [taxonomy, description, host, distribution: 8]; MacGil1921 [taxonomy, description, host, distribution: 451]; Mamet1958a [taxonomy, description, illustration, host, distribution: 374-376]; Marlat1908 [taxonomy, host, distribution: 136]; Maskel1897 [taxonomy, description, illustration, host, distribution: 297]; Prinsl1983 [distribution, biological control: 27]; Trjapi1989 [biological control: 293].

Selenaspidus eritreae Mamet

Selenaspidus eritreae Mamet, 1958a: 376. Type data: ERITREA: Dogali surroundings, on *Cadaba rotundifolia*; collected by G. De Lotto, 14.ii.1948. Holotype. Type depository: Paris: Muséum national d'Histoire naturelle, France.

SCALE COVER: Scale of female, circular, straw-coloured to pale brownish, low convex; exuviae central, pale golden yellow, obscured by secretion; diameter about 1.5 mm. Scale of male not observed (Mamet 1958a).

HOST PLANTS: Capparidaceae: *Cadaba rotundifolia* [Mamet1958a].

DISTRIBUTION: Afrotropical: Eritrea [Mamet1958a, Borchs1966].

GENERAL: Description and illustration of adult female by Mamet (1958a).

KEYS: Mamet 1958a: 365-366 (female) [World].

CITATIONS: Borchs1966 [catalogue: 254]; Mamet1958a [taxonomy, description, illustration, host, distribution: 376-378].

Selenaspidus euphorbiae (Newstead)

Aspidiotus (*Selenaspidus*) *euphorbiae* Newstead, 1912: 18. Type data: SOUTH AFRICA: Riet Tinkas, on *Euphorbia* sp., near *virosa* Willd. Holotype. Type depository: Paris: Muséum national d'Histoire naturelle, France.

Aspidiotus euphorbiae; Sasscer, 1912: 93. Change of combination.

Entaspidiotus euphorbiae; MacGillivray, 1921: 378. Change of combination.

Selenaspidus euphorbiae; Lindinger, 1943b: 264. Change of combination.

Selenaspidus eritreae; Mamet, 1958a: 379. Misspelling of species name. Notes: *Selenaspidus eritreae* Mamet in the caption on page 379 is a misspelling of *Selenaspidus euphorbiae* Newstead.

SYSTEMATICS: Mamet (1958a, p. 378) redescribed and illustrated this species, while referring to 'Plate 7'. However, no Plate 7 presented in Mamet (1958a), whereas Plate 6 was published twice, on page 377 and on page 379, each bearing the same caption, i.e. "Plate 6. - *Selenaspidus eritreae* Mamet, sp.n.: Adult female". We have compared the information for *Selenaspidus euphorbiae* (in redescription and key to species) with the illustration (bearing the caption "Plate 6. - *Selenaspidus eritreae* Mamet, sp.n.: Adult female" on page 379) and conclude that the latter is a *lapsus calami* of *Selenaspidus euphorbiae* Newstead, and that the illustration on page 379 should be regarded as the illustration of the latter.

SCALE COVER: Female scale, circular, thick, opaque; exuviae central or subcentral; larval pellicle golden brown to golden yellow, covered with secretion similar to that of larva; secretionary portion in two equal zones of pale ochreous and white, latter marginal; diameter about 1.75 to 2 mm (Newstead, 1912).

HOST PLANTS: **Euphorbiaceae**: *Euphorbia* [Newste1912, Mamet1958a, Borchs1966], *Euphorbia virosa* [Brain1918].

DISTRIBUTION: **Afrotropical**: South Africa [Newste1912, Mamet1958a, Borchs1966].

GENERAL: Description and illustration of adult female by Newstead (1912) and by Mamet (1958a). See discussion in Systematics above.

KEYS: Brain 1918: 130 (female) [South Africa].

CITATIONS: Borchs1966 [catalogue: 254]; Brain1918 [taxonomy, description, illustration, host, distribution: 132-133]; Fernal1903b [catalogue: 285]; Ferris1941e [taxonomy: 43]; Lindin1943b [taxonomy: 264]; MacGil1921 [taxonomy, description, host, distribution: 455]; Mamet1958a [taxonomy, description, illustration, host, distribution: 378]; Newste1912 [taxonomy, description, host, distribution: 18]; Sassce1912 [taxonomy, host, distribution: 93].

Selenaspidus euphorbiarum (Lindinger)
Aspidiotus euphorbiae Sasaki, I., 1935: 865. Type data: SOUTH AFRICA: East London, intercepted at Japan, Port Kobe, on *Euphorbia* sp. Syntypes, female. Homonym of *Aspidiotus (Selenaspidus) euphorbiae* Newstead, 1912. Notes: Depository unknown.
Aspidiotus luphorbiae; Sasaki, 1935: 865. Misspelling of species name.
Selenaspidus euphorbiae; Lindinger, 1943b: 207. Change of combination.
Aspidiotus euphorbiarum Lindinger, 1957: 545. Replacement name for *Aspidiotus euphorbiae* Sasaki, I. 1935.
Selenaspidus sasakii Mamet, 1958a: 400. Unjustified replacement name for *Selenaspidus euphorbiarum* Lindinger, 1957. Notes: Unnecessary replacement name, because it was predated by *Aspidiotus euphorbiarum* Lindinger, 1957.
Aspidiotus luphorbiae; Borchsenius, 1966: 254. Misspelling of species name.
Selenaspidus euphorbiarum; Borchsenius, 1966: 254. Change of combination.

SCALE COVER: Female scale subcircular, flat, brownish; first larval skin dark brown at centre; second larval skin reddish brown; about 2.8-3 mm. Male scale resembles that of female, but much smaller (Sasaki, 1935, translated by R. Takahashi, in Mamet, 1958a).

HOST PLANTS: **Euphorbiaceae**: *Euphorbia* [Sasaki1935, Mamet1958a].

DISTRIBUTION: **Afrotropical**: South Africa [Sasaki1935, Mamet1958a].

GENERAL: Description of adult female by Sasaki (1935, translated by R. Takahashi, in Mamet, 1958a). Mamet (1958a) noted that no specimens of this species available for his revision, but suggested that it is closely related to *Selenaspidus kamerunicus*.

CITATIONS: Borchs1966 [catalogue: 254]; Ferris1941e [taxonomy: 43,45]; Lindin1943b [taxonomy: 207]; Lindin1957 [taxonomy: 545]; Mamet1958a [taxonomy, description, host, distribution: 400]; Sasaki1935 [taxonomy, description, illustration, host, distribution: 865-866].

Selenaspidus eurylobus Munting

Selenaspidus eurylobus Munting, 1969a: 294. Type data: NIGERIA: taken in quarantine at Hawaii, on seeds of *Cycas* sp.; collected 3.xii.1962. Holotype female. Type depository: Washington: United States National Entomological Collection, U.S. National Museum of Natural History, District of Columbia, USA; type no. 3415/1.

SCALE COVER: Female scale sub-circular about 1.2 mm in diameter; brownish-yellow in colour. Male scale not seen (Munting, 1969a).

HOST PLANTS: **Cycadaceae**: *Cycas* [Muntin1969a].

DISTRIBUTION: **Afrotropical**: Nigeria [Muntin1969a].

BIOLOGY: Collected from seeds of *Cycas* sp. (Munting, 1969a).

GENERAL: Description and illustration of adult female by Munting (1969a).

CITATIONS: Muntin1969a [taxonomy, description, illustration, host, distribution: 293-294].

Selenaspidus ferox Lindinger

Selenaspidus ferox Lindinger, 1909d: 7. Type data: GHANA: vicinity of Wute, on an Euphorbiaceae plant resembling *Plumeria*. Holotype. Type depository: Hamburg: Zoologisches Institut und Zoologishces Museum, Universität von Hamburg, Germany.

Pseudaonidia ferox; Sanders, 1909a: 54. Change of combination.

Aspidiotus (*Selenaspidus*) *ferox*; Vayssière, 1913: 431. Change of combination.

Stringaspidiotus celastri; MacGillivray, 1921: 451. Misidentification; discovered by Mamet, 1958a: 380.

Selenaspidus ferox; Borchsenius, 1966: 255. Change of combination.

SCALE COVER: Scale of female clear brownish-grey, circular, flat; exuviae central, pale yellow, with obscured border; diameter of largest observed scale 3 mm (Lindinger, 1909d).

HOST PLANTS: **Euphorbiaceae** [Lindin1909d, Vayssi1913, Mamet1958a, Borchs1966].

DISTRIBUTION: **Afrotropical**: Côte d'Ivoire (=Ivory Coast) [Vayssi1913]; Ghana [Lindin1909d, Mamet1958a, Borchs1966].

GENERAL: Description and illustration of adult female by Lindinger (1909d). Mamet (1958a) noted that no specimens of this species were available for his revision of the *Selenaspidus* complex.

KEYS: Mamet 1962: 158 (female) [Afrotropical]; Lindinger 1909d: 4-5 (female) [Genus *Selenaspidus*].

CITATIONS: Borchs1966 [catalogue: 255]; Ferris1941e [taxonomy: 43]; Laing1929a [taxonomy: 498]; Lindin1909d [taxonomy, description, illustration, host, distribution: 7-8]; MacGil1921 [taxonomy, description, host, distribution: 380,451]; Mamet1958a [taxonomy, description, host, distribution: 378-380]; Sander1909a [taxonomy, host, distribution: 54]; Vayssi1913 [host, distribution: 431]; WeidneWa1968 [taxonomy: 178].

Selenaspidus griqua (Brain)

Aspidiotus (Selenaspidus) griqua Brain, 1918: 133. Type data: SOUTH AFRICA: Cape Province, Belmont, Griqualand East, on *Arthrosolen polycephalus*; collected by Mr. Thomsen, May 1915. Lectotype female, by subsequent designation Munting, 1970a: 38. Type depository: Pretoria: South African National Collection of Insects, South Africa; type no. 277/1.

Entaspidiotus griqus; MacGillivray, 1921: 456. Misspelling of species name.

Selenaspidus griqua; Lindinger, 1943b: 264. Change of combination.

SCALE COVER: Female scale 1.5 mm long, circular to elongate; colour buff to brownish, matt, smooth, shell-like; exuviae covered; first exuviae central and usually covered with a small circular area of secretion which is paler, or almost white; in rubbed specimens first exuviae brown; second exuviae covered with a layer of secretion similar to that of scale, but their outer margin is often indicated by a line of paler colour. Male scale similar, but smaller and more elongate, with first exuviae near one end, covered, yellow to orange when rubbed (Brain, 1918).

HOST PLANTS: **Thymelaeaceae**: *Arthrosolen polycephalus* [Brain1918, Mamet1958a, Borchs1966].

DISTRIBUTION: **Afrotropical**: South Africa [Brain1918, Mamet1958a].

GENERAL: Description and illustration of adult female by Brain (1918).

KEYS: Brain 1918: 130 (female) [South Africa].

CITATIONS: Borchs1966 [catalogue: 255]; Brain1918 [taxonomy, description, illustration, host, distribution: 133-134]; Ferris1941e [taxonomy: 44]; Lindin1943b [taxonomy: 264]; MacGil1921 [taxonomy, description, host, distribution: 456]; Mamet1958a [taxonomy, description, illustration, host, distribution: 380]; Muntin1970a [taxonomy: 38].

Selenaspidus incisus Lindinger

Selenaspidus silvaticus incisus Lindinger, 1913b: 100. Type data: SOMALIA: Osthorn, on upper surface of leaves of *Osyris abyssinica*. Holotype. Type depository: Hamburg: Zoologisches Institut und Zoologishces Museum, Universität von Hamburg, Germany.

Selenaspidus incisus; Mamet, 1958: 365. Change of status.

SYSTEMATICS: Material of this species not available to Mamet (1958a) for his revision of the *Selenaspidus* complex. In addition, Mamet (1958a) commented that its proper generic assignment doubtful, but suggested that until further information would become available, species should be retained in *Selenaspidus*.

HOST PLANTS: **Santalaceae**: *Osyris abyssinica* [Lindin1913b, Mamet1958a, Borchs1966].

DISTRIBUTION: **Afrotropical**: Somalia [Lindin1913b, Mamet1958a, Borchs1966].

CITATIONS: Borchs1966 [catalogue: 255]; Lindin1913b [taxonomy, description, host, distribution: 100]; Mamet1958a [taxonomy, description, host, distribution: 382]; WeidneWa1968 [taxonomy: 179].

Selenaspidus kamerunicus Lindinger

Selenaspidus kamerunicus Lindinger, 1909d: 7. Type data: CAMEROON: on an undetermined plant. Holotype. Type depository: Hamburg: Zoologisches Institut und Zoologishces Museum, Universität von Hamburg, Germany.

Pseudaonidia kamerunica; Sanders, 1909a: 54. Change of combination requiring emendation of species name for agreement in gender.

Aspidiotus (*Selenaspidus*) *kamerunicus*; Vayssière, 1913: 431. Change of combination.

Pseudoaonidia kamerunica; Leonardi, 1914: 202. Misspelling of genus name.

Selenaspidus kamerunicus; Borchsenius, 1966: 255. Revived combination.

SCALE COVER: Scale of female subcircular, pale yellowish to pale buff, thin, more or less transparent, flat; exuviae subcentral, yellowish; diameter of scale, about 2 mm. Scale of male not observed (Mamet 1958a).

HOST PLANTS: **Leguminosae**: *Cassia spectabilis* [MatileNo1984]. **Musaceae**: *Musa* [MatileNo1984]. **Sterculiaceae**: *Cacao* [Mamet1958a, Borchs1966], *Theobroma cacao* [Mamet1958a, Borchs1966].

DISTRIBUTION: **Afrotropical**: Angola [Almeid1973b]; Cameroon [Vayssi1913, MatileNo1984]; Ghana [Mamet1958a, Borchs1966]; Guinea [Leonar1914]; Sierra Leone [Hargre1937]; Uganda [Mamet1958a, Borchs1966]; Zaire [Mamet1958a, Borchs1966].

GENERAL: Description and illustration of adult female by Mamet (1958a).

KEYS: Mamet 1962: 158 (female) [Afrotropical]; Lindinger 1909d: 4-5 (female) [Genus *Selenaspidus*].

CITATIONS: Almeid1973b [host, distribution: 12]; Borchs1966 [catalogue: 255]; Ferris1941e [taxonomy: 44]; Hargre1937 [host, distribution, economic importance: 505-520]; Leonar1914 [taxonomy, host, distribution: 202]; Lepesm1947 [host, distribution: 214]; Lindin1909c [taxonomy, host, distribution: 451]; Lindin1909d [taxonomy, description, illustration, host, distribution: 7]; MacGil1921 [taxonomy, description, host, distribution: 452]; Mamet1958a [taxonomy, description, illustration, host, distribution: 382]; MatileNo1984 [host, distribution: 67]; Newste1920 [taxonomy: 198]; Sander1909a [taxonomy, host, distribution: 54]; Vayssi1913 [host, distribution: 431]; WeidneWa1968 [taxonomy: 178].

Selenaspidus latus Munting

Selenaspidus latus Munting, 1967a: 266. Type data: SOUTH AFRICA: Transvaal, Barberton district, Hemlock, on *Euphorbia ingens*; collected by J. Munting, 14.vi.1965. Holotype female. Type depository: Pretoria: South African National Collection of Insects, South Africa; type no. 1842/10.

SCALE COVER: Female scale subcircular, large, 3 mm in diameter, whitish with dark central exuviae. Male scale ovate, about 1 mm in length (Munting, 1967a).

HOST PLANTS: **Euphorbiaceae**: *Euphorbia ingens* [Muntin1967a].

DISTRIBUTION: **Afrotropical**: South Africa [Muntin1967a].

GENERAL: Description and illustration of adult female by Munting (1967a).

CITATIONS: Muntin1967a [taxonomy, description, illustration, host, distribution: 265-266].

Selenaspidus littoralis Mamet

Selenaspidus littoralis Mamet, 1954: 81. Type data: MADAGASCAR: Montagne d'Ambre, on *Dracaena* sp., Fort-Dauphin, on *Cycas thouarsi*. Syntypes. Type depository: Paris: Muséum national d'Histoire naturelle, France.

SCALE COVER: Female scale flat, buff-coloured; fragile texture, translucent; exuviae subcentral, somewhat shiny, darker than rest of scale; larval exuvia with a white, ring-like elevation at centre. Scale of male not observed (Mamet 1958a).

HOST PLANTS: **Agavaceae**: *Dracaena* [Mamet1954, Mamet1958a, Borchs1966]. **Cycadaceae**: *Cycas thouarsi* [Mamet1954, Mamet1958a, Borchs1966].

DISTRIBUTION: **Afrotropical**: Madagascar [Mamet1954, Mamet1958a, Mamet1959a, Borchs1966].

GENERAL: Description and illustration of adult female by Mamet (1953, 1958a).

KEYS: Mamet 1962: 158 (female) [Afrotropical].

CITATIONS: Borchs1966 [catalogue: 255]; Mamet1954 [taxonomy, description, illustration, host, distribution: 21,81]; Mamet1958a [taxonomy, description, illustration, host, distribution: 384]; Mamet1959a [host, distribution: 390].

Selenaspidus louisiae Mamet

Selenaspidus louisiae Mamet, 1958a: 384. Type data: SOUTH AFRICA: Durban, in garden between the beach and Marine Parade, on succulent plant. Holotype. Type depository: Paris: Muséum national d'Histoire naturelle, France.

SCALE COVER: Scale of female circular, whitish to pale buff, flat; exuviae central, pale golden yellow; diameter about 2.5 mm. Scale of male not observed (Mamet 1958a).

DISTRIBUTION: **Afrotropical**: South Africa [Mamet1958a, Borchs1966].

GENERAL: Description and illustration of adult female by Mamet (1958a).

KEYS: Mamet 1962: 158 (female) [Afrotropical].

CITATIONS: Borchs1966 [catalogue: 255]; Mamet1958a [taxonomy, description, illustration, host, distribution: 384,387].

Selenaspidus magnospinus (Newstead)

Aspidiotus (*Selenaspidus*) *articulatus magnospinus* Newstead, 1920: 197. Type data: UGANDA: Bufumira Isl., Sesse Islands, Lake Victoria, on the leaves of an unknown plant. Syntypes. Type depository: London: The Natural History Museum, UK.

Selenaspidus magnospinus; MacGillivray, 1921: 452. Change of combination and rank.

Aspidiotus (*Selenaspidus*) *magnospinus*; Ferris, 1941e: 45. Change of combination.

Selenaspidus magnospinus; Borchsenius, 1966: 255. Revived combination.

SCALE COVER: Scale of female subcircular, flat, pale brown; exuviae subcentral; diameter about 2.5 mm. Scale of male not observed (Mamet 1958a).

HOST PLANTS: **Myrsinaceae**: *Ardisia crenata* [Mamet1943a, Mamet1949, Mamet1958a, Borchs1966]. **Rubiaceae**: *Coffea* [Almeid1973b], *Coffea robusta* [Mamet1943a, Mamet1958a, Borchs1966].

DISTRIBUTION: **Afrotropical**: Angola [Almeid1973b]; Mauritius [Mamet1943a, Mamet1949, Mamet1958a, Borchs1966]; Uganda [Newste1920, Mamet1958a, Borchs1966].

GENERAL: Description and illustration of adult female by Newstead (1920) and by Mamet (1958a)
KEYS: Mamet 1962: 158 (female) [Afrotropical].
CITATIONS: Almeid1973b [host, distribution: 11]; Borchs1966 [catalogue: 255]; Ferris1941e [taxonomy: 45]; MacGil1921 [taxonomy, description, host, distribution: 451]; Mamet1943a [catalogue: 168]; Mamet1949 [catalogue: 65]; Mamet1958a [taxonomy, description, illustration, host, distribution: 386-388]; Newste1920 [taxonomy, description, illustration, host, distribution: 197-198].

Selenaspidus malzyi Balachowsky
Selenaspidus malzyi Balachowsky, 1954: 58. Type data: CAMEROON: 10km North of Guider and 100km South of Maroua, on *Tamarindus indica*. Holotype female. Type depository: Paris: Muséum national d'Histoire naturelle, France.
SCALE COVER: Scale of female, circular, a little convex, pure white, with larval exuviae central, brownish-mahogany-red; diameter about 2.1 mm. Scale of male not observed (Balachowsky, 1954; Mamet 1958a).
HOST PLANTS: **Anacardiaceae**: *Mangifera indica* [Mamet1958a, Borchs1966]. **Apocynaceae**: *Nerium oleander* [BalachMa1970]. **Leguminosae**: *Tamarindus indica* [Balach1954, Mamet1958a, Borchs1966].
DISTRIBUTION: **Afrotropical**: Cameroon [Balach1954, Mamet1958a, Borchs1966]; Mauritania [BalachMa1970]; Sudan [Balach1954, Mamet1958a, Borchs1966].
GENERAL: Description and illustration of adult female by Balachowsky (1954) and by Mamet (1958a).
CITATIONS: Balach1954 [taxonomy, description, illustration, host, distribution: 58-60]; BalachMa1970 [host, distribution: 1080]; Borchs1966 [catalogue: 255]; Mamet1958a [taxonomy, description, illustration, host, distribution: 389-390].

Selenaspidus perspinosus (Leonardi)
Pseudoaonidia ferox perspinosa Leonardi, 1914: 201. Type data: GUINEA: Conakry, on undetermined plant. Holotype female. Type depository: Portici: Dipartimento de Entomologia e Zoologia Agraria di Portici, Università di Napoli Federico II, Italy.
Pseudoaonidia ferox perspinosa; Leonardi, 1914: 201. Misspelling of genus name.
Selenaspidus perspinosus; Mamet, 1958a: 390. Change of combination and rank.
Selenaspidus perspinosus; Mamet, 1958a: 390. Change of combination requiring emendation of species name for agreement in gender.
SYSTEMATICS: According to Leonardi (1914), this species, which was described from a single specimen, differs from *Selenaspidus ferox* Lindinger by shape of the projections occurring on cephalic margin and by more numerous perivulvar pores. Mamet (1958a) noted that no specimens of this species available for his revision, and did not give an opinion on status of *S. perspinosus*.
SCALE COVER: Leonardi (1914) did not describe scale cover.
DISTRIBUTION: **Afrotropical**: Guinea [Leonar1914, Mamet1958a, Borchs1966].
GENERAL: Description and illustration of adult female by Leonardi (1914).

CITATIONS: Borchs1966 [catalogue: 255]; Leonar1914 [taxonomy, description, illustration, host, distribution: 201-202]; Lindin1943b [taxonomy: 264]; Mamet1958a [taxonomy, host, distribution: 392].

Selenaspidus pertusus (Brain)

Aspidiotus (*Selenaspidus*) *pertusus* Brain, 1918: 135. Type data: SOUTH AFRICA: East London, on *Euphorbia* sp. and *Mimusops* sp.; Komgha, Cape Province, on "laurel". Syntypes, female. Type depository: Pretoria: South African National Collection of Insects, South Africa. Notes: Munting (1970a) noted that the available type material was in too poor condition for lectotype designation.
Selenaspidus pertusus; MacGillivray, 1921: 452. Change of combination.
SCALE COVER: Female scale may reach 2.6 mm in diameter, circular, flat, dull rusty brown, with margins slightly paler. Exuviae yellowish or reddish brown. Male scale smaller, more elongate and paler in colour, yellowish brown (Brain, 1918).
HOST PLANTS: **Apocynaceae**: *Carissa grandiflora* [Mamet1958a, Borchs1966]. **Euphorbiaceae**: *Euphorbia* [Brain1918, Mamet1958a, Borchs1966]. **Oleaceae**: *Olea chrysophylla* [Mamet1958a, Borchs1966]. **Sapotaceae**: *Mimusops* [Brain1918, Mamet1958a, Borchs1966].
NATURAL ENEMIES: HYMENOPTERA **Encyrtidae**: *Habrolepis aspidioti* Compere & Annecke [Prinsl1983].
DISTRIBUTION: **Afrotropical**: Eritrea [Mamet1958a, Borchs1966]; South Africa [Brain1918, Mamet1958a, Borchs1966].
GENERAL: Description and illustration of adult female by Brain (1918) and by Mamet (1958a).
KEYS: Mamet 1962: 158 (female) [Afrotropical]; Brain 1918: 131 (female) [South Africa].
CITATIONS: Borchs1966 [catalogue: 255]; Brain1918 [taxonomy, description, illustration, host, distribution: 135-136]; DanzigPe1998 [catalogue: 355]; Ferris1941e [taxonomy: 47]; MacGil1921 [taxonomy, description, host, distribution: 452]; Mamet1958a [taxonomy, description, illustration, host, distribution: 391-392]; Muntin1970a [taxonomy: 39]; Prinsl1983 [distribution, biological control: 27]; Schmut1959 [taxonomy, description, host, distribution: 114,119]; Trjapi1989 [biological control: 293].

Selenaspidus podocarpi Mamet

Selenaspidus podocarpi Mamet, 1958a: 392. Type data: SOUTH AFRICA: Tzaneen, Transvaal, on *Podocarpus latifolius*. Holotype. Type depository: Paris: Muséum national d'Histoire naturelle, France.
SCALE COVER: Female scale, pale straw-coloured to dirty white, sometimes pale greyish through accumulation of dirt, flat to slightly convex; exuviae subcentral, golden-yellow to light buff, obscured by a layer of whitish secretion; length of scale about 3.5 mm. Scale of male, smaller than female, elongate, white to pale straw-coloured; exuviae towards one side, straw-coloured (Mamet 1958a).
HOST PLANTS: **Podocarpaceae**: *Podocarpus latifolius* [Mamet1958a, Borchs1966].
DISTRIBUTION: **Afrotropical**: South Africa [Mamet1958a, Borchs1966].

GENERAL: Description and illustration of adult female by Mamet (1958a).
CITATIONS: Borchs1966 [catalogue: 255-256]; Mamet1958a [taxonomy, description, illustration, host, distribution: 392-394].

Selenaspidus portulacariae Hall

Selenaspidus portulacariae Hall, 1939: 98. Type data: SOUTH AFRICA: Cape Province, Fort Beaufort, on *Portulacaria afra*. Syntypes, female. Type depository: London: The Natural History Museum, England, UK.
SCALE COVER: Female scale circular, 1.5 mm diameter; low convex, semitransparent; pale smoky brown; larval exuviae golden, nymphal exuviae pale brown to golden; in some specimens larval exuviae pale brown to golden; in some specimens the larval exuviae exhibit a dark median longitudinal stripe (Hall, 1939).
HOST PLANTS: **Portulacaceae**: *Portulacaria afra* [Hall1939, Mamet1958a].
DISTRIBUTION: **Afrotropical**: South Africa [Hall1939, Mamet1958a].
GENERAL: Description and illustration of adult female by Hall (1939) and by Mamet (1958a).
CITATIONS: Borchs1966 [catalogue: 256]; Hall1939 [taxonomy, description, illustration, host, distribution: 98-100]; Mamet1958a [taxonomy, description, illustration, host, distribution: 394-395].

Selenaspidus pumilus (Brain)

Aspidiotus (*Selenaspidus*) *pumilus* Brain, 1918: 133. Type data: SOUTH AFRICA: Natal, Stanger, on *Phormium tenax*; collected by C. Fuller, 11.xii.1903. Lectotype female, by subsequent designation Munting, 1970a: 39. Type depository: Pretoria: South African National Collection of Insects, South Africa; type no. 224/1.
Entaspidiotus pumilus; MacGillivray, 1921: 456. Change of combination.
Selenaspidus pumilus; Mamet, 1958a: 396. Change of combination.
SCALE COVER: Female scale about 1.6 mm in diameter, almost circular, margins flat, depressed, centre roundly and flatly conical, dull, grey brown, with yellow exuviae. Male scale smaller, narrower and elongate, of same colour and texture as female scale (Brain, 1918).
HOST PLANTS: **Agavaceae**: *Phormium tenax* [Brain1918, Mamet1958a, Borchs1966]. **Euphorbiaceae**: *Euphorbia* [Green1926a].
DISTRIBUTION: **Afrotropical**: South Africa [Brain1918, Mamet1958a, Borchs1966, Muntin1970a]; Sudan [Mamet1958a, Borchs1966]; Zimbabwe [Hall1931, Mamet1958a]. **Palaearctic**: United Kingdom (Scotland [Green1926a]).
GENERAL: Description and illustration of adult female by Brain (1918), Green (1926a) and by Mamet (1958a).
KEYS: Brain 1918: 130 (female) [South Africa].
CITATIONS: Borchs1966 [catalogue: 256]; Brain1918 [taxonomy, description, illustration, host, distribution: 133]; DanzigPe1998 [catalogue: 356]; Ferris1941e [taxonomy: 47]; Green1926a [taxonomy, description, illustration, host, distribution: 180-181]; Hall1931 [host, distribution: 288]; MacGil1921 [taxonomy, description, host, distribution: 456]; Mamet1958a [taxonomy, description, illustration, host, distribution: 396-397]; Muntin1970a [taxonomy: 39]; Schmut1959 [taxonomy, description, host, distribution: 114,117].

Selenaspidus quadrilobus Munting
Selenaspidus quadrilobus Munting, 1969a: 294. Type data: SOUTH AFRICA: Cape Province, Cederberg mountains, Uitkyk Pass, on *Serruria* sp.; collected by J. Munting, 12.v.1962. Holotype female. Type depository: Pretoria: South African National Collection of Insects, South Africa; type no. 1126/8.
SCALE COVER: Female scale subcircular, convex, sometimes flat; 1.5 mm in diameter; chocolate brown but with a pure white ventral skin. Male scale oval, about 1 mm long, slightly paler in colour than female (Munting, 1969a).
HOST PLANTS: **Proteaceae**: *Serruria* [Muntin1969a].
DISTRIBUTION: **Afrotropical**: South Africa [Muntin1969a].
GENERAL: Description and illustration of adult female by Munting (1969a).
CITATIONS: Muntin1969a [taxonomy, description, illustration, host, distribution: 294-296].

Selenaspidus rubidus McKenzie
Selenaspidus rubidus McKenzie, 1953a: 57. Type data: U.S.A.: California, San Diego, San Diego County, on *Euphorbia fasciculata*. Holotype female. Type depository: Washington: United States National Entomological Collection, U.S. National Museum of Natural History, District of Columbia, USA.
COMMON NAME: brownish euphorbia scale [McKenz1956].
SCALE COVER: Female scale approximately 1.5 mm in diameter, essentially circular, of a uniform brownish colour, with subcentral exuviae generally darker. Male scale not observed (McKenzie, 1953a).
HOST PLANTS: **Cactaceae**: *Cactus* [Mamet1958a, Borchs1966]. **Euphorbiaceae**: *Euphorbia* [Mamet1958a, Borchs1966], *Eup. captiosa* [McKenz1956, Mamet1958a, Borchs1966], *Eup. caput-medusae* [Mamet1958a, McKenz1956, Borchs1966], *Eup. fasciculata* [McKenz1953a, McKenz1956, Mamet1958a, Borchs1966], *Eup. mammillaris* [McKenz1956, Mamet1958a, Borchs1966], *Eup. tuberculata* [McKenz1956, Mamet1958a, Borchs1966],. **Proteaceae**: *Protea repens* [Muntin1969a].
DISTRIBUTION: **Afrotropical**: South Africa [Muntin1969a]. **Nearctic**: U.S.A.: (California [McKenz1953a, McKenz1956, Mamet1958a, Borchs1966]). **Oriental**: Singapore [Mamet1958a, Borchs1966]. **Palaearctic**: China (People's Republic) [Mamet1958a, Borchs1966]; Germany (United) [Mamet1958a, Borchs1966].
GENERAL: Description and illustration of adult female by McKenzie (1953a), Mamet (1958a) and by Gill (1997).
KEYS: Gill 1997: 256 (female) [Species of California]; McKenzie 1956: 26 (female) [U.S.A.: California].
CITATIONS: Borchs1966 [catalogue: 256]; DanzigPe1998 [catalogue: 356]; Gill1997 [host, distribution, taxonomy, illustration: 259-260]; Mamet1958a [taxonomy, description, illustration, host, distribution: 396-398]; McKenz1953a [taxonomy, description, illustration, host, distribution: 57-58]; McKenz1956 [taxonomy, description, illustration, host, distribution: 85-87]; Muntin1969a [host, distribution: 296]; Nakaha1982 [host, distribution: 82]; Schmut1959 [taxonomy, description, host, distribution: 114,119]; Tao1999 [taxonomy, host, distribution: 116-117].

Selenaspidus schultzei (Newstead)

Aspidiotus (*Selenaspidus*) *schultzei* Newstead, 1912: 18. Type data: SOUTH AFRICA: Kamagagas, Little Namaqualand, on a succulent plant. Holotype. Type depository: London: The Natural History Museum, England, UK.

Aspidiotus schultzei; Sasscer, 1912: 94. Change of combination.

Aspidiotus (*Selenaspidus*) *schultzei*; Brain, 1918: 132. Revived combination.

Entaspidiotus schultzei; MacGillivray, 1921: 456. Change of combination.

Selenaspidus schultzei; Lindinger, 1943b: 264. Change of combination.

SCALE COVER: Female scale circular, smooth, and rather thin; exuviae central, yellow to yellowish-brown; secreted part straw-coloured to ochreous white; diameter 1.5-2 mm. Scale of male unknown (Newstead, 1912).

DISTRIBUTION: **Afrotropical**: South Africa [Newste1912, Brain1918, Mamet1958a, Borchs1966].

GENERAL: Description and illustration of adult female by Newstead (1912) and by Mamet (1958a).

KEYS: Brain 1918: 130 (female) [South Africa].

CITATIONS: Borchs1966 [catalogue: 256]; Brain1918 [taxonomy, description, illustration, host, distribution: 132]; Ferris1941e [taxonomy: 48]; Lindin1943b [taxonomy: 264]; MacGil1921 [taxonomy, description, host, distribution: 456]; Mamet1958a [taxonomy, description, illustration, host, distribution: 399-400]; Newste1912 [taxonomy, description, illustration, host, distribution: 18]; Sassce1912 [taxonomy, host, distribution: 94].

Selenaspidus spinosus Laing

Selenaspidus spinosus Laing, 1929a: 497. Type data: SIERRA LEONE: Kogbotana, on banana. Holotype female. Type depository: London: The Natural History Museum, England, UK.

SCALE COVER: Female scale not observed (Laing, 1929a).

HOST PLANTS: **Euphorbiaceae**: *Gelonium procerum* [Mamet1958a, Borchs1966]. **Musaceae**: *Musa* [Laing1929a, Mamet1958a, Borchs1966]. **Sapotaceae**: *Pachystela brevipes* [Mamet1958a, Borchs1966].

NATURAL ENEMIES: HYMENOPTERA **Encyrtidae**: *Coccidencyrtus* sp. [Prinsl1983].

DISTRIBUTION: **Afrotropical**: Kenya [Mamet1958a, Borchs1966]; Sierra Leone [Laing1929a, Mamet1958a, Borchs1966]; Zaïre [Mamet1958a, Borchs1966].

GENERAL: Description and illustration of adult female by Laing (1929a) and by Mamet (1958a).

KEYS: Mamet 1962: 158 (female) [Afrotropical].

CITATIONS: Balach1948b [taxonomy: 269]; Borchs1966 [catalogue: 256]; ChuaWo1990 [host, distribution, economic importance: 548]; Hargre1937 [host, distribution, economic importance: 505-520]; Laing1929a [taxonomy, description, illustration, host, distribution: 497-498]; Lindin1943b [taxonomy: 264]; Mamet1958a [taxonomy, description, illustration, host, distribution: 401-402]; Prinsl1983 [distribution, biological control: 27].

Selenaspidus taizi Mamet
Selenaspidus taizi Mamet, 1958a: 404. Type data: YEMEN: Taiz, close to city walls, altitude 4500 feet, on a succulent *Euphorbia* sp. Holotype. Type depository: Paris: Muséum national d'Histoire naturelle, France.
SCALE COVER: Female scale almost circular; pure white; flat; exuviae subcentral, dark-brown, somewhat obscured by a layer of whitish secretion; diameter about 2 mm. Scale of male smaller than female, elongate-oval, white, with exuvia towards one end, brownish (Mamet, 1958a).
HOST PLANTS: **Euphorbiaceae**: *Euphorbia* [Mamet1958a, Borchs1966], *Euphorbia fractiflexa* [Matile1988].
DISTRIBUTION: **Afrotropical**: Yemen (South Yemen [Mamet1958a, Borchs1966]). **Palaearctic**: Saudi Arabia [Matile1988].
GENERAL: Description and illustration of adult female by Mamet (1958a).
CITATIONS: Borchs1966 [catalogue: 256]; DanzigPe1998 [catalogue: 356]; Mamet1958a [taxonomy, description, illustration, host, distribution: 403-404]; Matile1988 [host, distribution: 26].

Selenediella Mamet
Selenediella Mamet, 1958a: 424. Type species: *Hemiberlesia mckenziei* Takahashi, by original designation.
SYSTEMATICS: *Selenediella* Mamet differs from other genera of the *Selenaspidus* complex in having prosoma produced postero-laterally to form well-developed lobes (Mamet, 1958a).
GENERAL: Definition and characters by Mamet (1958a).
KEYS: Mamet 1958a: 362 (female) [*Selenaspidus* complex].
CITATIONS: Borchs1966 [catalogue: 257]; Mamet1958a [taxonomy, description: 362,424-426]; MorrisMo1966 [taxonomy, catalogue: 182].

Selenediella mckenziei (Takahashi)
Hemiberlesia mckenziei Takahashi, 1951b: 111. Type data: INDONESIA: Riouw Islands, Rempang, on undetermined plant. Holotype. Type depository: Sapporo: Entomological Institute, Faculty of Agriculture, Hokkaido University, Japan.
Selenediella mckenziei; Mamet, 1958a: 426. Change of combination.
SCALE COVER: Scale of female circular, pale yellowish brown on secretion, dark reddish brown on exuviae. Male scale unknown (Takahashi, 1951b).
DISTRIBUTION: **Oriental**: Indonesia [Takaha1951b, Mamet1958a, Borchs1966].
GENERAL: Description and illustration of adult female by Takahashi (1951b) and by Mamet (1958a).
CITATIONS: Borchs1966 [catalogue: 257]; Mamet1958a [taxonomy, description, illustration, host, distribution: 425-426]; Takaha1951b [taxonomy, description, illustration, host, distribution : 111-112].

Selenomphalus Mamet
Selenomphalus Mamet, 1958a: 426. Type species: *Aspidiotus euryae* Takahashi, by monotypy and original designation.

SYSTEMATICS: *Selenomphalus* was assigned by Mamet (1958a) to the *Selenaspidus* complex, because of spur-like shape of third lobe. Takagi (1984) suggested that it is related to *Metaspidiotus* and *Crassaspidiotus*, in arrangement of dorsal macroducts and other characters, but *Selenomphalus* is readily distinguished from the latter in shape of the third lobe.
GENERAL: Definition and characters by Mamet (1958a) and by Takagi (1969a).
KEYS: Chou 1985: 257 (female) [China]; Mamet 1958a: 362 (female) [*Selenaspidus* complex].
CITATIONS: Borchs1966 [catalogue : 257]; Chou1985 [taxonomy, description: 259]; DanzigPe1998 [catalogue: 356]; Kawai1980 [taxonomy: 225]; Mamet1958a [taxonomy, description: 362,426-428]; MorrisMo1966 [taxonomy, catalogue: 183]; Takagi1969a [taxonomy, description: 94]; Tao1999 [taxonomy: 117].

Selenomphalus distylii Takagi
Selenomphalus distylii Takagi, 1959a: 112. Type data: JAPAN: Honsyu, Wakayama-ken, Minabe, on *Distylium racemosum*. Syntypes, female. Type depository: Sapporo: Entomological Institute, Faculty of Agriculture, Hokkaido University, Japan.
SCALE COVER: Female scale subcircular, slightly convex, dorsally, and brown in colour (Takagi, 1959a).
HOST PLANTS: **Hamamelidaceae**: *Distylium racemosum* [Takagi1959a].
DISTRIBUTION: **Palaearctic**: Japan [Kawai1980] (Honshu [Takagi1959a]).
GENERAL: Description and illustration of adult female by Takagi (1959a).
CITATIONS: Borchs1966 [catalogue: 257]; DanzigPe1998 [catalogue: 356]; Kawai1980 [taxonomy, description, host, distribution: 225]; Muraka1970 [host, distribution: 78]; Takagi1959a [taxonomy, description, illustration, host, distribution: 112-114].

Selenomphalus euryae (Takahashi)
Aspidiotus euryae Takahashi, 1931b: 383. Type data: TAIWAN: Taihezan, Suisha, on *Eurya* sp. Syntypes, female. Type depository: Taichung: Entomology Collection, Taiwan Agricultural Research Institute, Wu-feng, Taichung, Taiwan.
Selenaspidus euryae; Lindinger, 1943b: 207. Change of combination.
Selenomphalus euryae; Mamet, 1958a: 428. Change of combination.
SCALE COVER: Female scale nearly circular or somewhat irregular in shape, flattened, about 1.7-2.0 mm in diameter; larval exuviae pale yellow, small, near margin of scale; secretion pale brown, very thin, semi-transparent (Takahashi, 1931b).
HOST PLANTS: **Loranthaceae**: *Loranthus parasiticus* [Mamet1958a, Borchs1966, Takagi1969a]. **Theaceae**: *Camellia sinensis* [Takagi1969a], *Eurya* [Takaha1931b, Takaha1932a, Takaha1933, Mamet1958a, Borchs1966, Takagi1969a], *Eurya japonica* [Takagi1969a], *Gordonia axillaris* [Takagi1969a].
DISTRIBUTION: **Oriental**: China (People's Republic) (Guangxi (Kwangsi) [Mamet1958a, Borchs1966]); Taiwan [Takaha1931b, Takaha1932a, Takaha1933, Mamet1958a, Borchs1966, Takagi1969a]. **Palaearctic**: China (People's Republic) [Takagi1969a].

GENERAL: Description and illustration of adult female by Takahashi (1931b), Mamet (1958a), Takagi (1969a) and by Chou (1985, 1986).
CITATIONS: Borchs1966 [catalogue: 257]; Chou1985 [taxonomy, description, host, distribution: 259-260]; Chou1986 [taxonomy, illustration: 658]; DanzigPe1998 [catalogue: 356-357]; Ferris1941e [taxonomy: 43]; Lindin1943b [taxonomy: 207]; Mamet1958a [taxonomy, description, illustration, host, distribution: 427-428]; Takagi1969a [taxonomy, description, illustration, host, distribution: 93,95,104]; Takaha1931b [taxonomy, description, illustration, host, distribution: 383-384]; Takaha1932a [host, distribution: 103]; Takaha1933 [host, distribution: 28,62]; Tao1999 [taxonomy, host, distribution: 117].

Semelaspidus MacGillivray

Semelaspidus MacGillivray, 1921: 393. Type species: *Aspidiotus cistuloides* Green, by original designation.
Partargionia MacGillivray, 1921: 394. Type species: *Aspidiotus artocarpi* Green, by monotypy and original designation. Synonymy by Lindinger, 1937: 192.
Protargionia Ferris, 1937c: 52. Synonymy by Borchsenius, 1966: 235.
Scmelaspidus; Chou, 1985: 253. Misspelling of genus name.
SYSTEMATICS: This genus belongs to group of genera centered around *Pseudaonidia* Cockerell, close to *Duplaspidiotus* MacGillivray, but differs in absence of plates between second and third lobes, and in pattern of the reticulation on dorsum of pygidium. Williams (1957a) also suggested that heavy sclerotization of pygidial margin anterior to fourth lobes was a diagnostic character of *Semelaspidus*.
GENERAL: Definition and characters by Williams (1957a).
KEYS: Chou 1985: 248 (female) [China]; Beardsley 1966: 502-504 (female) [Federated States of Micronesia]; Williams 1957a: 42 (female) [world].
CITATIONS: Beards1966 [taxonomy: 524]; Borchs1966 [catalogue: 235]; Chou1985 [taxonomy, description: 253]; Ferris1937c [taxonomy: 52]; Ferris1937d [taxonomy: 106]; Ferris1938 [taxonomy, description: 43,53]; Ferris1941d [taxonomy: 347]; Kozar1990f [distribution: 142]; Lindin1937 [taxonomy: 192,195]; MacGil1921 [taxonomy, description: 393-394,451,458]; McKenz1939 [taxonomy: 53]; MorrisMo1966 [taxonomy, catalogue: 183]; Rao1951b [taxonomy: 261]; Takaha1939d [taxonomy: 341,343]; Tao1999 [taxonomy: 117]; Willia1957a [taxonomy, description: 33-42].

Semelaspidus ambalangoda (Green)

Aspidiotus ambalangoda Green, 1922a: 1007. Type data: SRI LANKA: Ambalangoda, on upper surface of leaves of an undetermined shrub. Syntypes, female. Type depository: London: The Natural History Museum, England, UK.
Pseudischnaspis ambalangoda; Lindinger, 1937: 194. Change of combination.
Pseudaonidia ambalangoda; Green, 1937: 334. Change of combination.
Semelaspidus ambalangoda; Williams, 1957a: 35. Change of combination.
SCALE COVER: Green (1922a) described scale cover as: "Female scale flattish, oblong, much longer than broad, usually tapering to each extremity, 2.25 mm long,

1.5 mm wide; colour brownish black; larval exuvia reddish, surrounded by a narrow pale ring; nymphal exuvia concealed". Williams (1957a) described the scale cover: "Female scale varying in shape depending on position, but usually oval, flat, 2.5 mm by 1.5 mm, with reddish-brown exuviae of first stage larva situated towards one end. Male scale black, oval, measuring 1.25 mm by 0.75 mm, reddish-brown exuviae of first stage larva near centre of scale".

DISTRIBUTION: **Oriental**: Sri Lanka [Green1922a, Green1937, Willia1957a].

BIOLOGY: On leaves, especially on midribs and main veins (Williams, 1957a).

GENERAL: Description and illustration of adult female by Green (1922a) and by Williams (1957a).

KEYS: Williams 1957a: 42 (female) [World].

CITATIONS: Borchs1966 [catalogue: 235]; Ferris1941e [taxonomy: 40]; Green1922a [taxonomy, description, illustration, host, distribution: 1007]; Green1937 [taxonomy, host, distribution: 334]; Lindin1937 [taxonomy: 194]; Lindin1957 [taxonomy: 551]; Willia1957a [taxonomy, description, illustration, host, distribution: 35-37].

Semelaspidus artocarpi (Green)

Aspidiotus artocarpi Green, 1896c: 200. Type data: INDIA: Bombay, on leaves of Jak tree [=*Artocarpus integrifolia*]. Syntypes, female. Type depository: London: The Natural History Museum, England, UK.

Aspidiotus (*Mycetaspis ?*) *artocarpi*; Cockerell, 1897i: 27. Change of combination.

Targionia artocarpi; Leonardi, 1900: 315. Change of combination.

Aspidiotus (*Chrysomphalus*) *cistuloides* Green, 1905a: 342. Type data: SRI LANKA: Paradeniya, on *Cinnamomum* sp. Syntypes, female. Type depository: London: The Natural History Museum, England, UK. Synonymy by Rao, 1951b: 261.

Chrysomphalus cistuloides; Sanders, 1906: 15. Change of combination.

Aspidiotus (*Chrysomphalus*) *triglandulosus* Green, 1908a: 33. Type data: INDIA: Bombay, Mahableshwar, on undetermined tree. Syntypes, female. Type depository: London: Natural History Museum, England, UK. Synonymy by Rao, 1951b: 261.

Aspidiotus triglandulosus; Ramakrishna, 1919a: 21. Change of combination.

Semelaspidus cistuloides; MacGillivray, 1921: 451. Change of combination.

Semelaspidus triglandulosa; MacGillivray, 1921: 451. Change of combination requiring emendation of species name for agreement in gender.

Partargionia artocarpi; MacGillivray, 1921: 458. Change of combination.

Chrysomphalus (*Aspidiotus*) *triglandulosus*; Ramakrishna, 1924: 344. Change of combination.

Aspidiotus cistuloides; Ferris, 1937c: 52. Change of combination.

Melanaspis cistuloides; Lindinger, 1943a: 147. Change of combination.

Melanaspis triglandulosa; Lindinger, 1943a: 147. Change of combination.

Semelaspidus artocarpi; Rao, 1951: 261. Change of combination.

SCALE COVER: Scale of female moderately convex, oval, 2.5 mm X 1.5 mm; ventral scale elevated at one end and overlapping dorsal scale; exuviae of first stage larva reddish-brown and situated towards one end. Male scale black, elongate, 1.25 mm x 0.75 mm (Williams, 1957a).

HOST PLANTS: **Lauraceae**: *Cinnamomum* [Green1905a, Sander1906, Green1922, Green1937, Willia1957a]. **Moraceae**: *Artocarpus* [Ramakr1921a], *Artocarpus integrifolia* [Green1896c, Leonar1900, Willia1957a]. **Piperaceae**: *Piper nigrum* [Green1905a].

DISTRIBUTION: **Oriental**: India [Green1896c, Leonar1900, Green1908a, Ramakr1921a, Willia1957a]; Indonesia (Java [Green1905a]); Sri Lanka [Green1905a, Sander1906, Ramakr1921a, Green1922, Green1937, Willia1957a].

GENERAL: Description and illustration of adult female by Green (1905a) and by Williams (1957a).

KEYS: Williams 1957a: 42 (female) [World].

CITATIONS: Borchs1966 [catalogue: 235-236]; Cocker1897i [taxonomy, description, host, distribution: 27]; Cocker1899a [taxonomy: 395]; Fernal1903b [catalogue: 296]; Ferris1937c [taxonomy: 52]; Ferris1938 [taxonomy: 43,53]; Ferris1941e [taxonomy: 41-42,49]; Ferris1943a [taxonomy: 85]; Green1896c [taxonomy, description, illustration, host, distribution: 200-201]; Green1905a [taxonomy, description, illustration, host, distribution: 342-343]; Green1908a [taxonomy, description, host, distribution: 33-34]; Green1922 [host, distribution: 462]; Green1937 [host, distribution: 332]; Leonar1900 [taxonomy, description, illustration, host, distribution: 315-316]; Lindin1943a [taxonomy: 147]; MacGil1921 [taxonomy, description, host, distribution: 393,394,451,458]; McKenz1939 [taxonomy: 53,55]; Newste1917 [taxonomy: 375]; Newste1917b [host, distribution: 132]; Ramakr1919a [taxonomy, description, host, distribution: 21]; Ramakr1921a [host, distribution: 356,357]; Ramakr1924 [taxonomy, host, distribution: 344]; Ramakr1930 [taxonomy, host, distribution: 26]; Ramakr1938a [host, distribution: 341-351]; Rao1951b [taxonomy: 261]; Ruther1915a [taxonomy, description, host, distribution: 104]; Sander1906 [host, distribution: 15]; Sander1909a [taxonomy, host, distribution: 55]; UsmanPu1955 [host, distribution: 48]; Willia1957a [taxonomy, description, illustration, host, distribution: 37-39].

Semelaspidus mangiferae Takahashi

Semelaspidus mangiferae Takahashi, 1939d: 341. Type data: PHILIPPINES: on *Mangifera indica*. Syntypes, female. Type depository: Taichung: Entomology Collection, Taiwan Agricultural Research Institute, Wu-feng, Taichung, Taiwan.

SCALE COVER: Female scale black, slightly brownish, not shiny; thick; subcircular or elliptic, 1.0 mm long; convex; marginal area of venter attached to host; larval skins at center, or at apex of convex dorsum (Takahashi, 1939d) .

HOST PLANTS: **Anacardiaceae**: *Mangifera indica* [Takaha1939d, Willia1957a]. **Moraceae**: *Ficus* [Beards1966].

DISTRIBUTION: **Oriental**: Philippines [Takaha1939d, Willia1957a]; Taiwan [Tao1999].

BIOLOGY: On upper surfaces of leaves along midrib (Takahashi, 1939b).

GENERAL: Description and illustration of adult female by Takahashi (1939d). Mamet (1958b) did not study this species in the revision of the *Selenaspidus* complex.

KEYS: Williams 1957a: 42 (female) [World].

CITATIONS: Borchs1966 [catalogue: 236]; Chou1985 [taxonomy, host, distribution: 253-254]; KondoKa1995a [host, distribution: 97-98]; Takaha1939d [taxonomy, description, illustration, host, distribution: 341-343]; Tao1999 [taxonomy, distribution: 117]; Willia1957a [taxonomy, host, distribution: 39-40].

Semelaspidus theobromae Williams
Semelaspidus theobromae Williams, 1957a: 40. Type data: MALAYSIA: Malaya, Johore, Niyor, on *Theobroma cacao*. Holotype female. Type depository: London: The Natural History Museum, England, UK.
SCALE COVER: Female scale black; slightly convex; subcircular, about 2 mm in diameter; shape and size depending on position along leaf veins; exuviae of first stage larva reddish-brown, situated toward one edge (Williams, 1957a).
HOST PLANTS: **Sterculiaceae**: *Theobroma cacao* [Willia1957a].
DISTRIBUTION: **Oriental**: Malaysia (Malaya [Willia1957a]).
BIOLOGY: Occurring on both sides of leaves (Williams, 1957a).
GENERAL: Description and illustration of adult female by Williams (1957a).
KEYS: Williams 1957a: 42 (female) [World].
CITATIONS: Borchs1966 [catalogue: 236]; Willia1957a [taxonomy, description, illustration, host, distribution: 40-41].

Separaspis MacGillivray

Separaspis MacGillivray, 1921: 390. Type species: *Furcaspis proteae* Brain, by monotypy and original designation.
Truncaspidiotus MacGillivray, 1921: 390. Type species: *Lecanium capense* Walker, by monotypy and original designation. Synonymy by Lindinger, 1937: 197.
SYSTEMATICS: The genus *Separaspis,* together with several other genera, belongs in sub-tribe Furcaspidina (sensu Balachowsky, 1958b), and differ from other aspidiotine genera by: multisetose antennal tubercle, presence of perispiracular pores at anterior spiracles, and presence of glandular tubercles on venter of abdomen and thorax. It comes close to *Furcaspis* Lindinger, differing from it in: absence of perivulvar pores, and especially in presence of short, conical spines disposed submarginally on venter of abdominal segments.
GENERAL: Definition and characters by Lupo (1954) and by Balachowsky (1958b).
CITATIONS: Balach1958b [taxonomy, description: 250]; Borchs1966 [catalogue: 242]; Ferris1937c [taxonomy: 52]; Ferris1938 [taxonomy: 43]; Ferris1938b [taxonomy: 75]; Lindin1937 [taxonomy: 197]; Lupo1954 [taxonomy, description: 1-2]; MacGil1921 [taxonomy, description: 390,427]; MorrisMo1966 [taxonomy, catalogue: 183,199].

Separaspis capensis (Walker)
Lecanium capense Walker, 1852: 1079. Type data: SOUTH AFRICA: Cape Province, Algoa Bay, host plant not indicated. Syntypes, female. Type depository: London: The Natural History Museum, England, UK.

Aspidiotus (Aonidiella) capensis; Green, 1904b: 375. Change of combination requiring emendation of species name for agreement in gender.

Furcaspis capensis; Lindinger, 1908c: 99. Change of combination.

Aspidiotus reticulatus Newstead, 1912: 17. Type data: NAMIBIA: Klein-Namaland, Steinkopf, on undetermined plant. Syntypes, female. Type depository: London: The Natural History Museum, England, UK. Synonymy by Borchsenius, 1966: 242.

Spinaspidiotus reticulatus; MacGillivray, 1921: 430. Change of combination.

Truncaspidiotus capensis; MacGillivray, 1921: 431. Change of combination.

Separaspis capensis; Balachowsky, 1958b: 252. Change of combination.

SCALE COVER: Female scale sub-circular or oval; 2.5-3.5 mm in diameter; moderately convex, very dense and tough; highest point towards one side, from which, as a centre, concentric ridges or corrugations extend to scale margin; colour of scale varies with age; when young rich pale brown, but becomes more reddish later. Male scale small, linear, tapering slightly to posterior extremity, brown; exuviae at front end and hind end paler; length about 1.2 mm (Brain, 1918).

HOST PLANTS: **Liliaceae**: *Aloe* [Brain1918, Balach1958b, RosenDe1979], *Al. dichotoma* [Muntin1969], *Al. marlothii* [Balach1958b], *Al. ramosissima* [Muntin1969], *Al. rupestris* [Brain1918]. **Rutaceae**: *Teclea simplicifolia* [DeLott1967a].

NATURAL ENEMIES: HYMENOPTERA **Aphelinidae**: *Aphytis africanus* Quednau [RosenDe1979]. **Encyrtidae**: *Adelencyrtus* [Prinsl1983], *Habrolepis obscura* Compere & Annecke [AnneckIn1971], *Metaphycus aloinus* Annecke & Mynhardt [Prinsl1983].

DISTRIBUTION: **Afrotropical**: Kenya [DeLott1967a]; Namibia (Southwest Africa) [CockerRo1915a]; South Africa [Brain1918, Balach1958b, Muntin1969].

GENERAL: Description and illustration of adult female by Brain (1918) and by Balachowsky (1958b).

KEYS: Balachowsky 1958b: 250 (female) [Africa].

CITATIONS: AnneckIn1971 [host, distribution, biological control: 14]; Balach1958b [taxonomy, description, illustration, host, distribution: 251-254]; Borchs1966 [catalogue: 242]; Brain1918 [taxonomy, description, illustration, host, distribution: 137-138]; Cocker1919 [taxonomy: 116]; DeLott1967a [host, distribution: 115]; Fernal1903b [catalogue: 328]; Ferris1937c [taxonomy, illustration: 52]; Ferris1941e [taxonomy: 41,47]; Green1904b [taxonomy, description, illustration, host, distribution: 375-377]; Lindin1908b [taxonomy, host, distribution: 99]; Lindin1943b [taxonomy: 220]; MacGil1921 [taxonomy, description, host, distribution: 429-431]; Muntin1969 [taxonomy, host, distribution: 142]; Newste1912 [taxonomy, description, host, distribution: 17-18]; Nur1990a [taxonomy, structure, chromosomes: 184-185]; Prinsl1983 [distribution, biological control: 27]; RosenDe1979 [host, distribution, biological control: 542-545]; Signor1877 [taxonomy: 612]; Walker1852 [taxonomy, description, distribution: 1079].

Separaspis proteae (Brain)

Furcaspis proteae Brain, 1918: 139. Type data: SOUTH AFRICA: Buffelspoort, on *Faurea saligna*; collected by F. Thomsen, September 1909. Lectotype female, by subsequent designation Munting, 1970a: 39. Type depository: Pretoria: South African National Collection of Insects, South Africa; type no. 241/1.

Separaspis proteae; MacGillivray, 1921: 427. Change of combination.

SCALE COVER: Female scale similar to that of *Separaspis capensis*, but smaller and more brown than red; average size 1.6 mm in diameter. Male scale comparatively larger and wider than male of *S. capensis* (Brain, 1918).

HOST PLANTS: **Proteaceae**: *Faurea saligna* [Muntin1970a], *Protea* [Brain1918, Balach1958b, Muntin1970a].

NATURAL ENEMIES: HYMENOPTERA **Aphelinidae**: *Azotus separaspidis* Annecke & Insley [AnneckIn1970]. **Encyrtidae**: *Habrolepis* [Prinsl1983].

DISTRIBUTION: **Afrotropical**: South Africa [Brain1918, Balach1958b, Muntin1970a].

GENERAL: Description and illustration of adult female by Brain (1918) and by Balachowsky (1958b).

KEYS: Balachowsky 1958b: 250 (female) [Africa].

CITATIONS: AnneckIn1970 [host, distribution, biological control: 241-242]; Balach1958b [taxonomy, description, illustration, host, distribution: 253-255]; Borchs1966 [catalogue: 242]; Brain1918 [taxonomy, description, illustration, host, distribution: 139]; Ferris1937c [taxonomy: 52]; Ferris1938 [taxonomy: 43,55]; MacGil1921 [taxonomy, description, host, distribution: 427]; Muntin1970a [taxonomy: 39]; Prinsl1983 [distribution, biological control: 27].

Spinaspidiotus MacGillivray

Spinaspidiotus MacGillivray, 1921: 390. Type species: *Aspidiotus fissidens* Lindinger, by original designation.

SYSTEMATICS: Recent workers (Balachowsky, 1958b; Borchsenius, 1966; Morrison & Morrison, 1966) have accepted this genus as valid. It differs from *Hemiberlesia* Cockerell in: reduced size of scale cover, and characteristic sclerotization of margin of prosoma. It has affinity to *Schizaspis* Cockerell & Robinson, from which it differs mainly in absence of constrictions on prosoma.

GENERAL: Definition and characters by Balachowsky (1958b).

KEYS: Balachowsky 1958b: 228 (female) [*Aspidiotina* of Africa].

CITATIONS: Balach1958b [taxonomy, description: 218-219]; Borchs1966 [catalogue: 312]; Ferris1937c [taxonomy: 52]; Lindin1937 [taxonomy: 196]; MacGil1921 [taxonomy, description: 390,428-429]; Mamet1949 [taxonomy: 65]; MorrisMo1966 [taxonomy, catalogue: 187].

Spinaspidiotus fissidens (Lindinger)

Aspidiotus fissidens Lindinger, 1909e: 14. Type data: CAMEROON: Bipinde, Urwaldgebiet, on *Parinarium gabunense*, on *Paxia scandens* and on *Strychnos cinnabarina*. Syntypes, female. Type depository: Hamburg: Zoologisches Institut und Zoologishces Museum, Universität von Hamburg, Germany.

Aspidiotus fissidens pluridentatus Lindinger, 1910b: 35. Type data: TANZANIA: Kilimandjaro, Schira, altitude 1450 meters, on *Bosquiea cerasiflora*, Muoa mangroves, Usambara, on *Sideroxylon inerme*; MOZAMBIQUE: Quilimane, on *Chrysophyllum stuhlmanni*. Syntypes, female. Type depository: Hamburg: Zoologisches Institut und Zoologishces Museum, Universität von Hamburg, Germany. Synonymy by Ferris, 1941e: 47.

Aspidiotus fissus Lindinger, 1910b: 35. Type data: ETHIOPIA: near Harrar, on *Euphorbia* sp. Syntypes, female. Type depository: Hamburg: Zoologisches Institut und Zoologishces Museum, Universität von Hamburg, Germany. Synonymy by Borchsenius, 1966: 312.

Aspidiotus gowdeyi Newstead, 1913: 77. Type data: UGANDA: Entebbe, on *Annona muricata*; collected by C.G. Gowdey, 13.viii.1911 . Syntypes, female. Type depository: London: The Natural History Museum, England, UK. Synonymy by Ferris, 1941e: 44.

Hemiberlesia fissidens constricta Malenotti, 1916a: 340. Type data: SOMALIA: Giumbo, on leaves of *Rhizophora mucronata* (collected 13.vi.1913), and El Sai, on *Hyphaene pyrifera* (collected 21.vi.1913). Syntypes, female. Synonymy by Ferris, 1941e: 42.

Neosignoretia gowdeyi; MacGillivray, 1921: 425. Change of combination.

Spinaspidiotus fissidens; MacGillivray, 1921: 428. Change of combination.

Spinaspidiotus fissus; MacGillivray, 1921: 429. Change of combination.

Aspidiotus (*Aonidiella*) *gowdeyi*; Hall, 1929: 356. Change of combination.

Aspidiotus pluridentatus; Ferris, 1941e: 47. Change of status.

Hemiberlesia constricta; Ferris, 1941e: 47. Change of status.

Spinaspidiotus fissidens; Borchsenius: 1966: 312. Revived combination.

SCALE COVER: Lindinger (1909e) described scale cover as: "Scale pyriform, with rounded top, more or less circular, 1.4-1.8 mm long, 1.3-1.65 mm wide; exuviae in rounded part, yellow; colour brown with whitish or grey rims". Balachowsky (1958b) described the scale cover as: "Female scale circular, 0.6 mm in diameter; convex and suddenly truncate on top at margin of larval exuviae; larval exuviae central; margin circular; dark brown; outer margin paler, upper margin orange-brown to pale castaneous; exuviae completely hidden beneath a glistening white secretion, which is perfectly flat, quite circular in outline, and not raised above upper truncate margin of secretionary covering of scale; ventral vellum thin. Male scale oval, dark in center, brighter marginally; diameter, 0.4-0.5 mm". Illustration of scale cover by Balachowsky (1958b).

HOST PLANTS: **Annonaceae**: *Annona muricata* [Newste1913, Leonar1914, Balach1958b]. **Chrysobalanaceae**: *Parinari gabunense* [Lindin1909e, Balach1958b], *Parinari macrophylla* [Balach1958b]. **Connaraceae**: *Paxia scandans* [Lindin1909e, Balach1958b]. **Euphorbiaceae**: *Euphorbia* [Balach1958b]. **Guttiferae**: *Garcinia livingstoni* [Almeid1971]. **Leguminosae**: *Afzelia africana* [Balach1958b]. **Moraceae**: *Bosqueia cerasiflora* [Lindin1910b, Balach1958b], *Ficus* [Balach1958b]. **Palmae**: *Hyphaene pyrifera* [Balach1958b]. **Rhizophoraceae**: *Rhizophora micronata* [Balach1958b]. **Rubiaceae**: *Gardenia* [Balach1958b]. **Sapotaceae**: *Butyrospermum parkii* [Balach1958b], *Chrysophyllum agyrophyllum* [Hall1929, Balach1958b], *Chrysophyllum stuhlmanni* [Lindin1910b, Almeid1971],

Manilkara multinervis [Balach1958b], *Mimusops* [Almeid1971], *Pachystela brevipes* [Balach1958b], *Sideroxylon inerme* [Lindin1910b]. **Strychnaceae**: *Strychnos cinnabarina* [Lindin1909e, Balach1958b]. **Ulmaceae**: *Chaetacme aristata* [Brain1918, Balach1958b].

NATURAL ENEMIES: HYMENOPTERA **Encyrtidae**: *Habrolepis* sp. [Prinsl1983].

DISTRIBUTION: **Afrotropical**: Benin [Leonar1914]; Cameroon [Lindin1909e, Vayssi1913, MatileNo1984]; Eritrea [Balach1958b]; Ethiopia [Balach1958b]; Guinea [Balach1958b]; Kenya [Balach1958b]; Mozambique [Lindin1910b, Almeid1971]; Sierra Leone [Balach1958b]; Somalia [Balach1958b]; South Africa [Brain1918, Balach1958b]; Tanzania [Lindin1910b]; Uganda [Newste1913, Gowdey1917]; Zimbabwe [Hall1929, Balach1958b].

GENERAL: Description and illustration of adult female by Lindinger (1909e) and by Balachowsky (1958b).

KEYS: Brain 1918: 118 (female) [South Africa].

CITATIONS: Almeid1971 [host, distribution: 15-16]; Balach1958b [taxonomy, description, illustration, host, distribution: 219-222]; Borchs1966 [catalogue: 312]; Brain1918 [taxonomy, description, illustration, host, distribution: 118,123]; Ferris1937c [taxonomy: 52]; Ferris1941e [taxonomy: 43-44,46]; Gowdey1913 [host, distribution: 247-249]; Gowdey1917 [host, distribution: 189]; Hall1929 [taxonomy, host, distribution: 356-357]; Hall1946a [taxonomy: 536]; Hargre1927 [host, distribution, economic importance: 113-128]; Hargre1937 [host, distribution, economic importance: 505-520]; Laing1929a [taxonomy, host, distribution: 486]; Leonar1914 [taxonomy, host, distribution: 196]; Lepesm1947 [host, distribution: 196]; Lindin1909e [taxonomy, description, illustration, host, distribution: 14-15]; Lindin1910b [taxonomy, description, host, distribution: 35-36]; Lindin1957 [taxonomy: 545]; MacGil1921 [taxonomy, description, host, distribution: 425,428,429]; Maleno1916a [taxonomy, description, illustration, host, distribution: 340-342]; Mallam1954 [distribution: 24-60]; MatileNo1984 [host, distribution: 67]; McKenz1938 [taxonomy: 3]; Newste1913 [taxonomy, description, illustration, host, distribution: 77-78]; Prinsl1983 [distribution, biological control: 27]; Sassce1911 [taxonomy: 69]; Sassce1912 [taxonomy, host, distribution: 93]; Sassce1915 [taxonomy, host, distribution: 34]; Vayssi1913 [host, distribution: 430]; WeidneWa1968 [taxonomy: 172].

Spinaspidiotus maeandrius (Lindinger)

Aspidiotus maeandrius Lindinger, 1909e: 15. Type data: CAMEROON: Bipinde, Urwaldgebiet, on *Dichapetalum* sp. Syntypes, female. Type depository: Hamburg: Zoologisches Institut und Zoologishces Museum, Universität von Hamburg, Germany.

Spinaspidiotus maeandrius; MacGillivray, 1921: 430. Change of combination.

Aspidiotus meandrinus; Balachowsky, 1958b: 220. Misspelling of species name.

Aspidiotus maeandricus; Weidner & Wagner, 1968: 173. Misspelling of species name.

SCALE COVER: Female scale 2 mm long, 1-1.5 mm wide; white grey, with yellow brown exuviae; entire colour brown grey. Male scale whitish, 1.1 mm long,

0.7 mm wide; larval skin slightly subcentral towards to anterior end (Lindinger, 1909e).
HOST PLANTS: **Dichapetalaceae**: *Dichapetalum* [Lindin1909e].
DISTRIBUTION: **Afrotropical**: Cameroon [Lindin1909e, Vayssi1913].
GENERAL: Description and illustration of adult female by Lindinger (1909e).
CITATIONS: Balach1958b [taxonomy: 220]; Borchs1966 [catalogue: 312]; Ferris1941e [taxonomy: 45]; Lindin1909e [taxonomy, description, illustration, host, distribution: 15-17]; MacGil1921 [taxonomy, description, host, distribution: 430]; Sassce1911 [taxonomy: 69]; Vayssi1913 [host, distribution: 430]; WeidneWa1968 [taxonomy: 173].

Stringaspidiotus MacGillivray

Stringaspidiotus MacGillivray, 1921: 393. Type species: *Aspidiotus* (*Pseudaonidia*) *curculiginis* Green, by original designation.
SYSTEMATICS: Lindinger (1937) and Ferris (1938a) regarded this genus a synonym of *Furcaspis* Lindinger, while Balachowsky (1958b) and Borchsenius (1966) accepted it as valid. The genus *Stringaspidiotus* differs from *Furcaspis* in having unisetose antennal tubercle and second and third lobes being much smaller than median lobes (Balachowsky, 1958b).
GENERAL: Definition and characters by MacGillivray (1921) and by Balachowsky (1958b).
CITATIONS: Balach1958b [taxonomy: 250]; Borchs1966 [catalogue: 243]; Ferris1937c [taxonomy: 52]; Lindin1937 [taxonomy: 196]; MacGil1921 [taxonomy, description: 393,451]; MorrisMo1966 [taxonomy, catalogue: 190].

Stringaspidiotus curculiginis (Green)

Aspidiotus (*Pseudaonidia*) *curculiginis* Green, 1904a: 208. Type data: INDONESIA: Java, Buitenzorg, on both surfaces of leaves of *Curculigo recurvata*. Syntypes, female. Type depository: London: Natural History Museum, UK.
Pseudaonidia curculiginis; Marlatt, 1908: 137. Change of combination.
Furcaspis curculiginis; Lindinger, 1909b: 110. Change of combination.
Pseudaonidia circuliginis; Robinson, 1917: 33. Misspelling of species name.
Stringaspidiotus curculiginis; MacGillivray, 1921: 451. Change of combination.
Aspidiotus curculiginis; Ferris, 1937c: 52. Change of combination.
Stringaspidiotus curculiginis; Borchsenius, 1966: 243. Revived combination.
COMMON NAME: buli scale [VelasqRi1969].
SCALE COVER: Female scale elliptical, length 2-2.5 mm, breadth 1.25-1.50 mm; flattish, dark blackish-brown; Male scale similar, but smaller; exuviae nearer anterior extremity; length 1.5 mm; breadth 0.8 mm (Green, 1904a).
HOST PLANTS: **Hypoxidaceae**: *Curculigo recurvata* [Green1904a, Sander1906, Marlat1908]. **Palmae**: *Corypha elata* [Robins1917]. **Rubiaceae**: *Gardenia* [Takaha1942b].
DISTRIBUTION: **Oriental**: Indonesia (Java [Green1904a, Sander1906, Marlat1908]); Malaysia (Malaya [Takaha1942b]); Mongolia [DanzigKo1990]; Philippines (Luzon [Robins1917]); Thailand [Takaha1942b].

GENERAL: Description and illustration of adult female by Green (1904a).
KEYS: Robinson 1917: 33 (female) [Philippines]; Marlatt 1908: 135 (female) [World].
CITATIONS: Borchs1966 [catalogue: 243]; CockerRo1915 [taxonomy, host, distribution: 109]; DanzigKo1990 [host, distribution: 51]; Ferris1937c [taxonomy, illustration: 52,98]; Ferris1941e [taxonomy: 42]; Green1904a [taxonomy, description, illustration, host, distribution: 208-209]; Hunt1939 [host, distribution: 548-566]; Laing1929 [taxonomy: 498]; Lindin1909b [taxonomy: 110]; MacGil1921 [taxonomy, description, host, distribution: 451]; Marlat1908 [host, distribution: 135,137]; Robins1917 [taxonomy, description, host, distribution: 34]; Sander1906 [taxonomy, host, distribution: 15]; Takaha1942b [host, distribution: 49]; VelasqRi1969 [host, distribution: 195-208].

Sudanaspis Chou

Sudanaspis Chou, 1985: 278. Type species: *Aspidiotus* (*Hemiberlesia*) *vuilleti* Marchal, by monotypy and original designation.
SYSTEMATICS: This genus is very closely related to *Morganella* Cockerell, 1897, differing from the latter in shape of median lobes and shape of plates of pygidium.
GENERAL: Definition and characters by Chou (1985).
CITATIONS: Chou1985 [taxonomy, description: 278].
KEYS: Chou 1985: 260 (female) [genera of China].

Sudanaspis vuilleti (Marchal)

Aspidiotus (*Hemiberlesia*) *vuilleti* Marchal, 1909a: 587. Type data: SENEGAL: Bamako, on *Balanites* sp.; collected by Vuillet. Syntypes, female. Type depository: Paris: Muséum national d'Histoire naturelle, France.
Aspidiotus vuilleti; Sanders, 1909a: 53. Change of combination.
Hendaspidiotus vuilleti; MacGillivray, 1921: 440. Change of combination.
Aspidiotus (*Hemiberlesea*) *vuilleti*; Balachowsky, 1936a: 97. Misspelling of genus name.
Morganella vuilleti; Balachowsky, 1948b: 295. Change of combination.
Morganella vuilletti; Chou, 1985: 278. Misspelling of species name.
Sudanaspis vuilleti; Chou, 1985: 278. Change of combination.
SCALE COVER: Female scale circular, small, 0.8-1 mm in diameter; highly convex; dark, brown in median area; posterior margin curved in shape of a wooden-shoe; exuviae central; ventral vellum robust, attached to host (Balachowsky, 1948b).
HOST PLANTS: **Balanitaceae**: *Balanites* [Marcha1909a, Sander1909a], *Balanites aegyptiaca* [Balach1936a].
DISTRIBUTION: **Afrotropical**: Mauritania [BalachMa1970]; Senegal [Marcha1909a, Sander1909a]; Sudan [Balach1936a].
GENERAL: Description and illustration of adult female by Balachowsky (1936a, 1948b, 1956).
KEYS: Balachowsky 1956: 124 (female) [Africa]; Balachowsky 1948b: 293 (female) [Mediterranean].

CITATIONS: Balach1936a [taxonomy, description, illustration, host, distribution: 97-100]; Balach1948b [taxonomy, description, illustration, host, distribution: 295-297]; Balach1956 [taxonomy, description, illustration, host, distribution: 128-131]; Balach1958a [host, distribution: 35]; BalachMa1970 [host, distribution: 1081]; Borchs1966 [catalogue: 278]; Chou1985 [taxonomy: 278]; Ferris1941e [taxonomy: 49]; Lindin1957 [taxonomy: 546]; MacGil1921 [taxonomy, host, distribution: 440]; Marcha1909a [taxonomy, description, host, distribution: 587-588]; Marcha1909d [taxonomy, description, host, distribution: 178-179]; Sander1909a [taxonomy, host, distribution: 53]; Vayssi1913 [host, distribution: 431].

Taiwanaspidiotus Takagi

Taiwanaspidiotus Takagi, 1969a: 72. Type species: *Aspidiotus shakunagi* Takahashi, by original designation.

SYSTEMATICS: The type species of *Taiwanaspidiotus* Takagi is close to species of *Aspidiotus*, but cannot be retained in the latter owing to narrower body, evenly sclerotized dorsal derm of the pygidium, and reduction of plates on fifth abdominal segment (Takagi, 1969a).

GENERAL: Definition and characters by Takagi (1969a).

CITATIONS: Takagi1969a [taxonomy, description: 72]; Tao1999 [taxonomy: 119].

Taiwanaspidiotus shakunagi (Takahashi)

Aspidiotus shakunagi Takahashi, 1935: 32. Type data: TAIWAN: Taito Province, Chipponsan, on *Rhododendron* sp. Holotype female. Type depository: Sapporo: Entomological Institute, Faculty of Agriculture, Hokkaido University, Japan.

Taiwanaspidiotus shakunagi; Takagi, 1969a: 72. Change of combination.

SCALE COVER: Scale of the adult female pale whitish; thin; slightly convex dorsally; nearly circular, semitransparent, beneath the epidermis on the lower side of leaf, about 0.8 mm in diameter; exuviae pale yellow (Takahashi, 1935).

HOST PLANTS: **Ericaceae**: *Rhododendron* [Takaha1935, Takagi1969a].

DISTRIBUTION: **Oriental**: Taiwan [Takaha1935, Takagi1969a].

BIOLOGY: Insects develop beneath epidermis on lower side of leaf (Takahashi, 1935).

GENERAL: Description and illustration of adult female by Takahashi (1935) and by Takagi (1969a).

KEYS: Chou 1985: 262-263 (female) [Species of China].

CITATIONS: Borchs1966 [catalogue: 267]; Chou1985 [taxonomy, description, host, distribution: 272-273]; Ferris1941e [taxonomy: 48]; FoxWil1939 [host, distribution, economic importance: 2296]; Takagi1969a [taxonomy, description, illustration, host, distribution: 72-75]; Takaha1935 [taxonomy, description, illustration, host, distribution: 4,32-33]; Tao1999 [taxonomy, distribution: 119].

Taiwanaspidiotus yiei Takagi

Taiwanaspidiotus yiei Takagi, 1969a: 75. Type data: TAIWAN: Fen-chi-hu, on leaves of *Castanopsis kusanoi*. Holotype female. Type depository: Sapporo: Entomological Institute, Faculty of Agriculture, Hokkaido University, Japan.
Aspidiotus yiei; Chou, 1985: 399. Change of combination.
Taiwanaspidiotus yei; Tao, 1999: 119. Revived combination and misspelling of species name.
SCALE COVER: Takagi (1969a) did not describe scale cover.
HOST PLANTS: **Fagaceae**: *Castanopsis kusanoi* [Takagi1969a].
DISTRIBUTION: **Oriental**: Taiwan [Takagi1969a].
GENERAL: Description and illustration of adult female by Takagi (1969a) and by Chou (1985, 1986).
CITATIONS: Chou1985 [taxonomy, description, host, distribution: 399]; Chou1986 [taxonomy, illustration: 663]; ShiLi1991 [host, distribution: 166]; Takagi1969a [taxonomy, description, illustration, host, distribution: 74-75]; Tao1999 [taxonomy, host, distribution: 119].

Targionia Signoret

Targionia Signoret, 1869: 862. Nomen nudum.
Kermesoides Signoret, 1868: 862. Nomen oblitum. Type species: *Targionia nigra* Signoret, by implication. Notes: Stated by Signoret (1868) as "*Targ. nigra* Signoret, nov. spec. (*Kermesoides* id., olim)".
Targionia Signoret, 1869b: 99. Type species: *Targionia nigra* Signoret, by monotypy.
Aspidiotus (*Targionia*); Cockerell, 1897i: 14. Change of status.
Tagionia; Green, 1904: 66. Misspelling of genus name.
Targaspidiotus MacGillivray, 1921: 392. Type species: *Aspidiotus yuccarum* Cockerell, by original designation. Synonymy by Danzig, 1993: 221.
Schizotargionia Balachowsky, 1951: 644. Type species: *Aspidiotus arthrophyti* Archangelskaya, by monotypy and original designation. Synonymy by Danzig, 1993: 221.
Pseudomelanaspis Borchsenius, 1952: 262. Type species: *Pseudomelanaspis minima* Borchsenius, by monotypy and original designation. Synonymy by Danzig, 1993: 221.
Fisanotargionia Kaussari & Balachowsky, 1953b: 277. Type species: *Fisanotargionia quadrilobata* Kaussari & Balachowsky, by monotypy and original designation. Synonymy by Danzig, 1993: 221.
SYSTEMATICS: *Targionia* Signoret is close to *Rhizaspidiotus* MacGillivray, differing in form and distribution of dorsal macroducts on pygidium. In *Rhizaspidiotus* ducts are short and distributed sporadically, not in well-defined furrows, while in *Targionia* ducts are thin and long and disposed in distinct furrows (Balachowsky, 1951, 1958b; Danzig, 1993). *Pseudomelanaspis* Borchsenius (1952) was regarded a subjective synonym of *Targionia*. However, Borchsenius & Williams (1963) considered it valid and allied to *Melanaspis* Cockerell. Ferris (1943a) discussed in great details the nomenclature of *Targionia*. Ferris (1943a) and

Munting (1965b) discussed *Targionia* and *Targaspidiotus* MacGillivray, and indicated that the latter is available should it ever seem desirable to divide *Targionia*. Signoret (1868: 862) first published the generic name *Kermesoides* as "*Targ. nigra* Signoret, nov. spec. (*Kermesoides* id., olim)". Lindinger (1933e) and Ferris (1943a) accepted *Kermesoides* as an alternate specific name for *nigra*, whereas Morrison & Morrison (1966) considered the latter to be a generic name. The name *Kermesoides* was not used again since it was first introduced over 130 years ago, therefore we regard it a synonym of *Targionia,* as suggested by Morrison & Morrison (1966).

GENERAL: Definition and characters by Signoret (1870), Ferris (1938a, 1943a), Borchsenius (1950b, 1952), Kaussari & Balachowsky (1953b), Lupo (1957), Balachowsky (1951, 1958b), Gómez-Menor Ortega (1959), Borchsenius & Williams (1963), Bazarov & Shmelev (1971), Kosztarab & Kozár (1978), Danzig (1993) and by Yaşar (1995).

KEYS: Gill 1997: 24-26 (female) [Genera of California]; Danzig 1993: 221-222 (female) [Europe]; Tereznikova 1986: 78 (female) [Ukraine]; Kosztarab & Kozár 1978: 144-147 (female) [Hungary]; Bazarov & Shmelev 1971: 171 (female) [Central Asia]; Ezzat & Afifi 1966: 371-372 (female) [Egypt]; Danzig 1964: 646 (female) [Europe]; Balachowsky 1958b: 281 (female) [*Targionina* of Africa]; Ezzat 1958: 237-239 (female) [Egypt]; Gómez-Menor Ortega 1956: 7-8 (female) [Spain]; McKenzie 1956: 23 (female) [U.S.A.: California]; Balachowsky 1951: 632 (female) [Mediterranean]; Borchsenius 1950b: 168 (female) [USSR]; Gómez-Menor Ortega 1946: 59-61 (female) [Spain]; Ruiz Castro 1944: 57 (female) [Spain]; Ferris 1942: 27 (female) [North America]; Ferris 1942: 40 (female) [species North America]; Borchsenius 1937: 100 (female) [USSR]; Borchsenius 1937a: 32-33 (female) [Palaearctic Region]; Hollinger 1923: 6-7 (female) [U.S.A.: Missouri]; Leonardi 1920: 27 (female) [Italy]; Leonardi 1920: 104-105 (female) [Species of Italy]; Lawson 1917: 206 (female) [U.S.A.: Kansas]; Lawson 1917: 246 (female) [species U.S.A.: Kansas].

CITATIONS: Ashmea1891 [taxonomy: 101]; Atkins1886 [taxonomy: 273]; Balach1951 [taxonomy, description: 632-633,644-645]; Balach1958b [taxonomy, description: 290,292]; BazaroSh1971 [taxonomy, description: 181-182]; BerlesLe1898a [taxonomy: 10]; BlayGo1993 [taxonomy, description: 399,409]; Bodenh1924 [taxonomy: 21]; Bodenh1949 [taxonomy, description: 38]; Bodenh1952 [taxonomy: 330]; Borchs1937 [taxonomy, description: 100]; Borchs1937a [taxonomy: 33,67]; Borchs1949d [taxonomy: 195,250]; Borchs1950b [taxonomy, description: 168,230,233]; Borchs1952 [taxonomy, description: 262]; Borchs1966 [catalogue: 248,250-251,353]; BorchsWil963 [taxonomy, description: 389]; Cocker1893d [taxonomy: 8]; Cocker1897i [taxonomy: 14]; Cocker1899a [taxonomy: 395]; Cocker1905b [taxonomy: 200]; Danzig1964 [taxonomy: 654]; Danzig1993 [taxonomy, description: 221-222]; DanzigPe1998 [catalogue: 359]; Ezzat1958 [taxonomy: 238]; Fernal1903b [catalogue: 295]; Ferris1920b [taxonomy: 56]; Ferris1921b [taxonomy: 94]; Ferris1937c [taxonomy: 52]; Ferris1937e [taxonomy: 528]; Ferris1938 [taxonomy: 44]; Ferris1938a [taxonomy, description: 266]; Ferris1942 [taxonomy: 27]; Ferris1943a [taxonomy, description: 82-94]; Gill1997 [taxonomy: 267]; GomezM1937 [taxonomy, description: 121];

GomezM1946 [taxonomy: 60]; GomezM1956 [taxonomy: 8]; GomezM1959 [taxonomy, description: 160-161]; GomezM1965 [taxonomy, description: 99-100]; Green1904 [taxonomy: 66]; Hadzib1983 [taxonomy: 216]; Hollin1923 [taxonomy: 7,68]; Kaussa1952 [taxonomy: 181]; KaussaBa1953b [taxonomy, description: 277]; KosztaKo1978 [taxonomy, description: 176-177]; Kozar1990f [distribution: 142,143]; Lawson1917 [taxonomy : 206,246]; Lawson1917 [taxonomy, description: 246]; Leonar1897 [taxonomy: 284]; Leonar1897a [taxonomy: 375]; Leonar1897b [taxonomy: 109,111]; Leonar1900 [taxonomy, description: 302]; Leonar1920 [taxonomy, description: 27,104-105]; Lindin1908b [taxonomy: 98]; Lindin1911 [taxonomy: 382]; Lindin1932f [taxonomy: 194]; Lindin1937 [taxonomy: 197]; Lindin1943a [taxonomy: 152]; Low1882c [taxonomy: 521]; Lupo1957 [taxonomy, description: 54]; MacGil1921 [taxonomy, description: 392-393,446-448]; Maskel1887a [taxonomy, description: 40]; McKenz1939 [taxonomy: 54]; McKenz1956 [taxonomy, description: 23]; Miller1990 [taxonomy: 169-178]; Morgan1888b [taxonomy: 118]; MorrisMo1966 [taxonomy, catalogue: 79,166,180,192,193]; Muntin1965b [taxonomy: 209-211]; RuizCa1944 [taxonomy: 57]; Signor1869 [taxonomy: 862]; Signor1869b [taxonomy: 99]; Signor1870 [taxonomy, description: 105]; Silves1902 [taxonomy: 102]; ThiemGe1934a [taxonomy: 232]; Yasar1995a [taxonomy, description: 130].

Targionia anabasidis (Borchsenius)

Targaspidiotus anabasidis Borchsenius, 1952: 263. Type data: IRAN: Bender-Abbas, Oman bay, on *Anabasis aphylla*. Lectotype female, by subsequent designation Danzig, 1993: 224. Type depository: St. Petersburg: (= Leningrad) Zoological Museum, Academy of Science, Russia; type no. 222-48. Notes: Type data of this species is same as that of *Pseudomelanaspis minima* Borchsenius, 1952.

Pseudomelanaspis minima Borchsenius, 1952: 262. Type data: IRAN: Bender-abbas, Oman bay, on *Anabasis aphylla*. Lectotype female, by subsequent designation Danzig, 1993: 224. Type depository: St. Petersburg: (= Leningrad) Zoological Museum, Academy of Science, Russia. Synonymy by Danzig, 1993: 224. Synonymy by Danzig, 1993: 224. Notes: Type data of this species is same as that of *Targaspidiotus anabasidis* Borchsenius, 1952.

Schizotargionia anabasidis; Borchsenius, 1966: 251. Change of combination.

Targionia anabasidis; Danzig, 1993: 221. Change of combination.

SCALE COVER: Female scale circular, about 1.5 mm in diameter; slightly convex; colour light brown; exuviae central, covered with white secretionary matter (Borchsenius, 1952).

HOST PLANTS: Chenopodiaceae: *Anabasis aphylla* [Borchs1952, Matile1988], *Hammada salicornica* [Matile1988].

DISTRIBUTION: Palaearctic: Iran [Borchs1952, Danzig1993]; Saudi Arabia [Matile1988, Danzig1993].

GENERAL: Description of adult female by Borchsenius (1952) and by Borchsenius & Williams (1963) as *Pseudomelanaspis minima* Borchsenius.

KEYS: Danzig 1993: 221-222 (female) [Europe].

CITATIONS: Borchs1952 [taxonomy, description, host, distribution: 262-263]; Borchs1966 [catalogue: 251,353]; BorchsWi1963 [taxonomy, description, illustration: 389-390]; Danzig1993 [taxonomy, description, host, distribution: 224]; DanzigPe1998 [catalogue: 359-360]; Matile1988 [host, distribution: 25].

Targionia arthrophyti (Archangelskaya)

Aspidiotus (*Aonidiella*) *arthrophytoni* Archangelskaya, 1930: 85. Nomen nudum.

Aspidiotus (*Aonidiella*) *arthrophyti* Archangelskaya, 1931: 83. Type data: TURKMENISTAN: in sandy places near Repetek, on stems and branches of *Arthrophytum (Haloxylon) ammodendron*. Lectotype female, by subsequent designation Danzig, 1993: 224. Type depository: St. Petersburg: (= Leningrad) Zoological Museum, Academy of Science, Russia.

Aspidiotus arthrophyti; Borchsenius, 1937: 135. Change of combination.

Aonidiella arthrophyti; Archangelskaya, 1937: 95,97. Change of combination.

Targionia arthrophyti; Ferris, 1943a: 85. Change of combination.

Targaspidiotus arthrophyti; Borchsenius, 1950b: 232. Change of combination.

Schizotargionia arthrophyti; Balachowsky, 1951: 645. Change of combination.

Targionia arthrophyti; Danzig, 1993: 221. Revived combination.

SCALE COVER: Female scale white to black reddish; convex; diameter 1 mm; ventral scale thin, retained adherent to bark; exuviae brown, placed centrally; generally covered with secretion (Archangelskaya, 1931).

HOST PLANTS: **Chenopodiaceae**: *Arthrophytum ammodendron* [Archan1931, Balach1951, BazaroSh1971].

NATURAL ENEMIES: HYMENOPTERA **Encyrtidae**: *Comperiella schizotargioniae* Myartseva [Trjapi1989].

DISTRIBUTION: **Palaearctic**: Iran [BazaroSh1971]; Tajikistan [BazaroSh1971]; Turkmenistan [Archan1931, BazaroSh1971]; Uzbekistan [BazaroSh1971].

GENERAL: Description and illustration of adult female by Balachowsky (1951), Bazarov & Shmelev (1971) and by Danzig (1993).

KEYS: Danzig 1993: 221-222 (female) [Europe]; Bazarov & Shmelev 1971: 182 (female) [Central Asia]; Kaussari 1952: 182 (female) [Iran]; Balachowsky 1951: 645 (female) [Mediterranean].

CITATIONS: Archan1930 [taxonomy, host, distribution: 85]; Archan1931 [taxonomy, description, illustration, host, distribution: 83-85]; Archan1937 [taxonomy, description, host, distribution: 95,97]; Balach1951 [taxonomy, description, illustration, host, distribution: 645-648]; BazaroSh1971 [taxonomy, description, illustration, host, distribution: 182-184]; Borchs1937 [taxonomy, description, illustration, host, distribution: 134-135]; Borchs1939 [taxonomy, description, host, distribution: 11,41]; Borchs1950b [taxonomy, description, host, distribution: 230,232]; Borchs1952 [taxonomy: 263]; Borchs1966 [catalogue: 251]; DanzigPe1998 [catalogue: 360]; Ferris1941e [taxonomy: 41]; Ferris1943a [taxonomy, host, distribution: 85,87]; Kaussa1952 [taxonomy: 181]; Lindin1957 [taxonomy: 545]; MillerDa1990 [host, distribution, economic importance: 305]; SchmutKlLu1957 [host, distribution, economic importance: 493]; Trjapi1989 [biological control: 297].

Targionia balachowskyi (Kaussari)

Schizotargionia balachowskyi Kaussari, 1952: 182. Type data: IRAN: Beluchistan, near Zahedan, on *Tamarix* sp. Syntypes, female. Type depository: Paris: Muséum national d'Histoire naturelle, France.

Targionia balachowskyi; Danzig, 1993: 221. Change of combination.

SCALE COVER: Female scale circular, 2-2.2 mm in diameter; convex; grey with brown reflection; larval exuviae central, brown; ventral scale present; scale usually covered with particles of bark (Kaussari, 1952).

HOST PLANTS: **Tamaricaceae**: *Tamarix* [Kaussa1952].

DISTRIBUTION: **Palaearctic**: Iran [Kaussa1952, Kaussa1955].

GENERAL: Description and illustration of adult female by Kaussari (1952).

KEYS: Danzig 1993: 221-222 (female) [Europe]; Kaussari 1952: 182 (female) [Iran].

CITATIONS: Borchs1966 [catalogue: 251]; DanzigPe1998 [catalogue: 360]; Kaussa1952 [taxonomy, description, illustration, host, distribution: 181-184]; Kaussa1955 [host, distribution: 17].

Targionia bigeloviae (Cockerell), revived combination

Aspidiotus (*Hemiberlesia ?*) *bigeloviae* Cockerell, 1897i: 20. Type data: U.S.A.: California, Los Angeles, on *Bigelovia brachylepis*; collected by D.W. Coquillett. Syntypes, female. Type depository: U.S. National Entomological Collection, National Museum of Natural History, Washington, D.C., USA; type no. 4973.

Aspidiotus bigeloviae; Cockerell & Parrott, 1899: 278. Change of combination.

Aspidiotus (*Targionia*) *bigeloviae*; Cockerell, 1899a: 395. Change of combination.

Targionia bigeloviae; Leonardi, 1900: 343. Change of combination.

Leonardianna bigeloviae; MacGillivray, 1921: 450. Change of combination.

Targaspidiotus bigeloviae; Borchsenius, 1952: 263. Change of combination.

SYSTEMATICS: This species is re-assigned to *Targionia,* following the synonymy of *Targaspidiotus* with *Targionia* by Danzig (1993).

COMMON NAMES: bigelovia scale [McKenz1956].

SCALE COVER: Cockerell (1897i) described scale cover as: "Size and shape of *Aspidiotus rapax,* but dull greyish-brown; exuviae placed to one side as in *rapax*; when rubbed shining black, but more or less covered by a film of white secretion; removed from twig the scales leave a white patch". Ferris (1938a) described scale cover as: "Scale of the female circular, rather flat, exuviae subcentral, the scale bright reddish brown except for the area over the exuviae which is white. Scale of the male of similar colour or slightly paler, elongate, exuviae near one end".

HOST PLANTS: **Compositae**: *Bigelowia brachylepis* [Cocker1897i, Leonar1900, McKenz1956], *Gutierrezia* [Ferris1938a, McKenz1956], *Lepidospartum californicum* [Ferris1938a], *Lepidospartum squamatum* [McKenz1956].

DISTRIBUTION: **Nearctic**: Mexico [Nakaha1982]; United States of America (California [Cocker1897i, Ferris1938a, McKenz1956], Texas [Nakaha1982]).

BIOLOGY: Occurring on stems and crowns (Ferris, 1938a). Reported to have uniparental as well as biparental populations (Brown, 1965).

GENERAL: Description and illustration of adult female by Ferris (1938a, 1943a), McKenzie (1956) and by Gill (1997).

KEYS: McKenzie 1956: 26 (female) [U.S.A.: California]; Ferris 1943a: 94 (female) [World]; Ferris 1942: 40 (female) [North America].
CITATIONS: BenDov1990c [taxonomy: 115]; Borchs1952 [taxonomy: 263]; Borchs1966 [catalogue: 251]; Cocker1897i [taxonomy, description, host, distribution: 20]; Cocker1899a [taxonomy: 395]; CockerPa1899 [taxonomy, illustration: 278,282]; Fernal1903b [catalogue: 296]; Ferris1938a [taxonomy, description, illustration, host, distribution: 267]; Ferris1941e [taxonomy: 41]; Ferris1942 [taxonomy: 446:20]; Ferris1943a [taxonomy, description, illustration, host, distribution: 85,87-88,101]; Gerson1990 [taxonomy: 130]; Gill1997 [host, distribution, taxonomy, description, illustration, economic importance: 267,269]; Leonar1900 [taxonomy, host, distribution: 343]; MacGil1921 [taxonomy, description, host, distribution: 450]; McKenz1956 [taxonomy, description, illustration, host, distribution: 87-88]; Nakaha1982 [host, distribution: 83-84]; Newell1899 [taxonomy, description, host, distribution: 25]; Nur1990b [taxonomy, life history: 196].

Targionia fabianae Leonardi

Targionia fabianae Leonardi, 1911: 278. Type data: ARGENTINA: Cacheuta, *Fabiana denudata*. Lectotype female, by subsequent designation Claps, 2000: 94. Type depository: Portici: Dipartimento de Entomologia e Zoologia Agraria di Portici, Università di Napoli Federico II, Italy.
SCALE COVER: Female scale oval, 1.5 mm long, 1.4 mm wide; highly convex; exuviae small, eccentric, rust coloured; secreted part of scale robust; ochreous white; ventral vellum white, remains attached to host plant (Leonardi, 1911).
HOST PLANTS: **Solanaceae**: *Fabiana denudata* [Leonar1911, ClapsWoGo2001].
Verbenaceae: *Verbena aphylla* [ClapsWoGo2001].
DISTRIBUTION: **Neotropical**: Argentina [Leonar1911] (Mendoza [ClapsWoGo2001]).
GENERAL: Description and illustration of adult female by Leonardi (1911).
CITATIONS: Borchs1966 [catalogue: 249]; Claps2000 [taxonomy, description, illustration, host, distribution: 93-95]; ClapsWoGo2001 [host, distribution: 252]; Ferris1943a [taxonomy: 85]; Leonar1911 [taxonomy, description, illustration, host, distribution: 278-280]; MacGil1921 [taxonomy, description, host, distribution: 249]; Sassce1912 [taxonomy, host, distribution: 94].

Targionia halophila (Balachowsky)

Aspidiotus (*Aonidiella*) *halophilus* Balachowsky, 1928c: 277. Type data: ALGERIA: 40 km south of Constantine, on *Halocnemum strobilaceum*. Holotype female. Type depository: Paris: Muséum national d'Histoire naturelle, France.
Aonidiella halophila; Balachowsky, 1932d: xiii. Change of combination requiring emendation of species name for agreement in gender.
Targionia halophila; Balachowsky, 1932d: xiii. Change of combination.
Aspidiotus halophilus; Ferris, 1941e: 44. Change of combination.
Schizotargionia halophila; Balachowsky, 1951: 648. Change of combination.
Targaspidiotus halophilus; Borchsenius, 1952: 262. Change of combination.
Targionia halophila; Danzig, 1993: 222. Revived combination.

SCALE COVER: Female scale circular or subcircular; larval exuviae central or slightly eccentric; brown black, slightly covered with white secretion; 1.8 mm; ventral vellum attached to host plant (Balachowsky, 1951).

HOST PLANTS: **Chenopodiaceae**: *Halocnemum strobilaceum* [Balach1928c, Balach1932d, Ferris1943a], *Salicornia* [BlayGo1993], *Salsola longifolia* [GomezM1965, Martin1983], *Salsola vermiculata* [BlayGo1993], *Salsola weebi* [GomezM1965, Martin1983, BlayGo1993], *Suaeda fructicosa* [BlayGo1993].

DISTRIBUTION: **Palaearctic**: Algeria [Balach1928c, Balach1932d, Ferris1943a]; Spain [GomezM1965, Martin1983, BlayGo1993].

GENERAL: Description and illustration of adult female by Balachowsky (1928c, 1951).

KEYS: Danzig 1993: 221-222 (female) [Europe]; Kaussari 1952: 181 (female) [Iran]; Balachowsky 1951: 645 (female) [Mediterranean]; Ferris 1943a: 94 (female) [World].

CITATIONS: Balach1928c [taxonomy, description, illustration, host, distribution: 277-279]; Balach1932d [taxonomy, host, distribution: XIII]; Balach1951 [taxonomy, description, illustration, host, distribution: 648-650]; BlayGo1993 [taxonomy, description, illustration, host, distribution: 410-413]; Borchs1952 [taxonomy: 262]; Borchs1966 [catalogue: 251]; Danzig1993 [taxonomy: 222]; DanzigPe1998 [catalogue: 360]; Ferris1941e [taxonomy: 44]; Ferris1943a [taxonomy, host, distribution: 86,88]; GomezM1965 [taxonomy, description, illustration, host, distribution: 100-102]; GomezM1968 [host, distribution: 542]; Kaussa1952 [taxonomy: 181]; Lindin1957 [taxonomy: 545]; Martin1983 [taxonomy, host, distribution: 69].

Targionia haloxyloni Hall
Targionia haloxyloni Hall, 1926a: 27. Type data: EGYPT: Eastern Desert, Wadi Askhar South, Wadi Araba, Wadi Sennur, on *Haloxylon schweinfurthii*. Syntypes, female. Type depository: London: The Natural History Museum, England, UK.
Targionia haloxyli Lindinger, 1932f: 198. Unjustified emendation.
Targionia halophila; Rungs, 1934: 275. Misidentification; discovered by Rungs, 1935: 275.

SCALE COVER: Hall (1926a) described scale cover as: "Scale of adult female small, irregularly circular, diameter 1.25-1.75 mm; convex; dead white in colour owing to a thick covering of white secretionary matter; first exuvia straw-coloured when denuded and the second pellicle large and black; white secretionary matter is easily knocked off, coming away in one piece and revealing the black nymphal pellicle; ventral scale well developed but usually remaining attached to the host plant". Ferris (1943a) described scale cover as: "The scale of the female is described as being circular, convex, white, with the second exuvia black, ventral scale well developed but remaining attached to the host plant. Scale of the male not described".

HOST PLANTS: **Chenopodiaceae**: *Haloxylon schweinfurthii* [Hall1926a, Ferris1943a], *Haloxylon scoparium* [Rungs1942, Rungs1948], *Haloxylon tamariscifolium* [Rungs1935, Ferris1943a], *Salsola webbii* [Rungs1948], *Suaeda* [Balach1951]. **Cistaceae**: *Helianthemum lippii* [Ferris1943a]. **Fagaceae**: *Quercus pubescens* [Balach1951]. **Salsolaceae** [BalachMa1970].

DISTRIBUTION: **Afrotropical**: Mauritania [Rungs1942, Balach1958b, BalachMa1970]. **Palaearctic**: Egypt [Hall1926a, Ferris1943a, Ezzat1958]; Morocco [Rungs1935, Rungs1948, Ferris1943a]; Western Sahara [Rungs1942].

BIOLOGY: Occurring mainly on subterranean parts of plant (Ferris, 1943a).

GENERAL: Description and illustration of adult female by Hall (1926a), Ferris (1943a) and by Balachowsky (1951, 1958b).

KEYS: Danzig 1993: 221-222 (female) [Europe]; Balachowsky 1958b: 292 (female) [Africa]; Ezzat 1958: 242 (female) [Egypt]; Balachowsky 1951: 634 (female) [Mediterranean]; Ferris 1943a: 94 (female) [World].

CITATIONS: Balach1951 [taxonomy, description, illustration, host, distribution, economic importance: 637-640]; Balach1958a [host, distribution: 39]; Balach1958b [taxonomy, description, illustration, host, distribution: 292-294]; BalachMa1970 [host, distribution: 1081]; Borchs1966 [catalogue: 249]; DanzigPe1998 [catalogue: 360]; Ezzat1958 [distribution: 242]; EzzatAf1966 [taxonomy, description, illustration, host, distribution: 381-383]; EzzatNa1987 [distribution: 88]; Ferris1943a [taxonomy, description, illustration, host, distribution: 86-89,102]; Hall1926a [taxonomy, description, illustration, host, distribution: 27-28]; Hall1927b [taxonomy, description, illustration, host, distribution: 153-154]; Hosny1939 [taxonomy, host, distribution: 15]; Lindin1932f [taxonomy: 198]; Rungs1935 [host, distribution: 275]; Rungs1942 [host, distribution: 107]; Rungs1948 [host, distribution: 112].

Targionia kermesoides nomen nudum

Targionia kermesoides Signoret, 1868a: 862. Nomen nudum.
Targionia kermesoides Lindinger, 1933e: 68. Nomen nudum.
Targionia kermesoides Ferris, 1943a: 86. Nomen nudum.
Targionia kermesoides Borchsenius, 1966: 377. Nomen nudum.

Targionia kermoides nomen nudum

Targionia kermoides Lindinger, 1936: 166. Nomen nudum.

Targionia nigra Signoret

Targionia nigra Signoret, 1869: 862. Nomen nudum.
Targionia nigra Signoret, 1870: 106. Type data: FRANCE: Le Midi [=South East France], Cannes, on *Cineraria maritima*. Syntypes, both sexes. Type depository: Vienna: Naturhistorisches Museum Wien, Austria.
Aspidiotus signoreti Comstock, 1883: 82. Unjustified replacement name for *Targionia nigra* Signoret, 1870; discovered by Cockerell, 1897i: 19.
Aspidiotus (*Targionia*) *signoreti*; Cockerell, 1897i: 19. Change of combination.
Targionia nigra; Fernald, 1903b: 298. Revived combination.
Targionia deserti Balachowsky, 1927: 194. Type data: ALGERIA: in area of an oasis, 10 km southwest of Fort Mac-Mahon, on *Retama roetam*. Holotype female. Type depository: Paris: Muséum national d'Histoire naturelle, France. Synonymy by Ferris, 1943a: 85.
Targonia nigra; Balachowsky, 1932d: XII. Misspelling of genus name.
Aspidiotus deserti; Ferris, 1943a: 85. Change of combination.

Schizotargionia limonii Bazarov & Shmelev, 1967: 60. Type data: TURKMENISTAN: Kugi-Tang Ridge, on *Limonium suffruticosum*. Syntypes, female. Type depository: St. Petersburg: (= Leningrad) Zoological Museum, Academy of Science, Russia. Synonymy by Danzig, 1993: 225.

SCALE COVER: Ferris (1943a) described scale cover as: "Female scale highly convex, somewhat oval, dark brown or black, and quite rough because of transverse ridges. The ventral scale is very thick, almost as much so as the dorsal, the appearance being much that of a bivalve mollusk. Male scale is not represented in the material at hand, but the scale is described as being elongate-oval, apparently similar to that of the female in colour. Descriptions indicated that the insects occur on the stems and even on the roots of the host. It has been noted by various authors that specimens which seem on morphological grounds to belong to this species may present a wide range of variation in the appearance of the scales. In the three lots of material at hand this variation appears. Specimens from *Thymelaea hirsuta* from Tunis have the scales of the female white or slightly brown and almost porcellaneous in texture. Specimens from *Alhagi* in Egypt are pure white and of a slightly felted texture, and in these the ventral scale is quite thin".

HOST PLANTS: **Boraginaceae**: *Heliotropium luteum* [Ferris1943a]. **Chenopodiaceae**: *Salicornia* [GomezM1957, BlayGo1993], *Salsola* [Danzig1993], *Salsola longifolia* [GomezM1957, BlayGo1993], *Suaeda fructicosa* [Martin1983, BlayGo1993], *Suaeda vermiculata* [Balach1930c, Balach1932d, Ferris1943a]. **Compositae**: *Artemisia* [Bodenh1937, Danzig1993], *Artemisia judaica* [Bodenh1935, Ferris1943a], *Cineraria* [Danzig1993], *Cineraria maritima* [Signor1870, Leonar1918, Leonar1920, Balach1933a], *Helichrysum angustifolium* [Balach1932d, Ferris1943a], *Helichrysum italicum* [Leonar1920, Ferris1943a, Bachma1953], *Launaea spinosa* [Hall1926a, Ferris1943a], *Phagnalon* [Balach1935b, Martin1983], *Phagnalon sexatile* [Balach1933a], *Santolina chamaecyparissus incana* [GomezM1948, Martin1983, BlayGo1993], *Senecio* [Balach1930a, Ferris1943a], *Senecio cinerarea* [Balach1930, Balach1932d, Balach1933e, Ferris1943a, Foldi2000], *Senecio kleinia* [Balach1951]. **Cruciferae**: *Farsetia* [Balach1951], *Farsetia aegyptiaca* [Hall1925, Hall1926a, Bellio1929b, Ferris1943a], *Moricandia* sp. [Balach1958b], *Moricandia suffruticosa* [Rungs1935, Ferris1943a], *Zilla spinosa* [Ferris1943a]. **Cucurbitaceae**: *Citrullus colocynthus* [Hall1926a, Ferris1943a]. **Labiatae**: *Teucrium* [Balach1958b, Danzig1993], *Teucrium polium* [Rungs1935, Ferris1943a]. **Leguminosae**: *Alhagi maurorum* [Hall1923, Ferris1943a], *Glycyrrhiza* [Danzig1993], *Retama roetam* [Balach1930c, Balach1932d, Ferris1943a]. **Liliaceae**: *Asparagus horridus* [Balach1932d]. **Plumbaginaceae**: *Limonium suffruticosum* [BazaroSh1971, Danzig1993]. **Resedaceae**: *Ochradenus baccatus* [Hall1926a, Ferris1943a]. **Scrophulariaceae**: *Antirrhinum* [Balach1958b], *Antirrhinum ramosissimum* [Rungs1935, Ferris1943a]. **Thymelaeaceae**: *Thymelaea* [Bodenh1937], *Thymelaea hirsuta* [Balach1930c, Balach1932d, Bodenh1935, Ferris1943a].

DISTRIBUTION: **Palaearctic**: Algeria [Balach1930c, Balach1932d]; Canary Islands [Balach1951]; Corsica [Balach1931a, Balach1932d]; Egypt [Hall1923, Bodenh1924a, Hall1925, Hall1926a, Balach1951, Ezzat1958]; France [Signor1870, Balach1930, Balach1930a, Balach1932d, Balach1933e, Foldi2000]; Iran

[Balach1951, Kaussa1955]; Israel [Bodenh1924, Bodenh1927a, Bodenh1937]; Italy [Leonar1918, Leonar1920, LongoMaPe1995]; Libya [Danzig1993]; Morocco [Rungs1935]; Spain [GomezM1948, Martin1983, BlayGo1993, Danzig1993]; Tunisia [Balach1932d]; Turkmenistan [BazaroSh1971]; Yugoslavia [Balach1951, Bachma1953].

GENERAL: Description and illustration of adult female by Ferris (1943a), Balachowsky (1951, 1958b), Bazarov & Shmelev (1971) (as *Schizotargionia limonii*) and by Danzig (1993).

KEYS: Danzig 1993: 221-222 (female) [Europe]; Bazarov & Shmelev 1971: 182 (female) [Central Asia]; Balachowsky 1958b: 292 (female) [Africa]; Ezzat 1958: 242 (female) [Egypt]; Balachowsky 1951: 634 (female) [Mediterranean]; Ferris 1943a: 94 (female) [World]; Leonardi 1920: 104-105 (female) [Italy].

CITATIONS: Bachma1953 [host, distribution: 177]; Balach1927 [taxonomy, description, illustration, host, distribution: 194-197]; Balach1930 [host, distribution: 312]; Balach1930a [host, distribution: 179]; Balach1930c [host, distribution: 119]; Balach1931a [host, distribution: 98]; Balach1932d [taxonomy, host, distribution: XIII, XLIX]; Balach1933a [host, distribution: 38]; Balach1933e [host, distribution: 3]; Balach1935b [host, distribution: 260]; Balach1951 [taxonomy, description, illustration, host, distribution: 634-637]; Balach1958a [host, distribution: 38]; Balach1958b [taxonomy, description, illustration, host, distribution: 293-296]; BazaroSh1971 [taxonomy, description, illustration, host, distribution: 184-185]; Bellio1929b [taxonomy, description, illustration, host, distribution: 187-194]; BlayGo1993 [taxonomy, description, illustration, host, distribution: 400-403]; Bodenh1924 [taxonomy, host, distribution: 38]; Bodenh1924a [host, distribution: 122]; Bodenh1927a [host, distribution: 177]; Bodenh1935 [host, distribution: 247]; Bodenh1937 [host, distribution: 217]; Borchs1966 [catalogue: 249-250]; Cocker1896b [taxonomy, distribution: 333]; Cocker1897i [taxonomy, description, host, distribution: 19]; Comsto1881a [taxonomy, description, host, distribution: 543]; Comsto1883 [taxonomy, description, host, distribution: 82-83]; DanzigPe1998 [catalogue: 360-361]; ErlerKoTu1996 [host, distribution: 53-59]; Ezzat1958 [distribution: 242]; Fernal1903b [catalogue: 298]; Ferris1937c [taxonomy, illustration: 52,99]; Ferris1941e [taxonomy: 46,48]; Ferris1943a [taxonomy, description, illustration, host, distribution: 85-86,89-91,103]; Foldi2000 [host, distribution: 84]; Foldi2001 [distribution: 303-308]; Foldi2002 [host, distribution: 247]; GomezM1937 [taxonomy, description, illustration, host, distribution: 129-130]; GomezM1948 [taxonomy, description, illustration, host, distribution: 74-77]; GomezM1957 [host, distribution: 48]; GomezM1958a [host, distribution: 6]; GomezM1968 [host, distribution: 546]; Hall1923 [taxonomy, description, host, distribution: 29]; Hall1925 [host, distribution: 22]; Hall1926a [host, distribution: 32]; Hall1927b [taxonomy, description, illustration, host, distribution: 170-173]; Kaussa1955 [host, distribution: 16]; Leonar1897 [taxonomy: 286]; Leonar1918 [host, distribution: 193]; Leonar1920 [taxonomy, description, illustration, host, distribution: 111-114,304]; Lindin1910a [taxonomy: 438]; Lindin1911 [taxonomy: 382]; Lindin1912b [taxonomy, description, host, distribution: 93,94,104,150,162]; Lindin1933e [taxonomy: 68]; Lindin1935 [taxonomy: 146,147]; Lindin1936 [taxonomy: 166]; LongoMaPe1995 [distribution: 129]; Lupo1957 [taxonomy, description, illustration, host, distribution: 61-66]; MacGil1921 [taxonomy, description, host, distribution: 448]; Martin1983 [taxonomy, host, distribution: 69]; Rungs1933 [taxonomy: 116]; Rungs1935 [host, distribution: 275]; Signor1869 [taxonomy: 862]; Signor1869b [taxonomy: 100]; Signor1870 [taxonomy, description, illustration, host, distribution: 106].

Targionia parayuccarum Munting

Targionia parayuccarum Munting, 1965b: 211. Type data: SOUTH AFRICA: Cape Province, Cape Point Nature Reserve, on roots of *Cliffortia falcata*; collected J. Munting. Holotype female. Type depository: Pretoria: South African National Collection of Insects, South Africa; type no. 1605/1.
SCALE COVER: Female scale subcircular, about 1.5 mm in diameter; light brown except above exuviae which are covered with an off-white secretion; when this is rubbed off shiny, black exuviae exposed. Male scale oval, about 1.2 mm long, exuvia at anterior end, shiny black, covered becoming whiter towards flattened posterior end (Munting, 1965b).
HOST PLANTS: **Rosaceae**: *Cliffortia falcata* [Muntin1965b].
DISTRIBUTION: **Afrotropical**: South Africa [Muntin1965b].
GENERAL: Description and illustration of adult female by Munting (1965b)
CITATIONS: Muntin1965b [taxonomy, description, illustration, host, distribution: 210-211].

Targionia porifera (Borchsenius)

Rhizaspidiotus porifera Borchsenius, 1949b: 350. Type data: ARMENIA: near Erevan, on *Salsola* sp. Syntypes, female. Type depository: St. Petersburg: (= Leningrad) Zoological Museum, Academy of Science, Russia.
Fisanotargionia quadrilobata Kaussari & Balachowsky, 1953b: 277. Type data: IRAN: Yezd province, on *Seidlitzia* sp. Syntypes, female. Type depository: Paris: Muséum national d'Histoire naturelle, France. Synonymy by Danzig, 1993: 222.
Targionia porifera; Danzig, 1993: 222. Change of combination.
SCALE COVER: Female scale elliptical, diameter 1.2-1.4 mm; convex; white or white yellow; exuviae dark brown or black (Borchsenius, 1949b). Female scale of junior synonym *Fisanotargionia quadrilobata* subcircular, convex; exuviae central or subcentral, dark; secreted part bright grey, covered with powdery secretion; diameter 2-2.1 mm (Kaussari & Balachowsky, 1953b).
HOST PLANTS: **Chenopodiaceae**: *Kochia* [Danzig1993], *Salsola* [Borchs1949b, Danzig1993], *Seidlitzia* [KaussaBa1953b, Kaussa1957, Danzig1993]. **Zygophyllaceae**: *Zygophyllum* [Danzig1993].
DISTRIBUTION: **Palaearctic**: Armenia [Borchs1949b, Danzig1993]; Georgia [Danzig1993]; Iran [KaussaBa1953b, Kaussa1957]; Turkmenistan [Danzig1993].
GENERAL: Description and illustration of adult female by Borchsenius (1949b), Kaussari & Balachowsky (1953b) (as *Fisanotargionia quadrilobata*) and by Danzig (1993).
KEYS: Danzig 1993: 221-222 (female) [Europe].
CITATIONS: Borchs1949b [taxonomy, description, illustration, host, distribution: 350-351]; Borchs1949d [taxonomy, description, host, distribution: 250];

Borchs1950b [taxonomy, description, host, distribution: 233]; Borchs1966 [catalogue: 248,250]; Danzig1993 [taxonomy, description, illustration, host, distribution: 221-223]; DanzigPe1998 [catalogue: 361]; Kaussa1957 [host, distribution: 1]; KaussaBa1953b [taxonomy, description, illustration, host, distribution: 277-280].

Targionia prionota (Green & Laing)

Aspidiotus (*Targionia*) *prionota* Green & Laing, 1923: 128. Type data: TANZANIA: Ngerengere, on bark of an undetermined forest tree. Syntypes, female. Type depository: London: The Natural History Museum, England, UK.

Targionia prionota; Lindinger, 1937: 197. Change of combination.

Aspidiotus prionota; Ferris, 1943a: 86. Change of combination.

Targionia prionota; Borchsenius, 1966: 250. Revived combination.

SCALE COVER: Female scale low convex, subcircular, brownish black to black around margins, which are hard and thick, with centre pale yellowish grey, of a thin papery consistency (Green & Laing, 1923).

DISTRIBUTION: **Afrotropical**: Tanzania [GreenLa1923, Balach1958b].

GENERAL: Description and illustration of adult female by Green & Laing (1923) and by Balachowsky (1958b).

KEYS: Balachowsky 1958b: 292 (female) [Africa].

CITATIONS: Balach1958b [taxonomy, description, illustration, host, distribution: 295-296]; Borchs1966 [catalogue: 250]; Ferris1941e [taxonomy: 47]; Ferris1943a [taxonomy: 86]; GreenLa1923 [taxonomy, description, illustration, host, distribution: 128]; Lindin1937 [taxonomy: 197].

Targionia stoebae Munting

Targionia stoebae Munting, 1965b: 213. Type data: SOUTH AFRICA: Cape Province, Clanwilliam district, Cedarberg mountains near Uitkyk Pass, on *Stoebe plumosa*; collected by J. Munting. Holotype female. Type depository: Pretoria: South African National Collection of Insects, South Africa; type no. 1652/1.

SCALE COVER: Female scale subcircular, about 1.5 mm in diameter, with exuviae towards one end; blackish-brown; male scale more or less oval, up to 1 mm long; dark brown at anterior end and lighter at posterior extremity (Munting, 1965b).

HOST PLANTS: **Compositae**: *Stoebe plumosa* [Muntin1965b].

DISTRIBUTION: **Afrotropical**: South Africa [Muntin1965b].

GENERAL: Description and illustration of adult female by Munting (1965b).

CITATIONS: Muntin1965b [taxonomy, description, illustration, host, distribution: 212-213].

Targionia vitis (Signoret)

Aspidiotus vitis Signoret, 1876b: lii. Type data: FRANCE: near Nice, on "vigne" [=*Vitis vinifera*]. Syntypes, female. Type depository: Vienna: Naturhistorisches Museum Wien, Austria.

Diaspis blankenhornii Targioni Tozzetti, 1879: 17. Type data: ITALY: Novarese, on grapevine. Synonymy by Targioni Tozzetti, 1885: 109. Notes: Type material lost, Giuseppina Pellizzari, private communication to Yair Ben-Dov, 1999.

Diaspis blanckenhorni Targioni Tozzetti, 1885: 109. Misspelling of species name.
Aspidiotus (*Diaspidiotus*) *vitis*; Cockerell, 1897i: 19. Change of combination.
Targionia vitis; Leonardi, 1900: 304. Change of combination.
Targionia vitis suberi Leonardi, 1907b: 166. Type data: ITALY: Sardinia, Tempio, on *Quercus suber*. Syntypes, female. Type depository: Portici: Dipartimento de Entomologia e Zoologia Agraria di Portici, Università di Napoli Federico II, Italy. Synonymy by Ferris, 1943a: 86.
Targionia vitis arbutus Leonardi, 1909: 123. Type data: ITALY: Brindisi, on *Arbutus unedo*. Syntypes, female. Type depository: Portici: Dipartimento de Entomologia e Zoologia Agraria di Portici, Università di Napoli Federico II, Italy. Synonymy by Ferris, 1943a: 85.
Aspidiotus (*Targionia*) *vitis*; Feytaud, 1916: 11. Change of combination.
Targionia arbutus; Ferris, 1943a: 85. Change of status.
Targionia suberi; Ferris, 1943a: 86. Change of status.
SCALE COVER: Signoret (1876b) described scale cover as: "Female scale circular; black grey; exuviae central, colour brown shiny. Male scale oval very elongated". Balachowsky (1951) described scale cover as: "Female scale cover varies from ash-grey (on vine) to black (on *Quercus*); subcircular; size varies 1.8-2.4 mm; flat; matt; larval exuviae central or eccentric, colour brown red; ventral vellum white; attached to the host plant. Male scale similar in structure, oval 1.4-2 mm".
HOST PLANTS: **Ericaceae**: *Arbutus* [Ferris1943a, Danzig1993], *Arbutus unedo* [Leonar1920, Bodenh1949, Martin1983, BlayGo1993]. **Fagaceae** [Ferris1943a], *Castanea* [Borchs1936, Danzig1993], *Castanea sativa* [Borchs1934, Bodenh1949], *Fagus* [Borchs1936], *Fagus sylvatica* [Zahrad1972], *Quercus* [Leonar1909, Leonar1920, Borchs1934, Borchs1936, Bodenh1937, Bodenh1949, Bodenh1952, Hadzib1983], *Qu. cerris* [Zahrad1972], *Qu. coccifera* [Martin1983, BlayGo1993], *Qu. dentata* [Borchs1934], *Qu. ilex* [Leonar1920, Balach1930, Balach1932d, Balach1933e, Bachma1953, Zahrad1972, Martin1983, Foldi2000], *Qu. lanuginosa* [Zahrad1972], *Qu. pedunculata* [Borchs1934], *Qu. petraea* [Zahrad1972], *Qu. pubescens* [Zahrad1972], *Qu. sessiliflora* [Zahrad1972], *Qu. suber* [Leonar1907b, Leonar1909, Sander1909a, Balach1931a, Martin1983, BlayGo1993]. **Platanaceae**: *Platanus* [Danzig1993], *Platanus orientalis* [Bodenh1949, Zahrad1972]. **Salicaceae**: *Salix* [Danzig1993]. **Vitaceae**: *Vitis* [Danzig1993], *Vitis vinifera* [Signor1876b, Leonar1900, Balach1932d, Bodenh1949, ErlerTu2001].
NATURAL ENEMIES: ACARI **Phytoseiidae**: *Phytoseius finitimus* Ribaga [ErlerTu2001]. COLEOPTERA **Nitidulidae**: *Cybocephalus fodori* [StathaKo2001]. HYMENOPTERA **Aphelinidae**: *Aphytis abnormalis* Howard [Zahrad1972], *Azotus matritensis* Mercet [GomezM1959, Zahrad1972, Viggia1990b], *Coccophagoides moeris* (Walker) [GomezM1959, Zahrad1972, Gordh1979, Viggia1990a], *Pteroptrix dimidiata* Westwood [GomezM1959, Zahrad1972]. **Encyrtidae**: *Habrolepis pascuorum* Mercet [Zahrad1972, Trjapi1989], *Oencyrtus azureus* Mercet [Zahrad1972]. **Eulophidae**: *Chrysocharidia fimbriata* Erdos [Zahrad1972]. THYSANOPTERA **Thripidae**: *Karnyothrips flavipes* Jones [ErlerTu2001].
DISTRIBUTION: **Palaearctic**: Algeria [Leonar1900, Balach1927, Balach1932d, Ferris1943a]; Armenia [Danzig1993]; Azerbaijan (Azerbaijan [Borchs1936]); Corsica [Balach1931a, Balach1932d]; Czech Republic [Zahrad1977]; France

[Signor1876b, Balach1930, Balach1932d, Balach1933e, Foldi2000]; Georgia (Abkhaz ASSR [Borchs1934, Borchs1936], Georgia [Borchs1936, Hadzib1983]); Greece [Korone1934, Ferris1943a, StathaKo2001]; Hungary [Danzig1993]; Iran [Danzig1993]; Iraq [Danzig1993]; Israel [Bodenh1927a, Bodenh1937, Danzig1993]; Italy [Leonar1920, Ferris1943a, LongoMaPe1995]; Malta [Danzig1993]; Morocco [Balach1932d]; Portugal [Seabra1941, Danzig1993]; Romania [Danzig1993]; Russia (Caucasus [Borchs1934, Borchs1936], Dagestan AR [Borchs1936, Ferris1943a]); Sardinia [Leonar1907b, Leonar1909, Sander1909a]; Spain [Balach1935b, GomezM1937, Ferris1943a, Martin1983, BlayGo1993]; Turkey [Bodenh1949, Bodenh1952, Danzig1993, ErlerTu2001]; Ukraine (Krym (= Crimea) Oblast [Ferris1943a]); Yugoslavia [Balach1951, Bachma1953, Danzig1993].
BIOLOGY: Occurring on bark (Ferris, 1943a).
ECONOMIC IMPORTANCE: Common in Europe, Middle East and Armenia (see Distribution). Generally its populations are controlled below economic threshold by natural enemies (Zahradník, 1990).
GENERAL: Description and illustration of adult female by Ferris (1943a), Balachowsky (1951), Gómez-Menor Ortega (1959), Tereznikova (1986), Danzig (1993) and by Yaşar (1995a).
KEYS: Danzig 1993: 221-222 (female) [Europe]; Kosztarab & Kozár 1978: 177 (female) [Hungary]; Balachowsky 1951: 633 (female) [Mediterranean]; Ferris 1943a: 94 (female) [World]; Leonardi 1920: 104-105 (female) [Italy].
CITATIONS: Arras1976 [host, distribution, economic importance, chemical control: 15-19]; Bachma1953 [host, distribution: 177]; Balach1927 [host, distribution: 178]; Balach1930 [host, distribution: 312]; Balach1931a [host, distribution: 98]; Balach1932d [taxonomy, host, distribution: XIII, XLVIII]; Balach1933e [host, distribution, biological control: 3]; Balach1934 [taxonomy: 36]; Balach1935b [host, distribution: 260]; Balach1951 [taxonomy, description, host, distribution: 633,640-644]; Bodenh1927a [host, distribution: 177]; Bodenh1928 [host, distribution: 191]; Bodenh1937 [host, distribution: 217]; Bodenh1949 [taxonomy, description, illustration, host, distribution: 81-83]; Bodenh1952 [host, distribution: 348]; Borchs1934 [host, distribution: 31]; Borchs1935a [taxonomy, description, host, distribution: 36]; Borchs1936 [host, distribution: 138]; Borchs1937 [taxonomy, description, illustration, host, distribution: 136]; Borchs1937a [taxonomy, description, host, distribution: 67]; Borchs1939 [taxonomy, description, host, distribution: 11,40]; Borchs1939a [taxonomy, distribution: 43]; Borchs1949d [taxonomy, description, host, distribution: 250]; Borchs1950b [taxonomy, description, illustration, host, distribution: 231,233-234]; Borchs1966 [catalogue: 250]; BorianNi1995 [chemical control: 43]; ClapsWoGo2001a [taxonomy, host, distribution: 27]; Cocker1896b [distribution: 333]; Cocker1897i [taxonomy, description, host, distribution: 19]; Comsto1883 [taxonomy, description, host, distribution: 84]; Danzig1964 [taxonomy, host, distribution: 654]; Danzig1972 [taxonomy, host, distribution, economic importance: 221]; Danzig1993 [taxonomy, description, illustration, host, distribution: 222,226-227]; DanzigPe1998 [catalogue: 361]; Egger1990 [economic importance, chemical control: 27-28]; ErlerTu2001 [host, distribution, biological control: 303]; Fernal1903b [catalogue: 228,298]; Ferrar1987 [chemical control, biological control:

77-91]; Ferris1941e [taxonomy: 49]; Ferris1943a [taxonomy, description, illustration, host, distribution: 85-86,92-93,104]; Feytau1916 [taxonomy: 11]; Foldi1990 [structure: 43-54]; Foldi2000 [host, distribution: 84]; Foldi2001 [distribution: 303-308]; Garcia1930 [host, distribution, biological control]; Gavalo1931 [host, distribution: 8]; GomezM1937 [taxonomy, description, illustration, host, distribution: 124-127]; GomezM1958a [host, distribution: 6-7]; GomezM1959 [taxonomy, description, illustration, host, distribution, biological control: 161-165]; GomezM1960O [host, distribution: 170]; GomezM1968 [host, distribution: 546]; Gordh1979 [biological control: 901]; GuarioBaMe1996 [host, distribution, life history, economic importance, chemical control: 21,51-54]; GuarioLa1996 [host, distribution, economic importance, chemical control: 5,31-40]; Hadzib1983 [taxonomy, description, host, distribution, life history, biological control: 216-217]; Kiritc1932a [taxonomy: 268]; Korone1934 [taxonomy, description, illustration, host, distribution: 24]; KosztaKo1978 [taxonomy, description, host, distribution: 177]; Leonar1897 [taxonomy: 286]; Leonar1900 [taxonomy, description, illustration, host, distribution: 304-305]; Leonar1907b [taxonomy, description, illustration, host, distribution: 166]; Leonar1909 [taxonomy, description, host, distribution: 122-123]; Leonar1920 [taxonomy, description, illustration, host, distribution: 105-108]; Lindin1911 [taxonomy: 382]; Lindin1912b [taxonomy, description, host, distribution: 73,278,340]; Lindin1932c [taxonomy: 204]; Lindin1935 [taxonomy: 133]; Lombar1938 [taxonomy, description, host, distribution, life history: 117-138]; LongoMaPe1995 [distribution: 129]; Lupo1957 [taxonomy, description, illustration, host, distribution: 55-61]; MacGil1921 [taxonomy, description, host, distribution: 447,448]; Martel1913 [chemical control: 1-28]; Melis1949 [host, distribution: 17-25]; MillerDa1990 [host, distribution, economic importance: 305]; MoleasBa1994 [host, distribution, life history: 211-218]; Peleka1962 [host, distribution: 62]; Priore1964 [host, distribution: 131-178]; Priore1965 [host, distribution: 101-145]; Sander1909a [taxonomy, host, distribution: 55]; Sassce1911 [taxonomy: 71]; SchmutKlLu1957 [host, distribution, economic importance: 493]; Seabra1941 [distribution: 8]; Signor1876b [taxonomy, description, host, distribution: lii-liii]; Signor1877 [taxonomy, description, host, distribution: 601-603]; Silves1902 [taxonomy, description, host, distribution: 102]; Souzad1906 [taxonomy, description, host, distribution: 92]; StathaKo2001 [host, distribution, life history, economic importance, biological control: 134-139]; Targio1879 [taxonomy, description, host, distribution: 17,32]; Targio1885 [taxonomy: 109]; Terezn1986 [taxonomy, description, illustration, host, distribution: 81-82]; TranfaVi1987a [economic importance: 215-221]; Trjapi1989 [biological control: 293]; Viggia1990a [biological control: 125]; Viggia1990b [biological control: 178]; Yasar1995a [taxonomy, description, illustration, host, distribution: 130-132]; Zahrad1972 [taxonomy, description, host, distribution: 440]; Zahrad1977 [taxonomy, distribution: 121]; Zahrad1990a [host, distribution, description: 648-649].

Targionia yuccarum (Cockerell)

Aspidiotus yuccarum Cockerell, 1898m: 25. Type data: U.S.A.: New Mexico, Mesilla park, a short distance east of the Agricultural College, at bases of leaves of *Yucca elata*; collected May 1898. Syntypes, female. Type depository: Washington: United States National Entomological Collection, U.S. National Museum of Natural History, District of Columbia, USA.

Aspidiotus (Targionia) yuccarum; Cockerell, 1899a: 395. Change of combination.

Hemiberlesia yuccarum; Leonardi, 1900: 339. Change of combination.

Chrysomphalus (Melanaspis) tonilensis Cockrell, 1902t: 470. Type data: MEXICO: Jalisco, Tonila, on stalk, branches and root of a low bush of the sage family. Syntypes, female. Type depository: Washington: United States National Entomological Collection, U.S. National Museum of Natural History, District of Columbia, USA. Synonymy by Ferris, 1943a: 87.

Targionia yuccarum; Fernald, 1903b: 299. Change of combination.

Chrysomphalus covilleae Ferris, 1919a: 66. Type data: U.S.A.: Arizona, east of Phoenix, Mormon Flat, on *Covillea glutinosa*. Syntypes, female. Type depository: Davis: The Bohart Museum of Entomology, University of California, California, USA. Synonymy by Ferris, 1920a: 64.

Targionia covilleae; Ferris, 1919a: 68. Change of combination.

Targaspidiotus yuccarum; MacGillivray, 1921: 447. Change of combination.

Targionia tonilensis; McKenzie, 1939: 55. Change of combination.

Targionia yuccarum; Ferris, 1942: 446-40. Revived combination.

SCALE COVER: Female scale circular or suboval, about 2.75 mm in diameter; slightly convex; dark brown; rough and concentrically wrinkled; blackish towards middle; but central part, covering exuviae, covered by a large round patch of white secretion; when this is rubbed off, exuviae are exposed, shining black subcentral; thick ventral scale present; young female scales appear entirely white, or, when rubbed, white with a black spot. Male scale elongate, brown, with exuvia at one end covered by white secretion (Cockerell, 1898m).

HOST PLANTS: **Agavaceae**: *Dasylirion* [Ferris1943a], *Dasylirion wheeleri* [Ferris1938a], *Nolina* [Ferris1938a]. **Chenopodiaceae**: *Atriplex* [Ferris1921]. **Compositae**: *Baccharis* [Ferris1938a], *Chrysothamnus pulchellus* [McDani1970], *Gutierrezia* [Ferris1938a], *Gymnolomia* [Ferris1943a], *Gymnolomia tenuifolia* [Ferris1938a], *Isocoma* [Ferris1943a], *Isocoma heterophylla* [Ferris1938a]. **Liliaceae**: *Yucca elata* [Cocker1898m, Leonar1900]. **Onagraceae**: *Meriolix* [Ferris1943a], *Meriolix serrulata* [Ferris1938a]. **Polemoniaceae**: *Gilia* [Ferris1938a, McDani1970]. **Rhamnaceae**: *Ziziphus* [Ferris1943a], *Ziziphus lycioides* [Ferris1938a, McDani1970]. **Rubiaceae**: *Bigelovia* [Ferris1943a], *Bigelovia wrightii* [Ferris1938a]. **Zygophyllaceae**: *Covillea glutinosa* [Ferris1919a].

DISTRIBUTION: **Nearctic**: Mexico (Baja California [Ferris1921, Leonar1900], Colima [Ferris1938a], Jalisco [Cocker1902t]); United States of America (Arizona [Ferris1919a], New Mexico [Cocker1898m], Texas [Ferris1938a, McDani1970]).

BIOLOGY: Occurring on stems or frequently on roots of host (Ferris, 1938a).

GENERAL: Description and illustration of adult female by Ferris (1919a, 1938a).

KEYS: Ferris 1943a: 94 (female) [World]; Ferris 1942: 40 (female) [North America]; Cockerell 1905b: 201 (female) [U.S.A.: Colorado].

CITATIONS: Borchs1952 [taxonomy: 263]; Borchs1966 [catalogue: 251-252]; Cocker1898m [taxonomy, description, host, distribution: 25-26]; Cocker1899a [taxonomy: 395]; Cocker1902t [taxonomy, description, host, distribution: 470-471]; Cocker1905b [taxonomy: 201]; CockerPa1899 [taxonomy: 278,279]; Fernal1903b [catalogue: 294,299]; Ferris1919a [taxonomy, host, distribution: 66]; Ferris1920a [taxonomy: 64]; Ferris1921 [host, distribution: 132]; Ferris1921b [taxonomy: 94]; Ferris1937c [taxonomy: 52,56,100]; Ferris1938a [taxonomy, description, illustration, host, distribution: 268]; Ferris1941e [taxonomy: 49]; Ferris1942 [taxonomy: 446:40]; Ferris1943a [taxonomy, description, illustration, host, distribution: 85,86,93-94,105]; Leonar1900 [taxonomy, host, distribution: 339]; MacGil1921 [taxonomy, description, host, distribution: 447]; McDani1970 [taxonomy, illustration, host, distribution: 439-440]; McKenz1939 [taxonomy: 55]; Nakaha1982 [host, distribution: 84].

Tollaspidiotus MacGillivray

Tollaspidiotus MacGillivray, 1921: 389. Type species: *Aspidiotus mauritanus* Newstead, by original designation.
Tallaspidiotus; Balachowsky, 1958b: 249. Misspelling of genus name.
SYSTEMATICS: Lindinger (1937) considered this name a synonym of *Furcaspis* Lindinger, whereas McKenzie (1939), Ferris (1941e), Mamet (1949) and Borchsenius (1966) accepted the genus as valid. The genus differs from other genera of Furcaspidina, in circular shape of body, and presence of projecting lobes on prepygidial segments.
GENERAL: Definition and characters by MacGillivray (1921).
CITATIONS: Balach1958b [taxonomy: 249]; Borchs1966 [catalogue: 243]; Ferris1937c [taxonomy: 52]; Lindin1937 [taxonomy: 197]; MacGil1921 [taxonomy, description: 389,425-426]; Mamet1949 [taxonomy: 65]; McKenz1939 [taxonomy: 54]; MorrisMo1966 [taxonomy, catalogue: 192,196].

Tollaspidiotus mauritianus (Newstead)
Aspidiotus (Chrysomphalus?) mauritianus Newstead, 1917: 374. Type data: MAURITIUS: Botanic Gardens, on palm trees; collected by de Charmoy, 1915. Syntypes. Type depository: London: The Natural History Museum, England, UK.
Tollaspidiotus mauritianus; MacGillivray, 1921: 426. Change of combination.
Furcaspis mauritana; Lindinger, 1932e: 200. Change of combination requiring emendation of species name for agreement in gender.
Tollaspidiotus mauritianus; Borchsenius, 1966: 243. Revived combination.
SCALE COVER: Female scale subcircular, 0.8-1 mm long; anterior end slightly produced; posterior end strongly uptilted by ventral scale, which is markedly thickened and tongue-shaped, but does not project beyond dorsal scale; thickened portion rests upon a thinner scale of secretionary matter, so that scale resembles partly open bivalve shell of a mollusk, when examined in profile; larval exuviae subcentral, very prominent, and somewhat hemispherical; bright yellowish-buff or reddish-buff, with generally distinct concentric bands of a darker colour (Newstead, 1917). Illustration of female scale cover by Newstead (1917).

HOST PLANTS: **Palmae** [Newste1917], *Dictyosperma alba* [Mamet1943a, Mamet1949, Borchs1966], *Mascarena revaughanii* [Mamet1949, Borchs1966].
DISTRIBUTION: **Afrotropical**: Mauritius [Newste1917, Mamet1949].
GENERAL: Description and illustration of adult female by Newstead (1917).
CITATIONS: Borchs1966 [catalogue: 243]; Ferris1937c [taxonomy: 52]; Ferris1941e [taxonomy: 45]; Lepesm1947 [host, distribution: 195]; Lindin1932e [taxonomy: 200]; MacGil1921 [taxonomy, description, host, distribution: 426]; Mamet1943a [catalogue: 160]; Mamet1949 [catalogue: 65]; McKenz1939 [taxonomy: 54]; MoutiaMa1947 [distribution]; Newste1917 [taxonomy, description, illustration, host, distribution: 374-375].

Tsimanaspis Mamet

Tsimanaspis Mamet, 1959a: 477. Type species: *Tsimanaspis euphorbiae* Mamet, by monotypy and original designation.
SYSTEMATICS: Mamet (1959a) noted that affinities of *Tsimanaspis* are doubtful, but referred it to Aspidiotinae on account of length and slenderness of ducts and morphology of male scale. Mamet also suggested relation to *Comstockiella*, from which it differs by strongly sclerotized margin of pygidium, which bears well-developed paraphyses.
GENERAL: Description and definition by Mamet (1959a).
CITATIONS: Borchs1966 [catalogue: 357]; Mamet1959a [p. 477]; MorrisMo1966 [taxonomy, catalogue: 198].

Tsimanaspis euphorbiae Mamet

Tsimanaspis euphorbiae Mamet, 1959a: 478. Type data: MADAGASCAR: Lake Tsimanampetsotsa, on *Euphorbia* sp. Holotype. Type depository: Paris: Muséum national d'Histoire naturelle, France.
SCALE COVER: Female scale circular; conical; pale to dark bordeaux-red, with marginal area paler, sometimes obscured by a layer of white powdery wax; exuviae subcentral of same colour as scale; larval exuvia dot-like (Mamet, 1959a).
HOST PLANTS: **Euphorbiaceae**: *Euphorbia* [Mamet1959a, Borchs1966].
DISTRIBUTION: **Afrotropical**: Madagascar [Mamet1959a, Borchs1966].
GENERAL: Description and illustration of adult female by Mamet (1959a).
CITATIONS: Borchs1966 [catalogue: 357]; Mamet1959a [taxonomy, description, illustration, host, distribution: 478].

Unaspidiotus MacGillivray

Unaspidiotus MacGillivray, 1921: 387. Type species: *Aspidiotus corticispini* Lindinger, by original designation.
Japaspidiotus Takagi & Kawai, 1966: 118. Type species: *Japaspidiotus cedricola* Takagi & Kawai (= *Aspidiotus corticispini* Lindinger), by original designation. Synonymy by Takagi, 1967: 55.

SYSTEMATICS: This genus comes close to *Acanthaspidiotus* Borchsenius & Williams, differing in having median lobes set close to each other and lack spines between them, arrangement of dorsal macroducts of pygidium and the acute shape of pygidium (Takagi & Kawai, 1966).
GENERAL: Definition and characters by Takagi & Kawai (1966).
CITATIONS: Borchs1966 [catalogue: 367]; DanzigPe1998 [catalogue: 366]; Ferris1937c [taxonomy: 52]; Kawai1980 [taxonomy: 230-231]; Lindin1937 [taxonomy: 197]; MacGil1921 [taxonomy, description: 387,405-406]; MorrisMo1966 [taxonomy, catalogue: 201]; Takagi1967 [taxonomy: 54-55]; TakagiKa1966 [taxonomy, description: 118-119].

Unaspidiotus corticispini (Lindinger)

Aspidiotus (*Morganella*) *corticis-pini* Lindinger, 1909c: 448. Type data: JAPAN: Yokohama, on *Pinus densiflora*. Lectotype female, by subsequent designation Takagi, 1967: 55. Type depository: Hamburg: Zoologisches Institut und Zoologishces Museum, Universität von Hamburg, Germany.
Unaspidiotus corticis-pini; MacGillivray, 1921: 406. Change of combination.
Morganella corticis-pini; Lindinger, 1957: 544. Change of combination.
Japaspidiotus cedricola Takagi & Kawai, 1966: 118. Type data: JAPAN: Tokyo (Akisima, Ogoti, Okutama) on *Cedrus deodara* and *Tsuga diversifolia*; Simazima, Nagano-ken, on *Pinus densiflora*. Syntypes, female. Type depository: Sapporo: Entomological Institute, Faculty of Agriculture, Hokkaido University, Japan. Synonymy by Takagi, 1967: 55.
Unaspidiotus corticispini; Borchsenius, 1966: 367. Justified emendation.
SYSTEMATICS: Takagi & Kawai (1966) suggested that *Japaspidiotus cedricola* Takagi & Kawai, 1966 was identical with *Aspidiotus corticispini* Lindinger, 1909. Takagi (1967) confirmed synonymy after studying type series of latter.
SCALE COVER: Lindinger (1909c) did not describe scale cover of *Aspidiotus corticispini*, nor did Takagi & Kawai (1966) for junior synonym *Japaspidiotus cedricola*.
HOST PLANTS: **Pinaceae**: *Abies firma* [Kawai1977], *Cedrus deodara* [TakagiKa1966], *Pinus* [Kawai1977], *Pinus densiflora* [Sander1909a, TakagiKa1966, Takagi1967], *Tsuga diversifolia* [TakagiKa1966].
DISTRIBUTION: **Palaearctic**: Germany [Takagi1967]; Japan [Kuwana1917a, TakagiKa1966, Takagi1967, Kawai1977, Kawai1980].
GENERAL: Description and illustration of adult female by Takagi & Kawai (1966).
KEYS: Kuwana 1933b: 49 (female) [Japan].
CITATIONS: Borchs1966 [catalogue: 367]; DanzigPe1998 [catalogue: 366]; Ferris1937c [taxonomy: 52]; Ferris1941e [taxonomy: 42]; Kawai1977 [host, distribution, economic importance: 157]; Kawai1980 [taxonomy, description, host, distribution: 230-231]; Kuwana1917a [taxonomy, distribution: 174]; Lindin1909c [taxonomy, description, host, distribution: 448-449]; Lindin1911 [taxonomy, host, distribution: 86]; Lindin1932f [taxonomy: 200]; Lindin1957 [taxonomy: 544,545]; MacGil1921 [taxonomy, description, host, distribution: 406]; Muraka1970 [host, distribution: 78]; Sander1909a [taxonomy, host, distribution: 52]; Takagi1967

[taxonomy, host, distribution: 54-55]; TakagiKa1966 [taxonomy, description, illustration, host, distribution: 117-119]; WeidneWa1968 [taxonomy: 172].

Varicaspis MacGillivray

Varicaspis MacGillivray, 1921: 390. Type species: *Aspidiotus fiorineides* Newstead, by monotypy and original designation.

SYSTEMATICS: McKenzie (1947b) established *Africonidia* for the species *Africonidia halli*. However, Balachowsky (1954c; 1958b) showed that *A. halli* was a junior synonym of *Gymnaspis africana* Newstead, 1913, and transferred latter to *Africonidia*. Consequently, genus *Hallaspidiotus* Mamet, 1951 (type-species: *Gymnaspis africana* Newstead, became a junior objective synonym of *Africonidia*. Borchsenius (1966) did not accept validity of *Africonidia* and resurrected genus *Varicaspis* MacGillivray, 1921 (type-species: *Aspidiotus fiorineides* Newstead, 1920). Borchsenius' interpretation will doubtlessly be acceptable as soon as it can be shown that *A. fiorineides* - which at present is known only from inadequate type material (Balachowsky, 1958b) and a poor description - is in fact congeneric with *Africonidia africana*. Until these two genera have been revised, *Varicaspis* is restricted to its type species, while five species are placed in *Africonidia*, namely *africana, carreti, macdanieli, mkuzensis, subsimplex.*

GENERAL: Definition and characters by MacGillivray (1921).

CITATIONS: BenDov1974c [taxonomy: 19]; Borchs1966 [catalogue: 317]; Ferris1937c [taxonomy: 52]; Ferris1938b [taxonomy: 75]; Lindin1937 [taxonomy: 197]; MacGil1921 [taxonomy: 390,431]; MorrisMo1966 [catalogue: 202].

Varicaspis fiorineides (Newstead)

Aspidiotus fiorineides Newstead, 1920: 199. Type data: UGANDA: Jana Isl., Sesse Islands, Lake Victoria, on *Coffea robusta*. Syntypes, female and first instar. Type depository: London: The Natural History Museum, England, UK.

Varicaspis fiorineides; MacGillivray, 1921: 431. Change of combination.

Africonidia fiorineides; Balachowsky, 1954c: 78. Change of combination.

Varicaspis fiorineides; Borchsenius, 1966: 317. Revived combination.

SCALE COVER: Female scale attached to edge of leaf, with equal portions on both sides; very elongate; sides compressed; middle line of dorsum rather sharply keeled; exuviae central, bright orange-yellow or pale castaneous; secretionary portion very broad, thin, semi-opaque, dusky white; length 2-2.2 mm. Male scale similar to female, but much smaller (Newstead, 1920).

HOST PLANTS: **Myrtaceae**: *Syzygium guineense* [Almeid1973b]. **Oleaceae**: *Jasminum* [DeLott1967a]. **Rubiaceae**: *Coffea robusta* [Newste1920, DeLott1967a].

DISTRIBUTION: **Afrotropical**: Angola [Almeid1973b]; Kenya [DeLott1967a].

ECONOMIC IMPORTANCE: Reported as a very serious pest of Robusta coffee, *Coffea canephora* in Uganda (Chua & Wood, 1990).

GENERAL: Description and illustration of adult female by Newstead (1920).

CITATIONS: Almeid1973b [host, distribution: 8]; Balach1954c [taxonomy: 78]; Balach1958b [taxonomy: 150]; Borchs1966 [taxonomy, catalogue: 317]; ChuaWo1990 [host, distribution, economic importance: 549]; DeLott1967a [host, distribution: 112]; Ferris1937c [taxonomy: 52]; Ferris1941e [taxonomy: 43]; MacGil1921 [taxonomy, description, host, distribution: 431]; Newste1920 [taxonomy, description, illustration, host, distribution: 199-200].

Subfamily Comstockiellinae

Comstockiella **Cockerell**

Comstockiella Cockerell, 1896b: 320. Type species: *Aspidiotus sabalis* Comstock, by monotypy and original designation.

Constockiella; MacGillivray, 1921: 423. Misspelling of genus name.

SYSTEMATICS: Ferris (1938a) assigned this genus to Aspidiotinae, while pointing out differences in features from typical aspidiotine genera. Borchsenius (1965, 1966) placed it in a separate tribe Comstockiellini of the Aspidiotinae. Here it is regarded as the only genus in the Comstockiellinae. Brown & McKenzie, 1962) listed the distinguishing features of this monotypic genus: female pygidium without lobes and plates; adult female with paraspiracular pores; tubular ducts neither of the one-barred (aspidiotine) nor two-barred (diaspidine) type; first instar crawler with a six-segmented antennae.

GENERAL: Definition and characters by Ferris (1938a), Borchsenius (1965) and by Brown & McKenzie (1962).

KEYS: Gill 1997: 24-26 (female) [Genera of California]; McKenzie 1956: 22 (female) [U.S.A.: California]; Ferris 1942: 25 (female) [North America]; Ferris 1942: 32 (female) [species North America]; Green 1896e: 37 (female) [Sri Lanka].

CITATIONS: Borchs1966 [catalogue: 229]; BrownMc1962 [taxonomy, description: 141,166]; Cocker1896b [taxonomy: 320,335]; Cocker1897b; Cocker1899a [taxonomy: 396]; Fernal1903b [catalogue: 282]; Ferris1937c [taxonomy: 51]; Ferris1938a [taxonomy, description: 212]; Ferris1942; Gill1997 [taxonomy: 106]; Green1896e [taxonomy: 37]; Kozar1990f [distribution: 142]; Lepesm1947 [taxonomy: 213]; Lindin1908b [taxonomy: 98]; Lindin1937; MacGil1921 [taxonomy, description: 388,423]; McKenz1956; Miller1990 [taxonomy: 169-178]; MorrisMo1966 [taxonomy, catalogue: 44]; Sander1904a; Willia1969a [taxonomy: 323].

Comstockiella sabalis (Comstock)

Aspidiotus ? sabalis Comstock, 1883: 67. Type data: USA: Florida, Ft. George and Sanford, on leaves of palmetto [*Sabal*]. Syntypes, female. Type depository: Washington: United States National Entomological Collection, U.S. National Museum of Natural History, District of Columbia, USA.

Comstockiella sabalis; Cockerell, 1896b: 335. Change of combination.

Comstockiella sabalis mexicana Cockerell, 1897u: 267. Type data: MEXICO: probably from Maratlan, intercepted at USA, California, San Francisco, on palms. Syntypes, female. Type depository: Washington: United States National Entomological Collection, U.S. National Museum of Natural History, District of Columbia, USA. Synonymy by Ferris, 1938a: 213.

Comstockiella sabilis; Cockerell, 1899a: 396. Misspelling of species name.

Comstockiella mexicana; Lepesme, 1947: 214. Change of status.

Comstockiella sabialis; Schmutterer et al., 1957: 474. Misspelling of species name.

COMMON NAME: palmetto scale [Comsto1883, MerrilCh1923, Merril1953, McKenz1956, Dekle1965c].

SCALE COVER: Comstock (1883) described scale cover as: "female is snowy white; irregular in outline, but approximately circular; exuviae vary in position from central to marginal, they are covered and their position is indicated by a tubercle which is of a deeper white than the remainder of the scale. The scale of the male resembles that of the female, except that it is smaller and more elongated". Ferris (1938a) described scale cover as: "female white, oval, the exuviae subcentral, and concealed by wax; scale of the male similar in form and texture". Colour photograph by Gill (1997).

HOST PLANTS: **Globulariaceae**: *Globularia salicina* [Lindin1911]. **Palmae** [BesheaTiHo1973], *Cocos* [BesheaTiHo1973], *Erythea* [Borchs1966], *Sabal* [Comsto1883, MerrilCh1923, Ferris1938a, McKenz1956, Borchs1966, BesheaTiHo1973], *Sa. adamsonis* [Wilson1917], *Sa. minor* [BesheaTiHo1973], *Sa. palmetto* [Wilson1917, Dekle1965c, TippinBe1970, BesheaTiHo1973], *Serenoa repens* [Dekle1965c, TippinBe1970, BesheaTiHo1973]. **Umbelliferae**: *Washingtonia robusta* [Wilson1917].

NATURAL ENEMIES: HYMENOPTERA **Aphelinidae**: *Aphytis diaspidis* (Howard) [EvansPe1997], *Ap. mytilaspidis* (Le Baron) [RosenDe1978], *Coccobius donatellae* Pedata & Evans [EvansPe1997], *Encarsia citrina* (Howard) [EvansPe1997], *Encarsia portoricensis* Howard [RosenDe1978], *Physcus* [RosenDe1978]. THYSANOPTERA **Phlaeothripidae**: *Aleurodothrips fasciapennis* (Franklin) [RosenDe1978, PalmerMo1990].

DISTRIBUTION: **Nearctic**: Mexico [Cocker1899n, Merril1953, Borchs1966] (Baja California [Ferris1938a], Oaxaca [Ferris1938a], Sonora [Ferris1938a]); United States of America (California [MerrilCh1923, Ferris1938a, Merril1953, McKenz1956], Florida [Comsto1883, Comsto1916, Wilson1917, MerrilCh1923, Ferris1938a, Merril1953, Dekle1965c], Georgia [TippinBe1970, BesheaTiHo1973], Louisiana [Ferris1938a, Merril1953], Mississippi [Ferris1938a, Merril1953], North Carolina [Nakaha1982], South Carolina [BesheaTiHo1973], Texas [Ferris1938a, Merril1953, McDani1969]). **Neotropical**: Bahamas [Merril1953]; Bermuda [Merril1953, RosenDe1978]; Cuba [Merril1953]; Guadeloupe [MerrilCh1923, Ferris1938a, Merril1953]. **Palaearctic**: Canary Islands [Lindin1911]; USSR [Borchs1966].

BIOLOGY: The scale infests the leaves, frequently in great numbers (Ferris, 1938a).

ECONOMIC IMPORTANCE: A pest of palms and palmettos in southern and southwestern United States, west coast of Mexico, West Indies and Bermuda (Rosen & DeBach, 1978).

GENERAL: Description and illustration of adult female by Ferris (1938a), McKenzie (1956) and by Gill (1997). Intraspecific variation in number and grouping of perivulvar pores discussed and illustrated by Tippins & Beshear (1968).

KEYS: McKenzie 1956: 24 (female) [U.S.A.: California]; Comstock 1883: 55-57 (female) [North America].

CITATIONS: BeardsDaHo1976 [economic importance: 106]; BesheaTiHo1973 [host, distribution: 5-6]; Borchs1966 [catalogue: 229]; Brown1957 [taxonomy, structure: 362-363]; Brown1963 [taxonomy, structure: 360-406]; Cock1985a [biological control: 3]; Cocker1896b [taxonomy: 335]; Cocker1897i [taxonomy: 9]; Cocker1897u [taxonomy, description, host, distribution: 267]; Cocker1899a [taxonomy: 396]; Cocker1899n [host, distribution: 27]; Cocker1899s; Comsto1883 [taxonomy, description, illustration, host, distribution: 67-69]; Comsto1916 [taxonomy, description, illustration, host, distribution: 516,528-530]; Creigh1942 [host, distribution, control: 219-233]; Dekle1965c [taxonomy, description, host, distribution: 47]; Dekle1976 [taxonomy, description, host, distribution, economic importance: 66]; EvansPe1997 [host, distribution, biological control: 328-334]; Fernal1903b [catalogue: 282-283]; Ferris1937c [taxonomy, illustration: 51,67]; Ferris1938a [taxonomy, description, illustration, host, distribution: 213]; Ferris1941e [taxonomy: 48]; Giliom1990 [taxonomy, structure: 21]; Gill1997 [host, distribution, taxonomy, description, illustration, economic importance: 105-107]; HerricSe1999 [genetics: 41-71]; Howard1991 [host, distribution, life history, ecology, control: 217-225]; Howell1979 [taxonomy, description, illustration, host, distribution, life history: 556-558]; HowellTi1990 [taxonomy, structure, description, illustration: 34]; Kitchi1970 [structure, chemistry, anatomy, chromosome: 165-197]; Lepesm1947 [taxonomy, description, host, distribution, life history: 213-214]; Lindin1911 [taxonomy, description, illustration, host, distribution: 9-12]; Lobdel1937; MacGil1921 [taxonomy, description, host, distribution: 423]; Maskel1891 [taxonomy: 9]; McDani1969 [taxonomy, illustration, host, distribution: 89-90]; McKenz1956 [taxonomy, description, illustration, host, distribution: 56,59]; Merril1953 [taxonomy, description, host, distribution: 41]; MerrilCh1923 [taxonomy, description, host, distribution, economic importance: 227]; MillerDa1990 [host, distribution, economic importance: 301]; Nakaha1982 [host, distribution: 26]; Nur1990a [taxonomy, structure, chromosomes: 183-184]; PalmerMo1990 [biological control: 67-76]; RosenDe1978 [economic importance, biological control, life history, host, distribution: 106]; Sander1904a; SchmutKlLu1957 [host, distribution, economic importance: 474]; SchmutKlLu1959 [taxonomy: 374]; TippinBe1968 [taxonomy, description, structure: 67-69]; TippinBe1970 [host, distribution: 8]; Wilson1917; Wilson1917 [host, distribution: 45-46].

Subfamily Odonaspidinae

Berlesaspidiotus **MacGillivray**

Anoplaspis Leonardi, 1900: 344. Type species: *Aspidiotus* (*Odonaspis*) *bambusarum* Cockerell, by monotypy and original designation. Homonym of *Anoplaspis* Leonardi, 1898: 47.

Berlesaspidiotus MacGillivray, 1921: 389. Replacement name.

SYSTEMATICS: *Berlesaspidiotus* differs from other genera of Odonaspidinae, in possessing small cuticular projections on head and thorax, and antennal tubercle placed in cylindrical invagination in cuticle. Although *Anoplaspis* was first described by Leonardi (1898) as genus in subfamily Diaspidinae, he (Leonardi, 1900) rejected his 1898 usage of this name, and suggested it for an entirely different species *Aspidiotus* (*Odonaspis*) *bambusarum*) Cockerell, in Odonaspidinae. Leonardi's (1900) action is unacceptable according to Article 12(b)(5) of ICZN, and *Anoplaspis* Leonardi, 1898 is a valid genus in Diaspidinae. *Anoplaspis* Leonardi, 1900 is an homonym of *Anoplaspis* Leonardi, 1898. This homonymy was settled after MacGillivray (1921) introduced *Berlesaspidiotus*. For further discussion refer to Ben-Dov (1988b).

GENERAL: Definition and characters by Kuwana (1933) and by Ben-Dov (1988b).

KEYS: Ben-Dov 1988b: 15 (female) [species World]; Ben-Dov 1988b: 14 (female) [World]; Kuwana 1933a: 43-45 (female) [Japan].

CITATIONS: BenDov1988b [taxonomy, description: 15]; Borchs1966 [catalogue: 227]; Fernal1903b [catalogue: 299]; Ferris1937a [taxonomy: 33,36]; Kuwana1933 [taxonomy, description: 37]; Leonar1898 [taxonomy: 47]; Leonar1900 [taxonomy, description: 344]; MacGil1921 [taxonomy, description: 389]; MorrisMo1966 [catalogue: 11,23].

Berlesaspidiotus bambusarum (Cockerell)

Aspidiotus (*Odonaspis*) *bambusarum* Cockerell, 1898b: 191. Type data: JAPAN: on stalks of bamboo from Japan, intercepted at San Francisco, California, USA. Lectotype (examined), by subsequent designation Ben-Dov, 1988b: 16. Type depository: Washington: United States National Entomological Collection, U.S. National Museum of Natural History, District of Columbia, USA.

Anoplaspis bambusarum; Leonardi, 1900: 345. Change of combination.

Spatheaspis bambusarum; Cockerell, 1900f: 72. Change of combination.

Odonaspis bambusarum; Fernald, 1903b: 299. Change of combination.

Berlesaspidiotus bambusarum; MacGillivray, 1921: 424. Change of combination.

Dycryptaspis bambusarum; Lindinger, 1937: 184. Change of combination.

Froggattiella bambusarum; Borchsenius, 1966: 227. Change of combination.

Berlesaspidiotus bambusarum; Ben-Dov, 1988b: 15. Revived combination.

SCALE COVER: Female scale 2 mm diameter, very dark sepia brown, almost black, convex; dull; exuviae sub-central; first skin exposed, light orange; second, brown; with ventral scale (Cockerell, 1898b). Colour photograph by Kawai (1980).

HOST PLANTS: **Gramineae**: *Arundinaria simonii* [BenDov1988b], *Bambusa* [Takagi1959a], *Pleioblastus* [Kawai1977].
NATURAL ENEMIES: HYMENOPTERA **Encyrtidae**: *Anabrolepis japonica* Ishii [BenDov1988b, Trjapi1989].
DISTRIBUTION: **Palaearctic**: Japan [Kuwana1907, Kuwana1917a, Kuwana1933, Takagi1959a, Kawai1977, Kawai1980] (Kyushu [BenDov1988b]).
GENERAL: Description and illustration of adult female by Kuwana (1933) and by Ben-Dov (1988b).
KEYS: Ben-Dov 1988b: 15 (female) [World]; Kawai 1980: 199 (female) [Japan]; Takagi 1959a: 93-94 (female) [Japan]; Kuwana 1933: 37 (female) [Japan]; Kuwana 1933b: 50 (female) [Japan].
CITATIONS: BenDov1988b [taxonomy, description, illustration, distribution, host, biological control: 15-16,72,95]; BenDov1990e [host, distribution: 657]; Borchs1966 [catalogue: 227]; Cocker1898b [taxonomy, description, host, distribution: 191]; Cocker1900f [taxonomy: 72]; DanzigPe1998 [catalogue: 265]; Fernal1903b [catalogue: 299]; Ferris1937a [taxonomy, illustration: 33,36]; Ferris1941e [taxonomy: 41]; Ferris1955c [taxonomy: 33]; Ishii1928 [host, distribution, biological control: 79]; Ishii1932a [host, distribution, biological control: 161]; Kawai1977 [host, distribution, economic importance: 163]; Kawai1980 [taxonomy, description, host, distribution: 200-201]; Kuwana1907 [host, distribution: 196]; Kuwana1917a [taxonomy, distribution: 176]; Kuwana1933 [taxonomy, description, illustration, host, distribution: 37-38,xi]; Leonar1900 [taxonomy: 345]; MacGil1921 [taxonomy, description, host, distribution: 424]; Muraka1970 [host, distribution: 68]; Takagi1959a [taxonomy, host, distribution: 94-95]; Trjapi1989 [biological control: 293]; Willia1985a [taxonomy: 232].

Berlesaspidiotus crenulatus Ben-Dov
Berlesaspidiotus crenulatus Ben-Dov, 1988b: 16. Type data: INDIA: Madras, Pulney, on *Arundinaria walkeriana*; collected 18.xi.1913. Holotype female. Type depository: Washington: United States National Entomological Collection, U.S. National Museum of Natural History, District of Columbia, USA.
SCALE COVER: Ben-Dov (1988b) noted that scale cover of specimens from type-series were not available. However, scale cover of female from Philippines, circular, 3 mm in diameter; dorsal and ventral scales white; exuviae yellow, placed subcentrally (Ben-Dov, 1988b).
HOST PLANTS: **Gramineae**: *Arundinaria walkeriana* [BenDov1988b], *Bambusa* [BenDov1988b], *Schizostachyum lumampao* [BenDov1988b].
DISTRIBUTION: **Oriental**: India [BenDov1988b]; Philippines (Luzon [BenDov1988b]).
GENERAL: Description and illustration of adult female by Ben-Dov (1988b).
KEYS: Ben-Dov 1988b: 15 (female) [World].
CITATIONS: BenDov1988b [taxonomy, description, illustration, host, distribution: 16-17,96]; BenDov1990e [host, distribution: 657].

Circulaspis MacGillivray

Circulaspis MacGillivray, 1921: 393. Type species: *Odonaspis canaliculata* Green, by original designation.

SYSTEMATICS: This genus is distinguished from other genera of Odonaspidinae in possessing one invaginated tube at apex of the pygidium (Ben-Dov, 1988b).

GENERAL: Definition and characters by MacGillivray (1921), Ferris (1938a) and by Ben-Dov (1988b).

KEYS: Ben-Dov 1988b: 18 (female) [species World]; Ben-Dov 1988b: 14 (female) [World]; McDaniel 1974: 417-418 (female) [species U.S.A.: Texas]; Ferris 1942: 64 (female) [North America]; Ferris 1942: 65 (female) [species North America].

CITATIONS: Balach1948 [taxonomy: 24]; Balach1948b [taxonomy: 264]; Balach1949 [taxonomy: 109]; Balach1953g [taxonomy: 727]; Balach1958b [taxonomy: 298]; BenDov1980a [taxonomy, description: 76]; BenDov1988b [taxonomy, description: 18]; Borchs1966 [catalogue: 226]; Ferris1937a [taxonomy: 33,37]; Ferris1938 [taxonomy: 46]; Ferris1938a [taxonomy, description: 156]; Ferris1942 [taxonomy: 64]; Kozar1990f [distribution: 142]; Lindin1937 [taxonomy: 182]; MacGil1921 [taxonomy, description: 393,449]; MorrisMo1966 [catalogue, taxonomy: 38].

Circulaspis bibursella Ferris

Circulaspis bibursella Ferris, 1938a: 158. Type data: MEXICO: State of Colima, Manzanillo, on undetermined coarse grass. Holotype female. Type depository: Davis: Bohart Museum of Entomology, University of California, California, USA.

SCALE COVER: Scale of female brown, sometimes slightly fluffy from unconsolidated wax threads, broadly oval; scale of male similar in colour, slender (Ferris, 1938a).

HOST PLANTS: **Gramineae**: [Ferris1938a].

DISTRIBUTION: **Nearctic**: Mexico (Colima [Ferris1938a, BenDov1988b], Nayarit [Ferris1938a]).

BIOLOGY: Occurring exposed on leaves and beneath leaf sheaths.

GENERAL: Description and illustration of adult female by Ferris (1938a) and by Ben-Dov (1988b).

KEYS: Ben-Dov 1988b: 18 (female) [World]; Ferris 1942: 65 (female) [North America].

CITATIONS: BenDov1988b [taxonomy, description, illustration, host, distribution: 18-19,97]; Borchs1966 [catalogue: 226]; Ferris1938a [taxonomy, description, illustration, host, distribution : 158]; Ferris1942 [taxonomy: 65].

Circulaspis canaliculata (Green)

Odonaspis canaliculatus Green, 1900a: 72. Type data: SRI LANKA: Nuwara Eliya, on various kinds of bamboo. Syntypes, female. Type depository: London: The Natural History Museum, England, UK.

Aspidiotus canaliculata; MacGillivray, 1921: 226. Change of combination.

Circulaspis canaliculata; MacGillivray, 1921: 449. Change of combination.

Dycryptaspis canaliculata; Lindinger, 1937: 184. Change of combination.

Circulaspis canaliculata; Borchsenius, 1966: 226. Revived combination.
SCALE COVER: Female scale very dense and compact: irregularly circular: black: exuviae yellowish, second often concealed beneath black secretionary layer. Diameter 1.50-1.75 mm. Male scale elongate: outline often sinuous: both exuvia and secretionary area black; exuvia situated at anterior margin. Length 1.50 mm. Breadth 0.50 mm (Green, 1900a).
HOST PLANTS: **Gramineae**: *Bambusa* [Green1900a, Green1922, BenDov1988b].
DISTRIBUTION: **Oriental**: India [Ramakr1921a]; Sri Lanka [Green1900a, Ramakr1921a, Green1922, Green1937, BenDov1988b]. **Palaearctic**: China (People's Republic) [BenDov1988b].
GENERAL: Description and illustration of adult female by Green (1900a) and by Ben-Dov (1988b).
KEYS: Ben-Dov 1988b: 18 (female) [World].
CITATIONS: BenDov1988b [taxonomy, description, illustration, host, distribution: 19-20,98]; BenDov1990e [host, distribution: 657]; Borchs1966 [catalogue: 226]; DEDAC1923 [host, distribution]; Fernal1903b [catalogue: 299]; Ferris1937a [taxonomy, illustration: 33,37]; Green1900a [taxonomy, description, illustration, host, distribution: 72]; Green1922 [host, distribution: 463]; Green1937 [host, distribution: 335]; Lindin1937 [taxonomy: 184]; MacGil1921 [taxonomy, description, host, distribution: 393,449]; Ramakr1921a [host, distribution: 358]; WangVaXu1998 [host, distribution: 86].

Circulaspis fistulata (Ferris)
Odonaspis fistulata Ferris, 1921: 121. Type data: MEXICO: Baja California, Punta Palmilla, near San Jose del Cabo, on *Distichlis spicata*. Holotype female. Type depository: Davis: The Bohart Museum of Entomology, University of California, California, USA.
Dycryptaspis fistulata; Lindinger, 1937: 184. Change of combination.
Circulaspis fistulata; Ferris, 1938a: 159. Change of combination.
SCALE COVER: Scale of female brown, broadly elongate; scale of male white, slender, elongate (Ferris, 1938a).
HOST PLANTS: **Gramineae**: *Distichlis spicata* [Ferris1921, BenDov1988b], *Monanthochloe* [BenDov1988b], *Sporobolus* [Ferris1938a].
DISTRIBUTION: **Nearctic**: Mexico (Baja California [Ferris1921]); United States of America (Texas [Ferris1938a, McDani1974]).
BIOLOGY: Occurring on upper surface of strongly revolute leaves of host.
GENERAL: Description and illustration of adult female by Ferris (1921, 1938a), McDaniel (1974) and by Ben-Dov (1988b).
KEYS: Ben-Dov 1988b: 18 (female) [World]; McDaniel 1974: 418 (female) [U.S.A.: Texas]; Ferris 1942: 65 (female) [North America].
CITATIONS: BenDov1988b [taxonomy, description, illustration, host, distribution: 20-21,99]; Borchs1966 [catalogue: 226-227]; Ferris1921 [taxonomy, description, illustration, host, distribution: 121-123]; Ferris1938a [taxonomy, description, illustration, host, distribution: 159]; Ferris1942 [taxonomy: 65]; Lindin1937 [taxonomy: 184]; McDani1974 [taxonomy, description, illustration, host, distribution: 419]; Nakaha1982 [host, distribution: 23].

Circulaspis fistulella Ferris

Circulaspis fistulella Ferris, 1938a: 160. Type data: U.S.A.: Texas, Point Isabel, on *Distichlis spicata*. Holotype female. Type depository: Davis: The Bohart Museum of Entomology, University of California, California, USA.

COMMON NAME: salt grass scale [Dekle1976].

SCALE COVER: Scale of female elongate oval, tapering posteriorly, brownish; that of male slender, similar in colour to female (Ferris, 1938a).

HOST PLANTS: **Cyperaceae**: *Cladium jamaicense* [BenDov1988b]. **Gramineae**: *Cynodon dactylon* [BenDov1988b], *Distichlis spicata* [Ferris1938a], *Spartina* [TippinBe1970, BenDov1988b], *Spartina bakeri* [BenDov1988b], *Spartina patens* [BesheaTiHo1973, BenDov1988b], *Sporobolus* [Ferris1938a], *Sporobolus indicus* [BenDov1988b].

NATURAL ENEMIES: HYMENOPTERA **Signiphoridae**: *Signiphora* [BenDov1988b].

DISTRIBUTION: **Nearctic**: United States of America (Florida [BesheaTiHo1973, Dekle1976, BenDov1988b], Georgia [TippinBe1970, BesheaTiHo1973, BenDov1988b], Texas [Ferris1938a, McDani1974, BenDov1988b]).

BIOLOGY: Occurring on upper surface of strongly revolute leaves of host.

GENERAL: Description and illustration of adult female by Ferris (1938a) and by Ben-Dov (1988b).

KEYS: Ben-Dov 1988b: 18 (female) [World]; McDaniel 1974: 418 (female) [U.S.A.: Texas]; Ferris 1942: 65 (female) [North America].

CITATIONS: BenDov1988b [taxonomy, description, illustration, host, distribution, biological control: 21-22,72,100]; BesheaTiHo1973 [host, distribution: 14]; Borchs1966 [catalogue: 227]; Dekle1976 [taxonomy, description, host, distribution: 62]; Ferris1938a [taxonomy, description, illustration, host, distribution: 160]; Ferris1942 [taxonomy: 65]; McDani1974 [taxonomy, description, illustration, host, distribution: 419-420]; Nakaha1982 [host, distribution: 86]; TippinBe1970 [host, distribution: 8].

Dicirculaspis Ben-Dov

Dicirculaspis Ben-Dov, 1988b: 23. Type species: *Dicirculaspis philippina* Ben-Dov, by original designation.

SYSTEMATICS: This genus differs form other genera of Odonaspidinae in possessing two invaginated tubes at apex of pygidium (Ben-Dov, 1988b).

GENERAL: Definition and characters by Ben-Dov (1988b).

KEYS: Ben-Dov 1988b: 14 (female) [World]; Ben-Dov 1988b: 23 (female) [species World].

CITATIONS: BenDov1988b [taxonomy, description: 14,23].

Dicirculaspis bibursa (Ferris)

Circulaspis bibursa Ferris, 1938a: 157. Type data: U.S.A.: Texas, cliffs at Fort Davis, on grass. Lectotype female, by subsequent designation Ben-Dov, 1988b: 24. Type depository: Davis: The Bohart Museum of Entomology, University of California, California, USA; type no. T/695.

Dicirculaspis bibursa; Ben-Dov, 1988b: 23. Change of combination.
SCALE COVER: Scale of female white, elongate, oval, tapering posteriorly; that of male white, slender, elongate (Ferris, 1938a).
HOST PLANTS: **Gramineae** [Ferris1938a], *Stipa* [BenDov1988b], *Triodia pilosa* [Ferris1938a, McDani1974, BenDov1988b].
DISTRIBUTION: **Nearctic**: United States of America (Texas [Ferris1938a, McDani1974, BenDov1988b]).
BIOLOGY: Occurring beneath sheathing bases of leaves.
GENERAL: Description and illustration of adult female by Ferris (1938a) and by Ben-Dov (1988b).
KEYS: Ben-Dov 1988b: 23 (female) [World]; McDaniel 1974: 417-418 (female) [U.S.A.: Texas]; Ferris 1942: 65 (female) [North America].
CITATIONS: BenDov1988b [taxonomy, description, illustration, host, distribution: 23-24,101]; BenDov1990a [taxonomy: 88]; Borchs1966 [catalogue: 226]; Ferris1938a [taxonomy, description, illustration, host, distribution: 157]; Ferris1942 [taxonomy: 65]; McDani1974 [taxonomy, description, illustration, host, distribution: 418]; Nakaha1982 [host, distribution: 23].

Dicirculaspis philippina Ben-Dov
Dicirculaspis philippina Ben-Dov, 1988b: 24. Type data: PHILIPPINES: Luzon Island, Bikal, on *Schizostachyum* sp.; collected October 1925. Holotype female. Type depository: Washington: United States National Entomological Collection, U.S. National Museum of Natural History, District of Columbia, USA.
SCALE COVER: No information is available on female scale cover; male unknown (Ben-Dov, 1988b).
HOST PLANTS: **Gramineae**: *Bambusa* [BenDov1988b], *Schizostachyum* [BenDov1988b].
DISTRIBUTION: **Oriental**: Philippines (Luzon [BenDov1988b]).
GENERAL: Description and illustration of adult female by Ben-Dov (1988b).
KEYS: Ben-Dov 1988b: 23 (female) [World].
CITATIONS: BenDov1988b [taxonomy, description, illustration, host, distribution: 24-25,102]; BenDov1990e [host, distribution: 658].

Froggattiella Leonardi

Targionia (*Froggattiella*) Leonardi, 1900: 300. Type species: *Aspidiotus inusitatus* Green, by monotypy.
Targionia (*Froggattiella*); Cockerell, 1900f: 72. Notes: Authorship of genus name credited to Leonardi.
Froggattiella; MacGillivray, 1921: 393. Change of status.
SYSTEMATICS: *Froggattiella* was raised to generic rank by MacGillivray (1921) and interpreted as such by Ferris (1937a), although later (Ferris, 1938a) considered there was no ground to separate it from *Odonaspis*. Balachowsky (1953g) and Takagi (1959, 1969a) concurred with Ferris and regarded *Froggattiella* as a synonym of *Odonaspis*. However Ben-Dov (1988b) regarded it as a distinct genus.

GENERAL: Definition and characters by Williams & Watson (1988) and by Ben-Dov (1988b).

KEYS: Gill 1997: 31 (female) [Genera of California]; Ben-Dov 1988b: 26 (female) [species World]; Ben-Dov 1988b: 14 (female) [World]; Williams & Watson 1988: 19 (female) [Tropical South Pacific]; Chou 1985: 328 (female) [Genera of China]; Chou 1985: 331 (female) [Species of China].

CITATIONS: Balach1949 [taxonomy: 109]; Balach1953g [taxonomy: 727]; BenDov1988b [taxonomy, description: 26]; Borchs1966 [catalogue: 227]; Chou1985 [taxonomy, description: 330-331]; Cocker1900f [taxonomy: 72]; Danzig1993 [taxonomy: 133]; DanzigPe1998 [catalogue: 264-265]; Fernal1903b [taxonomy: 299]; Ferris1937a [taxonomy: 33,38]; Ferris1943a [taxonomy: 82]; Gill1997 [taxonomy: 150]; Kozar1990f [distribution: 142]; Leonar1900 [taxonomy, description: 300]; Lindin1937 [taxonomy: 185]; MacGil1921 [taxonomy, description: 393,450]; MorrisMo1966 [taxonomy: 81]; Takagi1959 [taxonomy: 92]; Takagi1969a [taxonomy: 60]; Tao1999 [taxonomy: 89]; WilliaWa1988 [taxonomy, description: 123].

Froggattiella inusitata (Green)

Aspidiotus inusitatus Green, 1896e: 66. Type data: SRI LANKA: on *Arundinaria* sp. Lectotype female (examined), by subsequent designation Ben-Dov, 1988b: 27. Type depository: London: The Natural History Museum, England, UK.

Aspidiotus (*Odonaspis*) *inusitatus*; Cockerell, 1899a: 395. Change of combination.

Targionia (*Froggattiella*) *inusitata*; Leonardi, 1900: 300. Change of combination requiring emendation of species name for agreement in gender.

Odonaspis inusitata; Green, 1900a: 71. Change of combination.

Aspidiotus inucitata; Kuwana, 1907: 193. Misspelling of species name.

Odonaspis (*Aspidiotus*) *inusitatus*; Ramakrishna Ayyar, 1921: 357. Change of combination.

Froggattiella inusitata; MacGillivray, 1921: 450. Change of combination.

Dycryptaspis inusitata; Lindinger, 1937: 184. Change of combination.

Froggattiella inusitata; Ben-Dov, 1988b: 26. Revived combination.

Odonaspis inusittatus; Wang, Varma & Xu, 1998: 86. Misspelling of species name.

SCALE COVER: Female scale large; flat; oval in front, elongated posteriorly; supplemental portion usually narrower than other; exuviae yellow, always concealed by secretion; approximately central in early adult stage, subsequently eccentric by backward extension of scale; ventral scale as developed as dorsal, and bearing what appear to be the ventral halves of exuviae; dorsal and ventral scale so much alike that it is often difficult to decide which dorsal and which ventral after it has been detached from stem; brownish-white or brownish-fulvous; length 3.5-7.5 mm; greatest breadth about 3.5 mm; male scale not known (Green, 1896e). Colour illustration of female scale cover by Green (1896e).

HOST PLANTS: **Gramineae**: *Arundinaria* [Green1896e, Green1937, BenDov1988b], *Bambusa* [Green1900a].

DISTRIBUTION: **Oriental**: Sri Lanka [Green1896e, Green1900a, Ramakr1921a, Green1937, BenDov1988b]; Taiwan [Green1937]. **Palaearctic**: Japan [Kuwana1917a] (Kyushu [Kuwana1902, Kuwana1907]).

GENERAL: Description and illustration of adult female by Green (1896e), Chou (1985, 1986) and by Ben-Dov (1988b).

KEYS: Ben-Dov 1988b: 26 (female) [World]; Chou 1985: 331 (female) [Species of China]; Green 1896e: 40 (female) [Sri Lanka].

CITATIONS: BenDov1988b [taxonomy, description, illustration, host, distribution: 26-28,103]; BenDov1990e [host, distribution: 658]; Borchs1966 [catalogue: 227]; Chou1985 [taxonomy, description, host, distribution: 331-333]; Chou1986 [taxonomy, illustration: 703]; Cocker1897i [taxonomy, description, host, distribution: 27-28]; Cocker1899a [taxonomy: 395]; Cocker1900f [taxonomy: 72]; DanzigPe1998 [catalogue: 265]; DEDAC1923 [host, distribution]; FangWuXu2001 [host, distribution: 108]; Fernal1903b [catalogue: 299]; Ferris1937a [taxonomy, illustration: 33,38]; Ferris1941e [taxonomy: 44]; Ferris1943a [taxonomy: 86]; Green1896e [taxonomy, description, illustration, host, distribution: 66-67]; Green1900a [taxonomy, host, distribution: 71-72]; Green1922 [taxonomy: 460]; Green1937 [host, distribution: 335]; Kuwana1902 [host, distribution: 65]; Kuwana1907 [taxonomy, host, distribution: 193]; Kuwana1917a [taxonomy, distribution: 176]; Leonar1900 [taxonomy, description, illustration, host, distribution: 300-302]; Lindin1937 [taxonomy: 184]; MacGil1921 [taxonomy, description, host, distribution: 450]; Ramakr1921a [host, distribution, taxonomy: 357]; Tao1999 [taxonomy, host, distribution: 89]; WangVaXu1998 [host, distribution: 86].

Froggattiella mcclurei Ben-Dov

Froggattiella mcclurei Ben-Dov, 1988b: 28. Type data: CHINA: Kwangtung, Yunq Hui, Wuchow, Kwanqs, on *Bambusa* sp.; collected xii.1928. Holotype female. Type depository: Washington: United States National Entomological Collection, U.S. National Museum of Natural History, District of Columbia, USA.

Froggattiella mccliurei; Tao, 1999: 89. Misspelling of species name.

SCALE COVER: Only slide-mounted specimens of females were available (Ben-Dov, 1988b).

HOST PLANTS: **Gramineae**: *Bambusa* [BenDov1988b], *Bambusa nana* [BenDov1988b].

DISTRIBUTION: **Oriental**: China (People's Republic) (Guangdong (Kwangtung) [BenDov1988b]); Hong Kong [BenDov1988b]; Indonesia (Java [BenDov1988b]); Philippines [BenDov1988b].

GENERAL: Description and illustration of adult female by Ben-Dov (1988b).

KEYS: Ben-Dov 1988b: 26 (female) [World].

CITATIONS: BenDov1988b [taxonomy, description, illustration, host, distribution: 28-29,104]; BenDov1990e [host, distribution: 658]; Tao1999 [taxonomy, host, distribution: 89].

Froggattiella penicillata (Green)

Odonaspis penicillata Green, 1905a: 346. Type data: SRI LANKA: Peradeniya, on *Gigantochloa aspera*. Lectotype female, by subsequent designation Ben-Dov, 1988b: 29. Type depository: London: The Natural History Museum, England, UK.

Froggattiella penicillata; Rutherford, 1915: 104. Change of combination.

Anoplaspis penicillata; Kuwana, 1933: 38. Change of combination.

Dycryptaspis penicillata; Lindinger, 1937: 184. Change of combination.

Odonaspis penicillata; Ferris, 1938a: 164. Revived combination.

Froggattiella penicillata; Borchsenius, 1966: 227. Revived combination.

Odonaspis penicillata; Takagi, 1969a: 62. Revived combination.

Froggattiella penicillata; Ben-Dov, 1988b: 29. Revived combination.

COMMON NAME: penicillate scale [McKenz1956, Dekle1965c, Borchs1966].

SCALE COVER: Female scale very pale fulvous; exuviae orange, usually concealed beneath whitish secretion, situated at anterior extremity; very firm and compact; ventral scale as dense as dorsal; two scales so firmly adherent that it is difficult to extract insect uninjured; elongate, broadest immediately behind exuviae, tapering posteriorly; flattened beneath; strong convex in front, depressed towards hinder extremity; length 1.5-2 mm, greatest breadth 1-1.1 mm. Male scale similar, but smaller, narrower and paler, length 1 mm (Green, 1905a). Female scale brown, elongate, broad anteriorly and tapering posteriorly. Male scale brown, elongate and slender (Ferris, 1938a). Colour photograph by Gill (1997).

HOST PLANTS: **Gramineae** [BesheaTiHo1973], *Arundinaria* [BenDov1988b], *Arundinaria hindsii* [BenDov1988b], *Bambusa* [McKenz1956, Takagi1959a, Dekle1965c, Muntin1965b, Takagi1969a, Hadzib1983, BenDov1988b, Danzig1993], *Ba. argentea-striata* [BenDov1988b], *Ba. nana* [BenDov1988b], *Ba. pervariabilis* [BenDov1988b], *Ba. stenostachya* [Takaha1929, Ferris1921a, McKenz1956, BenDov1988b], *Ba. vulgaris* [BenDov1988b], *Dendrocalamus* [Danzig1993], *De. latiflorus* [Takaha1929], *De. merrilliana* [BenDov1988b], *Gigantochloa* [Danzig1993], *Gigantochloa aspera* [Green1905a, Sander1906, Green1937, McKenz1956, Takagi1969a], *Ochlandra* [Danzig1993], *Ochlandra travancorica* [McKenz1956], *Panicum* [TippinBe1970], *Phyllostachys* [BenDov1988b], *Phyllostachys aurea* [BenDov1988b], *Phyllostachys bambusoides* [BesheaTiHo1973], *Phyllostachys pubescens* [BenDov1988b], *Schizostachyum glaucifolium* [WilliaWa1988], *Spartina* [TippinBe1970].

NATURAL ENEMIES: HYMENOPTERA **Encyrtidae**: *Adelencyrtus* [BenDov1988b], *Caenohomalopoda shikokuensis* (Tachikawa) [Prinsl1983, BenDov1988b, Trjapi1989].

DISTRIBUTION: **Afrotropical**: South Africa [Muntin1965b, BenDov1988b]. **Australasian**: Fiji [WilliaWa1988]; Hawaiian Islands (Hawaii [Takagi1969a, BenDov1988b]); Palau [Beards1966]. **Nearctic**: Mexico [BenDov1988b]; United States of America (Alabama [BenDov1988b], California [McKenz1956, BenDov1988b], Florida [Dekle1965c, BesheaTiHo1973, BenDov1988b], Georgia [TippinBe1970, BesheaTiHo1973, BenDov1988b], Louisiana [BenDov1988b], Mississippi [BenDov1988b], Texas [BenDov1988b]). **Neotropical**: Guyana [BenDov1988b]; Jamaica [BenDov1988b]; Puerto Rico & Vieques Island (Puerto Rico [BenDov1988b]). **Oriental**: Hong Kong [BenDov1988b]; India [Takagi1969a, BenDov1988b] (Tamil Nadu [Green1919c]); Pakistan [BenDov1988b]; Philippines [Takagi1969a, BenDov1988b]; Sri Lanka [Green1922, Green1937, Takagi1969a, BenDov1988b]; Taiwan [Ferris1921a, Takaha1929, Takagi1969a, BenDov1988b]. **Palaearctic**: Algeria [Takagi1969a]; China (People's Republic) [BenDov1988b] (Henan (Honan) [Shen1993]); Georgia [Hadzib1983, BenDov1988b, Danzig1993];

Iran [Kaussa1955, BenDov1988b]; Japan [Takagi1959a, Takagi1969a, Kawai1980, BenDov1988b].

BIOLOGY: Under sheathing bases of leaves; specimens observed being attached to stem and not to the leaf and showing no tendency to burrow beneath epidermis.

ECONOMIC IMPORTANCE: A minor pest of bamboos (Ben-Dov, 1990e; Gill, 1997).

GENERAL: Description and illustration of adult female by Green (1905a), Kuwana (1933), Ferris (1938a), McKenzie (1956), Takagi (1969a), Chou (1985, 1986), Ben-Dov (1988b), Gill (1997) and by Colon-Ferrer & Medina-Gaud (1998).

KEYS: Ben-Dov 1988b: 26 (female) [World]; Chou 1985: 331 (female) [Species of China]; Kawai 1980: 199 (female) [Japan]; Beardsley 1966: 525 (female) [Federated States of Micronesia]; Takagi 1959a: 93 (female) [Japan]; McKenzie 1956: 35 (female) [U.S.A.: California]; Balachowsky 1953g: 732 (female) [Mediterranean]; Ferris 1942: 65 (female) [North America]; Kuwana 1933: 37 (female) [Japan]; Kuwana 1933: 37 (female) [Japan]; Kuwana 1933b: 50 (female) [Japan].

CITATIONS: Balach1953g [taxonomy, description, illustration, host, distribution: 736-739]; BenDov1988b [taxonomy, description, illustration, host, distribution, biological control: 29-31,72,105]; BenDov1990c [taxonomy: 121]; BenDov1990e [host, distribution: 658]; BesheaTiHo1973 [host, distribution: 14]; Borchs1966 [catalogue: 227-228]; Brown1963 [taxonomy, structure: 360-406]; Chou1985 [taxonomy, description, host, distribution: 331-332]; Chou1986 [taxonomy, illustration: 704,705]; ColonFMe1998 [taxonomy, description, illustration, host, distribution: 133]; Danzig1972 [taxonomy, host, distribution, economic importance: 218]; Danzig1993 [taxonomy, description, illustration, host, distribution: 133-134]; DanzigPe1998 [catalogue: 265]; DEDAC1923 [host, distribution]; Dekle1965c [taxonomy, description, host, distribution: 99]; Dekle1976 [taxonomy, description, host, distribution, economic importance: 120]; FangWuXu2001 [host, distribution: 108]; Ferris1921a [host, distribution: 219]; Ferris1938a [taxonomy, description, illustration, host, distribution: 164]; Ferris1942 [taxonomy, host, distribution: 65]; Foldi1990 [structure: 43-54]; Gill1997 [host, distribution, taxonomy, description, illustration, economic importance: 150,152]; Green1905a [taxonomy, description, illustration, host, distribution: 346-347]; Green1919c [host, distribution: 439]; Green1922 [host, distribution: 463]; Green1937 [taxonomy, host, distribution: 335]; Hadzib1983 [host, distribution: 215]; Kaussa1955 [host, distribution: 17]; Kawai1980 [taxonomy, description, illustration, host, distribution: 200]; Kuwana1933 [taxonomy, description, host, distribution: 38-39]; Lindin1937 [taxonomy: 184]; Lindin1957 [taxonomy: 544]; Lobdel1937 [taxonomy, structure: 78]; MacGil1921 [taxonomy, description, host, distribution: 450]; McKenz1956 [taxonomy, description, illustration, host, distribution: 162-165]; Merril1953 [taxonomy, description, host, distribution: 63-64]; Muntin1965b [host, distribution: 193]; Muraka1970 [host, distribution: 68]; Nakaha1982 [host, distribution: 62]; Nur1990a [taxonomy, structure, chromosomes: 184-185]; Prinsl1983 [distribution, biological control: 26]; Ramakr1919a [taxonomy, host, distribution: 22]; Ramakr1930 [taxonomy, host, distribution: 27]; Ruther1915a [taxonomy, description, host, distribution: 104]; Sander1906 [taxonomy, host, distribution: 16]; Sassce1923 [host, distribution: 152-158]; Shen1993 [host, distribution: 59];

Takagi1959a [taxonomy, host, distribution: 94]; Takagi1969a [taxonomy, description, illustration, host, distribution: 61-62]; Takaha1929 [host, distribution: 78]; Tang1977 [taxonomy, description, illustration, host, distribution: 114-115]; Tao1999 [taxonomy, host, distribution: 89]; TippinBe1970 [host, distribution: 10-11]; WangVaXu1998 [host, distribution: 86]; WilliaWa1988 [taxonomy, illustration, host, distribution: 124-125].

Odonaspis Leonardi

Odonaspis Leonardi, 1897: 286. Type species: *Aspidiotus secretus* Cockerell, by monotypy.

Aspidiotus (*Dycryptaspis*) Cockerell in Leonardi, 1897a: 375. Type species: *Aspidiotus secretus* Cockerell, by original designation. Synonymy by Morrison & Morrison, 1966: 64.

Spatheaspis Leonardi, 1897b: 109. Unjustified replacement name.

Aspidiotus (*Odonaspis*); Cockerell, 1897i: 14. Change of status.

Odonaspis; Fernald, 1903b: 299. Revived rank.

Bakeraspis MacGillivray, 1921: 395. Type species: *Odonaspis schizostachyi* Cockerell & Robinson, by monotypy and original designation. Synonymy by Ferris, 1937a: 33.

Ligulaspis MacGillivray, 1921: 388. Type species: *Odonaspis janeirensis* Hempel (= *Odonaspis saccharicaulis* (Zehntner)), by monotypy and original designation. Synonymy by Lindinger, 1937: 188.

Dycryptaspis; Lindinger, 1937: 184. Change of status.

Odonaspis; Ben-Dov, 1988b: 32. Revived status.

SYSTEMATICS: *Odonaspis* was introduced by Leonardi as a new genus of the Aspidioti in an article Issued March 1897. A month later Leonardi (1897a) published a note in which he quoted the name "*Dycryptaspis* Cockerell," mentioned to him in a letter from Cockerell, with *Aspidiotus secretus* Cockerell as type species. Since *Odonaspis* Leonardi antedated *Dycryptaspis* Cockerell, the synonymy of the latter with *Odonaspis* is evident. In a third paper that year Leonardi (1897b) presumed that *Odonaspis* was preoccupied by *Odontaspis* Agassiz, and he Introduced *Spatheaspis* as a replacement name. According to Article 56 (a) of the International Code, *Odonaspis* Leonardi is not a homonym; therefore, *Spatheaspis* is an objective synonym of *Odonaspis*. Leonardi's error was raised again by Lindinger (1937), who considered *Odonaspis* Leonardi a preoccupied name, replaced it with *Dycryptaspis*, and transferred to the latter all the odonaspidine species described to that date. Lindinger's conclusions do not conform to ICZN and were rejected by earlier authors. *Odonaspis* differs from other genera in Odonaspidinae in: absence of pygidial invaginated tubes; apex of pygidium without a tuft of gland spines; antennal tubercles not placed in cylindrical invagination; cuticular striation without minute projections.

GENERAL: Definition and characters by Robinson (1917), Fullaway (1932), Kuwana (1933), Ferris (1937e, 1938a), Balachowsky (1948b, 1953g, 1958b), Borchsenius (1950b), Bustshik (1958), Almeida (1969), Takagi (1969a), Ben-Dov (1980a, 1988b) and by Williams & Watson (1988).

KEYS: Gill 1997: 31 (female) [Genera of California]; Gill 1997: 208 (female) [Species of California]; Zahradník 1990b: 73 (female) [Czech Republic]; Ben-Dov 1988b: 33-35 (female) [species World]; Ben-Dov 1988b: 14 (female) [World]; Williams & Watson 1988: 19 (female) [Tropical South Pacific]; Chou 1985: 328 (female) [Genera of China]; Chou 1985: 328 (female) [Species of China]; Beardsley 1966: 525 (female) [species Federated States of Micronesia]; Beardsley 1966: 502-504 (female) [Federated States of Micronesia]; Ezzat & Afifi 1966: 402 (female) [Egypt]; Danzig 1964: 645 (female) [Europe]; Takagi 1959a: 92 (female) [Japan]; Balachowsky 1958b: 299-300 (female) [species Africa]; Ezzat 1958: 247 (female) [Egypt]; Balachowsky 1953g: 733 (female) [species Mediterranean basin]; Borchsenius 1950b: 166 (female) [USSR]; Ferris 1942: 64 (female) [North America]; Ferris 1942: 65 (female) [species North America]; Borchsenius 1937: 100 (female) [USSR]; Borchsenius 1937a: 32-33 (female) [Palaearctic Region]; Kuwana 1933a: 43-45 (female) [Japan]; Fullaway 1932: 97-98 (female) [Hawaii]; Robinson 1917: 16-17 (female) [Philippines].

CITATIONS: Almeid1969 [taxonomy, description: 157-158]; Balach1942 [taxonomy: 47]; Balach1948b [taxonomy, description: 264]; Balach1949 [taxonomy: 109]; Balach1953g [taxonomy, description: 729-731]; Balach1958b [taxonomy, description: 298-299]; Beards1966 [taxonomy: 525]; BenDov1980a [taxonomy, description: 77]; BenDov1988b [taxonomy, description: 32-33]; BerlesLe1898 [taxonomy: 131,289]; Borchs1937a [taxonomy: 54,100]; Borchs1950b [taxonomy, description: 211]; Borchs1966 [catalogue: 223]; Bustsh1958 [taxonomy, description: 212-213]; Chou1985 [taxonomy, description: 328]; Cocker1897i [taxonomy: 14,31]; Cocker1899a [taxonomy: 395]; ColonFMe1998 [taxonomy, description: 132]; Danzig1964 [taxonomy: 650]; Danzig1993 [taxonomy: 132]; DanzigPe1998 [catalogue: 317]; Ezzat1958 [taxonomy: 247]; Fernal1903b [catalogue: 299]; Ferris1936a [taxonomy: 18]; Ferris1937a [taxonomy: 33-34,39]; Ferris1937e [taxonomy: 527]; Ferris1938a [taxonomy, description: 161]; Ferris1938b [taxonomy: 75]; Ferris1941d [taxonomy: 345]; Ferris1942 [taxonomy: 64,65]; Fullaw1932 [taxonomy, description: 108]; Gill1997 [taxonomy: 207]; Hadzib1983 [taxonomy: 214-215]; Kawai1980 [taxonomy: 199]; Kuwana1933 [taxonomy, description: 34-35,43]; Leonar1897 [taxonomy, description: 284-286]; Leonar1897a [taxonomy: 375]; Leonar1897b [taxonomy: 109,115]; Lepage1938 [taxonomy: 412]; Lindin1908 [taxonomy: 98]; Lindin1937 [taxonomy: 188,191,196]; MacGil1921 [taxonomy, description: 388,395,423,459]; Mamet1949 [taxonomy: 51]; McKenz1956 [taxonomy: 35]; Miller1990 [taxonomy: 169-178]; MorrisMo1966 [catalogue, taxonomy: 64,137,186]; Ramakr1930 [taxonomy: 27]; Robins1917 [taxonomy, description: 16-17]; Takagi1969a [taxonomy: 332]; Tao1999 [taxonomy: 102]; Willia1969a [taxonomy: 332]; WilliaWa1988 [taxonomy, description: 197]; Zimmer1948 [taxonomy: 426].

Odonaspis anneckei Ben-Dov

Odonaspis anneckei Ben-Dov, 1988b: 35. Type data: SOUTH AFRICA: Cape Province, Tradouw Pass, on *Ehrharta ramosa*; collected by J. Munting, 11.I.1969. Holotype female. Type depository: Pretoria: South African National Collection of Insects, South Africa.

SCALE COVER: Female scale oval and narrow, about 2 mm long, 1mm wide; white yellowish; larval exuviae brown, placed at anterior part of scale. Male scale oval and narrow, about 1 mm long, 0.3 mm wide; larval exuviae brown placed at anterior part of scale (Ben-Dov, 1988b).

HOST PLANTS: **Gramineae**: *Ehrharta ramosa* [BenDov1988b].

DISTRIBUTION: **Afrotropical**: South Africa [BenDov1988b].

GENERAL: Description and illustration of adult female by Ben-Dov (1988b).

KEYS: Ben-Dov 1988b: 34 (female) [World].

CITATIONS: BenDov1988b [taxonomy, description, illustration, host, distribution: 35,106].

Odonaspis arcusnotata Ben-Dov

Odonaspis arcusnotata Ben-Dov, 1988b: 35. Type data: TAIWAN: Suisha, on *Bambusa* sp.; collected 31.I.1930. Holotype female. Type depository: Washington: United States National Entomological Collection, U.S. National Museum of Natural History, District of Columbia, USA.

SCALE COVER: Only slide-mounted females available for description by Ben-Dov (1988b).

HOST PLANTS: **Gramineae**: *Arundinaria hindsii* [BenDov1988b], *Bambusa* [BenDov1988b], *Pleioblasthus* [Takagi2002].

DISTRIBUTION: **Oriental**: Taiwan [BenDov1988b]. **Palaearctic**: China (People's Republic) [BenDov1988b]; Japan [BenDov1988b, Takagi2002].

GENERAL: Description and illustration of adult female by Ben-Dov (1988b). The unnamed species, *Odonaspis* sp. listed by Takagi (1969a: 62) very likely belongs to *Odonaspis arcusnotata* Ben-Dov, 1988b (Y. Ben-Dov, personal information, 2000; see also Takagi (2002).

KEYS: Ben-Dov 1988b: 34 (female) [World].

CITATIONS: BenDov1988b [taxonomy, description, illustration, host, distribution: 35-36,107]; BenDov1990e [host, distribution: 658]; FangWuXu2001 [host, distribution: 108]; Takagi2002 [taxonomy, illustration, host, distribution: 59,96,97]; Tao1999 [taxonomy, host, distribution: 102].

Odonaspis aristidae Ben-Dov

Odonaspis aristidae Ben-Dov, 1988b: 36. Type data: SOUTH AFRICA: Cape Province, 70 miles northwest of Vryburg, on *Aristida diffusa* var. *burkei*; collected by J. Munting, 12.IX.1968. Holotype female. Type depository: Pretoria: South African National Collection of Insects, South Africa.

SCALE COVER: Female scale elliptical in outline, 2.0-3.5 mm long, 1.0-1.3 mm wide; white brownish; larval exuviae brown, placed at anterior part of scale. Male scale oval and narrow, 0.8-1 mm long, 0.4-0.5 mm wide; white; larval exuviae brown, placed at anterior part of scale (Ben-Dov, 1988b).

HOST PLANTS: **Gramineae**: *Aristida diffusa burkei* [BenDov1988b].
DISTRIBUTION: **Afrotropical**: South Africa [BenDov1988b].
GENERAL: Description and illustration of adult female by Ben-Dov (1988b).
KEYS: Ben-Dov 1988b: 34 (female) [World].
CITATIONS: BenDov1988b [taxonomy, description, illustration, host, distribution: 36-37,108].

Odonaspis australiensis Ben-Dov

Odonaspis australiensis Ben-Dov, 1988b: 37. Type data: AUSTRALIA: South Australia, Glen Osmond, Waite Agricultural Research Institute, on stolons of *Cynodon dactylon*; collected June 1952. Holotype female. Type depository: Canberra: Australian National Insect Collection, CSIRO Entomology, Australia.
SCALE COVER: Only slide-mounted females available for description by Ben-Dov (1988b).
HOST PLANTS: **Gramineae**: *Cynodon dactylon* [BenDov1988b].
DISTRIBUTION: **Australasian**: Australia (New South Wales [BenDov1988b], South Australia [BenDov1988b]).
GENERAL: Description and illustration of adult female by Ben-Dov (1988b).
KEYS: Ben-Dov 1988b: 34 (female) [World].
CITATIONS: BenDov1988b [taxonomy, description, illustration, host, distribution: 37-38,109].

Odonaspis benardi Balachowsky

Odonaspis benardi Balachowsky, 1957c: 203. Type data: MARTINIQUE: Sainte-Marie, on *Paspalum* sp. Holotype female. Type depository: Paris: Muséum national d'Histoire naturelle, France.
SCALE COVER: Female scale circular, 2 mm in diameter; wrinkled; bright brown; larval exuviae slightly subcentral. Male unknown (Balachowsky, 1957c).
HOST PLANTS: **Crassulaceae**: *Echeveria* [BenDov1988b]. **Gramineae**: *Andropogon* [BenDov1988b], *Paspalum* [Balach1957c, Borchs1966, BenDov1988b], *Paspalum notatum* [BenDov1988b]. **Marantaceae**: *Maranta* [BenDov1988b].
DISTRIBUTION: **Nearctic**: Mexico (Veracruz [BenDov1988b]); United States of America (Texas [BenDov1988b]). **Neotropical**: Costa Rica [BenDov1988b]; Cuba [BenDov1988b]; Guatemala [BenDov1988b]; Honduras [BenDov1988b]; Martinique [Balach1957c, Borchs1966, BenDov1988b]; Panama [BenDov1988b].
GENERAL: Description and illustration of adult female by Balachowsky (1957c) and by Ben-Dov (1988b).
KEYS: Ben-Dov 1988b: 34 (female) [World].
CITATIONS: Balach1957c [taxonomy, description, illustration, host, distribution: 203-206]; Balach1958b [taxonomy: 304]; BenDov1988b [taxonomy, description, illustration, host, distribution: 38-39,110]; BenDov1990e [host, distribution: 658]; Borchs1966 [catalogue: 224].

Odonaspis bromeliae Ben-Dov

Odonaspis bromeliae Ben-Dov, 1988b: 39. Type data: HONDURAS: on leaf of bromeliad (Bromeliaceae); collected 13.VIII.1975. Holotype female. Type depository: Washington: United States National Entomological Collection, U.S. National Museum of Natural History, District of Columbia, USA.

SCALE COVER: Only slide-mounted females were available for the description by Ben-Dov (1988b).

HOST PLANTS: **Bromeliaceae** [BenDov1988b], *Tillandsia* [BenDov1988b], *Tillandsia circinnata* [BenDov1988b].

DISTRIBUTION: **Neotropical**: Guatemala [BenDov1988b]; Honduras [BenDov1988b].

GENERAL: Description and illustration of adult female by Ben-Dov (1988b).

KEYS: Ben-Dov 1988b: 34 (female) [World].

CITATIONS: BenDov1988b [taxonomy, description, illustration, host, distribution: 39-40,111].

Odonaspis floridana Ben-Dov

Odonaspis floridana Ben-Dov, 1988b: 40. Type data: U.S.A.: Florida, Highlands County, Archbold Biological Station, on large grass; collected 28.IV.1975. Holotype female. Type depository: Washington: United States National Entomological Collection, U.S. National Museum of Natural History, District of Columbia, USA.

SCALE COVER: Only slide-mounted females available for description by Ben-Dov (1988b).

HOST PLANTS: **Gramineae**: *Spartina patens* [BenDov1988b].

DISTRIBUTION: **Nearctic**: United States of America (Florida [BenDov1988b]).

GENERAL: Description and illustration of adult female by Ben-Dov (1988b).

KEYS: Ben-Dov 1988b: 34 (female) [World].

CITATIONS: BenDov1988b [taxonomy, description, illustration, host, distribution: 40-41,112].

Odonaspis galapagoensis Ben-Dov

Odonaspis galapagoensis Ben-Dov, 1988b: 41. Type data: ECUADOR: Galapagos, Santa Cruz Island, Academy Bay, on *Sporobolus virginicus*; collected 24.VI.1977. Holotype female. Type depository: Washington: United States National Entomological Collection, U.S. National Museum of Natural History, District of Columbia, USA.

SCALE COVER: Only slide-mounted females available for description by Ben-Dov (1988b).

HOST PLANTS: **Gramineae**: *Sporobolus virginicus* [BenDov1988b].

DISTRIBUTION: **Neotropical**: Ecuador [BenDov1988b].

GENERAL: Description and illustration of adult female by Ben-Dov (1988b).

KEYS: Ben-Dov 1988b: 34 (female) [World].

CITATIONS: BenDov1988b [taxonomy, description, illustration, host, distribution: 41-42,113].

Odonaspis graminis Bremner
Odonaspis graminis Bremner, 1907: 368. Type data: U.S.A.: California, San Francisco, Presidio Hills, on roots of grass; collected by E.M. Ehrhorn. Syntypes, female. Type depository: Davis: The Bohart Museum of Entomology, University of California, California, USA.
Rugaspidis graminis; MacGillivray, 1921: 449. Change of combination.
Rugaspidis graminis; MacGillivray, 1921: 449. Misspelling of genus name.
Dycryptaspis graminis; Lindinger, 1937: 184. Change of combination.
Odonaspis graminis; Ben-Dov, 1988b: 42. Revived combination.
COMMON NAME: grass root scale [McKenz1956, Borchs1966].
SCALE COVER: Bremner (1907) described female scale as: "... the appearance of a clam, ranging in form from mytiliform to round, 1-1.5 mm in size, dirty-white in colour; the exuvia is at one side, and at the anterior extremity is glossy straw-coloured; the ventral scale is nearly as well developed as the dorsal, and has what appears to be the ventral half of the exuvia at the anterior end". Ferris (1938a) described scale cover as: "Female scale brown, broadly oval, exuviae apical. Scale of the male not seen".
HOST PLANTS: **Gramineae** [Bremne1907], *Danthonia* [Borchs1966], *Danthonia americana* [BenDov1988b], *Danthonia californica* [Ferris1938a, McKenz1956]. **Juncaceae**: *Juncus* [BenDov1988b].
DISTRIBUTION: **Nearctic**: Mexico [BenDov1988b] (District federal [Ferris1942]); U.S.A. (California [Bremne1907, Ferris1938a, McKenz1956, BenDov1988b]).
BIOLOGY: Occurring beneath sheathing bases of leaves and among scales on rootstocks.
GENERAL: Description and illustration of adult female by Ferris (1920b, 1938a) McKenzie (1956), Ben-Dov (1988b) and by Gill (1997).
KEYS: Gill 1997: 208 (female) [Species of California]; Ben-Dov 1988b: 33 (female) [World]; McKenzie 1956: 37 (female) [U.S.A.: California]; Ferris 1942: 65 (female) [North America].
CITATIONS: Balach1957c [taxonomy: 206]; BenDov1988b [taxonomy, description, illustration, host, distribution: 42,114]; Borchs1966 [catalogue: 224]; Bremne1907 [taxonomy, description, illustration, host, distribution: 368]; Ferris1920b [taxonomy, description, illustration, host, distribution: 57]; Ferris1938a [taxonomy, description, illustration, host, distribution: 162]; Ferris1942 [host, distribution: 8]; Gill1997 [host, distribution, taxonomy, description, illustration, economic importance: 208-209]; Lindin1937 [taxonomy: 184]; MacGil1921 [taxonomy, description, host, distribution: 449]; McKenz1956 [taxonomy, description, illustration, host, distribution: 37,162-163]; Nakaha1982 [host, distribution: 62]; Sander1909a [taxonomy, host, distribution: 55].

Odonaspis greenii Cockerell

Aspidiotus secretus; Green, 1896a: 64. Misidentification.

Odonaspis secretus greenii Cockerell, 1902a: 25. Type data: SRI LANKA (=CEYLON): Pundaluoya, on *Arundinaria* sp.; collected by E.E. Green, iv.1995. Lectotype, by subsequent designation Ben-Dov, 1988b: 43. Type depository: London: The Natural History Museum, UK.

Odonaspis greenii; Zimmerman, 1948: 428. Change of status.

Odonaspis greeni; Ben-Dov, 1988b: 42. Misspelling of species name.

COMMON NAME: Green's scale [Zimmer1948].

SCALE COVER: In original description of this species, Cockerell (1902a: 25) clearly referred to the description and illustration by Green (1896e: 64). Colour illustration of scale cover by Green (1896e).

HOST PLANTS: **Gramineae**: *Arundinaria* [MacGil1921, Ferris1950, BenDov1988b], *Bambusa* [BenDov1988b, Porcel1992], *Bambusa dissemulator* [BenDov1988b], *Bambusa nana* [BenDov1988b], *Bambusa pervariabilis* [BenDov1988b], *Bambusa vulgaris* [BenDov1988b, WilliaWa1988], *Dinochloa scandens* [BenDov1988b], *Gigantochloa verticillata* [BenDov1988b], *Miscanthus sinensis* [Takagi1990].

NATURAL ENEMIES: HYMENOPTERA **Encyrtidae**: *Caenohomalopoda guamensis* (Fullaway) [BenDov1988b].

DISTRIBUTION: **Australasian**: Hawaiian Islands (Hawaii [MacGil1921, Beards1966, BenDov1988b]); Samoa [BenDov1988b]; Western Samoa [BenDov1988b, WilliaWa1988]. **Nearctic**: United States of America (California [BenDov1988b]). **Neotropical**: Guadeloupe [BenDov1988b]; Guyana [BenDov1988b]; Martinique [BenDov1988b]; Saint Lucia [BenDov1988b]; Suriname [BenDov1988b]. **Oriental**: China (People's Republic) (Guangdong (Kwangtung) [Ferris1950, BenDov1988b], Yunnan [Ferris1950]); Hong Kong [BenDov1988b]; India [BenDov1988b]; Indonesia [MacGil1921, BenDov1988b]; Philippines [BenDov1988b]; Sri Lanka [MacGil1921, BenDov1988b]; Taiwan [BenDov1988b]; Thailand [BenDov1988b]; Vietnam [BenDov1988b]. **Palaearctic**: Czech Republic [BenDov1988b]; Italy [Porcel1992]; Japan [MacGil1921, BenDov1988b, Takagi1990]; Tunisia [BenDov1988b].

BIOLOGY: The young insect takes up its position upon the stem of plant beneath sheath; after completion of second moult the adult female insinuates itself into and between layers of sheath, excavating a cell for itself, and leaving the exuviae exposed upon surface (Green, 1896e).

GENERAL: Description and illustration of adult female by Green (1896e), Chou (1985, 1986), Ben-Dov (1988b), Zahradník (1990) and by Porcelli (1992).

KEYS: Ben-Dov 1988b: 34 (female) [World]; Williams & Watson 1988: 198 (female) [Tropical South Pacific]; Chou 1985: 328 (female) [Species of China]; Beardsley 1966: 525 (female) [Federated States of Micronesia].

CITATIONS: Balach1954e [description: 14]; Beards1966 [taxonomy, host, distribution: 528]; BenDov1988b [taxonomy, description, illustration, host, distribution, biological control: 42-43,72,115]; BenDov1990e [host, distribution: 658]; Borchs1966 [catalogue: 224]; Chou1985 [taxonomy, description, host, distribution: 328,330]; Chou1986 [taxonomy, illustration: 701]; Cocker1902a [taxonomy: 25,26]; DanzigPe1998 [catalogue: 317]; FangWuXu2001 [host, distribution: 108]; Fernal1903b [catalogue: 300]; Ferris1937a [taxonomy, illustration: 33,39]; Ferris1950 [taxonomy, description, host, distribution: 75]; Green1896e [taxonomy: 40,64]; Lindin1907a [taxonomy: 19]; LongoMaPe1995 [distribution: 128]; MacGil1921 [taxonomy, description, host, distribution: 423]; Porcel1992 [taxonomy, description, illustration, host, distribution: 33-36]; SoriaEsVi1998a [host, distribution: 337-342]; Takagi1990 [taxonomy, structure: 82,99,111-112]; Tao1999 [taxonomy, host, distribution: 102]; Willia1985a [taxonomy: 234]; WilliaWa1988 [taxonomy, host, distribution: 198]; Zahrad1990b [taxonomy, description, illustration, host, distribution: 117-119]; Zimmer1948 [taxonomy, description, host, distribution: 428].

Odonaspis lingnani Ferris

Odonaspis lingnani Ferris, 1955c: 33. Type data: CHINA: Kwangtung, Canton, Lingnan University, Old Bamboo Garden, on *Bambusa multiplex*. Lectotype female, by subsequent designation Ben-Dov, 1988b: 45. Type depository: Washington: United States National Entomological Collection, U.S. National Museum of Natural History, District of Columbia, USA.
SCALE COVER: Scale of female white (Ferris, 1955c).
HOST PLANTS: **Gramineae**: *Bambusa multiplex* [Ferris1955c, Borchs1966, BenDov1988b].
DISTRIBUTION: **Oriental**: China (People's Republic) (Guangdong (Kwangtung) [Ferris1955c, Borchs1966, BenDov1988b]); Indonesia [BenDov1988b].
BIOLOGY: Occurring massed in closely crowded axils of branches, so crowded that impossible to secure a scale by itself alone without destroying it and consequently no habit sketch given (Ferris, 1955c).
GENERAL: Description and illustration of adult female by Ferris (1955c), Chou (1985, 1986) and by Ben-Dov (1988b).
KEYS: Ben-Dov 1988b: 33 (female) [World]; Chou 1985: 328 (female) [Species of China].
CITATIONS: BenDov1988b [taxonomy, description, illustration, host, distribution: 45,116]; BenDov1990e [host, distribution: 658]; Borchs1966 [catalogue: 224]; Chou1985 [taxonomy, description, host, distribution: 328,330]; Chou1986 [taxonomy, illustration: 702]; FangWuXu2001 [host, distribution: 108]; Ferris1955c [taxonomy, description, illustration, host, distribution: 33]; Tao1999 [taxonomy, host, distribution: 102].

Odonaspis litorosa Ferris

Odonaspis litorosa Ferris, 1921: 119. Type data: MEXICO: Baja California, La Rivera, Eureca Ranch, on *Rhachidospermum mexicanum*; collected July 1919. Holotype female. Type depository: Davis: The Bohart Museum of Entomology, University of California, California, USA.
Dycryptaspis litorosa; Lindinger, 1937: 184. Change of combination.
Odonaspis litorosa; Borchsenius, 1966: 224. Revived combination.
SCALE COVER: Scale of female white, broadly oval, tapering posteriorly. Male scale not seen (Ferris, 1938a).

HOST PLANTS: **Gramineae**: *Arundo?* [Borchs1966], *Milium* [Borchs1966], *Rhachidospermum mexicanum* [Feris1938a, Borchs1966, BenDov1988b], *Sporobolus airoides* [Borchs1966, BenDov1988b], *Sp. wrightii* [Ferris1938a].
DISTRIBUTION: **Nearctic**: Mexico (Baja California [Ferris1921, Ferris1938a, BenDov1988b], Colima [Ferris1942], Guerrero [Ferris1942], Nayarit [Ferris1942], Oaxaca [Ferris1942], Sinola [Ferris1942], Veracruz [Ferris1942]); United States of America (Arizona [BenDov1988b], Colorado [BenDov1988b], Texas [Ferris1938a, BenDov1988b]). **Neotropical**: Panama [Borchs1966]. **Palaearctic**: Madeira Islands [Borchs1966].
BIOLOGY: Occurring beneath sheathing bases of leaves.
GENERAL: Description and illustration of adult female by Ferris (1921, 1938a) and by Ben-Dov (1988b).
KEYS: Ben-Dov 1988b: 34 (female) [World]; Ferris 1942: 65 (female) [North America].
CITATIONS: Balach1953g [taxonomy: 742,745]; Balach1957c [taxonomy: 206]; Balach1958b [taxonomy: 302,304]; BenDov1988b [taxonomy, description, illustration, host, distribution: 45-46,117]; Borchs1966 [catalogue: 224-225]; Ferris1921 [taxonomy, description, illustration, host, distribution: 119-121]; Ferris1938a [taxonomy, description, illustration, host, distribution: 163]; Ferris1942 [host, distribution: 8]; Lindin1937 [taxonomy: 184]; Nakaha1982 [host, distribution: 62].

Odonaspis minima Howell & Tippins
Odonaspis minima Howell & Tippins, 1978: 762. Type data: U.S.A.: Georgia, Charlton County, State Route 94, 8 km West of Saint George, on *Aristida* sp. Holotype female. Type depository: Washington: United States National Entomological Collection, U.S. National Museum of Natural History, District of Columbia, USA.
SCALE COVER: Female scale white, narrowing posteriorly, about 1 mm long, 0.5 mm wide; exuviae terminal, yellow. Male scale similar but smaller (Howell & Tippins, 1978).
HOST PLANTS: **Gramineae**: *Aristida* [HowellTi1978, BenDov1988b].
DISTRIBUTION: **Nearctic**: United States of America (Georgia [HowellTi1978, BenDov1988b]).
BIOLOGY: Adult female on outer surface of leaf, usually below soil surface (Howell & Tippins, 1978).
GENERAL: Description and illustration of adult female by Howell & Tippins (1978) and by Ben-Dov (1988b).
KEYS: Ben-Dov 1988b: 34 (female) [World].
CITATIONS: BenDov1988b [taxonomy, description, illustration, host, distribution: 46,118]; HowellTi1978 [taxonomy, description, illustration, host, distribution: 762-766]; HowellTi1990 [taxonomy, structure: 39]; Nakaha1982 [host, distribution: 62].

Odonaspis morrisoni Beardsley

Odonaspis morrisoni Beardsley, 1966: 525. Type data: FEDERATED STATES OF MICRONESIA: Truck Atoll, Dublon, on *Cynodon dactylon*; collected 17.X.1952. Holotype female. Type depository: Honolulu: Bernice P. Bishop Museum, Department of Entomology Collection, Hawaii, USA.

SCALE COVER: Only slide-mounted females available for description by Beardsley (1966).

HOST PLANTS: **Gramineae** [WilliaWa1988], *Cynodon dactylon* [Beards1966, BenDov1988b], *Distichlis* [BenDov1988b], *Zoysia matrella* [Beards1966, BenDov1988b]. **Rutaceae**: *Citrus limon* [WilliaWa1988].

DISTRIBUTION: **Australasian**: Federated States of Micronesia (Truk Islands [Beards1966, BenDov1988b]); Fiji [BenDov1988b, WilliaWa1988]. **Oriental**: Hong Kong [BenDov1988b]; Philippines [Beards1966, BenDov1988b].

GENERAL: Description and illustration of adult female by Beardsley (1966) and by Ben-Dov (1988b).

KEYS: Ben-Dov 1988b: 33 (female) [World]; Williams & Watson 1988: 198 (female) [Tropical South Pacific]; Beardsley 1966: 525 (female) [Federated States of Micronesia].

CITATIONS: Beards1966 [taxonomy, description, illustration, host, distribution: 525-527]; BenDov1988 [taxonomy, description, illustration, host, distribution: 46-47,119]; WilliaWa1988 [taxonomy, host, distribution: 198].

Odonaspis oshimaensis Kuwana

Odonaspis oshimaensis Kuwana, 1933: 35. Type data: JAPAN: Amami-Oshima, on *Panicum sanguinale*. Holotype female. Type depository: Davis: The Bohart Museum of Entomology, University of California, California, USA.

Dycryptaspis oshimaensis; Lindinger, 1937: 184. Change of combination.

Odonaspis oshimaensis; Borchsenius, 1966: 225. Revived combination.

SCALE COVER: Kuwana (1933) did not describe scale cover. Photograph of scale cover by Kawai (1980).

HOST PLANTS: **Gramineae** [Takagi1959], *Panicum sanguinale* [Kuwana1933, Balach1958b, BenDov1988b].

DISTRIBUTION: **Palaearctic**: Japan [Kuwana1933, Takagi1959, Borchs1966, Kawai1980, BenDov1988b].

GENERAL: Description and illustration of adult female by Kuwana (1933) and by Ben-Dov (1988b).

KEYS: Ben-Dov 1988b: 33 (female) [World]; Kawai 1980: 199 (female) [Japan]; Takagi 1959a: 93 (female) [Japan]; Kuwana 1933: 35 (female) [Japan]; Kuwana 1933b: 49 (female) [Japan].

CITATIONS: Balach1957c [taxonomy: 206]; Balach1958b [host, distribution: 304]; BenDov1988b [taxonomy, description, illustration, host, distribution: 47,120]; Borchs1966 [catalogue: 225]; DanzigPe1998 [catalogue: 317-318]; Kawai1980 [taxonomy, description, host, distribution: 202]; Kuwana1933 [taxonomy, description, illustration, host, distribution: 35]; Lindin1937 [taxonomy: 184]; Muraka1970 [host, distribution: 68]; Takagi1959 [taxonomy, host, distribution: 94].

Odonaspis pacifica Ben-Dov
Odonaspis pacifica Ben-Dov, 1988b: 48. Type data: GUAM: Guam, on bamboo; collected 7.v.1946. Holotype female. Type depository: Washington: United States National Entomological Collection, U.S. National Museum of Natural History, District of Columbia, USA.
SCALE COVER: Only slide-mounted females available for description by Ben-Dov (1988b).
HOST PLANTS: **Gramineae**: *Bambusa* [BenDov1988b].
DISTRIBUTION: **Australasian**: Guam [BenDov1988b].
GENERAL: Description and illustration of adult female by Ben-Dov (1988b).
KEYS: Ben-Dov 1988b: 33 (female) [World].
CITATIONS: BenDov1988b [taxonomy, description, illustration, host, distribution: 48,121]; BenDov1990e [host, distribution: 658].

Odonaspis panici Hall
Odonaspis panici Hall, 1926a: 26. Type data: EGYPT: Abu Sueir, on *Panicum turgidum*; collected 20.VI.1925. Holotype female. Type depository: London: The Natural History Museum, England, UK.
Aonidia panici; Lindinger, 1932f: 189. Change of combination.
Odonaspis panici; Borchsenius, 1966: 225. Revived combination.
SCALE COVER: Female scale, elongate, 2.5-3 mm long, 1.5-1.75 mm wide; broadened in front and somewhat narrowed behind; flattish; white; exuviae near one end naked and orange; ventral scale entire, dense white and continuous with dorsal scale. Male scale smaller, elongate with subparallel sides; exuviae orange; width of scale more than width of exuvia; length of scale of male 1 mm, breadth 0.5 mm (Hall, 1926a).
HOST PLANTS: **Gramineae**: *Panicum* [Ramakr1921a], *Panicum turgidum* [Hall1926a, Balach1953g, Borchs1966, BenDov1988b], *Pennisetum ciliare* [Rungs1937, Balach1953g, Borchs1966, BenDov1988b].
NATURAL ENEMIES: HYMENOPTERA **Aphelinidae**: *Physcus* sp. [BenDov1988b].
DISTRIBUTION: **Afrotropical**: Angola [Almeid1969]. **Oriental**: Sri Lanka [Ramakr1921a]. **Palaearctic**: Egypt [Hall1926a, Balach1953g, Ezzat1958, Borchs1966, BenDov1988b]; Iran [Borchs1966, BenDov1988b]; Israel [BenDov1988b]; Morocco [Rungs1937, Borchs1966, BenDov1988b].
GENERAL: Description and illustration of adult female by Hall (1926a) and by Ben-Dov (1988b).
KEYS: Ben-Dov 1988b: 35 (female) [World]; Ezzat 1958: 247 (female) [Egypt]; Balachowsky 1953g: 733 (female) [Mediterranean].
CITATIONS: Almeid1969 [taxonomy, description, host, distribution, biological control: 158]; Balach1953g [taxonomy, description, illustration, host, distribution: 746-748]; Balach1957c [taxonomy: 206]; BenDov1988b [taxonomy, description, illustration, host, distribution, biological control: 49-50,72,122]; Borchs1966 [catalogue: 225]; DanzigPe1998 [catalogue: 318]; Ezzat1958 [taxonomy, distribution, host: 247]; EzzatAf1966 [taxonomy, description, illustration, host, distribution: 402-404]; EzzatNa1987 [distribution: 87]; Hall1926a [taxonomy,

description, illustration, host, distribution: 26-27]; Hall1927b [taxonomy, description, host, distribution: 148-149]; Kaussa1955 [host, distribution: 17]; Lindin1932f [p. 189]; Ramakr1921a [host, distribution: 357]; Rungs1937 [host, distribution: 333].

Odonaspis paucipora Ben-Dov

Odonaspis paucipora Ben-Dov, 1988b: 49. Type data: INDIA: Calcutta, Royal Botanic Gardens, on *Bambusa tulda*; collected 26.V.1909. Holotype female. Type depository: Washington: United States National Entomological Collection, U.S. National Museum of Natural History, District of Columbia, USA.

SCALE COVER: Female scale oval, 1-2 mm long, 0.8-1.5 mm wide; white; larval exuviae yellow brown placed at anterior end of scale. Male scale elongate, parallel-sided, about 1 mm long, 0.5 mm wide; white; larval exuviae yellow brown placed at anterior end of scale (Ben-Dov, 1988b).

HOST PLANTS: **Gramineae**: *Bambusa tulda* [BenDov1988b].

DISTRIBUTION: **Neotropical**: Guyana [BenDov1988b]. **Oriental**: India [BenDov1988b].

GENERAL: Description and illustration of adult female by Ben-Dov (1988b).

KEYS: Ben-Dov 1988b: 34 (female) [World].

CITATIONS: BenDov1988b [taxonomy, description, illustration, host, distribution: 49-50,123]; BenDov1990e [host, distribution: 658].

Odonaspis phragmitis Hall

Odonaspis phragmitis Hall, 1935a: 221. Type data: ZIMBABWE: Inyazura, on *Phragmites communis*. Syntypes, female. Type depository: London: The Natural History Museum, England, UK.

Dycryptaspis phragmitis; Lindinger, 1937: 184. Change of combination.

Odonaspis phragmitis; Borchsenius. 1966: 225. Revived combination.

SCALE COVER: Female scale large, oval, 3-3.5 mm long, 0.7 mm wide; flat, oval and white in colour; texture of dorsal scale thin but tough. Male scale oblong, 1 mm long, 0.7 mm wide; tapering somewhat posteriorly, white with shiny brown naked exuvia (Hall, 1935a).

HOST PLANTS: **Gramineae**: *Cynodon dactylon* [Balach1958b], *Hemarthria altissima* [BenDov1988b], *Panicum viride* [Balach1958b], *Phragmites* [BenDov1988b], *Phragmites communis* [Hall1935a, Balach1958b, Borchs1966, BenDov1988b].

DISTRIBUTION: **Afrotropical**: Malawi [BenDov1988b]; Mauritius [Balach1958b]; Mozambique [Almeid1973]; South Africa [BenDov1988b]; Zimbabwe [Hall1935a, Balach1958b, Borchs1966, BenDov1988b].

GENERAL: Description and illustration of adult female by Hall (1935a), Balachowsky (1958b) and by Ben-Dov (1988b).

KEYS: Ben-Dov 1988b: 34 (female) [World]; Balachowsky 1958b: 300 (female) [Africa].

CITATIONS: Almeid1973 [host, distribution: 4]; Balach1958b [taxonomy, description, illustration, host, distribution: 300-301]; BenDov1988b [taxonomy, description, illustration, host, distribution: 50-51,124]; Borchs1966 [catalogue: 225];

Hall1935a [taxonomy, description, illustration, host, distribution: 221-222]; Lindin1937 [taxonomy: 184].

Odonaspis ruthae Kotinsky
Odonaspis ruthae Ehrhorn, 1907: 26. Nomen nudum.
Odonaspis graminis; Kotinsky, 1910: 129. Misidentification; discovered by McKenzie, 1956: 165.
Odonaspis ruthae Kotinsky, 1915: 102. Type data: HAWAII: Honolulu, on Bermuda grass; collected July, 1905. Lectotype female, by subsequent designation Ben-Dov, 1988b: 52. Type depository: Washington: United States National Entomological Collection, U.S. National Museum of Natural History, District of Columbia, USA.
Aonidia ruthae; Lindinger, 1932f: 189. Change of combination.
Dycryptaspis ruthae; Lindinger, 1943a: 150. Change of combination.
Odonaspis pseudoruthae Mamet, 1954: 67. Type data: MADAGASCAR: Mananjary, on a climbing Gramineae. Holotype. Type depository: Paris: Muséum national d'Histoire naturelle, France. Synonymy by Ben-Dov, 1988b: 51.
Odonaspis ruthae; Borchsenius, 1966: 225. Revived combination.
COMMON NAME: Bermuda grass scale [McKenz1956, Borchs1966].
SCALE COVER: Female scale oyster shaped or mytiliform when fully grown; about 1.75 mm long, 0.75 mm wide; chalk white; exuvia, straw-coloured, at elevated end, partly or entirely covered with white waxy dust which rubs off easily; ventral scale well developed, with dorsal scale completely enclosing and sealing insects. Male scale same shape, but only about half size of female (Kotinsky, 1915). Scale of female white, broad and elongate, tapering somewhat posteriorly; that of male white, elongate, and slender (Ferris, 1938a). Colour photograph by Gill (1997).
HOST PLANTS: **Arecaceae**: *Nypa fruticans* [WilliaWa1988]. **Euphorbiaceae**: *Euphorbia heterophylla* [BenDov1988b, WilliaWa1988]. **Gramineae** [Mamet1941a, BesheaTiHo1973], *Andropogon* [BesheaTiHo1973, BenDov1988b], *Axonopus compressus* [BenDov1988b], *Bambusa* [BenDov1988b], *Brachypodium distachyum* [BenDov1988b], *Cenchrus pauciflorus* [BenDov1988b], *Chloris* [WilliaBu1987, BenDov1988b, WilliaWa1988], *Chloris gayana* [BenDov1988b], *Cortaderia selloana* [BenDov1988b], *Cymbopogon citratus* [BenDov1988b], *Cynodon* [Bodenh1937, BesheaTiHo1973], *Cynodon dactylon* [Hall1925, Bodenh1935, Ferris1938a, Lepage1938, Hosny1939, Ferris1942, Mamet1949, McKenz1956], *Cynodon dactylon* [Dekle1965c], *Cynodon transvaalensis* [BenDov1988b], *Digitaria milanjiana* [BenDov1988b], *Digitaria sanguinalis* [BenDov1988b], *Distichlis* [BenDov1988b], *Eleusine indica* [BenDov1988b], *Eremochloa ophiuroides* [BesheaTiHo1973, BenDov1988b], *Festuca* [BesheaTiHo1973], *Lepturus repens* [BenDov1988b], *Panicum* [BesheaTiHo1973, BenDov1988b, WilliaWa1988], *Panicum maximum* [BenDov1988b], *Panicum viride* [Hall1925], *Paspalum distichum* [BenDov1988b], *Setaria geniculata* [BenDov1988b], *Sorghum halepense* [McKenz1956, BesheaTiHo1973, BenDov1988b], *Sorghum vulgare sudanense* [McKenz1956], *Spartina patens* [BesheaTiHo1973, BenDov1988b], *Sporobolus* [BesheaTiHo1973], *Sporobolus indicus* [BenDov1988b], *Stenotaphrum dimidiatum* [Mamet1957], *Stenotaphrum*

secundatum [BenDov1988b]. **Juncaceae**: *Juncus* [McKenz1956]. **Malvaceae**: *Hibiscus* [WilliaWa1988]. **Orchidaceae**: *Cattleya* [BenDov1988b].
NATURAL ENEMIES: COLEOPTERA **Nitidulidae**: *Cybocephalus pullus* Enrody-Younga [BenDov1988b]. HYMENOPTERA **Encyrtidae**: *Adelencyrtus odonaspidis* Fullaway [BenDov1988b, Trjapi1989]. THYSANOPTERA **Phlaeothripidae**: *Podothrips semiflavus* Hood [BenDov1988b, PalmerMo1990].
DISTRIBUTION: **Afrotropical**: Ethiopia [BenDov1988b]; Kenya [BenDov1988b]; Madagascar [BenDov1988b]; Mauritius [Mamet1941a, Mamet1943a, Mamet1949, Borchs1966, BenDov1988b]; Réunion [Mamet1957]; South Africa [BenDov1988b]; Tanzania [BenDov1988b]; Zimbabwe [Hall1935a, Green1937, BenDov1988b]. **Australasian**: Australia [Hall1925, Green1937] (New South Wales [BenDov1988b], Queensland [BenDov1988b], South Australia [BenDov1988b], Western Australia [BenDov1988b]); Christmas Island [BenDov1988b]; Cook Islands [BenDov1988b, WilliaWa1988]; French Polynesia (Society Islands [WilliaWa1988], Tahiti [BenDov1988b, WilliaWa1988]); Hawaiian Islands (Hawaii [Hall1925, Green1937, Ferris1938a, Merril1953, BenDov1988b]); Kiribati [WilliaWa1988] (Gilbert Island [BenDov1988b]); New Caledonia [Cohic1958]; Papua New Guinea [WilliaWa1988]; Tuvalu [BenDov1988b, WilliaWa1988]; Vanuatu [WilliaBu1987, WilliaWa1988] [WilliaBu1987, WilliaWa1988]. **Nearctic**: Mexico [Merril1953, BenDov1988b]; United States of America (Alabama [Nakaha1982], Arizona [BenDov1988b], Arkansas [Nakaha1982], California [Ferris1938a, McKenz1956, Merril1953, BenDov1988b], Florida [Merril1953, Dekle1965c, BesheaTiHo1973, BenDov1988b], Georgia [TippinBe1972, BesheaTiHo1973, BenDov1988b], Louisiana [Ferris1938a, Merril1953, BenDov1988b], Mississippi [BenDov1988b], North Carolina [BenDov1988b], South Carolina [BenDov1988b], Texas [Ferris1938a, Merril1953, BenDov1988b]). **Neotropical**: Argentina (Chaco [BenDov1988b]); Bermuda [BenDov1988b]; Bolivia [BenDov1988b]; Brazil (Bahia [BenDov1988b], Distrito federal (Brasilia) [Lepage1938]); Chile [GonzalCh1968, BenDov1988b]; Cuba [BenDov1988b]; Guadeloupe [BenDov1988b]; Panama [BenDov1988b]; Peru [BenDov1988b]; Puerto Rico & Vieques Island [BenDov1988b]; U.S. Virgin Islands [Nakaha1983, BenDov1988b]. **Oriental**: India [Hall1925]; Pakistan [BenDov1988b]; Sri Lanka [Green1922, Hall1925, Ferris1938a, Merril1953, BenDov1988b]. **Palaearctic**: Egypt [Hall1925, Green1937, Hosny1939, Ferris1938a, Merril1953, BenDov1988b]; Israel [Bodenh1937, BenDov1988b].
BIOLOGY: Occurring beneath sheathing bases of leaves and on rootstocks.
ECONOMIC IMPORTANCE: Most polyphagous species among Odonaspidinae, and reported to damage lawn grasses in Israel (Halperin, 1971; Berlinger & Barak, 1981) and to be a common pest of forage and turf grasses in Southern United States (Dale and McCoy, 1964; Tippins & Martin, 1982).
GENERAL: Description and illustration of adult female by Ferris (1938a), McKenzie (1956), Ben-Dov (1988b), Gill (1997) and by Colon-Ferrer & Medina-Gaud (1998).
KEYS: Gill 1997: 208 (female) [Species of California]; Ben-Dov 1988b: 34 (female) [World]; Ezzat 1988: 247 (female) [Egypt]; Williams & Watson 1988: 198 (female) [Tropical South Pacific]; Beardsley 1966: 525 (female) [Federated States of

Micronesia]; Balachowsky 1958b: 299 (female) [Africa]; McKenzie 1956: 37 (female) [U.S.A.: California]; Balachowsky 1953g: 732 (female) [Mediterranean]; Ferris 1942: 65 (female) [North America]; Fullaway 1932: 95-97, 108 (female) [Hawaii].

CITATIONS: AbouEl2001 [host, distribution, biological control: 185-195]; Balach1953g [taxonomy, description, illustration, host, distribution: 743-746]; Balach1958b [taxonomy, description, illustration, host, distribution: 299]; BeardsDaHo1976 [economic importance: 103]; BenDov1988b [taxonomy, description, illustration, host, distribution, economic importance, biological control: 51-55,67,71,139]; BenDov1990e [host, distribution: 658]; BerlinBa1981 [host, distribution, life history: 62-67]; BesheaTiHo1973 [host, distribution: 15]; Bodenh1935 [host, distribution: 247]; Bodenh1937 [host, distribution: 217]; Borchs1966 [catalogue: 225]; ClapsWoGo2001a [taxonomy, host, distribution: 22]; Cock1985a [biological control: 3]; Cohic1958 [host, distribution: 19]; ColonFMe1998 [taxonomy, description, illustration, host, distribution: 133-134]; CostaCoHa2000 [host, distribution, control]; CostaL1934 [host, distribution: 136]; DaleMc1964 [host, distribution, economic importance: 228]; DanzigPe1998 [catalogue : 318]; Dekle1965c [taxonomy, description, host, distribution: 100]; Dekle1976 [taxonomy, description, host, distribution, economic importance: 121]; Ehrhor1907 [taxonomy, host, distribution: 26]; Ezzat1958 [taxonomy, distribution: 247]; EzzatNa1987 [distribution: 87]; FDACSB1987 [host, distribution: 4-7]; Ferris1938a [taxonomy, description, illustration, host, distribution: 165]; Ferris1942 [description, host, distribution: 446]; Fullaw1932 [taxonomy: 95,108]; Gill1997 [host, distribution, taxonomy, description, illustration, economic importance: 208,210-211]; GonzalCh1968 [host, distribution: 111]; Green1922 [host, distribution: 463]; Green1937 [host, distribution: 335]; Hall1925 [taxonomy, description, host, distribution: 15]; Hall1935a [host, distribution: 222]; Hosny1939 [description, host, distribution: 15]; Kotins1910 [taxonomy: 129]; Kotins1915 [taxonomy, description, host, distribution: 102]; Lepage1938 [catalogue: 412]; Lindin1932f [taxonomy: 189]; Lindin1943a [taxonomy: 150]; Mamet1941a [host, distribution: 40]; Mamet1943a [catalogue: 164]; Mamet1949 [catalogue: 51]; Mamet1954 [taxonomy: 67]; Mamet1957 [host, distribution: 370,377]; Mamet1959a [host, distribution: 385]; McKenz1956 [taxonomy, description, illustration, host, distribution: 164-165]; Merril1953 [taxonomy, description, host, distribution: 64]; MerrilCh1923 [host, distribution: 246]; MillerDa1990 [host, distribution, economic importance: 304]; Nakaha1982 [host, distribution: 63]; Nakaha1983 [host, distribution: 13]; Nur1990b [taxonomy, life history: 196]; PalmerMo1990 [biological control: 67-76]; RiherdCh1952 [host, distribution, economic importance, control: 1-5]; Shetla1994 [host, distribution, economic importance]; TippinBe1968a [host, distribution: 134-136]; TippinBe1972 [host, distribution: 287]; TippinMa1982 [economic importance, life history, host, distribution: 319-321]; Trjapi1989 [biological control: 291]; WilliaBu1987 [host, distribution: 95]; WilliaWa1988 [taxonomy, illustration, host, distribution: 198-201]; Zimmer1948 [taxonomy, description, host, distribution: 426,428].

Odonaspis sabulincola Ben-Dov

Odonaspis sabulincola Ben-Dov, 1988b: 55. Type data: NAMIBIA [= SOUTH WEST AFRICA]: Namib Desert, Gobabeb, on *Stipagrostis* sp.; collected Y. Ben-Dov, 28.II.1975. Holotype female. Type depository: Pretoria: South African National Collection of Insects, South Africa.

SCALE COVER: Scale cover of young female circular, elongate oval in fully grown female, up to 4 mm long in reproducing female; white; larval exuviae yellow brown placed centrally. Male scale similar in colour and shape but smaller (Ben-Dov, 1988b).

HOST PLANTS: **Gramineae**: *Stipagrostis* [BenDov1988b], *Stipagrostis namaquensis* [BenDov1988b].

DISTRIBUTION: **Afrotropical**: Namibia (Southwest Africa) [BenDov1988b].

GENERAL: Description and illustration of adult female by Ben-Dov (1988b).

KEYS: Ben-Dov 1988b: 33 (female) [World].

CITATIONS: BenDov1988b [taxonomy, description, illustration, host, distribution: 55-56,126].

Odonaspis saccharicaulis (Zehntner)

Aspidiotus sacchari-caulis Zehntner, 1897a: 559. Type data: INDONESIA: Java, Pasoeroean, on "Glonggong" [a common name for *Saccharum arundinaceum* and *Saccharum spontaneum*]. Neotype female, by subsequent designation Ben-Dov, 1988b: 56-58. Type depository: Washington: United States National Entomological Collection, U.S. National Museum of Natural History, District of Columbia, USA.

Odonaspis secretus saccharicaulis; Cockerell, 1899: 274. Change of combination and rank and justified emendation.

Aspidiotus (*Odonaspis*) *janeirensis* Hempel, 1900a: 500. Type data: BRAZIL: Ilha das Flores, Bahia do Rio de Janeiro, on grass. Syntypes, female. Type depositories: São Paulo: Museu de Zoologia, Universidade de São Paulo, Brazil, and Curitiba: Departamento de Zoologia, Setor de Ciencias Biologicas, Universidade Federal do Parana, Brazil; type no. 1540. Synonymy by Ben-Dov, 1988b: 56.

Odonaspis janeirensis; Cockerell, 1902p: 256. Change of combination.

Odonaspis secreta saccharicaulis; Fernald, 1903b: 300. Change of combination.

Ligulaspis janeirensis; MacGillivray, 1921: 423. Change of combination.

Odonaspis saccharicaulis; Green & Laing, 1923: 128. Revived combination.

Dycryptaspis janeirensis; Lindinger, 1937: 184. Change of combination.

Dycryptaspis saccharicaulis; Lindinger, 1937: 184. Change of combination.

Odonaspis saccharicaulis; Ben-Dov, 1988b: 56. Revived combination.

COMMON NAME: paragrass scale [Dekle1965c].

SCALE COVER: Female scale circular to oval, 1.6-2.6 mm long, about 1 mm wide; white; larval exuviae yellow brown placed subcentrally. Male unknown (Ben-Dov, 1988b).

HOST PLANTS: **Gramineae** [Lepage1938], *Anadelphia arrecta* [BenDov1988b], *Bambusa* [Cocker1899j], *Chloris gayana* [BenDov1988b], *Cymbopogon citratus* [BenDov1988b], *Digitaria decumbens* [BenDov1988b], *Heliconia* [BenDov1988b], *Merostachys* [ClapsWoGo2001], *Panicum barbinode* [BenDov1988b], *Panicum purpurascens* [Dekle1965c], *Paspalum dissectum* [BenDov1988b], *Pennisetum*

[Balach1958b, Borchs1966, BenDov1988b], *Pennisetum purpureum* [Mamet1943a, Mamet1949, Dekle1965c, Borchs1966, BenDov1988b, ClapsWoGo2001], *Phragmites communis* [BenDov1988b], *Phragmites karka* [BenDov1988b], *Saccharum* [MatileNo1984, BenDov1988b, WilliaWa1988], *Saccharum arundinaceum* [Borchs1966, BenDov1988b], *Saccharum officinarum* [Lepage1938, BenDov1988b, ClapsWoGo2001], *Schizacirium* [BenDov1988b], *Scleria canescens* [BenDov1988b], *Sorghum* [MatileNo1984], *Sorghum halepense* [Balach1953g, BenDov1988b], *Sorghum vulgare* [Borchs1966, BenDov1988b], *Uniola paniculata* [BenDov1988b], *Zoysia matrella* [BenDov1988b].

DISTRIBUTION: **Afrotropical**: Cameroon [MatileNo1984, BenDov1988b]; Cape Verde Islands [SchmutPiKl1978]; Guinea [BenDov1988b]; Mauritius [Mamet1943a, Mamet1949, Borchs1966, BenDov1988b]; Mozambique [BenDov1988b]; São Tomé and Príncipe (São Tomé) [GreenLa1923, Balach1958b]); Somalia [BenDov1988b]; South Africa [BenDov1988b]; Tanzania [BenDov1988b]; Uganda [Nakaha1982]. **Australasian**: Australia [BenDov1988b]; Fiji [WilliaWa1988]; Hawaiian Islands (Hawaii [Cocker1899j, BenDov1988b]); Palau [Beards1966]; Papua New Guinea [WilliaWa1988]. **Nearctic**: United States of America (Alabama [Nakaha1982], Florida [Merril1953, Dekle1965c, BenDov1988b], Maryland [Nakaha1982], Texas [BenDov1988b]). **Neotropical**: Bahamas [BenDov1988b]; Brazil [BenDov1988b] (Bahia [Hempel1900a, Lepage1938], Rio de Janeiro [Lepage1938, ClapsWoGo2001], São Paulo [Lepage1938, ClapsWoGo2001]); Costa Rica [BenDov1988b]; Cuba [BenDov1988b]; Guatemala [Nakaha1982]; Honduras [Nakaha1982]; Puerto Rico & Vieques Island (Puerto Rico [BenDov1988b, ColonFMe1998]); U.S. Virgin Islands [Nakaha1983, BenDov1988b]; Venezuela [BenDov1988b]. **Oriental**: India [BenDov1988b]; Indonesia (Java [Merril1953, BenDov1988b], Sumatra [Green1930c]); Malaysia (Sabah [BenDov1988b]); Pakistan [BenDov1988b]; Philippines [VelasqRi1969, BenDov1988b]; Sri Lanka [Cocker1899j]; Thailand [BenDov1988b]. **Palaearctic**: Japan [Cocker1899j]; Madeira Islands [Balach1938a, BenDov1988b].

ECONOMIC IMPORTANCE: Occurring in major sugarcane growing areas of the world, but recorded as a pest of this crop only from India and the Philippines (Rao and Sankaran, 1969; Kalshoven, 1981). Reported to be common in southern Africa on sugarcane as well as on other grasses (Ben-Dov, 1988b); however, Carnegie et al. (1974) did not record it among coccoids that infest sugarcane in southern Africa.

GENERAL: Description and illustration of adult female by Balachowsky (1953g, 1958b), Ben-Dov (1988b) and by Colon-Ferrer & Medina-Gaud (1998).

KEYS: Ben-Dov 1988b: 34 (female) [World]; Williams & Watson 1988: 198 (female) [Tropical South Pacific]; Beardsley 1966: 525 (female) [Federated States of Micronesia]; Balachowsky 1958b: 299 (female) [Africa]; Balachowsky 1958b: 299 (female) [Africa]; Balachowsky 1953g: 732 (female) [Mediterranean].

CITATIONS: AgarwaSi1964 [host, distribution, economic importance: 149]; Balach1938a [taxonomy, host, distribution: 152]; Balach1953g [taxonomy, description, illustration, host, distribution: 739-743]; Balach1957c [taxonomy, host, distribution: 204]; Balach1958b [taxonomy, description, illustration, host, distribution: 302-305]; Beards1966 [host, distribution: 528]; BenDov1988b [taxonomy, description, illustration, host, distribution, economic importance: 56,67,127]; BenDov1990e [host, distribution: 658]; Borchs1966 [catalogue: 225]; Box1953 [host, distribution, biological control: 52]; Claps1993 [taxonomy: 5,10]; ClapsWoGo2001 [host, distribution: 253-254]; ClapsWoGo2001a [taxonomy, host, distribution: 22]; Cocker1899j [taxonomy, host, distribution: 274]; Cocker1902p [taxonomy: 256]; ColonFMe1998 [taxonomy, description, illustration, host, distribution: 134-135]; DanzigPe1998 [catalogue: 318-319]; Dekle1965c [taxonomy, description, host, distribution: 101]; Dekle1976 [taxonomy, description, host, distribution, economic importance: 122]; Fernal1903b [catalogue: 300]; Ferris1937a [taxonomy: 33]; Ferris1941e [taxonomy: 44,48]; Green1930c [host, distribution: 281]; GreenLa1923 [taxonomy, description, host, distribution: 129]; GreenLa1923 [taxonomy: 129]; Hazelh1929 [host, distribution: 1-8]; Hempel1900a [taxonomy, description, host, distribution: 500-501]; Hempel1901a [taxonomy, host, distribution: 106]; Kirkal1905 [taxonomy: 78]; Koning1908 [host, distribution: 1-7]; Lepage1938 [catalogue: 412]; Lindin1937 [taxonomy: 184]; MacGil1921 [taxonomy, description, host, distribution: 423]; Mamet1943a [catalogue: 164]; Mamet1949 [catalogue: 51,52]; MatileNo1984 [host, distribution: 66]; Merril1953 [taxonomy, description, host, distribution: 64-65]; Merril1953 [taxonomy, description, host, distribution: 64-65]; MillerDa1990 [host, distribution, economic importance: 304]; Nakaha1982 [host, distribution: 63]; Nakaha1983 [host, distribution: 13]; Nur1990a [taxonomy, structure, chromosomes: 181,184-185]; SchmutKlLu1957 [host, distribution, economic importance: 499]; SchmutPiKl1978 [host, distribution, economic importance: 330]; Tao1999 [taxonomy, host, distribution: 102]; VelasqRi1969 [host, distribution: 195-208]; WilliaGr1990 [host, distribution, economic importance, biological control: 563-578]; WilliaGr1990 [host, distribution, economic importance, biological control: 563-578]; WilliaWa1988 [taxonomy, host, distribution: 201]; Zehntn1897a [taxonomy: 559]; Zehntn1897b [taxonomy, description, host, distribution, biology: 1-10].

Odonaspis schizostachyi Cockerell & Robinson

Odonaspis schizostachyi Cockerell & Robinson, 1914: 327. Type data: PHILIPPINES: Luzon, Los Banos, on *Schizostachyum* sp.; collected December 1913. Lectotype female, by subsequent designation Ben-Dov, 1988b: 59. Type depository: Washington: United States National Entomological Collection, U.S. National Museum of Natural History, District of Columbia, USA.

Bakeraspis schizostachyi; MacGillivray, 1921: 459. Change of combination.

Dycryptaspis schizostachyi; Lindinger, 1937: 184. Change of combination.

Odonaspis schizostachyi; Borchsenius, 1966: 225. Revived combination.

COMMON NAME: anos scale [VelasqRi1969].

SCALE COVER: Female scale circular, little over 1 mm diameter; dull white; concentrically wrinkled; large first exuvia very pale yellowish (Cockerell & Robinson, 1914).

HOST PLANTS: **Gramineae**: *Shizostachyum* [CockerRo1914, MacGil1921, Ferris1937a, Borchs1966, BenDov1988b].

DISTRIBUTION: **Oriental**: Philippines (Luzon [CockerRo1914, MacGil1921, Ferris1937a, Borchs1966, BenDov1988b]).

GENERAL: Description and illustration of adult female by Cockerell & Robinson (1914) and by Ben-Dov (1988b).
KEYS: Ben-Dov 1988b: 33 (female) [World].
CITATIONS: BenDov1988b [taxonomy, description, illustration, host, distribution: 58-59,128]; BenDov1990a [taxonomy: 89]; BenDov1990e [host, distribution: 658]; Borchs1966 [catalogue: 225-226]; CockerRo1914 [taxonomy, description, illustration, host, distribution: 327-328]; Ferris1937a [taxonomy, illustration: 33,35]; Lindin1937 [taxonomy: 184]; MacGil1921 [taxonomy, description, host, distribution: 459]; Robins1917 [taxonomy, description, host, distribution: 17-18]; Sassce1915 [taxonomy, host, distribution: 35]; VelasqRi1969 [host, distribution: 195-208]; Willia1985a [taxonomy: 238].

Odonaspis secreta (Cockerell)
Aspidiotus secretus Cockerell, 1896h: 20. Type data: JAPAN: Tokyo, on bamboo. Lectotype female, by subsequent designation Ben-Dov, 1988b: 60. Type depository: Washington: United States National Entomological Collection, U.S. National Museum of Natural History, District of Columbia, USA; type no. 5944.
Odonaspis secreta; Leonardi, 1897: 286. Change of combination.
Aspidiotus secretus lobulata Maskell, 1897a: 241. Type data: JAPAN: on *Bambusa* sp.; No. K1513. Syntypes, female. Type depository: Auckland: New Zealand Arthropod Collection, Landcare Research, New Zealand. Synonymy by Borchsenius, 1966: 226. Notes: Described again as *Aspidiotus secretus* Cockerell, var. *lobulatus*, var. nov., by Maskell, 1898: 224.
Aspidiotus (*Dycryptaspis*) *secreta*; Leonardi, 1897a: 375. Change of combination requiring emendation of species name for agreement in gender.
Spatheaspis secreta; Leonardi, 1897b: 115. Change of combination.
Aspidiotus (*Odonaspis*) *secretus*; Cockerell, 1897i: 14. Change of combination.
Spatheaspis secreta lobulata; Leonardi, 1900: 338. Change of combination requiring emendation of species name for agreement in gender.
Odonaspis secreta; Fernald, 1903b: 300. Change of combination.
Parlatoria zeylanica Rutherford, 1915: 113. Type data: SRI LANKA: Paradeniya, on a 'small bamboo'; collected August, 1914. Syntypes, immature. Type depository: Paradeniya: Horticultural Crop Research and Development Institute, Sri Lanka. Synonymy by Takagi, 1987: 8. Notes: Rutherford (1915, page 114) also described another new species that was named *Parlatoria zeylanica*. The latter was a homonym and was replaced by *Parlatoria rutherfordi* Green, 1922a, in the genus *Parlatoria*.
Dycryptaspis secreta; Lindinger, 1937: 184. Change of combination.
Aspidiotus lobulatus; Ferris, 1941e: 45. Change of combination and rank.
Odonaspis secreta; Borchsenius, 1966: 226. Revived combination.
Odonaspis senireta; Wang, Varma & Xu, 1998: 86. Misspelling of species name.
COMMON NAME: skritaya bambukovaya shitovka [Borchs1966].
SYSTEMATICS: Takagi (1987, p.8-9) suggested that *Parlatoria zeylanica* Rutherford 1915 (page 113) (supposed by Rutherford to have been described as an adult female), was the second-instar male of an *Odonaspis* species (*O. secreta* or *O. greeni*). Besides other characters described by Rutherford (1915), three features -

absence of perivulvar pores, host plant bamboo, as well as "Associated with *Chionaspis simplex* Gr., and *Aspidiotus secretus* ... " - seem to be good reasons to accept Takagi's conclusion. Dr. Takagi (in letter, 30 August 2002, to Y. Ben-Dov), noted that Rutherford collection probably was lost, and re-iterated the conclusion that *Parlatoria zeylanica* is a synonym of *Odonaspis secreta* (Cockerell).

SCALE COVER: Female scale white, flat, broadly elongate, exuviae apical; scale of the male white, elongate, slender (Ferris, 1938a). Colour photograph of scale cover by Wong *et al.* (1999).

HOST PLANTS: **Gramineae**: *Arundinaria* [Green1896e, MacGil1921, Kuwana1933, Borchs1934, Borchs1937, BenDov1988b, Danzig1993], *Arundinaria japonica* [Kuwana1933, BenDov1988b], *Arundinaria simonii* [Beards1966, KawaiMaUm1971, BenDov1988b], *Bambusa* [Maskel1897a, Borchs1934, Takagi1959a, Takagi1970, BenDov1988b, WilliaWa1988, Danzig1993], *Bambusa fastuosa* [Balach1930a, Balach1932d, Balach1951, BenDov1988b], *Bambusa metake* [Lindin1912b, BenDov1988b], *Bambusa spinosa* [Lindin1912b], *Bambusa stenostachya* [Takaha1929, Takagi1970], *Bambusa tessellata* [BenDov1988b], *Bambusa veitchi-argenta* [BenDov1988b], *Miscanthus* [Takagi1970], *Phyllostachys hindsii* [BenDov1988b], *Pleioblastus* [Kawai1977], *Pleioblastus chino* [KawaiMaUm1971], *Sasa* [Borchs1950b, Danzig1993], *Sasa albo-marginata* [TakahaTa1956], *Sasa paniculata* [BenDov1988b], *Schizostachyum glaucifolium* [WilliaWa1988].

NATURAL ENEMIES: HYMENOPTERA **Aphelinidae**: *Aphelosoma plana* Nikolskaya [BenDov1988b, Viggia1990a], *Encarsia citrina* (Crawford) [BenDov1988b], *Physcus odonaspidis* Tachikawa [BenDov1988b], *Proaphelinoides bendovi* Tachikawa [BenDov1988b], *Proaphelinoides elongatiformis* Girault [BenDov1988b], *Proaphelinoides mirus* (Nikolskaya) [BenDov1988b]. **Encyrtidae**: *Caenohomalopoda koreana* Tachikawa, Paik & Paik [BenDov1988b, Trjapi1989], *Caenohomalopoda shikokuensis* (Tachikawa) [Trjapi1989]. THYSANOPTERA **Phlaeothripidae**: *Podothrips odonaspicola* (Kurosawa) [BenDov1988b, PalmerMo1990], *Podothrips sasacola* Kurosawa [BenDov1988b, PalmerMo1990].

DISTRIBUTION: **Australasian**: Australia [Leonar1897b]; Bonin Islands (= Ogasawara-Gunto) [Kuwana1909a, TakahaTa1956, KawaiMaUm1971, Kawai1987]; Hawaiian Islands (Hawaii [MacGil1921, Fullaw1932, Kuwana1933, Borchs1937, Ferris1938a]); Samoa [DoaneFe1916, Ferris1938a, Takagi1970]; Western Samoa [Laing1927]. **Nearctic**: United States of America (Louisiana [BenDov1988b], New Jersey [BenDov1988b]). **Neotropical**: Cuba [Houser1918, Ferris1938a]; Guadeloupe [Balach1957c]. **Oriental**: China (People's Republic) (Yunnan [Ferris1953]); Indonesia (Java [MacGil1921, Borchs1937]); Pakistan [BenDov1988b]; Sri Lanka [Green1896e, Leonar1897b, MacGil1921, Kuwana1933, Borchs1937, Ferris1938a, Takagi1970]; Taiwan [Takaha1929, TakahaTa1956, Takagi1970, WongChCh1999]. **Palaearctic**: Algeria [Lindin1912b, Borchs1937, Takagi1970]; China (People's Republic) (Xizang (Tibet) [BenDov1988b]); Corsica [Balach1932d]; Czech Republic [Zahrad1965, Zahrad1977]; France [Balach1930a, Balach1932d, Takagi1970, BenDov1988b]; Georgia (Abkhaz ASSR [Borchs1936, Borchs1937, Hadzib1983], Adzhar ASSR [Borchs1934, Borchs1936, Borchs1937,

Hadzib1983]); Iran [Kaussa1955]; Japan [Cocker1896h, Maskel1897a, Leonar1897b, Cocker1899j, Cocker1900f, Kuwana1933, Ferris1938a, Takagi1970] (Honshu [TakahaTa1956, Takagi1959a], Kyushu [TakahaTa1956], Shikoku [TakahaTa1956]); Russia (Caucasus [Borchs1950b, Danzig1993], St. Petersberg (= Leningrad) Oblast [Bustsh1958]); Ukraine (Krym (= Crimea) Oblast [Danzig1993]).

BIOLOGY: Occurring beneath sheathing bases of leaves, usually attached to inner surface of leaf rather than to stem and very commonly buried within tissues of leaf.

GENERAL: Description and illustration of adult female by Kuwana (1933), Ferris (1938a), Balachowsky (1953g), Chou (1985, 1986), Tereznikova (1986) and by Ben-Dov (1988b). Description and illustration of adult male by Bustshik (1958).

KEYS: Ben-Dov 1988b: 34 (female) [World]; Williams & Watson 1988: 2 (female) [Tropical South Pacific]; Chou 1985: 328 (female) [Species of China]; Kawai 1980: 201 (female) [Japan]; Beardsley 1966: 525 (female) [Federated States of Micronesia]; Takagi 1959a: 93 (female) [Japan]; Balachowsky 1953g: 733 (female) [Mediterranean]; Ferris 1942: 65 (female) [North America]; Kuwana 1933: 35 (female) [Japan]; Kuwana 1933b: 49 (female) [Japan]; Fullaway 1932: 95-97, 108 (female) [Hawaii]; Green 1896e: 40 (female) [Sri Lanka].

CITATIONS: Balach1930a [host, distribution: 179]; Balach1932d [taxonomy, host, distribution: XLVIII]; Balach1953a [taxonomy: 20]; Balach1953g [taxonomy, description, illustration, host, distribution: 733-736]; Balach1957c [host, distribution: 202]; Balach1958b [taxonomy: 304]; Beards1966 [host, distribution: 528]; BenDov1988b [taxonomy, description, illustration, host, distribution, biological control: 60-61,71-72,129]; BenDov1990c [taxonomy: 122]; BenDov1990e [host, distribution: 658]; Borchs1934 [host, distribution: 30]; Borchs1935a [taxonomy: 32]; Borchs1936 [host, distribution: 135]; Borchs1937 [taxonomy, description, illustration, host, distribution: 135]; Borchs1937a [taxonomy, description, host, distribution: 54-55]; Borchs1950b [taxonomy, description, illustration, host, distribution: 211,219]; Borchs1966 [catalogue: 198,199,226]; Box1953 [host, distribution, biological control: 52]; Bustsh1958 [taxonomy, description, illustration, host, distribution: 213]; Chou1985 [taxonomy, description, host, distribution: 328-330]; Chou1986 [taxonomy, illustration: 700]; Cocker1896b [distribution: 333]; Cocker1896h [taxonomy, description, host, distribution: 20]; Cocker1896i [taxonomy: 51]; Cocker1897i [taxonomy, description, host, distribution: 20]; Cocker1899j [distribution: 274]; Cocker1900f [description, host, distribution: 72]; Danzig1964 [taxonomy, host, distribution: 650]; Danzig1972 [taxonomy, host, distribution, economic importance: 218]; Danzig1993 [taxonomy, description, illustration, host, distribution: 132-133]; DanzigPe1998 [catalogue: 319]; DEDAC1923 [host, distribution]; DeitzTo1980 [taxonomy: 39]; DoaneFe1916 [host, distribution: 401]; FangWuXu2001 [host, distribution: 108]; Fernal1903b [catalogue: 300]; Ferris1937a [taxonomy: 33,39]; Ferris1938a [taxonomy, description, illustration, host, distribution: 166]; Ferris1941e [taxonomy: 45,48]; Ferris1942 [description, distribution: 446: 65]; Ferris1950a [taxonomy: 75]; Ferris1953 [host, distribution: 65]; Foldi2001 [distribution: 303-308]; Fullaw1932 [taxonomy, distribution: 95,108]; Green1896a [taxonomy: 84]; Green1896e [taxonomy, description, illustration, host, distribution: 64-65]; Green1922 [taxonomy: 460]; Green1922a [taxonomy: 1020]; Green1937 [host, distribution:

335]; Hadzib1983 [host, distribution, life history, biological control, economic importance: 215]; Houser1918 [description, host, distribution: 169]; HowardAs1895 [biological control: 633]; HowellTi1977 [taxonomy, description: 123]; Kaussa1955 [host, distribution: 17]; Kawai1977 [host, distribution, economic importance: 163]; Kawai1980 [taxonomy, description, host, distribution: 201]; Kawai1987 [host, distribution: 78]; KawaiMaUm1971 [host, distribution: 18]; Kozar1990a [life history, economic importance: 341-347]; Kuwana1902 [host, distribution: 66]; Kuwana1907 [host, distribution: 193-194]; Kuwana1909a [host, distribution: 160]; Kuwana1917a [taxonomy, distribution: 176]; Kuwana1933 [taxonomy, description, illustration, host, distribution: 36-37]; Leonar1897 [taxonomy: 286]; Leonar1897a [taxonomy: 375]; Leonar1897b [taxonomy, description, illustration, host, distribution: 115-117]; Leonar1900 [taxonomy: 338]; Lindin1912b [description, host, distribution: 81]; Lindin1935 [taxonomy: 141]; Lindin1937 [taxonomy: 184]; MacGil1921 [taxonomy, description, host, distribution: 423]; Maskel1897a [taxonomy, host, distribution: 241]; Maskel1898 [taxonomy, description, host, distribution: 224]; McKenz1945 [taxonomy: 73]; MillerDa1990 [host, distribution, economic importance: 304]; Morley1909 [host, distribution, biological control: 254-257]; Muraka1970 [host, distribution: 69]; Nakaha1982 [host, distribution: 63-64]; PalmerMo1990 [biological control: 67-76]; Ramakr1921a [host, distribution: 357]; Ruther1915 [taxonomy, description, host, distribution: 113-114]; SchmutKlLu1957 [host, distribution, economic importance: 499]; Takagi1959a [taxonomy, host, distribution: 94]; Takagi1970 [taxonomy, host, distribution: 138-139]; Takagi1987 [taxonomy: 8-9]; Takagi1990b [taxonomy, structure: 19-20]; Takaha1929 [host, distribution: 78]; TakahaTa1956 [host, distribution: 17]; Tao1999 [taxonomy, host, distribution: 102]; Terezn1986 [taxonomy, description, illustration, host, distribution: 118-119]; Trjapi1989 [biological control: 295]; Viggia1990a [biological control: 123]; WangVaXu1998 [taxonomy, host, distribution: 86]; Willia1985a [taxonomy: 238]; WilliaWa1988 [taxonomy, host, distribution: 201]; WongChCh1999 [taxonomy, description, host, distribution: 29,71]; Zahrad1965 [taxonomy, description, host, distribution: 305]; Zahrad1977 [taxonomy, distribution: 120].

Odonaspis serrata Ben-Dov

Odonaspis serrata Ben-Dov, 1988b: 61. Type data: SRI LANKA: Punduloya, on *Arundinaria* sp.; collected March, 1897. Holotype female. Type depository: Washington: United States National Entomological Collection, U.S. National Museum of Natural History, District of Columbia, USA.
SCALE COVER: Only slide-mounted females available for description by Ben-Dov (1988b).
HOST PLANTS: **Gramineae**: *Arundinaria* [BenDov1988b], *Bambusa* [BenDov1988b].
DISTRIBUTION: Oriental: Sri Lanka [BenDov1988b]; Vietnam [BenDov1988b].
GENERAL: Description and illustration of adult female by Ben-Dov (1988b).
KEYS: Ben-Dov 1988b: 33 (female) [World].
CITATIONS: BenDov1988b [taxonomy, description, illustration, host, distribution: 61-62,130]; BenDov1990e [host, distribution: 658].

Odonaspis siamensis (Takahashi)

Froggattiella siamensis Takahashi, 1942b: 50. Type data: THAILAND: Bangkok, on bamboo. Lectotype female, by subsequent designation Ben-Dov, 1988b: 62. Type depository: Taichung: Entomology Collection, Taiwan Agricultural Research Institute, Wu-feng, Taichung, Taiwan

Odonaspis siamensis; Ferris, 1955c: 33. Change of combination.

Troggattiella siamensis; Lindinger, 1957: 549. Misspelling of genus name.

SCALE COVER: Scale white, with yellowish larval skins (Takahashi, 1942b).

HOST PLANTS: **Gramineae**: *Bambusa* [Takaha1942b, Ferris1955c, BenDov1988b], *Bambusa nana* [BenDov1988b], *Dendrocalamus* [BenDov1988b], *Dendrocalamus latiflorus* [BenDov1988b].

DISTRIBUTION: **Oriental**: China (People's Republic) (Guangdong (Kwangtung) [Ferris1955c, BenDov1988b]); Philippines [BenDov1988b]; Thailand [Ferris1955c, Takaha1942b, Ferris1955c, Borchs1966, BenDov1988b].

GENERAL: Description and illustration of adult female by Takahashi (1942b), Chou (1985, 1986) and by Ben-Dov (1988b).

KEYS: Ben-Dov 1988b: 33 (female) [World]; Chou 1985: 328 (female) [Species of China].

CITATIONS: BenDov1988b [taxonomy, description, illustration, host, distribution: 62-63,131]; BenDov1990e [host, distribution: 658]; Borchs1966 [catalogue: 228]; Chou1985 [taxonomy, description, host, distribution: 328,333]; Chou1986 [taxonomy, illustration: 706]; FangWuXu2001 [host, distribution: 108]; Ferris1955c [taxonomy, host, distribution: 33]; Lindin1957 [taxonomy: 549]; Takaha1942b [taxonomy, description, illustration, host, distribution: 50-51]; Tao1999 [taxonomy, host, distribution: 102].

Odonaspis stipagrostis Ben-Dov

Odonaspis stipagrostis Ben-Dov, 1988b: 63. Type data: SOUTH AFRICA: Cape Province, 65 km west of Upington, at highway to Karasburg, on *Stipagrostis namaquensis*; collected Y. Ben-Dov, 10.IX.1974. Holotype female. Type depository: Pretoria: South African National Collection of Insects, South Africa.

SCALE COVER: Female scale oval, 1.0-1.6 mm long, 0.8-1.1 mm wide; white; larval exuviae yellow brown placed centrally. Scale of male similar in shape and colour to female but smaller (Ben-Dov, 1988b).

HOST PLANTS: **Gramineae**: *Stipagrostis namaquensis* [BenDov1988b].

DISTRIBUTION: **Afrotropical**: South Africa [BenDov1988b].

GENERAL: Description and illustration of adult female by Ben-Dov (1988b).

KEYS: Ben-Dov 1988b: 33 (female) [World].

CITATIONS: BenDov1988b [taxonomy, description, illustration, host, distribution: 63,132].

Odonaspis texana Ben-Dov

Odonaspis texana Ben-Dov, 1988b: 63. Type data: U.S.A.: Brewster, 8 miles north of Marathon, Texas, on *Bouteloua* sp.; collected 12.v.1976. Holotype female. Type depository: Washington: United States National Entomological Collection, U.S. National Museum of Natural History, District of Columbia, USA.

SCALE COVER: Only slide-mounted females available for description by Ben-Dov (1988b).
HOST PLANTS: **Gramineae**: *Bouteloua* [BenDov1988b].
DISTRIBUTION: **Nearctic**: United States of America (Texas [BenDov1988b]).
GENERAL: Description and illustration of adult female by Ben-Dov (1988b).
KEYS: Ben-Dov 1988b: 33 (female) [World].
CITATIONS: BenDov1988b [taxonomy, description, illustration, host, distribution: 63-64,133].

Odonaspis transkeiensis Ben-Dov
Odonaspis transkeiensis Ben-Dov, 1988b: 64. Type data: SOUTH AFRICA: Transkei, Port St. Johns, on *Dactyloctenium australe*; collected Y. Ben-Dov, 2.VII.1973. Holotype female. Type depository: Pretoria: South African National Collection of Insects, South Africa.
SCALE COVER: Female scale elongate oval; about 2 mm long, 1 mm wide; white greyish; larval exuviae brown placed at anterior part of scale. Male scale oval and narrow, about 1 mm long, 0.4 mm wide; white greyish; larval exuviae brown placed at anterior part of scale (Ben-Dov, 1988b).
HOST PLANTS: **Gramineae**: *Dactyloctenium australe* [BenDov1988b], *Digitaria* [BenDov1988b], *Stenotaphrum secundatum* [BenDov1988b].
DISTRIBUTION: **Afrotropical**: South Africa [BenDov1988b].
GENERAL: Description and illustration of adult female by Ben-Dov (1988b).
KEYS: Ben-Dov 1988b: 34 (female) [World].
CITATIONS: BenDov1988b [taxonomy, description, illustration, host, distribution: 64-65,134].

Odonaspis tsinjoarivensis Mamet
Odonaspis tsinjoarivensis Mamet, 1954: 19. Type data: MADAGASCAR: Tsinjoarivo, on climbing bamboo. Holotype. Type depository: Paris: Muséum national d'Histoire naturelle, France.
SCALE COVER: Scale cover not described by Mamet (1954).
HOST PLANT: **Gramineae**: *Bambusa* [Mamet1954, Borchs1966, BenDov1988b].
DISTRIBUTION: **Afrotropical**: Madagascar [Mamet1954, Borchs1966, BenDov1988b].
GENERAL: Description and illustration of adult female by Mamet (1954) and by Ben-Dov (1988b).
KEYS: Ben-Dov 1988b: 34 (female) [World].
CITATIONS: BenDov1988b [taxonomy, description, illustration, host, distribution: 65,135]; BenDov1990e [host, distribution: 658]; Borchs1966 [catalogue: 226]; Mamet1954 [taxonomy, description, illustration, host, distribution: 19,67-70].

TAXA TRANSFERRED FROM FAMILY

Aspidiotus bicarinatus Walker
Aspidiotus bicarinatus Walker, 1858: 306.
Current status: Following the interpretation of Fernald (1903b: 330) and Borchsenius (1966: 380), this species is regarded a larva of Lepidoptera.

Aspidiotus cryptus Signoret
Aspidiotus cryptus Signoret, 1869: 850.
Current status: Signoret (1870: 108) suggested that his 1869 species was an hymenopteron. Fernald (1903b: 330) formally removed it from the family. This binomen is accepted as an unrecognizable taxon that has been most likely described from eggs of an insect or a mite.

Aspidiotus gossypii Fitch
Aspidiotus gossypii Fitch, 1857f: 332.
Current status: *Aleyrodes gossypii* (Fitch, 1857) in the whitefly family, Aleyrodidae; see Borchsenius (1966) and Mound & Halsey (1978).

Aspidiotus hordeolum Walker
Aspidiotus hordeolum Walker, 1852: 1068.
Current status: *Coccus hordeolum* (Dalman, 1826) in the family Coccidae.

Aspidiotus luzulae Dufour
Aspidiotus luzulae Dufour, 1864: 208.
Current status: *Luzulaspis luzulae* (Dufour, 1864) in the family Coccidae.

REFERENCES

Abai, M. 1995. The report of *Nuculaspis abietis* (Schrank) from Mazandaran, Iran. *Applied Entomology and Phytopathology* **62**: 26, 108-109. [Abai1995]

Abbas, M.S.T. 1992. Comparative rates of infestation by four scale insects on citrus trees with special reference to rates of parasitism on the purple scale *Lepidosaphes beckii*. *Egyptian Journal of Agricultural Research* **70**: 477-485. [Abbas1992]

Abbott, W.S. 1926. Determining the effectiveness of dormant treatments against the San José scale. *Journal of Economic Entomology* **19**: 858-860. [Abbott1926]

Abbott, W.S., Culver, J.J. & Morgan, W.J. 1926. Effectiveness against the San José scale of the dry substitutes for liquid lime-sulphur. *United States Department of Agriculture Bulletin* **1371**: 1-26. [AbbottCuMo1926]

Abd El-Kareim, A.I., Darvas, B. & Kozár, F. 1988. Effects of the juvenoids fenoxycarb, hydroprene, kinoprene and methoprene on first instar larvae of *Epidiaspis leperii* Sign. (Hom., Diaspididae) and on its ectoparasitoid, *Aphytis mytilaspidis* (Le Baron) (Hym., Aphelinidae). *Journal of Applied Entomology* **106**: 270-275. [AbdElKDaKo1988]

Abd El-Kareim, A.I. & Kozár, F. 1988a. [Host plants in relation to the morphology and reproductive biology of *Aspidiotus nerii*, with description of the hypersensitive reaction of the host plant.].[In Hungarian]. *Kertgazdasag* **20(1)**: 55-60. [AbdElKKo1988a]

Abd-Rabou, S. 2001a. An annotated list of the Hymenopterous parasitoids of the Diaspididae (Hemiptera: Coccoidea) in Egypt, with new records. *Entomologica* **33(1999)**: 173-177. [AbdRab2001a]

Abdel-Fattah, M.I., El-Minshawy, A.M. & Darwish, E.T. 1978a. The seasonal abundance of two scale insects *Lepidosaphes beckii* (New.) and *Aonidiella aurantii* Mask. infesting citrus trees in Egypt. *Proceedings of the 4th Conference on Pest Control. Cairo* **1**: 74-84. [AbdelFElDa1978a]

Abdel-Rahman, A.G. 1995. Seasonal abundance of some pests attacking olives and their control under El-Qasr conditions, Matrouh Governorate. *Annals of Agricultural Science (Moshtohor)* **33**: 1553-1564. [AbdelR1995]

Abdelrahman, I. 1973. Toxicity of malathion to the natural enemies of California red scale, *Aonidiella aurantii* (Mask.) (Hemiptera: Diaspididae). *Australian Journal of Agricultural Research* **24**: 119-133. [Abdelr1973]

Abdelrahman, I. 1974. The effect of extreme temperatures on California red scale, *Aonidiella aurantii* (Mask.) (Hemiptera, Diaspididae), and its natural enemies. *Australian Journal of Zoology. Melbourne* **22**: 203-212. [Abdelr1974]

Abdelrahman, I. 1974a. Growth, development and innate capacity for increase in *Aphytis chrysomphali* Mercet and *A. melinus* DeBach, parasites of California red scale, *Aonidiella aurantii* (Mask.), in relation to temperature. *Australian Journal of Zoology. Melbourne* **22**: 213-230. [Abdelr1974a]

Abdelrahman, I. 1974b. Studies in ovipositional behaviour and control of sex in *Aphytis melinus* DeBach, a parasite of California red scale, *Aonidiella aurantii* (Mask.). *Australian Journal of Zoology. Melbourne* **22**: 231-247. [Abdelr1974b]

Ables, J.R. & Ridgway, R.L. 1981. Augmentation of entomophagous arthropods to control pest insects and mites. Pages 273-303. *in*: Beltsville Symposia in Agricultural Research. 5. Biological control in crop production. Totowa, N.J.: Allanheld, Osmun & Co.. 461 pp. [AblesRi1981]

Abou-Elkhair, S. 2001. Scale insects (Hemiptera: Coccidae) and their parasitoids on ornamental plants in Alexandria, Egypt. *Entomologica* **33(1999)**: 185-195. [AbouEl2001]

Ackerman, A.J. 1923. Preliminary report on control of San José scale with lubricating-oil emulsion. *Circular (United States Department of Agriculture)* **No. 263**: 18 pp. [Ackerm1923]

Adachi, M.S. & Fullaway, D.T. 1953. Two new diaspidid scales on *Auracaria*. *Proceedings of the Hawaiian Entomological Society* **15**: 87-91. [AdachiFu1953]

Agarwala, S.B.D. 1956. *Melanaspis glomerata* (Green) - A new coccid pest of sugarcane. *Proceedings of the Bihar Academy of Agricultural Sciences* **5**: 24-31. [Agarwa1956]

Agarwal, R.A. 1969. Morphological characteristics of sugarcane and insect resistance. *Entomologia Experimentalis et Applicata* **12**: 767-776. [Agarwa1969]

Agarwal, R.A. & Sharma, D.P. 1960. Studies on some epidermal characters and hardiness of sugarcane stem in relation to the incidence of scale insect; *Melanaspis glomerata* (Green). *Indian Journal of Entomology* **22**: 197-203. [AgarwaSh1960]

Agarwal, R.A., Sharma, D.P. & Kandazwamy, P.A. 1959. Feeding habit of sugarcane scale (*Targionia glomerata* Green). *Current Science (India)* **28**: 462-463. [AgarwaShKa1959]

Agarwal, R.A. & Siddiqi, Z.A. 1964. Sugarcane pests. Pages 149-186 *in*: Pant, N.C., Chief Editor. Entomology in India ["A special number of the Indian Journal of Entomology"]. New Delhi: The Entomological Society of India. 529 pp. [AgarwaSi1964]

Aguilar F., P.G. 1980. Apuntes sobre el control biológico y el control integrado de las plagas agrícolas en el Perú: introducción. *Revista Peruana de Entomología* **23**: 83-91. [Aguila1980]

Aguilar F., P.G., Salazar T., J. & Núñez, E. 1980. Apuntes sobre el control biológico y el control integrado de las plagas agrícolas en el Perú: el cultivo de cítricos. *Revista Peruana de Entomología* **23**: 97-100. [AguilaSaNu1980]

Aguilera P., A. 1970. Observaciones sobre *Melanaspis sitreara* (Hempel) (Hom.: Diaspididae). *Idesia* **1**: 111-133. [Aguile1970]

Aguilera P., A., Dias P., G. & Graña S., F. 1981. Nivel de ataque de las escamas blancas del olivo (Homoptera: Diaspididae en el valle de Azape (Arica, Chile). *Revista Peruana de Entomología* **24(1)**: 175-178. [AguileDiGr1981]

Aguilera P., A.., Mendoza, M.R., Vargas, C.H. & Diaz, P.G. 1984. [New contribution on predatory activity of *Coccidophilus citricola* (Coleoptera: Coccinellidae).].[In Spanish]. *Idesia* **8**: 47-54. [AguileMeVa1984]

Ahmad, R. & Ghani, M.A. 1966. Biology of *Chilocorus infernalis* Muls. *Technical Bulletin of the Commonwealth Institute of Biological Control* **7**: 101-106 [AhmadGh1966]

Aisagbonhi, C.I., Nwana, I.E. & Agwu, S.I. 1985. Preliminary analysis of a field population of *Aspidiotus destructor* Signoret (Homoptera: Diaspididae) and some soft scales on coconut palms. *Nigerian Journal of Entomology* **6**: 24-32. [AisagbNwAg1985]

Akinlosotu, T.A. & Kogbe, J.O.S. 1988. Studies on the incidence of yam scale, *Aspidiella hartii* on *Dioscorea* spp. and its chemical control. *Journal of Root Crops* **14(2)**: 21-23. [AkinloKo1988]

Alam, S.M. 1957. The taxonomy of some British encyrtid parasites (Hymenoptera) on scale insects (Coccoidea). *Transactions of the Royal Entomological Society of London* **109**: 421-466. [Alam1957]

Alam, M.Z. & Sattar, A. 1965. On the biology of citrus yellow scale *Aonidiella citrina* Coquillett in East Pakistan. Pages 161-171 *in*: Alam, M.Z., et al., Eds. A Review of Research Division of Entomology (1947-1964). Dacca: Agr. Inform. Serv.. [AlamSa1965]

Alden, C.H. 1925. San José scale control with lubricating oil emulsion on peach trees in the south. *Journal of Economic Entomology* **18(2)**: 253-257. [Alden1925]

Alderdice, M., Spino, C. & Weiler, L. 1984. Synthesis of the three isomeric components of San José scale pheromone. *Tetrahedron Letters* **25(16)**: 1643-1646. [AlderdSpWe1984]

Aldrich, J.M. 1899. The San José scale in Idaho. *Bulletin of the Idaho Agricultural Experiment Station* **16**: 1-16. [Aldric1899]

Aldrich, J.R. 1996. Sex pheromones in Homoptera and Heteroptera. Pages 199-233 *in*: Schaefer, C.W., Ed. Thomas Say Publications in Entomology. Proceedings. Studies on Hemipteran Phylogeny. Lanham, MD: Entomological Society of America. [Aldric1996]

Aleksidze, G. 1995. Armored scale insects (Diaspididae), pests of fruit orchards and their control in Republic of Georgia. *Israel Journal of Entomology* **29**: 187-190. [Aleksi1995]

Alencar, J.A. de 2000. Pragas potenciais presentes no agroecossistema do coqueiro anao irrigado no submedio do vale do Sao Francisco. *Documentos da Embrapa Semi Arido* **No. 152**: 12 pp. [Alenca2000]

Alexandrakis, V. 1980. Essai d'appréciation des dégâts provoqués sur oranger en Crète par la présence d'*Aonidiella aurantii* (Mask.) (Hom. Diaspididae). *Fruits* **35**: 555-560. [Alexan1980]

Alexandrakis, V. 1983. Données biologiques sur *Aonidiella aurantii* Mask. (Hom. Diaspididae) sur agrumes en Crète. *Fruits* **38**: 831-838. [Alexan1983]

Alexandrakis, V. 1990. Effect of *Dacus* control sprays, by air or ground, on the ecology of *Aspidiotus nerii* Bouché (Hom. Diaspididae). *Acta Horticulturae* **(286)**: 339-342. [Alexan1990]

Alexandrakis, V. & Bénassy, C. 1979. Essai d'appréciation des dégâts provoqués sur olivier en Crète par la présence d'*Aspidiotus nerii* Bouché (Homoptera, Diaspididae). *Revue de Zoologie Agricole et de Pathologie Végétale* **78**: 49-56. [AlexanBe1979]

Alexandrakis, V. & Bénassy, C. 1981. Essai de lutte biologique sur olivier en Crète par utilisation d'*Aphytis melinus* parasite d'*Aspidiotus nerii*. *Acta Oecologica Applicata* **2**: 13-25. [AlexanBe1981]

Alexandrakis, V. & Bénassy, C. 1982. [Influence of the host plant (olive) on the population dynamics of *Aspidiotus nerii* (Homoptera, Diaspididae).]. *Agronomie* **2**: 843-850. [AlexanBe1982]

Alexandrakis, V. & Michelakis, S. 1982. Distribution d'*Aonidiella aurantii* (Mask.) (Hom. Diaspididae) en fonction de son emplacement sur l'arbre et de la variété d'agrumes en Crète. *Fruits* **35**: 639-644. [AlexanMi1982]

Alexandrakis, V. & Neuenschwander, P. 1979. Influence de la poussière des chemins sur *Aspidiotus nerii* Bouché) Hom. Diaspididae) et son parasite principal, *Aphytis chilensis* How. (Hym., Aphelinidae), observés sur Olivier.*Annales de Zoologie - Ecologie Animale* **11**: 171-184. [AlexanNe1979]

Alexandrakis, V. & Neuenschwander, P. 1980. Le rôle d'*Aphytis chilensis*, parasite d'*Aspidiotus nerii* sur olivier en Crète. *Entomophaga* **25**: 61-71. [AlexanNe1980]

Alfieri, A. 1929. Les principaux insectes nuisibles infestant le jardin de Nouzha. *Bulletin de la Société Royale Entomologique d'Egypte* **13**: 7-9. [Alfier1929]

Ali, S.M. 1962. Coccids affecting sugarcane in Bihar (Coccidae: Hemiptera). *Indian Journal of Sugarcane Research and Development* **6**: 72-75. [Ali1962]

Ali, S.M. 1967a. Description of a new and records of some known coccids (Homoptera) from Bihar, India. *Oriental Insects. New Delhi* **1**: 29-43. [Ali1967a]

Ali, S.M. 1968. Coccids (Coccoidea: Hemiptera: Insecta) affecting fruit plants in Bihar (India). *Journal of the Bombay Natural History Society* **65**: 120-137. [Ali1968]

Almeida, D.M. de. 1969. Contribuicao para o estudo da quermofauna de Angola. *Boletim do Instituto de Investigacao Cientifica de Angola, Luanda* **6**: 129-164. [Almeid1969]

Almeida, D.M. de. 1971. Diaspididae (Homoptera - Coccoidea) of Mozambique, with the description of a new species. *Novos Taxa Entomológicos* **90**: 3-17. [Almeid1971]

Almeida, D.M. de. 1973. *Felixiella* n. gen. and other Diaspididae (Homoptera: Coccoidea) of Mozambique. *Novos Taxa Entomológicos* **100**: 3-8. [Almeid1973]

Almeida, D.M. de. 1973b. Coccoidea de Angola. 1- Revisao das especies conhecidas. *Boletim do Instituto de Investigacao Cientifica de Angola, Luanda* **10**: 1-23. [Almeid1973b]

Alstad, D.N. 1998. Population structure and the conundrum of local adaptation. Pages 3-21. *in*: Mopper, S. & Strauss, S.Y., Eds. Genetic Structure and Local Adaptation in Natural Insect Populations: Effects of Ecology, Life History, and Behavior. New York: Chapman & Hall. xix + 449 pp. [Alstad1998]

Alstad, D.N. & Edmunds, G.F. 1983. Selection, outbreeding depression, and the sex ratio of scale insects. *Science* **220**: 93-95. [AlstadEd1983]

Alstad, D.N. & Edmunds, G.F. 1983a. Adaptation, host specificity and gene flow in the black pineleaf scale. Pages 413-426 *in*: Denno, Robert F. & McClure, Mark S., eds. Variable Plants and Herbivores in Natural and Managed Systems. New York: Academic Press. [AlstadEd1983a]

Alstad, D.N. & Edmunds, G.F. 1987. Black pineleaf scale (Homoptera: Diaspididae) population density in relation to interdemic mating (Hemiptera: Diaspididae). *Annals of the Entomological Society of America* **80(5)**: 652-654. [AlstadEd1987]

Alstad, D.N. & Edmunds, G.F. 1989. Haploid and diploid survival differences demonstrate selection in scale insect demes. *Evolutionary Ecology* **3**: 253-263. [AlstadEd1989]

Alstad, D.N., Edmunds, G.F. & Johnson, S.C. 1980. Host adaptation, sex ratio, and flight activity in male black pineleaf scale. *Annals of the Entomological Society of America* **73**: 665-667. [AlstadEdJo1980]

Alvarado, J.A. 1939. Los insectos dañinos y los insectos auxiliares de la agricultura en Guatemala. Guatemala, C.A.. 285+ pp. [Alvara1939]

Alvarez, J.M. & Van Driesche, R. 1998. Biology of *Cybocephalus* sp. nr. *nipponicus* (Coleoptera: Cybocephalidae), a natural enemy of euonymus scale (Homoptera: Diaspididae). *Environmental Entomology* **27**: 130-136. [AlvareVa1998]

Alwood, W.B. 1896. The San José or pernicious scale (*Aspidiotus perniciosus*). *Bulletin of the Virginia Agricultural Experiment Station* **62**: 33-44. [Alwood1896]

Amin, A.H. 1981. An estimation of the population density of California red scale, *Aonidiella aurantii* (Mask.). *Research Bulletin, Ain Shams University, Faculty of Agriculture* **1537**: 1-11. [Amin1981]

Amin, A.H., Risk Madiha, A. & Sakr, H.E.A. 2001. Factors responsible for the extinction of *Chrysomphalus aonidum* (L.) from citrus orchards in Egypt. *Entomologica* **33(1999)**: 441. [AminRiSa2001]

Amos, J.M. 1933. A list of the Coccidae of Tippecanoe County and their hosts (Homoptera). *Proceedings of Indiana Academy of Science* **42**: 205-208. [Amos1933]

Amos, J.M. 1933a. Descriptions of Coccidae heretofore unreported from Indiana (Homoptera). *Proceedings of Indiana Academy of Science* **42**: 208-211. [Amos1933a]

Anderson, R.J., Adams, K.G., Chinn, H.R. & Henrick, C.A. 1980. Synthesis of the optical isomers of 3-methyl-6-isopropenyll-9-decen-1-yl acetate, a component of the California red scale pheromone. *Journal of Organic Chemistry* **45**: 2229-2236. [AndersAdCh1980]

Anderson, R.J., Chinn, H.R., Gill, K. & Henrick, C.A. 1979. Synthesis of 7-methyl-s-methylene-7-octen-1-yl propanoate and (Z)-3,7-dimethyl-2,7-octadien-1-yl propanoate, components of the sex pheromone of the San José scale. *Journal of Chemical Ecology* **5(6)**: 919-927. [AndersChGi1979]

Anderson, R.J., Gieselmann, M.J., Chinn, H.R., Adams, K.G., Henrick, C.A., Rice, R.E. & Roelofs, W.L. 1981. Synthesis and identification of a third component of the San José scale sex pheromone. *Journal of Chemical Ecology* **7**: 695-706. [AndersGiCh1981]

Anderson, R.J. & Henrick, C.A. 1979. Synthesis of (*E*)-3,9-dimethyl-6-isopropy-5,8-decadien-1-yl acetate, the sex pheromone of the yellow scale. *Journal of Chemical Ecology* **5(5)**: 773-779. [AndersHe1979]

André, M. 1942. Sur l'*Hemisarcoptes alus* Shimer (=*Coccisugus* Lignières) (Acarien). *Bulletin du Muséum d'Histoire Naturelle* **14(93)**: 173-180. [Andre1942]

Andries, H.E. [Ed.] 1932. Controlling Plant Pests in Southern Africa. Johannesburg: Cooper& Nephews. [215] pp. [Andrie1932]

Angerilli, N.P.D. 1990. 3.5.3 Chemical Control of Males. Pages 409-411 *in*: Rosen, D. (Ed.). Armored Scale Insects, Their Biology, Natural Enemies and Control [Series title: World Crop Pests, Vol. 4B]. Amsterdam, the Netherlands: Elsevier. 688 pp. [Angeri1990]

Angerilli, N.P.D., Gaunce, A.P. & Logan, D.M. 1986. Some effects of post-harvest fumigation, controlled-atmosphere storage, and cold storage on San José scale (Homoptera: Diaspididae) survival on two varieties of apples. *Canadian Entomologist* **118**: 493-497. [AngeriGaLo1986]

Angerilli, N.P.D. & Logan, D.M. 1985. Early season apple pest management: control of two species of scales (Homo.: Diaspididae) and Bruce Spanworm (Lep.: Geometridae) with methidathion. *Journal of the Entomological Society of British Columbia* **82**: 31-35. [AngeriLo1985]

Angerilli, N.P.D. & Logan, D.M. 1986. The use of pheromone and barrier traps to monitor San José scale (Homoptera: Diaspididae) phenology in the Okanagan Valley of British Columbia. *Canadian Entomologist* **118**: 767-774. [AngeriLo1986]

Annecke, D.P. 1963. Observations on some citrus pests in Moçambique and southern Rhodesia. *Journal of the Entomological Society of Southern Africa* **26**: 195-225. [Anneck1963]

Annecke, D.P. 1969. Recent developments in biological and integrated control of citrus pests in South Africa. Pages 849-854. *in*: Chapman, H.D. Proceedings of the First International Citrus Symposium, Vol. II. University of California Riverside. [Anneck1969]

Annecke, D.P. & Insley, H.P. 1970. New and little known African species of *Coccophagus* Westwood (Hymenoptera: Aphelinidae). *Journal of the Entomological Society of Southern Africa* **33(2)**: 227-237. [AnneckIn1970]

Annecke, D.P. & Insley, H.P. 1971. Catalogue of Ethiopian Encyrtidae and Aphelinidae (Hymenoptera: Chalcidoidea). *Entomology Memoir (Department of Agricultural Technical Services, Republic of South Africa)* **No. 23**: 53 pp. [AnneckIn1971]

Ansari, A.R. 1942. Short notes and exhibits. Occurrence of *Bourbon aspidiotus* (*Aspidiotus destructor* Sign.) in the Punjab. *Indian Journal of Entomology* **4(2)**: 233. [Ansari1942]

Ansari, M.A., Pawar, A.D., Murthy, K.R.K. & Ahmed, S.N. 1989. Sugarcane scale, *Melanaspis glomerata* Green and its biocontrol prospects in Karnataka. *Plant Protection Bulletin (Faridabad)* **41(1-2)**: 21-23. [AnsariPaMu1989]

Anthon, E.W. 1960. Insecticidal control of San José scale on stone fruits. *Journal of Economic Entomology* **53(6)**: 1085-1087. [Anthon1960]

Antongiovanni, E. 1937. [A new parasite of *Chrysomphalus dictyospermi* Morg.] [In Italian]. *Bollettino della Società Entomologica Italiana. Firenze* **69**: 44-46. [Antong1937]

Antongiovanni, E. 1954. Prove sull'efficacia di alcuni prodotti a base di Parathion nella lotta invernale contro le cocciniglie delle piante da frutto. Società Generale per l'Industria Mineraria e Chimica. 23 pp. [Antong1954]

Antropoli, A., Faccioli, G., Paswualini, E., Basaglia, M., Ermini, P. & Tosi, C. 1990. Valutazione dell'efficacia di fenoxycarb nella lotta a *Quadraspidiotus perniciosus*. *L'Informatore Agrario* **46**: 65-67. [AntropFaPa1990]

Apstein, S. 1915. Nomina conservanda. Unter Mitwirkung zahlreicher Spezialisten herausgegeben. *Sitzungsbericht der Gesellschaft naturforschender Freunde zu Berlin.* **No. 5**: 119-201. [Apstei1915]

Arancibia O., C., Sazo R., L. & Charlin C., R. 1990. Observaciones de la biologia de la escama del acacio *Diaspidiotus ancylus* (Putnam) en acacia blanca (*Robinia pseudoacacia*). *Simiente* **60(2)**: 106-108. [AranciSaCh1990]

Archangelskaya, A.D. 1929. [List of scale insects (Coccidae) collected in hothouses of the botanical gardens in Moscow and Leningrad in February 1929]. [In Russian]. *Bolezni Rastenii. Leningrad* **18**: 188-201. [Archan1929]

Archangelskaya, A.D. 1930. List of the scale insects (Coccidae) of Turkmenistan. [In Russian]. *Report of the Plant Protection Station for 1926 to 1929, Ashkhabad, Turkmenia* : 75-85. [Archan1930]

Archangelskaya, A.D. 1931. New species of scale insects, Coccidae, from Central Asia. [In Russian]. *Zashchita Rastenii, Leningrad* **7**: 69-85. [Archan1931]

Archangelskaya, A.D. 1937. The Coccidae of Middle Asia. [In Russian]. *Izdatelstvo Komiteta Nauk UZSSR, Tashkent.* 158 pp. [Archan1937]

Argov, Y., Zchori-Fein, E. & Rosen, D. 1995. Biosystematic studies in the *Aphytis lingnanensis* complex. *Israel Journal of Entomology* **29**: 315-320. [ArgovZcRo1995]

Argyriou, L.C. 1969. Biological control of citrus insects in Greece. Pages 817-822. *in*: Chapman, H.D. (Ed.). Proceedings First International Citrus Symposium. Vol. 2. Riverside: University of California. [Argyri1969]

Argyriou, L.C. 1970. [Scale insects on citrus in Greece.] Les cochenilles des citrus en Grèce. *Al Awamia* **37**: 57-65. [Argyri1970]

Argyriou, L.C. 1974. Data on the biological control of Citrus scales in Greece. *Bulletin SROP* **No. 3**: 89-94. [Argyri1974]

Argyriou, L.C. 1976. Some data on biology, ecology and distribution of *Aspidiotus nerii* Bouché (Homoptera: Diaspididae) in Greece. *Annales de l'Institut Phytopathologique Benaki* **11**: 209-218. [Argyri1976]

Argyriou, L.C. 1977a. Recherches sur un programme de lutte biologique et intégrée contre les ravageurs des agrumes en Grèce. *Fruits* **32(10**: 630-634. [Argyri1977a]

Argyriou, L.C. 1979a. The present status of integrated control of citrus pests in Greece. Pages 517-520. *in*: Proceedings: Internationales Symposium der IOBC/WPRS über Integrierten Pflanzenschutz in der Landund Forstwirtschaft. Wien: OILB/SROP. 648 pp. [Argyri1979a]

Argyriou, L.C. 1981. Establishment of the imported parasite *Prospaltella perniciosi* (Hym.: Aphelinidae) on *Quadraspidiotus perniciosus* (Hom.: Diaspididae) in Greece. *Entomophaga* **26(2)**: 125-129. [Argyri1981]

Argyriou, L.C. 1986. Integrated pest control in citrus in Greece. Pages 545-548. *in*: R. Cavalloro, E. Di Martino. Integrated Pest Control in Citrus Groves: Proceedings of the Experts' Meeting. Rotterdam, Netherlands: A.A. Balkema. 545-548. [Argyri1986]

Argyriou, L.C. 1990. 3.9.5 Olive. Pages 579-583 *in*: Rosen, D. (Ed.). Armored Scale Insects, Their Biology, Natural Enemies and Control [Series title: World Crop Pests, Vol. 4B]. Amsterdam, the Netherlands: Elsevier. 688 pp. [Argyri1990]

Argyriou, L.C. & Kourmadas, A.L. 1980. The phenology and natural enemies of *Aspidiotus nerii* Bouché in central Greece. *Fruits* **35**: 633-638. [ArgyriKo1980]

Argyriou, L.C. & Kourmadas, A.L. 1981. Contribution to the timing for the control of Diaspididae scales of olive trees. *Annales de l'Institut Phytopathologique Benaki* **13**: 65-72. [ArgyriKo1981]

Argyriou, L.C. & Mourikis, P.A. 1981. Current status of citrus pests in Greece. *Proceedings of the International Society of Citriculture* **2**: 623-627. [ArgyriMo1981]

Argyriou, L.C., Stavraki, H.G. & Mourikis, P.A. 1976. A list of recorded entomophagous insects of Greece. Athens. 73 pp. [ArgyriStMo1976]

Armoured scales and soft scales. 1991. *Zashchita rastenii. Moscow* **No. 6**: 31-33. [ArmourScan1991]

Arras, G. 1976. Fitofarmaci a confronto nella lotta contro la cocciniglia nera della vite *Targionia vitis* Sign. *Informatore Fitopatologico* 26: 15-19 [Arras1976].

Arruda, G.P. de 1972. [A new scale insect (Homoptera, Diaspididae), found on cashew in the state of Pernambuco.]. [In Portuguese]. *Anais do Instituto de Ciencias Biologia, Universidad Federal Rural de Pernambuco* **2**: 13-17. [Arruda1972]

Arruda, G.P. de 1976. A new species of scale insect (Homoptera, Diaspididae), found in Pernambuco. [In Portuguese]. *Anais do Instituto de Ciencias Biologia, Universidad Federal Rural de Pernambuco* **3**: 21-26. [Arruda1976]

Ashmead, W.H. 1880. On the red or circular scale of the orange *(Chrysomphalus ficus* Riley Ms.). *The American Entomologist* **3**: 267-269. [Ashmea1880]

Ashmead, W.H. 1891. A generic synopsis of the Coccidae. Family X. - Coccidae. *Transactions of the Entomological Society of America* **18**: 92-102. [Ashmea1891]

Asplanato, G. & Garcia Mari, F. 1998. Distribución del piojo rojo de California *Aonidiella aurantii* (Maskell) (Homoptera: Diaspididae) en árboles de naranjo. *Boletín de Sanidad Vegetal, Plagas* **24**: 3, 637-646. [AsplanGa1998]

Asplanato, G. & Garcia Mari, F. 2001. [Seasonal history of the California red scale, *Aonidiella aurantii* (Maskell) (Homoptera: Diaspididae) on orange trees of southern Uruguay.]. [In Spanish with summary in English.] *Agrociencia Montevideo* **5(1)**: 54-67. [AsplanGa2001]

Asquith, D., Croft, B.A., Hoyt, S.C., Glass, E.H. & Rice, R.E. 1980. The systems approach and general accomplishments toward better insect control in pome and stone fruits. Pages 249-317 *in*: Huffaker, C.B. (Ed.). New Technology of Pest Control. New York: Wiley-Interscience. 500 pp. [AsquitCrHo1980]

Atkinson, E.T. 1886. Insect-pests belonging to the homopterous family Coccidae. *Journal of the Asiatic Society of Bengal. Natural History* **55**: 267-298. [Atkins1886]

Atkinson, P.R. 1977. Preliminary analysis of a field population of citrus red scale, *Aonidiella aurantii* (Maskell), and the measurement and expression of stage duration and reproduction for life tables. *Bulletin of Entomological Research* **67**: 65-87. [Atkins1977]

Atkinson, P.R. 1983. Estimates of natural mortality related to environmental factor in a population of citrus red scale, *Aonidiella aurantii* (Maskell) (Hemiptera: Diaspididae). *Bulletin of Entomological Research* **73**: 239-258. [Atkins1983]

Atkinson, P.R. 1983a. Environmental factors associated with fluctuations in numbers of natural enemies of a population of citrus red scale, *Aonidiella aurantii* (Maskell) (Hemiptera: Diaspididae). *Bulletin of Entomological Research* **73**: 417-426. [Atkins1983a]

Aung, L.H., Leesch, J.G., Jenner, J.F. & Grafton Cardwell, E.E. 2001. Effects of carbonyl sulfide, methyl iodide, and sulfuryl fluoride on fruit phytotoxicity and insect mortality. *Annals of Applied Biology* **139(1)**: 93-100. [AungLeJe2001]

Autran, E. 1907. Las Cochinillas Argentinas. *Boletín de la Argentina Ministerio de Agricultura* **7**: 145-200. [Autran1907]

Avidov, Z. 1960. [Life history of the Egyptian black scale (*Chrysomphalus aonidum* L.) in the coastal plain of Israel. B. On citrus leaves. [In Hebrew]. *Bulletin of the National Institute of Agriculture* **65**: 17-31. [Avidov1960]

Avidov, Z. 1970. Biology of natural enemies of citrus scale insects and the development of methods for their mass production. Rehovot, Israel: Hebrew University of Jerusalem. 241 pp. [Avidov1970]

Avidov, Z., Balshin, M. & Gerson, U. 1970. Studies on *Aphytis coheni*, a parasite of the California red scale, *Aonidiella aurantii*, in Israel. *Entomophaga* **15**: 191-207. [AvidovBaGe1970]

Avidov, Z. & Gabai, N.M. 1960. [Life history of the Egyptian black scale (*Chrysomphalus aonidum*) in the coastal plain of Israel. A. On citrus fruits.]. [In Hebrew]. *Bulletin (National and University Institute of Agriculture, Rehovot, Israel)* **No. 311**: 5-16; V-VIII. [AvidovGa1960]

Avidov, Z., Rosen, D. & Gerson, U. 1963. A comparative study on the effects of aerial versus ground spraying of poisoned baits against the Mediterranean fruit fly on the natural enemies of scale insects in citrus groves. *Entomophaga* **8**: 205-212. [AvidovRoGe1963]

Ayoutantis, A. 1940. Scale insects observed on citrus in the island of Crete. *International Bulletin of Plant Protection* **14(1)**: 2M-4M. [Ayouta1940]

Aytas, M., Yumruktepe, R. & Mart, C. 2001. [Using pheromone traps to control California Red Scale *Aonidiella aurantii* (Maskell) (Hom.: Diaspididae) in the Eastern Mediterranean region.]. [In Turkish]. *Turkish Journal of Agriculture and Forestry* **25(2)**: 97-110. [AytasYuMa2001]

Azevedo, A. de 1923. As cochonilhas das laranjeiras. *Correio Agricola* **1**: 86-90. [Azeved1923A]

Azevedo Marques, L.A. de 1923. Pragas da videira. *Chacaras e Quintaes* **28**: 422. [Azeved1923LA]

Azevedo, A. de 1925. Insectos e fungos. *Correio Agricola* **3**: 85-86. [Azeved1925]

Azevedo, A. de 1929. Subsidios para o estudo da sanidade vegetal no Estado da Bahia. *Correio Agricola* **12**: 113-115. [Azeved1929]

Azevedo, A. de 1929a. Subsidios para o estudo da sanidade vegetal no Estado da Bahia. *Correio Agricola* **7**: 126-128. [Azeved1929a]

Azim, A. 1961. Mass production of *Chrysomphalus bifasciculatus* Ferris and its Hymenopterous parasites. *Mushi* **35(14)**: 97-109. [Azim1961]

Babaian, G.A. 1987. Scale - insects of stone fruit crops and control measures against them. *Bollettino del Laboratorio di Entomologia Agraria 'Filippo Silvestri'* **43** (Suppl.): 133-138. [Babaia1987]

Babenko, V.O. & Yanovskii, Y.P. 1997. [Disinfestation of apple seedlings by the Californian scale in nurseries in the northern Dnieper region.]. [In Ukrainian with summaries in English and Russian.] *10th International Conference of Plant Protection* **42**: 64-68. [BabenkYa1997]

Bachmann, F. 1952. Beitrag zur Biologie einheimischer Deckelschildläuse. *Schweizerische Entomologische Gesellschaftliche Mitteilungen* **25**: 144. [Bachma1952]

Bachmann, F. 1952a. *Quadraspidiotus schneideri* n. sp. (Homopt. Diaspidoid.), Eine neue Schildlausart. *Schweizerische Entomologische Gesellschaftliche Mitteilungen* **25**: 357. [Bachma1952a]

Bachmann, F. 1953. Beitrag zur Kenntnis der Jugoslawischen Schildlausfauna. *Srpska Akad. Nauk, Zbornik Radova* **30**: 175-184. [Bachma1953]

Bachmann, F. 1953a. Untersuchungen an den gelben Obstbaumschildläusen *Quadraspidiotus piri* Licht. Und *Quadraspidiotus schneideri* n. sp. *Zeitschrift für Angewandte Entomologie* **34**: 357-404. [Bachma1953a]

Bachmann, F. 1955. *Quadraspidiotus schneideri* Bachmann - Art oder rasse. *Zeitschrift für Angewandte Entomologie* **37**: 122-124. [Bachma1955]

Bachmann, F. 1956. Beitrag zur Kenntnis der Jugoslawischen Schildlaus-fauna. 2. Mitteilung. *Bioloski Glasnik* **8**: 99-102. [Bachma1956]

Bachmann, F. 1974. Theoretical and practical aspects of the control of armoured scales (Diaspididae). *Mededelingen Faculteit Landbouwkundige en Toegepaste Biologische Wetenschappen Universiteit Gent* **39(2)**: 755-760. [Bachma1974]

Bachmann, F. & Geier, P. 1950. Einige für die Schweiz neue oder wenig bekannte Cocciden aus der Unterfamilie der Diaspidinae. *Schweizerische Entomologische Gessell. Mitteilungen* **23**: 117-119. [BachmaGe1950]

Badawi, A. & Al-Ahmed, A.M. 1990. The population dynamics of the oriental scale insect, *Aonidiella orientalis* (Newstead) and factors affecting its seasonal abundance. *Arab Gulf Journal of Scientific Research* **8(3)**: 81-89. [BadawiAl1990]

Badenes-Perez, F.R., Zalom, F.G. & Bentley, W.J. 2002. Are San José scale (Hom., Diaspididae) pheromone trap captures predictive of crawler densities?. *Journal of Applied Entomology* **126(10)**: 545-549. [BadeneZaBe2002]

Badenes-Perez, F.R., Zalom, F.G. & Bentley, W.J. 2002a. Effects of dormant insecticide treatments on the San José scale (Homoptera: Diaspididae) and its parasitoids *Encarsia perniciosi* and *Aphytis* spp. (Hymenoptera : Aphelinidae). *International Journal of Pest Management* **48(4)**: 291-296. [BadeneZaBe2002a]

Baerensprung, F.V. 1849. Beobachtungen über einige einheimische Arten aus der Familie der Coccinen. *Zeitung für Zoologie, Zootomie und Palaeozoologie* **1**: 165-170, 173-176. [Baeren1849]

Baerg, W.J. 1921. Spraying for San José scale. *Bulletin (Agricultural Experiment Station, University of Arkansas)* No. 177. [Baerg1921]

Baeta Neves, C.M. 1947. Alguns insectos prejudiciais nos arvoredos de Sintra. *Revista Agronomica. Lisbon* **35**: 128-142. [BaetaN1947]

Baggiolini, M., Geier, P. & Mathys, G. 1951. Contaminabilité par le pou de San José (*Quadraspidiotus perniciosus* Comst.) des végétaux ligneux les plus connus en Suisse. *Ann. agr. de la Suisse* **52**: 931-937. [BaggioGeMa1951]

Baghel, C.L. & Dutta, S. 2001. Morphological study of mature male *Aonidiella orientalis* (Newstead). *Flora and Fauna (Jhansi)*. 7(2): 76-78. [BaghelDu2001]

Baicu, T. (Co-ordinator) 1982. [Tests of pesticides Vol. VII.] Testarea pesticidelor Vol. VII. Bucharest: Institutul de Cercetari Pentru Protectia Plantelor. 208 pp. [Baicu1982]

Baiocchi, A. & Galli, G. 1991. Ricerche innovative nella produzione integrata di mele in Valtellina. *Informatore Agrario* **47**: 44, 104-106. [BaioccGa1991]

Baker, H. 1933. The obscure scale on the pecan and its control. *United States Department of Agriculture Circular* **295**: 1-19. [Baker1933]

Baker, W.L. 1972. Eastern forest insects. *United States Department of Agriculture, Forest Service, Miscellaneous Publications* **1175**: 1-642. [Baker1972]

Baker, J.L. 1976. Determinants of host selection for species of *Aphytis* (Hymenoptera: Aphelinidae), parasites of Diaspine scales. *Hilgardia* **44**: 1-25. [Baker1976]

Balachowsky, A.S. 1926. Les principales cochenilles dont il faut redouter l'introduction en Algérie. *Revue Agricole de l'Afrique du Nord, Alger* **1926**: 1-8. [Balach1926]

Balachowsky, A.S. 1926a. Note sur un coccide de la faune neo-tropicale récemment acclimaté et nuisible au figuier en Algérie. *Bulletin de la Société d'Histoire Naturelle de l'Afrique du Nord* **17**: 63-69. [Balach1926a]

Balachowsky, A.S. 1927. Contribution à l'étude des coccides de l'Afrique mineure (1re note). *Annales de la Société Entomologique de France* **96**: 175-207. [Balach1927]

Balachowsky, A.S. 1927a. Les insectes nuisibles au figuier en Algérie et leurs traitements. *Bulletin de la Société d'Histoire Naturelle de l'Afrique du Nord* **17**: 69-75. [Balach1927a]

Balachowsky, A.S. 1928a. Contribution à l'étude des coccides de l'Afrique mineure (2e note). *Bulletin de la Société d'Histoire Naturelle de l'Afrique du Nord* **19**: 121-144. [Balach1928a]

Balachowsky, A.S. 1928b. Contribution à l'étude des coccides de L'Afrique mineure (3e note). *Chrysomphalus aonidum* L. - biologie - traitement. *Bulletin de la Société d'Histoire Naturelle de l'Afrique du Nord* **19**: 156-180. [Balach1928b]

Balachowsky, A.S. 1928c. Contribution à l'étude des Coccides de l'Afrique mineure (4e note). Nouvelle liste de Coccides nord-africains avec description d'espèces nouvelles. *Bulletin de la Société Entomologique de France* **1928**: 273-279. [Balach1928c]

Balachowsky, A.S. 1928d. Observations biologiques sur les parasites des coccides du Nord-Africain. *Annales des Epiphyties* **14(4)**: 280-312. [Balach1928d]

Balachowsky, A.S. 1929. Contribution à l'étude de la faune du Congo Belge. Diaspines nuisibles au caféier et au cacaoyer. *Revue de Pathologie Végétale et d'Entomologie Agricole de France* **16**: 141-145. [Balach1929]

Balachowsky, A.S. 1929a. Contribution à l'étude des coccides de l'Afrique mineure (6e note). Faune du Hoggar. *Annales de la Société Entomologique de France* **98**: 301-322. [Balach1929a]

Balachowsky, A.S. 1930. Contribution à l'étude des coccides de France (1re note). Faunule des Iles d'Hyères (Port-Cros et Levant). *Bulletin de la Société Entomologique de France* **1929**: 311-317. [Balach1930]

Balachowsky, A.S. 1930a. Contribution à l'étude des coccides de France (3e note). Coccides nouveaux ou peu connus de la faune de France. *Bulletin de la Société Entomologique de France* **1930**: 178-184. [Balach1930a]

Balachowsky, A.S. 1930c. Contribution à l'étude des coccides de l'Afrique mineure (9me note). Addition à la faune du nord-africain avec description de trois espèces nouvelles. *Bulletin de la Société d'Histoire Naturelle de l'Afrique du Nord* **21**: 119-125. [Balach1930c]

Balachowsky, A.S. 1931a. Contribution à l'étude des coccides de France (5e note). Faune de Corse. *Bulletin de la Société Entomologique de France* **1931**: 96-102. [Balach1931a]

Balachowsky, A.S. 1932. Contribution à l'étude des coccides de l'Afrique mineure [11me note]. Sur une diaspine nouvelle récoltée par M. P. Vayssière dans le haut-atlas. *Bulletin de la Société Entomologique de France* **37**: 18-20. [Balach1932]

Balachowsky, A.S. 1932b. La théorie des "Places Vides" et le peuplement des végétaux par les Coccidae. Pages 517-522 *in*: Livre du Centenaire Société de la Entomologique de France Paris. [Balach1932b]

Balachowsky, A.S. 1932d. Etude biologique des coccides du bassin occidental de la Méditerranée. Encyclopédie Entomologique Paris: P. Lechevalier & Fils. 214 pp + LXVII. [Balach1932d]

Balachowsky, A.S. 1932e. Contribution à l'étude des coccides de France supplément á la liste des coccides des Alpes-Maritimes et du var avec description d'un *Eriococcus* nouveau. *Bulletin de la Société Entomologique de France* **37**: 233-238. [Balach1932e]

Balachowsky, A.S. 1932f. Contribution à l'étude des coccides des colonies françaises. Sur une diaspine nouvelle du Tibesti. *Bulletin de la Société d'Histoire Naturelle de l'Afrique du Nord* **23**: 228-231. [Balach1932f]

Balachowsky, A.S. 1932g. Contribution à l'étude des coccides de France (12e note) sur la présence d'*Aspidiotus bavaricus* Ldgr. [Hem. Coccidae] dans la forêt de Fontainebleau. *Bulletin de l'Association des Naturalistes de la Vallée du Loing* **15**: 102-106. [Balach1932g]

Balachowsky, A.S. 1932j. Le pou de San José et l'importation des fruits frais. *Revue d'Horticulture et d'Agriculture de l'Afrique du Nord* **36**: 171-176. [Balach1932j]

Balachowsky, A.S. 1933a. Contribution à l'étude des coccides de France (14e note). *Annales de la Société Entomologique de France* **102**: 35-50. [Balach1933a]

Balachowsky, A.S. 1933b. Contribution à l'étude des coccides de l'Afrique mineure [13e note]. Sur une nouvelle diaspine récoltée par le Dr. R. Maire dans le Moyen-Atlas. *Bulletin de la Société Entomologique de France* **38**: 245-248. [Balach1933b]

Balachowsky, A.S. 1933e. Les coccides des Iles d'Hyères (Contribution à l'étude des coccides de France, 16e note). *Annales de la Société d'Histoire Naturelle de Toulon* **17**: 78-84. [Balach1933e]

Balachowsky, A.S. 1934. Recherches biogéographiques sur la faune des coccides de Corse. *Compte Rendu de la Société de Biogeographie. Paris* **11**: 34-39. [Balach1934]

Balachowsky, A.S. 1934a. Contribution à l'étude des coccides de France (17e note). Recherches complémentaires sur la faune de Corse .*Bulletin de la Société Entomologique de France* **39**: 67-72. [Balach1934a]

Balachowsky, A.S. 1934d. Les coccides du Sahara central. Mission du Hoggar. III. (Février à Mai 1928) *In* Seurat, L.-G., Études Zoologiques sur le Sahara Central. *Mémoires de la Société d'Histoire Naturelle de l'Afrique du Nord* **4**: 145-157. [Balach1934d]

Balachowsky, A.S. 1934e. Cochenilles nouvelles et peu connues du Maroc. [Contribution à l'étude des coccides du nord Africain, 14e note .[*Compte-Rendu de Association Française pour l'Avancement des Sciences* **58**: 160-163. [Balach1934e]

Balachowsky, A.S. 1935. Sur une diaspine nouvelle du sud de l'Espagne. *Bulletin de la Société Entomologique de France* **40**: 233-236. [Balach1935]

Balachowsky, A.S. 1935b. Les cochenilles de l'Espagne. *Revue de Pathologie vegetale et d'Entomologie Agricole de France* **22**: 255-269. [Balach1935b]

Balachowsky, A.S. 1936a. Note sur *Aspidiotus (Hemiberlesia) vuilleti* Marchal [Hem. Coccidae]. [Contribution à l'étude des coccides des colonies Françaises, 3e note.]. *Bulletin de la Société Entomologique de France* **41**: 97-100. [Balach1936a]

Balachowsky, A.S. 1937. Sur quelques cochenilles récoltées au cours du Congrès d'Avignon. [Contribution à l'étude des coccides de France, 21e note.]. *Bulletin de la Société Entomologique de France* **41**: 339-340. [Balach1937]

Balachowsky, A.S. 1937a. Sur une cochenille nouvelle de la forêt humide de Madère. [Contribution à l'étude des coccides du Nord-Africain, 16e note.]. *Bulletin de la Société Entomologique de France* **42**: 110-112. [Balach1937a]

Balachowsky, A.S. 1937c. Les cochenilles de Seine-et-Oise. (Contribution à l'étude des coccides de France, 23me note). *Bulletin de la Société des Sciences Naturelles de Seine-et-Oise (ser. 3)* **5**: 1-6. [Balach1937c]

Balachowsky, A.S. 1938a. Les cochenilles de Madère. I. Diaspididae - Asterolecaniinae. [Contribution à l'étude des coccides du nord Africain, 19e note]. *Revue de Pathologie Végétale et d'Entomologie Agricole de France* **25**: 144-155. [Balach1938a]

Balachowsky, A.S. 1941. Sur un *Chortinaspis* Ferris nouveau d'Asie mineure [Hem. Coccoidea, Diaspididae.]. *Bulletin de la Société Entomologique de France* **46**: 9-11. [Balach1941]

Balachowsky, A.S. 1942. Essai sur la classification des cochenilles (Homoptera - Coccoidea). *Annales de l'École Nationale d'Agriculture de Grignon* **3**: 34-48. [Balach1942]

Balachowsky, A.S. 1946. Etude biogéographique des Coccoidea des Iles Atlantides (Canaries et Madère). *Mémoires de la Société Biogéographique* **8**: 209-218. [Balach1946]

Balachowsky, A.S. 1948. Remarques sur deux cochenilles forestières du groupe de *Quadraspidiotus ostreaeformis* Curtis nouvelles pour la faune de France. [Contribution à l'étude des cochenilles en France, 26e note.]. *Revue de Pathologie Végétale et d'Entomologie Agricole de France* **27**: 14-24. [Balach1948]

Balachowsky, A.S. 1948a. Sur le statut et la biologie de *Quadraspidiotus ostreaeformis* Curtis et de *Quadraspidiotus pyri* Licht. [Contribution à l'étude des cochenilles de France, 27e note.]. *Revue de Pathologie Végétale et d'Entomologie Agricole de France* **27**: 89-97. [Balach1948a]

Balachowsky, A.S. 1948b. Les cochenilles de France, d'Europe, du nord de l'Afrique et du bassin Méditerranéen. IV. Monographie des Coccoidea, classification - Diaspidinae (Première partie). *Actualités Scientifiques et Industrielles* **1054**: 243-394. [Balach1948b]

Balachowsky, A.S. 1949. Sur un *Rugaspidiotus* (Coccoidea-Odonaspidini) nouveau d'Oranie. [Contribution à l'étude des Coccoidea du Nord-Africain, 27e note.]. *Bulletin de la Société d'Histoire Naturelle de l'Afrique du Nord* **40**: 107-110. [Balach1949]

Balachowsky, A.S. 1949a. Remarques complémentaires sur le genre *Murataspis* Balachw. et Rich. (Hom. Coccoidea). [Contribution á l'étude des Coccides du Nord-Africain, 24e note.]. *Bulletin de la Société des Sciences Naturelles du Maroc* **25-27**: 163-165. [Balach1949a]

Balachowsky, A.S. 1949b. Sur deux *Aonidia* (Hom. Coccoidea) nouveaux du cyprès et du laurier noble dans leurs stations spontanées du Haut-Atlas et du Moyen-Atlas (Maroc). *Revue de Pathologie Végétale et d'Entomologie Agricole de France* **28**: 112-117. [Balach1949b]

Balachowsky, A.S. 1949c. Sur un genre nouveau d' Aspidiotini [Hem. Coccoidea] du sud de l'Anti-Atlas (Maroc méridional). [Contribution á l'étude des Coccoidea du Nord de l'Afrique, 25e note. *Bulletin de la Société Entomologique de France* **54**: 74-76. [Balach1949c]

Balachowsky, A.S. 1950b. Les cochenilles de France, d'Europe, du Nord de l'Afrique et du Bassin Méditerranéen. V. - monographie des Coccoidea; Diaspidinae (deuxième partie) Aspidiotini. *Actualités Scientifique et Industrielles* **1087**: 397-557. [Balach1950b]

Balachowsky, A.S. 1950e. La destruction des insectes auxiliaires entomophages par les traitements insecticides et ses conséquences. *Comptes Rendus de l'Academie d'Agriculture de France* **36(6)**: 220-223. [Balach1950e]

Balachowsky, A.S. 1951. Les cochenilles de France, d'Europe, du Nord de l'Afrique et du bassin Méditerranéen. VI. - monographie des Coccoidea; Diaspidinae (Troisième partie) Aspidiotini (fin). *Actualités Scientifiques et Industrielles* **1127**: 561-720. [Balach1951]

Balachowsky, A.S. 1952a. Sur deux Diaspidinae [Hom. Coccoidea] nouveaux de Moyenne Guinée (A.O.F.) [Contribution à l'étude des Coccoidea de la France d'Outre-Mer, 5e note.]. *Bulletin de la Société Entomologique de France* **57**: 98-104. [Balach1952a]

Balachowsky, A.S. 1953a. Sur un nouveau genre de Parlatorini (Hom.-Coccoidea) originaire du massif du Béna (Moyenne-Guinée) (A.O.F.) [Contribution à l'étude des Coccoidea de la France d'Outre-Mer, 7e note.]. *Revue de Pathologie Végétale et d'Entomologie Agricole de France* **32**: 19-23. [Balach1953a]

Balachowsky, A.S. 1953b. Sur une cinquième espèce d'*Aonidiella* Berl. et Leon. (Hom. Coccoidea) nouvelle du Congo Belge. [Contribution à l'étude des Coccoidea du Congo Belge. III).]. *Revue de Zoologie et de Botanique Africaine* **47**: 77-80. [Balach1953b]

Balachowsky, A.S. 1953c. Sur un *Lindingaspis* MacGill. nouveau du massif du Béna (Moyenne-Guinée) [Hom. Coccoidea]. [Contribution à l'étude des Coccoidea de la France d'Outre-mer, 11e note.]. *Bulletin de la Société Entomologique de France* **58**: 109-112. [Balach1953c]

Balachowsky, A.S. 1953e. Sur un *Melanaspis* Ckll. nouveau vivant sur *Parinarium* en Moyenne Guinée. (Coccoidea, Diaspidinae). [Contribution à l'étude des cochenilles de la France d'Outre-Mer, 10e note.]. *Revue Française d'Entomologie* **20**: 146-149. [Balach1953e]

Balachowsky, A.S. 1953g. Les cochenilles de France d'Europe, du Nord de l'Afrique, et du bassin Méditerranéen. VII. - Monographie des Coccoidea; Diaspidinae-IV, Odonaspidini-Parlatorini. *Actualités Scientifiques et Industrielles* **1202**: 725-929. [Balach1953g]

Balachowsky, A.S. 1953i. Deux *Pseudaonidia* Ckll. (Hom. Coccoidea-Diaspidinae) nouveaux du massif du Béna (Moyenne Guinée) A.O.F. *Bulletin del l'Institut Français d'Afrique Noire* **15**: 1512-1522. [Balach1953i]

Balachowsky, A.S. 1953j. I. Sur un Pseudococcini (Hom. Coccoidea) nouveau du Cap Bon (Tunisie). II. Présence de *Rhizaspidiotus bivalvatus* Goux (Coccoidea, Diaspidinae) en Afrique du Nord. *Bulletin de la Société des Sciences Naturelles de Tunisie* **6**: 227-253. [Balach1953j]

Balachowsky, A.S. 1953k. Sur un *Hemiberlesia* Ckll. nouveau des montagnes du Cameroun. (Homoptera; Coccoidea). [Contribution à l'étude des Coccoidea de la France d'Outre-Mer, 8e note.]. *Beitrage zur Entomologie. Berlin* **3**: 111-115. [Balach1953k]

Balachowsky, A.S. 1954. Sur deux *Selenaspidus* Ckll. nouveaux (Coccoidea-Diaspididae) de Guinée Française et du Haut Cameroun. *Revue de Pathologie Végétale et d'Entomologie Agricole de France* **33**: 57-63. [Balach1954]

Balachowsky, A.S. 1954b. Etude comparative des cochenilles du cèdre au Liban et en Afrique du Nord. *Revue de Pathologie Végétale et d'Entomologie Agricole de France* **33**: 108-114. [Balach1954b]

Balachowsky, A.S. 1954c. Remarques sur le g. *Africonidia* McKenzie (Coccoidea-Diaspidinae) avec description d'une espèce nouvelle du Cameroun Méridional. *EOS* **30**: 77-80. [Balach1954c]

Balachowsky, A.S. 1954d. Sur deux *Aonidiella* Berl. et Leon. (Hom. Coccoidea) nouveaux du Congo Belge. *Annales du Musée Royal du Congo Belge. Tervuren* **1**: 296-300. [Balach1954d]

Balachowsky, A.S. 1954e. Les cochenilles paléarctiques de la tribu des Diaspidini. Paris: Institut Pasteur Mémoires Scientifiques. 450 pp. [Balach1954e]

Balachowsky, A.S. 1954f. Sur l'origine et le développement des insectes nuisibles aux plantes cultivées dans les oasis du Sahara Français. Pages 90-103 *in*: Cloudsley-Thompson, J.L. (ed.). Biology of Deserts Proceedings of Symposium Inst. Biol. [Balach1954f]

Balachowsky, A.S. 1955. Contribution à l'étude de la faune entomologique du Ruanda Urundi (mission P. Basilewsky 1935). Pt I. Homoptera Coccoidea. Sur un *Aspidiotus* Bouché (Diaspididae-Aspidiotini) nouveau du Ruanda (Contribution à l'étude des cochenilles du Congo Belge, 4ᵉ note. *Annales du Musée Royal du Congo Belge. Tervuren* **36**: 390-393. [Balach1955]

Balachowsky, A.S. 1956. Les cochenilles du continent Africain Noir. V. 1 - Aspidiotini (1ère partie). *Annales du Musée Royal du Congo Belge (Sciences Zoologiques). Tervuren* **3**: 1-142. [Balach1956]

Balachowsky, A.S. 1957. Sur deux Diaspidinae nouveaux du Thaïland (Coccoidea). *Tijdschrift voor Entomologie. Amsterdam* **100**: 29-34. [Balach1957]

Balachowsky, A.S. 1957c. Les cochenilles de la Guadeloupe et de la Martinique (première liste). *Revue de Pathologie Végétale et d'Entomologie Agricole de France* **36**: 198-208. [Balach1957c]

Balachowsky, A.S. 1958. Sur l'origine et la nocivité des insectes nuisibles aux plantes cultivées dans les oasis du Sahara Francais. *Travaux de l'Institut de Recherches Sahariennes* **3**: 7-29. [Balach1958]

Balachowsky, A.S. 1958a. Les cochenilles du Sahara Francais. 1ère partie. *Travaux de l'Institut de Recherches Sahariennes* **3**: 31-53. [Balach1958a]

Balachowsky, A.S. 1958b. Les cochenilles du continent Africain Noir. v. 2 Aspidiotini (2me partie), Odonaspidini et Parlatorini. *Annales du Musée Royal du Congo Belge (Sciences Zoologiques). Tervuren* **4**: 149-356. [Balach1958b]

Balachowsky, A.S. 1959. Nuevas cochinillas de Colombia. *Revista de la Academia Colombiana de Ciencias Exactas, Físicas y Naturales* **10**: 337-361. [Balach1959]

Balachowsky, A.S. 1959a. Otras cochinillas nuevas de Colombia. *Revista de la Academia Colombiana de Ciencias Exactas, Físicas y Naturales* **10**: 362-366. [Balach1959a]

Balachowsky, A.S. 1959b. Sur un nouvel Aspidiotini (Coccoidea) de l'Iran. *Revue de Pathologie Végétale et d'Entomologie Agricole de France* **38**: 211-212. [Balach1959b]

Balachowsky, A.S. 1968. Sur un *Quadraspidiotus* [Coccoidea-Aspidiotini] nouveau du centre de l'Ile de Ténériffe (Canaries). *Annales de la Société Entomologique de France* **(n.s.) 4**: 75-79. [Balach1968]

Balachowsky, A.S. & Ferrero, D. 1965. Sur un *Selenaspidus* Ckll. [Coccoidea-Aspidiotini] nouveau de République Centre-Africaine. *Annales de la Société Entomologique de France* **(n.s.) 1**: 711-715. [BalachFe1965]

Balachowsky, A.S. & Ferrero, D. 1967c. Sur un *Quadraspidiotus* McGill nouveau de l'Afrique équatoriale forestière [Hom. Coccoidea-Aspidiotini]. [Contribution à l'étude des Coccoidea de l'Afrique tropicale et équatoriale (11e note)]. *Bulletin de la Société Entomologique de France* **72**: 218-220. [BalachFe1967c]

Balachowsky, A.S. & Kaussari, M. 1951. Coccoidea-Diaspinae nouveaux du sud-est de l'Iran. *Bulletin de la Société Fouad 1er d'Entomologie* **35**: 1-15. [BalachKa1951]

Balachowsky, A.S. & Kaussari, M. 1953. Sur un *Melanaspis* Ckll. nouveau (Coccoidea-Diaspidinae) vivant sur chêne en Iran. *Revue de Pathologie Végétale et d'Entomologie Agricole de France* **32**: 28-31. [BalachKa1953]

Balachowsky, A.S. & Matile-Ferrero, D. 1970. Les cochenilles (Hom. Coccoidea) de la République Islamique de Mauritanie. Contribution à l'étude des Coccoidea de l'Afrique tropicale et équatoriale (14-e note). *Bulletin de l'Institut Fondamental d'Afrique Noire (Ser A)* **32**: 1078-1087. [BalachMa1970]

Balachowsky, A.S. & Richardeau, D. 1942. Sur un nouveau genre de Coccoidea vivant sur *Tamarix* dans le Sahara Nord-Africain avec description d'une espèce nouvelle. *Bulletin de la Société Entomologique de France* **47**: 100-103. [BalachRi1942]

Baldanza, F., Gaudio, L. & Viggiani, G. 1999. Cytotaxonomic studies of *Encarsia* Förster (Hymenoptera: Aphelinidae). *Bulletin of Entomological Research* **89(3)**: 209-215. [BaldanGaVi1999]

Ballou, H.A. 1912. Insect Pests of the Lesser Antilles. Bridgetown, Barbados: Advocate. 210 pp. [Ballou1912]

Ballou, H.A. 1912a. Report on the prevalence of some pests and diseases in the West Indies, for 1910 and 1911. *West Indian Bulletin* **12**: 412-425. [Ballou1912a]

Ballou, H.A. 1913. Notes on insect pests in Antigua. *Bulletin of Entomological Research* **4**: 61-65. [Ballou1913]

Ballou, H.A. 1915. Report on the prevalence of some pests and diseases in the West Indies during 1914. *West Indian Bulletin* **15(2)**: 121-147. [Ballou1915]

Ballou, H.A. 1922a. Report on the prevalence of some pests and diseases of crops in the West Indies during 1920. *West Indian Bulletin* **19**: 239-271. [Ballou1922a]

Banks, C.S. 1906. New Philippine insects. *Philippine Journal of Science* **1**: 229-236. [Banks1906]

Banks, C.S. 1906b. The principal insects attacking the coconut palm (Part II). *Philippine Journal of Science* **1**: 211-228. [Banks1906b]

Banks, H.J. 1990. 1.3.5 Physiology and biochemistry. Pages 267-274 *in*: Rosen, D. (Ed.). Armored Scale Insects, Their Biology, Natural Enemies and Control, [Series title: World Crop Pests, Vol. 4A]. Amsterdam, The Netherlands: Elsevier. 384 pp. [Banks1990]

Banti, A. 1893. Descrizione e figure dello *Aspidiotus ceratoniae* Colv. *Rivista di Patologia Vegetale. Firenze* **2**: 12-21. [Banti1893]

Bao, W. 1993. Studies on *Aphytis vandenboschi* a parasitoid of *Hemiberlesia pitysophila* (Homoptera: Diaspididae). *Natural Enemies of Insects* **15**: 1-5. [Bao1993]

Bar-Joseph, M. & Gerson, U. 1973. Avocado bark pitting associated with the latania scale. Yearbook 1972-73 of the Calif. Avocado Soc. 146-147 pp. [BarJosGe1973]

Bar-Zakay, I., Peleg, B.A. & Hefetz, A. 1989. Mating disruption of the California Red Scale, *Aonidiella aurantii* (Homoptera; Diaspididae). *Hassadeh* **69(7)**: 1228-1231. [BarZakPeHe1989]

Baranyovits, F. 1953. Some aspects of the biology of armoured scale insects. *Endeavour* **12**: 202-209. [Barany1953]

Barber, C.L. 1893. Leeward Island Coccidae. *Insect Life* **6(1)**: 50-51. [Barber1893]

Barbey, A. 1925. Traité d'entomologie forestière à l'usage des sylviculteurs des reboiseurs, des propriétaires de bois et des biologistes. Paris: Berger-Levrault. 749 pp. [Barbey1925]

Bardia Bardia, R. 1967. El "Piojo de San Jose" ensayos de tratamientos. *Boletín de Agron-Pecuario* 3-23. [Bardia1967]

Barnes, H.F. 1930. Gall midges (Cecidomyidae) as enemies of the Tingidae, Psyllidae, Aleyrodidae and Coccidae. *Bulletin of Entomological Research* **5**: 319-329. [Barnes1930]

Baroffio, C. 1993. [Description of the developmental stages of *Encarsia perniciosi* (Tower) using different techniques.] [In French]. *Mitteilungen der Schweizerischen Entomologischen Gesellschaft* **66(3-4)**: 371-378. [Baroff1993]

Baroffio, C. 1997. [Some aspects of the biology of *Encarsia perniciosi* (Tower) in its host, *Quadraspidiotus perniciosus* (Comstock), and application of the results in a biological control program in central Switzerland (Canton Zug).]. [In French]. *Mitteilungen der Schweizerischen Entomologischen Gesellschaft* **70**: 323-333. [Baroff1997]

Barritt, N.W. 1929. A new spray for scale insects on citrus in Egypt. *Bulletin of Entomological Research* **20**: 44. [Barrit1929]

Bartlett, B.R. 1953a. Retentive toxicity of field-weathered insecticide residues to entomophagous insects associated with citrus pests in California. *Journal of Economic Entomology* **46(4)**: 565-569. [Bartle1953a]

Bartlett, B.R. 1959. The biological control of black scale. *Sunkist Pest Control Circular* **No. 259**: 2 pp. [Bartle1959]

Bartlett, B.R. 1963. The contact toxicity of some pesticide residues to Hymenopterous parasites and Coccinellid predators. *Journal of Economic Entomology* **56(5)**: 694-698. [Bartle1963]

Bartlett, B.R. 1964. Integration of chemical and biological control. Pages 489-511 *in*: DeBach, P. & Schlinger, E.I., Eds. Biological Control of Insect Pests and Weeds. London: Chapman & Hall. 844 pp. [Bartle1964]

Bartlett, B. & DeBach, P. 1952. New natural enemies of avocado pests. *Citrus leaves* **32**: 16-17. [BartleDe1952]

Bartlett, B.R. & Fisher, T.W. 1950. Laboratory propagation of *Aphytis chrysomphali* for release to control California red scale. *Journal of Economic Entomology* **43**: 802-806. [BartleFi1950]

Bartlett, B.R. & Van den Bosch, R. 1964. Foreign exploration for beneficial organisms. Pages 283-304. *in*: DeBach, P. & Schlinger, E.I., Eds. Biological Control of Insect Pests and Weeds. London: Chapman & Hall. 844 pp. [BartleVa1964]

Bartra, C.E. 1974. Biología de *Selenaspidus articulatus* Morgan y sus principales controladores biológicos. *Revista Peruana de Entomología* **17**: 60-68. [Bartra1974]

Bartra, P.C.E. 1977. Observaciones biológicas sobre la 'quersa de laurel (*Aspidiotus hederae* Vallot, Homoptera: Diaspididae). *Revista Peruana de Entomología* **19**: 43-48. [Bartra1977]

Bassino, J.-P., Pelissier, P., Tisseur, M. & Codou, S. 1975. Les résultats de l'utilisation du parasite *Prospaltella perniciosi* Tower pour lutter contre le Pou de San José *Quadraspidiotus perniciosus* Comst. cochenille nuisible aux arbres frutiers. *La Défense des végétaux* **29(171)**: 21-23,26-42. [BassinPeTi1975]

Basu, A.C., Nath, D.K. & Chaterjee, P.B. 1969. Insects occurring on the orange plant (*Citrus reticulata* Blanco) in Darjeeling District, West Bengal, India. *Proceedings of the Zoological Society (Calcutta)* **11**: 169-178. [BasuNaCh1969]

Batra, R.C., Sandhu, G.S. & Sohi, A.S. 1989. Outbreak of California Red Scale on citrus and suppression through Coccinellid predators in the Punjab. *Bulletin of Entomology* **28(2)**: 161-162. [BatraSaSo1989]

Battaglia, D. 1989. Notizie biologiche sul *Coccidencyrtus dynaspidioti* sp. n. (Hymenoptera: Encyrtidae), parassitoide di *Dynaspidiotus britannicus* (Newstead) (Homoptera: Diaspididae). *Bollettino del Laboratorio di Entomologia Agraria 'Filippo Silvestri'* **45**: 149-165. [Battag1989]

Battaglia, D. & Viggiani, G. 1985. Primi reperti bioetologici sull'*Encarsia aspidioticola* (Mercet), nuovo parassita di *Dynaspidiotus britannicus* (Newst.) in Italia. *Atti XIV Congresso Nazionale Italiano di Entomologia, Palermo, 1985* 337-340. [BattagVi1985]

Battaglia, D. & Viggiani, G. 1987. Natural enemies of the holly scale *Dynaspidiotus britannicus* (Newstead) in Italy. *Bollettino del Laboratorio di Entomologia Agraria 'Filippo Silvestri'* **43** (Suppl.): 139-142. [BattagVi1987]

Bazarov, B.B. & Shmelev, G.P. 1970. [Hard scales brought in with plants and fruit.]. [In Russian]. *Izvestiya Akademii Nauk Tadzhikskoi SSR. Otdelenie Biologicheskikh Nauk* **41**: 109-111. [BazaroSh1970]

Bazarov, B.B. & Shmelev, G.P. 1971. [Scale insects (Homoptera, Coccoidea) of Tadzhikistan and the adjoining regions of middle Asia]. [In Russian]. *In* "Fauna of Tadzhikistan SSR.". Akad. Nauk Tadzh. SSR Inst. Zool. Parazitol. 11: 238 pp. [BazaroSh1971]

Beardsley, J.W. 1965. Notes and exhibitions. *Proceedings of the Hawaiian Entomological Society* **19**: 12, 14, 38. [Beards1965]

Beardsley, J.W. 1966. Insects of Micronesia. Homoptera: Coccoidea. *Insects of Micronesia* **6**: 377-562. [Beards1966]

Beardsley, J.W. 1970. *Aspidiotus destructor* Signoret, an armored scale pest new to the Hawaiian Islands. *Proceedings of the Hawaiian Entomological Society* **20**: 505-508. [Beards1970]

Beardsley, J.W., Davidson, J.A., Howell, J.O., Kosztarab, M., Miller, D.R., Nakahara, S. & Stoetzel, M. 1976. Syllabus for workshop on scale insect identification. 115 [BeardsDaHo1976]

Beardsley, J.W. & Gonzalez, R. H. 1975. The biology and ecology of armored scales. *Annual Review of Entomology* **20**: 47-73. [BeardsGo1975]

Beattie, G.A.C. 1984. Parasite lost, predator found - the ecology of citrus red scale in China. *Rural Newsletter* **98**: 25-29. [Beatti1984]

Beattie, G.A.C. & Gellatley, J.G. 1983. Citrus scale insects. *Agfacts (Department of Agriculture, New South Wales)* **220-226**: 6 pp. [BeattiGe1983]

Beattie, G.A.C. & Huang, M.D. 1988. The role of natural enemies in the control of citrus red scale, *Aonidiella aurantii* (Maskell) (Hemiptera: Diaspididae). *Proceedings of the 18th International Congress of Entomology* 386. [BeattiHu1988]

Beatty, H.A. 1944. The insects of St. Croix, V.I. *J. Agric. Univ.* **28(3/4)**: 114-172. [Beatty1944]

Beauchard, J. 1964. [The eradication of San José scale in the southwest of France.] L'éradication d'un foyer de pou de San José dans le sud-ouest de la France]. [In French]. *Phytoma* **16**: 45-49. [Beauch1964]

Beccari, F. 1971. [Contribution to the knowledge of the entomofauna of Saudi Arabia. First list of insects, mites and nematodes]. [In Italian]. *Rivista di Agricoltura Subtropicale e Tropicale. Firenze* **65**: 178-211. [Beccar1971]

Beckwith, C.S. 1945. The insects of the cultivated blueberry. *Circular (New Jersey Department of Agriculture)* **No. 356**: 43-54. [Beckwi1945]

Bederski, K.A. 1969. Insect population surveys and oil rotenone sprays: their use in Peruvian citrus orchards. Pages 933-940. *in*: Chapman, H.D. (Ed.). Proceedings First International Citrus Symposium. Vol. 2. Riverside: University of California. [Beders1969]

Bedford, E.C.G. 1968a. The biological control of red scale, *Aonidiella aurantii* (Mask.), on citrus in South Africa. *Journal of the Entomological Society of Southern Africa* **31**: 1-15. [Bedfor1968a]

Bedford, E.C.G. 1968b. Biological and chemical control of citrus pests in the Western Transvaal: An integrated spray programme. *South African Citrus Journal* **417**: 9,11,13,15,17,19,21-28. [Bedfor1968b]

Bedford, E.C.G. 1969. The selection of insecticides for the integration of biological and chemical control of citrus pests. *South African Citrus Journal* **431**: 3,5,7,9-10. [Bedfor1969]

Bedford, E.C.G. 1973. Citrus scale insects: Biological control proves successful. *Citrus & Subtropical Fruit Journal* **No. 470**: 5,7,9,11. [Bedfor1973]

Bedford, E.C.G. 1978. Other scale insects of minor importance. Pages 129-133 *in*: Bedford, E.C.G. (Ed.). Citrus Pests in the Republic of South Africa [Science Bulletin, No. 391]. Department of Agricultural Technical Services, South Africa. 253 pp. [Bedfor1978]

Bedford, E.C.G. 1981. Integrated citrus spray program for the Lowveld. Farming in South Africa. Citrus H.47.1 Pretoria, South Africa: Dept. of Agriculture & Water Supply. 6 pp. [Bedfor1981]

Bedford, E.C.G. 1989. The biological control of circular purple scale, *Chrysomphalus aonidum* (L.), on citrus in South Africa. Technical Communication No. 218. Pretoria, South Africa: Dept. of Agriculture & Water Supply. 16 pp. [Bedfor1989]

Bedford, E.C.G. 1989a. Red Scale on Citrus. Farming in South Africa, Citrus H.2. Pretoria, South Africa: Dept. of Agriculture & Water Supply. 4 pp. [Bedfor1989a]

Bedford, E.C.G. 1990. 3.7 Mechanical Control: High-pressure Rinsing of Fruit. Pages 507-513 *in*: Rosen, D. (Ed.). Armored Scale Insects, Their Biology, Natural Enemies and Control [Series title: World Crop Pests, Vol. 4B]. Amsterdam, the Netherlands: Elsevier. 688 pp. [Bedfor1990]

Bedford, E.C.G. 1998. Family Diaspididae: Armoured scale insects: Red scale: *Aonidiella aurantii* (Maskell). Pages 132-144. *in*: Bedford, E.C.G., Van den Berg, M.A. & De Villiers, E.A., Eds. Citrus Pests in the Republic of South Africa. 2nd ed. Nelspruit: Institute for Tropical and Subtropical Crops. 288 pp. [Bedfor1998]

Bedford, E.C.G. & Bohler, D. 1981. Integrated pest-management program for citrus - red scale under biological control: 1981-82. *Farming in South Africa* **H-47**: 1-15. [BedforBo1981]

Bedford, E.C.G. & Cilliers, C.J. 1994. The role of *Aphytis* in the biological control of armored scale insects on citrus in South Africa. Pages 143-179 *in*: Rosen, D., ed. Advances in the Study of Aphytis (Hymenoptera: Aphelinidae). Andover, UK: Intercept Limited. [BedforCi1994]

Bedford, E.C.G. & Deacon, V. 1984. An integrated pest control program for citrus in the Transvaal Lowveld. Farming in South Africa. Citrus H.47.1. Pretoria, South Africa: Dept. of Agriculture & Water Supply. 7 pp. [BedforDe1984]

Bedford, E.C.G. & Dorey, H. 1977. H.47 Integrated pest management programme for citrus, with red scale under biological control: 1977/78. *Farming in South Africa* **H.47**: 1-10. [BedforDo1977]

Bedford, E.C.G. & Georgala, M.B. 1978. Red scale *Aonidiella aurantii* (Mask.). Pages 109-118 *in*: Bedford, E.C.G. (Ed.). Citrus Pests in the Republic of South Africa [Science Bulletin, No. 391]. Department of Agricultural Technical Services, South Africa. 253 pp. [BedforGe1978]

Bedford, E.C.G. & Grobler, J.H. 1981. The current status of the biological control of red scale, *Aonidiella aurantii* (Mask.), on citrus in South Africa. *Proceedings of the International Society of Citriculture* 616-620. [BedforGr1981]

Bedford, E.C.G., Van den Berg, M.A. & De Villiers, E.A. 1998. Citrus Pests in the Republic of South Africa 2nd ed. Nelspruit: Institute for Tropical and Subtropical Crops. 288 pp. [BedforVaDe1998]

Bedford, E.C.G., Vercueil, S.W. & Deacon, V. 1985. A general guide to the integrated control of citrus pests with red scale under biological control. *Farming in South Africa* **H.47**: 1-16. [BedforVeDe1985]

Bedford, E.C.G., Vegoes-Houwens, G., Kok, I.B. & Vercueil, S.W. 1992. Impact of 46 pesticides on non-target pests under biological control in citrus orchards. Pretoria, South Africa: Citrus and Subtropical Fruit Research Institute. 112 pp. [BedforVeKo1992]

Beffa, G.D. 1937. Cronaca delle malattie e dei parassiti osservati nell'anno 1937. *Bollettino del Laboratorio sperimentale e R. Osservatorio di fitopatologia. Torino* **14**: 63-70. [Beffa1937]

Beglyarov, G.A. & Smetnik, A.I. 1977. Seasonal colonization of entomophages in the U.S.S.R. Pages 283-328. *in*: Ridgway, R.L. & Vinson, S.B. (Eds.). Biological Control by Augmentation of Natural Enemies: Insect and Mite Control with Parasites and Predators. New York: Plenum Press. 480 pp. [BeglyaSm1977]

Beingolea G., O.D. 1967. Control biológico de las plagas de los cítricos en el Perú. *Revista Peruana de Entomología* **10**: 67-81. [Beingo1967]

Beingolea G., O.D. 1969a. Notas sobre la biología de *Selenaspidus articulatus* Morgan (Hom.: Diaspididae). *Revista Peruana de Entomología* **12**: 119-129. [Beingo1969a]

Beingolea G., O.D. 1969d. Biological control of citrus pests in Peru. Pages 827-838. *in*: Chapman, H.D. (Ed.). Proceedings First International Citrus Symposium. Vol. 2. Riverside: University of California. [Beingo1969d]

Beingolea G., O.D. 1977. Culmina con éxito Proyecto de Control Biológico de la queresa redonda de los cítricos, *Selenaspidus articulatus* Morgan, en el Perú. *Noticiero Entomologico Soc. Entomol. del Peru* **1**: 2 pp. [Beingo1977]

Bellio, G. 1929b. Risultati zoologici della missione inviata dalla R. Societá Geografica Italiana per l'esplorazione dell' oasi di giarabub (1926-1927). Coccidae (Hemiptera). *Annali della Genoa Museo Civ. Storia Natur.* **53**: 187-194. [Bellio1929b]

Bellio, G. 1932. Esperienze col calore contro la bianca-rossa o crisomfalo degli agrumi (Chrysomphalus dictyospermi (Morg. Leon) ed altre cocciniglie. Messina: S.A. Industrie Grafiche meridionale. 22 pp. [Bellio1932]

Bellio, G. 1939. Hemiptera, Coccidae. *Missione Biologica nel Paese dei Borana. Raccolte Zoologiche. Roma* **3**: 225-239. [Bellio1939]

Bellows, T.S. 2001. Restoring population balance through natural enemy introductions. *Biological Control* 21(3): 199-205. [Bellow2001]

Belyavskaya, A.F. 1938. [Fumigation of fresh fruits by hydorcyanic acid gas with a view to destroying *Aspidiotus perniciosus* Comst. Pages 202-221 *in:* Kiritchenko, A.N. (Ed.), Works of Quarantine Laboratories, Leningrad, Sel'khozgiz. 272 pp. [Belyav1938]

Ben David, T. & Greenberg, Y. 2000. The control of California red scale (*Aonidiella aurantii* Mask.) by drench application of imidacloprid on lemon trees. [In Hebrew with summary in English.] *Alon Hanotea* **54**: 11-12, 423-425. [BenDavGr2000]

Ben-Dov, Y. 1972. A new species of *Pseudotargionia* Lindinger (Homoptera: Diaspididae) from South Africa. *Journal of the Entomological Society of Southern Africa* **35**: 309-312. [BenDov1972]

Ben-Dov, Y. 1974b. On a new genus and species and on some described armoured scale insects (Homoptera: Diaspididae) mainly from southern Africa. *Journal of the Entomological Society of Southern Africa* **37**: 315-325. [BenDov1974b]

Ben-Dov, Y. 1974c. New species of armoured scale insects (Homoptera: Diaspididae) from South Africa. *Phytophylactica* **6**: 19-30. [BenDov1974c]

Ben-Dov, Y. 1976. A new species of *Chortinaspis* Ferris from the Sinai peninsula (Homoptera: Diaspididae). *Revue de Zoologie Africaine* **90**: 204-208. [BenDov1976]

Ben-Dov, Y. 1980. Observations on scale insects (Homoptera: Coccoidea) of the Middle East. *Bulletin of Entomological Research* **70**: 261-271. [BenDov1980]

Ben-Dov, Y. 1980a. On the abdominal segmentation in the adult female of armoured scale insects of the Odonaspidini and Antakaspidini (Diaspididae). *Israel Journal of Entomology* **14**: 71-79. [BenDov1980a]

Ben-Dov, Y. 1981c. A catalogue of the Conchaspididae (Insecta, Homoptera, Coccoidea) of the world. *Annales de la Société Entomologique de France* **17**: 143-156. [BenDov1981c]

Ben-Dov, Y. 1985. Coccoidea. *In* "Insects of Southern Africa," edited by Scholtz, C.H. & Holm, E. Butterworths,. Durban. 502 pp. (168-175). [BenDov1985]

Ben-Dov, Y. 1988. Manna scale, *Trabutina mannipara* (Hemprich & Ehrenberg) (Homoptera: Coccoidea: Pseudococcidae). *Systematic Entomology* **13**: 387-392. [BenDov1988]

Ben-Dov, Y. 1988b. A taxonomic analysis of the armored scale tribe Odonaspidini of the world (Homoptera: Coccoidea: Diaspididae). *United States Department of Agriculture Technical Bulletin* No. 1723, 142 pp. [BenDov1988b]

Ben-Dov, Y. 1990a. 1.1.4.2 Taxonomic characters. Pages 85-91 *in*: Rosen, D. (Ed.). Armored Scale Insects, Their Biology, Natural Enemies and Control [Series title: World Crop Pests, Vol. 4A]. Amsterdam, The Netherlands: Elsevier. 384 pp. [BenDov1990a]

Ben-Dov, Y. 1990c. 1.1.4.4 Classification of diaspidoid and related Coccoidea. Pages 97-128 *in*: Rosen, D. (Ed.). Armored Scale Insects, Their Biology, Natural Enemies and Control [Series title: World Crop Pests, Vol. 4A]. Amsterdam, The Netherlands: Elsevier. 384 pp. [BenDov1990c]

Ben-Dov, Y. 1990d. 1.4.8 Relationships with ants. Pages 339-343 *in*: Rosen, D., Ed. Armored Scale Insects, Their Biology, Natural Enemies and Control [Series title: World Crop Pests, Vol. 4A]. Amsterdam, The Netherlands: Elsevier. 384 pp. [BenDov1990d]

Ben-Dov, Y. 1990e. 3.9.10 Bamboo. Pages 655-660 *in*: Rosen, D. (Ed.). Armored Scale Insects, Their Biology, Natural Enemies and Control [Series title: World Crop Pests, Vol. 4B]. Amsterdam, The Netherlands: Elsevier. 688 pp. [BenDov1990e]

Ben-Dov, Y. 1990h. 1.1.4 Systematics. 1.1.4.1 Status of our knowledge of diaspidoid systematics. Pages 81-84 *in*: Rosen, D. (Ed.). Armored Scale Insects, Their Biology, Natural Enemies and Control [Series title: World Crop Pests, Vol. 4A]. Amsterdam, The Netherlands: Elsevier. 384 pp. [BenDov1990h]

Ben-Dov, Y. 1993. A systematic catalogue of the soft scale insects of the world (Homoptera: Coccoidea: Coccidae) with data on geographical distribution, host plants, biology and economic importance. Flora & Fauna Handbook, No. 9.. Sandhill Crane Press, Gainesville, FL. 536 pp. [BenDov1993]

Ben-Dov, Y. 1994. A systematic catalogue of the mealybugs of the world (Insecta: Homoptera: Coccoidea: Pseudococcidae and Putoidae) with data on geographical distribution, host plants, biology and economic importance.. Intercept Limited, Andover, UK. 686 pp. [BenDov1994]

Ben-Dov, Y. 2001. [A note on quarantine interceptions of the San José scale.]. [In Hebrew]. *Alon Hanotea* **55**: 141-142. [BenDov2001]

Ben-Dov, Y. 2001c. Taxonomy of *Aspidiotus aharonii* Bodenheimer, 1924 (Hem. Coccoidea, Diaspididae) with new synonymy. *Entomologist's Monthly Magazine* **137**: 161. [BenDov2001c]

Ben-Dov, Y. & Matile-Ferrero, D. 1984. On the association of ants, genus *Melissotarsus* (Formicidae), with armoured scale insects (Diaspididae) in Africa. *Proceedings of the 10th International Symposium of Central European Entomofaunistics, Budapest, 15-20 August 1983* 278-380. [BenDovMa1984]

Ben-Dov, Y. & Matile-Ferrero, D. 1999. Nomenclature of the oleander scale (Hemiptera, Coccoidea, Diaspididae). *Revue Française d'Entomologie* **21(1)**: 5-8. [BenDovMa1999]

Ben-Dov, Y. & Marotta, S. 2001a. Stabilizing the name *Aspidiotus nerii* Bouché1833 , (Hem., Coccoidea, Diaspididae). *Bulletin de la Société Entomologique de France* **106(2)**: 181-191. [BenDovMa2001Sa]

Ben-Dov, Y. & Marotta, S. 2001b. Addenda to "Stabilizing the name *Aspidiotus nerii* Bouché1833 , (Hem., Coccoidea, Diaspididae)". *Bulletin de la Société Entomologique de France* **106(4)**: 426-427. [BenDovMa2001Sb]

Ben-Dov, Y. & Rosen, D. 1969. Efficacy of the California red scale on citrus in Israel. *Journal of Economic Entomology* **62(5)**: 1057-1060. [BenDovRo1969]

Ben-Dov, Y. & Venezian, A. 1983. New scale insects (Coccoidea) - orchard pests in the Arava. *Phytoparasitica* **11**: 124-125. [BenDovVe1983]

Ben-Dov, Y. & Williams, D.J. 2003. The identity of *Aspidiotus guianensis* Lindinger, 1957 (Hem., Coccoidea, Diaspididae). *Bulletin de la Société Entomologique de France* **108(2)**: 166-167. [BenDovWi2003]

Ben-Dov, Y. & Wysoki, M. 1990. *Aonidiella orientalis* (Maskell) - a mango pest in Israel. *Hassadeh* **70**: 1242. [BenDovWy1990]

Bénassy, C. 1954. Essai expérimental de constitution d'un foyer de *Prospaltella berlesei* How. dans le Lyonnais. *Comptes Rendus des Séances de l'Académie d'Agriculture de France* **40**: 528-530. [Benass1954]

Bénassy, C. 1957. Influence du facteur "exposition" sur la répartition des microhyménoptères parasites de Coccoidea-Diaspididae. *Entomophaga* **2(4)**: 283-291. [Benass1957]

Bénassy, C. 1958b. Remarques sur l'écologie de *Quadraspidiotus perniciosus* Comst. dans le midi Méditerranéen (Hom. Diaspidinae). *Entomophaga* **3(2)**: 93-108. [Benass1958b]

Bénassy, C. 1958d. Influence de l'hôte dans la croissance endoparasitaire de quelques Hyménoptères Chalcidiens, parasites de cochenilles Diaspines. *Compte Rendu de l'Académie des Sciences. Paris* **246(1)**: 179-181. [Benass1958d]

Bénassy, C. 1959b. Emplois complémentaires de la lutte biologique et de la lutte chimique contre les Cochenilles-Diaspines *Pseudaulacaspis pentagona* Targ. et *Quadraspidiotus perniciosus* Comst. *IVe Congrès International de Phytopharmacie Hambourg* **1**: 867-872. [Benass1959b]

Bénassy, C. 1961a. Contribution à l'étude de l'influence de quelques facteurs écologiques sur la limitation des pullulations de cochenilles-diaspines. Institut National de la Recherche Agronomique. Ser. A, No. 3747. 165 pp. [Benass1961a]

Bénassy, C. 1961b. Contribution a l'étude de l'influence de quelques facteurs écologiques sur la limitation des pullulations de cochenilles-diaspines. *Annales des Epiphyties* **12**: 1-157. [Benass1961b]

Bénassy, C. 1965a. Les cochenilles des agrumes dans le bassin méditerranéen méthodes de lutte .*Comptes Rendus (Journées Phytiatrie et Phytopharmacie circum-méditerranéennes)* 1:112-125. [Benass1965a]

Bénassy, C. 1967. Note sur une diaspine peu connue de l'olivier au Maroc: *Quadraspidiotus maleti* [Hom. Coccoidea]. *Annales de la Société Entomologique de France* **3**: 1133-1139. [Benass1967]

Bénassy, C. 1969. The biological control against coccids of citrus in the country of the French zone. Pages 793-799. *in*: Chapman, H.D. (Ed.). Proceedings First International Citrus Symposium. Vol. 2. Riverside: University of California. [Benass1969]

Bénassy, C. 1969a. La lutte biologique contre le pou de San José :Travaux menés par le Groupe de travail de l'OILB sur le pou de San José. *Publications de l'OEPP (Ser. A)* **No. 48**: 33-39. [Benass1969a]

Bénassy, C. 1970. Essai d'utilisation du pratique parasite spécifique du pou de San José en France. *Annales de Zoologie - Ecologie Animale* 45-50. [Benass1970]

Bénassy, C. 1971. La lutte biologique et son application pratique en arboriculture fruitière. *La Pomologie Française* **13(1)**: 9-23. [Benass1971]

Bénassy, C. 1977. Sobre algunos coccidos diaspinos de los citricos (*C. dictyospermi* Morg.; *L. beckii* Newm.; *U. yanonensis* Kuw.). *Boletín del Servicio de Defensa Contra Plagas Inspección Fitopatología* 3(1-2), 1-20. [Benass1977]

Bénassy, C. & Alexabdrakis, V. & Lionakis, S. 1984. Premier inventaire des cochenilles du kiwi en Crète. *Fruits* **39**: 325-327. [BenassAlLi1984]

Bénassy, C. & Bianchi, H. 1960. Sur l'écologie de *Prospaltella perniciosi* Tower (Hym. Aphelinidae), parasite spécifique importé de *Quadraspidiotus perniciosus* Comst. (Hom. Diaspidinae). *Entomophaga* **5(2)**: 165-191. [BenassBi1960]

Bénassy, C. & Bianchi, H. 1967. Note sur la faune des diaspines agrumicoles du littoral sud-est de la France. *Annales de la Société Entomologique de France* **3**: 247-256. [BenassBi1967]

Bénassy, C. & Bianchi, H. 1974. Observations sur *Aonidiella aurantii* Mask. et son parasite indigène *Comperiella bifasciata* How. (Hymenoptera, Encyrtidae). *Bulletin SROP* **1974/3**: 39-50. [BenassBi1974]

Bénassy, G., Bianchi, H. & Milaire, H. 1960a. Incidence des traitements insecticides sur les parasites de coccides: action des traitements "d'été "contre *Pseudaulacaspis pentagona* Targ. sur son parasite spécifique: *Prospaltella berlesei* How. *Phytiatrie-Phytopharmacie* **9**: 29-36. [BenassBiMi1960a]

Bénassy, C., Bianchi, H. & Milaire, H. 1962. État actuel des recherches en cours sur la "lutte intégrée" dans le cas des coccides en France. *Entomophaga* **7(1)**: 59-68. [BenassBiMi1962]

Bénassy, C., Bianchi, H. & Milaire, H. 1964. Expérimentation en matière de lutte intégrée dans deux vergers français. *Entomophaga* **9**: 271-280. [BenassBiMi1964]

Bénassy, G., Bianchi, H. & Milaire, H. 1964a. Recherches sur l'utilisation de *Prospaltella perniciosi* Tow. en France. *Annales des Epiphyties* **15(4)**: 457-473. [BenassBiMi1964a]

Bénassy, G., Bianchi, H. & Milaire, H. 1964b. Observations sur l'incidence de quelques produits insecticides et fongicides sur l'association pou de San José et parasite spécifique. *Revue de Zoologie Agricole* **63**: 27-37. [BenassBiMi1964b]

Bénassy, C., Bianchi, H. & Milaire, H. 1967. Lutte biologique contre le Pou de San José. *[Publication] Ministère de l'Agriculture, Institut National de la Recherche Agronomique, Service de la Protection des Végétaux,* 5 pp. [BenassBiMi1967]

Bénassy, C., Bianchi, H. & Milaire, H. 1967a. Note sur l'efficacité en France de *Prospaltella perniciosi* Tow. *Entomophaga* **12**: 241-255. [BenassBiMi1967a]

Bénassy, C., Bianchi, H. & Milaire, H. 1971. La température, facteur prévisionnel du cycle évolutif de *Prospaltella perniciosi* Tow. *Entomophaga* **16**: 4, 411-420. [BenassBiMi1971]

Bénassy, C. & Euverte, G. 1966. Premières applications de la lutte biologique contre *Aonidiella aurantii* au Maroc. *Al Awamia* **21**: 19-26. [BenassEu1966]

Bénassy, C. & Euverte, G. 1967. Perspectives nouvelles dans la lutte contre *Aonidiella aurantii* au Maroc (Hom. Diaspididae). *Entomophaga* **12(5)**: 449-459. [BenassEu1967]

Bénassy, C. & Euverte, G. 1967a. Notes sur *Chrysomphalus dictyospermi* au Maroc. *Al Awamia* **24**: 95-111. [BenassEu1967a]

Bénassy, C. & Euverte, G. 1968. Essai d'utilisation pratique de la lutte biologique contre le Pou de Californie (*Aonidiella aurantii*) au Maroc. *Al Awamia* **28**: 1-60. [BenassEu1968]

Bénassy, C. & Euverte, G. 1970. La lutte biologique contre *A. aurantii* au Maroc. *Al Awamia* **37**: 95-101. [BenassEu1970]

Bénassy, C. & Euverte, G. 1970a. Note sur l'action de deux espèces du genre *Aphytis* en tant qu'agents de lutte biologique contre deux coccides des *citrus* (*Aonidiella aurantii* Mask. et *Chrysomphalus dictyospermi* Morg.) au Maroc. *Annales de Zoologie - Ecologie Animale* **2(3)**: 357-372. [BenassEu1970a]

Bénassy, C., Mathys, G., Neuffer, G., Milaire, H., Bianchi, H. & Guignard, E. 1968. L'utilisation pratique de *Prospaltella perniciosi* Tow., parasite du pou San José *Quadraspidiotus perniciosus* Comst. *Entomophaga* **(No. 4)**: 1-28. [BenassMaNe1968]

Bénassy, C. & Soria, F. 1964. Observations écologiques sur les cochenilles diaspines nuisibles aux agrumes en Tunisie. *Annales de l'Institut National Agronomique de Tunisie* **37**: 193-222. [BenassSo1964]

Bénassy, C., Viggiani, G. & Rosen, D. 1979. La lutte intégrée en agrumiculture. Pages 281-287. *in*: Proceedings: Internationales Symposium der IOBC/WPRS über Integrierten Pflanzenschutz in der Landund Forstwirtschaft. Wien, Austria: OILB/SROP. 648 pp. [BenassViRo1979]

Bene, G. del, Gargani, E. & Landi, S. 2000. Evaluation of plant extracts for insect control. *Journal of Agriculture and Environment for International Development* **94(1)**: 43-61. [BeneGaLa2000]

Benfatto, D. & Carroccio, A. 1996. Cocciniglia rossa forte degli agrumi: quando intervenire?. *Informatore Agrario* **52**: 38, 73-76. [BenfatCa1996]

Benfatto, D. & Carroccio, A. 2002. Piu semplice il monitoraggio della cocciniglia rossa forte degli agrumi. (In Italian.) *Informatore Agrario* **58(21)**:73-75. [BenfatCa2002]

Benfatto, D., Conti, F. & Tumminelli, R. 1998. Lotta biologica e chimica alla cocciniglia rossa forte degli agrumi, *Aonidiella aurantii* Mask., in Sicilia. Pages 217-222 *in* [Proceedings of Giornate Fitopatologiche, held at Sicily, Ragusa, Italy, from 3-7 May 1998. [BenfatCoTu1998]

Bennett, F.D. 1974. Criteria for determination of candidate hosts and for selection of biotic agents. Pages 87-96 *in*: Maxwell, F.G. & Harris, F.A. (Eds.). Proceedings of the Summer Institute on Biological Control of Plant Insects and Diseases. Jackson: University Press of Mississippi. 647 pp. [Bennet1974]

Bennett, F.D., Rosen, D., Cochereau, P. & Wood, B.J. 1976. Biological control of pests of tropical fruits and nuts. Pages 359-395. *in*: Huffaker, C.B. & Messenger, P.S. (Eds.). Theory and Practice of Biological Control. New York: Academic Press. 788 pp. [BennetRoCo1976]

Bentley, G.M. 1913. The San José scale in Tennessee with methods for its control. Bulletin of the Agricultural Experiment Station of the University of Tennessee (Bull. No. 98) 10(4): 39-59. [Bentle1913]

Bentley, G.M. & Bartlett, I.L. 1931. Insects and allied pests of greenhouse plants with recommendations for their control. *Bulletin (Division of Plant Disease Control, Tennessee, State Department of Agriculture)* **No. 57**: [61 pp.]. [BentleBa1931]

Bentley, W.J., Coates, W.W., Hasey, J., Hendricks, C.L., Olson, W.H., Pickel, C., Sibbett, G.S. & Van Steenwyk, R.A. 2000. Pests of walnut. *UC Pest Management Guidelines* . [BentleCoHa2000]

Bentley, W.J., Hendricks, L., Duncan, R., Silvers, C., Martin, L., Gibbs, M. & Stevenson, M. 2001. BIOS and conventional almond orchard management compared. *California Agriculture* **55(5)**: 12-19. [BentleHeDu2001]

Bentley, W., Rice, R.E., Brazzle, J. & Day, K. 2000. Pests of nectarine. *UC Pest Management Guidelines* . [BentleRiBr2000]

Bentley, W.J., Rice, R.E. & Day, K.R. 2000. Pests of plum. *UC Pest Management Guidelines* . [BentleRiDa2000]

Benvenga, S.R., Gravena, S. & Bortoli, S.A. de 1997. Predaçao da cochinilha *Selenaspidus articulatus* Morgan (Hemiptera: Diaspididae) por *Pentilia egena* Mulsant (Coleoptera: Coccinellidae). Pages 105-106 *in*: Bento, J.M.S. & Delalibera, I. (Eds.). [16th Brazilian Congress of Entomology] 16o Congresso Brasileiro de Entomologia. Salvador: Sociedade Entomológica do Brasil/EMBRAPA-CNPMF. 400 pp. [BenvenGrBo1997]

Beran, F. 1962. Bemerkenswerte Nebenwirkung von Sevin gegen San José-Schildlaus. *Pflanzenarzt* **15**: 135-136. [Beran1962]

Beran, F. 1963. San José scale control in Austria. *Publications de l'OEPP. Serie A (European and Mediterranean Plant Protection Organisation)* **34**: 94-96. [Beran1963]

Berger, E.W. 1921. Natural enemies of scale insects and whiteflies in Florida. *Quarterly Bulletin of the Florida State Plant Board* **5**: 141-154. [Berger1921]

Berger, E.W. 1921a. Natural enemies of scale insects and whiteflies in Florida. *Review of Applied Entomology. Series A: Agricultural* **9**: 333-334. [Berger1921a]

Berger, E.W. 1932. The latest concerning natural enemies of citrus insects. *Proceedings of the Florida State Horticultural Society* **45**: 131-136. [Berger1932]

Berger, E.W. 1942a. Status of the friendly fungus parasites of armored scale-insects. *Florida Entomologist* **25**: 26-29. [Berger1942a]

Bergmann, E.C., Stradioto, M.F. & Brisolla, A.D. 1988. Ocorrencia de *Selenaspidus articulatus* (Morgan, 1988) em cultura de Seringueira (*Hevea brasiliensis* Muell. Arg. em Olimpia, Sao Paulo. *O Biológico* **54(1-6)**: 27-28. [BergmaStBr1988]

Berlese, A. 1895a. Le cocciniglie Italiane viventi sugli argumi. Parte III. I Diaspiti. [Cap. I. Note di sistematica e descrizione delle specie.]. *Rivista di Patologia Vegetale. Firenze* **4**: 74-170. [Berles1895a]

Berlese, A. 1895c. La cocciniglie Italiane viventi sugli agrumi. Parte III. I Diaspiti (cap. II. Osservazioni anatomiche. cap. III. Cenni di biologia e danni che I diaspiti recano arli agrumi.). *Rivista di Patologia Vegetale. Firenze* **4**: 195-292. [Berles1895c]

Berlese, A. 1896. Le cocciniglie Italiane viventi sugli agrumi. Parte II. I. *Lecanium. Rivista di Patologia Vegetale. Firenze* **3**: 49-100. [Berles1896]

Berlese, A. 1896b. Le cocciniglie italiane viventi sugli agrumi. Capitolo 2. Organi della digestione. *Rivista di Patologia Vegetale. Firenze* **5**: 3-73. [Berles1896b]

Berlese, A. 1911. Come progredisce la *Prospaltella berlesei* in Italia. *Redia* **7**: 436-461. [Berles1911]

Berlese, A. & Leonardi, G. 1896. Diagnosi di cocciniglie nuove. (Cont.). *Rivista di Patologia Vegetale. Firenze* **4**: 345-352. [BerlesLe1896]

Berlese, A. & Leonardi, G. 1898. Chermotheca Italica ... Cocciniglie Raccolte in Italia. 3 ed. Portici: Vesuviano. 51-75. [BerlesLe1898]

Berlese, A. & Leonardi, G. 1898a. Notizie intorno alle cocciniglie Americane che Minacciano la frutticultura Europea. *Annali di Agricoltura. (Ministero di Agricoltura, Industria e Commercio). Firenze, Roma* **2**: 1-142. [BerlesLe1898a]

Berlese, A. & Leonardi, G. 1899. Cocciniglie che minacciano la Frutticultura Europea: 4. Storia della scoperta e diffusione. *Rivista di Patologia Vegetale. Firenze* **7**: 253-273. [BerlesLe1899]

Berlese, A. & Paoli, G. 1916. Un endofago esotico efficace contro il *Chrysomphalus dictyospermi* Morg. *Redia* **11**: 305-307. [BerlesPa1916]

Berlinger, M.J. & Barak, R. 1981. The phenology of *Antonina graminis* (Homoptera, Pseudococcidae) and *Odonaspis ruthae* (Homoptera, Diaspididae) on lawn grasses in Israel. *Zeitschrift für Angewandte Entomologie* **91**: 62-67. [BerlinBa1981]

Berlinger, M.J., Segre, L., Podoler, H. & Taylor, R.A.J. 1999. Distribution and abundance of the oleander scale (Homoptera : Diaspididae) on jojoba. *Journal of Economic Entomology* **92(5)**: 1113-1119. [BerlinSePo1999]

Berro, J. 1927. Estación de fitopatología agricola de Almería. *Boletín de Patología Vegetal y Entomología Agrícola. Madrid* **2**: 55-59. [Berro1927]

Berry, J.A. 1991. Hymenoptera primary types in the New Zealand Arthropod Collection. *New Zealand Journal of Zoology* **18(3)**: 323-341. [Berry1991]

Berry, J.A., Morales, C.F., Hill, M.G., Lofroth, B.J. & Allan, D.J. 1989. The incidence of three diaspid scales on kiwifruit in New Zealand. *Proceedings of the Forty Second New Zealand Weed and Pest Control Conference. Palmerston North, New Zealand (Fruit Crops II)* : 182-186. [BerryMoHi1989]

Bertels M., A. 1956. Entomologia Agrícola Sul -- Brasileira. Série Didáctica, No. 16. Rio de Janeiro: Ministerio da Agricultura, Serviço de Informaçao Agrícola. 458 pp. [Bertel1956]

Bertels, A. & Bauke, O. 1966. Segunda relação das pragas das plantas cultivadas no Rio Grande do Sul. *Pesquisa Agropecuaria Brasileira* **1**: 17-46. [BertelBa1966]

Beshear, R.J. 1975. *Aleurodothrips fasciapennis* (Thysanoptera: Tubulifera) a predator of *Aspidiotus nerii* (Homoptera: Diaspididae). *Journal of the Georgia Entomological Society* **10**: 223-224. [Beshea1975]

Beshear, R.J. & Tippins, H.H. 1977. Collection records for the Hoke species of diaspidids. *Journal of the Georgia Entomological Society* **12**: 180-182. [BesheaTi1977]

Beshear, R.J., Tippins, H.H. & Howell, J.O. 1973. The armored scale insects (Homoptera: Diaspididae) of Georgia and their hosts. *Research Bulletin of the University of Georgia Agricultural Experiment Station* **146**: 1-15. [BesheaTiHo1973]

Bichina, T.I. Braiko, Ya.P. & Vaganova, L.D. 1985. [Quarantine coccids and methods of controlling them.]. [In Russian]. Kishinev, USSR: Kartya Moldovenyaske. 96 pp. [BichinYa1985]

Bichna, T.I. & Vaganova, L.D. 1985. [Ensuring the healthiness of planting material.]. [In Russian]. *Zashchita rastenii* **No. 11**: 31. [BichnaVa1985]

Biezanko, C.M. & Freitas, R.G. 1939. Catalogo dos insetos encontrados em Pelotas e seus arredores: Fasciculao II Homopteros. *Boletim (Escola de Agronomia "Eliseu Magiel", Pelotas, Rio Grande do Sul, Brazil)* **No. 26**: 1-18. [BiezanFr1939]

Biezanko, C.M. & Seta, F.D. 1940. Catalogo dos insetos encontrados no Rio Grande do Sul e seus arredores -- Homopteros (Homoptera). *O Campo* **11**: 67-68. [BiezanSe1940]

Biliotti, E. 1977. Augmentation of natural enemies in western Europe. Pages 341-347. *in*: Ridgeway, R.L. & Vinson, S.B. (Eds.). Biological Control by Augmentation of Natural Enemies: Insect and Mite Control with Parasites and Predators. New York: Plenum Press. 480 pp. [Biliot1977]

Biliotti, E., Bénassy, C., Bianchi, H. & Milaire, H. 1960. Premiers essais expérimentaux d'acclimatation en France de *Prospaltella perniciosi* Tower, parasite spécifique importé de *Quadraspidiotus perniciosus* Comst. *Comptes rendus des séances de l'Académie d'agriculture de France* **46**: 707-711. [BiliotBeBi1960]

Bilog Obra, G.P. & Morallo Rejesus, B. 2000. Biological studies of *Aphytis* sp. nr. *chrysomphali* (Hymenoptera: Aphelinidae). *Philippine Entomologist* **14(2)**: 137-147. [BilogOMo2000]

Bindra, O.S. & Varma, G.C. 1972. Pests of date-palm. *Punjab Horticultural Journal* **12(1)**: 14-24. [BindraVa1972]

Blahutiak, A. 1990. [Annotated list of *Quadraspidiotus perniciosus* Comstock entomophages.]. [In Slovensk]. *Entomol Probl* **20**: 9-19. [Blahut1990]

Blanchard, E. 1840. Septième famille. - Cocciniens; gallinsectes, Latr. *Histoire Naturelle des Insectes* **3**: 210-215. [Blanch1840]

Blanchard, E.E. 1922. Principales cochinillas de los citrus en Argentina. Primera parte: Cóccidos protegidos. *Boletín de la Argentina Ministerio de Agricultura* **27**: 387-98. [Blanch1922]

Blanchard, E.E. 1940. Apuntes sobre encirtidos Argentinos. *Anales de la Sociedad Científica Argentina* **130**: 106-128. [Blanch1940]

Blank, R.H., Gill, G.S.C. & Dow, B.W. 1997. Determining armoured scale distribution within kiwifruit blocks. Pages 293-297 *in*: Proceedings of the Fiftieth New Zealand Plant Protection Conference. Rotorua, New Zealand: New Zealand Plant Protection Society. [BlankGiDo1997]

Blank, R.H., Gill, G.S.C. & Dow, B.W. 1999. Armoured scale (Hemiptera : Diaspididae) distribution in kiwifruit blocks with reference to shelter. *New Zealand Journal of Crop and Horticultural Science* **27(1)**: 1-12. [BlankGiDo1999]

Blank, R.H., Gill, G.S.C. & Kelly, J.M. 2000. Development and mortality of greedy scale (Homoptera : Diaspididae) at constant temperatures. *Environmental Entomology* **29(5)**: 934-942. [BlankGiKe2000]

Blank, R.H., Gill, G.S.C., McKenna, C.E. & Stevens, P.S. 2000. Enumerative and binomial sampling plans for armored scale (Homoptera: Diaspididae) on kiwifruit leaves. *Journal of Economic Entomology* **93(6)**: 1752-1759. [BlankGiMc2000]

Blank, R.H., Gill, G.S.C. & Olson, M.H. 1994. Relationship between armoured scale infestations on kiwifruit leaves and fruit. Pages 304-309 *in*: Popay, A. J., ed. Proceedings of the Forty Seventh New Zealand Plant Protection Conference. Waitangi Hotel, New Zealand: Aug. 9-11, 1994. Rotorua, New Zealand: New Zealand Plant Protection Society. [BlankGiOl1994]

Blank, R.H., Gill, G.S.C., Olson, M.H. & Upsdell, M.P. 1995. Greedy scale (Homoptera: Diaspididae) phenology on taraire based on Julian day and degree-day accumulations. *Population Ecology* **24**: 1569-1575. [BlankGiOl1995]

Blank, R.H., Gill, G.S.C. & Olson, M.H. 1995a. Seasonal abundance of greedy scale (Homoptera: Diaspididae) and associated parasitoids on Taraire (*Beilschmiedia tarairi*). *Journal of Economic Entomology* **88**: 1634-1640. [BlankGiOl1995a]

Blank, R.H., Gill, G.S.C. & Stannard, K. 2000. Armoured scale infestation of fruit of Hort16A kiwifruit. *New Zealand Plant Protection* **53**: 205-210. [BlankGiSt2000]

Blank, R.H., Gill, G.S.C. & Upsdell, M.P. 1996. Greedy scale, *Hemiberlesia rapax* (Hemiptera: Diaspididae), phenology on kiwifruit leaves and wood. *New Zealand Journal of Crop and Horticultural Science* **24**: 239-248. [BlankGiUp1996]

Blank, R.H., Holland, P.T., Gill, G.S.C., Olson, M.H. & Malcolm, C.P. 1995. Efficacy and persistence of insecticide residues on fruit of kiwifruit to prevent greedy scale (Hemiptera: Diaspididae) crawler settlement. *New Zealand Journal of Crop and Horticultural Science* **23**: 13-23. [BlankHoGi1995]

Blank, R.H., Lo, P.L., Gill, G.S.C. & Olson, M.H. 1992. A residual bioassay technique to investigate scale crawler settlement on kiwifruit. *Proceedings of the Forty Fifth New Zealand Plant Protection Conference. Popay, A. J., ed. Wellington, New Zealand: August 11-13, 1992* : 174-179. [BlankLoGi1992]

Blank, R.H. & Olson, M.H. 1989. Lime sulphur for armoured scale and lichen control on kiwifruit. *Proceedings of the Forty Second New Zealand Weed and Pest Control Conference. Palmerston North, New Zealand (Fruit Crops II)* : 191-194. [BlankOl1989]

Blank, R.H. & Olson, M.H. 1989a. Toxicity of lime sulphur to armoured scale. *Proceedings of the Forty Second New Zealand Weed and Pest Control Conference. Palmerston North, New Zealand (Fruit Crops II)* : 187-190. [BlankOl1989a]

Blank, R.H. & Olson, M.H. 1990. Influence of temperature on toxicity of Diazinon to greedy scale. *Proceedings of the Forty Third New Zealand Weed and Pest Control Conference. Palmerston North, New Zealand* : 240-242. [BlankOl1990]

Blank, R.H. & Olson, M.H. 1990a. The toxicity of Methidathion against armoured scale pests on kiwifruit. *Proceedings of the Forty Third New Zealand Weed and Pest Control Conference. Palmerston North, New Zealand* : 243-246. [BlankOl1990a]

Blank, R.H., Olson, M.H. & Bell, D.S. 1987. Invasion of greedy scale crawlers (*Hemiberlesia rapax*) onto kiwifruit from taraire trees. *New Zealand Entomologist* **10**: 127-130. [BlankOlBe1987]

Blank, R.H., Olson, M.H., Clark, J.B. & Gill, G.S.C. 1993. Investigating two bee-safe materials for controlling latania scale on avocados during pollination. *Proceedings of the Forty Sixth New Zealand Plant Protection Conference. Christchurch, New Zealand: August 10-12, 1993; Rotorua, New Zealand* : 80-85. [BlankOlCl1993]

Blank, R.H., Olson, M.H. & Gill, G.S.C. 1992. Armoured scale, *Hemiberlesia lataniae* and *H. rapax* (Hemiptera: Diaspididae), infestation of kiwifruit rejected for export at two packhouses from 1987 to 1991. *New Zealand Journal of Crop and Hort Sci* **20**: 397-405. [BlankOlGi1992]

Blank, R.H., Olson, M.H. & Gill, G.S.C. 1993. An assessment of the quarantine risk of armoured scale (Hemiptera: Diaspididae) fruit infestations on kiwifruit. *New Zealand Journal of Crop and Hort Sci* **21**: 139-145. [BlankOlGi1993]

Blank, R.H., Olson, M. & Gill, G. 1994. Latania scale establish on avocados during pollination. *Orchardist of New Zealand* **67(4)**: 28-30. [BlankOlGi1994]

Blank, R.H., Olson, M.H. & Lo, P.L. 1990. Armoured scale (Hemiptera: Diaspididae) aerial invasion into kiwifruit orchards from adjacent host plants. *New Zealand Journal of Crop and Hort Science* **18(2-3)**: 81-87. [BlankOlLo1990]

Blank, R.H., Olson, M.H. & Lo, P.L. 1993. Mineral oil and diazinon to control armoured scale on kiwifruit. *Proceedings of the Forty Sixth New Zealand Plant Protection Conference. Christchurch, New Zealand: August 10-12, 1992; Rotorua, New Zealand* : 71-74. [BlankOlLo1993]

Blank, R.H., Olson, M.H., Tomkins, A.R., Greaves, A.J., Waller, J.E. & Pulford, W.M. 1994. Phytotoxicity investigations of mineral oil and diazinon sprays applied to kiwifruit in winter-spring for armoured scale control. *New Zealand Journal of Crop and Horticultural Science* **22**: 195-202. [BlankOlTo1994]

Blank, R.H., Olson, M.H. & Waller, J.E. 1991. Relative efficacy of chemicals for dormant season control of armoured scale on kiwifruit. *Proceedings of the Forty Fourth New Zealand Weed and Pest Control Conference. Palmerston North, New Zealand (Fruit Crops I)* : 75-79. [BlankOlWa1991]

Blay Goicoechea, M.A. 1993. [The Diaspididae Targioni-Tozzetti, 1868, from the Spanish peninsula and Baleares (Insecta: Hemiptera: Coccoidea)] La familia Diaspididae Targioni-Tozzetti, 1868 de España peninsular y Baleares (Insecta: Hemiptera: Coccoidea). Madrid: Editorial de la Universidad Complutense de Madrid. 736 pp. [BlayGo1993]

Bliss, C.I., Broadbent, B.M. & Watson, S.A. 1931. The life history of the California red scale *Chrysomphalus aurantii* Maskell: progress report. *Journal of Economic Entomology* **24**: 1222-1229. [BlissBrWa1931]

Bloesch, B. & Staubli, A. 1992. Le pou de San José, *Quadraspidiotus perniciosus* Comst. (Homoptera, Coccoidea). *Revue Suisse de Viticulture, d'Arboriculture et d'horticulture* **24(5)**: 305-310. [BloescSt1992]

Blumberg, D. 1971. Survival capacity of two species of *Cybocephalus* (Coleoptera: Cybocephalidae) under temperature and humidity extremes. *Entomologia Experimentalis et Applicata* **14**: 434-440. [Blumbe1971]

Blumberg, D. 1973. Field studies of *Cybocephalus nigriceps nigriceps* (J. Sahlberg) (Coleoptera: Cybocephalidae) in Israel. *Journal of Natural History* **7**: 567-571. [Blumbe1973]

Blumberg, D. 1973a. Survey and distribution of *Cybocephalidae* [*Coleoptera*] in Israel. *Entomophaga* **18(2)**: 125-131. [Blumbe1973a]

Blumberg, D. 1976. Adult diapause of *Cybocephalus nigriceps nigriceps* [*Col.: Cybocephalidae*]. *Entomophaga* **21(2)**: 131-139. [Blumbe1976]

Blumberg, D. 1990. 2.6.4 Host Resistance: Encapsulation of Parasites. Pages 221-228 *in*: Rosen, D. (Ed.). Armored Scale Insects, Their Biology, Natural Enemies and Control [Series title: World Crop Pests, Vol. 4B]. Amsterdam, the Netherlands: Elsevier. 688 pp. [Blumbe1990]

Blumberg, D. 1997. Parasitoid encapsulation as a defense mechanism in the Coccoidea (Homoptera) and its importance in biological control. *Biological Control* **8**: 225-236. [Blumbe1997]

Blumberg, D. & DeBach, P. 1979. Development of *Habrolepis rouxi* Compere (Hymenoptera: Encyrtidae) in two armoured scale hosts (Homoptera: Diaspididae) and parasite egg encapsulation by California red scale. *Ecological Entomology* **4**: 299-306. [BlumbeDe1979]

Blumberg, D., Ishaaya, I., Goldenberg, S., Tam, S. & Mendel, Z. 1994. Susceptibility to insect growth regulators (IGR's) of natural enemies of scale insects (Homoptera: Coccoidea). *Alon Hanotea* **48**: 434-440. [BlumbeIsGo1994]

Blumberg, D. & Luck, R.F. 1990. Differences in the rates of superparasitism between two strains of *Comperiella bifasciata* (Howard) (Hymenoptera: Encyrtidae) parasitizing California Red Scale (Homoptera: Diaspididae): an adaptation to circumvent encapsulation?. *Annals of the Entomological Society of America* **83(3)**: 591-597. [BlumbeLu1990]

Blumberg, D. & Swirski, E. 1974. The development and reproduction of Cybocephalid beetles on various foods. *Entomophaga* **19(4)**: 437-443. [BlumbeSw1974]

Blumberg, D. & Swirski, E. 1974a. Prey consumption and preying ability of three species of *Cybocephalus* (Coleoptera: Cybocephalidae). *Phytoparasitica* **2(1)**: 3-11. [BlumbeSw1974a]

Blumberg, D. & Swirski, E. 1982a. Comparative biological studies on two species of predatory beetles of the genus *Cybocephalus* [Col.: Cybocephalidae]. *Entomophaga* **27**: 67-76. [BlumbeSw1982a]

Blumberg, D., Swirski, E., Wysoki, M. & Izhar, Y. 1995. Biological control of pests in agriculture. *Hassadeh International* **2**: 33-37,44. [BlumbeSwWy1995]

Boboye, S.O. & Carman, G.E. 1975. Effects of insect growth regulators with juvenile hormone activity on the development of the California red scale. *Journal of Economic Entomology* **68**: 473-476. [BoboyeCa1975]

Bodenheimer, F.S. 1924. The Coccidae of Palestine. First report on this family. *Zionist Organization Institute of Agriculture and Natural History, Agricultural Experiment Station Bulletin* **1**: 1-100. [Bodenh1924]

Bodenheimer, F.S. 1924a. Observations about some scale-insects from El-Arish (Sinai) and Transjordania. *Bulletin de la Société Entomologique d'Egypte* **16**: 121-124. [Bodenh1924a]

Bodenheimer, F.S. 1926. Première note sur les cochenilles de Syrie. *Bulletin de la Société Entomologique de France* **1926**: 41-47. [Bodenh1926]

Bodenheimer, F.S. 1927a. Third note on the Coccidae of Palestine. *Agricultural Records, Tel Aviv* **2**: 177-186. [Bodenh1927a]

Bodenheimer, F.S. 1927b. First note on the zoocecidia of Palestine. *Bulletin de la Société Entomologique d'Egypte* **1926**: 64-88. [Bodenh1927b]

Bodenheimer, F.S. 1927c. □ber die für das Verbreitungsgebiet einer Art bestimmenden Faktoren. *Biologisches Zentralblatt* **47**: 25-44. [Bodenh1927c]

Bodenheimer, F.S. 1928. Eine kleine Cocciden-Ausbeute aus Griechenland. *Konowia* **7**: 191-192. [Bodenh1928]

Bodenheimer, F.S. 1929b. Die Coccidenfauna der Sinaihalbinsel. Pages 104-117. *in*: Bodenheimer, F.S. & Theodor, O. Ergebnisse der Sinai-expedition 1927 der Hebräischen Universität, Jerusalem. Leipzig: J.C. Hinrichs. 143 pp. [Bodenh1929b]

Bodenheimer, F.S. 1930. Contribution to the knowledge of citrus insects in Palestine. III. On the zoogeography and ecology of citrus insects, particularly those of Palestine and Mediterranean countries. *Hadar* **3**: 3-19. [Bodenh1930]

Bodenheimer, F.S. 1934. Contribution towards the knowledge of the red scale (*Chrysomphalus aurantii* Mask.) in Palestine. *Hadar* **7**: 139-148. [Bodenh1934]

Bodenheimer, F.S. 1935. Studies on the zoogeography and ecology of palearctic Coccidae I-III. *EOS* **10**: 237-271. [Bodenh1935]

Bodenheimer, F.S. 1935c. The Zoogeography of the Sinai Peninsula. *XII Congrès International de Zoologie,* Section V, Vol. **2**: 1138-1164. [Bodenh1935c]

Bodenheimer, F.S. 1936. Studies in animal populations. 1. Red scale populations in orange-groves of various ages in Palestine. *Hadar* **9**: 69,70,74. [Bodenh1936]

Bodenheimer, F.S. 1937. Prodromus Faunae Palestinae. Essai sur les Eléments Zoogéographiques et Historiques du Sud-Ouest du Sous-Règne Paléarctique. Le Caire Inst. Français d'Archéologie Orientale. T. 33, 286 pp. [Bodenh1937]

Bodenheimer, F.S. 1938. A decade of citrus entomology of Palestine. *Hadar* **11**: 45-50. [Bodenh1938]

Bodenheimer, F.S. 1943. A first survey of the Coccoidea of Iraq. *Government of Iraq, Ministry of Economics, Directorate General of Agriculture, Bulletin* **28**: 1-33. [Bodenh1943]

Bodenheimer, F.S. 1944. A conchaspid from Kurdistan and its importance in the phylogeny of the Diaspididae (Hemiptera). *Proceedings of the Royal Entomological Society of London* **13**: 4-5. [Bodenh1944]

Bodenheimer, F.S. 1944a. Additions to the Coccoidea of Iraq, with descriptions of two new species. *Bulletin de la Société Fouad 1er d'Entomologie* **28**: 81-84. [Bodenh1944a]

Bodenheimer, F.S. 1944b. Note on the Coccoidea of Iran, with description of new species. *Bulletin de la Société Fouad 1er d'Entomologie* **28**: 85-100. [Bodenh1944b]

Bodenheimer, F.S. 1949. The Coccidea of Turkey. Diaspididae. A monographic study. Ankara: Güney. 264 pp. [Bodenh1949]

Bodenheimer, F.S. 1951. Description of some new genera of Coccoidea. *Entomologische Berichten. Amsterdam* **13**: 328-331. [Bodenh1951]

Bodenheimer, F.S. 1951a. Citrus entomology in the Middle East with special reference to Egypt, Iran, Iraq, Palestine, Syria, Turkey. The Hague: W. Junk. 663 pp. [Bodenh1951a]

Bodenheimer, F.S. 1952. The Coccoidea of Turkey. I. *Revue de la Faculté des Sciences de l'Université d'Istanbul (Ser. B)* **17**: 315-351. [Bodenh1952]

Bodkin, G.E. 1913. Insects injurious to sugar cane in British Guiana, and their natural enemies. *Journal of the Board of Agriculture of British Guiana* **7**: 29-32. [Bodkin1913]

Bodkin, G.E. 1922. The scale insects of British Guiana. *Journal of the Board of Agriculture of British Guiana* **15**: 56-63. [Bodkin1922]

Bodkin, G.E. 1925. The fumigation of citrus trees in Palestine. *Bulletin of Entomological Research* **16(2)**: 143-149. [Bodkin1925]

Bogran, C.E., Heinz, K.M. & Ciomperlik, M.A. 2002. Interspecific competition among insect parasitoids: Field experiments with whiteflies as hosts in cotton. *Ecology* **83(3)**: 653-668. [BogranHeCi2002]

Böhm, H. 1951. Untersuchungen über die San José Schildlaus *Quadraspidiotus perniciosus* Comst. *Pflanzenschutzberichte. Wien* **6(5/6)**: 66-76. [Bohm1951]

Böhm, H. 1954. Auftreten von *Quadraspidiotus schneideri* n. sp. (Homopt. Diaspidoid.) in Österreich. *Pflanzenschutzberichte. Wien* **12 (pt 3-4)**: 55-57 [Bohm1954]

Böhm, H. 1955. 25 Jahr San José-Schildlaus (*Q. perniciosus* Comst.) in Österreich. Tatigkeitsbericht 1951-1955. *Bundesanstalt für Pflanzenschutz (Austria)* 245-266. [Bohm1955]

Böhm, H. 1963. Possibilities of biological control of San José scale in Austria. *Publications de l'OEPP. Serie A (European and Mediterranean Plant Protection Organisation)* **34**: 94-95. [Bohm1963]

Böhm, H. 1964. Bericht über den Stand der biologischen Bekämpfung der San José-Schildlaus in Österreich. *Pflanzenarzt (Wien)* **17**: 147-148. [Bohm1964]

Böhm, H. 1965. Einige weitere Bemerkungen zur Zucht und Einbürgerung der San José-Zehrwespe, *Prospaltella perniciosi* Tow., in» sterreich. *Pflanzenarzt (Wien)* **18**: 104. [Bohm1965]

Böhm, H. 1976. Immer wieder Schildläuse an Zimmerpflanzen. *Pflanzenarzt* **29**: 41-42. [Bohm1976]

Boisduval, J.B.A. 1867. Essai sur l'entomologie horticole. Paris: Donnaud. 648 pp. [Boisdu1867]

Boisduval, J.B.A. 1868. Conférence faite le 18 août par le Dr. Boisduval, au palais de l'industrie, sur les insectes qui ont ravagé les plantes exposées par Mm. Burel et Rivière. *Journal l'Insectologie Agricole* **2**: 266-283. [Boisdu1868]

Boisduval, J.B.A. 1868a. Note sur deux coccides nouvelles trouvées par M. Burel dans les serres chaudes à Paris. *Journal l'Insectologie Agricole* **2**: 266-283. [Boisdu1868a]

Bonafonte, P. 1979. Structure du sperme et migration des spermatozoides dans les voies genitales femelles de *Chrysomphalus ficus*. *Annales de la Société Entomologique de France* **15**: 505-512. [Bonafo1979]

Bonafonte, P. 1981. Contribution a l'étude de la reproduction chez les cochenilles diaspines. I. Influence de l'elevage au laboratoire et de l'accouplement differe sur la fecondite et le sex-ratio de *Chrysomphalus ficus* Ashmead (Homoptera - Diaspididae). *Annales de la Société Entomologique de France* **17**: 157-170. [Bonafo1981]

Bondar, G. 1914. Praga das laranjeiras e outras auranciaceas. *Boletim da Agricultura (São Paulo)* **15**: 1064-1106. [Bondar1914]

Bondar, G. 1915. Pragas das laranjeiras e outras auranciaceas. *Insectos Damninhos a' Agriculturea* **3**: 44-47. [Bondar1915]

Bondar, G. 1930. Insectos damninhos e molestias da batata doce no Brasil. *Correio Agricola* **8**: 343-348. [Bondar1930]

Bondar, G. 1930a. Insectos damninhos e molestias da batata doce no Brasil. *O Campo* **No. 9**: 17-19. [Bondar1930a]

Bonfanti, R. 1999. La cocciniglia di San Jose. *Informatore Agrario* **15(5)**: 24-26. [Bonfan1999]

Bonnemaison, L. 1936. Morphologie comparée du "pou de San José" (*Aonidiella perniciosa* Comst.) et de l'*Aspidiotus* des arbres fruitiers *Aspidiotus ostreaeformis* Curtis. *Revue de Pathologie Végétale et d'Entomologie Agricole de France* **23**: 230-243. [Bonnem1936]

Borang, A. 1996. Important insect pests of forest nurseries and plantations in Arunachal Pradesh. Pages 474-479. *in*: Nair, K.S.S., Sharma, J.K. & Varma, R.V. Impact of Diseases and Insect Pests in Tropical Forests. Kerala, India & Bangkok, Thailand: Kerala Forest Research Institute & Forestry Research Support Programme for Asia and the Pacific. 521 pp. [Borang1996]

Boratynski, K.L. 1953. Sexual dimorphism in the second instar of some Diaspididae (Homoptera: Coccoidea). *Transactions of the Royal Entomological Society of London* **104**: 451-479. [Boraty1953]

Borchsenius, N.S. 1934. [Survey of the coccids fauna of the Black Sea Coast of the Caucasus.]. [In Russian]. Abkhazia: Quarantine Station. 37 pp. [Borchs1934]

Borchsenius, N.S. 1935. [Five new species of scales (Coccidae) morphologically allied to San José scale (*Aspidiotus perniciosus* Comst.).]. [In Russian]. *Izvestiya Kursov Prikladoni Zoologii i Fitopathologii. Leningrad* **6**: 127-133. [Borchs1935]

Borchsenius, N.S. 1935a. [Detector of hard scales of the group Aspidiotini. Omis: Station Slavonik Azov-Black Sea Territory.] [In Russian]. 37 pp. [Borchs1935a]

Borchsenius, N.S. 1936. [On the fauna of scale insects (Coccidae) of the Caucasus.]. [In Russian]. *Trudy Krasnodarskogo sel'sko-Khozyaistvennogo Instituta. Krasnodar* **4**: 97-139. [Borchs1936]

Borchsenius, N.S. 1937. [Tables for the identification of coccids (Coccidae) injurious to cultivated plants and forests in the USSR.]. [In Russian]. Leningrad: Quarantine Regional Inspection. 148 pp. [Borchs1937]

Borchsenius, N.S. 1937a. [Coccidae of Quarantine Value for U.S.S.R. and their allied species.]. [In Russian]. Tbilisi, Georgia: Plant Quar. Sta.. 272 pp. + 72 pls. [Borchs1937a]

Borchsenius, N.S. 1938. [Observations on the Coccidae (Hem. Insecta) of the far eastern region.]. [In Russian]. *Vestnik DB Filiala Akademii Nauk SSSR* **29**: 131-146. [Borchs1938]

Borchsenius, N.S. 1939. [Systematic characteristics of second-instar larvae of the oysterlike Coccidae occurring in the USSR.]. [In Russian]. Leningrad Quarantine Laboratory. 47 pp. [Borchs1939]

Borchsenius, N.S. 1939a. [On the fauna of Coccidae in the Caucasus.]. [In Russian]. *Zashchita Rastenii. Leningrad* **18**: 43-51. [Borchs1939a]

Borchsenius, N.S. 1949. [Insects Homoptera. suborders mealybugs and scales (Coccoidea). Family mealybugs (Pseudococcidae).] [In Russian]. Vol. VII. *Fauna SSSR. Zoologicheskii Institut Akademii Nauk SSSR. N.S.* **38**: 1-382. [Borchs1949]

Borchsenius, N.S. 1949b. [A new genus and new species of hard and soft scales (Homoptera, Coccoidea) of USSR fauna.]. [In Russian]. *Entomologicheskoe Obozrenye* **30**: 334-353. [Borchs1949b]

Borchsenius, N.S. 1949c. [New genera of scale insects from the fauna of Central Asia (Insecta, Coccoidea, Diaspididae).]. [In Russian]. *Doklady Akademii Nauk SSSR. Moscow (n.s.)* **64**: 735-738. [Borchs1949c]

Borchsenius, N.S. 1949d. [Identification of the soft and armored scales of Armenia.]. [In Russian]. *Doklady Akademii Nauk Armyanoskoi SSSR* **1949**: 1-271. [Borchs1949d]

Borchsenius, N.S. 1950b. [Mealybugs and Scale Insects of USSR (Coccoidea).]. [In Russian]. Moscow: Akademii Nauk SSSR, Zoological Institute. 32: 250 pp. [Borchs1950b]

Borchsenius, N.S. 1952. [A new genus and new species of hard scales from Iran (Homoptera, Coccoidea).]. [In Russian]. *Entomologicheskoe Obozrenye* **32**: 261-263. [Borchs1952]

Borchsenius, N.S. 1955b. [New species of armored scales from the maritime region of the USSR and Northern Korea (Homoptera, Coccoidea).]. [In Russian]. *Trudy Zoologicheskogo Instituta. Leningrad* **21**: 247-252. [Borchs1955b]

Borchsenius, N.S. 1958. [Notes on the Coccoidea of China. 2. Descriptions of some new species of Pseudococcidae, Aclerdidae and Diaspididae (Homoptera, Coccoidea).]. [In Russian]. *Entomologicheskoe Obozrenye* **37**: 156-173. [Borchs1958]

Borchsenius, N.S. 1960b. [Contribution to the coccid fauna KNR [Chinese Peoples' Republic]. 4. Hard and soft scales, harmful fruit and grape culture in northeast and east KNR.]. [In Russian]. *Acta Entomologica Sinica* **10**: 214-218. [Borchs1960b]

Borchsenius, N.S. 1962b. [Descriptions of some new genera and species of Diaspididae (Homoptera, Coccoidea).]. [In Russian]. *Entomologicheskoe Obozrenye* **41**: 861-871. [Borchs1962b]

Borchsenius, N.S. 1964. [New genera and species of scale insects (Homoptera, Coccoidea, Diaspididae) from Transcaucasia, Middle and Eastern Asia.]. [In Russian]. *Entomologicheskoe Obozrenye* **34**: 152-168. [Borchs1964]

Borchsenius, N.S. 1965. [Essay on the classification of the armored scale insects (Homoptera, Coccoidea, Diaspididae).]. [In Russian]. *Entomologicheskoe Obozrenye* **44**: 208-214. [Borchs1965]

Borchsenius, N.S. 1966. A catalogue of the armoured scale insects (Diaspidoidea) of the world. Moscow & Leningrad: Nauka. 449 pp. [Borchs1966]

Borchsenius, N.S. & Williams, D.J. 1963. A study of the types of some little-known genera of Diaspididae with descriptions of new genera (Hemiptera: Coccoidea). *British Museum (Natural History) Entomological Bulletin* **13**: 353-394. [BorchsWi1963]

Bordage, E. 1914. Notes biologique recueillies à l'Ile de la Réunion. *Bulletin Biologique de la France et de la Belgique* 47. [Bordag1914]

Borer, E.T. 2002. Intraguild predation in larval parasitoids: implications for coexistence. *Journal of Animal Ecology* **71(6)**: 957-965. [Borer2002]

Borg, P. 1919. The scale insects of the Maltese Islands. 71 pp. [Borg1919]

Borg, J. 1922. Cultivation and diseases of fruit trees. Malta: Government Printing Office. 622 pp. [Borg1922]

Borges, M.L.V., Vianna e Silva, M. & Sá, N. 1986. National Agricultural Station, 50 years of activities, 1936-1986. Oeiras, Portugal: Ministério da Agricultura, Pescas e Alimentaçao Agrária. 41+ pp. [BorgesViSa1986]

Boriani, L. & Nicoli, G. 1995. Il polisolfuro di calcio tra passato e futuro. *Informatore Agrario* **51**: 43, 73-76. [BorianNi1995]

Bouché, P.F. 1833. Naturgeschichte der Schädlichen und Nützlichen Garteninsekten und die bewährtesten Mittel. Berlin: Nicolai. 176 pp. [Bouche1833]

Bouché, P.F. 1834. Naturgeschichte der Insekten, besonders in Hinsicht ihrer ersten Zustände als Larven und Puppen. Berlin: Nicolai. 216 pp. [Bouche1834]

Bouché, P.F. 1844. Beiträge zur Naturgeschichte der Scharlachläuse (Coccina). *Entomologische Zeitung, Stettin* **5**: 293-302. [Bouche1844]

Bouché, P.F. 1851. Neue Arten der Schildlausfamilie. *Entomologische Zeitung, Stettin* **12**: 110-112. [Bouche1851]

Bouhélier, R. 1935. Observations sur quelques Coccinelles coccidiphages au Maroc. *Revue de Zoologie Agricole et Appliquée* **34**: 17-20. [Bouhel1935]

Bouhélier, R., Defrance, P., de Francolini, J. & Rungs, C. 1932. La lutte contre les cochenilles nuisibles aux Aurantiacées. *Defense des Cultures. Direction Générale de l'Agriculture du Commerce et de la Colonisation. (Rabat, Morocco)* **No. 6**: 1-60 pp. [BouhelDeDe1932]

Bourijate, M. & Bonafonte, P. 1982. Influence de l'accouplement différé sur la fécondité ,la sexratio, l'oviposition, la formation du bouclier et le comportement chez quatre espèces de cochenilles diaspines (Hom., Diaspididae). *Annales de la Société Entomologique de France* **18**: 303-315. [BourijBo1982]

Bourne, B.A. 1921. Report on the Department of Agriculture, Barbados, for the financial year 1919-1920. *Report (Department of Agricultural, Barbados)* . [Bourne1921]

Bourne, B.A. 1923. Report on the Department of Agriculture, Barbados, 1922-1923. *Report (Barbados Department of Agriculture)* 25 pp. [Bourne1923]

Bovey, P. & Geier, P. 1946. Le pou de San-José *(Quadraspidiotus perniciosus* Comst.) Menace nos cultures fruitières. *Revue Horticole Suisse* **19**: 201-211. [BoveyGe1946]

Bower, C.C. 1987. Control of San José scale (*Comstockaspis perniciosus* (Comstock) (Hemiptera; Diaspididae)) and woolly aphid (*Eriosoma lanigerum* (Hausmann) (Hemiptera: Pemphigidae)) in an integrated mite control programme. *Plant Protection Quarterly* **2(2)**: 55-58. [Bower1987]

Bower, C.C. 1989. The phenology of *Comstockaspis perniciosus* (Comstock) (Hemiptera: Diaspididae) in an apple orchard at Orange, New South Wales. *Journal of the Australian Entomological Society* **28**: 239-245. [Bower1989]

Bower, C.C., Penrose, L.J. & Dodds, K. 1993. A practical approach to pesticide reduction on apple crops using supervised pest and disease control - preliminary results and problems. *Plant Protection Quarterly* **8**: 57-62. [BowerPeDo1993]

Box, H.E. 1953. Pages 51-65 *in*: List of Sugar-cane Insects. London: Commonwealth Institute of Entomology. 101 pp. [Box1953]

Boyce, A.M. 1928. Studies on the resistance of certain insects to hydrocyanic acid. *Journal of Economic Entomology* **21(95)**: 715-720. [Boyce1928]

Boyce, A.M. 1948. Insects and mites and their control. Pages 665-812 *in*: Batchelor, L.D. & Webber, H.J. (Eds.). The Citrus Industry. Berkeley & Los Angeles: University of California Press. 933 pp. [Boyce1948]

Boyce, A.M. 1950. Entomology of citrus and its contributions to entomological principles and practices. *Journal of Economic Entomology* **43(6)**: 741-766. [Boyce1950]

Boyce, A.M., Kagy, J.F., Persing, C.O. & Hansen, J.W. 1939. Studies with Dinitro-O-Cyclohexylphenol. *Journal of Economic Entomology* **32**: 432-450. [BoyceKaPe1939]

Boyer, F.D. & Ducrot, P.H. 1999. Total synthesis of the enantiomers of *Aspidiotus nerii* sex pheromone. *Comptes Rendus de l'Academie des Sciences Serie II Fascicule C-Chimie* **2(1)**: 29-33. [BoyerDu1999]

Boyer, F.D. & Ducrot, P.H. 1999a. Syntheses of cyclobutane derivatives: Total synthesis of (+) and (-) enantiomers of the oleander scale *Aspidiotus nerii* sex pheromone. *European Journal of Organic Chemistry* **(5)**: 1201-1211. [BoyerDu1999a]

Boyero, J.R., Antunez, E. & Moreno, R. 2000. [System to support tactical decision making for plant protection of orange orchards. I: Establishment of the sample unit of the tree.]. [In Spanish with summary in English.] *Boletín de Sanidad Vegetal, Plagas* **26(4)** Suppl.: 673-688. [BoyeroAnMo2000]

Boynton, M.F. 1901. Technical study of four species of *Aspidiotus*. *Bulletin of the New York State Museum (of Natural History)* **46**: 343-354. [Boynto1901]

Brain, C.K. 1918. The Coccidae of South Africa - II. *Bulletin of Entomological Research* **9**: 107-139. [Brain1918]

Brain, C.K. 1919. The Coccidae of South Africa - III. *Bulletin of Entomological Research* **9**: 197-239. [Brain1919]

Brain, C.K. & Kelly, A.E. 1917. The status of introduced coccids in South Africa in 1917. *Bulletin of Entomological Research* **8**: 181-185. [BrainKe1917]

Brandt, H. & Bollow, H. 1948. Schildläuse. Biologie, bekämpfung, und bestimmung der wichtigsten an obstgewächsen und früchten vorkommenden arten. *Pflanzenschutz* **1**: 14-18. [BrandtBo1948]

Bremner, O.E. 1907. New Coccidae from California. *Canadian Entomologist* **39**: 366-368. [Bremne1907]

Brewer, R.H. 1971. The influence of the parasite *Comperiella bifasciata* How. on the populations of two species of armoured scale insects, *Aonidiella aurantii* (Mask.) and *A. citrina* (Coq.), in South Australia. *Australian Journal of Zoology. Melbourne* **19**: 53-63. [Brewer1971]

Brick, C. 1909. X. Bericht über die Tätigkeit der Abteilung für Pflanzenschutz für die Zeit vom 1. Juli 1907 bis 30. Juni 1908. *Hamburg Bot. Staatsint., Abt. Planzenschutz* **10(1907/1908)**: 1-21. [Brick1909]

Brick, C. 1910. Bericht über die tätigkeit der abteilung für Pflanzenschutz für die zeit vom 9 Juli 1909 Bis 30 Juni 1910. *Jahrbuch der Hamburgischen Wissenschaftlichen Anstalten* **27**: 499-519. [Brick1910]

Brick, C. 1912. [Examination of fresh fruit.] Untersuchung des frischen Obstes. *Bericht über die Tätigkeit der Abteilung für Pflanzenschutz* **14**: 1-22. [Brick1912]

Brimblecombe, A.R. 1953. Studies of the Coccoidea. 1. New species of *Neoleonardia*.. *Queensland Journal of Agricultural Science* **10**: 161-166. [Brimbl1953]

Brimblecombe, A.R. 1954. Studies of Coccoidea. 2. Revision of some of the Australian Aspidiotini described by Maskell. *Queensland Journal of Agricultural Science* **11**: 149-160. [Brimbl1954]

Brimblecombe, A.R. 1955. Studies of the Coccoidea. 3. The genera *Chentraspis, Clavaspis, Lindingaspis* and *Morganella* in Queensland. *Queensland Journal of Agricultural Science* **12**: 39-56. [Brimbl1955]

Brimblecombe, A.R. 1956. Studies of the Coccoidea. 4. New species of Aspidiotini. *Queensland Journal of Agricultural Science* **13**: 107-122. [Brimbl1956]

Brimblecombe, A.R. 1957. Studies of the Coccoidea. 6. New genera and new species of Aspidiotini. *Queensland Journal of Agricultural Science* **14**: 261-191. [Brimbl1957]

Brimblecombe, A.R. 1958. Studies of the Coccoidea. 7. New designations of some Australian Diaspididae. *Queensland Journal of Agricultural Science* **15**: 59-94. [Brimbl1958]

Brimblecombe, A.R. 1959. Studies of the Coccoidea. 8. Three new genera and sixteen new species of Aspidiotini. *Queensland Journal of Agricultural Science* **16**: 121-156. [Brimbl1959]

Brimblecombe, A.R. 1959b. Studies of the Coccoidea. 10. New species of Diaspididae. *Queensland Journal of Agricultural Science* **16**: 381-407. [Brimbl1959b]

Brimblecombe, A.R. 1962. Studies of the Coccoidea. 12. Species occurring on deciduous fruit and nut trees in Queensland. *Queensland Journal of Agricultural Science* **19**: 219-229. [Brimbl1962]

Brimblecombe, A.R. 1962a. Studies of the Coccoidea. 13. The genera *Aonidiella, Chrysomphalus* and *Quadraspidiotus* in Queensland. *Queensland Journal of Agricultural Science* **19**: 403-423. [Brimbl1962a]

Brimblecombe, A.R. 1968. Studies of the Coccoidea. 14. The genera *Aspidiotus, Diaspidiotus* and *Hemiberlesia* in Queensland. *Queensland Journal of Agricultural and Animal Sciences* **25**: 39-56. [Brimbl1968]

Britton, W.E. 1901. The San José scale-insect: its appearance and spread in Connecticut. *Bulletin of the Connecticut Agricultural Experiment Station, Entomology Series* **5**: 3-14. [Britto1901]

Britton, W.E. 1901a. Preliminary experiments in spraying to kill the San José scale-insect: season of 1901. *Bulletin (Connecticut Agricultural Extension Service)* **No. 136**: 1-12 [Entom. Ser. no. 6]. [Britto1901a]

Britton, W.E. 1909. The San José scale and method of controlling it. *Bulletin of the Connecticut Agricultural Experiment Station* **No. 165**: 4 pp. [Britto1909]

Britton, W.E. 1919. Inspection of nurseries. *Bulletin of the Connecticut Agricultural Experiment Station* **211**: 255-261. [Britto1919]

Britton, W.E. 1920a. Inspection of nurseries. *Bulletin of the Connecticut Agricultural Experiment Station* **218**: 119-125. [Britto1920a]

Britton, W.E. 1920b. Inspection of nurseries. *Bulletin of the Connecticut Agricultural Experiment Station* **226**: 140-143. [Britto1920b]

Britton, W.E. 1923. The Aleyrodidae and Coccidae of Connecticut. Family Coccidae. *Connecticut Geological and Natural History Survey, Bulletin* **34**: 346-382. [Britto1923]

Britton, W.E. 1923a. Tests of sprays to control the San José scale. *Bulletin of the Connecticut Agricultural Experiment Station* **247**: 329-331. [Britto1923a]

Britton, W.E. 1923b. The Aleyrodidae and Coccidae of Connecticut. New Haven, Connecticut: Connecticut Geological and Natural History Survey. 382 pp. [Britto1923b]

Britton, W.E. 1924. Connecticut State Entomologist, Thirty-third report. *Connecticut Agricultural Experiment Station, Bulletin* **360**: 387-404. [Britto1924]

Britton, W.E. 1924a. Entomological features of 1923. *Bulletin of the Connecticut Agricultural Experiment Station* No. 256. [Britto1924a]

Britton, W.E. 1924b. Some of the principal insects attacking shade trees in Connecticut. *Bulletin of the Connecticut Agricultural Experiment Station* **No. 263**: 156-170. [Britto1924b]

Britton, W.E. 1926. Entomological features of 1925. *Bulletin of the Connecticut Agricultural Experiment Station* **275**: 219-245. [Britto1926]

Britton, W.E. 1928. Entomological features of 1927. *Bulletin of the Connecticut Agricultural Experiment Station* **No. 294**: 198-209. [Britto1928]

Britton, W.E. 1929. 28th Report of the State Entomologist of Connecticut. *Bulletin of the Connecticut Agricultural Experiment Station* **No. 305**: 669-688. [Britto1929]

Britton, W.E. 1930. Entomological features of 1929. *Bulletin of the Connecticut Agricultural Experiment Station* **315**: 490-501. [Britto1930]

Britton, W.E. 1931. Entomological features of 1930. *Bulletin of the Connecticut Agricultural Experiment Station* **327**: 461-473. [Britto1931]

Britton, W.E. 1932. Entomological features of 1931. *Bulletin of the Connecticut Agricultural Experiment Station* **338**: 499-515. [Britto1932]

Britton, W.E. 1937. Connecticut State Entomologist thirty-sixth report. Entomological features of 1936. *Bulletin of the Connecticut Agricultural Experiment Station* **396**: 293-311. [Britto1937]

Britton, W.E. 1938. Connecticut State Entomologist thirty-seventh report. *Bulletin of the Connecticut Agricultural Experiment Station* **408**: 137-152. [Britto1938]

Britton, W.E. & Walden, B.H. 1903. Fighting the San José scale-insect in 1903. *Bulletin (Connecticut Agricultural Extension Service)* **no. 144**: 1-26 [Entom. Ser. no. 10]. [BrittoWa1903]

Britton, W.E. & Walden, B.H. 1904. San José scale-insect experiments in 1904. *Bulletin (Connecticut Agricultural Extension Service)* **No. 146**: 1-32 [Entom. Ser. no. 11]. [BrittoWa1904]

Britton, W.E. & Zappe, M.P. 1927a. Some insect pests of nursery stock in Connecticut. *Bulletin of the Connecticut Agricultural Experiment Station* **No. 292**: 1-173. [BrittoZa1927a]

Britton, W.E. & Zappe, M.P. 1929a. Inspection of nurseries in 1928. *Bulletin of the Connecticut Agricultural Experiment Station* **305**: 689-700. [BrittoZa1929a]

Britton, W.E. & Zappe, M.P. 1930. Inspection of nurseries in 1929. *Bulletin of the Connecticut Agricultural Experiment Station* **315**: 502-515. [BrittoZa1930]

Brock, A.A. 1925. The scale situation in Orange country. *California Citrograph* **10(10)**: 349, 366. [Brock1925]

Brock, W.S. & Flint, W.P. 1919. Field Experiments in spraying for control of San José scale, 1919. *Circular (University of Illinois, Agricultural Experiment Station and Extension)* **No. 239**: 4 pp. [BrockFl1919]

Broodryk, S.W. 1964. Biological control of circular purple scale. *The South African Citrus Journal* **No. 372**: 7-13. [Broodr1964]

Brookes, H.M. & Hudson, N.M. 1968. The identification and distribution of *Quadraspidiotus* species (Homoptera: Diaspididae) on pome and stone fruits in Australia. *Journal of the Australian Entomological Society* **7**: 90-100. [BrookeHu1968]

Brookes, H.M. & Hudson, N.M. 1969. The distribution and host-plants of the species of *Quadraspidiotus* (Homoptera: Diaspididae) in Australia. *Australian Journal of Experimental Agriculture and Animal Husbandry* **9**: 228-233. [BrookeHu1969]

Brooks, R.F. 1969. Developing pesticide application equipment for citrus. Pages 923-932. *in*: Chapman, H.D. (Ed.). Proceedings First International Citrus Symposium. Vol. 2. Riverside: University of California. [Brooks1969]

Brooks, R.F. 1977. *Chrysomphalus ficus, Unaspis citri, Insulaspis gloverii*. In chapter: Leafhoppers, whiteflies, aphids, scale insects, etc.; *In* "Diseases, Pests and Weeds in Tropical Crops". Berlin, and Hamburg: Kranz, J., Schmutterer H., & Koch, W. Verlag, Paul. 666 pp. [Brooks1977]

Brooks, R.F. & Bullock, R.C. 1966. Control of yellow scale, *Aonidiella citrina*, on Florida citrus. *Florida Entomologist* **49(3)**: 185-188. [BrooksBu1966]

Brooks, R.F. & Thompson, W.L. 1963. Investigations of new scalicides for Florida citrus. *Florida Entomologist* **46(4)**: 279-284. [BrooksTh1963]

Brown, K.B. 1916. The specific effects of certain leaf-feeding Coccidae and Aphididae upon the pines. *Annals of the Entomological Society of America* **9**: 414-422. [Brown1916]

Brown, S.W. 1957. Chromosome behavior in *Comstockiella sabalis* (Comst.) (Coccoidea-Diaspididae). *Genetics* **42**: 362-363. [Brown1957]

Brown, S.W. 1960a. Chromosome aberration in two aspidiotine species of the armored scale insects (Coccoidea-Diaspididae). *Nucleus. Calcutta* **3**: 135-160. [Brown1960a]

Brown, S.W. 1963. The *Comstockiella* system of chromosome behavior in the armored scale insects (Coccoidea: Diaspididae). *Chromosoma. Berlin* **14**: 360-406. [Brown1963]

Brown, S.W. & De Lotto, G. 1959. Cytology and sex ratios of an African species of armored scale insect (Coccoidea-Diaspididae). *American Naturalist* **93**: 369-379. [BrownDe1959]

Brown, L.R. & Eads, C.O. 1967. Insects affecting ornamental conifers in southern California. *Bulletin of the California Agricultural Experiment Station* **834**: 1-72. [BrownEa1967]

Brown, S.W. & McKenzie, H.L. 1962. Evolutionary patterns in the armored scale insects and their allies (Homoptera: Coccoidea: Diaspididae, Phoenicococcidae, and Asterolecaniidae). *Hilgardia* **33**: 141-170. [BrownMc1962]

Brown, G.C. & Potter, D.A. 1990. 3.8.3 The Systems Approach to Integrated Pest Management with Emphasis on the Armored Scale Insects. Pages 527-533 *in*: Rosen, D. (Ed.). Armored Scale Insects, Their Biology, Natural Enemies and Control [Series title: World Crop Pests, Vol. 4B]. Amsterdam, the Netherlands: Elsevier. 688 pp. [BrownPo1990]

Browning, H.W. 1994a. Early classical biological control on citrus. Pages 27-49 *in*: Rosen, D., Bennett, F. D. & Capinera, J. L., eds. Pest Management in the Subtropics: Biological Control: A Florida Perspective. Andover, UK: Intercept Limited. [Browni1994a]

Bruner, S.C. 1930. Insectos útiles como enemigos de otros insectos en Cuba. *Revista de Agricultura, Comercio y Trabajo* 11-18. [Bruner1930]

Bruneteau, J. 1943. La lutte contre le pou de San José aux tats Unis d'Amérique et en Europe centrale. *Annales des Epiphyties* **IX**: 221-233. [Brunet1943]

Brüssel, E.W. van & Bhola, B. 1970. Biology and control of the red citrus scale *Chrysomphalus ficus* Ashm. *Surinaamse Land* **18**: 64-76. [BrusseBh1970]

Bruwer, I.J. & De Villiers, E.A. 1988. Biologiese beheer van die rooidopluis, *Aonidiella aurantii*, op die messinaproefplaas. *Citrus and Subtropical Fruit Research Institute Information Bulletin (South Africa)* **186**: 12-16. [BruwerDe1988]

Bryan, W.A. 1915. Natural history of Hawaii, being an account of the Hawaiian people, the geology and geography of the islands, and the native and introduced plants... Honolulu: [Printed for the author by] the Hawaiian Gazette. 596 pp. [Bryan1915]

Buhroo, A.A., Chishti, M.Z. & Masoodi, M.A. 2000. Degree-day phenology of San José scale, *Quadraspidiotus perniciosus* Comstock and assessment of its predator, *Chilocorus bijugus* Mulsant in Kashmir orchard ecosystem. *Indian Journal of Plant Protection* **28(2)**: 117-123. [BuhrooChMa2000]

Buchner, P. 1953. Endosymbiose der Tiere mit pflanzlichen Mikroorganismen. Pages 211-249. Basel/Stuttgart: Birkhäuser. 771 pp. [Buchne1953]

Buckley, R. 1987. Ant-Plant-Homoptera ineractions. *Advances in Ecological Research* **16**: 53-85. [Buckle1987]

Bünzli, G.H. 1935. Untersuchungen über coccidophile Ameisen aus dan Kaffeefeldern von Surinam. *Mitteilungen der Schweizerischen Entomologischen Gesellschaft* **16**: 455-581. [Bunzli1935]

Burditt, A.K. & Silhime, A.G. 1963. Experimental materials to control citrus scale insects in Florida. *Florida Entomologist* **46(3)**: 223-227. [BurditSi1963]

Burger, H.C. & Ulenberg, S.A. 1990. 3.2 Quarantine Problems and Procedures. Pages 313-327 *in*: Rosen, D. (Ed.). Armored Scale Insects, Their Biology, Natural Enemies and Control [Series title: World Crop Pests, Vol. 4B]. Amsterdam, the Netherlands: Elsevier. 688 pp. [BurgerU11990]

Burke, H.E. 1930. Which insects are the important enemies of shade, park and ornamental trees in the Pacific states. *Journal of Economic Entomology* **23**: 783-785. [Burke1930]

Burmeister, H. 1835. Scharlachlause. Schildläuse. Coccina. (Gallinsecta 1.). Pages 61-83 *in*: H. Burmeister (ed.). Handbuch der Entomologie. Berlin: T.F. Inslin. [Burmei1835]

Bustshik, T.N. 1958. [A contribution to the comparative morphology of the males of the scale insects (Homoptera, Coccoidea, Diaspididae).]. [In Russian]. *Trudy Vsesoyuznogo Entomologicheskogo Obshchestva. Akademiya Nauk SSSR. Moscow* **46**: 162-269. [Bustsh1958]

Bustshik, T.N. 1960. [Coccid fauna of western Kopet-Dagh.]. [In Russian]. *Trudy Akademii Nauk SSR Zoologicheskogo Instituta. St. Petersburg* **27**: 167-182. [Bustsh1960]

Butani, R.K. 1974. Les insectes parasites du palmier-dattier en Inde et leur contrôle. *Fruits* **29(10)**: 689-691. [Butani1974]

Butani, D.K. 1979. Insect pests of fruit crops and their control: jackfruit. *Pesticides* **13(11)**: 36-40. [Butani1979]

Buxton, P.A. 1920. Insect pests of dates and the date palm in Mesopotamia and elsewhere. *Bulletin of Entomological Research* **11**: 287-303. [Buxton1920]

Bytinski-Salz, H. & Sternlicht, M. 1967. Insects associated with oaks (*Quercus*) in Israel. *Israel Journal of Entomology* **2**: 107-143.

CAB International. 1951. *Chrysomphalus dictyospermi* (Morg.). *Distribution Maps of Pests, Series A, Agricultural* **Map no. 3**: 2 pp. [CABI1951]

CAB International. 1966. *Aspidiella hartii* (Ckll.). *Distribution Maps of Pests, Series A, Agricultural* **Map no. 217**: 2 pp. [CABI1966]

CAB International. 1966a. *Aspidiotus destructor* (Sign.). *Distribution Maps of Pests, Series A, Agricultural* **Map no. 218**: 2 pp. [CABI1966a]

CAB International. 1968. *Aonidiella aurantii* (Mask.). *Distribution Maps of Pests, Series A, Agricultural* **Map no. 2 (rev.)**: 2 pp. [CABI1968]

CAB International. 1970. *Aspidiotus nerii* Bch. *Distribution Maps of Pests, Series A, Agricultural* **Map no. 268**: 2 pp. [CABI1970]

CAB International. 1975. *Aonidiella citrina* (Coquillet). *Distribution Maps of Plant Pests, Series A, Agricultural* **Map no. 349 (1st Revision)**: 3 pp. [CABI1975]

CAB International. 1976. *Hemiberlesia lataniae* (Sign.). *Distribution Maps of Pests, Series A, Agricultural* **Map no. 360**: 2 pp. [CABI1976]

CAB International 1978b. *Aonidiella orientalis* (Newstead). *Distribution Maps of Pests, Series A, Agricultural* **Map no. 386**: 2 pp. [CABI1978b]

CAB International. 1981b. *Selenaspidus articulatus* (Morg.). *Distribution Maps of Pests, Series A, Agricultural* **Map no. 419**: 2 pp. [CABI1981b]

CAB International. 1986. *Quadraspidiotus perniciosus* (Comstock). *Distribution Maps of Pests, Series A, Agricultural* **Map no. 7 (rev.)**: 3 pp. [CABI1986]

CAB International. 1987a. *Hemiberlesia rapax* (Comstock). *Distribution Maps of Pests, Series A, Agricultural* **Map no. 484**: 3 pp. [CABI1987a]

CAB International. 1988. *Chrysomphalus aonidum* (Linnaeus). *Distribution Maps of Pests, Series A, Agricultural* **Map no. 4 (rev.)**: 3 pp. [CABI1988]

CAB International. 1996. *Aonidiella aurantii*. *Distribution Maps of Pests, Series A, Agricultural* **Map no. 2 (rev.)**: 5 pp. [CABI1996]

CAB International. 1997a. *Aonidiella citrina* (Coquillett). *Distribution Maps of Pests, Series A, Agricultural* **Map no. 349 (1st rev.)**: 3 pp. [CABI1997a]

Cabido Garcia, R. 1949. Contribuicao para o estudo da sistematica morfologia, biologia e ecologia da cochonilha amarela *(Chrysomphalus dictyospermi,* Morg.). *Boletim da Junta Nacional das Frutas* **9**: 374-465. [Cabido1949]

Cadahia, D. 1989. *Hemiberlesia pitysophila* Takagi (Homoptera, Diaspididae) plaga letal de *Pinus massoniana* Lamb. en China. *Boletín de Sanidad Vegetal, Plagas* **15(4)**: 343-363. [Cadahi1989]

Caesar, L. 1914. The San José and oyster-shell scales. *Bulletin of the Ontario Department of Agriculture* **219**: 1-30. [Caesar1914]

Cai, Z.J., Luo, Y.F., Wen, S.X., Zheng, Y.Q. & Wei, Z.P. 2001. [Comprehensive study on applying Sunspray oil year round to control citrus pests.]. [In Chinese with summary in English.] *Fujian Journal of Agricultural Science* **16(1)**: 28-32. [CaiLuWe2001]

California State Commission of Horticulture. 1914. Insect notes. *Monthly Bulletin, California (State) Commission of Horticulture* **3**: 189. [CSCSH1914]

Calkins, C.O. 1983. Research on exotic insects. Pages 321-359 *in*: Wilson, C.L. & Graham, C.L. (Eds.). Exotic Plant Pests and North American Agriculture. New York: Academic Press. 522 pp. [Calkin1983]

Caltagirone, L.E. 1985. Identifying and discriminating among biotypes of parasites and predators. Pages 189-200. *in*: Hoy, M.A. & Herzog, D.C. Biological Control in Agricultural IPM Systems. Orlando, FL: Academic Press. 589 pp. [Caltag1985]

Caltagarone, L.E. 1999. Adaptations of insects to modes of life. Pages 201-230. *in*: Huffaker, C.R. & Gutierrez, A.P., Eds. Ecological Entomology. 2nd ed. New York: John Wiley & Sons. 756 pp. [Caltag1999]

Cameron, J.W., Garman, G.E. & Soost, R.K. 1969. Differential resistance of Citrus species hybrids to infestation by the California red scale, *Aonidiella aurantii* (Mask.). *Journal American Society Horticultural Sciences* **94**: 694-696. [CameroGaSo1969]

Campbell, M.M. 1972. Observations on the effects of high density populations of spiders. Principally an *Ixeuticus* sp. on the control of *Aonidiella aurantii* Mask. by *Aphytis melinus* DeBach. *Experimental Record (South Australia. Dept. of Agriculture)* **62(11)**: 43. [Campbe1972]

Campbell, M.M. 1975. Duration of toxicity of residues of malathion and spray oil on citrus foliage in south Australia to adults of a California red scale parasite *Aphytis melinus* DeBach (Hymenoptera: Aphelinidae). *Journal of the Australian Entomological Society* **14**: 161-164. [Campbe1975]

Campbell, M.M. 1976. Colonisation of *Aphytis melinus* DeBach (Hymenoptera, Aphelinidae) in *Aonidiella aurantii* (Mask.) (Hemiptera, Coccidae) on citrus in South Australia. *Bulletin of Entomological Research* **65**: 659-668. [Campbe1976MM]

Canales Canales, A. & Valdivieso Jara, L. 1999. Manual de control biológico para la conducción del cultivo del olivo. Jesus María, Peru: Servicio Nacional de Sanidad Agraria. 37 pp. [CanaleVa1999]

Cangardel, H. 1960. Essais de différents traitements sur le "pou rouge de Californie" (*Aonidiella aurantii* Mask.). *Cahiers de la Recherche Agronomique. Morocco* **10**: 69-77. [Cangar1960]

Caprile, J., Varela, L., Pickel, C., Coates, W.W., Bentley, W.J. & Vossen, P.M. 2000. Pests of apple. *UC Pest Management Guidelines* . [CaprilVaPi2000]

Carbonell Bruhn, J. & Briozzo Beltrame, J. 1975. San José scale, *Quadraspidiotus perniciosus*. *Boletín Divulgación Ministerio de Ganadería y Agricultura Central Invest. Agric. Alberto Boer Ger.* **30**: 1-11. [CarbonBr1975]

Carimini, M. 1930. Una varietà di *"Aspidiotus"*. *Aspidiotus hederae* var. *unipectinata carmini.. Redia* **18**: 121-123. [Carimi1930]

Carman, G.E. 1948. Provisionary suggestions for the use of DDT to control red scale on citrus. *News Letter (University of California, Citrus Experiment Station)* **No. 36**: 1-8. [Carman1948]

Carman, G.E. 1953. Citrus scale treatments in California. *California Citrograph* **38**: 307-308. [Carman1953]

Carman, G.E. 1956. Field evaluation of malathion for control of California red scale on citrus. *Journal of Economic Entomology* **49**: 103-111. [Carman1956]

Carman, G.E. 1981. Comments on the red scale situation in South Africa. *Journal of Citrus and Subtropical Fruit* **575**: 7,9,11. [Carman1981]

Carman, G.E., Elmer, H.S. & Ewart, W.H. 1954. Information concerning malathion for the control of California red scale, purple scale, black scale and soft scale on citrus in California. *News Letter (University of California, Citrus Experiment Station)* **No. 64**: 11 pp. [CarmanElEw1954]

Carman, G.E. & Ewart, W.H. 1950. Present status of information on parathion for control of several pests on citrus in California. *Citrus leaves* **30**: 15A-16A, 31A [CarmanEw1950]

Carman, G.E. & Ewart, W.H. 1950a. Present status of information on parathion for the control of several pests on citrus in California. *Newsletter (Citrus Experiment Station - Division of Entomology, University of California, Riverside)* No. 48. [CarmanEw1950a]

Carman, G.E., Ewart, W.H. & Jeppson, L.R. 1951. Information concerning parathion for the control of several pests on citrus in southern California. *News Letter (University of California, Citrus Experiment Station)* **No. 52**: 16 pp. [CarmanEwJe1951]

Carman, G.E., Ewart, W.H., Jeppson, L.R., Riehl, L.A., Atkins, E.L., Ortega, J.C. & Klotz, L.J. 1956. 1956 Spray Program for California Citrus Fruits. *University of California Citrus Experiment Station at Riverside and Agricultural Experiment Station* . [CarmanEwJe1956]

Carman, G.E., Ewart, W.H., Jeppson, L.R., Riehl, L.A., Atkins, E.L. & Ortega, J.C. 1957. 1957 Spray Program for California Citrus Fruits. *University of California Citrus Experiment Station at Riverside and Agricultural Experiment Station* . [CarmanEwJe1957]

Carman, G.E., Ewart, W.H., Jeppson, L.R., Riehl, L.A., Atkins, E.L. & Ortega, J.C. 1958. 1958 Spray Program for California Citrus Fruits. *University of California Citrus Experiment Station at Riverside and Agricultural Experiment Station* . [CarmanEwJe1958]

Carman, G.E., Ewart, W.H., Jeppson, L.R., Riehl, L.A., Ortega, J.C., Atkins, E.L. & Elmer, H.S. 1959. 1959 Spray Program for California Citrus Fruits. *University of California Citrus Experiment Station at Riverside and Agricultural Experiment Station* . [CarmanEwJe1959]

Carman, G.E., Ewart, W.H., Jeppson, L.R., Riehl, L.A., Ortega, J.C., Atkins, E.L. & Elmer, H.S. 1960. 1960 Spray Program for California Citrus Fruits. *University of California Citrus Experiment Station at Riverside and Agricultural Experiment Station* . [CarmanEwJe1960]

Carman, G.E., Ewart, W.H., Jeppson, L.R., Riehl, L.A., Ortega, J.C., Atkins, E.L. & Elmer, H.S. 1961. 1961 Spray Program for California Citrus Fruits. *University of California Citrus Experiment Station at Riverside and Agricultural Experiment Station* . [CarmanEwJe1961]

Carman, G.E., Ewart, W.H., Jeppson, L.R., Riehl, L.A., Ortega, J.C., Atkins, E.L. & Elmer, H.S. 1962. 1962 Spray Program for California Citrus Fruits. *University of California Citrus Experiment Station at Riverside and Agricultural Experiment Station* . [CarmanEwJe1962]

Carman, G.E. & Lindgren, D.L. 1956. Evaluation of parathion in California red scale and yellow scale eradication programs. *Journal of Economic Entomology* **49(4)**: 534-539. [CarmanLi1956]

Carmin, J. 1936. Do fungi help to exterminate red scale in Palestine?. *Hadar* **9**: 173-175. [Carmin1936]

Carmin, J. & Scheinkin, D. 1934. Red scale in Palestine. (Hemiptera: Coccidae). *Bulletin de la Société Entomologique d'Egypte* **18**: 242-274. [CarminSc1934]

Carnes, E.K. 1907. The Coccidae of California. *Second Biennial Report of the Commissioner of Horticulture of the State of California* **(1905-1906)**: 155-222. [Carnes1907]

Carpenter, J.B. & Elmer, H.S. 1978. Pests and diseases of the date palm. *Agriculture Handbook (Agriculture Research Service, United States Department of Agriculture)* **No. 527**: 1-42. [CarpenEl1978]

Carroll, D.P. 1979. Within-tree distribution and host substrate influences on California red scale, *Aonidiella aurantii* (Mask.); density, survival, reproduction and parasitization. Riverside: Ph.D. Diss., Univ. Calif.. 1-145. [Carrol1979]

Carroll, D.P. & Luck, R.F. 1984. Bionomics of California red scale. *Aonidiella aurantii* (Maskell) (Homoptera: Diaspididae), on orange fruits, leaves and wood in California's San Joaquin Valley. *Environmental Entomology* **13**: 847-853. [CarrolLu1984]

Carroll, D.P. & Luck, R.F. 1984a. Within-tree distribution of California red scale, *Aonidiella aurantii* (Maskell) (Homoptera: Diaspididae), and its parasitoid *Comperiella bifasciata* Howard (Hymenoptera: Encyrtidae) on orange trees in the San Joaquin Valley. *Environmental Entomology* **13**: 179-183. [CarrolLu1984a]

Carvalho, J.P. & Aguiar, A.M.F. 1997. [Citrus pests in the Island of Madeira] Pragas dos citrinos na Ilha da Madeira Madeira: Secretaria Regional de Agricultura, Florestas e Pescs. 411 pp. [CarvalAg1997]

Casas, J., Nisbet, R.M., Swarbrick, S. & Murdoch, W.W. 2000. Eggload dynamics and oviposition rate in a wild population of a parasitic wasp. *Journal of Animal Ecology* **69(2)**: 185-193. [CasasNiSw2000]

Castel-Branco, A.J.F. 1951a. Himenópteros parasitas das cochonilhas na Africa oriental portuguesa. *Anais da Junta de Investigações Colonias* **6(4)**: 95-98. [Castel1951a]

Castel-Branco, A.J.F. 1956a. A cochonilha dos coqueiros na ilha do Príncipe (*Aspitiotus destructor* Sign.). *García de Orta* **4**: 225-238. [Castel1956a]

Castel-Branco, A.J.F. 1959. Lutte biologique contre *Aspidiotus destructor* Sign. a l'ile Principe (Afrique-Occidentale Portugaise). *Revue de Pathologie Végétale et d'Entomologie Agricole de France* **37**: 235-239. [Castel1959]

Castel-Branco, A.J.F. 1963. *Aspidiotus destructor* Sign. a outras cochonilhas do coqueiro (*Cocos nucifera* L.) e da palmeira do oleo *(Elaeis guineensis* Jacq.) nas Ilhas de S. Tome e Principe. *Memorias da Junta de Investigacoes do Ultramar. Lisboa (2a series)* **43**: 131-175. [Castel1963]

Casu, A.P., Onorato, M. & Gerardi, M.S. 1999. Trappole a feromoni per l'*Aonidiella aurantii. Informatore Agrario* **55(32)**: 69-72. [CasuOnGe1999]

Catchings, T.F. & Whitcomb, W.D. 1924. Notes on winter mortality of the three coccids *Pseudaonidia duplex* (Ckll.), *Chrysomphalus aonidum* (Linn.), and *Chrysomphalus dictyospermi* (Morg.), at New Orleans, Louisiana. *Journal of Economic Entomology* **17**: 604-606. [CatchiWh1924]

Catling, H.D. 1971. Studies on the citrus red scale, *Aonidiella aurantii* (Mask.), and its biological control in Swaziland. *Journal of the Entomological Society of Southern Africa* **34**: 393-411. [Catlin1971]

Catling, H.D. 1971a. Biological control of red scale: are we exploiting this approach sufficiently?. *The South African Citrus Journal* **No. 450**: 5-9. [Catlin1971a]

Cave, R.D. & Márquez C., G. 1994. Parasitoides de Diaspididae, Coccidae and Aleyrodidae atancando cítricos en Honduras. *CEIBA* **35(1)**: 3-8. [CaveMa1994]

Cendana, S.M. 1937. Studies on the biology of *Coccophagus* (Hymenoptera), a genus parasitic on nondiaspidine Coccidae. *University of California Publications in Entomology* **6**: 337-339. [Cendan1937]

Cevik, T., Okul, A., Soylu, O.Z., Bulut, H. & Zeki, C. 1996. [Summer chemical test against San-José scale (*Quadraspidiotus perniciosus* Comst.) harmful on fruit trees in Central Anatolia.]. [In Turkish]. *Zirai-Mucadele-Arastirma-Yilligi* No. 28-29, 61-62. [CevikOkSo1996]

Chambers, E.L. 1925. Controlling the pests of greenhouse and ornamental plants. *Annual Report (Wisconsin Horticultural Society)* **4**: 149-165. [Chambe1925EL]

Chambers, E.L. 1926. Nursery inspection work expands. *Biennial Report of the Wisconsin Department of Agriculture* **No. 82**: 55-63. [Chambe1926]

Chamberlin, J.C. 1927. Status and synonymy of the dictyospermum scale. *Monthly Bulletin, California Department of Agriculture* **16**: 484-491. [Chambe1927]

Chander, R. & Kakar, K.L. 1994. Requirement of dormant spray oil for the suppression of San José scale, *Quadraspidiotus perniciosus* (Comstock) on apple. *Journal of Insect Science* **7**: 222-223. [ChandeKa1994]

Chandler, S.C. 1950. Forbes scale as a major pest of peach. *Journal of Economic Entomology* **43(3)**: 398. [Chandl1950]

Chandra, J. & Avasthy, P.N. 1986. Biology of *Sticholotis madagassa* Weise on *Melanaspis glomerata* (Green) infesting *Erianthus munja*. *Bulletin of Entomology* **27(2)**: p. 194-196. [ChandrAv1986]

Chang, G.B., Dang, Z.Y. & Yu, C.Y. 1989. [Studies on the spatial distribution pattern of *Quadraspidiotus perniciosus*.]. [In Chinese]. *Forest Pest and Disease* **No. 3**: 29-30. [ChangDaYu1989]

Chapman, P.J., Avens, A.W. & Pearce, G.W. 1944. San José scale control experiments. *Journal of Economic Entomology* **37(2)**: 305-307. [ChapmaAvPe1944]

Charles, J.G. 1998. The settlement of fruit crop arthropod pests and their natural enemies in New Zealand: an historical guide to the future. *Biocontrol News and Information* **19(2)**: 47N-58N. [Charle1998]

Charles, J.G., Allan, D.J., Wearing, C.H., Burnip, G.M. & Shaw, P.W. 1998. Releases of *Hemisarcoptes coccophagus* Meyer (Acari: Hemisarcoptidae), a predator of armoured scale insects, in the South Island. *New Zealand Entomologist* **21**: 93-98. [CharleAlWe1998]

Charles, J.G., Hill, M.G. & Allan, D.J. 1995. Releases and recoveries of *Chilocorus* spp. (Coleoptera: Coccinellidae) and *Hemisarcoptes* spp. (Acari: Hemisarcoptidae) in kiwifruit orchards: 1987-93. *New Zealand Journal of Zoology* **22**: 319-324. [CharleHiAl1995]

Charles, J.G., Hill, M.G. & Allan, D.J. 1995a. Persistence of the predatory mite, *Hemisarcoptes coccophagus* Meyer (Hemisarcoptidae), on low populations of *Hemiberlesia lataniae* Signoret) (Diaspididae) in New Zealand. *Israel Journal of Entomology* **29**: 297-300. [CharleHiAl1995a]

Charles, J.G. & Walker, J.T.S. 1981. Mealybug control in apples. *The Orchardist of New Zealand* **54**: 252-253. [CharleWa1981]

Charmoy, D. d'Emmerez de. 1899. Notes sur les Cochenilles. *Planters and Commercial Gazette (Société 'Amicale Scientifique)* **1899**: 1-19. [Charmo1899]

Charmoy, D. d'Emmerez de & Gebert, S. 1921. Insect pests of various minor crops and fruit trees in Mauritius. *Bulletin of Entomological Research* **12(2)**: 181-190. [CharmoGe1921]

Chatterjee, N.C. & Bose, M. 1934. Entomological investigations on the spike disease of sandal. Coccinellidae. Supplementary data. *Indian Forest Records* **19**: 1-10. [ChatteBo1934]

Chazeau, J. 1981. La lutte biologique contre la cochenille transparente du cocotier *Temnaspidiotus destructor* aux Nouvelles-Hébrides. *Cahiers ORSTOM* **No. 44**: 11-22. [Chazea1981]

Chazeau, J. 1984. *Telsimia* de Nouvelle-Guinée (Col. Coccinellidae). *Bulletin de la Société Entomologique de France* **89**: 9-10, I-IX. [Chazea1984]

Cheah, L.H. & Irving, D.E. 1997. Kiwifruit. Pages 209-227. *in*: Mitra, S.K., Ed. Postharvest Physiology and Storage of Tropical and Subtropical Fruits. Wallingford, England, UK: CAB International. 423 pp. [CheahIr1997]

Chen, F.G. 1936. Notes on the scale insects of citrus in several districts of East Chekiang, with description of one new species. *Entomology and Phytopathology. Hangchow* **4**: 208-228. [Chen1936]

Chen, Y.G. & Hu, D.X. 1998. [Interspecific interaction between *Hemiberlesia pitysophila* Takagi (Homoptera: Diaspididae) and its parasite *Coccobius azumai* Tachikawa (Hymenoptera: Aphelinidae).]. [In Chinese]. *Natural Enemies of Insects* **20**: 136-142. [ChenHu1998]

Chen, F.G. & Wong, F.P. 1936. [A list of the known fruit insects of China.]. [In Chinese]. *Bur. Ent. Yearbook, Hangchow* **5**: 82-140. [ChenWo1936]

Cherkasova, T.F. & Duchak, A.N. 2000. [Efficacy of Efal and Fufanon.]. [In Russian.] *Zashchita i Karantin Rastenii* **No. 8**: 26. [CherkaDu2000]

Chi, D.F., Miao, J.C., Qu, H., Xiang, W.J., Li, C.Y., Lu, C.J. & Yi, L.S. 1997. [Control of *Quadraspidiotus gigas* and *Lepidosaphes salicina* using insect growth regulator, RH-5849.]. [In Chinese]. *Journal of Northeast Forestry University* **25**: 10-14. [ChiMiQu1997]

Chi, D.F., Zhang, F.B., Hu, Y.Y. & Sun, Y. 1997. [The influence of the kairomone of *Quadraspidiotus gigas* and oviposition deterring pheromone in parasitoids on the control ability of these parasitoids.]. [In Chinese]. *Journal of Northeast Forestry University* **5**: 15-21. [ChiZhHu1997]

Chiesa Molinari, O. 1938. Un nuevo Coccidae (Cochinilla) para nuestra olivicultura *(Aspidiotus latastei* Cockerell). *Boletín de Agricultura Ganadería Cordoba, Argentina* **166**: 5-13. [Chiesa1938]

Chiesa Molinari, O. 1938a. Coccidae (Cochinillas) que atacan el olivo en la Argentina. *Boletín de Agricultura Ganadería Cordoba, Argentina* **No. 167**: 1-21. [Chiesa1938a]

Chiesa Molinari, O. 1939. El piojo o cochinilla de San José *(Comstockaspis perniciosa* Comstock). *Boletín de Agricultura Ganadería Cordoba, Argentina* **171**: 27-44. [Chiesa1939]

Chiesa Molinari, O. 1948. Las plagas de la agricultura, manual práctico de procedimientos modernos para combatirlas. Buenos Aires: "El Ateneo". 497 pp. [Chiesa1948]

Chiesa Molinari, O.C. 1948a. Las plagas de la huerta y el jardín y modo de combatirlas. Buenos Aires: S.A. Editorial Bell. 205 pp. [Chiesa1948a]

Chiesa Molinari, O. 1963. *Abgrallaspis corporifuscus* especie nueva (Coccoidea: Diaspididae). *Revista de la Sociedad Entomológica Argentina* **26**: 159-162. [Chiesa1963]

Chiu, S., Liu, X., Huang, Z., Chen, W. & Wei, X. 1993. [The chemical control of the pine armoured scale *Hemiberlesia pitysophila* Takagi.]. [In Chinese]. *Acta Entomologica Sinica* **36(2)**: 177-184. [ChiuLiHu1993]

Chorley, J.K. 1939. Report of the Division of Entomology for the year ending 31st December, 1938. *Rhodesia Agricultural Journal* **36(8)**: 598-622. [Chorle1939]

Chou, I. 1938. Ridescrizione dell' *Aspidiotus destructor*. Sign. [Homoptera, Coccidae]. *Bollettino del R. Istituto Superior Agraria Laboratorio di Zoologia Generale e Agraria* **30**: 240-249. [Chou1938]

Chou, I. 1946. Descrizione di un nuovo *Chrysomphalus* (Coccidae Homoptera) delle Cina. *Insecta Sinensium. Shensi* **1**: 3-8. [Chou1946]

Chou, I. 1947. [A study on gen. *Chrysomphalus* of China (Hemip., Homop., Coccidae).]. [In Chinese]. *Entomologia Sinica. Shensi* **2**: 9-24. [Chou1947]

Chou, I. 1947a. [The scale insects of citrus and their cyanide fumigation.]. [In Chinese]. *Entomologia Sinica. Shensi* **2**: 25-39. [Chou1947a]

Chou, I. 1985. [Monograph of the Diaspididae of China.] Vol. 2. [In Chinese]. Shanxi Publ. House of Science & Technology. 196-432 +9. [Chou1985]

Chou, I. 1986. [Monograph of the Diaspididae of China.] Vol. 3. [In Chinese]. Shanxi Publ. House of Science & Technology. 433-771 + 9. [Chou1986]

Chu, C.L. 1992. Postharvest control of San José scale on apples by controlled atmosphere storage. *Postharvest Biology and Technology* **1(4)**: 361-369. [Chu1992]

Chua, T.H. & Wood, B.J. 1990. 3.9.2 Other Tropical Fruit Trees and Shrubs. Pages 543-552 *in*: Rosen, D. (Ed.). Armored Scale Insects, Their Biology, Natural Enemies and Control [Series title: World Crop Pests, Vol. 4B]. Amsterdam, the Netherlands: Elsevier. 688 pp. [ChuaWo1990]

Chumakova, B.M. 1957. [Parasites of the coccids in the Maritime Territory.]. [In Russian]. *Zoologicheskii Zhurnal. Moscow* **36**: 533-547. [Chumak1957]

Chumakova, B.M. 1961. [Parasites of injurious scale-insects from Kabardino-Balkaria (Hymenoptera, Chalcidoidea).]. [In Russian]. *Entomologicheskoe Obozrenye* **40**: 313-338. [Chumak1961]

Chumakova, B.M. 1964. [The San José scale *Diaspidiotus perniciosus* Comst. (Coccoidea, Diaspididae) and its parasites in the Soviet Far East.]. [In Russian]. *Entomologicheskoe Obozrenye* **43**: 535-552. [Chumak1964]

Chumakova, B.M. 1969. [Biology, ecology and species composition of the natural enemies of San José scale, *Quadraspidiotus perniciosus* Comst. (Homoptera, Coccoidea) in Sakhalin.]. [In Russian]. *Entomologicheskoe Obozrenye* **48**: 247-254. [Chumak1969]

Chumakova, B.M. & Goryunova, Z.S. 1963. [Development of males of *Prospaltella perniciosi* Tow. (Hymenoptera, Aphelinidae), parasite of San José scale (Homoptera, Coccoidea).]. [In Russian]. *Entomological Review* **42**: 178-181. [ChumakGo1963]

Chumakova, B.M. & Murasevszkaja, Z. Sz. 1975. [Utilization of natural enemies for the control of San José scale.] [In Russian]. *Biológiai növényvédelem* 174-184. [ChumakMu1975]

Cilliers, C.J. 1971. Observations on circular purple scale *Chrysomphalus aonidum* (Linn.), and two introduced parasites in Western Transvaal citrus orchards. *Entomophaga* **16**: 269-284. [Cillie1971]

Cilliers, C.J. 1978b. Circular purple scale (Chrysomphalus aonidum (L.) (=*Chrysomphalus ficus* Ashm.). Pages 119-122. *in*: Bedford, E.C.G. (Ed.). Citrus Pests in the Republic of South Africa, Scientific Bulletin No. 391. Department of Agricultural Technical Services, South Africa. 253 pp. [Cillie1978b]

Cilliers, C.J. 1998a. Circular purple scale: *Chrysomphalus aonidum* (L.) (=Chrysomphalus ficus Ashmead). Pages 145-149. *in*: Bedford, E.C.G., Van den Berg, M.A. & De Villiers, E.A., Eds. Citrus Pests in the Republic of South Africa. 2nd ed. Nelspruit: Institute for Tropical and Subtropical Crops. 288 pp. [Cillie1998a]

Cirio, U. 1979. Possibilities for integrated pest control in olive cultures. Pages 297-303. *in*: Proceedings: Internationales Symposium der IOBC/WPRS über Integrierten Pflanzenschutz in der Landund Forstwirtschaft. Wien, Austria: OILB/SROP. 648 pp. [Cirio1979]

Cividanes, F.J. & Gutierrez, A.P. 1996. Modeling the age-specific per capita growth and reproduction of *Rhizobius lophanthae* (Col. Coccinellidae). *Entomophaga* **41**: 257-266. [CividaGu1996]

Clancy, D.W., Selhime, A.G. & Muma, M.H. 1963. Establishment of *Aphytis holoxanthus* as a parasite of Florida red scale in Florida. *Journal of Economic Entomology* **56(5)**: 603-605. [ClancySeMu1963]

Claps, L.E. 1993. Lista de tipos y cotipos de Coccoidea depositados en colecciones entomologicas de instituciones de Brasil. *The Scale* **18**: 2-12. [Claps1993]

Claps, L.E. 2000. Redescripción de cinco especies de Diaspididae (Hemiptera, Coccoidea) de la región neotropical. *Revista Brasileira de Entomologia* **44(3/4)**: 91-95. [Claps2000]

Claps, L.E. & Terán, A.L. 2001. Diaspididae (Hemiptera: Coccoidea) asociadas a cítricos en la provincia de Tucumán (República Argentina). *Neotropical Entomology* **30(3)**: 391-402. [ClapsTe2001]

Claps, L.E., Wolff, V.R.S. & González, R.H. 2001. Catálogo de las especies de Diaspididae (Hemiptera: Coccoidea) nativas de Argentina, Brasil y Chile. *Insecta Mundi* **13(3/4)**: 239-256. [ClapsWoGo2001]

Claps, L.E., Wolff, V.R.S. & González, R.H. 2001a. Catálogo de las Diaspididae (Hemiptera: Coccoidea) exóticas de la Argentina, Brasil y Chile. *Revista de la Sociedad Entomológica Argentina* **60(1/-4)**: 9-34. [ClapsWoGo2001a]

Clark, S.W. & Friend, W.H. 1932. California red scale and its control in the lower Rio Grande Valley of Texas. *Bulletin (Texas Agricultural Experiment Station)* **No. 455**: 1-35. [ClarkFr1932]

Clarke, W.T. 1906. Two important scale insects and their control. *Circular (Agricultural Experiment Station of the Alabama Polytechnic Institute)* **No. 1**: 1-8. [Clarke1906]

Claus, C. 1864. Beobachtungen über die Bildung des insekteneies. *Zeitschrift für Wissenschaftliche Zoologie* **14**: 42-54. [Claus1864]

Clausen, C.P. 1940. Entomophagous Insects. New York: Hafner Publishing Co.. 688 pp. [Clause1940]

Clausen, C.P. 1956. Biological control of insect pests in the continental United States. *United States Department of Agriculture Technical Bulletin* **1139**: 1-151. [Clause1956]

Clausen, C.P. 1958. Biological control of insect pests. *Annual Review of Entomology* **3**: 291-310. [Clause1958]

Clausen, C.P. 1958a. The biological control of insect pests in the continental United States. Pages 443-447 *in*: Baecker, E.C., (Ed.). Proceedings of the Tenth International Congress of Entomology. Vol. 4. Ottawa. 1115 pp. [Clause1958a]

Clift, A.D. & Beattie, G.A.C. 1993. SCALEMAN: a computer program to help citrus growers manage red scale, *Aonidiella aurantii* (Homoptera: Diaspididae). Pages 470-472 *in*: Corey, S.A., Dall, D.J. & Milne, W.M., Eds. Pest Control and Sustainable Agriculture. Canberra: CSIRO Division of Entomology. [CliftBe1993]

Cochereau, P. 1965. Contrôle biologique d'*Aspidiotus destructor* Signoret (Homoptera-Diaspinae) dans l'Ile Vaté) Nouvelles Hébrides) au moyen de *Lindorus lophantae* Blaisd (Coleoptera-Coccinellidae). *Compte Rendu de l'Academie d'Agriculture de France* **51**: 318-321. [Cocher1965]

Cochereau, P. 1965a. Contre un ravageur du cocotier aux Nouvelles-Hébrides: Contrôle biologique d'*Aspidiotus destructor* Signoret (Homoptera-Diaspinae) par *Lindorus lophantae* Blaisd. (Coleoptera-Coccinellidae), Ile Vaté. *Oleagineux* **20(8-9)**: 507-512. [Cocher1965a]

Cochereau, P. 1969. Contrôle biologique d'*Aspidiotus destructor* Signoret (Homoptera, Diaspinae) dans l'île Vaté Nouvelles Hébrides) au moyen de *Rhizobius pulchellus* Montrouzier (Coleoptera, Coccinellidae). *Cahiers ORSTOM* **No. 8**: 57-100. [Cocher1969]

Cochereau, P. 1972. La lutte biologique dans le Pacifique. *Cahiers ORSTOM* **No. 16**: 89-104. [Cocher1972]

Cock, M.J.W. (Ed.) 1985a. A review of biological control of pests in the Commonwealth Caribbean and Bermuda up to 1982. *Technical Communication (Commonwealth Institute of Biological Control)* **No. 9**: 218 pp. [Cock1985a]

Cockerell, T.D.A. 1891. Institute of Jamaica. Notes from the museum no. 2. Some interesting scale insects. *Jamaica Post (Dec. 14)* . [Cocker1891]

Cockerell, T.D.A. 1892a. Additions to the museum. *Journal of the Institute of Jamaica* **1**: 54-56. [Cocker1892a]

Cockerell, T.D.A. 1892b. List of Coccidae observed in Jamaica. *Insect life* **4**: 333-334. [Cocker1892b]

Cockerell, T.D.A. 1892c. The West Indian rufous scale. *(Aspidiotus articulatus*, Morgan.). *Insect Life* **4**: 380-382. [Cocker1892c]

Cockerell, T.D.A. 1892e. Coccidae, or scale insects. *Bulletin of the Botanical Department, Jamaica* **36**: 5-9. [Cocker1892e]

Cockerell, T.D.A. 1892i. [no title]. *Proceedings of the Entomological Society of London* **1**: X. [Cocker1892i]

Cockerell, T.D.A. 1893bb. *Aspidiotus bowreyi*, n. sp. *Journal of the Institute of Jamaica* **1**: 383. [Cocker1893bb]

Cockerell, T.D.A. 1893cc. The distribution of Coccidae. *Insect Life* **6**: 99-103. [Cocker1893cc]

Cockerell, T.D.A. 1893d. Coccidae, or scale insects. - II. *Bulletin of the Botanical Department, Jamaica* **40**: 7-9. [Cocker1893d]

Cockerell, T.D.A. 1893e. West Indian Coccidae. *Entomologist's Monthly Magazine* **29**: 38-41, 80. [Cocker1893e]

Cockerell, T.D.A. 1893ff. Two new forms of Diaspinae. *Psyche* **6**: 571-573. [Cocker1893ff]

Cockerell, T.D.A. 1893j. A list of West Indian Coccidae. *Journal of the Institute of Jamaica* **1**: 252-256. [Cocker1893j]

Cockerell, T.D.A. 1893k. Coccidae, or scale insects, which live on orchids. *Gardeners' Chronicle and Agricultural Gazette* **13**: 548. [Cocker1893k]

Cockerell, T.D.A. 1893o. XIII.--Notes on some Mexican Coccidae. *Annals and Magazine of Natural History (Ser. 6)* **12**: 47-53. [Cocker1893o]

Cockerell, T.D.A. 1894b. A new scale insect on agave. *Entomological News* **5**: 59-60. [Cocker1894b]

Cockerell, T.D.A. 1894g. Notes on some scale insects of the sub-family Diaspinae. *Canadian Entomologist* **26**: 127-132. [Cocker1894g]

Cockerell, T.D.A. 1894l. Further notes on scale insects. (Coccidae). *Canadian Entomologist* **26**: 189-191. [Cocker1894l]

Cockerell, T.D.A. 1894n. The twentieth Neotropical *Aspidiotus*. *Actes de la Société Scientifique du Chili. Santiago* **4**: 35-36. [Cocker1894n]

Cockerell, T.D.A. 1895b. On a new scale insect found on plum. *Canadian Entomologist* **27**: 16-19. [Cocker1895b]

Cockerell, T.D.A. 1895w. IV. New species of Coccidae. *Psyche* **7**: 7-8. [Cocker1895w]

Cockerell, T.D.A. 1895y. A new scale insect infesting yam roots. *Miscellaneous Information Bulletin Royal Botanical Garden. Trinidad* **2**: 85-86. [Cocker1895y]

Cockerell, T.D.A. 1896b. A check list of the Coccidae. *Bulletin of the Illinois State Laboratory of Natural History* **4**: 318-339. [Cocker1896b]

Cockerell, T.D.A. 1896f. Notes and descriptions of the new Coccidae collected in Mexico by Prof. C.H.T. Townsend. *Bulletin, United States Department of Agriculture, Division of Entomology, Technical Series* **4**: 31-39. [Cocker1896f]

Cockerell, T.D.A. 1896h. Preliminary diagnoses of new Coccidae. *Psyche, Supplement* **7**: 18-21. [Cocker1896h]

Cockerell, T.D.A. 1896i. Some new species of Japanese Coccidae, with notes. *Bulletin, United States Department of Agriculture, Division of Entomology, Technical Series* **4**: 47-56. [Cocker1896i]

Cockerell, T.D.A. 1896k. On the Coccidae (scale insects) of Trinidad. *Bulletin Trinidad Royal Botanic Gardens, Miscellaneous Information* **2**: iii-v. [Cocker1896k]

Cockerell, T.D.A. 1896m. New Coccidae from Massachusetts and New Mexico. *Canadian Entomologist* **28**: 222-226. [Cocker1896m]

Cockerell, T.D.A. 1897a. Descriptive notes on two Coccidae. *The Entomologist* **30**: 12-14. [Cocker1897a]

Cockerell, T.D.A. 1897b. The palmetto scale. *Garden and Forest* **464**: 19. [Cocker1897b]

Cockerell, T.D.A. 1897i. The San José scale and its nearest allies. *United States Department of Agriculture, Division of Entomology, Technical Series* **6**: 1-31. [Cocker1897i]

Cockerell, T.D.A. 1897k. New and little-known Coccidae from Florida. I. Determinations and descriptions, including a new genus. *Psyche* **8**: 89-90. [Cocker1897k]

Cockerell, T.D.A. 1897l. Coccidae or scale insects. - XI. *Bulletin of the Botanical Department, Jamaica* **4**: 149-151. [Cocker1897l]

Cockerell, T.D.A. 1897q. The Coccidae of Ceylon by E.E. Green. *American Naturalist* **31**: 701-704. [Cocker1897q]

Cockerell, T.D.A. 1897s. Further notes on Coccidae from Brazil. *Revista do Museo Paulista* **2**: 383-384. [Cocker1897s]

Cockerell, T.D.A. 1897u. Some new and little-known Coccidae collected by Prof. C.H.T. Townsend in Mexico. *Canadian Entomologist* **29**: 265-271. [Cocker1897u]

Cockerell, T.D.A. 1898b. Two new scale-insects quarantined at San Francisco. *Psyche* **8**: 190-191. [Cocker1898b]

Cockerell, T.D.A. 1898bb. The San José scale. *Entomological News* **9**: 95. [Cocker1898bb]

Cockerell, T.D.A. 1898d. Two new scale insects. *The Entomologist* **31**: 65-66. [Cocker1898d]

Cockerell, T.D.A. 1898e. Three new Coccidae of the subfamily Diaspinae. *Psyche* **8**: 201-202. [Cocker1898e]

Cockerell, T.D.A. 1898i. A new scale-insect of the genus *Lecanium*. *Entomological News* **9**: 145-146. [Cocker1898i]

Cockerell, T.D.A. 1898j. New Coccidae from Mexico. *Annals and Magazine of Natural History (Ser. 7)* **1**: 426-440. [Cocker1898j]

Cockerell, T.D.A. 1898m. Some new Coccidae. *Annals and Magazine of Natural History (Ser. 7)* **2**: 24-27. [Cocker1898m]

Cockerell, T.D.A. 1898n. Note on *Aspidiotus greenii.. Entomologist's Monthly Magazine* **34**: 184-185. [Cocker1898n]

Cockerell, T.D.A. 1898q. New North American insects. *Annals and Magazine of Natural History (n.s.)* **2**: 321-331. [Cocker1898q]

Cockerell, T.D.A. 1898r. The Coccidae of the Sandwich Islands. *The Entomologist* **31**: 239-240. [Cocker1898r]

Cockerell, T.D.A. 1898s. A new scale insect found on Bearberry. *Canadian Entomologist* **30**: 266-267. [Cocker1898s]

Cockerell, T.D.A. 1899a. Article VII. - First supplement to the check-list of the Coccidae. *Bulletin of the Illinois State Laboratory of Natural History* **5**: 389-398. [Cocker1899a]

Cockerell, T.D.A. 1899d. Notes on Central American Coccidae with descriptions of three new species. *Annals and Magazine of Natural History* **3**: 167-170. [Cocker1899d]

Cockerell, T.D.A. 1899i. Four new diaspine Coccidae. *Canadian Entomologist* **31**: 105-107. [Cocker1899i]

Cockerell, T.D.A. 1899j. Some notes on Coccidae. *Proceedings of the Academy of Natural Sciences of Philadelphia* **1899**: 259-275. [Cocker1899j]

Cockerell, T.D.A. 1899k. Notes on Australian Coccidae. 3. - Two new species and a new variety. *Victorian Naturalist* **16**: 88-89. [Cocker1899k]

Cockerell, T.D.A. 1899n. Rhynchota, Hemiptera - Homoptera. [Aleurodidae and Coccidae]. *Biologia Centrali-Americana* **2**: 1-37. [Cocker1899n]

Cockerell, T.D.A. 1899p. Some synonymy. *Psyche* **8**: 311-312. [Cocker1899p]

Cockerell, T.D.A. 1899q. The Coccidae of the Sandwich Islands. *The Entomologist* **32**: 93,164. [Cocker1899q]

Cockerell, T.D.A. 1899r. The Coccidae of Mauritius. *American Naturalist* **33**: 899-901. [Cocker1899r]

Cockerell, T.D.A. 1899s. New records of Coccidae. *Journal of the New York Entomological Society* **7**: 257-259. [Cocker1899s]

Cockerell, T.D.A. 1900d. Four new Coccidae from Arizona. *Canadian Entomologist* **32**: 129-132. [Cocker1900d]

Cockerell, T.D.A. 1900f. Some Coccidae quarantined at San Francisco. *Psyche* **9**: 70-72. [Cocker1900f]

Cockerell, T.D.A. 1900h. Note on *Chrysomphalus dictyospermi,* a scale insect from Cannes. *Entomologist's Monthly Magazine* **36**: 157. [Cocker1900h]

Cockerell, T.D.A. 1900k. Table to separate the commoner scales (Coccidae) of the orange. *Memorias de la Sociedad Científica "Antonio Alazte"* **13**: 349-351. [Cocker1900k]

Cockerell, T.D.A. 1901. The Coccidae of Brazil [Review of Hempel's "As Coccidas Brazileiras."]. *American Naturalist* **35**: 63-64. [Cocker1901]

Cockerell, T.D.A. 1901c. Notes on some Coccidae of the earlier writers. *The Entomologist* **34**: 90-93. [Cocker1901c]

Cockerell, T.D.A. 1901d. Contributions from the New Mexico biological station. - XI. New and little-known insects from New Mexico. *Annals and Magazine of Natural History* **7**: 333-335. [Cocker1901d]

Cockerell, T.D.A. 1901l. South African Coccidae. *The Entomologist* **34**: 223-227, 248-250. [Cocker1901l]

Cockerell, T.D.A. 1902a. New genera and species of Coccidae, with notes on known species. *Annals and Magazine of Natural History (Ser. 7)* **9**: 20-26. [Cocker1902a]

Cockerell, T.D.A. 1902c. A new gall-making coccid. *Canadian Entomologist* **34**: 75. [Cocker1902c]

Cockerell, T.D.A. 1902k. A contribution to the knowledge of the Coccidae. Appendix. Some Brazilian Coccidae. *Annals and Magazine of Natural History (Ser. 7)* **9**: 450-456. [Cocker1902k]

Cockerell, T.D.A. 1902l. A new *Aspidiotus* from *Pinus sylvestris. Ohio Naturalist* **2**: 287-288. [Cocker1902l]

Cockerell, T.D.A. 1902p. A catalogue of the Coccidae of South America. *Revista Chilena de Historia Natural* **6**: 250-257. [Cocker1902p]

Cockerell, T.D.A. 1902t. Some Coccidae from Mexico. *Annals and Magazine of Natural History (Ser. 7)* **10**: 465-472. [Cocker1902t]

Cockerell, T.D.A. 1905. A table to facilitate the determination of the Mexican scale-insects of the genus *Aspidiotus* (sens. latiss.). *American Naturalist* **39**: 45-46. [Cocker1905]

Cockerell, T.D.A. 1905b. Tables for the identification of Rocky Mountain Coccidae (Scale insects and mealybugs). *Colorado University Studies* **2**: 189-203. [Cocker1905b]

Cockerell, T.D.A. 1905d. Three new South American Coccidae. *Entomological News* **16**: 161-163. [Cocker1905d]

Cockerell, T.D.A. 1905f. Some Coccidae from the Philippine Islands. *Proceedings of the Davenport Academy of Sciences. Davenport, Iowa* **10**: 127-136. [Cocker1905f]

Cockerell, T.D.A. 1919. A new coccid on the coconut palm. *Canadian Entomologist* **51**: 116. [Cocker1919]

Cockerell, T.D.A. 1920. A new scale insect on *Rhizophora. Philippine Journal of Science* **15**: 385-386. [Cocker1920]

Cockerell, T.D.A. 1920a. *Furcaspis biformis* (Homop., Coccidae). *Entomological News* **31**: 109. [Cocker1920a]

Cockerell, T.D.A. 1922a. Some Coccidae found on orchids (Hom.). *Entomological News* **33**: 149. [Cocker1922a]

Cockerell, T.D.A. & Bueker, E.D. 1930. New records of Coccidae (Homoptera). *American Museum Novitates (New York)* **424**: 1-8. [CockerBu1930]

Cockerell, T.D.A. & Parrott, P.J. 1899. Contribution to the knowledge of the Coccidae. *The Industrialist* **25**: 159-165, 227-237, 276-284. [CockerPa1899]

Cockerell, T.D.A. & Robbins, W.W. 1909a. Some new and little known Coccidae. *Journal of the New York Entomological Society* **17**: 104-107. [CockerRo1909aWW]

Cockerell, T.D.A. & Robinson, E. 1914. Descriptions and records of Coccidae. I. Subfamily Diaspinae. II. Non-Diaspine subfamilies. *Bulletin of the American Museum of Natural History* **33**: 327-335. [CockerRo1914]

Cockerell, T.D.A. & Robinson, E. 1915. Descriptions and records of Coccidae. *Bulletin of the American Museum of Natural History* **34**: 105-113. [CockerRo1915]

Cockerell, T.D.A. & Robinson, E. 1915a. Descriptions and records of Coccidae. *Bulletin of the American Museum of Natural History* **34**: 423-428. [CockerRo1915a]

Cohen, I. 1969. Biological control of citrus pests in Israel. Pages 769-772. *in*: Chapman, H.D. (Ed.). Proceedings First International Citrus Symposium, Vol. 2. Riverside: University of California. [Cohen1969]

Cohen, I. 1975. From biological to integrated control of citrus pest in Israel. *Technical Monograph, Ciba-Geigy Limited. Basie, Switzerland* **No. 4**: 38-41. [Cohen1975]

Cohen, E., Podoler, H. & El-Hamlauwi, M. 1987. Effects of the malathion-bait mixture used on citrus to control *Ceratitis capitata* (Wiedemann) (Diptera: Tephritidae) on the Florida red scale, *Chrysomphalus aonidum* (L.) (Hemiptera: Diaspididae), and its parasitoid *Aphytis holoxanthus* DeBach. *Bulletin of Entomological Research* **77**: 303-307. [CohenPoEl1987]

Cohen, E., Podoler, H. & El-Hamlauwi, M. 1988. Effect of malathion-bait mixture on two parasitoids of the Florida red scale, *Chrysomphalus aonidum* (L.). *Crop Protection* **7**: 91-95. [CohenPoEl1988]

Cohen, E., Podoler, H. & El-Hamlauwi, M. 1994. Delayed effects of malathion on hymenopteran parasites: the role of the scale cover of diaspidid insects. Pages 183-190 *in*: Rosen, D., ed. Advances in the Study of Aphytis (Hymenoptera: Aphelinidae). Andover, UK: Intercept Limited. [CohenPoEl1994]

Cohic, F. 1956. Parasites animaux des plantes cultivées on Nouvelle-Calédonie et dépendances. *Noumea, Inst. Franc. d'Oceanie* 1-92. [Cohic1956]

Cohic, F. 1958. Contribution à l'étude des cochenilles d'intérêt économique de Nouvelle-Calédonie et dépendances. *Documents Techniques de la Commission du Pacifique Sud* **116**: 1-35. [Cohic1958]

Coleman, G.A. 1903. Coccidae of the Coniferae, with the description of ten new species from California. *Journal of the New York Entomological Society* **11**: 61-85. [Colema1903]

Coles, R.B. & Talbot, P.H.B. 1976. *Septobasidium clelandii* and its conidial state *Harpographium corynelioides*. *Kew Bulletin* **31**: 481-488. [ColesTa1976]

Collard, J.W. 1918. Citrus-fruit culture in New Zealand. *New Zealand Journal of Agriculture* **46**: 154-162. [Collar1918]

Collier, T.R. 1995. Host feeding, egg maturation, resorption, and longevity in the parasitoid *Aphytis melinus* (Hymenoptera: Aphelinidae). *Annals of the Entomological Society of America* **88(2)**: 206-214. [Collie1995]

Collier, T.R., Murdoch, W.W. & Nisbet, R.M. 1994. Egg load and the decision to host-feed in the parasitoid, *Aphytis melinus*. *Journal of Animal Ecology* **63**: 299-306. [CollieMuNi1994]

Collins, P.B. 1950. Scale insects. *Discovery* **11**: 158-160. [Collin1950]

Collins, P.J., Lambkin, T.M. & Bodnaruk, K.P. 1994. Suspected resistance to Methidathion in *Aonidiella aurantii* (Maskell) (Hemiptera: Diaspididae) from Queensland. *Journal of the Australian Entomological Society* **33(4)**: 325-326. [CollinLaBo1994]

Collyer, E. & van Geldermalsen, M. 1975. Integrated control of apple pests in New Zealand. 1. Outline of experiment and general results. *New Zealand Journal of Zoology* **2(1)**: 101-134. [CollyeVa1975]

Colón-Ferrer, M. & Medina-Gaud, S. 1998. Contribution to the Systematics of the Diaspidids (Homoptera: Diaspididae) of Puerto Rico. Río Piedras, Puerto Rico: University of Puerto Rico, Mayagüez Campus, Agric. Experiment Station, Dept. of Crop Protection. 258 pp. [ColonFMe1998]

Colvée, P. 1880. Ensayo sobre una nueva enfermedad del olivo, producida por una nueva especie del genero *Aspidiotus. Gaceta Agrícola del Ministerio de Fomento* **14**: 21-41. [Colvee1880]

Colvée, P. 1881. Estudios sobre algunos insectos de la familia de los coccidos. Valencia: Nicasio Rius Monfort. 40 pp. [Colvee1881]

Colvée, P. 1882. Nuevos estudios sobre algunos insectos de la familia de los coccidos. Nuevos estudios sobre algunos insectos de la familia de los coccidos. Valencia.. 16 pp. [Colvee1882]

Compere, H. 1926. New coccid-inhabiting parasites (Encyrtidae, Hymenoptera) from Japan and California. *University of California Publications in Entomology* **4**: 33-50. [Comper1926]

Compere, H. 1928. New coccid-inhabiting chalcidoid parasites from Africa and California. *University of California Publications in Entomology* **4**: 209-230. [Comper1928]

Compere, H. 1936. Notes on the classification of the Aphelinidae with descriptions of new species. *University of California Publications in Entomology* **6**: 277-321. [Comper1936]

Compere, H. 1936a. A new species of *Habrolepis* parasitic in *Chrysomphalus aurantii*, Mask. *Bulletin of Entomological Research* **27(3)**: 493-496. [Comper1936a]

Compere, H. 1940a. The African species of *Metaphycus* Mercet. *Bulletin of Entomological Research* **31(1)**: 7-33. [Comper1940a]

Compere, H. 1953. An appraisal of Silvestri's work in the orient for the University of California, some misidentifications corrected, and two forms of *Casca* described as new species. *Bollettino del Laboratorio di Zoologia Generale e Agraria (Facolta di Sci. Agraria, Portici, Naples University)* **33**: 35-46. [Comper1953]

Compere, H. 1955. A systematic study of the genus *Aphytis* Howard (Hymenoptera, Aphelinidae) with descriptions of new species. *University of California Publications in Entomology* **10(4)**: 271-319. [Comper1955]

Compere, H. 1961. The red scale and its insect enemies. *Hilgardia* **31**: 173-278. [Comper1961]

Compere, H. 1961a. Descriptions of parasitic Hymenoptera and comments (Hymenopt.: Aphelinidae, Encyrtidae, Eulophidae). *Journal of the Entomological Society of Southern Africa* **24(1)**: 17-71. [Comper1961a]

Compere, H. 1969. Changing trends and objectives in biological control. Pages 755-764 *in*: Chapman, H.D. (Ed.). Proceedings First International Citrus Symposium. Vol. 2. Riverside: University of California. [Comper1969]

Compere, H. 1969a. The role of systematics in biological control: a backward look. *Israel Journal of Entomology* **4**: 5-10. [Comper1969a]

Compere, H. & Annecke, D.P. 1961. Descriptions of parasitic Hymenoptera and comments (Hymenopt.: Aphelinidae, Encyrtidae, Eulophidae). *Journal of the Entomological Society of Southern Africa* **24(1)**: 17-17. [ComperAn1961]

Compere, H., Flanders, S. & Smith, H.S. 1941. Use air transport from China for the introduction into California of a red scale inhabiting *Comperiella. California Citrograph* **26**: 291,301. [ComperFlSm1941]

Compere, H. & Smith, H.S. 1927. Note on the life-history of two oriental chalcidoid parasites of *Chrysomphalus. University of California Publications in Entomology* **4(4)**: 63-73. [ComperSm1927]

Compton, C.C. 1924. The use of lubricating oil emulsion on greenhouse scale insects. *Journal of Economic Entomology* **17**: 222-225. [Compto1924]

Comstock, J.H. 1881. Notes on Coccidae. *Canadian Entomologist* **13**: 8-9. [Comsto1881]

Comstock, J.H. 1881a. Report of the Entomologist. *Report of the Commissioner of Agriculture, United States Department of Agriculture* **1880/1881**: 276-349. [Comsto1881a]

Comstock, J.H. 1883. Second report on scale insects, including a monograph of the subfamily Diaspinae of the family Coccidae and a list, with notes of the other species of scale insects found in North America. *Department of Entomology Report, Cornell University Agricultural Experiment Station* **2**: 47-142. [Comsto1883]

Comstock, J.H. 1916. Report of the entomologist, United States Department of Agriculture. *Bulletin of the Cornell Agricultural Experiment Station, Entomology Division* **372**: 501-507. [Comsto1916]

Comstock, J.H. 1916a. Reports on scale insects. *Bulletin of the Cornell Agricultural Experiment Station* **372**: 425-603. [Comsto1916a]

Conway, T. 1951. Pests and diseases of fruit trees in New Zealand. *New Zealand Journal of Agriculture* **82**: 159-164. [Conway1951]

Cook, A.J. 1909. The red scale *(Chrysomphalus aurantii* Mask). *Pomona College Journal of Entomology* **1**: 15-21. [Cook1909]

Cooley, R.A. 1898b. Notes on some Massachusetts Coccidae. *Bulletin of the United States Department of Agriculture. Bureau of Entomology. Washington* **17**: 61-63. [Cooley1898b]

Cooremen, J. 1951. Sur quelques acariens vivants parmi les colonies de coccoides au Maroc. *Revue de Pathologie Végétale et d'Entomologie Agricole de France* **30(1)**: 30-34. [Coorem1951]

Coquillett, D.W. 1888. Report on various methods for destroying scale-insects. *Report of the Entomologist, Annual Report of the United States Department of Agriculture* pp. 123-133. [Coquil1888]

Coquillett, D.W. 1890a. Report on various methods for destroying the red scale of California. Letter of transmittal. *Bulletin (United States Department of Agriculture)* **22**: 9-? [Coquil1890a]

Coquillett, D.W. 1891a. Report on various methods for destroying scale insects. Letter of submittal. *Bulletin (United States Department of Agriculture)* **23**: 19-26. [Coquil1891a]

Corbett, G.H. 1932. Insects of Coconuts in Malaya. Kuala Lumpur: Department of Agriculture Straits Settlements and Federated Malay States.. 106 pp. [Corbet1932]

Coronado Blanco, J.M., Ruíz Cancino, E. & Monreal Hernández, L.S. 1997. Escamas armadas en cítricos. *Revista de la Universidad Autónoma de Tamaulipas* **52**: 38-41. [CoronaRuMo1997]

Corrêa, A.L., Anami, M.A.S. de A. & Lima, A.F. de 1997. Monitoramento de insetos pragas e predadores associados à culura do limon (*Citrus limonia*) em sistema orgánico de produçao no estado do Rio de Janeiro, Resultados parciais. Pages 312-313 *in*: Bento, J.M.S. & Delalibera, I. (Eds.). [16th Brazilian Congress of Entomology] 16o Congresso Brasileiro de Entomologia. Salvador: Sociedade Entomológica do Brasil/EMBRAPA-CNPMF. 400 pp. [CorreaAnLi1997]

Costa, F.L. 1997. Ratio *Pentilla egena: Selenaspidus articulatus* (Col.: Coccinellidae; Hem.: Diaspididae) on citrus; the effect of chlorpyriphos.]. Page 294 *in*: Bento, J.M.S. & Delalibera, I. (Eds.). [16th Brazilian Congress of Entomology] 16o Congresso Brasileiro de Entomologia. Salvador: Sociedade Entomológica do Brasil/EMBRAPA-CNPMF. 400 pp. [Costa1997]

Costa, H., Cowles, R., Hartin, J., Kido, K. & Kaya, H. 2000. Pests of turfgrass. *UC Pest Management Guidelines* . [CostaCoHa2000]

Costa Lima, A. 1922a. Relatorio da vigem feita ao Rio Grande do Sul pelo Chefe do Serviço de Vigilancia Sanitaria Vegetal, acompanhado do auxiliar do mesmo serviço, aum de averignar a existencia do piolho de S. José *Aspidiotus) (Diaspidiotus perniciosus*) e respectiva área de disseminaçao. *Boletim (Ministerio da Agricultura, Industria e Commercio,* **10(3)**: 37-45. [CostaL1922a]

Costa Lima, A. 1924. Sobre insectos parasitas da videira. *Almanak Agricola Brasileiro* **1924**: 135-141. [CostaL1924]

Costa Lima, A. 1924a. Sobre duas espécies de coccídeos do gênero *"Aonidiella"* ainda não assignaladas do Brasil. *Egatea* **9**: 195-199. [CostaL1924a]

Costa Lima, A. 1934. Sobre alguns coccideos. *Archivos do Instituto de Biologia Vegetal* **1**: 131-138. [CostaL1934]

Costa Lima, A. 1934a. Um novo coccídeos da cana de açúcar. *O Campo* **5**: 25. [CostaL1934a]

Costa Lima, A. 1942. Insectos do Brasil. Vol. 3 (Homopteros). Rio de Janeiro: Publicad. da Escola Nacional de Agronomia. 327 pp. [CostaL1942]

Costa Lima, A. 1949. Les possibilités de l'Amérique du Sud dans l'approvisionnement international en entomophages. *Union Internat. des Sci. Biol. Ser B.* **5**: 65-87. [CostaL1949]

Costa Lima, A. 1951. Uma nova espécie de *Aonidiella* (Homoptera, Coccoidea, Diaspididae). *Archivos do Instituto Biologico. São Paulo* **20**: 173-176. [CostaL1951]

Costa Lima, A. & Rangel, E. 1922. As pragas e molestias das plantas de cultura, no Brazil. *A Lavoura* **26**: 1101-13. [CostaLRa1922]

Costantino, G. 1938. Cocciniglie raccolte nel R. Orto Botanico di Catania. *Annali della Real Stazione Sperimentale di Frutticoltura e di Agrumicoltura* **15**: 25-44. [Costan1938]

Costantino, G. 1956a. Gli insetti dannosi agli agrumi. *Fitopatologia* **6(5)**: 74-79. [Costan1956a]

Costilla, M.A., Osores, V.M. & Basco, H.J. 1970. Biología y daño de la cochinilla roja Australiana *Aonidiella aurantii* (Mask.) bajo las condiciones de Tucumán. *Revista Industrial y Agrícola de Tucumán* **47**: 57-65. [CostilOsBa1970]

Cotte, H.J. 1912. Recherches sur les galles de Provence. Ph.D. Thesis. Université de Paris, Ecole Supérieure de Pharmacie. 240 pp. [Cotte1912]

Cottier, W. 1956. Insect Pests [Part 5]. Pages 209-481 *in*: Atkinson, J.D., Chamberlain, E.E., Dingley, J.M., Reid, W.D., Brien, R.M., Cottier, W., Jacks, H. & Taylor, G.G. Plant Protection in New Zealand. Wellington, New Zealand: [Technical Correspondence School, New Zealand Dept. of Education]. 699 pp. [Cottie1956]

Couch, J.N. 1931. The biological relationship between *Septobasidium retiforme* (B&C.) Pat. and *Aspidiotus osborni* New. and Ckll. *Quarterly Journal of Microscopical Science. London* **74**: 383-437. [Couch1931]

Coutin, R. 1997. Acariens et Insectes du Lierre. *Fiche Pédagogique* 4 pp. [Coutin1997]

Couturier, G., Matile-Ferrero, D. & Richard, C. Sur les cochenilles de la région de Tai (Cote d'Ivoire), recensées dans les cultures et en foret dense (Homoptera, Coccoidea). *Revue Française d'Entomologie (N.S.)* **7(5)**: 273-286. [CouturMaRi1985]

Cox, J.A. 1942. Effect of dormant sprays on parasites of the San José and terrapin scales. *Journal of Economic Entomology* **35(5)**: 698-701. [Cox1942]

Coy, J.P. 1938. *Comperiella bifasciata* in San Bernardino County. *Bulletin (California Department of Agriculture)* **27**: 445-446. [Coy1938]

Cravedi, P. 2000. Integrated peach production in Italy: objectives and criteria. *Pflanzenschutz Nachrichten Bayer* **53(2-3)**: 177-197. [Craved2000]

Cravedi, P. & Molinari, F. 1993. Synthetic pheromones in integrated pest management in peach and plum orchards in Italy. *IOBC/WPRS Bulletin* **16(10)**: 170-173. [CravedMo1993]

Cravedi, P. & Molinari, F. 1995. [Insect pheromones in the protection of peaches.] Feromoni degli insetti nella protezione dei pescheti. (In Italian.) *Informatore Agrario* **51**: 8, 115-121. [CravedMo1995]

Craw, A. 1891. Destructive insects, their natural enemies, remedies and recommendations. II. Scale insects. Description, history, and remedies for their destruction. *California State Board of Horticulture, Division of Entomology. Sacramento* **1891**: 7-15. [Craw1891]

Craw, A. 1896. Injurious insect pests found in the trees and plants from foreign countries. *Biennial Report of the California State Board of Horticulture* **5**: 33-47. [Craw1896]

Craw, A. 1906. Report of Superintendent of Entomology and Inspector. *Report (Division of Entomology, Hawaii Board of Agriculture and Forestry)* 139-158. [Craw1906]

Creighton, J.T. 1942. The control of some of Florida's shade tree pests. *Proceedings of Annual Meeting (International Shade Tree Conference)* **18th**: 219-233. [Creigh1942]

Cressman, A.W. 1933. Biology and control of *Chrysomphalus dictyospermi* (Morg.). *Journal of Economic Entomology* **26**: 696-706. [Cressm1933]

Cressman, A.W. 1941. Susceptibility of resistant and nonresistant strains of the California red scale to sprays of oil and cube resins. *Journal of Economic Entomology* **34(6)**: 859. [Cressm1941]

Cressman, A.W. 1943. Effectiveness against the California red scale of cube resins and nicotine in petroleum spray oil. *Journal of Agricultural Research* **67(1)**: 17-26. [Cressm1943]

Cressman, A.W. 1943a. Effectiveness against the California red scale of cube resins in light-medium and heavy spray oils. *Journal of Agricultural Research* **66(11)**: 413-419. [Cressm1943a]

Cressman, A.W. 1954. Concentrations of oil and parathion for control of California red scale. *Journal of Economic Entomology* **47(1)**: 100-102. [Cressm1954]

Cressman, A.W. 1955. Effect of oil and parathion sprays on soluble solids of oranges. *Journal of Economic Entomology* **48(2)**: 216-217. [Cressm1955]

Cressman, A.W. 1956. Timing of parathion sprays on oranges. *California Citrograph* **41**: 406,408. [Cressm1956]

Cressman, A.W. 1957. Light oil with parathion or malathion for control of black scale on oranges. *Journal of Economic Entomology* **50(5)**: 593-595. [Cressm1957]

Cressman, A.W. 1958. Effectiveness of different formulations of insecticides against California red scale. *Journal of Economic Entomology* **51(6)**: 911-912. [Cressm1958]

Cressman, A.W., Bliss, C.I., Kessels, L.T. & Dumestre, J.O. 1935. Biology of the camphor scale and a method for predicting the time of appearance of stages in the field. *Journal of Agricultural Research* **50**: 267-283. [CressmBlKe1935]

Cressman, A.W. & Broadbent, B.M. 1943. Effectiveness of cube and derris resins in a tank mix and an emulsive oil against California red scale. *Journal of Economic Entomology* **36(3)**: 439-441. [CressmBr1943]

Cressman, A.W. & Broadbent, B.M. 1944. Changes in California red scale populations following sprays of oils with and without derris resins. *Journal of Economic Entomology* **37(6)**: 809-813. [CressmBr1944]

Cressman, A.W. & Broadbent, B.M. 1953. Susceptibility of resistant and nonresistant strains of the California red scale to oil and to Parathion. *Journal of Economic Entomology* **46(5)**: 907. [CressmBr1953]

Cressman, A.W., Broadbent, B.M. and Munger, F. 1954. Control of red scale with parathion. *California Citrograph* **39**: 379,392,395. [CressmBrMu1954]

Cressman, A.W., Broadbent, B.M. and Munger, F. 1957. Screening tests of some organic insecticides against the California red scale. *RS-33 (United States. Agricultural Research Service.)* **40**: 1-7. [CressmBrMu1957]

Cressman, A.W. & Dawsey, L.H. 1936. The comparative insecticidal efficiency against the camphor scale of spray oils with different unsulphonatable residues. *Journal of Agricultural Research* **52(11)**: 865-878. [CressmDa1936]

Cressman, A.W. & Kessels, L.T. 1933. Winter mortality of the camphor scale and dictyosperma scale in 1933 at New Orleans, La. *Journal of Economic Entomology* **26**: 1177-1179. [CressmKe1933]

Cressman, A.W., Munger, F. & Broadbent, B.M. 1949. Tests with parathion for California red scale control. *California Citrograph* **34**: 332,350,351. [CressmMuBr1949]

Cressman, A.W., Munger, F. & Broadbent, B.M. 1950. Effectiveness of different concentrations of parathion alone and of oil with parathion to control California red scale. *Journal of Economic Entomology* **43(5)**: 610-614. [CressmMuBr1950]

Cressman, A.W., Munger, F. & Broadbent, B.M. 1953. Some factors influencing the effectiveness of parathion against California red scale. *Journal of Economic Entomology* **46(6)**: 1070-1074. [CressmMuBr1953]

Cressman, A.W., Munger, F. & Broadbent, B.M. 1954. Oil-parathion sprays for red scale on citrus. *California Citrograph* **39**: 424-440,441,444. [CressmMuBr1954]

Croft, B.A. 1982. Apple pest management. Pages 465-498 *in*: Metcalf, R.L. & Luckmann, W.H. (Eds.). Introduction to Insect Pest Management. 2nd ed. New York: John Wiley & Sons. 577 pp. [Croft1982]

Croft, B.A. & Bode, W.M. 1983. Tactics for deciduous fruit IPM. Pages 219-270 *in*: Croft, B.A. & Hoyt, S.C. (Eds.). Integrated Management of Insect Pests of Pome and Stone Fruits. New York: Wiley-Interscience. 454 pp. [CroftBo1983]

Croft, B.A. & Hull, L.A. 1983. The orchard as an ecosystem. Pages 19-42 *in*: Croft, B.A. & Hoyt, S.C. (Eds.). Integrated Management of Insect Pests of Pome and Stone Fruits. New York: Wiley-Interscience. 454 pp. [CroftHu1983]

Crouzel, I.S. de 1971. Studies on the biological control of Diaspidid scales on citrus in Argentina. *Record of Proceedings (Australian Academy of Science)* Vol. **1** (Abstracts of Papers): 200. [Crouze1971]

Crouzel, S. 1973. Estudios sobre control biológico de cochinillas Diaspididae que atacan cítricos en la República Argentina. *Idia* **304**: 15-39. [Crouze1973]

Crouzel, I.S., Bimboni, H.G., Zanelli, M. & Botto, E.N. 1973. Lucha biológica contra la "cochinilla roja australiana" *Aonidiella aurantii* (Maskell) (Hom. Diaspididae) en cítricos. *Revista de Investigaciones Agropecuarias Serie 5 Patología Vegetal* **10**: 251-318. [CrouzeBiZa1973]

Cui, W. & Hang, D.Z. 1987. [A preliminary study on *Aspidiotus nerii* Bouché.]. [In Chinese]. *Insect Knowledge* **24(6)**: 332-335. [CuiHa1987]

Curtis, J. (Ruricola) 1843c. *Aspidiotus ostreaeformis* (the pear-tree oyster scale). *Gardeners' Chronicle* **46**: 805. [Curtis1843c]

Curtis, J. (Ruricola) 1843d. Entomology -- No. XLVII. The small white-scale, or oleander shield-bearer, *Aspidiotus nerii* (Bouché). *Gardeners' Chronicle* **34**: 588. [Curtis1843d]

Division of Entomology, Department of Agriculture, Ceylon 1923. A preliminary list of the pests of cultivated plants in Ceylon. *Bulletin (Department of Agriculture, Ceylon)* **No. 67**: 1-68. [DEDAC1923]

Dahms, E.C. & Smith, D. 1994. The *Aphytis* fauna of Australia. Pages 245-255 *in*: Rosen, D., ed. Advances in the Study of Aphytis (Hymenoptera: Aphelinidae). Andover, UK: Intercept Limited. [DahmsSm1994]

Dale, J.L. & McCoy, C.E. 1964. The relationship of a scale insect to death of Bermuda grass in Arkansas. *Plant Disease Reporter* **48**: 228. [DaleMc1964]

Daneel, M., Merwe, S.v.d. & Jager, K.d. 1994. Sporadic scales affecting mango. *South African Mango Growers' Association Yearbook* **14**: 72-74. [DaneelMeJa1994]

Danzig, E.M. 1959. [On the scale insect fauna (Homoptera, Coccoidea) of the Leningrad region.]. [In Russian]. *Entomologicheskoe Obozrenye* **38**: 443-455. [Danzig1959]

Danzig, E.M. 1962. [Addition to the scale insect fauna (Homoptera, Coccoidea) of the Leningrad region.]. [In Russian]. *Trudy Akademii Nauk SSR Zoologicheskogo Instituta. St. Petersburg* **31**: 22-24. [Danzig1962]

Danzig, E.M. 1962b. [A short analysis of the fauna of the scale insects (Homoptera, Coccoidea) and their distribution in the Leningrad region.]. [In Russian]. *Trudy Akademii Nauk SSR Zoologicheskogo Instituta. St. Petersburg* **31**: 25-32. [Danzig1962b]

Danzig, E.M. 1964. [5. Suborder Coccinea -- Coccids or mealybugs and scale insects.]. [In Russian]. *Akademii Nauk SSR Zoologicheskogo Instituta.* **1**: 616-654. [Danzig1964]

Danzig, E.M. 1970. [Synonymy of some polymorphous species of coccids (Homoptera, Coccoidea).]. [In Russian]. *Zoologicheskii Zhurnal* **49**: 1015-1024. [Danzig1970]

Danzig, E.M. 1972. [Insects and ticks. *In* "Pests of Forest.".]. [In Russian]. *Akademii Nauk (SSR) Zoologicheskogo Instituta Leningrad* **1972**: 189-221. [Danzig1972]

Danzig, E.M. 1972b. [Contribution to the fauna of the white flies and scale insects (Homoptera: Aleyrodoidea, Coccoidea) of Mongolia.]. [In Russian]. *Insects of Mongolia* **1**: 325-348. [Danzig1972b]

Danzig, E.M. 1972c. [Contributions to the knowledge of the scale insects fauna (Homoptera, Coccoidea) of Afghanistan.]. [In Russian]. *Entomologicheskoe Obozrenye* **51**: 581-584. [Danzig1972c]

Danzig, E.M. 1977. [On the nomenclature and distribution of some injurious scale insects (Homoptera, Coccoidea).]. [In Russian]. *Entomologicheskoe Obozrenye* **56**: 99-102. [Danzig1977]

Danzig, E.M. 1977b. [An ecological and geographical review of the scale insects (Homoptera, Coccoidea) of the south of the Far East.]. Pages 37-60 *in*: Insect fauna of the Far East. Leningrad: Akademiya Nauk SSSR Zoologicheskii Institut. [Danzig1977b]

Danzig, E.M. 1978. [Scale insect fauna of South Sakhalin and Kunashir.]. [In Russian]. *Trudy Biologo-Pochvennogo, Akademii Nauk SSR, Vladivostok* **50**: 3-23. [Danzig1978]

Danzig, E.M. 1978a. [Fauna of scale insects (Homoptera, Coccoidea).]. [In Russian]. Pages 71-78 *in*: Ecological and faunistic investigations of the insect pests of Yakut. Yakutsk: Akademiya Nauk SSSR. [Danzig1978a]

Danzig, E.M. 1980b. [Coccoids of the Far East USSR (Homoptera, Coccinea) with phylogenetic analysis of scale insects fauna of the world.]. [In Russian]. Leningrad: Nauka. 367 pp. [Danzig1980b]

Danzig, E.M. 1983. [New and little known species of scale insects (Homoptera, Coccinea) of the fauna of the USSR.]. [In Russian]. *Entomologicheskoe Obozrenye* **62**: 514-523. [Danzig1983]

Danzig, E.M. 1985. [Contribution to the scale insect fauna (Homoptera, Coccinea) of Teberda State Reserve.]. [In Russian]. *Entomologicheskoe Obozrenye* **64**: 110-123. [Danzig1985]

Danzig, E.M. 1988. [Suborder Coccinea.]. [In Russian]. Pages 686-726 *in*: P.A. Lehr (ed.). Keys to insects of the Far East of the USSR. Leningrad: Nauka. [Danzig1988]

Danzig, E.M. 1990. [New species of coccids (Homoptera, Coccinea) from Iran, Mongolia and Vietnam.]. [In Russian]. *Entomologicheskoe Obozrenye* **69**: 373-376. [Danzig1990]

Danzig, E.M. 1993. [Fauna of Russia and neighbouring countries. Rhynchota, Volume X: suborder scale insects (Coccinea): families Phoenicococcidae and Diaspididae.]. [In Russian]. St. Petersburg: 'Nauka' Publishing House. 452 pp. [Danzig1993]

Danzig, E.M. 1995. Intraspecific variation in the scale insects (Homoptera: Coccinea). *Israel Journal of Entomology* **29**: 19-24. [Danzig1995]

Danzig, E.M. 2000. A new species of gall-forming armored scale insect from Israel (Homoptera, Coccinea: Diaspididae). *Zoosystematica Rossica* **8(2)**: 287-289. [Danzig2000]

Danzig, E.M. & Konstantinova, G.M. 1990. [On coccid (Homoptera, Coccinea) fauna of Vietnam.]. [In Russian]. *Trudy Zoologicheskogo Instituta Akademiya Nauk SSSR. Leningrad* **209**: 38-52. [DanzigKo1990]

Danzig, E.M. & Konstantinova, G.M. 1991. [Application of pheromones: Identification of male coccids in the southern European part of the USSR.]. [In Russian]. *Quarantine Pests, Diseases and Weeds* **1**: 1-15. [DanzigKo1991]

Danzig, E.M. & Pellizzari, G. 1998. Diaspididae. Pages 172-370 *in*: Kozár, F., Ed. Catalogue of Palaearctic Coccoidea. Budapest, Hungary: Plant Protection Institute, Hungarian Academy of Sciences. 526 pp. [DanzigPe1998]

Darling, D.C. & Johnson, N.F. 1984. Synopsis of nearctic Azotinae (Hymenoptera: Aphelinidae). *Proceedings of the Entomological Society of Washington* **86(3)**: 555-562. [DarlinJo1984]

Darvas, B., Farag, A.I., Kozár, F. & Darwish, E.T.E. 1985. Residual activity of Precocene II and Hydroprene against first instar larvae of *Quadraspidiotus perniciosus*. *Acta Phytopathologica Academiae Scientiarum Hungaricae* **20**: 347-350. [DarvasFaKo1985]

Darvas, B. & Szabo, L. 1987. Effect of moulting inhibitors (Buprofezin and Diflubenzuron) on Homoptera: *Acyrthosiphon pisum* (Aphididae), *Planococcus citri* (Pseudococcidae) and *Quadraspidiotus perniciosus* (Diaspididae). *Növényvédelem* **23**: 343-351. [DarvasSz1987]

Darvas, B. & Varjas, L. 1990. 3.5.2 Insect Growth Regulators. Pages 393-408 *in*: Rosen, D. (Ed.). Armored Scale Insects, Their Biology, Natural Enemies and Control [Series title: World Crop Pests, Vol. 4B]. Amsterdam, the Netherlands: Elsevier. 688 pp. [DarvasVa1990]

Darvas, B. & Virág, E.J. 1983. The effectivity of kinopren and hidropren against scale insects (Pseudococcidae, Coccidae, Diaspididae). *Növényvédelem* **19**: 455-463. [DarvasVi1983]

Das, S.C. 1988. Studies on *Aphytis* sp. ? *chrysomphali* (Mercet) - a parasite of black scale, *Chrysomphalus aonidum* (=*C. ficus*) Ashm. *Two and a Bud* **35(1-2)**: 44-45. [Das1988]

Das, S.C., Borthakur, M. & Gope, B. 1988. Bio-ecological studies on *Chilocorus circumdatus* Sch. an efficient predator of Black Scale, *Chrysomphalus aonidum* (=*C. ficus*) Ashm. *Two and a Bud* **35(1-2)**: 15-18. [DasBoGo1988]

Das, G.M. & Ganguli, R.N. 1961. Coccoids on tea in North-East India. *Indian Journal of Entomology* **23**: 245-256. [DasGa1961]

Dastgheyb-Beheshti, N., Behdad, E. & Barooti, S. 1988. Some biological studies on *Diaspidiotus prunorum* in Esfahan. *Entomologie et Phytopathologie Appliquées* **55(1/2)**: 31-32. [DastghBeBa1988]

David, H. & Easwaramoorthy, S. 1986. Biological control. Pages 383-421 *in*: David, H., Easwaramoorthy, S. & Jayanthi, R., eds. Sugarcane Entomology in India. Coimbatore, India: Sugarcane Breeding Institute. [DavidEa1986]

Davidson, J.A. 1964. The genus *Abgrallaspis* in North America (Homoptera: Diaspididae). *Annals of the Entomological Society of America* **57**: 638-643. [Davids1964]

Davidson, J.A. 1970. A new *Melanaspis* from Jamaica on bromeliad (Homoptera: Diaspididae). *Proceedings of the Entomological Society of Washington* **72**: 33-36. [Davids1970]

Davidson, J.A. 1970a. A new *Crenulaspidiotus* from Arizona (Homoptera: Diaspididae). *Proceedings of the Entomological Society of Washington* **72**: 500-503. [Davids1970a]

Davidson, N.A., Dibble, J.E., Flint, M.L., Marer, P.J. & Guye, A. 1991. Managing insects and mites with spray oils. *Publication (IPM Education and Publications, University of California)* **3347**: 47 pp. [DavidsDiFl1991]

Davidson, J.A. & Miller, D.R. 1977. A taxonomic study of *Hemigymnaspis* (Lindinger) (Homoptera: Diaspididae) including descriptions of four new species. *Proceedings of the Entomological Society of Washington* **79**: 499-517. [DavidsMi1977]

Davidson, J.A. & Miller, D.R. 1990. 3.9.8 Ornamental Plants. Pages 603-632 *in*: Rosen, D. (Ed.). Armored Scale Insects, Their Biology, Natural Enemies and Control [Series title: World Crop Pests, Vol. 4B]. Amsterdam, the Netherlands: Elsevier. 688 pp. [DavidsMi1990]

Davidson, J.A. & Raupp, M.J. 1999. Landscape IPM: Guidelines for Integrated Pest Management of Insect and Mite Pests on Landscape Trees and Shrubs [Bulletin 350]. 3rd ed. College Park, MD: Maryland Cooperative Extension, University of Maryland. 109 pp. [DavidsRa1999]

Davies, R.A.H. & McLaren, I.W. 1977. Tolerance of *Aphytis melinus* DeBach (Hymenoptera: Aphelinidae) to 20 orchard chemical treatments in relation to integrated control of red scale, *Aonidiella aurantii* (Maskell) (Homoptera: Diaspididae). *Australian Journal of Experimental Agriculture and Animal Husbandry* **17**: 323-328. [DaviesMc1977]

Davis, J.J. 1924. Estimating the abundance and damage by the San José scale. *Journal of Economic Entomology* **17**: 192-195. [Davis1924]

Davis, J.J. 1924a. Comparative tests with dormant sprays for San José scale control. *Journal of Economic Entomology* **17**: 285-289. [Davis1924a]

Davis, J.J. 1929. Insects of Indiana for 1928. *Proceedings of Indiana Academy of Science* **38**: 299-314. [Davis1929]

Davis, J.J. 1935. Insects of Indiana for 1934. *Proceedings of Indiana Academy of Science* **44**: 198-206. [Davis1935]

Davis, J.J. 1936. Insects of Indiana for 1935. *Proceedings of Indiana Academy of Science* **45**: 257-268. [Davis1936]

Davis, J.J. 1937. Insects of Indiana for 1936. *Proceedings of Indiana Academy of Science* **46**: 230-239. [Davis1937]

Davis, C.J. 1972. Recent introductions for biological control in Hawaii, XVII. *Proceedings of the Hawaiian Entomological Society* **21(2)**: 187-190. [Davis1972]

DeBach, J. 1948. The establishment of the Chinese race of *Comperiella bifasciata* on *Aonidiella aurantii* in southern California. *Journal of Economic Entomology* **41(6)**: 985. [DeBach1948]

DeBach, P. 1951a. The necessity for an ecological approach to pest control on citrus in California. *Journal of Economic Entomology* **44(4)**: 443-447. [DeBach1951a]

DeBach, P. 1958. The role of weather and entomophagous species in the natural control of insect populations. *Journal of Economic Entomology* **51(4)**: 474-484. [DeBach1958]

DeBach, P. 1958a. Selective breeding to improve adaptations of parasitic insects. Pages 759-768 *in*: Baecker, E.C., (Ed.). Proceedings of the Tenth International Congress of Entomology. Vol. 4. Ottawa. 1115 pp. [DeBach1958a]

DeBach, P. 1958b. Application of ecological information to control of citrus pests in California. Pages 187-194 *in*: Baecker, E.C., (Ed.). Proceedings of the Tenth International Congress of Entomology. Vol. 3. Ottawa. 895 pp. [DeBach1958b]

DeBach, P. 1960. The importance of taxonomy to biological control as illustrated by the cryptic history of *Aphytis holoxanthus* n. sp. (Hymenoptera: Aphelinidae), a parasite of *Chrysomphalus aonidum*, and *Aphytis coheni* n. sp., a parasite of *Aonidiella aurantii*. *Annals of the Entomological Society of America* **53**: 701-705. [DeBach1960]

DeBach, P. 1962a. Biological control of the California red scale, *Aonidiella aurantii* (Mask.), on citrus around the world. *XI Internationaler Kongress für Entomologie* 2(7,14) 749-753. [DeBach1962a]

DeBach, P. 1962b. Species of *Aphytis* for possible use in the biological control of *Chrysomphalus dictyospermi*. *FAO Plant Protection Bulletin* **10(2)**: 29. [DeBach1962b]

DeBach, P. 1964. Assistant Ed.: E.I. Schlinger. Biological Control of Insect Pests and Weeds London: Chapman & Hall. 844 pp. [DeBach1964]

DeBach, P. 1964b. Successes, trends, and future possibilities. Pages 673-713 *in*: DeBach, P. & Schlinger, E.I., Eds. Biological Control of Insect Pests and Weeds. London: Chapman & Hall. 844 pp. [DeBach1964b]

DeBach, P. 1964c. Some ecological aspects of insect eradication. *Bulletin of the Entomological Society of America* **10**: 221-224. [DeBach1964c]

DeBach, P. 1964d. Some species of *Aphytis* Howard (Hymenoptera: Aphelinidae) in Greece. *Annales de l'Institut Phytopathologique Benaki* **7(1)**: 5-18. [DeBach1964d]

DeBach, P. 1965. Weather and the success of parasites on population regulation. *Canadian Entomologist* **97**: 848-863. [DeBach1965]

DeBach, P. 1966. The competitive displacement and coexistence principles. *Annual Review of Entomology* **11**: 183-212. [DeBach1966]

DeBach, P. 1969. Biological control of diaspine scale insects on citrus in California. *Proceedings of the First International Citrus Symposium* **2**: 801-815. [DeBach1969]

DeBach, P. 1969a. Uniparental, sibling and semi-species in relation to taxonomy and biological control. *Israel Journal of Entomology* **4**: 11-28. [DeBach1969a]

DeBach, P. 1971. Fortuitous biological control from ecesis of natural enemies. Pages 293-307 *in*: Entomological essays to commemorate the retirement of K. Yasumatsu Tokyo: Hokuryukan Pub. Co.. 389 pp. [DeBach1971]

DeBach, P. 1972. The use of imported natural enemies in insect management ecology. Pages 211-233. *in*: Tall Timbers Conference on Ecological Animal Control by Habitat Management. Proceedings. Tallahassee. [DeBach1972]

DeBach, P. 1974. Biological Control by Natural Enemies. 1st ed. London: Cambridge University Press. 323 pp. [DeBach1974]

DeBach, P. 1979. *Aphytis riyadhi* n. sp. [Hym.: Aphelinidae], a parasite of *Aonidiella* spp. [Hom.: Diaspididae]. *Entomophaga* **24**: 131-138. [DeBach1979]

DeBach, P. & Argyriou, L.C. 1967. The colonization and success in Greece of some imported species (Hym., Aphelinidae) parasitic on citrus scale insects (Hom., Diaspididae). *Entomophaga* **12**: 325-342. [DeBachAr1967]

DeBach, P. & Bartlett, B.R. 1951. Effects of insecticides on biological control of insect pests of citrus. *Journal of Economic Entomology* **44(3)**: 372-383. [DeBachBa1951]

DeBach, P. & Bartlett, B.R. 1964. Methods of colonization, recovery and evaluation. Pages 402-426 *in*: DeBach, P. & Schlinger, E.I., Eds. Biological Control of Insect Pests and Weeds. London: Chapman & Hall. 844 pp. [DeBachBa1964]

DeBach, P., Dietrick, E.J. & Fleschner, C.A. 1949. A new technique for evaluating the efficiency of entomophagous insects in the field. *Journal of Economic Entomology* **42(3)**: 546-547. [DeBachDiFl1949]

DeBach, P., Dietrick, E.J., Fleschner, C.A. & Fisher, T.W. 1950. Periodic colonization of *Aphytis* for control of the California red scale. Preliminary tests, 1949. *Journal of Economic Entomology* **43(6)**: 783-802. [DeBachDiFl1950]

DeBach, P., Dietrick, E.J. & Fleschner, C.A. 1951. Ants vs. biological control of citrus pests. *California Citrograph* **36**: 312,347-348. [DeBachDiFl1951]

DeBach, P. & Fisher, T.W. 1956. Experimental evidence for sibling species in the oleander scale, *Aspidiotus hederae* (Vallot). *Annals of the Entomological Society of America* **49**: 235-239. [DeBachFi1956]

DeBach, P., Fisher, T.W. & Landi, J. 1955. Some effects of meteorological factors on all stages of *Aphytis lingnanensis*, a parasite of the California red scale. *Ecology* **36(4)**: 742-53. [DeBachFiLa1955]

DeBach, P., Fleschner, C.A. & Dietrick, E.J. 1949. California red scale studies in possible control by employment of natural enemies. *California Agriculture* **3(3)**: 12, 14. [DeBachFlDi1949]

DeBach, P. & Gordh, G. 1974. A new species of *Aphytis* that attacks important armored scale insects. *Entomophaga* **19**: 259-265. [DeBachGo1974]

DeBach, P., Hendrickson, R.M. & Rose, M. 1978. Competitive displacement: extinction of the yellow scale, *Aonidiella citrina* (Coq.) (Homoptera: Diaspididae), by its ecological homologue, the California red scale, *Aonidiella aurantii* (Mask.) in southern California. *Hilgardia* **46**: 1-35. [DeBachHeRo1978]

DeBach, P. & Huffaker, C.B. 1971. Experimental techniques for evaluation of the effectiveness of natural enemies. Pages 113-140 *in*: Huffaker, C.B., Ed. Biological control. New York & London: Plenum Press. 511 pp. [DeBachHu1971]

DeBach, P., Huffaker, C.B. & MacPhee, A.W. 1976. Evaluation of the impact of natural enemies. Pages 255-285 *in*: Huffaker, C.B. & Messenger, P.S. (Eds.). Theory and Practice of Biological Control. New York: Academic Press. 788 pp. [DeBachHuMa1976]

DeBach, P. & Landi, J. 1959. New parasites of California red scale. *California Citrograph* **44**: 301-304. [DeBachLa1959]

DeBach, P., Landi, J. & Jeppson, L. 1959. Integrated control measures: The integration of chemical control of mites with biological control of the California red scale. *California Agriculture* **13(7)**: 12,15. [DeBachLaJe1959]

DeBach, P., Landi, J.H. & White, E.B. 1955. Biological control of red scale. *California Citrograph* **40**: 254,271-2,274-275. [DeBachLaWh1955]

DeBach, P., Landi, P.J. & White, E.B. 1962. Parasites are controlling red scale in southern California citrus. *California Agriculture* **16**: 2-3. [DeBachLaWh1962]

DeBach, P. & Rosen, D. 1976a. Twenty new species of *Aphytis* (Hymenoptera: Aphelinidae) with notes and new combinations. *Annals of the Entomological Society of America* **69**: 541-545. [DeBachRo1976a]

DeBach, P. & Rosen, D. 1991. Biological Control by Natural Enemies. 2nd ed. Cambridge: Cambridge University Press. 440 pp. [DeBachRo1991]

DeBach, P., Rosen, D. & Kennett, C.E. 1971. Biological control of coccids by introduced natural enemies. Pages 165-194. *in*: Huffaker, C.B., Ed. Biological Control. New York and London: Plenum Press. 511 pp. [DeBachRoKe1971]

DeBach, P. & Sisojevic, P. 1960. Some effects of temperature and competition on the distribution and relative abundance of *Aphytis lingnanensis* and *A. chrysomphali* (Hymenoptera: Aphelinidae). *Ecology* **41(1)**: 153-160. [DeBachSi1960]

DeBach, P. & Sundby, R.A. 1963. Competitive displacement between ecological homologues. *Hilgardia* **34(5)**: 105-66. [DeBachSu1963]

DeBach, P. & White, E.B. 1960. Commercial mass culture of the California red scale parasite *Aphytis lingnanensis*. *Bulletin of the California Agricultural Experiment Station* **770**: 4-58. [DeBachWh1960]

de Freitas, S. & Mori, L.A.H. 1998. [Evaluation of the population of *Selenaspidus articulatus* (Morgan) (Hemiptera; Diaspididae) under two methods of water supply for citrus plants.]. [In Portuguese with summary in English.] *Científica (Jaboticabal)* **26(1-2)**: 19-24. [deFreiMo1998]

De Gregorio, A. 1915. Caratteri e biologia del *Chrysomphalus dictyospermi* Morg. auctorum (an potius *Aspidiotus agrumincola* De Greg.?) e del suo parassita distruttore *Aphelinus chrysomphali* Gar. Merc. var. *silvestrii* De Greg. con cenni di due ragni submicroscopici (Li-cosa). *Il Naturalista Siciliano* **22**: 125-190. [DeGreg1915]

DeJean, A. & Mony, R. 1991. Attaques d'arbres fruitiers tropicaux par les fourmis du genre *Melissotarsus* (Emery) (Hymenoptera, Formicidae) associées aux Homoptères Diaspididae. *Actes des Colloques Insectes Sociaux* **7**: 179-187. [DeJeanMo1991]

De Lillo, E. & Porcelli, F. 1993. *Pyemotes herfsi* (Oud.) (Acari, Pyemotidae) antagonista di *Melanaspis inopinata* (Leon.) (Coccoidea, Diaspididae) in Puglia. *Entomologica (Bari)* **27**: 117-124. [DeLillPo1993]

De Lotto, G. 1957. New Aspidiotini (Hom.: Coccoidea: Diaspididae) from Kenya. *Annals and Magazine of Natural History* **10**: 225-231. [DeLott1957]

De Lotto, G. 1963. The authorship and type of the genus *Diaspidiotus* (Homoptera: Diaspididae). *Journal of the Entomological Society of Southern Africa* **26**: 144-145. [DeLott1963]

De Lotto, G. 1967a. A contribution to the knowledge of the African Coccoidea (Homoptera). *Journal of the Entomological Society of southern Africa* **29**: 109-120. [DeLott1967a]

De Lotto, G. & Nastasi, V. 1955. Gli insetti dannosi alle piante coltivate e spontanee dell'Eritrea. *Rivista di Agricoltura Subtropicale e Tropicale. Firenze* **49**: 53-60. [DeLottNa1955]

deOng, E.R., Knight, H. & Chamberlin, J.C. 1927. A preliminary study of petroleum oil as an insecticide for citrus trees. *Hilgardia* **2(9)**: 351-384. [deOngKnCh1927]

De Santis, L. 1935. Un Himenóptera, parásito de la Cochinilla Roja de los Citrus, nuevo para la fauna argentina (*Aphytis chrysomphali*, G. Mercet). *Revista de la Facultad de Agronomía la Plata (Universidad Nacional de la Plata)* **20**: 262-271. [DeSant1935]

De Santis, L. 1940. Sinopsis del genero *Physcus* Howard con descripción de una especie nueva (Hym., Chalcidoidea). *Revista de la Facultad de Agronomía de La Plata* **24**: 29-44. [DeSant1940]

De Santis, L. 1941a. Lista de himenópteros parasitos y predatores de los insectos de la República Argentina. *Revista Agronomica. Lisbon* **4**: 21-24; 5: 120-123. [DeSant1941a]

De Santis, L. 1979. Catálogo de los himenópteros calcidoideos de américa al sur de los estadis unidos. Buenos Aires: La Plata, Prepared under the auspices of the Comisión de Investigación Científica. 488 pp. [DeSant1979]

De Stefani, T. 1910. *Chrysomphalus dictyospermi* var. *pinnulifera* Mask. negli agrumeti siciliani. *Bollettino del R. Orto Botanico e Giardino Coloniale di Palermo* **8(4)**: 189-196. [DeStef1910]

Dean, H.A. 1955. Factors affecting biological control of scale insects on Texas citrus. *Journal of Economic Entomology* **48**: 444-447. [Dean1955]

Dean, H.A. 1982. Reduced pest status of the Florida Red Scale on Texas citrus associated with *Aphytis holoxanthus. Journal of Economic Entomology* **75(1)**: 147-149. [Dean1982]

Debnath, S. & Handique, R. 1991. *Hedera helix* L., a new host (collateral host) for *Chrysomphalus aonidum* (*=ficus*) Ashmead -- a scale insect pest of tea. *Two and a Bud* **38(1-2)**: 39-40. [DebnatHa1991]

Decaux, M.F. 1899. Destruction rationnelle des insectes qui attaquent les arbres fruitièrs par l'emploi simultané des insecticides, des insectes auxiliaires, et par la protection et l'elevage de leurs ennemis naturel les parasites. *Journal de Société Nationale d'Horticulture de France* **22**: 158-184. [Decaux1899]

Deitz, L.L. & Davidson, J.A. 1986. Synopsis of the armored scale genus *Melanaspis* in North America (Homoptera : Diaspididae). Tech. Bull. 279. Raleigh, NC: North Carolina Agricultural Research Service. 91 pp. [DeitzDa1986]

Deitz, L.L. & Tocker, M.F. 1980. W.M. Maskell's Homoptera: Species-group names and type-material. *New Zealand Department of Scientific and Industrial Research, Information Series* **146**: 1-76. [DeitzTo1980]

Dekle, G.W. 1954. Some lychee insects of Florida. *Proceedings of the Florida State Horticultural Society* **67**: 226-228. [Dekle1954]

Dekle, G.W. 1957. Special projects, Asiatic red scale. *Biennial Report, Florida Department of Agriculture (21st)* 71-72. [Dekle1957]

Dekle, G.W. 1962. Camellia mining scale (*Pseudaonidia clavigera* (Ckll.). *Entomology Circular, Florida Department of Agriculture and Consumer Services Division of Plant Industry* **1**: 1. [Dekle1962]

Dekle, G.W. 1964. An armored scale (*Morganella longispina* (Morgan)) on Haitian citrus (Homoptera: Coccidae). *Entomology Circular, Florida Department of Agriculture and Consumer Services Division of Plant Industry* **23**: 1. [Dekle1964]

Dekle, G.W. 1964a. Camphor scale (*Pseudaonidia duplex* (Cockerell)) (Homoptera: Coccidae). *Entomology Circular, Florida Department of Agriculture and Consumer Services Division of Plant Industry* **28**: 1-2. [Dekle1964a]

Dekle, G.W. 1965c. Florida armored scale insects. Arthropods of Florida and Neighboring Land Areas. Gainesville, Florida: Fla. Dept. Agr. Div. Plant Indus.. V. 3: 265 pp. [Dekle1965c]

Dekle, G.W. 1966. Aglaonema scale (*Temnaspidiotus excisus* (Green)) (Homoptera: Diaspididae). *Entomology Circular, Florida Department of Agriculture and Consumer Services Division of Plant Industry* **49**: 1-2. [Dekle1966]

Dekle, G.W. 1976. Florida armored scale insects. *In* "Arthropods of Florida and neighboring land areas.". Fla. Dept. Agric. Consumer Serv. Div. Plant Ind.. 3: 345 pp. [Dekle1976]

Dekle, G.W. & Kuitert, L.C. 1975. Camellia mining scale, *Duplaspidiotus claviger* (Cockerell) (Diaspididae: Homoptera). *Entomology Circular, Florida Department of Agriculture and Consumer Services Division of Plant Industry* **152**: 1-4. [DekleKu1975]

Del Bene, G. 1984. [Observations on the biology of *Dynaspidiotus britannicus* (Newstead) (Homoptera: Diaspididae) and their natural enemies in Toscana.]. [In Italian]. *Redia* **67**: 323-336. [DelBen1984]

Del Estal, P., Soria, S. & Vinuela, E. 1994. Localizacion y ciclo biologico de *Nuculaspis regnieri* Balachw 1928 (Homoptera Diaspididae) en la zona centro de Espana. *Bol. Sanid. Veg. Plagas* **20(2)**: 477-486. [DelEstSoVi1994]

Del Guercio, G. 1894. Cocciniglie nuove, note e poco note. *Il Naturalista Siciliano* **13**: 141-158. [DelGue1894]

Del Guercio, G. 1906. Le cocciniglie degli agrumi. *Bollettino Ufficiale del Ministero Industriale e Commerciale (Roma)* **Anno V, Vol III, Fasc. 3**: 257-269. [DelGue1906]

Del Guercio, G. & Malenotti, E. 1915. Ricerche ed esperienze nuove contro la bianca-rossa degli agrumi in Sicilia nel 1914. *Redia* **11**: 1-127. [DelGueMa1915]

Delage-Darchen, B., Matile-Ferrero, D. & Balachowsky, A. 1972. Sur un cas aberrant de symbiose Cochenilles x Fourmis. *Compte Rendu de l'Academie des Sciences. Paris* **275**: 2359-2361. [DelageMaBa1972]

Delinska, N. 1989. A method of determining commencement of hatching of nymphs of San José scale *Quadraspidiotus perniciosus*. [In Russian]. *Rausteniev'dni Nauki (Plant Science)* **26(3)**: 102-105. [Delins1989]

Delucchi, V. 1965. Notes sur le pou de Californie (*Aonidiella aurantii* Maskell) au Maroc [Hom. Coccoidea]. *Annales de la Société Entomologique de France* **1(4)**: 739-788. [Delucc1965]

Delucchi, V. 1969. Biological control of citrus insects in the Mediterranean area excluding Maghreb countries and Israel. Pages 871-874. *in*: Chapman, H.D. (Ed.). Proceedings First International Citrus Symposium. Vol. 2. Riverside: University of California. [Delucc1969]

Delucchi, V., Rosen, D. & Schlinger, E.I. 1976. Relationship of systematics to biological control. Pages 81-91 *in*: Huffaker, C.B. & Messenger, P.S. (Eds.). Theory and Practice of Biological Control. New York: Academic Press. 788 pp. [DeluccRoSc1976]

Delucchi, V. & Traboulsi, R. 1965. *Habrolepis fanari* n. sp. [Chalc., Encyrtidae], parasite de *Chrysomphalus ficus* Ahmed en Afrique du nord et au moyen-orient. *Annales de la Société Entomologique de France* **1(2)**: 495-500. [DeluccTr1965]

Deng, H.J. & Nan, G.M. 2001. [The main diseases and pests of Korla's Xiangli pear variety and their control. [In Chinese.] *China Fruits* **No. 5**: 37-40. [DengNa2001]

Denno, R.F., McClure, M.S. & Ott, J.R. 1995. Interspecific interactions in phytophagous insects: competition reexamined and resurrected. *Annual Review of Entomology* **40**: 297-331. [DennoMcOt1995]

Devasahayam, S. 1992. Insect pests of black pepper and their control. *Planters Chronicle* **155**: 153. [Devasa1992]

Devasahayam, S. & Kozár, F. 1993. Additions to the insect fauna associated with tree spices. *Entomon* **18(1-2)**: 101-102. [DevasaKo1993]

Devasahayam, S., Koya, K.M.A. & Verghese, A. 1998. IPM in spices - challenges for the future. *Advances in IPM for horticultural crops: Environmental implications and thrusts.* 157-164. [DevasaKoVe1998]

Dhileepan, K. 1991. Insects associated with oil palm in India. *FAO Plant Protection Bulletin* **39(2/3)**: 94-99. [Dhilee1991]

Dhileepan, K. 1996. Parasitoids and predators of insects associated with oil palm (*Elaeis guineensis* Jacq.) in India. *Elaeis* **8**: 64-74. [Dhilee1996]

Dickson, R.C. 1941. Inheritance of resistance to hydrocyanic acid fumigation in the California red scale. *Hilgardia* **13(9)**: 515-522. [Dickso1941]

Dickson, R.C. 1951. Construction of the scale covering of *Aonidiella aurantii* (Mask.). *Annals of the Entomological Society of America* **44**: 596-602. [Dickso1951]

Dickson, R.C. & Fleschner, C.A. 1955. Homopterous insects attacking avocado in Central America and Mexico. *Journal of Economic Entomology* **48**: 614-615. [DicksoFl1955]

Dickson, R.C. & Lindgren, D.L. 1947. New data on California red scale. *Citrus Leaves* 6-7, 34. [DicksoLi1947]

Dickson, R.C. & Lindgren, D.L. 1947a. The California Red Scale. *California Citrograph* **32**: 524,542-544. [DicksoLi1947a]

Dietrick, E.J. 1981. Commercial production of entomophagous insects and their successful use in agriculture. Pages 151-160. *in*: Beltsville Symposia in Agricultural Research. 5. Biological control in crop production. Totowa, N.J.: Allanheld, Osmun & Co.. 461 pp. [Dietri1981]

Dietz, H.F. & Morrison, H. 1916. *Phenacaspis spinicola* n. sp.; an apparently new coccid from Indiana (Hem., Hom.). *Entomological News* **27**: 101-102. [DietzMo1916]

Dietz, H.F. & Morrison, H. 1916a. The Coccidae or scale insects of Indiana. *Indiana State Entomologist Eighth Annual Report (1914-1915)* **8**: 195-321. [DietzMo1916a]

Dikov, I. 1965. [Control of the Californian scale should not be abated.]. [In Russian]. *Rastitelna Zashchita* **13**: 8-9. [Dikov1965]

Dimitrov, T. 1935. [Contribution to the study of damage caused by insects and fungi parasites in the Bulgarian forests.]. [In Bulgarian]. *Annals of the University of Sofia* **13**: 220-252. [Dimitr1935]

Dinabandhoo, C.L. & Bhalla, O.P. 1981. Bionomics of the San José scale, *Quadraspidiotus perniciosus* (Hemiptera: Homoptera: Diaspididae) infesting temperate fruits. *Journal of Entomological Research. New Delhi* **4**: 63-67. [DinabaBh1981]

Ding, D., Pan, W., Tang, Z., Xie, G. Lian, J. & Weng, J. 1992. [Host preferences of *Coccobius azumai* (Hymenoptera: Aphelinidae), a parasitoid of *Hemiberlesia pitysophila* (Homoptera: Diaspididae).]. [In Chinese]. *Contributions of the Shanghai Institute of Entomology* **11**: 35-42. [DingPaTa1992]

Ding, D.C., Tang, X.H., Du, J.W., Pan, W.Y. & Lian, J.H. 1988. [Mating behaviour of *Hemiberlesia pitysophila* Takagi (Homoptera: Diaspididae).]. [In Chinese]. *Contributions of the Shanghai Institute of Entomology* **8**: 88. [DingTaDu1988]

Dingler, M. 1924. Biologische Notizen über verschiedene Cocciden. *Zeitschrift für Angewandte Entomologie* **10**: 364-386. [Dingle1924]

Dinther, J.B.M. van 1958. Insects of the coconut palm in Surinam. Page 421 *in*: Baecker, E.C., (Ed.). Proceedings of the Tenth International Congress of Entomology. Vol. 3. Ottawa. 895 pp. [Dinthe1958]

Disselkamp, C. 1954. Die Schildbildung der San José-Schildlaus *(Quadraspidiotus perniciosus* Comst. 1881). *Höfchen-Briefe (Bayer Pflanzenschutz- Nachrichten)* **7**: 109-155. [Dissel1954]

Dixon, A.F.G. 1997. Patch quality and fitness in predatory ladybirds. *Ecological Studies* **130**: 205-223. [Dixon1997]

Doane, R.W. 1931. Common Pests: How to Control Some of the Pests that affect Man's Health, Happiness and Welfare. Springfield, IL, Baltimore, MD: Charles C. Thomas. 397 pp. [Doane1931]

Doane, R.W. & Ferris, G.F. 1916. Notes on Samoan Coccidae with descriptions of three new species. *Bulletin of Entomological Research* **6**: 399-402. [DoaneFe1916]

Doane, R.W. & Hadden, E. 1909. Coccidae from the Society Islands. *Canadian Entomologist* **41**: 296-300. [DoaneHa1909]

Dohanian, S.M. 1937. The importation of Coccinellid enemies of Diaspine scales into Puerto Rico. *Journal of Agriculture of the University of Puerto Rico* **21(2)**: 243-247. [Dohani1937]

Domenici, M.G., Baldin, E.L.L. & Scomparin, C.H.J. 1997. Preferência da fixação de *Selenaspidus articulatus* (Morgan, 1889) (Hemiptera, Diaspididae) na folha de citros, em relação a sua localização. Page 227 *in*: Bento, J.M.S. & Delalibera, I. (Eds.). [16th Brazilian Congress of Entomology] 16o Congresso Brasileiro de Entomologia. Salvador: Sociedade Entomológica do Brasil/EMBRAPA-CNPMF. 400 pp. [DomeniBaSc1997]

Don, D., Miyanoshita, A. & Tatsuki, S. 1995. Host-associated differences in *Aspidiotus cryptomeriae* Kuwana (Homoptera: Coccoidea: Diaspididae): II. Morphology of *Torreya nucifera* race reared on non-host plant, *Cryptomeria japonica.. Japanese Journal of Applied Entomology and Zoology* **39(2)**: 159-162. [DonMiTa1995]

Donli, P.O., Buahin, G.K.A. & Aminu Kano, M. 1998. Management of oriental yellow scale insects (*Aonidiella orientalis* Newstead)(Homoptera: Diaspididae) on neem (*Azadirachta indica* A. Juss)(Meliaceae) in West Africa. Pages 1-10 *in* Lale, N.E.S., Molta, N.B., Donli, P.O. & Dike, M.C. (Eds.), Entomology in the Nigerian economy. Research focus in the 21st century. Maiduguri, Nigeria: Entomological Society of Nigeria. [DonliBuAm1998]

Donli, P.O., Gopal, B.V., Alkali, M. & Aminu Kano, M. 1998. Anatomy of neem (*Azadirachta indica* A. Juss) (Meliaceae) leaves in relation to the Oriental yellow scale insect (*Aonidiella orientalis* Newstead) (Homoptera: Diaspididae) infestation. Pages 39-44 *in* Lale, N.E.S., Molta, N.B., Donli, P.O. & Dike, M.C. (Eds.), Entomology in the Nigerian economy. Research focus in the 21st century. Maiduguri, Nigeria: Entomological Society of Nigeria. [DonliGoAl1998]

Donli, P.O., Sule, B.S. & Aminu Kano, M. 1998. Host range of Oriental yellow scale insect (*Aonidiella orientalis* Newstead) attacking neem as a background to its control. Pages 23-31 *in* Lale, N.E.S., Molta, N.B., Donli, P.O. & Dike, M.C. (Eds.), Entomology in the Nigerian economy. Research focus in the 21st century. Maiduguri, Nigeria: Entomological Society of Nigeria. [DonliSuAm1998]

Donli, P.O., Wabulari, D. & Aminu Kano, M. 1998. Effect of the scale insect (*Aonidiella orientalis* Newstead) (Homoptera: Diaspididae) on neem seed production and germinability in north-eastern Nigeria. Pages 55-61 *in* Lale, N.E.S., Molta, N.B., Donli, P.O. & Dike, M.C. (Eds.), Entomology in the Nigerian economy. Research focus in the 21st century. Maiduguri, Nigeria: Entomological Society of Nigeria. [DonliWaAm1998]

Dorge, S.K., Dalaya, V.P. & Pradhan, A.G. 1972. Studies on two predatory coccinellid beetles, *Pharoscymnus horni* Weise and *Chilocorus nigritus* F., feeding on sugar-cane scale *Aspidiotus glomeratus* G. *Labdev Journal of Science and Technology, B - Life Sciences* **10**: 3-4, 138-141. [DorgeDaPr1972]

Douglas, J.W. 1886. Notes on some British Coccidae (No. 2). *Entomologist's Monthly Magazine* **22**: 243-250. [Dougla1886]

Douglas, J.W. 1886c. Notes on some British Coccidae (No. 5). *Entomologist's Monthly Magazine* **23**: 150-155. [Dougla1886c]

Douglas, J.W. 1887. Note on some British Coccidae (No. 6). *Entomologist's Monthly Magazine* **23**: 239-243. [Dougla1887]

Douglass, B.W. 1912. Scale insects of Indiana. *Annual Report of the Indiana State Entomologist* **4**: 145-226. [Dougla1912]

Doutt, R.L. 1950. The parasite complex of *Furcaspis oceanica* Lindinger. *Annals of the Entomological Society of America* **43**: 501-507. [Doutt1950]

Doutt, R.L. 1959. The biology of parasitic Hymenoptera. *Annual Review of Entomology* **4**: 161-182. [Doutt1959]

Doutt, R.L., Annecke, D.P. & Tremblay, E. 1976. Biology and host relationships of parasitoids. Pages 143-185 *in*: Huffaker, C.B. & Messenger, P.S. (Eds.). Theory and Practice of Biological Control. New York: Academic Press. 788 pp. [DouttAnTr1976]

Downing, R.S. & Logan, D.M. 1977. A new approach to San José scale control. *Canadian Entomologist* **109**: 1249-1252 [DowninLo1977]

Dowson, V.H.W. 1935. Entomological notes. A list of date scale insects and a short note on their occurrence in Basrah. *Tropical Agriculture* **12**: 225. [Dowson1935]

Dozier, H.L. 1923. The Japanese camphor scale. *Gulf Coast Grower* **75**: 1-15. [Dozier1923]

Dozier, H.L. 1926. Notes on Porto Rican scale parasites. *Journal of the Department of Agriculture of Puerto Rico* **10(3/4)**: 267-277. [Dozier1926]

Dozier, H.L. 1933. Miscellaneous notes and descriptions of Chalcidoid parasites (Hymenoptera). *Proceedings of the Entomological Society of Washington* **35**: 85-100. [Dozier1933]

Dozier, H.L. 1937. Descriptions of miscellaneous Chalcidoid parasites from Puerto Rico (Hymenoptera). *Journal of Agriculture of the University of Puerto Rico* **21(2)**: 121-135. [Dozier1937]

Drake, C.J. 1935. Report of the State Entomologist. *Thirty-sixth Annual Year Book (Iowa)* **Part 4**: 83-91. [Drake1935]

Drea, J.J. 1990. 2.2.2 Other Coleoptera. Pages 41-49 *in*: Rosen, D. (Ed.). Armored Scale Insects, Their Biology, Natural Enemies and Control [Series title: World Crop Pests, Vol. 4B]. Amsterdam, the Netherlands: Elsevier. 688 pp. [Drea1990]

Dreistadt, S.H., Clark, J.K. & Flint, M.L. 1994. Pests of Landscape Trees and Shrubs: An Integrated Pest Management Guide. *Publication (Division of Agriculture and Natural Resources, University of California)* **3359**: 327 pp. [DreistClFl1994]

Drgon, S., Huba, A. & Beros, D. 1962. [The possible ways of using organophosphorus insecticides in combination with mineral oils for the control of the San José scale.]. [In Czech]. *Prace Lab. Ochr. Rast.* 221-232. [DrgonHuBe1962]

Du, J.W., Ding, D.C., Tang, X.H., Xu, S.F. & Wu, H. 1991. [Preliminary studies on sex pheromone of pine needles Hemiberlesian scale, *Hemiberlesia pitysophila* (Coccoidea: Diaspididae) I. Activity of crude extract and its separation methods.] [In Chinese]. *Contributions of the Shanghai Institute of Entomology* 45-50. [DuDiTa1991]

Dumas, P. & Vasseur, R. 1950. Les traitements coccicides des arbres fruitièrs. *Bulletin technique d'information: BTI/Ministère de l'agriculture.* **No. 49**: 235-245. [DumasVa1950]

Dumbleton, L.J. 1954. A list of insect pests recorded in South Pacific Territories. *Technical Paper of the South Pacific Commission* **79**: 1-202. [Dumble1954]

Dunkelblum, E. 1999. Scale Insects. Pages 251-276 *in*: Hardie, J. & Minks, A.K., Eds. Pheromones of Non-Lepidopteran Insects Associated with Agricultural Plants. Oxon, UK & New York: CABI Publishing. [Dunkel1999]

Dupont, P.R. 1931. Annual report of the Department of Agriculture for the year 1930. *Annual Report (Seychelles. Department of Agriculture)* 1-18. [Dupont1931]

Dürr, H.J.R. 1951. The effect of parathion on the pernicious scale *Aspidiotus perniciosus* Comst. *Journal of the Entomological Society of Southern Africa* **14(2)**: 200-201. [Durr1951]

Dušková, F. 1952. Vergleich der Morphologischen hauptmerkmale der Schildläuse *Quadraspidiotus piri* (Lichtenstein) und *Quadraspidiotus marani* Zahradnik (Coccoidea: Diaspididae). *Beiträge zur Entomologie. Berlin* **2**: 452-455. [Duskov1952]

Dušková, F. 1953. Morfologie *Quadraspidiotus piri* (Lichtenstein) (Homoptera, Coccoidae). *Acta Ceskoslov. Spolec. Zool. Vest....* **17**: 8-36. [Duskov1953]

Dušková, F. 1953a. Die Morphologischen Merkmale und ökologische Bemerkungen zu den Weibchen der Schildläuse *Quadraspidiotus piri* (Licht.), *Q. marani* Zahr., *Q. ostreaeformis* Curt. und *Q. pernicious* (Comst.) (Homoptera, Coccoidea). *Acta Ceskoslov. Spolec. Zool. Vest....* **17**: 229-250. [Duskov1953a]

Dutta, S. 1990. Contribution towards the studies of scale insects (Homoptera: Coccoidea: Diaspididae) of North India. *Bulletin of Entomology* **31**: 152-163. [Dutta1990]

Dutta, S.K. 1992. Response of certain sugarcane varieties to scale insect and association of plant character with infestation. *Cooperative Sugar* **23(5)**: 337-342. [Dutta1992]

Dutta, S. & Baghel, C.L. 1991. On the morphology of mature female *Aonidiella orientalis* (Newstead). *Uttar Pradesh Journal of Zoology* **11(1)**: 31-35. [DuttaBa1991]

Dutta, S.K. & Devaiah, M.C. 1987. Morphology of the crawler of Sugarcane Scale Insect, *Melanaspis glomerata* (Green) (Diaspididae: Homoptera). *Journal of Research Assam Agriculture University* **8(1-2)**: 30-35. [DuttaDe1987]

Dutta, S.K. & Devaiah, M.C. 1987a. Preliminary observations on parasitic habit of *Azotus* sp. (Aphelinidae: Hymenoptera). *Journal of Research Assam Agriculture University* **8(1-2)**: 54-56. [DuttaDe1987a]

Dutta, S.K. & Devaiah, M.C. 1990. Infestation of sugarcane scale, *Melanaspis glomerata* Green on some popular and promising sugarcane varieties in Karnataka. *Indian Sugar* **40(3)**: 161-163. [DuttaDe1990]

Dutta, S.K. & Devaiah, M.C. 1994. External morphology of the head of adult *Anabrolepis mayurae* Subba Rao (Hymenoptera: Encyrtidae). *Journal of the Agricultural Science Society of North East India* **7**: 19-21. [DuttaDe1994]

Dutta, S.K. & Devaiah, M.C. 1995. Morphology of thorax and gaster of *Adelencyrtus (Anabrolepis) mayurai* (Subba Rao) (Hymenoptera: Encyrtidae). *Plant Health* **1**: 39-47. [DuttaDe1995]

Dutta, S.K. & Devaiah, M.C. 1995a. External morphology of the pupa of *Adelencyrtus mayurai* (Subba Rao)(Hymenoptera: Encyrtidae). *Bioved* **6**: 75-78. [DuttaDe1995a]

Dutta, S. & Singh, R.V. 1990. Description of new species *Abgrallaspis narainus* (Homoptera: Coccoidea: Diaspididae) collected from Firozabad, U.P. (India). *Bulletin of Entomology* **31**: 1-6. [DuttaSi1990]

du Toit, W.J. 1996. Integrated pest management of citrus in the eastern Cape: a review. *Proceedings of the International Society of Citriculture* **I**: 526-529. [duToit1996]

Düzgünes, Z. 1969. Position in Turkey of the San José scale (*Quadraspidiotus perniciosus* Comst.). Rapport de la Conférence Internationale OEPP-OILB sur le Pou de San José (Milan-Vérone, 1968). Paris: European Mediter. Plant Protect. Organ.. 111 pp. [Duzgun1969]

Dymock, J.J. & Holder, P.W. 1996. Nationwide survey of arthropods and molluscs on cut flowers in New Zealand. *New Zealand Journal of Crop and Horticultural Science* **24**: 249-257. [DymockHo1996]

Dzhashi, V.S. 1971. Tea plant pests in the Soviet subtropics. *Proceedings of the 13th International Congress of Entomology, Moscow 1968* **2**: 325-326. [Dzhash1971]

Dzhashi, V.S. 1988. [Results of a study of the destructive shield bug.]. *Subtropicheskie Kul'tury* No. 3, 134-143. [Dzhash1988]

Dzhashi, V.S. 1989. [The Japanese camellia scale - a specialised pest of the branches and shoots of the tea plant.]. [In Russian]. *Subtropicheskie Kul'tury* No. 3: 93-101 [Dzhash1989]

Dziedzicka, A. 1989. Scale insects (Coccinea) occurring in Polish greenhouses I. Diaspididae. *Acta Biologica Cracoviensia; Series: Zoologia* **31**: 93-114. [Dziedz1989]

Dziedzicka, A. & Karnkowski, W. 1990. The contribution to knowledge of *Selenaspidus articulatus* (Morgan) (Homoptera, Coccinea, Diaspididae). *Acta Biologica Cracoviensia; Series: Zoologia* **32**: 39-43. [DziedzKa1990]

Entomologist of the Hawaiian Government. 1897. Report of the Entomologist of the Hawaiian Government. *Planters Monthly* **16**: 67-85. [EHG1897]

European and Mediterranean Plant Protection Organisation. 1955. Rapport du Groupe de travail pour l'étude du pou de San José. Paris. 19 pp. [EMPPO1955]

European and Mediterranean Plant Protection Organisation. Working Party on Phytosanitary Regulations. 1962. International Conference on *Ceratitis capitata* and *Quadraspidiotus perniciosus*. *Report (European and Mediterranean Plant Protection Organization (EPPO))* **(Ser. A) No.3 4**: 7-27. [EMPPO1962]

European and Mediterranean Plant Protection Organisation. 1965. [*Quadraspidiotus perniciosus* in 1964 (San José scale).] *Quadraspidiotus perniciosus* en 1964 (pou de San José). *EPPO Publications (Series B)* **No. 55**: 8 pp. [EMPPO1965]

Early, J.W. 1984. Parasites and predators. Pages 271-308 *in*: Scott, R.R. (Ed.). New Zealand Pest and Beneficial Insects. Canterbury: Lincoln University College of Agriculture. 373 pp. [Early1984]

Easwaramoorthy, S. 1986. Ecology of sugarcane pests. Pages 31-67 *in*: David, H., Easwaramoorthy, S. & Jayanthi, R., eds. Sugarcane Entomology in India. Coimbatore, India: Sugarcane Breeding Institute. [Easwar1986]

Easwaramoorthy, S. 1986a. New approaches in sugarcane pest control. Pages 437-457 *in*: David, H., Easwaramoorthy, S. & Jayanthi, R., eds. Sugarcane Entomology in India. Coimbatore, India: Sugarcane Breeding Institute. [Easwar1986a]

Easwaramoorthy, S., David, H. & Bai, K.S. 1994. Relative toxicity of certain insecticides to *Adelencyrtus mayurai* (Subba Rao), a parasitoid of sugarcane scale insect, *Melanaspis glomerata* (Green). *Journal of Biological Control* **8(1)**: 14-17. [EaswarDaBa1994]

Easwaramoorthy, S., David, H. & Gai, K.S. 1996. Studies on *Botryoideclava bharatiya* Subba Rao, a parasite of sugarcane scale insect, *Melanaspis glomerata* (Green). *Entomon* **21**: 55-64. [EaswarDaGa1996]

Easwaramoorthy, S. & Kurup, N.K. 1986. The scale insect, *Melanaspis glomerata* (Green). Pages 233-258 *in*: David, H., Easwaramoorthy, S. & Jayanthi, R., eds. Sugarcane Entomology in India. Coimbatore, India: Sugarcane Breeding Institute. [EaswarKu1986]

Easwaramoorthy, S. & Rao, T.C.R. 1983. A reliable field evaluation technique for sugarcane scale insect (*Melanaspis glomerata* Green) resistance. *National Seminar on Breeding Crop Plants for Resistance to Pests and Diseases* 64-65. [EaswarRa1983]

Ebeling, W. 1931. Method for determination of the efficiency of sprays and HCN gas used in the control of red scale. *Monthly Bulletin, California Department of Agriculture* **10**: 669-672. [Ebelin1931]

Ebeling, W. 1932. Experiments with oil sprays used in the control of the California red scale, *Chrysomphalus aurantii* (Mask.) (Homoptera: Coccidae) on lemons. *Journal of Economic Entomology* **25**: 1007-1012. [Ebelin1932]

Ebeling, W. 1933. Variation in the population density of the California red scale, *Aonidiella aurantii* Mask., in a hilly lemon grove. *Journal of Economic Entomology* **26**: 851-854. [Ebelin1933]

Ebeling, W. 1934. A fungus found attacking the red scale in groves. *California Citrograph* **19**: 362-363. [Ebelin1934]

Ebeling, W. 1936. Effect of oil spray on California red scale at various stages of development. *Hilgardia* **10(4)**: 95-125. [Ebelin1936]

Ebeling, W. 1940. Toxicants and solids added to spray oil in control of California red scale. *Journal of Economic Entomology* **33(1)**: 92-102. [Ebelin1940]

Ebeling, W. 1945. DDT and rotenone used in oil to control the California red scale. *Journal of Economic Entomology* **38(5)**: 556-563. [Ebelin1945]

Ebeling, W. 1947. DDT preparations to control certain scale insects on citrus. *Journal of Economic Entomology* **40(5)**: 619-632. [Ebelin1947]

Ebeling, W. 1949. Subtropical Entomology San Francisco: Lithotype Process Co. 747 pp. [Ebelin1949]

Ebeling, W. 1959. Subtropical fruit pests. Los Angeles: Division of Agricultural Sciences, Univ. of Calif.. 436 pp. [Ebelin1959]

Ebeling, W. 1975. Urban Entomology. Richmond, CA: University of California. 695 pp. [Ebelin1975]

Ebeling, W. & Pence, R.J. 1953. Avocado pests. *Circular (California Agricultural Experiment Station)* **428**: 1-35. [EbelinPe1953]

Eblakoba, A.A. 1938. Experiments on the control of *Ceroplastes sinensis* De Guer. with the fungus *Cephalosporium lecanii* Zimm. *Summary of the Scientific Research Work of the Institute of Plant Protection for the year 1936* Part **III**: 75-77. [Eblako1938]

Edland, T. 1990. Utanlandske skadedyr (3). San José skjoldlus *Quadraspidiotus perniciosus* (Comstock) - ein farleg skadegjerar pa mange planter. *Gartneryrket* 80(8-9) 19-22. [Edland1990]

Edmundson, W.C. 1918. Sprays for the control of San José scale. *Bulletin (Idaho Agricultural Experiment Station)* **108**: 16 pp. [Edmund1918]

Edmunds, G.F. 1973. Ecology of black pineleaf scale (Homoptera: Diaspididae). *Environmental Entomology* **2**: 765-777. [Edmund1973]

Edmunds, G.F. & Allen, R.K. 1958. Comparison of black pine leaf scale population-density on normal ponderosa pine and those weakened by other agents. Pages 391-392 *in*: Baecker, E.C., (Ed.). Proceedings of the Tenth International Congress of Entomology. Vol. 4. Ottawa. 1115 pp. [EdmundAl1958]

Edmunds, G.F. & Alstad, D.N. 1981. Responses of black pineleaf scales to host plant variability. Denno, R.F. & Dingle, H. Insect Life History Patterns: Habitat and Geographic Variation. New York: Springer Verlag. 225 pp. [EdmundAl1981]

Edmunds, G.F. & Alstad, D.N. 1984. High summer mortality of black pineleaf scale (Homoptera: Diaspididae). *Pan-Pacific Entomologist* **60**: 267-268. [EdmundAl1984]

Edmunds, G.F. & Alstad, D.N. 1985. Malathion-induced sex ratio changes in black pineleaf scale (Hemiptera: Diaspididae). *Annals of the Entomological Society of America* **78**: 403-405. [EdmundAl1985]

Edwards, W.H. 1936. Pests attacking citrus in Jamaica. *Bulletin of Entomological Research* **27**: 335-337. [Edward1936]

Efimoff, A. 1937. [List of Insect Pests in Spain and Portugal.]. [In Russian]. Moscow: Central Laboratory of Plant Quarantine Admin.. 110 pp. [Efimof1937]

Egger, E. 1990. Protezione fitosanitaria vite 1990. *Informatore Agrario* **46(16)**: 27-135. [Egger1990]

Ehler, L.E. 1995. Biological control of obscure scale (Homoptera: Diaspididae) in California: an experimental approach. *Environmental Entomology* **24**: 779-795. [Ehler1995]

Ehler, L.E. 1996. Structure and impact on Natural enemy guilds in biological control of insect pests. Pages 337-342. *in*: Polis, G.A. & Winemiller, K.O., Eds. Food webs; Integration of patterns and dynamics. New York, Albany: Chapman and Hall. 472 pp. [Ehler1996]

Ehler, L.E. 1997. Obscure scale declines after parasitic wasp introduced. *California Agriculture* **6**: 29-32. [Ehler1997]

Ehler, L.E. & Endicott, P.C. 1984. Effect of malathion-bait sprays on biological control of insect pests of olive, citrus, and walnut. *Hilgardia* **52(5)**: 1-47. [EhlerEn1984]

Ehrhorn, E.M. 1907. Annual report of Deputy Commissioner of Horticulture. Pages 24-27 *in*: Coper, E. Second Biennial Report of the Commissioner of Horticulture of the State of California for 1905-1906. Sacramento. [Ehrhor1907]

Ehrhorn, E. 1913. Report of the Superintendent of Entomology. *Report of the Hawaii Board of Commissioner of Agriculture and Forestry* 101-151. [Ehrhor1913]

Ehrhorn, E.M. 1925. Report of Chief Plant Inspector, December, 1924. *Hawaiian Forest and Agriculture* **22**: 18-20. [Ehrhor1925]

Ehrhorn, E.M. 1925a. Report of Chief Plant Inspector, January, 1925. *Hawaiian Forest and Agriculture* **22**: 20-21. [Ehrhor1925a]

Ehrhorn, E.M. 1925g. Report of Chief Plant Inspector, July 1925. *Hawaiian Forest and Agriculture* **22**: 101. [Ehrhor1925g]

Ehrhorn, E.M. 1926c. Report of Chief Plant Inspector, February, 1926. *Hawaiian Forest and Agriculture* **23**: 20. [Ehrhor1926c]

Ehrhorn, E.M., Fullaway, D.T. & Swezey, O.H. 1913. Report of Committee on Common Names of Economic Insects in Hawaii. *Proceedings of the Hawaiian Entomological Society* **2**: 295-300. [EhrhorFuSw1913]

Einhorn, J., Guerrero, A., Ducrot, P.H., Boyer, F.D. & Gieselmann, M. 1998. Sex pheromone of the oleander scale, *Aspidiotus nerii*: structural characterization and absolute configuration of an unusual functionalized cyclobutane. *Proceedings of the National Academy of Science. USA* **95**: 9867-9872. [EinhorGuDu1998]

El-Minshawy, A.M., El-Sawaf, S.K., Hammad, S.M. & Donia, A. 1971a. The biology of *Hemiberlesia lataniae* (Sig.) in Alexandria district (Hemiptera-Homoptera-Diaspididae). *Bulletin de la Société Entomologique d'Egypte* **55**: 461-467. [ElMinsElHa1971a]

El-Minshawy, A.M., El-Sawaf, S.K., Hammad, S.M. & Donia, A. 1974. Survey of the scale insects attacking fruit trees in Alexandria district (Part 1: Fam. Diaspididae; subfam. Diaspidinae, tribe Aspidiotini). *Alexandria Journal of Agricultural Research* **22**: 223-232. [ElMinsElHa1974]

El-Minshawy, A.M. & Osman, M. 1973. The morphology of the male scale insect *Mycetaspis personata* (Comstock) (Coccoidea: Diaspididae). *Bollettino del Laboratorio di Entomologia Agraria 'Filippo Silvestri'. Portici* **30**: 169-173. [ElMinsOs1973]

El-Minshawy, A.M. & Osman, M. 1974. Biological and ecological studies on the masked scale insect *Mycetaspis personata* (Comstock) in Alexandria area (Coccoidea: Diaspididae). *Bollettino del Laboratorio di Entomologia Agraria 'Filippo Silvestri'. Portici* **31**: 152-170. [ElMinsOs1974]

Elder, R.J. & Bell, K.L. 1998. Establishment of *Chilocorus* spp. (Coleoptera : Coccinellidae) in a *Carica papaya* L. orchard infested by *Aonidiella orientalis* (Newstead) (Hemiptera : Diaspididae). *Australian Journal of Entomology* **37(4)**: 362-365. [ElderBe1998]

Elder, R. J., Gultzow, D., Smith, D. & Bell, K.L. 1997. Oviposition by *Comperiella lemniscata* Compere and Annecke (Hymenoptera: Encyrtidae) in *Aonidiella orientalis* (Newstead) (Hemiptera: Diaspididae). *Australian Journal of Entomology* **36(3)**:299-301. [ElderGuSm1997]

Elder, R.J. & Smith, D. 1995. Mass rearing of *Aonidiella orientalis* (Newstead) (Hemiptera: Diaspididae) on butternut gramma. *Journal of the Australian Entomological Society* **34**: 253-254. [ElderSm1995]

Elder, R.J., Smith, D. & Bell, K.L. 1998. Successful parasitoid control of *Aonidiella orientalis* (Newstead) (Hemiptera: Diaspididae) on *Carica papaya* L. *Australian Journal of Entomology* **37**: 74-79. [ElderSmBe1998]

Elgueta, N. 1932. Un enemigo de la "escama roja". *Revista Chilena de Historia Natural* **36**: 85. [Elguet1932]

Eliraz, A. & Rosen, D. 1978. II. Larval criteria in the systematics of *Aphytis*. *Hilgardia* **46(3)**: 96-101. [ElirazRo1978]

Elkins, R.B., Van Steenwyk, R.A., Varela, L.G. & Pickel, C. 2000. Pests of pear. *UC Pest Management Guidelines* . [ElkinsVaVa2000]

Elmer, H.S. & Brawner, O.L. 1982. Seven-year study of effects of California red scale (Homoptera: Diaspididae) on navel orange production in California's San Joaquin Valley. *Journal of Economic Entomology* **75**: 699-700. [ElmerBr1982]

Elmer, H.S., Brawner, O.L. & Ewart, W.H. 1980. California red scale predator may create citricola control dilemma. *California Agriculture* **34**: 20-21. [ElmerBrEw1980]

Elmer, H.S., Ewart, W.H. & Carman, G.E. 1951. Abnormal increase of *Coccus hesperidum* in citrus groves treated with parathion. *Journal of Economic Entomology* **44(4)**: 593-597. [ElmerEwCa1951]

Elwan, A.A. 2000. Survey of the insect and mite pests associated with date palm trees in Al-Dakhliya region, Sultanate of Oman. *Egyptian Journal of Agricultural Research* **78(2)**: 653-664. [Elwan2000]

Embleton, A.L. 1902. On the economic importance of the parasites of Coccidae. *Transactions of the Entomological Society of London* **35**: 219-229. [Emblet1902]

Emms, L.M. & McClaren, G.F. 1984. San José scale *Quadraspidiotus perniciosus* (Comstock) (Homoptera: Diaspididae) in central Otago. *New Zealand Entomologist* **8**: 81-82. [EmmsMc1984]

English, L.L. & Decker, G.C. 1954. Spray mixtures to control Putnam scale and bark beetles on elms. *Journal of Economic Entomology* **47**: 624-627. [EnglisDe1954]

English, L.L. & Turnipseed, G.F. 1933. A method for timing sprays for the control of scale insects on citrus. *Journal of Economic Entomology* **26(5)**: 987-989. [EnglisTu1933]

English, L.L. & Turnipseed, G.F. 1940. Insect pests of azaleas and camellias and their control. *Circular (Agricultural Experiment Station of the Alabama Polytechnic Institute)* **84**: 3-18. [EnglisTu1940]

Erichsen, C., Samways, M.J. & Hattingh, V. 1991. Reaction of the ladybird *Chilocorus nigritus* (F.) (Col., Coccinellidae) to a doomed food resource. *Journal of Applied Entomology* **112**: 493-498. [ErichsSaHa1991]

Erler, F., Kozár, F. & Tunç, I. 1996. A preliminary study on the armoured scale insect (Homoptera, Coccoidea: Diaspididae) fauna of Antalya. *Acta Phytopathologica et Entomologica Hungarica* **31**: 53-59. [ErlerKoTu1996]

Erler, F. & Tunç, I. 2001. A survey (1992-1996) of natural enemies of Diaspididae species in Antalya, Turkey. *Phytoparasitica* **29(4)**: 299-305. [ErlerTu2001]

Ervin, R.T. 1982. A history and economic appraisal of an insect monitoring program: California red scale. Riverside: University of California. 174 pp. [Ervin1982]

Ervin, R.T., Gardner, P.D., Baritelle, J.L. & Moreno, D.S. 1986. A history of red scale districts in central California. *Citrograph* **71**: 71-77. [ErvinGaBa1986]

Esaki, T. 1940a. [Fauna of injurious insects of the Japanese Mandated South Sea islands and their control.]. [In Japanese]. *Bot. Zool. Tokyo* **8(1)**: 274-280. [Esaki1940a]

Essig, E.O. 1913c. Insect notes. *Monthly Bulletin, California (State) Commission of Horticulture* **2(7)**: 597. [Essig1913c]

Essig, E.O. 1916a. Two newly established scale insects. Order - Hemiptera. Family - Coccidae. *Monthly Bulletin, California (State) Commission of Horticulture* **5**: 192-197. [Essig1916a]

Essig, E.O. 1920. The grape scale in California. *Monthly Bulletin, California Department of Agriculture* **9**: 37-39. [Essig1920]

Essig, E.O. 1928. Some insects of the Yosemite National Park. *Pan-Pacific Entomologist* **5**: 76-78. [Essig1928]

Essig, E.O. 1949. Insect surveys in relation to quarantine and control of insect pests. *Journal of Economic Entomology* **41**: 673-677. [Essig1949]

Essig, E.O. & Chapman, C.C. 1909. The purple and red scales. *Bulletin (Claremont Pomological Club)* **No. 2**: 14 pp. [EssigCh1909]

Etienne, J. & Matile-Ferrero, D. 1993. Les cochenilles de Casamance (Senegal) (Hemiptera, Coccoidea). *Journal of African Zoology* **107**: 253-267. [EtiennMa1993]

Euverte, G. 1967. L'insectarium de lutte biologique production massive d'A*phytis* parasites de cochenilles. *Al Awamia* **23**: 59-100. [Euvert1967]

Evans, J.W. 1942. Scale insects. *The Tasmanian Journal of Agriculture* **13**: 156-159. [Evans1942]

Evans, J.W. 1943. Insect Pests and their Control. Tasmania: H.H. Pimblett, Government Printer. 178 pp. [Evans1943]

Evans, H.H. 1975. The Comstock heritage. Pages xvii-xxv *in*: Pimentel, D. (Ed.). Insects, Science, & Society New York: Academic Press. 284 pp. [Evans1975]

Evans, G.A. & Pedata, P.A. 1997. Parasitoids of *Comstockiella sabalis* (Homoptera: Diaspididae) in Florida and description of a new species of the genus *Coccobius* (Hymenoptera: Aphelinidae). *Florida Entomologist* **80**: 328-334. [EvansPe1997]

Evans, H.C. & Prior, C. 1990. 2.1 Entomopathogenic fungi. Pages 3-17 *in*: Rosen, D. (Ed.). Armoured Scale Insects, Their Biology, Natural Enemies and Control [Series title: World Crop Pests, Vol. 4B]. Amsterdam, the Netherlands: Elsevier. 688 pp. [EvansPr1990]

Ewart, W.H. 1969. Factors involved in the development of control treatments for unarmored scales. Pages 879-880. *in*: Chapman, H.D. (Ed.). Proceedings First International Citrus Symposium. Vol. 2. Riverside: University of California. [Ewart1969]

Ewart, W.H. & Carman, G.E. 1951. Control programs for citrus insects in central California for 1951. *News Letter (University of California, Citrus Experiment Station)* **No. 50**: 10 pp. [EwartCa1951]

Ewart, W.H., Carman, G.E. & Jeppson, L.R. 1952. Information on the use of the newer insecticides for the control of insect and mite pests of citrus in central California. *News Letter (University of California, Citrus Experiment Station)* **No. 54**: 6 pp. [EwartCaJe1952]

Ewart, W.H., Carman, G.E. & Jeppson, L.R. 1953. The use of new pesticides for the control of citrus pests in central California in 1953. *News Letter (University of California, Citrus Experiment Station)* **No. 60**: 7 pp. [EwartCaJe1953]

Ewart, W.H., Carman, G.E., Jeppson, L.R. & Elmer, H.S. 1954. New pesticide treatments for citrus in central California in 1954. *Newsletter (Citrus Experiment Station - Division of Entomology, University of California, Riverside)* **No. 62**: 11 pp. [EwartCaJe1954]

Ewing, B., Yandell, B.S., Barbieri, J.F., Luck, R.F. & Forster, L.D. 2002. Event-driven competing risks. *Ecological Modelling* **158(1-2)**: 35-50. [EwingYaBa2002]

Ezzat, Y.M. 1958. Classification of the scale insects, family Diaspididae, as known to occur in Egypt [Homoptera: Coccoidea]. *Bulletin de la Société Entomologique d'Egypte* **42**: 233-251. [Ezzat1958]

Ezzat, Y.M. & Afifi, S. 1966. Redescription and classification of the scale insects of the family Diaspididae, originally described by W.J. Hall from Egypt (Homoptera: Coccoidea). *Bulletin de la Société Entomologique d'Egypte* **49**: 367-409. [EzzatAf1966]

Ezzat, Y.M. & Nada, S.M.A. 1987. List of Superfamily Coccoidea as known to exist in Egypt. *Bollettino del Laboratorio di Entomologia Agraria 'Filippo Silvestri'* **43** (Suppl.): 85-90. [EzzatNa1987]

Ezzat, Y.M. & Rahwy, S.H. 1966. Control of certain scale insects infesting citrus trees in U.A.R. Ministry of Agriculture, Plant Protection Department, United Arab Republic. 46 pp. [EzzatRa1966]

Fain, A. & Ripka, G. 1998. A new species of *Hemisarcoptes* Lignieres, 1893 (Acari: Hemisarcoptidae) from ornamental trees in Hungary. *International Journal of Acarology* **24(1)**: 33-39. [FainRi1998]

Fang, Z.G., Wu, S.A. & Xu, H.C. 2001. [A list of bamboo scale insects in China. (Homoptera: Coccoidea).]. [In Chinese]. *Journal of Zhejiang Forestry College* **18(1)**: 102-110. [FangWuXu2001]

Fargerlund, J. & Moreno, D.S. 1974. A method of handling card traps in mealybug, scale surveys. *Citrograph* **60**: 26-28. [FargerMo1974]

Fari, L. & Fari, L. 1935. Sur le traitement des arbres fruitiers avec les Carbolineums et sur la lutte contre la cochenille de San José. [In Hungarian]. *Review of Applied Entomology* 23: 339. [FariFa1935].

Fawcett, H.S. 1948. Biological control of citrus insects by parasitic fungi and bacteria. Pages 627-664 *in*: Batchelor, L.D. & Webber, H.J. (Eds.). The Citrus Industry. Vol. II. Production of the Crop. Berkeley & Los Angeles: University of California Press. 933 pp. [Fawcet1948]

Felt, E.P. 1901. Scale insects of importance and list of species in New York State. *Bulletin of the New York State Museum (of Natural History)* **46**: 289-377. [Felt1901]

Felt, E.P. 1924. Manual of Tree and Shrub Insects. New York: MacMillan. 382 pp. [Felt1924]

Felt, E.P. 1925. Key to gall midges (A resumé of studies I-VII, Itonidae). *New York State Museum Bulletin* **No. 257**: 239 pp. [Felt1925]

Felt, E.P. 1929. Protection of trees from insects. *Report of the Quebec Society for the Protection of Plants* 43-48. [Felt1929]

Fenton, F.A. 1939. The insect record for Oklahoma, 1938. *Proceedings of the Oklahoma Academy of Science* **19**: 71-78. [Fenton1939]

Ferguson, A.M. 1974. Biology of greedy scale, *Hemiberlesia rapax* (Comstock) on kiwifruit, *Actinidia chinensis. Proceedings of Kiwifruit Seminar held at Tauranga, September 19, 1974* 20-26. [Fergus1974]

Ferguson, A.M. & Fletcher, J.D. 1991. Greedy scale and the fungus *Fusarium stilboides. Proceedings of the Forty Fourth New Zealand Weed and Pest Control Conference. Palmerston North, New Zealand (Fruit Crops I)* : 260-261. [FergusFl1991]

Ferguson, A.M. & Stratton, A.E. 1978. Insect control on kiwifruit, Part 1. *Proceedings of the New Zealand Weed and Pest Control Conference* **31st**: 135-139. [FergusSt1978]

Ferguson, A.M. & Stratton, A.E. 1978a. Insect control on kiwifruit, Part II. *Proceedings of the New Zealand Weed and Pest Control Conference* **31st**: 140-146. [FergusSt1978a]

Ferguson, A.M. & Wallace, A.R. 1979. A laboratory technique to evaluate pesticides for control of greedy scale on kiwifruit. *Proceedings of the New Zealand Weed and Pest Control Conference* **32nd**: 220-224. [FergusWa1979]

Fernald, M.E. 1903b. A catalogue of the Coccidae of the world. *Bulletin of the Hatch Experiment Station of the Massachusetts Agricultural College* **88**: 1-360. [Fernal1903b]

Fernandes, I.M. 1972. Contribuicao para o conhecimento de alguns Homoptera Coccoidea do arquipelago de Cabo Verde (2a parte). *Garcia de Orta. Serie de Zoologia* **1**: 11-16. [Fernan1972]

Fernandes, I.M. 1973a. Alguns Coccoidea (Homoptera) do Arquipelago de Cabo Verde. *Memorias. Junta de Investigacoes do Ultramar. Lisboa* **58**: 251-269. [Fernan1973a]

Fernandes, I.M. 1981. Contribuiçao para o conhecimento da quermofauna do arquipelago dos Açores. *Garcia de Orta. Serie de Zoologia. Lisboa* **10**: 47-50. [Fernan1981]

Fernandes, I.M. 1987a. Contribuicao para o conhecimento da quermofauna da Guine-Bissau. *Garcia de Orta, Ser. Zool.* **14(1)**: 31-37. [Fernan1987a]

Fernandes, I.M. 1989. Contribuicao para o conhecimento de Coccoidea (Homoptera) de Angola. *Garcia de Orta, Ser. Zool.* **15(1)**: 129-134. [Fernan1989]

Fernandes, I.M. 1992. Contribuiçao para o conhecimento de Coccoidea (Homoptera) de Portugal. I - Lista anotada de cochonilhas do jardim do Centro de Zoologia. *Garcia de Orta, Serie de Zoologia, Lisboa (1990)* **17(1-2)**: 59-63. [Fernan1992]

Fernandes, I.M. 1993. Novos dados para o conhecimento da quermofauna das ilhas de S. Tome e Principe. *Garcia de Orta, Ser. Zool.* **18(1-2)**: 111-113. [Fernan1993]

Fernandes, I.M. 1999. Novos dados para o conhecimento da quermofauna do Arquipelago de Cabo Verde. *García de Orta, Serie de Zoologia. Lisboa* **23(1)**: 85-89. [Fernan1999]

Fernando, L.C.P. & Walter, G.H. 1997. Species status of two host-associated populations of *Aphytis lingnanensis* (Hymenoptera: Aphelinidae) in citrus. *Bulletin of Entomological Research* **87**: 137-144. [FernanWa1997]

Fernando, L.C.P. & Walter, G.H. 1999. Activity patterns and oviposition rates of *Aphytis lingnanensis* females, a parasitoid of California red scale *Aonidiella aurantii*: implications for successful biological control. *Ecological Entomology* **24(4)**: 416-425. [FernanWa1999]

Ferrari, R. 1976. [Tests on the survival and fecundity of the apple infestant *Quadraspidiotus perniciosus* Comst. after controlled atmosphere storage.]. [In Italian]. *Redia* **56**: 391-400. [Ferrar1976]

Ferrari, R. 1987. *Targionia vitis* Sign.: ricerche di biologia e lotta. *Redia* **70**: 77-91. [Ferrar1987]

Ferrière, C. 1927. Chalcidiens parasites de la Cochenille du Pin (*Leucaspis pini* Hart.). *Revue Suisse de Zoologie* **34(2)**: 55-67. [Ferrie1927]

Ferris, G.F. 1919a. A contribution to the knowledge of the Coccidae of Southwestern United States. *Stanford University Publications, University Series. Palo Alto* **1919**: 1-68. [Ferris1919a]

Ferris, G.F. 1920a. Notes on Coccidae-VI. (Hemiptera). *Canadian Entomologist* **52**: 61-65. [Ferris1920a]

Ferris, G.F. 1920b. Scale insects of the Santa Cruz Peninsula. *Stanford University Publications, Biological Sciences. Palo Alto* **1**: 1-57. [Ferris1920b]

Ferris, G.F. 1921. Report upon a collection of Coccidae from Lower California. *Stanford University Publications, Biological Sciences. Palo Alto* **1**: 61-132. [Ferris1921]

Ferris, G.F. 1921a. Some Coccidae from Eastern Asia. *Bulletin of Entomological Research* **12**: 211-220. [Ferris1921a]

Ferris, G.F. 1921b. Notes on Coccidae - VII, VIII (Hemiptera). *Canadian Entomologist* **53**: 57-61, 91-95. [Ferris1921b]

Ferris, G.F. 1935. Scale insects (Hemiptera: Coccoidea) from the Marquesas. *Pacific Entomological Survey, Pub. 8, Art. 9. (Bound in Bulletin of the Bernice P. Bishop Museum, Honolulu, (1939)* **142**: 125-131. [Ferris1935]

Ferris, G.F. 1936. Contributions to the knowledge of the Coccoidea (Homoptera). *Microentomology* **1**: 1-16. [Ferris1936]

Ferris, G.F. 1936a. Contributions to the knowledge of the Coccoidea (Homoptera). II. (Contribution no. 2). *Microentomology* **1**: 17-92. [Ferris1936a]

Ferris, G.F. 1937a. Contributions to the knowledge of the Coccoidea (Homoptera). IV. (Contribution no. 5). *Microentomology* **2**: 1-45. [Ferris1937a]

Ferris, G.F. 1937c. Contributions to the knowledge of the Coccoidea (Homoptera). V. (Contribution no. 6). *Microentomology* **2**: 47-101. [Ferris1937c]

Ferris, G.F. 1937d. Contributions to the knowledge of the Coccoidea (Homoptera). VI. (Contribution no. 7). *Microentomology* **2**: 103-122. [Ferris1937d]

Ferris, G.F. 1937e. On nomenclatorial and other problems in the systematics of the Coccoidea (Insecta: Homoptera). *Annals and Magazine of Natural History* **20**: 525-530. [Ferris1937e]

Ferris, G.F. 1938. Contributions to the knowledge of the Coccoidea (Homoptera). VII. (Contribution no 9). *Microentomology* **3**: 37-56. [Ferris1938]

Ferris, G.F. 1938a. Atlas of the scale insects of North America. Series 2. Palo Alto, California: Stanford University Press. [Ferris1938a]

Ferris, G.F. 1938b. Contributions to the knowledge of the Coccoidea (Homoptera). VIII. (Contribution no. 10). *Microentomology* **3**: 57-75. [Ferris1938b]

Ferris, G.F. 1941d. Atlas of the scale insects of North America. Series 3. Palo Alto, California: Stanford University Press. [Ferris1941d]

Ferris, G.F. 1941e. The genus *Aspidiotus* (Homoptera; Coccoidea; Diaspididae) (Contribution no. 28). *Microentomology* **6**: 33-69. [Ferris1941e]

Ferris, G.F. 1941f. Contributions to the knowledge of the Coccoidea (Homoptera) X: Illustrations of eleven genotypes of the Diaspididae. (Contribution no. 25.). *Microentomology* 6(1) 11-24. [Ferris1941f]

Ferris, G.F. 1942. Atlas of the scale insects of North America. Series 4. Palo Alto, California: Stanford University Press. [Ferris1942]

Ferris, G.F. 1943. Additions to the knowledge of the Diaspididae (Homoptera: Coccoidea). (Contribution no. 41). *Microentomology* **8**: 58-79. [Ferris1943]

Ferris, G.F. 1943a. The genus *Targionia* Signoret and some of its allies (Homoptera: Coccoidea: Diaspididae). (Contribution no. 42). *Microentomology* **8**: 81-111. [Ferris1943a]

Ferris, G.F. 1946. Information concerning the genera *Chortinaspis* and *Aspidiotus* (Homoptera: Coccoidea: Diaspididae). (Contribution no. 51). *Microentomology* **11**: 37-49. [Ferris1946]

Ferris, G.F. 1950. Report upon scale insects collected in China (Homoptera: Coccoidea). Part I. (Contribution no. 66). *Microentomology* **15**: 1-34. [Ferris1950]

Ferris, G.F. 1950a. Report upon scale insects collected in China (Homoptera: Coccoidea). Part II. (Contribution no. 68). *Microentomology* **15**: 69-97. [Ferris1950a]

Ferris, G.F. 1952a. Report upon scale insects collected in China (Insecta: Homoptera). Part III. (Contribution No. 76). *Microentomology* **17**: 6-16. [Ferris1952a]

Ferris, G.F. 1953. Report upon scale insects collected in China (Homoptera: Coccoidea). Part IV. (Contribution No. 84). *Microentomology* **18**: 59-84. [Ferris1953]

Ferris, G.F. 1954. New species of Diaspididae from Florida and the Caribbean Islands (Homoptera; Coccoidea). (Contribution No. 88). *Microentomology* **19**: 41-50. [Ferris1954]

Ferris, G.F. 1955b. Some miscellaneous Coccoidea (Insecta: Homoptera). (Contribution No. 91). *Microentomology* **20**: 22-29. [Ferris1955b]

Ferris, G.F. 1955c. Report upon a collection of scale insects from China. Part VI. (Insecta: Homoptera: Coccoidea). (Contribution no. 92). *Microentomology* **20**: 30-40. [Ferris1955c]

Ferris, G.F. & Kelly, J.B. 1923. XIV expedition of the California Academy of Sciences to the Gulf of California in 1921. Some Coccidae from about the Gulf of California. *Proceedings of the California Academy of Sciences* **12**: 315-318. [FerrisKe1923]

Feytaud, J. 1916. Les cochenilles de la vigne. *Société d'étude et Vulgarisation de la Zoologie agricole* **15**: 1-90. [Feytau1916]

Fidalgo, A.P. 1983. Los parasitoides y predatores de *Melanaspis paulistus* Hempl. (Hom.: Coccoidea) en el norte Argentino. *Acta Zoologica Lilloana* **37**: 119-125. [Fidalg1983]

Figueroa Potes, A. 1952. Catálogo de los artrópodos de las clases Arachnida e Insecta encontrados en el hombre, los animales y las plantas de la República de Colombia - II. *Acta Agronomica* **2**: 199-223. [Figuer1952]

Filho, O. 1931. Combate aos parasitas da laranjeira. *Chacaras e Quintaes* **43**: 606. [Filho1931]

Finney, G.L. & Fisher, T.W. 1964. Culture of entomophagous insects and their hosts. Pages 328-355 *in*: DeBach, P. & Schlinger, E.I., Eds. Biological Control of Insect Pests and Weeds. London: Chapman & Hall. 844 pp. [FinneyFi1964]

Fisher, F.E. 1950. Entomogenous fungi attacking scale insects and rust mites on citrus in Florida. *Journal of Economic Entomology* 305-309. [Fisher1950]

Fisher, T.W. 1964. Quarantine handling of entomophagous insects. Pages 305-325. *in*: DeBach, P. & Schlinger, E.I., Eds. Biological Control of Insect Pests and Weeds. London: Chapman & Hall. 844 pp. [Fisher1964]

Fisher, T.W. & De Bach, P. 1976. Principles, strategies, and tactics of pest population regulation and control in citrus ecosystems. *Proceedings of the Tall Timbers Conference on Ecology and Animal Control by Habitat Management* **6**: 43-50. [FisherDe1976]

Fisher, F.E. & Griffiths, J.T. 1950. The fungicidal effect of sulfur on entomogenous fungi attacking purple scale. *Journal of Economic Entomology* **43(5)**: 712-718. [FisherGr1950]

Fisher, F.E., Thompson, W.L. & Griffiths, J.T. 1949. Progress report on the fungus diseases of scale insects attacking citrus in Florida. *Florida Entomologist* **32(1)**: 1-11. [FisherThGr1949]

Fitch, A. 1857. Pear bark-louse, *Lecanium pyri*, Schrank. 54 - Scurfy bark-louse, *Aspidiotus furfurus*, new species. (Homoptera, Coccidae) *In* Third report on noxious and other insects of the state of New York, 53. *Transactions of the New York State Agricultural Society with an Abstract of the Proceedings of the County Agricultural Societies* **16**: 352-353. [Fitch1857]

Fitch, A. 1857a. 74. - Cherry bark-louse, *Lecanium cerasifex*, new species. 75. - Cherry scale insect, *Aspidiotus cerasi*, new species. (Homoptera. Coccidae) *In* Third report on noxious and other insects of the state of New York. *Transactions of the New York State Agricultural Society with an Abstract of the Proceedings of the County Agricultural Societies* **16**: 368. [Fitch1857a]

Fitch, A. 1857b. 139. - Circular bark louse, *Aspidiotus circularis*, new species; 140. - Currant bark louse, *Lecanium ribis*, new species. (Homoptera. Coccidae) *In* Third report on noxious and other insects of the state of New York. *Transactions of the New York State Agricultural Society with an Abstract of the Proceedings of the County Agricultural Societies* **16**: 426-427. [Fitch1857b]

Fjelddalen, J. 1996. Skjoldlus (Coccinea, Hom.) i Norge. *Insekt-Nytt* **21**: 4-24. [Fjeldd1996]

Flanders, S.E. 1934. Insects on rutaceous trees native to subtropical Australia. *Pan-Pacific Entomologist* **10**: 145-150. [Flande1934]

Flanders, S.E. 1934a. The life histories of three newly imported predators of the red scale. *Journal of Economic Entomology* **27**: 723-724. [Flande1934a]

Flanders, S.E. 1936a. *Coccidophilus citricola* Brèthes, a predator enemy of red and purple scales. *Journal of Economic Entomology* **29(5)**: 1023-1024. [Flande1936a]

Flanders, S.E. 1936b. A biological phenomenon affecting the establishment of Aphelinidae as parasites. *Annals of the Entomological Society of America* **29**: 251-255. [Flande1936b]

Flanders, S.E. 1937. Ovipositional instincts and developmental sex differences in the genus *Coccophagus*. *University of California Publications in Entomology* **6(15)**: 401-422. [Flande1937]

Flanders, S.E. 1940b. Environmental resistance to the establishment of parasitic Hymenoptera. *Annals of the Entomological Society of America* **33**: 245-253. [Flande1940b]

Flanders, S.E. 1942c. Abortive development in parasitic Hymenoptera, induced by the food-plant of the insect host. *Journal of Economic Entomology* **35**: 834-835. [Flande1942c]

Flanders, S.E. 1943a. Mass production of the California red scale, and its parasite *Comperiella bifasciata*. *Journal of Economic Entomology* **36(2)**: 233-235. [Flande1943a]

Flanders, S.E. 1943b. Red scale parasites: Comparison of oriental races of *Comperiella bifasciata* in California. *California Citrograph* **28(3)**: 78. [Flande1943b]

Flanders, S.E. 1944a. Observations on *Comperiella bifasciata*, an endoparasite of diaspine coccids. *Annals of the Entomological Society of America* **37**: 365-371. [Flande1944a]

Flanders, S.E. 1944b. Observations on *Prospaltella perniciosi* and its mass production. *Journal of Economic Entomology* **37**: 105. [Flande1944b]

Flanders, S.E. 1945a. Coincident infestations of *Aonidiella citrina* and *Coccus hesperidum*, a result of ant activity. *Journal of Economic Entomology* **38(6)**: 711-712. [Flande1945a]

Flanders, S.E. 1945b. The bisexuality of uniparental Hymenoptera, a function of the environment. *American Naturalist* **79**: 122-141. [Flande1945b]

Flanders, S.E. 1945c. Uniparentalism in the Hymenoptera and its relation to polyploidy. *Science* **100 (2591)**: 168=169. [Flande1945c]

Flanders, S.E. 1947. Use of potato tubers in mass culture of Diaspine scale insects. *Journal of Economic Entomology* **40(5)**: 746-747. [Flande1947]

Flanders, S.E. 1948a. Biological control of the yellow scale. *California Citrograph* **34**: 56, 76-77. [Flande1948a]

Flanders, S.E. 1949. George Compere--a pioneer in the biological control of red scale. *California Citrograph* **34**: 160-162. [Flande1949]

Flanders, S.E. 1949a. Culture of entomophagous insects. *Canadian Entomologist* **81(11)**: 257-274. [Flande1949a]

Flanders, S.E. 1950a. Races of apomictic parasitic Hymenoptera introduced into California. *Journal of Economic Entomology* **43**: 719-720. [Flande1950a]

Flanders, S.E. 1951b. The role of the ant in the biological control of homopterous insects. *Canadian Entomologist* **83**: 93-98. [Flande1951b]

Flanders, S.E. 1953. Hymenopterous parasites of three species of oriental scale insects. *Bollettino del Laboratorio di Zoologia Generale e Agraria della Facolta Agraria in Portici* **33**: 10-28. [Flande1953]

Flanders, S.E. 1953a. Variations in susceptibility of citrus-infesting coccids to parasitization. *Journal of Economic Entomology* **46(2)**: 266-269. [Flande1953a]

Flanders, S.E. 1954. Casca's elusive husband. *California Citrograph* **39**: 343-352. [Flande1954]

Flanders, S.E. 1958a. The role of the ant in the biological control of scale insects in California. Pages 579-584 *in*: Baecker, E.C., (Ed.). Proceedings of the Tenth International Congress of Entomology. Vol. 4. Ottawa. 1115 pp. [Flande1958a]

Flanders, S.E. 1959b. Differential host relations of the sexes in parasitic Hymenoptera. *Entomologia Experimentalis et Applicata* **2**: 125-142. [Flande1959b]

Flanders, S.E. 1960. The status of San José scale parasitization (including biological notes). *Journal of Economic Entomology* **53(5)**: 757-759. [Flande1960]

Flanders, S.E. 1962. The parasitic Hymenoptera: specialists in population regulation. *Canadian Entomologist* **94(11)**: 1133-1147. [Flande1962]

Flanders, S.E. 1966. Unique biological aspects of the genus *Casca* and a description of a new species (Hymenoptera: Aphelinidae). *Annals of the Entomological Society of America* **59(1)**: 79-82. [Flande1966]

Flanders, S.E. 1969. An historical account of *Casca smithi* and its competitor *Aphytis holoxanthus*, parasites of Florida red scale. *Israel Journal of Entomology* **4**: 29-33. [Flande1969]

Flanders, S.E. 1971. Multiple parasitism of armored coccids (Homoptera) by host-regulative aphelinids (Hymenoptera); ectoparasites versus endoparasites. *Canadian Entomologist* **103**: 857-872. [Flande1971]

Flanders, S.E., Gressitt, J.L. & DeBach, P. 1950. Parasite of Glover's scale established in California. *California Citrograph* **35**: 254-255. [FlandeGrDe1950]

Flanders, S.E., Gressit, J.L. & Fisher, T.W. 1958. *Casca chinensis*, an internal parasite of California red scale. *Hilgardia* **29(3)**: 65-91. [FlandeGrFi1958]

Fleschner, C.A. 1960. Parasites and predators for pest control. Pages 183-208 *in*: Reitz, L.P. (Ed.). Biological and Chemical Control of Plant and Animal Pests. Washington, D.C.: American Association for the Advancement of Science (Publication No. 61). 273 pp. [Flesch1960]

Fletcher, T.B. 1919. Annotated list of Indian crop-pests. *Proceedings of the 3rd Entomological Meeting. Pusa, India* **1**: 33-314. [Fletch1919]

Fletcher, W.A. 1951. Citrus pests and diseases in New Zealand. *Bulletin (New Zealand Department of Agriculture)* **20**: 1-24. [Fletch1951]

Fletcher, J. & Gibson, A. 1908. Entomological record, 1907. *Report of the Entomological Society of Ontario* **19**: 113-133. [FletchGi1908]

Fleury, A.C. 1934. Bureau of Plant Quarantine. *Monthly Bulletin, Department of Agriculture, State of California* **23**: 483-500. [Fleury1934]

Fleury, A.C. 1934a. Sixteenth annual conference Western Plant Quarantine Board. *Monthly Bulletin, California Department of Agriculture* **23**: 278-289. [Fleury1934a]

Fleury, A.C. 1935. Bureau of Plant Quarantine. *Bulletin of the California Department of Agriculture* **24**: 504-527. [Fleury1935]

Flint, W.P. 1920. Further tests of dry sulfur compounds for the control of the San José scale. *Bulletin (State of Illinois, Dept. of Registration and Education, Division of Natural History Survey)* **13(13)**: 339-343. [Flint1920]

Flint, W.P. 1923. Shall we change our recommendations for San José scale control. *Journal of Economic Entomology* **16**: 209-215. [Flint1923]

Flint, M.L. & van den Bosch, R. 1981. Introduction to Integrated Pest Management. New York: Plenum Press. 240 pp. [FlintVa1981]

Florida Department of Agriculture and Consumer Services, Bureau of Entomology. 1982. Plant disease report. *Tri-ology* **12(11)**: 5-11. [FDACSB1982]

Florida Department of Agriculture and Consumer Services, Bureau of Entomology 1983. Insects affecting. *Tri-ology* **22(2)**: 6-8. [FDACSB1983]

Florida Department of Agriculture and Consumer Services, Bureau of Entomology. 1987. Technical report. *Tri-ology* **26(5)**: 4-7. [FDACSB1987]

Florida Department of Agriculture and Consumer Services, Bureau of Entomology 1988. Special insects of regional importance. *Tri-ology* **27(11)**: 5-6. [FDACSB1988]

Foldi, I. 1982. Étude structurale et expérimentale de la formation du bouclier chez les cochenilles diaspines (Hom., Coccoidea, Diaspididae). *Annales de la Société Entomologique de France* **18**: 317-330. [Foldi1982]

Foldi, I. 1990. Morphologie des stades larvaires et imaginal du male d'*Eurhizococcus brasiliensis* (Hempel in wille, 1922) (Homoptera: Coccoidea: Margarodidae). *Nouvelle Revue d'Entomologie* **7(4)**: 405-418. [Foldi1990]

Foldi, I. 1990c. 1.2.3 Gametogenesis. 1.2.3.1 Oocytes and spermatozoa formation. Pages 199-204 *in*: Rosen, D., (Ed.). Armored Scale Insects, Their Biology, Natural Enemies and Control [Series title: World Crop Pests, Vol. 4A]. Amsterdam, the Netherlands: Elsevier. 384 pp. [Foldi1990c]

Foldi, I. 1990d. 1.3.4 Moulting and scale-cover formation. Pages 257-265 *in*: Rosen, D. (Ed.). Armored Scale Insects, Their Biology, Natural Enemies and Control [Series title: World Crop Pests, Vol. 4A]. Amsterdam, the Netherlands: Elsevier. [Foldi1990d]

Foldi, I. 2000. Diversité et modification des peuplements de cochenilles des îles d'Hyères en milieux naturels et anthropisés (Hemiptera: Coccoidea). *Annales de la Société Entomologique de France* **36(1)**: 75-94. [Foldi2000]

Foldi, I. 2001. Liste des cochenilles de France (Hemiptera, Coccoidea). *Bulletin de la Société Entomologique de France* **106(3)**: 303-308. [Foldi2001]

Foldi, I. 2002. Cochenilles, pucerons, aleurodes et psylles de la Réserve Nationale de Camargue (Hemiptera, Sternorrhyncha). *Bulletin de la Société Entomologique de France* **107(3)**: 243-251. [Foldi2002]

Foldi, I. & Delplanque, A. 1998. Les Homoptères: Sternorhyncha. II. Les Cochenilles. Pages 197-203 *in*: Delplanque, A., (Editor). Les insectes associés aux peupliers. Editions Memor. 350 pp. + 71 plates. [FoldiDe1998]

Foldi, I. & Pesson, P. 1978. Étude ultrastructurale de la spermiogenèse chez une cochenille *Aonidiella aurantii* Mask. (Homoptera, Coccoidea, Diaspididae). *Annales des Sciences Naturelles. Zoologie et Biologie Animale* **20**: 321-337. [FoldiPe1978]

Foldi, I. & Soria, S.J. 1989. Les cochenilles nuisibles a la vigne en amerique du sud (Homoptera: Coccoidea). *Annales de la Société entomologique de France (N.S)* **25(4)**: 411-430. [FoldiSo1989]

Folkina, M.Y. 1978. [A case of mass extermination of the San José scale *Quadraspidiotus perniciosus* by an ant *Crematogaster subdentata*.]. [In Russian]. *Zoologicheskii Zhurnal* **57**: 301. [Folkin1978]

Folkina, M.Y. 1980. [The San José scale *Quadraspidiotus perniciosus* (Comst.) (Homoptera, Diaspididae) in the south of the Chimkent region.]. [In Russian]. *Akademii Nauk Kazakhskoi SSR, Instituta Zoologicheskogo (Trudy)* **39**: 24-31. [Folkin1980]

Fonseca, J.P. da. 1934a. Relaçao das principais pragas observadas nos anos de 1931, 1932 E 1933, nas plantas de maior cultivo no estado de S. Paulo. *Arquivos do Instituto Biológico. Sao Paulo* **5**: 263-289. [Fonsec1934a]

Fonseca, J.P. da. 1962. Contribuiçao ao conhecimento dos coccideos do Brasil (Homoptera - Coccoidea). *Arquivos do Instituto Biologico, Sao Paulo* **29**: 13-28. [Fonsec1962]

Fonseca, J.P. da 1963. Uma nova praga da mangueira recentemente introduzida no Brasil. *O Biológico* **29(2)**: 32-35. [Fonsec1963]

Fonseca, J.P. da 1964. Uma nova praga da mangueira recentemente introduzida no Brasil. *A Rural* **44**: 515. [Fonsec1964]

Fonseca, J.P. da. 1969. Contribuiçao ao conhecimento dos coccideos do Brasil (Homoptera - Coccoidea). *Arquivos do Instituto Biologico, Sao Paulo* **36**: 9-40. [Fonsec1969]

Fonseca, J.P. da & Autuori, M. 1932a. Lista dos principaes insectos que atacam plantas citricas no Brasil. *Revista de Entomologia (Sao Paulo)* **2**: 202-214. [FonsecAu1932a]

Fontenla Rizo, J.L., Rodriguez, R. & Suri, M. 1987. Estructura y organizacion de dos comunidades de Coccoidea (Insecta: Homoptera) en dos cultivares de citricos). *Reporte de Investigacion del Instituto de Ecologia y Sistematica (Academia de Ciencias de Cuba, Havana)* **(45)**: 28 pp. [FontenRoSu1987]

Food and Agriculture Organization (FAO). 1976. New records. *Quarterly Newsletter, FAO Plant Protection Commission of Southeast Asia and Pacific Region* **19**: 6-8. [FAO1976]

Forbes, S.A. 1898. The San José scale in Illinois. (*Aspidiotus perniciosus* Comstock.). *Annual Report of the Illinois State Entomologist* **20**: 1-25. [Forbes1898]

Forbes, S.A. 1915. Observations and experiments on the San José scale. *Bulletin (Agricultural Experiment Station of the University of Illinois)* **No. 180**: 545-561. [Forbes1915]

Forbes, S.A. 1915a. The San José scale. *28th Report of the State Entomologist of Illinois* 63-79. [Forbes1915a]

Forster, L.D. & Luck, R.F. 1996. The role of natural enemies of California red scale in an IPM program in California citrus. *Proceedings of the International Society of Citriculture* **I**: 504-507. [ForsteLu1996]

Forster, L.D., Luck, R.F. & Grafton-Cardwell, E.E. 1995. Life stages of California red scale and its parasitoids. *Publication (Division of Agriculture and Natural Resources, University of California)* No. 21529. [ForsteLuGr1995]

Fotidar, M.R. & Raina, J.L. 1936. San José scale in Kashmir and its control. *Review of Applied Entomology* **24**: 692-693. [FotidaRa1936]

Foua-Bi, K. 1982. Effet des piqûres d'*Aspidiella hartii* Ckll. (Homoptère, Coccidae) sur la levée, le développement de l'appareil végétatif, et la productivité de l'igname. Pages 265-273 *in*: Miege & Lyonga. Yams - Ignames. Oxford: Clarendon Press. [FouaBi1982]

Fox-Wilson, G. 1939. Insect pests of the genus *Rhododendron*. *Int. Kongr. Entom. (Berlin)* **7**: 2296-2323. [FoxWil1939]

Frank, A.B. & Kruger, F. 1898. Die europäischen Verwandten der San José Schildlaus. *Gartenflora* **47**: 393-400. [FrankKr1898]

Frank, A.B. & Kruger, F. 1900. Schildlausbuch. Beschreibung und Bekämpfung der für den Deutschen Obst-und Weinanbau wichtigsten Schildläuse. Berlin: Parey. 120 pp. [FrankKr1900]

Franz, J.M. 1958. Biological control in Germany. Pages 461-465 *in*: Baecker, E.C., (Editor). *Proceedings of the Tenth International Congress of Entomology.* Vol. 4. Ottawa. 1115 pp. [Franz1958]

Frauenfeld, G. von 1868. Zoologische Miscellen XV. *Verhandlungen der Zoologisch-Botanischen Gesellschaft in Wien* **18**: 885-899. [Frauen1868]

Freitas, A. de. 1962. Distribuicao e hospedeiros da cochonilha-de-Sao-José (*Quadraspidiotus perniciosus* (Comst.)) em Portugal continental. *Agronomia Lusitana* **24**: 21-32. [Freita1962]

Freitas, A. de. 1966. O comportamento bio-ecologico da cochonilha-de-Sao-José *Quadraspidiotus perniciosus* (Comst.)] em Portugal continental. *Agronomia Lusitana* **26**: 289-335. [Freita1966]

Freitas, A. de. 1975. [The bio-ecological behaviour of the San José scale (*Quadraspidiotus perniciosus*) (Comst.)) in continental Portugal. II. Natural mortality and parasitism on apple.]. [In Portuguese]. *Agronomia Lusitana* **36**: 235-285. [Freita1975]

Frey, J.E. & Frey, B. 1995. Molecular identification of six species of scale insects (*Quadraspidiotus perniciosus*) by RAPD-PCR: assessing the field-specificity of pheromone traps. *Molecular Cell Biology* **4**: 777-780. [FreyFr1995]

Frey, J.E. & Frey, B. 1995a. Species identification with RAPD-PCR: a determination key for six species of the genus, *Quadraspidiotus* MacGillivray (Diaspididae) (Abstract only). *Israel Journal of Entomology* **29**: 100. [FreyFr1995a]

Froggatt, W.W. 1914. Descriptive catalogue of the scale insects ("Coccidae") of Australia (Part I.). *Agricultural Gazette of New South Wales* **25**: 127-136. [Frogga1914]

Froggatt, W.W. 1915. A descriptive catalogue of the scale insects ('Coccidae') of Australia. *Agricultural Gazette of New South Wales* **26**: 411-423, 511-516, 603-615, ... [Frogga1915]

Froggatt, W.W. 1933. The Coccidae of the casuarinas. *Proceedings of the Linnean Society of New South Wales* **58**: 363-374. [Frogga1933]

Fryer, J.C.S. 1936. Report on insect pests of crops in England and Wales: 1932-1934. *Bulletin of the Ministry of Agriculture and Forestry* **No. 99**: 50 pp. [Fryer1936]

Fullaway, D.T. 1932. Synopsis of the Hawaiian diaspinae (Coccidae). *Proceedings of the Hawaiian Entomological Society* **8**: 93-111. [Fullaw1932]

Fuller, C. 1897. A gall-making diaspid. *Agricultural Gazette of New South Wales* **8**: 579-580. [Fuller1897]

Fuller, C. 1897b. Some Coccidae of Western Australia. *Journal of Western Australia Bureau of Agriculture* **4**: 1344-1346. [Fuller1897b]

Fuller, C. 1897c. Notes on Coccidae. I. Some Coccidae of Western Australia. II. Coccid Literature. Perth, Western Australia: (privately published). 15 pp. [Fuller1897c]

Fuller, C. 1899. XIV. Notes and descriptions of some species of Western Australian Coccidae. *Transactions of the Entomological Society of London* **1899**: 435-473. [Fuller1899]

Fuller, C. 1907. The scale insects, bark lice and mealy bugs. *Natal Agricultural Journal* **3**: 1031-1055. [Fuller1907]

Fulton, R.A. & Busbey, R.L. 1943. Apparatus for laboratory fumigation of the California red scale. *Journal of Economic Entomology* **36(4)**: 628-629. [FultonBu1943]

Furness, G.O., Buchanan, G.A., George, R.S. & Richardson, N.L. 1983. A history of the biological and integrated control of red scale, *Aonidiella aurantii* on citrus in the Lower Murray Valley of Australia. *Entomophaga* **28(3)**: 199-212. [FurnesBuGe1983]

Furniss, R.L. & Carolin, V.M. 1977. Western Forest Insects. *Miscellaneous Publications United States Department of Agriculture* **No. 1339**: 1-654. [FurnisCa1977]

Fursch, H. 1985. The southern African species of *Pharoscymnus* Bedel and *Paropsis Casey* (Coleoptera: Coccinellidae). *Journal of the Entomological Society of Southern Africa* **48**: 223-231. [Fursch1985]

Fursch, H. 1987. Neue afrikanische *Scymnini*-Arten (Coleoptera Coccinellidae) als FreBfeinde von Manihot-Schädlingen. *Revue de Zoologie Africaine* **100(4)**: 387-394. [Fursch1987]

Gabritschevsky, E. 1923. Postembryonale Entwicklung, Parthenogenese und "Pedogamie" bei den Schildläusen (Coccidae). *Revue Zoologique Russe (Ruskii Zoologicheskii Zhurnal)* **3**: 295-332. [Gabrit1923]

Gahan, A.B. 1925. Some new parasitic Hymenoptera with notes on several described forms. *Proceedings of the United States National Museum. Washington* **4**: 1-23. [Gahan1925]

Gahan, A.B. 1927a. On some Chalcidoid scale parasites from Java. *Bulletin of Entomological Research* **18**: 149-153. [Gahan1927a]

Gambaro, P. 1947. Il ciclo biologico dell' *Aspidiotus perniciosus* Comst. nel veronese. *Memorie della Società Entomologica Italiana* **26**: 45-58. [Gambar1947]

Gambaro, P. 1955. L'ibernazioni di *Quadraspidiotus perniciosus* Comst. E i suoi rapporti con il clima. *Bollettino del Laboratorio di Zoologia Generale e Agraria della Facolta Agraria in Portici* **33**: 255-272. [Gambar1955]

Gambaro, P.I. 1963a. Prove di acclimatazione di *Prospaltella perniciosi* Tow. nel Veronese. *Atti del'Istituto Veneto di scienze letiere ed arte* **122**: 403-406. [Gambar1963a]

Gambaro, P.I. 1965. Osservazioni sulla biologia della *Prospaltella perniciosi* Tow. nella valle Padana. *Entomophaga* **10**: 373-376. [Gambar1965]

Gamzaev, I.M. 2002. [Natural enemies of the California scale]. [In Russian.] *Zashchita i Karantin Rastenii* **(No. 1)**: 27-28. [Gamzae2002]

Gan, Z.Y., Liu, X.Q. & Zhang, S.J. 1993. [Occurrence and forecasting of the orange brown scale, *Chrysomphalus aonidum* (L.)]. [In Chinese]. *Entomological Knowledge* **30**: 6, 347-348. [GanLiZh1993]

Gaprindashvili, N.K. 1954. [Results of a study of *Lindorus lophanthae* as an entomophage in control of several species of coccids of the Black seacoast of Adzharian SSR.]. [In Russian]. *Zoologicheskii Zhurnal* **33(3)**: 587-597. [Gaprin1954]

Gaprindashvili, N.K. 1956. [Investigation results of the species composition and efficiency of coccid and aleyrodid entomophages on subtropical plans in Adjaria.]. [In Russian]. *Akademii Nauk Gruzinskoi SSR Instituta Zashchitii Rastenii Trudy* **11**: 103-137. [Gaprin1956]

Gaprindashvili, N.K. 1975. [Biological control of the main pests on tea plantations in the Georgian SSR.]. [In Russian]. *VIII International Plant Protection Congress* **3**: 29-33. [Gaprin1975]

García Mercet, R. 1916. Los parásitos del <<poll-roig>>. *Revista de la Academia de Ciencias Exactas, Físicas y Naturales de Madrid* **14**: 776-788. [Garcia1916]

García Mercet, R. 1921. Himenópteros Fam. Encírtidos. Madrid: Museo Nacional de Ciencias Naturales. 732 pp. [Garcia1921]

García Mercet, R. 1922. El género *Azotus* Howard (Him. Calcídidos). *Boletín de la Real Sociedad Española de Historia Natural. Madrid (Sección Biológica)* **22(4)**: 196-200. [Garcia1922]

García Mercet, R. 1930. Los afelinidos de España. *Revista de Biología Forestal y Limnología* **BG**: 29-106. [Garcia1930]

García Mercet, R. 1931a. Afelinidos paleárticos (Hym. Chalc.) 8a nota. *Boletín de la Sociedad Española de Historia Natural* **31(9)**: 659-669. [Garcia1931a]

Garcia, R.C. 1949. Contribuiçao para o estudo da sistematica morfologia, biologia e ecologia da cochonilha amarela. (*Chrysomphalus dictyospermi* Morg.). *Boletim da Junta Nacional das Frutas* **9**: 374-465. [Garcia1949]

Garcia, J.N., Pinto, R.A. & Gravena, S. 1997. Diferentes doses de Aldicarb no control de pragas da laranjeira doce. Page 160 *in*: Bento, J.M.S. & Delalibera, I. (Eds.). [16th Brazilian Congress of Entomology] 16o Congresso Brasileiro de Entomologia. Salvador: Sociedade Entomológica do Brasil/EMBRAPA-CNPMF. 400 pp. [GarciaPiGr1997]

García-Marí, F. & Rodrigo, E. 1995. Life cycle of the diaspidids *Aonidiella aurantii, Lepidosaphes beckii* and *Parlatoria pergandii* in an orange grove in Valencia (Spain). *SROP Bulletin* **18**: 118-125. [GarciaRo1995]

Gardner, R. 1958. Biological control of insect and plant pests in the Trust Territory and Guam. Pages 465-471 *in*: Baecker, E.C., (Ed.). Proceedings of the Tenth International Congress of Entomology. Vol. 4. Ottawa. 1115 pp. [Gardne1958]

Gardner, P.D., Ervin, R.T., Moreno, D.S. & Baritelle, J.L. 1983. California red scale (Homoptera: Diaspididae): cost analysis of a pheromone monitoring program. *Journal of Economic Entomology* **76(3)**: 601-604. [GardneErMo1983]

Gardosik, S.M. 2001. *Aspidiotus cryptomeriae* Kuwana, an armored scale pest of conifers (Homoptera: Diaspididae). *Regulatory Horticulture (PA Dept. of Agric.)* **27(200)**: 23-25. [Gardos2001]

Garonna, A.P. 1991. Aspetti biologici di *Comperiella lemniscata* Compere & Annecke (Hymenoptera: Encyrtidae), parassitoide endofago di *Chrysomphalus dictyospermi* (Morgan) (Homoptera: Diaspididae). *Atti XVI Congresso Nazionale Italiano di Entomologia* **16**: 363-366. [Garonn1991]

Garonna, A.P. 1994. On the occurrence in Italy of *Aphytis acrenulatus* Rosen & DeBach (Hymenoptera: Aphelinidae) parasitic on *Rhizaspidiotus donacis* Leonardi (Homoptera: Diaspididae). *Bollettino del Laboratorio di Entomologia Agraria 'Filippo Silvestri'* **49**: 53-59. [Garonn1994]

Garonna, A.P. & Viggiani, G. 1989. Notizie preliminari sulla *Comperiella lemniscata* Compere & Annecke (Hymenoptera: Encyrtidae), parassitoide di *Chrysomphalus dictyospermi* (Morg.) in Italia. *Redia* **72(2)**: 523-527. [GaronnVi1989]

Garonna, A.P. & Viggiani, G. 1993. Ulteriori osservazioni morfo-biologiche di laboratorio du *Comperiella lemniscata* Compere & Annecke (Hym.: Encyrtidae), parassitoide endofago di *Chrysomphalus dictyospermi* (Morgan) (Hom.: Diaspididae). *Bollettino del Laboratorio di Entomologia Agraria 'Filippo Silvestri'* **48**: 117-124. [GaronnVi1993]

Gatina, E.S. 1973. [Plum in the flood plain of the Dnester Selekt.]. [In Russian]. *I Sortoizuch. I Yagod. Kul'tur. Kishinev, Moldavian SSR* 121-192. [Gatina1973]

Gauthier, N.L. 1993. Scale control on highbush blueberry. *The Grower: Vegetable and Small Fruit Newsletter* **93(2)**: 4-5. [Gauthi1993]

Gavalov, I. 1932b. [Materials for the knowledge of pests and diseases of agricultural plants of the Adugei region.]. [In Russian]. Krasyudar: Izdanie Adigeiskoi Oblactnoi Planovoi Komissii. 33 pp. [Gavalo1932b]

Gavalov, I. 1936. [Important coccid pests (Coccidae) of garden plants.]. [In Russian]. *Proceedings of the Agricultural Institute of Krasnodar* **4**: 58-93. [Gavalo1936]

Geier, P. 1950. Note préliminaire sur l'hivernage de *Quadraspidiotus perniciosus* Comst. (Hemipt. Diaspididae). *Mitteilungen der Schweizerischen Entomologischen Gesellschaft* **23**: 329-336 [Geier1950]

Gendrier, J.P. 1999. La lutte biologique en arboriculture fruitière. *Les Dossiers de l'Environnement* **No. 19**: 1-6. [Gendri1999]

Gennadius, P. 1881. Sur une nouvelle espèce de cochenille du genre *Aspidiotus (Aspidiotus coccineus)*. *Annales de la Société Entomologique de France* **1**: 189-192. [Gennad1881]

Gentile, A.G. & Summers, F.M. 1958. The biology of San José scale on peaches with special reference to the behavior of males and juveniles. *Hilgardia* **27**: 269-285. [GentilSu1958]

Gentry, J.W. 1965. Crop insects of northeast Africa-southwest Asia. *Agriculture Handbook (Agriculture Research Service, United States Department of Agriculture)* **No. 273**: 210 pp. [Gentry1965]

Georgala, M.B. 1963a. Prospect of a new approach to red scale control. *South African Citrus Journal* **April 1963**:5,7. [Georga1963a]

Georgala, M.B. 1964a. Comments on the corrective control of red scale. *The South African Citrus Journal* **371**: 3,5,7,9,11,15. [Georga1964a]

Georgala, M.B. 1967. Red scale: recent experiments with corrective control measures. *South African Citrus Journal* **407**: 3,5. [Georga1967]

Georgala, M.B. 1984. Analysis of the requirements for red scale control. *Citrus & Subtropical Fruit Journal* **606**: 13-4, 16-18. [Georga1984]

Georgala, M.B. 1984a. Preventive programme options for the control of red scale. *Citrus & Subtropical Fruit Journal* **606**: 9-12. [Georga1984a]

Georgala, M.B., Buitendag, C.H. & Hofmeyr, J.H. 1972. Recommendations for the control of major citrus pests on bearing trees during the 1972-73 season. *The Citrus Grower and Sub-Tropical Fruit Journal* **464**: 16-22. [GeorgaBuHo1972]

Georghiou, G.P. & Mellon, R.B. 1983. Pesticide resistance in time and space. Pages 1-46 *in*: Georghiou, G.P. & Saito, T. (Eds.). Pest Resistance to Pesticides. New York: Plenum Press. 809 pp. [GeorghMe1983]

Georghiou, G.P. & Taylor, C.E. 1976. Pesticide resistance as an evolutionary phenomenon. Pages 759-785 *in*: Packer, J.S. & White, D. (Eds.). Proceedings of XV International Congress of Entomology (Washington, D.C., August 19-27, 1976). College Park, MD: Entomological Association of America. 824 pp. [GeorghTa1976]

Gerasimova, A.A. 1938. [Bio-ecology of San Jose scale in Adjaristan. Pages 47-65 *in*: Kiritchenko, A.N. (Ed.), Works of Quarantine Laboratories, Leningrad, Sel'khozgiz. 272 pp. [Gerasi1938]

Gerhardt, P.D. & Lindgren, D.L. 1952. Dictyospermum scale in California. *Journal of Economic Entomology* **45**: 874-877. [GerharLi1952]

Gerling, D. & Bar, D. 1971. Reciprocal host-parasite relations as exemplified by *Chrysomphalus aonidum* (Homoptera: Diaspididae) and *Pteroptrix smithi* (Hymenoptera: Aphelinidae). *Entomophaga* **16**: 37-44. [GerlinBa1971]

Germain, J.-F. & Bertaux, F. 2002. *Aonidiella citrina* maintenant présente en France. *Phytoma* **No. 550**: 49-51. [GermaiBe2002]

Gerson, U. 1967c. Observations on *Hemisarcoptes coccophagus* Meyer (Astigmata: Hemisarcoptidae), with a new synonym. *Acarologia* **9**: 632-638. [Gerson1967c]

Gerson, U. 1990. 1.1.4.5 Biosystematics. Pages 129-134 *in*: Rosen, D. (Ed.). Armored Scale Insects, Their Biology, Natural Enemies and Control [Series title: World Crop Pests, Vol. 4A]. Amsterdam, The Netherlands: Elsevier. 384 pp. [Gerson1990]

Gerson, U. & Davidson, J.A. 1974. Resurrection of *Greenoidea* MacGillivray and description of the related *Avidovaspis*, new genus (Homoptera: Diaspididae). *Proceedings of the Entomological Society of Washington* **76**: 156-162. [GersonDa1974]

Gerson, U. & Hazan, A. 1979. A biosystematic study of *Aspidiotus nerii* Bouché (Homoptera: Diaspididae), with the description of one new species. *Journal of Natural History* **13**: 275-284. [GersonHa1979]

Gerson, U. & Izraylevich, S. 1997. A review of host utilization by *Hemisarcoptes* (Acari: Hemisarcoptidae) parasitic on scale insects. *Systematic and Applied Acarology* **2**: 33-42. [GersonIz1997]

Gerson, U., O'Connor, B.M. & Houck, M.A. 1990. 2.2.6 Acari. Pages 77-97 *in*: Rosen, D., ed. Armoured Scale Insects, Their Biology, Natural Enemies and Control, Vol. 4B. World Crop Pests. Amsterdam, the Netherlands: Elsevier. [GersonOcHo1990]

Gerson, U. & Zor, Y. 1973. The armoured scale insects (Homoptera: Diaspididae) of avocado trees in Israel. *Journal of Natural History* **7**: 513-533. [GersonZo1973]

Gertsson, C.A. 2000. Nya arter och nya landskapsfynd av sköldlöss från Sverige fram till år 2000. *Entomologisk Tidskrift. Stockholm* **121(4)**: 147-153. [Gertss2000]

Gertsson, C.A. 2001. An annotated checklist of the scale insects (Homoptera: Coccoidea) of Sweden. *Entomologisk Tidskrift. Stockholm* **122(3)**: 123-130. [Gertss2001]

Ghabbour, M.W. 1995. Description of the first instars of the Egyptian species in the genus *Aonidiella* (Diaspididae - Coccoidea, Homoptera). *Egyptian Journal of Agricultural Research* **73**: 379-387. [Ghabbo1995]

Ghauri, M.S.K. 1962. The morphology and taxonomy of male scale insects (Homoptera: Coccoidea). London: British Museum (Natural History). 221 pp. [Ghauri1962]

Ghesquière, J. 1927. Notes sur les coccides parasites des agrumes au Congo Belge. *Revue de Zoologie Africaine* **14(3)**: 310-316. [Ghesqu1927]

Ghesquière, J. 1932. Quelques coccides Congolais. *Bulletin du Cercle Botanique Congolais. Bruxelles* **9**: 58-63. [Ghesqu1932]

Giannotti, O. 1942. Duas novas especies de coccideos do Brasil (Homoptera-Coccoidea). *Archivos do Instituto Biologico. São Paulo* **13**: 213-216. [Gianno1942]

Gibson, A. & Ross, W.A. 1922. Insects affecting greenhouse plants. *Bulletin (Dominion of Canada, Department of Agriculture)* **No. 7**: 1-36. [GibsonRo1922]

Gieselmann, M.J. 1990. 1.2.4 Pheromones. 1.2.4.1 Pheromones and mating behaviour. Pages 221-224 *in*: Rosen, D. (Ed.). Armored Scale Insects, Their Biology, Natural Enemies and Control [Series title: World Crop Pests, Vol. 4A]. Amsterdam, The Netherlands: Elsevier. 384 pp. [Giesel1990]

Gieselmann, M.J. 1990a. 1.2.4.2 Pheromone identification and chemistry. Pages 225-232 *in*: Rosen, D. (Ed.). Armored Scale Insects, Their Biology, Natural Enemies and Control [Series title: World Crop Pests, Vol. 4A]. Amsterdam, The Netherlands: Elsevier. 384 pp. [Giesel1990a]

Gieselmann, M.J., Henrick, C.A., Anderson, R.J., Moreno, D.S. & Roelofs, W.L. 1980. Responses of male California red scale to sex pheromone isomers. *Journal of Insect Physiology* **26**: 179-182. [GieselHeAn1980]

Gieselmann, M.J., Moreno, D.S., Fargerlund, J. Tashiro, H. & Roelofs, W.L. 1979. Identification of the sex pheromone of the yellow scale. *Journal of Chemical Ecology* **5**: 27-33. [GieselMoFa1979]

Gieselmann, M.J. & Rice, R.E. 1990. 3.4.3 Use of Pheromone Traps. Pages 349-352 *in*: Rosen, D. (Ed.). Armored Scale Insects, Their Biology, Natural Enemies and Control [Series title: World Crop Pests, Vol. 4B]. Amsterdam, the Netherlands: Elsevier. 688 pp. [GieselRi1990]

Gieselmann, M.J., Rice, R.E., Jones, R.A. & Roelofs, W.L. 1979. Sex pheromone of the San José scale. *Journal of Chemical Ecology* **5**: 891-900. [GieselRiJo1979]

Giliomee, J.H. 1990. 1.1.2.2 The adult male. Pages 21-28 *in*: Rosen, D. (Ed.). Armored Scale Insects, Their Biology, Natural Enemies and Control [Series title: World Crop Pests, Vol. 4A]. Amsterdam, The Netherlands: Elsevier. 384 pp. [Giliom1990]

Gill, R.J. 1990. 3.3 Eradication. Pages 329-333 *in*: Rosen, D. (Ed.). Armored Scale Insects, Their Biology, Natural Enemies and Control [Series title: World Crop Pests, Vol. 4B]. Amsterdam, the Netherlands: Elsevier. 688 pp. [Gill1990]

Gill, R.J. 1997. The Scale Insects of California: Part 3. The Armored Scales (Homoptera: Diaspididae). Sacramento, CA: California Dept. of Food & Agriculture. 307 pp. [Gill1997]

Gill, S., Clement, D.L. & Dutky, E. 1999. Pests & Diseases of Herbaceous Perennials: the Biological Approach. Batavia: Ball Publishing (GrowerTalks Bookshelf). xvi + 304 pp. [GillClDu1999]

Gillette, C.P. & List, G.M. 1918. Ninth annual report of the State Entomologist of Colorado for the year 1917. *Circular of the Colorado State Entomologist* **26**: 3-52. [GilletLi1918]

Gillette, C.P. & List, G.M. 1922. Circular 36. Thirteenth annual report of the State Entomologist of Colorado. Fort Collins, Colorado: Office of the State Entomol., Colo. Agric. College. 64 pp. [GilletLi1922]

Gillette, C.P. & List, G.M. 1923. Notes on insect and mite pests. Pages 12-18 *in*: Circular 38. Thirteenth annual report of the State Entomologies. Fort Collins, Colorado: Office of the State Entomol. Colo. Agric. College. [GilletLi1923]

Giraldi, G. & Rui, D. 1962. Ricerche sulla possibilità di acclimatamento nel venuto della cocciniglia del cocco e del banano (*Aspidiotus destructor* Sign.) (Hemip., Coccidae). *Relazioni e Monografie Agrarie Subtropicali e Tropicali* **No. 83**: 121-144. [GiraldRu1962]

Girault, A.A. 1911a. Notes on the Hymenoptera Chalcidoidea, with descriptions of several new genera and species. *Journal of the New York Entomological Society* **19**: 175-189. [Giraul1911a]

Girault, A.A. 1916. Notes on described Chalcidoid Hymenoptera with new genera and species. *Societas Entomologica (Stuttgart)* **31(9)**: 42-44. [Giraul1916]

Girola, C.D. & Araujo, J.J. 1925. Enfermedades de las plantas: lista de las observadas en la República Argentina en los años de 1918 a 1923. *Publicación del Museo Agrícola S.R.A.* No. 46. [GirolaAr1925]

Glaser, L. 1877. Ueber einige in der Rheingegend aufretende Schnabelkerfe. *Zool. Garten* **18**: 45-51. [Glaser1877]

Glenn, P.A. 1915. The San José scale (*Aspidiotus perniciosus* Comstock.). *Annual Report of the Illinois State Entomologist* **28**: 87-106. [Glenn1915]

Glenn, P.A. 1931. Use of temperature accumulations as an index to the time of appearance of certain insect pests during the season. *Transactions of the Illinois State Academy of Science* **24**: 167-180. [Glenn1931]

Glick, P.A. 1922. Insects injurious to Arizona crops during 1922. (Part III). *Annual report (Arizona Commission of Agriculture and Horticulture)* **14th**: 55-77. [Glick1922]

Glover, P.M. 1933. *Aspidiotus (Furcaspis) orientalis* Newstead (Coccidae), its economic importance in lac cultivation and its control. *Bulletin of the Indian Lac Research Institute* **No. 16**: 1-23. [Glover1933]

Glover, P.M. 1935. An account of the occurrence of *Chrysomphalus aurantii* Mask. and *Laccifer lacca*, Kerr on grape fruit in Ranchi District, Chota Nagpur, with a note on the Chalcidoid parasites of *Aspidiotus orientalis* Newst. *Journal, Bombay Natural History Society* **38**: 151-153. [Glover1935]

Goberdhan, L.C. 1962. Scale insects of the coconut palm with special reference to *Aspidiotus destructor*. *Journal of the Agricultural Society of Trinidad and Tobago* **62**: 49-71. [Goberd1962]

Goethe, R. 1884. Beobachtungen über Schildläuse und deren Feinde, angestellt an Obstbäumen und Reben im Rheingau. *Jahrbücher des Nassauischen Vereins für Naturkunde* **37**: 107-131. [Goethe1884]

Goethe, R. 1899. Tierische Feinde. (Beobachtungen über Schildläuse.). *Königlichen Lehranstalt für Obst-, Wein- und Gartenbau. Berlin* **1898/99**: 16-22. [Goethe1899]

Goidanich, A. 1956. Sulla permeabilita' del follicolo sericeo in alcuni Diaspididi (Hem. Cocc. Diaspididae). *Proceedings of the 10th International Congress of Entomology. Montreal* 207-224. [Goidan1956]

Goidanich, A. 1958. Sulla permeabilità del follicolo sericeo in alcuni Diaspididi (Hem. Cocc. Diaspididae). Page 387 *in*: Baecker, E.C., (Ed.). Proceedings of the Tenth International Congress of Entomology. Vol. 2. Ottawa. 1055 pp. [Goidan1958]

Golan, K. & Lagowska, B. 1998. [Occurrence of scale insects on ornamental conifers.] Wystepowanie czerwcow na ozdobnych roslinach iglastych. *Ochrona Roslin* **42**: 7, 4-6. [GolanLa1998]

Golan, K., Lagowska, B. & Jaskiewicz, B. 2001. Scale insects (Hemiptera, Coccoidea) of the Kazimierz Landscape Park in Poland. *Fragmenta Faunistica* **44**: 229-249. [GolanLaJa2001]

Gomes Costa, R. 1940. Como combater as Cochonilhas. *Anais do II Congresso Riograndense de Agronomía* 329-354. [GomesC1940]

Gomes Costa, R. & Redaelli, D.C. 1947. Cochonilhas ou Coccideas do Rio Grande do Sul. *Revista Agronomica. Lisbon* **11**: 5-8, 65-68, 123-126, 170-175... [GomesCRe1947]

Gomez, J. 1936. Novos hospedeiros e novas regioes de alguns insetos do Brasil. *O Campo* **7 (82)**: 42-43. [Gomez1936]

Gomez Clemente, F. 1943. Cochinillas que atacan a los agrios en la región de Levante. *Boletín de Patología Vegetal y Entomología Agrícola. Madrid* **12**: 299-328. [GomezC1943]

Gómez Clemente, F. 1950. Los insectos auxiliares en la lucha contra los nocivos a los agrios. *Boletín de Patología Vegetal y Entomología Agrícola. Madrid* **19**: 1-18. [GomezC1950]

Gómez Clemente, F. 1954a. Estado actual de la lucha biológica contra algunas cochinillas de los agrios. *Boletín de Patología Vegetal y Entomología Agrícola. Madrid* **19**: 19-35. [GomezC1954a]

Gómez-Menor Guerrero, J.M. 1962. Diaspidos de gran Canaria y Tenerife (Islas Canarias) (Hom. Coccoidea). *Anales de Estudios Atlánticos* **8**: 147-214. [GomezM1962]

Gómez-Menor Ortega, J. 1927a. Algunos cóccidos nuevos de España (Hem. Cocc.). *EOS* **3**: 288-298. [GomezM1927a]

Gómez-Menor Ortega, J. 1937. Cóccidos de España. Madrid: Instituto de Investigaciones Agronómicas, Estación. 432 pp. [GomezM1937]

Gómez-Menor Ortega, J. 1941. Cóccidos de la República Dominicana (Hom. Cocc.). *EOS* **16**: 125-143. [GomezM1941]

Gómez-Menor Ortega, J. 1946. Adiciones a los "Cóccidos de España" (1a nota). *EOS* **22**: 59-106. [GomezM1946]

Gómez-Menor Ortega, J. 1948. Adiciones a los "Cóccidos de España" (2a nota). *EOS* **24**: 73-121. [GomezM1948]

Gómez-Menor Ortega, J. 1954. Adiciones a los "Cóccidos de España" (3a nota). *EOS* **30**: 119-148. [GomezM1954]

Gómez-Menor Ortega, J. 1956. Cochinillas que atacan a los frutales (Homoptera, Coccoidea: 1. Familia Diaspididae). *Boletín de Patología Vegetal y Entomología Agrícola. Madrid* **22**: 1-105. [GomezM1956]

Gómez-Menor Ortega, J. 1956a. Datos sobre cóccidos de Espana. *Bollettino del Laboratorio di Zoologia Generale e Agraria della Facolta Agraria in Portici* **33**: 611-620. [GomezM1956a]

Gómez-Menor Ortega, J. 1956b. Quelques nouveaux coccides d'espagne. *Proceedings of the 14th International Congress of Zoology. Copenhagen* 482-386. [GomezM1956b]

Gómez-Menor Ortega, J. 1957. Adiciones a los cóccidos de España (4a nota). *EOS* **33**: 39-86. [GomezM1957]

Gómez-Menor Ortega, J. 1958a. Distribución geográfica y ensayo de la ecológica de los cóccidos en España .*Publicado del Instituto Biologia Aplicata Barcelona* **27**: 5-15. [GomezM1958a]

Gómez-Menor Ortega, J. 1958c. Homopteros Sternorrhyncha de la Provincia de Granada. *Boletín de la Real Sociedad Española de Historia Natural. Madrid (Sección Biológica)* **55**: 403-408. [GomezM1958c]

Gómez-Menor Ortega, J. 1959. Homopteros "Sternorrhyncha" que atacan a la encina, II Parte. *Revista de Entomologia de España* **17**: 141-201. [GomezM1959]

Gómez-Menor Ortega, J. 1960. Adiciones a los "Cóccidos de España". V. (Superfamilia Coccoidea). *EOS* **36**: 157-204. [GomezM1960O]

Gómez-Menor Ortega, J .1965 .Adiciones a los "Cóccidos de España", (6a nota). Especies del genero *Evallaspis* con su distribución geográfica en la Peninsula a las Islas Baleares. *EOS* **41**: 87-114. [GomezM1965]

Gómez-Menor Ortega, J. 1967. Lista de Coccoidea de las Islas Canarias (adiciones) (Hemiptera, Homoptera). *EOS* **43**: 131-134. [GomezM1967O]

Gómez-Menor Ortega, J. 1968. Adiciones a los "Coccidos de España" VII nota (Hem. Homoptera). *EOS* **43**: 541-563. [GomezM1968]

Gonzalez, C., Hernandez, D., Sibat, R. & Rodriguez, J.L. 1991. [Citrus coccids and their natural enemies in Cuba.]. [In Spanish]. Pages 433-441. *in*: Pavis, C. and A. Kermarrec (Eds.). Caribbean Meetings on Biological Control. Guadeloupe, France, Nov. 5-7, 1990. Paris, France: Institut National de la Recherche Agronomique (INRA). 569 pp. [GonzalHeSi1991]

González, R.H. 1956. Biologia y control de la escama de San Jose y su control con insecticidas fosforados. *Agronomia (Chile)* **1**: 9-13 [Gonzal1956]

Gonzáles, R.H. 1969. Biological control of citrus pests in Chile. Pages 839-847. *in*: Chapman, H.D. Proceedings First International Citrus Symposium. Vol. 2. Riverside: University of California. [Gonzal1969]

González, R.H. 1973. Integrated control strategies on deciduous fruit trees. *FAO Plant Protection Bulletin* **21(3)**: 56-63. [Gonzal1973]

González, R.H. 1975. Integrated pest control in orchards in Chile and perspectives in South America. *C.R. 5e Symposium Lutte integree en vergers OILB/SROP* 135-145. [Gonzal1975]

González, R.H. 1981. Biología, ecología y control de la escama de San José en Chile, *Quadraspidiotus perniciosus* (Comst.). *Publicaciones en Ciencia Agricultura, Universidad de Chile, Facultad Ciencia Agrarias, Veterinarias, y Forestales* **9**: 1-64. [Gonzal1981]

González, R.H. 1986. Plagas del kiwi en Chile. *Fruticola* **7(1)**: 13-27. [Gonzal1986]

González, R.H. 1989. Insectos y acaros de importancia agricola y cuarentenaria en Chile. Santiago: Universidad de Chile. 310 pp. [Gonzal1989]

González, R.H. 1989a. Manejo de plagas del kiwi en chile: 1. Degradacion de residuos de los insecticidas *Chlorpyrifos y Phosmet*. *Revista Fruticola* **10(2)**: 35-43. [Gonzal1989a]

González, R.H. & Charlin, R. 1968. Nota preliminar sobre los insectos cóccoideos de Chile. *Revista Chilena de Entomología* **6**: 109-113. [GonzalCh1968]

González, R.H. & Curkovic, S.T. 1994. Manejo de plagas y degradación de residuos de pesticidas en kiwi. *Revista Fruticola* **15**: 5-20. [GonzalCu1994]

González, R.H. & Rojas P., S. 1967. Estudio analítico del control biológico de plagas agrícolas en Chile. *Agricultura Técnica. Chile* **26(4)**: 133-147. [GonzalRo1967]

Goot, P. van der 1928. Ziekten en plagen der cultuurgewassen in Nederlandsch-indie in 1927. *Mededeelingen van het Intituut vor Plantenziekten* **No. 74**: 1-85. [Goot1928]

Gordh, G. 1977. Biosystematics of natural enemies. Pages 125-148. *in*: Ridgway, R.L. & Vinson, S.B. Biological Control by Augmentation of Natural Enemies: Insect and Mite Control with Parasites and Predators. New York: Plenum Press. 480 pp. [Gordh1977]

Gordh, G. 1979. Family Encyrtidae. Pages 890-967 *in*: Krombein, K.V., Hurd, P.D., Jr., Smith, D.R. & Burks, B.D., Eds. Catalog of Hymenoptera in America North of Mexico. Washington, D.C.: Smithsonian Institution Press. 1198 pp. [Gordh1979]

Gordh, G. 1994. A biographical account of Compere, Harold (1896-1978), Biological-control foreign explorer. *Pan-Pacific Entomologist* **70(3)**: 188-205. [Gordh1994]

Gordon, F.C. & Potter, D.A. 1988. Seasonal biology of the walnut scale, *Quadraspidiotus juglansregiae* (Homoptera: Diaspididae), and associated parasites on red maple in Kentucky. *Journal of Economic Entomology* **81(4)**: 1181-1185. [GordonPo1988]

Gordon, R.D. 1978. West Indian Coccinellidae. II. (Coleoptera): Some scale predators with keys to genera and species. *The Coleopterists Bulletin* **32(3)**:205-218. [Gordon1978]

Gordon, R.D. 1980. The tribe Azyini (Coleoptera: Coccinellidae): historical review and taxonomic revision. *Transactions of the American Entomological Society* **106(2)**: 149-203. [Gordon1980]

Goryunova, Z.S. 1964. [Intraspecific differentiation in Prospaltella (*Prospatella perniciosi* Tow.) the parasite of the San José scale.]. [In Russian]. *Trudy Vsesoiuznogo nauchno-issledovatel'skogo instituta zashchity rastenii [Proceedings of the All-Union Scientific-Research Inst. for Plant Prot.]* **21**: 40-55. [Goryun1964]

Gough, L. 1914. The fumigation of citrus trees. *Agricultural Journal of Egypt* **4(1)**: 17-19. [Gough1914]

Goux, L. 1937c. Notes sur les coccides [Hem. Coccidae] de la France. (19e note). Etude d'un *Rhizaspidiotus* et d'un *Aspidiotus* nouveaux. *Bulletin de la Société Zoologique de France* **62**: 341-349. [Goux1937c]

Goux, L. 1941a. Notes sur les *Trionymus* de la France et sur quelques espèces nouvelles pour la faune Française (Hem. Coccidae) (Notes sur les coccides (Hem. Coccidae). de la France (30e note). *Bulletin de la Société d'Histoire Naturelle de l'Afrique du Nord* **32**: 31-44. [Goux1941a]

Goux, L. 1945. Notes sur les coccides (Hem. Coccoidea) de la France (34me note). Description d'un *Trionymus* nouveau et de sa larve néonate et remarques sur quelques espèces nouvelles pour la faune française. *Bulletin du Muséum d'Histoire Naturelle de Marseille* **5**: 30-38. [Goux1945]

Goux, L. 1949. Notes sur les coccids (Hem. Coccoidea) de la France. (40me note). Étude d'un *Quadraspidiotus* nouveau parasite de l'olivier et remarques sur une espèce nouvelle pour la faune Française. *Bulletin de la Société Linnéenne de Provence* **17**: 27-32. [Goux1949]

Goux, L. 1951. Notes sur les coccids (Hem. Coccoidea) de la France (41e note); étude d'un *Diaspidiotus* nouveau des environs de Marseille. *Bulletin de la Société Entomologique de France* **56**: 10-14. [Goux1951]

Gowdey, C.C. 1913. A list of Uganda Coccidae and their food-plants. *Bulletin of Entomological Research* **4**: 247-249. [Gowdey1913]

Gowdey, C.C. 1917. A list of Uganda Coccidae, their food-plants and natural enemies. *Bulletin of Entomological Research* **8**: 187-189. [Gowdey1917]

Gowdey, C.C. 1921. The Coccidae of Jamaica. *Entomological Bulletin, Jamaican Department of Science and Agriculture* **1**: 1-46. [Gowdey1921]

Graebner, L., Moreno, D.S. & Baritelle, J.L. 1984. The Fillmore Citrus Protective District: a success story in integrated pest management. *Bulletin of the Entomological Society of America* **30**: 27-33. [GraebnMoBa1984]

Grafton-Cardwell, B. 1994. Resistance of California red and yellow scale in the San Joaquin valley of California. *Resistant Pest Management* **6(1)**: 7-9. [Grafto1994]

Grafton-Cardwell, E.E., Millar, J.G., O'Connell, N.V. & Hanks, L.M. 2000. Sex pheromone of yellow scale, *Aonidiella citrina* (Homoptera: Diaspididae): Evaluation as an IPM tactic. *Journal of Agricultural and Urban Entomology* **17(2)**: 75-88. [GraftoMiOC2000]

Grafton-Cardwell, E.E., Morse, J.G., O'Connell, N.V. & Phillips, P.A. 2000. Pests of citrus. *UC Pest Management Guidelines* . [GraftoMoOC2000]

Grafton-Cardwell, E.E. & Ouyang, Y. 1993. Toxicity of four insecticides to various populations of the predacious mite, *Euseius tularensis* Congdon (Acarina: Phytoseiidae) from San Joaquin Valley California citrus. *Journal of Agricultural Entomology* **10(1)**: 21-29. [GraftoOu1993]

Grafton-Cardwell, E.E., Ouyang, Y. & Salse, J. 1998. Insecticide resistance and esterase enzyme variation in the California red scale (Homoptera: Diaspididae). *Journal of Economic Entomology* **9**: 812-819. [GraftoOuSa1998]

Grafton-Cardwell, E.E., Ouyang, Y., Striggow, R. & Vehrs, S. 2001. Armored scale insecticide resistance challenge San Joaquin Valley citrus growers. *California Agriculture* **55(5)**: 20-25. [GraftoOuSt2001]

Grafton-Cardwell, E.E. & Reagan, C.A. 1995. Selective use of insecticides for control of armored scale (Homoptera: Diaspididae) in San Joaquin Valley California citrus. *Journal of Economic Entomology* **88**: 1717-1725. [GraftoRe1995]

Grafton-Cardwell, E.E., Striggow, R.A. & Ouyang, Y. 1996. Pesticide resistance in armoured scale influences citrus IPM in California. *Proceedings of the International Society of Citriculture* **I**: 553-555. [GraftoStOu1996]

Grafton-Cardwell, E.E. & Vehrs, S.L.C. 1995. Monitoring for organophosphate- and carbamate-resistant armored scale (Homoptera: Diaspididae) in San Joaquin Valley citrus. *Journal of Economic Entomology* **88**: 495-504. [GraftoVe1995]

Graham, J.C. & Streibert, H.P. 1996. Resistance management for California red scale, *Aonidiella aurantii* (Mask.): insecticide resistance action committee guidelines. *Proceedings of the International Society of Citriculture* **I**: 649-651. [GrahamSt1996]

Grandpré, A. Daruty & Charmoy, d'Emmerz de. 1899. Liste raisonnée des cochenilles de l'Île Maurice. *Publications de la Société Amicale Scientifique, Maurice 24 Mars* **1899**: 1-48. [GrandpCh1899]

Gravena, S., Fernandes, C.D., Santos, A.C., Pinto, A.S. & Paiva, P.S.B. 1992. Efeito de Buprofezin e Abamectin sobre *Pentilia egena* (Muls.) (Coleoptera: Coccinellidae) e crispideos em citros. *Anais da Sociedade Entomologica do Brasil* **21(3)**: 215-222. [GravenFeSa1992]

Gravena, S. & Fornasieri, J.L. 1979. [Population fluctuations of some scale insects and entomophagous predators in citrus groves and the influence of meteorological factors.]. [In Portuguese]. *Científica* **7**: 109-113. [GravenFo1979]

Gravena, S., Leao-Neto, R.D.R., Moretti, F.C. & Tozatti, G. 1988. [Efficacy of insecticides on Selenaspidus articulatus (Morgan) (Homoptera, Diaspididae) and effect on natural enemies in citrus crop.]. [In Portuguese]. *Científica (Jaboticabal)* **16(2)**: 209-218. [GravenLeMo1988]

Gravena, S., Paiva, P.S.B. & Yamamoto, H. 1993. Impact of methidathion on citrus entomofauna: I. Perspective of ecological selectivity. *Bulletin OILB/SROB* **16(7)**: 16-25. [GravenPaYa1993]

Gravena, S., Yamamoto, P.T., Fernandes, O.D., & Benetoli, I. 1992. [Effect of ethion and aldicarb against *Selenaspidus articulatus* (Morgan), *Parlatoria ziziphus* (Lucas) (Hemiptera: Diaspididae) and influence on beneficial fungus.]. [In Portuguese]. *Anais da Sociedade Entomologica do Brasil* **21(2)**: 101-111. [GravenYaFe1992]

Gray, G.P. & Kirkpatrick, A.F. 1929. The protective stupefaction of certain scale insects by hydrocyanic acid vapor. *Journal of Economic Entomology* **22**: 878-892. [GrayKi1929]

Greany, P.D., Vinson, S.B. & Lewis, W.J. 1984. Insect parasitoids: finding new opportunities for biological control. *Bioscience* **34(11)**: 690-696. [GreanyViLe1984]

Greathead, D.J. 1971. A review of biological control in the Ethiopian region. *Technical Communication (Commonwealth Institute of Biological Control)* **5**: 162 pp. [Greath1971]

Greathead, D.J. 1973. A review of introductions of *Lindorus lophanthae* (Blaisd.) (Col: Coccinellidae) against hard scales (Diaspididae). *Technical Bulletin of the Commonwealth Institute of Biological Control* **16**: 29-33. [Greath1973]

Greathead, D.J. (Editor) 1976. A Review of Biological Control in Western and Southern Europe. Farnham Royal, Slough: Commonwealth Agricultural Bureaux. 182 pp. [Greath1976]

Greathead, D.J. 1986. Parasitoids in classical biological control. Pages 289-318 *in*: Waage, J. & Greathead, D. (Eds.). Insect Parasitoids. London: Academic Press. 389 pp. [Greath1986]

Greathead, D.J. 1989. Biological control as an introduction phenomenon: a preliminary examination of progrmmes against Homoptera. *The Entomologist* **108(1/2)**: 28-37. [Greath1989]

Greathead, D.J. 1990. 1.4.3 Crawler behaviour and dispersal. Pages 305-308 *in*: Rosen, D. (Ed.). Armored Scale Insects, Their Biology, Natural Enemies and Control [Series title: World Crop Pests, Vol. 4A]. Amsterdam, The Netherlands: Elsevier. 384 pp. [Greath1990]

Greathead, D.J. & Greathead, A.H. 1992. Biological control of insect pests by insect parasitoids and predators: the BIOCAT database. *Biocontrol News and Information* **13(4)**: 61N-68N. [GreathGr1992]

Greaves, A.J., Davys, J.W., Dow, B.W., Tomkins, A.R., Thomson, C. & Wilson, D.J. 1994. Seasonal temperatures and the phenology of greedy scale populations (Homoptera: Diaspididae) on kiwifruit vines in New Zealand. *New Zealand Journal of Crop and Horticultural Science* **22**: 7-16. [GreaveDaDo1994]

Greaves, A.J., Tomkins, A.R., Wilson, D.J. & Thomson, C. 1992. Abamectin to control armoured scales (Hemiptera: Diaspididae) on kiwifruit. *New Zealand Journal of Crop and Hort Sci* **20**: 79-83. [GreaveToWi1992]

Green, E.E. 1890. Insect pests of the tea plant. Part 1. 104 pp. [Green1890]

Green, E.E. 1895. Notes on coccids from Kent. *Entomologist's Monthly Magazine* **31**: 229-233. [Green1895]

Green, E.E. 1896. Catalogue of Coccidae collected in Ceylon. *Indian Museum Notes* **4**: 2-10. [Green1896]

Green, E.E. 1896a. On an intermediate "Aonidiform" stage in *Aspidiotus*. *Entomologist's Monthly Magazine* **32**: 84. [Green1896a]

Green, E.E. 1896c. New Indian coccids. *Entomologist's Monthly Magazine* **32**: 199-201. [Green1896c]

Green, E.E. 1896e. The Coccidae of Ceylon. Part I. London: Dulau. 103 pp. [Green1896e]

Green, E.E. 1897. Notes on Coccidae from the Royal Gardens, Kew. [with additions by R. Newstead.]. *Entomologist's Monthly Magazine* **33**: 68-77. [Green1897]

Green, E.E. 1899c. Observations on *Aspidiotus lataniae*, Sign. *Entomologist's Monthly Magazine* **35**: 181-183. [Green1899c]

Green, E.E. 1900a. Supplementary notes on the Coccidae of Ceylon. *Journal of the Bombay Natural History Society* **13**: 66-76, 252-257. [Green1900a]

Green, E.E. 1900c. Remarks on Indian scale insects (Coccidae), with descriptions of new species. *Indian Museum Notes* **5**: 1-13. [Green1900c]

Green, E.E. 1901a. Note on the genus *Lecaniodiaspis*, Targ. *Entomologist's Monthly Magazine* **37**: 293-295. [Green1901a]

Green, E.E. 1903a. Remarks on Indian scale insects (Coccidae). With descriptions of new species. Pt. II. *Indian Museum Notes* **5**: 93-103. [Green1903a]

Green, E.E. 1904. Descriptions of some new Victorian Coccidae. *The Victorian Naturalist* **21**: 65-69. [Green1904]

Green, E.E. 1904a. On some Javanese Coccidae: with descriptions of new species. *Entomologist's Monthly Magazine* **40**: 204-210. [Green1904a]

Green, E.E. 1904b. On some Coccidae in the collection of the British Museum. *Annals and Magazine of Natural History* **14**: 373-378. [Green1904b]

Green, E.E. 1905. On some Javanese Coccidae: With descriptions of new species. *Entomologist's Monthly Magazine* **41**: 28-33. [Green1905]

Green, E.E. 1905a. Supplementary notes on the Coccidae of Ceylon. *Journal of the Bombay Natural History Society* **16**: 340-357. [Green1905a]

Green, E.E. 1905b. Some new Victorian Coccidae. *Victorian Naturalist* **22**: 3-8. [Green1905b]

Green, E.E. 1907. XII. Notes on the Coccidae collected by the Percy Sladen Trust Expedition to the Indian Ocean: supplemented by a collection received from Mr. R. Dupont, Director of Agriculture, Seychelles. *Transactions of the Linnean Society of London, Zoology* **12**: 197-207. [Green1907]

Green, E.E. 1908a. Remarks on Indian scale insects (Coccidae), Part III. With a catalogue of all species hitherto recorded from the Indian continent. *Memoirs of the Department of Agriculture in India, Entomology Series* **2**: 15-46. [Green1908a]

Green, E.E. 1910a. Remarks on Coccidae from Uganda. *Bulletin of Entomological Research* **1**: 201-202. [Green1910a]

Green, E.E. 1911. On some Coccidae affecting rubber trees in Ceylon with descriptions of new species. *Journal of Economic Biology* **6**: 27-37. [Green1911]

Green, E.E. 1914c. Remarks on a small collection of Coccidae from Northern Australia. *Bulletin of Entomological Research* **5**: 231-234. [Green1914c]

Green, E.E. 1914d. On some coccid pests from the Seychelles. *Journal of Economic Biology. London* **9**: 47-48. [Green1914d]

Green, E.E. 1915c. Notes on Coccidae collected by F.P. Jepson, Government Entomologist, Fiji. *Bulletin of Entomological Research* **6**: 44. [Green1915c]

Green, E.E. 1915d. New species of Coccidae from Australia. *Bulletin of Entomological Research* **6**: 45-53. [Green1915d]

Green, E.E. 1915e. On some animal pests of the hevea rubber tree. *Transactions of the Third International Congress of Tropical Agriculture* **1**: 608-636. [Green1915e]

Green, E.E. 1916. On two new British Coccidae, with notes on some other British species. *Entomologist's Monthly Magazine* **52**: 23-31. [Green1916]

Green, E.E. 1916e. Remarks on Coccidae from Northern Australia - II. *Bulletin of Entomological Research* **7**: 53-65. [Green1916e]

Green, E.E. 1916f. Notes on Coccidae occurring in the Seychelles Islands, with descriptions of new species. *Bulletin of Entomological Research* **7**: 193-196. [Green1916f]

Green, E.E. 1918. A list of Coccidae affecting various genera of plants. *Annals of Applied Biology* **4**: 228-239; 5: 143-156, (1919). [Green1918]

Green, E.E. 1919c. Notes on Indian Coccidae of the subfamily Diaspidinae, with description of new species. *Records of the Indian Museum* **16**: 433-449. [Green1919c]

Green, E.E. 1920. Observations on British Coccidae. No. V. *Entomologist's Monthly Magazine* **56**: 114-130. [Green1920]

Green, E.E. 1922. The Coccidae of Ceylon, Part V. Pages 345-472. London: Dulau & Co.. [Green1922]

Green, E.E. 1922a. Supplementary notes on the Coccidae of Ceylon. Part IV. *Journal of the Bombay Natural History Society* **28**: 1007-1037. [Green1922a]

Green, E.E. 1923b. Observations on the Coccidae of the Madeira Islands. *Bulletin of Entomological Research* **14**: 87-99. [Green1923b]

Green, E.E. 1925. Observations on British Coccidae. - IX. *Entomologist's Monthly Magazine* **61**: 34-44. [Green1925]

Green, E.E. 1925b. Notes on the Coccidae of Guernsey (Channel Islands), with descriptions of some new species. *Annals and Magazine of Natural History* **16**: 516-527. [Green1925b]

Green, E.E. 1926. On some new genera and species of Coccidae. *Bulletin of Entomological Research* **17**: 55-65. [Green1926]

Green, E.E. 1926a. Observations on British Coccidae. - X. *Entomologist's Monthly Magazine* **62**: 172-183. [Green1926a]

Green, E.E. 1927. A brief review of the indigenous Coccidae of the British Islands, with emendations and additions. *Entomologist's Record and Journal of Variation. London* **39**: 1-4. [Green1927]

Green, E.E. 1928. A brief review of the indigenous Coccidae of the British Islands, with emendations and additions (to February, 1928). *Entomologist's Record and Journal of Variation* **40**: 1-14. [Green1928]

Green, E.E. 1928c. On a racial form of *Aspidiotus lenticularis*, Lindgr. with some remarks upon the Leonardi classification of the Aspidioti. *Annals and Magazine of Natural History* **1**: 374-376. [Green1928c]

Green, E.E. 1929. Some Coccidae collected by Dr. J.G. Myers in New Zealand. *Bulletin of Entomological Research* **19**: 369-389. [Green1929]

Green, E.E. 1930. Observations on British Coccidae. - XII. *Entomologist's Monthly Magazine* **66**: 9-17. [Green1930]

Green, E.E. 1930b. Notes on some Coccidae collected by Dr. Julius Melzer, at Sao Paulo, Brazil (Rhynch.). *Stettiner Entomologische Zeitung* **91**: 214-219. [Green1930b]

Green, E.E. 1930c. Fauna Sumatrensis (Bijdrag Nr. 65). Coccidae. *Tijdschrift voor Entomologie* **73**: 279-297. [Green1930c]

Green, E.E. 1931a. Observations on British Coccidae. - XIII. *Entomologist's Monthly magazine* **67**: 99-106. [Green1931a]

Green, E.E. 1934a. On a new species of *Targionia* (Coccidae) from Russia. *Stylops* **3**: 95-96. [Green1934a]

Green, E.E. 1934d. Observations on British Coccidae. - XIV. *Entomologist's Monthly Magazine* **70**: 108-114. [Green1934d]

Green, E.E. 1937. An annotated list of the Coccidae of Ceylon, with emendations and additions to date. *Ceylon Journal of Science Section B. Zoology and Geology* **20**: 277-341. [Green1937]

Green, E.E. & Laing, F. 1921. Coccidae from the Seychelles. *Bulletin of Entomological Research* **12**: 125-128. [GreenLa1921]

Green, E.E. & Laing, F. 1923. Descriptions of some new species and some new records of Coccidae. I. Diaspidinae. *Bulletin of Entomological Research* **14**: 123-131. [GreenLa1923]

Green, E.E. & Mann, H.H. 1907. The Coccidae attacking the tea plant in India and Ceylon. *Memoirs of the Department of Agriculture in India, Entomological Series* **1**: 337-355. [GreenMa1907]

Greig, A.M.W. 1944. Recognition and control of citrus diseases in New Zealand. *Bulletin (New Zealand Department of Agriculture)* No. 20. [Greig1944]

Gressitt, J.L. 1958. Entomology in the Pacific area with special reference to agriculture. Pages 395-398 *in*: Baecker, E.C., (Ed.). Proceedings of the Tenth International Congress of Entomology. Vol. 3. Ottawa. 895 pp. [Gressi1958]

Gressitt, J.L. & Flanders, S.E. 1949. New developments in the transport of beneficial insects. *Journal of Economic Entomology* **42(1)**: 150. [GressiFl1949]

Griffith, J.G. 1919. Biology. *Annual Report (New Mexico Agricultural Experiment Station)* **30th**: 14-18. [Griffi1919]

Griffiths, J.T. 1951. Possibilities for better citrus insect control through the study of the ecological effects of spray programs. *Journal of Economic Entomology* **44(4)**: 464-468. [Griffi1951]

Griffiths, J.T. & Fisher, F.E. 1949. Residues on citrus trees in Florida. I. Changes in insect populations following the use of different chemical materials used at the same amounts of residue per 100 gallons of spray. *Journal of Economic Entomology* **42(5)**: 829-833. [GriffiFi1949]

Griffiths, H.J., Hoffman, R., Luck, R. & Moreno, D. 1985. New attempt at managing red scale in the central valley. *Newsletter, Association of Applied Insect Ecologists* **5**: 1-6. [GriffiHoLu1985]

Griffiths, J.T. & Stearns, C.R. 1947. A further account of the effects of DDT when used on citrus trees in Florida. *Florida Entomologist* **30(1/2)**: 2-8. [GriffiSt1947]

Griffiths, J.T. & Thompson, W.L. 1947. The use of DDT on citrus trees in Florida. *Journal of Economic Entomology* **40(3)**: 386-388. [GriffiTh1947]

Griffiths, J.T. & Thompson, W.L. 1949. A preliminary report on the possibilities for forecasting periods of oviposition activity for purple and Florida red scales. *Proceedings of the Florida State Horticultural Society* **61**: 101-109. [GriffiTh1949]

Griffiths, J.T. & Thompson, W.L. 1949a. Identification of Florida red and purple scales on citrus trees in Florida. *Florida Agricultural Experiment Station Circular* **S-5**: 1-13. [GriffiTh1949a]

Griffiths, J.T. & Thompson, W.L. 1957. Insects and mites found on Florida citrus. *Bulletin of the Florida Agricultural Experiment Station* **591**: 5-30. [GriffiTh1957]

Griswold, G.H. 1926. Notes on some feeding habits of two chalcid parasites. *Annals of the Entomological Society of America* **19**: 331-334. [Griswo1926]

Grout, T.G., Du Toit, W.J., Hofmeyr, J.H. & Richards, G.I. 1989. California Red Scale (Homoptera: Diaspididae) phenology on citrus in South Africa. *Journal of Economic Entomology* **82(3)**: 793-798. [GroutDuHo1989]

Grout, T.G. & Richards, G.I. 1989. The multiple cohort structure in populations of red scale, *Aonidiella aurantii* (Maskell) (Homoptera: Diaspididae), on citrus in South Africa. *Journal of the Entomological Society of Southern Africa* **52(2)**: 277-283. [GroutRi1989]

Grout, T.G. & Richards, G.I. 1989a. Three methods of monitoring male red scale, *Aonidiella aurantii* (Maskell), with synthetic pheromone. *Citrus & Subtropical Fruit Journal* **645**: 11-13. [GroutRi1989a]

Grout, T.G. & Richards, G.I. 1991. Effect of Buprofezin applications at different phenological times on California Red Scale (Homoptera: Diaspididae). *Journal of Economic Entomology* **84(6)**: 1802-1805. [GroutRi1991]

Grout, T.G. & Richards, G.I. 1991a. Value of pheromone traps for predicting infestations of red scale, *Aonidiella aurantii* (Maskell) (Hom., Diaspididae), limited by natural enemy activity and insecticides used to control citrus thrips, *Scirtothrips aurantii* Faure (Thys., Thripidae). *Journal of Applied Entomology* **111**: 20-27. [GroutRi1991a]

Grout, T.G. & Richards, G.I. 1992. Organophosphate resistance in California red scale (Homoptera: Diaspididae) on citrus in the eastern Cape and the effect of oil as an organophosphate synergist. *Journal of the Entomological Society of Southern Africa* **55(1)**: 1-7. [GroutRi1992]

Grout, T.G., Richards, G.I. & Stephen, P.R. 1992. The possibility of reducing spray volumes used for the control of red scale *Aonidiella aurantii* (Mask.). *Citrus Journal* **2(3)**: 34-36. [GroutRiSt1992]

Gruys, P. 1979. Significance and practical application of selective pesticides. Pages 107-112. *in*: Proceedings: Internationales Symposium der IOBC/WPRS über Integrierten Pflanzenschutz inder Landund Forstwirtschaft. Wien, Austria: OILB/SROP. 648 pp. [Gruys1979]

Gu, D.X. & Chen, Y.G. 1998. Life tables of *Hemiberlesia pitysophila* Takagi and its parasitism rate by *Coccobius azumai* Tachikawa. *Natural Enemies of Insects* **20**: 4, 156-163. [GuCh1998]

Gu, D. & Murakami, Y. 1990. Ecological studies on the pine needle Hemiberlesian scale, *Hemiberlesia pitysophila* Takagi (Homoptera: Diaspididae) and its parasitoid *Coccobius azumai* Tachikawa (Hymenoptera: Aphelinidae). *Science Bulletin Faculty Agric. Kyushu Univ.* **45(1-2)**: 31-36. [GuMu1990]

Guario, A., Baldacchino, F. & Merlino. S. 1996. La cocciniglia nera della vite *Targionia vitis* (Sign.). *Informatore Agrario* **52**: 21, 51-54. [GuarioBaMe1996]

Guario, A. & Laccone, G. 1996. La difesa dell'uva da tavola dai fitofagi. *Informatore Agrario* **Suppl.**: 52: 50, 31-40. [GuarioLa1996]

Guerrieri, E. & Noyes, J.S. 2000. Revision of European species of the genus *Metaphycus* Mercet (Hymenoptera: Chalcidoidea: Encyrtidae), parasitoids of scale insects (Homoptera: Coccoidea). *Systematic Entomology* **25(2)**: 147-222. [GuerriNo2000]

Gulmahamad, H. & DeBach, P. 1978. Biological control of the San José scale *Quadraspidiotus perniciosus* (Comstock) (Homoptera: Diaspididae) in southern California. *Hilgardia* **46**: 205-238. [GulmahDe1978]

Gulmahamad, H. & DeBach, P. 1978a. Biological studies on *Aphytis aonidiae* (Mercet) (Hymenoptera: Aphelinidae), an important parasite of the San José Scale. *Hilgardia* **46**: 239-256. [GulmahDe1978a]

Gumus, M. & Uygun, N. 1992. Determining a better sampling scheme of the *Aonidiella aurantii* (Maskell) (Homoptera: Diaspididae) which is an important pest of citrus. *Turkiye Entomoloji Dergisi* **16(4)**: 209-216. [GumusUy1992]

Gupta, A.P. 1978. Scale insects *Melanaspis glomerata*: a potential danger for sugar industry in eastern Uttar Pradesh. *Proceedings of the 42nd Annual Convention of Sugar Technology Association of India* 67-74. [Gupta1978]

Gupta, B.P. & Singh, Y.P. 1988. Mango scale insects -- occurrence in western Uttar Pradesh and their control. *Progressive Horticulture* **20(3-4)**: 357-361. [GuptaSi1988]

Haas, A.R.C. 1934. Relation between the chemical composition of citrus scale insects and their resistance to hydrocyanic acid fumigation. *Journal of Agricultural Research* **19(6)**: 477-492. [Haas1934]

Habib, A. & Atallah, Y.H. 1960. Population studies on the black scale, *Chrysomphalus ficus* Ashmead. III. The build up of the population on different kinds of citrus. *Bulletin de la Société Entomologique d'Egypte* **44**: 353-365. [HabibAt1960]

Habib, A., El-Kady, E.A., Rawhy,, S.H. & Soliman, A.F. 1973. Control of the citrus waxy scale insect, *Ceroplastes floridensis* Comstock and the purple scale insect, *Lepidosaphes beckii* (Newman). *Bulletin de la Société Entomologique d'Egypte* **7**: 133-141. [HabibElRa1973]

Habib, A., Ezzat, Y.M. & Atallah, Y.H. 1960. Sexual dimorphism in the second instar of *Chrysomphalus ficus* Ashmead (Homoptera: Coccoidea-Diaspididae). *Bulletin de la Société Entomologique d'Egypte* **44**: 329-336. [HabibEzAt1960]

Habib, A., Salama, H.S. & Amin, A.H. 1971. Population studies on scale insects infesting citrus trees in Egypt. *Zeitschrift für Angewandte Entomologie* **69**: 318-330. [HabibSaAm1971]

Habib, A., Salama, H.S. & Amin, A.H. 1972. Population of *Aonidiella aurantii* on citrus varieties in relation to their physical and chemical characteristics. *Entomologia Experimentalis et Applicata* **15**: 324-328. [HabibSaAm1972]

Habib, A., Salama, H.S. & Amin, A.H. 1972a. The build up of population of the red scale *Aonidiella aurantii* (Maskell) on citrus trees in Egypt. *Zeitschrift für Angewandte Entomologie* **70**: 378-385. [HabibSaAm1972a]

Habibian, A. 1980. [Biology and control of San Jose scale in Guilan Province *Quadraspidiotus perniciosus* (Comstock)]. [In Persian]. *Entomologie et Phytopathologie Appliquées* 48: 127-134. [Habibi1980]

Hadzibejli, Z.K. 1957a. [Regionalization of landscape subdivisions of scale insects of Georgia.]. [In Russian]. *Third Conference of the Soviet Entomology Society* **2**: 100-102. [Hadzib1957a]

Hadzibejli, Z.K. 1983. [Coccids of the subtropical zone of Gruzia.]. [In Russian]. Tbilisi: Metsniereba. 293 pp. [Hadzib1983]

Hafez, M., Tawfik, M.F.S. & Raouf, A. 1970. Fluctuations of the population densities of the black scale, *Chrysomphalus ficus* Ashm., at certain localities in U.A.R. On the population dynamics of the black scale *Chrysomphalus ficus* Ashm. *Technical Bulletin, U.A.R. Min. Agric & Agrar. Agrar. Reform U.A.R, Plant Protection Department* **2**: 3-16. [HafezTaRa1970]

Hagen, K.S. 1974. The significance of predaceous Coccinellidae in biological and integrated control of insects. *Entomophaga* **Mem. H.S. 7**: 25-44. [Hagen1974]

Hagen, K.S., Bombosch, S. & McMurtry, J.A. 1976. The biology and impact of predators. Pages 93-142 *in*: Huffaker, C.B. & Messenger, P.S. (Eds.). Theory and Practice of Biological Control. New York: Academic Press. 788 pp. [HagenBoMc1976]

Hagen, K.S., van den Bosch, R. & Dahlsten, D.L. 1971. The importance of naturally-occurring biological control in the western United States. Pages 253-293. *in*: Huffaker, C.B. (Ed.). Biological Control. New York: Plenum Press. 511 pp. [HagenVaDa1971]

Hakkonen, H. & Pimentel, D. 1984. New approach for selecting biological control agents. *Canadian Entomologist* **116**: 1109-1121. [HakkonPi1984]

Halbert, S.E. 1995. Entomology Section. *Tri-ology* **34**: 6-9. [Halber1995]

Halbert, S.E. 1996. Entomology section. *Tri-ology* **35(3)**: 4-10. [Halber1996]

Halbert, S.E. 1996a. Entomology section. *Tri-ology* **35(4)**: 4-8. [Halber1996a]

Halbert, S.E. 2000. Entomology section. *Tri-ology* **39(1)**: 3-4. [Halber2000]

Hall, I.M. 1969. Diseases of citrus insects and mites. Pages 823-826. *in*: Chapman, H.D. (Ed.). Proceedings First International Citrus Symposium. Vol. 2. Riverside: University of California. [Hall1969]

Hall, W.J. 1922. Observations on the Coccidae of Egypt. *Bulletin, Ministry of Agriculture, Egypt, Technical and Scientific Service* **22**: 1-54. [Hall1922]

Hall, W.J. 1923. Further observations on the Coccidae of Egypt. *Bulletin, Ministry of Agriculture, Egypt, Technical and Scientific Service* **36**: 1-61. [Hall1923]

Hall, W.J. 1924a. The insect pests of citrus trees in Egypt. *Bulletin, Ministry of Agriculture, Egypt, Technical and Scientific Service* **45**: 1-29. [Hall1924a]

Hall, W.J. 1925. Notes on Egyptian Coccidae with descriptions of new species. *Bulletin, Ministry of Agriculture, Egypt, Technical and Scientific Service* **64**: 1-31. [Hall1925]

Hall, W.J. 1926a. Contribution to the knowledge of the Coccidae of Egypt. *Bulletin, Ministry of Agriculture, Egypt, Technical and Scientific Service* **72**: 1-41. [Hall1926a]

Hall, W.J. 1927. On a small collection of Coccidae from Palestine. *Bulletin de la Société Entomologique d'Egypte* **1926**: 107-109. [Hall1927]

Hall, W.J. 1927b. Notes on the Coccidae of the eastern desert of Egypt. *Bulletin de la Société Entomologique d'Egypte* **1926**: 118-177. [Hall1927b]

Hall, W.J. 1927d. Miscellaneous notes on Egyptian Coccidae with descriptions of three new species. *Bulletin de la Société Entomologique d'Egypte* **1926**: 267-287. [Hall1927d]

Hall, W.J. 1928. Observations on the Coccidae of southern Rhodesia. - I. *Bulletin of Entomological Research* **19**: 271-292. [Hall1928]

Hall, W.J. 1929. Observations on the Coccidae of southern Rhodesia. - II. *Bulletin of Entomological Research* **20**: 345-358. [Hall1929]

Hall, W.J. 1929a. Observations on the Coccidae of southern Rhodesia. - III. *Bulletin of Entomological Research* **20**: 359-376. [Hall1929a]

Hall, W.J. 1931. Observations on the Coccidae of southern Rhodesia. *Transactions of the Entomological Society of London* **79**: 285-303. [Hall1931]

Hall, W.J. 1935a. Observations on the Coccidae of southern Rhodesia. - VII. *Stylops* **4**: 217-227. [Hall1935a]

Hall, W.J. 1937. Observations on the Coccidae of southern Rhodesia. *Transactions of the Royal Entomological Society of London* **86**: 119-134. [Hall1937]

Hall, W.J. 1939. A new genus and four apparently new species of Coccidae (Homoptera) from the Union of South Africa. *Journal of the Entomological Society of Southern Africa* **2**: 93-100. [Hall1939]

Hall, W.J. 1941. On some new species and two new genera of Coccidae (Homoptera) from southern Rhodesia. *Journal of the Entomological Society of southern Africa* **4**: 221-239. [Hall1941]

Hall, W.J. 1943. Notes on some Coccidae (Homoptera) from southern Rhodesia, with descriptions of two new species. *Journal of the Entomological Society of Southern Africa* **6**: 1-6. [Hall1943]

Hall, W.J. 1946. New or little known species of Diaspididae (Coccoidea) from Africa. *Transactions of the Royal Entomological Society of London* **97**: 55-73. [Hall1946]

Hall, W.J. 1946a. On the Ethiopian Diaspidini (Coccoidea). *Transactions of the Royal Entomological Society of London* **97**: 497-592. [Hall1946a]

Hall, W.J. & Ford, W.K. 1933. Note on some citrus insects of southern Rhodesia. *Mazoe Citrus Experiment Station. British South Africa Company.* **No. 2**: 55 pp. [HallFo1933]

Hall, W.J. & Williams, D.J. 1962. New Diaspididae (Homoptera: Coccoidea) from the Indo-Malayan region. *Bulletin of the British Museum (Natural History) Entomology* **13**: 21-43. [HallWi1962]

Hambleton, E.J. 1947. [Plague of insects that attack coffee in Latin America.] Plaga de insectos que atacan el café en la America Latina. *Café de El Salvador* **17**: 949-956. [Hamble1947]

Hamilton, C.C. 1936. The identification of insect injury. *National Shade Tree Conference* **12th**: 150-160. [Hamilt1936]

Hamlen, R.A. 1974. Populations of economically important insects and mites on Florida grown tropical foliage crops. *Florida Foliage Grower* **11(5)**: 6-8. [Hamlen1974]

Hammad, S.M., Kadous, A.A. & Ramadan, M.M. 1981. Insects and mites attacking date palm in the eastern province of Saudi Arabia. *Proceedings, Fifth Conference on the Biological Aspects of Saudi Arabia* pp. 252-268. [HammadKaRa1981]

Hamon, A.B. 1980d. *Pseudaonidia trilobitiformis* (Green) (Homoptera: Diaspididae). *33rd Biennial Report, Florida Department of Agriculture and Consumer Services, Division of Plant Industry,* pp. 38-39. [Hamon1980d]

Hamon, A.B. 1981. Plumose scale, *Morganella longispina* (Morgan) (Homoptera: Coccoidea: Diaspididae). *Entomology Circular, Florida Department of Agriculture and Consumer Services Division of Plant Industry,* **No. 226**: 1-2. [Hamon1981]

Hamon, A.B. 1983. *Morganella longispina* (Morgan) (Homoptera: Coccoidea: Diaspididae). *34th Biennial Report, Florida Department of Agriculture and Consumer Services, Division of Plant Industry,* p. 45. [Hamon1983]

Hamon, A.B. 1985. *Oceanaspidiotus araucariae* (Adachi and Fullaway) (Homoptera: Coccoidea: Diaspididae). *Entomology Circular, Florida Department of Agriculture and Consumer Services Division of Plant Industry* **No. 276**: 1-2. [Hamon1985]

Handa, S. & Dahiya, K.K. 1999. Chemical management of mango scales. *Agricultural Science Digest* **19(2)**: 112-114. [HandaDa1999]

Hanks, L.M. & Denno, R.F. 1998. Dispersal and adaptive deme formation in sedentary coccoid insects. Pages 239-262. *in*: Mopper, S. & Strauss, S.Y., Eds. Genetic Structure and Local Adaptation in Natural Insect Populations: Effects of Ecology, Life History, and Behavior. New York: Chapman & Hall. xix + 449 pp. [HanksDe1998]

Hansen, J.D. 2001. Ultrasound treatments to control surface pests of fruit. *HortTechnology* **11(2)**: 186-188. [Hansen2001]

Hardie, J. & Minks, A.K. (Editors.) 1999. Pheromones of Non-Lepidopteran Insects Associated with Agricultural Plants. Oxon, U.K. & New York: CABI Publishing. 466 pp. [HardieMi1999]

Hardison, A.C. 1941. Controlling scale cooperatively. *California Citrograph* **26**: 283, 308-309. [Hardis1941]

Hardman, N.F. & Craig, R. 1941. A physiological basis for the differential resistance of the two races of red scale to HCN. *Science* **94(2434)**: 187. [HardmaCr1941]

Hare, J.D. 1996. Priming *Aphytis*: behavioral modification of host selection by exposure to a synthetic contact kairomone. *Entomologia Experimentalis et Applicata* **78**: 263-269. [Hare1996]

Hare, J.D. & Luck, R.F. 1991. Indirect effects of citrus cultivars on life history parameters of a parasitic wasp. *Ecology* **72(5)**: 1576-1585. [HareLu1991]

Hare, J.D., Millar, J.G. & Luck, R.F. 1993. A caffeic acid ester mediates host recognition by a parasitic wasp. *Naturwissenschaften* **80(2)**: 92-94. [HareMiLu1993]

Hare, J.D. & Morgan, D.J.W. 1997. Mass-priming *Aphytis*: Behavioral improvement of insectary-reared biological control agents. *Biological Control* **10(3)**: 207-214. [HareMo1997]

Hare, J.D. & Morgan, D.J.W. 2000. Chemical conspicuousness of an herbivore to its natural enemy: Effect of feeding site selection. *Ecology* **81(2)**: 509-519. [HareMo2000]

Hare, J.D., Morgan, D.J.W. & Nguyun, T. 1997. Increased parasitization of California red scale in the field after exposing its parasitoid, *Aphytis melinus*, to a synthetic kairomone. *Entomologia Experimentalis et Applicata* **82**: 73-81. [HareMoNg1997]

Hare, J.D., Yu, D.S. & Luck, R.F. 1990. Variation in life history parameters of California red scale on different citrus cultivars. *Ecology* **71(4)**: 1451-1460. [HareYuLu1990]

Hargreaves, E. 1927. Some insect pests of Sierra Leone. Pages 113-128 *in*: Proceedings of the First West African Agricultural Conference Lagos, Nigeria. 196 pp. [Hargre1927]

Hargreaves, E. 1937. Some insects and their food-plants in Sierra Leone. *Bulletin of Entomological Research* **28(3)**: 505-520. [Hargre1937]

Hargreaves, H. 1948. List of Recorded Cotton Insects of the World. London: Commonwealth Institute of Entomology. 50 pp. [Hargre1948]

Hariri, G. 1971. A list of recorded insect fauna of Syria Part 2. Faculty of Agriculture, University of Aleppo, Aleppo, Syria. 306 pp. [Hariri1971]

Harned, R.W. 1928. Annual report of the Department of Zoology and Entomology: Scale insect project (Adams Fund). *Annual Report (Mississippi Agricultural Experiment Station)* 23-24. [Harned1928]

Harpaz, I. & Rosen, D. 1971. Development of integrated control programs for crop pests in Israel. Pages 458-468 *in*: Huffaker, C.B., Ed. Biological Control. New York & London: Plenum Press. 511 pp. [HarpazRo1971]

Harrison, J.W.H. 1916. Coccidae and Aleyrodidae in Northumberland, Durham, and north-east Yorkshire. *The Entomologist* **49**: 172-174. [Harris1916]

Harris, K.M. 1990. 2.2.4 Cecidomyiidae and other Diptera. Pages 61-66 *in*: Rosen, D. (Ed.). Armored Scale Insects, Their Biology, Natural Enemies and Control [Series title: World Crop Pests, Vol. 4B]. Amsterdam, the Netherlands: Elsevier. 688 pp. [Harris1990]

Harris, K.M. 1997. Cecidomyiidae and other Diptera. Pages 61-68. *in*: Ben-Dov, Y. & Hodgson, C.J. (Eds.). Soft Scale Insects - Their Biology, Natural Enemies and Control [Vol. 7B]. Amsterdam & New York: Elsevier. 442 pp. [Harris1997]

Harris, W.V. 1937. Annotated list of insects injurious to native food crops in Tanganyika. *Bulletin of Entomological Research* **28(3)**: 483-488. [Harris1937]

Harris, W.V. 1937a. Annual report of the entomologist, 1936. *Annual report (Tanganyika Department of Agriculture)* 88-94. [Harris1937a]

Hart, J.H. 1896. Coccidae, new sp. & var. *Miscellaneous Information Bulletin Royal Botanical Garden. Trinidad* **2**: 156. [Hart1896]

Hart, W.G. 1990. 3.4.4 Remote Sensing. Pages 353-356 *in*: Rosen, D. (Ed.). Armored Scale Insects, Their Biology, Natural Enemies and Control [Series title: World Crop Pests, Vol. 4B]. Amsterdam, the Netherlands: Elsevier. 688 pp. [Hart1990]

Hartig, T. 1839. Jahresberichte über die fortschritte der fortwissenschaft und forstlichen naturkunde nebst original abhandlungen aus dem gebiete dieser wissenschaften. 646 pps. [Hartig1839]

Haseman, L. & Sullivan, K.C. 1923. Controlling San José scale with lubricating oil emulsion. *Circular (Agricultural Experiment Station, College of Agriculture, University of Missouri, Columbia, MO)* **No. 109**: 4 pp. [HasemaSu1923]

Hasey, J., Olson, W.H., Van Steenwyk, R. & Beede, R. 1999. Pests of kiwifruit. *UC Pest Management Guidelines* . [HaseyOlVa1999]

Hassan, T.A. 1999. Joint action of certain insecticides on the California Red Scale Insect, *Aonidiella aurantii* (Mask). *Journal of Natural and Applied Sciences (University of Aden)* **3(2)**: 23-29. [Hassan1999]

Hassan, E. & Summers, R.G. 1997. Testing the toxicity effects on California red scale parasitoid (*Aphytis lingnanensis* Compere) of two insecticides used to control California red scale (*Aonidiella aurantii* Mask.) on citrus in the laboratory. *Zeitschrift für Pflanzenkrankheiten und Pflanzenschutz* **104**: 415-418. [HassanSu1997]

Hassanein, F.A. & Hamed, A.R. 1985. On the population dynamics of *Hemiberlesia lataniae* (Signoret) and its parasite *Habrolepis aspidioti* Compare (Compere?) & Annecke in Egypt (Homoptera: Diaspididae: Hymenoptera: Encyrtidae). *Bulletin of the Entomological Society of Egypt* **14**: 63-72. [HassanHa1985]

Hassell, M.P. & Waage, J.K. 1984. Host-parasitoid population interactions. *Annual Review of Entomology* **29**: 89-114. [HasselWa1984]

Hattingh, V. 1996. The use of insect growth regulators in integrated pest management of citrus in southern Africa. *Citrus Journal* **6**: 14-17. [Hattin1996]

Hattingh, V. & Samways, M.J. 1990. Absence of intraspecific interference during feeding by the predatory ladybirds *Chilocorus* spp. (Coleoptera: Coccinellidae). *Ecological Entomology* **15(4)**: 385-390. [HattinSa1990]

Hattingh, V. & Samways, M.J. 1991. Determination of the most effective method for field establishment of biocontrol agents of the genus *Chilocorus* (Coleoptera: Coccinellidae). *Bulletin of Entomological Research* **81**: 169-174. [HattinSa1991]

Hattingh, V. & Samways, M.J. 1991a. A forced change in prey type during field introductions of Coccidophagous biocontrol agents *Chilocorus* species (Coleoptera Coccinellidae); is it an important consideration in achieving establishment?. Pages 143-148 *in*: Dixon, L., Polgar, R. J., Chambers, A. F. G. & Hodek, I., eds. Behaviour and Impact of Aphidophaga. Godollo, Hungary: September 1990. Hague, the Netherlands: SPB Academic Publishing. [HattinSa1991a]

Hattingh, V. & Samways, M.J. 1992. Prey choice and substitution in *Chilocorus* spp. (Coleoptera: Coccinellidae). *Bulletin of Entomological Research* **82**: 327-334. [HattinSa1992]

Hattingh, V. & Samways, M.J. 1993. Evaluation of artificial diets and two species of natural prey as laboratory food for *Chilocorus* spp. *Entomologia Experimentalis et Applicata* **69(1)**: 13-20. [HattinSa1993]

Hattingh, V. & Tate, V. 1995. Effects of field-weathered residues of insect growth regulators on some Coccinellidae (Coleoptera) of economic importance as biocontrol agents. *Bulletin of Entomological Research* **85**: 489-493. [HattinTa1995]

948 REFERENCES

Hattingh, V. & Tate, B.A. 1996. The effects of insect growth regulator use on IPM in southern African citrus. *Proceedings of the International Society of Citriculture* I: 523-525. [HattinTa1996]

Hattingh, V. & Tate, B.A. 1996a. The pest status of mealybugs on citrus in Southern Africa. *Proceedings of the International Society of Citriculture* I: 560-563. [HattinTa1996a]

Havron, A., Kenan, G. & Rosen, D. 1991. Selection for pesticide resistance in *Aphytis*. II. *A. lingnanensis* a parasite of the California red scale. *Entomologia Experimentalis et Applicata* **61**: 229-235. [HavronKeRo1991]

Havron, A. & Rosen, D. 1992. [Selection of two species of *Aphytis* for resistance to cotnion (azinphos-methyl)]. *Hassadeh* **72**: 8, 984-987. [HavronRo1992]

Havron, A. & Rosen, D. 1994. Selection for organophosphorus pesticide resistance in two species of *Aphytis*. Pages 209-220 *in*: Rosen, D., Ed. Advances in the Study of Aphytis (Hymenoptera: Aphelinidae). Andover, U.K.: Intercept. 362 pp. [HavronRo1994]

Havron, A., Rosen, D., Prag, H. & Rossler, Y. 1991. Selection for pesticide resistance in *Aphytis*: I.*A. holoxanthus*, a parasite of the Florida red scale. *Entomologia Experimentalis et Applicata* **61**: 221-228. [HavronRoPr1991]

Havron, A., Rosen, D. & Rubin, A. 1995. Release of pesticide-resistant *Aphytis* strains in Israeli citrus orchards. *Israel Journal of Entomology* **29**: 309-313. [HavronRoRu1995]

Hawkins, B.A. 1994. Pattern and Process in Host-parasitoid Interactions. Cambridge: University Press. 190 pp. [Hawkin1994]

Hayat, M. 1989. A revision of the species of *Encarsia* Foerster (Hymenoptera: Aphelinidae) from India and the adjacent countries. *Oriental Insects. New Delhi* **23**: 1-131. [Hayat1989]

Hayward, K.J. 1939. [The main parasites of citrus and means of control.] Principales parasitos de los citrus y forma de combatirlos. *Boletín de Frutas y Hortalizas (Buenos Aires)* **4(39)**: 288 pp. [Haywar1939]

Hayward, K.J. 1944. [First list of pest insects of Tucuman.] Primera list de insectos tucumanos perjudiciales. *Publicaciones Misceláneas de la Estación Experimental Agrícola de Tucumán* **4**: 1-32. [Haywar1944]

Hazelhoff, E.H. 1929. Insect pests of sugar cane in Java. *Bulletin (3rd Congress of the International Society of Sugarcane Technologies)* **4**: 1-8. [Hazelh1929]

He, G.F., Bao, W.M., Lu, A.P. & Zhang, G.X. 1998. A biological study of armored scale, *Abgrallaspis cyanophylli* with emphasis on temperature and humidity relations. *Chinese Journal of Biological Control* **14**: 1-3. [HeBaLu1998]

He, G., Cui, B., Bao, W. & Pu, Z. 1991. [The effect of temperature on the survival, development and fecundity of three armored scales, *Abgrallaspis cyanophylli, Aonidiella aurantii* and *Chrysomphalus* sp.]. [In Chinese]. *Natural Enemies of Insects* **13(2)**: 58-60. [HeCuBa1991]

Headlee, T.J. 1927. Pages 112-113 *in*: Report of the Department of Entomology of the New Jersey Agricultural Experiment Station for the Year Ended June 30, 1927. New Brunswick, N.J. [Headle1927]

Headlee, T.J. 1929. Forty-ninth annual report of the New Jersey State Agricultural Report of the Department of Entomology. *Report (New Jersey Agricultural Experiment Station)* (year ending June 30, **1928)**: 125-? [Headle1929]

Hecht, O. 1936. Studies on the biology of *Chilocorus bipustulatus* (Coleoptera-Coccinellidae); an enemy of the red scale *Chrysomphalus aurantii*. *Bulletin de la Société Royale Entomologique d'Egypte* **Séance du 16 Décembre 1936**: 299-326 [Hecht1936]

Hefetz, A., Kronenberg, S., Peleg, B.A. & Bar-Zakay, I. 1988. Mating disruption of the California Red Scale *Aonidiella aurantii* (Homoptera: Diaspididae). Pages 1121-1127 *in*: Goren, R. & Mendel, K., eds. Citriculture: Proceedings of the Sixth International Citrus Congress: Middle-East. Tel Aviv: March 6-11, 1988. Rehovot, Israel: Balaban. [HefetzKrPe1988]

Heimpel, G.E. & Rosenheim, J.A. 1995. Dynamic host feeding by the parasitoid *Aphytis melinus*: the balance between current and future reproduction. *Journal of Animal Ecology* **64(2)**: 153-167. [HeimpeRo1995]

Heimpel, G.E. & Rosenheim, J.A. 1998. Egg limitation in parasitoids: a review of the evidence and a case study. *Biological Control* **11**: 160-168. [HeimpeRo1998]

Heimpel, G.E., Rosenheim, J.A. & Kattari, D. 1997. Adult feeding and lifetime reproductive success in the parasitoid *Aphytis melinus*. *Entomologia Experimentalis et Applicata* **83**: 305-315. [HeimpeRoKa1997]

Heimpel, G.E., Rosenheim, J.A. & Mangel, M. 1996. Egg limitation, host quality, and dynamic behavior by a parasitoid in the field. *Ecology* **77**: 2410-2420. [HeimpeRoMa1996]

Hellén, W. 1921. Veränderungen in der Kenntnis der Insektenfauna Finnlands bis zum Jahr 2021. *Notulae Entomologicae* **I**: 120-128. [Hellen1921]

Helmy, E.I., Hanafy, H.A., Hassan, N.A., El-Imery, S.M. & Mohamed, F.A. 1992. New approach to control scale insects by using five Egyptian miscible oils on orange trees in Egypt. *Egyptian Journal of Agricultural Research* **70**: 763-771. [HelmyHaHa1992]

Helmy, E.I., Hindy, M.A., Hassan, N.A. & El-Imery, S.M. 1997. Comparison between aerial and ground spraying against the California red scale, *Aonidiella aurantii* (Mask.) and *Dialeurodes citri* (Ashmead) on citrus trees. *Egyptian Journal of Agricultural Research* **75**: 601-609. [HelmyHiHa1997]

Hempel, A. 1900a. As coccidas Brasileiras. *Revista do Museu Paulista. São Paulo* **4**: 365-537. [Hempel1900a]

Hempel, A. 1901a. Descriptions of Brazilian Coccidae. *Annals and Magazine of Natural History* **8**: 62-72, 100-111. [Hempel1901a]

Hempel, A. 1904. Resultado do exame de diversas collecçoes de coccidas enviadas ao Instituto Agronomico pelo Sr. Carlos Moreira, do Museu Nacional, Rio de Janeiro. *Boletim da Agricultura (São Paulo)* **1**: 311-323. [Hempel1904]

Hempel, A. 1918. Descripção de sete novas espécies de coccidas. *Revista do Museu Paulista. São Paulo* **10**: 193-208. [Hempel1918]

Hempel, A. 1920. Coccidas que infestam as nossas árvores fructíferas. *Revista do Museu Paulista. São Paulo* **12**: 109-143. [Hempel1920]

Hempel, A. 1922. As pragas da lavoura. *Boletim da Agricultura (São Paulo)* **23**: 133-144. [Hempel1922]

Hempel, A. 1932. Descripção o de vinte a duas espécies novas de coccideos (Hemiptera - Homoptera). *Revista de Entomologia* **2**: 310-339. [Hempel1932]

Hempel, A. 1937. Novas espécies de coccídeos (Homoptera) do Brasil. *Archivos do Instituto Biologico. São Paulo* **8**: 5-36. [Hempel1937]

Henderson, R.C. 2001a. New synonymies for two armoured scale insects (Hemiptera: Coccoidea: Diaspididae) on *Dysoxylum spectabile*. *New Zealand Entomologist* **24**: 89-90. [Hender2001a]

Hendricks, H.J. & Williams, M.L. 1992. Life history of *Melanaspis obscura* (Homoptera: Diaspididae) infesting pin oak in Alabama. *Annals of the Entomological Society of America* **85(4)**: 452-457. [HendriWi1992]

Henrik, C.A., Carney, R.L. & Anderson, R.J. 1982. Some aspects of the synthesis of insect sex pheromones. *Symposium Series of the American Chemical Society* **190**: 27-60. [HenrikCaAn1982]

Heraty, J.M. & Schauff, M.E. 1998. Mandibular teeth in Chalcidoidea: function and phylogeny. *Journal of Natural History* **32**: 1227-1244. [HeratySc1998]

Herbert, F.B. 1919a. Insect problems of western shade trees. *Journal of Economic Entomology* **12**: 333-337. [Herber1919a]

Heriot, A.D. 1934. The renewal and replacement of the stylets of sucking insects during each stadium, and the method of penetration. *Canadian Journal of Research* **11**: 602-612. [Heriot1934]

Herrera, J.M. 1964. Ciclos biológicos de las queresas de los cítricos en la costa central. Métodos para su control. *Revista Peruana de Entomología* **7**: 1-8. [Herrer1964]

Herrick, G.W. 1910. A new species of *Aspidiotus*. *Entomological News* **21**: 22. [Herric1910]

Herrick, G.W. 1910a. *Targionia celtis*. *Canadian Entomologist* **42**: 373-374. [Herric1910a]

Herrick, G.W. 1911. Some scale insects of Mississippi with notes on certain species from Texas. *Mississippi Agricultural Experiment Station Technical Bulletin* **2**: 1-78. [Herric1911]

Herrick, G.W. 1920. Insects of Economic Importance: Outlines of Lectures in Economic Entomology. New York: MacMillan. 172 pp. [Herric1920]

Herrick, G.W. 1925. Manual of Injurious Insects. New York: Henry Holt. 489 pp. [Herric1925]

Herrick, G. & Seger, J. 1999. Imprinting and paternal genome elimination in insects. Pages 41-71 *in*: Ohlsson, R., (Ed.). Genomic Imprinting: An Interdisciplinary Approach. Berlin: Springer. [HerricSe1999]

Hewitt, J.L. 1943. Nursery service. *Bulletin (California Department of Agriculture)* **32(4)**: 266-274. [Hewitt1943]

Hickel, E.R. & Ducroquet, J.P.H.J. 1995. [Scales (Homoptera: Coccoidea) pests of feijoa.]. [In Portuguese]. *Anais da Sociedade Entomologica do Brasil* **24(3)**: 665-668. [HickelDu1995]

Hill, D.S. 1975. Agricultural Insect Pests of the Tropics and their Control. Cambridge, London, New York, Melbourne: Cambridge University Press. 516 pp. [Hill1975]

Hill, M.G. 1989a. Diaspididae, armoured scales (Homoptera). Pages 177-182 *in*: Cameron, P.J., Hill, R.L., Bain, J. & Thomas, W.P. (Eds.). A Review of biological control of invertebrate pests and weeds in New Zealand 1874 to 1987. Wallingford Oxon, U.K.: CAB International Institute of Biological Control. 424 pp. [Hill1989a]

Hill, M.G., Allan, D.J., Henderson, R.C. & Charles, J.G. 1993. Introduction of armoured scale predators and establishment of the predatory mite *Hemisarcoptes coccophagus* (Acari: Hemisarcoptidae) on Latania scale, *Hemiberlesia lataniae* (Homoptera: Diaspididae) in kiwifruit shelter trees in New Zealand. *Bulletin of Entomological Research* **83(3)**: 369-376. [HillAlHe1993]

Hillman, F.H. 1895. The San José Scale. Bulletin No. 28. Reno, NV: Nevada St. Univ., Agricultural Experiment Station. 8 pp. [Hillma1895]

Hinckley, A.D. 1963. Trophic records of some insects mites, and ticks in Fiji. *Bulletin (Department of Agriculture, Fiji)* **No. 45**: 1-116. [Hinckl1963]

Hindi, A., Amer, A., Rofail, F., Rawhy, S. & Madkour, A. 1964. The effect of formulations of local mineral oils used alone, or in combination with phosphorous compounds on the black scale, *Chrysomphalus ficus*, and other scale insects. *Agricultural Research Review* **42(2)**: 27-43. [HindiAmRo1964]

Hippe, C. & Mani, E. 1995. Flight monitoring of the San José scale, *Quadraspidiotus perniciosus* (Comstock) (Diaspididae) and of its parasitoid *Encarsia perniciosi* (Tower) (Aphelinidae) in northwest Switzerland (Abstract only). *Israel Journal of Entomology* **29**: 184. [HippeMa1995]

Hippe, C., Schwaller, F., Mani, E., Kull, H. & Kozár, F. 1995. Bekampfung einheimischer Austernschildlause. *Obst- und Weinbau* **131**: 4, 84-85. [HippeScMa1995]

Hirayama, Y. & Nogami, Y. 1975. [Ecology and control of *Comstockaspis macroporana*, the insect pest of chestnut trees.]. [In Japanese]. *Shokubutsu Boeki* [Plant Protection] **29**: 2-6. [HirayaNo1975]

Hirose, Y., Nakamura, T. & Takagi, M. 1990. Successful biological control: a case study of parasitoid aggregation. Pages 171-183 *in*: Mackkauer, M., Ehler, L.E. & Roland, J. (Eds.). Critical Issues in Biological Control. Andover, Hants: Intercept. 330 pp. [HiroseNaTa1990]

Hix, R.L., Pless, C.D., Deyton, D.E. & Sams, C.E. 1999. Management of San José scale on apple with soybean-oil dormant sprays. *HortScience* **34(1)**: 106-108. [HixPlDe1999]

Hodek, I. 1973. Biology of Coccinellidae [with Keys for Identification of larvae by co-authors. The Hague and Prague: Junk and Academia, Publishing House of the Czechoslovak Academy of Sciences. 260 pp. [Hodek1973]

Hodgkiss, H.E. 1904. The life history and treatment of a common palm scale *(Chrysomphalus dictyospermi* Morgan.). *Annual Report of the Massachusetts Agricultural College* **41**: 97-106. [Hodgki1904]

Hodgkiss, H.E. & Parrott, P.J. 1914. The parasites of the San José scale in New York. *Journal of Economic Entomology* **7**: 227-228. [HodgkiPa1914]

Hodgson, C.J. 1969c. Pests of citrus and their control *in* Citrus in Rhodesia. *Rhodesia Agricultural Journal Technical Bulletin* **No. 7**: 18-29. [Hodgso1969c]

Hofer, J. 1903. Beitrag zur Cocciden-Fauna der Schweiz. *Mitteilungen der Schweizerischen Entomologischen Gesellschaft* **10**: 474-483. [Hofer1903]

Hoffman, R.W. & Kennett, C.E. 1985. Tracking CRS (California red scale) development by degree-days. *California Agriculture* **39**: 19-20. [HoffmaKe1985]

Hoke, G. 1927. Some undescribed diaspines from Mississippi. *Annals of the Entomological Society of America* **20**: 349-358. [Hoke1927]

Hole, U.B. & Salunkhe, G.N. 1997. Field efficacy of some insecticides against red scale, *Aonidiella aurantii* Maskell on rose. *Plant Protection Bulletin (Faridabad)* **49(1/4)**: 12-13. [HoleSa1997]

Hole, U.B. & Salunkhe, G.N. 1998. Evaluation of rose cultivars against red scale *(Aonidiella aurantii* Maskell). *Journal of Maharashtra Agricultural Universities* **22**: 199-201. [HoleSa1998]

Hole, U.B. & Salunkhe, G.N. 1999. Relationship between the population build up of *Aonidiella aurantii* (Maskell) on rose and weather parameters. *Indian Journal of Agricultural Research* **33(2)**: 93-102. [HoleSa1999]

Hollinger, A.H. 1923. Scale insects of Missouri. *Research Bulletin. Missouri Agricultural Experiment Station. Columbia* **58**: 1-71. [Hollin1923]

Holman, J. 2002. [The natural enemies of San José scale *(Quadraspidiotus perniciosus* Coms.) from the order Hymenoptera in the Czech Republic.]. [In Czech with summary in English.] *Acta Universitatis Agriculturae et Silviculturae Mendelianae Brunensis* **50(3)**: 41-54. [Holman2002]

Holmes, J.J. & Davidson, J.A. 1984. Integrated pest management for arborists: implementation of a pilot program. *Journal of Arboriculture* **10(3)**: 65-70. [HolmesDa1984]

Holzapfel, M. 1932. Die Gewächshausfauna des Berner Botanischen Gartens. *Revue Suisse de Zoologie* **39**: 325-374. [Holzap1932]

Honda, J.Y. & Luck, R.F. 1995. Scale morphology effects on feeding behavior and biological control potential of *Rhyzobius lophanthae* (Coleoptera: Coccinellidae). *Annals of the Entomological Society of America* **88(4)**: 441-450. [HondaLu1995]

Honiball, F. 1975. [Prospects for biological control of red scale *Aonidiella aurantii* (Mask.) in the Citrusdal area.]. [In Afrikaans]. Pages 217-220 *in*: Durr, H.J.R., Giliomee, J.H. & Neser, S., (Eds.). Proceedings of the First Congress of the Entomological Society of Southern Africa. Pretoria: Entomological Society of Southern Africa. 273 pp. [Honiba1975]

Honiball, F., Giliomee, J.H. & Randall, J.H. 1979. Mechanical control of red scale *Aonidiella aurantii* (Mask.) on harvested oranges. *Citrus & Subtropical Fruit Journal* **549**: 17-18. [HonibaGiRa1979]

Horn, D.J. 1988. Ecological Approach to Pest Management. New York: Guilford Press. 285 pp. [Horn1988]

Hornok, L. & Kozár, F. 1984. Fungi associated with a scale insect, *Quadraspidiotus ostreaeformis* (Curtis, 1843) (Homoptera, Coccoidea: Diaspididae). *Acta Phytopathologica Academiae Scientiarum Hungaricae* **19**: 9-11. [HornokKo1984]

Horticultural Inspection Notes. 1923. *Journal of Economic Entomology* **16**: 236-239. [HorticInNo1923]

Horticultural Inspection Notes. 1923a. *Journal of Economic Entomology* **16**: 237-239. [HorticInNo1923a]

Horton, J.R. 1918. The Argentine ant in relation to citrus groves. *Bulletin (United States Department of Agriculture)* **No. 647**: 1-74. [Horton1918]

Horvath, G. 1897. Description d'hémiptères nouveaux et notes diverses. *Revue d'Entomologie* **16**: 81-97. [Horvat1897]

Hosny, M. 1939. On coccids found on roots of plants in Egypt. *Bulletin, Ministry of Agriculture, Egypt, Technical and Scientific Service* **237**: 1-21. [Hosny1939]

Hosny, M.M. 1968. Classification of certain insect pests of Egyptian agriculture into three types of time-relative population trends. *Bulletin de la Societe Entomologique d'Egypte* **52**: 179-182. [Hosny1968]

Hosny, M.M., Amin, A.H. & El-Saadany, G.B. 1972. The damage threshold of the red scale, *Aonidiella aurantii* (Maskell) infesting mandarin trees in Egypt. *Zeitschrift für Angewandte Entomologie* **71**: 286-296. [HosnyAmEl1972]

Hosny, M. & Ezzat, Y.M. 1957. Further additions of the Coccoidea of Egypt. *Bulletin de la Société Entomologique d'Egypte* **41**: 331-333. [HosnyEz1957]

Houard, C. 1913. Les Zoocécidies des Plantes d'Europe et du Bassin de la Méditerranée. Vol. 3. Paris: Librairie Scientifique A. Hermann et Fils. 1560 pp. [Houard1913]

Houck, M.A. 1999. Phoresy by *Hemisarcoptes* (Acari: Hemisarcoptidae) on *Chilocorus* (Coleoptera: Coccinellidae): influence of subelytral ultrastructure. *Experimental & Applied Acarology* **23(2)**: 97-118. [Houck1999]

Houck, L.G., Jenner, J.F., Moreno, D.S. & Mackey, B.E. 1989. Permeability of polymer film wraps for citrus fruit fumigated with hydrogen cyanide to control California red scale. *Journal of the American Society for Horticultural Science* **114(2)**: 287-292. [HouckJeMo1989]

Houck, M.A. & O'Connor, B.M. 1996. Temperature and host effects on key morphological characters of *Hemisarcoptes cooremani* and *Hemisarcoptes malus* (Acari: Hemisarcoptidae). *Experimental & Applied Acarology* **20**: 667-682. [HouckOc1996]

Hough, W.S. 1932. The efficiency of tar distillate sprays in controlling San José and scurfy scales in 1931. *Journal of Economic Entomology* **25**: 613-617. [Hough1932]

Hough, W.S. 1933. The efficiency of tar distillate sprays in controlling San José scale in 1932. *Journal of Economic Entomology* **26(3)**: 470-473. [Hough1933]

Houser, J.S. 1908. The more important insects affecting Ohio trees. *Bulletin of the Ohio Agricultural Experiment Station* **194**: 173-181. [Houser1908]

Houser, J.S. 1918. The Coccidae of Cuba. *Annals of the Entomological Society of America* **11**: 157-172. [Houser1918]

Houser, J.S. 1920. Recent tests of materials to control San José scale. *Monthly Bulletin (Ohio Agricultural Experiment Station)* **5**: 49-51. [Houser1920]

Houser, J.S. 1926. The care of samples used in scoring the results of dormant sprays for the San José scale. *Journal of Economic Entomology* **19**: 94-95. [Houser1926]

Houska, J.B. 1926. The care of samples used in scoring the results of dormant sprays for the San José scale. *Journal of Economic Entomology* **19**: 94-95. [Houska1926]

Houston, K.J. 1991. *Chilocorus circumdatus* Gyllenhal newly established in Australia and additional records for *Coccinella undecimpunctata* L. (Coleoptera: Coccinellidae). *Journal of the Australian Entomological Society* **30(4)**: 341-342. [Housto1991]

Howard, C.W. 1908. The scale insects of citrus trees. *Transvaal Agricultural Journal* **6**: 265-277. [Howard1908]

Howard, F.W. 1991. Ecology and control of hemipterous pests of cultivated palms. *American Entomologist* (**Winter**): 217-225. [Howard1991]

Howard, L.O. 1894. The hymenopterous parasites of the California red scale. *Insect Life* **6**: 227-236. [Howard1894]

Howard, L.O. 1894c. Two parasites of important scale-insects. *Insect Life* **7**: 5-8. [Howard1894c]

Howard, L.O. 1895e. Revision of the Aphelininae of North America. *Technical Series (U.S. Department of Agriculture, Department of Entomology)* **1**: 1-44. [Howard1895e]

Howard, L.O. 1897b. Australian and New Zealand Coccidae. *Bulletin of the United States Department of Agriculture. Bureau of Entomology. Washington* **7**: 81-82. [Howard1897b]

Howard, L.O. 1898a. The San José scale in 1896-1897. *Bulletin of the United States Department of Agriculture.* **12**: 1-31. [Howard1898a]

Howard, L.O. 1898b. On some parasites of Coccidae, with descriptions of two new genera of Aphelininae. *Proceedings of the Entomological Society of Washington* **4**: 133-139. [Howard1898b]

Howard, L.O. 1907. New genera and species of Aphelininae, with a revised table of genera. *Technical Series (U.S. Department of Agriculture, Department of Entomology)* **Ser. 12, Pt. 4**: 69-88. [Howard1907]

Howard, L.O. 1911. A new species of *Coccophagus* with a table of the host relations of those species of the genus known to the writer. *Journal of Economic Entomology* **4**: 276-277. [Howard1911]

Howard, L.O. & Ashmead, W.H. 1895. On some reared parasitic hymenopterous insects from Ceylon. *Proceedings of the United States National Museum. Washington* **18**: 633-648. [HowardAs1895]

Howard, L.O. & Marlatt, C.L. 1899. The original home of the San José scale. *United States Department of Agriculture, Division of Entomology, Bulletin* **20**: 36-39. [HowardMa1899]

Howell, J.F. & George, D.A. 1984. Efficacy and persistence of chlorpyrifos residues on peaches for control of San José scale (Homoptera: Diaspididae). *Journal of Economic Entomology* **77**: 534-536. [HowellGe1984]

Howell, J.O. 1979. The adult male of *Comstockiella sabalis*; morphology and systematic significance. *Annals of the Entomological Society of America* **72**: 556-558. [Howell1979]

Howell, J.O. & Tippins, H.H. 1975. A new species of *Clavaspis* (Homoptera: Diaspididae) from pecan. *Annals of the Entomological Society of America* **68**: 338-340. [HowellTi1975]

Howell, J.O. & Tippins, H.H. 1977. Descriptions of first instars of nominal type-species of eight diaspidid tribes. *Annals of the Entomological Society of America* **70**: 119-135. [HowellTi1977]

Howell, J.O. & Tippins, H.H. 1978. Morphology and systematics of *Odonaspis minima*, n. sp. *Annals of the Entomological Society of America* **71**: 762-766. [HowellTi1978]

Howell, J.O. & Tippins, H.H. 1990. 1.1.2.3 The immature stages. Pages 29-54 *in*: Rosen, D. (Ed.). Armored Scale Insects, Their Biology, Natural Enemies and Control [Series title: World Crop Pests, Vol. 4A]. Amsterdam, The Netherlands: Elsevier. 384 pp. [HowellTi1990]

Hoy, M.A. 1990. 3.6.3.2 Genetic Improvement of Natural Enemies of Armored Scale Insects. Pages 441-451 *in*: Rosen, D. (Ed.). Armored Scale Insects, Their Biology, Natural Enemies and Control [Series title: World Crop Pests, Vol. 4B]. Amsterdam, the Netherlands: Elsevier. 688 pp. [Hoy1990]

Hoy, M.A. & Herzog, D.C. 1985. Biological Control in Agricultural IPM Systems. Orlando, FL: Academic Press. 589 pp. [HoyHe1985]

Hoyt, S.C. & Caltagirone, L.E. 1971. The developing programs of integrated control of pests of apples in Washington and peaches in California. Pages 395-421 *in*: Huffaker, C.B., Ed. Biological Control. New York & London: Plenum Press. 511 pp. [HoytCa1971]

Hoyt, S.C., Leeper, J.R., Brown, G.C. & Croft, B.A. 1983. Basic biology and management components for insect IPM. Pages 93-151 *in*: Croft, B.A. & Hoyt, S.C. (Eds.). Integrated Management of Insect Pests of Pome and Stone Fruits. New York: Wiley-Interscience. 454 pp. [HoytLeBr1983]

Hoyt, S.C., Westigard, P.H. & Rice, R.E. 1983. Development of pheromone trapping techniques for male San José scale (Homoptera: Diaspididae). *Environmental Entomology* **12**: 371-375. [HoytWeRi1983]

Hoz, M.T. de la 1983. Desarrollo posembrionario de *Selenaspidus articulatus* (Homoptera: Diaspididae) en condiciones ambientales. *Agrotécnica de Cuba* **15**: 21-33. [Hoz1983]

Hsu, S.K.T. 1935. A list of Chinese insects intercepted in other countries. *Entomology and Phytopathology. Hangchow* **3**: 578-590. [Hsu1935]

Hu, Y.Y., Dai, H.G & Hu, C.S. 1982. A preliminary study on poplar scale-insect *Quadraspidiotus gigas* (Thiem et Gerneck). *Scientia Silvae Sinicae* **18**: 160-169. [HuDaHu1982]

Huba, A. 1957. [The natural enemies of the San José scale (*Quadraspidiotus perniciosus* Comst.) in Slovakia and opportunities for practical exploitation.]. [In Slovakian]. *Pol'nohospodárstvo. Bratislava* **4(2)**: 306-353. [Huba1957]

Huba, A. 1958. Effektivität einer Introduktion von Parasiten der San José Schildlaus in der Tschechoslowakei. *Trans. I. Int. Conf. Insect Pathology and Biol. Control, Praha* 395-403. [Huba1958]

Huba, A. 1960. A contribution to the differentiation of the Diaspididae. *Pol'nohospodárstvo. Bratislava* **7**: 39-50. [Huba1960]

Huba, A. 1962. The prognosis of the spread and harmfulness of the San José scale (*Quadraspidiotus perniciosus* Comst.) in Europe. *Pol'nohospodárstvo. Bratislava* **9**: 415-422. [Huba1962]

Huffaker, C.B. 1990. 2.6.3 Effects of Environmental Factors on Natural Enemies of Armored Scale Insects. Pages 205-220 *in*: Rosen, D. (Ed.). Armored Scale Insects, Their Biology, Natural Enemies and Control [Series title: World Crop Pests, Vol. 4B]. Amsterdam, the Netherlands: Elsevier. 688 pp. [Huffak1990]

Huffaker, C.B. & Caltagirone, L.E. 1986. The impact of biological control on the development of the Pacific. *Agriculture, Ecosystems & Environment* **15**: 95-107. [HuffakCa1986]

Huffaker, C.B. & Doutt, R.L. 1965. Establishment of the Coccinellid, *Chilocorus bipustulatus* Linnaeus, in California olive groves. *Pan-Pacific Entomologist* **41**: 61-63. [HuffakDo1965]

Huffaker, C.B. & Gutierrez, A.P. 1990a. 3.6.5 Evaluation of Efficiency of Natural Enemies in Biological Control. Pages 473-495 *in*: Rosen, D. (Ed.). Armored Scale Insects, Their Biology, Natural Enemies and Control [Series title: World Crop Pests, Vol. 4B]. Amsterdam, the Netherlands: Elsevier. 688 pp. [HuffakGu1990a]

Huffaker, C.B., Messenger, P.S. & DeBach, P. 1971. The natural enemy component in natural control and the theory of biological control. Pages 16-67 *in*: Huffaker, C.B., Ed. Biological Control. New York & London: Plenum Press. 511 pp. [HuffakMeDe1971]

Huffaker, C.B., Simmonds, F.J. & Laing, J.E. 1976. The theoretical and empirical basis of biological control. Pages 41-78. *in*: Huffaker, C.B. & Messenger, P.S. (Eds.). Theory and Practice of Biological Control. New York: Academic Press. 788 pp. [HuffakSiLa1976]

Huffaker, C.B. & Smith, R.F. 1980. Rationale, organization, and development of a national integrated pest management project. Pages 1-24 *in*: Huffaker, C.B. (Ed.). New Technology of Pest Control. New York: Wiley-Interscience. 500 pp. [HuffakSm1980]

Huffaker, C.B. & Stinner, R.E. 1971. The role of natural enemies in pest control programs. Pages 333-350 *in*: Entomological essays to commemorate the retirement of K. Yasumatsu Tokyo: Hokuryukan Pub. Co.. 389 pp. [HuffakSt1971]

Hughes-Schrader, S. 1957. Differential polyteny and polyploidy in diaspine coccids (Homoptera: Coccoidea). *Chromosoma. Berlin* **8**: 709-718. [Hughes1957]

Hulsenberg, 1928. Versuche mit Calciumcyanid zur Bekämpfung von Gewächshausschädlingen. *Zeitschrift für Angewandte Entomologie* **14**: 285-315. [Hulsen1928]

Hungerford, H.B. 1944. Report on some insects of interest to the horticulturist. *Biennial Report of the Kansas State Horticultural Society* **47**: 52-55. [Hunger1944]

Hunt, H.A. 1939. Plant quarantine. *Annual Report California Agricultural Experiment Station* **28**: 548-566. [Hunt1939]

Hunter, S.J. 1899. The Coccidae of Kansas. *Kansas University Quarterly* **8**: 1-15. [Hunter1899]

Hunter, M.S. & Woolley, J.B. 2001. Evolution and behavioral ecology of heteronymous aphelinid parasitoids. *Annual Review of Entomology* **46**: 251-290. [HunterWo2001]

Hurt, R.H. 1933. Tar oil distillates as dormant spray materials for fruit trees. *Bulletin (Virginia Agricultural Experiment Station)* **293**: 3-15. [Hurt1933]

Hussain, A.A. 1974. Date palms and dates with their pests in Iraq. Pages 55-68. 1st. ed. Baghdad, Iraq: University of Baghdad. 166 pp. [Hussai1974]

Hutson, J.C. 1916. Pests of sugarcane in British Guiana II. *The Agricultural News (Barbados)* **15**: 426. [Hutson1916]

Hutson, J.C. 1933. The coconut scale (*Aspidiotus destructor*). *Tropical Agriculturist* **80(4)**: 254-256. [Hutson1933]

Il'ina, A.I. 1938. [Pathologo-anatomical modification of apple trees damaged by San Jose scale. Pages 133-146 *in:* Kiritchenko, A.N. (Ed.), Works of Quarantine Laboratories, Leningrad, Sel'khozgiz. 272 pp. [Ilina1938]

Institut français de recherches fruitières d'outre-mer (IFAC) 1962d. [Citrus at S.E.A.] Les agrumes a la S.E.A. *Rapport Annuel (IFAC, Paris)* **1962**: 140-151. [IFAC1962d]

Integrated Pest Management for Walnuts 1987. 2nd ed. Berkely, California: Univ. of Calif., Statewide Integrated Pest Management Project, Division of Agric. Natural Resources. 96 pp. [IPMW1987]

Ibarra-Nunez, G. 1990. Arthropods associated with coffee trees in a mixed plantation in Soconusco, Chiapas, Mexico. I. Variety and abundance. *Folia Entomologica Mexicana* **(79)**: 207-231. [Ibarra1990]

Imenes, S.D.L., Bergmann, E.C. & Wolff, V.R.S. 1999. Primeira ocorrencia de *Hemiberlesia lataniae* (Signoret, 1869) (Hemiptera, Diaspididae) em *Syngonium podophyllum* Schott (Araceae), no Brasil. *Arquivos do Instituto Biológico. Sao Paulo* **66(1)**: 117-119. [ImenesBeWo1999]

Imms, A.D. 1931. Biological control. I. Insect pests. *Tropical Agriculture* **8**: 98-102. [Imms1931]

Ingram, B.F. & Nimmo, P.R. 1991. Control of San José scale *Comstockaspis perniciosus* (Comstock) on apples in the Stanthorpe district, Queensland. *General and Applied Entomology* **23**: 53-58. [IngramNi1991]

Innerhofer, J., Redl, H., Weindlmayr, J. & Vigl, J. 1985. Prufung moderner Bekampfungsmittel gegen die San José-Schildlaus unter Zuhilfenahme einer neuen Bonitierungsmethode. *Mitteilungen Klosterneuburg Rebe und Wein, Obstbau und Fruchteverwertung* **35(6)**: 261-268. [InnerhReWe1985]

Inserra, S. 1966. Introduzione ed acclimatazione di due *Aphytis* (*A. melinus* De Bach ed *A. lignanensis* Compere) parassiti ectofagi di alcune cocciniglie degli agrumi. *Tecnica Agricola. Catania* **18(2)**: 176-186. [Inserr1966]

Inserra, S. 1966a. La cocciniglia rossa forte degli agrumi (*Aonidiella aurantii* Maskell) in Sicilia. *Bollettino del Laboratorio di Entomologia Agraria 'Filippo Silvestri'. Portici* **27**: 1-26. [Inserr1966a]

Inserra, S. 1968. Prove di lotta integrata contro l'*Aonidiella aurantii* Mask. ed altre cocciniglie degli agrumi in Sicilia. *Entomologica* **4**: 45-77. [Inserr1968]

Inserra, S. 1969. La cocciniglia rossa forte degli agrumi (*Aonidiella aurantii* Maskell) in Sicilia. *Bollettino del Laboratorio di Entomologia Agraria 'Filippo Silvestri'. Portici* **27**: 1-26. [Inserr1969]

Inserra, S. 1970a. Acclimatation, diffusion et notes sur la biologie d'*Aphytis melinus* De Bach en Sicile. *Al Awamia* **37**: 39-46. [Inserr1970a]

Inserra, S. & Calabretta, C. 1987. Research on scale insects (Homoptera : Coccoidea : Diaspididae) living on *Ceratonia siliqua* L., *Pistacia vera* L. and *Pistacia lentiscus* L. in Sicily. Part II. *Bollettino del Laboratorio di Entomologia Agraria 'Filippo Silvestri'* **43** (Suppl.): 43 (Suppl.): 91-95. [InserrCa1987]

Integrated Pest Management for Citrus 1991. Integrated Pest Management for Citrus. *Publication (Division of Agriculture and Natural Resources, University of California)* **3303**: 144 pp. [IntegrPeMa1991]

Iperti, G. 1961. Les coccinelles leur utilisation en agriculture. *Revue de Zoologie Agricole* **1/3**: 14-30. [Iperti1961]

Iperti, G. & Brun, J. 1969. Rôle d'une quarantaine pour la multiplication des Coccinellidae coccidiphages destines a combattre la cochenille du Palmier-Dattier (*Parlatoria blanchardii* Targ.) en Adrar mauritanien. *Entomophaga* **14(2)**: 149-157. [IpertiBr1969]

Iperti, G. & Laudého, Y. 1968. [Bio-ecological intervention intended for control of the date palm scale: *Parlatoria blanchardi* Targ. (Coccoidea - Diaspididae) in the Mauritanian Adrar.]. [In French]. *Fruits* **23(10)**: 543-552. [IpertiLa1968]

Iren, S. 1970. [A study on the species of *Septobasidium* and scale insects on apple and pear in Tirebolu (Turkey) and the relations between them.]. [In Turkish]. *Tarim Bakanligi, Zirai Mucadele ve Zirai Karantina Genel Mudurlugu, Ankara* 21 pp. [Iren1970]

Isayev, V.V. 1975. Significance of quarantine measures for preventing spread of San José scale [*Quadraspidiotus perniciosus* (Comst.)] and its eradication in northern Caucasus. *VIII International Plant Protection Congress* **3**: 181-187. [Isayev1975]

Isely, D. 1924. Lubricating oil emulsion for the control of San José scale. *Extension Circular (College of Agriculture, University of Arkansas)* **No. 164**: 4 pp. [Isely1924]

Ishaaya, I. 1971. Observations on the phenoloxidase system in the armored scales, *Aonidiella aurantii* and *Chrysomphalus aonidum*. *Comparative Biochemistry and Physiology B* **39**: 935-943. [Ishaay1971]

Ishaaya, I., Blumberg, D. & Yarom, I. 1989. Buprofezin -- a novel IGR for controlling whiteflies and scale insects. *Mededelingen van de Faculteit Landbouwwetenschappen, Rijksuniversiteit Gent* **54(3b)**: 1003-1008. [IshaayBlYa1989]

Ishaaya, I., Mendel, Z. & Blumberg, D. 1992. Effect of buprofezin on California Red Scale *Aonidiella aurantii* (Maskell), in a citrus orchard. *Israel Journal of Entomology* **25-26**: 67-71. [IshaayMeBl1992]

Ishaaya, I. & Swirski, E. 1970. A rapid laboratory test for determining death in some armored scale species (Coccoidea: Diaspididae). *Entomologia Experimentalis et Applicata* **13**: 37-42. [IshaaySw1970]

Ishaaya, I. & Swirski, E. 1970a. Invertase and amylase activity in the armoured scales *Chrysomphalus aonidum* and *Aonidiella aurantii*. *Journal of Insect Physiology* **16**: 1599-1605. [IshaaySw1970a]

Ishaaya, I. & Swirski, E. 1990. 1.5.2 Iodine test for determining live and dead scale insects. Pages 353-356. *in*: Rosen, D. (Ed.). Armored Scale Insects, Their Biology, Natural Enemies and Control [Series title: World Crop Pests, Vol. 4A]. Amsterdam, The Netherlands: Elsevier. 384 pp. [IshaaySw1990]

Ishaaya, I., Swirski, E. & Neubauer, I. 1980. Digestive enzymes and trehalase activity in some hemipterous insects and their relation to insect hosts compatibility and insect behavior (abstract). *Abstracts of the 16th International Congress of Entomology. Kyoto,* p. 212. [IshaaySwNe1980]

Ishchenko, R.I., Bichina, T.I. & Kovalev, B.G. 1988. [Attractiveness of 3,7-dimethyl-2,7 octadienyl propionate and its positional isomer for San José scale males.]. [In Russian]. *Khemoretseptsiya Nasekomykh* **(No. 10)**: 99-101. [IshcheBiKo1988]

Ishii, T. 1923. Observations on the Hymenopterous parasites of *Ceroplastes rubens* Mask., with descriptions of new genera and species of the subfamily Encyrtinae. *Imperial Plant Quarantine Station Bulletin (Department of Agriculture and Commerce, Japan)* **No. 3**: 69-114. [Ishii1923]

Ishii, T. 1926. [On Hymenoptera parasitic upon Coccidae.]. [In Japanese]. *Kontyû* **1(1)**: 31-36. [Ishii1926]

Ishii, T. 1928. The Encyrtinae of Japan. *Bulletin of the Imperial Central Agricultural Experiment Station, Japan* **3**: 79-160. [Ishii1928]

Ishii, T. 1932a. The Encyrtinae of Japan II. Studies on morphology and biology. *Bulletin of the Imperial Central Agricultural Experiment Station, Japan* **3(3)**: 161-202. [Ishii1932a]

Ivanova, A.N. & Pavlyuchuk, M.V. 1988. [Phytohormones.]. [In Russian]. *Zashchita rastenii. Moscow* **No. 11**: 28-29. [IvanovPa1988]

Izraylevich, S. & Gerson, U. 1993. Mite parasitization on armored scale insects: host suitability. *Experimental and Applied Acarology* **17(12)**: 861-875. [IzraylGe1993]

Izraylevich, S. & Gerson, U. 1993a. Population dynamics of *Hemisarcoptes coccophagus* Meyer (Astigmata: Hemisarcoptidae) attacking three species of armored scale insects (Homoptera: Diaspididae). *Experimental and Applied Acarology* **17(12)**: 877-888. [IzraylGe1993a]

Izraylevich, S. & Gerson, U. 1995. Sex ratio of *Hemisarcoptes coccophagus*, a mite parasitic on insects: density-dependent processes. *Oikos* **74**: 439-446. [IzraylGe1995]

Izraylevich, S. & Gerson, U. 1995b. Spatial patterns of the parasitic mite *Hemisarcoptes coccophagus* (Astigmata: Hemisarcoptidae): host effect, density-dependence of aggregation, and implications for biological control. *Bulletin of Entomological Research* **85**: 235-240. [IzraylGe1995b]

Izraylevich, S., Gerson, U. & Hasson, O. 1996. Numerical response of a parasitic mite: host effect and mechanism. *Environmental Entomology* **25**: 390-395. [IzraylGeHa1996]

Izraylevich, S., Hasson, O. & Gerson, U. 1995. Frequency-dependent host selection by parasitic mites: a model and a case study. *Oecologia* **102(2)**: 138-145. [IzraylHaGe1995]

Jaap, O. 1914. Verzeichnis der bei Triglitz in der Prignitz beobachteten Cocciden. *Verhandlungen des Botanischen Vereins Brandenburg* **56**: 135-142. [Jaap1914]

Jackson, B.D. 1913. Pages 1-48 *in*: Catalogue of the Linnean Specimens of Amphibia, Insecta and Testacea, noted by Carl von Linné. London: Linnean Society. [Jackso1913]

Jahn, S. & Polesny, F. 1999. Population dynamics of San José scale and San José scale parasitoids in three different sites in Austria. *IOBC-WPRS Bulletin* **22(7)**: 201-202. [JahnPo1999]

Jalali, S.K. & Singh, S.P. 1995. Effect of pesticide on mortality and parasitizing ability of parasitoid *Aphytis* species of San José scale (*Quadraspidiotus perniciosus*). *Indian Journal of Agricultural Sciences* **65**: 617-620. [JalaliSi1995]

Jalaluddin, M. & Mohanasundaram, M. 1989. Control of the coconut scale *Aspidiotus destructor* Sign. in the nursery. *Entomon* **14(3-4)**: 203-206. [JalaluMo1989]

Jalaluddin, M. & Mohanasundaram, M. 1989a. Residual toxicity of four insecticides recommended for control of coconut coccids on the parasitoid fauna of *Opisina arenosella* Wlk. *Entomon* **14(3-4)**: 199-202. [JalaluMo1989a]

Jalaluddin, S.M., Mohanasundaram, M. & Sundara Babu, P.C. 1992. Influence of weather on fecundity potential of coconut scale *Aspidiotus destructor* Sing. *Indian Coconut Journal* **22(12)**: 5-7. [JalaluMoSu1992]

Jalaluddin, S.M., Thirumoorthy, S., Mohanasundaram, M., Chinniah, C. & Chinnaswami, K.N. 1991. Coccid complex of coconut in Tamil Nadu. *Indian Coconut Journal* **22(3)**: 17. [JalaluThMo1991]

Jalaluddin, S.M., Sadakathulla, S. & Manuel, W.W. 2001. New record of yellow hard scale *Aspidiotus destructor* Sign. (Homoptera: Diaspididae) on betelvine. *Entomon* **26(3-4)**: 347-348. [JalaluSaMa2001]

James, D.G., Stevens, M.M. & O'Malley, K.J. 1997. The impact of foraging on populations of *Coccus hesperidum* L. (Hem., Coccidae) and *Aonidiella aurantii* (Maskell) (Hem., Diaspididae) in an Australian citrus grove. *Journal of Applied Entomology* **121**: 257-259. [JamesStOM1997]

Jamieson, L.E., Dobson, S., Cave, J. & Stevens, P.S. 2002. A survey of armoured scale insects on kiwifruit shelter. *New Zealand Plant Protection* **55**: 354-360. [JamiesDoCa2002]

Jancke, G.D. 1955. Zur Morphologie der männlichen Cocciden. *Zeitschrift für Angewandte Entomologie* **37**: 265-314. [Jancke1955]

Janjua, N.A. 1959. Insects of Baluchistan and their distribution. Appendix H. *Papers of the Peabody Museum of American Archaeology and Ethnology* **52**: 231-264. [Janjua1959]

Jannone, G. 1940. Principali cause di natura animale riscontrate dannose all'agricoltura dell'Africa Orientale Italiana durante il 1939. *L'Agricoltura Coloniale* **34(6)**: 241-253. [Jannon1940]

Jannone, G., Capra, F. & Binaghi, G. 1962. Ricerche svolte a Genova sull *Aspidiotus destructor* (Sign.) (Hemip., Coccidae) vivente sulle banane provenienti dalla Somalia. *Relazioni e Monografie Agrarie Subtropicali e Tropicali* **No. 83**: 5-126. [JannonCaBi1962]

Jansen, M.G.M. 2001. An annotated list of the scale insects (Hemiptera: Coccoidea) of the Netherlands. *Entomologica* **33(1999)**: 197-206. [Jansen2001]

Jarraya, A. 1970. État phytosanitaire des agrumes de Tunisie et perspectives de lutte contre leurs principaux ravageurs. *Al Awamia* **37**: 85-89. [Jarray1970]

Jarvis, C.D. 1907. Petroleum emulsion for the San José scale. *Bulletin (Agricultural Experiment Station, Connecticut Agricultural College, Storrs, CT)* **No. 49**: 5-12. [Jarvis1907]

Jarvis, C.D. 1908. Proprietary and home-made miscible oils for the control of the San José scale. *Bulletin (Agricultural Experiment Station, Connecticut Agricultural College, Storrs, CT)* **No. 54**: 169-197. [Jarvis1908CD]

Jarvis, T.D. 1908. A preliminary list of the scale insects of Ontario. *Annual Report of the Entomological Society of Ontario* **38**: 50-72. [Jarvis1908TD]

Jarvis, H. 1927. The San José scale (*Aspidiotus perniciosus* Comstock). *Queensland Agricultural Journal* **27**: 513-517. [Jarvis1927]

Jaszai, E.V. & Darvas, B. 1983. Effectiveness of kinoprene and hydroprene on pests of greenhouse ornamentals; mealybugs (Pseudococcidae), soft scales (Coccidae) and armored scales (Diaspididae). Pages 4:198-202 *in*: Darvas, B., Toth, M. & Vajna, L., Eds. Proceedings of the International Conference on Integrated Plant Protection Budapest: Horticultural University Press. [JaszaiDa1983]

Jayanthi, R. 1999. New records of scale insects on sugarcane in Tamil Nadu, India (Hemiptera: Coccoidea). *Bionotes* **1(4)**: 76. [Jayant1999]

Jayanthi, R. & David, H. 1986. Varietal resistance. Pages 363-381 *in*: David, H., Easwaramoorthy, S. & Jayanthi, R., eds. Sugarcane Entomology in India. Coimbatore, India: Sugarcane Breeding Institute. [JayantDa1986]

Jayanthi, R., David, H. & Goud, Y.S. 1994. Physical characters of sugarcane plant in relation to infestation by *Melanaspis glomerata* (G.) and *Saccharicoccus sacchari* (Ckll.). *Journal of Entomological Research. New Delhi* **18**: 305-314. [JayantDaGo1994]

Jayanthi, R., David, H. & Goud, Y.S. 1996. Behaviour of neonate crawlers of sugarcane coccoids, *Melanaspis glomerata* and *Saccharicoccus sacchari*. *Bulletin of Entomology* **37(1-2)**: 7-9. [JayantDaGo1996]

Jayanthi, R. & Goud, Y.S. 2000. Feeding sites of the scale insect, *Melanaspis glomerata*, and mealy bug, *Saccharicoccus sacchari*, in the sugarcane stem. *Bionotes* **2(1)**: 11. [JayantGo2000]

Jenkins, C.F.H. 1945. The citrus red scale. *Journal of Agriculture (Western Australia Dept. of Agriculture)* **22(1)**: 10-16. [Jenkin1945]

Jenser, G. & Sheta, I.B. 1972. Importance of male control in preventing damage done by the San José scale (*Quadraspidiotus perniciosus* Comst.). *Acta Agronomica Academiae Scientiarum Hungaricae* **12(1/2)**: 119-124. [JenserSh1972]

Jeppson, L.R. 1969. Impact of insect and mite resistance on chemical control programs on citrus. Pages 917-921. *in*: Chapman, H.D. (Ed.). Proceedings First International Citrus Symposium. Vol. 2. Riverside: University of California. [Jeppso1969]

Jermini, M. & Brunetti, R. 1994. Phytoxicite du fenoxycarb et d'huile minerale utilises en hiver pour lutter contre le pou de San José *(Quadraspidiotus perniciosus* Comst.) en pepiniere de plantes ornementales: deux années d'essai au Tessin. *Revue Horticole Suisse* **67**: 5-6, 115-119. [JerminBr1994]

Jhansi, K. & Babu, P.R. 2002. Judicious use of agronomical practice and insecticides in management of sugarcane scale, *Melanaspis glomerata* (Green). *Indian Sugar* **51(9)**: 615-618. [JhansiBa2002]

Ji, L., Izraylevich, S., Gazit, S. & Gerson, U. 1996. A sex-specific tri-trophic-level effect in a phoretic association. *Experimental & Applied Acarology* **20**: 503-509. [JiIzGa1996]

Ji, L.Z. & Yang, J.K. 1990. [Research progress on the predacious mite, *Hemisarcoptes* spp. (Astigmata: Hemisarcoptidae), associated with armoured scale insects (Coccoidea)]. [In Chinese]. *Chinese Journal of Biological Control* **6(3)**: 134-136. [JiYa1990]

Jimenez J., E. 1969. Control biológico de algunas plagas de los cítricos en México. Pages 781-784. *in*: Chapman, H.D. (Ed.). Proceedings First International Citrus Symposium. Vol. 2 Riverside: University of California. [Jimene1969]

John, O. 1930. Material zur Insektenfauna Lettlands: 1. Zur Coccidenfauna Lettlands. *Folia Zoologica et Hydrobiologica* **2**: 3-7. [John1930]

Johnson, D.T., Lewis, B.A. & Whitehead, J.D. 1999. Grape scale (Homoptera: Diaspididae) biology and management on grapes. *Journal of Entomological Science* **34(2)**: 161-170. [JohnsoLeWh1999]

Johnson, R.H., Young, B.L. & Alstad, D.N. 1997. Responses of ponderosa pine growth and volatile terpene concentrations to manipulation of soil water and sunlight availability. *Canadian Journal of Forest Research* **27**: 1794-1804. [JohnsoYoAl1997]

Johnson, W.G. 1896. Preliminary notes on five new species of scale insects. *Entomological News* **7**: 150-152. [Johnso1896]

Johnson, W.G. 1898. Notes on the external characters of the San José scale, cherry scale and Putnam's scale. *Canadian Entomologist* **30**: 82-83. [Johnso1898]

Johnson, W.G. 1899. The odour of Coccidae. *Canadian Entomologist* **31**: 87-88. [Johnso1899]

Johnson, W.G. 1900. *Aphelinus fuscipennis* an important parasite upon the San José scale in eastern United States. *Bulletin of the United States Department of Agriculture. Bureau of Entomology. Washington* **26**: 73-75. [Johnso1900]

Johnson, W.T. & Lyon, H.H. 1991. Insects that Feed on Trees and Shrubs. 2nd ed. Ithaca, N.Y.: Comstock Pub. Associates. 560 pp. [JohnsoLy1991]

Johnston, J.R. 1915. The entomogenous fungi of Porto Rico. *Bulletin (Board of Commissioners of Agriculture, Government of Porto Rico)* **No. 10**: 1-33. [Johnst1915]

Jones, E.P. 1936. The bionomics and ecology of red scale - *Aonidiella aurantii* Mask. - in southern Rhodesia. *British South Africa Co., Mazoe Citrus Experiment Station Publication* **5**: 11-52. [Jones1936]

Jorgensen, D.D. 1934. A study of some Utah Coccidae (scale insects). *Proceedings of the Utah Academy of Sciences, Arts, and Letters* **11**: 273-284. [Jorgen1934]

Jorgensen, C.D., Rice, R.E., Hoyt, S.C. & Westigard, P.H. 1981. Phenology of the San José scale (Homoptera: Diaspididae). *Canadian Entomologist* **113**: 149-159. [JorgenRiHo1981]

Joséph, E. 1984. Programme d'action pour l'amélioration de la protection des végétaux. Algérie. Deuxième consultation sur le contrôle phytosanitaire 26 juillet-11 août 1982. Algeria. Second meeting on plant health surveillance 26 July-11. August 1982. 32 pp. [Joseph1984]

Jourdheuil, P. 1979. Progrès réalisés dans l'utilisation des entomophages. Pages 75-79. *in*: Proceedings: Internationales Symposium der IOBC/WPRS über Integrierten Pflanzenschutz in der Landund Forstwirtschaft. Wien, Austria: OILB/SROP. 648 pp. [Jourdh1979]

Jura, C. 1959. The early developmental stages of the ovoviviparous scale insect (*Quadraspidiotus ostreaeformis*) (Curt.) (Homoptera, Coccidae, Aspidiotini). *Zoologica Poloniae. Krakow* **9**: 17-34. [Jura1959]

Kagy, J.F. 1936. Laboratory method of comparing the toxicity of substances to San José scale. *Journal of Economic Entomology* **29**: 393-395. [Kagy1936]

Kalshoven, L.G.E. 1981. Pests of crops in Indonesia. Pages 160-193. Jacarta, Indonesia: Ichtiar Baru. 701 pp. [Kalsho1981]

Kamath, M.K. 1979. A review of biological control of insect pests and noxious weeds in Fiji (1969-1978). *Fiji Agricultural Journal* **41**: 55-72. [Kamath1979]

Kansu, L.A. & Uygun, N. 1979. Integrierte Schädlingsbekämpfung im Citrusanbau in der Türkei. Pages 565-567. *in*: Proceedings: Internationales Symposium der IOBC/WPRS über Integrierten Pflanzenschutz in der Land und Forstwirtschaft. Wien, Austria: OILB/SROP. 648 pp. [KansuUy1979]

Kapur, A.P. 1942. Bionomics of some Coccinellidae predaceous on aphids and coccids in north India. *Indian Journal of Entomology* **4(1)**: 49-66. [Kapur1942]

Kapur, A.P. 1956. Systematic and biological notes on the lady-bird beetles predacious on the San José scale in Kashmir with description of a new species (Coleoptera: Coccinellidae). *Records of the Indian Museum* **52**:257-274. [Kapur1956]

Karaca, I. 1998. Parasitization efficacy of *Aphytis melinus* DeBach (Hymenoptera: Aphelinidae) as affected by host size and size distribution of *Aonidiella aurantii* (Maskell) (Homoptera: Diaspididae) in a lemon orchard. *Turkiye Entomoloji Dergisi* **22**: 2, 101-108. [Karaca1998]

Karaca, I., Senal, D., Colkesen, T. & Ozgokce, S. 1998. Observations on the oleander scale, *Aspidiotus nerii* Bouché (Hemiptera: Diaspididae) and its natural enemies on blueleaf wattle in the Adana province of Turkey. Page 23 *in*: VIIIth International Symposium on Scale Insect Studies. 41 pp. [KaracaSeCo1998]

Karaca, I., Senal, D., Colkesen, T. & Özgökçe, M.S. 2001. Observations on the oleander scale, *Aspidiotus nerii* Bouché (Hemiptera: Diaspididae) and its natural enemies on blueleaf wattle in Adana Province, Turkey. *Entomologica* **33(1999)**: 407-412. [KaracaSeCo2001]

Karaca, I., Sekeroglu, E. & Uygun, N. 1987. [Life tables of the California Red Scale, *Aonidiella aurantii* (Mask.) (Homoptera: Diaspididae) in laboratory conditions.] Kirmizi kabuklu bit (*Aonidiella aurantii* (Mask.) (Homoptera: Diaspididae)) in laboratuvar kosullarinda yasam cizelgesi. *Turkiye I. Entomoloji Kongresi* : 129-138. [KaracaSeUy1987]

Karaca, I. & Uygun, N. 1990. [Natural enemies of *Aonidiella aurantii* (Maskell) (Homoptera, Diaspididae) in East Mediterranean citrus areas, and their population development on different citrus varieties.]. [In Turkish]. *Proceedings of the Second Turkish National Congress of Entomology. Izmir, Turkey: 26-29 Eylul 1990* : 97-108. [KaracaUy1990]

Karaca, I. & Uygun, N. 1992. Population development of *Aonidiella aurantii* (Maskell) (Homoptera: Diaspididae) on different citrus species and varieties. *Proceedings of the Second Turkish National Congress of Entomology. Izmir, Turkey* : 9-19. [KaracaUy1992]

Karaca, I. & Uygun, N. 1993. [Life tables of oleander scale, *Aspidiotus nerii* Bouché on different host plants.]. [In Turkish]. *Turkiye Entomoloji Dergisi* **17(4)**: 217-224. [KaracaUy1993]

Karaca, I., Uygun, N., Elekcioglu, N.Z. & Senal, D. 2002. Population development of *Aonidiella aurantii* (Maskell) and *Parlatoria pergandii* Comstock (Homoptera: Diaspididae) in Cukurova Region of Turkey. *Bollettino di Zoologia Agraria e di Bachicoltura (Milano)* **33(3)**: 313-317. [KaracaUyEl2002]

Kasargode, R.S. 1914. A preliminary account of the Coccidae of western India. *Journal of the Bombay Natural History Society* **23**: 133-137. [Kasarg1914]

Kaston, B.J. 1938. Checklist of elm insects. *Bulletin of the Connecticut Agricultural Experiment Station* **No. 408**: 235-242. [Kaston1938]

Kathiresan, K. 1993. Dangerous pest on nursery seedlings of *Rhizopora*. *Indian Forester* **119(12)**: 1026. [Kathir1993]

Kato, T. 1968. [Predacious behaviour of coccidophagus coccinellid *Chilocorus kuwanae* Silvestri in the hedge of *Euonymus japonicus* Thunberg]. [In Japanese]. *Kontyû* 29-38. [Kato1968]

Katsoyannos, P. 1992. Olive pests and their control in the Near East. FAO Plant Production and Protection Paper No. 115. Rome, Italy: FAO (United Nations). 178 pp. [Katsoy1992]

Katsoyannos, P. 1993. IPM for citrus insect pests in northern Mediterranean countries. *FAO Plant Protection Bulletin* **41(3-4)**: 177-198. [Katsoy1993]

Katsoyannos, P. 1996. Integrated insect pest management for citrus in Northern Mediterranean countries. Athens, Greece: Benaki Phytopathologique Institute.. 110 pp. [Katsoy1996]

Katsoyannos, P.I. & Argyriou, L.C. 1985. The phenology of the San José scale *Quadraspidiotus perniciosus* (Hom.: Diaspididae) and its association with its natural enemies on almond trees in northern Greece. *Entomophaga* **30(1)**: 3-11. [KatsoyAr1985]

Katsoyannos, P. & Stathas, G.J. 1997. Phenology of *Melanaspis inopinata* on pistachio trees in Greece. *Phytoparasitica* **25**: 331-332. [KatsoySt1997]

Kattari, D., Heimpel, G.E., Ode, P.J. & Rosenheim, J.A. 1999. Hyperparasitism by *Ablerus clisiocampae* Ashmead (Hymenoptera : Aphelinidae). *Proceedings of the Entomological Society of Washington* **101(3)**: 640-644. [KattarHeOd1999]

Kaussari, M. 1952. Sur une cochenille nouvelle du sud-est de l'Iran. *Revue de Pathologie Végétale et d'Entomologie Agricole de France* **31**: 181-184. [Kaussa1952]

Kaussari, M. 1954. Sur un *Dynaspidiotus* Thiem et Gerneck (Hem. Coccidae) nouveau de l'ouest de l'Iran. *Revue de Pathologie Végétale et d'Entomologie Agricole de France* **33**: 80-83. [Kaussa1954]

Kaussari, M. 1955. La première liste des cochenilles de l'Iran. *Ministère de l'Agriculture, Entomologie et Phytopathologie Appliquées, Publication du Département Général de la Protection Plantes. Tehran* **15**: 14-20. [Kaussa1955]

Kaussari, M. 1955a. Sur trois Coccoidea Diaspidinae nouveaux de l'Iran. *Revue de Pathologie Végétale et d'Entomologie Agricole de France* **34**: 229-237. [Kaussa1955a]

Kaussari, M. 1956. Sur un *Dynaspidiotus* Thiem et Gerneck (Coccoidea-Aspidiotini) nouveau des environs de Chiraz (Iran). *Revue de Pathologie Végétale et d'Entomologie Agricole de France* **35**: 102-103. [Kaussa1956]

Kaussari, M. 1957. La deuxième liste des cochenilles de l'Iran. *Ministère de l'Agriculture, Entomologie et Phytopathologie Appliquées, Publication du Département Général de la Protection Plantes. Tehran* **16-17**: 1-3. [Kaussa1957]

Kaussari, M. 1958. Sur deux Aspidiotini-Aspidiotina (Coccoidea) du sud de l'Iran. *Revue de Pathologie Végétale et d'Entomologie Agricole de France* **37**: 229-234. [Kaussa1958]

Kaussari, M. & Balachowsky, A. 1953. Sur un *Diaspidiotus* (Hom. Coccoidea) nouveau vivant sur *Tamarix* dans le sud de l'Iran. *Revue de Pathologie Végétale et d'Entomologie Agricole de France* **32**: 24-26. [KaussaBa1953]

Kaussari, M. & Balachowsky, A. 1953a. Sur un *Aspidaspis* Ferris (Coccoidea-Aspidiotini) nouveau du sud de l'Iran. *Revue de Pathologie Végétale et d'Entomologie Agricole de France* **32**: 99-102. [KaussaBa1953a]

Kaussari, M. & Balachowsky, A. 1953b. Sur un nouveau genre d'Aspidiotini *Targionina* (Coccoidea-Diaspidinae) du centre de l'Iran. *Revue de Pathologie Végétale et d'Entomologie Agricole de France* **32**: 276-280. [KaussaBa1953b]

Kawai, S. 1977. Changes of Coccid-Fauna with urbanization in Tokyo. Pages 148-172 *in*: Numata, M., Ed. "Tokyo Project" Interdisciplinary Studies of Urban Ecosystem in the Metropolis of Tokyo [Kawai1977]

Kawai, S. 1980. Scale insects of Japan in colors. Tokyo: National Agricultural Education Association. 455 pp. [Kawai1980]

Kawai, S. 1987. [The coccid fauna of the Ogasawara (Bonin) Islands.]. Pages 75-86. *in*: [Development, Technique and Cooperation in Agriculture: A Collection of Papers to Commemorate the 30th Anniversary.] Tokyo University of Agriculture. [Kawai1987]

Kawai, S., Matsubara, Y. & Umesawa, K. 1971. A preliminary revision of the Coccoidea-fauna of the Ogasawara (Bonin) Islands (Homoptera: Coccoidea). *Applied Entomology and Zoology, Tokyo* **6**: 11-26. [KawaiMaUm1971]

Kawecki, Z. 1935. Czerwce (Coccidae) województwa krakowskiego I kieleckiego zebrane w latach 1933-1934. (Die schildläuse (Coccidae) der wojewodschaften Kraków und kielce die in den jahren 1933-1934 gesammelt wurden.). *Sprawozdanie Komisji Fizjograficznej Polskiej Akad. Umiejetnosci (Annual Report of the Physiographical Commission of Polish Acad. of Sciences & Letter* **68**: 73-90. [Kaweck1935]

Kawecki, Z. 1948. Some Coccidae from Poland. *Materialy do Fizjografii Kraju Documenta Physiographica Poloniae* **10**: 1-10. [Kaweck1948]

Kawecki, Z. 1950. Tarcznik niszczyciel (tarcznik San José) - *Quadraspidiotus (Aspidiotus) perniciosus* Comst. w. Europie i jego pojawienie sie W. polsce. (San José Scale - *Quadraspidiotus (Aspidiotus) perniciosus* Comst. in Europe and its appearance in Poland.). *Polska Akademia Umiejetnosci, Prace Rolniczo-Lesne* **55**: 1-54. [Kaweck1950]

Kaydan, M.B., Ülgentürk, S., Toros, S. & Kozár, F. 2002. Scale insects (Homoptera: Coccoidea) of natural and agriculture areas in Kapadokya, Turkey. *Bollettino di Zoologia Agraria e di Bachicoltura (Milano)* **33(3)**: 253-257. [KaydanUlTo2002]

Kazandjiev, V., Tzalev, M. & Staneva, E. 1995. Reproduction of San José scale (*Quadraspidiotus perniciosus* Comst.; Homoptera: Diaspididae) in Bulgaria in dependence of the thermal conditions. *Bulgarian Journal of Agricultural Science* **1**: 247-252. [KazandTzSt1995]

Keen, F.P. 1952. Insect enemies of western forests. *Miscellaneous Publications United States Department of Agriculture* **No. 273**: 1-280. [Keen1952]

Keetch, D.P. & Whitnall, A.B.M. 1972. Ecology of the citrus red mite, Panonychus citri (McGregor), (Acarina: Tetranychidae) in South Africa. 4. The influence of red scale sprays on the population density of *P. citri. Journal of the Entomological Society of Southern Africa* **35(2)**: 253-263. [KeetchWh1972]

Kehat, M., Blumberg, D. & Greenberg, S. 1968. Tests with azinphosmethyl for controlling the red scale (*Aonidiella aurantii*) Mask. on greenhouse roses. *International Pest Control* **10**: 28-29. [KehatBlGr1968]

Kervegant, D. 1932. Le caféier a la Martinique. *Bulletin de l'Agence Générale dés Colonies* **28**: 1653-1686. [Kerveg1932]

Kfir, R. & Rosen, D. 1981. Biology of the hyperparasite *Marietta javensis* Howard) (Hymenoptera: Aphelinidae) reared on *Microterys flavus* (Howard) in brown soft scale. *Journal of the Entomological Society of Southern Africa* **44**: 141-150. [KfirRo1981]

Khajuria, D.R. & Sharma, H.K. 1997. Use of miscible oil and insecticidal combinations in management of San José scale (*Quadraspidiotus perniciosus*) on apple (*Malus domestica*). *Indian Journal of Agricultural Sciences* **67**: 488-489. [KhajurSh1997]

Khalaf, J. & Sokhansanj, M. 1993. Bioecological studies on Orientalis yellow scale (*Aonidiella orientalis* New.) and its control by integrated methods in Fars Province. *Applied Entomology and Phytopathology* **60(1/2)**: 11-12. [KhalafSo1993]

Khalifa, A. & Habib, A. 1958. Population studies on the black scale, *Chrysomphalus ficus* Ashmead: II. Interactions of factors affecting distribution of population. *Bulletin de la Societe Entomologique d'Egypte* **42**: 449-459. [KhalifHa1958]

Khanna, K.I. 1957. On the coccid *Melanaspis glomerata* Green in Bihar. *Indian Journal of Sugar Research and Development* **2**: 19-22. [Khanna1957]

Kidd, N.A.C. & Jervis, M.A. 1996. Population dynamics. Pages 293-374 *in*: Jervis, M. & Kidd, N. (Ed.). Insect Natural Enemies: Practical Approaches to their Study and Evaluation. London: Chapman & Hall. 491 pp. [KiddJe1996]

Killington, F.J. 1932. Coccidae (Hemipt.) taken in Hanta in 1930 and 1931. *Journal (Society for British Entomology)* **1**: 17-18. [Killin1932]

Killington, F.J. 1936. Further records of Coccidae from Hants. *Journal of the Society for British Entomology* **1**: 119. [Killin1936]

Kim, K.C. 1983. How to detect and combat exotic pests. Pages 261-319 *in*: Wilson, C.L. & Graham, C.L. (Eds.). Exotic Plant Pests and North American Agriculture. New York: Academic Press. 522 pp. [Kim1983]

Kimouilo, H. & Costa, A.S. 1987. [Symptoms of yellow net of grapevine associated with scale insect feeding.]. [In Portuguese]. *Fitopatologia Brasileira* **12(1)**: 104-106. [KimouiCo1987]

Kinawy, M.M. 1991. Biological control of the coconut scale insect (*Aspidiotus destructor* Sign, Homoptera: Diaspididae) in the southern region of Oman (Dhofar). *Tropical Pest Management* **37(4)**: 387-389. [Kinawy1991]

King, G.B. 1899d. Contributions to the knowledge of Massachusetts Coccidae. - IV. *Canadian Entomologist* **31**: 251-255. [King1899d]

King, G.B. 1899h. Two new coccids from Bermuda. *Psyche* **8**: 350. [King1899h]

King, G.B. 1900b. The Coccidae of the ivy. *Canadian Entomologist* **32**: 214-215. [King1900b]

King, G.B. 1903b. Some new records of Coccidae. *Canadian Entomologist* **35**: 191-197. [King1903b]

King, E.G. & Leppla, N.C. 1984. Advances and Challenges in Insect Rearing. New Orleans: Agricultural Research Service, U.S.D.A. 306 pp. [KingLe1984]

King, E.G. & Morrison, R.K. 1984. Some systems for production of eight entomophagous arthropods. Pages 206-222 *in*: King, E.G. & Leppla, N.C. Advances and Challenges in Insect Rearing. New Orleans: Agricultural Research Service, U.S.D.A. 306 pp. [KingMo1984]

Kinjo, M., Nakasone, F., Higa, Y., Nagamine, M., Kawai, S. & Kondo, T. 1996. Scale insects on mango in Okinawa Prefecture. *Proceedings of the Association for Plant Protection of Kyushu* **42**: 125-127. [KinjoNaHi1996]

Kiritchenko, A.N. 1929. A key for identification of scales injurious to citrus. *Subtropics* **5-6**: 170-174. [Kiritc1929]

Kiritchenko, A.N. 1931. [Second contribution to the coccid fauna of Ukraine and Crimea.]. [In Ukrainian]. *Zashchita Rastenii. Leningrad* **7**: 307-321. [Kiritc1931]

Kiritchenko, A.N. 1932a. [Suborder Coccoidea. Scale insects.]. [In Ukrainian]. *Trudy po Zashchite Rastenii, Series Entomologya* **5**: 63-66. [Kiritc1932a]

Kiritchenko, A.N. 1938b. [The San José scale in conditions of USSR. Pages 222-257 *in:* Kiritchenko, A.N. (Ed.), Works of Quarantine Laboratories.] [In Russian]. Leningrad, Sel'khozgiz. 272 pp. [Kiritc1938b]

Kiritchenko, A.N. 1940. [Third report on the coccid fauna of USSR.]. [In Ukrainian]. *Trudy Zoologicheskogo Instituta Akademii Nauk SSSR* **1940**: 115-137. [Kiritc1940]

Kiriukhin, G. 1947. Note complémentaire sur deux coccides nuisibles au pistacier. *Ministère de l'Agriculture, Entomologie et Phytopathologie Appliquées. Tehran* **3**: 1 p. (Persian); p. 21 (French). [Kiriuk1947]

Kiriukhin, G. & Taghi-Zadeh, F. 1947. Les coccides du genre *Aonidiella* dans les régions subtropicales du nord de l'Iran. *Ministère de l'Agriculture, Entomologie et Phytopathologie Appliquées. Tehran* **3**: 4 pp. (Persian); p. 19 (French). [KiriukTa1947]

Kirkaldy, G.W. 1902. Hemiptera. *Fauna Hawaiiensis* **3**: 93-174. [Kirkal1902]

Kirkaldy, G.W. 1905. Bibliographical notes on the Hemiptera [Fam. Coccidae]. *The Entomologist* **38**: 78. [Kirkal1905]

Kislyi, A. 1965. [The yellow pear scale.]. [In Russian].*Zashchita Rastenii of Vreditelei i Boleznei* **6**: 31-32. [Kislyi1965]

Kitchin, R.M. 1970. A radiation analysis of a *Comstockiella* chromosome system: destruction of heterochromatic chromosomes during spermatogenesis in *Parlatoria oleae* (Coccoidea: Diaspididae). *Chromosoma. Berlin* **31**: 165-197. [Kitchi1970]

Klein, H.Z. 1935. On the biology of the red scale (*Chrysomphalus aurantii* Mask.) in the Jordan Valley. *Hadar* **8(3)**: 71-73; 8(4): 115-116. [Klein1935]

Klein, H.Z. 1940. Citrus pests in Palestine. Sifriat Hassadeh Publishers. 215 pp. [Klein1940]

Klett, W. 1963. La lutte biologique contre le pou de San José: Possibilités et avantages. *OEPP/EPPO Public. Ser. A* **34**: 86-91. [Klett1963]

Knight, H. 1932. Some notes on scale resistance and population density. *Journal of Entomology and Zoology* **24(1)**: 1. [Knight1932]

Knight, A.L., Christianson, B.A. & Unruh, T.R. 2001. Impacts of seasonal kaolin particle films on apple pest management. *Canadian Entomologist* **133(3)**: 413-428. [KnightChUn2001]

Knight, R.L. & Alston, F.H. 1974. Pest resistance in fruit breeding. Pages 73-86 *in*: Jones, D.P. & Solomon, M.E., Eds. Biology in Pest and Disease Control. Oxford: Blackwell. [KnightAl1974]

Knowlton, G.F. 1932. Notes on injurious Utah insects -- 1931. *Proceedings of the Utah Academy of Sciences, Arts, and Letters* **9**: 79-83. [Knowlt1932]

Knowlton, G.F. & Smith, C.F. 1936. Rose insects. *Proceedings of the Utah Academy of Sciences, Arts, and Letters* **13**: 263-267. [KnowltSm1936]

Kobakhidze, D.N. 1965. Some results and prospects of the utilization of beneficial entomophagous insects in the control of insect pests in Georgian SSR (USSR). *Entomophaga* **10(4)**: 323-330. [Kobakh1965]

Kocourek, F., Berankova, J. & Streinz, L. 1999. San José scale (*Quadraspidiotus perniciosus* Comst.) male flight pattern monitoring by means of pheromone traps. *IOBC-WPRS Bulletin* **23(7)**: 407-412. [KocourBeSt1999]

Koebele, A. 1890. Report of a trip to Australia made under direction of the entomologist to investigate the natural enemies of the fluted scale. *United States Department of Agriculture, Division of Entomology, Bulletin* **No. 21**: 1-32. [Koebel1890]

Koebele, A. 1893. Studies of parasitic and predaceous insects in New Zealand, Australia, and adjacent islands. Washington, D.C.: US Dept. of Agriculture. 39 pp. [Koebel1893]

Koebele, A. 1907. Parasites and predatory insects forwarded for *Pseudococcus nipae*. *Report of the Hawaii Board of Commissioner of Agriculture and Forestry* 159-164. [Koebel1907]

Koebele, A. & Marsden, J. 1897. Koebele and his Work. *Planters' Monthly* **16**: 65-67. [KoebelMa1897]

Koehler, C.S. 1964. Insect Pest Management Guidelines for California landscape Ornamentals. Oakland, California: Cooperative Extension, University of California, Division of Agriculture and Natural Resources. 82 pp. [Koehle1964]

Koller, F.F.C., Costa, M.K.M. & Corseuil, E. 1997. [Occurrence of scales and their natural enemies in two citrus orchards in Vamo, RS.]. [In Portuguese]. Page 122 *in*: Bento, J.M.S. & Delalibera, I. (Eds.). [16th Brazilian Congress of Entomology] 16o Congresso Brasileiro de Entomologia. Salvador: Sociedade Entomológica do Brasil/EMBRAPA-CNPMF. 400 pp. [KollerCoCo1997]

Komemoto, K.I. 1982. [Spatial distribution of coccoids on shoots of *Miscanthus sinensis*.]. [In Japanese]. *Nanki-Seibutsu* **24**: 26-30. [Komemo1982]

Komosinska, H. 1961. On some scale-insects (Homoptera, Coccoidea) living in greenhouses in Poland. *Fragmenta Faunistica* **9**: 221-232. [Komosi1961]

Komosinska, H. 1964. Armored scale insects (Homoptera, Coccoidea, Diaspididae) of citrus fruits imported to Poland. *Fragmenta Faunistica* **11**: 207-231. [Komosi1964]

Komosinska, H. 1965. *Reclavaspis* gen. n. and *Reclavaspis australicus* sp. n. from Australia (Homoptera, Coccoidea, Diaspididae). *Frustula Entomologica, Pisa (N.S.)* **8**: 1-4. [Komosi1965]

Komosinska, H. 1965a. A new species of *Abgrallaspis* Balach. (Homoptera, Coccoidea, Diaspididae) from greenhouses in Poland. *Frustula Entomologica, Pisa (N.S.)* **8**: 1-6. [Komosi1965a]

Komosinska, H. 1968. Investigations on the scale insects (Homoptera, Coccoidea, Diaspididae) living in greenhouses in Poland. Part I. *Polskie Pismo Entomologiczne* **38**: 205-208. [Komosi1968]

Komosinska, H. 1968a. [*Quadraspidiotus gigas* (Thiem et Gerneck) a new species of scale insects (Homoptera Coccoidea) for the fauna of Poland.]. [In Polish]. *Sylwan* **1**: 51-54. [Komosi1968a]

Komosinska, H. 1969. Studies on the genus *Abgrallaspis* Balachowsky, 1948, (Homoptera, Coccoidea, Diaspididae). *Acta Zoologica Cracoviensia* **14**: 43-85. [Komosi1969]

Komosinska, H. 1974a. [Physiographical and ecological investigations upon armored scale insects (Homoptera, Coccoidea, Diaspididae) of Poland.]. [In Polish]. *Zeszyty Naukowe Akademii Rolniczej w Warszawie, Rozprawy Naukowe* **43**: 1-84. [Komosi1974a]

Komosinska, H. 1986. Occurrence of scale insects (Homoptera, Coccoidea) on avenue trees in Warsaw. *Annals of Warsaw Agricultural University* **No. 20**: 3-12. [Komosi1986]

Komosinska, H. 1986a. Occurrence of scale insects (Homoptera) on trees and shrubs of the Warsaw housing settlements. *Annals of Warsaw Agricultural University* **No. 20**: 13-20. [Komosi1986a]

Komosinska, H. 1987. Occurrence of scale insects (Homoptera, Coccoidea) on trees and shrubs of the Warsaw parks. *Annals of Warsaw Agricultural University - SGGW-AR* **(21)**: 95-103. [Komosi1987]

Komosinska, H. 1987a. Occurrence of scale insects (Homoptera, Coccoidea) on trees and shrubs of forests in the Warsaw environs. *Annals of Warsaw Agricultural University - SGGW-AR* **(21)**: 105-116. [Komosi1987a]

Konar, A. & Ghosh, M.R. 1994. Incidence pattern of scale insects on orange, *Citrus reticulate* Blanco in different orchards of Darjeeling district, West Bengal. *Annals of Entomology* **12**: 9-14. [KonarGh1994]

Kondo, T. 2001. Las cochinillas de Colombia (Hemiptera: Coccoidea).]. *Biota Colombiana* **2(1)**: 31-48. [Kondo2001]

Kondo, T. & Kawai, S. 1995. Scale insects (Homoptera: Coccoidea) on mango in Colombia. *Japanese Journal of Tropical Agriculture* **39**: 97-98. [KondoKa1995]

Kondo, T. & Kawai, S. 1995a. The scale insects (Homoptera: Coccoidea) on mango in India. *Japanese Journal of Tropical Agriculture* **39**: 97-98. [Extra Issue 2] [KondoKa1995a]

Koningsberger, J.C. 1908. Tweede Overzicht der Schadelijke en Nuttige Insecten van Java. *Mededeelingen witgaande van het Department van Landbow* **6**: 1-7. [Koning1908]

Konstantinova, G.M. 1976. [Coccids - pests of apple.]. [In Rusian]. *Zashchita Rastenii. Leningrad* **12**: 49-50. [Konsta1976]

Konstantinova, G.M. 1988. [The California scale.]. [In Rusian]. *Zashchita rastenii. Moscow* No. 9: 50-51. [Konsta1988]

Konstantinova, G.M. & Gura, N.A. 1987. Harmful coccids (Homoptera : Coccinea) and their quarantine importance. *Bollettino del Laboratorio di Entomologia Agraria 'Filippo Silvestri'* **43 (Suppl.)**: 161-165. [KonstaGu1987]

Konstantinova, G.M., Kudina, Zh. D. & Derevyanko, N.M. 1985. [An express method.]. [In Rusian]. *Zashchita rastenii. Moscow* No. 4, 29-30. [KonstaKuDe1985]

Konstantinova, G.M., Maximova, V.H. & Kozár, F. 1982. [Integrated protection of groves and orchards from the San José scale *Aspidiotus perniciosus*, fruit pest, USSR.]. [In Rusian]. *Zashchita rastenii. Moscow* **3**: 24-25. [KonstaMaKo1982]

Konstantinova, G.M. & Mordkovich, Ya. B. 1982. [The California scale.]. [In Rusian]. *Zashchita rastenii. Moscow* **No. 7**: 63. [KonstaMo1982]

Korchagin, V.N. 1987. [Diaspidids and coccids.]. [In Rusian]. *Zashchita rastenii. Moscow* **(No. 3)**: 58-59. [Korcha1987]

Korecka, T. 1974. The haemolymph in *Quadraspidiotus ostreaeformis* (Curt.) (Homoptera, Coccoidea, Aspidiotini). *Acta Biologica Cracoviensia Series Zoologia* **17**: 85-93. [Koreck1974]

Koroneos, J. 1934. Les Coccidae de la Grèce surtout du Pélion (Thessalie). I. Diaspinae. Athens. 95 pp. [Korone1934]

Kosztarab, M. 1955. Revision und Ergänzung der in der "Fauna regni Hungariae" angeführten Cocciden. *Annales Historico-Naturales Musei Nationalis Hungarici. Budapest* **6**: 371-385. [Koszta1955]

Kosztarab, M. 1956. Eine neue Schildlaus-art-*Diaspidiotus hungaricus* n. sp. (Coccoidea, Diaspididae) aus Ungarn. *Annales Historico-Naturales Musei Nationalis Hungarici. Budapest* **7**: 435-438. [Koszta1956]

Kosztarab, M. 1963. The armored scale insects of Ohio (Homoptera: Coccoidea: Diaspididae). *Bulletin of the Ohio Biological Survey* **2 (n.s.)**: 120 pp. [Koszta1963]

Kosztarab, M. 1987. Everything unique or unusual about scale insects (Homoptera: Coccoidea). *Bulletin of the Entomological Society of America* **2**: 215-220. [Koszta1987]

Kosztarab, M. 1990. 3.1.2 Economic Importance. Pages 307-311 *in*: Rosen, D. (Ed.). Armored Scale Insects, Their Biology, Natural Enemies and Control [Series title: World Crop Pests, Vol. 4B]. Amsterdam, the Netherlands: Elsevier. 688 pp, [Koszta1990]

Kosztarab, M. 1996. Scale insects of Northeastern North America. Identification, biology, and distribution. Martinsburg, Virginia: Virginia Museum of Natural History. 650 pp. [Koszta1996]

Kosztarab, M., Ben-Dov, Y. & Kosztarab, M. 1986. An annotated list of generic names of the scale insects (Homoptera: Coccoidea) Third Supplement. *Virginia Polytechnic Institute & State University, Blacksburg, Agricultural Experiment Station Bulletin* **86-2**: 1-34. [KosztaBeKo1986]

Kosztarab, M. & Kosztarab, M.P. 1988. A selected bibliography of the Coccoidea (Homoptera) Third Supplement (1970-1985). *Virginia Polytechnic Institute and State University, Agricultural Experiment Station Bulletin* **88-1**: 1-252. [KosztaKo1988MP]

Kosztarab, M. & Kozár, F. 1978. Scale insects - Coccoidea. *Fauna Hungariae, Akadémiai Kiadó, Budapest* **17**: 1-192. [KosztaKo1978]

Koteja, J. 1990b. 1.3. Developmental Biology and Physiology; 1.3.1. Embryonic Development; Ovipary and Vivipary. Pages 233-242 *in*: Rosen, D. (Ed.). Armored Scale Insects, their Biology, Natural Enemies and Control [Series title: World Crop Pests, Vol. 4A]. Amsterdam, The Netherlands: Elsevier. 384 pp. [Koteja1990b]

Koteja, J. 1990c. 1.3.2. Life history. Pages 243-254 *in*: Rosen, D. (Ed.). Armored Scale Insects, their Biology, Natural Enemies and Control [Series title: World Crop Pests, Vol. 4A]. Amsterdam, The Netherlands: Elsevier. 384 pp. [Koteja1990c]

Koteja, J. 2000a. [Scale insects (Hemiptera: Sternorryncha: Coccinea).] Czerwce (Hemiptera: Sternorryncha: Coccinea). *Flora i Fauna Pienin -- Monografie Pininskie* **1**: 169-173. [Koteja2000a]

Koteja, J. 2000c. Advances in the study of fossil coccids (Hemiptera: Coccinea). *Polskie Pismo Entomologiczne* **69**: 187-218. [Koteja2000c]

Koteja, J. & Ben-Dov, Y. 2003. Notes on the fossil armoured scale insect *Aspidiotus crenulatus* (Pampaloni) (Hem., Coccoidea, Diaspididae). *Bulletin de la Société Entomologique de France* **108(2)**: 165-166. [KotejaBe2003]

Kotikal, Y.K. & Kulkarni, K.A. 2000. Incidence of insect pests of turmeric (*Curcuma longa* L.) in northern Karnataka, India. *Journal of Spices & Aromatic Crops* **9(1)**: 51-54. [KotikaKu2000]

Kotinsky, J. 1903. The first North American leaf-gall diaspine. *Proceedings of the Entomological Society of Washington* **5**: 149-150. [Kotins1903]

Kotinsky, J. 1907. Insects affecting rubber plants. *Hawaiian Forest and Agriculture* **4(10**:304-308. [Kotins1907]

Kotinsky, J. 1908. Some Coccidae from Singapore collected by F. Muir. *Proceedings of the Hawaiian Entomological Society* **1**: 167-171. [Kotins1908]

Kotinsky, J. 1909. Report of Superintendent of Entomology. *Report (Division of Entomology, Hawaii Board of Agriculture and Forestry)* 97-113. [Kotins1909]

Kotinsky, J. 1910. Coccidae not hitherto recorded from these islands (Hemiptera-Homoptera). *Proceedings of the Hawaiian Entomological Society* **2**: 127-131. [Kotins1910]

Kotinsky, J. 1915. The Bermuda grass *Odonaspis*. *Proceedings of the Entomological Society of Washington* **17**: 101-104. [Kotins1915]

Kotter, E. 1977. Citrus pest surveys to guide pest management in the Jordan Valley. *Hassadeh* **57**: 1805, 1807-1815. [Kotter1977]

Koya, K.M.A., Devasahayam, S., Selvakumaran, S. & Kallil, M. 1996. Distribution and damage caused by scale insects and mealy bugs associated with black pepper (*Piper nigrum* Linnaeus) in India. *Journal of Entomological Research. New Delhi* **20**: 129-136. [KoyaDeSe1996]

Kozár, F. 1990. 3.4.1 Forecasting. Pages 335-340 *in*: Rosen, D. (Ed.). Armored Scale Insects, Their Biology, Natural Enemies and Control [Series title: World Crop Pests, Vol. 4B]. Amsterdam, the Netherlands: Elsevier. 688 pp. [Kozar1990]

Kozár, F. 1990a. 3.4.2 Sampling and census-taking. Pages 341-347 *in*: Rosen, D. (Ed.). Armored Scale Insects, Their Biology, Natural Enemies and Control [Series title: World Crop Pests, Vol. 4B]. Amsterdam, the Netherlands: Elsevier. 688 pp. [Kozar1990a]

Kozár, F. 1990c. 3.9.7 Deciduous fruit trees. Pages 593-602 *in*: Rosen, D. (Ed.). Armored Scale Insects, Their Biology, Natural Enemies and Control [Series title: World Crop Pests, Vol. 4B]. Amsterdam, the Netherlands: Elsevier. 688 pp. [Kozar1990c]

Kozár, F. 1990f. 1.1.5 Zoogeographical considerations. Pages 135-148 *in*: Rosen, D. (Ed.). Armored Scale Insects, Their Biology, Natural Enemies and Control [Series title: World Crop Pests, Vol. 4A]. Amsterdam, the Netherlands: Elsevier. 384 pp. [Kozar1990f]

Kozár, F. 1995b. [Investigations on the morphology of male scale insects belonging to the genus *Quadraspidiotus*.] *Quadraspidiotus* nembe tartozó pajjjzstetü fajok hímjeinek morfológiai vizsgálata (Homoptera: Coccoidea). *Növényvédelmi Tudományos Napok [Plant Protection Science Days]* p. 51. [Kozar1995b]

Kozár, F. 1999. Activity and selectivity of pheromone traps of Diaspididae (Homoptera: Coccoidea). Page 125 *in*: 51st International Symposium on Crop Protection [Kozar1999]

Kozár, F. 1999a. Data to the scale insect (Homoptera: Coccoidea) fauna of the Aggtelek National Park. *The Fauna of the Aggtelek National Park* 137-142. [Kozar1999a]

Kozár, F., Guignard, E., Bachmann, F., Mani, E. & Hippe, C. 1994. The scale insect and whitefly species of Switzerland (Homoptera: Coccoidea and Aleyrodoidea). *Mitteilungen der Schweizerischen Entomologischen Gesellschaft/Bulletin of the Societe Entomologique Suisse* **67(1-2)**: 151-161. [KozarGuBa1994]

Kozár, F. & Hippe, C. 1996. A new species from the genus *Greenisca* Borchsenius, 1948 and additional data on the occurrence of scale insects (Homoptera: Coccoidea) in Switzerland. *Folia Entomologica Hungarica* **62**: 91-96. [KozarHi1996]

Kozár, F., Hippe, C. & Mani, E. 1996. Morphometric analyses of the males of *Quadraspidiotus* species (Hom., Diaspididae) found in European orchards or their vicinity . *Journal of Applied Entomology* **120**: 433-437. [KozarHiMa1996]

Kozár, F., Hippe, C. & Mani, E. 1997. [Key to determine *Quadraspidiotus* (Homoptera: Coccoidea) males and to distinguish them from males of other genera.] Kulcs a *Quadraspidiotus* (Homoptera: Coccoidea) pajzstetu nembe tartozo himek meghatarozasahoz. *Növényvédelem* **33**: 321-327. [KozarHiMa1997]

Kozár, F. & Kienitz, F.K. 1979. The occurrence of scale insects (Homoptera, Coccoidea) in tropical fruit shipments. *Növényvédelem* **15**: 246-250. [KozarKi1979]

Kozár, F. & Konstantinova, G.M. 1981a. San José scale in deciduous fruit orchards of some European countries (Survey of scale insect infestations in European orchards No. IV). *EPPO (European and Mediterranean Plant Protection Organization) Bulletin* **11**: 127-133. [KozarKo1981a]

Kozár, F. & Konczné Benedicty, Z. 1996. [New data on the swarming of San José scale [*Quadraspidiotus perniciosus* (Comstock, 1881) Hom.: Coccoidea] males and scale parasitoids.]. *Növényvédelem* **32(10)**: 499-506. [KozarKo1996]

Kozár, F. & Matile-Ferrero, D. 1983. Two new species of armoured scale insects from Hungary (Homoptera, Coccoidea: Diaspididae). *Acta Zoologica Academiae Scientiarum Hungaricae* **29**: 389-395. [KozarMa1983]

Kozár, F. & Walter, J. 1985. Check-list of the Palaearctic Coccoidea (Homoptera). *Folia Entomologica Hungarica* **46**: 63-110. [KozarWa1985]

Kozarzhevskaya, E.F. & Konstantinova, G.M. 1972. Records on pseudo-Californian scale. *Akademii Nauk SSSR Glav. Bot. Sad. Zashchita Rastenii Vreditelei i Bodez.* **1**: 42-46. [KozarzKo1972]

Krambias, A. 1977. Rapport préliminaire sur le pou de Californie (*Aonidiella aurantii*) des citrus a Chypre. *Fruits* **32**: 351-353. [Krambi1977]

Krassilstschik, J. 1893. Zur Entwicklungsgeschichte der Phytophthires. (Über Viviparität mit geschlechtlicher Fortpflanzung bei den Cocciden.). *Zoologischer Anzeiger. Jena* **16**: 69-76. [Krassi1893]

Krause, G. 1950. Erkennung der San José-schildlaus und anderer deckelschildläuse auf einheimischem und importiertem obst. *Zeitschrift für Pflanzenbrau und Pflanzenschutz* **1950**: 1-36. [Krause1950]

Krishnamoorthy, A. & Rajagopal, D. 1995. Comparative toxicity of pesticides to natural enemies of California red scale, *Aonidiella aurantii* Maskell). *Pest Management in Horticultural Ecosystems* **1**: 71-79. [KrishnRa1995]

Krishnamoorthy, A. & Rajagopal, D. 1998. Effect of insecticides on the California red scale, *Aonidiella aurantii* (Maskell) and its natural enemies. *Pest Management in Horticultural Ecosystems* **4**: 2, 83-88. [KrishnRa1998]

Krishnamoorthy, A. & Rajagopal, D. 1999. Selection of host for mass production of California red scale. *Insect Environment* **5(2)**: 86-87. [KrishnRa1999]

Krnjajic, S., Injac, M., Peric, P., Dulic, K., Stamenov, M. & Graora, D. 1993. [Monitoring the flight of apple pests with pheromones.] Pracenje leta stetocina jabuke feremonima. *Zastita bilja* **44(1)**: 63-71. [KrnjajInPe1993]

Krüger, L. 1899. Insektenwanderungen zwischen Deutschland und den Vereinigten Staaten Berlin: Stettin. 171+ pp. [Kruger1899]

Krzysztofowicz, A. 1957. L'addition a la faune d'aspidiotini de la Pologne (Homoptera, Coccoidea, Aspidiotini). *Zeszyty Naukowe Uniwersytetu Jagiellonskiego* **10**: 223-240. [Krzysz1957]

Ksiazkiewica, M. 1980. Ultrastructure of the trophic chamber and nutritive cord of *Aspidiotus hederae* (Homoptera, Coccoidea). *Cell and Tissue Research* **213**: 149-157. [Ksiazk1980]

Ksiazkiewica-Kapralska, M. 1984. Ultrastructure of the imaginal oenocytes in *Aspidiotus hederae* (Homoptera: Coccoidea). *Acta Biologica Cracoviensia Series Zoologia* **26**: 1-6. [Ksiazk1984]

Kuhmina, A.V. 1970. *Ceroplastes rusci* L., *Quadraspidiotus perniciosus* Comst., *Pseudococcus citriculus* Green, *Lopholeucaspis japonica* Ckll., *Unaspis yanonensis* Kuw., *Unaspis citri* Comst. Pages 45-96 *in*: Shutova, N.N., Ed. Handbook of Quarantine and Other Dangerous Pests, Diseases and Weedy Plants. Moscow: Kolos. [Kuhmin1970]

Kuwana, S.I. 1901a. Illustrated treatise on scale insects. *Insect World* **5**: 1-98. [Kuwana1901a]

Kuwana, S.I. 1902. Coccidae (scale insects) of Japan. *Proceedings of the California Academy of Sciences* **3**: 43-98. [Kuwana1902]

Kuwana, S.I. 1902a. Coccidae from the Galapagos Islands. *Journal of the New York Entomological Society* **10**: 28-33. [Kuwana1902a]

Kuwana, S.I. 1904. The San José scale in Japan. Tokyo: Imperial Agricultural Experiment Station Nishigaha. 33 pp. [Kuwana1904]

Kuwana, S.I. 1907. Coccidae of Japan, I. A synoptical list of Coccidae of Japan with descriptions of thirteen new species. *Bulletin of the Imperial Central Agricultural Experiment Station, Japan* **1**: 177-212. [Kuwana1907]

Kuwana, S.I. 1909. Coccidae of Japan (III). First supplemental list of Japanese Coccidae, or scale insects, with description of eight new species. *Journal of the New York Entomological Society* **17**: 150-158. [Kuwana1909]

Kuwana, S.I. 1909a. Coccidae of Japan (IV). A list of Coccidae from the Bonin Islands (Ogasawarajima), Japan. *Journal of the New York Entomological Society* **17**: 158-164. [Kuwana1909a]

Kuwana, S.I. 1917a. A check list of the Japanese Coccidae. Pages 163-182 *in*: K. Nagano (ed.). A collection of essays for Mr. Yasushi Nawa, written in commemoration of his 60 birthday. Japan: Gifu. [Kuwana1917a]

Kuwana, S.I. 1927. A list of Coccidae (scale insects) known from China. *Lingnaam Agricultural Review* **4**: 70-72. [Kuwana1927]

Kuwana, S.I. 1931b. Scale insects of Amami-Oshima, with descriptions of four new species. *Dobutsugaku Zasshi (Journal of the Zoological Society of Japan). Tokyo* **43**: 163-171. [Kuwana1931b]

Kuwana, S.I. 1932. Two new conifer-infesting scale insects from Japan. *Philippine Journal of Science* **48**: 51-54. [Kuwana1932]

Kuwana, S.I. 1933. The diaspine Coccidae of Japan, VII. *Scientific Bulletin (Ministry of Agriculture and Forestry, Japan)* **3**: 1-42. [Kuwana1933]

Kuwana, S.I. 1933a. Key to genera of the diaspine Coccidae of Japan. *Scientific Bulletin (Ministry of Agriculture and Forestry, Japan)* **3**: 43-45. [Kuwana1933a]

Kuwana, S.I. & Muramatsu, K. 1931. A new scale insect of the genus *Aspidiotus* found on *Cymbidium faberi* from China. *Journal of Plant Protection, Nippon Plant Protection Society (Byochugai Zasshi)* **18**: 335-338. [KuwanaMu1931]

Kuwana, S.I. & Muramatsu, K. 1931a. New scale insects and white fly found upon plants entering Japanese ports. *Dobutsugaku Zasshi (Journal of the Zoological Society of Japan). Tokyo* **43**: 647-660. [KuwanaMu1931a]

Kuwana, S.I. & Muramatsu, K. 1932a. Three new species of the diaspine coccids from Japan. *Journal of Plant Protection, Nippon Plant Protection Society (Byochugai Zasshi)* **19**: 95-100. [KuwanaMu1932a]

Kuznetsov, N.N. 1976. [A new species of armored scale insects (Homoptera, Coccoidea, Diaspididae) from the Crimea.]. [In Russian]. *Entomologicheskoe Obozrenye* **55**: 77-79. [Kuznet1976]

Kyparissoudas, D.S. 1987. The occurrence of *Encarsia perniciosi* in areas of northern Greece as assessed by sex pheromone traps of its host *Quadraspidiotus perniciosus*. *Entomologia Hellenica* **5**: 7-12. [Kypari1987]

Kyparissoudas, D.S. 1987a. Flight of San José scale *Quadraspidiotus perniciosus* males and time of crawler appearance in orchards of northern Greece. *Entomologia Hellenica* **5(2)**: 75-80. [Kypari1987a]

Kyparissoudas, D.S. 1990. Determination of spray dates of the control of the first generation of *Quadraspidiotus perniciosus* in Northern Greece. *Entomologia Hellenica* **8**: 5-9. [Kypari1990]

Labanowski, G. 1999. [Scale insects - dangerous pests of *Dracaena*.] Tarczniki - grozne szkodniki draceny. *Ochrona Roslin* **43(6)**: 14-15. [Labano1999]

Lacroix, D.S. 1926. Miscellaneous observations on a cranberry scale *Targionia dearnessi* (Ckll.) (Homop.: Coccidae). *Entomological News* **37**: 249-251. [Lacroi1926]

Lagowska, B. 1990. *Rhizaspidiotus canariensis* (Lindinger) (Homoptera Diaspididae) new to the Polish fauna. *Polskie Pismo Entomologiczne [Bulletin Entomologique]* **60**: 261-264. [Lagows1990]

Lagowska, B. 1995. The biological control perspective of scale insects (Homoptera, Coccinea) on ornamental plants in glasshouses. *Wiadomosci Entomologiczne* **14**: 5-10. [Lagows1995]

Lagowska, B. 1995a. [San José scale - a dangerous quarantine pest.] Tarcznik niszczyciel, grozny szkodnik kwarantannowy. *Ochrona Roslin* **39**: 2, 8. [Lagows1995a]

Lagowska, B. 1998a. [The occurrence of scale insects (Homoptera, Coccinea) on trees and shrubs in urban habitat.] Wystepowanie czerwców (Homoptera, Coccinea) na drzewach i krzewach w srodowisku miejskim. Pages 63-71. *in*: Barczak, T. & Indykiewicz, P. (Eds.). [Urban fauna.] Fauna miast. Bydgoszcz: Wyd. ATR. [Lagows1998a]

Lagowska, B. 2002a. [New data on the occurrence of *Diaspidiotus alni* (Marchal, 1909) (Hemiptera: Coccoidea: Diaspididae) in Poland.] [In Polish.] *Wiadomosci Entomologiczne* **20(3-4)**: 184. [Lagows2002a]

Lagowska, B. & Filipowicz, A 1995. Occurrence of scale insects (Homoptera, Coccinea) on pot ornamental plants in greenhouses. *Materialy Ogolnopolskiej Konferencji Naukowej "Nauka Praktyce Ogrodniczej" Akademia Rolnicza (Lublin)* 375-378. [LagowsFi1995]

Lagowska, B. & Golan, K. 1998. [The harmfulness of scale insects in orchards.] Szkodliwosc czerwcow w sadach. *Ochrona Roslin* **42**: 6, 21-23. [LagowsGo1998]

Lagowska, B. & Koteja, J. 1996. [Scale insects (Homoptera, Coccinea) of Roztocze.] Czerwce (Homoptera, Coccinea) Roztocza. *Fragmenta Faunistica* **39**: 29-42. [LagowsKo1996]

Laing, F. 1925a. Descriptions of some new genera and species of Coccidae. *Bulletin of Entomological Research* **16**: 51-66. [Laing1925a]

Laing, F. 1927. Coccidae, Aphididae and Aleyrodidae. *Insects of Samoa. London* **2**: 35-45. [Laing1927]

Laing, F. 1929. Report on Australian Coccidae. *Bulletin of Entomological Research* **20**: 15-37. [Laing1929]

Laing, F. 1929a. Descriptions of new, and some notes on old species of Coccidae. *Annals and Magazine of Natural History* **4**: 465-501. [Laing1929a]

Laing, F. 1931. A new coccid injurious to fruit trees in Baluchistan. *Memoirs of the Department of Agriculture in India, Entomology Series* **12**: 99-100. [Laing1931]

Laing, F. 1932. On a small collection of Coccidae from the Belgian Congo. *Revue de Zoologie et de Botanique Africaine* **23**: 61-69. [Laing1932]

Laing, F. 1933. The Coccidae of New Caledonia. *Annals and Magazine of Natural History* **11**: 675-678. [Laing1933]

Lale, N.E.S. 1998. Neem in the Conventional Lake Chad Basin area and the threat of Oriental yellow scale insect (*Aonidiella orientalis* Newstead) (Homoptera: Diaspididae). *Journal of Arid Environments* **40**: 2, 191-197. [Lale1998]

Lali, T.S. & Muniappan, R. 1996. Distribution of the red coconut scale, *Furcaspis oceanica* Lindinger (Homoptera : Diaspididae) and its introduced parasitoid, *Adelencyrtus oceanicus* Doutt (Hymenoptera : Encyrtidae) in Guam. *Journal of Biological Control* **10**: 15-20. [LaliMu1996]

Lanchester, H.P. & Dean, F.P. 1957. Control of San José scale on fruit trees during the prebloom period of pears. *Journal of Economic Entomology* **50(1)**: 14-15. [LancheDe1957]

Landi, S. & Del Bene, G. 1994. Osservazioni bio-ecologiche su *Aonidia lauri* (Bouché) (Homoptera Diaspididae) [Bio-ecological observations on *Aonidia lauri* (Bouché)]. *Redia* **77**: 33-45. [LandiDe1994]

Langford, G.S. & Cory, E.N. 1939. Common insects of lawns, ornamental shrubs and shade trees. *Bulletin (University of Maryland Extension Service)* **84**: 3-54. [LangfoCo1939]

Laporte, M.L. 1947. Les parasites du pou de San José en Algérie. *Annales de l'Institut Agricole d'Algérie* **2**: 209-212. [Laport1947]

Laporte, M.L. 1948. Hôtes nouveaux d'*Aspidiotus citrinus* How. (Hym. Chalcididae) parasite de cochenilles en Algérie. *Revue de Pathologie Végétale et d'Entomologie Agricole de France* **27(1)**: 35-37. [Laport1948]

Laporte, M.L. 1949. Les parasites de *Chrysomphalus ficus* Ashm. (Hom. Coccoidea) en Algérie. *Revue de Pathologie Végétale et d'Entomologie Agricole de France* **28**: 150-158. [Laport1949]

Larew, H.G. 1990. 1.4.2.2 Gall formation. Pages 293-300 *in*: Rosen, D. (Ed.). Armored Scale Insects, Their Biology, Natural Enemies and Control [Series title: World Crop Pests, Vol. 4A]. Amsterdam, The Netherlands: Elsevier. 384 pp. [Larew1990]

Larter, L.N.H. 1937. Report of the Government Botanist. *Annual Report of the Department of Science and Agriculture (Jamaica)* 65-72. [Larter1937]

Lawson, J.H. & Hassell, M.P. 1984. Interspecific competition in insects. Pages 451-495. *in*: Huffaker, C.B. & Rabb, R.L. (Eds.). Ecological Entomology. New York: Wiley-Interscience. 844 pp. [LawsonHa1984]

Lawson, P.B. 1917. Scale insects injurious to fruit and shade trees. The Coccidae of Kansas. *Bulletin of The University of Kansas. Biological Series* **18**: 161-279. [Lawson1917]

Le Pelley, R.H. 1959. Agricultural insects of East Africa. Pages 33-48. Nairobi, Kenya: East Africa High Comm.. 307 pp. [LePell1959]

Lee, H.S. & Wen, H.C. 1977. [Seasonal occurrence and control of the papaya red scale, *Aonidiella inornata* McKenzie (Homoptera: Diaspididae).]. [In Chinese]. *Plant Protection Bulletin* **19(3)**: 196-201. [LeeWe1977]

Leefmans, S. 1929. [Diseases and pests of cultivated plants in the Netherlands-Indies in 1928.] Ziekten en plagen der cultuurgewassen in Nederlandsch-indie in 1928. *Mededeelingen van het Intituut vor Plantenziekten* **No. 75**: 1-96. [Leefma1929]

Lelláková-Dusková, F. 1959. Männchen der Schildläuse *Diaspidiotus wünni* (Lindinger) (Coccoidea, Diaspididae). *Casopis Ceské (Ceskoslovenské) Spolecnosti Entomologické (Acta Societatis Entomologicae Bohemiae)* **56**: 338-342. [Lellak1959]

Lelláková-Dusková, F. 1963. The morphology, metamorphosis, and the life-cycle of the scale insects *Quadraspidiotus gigas* (Thiem et Gerneck) (Homoptera, Coccoidea, Diaspididae). *Acta Entomologica Musei Nationalis Pragae* **35**: 611-648. [Lellak1963]

Lelláková-Dusková, F. 1965. *Quadraspidiotus marani* Zahr., the morphology of developmental stages in the male (Homoptera, Coccoidea, Diaspididae). *Acta Entomologica Bohemoslovaca* **62**: 202-209. [Lellak1965]

Lelláková-Dusková, F. 1966. Die auf Weidengewächsen (Salicales) lebenden Schildläuse (Coccoidea) in der Tschechoslowakei. *Zeitschrift für Angewandte Entomologie* **57**: 294-309. [Lellak1966]

Lenteren, J.C. van 1994. Oviposition behavior of *Aphytis* (Hymenoptera, Chalcidoidea, Aphelinidae), parasitoids of armored scale insects (Homoptera, Coccoidea, Diaspididae). Pages 13-39 *in*: Rosen, D., ed. Advances in the Study of Aphytis (Hymenoptera: Aphelinidae). Andover, UK: Intercept Limited. [Lenter1994]

Lenteren, J.C. van & DeBach, P. 1981. Host discrimination in three ectoparasites (*Aphytis coheni, A. lingnanensis* and *A. melinus*) of the oleander scale (*Aspidiotus nerii*). *Netherlands Journal of Zoology* **31**: 504-532. [LenterDe1981]

Leonardi, G. 1897. Monografia de genere *Aspidiotus* (nota preventiva). *Rivista di Patologia Vegetale. Firenze* **5**: 283-286. [Leonar1897]

Leonardi, G. 1897a. Intorno al genere *Aspidiotus*. *Rivista di Patologia Vegetale. Firenze* **5**: 375. [Leonar1897a]

Leonardi, G. 1897b. Generi e specie diaspiti. Saggio di sistematica degli *Aspidiotus*. *Rivista di Patologia Vegetale. Firenze* **6**: 102-134. [Leonar1897b]

Leonardi, G. 1898. Monografia del genere *Mytilaspis*. Nota preventiva. *Rivista di Patologia Vegetale. Firenze* (1897) **6**: 45-47. [Leonar1898]

Leonardi, G. 1898a. Generi e specie di diaspiti. Saggio di sistematica degli *Aspidiotus*.. *Rivista di Patologia Vegetale. Firenze* **6**: 48-78. [Leonar1898a]

Leonardi, G. 1898c. Generi e specie di diaspiti saggio di sistematica degli *Aspidiotus*.. *Rivista di Patologia Vegetale. Firenze* **7**: 38-86. [Leonar1898c]

Leonardi, G. 1899. Generi e specie di diaspiti. Saggio di sistematica degli *Aspidiotus*.. *Rivista di Patologia Vegetale. Firenze* **7**: 173-225. [Leonar1899]

Leonardi, G. 1900. Generi e specie di diaspiti. Saggio di sistematica degli *Aspidiotus*. *Rivista di Patologia Vegetale. Firenze* **8**: 298-363. [Leonar1900]

Leonardi, G. 1903. Generi e specie di diaspiti. Saggio di sistematica delle *Mytilaspides*. *Annali della R. Scuola Superiore di Agricoltura. Portici* **5**: 1-114. [Leonar1903]

Leonardi, G. 1903a. Generi e specie di diaspiti. Saggio di sistematica delle parlatoriae. *Annali della R. Scuola Superiore di Agricoltura. Portici* **5**: 1-59. [Leonar1903a]

Leonardi, G. 1906. Diagnosi di coccinglie nuove. *Redia* **3**: 1-7. [Leonar1906]

Leonardi, G. 1907. Notizie sopra alcune cocciniglie dell'isola di giava raccolte dal Prof. O. Penzig. *Annali della R. Scuola Superiore di Agricoltura. Portici* **7**: 1-22. [Leonar1907]

Leonardi, G. 1907a. Notizie sopra una cocciniglia degli agrumi nuova per l'Italia (*Aonidiella aurantii* Mask.). *Bollettino del Laboratorio di Zoologia Generale e Agraria della R. Scuola Superior Agricoltura. Portici* **1**: 117-134. [Leonar1907a]

Leonardi, G. 1907b. Contribuzione alla conoscenza delle cocciniglie Italiane. *Bollettino del Laboratorio di Zoologia Generale e Agraria della R. Scuola Superior Agricoltura. Portici* **1**: 135-169. [Leonar1907b]

Leonardi, G. 1908a. Seconda contribuzione alla conoscenza della cocciniglie Italiane. *Bollettino del R. Laboratorio di Entomologia Agraria di Portici* **3**: 150-191. [Leonar1908a]

Leonardi, G. 1909. Chermotheca Italica...Cocciniglie Raccolte in Italia. Pages 5:101-125. Portici: Vesuviano. 125 pp. [Leonar1909]

Leonardi, G. 1910. Su due cocciniglie dannose agli agrumi di recento introduzione in Italia. *Annali della R. Scuola Superiore di Agricoltura. Portici* **4**: 1-19. [Leonar1910]

Leonardi, G. 1911. Contributo alla conoscenza delle coccinglie della Repúbblica Argentina. *Bollettino del R. Laboratorio di Entomologia Agraria di Portici* **5**: 237-284. [Leonar1911]

Leonardi, G. 1911a. Contributo alla conoscenza delle coccinglie della Repúbblica Argentina. *Bollettino della Real Scuola Superiore di Agricoltura. Portici* (**ser 2**) **10**: 3-50. [Leonar1911a]

Leonardi, G. 1913b. Nuove specie di coccinglie raccolte in Italia. *Bollettino del Laboratorio di Zoologia Generale e Agraria della R. Scuola Superior Agricoltura. Portici* **7**: 59-65. [Leonar1913b]

Leonardi, G. 1913c. Nuove specie di diaspiti viventi sull'olivo. *Bollettino del Laboratorio di Zoologia Generale e Agraria della R. Scuola Superior Agricoltura. Portici* **7**: 66-71. [Leonar1913c]

Leonardi, G. 1914. Contributo alla conoscenza delle coccinglie dell'Africa occidentale e meridionale. *Bollettino del Laboratorio di Zoologia Generale e Agraria della R. Scuola Superior Agricoltura. Portici* **8**: 187-224. [Leonar1914]

Leonardi, G. 1918. Terza contribuzione alla conoscenza delle cocciniglie Italiane. *Bollettino del R. Laboratorio di Entomologia Agraria di Portici* **12**: 188-216. [Leonar1918]

Leonardi, G. 1920. Monografia delle cocciniglie Italiane. Portici: Della Torre. 555 pp. [Leonar1920]

Lepage, H.S. 1938. Catálogo dos coccídeos do Brasil. *Revista do Museu Paulista. São Paulo* **23**: 327-491. [Lepage1938]

Lepage, H.S. 1940. Notas sobre coccoideos do Brasil (com descricao de especie nova) (Homoptera-Coccoidea). *Papéis Avulsos do Departamento de Zoologia, Secretaria da Agricultura. São Paulo* **1**: 69-78. [Lepage1940]

Lepage, H.S. 1942. Descriçao de onze espécies novas de coccídeos do Brasil (Homoptera-Coccoidea). *Archivos do Instituto Biologico. São Paulo* **13**: 173-189. [Lepage1942]

Lepage, H.S. & Figueiredo, E.R. 1947. As pragas das orquidáceas. Ostrensky: Inst. Biol. (*São* Paulo). 49 pp. [LepageFi1947]

Lepage, H.S. & Giannotti, O. 1942. Contribuição para o levantamento fitossanitário do nordeste Brasileiro. Coccoideos assinalados nos estados do Ceara, Paraiba do Norte e Rio Grande do Norte. *Boletim da Sociedade Brasileira de Agron.* **5**: 444-458. [LepageGi1942]

Lepage, H.S. & Giannotti, O. 1943. Notas coccidológicas (Homoptera-Coccoidea). *Arquivos do Instituto Biologico. São Paulo* **14**: 331-350. [LepageGi1943]

Lepage, H.S. & Giannotti, O. 1944. Algumas espécies novas de coccideos do Brasil (Homoptera - Coccoidea). *Arquivos do Instituto Biologico. São Paulo* **15**: 299-306. [LepageGi1944]

Lepesme, P. 1947. Les Insectes des palmiers. Paris: André903 . pp. [Lepesm1947]

Lepiney, J. 1928. Les insectes nuisible du chêne liège dans les forêts du Maroc. *Annales des Epiphyties* **14**: 313-321. [Lepine1928]

Lever, R.J.A.W. 1969. Les Ravageurs du Cocotier. (Études agricoles de la FAO, No. 77). Rome: FAO. U.N. 196 pp. [Lever1969]

Levitin, E. & Cohen, E. 1998. The involvement of acetylcholinesterase in resistance of the California red scale *Aonidiella aurantii* to organophosphorus pesticides. *Entomologia Experimentalis et Applicata* **88**: 115-121. [LevitiCo1998]

Lewis, C. & Campbell, J.D. (Editors). 1958. The Oxford Atlas. Readers Union, Oxford University Press. 96 pp. + xxvi + 92 pp.

Li, C.D. 1996. [Two new species of Aphelinidae (Hymenoptera: Chalcidoidea) from northeastern region of China.]. [In Chinese]. *Journal of Northeast Forestry University* **24**: 98-101. [Li1996]

Li, C. & Liao, D. 1990. [Methods of mass rearing four citrus scale insects.]. [In Chinese]. *Chinese Journal of Biological Control* **6(2)**: 68-70. [LiLi1990]

Li, C.D., Lou, Y.M., Cao, C.W., Yu, C.J. & Yang, X.G. 1997. [Species of *Encarsia* Foerster (Hymenoptera: Aphelinidae) from northeastern China.]. [In Chinese]. *Journal of Forestry Research* **8(4)**: 225-226. [LiLoCa1997]

Lian, J.H. & Tang, Z.Y. 1985. [Experimental control of *Hemiberlesia pitysophila* Takagi with 'Emulsion 82'.]. [In Chinese]. *Forest Science and Technology* **No. 9**: 29-31. [LianTa1985]

Liang, C.F. 1988. [Study status of introduced parasite [*Coccobius* sp.] to control *Hemiberlesia pitysophila* Takagi (Homop.: Diaspididae).]. [In Chinese]. *Natural Enemies of Insects* **10(1)**: 62. [Liang1988]

Liang, T., Ai, S.J. & Zhang, Q.X. 1997. [Bionomics of *Quadraspidiotus perniciosus* (Comstock) and its control.]. [In Chinese]. *Entomological Knowledge* **34**: 152-153. [LiangAiZh1997]

Liang, T., Ai, S.J., Zhang, Q.X., Chen, H.M., Tuo, H.T. & Li, Z.G. 1999. [Study on the biological characteristics of scales, their dominant natural enemies and control on pear trees in Xinjiang.]. [In Chinese]. *China Fruits* **(No. 2)**: 29-30. [LiangAiZh1999]

Liang, G. & Chen, Z. 1990. [A preliminary survey of parasitoid wasps of *Hemiberlesia pitysophila*.]. [In Chinese]. *Natural Enemies of Insects* **12(1)**: 1-6. [LiangCh1990]

Lichtenstein, J. 1881. [*Aspidiotus pyri* Licht.]. *Annales de la Société Entomologique de France* **1**: li-liii. [Lichte1881]

Lidgett, J. 1898. Notes and observations on some Victorian Coccidae. *Wombat* **3**: 80-95. [Lidget1898]

Lidgett, J. 1898a. Description of two new Australian coccids. *Wombat. Geelong* **4**: 13-15. [Lidget1898a]

Lidgett, J. 1902. *Aspidiotus hederae* in Australia. *Entomological News* **13**: 43-45. [Lidget1902]

Liegeois, P. 1944. La culture du cacaoyer au Congo Belge. *Bulletin agricole du Congo belge* **35(1-4)**: 147-173. [Liegeo1944]

Lindblom, A. 1938. [Pests of Sweden.] Skadedjur i Sverige ar. *Statens Vaxtskyddsanstalt* **26**: 1-102. [Lindbl1938]

Lindgren, D.L. 1938. The stupefaction of red scale, *Aonidiella aurantii*, by hydrocyanic acid. *Hilgardia* **11(5)**: 213-225. [Lindgr1938]

Lindgren, D.L. 1941. Factors influencing the results of fumigation of the California red scale. *Hilgardia* **13(9)**: 49-511. [Lindgr1941]

Lindgren, D.L. & Boyce, A.M. 1944. Results with dichloro-diphenyl-tricloroethane in control of the California red scale. *Journal of Economic Entomology* **37(1)**: 123-124. [LindgrBo1944]

Lindgren, D.L. & Dickson, R.C. 1942. Spray-fumigation experiments on California red scale. *Journal of Economic Entomology* **35(6)**: 827-829. [LindgrDi1942]

Lindgren, D.L. & Dickson, R.C. 1945. Repeated fumigation with HCN and the development of resistance in the California red scale. *Journal of Economic Entomology* **38(3)**: 296-299. [LindgrDi1945]

Lindgren, D.L. & Gerhardt, P.D. 1947. The response of California red scale to fumigation with ethylene dibromide and ethylene dibromide-HCN. *Journal of Economic Entomology* **40**: 680-682. [LindgrGe1947]

Lindgren, D.L., LaDue, J.P. & Dickson, R.C. 1945. Laboratory studies with rotenone oil in sprays to control the California red scale. *Journal of Economic Entomology* **38(5)**: 567-572. [LindgrLaDi1945]

Lindgren, D.L., LaDue, J.P. & Dow, D. 1944. Mystery insecticide studies: Remarkable results obtained in DDT experiments on red scale. *Citrus leaves* **24(4)**: 6-7,30. [LindgrLaDo1944]

Lindgren, D.L., LaDue, J.P. & Dow, D. 1944a. Preliminary studies with DDT in control of the California red scale. *California Citrograph* **29(7)**: 180-181. [LindgrLaDo1944a]

Lindgren, D.L. & Sinclair, W.B. 1944. Relation of mortality to amounts of hydrocyanic acid recovered from fumigated resistant and nonresistant citrus scale insects. *Hilgardia* **16(6)**: 303-315. [LindgrSi1944]

Lindinger, L. 1906. Die Schildlausgattung *Leucaspis*. *Jahrbuch der Hamburgischen Wissenschaftlichen Anstalten. Hamburg* (1905) **3**: 1-60. [Lindin1906]

Lindinger, L. 1907. Bestimmungstafel der Deutschen Diaspinen. *Entomologische Blätter* **3**: 4-6. [Lindin1907]

Lindinger, L. 1907a. Betrachtungen über die Cocciden-Nomenklatur. *Entomologisches Wochenblatt. Stuttgart* **24**: 19-20, 22-23. [Lindin1907a]

Lindinger, L. 1908. Nomenklaturbetrachtungen. *Berliner Entomologische Zeitschrift* **52**: 83-95. [Lindin1908]

Lindinger, L. 1908b. Coccidenstudien. I. Zur Systematik der Diaspinen. II. Kritische Notizen. *Berliner Entomologische Zeitschrift* **52**: 96-106. [Lindin1908b]

Lindinger, L. 1908e. Literatur-referate. - Marlatt, C.L. new species of diaspine scale insects. *Entomologische Blätter* **4**: 240-241. [Lindin1908e]

Lindinger, L. 1909. Die Schildlausgattung *Gymnaspis*. *Zeitschrift für Deutsche Entomologen* **1909**: 148-153. [Lindin1909]

Lindinger, L. 1909a. Zwei Lorbeerschädlinge aus der Familie der Schildläuse. *Zeitschrift für Pflanzenkrankheiten* **18**: 321-336. [Lindin1909a]

Lindinger, L. 1909b. Beiträge zur Kenntnis der Schildläuse und ihrer Verbreitung. *Zeitschrift für Wissenschaftliche Insektenbiologie* **5**: 105-110; 147-152; 220-226. [Lindin1909b]

Lindinger, L. 1909c. Bemerkenswerte Schildläuse auf den im Berichtsjahr untersuchten Pflanzen. *Jahrbuch der Hamburgischen Wissenschaftlichen Anstalten. Hamburg* **26**: 448-451, 464-466. [Lindin1909c]

Lindinger, L. 1909d. Die Schildläusgattung *Selenaspidus*. *Jahrbuch der Hamburgischen Wissenschaftlichen Anstalten. Hamburg* **3**: 1-12. [Lindin1909d]

Lindinger, L. 1909e. Afrikanische Schildläuse. I. (Diaspinen aus kamerun.) und II. (Eine der San-José Schildlaus ähnliche südafrikanische Diaspine.). *Jahrbuch der Hamburgischen Wissenschaftlichen Anstalten. Hamburg* **3**: 13-46. [Lindin1909e]

Lindinger, L. 1910. Die Cocciden-Literatur des Jahres 1908. *Zeitschrift für Wissenschaft Insektenbiologie* **6**: 123-124, 151-156, 190-192, 258-262, 332-330. [Lindin1910]

Lindinger, L. 1910a. Die Schildlausgattung *Gymnaspis*. (Hemipt.) II. *Zeitschrift für Deutsche Entomologen* **4**: 437-440. [Lindin1910a]

Lindinger, L. 1910b. Afrikanische Schildläuse. III. (Cocciden des ostlichen Afrikas.). *Jahrbuch der Hamburgischen Wissenschaftlichen Anstalten. Hamburg* **27**: 33-48a. [Lindin1910b]

Lindinger, L. 1910c. Beiträge zur Kenntnis der Schildläuse und ihrer Verbreitung II. *Zeitschrift für Wissenschaftliche Insektenbiologie* **6**: 371-376, 437-441. [Lindin1910c]

Lindinger, L. 1911. Beiträge zur Kenntnis der Schildläuse und ihre Verbreitung II. *Zeitschrift für Wissenschaftliche Insektenbiologie* **7**: 9-12, 86-90, 126-130, 172-177... [Lindin1911]

Lindinger, L. 1911a. Afrikanische Schildläuse. IV. Kanarische Cocciden. Ein Beitrag zur Fauna der Kanarischen Inseln. *Jahrbuch der Hamburgischen Wissenschaftlichen Anstalten* **28**: 1-38. [Lindin1911a]

Lindinger, L. 1912. Nachtrag zu den Beiträgen zur Kenntnis der Schildläuse usw. II. *Zeitschrift für Wissenschaftliche Insektenbiologie* **8**: 31. [Lindin1912]

Lindinger, L. 1912b. Die Schildläuse (Coccidae) Europas, Nordafrikas und Vorder-Asiens, einschliesslich der Azoren, der Kanaren und Madeiras. Stuttgart: Ulmer. 388 pp. [Lindin1912b]

Lindinger, L. 1913. Afrikanische Schildläuse. V. Die Schildläuse Deutsche-ostafrikas. *Jahrbuch der Hamburgischen Wissenschaftlichen Anstalten* **30**: 59-95. [Lindin1913]

Lindinger, L. 1913b. Einige Cocciden aus dem ausserdeutschen Ostafrika. [Supplement to Afrikanische Schildläuse V.]. *Jahrbuch der Hamburgischen Wissenschaftlichen Anstalten. Hamburg* **30**: 96-100. [Lindin1913b]

Lindinger, L. 1914. Die Cocciden-Literatur des Jahres 1909. *Zeitschrift für Wissenschaftliche Insektenbiologie* **10**: 114-120, 155-160, 243-249. [Lindin1914]

Lindinger, L. 1921. Tätigkeitsbericht der Schädlingsabteilung des Instituts für angewandte Botanik zu Hamburg für die Zeit vom 14. Februar bis zum 30. Juni 1920. *Zeitschrift für Angewandte Entomologie* **7**: 424-440. [Lindin1921]

Lindinger, L. 1923. Einführung in die Kenntnis der deutschen Schildläuse. *Entomologisches Jahrbuch* **32**: 138-152. [Lindin1923]

Lindinger, L. 1924. Die Schildläuse der mitteleuropäischen Gewächshäuser. *Entomologisches Jahrbuch* **33/34**: 167-191. [Lindin1924]

Lindinger, L. 1928. Bericht über die Tätigkeit der Abteilung für Pflanzenschutz. A. berwachung der Ein- und Ausfuhr von Obst und Pflanzen (amtliche Pflanzenbeschau.). *Hamburgerischen Institut für Angewandte Botanie. Jahresberichte* : 94-110. [Lindin1928]

Lindinger, L. 1931. Bericht über die Tätigkeit der Abteilung für Pflanzenschutz... A. die berwachung der Ein- und Ausfuhr von Obst, Pflanzen und Pflanzenteilen (amtliche pflanzenbeschau). *Jahresbericht, Institut für Angewandte Botanik. Hamburg* **1930**: 102-125. [Lindin1931]

Lindinger, L. 1931a. Beiträge zur Kenntnis der Schildläuse III. *Entomologische Rundschau. Stuttgart* **48**: 8-180. [Lindin1931a]

Lindinger, L. 1932. Die Synonymie von Walkers Cocciden (Hem.). *Mitteilungen der Deutschen Entomologischen Gesellschaft* **3**: 26-27. [Lindin1932]

Lindinger, L. 1932a. Beiträge zur Kenntnis der Schildläuse III. (Hemipt. Cocc.). *Entomologische Rundschau. Stuttgart* **49**: 79. [Lindin1932a]

Lindinger, L. 1932c. Beiträge zur Kenntnis der Schildläuse (Coccidae). Wissenschaftl. Ballast: MS-Namen ohne Beschreibung u. dgl. *Entomologische Rundschau. Stuttgart* **49**: 203-205. [Lindin1932c]

Lindinger, L. 1932e. Randbemerkungen. *Entomologische Rundschau. Stuttgart* **49**: 222-223. [Lindin1932e]

Lindinger, L. 1932f. Beiträge zur Kenntnis der Schildläuse. *Konowia* **11**: 177-205. [Lindin1932f]

Lindinger, L. 1932g. Beiträge zur Kenntnis der Schildläuse (Hemiptera - Homoptera, Coccidae). *Wiener Entomologische Zeitung* **49**: 217-225. [Lindin1932g]

Lindinger, L. 1933e. Beiträge zur Kenntnis der Schildläuse (Coccidae). Die richtige Benennung von *Aspidiotus niger* Sign. und *Targionia nigra* Sign. *Entomologische Zeitschrift* **47**: 68. [Lindin1933e]

Lindinger, L. 1934. Beiträge zur Kenntnis der Schildläuse (Hemipt. - Homopt., Coccid). *Entomologischer Anzeiger* **14**:15-64. [Lindin1934]

Lindinger, L. 1934c. *Melanaspis eugeniae* sp. nov. aus Porto Rico. (Homopt. Coccoidea). *Entomologische Rundschau. Stuttgart* **51**: 45-46. [Lindin1934c]

Lindinger, L. 1934e. Die Schildlaus-Arten P. Fr. Bouchés und ihre Deutung. *Entomologisches Jahrbuch* **43**: 153-169. [Lindin1934e]

Lindinger, L. 1935. Die nunmehr gültigen Namen der Arten in meinem 'Schildläusebuch' und in den 'Schildläusen der Mitteleuropäischen Gewächshäuser'. *Arbeiten über physiologische und angewandte Entomologie* **44**: 127-149. [Lindin1935]

Lindinger, L. 1936. Neue Beiträge zur Kenntnis der Schildläuse (Coccidae). *Entomologisches Jahrbuch* **45**: 148-167. [Lindin1936]

Lindinger, L. 1936b. Wie neue Schildlaus-arten gemacht werden. *Entomologische Zeitschrift* **50**: 285-286. [Lindin1936b]

Lindinger, L. 1936c. Über einige Schildläuse des Berliner botanischen Gartens (Coccidae). *Arbeiten über physiologische und angewandte Entomologie* **3**: 153-155. [Lindin1936c]

Lindinger, L. 1937. Verzeichnis der Schildlaus-Gattungen. (Homoptera - Coccoidea Handlirsch, 1903). *Entomologischen Jahrbuch* **46**: 178-198. [Lindin1937]

Lindinger, L. 1939. Nachtrag zur Schildlaus-fauna Nordwestdeutschlands. *Bombus* **10**: 37-38. [Lindin1939]

Lindinger, L. 1943a. Die Schildlausnamen in Fulmeks Wirtindex 1943. *Arbeiten über Morphologische und Taxonomische Entomologie aus Berlin-Dahlem* **10**: 145-152. [Lindin1943a]

Lindinger, L. 1943b. Verzeichnis der Schildlaus-Gattungen, 1. Nachtrag. (Homoptera: Coccoidea). *Zeitschrift der Wiener Entomologischen Gesellschaft* **28**: 205-208, 217-224, 264-265. [Lindin1943b]

Lindinger, L. 1949. Einige Mitteilungen über Schildläuse (Homopt., Coccoidea). *Zeitschrift für die Gesam. Insektenkunde Entomologischen Internationaler* **1**: 210-213. [Lindin1949]

Lindinger, L. 1957. Ein weiterer Beitrag zur Synonymie der Cocciden. *Beiträge zur Entomologie. Berlin* **7**: 543-553. [Lindin1957]

Lindner, G. 1954. Zur Zytologie von *Aspidiotus perniciosus* Comst. (Homoptera-Coccidae). *Biologisches Zentralblatt. Erlangen and Leipzig* **73**: 456-458. [Lindne1954]

Linnaeus, C. 1758. Insecta. Hemiptera. Coccus. Systema naturae. X ed. Holmiae: Salvii. 823 pp. [Linnae1758]

Lintner, J.A. 1895. The San José scale, *Aspidiotus perniciosus* and some other destructive scale-insects of the state of New York. *Bulletin of the New York State Museum* **3(13)**: 263-305. [Lintne1895]

Lintner, J.A. 1896. Notes on some of the insects of the year in the state of New York. *Bulletin of the United States Department of Agriculture. Bureau of Entomology. Washington* **6**: 54-61. [Lintne1896]

Liotta, G. 1970. Diffusion des cochenilles des agrumes en Sicile et introduction d'une nouvelle espèce en Sicile occidentale. *Al Awamia* **37**: 33-38. [Liotta1970]

Liotta, G. 1974. Effeti secondari dei piu' comuni fitofarmaci adopterati contro i diaspini degli agrumi in Sicilia su *Aphytis chilensis* (How.) (Hym. Aphelinidae). *Bollettino dell' Istituto di Entomologia Agraria e dell' Osservatorio di Fitopatologia. Palermo* **9**: 175-186. [Liotta1974]

Liotta, G. 1974a. Effetti secondari dei piu' comuni fitofarmaci usati contro i diaspini degli agrumi in Sicilia su *Aspidiotiphagus citrinus* (Craw.) (Hym. - Aphelinidae). *Bollettino dell' Istituto di Entomologia Agraria e dell' Osservatorio di Fitopatologia. Palermo* **9**: 187-194. [Liotta1974a]

Liotta, G. 1974b. Essais d'élevage d'*Aphytis chilensis* How. (Hym. Aphelinidae). *Bulletin OILB/SROP (Sect. Reg. Ouest Paléarctique)* **No. 3**: 83-88. [Liotta1974b]

Liotta, G. 1980. Rapporti tra *Aphytis chilensis* (How.) e *Aspidiotiphagus citrinus* (Craw.) (Hym. Aphelinidae) sull'ospite *Aspidiotus nerii* Bouche (Hom. Diaspididae). *Fruits* **35**: 695-699. [Liotta1980]

Liotta, G., Bissanti, G. & Lombardo, A. 1985. Dinamica di popolazione di *Aspidiotus nerii* Bouche (Hom. Diaspididae) su limone in Sicilia. *Atti XIV Congresso Nazionale Italiano di Entomologia, Palermo, 1985*, pp. 591-598. [LiottaBiLo1985]

Liotta, G., Burgio, S. & Montagna, M.G. 1985. Biocenosi parassitaria di *Aspidiotus nerii* Bouche (Hom. Diaspididae) in Sicilia. *Atti XIV Congresso Nazionale Italiano di Entomologia* 833-840. [LiottaBuMo1985]

Liotta, G. & Maniglia, G. 1974. Valutazione dell'efficacia dei fitofarmaci comunemente adoperati contro *Aspidiotus hederae* (Vall.) (Hom.-Diaspididae) in Sicilia. *Bollettino dell' Istituto di Entomologia Agraria e dell' Osservatorio di Fitopatologia. Palermo* **9**: 207-213. [LiottaMa1974]

Liotta, G. & Mineo, G. 1974. [Studies on the chemical control of *Aonidiella aurantii* (Mask.).]. [In French]. *EPPO (European and Mediterranean Plant Protection Organization) Bulletin* **4(3)**: 381-385. [LiottaMi1974]

Lit, I.L. 1997b. First report of the family Lecanodiaspididae and other new records and notes on Philippine scale insects (Coccoidea, Hemiptera). *Philippine Entomologist* **11**: 87-95. [Lit1997b]

Lit, I.L. & Rimando, L.C. 1993. A review of Velasquez's works on Philippine Diaspididae (Coccoidea, Hemiptera). *Philippine Entomologist* **9(2)**: 163-167. [LitRi1993]

Liu, J.X., Liu, K.Y., Yan, S.C., Lin, T., Chi, D.F., Wu, Q.Y., Li, F.R. & Li, C.S. 1997. [Regulations of the outbreak of *Quadraspidiotus gigas*.]. [In Chinese]. *Journal of Northeast Forestry University* **25**: 5-9. [LiuLiYa1997]

Lizer y Trelles, C.A. 1916a. Un cóccido asiático nuevo para la República Argentina: *"Chrysomphalus dictyospermi pinnulifera"* (Mask.) (Hem. Hom.). *Physis. Buenos Aires* **2**: 177. [Lizery1916a]

Lizer y Trelles, C.A. 1916c. Sobre la presencia del *"Chrysomphalus paulistus"* Hemp. en el delta del Paraná. *Physis. Buenos Aires* **2**: 432-433. [Lizery1916c]

Lizer y Trelles, C.A. 1917b. Une nouvelle variété de *"Chrysomphalus obscurus"* Comst. (Coccidae) *(Chrysomphalus obscurus* var. *lahillei* nov.). *Physis. Buenos Aires* **3**: 242-244. [Lizery1917b]

Lizer y Trelles, C.A. 1936. Algunas cochinillas nuevas para la fauna de la República Argentina. *Physis. Buenos Aires* **12**: 113-116. [Lizery1936]

Lizer y Trelles, C.A. 1939. Catálogo sistemático razonado de los cóccidos (Hom. Sternor) vernáculos de la Argentina. *Physis. Buenos Aires* **17**: 157-210. [Lizery1939]

Lizer y Trelles, C.A. 1942. La colección coccidológica de Pedro Jorgensen. *Notas del Museo de la Plata, Zoologia* **7**: 69-80. [Lizery1942]

Lizer y Trelles, C.A. 1942a. Apuntaciones coccidológicas. I. *Revista de la Sociedad Entomológica Argentina* **11**: 319-335. [Lizery1942a]

Lizer y Trelles, C.A. 1942c. Cochinillas halladas por primera vez en la Argentina (Hom. Sternor.). *Revista de la Sociedad Entomológica Argentina* **11**: 230-236. [Lizery1942c]

Lizer y Trelles, C.A. 1943a. Apuntaciones coccidológicas. I. *Revista de la Sociedad Entomológica Argentina* **11**: 319-335. [Lizery1943a]

Lizer y Trelles, C.A. 1954. A new species of *Duplaspidiotus* (Hom., Coccoidea) found in South America. *Entomologist's Monthly Magazine* **90**: 192-193. [Lizery1954]

Lizzio, S., Siscaro, G. & Longo, S. 1998. Analisi dei principali fattori di mortalita di *Aonidiella aurantii* (Maskell) in agrumeti della Sicilia. *Bollettino di Zoologia Agraria e di Bachicoltura (Milano)* **30**: 2, 165-183. [LizzioSiLo1998]

Llorens Climent, J.M. 1990. Homoptera I, Cocchinillas de los cítricos y su control biológico. Alicante, Spain: Pisa Ediciones. 260 pp. [Lloren1990]

Lo, P.L. & Blank, R.H. 1989. A survey of armoured scale species (Hemiptera: Diaspididae) in kiwifruit orchards. *New Zealand Entomologist* **12**: 1-4. [LoBl1989]

Lobdell, G. H. 1937. Two-segmented tarsi in coccids; other notes (Homoptera). *Annals of the Entomological Society of America* **30**: 75-80. [Lobdel1937]

Lochhead, W. 1900. The San José and other scale insects. *Bulletin of the Ontario Department of Agriculture* **1900**: 1-48. [Lochhe1900]

Logan, D.P. & Thomson, C. 2002. Temperature-dependent development of parasitoids on two species of armoured scale insects. *New Zealand Plant Protection* 55 361-367. [LoganTh2002]

Lombardi, D. 1938. Osservazioni sulla morfologia e biologia della *Targionia vitis* Sign. *Bollettino dell 'Istituto di Entomologia della Università (degli Studi) di Bologna* **10**: 117-138. [Lombar1938]

Lombardo, D.A. & Weedon, A.C. 1986. Photoenolisation of conjugated esters: Synthesis of a San José scale pheromone by partially regio-controlled photochemical deconjugation. *Tetrahedron Letters* **27(46)**: 5555-5558. [LombarWe1986]

Longo, S., Mazzeo, G., Palmeri, V., Benfatto, D., Maurello, S. & Di Leo, A. 2002. New remarks on the distribution and biology of *Aonidiella citrina* (Coquillett) (Hemiptera, Coccoidea) in Italy. *Bollettino di Zoologia Agraria e di Bachicoltura (Milano)* **33(3)**: 508-109. [LongoMaPa2002]

Longo, S., Marotta, S., Pellizzari, G., Russo, A. & Tranfaglia, A. 1995. An annotated list of the scale insects (Homoptera: Coccoidea) of Italy. *Israel Journal of Entomology* **29**: 113-130. [LongoMaPe1995]

Longo, S., Mazzeo, G., Russo, A. & Siscaro, G. 1994a. *Aonidiella citrina* (Coquillet) nuovo parassita degli agrumi in Italia. *Informatore Fitopatologia* **44**: 12; 19-25. [LongoMaRu1994a]

Longo, S., Mazzeo, G., Russo, A. & Siscaro, G. 1995a. Armoured scales injurious to citrus in Italy. *IOBC-WPRS Bulletin* **18**: 126-129. [LongoMaRu1995a]

Longo, S., Mazzeo, G., Russo, A. & Siscaro, G. 1996. Armoured scales injurious to Citrus in Italy. *IOBC-WPRS Bulletin* **18(5)**: 126-130. [LongoMaRu1996]

Longo, S., Russo, A. & Siscaro, G. 1989. Rilievi bio-etologici su *Quadraspidiotus perniciosus* (Homoptera: Diaspididae) in pescheti della Sicilia orientale. *Tecnica Agricola* **41(3)**: 3-11. [LongoRuSi1989]

Lorbeer, H. 1971. Integrated biological control in Fillmore citrus groves. *Citrograph* **April**: 199-201. [Lorbee1971]

Lord, H.V.S. 1922. Care of shade trees: requirements, diseases, insects and their control, and tree surgery. *Altoona Mirror* **August 26**: 77 pp. [Lord1922]

Loucif, Z. & Bonafonte, P. 1977. Observation des population du pou de San José, *Quadraspidiotus perniciosus* Comst. (Hom. Diaspididae) dans la pleine de la Mitidja (Algérie) d'Octobre 1975 a Mai 1976. *Fruits* **32**: 253-261. [LoucifBo1977]

Lounsbury, C.P. 1898. Notes and Studies on the Season's Pests. Pages 35-58. *in*: Report of the Government Entomologist for year 1898. Cape Town, South Africa: W.A. Richards & Sons. 65. [Lounsb1898]

Lounsbury, C.P. 1906. Report of the government entomologist for the year 1906. *Report (Cape of Good Hope, South Africa, Department of Agriculture)* 80-91. [Lounsb1906]

Lounsbury, C.P. 1914. Division of Entomology. Annual report, 1913-14. *Annual report (Division of Entomology, South Africa Union)* 1-26. [Lounsb1914]

Lounsbury, C.P. 1916. Appendix X. Division of Entomology. Annual report, 1915-16. *Annual report (Division of Entomology, South Africa Union)* 83-103. [Lounsb1916]

Lounsbury, C.P. 1921. Report No. V. Division of Entomology. *Annual report (Division of Entomology, South Africa Union)* 35-38. [Lounsb1921]

Lounsbury, C.P. 1922. Entomology. *Annual report (Division of Entomology, South Africa Union)* **No. 9**: 205-210. [Lounsb1922]

Loureiro Ferreira, M. 1992. Uma cochonilha perigosa (*Aonidiella perniciosa* Comst.). *Arquivos da Seccao de Biologia (Universidade de Coimbra. Museu Zoologico)* **2(1)**: 5-24. [Lourei1992]

Love, J.L. & Ferguson, A.M. 1977. Pesticide residues and greedy scale control on kiwifruit. *New Zealand Journal of Agricultural Research* **20**: 95-103. [LoveFe1977]

Löw, F. 1882. Zur Naturgeschichte des *Acanthococcus aceris* Sign. *Wiener Entomologische Zeitung* **1**: 81-85. [Low1882]

Löw, F. 1882c. Der Schild der Diaspiden. *Verhandlungen der Zoologisch-Botanischen Gesellschaft in Wien* **32**: 513-522. [Low1882c]

Lozzia, G.C. 1985. Su alcune cocciniglie rinvenute in Lombardia e zone Limitrofe su piante ornamentali. *Notiziario Malattie delle Piante* **106 (III series, no. 33)**: 122-124. [Lozzia1985]

Lu, Y.D. 1989. [Studies on experimental populations of citrus red scale using life tables and the role of parasites in the control of a natural population.]. [In Chinese]. Pages 207-217 *in*: Studies on the Integrated Management of Citrus Insect Pests. Guangzhou, Guangdong, China: Academic Book & Periodical Press. [Lu1989]

Lu, Y. 1989a. [The probability of interspecific competition between two citrus armoured scales: red scale and circular black scale.]. [In Chinese]. Pages 218-223 *in*: Studies on the Integrated Management of citrus insect pests. Guangzhou, Guangdong, China: Academic Book & Periodical Press. [Lu1989a]

Luck, R.F. 1986. Biological control of California red scale. Pages 69-84. *in*: Ecological Knowledge and Environmental Problem-Solving. Washington, D.C.: National Research Council. [Luck1986]

Luck, R.F. 1986a. The role of scale size in the biological control of California red scale. Pages 355-363. *in*: Cavalloro, R. & di Martino, E., Eds. Integrated Pest Control in Citrus-Groves. Proceedings of the Experts' Meeting. [Luck1986a]

Luck, R.F., Allen, J.C. & Baasch, D. 1980. A systems approach to research and decision-making in the citrus ecosystem. Pages 365-396. *in*: Huffaker, C.B. (Ed.). New Technology of Pest Control. New York: Wiley-Interscience. 500 pp. [LuckAlBa1980]

Luck, R.F., Forster, L.D. & Morse, J.G. 1996. An ecologically based IPM program for citrus in California's San Joaquin Valley using augmentative biological control. *Proceedings of the International Society of Citriculture* **I**: 499-503. [LuckFoMo1996]

Luck, R.F., Jiang, G. & Houck, I.A. 1999. A laboratory evaluation of the astigmatid mite *Hemisarcoptes cooremani* Thomas (Acari : Hemisarcoptidae) as a potential biological control agent for an armored scale, *Aonidiella aurantii* (Maskell) (Homoptera : Diaspididae). *Biological Control* **15(2)**: 173-183. [LuckJiHo1999]

Luck, R.F. & Podoler, H. 1985. Competitive exclusion of *Aphytis lingnanensis* by *A. melinus*; potential role of host size. *Ecology* **66**: 904-913. [LuckPo1985]

Luck, R.F., Podoler, H. & Kfir, R. 1982. Host selection and egg allocation behaviour by *Aphytis melinus* and *A. lingnanensis*: comparison of two facultatively gregarious parasitoids. *Ecological Entomology* **7**: 397-408. [LuckPoKf1982]

Luck, R.F. & Uygun, N. 1986. Host recognition and selection by *Aphytis* species: response to California red, oleander, and cactus scale cover extracts. *Entomologia Experimentalis et Applicata* **40**: 129-136. [LuckUy1986]

Luck, R.F., van den Bosch, R. & Garcia, R. 1977. Chemical insect control - a troubled pest management strategy. *Bioscience* **27(9)**: 606-611. [LuckVaGa1977]

Ludicke, M. 1950. Über biologische Besonderheiten der San José-Schildlaus im Zusammenhang mit der Wirkund von Phosphorsäureestern. *Hofchenbriefe* **3(2)**: 17-32. [Ludick1950]

Lugger, O. 1900. Bugs injurious to our cultivated plants. *Bulletin (University of Minnesota. Agricultural Experiment Station)* **No. 69**: 208-245. [Lugger1900]

Luna-Salas, J.F. & Martinez-Shio, M.E. 1995. Scale insects on citrus in central Tamaulipas, Mexico (Abstract only). *Israel Journal of Entomology* **29**: 265. [LunaSaMa1995]

Lupo, V. 1936. Revisione delle specie di *Aonidiella* Berl. et Leon. del gruppo *A. aurantii* (Mask). *Bollettino del Laboratorio di Zoologia Generale e Agraria della Facolta Agraria in Portici* **29**: 249-261. [Lupo1936]

Lupo, V. 1948. Revisione delle cocciniglie Italiane. VI. (*Aspidiotus, Quadraspidiotus, Diaspidiotus, Rhizaspidiotus, Nuculaspis*). *Bollettino del R. Laboratorio di Entomologia Agraria. Portici* **8**: 137-208. [Lupo1948]

Lupo, V. 1953. Revisione delle cocciniglie Italiane VII. (Gen. *Pseudaonidia, Chrysomphalus, Aspidiotus*). *Bollettino del R. Istituto Superior Agraria Laboratorio de Entomologia Agraria, "Filippo Silvestri"* **12**: 16-44. [Lupo1953]

Lupo, V. 1953a. Revisione delle cocciniglie Italiane. VIII. Diaspinae. Gen. *Hemiberlesia* Cockerell. *Bollettino del R. Istituto Superior Agraria Laboratorio de Entomologia Agraria, "Filippo Silvestri"* **12**: 74-97. [Lupo1953a]

Lupo, V. 1954. Revisione delle cocciniglie Italiane. IX. (*Separaspis, Gonaspidiotus, Chorizaspidiotus*). *Bollettino del R. Istituto Superior Agraria Laboratorio de Entomologia Agraria, "Filippo Silvestri"* **13**: 1-33. [Lupo1954]

Lupo, V. 1954a. Revisione delle cocciniglie Italiane. X. (*Pelomphala, Aonidiella, Comstockiaspis*). *Bollettino del R. Istituto Superior Agraria Laboratorio de Entomologia Agraria, "Filippo Silvestri"* **13**: 34-63. [Lupo1954a]

Lupo, V. 1957. Revisione delle coczinglie Italiane. 11. (Gen. *Targionia, Rhizaspidiotus, Aonidia.*). *Bollettino del Laboratorio di Entomologia Agraria 'Filippo Silvestri'. Portici* **15**: 54-84. [Lupo1957]

Lupo, V. 1957a. Precisazione sulla posizion e rottura delle esuvie larvali delle Diaspinae. *Bollettino dell' Accademia Gioenia di Scienze Nat. Catania* **(ser. 4) 3**: 419-429. [Lupo1957a]

Lurie, S., Fallik, E., Klein, J.D., Kozár, F. & Kovacs, K. 1998. Postharvest heat treatment of apples to control San José scale (*Quadraspidiotus perniciosus* Comstock) and Blue Mold (*Penicillium expansum* Link) and maintain fruit firmness. *Journal of the American Society for Horticultural Science* **123**: 110-114. [LurieFaKl1998]

Lyle, C. 1947. Achievements and possibilities in pest eradication. *Journal of Economic Entomology* **40**: 1-8. [Lyle1947]

Lyne, W.H. 1921. A talk on insects imported from the Orient. *Proceedings of the Entomological Society of British Columbia* **13/15**: 146-148. [Lyne1921]

Ma, H.Q., Chen, X.R. & Zhou, R.Z. 1995. [Homoptera: Coccoidea.]. [In Chinese]. Pages 117-119. *in*: Wu, H., Ed. Insects of Baishanzu Mountain, Eastern China. Beijing: China Forestry Publishing House. 586 pp. [MaChZh1995]

Ma, L., Li, C.D., Liu, J.Q., Sun, Y., Wang, X.L. & Ji, Y.J. 1997. [Predatory function of *Chilocorus kuwanae* Silvestri on *Quadraspidiotus gigas* (Thiem et Gerneck).]. [In Chinese]. *Journal of Northeast Forestry University* **25**: 64-67. [MaLiLi1997]

Ma, L., Li, C.D., Liu, J.Q., Sun, Y., Wang, X.O. & Ji, Y.J. 1997a. [The life course of *Chilocorus kuwanae* Silvestri.]. [In Chinese]. *Journal of Northeast Forestry University* **25**: 59-61. [MaLiLi1997a]

Ma, L. & Lin, T. 2001. Study on the biological characteristics of *Homalotylus flaminius*. *Journal of Forestry Research* **4**: 269-270. [MaLi2001]

Ma, L. & Lin, T. 2002. Bioecology of *Sticholotis cribellata* Sicard (Coleoptera: Coccinellidae), a potential predator of *Melanaspis glomerata* (Green) (Homoptera: Diaspididae). *Journal of Biological Control* **15(1)**: 21-26. [MaLi2002]

Ma, W.L., Shi, Y.Q. & Wan, J.J. 1985. [A preliminary study of *Quadraspidiotus perniciosus* on jujube.]. [In Chinese]. *Plant Protection* **11(1)**: 9-10. [MaShWa1985]

Maber, J., Holland, P.T. & Tomkins, A. 1986. Evaluation of low-volume spraying in kiwifruit. Pages 143-147. *in*: Proceedings of the New Zealand Weed and Pest Control Conference [MaberHoTo1986]

MacGillivray, A.D. 1921. The Coccidae. Tables for the Identification of the Subfamilies and Some of the More Important Genera and Species, together with Discussions . . . Urbana, Ill.: Scarab. 502 pp. [MacGil1921]

Mackie, D.B. 1931. A report of the coccids infesting avocados in California with special reference to *Chrysomphalus dictyospermi* (Morg.). *Bulletin of the California Department of Agriculture* **20**: 419-441. [Mackie1931]

Mackie, D.B. 1933. Entomology. *Monthly Bulletin, California Department of Agriculture* **22**: 457-472. [Mackie1933]

Mackie, D.B. 1934. Entomological service. *Monthly Bulletin, California Department of Agriculture* **23**: 396-418. [Mackie1934]

Mackie, D.B. 1934a. Report of pest treatment committee. *Monthly Bulletin, California Department of Agriculture* **23**: 268-269. [Mackie1934a]

Mackie, D.B. 1935. Entomological service. *Monthly Bulletin, California Department of Agriculture* **24**: 403-430. [Mackie1935]

Mackie, D.B. 1936. Entomological service. *Bulletin of the California Department of Agriculture* **25**: 455-481. [Mackie1936]

Madsen, H.F. & Morgan, C.V.G. 1970. Pome fruit pests and their control. *Annual Review of Entomology* **15**: 295-320. [MadsenMo1970]

Magagula, C.N. & Samways, M.J. 2000. Effects of insect growth regulators on *Chilocorus nigritus* (Coleoptera: Coccinellidae), a non-target natural enemy of citrus red scale, *Aonidiella aurantii* (Homoptera: Diaspididae), in southern Africa: evidence from laboratory and field trials. *African Entomology* **8(1)**: 47-56 [MagaguSa2000]

Mague, D.L. 1982. Biology of the San José scale (*Quadraspidiotus perniciosus* (Comstock)) (Homoptera: Diaspididae) in New York apple orchards; male flight phenology and crawler dispersal. *Dissertation Abstracts International* **(B) 43**: 975. [Mague1982]

Mague, D.L. & Reissig, W.H. 1983. Airborne dispersal of San José scale, *Quadraspidiotus perniciosus* (Comstock) (Homoptera: Diaspididae), crawlers infesting apple. *Environmental Entomology* **12**: 692-696. [MagueRe1983]

Mague, D.L. & Reissig, W.H. 1983a. The phenology of the San José scale (Homoptera: Diaspididae) in New York State apple orchards. *Canadian Entomologist* **115**: 717-722. [MagueRe1983a]

Maheswari, T.U. & Purushotham, K. 1999. Widespread occurrence of scale insect on papaya. *Insect Environment* **5(1)**: 17. [MaheswPu1999]

Mahmood, R. & Mohyuddin, A.I. 1986. Integrated control of mango pests. Islamabad: Pakistan Agricultural Research Council. 11 pp. [MahmooMo1986]

Mahmoud, F.A., Hamdy, M.K. & Hegazi, A.G. 2001. Antimicrobial activity of secretory materials of some scale insects. *Entomologica* **33(1999)**: 441. [MahmouHaHe2001]

Maksimova, V.I. 1973. [Characteristics of the San José scale (*Quadraspidiotus perniciosus* Comst.)]. [In Russian]. *Vestnik Zoologii. Institut Zoologii Akademiya Nauk Ukrainskoi SSR. Kiev* **7**: 81-84. [Maksim1973]

Maksimova, V.I. 1973a. [Colonization of San José scale (*Quadraspidiotus perniciosus* Comst.)]. [In Russian]. *Vestnik Zoologii. Institut Zoologii Akademiya Nauk Ukrainskoi SSR. Kiev* **7**: 78-80. [Maksim1973a]

Maksimova, V.I. 1974. [The effect of the ambient temperature on the development of San José scale (*Quadraspidiotus perniciosus* Comst.)]. [In Russian]. *Vestnik Zoologii. Institut Zoologii Akademiya Nauk Ukrainskoi SSR. Kiev* **No. 2**: 81-82. [Maksim1974]

Maksimova, V.I. 1976. [Sex ratio and female sterility in *Quadraspidiotus perniciosus* Comst.]. [In Russian]. V*estnik Zoologii. Institut Zoologii Akademiya Nauk Ukrainskoi SSR. Kiev* **No. 2**: 76-78. [Maksim1976]

Maksimova, V.I. 1978. [Respiration of the San José scale *Quadraspidiotus perniciosus* (Insecta, Coccoidea)]. [In Russian]. *Vestnik Zoologii. Institut Zoologii Akademiya Nauk Ukrainskoi SSR. Kiev* **No. 5**: 58-61. [Maksim1978]

Maksimova, V.I. 1978a. [Forecasting the numbers of the California scale.]. [In Russian]. *Zashchita Rastenii (Minsk, Byelorussian S.S.R.)* **3**: 46. [Maksim1978a]

Maksimova, V.I. 1991. [Methods of placing traps containing pheromone of the California scale.]. [In Russian]. *Zashchita rastenii. Moscow* **No. 1**: 42. [Maksim1991]

Maksimova, V.I. 1991a. [A food-trap for *Synanthedon myopaeformis*.]. [In Russian]. *Zashchita rastenii. Moscow* **No. 9**: 24. [Maksim1991a]

Malenotti, E. 1916. Specie nouve e critiche di diaspiti. *Redia* **11**: 309-321. [Maleno1916]

Malenotti, E. 1916a. Diaspiti raccolti nella Somalia Italiana Meridionale. *Redia* **11**: 321-358. [Maleno1916a]

Malenotti, E. 1916c. Sulle pretese varietà del *"Chrysomphalus dictyospermi"* (Morg.) Leon. *Redia* **12**: 109-123. [Maleno1916c]

Malenotti, E. 1916d. *Signiphora merceti*, Malen n. sp. *Redia* **12(1)**: 181-182. [Maleno1916d]

Malenotti, E. 1918. I nemici naturali della "Bianca-Rossa" (*Chrysomphalus dictyospermi*).]. *Redia* **113(1/2)**: 17-53. [Maleno1918]

Malenotti, E. 1917. Sopra un caso di endofagia dell'*Aspidiotiphagus citrinus* (Craw.). How. sul *Chrysomphalus dictyospermi* (Morg.) Leon. (In Italian.) *Redia* **12**: 195-196. [Maleno1917]

Malenotti, E. 1927. Il valore pratico dei follicoli nella diagnosi dei communi diaspiti. *Italia Agricola* **64**: 52-55. [Maleno1927]

Malenotti, E. 1946. Possono le pesche diffondere l'*Aspidiotus perniciosus*?. *Rivista della Società Toscana di Orticoltura* **71st year, 31(5/6)**: 3-8. [Maleno1946]

Malipatil, M.B., Dunn, K.L. & Smith, D. 2000. An Illustrated Guide to the Parasitic Wasps Associated with Citrus Scale Insects and Mealybugs in Australia. Victoria: Natural Resources and Environment, Agriculture Resources Conservation Land Management, Agriculture. 152 pp. [MalipaDuSm2000]

Mallamaire, M.A. 1954. [Catalog of the main pest insects, Nematodes, Myriapodes and Acariens of cultivated plants of French West Africa and Togo.]. Pages 24-60. *in*: Inspection Générale de l'Agriculture. Bulletin de la Protection des Végétaux, No. 1/2 Institut Français d'Outre-mer, Marseille. [Mallam1954]

Mallea, A.R., Macola, G.S., Garcia, S.J.G., Bahamondes, L.A. & Suarez, J.H. 1974. Observaciones sobre posturas de *Aspidiotus hederae* (Vallot) (Homoptera - Diaspididae). *Revista de la Facultad de Ciencias Agrarias. Argentina* **20**: 145-147. [MalleaMaGa1974]

Maltby, H.L., Jimenez-Jimenez, E. & DeBach P. 1968. Biological control of armored scale insects in Mexico. *Journal of Economic Entomology* **61**: 1086-1088. [MaltbyJiDe1968]

Malumphy, C. 1997. Laurel scale, *Aonidia lauri* (Bouché) (Homoptera: Coccoidea, Diaspididae), a pest of bay laurel, new to Britain. *Entomologist's Gazette* **48**: 195-198. [Malump1997]

Mamet, J.R. 1936. New species of Coccidae (Hemipt. Homopt.) from Mauritius. *Proceedings of the Royal Entomological Society of London (Ser. B)* **5**: 90-96. [Mamet1936]

Mamet, J.R. 1939b. Some new genera and species of Coccidae (Hemipt. Homopt.) from Mauritius. *Transactions of the Royal Entomological Society of London* **89**: 579-589. [Mamet1939b]

Mamet, J.R. 1941. On some Coccidae (Hemipt. Homopt.) described from Mauritius by de Charmoy. *Mauritius Institute Bulletin. Port Louis* **2**: 23-29. [Mamet1941]

Mamet, J.R. 1941a. Report on a few Coccidae (Homopt.) collected by Mr. P.O. Wiehe in the Chagos Archipelago. *Mauritius Institute Bulletin* **2**: 40. [Mamet1941a]

Mamet, J.R. 1942. On a few Coccidae (Homopt.) recently described from Mauritius. *Proceedings of the Royal Entomological Society of London, Series B: Taxonomy* **11**: 35-37. [Mamet1942]

Mamet, J.R. 1943a. A revised list of the Coccoidea of the islands of the western Indian Ocean, south of the equator. *Mauritius Institute Bulletin. Port Louis* **2**: 137-170. [Mamet1943a]

Mamet, J.R. 1946. On some species of Coccoidea recorded by de Charmoy from Mauritius. *Mauritius Institute Bulletin. Port Louis* **2**: 241-246. [Mamet1946]

Mamet, J.R. 1949. An annotated catalogue of the Coccoidea of Mauritius. *Mauritius Institute Bulletin. Port Louis* **3**: 1-81. [Mamet1949]

Mamet, J.R. 1950. Notes on the Coccoidea of Madagascar - I. *Mémoires de l'Institut Scientifique de Madagascar (Ser. A).* **4**: 17-38. [Mamet1950]

Mamet, J.R. 1951. Notes on the Coccoidea of Madagascar - II. *Mémoires de l'Institut Scientifique de Madagascar (Ser. A)* **5**: 213-254. [Mamet1951]

Mamet, J.R. 1953a. The authorship of the species of Coccoidea (Hemiptera) described from Mauritius in 1899. *Proceedings of the Royal Entomological Society of London (A)* **28**: 149-152. [Mamet1953a]

Mamet, J.R. 1954. Notes on the Coccoidea of Madagascar, III. *Mémoires de l'Institut Scientifique de Madagascar (Ser. E)* **4**: 1-86. [Mamet1954]

Mamet, J.R. 1954a. Miscellaneous notes on the Coccoidea (Homoptera) of the Mascarene Islands and of the Chagos Archipelago. *Mauritius Institute Bulletin. Port Louis* **3**: 260-265. [Mamet1954a]

Mamet, J.R. 1954b. A monograph of the Conchaspididae Green (Hemiptera: Coccoidea). *Transactions of the Royal Entomological Society of London* **105**: 189-239. [Mamet1954b]

Mamet, J.R. 1956. Miscellaneous coccid studies (Homoptera). *Naturaliste Malgache* **8**: 133-141. [Mamet1956]

Mamet, J.R. 1957. On some Coccoidea from Reunion Island (Homoptera). *Mémoires de l'Institut Scientifique de Madagascar (Ser. E)* **8**: 367-386. [Mamet1957]

Mamet, J.R. 1958a. The *Selenaspidus* complex (Homoptera, Coccoidea). *Annales du Musée Royal du Congo Belge. Zoologiques, Miscellanea Zoologica, Tervuren* **4**: 359-429. [Mamet1958a]

Mamet, J.R. 1959a. Notes on the Coccoidea of Madagascar IV. *Mémoires de l'Institut Scientifique de Madagascar (Ser. E)* **11**: 369-479. [Mamet1959a]

Mamet, J.R. 1960. On some Coccoidea from the Comoro archipelago. *Naturaliste Malgache* **12**: 155-162. [Mamet1960]

Mamet, J.R. 1962. Notes on the Coccoidea of Madagascar, V. (Homoptera). *Naturaliste Malgache* **13**: 153-202. [Mamet1962]

Mamet, J.R. 1974. Two new species of Diaspididae from Agalega (Indian Ocean) (Hom. Coccoidea). *Bulletin de la Société Entomologique de France* **79(5-6)**: 166-168. [Mamet1974]

Mani, E., Hippe, C. & Schwaller, F. 1993. [Increased incidence of oystershell scales in fruit orchards.] Vermehrtes Auftreten von Austernschildlausen in Obstanlagen. *Schweizerische Zeitschrift für Obst- und Weinbau* **129**: 12, 299-302. [ManiHiSc1993]

Mani, M. & Krishnamoorthy, A. 1996. *Aonidiella orientalis* (Newstead) (Diaspididae, Homoptera) and its natural enemies found on Sapota, Ber, Custard apple and banana. *Entomon* **21**: 273-274. [ManiKr1996]

Mani, E., Schwaller, F., Baroffio, C. & Hippe, C. 1995. Die San-Jose-Schildlaus in der deutschen Schweiz: Wo stehen wir heute?. *Obst- und Weinbau* **131**: 10, 264-267. [ManiScBa1995]

Mansfield, W. 1920. Insects injurious to economic crops in the Zanzibar protectorate. *Bulletin of Entomological Research* **10(2)**: 145-155. [Mansfi1920]

Maranhao, Z.C. 1946. Nomes vulgares inglêses de insetos. *Revista Agricola (Piracicaba)* **21**: 164-179. [Maranh1946]

Marchal, P. 1904. Sur la biologie du *Chrysomphalus dictyospermi* var. *minor* Berlese, et sur l'extension de cette cochenille dans le Bassin Méditerranéen. *Bulletin de la Société Entomologique de France* **16**: 246-249. [Marcha1904]

Marchal, P. 1909. Sur les cochenilles du midi de la France et de la Corse. *Compte Rendu de l'Academie des Sciences. Paris* **148**: 871-872. [Marcha1909]

Marchal, P. 1909a. Sur les cochenilles de l'Afrique occidentale. *Compte Rendu des Séances de la Société de Biologie* **66**: 586-588. [Marcha1909a]

Marchal, P. 1909b. Sur deux cochenilles nouvelles sur les *Ephedra*. *Bulletin de la Société Zoologique de France* **34**: 59-60. [Marcha1909b]

Marchal, P. 1909c. Cochenilles nouvelles de l'Afrique occidentale. *Bulletin de la Société Zoologique de France* **34**: 68-69. [Marcha1909c]

Marchal, P. 1909d. Contribution à l'étude des coccides de l'Afrique occidentale. *Mémoires de la Societe Zoologique de France* **22**: 165-182. [Marcha1909d]

Marchal, P. 1911. Sur une cochenille nouvelle d'Algérie (Hem. Coccidae). *Bulletin de la Société Entomologique de France* **4**: 71. [Marcha1911]

Marchal, P. 1911a. Sur une nouvelle cochenille cécidogéne. *Bulletin de la Société Zoologique de France* **36**: 150. [Marcha1911a]

Marco, R.I. 1959. Notes on the biological control of pests of agriculture in Chile. *FAO Plant Protection Bulletin* **8(3)**: 25-30. [Marco1959]

Mariau, D. 1998. Hemiptera (Insecta) on oil palm and coconut. *Insect Science and its Application* **18(4)**: 269-277. [Mariau1998]

Mariau, D. & Julia, J.F. 1977. Nouvelles recherches sur la Cochenille du cocotier *Aspidiotus destructor* (Sign.). *Oleagineux* **32(5)**: 217-224. [MariauJu1977]

Mark, E.L. 1877. Beiträge zur Anatomie und Histologie der Pflanzenläuse, insbesondere der Cocciden. *Archiv für Mikroskopische Anatomie* **13**: 31-81. [Mark1877]

Marlatt, C.L. 1899b. *Aspidiotus convexus*, Comst. - a correction. *Canadian Entomologist* **31**: 208-211. [Marlat1899b]

Marlatt, C.L. 1899d. An account of *Aspidiotus ostreaeformis*. *United States Department of Agriculture, Division of Entomology, Bulletin* **20**: 76-82. [Marlat1899d]

Marlatt, C.L. 1899e. Recent work on Coccidae. *Science (n.s.)* **10**: 657-660. [Marlat1899e]

Marlatt, C.L. 1900. *Aspidiotus diffinis*. Another scale insect of probable European origin recently found in North America. *Entomological News* **11**: 425-427. [Marlat1900]

Marlatt, C.L. 1900a. The European pear scale. *Diaspis piricola* (Del Guercio) Saccardo, 1895. *Entomological News* **11**: 590-594. [Marlat1900a]

Marlatt, C.L. 1902. Résumé of the search for the native home of the San José scale of Japan and China. *United States Department of Agriculture, Division of Entomology, Bulletin* **37**: 65-78. [Marlat1902]

Marlatt, C.L. 1902a. The San José scale: its native home and natural enemy. *United States Department of Agriculture Yearbook* **1902**: 155-174. [Marlat1902a]

Marlatt, C.L. 1903. Scale insects and mites on citrus trees. *Farmer's Bulletin (USDA)* **No. 172**: 1-43. [Marlat1903]

Marlatt, C.L. 1906. The San José or Chinese scale. *United States Department of Agriculture, Bureau of Entomology, Technical Series* **62**: 1-89. [Marlat1906]

Marlatt, C.L. 1908. The genus *Pseudaonidia*. *Proceedings of the Entomological Society of Washington* **9**: 131-141. [Marlat1908]

Marlatt, C.L. 1908b. *Aspidiotus ancylus* Putnam vs. *circularis* Fitch. *Entomological News* **19**: 309-311. [Marlat1908b]

Marlatt, C.L. 1908c. New species of diaspine scale insects. *United States Department of Agriculture, Bureau of Entomology, Technical Series* **16**: 11-32. [Marlat1908c]

Marlatt, C.L. 1911. A newly imported scale-pest on Japanese hemlock (Rhynch.). *Entomological News* **22**: 385-387. [Marlat1911]

Marlatt, C.L. 1915. Scale insects and mites on citrus trees. *Farmer's Bulletin (USDA)* No. 172. [Marlat1915]

Marlatt, C.L. 1921a. Pests collected from imported plants and plant products from January 1, 1921, to December 31, 1921, inclusive. *Annual Letter of Information (Federal Horticultural Board, United States Department of Agriculture)* **No. 35**: 1-36. [Marlat1921a]

Marr, V.G. 1949. Versuche über Sommerbekämpfung der San José-Schildlaus. *Hofchen-Briefe* **2(1)**: 18-25. [Marr1949]

Marshall, J. 1952. Applied entomology in the orchards of British Columbia, 1900-1951. *Entomological Society of British Columbia* **48**: 25-31. [Marsha1952]

Marshall, J. 1953. A decade of pest control in British Columbia orchards. *Proceedings of the Entomological Society of British Columbia* **49**: 7-11. [Marsha1953]

Marshall, J. 1958. A decade of concentrate spraying in deciduous orchards. Pages 249-254 *in*: Baecker, E.C., (Ed.). Proceedings of the Tenth International Congress of Entomology. Vol. 3. Ottawa. 895 pp. [Marsha1958]

Martelli, G. 1913. La lotta contro il *Crisonfalo pinnulifero* (*Chrysomphalus dictyospermi* var. *pinnulifera*), volg. Bianca-Rossa, durante il 1912. *Cattedra Ambulante di Agricoltura per la Provincia di Messina* 28 pp. [Martel1913]

Martin, H. 1954. Scale insects on citrus in Tripolitania. *FAO Plant Protection Bulletin* **11(8)**: 113-116. [Martin1954]

Martin, H. 1958. Pests and diseases of date palm in Libya. *FAO Plant Protection Bulletin* **6**: 120-123. [Martin1958]

Martin-Mateo, M.P. 1983. Inventario preliminar de los cóccidos de España. I. Diaspididae. *Graellsia, Revista de Entomólogos Ibéricos. Madrid* **39**: 47-71. [Martin1983]

Martorell, L.F. 1976. Annotated food plant catalog of the insects of Puerto Rico. Agric. Expt. Sta., Univ. Puerto Rico, Dept. Entomo. 303 pp. [Martor1976]

Marucci, P.E. 1966. Insects and their control. Pages 199-235. *in*: Eck, P. & Childers, N.F., Eds. Blueberry Culture. New Brunswick, NJ: Rutgers University Press. 378 pp. [Marucc1966]

Marutani, M. & Muniappan, R. 1989. Incidence of the red coconut scale, *Furcaspis oceanica* [Homoptera: Diaspididae] and its parasites in Micronesia. *Journal of Plant Protection in Tropics* **6(1)**: 61-66. [MarutaMu1989]

Marutani, M. & Muniappan, R. 1990. Use of *Adelencyrtus oceanicus* (Hym.: Encyrtidae) for controlling the red coconut scale, *Furcaspis oceanica* (Hom.: Diaspididae) in Guam. Page 1 p. *in*: FFT-NARC International Seminar: the Use of Parasitoids and Predators to Control Agricultural Pests. Tukuba-gun, Japan: National Agricultural Research Centre (NARC). [MarutaMu1990]

Masi, L. 1934. [Descriptions of some Chalcids from Morocco.] Descrizione di alcuni calcididi del Marocco. *Bollettino della Società Entomologica Italiana. Firenze* **66**: 97-102. [Masi1934]

Maskell, W.M. 1879. On some Coccidae in New Zealand. *Transactions and Proceedings of the New Zealand Institute* **11**: 187-228. [Maskel1879]

Maskell, W.M. 1882. Further notes on Coccidae in New Zealand, with descriptions of new species. *Transactions and Proceedings of the New Zealand Institute* **14**: 215-229. [Maskel1882]

Maskell, W.M. 1884. Art. V. - Further notes on Coccidae in New Zealand, with descriptions of new species. *Transactions and Proceedings of the New Zealand Institute* **16**: 120-144. [Maskel1884]

Maskell, W.M. 1885a. Further notes on Coccidae in New Zealand. *Transactions and Proceedings of the New Zealand Institute* **17**: 20-31. [Maskel1885a]

Maskell, W.M. 1887. Further notes on New Zealand Coccidae. *Transactions and Proceedings of the New Zealand Institute* **19**: 45-49. [Maskel1887]

Maskell, W.M. 1887a. An account of the insects noxious to agriculture and plants in New Zealand, the scale insects (Coccididae). Wellington: State Forests & Agricultural Department. 116 pp. [Maskel1887a]

Maskell, W.M. 1889. On some new South Australian Coccidae. *Transactions of the Royal Society of South Australia (1888)* **11**: 101-111. [Maskel1889]

Maskell, W.M. 1891. Further coccid notes: with descriptions of new species from New Zealand, Australia, and Fiji. *Transactions and Proceedings of the New Zealand Institute (1890)* **23**: 1-36. [Maskel1891]

Maskell, W.M. 1891a. Descriptions of new Coccidae. *Indian Museum Notes* **2**: 59-62. [Maskel1891a]

Maskell, W.M. 1892. Further coccid notes: with descriptions of new species, and remarks on coccids from New Zealand, Australia and elsewhere. *Transactions and Proceedings of the New Zealand Institute* **24**: 1-64. [Maskel1892]

Maskell, W.M. 1893b. Further coccid notes: with descriptions of new species from Australia, India, Sandwich Islands, Demerara, and South Pacific. *Transactions and Proceedings of the New Zealand Institute* **25**: 201-252. [Maskel1893b]

Maskell, W.M. 1894b. Further coccid notes with descriptions of several new species and discussion of various points of interest. *Transactions and Proceedings of the New Zealand Institute* **26**: 65-105. [Maskel1894b]

Maskell, W.M. 1895a. Synoptical list of Coccidae reported from Australia and the Pacific Islands up to December 1894. *Transactions and Proceedings of the New Zealand Institute* **27**: 1-35. [Maskel1895a]

Maskell, W.M. 1895b. Further coccid notes: with description of new species from New Zealand, Australia, Sandwich Islands, and elsewhere, and remarks upon many species already reported. *Transactions and Proceedings of the New Zealand Institute* **27**: 36-75. [Maskel1895b]

Maskell, W.M. 1896. *Aspidiotus perniciosus* Comstock, and *Aonidia fusca* Maskell: A question of identity or variation. *Canadian Entomologist* **28**: 14-16. [Maskel1896]

Maskell, W.M. 1896b. Further coccid notes, with descriptions of new species and discussions of questions of interest. *Transactions and Proceedings of the New Zealand Institute* **28**: 380-411. [Maskel1896b]

Maskell, W.M. 1897. Further coccid notes: with descriptions of new species and discussions of points of interest. *Transactions and Proceedings of the New Zealand Institute* **29**: 293-331. [Maskel1897]

Maskell, W.M. 1897a. On a collection of Coccidae, principally from China and Japan. *Entomologist's Monthly Magazine* **33**: 239-244. [Maskel1897a]

Maskell, W.M. 1898. Further coccid notes: with descriptions of new species, and discussion of points of interest. *Transactions and Proceedings of the New Zealand Institute* **30**: 219-252. [Maskel1898]

Maskew, F. 1914. Report for the month of February, 1914. *Monthly Bulletin, California (State) Commission of Horticulture* **3**: 193-194. [Maskew1914]

Maskew, F. 1914a. Report for the month of August, 1914. *Monthly Bulletin, California (State) Commission of Horticulture* **3**: 446-447. [Maskew1914a]

Maskew, F. 1916. Report for the month of June, 1916. *Monthly Bulletin, California (State) Commission of Horticulture* **5**: 308-309. [Maskew1916]

Masoodi, M.A., Bhat, A.M. & Koul, V.K. 1989. Toxicity of insecticide to adults of *Encarsia* (= *Prospaltella*) *perniciosi* (Hymenoptera: Aphelinidae). *Indian Journal of Agricultural Sciences* **59(1)**: 50-52. [MasoodBhKo1989]

Masoodi, M.A., Bhagat, K.C., Sofi, M.R. & Koul, V.K. 1995. Toxicity of some fungicides to adult parasitoids *Aphytis proclia* Walker and *Encarsia perniciosi* Tower of San José scale. *Journal of Insect Science* **8**: 154-156. [MasoodBhSo1995]

Masoodi, M.A., Bhagat, K.C. & Sofi, M.R. 1996. Preliminary observations on the natural enemies of San José scale infesting apple trees in Kashmir. *Shashpa* **3**: 77-78. [MasoodBhSo1996]

Masoodi, M.A., Sofi, A.R., Bhagat, K.C. & Koul, V.K. 1996. Relative susceptibility of apple cultivars to San José scale, *Quadraspidiotus perniciosus* (Comstock). *Pest Management and Economic Zoology* **4(1/2)**: 119-121. [MasoodSoBh1996]

Masoodi, M.A., Trali, A.R., Bhat, A.M., Tiku, R.K. & Nehru, R.K. 1989. Establishment of *Encarsia* (= *Prospaltella*) *Perniciosi*, a specific parasite of San José Scale on apple in Kashmir. *Entomophaga* **34(1)**: 39-43. [MasoodTrBh1989]

Masoodi, M.A., Trali, A.R., Bhat, A.M., Tiku, R.K. & Nehru, R.K. 1989a. Establishment of *Aphytis* sp. *Proclia* group on San José scale in Kashmir. *Indian Journal of Plant Protection* **17(1)**: 71-73. [MasoodTrBh1989a]

Mathis, W. 1941. Notes on the biology of the Florida red scale (*Chrysomphalus aonidum* (L.)). *Florida Entomologist* **24(1)**: 1-5. [Mathis1941]

Mathis, W. 1947. Biology of the Florida red scale in Florida. *Florida Entomologist* **29(2/3)**: 13-35. [Mathis1947]

Mathys, G. 1953. Observations sur la mobilité des larves néonates de *Quadraspidiotus perniciosus* Comst. et sur leur transport par le vent. *Landwirtschaftiches Jahrbuch der Schweiz* **2**: 981-984. [Mathys1953]

Mathys, G. 1959. Vers un tournant dans la lutte contre le pou de San-José?. *Revue Romande d'agriculture, de viticulture et d'arboriculture* **15(6)**: 53-56. [Mathys1959]

Mathys, G. 1966. Possibilités de lutte contre le pou de San José par la méthode biologique et intégrée. *Proceedings (FAO Symposium on Integrated Pest Control)* **3**: 53-64. [Mathys1966]

Mathys, G. & Guignard, E. 1961. L'efficacité e *Prospaltella perniciosi* Tow., parasite du pou de San-José *Quadraspidiotus perniciosus* Comst.): Essai de conciliation de la lutte biologique et de la lutte chimique en vergers exploités commercialement. *Revue Romande d'agriculture, de viticulture et d'arboriculture* **17(6)**: 53-56. [MathysGu1961]

Mathys, G. & Guignard, E. 1962. Un important allié dans la lutte contre le pou de San-José: *Prospaltella perniciosi* Tow. *Agriculture Romande* **1(5) Ser. A**: 59-61. [MathysGu1962]

Mathys, G. & Guignard, E. 1963. Appendix 18: *Prospaltella perniciosi* Tower, an important ally in the control of San José scale a Swiss experiment in integrated control. *Publications de l'OEPP. Serie A (European and Mediterranean Plant Protection Organisation)* **A 34**: 96-101. [MathysGu1963]

Mathys, G. & Guignard, E. 1965. Étude de l'efficacité de *Prospaltella perniciosi* Tow. en Suisse parasite du pou de San-José. *Entomophaga* **10(93)**: 193-220. [MathysGu1965]

Mathys, G. & Guignard, E. 1967. Quelques aspects de la lutte biologique contre le pou de San José *Quadraspidiotus perniciosus* Comst à l'aide de l'aphélinide *Prospaltella perniciosi* Tow. *Entomophaga* **12**: 223-224. [MathysGu1967]

Mathys, G. & Guignard, E. 1967a. Enseignements recueillis au cours de neuf ans de travaux avec *Prospaltella perniciosi* Tow., parasite du pou de San José *Quadraspidiotus perniciosus* Comst. *Entomophaga* **12(3)**: 212-222. [MathysGu1967a]

Mathys, G., Guignard, E. & Stahl, J. 1965. L'identification des cochenilles du genre *Quadraspidiotus* importantes en arboriculture. *Agriculture Romande* **4 (Ser. A)**: 65-68. [MathysGuSt1965]

Matile-Ferrero, D. 1976. La faune terrestre de l'Île de Sainte-Hélène. 7. Coccoidea. *Musée Royal de l'Afrique centrale, Tervuren, Belgique, Annales (Serie 8) Sciences Zoologiques* **215**: 292-318. [Matile1976]

Matile-Ferrero, D. 1978. Homoptères Coccoidea de l'Archipel des Comores *In* Faune Entomologique de l'Archipel des Comores, Matile, L., Ed. *Memoires du Museum National d'Histoire Naturelle (N.S.) Serie A, Zoologie* **109**: 39-70. [Matile1978]

Matile-Ferrero, D. 1984c. Insects of Saudi Arabia Homoptera: Subordo Coccoidea. *Fauna of Saudi Arabia* **6**: 219-228. [Matile1984c]

Matile-Ferrero, D. 1988. Sternorrhyncha: Suborder Coccoidea of Saudi Arabia (Part 2). *Fauna of Saudi Arabia* **9**: 23-38. [Matile1988]

Matile-Ferrero, D. & Balachowsky, A.S. 1972. Contribution à l'étude de la faune des Coccoidea des îles Canaries avec description de deux espèces nouvelles (Hom.). *Bulletin de la Société Entomologique de France* **77**: 106-114. [MatileBa1972]

Matile-Ferrero, D. & Balachowsky, A.S. 1973. Deux Aspidiotini (Homoptera, Coccoidea - Diaspididae) nouveaux de Nouvelle-Calédonie. *Cahiers Pacifique* **No. 17**: 239-243. [MatileBa1973]

Matile-Ferrero, D. & Nonveiller, G. 1984. Coccoidea. Pages 62-70 *in*: G. Nonveiller (Editor). Catalogue commenté et illustré des Insectes du Cameroun d'intérêt agricole (apparitions, réparition, importance). Memoires. Beograd: Institut pour la Protection des Plantes.. 210 pp. [MatileNo1984]

Matile-Ferrero, D. & Oromí, P. 2001. Hemiptera. Coccoidea. Izquierdo, I., Martín, J.L., Zurita, N. & Arechavaleta, M. (Editors). [List of wild species from the Canaries (mushrooms, plants and land animals).] Tenerife: Consejería de Política Territorial y Medio Ambiente Gobierno de Canarias. 186-190, 193, 195-196. [MatileOr2001]

Matsuda, M. 1927. Studies on the rotatory movements necessary for the formation of the scale in *Chrysomphalus aonidum* L. *Transactions of the Natural History Society of Formosa. Taihoku* **17**: 391-417. [Matsud1927]

Matsuda, M. 1929. Studies on *Chrysomphalus aonidum* 1. in Formosa. *Report, Government Research Institute, Department of Agriculture, Formosa* **39**: 1-79. [Matsud1929]

Matsuda, M. 1935. Studies on *Chrysomphalus aonidum* Linné, IX. Seasonal life history in Kyoto. *Mushi* **8**: 29-36. [Matsud1935]

Matta, V.A. 1979. [Natural enemies of white diaspid scales from olive trees in the Azapa Valley, Arica, Chile.]. [In Spanish]. *Idesia* **(5)**: 231-242. [Matta1979]

Matvievskii, A.S. 1983. [The use of pesticides in orchards in the Ukraine.]. [In Russian]. *Zashchita rastenii. Moscow* (No. 6) 25. [Matvie1983]

Maxwell-Lefroy, H. 1903. The scale insects of the Lesser Antilles. Part II. *Imperial Department of Agriculture for the West Indies Pamphlet Series* **22**: 1-50. [Maxwel1903]

May, W. 1899a. Über die Larven einiger *Aspidiotus* Arten. *Mitteilungen aus dem Naturhistorischens Museum in Hamburg* **16**: 149-153. [May1899a]

Mayné, R. & Ghesquière, J. 1934. Hémiptères nuisibles aux végétaux du Congo belge. *Annales de Gembloux* **40((1)**: 3-38. [MayneGh1934]

Mazzeo, G., Longo, S., Benfatto, D., Palmeri, V. & Di Leo, A. 2002. Trials of biological control of *Aonidiella aurantii* Maskell (Hemiptera, Coccoidea) in citrus groves in Italy. *Bollettino di Zoologia Agraria e di Bachicoltura (Milano)* **33(3)**: 485-488. [MazzeoLoBe2002]

Mazzoni, E. 2001. Overwintering of the San José scale on stone fruit in northern Italy. *IOBC/WPRS Bulletin* **24(5)**: 201-206. [Mazzon2001]

Mazzoni, E. & Cravedi, P. 2002. Observations on the overwintering and winter mortality of the San scale in fruit orchards in Emilia Romagna (Northern Italy). *Bollettino di Zoologia Agraria e di Bachicoltura (Milano)* **33(3)**: 373-383. [MazzonCr2002]

Mazzoni, E. & Polesny, F. 2001. Overwintering of the San José Scale on stone fruit in Northern Italy. *IOBC/WPRS Bulletin* **24(5)**: 201-205. [MazzonPo2001]

McBeth, I.G. & Allison, J.R. 1941. Rotenone and oil sprays: some observations on use of toxic materials in control of scale pests. *California Citrograph* **26(10)**: 310-311. [McBethAl1941]

McCabe, T.L. & Johnson, L.M. 1980. Catalogue of the types in the New York State Museum insect collection. *Bulletin of the New York State Museum (of Natural History)* **No. 434**: 1-38. [McCabeJo1980]

McClain, D.C., Rock, G.C. & Stinner, R.E. 1990. San José scale (Homoptera: Diaspididae): Simulation of seasonal phenology in North Carolina orchards. *Environmental Entomology* **19(4)**: 916-925. [McClaiRoSt1990]

McClain, D.C., Rock, G.C. & Stinner, R.E. 1990a. Thermal requirements for development and simulation of the seasonal phenology of *Encarsia perniciosi* (Hymenoptera: Aphelinidae), a parasitoid of the San José scale (Homoptera: Diaspididae) in North Carolina orchards. *Environmental Entomology* **19(5)**: 1396-1402. [McClaiRoSt1990a]

McClain, D.C., Rock, G.C. & Woolley, J.B. 1990. Influence of trap color and San José scale (Homoptera: Diaspididae) pheromone on sticky trap catches of 10 aphelinid parasitoids (Hymenoptera). *Environmental Entomology* **19(4)**: 926-931. [McClaiRoWo1990]

McClure, M.S. 1978. Seasonal development of *Fiorinia externa, Tsugaspidiotus tsugae* (Homoptera: Diaspididae), and their parasite-host synchronism to the population dynamics of two scale pests of hemlock. *Environmental Entomology* **7**: 863-870. [McClur1978]

McClure, M.S. 1980a. Competition between exotic species: scale insects on hemlock. *Ecology* **61**: 1391-1401. [McClur1980a]

McClure, M.S. 1981. Effects of voltinism, interspecific competition and parasitism on the population dynamics of the hemlock scales, *Fiorinia externa* and *Tsugaspidiotus tsugae* (Homoptera: Diaspididae). *Ecological Entomology* **6**: 47-54. [McClur1981]

McClure, M.S. 1982. Distribution and damage of two *Pineus* species (Homoptera: Adelgidae) on red pine in New England. *Annals of the Entomological Society of America* **75**: 150-157. [McClur1982]

McClure, M.S. 1983b. Reproduction and adaptation of exotic hemlock scales (Homoptera: Diaspididae) on their new and native hosts. *Environmental Entomology* **12**: 1811-1815. [McClur1983b]

McClure, M.S. 1985. Patterns of abundance, survivorship, and fecundity of *Nuculaspis tsugae* (Homoptera: Diaspididae) on *Tsuga* species in Japan in relation to elevation. *Environmental Entomology* **14**: 413-415. [McClur1985]

McClure, M.S. 1986a. Population dynamics of Japanese hemlock scales: a comparison of endemic and exotic communities. *Ecology* **67**: 141-142. [McClur1986a]

McClure, M.S. 1988. The armored scales of hemlock. Pages 46-65 *in*: Berryman, Alan A., ed. Dynamics of Forest Insect Populations: Patterns, Causes, Implications. New York: Plenum Press. [McClur1988]

McClure, M.S. 1989. Importance of weather to the distribution and abundance of introduced Adelgid and scale insects. *Agricultural and Forest Meteorology* **47(2-4)**: 291-302. [McClur1989]

McClure, M.S. 1990a. 1.1.7 Coevolution of armored scale insects and their host plants: speciation processes. Pages 165-168 *in*: Rosen, D. (Ed.). Armored Scale Insects, Their Biology, Natural Enemies and Control [Series title: World Crop Pests, Vol. 4A]. Amsterdam, The Netherlands: Elsevier. 384 pp. [McClur1990a]

McClure, M.S. 1990b. 1.4 Ecology. 1.4.1 Habitats and hosts. Pages 285-288 *in*: Rosen, D. (Ed.). Armored Scale Insects, Their Biology, Natural Enemies and Control [Series title: World Crop Pests, Vol. 4A]. Amsterdam, The Netherlands: Elsevier. 384 pp. [McClur1990b]

McClure, M.S. 1990c. 1.4.2 Host relationships. 1.4.2.1 Impact on host plants. Pages 289-291 *in*: Rosen, D. (Ed.). Armored Scale Insects, Their Biology, Natural Enemies and Control [Series title: World Crop Pests, Vol. 4A]. Amsterdam, The Netherlands: Elsevier. 384 pp. [McClur1990c]

McClure, M.S. 1990d. 1.4.2.3 Patterns of host specificity. Pages 301-303 *in*: Rosen, D. (Ed.). Armored Scale Insects, Their Biology, Natural Enemies and Control [Series title: World Crop Pests, Vol. 4A]. Amsterdam, The Netherlands: Elsevier. 384 pp. [McClur1990d]

McClure, M.S. 1990e. 1.4.4 Patterns of temporal and spatial distribution. Pages 309-314 *in*: Rosen, D. (Ed.). Armored Scale Insects, Their Biology, Natural Enemies and Control [Series title: World Crop Pests, Vol. 4A]. Amsterdam, The Netherlands: Elsevier. 384 pp. [McClur1990e]

McClure, M.S. 1990f. 1.4.5 Seasonal history. Pages 315-318 *in*: Rosen, D. (Ed.). Armored Scale Insects, Their Biology, Natural Enemies and Control [Series title: World Crop Pests, Vol. 4A]. Amsterdam, The Netherlands: Elsevier. 384 pp. [McClur1990f]

McClure, M.S. 1990g. 1.4.6 Influence of environmental factors. Pages 319-330 *in*: Rosen, D. (Ed.). Armored Scale Insects, Their Biology, Natural Enemies and Control [Series title: World Crop Pests, Vol. 4A]. Amsterdam, The Netherlands: Elsevier. 384 pp. [McClur1990g]

McClure, M.S. 1990h. 1.4.7 Life-tables, models and population dynamics. Pages 331-337 *in*: Rosen, D. (Ed.). Armored Scale Insects, Their Biology, Natural Enemies and Control [Series title: World Crop Pests, Vol. 4A]. Amsterdam, The Netherlands: Elsevier. 384 pp. [McClur1990h]

McClure, M.S. 1991. Adelgid and scale insect guilds on hemlock and pine. *General Technical Report, (U.S.D.A., Forest Service)* **(NE-153)**: 256-270. [McClur1991]

McClure, M.S. & Fergione, M.B. 1977. *Fiorinia externa* and *Tsugaspidiotus tsugae* (Homoptera: Diaspididae): distribution, abundance, and new hosts of two destructive scale insects of eastern hemlock in Connecticut. *Environmental Entomology* **6**: 807-811. [McClurFe1977]

McClure, M.S. & Hare, J.D. 1984. Foliar terpenoids in *Tsuga* species and the fecundity of scale insects. *Oecologia* **63**: 185-193. [McClurHa1984]

McCoy, C.W. 1985. Citrus: Current status of biological control in Florida. Pages 481-499. *in*: Hoy, M.A. & Herzog, D.C. (Eds.). Biological Control in Agricultural IPM Systems Orlando, FL: Academic Press. 589 pp. [McCoy1985]

McDaniel, B. 1968. The armored scale insects of Texas (Homoptera: Coccoidea: Diaspididae). *Southwestern Naturalist* **13**: 201-242. [McDani1968]

McDaniel, B. 1969. The armored scale insects of Texas (Homoptera: Coccoidea: Diaspididae). Part II. *Southwestern Naturalist* **14**: 89-113. [McDani1969]

McDaniel, B. 1970. The armored scale insects of Texas (Homoptera: Coccoidea: Diaspididae). Part III. *Southwestern Naturalist* **14**: 411-440. [McDani1970]

McDaniel, B. 1974. The armored scale insects of Texas (Homoptera: Coccoidea: Diaspididae). Part VII. *Southwestern Naturalist* **18**: 417-442. [McDani1974]

McIntire, S.J. 1889. Further notes upon some remarkable Coccidae from British Guiana. *Journal of the Quekett Microscopical Club, Ser. II* **4**: 22-25. [McInti1889]

McKenna, C.E. & Retamales, J. 1999. Evaluation of vegetable oils for armoured scale control in kiwifruit orchards. *Acta Horticulturae* **(No. 498)**: 365-370. [McKennRe1999]

McKenzie, H.L. 1935. Biology and control of avocado insects and mites. *Bulletin (University of California, College of Agriculture, Agricultural Experiment Station)* **No. 592**: 1-48. [McKenz1935]

McKenzie, H.L. 1937. Morphological differences distinguishing California red scale, yellow scale, and related species (Homoptera, Diaspididae). *University of California Publications in Entomology* **6**: 323-335. [McKenz1937]

McKenzie, H.L. 1937a. Generic characteristics of *Aonidiella* Berlese and Leonardi, and a description of a new species from Australia (Homoptera-Diaspididae). *Pan-Pacific Entomologist* **13**: 176-180. [McKenz1937a]

McKenzie, H.L. 1938. The genus *Aonidiella* (Homoptera; Coccoidea: Diaspididae). (Contribution number 8). *Microentomology* **3**: 1-36. [McKenz1938]

McKenzie, H.L. 1939. A revision of the genus *Chrysomphalus* and supplementary notes on the genus *Aonidiella* (Homoptera: Coccoidea: Diaspididae). *Microentomology* **4**: 51-77. [McKenz1939]

McKenzie, H.L. 1942b. Two new species related to red scale (Homoptera; Coccoidea; Diaspididae). *Bulletin of the California Department of Agriculture* **31**: 141-147. [McKenz1942b]

McKenzie, H.L. 1943. Miscellaneous diaspid studies including notes on *Chrysomphalus* (Homoptera; Coccoidea; Diaspididae). *Bulletin of the California Department of Agriculture* **32**: 148-162. [McKenz1943]

McKenzie, H.L. 1943a. The seasonal history of *Matsucoccus vexillorum* Morrison (Homoptera; Coccoidea; Margarodidae). (Contribution No. 39.). *Microentomology* **8**: 42-52. [McKenz1943a]

McKenzie, H.L. 1944. Miscellaneous diaspid scale studies (Homoptera; Coccoidea; Diaspididae). *Bulletin of the California Department of Agriculture* **33**: 53-57. [McKenz1944]

McKenzie, H.L. 1945. A revision of *Parlatoria* and closely allied genera. (Homoptera: Coccoidea: Diaspididae). *Microentomology* **10**: 47-121. [McKenz1945]

McKenzie, H.L. 1946. Supplementary notes on the genera *Aonidiella* and *Parlatoria* (Homoptera: Coccoidea: Diaspididae). *Microentomology* **11**: 29-36. [McKenz1946]

McKenzie, H.L. 1946a. General distribution of red scale, *Aonidiella aurantii* (Maskell) in California. *Bulletin of the California Department of Agriculture* **35**: 95-99. [McKenz1946a]

McKenzie, H.L. 1947. Diaspid scale studies with notes on California species (Homoptera; Coccoidea; Diaspididae). *Bulletin of the California Department of Agriculture* **36**: 31-36. [McKenz1947]

McKenzie, H.L. 1947b. Miscellaneous diaspid scale studies. Part V. (Homoptera; Coccoidea; Diaspididae). *Bulletin of the California Department of Agriculture* **36**: 107-114. [McKenz1947b]

McKenzie, H.L. 1950. The genera *Lindingaspis* MacGillivray and *Marginaspis* Hall (Homoptera; Coccoidea; Diaspididae). (Contribution No. 69). *Microentomology* **15**: 98-124. [McKenz1950]

McKenzie, H.L. 1951. Miscellaneous diaspidid scale studies (Homoptera; Coccoidea; Diaspididae). Scale studies - Part VII. *Bulletin of the California Department of Agriculture* **40**: 80-82. [McKenz1951]

McKenzie, H.L. 1953. A new scale insect from the Ryukyu Islands related to the red scale. (Homoptera; Coccoidea; Diaspididae). Scale studies - Part XI. *Bulletin of the California Department of Agriculture* **42**: 35-38. [McKenz1953]

McKenzie, H.L. 1953a. Two new *Selenaspidus* scales infesting *Euphorbia* in California. (Homoptera; Coccoidea; Diaspididae). Scale studies - Part XII. *Bulletin of the California Department of Agriculture* **42**: 53-58. [McKenz1953a]

McKenzie, H.L. 1956. The armored scale insects of California. *Bulletin of the California Insect Survey* **5**: 1-209. [McKenz1956]

McKenzie, H.L. 1957. A new armored scale insect on *Elaeagnus* from Texas (Homoptera: Coccoidea: Diaspididae). Scale studies pt. 13. *Bulletin of the California Department of Agriculture* **46**: 218-220. [McKenz1957]

McKenzie, H.L. 1963. Miscellaneous Diaspidid scale studies, including a new Asterolecaniid from Florida. (Homoptera; Coccoidea; Diaspididae; Asterolecaniidae). Scale studies -- Part XV. *Bulletin of the California Agricultural Experiment Station* **52**: 29-39. [McKenz1963]

McLaren, I.W. 1971. A comparison of the population growth potential in California red scale, *Aonidiella aurantii* (Maskell), and yellow scale, *A. citrina* (Coquillett), on citrus. *Australian Journal of Zoology. Melbourne* **19**: 189-204. [McLare1971]

McLaren, G.F. 1989. Control of oystershell scale *Quadraspidiotus ostreaeformis* (Curtis) on apples in Central Otago. *New Zealand Journal of Crop and Hort. Sci.* **17**: 221-227. [McLare1989]

McLaren, G.F. 1989a. Phenology of oystershell scale *Quadraspidiotus ostreaeformis* (Curtis) in Central Otago. *New Zealand Journal of Crop and Horticultural Science* **17(3)**: 215-219. [McLare1989a]

McLaren, I.W. & Buchanan, G.A. 1973. Parasitism by *Aphytis chrysomphali* Mercet and *A. melinus* DeBach of California red scale, *Aonidiella aurantii* (Maskell), in relation to seasonal availability of suitable stages of the scale. *Australian Journal of Zoology. Melbourne* **21**: 111-117. [McLareBu1973]

McLaren, G.F. & Fraser, J.A. 1992. Leafroller and codling moth flights in Central Otago. *Orchardist of New Zealand* **65(9)**: 21. [McLareFr1992]

Mead, F.W. 1983. Technical report: Bureau of Entomology. *Tri-ology* **22**: 1-5. [Mead1983]

Mead, F.W. 1985. Technical report: Bureau of Entomology. *Tri-ology* **24(7)**: 1-3. [Mead1985]

Mead, F.W. 1987. Technical report: Bureau of Entomology. *Tri-ology* **26(4)**: 2-4. [Mead1987]

Meerwarth, H. 1900. Die Randstructur des Leizten Hinterleibssegments von *Aspidiotus perniciosus* Comst. *Jahrbuch der Hamburgischen Wissenschaftlichen Anstalten. Hamburg* **3**: 1-15. [Meerwa1900]

Mehrnejad, M.R. & Ak, B.E. 2001. The current status of pistachio pests in Iran. *Cahiers Options Méditerranéennes* 56: 315-322.[MehrneAk2001]

Melander, A.L. 1914. Can insects become resistant to sprays?. *Journal of Economic Entomology* **7**: 167-173. [Meland1914]

Melander, A.L. 1915. Varying susceptibility of the San José scale to sprays. *Journal of Economic Entomology* **8**: 475-481. [Meland1915]

Melander, A.L. 1915a. The San José scale insect. *State Agriculture Experiment Station Pullman Washington Pop. Bulletin* **78**: 1-7. [Meland1915a]

Melander, A.L. 1922. San José scale. *Bulletin (Washington Agricultural Experiment Station)* **175**: 21-22. [Meland1922]

Melander, A.L. 1922a. San José scale. *Bulletin (Washington Agricultural Experiment Station)* **167**: 24-26. [Meland1922a]

Melander, A.L. 1923. Tolerance of San José scale to sprays. *Technical Paper (Agricultural Experiment Station, State College of Washington, Pullman, WA)* **Bulletin No. 174**: 1-52. [Meland1923]

Melander, A.L. 1926. Are insects becoming resistant?. *American Fruit Grower Magazine* **(Feb. 1926)**: 7, 40. [Meland1926]

Melis, A. 1930. Contribuzione alla conoscenza degli insetti dannosi alle piante agrarie e forestali della Sardegna. *Redia* **18**: 1-120. [Melis1930]

Melis, A. 1943. Contributo alla conoscenza dell' *"Aspidiotus perniciosus"* Comst. *Redia* **29**: 1-170. [Melis1943]

Melis, A. 1949. Elenco delle principali specie di Insetti che hanno prodotto infestazioni degne dinota in Italia durante l'anno 1949. *Redia* **34**: 17-25. [Melis1949]

Melis, A. 1951. Precisazioni morfo-biologiche sull' *Aspidiotus perniciosus* Comst. *Redia* **36**: 1-91. [Melis1951]

Melton, F.M. & Shives, S.A. 1939. Seasonal occurrence of Florida red scale, *Chrysomphalus aonidum* (L.), and two-spotted spider mite, *Tetranychus urticae* Koch, in Manatee County. *Proceedings of the Florida State Horticultural Society* 111: 37-38. [MeltonSh1939]

Mendel, Z., Blumberg, D. & Ishaaya, I. 1994. Effects of some insect growth regulators on natural enemies of scale insects (Hom.: Coccoidea). *Entomophaga* **39(2)**: 199-209. [MendelBlIs1994]

Mendel, Z., Podoler, H. & Rosen, D. 1990. 2.7.4 Analysis of the Gut Contents of Predators. Pages 289-290 *in*: Rosen, D., ed. Armoured Scale Insects, Their Biology, Natural Enemies and Control, Vol. 4B. World Crop Pests. Amsterdam, the Netherlands: Elsevier. [MendelPoRo1990]

Merkel, L. 1938. IV. Pflanzenschutz einschl. Obstbauberatung. A. Amtliche Pflanzenbeschau im Freihafen. Uberwachung der Ein- and Ausfurh von Obst, Pflanzen and Pflanzenteilen. *Jahresberichte Institut für Angewandte Botanik. Hamburg* **55**: 88-99. [Merkel1938]

Merrill, G.B. 1923. A new scale insect from Florida. *Quarterly Bulletin of the Florida State Plant Board* 7: 167-168. [Merril1923]

Merrill, G.B. 1953. A revision of the scale insects of Florida. *Bulletin of the Florida State Plant Board* 1: 1-143. [Merril1953]

Merrill, G.B. & Chaffin, J. 1923. Scale insects of Florida. *Quarterly Bulletin of the Florida State Plant Board* 7: 177-298. [MerrilCh1923]

Messenger, P.S., Biliotti, E. & van den Bosch, R. 1976. The importance of natural enemies in integrated control. Pages 543-563 *in*: Huffaker, C.B. & Messenger, P.S. (Eds.). Theory and Practice of Biological Control. New York: Academic Press. 788 pp. [MessenBiVa1976]

Messenger, P.S. & van den Bosch, R. 1971. The adaptability of introduced biological control agents. Pages 68-92 *in*: Huffaker, C.B. Biological Control. New York & London: Plenum Press. 511 [MessenVa1971]

Messenger, P.S., Wilson, F. & Whitten, M.J. 1976. Variation, fitness, and adaptability of natural enemies. Pages 209-231 *in*: Huffaker, C.B. & Messenger, P.S. (Eds.). Theory and Practice of Biological Control. New York: Academic Press. 788 pp. [MessenWiWh1976]

Metcalf, R.L. 1982. Insecticides in pest management. Pages 217-277 *in*: Metcalf, R.L. & Luckmann, W.H. (Eds.). Introduction to Insect Pest Management. 2ed ed. New York: John Wiley & Sons. 577 pp. [Metcal1982]

Metcalf, R.L., Carlson, R.B. & Murphy, U.E. 1949. Citrus pest insecticides screened by laboratory and field tests as new control chemicals are developed. *California Agriculture* **3(7)**: 5, 12. [MetcalCaMu1949]

Metcalf, R.L. & Metcalf, R.A. 1993. Destructive and Useful Insects: Their Habits and Control. 5th ed. New York: McGraw Hill. [MetcalMe1993]

Metschnikoff, E. 1866. Embryologische Studien an Insekten. *Zeitschrift für Wissenschaftliche Zoologie* **16**: 389-500. [Metsch1866]

Meyer, M.K.P. 1962. Two new mite predators of red scale (*Aonidiella aurantii*) in South Africa. *South African Journal of Agricultural Science* **5(3)**: 411-417. [Meyer1962]

Meyer, R.H. & Randall, R. 1973. San José scale control with superior oil. *Journal of Economic Entomology* **66(6)**: 1354. [MeyerRa1973]

Michaelides, R. & Previs, S. 1960. Control of citrus pests in Cyprus. *World Crops* **12**: 389-391. [MichaePr1960]

Michelakis, S. 1990. The influence of pests and diseases on the quantity and quality of olive oil production. *Olivae* **67(30)**: 38-40. [Michel1990]

Michelbacher, A.E. & Ortega, J.C. 1958. A technical study of insects and related pests attacking walnuts. *Bulletin (University of California, College of Agriculture, Agricultural Experiment Station)* **No. 764**: 1-86. [MichelOr1958]

Milaire, H. 1958. La situation du Pou de San José en France. *Phytoma* **96**: 15-18. [Milair1958]

Milaire, H. 1969. L'utilisation pratique de *Prospaltella perniciosi* Tow. dans la lutte contre le pou de San José dans les vergers français. *EPPO (European and Mediterranean Plant Protection Organization) Bulletin* **(Ser. A) no. 48**: 77-85. [Milair1969]

Milaire, H.G. & Audemard, H. 1979. La protection intégrée des arbres fruitiers a noyau: pêchers et pruniers. Pages 321-355. *in*: Proceedings: Internationales Symposium der IOBC/WPRS über Integrierten Pflanzenschutz in der Land und Forstwirtschaft. Wien, Austria: OILB/SROP. 648 pp. [MilairAu1979]

Milholland, R.D. & Meyer, J.R. 1984. Diseases and arthropod pests of blueberries. *Bulletin (North Carolina Agricultural Research Service)* **No. 468**: 1-33. [MilholMe1984]

Millar, J.G. & Hare, J.D. 1993. Identification and synthesis of a kairomone inducing oviposition by the parasitoid *Aphytis melinus* from California red scale covers. *Journal of Chemical Ecology* **19(8)**: 1721-1736. [MillarHa1993]

Miller, R.L. 1937. The control of scale insects on citrus. *Proceedings of the Florida State Horticultural Society* **50**: 100-106. [Miller1937RL]

Miller-Weeks, M. 1983. Current status of beech bark disease in New England and New York. Pages 21-23. *in*: Proceedings, I.U.F.R.O. Beech Bark Disease Working Party Conference, General Technical Report WO-37. [Washington, D.C.]: USDA Forest Service. 140 pp. [Miller1983]

Miller, D.R. 1990. 1.1.8 Phylogeny. Pages 169-178 *in*: Rosen, D. (Ed.). Armored Scale Insects, Their Biology, Natural Enemies and Control [Series title: World Crop Pests, Vol. 4A]. Amsterdam, The Netherlands: Elsevier. 688 pp. [Miller1990]

Miller, F. 1999. Take control of scale infestations. *American Nurseryman* **190(10)**: 14. [Miller1999]

Miller, D.R. & Davidson, J.A. 1981. A systematic revision of the armoured scale genus *Crenulaspidiotus* MacGillivray (Diaspididae, Homoptera). *Polskie Pismo Entomologiczne* **51**: 531-595. [MillerDa1981]

Miller, D.R. & Davidson, J.A. 1990. 3.1.1 A List of the Armored Scale Insect Pests. Pages 299-306 *in*: Rosen, D. (Ed.). Armored Scale Insects, Their Biology, Natural Enemies and Control [Series title: World Crop Pests, Vol. 4B]. Amsterdam, the Netherlands: Elsevier. 688 pp. [MillerDa1990]

Miller, D.R. & Davidson, J.A. 1998. A new species of armored scale (Hemiptera: Coccoidea: Diaspididae) previously confused with *Hemiberlesia diffinis* (Newstead). *Proceedings of the Entomological Society of Washington* **100**: 193-201. [MillerDa1998]

Miller, D.R., Davidson, J.A. & Stoetzel, M.B. 1984. A taxonomic study of the armored scale *Pseudischnaspis* Hempel (Homoptera: Coccoidea: Diaspididae). *Proceedings of the Entomological Society of Washington* **86**: 94-109. [MillerDaSt1984]

Miller, D.R. & Howard, F.W. 1981. A new species of *Abgrallaspis* (Homoptera: Coccoidea: Diaspididae) from Louisiana. *Annals of the Entomological Society of America* **74**: 164-166. [MillerHo1981]

Miller, D.R. & Gimpel, M.E. 2000. A Systematic Catalogue of the Eriococcidae Felt Scales) (Hemiptera: Coccoidea) of the World. Andover, U.K.: Intercept Ltd.. 589 pp. [MillerGi2000]

Miyanoshita, A., Kawai, S. & Fujii, K. 1993. Host-associated differences in *Aspidiotus cryptomeriae* Kuwana (Homoptera: Coccoidea: Diaspididae) I. Adult morphology and host preference. *Applied Entomology and Zoology* **28(1)**: 71-80. [MiyanoKaFu1993]

Miyanoshita, A. & Tatsuki, S. 1995. [Host-associated differences in *Aspidiotus cryptomeriae* Kuwana (Homoptera: Coccoidea: Diaspididae). II. Morphology of *Torreya nucifera* race reared on non-host plant, *Cryptomeria japonica* D. Don.]. *Japanese Journal of Applied Entomology and Zoology* **39**: 159-162. [MiyanoTa1995]

Miyanoshita, A. & Tatsuki, S. 2001. Role of sex pheromones in reproductive isolation between two host races in *Aspidiotus cryptomeriae* Kuwana (Homoptera: Diaspididae). *Applied Entomology and Zoology. Tokyo* **36(no. 2)**: 199-202. [MiyanoTa2001]

Miyanoshita, A., Tatsuki, S., Kusano, T. & Koichi, F. 1991. [Variation of esterase isozymes in *Aspidiotus cryptomeriae*.]. [In Japanese]. *Japanese Journal of Applied Entomology and Zoology* **35**: 317-321. [MiyanoTaKu1991]

Miyoshi, K. 1926. [List of injurious insects of citrus from Japan.]. [In Japanese]. *Insect World* **30(9)**: 303-326. [Miyosh1926]

Moffitt, L.J. & Baritelle, J.L. 1990. 3.4.5 Economic Thresholds. Pages 357-362 *in*: Rosen, D. (Ed.). Armored Scale Insects, Their Biology, Natural Enemies and Control [Series title: World Crop Pests, Vol. 4B]. Amsterdam, the Netherlands: Elsevier. 688 pp. [MoffitBa1990]

Mohammad, Z.K., Ghabbour, M.W. & Tawfik, M.H. 2001. Population dynamics of *Aonidiella orientalis* (Newstead) (Coccoidea: Diaspididae) and its parasitoid *Habrolepis aspidioti* Compere & Annecke (Hymenoptera: Encyrtidae). *Entomologica* **33(1999)**: 413-418. [MohammGhTa2001]

Moiseenkov, A.M., Ishchenko, R.I., Veselovskii, V.V., Odinokov, V.N., Polunin, E.V., Kovalev, B.G., Cheskis, B.A. & Tolstikov, G.A. 1989. Synthesis of alpha-neryl and alpha-geranyl propionates - components of the sex pheromone of San José Scale. *Chemistry of Natural Compounds* **25(3)**: 366-368. [MoiseeIsVe1989]

Moleas, T. & Baldacchino, F. 1994. Osservazioni bioetologiche su *Targionia vitis* (Sign.) (Rhynchota- Diaspididae) nel biotopo vite allevata a tendone. *Atti Giornato Fitopatologiche* 2: 211-218. [MoleasBa1994]

Monastero, S. 1958. Gli insetti piu nocivi agli agrumi e i metodi per compatterli. *Bollettino dell' Istituto di Entomologia Agraria Osservatorio Fitopatologia. Palermo* **2**: 131-165. [Monast1958]

Monastero, S. & Zaami, V. 1960. Le cocciniglie degli agrumi in Sicilia (*Chrysomphalus dictyospermi, Parlatoria pergandei, Aspidiotus hederae*. (II Nota). *Bollettino dell' Istituto di Entomologia Agraria dell Università Palmero* 3: 169-236. [MonastZa1960]

Monrreal Hdz., L.S., Coronado Blanco, J.M. & Ruiz Cancino, E. 1997. La escama roja de Florida *Chrysomphalus aonidum* (L.), plaga importante de la naranja en Mexico. *Revista de la Universidad Autónoma de Tamaulipas* **No. 55**: 43-48. [MonrreCoRu1997]

Monte, O. 1930. Os pulgoes dos vegetaes. *Boletim de Agricultura Zootech e Vet* **3**: 3-36. [Monte1930]

Monte, O. 1943. F.B. -- Capital -- Cochonilha (*Capulinia*) da jaboticabeira. *O Biológico* **9**: 133. [Monte1943]

Monteiro Guimarães, J.A. 1973. Catálogo das pragas das culturas em Portugal Continental. Vol. I 299 pp. [Montei1973]

Montgomery, J.H. 1921. Quarantine Department. *Quarterly Bulletin of the Florida State Plant Board* **2(2)**: 41-55. [Montgo1921]

Mony, R., Kenne, M., Orivel, J. & Dejean, A. 2002. Biology and ecology of pest ants of the genus *Melissotarsus* (Formicidae: Myrmicinae), with special reference to tropical fruit tree attacks. *Sociobiology* **40(3)**: 645-654. [MonyKeOr2002]

Moore, H.W.B. 1915. A list of the insects affecting sugarcane in British Guiana. *Timehri. Journal of the Royal Agricultural and Commercial Society of British Guiana (Georgetown)* **3**: 305-310. [Moore1915]

Moore, W. 1933. Studies of the "resistant" California red scale *Aonidiella aurantii* Mask. in California. *Journal of Economic Entomology* **26**: 1140-1161. [Moore1933]

Moore, W. 1934. Fumigation experiments with the California red scale under orchard conditions. *Journal of Economic Entomology* **28**: 1042-1055. [Moore1934]

Moore, W. 1936. Differences between resistant and non-resistant red scale in California. *Journal of Economic Entomology* **19(1)**: 65-78. [Moore1936]

Morales, C.F. 1988. The occurrence of latania scale, *Hemiberlesia lataniae* (Signoret) (Hemiptera: Diaspididae), in New Zealand. *New Zealand Journal of Experimental Agriculture* **16(1)**: 77-82. [Morale1988]

Mordkovich, Y.B. & Chernei, L.B. 1993. [Phostoxin for fumigation of planting stock]. [In Russian]. *Zashchita rastenii. Moscow* No. 12, 30. [MordkoCh1993]

Moreira, C. 1899. Contra os inimigos. *A Lavoura* **3**: 89-90. [Moreir1899]

Moreira, C. 1921a. Insectos parasitas da videira e do caffeeiro. *Entomologia Agricola* **1**: 124-126. [Moreir1921a]

Moreira, C. 1921b. Entomologia agricola Brasileira. *Boletim (Ministerio da Agricultura, Industria e Commercio, Instituto Biologico de Defesa Agricola, Rio de Janeiro)* **No. 1**: 182. [Moreir1921b]

Moreira, C. 1929. Legislative and administrative measures. *International Bulletin of Plant Protection* **3**: 85-89. [Moreir1929]

Moreira, C. 1929a. Entomologia Agricola Brasileira [Boletim No. 1]. Rio de Janeiro: Min. da Agric., Ind. e Comercio. 274 pp. [Moreir1929a]

Moreno, D.S. 1972. Location of the site of production of the sex pheromone in the yellow scale and the California red scale. *Annals of the Entomological Society of America* **65**: 1283-1286. [Moreno1972]

Moreno, D.S. 1983. Efficiency of pheromone traps in citrus pest detection. *Citrograph* **68**: 77-79. [Moreno1983]

Moreno, D.S., Carman, G.E., Fargerlund, J. & Shaw, J.G. 1974. Flight and dispersal of the adult male yellow scale. *Annals of the Entomological Society of America* **67**: 15-20. [MorenoCaFa1974]

Moreno, D.S., Carman, G.E., Rice, R.E., Shaw, J.G. & Bain, N.S. 1972. Demonstration of a sex pheromone of the yellow scale, *Aonidiella citrina. Annals of the Entomological Society of America* **65**: 443-446. [MorenoCaRi1972]

Moreno, D.S. & Fargerlund, J. 1975. Histology of the sex pheromone glands in the yellow scale and California red scale. *Annals of the Entomological Society of America* **68**: 425-428. [MorenoFa1975]

Moreno, D.S., Fargerlund, J. & Shaw, J.G. 1973. California red scale: captures of males in modified pheromone traps. *Journal of Economic Entomology* **66**: 1333. [MorenoFaSh1973]

Moreno, D.S., Fargerlund, J. & Shaw, J.G. 1976. California red scale: development of a method for testing juvenoids. *Journal of Economic Entomology* **69**: 292-297. [MorenoFaSh1976]

Moreno, D.S. & Kennett, C.E. 1983. Monitoring insect pest populations by trapping in California citrus orchards. *Proceedings of the International Society of Citriculture* **2**: 684-689. [MorenoKe1983]

Moreno, D.S. & Kennett, C. 1985. Predictive year-end California red scale (Homoptera: Diaspididae) orange fruit infestations based on catches of males in the San Joaquin Valley. *FORUM: Journal of Economic Entomology* **78**: 1-9. [MorenoKe1985]

Moreno, D.S. & Luck, R.F. 1992. Augmentative releases of *Aphytis melinus* (Hymenoptera: Aphelinidae) to suppress California Red Scale (Homoptera: Diaspididae) in southern California lemon orchards. *Journal of Economic Entomology* **85(4)**: 1112-1119. [MorenoLu1992]

Moreno, D.S., Rice, R.E. & Carman, G.E. 1972. Specificity of the sex pheromones of female yellow scales and California red scales. *Journal of Economic Entomology* **65**: 698-701. [MorenoRiCa1972]

Morgan, A.C.F. 1887. Observations upon *Aspidiotus rapax* Comstock and *A. camelliae* (Boisd.) Signoret: two allied species of Coccidae. *Entomologist's Monthly Magazine* **24**: 79-82. [Morgan1887]

Morgan, A.C.F. 1888. *Aspidiotus zonatus*, Frauenfeld. *Entomologist's Monthly Magazine* **24**: 205-208. [Morgan1888]

Morgan, A.C.F. 1888a. Observations on Coccidae (No. 1). *Entomologist's Monthly Magazine* **25**: 42-48. [Morgan1888a]

Morgan, A.C.F. 1888b. Observations on Coccidae (No. 2). *Entomologist's Monthly Magazine* **25**: 118-120. [Morgan1888b]

Morgan, A.C.F. 1889a. Observations on Coccidae (No. 5). *Entomologist's Monthly Magazine* **25**: 349-353. [Morgan1889a]

Morgan, A.C.F. 1890. Observations on Coccidae (No. 6). *Entomologist's Monthly Magazine* **26**: 42-45. [Morgan1890]

Morgan, A.C.F. 1893. *Aspidiotus palmae*, n. sp. *Entomologist's Monthly Magazine* **29**: 40-41. [Morgan1893]

Morgan, C.V.G. 1967. Fate of the San José scale and the European fruit scale (Homoptera: Diaspididae) on apples and prunes held in standard cold storage and controlled atmosphere storage. *Canadian Entomologist* **99**: 650-659. [Morgan1967]

Morgan, C.V.G. & Angle, B.J. 1968. Notes on the habits of the San José scale and the European fruit scale (Homoptera: Diaspididae) on harvested apples in British Columbia. *Canadian Entomologist* **100**: 499-503. [MorganAn1968]

Morgan, C.V.G. & Angle, B.J. 1969. Distribution and development of the San José scale (Homoptera: Diaspididae) on the leaves, bark, and fruit of some orchard and ornamental trees in British Columbia. *Canadian Entomologist* **10**: 983-989. [MorganAn1969]

Morgan, C.V.G. & Arrand, J.C. 1971. San José scale and European fruit scale in interior British Columbia. *Journal of the Entomological Society of British Columbia* **67(1)**: 1-12. [MorganAr1971]

Morgan, D.J.W. & Hare, J. D. 1997. Uncoupling physical and chemical cues: the independent roles of scale cover size and kairomone concentration on host selection by *Aphytis melinus* DeBach (Hymenoptera: Aphelinidae). *Journal of Insect Behavior* **10**: 679-684. [MorganHa1997]

Morgan, D.J.W. & Hare, J.D. 1998. Innate and learned cues: scale cover selection by *Aphytis melinus* (Hymenoptera: Aphelinidae). *Journal of Insect Behavior* **11**: 463-479. [MorganHa1998]

Morgan, D.J.W. & Hare, J.D. 1998a. Volatile cues used by the parasitoid, *Aphytis melinus*, for host location: California red scale revisited. *Entomologia Experimentalis et Applicata* **88(3)**: 235-245. [MorganHa1998a]

Morgan, H.A. 1894. Report of entomologist. Part IV. *Bulletin of the Louisiana Experiment Station* **28**: 982-1006. [Morgan1894]

Mori, K. 1981. Recent progress in the synthesis of optically active pheromones. *Colloques INRA* **No. 7**: 41-53. [Mori1981]

Mori, K. and Kuwahara, S. 1982. Synthesis of optically active forms of (E)-6-isopropyl-3,9-dimethyl-5,8-decadienyl acetate, the pheromone of the yellow scale. *Tetrahedron* **38**: 521-525. [MoriKu1982]

Morley, C. 1909. On the Hymenopterous parasites of Coccidae. *The Entomologist* **42(558)**: 254-257; 276-278. [Morley1909]

Morrill, A.W. 1911. The rufous scale at Key West, Florida. *Journal of Economic Entomology* **4**: 277. [Morril1911]

Morrison, H. 1923. A report on a collection of Coccidae from Argentine II. (Hemiptera Coccidae). *Proceedings of the Entomological Society of Washington* **25**: 122-127. [Morris1923]

Morrison, H. & Morrison, E.R. 1922. A redescription of the type species of the genera of Coccidae based on species originally described by Maskell. *Proceedings of the United States National Museum. Washington* **60**: 1-120. [MorrisMo1922]

Morrison, H. & Morrison, E.R. 1965. A selected bibliography of the Coccoidea. First Supplement. *Miscellaneous Publication United States Department of Agriculture* 987: 1-44. [MorrisMo1965]

Morrison, H. & Morrison, E.R. 1966. An annotated list of generic names of the scale insects (Homoptera: Coccoidea). *Miscellaneous Publication United States Department of Agriculture* **1015**: 1-206. [MorrisMo1966]

Morrison, H. & Renk, A.V. 1957. A selected bibliography of the Coccoidea. *Miscellaneous Publications United States Department of Agriculture* 734: 1-222. [MorrisRe1957]

Morrison, R.K. & King, E.G. 1977. Mass production of natural enemies. Pages 183-217. *in*: Ridgway, R.L. & Vinson, S.B. (Eds.). Biological Control of Augmentation of Natural Enemies: Insect and Mite Control with Parasites and Predators. New York: Plenum Press. 480 pp. [MorrisKi1977]

Morse, J.G., Arbaugh, M.J. & Moreno, D.S. 1985. California red scale - computer simulation on CRS population. *California Agriculture* **39**: 8-10. [MorseArMo1985]

Morse, S., Acholo, M., McNamara, N. & Oliver, R. 2000. Control of storage insects as a means of limiting yam tuber fungal rots. *Journal of Stored Products Research* **36 (1)**: 37-45. [MorseAcMc2000]

Mottareale, G. 1915. Su d'un caso di morte naturale della Bianco-Rossa (*Chrysomphalus dictyospermi* v. *pinnulifera* Mask.) in provincia di Reggio Calabria. *Bollettino della Societa dei naturalisti in Napoli* **66(Ser. 6)**: 29-31. [Mottar1915]

Moulton, D. 1931. Division of Entomology and pest control. *Monthly Bulletin, California Department of Agriculture* **20**: 745-76. [Moulto1931]

Mourad, A.K., Mesbah, H.A., Fata Aziza, A.S., Moursi Khadiga, S. & Abdel Razak Soad, I. 2001. Survey of scale insects of ornamental plants in Alexandria Governorate, Egypt. *Mededelingen Faculteit Landbouwkundige en Toegepaste Biologische Wetenschappen Universiteit Gent* **66(2B)**: 571-580. [MouradMeFa2001]

Moussa, M.E. 1986. Bionomics, relative abundance and natural mortality rates of the oriental red scale, *Aonidiella orientalis* (Newstead) in Riyadh, Saudi Arabia (Homoptera, Diaspididae). *Journal of the College of Agriculture, King Saud University* **8(1)**: 227-234. [Moussa1986]

Moutia, L.A. & Mamet, R.J. 1946. A review of twenty-five years of economic entomology in the island of Mauritius. *Bulletin of Entomological Research* **36**: 439-472. [MoutiaMa1946]

Moutia, L.A. & Mamet, R.J. 1947. An annotated list of insects and Acarina of economic importance in Mauritius. *Bulletin (Department of Agriculture, Mauritius)* **No. 29**: 43 pp. [MoutiaMa1947]

Moznette, G.F. 1920. The dictyospermum scale on the avocado and how it may be controlled. *Quarterly Bulletin of the Florida State Plant Board* **5**: 5-11. [Moznet1920]

Moznette, G.F. 1922. Insects injurious to the mango in Florida and how to combat them. *Farmer's Bulletin (USDA)* No. 1257. [Moznet1922]

Moznette, G.F. 1922a. The avocado: Its insect enemies and how to combat them. *Farmer's Bulletin (USDA)* **1261**: 31 pp. [Moznet1922a]

Mueller, F.P. & Eisenschmidt, H. 1954. *Quadraspidiotus schneideri* Bachmann (= *Marani* Zahradnik), eine der San Jose-Schildlaus ähnliche Deckelschildlaus. *Nachrichtenblatt für den Deutschen Pflanzenschutzdienst. Berlin* **8**: 151-153. [MuelleEi1954]

Muma, M.H. 1948. Insect parasitism and related biological factors as concerned with citrus insect and mite control. *Florida Agricultural Experiment Station Annual Report* **1957**: 1993-194. [Muma1948]

Muma, M.H. 1955. Factors contributing to the natural control of citrus insects and mites in Florida. *Journal of Economic Entomology* **48(4)**: 432-438. [Muma1955]

Muma, M.H. 1959. Natural control of Florida red scale on citrus in Florida by predators and parasites. *Journal of Economic Entomology* **52**: 577-586. [Muma1959]

Muma, M.H. 1969. Biological control of various insects and mites on Florida citrus. Pages 863-870. *in*: Chapman, H.D. (Ed.). Proceedings First International Citrus Symposium. Vol. 2. Riverside: University of California. [Muma1969]

Muma, M.H. 1971. Parasitism of armored scale insects in commercial Florida citrus groves. *Florida Entomologist* **54**: 139-150. [Muma1971]

Muma, M.H. & Clancy, D.W. 1961. Parasitism of purple scale and Florida red scale in Florida citrus groves. *Proceedings of the Florida State Horticultural Society* **74**:29-32. [MumaCl1961]

Muma, M.H., Selhime, A.G. & Denmark, H.A. 1961. An annotated list of predators and parasites associated with insects and mites of Florida citrus. *Bulletin (University of Florida, Agricultural Experiment Stations)* **No. 634**: 1-39. [MumaSeDe1961]

Munger, F. 1948. Body-temperature measurements of the California red scale. *Journal of Economic Entomology* **41**: 422-423. [Munger1948]

Munger, F. & Cressman, A.W. 1948. Effect of constant and fluctuating temperatures on the rate of development of California red scale. *Journal of Economic Entomology* **41**: 424-427. [MungerCr1948]

Muniappan, R. & Marutani, M. 1989. Biology and biological control of the red coconut scale, *Furcaspis oceanica* (Lindinger). Pages 17-18 *in*: Tropical and Subtropical Agricultural Research Under PL 89-106. Special Research Grants. Progress and Achievements. The Pacific Basin Group. Honolulu, Hawaii: University of Hawaii at Manda. [MuniapMa1989]

Muñoz Ginarte, B. 1937. Cultivo del cocotero. *Revista de Agricultura (Havana, Cuba)* **2(1)**: 3-9. [MunozG1937]

Munting, J. 1965. New and little known armoured scales (Homoptera: Diaspididae) from South Africa -- 1. *Journal of the Entomological Society of Southern Africa* **27**: 230-238. [Muntin1965]

Munting, J. 1965a. Lac insects (Homoptera: Lacciferidae) from South Africa. *Journal of the Entomological Society of Southern Africa* **28**: 32-43. [Muntin1965a]

Munting, J. 1965b. New and little known armoured scales (Homoptera: Diaspididae) from South Africa -- 2. *Journal of the Entomological Society of Southern Africa* **28**: 179-216. [Muntin1965b]

Munting, J. 1967a. New and little known armoured scales (Homoptera: Diaspididae) from South Africa -- 3. *Journal of the Entomological Society of Southern Africa* **30**: 251-274. [Muntin1967a]

Munting, J. 1969. Observations on some armoured scale insects (Homoptera: Coccoidea: Diaspididae) from south west Africa and neighbouring territories. *Cimbebasia* **1 (ser. A)**: 115-161. [Muntin1969]

Munting, J. 1969a. Three new species of the subtribe Selenaspidina (Coccoidea: Diaspididae) from Africa. *Journal of the Entomological Society of Southern Africa* **32**: 291-296. [Muntin1969a]

Munting, J. 1970a. Lectotype designation for some South African armoured scale insects. *Revue de Zoologie et de Botanique Africaine* **82**: 35-40. [Muntin1970a]

Munting, J. 1971. On the identity of the grey scale *Diaspidiotus africanus* Marlatt and some related species from southern Africa (Homoptera: Diaspididae). *Journal of the Entomological Society of Southern Africa* **34**: 119-143. [Muntin1971]

Munting, J. 1971a. Notes on the genus *Aspidiotus* Bouché (Homoptera: Diaspididae). *Journal of the Entomological Society of Southern Africa* **34**: 305-317. [Muntin1971a]

Murakami, Y. 1970. A review of biology and ecology of Diaspine scales in Japan (Homoptera, Coccoidea). *Mushi* **43**: 65-114. [Muraka1970]

Murakami, Y., Abe, N. & Cosenza, G.W. 1984. Parasitoids of scale insects and aphids on citrus in the Cerrados region of Brazil (Hymenoptera: Chalcidoidea). *Applied Entomology and Zoology. Tokyo* **19**: 237-244. [MurakaAbCo1984]

Murdoch, W.W. 1990. The relevance of pest-enemy models to biological control. Pages 1-24 *in*: Mackauer, M., Ehler, L.E. & Roland, J. (Eds.). Critical Issues in Biological Control. Andover, Hants: Intercept. 330 pp. [Murdoc1990]

Murdoch, W.W., Briggs, C.J. & Nisbet, R.M. 1996. Competitive displacement and biological control in parasitoids: a model. *American Naturalist* **148**: 807-826. [MurdocBrNi1996]

Murdoch, W.W., Chesson, J. & Chesson, P.L. 1985. Biological control in theory and practice. *American Naturalist* **125(3)**: 344-366. [MurdocChCh1985]

Murdoch, W.W., Luck, R.F., Swarbrick, S.L., Walde, S.J., Yu, D.S. & Reeve, J.D. 1995. Regulation of an insect population under biological control. *Ecology* **76(1)**: 206-217. [MurdocLuSw1995]

Murdoch, W.W., Luck, R.F., Walde, S.J., Reeve, J.D. & Yu, D.S. 1989. A refuge for red scale under control by *Aphytis*: structural aspects. *Ecology* **70(6)**: 1707-1714. [MurdocLuWa1989]

Murdoch, W.W., Nisbet, R.M., Luck, R.F., Godfray, H.C.J. & Gurney, W.S.C. 1992. Size-selective sex-allocation and host feeding in a parasitoid-host model. *Journal of Animal Ecology* **61**: 533-541. [MurdocNiLu1992]

Murdoch, W.W., Swarbrick, S.L., Luck, R.F., Walde, S. & Yu, D.S. 1996. Refuge dynamics and metapopulation dynamics: an experimental test. *American Naturalist* **147**: 424-444. [MurdocSwLu1996]

Myartseva, S.N. 1972. [The fauna of Coccidae and Aleurodidae in South Turkmenia and the biology of the principal species.]. [In Russian]. *Izvestiya Akademii Nauk Turkmenskoi SSR Biologii Nauk* **1**: 53-60. [Myarts1972]

Myartseva, S.N. 1982. [New species of parasitoids Hymenoptera (Chalcidoidea, Encyrtidae) from central Asia and Afghanistan.]. [In Russian]. *Izvestiya Akademii Nauk Turkmenskoi SSR Biologii Nauk* **6**: 39-46. [Myarts1982]

Myartseva, S.N. 1984. [Natural enemies of the family Encyrtidae (Hymenoptera, Chalcidoidea) in Turkmenistan and scientific bases for their use in plant protection.]. [In Russian]. *Izvestiya Akademii Nauk Turkmenskoi SSR Biologii Nauk* **No. 2**: 23-30. [Myarts1984]

Myartseva, S.N. & Ruíz-Cancino, E. 2000. Annotated checklist of the Aphelinidae (Hymenoptera: Chalcidoidea) of México. *Folia Entomologica Mexicana* **109**: 7-33. [MyartsRu2000]

Myers, L.E. 1925. The generic types of the Diaspididae (Hemiptera). Part I. The genotypes of *Diaspis* and *Aspidiotus*. *Bulletin of Entomological Research* **16**: 164-167. [Myers1925]

Myers, L.E. 1927. The generic types of the Diaspididae (Hemiptera). Part II. *Bulletin of Entomological Research* **17**: 341-346. [Myers1927LE]

National Research Council (U.S.). Committee on Plant and Animal Pests. Subcommittee on Insect Pests. 1969. Insect-Pest Management and Control. Vol. 3. Publication 1695. Washington, D.C.: National Academy of Sciences. 508 pp. [NRC1969]

New South Wales. Dept. of Agriculture. Entomology Branch. 1963. Insect Pest Survey for the Year Ending 30th June, 1962. [n.p.]. 46 pp. [NSWDAE1963]

Nadel, D.J. & Biron, S. 1964. Laboratory studies and controlled mass rearing of *Chilocorus bipustulatus* Linn., a citrus scale predator in Israel. *Rivista di Parassitologia* **25(3)**: 195-206. [NadelBi1964]

Nafus, D.M. 1996. An insect survey of the Marshall Islands. *Technical Paper of the South Pacific Commission* **No. 208**: 1-35. [Nafus1996]

Nagaraja, H. & Hussainy, S.U. 1968. A study of six species of *Chilocorus* (Coleoptera: Coccinellidae) predaceous on San José and other scale insects. *Oriental Insects. New Delhi* **1(3/4)**: 249-256. [NagaraHu1968]

Nagarkatti, S. & Sankaran, T. 1990. 3.9.3 Tea. Pages 553-562 *in*: Rosen, D. (Ed.). Armored Scale Insects, Their Biology, Natural Enemies and Control [Series title: World Crop Pests, Vol. 4B]. Amsterdam, the Netherlands: Elsevier. 688 pp. [NagarkSa1990]

Nair, M.R.G.K. 1964. Pests of coconut. Pages 72-82. *in*: Pant, N.C., Chief Editor. Entomology in India ["A special number of the Indian Journal of Entomology"]. New Delhi: The Entomological Society of India. pp. 529 [Nair1964]

Nair, R. Balakrishnan & Menon, R. 1963. Major and minor pests in the arecanut crop *Areca catechu* Linn. *Arecanut Journal* **14**: 139-147. [NairMe1963]

Nakahara, S. 1982. Checklist of the armored scales (Homoptera: Diaspididae) of the conterminous United States. U.S. Dept. Agric. Animal & Plant Health Insp. Serv. 110 pp. [Nakaha1982]

Nakahara, S. 1984. A new genus and two new species of armored scales from Mexico (Homoptera: Diaspididae). *Proceedings of the Entomological Society of Washington* **86**: 935-941. [Nakaha1984]

Nakao, S., Takagi, S., Tachikawa, T. & Wongsiri, T. 1977. Scale insects collected on citrus and other plants and their hymenopterous parasites in Thailand. *Insecta Matsumurana (n.s.)* **11**: 61-68. [NakaoTaTa1977]

Nakayama, S. 1912. *Chilocorus similis* Rossi and its relation to scale insects in Japan. *Monthly Bulletin, California (State) Commission of Horticulture* **1(13)**: 932-936. [Nakaya1912]

Nandakumar, C., Reghunath, P. & Sheela, K.R. 1988. On control of the red scale *Aonidiella aurantii* Maskell on rose. *Entomon* **13(3 & 4)**: 275-277. [NandakReSh1988]

Narayanan, E.S., Subba Rao, B.R. & Kaur, R.B. 1957. Some known and new records of parasites of sugarcane scale insects from India. *Indian Journal of Entomology* **19 (Part II)**: 144-146. [NarayaSuKa1957]

Nassar, S.G., Stall, G.B. & Martin, J.W. 1972. The biological effect of juvenile hormone analogs on the development of latania scale, *Hemiberlesia lataniae*. 56th Annual Meeting Pacific Branch Entomological Society of America 1972: 26.

Nath, K. 1972. Studies on the citrus inhabiting coccids (Coccoidea: Hemiptera) of Darjeeling District, West Bengal. *Bulletin of Entomology* **13**: 1-10. [Nath1972]

Navrozidis, E.I., Zartaloudis, Z.D., Papadopoulou, S.H. & Karayiannis, I. 1999. Biology and control of San José scale, *Quadraspidiotus perniciosus* (Comstock) (Hemiptera, Diaspididae) on apricot trees in northern Greece. *Acta Horticulturae* **(No. 488)**: 695-698. [NavrozZaPa1999]

Neiswander, R.B. 1966. Insect and mite pests of trees and shrubs. *Research Bulletin (Ohio Agricultural Research and Development Center -- Wooster, Ohio)* **983**: 1-54. [Neiswa1966]

Nekrasova, Z.N. 1938. Field tests of hydrogen sulfide as a control of the San José scale. *Review of Applied Entomology* **26**: 481. [Nekras1938]

Nel, R.G. 1933. A comparison of *Aonidiella aurantii* and *Aonidiella citrina*, including a study of the internal anatomy of the latter. *Hilgardia* **7**: 417-466. [Nel1933]

Nel, J.J.C., De Lange, L. & Van Ark, H. 1979. Resistance of citrus red scale, *Aonidiella aurantii* (Mask.), to insecticides. *Journal of the Entomological Society of Southern Africa* **42**: 275-281. [NelDeVa1979]

Nepveu, P. 1943. L'affaiblissement végétatif des pêchers attaqués par le pou de San-José: une méthode de mesure. *Compte Rendu de l'Academie d'Agriculture de France* **19(16)**: 461-462. [Nepveu1943]

Nepveu, P. & Vasseur, R. 1946. Sue la contaminabilité des plantes horticoles et spontanées par le pou de San-José *Quadraspidiotus perniciosus* Comst.). *Compte Rendu de l'Academie d'Agriculture de France* **32(10)**: 415-418. [NepveuVa1946]

Nepveu, P. & Vasseur, R. 1948. La contaminabilité de la flore horticole française par le pou de San-José *Quadraspidiotus perniciosus* Comst.). (Document Phytosanitaire, No. 5). Paris: Ministère de l'Agric., Inst. National de la Recherche Agron. et Serv. de la Protection des Végétaux. 41 pp. [NepveuVa1948]

Neuenschwander, P., Michelakis, S. & Alexandrakis, V. 1977. Biologie et écologie d'*Aspidiotus nerii* Bouché (Hom. Diaspididae) sur olivier en Crête Occidentale (Grèce). *Fruits* **32**: 418-427. [NeuensMiAl1977]

Neuffer, G. 1962. Zur Zucht und Verbreitung von *Prospaltella perniciosi* Tower (Hymenoptera, Aphelinidae) und anderen Parasiten der San-José-Schildlaus (*Quadraspidiotus perniciosus* Comstock; Homoptera, Diaspidinae) in Baden-Würtenburg. *Nachrichtenblatt für den Deutschen Pflanzenschutzdienst. Berlin* **7**: 97-101. [Neuffe1962]

Neuffer, G. 1964. Bemerkungen zur Parasitenfauna von *Quadraspitiotus perniciosus* Comst. and zur Zucht bisexueller *Prospaltella perniciosi* Tow. im Insektarium. *Zeitschrift für Pflanzenkrankheiten und Pflanzenschutz* **71(1)**: 1-11. [Neuffe1964]

Neuffer, G. 1964a. Zu den aussetzversuchen von *Prospaltella perniciosi* Tower (Hymenoptera, Aphelinidae) gegen die San José schildlaus (*Quadraspidiotus perniciosus* Comstock) in Baden-Württemberg. *Entomophaga* **9**: 131-136. [Neuffe1964a]

Neuffer, G. 1966. Zur parasitenfauna von *Quadraspidiotus perniciosus* Comstock (Hom., Diaspidinae) unter Besonderer Berücksichtigung der Importierten *Prospaltella perniciosi* Tower (Hym., Aphelinidae). *Entomophaga* **11(4)**: 383-393. [Neuffe1966]

Neuffer, G. 1967. Erfahrungen in der Massenzucht von *Prospaltella perniciosi* Tow. im Veränderten Stuttgarter Insektarium. *Entomophaga* **12**: 235-239. [Neuffe1967]

Neuffer, G. 1968. Die Wirksamfeit der Aphelinide *Prospaltella perniciosi* Tower im südwestdeutschen Befallsgebiet der San-José-Schildlaus *Quadraspidiotus perniciosus* Comstock. *Anzeiger für Schädlingskunde vereinigt mit Schädlingsbekämpfung* **41**: 97-101. [Neuffe1968]

Neuffer, G. 1969. Biological control of the San José scale with *Prospaltella perniciosi* Tow. in south-western Germany. *OEPP/EPPO Public. Ser. A* **No. 48**: 49-55. [Neuffe1969]

Neuffer, G. 1969a. Use of climate-diagrams to characterize the climatic conditions of areas infested by the San José scale (*Quadraspidiotus perniciosus* Comst.). *OEPP/EPPO Public. Ser. A* **No. 48**: 63-68. [Neuffe1969a]

Neuffer, G. 1990. Zur Abundanz und Gradation der San-José-Schildlaus *Quadraspidiotus perniciosus* Comst. und deren Gegenspieler *Prospaltella perniciosi* Tow. Eine Zusammenfassung 30 jahriger Untersuchungen. *Gesunde Pflanzen* **42**: 89-96. [Neuffe1990]

Neves, M. 1935. [Contribution to the study of Portuguese coccids.] Contribution à l'étude des coccides du Portugal.]. [In French]. *Bulletin de la Société Portugaise des Sciences Naturelles* **12**: 59-61. [Neves1935]

Newcomer, E.J. & Yothers, M.A. 1929. Sterility in the San José scale. *Journal of Economic Entomology* **22**: 821-822. [NewcomYo1929]

Newcomer, E.J. & Yothers, M.A. 1931. Experiments for the control of the San José scale with lubricating-oil emulsions in the Pacific northwest. *Circular (United States Department of Agriculture)* **No. 175**: 1-12. [NewcomYo1931]

Newell, W. 1899. On the North American species of the subgenera *Diaspidiotus* and *Hemiberlesia*, of the genus *Aspidiotus*. *Contributions of the Department of Zoology and Entomology, Iowa State College* **3**: 1-31. [Newell1899]

Newell, W. 1923. Tropical and sub-tropical quarantines: Summary. *Journal of Economic Entomology* **16**: 263-266. [Newell1923]

Newell, W. & Cockerell, T.D.A. 1898. [Descriptions of new coccids.]. *In* Osborn, H., Notes on Coccidae Occurring in Iowa. *Proceedings of the Iowa Academy of Sciences* **5**: 224-31. [NewellCo1898]

Newell, W. & Rosenfeld, A.H. 1908. A brief summary of the more important injurious insects of Louisiana. *Journal of Economic Entomology* **1**: 150-155. [NewellRo1908]

Newstead, R. 1893d. Observations of Coccidae (No. 5). *Entomologist's Monthly Magazine* **29**: 185-188. [Newste1893d]

Newstead, R. 1893f. Observations on Coccidae (No. 7). *Entomologist's Monthly Magazine* **29**: 279-280. [Newste1893f]

Newstead, R. 1894. Observations on Coccidae (No. 8). *Entomologist's Monthly Magazine* **30**: 179-183. [Newste1894]

Newstead, R. 1894c. Scale insects in Madras. *Indian Museum Notes* **3**: 21-32. [Newste1894c]

Newstead, R. 1896. Observations on Coccidae (no. 14). *Entomologist's Monthly Magazine* **32**: 57-60. [Newste1896]

Newstead, R. 1896a. Observations on Coccidae (No. 15). *Entomologist's Monthly Magazine* **32**: 132-134. [Newste1896a]

Newstead, R. 1897a. New Coccidae collected in Algeria by the Rev. Alfred E. Eaton. *Transactions of the Entomological Society of London* **1897**: 93-103. [Newste1897a]

Newstead, R. 1898. Observations on Coccidae (No. 17). *Entomologist's Monthly Magazine* **34**: 92-99. [Newste1898]

Newstead, R. 1901a. Observations on Coccidae (no. 19). *Entomologist's Monthly Magazine* **37**: 81-86. [Newste1901a]

Newstead, R. 1901b. Monograph of the Coccidae of the British Isles. London: Ray Society. 220 pp. [Newste1901b]

Newstead, R. 1906a. Report on insects sent from Der Kaiserliche Biologische Anstalt fur Land- und Forstwirtschaft Dahlem, Berlin. *Quarterly Journal. Institute of Commercial Research in the Tropics. University of Liverpool* **1**: 73-74. [Newste1906a]

Newstead, R. 1908b. On a collection of Coccidae and other insects affecting some cultivated and wild plants in Java and in Tropical Western Africa. *Journal of Economic Biology. London* **3**: 33-42. [Newste1908b]

Newstead, R. 1910a. On scale insects (Coccidae) from the Uganda Protectorate. *Bulletin of Entomological Research* **1**: 64-69. [Newste1910a]

Newstead, R. 1910c. Some further observations on the scale insects (Coccidae) of the Uganda Protectorate. *Bulletin of Entomological Research* **1**: 185-199. [Newste1910c]

Newstead, R. 1911. Observations on African scale insects (Coccidae). (No. 3). *Bulletin of Entomological Research* **2**: 85-104. [Newste1911]

Newstead, R. 1911a. On a collection of Coccidae and Aleurodidae, chiefly African, in the collection of the Berlin Zoological Museum. *Mitteilungen aus dem Zoologischen Museum in Berlin* **5**: 155-174. [Newste1911a]

Newstead, R. 1912. On a collection of African Coccidae collected by Mr. L. Schultze in South and West Africa. *Zool. Anthropol. Ergeb. Forschungsreise West. und Zent. Südafrika* **5**: 15-20. [Newste1912]

Newstead, R. 1913. Notes on scale-insects (Coccidae). - Part I. *Bulletin of Entomological Research* **4**: 67-81. [Newste1913]

Newstead, R. 1914. Notes on scale-insects (Coccidae). Part II. *Bulletin of Entomological Research* **4**: 301-311. [Newste1914]

Newstead, R. 1917. Observations on scale-insects (Coccidae) - III. *Bulletin of Entomological Research* **7**: 343-380. [Newste1917]

Newstead, R. 1917a. Observations on scale-insects (Coccidae) - IV. *Bulletin of Entomological Research* **8**: 1-34. [Newste1917a]

Newstead, R. 1917b. Observations on scale-insects (Coccidae) - V. *Bulletin of Entomological Research* **8**: 125-134. [Newste1917b]

Newstead, R. 1920. Observations on scale-insects (Coccidae) - VI. *Bulletin of Entomological Research* **10**: 175-207. [Newste1920]

Nicolas, J. 1979. [The San José scale: Periods of swarming and chemical control.]. *Bulletin technique des Pyrénées-Orientales* **93**: 143-148. [Nicola1979]

Noel, P. 1894. Un insecte nouveau pour la faune Francaise. L'*Aspidiotus ostreaeformis*. *Bulletin de la Société des Amis des Sciences Naturelles de Rouen* **29**: 67-72. [Noel1894]

Nonell Comas, D.J. 1932. Trabajos de las estaciones de fitopatología agrícola en el año 1931. Estación de fitopatología agrícola de Barcelona. *Boletín de Patología Vegetal y Entomología Agrícola. Madrid* **6**: 183-190. [Nonell1932]

Nonell Comas, D.J. 1935. Estación de Fitopatología Agrícola de Barcelona. *Plagas del Campo, Memoria del Servicio Fitopatológico Agrícola (Madrid)* **3**: 281-287. [Nonell1935]

Normant, J.F. & Alexakis, A. 1982. Reactifs organo-cuivreux: nouveaux agents de synthèse de phéromones sexuelles d'insectes. *Colloques de l'INRA* **(no. 7)**: 33-40. [NormanAl1982]

Norris, D.M. & Kogan, M. 1 1980. Biochemical and morphological bases of resistance. Pages 22-61 *in*: Maxwell, F.G. & Jennings, P.R. (Eds.). Breeding Plants Resistant to Insects. New York: John Wiley. 683 pp. [NorrisKo1980]

Notz P., A. 1974. Reconocimiento de predatores de escamas (Homoptera, Coccoidea) en las zonas de Maracay y sus alrededores, Estado Aragua, Venezuela. *Revista Facultad de Agronomía. Maracay* **8**: 127-143. [NotzP1974]

Nour El-Dine, A.M. & Rizkallah, R. 1970. Unrecorded insects and pests injurious to some ornamental plants in Egypt. *Bulletin de la Société Entomologique d'Egypte* **54**: 123-127. [NourElRi1970]

Novartis Agribusiness Chile SA. 2000. [Integrated pest management programme. Pome fruit orchards.] Programa manejo integrado de plagas. 31 pp. [NAC2000]

Novitsky, S. 1961. Synonymic notes on the San José scale parasite *Pteroptrix dimidiata* Westw. (Hym., Chalcidoidea, Aphelinidae). *Entomologist's Monthly Magazine* **97**: 193-194. [Novits1961]

Noyes, J.S. 1990a. 2.4.2 Encyrtidae. Pages 133-166 *in*: Rosen, D. (Ed.). Armored Scale Insects, Their Biology, Natural Enemies and Control [Series title: World Crop Pests, Vol. 4B]. Amsterdam, the Netherlands: Elsevier. 688 pp. [Noyes1990a]

Nucifora, S. 1991. *Rungaspis capparidis* (Bodenheimer) su genistinae e capparidaceae in Pantelleria e in Sicilia (Homoptera: Diaspididae). *Atti XVI Congresso Nazionale Italiano di Entomologia. Bari - Martina Franca (Ta): 23/28 September. Italy* **16**: 533-536. [Nucifo1991]

Nucifora, S. & Watson, G.W. 2001. Armoured scale insects (Hemiptera: Coccoidea: Diaspididae) new to Sicily: records and observations. *Entomologica* **33(1999)**: 207-211. [NucifoWa2001]

Nur, U. 1990a. 1.2 Reproductive biology and genetics. 1.2.1 Chromosomes, sex-ratios, and sex determination. Pages 179-190 *in*: Rosen, D. (Ed.). Armored Scale Insects, Their Biology, Natural Enemies and Control [Series title: World Crop Pests, Vol. 4A]. Amsterdam, The Netherlands: Elsevier. 384 pp. [Nur1990a]

Nur, U. 1990b. 1.2.2 Parthenogenesis. Pages 191-197 *in*: Rosen, D. (Ed.). Armored Scale Insects, Their Biology, Natural Enemies and Control [Series title: World Crop Pests, Vol. 4A]. Amsterdam, The Netherlands: Elsevier. 384 pp. [Nur1990b]

Obra, G.P.B. & Rejesus, B.M. 2000. Biological studies of *Aphytis* sp. nr. *chrysomphali* (Hymenoptera: Aphelinidae). *Philippine Entomologist* **14(2)**: 137-147. [ObraRe2000]

Obretenchev, D. & Stanev, M. 1992. [Methyl-bromide gassing of planting fruit stock against the San José scale *Quadraspidiotus perniciosus*.]. [In Bulgarian]. *Rausteniev'dni Nauki (Plant Science)* **29(3/6)**: 118-123. [ObreteSt1992]

Odinokov, V.N., Kukovinets, O.S., Zainullin, R.A., Tsyglintseva, E.Yu., Sultanmuratova, V.R., Veselovskii, V.V., Dragan, V.A. & Rubinskaya, T.Ya. 1989. Insect pheromones and their analogs. XIX. Preparation of alpha-geranyl propionate - the main component of the sex pheromone of San José scale. *Chemistry of Natural Compounds* **25(3)**: 364-366. [OdinokKuZa1989]

Ody, M., Mathys, G. & Huggel, H. 1969. Contribution à l'étude anatomique du pou de San José: *Aspidiotus perniciosus* Comst. (Cochenille, Diaspines). *Schweizerische Entomologische Gesellschaftliche Mitteilungen* **42**: 38-45. [OdyMaHu1969]

Ofek, G., Huberman, G., Yzhar, Y., Wysoki, M., Kuzlitzky, W. & Reneh, S. 1997. The control of the oriental red scale, *Aonidiella orientalis* Newstead and the California red scale, *A. aurantii* (Maskell) (Homoptera: Diaspididae) in mango orchards in Hevel Habsor (Israel). *Alon Hanotea* **51**: 212-218. [OfekHuYz1997]

Ohgushi, R., Nishino, T., Tazoe, S. & Gamo, Y. 1967. Ecological observations and control of California red scale, *Aonidiella aurantii* Maskell. *Proceedings of the Kyushu Association of Plant Protection* **13**: 118-121. [OhgushNiTa1967]

Oil palm and coconut pests in West Africa. 1981. Oil palm and coconut pests in West Africa. *Oleagineux* **36**: 168-228. [OilPaCo1981]

O'Kane, W.C. & Conklin, J.G. 1930. Lime sulphur in relation to San Jose scale and oyster shell scale. Studies of contact insecticides. *Technical Bulletin New Hampshire, Agricultural Experiment Station* 40: 1-15. [OkaneCo1930]

Omer-Cooper, M., Jones, C.M. & Whitehead, G.B. 1946. Predators of mussel scale. *The Citrus Grower* No. 154. [OmerCoJoWh1946]

Omer-Cooper, J. & Whitehead, G.B. 1950. Studies on the biological control of red scale in the Eastern Cape Province. *The Citrus Grower* **195**: 3,12. [OmerCoWh1950]

Omer-Cooper, J., Whitnall, A.B.M. & Glaholm, J. 1946. Natural control of red scale, *Aonidiella aurantii* Mask. *The Citrus Grower* **145**: 5-8. [OmerCoWhGl1946]

Onder, E. P. 1982. Investigations on the biology, food-plants, damage and factors affecting seasonal population fluctuations of *Aonidiella* species (Homoptera: Diaspididae) injurious to citrus trees in Izmir and its surroundings. *Arastirma Eserleri Serisi* **No. 43**: 1-171. [Onder1982]

O'Neil, W.J. & Hantsbarger, W.M. 1952. Apple and pear insect investigations at the Tree Fruit Experiment Station. *Proceedings Washington State Horticulture Association* 488: 170-178. [OneilHa1950]

Ordish, G. 1967. Biological Methods in Crop Pest Control. London: Constable and Company, Ltd.. 242 pp. [Ordish1967]

Ordogh, G. & Takacs, A. 1983. [The effectivity of Pyrotox against larvae of scale insects (*Pseudococcus maritimus, Saissetia coffeae, Aspidiotus nerii*) damaging potted ornamentals.]. [In Hungarian]. *Növényvédelem* **19(9)**:417-419. [OrdoghTa1983]

Orlinskii, A.D. 1989a. [Development of the bio-method in Bulgaria.]. [In Russian]. *Zashchita Rastenii. Moscow* **No. 12**: 54. [Orlins1989a]

Orphanides, G.M. 1983. Biology of the California red scale, *Aonidiella aurantii* (Maskell) (Homoptera, Diaspididae), and its seasonal availability for parasitization by *Aphytis* spp. in Cyprus. *Bollettino del Laboratorio di Entomologia Agraria 'Filippo Silvestri'. Portici* **39**: 203-212. [Orphan1983]

Orphanides, G.M. 1984. Population dynamics of the California red scale, *Aonidiella aurantii* (Maskell) (Homoptera: Diaspididae) on citrus in Cyprus. *Bollettino del Laboratorio di Entomologia Agraria 'Filippo Silvestri'. Portici* **41**: 195-209. [Orphan1984]

Orphanides, G.M. 1984a. Competitive displacement between *Aphytis* spp. [Hym. Aphelinidae] parasites of the California red scale in Cyprus. *Entomophaga* **29**: 275-281. [Orphan1984a]

Orphanides, G.M. 1996. Establishment of *Comperiella bifasciata* (Hym.: Encyrtidae) on *Aonidiella aurantii* (Hom.: Diaspididae) in Cyprus. *Entomophaga* **41**: 53-57. [Orphan1996]

Ortu, S. & Acciaro, M. 1999. [Preliminary observations on the use of developmental degree-day accumulation for the control of *Aonidiella aurantii* (Maskell) in Sardinia.]. [In Italian]. *Frustula Entomologica, Pisa (N.S.)* **22(35)**: 127-131. [OrtuAc1999]

Osborn, H. 1898. Notes on Coccidae occurring in Iowa. *Proceedings of the Iowa Academy of Sciences* **5**: 224-231. [Osborn1898]

Osburn, M.R. 1939. Control of the purple scale and Florida red scale. *Journal of Economic Entomology* **32(5)**: 688-690. [Osburn1939]

Osburn, M.R. & Mathis, W. 1944. Oil sprays with or without derris resins to control Florida Red Scale. *Journal of Economic Entomology* **37(4)**: 516-519. [OsburnMa1944]

Osburn, M.R. & Mathis, W. 1946. Effect of cultivation on Florida red scale populations. *Journal of Economic Entomology* **39(5)**: 571-574. [OsburnMa1946]

Osburn, M.R. & Mathis, W. 1948. Effect of DDT on Florida red scale populations. *Journal of Economic Entomology* **41(3)**: 454-456. [OsburnMa1948]

Osburn, M.R. & Pierce, W.C. 1963. Controlling insects and diseases of the pecan. *Agriculture Handbook (Agriculture Research Service, United States Department of Agriculture)* **No. 240**: 1-52. [OsburnPi1963]

Osburn, M.R. & Spencer, H. 1938. Effect of spray residues on scale insect populations. *Journal of Economic Entomology* **31(6)**: 731-732 [OsburnSp1938]

Otero, A.R. 1935. Insectos del Guayabo en Cuba. *Circular (Estación Experimental Agronómica, Santiago de Las Vegas, Cuba)* **No. 78**: 1-26. [Otero1935]

Otero, O., Castellanos, A., Mora, J., Rodrigeuz, N., Reinaldo, I., Caarera, C. González, Arteaga, E., Broch, R., Montes, M., Fernández, O. & Borges, M 1996. Integrated pest management in Cuban citrus groves. *Proceedings of the International Society of Citriculture* **I**: 530-535. [OteroCaMo1996]

Pace, T. 1939. Doenças e pragas da laranjeira. *Chacaras e Quintaes* **59**: 664-665. [Pace1939]

Pagliano, T. 1929. Les insectes et autres parasitres de l'olivier. Pages 274-307 *in*: [9th International Congress of Olive Culture.] IXe Congrès international d'oléiculture. Vol. II. Tunis: Régence de Tunis - Protectorat Français. Direction Gén. de l'agriculture, du Commerce/Colonisation. [Paglia1929]

Paik, W.H. 1958. [Citrus insect problem in Quelpart Island.]. [In Korean]. *Journal of Applied Zoology* **1(1)**: 31-35. [Paik1958]

Paik, W.H. 1972. [Scale insects found in the greenhouses in Korea.]. [In Korean]. *Korean Journal of Plant Protection* **11**: 1-4. [Paik1972]

Painter, R.H. 1951. Insect Resistance in Crop Plants. New York: The Macmillan Company. 520 pp. [Painte1951]

Paiva, P.E.B., Pinto, R.A., Silva, J.L. & Gravena, S. 1997. Efeito de abamectin associado a óleo vegetal sobre a cochonilha pardinha {Selenaspidus articulatus} e ácaro de ferrugem *Phyllocoptruta oleivora* em citros. Page 161 *in*: Bento, J.M.S. & Delalibera, I. (Eds.). [16th Brazilian Congress of Entomology] 16o Congresso Brasileiro de Entomologia. Salvador: Sociedade Entomológica do Brasil/EMBRAPA-CNPMF. 400 pp. [PaivaPiSi1997]

Palaro, M.R., Silva, J.L., Paiva, P.E.B. & Gravena, S. 1997. Efeito de ethion gel no controle da cocho-nilha pardinha *Selenaspidus articulatus* em citros. Page 161 *in*: Bento, J.M.S. & Delalibera, I. (Eds.). [16th Brazilian Congress of Entomology] 16o Congresso Brasileiro de Entomologia. Salvador: Sociedade Entomológica do Brasil/EMBRAPA-CNPMF. 400 pp. [PalaroSiPa1997]

Palmer, J.M. & Mound, L.A. 1990. 2.2.5 Thysanoptera. Pages 67-76 *in*: Rosen, D. (Ed.). Armored Scale Insects, Their Biology, Natural Enemies and Control [Series title: World Crop Pests, Vol. 4B]. Amsterdam, the Netherlands: Elsevier. 688 pp. [PalmerMo1990]

Paloukis, S. 1987. Evaluation of two new insecticides for the control of scale insects on fruit trees in northern Greece. *Bollettino del Laboratorio di Entomologia Agraria 'Filippo Silvestri'* **43 (Suppl.)**: 179-183. [Palouk1987]

Paloukis, S.S. & Navrozidis, E.I. 1995. Effectiveness of a new insecticide (Diofenolan) for control of San José scale, *Quadraspidiotus perniciosus* (Comstock) (Diaspididae), on peach trees in northern Greece. *Israel Journal of Entomology* **29**: 285-286. [PaloukNa1995]

Pampaloni, L. 1902. I resti organici nel disodile dimelilli in Sicilia. *Palaeontographia Italica. Memorie di Paleontologia* **8**: 121-130. [Pampal1902]

Pampaloni, L. 1903. Microflora and microfauna in "disodile" in Melilli, Sicily. *Bollettino dell' Accademia Gioenia di Scienze Nat. Catania* **67**: 248-253. [Pampal1903]

Pan, W.Y., Chen, S.L., Lian, J.H., Qiu, H.Z. & Lan, G. 1989. [A preliminary report on control of *Hemiberlesia pitysophila* using *Cladosporium cladosporioides*.]. [In Chinese]. *Forest Pest and Disease* **No. 3**: 22-23. [PanChLi1989]

Pan, W.Y., Tang, Z.Y., Chen, Z.F., Yang, Z.X. & Lian, J.H. 1989. [Studies on bionomics of *Hemiberlesia pitysophila* and its control.]. [In Chinese].*Forest Pest and Disease* **No. 1**: 1-6. [PanTaCh1989]

Pan, W.Y., Tang, Z.Y., Xia, G.L., Lian, J.H. & Hu, J.L. 1987. [Studies of a new destructive forest pest in South China: pine needle Hemiberlesian scale (Coccoidea: Diaspididae).]. [In Chinese]. *Contributions from Shanghai Institute of Entomology* **7**: 177-189. [PanTaXi1987]

Paoli, G. 1915. Contributo alla conoscenze della cocciniglie della Sardegna. *Redia (1915)* **11**: 239-268. [Paoli1915]

Paoli, G. 1922. [A mission to collect parasites of *Chrysomphalus dictyospermi* Morgan in Madera Island.]. [In Italian]. *Nuovi annali del Ministero per l'Agricoltura* **2**: 407-416. [Paoli1922]

Paoli, G. 1927a. Casi fitopatologici osservati in Liguria nella primavera-estate 1927. *Bollettino della stazione di patologia vegetale* **7**: 382-387. [Paoli1927a]

Paolo, L. 1920. Coleotteri esotici utili e dannosi alle piante importati in italia e rinvenuti nel Lazio. *Atti Pont Acad. Nuovi Lincei Rome* **74**: 28-31. [Paolo1920]

Pappas, J.L. & Carman, G.E. 1952. Laboratory studies with several organic phosphorous materials on California red scale. *Paper presented at the Pacific Branch meeting of the American Association of Economic Entomologists, Santa Barbara, California, June 25, 1952.* . [PappasCa1952]

Parrott, P.J. 1899. *Aspidiotus fernaldi* (Ckll.) sub-sp., *cockerelli*, sub-sp. nov. *Canadian Entomologist* **31**: 10-11. [Parrot1899]

Parrott, P.J. 1899b. New coccids from Kansas. *Canadian Entomologist* **31**: 280-282. [Parrot1899b]

Parsana, G.J., Malavia, D.D. & Koshiya, D.J. 1994. Effect of drip irrigation on incidence of insect pests of sugarcane. *Gujarat Agricultural University, Research Journal. India* **20**: 15-17. [ParsanMaKo1994]

Pasqualini, E. & Civolani, S. 2000. La difesa dalla cocciniglia di San Jose. *Informatore Agrario* 56(6) 93-95. [PasquaCi2000]

Patil, A.P., Thakur, S.G. & Mohalkar, P.R. 1988. Incidence of pests of turmeric and ginger in Western Maharashtra. *Indian Cocoa, Arecanut & Spices Journal* **12(1)**: 8-9. [PatilThMo1988]

Payne, J.A., Malstrom, H.L. & KenKnight, G.E. 1979. Insect pests and diseases of the pecan. *Agricultural Reviews and Manual (USDA, Science and Education Administration)* ARM-S-5, 43 pp. [PayneMaKe1979]

Peck, S.B., Heraty, J., Landry, B. & Sinclair, B.J. 1998. Introduced insect fauna of an oceanic archipelago: the Galápagos Islands, Ecuador. *American Entomologist* **44(4)**: 219-237. [PeckHeLa1998]

Pedata, P.A. 1999. Description of a new species of *Coccobius* (Hymenoptera: Aphelinidae) parasitoid of *Ephedraspis ephedrarum*, with complementary notes on *Coccobius reticulatus* (Compere et Annecke). *Bollettino del Laboratorio di Entomologia Agraria 'Filippo Silvestri'. Portici* **55**: 45-51. [Pedata1999]

Pegazzano, F. 1948. Ricerche biologiche sulla cocciniglia S. José *Comstockaspis (=Aspidiotus) perniciosa* Comst. *Annali della Facolta di Agraria. Pisa* **9**: 178-188. [Pegazz1948]

Pegazzano, F. 1949. Un diaspino nuovo per la fauna Paleartica: *Pseudaonidia poeoniae* [!] Cockll. (Hemiptera, Coccidae). *Redia* **34**: 233-235. [Pegazz1949]

Pegazzano, F. 1953. Contributo alla conoscenza della *Pseudaonidia paeoniae* Cockll. (Hemiptera, Coccidae, Diaspinae). *Redia* **38**: 281-315. [Pegazz1953]

Pegazzano, F. 1955. *Quadraspidiotus marani* Zahradnik *F. olivicola* F.N. (Hem. Hom. Coccidae, Diaspini) nuovo ospite dell'olivo. *Redia* **40**: 311-324. [Pegazz1955]

Peleg, B.A. 1982. Effect of a new insect growth regulator RO-13-5223 on scale insects. *Phytoparasitica* **10**: 27-32. [Peleg1982]

Peleg, B.A. 1983. Effect of a new insect growth regulator, RO 13-5223, on hymenopterous parasites of scale insects. *Entomophaga* **28**: 367-372. [Peleg1983]

Peleg, B.A. 1988. Effect of a new phenoxy juvenile hormone analog on California Red Scale (Homoptera: Diaspididae), Florida wax scale (Homoptera: Coccidae) and the ectoparasite *Aphytis holoxanthus* DeBach (Hymenoptera: Aphelinidae). *Journal of Economic Entomology* **81**: 88-92. [Peleg1988]

Peleg, B.A. & Bar-Zakay, I. 1995. The pest status of citrus scale insects in Israel (1984-1994). *Israel Journal of Entomology* **29**: 261-264. [PelegBa1995]

Peleg, B.A. & Gothilf, S. 1981. Effect of the insect growth regulators Diflubenzuron and Methoprene on scale insects. *Journal of Economic Entomology* **74**: 124-126. [PelegGo1981]

Peleg, B.A. & Nadel, D.J. 1966. A method of P32 labeling of the armored and soft scale predator *Chilocorus bipustulatus* L. *Israel Journal of Agricultural Research* **16(2)**: 97-98. [PelegNa1966]

Pelekassis, C.E.D. 1962. A catalogue of the more important insects and other animals harmful to the agricultural crops of Greece during the last thirty-year period. *Annales de l'Institut Phytopathologique Benaki* **5(1)**: 104 pp. [Peleka1962]

Pelekassis, C.D. 1974. Historical review of biological control of citrus scale insects in Greece. *Bulletin SROP* **1974/3**: 14-20. [Peleka1974]

Pellizzari-Scaltriti, G. 1987. New data on distribution of some scale insects in Italy. *Bollettino del Laboratorio di Entomologia Agraria 'Filippo Silvestri'* **43 (Suppl.)**: 117-125. [Pelliz1987]

Pellizzari-Scaltriti, G. 1994a. [Homoptera Coccoidea new for Italy.]. [In Italian]. *Bollettino di Zoologia Agraria e di Bachicoltura (Milano)* **26(2)**: 271-274. [Pelliz1994a]

Pellizzari-Scaltriti, G. & Camporese, P. 1991a. Contributo alla conoscenza delle cocciniglie (Homoptera, Coccoidea) delle querce in Italia. Pages 193-209. *in*: [Atti del Convegno: Problematiche fitopatologiche del genere Quercus in Italia. Florence, Italy. [PellizCa1991a]

Pelot, B.L. 1950. Scales *on Chrysalidocarpus*. *Proceedings of the Hawaiian Entomological Society* **14(1)**: 17. [Pelot1950]

Peña, J.E. & Duncan, R. 1999. Seasonality and control of arthropods on carambola cultivars in Southern Florida. *Proceedings of the 112th Annual Meeting of the Florida State Horticultural Society* **No. 112**: 210-212. [PenaDu1999]

Penman, D.R. 1984. Deciduous tree fruit pests. Pages 33-50 *in*: Scott, R.R. (Ed.). New Zealand Pest and Beneficial Insects. Canterbury: Lincoln University College of Agriculture. 373 pp. [Penman1984]

Penzig, O. 1887. Studi Botanici sugli Agrumi e sulle Piante Affini. Rome: Annali di Agricoltura. 590 pp. [Penzig1887]

Peral L., A. 1968. Métodos de cultivo másivo de algunas especies de escamas para la reproducción de parásitos en gran escala. *Fitofilo* **21(59)**: 22-29. [PeralL1968]

Pereira, C.H., Sampaio, M.V., Assunção, E.D., Silva, M.J.P.R. & Cassino, P.C.R. 1997. Flutuação populacional de *Selenaspidus articulatus* no parque da Gleba E, barra da Tijuca, RJ. Page 223 *in*: Bento, J.M.S. & Delalibera, I. (Eds.). [16th Brazilian Congress of Entomology] 16o Congresso Brasileiro de Entomologia] Salvador: Sociedade Entomológica do Brasil/EMBRAPA-CNPMF. 400 pp. [PereirSaAs1997]

Pérez, E.Q. 1972. The biology of the California red scale, *Aonidiella aurantii* (Maskell) (Diaspididae: Homoptera). *Philippine Entomologist* **2**: 227-245. [Perez1972]

Pérez Guerra, G. & Carnero Hernandez, A. 1987. Coccids of horticultural crops in the Canary Islands. *Bollettino del Laboratorio di Entomologia Agraria 'Filippo Silvestri'* **43** (Suppl.): 127-130. [PerezGCa1987]

Perkins, J.H. 1982. Insects, Experts, and the Insecticide Crisis: The Quest for New Pest Management Strategies. New York: Plenum Press. 304 pp. [Perkin1982]

Pernicious (San Jose) scale (Quadraspidiotus perniciosus). 1971. Pernicious (San Jose) scale (*Quadraspidiotus perniciosus*). *Rhodesian Farmer* **42**: 27. [PernicSaJo1971]

Perruso, J.C. & Cassino, P.C.R. 1993. Flutuacao populacional de *Selenaspidus articulatus* (Morg.) (Homoptera: Diaspididae) em *Citrus sinensis* (L.) no estado do Rio de Janeiro. *Anais da Sociedade Entomologica do Brasil* **22(2)**: 401-404. [PerrusCa1993]

Perruso, J.C. & Cassino, P.C.R. 1997. [Presence-absence sampling plan for *Selenaspidus articulatus* (Morg.) (Homoptera: Diaspididae) on citrus.]. [In Portuguese]. *Anais da Sociedade Entomologica do Brasil* **26**: 321-326. [PerrusCa1997]

Perruso, J.C., Cassino, P.C.R., Soares, M.A. & Azevedo, O.R.F. 1997. Aspectos ecológicos de *Selenaspidus articulatus* (Homoptera, Diaspididae) na cultura da laranja no Rio de Janeiro. Page 241 *in*: Bento, J.M.S. & Delalibera, I. (Eds.). [16th Brazilian Congress of Entomology] 16o Congresso Brasileiro de Entomologia. Salvador: Sociedade Entomológica do Brasil/EMBRAPA-CNPMFv. 400 pp. [PerrusCaSo1997]

Pesson, P. 1950. Sur un phénomène de phorésie des spermatozoides par des cellules oviductaires, chez *Aspidiotus ostreaeformis* Curt. (Hemiptera-Homoptera-Coccoidea). *Proceedings of the 8th International Congress of Entomology* 566-570. [Pesson1950]

Pesson, P. & Foldi, I. 1978. Fine structure of the tegumentary glands secreting the protective "shield" in a sessile insect (Homoptera, Diaspididae). *Tissue and Cell* **10**: 389-399. [PessonFo1978]

Petch, T. 1921. Fungi parasitic on scale insects. *Transactions of the British Mycological Society* **7**: 18-40. [Petch1921]

Petch, T. 1921a. Studies in entomogenous fungi. 1. The nectriae parasitic on scale insects. *Transactions of the British Mycological Society* **7**: 89-167. [Petch1921a]

Petschen, I., Borsh, M.P. & Guerrero, A. 2000. Enzyme-catalyzed synthesis and absolute configuration of (1S,2R,5S)- and (1R,2S,5R)-2-(1-hydroxyethyl)-1-(methoxymethyloxyethyl)cyclobutane- 1-carbonitri le, key intermediates for the preparation of chiral cyclobutane-containing pheromones. *Tetrahedron: Assymetry* **11(8)**: 1691-1695. [PetschBoGu2000]

Petschen, I., Parrilla, A., Bosch, M.P., Amela, C. & Botar, A.A. 1999. First total synthesis of the sex pheromone of the oleander scale *Aspidiotus nerii*: An unusual sesquiterpenic functionalized cyclobutane. *Chemistry - A European Journal* **5(11)**: 3299-3309. [PetschPaBo1999]

Pettit, R.H. 1900. Some insects of the year 1899. *Bulletin of the Michigan Agricultural Experiment Station (Entomology Department)* **180**: 1-141. [Pettit1900]

Pfeiffer, D.G. 1985. Pheromone trapping of males and prediction of crawler emergence of San José scale (Homoptera: Diaspididae) in Virginia apple orchards. *Journal of Entomological Science* **20**: 351-353. [Pfeiff1985]

Philipp, W. 1965. Ist die San-José-Schildlaus noch ein Problem?. *Gesunde Pflanzen* **17**: 218-219. [Philip1965]

Picart, J.L. & Matile-Ferrero, D. 2000. Cochenilles en serres de collections. *Phytoma* **524**: 44-46. [PicartMa2000]

Pickel, D.B. 1928. Uma rica collecçao de preparaçooes microscopicas. *Chacaras e Quintaes* **37**: 55. [Pickel1928]

Pickel, C., Bentley, W., Rice, R.E., Brazzle, J., Hasey, J. & Day, K. 2000. Pests of peach. *UC Pest Management Guidelines* . [PickelBeRi2000]

Pickel, C., Olson, W.H., Zalom, F. & Reil, W. 2000. Pests of prune. *UC Pest Management Guidelines* . [PickelOlZa2000]

Pierce, W.C. 1938. Control of the obscure scale on pecan with low concentrations of lubricating oil emulsions. *Journal of Economic Entomology* **31(6)**: 722-724. [Pierce1938]

Pietri-Tonelli, P. de, Biondi, G. & Barontini, A. 1961. [Control experiments against some species of citrus scales.]. [In Italian]. *Contributi (Istituto di ricerche agrarie)* **5**: 21-46. [PietriBiBa1961]

Pietri-Tonelli, P. de, Biondi, G. & Corradini, V. 1969. Development and evaluation of non-oil chemical control treatments. Pages 909-915. *in*: Chapman, H.D. (Ed.). Proceedings First International Citrus Symposium. Vol. 2. Riverside: University of California. [PietriBiCo1969]

Pimentel, D. 1963. Introducing parasites and predators to control native pests. *Canadian Entomologist* **95(8)**: 785-792. [Piment1963]

Pinto, A. de S., Busoli, A.C. & Gravena, S. 1997. Flutuação populacional de *Selenaspidus articulatus* (Morgan) (Hemiptera: Diaspididae) em citros e ocorrência de inimigos naturais no município de Taquaritinga, SP. Page 239 *in*: Bento, J.M.S. & Delalibera, I. (Eds.). [16th Brazilian Congress of Entomology] 16o Congresso Brasileiro de Entomologia. Salvador: Sociedade Entomológica do Brasil/EMBRAPA-CNPMF. 400 pp. [PintoBuGr1997]

Pitt, D.T. 1945. The cultivated blueberry industry in New Jersey, 1944. *Circular (New Jersey Department of Agriculture)* **No. 356**: 3-42. [Pitt1945]

Pizzamiglio, M.A. 1977. Population studies on *Aspidiotus nerii* Bouché (Homoptera: Diaspididae) and two of its natural enemies on California bay trees. *M.S. Thesis, University of California, Berkeley* 96 pp. [Pizzam1977]

Pless, C.D., Deyton, D.E. & Sams, C.E. 1995. Control of San José scale, terrapin scale, and European red mite on dormant fruit trees with soybean oil. *HortScience* **30(1)**: 94-97. [PlessDeSa1995]

Podsiadlo, E. 1995. Interrelations of *Quadraspidiotus zonatus* (Frauenfeld) and its encyrtid parasite of the genus *Metaphycus* Mercet. *Israel Journal of Entomology* **29**: 237-238. [Podsia1995]

Podsiadlo, E. 1997. Morphology of the first instar of *Quadraspidiotus zonatus* (Frauenfeld) (Homoptera, Coccinea, Diaspididae). *Annales Zoologici (Warsaw)* **46**: 193-200. [Podsia1997]

Podsiadlo, E. 1998. Morphology of some developmental stages of *Quadraspidiotus zonatus* (Frauenfeld). Page 32 *in*: VIIIth International Symposium on Scale Insect Studies. 41 [Podsia1998]

Podsiadlo, E. 2000. Morphology of the second instar larva of *Quadraspidiotus zonatus* (Hemiptera: Coccinea: Diaspididae). *Polskie Pismo Entomologiczne* **69**: 397-404. [Podsia2000]

Podsiadlo, E. 2001. Sexual dimorphism in the first- and second-nymphal instars of *Quadraspidiotus zonatus* (Frauenfeld) (Hemiptera: Coccinea: Diaspididae). *Entomologica* **33(1999)**: 139-140. [Podsia2001]

Podsiadlo, E. 2002. A note on the position of thoracic spiracles in Diaspididae (Hemiptera: Coccinea). *Polskie Pismo Entomologiczne* **71**: 159-164. [Podsia2002]

Pokrovskii, E.A., Unterberger, V.K. & Deniskina, G.P. 1960. [Insecticides to control the San José scale.]. [In Russian]. *Zashchita Rastenii of Vreditelei i Boleznei* **1**: 27. [PokrovUnDe1960]

Polavarapu, S., Davidson, J.A. & Miller, D.R. 2000. Life history of the putnam scale, *Diaspidiotus ancylus* (putnam) (Hemiptera: Coccoidea: Diaspididae) on blueberries (*Vaccinium corymbosum*, Ericaceae) in New Jersey, with a world list of scale insects on blueberries. *Proceedings of the Entomological Society of Washington* **102(3)**: 549-560. [PolavaDaMi2000]

Pollini, A. & Bariselli, M. 1998. Stratégie de lutte intégrée contre la migration des larves de pou de San José (*Comstockaspis perniciosa*) en vergers de poiriers en Emilia-Romagna (Italie). Pages 79-80. *in*: First transnational workshop on biological, integrated and rational control: status and perspectives with regard to regional and European experiences Loos-en-Gohelle: Service Regional de la Protection des Végétaux, Nord Pas-De-Calais. [PollinBa1998]

Ponnamma, K.N. 1999. Coccoids associated with oil palm in India - a review. *Planter* **75(882)**: 445-451. [Ponnam1999]

Ponsonby, D.J. & Copland, M.J.W. 2000. Maximum feeding potential of larvae and adults of the scale insect predator, *Chilocorus nigritus* with a new method of estimating food intake. *Biocontrol Dordecht* **45(3)**: 295-310. [PonsonCo2000]

Pope, R.D. 1981. *Rhyzobius ventralis* (Coleoptera: Coccinellidae), its constituent species, and their taxonomy and historical roles in biological control. *Bulletin of Entomological Research* **71**: 19-31. [Pope1981]

Popova, A.I. 1962. [Biology of *Prospaltella perniciosi* Tow., a parasite of San José scale, and its use on the Black seacoast of the Krasnodar area.]. [In Russian]. Pages 147-175 *in*: Zimina, L.C. & Shchepetilinkovoi, V.A. (Eds.). [Biological Method of Controlling Pests and Diseases of Agricultural Crops.] Moscow: Izdatelstvo Selckokhozyaistvennoi Literaturi, Jhurnal i Plakatov. [Popova1962]

Popova, A. I. 1964. [Method of mass rearing a parasite of San José scale.]. [In Russian]. *Trudy Vsesoiuznogo nauchno-issledovatel'skogo instituta zashchity rastenii [Proceedings of the All-Union Scientific-Research Inst. for Plant Prot.]* **20(I)**: 61-64. [Popova1964]

Popova, A.I. 1971. [Chilocoruses in controlling the San José scale.]. [In Russian]. *Zashchita rastenii* **16(2)**: 42-43. [Popova1971]

Popova, A.I. 1979. [Parasites of the males of *Diaspidiotus perniciosus* Comst. (Homoptera, Coccoidea).]. [In Russian]. *Entomologicheskoe Obozrenye* **58**: 538-547. [Popova1979]

Porath, A. 1969. On the establishment and distribution of *Casca smithi* Comp. a parasite of *Chrysomphalus aonidum* L. in Israel. *Israel Journal of Entomology* **4**: 41-43. [Porath1969]

Porcelli, F. 1992. Cocciniglie nuove per l'Italia. *Frustula Entomologica* **13(26)**: 31-38. [Porcel1992]

Porcelli, F. 1995. Antennal sensilla of male Diaspididae (Homoptera): comparative morphology and functional interpretation. *Israel Journal of Entomology* **29**: 25-45. [Porcel1995]

Porcelli, F. & Pizza, M. 1993. [*Quadraspidiotus marani* Zahradník (Homoptera, Diaspididae).]. [In Italian]. *Olivicoltura* **4**: 13-16. [PorcelPi1993]

Potter, D.A., Jensen, M.P. & Gordon, F.C. 1989. Phenology and degree-day relationships of the obscure scale (Homoptera: Diaspididae) and associated parasites on pin oak in Kentucky. *Journal of Economic Entomology* **82(2)**: 551-55. [PotterJeGo1989]

Poutiers, R. 1922. L'acclimatation de *Cryptolaemus montrouzieri* Muls. dans le Midi de la France. *Annales des Epiphyties* **8**: 3-28. [Poutie1922]

Poutiers, R. 1924. Les parasites due *Chrysomphalus dictyospermi* Morg. en France. *Comptes rendus des séances (Académie d'agriculture de France)* **10**: 490-496. [Poutie1924]

Poutiers, R. 1928. Observations sur quelques Hyménoptères parasites de Coccides sur le littoral méditerranéen. *Revue de Pathologie Végétale et d'Entomologie Agricole de France* **15(7)**: 267-270. [Poutie1928]

Poutiers, R. 1948. Le pou de San-José. *Phytoma* **2**: 23-26. [Poutie1948]

Powell, W. 1986. Enhancing parasitoid activity in crops. Pages 319-340 *in*: Waage, J. & Greathead, D. (Eds.). Insect Parasitoids. London: Academic Press. 389 pp. [Powell1986]

Powell, J.A. & Hogue, C.L. 1979. California Insects. (California Natural History Guides, No. 44) Berkeley: University of California Press. 388 pp. [PowellHo1979]

Prasada Rao, V.L.V., Venugopala Rao, N. & Muralidhara Rao, G. 1991. Effect of trashing and insecticidal application on control of scale insect (*Melanaspis glomerata*) of sugarcane (*Saccharum officinarum*). *Indian Journal of Agricultural Sciences* **61(7)**: 516-520. [PrasadVeMu1991]

Pratt, R.M. 1956. Long range relationships between weather factors and scale insect populations. *Proceedings of the Florida State Horticultural Society* **69**: 87-93. [Pratt1956]

Pratt, R.M. 1958. Florida Guide to Citrus Insects, Diseases and Nutritional Disorders in Color. Gainesville, Florida: Agricultural Experiment Station. 191 pp. [Pratt1958]

Pree, D.J., Menzies, D.R., Marshall, D.B. & Fisher, R.W. 1985. Influence of spray volumes on control of San José scale (Homoptera: Diaspididae). *Proceedings of the Entomological Society of Ontario* **115**: 19-23. [PreeMeMa1985]

Price, P.W. 1984. The Concept of the Ecosystem. Pages 19-52. *in*: Huffaker, C.B. & Rabb, R.L. (Eds.). Ecological Entomology. New York: Wiley-Interscience. 844 pp. [Price1984]

Priesner, H. 1931. On the biology of *Chrysomphalus ficus* Ril. (Hem., Cocc.) with suggestions on the control of this species in Egypt. *Bulletin, Ministry of Agriculture, Egypt, Technical and Scientific Service* **117**: 1-19. [Priesn1931]

Priesner, H. & Hosny, M. 1932. The "masked scale" *Chrysomphalus personatus* in Egypt. *Bulletin de la Société Entomologique d'Egypte* **16**: 92-96. [PriesnHo1932]

Priesner, H. & Hosny, M. 1940. Notes on parasites and predators of Coccidae and Aleurodidae in Egypt. *Bulletin de la Société Fouad 1er d'Entomologie* **24**: 58-70. [PriesnHo1940]

Prinsloo, G.L. 1983. A parasitoid-host index of Afrotropical Encyrtidae (Hymenoptera: Calcidoidea). *Entomology Memoirs, Department of Agricultural Technical Services, Republic of South Africa. Pretoria* **60**: 1-35. [Prinsl1983]

Prinsloo, G.L. 1984. An illustrated guide to the parasitic wasps associated with citrus pests in the Republic of South Africa. *Science Bulletin (Department of Agriculture, Republic of South Africa)* **(402)**: 1- 19. [Prinsl1984]

Prinsloo, G.L. & Annecke, D.P. 1976. New Encyrtidae (Hymenoptera) from south west Africa. *Journal of the Entomological Society of Southern Africa* **39(2)**: 185-199. [PrinslAn1976]

Prints, E.Y. 1965. [On the susceptibility of different pear varieties to damage by the San José scale.]. *Vrediya I Poleznaya Fauna Besnozvonochiix Moldavii* **2**: 49-52. [Prints1965]

Priore, R. 1964. Reperti sugli stadi di ibernamento dei doccidi più diffusi e dannosi alle piante, in particolare ai fruttiferi, in Campania. *Bollettino del Laboratorio di Entomologia Agraria 'Filippo Silvestri'. Portici* **22**: 131-178. [Priore1964]

Priore, R. 1964a. Sperimentazione di lotta chimica invernale contro le principali cocciniglie diaspini dei fruttiferi nelle province di Napoli e Caserta Anni 1961-1964. *Bollettino del Laboratorio di Entomologia Agraria. Portici* **22**: 179-204. [Priore1964a]

Priore, R. 1965. Reperti sugli stadi di ibernamento dei doccidi più diffusi e dannosi alle piante, in particolare ai fruttiferi, in Campania. *Annali della Facolta di Scienze Agrarie della Universita degli Studi di Torino* **30(1964-1965)**: 101-145. [Priore1965]

Pritts, M.P. & Hancock, J.F. (Eds.) 1992. Highbush Blueberry Production Guide Ithaca, N.Y.: Northeast Regional Agricultural Engineering Service. 200 pp. [PrittsHa1992]

Provencher, L. 2002. Is *Aspidiotus nerii* Bouché a complex of multiple biotypes of sexual and asexual lineages within a single species?. *Bollettino di Zoologia Agraria e di Bachicoltura (Milano)* **33(3)**: 512. [Proven2002]

Pruna, P.M. & Alayo, R. 1973. Muestreo de Coccoideos y Aleyródidos en los cítricos de Isla de Pinos. *Poeyana (Instituto de Zoologia, Academia de Ciencias de Cuba)* **107**: 1-12. [PrunaAl1973]

Pruthi, H.S. & Batra, H.N. 1960. Important Fruit Pests of North-west India. Delhi: The Indian Council of Agricultural Research (New Delhi). 113 pp. [PruthiBa1960]

Pruthi, H.S. & Mani, M.S. 1945. Our knowledge of the insect and mite pests of citrus in India and their control. *Scientific Monograph (The Imperial Council of Agricultural Research, India)* **No. 16**: 42. [PruthiMa1945]

Pruthi, H.S. & Rao, V.P. 1942. Coccids attacking sugarcane in India. *Indian Journal of Entomology* **4**: 87-88. [PruthiRa1942]

Pruthi, H.S. & Rao, V.P. 1951. San José scale in India. *Bulletin of the Indian Council of Agricultural Research* **71**: 1-48. [PruthiRa1951]

Pulselli, A. 1927. Microcera coccophila. Desm. (1848). *Bollettino della Stazione di Patologia Vegetale* **7**: 300-327. [Pulsel1927]

Putnam, J.D. 1878. Report on maple bark louse. *Annual Report of the Iowa State Horticultural Society* **12**: 317-324. [Putnam1878]

Putnam, J.D. 1880. Biological and other notes on Coccidae. *Proceedings of the Davenport Academy of Natural Sciences (1879-1880)* **2**: 293-347. [Putnam1880]

Puttarudriah, M. & Channabasavanna, G.P. 1953a. Beneficial coccinellids of Mysore - I. *Indian Journal of Entomology* **15**: 87-95. [PuttarCh1953a]

Qin, H.Z., Tang, S.J. & Feng, J.R. 1997. [Preliminary study on *Aonidiella taxus* Leonardi.]. [In Chinese]. *Journal of Shanghai Agricultural College* **15**: 109-113. [QinTaFe1997]

Quaintance, A.L. 1909. Fumigation of apples for the San José scale. *Bulletin (Bureau of Entomology, USDA)* **No. 84**: 33 pp. [Quaint1909]

Quaintance, A.L. 1915. The San José scale and its control. *Farmer's Bulletin (USDA)* **No. 650**: 27 pp. [Quaint1915]

Quaintance, A.L. & Scott, W.M. 1912. The more important insect and fungous enemies of the fruit and foliage of the apple. *Farmer's Bulletin (USDA)* **492**: 48 pp. [QuaintSc1912]

Quayle, H.J. 1910. *Aphelinus diaspidis* Howard. *Journal of Economic Entomology* **3**: 398-401. [Quayle1910]

Quayle, H.J. 1911. Locomotion of certain young scale insects. *Journal of Economic Entomology* **4**: 301-306. [Quayle1911]

Quayle, H.J. 1911a. The red or orange scale *Chrysomphalus aurantii* Mask. *Bulletin of the California Agricultural Experiment Station* **222**: 99-150. [Quayle1911a]

Quayle, H.J. 1911d. Citrus fruit insects. *Bulletin (University of California, College of Agriculture, Agricultural Experiment Station)* **No. 214**: [443]-512. [Quayle1911d]

Quayle, H.J. 1911e. Scale insect parasitism in California. *Journal of Economic Entomology* **5**: 510-515. [Quayle1911e]

Quayle, H.J. 1916. Dispersion of scale insects by the wind. *Journal of Economic Entomology* **9**: 486-492. [Quayle1916]

Quayle, H.J. 1916a. Are scales becoming resistant to fumigation?. *The U.C. Journal of Agriculture* **3**: 333-334,358. [Quayle1916a]

Quayle, H.J. 1920. Recent fumigation experiments. *California Citrograph* **5(6)**: 188,189,193. [Quayle1920]

Quayle, H.J. 1922. Resistance of certain scale insects in certain localities to hydrocyanic acid fumigation. *Journal of Economic Entomology* **15(6)**: 400-405. [Quayle1922]

Quayle, H.J. 1927. Spray and fumigation combination for resistant red scale. *Journal of Economic Entomology* **20**: 667-673. [Quayle1927]

Quayle, H.J. 1932. Biology and control of citrus insects and mites. *Bulletin (University of California, College of Agriculture, Agricultural Experiment Station)* **No. 542**: 1-87. [Quayle1932]

Quayle, H.J. 1938. The development of resistance to hydrocyanic acid in certain scale insects. *Hilgardia* **11**: 183-210. [Quayle1938]

Quayle, H.J. 1938a. Insects of Citrus and Other Subtropical Fruits. Ithaca, New York: Comstock Publishing Co.. 583 pp. [Quayle1938a]

Quayle, H.J. 1942. A physiological difference in the two races of red scale and its relation to tolerance to HCN. *Journal of Economic Entomology* **35**: 813-816. [Quayle1942]

Quayle, H.J. & Ebeling, W. 1934. Spray-fumigation treatment for resistant red scale on lemons. *Bulletin of the Florida Agricultural Experiment Station* **583**: 1-22. [QuayleEb1934]

Quayle, H.J. & Rohrbaugh, P.W. 1934. Temperature and humidity in relation to HCN fumigation for the red scale. *Journal of Economic Entomology* **28**: 1083-1095. [QuayleRo1934]

Quednau, F.W. 1964. Experimental evidence of differential fecundity on red scale [*Aonidiella aurantii*)] in six species of *Aphytis* (Hymenoptera, Aphelinidae). *South African Journal of Agricultural Science* **7**: 335-340. [Quedna1964]

Quednau, F.W. 1964a. An evaluation of fecundity, host-mutilation and longevity on three species of diaspine scales in *Aphytis lingnanensis* Compere (Hymenoptera, Aphelinidae). *South African Journal of Agricultural Science* **7**: 521-530. [Quedna1964a]

Quednau, F.W. 1964b. A contribution on the genus *Aphytis* Howard in South Africa (Hymenoptera: Aphelinidae). *Journal of the Entomological Society of Southern Africa* **27(1)**: 86-116. [Quedna1964b]

Quednau, F.W. 1965. A technique for identifying mixed populations of six species of *Aphytis* parasitic on red scale [*Aonidiella aurantii* (Mask.)], for recognition after recovery from scales collected in citrus orchards (Hymenoptera, Aphelinidae). *South African Journal of Agricultural Science* **8**: 43-56. [Quedna1965]

Quednau, F.W. & Annecke, D.P. 1963. Biological control of citrus red scale. *The South African Citrus Journal* **No. 357**: 11-18. [QuednaAn1963]

Quelch, J.J. (Ed.) 1890. Scale insects *In* Occasional Notes. *Timehri. Journal of the Royal Agricultural and Commercial Society of British Guiana (Georgetown)* **4 (N.S.)**: 369-370. [Quelch1890]

Quezada, J.R., Cornejo, C., Diaz de Mira, A. & Hidalgo, F. 1972. [The main species of insects associated with citrus crops in El Salvador.] [In Spanish]. San Salvador, E. Salvador: Ministerio de Agricultura y Ganaderia. 49 pp. [QuezadCoDi1972]

Quilis Perez, M. 1935. Influencia de los factores climáticos en el cálculo de los ciclos biológicos de los insectos. *VI Congreso Intern. Entom.* **2**: 621-633. [Quilis1935]

Quiroga, D., Arretz, P. & Araya, J.E. 1991. Sucking insects damaging jojoba, *Simmondsia chinensis* (Link) Schneider, and their natural enemies, in the North Central and Central regions of Chile. *Crop Protection* **10(6)**: 469-472. [QuirogArAr1991]

Quisumbing, A.R. & Kydonieus, A.F. 1989. Plastic laminate dispensers. Pages 149-171 *in*: Jutsum, A.R. & Gordon, R.F.S. (Eds.). Insect Pheromones in Plant Protection. Chichester: John Wiley & Sons. 369 pp. [QuisumKy1989]

Royal Legation of Egypt, Rome. 1935. Discoveries and current events: Egypt. *International Bulletin of Plant Protection* **9(3)**: 60-62. [RLER1935]

R. Stazione di entomologia agraria. Florence (Italy). 1915. [Agronomic Entomology: Manual of pest insects on cultivated and native plants and its produce, and control techniques.] [In Italian]. Florence: Tipografia di M. Ricci. 483 pp. [RSEA1915]

Rabb, R.L., DeFoliart, G.K. & Kennedy, G.G. 1984. An ecological approach to managing insect populations. Pages 697-728 *in*: Huffaker, C.B. & Rabb, R.L. (Eds.). Ecological Entomology. New York: Wiley-Interscience. 844 pp. [RabbDeKe1984]

Raciti, E. & Saraceno, F. 2001. Allevamento di *Aphytis melinus* per il controllo biologico di *Aonidiella aurantii*. *Informatore Agrario* **57(19)**: 39-40. [RacitiSa2001]

Raciti, E., Tumminelli, R., Marano, G., Conti, F., Barraco, D., Dinatale, A., Fisicaro, R. & Schillirò, E. 1996. A strategy of integrated pest management in eastern Sicily citrus: first results and economic evaluation. *Proceedings of the International Society of Citriculture* **I**: 652-658. [RacitiTuMa1996]

Rafal'-skii, A.K. & Nishenko, A.A. 1988. [The California scale in the southern steppe of the Ukraine.]. [In Russian]. *Zashchita rastenii. Moscow* **No. 9**: 40-42. [RafalsNi1988]

Ragusa, S. & Russo, A. 1989. Gli artropodi dell'annona in Calabria. *L'informatore* **45(29)**: 71-74. [RagusaRu1989]

Rahman, K.A. & Ansari, A.R. 1941. Scale insects of the Punjab and north-west frontier province usually mistaken for San José scale (with descriptions of two new species). *Indian Journal of Agricultural Sciences* **11**: 816-830. [RahmanAn1941]

Rahman, K.A. & Kalra, A.N. 1940. Volant animals which act as carriers of San José scale. *Review of Applied Entomology* **29**: 272. [RahmanKa1940]

Rajagopal, D. & Krishnamoorthy, A. 1996. Bionomics and management of oriental yellow scale, *Aonidiella orientalis* (Newstead) (Homoptera: Diaspididae): an over view. *Agricultural Research* **17**: 139-146. [RajagoKr1996]

Raju, A.K. & Rao, K.K.P. 1983. Control of the sugarcane scale insect *Melanaspis glomerata* Green with some insecticides and its natural enemies. *Cooperative Sugar* **14**: 571-572. [RajuRa1983]

Raju, K. & Seshagiri, R.G. 1982. A note on the biology of sugarcane scale insect *Melanaspis glomerata* (Green) in Andhra Pradesh. *Indian Sugar* **32**: 19-20. [RajuSe1982]

Ramakrishna Ayyar, T.V. 1919. Some south Indian coccids of economic importance. *Journal of the Bombay Natural History Society* **26**: 621-628. [Ramakr1919]

Ramakrishna Ayyar, T.V. 1919a. A contribution to our knowledge of South Indian Coccidae. *Bulletin of the Agricultural Research Institute, Pusa, India* **87**: 1-50. [Ramakr1919a]

Ramakrishna Ayyar, T.V. 1919b. Notes on new and unrecorded species of Indian Coccidae. *Bulletin of the Agricultural Research Institute. Pusa* **89**: 91-99. [Ramakr1919b]

Ramakrishna Ayyar, T.V. 1921. Short notes on new and known insects from south India. *Proceedings of the Entomology Meetings. India* **4**: 29-40. [Ramakr1921]

Ramakrishna Ayyar, T.V. 1921a. A check list of the Coccidae of the Indian region. *Proceedings of the Entomology Meetings. India* **4**: 336-362. [Ramakr1921a]

Ramakrishna Ayyar, T.V. 1924. A further contribution to our knowledge of south Indian Coccidae. *Proceedings of the Entomology Meetings. India* **5**: 339-344. [Ramakr1924]

Ramakrishna Ayyar, T.V. 1926. Recent additions to the Indo-Ceylonese coccid fauna, with notes on known and new forms. *Journal of the Bombay Natural History Society* **31**: 450-457. [Ramakr1926]

Ramakrishna Ayyar, T.V. 1930. A contribution to our knowledge of South Indian Coccidae (Scales and Mealybugs). *Bulletin of the Imperial Institute of Agricultural Research, Pusa, India* **197**: 1-73. [Ramakr1930]

Ramakrishna Ayyar, T.V. 1938a. An annotated conspectus of the insects affecting fruit crops in S. India. *Madras Agricultural Journal* **26(9)**: 341-351. [Ramakr1938a]

Ramos, J.A. 1946. The insects of Mona Island (West Indies). *Journal of Agriculture of the University of Puerto Rico* **30**: 1-74. [Ramos1946]

Rangel, J.F. & Gomes, J.G. 1945. Guia para reconhecimento e combate das principais doencas e pragas da laranjeira.. *Publicaçao (Serviço de documentaçao, Ministerio da Agricultura, Rio de Janeiro, Brasil)* **No. 11**: 1-44. [RangelGo1945]

Rao, B.R.S. 1957. Some new species of Indian Hymenoptera. *Proceedings of the Indian Academy of Sciences* **46**: 376-390. [Rao1957]

Rao, V.P. 1943. The San José scale - *Quadraspidiotus perniciosus* (Comstock) in South India. *Indian Journal of Entomology* **5**: 245-246. [Rao1943VPa]

Rao, V.P. 1951b. Synonymy of *Semelaspidus artocarpi* (Green). *Proceedings of the Entomological Society of Washington* **53**: 261-262. [Rao1951b]

Rao, V.P. 1969. India as a source of natural enemies of pests of citrus. Pages 785-792. *in*: Chapman, H.D. (Ed.). Proceedings First International Citrus Symposium. Vol. 2. Riverside: University of California. [Rao1969]

Rao, V.P. & Chatterjee, S.N. 1950. The San José scale and other scale insects usually mistaken there for in India. *Indian Journal of Entomology* **10**: 5-29. [RaoCh1950]

Rao, V.P., Ghani, M.A., Sankaran, T. & Mathur, K.C. 1971. A Review of the Biological Control of Insects and Other Pests in South-East Asia and the Pacific Region. Farnam Royal, Slough, England: Commonwealth Agricultural Bureaux. 149 pp. [RaoGhSa1971]

Rao, V.P. & Sankaran, T. 1969. The scale insects of sugar cane. Pages 325-341. Amsterdam, London, New York.: Elsevier. 568 pp. [RaoSa1969]

Rao, V.L.V.P. & Rao, N.V. 1990. Integrated control of scale insect (*Melanaspis* Green) on sugarcane. *Bharatiya Sugar* **15(10)**: 60. [RaoRa1990]

Rao, V.L.V.P. & Rao, N.V. 1990a. Utility of trashing and insecticidal application in the control of the scale insect (*Melanaspis glomerata* G.) on sugarcane. *Cooperative Sugar* **22(3)**: 173-175. [RaoRa1990a]

Rao, V.L.V.P. & Rao, N.V. 1992. Efficacy of soil application of some insecticides in controlling scale insect (*Melanaspis glomerata* Green) on sugarcane ratoon. *Indian Journal of Plant Protection* **20(2)**: 125-128. [RaoRa1992]

Rao, A.V., Rao, T.R., Satyanarayanamurthy, M., Seshagirirao, C. & Rao, B.H.K. 1983. Chemical control of the scale insect (*Melanaspis glomerata* Green) on sugarcane. *Cooperative Sugar* **15**: 11-18. [RaoRaSa1983]

Rauch, F. & Lhoste, J. 1964. Étude sur la phytotoxicité de certains coccicides sur agrumes au Maroc. *Al Awamia* **12(1)**: 61-73. [RauchLh1964]

Raupp, M.J., Holmes, J.J., Sadof, C., Shrewsbury, P. & Davidson, J.A. 2001. Effects of cover sprays and residual pesticides on scale insects and natural enemies in urban forests. *Journal of Arboriculture* **27(4)**: 203-214. [RauppHoSa2001]

Ravindranath, K. & Subbaratnam, G.V. 1994. Management of sugarcane scale insects *Melanaspis glomerata* (Green). *Indian Sugar* **44**: 333-343. [RavindSu1994]

Ravindranath, K. & Subbaratnam, G.V. 1995. Natural enemies of sugarcane scale insects *Melanaspis glomerata* in Andhra Pradesh. *Bulletin of Entomology* **36**: 117-119. [RavindSu1995]

Ravindranath, K. & Subbaratnam, G.V. 1998. Relative susceptibility of some varieties of sugarcane to the scale insect, *Melanaspis glomerata* (Green). *Pest Management and Economic Zoology* **6(2)**: 159-161 [RavindSu1998]

Rawat, U.S. & Pawar, A.D. 1992. Biocontrol of San José scale, *Quadraspidiotus perniciosus* (Comstock) by predatory beetle, *Chilocorus bijugus* Mulsant in Himachal Pradesh. *Plant Protection Bulletin (Faridabad)* **44(4)**: 7-10. [RawatPa1992]

Rawat, U.S. & Sangal, S.K. 1993. Influence of quality of food and host density on the ovipositional behaviour and longevity of predatory beetle, *Chilocorus bijugus* Mulsant. *Journal of Insect Science* **6(1)**: 109-110. [RawatSa1993]

Rawat, U.S., Thakur, J.N. & Pawar, A.D. 1988. Introduction and establishment of *Chilocorus bijugus* Mulsant and *Pharoscymnus flexibilis* Mulsant, predatory beetles of San José scale at Thanedhar areas in Himachal Pradesh. *Current Science* **57(22)**: 1250-1251. [RawatThPa1988]

Reboul, J.L. 1976. Principaux parasites et maladies des plantes cultivées en Polynésie Française. *Service de l'Economic Rurale, Recherche Agronomique* **No. 129/ER/RA**: 58 pp. [Reboul1976]

Reeve, J.D. & Murdoch, W.W. 1985. Aggregation by parasitoids in the successful control of the California red scale: a test of theory. *Journal of Animal Ecology* **54**: 797-816. [ReeveMu1985]

Reh, L. 1900. Zucht-ergebnisse mit *Aspidiotus perniciosus* Comst. *Jahrbuch der Hamburgischen Wissenschaftlichen Anstalten. Hamburg* **17**: 237-257. [Reh1900]

Reh, L. 1900a. Über *Aspidiotus ostreaeformis* Curt. und verwandte Formen. *Jahrbuch der Hamburgischen Wissenschaftlichen Anstalten. Hamburg* **17**: 259-271. [Reh1900a]

Reh, L. 1900b. Über *Aspidiotus ostraeformis* Curt. und *A. pyri* Licht. *Zoologischer Anzeiger. Jena* **23**: 497-499. [Reh1900b]

Reh, L. 1900c. Über Schildbildung und häutung bei *Aspidiotus perniciosus* Comst. *Zoologischer Anzeiger. Jena* **23**: 502-504. [Reh1900c]

Reh, L. 1903. Zur Naturgeschichte Mittel-und Nordeuropäischer Schildläuse. *Allgemeine Zeitschrift für Entomologie* **8**: 301-469. [Reh1903]

Reh, L. 1926. Pflanzenschädliche Insekten. *Schröder's Handbuch der Entomologie* **2**: 308-328. [Reh1926]

Reh, L. 1933. Witterung und Insekten. *Anzeiger für Schädlingskunde* **9**: 109-112. [Reh1933]

Rehman, S.U. 1996. Residual effects of pesticides on parasitism by Aphytis holoxanthus DeBach (Hymenoptera: Aphelinidae) on Florida red scale Chrysomphalus aonidum. 88 pp. [Rehman1996]

Rehman, S.U., Browning, H.W., Nigg, H.N. & Harrison, J.M. 1999. Residual effects of carbaryl and dicofol on *Aphytis holoxanthus* DeBach (Hymenoptera : Aphelinidae). *Biological Control* **16(3)**: 252-257. [RehmanBrNi1999]

Rehman, S.U., Browning, H.W., Nigg, H.N. & Harrison, J.M. 2000. Increases in Florida red scale populations through pesticidal elimination of *Aphytis holoxanthus* DeBach in Florida citrus. *Biological Control* **18(2)**: 87-93. [RehmanBrNi2000]

Rehman, S.U., Browning, H.W., Salyani, M. & Nigg, H.N. 1999a. Effects of pesticide deposit pattern on rate of contact and mortality of *Aphytis holoxanthus* (Hymenoptera: Aphelinidae). *Florida Entomologist* **82(1)**: 28-33. [RehmanBrSa1999a]

Rehman, M.H., Ghani, M.A. & Kazimi, S.K. 1961. Introduction of exotic natural enemies of San José scale into Pakistan. *Technical Bulletin of the Commonwealth Institute of Biological Control* **1**: 165-177. [RehmanGhKa1961]

Reiderne S., K. & Kozár, F. 1994. [Recently appeared scale insect species (Homoptera, Coccoidea: Diaspididae) on glasshouse ornamentals in Hungary.]. [In Hungarian]. *Növényvédelem* **30(9)**: 423-427. [ReiderKo1994]

Reis, P.R. & Melo, L.A.S. 1984. Pragas da videira. *Informe Agropecuario Belo Horizonte* **10**:(117): 68-72. [ReisMe1984]

Reissig, W.H., Weires, R.W., Onstad, D.W., Stanley, B.H. & Stanley, D.M. 1985. Timing and effectiveness of insecticide treatments against the San José scale (Homoptera: Diaspididae). *Journal of Economic Entomology* **78**: 238-248. [ReissiWeOn1985]

Rekk, G.F. 1938. [On the injuriousness of San José scale in Georgia. Pages 66-93 *in:* Kiritchenko, A.N. (Ed.), Works of Quarantine Laboratories.] [In Russian]. Leningrad, Sel'khozgiz. 272 pp. [Rekk1938]

Reyne, A. 1947. Notes on the biology of *Comperiella unifasciata* Ishii and its host *Aspidiotus destructor rigidus* nov. subspec. *Tijdschrift voor Entomologie. Amsterdam* **88**: 294-302. [Reyne1947]

Reyne, A. 1948. Studies on a serious outbreak of *Aspidiotus destructor rigidus* in the cocoanut-palms of Sangi (North Celebes). *Tijdschrift voor Entomologie. Amsterdam* **89**: 83-123. [Reyne1948]

Reyne, A. 1949. Nederlandse Coccidae. *Tijdschrift voor Entomologie, Amsterdam* **91**: lxix-lxxi. [Reyne1949]

Reyne, A. 1957. Snavelinsecten - Rhynchota, I, Nederlandse schildluizen (Coccidae). *Wetenschappelijke Mededlingen van de Koninklijke Nederlandes Natuurhistorische Vereniging* **22**: 1-44. [Reyne1957]

Ribaga, C. 1902. Attività del *Novius cardinalis* Muls. contro l'*Icerya purchasi* Mask. in Italia. *Rivista di Patologia Vegetale. Firenze* **10**: 299-323. [Ribaga1902]

Rice, R.E. 1974. San José scale: Field study with a sex pheromone. *Journal of Economic Entomology* **67**: 561-562. [Rice1974]

Rice, R.E., Atterholt, C.A., Delwiche, M.J. & Jones, R.A. 1997. Efficacy of mating disruption pheromones in paraffin emulsion dispensers. *Bulletin OILB/SROP (Sect. Reg. Ouest Paléarctique)* **20**: 151-161. [RiceAtDe1997]

Rice, R.E., Dibble, J.E., Jones, R.A. & La Rue J.H. 1974. Insecticides and timing sprays for control of San José scale. *California Agriculture* **28(4)**: 3. [RiceDiJo1974]

Rice, R.E., Flaherty, D.L. & Jones, R.A. 1982. Monitoring and modeling San José scale. *California Agriculture* **36**: 13-14. [RiceFlJo1982]

Rice, R.E. & Hoyt, S.C. 1980. Response of San José scale to natural and synthetic sex pheromones. *Environmental Entomology* **9**: 190-194. [RiceHo1980]

Rice, R.E. & Jones, R.A. 1977. Monitoring flight patterns of San José scale (Homoptera: Diaspididae). *Canadian Entomologist* **109**: 1403-1404. [RiceJo1977]

Rice, R.E. & Jones, R.A. 1982. Collections of *Prospaltella perniciosi* Tower (Hymenoptera: Aphelinidae) on San José scale (Homoptera: Diaspididae) pheromone traps. *Environmental Entomology* **11(4)**: 876-880. [RiceJo1982]

Rice, R.E. & Moreno, D.S. 1969. Marking and recapture of California red scale for field studies. *Annals of the Entomological Society of America* **62**: 558-560. [RiceMo1969]

Rice, R.E. & Moreno, D.S. 1969a. Comparative production of pheromone by the California red scale reared on lemons and potatoes. *Journal of Economic Entomology* **62**: 958. [RiceMo1969a]

Rice, R.E. & Moreno, D.S. 1970. Flight of male California red scale. *Annals of the Entomological Society of America* **63**: 91-96. [RiceMo1970]

Richards, A.M. 1960. Scale insect survey on apples 1959-60. *New Zealand Journal of Agricultural Research* **3**: 693-698. [Richar1960AM]

Ridgway, R.L., King, E.G. & Carrillo, J.L. 1977. Augmentation of natural enemies for control of plant pests in the western hemisphere. Pages 379-416. *in*: Ridgeway, R.L. & Vinson, S.B. Biological Control by Augmentation of Natural Enemies: Insect and Mite Control with Parasites and Predators. New York: Plenum Press. 480 pp. [RidgwaKiCa1977]

Riehl, L.A. 1969. Advances relevant to narrow-range spray oils for citrus pest control. Pages 897-907. *in*: Chapman, H.D. (Ed.). Proceedings First International Citrus Symposium. Vol. 2. Riverside: University of California. [Riehl1969]

Riehl, L.A. 1990. 3.5.1 Control Chemicals. Pages 365-392 *in*: Rosen, D. (Ed.). Armored Scale Insects, Their Biology, Natural Enemies and Control [Series title: World Crop Pests, Vol. 4B]. Amsterdam, the Netherlands: Elsevier. 688 pp. [Riehl1990]

Riehl, L.A., Brooks, R.F., McCoy, C.W., Fisher, T.W. & Dean, H.A. 1980. Accomplishments toward improving integrated pest management for citrus. Pages 319-363 *in*: Huffaker, C.B. (Ed.). New Technology of Pest Control. New York: Wiley-Interscience. 500 pp. [RiehlBrMc1980]

Riehl, L.A. & Carman, G.E. 1953. Narrow-cut petroleum fractions of naphthenic and paraffinic composition for control of California red scale. *Journal of Economic Entomology* **46(6)**: 1007-1013. [RiehlCa1953]

Riehl, L.A., Garber, M.J., LaDue, J.P., Rodriguez, J.L. & Wilson, E.L. 1964. Hydrogenation refining vs. efficiencies of spray oils against citrus red mite eggs and California red scale. *Journal of Economic Entomology* **57(4)**: 522-525. [RiehlGaLa1964]

Riehl, L.A. & LaDue, J.P. 1950. Insecticidal properties of hydrocarbon compositions. Part II - Evaluation of petroleum fractions against California red scale and citrus red mite. *Joint Symposium on the Agricultural Applications of Petroleum Products* 29-43. [RiehlLa1950]

Riehl, L.A. & LaDue, J.P. 1952. Evaluation of petroleum fractions against California red scale and citrus red mite. *Advances in Chemistry Series* **7**: 25-36. [RiehlLa1952]

Riehl, L.A., LaDue, J.P. & Rodriguez, J.L. 1958. Evaluation of representative California spray oils against citrus red mite and California red scale. *Journal of Economic Entomology* **51(2)**: 193-195. [RiehlLaRo1958]

Riehl, L.A., LaDue, J.P. & Rodriguez, J.L. 1959. Efficiency of ethion in oil spray against California red scale and citrus red mite. *Journal of Economic Entomology* **52**: 857-860. [RiehlLaRo1959]

Riehl, L.A., LaDue, J.P. & Rodriguez, J.L. 1965. Efficiency of a re-formed oil against citrus red mite eggs and California red scale. *Journal of Economic Entomology* **58(5)**: 907-909. [RiehlLaRo1965]

Riherd, P.T. & Chada, H.L. 1952. Some scale insects attacking grasses in Texas. *Progress Report (Texas Agricultural Experiment Station)* **1461**: 1-5. [RiherdCh1952]

Rivnay, E. 1945. Notes on Encyrtidae from Palestine with the description of a new species. *Journal of the Entomological Society of Southern Africa* **8**: 117-122. [Rivnay1945]

Rivnay, E. 1968. Biological control of pests in Israel (a review 1905 - 1965). *Israel Journal of Entomology* **3(1)**: 1-156. [Rivnay1968]

Roaf, J.R. & Mote, D.C. 1935. The holly scale, *Aspidiotus britannicus* Newstead, and other insect pests of English holly in Oregon. *Journal of Economic Entomology* **28**: 1041-1049. [RoafMo1935]

Robb, R.L., Costa, H.S., Bethke, J., Cowles, R. & Parrella, M.P. 2001. Pests of floricultural and ornamental nurseries. *UC Pest Management Guidelines* . [RobbCoBe2001]

Roberts, F.S. 1958. Insects affecting banana production in Central America. Pages 411-415 *in*: Baecker, E.C., (Ed.). Proceedings of the Tenth International Congress of Entomology. Vol. 3. Ottawa. 895 pp. [Robert1958]

Robertson, G. 1971. Marking red-scale with carbon-14 and its effect on reproduction and survival. *Entomologia Experimentalis et Applicata* **14**: 449-456. [Robert1971]

Robertson, G. 1973. The sensitivity of scale insects to carbon-14 particles and to cobalt-60 gamma rays. *International Journal of Radiation Biology* **24**: 313-323. [Robert1973]

Robertson, G. & Slowiak, D. 1974. The distribution of carbon-14 in plants labelled as a source of food for red scale. *Entomologia Experimentalis et Applicata* **17**: 163-175. [RobertSl1974]

Robinson, E. 1917. Coccidae of the Philippine Islands. *Philippine Journal of Science (Ser. D.)* **12**: 1-47. [Robins1917]

Robinson, E. 1918. Descriptions and records of Philippine Coccidae. *Philippine Journal of Science* **13**: 145-147. [Robins1918]

Robison, W.G. 1990. 1.2.3.2 Sperm ultrastructure, behaviour, and evolution. Pages 205-220 *in*: Rosen, D., (Ed.). Armored Scale Insects, Their Biology, Natural Enemies and Control [Series title: World Crop Pests, Vol. 4A]. Amsterdam, the Netherlands: Elsevier. 384 pp. [Robiso1990]

Rock, G.C. & McClain, D.C. 1990. Effects of constant photoperiods and temperatures on the hibernating life stages of the San José Scale (Homoptera: Diaspididae) in North Carolina. *Journal of Entomological Science* **25(4)**: 615-621. [RockMc1990]

Rodrigo, E. & Garcia-Mari, F. 1990. Comparacion del ciclo biologico de los diaspinos *Parlatoria pergandii, Aonidiella aurantii* y *Lepidosaphes beckii* (Homoptera, Diaspididae) en citricos. *Boletín de Sanidad Vegetal, Plagas* **16(1)**: 25-35. [RodrigGa1990]

Rodrigo, E. & Garcia-Mari, F. 1992. Ciclo biologico de los diaspinos de citricos *Aonidiella aurantii* (Mask.), *Lepidosaphes beckii* (Newm.) y *Parlatoria pergandei* (Comst.) en 1990. *Boletín de Sanidad Vegetal, Plagas* **18(1)**: 31-44. [RodrigGa1992]

Rodrigo, E. & Garcia-Mari, F. 1994. Estudio de la abundancia y distribucion de algunos coccidos diaspididos de citricos. *Bol. Sanid. Veg. Plagas* **20(1)**: 151-164. [RodrigGa1994]

Rodrigo, E., Troncho, P. & Garcia-Mari, F. 1996. Parasitoids (Hym.: Aphelinidae) of three scale insects (Hom.: Diaspididae) in a citrus grove in Valencia, Spain. *Entomophaga* **41**: 77-94. [RodrigTrGa1996]

Rodrigues, A.N., Torres, L.M. & Polesny, F. 2001. Phenology of San José Scale, *Quadraspidiotus perniciosus* (Comstock) on apple in Guarda region (central eastern Portugal). *IOBC/WPRS Bulletin* **24(5)**: 195-199. [RodrigToPo2001]

Rodrigues, W.C., Gouvea, A., Azevedo, O.R.F., Soares, M.A. & Cassino, P.C.R. 1997. Avaliação do grau de parasitismo de *Aphytis* sp. sobre *Selenaspidus articulatus* na Baixada Litoránea do estado do Rio de Janeiro. Page 132 *in*: Bento, J.M.S. & Delalibera, I. (Eds.). [16th Brazilian Congress of Entomology] 16o Congresso Brasileiro de Entomologia. Salvador: Sociedade Entomológica do Brasil/EMBRAPA-CNPMF. 400 pp. [RodrigGoAz1997]

Roelofs, W.L., Gieselmann, M.J., Cardé, A.M., Tashiro, H., Moreno, D.S., Henrick, C.A. & Anderson, R. 1977. Sex pheromone of the California red scale, *Aonidiella aurantii. Nature (London)* **267**: 698-699. [RoelofGiCa1977]

Roelofs, W.L., Gieselmann, M.J., Cardé, A.M., Tashiro, H., Moreno, D.S., Henrick, C.A. & Anderson, R. 1978. Identification of the California red scale sex pheromone. *Journal of Chemical Ecology* **4**: 211-224. [RoelofGiCa1978]

Roelofs, W.L., Gieselmann, M.J., Mori, K. & Moreno, D.S. 1982. Sex pheromone chirality comparison between sibling species -- California red scale and yellow scale. *Naturwissenschaften* **69**: 348. [RoelofGiMo1982]

Rolfs, P.H. 1897. The fungus disease of the San José Scale (*Sphaerostilbe coccophila*, Tul). *Bulletin of the Florida Agricultural Experiment Station* **41**: 519-536. [Rolfs1897]

Rong, I.H. & Grobbelaar, E. 1998. South African records of associations between fungi and arthropods. *African Plant Protection* **4**: 1, 43-63. [RongGr1998]

Ronna, E. 1928. Os Insectos do Brasil. Sao Paulo: Chacaras e Quintaes. 175 pp. [Ronna1928]

Rose, M. 1990. 2.7.1 Sampling for Natural Enemies. Pages 229-234 *in*: Rosen, D. (Ed.). Armored Scale Insects, Their Biology, Natural Enemies and Control [Series title: World Crop Pests, Vol. 4B]. Amsterdam, the Netherlands: Elsevier. 688 pp. [Rose1990]

Rose, M. 1990a. 2.7.3 Rearing and Mass Rearing of Natural Enemies. Pages 263-287 *in*: Rosen, D. (Ed.). Armored Scale Insects, Their Biology, Natural Enemies and Control [Series title: World Crop Pests, Vol. 4B]. Amsterdam, the Netherlands: Elsevier. 688 pp. [Rose1990a]

Rose, M. 1990b. 3.6.3.1 Periodic Colonization of Natural Enemies. Pages 433-440 *in*: Rosen, D. (Ed.). Armored Scale Insects, Their Biology, Natural Enemies and Control [Series title: World Crop Pests, Vol. 4B]. Amsterdam, the Netherlands: Elsevier. 688 pp. [Rose1990b]

Rose, M. 1990c. 3.9.1 Citrus. Pages 535-541 *in*: Rosen, D. (Ed.). Armored Scale Insects, Their Biology, Natural Enemies and Control [World Crop Pests, Vol. 4B]. Amsterdam, the Netherlands: Elsevier. 688 pp. [Rose1990c]

Rose, M. 1990d. 1.5.3 Rearing and mass rearing. Pages 357-365 *in*: Rosen, D. (Ed.). Armored Scale Insects, Their Biology, Natural Enemies and Control [World Crop Pests, Vol. 4A]. Amsterdam, the Netherlands: Elsevier. 384 pp. [Rose1990d]

Rose, M. 1992. Biological control by natural enemies in the interior plantscape. College Station, TX: Dept. of Entomology, Texas A & M. [Rose1992]

Rose, M. & DeBach, P. 1990a. 3.6.4 Conservation of Natural Enemies. Pages 461-472 *in*: Rosen, D. (Ed.). Armored Scale Insects, Their Biology, Natural Enemies and Control [Series title: World Crop Pests, Vol. 4B]. Amsterdam, the Netherlands: Elsevier. 688 pp. [RoseDe1990a]

Rosen, D. 1965. The hymenopterous parasites of citrus armored scales in Israel (Hymenoptera: Chalcidoidea). *Annals of the Entomological Society of America* **58**: 388-396. [Rosen1965]

Rosen, D. 1966. Keys for the identification of the hymenopterous parasites of scale insects, aphids and aleyrodids on citrus in Israel. *Scripta Hierosalym* **18**: 43-79. [Rosen1966]

Rosen, D. 1967b. Biological and integrated control of citrus pests in Israel. *Journal of Economic Entomology* **60(3)**: 1422-1427. [Rosen1967b]

Rosen, D. 1969. The parasites of coccids, aphids and aleyrodids on citrus in Israel: some zoogeographical considerations. *Israel Journal of Entomology* **4**: 45-53. [Rosen1969]

Rosen, D. 1973. Methodology for biological control of armored scale insects. *Phytoparasitica* **1**: 47-54. [Rosen1973]

Rosen, D. 1979. Integrated control of citrus pests in Israel. Pages 289-292. *in*: Proceedings: Internationales Symposium der IOBC/WPRS über Integrierten Pflanzenschutz in der Landund Forstwirtschaft. Wien, Austria: IOLB/SROP. 648 pp. [Rosen1979]

Rosen, D. 1986. The role of taxonomy in effective biological control programs. *Agriculture, Ecosystems & Environment* **15**: 121-129. [Rosen1986]

Rosen, D. 1987. Natural enemies of the Diaspididae, and their utilization in biological control. *Bollettino del Laboratorio di Entomologia Agraria 'Filippo Silvestri'* **43 (Suppl.)**: 189-194. [Rosen1987]

Rosen, D. 1990. 3.6.1 Biological Control: Introduction. Pages 413-415 *in*: Rosen, D. (Ed.). Armored Scale Insects, Their Biology, Natural Enemies and Control [Series title: World Crop Pests, Vol. 4B]. Amsterdam, the Netherlands: Elsevier. 688 pp. [Rosen1990]

Rosen, D. 1990a. 3.6.6 Biological Control: Selected Case Histories. Pages 497-505 *in*: Rosen, D. (Ed.). Armored Scale Insects, Their Biology, Natural Enemies and Control [Series title: World Crop Pests, Vol. 4B]. Amsterdam, the Netherlands: Elsevier. 688 pp. [Rosen1990a]

Rosen, D. 1993. Parasitic Hymenoptera in biological control of the genus *Aphytis*. Pages 411-416 *in*: Advances in Parasitic Hymenoptera Research: Proceedings of the II Conference on the Taxonomy and Biology of Parasitic Hymenoptera. [Rosen1993]

Rosen, D. & DeBach, P. 1973. Systematics, morphology and biological control. *Entomophaga* **18(3)**: 215-222. [RosenDe1973]

Rosen, D. & DeBach, P. 1978. Diaspididae. Pages 78-128. *in*: Clausen, C.P. Introduced Parasites and Predators of Arthropod Pests and Weeds: a World Review. Washington, D.C.: Agricultural Research Service, United States Department of Agriculture. 545 pp. [RosenDe1978]

Rosen, D. & DeBach, P. 1979. Species of Aphytis of the world (Hymenoptera: Aphelinidae). (Series Entomologica: vol. 17). The Hague, Boston, London: Dr. W. Junk. 801 pp. [RosenDe1979]

Rosenheim, J.A. & Heimpel, G.E. 1994. Sources of intraspecific variation in oviposition and host-feeding behaviour. Pages 41-78 *in*: Rosen, D., Ed. Advances in the Study of Aphytis (Hymenoptera: Aphelinidae). Andover, U.K.: Intercept. 362 pp. [RosenhHe1994]

Rosenheim, J.A. & Rosen, D. 1991. Foraging and oviposition decisions in the parasitoid *Aphytis lingnanensis*: distinguishing the influences of egg load and experience. *Journal of Animal Ecology* **60**: 873-893. [RosenhRo1991]

Rosenheim, J.A. & Rosen, D. 1992. Influence of egg load and host size on host-feeding behaviour of the parasitoid *Aphytis lingnanensis. Ecological Entomology* **17**: 263-272. [RosenhRo1992]

Rossler, Y. & Rosen, D. 1990. 3.8.2 A Case History: IPM in Citrus in Israel. Pages 519-526 *in*: Rosen, D. (Ed.). Armored Scale Insects, Their Biology, Natural Enemies and Control [Series title: World Crop Pests, Vol. 4B]. Amsterdam, the Netherlands: Elsevier. 688 pp. [RossleRo1990]

Rota, P. 1953. Contributo alla conoscenza degli stadi ibernanti di *Quadraspidiotus (Aspidiotus) perniciosus* Comst. *Bollettino di Zoologia Agraria e di Bachicoltura (Milano)* **19**: 77-84. [Rota1953]

Rota, P. 1955. Contributo alla conoscenza della ghiandole sericipare negli Aspidiotini. *Bollettino di Zoologia Agraria e di Bachicoltura (Milano)* **21**: 231-252. [Rota1955]

Rothschild, M., Euw, J. von & Reichstein, T. 1973. Cardiac glycosides in a scale insect (*Aspidiotus*), a ladybird (*Coccinella*) and a lacewing (*Chrysopa*). *Journal of Entomology (A)* **48**: 89-90. [RothscEuRe1973]

Rübsaamen, E.H. 1902. Über Zoocecidien von den Kanarischen Inseln und Madeira. *Marcellia (Rivista Internazionale di Cecidologia)* **1**: 60-65. [Rubsaa1902]

Rübsaamen, E.H. 1907. Beiträge zur Kenntnis aussereuropäischer Zoocecidien. I. Beitrag: Gallen vom Bismarck-Archipel. II. Beitrag: Gallen aus Brasilien und Peru. *Marcellia (Rivista Internazionale di Cecidologia)* **4**: 5-25, 65-85. [Rubsaa1907]

Rubtsov, I.A. 1947a. [New effective biological agents in controlling the San José scale.]. [In Russian]. *Priroda* **2**: 61-63. [Rubtso1947a]

Rubtsov, I.A. 1949. [New effective entomophages effective against Diaspidinae scale insects.]. [In Russian]. *Priroda* **No. 10**: 74-75. [Rubtso1949]

Rubtsov, I.A. 1952a. *Lindorus* - an effective predator of diaspine scales. *Review of Applied Entomology* **(A)43**: 109. [Rubtso1952a]

Ruhl, M. 1913. Liste neuerdings beschriebener oder gezogener Parasiten und ihrer Wirte. *Societas entomologica (Stuttgart)* **28**: 79-80. [Ruhl1913]

Ruiz Castro, A. 1944. Un coccido ampelófago, nuevo en España *[Diaspidiotus uvae* (Comstock)]. *Boletín de Patología Vegetal y Entomología Agrícola. Madrid* **13**: 55-74. [RuizCa1944]

Rungs, C. 1933. Une diaspine nouvelle du Maroc: *Targionia regnieri*, n. sp. *Bulletin de la Société d'Histoire Naturelle de l'Afrique du Nord* **24**: 114-117. [Rungs1933]

Rungs, C. 1935. Coccidae du Maroc (3me note). *Revue de Pathologie Végétale et d'Entomologie Agricole de France* **22**: 270-282. [Rungs1935]

Rungs, C. 1936. Diaspines [Hem. Coccidae] nouvelles du Maroc. [Coccides du Maroc, 4 note.]. *Bulletin de la Société Entomologique de France* **41**: 50-56. [Rungs1936]

Rungs, C. 1937. Une diaspine nouvelle du Maroc. *Aspidiotus braunschvigi* n. sp. [Hem. Coccidae]. *Bulletin de la Société Entomologique de France* **41**: 329-335. [Rungs1937]

Rungs, C. 1939. Une diaspine nouvelle du Maroc. (Coccidae du Maroc, 6e note). (*Hemiberlessa jourdani*) n. sp. [Hem. Coccidae]. *Bulletin de la Société Entomologique de France* **43**: 231-234. [Rungs1939]

Rungs, C. 1941. Une diaspine nouvelle capturée au Portugal: *Aonidiella mimeuri* nov. sp. (Hémipt. Coccidae). *Bulletin de la Société d'Histoire Naturelle de l'Afrique du Nord* **32**: 27-30. [Rungs1941]

Rungs, C. 1942. Récoltes de la mission d'étude des acridiens E. Morales Agacion et Ch. Rungs au Sahara Espagnol (Hem. Cocc.). *EOS* **18**: 107-111. [Rungs1942]

Rungs, C. 1948. Contribution au catalogue des Coccidae du Maroc. [Coccoidea du Maroc, 11 note.]. *Revue de Pathologie Végétale et d'Entomologie Agricole de France* **27**: 110-117. [Rungs1948]

Rungs, C. 1950. Sur l'extension spontanée au Maroc du *Rhizobius (Lindorus) lophanthae* Blaisd. Col. Coccinellidae. *Bulletin de la Société Entomologique de France* **55**: 9-11. [Rungs1950]

Rungs, C. 1952. Contribution a la connaissance des ennemis de l'arganier *Argania spinosa* (L.). *Bulletin de la Société des Sciences naturelles du Maroc* **32**: 61-76. [Rungs1952]

Rungs, C.E.E. 1970. Problèmes phytosanitaires posés a l'agrumiculture marocaine. *Al Awamia* **37**: 91-94. [Rungs1970]

Russell, L.M., Kosztarab, M. & Kosztarab, M.P. 1974. A selected bibliography of the Coccoidea. Second supplement. *Miscellaneous Publications United States Department of Agriculture* **1281**: 1-122. [RusselKoKo1974]

Russo, G. 1929. Los insectos daninos en la economía agricola. *Revista de Agricultura (Republica Dominicana)* **20(1)**: 2-3. [Russo1929]

Russo, G. 1951. La lotta biologica contro i parassiti animali dei vegetali in Italia. *Bollettino del Laboratorio di Entomologia Agraria. Portici* **10**: 86-97. [Russo1951]

Russo, G. 1958. Problemi italiani di lotta biologica dei fitofagi. *Bollettino del Laboratorio di Entomologia Agraria. Portici* **16**: 141-147. [Russo1958]

Russo, G. 1959. La cocciniglie degli agrume e mezzi di lotta. *Tecnica Agricola. Catania* **11**: 354-387. [Russo1959]

Russo, A. 1987. Remarks on the biological behaviour of *Quadraspidiotus perniciosus* Comst. in Sicily. *Bollettino del Laboratorio di Entomologia Agraria 'Filippo Silvestri'* **43** (Suppl.): 203-208. [Russo1987]

Russo, A., Mazzeo, G. & Suma, P. 1999. Sulla presenza di *Entaspidiotus lounsburyi* (Marlatt, 1908) su Mesembryanthemaceae in Italia (Homoptera, Coccoidea)]. *Redia* **82**: 83-87. [RussoMaSu1999]

Rutherford, A. 1914. Some Ceylon Coccidae. *Bulletin of Entomological Research* **5**: 259-268. [Ruther1914]

Rutherford, A. 1915. Some new Ceylon Coccidae. *Journal of the Bombay Natural History Society* **24**: 111-118. [Ruther1915]

Rutherford, A. 1915a. Notes on Ceylon Coccidae. *Spolia Zeylanica* **10**: 103-115. [Ruther1915a]

Rutherford, D.M. 1948. Red scale is no problem in some citrus orchards. *Citrograph* **34(2)**: 55,68 [Ruther1948]

Ryan, H.J. 1946. Some Los Angeles county experiences with new insect pests and insect eradication projects. *Bulletin (California Department of Agriculture)* **35(3)**: 124-125. [Ryan1946]

Rzaeva, L.M. & Yaminova, G.A. 1985. [Fauna of aphelinids (Hymenoptera, Aphelinidae) -- coccid parasites in the Kuba-Khachmass zone of the Azerbaijan SSR (USSR).]. [In Russian]. *Izvestiya Akademii Nauk Azerbaidzan SSR, Biol. Nauk* **No. 6**: 55-58. [RzaevaYa1985]

Saakyan-Baranova, A.A. 1954. [Pests of greenhouse plants.]. [In Russian]. *Akademii Nauk SSSR Glav. Bot. Sad. Zashchita Rastenii Vreditelei i Bodez.* **4**: 7-41. [Saakya1954]

Sachtleben, H. 1944. Über einen Rest der Sammlung Bouche's und die in ihm enthaltenen Cocciden. (Diptera, Hymenoptera & Hemiptera - Homoptera: Coccoidea). *Arbeiten über Morphologische und Taxonomische Entomologie aus Berlin-Dahlem* **11**: 65-76. [Sachtl1944]

Sadakathulla, S. 1993. Technique of mass production of the predatory coccinellid, *Chilocorus nigritus* (Fabricius) on coconut scale, *Aspidiotus destructor* Sign. *Indian Coconut Journal* **23(9)**: 12-13. [Sadaka1993]

Sahai, B. & Joshi, L.D. 1965. Bionomics and biological control of San José scale (*Quadraspidiotus perniciosus* Comstock) in U.P. *Punjab Horticultural Journal* **5(1)**: 37-43. [SahaiJo1965]

Sailer, R.I. 1973. A look at USDA's biological control in insect pests: 1888 to present. *Agricultural Science Review* **10(4)**: 15-27. [Sailer1973]

Sailer, R.I. 1976. Future role of biological control in management of pests. *Proceedings of the Tall Timbers Conference on Ecology and Animal Control by Habitat Management* **6**: 195-209. [Sailer1976]

Sailer, R.I. 1981. Elements of opportunity in biological control. Pages 419-432 *in*: Beltsville Symposia in Agricultural Research. 5. Biological control in crop production. Totowa, N.J.: Allanheld, Osmun & Co.. 461 pp. [Sailer1981]

Sailer, R.I. 1983. History of insect introductions. Pages 15-38 *in*: Wilson, C.L. & Graham, C.L. (Eds.). Exotic Plant Pests and North American Agriculture. New York: Academic Press. 522 pp. [Sailer1983]

Sakai, K. 1939. Comparison of the red scale (*Aonidiella aurantii* (Mask.)) and its allied species in Japan. *Oyo-Kontyu. Tokyo* **2**: 45-62. [Sakai1939]

Salama, H.S. 1970. Ecological studies on the scale insect, *Chrysomphalus dictyospermi* (Morgan) in Egypt. *Zeitschrift für Angewandte Entomologie* **65**: 427-430. [Salama1970]

Salama, H.S. 1970a. Population dynamics of the scale insect *Mycetaspis personatus* (Comstock) in Egypt (Homoptera - Coccoidea). *Zeitschrift für Angewandte Entomologie* **66**: 42-46. [Salama1970a]

Salama, H.S. 1972. On the population density and bionomics of *Parlatoria blanchardi* (Targ.) and *Mycetaspis personatus* (Comstock) (Homoptera - Coccoidea). *Zeitschrift für Angewandte Entomologie* **70**: 403-407. [Salama1972]

Salama, H.S., Amin, A.H. & Hawash, M. 1972. Effect of nutrients supplied to citrus seedlings on their susceptibility to infestation with the scale insects, *Aonidiella aurantii* (Maskell) and *Lepidosaphes beckii* (Newman) (Coccoidea). *Zeitschrift für Angewandte Entomologie* **71**: 395-405. [SalamaAmHa1972]

Salama, H.S. & Hamdy, M.K. 1974. Studies on populations on two scale insects infesting fig trees in Egypt (Coccoidea). *Zeitschrift für Angewandte Entomologie* **75**: 200-204. [SalamaHa1974]

Salama, H.S. & Hamdy, M.K. 1974a. Zum Auftreten von *Aspidiotus hederae* (Vallot) (Coccoidea, Diaspididae) an Cassia-Bäumen in. gypten. *Anzeiger Schädlingskunde, Pflanzen(schutz) und Umweltschutz* **47**: 138-139. [SalamaHa1974a]

Salama, H.S. & Saleh, M.R. 1972. Population of the scale insect *Mycetaspis personatus* (Comstock) on different varieties of *Mangifera indica* L. *Zeitschrift für Angewandte Entomologie* **70**: 328-331. [SalamaSa1972]

Salama, H.S. & Saleh, M.R. 1984. Components of the essential oil of three citrus species in correlation to their infestation with scale insects. *Zeitschrift für Angewandte Entomologie* **97**: 393-398. [SalamaSa1984]

Salas, H. & Goane, L. 2001. [Monitoring of the principal pests of lemon in Tucumán.] Monitoreo de las principales plagas del limon en Tucumán. [In Spanish.] *Avance Agroindustrial* **22(3)**: 27-30. [SalasGo2001]

Salazar Torres, J.C. 1989. [Scales (Homoptera: Coccoidea) occurring in four fruit species of the Rosaceae family in Zacatlan, Puebla.] Chapingo, Mexico. 93 pp. [Salaza1989]

Salazar Torres, J.C. & Solis Aguilar, J.F. 1990. [Scale insects (Homoptera: Coccoidea) present on four species of fruit trees of the family Rosaceae in Zacatlan, Puebla.] Escamas (Homoptera: Coccoidea) presentes en cuatro especies frutales de la familia Rosaceae en Zacatlan, Puebla. *Revista Chapingo* **15(67-68)**: 135-137. [SalazaSo1990]

Salem, M. & El-Saadany, G.B. 1979. Pheromone traps as monitors of the population of the red scale insect, *Aonidiella aurantii* (Mask.) in lemon orchards at Sharkeya, Egypt (Homoptera: Diaspididae). *Proceedings of the Conference of the Agricultural Development Research. Faculty of Agriculture, Ain Shams University: Dec. 17-19* : 129-138. [SalemEl1979]

Salem, S.A. & Saleh, M.R. 1992. Ecological studies on some scale insects infesting citrus trees in El-Mollake region, Sharkeia Governorate. *Bulletin of the Entomological Society of Egypt* **70**: 209-216. [SalemSa1992]

Sampaio, T. 1898. Os piolhos vegetaes. *Revista Agricola (Sao Paolo)* **3**: 73-80. [Sampai1898]

Samways, M.J. 1981. Comparison of ant community structure (Hymenoptera: Formicidae) in citrus orchards under chemical and biological control of red scale, *Aonidiella aurantii* (Maskell) (Hemiptera: Diaspididae). *Bulletin of Entomological Research* **71**: 663-670. [Samway1981]

Samways, M.J. 1981a. Biological Control of Pests and Weeds (Studies in Biology, The Institute of Biology, no. 132). London: Edward Arnold. 58 pp. [Samway1981a]

Samways, M.J. 1985. Relationship between red scale, *Aonidiella aurantii* (Maskell) (Hemiptera: Diaspididae), and its natural enemies in the upper and lower parts of citrus trees in South Africa. *Bulletin of Entomological Research* **75**: 379-393. [Samway1985]

Samways, M.J. 1986. Combined effect of natural enemies (Hymenoptera: Aphelinidae & Coleoptera: Coccinellidae) with different niche breadths in reducing high populations of red scale, *Aonidiella aurantii* (Maskell) (Hemiptera: Diaspididae). *Bulletin of Entomological Research* **76**: 671-683. [Samway1986]

Samways, M.J. 1986a. Spatial and temporal population patterns of *Aonidiella aurantii* Hemiptera Homoptera Diaspididae parasitoids Hymenoptera Aphelinidae and Encyrtidae caught on yellow sticky traps in citrus. *Bulletin of Entomological Research* **76**: 265-274. [Samway1986a]

Samways, M.J. 1989. Climate diagrams and biological control: an example from the areography of the ladybird *Chilocorus nigritus* (Fabricius, 1798) (Insecta, Coleoptera, Coccinellidae). *Journal of Biogeography* **16**: 345-351. [Samway1989]

Samways, M.J., Grout, T.G. & Prins, A.J. 1998. Ants as citrus pests. Pages 234-242. *in*: Bedford, E.C.G., Van den Berg, M.A. & de Villiers, E.A., Eds. Citrus Pests in the Republic of South Africa. 2nd ed. Nelspruit: Institute for Tropical and Subtropical Crops. 288 pp. [SamwayGrPr1998]

Samways, M.J. & Mapp, J. 1983. A new method for the mass-introduction of *Chilocorus nigritus* (F.) (Coccinellidae) into citrus orchards. *Citrus and Subtropical Fruit Journal* **598**: 4-6. [SamwayMa1983]

Samways, M.J., Nel, M. & Prins, A.J. 1982. Ants (Hymenoptera: Formicidae) foraging in citrus trees and attending honeydew-producing Homoptera. *Phytophylactica* **14**: 155-157. [SamwayNePr1982]

Samways, M.J., Osborn, R., Hastings, H. & Hattingh, V. 1999. Global climate change and accuracy of prediction of species' geographical ranges: establishment success of introduced ladybirds (Coccinellidae, *Chilocorus* spp.) worldwide. *Journal of Biogeography* **26**: 795-812. [SamwayOsHa1999]

Sanders, J.G. 1904. Three new scale insects from Ohio. *Ohio Naturalist* **4**: 94-98. [Sander1904]

Sanders, J.G. 1904a. The Coccidae of Ohio. *Proceedings of the Ohio Academy of Science* **4**: 25-92. [Sander1904a]

Sanders, J.G. 1906. Catalogue of recently described Coccidae. *United States Department of Agriculture, Bureau of Entomology, Technical Series* **12**: 1-18. [Sander1906]

Sanderson, E.D. 1908. The San José scale. *Circular (New Hampshire Agricultural Experiment Station)* 12 pp. [Sander1908]

Sanders, J.G. 1909a. Catalogue of recently described Coccidae - II. *United States Department of Agriculture, Bureau of Entomology, Technical Series* **16**: 33-60. [Sander1909a]

Sanders, J.G. 1910. A review of the Coccidae described by Dr. Asa Fitch. *Proceedings of the Entomological Society of Washington* **12**: 56-61. [Sander1910]

Sankaran, T. 1984. Survey for Natural Enemies of Diaspine Scale Insects in South India: Final Technical Report for the Period November 5, 1980 to November 4, 1983. Bangalore, India: Commonwealth Institute of Biological Control. 87 pp. [Sankar1984]

Santos, A.C. & Gravena, S. 1995. [Control of the Rufous Scale *Selenaspidus articulatus* Morgan (Homoptera: Diaspididae) with mineral oil and dimethoate.] Controle da cochonilha pardinha *Selenaspidus articulatus* Morgan (Homoptera: Diaspididae) com óleo mineral e dimetoato. *Anais da Sociedade Entomologica do Brasil* **24**: 411-414. [SantosGr1995]

Sasaki, C. 1901. On the Japanese species allied to the San José scale in America. *Annotationes Zoologicae Japonenses. Tokyo* **3**: 165-173. [Sasaki1901]

Sasaki, I. 1935. [On three new species of *Aspidiotus* and *Chrysomphalus* discovered in plant inspection.]. [In Japanese]. *Journal of Plant Protection, Nippon Plant Protection Society (Byochugai Zasshi)* **22**: 864-867. [Sasaki1935]

Sasscer, E.R. 1911. Catalogue of recently described Coccidae - III. *Technical Series. Bureau of Entomology, United States Department of Agriculture. Washington* **16**: 61-74. [Sassce1911]

Sasscer, E.R. 1912. Catalogue of recently described Coccidae - IV. *Technical Series. Bureau of Entomology, United States Department of Agriculture. Washington* **16**: 83-97. [Sassce1912]

Sasscer, E.R. 1915. Catalogue of recently described Coccidae - V. *Proceedings of the Entomological Society of Washington* **17**: 25-38. [Sassce1915]

Sasscer, E.R. 1918. Important foreign insect pests collected on imported nursery stock in 1917. *Journal of Economic Entomology* **11**:125-129. [Sassce1918]

Sasscer, E.R. 1923. Important foreign insects collected on imported nursery stock in 1922. *Journal of Economic Entomology* **16**: 152-158. [Sassce1923]

Sasscer, E.R. & Weigel, C.A. 1923. Further data on fumigation with hydrocyanic acid gas in greenhouses on a commercial basis. *Journal of Economic Entomology* **16**: 84-87. [SassceWe1923]

Sasscer, E.R. & Weigel, C.A. 1924. Recent developments in greenhouse fumigation with hydrocyanic-acid gas. *Journal of Economic Entomology* **17**: 214-222. [SassceWe1924]

Savastano, L. 1930. Della biancarossa *(Crysomphalus dictyospermi* Morg.) negli agrumi e in altre species ospitanti nell'Italia. Studio di fitopatologia arborea. *Annali della Real Stazione Sperimentale di Frutticoltura e di Agrumicoltura* **10**: 1-77. [Savast1930]

Savescu, A.D. 1953. Paduchi testosi la prun si Combaterea lor. *Repub. Pop. Româna Min. Agr. Inst. de Cercetari Agron. Indrumari Teh.* **36**: 3-46. [Savesc1953]

Savescu, A. (Editor). 1982. Tratat de Zoologie Agricola: Daunatorii plantelor cultivate, Vol. II. Romania: Editura Academiei Republicii Socialiste. 353 pp. [Savesc1982]

Schaub, L., Bloesch, B., Hippe, C., Keimer, C., Schmid, A. & Brunetti, R. 1999. Validation d'un modele de la phénologie du pou de San José. *Revue Suisse de Viticulture, d'Arboriculture et d'Horticulture* **31(5)**: 253-257. [SchaubBlHi1999]

Schaub, L.P., Mani, E., Bloesch, B. & Schwaller. 1995. Distribution of *Quadraspidiotus perniciosus* (San José scale) in Switzerland based on interpolated pheromone trap data. *Bulletin OEPP* **25**: 631-636. [SchaubMaBl1995]

Schausberger, P. 1998. Survival, development and fecundity in *Euseius finlandicus, Typhlodromus pyri* and *Kampimodromus aberrans* (Acari, Phytoseiidae) feeding on the San José scale *Quadraspidiotus perniciosus* (Coccina, Diaspididae). *Journal of Applied Entomology* **122**: 53-56. [Schaus1998]

Scherney, F. 1954. 1st *Quadraspidiotus schneideri* Bachmann eine Art?. *Zeitschrift für Angewandte Entomologie* **36**: 225-231. [Schern1954]

Scheurer, R. & Ruzette, M.A. 1974. Effects of insect growth regulators on the oleander scale (*Aspidiotus nerii*) and the European fruit lecanium (*Parthenolecanium corni*). *Zeitschrift für Angewandte Entomologie* **77**: 218-222. [ScheurRu1974]

Schilder, F.A. & Schilder, M. 1928. Die Nahrung der Coccinelliden und ihre Beziehung zur Verwandtschaft der Arten. *Arbeiten aus der Biologischen Reichsanstalt für Land-und Forstwirtschaft* **2**: 213-282. [SchildSc1928]

Schlinger, E.I. & Doutt, R.L. 1964. Systematics in relation to biological control. Pages 247-280 *in*: DeBach, P. & Schlinger, E.I., Eds. Biological Control of Insect Pests and Weeds. London: Chapman & Hall. 844 pp. [SchlinDo1964]

Schmidt, O. 1883. Metamorphose und Anatomie des männlichen *Aspidiotus nerii. Archiv für Naturgeschichte. Berlin* **51**: 169-200. [Schmid1883]

Schmutterer, H. 1951. Zur Lebensweise der Nadelholz-Diaspidinen (Homoptera, Coccoidea, Diaspididae, Diaspidinae) und ihrer Parasiten in den Nadelwäldern Frankens. *Zeitschrift für Angewandte Entomologie* **33**: 111-136. [Schmut1951]

Schmutterer, H. 1952. Die Ökologie der Cocciden (Homoptera, Coccoidea) Frankens. 2. Abschnitt. *Zeitschrift für Angewandte Entomologie* **33**: 369-420, 544-584; **34**: 65-100. [Schmut1952]

Schmutterer, H. 1957a. Untersuchungen über die Schildlausfauna einiger botanischer Gärten in Westdeutschland. *Berichte der Oberhessischen Gesellschaft für Natur- und Heilkunde. Giessen* **28**: 133-140. [Schmut1957a]

Schmutterer, H. 1957b. Die Mermale der deutschen Deckelschildläuse und ihre Bedeutung für die Artbestimmung. Ber. über die 8. Pages 148-150. Munchen: Wanderversamml. Deut. Ent.. [Schmut1957b]

Schmutterer, H. 1959. Schildläuse oder Coccoidea. 1. Deckelschildläuse oder Diaspididae. Die Tierwelt Deutschlands und der angrenzenden Meeresteile. Jena: Fischer. 260 pp. [Schmut1959]

Schmutterer, H. 1964. Zur Kenntnis der Schädlinge und Krankheiten der Kulturpflanzen in Südsomalia. *Anzeiger für Schädlingskunde* **37(7)**: 102-106. [Schmut1964]

Schmutterer, H. 1969. Pests of crops in Northeast and Central Africa with particular reference to Sudan. Pages 97-118. Portland: Gustav Fischer Verlag. 269 pp. [Schmut1969]

Schmutterer, H. 1990a. Observations on pests of *Azadirachta indica*(neem tree) and of some *Melia* spp. *Journal of Applied Entomology* **109(4)**: 390-400. [Schmut1990a]

Schmutterer, H. 1998. Some arthropod pests and a semi-parasitic plant attacking neem (*Azadirachta indica*) in Kenya. *Anzeiger Schädlingskunde, Pflanzen(schutz) und Umweltschutz* **71**: 36-38. [Schmut1998]

Schmutterer, H. 2001. The scale insects, whiteflies, aphids and psyllids of the neem tree, *Azadirachta indica* (Meliaceae). *Entomologica* **33(1999)**: 339-345. [Schmut2001]

Schmutterer, H., Kloft, W. & Lüdicke, M. 1957. Coccoidea, Schildläuse, scale insects, cochenilles. Tierische Schädlinge an Nutzpflanzen. 2. Teil, Vierte Lieferung. Homoptera II. Teil. Pages 403-520 *in*: Sorauer, P. Handb. der Pflanzenkrankheiten. V. 5. Berlin: Paul Parey. [SchmutKlLu1957]

Schmutterer, H., Kloft, W. & Lüdicke, M. 1959. Kritische Bemerkungen zu zwei Veröffentlichungen von L. Lindinger. *Beiträge zur Entomologie. Berlin* **9**: 373-375. [SchmutKlLu1959]

Schmutterer, H., Pires, A. & Klein Koch, C. 1978. Zur Schädlingsfauna der Kapverdischen Inseln. *Zeitschrift für Angewandte Entomologie* **86**: 320-336. [SchmutPiKl1978]

Schoonees, J. & Giliomee, J.H. 1982. The toxicity of methidathion to parasitoids of red scale, *Aonidiella aurantii* (Hemiptera: Diaspididae). *Journal of the Entomological Society of Southern Africa* **45**: 261-273. [SchoonGi1982]

Schrader, F. 1929a. Notes on reproduction in *Aspidiotus hederae* (Coccidae). *Psyche* **26**: 232-236. [Schrad1929a]

Schrank, F. 1776. Beyträge zur Naturgeschichte. Augsburg: Veith. 137 pp. [Schran1776]

Schrank, F. 1781. Coccus, Gallinsect. numeratio insectorum Austriae indigenorum. Augustae Vindelicorum. 552 pp. [Schran1781]

Schrank, F. 1801. Fauna Boica. Nurnberg. 374 pp. [Schran1801]

Schread, J.C. 1970. Controls of scale insects and mealybugs on ornamentals. *Bulletin of the Connecticut Agricultural Experiment Station* **No. 710**: 1-27. [Schrea1970]

Schreiner, I. 1989. Biological control introductions in the Caroline and Marshall Islands. *Proceedings of the Hawaiian Entomological Society. Honolulu, Hawaii: Nov. 30, 1989* **(29)**: 57-69. [Schrei1989]

Schuh, J. & Mote, D.C. 1948. Insect pests of nursery and ornamental trees and shrubs in Oregon. *Bulletin (Agricultural Experiment Station, Oregon State College)* **449**: 164 pp. [SchuhMo1948]

Schuhmann, G. 1954. Die Einwirkung des Diäthyl-p-nitrohenyl-Esters der Thiophosphorsäure auf die San José-Schildlaus (*Quadraspidiotus perniciosus* Comst.). *Zeitschrift für Angewandte Entomologie* **36(3)**: 284-303. [Schuhm1954]

Schweig, C. & Grunberg, A. 1936. The problem of black scale (*Chrysomphalus ficus* Ashm.) in Palestine. *Bulletin of Entomological Research* **27**: 677-714. [SchweiGr1936]

Scott, R.R. 1984a. Berry fruit pests. Pages 11-31 *in*: Scott, R.R. (Ed.). New Zealand Pest and Beneficial Insects. Canterbury: Lincoln University College of Agriculture. 373 pp. [Scott1984a]

Seabra, A.F. de. 1921. As doencas das plantacoes de cacau das Ilhas de S. Tomé e Principe. Os servicos tecnicos de combate contra as epiphytias. Lisboa: Inst. Internac. de Biol. Colon.. 142 pp. [Seabra1921]

Seabra, A.F. de. 1922. Insectes de S. Tomé Provenant de la mission d'étude du Professeur Sousa da Camera en 1920. Coimbra: Impresna da Universidade. 21 pp. [Seabra1922]

Seabra, A.F. de. 1925. Insectes de S. Tomé provenant de la mission d'étude du Professeur Sousa da Camara en 1920. *Anais do Instituto de Agronomia* **2**: 27-45. [Seabra1925]

Seabra, A.F. de. 1930. Subsidios para o conhecimento da fauna das matas nacionais. Conclusoes de estudos realizados durante os meses de Julho E. Agosto de 1925 na mata de leiria. *Arquivos da Seccao de Biologia e Parasitologia* **1**: 267-282. [Seabra1930]

Seabra, A.F. de. 1930a. Apontamentos para o estudo das cochonilhas de Portugal (Hemípteros-Homopteros). *Arquivos da Seccao de Biologia e Parasitologia* **1**: 143-148. [Seabra1930a]

Seabra, A.F. de. 1941. Contribuiçoes para o inventário da fauna Lusitânica insecta: Homoptera (Coccidae). *Memorias y Estudos deo Museo Zoologico da Universidade de Coimbra* **125**: 1-8. [Seabra1941]

Seabra, A.F. de. 1942. Contribuiçoes para o inventário da fauna Lusitânica. Insecta: Homoptera (Coccidae). 1. Aditamento. *Memorias y Estudos do Museo Zoologico da Universidade de Coimbra* **128**: 1-2. [Seabra1942]

Searle, C.M. 1964. A list of the insect enemies of citrus pests in southern Africa. *Technical Communication (Department of Agricultural Technical Services, Pretoria, South Africa)* No. 30: 1-18. [Searle1964]

Searle, C.M., Annecke, D.P. & Wiese, I.H. 1963. Insecticides, beneficial insects, and citrus. *The South African Citrus Journal* **349**: 3-9. [SearleAnWi1963]

Sefer, E. 1961. Catálogo dos insetos que atacam as plantas cultivadas da Amazonia. *Boletim Técnico do Instituto Agrónomico do Norte. Brazil* No. 43: 23-53. [Sefer1961]

Seghatoleslami, H. 1980. Some complementary studies on San José scale in Bandar Anzali (Guilan Province). *Entomologie et Phytopathologie Appliquées* **48**: 149-154. [Seghat1980]

Sekeroglu, E., Uygun, N. & Karaca, I. 1989. The effect of different irrigation systems on population dynamics of California red scale (Homoptera: Diaspididae) on lemon trees in Adana, Turkey. *Turkiye Entomoloji Dergisi* **13(3)**: 147-152. [SekeroUyKa1989]

Selhime, A.G. & Brooks, R.F. 1977. Biological control of some armored scale insects on citrus. *Proceedings of the International Society of Citriculture* **2**: 475-478. [SelhimBr1977]

Selhime, A.G., Muma, M.H., Simanton, W.A. & McCoy, C.W. 1969. Control of Florida red scale in Florida with the parasite *Aphytis holoxanthus*. *Journal of Economic Entomology* **62(4)**: 954-955. [SelhimMuSi1969]

Selvakumaran, S., Kallil, M. & Devasahayam, S. 1996. Natural enemies of two major species of scale insects infesting black pepper (*Piper nigrum* L.) in India. *Pest Management in Horticultural Ecosystems* **2(2)**: 79-83. [SelvakKaDe1996]

Senal, D., Karaca, I. & Undag, H. 2002. Storage possibilities of scale insect predator *Chilocorus bipustulatus* (L.) eggs at different temperatures. *Bollettino di Zoologia Agraria e di Bachicoltura (Milano)* **33(3)**: 427-433. [SenalKaUn2002]

Sengonca, C., Uygun, N., Karaca, I. & Schade, M. 1998. Primary studies on the parasitoid fauna of Coccoidea in cultivated and non-cultivated areas in the east Mediterranean region of Turkey. *Anzeiger Schädlingskunde, Pflanzen(schutz) und Umweltschutz* **71**: 7, 128-131. [SengonUyKa1998]

Seno, K.C.A. & Galli, J.C. 1999. Effect of aldicarbe and fosetyl-AL on *Selenaspidus articulatus* (Morgan) (Homoptera: Diaspididae) on citrus plants with and without variegated chlorosis symptoms. *Anais da Sociedade Entomologica do Brasil* **28(4)**: 715-720. [SenoGa1999]

Severin, H.C. & Severin, H.H.P. 1909. A preliminary list of the Coccidae of Wisconsin. *Journal of Economic Entomology* **2**: 296-298. [SeveriSe1909]

Shafik, M. & Husni, M. 1938. The ideal spray emulsion for the control of scale insects on citrus in Egypt. *Bulletin de la Société Fouad 1er d'Entomologie* **22**: 357-395. [ShafikHu1938]

Shah, A.H., Jhala, R.C. & Patel, C.B. 1988. Bioefficacy of Aldicarb and BPMC against mango scales and its residues on/in mango fruits. *Gujarat Agricultural University Research Journal* **13(2)**: 19-22. [ShahJhPa1988]

Shalaby, F. 1961. A preliminary survey of the insect fauna of Saudi Arabia. *Bulletin de la Société Entomologique d'Egypte* **45**: 211-228. [Shalab1961]

Sharipov, M. 1980. [*Anthemus aspidioti* (Hymenoptera, Encyrtidae): A parasite of scale insects (Homoptera, Diaspididae) in Central Asia and Kazakh SSR, USSR.]. [In Russian]. *Entomologicheskoe Obozrenye* **59(2)**: 381-384. [Sharip1980]

Sharma, D.C., Rawat, U.S. & Pawar, A.D. 1990. Effect of temperature and humidity on the development, longevity and predatory potential of *Pharoscymnus flexibilis* Muls. on San José scale. *Journal of Biological Control* **4(1)**: 11-14. [SharmaRaPa1990]

Sharoni, M. 1980. [Biological and ecological parameters characterizing *Aphytis holoxanthus* as an efficient parasite of the Florida red scale (*Chrysomphalus aonidum* L.).] [In Hebrew, English Abstract]. *M.Sc. Thesis, The Hebrew University of Jerusalem, Rehovot.* 59 pp. [Sharon1980]

Shaw, P.W., Bradley, S.J. & Walker, J.T.S. 2000. Efficacy and timing of insecticides for the control of San José scale on apple. *New Zealand Plant Protection* **53**: 13-17. [ShawBrWa2000]

Shaw, J.G., Moreno, D.S. & Fargerlund, J. 1971. Virgin female California red scales used to detect infestations. *Journal of Economic Entomology* **64**: 1305-1306. [ShawMoFa1971]

Shaw, J.G., Moreno, D.S., Heard, H.L. & Howie, R.M. 1971. Scale detection system for Coachella. *Citrograph* **57**: 67-69. [ShawMoHe1971]

Shaw, J.G., Sunderman, R.P., Moreno, D.S. & Fargerlund, J. 1973. California red scale: Females isolated by treatment with dichlorvos. *Environmental Entomology* **2**: 1062-1063. [ShawSuMo1973]

Shen, X.C. (Ed.) 1993. A Checklist of Insects from Henan. Beijing: China Agricultural Science and Technology Press. 353 pp. [Shen1993]

Shenefelt, R.D. & Benjamin, D.M. 1955. Insects of Wisconsin forests. *Circular (University of Wisconsin, College of Agriculture, Extension Service)* **No. 500**: 1-110. [ShenefBe1955]

Sheta, I.B. & Jenser, G. 1970. [Phenological studies of the San José scale, *Quadraspidiotus perniciosus* Comst.] Kaliforniai pajzstetün (*Quadraspidiotus perniciosus* Comst.) végzett fenológiai vizsgálatok. *Növényvédelem* **2**: 76-81. [ShetaJe1970]

Shetlar, D.J. 1994. Turfgrass insect and mite management. Pages 171-343. *in*: Watschke, T.L., Dernoeden, P.H. & Shetlar, D.J. Managing Turfgrass Pests (Advances in Turfgrass Science). Boca Raton; Ann Arbor; London; Tokyo: Lewis Publishers. 361 pp. [Shetla1994]

Shi, Y.L. & Liu, Y.S. 1991. [Material of scale insect fauna on Fagaceae plants.]. [In Chinese]. *Journal of Shandong Agricultural University* **22(2)**: 160-166. [ShiLi1991]

Shi, G.L., Liu, X.Q., Li, J., Li, L.C. & Yang, F.D. 1997. [Studies on the bionomics of *Quadraspidiotus perniciosus* (Comstock) and its occurrence prediction.]. [In Chinese]. *Scientia Silvae Sinicae* **33**: 161-167. [ShiLiLi1997]

Shiao, S.N. 1979. [Morphology, life history and bionomics of the palm scale *Hemiberlesia cyanophylli* (Homoptera, Diaspididae) in northern parts of Taiwan.]. [In Chinese]. *Bulletin of Plant Protection. Taiwan* **21**: 267-276. [Shiao1979]

Shidrawi, G.R. 1990. A WHO global programme for monitoring vector resistance to pesticides. *WHO Bulletin OMS* **68(4)**: 403-408. [Shidra1990]

Shiraki, T. 1912. Injurious Insects of Formosa, Vol. 1. Agricultural Experiment Station, Government of Formosa, Japan. [Shirak1912]

Shiroma, E.S. 1969. Notes and exhibitions. *Proceedings of the Hawaiian Entomological Society* **20**: 283. [Shirom1969]

Shoemaker, C.A. 1980. The role of systems analysis in integrated pest management. Pages 25-49 *in*: Huffaker, C.B. (Ed.). New Technology of Pest Control. New York: Wiley-Interscience. 500 pp. [Shoema1980]

Shoemaker, C.A., Huffaker, C.B. & Kennett, C.E. 1978. Integrated pest management for olives. *California Agriculture* **32**: 16-17. [ShoemaHuKe1978]

Shukla, G.S. & Tripathi, N. 1979. Studies on the preference of crawlers of sugarcane scale insect, *Melanaspis glomerata* (Green (Hemiptera: Coccidae) in relation to the age of the internodes of sugarcane. *Indian Sugar* **29**: 535-536. [ShuklaTr1979]

Shukla, G.S. & Tripathi, N. 1981. Studies on the survival of sugarcane scale insect, *Melanaspis glomerata* (Green) crawlers after emergence. *Indian Sugar* **31**: 101-102. [ShuklaTr1981]

Shukla, G.S. & Tripathi, N. 1983. Life-history and seasonal history of sugarcane scale-insect, *Melanaspis glomerata* (Green) (Homoptera: Coccidae), in Eastern Uttar Pradesh. *Indian Journal of Agricultural Sciences* **53**: 160-162. [ShuklaTr1983]

Sibbett, G.S., Van Steenwyk, R.A. & Ferguson, L. 2000. Pests of olive. *UC Pest Management Guidelines* . [SibbetVaFe2000]

Siddiqi, Z.A. 1966. Crop pests in Afghanistan. *Publication (Plant Protection Association of Afghanistan)* **No. 1**: 4-5. [Siddiq1966]

Siddiqi, P.M. 1981. Pages 172-180 *in*: Major Agriculture Pests in Afghanistan and Methods for their Control. Kabul: Faculty of Agriculture, Kabul University. [Siddiq1981]

Siegler, E.H. & Baker, H. 1924. Parasitism of scales -- San José and oyster shell. *Journal of Economic Entomology* **17**: 497-499. [SiegleBa1924]

Signoret, V. 1869. Essai sur les cochenilles ou gallinsectes (Homoptères - Coccides) 2e partie. *Annales de la Société Entomologique de France (serie 4)* **8**: 829-876. [Signor1869]

Signoret, V. 1869b. Essai sur les cochenilles ou gallinsectes (Homoptères - Coccides), 3e partie. *Annales de la Société Entomologique de France (serie 4)* **9**: 97-104. [Signor1869b]

Signoret, V. 1869c. Essai sur les cochenilles ou gallinsectes (Homoptères - Coccides), 4e partie. *Annales de la Société Entomologique de France* **9**: 109-138. [Signor1869c]

Signoret, V. 1869d. Essai sur les cochenilles ou gallinsectes (Homoptères - Coccides), 5e partie. *Annales de la Société Entomologique de France* **9**: 431-452. [Signor1869d]

Signoret, V. 1870. Essai sur les cochenilles ou gallinsectes (Homoptères-Coccides), 6e partie. *Annales de la Société Entomologique de France, (serie 4)* **10**: 91-110. [Signor1870]

Signoret, V. 1876b. Note sur une nouvelle espèce d'*Aspidiotus*. *Annales de la Société Entomologique de France, Bulletin Entomologique (serie 5)* **6**: lii-liii. [Signor1876b]

Signoret, V. 1877. Essai sur les cochenilles ou gallinsectes (Homoptères - Coccides), 18e et dernière partie. *Annales de la Société Entomologique de France (serie 5)* **6**: 591-676. [Signor1877]

Signoret, V. 1882b. Diverses notes sur plusieurs Hemiptères, *Spondyliaspis* Sign. = *Inglina*. *Annales de la Société Entomologique de France, Bulletin Entomologique (serie 6* :**2** (clxxxiii-clxxxv. [Signor1882b[

Sigwalt, B. 1971. Demographic studies of Diaspine scales. Applications to three species noxious for orange trees in Tunisia. Particular case of a species with overlapping generations *Parlatoria ziziphi* Lucas. *Annales de Zoologie - Ecologie Animale* **3**: 5-15. [Sigwal1971]

Sikes, J.H. 1931. The effects of contact insecticides upon the obscure scale, *Chrysomphalus obscurus* (Comstock) [microform]. *[microform]* 40 leaves. [Sikes1931]

Silva, P. 1944. Insect pests of cacao in the state of Bahia, Brazil. *Tropical Agriculture* **21**: 8-14. [Silva1944]

Silva, J.L., Paiva, P.E.B., Pinto, R.A., Yamamoto, P.T. & Gravena, S. 1997. Effect of Chlorpyriphos on the control of rufous scale *Selenaspidus articulatus* on citrus. Page 161 *in*: Bento, J.M.S. & Delalibera, I. (Eds.). [16th Brazilian Congress of Entomology] 16o Congresso Brasileiro de Entomologia. Salvador: Sociedade Entomológica do Brasil/EMBRAPA-CNPMF. 400 pp. [SilvaPaPi1997]

Silvestri, F. 1902. Coccidei parassiti della vite. *Bollettino di Entomologica e Agraria Patologia Vegetale* **9**: 74-149. [Silves1902]

Silvestri, F. 1915. Contributo alla conoscenza degli insetti dell'olivo dell'Eritrea e dell'Africa meridionale. *Bollettino del Laboratorio di Zoologia Generale e Agraria della R. Scuola Superiore d'Agricoltura in Portici* **9**: 249-263. [Silves1915]

Silvestri, S. 1921. Il crisomfalo o cocciniglia rossa degli agrumi. *Bollettino del Laboratorio di Entomologia Agraria 'Filippo Silvestri'. Portici* **2**: 1-11. [Silves1921]

Silvestri, F. 1926a. Lotta contro alcune cocciniglie degli agrumi. *Nuovi annali dell'agricoltura* **6**: 97-101. [Silves1926a]

Silvestri, F. 1929. Preliminary report on the citrus scale-insects of China. Pages 897-904 *in*: International Congress of Entomology (4th : 1928 : Ithaca); Jordan, K. & Horn, W. (Eds.). Fourth International Congress of Entomology, Ithaca, August 1928. Vol. II. Naumburg a/Saale, G. Pätz. 2 v. in 1. [Silves1929]

Simanton, W.A. 1960a. Seasonal populations of citrus insects and mites in commercial groves. *Florida Entomologist* **43(1)**: 49-57. [Simant1960a]

Simanton, W.A. 1962. Preference and use of pesticides on citrus during a ten year period. *Proceedings of the Florida State Horticultural Society* **75**: 99-103. [Simant1962]

Simanton, W.A. 1962a. Operation of an ecological survey for Florida citrus pests. *Journal of Economic Entomology* **55(!)**: 105-112. [Simant1962a]

Simanton, W.A. 1962b. Losses and production costs attributable to insects and related arthropods attacking citrus in Florida in 1961. *Cooperative Economic Insect Report (USDA)* **12(45)**: 1182. [Simant1962b]

Simanton, W.A. 1969. Sixteen years of pesticide preferences and use on citrus. *Proceedings of the Pest Control Conference* **3**: 18-25. [Simant1969]

Simanton, W.A. 1969a. Use and value of a continuous field survey in the control of citrus pests. Pages 889-896. *in*: Chapman, H.D. (Ed.). Proceedings First International Citrus Symposium. Vol. 2. Riverside: University of California. [Simant1969a]

Simanton, W.A. 1976. Occurrence of insect and mite pests of citrus, their predators and parasitism in relation to spraying operations. *Proceedings of the Tall Timbers Conference on Ecology and Animal Control by Habitat Management* **6**: 135-164. [Simant1976]

Simmonds, H.W. 1944. The effect of the host fruit upon the scale *Aonidiella aurantii* Mask. in relation to its parasite *Comperiella bifasciata* How. *Journal of the Australian Institute of Agricultural Science* **10**: 38-39. [Simmon1944]

Simmonds, F.J. 1958b. Recent work on biological control in the British West Indies. Pages 475-478 *in*: Baecker, E.C., (Ed.). Proceedings of the Tenth International Congress of Entomology. Vol. 4. Ottawa. 1115 pp. [Simmon1958b]

Simmonds, F.J. 1959a. Biological control -- past, present and future. *Journal of Economic Entomology* **52(6)**: 1099-1102. [Simmon1959a]

Simmonds, F.J. 1960. Biological control of the coconut scale, *Aspidiotus destructor* Sign., in Principe, Portuguese West Africa. *Bulletin of Entomological Research* **51**: 223-237. [Simmon1960]

Simmonds, F.J. 1969a. The work of the Commonwealth Institute of Biological Control relating to citrus insects. Pages 765-767. *in*: Chapman, H.D. (Ed.). Proceedings First International Citrus Symposium. Vol. 2. Riverside: University of California. [Simmon1969a]

Simmonds, F.J. & Bennett, F.D. 1976. Biological control of agricultural pests. Pages 46-472 *in*: Packer, J.S. & White, D. (Eds.). Proceedings of XV International Congress of Entomology (Washington, D.C., August 19-27, 1976). College Park, MD: Entomological Society of America. 824 pp. [SimmonBe1976]

Simmonds, F.J., Franz, J.M. & Siler, R.I. 1976. History of biological control. Pages 17-39. *in*: Huffaker, C.B. & Messenger, P.S. (Eds.). Theory and Practice of Biological Control. New York: Academic Press. 788 pp. [SimmonFrSi1976]

Simmonds, F.J. & Greathead, D.J. 1977. Introductions and pest and weed problems. Pages 109-124. *in*: Cherrett, J.M. & Sagar, G.R. (Eds.). Origins of Pest, Parasite, Disease and Weed Problems. Oxford: Blackwell Scientific. 413 pp. [SimmonGr1977]

Singh, C. 1963. San José scale – a menace of hill orchards. *Indian Horticulture* **7**: 13. [Singh1963]

Singh, C. 1964. Temperate fruits pests. Pages 213-226 *in*: Pant, N.C., Chief Editor. Entomology in India ["A special number of the Indian Journal of Entomology"]. New Delhi: The Entomological Society of India. 529 pp. [Singh1964]

Sinha, P.K. & Dinesh, D.S. 1984. A report on the coccids (Hemiptera: Coccoidea), their host plants and natural enemies at Bhagalpur. *Biological Bulletin of India* **6**: 7-13. [SinhaDi1984]

Siscaro, G., Longo, S. & Lizzio, S. 1999. Ruolo degli entomofagi di *Aonidiella aurantii* (Maskell) (Homoptera, Diaspididae) in agrumeti siciliani. *Phytophaga. Palermo* **9(Suppl.)**: 41-52. [SiscarLoLi1999]

Skiba, N.S. & Parii, I.F. 1989. [Pests and diseases of cherry.]. [In Russian]. *Zashchita rastenii. Moscow* **No. 8**: 48-51. [SkibaPa1989]

Skorkin, L.A. 1938. [Importance of fruits as a possible means of carriage of San José scale. Pages 147-186 *in:* Kiritchenko, A.N. (Ed.), Works of Quarantine Laboratories.] [In Russian]. Leningrad, Sel'khozgiz. 272 pp. [Skorki1938]

Smetnik, A.I. 1991. [Pheromones of scales.]. [In Russian]. *Quarantine Pests, Diseases and Weeds* **I**: 92-129. [Smetni1991]

Smetnik, A.I. & Konstantinova,G.M. 1983. [The use of synthetic sex pheromones in plant quarantine in the USSR.]. [In Russian]. Page 882 *in*: 10th International Congress of Plant Protection: Plant protection for human welfare. Vol. 2. Croydon, UK: British Crop Protection Council. [SmetniKo1983]

Smetnik, A.I., Konstantinova, G.M., Kovalev, B.G., Bychina, T.I., Rotar', M.G. & Maksimova, V.I. 1982. [Testing of pheromone traps for San José scale *Aspidiotus perniciosus*, fruit pest control, USSR.]. [In Russian]. *Zashchita rastenii. Moscow. "Kolos"* (1) 39. [SmetniKoKo1982]

Smetnik, A.I., Konstantinova, G.M., Maksimova, V.I., Rozinskaya, E.M. & Shelukhin, V.I. 1987. Application of Coccids sex pheromones in the practice of the USSR Quarantine Services. *Bollettino del Laboratorio di Entomologia Agraria 'Filippo Silvestri'* **43 (Suppl.)**: 209-213. [SmetniKoMa1987]

Smetnik, A.I., Maksimova, V.I., Konstantinova, G.M., Ishchenko, R.I. & Shelukhin, V.I. 1983. [Trails with the sex pheromone of the San José scale [*Aspidiotus perniciosus*, genetic control, USA, USSR]. [In Russian]. *Zashchita rastenii. Moscow. "Kolos"* (9) 38. [SmetniMaKo1983]

Smirnoff, W.A. 1950a. Sur la biologie au Maroc de *Rhizobius (Lindorus) lophantae* Blaisd. (Col. Coccinellidae). *Revue de Pathologie Végétale et d'Entomologie Agricole de France* **29(4)**: 190-194. [Smirno1950a]

Smirnoff, W.A. 1951. La cochenille "Pou rouge" dans les cultures d'agrumes au Maroc. *Terre Marocaine* **No. 264**: 3-7. [Smirno1951]

Smirnoff, W.A. 1951a. Aperçu sur le développement de quelques cochenilles parasites des agrumes au Maroc. *Travaux originaux (Service de la Défense des Végétaux, Morroco)* **1**: 1-29. [Smirno1951a]

Smirnoff, W.A. 1952. *Aspidiotiphagus lounsburyi*, parasite de certaines espèces de Diaspididae au Maroc. *Revue de Pathologie Végétale et d'Entomologie Agricole de France* **31**: 63-69. [Smirno1952]

Smirnoff, W.A. 1957a. [*Aonidiella aurantii* Mask. (California red scale): citrus pest in Morocco and how to control them.] *Aonidiella aurantii* Mask. (pou rouge de Californie): parasite des agrumes au Maroc et les moyens de le combattre. *Fruits et Primeurs de l'Afrique du Nord: La revue française de l'oranger de l'arboriculture fruitière et des cultures irriguées* **No. 287**: 47-55. [Smirno1957a]

Smirnov, B.A. 1938. [Effect of injury by *Aspidiotus perniciosus* Comst. on the marketability of fresh fruits. Pages 187-201 *in:* Kiritchenko, A.N. (Ed.), Works of Quarantine Laboratories.] [In Russian]. Leningrad, Sel'khozgiz. 272 pp. [Smirno1938]

Smit, B. 1964. Insects in Southern Africa: How to Control Them. Cape Town: Oxford University Press. 399 pp. [Smit1964]

Smith, A.D.M. & Maelzer, D.A. 1986. Aggregation of parasitoids and density-independence of parasitism in field populations of the wasp *Aphytis melinus* and its host, the red scale *Aonidiella aurantii*. *Ecological Entomology* **11**: 425-434. [SmithMa1986]

Smith, D. 1978a. Biological control of scale insects on citrus in southeastern Queensland. II. Control of circular black scale *Chrysomphalus ficus* Ashmead by the introduced parasite, *Aphytis holoxanthus* De Bach. *Journal of the Australian Entomological Society* **17**: 373-377. [Smith1978a]

Smith, D., Beattie, G.A.C. & Broadley, R.H. 1997. Citrus Pests and their Natural Enemies: Integrated Pest Management in Australia. Brisbane, Australia: State of Queensland, Dept. of Primary Industries, and Horticultural Research and Development Corp.. 263+ pp. [SmithBeBr1997]

Smith, D., Freebairn, C.G. & Papacek, D.F. 1996. The effect of host density and parasitoid inoculum size on the mass production of *Leptomastix dactylopii* Howard (Hymenoptera: Encyrtidae) and *Aphytis lingnanensis* Compere (Hymenoptera: Aphelinidae) in Queensland. *General and Applied Entomology* **27**: 57-64. [SmithFrPa1996]

Smith, D., Smith, N.J. & Smith, K.M. 1998. Effect of abamectin on citrus rust mite *Phyllocoptruta oleivora* and brown citrus rust mite *Tegolophus australis* and the scale natural enemies *Aphytis lingnanensis* and *Chilocorus circumdatus* on oranges. *Plant Protection Quarterly* **13**: 136-139. [SmithSmSm1998]

Smith, H.S. 1926. The present status of biological control work in California. *Journal of Economic Entomology* **19**: 294-302. [Smith1926]

Smith, H.S. 1934. University of California to continue search for red scale parasites. *California Citrograph* **19**: 280-282. [Smith1934]

Smith, H.S. 1941. Status of biological control of scale pests. *California Citrograph* **26(3)**: 58, 76-77. [Smith1941]

Smith, H.S. 1948. Biological control of insect pests. Pages 597-664 *in:* Batchelor, L.D. & Webber, H.J. (Eds.). The Citrus Industry. Vol. II. Production of the Crop Berkeley & Los Angeles: University of California Press. 933 pp. [Smith1948]

Smith, H.S. 1948a. Quarantine and quarantine service. Pages 813-830 *in:* Batchelor, L.D. & Webber, H.J. (Eds.). The Citrus Industry. Vol. II. Production of the Crop. Berkeley & Los Angeles: University of California Press. 933 pp. [Smith1948a]

Smith, H.S. & Compere, H. 1931a. An imported parasite attacks the yellow scale. *California Citrograph* **16**: 328. [SmithCo1931a]

Smith, H.S., Essig, E.O., Fawcett, H.S., Peterson, G.M., Qualyle, H.J., Smith, R.E. & Tolley, H.R. 1933. Efficacy and economic effects of plant quarantines in California. *Bulletin (Connecticut Agricultural Extension Station)* No. 553. [SmithEsFa1933]

Smith, H.S. & Flanders, S.E. 1948. Search for parasites of red scale is reoriented. *California Citrograph* **34**: 4, 17-18, 20. [SmithFl1948]

Smith, H.S. & Flanders, S.E. 1950. The search for natural enemies of citrus pests. *California Citrograph* **35**: 362,376,378. [SmithFl1950]

Smith, J.B. 1898. The distribution of the San José or pernicious scale in New Jersey. *United States Department of Agriculture, Division of Entomology, Bulletin* **17**: 32-39. [Smith1898]

Smith, J.M. 1957. Effects of the food plant of California red scale, *Aonidiella aurantii* (Mask.) on reproduction of its Hymenopterous parasites. *Canadian Entomologist* **89**: 219-230. [Smith1957]

Smith, K.M., Smith, D. & Lisle, A.T. 1999. Effect of field-weathered residues of pyriproxyfen on the predatory coccinellids *Chilocorus circumdatus* Gyllenhal and *Cryptolaemus montrouzieri* Mulsant. *Australian Journal of Experimental Agriculture* **39**: 995-1000. [SmithSmLi1999]

Snitko, N.F. 1938. [The growth of shoots of fruit trees infested by San José scale. Pages 94-131 *in:* Kiritchenko, A.N. (Ed.), Works of Quarantine Laboratories.] [In Russian]. Leningrad, Sel'khozgiz. 272 pp. [Snitko1938]

Soares Brandao, J. 1942. As orquideas sao tambem parasitadas por pragas e doenças. *Sitios e Fazendas* **7**: 54-56. [Soares1942]

Soares Brandao, J. 1945. Os parasitos de algumas plantas floriferas. *Boletim do Ministerio da Agricultura. Brazil.* **34**: 49-81. [Soares1945]

Soares, M.A., Marchior, L.C. & Cassino, P.C.R. 1998. Variaçao sazonal de *Selenaspidus articulatus* (Morgan, 1889) (Homoptera, Diaspididae) em *Citrus reticulata* Blanco, no campus da UFRRJ. *Floresta e Ambiente* **5(1)**: 124-129. [SoaresMaCa1998]

Sokoloff, V.P. & Klotz, L.J. 1941. A bacterial pathogen of the citrus red scale. *Science* **94(2428)**: 40-41. [SokoloKl1941]

Sokoloff, V.P. & Klotz, L.J. 1942. Mortality of the red scale on citrus through infection with a spore-forming bacterium. *Phytopathology* **32**: 187-198. [SokoloKl1942]

Sokoloff, V.P. & Klotz, L.J. 1943. Susceptibility of California citrus red scale to bacterial infection in relation to nitrogen content of the substratum. *Citrus leaves* **23(11)**: 6-7. [SokoloKl1943]

Soria, S., Estal, P. del & Vinuela, E. 1996. Los coccidos del Tejo (*Taxus baccata* L.) en Espana. *Boletín de Sanidad Vegetal, Plagas* **22**: 2, 241-249. [SoriaEsVi1996]

Soria, S., Estal, P. del & Vinuela, E. 1998a. Presencia en España de *Odonaspis greeni* (Cockerell) y *Bambusaspis bambusae* Boisduval sobre plantas ornamentales de bambu. *Boletín de Sanidad Vegetal, Plagas* **24**: 2, 337-342. [SoriaEsVi1998a]

Soria, S., Moreno, M., Viñuela, E. & Estal, P. del. 2000. Principales cochinillas en los pinos españoles. *Boletín de Sanidad Vegetal, Plagas* **26(3)**: 335-348. [SoriaMoVi2000]

Souissi, R. & Panis, A. 1999. [Estimation of the fecundity of two scales of fruit trees *Aspidiotus nerii* Bouche and *Pseudaulacaspis pentagona* (Targioni-Tozzetti) (Hemiptera: Diaspididae) using a micro-cage technique.]. [In French]. *Annales de la Société Entomologique de France* **35(Suppl. S)**: 87-92. [SouissPa1999]

Souza da Camara, M. de. 1906. Subsidios para o estudo das cochonilhas Portuguezas. *Revista Agronomica. Lisbon* **4**: 92-95, 148-149. [Souzad1906]

Spencer, H. & Osborn, M.R. 1938. Citrus insect projects of the Orlando laboratory. *Journal of Economic Entomology* **31(6)**: 728-730. [SpenceOs1938]

Spiller, D. 1952. Truncated lognormal distribution of red scale *Aonidiella aurantii* Mask.) on citrus leaves. *New Zealand Journal of Science and Technology* **33**: 483-487. [Spille1952]

Spollen, K.M. & Hoy, M.A. 1993. Laboratory-selected California red scale parasite is resistant to Sevin. *California Agriculture* **47(1)**: 16-19. [SpolleHo1993]

Spollen, K.M. & Hoy, M.A. 1993a. Residual toxicity of five citrus pesticides to a carbaryl-resistant and a wild strain of the California red scale parasite *Aphytis melinus* DeBach (Hymenoptera: Aphelinidae). *Journal of Economic Entomology* **86**: 195-204. [SpolleHo1993a]

Spollen, K.M. & Hoy, M.A. 1993b. Genetic improvement of an arthropod natural enemy: relative fitness of a carbaryl-resistant strain of the California red scale parasite *Aphytis melinus* DeBach. *Biological Control* **2**: 87-94. [SpolleHo1993b]

Spollen, K.M., Rosenheim, J.A. & Hoy, M.A. 1994. Intraspecific variation in response to pesticides in *Aphytis melinus* DeBach from California citrus: results of natural and artificial selection between 1984 and 1991. Pages 189-208 *in*: Rosen, D., Ed. Advances in the Study of Aphytis (Hymenoptera: Aphelinidae). Andover, U.K.: Intercept. 362 pp. [SpolleRoHo1994]

Sproul, A.N. 1981. Citrus red scale, from sprays to parasites. *Journal of Agriculture of Western Virginia* **5**: 50. [Sproul1981]

Stafford, E.W. 1915. Studies in diaspinine pygidia. *Annals of the Entomological Society of America* **8**: 67-73. [Staffo1915]

Stanev, M.Ts. 1963. [Studies on the biology and on control of *Quadraspidiotus perniciosus* Comst. in Bulgaria.]. [In Bulgarian]. *Bulletin of the Institute of Plant Protection* **5**: 5-27. [Stanev1963]

Stanev, M. & Delinska, N. 1989. [Flight of males of San José scale, *Quadraspidiotus perniciosus* Comst. (Homoptera; Diaspididae).]. [In Bulgarian]. *Rasteniev'dni Nauki* **26(6)**: 85-91. [StanevDe1989]

Stannard, L.J. 1965. Polymorphism in the Putnam's scale, *Aspidiotus ancylus* (Homoptera: Coccoidea). *Annals of the Entomological Society of America* **58**: 573-576. [Stanna1965]

Starler, N.H. & Ridgway, R.L. 1977. Economic and social considerations for the utilization of augmentation of natural enemies. Pages 431-448. *in*: Ridgeway, R.L. & Vinson, S.B. (Eds.). Biological Control by Augmentation of Natural Enemies: Insect and mite Control with Parasites and Predators. New York: Plenum Press. 480 pp. [StarleRi1977]

Starnes, H.N. 1897. The San José and other scales in Georgia. *Bulletin of the Georgia Experiment Station* **36**: 1-31. [Starne1897]

Stathas, G.J. 2000. *Rhyzobius lophanthae* prey consumption and fecundity. *Phytoparasitica* **28(3)**: 203-211. [Statha2000]

Stathas, G.J. 2000a. The effect of temperature on the development of the predator *Rhyzobius lophanthae* and its phenology in Greece. *Biocontrol Dordecht* **45(4)**: 439-451. [Statha2000a]

Stathas, G.J. 2001b. Studies on morphology and biology of immature stages of the predator *Rhyzobius lophanthae* Blaisdell (Col.: Coccinellidae). *Anzeiger für Schädlingskunde* **74(5)**: 113-116. [Statha2001b]

Stathas, G.J. & Eliopoulos, P.A. 2001. Prey consumption of the predator *Chilocorus bipustulatus* Linnaeus on *Aspidiotus nerii* Bouché. *Annales de l'Institut Phytopathologique Benaki* **(N.S.) 19**: 125-133. [StathaEl2001]

Stathas, G.J., Eliopoulos, P.A., Kontodimas, D.C. & Siamos, D.T. 2002. Adult morphology and life cycle under constant temperatures of the predator *Rhyzobius lophanthae* Blaisdell (Col., Coccinellidae). *Anzeiger für Schädlingskunde* **75(4)**: 105-109. [StathaElKo2002]

Stathas, G.J. & Kontodimas, D.C. 2001. Ecological data of the scale *Targionia vitis* on grapes in southern Greece. *Annales de l'Institut Phytopathologique Benaki* **(N.S.) 19**: 134-139. [StathaKo2001]

Stavraki, H.G., Argyriou, LC. & Yamvrias, C. 1979. Sur un programme de lutte intégrée contre les ennemis de l'olivier en Grèce. Pages 574-577. *in*: Proceedings: Internationales Symposium der IOBC/WPRS über Integrierten Pflanzenschutz in der Land und Forstwirtschaft. Wien: OILB/SROP. 648 pp. [StavraArYa1979]

Stehr, F.W. 1974. Release, establishment and evaluation of parasites and predators. Pages 124-136 *in*: Maxwell, F.G. & Harris, F.A. (Eds.). Proceedings of the Summer Institute on Biological Control of Plant Insects and Diseases. Jackson: University Press of Mississippi. 647 pp. [Stehr1974]

Steinberg, S., Podoler, H. and Rosen, D. 1987. Competition between two parasites of the Florida red scale in Israel. *Ecological Entomology* **12**: 299-310. [SteinbPoRo1987]

Steinberg, S., Podoler, H. & Rosen, D. 1987a. Biological control of the Florida red scale, *Chrysomphalus aonidum*, in Israel by two parasite species: current status in the coastal plain. *Phytoparasitica* **14**: 199-204. [SteinbPoRo1987a]

Steinberg, S., Podoler, H. & Rosen, D. 1994. r and K strategies in parasitoids of diaspidid scale insects: *Aphytis* vs. endoparasitoids. Pages 79-91 *in*: Rosen, D., ed. Advances in the Study of Aphytis (Hymenoptera: Aphelinidae). Andover, UK: Intercept Limited. [SteinbPoRo1994]

Steiner, M. 1987. Mealybugs and scales in greenhouses and interior plantscapes. *Ohio Florists' Association Bulletin* **No. 694**: 1-6. [Steine1987]

Steinitz, H. 1938. Experiments on citrus varieties as host-plants of the *Aspidiotus hederae* Vall. in Palestine. *Hadar* **11**: 160-163. [Steini1938]

Steinweden, J.B. 1948. Identification and control of orchid pests. *Orchid Digest* **May-June**: 105-111. [Steinw1948]

Stephani Perez, T. de. 1909. *Chrysomphalus dictyospermi* var. *pinnulifera* Mask. negli agrumeti Siciliani. *Bollettino R. Orthno Botanico e Giard. Colon.* **8**: 189-196. [Stepha1909]

Stern, V.M., Adkisson, P.L., Beingolea G., O. & Viktorov, G.A. 1976. Cultural controls. Pages 593-613 *in*: Huffaker, C.B. & Messenger, P.S. (Eds.). Theory and Practice of Biological Control. New York: Academic Press. 788 pp. [SternAdBe1976]

Sternlicht, M. 1973a. The sex attractants of the flower moth *Prays citri* and of the California red scale *Aonidiella aurantii* as a help in controlling these pests. *I Congreso Mundial de Citricultura (Murcia-Valencia, 29 abril-10 mayo 1973)* **2**: 513-519. [Sternl1973a]

Steven, D., Blank, R.H. & Tomkins, A.R. 1991. Monitoring scale to reduce sprays in kiwifruit. *New Zealand Agric. Sci.* **26**: 84. [StevenBlTo1991]

Stevens, P.S., McKenna, C.E., Blank, R.H., Tomkins, A.R. & Steven, D. 1997. Comparison of armoured scale spray thresholds in kiwifruit. Pages 288-292. *in*: Proceedings of the Fiftieth New Zealand Plant Protection Conference. Roturua, New Zealand: New Zealand Plant Protection Society. [StevenMcBl1997]

Steven, D. & Retamales, J. 1999. Integrated and organic production of kiwifruit. *Acta Horticulturae* **(No. 498)**: 345-354. [StevenRe1999]

Steven, D., Tomkins, A.R., Blank, R.H. & Charles, J.G. 1994. A first-stage integrated pest management system for kiwifruit. Pages 135-142 *in*: Proceedings -- Brighton Crop Protection Conference Pests and Diseases. Brighton, UK: November 21-24, 1994. Farnham, UK: British Crop Protection Council. [StevenToBl1994]

Steven, D., Valenzuela, L. & Gonzalez, R.H. 1997. Kiwifruit pests in Chile. *Acta Horticulturae* **444**: 773-777. [StevenVaGo1997]

Steyn, J.J. 1951. The effect of low calcium, phosphorus or nitrogen on the life-cycle of red scale *(Aonidiella aurantii* Mask.). *Journal of the Entomological Society of Southern Africa* **14**: 165-170. [Steyn1951]

Steyn, J.J. 1954. The effect of the cosmopolitan brown house ant (*Pheidole megacephala* F.) on citrus red scale (*Aonidiella aurantii* Mask.) at Letaba. *Journal of the Entomological Society of Southern Africa* **17**: 252-264. [Steyn1954]

Steyn, J.J. 1954a. The pugnacious ant (*Anoplolepis custodiens* Smith) and its relation to the control of citrus scales at Letaba. *Memoirs of the Entomological Society of Southern Africa* **3**: 1-96. [Steyn1954a]

Steyn, J.J. 1955. The effect of mixed ant populations on red scale (*Aonidiella aurantii* Mask.) on citrus at Letaba. *Journal of the Entomological Society of Southern Africa* **18(1)**: 93-105. [Steyn1955]

Steyn, J.J. 1958. The effect of ants on citrus scales at Letaba, South Africa. Pages 589-594 *in*: Baecker, E.C., (Ed.). Proceedings of the Tenth International Congress of Entomology. Vol. 4. Ottawa. 1115 pp. [Steyn1958]

Steyn, J.J. 1959. Ant protection and citrus scales at Letaba, South Africa. *Proceedings of the IVth International Congress of Crop Protection (Hamburg, September 1957)* **1**: 899-902. [Steyn1959]

Stimmel, J.F. 1976. Putnam scale, *Diaspidiotus ancylus* (Putnam) (Homoptera: Diaspididae). *Regulatory Horticulture (PA Dept. of Agric.)* **(Ent. Circ. No. 11**: 19-20). [Stimme1976]

Stimmel, J.F. 1980a. Obscure scale, *Melanaspis obscura* Comstock (Homoptera: Diaspididae). *Regulatory Horticulture (PA Dept. of Agric.)* **6(1)**: 11-12 (Ent. Circ. No. 46). [Stimme1980a]

Stimmel, J.F. 1986. *Aspidiotus cryptomeriae* Kuwana, an armored scale pest of conifers. *Regulatory Horticulture (PA Dept. of Agric.)* **12(2)**: 21-22 (Ent. Circ. No. 108). [Stimme1986]

Stimmel, J.F. 2000. Hemlock scale, *Abgrallaspis ithacae* (Ferris) Homoptera: Diaspididae. *Regulatory Horticulture (PA Dept. of Agric.)* **26**: 15-17. [Stimme2000]

Stimmel, J.F. 2002. *Nuculaspis pseudomeyeri* (Kuwana), a scale insect on evergreen conifers (Homoptera: Diaspididae).*Regulatory Horticulture (PA Dept. of Agric.)* **28**: 27-29. [Stimme2002]

Stoetzel, M.B. & Davidson, J.A. 1971. Biology of the obscure scale, *Melanaspis obscura* (Homoptera: Diaspididae), on pin oak in Maryland. *Annals of the Entomological Society of America* **64**: 45-50. [StoetzDa1971]

Stoetzel, M.B. & Davidson, J.A. 1973. Life history of the obscure scale (Homoptera: Diaspididae) on pin oak and white oak in Maryland. *Annals of the Entomological Society of America* **66**: 308-311. [StoetzDa1973]

Stoetzel, M.B. & Davidson, J.A. 1974. Sexual dimorphism in all stages of the Aspidiotini (Homoptera: Diaspididae). *Annals of the Entomological Society of America* **67**: 138-140. [StoetzDa1974]

Stoetzel, M.B. & Davidson, J.A. 1974a. Biology, morphology and taxonomy of immature stages of nine species in the Aspidiotini (Homoptera: Diaspididae). *Annals of the Entomological Society of America* **67**: 475-509. [StoetzDa1974a]

Stofberg, F.J. 1937a. The citrus red scale (*Aonidiella aurantii* Mask.). *Science Bulletin, Union of South Africa, Department of Agriculture and Forestry* **167**: 1-24. [Stofbe1937a]

Stoliarova, F.A., Okolelova, V.A. & Chuguniaeva, A.T. 1977. [Entomophagous insects of San José scale *Quadraspidiotus perniciosus*, pest of fruit trees, biological control.]. *Zashchita Rastenii* **3**: 54-55. [StoliaOkCh1977]

Stouthamer, R. & Luck, R.F. 1991. Transition from bisexual to unisexual cultures in *Encarsia perniciosi* (Hymenoptera, Aphelinidae) - new data and a reinterpretation. *Annals of the Entomological Society of America* **84(2)**: 150-157. [StouthLu1991]

Streibert, H.P., Frischknecht, M.L. & Karrer, F. 1994. Diofenolan - a new insect growth regulator for the control of scale insects and important lepidopterous pests in deciduous fruit and citrus. *Proceedings (Brighton Crop Protection Conference, Pests and Diseases)* **1**: 23-30. [StreibFrKa1994]

Strong, L.A. 1922. A synopsis of work for the months of March, April, May, June and July, 1922. *Monthly Bulletin, Department of Agriculture, State of California* **11**: 775-780. [Strong1922]

Struble, G.R. & Johnson, P.C. 1964. Black pine-leaf scale. *Forest Pest Leaflet (U.S. Dept. of Agriculture)* **91**: 1-6. [StrublJo1964]

Su, T.H. & Wang, C.M. 1988. [Life history and control measures of the citrus mealybug and the latania scale insects on grapevine.]. [In Chinese]. *Plant Protection Bulletin. Taipei* **30(3)**: 279-288. [SuWa1988]

Sudha Rao, V. & Rao, V.P. 1960. Introduction of *Prospaltella perniciosi* for control of San José scale in India and Pakistan. *FAO Plant Protection Bulletin* **10(8)**: 120-123. [SudhaRRa1960]

Sudoi, V. 1995. Effect of spraying petroleum white oil (Murphoil) on control of scale insects *Aspidiotus* sp. and their effect on natural enemies *Aphytis* sp. *Tea* **16**: 119-123. [Sudoi1995]

Sugimoto, S. 1994. Scale insects intercepted on banana fruits from Mindanao Is., the Philippines (Coccoidea: Homoptera). *Research Bulletin of the Plant Protection Service Japan* **30**: 115-121. [Sugimo1994]

Sugimoto, S., Kadoi, M. & Tasaka, E. 1996. Miscellaneous notes on the scale insects (Homoptera: Coccoidea) intercepted at quarantine inspection on bananas. *Research Bulletin of the Plant Protection Service Japan* **32**: 99-101. [SugimoKaTa1996]

Šulc, K. 1934. [*Thymaspis fusca* n. gen. n. sp. *Aspidiotus baudysi* n. sp. (Coccidae, sf. Diaspinae). Patria Moravia. CSR.]. [In Czech]. *Prace Moravské Prírodovedecké Spolecnosti (Acta Societatis Scientiarum Naturalium Moravicae)* **9**: 1-21. [Sulc1934]

Sullivan, K.C. 1930. State Plant Division. *Biennial Report (Missouri State Board of Agriculture)* **28**: 51-59. [Sulliv1930]

Sumaroka, A.F. 1967. Factors affecting the sex ratio of *Aphytis proclia* Wlk., external parasite of the San José scale. *Entomological Review* **46**: 179-185. [Sumaro1967]

Summerland, S.A. & Hamilton, D.W. 1951. *Hemisarcoptes malus*, a predator of Forbes scale. *Journal of Economic Entomology* **44(5)**: 818. [SummerHa1951]

Sunil, J., Ballal, C.R., Rao, N.S. & Joshi, S. 2001. Influence of temperature on biological, predatory and reproductive attributes of *Sticholotis cribellata* Sicard (Coleoptera: Coccinellidae) on *Melanaspis glomerata* Green. *Annals of Plant Protection Sciences* **9(1)**: 26-31. [SunilBaRa2001]

Sunil, J., Poorani, J. & Singh, S.P. 2002. Bioecology of *Sticholotis cribellata* Sicard (Coleoptera: Coccinellidae), a potential predator of *Melanaspis glomerata* (Green) (Homoptera: Diaspididae). *Journal of Biological Control* **15(1)**: 21-26. [SunilPoSi2002]

Suresh, S. & Mohanasundaram, M. 1995. Occurrence of *Morganella cueroensis* (Cockerell) (Diaspididae: Coccoidea: Homoptera) in South India. *Journal of the Bombay Natural History Society* **92**: 429-430. [SureshMo1995]

Suresh, S. & Mohanasundaram, M. 1996. Coccoid (Coccoidea: Homoptera) fauna of Tamil Nadu, India. *Journal of Entomological Research. New Delhi* **20**: 233-274. [SureshMo1996]

Suris, M. 1999. Disposición espacial de *Selenaspidus articulatus* Morg. (Coccoidea: Diaspididae) en naranjo Valencia (*Citrus sinensis* L.). *Revista de Protección Vegetal* **14(1)**: 17-22. [Suris1999]

Suris Campos, M. 1984. [Population fluctuation of *Selenaspidus articulatus* Morg. (Coccoidea: Diaspididae) in mature groves of Valencia orange (in Cuba)]. [In Spanish]. *Centro Agrícola* **11(3)**: 110. [SurisC1984]

Suris Campos, M. 1993. Guía para la identificación de los principales coccidos de cítricos en fase migrante. *Revista de Protección Vegetal* **8**: 2, 121-127. [SurisC1993]

Suris, M. & Castillo, N. 1987. Variaciones en la distribucion de diaspididos segun variedad y organo infestado de citrico en la provincia Matanzas. *Revista de Proteccion Vegetal* **2(1)**: 27-31. [SurisCa1987]

Suris, M., Hernandez, N. & Varona, I. 1989. Analisis de un procedimiento de muestreo para *Selenaspidus articulatus* Morg. (Homoptera: Diaspididae) en citricos. *Revista de Proteccion Vegetal* **4(1)**: 1-8. [SurisHeVa1989]

Suris, M. & Varonna, I. 1988. Distribucion espacial de *Selenaspidus articulatus* (Coccoidea: Diaspididae) en una plantacion de naranjo Valencia. *Revista de Proteccion Vegetal* **3(1)**: 38-44. [SurisVa1988]

Swailem, S.M., Awadallah, K.T. & Shaheen, A.A. 1976. Abundance of *Lindingaspis rossi* Mask. on ornamental host plants in Giza and Zagazig region, Egypt (Hemiptera-Homoptera: Diaspididae). *Bulletin de la Société Entomologique d'Egypte* **60**: 257-263. [SwaileAwSh1976]

Sweetman, H.L. 1958. Successful biological control against animals. Pages 449-459 *in*: Baecker, E.C., (Ed.). Proceedings of the Tenth International Congress of Entomology. Vol. 4. Ottawa. 1115 pp. [Sweetm1958]

Swezey, O.H. 1945. Insects associated with orchids. *Proceedings of the Hawaiian Entomological Society* **12**: 343-403. [Swezey1945]

Swiderski, Z. 1980. The fine structure of the sperm bundles of the coccid *Aspidiotus perniciosus* and sperm movement. *International Journal of Invertebrate Reproduction* **295**: 331-339. [Swider1980]

Swingle, H.S. & Snapp, O.I. 1931. Petroleum oils and oil emulsions as insecticides, and their use against the San José scale on peach trees in the south. *Technical Bulletin (U.S.D.A.)* **No. 253**: 1-48. [SwinglSn1931]

Swirski, E. 1976b. Pests of subtropical trees. *the Encyclopedia of Agriculture* **3**: 555-559. [Swirsk1976b]

Swirski, E. 1989. Biological control of pests by arthropod natural enemies (insects and mites). *Israel Agresearch* **3**: 11-44. [Swirsk1989]

Swirski, E. & Arenstein, Z. 1971. List of common insects found in subtropical orchards. *Publication of the Volcani Institute of Agriculture Research (Bet Dagan, Israel)* 12-15. [SwirskAr1971]

Swirski, E., Wysoki, M. & Izhar, Y. 2002. Subtropical Fruits Pests in Israel. Tel Aviv: Fruit Board of Israel. 284 pp. [SwirskWyIz2002]

Symons, T.B., Cory, E.N. & Babcock, O.G. 1911. Treatment for the San José scale and terrapin scale insects. *Bulletin of the Maryland Agriculture Experiment Station* **161**: 221-234. [SymonsCoBa1911]

Symons, T.B. & Weldon, G.P. 1907. Spraying for San José scale. *Bulletin (Maryland Agricultural Experiment Station)* **(123)**: 139-152. [SymonsWe1907]

Szelényi, G.I. 1936. [Research on the development and epidemiology of *Aspidiotus piri* (Licht.) Reh, on the generation of the year 1935.]. *Hungary Ackerbau Min., Kisérlet. Közlem. Rec. Hungarian Agr. Expt. Sta.* **39**: 159-170. [Szelen1936]

Szent-Ivany, J.J.H. 1963. Non-coleopterous insects of *Cocos nucifera* on South Pacific Islands, with special reference to the territory of Papua and New Guinea. *Proceedings of the 9th Pacific Science Congress. Bangkok* **9**: 67-71. [SzentI1963]

Szklarzewicz, T. & Bilinski, S.M. 1995. Structure of ovaries in ensign scale insects, the most primitive representatives of Coccomorpha (Insecta, Hemiptera). *Journal of Morphology* **224**: 23-29. [SzklarBi1995]

Szulczewski, J.W. 1926. [Scale records from the city of Poznan.] Materjaly do fauny czerwców miasta Poznania.]. [In Polish]. *Polskie Pismo Entomologiczne* **5**: 137-143. [Szulcz1926]

Szulczewski, J.W. 1931. [Entomological and zoocecidiological notes in the Lublin county in Upper Silesia.] Notatki entomologiczne i zoocecidiologiczne z powiatu lublinieckiego na Górnym Slasku. *Polskie Pismo Entomologiczne* **10**: 124-135. [Szulcz1931]

Szulczewski, J.W. 1949. [Contribution to the Coccidae fauna in the district of "Ziemia lubuska".] Przyczynek do fauny czerwcow (Coccinae) Ziemi Lubuskiej. *Bad. Fizjograf. Pol. Zach. Pozn. TPN* **2**: 219-224. [Szulcz1949]

Tabibullah, M. & Gabriel, B.F. 1973. Biological study of *Aspidiotus destructor* Signoret in different coconut varieties and other host plants. *Philippine Entomologist* **2**: 409-426. [TabibuGa1973]

Tachikawa, T. 1973. Discovery of *Aspidiotus beilschmiediae* Takagi (Homoptera: Diaspididae) from Japan and its aphelinid parasite (Hymenoptera). *Transactions of the Shikoku Entomological Society* **11**: 137. [Tachik1973]

Tachikawa, T. 1988. A new and economically important species of *Coccobius* (Hymenoptera: Aphelinidae) parasitic on *Hemiberlesia pitysophila* Takagi (Homoptera: Diaspididae in Okinawa, Japan. *Transactions of the Shikoku Entomological Society* **19(1-2)**: 69-71. [Tachik1988]

Tachikawa, T. & Valentine, E.W. 1969. A new species of *Aphycomorpha* (Hymenoptera: Encyrtidae) parasitic on a diaspidine scale from New Zealand. *New Zealand Journal of Science* **12**: 535-540. [TachikVa1969]

Tadic, M. 1961. [The indigenous natural enemies of the San José scale in Yugoslavia.] Les insectes entomophages autochtones du Pou de San-José *Quadraspidiotus perniciosus* Comst.) en Yougoslavie. *Arhiv za poljoprivredne nauke (Journal for scientific agricultural research)* **46**: 130-132. [Tadic1961]

Tadic, M. 1967. [Acclimatization of *Prospaltella perniciosi* Tow. in Yugoslavia.] Odomacivanje *Prospaltella perniciosi* Tow. u Jugoslaviji. *Zastita Bilja. Beograd* **93/97**: 147-154. [Tadic1967]

Takagi, S. 1956. Notes on the Japanese species of the genus *Pseudaulacaspis* MacGillivray, with description of a new species. *Insecta Matsumurana* **19**: 113-116. [Takagi1956]

Takagi, S. 1956b. Four new species of *Diaspidiotus* and *Quadraspidiotus* (Homoptera, Coccoidea). *Insecta Matsumurana* **20**: 83-89. [Takagi1956b]

Takagi, S. 1957. A revision of the Japanese species of the genus *Aspidiotus*, with description of a new genus and a new species. *Insecta Matsumurana* **21**: 31-40. [Takagi1957]

Takagi, S. 1958. New or little known scale insects of the tribe Aspidiotini, with a list of the genera and species occurring in Japan. *Insecta Matsumurana* **21**: 121-131. [Takagi1958]

Takagi, S. 1959. Notes on the scale insects of the tribe Odonaspidini occurring in Japan (Homoptera, Coccoidea). *Insecta Matsumurana* **22**: 92-95. [Takagi1959]

Takagi, S. 1959a. A new species of the genus *Selenomphalus* Mamet from Japan. *Insecta Matsumurana* **22**: 112-114. [Takagi1959a]

Takagi, S. 1961. A contribution to the knowledge of the Diaspidini of Japan (Homoptera: Coccoidea) Pt. II. *Insecta Matsumurana* **24**: 4-42. [Takagi1961]

Takagi, S. 1962b. Records of some Diaspididae of Japan. *Insecta Matsumurana* **25**: 52. [Takagi1962b]

Takagi, S. 1962c. *Aonidiella comperei* McKenzie from Formosa. *Insecta Matsumurana* **25**: 52. [Takagi1962c]

Takagi, S. 1963a. Discovery of *Duplaspidiotus claviger* in Japan. *Insecta Matsumurana* **25**: 123. [Takagi1963a]

Takagi, S. 1967. Examinations of the type slides of three Diaspididae described from Japan (Homoptera: Coccoidea). *Insecta Matsumurana* **30**: 52-55. [Takagi1967]

Takagi, S. 1969a. Diaspididae of Taiwan based on material collected in connection with the Japan-U.S. Co-operative Science Programme, 1965 (Homoptera: Coccoidea). Part I. *Insecta Matsumurana* **32**: 1-110. [Takagi1969a]

Takagi, S. 1970. Diaspididae of Taiwan based on material collected in connection with the Japan-U.S. Cooperative Science Programme, 1965 (Homoptera: Coccoidea). Pt. II. *Insecta Matsumurana* **33**: 146 pp. [Takagi1970]

Takagi, S. 1974. An approach to the *Hemiberlesia* problem (Homoptera: Coccoidea). *Insecta Matsumurana* **(n.s.) 3**: 1-33. [Takagi1974]

Takagi, S. 1975. Coccoidea collected by the Hokkaido University expedition to Nepal, Himalaya, 1968 (Homoptera). *Insecta Matsumurana (New Series)* **(n.s.)6**: 1-33. [Takagi1975]

Takagi, S. 1984. Some aspidiotine scale insects with enlarged setae on the pygidial lobes (Homoptera: Coccoidea: Diaspididae). *Insecta Matsumurana* **28**: 1-69. [Takagi1984]

Takagi, S. 1987. Two new Parlatoriine scale insects with Odonaspidine characters: the other side of the coin (Homoptera: Coccoidea: Diaspididae). *Insecta Matsumurana* **37**: 1-25. [Takagi1987]

Takagi, S. 1990. Disc pores of Diaspididae: microstructure and taxonomic value (Homoptera: Coccoidea). *Insecta Matsumurana* **New Series 44**: 81-112. [Takagi1990]

Takagi, S. 1990a. SEM observations on the tests of some Diaspididae. *Insecta Matsumurana* **New Series 44**: 17-80. [Takagi1990a]

Takagi, S. 1990b. 1.1.2 External morphology. 1.1.2.1 The adult female. Pages 5-20 *in*: Rosen, D. (Ed.). Armored Scale Insects, Their Biology, Natural Enemies and Control [Series title: World Crop Pests, Vol. 4A]. Amsterdam, The Netherlands: Elsevier. 384 pp. [Takagi1990b]

Takagi, S. 1990c. 1.1.2.6 Polymorphism. Pages 59-64 *in*: Rosen, D. (Ed.). Armored Scale Insects, Their Biology, Natural Enemies and Control [Series title: World Crop Pests, Vol. 4A]. Amsterdam, The Netherlands: Elsevier. 384 pp. [Takagi1990c]

Takagi, S. 2000. Four extraordinary diaspidids (Homoptera: Coccoidea). *Insecta Matsumurana* **57**: 39-87. [Takagi2000]

Takagi, S. 2002. One new subfamily and two new tribes of the Diaspididae (Homoptera: Coccoidea). *Insecta Matsumurana* **59**: 55-100. [Takagi2002]

Takagi, S. & Kawai, S. 1966. Some Diaspididae of Japan (Homoptera: Coccoidea). *Insecta Matsumurana* **28**: 93-119. [TakagiKa1966]

Takagi, K. & Rosen, D. 1981. Note on the species of *Aphytis* Howard (Hymenoptera, Aphelinidae) occurring in Japan. *Kontyû* **49(2)**: 314-321. [TakagiRo1981]

Takagi, S. & Tippins, H.H. 1972. Two new species of the Diaspididae occurring on Spanish moss in North America (Homoptera: Coccoidea). *Kontyû* **40**: 180-186. [TakagiTi1972]

Takagi, S. & Yamamoto, M. 1974. Two new banana-infesting scale insects of *Hemiberlesia* or *Abgrallaspis* from Ecuador (Homoptera: Coccoidea). *Insecta Matsumurana* **(n.s.)** **3**: 35-42. [TakagiYa1974]

Takahashi, R. 1929. Observations on the Coccidae of Formosa. - 1. *Report, Government Research Institute, Department of Agriculture, Formosa* **40**: 1-82. [Takaha1929]

Takahashi, R. 1929a. Aphididae and Coccidae of the Pescadores. *Transactions of the Natural History Society of Formosa* **19**: 425-431. [Takaha1929a]

Takahashi, R. 1931. Some Coccidae of Formosa. *Transactions of the Natural History Society of Formosa* **21**: 1-5. [Takaha1931]

Takahashi, R. 1931b. Records and descriptions of the Coccidae from Formosa. Part I. *Journal of the Society of Tropical Agriculture. Taiwan* **3**: 377-385. [Takaha1931b]

Takahashi, R. 1932a. New food plants of the Coccidae in Formosa. *Transactions of the Formosa Natural History Society* **22**: 102-105. [Takaha1932a]

Takahashi, R. 1933. Observations on the Coccidae of Formosa. III. *Report. Government Research Institute. Department of Agriculture. Formosa* **60**: 1-64. [Takaha1933]

Takahashi, R. 1934. Observations on the Coccidae of Formosa. Part IV. *Report, Government Research Institute, Department of Agriculture, Formosa* **63**: 1-38. [Takaha1934]

Takahashi, R. 1935. Observations on the Coccidae of Formosa. V. *Report Department of Agriculture Government Research Institute, Formosa* **66**: 1-37. [Takaha1935]

Takahashi, R. 1936. Two interesting scale insects attacking the Lauraceae in Formosa (Hemiptera). *Transactions of the Formosa Natural History Society* **26**: 80-83. [Takaha1936]

Takahashi, R. 1936b. A new scale insect causing galls in Formosa (Homoptera). *Transactions of the Formosa Natural History Society* **26**: 426-428. [Takaha1936b]

Takahashi, R. 1936c. Some Aleyrodidae, Aphididae, Coccidae (Homoptera), and Thysanoptera from Micronesia. *Tenthredo. (Acta Entomologica. Takeuchi Entomological Laboratory)* **1**: 109-120. [Takaha1936c]

Takahashi, R. 1936d. Some Coccidae from Formosa and Japan (Homoptera), I. *Mushi* **9**: 1-8. [Takaha1936d]

Takahashi, R. 1939b. Some Aleyrodidae, Aphididae, and Coccidae from Micronesia (Homoptera). *Tenthredo. (Acta Entomologica. Takeuchi Entomological Laboratory)* **2**: 234-272. [Takaha1939b]

Takahashi, R. 1939c. Some Coccidae from Formosa and Japan (Homoptera), IV. *Mushi* **12**: 86-89. [Takaha1939c]

Takahashi, R. 1939d. Two new diaspine Coccidae from India and the Philippines (Hemiptera). *Tenthredo. (Acta Entomologica. Takeuchi Entomological Laboratory)* **2**: 339-343. [Takaha1939d]

Takahashi, R. 1940. Some Coccidae from Formosa and Japan (Homoptera), V. *Mushi, Fukuoka Entomological Society* **13**: 18-28. [Takaha1940]

Takahashi, R. 1940a. Insects of the Sternorrhyncha (Hemiptera) of Daito Jima, the Loochoo Islands. *Transactions of the Natural History Society of Formosa* **30**: 327-332. [Takaha1940a]

Takahashi, R. 1941b. Some species of Aleyrodidae, Aphididae, and Coccidae from Micronesia (Homoptera). *Tenthredo. (Acta Entomologica. Takeuchi Entomological Laboratory)* **3**: 208-220. [Takaha1941b]

Takahashi, R. 1942. Some Coccidae from Malaya and Hongkong (Homoptera). *Transactions of the Formosa Natural History Society* **32**: 63-68. [Takaha1942]

Takahashi, R. 1942b. Some injurious insects of agricultural plants and forest trees in Thailand and Indo-China. II. Coccidae. *Report. Government Research Institute. Department of Agriculture. Formosa* **81**: 1-56. [Takaha1942b]

Takahashi, R. 1942d. Some species of Aleyrodidae, Aphididae and Coccidae in Micronesia (Homoptera). *Tenthredo. (Acta Entomologica. Takeuchi Entomological Laboratory)* **3**: 349-358. [Takaha1942d]

Takahashi, R. 1951b. Some species of Coccidae from the Riouw Islands - Part II. *Insecta Matsumurana* **17**: 103-112. [Takaha1951b]

Takahashi, R. 1952a. Descriptions of five new species of Diaspididae from Japan, with notes on dimorphism in *Chionaspis* or *Phenacaspis* (Coccoidea, Homoptera). *Miscellaneous Reports of the Research Institute on Natural Resources* **27**: 7-15. [Takaha1952a]

Takahashi, R. 1953a. [Nomenclature of the scale insects attacking the citrus.]. *Kankitsu* **5**: 10-13. [Takaha1953a]

Takahashi, R. 1955e. *Lepidosaphes* of Japan. (Diaspididae, Coccoidea, Homoptera). *Bulletin of the Osaka Perfecture* **5**: 67-78. [Takaha1955e]

Takahashi, R. 1955f. Some scale insects of the Loochoo Islands (Homoptera). *Bulletin of the Biogeographical Society of Japan* **16-19**: 238-242. [Takaha1955f]

Takahashi, R. 1956b. Three new genera and a new species of Diaspididae from Japan. *Insecta Matsumurana* **20**: 23-28. [Takaha1956b]

Takahashi, R. & Tachikawa, T. 1956. Scale insects of Shikoku (Homoptera: Coccoidea). *Transactions of the Shikoku Entomological Society* **5**: 1-17. [TakahaTa1956]

Takahashi, R. & Takagi, S. 1957. A new genus of Diaspididae from Japan (Coccoidea, Homoptera). *Kontyû* **25**: 102-105. [TakahaTa1957]

Talhouk, A.M.S. 1950. A list of insects observed on economically important plants and plant products in Lebanon. *Bulletin de la Société Fouad 1er d'Entomologie d'Egypte* **34**: 133-141. [Talhou1950]

Talhouk, A.M.S. 1969. Insects and mites injurious to crops in Middle Eastern countries. *Monographien zur Angewandte Entomologie* **21**: 1-239. [Talhou1969]

Talhouk, A.M.S. 1975. Citrus pests throughout the world. *Ciba-Geigy Agrochemicals, Technical Monograph* **4**: 21-23. [Talhou1975]

Tamaki, Y. 1997. 1.1.2.5 Chemistry of the Test Cover. Pages 55-72. *in*: Ben-Dov, Y. & Hodgson, C.J., Eds. Soft Scale Insects: Their Biology, Natural Enemies and Control [Vol. 7A]. Amsterdam & New York: Elsevier. 55-72. [Tamaki1997]

Tanaka, M. 1966. [Fundamental studies on the utilization of natural enemies in the citrus grove in Japan. I. The bionomics of natural enemies of the most serious pests. II. *Stethorus japonicus* H. Kamiya, a predator of the citrus red mites, *Panonychus citri* McG.]. *Bulletin of the Horticultural Research Station (Ministry of Agriculture and Forestry) Series D* **4**: 1-42. [Tanaka1966]

Tang, F.T. 1977. [The scale insects of horticulture and forest of China. Vol. I.]. [In Chinese]. Liaoning, China: The Institute of Gardening. 259 pp. [Tang1977]

Tang, F.T. 1984. [The scale insects of horticulture and forests of China.]. [In Chinese]. *Shanxi Agricultural University Press Research Publication* **2**: 1-115. [Tang1984]

Tang, F.T. & Chu, C.Y. 1983. [Three new species of Diaspididae from *Keteleeria evelyniana* in vicinity of Kunming, Yunnan.]. [In Chinese]. *Acta Zootaxonomica Sinica* **8**: 301-306. [TangCh1983]

Tang, F.T., Hao, J., Shi, G. & Tang, Y. 1991. [On a newly found genus and three new species of Diaspididae from China (Homoptera).]. [In Chinese]. *Acta Entomologica Sinica* **34(4)**: 458-464. [TangHaSh1991]

Tang, S., Qin, H. & Wang, D. 1990. [Study on *Aspidiotus chinensis* Kuw. et Mur.]. [In Chinese]. *Journal of Shanghai Agricultural College* **8(3)**: 187-194. [TangQiWa1990]

Tanquary, M.C. & Hayes, M.E. 1920. Commercial-sulphur products as dormant sprays for control of the San José scale. *Circular (Texas Agricultural Experiment Station)* **No. 24**: 3-7. [TanquaHa1920]

Tao, C.C.C. 1999. List of Coccoidea (Homoptera) of China. *Special Publication (Taiwan Agric. Res. Inst.)* **No. 78**: 1-176. [Tao1999]

Targioni Tozzetti, A. 1867. Studii sulle Cocciniglie. *Memorie della Società Italiana di Scienze Naturali. Milano* **3(3)**: 1-87. [Targio1867]

Targioni Tozzetti, A. 1868. Introduzione alla seconda memoria per gli studi sulle cocciniglie, e catalogo dei generi e delle specie della famiglia dei coccidi. *Atti della Società italiana di scienze naturali* **11**: 721-738. [Targio1868]

Targioni Tozzetti, A. 1879. [*Diaspis blankenhornii.*] *Bollettino della Società Entomologica Italiana. Resoconti delle Aduanze.* Firenze 1879: 32. [Targio1879]

Targioni Tozzetti, A. 1881. Relazione intorno ai lavori della R. Stazione di Entomologia Agraria di Firenze per gli anni 1877-78. Parte scientifica. Fam. coccidi. *Annali di Agricoltura. (Ministero di Agricoltura, Industria e Commercio). Firenze, Roma* **1881**: 134-161. [Targio1881]

Targioni Tozzetti, A. 1884. Relazione intorno ai lavori della R. Stazione di Entomologia Agraria di Firenze per gli anni 1879-80. Article V. - omotteri. *Annali di Agricoltura. (Ministero di Agricoltura, Industria e Commercio). Firenze, Roma* **1884**: 383-414. [Targio1884]

Targioni Tozzetti, A. 1885. Note sopra alcune cocciniglie (coccidei). *Bollettino della Società Entomologica Italiana* **17**: 100-120. [Targio1885]

Targioni Tozzetti, A. 1888. Relazione intorno ai lavori della R. Stazione di Entomologia Agraria di Firenze per gli anni 1883-84-85. Article VI. Ord. eterotteri-omotteri. Hemiptera, 1. pars, Rhyngota, F. Rhyncota, Fieb. (sottord. coccidi). *Annali di Agricoltura. (Ministero di Agricoltura, Industria e Commercio). Firenze, Roma* **1888**: 415-437. [Targio1888]

Targioni Tozzetti, A. 1892. *Aonidia blanchardi*, nouvelle espèce de cochenille du dattier du Sahara. *Memoires de la Société Zoologique de France* **5**: 69-82. [Targio1892]

Tashiro, H. 1966. Improved laboratory techniques for rearing California red scale on lemons. *Annals of the Entomological Society of America* **59**: 604-608. [Tashir1966]

Tashiro, H. & Beavers, J.B. 1968. Growth and development of the California red scale, *Aonidiella aurantii. Annals of the Entomological Society of America* **61**: 1009-1014. [TashirBe1968]

Tashiro, H., Beavers, J.B. & Moreno, D. 1969. Comparative response of two strains of California red scale, *Aonidiella aurantii* males to pheromone extract and to females of the reciprocal strain. *Annals of the Entomological Society of America* **62**: 279-280. [TashirBeMo1969]

Tashiro, H. & Chambers, D.L. 1967. Reproduction in the California red scale, *Aonidiella aurantii* (Homoptera: Diaspididae). I. Discovery and extraction of a female sex pheromone. *Annals of the Entomological Society of America* **60**: 1166-1170. [TashirCh1967]

Tashiro, H., Chambers, D.L., Moreno, D. & Beavers, J.B. 1969. Reproduction in the California red scale, *Aonidiella aurantii*. III. Development of an olfactometer for bioassay of the female sex pheromone. *Annals of the Entomological Society of America* **62**: 935-940. [TashirChMo1969]

Tashiro, H. & Moffitt, C. 1968. Reproduction in the California red scale, *Aonidiella aurantii*. II. Mating behavior and postinsemination female changes. *Annals of the Entomological Society of America* **61**: 1014-1020. [TashirMo1968]

Tawfik, M.H. & Ghabbour, M.W. 2001. Monitoring California red scale and its aphelinid parasitoid using yellow sticky traps. *Entomologica* **33(1999)**: 267-272. [TawfikGh2001]

Tawfik, M.H. & Mohammad, Z.K. 2002. Ecological studies of two scale insects (Hemiptera, Coccoidea) on Morus alba in Egypt. *Bollettino di Zoologia Agraria e di Bachicoltura (Milano)* **33(3)**: 267-273. [TawfikMo2002]

Taylor, T.H.C. 1935. The campaign against *Aspidiotus destructor*, Sign., in Fiji. *Bulletin of Entomological Research* **26**: 1-102. [Taylor1935]

Ter-Grigorian, M.A. 1956. [Coccids of cultivated fruit in Armenia.]. [In Russian]. *Sbornik, Institut Zoologii, Akademiya Nauk Armianskoi SSR (Zoological Papers, Academy of Sciences of Armenian SSR, Zoological Institute)* **9**: 33-57. [TerGri1956]

Ter-Grigorian, M.A. 1962. [The Coccoidea of the woods of Armenia (Insecta, Homoptera, Coccoidea).]. [In Russian]. *Zoological Papers, Academy of Sciences of Armenian SSR, Zoological Institute* **12**: 125-161. [TerGri1962]

Tereznikova, E.M. 1986. Coccoids. *Fauna Ukraini*. [In Ukrainian]. *Akademiya Nauk Ukrainskoi RSR. Institut Zoologii. Kiev* **20**: 1-131. [Terezn1986]

Tewari, G.C. & Tripathi, G.M. 1979. Sugarcane scale insect and its control. *Indian Farming* **29(2)**: 19-21. [TewariTr1979]

Thakur, M.L. 1999. Insect pest status of poplars in India. *Indian Forester* **9**: 866-872. [Thakur1999]

Thakur, J.N., Pawar, A.D. & Rawat, U.S. 1993. Introduction, colonisation and new records of some biocontrol agents of San José scale, *Quadraspidiotus perniciosus* Comstock (Hemiptera: Coccidae) in Kullu Valley, H. P., India. *Journal of Biological Control* **7(2)**: 99-101. [ThakurPaRa1993]

Thakur, J.N., Rawat, U.S. & Pawar, A.D. 1988. Effects of commonly used fungicides on longevity and mortality of *Encarsia perniciosi* (Tower) and *Aphytis* sp. *proclia* group (Hymenoptera: Aphelinidae). *Journal of Biological Control* **2(2)**: 72-73. [ThakurRaPa1988]

Thakur, J.N., Rawat, U.S. & Pawar, A.D. 1989. Investigations on the occurrence of natural enemies of San José scale, *Quadraspidiotus perniciosus* Comstock (Hemiptera: Coccidae) in Jummu & Kashmir and Himachal Pradesh. *Entomon* **14(1/2)**: 143-146. [ThakurRaPa1989]

Theron, J.G. 1958. Comparative studies on the morphology of male scale insects (Hemiptera: Coccoidea). *Annals of the University of Stellenbosch (Section A)* **34**: 1-71. [Theron1958]

Thiem, H. & Gerneck, R. 1934. Verbreitung, Entwicklung und Bestimmung der bisher in Deutschland aufgefundenen Austernschildläuse (Aspidotini) unter Einschluss der roten Austernschildlaus *(Epidiaspis betulae)* und der San José-Schildlaus *(Aspidiotus perniciosus)*. *Zeitschrift für Pflanzenkrankheiten* **44**: 529-555. [ThiemGe1934]

Thiem, H. & Gerneck, R. 1934a. Untersuchungen an Deutschen Austernschildläusen (Aspidiotini) im vergleich mit der San José-Schildlaus *(Aspidiotus perniciosus* Comst.). *Arbeiten über physiologische und angewandte Entomologie* **1**: 130-158; 208-238. [ThiemGe1934a]

Thompson, W.L. 1935. Lime-sulfur sprays for the combined control of purple scale and rust mites. *Bulletin of the California Agricultural Experiment Station* **282**: 1-38. [Thomps1935]

Thompson, W.L. 1938. Results of different methods of oil application for the control of scale insects on citrus. *Proceedings of the Florida State Horticultural Society* **51**: 109-114. [Thomps1938]

Thompson, W.L. 1938a. Some possible reasons for the increases of purple scale infestations. *Citrus Industry* **19(12)**: 6-7,17,20. [Thomps1938a]

Thompson, W.L. 1939. Cultural practices and their influence upon citrus pests. *Journal of Economic Entomology* **32(6)**: 782-789. [Thomps1939]

Thompson, W.L. 1939a. Factors influencing the development and control of scale insects on citrus. *Proceedings of the Florida State Horticultural Society* **52nd (1939)**: 104-111. [Thomps1939a]

Thompson, W.L. 1942a. Some problems of control of scale insects on citrus. *Proceedings of the Florida State Horticultural Society* **55**: 51-59. [Thomps1942a]

Thompson, W.L. 1950. Combined control of scale insects and mites on citrus. *Annual Report (Florida Agricultural Experiment Station)* **1950**: 155. [Thomps1950]

Thompson, W.R. 1958. Biological control in some Commonwealth countries. Pages 479-482 *in*: Baecker, E.C., (Ed.). Proceedings of the Tenth International Congress of Entomology. Vol. 4. Ottawa. 1115 pp. [Thomps1958]

Thompson, W.L. & Griffiths, J.T. 1949. Purple scale and Florida red scale as insect pests of citrus in Florida. *Bulletin of the Florida Agricultural Experiment Station* **462**: 1-40. [ThompsGr1949]

Thomson, C., Tomkins, A.R. & Wilson, D.J. 1996. Effect Of insecticides on immature and mature stages of *Encarsia citrina* an armoured scale parasitoid. *Proceedings of the New Zealand Plant Protection Conference* **49th**: 1-5. [ThomsoToWi1996]

Thorpe, W.H. 1930. Biological races in insects and allied groups. *Biological Reviews of the Cambridge Philosophical Society* **5**: 177-212. [Thorpe1930WH]

Tian, M. & Chen, S. 1991. [Studies on the bionomics of *Aleurodothrips fasciapennis* (Thy. Phlaeothripidae), a predator of diaspid scales on citrus.] [In Chinese]. *Chinese Journal of Biological Control* **7(2)**: 64-66. [TianCh1991]

Timberlake, P.H. 1916. Revision of the parasitic hymenopterous insects of the genus *Aphycus* Mayr, with notice of some related genera. *Proceedings of the United States National Museum. Washington* **50**: 561-640. [Timber1916]

Timberlake, P.H. 1924. Records of the introduced and immigrant chalcid-flies of the Hawaiian Islands. *Proceedings of the Hawaiian Entomological Society* **5**: 418-449. [Timber1924]

Timlin, J.S. 1964a. The distribution and relative importance of some armoured scale insects on pip fruit in the Nelson/Marlborough orchards during 1959/60. *New Zealand Journal of Agricultural Research* **7**: 531-535. [Timlin1964a]

Timofeeva, T.V. 1938. [Bio-ecology of San José scale in conditions of East Georgia. Pages 5-46 *in:* Kiritchenko, A.N. (Ed.), Works of Quarantine Laboratories.] [In Russian]. Leningrad, Sel'khozgiz. 272 pp. [Timofe1938]

Tippins, H.H. & Beshear, R.J. 1968. Variations in the number and groupings of disc pores in *Comstockiella sabalis* (Comst.) (Homoptera: Diaspididae). *Journal of the Georgia Entomological Society* **3**: 67-69. [TippinBe1968]

Tippins, H.H. & Beshear, R.J. 1968a. Scale insects (Homoptera: Coccoidea) from grasses in Georgia. *Journal of the Georgia Entomological Society* **3**: 134-136. [TippinBe1968a]

Tippins, H.H. & Beshear, R.J. 1970. The armored scale insects (Homoptera: Diaspididae) of Cumberland Island, Georgia. *Journal of the Georgia Entomological Society* **5**: 7-12. [TippinBe1970]

Tippins, H.H. & Beshear, R.J. 1971. *Aspidiotus* n. sp. (Homoptera: Diaspididae) from saw grass, *Mariscus jamaicensis*. *Journal of the Georgia Entomological Society* **6**: 85-86. [TippinBe1971]

Tippins, H.H. & Beshear, R.J. 1972. Additional scale insects from grasses. *Journal of the Georgia Entomological Society* **7**: 286-288. [TippinBe1972]

Tippins, H.H. & Beshear, R.J. 1978. Some new records of scale insects from grasses in Georgia. *Journal of the Georgia Entomological Society* **13**: 12-14. [TippinBe1978]

Tippins, H.H., Clay, H. & Barry, R.M. 1977. Peony scale: a new host and biological information. *Journal of the Georgia Entomological Society* **12**: 68-71. [TippinClBa1977]

Tippins, H.H., Howell, J.O. & Beshear, R.J. 1981. Some immature stages of *Pseudaonidia paeoniae* (Homoptera: Coccoidea, Diaspididae). *Journal of the Georgia Entomological Society* **16**: 356-361. [TippinHoBe1981]

Tippins, H.H. & Martin, P.B. 1982. Seasonal occurrence of Bermudagrass scale. *Journal of the Georgia Entomological Society* **17**: 319-321. [TippinMa1982]

Toledo, A.A., de. 1940. Notas sobre a biologia do "*Chrysomphalus aonidum* L. 1758" (Homoptera) no estado de S. Paulo - Brasil. *Archivos do Instituto Biologico. São Paulo* **11**: 559-578. [Toledo1940]

Tomkins, A.R. 1992. Teflubenzuron and Phosalone alone and in combination for pest control on kiwifruit: I. Armoured scale insects. *Proceedings of the Forty Fifth New Zealand Plant Protection Conference. Popay, A. J., ed. Wellington, New Zealand: August 11-13, 1992* : 151-155. [Tomkin1992]

Tomkins, A.R. 1992a. Evaluation of an insect growth regulator, fenoxycarb, for greedy scale control on kiwifruit. *Acta Horticulturae* **297**: 517-522. [Tomkin1992a]

Tomkins, A.R., Allison, P.A., Thomson, C. & Wilson, D.J. 1997. Development of a model to predict the phenology of oleander scale (*Aspidiotus nerii*) infesting kiwifruit. *Acta Horticulturae* **444**: 791-795. [TomkinAlTh1997]

Tomkins, A.R., Allison, P.A., Wilson, D.J. & Thomson, C. 1996. 3D imaging shows within-plant distribution of pests on unsprayed New Zealand kiwifruit vines. *Acta Horticulturae* **444**: 785-789. [TomkinAlWi1996]

Tomkins, A.R., Follas, G., Thomson, C. & Wilson, D.J. 1994. A new insect growth regulator, CGA 59205, with promising activity against scale insects. Pages 337-340 *in:* Popay, A. J., ed. Proceedings of the Forty Seventh New Zealand Plant Protection Conference. Waitangi Hotel, New Zealand: Aug. 9-11, 1994. Rotorua, New Zealand: New Zealand Plant Protection Society. [TomkinFoTh1994]

Tomkins, A.R., Greaves, A.J., Wilson, D.J. & Thomson, C. 1992. Field evaluation of Avermectin B1 for pest control on kiwifruit. *Proceedings of the Forty Fifth New Zealand Plant Protection Conference. Popay, A. J., ed. Wellington, New Zealand: August 11-13, 1992* : 146-150. [TomkinGrWi1992]

Tomkins, A.R., Greaves, A.J., Wilson, D.J. & Thomson, C. 1996. Evaluation of a mineral oil for pest control on kiwifruit. Pages 12-16. *in*: O'Callaghan, M., Ed. Proceedings of the Forty Ninth New Zealand Plant Protection Conference, Quality Hotel Rutherford, Nelson, New Zealand, 13-15 August, 1996. Rotorua, New Zealand: New Zealand Plant Protection Society. [TomkinGrWi1996]

Tomkins, A.R., Thomson, C., Wilson, D.J. & Greaves, A.J. 1992. Armoured scale insects (Hemiptera: Diaspididae) on unsprayed kiwifruit vines in the Waikato. *New Zealand Entomologist* **(15)**: 58-63. [TomkinThWi1992]

Tomkins, A.R., Thomson, C. & Wilson, D.J. 1995. Effect of tebufenozide on *Encarsia citrina*, an armoured scale parasitoid. *Proceedings of the Forty Eighth New Zealand Plant Protection Conference* 139-142. [TomkinThWi1995]

Tomkins, A.R., Wilson, D.J., Thomson, C. & Allison, P. 2000. Incidence of armoured scale insects on persimmons. *New Zealand Plant Protection* **53**: 211-215. [TomkinWiTh2000]

Tong, C.J., Tang, Z.Y. & Pan, W.Y. 1988. [Preliminary study on the fluctuation of natural populations of *Hemiberlesia pitysophila*.] [In Chinese]. *Forest Science and Technology* **(No. 2)**: 6-11. [TongTaPa1988]

Torres, A.F.M. 1922. Defesa agricola. *Boletim (Ministerio da Agricultura, Industria e Commercio, Instituto Biológico de Defesa Agricola, Rio de Janeiro)* **11**: 72-90. [Torres1922]

Torres, L.M., Rodrigues, A.N., Avilla, J. & Polesny, F. 2001. Chemical control of *Quadraspidiotus perniciosus* (Comstock) (Homoptera: Diaspididae) in apples and side effects on phytoseiid mites (Acari: Phytoseiidae). *IOBC/WPRS Bulletin* **24(5)**: 207-212. [TorresRoAv2001]

Toumey, J.W. 1895. Notes on scale insects in Arizona. *Bulletin (Arizona Agricultural Experiment Station)* **No.14**: 29-56. [Toumey1895]

Tourneur, J.C. 1970. L'utilisation des coccinelles prédatrices en lutte biologique. *Fruits* **25(2)**: 97-107. [Tourne1970]

Tower, W.V. 1911. Insects injurious to citrus fruits and methods for combating them. *Bulletin (Porto Rico Agricultural Experiment Station)* **10**: 1-35. [Tower1911]

Tower, D.G. 1913. A new hymenopterous parasite of *Aspidiotus perniciosus* Comst. *Annals of the Entomological Society of America* **6**: 125-126. [Tower1913]

Tower, D.G. 1914. Notes on the life history of *Prospaltella perniciosi* Tower. *Journal of Economic Entomology* **7**: 422-432. [Tower1914]

Townsend, C.H.T. 1897. Locality and food plant catalogue of Mexican Coccidae. *Journal of the New York Entomological Society* **5**: 178-190. [Townse1897]

Townsend, C.H.T. & Cockerell, T.D.A. 1898. Coccidae collected in Mexico by Messrs. Townsend and Koebele in 1897. *Journal of the Entomological Society of New York* **6**: 165-180. [TownseCo1898]

Trabut, L. 1911. Catalogue des cochenilles observées en Algérie. *Bulletin de la Société d'Histoire Naturelle de l'Afrique du Nord* **3**: 51-64. [Trabut1911]

Tranfaglia, A. & Viggiani, G. 1987. Scale insects of economic importance and their control in Italy. *Bollettino del Laboratorio di Entomologia Agraria 'Filippo Silvestri'* **43** (Suppl.): 215-221. [TranfaVi1987a]

Treherne, R.C. 1914. A review of applied entomology in British Columbia. *Proceedings of the Entomological Society of British Columbia* **4**: 67-71. [Treher1914]

Treherne, R.C. 1916a. Some orchard insects of economic importance in British Columbia. *Proceedings of the Entomological Society of British Columbia* **9**: 66-83. [Treher1916a]

Tremblay, E. 1990a. 1.3.6 Endosymbionts. Pages 275-283 *in*: Rosen, D., (Ed.). Armored Scale Insects, Their Biology, Natural Enemies and Control [Series title: World Crop Pests, Vol. 4A]. Amsterdam, the Netherlands: Elsevier. 384 pp. [Trembl1990a]

Tremblay, E. 1999. [Quarantine insects and the risks of their introduction into Italy.]. [In Italian]. *Notiziario sulla Protezione delle Piante* **No. 10**: 19-28. [Trembl1999]

Tremblay, E. & Rotundo, G. 1975. The pheromones of the Coccoidea and their possible use in integrated control. *Proceedings of the 8th International Plant Protection Congress (Moscow). Reports & Information Section V - Biological & Genetic Control.* 195-200. [TremblRo1975]

Trimble, F.M. 1929. Scale insects injurious in Pennsylvania. *Bulletin of the Pennsylvania Department of Agriculture* **12**: 1-21. [Trimbl1929]

Tripathi, N. & Shukla, G.S. 1982. Effect of temperature on the establishment and development of crawlers of sugarcane scale insect, *Melanaspis glomerata* (Green) (Hemiptera: Coccidae). *Comparative Physiology and Ecology* **7**: 83-84. [TripatSh1982]

Tripathi, S.R. & Tewary, P. 1984. Biology of the sugarcane scale insect, *Melanaspis glomerata* (Diaspididae: Coccoidea). *Journal of Advanced Zoology* **5**: 68-78. [TripatTe1984]

Trjapitsin, V.A. 1989. [Parasitic Hymenoptera of the Fam. Encyrtidae of Palaearctics.] [In Russian]. Leningrad: Akademia Nauk SSSR. Zoologicheskii. 487 pp. [Trjapi1989]

Trjapitsin, V.A. & Myartseva, S.N. 2001. A new species of the genus *Adelencyrtus* from Malaysia (Hymenoptera: Encyrtidae). *Zoosystematica Rossica* 10: 163-165. [TrjapiMy2001]

Troncho, P., Rodrigo, E., Garcia-Mari, F. & Troncho, P. 1992. Ciclo biológico de los diaspinos de cítricos *Aonidiella aurantii* (Mask.), *Lepidosaphes beckii* (Newm.) y *Parlatoria pergandi* (Comstock) en una parcela de naranjo. *Boletín de Sanidad Vegetal, Plagas* **18(1)**: 11-30. [TronchRoGa1992]

Trouvelot, B. & Vezin, C. 1942. Le pou de San-José sur les cultures fruitières en France. *Comptes Rendus de Séances de l'Academie d'Agriculture de France* **28(1)**: 1-12. [TrouveVe1942]

Trujillo Peluffo, A. 1942. Insectos y otros parásitos de la agricultura y sus productos en el Uruguay. Montevideo. 323 pp. [Trujil1942]

Tryon, H. 1889. Report on Insect and Fungus Pests, No. 1. Brisbane: Department of Agriculture, Queensland. 238 pp. [Tryon1889]

Tschorbadjiew, P. 1939. [Short note on scale insects (Coccidae, Rhynchota) in Bulgaria.] [In Bulgarian]. *Izvestiya na Bulgarskoto Entomologichno Druzhestvo. Sofiya* **10**: 88-90. [Tschor1939]

Tsolakis, H. & Sinacori, A. 1995. Prove in campo sull'efficacia di buprofezinnei confronti di *Aonidiella aurantii* (Maskell). *Phytophaga. Palermo* **6**: 223-229. [TsolakSi1995]

Tsukamoto, M. 1983. Methods of genetic analysis of insecticide resistance. Pages 71-98 *in*: Georghiou, G.P. & Saito, T. (Eds.). Pest Resistance to Pesticides. New York: Plenum Press. 809 pp. [Tsukam1983]

Tullgren, A. 1906. Om sköldlöss. *Entomologisk Tidskrift. Stockholm* **27**: 69-95. [Tullgr1906]

Tumminelli, R., Amico, C., Conti, F., Fisicaro, R., Saraceno-F. & Mazzone, A. 2001. Resultati di quattro anni di sperimentazione: gestione della cocciniglia rossa forte degli agrumi. *Informatore Agrario* **57(19)**: 31-36. [TumminAmCo2001]

Tumminelli, R., Conti, F., Saraceno, F., Raciti, E. & Schilirò. 1996 Seasonal development of California red scale (Homoptera: Diaspididae) and *Aphytis melinus* De Bach (Hymenoptera: Aphelinidae) on citrus in eastern Sicily. *Proceedings of the International Society of Citriculture* **I**: 493-503. [TumminCoSa1996]

Tuncyürek, M. 1970. [Natural enemies of citrus and fig coccids in western Turkey.]. [In Turkish]. *Bitki Koruma Bulteni. Ankara (Plant Protection Bulletin)* **10**: 30-52. [Tuncyu1970]

Tuncyürek, M. 1970a. Les cochenilles nuisibles aux *Citrus* en Turquie. *Al Awamia* **37**: 67-80. [Tuncyu1970a]

Tuncyürek, M. 1976. The list of natural enemies of some insect pests in Turkey -- Part I. *Plant Protection Bulletin* **16(1)**: 32-45. [Tuncyu1976]

Tzalev, M. 1964b. [New scale insects established on plants in greenhouses in this country [Bulgaria].]. [In Russian]. *Zashchita Rastenii. Leningrad* **12**: 15-20. [Tzalev1964b]

Tzalev, M. & Valcheva, M. 1965. [The *Quadraspidiotus marani* Zahr.]. [In Russian]. *Zashchita Rastenii. Leningrad* **13**: 22-25. [TzalevVa1965]

Uematsu, H. 1972. [Studies on *Marietta carnesi* (Hymenoptera: Aphelinidae), a hyperparasite of diaspine scales (Homoptera: Diaspididae). I. Host relation.]. [In Japanese]. *Japanese Journal of Applied Entomology and Zoology* **16**: 187-192. [Uemats1972]

Uematsu, H. 1974. [Studies on *Marietta carnesi* (Hymenoptera: Aphelinidae), a hyperparasite of diaspine scales (Homoptera: Diaspididae). II. Bionomics.]. [In Japanese]. *Japanese Journal of Applied Entomology and Zoology* **18**: 177-182. [Uemats1974]

Uematsu, H. 1976. [Studies on *Marietta carnesi* (Hymenoptera: Aphelinidae), a hyperparasite of diaspine scales (Homoptera: Diaspididae). III. Reproductive capacity.]. [In Japanese]. *Japanese Journal of Applied Entomology and Zoology* **29**: 115-119. [Uemats1976]

Uematsu, H. 1978. [Bionomics of *Aonidiella taxus* Leonardi (Homoptera: Diaspididae).]. [In Japanese]. *Science Bulletin of the Faculty of Agriculture, Kyushu University* **33**: 25-31. [Uemats1978]

Uematsu, H. 1978a. [Studies on *Marietta carnesi* (Hymenoptera: Aphelinidae), a hyperparasite of Diaspine scales (Homoptera; Diaspididae). IV. Host discrimination.]. [In Japanese]. *Japanese Journal of Applied Entomology and Zoology* **22**: 135-140. [Uemats1978a]

Uematsu, H. 1979. [Studies on life table for an armored scale insect, *Aonidiella taxus* Leonardi (Homoptera: Diaspididae).]. [In Japanese]. *Science Bulletin of the Faculty of Agriculture, Kyushu University* **33**: 79-86. [Uemats1979]

Ugolini, A. 1979. Supervised control in apple orchards in Piedmont. Pages 597-605. *in*: Proceedings: Internationales Symposium der IOBC/WPRS über Integrierten Pflanzenschutz in der Landund Forstwirtschaft. Wien, Austria: OILB/SROP. 648 pp. [Ugolin1979]

Ülgentürk, S. 1996. [Diaspididae (Homoptera: Coccoidea) species from ornamental trees in Ankara province]. [In Turkish]. *Proceedings of the Third Turkish National Congress of Entomology,* pp. 541-548. [Ulgent1996]

Ülgentürk, S. & Toros, S. 2000.]Preliminary studies on parasitoids and predators of Diaspididae (Homoptera: Coccoidea) species on park plants.] Park bitkilerinde s [In Turkish]. *Tarim Bilimleri Dergisi* 6: 106-110. [UlgentTo2000]

Ullyett, G.C. 1946. Red scale on citrus and its control by natural factors. *Farming in South Africa* **21**: 333-337. [Ullyet1946]

Unterberger, V.K. 1964. [The penetration of petroleum oil under the field of California scale, *Aspidiotus perniciosus* Comst.]. [In Russian]. *Khimiia v sel'skom khoziaistve* **1**: 26-27. [Unterb1964]

Unterberger, V.K. 1964a. [Use of phosphorous insecticides in the form of oil emulsions against the San José scale.]. [In Russian]. *Khimiia v sel'skom khoziaistve* **7**: 34-41. [Unterb1964a]

Upadhyay, S. & Vaidya, M.S. 1993. Loss estimation in sugarcane due to scale insect *Melanaspis glomerata* (Green) in northwestern Madhya Pradesh. *Research and Development Reporter* **10**: 26-29. [UpadhyVa1993]

Usman, S. & Puttarudraiah, M. 1955. A list of the insects of Mysore including the mites. *Entomology Series, Department of Agriculture, Mysore State* Bulletin No. 16. [UsmanPu1955]

Uygun, N. & Elekcioglu, N.Z. 1998. Effect of three Diaspididae prey species on development and fecundity of the ladybeetle *Chilocorus bipustulatus* in the laboratory. *Biocontrol* **43(2)**: 153-162. [UygunEl1998]

Uygun, N., Karaca, I., Sekeroglu, E. & Ulyusoy, M.R. 1992. [Integrated pest management studies in a newly established citrus orchard in Cukurova.] Cukurova'da yeni kurulan bir turuncgil bahcesinde zararliara karsi integre savas calismalari. *Proceedings of the Second Turkish National Congress of Entomology. Izmir, Turkey* : 171-182. [UygunKaSe1992]

Uygun, N., Karaca, I. & Sekeroglu, E. 1995. Population dynamics of *Aonidiella aurantii* (Maskell) (Homoptera: Diaspididae) and its natural enemies on citrus in the Mediterranean region of Turkey from 1976 to 1993. *Israel Journal of Entomology* **29**: 239-246. [UygunKaSe1995]

Uygun, N., Karaca, I., Ulusoy, M.R. & Tekeli, N.Z. 1996. Status of citrus pests and their control in Turkey. *IOBC-WPRS Bulletin* **18(5)**: 171-183. [UygunKaUl1996]

Uygun, N., Sengonca, C., Erkilic, L. & Schade, M. 1998. The Coccoidea fauna and their host plants in cultivated and non-cultivated areas in the East Mediterranean region of Turkey. *Acta Phytopathologica Academiae Scientiarum Hungaricae* **33**: 1-2, 183-191. [UygunSeEr1998]

Vacante, V. 1985. Acari simbionti di Diaspididae (Rynchota, Homoptera) in Sicilia: rilevamento delle specie. Io contributo. *Atti XIV Congresso Nazionale Italiano di Entomologia, Palermo, 1985* 741-747. [Vacant1985]

Vacante, V. 1985a. Acari simbionti di Diaspididae in Sicilia: *Hemisarcoptes malus* (Shimer). IIo contributo. *Atti XIV Congresso Nazionale Italiano di Entomologia, Palermo, 1985* 749-758. [Vacant1985a]

Vacante, V. & Gerson, U. 1987. Three species of *Eryngiopus* (Acari: Stigmaeidae) from Italy, with key to species and summary of habitats. *Redia* **70**: 385-401. [VacantGe1987]

Valentine, E.W. 1963. New records of *Hymenopterous* parasites of Homoptera in New Zealand. *New Zealand Journal of Science* **6**: 6-13. [Valent1963]

Valentine, E.W. 1967. A list of the hosts of entomophagous insects of new Zealand. *New Zealand Journal of Science* **10**: 1100-1210. [Valent1967]

Valles, E. 1965. [Aspects of biological control in general and in the special case of the coconut cochineal (*Aspidiotus destructor* Sign.)]. [In Portuguese]. *Jornadas Silvo-Agronómicas (Angola)* **2**: 259-279. [Valles1965]

Van Der Merwe, C.P. 1962. A note on the introduction and spread of pernicious scale, *Aspidiotus (Quadraspidiotus) perniciosus* (Comstock), in South Africa. *Journal of the Entomological Society of Southern Africa* **25**: 328-331. [VanDer1962]

Van Dine, D.L. 1913. The insects affecting sugar cane in Porto Rico. *Journal of Economic Entomology* **6(2)**: 251-257. [VanDin1913]

Van den Bosch, R. & Telford, A.D. 1964. Environmental modification and biological control. Pages 459-488 *in*: DeBach, P. & Schlinger, E.I., Eds. Biological Control of Insect Pests and Weeds. London: Chapman & Hall. 844 pp. [VandenTe1964]

van Emden, H.F. 1990. Plant diversity and natural enemy efficiency in agroecosystems. Pages 63-80 *in*: Mackauer, M., Ehler, L.E. & Roland, J. (Eds.). Critical Issues in Biological Control. Andover, Hants: Intercept. 330 pp. [vanEmd1990]

Vargas M., R. 1987. Control integrado de escama de San José. *Investigación y Progreso Agropecuario, La Platina (Instituto de Investigaciones Agropecuarias, Estación Experimental La Platina)* **44**: 44-48. [Vargas1987]

Varun, C.L. 1993. Cultural control of sugarcane scale insect (*Melanaspis glomerata* Green) in Bhat soil. *Indian Journal of Entomology* **55**: 219-220. [Varun1993]

Vasquez, J., Delgado, C., Couturier, G. & Matile-Ferrero, D. 2002. Les insectes nuisibles au goyavier (*Psidium guajava* L. Myrtaceae) en Amazonie péruvienne. *Fruits* **57(5-6)**: 323-334. [VasqueDeCo2002]

Vasseur, R. 1949. Quelques données sur la biologie du pou de San José *Quadraspidiotus perniciosus* Comst. dans la Région Lyonnaise: Perspectives nouvelles de la lutte chimique. *La Pomologie Française* **76**: 47-51. [Vasseu1949]

Vasseur, R. & Bénassy, C. 1953. Sur la faune parasitaire du pou de San José *Quadraspidiotus perniciosus* Comst. dans la région lyonnaise et ses relations avec les conditions climatiques. *Annales des Epiphyties* **4**: 283-290. [VasseuBe1953]

Vasseur, R. & Bianchi, H. 1949. Quelques données nouvelles sur la lutte chimique contre le pou de San José *Quadraspidiotus perniciosus* Comst.). *Comptes Rendus des Séances de l'Académie d'Agriculture de France* **35(7)**: 280-282. [VasseuBi1949]

Vasseur, R. & Bianchi, H. 1953. Sur l'efficacité de diverses formules insecticides utilisées en traitement d'hiver contre les cochenilles des arbres fruitiers. *Revue de Zoologie Agricole* **no. 7-9**: 76-83. [VasseuBi1953]

Vasseur, R. & Bianchi, H. 1953a. Sur l'efficacité de produits insecticides divers utilisés contre les cochenilles diaspines des arbres fruitiers en traitements d'hiver. *Annales des Epiphyties* **4**: 45-58. [VasseuBi1953a]

Vasseur, R. & Schvester, D. 1957. Biologie et écologie du pou de San José (*Quadraspidiotus perniciosus* Comst.) en France. *Annales des Epiphyties* **8**: 5-66. [VasseuSc1957]

Vasseur, R., Schvester, D. & Bianchi, H. 1952. Sur l'effet aphicide de certains traitements contre le pou de San José *Quadraspidiotus perniciosus* (Comst.). *Annales des Epiphyties* **3**: 339-350. [VasseuScBi1952]

Vasseur, R., Schvester, D. & Bianchi, H. 1957. Observations sur les traitements en cours de végétation contre le pou de San José (*Quadraspidiotus perniciosus* Comst.). *Annales des Epiphyties* **8**: 101-110. [VasseuScBi1957]

Vayssière, P. 1913. Note sur les coccides de l'Afrique occidentale. *Annales du Service des Epiphyties* **1**: 424-432. [Vayssi1913]

Vayssière, P. 1914a. Note sur quelques coccides nouveaux ou peu connus. *Bulletin de la Société Entomologique de France* **1914**: 206-208. [Vayssi1914a]

Vayssière, P. 1915. Note sur quelques coccides reçues à la Station Entomologique de Paris en 1913. *Annales du Service des Epiphyties* **2**: 288-301. [Vayssi1915]

Vayssière, P. 1920. Les insectes nuisibles aux cultures du Maroc (2e note). *Bulletin de la Société Entomologique de France* **15**: 256-259. [Vayssi1920]

Vayssière, P. 1930. Les Insectes Nuisibles au Cotonnier dans les Colonies Françaises. Paris: Société d'Editions Géographiques, Maritimes et Coloniales. 438 pp. [Vayssi1930]

Vayssière, P. 1932a. Insectes utiles ou nuisibles acclimatés en France ou dans les colonies depuis un siècle. Pages 629-648. *in*: Livre du Centenaire Paris: Société Entomologique de France. [Vayssi1932a]

Vazirani, T.G. 1982. *Sukunahikona popei* sp.n. (Coleoptera: Coccinellidae) feeding on scale insects (Hemiptera: Diaspididae) infesting coconut palm in Gujarat, India. *Bulletin of Entomological Research* **72**: 29-32. [Vazira1982]

Vehrs, S.L.C. & Grafton-Cardwell, E.E. 1994. Chlorpyrifos effect on armored scale (Homoptera: Diaspididae) populations in San Joaquin Valley citrus. *Journal of Economic Entomology* **87(4)**: 1046-1057. [VehrsGr1994]

Velasquez, F.J. 1971. Some Philippine armored scale insects of the tribe Aspidiotini (Diaspididae, Homoptera). *Philippine Entomologist* **2(2)**: 89-153. [Velasq1971]

Velasquez, F.J. & Rimando, L. 1969. A check list and host index of the armored scale insects of the Philippines (Diaspididae: Homoptera). *Philippine Entomologist* **1**: 195-208. [VelasqRi1969]

Vernalha, M.M. 1953. Coccideos da colecao I.B.P.T. *Arquivos de Biologia e Tecnologia, Curitiba* **8**: 111-304. [Vernal1953]

Vernalha, M.M. 1957. Contribuicao para o conhecemiento dos coccideos (Homoptera, Coccoidea) de Ilex sp. no Estado do Parana. *Boletim Instituto de Biologia e Pesquisas Tecnologicas. Curitiba* **39**: 1-52. [Vernal1957]

Vernalha, M.M., Gabardo, J.C. & Rocha, M.A.L. da. 1970. [New pest of *Ficus elastica* Roxb.]. [In Portuguese]. *Arquivos de Biología e Tecnología. Curitiba* **14**: 41-43. [VernalGaRo1970]

Vernalha, M.M., Silva, R.P., Gabardo, J.C. & Novacki, M.J. 1985. [*Quadraspidiotus perniciosus*(Comstock, 1881) (Homoptera Diaspididae), the Sao José louse.]. [In Portuguese]. *Acta Biológica Paranaense. Curitiba* **14(1-4)**: 193-206. [VernalSiGa1985]

Vesey-Fitzgerald, D. 1940. The control of Coccidae on coconuts in Seychelles. *Bulletin of Entomological Research* **31**: 253-285. [VeseyF1940]

Vesey-Fitzgerald, D. 1941. Progress of the control of coconut-feeding Coccidae in Seychelles. *Bulletin of Entomological Research* **32**: 161-164. [VeseyF1941]

Vesey-Fitzgerald, D. 1953. Review of the biological control of coccids on coconut palms in the Seychelles. *Bulletin of Entomological Research* **44(3)**: 405-413. [VeseyF1953]

Viel, G. 1946. Étude sur la valeur insecticide de produits divers dans la lutte contre le pou de San-José}) Quadraspidiotus perniciosus} Comst.). *Annales des Epiphyties* **12(4)**: 401-413. [Viel1946]

Viggiani, G. 1970a. Les cochenilles des agrumes en Italie et les problèmes se rapportant aux moyens de les combattre. *Al Awamia* **37**: 47-55. [Viggia1970a]

Viggiani, G. 1974. Recherches sur les cochenilles des agrumes. *Bulletin OILB/SROP (Sect. Reg. Ouest Palearctique)* **1974/3**: 117-120. [Viggia1974]

Viggiani, G. 1978. Current state of biological control of olive scales. *Bollettino del Laboratorio di Entomologia Agraria 'Filippo Silvestri'. Portici* **35**: 30-38. [Viggia1978]

Viggiani, G. 1984. Bionomics of the Aphelinidae. *Annual Review of Entomology* **19**: 257-276. [Viggia1984]

Viggiani, G. 1985. [Phenology, natural enemies and control of *Quadraspidiotus ostraeformis* Curtis (Homoptera: Diaspididae) on filbert.] [In Italian]. *Bollettino del Laboratorio di Entomologia Agraria 'Filippo Silvestri'. Portici* **42**: 3-9. [Viggia1985]

Viggiani, G. 1987. Le specie italiane del genere *Encarsia* Forster (Hymnoptera: Aphelinidae. *Bollettino del Laboratorio di Entomologia Agraria 'Filippo Silvestri'. Portici* **44**: 121-179. [Viggia1987]

Viggiani, G. 1989. Su alcuni insetti fitofagi di attualita. *Informatore Agrario* **45(28)**: 63-64. [Viggia1989]

Viggiani, G. 1990a. 2.4 Endoparasites. Pages 121-132 *in*: Rosen, D. (Ed.). Armored Scale Insects, Their Biology, Natural Enemies and Control [Series title: World Crop Pests, Vol. 4B]. Amsterdam, the Netherlands: Elsevier. 688 pp. [Viggia1990a]

Viggiani, G. 1990b. 2.5 Hyperparasites. Pages 177-181 *in*: Rosen, D. (Ed.). Armored Scale Insects, Their Biology, Natural Enemies and Control [Series title: World Crop Pests, Vol. 4B]. Amsterdam, the Netherlands: Elsevier. 688 pp. [Viggia1990b]

Viggiani, F. & Iannaccone, F. 1973. Osservazioni sulla biologia e sui parassiti del Diaspini *Chrysomphalus* ... *Bollettino del Laboratorio di Entomologia Agraria 'Filippo Silvestri'. Portici* **30**: 104-116. [ViggiaIa1973]

Vikberg, V. 1991. *Diaspidiotus bavaricus* Lindinger (Diaspididae) and *Acanthococcus baldonensis* Rasina (Eriococcidae) new to the Finnish fauna (Homoptera, Coccoidea). *Entomologica Fennica* **2(1)**: 4. [Vikber1991]

Vinson, S.B. 1977. Behavioral chemicals in the augmentation of natural enemies. Pages 237-279. *in*: Ridgway, R.L. & Vinson, S.B. (Eds.). Biological Control by Augmentation of Natural Enemies: Insect and Mite Control with Parasites and Predators. New York: Plenum Press. 480 pp. [Vinson1977]

Vinson, S.B. 1990. 3.6.3.3. Potential for Semiochemical Manipulation of Beneficial Insects used to Control Scale Insects. Pages 453-459 *in*: Rosen, D. (Ed.). Armored Scale Insects, Their Biology, Natural Enemies and Control [Series title: World Crop Pests, Vol. 4B]. Amsterdam, the Netherlands: Elsevier. 688 pp. [Vinson1990]

Vogel, V.W. 1960. Einige mikroskopische Merkmale der Weibchen der San-Jose-Schildlaus, *Quadraspidiotus perniciosus. Schweizerische Zeitschrift für Obst- und Weinbau* **69**: 265-267. [Vogel1960]

Wadhi, S.R. & Batra, H.N. 1964. Pests of tropical and subtropical fruit trees. Pages 227-260 *in*: Pant, N.C., Chief Editor. Entomology in India ["A special number of the Indian Journal of Entomology"]. New Delhi: The Entomological Society of India. 529 pp. [WadhiBa1964]

Wahl, B. 1933. Die San José-Laus *(Aspidiotus perniciosus)* und deren Auftreten in Europa. *Proceedings 5e Congress International d'Entomologie (Paris 1932)* **5**: 693-698. [Wahl1933]

Wailes, G. 1859. Rhododendrons and their Enemies. *Proceedings of the Entomological Society of London* **5**: 85. [Wailes1859]

Walde, S.J., Luck, R.F., Yu, D.S. & Murdoch, W.W. 1989. A refuge for red scale: the role of size-selectivity by a parasitoid wasp. *Ecology* **70(6)**: 1700-1706. [WaldeLuYu1989]

Waldner, W. 1988. Was bringt Insegar im Sudtiroler Obstbau?. *Obstbau Weinbau* **25(6)**: 186-188. [Waldne1988]

Walker, F. 1852. List of the specimens of homopterous insects in the collection of the British Museum, Part IV. London: British Museum (Natural History). 1188 pp. [Walker1852]

Walker, G.P. & Aitken, D.C.G. 1996. Non-target effect of sprays to control California red scale (*Aonidiella aurantii* Maskell)), Hom., Diaspididae) on citrus red mite (*Panonychus citri* (McGregor), Acari, Tetranychidae). *Journal of Applied Entomology* **120**: 175-180. [WalkerAi1996]

Walker, G.P., Aitken, D.C.G. & O'Connell, N.V. 1990. Using phenology to time insecticide applications for control of California Red Scale (Homoptera: Diaspididae) on citrus. *Journal of Economic Entomology* **83(1)**: 189-196. [WalkerAiOC1990]

Walker, G.P. & Bednar, L.F. 1986. Behavioral response of California red scale crawlers (Homoptera: Diaspididae) to solutions behind artificial membranes. *Annals of the Entomological Society of America* **79**: 549-553. [WalkerBe1986]

Walker, G.P., Zareh, N. & Arpaia, M.L. 1999. Effect of pressure and dwell time on efficiency of a high-pressure washer for postharvest removal of California red scale (Homoptera : Diaspididae) from citrus fruit. *Biocontrol Dordecht* **44(1)**: 47-58. [WalkerZaAr1999]

Walker, G.P., Zareh, N. & Arpaia, M.L. 1999a. Effect of pressure and dwell time on efficiency of a high-pressure washer for postharvest removal of California red scale (Homoptera: Diaspididae) from citrus fruit. *Journal of Economic Entomology* **92(4)**: 906-914. [WalkerZaAr1999a]

Wang, C.N. 1936. [A tentative list of insects of Hunan Province I.]. [In Chinese]. *Entomology and Phytopathology. Hangchow* **4(19)**: 382-389. [Wang1936]

Wang, C. & Su, T. 1989. [The effect of temperature on population parameters of the Latania scale, *Hemiberlesia lataniae* Signoret.]. [In Chinese]. *Chinese Journal of Biological Control* **9(2)**: 151-156. [WangSu1989]

Wang, H.J., Varma, R.V. & Xu, T.S. 1998. Insect Pests of Bamboos in Asia. New Delhi, India: International Network for Bamboo and Rattan. 200 pp. [WangVaXu1998]

Wang, T.C. & Zhang, X.J. 1994a. [A new species of the genus *Diaspidiotus* Cockerell (Homoptera: Coccoidea: Diaspididae).] [In Chinese]. *Acta Zootaxonomica Sinica* **19(3)**: 326-328. [WangZh1994a]

Ward, R. 1890. Notes on scale and other parasitical insects. *Timehri. Journal of the Royal Agricultural and Commercial Society of British Guiana (Georgetown)* **4 (n.s.)**: 302-311. [Ward1890]

Ware, A.B. & Pasques, B.P. 1996. *Aphytis* augmentation in the southern African Lowveld. *Proceedings of the International Society of Citriculture* **I**: 511-514. [WarePa1996]

Warthen, J.D., Rudrum, M., Moreno, D.S. & Jacobson, M. 1970. *Aonidiella aurantii* sex pheromone isolation. *Journal of Insect Physiology* **16**: 2207-2209. [WartheRuMo1970]

Washington, J.R. & Walker, G.P. 1990. Histological studies of California Red Scale (Homoptera: Diaspididae) feeding on citrus. *Annals of the Entomological Society of America* **83(5)**: 939-948. [WashinWa1990]

Watanabe, M.A. 1997. Ocorréncia de cochonilha pardinha *Selenaspidus articulatus* e seus parasitóides nativos na região de *Limeira* sp. Page 103 *in*: Bento, J.M.S. & Delalibera, I. (Eds.). [16th Brazilian Congress of Entomology] 16o Congresso Brasileiro de Entomologia. Salvador: Sociedade Entomológica do Brasil/EMBRAPA-CNPMF. 400 pp. [Watana1997]

Watanabe, M.A., Tambasco, F.J., Costa, V.A., Nardo, E.A.B. de, Facanali, R. & de Nardo, E.A.B. 2000. Flutuaçao populacional de cochonilhas de carapaca na cultura de citros em municipios paulistas. *Laranja* **21(1)**: 49-64. [WatanaTaCo2000]

Watanabe, M.A., Tambasco, F.J., Nardo, E.A.B. de, Viana, R.I. & Pereira, G.D. 2000. [Competition between *Selenaspidus articulatus* and *Parlatoria ziziphi* scales in orchards in the Citrus region of Sao Paulo State.]. [In Portuguese]. *Laranja* 21(1) 81-97. [WatanaTaNa2000]

Watanabe, M.A. & Yoshii, C. 1992. Parasitismo em cochonilha pardinha *Selenaspidus articulatus* (Hemiptera-Homoptera, Diaspididae) por *Aphytis* sp. (Hymenoptera, Aphelinidae). *Revista de Agricultura Piracicaba* **67**: 1, 63-65. [WatanaYo1992]

Watanabe, M.A., Yoshii, C. & Siloto, R.C. 1994. Parasitismo em cochonilha *Selenaspidus articulatus* (Morgan, 1889) (Hemiptera/Homoptera, Diaspididae) em citros nas regioes de Jaguariuna e *Limeira*, sp. *Revista de Agricultura (Piracicaba)* **69(2)**: 193-200. [WatanaYoSi1994]

Waterworth, H.E. 1981. Control of plant diseases by exclusion: quarantines and disease-free stocks. Pages 269-296. *in*: Pimentel, D. (Ed.). Handbook of Pest Management in Agriculture. Vol. I. (CRC Series in Agriculture). Boca Raton, Florida: CRC Press. 597 pp. [Waterw1981]

Watson, D.M., Du, T.Y., Li, M., Xiong, J.J., Liu, D.G., Huang, M.D., Rae, D.J. & Beattie, G.A.C. 2000. Life history and feeding biology of the predatory thrips, *Aleurodothrips fasciapennis* (Thysanoptera: Phlaeothripidae). *Bulletin of Entomological Research* 88: 351-357. [WatsonDuLi2000]

Watson, D.M., Du, T.Y., Li, M. Xiong, J.J., Liu, D.G., Huang, M.D., Rae, D.J. & Beattie, GAC. 2000. The effect of two prey species, *Chrysomphalus aonidum* and *Corcyra cephalonica*, on the quality of the predatory thrips, *Aleurodothrips fasciapennis*, reared in the laboratory. *Biocontrol Dordecht* **45(1)**: 45-61. [WatsonDuLi2000]

Watson, D.M., Du, T.Y., Li, M., Xiong, J.J., Liu, D.G., Huang, M.D., Rae, D.J. & Beattie, G.A.C. 2000a. Functional responses of, and mutual interference in *Aleurodothrips fasciapennis* (Franklin) (Thysanoptera: Phlaeothripidae) and implications for its use as a biocontrol agent. *General and Applied Entomology* **29**: 31-37. [WatsonDuLi2000a]

Watson, J.R. 1918. Insects of a citrus grove. *Bulletin (Agricultural Experiment Station, University of Florida)* **148**: 165-267. [Watson1918]

Watson, J.R. 1926. Citrus insects and their control. *Bulletin of the Florida Agricultural Experiment Station* **148**: 293-423. [Watson1926]

Watson, J.R. & Berger, E.W. 1937. Citrus insects and their control. *Bulletin (Agricultural Extension Service, Florida)* **88**: 135 pp. [WatsonBe1937]

Wearing, C.H. 1976. San José scale, *Quadraspidiotus perniciosus* (Comstock), life cycle. *Information Series of the (New Zealand) Department of Scientific and Industrial Research (DSIR)* **No. 105/121**: 1-2. [Wearin1976]

Webber, H.J. 1897. The new method of warfare against scale insects. *Proceedings of the Florida State Horticultural Society* **10**: 53-58. [Webber1897]

Webster, F.M. 1899. Odour of San José scale, *Aspidiotus perniciosus*. *Canadian Entomologist* **31**: 4. [Webste1899]

Webster, F.M. 1900. The native home of the San José scale. *Annual Report of the Entomological Society of Ontario* **30**: 55-56. [Webste1900]

Weglarska, B. 1962. Oogenesis in the ovoviviparous scale insect *Quadraspidiotus ostreaeformis* (Curt.) (Homoptera, Coccidae, Aspidiotini). Part I. Morphological and cytological investigations. *Zoologica Poloniae. Krakow* 267-294. [Weglar1962]

Weglarska, B. 1962a. Oogenesis in the ovoviviparous scale insects *Quadraspidiotus ostreaeformis* (Curt.), (Homoptera, Coccidae, Aspidiotini). Part II. Histochemical investigations. *Zoologica Poloniae. Krakow* **12**: 41-68. [Weglar1962a]

Weglarska, B. 1966. Development of testes and spermatogenesis in *Quadraspidiotus ostreaeformis* (Curt.) (Homoptera, Coccoidea, Aspidiotini). Development of testes including sperm bundles formation. *Zeszyty Naukowe Uniwersytetu Jagiellonskiego Prace Zool.* **12**: 59-98. [Weglar1966]

Weglarska, B. 1968. Spermatogenesis in *Quadraspidiotus ostreaeformis* (Curt.) (Homoptera, Coccoidea, Aspidiotini). *Zeszyty Naukowe Uniwersytetu Jagiellonskiego Prace Zool.* **14**: 63-82. [Weglar1968]

Weidner, V.H. & Wagner, W. 1968. Die Entomologischen Sammlungen des Zoologischen Staatsinstituts und Zoologischen Museums Hamburg VII. Teil. Insecta IV. *Mitteilungen des Hamburgerischen Zoologisches Museums und Instituts* **65**: 123-179. [WeidneWa1968]

Weigel, C.A. & Broadbent, B.M. 1924. Lubricating-oil emulsion as a control for *Chrysomphalus aonidum* in greenhouses. *Journal of Economic Entomology* **17**: 386-389. [WeigelBr1924]

Wentzel, P.C. 1970. [The mass-rearing of red scale, *Aonidiella aurantii* (Maskell).]. *Phytophylactica* **2**: 195-198. [Wentze1970]

Wester, P.J. 1918. The coconut, its culture and uses. *Philippine Agricultural Review* **1(1)**: 5-57. [Wester1918]

Wester, P.J. 1920. The mango. *Bulletin (Bureau of Agriculture, the Philippines)* **No. 18**: 70 pp. [Wester1920]

Westwood, J.O. 1840. An introduction to the modern classification of insects; founded on the natural habits and corresponding organization of different families. Vol. II. London: Longman, Orme, Brown, Green and Longmans. 587 pp. [Westwo1840]

Whitcomb, W.H. 1974. Natural populations of entomophagous arthropods and their effect on the agroecosytem. Pages 150-169 *in*: Maxwell, F.G. & Harris, F.A. (Eds.). Proceedings of the Summer Institute on Biological Control of Plant Insects and Diseases. Jackson: University Press of Mississippi. 647 pp. [Whitco1974]

Whitehead, J.D. 1963. The Biology and Control of the Grape Scale, Aspidiotus uvae (Const.) in Northeastern Arkansas. The University of Arkansas. 98 pp. [Whiteh1963]

Whiting, D.C., Hoy, L.E., Connolly, P.G., McDonald, R.M. & O'Callaghan, M. 1998. Effects of high-pressure water jets on armoured scale insects and other contaminants of harvested kiwifruit. *Proceedings of the 51st New Zealand Plant Protection Conference* 211-215. [WhitinHoCo1998]

Whitnall, B. 1959. Red scale on citrus. *The Citrus Grower* **305**: 5-9. [Whitna1959]

Whitney, L.A. 1933. Report of Associate plant inspector, 1931-1932. *Hawaiian Forest and Agriculture* **30**: 59-70. [Whitne1933]

Willard, J.R. 1972. Studies on the rates of development and reproduction of California red scale, *Aonidiella aurantii* (Mask.) (Homoptera: Diaspididae) on citrus. *Australian Journal of Zoology. Melbourne* **20**: 37-47. [Willar1972]

Willard, J.R. 1972a. The rhythm of emergence of crawlers of California red scale, *Aonidiella aurantii* (Mask.), (Homoptera: Diaspididae). *Australian Journal of Zoology. Melbourne* **20**: 49-65. [Willar1972a]

Willard, J.R. 1973. Wandering time of the crawlers of California red scale, *Aonidiella aurantii* (Mask.) (Homoptera: Diaspididae), on citrus. *Australian Journal of Zoology. Melbourne* **21**: 217-229. [Willar1973]

Willard, J.R. 1973a. Survival of crawlers of California red scale, *Aonidiella aurantii* (Mask.) (Homoptera: Diaspididae). *Australian Journal of Zoology. Melbourne* **21**: 567-573. [Willar1973a]

Willard, J.R. 1974. Horizontal and vertical dispersal of California red scale, *Aonidiella aurantii* (Mask.) (Homoptera: Diaspididae). *Australian Journal of Zoology. Melbourne* **22**: 531-548. [Willar1974]

Willard, J.R. 1976. Leaf disc method for rearing California red scale *Aonidiella aurantii* (Maskell) (Homoptera: Diaspididae). *Journal of the Australian Entomological Society* **15**: 7-11. [Willar1976]

Wille, J.E. 1941. Tres informes de observaciones entomológicas en la costa en 1940. *Informe (Ministerio de Fomento, Dirección de Agricultura y Ganadería, Lima, Peru)* **No. 53**: 3-26. [Wille1941]

Williams, D.J. 1957a. The genus *Semelaspidus* MacGillivray (Aspidiotini, Coccoidea: Hemiptera). *Proceedings of the Royal Entomological Society of London, Series B: Taxonomy* **26**: 33-42. [Willia1957a]

Williams, D.J. 1960d. A new species of *Greeniella* Cockerell (Aspidiotini: Coccoidea: Homoptera) from Malaya. *Proceedings of the Royal Entomological Society of London, Series B: Taxonomy* **29**: 153-154. [Willia1960d]

Williams, D.J. 1962a. Two new South African Diaspididae (Homoptera). *Journal of the Entomological Society of Southern Africa* **25**: 128-132. [Willia1962a]

Williams, D.J. 1963a. A new species of *Entaspidiotus* MacGillivray (Homoptera: Coccoidea: Diaspididae) from South Africa. *Proceedings of the Royal Society of London* **32**: 44-46. [Willia1963a]

Williams, D.J. 1963b. Synoptic revisions of I. *Lindingaspis* and II. *Andaspis* with two new allied genera (Hemiptera: Coccoidea). *Bulletin of the British Museum (Natural History) Entomology* **15**: 1-31. [Willia1963b]

Williams, D.J. 1969a. The family-group names of the scale insects (Hemiptera: Coccoidea). *Bulletin of the British Museum (Natural History) Entomology* **23**: 315-341. [Willia1969a]

Williams, D.J. 1970. The mealybugs (Homoptera, Coccoidea, Pseudococcidae) of sugarcane, rice and sorghum. *Bulletin of Entomological Research* **60**: 109-188. [Willia1970DJ]

Williams, D.J. 1971a. A new species of *Abgrallaspis* Balachowsky (Homoptera, Coccoidea, Diaspididae) from a nursery in England causing leaf-spotting on orchids. *Bulletin of Entomological Research* **60**: 453-455. [Willia1971a]

Williams, D.J. 1974. The type-species of *Lindingaspis* MacGillivray (Homoptera: Coccoidea: Diaspididae). *Journal of Entomology (B)* **42**: 217-220. [Willia1974]

Williams, D.J. 1985a. T.D.A. Cockerell's scale insects (Homoptera: Coccoidea) in the British Museum (Natural History). *Folia Entomologica Hungarica* **46**: 215-240. [Willia1985a]

Williams, D.J. 1985e. Some South American scale insects (Homoptera: Coccoidea) on *Nothofagus. Journal of Natural History* **19**: 249-258. [Willia1985e]

Williams, D.J. & Butcher, C.F. 1987. Scale insects (Hemiptera: Coccoidea) of Vanuatu. *New Zealand Entomologist* **9**: 88-99. [WilliaBu1987]

Williams, J.R. & Greathead, D.J. 1990. 3.9.4 Sugar Cane. Pages 563-578 *in*: Rosen, D. (Ed.). Armored Scale Insects, Their Biology, Natural Enemies and Control [Series title: World Crop Pests, Vol. 4B]. Amsterdam, the Netherlands: Elsevier. 688 pp. [WilliaGr1990]

Williams, D.J. & Watson, G.W. 1988. The scale insects of the tropical South Pacific region. Pt. 1: The armoured scales (Diaspididae). London: CAB International Institute of Entomology. 290 pp. [WilliaWa1988]

Willis, J.C. 1988. A dictionary of the flowering plants & ferns. 8th Edition. Cambridge: Cambridge University Press. 1245 pp. [Willis1988].

Wilson, C.E. 1917. Some Florida scale-insects. *Quarterly Bulletin of the Florida State Plant Board* **2**: 2-65. [Wilson1917]

Wilson, C.E. 1921. Report of the entomologist. *Report (Virgin Islands Agricultural Experiment Station)* 20-34. [Wilson1921]

Wilson, K. & Goldsmid, J.M. 1962. Rhodesian citrus pests and their control. *Rhodesia Agricultural Journal* **59**: 41-61. [WilsonGo1962]

Woglum, R.S. 1906. Two new scale insects. *Canadian Entomologist* **38**: 73-75. [Woglum1906]

Woglum, R.S. 1923. Fumigation of citrus trees for control of insect pests. *Farmer's Bulletin (USDA)* **No. 1321**: 1-59. [Woglum1923]

Woglum, R.S. 1925. Observations on insects developing immunity to insecticides. *Journal of Economic Entomology* **18**: 593-597. [Woglum1925]

Woglum, R.S. 1925a. Value of sprays and fumigation for resistant black scale control. *Bulletin (California Fruit Growers Exchange Bulletin)* No. 2. [Woglum1925a]

Woglum, R.S. 1925b. Is scale becoming resistant?. *California Citrograph* **10(5)**: 178. [Woglum1925b]

Woglum, R.S. 1926. The use of oil spray on citrus trees. *Journal of Economic Entomology* **19**: 723-733. [Woglum1926]

Woglum, R.S. 1942. The commercial value during 1941-42 season of toxic oil sprays for red scale. *The Exchange: Pest Control Circular* **No. 90**: 1-3. [Woglum1942]

Woglum, R.S. & LaFollette, J.R. 1926. Recent investigations of resistant red scale. *California Citrograph* **11(10)**: 396-397. [WoglumLa1926]

Woglum, R.S., LaFollette, J.R., Landon, W.E. & Lewis, H.C. 1947. The effect of field-applied insecticides on beneficial insects of citrus in California. *Journal of Economic Entomology* **40(6)**: 818-820. [WoglumLaLa1947]

Wolcott, G.N. 1958. The evanescence of perfect biological control. Pages 511-513 *in*: Baecker, E.C., (Ed.). Proceedings of the Tenth International Congress of Entomology. Vol. 4. Ottawa. 1115 pp. [Wolcot1958]

Wolfe, H.S., Toy, L.R. & Stahl, A.L. 1934. Avocado production in Florida. *Bulletin of the Florida Agricultural Experiment Station* **272**: 1-96. [WolfeToSt1934]

Wolfenbarger, D.O. 1951. Dictyospermum scale control on avocados. *Florida Entomologist* **34(2)**: 54-58. [Wolfen1951]

Wolfenbarger, D.O. 1955. Mango insect pest control. *Proceedings of the Florida Mango Forum, Miami, September 21, 1955* 27-34. [Wolfen1955]

Wolff, M. 1911. Forstlich wichtige Schildläuse. *Kaiser-Wilhelm Institut für Landwirtschaft, Vorträge über Pflanzenschutz, Forstschutz* **2**: 49-94. [Wolff1911]

Wolff, V.R.S. & Claps, L.E. 1999. Uma nova espécie do género {Dynaspidiotus} Thiem & Gerneck (Coccoidea: Diaspididae) ocorrendo em {Araucaria angustifolia} (Bertol.) *Arquivos do Instituto Biológico. Sao Paulo* **66(1)**: 21-25. [WolffCl1999]

Wolff, V.R.S. & Corseuil, E. 1993. Espécies de Diaspididae (Hom.: Coccoidea) ocorrentes em plantas cítricas no Rio Grande do Sul, Brasil: I - Aspidiotinae. *Biociências* **1(1)**: 25-60. [WolffCo1993]

Wolff, V.R.S. & Corseuil, E. 1993a. Diaspidídeos ocorrentes em mangueira no Brasil com caracterização e registro de {Aulacaspis tubercularis} Newst., 1906 (Hom.: Coccoidea) no Rio Grande do Sul. *Biociências* **1(1)**: 151-161. [WolffCo1993a]

Wong, C.Y., Chen, S.P. & Chou, L.Y. 1999. [Guidebook to Scale Insects of Taiwan.] [In Chinese]. Wufeng, Taichung, Taiwan: Taiwan Agricultural Research Institute. 98 pp. [WongChCh1999]

Wood, B.J. 1962. Natural enemies in the control of citrus pests in Cyprus. *Countryman Nicosia* 8-11. [Wood1962]

Woodhill, A.R. 1928. Spring versus autumn spraying for red scale of citrus. *Agricultural Gazette of New South Wales* **39**: 561. [Woodhi1928]

Woodruff, R.E., Beck, B.M., Skelley, P.E., Schotman, C.Y.L. & Thomas, M.C. 1998. Checklist and Bibliography of the Insects of Grenada and the Grenadines. Gainesville, FL: Center for Systematic Entomology. 286 pp. [WoodruBeSk1998]

Woodward, T.E., Evans, J.W. & Eastop, V.F. 1970. Hemiptera (Bugs, leafhoppers, etc.). Pages 387-457. *in*: Mackerras, I.M., Ed. The Insects of Australia: A Textbook for Students and Research Workers. Melbourne: Melbourne University Press. 1029 pp. [WoodwaEvEa1970]

Woodworth, C.W. 1909. California horticultural quarantine. *Journal of Economic Entomology* **2**: 359-360. [Woodwo1909]

Woolley, J.B. 1990. 2.4.3 Signiphoridae. Pages 167-176 *in*: Rosen, D. (Ed.). Armored Scale Insects, Their Biology, Natural Enemies and Control [Series title: World Crop Pests, Vol. 4B]. Amsterdam, the Netherlands: Elsevier. 688 pp. [Woolle1990]

Worthlet, X.N. 1932. Some tests of spray materials against San José scale at the Pennsylvania State College. *Pennsylvania State Horticulture Association News* **9(1)**: 22-26. [Worthl1932]

Wu, S.A. 1999b. [Homoptera: Coccoidea.]. [In Chinese]. Pages 231-235 *in*: Shen, X.C. & Pei, H.C. (Eds.). The Fauna and Taxonomy of Insects in Henan. Vol. 4. Insects of the Mountains Funiu and Dabie Regions. Beijing: China Agricultural Scientech Press. 415 pp. [Wu1999b]

Wu, W.J., Chen, S.P. & Wei, H.Y. 1989. [Functional response of *Chilocorus kuwanae* Silvestri (Coleop.: Coccinellidae) to *Hemiberlesia pitysophila* Takagi (Homop.: Diaspididae).]. [In Chinese]. *Natural Enemies of Insects* **11(1)**: 28-30. [WuChWe1989]

Wu, M.L. & Liu, X.Q. 1994. [Repellent effects of some plant oils on *Hemiberlesia pitysophila* Takagi.]. [In Chinese]. *Entomological Knowledge* **31**: 1, 28-30. [WuLi1994]

Wu, J.L., Pan, Y.Z., Zhao, Q. & Su, J.X. 1998. [Studies on the control threshold of the silk tree scale (*Aonidiella sotetsu*) (Takahashi).]. [In Chinese]. *Scientia Silvae Sinicae* **34**: 2, 51-57. [WuPaZh1998]

Wysoki, M. 1977. Insect pests of macadamia in Israel. *Phytoparasitica* **5**: 187-188. [Wysoki1977]

Wysoki, M. 1997. Present status of arthropod fauna in mango orchards in Israel. *Acta Horticulturae* No. 455: 805-811. [Wysoki1997]

Wysoki, M., Israeli, Y. & Rosen, D. 1995. The oriental red scale, *Aonidiella orientalis* (Newstead) (Diaspididae): biology, phenology, geographic distribution and natural enemies in Israel (Abstract only). *Israel Journal of Entomology* **29**: 267. [WysokiIsRo1995]

Xavier, A.L.Q., de Freitas, S. & Scomparin, C.H.J. 1997. Avaliaçao da capacidade de predaçao de *Chrysoperla externa* (Hagen, 1961) (Neuroptera, Chrysopidae) sobre a cochonilha *Selenaspidus articulatus* (Morgan, 1889) (Hemiptera, Diaspididae). Page 135 *in*: Bento, J.M.S. & Delalibera, I. (Eds.). [16th Brazilian Congress of Entomology] 16o Congresso Brasileiro de Entomologia. Salvador: Sociedade Entomológica do Brasil/EMBRAPA-CNPMF. 400 pp. [XavierDeSc1997]

Xie, Y.P. 1998. [The Scale Insects of the Forest and Fruit Trees in Shanxi of China.]. [In Chinese]. China Forestry Publishing House. 147 pp. [Xie1998]

Xie, G.L., Pan, W.Y., Tang Z.Y., Ding, D.C. & Lian, J.H. 1997. [Evaluation on the effective and stable control of *Hemiberlesia pitysophila* Takagi with *Coccobius azumai* Tachikawa.]. [In Chinese]. *Acta Entomologica Sinica* **40**: 135-144. [XiePaTa1997]

Xie, Y.P., Xue, J.L., Liu, X.Q., Li, J.P. & Tang, M. 1995. [The epidemic of city forest scales and its management strategy.]. [In Chinese]. *Forest Research* **8**: 114-118. [XieXuLi1995]

Yamamoto, A. & Ogawa, K. 1989. Chemistry and commercial production of pheromones and other behaviour-modifying chemicals. Pages 123-148 *in*: Jutsum, A.R. & Gordon, R.F.S. (Eds.). Insect Pheromones in Plant Protection. Chichester: John Wiley & Sons. 369 pp. [YamamoOg1989]

Yan, J.Y. 1981. Studies on the behaviour of males of California red scale *Aonidiella aurantii* (Maskell). Ph.D. dissertation. Adelaide, South Australia: University of Adelaide. [Yan1981]

Yan, J.Y. & Isman, M.B. 1986. Environmental factors limiting emergence and longevity of male California red scale, *Aonidiella aurantii* Homoptera Diaspididae. *Environmental Entomology* **15**: 971-975. [YanIs1986]

Yanovskii, Y.P. 2001. [The resistance of apple to San José scale.]. [In Russian]. *Sadovodstvo i Vinogradarstvo* **No. 1**: 13-14. [Yanovs2001]

Yanovskii, Y.P. & Larcheva, E.I. 2000. [New insecticides for apple pest control.]. (In Russian.) *Zashchita i Karantin Rastenii* No. 11, 25. [YanovsLa2000]

Yarom, I., Blumberg, D. & Ishaaya, I. 1988. Effects of Buprofezin on California Red Scale (Homoptera: Diaspididae) and Mediterranean Black Scale (Homoptera: Coccidae). *Journal of Economic Entomology* **81**: 1581-1585. [YaromBlIs1988]

Yaşar, B. 1995. [Two new species of armoured scale insects for the Turkish fauna (Homoptera: Coccoidea: Diaspididae).]. [In Turkish with summary in English.] *Journal of Agricultural Faculty (Turkey)* **(5)1**: 49-56. [Yasar1995]

Yaşar, B. 1995a. [Taxonomic studies on the fauna of Diaspididae (Homoptera: Coccoidea in Turkey]. [In Turkish]. 289 pp. Yüzüncü Yil Üniversitesi Matbaasi, Van, Turkey. [Yasar1995a]

Yasnosh, V.A. 1987. Integrated control of scale insects in citrus groves in USSR. *Bollettino del Laboratorio di Entomologia Agraria 'Filippo Silvestri'* **43 (Suppl.)**: 229-234. [Yasnos1987]

Yasnosh, V.A. 1994. *Aphytis* species occurring in the former USSR and their role in biological control. Pages 317-333 *in*: Rosen, D., ed. Advances in the Study of Aphytis (Hymenoptera: Aphelinidae). Andover, UK: Intercept Limited. [Yasnos1994]

Yasnosh, V.A. 1995. Coccids of economic importance and their control in the republic of Georgia. *Israel Journal of Entomology* **29**: 247-251. [Yasnos1995]

Yasnosh, V.A. & Myartseva, S.N. 1971a. [Two new aphelinid species (Chalcidoidea, Aphelinidae) - parasites of armored scales (Coccoidea, Diaspididae) in Central Asia.]. [In Russian]. *Bulletin of the Academy of Sciences (Turkmenian Soviet Republic, Ashkhabad, Biological Series)* **6**: 35-41. [YasnosMy1971a]

Yerushalmi, N. & Cohen, E. 2002. Acetylcholinesterase of the California red scale *Aonidiella aurantii* Mask.: Catalysis, inhibition, and reactivation. *Pesticide Biochemistry and Physiology* **72(3)**: 133-141. [YerushCo2002]

Yinon, U. 1969. Food consumption of the armored scale lady-beetle *Chilocorus bipustulatus* (Coccinellidae). *Entomologia Experimentalis et Applicata* **12**: 139-146. [Yinon1969]

York, H.H. 1905. A new *Aspidiotus* from *Aesculus glabra*. *Ohio Naturalist* **5**: 325-326. [York1905]

Yothers, W.W. 1917. The effects of the freeze of February 2-4, 1917 on the insect pests and mites on citrus. *Florida Buggist* **1**: 30-35,38-40. [Yother1917]

Yothers, W.W. 1922. Spraying for the control of the Florida red scale. *Proceedings of the Florida State Horticultural Society* **1922**: 63-67. [Yother1922]

Yothers, M.A. 1940. Females of the San José scale rendered unproductive by lime-sulfur. *Journal of Economic Entomology* **33(6)**: 890-892. [Yother1940]

Yothers, W.W. & Miller, R.L. 1933. Sprays for scale insects and whiteflies on citrus trees in Florida. *Proceedings of the Florida State Horticultural Society* **46**: 38-52. [YotherMi1933]

Young, B.L. 1986. New genera and species of *Diaspididae* (Coccoidea) from Yunnan and Guizhou. *Contributions of the Shanghai Institute of Entomology* **6**: 199-211. [Young1986]

Young, B.L., Johnson, R.H. & Alstad, D.N. 1993. Sex-ratio variation in scale insects on ponderosa pine: effects of needle age class. *Bulletin of the Ecological Society of America* **74** (2 suppl.): 497. [YoungJoAl1993]

Young, B.L. & Lu, D. 1988. A new scale insect on pine needles from Sichuan (Coccoidea: Diaspididae). *Contributions of the Shanghai Institute of Entomology* **8(8)**: 189-192. [YoungLu1988]

Yousef, A.E.T.A. & El Halawany, M.E. 1982. Effect of prey species on the biology of *Amblyseius gossipi* El Badry (Acari: Mesostigmata: Phytoseiidae). *Acarologia* **23(2)**: 113-117. [YousefEl1982]

Yu, J.J. 1986. [Biological control of agricultural insect pests in Egypt.]. *Chinese Journal of Biological Control* **2(3)**: 143-144. [Yu1986]

Yu, D.S. & Luck, R.F. 1988. Temperature-dependent size and development of California Red Scale (Homoptera: Diaspididae) and its effect on host availability for the ectoparasitoid, *Aphytis melinus* De Bach (Hymenoptera: Aphelinidae). *Environmental Entomology* **17**: 154-161. [YuLu1988]

Yu, D.S., Luck, R.F. & Murdoch, W.W. 1990. Competition, resource partitioning and coexistence of an endoparasitoid *Encarsia perniciosi* and an ectoparasitoid *Aphytis melinus* of the California red scale. *Ecological Entomology* **15(4)**: 469-480. [YuLuMu1990]

Yust, H.R. 1941. A morphological difference between the California red scale and the yellow scale. *Journal of Economic Entomology* **34**: 785-787. [Yust1941]

Yust, H.R. 1943. Productivity of the California red scale on lemon fruit. *Journal of Economic Entomology* **36**: 868-872. [Yust1943]

Yust, H.R., Busbey, R.L. & Howard, L.B. 1942. Laboratory fumigations of the California red scale with methyl bromide, alone and with hydrocyanic acid. *Journal of Economic Entomology* **35(4)**: 521-524. [YustBuHo1942]

Yust, H.R., Busbey, R.L. & Nelson, H.D. 1942. Reaction of resistant and nonresistant strains of California red scale to fumigation with HCN. *Journal of Economic Entomology* **35(6)**: 816-820. [YustBuNe1942]

Yust, H.R. & Cevallow, M.A. 1956. [Preliminary list of agricultural pests of Ecuador.] Lista preliminar de plagas de la agricultura del Ecuador. *Revista Ecuatoriana de Entomología y Parasitología* **2**: 425-442. [YustCe1956]

Yust, H.R., Fulton, R.A. & Nelson, H.D. 1951. Development and stability of resistance of California red scale to fumigation with hydrocyanic acid. *Journal of Economic Entomology* **44(6)**: 833-838. [YustFuNe1951]

Yust, H.R. & Howard, L.B. 1942. Factors influencing results of laboratory fumigation of California red scale with HCN. *Journal of Economic Entomology* **35(6)**: 821-824. [YustHo1942]

Yust, H.R., Nelson, H.D. & Busbey, R.L. 1942. The stupefaction of California red scale with sublethal dosages of hydrocyanic acid. *Journal of Economic Entomology* **35(3)**: 339-342. [YustNeBu1942]

Yust, H.R., Nelson, H.D. & Busbey, R.L. 1942a. Fumigation of yellow scale with HCN under laboratory conditions. *Journal of Economic Entomology* **35(6)**: 825-826. [YustNeBu1942a]

Yust, H.R., Nelson, H.D. & Busbey, R.L. 1943. Comparative susceptibility of two strains of California red scale to HCN, with special reference to the inheritance of resistance. *Journal of Economic Entomology* **36(5)**: 744-749. [YustNeBu1943]

Yust, H.R., Nelson, H.D. & Busbey, R.L. 1943a. The influence of repeated fumigation with HCN on the susceptibility of the California red scale. *Journal of Economic Entomology* **36(6)**: 872-874. [YustNeBu1943a]

Yust, H.R. & Shelden, F.F. 1952. A study of the physiology of resistance to hydrocyanic acid in the California red scale. *Annals of the Entomological Society of America* **45(2)**: 220-228. [YustSh1952]

Zagainyi, S.A. 1956. [Pests of greenhouse plants in Krasnodar territory.]. *Akademii Nauk SSSR Glav. Bot. Sad. Zashchita Rastenii Vreditelei i Bodez.* **26**: 85-90. [Zagain1956]

Zahradník, J. 1951. Contribution à la connaissance des cochenilles (Homopt. Coccoidea). *Acta Societatis Entomologicae Cechosloveniae* **48**: 198-200. [Zahrad1951]

Zahradník, J. 1952. Revision der Tschechoslovakischen Arten der Schildläuse aus der Unterfamilie der Diaspidinae. *Acta Entomologica Musei Nationalis Pragae* **27**: 89-200. [Zahrad1952]

Zahradník, J. 1952a. Eine neue Schildlausart - *Quadraspidiotus marani* n. sp. (Coccoidea: Diaspididae). *Beiträge zur Entomologie. Berlin* **2**: 449-451. [Zahrad1952a]

Zahradník, J. 1953. Première contribution à la connaissance des cochenilles vivant dans les serres du jardin des plantes à praha. *Acta Entomologica Musei Nationalis Pragae* **28**: 125-148. [Zahrad1953]

Zahradník, J. 1955. Zur Nomenclatur von *Quadraspidiotus marani* Zahradník *Quadraspidiotus schneideri* (Bachamann). *Beiträge zur Entomologie. Berlin* **5**: 88-90. [Zahrad1955]

Zahradník, J. 1955a. Über die wichtigsten artdifferenzierenden Merkmale der Schildlaus *Quadraspidiotus marani* Zahradník (=Qu. schneideri Bachmann). *Zeitschrift für Angewandte Entomologie* **37**: 125-127. [Zahrad1955a]

Zahradník, J. 1957. Neue Schildlausfunde in den Gewächshäusern der Tschechoslowakei. *Pflanzenschutzberichte. Wien* **19**: 45-52. [Zahrad1957]

Zahradník, J. 1959. *Borchseniaspis* novum genus, typus *Aspidiotus palmae* Morgan et Cockerell, 1893 (Homoptera, Diaspididae). *Sborník Faunistych Pracì Entomologica Oddeleni Národního Musea v Praze (Acta Faunistica Entomologica Musei Nationalis Pragae* **5**: 65-67 [Zahrad1959]

Zahradník, J. 1959b. Kritická bibliografie Cervcu Ceskoslovenska. *Acta Faunistica Entomologica Musei Nationalis Pragae. Supplementum* **1**: 7-69. [Zahrad1959b]

Zahradník, J. 1962. Diaspidides nouveaux ou peu connus dans la faune Tchécoslovaque. (Homoptera, Diaspidae). *Sborník Faunistych Pracì Entomologica Oddeleni Národního Musea v Praze (Acta Faunistica Entomologica Musei Nationalis Pragae)* **8**: 91-94. [Zahrad1962]

Zahradník, J. 1965. Sur des cochenilles nouvelles dans les serres en Tchécoslovaquie (Coccinea). *Acta Faunistica Entomologica Musei Nationalis Pragae (Sborník Faunistych Pracì Entomologického Oddeleni Národního Musea v Praze)* **11**: 303-306. [Zahrad1965]

Zahradník, J. 1972. Überfamilie Archaeococcoidea, Neococcoidea. Pages 391-446 *in*: Schwenk, W., Ed. Die Forstschädlinge Europas. Hamburg-Berlin: Paul Parey. [Zahrad1972]

Zahradník, J. 1977. Aleyrodinea - Coccinea. *In* "Enumeratio Insectorum Bohemoslovakiae". *Acta Faunistica Entomologicae Musei Nationalis Pragae, Supplement* **4**: 117-122. [Zahrad1977]

Zahradník, J. 1990. 3.9.9.1 Conifers. Pages 633-644 *in*: Rosen, D. (Ed.). Armored Scale Insects, Their Biology, Natural Enemies and Control [Series title: World Crop Pests, Vol. 4B]. Amsterdam, the Netherlands: Elsevier. 688 pp. [Zahrad1990]

Zahradník, J. 1990a. 3.9.9.2 Other forests. Pages 645-654 *in*: Rosen, D. (Ed.). Armored Scale Insects, Their Biology, Natural Enemies and Control [Series title: World Crop Pests, Vol. 4B]. Amsterdam, the Netherlands: Elsevier. 688 pp. [Zahrad1990a]

Zahradník, J. 1990b. Die Schildläuse (Coccinea) auf Gewächshaus- un Zimmerpflanzen in den Tschechischen Ländern. *Acta Universitatis Carolinae -- Biologica* **34**: 1-160. [Zahrad1990b]

Zalom, F.G., Van Steenwyk, R.A., Bentley, W.J., Coviello, R.L., Rice, R.E., Hendricks, L., Pickel, C. & Freeman, M.W. 2001. Pests of almond. *UC Pest Management Guidelines* . [ZalomVaBe2001]

Zchori-Fein, E., Faktor, M., Zeidan, Y., Gottlieb, H. Czosnek & Rosen, D. 1995. Parthenogenesis-inducing microorganisms in *Aphytis* (Hymenoptera: Aphelinidae). *Insect Molecular Biology* **4**: 173-178. [ZchoriFaZe1995]

Zchori-Fein, E., Rosen, D. & Roush, R.T. 1994. Microorganisms associated with thelytoky in *Aphytis lingnanensis (lingnanensis?* Compere (Hymenoptera: Aphelinidae). *International Journal of Insect Morphology and Embryology* **23(2)**: 169-172. [ZchoriRoRo1994]

Zehntner, L. 1897a. Overzicht van de Ziekten van het Suikerriet op Java. *Archief voor Java-Suikerindustrie* **5**: 525-575. [Zehntn1897a]

Zehntner, L. 1897b. De plantenluizen van het Suikerriet op Java. IV. *Aspidiotus sacchari caulis* n. sp. *Mededeelingen van het Proefstation Oost-Java* **39**: 1-10. [Zehntn1897b]

Zhang, S.Y. 1983. [Observations on the effectiveness of the natural control of *Chilocorus kuwanae* Silvestri (Col.: Coccinellidae) against two species of scales in the orchard.]. *Natural Enemies of Insects* **5(2)**: 86-88. [Zhang1983]

Zhang, X.L. 2000. [The extra effective insecticides for control of pomelo scales.]. *South China Fruits* **29(6)**: 23. [Zhang2000]

Zhang, X.W. & Gu, D. 1994. Experimental release of *Aphytis melinus* to control *Aonidiella aurantii* in a citrus orchard in Guangdong. *Chinese Journal of Biological Control* **10(3)**: 103-105. [ZhangGu1994]

Zhang, X.W., Gu, D.X. & Zhou, Z.M. 1992. [Studies on the growth, development and functional response on *Aonidiella aurantii* of *Aphytis melinus*. *Chinese Journal of Biological Control* **8(1)**: 45. [ZhangGuZh1992]

Zhao, J. 1990. An advanced study on the biology and ecology of the California Red Scale *Aonidiella aurantii* (Mask.) (Hemiptera: Diaspididae) and its natural enemy *Aphytis melinus* DeBach. Yangling, Shaanxi, China: Tianze Eldonejo. 245 pp. [Zhao1990]

Zhou, C.A., Zou, J.J. & Peng, J.C. 1993. [Bionomics of coconut scale - a main pest insect on Actinidia and its control.]. *Entomological Knowledge* **30**: 1, 18-20. [ZhouZoPe1993]

Zimmer, J.F. 1912. The grape scale. (*Aspidiotus [Diaspidotus] uvae* Comst.). *United States Department of Agriculture, Division of Entomology, Bulletin* **97**: 115-124. [Zimmer1912]

Zimmerman, E.C. 1948. Homoptera: Sternorrhyncha. *Insects of Hawaii* **5**: 1-464. [Zimmer1948]

INDEX TO GENERA

Myrtophila Brimblecombe, accepted valid name	*Myrtophila*
Mytrophila; Borchsenius, misspelling of genus name	*Myrtophila*
Neglectaspis Lindinger, junior synonym	*Chentraspis*
Neoclavaspis Brimblecombe, accepted valid name	*Neoclavaspis*
Neofurcaspis Green, accepted valid name	*Neofurcaspis*
Neoleonardia MacGillivray, accepted valid name	*Neoleonardia*
Neomorgania MacGillivray, accepted valid name	*Neomorgania*
Neoselenaspidus Mamet, accepted valid name	*Neoselenaspidus*
Nigridiaspis Ferris, accepted valid name	*Nigridiaspis*
Nuculaspis Ferris, junior synonym	*Dynaspidiotus*
Obtuaspis; Ben-Dov, misspelling of genus name	*Obtusaspis*
Obtusaspis MacGillivray, accepted valid name	*Obtusaspis*
Oceanaspidiotus Takagi, accepted valid name	*Oceanaspidiotus*
Octaspidiotus MacGillivray, accepted valid name	*Octaspidiotus*
Odonaspis Leonardi, accepted valid name	*Odonaspis*
Odonaspis; Fernald, revived rank	*Odonaspis*
Operculaspis Laing, accepted valid name	*Operculaspis*
Palinaspis Ferris, accepted valid name	*Palinaspis*
Paranewsteadia MacGillivray, accepted valid name	*Paranewsteadia*
Paraonidea; MacGillivray, misspelling of genus name	*Paraonidia*
Paraonidia MacGillivray, accepted valid name	*Paraonidia*
Paraonidiella MacGillivray, accepted valid name	*Paraonidiella*
Paraselenaspidus Mamet, accepted valid name	*Paraselenaspidus*
Paraspidiotus Thiem & Gerneck, junior synonym	*Diaspidiotus*
Parrottia MacGillivray, accepted valid name	*Parrottia*
Partargionia MacGillivray, junior synonym	*Semelaspidus*
Pelomphala MacGillivray, junior synonym	*Melanaspis*
Phaspis Ben-Dov, accepted valid name	*Phaspis*
Phaulaspis Leonardi, accepted valid name	*Phaulaspis*
Protargionia Ferris, junior synonym	*Semelaspidus*
Pseudaonidea; MacGillivray, misspelling of genus name	*Pseudaonidia*
Pseudaonidiella MacGillivray, junior synonym	*Pseudaonidia*
Pseudischnaspis Hempel, accepted valid name	*Pseudischnaspis*
Pseudoaonidia; Chou, misspelling of genus name	*Pseudaonidia*
Pseudoaonidia; Leonardi, misspelling of genus name	*Pseudaonidia*
Pseudoaonidia; Rungs, misspelling of genus name	*Pseudaonidia*
Pseudoaunidia; Minamikawa, misspelling of genus name	*Pseudaonidia*
Pseudomelanaspis Borchsenius, junior synonym	*Targionia*
Pseudoselenaspidus Fonseca, accepted valid name	*Pseudoselenaspidus*
Pseudotargionia Lindinger, accepted valid name	*Pseudotargionia*
Pygidiaspis MacGillivray, accepted valid name	*Pygidiaspis*
Qiadraspidiotus; Borchsenius, misspelling of genus name	*Diaspidiotus*
Quadraspidiotus MacGillivray, accepted valid name	*Diaspidiotus*
Reclavaspis Komosinska, accepted valid name	*Reclavaspis*
Remataspidiotus; Chou, misspelling of genus name	*Remotaspidiotus*
Remotaspidiotus MacGillivray, accepted valid name	*Remotaspidiotus*

INDEX TO SPECIES

Due to technical reasons, it was pertinent to use, in part of entries, the following **abbreviations** in the status description of some taxa:

acc. - accepted	**agree.** - agreement
& - and	**cha.** - change
comb. - combination	**epit.** - epithet
gend. - gender	**jun.** - junior
misspel. - misspelling	**nom.** - nomen
nud. - nudum	**replace.** - replacement
spe. - species	**syn.** - synonym
val. - valid	

aberrans Mamet, accepted valid name	*Chrysomphalus aberrans*
abieticola Koroneos, accepted valid name	*Dynaspidiotus abieticola*
abietina Hall & Williams, accepted valid name	*Aonidiella abietina*
abietis Comstock, junior homonym	*Abgrallaspis ithacae*
abietis pini; Cockerell, change in rank	*Dynaspidiotus californicus*
abietis Schrank, accepted valid name	*Dynaspidiotus abietis*
abietis Takagi & Kawai, accepted valid name	*Clavaspidiotus abietis*
abietis; Cockerell, misidentification	*Dynaspidiotus californicus*
abietoides Borchsenius, nomen nudum that is unplaced	*Aspidiotus abietoides*
abietoides Cockerell, nomen nudum that is unplaced	*Aspidiotus abietoides*
abietoides Ferris, nomen nudum that is unplaced	*Aspidiotus abietoides*
acaciae Morgan, junior synonym	*Neomorgania eucalypti*
acaciae Munting, accepted valid name	*Morganella acaciae*
acaciae propinqua Maskell, junior synonym	*Neomorgania eucalypti*
acephala Ferris, accepted valid name	*Pseudischnaspis acephala*
acuminatus Targioni Tozzetti, junior synonym	*Hemiberlesia rapax*
acuta Mamet, accepted valid name	*Acutaspis acuta*
acutus Borchsenius, accepted valid name	*Diaspidiotus acutus*
adiscus; this catalogue, change in rank	*Rhizaspidiotus adiscus*
adnatae Brimblecombe, accepted valid name	*Myrtophila adnatae*
adonidum; Ferris & Kelly, misspelling of spe. name	*Chrysomphalus aonidum*
aegyptiacus; Ferris, change of combination and rank	*Diaspidiotus pyri*
aesculi Johnson, accepted valid name	*Diaspidiotus aesculi*
aesculi solus Hunter, junior synonym	*Diaspidiotus ancylus*
affinis Leonardi, accepted valid name	*Marginaspis affinis*
affinis Newstead, junior homonym	*Hemiberlesia diffinis*

aristidae Ben-Dov, accepted valid name — *Odonaspis aristidae*
aristotelesi Lepage & Giannotti, accepted valid name — *Melanaspis aristotelesi*
aristotelis; Lindinger, misspelling of species name — *Melanaspis aristotelesi*
armeniacus Balachowsky, emendation unjustified — *Diaspidiotus armenicus*
armeniacus Borchsenius, emendation that is unjustified — *Diaspidiotus armenicus*
armenicus Borchsenius, accepted valid name — *Diaspidiotus armenicus*
armigera Ferris, accepted valid name — *Nigridiaspis armigera*
arnaldoi Azevedo Marques, nom. nudum that is placed — *Melanaspis arnaldoi*
arnaldoi Costa Lima, accepted valid name — *Melanaspis arnaldoi*
arnoldi; McKenzie, misspelling of species name — *Melanaspis arnaldoi*
arroyoi Balachowsky, accepted valid name — *Diaspidiotus arroyoi*
artemisiae adiscus Gómez-Menor Ortega, accepted valid name — *Rhizaspidiotus adiscus*
artemisiae Hall, junior synonym — *Rhizaspidiotus canariensis*
artemisiae Mamet, accepted valid name — *Melanaspis artemisiae*
arthrophyti Archangelskaya, accepted valid name — *Targionia arthrophyti*
arthrophytoni Archangelskaya, nom. nudum placed — *Targionia arthrophyti*
articulata; Malenotti, cha. spe. name gender agreement — *Selenaspidus articulatus*
articulatus celastri Maskell, accepted valid name — *Selenaspidus celastri*
articulatus magnospinus Newstead, acc. valid name — *Selenaspidus magnospinus*
articulatus Morgan, accepted valid name — *Selenaspidus articulatus*
articulatus rufescens Cockerell, junior synonym — *Selenaspidus articulatus*
articulatus simplex Grandpré & Charmoy, jun. syn. — *Selenaspidus articulatus*
artocarpi Green, accepted valid name — *Semelaspidus artocarpi*
artus Munting, accepted valid name — *Aspidiotus artus*
arundinariae Deitz & Davidson, accepted valid name — *Melanaspis arundinariae*
askleniae; Borchsenius, misspelling of species name — *Hemiberlesia lataniae*
aspidiotiformis Balachowsky, accepted valid name — *Neoselenaspidus aspidiotiformis*
aspleniae Sasaki, I., junior synonym — *Hemiberlesia lataniae*
asymmetrica Brimblecombe, accepted valid name — *Pseudotargionia asymmetrica*
atherospermae Maskell, junior synonym — *Aspidiotus nerii*
atherospinmae; Tao, misspelling of species name — *Aspidiotus nerii*
atlantica Balachowsky, accepted valid name — *Dynaspidiotus atlanticus*
atlantica Ferris, accepted valid name — *Aonidia atlantica*
atlanticus Lindinger, cha. spe. name gender agreement — *Aonidia atlantica*
atlanticus; Lindinger, change spe. name gender agree. — *Dynaspidiotus atlanticus*
atlantorum Matile-Ferrero & Balachowsky, accepted valid name — *Aonidiella atlantorum*
atomaria Hall, accepted valid name — *Aspidiotus atomarius*
atomariae; Borchsenius, misspelling species name — *Aspidiotus atomarius*
atomarius; Munting, cha. spe. name gender agreement — *Aspidiotus atomarius*
atripileus Munting, accepted valid name — *Aspidiotus atripileus*
aubrevillei Balachowsky & Ferrero, acc. valid name — *Selenaspidus aubrevillei*
aurantii citrina; Leonardi, cha. spe. name gender agree. — *Aonidiella citrina*

pectinata; MacGillivray, cha. spe. name gender agree.	*Diaspidiotus africanus*
pectinatus Lindinger, junior synonym	*Diaspidiotus africanus*
pedilanthi Ferris, accepted valid name	*Clavaspis pedilanthi*
pedina Munting, accepted valid name	*Melanaspis pedina*
pedroniformis Cockerell & Robinson, junior synonym	*Aonidiella orientalis*
pedronis Green, junior synonym	*Aonidiella orientalis*
pelagijae Bachmann, junior synonym	*Diaspidiotus pyri*
pemiciabilus; Tang & Zhang, misspelling spe. name	*Diaspidiotus perniciabilus*
penicillata Green, accepted valid name	*Froggattiella penicillata*
penniseti Hall, accepted valid name	*Lindingaspis penniseti*
perieri Goux, accepted valid name	*Diaspidiotus perieri*
perineti Mamet, accepted valid name	*Sakalavaspis perineti*
perniciabilus Wang & Zhang, accepted valid name	*Diaspidiotus perniciabilus*
perniciosa; Leonardi, cha. spe. name gender agreement	*Diaspidiotus perniciosus*
perniciosus albopunctatus; Cockerell, change of combination and rank	*Diaspidiotus perniciosus*
perniciosus andromelas; Cockerell, change of combination and rank	*Diaspidiotus perniciosus*
perniciosus Comstock, accepted valid name	*Diaspidiotus perniciosus*
perniciosus eucalypti Fuller, junior homonym	*Chrysomphalus rubribullatus*
perplexa Green, accepted valid name	*Aonidia perplexa*
perplexa Munting, accepted valid name	*Clavaspis perplexa*
perseae Comstock, accepted valid name	*Acutaspis perseae*
perseae; Ben-Dov & German, emendation justified	*Abgrallaspis perseae*
persearum Cockerell, junior synonym	*Oceanaspidiotus spinosus*
perseus Davidson, accepted valid name	*Abgrallaspis perseae*
personata; Leonardi, change spe. name gender agreem.	*Mycetaspis personata*
personates; Cockerell, misspelling of species name	*Mycetaspis personata*
personatus Comstock, accepted valid name	*Mycetaspis personata*
perspinosus; Mamet, change of combination and rank	*Selenaspidus perspinosus*
pertusus Brain, accepted valid name	*Selenaspidus pertusus*
phaneraspis Takagi, accepted valid name	*Hypaspidiotus phaneraspis*
phenax Cockerell, accepted valid name	*Melanaspis phenax*
philippiae Mamet, accepted valid name	*Melanaspis philippiae*
philippina Ben-Dov, accepted valid name	*Dicirculaspis philippina*
philippinensis Velasquez, accepted valid name	*Aspidiotus philippinensis*
phoenax; Balachowsky, misspelling of species name	*Melanaspis phenax*
phoenicis Gerson & Davidson, accepted valid name	*Avidovaspis phoenicis*
phormii Signoret, accepted valid name	*Aspidiotus phormii*
phormii Signoret, nomen nudum that is placed	*Aspidiotus phormii*
phragmitis Hall, accepted valid name	*Odonaspis phragmitis*
phragmitis Takahashi, accepted valid name	*Aspidiella phragmitis*
phyllanthi Green, accepted valid name	*Greenoidea phyllanthi*
picea; Lindinger, change spe. name gender agreement	*Lindingaspis piceus*
piceae Tang, Hao, Shi & Tang, accepted valid name	*Dynaspidiotus piceae*

rhododendronii Tang, accepted valid name — *Octaspidiotus rhododendronii*

rhois Lindinger, emendation that is unjustified — *Aonidia rhusae*
rhusae Brain, accepted valid name — *Aonidia rhusae*
rigida Ferris, accepted valid name — *Aspidiella rigida*
rigidus; Borchsenius, change in rank — *Aspidiotus rigidus*
rigidus; Ferris, change sp. name gender agreement — *Aspidiella rigida*
rigina; Borchsenius, misspelling of species name — *Aspidiella rigida*
riograndensis Wolff & Claps, accepted valid name — *Dynaspidiotus riograndensis*

ritchiei Green & Laing, junior synonym — *Melanaspis odontoglossi*
ritchiiei; Lindinger, misspelling of species name — *Melanaspis odontoglossi*
riverae Cockerell, accepted valid name — *Aspidiotus riverae*
riverai Lindinger, emendation that is unjustified — *Aspidiotus riverae*
robiniae Borchsenius, nomen nudum that is unplaced — *Aspidiotus robiniae*
robiniae Lindinger, nomen nudum that is unplaced — *Aspidiotus robiniae*
robiniae Targioni Tozzetti, nomen nudum is unplaced — *Aspidiotus robiniae*
robusta Grassi & Berlese, junior synonym — *Melanaspis inopinata*
robustus; McKenzie, cha. spe. name gender agreement — *Melanaspis inopinata*
rosei; Kasargode, misspelling of species name — *Lindingaspis rossi*
roseni Danzig, accepted valid name — *Diaspidiotus roseni*
rossi colae Laing, accepted valid name — *Lindingaspis colae*
rossi ferrandii Malenotti, junior synonym — *Lindingaspis opima*
rossi greeni Brain & Kelly, accepted valid name — *Lindingaspis greeni*
rossi Maskell, accepted valid name — *Lindingaspis rossi*
rossi musae Laing, accepted valid name — *Lindingaspis musae*
rossi musae; McKenzie, change in rank — *Lindingaspis musae*
rossi victoriae Cockerell, accepted valid name — *Lindingaspis victoriae*
rossi; Doane & Ferris, misidentification — *Lindingaspis similis*
rossi; Green, misidentification — *Lindingaspis greeni*
rotunda Brimblecombe, accepted valid name — *Mimeraspis rotunda*
rotunda Ferris, accepted valid name — *Melanaspis rotunda*
ruandensis Balachowsky, accepted valid name — *Aspidiotus ruandensis*
rubidus McKenzie, accepted valid name — *Selenaspidus rubidus*
rubribullata Froggatt, accepted valid name — *Chrysomphalus rubribullatus*

rubribullata Froggatt, replacement name — *Chrysomphalus rubribullatus*

rubribullatus; Ferris, change spe. name gender agree. — *Chrysomphalus rubribullatus*

rübsaameni Cockerell, accepted valid name — *Abgrallaspis ruebsaameni*
ruebsaameni Borchsenius, emendation that is justified — *Abgrallaspis ruebsaameni*
rufa Lindinger, accepted valid name — *Furcaspis rufa*
rufescens Cockerell, junior synonym — *Selenaspidus articulatus*
rufescens; Borchsenius, change of combination & rank — *Selenaspidus articulatus*
rufus; MacGillivray, cha. spe. name gender agreement — *Furcaspis rufa*